THE UNIVERSAL HISTORY OF

# NUMBERS

## FROM PREHISTORY TO THE INVENTION OF THE COMPUTER

GEORGES IFRAH

*Translated from the French*
*by David Bellos, E. F. Harding, Sophie Wood, and Ian Monk*

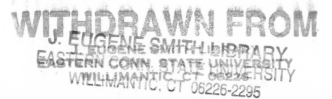

John Wiley & Sons, Inc.
New York • Chichester • Weinheim • Brisbane • Singapore • Toronto

*For you, my wife,*
*the admirably patient witness of the joys and agonies that this hard labour has procured me, or to which you have been subjected, over so many years.*
*For your tenderness and for the intelligence of your criticisms.*
*For you, Hanna, to whom this book and its author owe so much.*

*And for you, Gabrielle and Emmanuelle,*
*my daughters, my passion.*

* *

This book is printed on acid-free paper. ♾

Published by John Wiley & Sons, Inc., in 2000.
Published simultaneouly in Canada.

First published in France with the title *Histoire universelle des chiffres*
by Editions Robert Laffont, Paris, in 1994.

First published in Great Britain in 1998 by The Harvill Press Ltd



This translation has been published with the financial support of the European Commission and of the French Ministry of Culture.

All illustrations, with the exception of Fig. 1.30–36 and 2.10 by Lizzie Napoli, have been drawn, or recopied, by the author.



***Library of Congress Cataloging-in-Publication Data:***

Ifrah, Georges
[Histoire universelle des chiffres. English]
The universal history of numbers : from prehistory to the invention of the computer / Georges Ifrah ; [translated by David Bellos, E. F. Harding, Sophie Wood, and Ian Monk].
p. cm.
Includes bibliographical references and index.
ISBN 0-471-37568-3 (acid-free paper)
1. Numeration—History I. Title.

QA141.I3713 2000
513.2 21—dc21 99-045531

Printed in the United States of America

10 9 8 7 6 5 4 3 2 1

# SUMMARY TABLE OF CONTENTS

*Foreword* v

*List of Abbreviations* vi

*Introduction* xv
*Where "Numbers" Come From*

CHAPTER 1 Explaining the Origins: Ethnological and Psychological Approaches to the Sources of Numbers 3

CHAPTER 2 Base Numbers and the Birth of Number-systems 23

CHAPTER 3 The Earliest Calculating Machine – The Hand 47

CHAPTER 4 How Cro-Magnon Man Counted 62

CHAPTER 5 Tally Sticks: Accounting for Beginners 64

CHAPTER 6 Numbers on Strings 68

CHAPTER 7 Number, Value and Money 72

CHAPTER 8 Numbers of Sumer 77

CHAPTER 9 The Enigma of the Sexagesimal Base 91

CHAPTER 10 The Development of Written Numerals in Elam and Mesopotamia 96

CHAPTER 11 The Decipherment of a Five-thousand-year-old System 109

CHAPTER 12 How the Sumerians Did Their Sums 121

CHAPTER 13 Mesopotamian Numbering after the Eclipse of Sumer 134

CHAPTER 14 The Numbers of Ancient Egypt 162

CHAPTER 15 Counting in the Times of the Cretan and Hittite Kings 178

CHAPTER 16 Greek and Roman Numerals 182

CHAPTER 17 Letters and Numbers 212

CHAPTER 18 The Invention of Alphabetic Numerals 227

CHAPTER 19 Other Alphabetic Number-systems 240

CHAPTER 20 Magic, Mysticism, Divination, and Other Secrets 248

CHAPTER 21 The Numbers of Chinese Civilisation 263

CHAPTER 22 The Amazing Achievements of the Maya 297

CHAPTER 23 The Final Stage of Numerical Notation 323

CHAPTER 24 PART I Indian Civilisation: the Cradle of Modern Numerals 356

CHAPTER 24 PART II Dictionary of the Numeral Symbols of Indian Civilisation 440

CHAPTER 25 Indian Numerals and Calculation in the Islamic World 511

CHAPTER 26 The Slow Progress of Indo-Arabic Numerals in Western Europe 577

CHAPTER 27 Beyond Perfection 592

*Bibliography* 601

*Index of Names and Subjects* 616

# FOREWORD

The main aim of this two-volume work is to provide in simple and accessible terms the full and complete answer to all and any questions that anyone might want to ask about the history of numbers and of counting, from prehistory to the age of computers.

More than ten years ago, an American translation of the predecessor of *The Universal History of Numbers* appeared under the title *From One to Zero*, translated by Lowell Bair (Viking, 1985). The present book – translated afresh – is many times larger, and seeks not only to provide a historical narrative, but also, and most importantly, to serve as a comprehensive, thematic encyclopaedia of numbers and counting. It can be read as a whole, of course; but it can also be consulted as a source-book on general topics (for example, the Maya, the numbers of Ancient Egypt, Arabic counting, or Greek acrophonics) and on quite specific problems (the proper names of the nine mediaeval *apices*, the role of Gerbert of Aurillac, how to do a long division on a dust-abacus, and so on).

Two maps are provided in this first volume to help the reader find what he or she might want to know. The *Summary Table of Contents* above gives a general overview. The Index of names and subjects, from p. 616, provides a more detailed map to this volume.

The bibliography has been divided into two sections: sources available in English; and other sources. In the text, references to works listed in the bibliography give just the author name and the date of publication, to avoid unnecessary repetition. Abbreviations used in the text, in the captions to the many illustrations, and in the bibliography of this volume are explained below.

# LIST OF ABBREVIATIONS

*Where appropriate, cross-references to fuller information in the Bibliography are given in the form: "see:* AUTHOR*"*

| | | |
|---|---|---|
| AA | *The American Anthropologist* | Menasha, Wisconsin |
| AAN | *American Antiquity* | |
| AANL | *Atti dell'Accademia Pontificia de' Nuovi Lincei* | Rome |
| AAR | *Acta Archaeologica* | Copenhagen |
| AAS | *Annales archéologiques syriennes* | Damascus |
| AASOR | *Annual of the American School of Oriental Research* | Cambridge, MA |
| AAT | *Aegypten und altes Testament* | |
| ABSA | *Annual of the British School in Athens* | London |
| ACII | *Appendice al Corpus Inscriptionum Italicorum* | see: GAMURRINI |
| ACLHU | *Annals of the Computation Laboratory of Harvard University* | Cambridge, MA |
| ACT | *Astronomical Cuneiform Texts* | See: NEUGEBAUER |
| ACOR | *Acta Orientalia* | Batavia |
| ADAW | *Abhandlungen der deutschen Akademie der Wissenschaften zu Berlin* | |
| ADFU | *Ausgrabungen der deutschen Forschungsgemeinschaft in Uruk-Warka* | Berlin |
| ADO | *Annals of the Dudley Observatory* | Albany, NY |
| ADOGA | *Ausgrabungen der deutschen Orient-Gesellschaft in Abusir* | |
| ADP | *Archives de Psychologie* | Geneva |
| ADSM | *Album of Dated Syriac Manuscripts* | see: HATCH |
| AEG | *Aegyptus. Rivista italiana di egittologia e papirologia* | Milan |
| AESC | *Annales*. Economies, Sociétés, Civilisations | Paris |
| AFD | *Annales d'une famille de Dilbat* | see: GAUTIER |
| AFO | *Archiv für Orientforschung* | Graz |
| AGA | *Aegyptologische Abhandlungen* | Wiesbaden |
| AGM | *Abhandlungen zur Geschichte der Mathematik* | Leipzig |
| AGMNT | *Archiv für Geschichte der Mathematik, der Naturwissenschaften und der Technik* | |
| AGW | *Abhandlungen der Gesellschaft der Wissenschaften* | Göttingen |
| AHC | *Annals of the History of Computing* | IEEE, New York |
| AHES | *Archive for the History of the Exact Sciences* | |
| AI | *Arad Inscriptions* | see: AHARONI |
| AIEE | American Institute of Electrical Engineers | New York, NY |
| AIHS | *Archives internationales d'histoire des sciences* | Paris |
| AIM | *Artificial Intelligence Magazine* | |
| AJ | *Accountants Journal* | |
| AJA | *American Journal of Archaeology* | New York, NY |
| AJPH | *American Journal of Philology* | New York, NY |
| AJPS | *American Journal of Psychology* | New York, NY |
| AJS | *American Journal of Science* | New York, NY |
| AJSL | *American Journal of Semitic Languages and Literature* | Chicago, IL |
| AKK | *Akkadika* | Brussels |
| AKRG | *Arbeiten der Kaiserlichen Russischen Gesandschaft zu Peking* | Berlin |
| AM | *American Machinist* | |
| AMA | *Asia Major* | Leipzig & London |
| AMI | *Archaeologische Mitteilungen aus Iran* | Berlin |
| AMM | *American Mathematical Monthly* | |
| AMP | *Archiv der Mathematik und Physik* | |
| ANS | *Anatolian Studies* | London |
| ANTH | *Anthropos* | Göteborg |
| ANTHR | *Anthropologie* | Paris |
| AOAT | *Alter Orient und Altes Testament* | Neukirchen-Vluyn |
| AOR | *Analecta Orientalia* | Rome |

| | | |
|---|---|---|
| AOS | *American Oriental Series* | New Haven, CT |
| APEL | *Arabic Papyri in the Egyptian Library* | see: GROHMANN |
| ARAB | *Arabica. Revue d'études arabes* | Leyden |
| ARBE | *Annual Report of the American Bureau of Ethnology* | Washington, DC |
| ARBS | *Annual Report of the Bureau of the Smithsonian Institution* | Washington, DC |
| ARC | *Archeion* | Rome |
| ARCH | *Archeologia* | Rome |
| ARCHL | *Archaeologia* | London |
| ARCHN | *Archaeology* | New York, NY |
| ARM | *Armenia* | |
| ARMA | *Archives royales de Mari* | Paris |
| AROR | *Archiv Orientálni* | Prague |
| ARYA | | see: SHUKLA & SARMA |
| AS | *Automata Studies* | Princeton, NJ, 1956 |
| ASAE | *Annales du service de l'antiquité de l'Egypte* | Cairo |
| ASB | *Assyriologische Bibliothek* | Leipzig |
| ASE | *Archaeological Survey of Egypt* | London |
| ASI | *Archaeological Survey of India* | New Delhi |
| ASMF | *Annali di scienze matematiche e fisiche* | Rome |
| ASNA | *Annuaire de la société française de numismatique et d'archéologie* | Paris |
| ASOR | *American School of Archaeological Research* | Ann Arbor, MI |
| ASPN | *Annales des sciences physiques et naturelles* | Lyon |
| ASR | *Abhandlungen zum schweizerischen Recht* | Bern |
| ASS | *Assyriological Studies* | Chicago, IL |
| ASTP | *Archives suisses des traditions populaires* | |
| ASTRI | *L'Astronomie indienne* | see: BILLARD |
| AT | *Annales des Télécommunications* | Paris |
| ATU | *Archaische Texte aus Uruk* | see: FALKENSTEIN |
| ATU2 | *Zeichenliste der Archaischen Texte aus Uruk* | see: GREEN & NISSEN |
| AUT | *Automatisme* | Paris |
| BAB | *Bulletin de l'Académie de Belgique* | Brussels |
| BAE | *Bibliotheca Aegyptica* | Brussels |
| BAMNH | *Bulletin of the American Museum of Natural History* | New York, NY |
| BAMS | *Bulletin of the American Mathematical Society* | |
| BAPS | *Bulletin de l'Académie polonaise des Sciences* | Warsaw |
| BARSB | *Bulletin de l'Académie royale des sciences et belles-lettres de Bruxelles* | Brussels |
| BASOR | *Bulletin of the American School of Oriental Research* | Ann Arbor, MI |
| BCFM | *Bulletin du Club français de la médaille* | Paris |
| BCMS | *Bulletin of the Calcutta Mathematical Society* | Calcutta |
| BDSM | *Bulletin des sciences mathématiques* | Paris |
| BEFEO | *Bulletin de l'Ecole française d'Extrême-Orient* | Paris & Hanoi |
| BEPH | *Beiträge zur englischen Philologie* | Leipzig |
| BFT | *Blätter für Technikgeschichte* | Vienna |
| BGHD | *Bulletin de géographie historique et descriptive* | Paris |
| BHI | *Bulletin hispanique* | Bordeaux |
| BHR | *Bibliothèque d'humanisme et de Renaissance* | Geneva |
| BIFAO | *Bulletin de l'Institut français d'antiquités orientales* | Cairo |
| BIMA | *Bulletin of the Institute of Mathematical Applications* | |
| BJRL | *Bulletin of the John Rylands Library* | Manchester, UK |
| BLPM | *Bulletin de liaison des professeurs de mathématiques* | Paris |
| BLR | *Bell Laboratories Record* | Murray Hill, NJ |
| BMA | *Biblioteca Mathematica* | |
| BMB | *Bulletin for Mathematics and Biophysics* | |
| BMET | *Bulletin du Musée d'ethnologie du Trocadéro* | Paris |
| BMFRS | *Biographical Memoirs of Fellows of the Royal Society* | London |
| BMGM | *Bulletin of the Madras Government Museum* | Madras |
| BRAH | *Boletín de la Real Academia de la Historia* | Madrid |
| BSA | *Bulletin de la Société d'anthropologie* | Paris |
| BSC | *Bulletin scientifique* | Paris |

| | | |
|---|---|---|
| BSEIN | *Bulletin de la Société d'encouragement pour l'industrie nationale* | Paris |
| BSFE | *Bulletin de la Société française d'Egyptologie* | Paris |
| BSFP | *Bulletin de la Société française de philosophie* | Paris |
| BSPF | *Bulletin de la Société préhistorique française* | Paris |
| BSI | *Biblioteca Sinica* | Paris |
| BSM | *Bulletin de la Société mathématique de France* | Paris |
| BSMA | *Bulletin des sciences mathématiques et astronomiques* | Paris |
| BSMF | *Bollettino di bibliografia e di storia delle scienze matematiche e fisiche* | Rome |
| BSMM | *Bulletin de la Société de médecine mentale* | Paris |
| BSNAF | *Bulletin de la Société nationale des antiquaires de France* | Paris |
| BSOAS | *Bulletin of the School of Oriental and African Studies* | London |
| BST | *Bell System Technology* | Murray Hill, NJ |
| CAA | *Contributions to American Archaeology* | Washington, DC |
| CAAH | *Contributions to American Anthropology and History* | Washington, DC |
| CAH | *The Cambridge Ancient History* | Cambridge, UK, 1963 |
| CAPIB | *Corpus of Arabic and Persian Inscriptions of Bihar* | Patna |
| CAW | Carnegie Institution | Washington, DC |
| CDE | Chronique d'Egypte, in: *Bulletin périodique de la Fondation égyptienne de la reine Elisabeth* | Brussels |
| CENT | *Centaurus* | Copenhagen |
| CETS | *Comparative Ethnographical Studies* | |
| CGC | *A Catalogue of Greek Coins in the British Museum* | see: POOLE |
| CHR | *China Review* | |
| CIC | *Corpus des Inscriptions du Cambodge* | see: COEDÈS |
| CIE | *Corpus inscriptionum etruscarum* | 1970 |
| CIG | *Corpus inscriptionum graecarum* | see: BOECKH, FRANZ, CURTIUS & KIRKHOFF |
| CII | *Corpus inscriptionum Iudaicorum* | see: FREY |
| CIIN | *Corpus inscriptionum Indicarum* | London, Benares & Calcutta, 1888–1929 |
| CIL | *Corpus inscriptionum latinarum* | Leipzig & Berlin, 1861–1943 |
| CIS | *Corpus inscriptionum semiticarum* | Paris, 1889–1932 |
| CJW | *Coins of the Jewish War* | see: KADMAN |
| CNAE | *Contributions to North American Ethnology* | Washington, DC |
| CNP | *Corpus Nummorum Palaestiniensium* | Jerusalem |
| COWA | *Relative Chronologies in Old World Archaeology* | see: ERICHSEN |
| CPH | *Classical Philology* | Chicago, IL |
| CPIN | *Le Cabinet des poinçons de l'Imprimerie national* | Paris, 1963 |
| CR | *Classical Review* | |
| CRAI | *Comptes-rendus des séances de l'Académie des Inscriptions et Belles-Lettres* | Paris |
| CRAS | *Comptes-rendus des séances de l'Académie des Sciences* | Paris |
| CRCIM | *Comptes-rendus du Deuxième Congrès international de Mathématiques de Paris* | see: DUPORCK |
| CRGL | *Comptes-rendus du Groupe linguistique d'études hamito-sémitiques* | Paris |
| CRSP | *Comptes-rendus de la Société impériale orthodoxe de Palestine* | |
| CSKBM | *Catalogue of Sanskrit Buddhist Manuscripts in the British Museum* | London |
| CSMBM | *Catalogue of Syriac Manuscripts in the British Museum* | see: WRIGHT |
| CTBM | *Cuneiform Texts from Babylonian Tablets in the British Museum* | London, 1896 |
| D | *Le Temple de Dendara* | see: CHASSINAT |
| DAA | *Denkmäler aus Aegypten und Aethiopien* | see: LEPSIUS |

| | | |
|---|---|---|
| DAB | *Dictionnaire archéologique de la Bible* | Paris: Hazan, 1970 |
| DAC | *Dictionnaire de l'Académie française* | |
| DAE | *Deutsche Aksoum-Expedition* | Berlin |
| DAFI | *Cahiers de la Délégation archéologique française en Iran* | Paris |
| DAGR | *Dictionnaire des antiquités grecques et romaines* | see: DAREMBERG & SAGLIO |
| DAR | *Denkmäler des Alten Reiches im Museum von Kairo* | see: BORCHARDT |
| DAT | *Dictionnaire archéologique des techniques* | Paris: L'Accueil, 1963 |
| DCI | *Dictionnaire de la civilisation indienne* | see: FRÉDÉRIC |
| DCR | *Dictionnaire de la civilisation romaine* | see: FREDOUILLE |
| DG | *Demotisches Glossar* | see: ERICHSEN |
| DgRa | *De Gestis regum Anglorum libri* | see: MALMESBURY |
| DI | *Der Islam* | |
| DJD | *Discoveries in the Judaean Desert of Jordan* | Clarendon Press, Oxford |
| DMG | *Documents in Mycenean Greek* | see: VENTRIS & CHADWICK |
| DR | *Divination et rationalité* | Paris: Le Seuil, 1974 |
| DS | *Der Schweiz* | |
| DSB | *Dictionary of Scientific Biography* | see: GILLESPIE |
| DTV | *Dictionnaire de Trévoux* | Paris, 1771 |
| E | *Le Temple d'Edfou* | see: CHASSINAT |
| EA | *Etudes asiatiques* | Zürich |
| EBR | *Encyclopaedia Britannica* | London |
| EBOR | *Encyclopédie Bordas* | Paris |
| EC | *Etudes crétoises* | Paris |
| EE | *Epigrafia etrusca* | see: BUONAMICI |
| EEG | *Elementa epigraphica graecae* | see: FRANZ |
| EENG | *Electrical Engineering* | |
| EG | *Epigraphia greca* | see: GUARDUCCI |
| EI | *Epigraphia Indica* | Calcutta |
| EIS | *Encyclopédie de l'Islam* | Leyden, 1908–1938 |
| EJ | *Encyclopaedia Judaica* | Jerusalem |
| EMDDR | *Entwicklung der Mathematik in der DDR* | Berlin, 1974 |
| EMW | *Enquêtes du Musée de la vie wallone* | |
| ENG | *Engineering* | Paris |
| EP | *Encyclopédie de la Pléiade* | Paris |
| EPP | *L'Ecriture et la psychologie des peuples* | Paris: A. Colin, 1963 |
| ERE | *Encyclopaedia of Religions and Ethics* | Edinburgh & New York, 1908–1921 |
| ESIP | *Ecritures. Systèmes idéographiques et pratiques expressives* | see: CHRISTIN |
| ESL | *L'Espace et la lettre* | Paris: UGE, 1977 |
| ESM | *Encyclopédie des sciences mathématiques* | Paris, 1909 |
| EST | *Encyclopédie internationale des sciences et des techniques* | Paris, 1972 |
| EUR | *Europe* | |
| EXP | *Expedition* | Philadelphia, PA |
| FAP | *Fontes atque Pontes. Eine Festgabe für Helmut Brunner* | see: AAT 5 (1983) |
| FEHP | *Facsimile of an Egyptian Hieratic Papyrus* | see: BIRCH |
| FIH | *Das Mathematiker-Verzeichnis im Fihrist des Ibn Abi Jakub an Nadim* | see: SUTER |
| FMAM | *Field Museum of Natural History* | Chicago, IL |
| FMS | *Frühmitelalterliche Studien* | Berlin |
| GIES | Glasgow Institute of Engineers and Shipbuilders in Scotland | |
| GKS | *Das Grabdenkmal des Königs S'ahu-Re* | see: BORCHARDT |
| GLA | *De sex arithmeticae practicae specibus Henrici Glareani epitome* | Paris, 1554 |
| GLO | *Globus* | |
| GORILA | *Recueil des inscriptions en linéaire A* | see: GODART & OLIVIER |
| GT | *Ganitatilaka, by Shrîpati* | see: KAPADIA |
| GTSS | *Ganitasârasamgraha by Mahâvîra* | see: RANGACARYA |

| | | |
|---|---|---|
| HAN | *Hindu-Arabic Numerals* | see: SMITH & KARPINSKI |
| HESP | *Hesperis. Archives berbères et Bulletin de l'Institut des Hautes Etudes marocaines* | |
| HF | *Historical Fragments* | see: LEGRAIN |
| HG | *"Hommages à H. G. Güterboch" in Anatolian Studies* | Istanbul, 1974 |
| HGE | *Handbuch der griechischen Epigraphik* | see: LARFELD |
| HGS | *Histoire générale des sciences* | see: TATON |
| HLCT | *Haverford Library Collection of Cuneiform Tablets* | New Haven, CT |
| HMA | *Historia mathematica* | |
| HMAI | *Handbook of Middle American Indians* | Austin, TX |
| HNE | *Handbuch der Nordsemitischen Epigraphik* | see: LIDZBARSKI |
| HOR | *Handbuch der Orientalistik* | Leyden & Cologne |
| HP | *Hieratische Paläographie* | see: MÖLLER |
| HPMBS | *The History and Palaeography of Mauryan Brahmi Script* | see: UPASAK |
| HUCA | *Hebrew Union College Annual*, ed. S. H. Blank | |
| IA | *Indian Antiquary* | Bombay |
| IDERIC | Institut d'études et de recherches interethniques et interculturelles | Nice |
| IEJ | *Israel Exploration Journal* | Jerusalem |
| IESIS | *Indian Epigraphy and South Indian Scripts* | see: SIVARAMAMURTI |
| IHE | *Las Inscripciónes hebraïcas de España* | see: CANTERA & MILLAS |
| IHQ | *Indian Historical Quarterly* | Calcutta |
| IJES | *International Journal of Environmental Studies* | |
| IJHS | *Indian Journal of History of Science* | |
| IMCC | *Listes générales des Inscriptions et Monuments du Champa et du Cambodge* | see: COEDÈS & PARMENTIER |
| INEP | *Indian Epigraphy* | see: SIRCAR |
| INM | *Indian Notes and Monographs* | |
| INSA | *Die Inschriften Asarhaddons, König von Assyrien* | see: BORGER |
| IOS | *Israel Oriental Studies* | |
| IP | *Indische Palaeographie* | see: BUHLER |
| IR | *Inscription Reveal, Documents from the Time of the Bible, the Mishna and the Talmud* | Jerusalem, 1973 |
| ISCC | *Inscriptions sanskrites du Champa et du Cambodge* | see: BARTH & BERGAIGNE |
| IS | *Isis, revue d'histoire des sciences* | |
| JA | *Journal asiatique* | Paris |
| JAI | *Journal of the Anthropological Institute of Great Britain* | |
| JAOS | *Journal of the American Oriental Society* | Baltimore, MD |
| JAP | *Journal of Applied Psychology* | |
| JASA | *Journal of the American Statistical Association* | |
| JASB | *Journal of the Asiatic Society of Bengal* | Calcutta |
| JB | *Jinwen Bián* | see: RONG REN |
| JBRAS | *Journal of the Bombay branch of the Royal Asiatic Society* | Bombay |
| JCS | *Journal of Cuneiform Studies* | New Haven, CT |
| JEA | *Journal of Egyptian Archaeology* | London |
| JFI | *Journal of the Franklin Institute* | |
| JFM | *Jahrbuch über die Fortschritte der Mathematik* | |
| JHS | *Journal of Hellenic Studies* | London |
| JIA | *Journal of the Institute of Actuaries* | |
| JJS | *Journal of Jewish Studies* | London |
| JNES | *Journal of Near Eastern Studies* | Chicago, IL |
| JPAS | *Journal and Proceedings of the Asiatic Society of Bengal* | Calcutta |
| JRAS | *Journal of the Royal Asiatic Society* | London |
| JRASB | *Journal of the Royal Asiatic Society of Bengal* | |
| JRASI | *Journal of the Royal Asiatic Society of Great Britain and Ireland* | London |

| | | |
|---|---|---|
| JRSA | *Journal of the Royal Society of the Arts* | London |
| JRSS | *Journal of the Royal Statistical Society* | |
| JSA | *Journal de la société des américanistes* | Paris |
| JSI | *Journal of Scientific Instruments* | London |
| JSO | *Journal de la société orientale d'Allemagne* | |
| KAI | *Kanaanaïsche und Aramaïsche Inschriften* | see: DONNER & RÖLLIG |
| KAV | *The Kashmirian Atharva-Veda* | Baltimore, MD, 1901 |
| KR | *The Brooklyn Museum Aramaic Papyri* | see: KRAELING |
| KS | *Keilschriften Sargons, König von Assyrien* | see: LYON |
| LAA | *Annals of Archaeology and Anthropology* | Liverpool |
| LAL | *Lalitavistara Sûtra* | see: LAL LITRA |
| LAT | *Latomus* | Brussels |
| LAUR | *Petri Laurembergi Rostochiensis Institutiones arithmeticae* | Hamburg, 1636 |
| LBAT | *Late Babylonian Astronomical and Related Texts* | see: PINCHES & STRASMAIER |
| LBDL | *Late Old Babylonian Documents and Letters* | see: FINKELSTEIN |
| LEV | *Levant* | |
| LIL | *Lîlâvatî by Bhâskata* | see: DVIVEDI |
| LOE | *The Legacy of Egypt* | see: HARRIS |
| LOK | *Lokavibhâga* | see: ANONYMOUS |
| MA | *Mathematische Annalen* | |
| MAA | *Les Mathématiques arabes* | see: YOUSHKETVITCH |
| MACH | *Machriq* | Baghdad |
| MAF | *Mémorial de l'artillerie française* | Paris |
| MAGW | *Mitteilungen der Anthropologischen Gesellschaft in Wien* | Vienna |
| MAPS | *Memoirs of the American Philosophical Society* | Philadelphia, PA |
| MAR | *Die Mathematiker und Astronomen der Araber und ihre Werke* | see: SUTER |

| | | |
|---|---|---|
| MARB | *Mémoires de l'Académie royale de Bruxelles* | Brussels |
| MARI | *Mari. Annales de recherches interdisciplinaires* | Paris |
| MAS | *Memoirs of the Astronomical Society* | |
| MCM | *Memoirs of the Carnegie Museum* | Washington, DC |
| MCT | *Mathematical Cuneiform Texts* | see: NEUGEBAUER & SACHS |
| MDP | *Mémoires de la délégation archéologique en Susiane* (vols. 1–5), continued as: *Mémoires de la Délégation en Perse* (vols. 6–13), *Mémoires de la mission archéologique en Perse* (vols.14–30), *Mémoires de la mission archéologique en Iran* (vols.31–40), *Mémoires de la délégation archéologique en Iran* (vols.41–) | |
| MDT | *Mémoires de Trévoux* | |
| MFO | *Mélanges de la Faculté orientale* | Beirut |
| MG | Morgenländische Gesellschaft | |
| MGA | *Mathematical Gazette* | |
| MIOG | *Mitteilungen des Instituts für österreichische Geschichtsforschung* | Innsbruck |
| MM | *Mitteilungen für Münzsammler* | Frankfurt / Main |
| MMA | *Memoirs of the Museum of Anthropology* | Ann Arbor, MI |
| MMO | *Museum Monographs* | Philadelphia, PA |
| MNRAS | *Monthly Notes of the Royal Astronomical Society* | |
| MP | *Michigan Papyri* | Ann Arbor, MI |
| MPB | *Mathematisch-physikalische Bibliothek* | Leipzig |
| MPCI | *Mémoire sur la propagation des chiffres indiens* | see: WOEPKE |
| MSA | *Mémoires de la société d'anthropologie* | Paris |
| MSPR | *Mitteilungen aus der Sammlung der Papyrus Rainer* | |
| MT | *Mathematics Teacher* | |
| MTI | *Mathematik Tijdschrift* | |
| MUS | *Mélanges de l'université Saint-Joseph* | Beirut |

| | | |
|---|---|---|
| N | *Le Nabatéen* | see: CANTINEAU |
| NA | *Nature* | London |
| NADG | *Neues Archiv der Gesellschaft für ältere deutsche Geschichtskunde* | Hanover |
| NAM | *Nouvelles Annales de Mathématiques* | Paris |
| NAT | *La Nature* | Paris |
| NAW | *Nieuw Archief voor Wiskunde* | |
| NAWG | *Nachrichten der Akademie der Wissenschaften zu Göttingen* | Göttingen |
| NC | *Numismatic Chronicle* | London |
| NCEAM | *Notices sur les caractères étrangers anciens et modernes* | see: FOSSEY |
| NEM | *Notices et Extraits des Manuscrits de la Bibliothèque nationale* | Paris |
| NMM | *National Mathematics Magazine* | |
| NNM | *Numismatical Notes and Monographs* | New York |
| Nott | *Christophori Nottnagelii Professoris Wittenbergensis Institutionum mathematicarum* | Wittenberg, 1645 |
| NS | *New Scientist* | London |
| NYT | *New York Times* | |
| NZ | *Numismatische Zeitschrift* | Vienna |
| | | |
| OED | *Oxford English Dictionary* | |
| OIP | *Oriental Institute Publications* | Chicago, IL |
| OR | *Orientalia* | Rome |
| | | |
| PA | *Popular Astronomy* | |
| PEQ | *Palestine Exploration Quarterly* | London |
| PFT | *Persepolis Fortification Tablets* | see: HALLOCK |
| PGIFAO | *Papyrus grecs de l'Institut français d'Archéologie orientale* | Cairo |
| PGP | *Paläographie der griechischen Papyri* | see: SEIDER |
| PHYS | *Physis* | Buenos Aires |
| PI | *The Paleography of India* | see: OJHA |
| PIB | *Paleographia Iberica* | see: BURNAM |
| PLMS | *Proceedings of the London Mathematical Society* | London |
| PLO | *Porta Linguarum Orientalum* | Berlin |
| PM | *The Palace of Minos* | see: EVANS |
| PMA | *Periodico matematico* | |
| PMAE | *Papers of the Peabody Museum* | Cambridge, MA |
| PPS | *Proceedings of the Prehistoric Society* | |
| PR | *Physical Review* | |
| PRMS | *Topographical Bibliography* | see: PORTER and MOSS |
| PRS | *Proceedings of the Royal Society* | London |
| PRU | *Le Palais royal d'Ugarit* | see: SCHAEFFER |
| PSBA | *Proceedings of the Society of Biblical Archaeology* | London |
| PSREP | *Publications de la société royale égyptienne de papyrologie* | Cairo |
| PTRSL | *Philosophical Transactions of the Royal Society* | London |
| PUMC | *Papyri in the University of Michigan Collection* | see: GARETT-WINTER |
| | | |
| QSG | *Quellen und Studien zur Geschichte der Mathematik, Astronomie und Physik* | Berlin |
| | | |
| RA | *Revue d'Assyriologie et d'Archéologie orientale* | Paris |
| RACE | Real Academia de Ciencias Exactas, Físicas y Naturales | Madrid |
| RAR | *Revue archéologique* | Paris |
| RARA | *Rara Arithmetica* | see: D. E. SMITH |
| RB | *Revue biblique* | Saint-Etienne |
| RBAAS | *Report of the British Society for the Advancement of Science* | London |
| RCAE | *Report of the Cambridge Anthropological Expedition to the Torres Straits* | Cambridge, 1907 |
| RdSO | *Revista degli Studi Orientali* | Rome |
| RE | *Revue d'Egyptologie* | Paris |

| | | |
|---|---|---|
| REC | *Revue des Etudes Celtiques* | Paris |
| REG | *Revue des Etudes Grecques* | Paris |
| REI | *Revue des Etudes Islamiques* | Paris |
| RES | *Répertoire d'épigraphie sémitique* | Paris |
| RFCB | *Reproduccion fac similar* | see: SELER |
| RFE | *Recueil de facsimilés* | see: PROU |
| RH | *Revue historique* | Paris |
| RHA | *Revue de Haute-Auvergne* | Aurillac |
| RHR | *Revue de l'Histoire des Religions* | Paris |
| RHS | *Revue d'Histoire des Sciences* | Paris |
| RHSA | *Revue d'Histoire des Sciences et de leurs applications* | Paris |
| RMM | *Revue du Monde musulman* | Paris |
| RN | *Revue numismatique* | Paris |
| RRAL | *Rendiconti della Reale Accademia dei Lincei* | Rome |
| RSS | *Rivista di Storia della Scienza* | Florence |
| RTM | *The Rock Tombs of Meir* | see: BLACKMAN |
| S | *Aramäische Papyrus und Ostraka* | see: SACHAU |
| SAOC | *Studies in Ancient Oriental Civilizations* | Chicago, IL |
| SC | *Scientia* | |
| SCAM | *Scientific American* | New York |
| SE | *Studi etruschi* | Florence |
| SEM | *Semitica* | Paris |
| SGKIO | *Studien zur Geschichte und Kultur des islamischen Orients* | Berlin |
| SHAW | *Sitzungsberichte der Heidelberger Akademie der Wissenschaften* | Heidelberg |
| SHM | *Sefer ha Mispar* | see: SILBERBERG |
| SIB | *Scripta Pontificii Instituti Biblici* | Rome |
| SIP | *Elements of South Indian Paleography* | see: BURNELL |
| SJ | *Science Journal* | |
| SKAW | *Sitzungsberichte der kaiserlichen Akademie der Wissenschaften* | Vienna |
| SM1 | *Scripta Minoa*, 1 | see: EVANS |
| SM2 | *Scripta Minoa*, 2 | see: EVANS & MYRES |
| SMA | *Scripta Mathematica* | |
| SME | *Studi medievali* | Turin |
| SMS | *Syrio-Mesopotamian Studies* | Los Angeles, CA |
| SPA | *La scrittura proto-elamica* | see: MERIGGI |
| SPRDS | *Scientific Proceedings of the Royal Dublin Society* | Dublin |
| SS | *Schlern Schriften* | Innsbruck |
| STM | *Studia Mediterranea* | Pavia |
| SUM | *Sumer* | Baghdad |
| SVSN | *Mémoires de la société vaudoise des sciences naturelles* | Lausanne |
| SWG | *Schriften der Wissenschaftlichen Gesellschaft in Strassburg* | Strasburg |
| TA | *Tablettes Albertini* | see: COURTOIS, LESCHI, PERRAT & SAUMAGNE |
| TAD | *Türk Arkeoloji Dergisi* | |
| TAPS | *Transactions of the American Philosophical Society* | |
| TASJ | *Transactions of the Asiatic Society of Japan* | Yokohama |
| TCAS | *Transactions published by the Connecticut Academy of Arts and Sciences* | New Haven, CT |
| TDR | *Tablettes de Drehem* | see: GENOUILLAC |
| TEB | *Tablettes de l'époque babylonienne ancienne* | see: BIROT |
| TH | *Theophanis Chronographia* | Paris, 1655 |
| TIA | *Thesaurus Inscriptionum Aegypticum* | see: BRUGSCH |
| TLE | *Testimonia Linguae Etruscae* | 1968 |
| TLSM | *Transactions of the Literary Society of Madras* | Madras |
| TMB | *Textes mathématiques de Babylone* | see: THUREAU-DANGIN |
| TMIE | *Travaux et mémoires de l'Institut d'Ethnologie de Paris* | Paris |
| TMS | *Textes mathématiques de Suse* | see: BRUINS & RUTTEN |
| TRAR | *Trattati d'Aritmetica* | see: BONCOMPAGNI |

| | | |
|---|---|---|
| TRIA | *Transactions of the Royal Irish Academy* | Dublin |
| TSA | *Tablettes sumériennes archaïques* | see: GENOUILLAC |
| TSM | *Taylor's Scientific Memoirs* | London |
| TTKY | *Türk Tarih Kurumu Yayinlarindan* | Ankara |
| TUTA | *Tablettes d'Uruk* | see: THUREAU-DANGIN |
| TZG | *Trierer Zeitschrift zur Geschichte und Kunst des Trierer Landes* | Trier |
| UAA | *Urkunden des Aegyptischen Altertums* | see: STEINDORFF |
| UCAE | *University of California Publication of American Archaeology and Ethnology* | Berkeley, CA |
| UMN | *Unterrichtsblätter für Mathematik und Naturwissenschaften* | |
| URK | *Hieroglyphischen Urkunden der griechischen-römischen Zeit* | see: SETHE |
| URK.I | *Urkunden des Alten Reichs* | see: SETHE |
| URK.IV | *Urkunden der 18.ten Dynastie* | see: SETHE & HELCK |
| UVB | *Vorläufiger Bericht über die Ausgrabungen in Uruk-Warka* | Berlin |
| VIAT | *Viator. Medieval and Renaissance Studies* | Berkeley, CA |
| WKP | *Wochenschrift für klassische Philologie* | |
| WM | *World of Mathematics* | |
| YI | *Xiao dun yin xu wenzi: yi bián* | see: DONG ZUOBIN |
| YOS | *Yale Oriental Series* | New Haven, CT |
| ZA | *Zeitschrift für Assyriologie* | Berlin |
| ZAS | *Zeitschrift für Aegyptische Sprache und Altertumskunde* | Berlin |
| ZDMG | *Zeitschrift der Deutschen Morgenländischen Gesellschaft* | Wiesbaden |
| ZDP | *Zeitschrift des Deutschen Palästina-Vereins* | Leipzig & Wiesbaden |
| ZE | *Zeitschrift für Ethnologie* | Braunschweig |
| ZKM | *Zeitschrift für die Kunde des Morgenlandes* | Göttingen |
| ZMP | *Zeitschrift für Mathematik und Physik* | |
| ZNZ | *Zbornik za Narodni Zivot i Obicaje juznih Slavena* | Zagreb |
| ZOV | *Zeitschrift für Osterreichische Volkskunde* | Vienna |
| ZRP | *Zeitschrift für Romanische Philologie* | Tübingen |

# INTRODUCTION

*Where "Numbers" Come From*

## TEACHER LEARNS A LESSON

This book was sparked off when I was a schoolteacher by questions asked by children. Like any decent teacher, I tried not to leave any question unanswered, however odd or naive it might seem. After all, a curious mind often is an intelligent one.

One morning, I was giving a class about the way we write down numbers. I had done my own homework and was well-prepared to explain the ins and outs of the splendid system that we have for representing numbers in Arabic numerals, and to use the story to show the theoretical possibility of shifting from base 10 to any other base without altering the properties of the numbers or the nature of the operations that we can carry out on them. In other words, a perfectly ordinary maths lesson, the sort of lesson you might have once sat through yourself – a lesson taught, year in, year out, since the very foundation of secondary schooling.

But it did not turn out to be an ordinary class. Fate, or Innocence, made that day quite special for me.

Some pupils – the sort you would not like to come across too often, for they can change your whole life! – asked me point-blank all the questions that children have been storing up for centuries. They were such simple questions that they left me speechless for a moment:

"Sir, where do numbers come from? Who invented zero?"

Well, where do numbers come from, in fact? These familiar symbols seem so utterly obvious to us that we have the quite mistaken impression that they sprang forth fully formed, as gods or heroes are supposed to. The question was disconcerting. I confess I had never previously wondered what the answer might be.

"They come . . . er . . . they come from the remotest past," I fumbled, barely masking my ignorance.

But I only had to think of Latin numbering (those Roman numerals which we still use to indicate particular kinds of numbers, like sequences of kings or millionaires of the same name) to be quite sure that numbers have not always been written in the same way as they are now.

"Sir!" said another boy, "Can you tell us how the Romans did their sums? I've been trying to do a multiplication with Roman numerals for days, and I'm getting nowhere with it!"

"You can't do sums with those numerals," another boy butted in. "My dad told me the Romans did their sums like the Chinese do today, with an abacus."

That was almost the right answer, but one which I didn't even possess.

"Anyway," said the boy to the rest of the class, "if you just go into a Chinese restaurant you'll see that those people don't need numbers or calculators to do their sums as fast as we do. With their abacuses, they can even go thousands of times faster than the biggest computer in the world."

That was a slight exaggeration, though it is certainly true that skilled abacists can make calculations faster than they can be done on paper or on mechanical calculating machines. But modern electronic computers and calculators obviously leave the abacus standing.

I was fortunate and privileged to have a class of boys from very varied backgrounds. I learned a lot from them.

"My father's an ethnologist," said one. "He told me that in Africa and Australia there are still primitive people so stupid that they can't even count further than two! They're still cavemen!"

What extraordinary injustice in the mouth of a child! Unfortunately, there used to be plenty of so-called experts who believed, as he did, that "primitive" peoples had remained at the first stages of human evolution. However, when you look more closely, it becomes apparent that "savages" aren't so stupid after all, that they are far from being devoid of intelligence, and that they have extraordinarily clever ways of coping without numbers. They have the same potential as we all do, but their cultures are just very different from those of "civilised" societies.

But I did not know any of that at the time. I tried to grope my way back through the centuries. Before Arabic numerals, there were Roman ones. But does "before" actually mean anything? And even if it did, what was there before those numerals? Was it going to be possible to use an archaeology of numerals and computation to track back to that mind-boggling moment when someone first came up with the idea of counting?

Several other allegedly naive questions arose as a result of my pupils' curious minds. Some concerned "counting animals" that you sometimes see at circuses and fairs; they are supposed to be able to count (which is why some people claim that mathematicians are just circus artistes!) Other pupils put forward the puzzle of "number 13", alternately considered an omen of good luck and an omen of bad luck. Others wondered what was in the minds of mathematical prodigies, those phenomenal beings who can perform very complex operations in their heads at high speed – calculating the cube root of a fifteen-digit number, or reeling off all the prime numbers between seven million and ten million, and so on.

In a word, a whole host of horrendous but fascinating questions exploded in the face of a teacher who, on the verge of humiliation, took the full measure of his ignorance and began to see just how inadequate the teaching of mathematics is if it makes no reference to the history of the subject. The only answers I could give were improvised ones, incomplete and certainly incorrect.

I had an excuse, all the same. The arithmetic books and the school manuals which were my working tools did not even allude to the history of numbers. History textbooks talk of Hammurabi, Caesar, King Arthur, and Charlemagne, just as they mention the travels of Marco Polo and Christopher Columbus; they deal with topics as varied as the history of paper, printing, steam power, coinage, economics, and the calendar, as well as the history of human languages and the origins of writing and of the alphabet. But I searched them in vain for the slightest mention of the history of numbers. It was almost as if a conspiracy of obviousness aimed to make a secret, or, even worse, just to make us ignorant of one of the most fantastic and fertile of human discoveries. Counting is what allowed people to take the measure of their world, to understand it better, and to put some of its innumerable secrets to good use.

These questions had a profound impact on me, beginning with this lesson in modesty: my pupils, who were manifestly more inquisitive than I had been, taught me a lesson by spurring me on to study the history of a great invention. It turned out to be a history that I quickly discovered to be both universal and discontinuous.

## THE QUEST FOR THE MATHEMATICAL GRAIL

I could not now ever let go of these questions, and they soon drew me into the most fascinating period of learning and the most enthralling adventure of my life.

My desire to find the answers and to have time to think about them persuaded me, not without regrets, to give up my teaching job. Though I had only slender means, I devoted myself full-time to a research project that must have seemed as mad, in the eyes of many people, as the mediaeval quest for the Holy Grail, the magical vessel in which the blood of Christ on the cross was supposed to have been collected. Lancelot, Perceval, and Gawain, amongst many other valiant knights of Christendom, set off in search of the grail without ever completing their quest, because they were not pure enough or lacked sufficient faith or chastity to approach the Truth of God.

I couldn't claim to have chastity or purity either. But faith and calling led me to cross the five continents, materially or intellectually, and to glimpse horizons far wider than those that the cloistered world of mathematics usually allows. But the more my eyes opened onto the wider world, the more I realised the depth of my ignorance.

Where, when and how did the amazing adventure of the human intellect begin? In Asia? In Europe? Or somewhere in Africa? Did it take place at the time of Cro-Magnon man, about thirty thousand years ago, or in the Neanderthal period, more than fifty thousand years ago? Or could it have been half a million years ago? Or even – why not? – a million years ago?

What motives did prehistoric peoples have to begin the great adventure of counting? Were their concerns purely astronomical (to do with the phases of the moon, the eternal return of day and night, the cycle of the seasons, and so on)? Or did the requirements of communal living give the first impulse towards counting? In what way and after what period of time did people discover that the fingers of one hand and the toes of one foot represent the same concept? How did the need for calculation impose itself on their minds? Was there a chronological sequence in the discovery of the cardinal and ordinal aspects of the integers? In which period did the first attempts at oral numbering occur? Did an abstract conception of number precede articulated language? Did people count by gesture and material tokens before doing so through speech? Or was it the other way round? Does the idea of number come from experience of the world? Or did the idea of number act as a catalyst and make explicit what must have been present already as a latent idea in the minds of our most distant ancestors? And finally, is the concept of number the product of intense human thought, or is it the result of a long and slow evolution starting from a very concrete understanding of things?

These are all perfectly normal questions to ask, but most of the answers cannot be researched in a constructive way since there is no longer any trace of the thought-processes of early humans. The event, or, more probably, the sequence of events, has been lost in the depths of prehistoric time, and there are no archaeological remains to give us a clue.

However, archaeology was not necessarily the only approach to the problem. What other discipline might there be that would allow at least a stab at an answer? For instance, might psychology and ethnology not have some power to reconstitute the origins of number?

The Quest for Number? Or a quest for a wraith? That was the question. It was not easy to know which it was, but I had set out on it and was soon to conquer the whole world, from America to Egypt, from India to Mexico, from Peru to China, in my search for more and yet more numbers. But as I had no financial backer, I decided to be my own sponsor, doing odd jobs (delivery boy, chauffeur, waiter, night watchman) to keep body and soul together.

As an intellectual tourist I was able to visit the greatest museums in the world, in Cairo, Baghdad, Beijing, Mexico City, and London (the British Museum and the Science Museum); the Smithsonian in Washington, the Vatican Library in Rome, the libraries of major American universities (Yale, Columbia, Philadelphia), and of course the many Paris collections at the Musée Guimet, the Conservatoire des arts et métiers, the Louvre, and the Bibliothèque nationale. I also visited the ruins of Pompeii and Masada. And took a trip to the Upper Nile Valley to see Thebes, Luxor, Abu Simbel, Gizeh. Had a look at the Acropolis in Athens and the Forum in Rome. Pondered on time's stately march from the top of the Mayan pyramids at Quiriguá and Chichén Itzá. And from here and from there I gleaned precious information about past and present customs connected with the history of counting.

When I got back from these fascinating ethno-numerical and archaeo-arithmetical expeditions I buried myself in popularising and encyclopaedic articles, plunged into learned journals and works of erudition, and fired off thousands of questions to academic specialists in scores of different fields.

At the start, I did not get many replies. My would-be correspondents were dumbfounded by the banality of the topic.

There are of course vast numbers of oddballs forever pestering specialists with questions. But I had to persuade them that I was serious. It was essential for me to obtain their co-operation, since I needed to be kept up to date about new and recent discoveries in their fields, however apparently insignificant, and as an amateur I needed their help in avoiding misinterpretations. And since I was dealing with many specialists who were far outside the field of mathematics, I had not only to persuade them that I was an honest toiler in a respectable field, but also to get them to accept that "numbers" and "mathematics" are not quite the same thing. As we shall see . . .

All this work led me to two basic facts. First, a vast treasure-house of documentation on the history of numbers does actually exist. I owe a great deal to the work of previous scholars and mention it frequently throughout this book. Secondly, however, the articles and monographs in this store of knowledge each deal with only one specialism, are addressed to other experts in the same field, and are far from being complete or comprehensive accounts. There were also a few general works, to be sure, which I came across later, and which also gave me some help. But as they describe the state of knowledge at the time they were written, they had been long overtaken by later discoveries in archaeology, psychology, and ethnography.

No single work on numbers existed which covered the whole of the available field, from the history of civilisations and religions to the history of science, from prehistoric archaeology to linguistics and philology, from mythical and mathematical interpretation to ethnography, ranging over the five continents.

Indeed, how can one successfully sum up such heterogeneous material without losing important distinctions or falling into the trap of simplification? The history of numbers includes topics as widely divergent as the perception of number in mammals and birds, the arithmetical use of prehistoric notched bones, Indo-European and Semitic numbering systems, and number-techniques among so-called primitive populations in Australia, the Americas, and Africa. How can you catch in one single net things as different as finger-counting and digital computing? counting with beads and Amerindian or Polynesian knotted string? Pharaonic epigraphy and Babylonian baked clay tablets? How can you talk in the same way about Greek and Chinese arithmetic, astronomy and Mayan inscriptions, Indian poetry and mathematics, Arabic algebra and the mediaeval quadrivium? And all of that so as to obtain a coherent overall vision of the development through time and space of the defining invention of modern humanity, which is our present numbering system? And where do animals fit into what is already an enormously complex field? Not to mention human infants . . .

What I had set out to do was manifestly mad. The topic sat at the junction of all fields of knowledge and constituted an immense universe of human intellectual evolution. It covered a field so rich and huge that no single person could hope to grasp it alone.

Such a quest is by its nature unending. This book will occupy a modest place in a long line of outstanding treatises. It will not be the last of them, to be sure, for so many more things remain undiscovered or not yet understood. All the same, I think I have brought together practically everything of significance from what the number-based sciences, of the logical and historical kinds, have to teach us at the moment. Consequently, this is also probably the only book ever written that gives a more or less universal and comprehensive history of numbers and numerical calculation, set out in a logical and chronological way, and made accessible in plain language to the ordinary reader with no prior knowledge of mathematics.

And since research never stands still, I have been able to bring new solutions to some problems and to open up other, long-neglected areas of the universe of numbers. For example, in one of the chapters you will find a solution to the thorny problem of the decipherment of Elamite numbering, used nearly five thousand years ago in what is now Iran. I have also shown that Roman numbering, long thought to have been derived from the Greek system, was in fact a "prehistoric fossil", developed from the very ancient practice of notching. There are also some new contributions on Mesopotamian numbering and arithmetic, as well as a quite new way of looking at the fascinating and sensitive topic of how "our" numbers evolved from the unlikely conjunction of several great ideas. Similarly, the history of mechanical calculation culminating in the invention of the computer is entirely new.

## A VERY LONG STORY

If you wanted to schematise the history of numbering systems, you could say that it fills the space between One and Zero, the two concepts which have become the symbols of modern technological society.

Nowadays we step with careless ease from Zero to One, so confident are we, thanks to computer scientists and our mathematical masters, that the Void always comes before the Unit. We never stop to think for a moment that in terms of time it is a huge step from the invention of the number "one", the first of all numbers even in the chronological sense, to the invention of the number "zero", the last major invention in the story of numbers. For in fact the whole history of humanity is spread out backwards between the time when it was realised that the void was "nothing" and the time when the sense of "oneness" first arose, as humans became aware of their individual solitude in the face of life and death, of the specificity of their species as distinct from other living beings, of the singularity of their selves as distinct from others, or of the difference of their sex as distinct from that of their partners.

But the story is neither abstract nor linear, as the history of mathematics is sometimes (and erroneously) imagined to be. Far from being an impeccable sequence of concepts each leading logically to the next, the history of numbers is the story of the needs and concerns of enormously diverse social groupings trying to count the days in the year, to make deals and bargains, to list their members, their marriages, their bereavements, their goods and flocks, their soldiers, their losses, and even their prisoners, trying also to record the date of the foundation of their cities or of one of their victories.

Goatherds and shepherds needed to know when they brought their flocks back from grazing that none had been lost; people who kept stocks of tools or arms or stood guard over food supplies for a community needed to know whether the complement of tools, arms or supplies had remained the same as when they last checked. Or again, communities with hostile neighbours must have been concerned to know whether, after each military foray, they still had the same number of soldiers, and, if not, how many they had lost in the fight. Communities that engaged in trading needed to be able to "reckon" so as to be able to buy or barter goods. For harvesting, and also in order to prepare in time for religious ceremonies, people needed to be able to count and to measure time, or at the very least to develop some practical means of managing in such circumstances.

In a word, the history of numbers is the story of humanity being led by the very nature of the things it learned to do to conceive of needs that could only be satisfied by "number reckoning". And to do that, everything and anything was put in service. The tools were approximate, concrete, and empirical ones before becoming abstract and sophisticated, originally imbued with strange mystical and mythological properties, becoming disembodied and generalisable only in the later stages.

Some communities were utilitarian and limited the aims of their counting systems to practical applications. Others saw themselves in the infinite and eternal elements, and used numbers to quantify the heavens and the earth, to express the lengths of the days, months and years since the creation of the universe, or at least from some date of origin whose meaning had subsequently been lost. And because they found that they needed to represent very large numbers, these kinds of communities did not just invent more symbols, but went down a path that led not only towards the fundamental rule of position, but also onto the track of a very abstract concept that we call "zero", whence comes the whole of mathematics.

## THE FIRST STEPS

No one knows where or when the story began, but it was certainly a very long time ago. That was when people were unable to conceive of numbers as such, and therefore could not count. They were capable, at most, of the concepts of *one, two*, and *many*.

As a result of studies carried out on a wide range of beings, from crows to humans as diverse as infants, Pygmies, and the Amerindian inhabitants of Tierra del Fuego, psychologists and ethnologists have been able to establish the absolute zero of human number-perception. Like some of the higher animals, the human adult with no training at all (for example, learning to recognise the 5 or the 6 at cards by sight, through sheer practice) has direct and immediate perception of the numbers 1 to 4 only. Beyond that level, people have to learn to count. To do that they need to develop, firstly, advanced number-manipulating skills, then, for the purposes of memorisation and of communication, they need to develop a linguistic instrument (the names of the numbers), and, finally, and much later on, they need to devise a scheme for writing numbers down.

However, you do not have to "count" the way we do if what you want to do is to find the date of a ceremony, or to make sure that the sheep and the goats that set off to graze have all come back to the byre. Even in the complete absence of the requisite words, of sufficient memory, and of the abstract concepts of number, there are all sorts of effective substitute devices for these kinds of operation. Various present-day populations in Oceania, America, Asia, and Africa whose languages contain only the words for *one*, *two*, and *many*, but who nonetheless understand one-for-one parities perfectly well, use notches on bones or wooden sticks to keep a tally. Other populations use piles or lines of pebbles, shells, knucklebones, or sticks. Still others tick things off by the parts of their body (fingers, toes, elbows and knees, eyes, nose, mouth, ears, breasts, and chest).

## THE EARLIEST COUNTING MACHINES

Early humanity used more or less whatever came to hand to manage in a quantitative as well as a qualitative universe. Nature itself offered every cardinal model possible: birds with two wings, the three parts of a clover-leaf, four-legged animals, and five-fingered hands . . . But as everyone began counting by using their ten fingers, most of the numbering systems that were invented used base 10. All the same, some groups chose base 12. The Mayans, Aztecs, Celts, and Basques, looked down at their feet and realised that their toes could be counted like fingers, so they chose base 20. The Sumerians and Babylonians, however, chose to count on base 60, for reasons that remain mysterious. That is where our present division of the hour into 60 minutes of 60 seconds comes from, as does the division of a circle into 360 degrees, each of 60 minutes divided into 60 seconds.

The very oldest counting tools that archaeologists have yet dug up are the numerous animal bones found in western Europe and marked with one or more sets of notches. These tally sticks are between twenty thousand and thirty-five thousand years old.

The people using these bones were probably fearsome hunters, and, for each kill, they would score another mark onto the tally stick. Separate counting bones might have been used for different animals – one tally for bears, another for bison, another for wolves, and so on.

They had also invented the first elements of accounting, since what they were actually doing was writing numbers in the simplest notation known.

The method may seem primitive, but it turned out to be remarkably robust, and is probably the oldest human invention (apart from fire) still in use today. Various tallies found on cave walls next to animal paintings leave us in little real doubt that we are dealing with an animal-counting device. Modern practice is no different. Since time immemorial, Alpine shepherds in Austria and Hungary, just like Celtic, Tuscan, and Dalmatian herdsmen, have checked off their animals by scoring vertical bars, Vs and Xs on a piece of wood, and that is still how they do it today. In the eighteenth century, the same "five-barred gate" was used for the shelf marks of parliamentary papers at the British House of Commons Library; it was used in Tsarist Russia and in Scandinavia and the German-speaking countries for recording loans and for calendrical accounts; whereas in rural France at that time, notched sticks did all that present-day account books and contracts do, and in the open markets of French towns they served as credit "slates". Barely twenty years ago a village baker in Burgundy made notches in pieces of wood when he needed to tot up the numbers of loaves each of his customers had taken on credit. And in nineteenth-century Indo-China, tally sticks were used as credit instruments, but also as signs of exclusion and to prevent contact with cholera victims. Finally, in Switzerland, we find notched sticks used, as elsewhere, for credit reckonings, but also for contracts, for milk deliveries, and for recording the amounts of water allocated to different grazing meadows.

The long-lasting and continuing currency of the tally system is all the more surprising for being itself the source of the Roman numbering system, which we also still use alongside or in place of Arabic numerals.

The second concrete counting tool, the hand, is of course even older. Every population on earth has used it at one stage or another. In various places in Auvergne (France), in parts of China, India, Turkey, and the former Soviet Union, people still do multiplication sums with their fingers, as the numbers are called out, and without any other tool or device. Using joints and knuckles increases the possible range, and it allowed the Ancient Egyptians, the Romans, the Arabs and the Persians (not forgetting Western Christians in the Middle Ages) to represent concretely all the numbers from 1 to 9,999. An even more ingenious variety of finger-reckoning allowed the Chinese to count to 100,000 on one hand, and to one million using both hands!

But the story of numbers can be told in other ways too. In places as far apart as Peru, Bolivia, West Africa, Hawaii, the Caroline Islands, and Ryû-Kyû, off the Japanese coast, you can find knotted string used to represent numbers. It was with such a device that the Incas sorted the archives of their very effective administration.

A third system has a far from negligible role in the history of arithmetic – the use of pebbles, which really underlies the beginning of calculation. The pebble-method is also the direct ancestor of the abacus, a device still in wide use in China, Japan and Eastern Europe. But it is the very word *calculation* that sends us back most firmly to the pebble-method: for in Latin the word for pebble is *calculus*.

## THE FIRST NUMBERS IN HISTORY

The pebble-method actually formed the basis for the first written numbering system in recorded history. One day, in the fourth millennium BCE, in Elam, located in present-day Iran towards the Persian Gulf, accountants had the idea of using moulded, unbaked clay tokens in the place of ordinary or natural pebbles. The tokens of various shapes and sizes were given conventional values, each different type representing a unit of one order of magnitude within a numbering system: a stick shape for 1, a pellet for 10, a ball for 100, and so on. The idea must have been in the air for a long time, for at about the same period a similarly clay-based civilisation in Sumer, in lower Mesopotamia, invented an identical system. But since the Sumerians counted to base 60 (sexagesimal reckoning), their system was slightly different: a small clay cone stood for 1, a pellet stood for 10, a large cone for 60, a large perforated cone stood for 600, a ball meant 3,600, and so on.

These civilisations were in a phase of rapid expansion but remained exclusively oral, that is to say without writing. They relied on the rather limited potential of human memory. But the accounting system that was developed from the principles just explained turned out to be very serviceable. In the first development, the idea arose of enclosing the tokens in a spherical clay case. This allowed the system not only to serve for actual arithmetical operations, but also for keeping a record of inventories and transactions of all kinds. If a check on past dealings was needed, the clay cases could be broken open. But the second development was even more pregnant. The idea was to symbolise on the outside of the clay case the objects that were enclosed within it: one notch on the case signified that there was one small cone inside, a pellet was symbolised by a small circular perforation, a large cone by a thick notch, a ball by a circle, and so on. Which is how the oldest numbers in history, the Sumerian numerals, came into being, around 3200 BCE.

This story is obviously related to the origins of writing, but it must not be confused with it entirely. Writing serves not only to give a visual representation to thought and a physical form to memory (a need felt by all advanced societies), but above all to record articulated speech.

## THE COMMON STRUCTURE OF THE HUMAN MIND

It is extraordinary to see how peoples very distant from each other in time and space used similar methods to reach identical results.

All societies learned to number their own bodies and to count on their fingers; and the use of pebbles, shells and sticks is absolutely universal. So the fact that the use of knotted string occurs in China, in Pacific island communities, in West Africa, and in Amerindian civilisations does not require us to speculate about migrations or long-distance travellers in prehistory. The making of notches to represent number is just as widespread in historical and geographical terms. Since the marking of bone and wood has the same physical requirements and limitations wherever it is done, it is no surprise that the same kinds of lines, Vs and Xs are to be seen on armbones and pieces of wood found in places as far apart as Europe, Asia, Africa, Oceania and the Americas. That is also why these marks crop up in virtually identical form in civilisations as varied as those of the Romans, the Chinese, the Khâs Boloven of Indo-China, the Zuñi Indians of New Mexico, and amongst contemporary Dalmatian and Celtic herdsmen. It is therefore not at all surprising that some numbers have almost always been represented by the same figure: 1, for instance, is represented almost universally by a single vertical line; 5 is also very frequently, though slightly less universally, figured by a kind of V in one orientation or another, and 10 by a kind of X or by a horizontal bar.

Similarly, the Ancient Egyptians, the Hittites, the Greeks, and the Aztecs worked out written numbering systems that were structurally identical, even if their respective base numbers and figurations varied considerably. Likewise the common system of Sumerian, Roman, Attic, and South Arabian numbering. Several family groupings of the same kind can be found in other sets of unrelated cultures. There is no need to hypothesise actual contact between the cultures in order to explain the similarities between their numbering systems.

So it would seem that human beings possess, in all places and at all times, a permanent capacity to repeat an invention or discovery already made elsewhere, provided only that the society or individual involved encounters cultural, social, and psychological conditions similar to those that prevailed when the invention was first made.

This is what explains why in modern science, the same discovery is sometimes made at almost the same time by two different scientists working in complete isolation from each other. Famous examples of such coincidences of invention include the simultaneous development of analytical geometry by Descartes and Fermat, of differential calculus by Newton and Leibnitz, of the physical laws of gasses by Boyle and Mariotte, and of the principles of thermodynamics by Joule, Mayer, and Sadi Carnot.

## NUMBERS AND LETTERS

Ever since the invention of alphabetic writing by the Phoenicians (or at least, by a northwestern Semitic people) in the second millennium BCE, letters have been used for numbers. The simplicity and ingenuity of the alphabetic system led to its becoming the most widespread form of writing, and the Phoenician scheme is at the root of nearly every alphabet in the world today, from Hebrew to Arabic, from Berber to Hindu, and of course Greek, which is the basis of our present (Latin) lettering.

Given their alphabets, the Greeks, the Jews, the Arabs and many other peoples thought of writing numbers by using letters. The system consists of attributing numerical values from 1 to 9, then in tens from 10 to 90, then in hundreds, etc., to the letters in their original Phoenician order (an order which has remained remarkably stable over the millennia).

Number-expressions constructed in this way worked as simple accumulations of the numerical values of the individual letters. The mathematicians of Ancient Greece rationalised their use of letter-numbers within a decimal system, and, by adding diacritic signs to the base numbers, became able to express numbers to several powers of 10.

In poetry and literature, however, and especially in the domains of magic, mysticism, and divination, it was the sum of the number-values of the letters in a word that mattered.

In these circumstances, every word acquired a number-value, and conversely, every number was "loaded" with the symbolic value of one or more words that it spelled. That is why the number 26 is a divine number in Jewish lore, since it is the sum of the number-values of the letter that spell YAHWEH, the name of God:

ה ו ה י = 5 + 6 + 5 + 10

The Jews, Greeks, Romans, Arabs (and as a result, Persians and Muslim Turks) pursued these kinds of speculation, which have very ancient origins: Babylonian writings of the second millennium BCE attribute a numerical value to each of the main gods: 60 was associated with Anu, god of the sky; 50 with Enlíl, god of the earth; 40 with Ea, god of water, and so forth.

The device also allowed poets like Leonidas of Alexandria to compose quite special kinds of work. It is also the basis for the art of the chronogram (verses that express a date simultaneously in words and in numbers) that can be found amongst the poets and stone-carvers of North Africa, Turkey, and Iran.

From ancient times to the present, the device has given a rich field to cabbalists, Gnostics, magicians, soothsayers, and mystics of every hue, and innumerable speculations, interpretations, calculations and predictions have been built on letter-number equivalences. The Gnostics, for example, thought they could work out the "formula" and thus the true name of God, which would enable them to penetrate all the secrets of the divine. Several religious sects are based on beliefs of this kind (such as the Hurufi or "Lettrists" of Islam) and they still have many followers, some of them in Europe.

The Greeks and Jews who first established a number-coded alphabet certainly could not have imagined that fifteen hundred or two thousand years later a Catholic theologian called Petrus Bungus would churn out a seven-hundred page numerological treatise "proving" (subject to a few spelling improvements!) that the name of Martin Luther added up to 666. It was a proof that the "isopsephic" initiates knew how to read, since according to St John the Apostle, 666 was the number of the "Beast of the Apocalypse", that is to say the Antichrist. Bungus was neither the first nor the last to make use of these methods. In the late Roman Empire, Christians tried to make Nero's name come to 666; during World War II, would-be numerological prophets managed to "prove" that Hitler was the real "Beast of the Apocalypse". A discovery that many had already made without the help of numbers.

## THE HISTORY OF A GREAT INVENTION

Logic was not the guiding light of the history of number-systems. They were invented and developed in response to the concerns of accountants, first of all, but also of priests, astronomers, and astrologers, and only in the last instance in response to the needs of mathematicians. The social categories dominant in this story are notoriously conservative, and they probably acted as a brake on the development and above all on the accessibility of numbering systems. After all, knowledge (however rudimentary it may now appear) gives its holders power and privilege; it must have seemed dangerous, if not irreligious, to share it with others.

There were also other reasons for the slow and fragmentary development of numbers. Whereas fundamental scientific research is pursued in terms of scientists' own criteria, inventions and discoveries only get developed and adopted if they correspond to a perceived social need in a civilisation. Many scientific advances are ignored if there is, as people say, no "call" for them.

The stages of mathematical thought make a fascinating story. Most peoples throughout history failed to discover the rule of position, which was discovered in fact only four times in the history of the world. (The rule of position is the principle of a numbering system in which a 9, let's say, has a different magnitude depending on whether it comes in first, second, third . . . position in a numerical expression.) The first discovery of this essential tool of mathematics was made in Babylon in the second millennium BCE. It was then rediscovered by Chinese arithmeticians at around the start of the Common Era. In the third to fifth centuries CE, Mayan astronomers reinvented it, and in the fifth century CE it was rediscovered for the last time, in India.

Obviously, no civilisation outside of these four ever felt the need to invent zero; but as soon as the rule of position became the basis for a numbering system, a zero was needed. All the same, only three of the four (the Babylonians, the Mayans and the Indians) managed to develop this final abstraction of number: the Chinese only acquired it through Indian influences. However, the Babylonian and Mayan zeros were not conceived of as numbers, and only the Indian zero had roughly the same potential as the one we use nowadays. That is because it is indeed the Indian zero, transmitted to us through the Arabs together with the number-symbols that we call Arabic numerals and which are in reality Indian numerals, with their appearance altered somewhat by time, use and travel.

Our knowledge of the history of numbers is of course only fragmentary, but all the pieces converge inexorably towards the system that we now use and which in recent times has conquered the whole planet.

## COMPUTATION, FIGURES, AND NUMBERS

Arithmetic has a history that is by no means limited to the history of the figures we use to represent numbers. In this history of computation, figures arose quite late on; and they constitute only one of many possible ways of representing number-concepts. The history of numbers ran parallel to the history of computation, became part of it only when modern written arithmetic was invented, and then separated out again with the development of modern calculating machines.

Numbers have become so integrated into our way of thinking that they often seem to be a basic, innate characteristic of human beings, like walking or speaking. But that is not so. Numbers belong to human culture, not nature, and therefore have their own long history. For Plato, numbers were "the highest degree of knowledge" and constituted the essence of outer and inner harmony. The same idea was taken up in the Middle Ages by Nicholas Cusanus, for whom "numbers are the best means of approaching divine truths". These views all go back to Pythagoras, for whom "numbers alone allow us to grasp the true nature of the universe".

In truth, though, it is not numbers that govern the universe. Rather, there are physical properties in the world which can be expressed in abstract terms through numbers. Numbers do not come from things themselves, but from the mind that studies things. Which is why the history of numbers is a profoundly human part of human history.

## IN CONCLUSION

Once a person's curiosity, on any subject, is aroused it is surprising just how far it may lead him in pursuit of its object, how readily it overcomes every obstacle. In my own case my curiosity about, or rather my absolute fascination with, numbers has been well served by a number of assets with which I set out: a Moroccan by birth, a Jew by cultural heritage, I have been afforded a more immediate access to the study of the work of Arab and Hebrew mathematicians than I might have obtained as a born European. I could harmonise within myself the mind-set of Eastern metaphysics with the Cartesian logic of the West. And I was able to identify the basic rules of a highly complex system. Moreover I possessed a sufficient aptitude for drawing to enable me to make simple illustrations to help clarify my text. I hope that the reader will recognise in this History that numbers, far from being tedious and dry, are charged with poetry, are the very vehicle for traditional myths and legends – and the finest witness to the cultural unity of the human race.

# THE UNIVERSAL HISTORY OF NUMBERS

CHAPTER 1

# EXPLAINING THE ORIGINS

## *Ethnological and Psychological Approaches to the Sources of Numbers*

### WHEN THE SLATE WAS CLEAN

There must have been a time when nobody knew how to count. All we can surmise is that the concept of number must then have been indissociable from actual objects – nothing very much more than a direct apperception of the plurality of things. In this picture of early humanity, no one would have been able to conceive of a number as such, that is to say as an abstraction, nor to grasp the fact that sets such as "day-and-night", a brace of hares, the wings of a bird, or the eyes, ears, arms and legs of a human being had a common property, that of "being two".

Mathematics has made such rapid and spectacular progress in what are still relatively recent periods that we may find it hard to credit the existence of a time without number. However, research into behaviour in early infancy and ethnographic studies of contemporary so-called primitive populations support such a hypothesis.

### CAN ANIMALS COUNT?

Some animal species possess some kind of notion of number. At a rudimentary level, they can distinguish concrete quantities (an ability that must be differentiated from the ability to count numbers in abstract). For want of a better term we will call animals' basic number-recognition the *sense of number*. It is a sense which human infants do not possess at birth.

Humans do not constitute the only species endowed with intelligence: the higher animals also have considerable problem-solving abilities. For example, hungry foxes have been seen to "play dead" so as to attract the crows they intend to eat. In Kenya, lions that previously hunted alone learned to hunt in a pack so as to chase prey towards a prepared ambush. Monkeys and other primates, of course, are not only able to make tools but also to learn how to manipulate non-verbal symbols. A much-quoted example of the first ability is that of the monkey who constructed a long bamboo tube so as to pick bananas that were out of reach. Chimpanzees have been taught to use tokens of different shapes to obtain bananas, grapes, water, and so on, and some even ended up hoarding the tokens against future needs. However, we must be careful not to be taken in by the kind of "animal intelligence" that you can see at the circus and the fairground. Dogs that can "count" are examples of effective training or (more likely) of clever trickery, not of the intellectual properties of canine minds. However, there are some very interesting cases of number-sense in the animal world.

Domesticated animals (for instance, dogs, cats, monkeys, elephants) notice straight away if one item is missing from a small set of familiar objects. In some species, mothers show by their behaviour that they know if they are missing one or more than one of their litter. A sense of number is marginally present in such reactions. The animal possesses a natural disposition to recognise that a small set seen for a second time has undergone a numerical change.

Some birds have shown that they can be trained to recognise more precise quantities. Goldfinches, when trained to choose between two different piles of seed, usually manage to distinguish successfully between three and one, three and two, four and two, four and three, and six and three.

Even more striking is the untutored ability of nightingales, magpies and crows to distinguish between concrete sets ranging from one to three or four. The story goes that a squire wanted to destroy a crow that had made its nest in his castle's watchtower. Each time he got near the nest, the crow flew off and waited on a nearby branch for the squire to give up and go down. One day the squire thought of a trick. He got two of his men to go into the tower. After a few minutes, one went down, but the other stayed behind. But the crow wasn't fooled, and waited for the second man to go down too before coming back to his nest. Then they tried the trick with three men in the tower, two of them going down: but the third man could wait as long as he liked, the crow knew that he was there. The ploy only worked when five or six men went up, showing that the crow could not discriminate between numbers greater than three or four.

These instances show that some animals have a potential which is more fully developed in humans. What we see in domesticated animals is a rudimentary perception of equivalence and non-equivalence between sets, but only in respect of numerically small sets. In goldfinches, there is something more than just a perception of equivalence – there seems to be a sense of "more than" and "less than". Once trained, these birds seem to have a perception of intensity, halfway between a perception of quantity (which requires an ability to numerate beyond a certain point) and a perception of quality. However, it only works for goldfinches when the "moreness" or "lessness" is quite large; the bird will almost always confuse five and four, seven and five, eight and six, ten and six. In other words, goldfinches can recognise differences of intensity if they are large enough, but not otherwise.

Crows have rather greater abilities: they can recognise equivalence and non-equivalence, they have considerable powers of memory, and they can perceive the relative magnitudes of two sets of the same kind separated in time and space. Obviously, crows do not count in the sense that we do, since in the absence of any generalising or abstracting capacity they cannot conceive of any "absolute quantity". But they do manage to distinguish concrete quantities. They do therefore seem to have a basic number-sense.

## NUMBERS AND SMALL CHILDREN

Human infants have few innate abilities, but they do possess something that animals never have: a potential to assimilate and to recreate stage by stage the conquests of civilisation. This inherited potential is only brought out by the training and education that the child receives from the adults and other children in his or her environment. In the absence of permanent contact with a social milieu, this human potential remains undeveloped – as is shown by the numerous cases of *enfants sauvages*. (These are children brought up by or with animals in the wild, as in François Truffaut's film, *The Wild Child*. Of those recaptured, none ever learned to speak and most died in adolescence.)

We should not imagine a child as a miniature adult, lacking only judgement and knowledge. On the contrary, as child psychology has shown, children live in their own worlds, with distinct mentalities obeying their own specific laws. Adults cannot actually enter this world, cannot go back to their own beginnings. Our own childhood memories are illusions, reconstructions of the past based on adult ways of thinking.

But infancy is nonetheless the necessary prerequisite for the eventual transformation of the child into an adult. It is a long-drawn-out phase of preparation, in which the various stages in the development of human intelligence are re-enacted and reconstitute the successive steps through which our ancestors must have gone since the dawn of time.

According to N. Sillamy (1967), three main periods are distinguished: *infancy* (up to three years of age), *middle childhood* (from three to six or seven); and *late childhood*, which ends at puberty. However, a child's intellectual and emotional growth does not follow a steady and linear pattern. Piaget (1936) distinguishes five well-defined phases:

1. a *sensory-motor period* (up to two years of age) during which the child forms concepts of "object" out of fragmentary perceptions and the concept of "self" as distinct from others;
2. a *pre-operative stage* (from two to four years of age), characterised by egocentric and anthropomorphic ways of thinking ("look, mummy, the moon is following me!");
3. an *intuitive period* (from four to six), characterised by intellectual perceptions unaccompanied by reasoning; the child performs acts which he or she would be incapable of deducing, for example, pouring a liquid from one container into another of a different shape, whilst believing that the volume also changes;
4. a stage of *concrete operations* (from eight to twelve) in which, despite acquiring some operational concepts (such as class, series, number, causality), the child's thought-processes remain firmly bound to the concrete;
5. a period (around puberty) characterised by the emergence of *formal operations*, when the child becomes able to make hypotheses and test them, and to operate with abstract concepts.

Even more precisely: the new-born infant in the cradle perceives the world solely as variations of light and sound. Senses of touch, hearing and sight slowly grow more acute. From six to twelve months, the infant acquires some overall grasp of the space occupied by the things and people in its immediate environment. Little by little the child begins to make associations and to perceive differences and similarities. In this way the child forms representations of relatively simple groupings of beings and objects which are familiar both by nature and in number. At this age, therefore, the child is able to reassemble into one group a set of objects which have previously been moved apart. If one thing is missing from a familiar set of objects, the child immediately notices. But the abstraction of number – which the child simply feels, as if it were a feature of the objects themselves – is beyond the child's grasp. At this age babies do not use their fingers to indicate a number.

Between twelve and eighteen months, the infant progressively learns to distinguish between one, two and several objects, and to tell at a glance the relative sizes of two small collections of things. However, the infant's numerical capabilities still remain limited, to the extent that no clear distinction is made between the numbers and the collections that they represent. In other words, until the child has grasped the generic principle of the natural numbers ($2 = 1 + 1$; $3 = 2 + 1$; $4 = 3 + 1$, etc.), numbers remain nothing more than "number-groupings", not separable from the concrete nature of the items present, and they can only be recognised by the principle of *pairing* (for instance, on seeing two sets of objects lined up next to each other).

Oddly enough, when a child has acquired the use of speech and learned to name the first few numbers, he or she often has great difficulty in symbolising the number three. Children often count from one to two and then miss three, jumping straight to four. Although the child can recognise, visually and intuitively, the concrete quantities from one to four, at this stage of development he or she is still at the very doorstep of abstract numbering, which corresponds to *one, two, many*.

However, once this stage is passed (at between three and four years of age, according to Piaget), the child quickly becomes able to count properly. From then on, progress is made by virtue of the fact that the abstract concept of number progressively takes over from the purely perceptual aspect of a collection of objects. The road lies open which leads on to the acquisition of a true grasp of abstract calculation. For this reason, teachers call this phase the "pre-arithmetical stage" of intellectual development. The child will first learn to count up to ten, relying heavily on the use of fingers; then the number series is progressively extended as the capacity for abstraction increases.

## ARITHMETIC AND THE BODY

The importance of the hand, and more generally of the body in children's acquisition of arithmetic can hardly be exaggerated. Inadequate access to or use of this "counting instrument" can cause serious learning difficulties.

> In earliest infancy, the child plays with his or her fingers. It constitutes the first notion of the child's own body. Then the child touches everything in order to make acquaintance with the world, and this also is done primarily with the hands. One day, a well-intentioned teacher who wanted arithmetic to be "mental", forbade finger-counting in his class. Without realising it, the teacher had denied the children the use of their bodies, and forbidden the association of mathematics with their bodies. I've seen many children profoundly relieved to be able to use their hands again: their bodies were at last accepted [ . . . ] Spatio-temporal disabilities can likewise make learning mathematics very difficult. Inadequate grasp of the notions of "higher than" and "lower than" affect the concepts of number, and all operations and relations between them. The unit digits are written to the right, and the hundred digits are written to the left, so a child who cannot tell left from right cannot write numbers properly or begin an operation at all easily. Number skills and the whole set of logical operations of arithmetic can thus be seriously undermined by failure to accept the body. [L. Weyl-Kailey (1985)]

## NUMBERS AND THE PRIMITIVE MIND

A good number of so-called primitive people in the world today seem similarly unable to grasp number as an abstract concept. Amongst these populations, number is "felt" and "registered", but it is perceived as a *quality*, rather as we perceive smell, colour, noise, or the presence of a person or thing outside of ourselves. In other words, "primitive" peoples are affected only by changes in their visual field, in a direct subject-object relationship. Their grasp of number is thus limited to what their predispositions allow them to see in a single visual glance.

However, that does not mean that they have no perception of quantity. It is just that the plurality of beings and things is measured by them not in a quantitative but in a qualitative way, without differentiating individual items. Cardinal reckoning of this sort is never fixed in the abstract, but always related to concrete sets, varying naturally according to the type of set considered.

> A well-defined and appropriately limited set of things or beings, provided it is of interest to the primitive observer, will be memorised with all its characteristics. In the primitive's mental representation of it, the exact number of the things or beings involved is implicit: it resembles a quality by which this set is different from another group consisting of one or several more or fewer members. Consequently, when he sets eyes on the set for a second time, the primitive knows if it is complete or if it is larger or smaller than it was previously. [L. Lévy-Bruhl (1928)]

## ONE, TWO . . . MANY

In the first years of the twentieth century, there were several "primitive" peoples still at this basic stage of numbering: Bushmen (South Africa), Zulus (South and Central Africa), Pygmies (Central Africa), Botocudos (Brazil), Fuegians (South America), the Kamilarai and Aranda peoples in Australia, the natives of the Murray Islands, off Cape York (Australia), the Vedda (Sri Lanka), and many other "traditional" communities.

According to E. B. Tylor (1871), the Botocudos had only two real terms for numbers: one for "one", and the other for "a pair". With these lexical items they could manage to express three and four by saying something like "one and two" and "two and two". But these people had as much difficulty conceptualising a number above four as it is for us to imagine quantities of a trillion billions. For larger numbers, some of the Botocudos just pointed to their hair, as if to say "there are as many as there are hairs on my head".

A. Sommerfelt (1938) similarly reports that the Aranda had only two number-terms, *ninta* (one), and *tara* (two). Three and four were expressed as *tara-mi-ninta* (one and two) and *tara-ma-tara* ("two and two"), and the number series of the Aranda stopped there. For larger quantities, imprecise terms resembling "a lot", "several" and so on were used.

Likewise G. Hunt (1899) records the Murray islanders' use of the terms *netat* and *neis* for "one" and "two", and the expressions *neis-netat* (two + one) for "three", and *neis-neis* (two + two) for "four". Higher numbers were expressed by words like "a crowd of . . ."

Our final example is that of the Torres Straits islanders for whom *urapun* meant "one", *okosa* "two", *okosa-urapun* (two-one) "three", and *okosa-okosa* (two-two) "four". According to A. C. Haddon (1890) these were the only terms used for absolute quantities; other numbers were expressed by the word *ras*, meaning "a lot".

Attempts to teach such communities to count and to do arithmetic in the Western manner have frequently failed. There are numerous accounts of natives' lack of memory, concentration and seriousness when confronted with numbers and sums [see, for example, M. Dobrizhoffer (1902)]. It generally turned out much easier to teach primitive peoples the arts of music, painting, and sculpture than to get them to accept the interest and importance of arithmetic. This was perhaps not just because primitive peoples felt no need of counting, but also because numbers are amongst the most abstract concepts that humanity has yet devised. Children take longer to learn to do sums than to speak or to write. In the history of humanity, too, numbers have proved to be the hardest of these three skills.

## PARITY BEFORE NUMBER

These primitive peoples nonetheless possessed a fundamental arithmetical rule which if systematically applied would have allowed them to manipulate numbers far in excess of four. The rule is what we call the *principle of base 2* (or binary principle). In this kind of numbering, five is "two-two-one", six is "two-two-two", seven is "two-two-two-one", and so on. But primitive societies did not develop binary numbering because, as L. Gerschel (1960) reminds us, they possessed only the most basic degree of numeracy, that which distinguishes between the singular and the dual.

A. C. Haddon (1890), observing the western Torres Straits islanders, noted that they had a pronounced tendency to count things in groups of two or in couples. M. Codrington, in *Melanesian Languages*, noticed the same thing in many Oceanic populations: "The natives of Duke of York's Island count in couples, and give the pairings different names depending how many of them there are; whereas in Polynesia, numbers are used although it is understood that they refer to so many pairs of things, not to so many things." Curr, as quoted by T. Dantzig (1930), confirms that Australian aborigines also counted in this way, to the extent that "if two pins are removed from a set of seven the aborigines rarely notice it, but they see straight away if only one is removed".

These primitive peoples obviously had a stronger sense of parity than of number. To express the numbers three and four, numbers they did not grasp as abstracts but which common sense allowed them to see in a single glance, they had recourse only to concepts of *one* and *pair*. And so for them groups like "two-one" or "two-two" were themselves pairs, not (as for us) the abstract integers (or "whole numbers") "three" and "four". So it is easy to see why they never developed the binary system to get as far as five and six, since these would have required three digits, one more than the pair which was their concept of the highest abstract number.

## THE LIMITS OF PERCEPTION

The limited arithmetic of "primitive" societies does not mean that their members were unintelligent, nor that their innate abilities were or are lesser than ours. It would be a grave error to think that we could do better than a Torres Straits islander at recognising number if all we had to use were our natural faculties of perception.

In practice, when we want to distinguish a quantity we have recourse to our memories and/or to acquired techniques such as comparison, splitting, mental grouping, or, best of all, actual counting. For that reason it is rather difficult to get to our natural sense of number. There is an exercise that we can try, all the same. Looking at Fig. 1.1, which contains sets of objects *in line*, try to estimate the quantity of each set of objects in a single visual glance (that is to say, *without* counting). What is the best that we can do?

FIG. 1.1.

Everyone can see the sets of one, of two, and of three objects in the figure, and most people can see the set of four. But that's about the limit of our natural ability to numerate. Beyond four, quantities are vague, and our eyes alone cannot tell us how many things there are. Are there fifteen or twenty plates in that pile? Thirteen or fourteen cars parked along the street? Eleven or twelve bushes in that garden, ten or fifteen steps on this staircase, nine, eight or six windows in the façade of that house? The correct answers cannot be just seen. *We have to count to find out!*

The eye is simply not a sufficiently precise measuring tool: its natural number-ability virtually never exceeds four.

There are many traces of the "limit of four" in different languages and cultures. There are several Oceanic languages, for example, which distinguish between nouns in the singular, the dual, the triple, the quadruple, and the plural (as if in English we were to say *one bird, two birdo, three birdi, four birdu, many birds*).

In Latin, the names of the first four numbers (*unus, duos, tres, quatuor*) decline at least in part like other nouns and adjectives, but from five (*quinque*), Latin numerical terms are invariable. Similarly, Romans gave "ordinary" names to the first four of their sons (names like Marcus, Servius, Appius, etc.), but the fifth and subsequent sons were named only by a numeral: Quintus (the fifth), Sixtus (the sixth), Septimus (the seventh), and so on. In the original Roman calendar (the so-called "calendar of Romulus"), only the first four months had names (Martius, Aprilis, Maius, Junius), the fifth to tenth being referred to by their order-number: Quintilis, Sextilis, September, October, November, December.*

Perhaps the most obvious confirmation of the basic psychological rule of the "limit of four" can be found in the almost universal counting-device called (in England) the "five-barred gate". It is used by innkeepers keeping a tally or "slate" of drinks ordered, by card-players totting up scores, by prisoners keeping count of their days in jail, even by examiners working out the mark-distribution of a cohort of students:

| | | | | | |
|---|---|---|---|---|---|
| 1 | I | 6 | ~~IIII~~ I | 11 | ~~IIII~~ ~~IIII~~ I |
| 2 | II | 7 | ~~IIII~~ II | 12 | ~~IIII~~ ~~IIII~~ II |
| 3 | III | 8 | ~~IIII~~ III | 13 | ~~IIII~~ ~~IIII~~ III |
| 4 | IIII | 9 | ~~IIII~~ IIII | 14 | ~~IIII~~ ~~IIII~~ IIII |
| 5 | ~~IIII~~ | 10 | ~~IIII~~ ~~IIII~~ | 15 | ~~IIII~~ ~~IIII~~ ~~IIII~~ |

FIG. 1.2. *The five-barred gate*

* The original ten-month Roman calendar had 304 days and began with *Martius*. It was subsequently lengthened by the addition of two further months, *Januarius* and *Februarius* (our January and February). Julius Caesar further reformed the calendar, taking the start of the year back to 1 January and giving it 365 days in all. Later, the month of *Quintilis* was renamed *Julius* (our July) in honour of Caesar, and *Sextilis* became *Augustus* in honour of the emperor of that name.

Most human societies the world has known have used this kind of number-notation at some stage in their development and all have tried to find ways of coping with the unavoidable fact that beyond four (IIII) nobody can "read" intuitively a sequence of five strokes (IIIII) or more.

ARAMAIC (Egypt)
Elephantine script: 5th to 3rd centuries BCE

| 1 | 2 | 3 | 4 | 5 | 6 | 7 | 8 | 9 |
|---|---|---|---|---|---|---|---|---|

FIG. 1.3.

ARAMAIC (Mesopotamia)
Khatra script: First decades of CE

| 1 | 2 | 3 | 4 | 5 | 6 | 7 | 8 | 9 |
|---|---|---|---|---|---|---|---|---|

FIG. 1.4.

ARAMAIC (Syria)
Palmyrenean script: First decades of CE

| 1 | 2 | 3 | 4 | 5 | 6 | 7 | 8 | 9 |
|---|---|---|---|---|---|---|---|---|

FIG. 1.5.

CRETAN CIVILISATION
Hieroglyphic script: first half of second millennium BCE

| 1 | 2 | 3 | 4 | 5 | 6 | 7 | 8 | 9 |
|---|---|---|---|---|---|---|---|---|

FIG. 1.6.

CRETAN CIVILISATION
Linear script: 1700–1200 BCE

| 1 | 2 | 3 | 4 | 5 | 6 | 7 | 8 | 9 |
|---|---|---|---|---|---|---|---|---|

FIG. 1.7.

EGYPT
Hieroglyphic script: third to first millennium BCE

| 1 | 2 | 3 | 4 | 5 | 6 | 7 | 8 | 9 |
|---|---|---|---|---|---|---|---|---|

Fig. 1.8.

ELAM
"Proto-Elamite" script: Iran, first half of third millennium BCE

| 1 | 2 | 3 | 4 | 5 | 6 | 7 | 8 | 9 |
|---|---|---|---|---|---|---|---|---|

Fig. 1.9.

ETRUSCAN CIVILISATION
Italy, 6th to 4th centuries BCE

| 1 | 2 | 3 | 4 | 5 | 6 | 7 | 8 | 9 |
|---|---|---|---|---|---|---|---|---|

Fig. 1.10.

GREECE
Epidaurus and Argos, 5th to 2nd centuries BCE

| 1 | 2 | 3 | 4 | 5 | 6 | 7 | 8 | 9 |
|---|---|---|---|---|---|---|---|---|

Fig. 1.11.

GREECE
Taurian Chersonesus, Chalcidy, Troezen, 5th to 2nd centuries BCE

| 1 | 2 | 3 | 4 | 5 | 6 | 7 | 8 | 9 |
|---|---|---|---|---|---|---|---|---|

*π, initial of pente, five

Fig. 1.12.

GREECE
Thebes, Karistos, 5th to 1st centuries BCE

| 1 | 2 | 3 | 4 | 5 | 6 | 7 | 8 | 9 |
|---|---|---|---|---|---|---|---|---|

*π, initial of pente, five

Fig. 1.13.

INDUS CIVILISATION
2300–1750 BCE

| 1 | 2 | 3 | 4 | 5 | 6 | 7 | 8 | 9 |
|---|---|---|---|---|---|---|---|---|

Fig. 1.14.

HITTITE CIVILISATION
Hieroglyphic: Anatolia, 1500–800 BCE

| 1 | 2 | 3 | 4 | 5 | 6 | 7 | 8 | 9 |
|---|---|---|---|---|---|---|---|---|

Fig. 1.15.

LYCIAN CIVILISATION
Asia Minor, first half of first millennium BCE

| 1 | 2 | 3 | 4 | 5 | 6 | 7 | 8 | 9 |
|---|---|---|---|---|---|---|---|---|

Fig. 1.16.

LYDIAN CIVILISATION
Asia Minor, 6th to 4th centuries BCE

| 1 | 2 | 3 | 4 | 5 | 6 | 7 | 8 | 9 |
|---|---|---|---|---|---|---|---|---|

Fig. 1.17.

MAYAN CIVILISATION
Pre-Columbian Central America, 3rd to 14th centuries CE

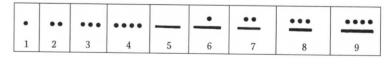

| 1 | 2 | 3 | 4 | 5 | 6 | 7 | 8 | 9 |
|---|---|---|---|---|---|---|---|---|

Fig. 1.18.

MESOPOTAMIA
Archaic Sumerian, beginning of third millennium BCE

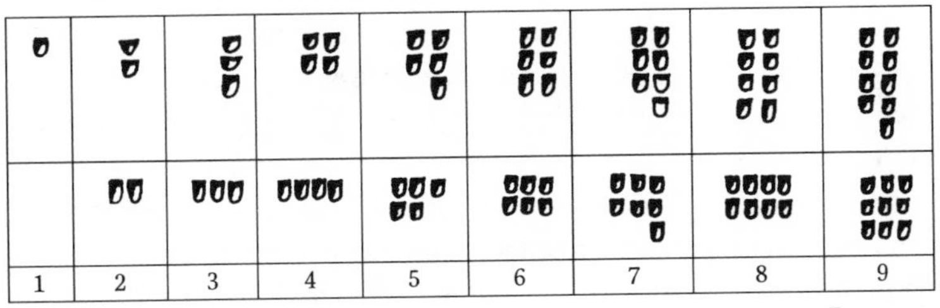

| 1 | 2 | 3 | 4 | 5 | 6 | 7 | 8 | 9 |
|---|---|---|---|---|---|---|---|---|

Fig. 1.19.

MESOPOTAMIA
Sumerian cuneiform, 2850–2000 BCE

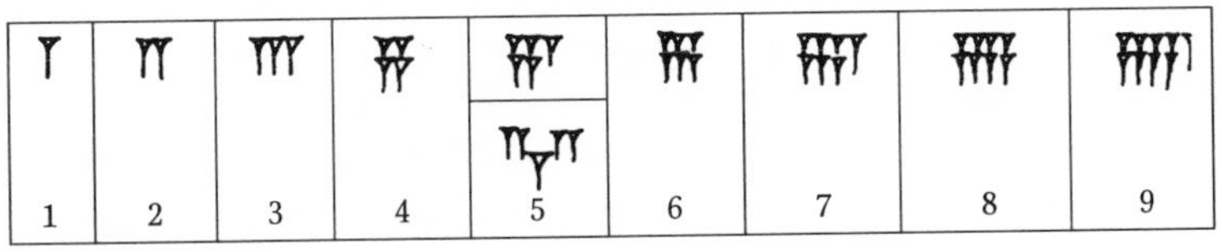

| 1 | 2 | 3 | 4 | 5 | 6 | 7 | 8 | 9 |
|---|---|---|---|---|---|---|---|---|

Fig. 1.20.

MESOPOTAMIA
Assyro-Babylonian cuneiform, second to first millennium BCE

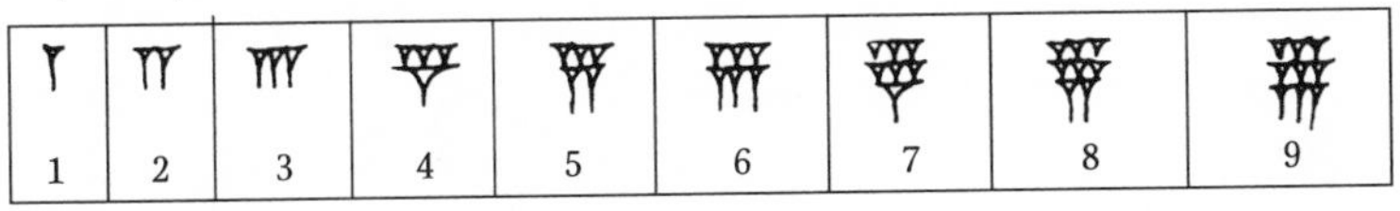

| 1 | 2 | 3 | 4 | 5 | 6 | 7 | 8 | 9 |
|---|---|---|---|---|---|---|---|---|

Fig. 1.21.

CIVILISATIONS OF MA'IN & SABA (SHEBA)
Southern Arabia, 5th to 1st centuries BCE

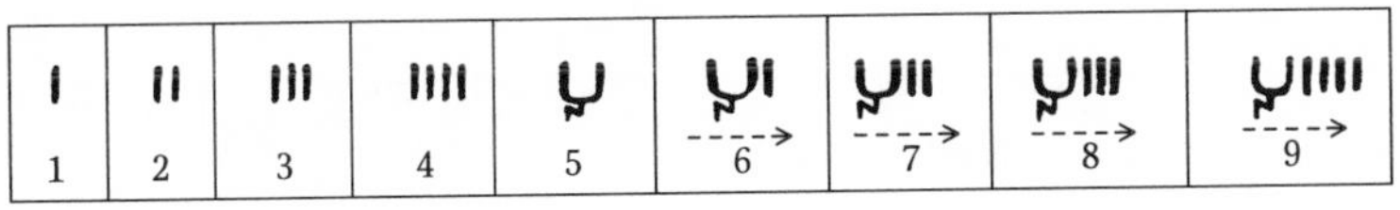

| 1 | 2 | 3 | 4 | 5 | 6 | 7 | 8 | 9 |
|---|---|---|---|---|---|---|---|---|

Fig. 1.22.

PHOENICIAN CIVILISATION
From 6th century BCE

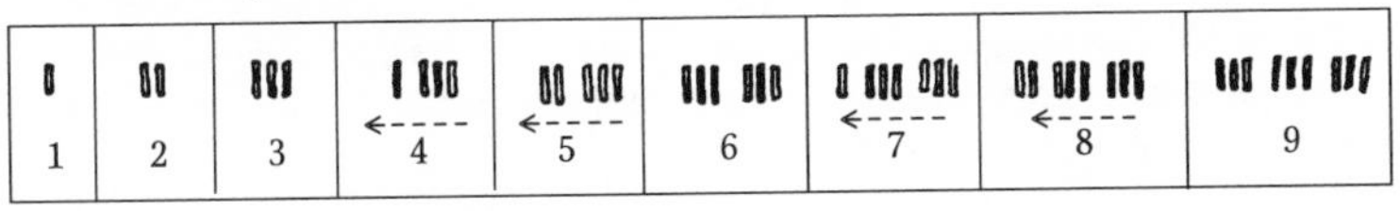

| 1 | 2 | 3 | 4 | 5 | 6 | 7 | 8 | 9 |
|---|---|---|---|---|---|---|---|---|

Fig. 1.23.

URARTU
Hieroglyphic script, Armenia, 13th to 9th centuries BCE

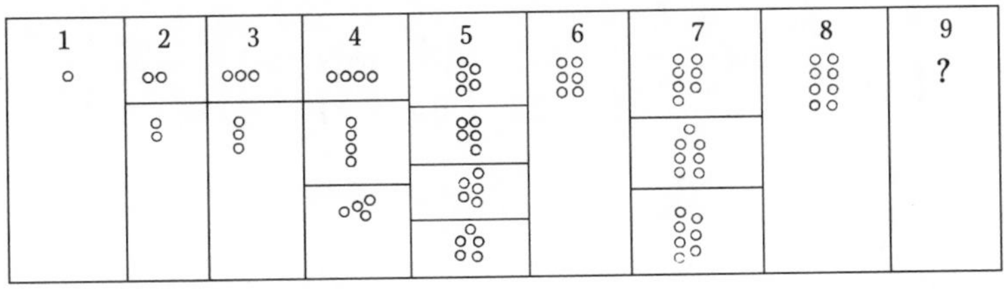

| 1 | 2 | 3 | 4 | 5 | 6 | 7 | 8 | 9 |
|---|---|---|---|---|---|---|---|---|
| | | | | | | | | ? |

Fig. 1.24.

To recapitulate: at the start of this story, people began by counting the first nine numbers by placing in sequence the corresponding number of strokes, circles, dots or other similar signs representing "one", more or less as follows:

| I | II | III | IIII | IIIII | IIIIII | IIIIIII | IIIIIIII | IIIIIIIII |
|---|---|---|---|---|---|---|---|---|
| 1 | 2 | 3 | 4 | 5 | 6 | 7 | 8 | 9 |

Fig. 1.25.

But because series of identical signs are not easy to read quickly for numbers above four, the system was rapidly abandoned. Some civilisations (such as those found in Egypt, Sumer, Elam, Crete, Urartu, and Greece) got round the difficulty by grouping the signs for numbers from five to nine to 9 according to a principle that we might call *dyadic representation:*

| I | II | III | IIII | III<br>II | III<br>III | IIII<br>III | IIII<br>IIII | IIIII<br>IIII |
|---|---|---|---|---|---|---|---|---|
| 1 | 2 | 3 | 4 | 5 | 6 | 7 | 8 | 9 |
| | | | | (3 + 2) | (3 + 3) | (4 + 3) | (4 + 4) | (5 + 4) |

Fig. 1.26.

Other civilisations, such as the Assyro-Babylonian, the Phoenician, the Egyptian-Aramaean and the Lydian, solved the problem by recourse to a *rule of three*:

| I | II | III | III<br>I | III<br>II | III<br>III | III<br>III<br>I | III<br>III<br>II | III<br>III<br>III |
|---|---|---|---|---|---|---|---|---|
| 1 | 2 | 3 | 4 | 5 | 6 | 7 | 8 | 9 |
| | | | (3 + 1) | (3 + 2) | (3 + 3) | (3 + 3 + 1) | (3 + 3 + 2) | (3 + 3 + 3) |

Fig. 1.27.

And yet others, like the Greeks, the Manaeans and Sabaeans, the Lycians, Mayans, Etruscans and Romans, came up with an idea (probably based on finger-counting) for a special sign for the number five, proceeding thereafter on a *rule of five* or quinary system (6 = 5 + 1, 7 = 5 + 2, and so on).

There really can be no debate about it now: *natural human ability to perceive number does not exceed four!*

So the basic root of arithmetic as we know it today is a very rudimentary numerical capacity indeed, a capacity barely greater than that of some animals. There's no doubt that the human mind could no more accede *by innate aptitude alone* to the abstraction of counting than could crows or goldfinches. But human societies have enlarged the potential of these very limited abilities by inventing a number of mental procedures of enormous fertility, procedures which opened up a pathway into the universe of numbers and mathematics . . .

## DEAD RECKONING

Since we can discriminate unreflectingly between concrete quantities only up to four, we cannot have recourse only to our natural sense of number to get to any quantity greater than four. We must perforce bring into play the device of abstract counting, the characteristic quality of "civilised" humanity.

But is it therefore the case that, in the absence of this mental device for counting (in the way we now understand the term), the human mind is so enfeebled that it cannot engage in any kind of numeration at all?

It is certainly true that without the abstractions that we call "one", "two", "three", and so on it is not easy to carry out mental operations. But it does not follow at all that a mind without numbers of our kind is incapable of devising specific tools for manipulating quantities in concrete sets. There are very good reasons for thinking that for many centuries people were able to reach several numbers without possessing anything like number-concepts.

There are many ethnographic records and reports from various parts of Africa, Oceania and the Americas showing that numerous contemporary "primitive" populations have numerical techniques that allow them to carry out some "operations", at least to some extent.

These techniques, which, in comparison to our own, could be called "concrete", enable such peoples to reach the same results as we would, by using *mediating objects* or *model collections* of many different kinds (pebbles, shells, bones, hard fruit, dried animal dung, sticks, the use of notched bones or sticks, etc.). The techniques are much less powerful and often more complicated than our own, but they are perfectly serviceable for establishing (for example) whether as many head of cattle have returned from grazing as went out of the cowshed. You do not need to be able to count by numbers to get the right answer for problems of that kind.

## ELEMENTARY ARITHMETIC

It all started with the device known as "one-for-one correspondence". This allows even the simplest of minds to compare two collections of beings or things, of the same kind or not, without calling on an ability to count in numbers. It is a device which is both the prehistory of arithmetic, and the dominant mode of operation in all contemporary "hard" sciences.

Here is how it works: You get on a bus and you have before you (apart from the driver, who is in a privileged position) two sets: a set of seats and a set of passengers. In one glance you can tell whether the two sets have "the same number" of elements; and, if the two sets are not equal, you can tell just as quickly which is the larger of the two. This ready-reckoning of number without recourse to numeration is more easily explained by the device of one-for-one correspondence.

If there was no one standing in the bus and there were some empty seats, you would know that each passenger has a seat, but that each seat does not necessarily have a passenger: therefore, there are fewer passengers than seats. In the contrary case – if there are people standing and all the seats are taken – you know that there are more passengers than seats. The third possibility is that there is no one standing and all seats are taken: as each seat corresponds to one passenger, there are as many passengers as seats. The last situation can be described by saying that there is a mapping (or a *biunivocal correspondence*, or, in terms of modern mathematics, a *bijection*) between the number of seats and the number of passengers in the bus.

At about fifteen or sixteen months, infants go beyond the stage of simple observation of their environment and become capable of grasping the principle of one-for-one correspondence, and in particular the property of mapping. If we give a baby of this age equal numbers of dolls and little chairs, the infant will probably try to fit one doll on each seat. This kind of play is nothing other than mapping the elements of one set (dolls) onto the elements of a second set (chairs). But if we set out more dolls than chairs (or more chairs than dolls), after a time the baby will begin to fret: it will have realised that the mapping isn't working.

FIG. 1.28. *Two sets map if for each element of one set there is a corresponding single element of the other, and vice versa.*

This mental device does not only provide a means for comparing two groups, but it also allows its user to manipulate several numbers without knowing how to count or even to name the quantities involved.

If you work at a cinema box-office you usually have a seating plan of the auditorium in front of you. There is one "box" on the plan for each seat in the auditorium, and, each time you sell a ticket, you cross out one of the boxes on the plan. What you are doing is: mapping the seats in the cinema onto the boxes on the seating plan, then mapping the boxes on the plan onto the tickets sold, and finally, mapping the tickets sold onto the number of people allowed into the auditorium. So even if you are too lazy to add up the number of tickets you've sold, you'll not be in any doubt about knowing when the show has sold out.

To recite the attributes of Allah or the obligatory laudations after prayers, Muslims habitually use a string of prayer-beads, each bead corresponding to one divine attribute or to one laudation. The faithful "tell their beads" by slipping a bead at a time through their fingers as they proceed through the recitation of eulogies or of the attributes of Allah.

FIG. 1.29. *Muslim prayer-beads (*subha *or* sebha *in Arabic) used for reciting the 99 attributes of Allah or for supererogatory laudations. This indispensable piece of equipment for pilgrims and dervishes is made of wooden, mother-of-pearl or ivory beads that can be slipped through the fingers. It is often made up of three groups of beads, separated by two larger "marker" beads, with an even larger bead indicating the start. There are usually a hundred beads on a string (33 + 33 + 33 + 1), but the number varies.*

Buddhists have also used prayer-beads for a very long time, as have Catholics, for reciting *Pater noster*, *Ave Maria*, *Gloria Patri*, etc. As these litanies must be recited several times in a quite precise order and number, Christian rosaries usually consist of a necklace threaded with five times ten small beads, each group separated by a slightly larger bead, together with a chain bearing one large then three small beads, then one large bead and a cross. That is how the litanies can be recited without counting but without omission – each small bead on the ring corresponds to one *Ave Maria*, with a *Gloria Patri* added on the last bead of each set of ten, and a *Pater noster* is said for each large bead, and so on.

The device of one-for-one correspondence has thus allowed these religions to devise a system which ensures that the faithful do not lose count of their litanies despite the considerable amount of repetition required. The device can thus be of use to the most "civilised" of societies; and for the completely "uncivilised" it is even more valuable.

Let us take someone with no arithmetical knowledge at all and send him to the grocery store to get ten loaves of bread, five bottles of cooking oil, and four bags of potatoes. With no ability to count, how could this person be trusted to bring back the correct amount of change? But in fact such a person is perfectly capable of carrying out the errand provided the proper equipment is available. The appropriate kit is necessarily based on the principle of one-for-one correspondence. We could make ten purses out of white cloth, corresponding to the ten loaves, five yellow purses for the bottles of cooking oil, and four brown purses, for the bags of potatoes. In each purse we could put the exact price of the corresponding item of purchase, and all the uneducated shopper needs to know is that a white purse can be exchanged for a loaf, a yellow one for a bottle of oil and a brown one for a bag of potatoes.

This is probably how prehistoric humanity did arithmetic for many millennia, before the first glimmer of arithmetic or of number-concepts arose.

Imagine a shepherd in charge of a flock of sheep which is brought back to shelter every night in a cave. There are fifty-five sheep in this flock. But the shepherd doesn't know that he has fifty-five of them since he does not know the number "55": all he knows is that he has "many sheep". Even so, he wants to be sure that all his sheep are back in the cave each night. So he has an idea – the idea of a concrete device which prehistoric humanity used for many millennia. He sits at the mouth of his cave and lets the animals in one by one. He takes a flint and an old bone, and cuts a notch in the bone for every sheep that goes in. So, without realising the mathematical meaning of it, he has made exactly fifty-five incisions on the bone by the time the last animal is inside the cave. Henceforth the shepherd can check whether any sheep in his flock are missing. Every time he comes back from grazing, he lets the sheep into the cave one by one, and moves his finger over one indentation in the tally stick for each one. If there are any marks left on the bone after the last sheep is in the cave, that means he has lost some sheep. If not, all is in order. And if meanwhile a new lamb comes along, all he has to do is to make another notch in the tally bone.

So thanks to the principle of one-for-one correspondence it is possible to manage to count even in the absence of adequate words, memory or abstraction.

One-for-one mapping of the elements of one set onto the elements of a second set creates an abstract idea, entirely independent of the type or nature of the things or beings in the one or other set, which expresses a property common to the two sets. In other words, mapping abolishes the distinction that holds between two sets by virtue of the type or nature of the elements that constitute them. This abstract property is precisely why one-for-one mapping is a significant tool for tasks involving enumeration; but in practice, the methods that can be based on it are only suitable for relatively small sets.

This is why *model collections* can be very useful in this domain. Tally sticks with different numbers of marks on them constitute so to speak a range of *ready-made mappings* which can be referred to independently of the type or nature of the elements that they originally referred to. A stick of ivory or wood with twenty notches on it can be used to enumerate twenty men, twenty sheep or twenty goats just as easily as it can be used for twenty bison, twenty horses, twenty days, twenty pelts, twenty kayaks, or twenty measures of grain. The only number technique that can be built on this consists of choosing the most appropriate tally stick from the ready-mades so as to obtain a one-to-one mapping on the set that you next want to count.

However, notched sticks are not the only concrete *model collection*s available for this kind of matching-and-counting. The shepherd of our example could also have used pebbles for checking that the same number of sheep come into the cave every evening as went out each morning. All he needs to do to use this device would be to associate one pebble with each head of sheep, to put the resulting pile of pebbles in a safe place, and then to count them out in a reverse procedure on returning from the pasture. If the last animal in matches the last pebble in the pile, then the shepherd knows for sure that none of his flock has been lost, and if a lamb has been born meanwhile, all he needs to do is to add a pebble to the pile.

All over the globe people have used a variety of objects for this purpose: shells, pearls, hard fruit, knucklebones, sticks, elephant teeth, coconuts, clay pellets, cocoa beans, even dried dung, organised into heaps or lines corresponding in number to the tally of the things needing to be checked. Marks made in sand, and beads and shells, strung on necklaces or made into rosaries, have also been used for keeping tallies.

Even today, several "primitive" communities use parts of the body for this purpose. Fingers, toes, the articulations of the arms and legs (elbow, wrist, knee, ankle . . . ), eyes, nose, mouth, ears, breasts, chest, sternum, hips and so on are used as the reference elements of one-for-one counting systems. Much of the evidence comes from the Cambridge Anthropological Expedition to Oceania at the end of the last century. According to Wyatt Gill, some Torres Straits islanders "counted visually" (see Fig. 1.30):

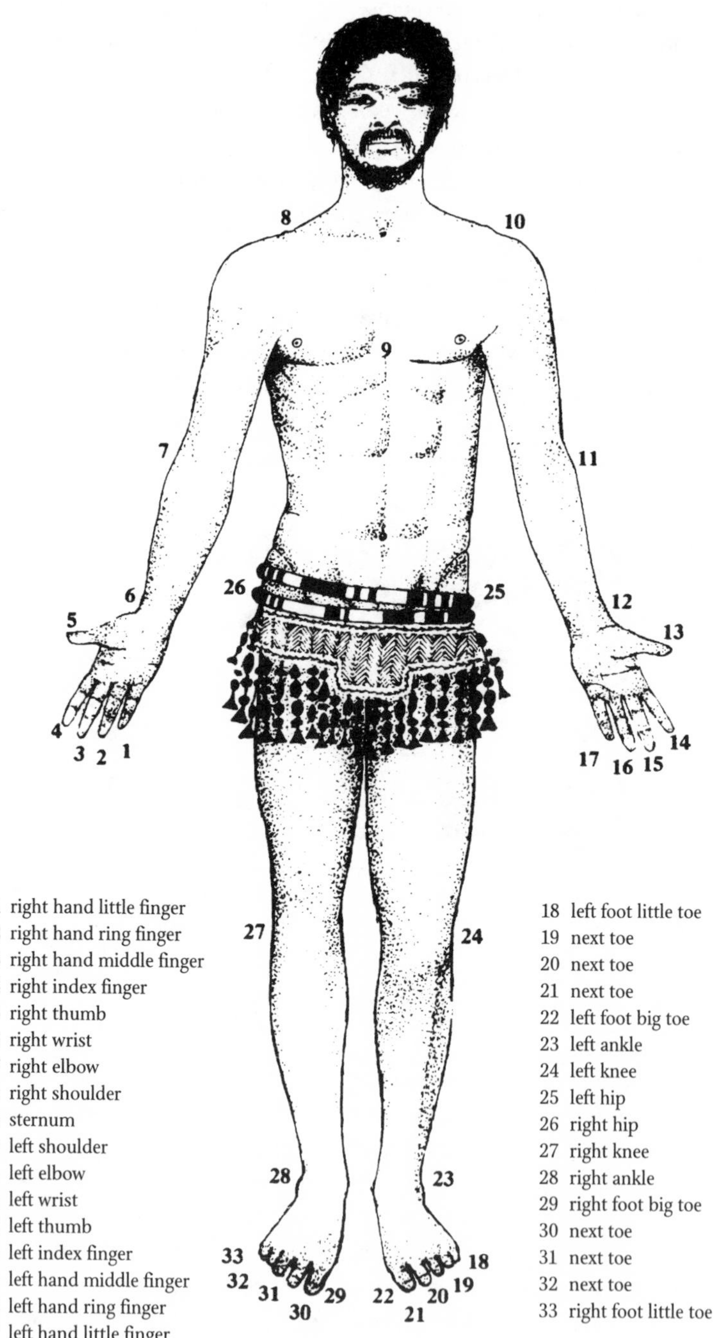


FIG. 1.30. *Body-counting system used by Torres Straits islanders*

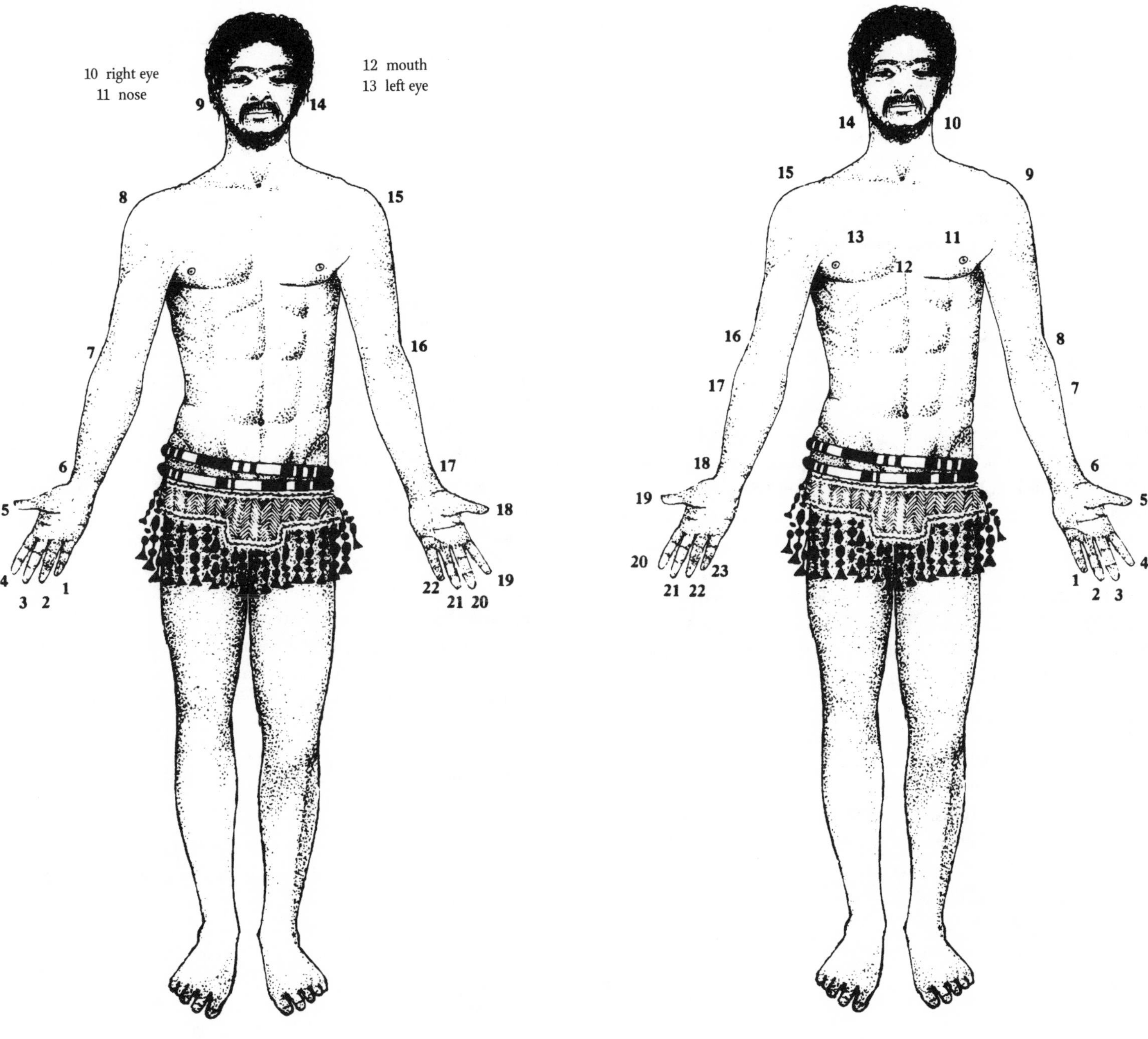


Fig. 1.31. *System used by Papuans (New Guinea)*

Fig. 1.32. *Body-counting system used by the Elema (New Guinea)*

They touch first the fingers of their right hand, one by one, then the right wrist, elbow and shoulder, go on to the sternum, then the left-side articulations, not forgetting the fingers. This brings them to the number seventeen. If the total needed is higher, they add the toes, ankle, knee and hip of the left then the right hand side. That gives 16 more, making 33 in all. For even higher numbers, the islanders have recourse to a bundle of small sticks. [As quoted in A. C. Haddon (1890)]

Murray islanders also used parts of the body in a conventional order, and were able to reach 29 in this manner. Other Torres Straits islanders used similar procedures which enabled them to "count visually" up to 19; the same customs are found amongst the Papuans and Elema of New Guinea.

## NUMBERS, GESTURES, AND WORDS

The question arises: is the mere enumeration of parts of the body in regular order tantamount to a true arithmetical sequence? Let us try to find the answer in some of the ethnographic literature relating to Oceania.

The first example is from the Papuan language spoken in what was British New Guinea. According to the report of the Cambridge Expedition to the Torres Straits, Sir William MacGregor found that "body-counting" was prevalent in all the villages below the Musa river. "Starting with the little finger on the right hand, the series proceeds with the right-hand fingers, then the right wrist, elbow, shoulder, ear and eye, then on to the left eye, and so on, down to the little toe on the left foot." Each of the gestures to these parts of the body is accompanied, the report continues, by a specific term in Papuan, as follows:

| NUMBER | NUMBER-GESTURE | GESTURE-WORD |
|---|---|---|
| 1 | right hand little finger | *anusi* |
| 2 | right hand ring finger | *doro* |
| 3 | right hand middle finger | *doro* |
| 4 | right hand index finger | *doro* |
| 5 | right thumb | *ubei* |
| 6 | right wrist | *tama* |
| 7 | right elbow | *unubo* |
| 8 | right shoulder | *visa* |
| 9 | right ear | *denoro* |
| 10 | right eye | *diti* |
| 11 | left eye | *diti* |
| 12 | nose | *medo* |
| 13 | mouth | *bee* |
| 14 | left ear | *denoro* |
| 15 | left shoulder | *visa* |
| 16 | left elbow | *unubo* |
| 17 | left wrist | *tama* |
| 18 | left thumb | *ubei* |
| 19 | left hand index finger | *doro* |
| 20 | left hand middle finger | *doro* |
| 21 | left hand ring finger | *doro* |
| 22 | left hand little finger | *anusi* |

The words used are simply the names of the parts of the body, and strictly speaking they are not numerical terms at all. *Anusi*, for example, is associated with both 1 and 22, and is used to indicate the little fingers of both the right and the left hands. In these circumstances how can you know which number is meant? Similarly the term *doro* refers to the ring, middle and index fingers of both hands and "means" either 2 or 3 or 4 or 19 or 20 or 21. Without the accompanying gesture, how could you possibly tell which of these numbers was meant?

However, there is no ambiguity in the system. What is spoken is the name of the part of the body, which has its rank-order in a fixed, conventional sequence within which no confusion is possible. So there is no doubt that the mere enumeration of the parts of the body does not constitute a true arithmetical sequence unless it is associated with a corresponding sequence of gestures. Moreover, the mental counting process has no direct oral expression – you can get to the number required without uttering a word. A conventional set of "number-gestures" is all that is needed.

In those cases where it is possible to recover the original meanings of the names given to numbers, it often turns out that they retain traces of body-counting systems like those we have looked at. Here, for example, are the number-words used by the Bugilai (former British New Guinea) together with their etymological meanings:

| | | |
|---|---|---|
| 1 | *tarangesa* | left hand little finger |
| 2 | *meta kina* | next finger |
| 3 | *guigimeta kina* | middle finger |
| 4 | *topea* | index finger |
| 5 | *manda* | thumb |
| 6 | *gaben* | wrist |
| 7 | *trankgimbe* | elbow |
| 8 | *podei* | shoulder |
| 9 | *ngama* | left breast |
| 10 | *dala* | right breast |

[Source: J. Chalmers (1898)]

E. C. Hawtrey (1902) also reports that the Lengua people of the Chaco (Paraguay) use a set of number-names broadly derived from specific number-gestures. Special words apparently unrelated to body-counting are used for 1 and 2, but for the other numbers they say something like:

| | |
|---|---|
| 3 | "made of one and two" |
| 4 | "both sides same" |
| 5 | "one hand" |
| 6 | "reached other hand, one" |
| 7 | "reached other hand, two" |
| 8 | "reached other hand, made of one and two" |
| 9 | "reached other hand, both sides same" |
| 10 | "finished, both hands" |
| 11 | "reached foot, one" |
| 12 | "reached foot, two" |
| 13 | "reached foot, made of one and two" |
| 14 | "reached foot, both sides same" |
| 15 | "finished, foot" |
| 16 | "reached other foot, one" |
| 17 | "reached other foot, two" |
| 18 | "reached other foot, made of one and two" |
| 19 | "reached other foot, both sides same" |
| 20 | "finished, feet" |

The Zuñi have names for numbers which F. H. Cushing (1892) calls "manual concepts":

| | | |
|---|---|---|
| 1 | *töpinte* | taken to begin |
| 2 | *kwilli* | raised with the previous |
| 3 | *kha'i* | the finger that divides equally |
| 4 | *awite* | all fingers raised bar one |
| 5 | *öpte* | the scored one |
| 6 | *topalïk'ye* | another added to what is counted already |
| 7 | *kwillik'ya* | two brought together and raised with the others |
| 8 | *khailïk'ya* | three brought together and raised with the others |
| 9 | *tenalïk'ya* | all bar one raised with the others |
| 10 | *ästem'thila* | all the fingers |
| 11 | *ästem'thila topayä'thl' tona* | all the fingers and one more raised |

and so on.

All this leads us to suppose that in the remotest past gestures came before any oral expression of numbers.

## CARDINAL RECKONING DEVICES FOR CONCRETE QUANTITIES

Let us now imagine a group of "primitive" people lacking any conception of abstract numbers but in possession of perfectly adequate devices for "reckoning" relatively small sets of concrete objects. They use all sorts of model collections, but most often they "reckon by eye" in the following manner: they touch each other's right-hand fingers, starting with the little finger, then the right wrist, elbow, shoulder, ear, and eye. Then they touch each others' nose, mouth, then the left eye, ear, shoulder, elbow, and wrist, and on to the little finger of the left hand, getting to 22 so far. If the number needed is higher, they go on to the breasts, hips, and genitals, then the knees, ankles and toes on the right then the left sides. This extension allows 19 further integers, or a total of 41.

The group has recently skirmished with a rebellious neighbouring village and won. The group's leader decides to demand reparations, and entrusts one of his men with the task of collecting the ransom. "For each of the warriors we have lost", says the chief, "they shall give us as many pearl necklaces as there are from the little finger on my right hand to my right eye, as many pelts as there are from the little finger of my right hand to my mouth, and as many baskets of food as there are from the little finger of my right hand to my left wrist." What this means is that the reparation for each lost soldier is:

*10 pearl necklaces*
*12 pelts*
*17 baskets of food*

In this particular skirmish, the group lost sixteen men. Of course none amongst the group has a notion of the number "16", but they have an infallible method of determining numbers in these situations: on departing for the fight, each warrior places a pebble on a pile, and on his return each surviving warrior picks a pebble out of the pile. The number of unclaimed pebbles corresponds precisely to the number of warriors lost.

One of the leader's envoys then takes possession of the pile of remaining pebbles but has them replaced by a matching bundle of sticks, which is easier to carry. The chief checks the emissaries' equipment and their comprehension of the reparations required, and sends them off to parley with the enemy.

The envoys tell the losing side how much they owe, and proceed to enumerate the booty in the following manner: one steps forward and says:

"Bring me a pearl necklace each time I point to a part of my body," and he then touches in order the little finger, the ring finger, the middle finger, the index finger and the thumb of his right hand. So the vanquished bring him one necklace, then a second, then a third and so on up to the fifth. The envoy then repeats himself, but pointing to his right wrist, elbow, shoulder, ear and eye, which gets him five more necklaces. So without having any concept of the number "10" he obtains precisely ten necklaces.

Another envoy proceeds in identical fashion to obtain the twelve pelts, and a third takes possession of the seventeen baskets of food that are demanded.

That is when the fourth envoy comes into the equation, for he possesses the tally of warriors lost in the battle, in the form of a bundle of sixteen sticks. He sets one aside, and the three other envoys then repeat their operations, allowing him to set another stick aside, and so on, until there are no sticks left in the bundle. That is how they know that they have the full tally, and so collect up the booty and set off with it to return to their own village.

As can be seen, "primitives" of this kind are not using body-counting in exactly the same way as we might. Since we know how to count, a conventional order of the parts of the body would constitute a true arithmetical sequence; each "body-point" would be assimilated in our minds to a cardinal (rank-order) number, characteristic of a particular quantity of things or beings. For instance, to indicate the length of a week using this system, we would not need to remember that it contained as many days as mapped onto our bodies from the right little finger to the right elbow, since we could just attach to it the "rank-order number" called "right elbow", which would suffice to symbolise the numerical value of any set of seven elements.

That is because we are equipped with *generalising abstractions* and in particular with number-concepts. But "primitive" peoples are not so equipped: they cannot *abstract* from the "points" in the numbering sequence: their grasp of the sequence remains embedded in the specific nature of the "points" themselves. Their understanding is in effect restricted to one-for-one mapping; the only "operations" they make are to add or remove one or more of the elements in the basic series.

Such people do not of course have any abstract concept of the number "ten", for instance. But they do know that by touching in order their little finger, ring finger, middle finger, index finger and thumb on the right hand, then their right wrist, elbow, shoulder, ear, and eye, they can "tally out" as many men, animals or objects as there are body-points in the sequence. And having done so, they remember perfectly well which body-point any particular tally of things or people reached, and are able to repeat the operation in order to reach exactly the same tally whenever they want to.

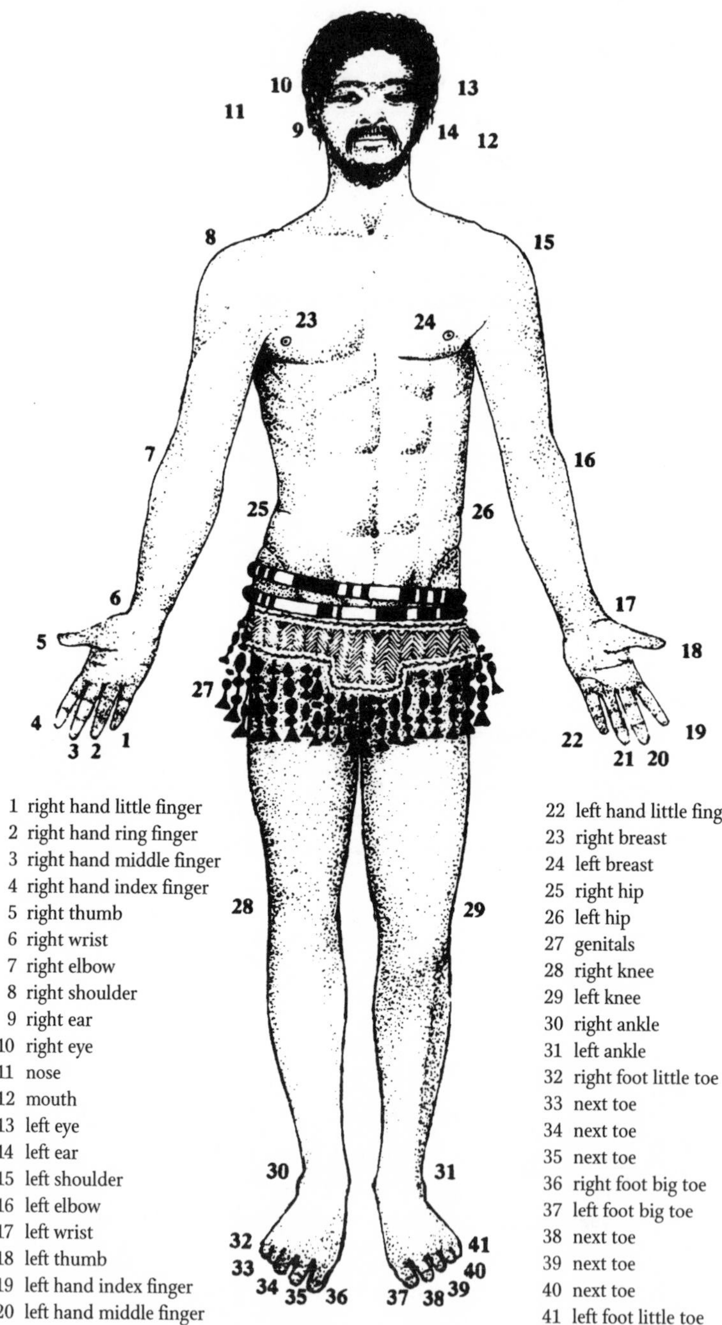


FIG. 1.33.

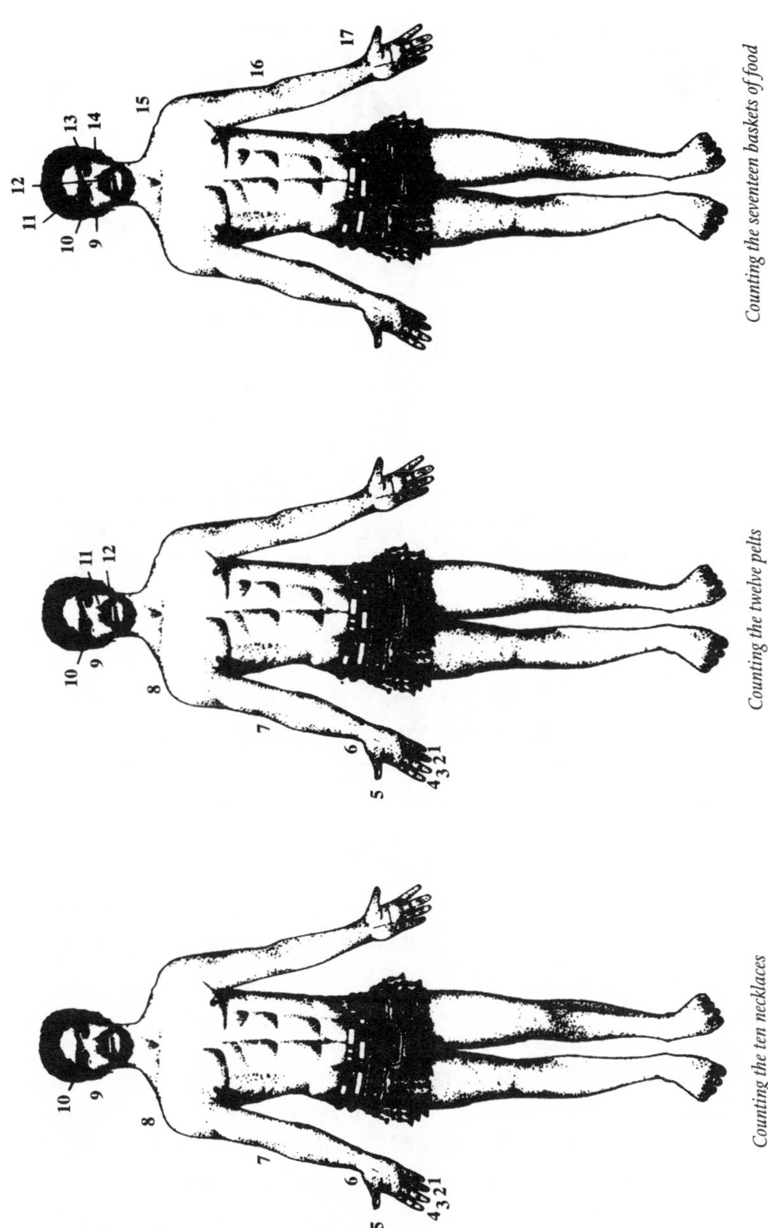


*Counting the ten necklaces*

*Counting the twelve pelts*

*Counting the seventeen baskets of food*

Fig. 1.34.

In other words, this procedure is a simple and convenient means of establishing ready-made mappings which can then be mapped one-to-one onto any sets for which a total is required. So when our imaginary tribe went to collect its ransom, they used only these notions, not any true number-concepts. They simply mapped three such ready-made sets onto a set of ten necklaces, a set of twelve pelts, and a set of seventeen baskets of food for each of the lost warriors.

These body-counting points are thus not thought of by their users as "numbers", but rather as the last elements of model sets arrived at after a regulated (conventional) sequence of body-gestures. This means that for such people the mere designation of any one of the points *is not sufficient to describe a given number of beings or things unless the term uttered is accompanied by the corresponding sequence of gestures*. So in discussions concerning such and such a number, no real "number-term" is uttered: instead, a given number of body-counting points will be enumerated, alongside the simultaneous sequence of gestures. This kind of enumeration therefore fails to constitute a genuine arithmetical series; participants in the discussion must also necessarily keep their eyes on the speaker!

All the same, our imaginary tribesmen have unknowingly reached quite large numbers, even with such limited tools, since they have collected:

$16 \times 10 = 160$ necklaces
$16 \times 12 = 192$ pelts
$16 \times 17 = 272$ baskets of food

or *six hundred and twenty-four* items in all! (see Fig. 1.34)

There is a simple reason for this: they had thought of associating easily manipulated material objects with the parts of the body involved in their counting operations. It is true that they counted out the necklaces, pelts and food-baskets by their traditional body-counting method, but the determining element in calculating the ransom to be paid (the number of men lost in the battle) was "numerated" with the help of pebbles and a bundle of sticks.

Let us now imagine that the villagers are working out how to fix the date of an important forthcoming religious festival. The shaman who that morning proclaimed the arrival of the new moon also announced in the following way, accompanying his words with quite precise gestures of his hands, that the festival will fall on the *thirteenth day of the eighth moon thereafter*: "Many suns and many moons will rise and fall before the festival. The moon that has just risen must first wax and then wane completely. Then it must wax as many times again as there are from the little finger on my right hand to the elbow on the same side. Then the sun will rise and set as many times as there are from the little finger on my right hand to my mouth. That is when the sun will next rise on the day of our Great Festival."

This community obviously has a good grasp of the lunar cycle, which is only to be expected, since, after the rising and the setting of the sun, the moon's phases constitute the most obvious regular phenomenon in the natural environment. As in all *empirical calendars*, this one is based on the observation of the first quarter after the end of each cycle. With the help of model collections inherited from forebears, many generations of whom must have contributed to their slow development, the community can in fact "mark time" and compute the date thus expressed without error, as we shall see.

On hearing the shaman's pronouncement, the chief of the tribe paints a number of marks on his own body with some fairly durable kind of colouring material, and these marks enable him to record and to recognise the festival date unambiguously. He first records the series of reappearances that the moon must make from then until the festival by painting *small circles* on his right-hand little finger, ring finger, middle finger, index finger, thumb, wrist, and elbow. Then he records the number of days that must pass from the appearance of the last moon by painting a *thin line*, first of all on each finger of his right hand, then on his right wrist, elbow, shoulder, ear, and eye, then on his nose and mouth. To conclude, he puts a *thick line* over his left eye, thereby symbolising the dawn of the great day itself.

The following day at sunset, a member of the tribe chosen by the chief to "count the moons" takes one of the ready-made ivory tally sticks with thirty incised notches, the sort used whenever it is necessary to reckon the days of a given moon in their order of succession (see Fig. 1.35). He ties a piece of string around the first notch. The next evening, he ties a piece of string around the second notch, and so on every evening until the end of the moon. When he reaches the penultimate notch, he looks carefully at the night sky, in the region where the sun has just set, for he knows that the new moon is soon due to appear.

On that day, however, the first quarter of the new moon is not visible in the sky. So he looks again the next evening when he has tied the string around the last notch on the first tally stick; and though the sky is not clear enough to let him see the new moon, he decides nonetheless that a new month has begun. That is when he paints a little circle on his right little finger, indicating that one lunar cycle has passed.

At dusk the following day, our "moon-counter" takes another similar tally stick and ties a string around the first notch. The day after, he or she proceeds likewise with the second notch, and so on to the end of the second month. But at that month's end the tally man knows he will not need to scan the heavens to check on the rising of the new moon. For in this tribe, the knowledge that moon cycles end alternately on the penultimate and last notches of the tally sticks has been handed down for generations. And this knowledge is only very slightly inaccurate, since the average length of a lunar cycle is 29 days and 12 hours.

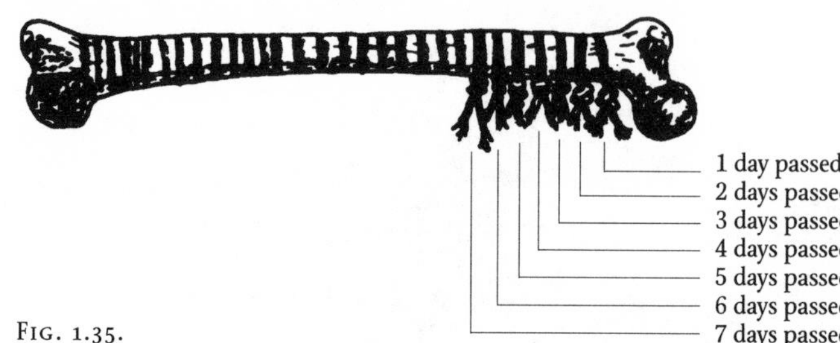


FIG. 1.35.

The moon-counter proceeds in this manner through alternating months of 29 and 30 days until the arrival of the last moon, when he paints a little circle on his right elbow. There are now as many circles on the counter's body as on the chief's: the counter's task is over: the "moon tally" has been reached.

The chief now takes over as the "day-counter", but for this task tally sticks are not used, as the body-counting points suffice. The community will celebrate its festival when the chief has crossed out all the *thin lines* from his little finger to his mouth and also the *thick line* over his left eye, that is to say on the thirteenth day of the eighth moon (Fig. 1.36)

This reconstitution of a non-numerate counting system conforms to many of the details observed in Australian aboriginal groups, who are able to reach relatively high numbers through the (unvocalised) numeration of parts of the body when the body-points have a fixed conventional order and are associated with manipulable model collections – knotted string, bundles of sticks, pebbles, notched bones, and so on.

Valuable evidence of this kind of system was reported by Brooke, observing the Dayaks of South Borneo. A messenger had the task of informing a number of defeated rebel villages of the sum of reparations they had to pay to the Dayaks.

> The messenger came along with some dried leaves, which he broke into pieces. Brooke exchanged them for pieces of paper, which were more convenient. The messenger laid the pieces on a table and used his fingers at the same time to count them, up to ten; then he put his foot on the table, and counted them out as he counted out the pieces of paper, each of which corresponded to a village, with the name of its chief, the number of warriors and the sum of the reparation. When he had used up all his toes, he came back to his hands. At the end of the list, there were forty-five pieces of paper laid out on the table.* Then

* Each finger is associated with one piece of paper and one village, in this particular system, and each toe with the set of ten fingers.

he asked me to repeat the message, which I did, whilst he ran through the pieces of paper, his fingers and his toes, as before.

"So there are our letters," he said. "You white folk don't read the way we do."

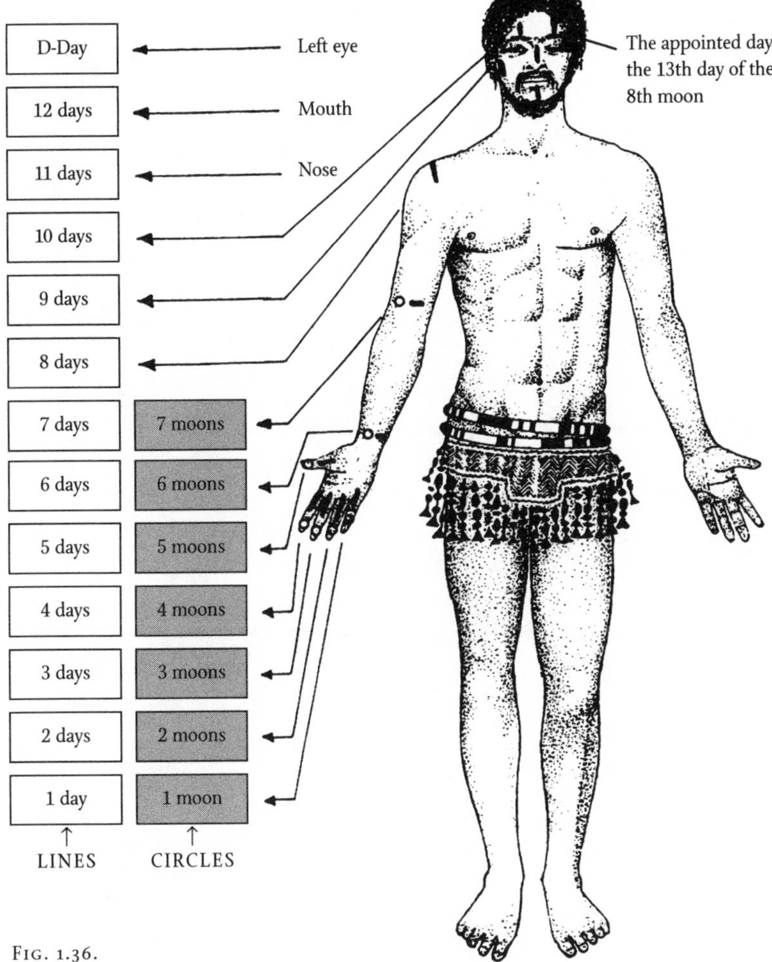


FIG. 1.36.

Later that evening he repeated the whole set correctly, and as he put his finger on each piece of paper in order, he said:

"So, if I remember it tomorrow morning, all will be well; leave the papers on the table."

Then he shuffled them together and made them into a heap. As soon as we got up the next morning, we sat at the table, and he re-sorted the pieces of paper into the order they were in the previous day, and repeated all the details of the message with complete accuracy. For almost a month, as he went from village to village, deep in the interior, he never forgot the different sums demanded. [Adapted from Brooke, *Ten Years in Sarawak*]

All this leads us to hypothesise the following evolution of counting systems:

*First stage*

Only the lowest numbers are within human grasp. Numerical ability remains restricted to what can be evaluated in a single glance. "Number" is indissociable from the concrete reality of the objects evaluated.* In order to cope with quantities above four, a number of concrete procedures are developed. These include finger-counting and other body-counting systems, all based on one-for-one correspondence, and leading to the development of simple, widely-available ready-made mappings. What is articulated (lexicalised) in the language are these ready-made mappings, accompanied by the appropriate gestures.

*Second stage*

By force of repetition and habit, the list of the names of the body-parts in their numerative order imperceptibly acquire abstract connotations, especially the first five. They slowly lose their power to suggest the actual parts of the body, becoming progressively more attached to the corresponding number, and may now be applied to any set of objects. (L. Lévy-Bruhl)

*Third stage*

A fundamental tool emerges: numerical nomenclature, or the names of the numbers.

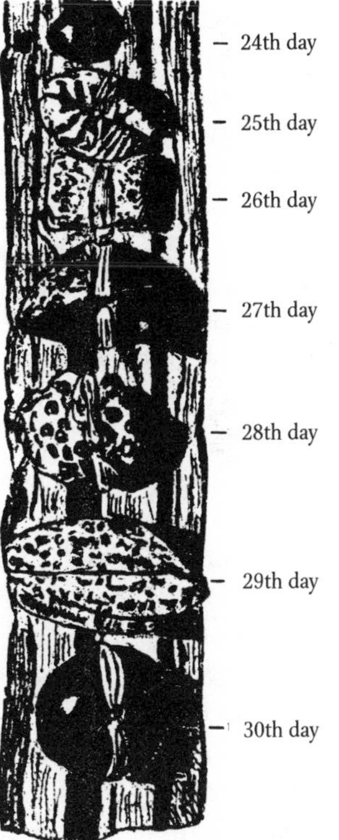


FIG. 1.37. *Detail from a "material model" of a lunar calendar formerly in use amongst tribal populations in former Dahomey (West Africa). It consists of a strip of cloth onto which thirty objects (seeds, kernels, shells, hard fruit, stones, etc.) have been sewn, each standing for one of the days of the month. (The fragment above represents the last seven days). From the Musée de l'Homme, Paris.*

* Thus as L. Lévy-Bruhl reports, Fijians and Solomon islanders have collective nouns for tens of arbitrarily selected items that express neither the number itself nor the objects collected into the set. In Fijian, *bola* means "a hundred dugouts", *koro* "a hundred coconuts", *salavo* "a thousand coconuts". Natives of Mota say *aka peperua* ("butterfly two dugout") for "a pair of dugouts" because of the appearance of the sails. See also Codrington, E. Stephen and L. L. Conant.

## COUNTING: A HUMAN FACULTY

The human mind, evidently, can only grasp integers as abstractions if it has fully available to it the notion of distinct units as well as the ability to "synthesise" them. This intellectual faculty (which presupposes above all a complete mastery of the ability to analyse, to compare and to abstract from individual differences) rests on an idea which, alongside mapping and classification, constitutes the starting point of all scientific advance. This creation of the human mind is called "hierarchy relation" or "order relation": it is the principle by which things are ordered according to their "degree of generality", from *individual*, to *kind*, to *type*, to *species*, and so on.

Decisive progress towards the art of abstract calculation that we now use could only be made once it was clearly understood that the integers could be classified into a *hierarchised system of numerical units* whose terms were related as kinds within types, types within species, and so on. Such an organisation of numerical concepts in an invariable sequence is related to the generic principle of "recurrence" to which Aristotle referred (*Metaphysics* 1057, a) when he said that an integer was a "multiplicity measurable by the one". The idea is really that integers are "collections" of abstract units obtained successively by the adjunction of further units.

*Any element in the regular sequence of the integers (other than 1) is obtained by adding 1 to the integer immediately preceding in the "natural" sequence that is so constituted* (see Fig. 1.38). As the German philosopher Schopenhauer put it, any natural integer presupposes its preceding numbers as the cause of its existence: for our minds cannot conceive of a number as an abstraction unless it subsumes all preceding numbers in the sequence. This is what we called the ability to "synthesise" distinct units. Without that ability, number-concepts remain very cloudy notions indeed.

But once they have been put into a natural sequence, the set of integers permits another faculty to come into play: numeration. To numerate the items in a group is to assign to each a symbol (that is to say, a word, a gesture or a graphic mark) corresponding to a number taken from the natural sequence of integers, beginning with 1 and proceeding in order until the exhaustion of that set (Fig. 1.40). The symbol or name given to each of the elements within the set is the name of its order number within the collection of things, which becomes thereby a sequence or procession of things. The order number of the last element within the ordered group is precisely equivalent to the number of elements in the set. Obviously the number obtained is entirely independent of the order in which the elements are numerated – whichever of the elements you begin with, you always end up with the same total.

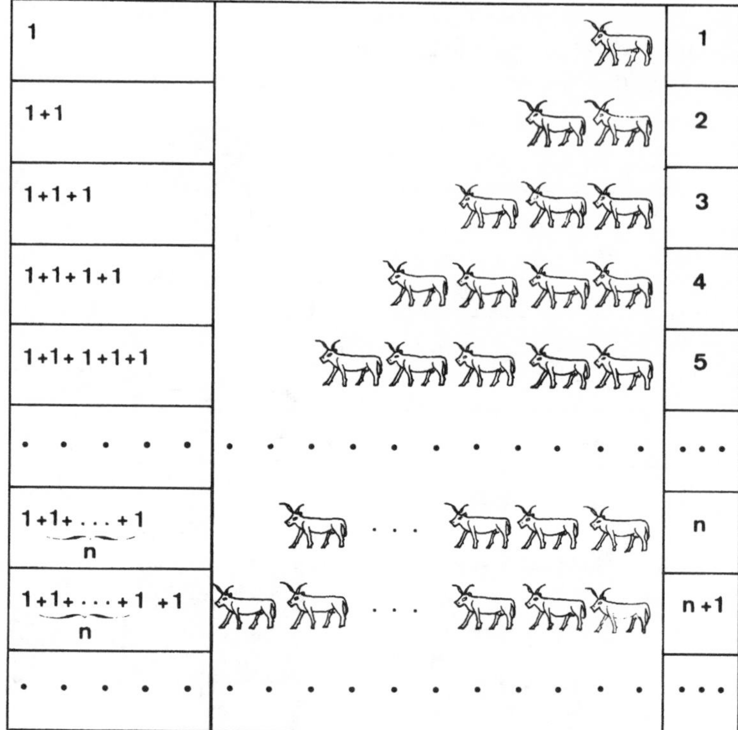


FIG. 1.38. *The generation of integers by the so-called procedure of recurrence*

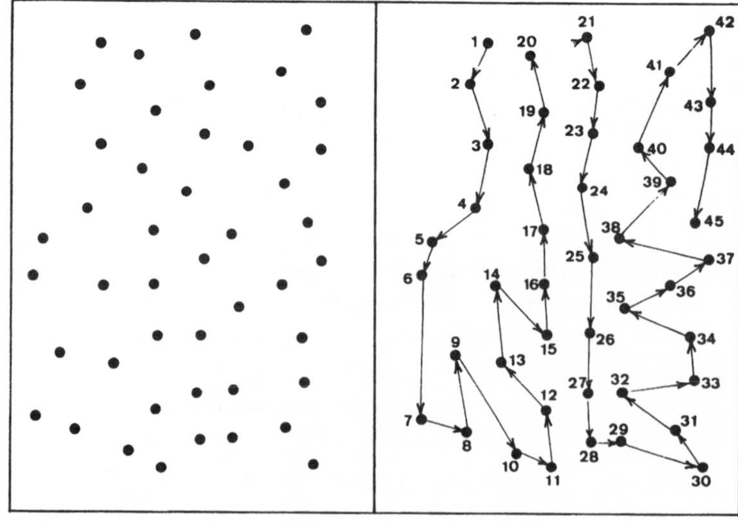

FIG. 1.39. *Numeration of a "cloud" of dots*

For example, let us take a box containing "several" billiard balls. We take out one at random and give it the "number" 1 (for it is the first one to come out of the box). We take another, again completely at random, and give it the "number" 2. We continue in this manner until there are no billiard balls left in the box. When we take out the last of the balls, we give it a specific number from the natural sequence of the integers. If its number is 20, we say that there are "twenty" balls in the box. Numeration has allowed us to transform a vague notion (that there are "several" billiard balls) into exact knowledge.

In like manner, let us consider a set of "scattered" points, in other words dots in a "disordered set" (Fig. 1.39). To find out how many dots there are, all we have to do is to connect them by a "zigzag" line passing through each dot once and no dot twice. The points then constitute what is commonly called a chain. We then give each point in the chain an order-number, starting from one of the ends of the chain we have just made. The last number, given therefore to the last point in the chain, provides us with the total number of dots in the set.

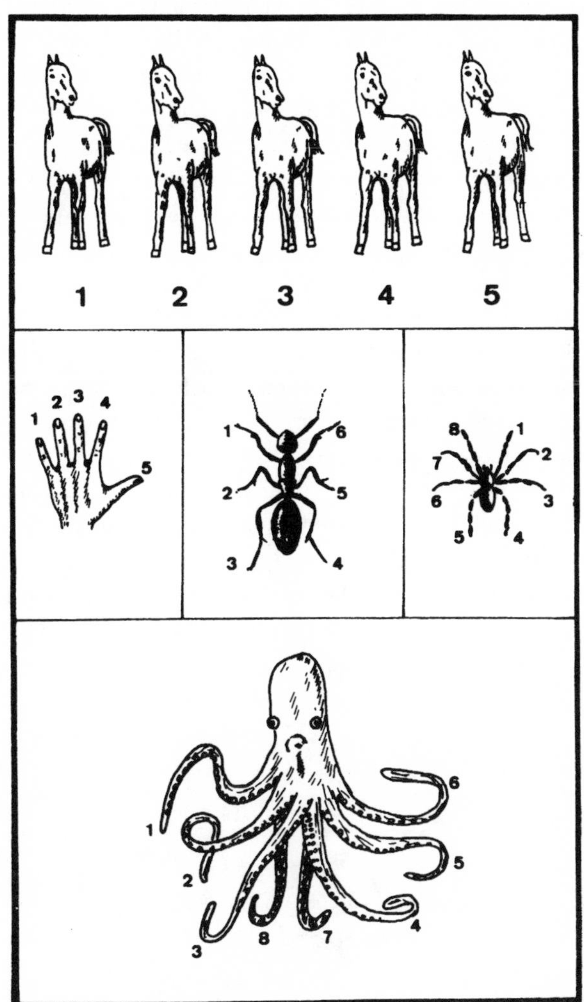


FIG. 1.40. *Numeration allowing us to advance from concrete plurality to abstract number*

So with the notions of succession and numeration we can advance from the muddled, vague and heterogeneous apperception of concrete plurality to the abstract and homogenous idea of "absolute quantity".

So the human mind can only "count" the elements in a set if it is in possession of all three of the following abilities:

1. the ability to assign a "rank-order" to each element in a procession;
2. the ability to insert into each unit of the procession the memory of all those that have gone past before;
3. the ability to convert a sequence into a "stationary" vision.

The concept of number, which at first sight seemed quite elementary, thus turns out to be much more complicated than that. To underline this point I should like to repeat one of P. Bourdin's anecdotes, as quoted in R. Balmès (1965):

> I once knew someone who heard the bells ring four as he was trying to go to sleep and who counted them out in his head, one, one, one, one. Struck by the absurdity of counting in this way, he sat up and shouted: "The clock has gone mad, it's struck one o'clock four times over!"

## THE TWO SIDES OF THE INTEGERS

The concept of number has two complementary aspects: cardinal numbering, which relies only on the principle of mapping, and ordinal numeration, which requires both the technique of pairing and the idea of succession.

Here is a simple way of grasping the difference. January has 31 days. The number 31 represents the total number of days in the month, and is thus in this expression a cardinal number. However, in expressions such as "31 January 1996", the number 31 is not being used in its cardinal aspect (despite the terminology of grammar books) because here it means something like "the thirty-first day" of the month of January, specifying not a total, but a rank-order of a specific (in this case, the last) element in a set containing 31 elements. It is therefore unambiguously an ordinal number.

> We have learned to pass with such facility from cardinal to ordinal number that the two aspects appear to us as one. To determine the plurality of a collection, i.e. its cardinal number, we do not bother any more to find a model collection with which we can match it – we *count*

it. And to the fact that we have learned to identify the two aspects of number is due our progress in mathematics. For whereas in practice we are really interested in the cardinal number, this latter is incapable of creating an arithmetic. The operations of arithmetic are based on the tacit assumption that *we can always pass from any number to its successor*, and this is the essence of the ordinal concept.

And so matching by itself is incapable of creating an art of reckoning. Without our ability to arrange things in ordered succession little progress could have been made. Correspondence and succession, the two principles which permeate all mathematics – nay, all realms of exact thought – are woven into the very fabric of our number-system. [T. Dantzig (1930)]

## TEN FINGERS TO COUNT BY

Humankind slowly acquired all the necessary intellectual equipment thanks to the ten fingers on its hands. It is surely no coincidence if children still learn to count with their fingers – and adults too often have recourse to them to clarify their meaning.

Traces of the anthropomorphic origin of counting systems can be found in many languages. In the Ali language (Central Africa), for example, "five" and "ten" are respectively *moro* and *mbouna*: *moro* is actually the word for "hand" and *mbouna* is a contraction of *moro* ("five") and *bouna*, meaning "two" (thus "ten" = "two hands").

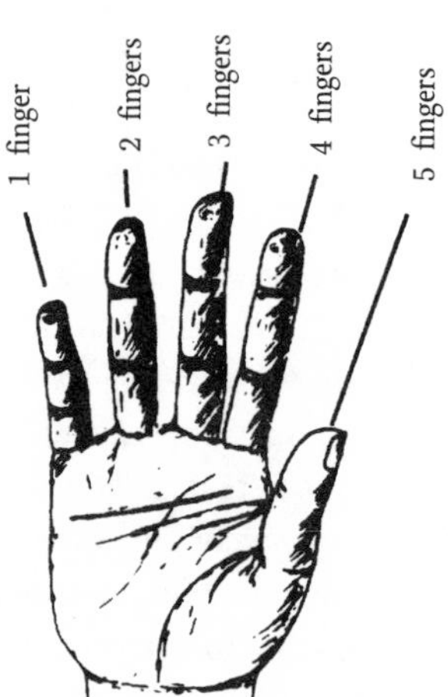


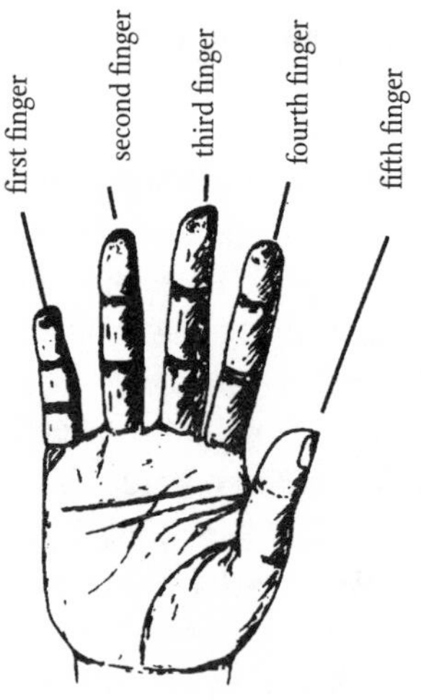


FIG. 1.41.

It is therefore very probable that the Indo-European, Semitic and Mongolian words for the first ten numbers derive from expressions related to finger-counting. But this is an unverifiable hypothesis, since the original meanings of the names of the numbers have been lost.

In any case, the human hand is an extremely serviceable tool and constitutes a kind of "natural instrument" well suited for acquiring the first ten numbers and for elementary arithmetic.

Because there are ten fingers and because each can be moved independently of the others, the hand provides the simplest "model collection" that people have always had – so to speak – to hand.

The asymmetric disposition of the fingers puts the hand in harmony with the normal limitation of the human ability to recognise number visually (a limit set at four). As the thumb is set at some distance from the index finger it is easy to treat it as being "in opposition" to the elementary set of four, and makes the first five numbers an entirely natural sequence. Five thus imposes itself as a basic unit of counting, alongside the other natural grouping, ten. And because each of the fingers is actually different from the others, the human hand can be seen as a true succession of abstract units, obtained by the progressive adjunction of one to the preceding units.

In brief, one can say that the hand makes the two complementary aspects of integers entirely intuitive. It serves as an instrument permitting natural movement between cardinal and ordinal numbering. If you need to show that a set contains three, four, seven or ten elements, you raise or bend *simultaneously* three, four, seven or ten fingers, using your hand as cardinal mapping. If you want to count out the same things, then you bend or raise three, four, seven or ten fingers *in succession*, using the hand as an ordinal counting tool (Fig. 1.41).

The human hand can thus be seen as the simplest and most natural counting machine. And that is why it has played such a significant role in the evolution of our numbering system.

CHAPTER 2

# BASE NUMBERS
## AND THE BIRTH OF NUMBER-SYSTEMS

### NUMBERS AND THEIR SYMBOLS

Once they had grasped abstract numbers and learned the subtle distinction between cardinal and ordinal aspects, our ancestors came to have a different attitude towards traditional "numbering tools" such as pebbles, shells, sticks, strings of beads, or points of the body. Gradually these simple mapping devices became genuine numerical symbols, which are much better suited to the tasks of assimilating, remembering, distinguishing and combining numbers.

Another great step forward was the creation of names for the numbers. This allowed for much greater precision in speech and opened the path towards real familiarity with the universe of abstract numbers.

Prior to the emergence of number-names, all that could be referred to in speech were the "concrete maps" which had no obvious connection amongst themselves. Numbers were referred to by intuitive terms, often directly appealing to the natural environment. For instance, 1 might have been "sun", "moon", or "penis"; for 2, you might have found "eyes", "breasts", or "wings of a bird"; "clover" or "crowd" for 3; "legs of a beast" for 4; and so on. Subsequently some kind of structure emerged from body-counting. At the start, perhaps, you had something like this: "the one to start with" for 1; "raised with the preceding finger" for 2; "the finger in the middle" for 3; "all fingers bar one" for 4; "hand" for 5; and so on. Then a kind of anatomical mapping occurred, so that "little finger" = 1, "ring finger" = 2, "middle finger" = 3, "index finger" = 4, "thumb" = 5, and so on. However, the need to distinguish between the number-symbol and the name of the object or image being used to symbolise the number led people to make an ever greater distinction between the two names, so that eventually the connection between them was entirely lost. As people progressively learned to rely more and more on language, the sounds superseded the images for which they stood, and the originally concrete models took on the abstract form of number-words. The idea of a natural sequence of numbers thus became ever clearer; and the very varied set of initial counting maps or model collections turned into a real *system of number-names*. Habit and memory gave a concrete form to these abstract ideas, and, as T. Dantzig says (p. 8), that is how "mere words became measures of plurality".

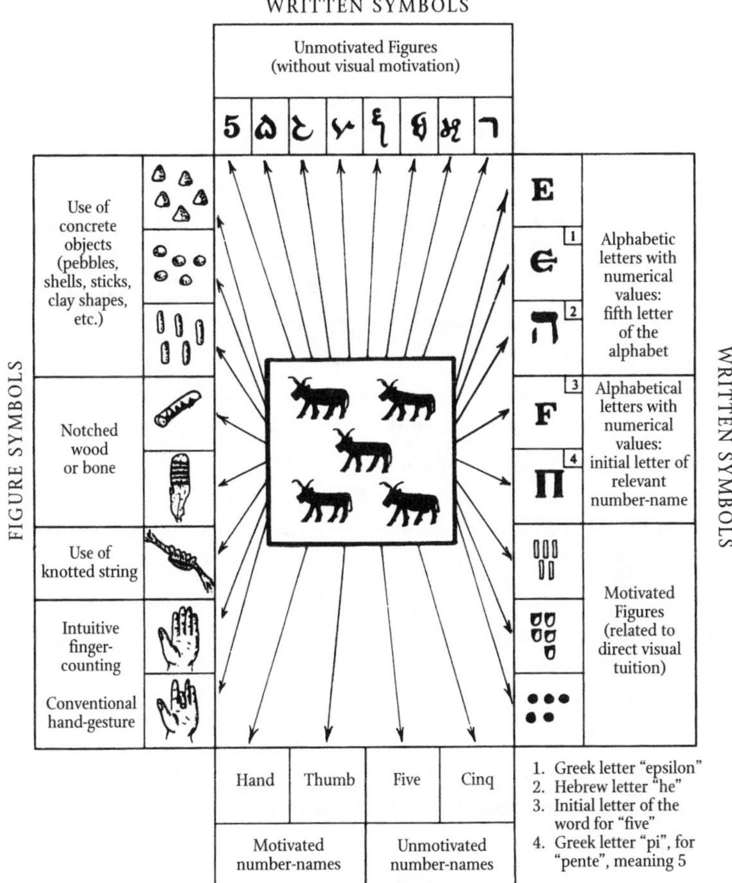


FIG. 2.1.

Of course, concrete symbols and spoken expressions were not the only devices that humanity possessed for mastering numbers. There was also writing, even if that did arise much later on. Writing involves *figures*, that is to say, graphic signs, of whatever kind (carved, drawn, painted, or scored on clay or stone; iconic signs, letters of the alphabet, conventional signs, and so on). We should note that figures are not numbers. "Unit", "pair of", "triad" are "numbers", whilst 1, 2, 3 are "figures", that is to say, conventional graphic signs that represent number-concepts. A figure is just one of the "dresses" that a number can have: you can change the way a number is written without changing the number-concept at all.

These were very important developments, for they allowed "operations" on things to be replaced by the corresponding operation on number-symbols. For numbers do not come from things, but from the laws of the human mind as it works on things. Even if numbers seem latent in the natural world, they certainly did not spring forth from it by themselves.

### THE DISCOVERY OF BASE NUMBERS

There were two fundamental principles available for constructing number-symbols: one that we might call a *cardinal* system, in which you adopt a standard sign for the unit and repeat it as many times as there are units in the number; and another that we could call an *ordinal* system, in which each number has its own distinctive symbol.

In virtue of the first principle, the numbers 2 to 4 can be represented by repeating the name of the number 1 two, three or four times, or by laying out in a line, or on top of each other, the appropriate number of "unit signs" in pebbles, fingers, notches, lines, or dots (see Fig. 2.2).

The second principle gives rise to representations for the first four numbers (in words, objects, gestures or signs) that are each different from the others (see Fig. 2.3).

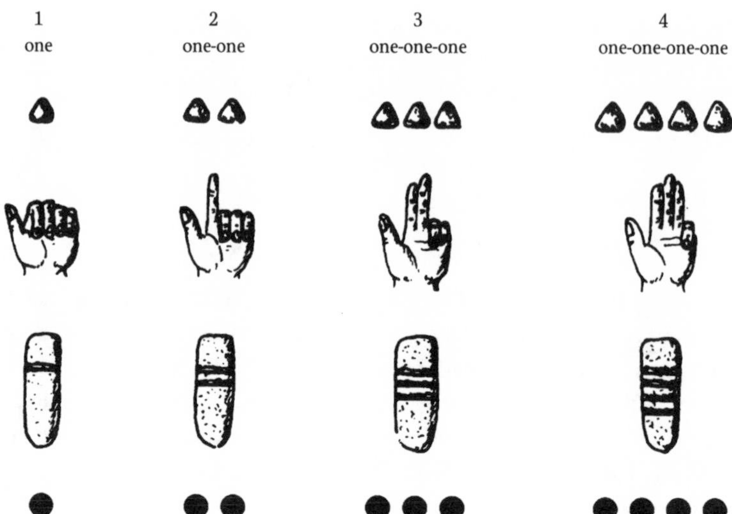


FIG. 2.2. *"Cardinal" representations of the first four numbers*

thumb one | index two | middle three | ring four

1 | 2 | 3 | 4

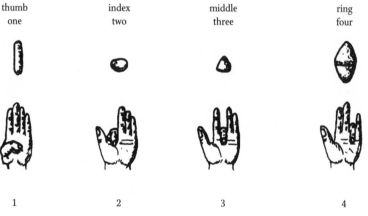

FIG. 2.3. *"Ordinal" representations of the first four numbers*

Either of these principles is an adequate basis for acquiring a grasp of ever larger sets – but the application of both principles quickly runs into difficulty. To represent larger numbers, you can't simply use more and more pebbles, sticks, notches, or knotted string; and the number of fingers and other counting points on the body is not infinitely extensible. Nor is it practicable to repeat the same word any number of times, or to create unique symbols for any number of numbers. (Just think how many different symbols you would need to say how many cents there are in a ten-dollar bill!)

To make any progress, people had first to solve a really tricky problem: What in practice is the smallest set of symbols in which the largest numbers can in theory be represented? The solution found is a remarkable example of human ingenuity.

The solution is to give one particular set (for example, the set of ten, the set of twelve, the set of twenty or the set of sixty) a special role and to classify the regular sequence of numbers in a hierarchical relationship to the chosen ("base") set. In other words, you agree to set up a *ladder* and to organise the numbers and their symbols on ascending *steps* of the ladder. On the first step you call them "first-order units", on the second step, "units of the second order", on the third step, "units of the third order", and so on. And that is all there was to the invention of a number-system that saves vast amounts of effort in terms of memorisation and writing-out. The system is called "the rule of position" (or "place-value system"), and its discovery marked the birth of numbering systems where the "base" is the number of units in the set that constitutes the unit of the next order. The place-value system can be applied to material "relays", to words in a language, or to graphic marks – producing respectively concrete, oral, and written numbering systems.

### WHY BASE 10?

Not so long ago shepherds in certain parts of West Africa had a very practical way of checking the number of sheep in their flocks. They would make the animals pass by one by one. As the first one went through the gate, the shepherd threaded a shell onto a white strap; as the second went through, he threaded another shell, and so on up to the ninth. When the tenth went through, he took the shells off the white strap and put one on a blue strap, which served for counting in tens. Then he began again threading shells onto the white strap until the twentieth sheep went through, when he put a second shell on the blue strap. When there were ten shells on the blue strap, meaning that one hundred sheep had now been counted, he undid the blue strap and threaded a shell onto a red strap, which was the "hundreds" counting device. And so he continued until the whole flock had been counted. If there were for example two hundred and fifty-eight head in the flock, the shepherd would have eight shells on the white strap, five on the blue strap and two on the red strap. There's nothing "primitive" about this method, which is in effect the one that we use now, though with different symbols for the numbers and orders of magnitude.

The basic idea of the system is the primacy of grouping (and of the rhythm of the symbols in their regular sequence) in "packets" of tens, hundreds (tens of tens), thousands (tens of tens of tens), and so on. In the shepherd's concrete technique, each shell on the white strap counts as a simple unit, each shell on the blue strap counts for ten, and each shell on the red strap counts for a hundred. This is what is called *the principle of base 10*. The shepherd's device is an example of a concrete decimal number-system.

Obviously, instead of using threaded shells and leather straps, we could apply the same system to words or to graphic signs, producing oral or written decimal numeration. Our current number system is just such, using the following graphic signs, often referred to as Arabic numerals:

1 2 3 4 5 6 7 8 9 0

The first nine symbols represent the simple units, or units of the first decimal order (or "first magnitude"). They are subject to the rule of position, or place-value, since their value depends on the place or position that they occupy in a written numerical expression (a 3, for instance, counts for three units, three tens, or three hundreds depending on its position in a three-digit numerical expression). The tenth symbol above represents what we call "zero", and it serves to indicate the absence of any unit of a particular decimal order, or order of magnitude. It also has the meaning of "nought" – for example, the number you obtain when you subtract a number from itself.

The base of ten, which is the first number that can be represented by two figures, is written as 10, a notation which means "one ten and no units".

The numbers from 11 to 99 are represented by combinations of two of the figures according to the rule of position:

11 "one ten, one unit"
12 "one ten, two units"
20 "two tens, no units"
21 "two tens, one unit"
30 "three tens, no units"
40 "four tens, no units"
50 "five tens, no units"

The *hundred*, equal to the square of the base, is written: 100, meaning "one hundred, no tens, no units", and is the smallest number that can be written with three figures.

Numbers from 101 to 999 are represented by combinations of three of the basic figures:

101 "one hundred, no tens, one unit"
358 "three hundreds, five tens, eight units"

There then comes the *thousand*, equal to the cube of the base, which is written : 1,000 ("one thousand, no hundreds, no tens, no units"), and is the smallest number that can be written with four figures. The following step on the ladder is the *ten thousand*, the base to the power of four, which is written 10,000 ("one ten thousand, no thousands, no hundreds, no tens, no units") and is the smallest number that can be written with five figures; and so on.

In oral (spoken) numeration constructed in the same way things proceed in very similar general manner, but with one difference that is inherent to the nature of language: all the numbers less than or equal to ten and also the several powers of ten (100, 1,000, 10,000, etc.) have individual names entirely unrelated to each other, whereas all other numbers are expressed by words made up of combinations of the various number-names.

In English, if we restrict ourselves for a moment to cardinal numbers, the system would proceed in theory as follows. For the first ten numbers we have individual names:

| *one* | *two* | *three* | *four* | *five* | *six* | *seven* | *eight* | *nine* | *ten* |
|---|---|---|---|---|---|---|---|---|---|
| 1 | 2 | 3 | 4 | 5 | 6 | 7 | 8 | 9 | 10 |

The first nine are "units of the first decimal order of magnitude" and the tenth constitutes the "base" of the system (and by definition is therefore the sign for the "unit of the second decimal order of magnitude"). To name the numbers from 11 to 19, the units are grouped in "packets" of ten and we proceed (in theory) by simple addition:

11 *one-ten* (= 1 + 10)
12 *two-ten* (= 2 + 10)
13 *three-ten* (= 3 + 10)
14 *four-ten* (= 4 + 10)

| | | |
|---|---|---|
| 15 | *five-ten* | (= 5 + 10) |
| 16 | *six-ten* | (= 6 + 10) |
| 17 | *seven-ten* | (= 7 + 10) |
| 18 | *eight-ten* | (= 8 + 10) |
| 19 | *nine-ten* | (= 9 + 10) |

Multiples of the base, from 20 to 90, are the "tens", or units of the second decimal order, and they are expressed by multiplication:

| | | |
|---|---|---|
| 20 | *two-tens* | (= 2 × 10) |
| 30 | *three-tens* | (= 3 × 10) |
| 40 | *four-tens* | (= 4 × 10) |
| 50 | *five-tens* | (= 5 × 10) |
| 60 | *six-tens* | (= 6 × 10) |
| 70 | *seven-tens* | (= 7 × 10) |
| 80 | *eight-tens* | (= 8 × 10) |
| 90 | *nine-tens* | (= 9 × 10) |

If the number of tens is itself equal to or higher than ten, then the tens are also grouped in packets of ten, constituting the "units of the third decimal order", as follows:

| | | |
|---|---|---|
| 100 | hundred | (= $10^2$) |
| 200 | two hundreds | (= 2 × 100) |
| 300 | three hundreds | (= 3 × 100) |
| 400 | four hundreds | (= 4 × 100) |
| . . . | . . . | . . . |

The hundreds are themselves then grouped into packets of ten, constituting "units of the fourth decimal order", or thousands:

| | | |
|---|---|---|
| 1,000 | one thousand | (=$10^3$) |
| 2,000 | two thousands | (2 × 1,000) |
| 3,000 | three thousands | (3 × 1,000) |
| . . . | . . . | . . . |

Then come the ten thousands, which used to be called *myriads,* corresponding to the "units of the fifth decimal order":

| | | |
|---|---|---|
| 10,000 | a myriad | (=$10^4$) |
| 20,000 | two myriads | (= 2 × 10,000) |
| 30,000 | three myriads | (= 3 × 10,000) |
| . . . | . . . | . . . |

Using only these words of the language, the names of all the other numbers are obtained by creating expressions that rely simultaneously on multiplication and addition in strict descending order of the powers of the base 10:

53,781

| five-myriads | three-thousands | seven-hundreds | eight-tens | one |
|---|---|---|---|---|
| (= 5 × 10,000 | + 3 × 1,000 | + 7 × 100 | + 8 × 10 | + 1) |

Such then are the general rules for the formation of the names of the cardinal numbers in the "base 10" system of the English language.

It must have taken a very long time for people to develop such an effective way of naming numbers, as it obviously presupposes great powers of abstraction. However, what we have laid out is evidently a purely theoretical naming system, which no language follows with absolute strictness and regularity. Particular oral traditions and the rules of individual languages produce a wide variety of irregularities: here are some characteristic examples from around the world.

### *Numbers in Tibetan*

[For sources and further details, see M. Lalou (1950), S. C. Das (1915); H. Bruce Hannah (1912). Information kindly supplied by Florence and Hélène Béquignon]

Tibetan has an individual name for each of the first ten numbers:

| *gcig* | *gnyis* | *gsum* | *bzhi* | *lnga* | *drug* | *bdun* | *brgyad* | *dgu* | *bcu* |
|---|---|---|---|---|---|---|---|---|---|
| 1 | 2 | 3 | 4 | 5 | 6 | 7 | 8 | 9 | 10 |

For numbers from 11 to 19, Tibetan uses addition:

| | | |
|---|---|---|
| 11 | *bcu-gcig* | (= 10 + 1) |
| 12 | *bcu-gnyis* | (= 10 + 2) |
| 13 | *bcu-gsum* | (= 10 + 3) |
| 14 | *bcu-bzhi* | (= 10 + 4) |
| 15 | *bcu-lnga* | (= 10 + 5) |
| 16 | *bcu-drug* | (= 10 + 6) |
| 17 | *bcu-bdun* | (= 10 + 7) |
| 18 | *bcu-brgyad* | (= 10 + 8) |
| 19 | *bcu-dgu* | (= 10 + 9) |

And for the tens, multiplication is applied:

| | | | |
|---|---|---|---|
| 20 | *gnyis-bcu* | "two-tens" | (= 2 × 10) |
| 30 | *gsum-bcu* | "three-tens" | (= 3 × 10) |
| 40 | *bzhi-bcu* | "four-tens" | (= 4 × 10) |
| 50 | *lnga-bcu* | "five-tens" | (= 5 × 10) |
| 60 | *drug-bcu* | "six-tens" | (= 6 × 10) |
| 70 | *bdun-bcu* | "seven-tens" | (= 7 × 10) |
| 80 | *brgyad-bcu* | "eight-tens" | (= 8 × 10) |
| 90 | *dgu-bcu* | "nine-tens" | (= 9 × 10) |

For a hundred (=$10^2$) there is the word *brgya*, and the corresponding multiples are obtained by the same principle of multiplication:

| | | | |
|---|---|---|---|
| 200 | *gnyis-brgya* | "two-hundreds" | (= 2 × 100) |
| 300 | *gsum-brgya* | "three-hundreds" | (= 3 × 100) |
| 400 | *bzhi-brgya* | "four-hundreds" | (= 4 × 100) |
| 500 | *lnga-brgya* | "five-hundreds" | (= 5 × 100) |
| 600 | *drug-brgya* | "six-hundreds" | (= 6 × 100) |
| 700 | *bdun-brgya* | "seven-hundreds" | (= 7 × 100) |
| 800 | *brgyad-brgya* | "eight-hundreds" | (= 8 × 100) |
| 900 | *dgu-brgya* | "nine-hundreds" | (= 9 × 100) |

There are similarly individual words for "thousand", "ten thousand" and so on, producing a very simple naming system for all intermediate numbers:

| | | |
|---|---|---|
| 21: | *gnyis-bcu rtsa gcig* | "two-tens and one" (= 2 × 10 + 1) |
| 560: | *lnga-brgya rtsa drug-bcu* | "five-hundreds and six-tens" (= 5 × 100 + 6 × 10) |

*Numbers in Mongolian*

[Source: L. Hambis (1945)]
Numbering in Mongolian is similarly decimal, but with some variations on the regular system we have seen in Tibetan. It has the following names for the first ten numbers:

| *ṇigän* | *qoyar* | *ɣurban* | *dörbän* | *tabun* | *ǰirɣu'an* | *dolo'an* | *naiman* | *yisün* | *arban* |
|---|---|---|---|---|---|---|---|---|---|
| 1 | 2 | 3 | 4 | 5 | 6 | 7 | 8 | 9 | 10 |

and proceeds in a perfectly normal way for the numbers from eleven to nineteen:

| | | |
|---|---|---|
| 11 | *arban nigän* | ("ten-one") |
| 12 | *arban qoyar* | ("ten-two") |
| . . . | . . . | . . . |

However, the tens are formed rather differently. Instead of using analytic combinations of the "two-tens", "three-tens" type, Mongolian has specific words formed from the names of the corresponding units, subjected to a kind of "declension" or alteration of the ending of the word:

| | | |
|---|---|---|
| 20 | *qorin* | (from *qoyar* = 2) |
| 30 | *ɣučin* | (from *ɣurban* = 3) |
| 40 | *döčin* | (from *dörbän* = 4) |
| 50 | *tabin* | (from *tabun* = 5) |
| 60 | *ǰirin* | (from *ǰirɣu'an* = 6) |
| 70 | *dalan* | (from *dolo'an* = 7) |
| 80 | *nayan* | (from *naiman* = 8) |
| 90 | *ǰarin* | (from *yisün* = 9) |

From one hundred, however, numbers are formed in a regular way based on multiplication and addition, as explained above:

| | | |
|---|---|---|
| 100 | *ǰa'un* | ("hundred") |
| 200 | *qoyar ǰa'un* | ("two-hundreds") |
| 300 | *ɣurban ǰa'un* | ("three-hundreds") |
| 400 | *dörbän ǰa'un* | ("four-hundreds") |
| . . . | . . . | . . . |
| 1,000 | *mingɣan* | ("thousand") |
| 2,000 | *qoyar mingɣan* | ("two-thousands") |
| 3,000 | *ɣurban mingɣan* | ("three-thousands") |
| . . . | . . . | . . . |
| 10,000 | *tümän* | ("myriad") |
| 20,000 | *qoyar tümän* | ("two-myriads") |
| . . . | . . . | . . . |

| | | | | |
|---|---|---|---|---|
| 20,541 | *qoyar tümän* | *tabun ǰa'un* | *döčin* | *nigän* |
| | "two myriads | five-hundreds | forty | one" |
| | (= 2 × 10,000 | + 5 × 100 | + 40 | + 1) |

*Ancient Turkish numbers*

[Source: A. K. von Gabain (1950)]
This section describes the numerals in spoken Turkish of the eighth century CE as deduced from Turkish inscriptions found in Mongolia. The system has some remarkable features.

The first nine numbers are as follows:

| *bir* | *iki* | *üč* | *tört* | *beš* | *altï* | *yěti* | *säkiz* | *tokuz* |
|---|---|---|---|---|---|---|---|---|
| 1 | 2 | 3 | 4 | 5 | 6 | 7 | 8 | 9 |

For the tens, the following set of names are used:

| | |
|---|---|
| 10 | *on* |
| 20 | *yegirmi* |
| 30 | *otuz* |
| 40 | *kïrk* |
| 50 | *ällig* |
| 60 | *altmïš* |
| 70 | *yětmiš* |

| | |
|---|---|
| 80 | *säkiz on* |
| 90 | *tokuz on* |

The tens from 20 to 50 do not seem to have any etymological relation with the corresponding units. However, *altmiš* (= 60) and *yetmiš*(= 70) derive respectively from *altï* (= 6) and *yeti* (= 7) by the addition of the ending *miš* (or *miš*). The words for 80 and 90 also derive from the names of 8 and 9, but by analytical combination with the word for 10, so that they mean something like "eight tens" and "nine tens".

The word for 50, however, is very probably derived from the ancient method of finger-counting, since *ällig* is clearly related to *äl* (or *älig*), the Turkish word for "hand". (Turkish finger-counting is still done in the following way: using one thumb, you touch in order on the other hand the tip of the little finger, the ring finger, the middle finger and the index finger, which gets you to 4; for 5, you raise the thumb of the "counted" hand; then you bend back the thumb and raise in order the index finger, the middle finger, the ring finger, the little finger, and finally the thumb again, so that for 10 you have all the fingers of the "counted" hand stretched out. This technique represents the trace of an even older system in which the series was extended by raising one finger of the other hand for each ten counted, so that one hand with all fingers stretched out meant 10, and the other hand with all fingers stretched out meant 50.)

The system then gives the special name of *yüz* to the number 100, and proceeds by multiplication for the names of the corresponding multiples of a hundred:

| | | | |
|---|---|---|---|
| 100 | *yüz* | | |
| 200 | *iki yüz* | "two-hundreds" | (= 2 × 100) |
| 300 | *üč yüz* | "three-hundreds" | (= 3 × 100) |
| 400 | *tört yüz* | "four-hundreds" | (= 4 × 100) |
| 500 | *beš yüz* | "five-hundreds" | (= 5 × 100) |
| 600 | *altï yüz* | "six-hundreds" | (= 6 × 100) |
| 700 | *yeti yüz* | "seven-hundreds" | (= 7 × 100) |
| 800 | *säkiz yüz* | "eight-hundreds" | (= 8 × 100) |
| 900 | *tokuz yüz* | "nine-hundreds" | (= 9 × 100) |

The word for a thousand is *bïng* (which in some Turkic dialects also means "a very large amount"), and the multiples of a thousand are similarly expressed by analytical combinations of the same type:

| | | | |
|---|---|---|---|
| 1,000 | *bïng* | | |
| 2,000 | *iki bïng* | "two-thousands" | (= 2 × 1,000) |
| 3,000 | *üč bïng* | "three-thousands" | (= 3 × 1,000) |
| 4,000 | *tört bïng* | "four-thousands" | (= 4 × 1,000) |
| 5,000 | *beš bïng* | "five-thousands" | (= 5 × 1,000) |
| 6,000 | *altï bïng* | "six-thousands" | (= 6 × 1,000) |
| 7,000 | *yeti bïng* | "seven-thousands" | (= 7 × 1,000) |
| 8,000 | *säkiz bïng* | "eight-thousands" | (= 8 × 1,000) |
| 9,000 | *tokuz bïng* | "nine-thousands" | (= 9 × 1,000) |

What is unusual about the ancient Turkish system is the way the numbers from 11 to 99 are expressed. In this range, what is given is first the unit, and then, not the multiple of ten already counted, but the multiple not yet reached, by a kind of "prospective account". This gives, for example:

| | | |
|---|---|---|
| 11 | *bir yegirmi* | literally: "one, twenty" |
| 12 | *iki yegirmi* | literally: "two, twenty" |
| 13 | *üč yegirmi* | literally: "three, twenty" |
| 21 | *bir otuz* | literally: "one, thirty" |
| 22 | *iki otuz* | literally: "two, thirty" |
| 53 | *üč altmiš* | literally: "three, sixty" |
| 65 | *beš yetmiš* | literally: "five, seventy" |
| 78 | *säkiz säkiz on* | literally: "eight, eighty" |
| 99 | *tokuz yüz* | literally: "nine, one hundred" |

What is involved is neither a multiplicative nor a subtractive principle but something like an ordinal device, as follows:

| | |
|---|---|
| 11 | "the first unit before twenty" |
| 12 | "the second unit before twenty" |
| 21 | "the first unit before thirty" |
| 23 | "the third unit before thirty" |
| 53 | "the third unit before sixty" |
| 87 | "the seventh unit before ninety" |
| 99 | "the ninth unit before a hundred" |

This way of counting is reminiscent of the way time is expressed in contemporary German, where, for "a quarter past nine" you say *viertel zehn*, meaning "a quarter of ten" (= "the first quarter before ten"), or for "half past eight" you say *halb neun*, meaning "half nine" (= "the first half before nine").

However, around the tenth century CE, under Chinese influence, which was very strong in the eastern Turkish-speaking areas, this rather special way of counting was "rationalised" (by the Uyghurs, first of all, who always have been close to Chinese civilisation). Using the Turkic stem *artuk*, meaning "overtaken by", the following expressions were created:

| | | |
|---|---|---|
| 11 | *on artukï bir* | ("ten overtaken by one") |
| 23 | *yegirmi artukï üč* | ("twenty overtaken by three") |
| 53 | *ällig artukï üč* | ("fifty overtaken by three") |
| 87 | *säkiz on artukï yeti* | ("eighty overtaken by seven") |

Whence come the simplified versions still in use today:

| | | |
|---|---|---|
| 11 | *on bir* | (= 10 + 1) |
| 23 | *yegirmi üč* | (= 20 + 3) |
| 53 | *ällig üč* | (= 50 + 3) |
| 87 | *säkiz on yeti* | (= 80 + 7) |

*Sanskrit numbering*

The numbering system of Sanskrit, the classical language of northern India, is of great importance for several related reasons. First of all, the most ancient written texts that we have of an Indo-European language are the Vedas, written in Sanskrit, from around the fifth century BCE, but with traces going as far back as the second millennium BCE. (All modern European languages with the notable exceptions of Finnish, Hungarian, Basque, and Turkish belong to the Indo-European group: see below). Secondly, Sanskrit, as the sacred language of Brahmanism (Hinduism), was used throughout India and Southeast Asia as a language of literary and scholarly expression, and (rather like Latin in mediaeval Europe) provided a means of communication between scholars belonging to communities and lands speaking widely different languages. The numbering system of Sanskrit, as a part of a written language of great sophistication and precision, played a fundamental role in the development of the sciences in India, and notably in the evolution of a place-value system.

The first ten numbers in Sanskrit are as follows:

| | |
|---|---|
| 1 | *eka* |
| 2 | *dvau, dva, dve, dvi* |
| 3 | *trayas, tisras, tri* |
| 4 | *catvaras, catasras, catvari, catur* |
| 5 | *pañca* |
| 6 | *ṣaṭ* |
| 7 | *sapta* |
| 8 | *aṣṭau, aṣṭa* |
| 9 | *náva* |
| 10 | *daśa* |

Numbers from 11 to 19 are then formed by juxtaposing the number of units and the number 10:

| | | | |
|---|---|---|---|
| 11 | *eka-daśa* | "one-ten" | (= 1 + 10) |
| 12 | *dva-daśa* | "two-ten" | (= 2 + 10) |
| 13 | *tri-daśa* | "three-ten" | (= 3 + 10) |
| 14 | *catvari-daśa* | "four-ten" | (= 4 + 10) |
| 15 | *pañca-daśa* | "five-ten" | (= 5 + 10) |
| 16 | *ṣaṭ-daśa* | "six-ten" | (= 6 + 10) |
| 17 | *sapta-daśa* | "seven-ten" | (= 7 + 10) |
| 18 | *aṣṭa-daśa* | "eight-ten" | (= 8 + 10) |
| 19 | *náva-daśa* | "nine-ten" | (= 9 + 10) |

For the following multiples of 10, Sanskrit has names with particular features:

| | |
|---|---|
| 20 | *viṃśati* |
| 30 | *triṃśati* |
| 40 | *catvāriṃśati* |
| 50 | *pañcāśat* |
| 60 | *ṣaṣti* |
| 70 | *sapti* |
| 80 | *aṣīṭi* |
| 90 | *návati* |

Broadly speaking, the names of the tens from 20 upwards are formed from a word derived from the name of the corresponding unit plus a form for the word for 10 in the plural.

One hundred is *śatam* or *śata*, and for multiples of 100 the regular formula is used:

| | | |
|---|---|---|
| 100 | *śatam, śata* | |
| 200 | *dviśata* | (= 2 × 100) |
| 300 | *triśata* | (= 3 × 100) |
| 400 | *caturśata* | (= 4 × 100) |
| 500 | *pañcaśata* | (= 5 × 100) |

For 1,000, the word *sahásram* or *sahásra* is used, in analytical combination with the names of the units, tens and hundreds to form multiples of the thousands, the ten thousands, and hundred thousands:

| | | |
|---|---|---|
| 1,000 | *sahásra* | |
| 2,000 | *dviṣahásra* | (= 2 × 1,000) |
| 3,000 | *trisahásra* | (= 3 × 1,000) |
| ... | ... | ... |
| 10,000 | *daśasahásra* | (= 10 × 1,000) |
| 20,000 | *viṃsatsahásra* | (= 20 × 1,000) |
| 30,000 | *triṃsatsahásra* | (= 30 × 1,000) |
| ... | ... | ... |

| | | |
|---|---|---|
| 100,000 | *śatasahásra* | (= 100 × 1,000) |
| 200,000 | *dviśatasahásra* | (= 200 × 1,000) |
| 300,000 | *triśatasahásra* | (= 300 × 1,000) |
| ... | ... | ... |

This gives the following expressions for intermediate numbers:

4,769:

| *nava* | *ṣaṣti* | *saptaśata* | *ca* | *catursahásra* |
|---|---|---|---|---|
| ("nine | sixty | seven-hundreds | and | four-thousands") |
| (= 9 | + 60 | + 7 × 100 | + | 4 × 1,000) |

Sanskrit thus has a decimal numbering system, like ours, but with combinations done "in reverse", that is to say starting with the units and then in *ascending* order of the powers of 10.

## WHAT IS INDO-EUROPEAN?

"Indo-European" is the name of a huge family of languages spoken nowadays in most of the European land-mass, in much of western Asia, and in the Americas. There has been much speculation about the geographical origins of the peoples who first spoke the language which has split into the many present-day branches of the Indo-European family. Some theories hold that the Indo-Europeans originally came from central Asia (the Pamir mountains, Turkestan); others maintain that they came from the flat lands of northern Germany, between the Elbe and the Vistula, and the Russian steppes, from the Danube to the Ural mountains. The question remains unresolved. All the same, some things are generally agreed. The Indo-European languages derive from dialects of a common "stem" spoken by a wide diversity of tribes who had numerous things in common. The Indo-Europeans were arable farmers, hunters, and breeders of livestock; they were patriarchal and had social ranks or castes of priests, farmers, and warriors; and a religion that involved the cult of ancestors and the worship of the stars. However, we know very little about the origins of these peoples, who acquired writing only in relatively recent times.

The Indo-European tribes began to split into different branches in the second millennium BCE at the latest, and over the following thousand years the following tribes or branches appear in early historical records: *Aryans*, in India, and *Kassites*, *Hittites*, and *Lydians* in Asia Minor; the *Achaeans, Dorians, Minoans*, and *Hellenes*, in Greece; then the *Celts* in central Europe, and the *Italics* in the Italian peninsula. Further migrations from the East occurred towards the end of the Roman Empire in the fourth to sixth centuries CE, bringing the *Germanic* tribes into western Europe.

The Indo-European language family is thus spread over a very wide area and is traditionally classified in the following branches, for each of which the earliest written traces date from different periods, but none from before the second millennium BCE:

- The *Indo-Aryan* branch: Vedic, classical Sanskrit, and their numerous modern descendants, of which there are five main groups:
  - the western group, including Sindhi, Gujurati, Landa, Mahratta, and Rajasthani
  - the central group, including Punjabi, Pahari, and Hindi
  - the eastern group, including Bengali, Bihari, and Oriya
  - the southern group (Singhalese)
  - so-called "Romany" or gypsy languages
- The *Iranian* branch, including ancient Persian (spoken at the time of Darius and Xerxes), Avestan (the language of Zoroaster), Median, Scythian, as well as several mediaeval and modern languages spoken in the area of Iran (Sogdian, Pahlavi, Caspian and Kurdish dialects, Ossetian (spoken in the Caucasus), Afghan, and Baluchi)
- A branch including the *Anatolian* language of the Hittite Empire as well as Lycian and Lydian
- The *Tokharian* branch. This language (with its two dialects, Agnaean and Kutchian) was spoken by an Indo-European population settled in Chinese Turkestan between the fifth and tenth centuries CE, but became extinct in the Middle Ages. As an ancient language related to Hittite as well as to Western branches of the Indo-European family (Greek, Latin, Celtic, Germanic), it is of great importance for historical linguists and is often used in tracing the etymologies of common Indo-European words
- The *Armenian* branch, with two dialects, western (spoken in Turkey) and eastern (spoken in Armenia)
- The *Hellenic* branch, which includes ancient dialects such as Dorian, Achaean, Creto-Minoan, as well as Homeric (classical) Greek, Koiné (the spoken language of ancient Greece), and Modern Greek
- The *Italic* branch, which includes ancient languages such as Oscan, Umbrian and Latin, and all the modern Romance languages (Italian, Spanish, Portuguese, Provençal, Catalan, French, Romanian, Sardinian, Dalmatian, Rhaeto-Romansch, etc.)
- The *Celtic* branch, which has two main groups:
  - "continental" Celtic dialects, including the extinct language of the Gauls
  - "island" Celtic, itself possessing two distinct subgroups, the Brythonic (Breton, Welsh, and Cornish) and the Gaelic (Erse, Manx, and Scots Gaelic)
- The *Germanic* branch, which has three main groups:
  - Eastern Germanic, of which the main representative is Gothic

– the Nordic languages (Old Icelandic, Old Norse, Swedish, Danish)
– Western Germanic languages, including Old High German and its mediaeval and modern descendants (German), Low German, Dutch, Friesian, Old Saxon, Anglo-Saxon, and its mediaeval and modern descendants (Old English, Middle English, contemporary British and American English)

- The *Slavic* branch, of which there are again three main groups:
  – Eastern Slavic languages (Russian, Ukrainian, and Belorussian)
  – Southern Slavic languages (Slovenian, Serbo-Croatian, Bulgarian)
  – Western Slavic languages (Czech, Slovak, Polish, Lekhitian, Sorbian, etc.)
- The *Baltic* branch, comprising Baltic, Latvian, Lithuanian, and Old Prussian
- *Albanian*, a distinct branch of the Indo-European family, with no "close relatives" and two dialects, Gheg and Tosk
- The *Thraco-Phrygian* branch, with traces found in the Balkans (Thracian, Macedonian) and in Asia Minor (Phrygian)
- And finally a few minor dialects with no close relatives, such as Venetian and Illyrian.

## INDO-EUROPEAN NUMBER-SYSTEMS

Sanskrit is thus a particular case of a very large "family" of languages (the Indo-European family) all of whose members use decimal numbering systems. The general rule that all these systems have in common is that the numbers from 1 to 9 and each of the powers of 10 have individual names, all other numbers being expressed as analytical combinations of these names.

Nonetheless, some of these languages have additional number-names that seem to have no etymological connection with the basic set of names: for example "eleven" and "twelve" in English, like the German *elf* and *zwölf*, have no obvious connection to the words for "ten" (*zehn*) and "one" (*eins*) or "ten" and "two" (*zwei*) respectively, whereas all the following numbers are formed in regular fashion:

| | ENGLISH | | GERMAN | |
|---|---|---|---|---|
| 13 | thirteen | (= three+ten) | *dreizehn* | (= *drei+zehn*) |
| 14 | fourteen | (= four+ten) | *vierzehn* | (= *vier+zehn*) |
| 15 | fifteen | (= five+ten) | *fünfzehn* | (= *fünf+zehn*) |
| 16 | sixteen | (= six+ten) | *sechszehn* | (= *sechs+zehn*) |
| 17 | seventeen | (= seven+ten) | *siebzehn* | (= *sieben+zehn*) |
| 18 | eighteen | (= eight+ten) | *achtzehn* | (= *acht+zehn*) |
| 19 | nineteen | (= nine+ten) | *neunzehn* | (= *neun+zehn*) |

The "additional" number-names in the range 11–19 in the Romance languages, on the other hand, are obvious contractions of the analytical Latin names (with the units in first position) from which they are all derived:

| | LATIN | | ITALIAN | FRENCH | SPANISH |
|---|---|---|---|---|---|
| 11 | *undecim* | ("one-ten") | *undici* | *onze* | *once* |
| 12 | *duodecim* | ("two-ten") | *dodici* | *douze* | *doce* |
| 13 | *tredecim* | ("three-ten") | *tredici* | *treize* | *trece* |
| 14 | *quattuordecim* | ("four-ten") | *quattordici* | *quatorze* | *catorce* |
| 15 | *quindecim* | ("five-ten") | *quindici* | *quinze* | *quince* |
| 16 | *sedecim* | ("six-ten") | *sedici* | *seize* | |
| 17 | *septendecim* | ("seven-ten") | | | |
| 18 | *octodecim* | ("eight-ten") | | | |
| 19 | *undeviginti* | ("one from twenty") | | | |

The remaining numbers before 20 are constructed analytically: *dix-sept, dix-huit, dix-neuf* (French), *dieci-sette, dieci-otto* (Italian), etc.

In the Germanic languages, the tens are constructed in regular fashion using an ending clearly derived from the word for "ten" on the stem of the word for the corresponding unit : in English, twenty = "two - ty", thirty = "three - ty", and so on, and in German, *zwanzig* = "*zwei - zig*", *dreissig* = *drei + sig*, and so on. In order to avoid confusion between the "teens" and the "tens" in Latin, multiples of 10, which similarly have the unit-name in first position, use the ending "*-ginta*", giving the following contractions in the Romance languages derived from it:

| | LATIN | ITALIAN | FRENCH | SPANISH |
|---|---|---|---|---|
| 30 | *triginta* | *trenta* | *trente* | *treinta* |
| 40 | *quadraginta* | *quaranta* | *quarante* | *cuarenta* |
| 50 | *quinquaginta* | *cinquanta* | *cinquante* | *cincuenta* |
| 60 | *sexaginta* | *sessanta* | *soixante* | *sessanta* |
| 70 | *septuaginta* | *settanta* | *septante** | *setenta* |
| 80 | *octoginta* | *ottanta* | *octante** | *ochenta* |
| 90 | *nonaginta* | *novanta* | *nonante** | *noventa* |

The French numerals marked with an asterisk are the "regular" versions found only in Belgium and French-speaking Switzerland; "standard" French uses irregular expressions for 70 (*soixante-dix*, "sixty-ten"), 80 (*quatre-vingts*, "four-twenty"), and 90 (*quatre-vingt-dix*, "four-twenty-ten"). In addition, of course, we have omitted from the table above the Latin and Romance names for the number 20, which seems to be a problem at first sight. In Latin it is *viginti*, a word with no relation to the words for "ten" (*decem*) or for "two" (*duo*); and its Romance derivatives, with the exception of Romanian, follow the irregularity (*venti* in Italian, *vingt* in French, *veinte* in Spanish). So where does the "Romance twenty" come from?

*Roots*

The richness of the descendance of the original Indo-European language means that, by comparison and deduction, it is possible to reconstruct the form that many basic words must have had in the "root" or "stem" language, even though no written trace remains of it. Indo-European root words, being hypothetical, are therefore always written with an asterisk. The original number-set is believed to have been this:

1 **oi-no*, **oi-ko*, **oi-wo*
2 **dwō*, **dwu*, **dwoi*
3 **tri* (and derivative forms: **treyes*, **tisores*)
4 **kwetwores*, **kwetesres*, **kwetwor*
5 **pénkwe*, **kwenkwe*
6 **seks*, **sweks*
7 **septm*
8 **oktṓ*, **oktu*
9 **néwn̥*
10 **dékm̥*

This helps us to see that despite their apparent difference, the words for "one" in Sanskrit, Avestan, and Czech, for example (respectively *eka*, *aeva* and *jeden*) are all derived from the same "root" or prototype, as are the Latin *unus*, German *eins* and Swedish *en*.

All trace has been lost of the concrete meanings that these Indo-European number-names might have had originally. However, Indo-European languages do bear the visible marks of that long-distant time when, in the absence of any number-concept higher than two, the word for all other numbers meant nothing more than "many".

The first piece of evidence of this ancient number-limit is the grammatical distinction made in several Indo-European languages between the *singular*, the *dual*, and the *plural*. In classical Greek, for example, *ho lukos* means "the wolf", *hoi lukoi* means "the wolves", but for "two wolves" a special ending, the mark of the "dual", is used: *tṓ luko*.

Another piece of the puzzle is provided by the various special meanings and uses of words closely associated with the name of the number 3. Anglo-Saxon *thria* (which becomes "three" in modern English) is related to the word *throp*, meaning a pile or heap; and words like *throng* are similarly derived from a common Germanic root having the sense of "many". In the Romance languages there are even more evident connections between the words for "three" and words expressing plurality or intensity: the Latin word *tres* (three) has the same root as the preposition and prefix *trans-* (with meanings related to "up until", "through", "beyond"), and in French, derivations from this common stem produce words like *très* ("very"), "*trop*" (too much), and even *troupe* ("troop"). It can be deduced from these and many other instances that in the original Indo-European stem language, the name of the number "three" (*tri*) was also the word for plurality, multiplicity, crowds, piles, heaps, and for the beyond, for what was beyond reckoning.

The number-systems of the Indo-European languages, which are all strictly decimal, have remained amazingly stable over many millennia, even whilst most other features of the languages concerned have changed beyond recognition and beyond mutual comprehension. Even the apparent irregularities within the system are for the most part explicable within the logic of the original decimal structure – for example, the problem mentioned above of the "Romance twenty". French *vingt*, Spanish *veinte*, etc. derive from Latin *viginti*, which is itself fairly self-evidently a derivative of the Sanskrit *viṃsati*. And Sanskrit "twenty" is not irregular at all, being a contraction of a strictly decimal *dvi-daśati* ("two-tens") ⇒ *visati* ⇒ *viṃsati*. Similar derivations can be found in other branches of the Indo-European family of languages. In Avestan, 20 is *visaiti*, formed from *baē*, "two", and *dása* (= 10); and in Tokharian A, where *wu* = 2 and *śäk* = 10, *wi-śäki* (= 2 × 10) became *wiki*, "twenty".

THE NAMES OF THE NUMBER 1

| Indo-European prototypes: | **oi-no*, **oi-ko*, **oi-wo* |
|---|---|
| SANSKRIT | *eka* |
| AVESTAN | *aēva* |
| GREEK | *hén* |
| EARLY LATIN<br>LATIN<br>ITALIAN<br>SPANISH<br>FRENCH<br>PORTUGUESE<br>ROMANIAN | *oinos, oinom*<br>*unus, unum*<br>*uno*<br>*uno*<br>*un*<br>*um*<br>*uno* |
| OLD ERSE<br>MODERN IRISH<br>BRETON<br>SCOTS GAELIC<br>WELSH | *oen*<br>*oin*<br>*eun*<br>*un*<br>*un* |
| GOTHIC<br>DUTCH<br>OLD ICELANDIC<br>SWEDISH<br>DANISH<br>OLD SAXON<br>ANGLO-SAXON<br>ENGLISH<br>OLD HIGH GERMAN<br>GERMAN | *ain (-s)*<br>*een*<br>*einn*<br>*en*<br>*en*<br>*en*<br>*an*<br>*one*<br>*ein, eins*<br>*ein* |
| CHURCH SLAVONIC<br>RUSSIAN<br>CZECH<br>POLISH | *inŭ*<br>*odin*<br>*jeden*<br>*jeden* |
| LITHUANIAN<br>BALTIC | *vienas*<br>*vienes* |

FIG. 2.4A.

THE NAMES OF THE NUMBER 2

| Indo-European prototypes: | **dwō*, **dwu*, **dwoi* |
|---|---|
| SANSKRIT | *dvau, dva, dvi* |
| AVESTAN | *baè* |
| HITTITE | *tā* |
| TOKHARIAN A | *wu, we* |
| ARMENIAN | *erku* |
| GREEK | *dùô* |
| LATIN<br>SPANISH<br>FRENCH<br>ROMANIAN | *duo, duae*<br>*dos*<br>*deux*<br>*doi* |
| OLD ERSE<br>MODERN IRISH<br>BRETON<br>SCOTS GAELIC<br>WELSH | *dáu, dő*<br>*da*<br>*diou*<br>*dow*<br>*dwy, dau* |
| GOTHIC<br>DUTCH<br>OLD ICELANDIC<br>SWEDISH<br>DANISH<br>OLD SAXON<br>ANGLO-SAXON<br>ENGLISH<br>OLD HIGH GERMAN<br>GERMAN | *twai, twa*<br>*twee*<br>*tveir*<br>*två*<br>*to*<br>*twene*<br>*twegen*<br>*two*<br>*zwene*<br>*zwei* |
| CHURCH SLAVONIC<br>RUSSIAN<br>POLISH | *dŭva, dŭvě*<br>*dva*<br>*dwa* |
| LITHUANIAN | *dù, dvi* |
| ALBANIAN | *dy, dyj* |

FIG. 2.4B.

THE NAMES OF THE NUMBER 3

| Indo-European prototypes: | **treyes*, **tisores*, **tri* |
|---|---|
| SANSKRIT | *trayas, tisras, tri* |
| AVESTAN | *thrayō, tisrō, tri* |
| HITTITE | *tri* |
| TOKHARIAN B | *trai* |
| ARMENIAN | *erekh* |
| GREEK | *treis* |
| OSCAN<br>LATIN<br>ITALIAN<br>SPANISH<br>FRENCH<br>ROMANIAN | *trís*<br>*trēs, tria*<br>*tre*<br>*tres*<br>*trois*<br>*trei* |
| OLD ERSE<br>WELSH | *téoir, trí*<br>*tri, tair* |
| GOTHIC<br>DUTCH<br>OLD ICELANDIC<br>SWEDISH<br>OLD SAXON<br>ANGLO-SAXON<br>ENGLISH<br>OLD HIGH GERMAN<br>GERMAN | *threis, thrija*<br>*drie*<br>*thrir*<br>*tre*<br>*thria*<br>*thri*<br>*three*<br>*dri*<br>*drei* |
| CHURCH SLAVONIC<br>RUSSIAN<br>POLISH | *trije, tri*<br>*tri*<br>*trzy* |
| LITHUANIAN | *trỹs* |
| ALBANIAN | *tre, tri* |

FIG. 2.4C.

THE NAMES OF THE NUMBER 4

| Indo-European prototypes: | **kwetwores*, **kwetesres*, **kwetwor* |
|---|---|
| SANSKRIT | *catvaras, catasras, catvari, catur* |
| AVESTAN | *čathwārō* |
| TOKHARIAN A<br>TOKHARIAN B | *śtwar*<br>*śtwer* |
| ARMENIAN | *čorkh* |
| ANCIENT GREEK | *téttares, téssares, tétores* |
| OSCAN<br>LATIN<br>ITALIAN<br>SPANISH<br>FRENCH<br>ROMANIAN | *pettiur, petora*<br>*quattuor*<br>*quattro*<br>*cuatro*<br>*quatre*<br>*patru* |
| OLD ERSE<br>BRETON<br>WELSH<br>SCOTS GAELIC | *cethir, cethoir*<br>*pevar*<br>*pedwar*<br>*peswar* |
| GOTHIC<br>OLD ICELANDIC<br>SWEDISH<br>OLD SAXON<br>ANGLO-SAXON<br>ENGLISH<br>OLD HIGH GERMAN<br>GERMAN | *fidwor*<br>*fjorer*<br>*fyra*<br>*fiuwar*<br>*foewer*<br>*four*<br>*vier*<br>*vier* |
| CHURCH SLAVONIC<br>RUSSIAN<br>CZECH | *četyre*<br>*četyre*<br>*ctyri* |
| POLISH | *cztery* |
| LITHUANIAN<br>BALTIC | *keturi*<br>*keturi* |

FIG. 2.4D.

THE NAMES OF THE NUMBER 5

| Indo-European prototypes: | *pénkwe, *kwenkwe |
|---|---|
| SANSKRIT | *pañča* |
| AVESTAN | *panča* |
| HITTITE | *panta* |
| TOKHARIAN A | *päñ* |
| TOKHARIAN B | *piš* |
| ARMENIAN | *hing* |
| GREEK | *pénte* |
| LATIN | *quinque* |
| SPANISH | *cinco* |
| FRENCH | *cinq* |
| ROMANIAN | *cinci* |
| OLD ERSE | *cóic* |
| MODERN IRISH | *coic* |
| WELSH | *pump* |
| BRETON | *pemp* |
| GOTHIC | *fimf* |
| DUTCH | *viif* |
| OLD ICELANDIC | *fimm* |
| SWEDISH | *fem* |
| OLD SAXON | *fif* |
| ANGLO-SAXON | *fif* |
| ENGLISH | *five* |
| OLD HIGH GERMAN | *finf* |
| GERMAN | *fünf* |
| CHURCH SLAVONIC | *petĭ* |
| RUSSIAN | *piat'* |
| POLISH | *piec* |
| LITHUANIAN | *penki* |
| ALBANIAN | *pęsë* |

FIG. 2.4E.

THE NAMES OF THE NUMBER 6

| Indo-European prototypes: | *seks, *sweks |
|---|---|
| SANSKRIT | *ṣaṭ* |
| AVESTAN | *xšvaš* |
| TOKHARIAN A | *ṣäk* |
| ARMENIAN | *vec* |
| ANCIENT GREEK | *wéks* |
| MODERN GREEK | *héx* |
| LATIN | *sex* |
| ITALIAN | *sei* |
| SPANISH | *seis* |
| FRENCH | *six* |
| ROMANIAN | *shase* |
| OLD ERSE | *sé* |
| MODERN IRISH | *se* |
| WELSH | *chwech* |
| BRETON | *c'houec'h* |
| GOTHIC | *saihs* |
| DUTCH | *zes* |
| OLD ICELANDIC | *sex* |
| SWEDISH | *sex* |
| OLD SAXON | *sehs* |
| ANGLO-SAXON | *six* |
| ENGLISH | *six* |
| OLD HIGH GERMAN | *sehs* |
| GERMAN | *sechs* |
| CHURCH SLAVONIC | *šestĭ* |
| RUSSIAN | *chest'* |
| POLISH | *szesc* |
| LITHUANIAN | *sesi* |
| ALBANIAN | *giashtë* |

FIG. 2.4F.

THE NAMES OF THE NUMBER 7

| Indo-European prototype: | *septṃ |
|---|---|
| SANSKRIT | *sapta* |
| AVESTAN | *hapta* |
| HITTITE | *sipta* |
| TOKHARIAN A | *ṣpät* |
| ARMENIAN | *ewhtn* |
| GREEK | *heptá* |
| LATIN | *septem* |
| SPANISH | *siete* |
| FRENCH | *sept* |
| ROMANIAN | *shapte* |
| OLD ERSE | *secht* |
| MODERN IRISH | *secht* |
| WELSH | *saith* |
| BRETON | *seiz* |
| GOTHIC | *sibun* |
| DUTCH | *zeven* |
| OLD ICELANDIC | *siau* |
| SWEDISH | *sju* |
| OLD SAXON | *sibun* |
| ENGLISH | *seven* |
| OLD HIGH GERMAN | *siben* |
| GERMAN | *sieben* |
| CHURCH SLAVONIC | *sedmĭ* |
| RUSSIAN | *sem'* |
| POLISH | *siedem* |
| LITHUANIAN | *septyni* |

FIG. 2.4G.

THE NAMES OF THE NUMBER 8

| Indo-European prototypes: | *októ, *oktu |
|---|---|
| SANSKRIT | *aṣṭ'á, aṣṭau* |
| AVESTAN | *asta* |
| TOKHARIAN B | *okt* |
| ARMENIAN | *uth* |
| GREEK | *okto* |
| LATIN | *octó* |
| SPANISH | *ochó* |
| FRENCH | *huit* |
| ROMANIAN | *opt* |
| OLD ERSE | *ocht* |
| MODERN IRISH | *ocht* |
| WELSH | *wyth* |
| BRETON | *eiz* |
| GOTHIC | *ahtau* |
| DUTCH | *acht* |
| OLD ICELANDIC | *atta* |
| SWEDISH | *åtta* |
| OLD SAXON | *ahto* |
| ANGLO-SAXON | *eahta* |
| ENGLISH | *eight* |
| OLD HIGH GERMAN | *ahto* |
| GERMAN | *acht* |
| CHURCH SLAVONIC | *osmi* |
| RUSSIAN | *vosem'* |
| POLISH | *osiem* |
| LITHUANIAN | *aštuoni* |

FIG. 2.4H.

THE NAMES OF THE NUMBER 9

Indo-European prototype: **néwn*

| | |
|---|---|
| SANSKRIT | *náva* |
| AVESTAN | *nava* |
| TOKHARIAN A<br>TOKHARIAN B | *ñu*<br>*ñu* |
| ARMENIAN | *inn* |
| GREEK | *en-néa* |
| LATIN<br>ITALIAN<br>SPANISH<br>FRENCH<br>ROMANIAN<br>PORTUGUESE | *novem*<br>*nove*<br>*nueve*<br>*neuf*<br>*noue*<br>*noue* |
| OLD ERSE<br>MODERN IRISH<br>WELSH<br>BRETON | *nóin*<br>*nói*<br>*naw*<br>*nao* |
| GOTHIC<br>DUTCH<br>OLD ICELANDIC<br>SWEDISH<br>OLD SAXON<br>ANGLO-SAXON<br>ENGLISH<br>OLD HIGH GERMAN<br>GERMAN | *nium*<br>*negon*<br>*nio*<br>*nio*<br>*nigun*<br>*nigon*<br>*nine*<br>*niun*<br>*neun* |
| CZECH<br>RUSSIAN<br>POLISH | *devet*<br>*deviat'*<br>*dziewiec* |
| LITHUANIAN | *devyni* |
| ALBANIAN | *nëndë* |

FIG. 2.4I.

THE NAMES OF THE NUMBER 10

Indo-European prototype: **dékm̥*

| | |
|---|---|
| SANSKRIT | *dáśa* |
| AVESTAN | *dasa* |
| TOKHARIAN A<br>TOKHARIAN B | *śäk*<br>*śak* |
| ARMENIAN | *tasn* |
| GREEK | *déka* |
| LATIN<br>ITALIAN<br>SPANISH<br>FRENCH<br>ROMANIAN<br>PORTUGUESE | *decem*<br>*dieci*<br>*diez*<br>*dix*<br>*zece*<br>*dez* |
| OLD ERSE<br>MODERN IRISH<br>WELSH<br>BRETON | *deich*<br>*deich*<br>*deg*<br>*dek* |
| GOTHIC<br>DUTCH<br>OLD ICELANDIC<br>SWEDISH<br>OLD SAXON<br>ANGLO-SAXON<br>ENGLISH<br>OLD HIGH GERMAN<br>GERMAN | *taikun*<br>*tien*<br>*tio*<br>*tio*<br>*techan*<br>*tyn*<br>*ten*<br>*zehan*<br>*zehn* |
| CZECH<br>RUSSIAN<br>POLISH | *deset*<br>*desiat'*<br>*dziesiec* |
| LITHUANIAN | *desimt* |
| ALBANIAN | *dietë* |

FIG. 2.4J.

| | LATIN | ITALIAN | FRENCH | SPANISH | ROMANIAN |
|---|---|---|---|---|---|
| 1 | *unus* | *uno* | *un* | *uno* | *uno* |
| 2 | *duo* | *due* | *deux* | *dos* | *doi* |
| 3 | *tres* | *tre* | *trois* | *tres* | *trei* |
| 4 | *quattuor* | *quattro* | *quatre* | *cuatro* | *patru* |
| 5 | *quinque* | *cinque* | *cinq* | *cinco* | *cinci* |
| 6 | *sex* | *sei* | *six* | *seis* | *shase* |
| 7 | *septem* | *sette* | *sept* | *siete* | *shapte* |
| 8 | *octo* | *otto* | *huit* | *ocho* | *opt* |
| 9 | *novem* | *nove* | *neuf* | *nueve* | *noue* |
| 10 | *decem* | *dieci* | *dix* | *diez* | *zece* |
| 11 | *undecim* | *undici* | *onze* | *once* | *un spree zece* |
| 12 | *duodecim* | *dodici* | *douze* | *doce* | *doi spree zece* |
| 20 | *viginti* | *venti* | *vingt* | *veinte* | *doua-zeci* |
| 30 | *triginta* | *trenta* | *trente* | *treinta* | *trei-zeci* |
| 40 | *quadraginta* | *quaranta* | *quarante* | *cuarenta* | *patru-zeci* |
| 50 | *quinquaginta* | *cinquanta* | *cinquante* | *cincuenta* | *cinci-zeci* |
| 60 | *sexaginta* | *sessanta* | *soixante* | *sesenta* | *shase-zeci* |
| 70 | *septuaginta* | *settanta* | *soixante-dix* | *setenta* | *shapte-zeci* |
| 80 | *octoginta* | *ottanta* | *quatre-vingts* | *ochenta* | *opt-zeci* |
| 90 | *nonaginta* | *novanta* | *quatre-vingt-dix* | *noventa* | *noua-zeci* |
| 100 | *centum* | *cento* | *cent* | *ciento* | *o suta* |
| 1,000 | *mille* | *mille* | *mille* | *mil* | *o mie* |

| | GOTHIC | OLD HIGH GERMAN | GERMAN | ANGLO-SAXON | ENGLISH |
|---|---|---|---|---|---|
| 1 | *ains* | *ein* | *eins* | *an* | *one* |
| 2 | *twa* | *zwene* | *zwei* | *twegen* | *two* |
| 3 | *preis* | *dri* | *drei* | *pri* | *three* |
| 4 | *fidwoor* | *vier* | *vier* | *feower* | *four* |
| 5 | *fimf* | *fünf* | *fünf* | *fif* | *five* |
| 6 | *saíhs* | *sehs* | *sechs* | *six* | *six* |
| 7 | *sibun* | *siben* | *sieben* | *seofou* | *seven* |
| 8 | *ahtaú* | *ahte* | *acht* | *eahta* | *eight* |
| 9 | *niun* | *niun* | *neun* | *nigon* | *nine* |
| 10 | *taíhun* | *zehan* | *zehn* | *tyn* | *ten* |
| 11 | *ain-lif* | *einlif* | *elf* | *endleofan* | *eleven* |
| 12 | *twa-lif* | *zwelif* | *zwölf* | *twelf* | *twelve* |
| 20 | *twai-tigjus* | *zwein-zug* | *zwanzig* | *twentig* | *twenty* |
| 30 | *threo-tigjus* | *driz-zug* | *dreißig* | *thritig* | *thirty* |
| 40 | *fidwor-tigjus* | *fior-zug* | *vierzig* | *feowertig* | *forty* |
| 50 | *fimf-tigjus* | *finf-zug* | *fünfzig* | *fiftig* | *fifty* |
| 60 | *saíhs-tigjus* | *sehs-zug* | *sechzig* | *sixtig* | *sixty* |
| 70 | *sibunt-ehund* | *sibun-zo* | *siebzig* | *hund-seofontig* | *seventy* |
| 80 | *ahtaút-ehund* | *ahto-zo* | *achtzig* | *hund-eahtatig* | *eighty* |
| 90 | *niunt-ehund* | *niun-zo* | *neunzig* | *hund-nigontig* | *ninety* |
| 100 | *taíhun-taíhund* | *zehan-zo* | *hundert* | *hund-teontig* | *hundred* |
| 1,000 | *thusundi* | *dusent* | *tausend* | *thusund* | *thousand* |

FIG. 2.5. *The decimal nature of Indo-European number-names*

## OTHER SOLUTIONS TO THE PROBLEM OF THE BASE

Not all civilisations came up with the same solution to the problem of the base. In other words, base 10 is not the only way of constructing a number-system.

There are many examples of numeration built on a base of 5. For example: Api, a language spoken in the New Hebrides (Oceania), gives individual names to the first five numbers only:

| | |
|---|---|
| 1 | *tai* |
| 2 | *lua* |
| 3 | *tolu* |
| 4 | *vari* |
| 5 | *luna* (literally, "the hand") |

and then uses compounds for the numbers from 6 to 10:

| | | |
|---|---|---|
| 6 | *otai* | (literally, "the new one") |
| 7 | *olua* | (literally, "the new two") |
| 8 | *otolu* | (literally, "the new three") |
| 9 | *ovari* | (literally, "the new four") |
| 10 | *lualuna* | (literally, "two hands") |

The name of 10 then functions as a new base unit:

| | | |
|---|---|---|
| 11 | *lualuna tai* | $(= 2 \times 5 + 1)$ |
| 12 | *lualuna lua* | $(= 2 \times 5 + 2)$ |
| 13 | *lualuna tolu* | $(= 2 \times 5 + 3)$ |
| 14 | *lualuna vari* | $(= 2 \times 5 + 4)$ |
| 15 | *toluluna* | $(= 3 \times 5)$ |
| 16 | *toluluna tai* | $(= 3 \times 5 + 1)$ |
| 17 | *toluluna lua* | $(= 3 \times 5 + 2)$ |

and so on. [Source: T. Dantzig (1930), p. 18]

Languages that use base 5 or have traces of it in their number-systems include Carib and Arawak (N. America); Guarani (S. America); Api and Houailou (Oceania); Fulah, Wolof, Serere (Africa), as well as some other African languages: Dan (in the Mande group), Bete (in the Kroo group), and Kulango (one of the Voltaic languages); and in Asia, Khmer. [See: M. Malherbe (1995); F. A. Pott (1847)].

Other civilisations preferred base 20 – the "vigesimal base" – by which things are counted in packets or groups of twenty. Amongst them we find the Tamanas of the Orinoco (Venezuela), the Eskimos or Inuits (Greenland), the Ainus in Japan and the Zapotecs and Maya of Mexico. The Mayan calendar consisted of "months" of 20 days, and laid out cycles of 20 years, 400 years ($= 20^2$) 8,000 years ($= 20^3$), 160,000 years ($= 20^4$), 3,200,000 years ($= 20^5$), and even 64,000,000 years ($= 20^6$).

Like all the civilisations of pre-Columbian Central America, the Aztecs and Mixtecs measured time and counted things in the same way, as shown in numerous documents seized by the conquistadors. The goods collected by Aztec administrators from subjugated tribes were all quantified in vigesimal terms, as Jacques Soustelle explains:

> For instance, *Toluca* was supposed to provide twice a year 400 loads of cotton cloth, 400 loads of decorated *ixtle* cloaks, 1,200 ($3 \times 20^2$) loads of white *ixtle* cloth . . . *Quahuacan* gave four yearly tributes of 3,600 ($9 \times 20^2$) beams and planks, two yearly tributes of 800 ($2 \times 20^2$) loads of cotton cloth and the same number of loads of *ixtle* cloth . . . *Quauhnahuac* supplied the Imperial Exchequer with twice-yearly deliveries of 3,200 ($8 \times 20^2$) loads of cotton cloaks, 400 loads of loin-cloths, 400 loads of women's clothing, 2,000 ($5 \times 20^2$) ceramic vases, 8,000 ($20^3$) sheaves of "paper" . . .
>
> [From the *Codex Mendoza*]

This is how the Aztec language gives form to a quinary-vigesimal base:

| | | | |
|---|---|---|---|
| 1 | *ce* | 11 | *matlactli-on-ce* (10 + 1) |
| 2 | *ome* | 12 | *matlactli-on-ome* (10 + 2) |
| 3 | *yey* | 13 | *matlactli-on-yey* (10 + 3) |
| 4 | *naui* | 14 | *matlactli-on-naui* (10 + 4) |
| 5 | *chica* or *macuilli* | 15 | *caxtulli* |
| 6 | *chica-ce* (5 + 1) | 16 | *caxtulli-on-ce* (15 + 1) |
| 7 | *chica-ome* (5 + 2) | 17 | *caxtulli-on-ome* (15 + 2) |
| 8 | *chica-ey* (5 + 3) | 18 | *caxtulli-on-yey* (15 + 3) |
| 9 | *chica-naui* (5 + 4) | 19 | *caxtulli-on-naui* (15 + 4) |
| 10 | *matlactli* | 20 | *cem-poualli* (1 × 20, "a score") |

| | | |
|---|---|---|
| 30 | *cem-poualli-on-matlactli* | (20 + 10) |
| 40 | *ome-poualli* | (2 × 20) |
| 50 | *ome-poualli-on-matlactli* | (2 × 20 + 10) |
| 100 | *macuil-poualli* | (5 × 20) |
| 200 | *matlactli-poualli* | (10 × 20) |
| 300 | *caxtulli-poualli* | (15 × 20) |
| 400 | *cen-tzuntli* | (1 × 400, "one four-hundreder") |
| 800 | *ome-tzuntli* | (2 × 400) |
| 1,200 | *yey-tzuntli* | (3 × 400) |
| 8,000 | *cen-xiquipilli* | (1 × 8,000, "one eight-thousander") |

FIG. 2.6.

There are many populations outside of America and Europe (for instance, the Malinke of Upper Senegal and Guinea, the Banda of Central Africa, the Yebu and Yoruba people of Upper Senegal and Nigeria, etc.) who continue to count in this fashion. Yebu numeration is as follows, according to C. Zaslavsky (1973):

| | | |
|---|---|---|
| 1 | *otu* | |
| 2 | *abuo* | |
| 3 | *ato* | |
| 4 | *ano* | |
| 5 | *iso* | |
| 6 | *isii* | |
| 7 | *asaa* | |
| 8 | *asato* | |
| 9 | *toolu* | |
| 10 | *iri* | |
| 20 | *ohu* | |
| 30 | *ohu na iri* | (= 20 +10) |
| 40 | *ohu abuo* | (= 20 × 2) |
| 50 | *ohu abuo na iri* | (= 20 × 2 +10) |
| 60 | *ohu ato* | (= 20 × 3) |
| . . . | . . . | . . . |
| 100 | *ohu iso* | (= 20 × 5) |
| 200 | *ohu iri* | (= 20 × 10) |
| . . . | . . . | . . . |
| 400 | *nnu* | (= $20^2$) |
| 8,000 | *nnu khuru ohu* | (= $20^3$ = "400 meets 20") |
| 160,000 | *nnu khuru nnu* | (= $20^4$ = "400 meets 400") |

The Yoruba, however, proceed in a quite special way, using additive and subtractive methods alternately [Zaslavsky (1973)]:

| | | |
|---|---|---|
| 1 | *ookan* | |
| 2 | *eeji* | |
| 3 | *eeta* | |
| 4 | *eerin* | |
| 5 | *aarun* | |
| 6 | *eeta* | |
| 7 | *eeje* | |
| 8 | *eejo* | |
| 9 | *eesan* | |
| 10 | *eewaa* | |
| 11 | *ookan laa* | (= 1 + 10: *laa* from *le ewa*, "added to 10") |
| 12 | *eeji laa* | (= 2 + 10) |
| 13 | *eeta laa* | (= 3 + 10) |
| 14 | *eerin laa* | (= 4 + 10) |
| 15 | *eedogun* | (= 20 – 5; from *aarun din ogun*,"5 taken from 20") |
| 16 | *erin din logun* | (= 20 – 4) |
| 17 | *eeta din logun* | (= 20 – 3) |
| 18 | *eeji din logun* | (= 20 – 2) |
| 19 | *ookan din logun* | (= 20 – 1) |
| 20 | *ogun* | |
| 21 | *ookan le loogun* | (= 20 + 1) |
| 25 | *eedoogbon* | (= 30 – 5) |
| 30 | *ogbon* | |
| 35 | *aarun din logoji* | (= (20 × 2) – 5) |
| 40 | *logoji* | (= 20 × 2) |
| 50 | *aadota* | (= (20 × 3) –10) |
| 60 | *ogota* | (= 20 × 3) |
| . . . | . . . | . . . |
| 100 | *ogorun* | (= 20 × 5) |
| . . . | . . . | . . . |
| 400 | *irinwo* | |
| 2,000 | *egbewa* | (= (20 × 10) × 10) |
| 4,000 | *egbaaji* | (= 2,000 × 2) |
| 20,000 | *egbaawaa* | (= 2,000 × 10) |
| 40,000 | *egbaawaa lonan meji* | (= (2,000 × 10) × 2) |
| 1,000,000 | *egbeegberun* | (literally: "1,000 × 1,000") |

The source of this bizarre vigesimal system lies in the Yorubas' traditional use of cowrie shells as money: the shells are always gathered in "packets" of 5, 20, 200 and so on.

According to Mann (JAI, 16), Yoruba number-names have two meanings – the number itself, and also the things that the Yoruba count most of all, namely cowries. "Other objects are always reckoned against an equivalent number of cowries . . ." he explains. In other words, Yoruba numbering retains within it the ancient tradition of purely cardinal numeration based on matching sets.

Various other languages around the world retain obvious traces of a 20-based (vigesimal) number-system. For example, Khmer (spoken in Cambodia) has some combinations based on an obsolete word for 20, and, according to F. A. Pott (1847), used to have a special word (*slik*) for 400 (= 20 × 20). Such features are of course also to be found in European languages, and nowhere more clearly than in the English word *score*. "Four score and seven years ago . . . " is the famous opening sentence of Abraham Lincoln's Gettysburg Address. Since *to score* also means to scratch, mark, or incise (wood, stone or paper), we can see the very ancient origin of its use for the number 20: a *score* was originally a counting stick "scored" with twenty notches.

French also has many traces of vigesimal counting. The number 80 is "four-twenties" (*quatre-vingts*) in modern French, and until the seventeenth century other multiples of twenty were in regular use. *Six-vingts* (6 x 20 = 120) can be found in Molière's *Le Bourgeois Gentilhomme* (Act III, scene iv); the seventeenth-century corps of the sergeants of the city of Paris, who numbered 220 in all, was known as the *Corps des Onze-Vingts* (11 x 20), and the hospital, originally built by Louis XI to house 300 blind veteran soldiers, was and still is called the *Hôpital des Quinze-Vingts* (15 x 20 = 300).

Danish also has a curious vigesimal feature. The numbers 60 and 80 are expressed as "three times twenty" (*tresindstyve*) and "four times twenty" (*firsindstyve*); 50, 70 and 90, moreover, are *halvtresindstyve*, *halvfirsindstyve*, and *halvfemsindstyve*, literally "half three times twenty", "half four times twenty", and "half five times twenty", respectively. The prefix "half" means that only half of the last of the multiples of 20 should be counted. This accords with the kind of "prospective account" that we observed in ancient Turkish numeration (see above, p. 000):

50 = $3 \times 20$ *minus* half of the third twenty = $3 \times 20 - 10$
70 = $4 \times 20$ *minus* half of the third twenty = $4 \times 20 - 10$
90 = $5 \times 20$ *minus* half of the third twenty = $5 \times 20 - 10$

Even clearer evidence of vigesimal reckoning is found in Celtic languages (Breton, Welsh, Irish). In modern Irish, for example, despite the fact that 100 and 1,000 have their own names by virtue of the decimality that is common to all Indo-European languages, the tens from 20 to 50 are expressed as follows:

| | | |
|---|---|---|
| 20 | *fiche* | ("twenty") |
| 30 | *deich ar fiche* | ("ten and twenty") |
| 40 | *da fiche* | ("two-twenty") |
| 50 | *deich ar da fiche* | ("ten and two-twenty") |

We can only presume that the Indo-European peoples who settled long ago in regions stretching from Scandinavia to the north of Spain, including the British Isles and parts of what is now France, found earlier inhabitants whose number-system used base 20, which they adopted for the commonest numbers up to 99, integrating these particular vigesimal expressions into their own Indo-European decimal system. Since all trace of the languages of the pre-Indo-European inhabitants of Western Europe has disappeared, this explanation, though plausible, is only speculation, but it is supported, if not confirmed, by the use of base 20 in the numbering system of the Basques, one of the few non-Indo-European languages spoken in Western Europe and whose presence is not accounted for by any recorded invasion or conquest.

| | IRISH | | WELSH | | BRETON | |
|---|---|---|---|---|---|---|
| 1 | *oin* | | *un* | | *eun* | |
| 2 | *da* | | *dau* | | *diou* | |
| 3 | *tri* | | *tri* | | *tri* | |
| 4 | *cethir* | | *petwar* | | *pevar* | |
| 5 | *coic* | | *pimp* | | *pemp* | |
| 6 | *se* | | *chwe* | | *chouech* | |
| 7 | *secht* | | *seith* | | *seiz* | |
| 8 | *ocht* | | *wyth* | | *eiz* | |
| 9 | *nói* | | *naw* | | *nao* | |
| 10 | *deich* | | *dec, deg* | | *dek* | |
| 11 | *oin deec* | 1 + 10 | *un ar dec* | 1 + 10 | *unnek* | 1 + 10 |
| 12 | *da deec* | 2 + 10 | *dou ar dec* | 2 + 10 | *daou-zek* | 2 + 10 |
| 13 | *tri deec* | 3 + 10 | *tri ar dec* | 3 + 10 | *tri-zek* | 3 + 10 |
| 14 | *cethir deec* | 4 + 10 | *petwar ac dec* | 4 + 10 | *pevar-zek* | 4 + 10 |
| 15 | *coic deec* | 5 + 10 | *hymthec* | 5 + 10 | *pem-zek* | 5 + 10 |
| 16 | *se deec* | 6 + 10 | *un ar hymthec* | 1 + 15 | *choue-zek* | 6 + 10 |
| 17 | *secht deec* | 7 + 10 | *dou ar hymthec* | 2 + 15 | *seit-zek* | 7 + 10 |
| 18 | *ocht deec* | 8 + 10 | *tri ar hymthec*[1] | 3 + 15 | *eiz-zek*[2] | 8 + 10 |
| 19 | *noi deec* | 9 + 10 | *pedwar ar hymthec* | 4 + 15 | *naou-zek* | 9 + 10 |
| 20 | *fiche* | 20 | *ugeint* | 20 | *ugent* | 20 |
| 30 | *deich ar fiche* | 10 + 20 | *dec ar ugeint* | 10 + 20 | *tregont* | |
| 40 | *da fiche* | $2 \times 20$ | *de-ugeint* | $2 \times 20$ | *daou-ugent* | $2 \times 20$ |
| 50 | *deich ar dafiche* | $10 + (2 \times 20)$ | *dec ar de-ugeint* | $10 + (2 \times 20)$ | *hanter-kant* | half-100 |
| 60 | *tri fiche* | $3 \times 20$ | *tri-ugeint* | $3 \times 20$ | *tri-ugent* | $3 \times 20$ |
| 70 | *dech ar tri fiche* | $10 + (3 \times 20)$ | *dec ar tri-ugeint* | $10 + (3 \times 20)$ | *dek ha tri-ugent* | $10 + (3 \times 20)$ |
| 80 | *ceithri fiche* | $4 \times 20$ | *pedwar-ugeint* | $4 \times 20$ | *pevar-ugent* | $4 \times 20$ |
| 90 | *deich ar ceithri fiche* | $10 + (4 \times 20)$ | *dec ar pedwar-ugeint* | $10 + (4 \times 20)$ | *dek ha pevar-ugent* | $10 + (4 \times 20)$ |
| 100 | *cet* | | *cant* | | *kant* | |
| 1,000 | *mile* | | *mil* | | *mil* | |

[1]alternatively, *deu naw* (= $2 \times 9$)   [2]alternatively, *tri-ouech* (= $3 \times 6$)

FIG. 2.7. *Celtic number-names*

Basque numbers are as follows:

| | | | | | | |
|---|---|---|---|---|---|---|
| 1 | *bat* | | 16 | *hamasei* | = 10 + 6 |
| 2 | *bi, biga, bida* | | 17 | *hamazazpi* | = 10 + 7 |
| 3 | *hiru, hirur* | | 18 | *hamazortzi* | = 10 + 8 |
| 4 | *lau, laur* | | 19 | *hemeretzi* | = 10 + 9 |
| 5 | *bost, bortz* | | 20 | *hogei* | = 20 |
| 6 | *sei* | | 30 | *hogeitabat* | = 10 + 20 |
| 7 | *zazpi* | | 40 | *berrogei* | =2 × 20 |
| 8 | *zortzi* | | 50 | *berrogei-tamar* | = (2 × 20) + 10 |
| 9 | *bederatzi* | | 60 | *hirurogei* | = 3 × 20 |
| 10 | *hamar* | | 70 | *hirurogei-tamar* | = (3 × 20) + 10 |
| 11 | *hamaika* | irregular | 80 | *laurogei* | = 4 × 20 |
| 12 | *hamabi* | = 10 + 2 | 90 | *laurogei-tamar* | = ( 4 × 20) + 10 |
| 13 | *hamahiru* | = 10 + 3 | 100 | *ehun* | |
| 14 | *hamalau* | = 10 + 4 | 1,000 | *mila* | |
| 15 | *hamabost* | = 10 + 5 | | | |

The mystery of Basque remains entire. As can be seen, it is a decimal system for numbers up to 19, then a vigesimal system for numbers from 20 to 99, and it then reverts to a decimal system for larger numbers. It may be that, like the Indo-European examples given above (Danish, French, and Celtic), it was originally a decimal system which was then "contaminated" by contact with populations using base 20; or, on the contrary, Basque may have been originally vigesimal, and subsequently "reformed" by contact with Indo-European decimal systems. The latter seems to be supported by the obviously Indo-European root of the words for 100 (not unlike "hundred") and 1,000 (almost identical to Romance words for "thousand"); but neither hypothesis about the origins of Basque numbering can be proven.

## THE COMMONEST BASE IN HISTORY: 10

Base 20, although quite widespread, has never been predominant in the history of numeration. Base 10, on the other hand, has always been by far the commonest means of establishing the rule of position. Here is a (non-exhaustive) alphabetical listing of the languages and peoples who have used or still use a numbering system built on base 10:

| | |
|---|---|
| AMORITES: | Northwestern Mesopotamia, founders of Babylon c. 1900 BCE, and of the first Babylonian dynasty |
| ARABS: | before and after the birth of Islam |
| ARAMAEANS: | Syria and northern Mesopotamia, second half of second millennium BCE |
| ASSYRIANS: | Mesopotamia, from the start of the second millennium BCE to c. 500 BCE |
| BAMOUNS : | Cameroon |
| BAOULE : | Ivory Coast |
| BERBERS : | Fair-skinned people settled in North Africa since at least Classical times |
| SHAN : | Indo-China, from second century CE |
| CHINESE : | from the origins |
| EGYPTIANS : | from the origins |
| ELAMITES: | Khuzestan, southwestern Iran, from fourth century BCE |
| ETRUSCANS : | probably from Asia Minor, settled in Tuscany from the late seventh century BCE |
| GOURMANCHES : | Upper Volta |
| GREEKS : | from the Homeric period |
| HEBREWS : | before and after the Exile |
| HITTITES: | Anatolia, from second millennium BCE |
| INCAS : | Peru, Ecuador, Bolivia, twelfth to sixteenth centuries CE |
| INDIA : | All civilisations of northern and southern India |
| INDUS CIVILISATION: | River Indus area, c. 2200 BCE |
| LYCIANS : | Asia Minor, first half of first millennium BCE |
| MALAYSIANS | |
| MALAGASY : | Madagascar |
| MANCHUS | |
| MINOANS : | Crete, second millennium BCE |
| MONGOLIANS | |
| NUBIANS : | Northeast Africa, since Pharaonic times |
| PERSIANS | |
| PHOENICIANS | |
| ROMANS | |
| TIBETANS | |
| UGARITIC PEOPLE : | Syria, second millennium BCE |
| URARTIANS : | Armenia, seventh century BCE |

In the world today, base 10 is used by a multitude of languages, including:

> Albanian; the Altaic languages (Turkish, Mongolian, Manchu); Armenian; Bamoun (Cameroon); Baoule (Ivory Coast); Batak; Chinese; the Dravidian languages (Tamil, Malayalam, Telugu); the Germanic languages (German, Dutch, Norwegian, Danish, Swedish, Icelandic, English); Gourmanche (Upper Volta); Greek; Indo-Aryan languages (Sindhi, Gujurati, Mahratta, Hindi, Punjabi, Bengali, Oriya, Singhalese); Indonesian; Iranian languages (Persian, Pahlavi, Kurdish, Afghan); Japanese; Javanese; Korean; Malagasy; Malay; Mon-Khmer languages (Cambodian, Kha); Nubian (Sudan); Polynesian languages (Hawaiian, Samoan, Tahitian, Marquesan); the Romance languages (French, Spanish, Italian, Portuguese, Romanian, Catalan, Provençal, Dalmatian); Semitic languages (Hebrew, Arabic, Amharic, Berber); the Slavic languages (Russian, Slovene, Serbo-Croat, Polish, Czech, Slovak); Thai languages (Laotian, Thai, Vietnamese); Tibeto-Burmese languages (Tibetan, Burmese, Himalayan dialects); Uralian (Finno-Ugrian) languages (Finnish, Hungarian).

These lists show, if it needed to be shown, just how successful base 10 has been and ever remains.

## ADVANTAGES AND DRAWBACKS OF BASE 10

The ethnic, geographical, and historical spread of base 10 is enormous, and we can say that it has become a virtually universal counting system. Is that because of its inherent practical or mathematical properties? Certainly not!

To be sure, base 10 has a distinct advantage over larger counting units such as 60, 30, or even 20: its magnitude is easily managed by the human mind, since the number of distinct names or symbols that it requires is quite limited, and as a result addition and multiplication tables using base 10 can be learned by rote without too much difficulty. It is far, far harder to learn the sixty distinct symbols of a base 60 system, even if large numbers can then be written with far fewer symbols; and the multiplication tables for even very simple Babylonian arithmetic require considerable feats of memorisation (sixty tables, each with sixty lines.)

At the other extreme, small bases such as 2 and 3 produce very small multiplication and addition tables to learn by heart; but they require very lengthy strings to express even relatively small numbers in speech or writing, a difficulty that base 10 avoids.

Let us look at a concrete alternative system, an English oral numbering system using base 2. Initially such a system would have only two number-names: "one" to express the unit, and "two" (let us call it "twosome") to express the base.

| 1 | 2 |
|---|---|
| *one* | *twosome* |

It would then acquire special names for each of the powers of the base: let us say "foursome" for $2^2$, "eightsome" for $2^3$, "sixteensome" for $2^4$, and so on. Analytical combinations would therefore produce a set of number-names something like this:

| | | | |
|---|---|---|---|
| 1 | one | 10 | eightsome twosome |
| 2 | twosome | 11 | eightsome twosome-one |
| 3 | twosome-one | 12 | eightsome foursome |
| 4 | foursome | 13 | eightsome foursome-one |
| 5 | foursome one | 14 | eightsome foursome twosome |
| 6 | foursome twosome | 15 | eightsome foursome twosome-one |
| 7 | foursome twosome-one | 16 | sixteensome |
| 8 | eightsome | 17 | sixteensome-one |
| 9 | eightsome one | | and so on. |

If our written number-system, using the rule of position, were constructed on base 2, then we would need only two digits, 0 and 1. The number two ("twosome"), which constitutes the base of the system, would be written 10, just like the present base "ten", but meaning "one twosome and no units"; three would be written 11 ("one twosome and one unit"), and so on:

| | | | |
|---|---|---|---|
| 1 | would be written | 1 | |
| 2 | would be written | 10 | |
| 3 | would be written | 11 | $= 1 \times 2 + 1$ |
| 4 | . . . | 100 | $= 1 \times 2^2 + 0 \times 2 + 0 \times 1$ |
| 5 | | 101 | $= 1 \times 2^2 + 0 \times 2 + 1 \times 1$ |
| 6 | | 110 | $= 1 \times 2^2 + 1 \times 2 + 0 \times 1$ |
| 7 | | 111 | $= 1 \times 2^2 + 1 \times 2 + 1 \times 1$ |
| 8 | | 1000 | $= 1 \times 2^3 + 0 \times 2^2 + 0 \times 2 + 0 \times 1$ |
| 9 | | 1001 | $= 1 \times 2^3 + 0 \times 2^2 + 0 \times 2 + 1 \times 1$ |
| 10 | | 1010 | $= 1 \times 2^3 + 0 \times 2^2 + 1 \times 2 + 0 \times 1$ |
| 11 | | 1011 | $= 1 \times 2^3 + 0 \times 2^2 + 1 \times 2 + 1 \times 1$ |
| 12 | | 1100 | $= 1 \times 2^3 + 1 \times 2^2 + 0 \times 2 + 0 \times 1$ |
| 13 | | 1101 | $= 1 \times 2^3 + 1 \times 2^2 + 0 \times 2 + 1 \times 1$ |
| 14 | | 1110 | $= 1 \times 2^3 + 1 \times 2^2 + 1 \times 2 + 0 \times 1$ |
| 15 | | 1111 | $= 1 \times 2^3 + 1 \times 2^2 + 1 \times 2 + 1 \times 1$ |
| 16 | | 10000 | $= 1 \times 2^4 + 0 \times 2^3 + 0 \times 2^2 + 0 \times 2 + 0$ |
| 17 | | 10001 | $= 1 \times 2^4 + 0 \times 2^3 + 0 \times 2^2 + 0 \times 2 + 1$ |

FIG. 2.8.

Now, whilst we now require only four digits to express the number two thousand four hundred and forty-eight (2,448) in a base 10 number system, a base 2 or binary system (which is in fact the system used by computers) requires no fewer than twelve digits:

$$100110010000$$
$$(= 1 \times 2^{11} + 0 \times 2^{10} + 0 \times 2^{9} + 1 \times 2^{8} + 1 \times 2^{7} + 0 \times 2^{6} + 0 \times 2^{5} + 1 \times 2^{4} + 0 \times 2^{3} + 0 \times 2^{2} + 0 \times 2 + 0)$$

Using these kinds of expressions would produce real practical problems in daily life: cheques would need to be the size of a sheet of A3 paper in order to be used to pay the deposit on a new house, for example; and it would take quite a few minutes just to say how much you think a second-hand Ferrari might be worth.

Nonetheless, there are several other numbers that could serve as base just as well as 10, and in some senses would be preferable to it.

There is nothing impossible or impracticable about changing the "steps on the ladder" and counting to a different base. Bases such as *7, 11, 12,* or even *13* would provide orders of magnitude that would be just as satisfactory as base 10 in terms of the human capacity for memorisation. As for arithmetical operations, they could be carried out just as well in these other bases, and in exactly the same way as we do in our present decimal system. However, we would have to lose our mental habit of giving a special status to 10 and the powers of 10, since the corresponding names and symbols would be just as useless in a 12-based system as they would in one based on 11.

If we were to decide one day on a complete reform of the number-system, and to entrust the task of designing the new system to a panel of experts, we would probably see a great battle engaged, as is often the case, between the "pragmatists" and the "theoreticians". "What we need nowadays is a system that is mathematically satisfactory," one of them would assert. "The best systems are those with a base that has the largest number of divisors," the pragmatist would propose. "And of all such bases, *12* seems to me to be by far the most suitable, given the limits of human memory. I don't need to remind you how serviceable base 12 was found to be by traders in former times – nor that we still have plenty of traces of the business systems of yore, such as the *dozen* and the *gross* (12 × 12), and that we still count eggs, oysters, screws and suchlike in that way. Base 10 can only be divided by 2 and 5; but 12 has 2, 3, 4, and 6 as factors, and that's precisely why a duodecimal system would be really effective. Just think how useful it would be to arithmeticians and traders, who would much more easily be able to compute halves, thirds, quarters, and even sixths of every quantity or sum. Such fractions are so natural and so common that they crop up all the time even without our noticing. And that's not the whole story! Just think how handy it would be for calculations of time: the number of months in the year would be equal to the base of the system; a day would be twice the base in terms of hours; an hour would be five times the base in minutes; and a minute the same number of seconds. It would be enormously helpful as well for geometry, since arcs and angles would be measured in degrees equal to five times the base in minutes, and minutes would be the same number of seconds. The full circle would be thirty times the base 12, and a straight line just fifteen times the base. Astronomers too would find it more than handy . . ."

"But those are not the most important considerations in our day and age," the theoretician would argue. "I've no historical example to support what I'm going to propose, but enough time has passed for my ideas to stand up on their own. The main purpose of a written number system – I'm sure everyone will agree – is to allow its users to represent all numbers simply and unambiguously. And I do mean *all* numbers – integers, fractions, rational and irrational numbers, the whole lot. So what we are looking for is a numbering system with a base that has no factor other than itself, in other words, a number system having a prime number as its base. The only example I'll give is base 11. This would be much more useful than base 10 or 12, since under base 11 most fractions are irreducible: they would therefore have one and only one possible representation in a system with base 11. For instance: the number which in our present decimal system is written 0.68 corresponds in fact to several other fractions – 68/100, 34/50, and 17/25. Admittedly, these expressions all refer to the same fraction, but there is an ambiguity all the same in representing it in so many different ways. Such ambiguities would vanish completely in a system using base 11 or 7 (or indeed, any system with a prime number as its base), since the irreducibility of fractions would mean that any number had one and only one representation. Just think of the mathematical advantages that would flow from such a reform . . ."

So, since it has only two factors and is not a prime number, base 10 would have no supporters on such a committee of experts!

Base 12 really has had serious supporters, even in recent times. British readers may recall the rearguard defence of the old currency – 12 pence (d) to the shilling, 20 shillings to the pound sterling – at the time it was abandoned in 1971: the benefits of teaching children to multiply and divide by 2, 3, 4, and 6 (for the smaller-value coins of 3d and 6d) and by 8 (for the "half-crown", worth 2s 6d) were vigorously asserted, and many older people in Britain continue to maintain that youngsters brought up on decimal coinage no longer "know how to count". In France, a civil servant by the name of Essig proposed a duodecimal system for weights and measures in 1955, but failed to persuade the nation that first universalised the metric system to all forms of measurement.

It seems quite unrealistic to imagine that we could turn the clock back now and modify the base number of both spoken and written number-systems. The habit of counting in tens and powers of 10 is so deeply ingrained in our traditions and minds as to be well-nigh indestructible. The best thing to do was to reform the bizarre divisions of older systems of weights and measures and to replace them with a unified system founded on the all-powerful base of 10. That is precisely what was done in France in the Revolutionary period: the Convention (a form of parliament) created the metric system and imposed it on the nation by the Laws of 18 Germinal Year III, in the revolutionary calendar (8 April 1795) and 19 Frimaire Year VIII (19 December 1799).

## A BRIEF HISTORY OF THE METRIC SYSTEM

Until the late eighteenth century, European systems of weights and measures were diverse, complicated, and varied considerably from one area to another. Standards were fixed with utter whimsicality by local rulers, and quite arbitrary objects were used to represent lengths, volumes, etc. From the late seventeenth century onwards, as the experimental sciences advanced and the general properties of the physical world became better understood, scholars strove to devise stable and coherent measuring systems based on permanent, universal and unmodifiable standards. The growth of trade throughout the eighteenth century also created a need for common measurements at least within each country, and a uniform system of weights and measures. Thus the metric system emerged towards the end of the eighteenth century. It is a fully consistent and coherent measurement system using base 10 (and therefore fully compatible with the place-value system of written numbering that the Arabs had brought to Europe in the Middle Ages, having themselves learned it from the Indians), which the French Revolution offered "to all ages and to all peoples, for their greater benefit". It produced astounding progress in applied areas, since it is perfectly adapted to numerical calculation and is extremely simple to operate in fields of every kind.

**Around 1660**: In order to harmonise measurement of time and length and also so as to compare the various standards used for measuring length around the world, the Royal Society of London proposed to establish as the unit of length the length of a pendulum that beats once per second. The idea was taken up by Abbé Jean Picard in *La Mesure de la Terre* ("The Measurement of the Earth") in 1671, by Christian Huygens in 1673, and by La Condamine in France, John Miller in England, and Jefferson in America.

**1670**: Abbé Gabriel Mouton suggested using the sexagesimal minute of the meridian (= 1/1000 of the nautical mile) as the unit of length. But this unit, of roughly 1.85 metres, was too long to be any practical use.

**1672**: Richer discovered that the length of a pendulum that beats once per second is less at Cayenne (near the Equator) than in Paris. The consequence of this discovery was that, because of the variation in length of the pendulum caused by the variation in gravity at different points on the globe, the choice of the location of the standard pendulum would be politically very tricky. As a result the idea of using the one-second pendulum as a unit of length was eventually abandoned.

**1758**: In *Observations sur les principes métaphysiques de la géometrie* ("Observations on the Metaphysical Principles of Geometry"), Louis Dupuy suggested unifying measurements of length and weight by fixing the unit of weight as that of a volume of water defined by units of length.

**1790**: 8 May: Talleyrand proposed, and the *Assemblée constituante* (Constituent Assembly) approved the creation of a stable, simple and uniform system of weights and measures. The task of defining the system was entrusted to a committee of the Academy of Sciences, with a membership consisting of Lagrange, Laplace, and Monge (astronomical and calendrical measurements), Borda (physical and navigational measurement), and Lavoisier (chemistry). The base unit initially chosen was the length of the pendulum beating once per second.

**1791**: 26 March: The committee decided to abandon the pendulum as the base unit and persuaded the Constituent Assembly to choose as the unit of length the ten-millionth part of one quarter of the earth's meridian, which can be measured exactly as a fraction of the distance from the pole to the Equator. At Borda's suggestion, this unit would be called the *metre* (Greek for "measure").

What the committee then had to do was to produce conventional equivalencies between the various units chosen so that all of them (except units of time) could be derived from the metre. So, for measuring surface area, the unit chosen was the *are*, a square with a side of 10 metres; for measuring weight, the *kilogram* was defined as the weight of a unit of volume (1 litre) of pure water at the temperature of melting ice, corrected for the effects of latitude and air pressure. All that now had to be done to set up the entire metric system was to make the key measurement, the distance from the pole to the Equator – a measurement that was all the more interesting at that time as Isaac Newton had speculated that the globe was an ellipsoid with flattened ends (contradicting Descartes, who believed it was a sphere with elongated or pointed ends).

**1792**: The "meridian expedition" began. A line was drawn from Dunkirk to Barcelona and measured out by triangulation points located thanks to

Borda's goniometer, with some base stretches measured out with greater precision on the ground. Under the direction of Méchain and Delambre, one team was in charge of triangulation, one was responsible for the standard length in platinum, and one for drafting the users' manuals of the new system. Physicists such as Coulomb, Haüy, Hassenfrantz, and Borda, and the mathematicians Monge, Lagrange, and Laplace were amongst the many scientists who collaborated on this project which was not fully completed until 1799.

**1793**: 1 August: The French government promulgated a decree requiring all measures of money, length, area, volume, and weight to be expressed in decimal terms : all the units of measure would henceforth be hierarchised according to the powers of 10. As it overturned all the measures in current use (most of them using base 12), the decimalisation decree required new words to be invented, but also created the opportunity for much greater coherence and accuracy in counting and calculation.

**1795**: 7 April: Law of 18 Germinal, Year III, which organised the metric system, gave the first definition of the metre as a fraction of the terrestrial meridian, and fixed the present nomenclature of the units (decimetre, centimetre, millimetre; are, deciare, centiare, hectare; gram, decigram, centigram, kilogram; franc, centime; etc.)

**1795**: 9 June: Lenoir fabricated the first legal metric standard, on the basis of the calculation made by La Caille of the distance between the pole and the Equator at 5,129,070 *toises de Paris* (in 1799, Delambre and Méchain obtained a different, but actually less accurate figure of 5,130,740 *toises de Paris*).

**1795**: 25 June: Establishment of the *Bureau des Longitudes* (Longitude Office) in Paris.

**1799**: First meeting in Paris of an international conference to discuss universal adoption of the metric system. The system was considered "too revolutionary" to persuade other nations to "think metric" at that time.

**1799**: 22 June: The definitive standard metre and kilogram, made of platinum, were deposited in the French National Archives.

**1799**: 10 December: Law of 10 Frimaire, Year VIII, which confirmed the legal status of the definitive standards, gave the second definition of the metre (the length of the platinum standard in the National Archives, namely 3 feet and 11.296 "lines" of the *toise de Paris*), and in theory made the use of the metric system obligatory. (In fact, old habits of using pre-metric units of measurement persisted for many years and were tolerated.)

**1840**: 1 January: With the growing spread of primary education in France, the law was amended to make the use of the metric system genuinely obligatory on all.

**1875**: Establishment of the International Bureau of Weights and Measures at Sèvres (near Paris). The Bureau created the new international standard metre, made of iridoplatinum.

**1876**: 22 April: The new international standard metre was deposited in the Pavillon de Breteuil, at Sèvres, which was then ceded by the nation to the International Weights and Measures Committee and granted the status of "international territory".

**1899**: The General Conference on Weights and Measures met and provided the third definition of the metre. The length of the meridian was abandoned as a basis of calculation. Henceforth, the metre was defined as the distance at 0°C of the axis of the three median lines scored on the international standard iridoplatinum metre.

**1950s**: The invention of the laser allowed significant advances in optics, atomic physics, and measurement sciences. Moreover, quartz and atomic clocks resulted in the discovery of variations in the length of the day, and put an end to the definition of units of time in terms of the earth's rotation on its axis.

**1960**: 14 October: Fourth definition of the metre as an optical standard (one hundred times more accurate than the metre of 1899): the metre now becomes equal to 1,650,763.73 wave-lengths of orange radiation in a void of krypton 86 (krypton 86 being one of the isotopes of natural krypton).

**1983**: 20 October: The XVIIth General Conference on Weights and Measures gives the fifth definition of the metre, based on the speed of light in space (299,792,458 metres per second): a metre is henceforth the distance travelled by light in space in 1/299,792,458 of a second. As for the second, it is defined as the duration of 9,192,631,770 periods of radiation corresponding to the transition between the two superfine levels of the fundamental state of an atom of caesium 133. At the same conference, definitions of the five other basic units (kilogram, amp, kelvin, mole, and candela) were also adopted, as well as the standards that constitute the current International Standards system (IS).*

## THE ORIGIN OF BASE 10

Well, then: where *does* base 10 come from?

In the second century CE, Nicomachus of Gerasa, a neo-Pythagorean from Judaea, wrote an *Arithmetical Introduction* which, in its many translations, influenced Western mathematical thinking throughout the Middle Ages. For Nicomachus, the number 10 was a "perfect" number, the number of the divinity, who used it in his creation, notably for human toes and fingers,

* For information contained in this section on the metric system I am indebted to Jean Dhombres, President of the French Association for the History of Science.

and inspired all peoples to base their counting systems on it. For many centuries, indeed, numbers were thought to have mystical properties; in Pythagorean thinking, 10 was held to be "the first-born of the numbers, the mother of them all, the one that never wavers and gives the key to all things".

Such attitudes to numbers, which had their place in a world-view which was itself mystical through and through, now seem as circular and self-defeating as the observation that God had the wisdom to cause rivers to flow through the middle of towns.

In fact, the almost universal preference for base 10 comes from nothing more obscure than the fact that we learn to count on our fingers, and that we happen to have ten of them. We would use base 10 even if we had no language, or were bound to a vow of total silence: for just like the North African shepherd and his shells and straps discussed on p. 24–25 above, we could use our raised fingers to count out the first ten in silence, a colleague could then raise one finger to keep count of the tens, and so on to 99, when (for numbers of 100 and more) the fingers of a third colleague would be needed. Fig. 2.9 shows the position of the three silent colleagues' hands at number 627.

| Helper No. 3 | | Helper No. 2 | | Helper No. 1 | |
|---|---|---|---|---|---|
| Left | Right | Left | Right | Left | Right |
| 6 | | 2 | | 7 | |
| 600 | | 20 | | 7 | |

Fig. 2.9.

The obvious practicality of such a non-linguistic counting system using only our own bodies shows that the idea of grouping numbers into packets of ten and powers of ten is based on the "accident of nature" that is the physiology of the human hand. Since that physiology is universal, base 10 necessarily occupies a dominant, not to say inexpugnable position in counting systems.

If nature had given us six fingers, then the majority of counting systems would have used base 12. If on the other hand evolution had brought us down to four fingers on each hand (as it has for the frog), then we would doubtless have long-standing habits and traditions of counting on base 8.

## THE ORIGINS OF THE OTHER BASES

The reason for the adoption of vigesimal (base 20) systems in some cultures can be seen by the basic idea of Aztec numbering as laid out in Fig. 2.6 above. In the language of the Aztecs

- the names of the first five numbers can be associated with the fingers of one hand;
- the following five numbers can be associated with the fingers of the other hand;
- the next five numbers can be associated with the toes of one foot;
- and the last five numbers can be associated with the toes on the other foot.

And so 20 is reached with the last toe of the second foot (see Fig. 2.10).

This is no coincidence. It is simply that some communities, because they realised that by leaning forward a little they could count toes as well as fingers, ended up using base 20.

One remarkable fact is that both the Inuit (Greenland) and the Tamanas (in the Orinoco basin) used the same expression for the number 53, literally meaning: "of the third man, three on the first foot".

According to C. Zaslavsky (1973), the Banda people in Central Africa express the number 20 by saying something like "a hanged man": presumably because when you hang a man you can see straight away all his fingers and toes. In some Mayan dialects, the expression *hun uinic*, which means 20, also means "one man". The Malinke (Senegal) express 20 and 40 by saying respectively "a whole man" and "a bed" – in other words, two bodies in a bed!

In the light of all this there can be no doubt at all that the origin of vigesimal systems lies in the habit of counting on ten fingers *and* ten toes . . .

The origin of base 5 is similarly anthropomorphic. Quinary reckoning is founded on learning to count using the fingers of one hand only.

The following finger-counting technique, which is found in various parts of Oceania and is also currently used by many Bombay traders for various specific purposes, is a good example of how a primitive one-hand counting system can give rise to more elaborate numbering. You use the five fingers of the left hand to count the first five units. Then, once this number is reached, you extend the thumb of the right hand, and go on counting to 10 with the fingers of the left hand; then you extend the index finger of the right hand and count again on the left hand from 11 to 15; and so on, up to 25. The series can be extended to 30 since the fingers of the left hand are usable six times over in all.

However, this obviously fails to resolve the basic mystery: why did base 5 – which must be considered the most natural base by far, since it is

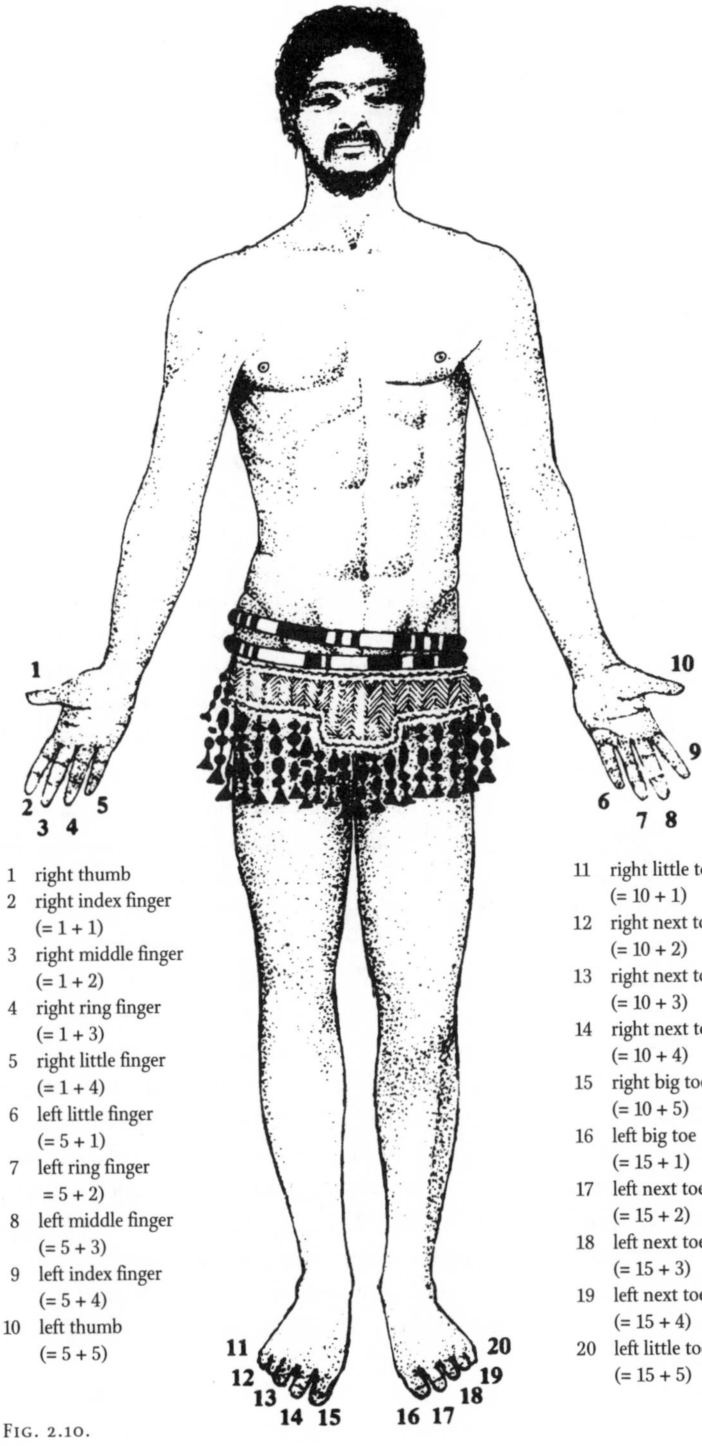

1 right thumb
2 right index finger (= 1 + 1)
3 right middle finger (= 1 + 2)
4 right ring finger (= 1 + 3)
5 right little finger (= 1 + 4)
6 left little finger (= 5 + 1)
7 left ring finger = 5 + 2)
8 left middle finger (= 5 + 3)
9 left index finger (= 5 + 4)
10 left thumb (= 5 + 5)
11 right little toe (= 10 + 1)
12 right next toe (= 10 + 2)
13 right next toe (= 10 + 3)
14 right next toe (= 10 + 4)
15 right big toe (= 10 + 5)
16 left big toe (= 15 + 1)
17 left next toe (= 15 + 2)
18 left next toe (= 15 + 3)
19 left next toe (= 15 + 4)
20 left little toe (= 15 + 5)

FIG. 2.10.

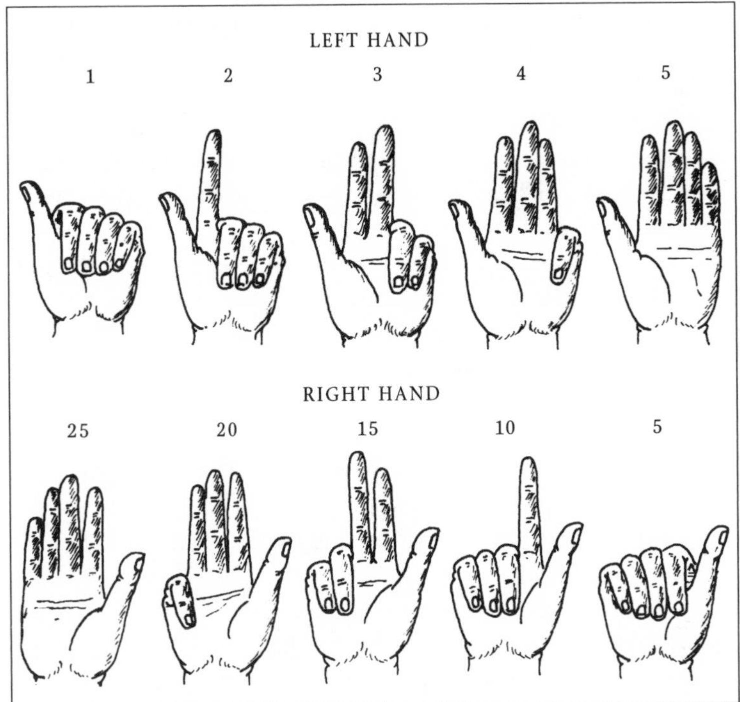


FIG. 2.11.

virtually dictated by the basic features of the human body and must be self-evident from the very moment of learning to count – why did base 5 not become adopted as the universal human counting tool? Why, in other words, was the apparently inevitable construction of quinary counting words generally avoided? Why did so many cultures go up to 10, to 20 or, in the case of the Sumerians, whom we will discuss again, as far as 60? Even more mysterious are those cultures which possessed a concept of number and knew how to count, but went back down to 4 for their numerical base.

L. L. Conant (1923) tackled the whole problem in detail without claiming to have found the final answer. The anthropologist Lévy-Bruhl, on the other hand, thought it was a false problem. In his view, we should not suppose that people ever invented number-systems *in order* to carry out arithmetical operations or devised systems that were intended to be best suited to operations that, prior to the devising of the system, could not be imagined. "Numbering systems, like languages, from which they can hardly be distinguished, are in the first place social phenomena, closely dependent on collective mentalities," he claimed. "The mentality of any society is completely bound up with its internal functioning and its institutions."

To conclude this chapter we shall return with Lévy-Bruhl to very primitive counting systems which do not yet clearly distinguish between the cardinal and ordinal aspects of number. In the kind of "body-counting" explained above and demonstrated in Fig. 1.30, 1.31 and 1.32, there are no "privileged" points or numbers, and therefore no concept of a base at all. Using Petitot's dictionary of the language of Dene-Dindjie Indians (Canada), Lévy-Bruhl explains how things are counted out in a system with no base:

> You hold out your left hand (always the left hand) with the palm turned towards your face, and bend your little finger, saying
>
> for 1 : *the end is bent*
> or *on the end*
>
> Then you bend your ring finger and say:
>
> for 2: *it's bent again*
>
> Then you bend your middle finger and say
>
> for 3: *the middle one is bent*
>
> Then you bend your index finger, leaving the thumb stretched out, and say:
>
> for 4: *there's only that left*
>
> Then you open out your whole hand and say:
>
> for 5: *it's OK on my hand*
> or *on a hand*
> or *my hand*
>
> Then you fold back three fingers together on your left hand, keeping the thumb and index stretched out, and touch the left thumb with the right thumb, saying:
>
> for 6: *there's three on each side*
> or: *three by three*
>
> Then you bend down four fingers on your left hand and touch your left thumb (still stretched out) with the thumb and index finger of your right hand, and say:
>
> for 7: *on one side there are four*
> or *there are still three bent*
> or *three on each side and one in the middle*
>
> Then you stretch out three fingers of your right hand and touch the outstretched thumb of your left hand, creating two groups of four fingers (bent and extended), and say:
>
> for 8: *four on four*
> or *four on each side*
>
> Then you show the little finger of your right hand, the only one now bent, and say:
>
> for 9: *there's still one down*
> or *one still short*
> or *the little finger's lying low*
>
> Then you start the gestures over again, saying "*one full plus one*", or "*one counted plus one*", "*one counted plus two*", "*one counted plus three*", and so on.

Lévy-Bruhl argues that in this system, which does not prevent the Dene-Dindjie from counting properly, there is no concept of a quinary base: 6 is not "a second one", 7 not "a new two", as we find in so many other numbering systems. On the contrary, he says, 6 here is "three and three" – which shows that finishing the count on one hand is in no way a "marker" or a "privileged number" in this system. The periodicity of numbers is not derived from the physical manner of counting, does not come from the series of movements made to indicate the sequence of the numbers.

In this view, numbering systems relate much more directly to the "mental world" of the culture or civilisation, which may be mythical rather than practical, attributing more significance to the four cardinal points of the compass, or to the four legs of an animal, than to the five fingers of the hand. We do not have to try and guess why this base rather than another was "chosen" by a given people for their numbering system, even if they do effectively use the five fingers of their hand for counting things out. Where a numbering system has a base, the base was never "chosen", Lévy-Bruhl asserts. It is a mistake to think of "the human mind" constructing a number system in order to count: on the contrary, people began to count, slowly and with great difficulty, long before they acquired the concept of number.

However, it is clear that the adoption of base 5 is related to the way we count on the fingers of our hands. But why did those cultures that adopted base 5 not extend it, like so many others, to the base 10 that corresponds to the fingers of both hands? Dantzig has speculated that it may have to do with the conditions of life in warrior societies, in which men rarely go about unarmed. If they want to count, they tuck their weapon under their left arm and count on the left hand, using the right hand as a check-off. The right hand remains free to seize the weapon if needed. This may explain why the left hand is almost universally used by right-handed people for counting, and vice versa [T. Dantzig (1930), p. 13].

However this may be, base numbers arise for many reasons, many of which have nothing at all to do with their suitability for counting or for arithmetical operations; and they may indeed have arisen long before any kind of abstract arithmetic was invented.

## CHAPTER 3

# THE EARLIEST CALCULATING MACHINE – THE HAND

That uniquely flexible and useful tool, the human hand, has also been the tool most widely used at all times as an aid to counting and calculation.

Greek writers from Aristophanes to Plutarch mention it, and Cicero tells us that its use was as common in Rome: *tuos digitos novi* – "I well know your skill at calculating on your fingers" (*Epistulae ad Atticum*, V, 21, 13); Seneca says much the same: "Greed was my teacher of arithmetic: I learned to make my fingers the servants of my desires" (*Epistles*, LXXXVII); and, later, Tertullian said: "Meanwhile, I have to sit surrounded by piles of papers, bending and unbending my fingers to keep track of numbers." The famous orator Quintilian stressed the importance of calculating on the fingers, especially in the context of pleading at law: "Skill with numbers is needed not only by the Orator, but also by the pleader at the Bar. An Advocate who stumbles over a multiplication, or who merely exhibits hesitation or clumsiness in calculating on his fingers, immediately gives a bad impression of his abilities;" and the digital techniques he referred to, which were in common use by the inhabitants of Rome, required very considerable dexterity (see Fig. 3.13). Pliny the Elder, in his *Natural History* (XVI), described how King Numa offered up to the God Janus (the god of the Year, of Age, and of Time) a statue whose fingers displayed the number of days in a year. Such practices were by no means confined to the Greeks and Romans. Archaeologists, historians, ethnologists, and philologists have come upon them at all times and in all regions of the world, in Polynesia, Oceania, Africa, Europe, Ancient Mesopotamia, Egypt under the Pharaohs, the Islamic world, China, India, the Americas before Columbus, and the Western world in the Middle Ages. We can conclude, therefore, that the human hand is the original "calculating machine". In the following we shall show how, once people had grasped the principle of the base, over the ages they developed the arithmetical potential of their fingers to an amazing degree. Indeed, certain details of this are evidence of contacts and influences between different peoples, which could never have been inferred in any other way.

## EARLY WAYS OF COUNTING ON FINGERS

The simplest method of counting on the fingers consists of associating an integer with each finger, in a natural order. This may be done in many ways. One may start with the fingers all bent closed, and count by successively straightening them; or with the fingers open, and successively close them. One may count from the left thumb along the hands to the right little finger, or from the little finger of the left hand through to the thumb of the right, or from the index finger to the little finger and finally the thumb (see Fig. 3.3). The last method was especially used in North Africa. It seems likely that at the time of Mohammed the Arabs used this method. One of the *Ḥadiths* tells how the Prophet showed his disciples that a month could have 29 days, showing "his open hands three times, but with one finger bent the third time". Also, a Muslim believer always raises the index finger when asserting the unity of Allah and expressing his faith in Islam, in performing the prayer of *Shahādah* ("witness").

FIG. 3.1. *Finger-counting among the Aztecs (Pre-Columbian Mexico). Detail of a mural by Diego Rivera. National Museum of Mexico*

Fig. 3.2. *Boethius (480–524 CE), the philosopher and mathematician, counting on his fingers. From a painting by Justus of Ghent (15th C.). See P. Dedron and J. Itard (1959).*

## A STRANGE WAY OF BARGAINING

There is a similar method, of very ancient origin, which persisted late in the East and was common in Asia in the first half of the twentieth century. It is a special way of finger-counting used by oriental dealers and their clients in negotiating their terms. Their very curious procedure was described by the celebrated Danish traveller Carsten Niebuhr in the eighteenth century, as follows:

> I have somewhere read, I think, that the Orientals have a special way of settling a deal in the presence of onlookers, which ensures that none of these becomes aware of the agreed price, and they still regularly make use of it. I dreaded having someone buy something on my behalf in this way, for it allows the agent to deceive the person for whom he is acting, even when he is watching. The two parties indicate what price is asked, and what they are willing to pay, by touching fingers or knuckles. In doing so, they conceal the hand in a corner of their dress, not in order to conceal the mystery of their art, but simply in order to hide their dealings from onlookers. . . . (*Beschreibung von Arabien*, 1772)

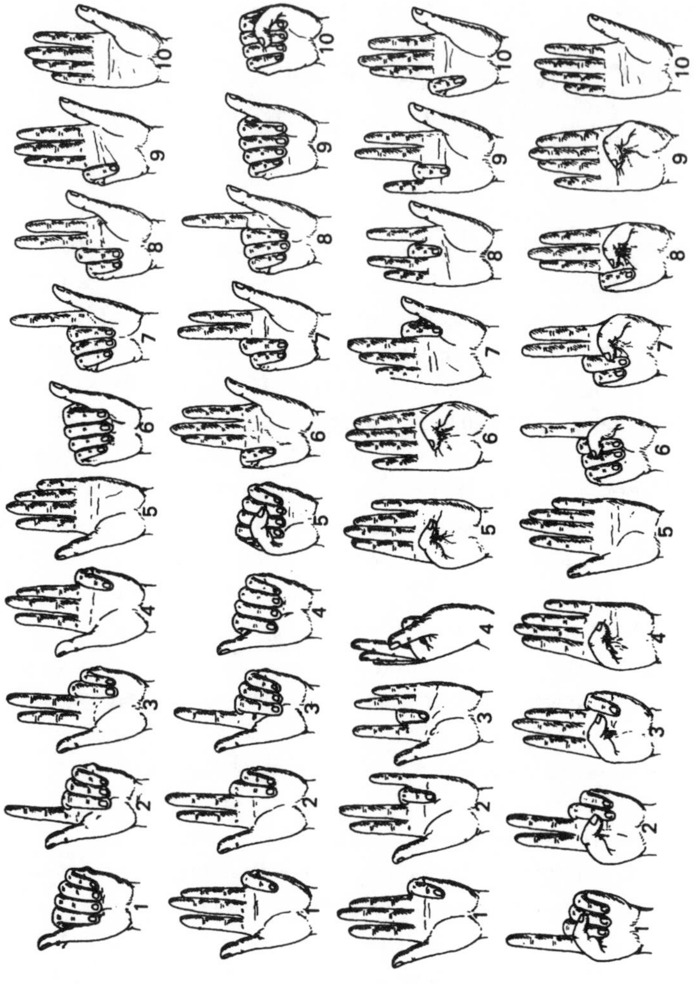


Fig. 3.3. *Variants of basic counting with the fingers*

To indicate the number 1, one of the negotiators takes hold of the index finger of the other; to indicate 2, 3 or 4, he takes hold of index and middle fingers, index, middle and fourth, or all four fingers. To indicate 5, he grasps the whole hand. For 6, he twice grasps the fingers for 3 (2 × 3), for 7, the fingers for 4 then the fingers for 3 (4 + 3), for 8, he twice grasps the fingers for

4 (2 × 4), and for 9, he grasps the whole hand, and then the fingers for 4 (5 + 4). For 10, 100, 1,000 or 10,000 he again takes hold of the index finger (as for 1); for 20, 200, 2,000 or 20,000 the index and middle fingers (as for 2), and so on (see Fig. 3.4). This does not, in fact, lead to confusion because the two negotiators will have agreed beforehand on roughly what the price will be (whether about 40 dinars, or 400, for example). Niebuhr does not tell that he himself saw such a deal take place, but J. G. Lemoine found traces of the method in Bahrain, a place famous for its pearl fishery, when he made a study of this topic at the beginning of this century. He gathered information from pearl dealers in Paris, who had often visited Bahrain and had occasion to employ this procedure in dealing with the Bahrainis. He states:

> The two dealers, seated face to face, bring their right hands together and, with the left hand, hold a cloth over them so that their right hands are concealed. The negotiation, with all its "discussions", takes place without a word being spoken, and their faces remain totally impassive. Those who have observed this find it extremely interesting, for the slightest visible sign could be taken to the disadvantage of one or other of the dealers.

Similar methods of negotiation have been reported from the borders of the Red Sea, from Syria, Iraq and Arabia, from India and Bengal, China and Mongolia, and – from the opposite end of the world – Algeria. P. J. Dols (1918), reporting on "Chinese life in Gansu province", describes how dealings were still being conducted in China and Mongolia in the early twentieth century.

> The buyer puts his hands into the sleeves of the seller. While talking, he takes hold of the seller's index finger, thereby indicating that he is offering 10, 100 or 1,000 francs. "No!" says the other. The buyer then takes the index and middle fingers together. "Done!" says the seller. The deal has been struck, and the object is sold for 20, or for 200, francs. Three fingers together means 30 (or 300 or 3,000), four fingers 40 (or 400 or 4,000). When the buyer takes the whole hand of the seller, it is 50 (or 500 or 5,000). Thumb and little finger signify 60 (note the difference from the Middle Eastern system described above). Placing the thumb in the vendor's palm means 70, thumb and index together 80. When the buyer, using his thumb and index finger together, touches the index finger of the seller, this indicates 90.

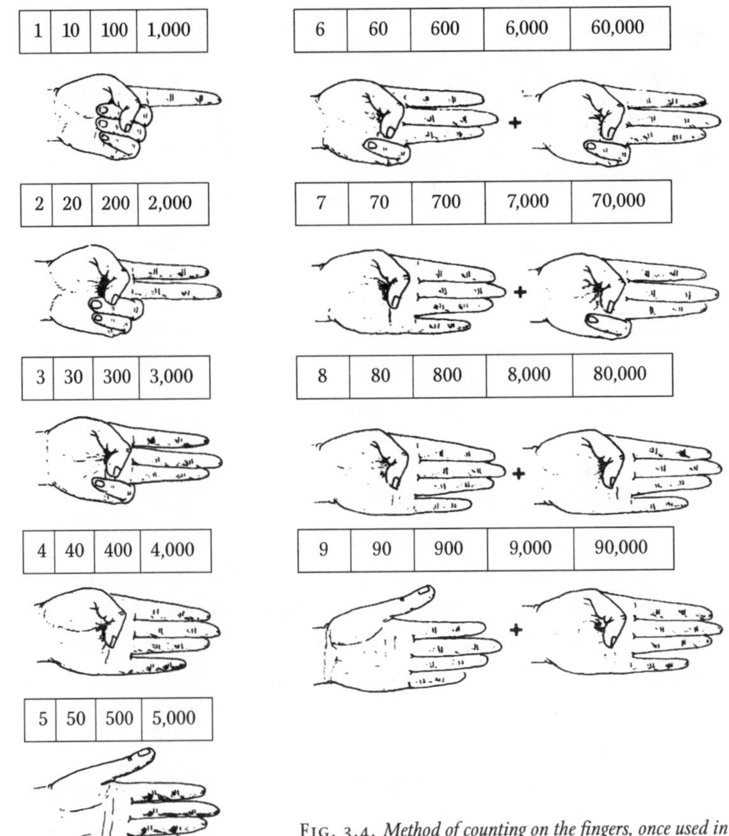


Fig. 3.4. *Method of counting on the fingers, once used in bargaining between oriental dealers*

## COUNTING ALONG THE FINGERS

There is more to fingers than a single digit: they have a knuckle, two joints, and three bones (but one joint and two bones for the thumbs). Amongst many Asiatic peoples, this more detailed anatomy has been exploited for counting. In southern China, Indo-China and India, for example, people have counted one for each joint, including the knuckle, working from base to tip of finger (or in reverse) and from little finger to thumb, pointing with a finger of the other hand. Thus each hand can count up to 14, and both hands up to 28 (see Fig. 3.5). A Chinaman from Canton once told me of a singular application of this method. Since a woman's monthly cycle lasts 28 days, his mother used to tie a thread around each joint as above for each day of her cycle, to detect early or late menstruation. The Venerable Bede (673–735 CE, a monk in the Monastery of Saints Peter and Paul at Wearmouth and Jarrow and author of the influential *De ratione temporum*, "Of the Division of Time"), applied similar counting methods for his calculations of time. To count the twenty-eight years of the solar cycle, beginning with a leap year, he started from the tip of the little finger and counted across the four fingers, winding back and forth and working down to the base of the fingers to count up to 12, then moving to the other hand to count up to 24, finally using the two thumbs to count up to 28 (see Fig. 3.6).

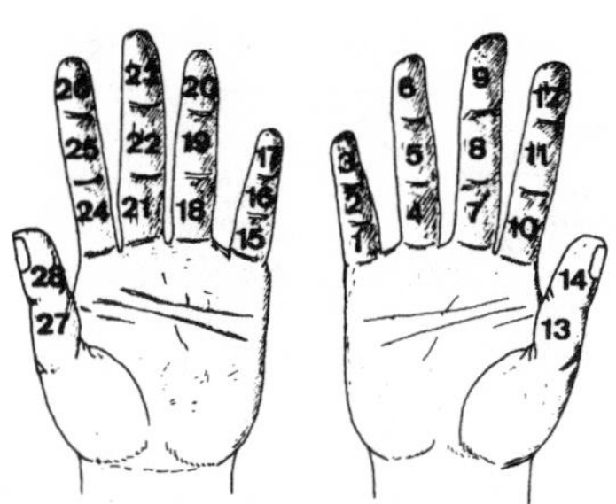

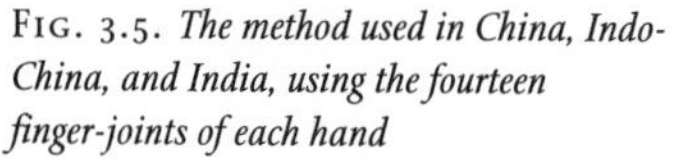

FIG. 3.5. *The method used in China, Indo-China, and India, using the fourteen finger-joints of each hand*

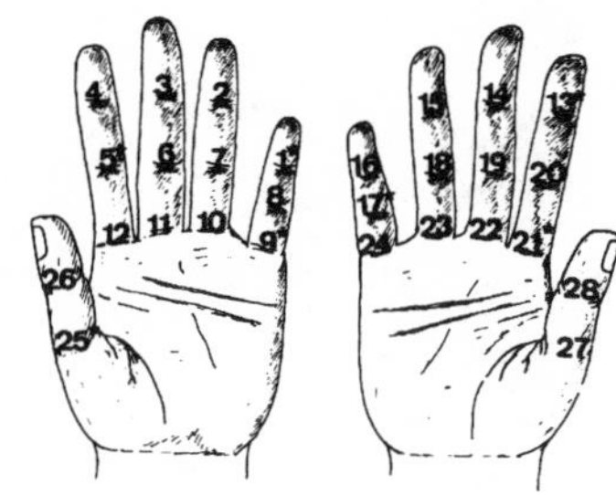

FIG. 3.6. *The Venerable Bede's method of counting the twenty-eight years of the solar cycle on the knuckles (7th C.). Leap years are marked with asterisks*

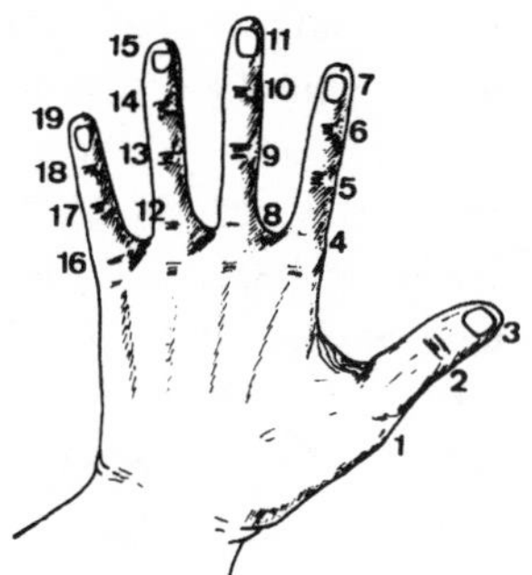

FIG. 3.7. *Bede's method of counting the nineteen years of the lunar cycle, using the knuckles and fingernails of the left hand*

FIG. 3.8. *The Indian and Bengali method, using the knuckles of fingers and thumb, and the ball of the thumb*

He used a similar method to count the nineteen years of the lunar cycle, counting up to 14 on the knuckles of the left hand, going on up to 19 by also counting the fingernails (see Fig. 3.7). The objective was to determine the date of Easter, the subject of complicated controversies in the early Church. In particular there was dispute between the British and Irish Churches, on the one hand, and the Roman Church on the other, regarding which lunar cycle to adopt for the date of Easter. Bede's calculations brought together the solar year and the solar and lunar cycles of the Julian calendar, and its leap years.

A different method of counting on the knuckles was long used in northeast India, and is still found in the regions of Calcutta and Dacca. It was reported by seventeenth- and eighteenth-century travellers, especially the Frenchman J. B. Tavernier (1712). According to N. Halhed (1778), the Bengalis counted along the knuckles from base to tip, starting with the little finger and ending with the thumb, using the ball of the thumb as well and thus counting up to 15 (Fig. 3.8).

This method of counting on the knuckles has given rise to the practice, common among Indian traders, of fixing a price by offering the hand under cover of a cloth; they then touch knuckles to raise or lower their proposed prices (Halhed).

There are 15 days in the Hindu month, the same number as can be counted on a hand; and, according to Lemoine, this is no coincidence. The Hindu year (360 days) consists of 12 seasons (*Nitus*) each of two "months" (*Masas*). One month of 15 days corresponds to one phase of the moon, and the following month to another phase. The first, waxing, phase is called *Rahu*, and the second, waning, phase is called *Ketu*. In this connection we may refer to the legend which tells how, before the raising up of the oceans, these two "faces" of the moon formed a single being, subsequently cut in two by Mohini (*Vishnu*). Such a system for counting on the hands is also found throughout the Islamic world, but mainly for religious purposes in this case. Muslims use it when reciting the 99 incomparable attributes of Allah or for counting in the litany of *subḥa* (which consists of the 33-fold repetition of each of the three "formulas"), which is recited following the obligatory prayer. To do either of these conveniently, a count of 33 must be achieved. This is done by counting the knuckles, from base to tip, of each finger and the thumb (including the ball of the thumb), first on the left hand and then on the right. In this way a count of 30 is attained, which is brought up to 33 by further counting on the tips of the little, ring, and middle fingers of the right hand.

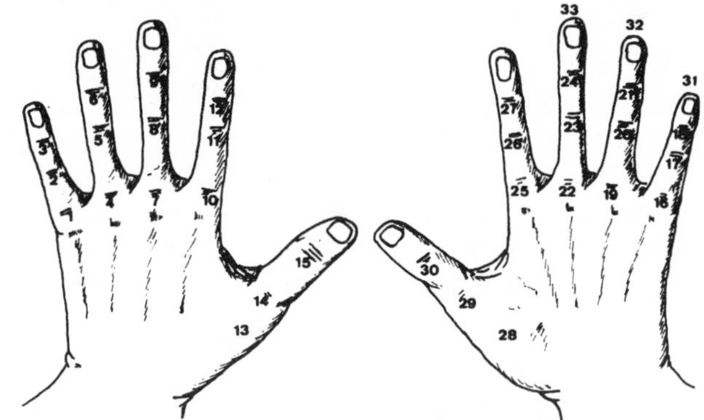

FIG. 3.9. *The Muslim method of counting up to 33, used for reciting the 99 (= 3 times 33) attributes of Allah, and for the 33-fold repetitions of the* subḥa

Nowadays, Muslims commonly adopt a rosary of prayer-beads for this purpose, but the method just described may still be adopted if the beads are not to hand. However, the hand-counting method is extremely ancient and undoubtedly pre-dates the use of the beads. Indeed, it finds mention in the oral tradition, in which the Prophet is described as admonishing women believers against the use of pearls or pebbles, and encouraging them to use their fingers to count the Praises of Allah. I. Godziher (1890) finds in this tradition some disapproval, by the Islamic authorities, of the use of the rosary subsequent to its emergence in the ninth century CE, which persisted until the fifteenth century CE. Abū Dawūd al Tirmidhī tells it as follows: "The Prophet of Allah has said to us, the women of Medina: Recite the *tasbīḥ*, the *tahlīl* and the *taqdīs*; and count these Praises on your fingers, for your fingers are for counting." This parallelism between Far-Eastern commercial practices and the ancient and widespread customs of Islamic religious tradition is extremely interesting.

## THE GAME OF MORRA

For light relief, let us consider the game of Morra, a simple, ancient and well-known game usually played between two players. It grew out of finger-counting. The two players stand face to face, each holding out a closed fist. Simultaneously, the two players open their fists; each extends as many fingers as he chooses, and at the same time calls out a number from 1 to 10. If the number called by a player equals the sum of the numbers of fingers shown by both players, then that player wins a point. (The players may also use both hands, in which case the call would be between 1 and 20.) The game depends not only on chance, but also on the quickness, concentration, judgment and anticipation of the players. Because the game is so well defined, and also of apparently ancient origin, it is very interesting for our purpose to follow its traces back into history, and across various peoples; and we shall come upon many signs of contacts and influences which will be important for us. It is still popular in Italy (where it is called *morra*), and is also played in southeast France (*la mourre*), in the Basque region of Spain, in Portugal, in Morocco and perhaps elsewhere in North Africa. As a child I played a form of it myself in Marrakesh, with friends, as a way of choosing "It". We would stand face to face, hands behind our backs. One of the two would bring forward a hand with a number of fingers extended. The other would call out a number from 1 to 5, and if this was the same as the number of fingers then he was "It"; otherwise the first player was "It". In China and Mongolia the game is called *hua quan* (approximately, "fist quarrel"). According to Joseph Needham, it is a popular entertainment in good circles. P. Perny (1873) says: "If the guests know each other well, their host will propose *qing hua quan* ('let us have a fist quarrel'). One of the guests is appointed umpire. For reasons of politeness, the host and one of the guests commence, but the host will soon give way to someone else. The one who loses pays the 'forfeit' of having to drink a cup of tea." The game of Morra was very popular, in Renaissance times, in France and Italy, amongst valets, pages and other servants to while away their idle hours. "The pages would play Morra at a flick of the fingers" (Rabelais, *Pantagruel* Book IV, Ch. 4); "Sauntering along the path like the servants sent to get wine, wasting their time playing at Morra" (Malherbe, *Lettres* Vol. II p. 10). Fifteen hundred years earlier, the Roman plebs took great delight in the game, which they called *micatio* (Fig. 3.10). Cicero's phrase for a man one could trust was: "You could play *micatio* with him in the dark." He says it was a common turn of phrase, which indicated the prevalence of the game in the popular culture.

FIG. 3.10. *Mural painting showing the game of Morra. Farnesina, Rome*

The game also served in the settling of disputes, legal or mercantile, when no other means prevailed, much as in "drawing the short straw", and was even forbidden by law in public markets (G. Lafaye, 1890). Vases and other Ancient Greek relics depict the game (Fig. 3.11). According to legend it was Helen who invented the game, to amuse her lover Paris.

Much earlier, the Egyptians had a similar game, as shown in the two funerary paintings reproduced in Fig. 3.12. The top is from a tomb at Beni

FIG. 3.11. *The game of Morra in Greek times. (Left) Painted vase in the Lambert Collection, Paris. (Right) Painted vase, Munich Museum. (DAGR, pp. 1889–90)*

FIG. 3.12. *Two Egyptian funeral paintings showing the game of Morra. (Above) Tomb no. 9 of Beni-Hassan (Middle Kingdom). See Newberry, ASE vol. 2 (1893), plate 7. (Below) Theban tomb no. 36 (Aba's tomb, XXVIth dynasty). See Wilkinson (1837), vol. 2, p. 55 (Fig. 307). See also photo no. 9037 by Schott at the Göttingen Institute of Egyptology.*

Hassan dating from the Middle Kingdom (21st–17th centuries BCE), and it shows two scenes. In the first scene, one man holds his hand towards the eyes of the other, hiding the fingers with his other hand. The other man holds his closed fist towards the first. The lower scene depicts similar gestures, but directed towards the hand. According to J. Yoyotte (in G. Posener, 1970), the hieroglyphic inscriptions on these paintings mean: *Left legend*: Holding the *íp* towards the forehead; *Right legend*: Holding the *íp* towards the hand. The Egyptian word *íp* means "count" or "calculate", so these paintings must refer to a game like Morra. The lower painting, from Thebes, is from the time of King Psammetichus I (seventh century BCE) and was (according to Leclant) copied from an original from the Middle Kingdom. This too shows two pairs of men, showing each other various combinations of open and closed fingers.

We may therefore conclude that the game of Morra, in one form or another, goes back at least to the Middle Kingdom of Pharaonic Egypt. In the world of Islam, Morra is called *mukhāraja* ("making it stick out"). At the start of the present century it was played in its classical form in remote areas of Arabia, Syria and Iraq. *Mukhāraja* was above all, however, a divination ritual amongs the Muslims and was therefore forbidden to the faithful (fortune-telling is proscribed by both Bible and Koran); so it was a much more serious matter than a mere game. An Arabian fortune-telling manual shows circular maps of the universe (*Zā' irjat al 'alam*), divided into sectors corresponding to the stars, where each star has a number. There are also columns of numbers which give possible "answers" to questions which might be asked. The *mukhāraja* was then used to establish a relationship between the two sets of numbers.

## COUNTING AND SIGN-LANGUAGE

There is a much more elaborate way of counting with the hand which, from ancient times until the present day, has been used by the Latins and can also be found in the Middle East where, apparently, it may go back even further in time. It is rather like the sign language used by the deaf and dumb. Using one or both hands at need, counting up to 9,999 is possible by this method. From two different descriptions we can reconstruct it in its entirety. These are given in parallel to each other in Fig. 3.13.

The first was written in Latin in the seventh century by the English monk Bede ("The Venerable") in his *De ratione temporum*, in the chapter *De computo vel loquela digitorum* ("Counting and talking with the fingers"). The other is to be found in the sixteenth-century Persian dictionary *Farhangi Djihangiri*. There is a most striking coincidence between these two descriptions written nine centuries apart and in such widely separated places.

With one hand (the left in the West and the right in the East), the little finger, fourth and middle fingers represented units, and either the thumb or the index finger (or both) was used for tens. With the other hand, hundreds and thousands were represented in the same way as the units and tens.

Both accounts also describe how to show numbers from 10,000 upwards. In the Eastern description: "for 10,000 bring the whole top joint of the thumb in contact with the top joint of the index finger and part of its second joint, so that the thumbnail is beside the nail of the index finger and the tip of the thumb is beside the tip of the index finger." For his part, Bede says: "For 10,000 place your left hand, palm outwards, on your breast, with the fingers extended backwards and towards your neck." Therefore the two descriptions diverge at this point.

Let us however follow Bede a little further.

> For 20,000 spread your left hand wide over your breast. For 30,000 the left hand should be placed towards the right and palm downwards, with the thumb towards the breastbone. For 50,000 similarly place the left hand at the navel. For 60,000 bring your left hand to your left thigh, inclining it downwards. For 70,000 bring your left hand to the

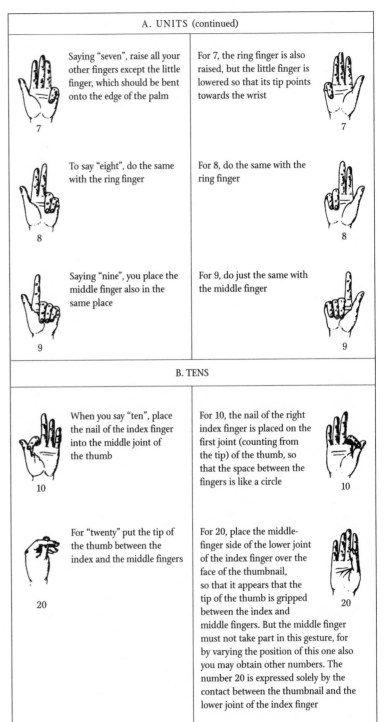

| WESTERN DESCRIPTION<br>*From the Latin of the Venerable Bede, seventh century* | EASTERN DESCRIPTION<br>*From a sixteenth-century Persian dictionary* |
|---|---|
| **A. UNITS** | |
|  1 When you say "one", bend your left little finger so as to touch the central fold of your palm | For 1, bend down your little finger  1 |
| 2 For "two", bend your next finger to touch the same spot | For 2, your ring finger must join your little finger  2 |
| 3 When you say "three", bend your third finger in the same way | For 3, bring your middle finger to join the other two  3 |
| 4 When you say "four", raise up your little finger from its place | For 4, raise the little finger (the other fingers should stay where they were before)  4 |
| 5 Saying "five", raise your second finger in the same way | For the number 5, also raise your ring finger  5 |
| 6 When you say "six", you also raise your third finger, but you must keep your ring finger in the middle of your palm | For the number 6, raise the middle finger, keeping your ring finger down (so that its tip is in the centre of the palm)  6 |

| A. UNITS (continued) | |
|---|---|
|  7 Saying "seven", raise all your other fingers except the little finger, which should be bent onto the edge of the palm | For 7, the ring finger is also raised, but the little finger is lowered so that its tip points towards the wrist  7 |
| 8 To say "eight", do the same with the ring finger | For 8, do the same with the ring finger 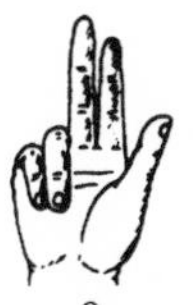 8 |
| 9 Saying "nine", you place the middle finger also in the same place | For 9, do just the same with the middle finger  9 |
| **B. TENS** | |
|  10 When you say "ten", place the nail of the index finger into the middle joint of the thumb | For 10, the nail of the right index finger is placed on the first joint (counting from the tip) of the thumb, so that the space between the fingers is like a circle  10 |
|  20 For "twenty" put the tip of the thumb between the index and the middle fingers | For 20, place the middle-finger side of the lower joint of the index finger over the face of the thumbnail, so that it appears that the tip of the thumb is gripped between the index and middle fingers. But the middle finger must not take part in this gesture, for by varying the position of this one also you may obtain other numbers. The number 20 is expressed solely by the contact between the thumbnail and the lower joint of the index finger  20 |

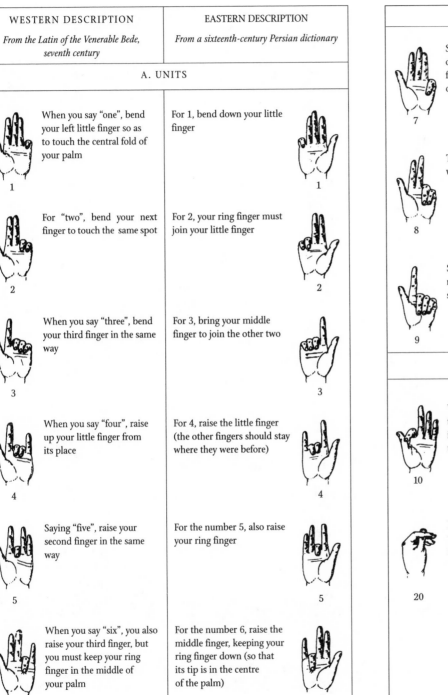

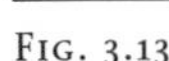

FIG. 3.13.

| B. TENS (continued) | |
|---|---|
|  30 For "thirty", touch thumb and index in a gentle kiss | For 30, hold the thumb straight and touch the tip of the thumbnail with the index finger, so that together they resemble the arc of a circle with its chord (if you need to bend the thumb somewhat, the number will be equally well indicated and no confusion should result)  30 |
|  40 For "forty", place the inside of the thumb against the side or the back of the index finger, keeping both of them straight | For 40, place the inside of the tip of the thumb on the back of the index finger, so that there is no space between the thumb and the edge of the palm  40 |
|  50 For "fifty", bend the thumb across the palm of the hand, with the top joint bent over, like the Greek letter Γ | For 50, hold the index finger straight up, but bend the thumb and place it in the palm of the hand, in front of the index finger  50 |
|  60 For "sixty", with the thumb bent as for fifty, the index finger is brought down to cover the face of the thumbnail | For 60, bend the thumb and place the second phalanx of the index finger on the face of the thumbnail  60 |
|  70 For "seventy", with the index finger as before, that is closely covering the thumbnail, raise the thumbnail across the middle joint of the index finger | For 70, raise the thumb and place the underside of the first joint of the index finger on the tip of the thumbnail so that the face of the thumbnail remains uncovered  70 |

FIG. 3.13. *(continued)*

| B. TENS (continued) | |
|---|---|
|  80 For "eighty", with the index raised as above, and the thumb straight, place the thumbnail within the bent middle phalanx of the index finger | For 80, hold the thumb straight and place the tip of the index finger on the curve of its top joint. (Note the discrepancy between the two accounts here) 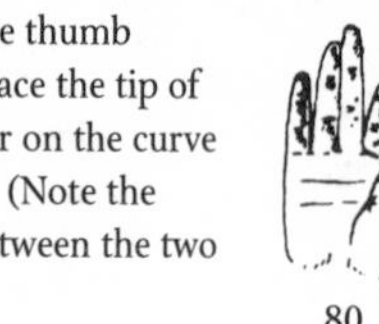 80 |
|  90 For "ninety", press the nail of the index finger against the root of the thumb | For 90, put the nail of the index finger over the joint of the second phalanx of the thumb (just as, for 10, you place it over the joint of the first phalanx) 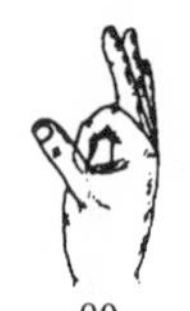 90 |

## C. HUNDREDS AND THOUSANDS

| | |
|---|---|
| 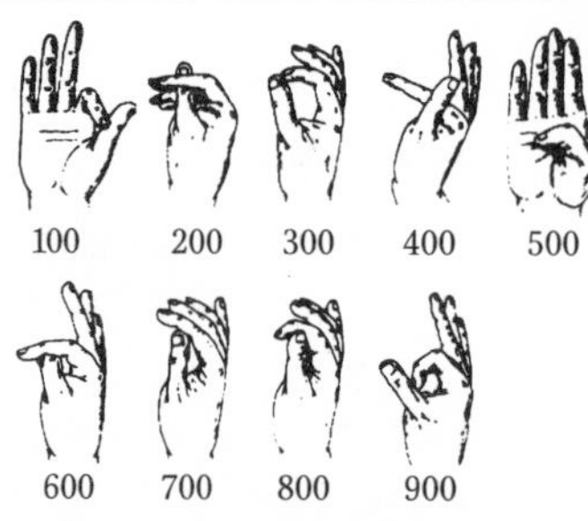 100 200 300 400 500 600 700 800 900 | 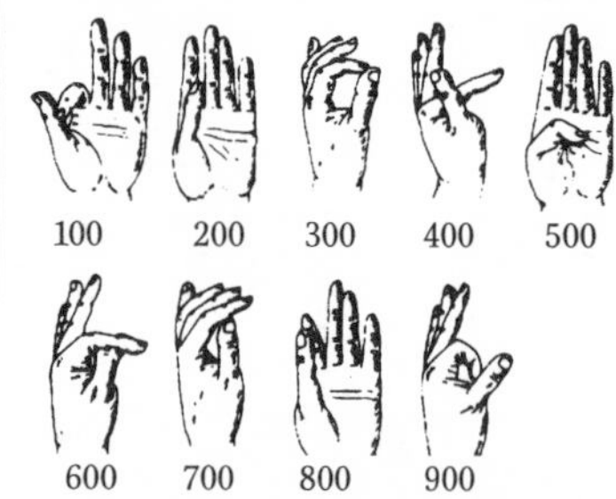 100 200 300 400 500 600 700 800 900 |
| When you say "a hundred", on your right hand do as for ten on the left hand; "two hundred" on the right hand is like twenty on the left; "three hundred" on the right like thirty on the left; and so on up to "nine hundred"<br>When you say "a thousand", with your right hand you do as for one with the left; "two thousand", on the right is like two on the left; "three thousand" on the right like three on the left, and so on up to "nine thousand" | Once you have mastered these eighteen numbers, the nine combinations of the little, ring and middle fingers as well as the nine combinations of the thumb and the index finger then you can readily understand that what serves on the right hand to show the units from 1 to 9 will on the left hand show from 1,000 to 9,000; and that what on the right hand shows the tens, on the left hand shows the hundreds from 100 to 900 |
| 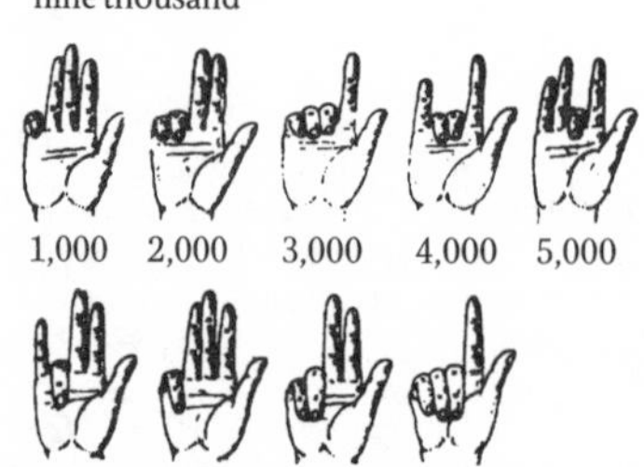 1,000 2,000 3,000 4,000 5,000 6,000 7,000 8,000 9,000 | 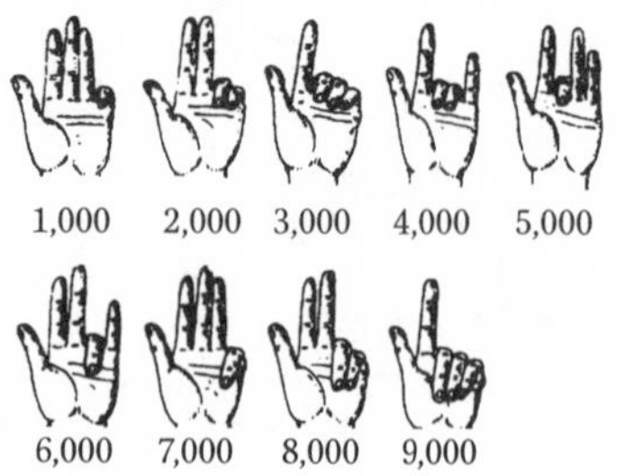 1,000 2,000 3,000 4,000 5,000 6,000 7,000 8,000 9,000 |

same place, but palm outwards. For 80,000 grasp your thigh with your hand. For 90,000 grasp your loins with the left hand, the thumb towards the genitals.

Bede continues by describing how, by using the same signs on the right-hand side of the body, and with the right hand, the numbers from 100,000 to 900,000 may be represented. Finally he explains that one million may be indicated by crossing the two hands, with the fingers intertwined.

## FINGER-COUNTING THROUGHOUT HISTORY

The method described above is extremely ancient. It is likely that it goes back to the most extreme antiquity, and it remained prominent until recent times in both the Western and Eastern worlds and, in the latter, persisted until recent times. In the Egypt of the Pharaohs it was in use from the Old Kingdom (2800–2300 BCE), as it would seem from a number of funeral paintings of the period. For example, Fig. 3.14 shows, from right to left, three men displaying numbers on their fingers according to the method just described. The first figure seems to be indicating 10 or 100, the fourth 6 or 6,000 and the sixth 7 or 7,000. According to traditions which have been repeated by various authors, Egypt clearly appears to have been the source of this system.

FIG. 3.14. *Finger-counting shown on a Egyptian monument of the Old Kingdom (Fifth Dynasty, 26th century BCE). Mastaba D2 at Saqqara. See Borchardt (1937), no. 1534A, plate 48.*

C. Pellat (1977) quotes two Arab manuscripts. One of these is at the University of Tunis (no. 6403) and the other is in the library of the Waqfs in Baghdad (*Majami'* 7071/9). The counting system in question is, in the first manuscript, attributed to "the Copts of Egypt"; the title of the second clearly suggests that it is of Egyptian origin. (*Treatise on the Coptic manner of counting with the hands*).

A *qasida* (poem in praise of a potential patron) attributed to Mawsili al Ḥanbali describes "the sign language of the Copts of Egypt, which expresses numbers by arranging the fingers in special ways". Ibn al Maghribi states, "See! I follow in the steps of every learned man. The spirit moves me to write something of this art and to compose a *Ragaz*, to be called *The Table of Memory*, which shall include the art of counting of the Copts." Finally, Juan Perez de Moya (*Alcala de Henares*, 1573) comes to the following conclusion: "No one knows who invented this method of counting, but since the Egyptians loved to be sparing of words (as Théodoret has said), it must be from them that it has come."

There is also evidence for its use in ancient Greece. Plutarch (*Lives of Famous Men*) has it that Orontes, son-in-law of Artaxerxes King of Persia, said: "Just as the fingers of one who counts are sometimes worth ten thousand and sometimes merely one, so also the favourites of the King may count for everything, or for nothing."

The method was also used by the Romans, as we know in the first instance from "number-tiles" discovered in archaeological excavations from several parts of the Empire, above all from Egypt, which date mostly from the beginning of the Christian era (Fig. 3.15). These are small counters or tokens, in bone or ivory, each representing a certain sum of money. The Roman tax collectors gave these as "receipts". On one side there was a representation of one of the numbers according to the sign system described above, and on the other side was the corresponding Roman numeral. (It would seem, however, that these numbers never went above 15 in these counters from the Roman Empire).

FIG. 3.15. *Roman numbered tokens* (tesserae) *from the first century CE. The token on the left shows on one face the number 9 according to a particular method of finger-counting; on the reverse face, the same value is shown in Roman numerals. British Museum. The token on the right shows a man making the sign for 15, according to the same system, on the fingers of his left hand. Bibliothèque nationale (Paris). Tessera no. 316. See Frohner (1884).*

We also know about this from the writings of numerous Latin authors. Juvenal (c. 55–135 CE) speaks thus of Nestor, King of Pylos, who lived, it is said, for more than a hundred years: "Fortunate Nestor who, having attained one hundred years of age, henceforth shall count his years on his right hand!" This tells us that the Romans counted tens and units on the left hand, and hundreds and thousands on the right hand. Apuleus (c. 125–170 CE) describes in his *Apologia* how, having married a rich widow, a certain Aemilia Pudentilla, he was accused of resorting to magic means to win her heart. He defended himself before the Pro-Consul

Claudius Maximus in the presence of his chief accuser Emilianus. Emilianus had ungallantly declared that Aemilia was sixty years old, whereas she was really only forty. Here is how Apuleus challenges Emilianus.

> How dare you, Emilianus, increase her true number of years by one half again? If you said thirty for ten, we might think that you had ill-expressed it on your fingers, holding them out straight instead of curved (Fig. 3.16). But forty, now that is easily shown: it is the open hand! So when you increase it by half again this is not a mistake, unless you allow her to be thirty years old and have doubled the consular years by virtue of the two consuls.

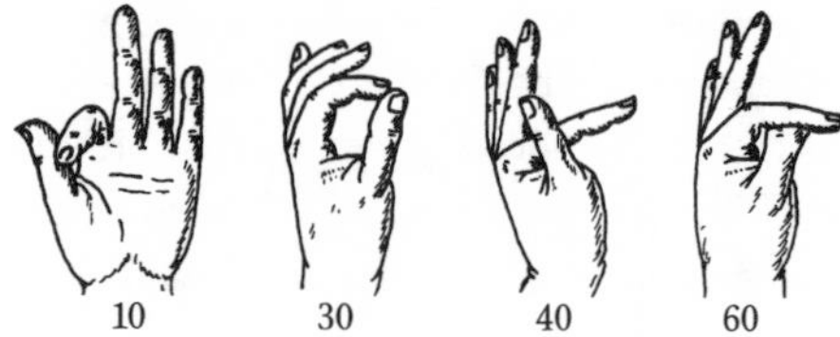


Fig. 3.16.

And we may cite Saint Jerome, Latin philologist of the time of Saint Augustine:

> One hundred, sixty, and thirty are the fruits of the same seed in the same earth. Thirty is for marriage, since the joining of the two fingers as in a tender kiss represents the husband and the wife. Sixty depicts the widow in sadness and tribulation. And the sign for one hundred (pay close attention, gentle reader), copied from the left to right with the same fingers, shows the crown of virginity (Fig. 3.17).

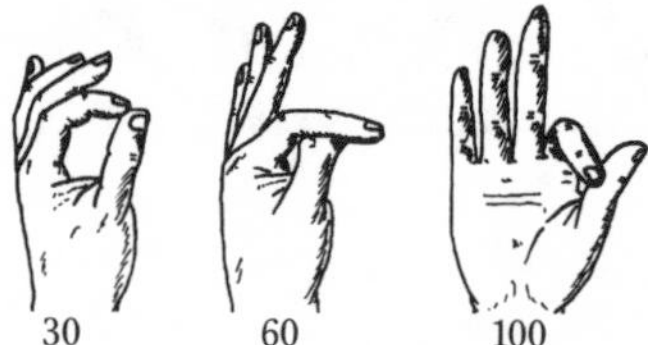


Fig. 3.17.

Again, the patriarch Saint Cyril of Alexandria (376–444) gives us the oldest known description of this system (*Liber de computo*, Chapter CXXXVIII: *De Flexibus digitorum*, III, 135). The description exactly matches a passage in a sixth-century Spanish encyclopaedia, *Liber etymologiarum*, which was the outcome of an enormous compilation instituted by Bishop Isidor of Seville (570–636). The Venerable Bede, in his turn, drew inspiration from it in the seventh century for his chapter *De computo vel loquela digitorum*.

One of the many reasons why this system remained popular was its secret, even mysterious, aspect. J. G. Lemoine (1932) says: "What a splendid method for a spy to use, from the enemy camp, to inform his general at a distance of the numerical force of the enemy, by a simple, apparently casual, gesture or pose." Bede also gives an example of such silent communication: "A kind of manual speech [*manualis loquela*] can be expressed by the system which I have explained, as a mental exercise or as an amusement." Having established a correspondence between the Latin letters and the integers, he says: "To say *Caute age* ('look out!') to a friend amongst doubtful or dangerous people, show him (the following finger gestures)" (Fig. 3.18).

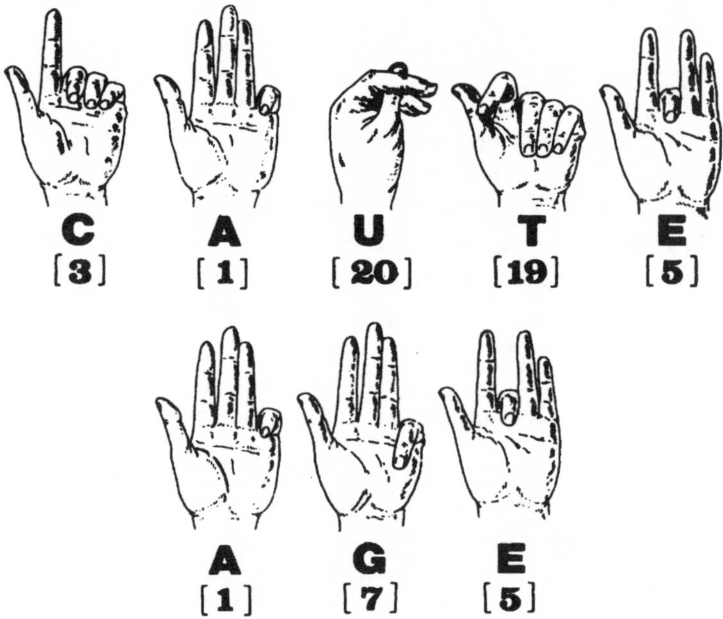


Fig. 3.18.

Following the fall of the Roman Empire, the same manual counting remained extraordinarily in vogue until the end of the Middle Ages (Fig. 3.19 to 3.21), and played a most important part in mediaeval education. The finger counting described in Bede's *De computo vel loquela digitorum* (cited above) was extensively used in the teaching of the *Trivium* of grammar, rhetoric and logic during the undergraduate years leading to the B.A. degree, which, with the *Quadrivium* (literally "crossroads", the meeting of the Four Ways of arithmetic, geometry, astronomy, and music) studied in the following years leading to the M.A. degree, made up the Seven Liberal Arts of the scholarly curriculum, from the sixth to the fifteenth centuries. Barely four hundred years ago, a textbook of arithmetic was not considered complete without detailed explanations of this system (Fig. 3.22). Only when written arithmetic became widespread, with the adoption of the use of Arabic numerals, did the practice of arithmetic on hands and fingers finally decline.

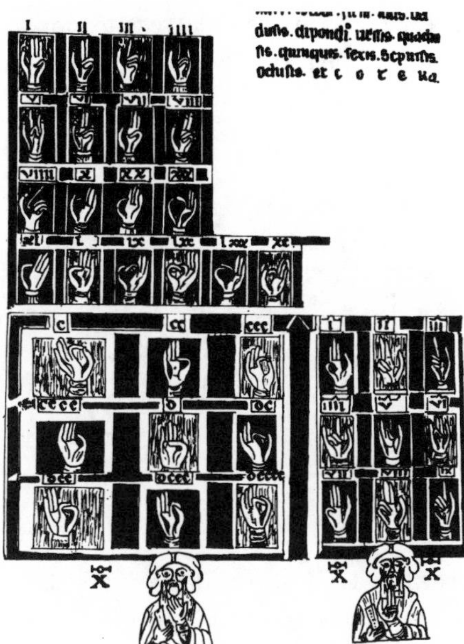

FIG. 3.19. *The system described in Fig. 3.13 illustrated in a manuscript by the Spanish theologian Rabano Mauro (780–856). Codex Alcobacense 394, folio 152 V. National Library of Lisbon. See Burnam (1912–1925), vol. 1, plate XIV.*

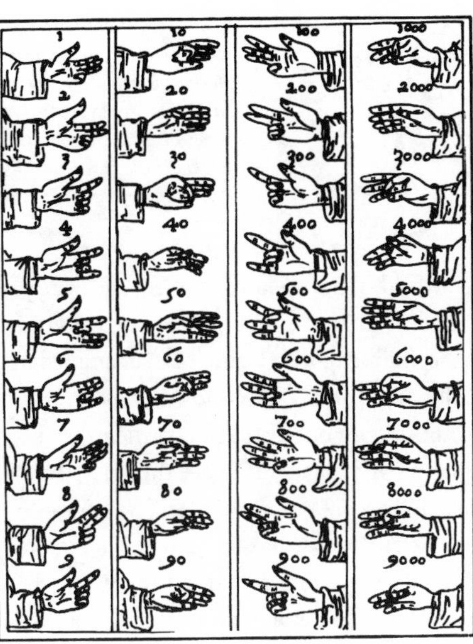


FIG. 3.21. *The same system yet again in a mathematical work published in Vienna in 1494. Extract from the work by Luca Pacioli:* Summa de Arithmetica, Geometrica, proportioni e proportionalita

FIG. 3.20. *The same system again, in a Spanish manuscript of 1130. Detail of a codex from Catalonia (probably from Santa Maria de Ripoll). National Library of Madrid, Codex matritensis A19, folio 3V. See Burnam (1912–1925), vol. 3, plate XLIII.*

FIG. 3.22. *The same system of signs in a work on arithmetic published in Germany in 1727: Jacob Leupold,* Theatrum Arithmetico-Geometricum

FIG. 3.23. *In the Arab-Persian system of number gestures, the number 93 is shown by placing the nail of the index finger right on the joint of the second phalanx of the thumb (which represents 90), and then bending the middle, ring and little fingers (which represents 3); and this, nearly enough, gives rise to a closed fist.*

In the Islamic world, the system was at least as widely spread as in the West, as recounted by many Arab and Persian writers from the earliest times. From the beginning of the Hegira, or Mohammedan era (dated from the flight of Mohammed from Mecca to Medina on 15 July 622 CE), we find an oblique allusion among these poets when they say that a mean or ungenerous person's hand "makes 93" (see the corresponding closed hand, symbol of avarice, in Fig. 3.23). One of them, Yaḥyā Ibn Nawfal al Yamānī (seventh century) says: "Ninety and three, which a man may show as a fist closed to strike, is not more niggardly than thy gifts, Oh Yazid." Another, Khalîl Ibn Aḥmad (died 786), grammarian and one of the founders of Arab poetry, writes: "Your hands were not made for giving, and their greed is notorious: one of them makes 3,900 (the mirror image of 93) and the other makes 100 less 7."

One of the greatest Persian poets, Abu'l Kassim Firdūsi, dedicated *Shah Namēh* (The Book of Kings) to Sultan Mahmūd le Ghaznavide but found himself poorly rewarded. In a satire on the Sultan's gross avarice, he wrote: "The hand of King Mahmūd, of noble descent, is nine times nine and three times four."

A *qasīda* of the Persian poet Anwari (died 1189 or 1191) praises the Grand Vizir Nizam al Mūlk for his precocity in arithmetic: "At the age when most children suck their thumbs, you were bending the little finger of your left hand" (implying that the Vizir could already count to a thousand) (Fig. 3.13C).

A dictum of the Persian poet Abu'l Majîd Sanāyî (died 1160) reminds us that by twice doing the same thing in one's life, one may take away from its value: "What counts for 200 on the left hand, on the right hand is worth no more than 20" (Fig. 3.24). The poet Khāqānī (1106–1200) exclaims: "If I could count the turns of the wheel of the skies, I would number them on my left hand!" and: "Thou slayest thy lover with the glaive of thy glances, so many as thou canst count on thy left hand" (the left hand counts the hundreds and the thousands).

Another quotation from Anwari: "One night, when the service I rendered thee did wash the face of my fortune with the water of kindness, you did give to me that number (50) which thy right thumb forms when it tries to hide its back under thy hand" (Fig. 3.25).

And some verses of Al Farazdaq (died 728) refer to forming the number 30 by opposing thumb and forefinger, in a description of crushing pubic lice (Fig. 3.26).

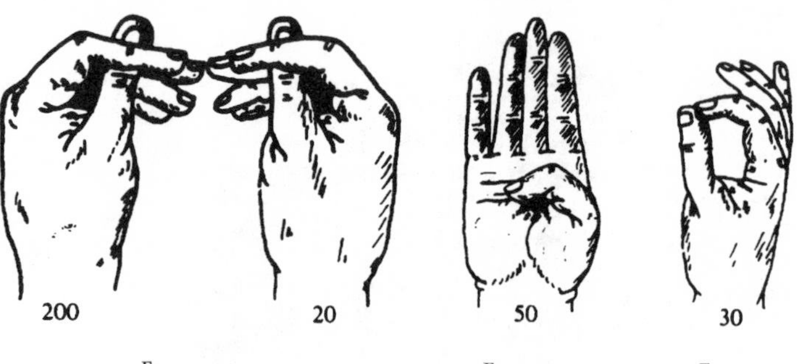


FIG. 3.24. FIG. 3.25. FIG. 3.26.

According to Levi della Vida (1920), one of the earliest datable references from the Islamic world to this numerical system can be found in the following quotation from Ibn Sa'ad (died c. 850): "Hudaifa Ibn al Yamān, companion of the Prophet, signalled the murder of Khalif 'Otman as one shows the number 10 and sighed: 'This will leave a void [forming a round between finger and thumb, Fig. 3.27] in Islam which even a mountain could hardly fill.'"

A poem attributed to Al Mawsilī al Ḥanbalī says: "If you place the thumb against the forefinger like – listen carefully – someone who takes hold of an arrow, then it means 60" (Fig. 3.28); and, in verses attributed to Abūl Ḥassan 'Alī: "For 60, bend your forefinger over your thumb, as a bowman grasps an arrow [Fig. 3.28] and for 70 do like someone who flicks a dinar to test it" (Fig. 3.29).

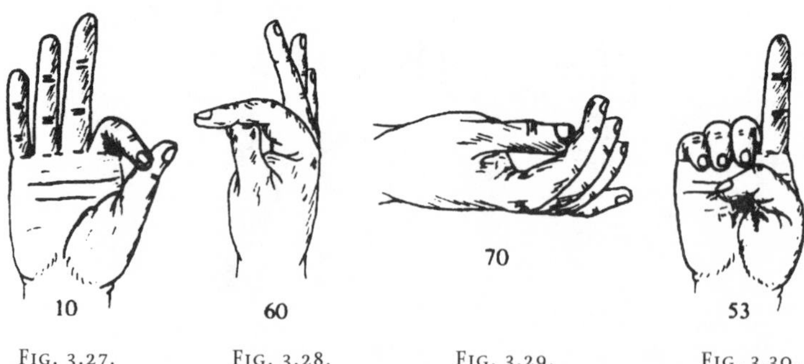


FIG. 3.27. FIG. 3.28. FIG. 3.29. FIG. 3.30.

Aḥmad al Barbīr al Tarābulusī (a writer on secular Arab and Persian texts), talking of what he calls counting by bending the fingers, says: "We know the traditionalists use it, because we find references; and it is the same with the *fuqahā*,* for these lawyers refer to it in relation to prayer in connection with the Confession of Faith;† they say that, according to the rule of tradition, he who prays should place his right hand on his thigh when he squats for the Tashahud, forming the number 53" (Fig. 3.30).

From the poet Khāqānī we have: "What struggle is this between Rustem and Bahrām? What fury and dispute is it that perturbs these two sons of noble lines? Why, they fight day and night to decide which army shall do a 20 on the other's 90."

This may seem obscure to the modern reader, unfamiliar with the finger signs in question. But look closely at the gestures that correspond to the numbers (Fig. 3.31): "90" undoubtedly represents the anus (and, by extension, the backside), as it commonly did in vulgar speech; while "to do a 20 on someone" is undoubtedly an insulting reference to the sexual act (apparently expressed as "to make a thumb" in Persian) and therefore (by extension in this military context) to "get on top of".

More obscenely, Aḥmād al Barbīr al Tarābulusī could not resist offering his pupils the following mnemonic for the gestures representing 30 and 90:

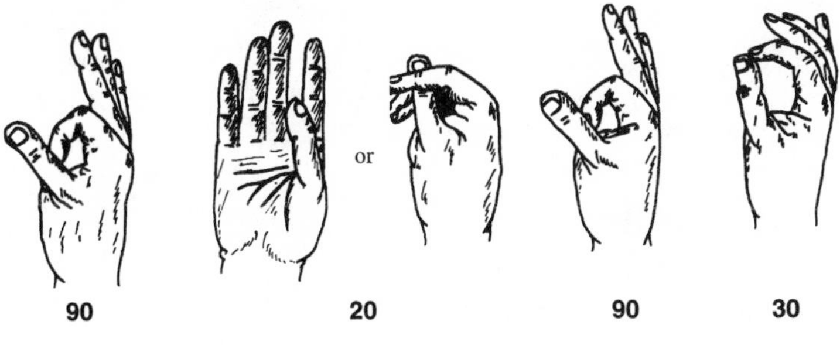


FIG. 3.31. FIG. 3.32.

"A poet most elegantly said, of a handsome young man: Khālid set out with a fortune of 90 dirhams, but had only one third of it left when he returned!" plainly asserting that Khālid was homosexual (Fig. 3.32), having started "narrow" (90) but finished "wide" (30).

These many examples amply show how numbers formed by the fingers served as figures of speech, no doubt much appreciated by the readers of the time. These ancient origins find etymological echo today, as in *digital computing*. There is no longer any question of literally counting on the fingers, but the Latin words *digiti* ("fingers") and *articuli* ("joints") came to represent "units" and "tens", respectively, in the Middle Ages, whence *digiti* in turn came to mean the signs used to represent the units of the decimal system. The English word *digit*, meaning a single decimal numeral, is derived directly from this. In turn, this became applied to computation, hence the term *digital* computing in the sense of "computing by numbers". With the development and recent enormous spread of "computers" (*digital* computers), the meaning of "digital computation" has been extended to include every aspect of the processing of information by machine in which any entity, numerical or not, and whether or not representing a variable physical quantity, is given a *discrete* representation (by which is meant that distinct representations correspond to different values or entities, there is a finite – though typically enormous – number of possible distinct representations, and different repesentations are encoded as sequences of symbols taken from a finite set of available symbols). In the modern digital computer, the primitive symbols are two in number and denoted by "0" and "1" (the *binary* system) and realised in the machine in terms of distinct physical states which are reliably distinguishable.

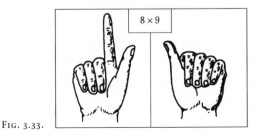

* Islamic lawyers who concern themselves with every kind of social or personal matter, with the order of worship and with ritual requirements

† Asserting that Allah is One, affirming belief in Mohammed, at the same time raising the index finger and closing the others

### HOW TO CALCULATE ON YOUR FINGERS

After this glance at the modern state of the art of digital information processing, let us see how the ancients coped with their "manual informatics".

The hand can be used not only for counting, but also for systematically performing arithmetical calculations. I used to know a peasant from the Saint-Flour region, in the Auvergne, who could multiply on his fingers, with no other aid, any two numbers he was given. In so doing, he was following in a very ancient tradition.

For example, *to multiply 8 by 9*, he closed on one hand as many fingers as the excess of 8 over 5, namely 3, keeping the other two fingers extended. On the other hand, he closed as many fingers as the excess of 9 over 5, namely 4, leaving the fifth finger extended (Fig. 3.33). He would then (mentally) multiply by 10 the total number (7) of closed fingers (70), multiply together the numbers of extended fingers on the two hands ($2 \times 1 = 2$), and finally add these two results together to get the answer (72). That is to say:

$$8 \times 9 = (3 + 4) \times 10 + (2 \times 1) = 72$$

FIG. 3.33.

Similarly, *to multiply 9 by 7*, he closed on one hand the excess of 9 over 5, namely 4, and on the other the excess of 7 over 5, namely 2, in total 6; leaving extended 1 and 3 respectively, so that by his method the result is obtained as

$$9 \times 7 = (4 + 2) \times 10 + (1 \times 3) = 63$$

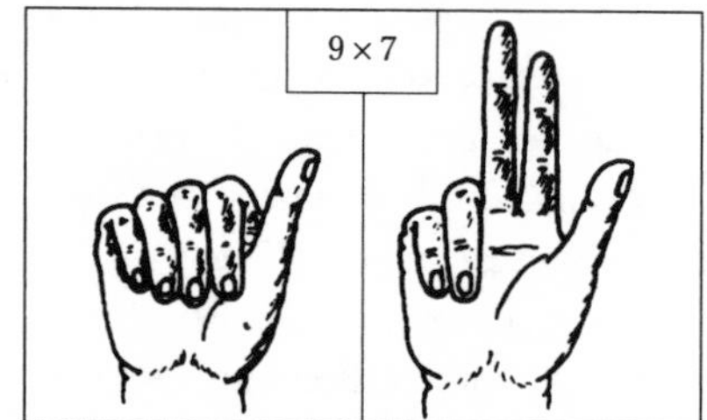


FIG. 3.34.

Although undoubtedly discovered by trial and error by the ancients, this method is infallible for the multiplication of any two whole numbers between 5 and 10, as the following proves by elementary (but modern) algebra. To multiply together two numbers $x$ and $y$ each between 5 and 10, close on one hand the excess $(x - 5)$ of $x$ over 5, and on the other the excess $(y - 5)$ of $y$ over 5; the total of these two is $(x - 5) + (y - 5)$, and 10 times this is

$$((x - 5) + (y - 5)) \times 10 = 10\,x + 10\,y - 100$$

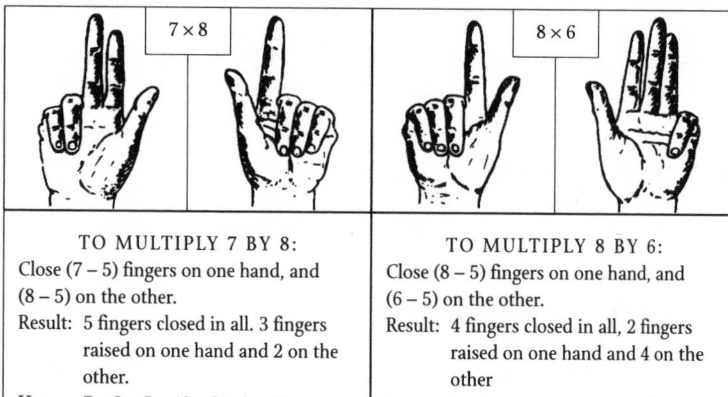


TO MULTIPLY 7 BY 8:
Close (7 – 5) fingers on one hand, and (8 – 5) on the other.
Result: 5 fingers closed in all. 3 fingers raised on one hand and 2 on the other.
Hence: $7 \times 8 = 5 \times 10 + 3 \times 2 = 56$

TO MULTIPLY 8 BY 6:
Close (8 – 5) fingers on one hand, and (6 – 5) on the other.
Result: 4 fingers closed in all, 2 fingers raised on one hand and 4 on the other
Hence: $8 \times 6 = 4 \times 10 + 2 \times 4 = 48$

FIG. 3.35.

The number of fingers remaining extended on the first hand is $5 - (x - 5) = 10 - x$, and on the other, similarly, $10 - y$. The product of these two is

$$(10 - x) \times (10 - y) = 100 - 10\,x - 10\,y + x\,y$$

Adding these two together, according to the method, therefore results in

$$(10\,x + 10\,y - 100) + 100 - 10\,x - 10\,y + xy = x\,y$$

namely the desired result of multiplying $x$ by $y$.

He had a similar way of multiplying numbers exceeding 9. For example, *to multiply 14 by 13*, he closed on one hand as many fingers as the excess of 14 over 10, namely 4, and on the other, similarly, 3, making in all 7. Then he mentally multiplied this total (7) by 10 to get 70, adding to this the product $(4 \times 3 = 12)$ to obtain 82, finally adding to this result $10 \times 10 = 100$ to obtain 182 which is the correct result.

By similar methods, he was able to multiply numbers between 15 and 20, between 20 and 25, and so on. It is necessary to know the squares of 10, 15, 20, 25 and so on, and their multiplication tables. The mathematical justifications of some of these methods are as follows.

*To multiply two numbers x and y between 10 and 15:*

$$10\,[(x - 10) + (y - 10)] + (x - 10) \times (y - 10) + 10^2 = x\,y$$

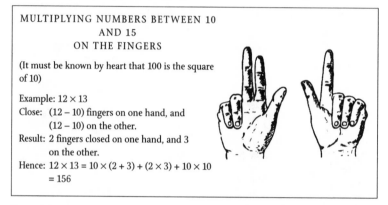

MULTIPLYING NUMBERS BETWEEN 10 AND 15 ON THE FINGERS

(It must be known by heart that 100 is the square of 10)

Example: $12 \times 13$
Close: (12 – 10) fingers on one hand, and (12 – 10) on the other.
Result: 2 fingers closed on one hand, and 3 on the other.
Hence: $12 \times 13 = 10 \times (2 + 3) + (2 \times 3) + 10 \times 10 = 156$

FIG. 3.36.

*To multiply two numbers x and y between 15 and 20:*

$$15\,[(x - 15) + (y - 15)] + (x - 15) \times (y - 15) + 15^2 = xy$$

MULTIPLYING NUMBERS BETWEEN 15 AND 20 ON THE FINGERS

(It must be known by heart that 225 is the square of 15)

Example: $18 \times 16$
Close: (18 – 15) fingers on one hand, and (16 – 15) on the other.
Result: 3 fingers closed on one hand, and 1 on the other
Hence: $15 \times (3 + 1) + (3 \times 1) + 15 \times 15 = 288$

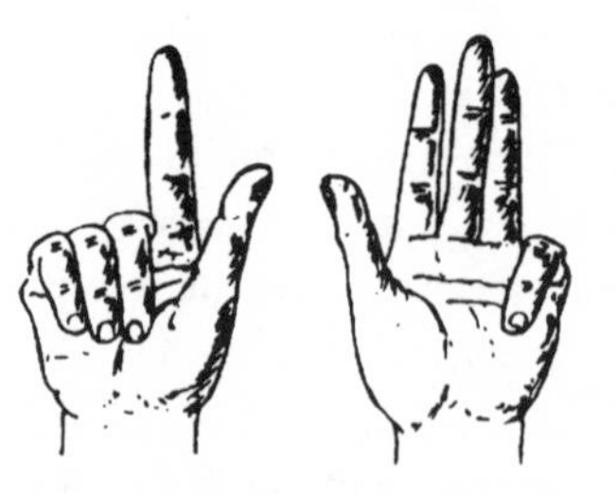

FIG. 3.37.

*To multiply two numbers x and y between 20 and 25:*

$$20\,[(x-20)+(y-20)]+(x-20)\times(y-20)+20^2=xy$$

and so on.*

It can well be imagined, therefore, how people who did not enjoy the facility in calculation which our "Arabic" numerals allow us were none the less able to devise, by a combination of memory and a most resourceful ingenuity in the use of the fingers, ways of overcoming their difficulties and obtaining the results of quite difficult calculations.

Fig. 3.38. *Calculating by the fingers shown in an Egyptian funeral painting from the New Kingdom. This is a fragment of a mural on the tomb of Prince Menna at Thebes, who lived at the time of the 18th Dynasty, in the reign of King Thutmosis, at the end of the 15th century BCE. We see six scribes checking while four workers measure out grain and pour bushels of corn from one heap to another. On the right, on one of the piles of grain, the chief scribe is doing arithmetic on his fingers and calling out the results to the three scribes on the left who are noting them down. Later they will copy the details onto papyrus in the Pharaoh's archives. (Theban tomb no. 69)*

## COUNTING TO THOUSANDS USING THE FINGERS

The method to be described is a much more developed and mathematically more interesting procedure than the preceding one. There is evidence of its use in China at any rate since the sixteenth century, in the arithmetical textbook *Suan fa tong zong* published in 1593. E. C. Bayley (1847) attests that it was in use in the nineteenth century, and Chinese friends of mine from Canton and Peking have confirmed that it is still in use.

In this method, each knuckle is considered to be divided into three parts: left knuckle, middle knuckle and right knuckle. There being three knuckles to a finger, there is a place for each of the nine digits from 1 to 9. Those on the little finger of the right hand correspond to the units, those on the fourth finger to the tens, on the middle finger to the hundreds, the forefinger to the thousands, and finally the right thumb corresponds to the tens of thousands. Similarly on the left hand, the left thumb corresponds to the hundreds of thousands, the forefinger to the millions, the middle finger to the tens of millions, and so on (Fig. 3.39); finally, therefore, on the little finger of the left hand we count by steps of thousands of millions, i.e. by billions.

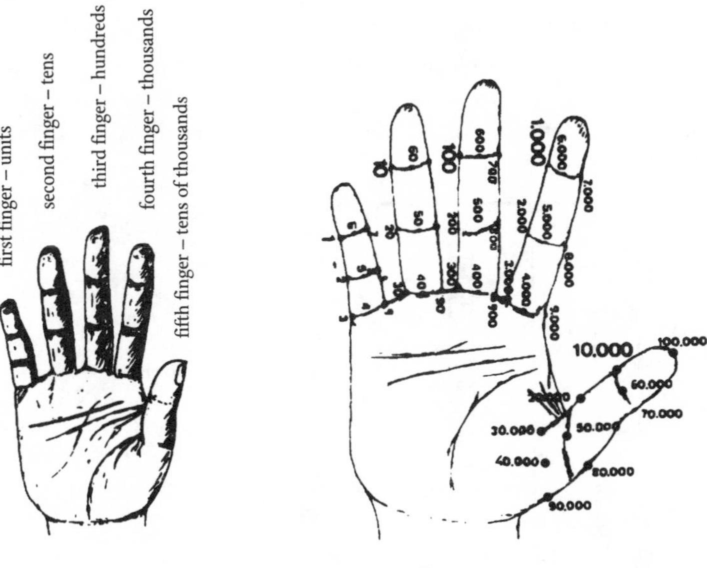


Fig. 3.39. Fig 3.40.

With the right hand palm upwards (Fig. 3.40), we count on the little finger from 1 to 3 by touching the "left knuckles" from tip to base; then from 4 to 6 by touching the "centre knuckles" from base to tip; and finally, from 7 to 9 by touching the "right knuckles" from tip to base. We count the tens similarly on the fourth finger, the hundreds on the middle finger, and so on.

In this way it is, in theory, possible to count up to 99,999 on one hand, and up to 9,999,999,999 with both: a remarkable testimony to human ingenuity.

* The general rule being:
$N((x-N)+(y-N))+(x-N)(y-N)+N^2=Nx+Ny-2N^2+xy-xN-yN+N^2+N^2=xy.$

CHAPTER 4

# HOW CRO-MAGNON MAN COUNTED

Among the oldest and most widely found methods of counting is the use of marked bones. People must have made use of this long before they were able to count in any abstract way.

The earliest archaeological evidence dates from the so-called Aurignacian era (35,000–20,000 BCE), and are therefore approximately contemporary with Cro-Magnon Man. It consists of several bones, each bearing regularly spaced markings, which have been mostly found in Western Europe (Fig. 4.1).

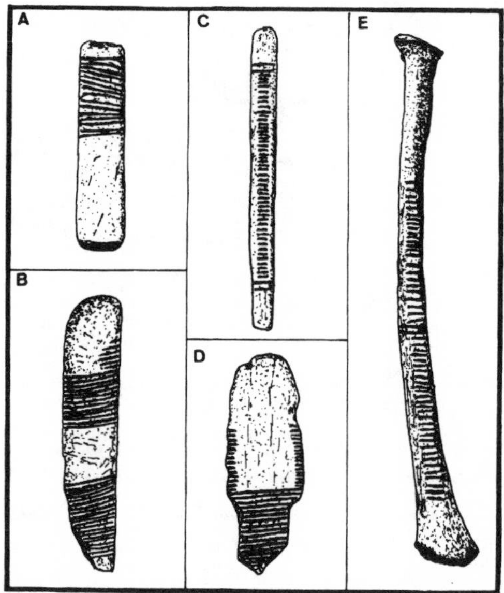


FIG. 4.1. *Notched bones from the Upper Palaeolithic age.*

*A and C: Aurignacian. Musée des Antiquités nationales, St-Germain-en-Laye. Bone C is from Saint-Marcel (Indre, France).*

*B and D: Aurignacian. From the Kulna cave (Czech Republic).*

*E: Magdalenian (19,000–12,000 BCE). From the Pekarna cave (Czech Republic). See Jelinek (1975), pp. 435–453.*

Amongst these is the radius bone of a wolf, marked with 55 notches in two series of groups of five. This was discovered by archaeologists in 1937, at Dolní Věstonice in Czechoslovakia, in sediments which have been dated as approximately 30,000 years old. The purpose of these notches remains mysterious, but this bone (whose markings are systematic, and not artistically motivated) is one of the most ancient arithmetic documents to have come down to us. It clearly demonstrates that at that time human beings were not only able to conceive number in the abstract sense, but also to represent number with respect to a base. For otherwise, why would the notches have been grouped in so regular a pattern, rather than in a simple unbroken series?

The man who made use of this bone may have been a mighty hunter. Each time he made a kill, perhaps he made a notch on his bone. Maybe he had a different bone for each kind of animal: one for bears, another for deer, another for bison, and so on, and so he could keep the tally of the larder. But, to avoid having to re-count every single notch later, he took to grouping them in fives, like the fingers of the hand. He would therefore have established a true graphical representation of the first few whole numbers, in base 5 (Fig. 4.2).

| I I I I I | I I I I I | I I I I I | I I I I I | . . . |
|---|---|---|---|---|
| 1 2 3 4 5 | 6 7 8 9 10 | 11 . . . 15 | 16 . . . 20 | |
| 1 hand | 2 hands | 3 hands | 4 hands | |

FIG. 4.2.

Also of great interest is the object shown in Fig. 4.3, a point from a reindeer's antler found some decades ago in deposits at Brassempouy in the Landes, dating from the Magdalenian era. This has a longitudinal groove which separates two series of transverse notches, each divided into distinct groups (3 and 7 on one side, 5 and 9 on the other). The longitudinal notch, which is much closer to the 9–5 series than to the 3–7 series, seems to form a kind of link or *vinculum* (as is sometimes used in Mathematics) joining the group of nine to the group of five.

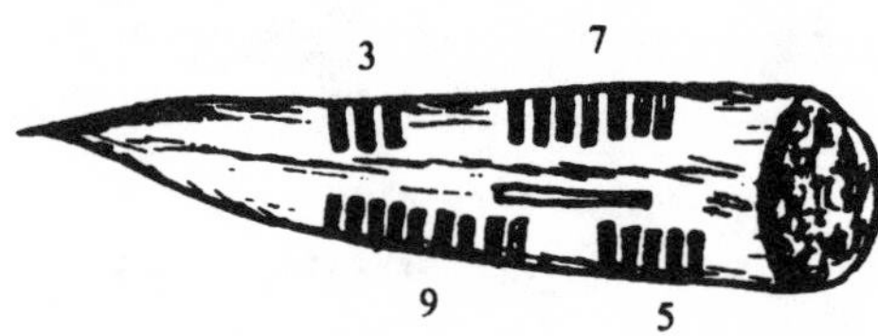


FIG. 4.3. *Notched bone from the Magdalenian era (19,000–12,000 BCE), found at Brassempouy (Landes, France). Bordeaux, Museum of Aquitaine*

Now what could this be for? Was it perhaps a simple tool, or a weapon, which had been grooved to stop it slipping in the hand? Unlikely. Anyway, what purpose would the longitudinal groove then serve? And even if this were the case, why do we not find such markings on similar prehistoric implements?

In fact, this object also bears witness of some activity with arithmetical connotations. The way the numbers 3, 5, 7, and 9 are arranged, and the frequency with which these numbers occur in many artefacts from the same period, suggest a possible explanation.

Let us suppose that the longitudinal groove represents unity, and that the transverse lines represent other odd numbers (which are all prime except for 9 which is the square of 3).

This spike from an antler with its grooves then makes a kind of arithmetical tool, showing a graphical representation of the first few odd numbers arranged in such a way that some of their simpler properties are exhibited (Fig. 4.4).

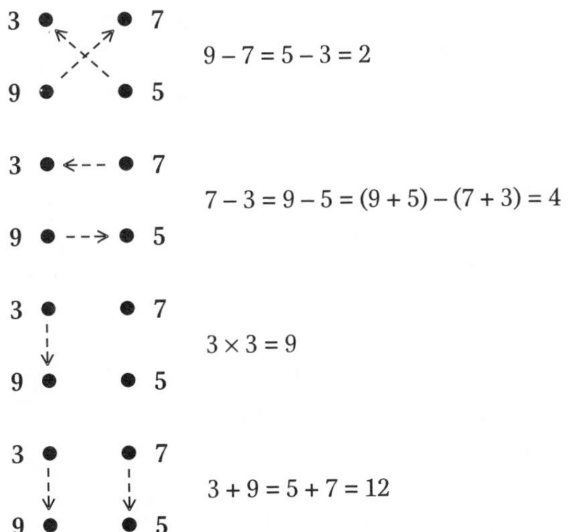


FIG. 4.4. *Some of the arithmetical properties of the groupings of the grooves on the bone shown in Fig. 4.3.*

As well as giving us concrete evidence for the memorisation and recording of numbers, the practice of making tally marks such as described is also a precursor of counting and book-keeping. We are therefore led to suppositions such as the following.

Our distant forefathers possibly used this piece of antler for taking count of people, things or beasts. It could perhaps have served a tool-maker to keep account of his own tools:

3 graters and 7 knives (in stone)
9 scrapers and 5 needles (in bone)

where the longitudinal groove linking the 5 and the 9 may, in this man's mind, have denoted the common material (bone) from which they were made.

Or perhaps a warrior might similarly keep count of his weapons:

3 knives and 7 daggers
9 spears with plain blade, and 5 with split blade

Or the hunter might record the numbers of different types of game brought back for the benefit of his people:

3 bison and 7 buffalo
9 reindeer and 5 stags

We can also imagine how a herdsman could count the beasts in his keeping, sheep and goats on the one hand, cattle on the other.

A messenger could use an antler engraved in this way to carry a promissory note to a neighbouring tribe:

In 3 moons and 7 days we will bring
9 baskets of food and 5 fur animals

We can also imagine it being used as a receipt for goods, or a delivery manifest, or for accounting for an exchange or distribution of goods.

Of course, these are only suppositions, since the true meaning has eluded the scholars. And in fact the true purpose of these markings will remain unknown for ever, because with this kind of symbolism the things themselves to which the operations apply are represented only by their quantity, and not by specific signs which depict the nature of the things.

Human kind was still unable to write. But, by representing as we have described the enumeration of this or that kind of unit, the owner of the antler, and his contemporaries, had nonetheless achieved the inventions of written number: in truth, they wrote figures in the most primitive notation known to history.

# CHAPTER 5

# TALLY STICKS
## ACCOUNTING FOR BEGINNERS

Notched sticks – tally sticks – were first used at least forty thousand years ago. They might seem to be a primitive method of accounting, but they have certainly proved their value. The technique has remained much the same through many centuries of evolutionary, historical, and cultural change, right down to the present day. Although our ancestors could not have known it, their invention of the notched stick has turned out to be amongst the most permanent of human discoveries. Not even the wheel is as old; for sheer longevity, only fire could possibly rival it.

Notch-marks found on numerous prehistoric cave-wall paintings alongside outlines of animals leave no doubt about the accounting function of the notches. In the present-day world the technique has barely changed at all.

For instance, in the very recent past, native American labourers in the Los Angeles area used to keep a tally of hours worked by scoring a fine line in a piece of wood for each day worked, with a deeper or thicker line to mark each week, and a cross for each fortnight completed.

More colourful users of the device in modern times include cowboys, who made notches in the barrels of their guns for each bison killed, and the fearsome bounty hunters who kept a tally in the same manner for every outlaw that they gunned down. And Calamity Jane's father also used the device for keeping a reckoning of the number of marriageable girls in his town.

On the other side of the world, the technique was in daily use in the nineteenth century, as we learn from explorer's tales:

> On the road, just before a junction with a smaller track, I came upon a heavy gate made of bamboo and felled tree trunks, and decorated with hexagonal designs and sheaves. Over the track itself was hung a small plank with a set of regularly-sized notches, some large, some small, on each side. On the right were twelve small notches, then four large ones, then another set of twelve small ones. This meant: *Twelve days march from here, any man who crosses our boundary will be our prisoner or will pay a ransom of four water-buffalo and twelve ticals* (rupees). On the left, eight large notches, eleven middle-sized ones, and nine small ones, meaning: *There are eight men, eleven women and nine children in our village*. [J. Harmand (1879): Laos]

> In Sumatra, the Lutsu declared war by sending a piece of wood scored with lines together with a feather, a scrap of tinder and a fish. Translation: they will attack with as many hundreds (or thousands) of men as there are scored lines; they will be as swift as a bird (the feather), will lay everything waste (the tinder = fire), and will drown their enemies (the fish). [J. G. Février (1959)]

Only a few generations ago, shepherds in the Alps and in Hungary, as well as Celtic, Tuscan, and Dalmatian herders, used to keep a tally of the number of head in their flocks by making an equivalent number of notches or crosses on wooden sticks or planks. Some of them, however, had a particularly developed and subtle version of the technique as L. Gerschel describes:

> On one tally-board from the Moravian part of Walachia, dating from 1832, the shepherd used a special form of notation to separate the milk-bearing sheep from the others, and within these, a special mark indicated those that only gave half the normal amount. In some parts of the Swiss Alps, shepherds used carefully crafted and decorated wooden boards to record various kinds of information, particularly the number of head in their flock, but they also kept separate account of sterile animals, and distinguished between sheep and goats . . . .
>
> We can suppose that shepherds of all lands cope with much the same realities, and that only the form of the notation varies (using, variously, knotted string or *quipu* [see Chapter 6 below], primitive notched sticks, or a board which may include (in German-speaking areas) words like *Küo* (cows), *Gallier* (sterile animals), *Geis* (goats) alongside their tallies. There is one constant: the shepherd must know how many animals he has to care for and feed; but he also has to know how many of them fall into the various categories – those that give milk and those that don't, young and old, male and female. Thus the counts kept are not simple ones, but threefold, fourfold or more *parallel tallies* made simultaneously and entered side by side on the counting tool.

FIG. 5.1. *Swiss shepherd's tally stick (Late eighteenth century, Saanen, Canton of Bern). From the Museum für Völkerkunde, Basel; reproduced from Gmür (1917)*

In short, shepherds such as these had devised a genuine system of accounting.

Another recent survival of ancient methods of counting can be found in the name that was given to one of the taxes levied on serfs and commoners in France prior to 1789: it was called *la taille,* meaning "tally" or "cut", for the simple reason that the tax-collectors totted up what each taxpayer had paid on a wooden tally stick.

In England, a very similar device was used to record payments of tax and to keep account of income and expenditure. Larger and smaller notches on wooden batons stood for one, ten, one hundred, etc., pounds sterling (see Fig. 5.2). Even in Dickens's day, the Treasury still clung on to this antiquated system! And this is what the author of *David Copperfield* thought of it:

> Ages ago, a savage mode of keeping accounts on notched sticks was introduced into the Court of Exchequer; the accounts were kept, much as Robinson Crusoe kept his calendar on the desert island. In course of considerable revolutions of time . . . a multitude of accountants, book-keepers, actuaries and mathematicians, were born and died; and still official routine clung to these notched sticks, as if they were pillars of the constitution, and still the Exchequer accounts continued to be kept on certain splints of elm wood called "tallies". Late in the reign of George III, some restless and revolutionary spirit originated the suggestion, whether, in a land where there were pens, ink and paper, slates and pencils, and systems of accounts, this rigid adherence to a barbarous usage might not border on the ridiculous? All the red tape in the public offices turned redder at the bare mention of this bold and original conception, and it took till 1826 to get these sticks abolished.
>
> [Charles Dickens (1855)]

Britain may be a conservative country, but it was no more backward than many other European nations at that time. In the early nineteenth century, tally sticks were in use in various roles in France, Germany, and Switzerland, and throughout Scandinavia. Indeed, I myself saw tally sticks in use as credit tokens in a country bakery near Dijon in the early 1970s. This is how it is done: two small planks of wood, called *tailles,* are both marked with a notch each time the customer takes a loaf. One plank stays with the baker, the other is taken by the customer. The number of loaves is totted up and payment is made on a fixed date (for example, once a week). No dispute over the amount owed is possible: both planks have the same number of notches, in the same places. The customer could not have removed any, and there's an easy way to make sure the baker hasn't added any either, since the two planks have to match (see Fig. 5.3).

The French baker's tally stick was described thus in 1869 by André Philippe, in a novel called *Michel Rondet:*

> The women each held out a piece of wood with file-marks on it. Each piece of wood was different – some were just branches, others were planed square. The baker had identical ones threaded onto a strap. He looked out for the one with the woman's name on it on his strap, and the file-marks tallied exactly. The notches matched, with Roman numerals – I, V, X – signifying the weight of the loaves that had been supplied.

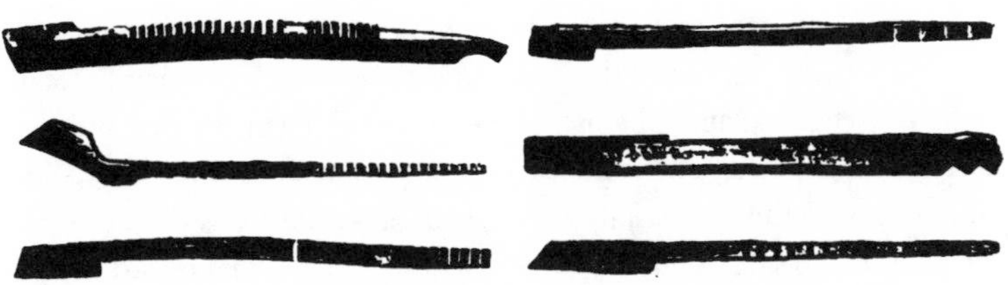

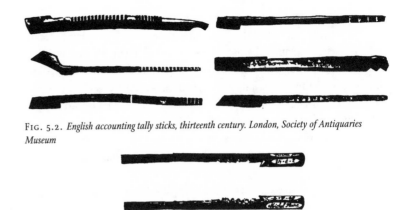

FIG. 5.2. *English accounting tally sticks, thirteenth century. London, Society of Antiquaries Museum*

FIG. 5.3. *French country bakers' tally sticks, as used in small country towns*

René Jonglet relates a very similar scene that took place in Hainault (French-speaking Belgium) around 1900:

> The baker went from door to door in his wagon, calling the housewives out. Each would bring her "tally" – a long and narrow piece of wood, shaped like a scissor-blade. The baker had a duplicate of it, put the two side by side, and marked them both with a saw, once for each six-pound loaf that was bought. It was therefore very easy to check what was owed, since the number of notches on the baker's and housewife's tally stick was the same. The housewife couldn't remove any from both sticks, nor could the baker add any to both.

The tally stick therefore served not just as a curious form of bill and receipt, but also as a wooden credit card, almost as efficient and reliable as the plastic ones with magnetic strips that we use nowadays.

French bakers, however, did not have a monopoly on the device: the use of twin tally sticks to keep a record of sums owing and to be settled can be found in every period and almost everywhere in the world.

The technique was in use by the Khâs Boloven in Indo-China, for example, in the nineteenth century:

> For market purchases, they used a system similar to that of country bakers: twin planks of wood, notched together, so that both pieces held the same record. But their version of this memory-jogger is much more complicated than the bakers', and it is hard to understand how they coped with it. Everything went onto the planks – the names of the

sellers, the names of the buyer or buyers, the witnesses, the date of delivery, the nature of the goods and the price. [J. Harmand (1880)]

As Gerschel explains, the use of the tally stick is, in the first place, to keep track of partial and successive numbers involved in a transaction. However, once this use is fully established, other functions can be added: the tally stick becomes a form of memory, for it can hold a record not just of the intermediate stages of a transaction, but also of its final result. And it was in that new role, as the record of a completed transaction, that it acquired an economic function, beyond the merely arithmetic function of its first role.

The *mark of ownership* was the indispensable additional device that allowed tally sticks to become economic tools. The mark symbolised the name of its owner: it was his or her "character" and represented him or her legally in any situation, much like a signature. Improper use of the mark of ownership was severely punished by the law, and references to it are found in French law as late as the seventeenth century.

The mark of ownership thus took the notched stick into a different domain. Originally, notched sticks had only notches on them: but now they also carried signs representing not numbers but names.

FIG. 5.4. *Examples of marks of ownership used over the ages. The signs were allocated to specific members of the community and could not be exchanged or altered.*

Here is how they were used amongst the Kabyles, in Algeria:

> Each head of cattle slaughtered by the community is divided equally between the members, or groups of members. To achieve this, each member gives the chief a stick that bears a mark; the chief shuffles the sticks and then passes them to his assistant, who puts a piece of meat on each one. Each member then looks for his own stick and thus obtains his share of the meat. This custom is obviously intended to ensure a fair share for everyone. [J. G. Février (1959)]

The mark of ownership probably goes back to the time before writing was invented, and it is the obvious ancestor of what we call a signature (the Latin verb *signare* actually means "to make a cross or mark"). So the mark, the "signature" of the illiterate, can be associated with the tally stick, the accounting device for people who cannot count.

But once you have signatures, you have contracts: which is how tally sticks with marks of ownership came to be used to certify all sorts of commitments and obligations. One instance is provided by the way the Cheremiss and Chuvash tribesfolk (central Russia) recorded loans of money in the nineteenth century. A tally stick was split in half lengthways, each half therefore bearing the same number of notches, corresponding to the amount of money involved. Each party to the contract took one of the halves and inscribed his personal mark on it (see Fig. 5.4), and then a witness made his or her mark on both halves to certify the validity and completeness of the transaction. Each party then took and kept the half with the *other's* signature or mark. Each thus retained a certified, legally enforceable and unalterable token of the amount of capital involved (indicated by the notched numbers on both tally sticks). The creditor could not alter the sum, since the debtor had the tally stick with the creditor's mark; nor could the debtor deny his debt, since the creditor had the tally stick with the debtor's mark on it.

According to A. Conrady (1920), notched sticks similarly constituted the original means of establishing pacts, agreements and transactions in preliterate China. They gave way to written formula only after the development of Chinese writing, which itself contains a trace of the original system: the ideogram signifying *contract* in Chinese is composed of two signs meaning, respectively, "notched stick" and "knife".

FIG. 5.5.

The Arabs (or their ancestors) probably had a similar custom, since a similar derivation can be found in Arabic. The verb-root *farada* means both "to make a notch" and "to assign one's share (of a contract or inheritance) to someone".

In France, tally sticks were in regular use up to the nineteenth century as waybills, to certify the delivery of goods to a customer. Article 1333 of the *Code Napoléon*, the foundation stone of the modern French legal system, makes explicit reference to tally sticks as the means of guaranteeing that deliveries of goods had been made.

In many parts of Switzerland and Austria, tally sticks constituted until recently a genuine social and legal institution. There were, first of all, the *capital tallies* (not unlike the tokens used by the Chuvash), which recorded loans made to citizens by church foundations and by local authorities. Then there were the *milk tallies*. According to L. Gerschel, they worked in the following ways:

> At Ulrichen, there was a single tally stick of some size on which was inscribed the mark of ownership of each farmer delivering milk, and opposite his mark, the quantity of milk delivered. At Tavetsch (according to Gmür), each farmer had his own tally stick, and marked on it the amount of milk he owed to each person whose mark of ownership was

on the stick; reciprocally, what was owed to him appeared under his mark of ownership on others' tally sticks. When the sticks were compared, the amount outstanding could be computed.

There were also *mole tallies*: in some areas, the local authorities held tallies for each citizen, marked with that citizen's mark of ownership, and would make a notch for each mole, or mole's tail, surrendered. At the year's end, the mole count was totted up and rewards paid out according to the number caught.

Tallies were also used in the Alpine areas for recording pasture rights (an example of such a tally, dated 1624, is said by M. Gmür to be in the Swiss Folklore Museum in Basel) and for water rights. It must be remembered that water was scarce and precious, and that it almost always belonged to a feudal overlord. That ownership could be rented out, sold and bequeathed. Notched planks were used to record the sign of ownership of the family, and to indicate how many hours (per day) of a given water right it possessed.

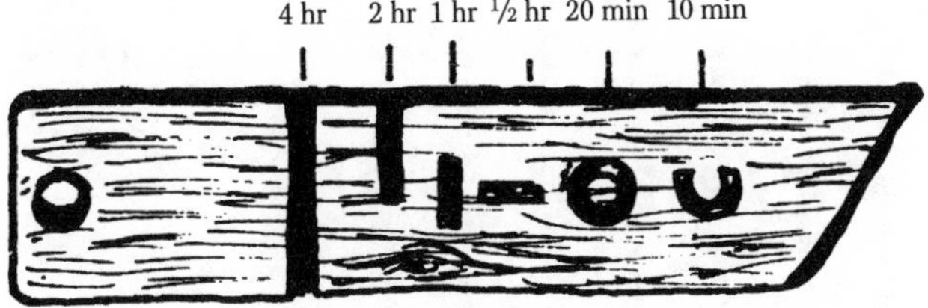


FIG. 5.6. *A water tally from Wallis (Switzerland). Basel, Museum für Völkerkunde. See Gmür, plate XXVI*

Finally, the Alpine areas also used *Kehrtesseln* or "turn tallies", which provided a practical way of fixing and respecting a duty roster within a guild or corporation (night watchmen, standard-bearers, gamekeepers, churchwardens, etc.).

In the modern world there are a few surviving uses of the notched-stick technique. Brewers and wine-dealers still mark their barrels with Xs, which have a numerical meaning; publicans still use chalk-marks on slate to keep a tally of drinks yet to be settled. Air Force pilots also still keep tallies of enemy aircraft shot down, or of bombing raids completed, by "notching" silhouettes of aircraft or bombs on the fuselage of their aircraft.

The techniques used to keep tallies of numbers thus form a remarkably unbroken chain over the millennia.

# CHAPTER 6

# NUMBERS ON STRINGS

Although it was certainly the first physical prop to help our ancestors when they at last learned to count, the hand could never provide more than a fleeting image of numerical concepts. It works well enough for representing numbers visually and immediately: but by its very nature, finger-counting cannot serve as a recording device.

As crafts and trade developed within different communities and cultures, and as communication between them grew, people who had not yet imagined the tool of writing nonetheless needed to keep account of the things that they owned and of the state of their exchanges. But how could they retain a durable record of acts of counting, short of inventing written numerals? There was nothing in the natural world that would do this for them. So they had to invent something else.

In the early years of the sixteenth century, Pizarro and his Spanish conquistadors landed on the coast of South America. There they found a huge empire controlling a territory more than 4,000 km long, covering an area as large as Western Europe, in what is now Bolivia, Peru, and Ecuador. The Inca civilisation, which went back as far as the twelfth century CE, was then at the height of its power and glory. Its prosperity and cultural sophistication seemed at first sight all the more amazing for the absence amongst these people of knowledge of the wheel, of draught animals, and even of writing in the strict sense of the word.

However, the Incas' success can be explained by their ingenious method of keeping accurate records by means of a highly elaborate and fairly complex system of knotted string. The device, called a *quipu* (an Inca word meaning "knot") consisted of a main piece of cord about two feet long onto which thinner coloured strings were knotted in groups, these pendant strings themselves being knotted in various ways at regular intervals (see Fig. 6.1).

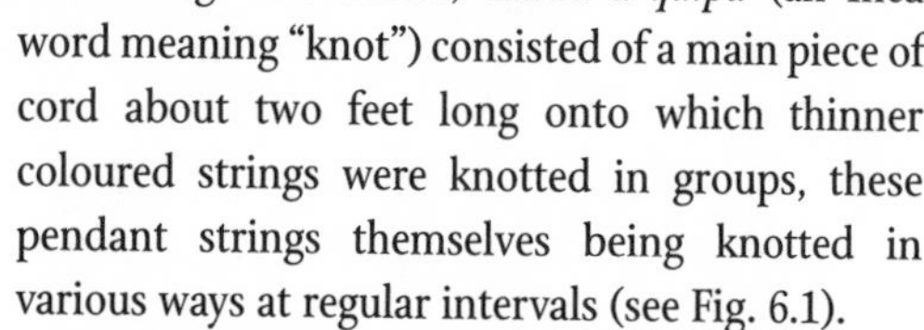

*Quipus*, sometimes incorrectly described as "abacuses", were actually recording devices that met the various needs of the very efficient Inca administration. They provided a means for representing liturgical, chronological, and statistical records, and could occasionally also serve as calendars and as messages. Some string colours had conventional meanings, including both tangible objects and abstract notions: white, for instance, meant either "silver" or "peace"; yellow signified "gold"; red stood for "blood" or "war"; and so on. *Quipus* were used primarily for book-keeping, or, more precisely, as a concrete enumerating tool. The string colours, the number and relative positions of the knots, the size and the spacing of the corresponding groups of strings all had quite precise numerical meanings (see Fig. 6.2, 6.3 and 6.4). *Quipus* were used to represent the results of counting (in a decimal verbal counting system, as previously stated) all sorts of things, from military matters to taxes, from harvest reckonings to accounts of animals slain in the enormous annual culls that were held, from delivery notes (see Fig. 6.5) to population censuses, and including calculations of base values for levies and taxes for this or that administrative unit of the Inca Empire, inventories of resources in men and equipment, financial records, etc.

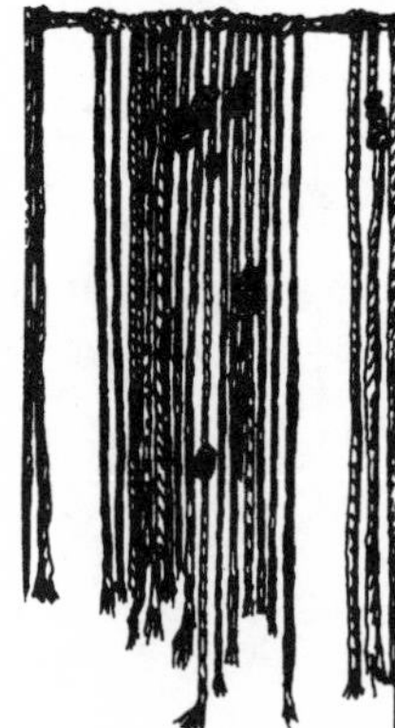

Fig. 6.1. *A Peruvian* quipu

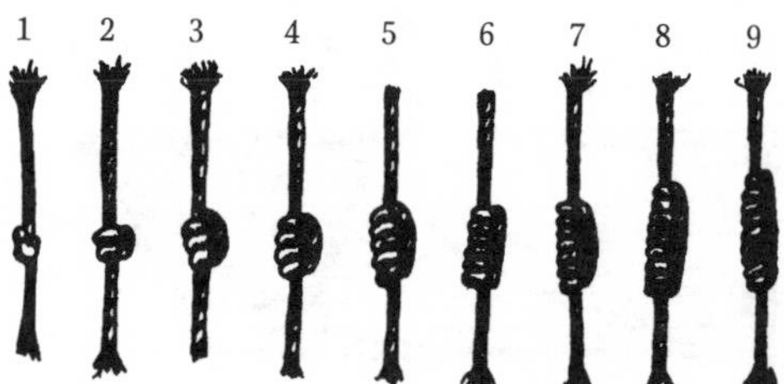


Fig. 6.2. *The first nine numbers represented on a string in the manner of the Inca* quipu

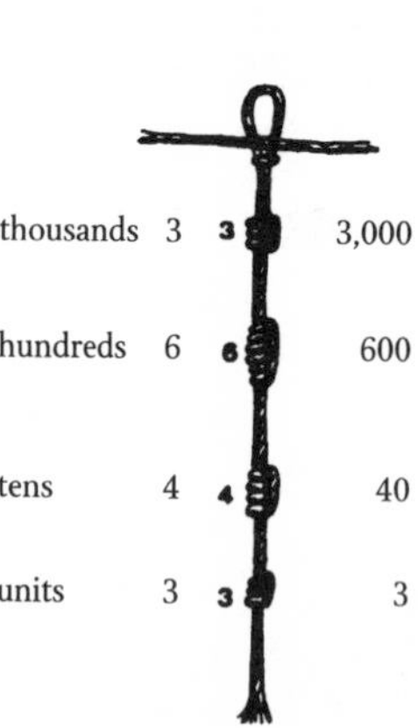


Fig. 6.3. *The number 3,643 as it would be represented on a string in the manner of a Peruvian* quipu

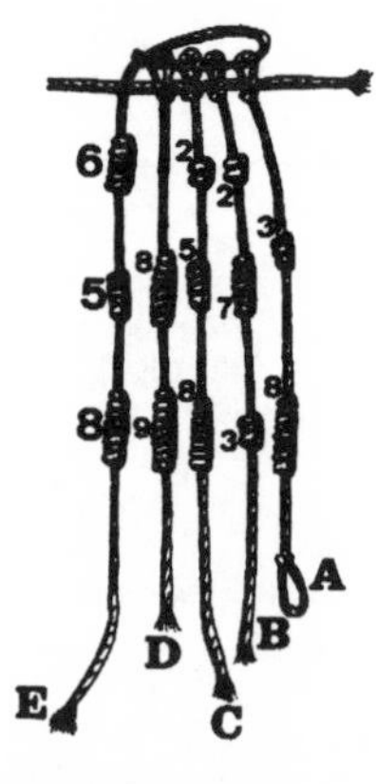


A = 38 B = 273 C = 258 D = 89
E = A + B + C + D = 658

Fig. 6.4. *Numerical reading of a bunch of knotted strings, from an Inca* quipu, *American Museum of Natural History, New York, exhibit B 8713, quoted in Locke (1924): the number 658 on string E equals the sum of the numbers represented on strings A, B, C and D.*

*Quipus* were based on a fairly simple, strictly decimal system of positions. Units were represented by the string being knotted a corresponding number of times around the first fixed position-point (counting from the end or bottom of the string), tens were represented similarly by the number of times the string was knotted around the second position-point, the third point served for recording hundreds, the fourth for thousands, etc. So to "write" the number 3,643 on Inca string (as shown in Fig. 6.3), you knot the string three times at the first point, four times at the second, six times at the third, and three times at the fourth position-point.

Officers of the king, called *quipucamayocs* ("keepers of the knots"), were appointed to each town, village and district of the Inca Empire with responsibility for making and reading *quipus* as required, and also for supplying the central government with whatever information it deemed important (see Fig. 6.5). It was they who made annual inventories of the region's produce and censuses of population by social class, recorded the results on string with quite surprising regularity and detail, and sent the records to the capital.

FIG. 6.5. *An Inca* quipucamayoc *delivering his accounts to an imperial official and describing the results of an inventory recorded on the* quipu. *From the Peruvian Codex of the sixteenth-century chronicler Guaman Poma de Ayala (in the Royal Library, Copenhagen), reproduced from* Le Quipucamayoc, *p. 335*

> One of the *quipucamayocs* was responsible for the revenue accounts, and kept records of the quantities of raw materials parcelled out to the workers, of the amount and quality of the objects each made, and of the total amount of raw materials and finished goods in the royal stores. Another kept the register of births, marriages and deaths, of men fit for combat, and other details of the population in the kingdom. Such records were sent in to the capital every year where they were read by officers learned in the art of deciphering these devices. The Inca government thus had at its disposal a valuable mass of statistical information: and these carefully stored collections of skeins of coloured string constituted what might have been called the Inca National Archives. [Adapted from W. H. Prescott (1970)]

*Quipus* are so simple and so valuable that they continued to be used for many centuries in Peru, Bolivia and Ecuador. In the mid-nineteenth century, for example, herdsmen, particularly in the Peruvian Altiplano, used *quipus* to keep tallies of their flocks [M. E. de Rivero & J. D. Tschudi (1859)]. They used bunches of white strings to record the numbers of their sheep and goats, usually putting sheep on the first pendant string, lambs on the second, goats on the third, kids on the fourth, ewes on the fifth, and so on; and bunches of green string to count cattle, putting the bulls on the first pendant string, dairy cows on the second, heifers on the third, and then calves, by age and sex, and so on (see Fig. 6.6).

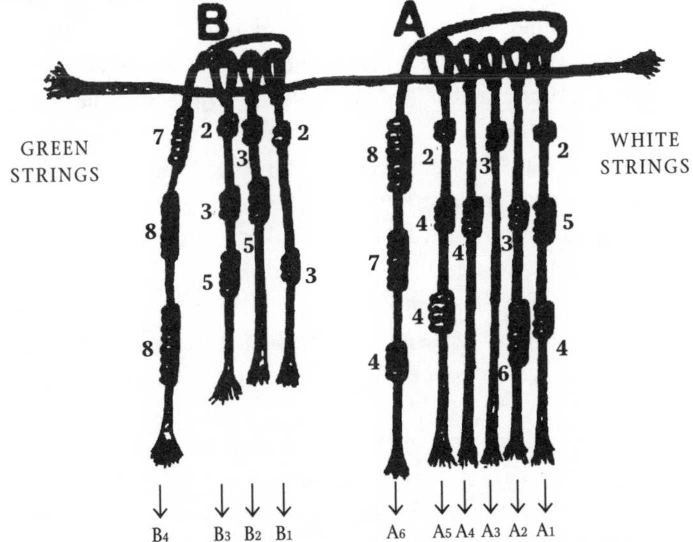


FIG. 6.6. *A livestock inventory on a nineteenth-century* quipu *from the Peruvian Altiplano. On bunch A (white string), small livestock: 254 sheep (string A1), 36 lambs (A2), 300 goats (A3), 40 kids (A4), 244 ewes (A5), total = 874 sheep and goats (A6). On bunch B (green string), cattle: 203 bulls (B1), 350 dairy cows (B2), 235 sterile cows (B3), total = 788 head of cattle (B4).*

Even today native Americans in Bolivia and Peru use a very similar device, the *chimpu*, a direct descendant of the *quipu*. A single string is used to represent units up to 9, with each knot on it indicating one unit, as on a *quipu*; tens are figured by the corresponding number of knots tied on two strings held together; hundreds in like manner on three strings, thousands on four strings, and so on. On *chimpus*, therefore, the magnitude of a number in powers of 10 is represented by the number of strings included in the knot – six knots may have the value of 6, 60, 600 or 6,000 according to whether it is tied on one, two, three, or four strings together.

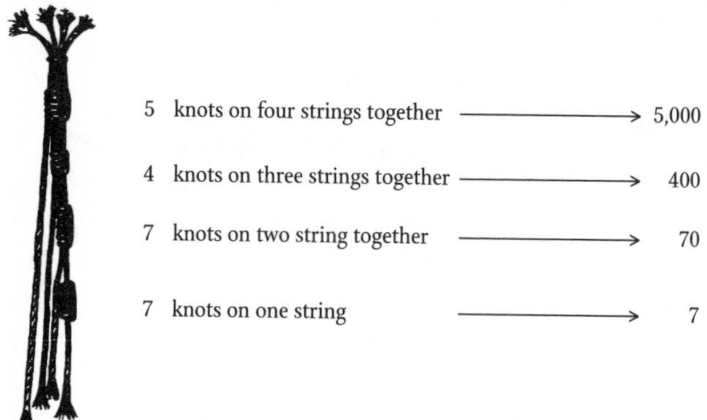


FIG. 6.7. *A* chimpu *(Bolivia and Peru)*

These remarkable systems are not however uniquely found in Inca or indeed South American civilisations. The use of knotted string is attested since classical times, and in various regions of the world.

Herodotus (485–425 BCE) recounts how, in the course of one of his expeditions, Darius, King of Persia (522–486 BCE) entrusted the rearguard defence of a strategically vital bridge to Greek soldiers, who were his allies. He gave them a leather strap tied into sixty knots, and ordered them to undo one knot each day, saying:

> "If I have not returned by the time all the knots are undone, take to your boats and return to your homes!"

In Palestine, in the second century CE, Roman tax-collectors used a "great cable", probably made up of a collection of strings, as their register. In addition, receipts for taxes paid took the form of a piece of string knotted in a particular way.

Arabs also used knotted string over a long period of time not only as a concrete counting device, but also for making contracts, for giving receipts, and for administrative book-keeping. In Arabic, moreover, the word *aqd*, meaning "knot", also means "contract", as well as any class of numbers constituted by the products of the nine units to any power of ten (several Arabic mathematicians refer to the *aqd* of the hundreds, the *aqd* of the thousands, and so on).

The Chinese were also probably familiar with knotted-string numbers in ancient times before writing was invented or widespread. The semi-legendary Shen Nong, one of the three emperors traditionally credited with founding Chinese civilisation, is supposed to have had a role in developing a counting system based on knots and in propagating its use for book-keeping and for chronicles of events. References to a system reminiscent of Peruvian *quipus* can be found in the *I Ching* (around 500 BCE) and in the *Tao Te Ching*, traditionally attributed to Lao Tse.

The practice is still extant in the Far East, notably in the Ryû-Kyû Islands. On Okinawa, workers in some of the more mountainous areas use plaited straw to keep a record of days worked, money owed to them, etc. At Shuri, moneylenders keep their accounts by means of a long piece of reed or bark to which another string is tied at the middle. Knots made in the upper half of the main "string" signify the date of the loan, and on the lower half, the amount. On Yaeyama, harvest tallies were kept in similar fashion; and taxpayers received, in lieu of a written "notice to pay", a piece of string so knotted as to indicate the amount due [J. G. Février (1959)].

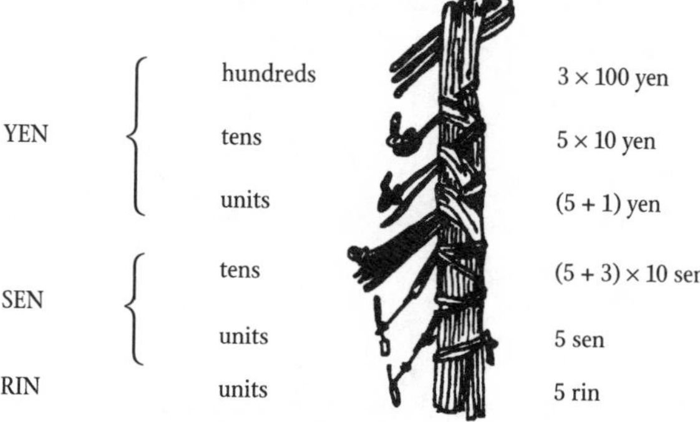


FIG. 6.8. *A sum of money as expressed in knotted string in the style used by workers on Okinawa and tax-collectors on Yaeyama. The figure shows 356 yen, 85 sen and 5 rin (1yen = 100 sen, 1 sen = 10 rin). The number 5 is represented by a knot at the end of the trailing straw. See also Chapter 25, Fig. 25.9.*

The same general device can be found in the Caroline Islands, in Hawaii, in West Africa (specifically amongst the Yebus, who live in the hinterland of Lagos (Nigeria)), and also at the other ends of the world, amongst native Americans such as the Yakima (eastern Washington State), the Walapai and the Havasupai (Arizona), the Miwok and Maidu (North & South Carolina), and of course amongst the Apache and Zuñi Indians of New Mexico.

A bizarre survival of the formerly wide role of knotted string was to be found as late as the end of the last century amongst German flour-millers, who kept records of their dealings with bakers by means of rope (see Fig. 6.9 below). Similarly, knotted-string rosaries (like their beaded and notched counterparts), for keeping count of prayers, are common to many religions. Tibetan monks, for example, count out the *one hundred and eight unities* (the number 108 is considered a sacred number) on a bunch of 108 knotted strings (or a string of 108 beads) whose colour varies with the deity to be invoked: yellow string (or beads) for prayers to *Buddha*; white string (or white beads made from shells) for *Bodhisattva*; red strings (or coral beads) for *the one who converted Tibet*; etc. A very similar practice was current only a few decades ago amongst various Siberian tribes (Voguls, Ostyaks, Tungus, Yakuts, etc.); and there is also a Muslim tradition, handed down by Ibn Sa'ad, according to which Fatima, Mohammed's daughter, counted out the 99 attributes of Allah and the supererogatory laudations on a piece of knotted string, not on a bead rosary.

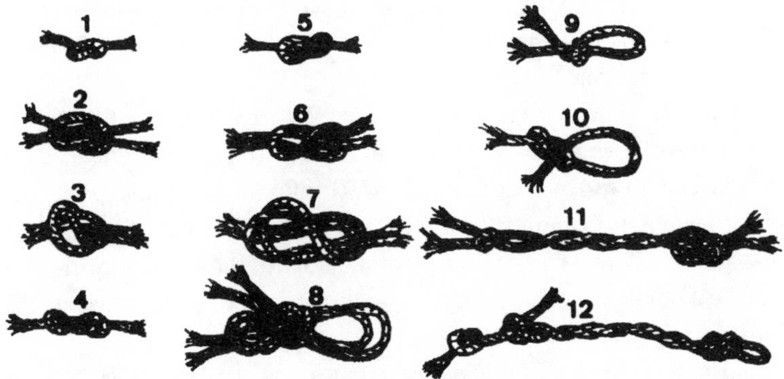


FIG. 6.9. *German millers' counting device using knotted rope (the system in force at Baden in the nineteenth century is illustrated)*

FIG. 6.10. *The bands and fringe of prayer-shawl*

For morning prayers (*Shaḥrit*) and other services in the synagogue, Jews wear a prayer-shawl (*talit*) adorned with fringes (*tsitsit*). Now, the four corner-threads of the fringe are always tied into a quite precise number of knots: 26 amongst "Eastern" (Sephardic) Jews, and 39 amongst "Western" (Ashkenazi) Jews. The number 26 corresponds to the numerical value of the Hebrew letters which make up the name of God, YHWH (see below, Chapters 17 and 20, for more detail on letter-counting systems), and 39 is the total of the number-values of the letters in the expression "God is One", YHWH EHD (see below). 39 is also the "value" of the Hebrew word meaning "morning dew" (*tal*), and rabbis have often commented that at prayer the religious Jew is able to hear the word of God "which falls from his mouth as morning dew falls on the grass".

| | | |
|---|---|---|
| | י ה ו ה | |
| YHWH | 5 + 6 + 5 + 10 | = 26 |
| *Yahwe*, "the Lord" | | |
| | י ה ו ה א ח ד | |
| YHWH EHD | 4 + 8 + 1 + 5 + 6 + 5 + 10 | = 39 |
| *Yahwe ehad*, "the Lord is One" | | |
| | ט ל | |
| ṬL | 30 + 9 | = 39 |
| *tal*, "morning dew" | | |

FIG. 6.11.

Knotted string has thus served not only as a device for concrete numeration, but also as a mnemotechnic tool (for recording numbers, maintaining administrative archives, keeping count of contracts, calendars, etc.). Although knotted string does not constitute a form of writing in the strict sense, it has performed all of writing's main functions – to preserve the past and to ensure the survival of contracts between members of the same society. Numbers on strings can therefore be considered for our purposes as a special form of written numbers.

CHAPTER 7

# NUMBER, VALUE AND MONEY

At a time when people lived in small groups, and could find what they needed in the nature around them, there would have been little need for different communities to communicate with each other. However, once some sort of culture developed, and people began to craft objects of use or desire, then, because the raw resources of nature are unequally distributed, trade and exchange became necessary.

The earliest form of commercial exchange was barter, in which people exchange one sort of foodstuff or goods directly for another, without making use of anything resembling our modern notion of "money". On occasion, if the two parties to the exchange were not on friendly terms, these exchanges took the form of silent barter. One side would go to an agreed place, and leave there the goods on offer. Next day, in their place or beside them, would be found the goods offered in exchange by the other side. Take it or leave it: if the exchange was considered acceptable, the goods offered in exchange would be taken away and the deal was done. However, if the offer was not acceptable then the first side would go away, and come back next day hoping to find a better offer. This could go on for several days, or even end without a settlement.

Among the Aranda of Australia, the Vedda of Ceylon, Bushmen and Pygmies of Africa, the Botocoudos in Brazil, in Siberia, in Polynesia – such transactions have been observed. But with growth in communication, and the increasing importance of trade, barter became increasingly inconvenient, depending as it does on the whims of individuals or on interminable negotiations.

The need grew, therefore, for a stable system of equivalences of value. This would be defined (much as numbers are expressed in terms of a base) in terms of certain fixed units or standards of exchange. With such a system it is not only possible to evaluate the transactions of trade and commerce, but also to settle social matters – such as "bride price" or "blood money" – so that, for instance, a woman would be worth so many of a certain good as a bride, the reparation for a robbery so many. In pre-Hellenic Greece, the earliest unit of exchange that we find is the ox. According to Homer's *Iliad* (XXIII, 705, 749–751; VI, 236; eighth century BCE), a "woman good for a thousand tasks" was worth four oxen, the bronze armour of Glaucos was worth nine, and that of Diomedes (in gold) was worth 100. And, in decreasing order of value, are given: a chased silver cup, an ox, and half a golden talent. The Latin word *pecunia* (money), from which we get "pecuniary", comes from *pecus*, meaning "cattle"; and the related word *peculium* means "personal property", from which we also get "peculiar". In fact the strict sense of *pecunia* is "stock of cattle". The English word *fee* has come to us partly from Old English *feoh* meaning both "cattle" and "property" which itself is believed to be derived via a Germanic root from *pecus* (compare modern German *Vieh*, "livestock"), and partly from Anglo-French *fee* which is probably also of similar Germanic derivation. Like the Sanskrit *rupa* (whence "rupee"), these words remind us of a time when property, recompense, offerings, and ritual sacrifices were evaluated in heads of cattle. In some parts of East Africa, the dowry of a bride is counted in cattle. The Latin *capita* ("head") has given us "capital". In Hebrew, *keseph* means both "sheep" and "money"; and the root-word made of the letters GML stands for both "camel" and "wages".

In ancient times, however, barter was a far from simple affair. It was surrounded by complicated formalities, which were probably associated with mysticism and magical practices, as is confirmed by ethnological study of contemporary "primitive" societies and by archaeological findings. We may imagine, therefore, that in pastoral societies the concept of the "ox standard" grew out of the "ox for the sacrifice" which itself depended on the intrinsic value attributed to the animal.

L. Hambis (1963–64), describing certain parts of Siberia, says "Buying and selling was still done by barter, using animal pelts as a sort of monetary unit; this system was employed by the Russian government until 1917 as a means of levying taxes on the people of these parts." In the Pacific islands, on the other hand, goods were valued not in terms of livestock but in terms of pearl or sea-shell necklaces. The Iroquois, Algonquin, and other north-east American Indians used strings of shells called *wampum*. Until recently, the Dogon of Mali used cowrie shells. One Ogotemmêli, interviewed by M. Griaule (1966), says "a chicken is worth three times eighty cowries, a goat or a sheep three times eight hundred, a donkey forty times eight hundred, a horse eighty times eight hundred, an ox one hundred and twenty times eight hundred." "But", continues Griaule, "in earlier times the unit of exchange was not the cowrie. At first, people bartered strips of cloth for animals or goods. The cloth was their money. The unit was the 'palm' of a strip of cloth twice eighty threads wide. So a sheep was worth eight cubits of three 'palms' . . . Subsequently, values were laid down in terms of cowries by Nommo the Seventh, Master of the Word."

With some differences of detail, practices were similar in pre-Columbian Central America. The Maya used also cotton, cocoa, bitumen, jade, pots, pearls, stones, jewels, and gold. For the Aztecs, according to J. Soustelle,

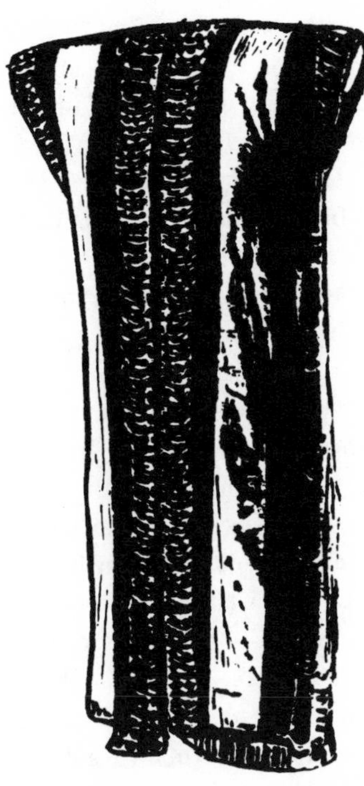

FIG. 7.1. *Tunic worn in the nineteenth century by members of the Tyal tribe in Formosa. More than 2,500 precious stones are attached in bands, at the edges and on either side of the centre line. Such tunics were used as "money" in buying livestock and in the trade in young women. The bands of precious stones could be detached separately to serve as pocket money for everyday purchases. New York, Chase Manhattan Bank Museum of Money*

"certain foodstuffs, goods or objects were employed as standards of value and as tokens of exchange: the *quachtli* (a piece of cloth) and 'the load' (20 *quachtlis*); the cocoa bean used as 'small change' and the *xiquipilli* (a bag of 8,000 beans); little T-shaped axes of copper; feather quills filled with gold." The same kind of economy was practised in China prior to the adoption of money in the modern sense. In the beginning, foodstuffs and goods were exchanged, their value being expressed in terms of certain raw materials, or certain necessities of life, which were adopted as standards. These might include the teeth and horns of animals, tortoise shells, sea-shells, hides, or fur pelts. Later, weapons and utensils were adopted as tokens of value: knives, shovels, etc. These would at first have been made of stone, but later, from the Shang Dynasty, of bronze (sixteenth to eleventh century BCE).

However, regular use of such kinds of items was cumbersome and not always easy. As a result, metal played an increasingly important role, in the form of blocks or ingots, or fashioned into tools, ornaments or weapons, until finally metal tokens were adopted as money in preference to other forms, for the purposes of buying and selling. The value of a merchandise was measured in terms of weight, with reference to a standard weight of one metal or another.

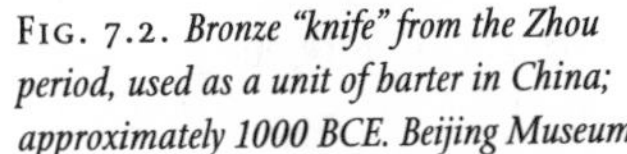

FIG. 7.2. *Bronze "knife" from the Zhou period, used as a unit of barter in China; approximately 1000 BCE. Beijing Museum*

FIG. 7.3. *Lance-head once used as money by Central Congo tribes. One of these would possibly buy a fowl; five or six would be the price of a slave. New York, Chase Manhattan Bank Museum of Money*

Thus it was that "When Abraham purchased the Makpelah Cave, he weighed out four hundred silver shekels for Ephron the Hittite."* Later on Saul, seeking his father's she-asses, sought the help of a seer for which he gave one quarter of a shekel of silver (I Samuel IX, 8). Similarly, the fines laid down in the Code of the Alliance were stipulated in shekels of silver, as also was the poll tax (Exodus XXX, 12–15) [A. Negev (1970)].

In the Egypt of the Pharaohs likewise, foodstuffs and goods were often valued, and paid for, with metal (copper, bronze, sometimes gold or silver) measured out in nuggets or in flakes, or given in the form of bars or rings which were measured by weight. The principal standard of weight was the *deben*, equivalent to 91 grams of our measure. For certain purchases, value was determined in certain fractions of the *deben*. For example, in the Old Kingdom (2780–2280 BCE) the *shât*, one twelfth of a *deben*, was used (equivalent, therefore, to 7.6 grams). In the New Kingdom (1552–1070 BCE) the *shât* gave way to the *qat*, one tenth of the *deben* or 9.1 grams.

In a contract from the Old Kingdom we can see how value was expressed in terms of the *shât*. According to this, the rent of a servant was to be paid as follows. the values being in *shâts* of bronze:

| | | | | |
|---|---|---|---|---|
| 8 | bags of grain | value | 5 | *shâts* |
| 6 | goats | value | 3 | *shâts* |
| | silver | value | 5 | *shâts* |
| | Total | value | 13 | *shâts* |

* The Old Testament shekel is equivalent to 11.4 grams of our measure.

As another example, the following account from the New Kingdom shows *debens* of copper being used as a standard of value.

> Sold to Hay by Nebsman the Brigadier:
> 1 ox, worth 120 *debens* of copper
> Received in exchange:
> 2 pots of fat, value 60 *debens*
> 5 loin-cloths in fine cloth, worth 25 *debens*
> 1 vestment of southern flax, worth 20 *debens*
> 1 hide, worth 15 *debens*

In this example we can see how goods could be used in payment as well as metal tokens in the marketplace of ancient times. That ox, for instance, cost 120 *debens* of copper, but not one piece of real metal had changed hands: 60 of the *debens* owing had been settled by handing over 2 pots of fat, 25 more with 5 loin-cloths, and so on.

Although goods had been exchanged for goods, therefore, this was not a straightforward barter. It in fact reflected a real monetary system. Thenceforth, by virtue of the metal standard, goods were no longer bartered at the whim of the dealers or according to arbitrary established practice, but in terms of their "market price".

There is a letter dating from around 1800 BCE which gives a vivid illustration of these matters. It comes from the Royal Archives of the town of Mari, and was sent by Iškhi-Addu, King of Qatna, to Išme-Dagan, King of Êkallâtim. Iškhi-Addu roundly reproaches his "brother" for sending a meagre "sum" in pewter, in payment for two horses worth several times that amount.

> Thus [speaks] Iškhi-Addu thy brother:
> This should not have to be said! But speak I must, to console my heart. . . . Thou hast asked of me the two horses that thou didst desire, and I did have them sent to thee. And see! how thou hast sent to me merely twenty rods of pewter! Didst thou not gain thy whole desire from me without demur? And yet thou dare'st send me so little pewter! . . . Know thou that here in Qatna, these horses are worth six hundred shekels of silver. And see, how thou hast sent me but twenty rods of pewter! What will they say of this, when they hear of it?

An understandable indignation, since a shekel of silver was worth three or four rods of pewter at the time.

It should not be thought, though, that "money", in the modern sense of the word, was used in payment in those times. It was not a "coinage" in the sense of pieces of metal, die-cast in a mint which is the prerogative of the State, and guaranteed in weight and value. The idea of a coinage sound in weight and alloy did not come about until the first millennium BCE, most probably with the Lydians. Until that time, only a kind of "base-weight" played a role in transactions and in legal deeds, acting as a unit of value in terms of which the prices of individual items of merchandise, or individual deeds, could be expressed. On this basis, this or that metal was first counted out in ingots, rings, or other objects, and then its weight, in units of the "base-weight", was determined, and in this way could be used as "salary", "fine", or "exchange".

Let us go back a few thousand years and, in the description of Maspero, observe a market from Egypt of the Pharaohs.

> Early in the morning endless streams of peasants come in from the surrounding country, and set up their stalls in the spots reserved for them as long as anyone can remember. Sheep, geese, goats and wide-horned oxen are gathered in the centre to await buyers. Market gardeners, fishermen, fowlers and gazelle hunters, potters and craftsmen squat at the roadside and beside the houses, their goods heaped in wicker baskets or on low tables, fruits and vegetables, fresh-baked bread and cakes, meats raw or variously prepared, cloths, perfumes, jewels, the necessities and the frivolities of life, all set out before the curious eyes of their customers. Low and middle class alike can provide for themselves at lower cost than in the regular shops, and take advantage of it according to their means.
>
> The buyers have brought with them various products of their own labours, new tools, shoes, mats, pots of lotion, flasks of drink, strings of cowrie shells or little boxes of copper or silver or even golden rings each weighing one *deben*,* which they will offer to exchange for the things they need.
>
> For purchase of a large beast, or of objects of great value, loud, bitter and protracted arguments take place. Not only the price, but in what species the price shall be paid, must be settled, so they draw up lists whereon beds, rods, honey, oil, pick-axes or items of clothing may make up the value of a bull or a she-ass.

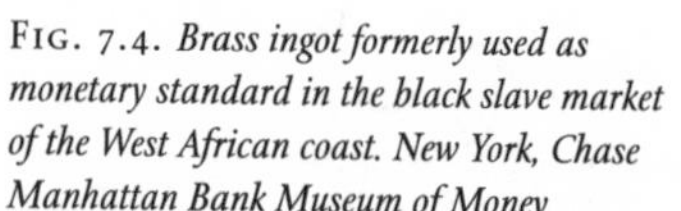

FIG. 7.4. *Brass ingot formerly used as monetary standard in the black slave market of the West African coast. New York, Chase Manhattan Bank Museum of Money*

* Maspero uses *tabnou*, here replaced by the the more precise term *deben*.

The retail trading does not involve so much complicated reckoning. Two townsmen have stopped at the same moment in front of a fellah with onions and corn displayed in his basket.* The first's liquid assets are two necklaces of glass pearls or coloured enamelled clay; the second one has a round fan with a wooden handle, and also one of those triangular fans which cooks use to boost the fire.

FIG. 7.5. *Market scenes in an Egyptian funeral painting of the Old Kingdom, Fifth or Sixth Dynasty (around 2500 BCE). The painting adorns the tomb of Feteka at the northern end of the necropolis of Saqqara (between Abusîr and Saqqara). See Lepsius (1854–59), vol. II, page 96 (Tomb no. 1), and Porter & Moss (1927–51), vol. 3 part 1, page 351.*

"This necklace would really suit you," calls the first, "it's just your style!"

"Here is a fan for your lady and a fan for your fire," says the other.

Still, the fellah calmly and methodically takes one of the necklaces to examine it:

"Let's have a look, I'll tell you what it's worth."

With one side offering too little, and the other asking for too much, they proceed by giving here and taking there, and finally agree on the number of onions or the amount of grain which will just match the value of the necklace or the fan.

Further along, a shopper wants some perfume in exchange for a pair of sandals and cries his wares heartily:

"Look, fine solid shoes for your feet!"

* Some of the scenes described can be seen on an Egyptian funeral painting from the Old Kingdom, reproduced here in Fig. 7.5.

But the merchant is not short of footwear just now, so he asks for a string of cowries for his little jars:

"See how sweet it smells when you put a few drops around!" he says winningly.

A woman passes two earthen pots, probably of ointment she has made, beneath the nose of a squatting man.

"This lovely scent will catch your fancy!"

Behind this group, two men argue the relative worth of a bracelet and a packet of fish-hooks; and a women with a small box in her hand is negotiating with a man selling necklaces; another woman is trying to get a lower price on a fish which the seller is trimming for her.

Barter against metal requires two or three more stages than simple barter. The rings or the folded sheets which represent *debens* do not always have the standard content of gold or silver, and may be of short weight. So they must be weighed for each transaction to establish their real value, which offers the perfect opportunity for those concerned to enter into heated dispute. After they have passed a quarter of an hour yelling that the scales do not work, that the weighing has been messed up, that they have to start all over again, they finally weary of the struggle and come to a settlement which roughly satisfies both sides.

However, sometimes someone cunning or unscrupulous will adulterate the rings by mixing their precious metal with as much false metal as possible short of making their trickery apparent. An honest trader who is under the impression that he received a payment of eight gold *debens*, who was in fact paid in metal which was one third silver, has unwittingly lost almost one third of his part. Fear of being cheated in this way held back the common use of *debens* for a long time, and caused the use of produce and artisanal objects in barter to be maintained.

At the end of the day, the use of *money* (in the modern sense of the term) became established once the metal was cast into small blocks or coins, which could be easily handled, had constant weight, and were marked with the official stamp of a public authority who had the sole right to certify good weight and sound metal.

This ideal system of exchange in commercial transactions was invented in Greece and Anatolia during the seventh century BCE. (In China the earliest similar usage occurred also at about the same time, apparently, around 600–700 BCE, during the Chow Dynasty.) Who might have first thought of it? Some consider that Pheidon, king of Argos in the Peloponnese, introduced the system in his own city and in Ægina, around 650 BCE. However, the majority of scholars agree that the honour of the invention should go to Asia Minor under the Greeks, most probably to Lydia.

Be that as it may, the many advantages of the use of coins led to its rapid adoption in Greece and Rome, and amongst many other peoples. The rest is another story.

FIG. 7.6. *Greek coins.* *Left: silver tetradrachma from Agrigento, around 415 BCE.* *Right: tetradrachma from Syracuse, around 310 BCE. Agrigento Museum*

By learning how to count in the abstract, grouping every kind of thing according to the principle of numerical base, people also learned how to *estimate, evaluate* and *measure* all sorts of magnitudes – weights, lengths, areas, volumes, capacities and so on. They likewise managed to conceive ever larger numbers, though they could not yet attain the concept of infinity. They worked out many technical procedures (mental, material and later written), and laid the early foundations of arithmetic which, at first, was purely practical and only later became abstract and led on to algebra.

The way also opened up for the devising of a calendar, for a systemisation of astronomy, and for the development of a geometry which was at first based straightforwardly on measurement of length, area and volume, before becoming theoretical and axiomatic. In short, the grasp of these fundamental data allowed the human race to attempt the measurement of its world, little by little to understand it better and better, to press into humanity's service some of their world's innumerable secrets, and to organise and to develop their economy.

# CHAPTER 8

# NUMBERS OF SUMER

## WRITING: THE INVENTION OF SUMER

Writing, as a system enabling articulated speech to be recorded, is beyond all doubt among the most potent intellectual tools of modern man. Writing perfectly meets the need (which every person in any advanced social group feels) for visual representation and the preservation of thought (which of its nature would otherwise evanesce). It also offers a remarkable method of expression and of preservation of communication, so that anyone can keep a permanent record of words long since spoken and flown. However, it is much more than a mere instrument.

> By recording speech in silent form, writing does not merely conserve it, but also stimulates thought such as, otherwise, would have remained latent. The simplest of marks made on stone or paper are not just a tool: they entomb old thoughts, but also bring them back to life. As well as fixing language, writing is also a new language, silent perhaps, which lays a discipline on thought and, in transcribing it, organises it. . . . Writing is not only a means of durable expression: it also gives direct access to the world of ideas. It faithfully represents the spoken word, but it also facilitates the understanding of thought and gives thought the means to traverse both space and time. [C. Higounet (1969)]

Writing, therefore, in revolutionising human life, is one of the greatest of all inventions. The earliest known writing appeared around 3000 BCE, not far from the Persian Gulf, in the land of Sumer, which lay in Lower Mesopotamia between the Tigris and Euphrates rivers. Here also were developed the earliest agriculture, the earliest technology, the first towns and cities, by the Sumerians, a non-Semitic people of still obscure origins.

As evidence of this we have numerous documents known as "tablets" which were used as a kind of "paper" by the inhabitants of this region. The oldest of these (which also carry the most archaic form of the writing) were discovered at the site of Uruk,* more precisely at the archaeological level designated as *Uruk IVa*.†

These tablets are, in fact, small plaques of dry clay, roughly rectangular

* The royal city of Uruk was situated south of Lower Mesopotamia on the Iraqi site of Warka (now about twenty km north of the Euphrates). It has given its name to the epoch in which, it is presumed, the Sumerian people first appeared in the region and in which writing was invented in Mesopotamia.

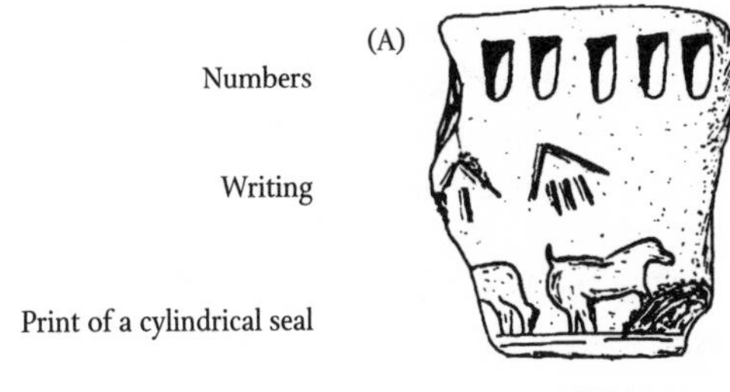

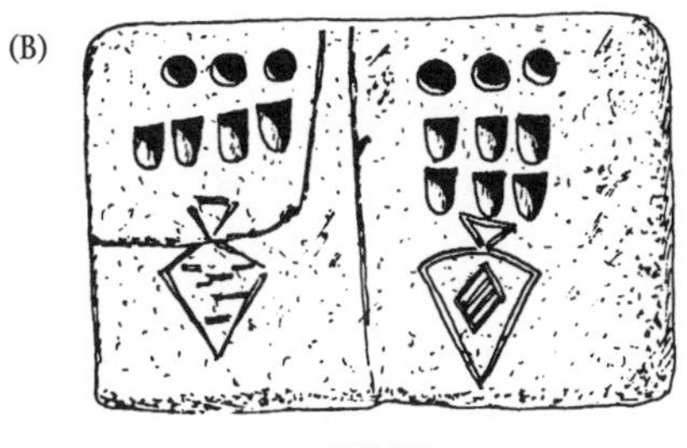

FIG. 8.1. *Archaic Sumerian tablets, discovered at Uruk (level IVa). They are among the earliest known instances of Sumerian writing. Several of these tablets are divided by horizontal and vertical lines into panels which contain numbers and signs representing writing (which already seem to follow a standard pattern). These indicate a degree of precise analytical thought, composed of separate elements brought together, as in articulate speech. The Iraqi Museum, Baghdad*

in outline and convex on their two faces (see Fig. 8.1). On one side, sometimes on both, they bear hollowed-out markings of various shapes and sizes. These marks were made on the clay while still soft by the pressure of a particular tool. As well as these hollow markings we may also find outline drawings made with a pointed tool, representing all kinds of things or

† The best known of the Sumerian archaeological sites, and the first to be excavated, Uruk has served to establish a "time scale" for this civilisation. In certain sectors deep excavation has revealed a series of strata to which archaeologists refer to determine approximate dates for their finds: the ordering of the different layers, from top to bottom, corresponds to the different stages in the history of the civilisation.

beings. The hollow markings correspond to the different units in the Sumerian sequence of enumeration (in the archaic graphology); they are, therefore, the most ancient "figures" known in history (see Fig. 8.2). The drawings are simply the characters in the archaic writing system of Sumeria (Fig. 8.3).

Some of these tablets also have symbolic motifs in relief, made by rolling cylindrical seals over the surface of the tablet, from one end to the other.

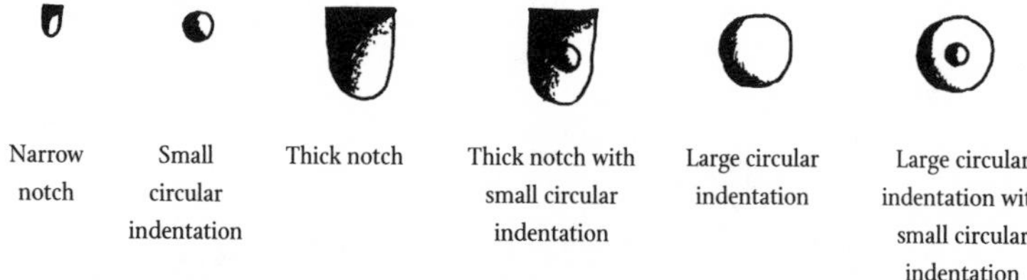


FIG. 8.2. *The shapes of archaic Sumerian numbers*

These tablets seem to have served as records of various quantities associated with different kinds of goods – invoices, as it were, for supplies, deliveries, inventories, or exchanges. Let us have a closer look at the drawings on these tablets, and try to discern the principal character of this writing system. Some of these drawings are very realistic and show the essential outlines of material objects, which may be quite complex (Fig. 8.3).

On occasion, the drawings are much simplified, but still strongly evoke their subject. For example, the heads of the ox, the ass, the pig, and the dog are drawn in a concrete though very stylised way, and the drawing of the animal's head stands for the animal itself.

More often, however, the original object is no longer directly recognisable; the part stands for the whole, and effect represents cause, in a stylised and condensed symbolism. A woman, for instance, is represented by a schematic drawing of the triangle of pubic hair (Fig. 8.3 F), and the verb to impregnate by a drawing of a penis (Fig. 8.3 E).

Generally speaking, as a result of these abbreviations and the subtly simplified relation between representation and object represented, the latter mostly eludes us. The symbols are simple geometric drawings, and the represented objects (where we can determine what they are, by semantic or palaeographic means) have little apparently in common with their representations. Consider the sign for a sheep, for example (Fig. 8.3 U): what might this drawing possibly represent, a circle surrounding a cross? A sheep-pen? A brand? We have no idea.

What is striking about these drawings is their constant and definite character* in which each particular symbol exhibits little variation of form.

* This means that the design has been finalised once and for all, so that "writing" implies choosing and setting up a repertoire of generally accepted and recognised symbols.

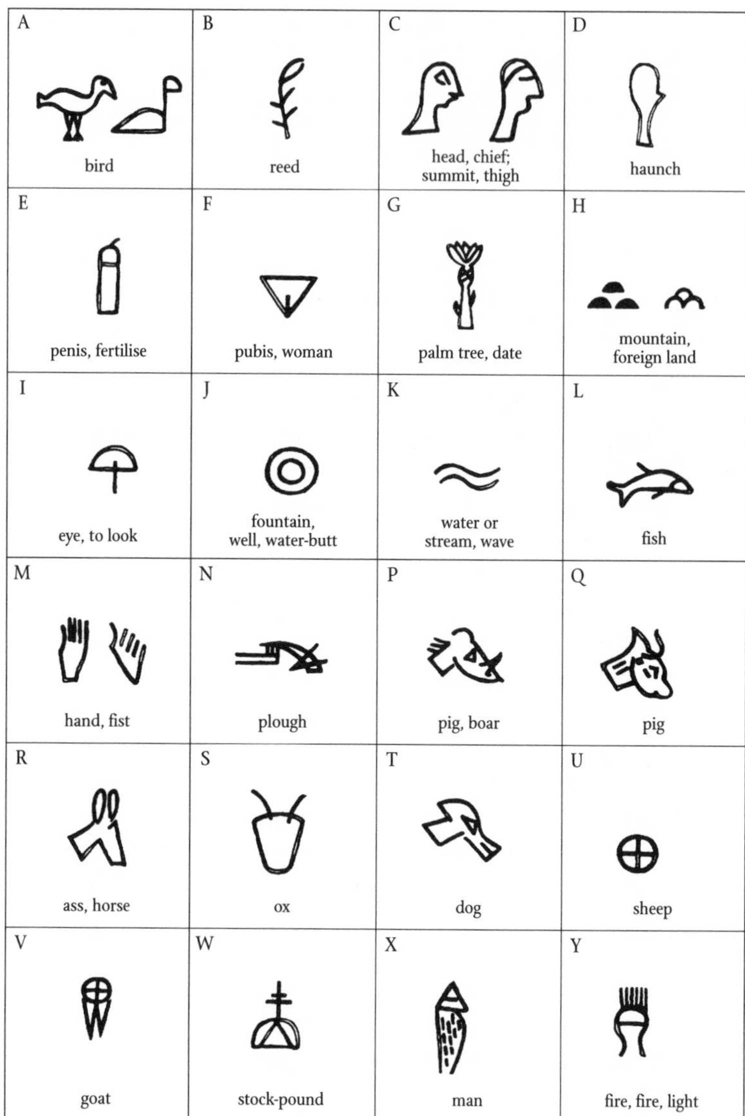


FIG. 8.3. *Pictograms from archaic Sumerian writing*

Comparing this with the number of variations which will emerge in subsequent periods, we are obliged to see in this constancy and regularity the mark of true writing – in the sense of a fully worked-out system which everyone has adopted – and therefore to consider that we are seeing the very origin of writing or, at any rate, its earliest stages, based no doubt on earlier usages but bearing this essential new feature of being a generally accepted uniform practice.

We find ourselves contemplating, therefore, a system of graphical

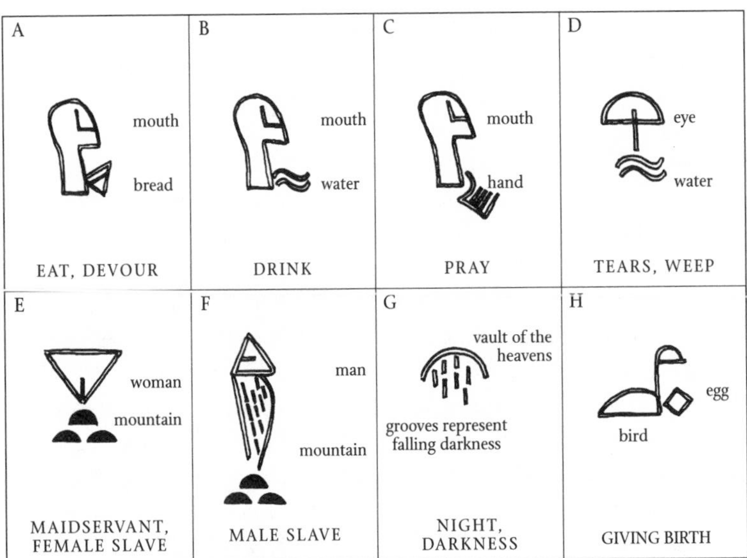


FIG. 8.4. *Some examples of the evocative "logical aggregates" used in archaic Sumerian writing*

symbols intended to express the precise thoughts which occur in speech. However, it is still not writing in the strict and full sense:* we are still in the "prehistory", or rather the "protohistory", of the development of writing (that is, in the pictographic stage).

All of these symbols, whether we know what they mean or not, are graphical representations of material objects.

But we should still not conclude that they can only represent material objects. Each object can be used to symbolise not only the activities or actions directly implied by the object, but also related concepts. The leg, for example, can also represent "walking", "going" or "standing up"; the hand can stand for "taking", "giving", "receiving" (Fig. 8.3 M); the rising sun for "day", "light" or "brightness"; the plough for "ploughing", "sowing", "digging" (Fig. 8.3 N) and, by extension, "ploughman", "farmer", and so on.

The scope of each ideogram can also be extended by a device which had already, at this time, been long applied to symbolism. Two parallel lines can represent the idea of "friend" and "friendship"; two lines crossing each other, the idea of "enemy" or "hostility". The Sumerians gave great place to this idea of enlarging the possible range of meanings of their drawings by combining two or more together to represent new ideas, or aspects of reality otherwise hard to express. The combination *mouth* + *bread* thereby expresses "eat", "devour"; *mouth* + *water* expresses "drink"; *mouth* + *hand* expresses "prayer" (in accordance with Sumerian ritual); and *eye* + *water* denotes "tears", "weeping".

In the same way, an *egg* beside *fowl* suggests the act of "giving birth", strokes underneath a semi-circle suggest darkness falling from the heavenly vault, "night", "the dark". In that flat, lowland country, where "mountain" was synonymous with foreign lands and enemy country (Fig. 8.3 H), the juxtaposition *woman* + *mountain* meant "foreign woman" (literally "woman from the mountains") and therefore, by extension, "female slave" or "maid-servant" (since women were brought to Sumer, bought or captured, to serve as slaves). The same association of ideas gave rise to the combination *man* + *mountain* to denote a male slave (Fig. 8.4 F).

Human thought could therefore be better expressed by this system of pictograms and ideograms than by a purely representational visual art. This system was a systematic attempt to express the whole of thought in the same way as it was represented and dissected in spoken language. But it was still far from perfect, being a long way yet from being able to denote with precision, and without ambiguity, everything that could be expressed in spoken language. Because it depended excessively on the material world of objects which could be drawn as pictures, it required a very large number of different symbols. In fact, the total number of symbols used in this first age of writing in Mesopotamia has been estimated to be about two thousand.

Furthermore, not only was this writing system difficult to manipulate, it was also seriously ambiguous. If, for example, *plough* can also mean "ploughman", how are we supposed to know which one is meant? Even for one and the same word, how can its various nuances be distinguished – nuances which language can meticulously encapsulate and which are essential to complete understanding of the thought (including such qualities as gender, singular and plural, quality, and the countless relationships between things in time and in space)? How can one distinguish the many ways in which actions vary with time?

This writing was certainly a step towards representing spoken language, but it was limited to what could be expressed in images, that is to say to the immediately representable aspects of objects and actions, or to their immediately cognate extensions. For such reasons the original Sumerian writing remains, and will no doubt always remain, undecipherable. Consider the *bull's head* in Fig. 8.1 D. Is it really "the head of a bull"? Or is it – more plausibly – "a bull", a unit of livestock ("one ox"), one of the many products one can obtain from cattle (leather, milk, horn, meat)? Or does it represent some person who may have had a name on the lines of "Mr Bull" (thus

* In the strict sense of the word, a mere visual representation of thought by means of symbols of material objects cannot be considered true writing, since it is more closely related to spoken words than to thought itself. For it to be considered true writing, it would in addition need to be a systematic representation of spoken language, since writing, like language, is a *system* and not a random sequence of items. Février (1959) says: "Writing is a system for human communication using well-defined conventional symbols for the representation of language, which can be transmitted and received, which can be equally well understood by both parties, and which are associated with the words of the spoken language."

being the equivalent of a signature)? Only the few people immediately implicated would be in a position to know what exactly was intended by the bull's head on this particular tablet.

In these circumstances, Sumerian writing at this stage of development is better thought of as an aide-mémoire than as a written record in the proper sense of the term: something which served to help people recall what they already exactly knew (possibly missing out some essential detail), rather than something which could exactly express this to someone who had never known it directly.

Such a scheme answered the purposes of the time well enough. Apart from a few "lists of symbols", all of the known archaic Sumerian tablets carry summaries of administrative actions or of exchanges, as we can see from the totalled numbers which can be found at the end of the document (or on its other side). All of these tablets are, therefore, accounts (in the financial sense of the term). Pure economic necessity therefore played, beyond doubt, a leading role in the story:* the emergence of this writing system was undoubtedly inspired by the necessities of accounting and stocktaking, which caused the Sumerians to become aware of the fact that the old order, which was still based on a purely oral tradition, was running out of steam and that a completely new approach to the organisation of work was called for.

As P. Amiet explains, "Writing was invented by accountants faced with the task of noting economic transactions which, in the rapidly developing Sumerian society, had become too numerous and too complex to be merely entrusted to memory. Writing bears witness to a radical transformation of the traditional way of life, in a novel social and political environment already heralded by the great constructions of the preceding era." At that time the temples were solely responsible for the economy of all Sumer, where continual over-production required a very centralised system of redistribution which became increasingly complicated, a situation which undoubtedly gave rise to the invention of writing. But accounting is simply the recording, by rote or by writing, of operations which have already taken place, and which concern solely the displacements of objects and of people. According to J. Bottero, archaic Sumerian writing is perfectly adapted to this function, which is the reason why its earliest form – which had a profound effect on later developments – was such as to serve above all as an aide-mémoire.

In order, however, to become completely intelligible, and above all in order to attain the status of "writing" in the true sense of the word (i.e. capable of recording unambiguously whatever could be expressed in language), this archaic picto-ideography was therefore obliged to make great advances not only in clarity and precision, but also in universality of reference.

This transition began to occur around 2800–2700 BCE, at which time Sumerian writing became allied to spoken language (which is the most developed way of analysing and communicating reality).

The idea at the root of this development was to use the picture-signs, no longer merely pictorially or ideographically, but phonetically, by relating them to spoken Sumerian, somewhat as in our picture-puzzles, where a phrase is punningly represented by objects whose names form parts of the sounds in the spoken phrase. For example, a picture showing a needle and thread being used to sew a bunch of thyme, a goalkeeper blocking a goal-kick, and the digit "9", could (in English) represent the saying "A stitch in time saves nine".

Thus a picture of an oven is at this time (2800–2700 BCE) no longer used to represent the object, but rather to represent the sound *ne*, which is the Sumerian word for "oven". Likewise, a picture of an arrow (in Sumerian *ti* stands for the sound *ti*; and since the word for "life" in Sumerian is also pronounced *ti* the arrow picture also stands for this word. As Bottero explains: "Using the pictogram of the arrow (*ti*) to denote something quite different which is also pronounced *ti* ('life') completely breaks the primary relation of the image to the object (arrow) and transfers it to the phoneme (*ti*); to something, that is, which is not situated in the material world but is inherent solely in spoken language, and has a more universal nature. For while the arrow, purely as a pictogram, can only refer to the object 'arrow' and possibly to a limited group of related things (weapon, shooting, hunting, etc.), the sound *ti* denotes precisely the phoneme, no matter where it may be encountered in speech and without reference to any material object whatever, and corresponds solely to this word, or to this part of a word (as in *ti-bi-ra*, 'blacksmith'). The sign of the arrow is therefore no longer a pictogram (it depicts nothing) but a *phonogram* (evoking a *phoneme*). The graphical system no longer serves to write things, but to write words, and it no longer communicates one single idea, but the whole of speech and language."

This represents an enormous advance, because such a system is now capable of representing the various grammatical parts of speech: pronouns, articles, prefixes, suffixes, nouns, verbs, and phrases, together with all the nuances and qualifications which can hardly, if at all, be represented in any other way. "As such," adds Bottero, "even if this now means that the reader must know the language of the writer in order to understand, the system can record whatever the spoken language expresses, exactly as it is

* Does the development of this writing have solely an economic explanation? Did not different needs (religious, divinatory, even literary) also play a part? Did people not communicate with each other at a distance in writing, for instance? There are those who think so; but so far no archaeological find has lent support to such possibilities.

expressed: the system no longer serves merely as a record to assist memory and recall, but can also inform and instruct."

It is not our business to go into the specific details of the language for which the Sumerians developed their graphical system, once they had reached the phonetic stage in the above way. But we may echo Bottero in saying that Sumerian writing (enormous advance though it was), because it was born of a pictography designed to aid and extend the memory, remained fundamentally a way of writing words: an aide-mémoire developed into a system, enhanced by the extension into phonetics, but not essentially transformed by it. (After the entry of the phonetic aspect, the Sumerians in fact kept many of their archaic ideograms of which each one continued to denote a word designating a specific entity or object, or even several words connected by more or less subtle relations of meaning, causality or symbolism.)

## THE SUMERIANS

(Adapted from G. Rachet's *Dictionnaire de l'archéologie*)

The geographical origins of the Sumerians remain a topic of controversy. Though some would have them originate from Asia Minor, it seems rather that they arrived in Lower Mesopotamia from Iran, having come from central Asia.

Their language, which remains imperfectly known, was agglutinative, like the early Asiatic (pre-Semitic and pre-Indo-European) languages, and the Caucasian and Turco-Mongolian languages of today. In any case, wherever they came from was mountainous, as is shown by two things which they brought with them to south Mesopotamia: the *ziggurat*, a relic of ancient mountain religions, and stone-carving; whereas the Mesopotamian region is bare of stone.

Their most likely date of arrival in Mesopotamia can be placed in the so-called *Uruk* period, during the second half of the fourth millennium BCE, either during the *Uruk IV* period, or that of *Uruk V*. Quite possibly they arrived gradually, in minor waves, thereby leaving no archaeological traces for the whole of the Uruk period. It certainly seems that this city, home of the epic hero Gilgamesh, had been the primordial centre of the culture they bore. And it is certain the so-called *Jemdet Nasr* period began under their initiative, at the end of the fourth millennium BCE, to be followed by the *pre-Sargonic* era or *Ancient Dynasty* which saw the first culmination of Sumerian civilisation.

These periods were marked by three cultural manifestations: the development of *glyptics* (where cylinders engraved with parades of animals, and various scenes of a religious nature are dominant among the tablets); the development of sculpture with relief on stone vases, animals and personages in the round, themes treated with great mastery and with a force which did not exclude elegance, the masterpiece of this period being the mask, known as the Lady of Warka, imbued with a delicate realism; finally, the emergence of writing which, if it has not given us annals, allows us to identify the gods to whom the temples were dedicated and to learn the names of certain personages, in particular those which have been found in the royal tombs of Ur.

The towns of the land of Sumer: Ur, Uruk, Lagaš, Umna, Adab, Mari, Kiš, Awan, Akšak, were constituted as city-states or, as Falkenstein has said, city-temples, which fought incessantly to exert a hegemony which they exercised more or less by turns. Up to the Archaic Dynasty II, we nowhere find a palace, since the king was in reality a priest, vicar of the god, who lived in the precincts of the temple, the Gir-Par, of which it seems we have an example in the edifice of Nippur.

The priest-king bore the title of EN, "Lord"; it is only during the Archaic Dynasty II that the title of king, *Lugal*, emerges, and at the same time the palace, witness to the separation of State and priesthood, and the emergence of a military monarchy. The earliest known palace is that of Tell A at Kiš, and the first personage who bore the title of Lugal was in fact a king of Kiš, Mebaragesi (around 2700 BCE). The furnishings of the tombs of Ur, which date from subsequent centuries, reveal the high level of material civilisation which the Sumerians had attained. The metallurgists had acquired a great mastery of their art and the sculptors had produced fine in-the-round works. We see a parallel development of urbanisation and of monumental building: the oval temple of Khafaje, the square temple of Tell Asmar, the temple of Ishtar at Mari, the temple of Inanna at Nippur. The expansion of the Sumerian cities was brusquely arrested in the twenty-fourth century BCE by the formation of the Semitic empire of Akkad. But the Akkadians assimilated the Sumerian culture and spread it beyond the land of Sumer. Savage tribes from the neighbouring mountains, Lullubi and Guti, put an end to the Akkadian Empire and ravaged the countryside until the king of Uruk, Utu-Hegal, overthrew the power of the Guti and captured their king, Tiriqan. Now an age of Sumerian renaissance began, with the hegemony of Lagaš and above all of Ur.

At the beginning of the second millennium BCE, the Sumerians were once again dominant with the dynasties of Isin and of Larsa, but after the triumph of Babylon, under Hammurabi, Sumer disappeared politically; but nevertheless the Sumerian language remained a language of priests, and many features of their civilisation, assimilated by the Babylonian Semites, were to survive across the Mesopotamian culture of Babylon.

## THE SEXAGESIMAL SYSTEM

Let us now pass to the numbers themselves. The Sumerians did not count in tens, hundreds and thousands, but adopted instead the numerical base 60, grouping things by sixties and by powers of 60.

We ourselves have vestiges of this base, visible in the ways we express time in hours, minutes and seconds, and circular measure in degrees, minutes and seconds. For instance, if we have to set a digital timepiece to

9; 08; 43

then we know that this corresponds to 9 hours, 8 minutes and 43 seconds, being time elapsed since midnight; and this can be expressed in seconds as follows:

$$9 \times 60^2 + 8 \times 60 + 43 = 32{,}923 \text{ seconds.}$$

Likewise, when a ship's officer determines the latitude of a position he will express it as, for instance: 25°; 36′; 07″, and everyone then knows that the position is

$$25 \times 60^2 + 36 \times 60 + 7 = 92{,}167''$$

north of the Equator.

With the Greeks, and later the Arabs, this was used as a scientific number-system, adopted by astronomers. Since the Greeks, however, with few and belated exceptions, this system has been used solely to express fractions (e.g. minutes and seconds as subdivisions of an hour). But in more distant times, as excavations in Mesopotamia have revealed, it gave rise to two quite separate number-systems which were used for whole numbers as well as fractions. One was the system used solely for scientific purposes by the Babylonian mathematicians and astronomers, later inherited by the Greeks who in turn passed it down to us by way of the Arabs. The other, more ancient yet and which we are about to discuss, was the number-system in common use amongst the Sumerians, predecessors of the Babylonians, and exclusively amongst them.

## THE SUMERIAN ORAL COUNTING METHOD

60 is certainly a large number to use as base for a number-system, placing considerable demands on the memory since – in principle at least – it requires knowledge of sixty different signs or words to stand for the numbers from 1 to 60. But the Sumerians overcame this difficulty by using 10 as an intermediary to lighten the burden on the memory, as a kind of stepping-stone between the different sexagesimal orders of magnitude (1, 60, $60^2$, $60^3$, etc.).

Ignoring sundry variants, the Sumerian names for the first ten numbers, according to Deimel, Falkenstein and Powell, were

| | | | |
|---|---|---|---|
| 1 | *geš* (or *aš* or *dič*) | 6 | *àš* |
| 2 | *min* | 7 | *imin* |
| 3 | *eš* | 8 | *ussu* |
| 4 | *limmu* | 9 | *ilimmu* |
| 5 | *iá* | 10 | *u* |

FIG. 8.5A.

They also gave a name to each multiple of 10 below 60 (so, up this point, it was a decimal system):

| | |
|---|---|
| 10 | *u* |
| 20 | *niš* |
| 30 | *ušu* |
| 40 | *nišmin* (or *nimin* or *nin*) |
| 50 | *ninnû* |
| 60 | *geš* (or *gešta*) |

FIG. 8.5B.

Apart from the case of 20 ( *niš* seems to be independent of *min* = 2 and of *u* = 10), these names are in fact compound words. The word for 30, therefore, is formed by combining the word for 3 with the word for 10:

$$30 = \textit{ušu} < {}^*\textit{eš.u} = 3 \times 10$$

(where the asterisk indicates that an intermediate word has been restored).

In the same way, the word for 40 is derived by combining the word for 20 with the word for 2:

$$40 = \textit{nišmin} = \textit{niš.min} = 20 \times 2\ .$$

The variants of this are simply contractions of *nišmin*:

$$40 = \textit{nin} < \textit{ni.(–m).in} = \textit{ni.(–š).min} < \textit{nišmin}.$$

The word for 50 comes from the following combination:

$$50 = \textit{ninnû} < {}^*\textit{nimnu} = \textit{niminu} = \textit{nimin.u} = 40 + 10.$$

In the words of F. Thureau-Dangin, the Sumerian names for the numbers 20, 40 and 50 seem like a sort of "vigesimal enclave" in this system. Note, by the way, that the word for 60 (*geš*) is the same as the word for unity. No doubt this was because the Sumerians thought of 60 as a large unity. Nevertheless, to avoid ambiguity, it was sometimes called *gešta*.

The number 60 represents a certain level, above which, in this oral numeration system, multiples of 60 up to 600 were expressed by using 60 as a new unit:

| | | | | | |
|---|---|---|---|---|---|
| 60 | *geš* | | 360 | *geš-àš* | (= 60 × 6) |
| 120 | *geš-min* | (= 60 × 2) | 420 | *geš-imin* | (= 60 × 7) |
| 180 | *geš-eš* | (= 60 × 3) | 480 | *geš-ussu* | (= 60 × 8) |
| 240 | *geš-limmu* | (= 60 × 4) | 540 | *geš-ilimmu* | (= 60 × 9) |
| 300 | *geš-iá* | (= 60 × 5) | 600 | *geš-u* | (= 60 × 10) |

Fig. 8.5c.

The next level is reached at 600, which is now treated as another new unit whose multiples were used up to 3,000:

| | | | | | |
|---|---|---|---|---|---|
| 600 | *geš-ŭ* | | 2,400 | *geš-u-limmu* | (= 600 × 4) |
| 1,200 | *geš-u-min* | (= 600 × 2) | 3,000 | *geš-u-iá* | (= 600 × 5) |
| 1,800 | *geš-u-eš* | (= 600 × 3) | 3,600 | *šàr* | (= $60^2$) |

Fig. 8.5d.

The number 3,600 (sixty sixties) is the next level, and it is given a new name (*šàr*) and in turn becomes yet another new unit:

| | | | | | |
|---|---|---|---|---|---|
| *šàr* | 3,600 | (= $60^2$) | *šàr-àš* | 21,600 | (= 3,600 × 6) |
| *šàr-min* | 7,200 | (= 3,600 × 2) | *šàr-imin* | 25,200 | (= 3,600 × 7) |
| *šàr-eš* | 10,800 | (= 3,600 × 3) | *šàr-ussu* | 28,800 | (= 3,600 × 8) |
| *šàr-limmu* | 14,400 | (= 3,600 × 4) | *šàr-ilimu* | 32,400 | (= 3,600 × 9) |
| *šàr-iá* | 18,000 | (= 3,600 × 5) | *šàr-u* | 36,000 | (= 3,600 × 10) |

Fig. 8.5e.

The following levels correspond to the numbers 36,000, 216,000, 12,960,000, and so on, proceeding in the same sort of way as above:

| | | | | | |
|---|---|---|---|---|---|
| 36,000 | *šàr-u* | (= $60^2$ × 10) | 144,000 | *šàr-u-limmu* | (= 36,000 × 4) |
| 72,000 | *šàr-u-min* | (= 36,000 × 2) | 180,000 | *šàr-u-iá* | (= 36,000 × 5) |
| 108,000 | *šàr-u-eš* | (= 36,000 × 3) | 216,000 | *šàrgal* | (= $60^3$) (literally: "big 3,600") |

Fig. 8.5f.

| | | | | | |
|---|---|---|---|---|---|
| 216,000 | *šàrgal* | (= $60^3$) | 1,296,000 | *šàrgal-aš* | (= 216,000 × 6) |
| 432,000 | *šàrgal-min* | (= 216,000 × 2) | 1,512,000 | *šàrgal-imin* | (= 216,000 × 7) |
| 1,080,000 | *šàrgal-iá* | (= 216,000 × 5) | 2,160,000 | *šàrgal-u* | (= 216,000 × 10) |

Fig. 8.5g.

| | | | | | |
|---|---|---|---|---|---|
| 2,160,000 | *šàrgal-u* | (= $60^3$× 10) | 8,640,000 | *šàrgal-u-limmu* | (= 2,160,000 × 4) |
| 4,320,000 | *šàrgal-u-min* | (= 2,160,000 × 2) | 10,800,000 | *šàrgal-u-iá* | (= 2,160,000 × 5) |
| 6,480,000 | *šàrgal-u-eš* | (= 2,160,000 × 3) | | | |
| 12,960,000 | *šàrgal-šu-nu-tag* | (= $60^4$) ("Unit greater than big šàr") | | | |

Fig. 8.5h.

## FROM THE ORAL TO THE WRITTEN NUMBER-SYSTEM

When, around 3200 BCE, the Sumerians devised a numerical notation, they gave a special graphical symbol to each of the units 1; 10; 60; 600 (= 60 × 10); 3,600 (= $60^2$); 36,000 (= $60^2 \times 10$), that is to say to each term in the sequence generated by the following schema:

$$1$$
$$10$$
$$10 \times 6$$
$$(10 \times 6) \times 10$$
$$(10 \times 6 \times 10) \times 6$$
$$(10 \times 6 \times 10 \times 6) \times 10$$

They therefore mimicked the names of the different units in their oral system which, as we have seen, used base 60 and proceeded by a system of levels constructed alternately on auxiliary bases of 6 and of 10 (Fig. 8.6).

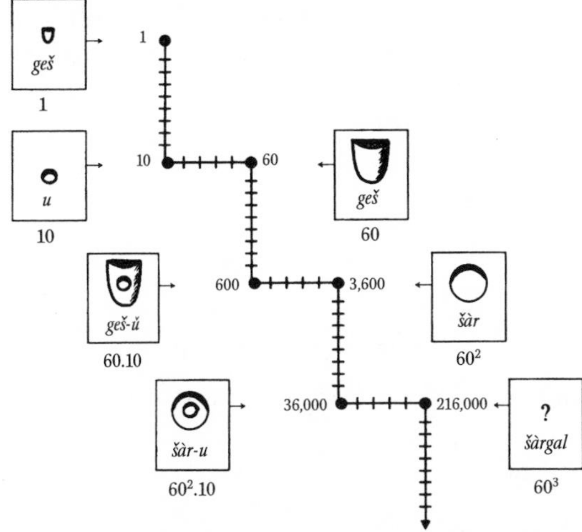


*Fig. 8.6. The structure of the Sumerian number-system, which was a sexagesimal system constructed upon a base of 10 alternating with a base of 6 (thus activating in turn two divisors of the base 60: 10 × 6 = 60)*

## THE VARIOUS FORMS OF SUMERIAN NUMBERS

In the archaic epochs, unity was represented by a small notch (sometimes elongated), 10 by a circular indentation of small diameter, 600 (= 60 × 10) by a combination of these two, 3,600 (= $60^2$) by a circular indentation of large diameter, and 36,000 (= 3,600 × 10) by the smaller circular indentation within the larger circular indentation (Fig. 8.2 and 8.6).

To start with, these symbols were impressed on the tablets in the following orientation:

FIG. 8.7.

However, starting in the twenty-seventh century BCE, these became rotated anticlockwise through 90°. Thus the non-circular symbols thenceforth no longer pointed from top to bottom but from left to right:

FIG. 8.8.

After the development of the cuneiform script, these number-symbols took on a completely new form, angular and with much sharper outlines.

- The number 1 was thereafter represented by a small vertical wedge (instead of a small cylindrical notch);
- the number 10 was represented by a chevron (instead of the small circular impression);
- the number 60 was represented by a larger vertical wedge (instead of a wide notch);
- the number 600 by this larger vertical wedge combined with the chevron of the number 10;
- the number 3,600 by a polygon formed by joining up four wedges (instead of the larger circle);
- the number 36,000 by the polygon for 3,600, with the wedge for 10 in its centre;

and, finally, the number 216,000 (the cube of 60, for which a special symbol was introduced into cuneiform script) was represented by combining the polygon for 3,600 with the wedge for 60 (see Fig. 8.9).

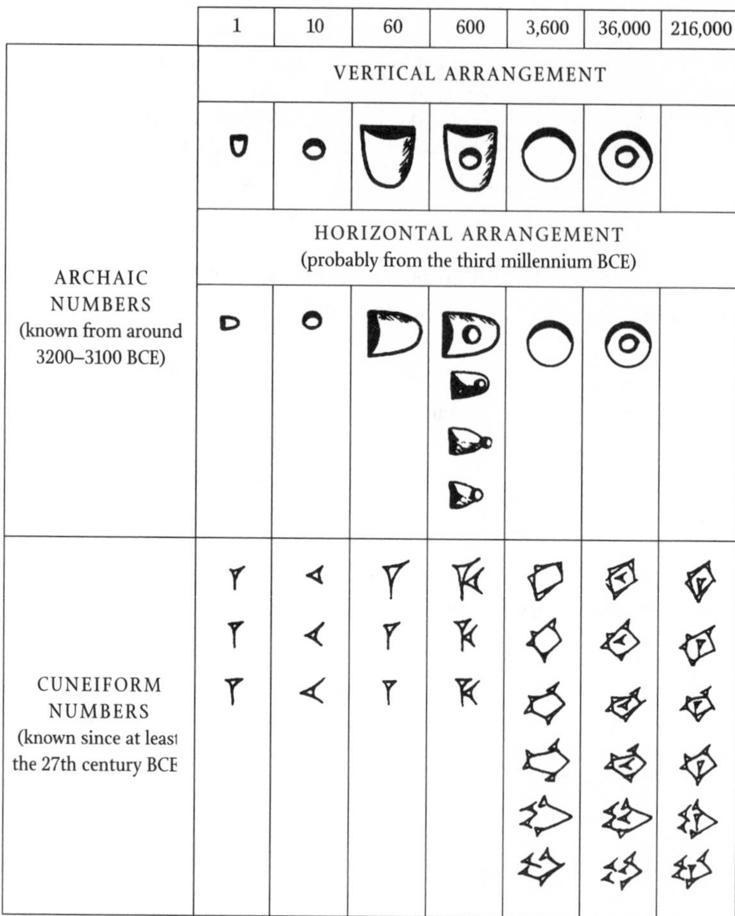


FIG. 8.9. *The development of the shapes of numbers originating in Sumeria. The change from the archaic to the cuneiform shapes resulted from the replacement of the "old stylus", which was cylindrical at one end, and pointed at the other, by the "flat" stylus shaped something like a modern ruler. This new writing instrument conduced its users to break the curves into a series of wedges or chevrons. See Deimel (1924, 1947) and Labat (1976 and in EPP).*

### *Clay as Mesopotamian "paper" and how to write on it*

In Mesopotamia, stone is rare; wood, leather and parchment are difficult to preserve, and the soil consists of alluvial deposits. The inhabitants of this region therefore took what came to hand for the purpose of expressing their thoughts or for recording the spoken word, and what they had to hand was clay. They had used this raw material since very early times for modelling figurines, for sculpture, and for glyptics*, and later most ingeniously put it

* Close examination of archaeological finds leads one to believe that the usages of clay held no secrets for the Mesopotamians, four thousand years BCE. This is an important consideration for the history of writing in this region, since it effectively implies that they were fully aware of the possibilities of the medium. The

to diverse uses, especially for the purpose of writing, for more than three thousand years, in more than a dozen languages†. To borrow a phrase from J. Nougayrol (1945), you might say that these people created "civilisations of clay".

The originality of Mesopotamian graphics directly reflects the nature of this material and the techniques available to work with it, and we have an interest in devoting some attention to this; what follows will allow us to better trace the evolution of the forms of figures and written characters which originated in Sumer.

We have seen how Sumerian figures were hollow marks of different shapes and sizes (Fig. 8.2), while the written characters were real drawings representing beings and objects of every kind (Fig. 8.3). Originally, therefore, there were fundamental differences of technique between the production of the one and the production of the other. The number-signs, like the motifs created using cylindrical or stamp-like seals, were produced by *impression*; the written characters on the other hand were *traced*.

For these purposes the Sumerians used a reed stem (or possibly a rod made of bone or ivory), which at one end was shaped into a cylindrical stylus, while the other end was sharpened to a point somewhat like a modern pen (Fig. 8.10).

The pictograms were made by pressing the pointed end quite deeply into the clay, still fresh, of the tablets (Fig. 8.11). To draw a line, the same pointed end was pressed in as before, and then drawn parallel to the surface through the required distance. Of course, this would often result in a wavy line, and could give rise to spillover on either side, because of the softness of the material.

FIG. 8.11. *How the archaic Sumerian pictograms were drawn on soft clay tablets*

For the numbers, on the other hand, the Sumerians made these by making an imprint on the soft damp clay with the other end of the instrument, the end shaped into a circular stylus. This was done with the stylus held at a certain angle to the surface of the clay. They had two styluses of different diameters: one about 4 mm, the other about 1 cm (Fig. 8.10). According to the angle at which they held the stylus, either a circular imprint, or a notch, would be obtained, and its size would depend on the diameter of the stylus (Fig. 8.12):

- a circular imprint of smaller or larger diameter if the stylus was held perpendicular to the surface of the clay;
- a notch, narrow or wide, if the stylus was held at an angle of 30° – 45° to the surface; the imprint would be more elongated if the angle was small.

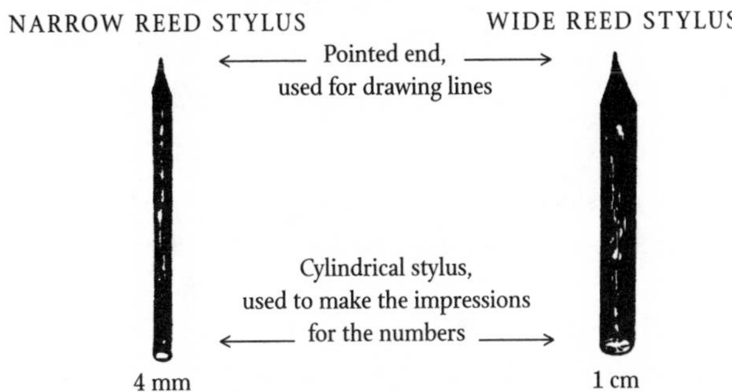


FIG. 8.10. *Reconstruction of the writing instruments of the Sumerian scribes (archaic era)*

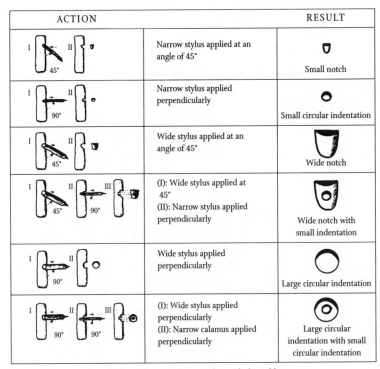

| ACTION | | RESULT |
|---|---|---|
| I 45° II | Narrow stylus applied at an angle of 45° | Small notch |
| I 90° II | Narrow stylus applied perpendicularly | Small circular indentation |
| I 45° II | Wide stylus applied at an angle of 45° | Wide notch |
| I 45° II 90° III | (I): Wide stylus applied at 45°<br>(II): Narrow stylus applied perpendicularly | Wide notch with small indentation |
| I 90° II | Wide stylus applied perpendicularly | Large circular indentation |
| I 90° II 90° III | (I): Wide stylus applied perpendicularly<br>(II): Narrow calamus applied perpendicularly | Large circular indentation with small circular indentation |

FIG. 8.12. *How the archaic numbers were impressed on soft clay tablets*

character, possibly religious but certainly symbolic, of the motifs appearing on these vases and jugs, their repeated occurrence and their systematic stylisation, must not only have accustomed their creators to express a number of thoughts and ideas in this way but also to subsume these into ever simpler and more concise designs.

† At the dawn of the second millennium BCE, at the time when writing emerged in Sumer, the use of clay for "tablets" intended to bear conventional signs was already widespread in the region. This point also is very important, for it clarifies one reason why Sumerian writing moved on to a systematic phase: considering the difficulty of sculpture, carving, and painting, and the fact that they demand time for their execution, the universal adoption of clay throughout Mesopotamia is readily explained by the ease with which it can be worked (compared with wood, bone, or stone, whether for engraving, embossing, impression, moulding, or cutting).

### *Why Sumerian writing changed direction*

In the very earliest times, the signs used in Sumerian writing were drawn on the clay tablets in the natural orientation of whatever they were supposed to represent: vases stood upright, plants grew upwards, living things were vertical, etc. Similarly, the non-circular figures for numbers were also vertical (the stylus being held sloping towards the bottom of the tablet).

These signs and figures were generally arranged on the tablets in horizontal rows which, in turn, were subdivided into several compartments or boxes (Fig. 8.1, tablet E). Within each box, the figures were generally at the top, starting from the right, while the drawings used for writing were at the very bottom, like this:

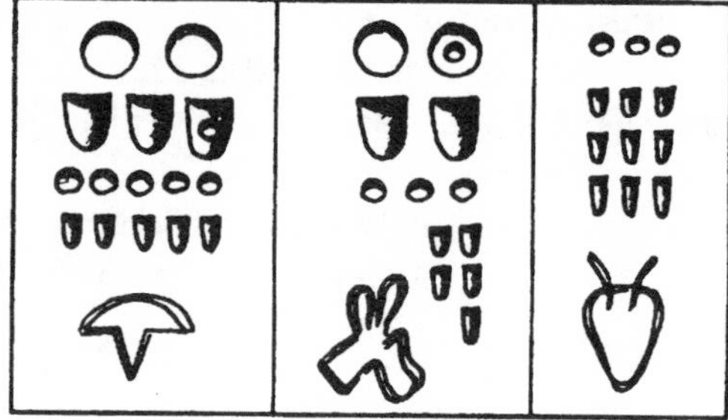

FIG. 8.13.

Now, if we examine the arrangement of figures and drawings on one of the tablets of the so-called Uruk period (around 3100 BCE), we find that where one of the boxes is not completely full the empty space is always on the left of the box (see the second box from the right in the top row of the tablet in Fig. 8.14).

This proves that the scribes of the earliest times wrote from right to left and from the top to the bottom. The non-circular figures were vertical, and the drawings had their natural orientations. In short, in the beginning Sumerian writing was read from right to left and from top to bottom.

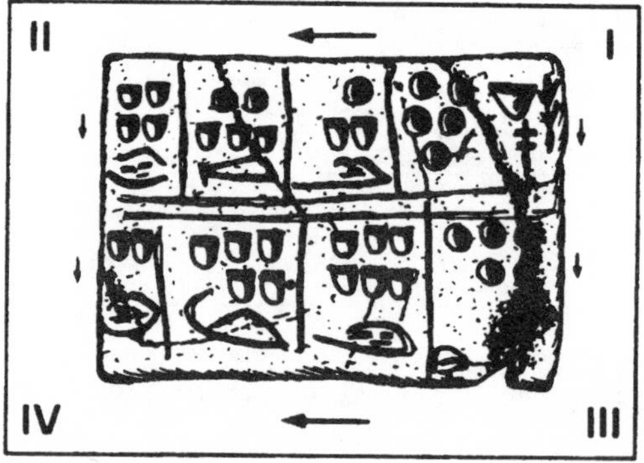


FIG. 8.14. *Sumerian tablet from Uruk, from around 3100 BCE. Iraqi Museum, Baghdad*

This arrangement long persisted on Mesopotamian stone inscriptions. It can be seen especially on the Stele of the Vultures (where the text is arranged in horizontal bands, and the boxes succeed each other from right to left and from top to bottom), in the celebrated Code of Hammurabi (whose inscription, which is read from right to left and from left to right, is arranged in vertical columns), and in several legends later than the seventeenth century BCE.

It went quite differently in the case of clay tablets, however: that is, in the case of everyday writings. Starting around the twenty-seventh century BCE, the signs used for writing, and the figures used for numbers, underwent a rotation through 90° anticlockwise.

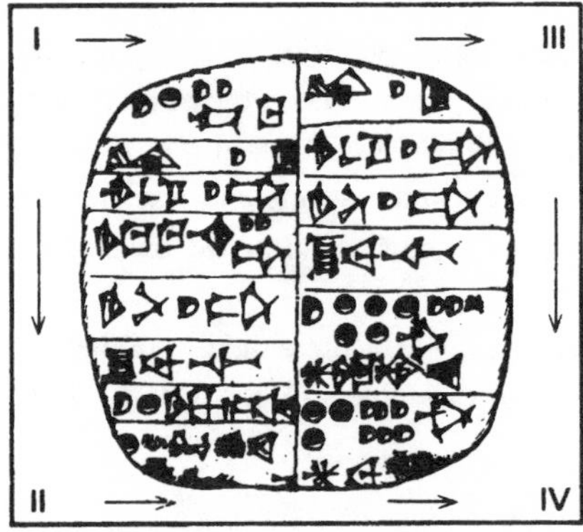


FIG. 8.15. *Sumerian tablet (Tello, about 3500 BCE). Bibliothèque nationale, Paris, Cabinet des Médailles (CMH 870 F). See de Genouillac (1909), plate IX*

To verify this, consider the tablet in Fig. 8.15, and look at in the direction I → II indicated by the long arrow in the Fig., after turning it 90° clockwise so that I → II is from right to left and at the top. Then we can see that if a compartment is not full up, the empty space is at the bottom, and not at the left. Likewise, in the original position of the tablet, the empty space is at the right.

According to C. Higounet (1969), this would be due to a change in the orientation with which the tablets were held.

With the small tablets of the earliest times, holding the tablet obliquely in the hand made it easier to trace drawings in columns from top to bottom. But, when the tablets became larger, the scribes had to place them upright in front of them, and the signs became horizontal, and the writing went in lines from left to right.

Be that as it may, thenceforth the drawings and the non-circular figures had an orientation 90° anticlockwise from their original one (Fig. 8.16);

"turned sideways, they became less pictorial, and therefore more liable to undergo a certain systematisation." [R. Labat]

FISH

LEGS

HEAD

OX

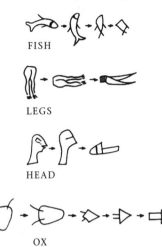

Fig. 8.16. *Anticlockwise rotation, through a quarter turn, of the Sumerian signs and numbers*

*The emergence of the cuneiform signs*

The radical transformation which the Sumerian characters underwent after the Pre-Sargonic era (2700–2600 BCE) is due simply to a change of implement.

While the drawings used in writing had originally been traced out with the pointed end of the stylus, this changed when they had the idea of using instead, for this, the method which had always been used for the figures denoting numbers, namely impressing the marks on the clay. Instead of using a pointed stylus for tracing lines, they preferred to use a reed stem (or a rod of bone or ivory) whose end was trimmed in such a way that its tip formed a straight edge, and no longer a circle or a point. This edge was then pressed into the clay, to achieve cleanly, at one stroke, a line segment of a certain length; this clearly was much more rapid than drawing it with a pointed tool.

Of course this new type of stylus made characters of quite a different shape, with sharper lines and an angular appearance; these signs are called *cuneiform* (from the Latin *cuneus*, "a wedge") (Fig. 8.17).

The angular shapes of the imprints made by such a stylus on the clay naturally led to greater stylisation of the shapes of the various signs. Curves were broken up, and where necessary were replaced by a series of line segments, so that a picture was reduced to a collection of broken lines. In this new form of Sumerian writing, a circle, for example, became a polygon, and curves were replaced by polygonal lines (Fig. 8.18).

This did not occur all at once, however. It is not seen at all around 2850 BCE. It begins to appear in the archaic tablets of Ur (2700–2600 BCE), and in those of Fara (Šuruppak), where the majority of the signs are made up of impressed lines, while many other tablets of the same period continue to show the curved lines traced by the older method.

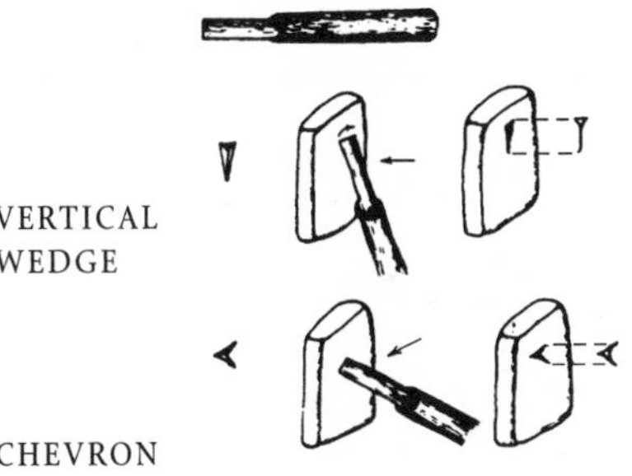


Fig. 8.17. *Impressing cuneiform signs on soft clay. The vertical wedge was made by pressing lightly on the clay with one of the corners of the "beak" of the stylus (the heavier the pressure, the larger the wedge).*

| | ARCHAIC SIGNS | | CUNEIFORM SIGNS | |
|---|---|---|---|---|
| | Uruk period (about 3100 BCE) | Jemdet-Nasr era (about 2850 BCE) | Pre-Sargonic Era (about 2600 BCE) | Third Ur Dynasty (about 2000 BCE) |
| STAR DIVINITY | | | | |
| EYE | | | | |
| HAND | | | | |
| BARLEY | | | | |
| LEG | | | | |
| FIRE TORCH | | | | |
| BIRD | | | | |
| HEAD SUMMIT CHIEF | | | | |

Fig. 8.18.

At the beginning of this change in form, the signs nevertheless remained very complex, since people wished to preserve as much as possible of the detail of the original drawings, and because in the majority of cases the objective was still to achieve the outline of a concrete object. But, after a long period of adaptation, from the end of the third millennium BCE the scribes only kept what was essential and therefore made their marks much more rapidly than before.

And this is how the signs in Sumerian writing finally lost all resemblance to the real objects which they were meant to represent in the first place.

## THE SUMERIAN WRITTEN COUNTING METHOD

Starting with these basic symbols, the first nine whole numbers were represented by repeating the sign for unity as often as required; the numbers 20, 30, 40, and 50 by repeating the sign for 10 as often as required, the numbers 120, 180, 240, etc. by repeating the symbol for 60, and so on.

Generally, since the system was based on the additive principle, a number was represented by repeating, at the level of each order of magnitude, the requisite symbol as often as required.

For example, a tablet dating from the fourth millennium BCE (Fig. 8.1, tablet C) carries the representation of the number 691 in the following form:

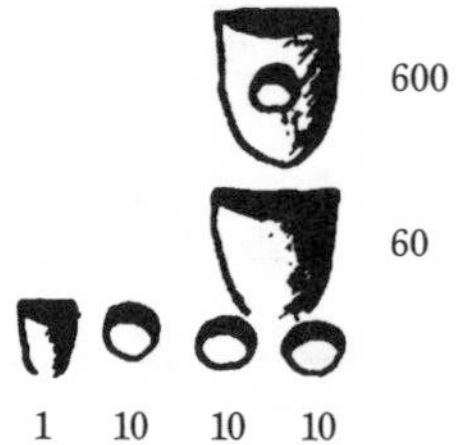


FIG. 8.19.

Likewise, on a tablet from Šuruppak, from around 2650 BCE, the number 164,571 is represented as follows (Fig. 8.20 and 12.1):

| | | | |
|---|---|---|---|
| 36,000 drawn 4 times over = | 36,000 × 4 = | 144,000 |
| 3,000 drawn 5 times over = | 3,600 × 5 = | 18,000 |
| 600 drawn 4 times over = | 600 × 4 = | 2,400 |
| 60 drawn 2 times over = | 60 × 2 = | 120 |
| 10 drawn 5 times over = | 10 × 5 = | 50 |
| 1 drawn 1 time = | = | 1 |
| | | 164,571 |

FIG. 8.20.

| | | | | | |
|---|---|---|---|---|---|
| | 30 8 | 60 50 7 | 180 40 1 | 240 40 1 | 120 10 9 |
| 4 | 38 | 117 | 221 | 281 | 139 |

FIG. 8.21A.

TRANSLATION

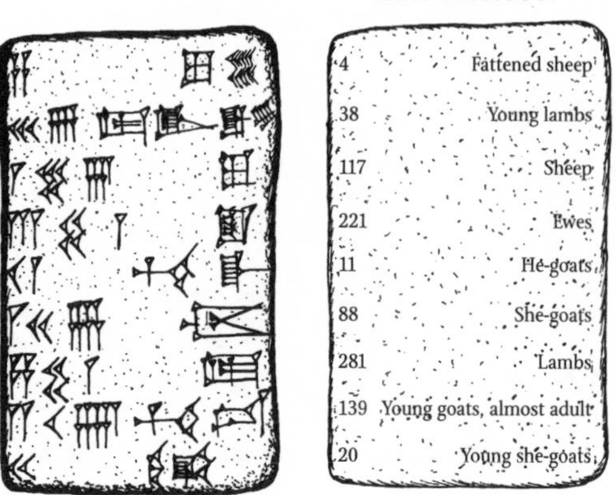

4 Fattened sheep
38 Young lambs
117 Sheep
221 Ewes
11 He-goats
88 She-goats
281 Lambs
139 Young goats, almost adult
20 Young she-goats


FIG. 8.21B. *Sumerian tablet from about 2000 BCE, giving a tally of livestock by means of cuneiform signs and numbers. Translation by Dominique Charpin. See de Genouillac (1911), plate V, no. 4691 F*

Similarly, for the cuneiform representation, on a tablet dating from the second dynasty of Ur (about 2000 BCE), found in a warehouse at Drehem (Ašnunak Patesi), various numbers are represented as shown on Figs. 8.21 A and B.

Finally, and in the same way, on a tablet contemporary with this last one, but from a clandestine excavation at Tello, we find the numbers 54,492 and 199,539 also expressed in cuneiform symbols:

| | |
|---|---|
| 54,492 | 36,000 drawn 1 time = 36,000<br>3,600 drawn 5 times over = 18,000<br>60 drawn 8 times over = 480<br>10 drawn 1 time = 10<br>1 drawn 2 times over = 2<br>54,492 |
| 199,539 | 36,000 drawn 5 times over = 180,000<br>3,600 drawn 5 times over = 18,000<br>600 drawn 2 times over = 1,200<br>60 drawn 5 times over = 300<br>10 drawn 3 times over = 30<br>1 drawn 9 times over = 9<br>199,539 |

FIG. 8.22. *Barton (1918), Table Hlb 24, no. 16*

We may observe in passing that the Sumerians grouped the identical repeated symbols in such a way as to facilitate the grasp, in one glance, of the values of the assemblages within each order of magnitude. Considering just the representations of the first nine numbers, these groupings were initially made according to a dyadic or binary principle (Fig. 8.23) and later according to a ternary principle in which the number 3 played a special role (Fig. 8.24).

ARCHAIC NUMBERS

| 1 | 2 | 3 | 4 | 5 | 6 | 7 | 8 | 9 |
|---|---|---|---|---|---|---|---|---|

CUNEIFORM NUMBERS

| 1 | 2 | 3 | 4 | 5 | 6 | 7 | 8 | 9 |
|---|---|---|---|---|---|---|---|---|

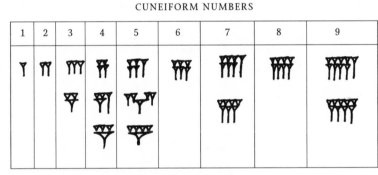

FIG. 8.23. *The dyadic (binary) principle of representing the nine units*

CUNEIFORM NUMBERS

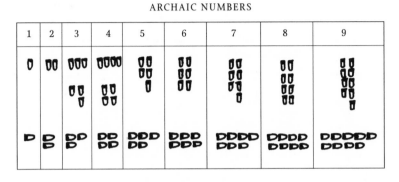

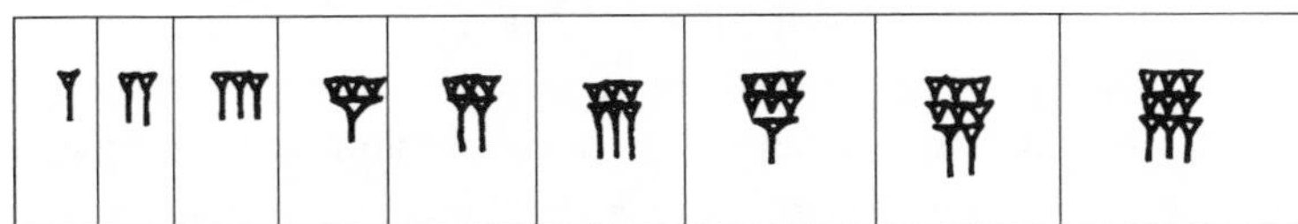

FIG. 8.24. *The ternary principle of representing the nine units*

Thus the Sumerian numbering system sometimes required inordinate repetitions of identical marks, since it placed symbols side by side to represent addition of their values. For example, the number 3,599 required a total of twenty-six symbols!

For this reason, the Sumerian scribes would seek simplification by often using a subtractive convention, writing numbers such as 9, 18, 38, 57, 2,360 and 3,110 in the form:

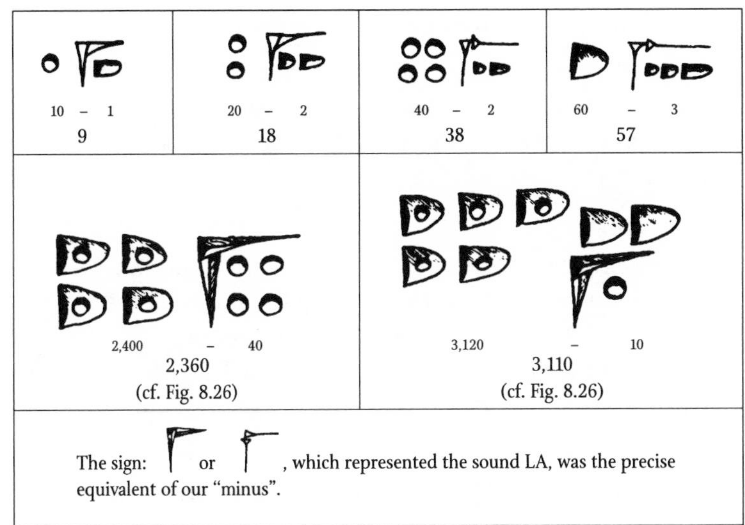


FIG. 8.25.

FIG. 8.26. *Sumerian tablet from Šuruppak (Fara), 2650 BCE. Istanbul Museum. See Jestin (1937), plate LXXXIV, 242 F*

From the pre-Sargonic era (about 2500 BCE), certain irregularities start to appear in the cuneiform representation of numbers. As well as the subtractive convention just described, the multiples of 36,000 can be found represented as shown in Fig. 8.27, instead of simply repeating the symbol for 36,000 once, twice, or three, four, or five times.

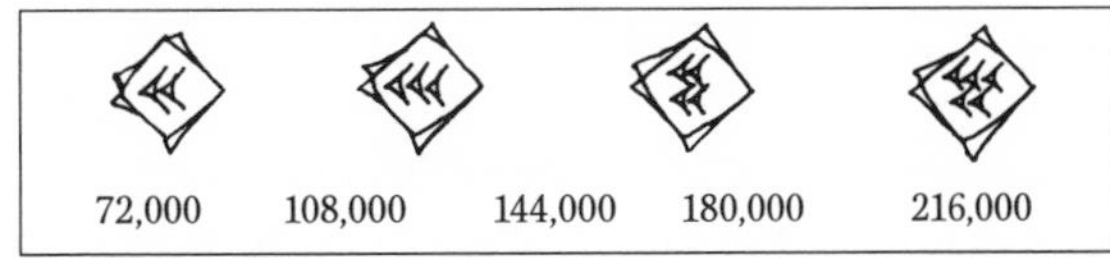


FIG. 8.27. *See Deimel*

These forms evidently correspond to the arithmetical formulae

72,000 = 3,600 × 20 (instead of 36,000 + 36,000)
108,000 = 3,600 × 30 (instead of 36,000 + 36,000 + 36,000)
144,000 = 3,600 × 40 (instead of 36,000 + 36,000 + 36,000 + 36,000)
180,000 = 3,600 × 50
(instead of 36,000 + 36,000 + 36,000 + 36,000 + 36,000).

In this, the Sumerians were doing nothing other than what we would today refer to as "expressing in terms of a common factor". Observing that the symbol for 3,600 is itself made up of the symbol for 360 with the symbol for 10, they also, after their fashion, made the number 3,600 a common factor so that, for instance, instead of representing 144,000 in the form

$$(3{,}600 \times 10) + (3{,}600 \times 10) + (3{,}600 \times 10) + (3{,}600 \times 10)$$

they used instead the simpler form

$$3{,}600 \times (10 + 10 + 10 + 10).$$

Another special point arising in the cuneiform notation concerned the two numbers 70 (= 60 + 10) and 600 (= 60 × 10), since both involved juxtaposing the symbol for 60 and the symbol for 10. This can clearly lead to ambiguity, since for 70 they are combined additively, and for 600 multiplicatively. This ambiguity was not present, however, in the archaic notation:

Fig. 8.28A. Fig. 8.28B.

They were, however, able to eliminate any possible confusion. In the case of 70, they placed a clear separation between the wedge (for 10) and the chevron (for 60) so as to indicate addition (Fig. 8.29 A), while for 600 they put them in contact so as to form an indivisible group, to represent multiplication (Fig. 8.29 B).

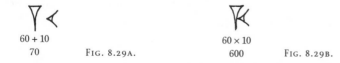

60 + 10
70

Fig. 8.29A.

60 × 10
600

Fig. 8.29B.

A different problem arose with the representation of the numbers 61, 62, 63, etc. In the beginning, the number 1 was represented by a small wedge, and the number 60 by a larger wedge, and so there was no ambiguity:

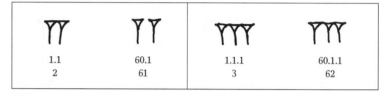

| 60 1 | 60 2 | 60 3 | 60 4 | 60 5 | 60 6 | 60 7 | 60 8 | 60 9 |
|---|---|---|---|---|---|---|---|---|
| 61 | 62 | 63 | 64 | 65 | 66 | 67 | 68 | 69 |

Fig. 8.30.

Later, however, 1 and 60 came to be represented by the same size of vertical wedge, and it was very difficult to distinguish between 2 and 61, or between 3 and 62, for example:

| 1.1 | 60.1 | 1.1.1 | 60.1.1 |
|---|---|---|---|
| 2 | 61 | 3 | 62 |

Fig. 8.31.

Therefore they had the idea of distinctly separating the unit symbols for the sixties from those for the units.

| 60 1 | 60 2 | 60 3 | 60 4 | 60 5 | 60 6 | 60 7 | 60 8 | 60 9 |
|---|---|---|---|---|---|---|---|---|
| 61 | 62 | 63 | 64 | 65 | 66 | 67 | 68 | 69 |

Fig. 8.32.

This particular problem with the cuneiform sexagesimal notation was the root of a most interesting simplification to which we shall return in Chapter 13.

For a long time, the cuneiform characters (known since at least twenty-seven centuries BCE) coexisted with the archaic numeral signs (Fig. 8.9). On certain tablets contemporary with the kings of the Akkad Dynasty (second half of the third millennium BCE), we see the cuneiform numbers side by side with their archaic counterparts. The intention, it seems, was to mark a distinction of rank between the people being enumerated: the cuneiform figures were for people of higher social standing, and the others for slaves or common people [M. Lambert, *personal communication*]. The cuneiform number-symbols did not definitively supplant the archaic ones until the third dynasty of Ur (2100–2000 BCE).

## CHAPTER 9

# THE ENIGMA OF THE SEXAGESIMAL BASE

In all of human history the Sumerians alone invented and made use of a *sexagesimal system* – that is to say, a system of numbers using 60 as a base. This invention is without doubt one of the great triumphs of Sumerian civilisation from a technical point of view, but it is nonetheless one of the greatest unresolved enigmas in the history of arithmetic. Although there have been many attempts to make sense of it since the time of the Greeks, we do not know the reasons which led the Sumerians to choose such a high base. Let us begin with a review of the explanations that have been put forward so far.

### THEON OF ALEXANDRIA'S HYPOTHESIS

Theon of Alexandria, a Greek editor of Ptolemaic texts, suggested in the fourth century CE that the Sumerians chose base 60 because it was the "easiest to use" as well as the lowest of "all the numbers that had the greatest number of divisors". The same argument also cropped up 1,300 years later in *Opera mathematica*, by John Wallis (1616–1703), and again, in a slightly different form, in 1910, when Löfler argued that the system arose "in priestly schools where it was realised that 60 has the property of having all of the first six integers as factors".

### FORMALEONI'S AND CANTOR'S HYPOTHESES

In 1789 a different approach was suggested by the Venetian scholar Formaleoni, and then repeated in 1880 by Moritz Cantor. They held that the Sumerian system derived from exclusively "natural" considerations: on this view, the number of days in a year, rounded down to 360, was the reason for the circle being divided into 360 degrees, and the fact that the chord of a sextant (one sixth of a circle) is equal to the radius gave rise to the division of the circle into six equal parts. This would have made 60 a natural unit of counting.

### LEHMANN-HAUPT'S HYPOTHESIS

In 1889, Lehmann-Haupt believed he had identified the origin of base 60 in the relationship between the Sumerian "hour" (*danna*), equivalent to two of our current hours, and the visible diameter of the sun expressed in units of time equivalent to two minutes by current reckoning.

### NEUGEBAUER'S HYPOTHESIS

In 1927 O. Neugebauer proposed a new solution which located the source of base 60 in terms of systems of weights and measures. This is how the proposal was explained by O. Becker and J. E. Hoffmann (1951):

> It arose from the combination of originally quite separate measurement units using base 10 and having (as in spoken language, and like the Egyptian systems) different symbols for 1, 10, and 100 as well as for the "natural fractions", 1/2, 1/3, and 2/3. The need to combine the systems arose particularly for measures of weight corresponding to measures of price or value. The systems were too disparate to be harmonised by simple equivalence tables, and so they were combined to give a continuous series such that the elements in the set of higher values (B) became whole multiples of elements in the set of lower values (A). Since both sets of values had the structure 1/1, 1/2, 2/3, 1, 2, 3 . . . 10, the relationship between the two sets A and B had to allow for division by 2 and by 3, which introduced factor 6. So from the decimal structure of the original number-system, the Sumerians ended up with 60 as the base element of the new (combined) system.

On the other hand, F. Thureau-Dangin (1929) took the view that this entirely theoretical explanation cannot be a correct account of the origin of Sumerian numbering, because it is "undoubtedly the case that base 60 only occurs in Sumerian weights and measures because it was already available in the number-system".

### OTHER SPECULATIONS

The Mesopotamians, according to D. J. Boorstin (1986), got to 60 by multiplying the number of planets (Mercury, Venus, Mars, Jupiter, and Saturn) by the number of months in the year: 5 × 12 is also a multiple of 6.

In 1910, E. Hoppe tried to refute, then to adapt Neugebauer's hypothesis: in this view, the Sumerians would have seen that base 30 provided for most of their needs, but chose the higher base of 60 because it was also divisible by 4. He subsequently proposed another explanation, based on geometry: the sexagesimal system, he argued, must have been in some relationship to the division of the circle into six equal parts instead of into four right angles, which made the equilateral triangle, instead of the square, the fundamental figure of Sumerian geometry. If the angle of an equilateral triangle is divided into 10 "degrees", in a decimal numbering system, then the circle would have 60 degrees, thus giving the origin of base 60 for the developed numbering system.

However, as was pointed out by the Assyriologist G. Kewitsch (1904), neither astronomy nor geometry can actually explain the origin of a number-system. Hoppe's and Neugebauer's speculations are far too theoretical, presupposing as they do that abstract considerations preceded concrete applications. They require us to believe that geometry and astronomy existed as fully-developed sciences before any of their practical applications. The historical record tells a very different story!

I once knew a professor of mathematics who likewise tried to persuade his students that abstract geometry was historically prior to its practical applications, and that the pyramids and buildings of ancient Egypt "proved" that their architects were highly sophisticated mathematicians. But the first gardener in history to lay out a perfect ellipse with three stakes and a length of string certainly held no degree in the theory of cones! Nor did Egyptian architects have anything more than simple devices – "tricks", "knacks" and methods of an entirely empirical kind, no doubt discovered by trial and error – for laying out their ground plans. They knew, for example, that if you took three pieces of string measuring respectively three, four, and five units in length, tied them together, and drove stakes into the ground at the knotted points, you got a perfect right angle. This "trick" *demonstrates* Pythagoras's theorem (that in a right-angled triangle the square on the hypotenuse equals the sum of the squares on the other two sides) with a particular instance in whole numbers $((3 \times 3) + (4 \times 4) = 5 \times 5)$, but it does not presuppose *knowledge* of the abstract formulation, which the Egyptians most certainly did not have.

All the same, the Sumerians' mysterious base 60 has survived to the present day in measurements of time, arcs, and angles. Whatever its origins, its survival may well be due to the specific arithmetical, geometrical and astronomical properties of the number.

## KEWITSCH'S HYPOTHESIS

Kewitsch speculated in 1904 that the sexagesimal system of the Sumerians resulted from the fusion of two civilisations, one of which used a decimal number-system, and the other base 6, deriving from a special form of finger-counting. This is not easily acceptable as an explanation, since there is no historical record of a base 6 numbering system anywhere in the world [F. Thureau-Dangin (1929)].

## BASE 12

On the other hand, *duodecimal systems* (counting to base 12) are widely attested, not least in Western Europe. We still use it for counting eggs and oysters, we have the words *dozen* and *gross* (= 12 × 12), and measurements of length and weight based on 12 were current in France prior the Revolution of 1789, in Britain until only a few years ago, and still are in the United States.

The Romans had a unit of weight, money, and arithmetic called the *as*, divided into 12 *ounces*. Similarly, one of the monetary units of pre-Revolutionary France was the *sol*, divided into 12 *deniers*. In the so-called Imperial system of weights and measures, in use in continental Europe prior to the introduction of the metric system (see above, pp. 42–3), length is measured in *feet* divided into 12 *inches* (and each inch into 12 *lines* and each line into 12 *points*, in the obsolete French version).

The Sumerians, Assyrians, and Babylonians used base 12 and its multiples and divisors very widely indeed in their measurements, as the following table shows:

| | | | |
|---|---|---|---|
| LENGTH | | | |
| 1 *ninda* | | 12 cubits | |
| 1 *ninni* | "perch" | 10 × 12 ells | |
| 1 *šu* | | 2/12ths of a cubit | |
| WEIGHT | | | |
| 1 *gín* | "shekel" | 3 × 12 *šu* | (8.416 grams) |
| AREA | | | |
| 1 *bùr* | | 150 × 12 *sar* | |
| 1 *sar* | | 12 × 12 square cubits | (35.29 centiares) |
| VOLUME | | | |
| 1 *gur* | | 25 × 12 *sìla* | |
| 1 *pi* | | 3 × 12 *sìla* | |
| 1 *baneš* | | 3 × 6 *sìla* | |
| 1 *bán* | | 6 *sìla* | |
| 1 *sìla* | | | (842 ml) |

The Mesopotamian day was also divided into twelve equal parts (called *danna*), and they divided the circle, the ecliptic, and the zodiac into twelve equal sectors of 30 degrees.

Moreover, there is clear evidence on tablets from the ancient city of Uruk [see Green & Nissen (1985); Damerov & Englund (1985)] of several different Sumerian numerical notations, which must have been used concurrently with the classical system (see Fig. 8.9, recapitulated in Fig. 9.1 below), amongst which there are the measures of length shown in Fig. 9.2.

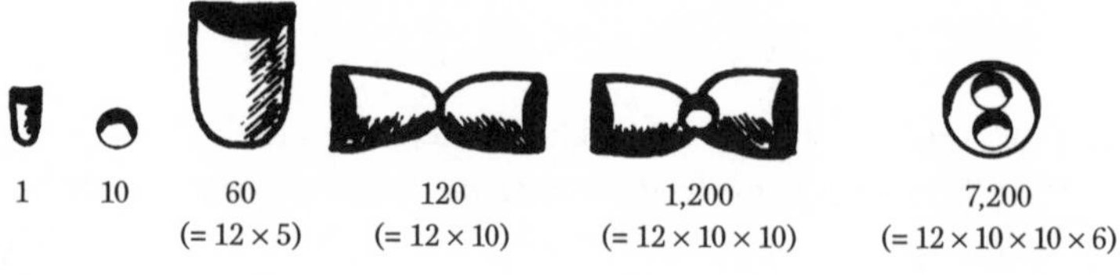


FIG. 9.1.

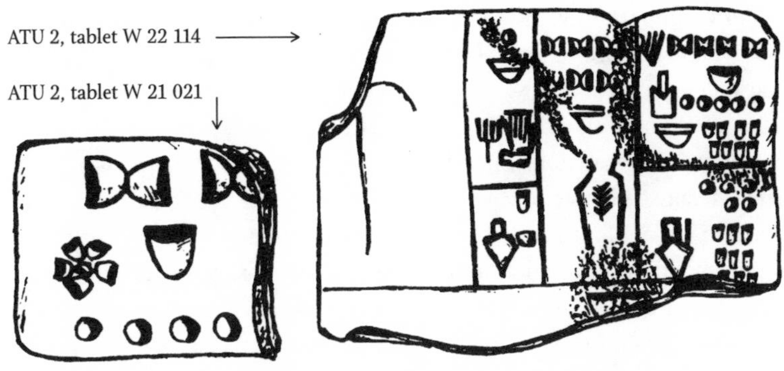


FIG. 9.2. *Archaic Sumerian tablets from Uruk, showing a numerical notation that is different from the standard one. (Numerous tablets of this kind prove that the Sumerians had several parallel systems). Date: c. 3000 BCE. Baghdad, Iraqi Museum. Source: Damerov & Englund (1985)*

To sum up, base 12 could well have played a major role in shaping the Sumerian number-system.

## AN ATTRACTIVE HYPOTHESIS

The major role given to base 10 in Sumerian arithmetic is similarly well-attested: as we saw in Chapter 8, it was used as an auxiliary unit to circumvent the main difficulty of the sexagesimal system, which in theory requires sixty different number-names or signs to be memorised. This is all the more interesting because the Sumerian word for "ten", pronounced *u*, means "fingers", strongly suggesting that we have a trace of an earlier finger-counting system of numerals.

This makes it possible to go back to Kewitsch's hypothesis and to give it a different cast: to suppose that the choice of base 60 was a learned solution to the union between two peoples, one of which possessed a decimal system and the other a system using base 12. For 60 is the lowest common multiple of 10 and 12, as well as being the lowest number of which all the first six integers are divisors.

Our hypothesis is therefore this: that Sumerian society had to begin with both decimal and duodecimal number-systems; and that its mathematicians, who reached a fairly advanced degree of sophistication (as we can see from the record of their achievements), subsequently devised a learned system that combined the two bases according to the principle of the LCM (lowest common multiple), producing a sexagesimal base, which had the added advantage of convenience for numerous types of calculation.

This is a very attractive and quite plausible hypothesis: but it fails as a historical explanation of origins on the obvious grounds that it presupposes too much intellectual sophistication. For we must not forget that most historically and ethnographically attested base numbers arose for reasons quite independent of arithmetical convenience, and that they were chosen very often without reference to a structure or even to the concepts of abstract numbers.

## ARE THERE MYSTICAL REASONS FOR BASE 60?

Sacred numbers played a major role in Mesopotamian civilisations; Sumerian mathematics developed in the context of number-mysticism; and so it is tempting to see some kind of religious or mystical basis for the sexagesimal system.

> Sumerian mathematics, like astrology, cannot be disentangled from numerology, with which it has reciprocal relations. From the dawn of the third millennium BCE, the number 50 was attributed to the temple of Lagaš, son of the earth-god, and this shows that from the earliest times numbers had "speculative" meanings. The Akkadians brought number-symbolism into Babylonian thought, making it an essential element of the Name, the Individual and the Work. Alongside their scientific or intellectual functions, numbers became part of the way the Mesopotamians conceived the structure of the world. For example, the numeral *šar* or *šaros* (= 3,600) is written in cuneiform as a sign which is clearly a deformation of the *circle* [see Fig. 8.9], and it also means "everything", "totality", "cosmos". In Sumerian cosmogony, two primordial entities, the "Upper Totality" or *An-Šar* and the "Lower Totality" or *Ki-Šar* came together to give birth to the first gods. Moreover, the full circle of 360° is divided into degrees, whose basic unit of 1/360 is called *Geš* – and the symbol for *Geš* is precisely what is used to signify "man" and thus for elaborating the names of masculine properties. The higher base unit or *sosse* (= 60) is also pronounced *Geš* [see Fig. 8.5], and its sign (with an added asterisk or star) is the figure of the "Upper God", or heaven, whose name is pronounced *An(u)*, by virtue of the ideogram that defines it as a divinity and as heaven. So the celestial god, 60, is the father of the earth-god, 50; the god of the Abyss is 40, two thirds of 60. The moon-god is 30 (it has been suggested, without any evidence, that the moon-god has this number in virtue of the number of days in the lunar cycle); and the sun-god has the number 20, which is also the determining number of "king" . . . [Adapted from M. Rutten (1970)]

It seems plausible, in this context, to think that base 60 commended itself to the mystic minds of Sumerians because of their cult of the "Upper God" *Anu*, whose number it was.

There are many attested examples, in Australia, Africa, the Americas, and Asia, of number-systems with a base (most often, base 4) that has mystical ramifications. However, the Sumerian system is much more developed than any of these, and presupposes complete familiarity with abstract number-concepts. For this reason it does not seems right to consider Sumerian mysticism as the origin of the Sumerian base 60. Things should rather be looked on the other way round: it is far more probable that 60 was the "number" of the Upper God Anu precisely because it was *already* the larger of the units of Sumerian arithmetic.

## THE PROBABLE ORIGIN OF THE SEXAGESIMAL SYSTEM

So where does base 60 come from? Here is what I believe to be the solution to this enigma.

It is necessary to suppose (without a great deal of material evidence) that the Mesopotamian basin had one or more *indigenous populations* prior to Sumerian domination. A second essential premise (but one that is not at all controversial) is that the Sumerians were immigrants, that they *came from somewhere else,* more than probably in the fourth millennium BCE. Though we know very little about the indigenous population, and almost nothing about the prior cultural connections of the Sumerians, who seem to have broken all ties with their previous environment, we can speculate with a fair degree of confidence that these two cultures possessed different counting systems, one of which was duodecimal, and the other quinary.

Let us look again at Sumerian number-names.

| 1 | 2 | 3 | 4 | 5 | 6 | 7 | 8 | 9 | 10 |
|---|---|---|---|---|---|---|---|---|---|
| *geš* | *min* | *eš* | *limmu* | *iá* | *àš* | *imin* | *ussu* | *ilimmu* | *u* |

*Geš* (1) is a word that also means "man", "male" and "erect phallus"; *min* (2) also means "woman"; and *eš* (3) is also the plural suffix in Sumerian (rather like *-s* in English). The symbolism of these number-names is both apparent and very ancient indeed, taking us back to "primitive" perceptions of man as vertical (in distinction to all other animals) and alone, of woman as the "complement" of a pair (man and woman, or woman and child), and of "the many" beginning at three. (In Pharaonic Egypt as in the Hittite Empire, plurals were indicated by writing the same hieroglyph three times over, or by adding three vertical bars after the sign; in classical Chinese, the ideogram for "forest" consists of three ideograms for "tree", whereas the concept "crowd" was represented by a triple repetition of the ideogram for "man".) So the semantic meanings of the names of the first three numbers of Sumerian is a trace of those lost ages when people had only the most rudimentary concepts of number, counting only "one, two, and many".

More importantly, however, the names of the numbers in spoken Sumerian also carry unmistakable traces of a quinary system. *Às,* six, looks like an elision of *iá* and *geš,* "five (and) one"; *imin,* seven, is more certainly a contraction of *iá* and *min,* "five and two"; *ilimmu* is clearly related to *iá* and *limmu,* "five and four". In other words, Sumerian number-names derive from a vanished system using base 5. We speculate therefore that one of the two populations involved had a quinary counting system, and that in contact with a civilisation using base 12, the sexagesimal system was invented or chosen, since $5 \times 12 = 60$.

As we have already seen, the quinary base is anthropomorphic and derives from learning to count on the fingers of one hand and using the other hand as a "marker" when counting beyond 5. However, the origin of base 12 is far less obvious. My own view is that it was probably also based on the human hand.

Each finger has three articulations (or phalanxes): and if you leave out the thumb (as you have to, since you use it to check off the phalanxes counted), you can get to 12 using only the fingers of one hand, as in Fig. 9.3 below:

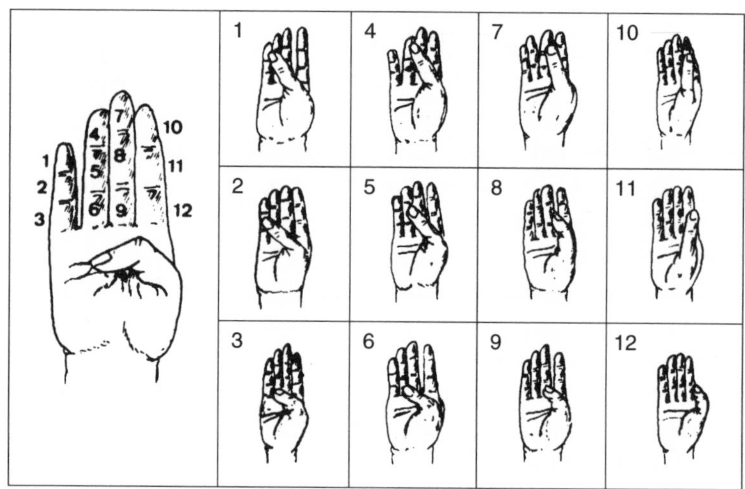


FIG. 9.3.

Repeating the device once over, you get from 13 to 24, then from 25 to 36, and so on. In other words, with a finger-counting device of this kind, base 12 seems the most natural for a numbering system.

This hypothesis is difficult to prove, but phalanx-counting of this type does exist and is in use today in Egypt, Syria, Iraq, Iran, Afghanistan, Pakistan, and some parts of India. Sumerians could therefore easily have used it at the dawn of their civilisation.

How then can we explain the fact that *u*, the Sumerian word for "10", means "fingers", and that there is no trace of a duodecimal system in spoken numbers, and no special word for the dozen ("12" is *u-min*, meaning "ten-two")?

My view is that spoken Sumerian numbers carry no trace of either base 12 or base 10: in other words, the name of the number 10 is not evidence of a lost decimal number-system, but merely the metaphoric expression of a universal human perception of human anatomy, the fact that there are ten fingers in all on the two hands.

At all events, my hypothesis has the advantage over all other speculations of giving a concrete explanation for the mysterious origin of base 60. As we saw in Chapter 3, basic finger-counting techniques, supplemented by mental effort (which quickly becomes quite "natural" once the principle of the base has been grasped), has often opened the way to arithmetical elaborations far superior to the original rudimentary system involved. From this, we can assert that the origin of base 60 could well have been connected to the finger-counting scheme shown in Fig. 9.4, currently in use across a broad band stretching from the Middle East to Indo-China.

This particular device makes 60 the principal base, with 5 and 12 serving as auxiliaries. This is how it is done:

Using your right hand, you count from 1 through 12 by pressing the tip of your thumb onto each of the three phalanxes (articulations) of the four opposing fingers. When you reach a dozen on the right hand, you check it off by folding the little finger of your left hand. You return to the right hand and count from 13 through 24 in similar fashion, then fold down the ring finger of your left hand, then count from 25 through 36 again on the right hand. The middle finger of the left hand is folded down to mark off 36, and you proceed to count from 37 to 48 on the right, then folding down the left index finger. Repeating the operation once more, you get to 60, and fold down the last remaining finger of the left hand (the thumb). As you can't count any higher numbers with this system, 60 is the obvious base.

LEFT HAND

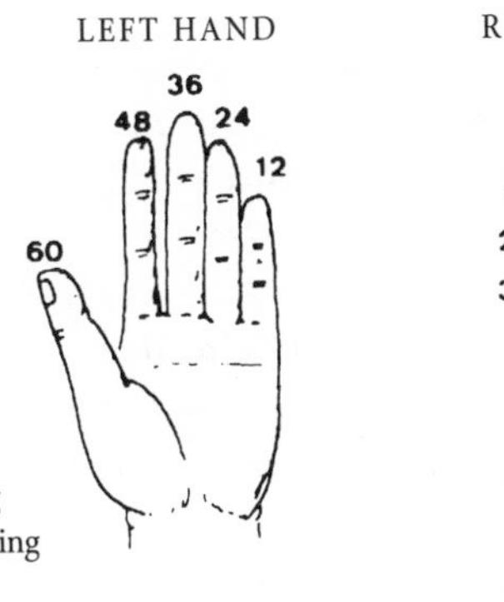


Finger-counting with each standing for a dozen

RIGHT HAND

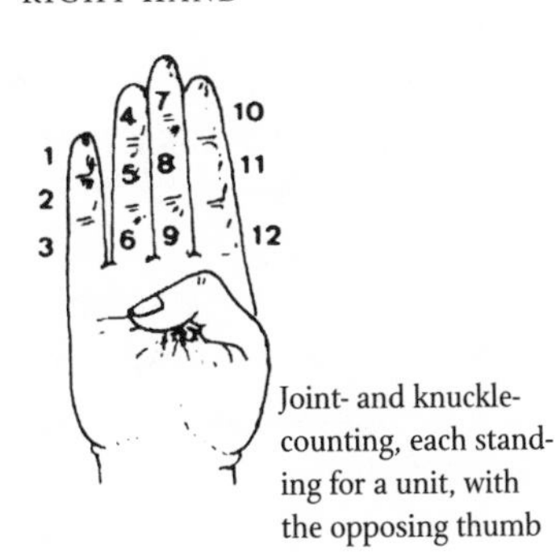


Joint- and knuckle-counting, each standing for a unit, with the opposing thumb

FIG. 9.4.

My hypothesis can therefore be told as a story. As a result of the symbiosis of two different cultures, one of which used a quinary finger-counting method, and the other a duodecimal base deriving from a system of counting the phalanxes with the opposing thumb, 60 was chosen as the new higher unit of counting as it represented the combination of the two prior bases.

Since 60 was a pretty large number to use as a base, arithmeticians looked for an intermediate number to use, so as to mitigate the difficulties that arise from people's limited capacity to memorise number-names. Base 5 was too small compared to 60 – it would have required very long number-strings to express intermediate numbers; so 10 was chosen, a number provided by nature, so to speak, and of an ideal magnitude for the task in hand. Why not base 12? It has many advantages over 10, but it would probably have disoriented those accustomed to the quinary base, for whom 10, being twice the number of fingers on one hand, must have seemed more natural.

Since 6 is the coefficient required to turn 10 into 60, the Sumerian system, by its own dynamic, or rather, because of the inherent properties of the numbers involved, became a kind of compromise between 6 and 10, which served as the alternate and auxiliary bases of the sexagesimal system. Only subsequently could it have been observed that the resulting base had very valuable arithmetical properties as well as advantages for astronomy and geometry, which could only have been discovered as mastery of the counting tool and of the applied sciences progressed. Those properties and advantages came to seem so considerable and numerous that the Sumerians gave the main units the names of their own gods.

That is, in my view, the most plausible explanation of base 60. All the same, it should only be taken as a story – a story for which no archaeological proof or even evidence exists, as far as I know. However, if it were the true story of the origin of the sexagesimal system, then it would give added support to the anthropomorphic origin of the other common and historically-attested bases (5, 10 and 20), and thus underline the huge importance of human fingers in the history of numbers and counting.

## CHAPTER 10

# THE DEVELOPMENT OF WRITTEN NUMERALS IN ELAM AND MESOPOTAMIA

As we have seen, by the fourth millennium BCE clay was already a traditional material, not only for building work, but also and above all as the *basic medium for the expression of human thought*. In this period, the Mesopotamian peoples were entirely at ease with clay in a wide range of applications, and they used it to throw earthenware and ceramic vessels and figurines, to mix mortar, to mould bricks, and to shape seals, beads, jewels, and so on. It is therefore not unreasonable to suppose that the inhabitants of Sumer, long before they devised their written numerals and writing system, made diverse kinds of clay or earthenware objects or tokens with conventional values in order to symbolise and to manipulate numbers.

### FROM PEBBLES TO ARITHMETIC

Concrete arithmetic (which as we shall see most certainly existed in the region of Mesopotamia) necessarily derived from the archaic "heap of pebbles" counting method, put to numerical use. The "pebbles" method is attested in every corner of the globe and clearly played a major role in the history of arithmetic – for pebbles first allowed people to learn how to perform arithmetical operations.

The English word *calculus*, like the French *calcul* (which has a more general meaning of "arithmetic", "counting operation", or "calculation") comes directly from the Latin *calculus*, which means . . . a pebble, and, by extension, a ball, token, or counter. The Latin word is related to *calx, calcis*, which, like the similar-sounding Greek word *khaliks*, means "rock" or "limestone", and which has numerous etymological derivations in modern European languages, from German *Kalkstein*, "limestone", to English *calcium* and French *calcul*, in the sense of "kidney stones".

Because the Greeks and Romans taught their children to count and to perform arithmetical operations with the help of pebbles, balls, tokens, and counters made of stone (probably limestone, which is lighter and easier to fashion than marble or granite) their word for "doing pebbles" (*calculation*) has come to refer to all and any of the elementary arithmetical operations – addition, subtraction, multiplication, and division.

Greek and Arabic both have their own independent etymological proofs of the origins of arithmetic in the manipulation of stones. Greek *pséphos* means both "stone" and "number"; Arabic *ḥaswa*, meaning "pebble", has the same root as *iḥsā'*, which means "a count (of things)" and "statistics".

At its simplest, the pebble method is extremely primitive: even more than the basic forms of notched-stick counting described in Chapter 1, it represents the "absolute zero" of number-techniques. It can only supply cardinal numbers, requires no memorisation and no abstraction, and uses exclusively the principle of one-for-one correspondence.

However, once abstract counting has been mastered, the pebble method is sufficiently adaptable to allow great strides to be made. In some African villages, accounts of marriageable girls (and of boys of military age) were kept until quite recently by this method. On reaching puberty each village girl gave a metal bangle to the local matchmaker, who threaded it onto a strap alongside other similar bangles; when her marriage was imminent, the girl would take back her bangle, leaving the matchmaker with an accurate and immediate account of the number of "matches" left to make. It is a most convenient way of performing a subtraction in the absence of any knowledge of arithmetic as such.

In Abyssinia (now Ethiopia) tribal warriors had a similar device. On leaving for a foray each warrior placed a pebble on a heap, and on return to the village, each survivor removed his pebble. The number of unclaimed pebbles provided the precise total of losses in the skirmish. Exactly the same device is portrayed in the opening sequence of Eisenstein's film, *Ivan The Terrible* (part I): we see each soldier in the army of Ivan IV Vassilievich, Tsar of all the Russias, placing a metal token on a tray before setting off for the siege of Kazan.

In the course of time it became apparent that the device could not be taken very far, and did not satisfy many perfectly ordinary requirements. For instance, you need to collect a thousand pebbles just to count up to 1,000! But once the principle of a base in a numbering system had arisen, pebbles could be used more imaginatively.

In some cultures, the idea arose of replacing natural pebbles with pieces of stone of various sizes and attributing a conventional value of a different order of magnitude to each size. So, in a decimal numbering system, a unit of the first order might be represented by a small pebble, a unit of the second order (tens) by a slightly larger pebble, a unit of the third order (hundreds) by a larger stone, a unit of the fourth order (thousands) by an even bigger one, and so on. To represent the other numbers in the series by this method, you just needed an appropriate number of pebbles of the appropriate size.

It was a practical device, but not yet quite serviceable, because it was hard to find a sufficiency of pebbles or stones of identical sizes and shapes.

For some societies there was an additional and quite crucial obstacle if they inhabited regions where stone was uncommon.

The pebble method was therefore perfected by recourse to malleable earth, a material far better suited to making regular counting tokens. That is what happened in Elam and Mesopotamia in prehistoric times, prior to the invention of writing and of written numerals.

## MESOLITHIC AND NEOLITHIC TOKENS IN THE MIDDLE EAST

In several archaeological sites in the Middle East, in places as far from each other as Anatolia, the Indus Valley, the shores of the Caspian Sea, and the Sudan (Fig. 10.2), researchers have unearthed thousands upon thousands of small objects in a wide variety of sizes and regular geometrical shapes, such as cones, discs, spheres, pellets, sticks, tetrahedrons, cylinders, and so on (Fig. 10.1). These are the objects to which we shall apply the generic term of *calculi*.

Some of these *calculi* are inscribed with parallel lines, crosses, and other similar patterns (Fig. 10.1 B, C, D, E, M, O). Others are decorated with carved or moulded figurines that are visible representations of different kinds of beings or things (jars, cattle, dogs, etc.). Finally, there are some that have neither pattern nor figurine (Fig. 10.1 A, F, G, H, I, J, L, N, P, Q, R, S, T, U).

The oldest *calculi* found so far, dating from the ninth to the seventh millennium BCE, come from Beldibi (Anatolia), Tepe Asiab (Mesopotamia), Ganj Dareh Tepe (Iran), Khartoum (Sudan), Jericho (West Bank)) and Abu Hureyra (Syria). The most recent, dating from the second millennium BCE, were found at Tepe Hissar (Iran), Megiddo (Israel), and Nuzi (Mesopotamia).

Most of the *calculi* were found scattered around at ground level. Others, however, were found inside or next to egg-shaped or spherical hollow clay balls or *bullae*. However, although *bullae* are not found prior to the fourth millennium BCE, hundreds of them have been unearthed at Tepe Yahya (Iran), Habuba Kabira (Syria), Uruk (Mesopotamia), Susa (Iran), Chogha Miš (Iran), Nineveh (Mesopotamia), Tall-i-Malyan (Iran) and Nuzi (Mesopotamia).

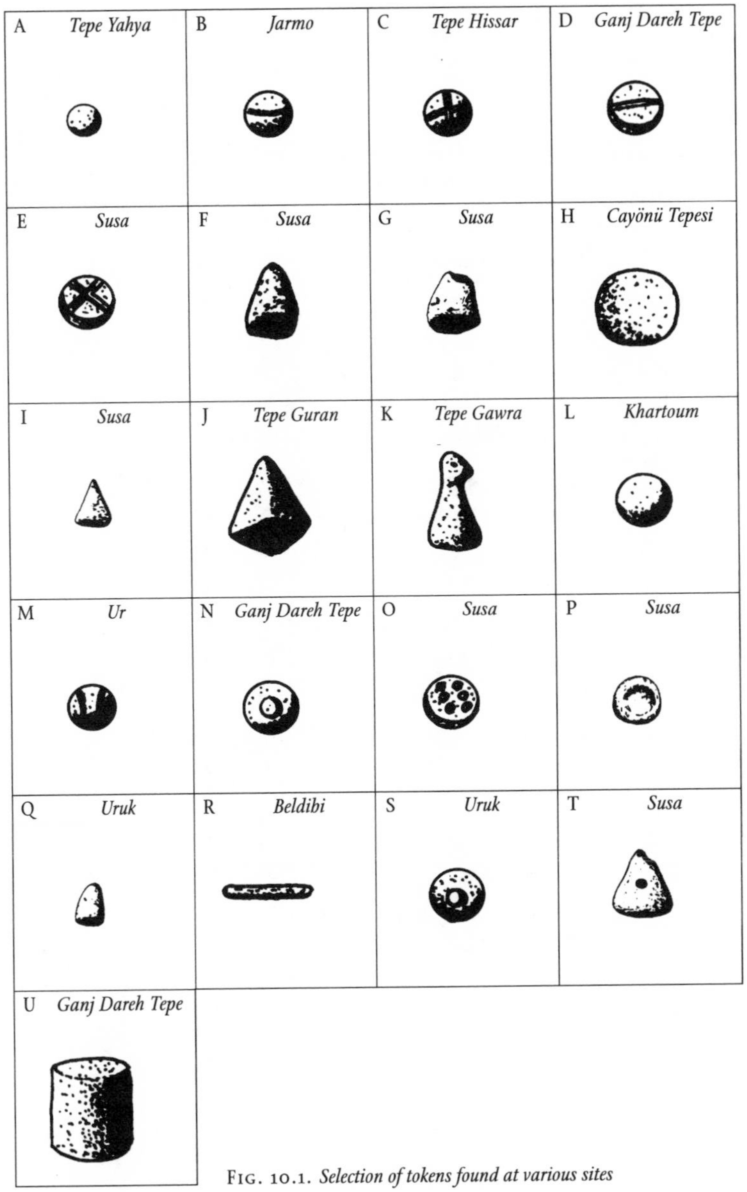

| | | | |
|---|---|---|---|
| A *Tepe Yahya* | B *Jarmo* | C *Tepe Hissar* | D *Ganj Dareh Tepe* |
| E *Susa* | F *Susa* | G *Susa* | H *Cayönü Tepesi* |
| I *Susa* | J *Tepe Guran* | K *Tepe Gawra* | L *Khartoum* |
| M *Ur* | N *Ganj Dareh Tepe* | O *Susa* | P *Susa* |
| Q *Uruk* | R *Beldibi* | S *Uruk* | T *Susa* |
| U *Ganj Dareh Tepe* | | | |

FIG. 10.1. *Selection of tokens found at various sites*

## WHAT DO THE COUNTING TOKENS MEAN?

Denise Schmandt-Besserat has put together all that is known about these tokens and argues that for the Middle Eastern civilisations of the ninth to the second millennium BCE they constituted three-dimensional *pictograms* of the specific goods or produce which they served to account for in commercial exchanges.

In other words, she believes that the tokens were shaped so as to represent or symbolise the very things that they "counted", using actual or schematic images of, for instance, pots, heads of cattle, and so on, and in some cases were marked with dots or lines to indicate their places in a numbering series (one rectangular plaque has 2 × 5 dots on it, for example, and one cow's head figurine has 2 × 3 dots marked on it).

| MILLENNIUM BCE | SITE | REGION | TYPE OF TOKEN: cylinders | discs | spheres and pellets | cones (various sizes) | sticks | hollow clay balls or bullae |
|---|---|---|---|---|---|---|---|---|
| 9th | Beldibi | Anatolia | ★ | ★ | ★ | ★ | ★ | |
| | Tepe Asiab | Mesopotamia | ★ | ★ | ★ | ★ | ★ | |
| 9th–8th | Ganj Dareh Tepe | Iran | ★ | ★ | ★ | ★ | ★ | |
| 8th | Khartoum | Sudan | ★ | ★ | ★ | | | |
| 8th–7th | Cayönü Tepesi | Anatolia | ★ | ★ | ★ | ★ | ★ | |
| 7th–6th | Jericho | West Bank | | | ★ | ★ | | |
| | Tell Ramad | Syria | ★ | ★ | ★ | ★ | | |
| | Ghoraife | Syria | | ★ | ★ | ★ | | |
| | Suberde | Anatolia | ★ | | ★ | ★ | ★ | |
| | Jarmo | Mesopotamia | ★ | ★ | ★ | ★ | ★ | |
| | Tepe Guran | Iran | | ★ | ★ | ★ | | |
| | Anau | Iran | | ★ | ★ | ★ | | |
| 6th | Tell As Sawwan | Mesopotamia | | | ★ | ★ | | |
| | Can Hasan | Anatolia | | ★ | ★ | ★ | | |
| | Tell Arpichiya | Mesopotamia | | | ★ | ★ | | |
| 6th–5th | Chaga Sefid | Iran | ★ | ★ | ★ | ★ | ★ | |
| | Tal-i-Iblis | Iran | | ★ | ★ | ★ | | |
| 4th | Tepe Yahya | Iran | | ★ | ★ | ★ | | ★ |
| | Habuba Kabira | Iran | | ★ | ★ | ★ | | ★ |
| | Warka (Uruk) | Mesopotamia | ★ | ★ | ★ | ★ | | ★ |
| | Susa | Iran | ★ | ★ | ★ | ★ | ★ | ★ |
| | Chogha Miš | Iran | ★ | ★ | ★ | ★ | | ★ |
| | Nineveh | Mesopotamia | | | | | | ★ |
| 4th–3rd | Tall-i-Malyan | Iran | ★ | | ★ | ★ | | ★ |
| | Tepe Gawra | Iran | | | | ★ | | |
| 3rd | Jemdet Nasr | Mesopotamia | | ★ | | ★ | | |
| | Kiš | Mesopotamia | | ★ | ★ | ★ | | |
| | Tello | Mesopotamia | ★ | ★ | ★ | ★ | | |
| | Fara | Mesopotamia | | ★ | ★ | ★ | | |
| 3rd–2nd | Tepe Hissar | Iran | | ★ | ★ | ★ | | |
| 2nd | Megiddo | Israel | ★ | | | | | |
| | Nuzi | Mesopotamia | | | | | | ★ |

FIG. 10.2. *Middle Eastern archaeological sites with finds of clay objects of various shapes and sizes, some of which are known to have been used for arithmetical operations and for accounts*

It is an attractive idea, and if it could be proved it would show that there was a very sophisticated accounting system in the Middle East in the earliest periods of the prehistoric record in that area.

However, it is only a hypothesis, and there is no solid evidence to support it. It presupposes the existence of a sufficiently complex market economy to have created the need felt for such an elaborate counting system. Schmandt-Besserat nonetheless takes her argument even further, and claims that this "three-dimensional symbolic system" is the origin of Sumerian pictograms and ideograms, that is to say is the source of the earliest writing system in the world.

Her conclusions derive from the discovery of a very large number of objects of various shapes (discs, spheres, cones, cylinders, and triangles) inscribed with exactly the same motifs – parallel lines, concentric circles, crossed lines – as are found on Sumerian tablets of the Uruk period, where a cross inside a circle stands for "sheep", three parallel lines inside a circle stands for "clothes", and so on (Fig. 10.3). The signs of Sumerian writing, she says, are simply two-dimensional reproductions of the three-dimensional tokens.

This important claim is nonetheless somewhat specious because it presupposes that there was a completely common and standardised set of traditions and conventions over a huge geographical area throughout a period of several thousand years – and what we know of the area and period suggests on the contrary that its cultures were very diverse. It is quite wrong to "explain" Sumerian pictograms by the shape of tokens found as far afield as Beldibi, Jericho, Khartoum, or Tepe Asiab, dating from eras as varied as the fourth, sixth, and ninth millennia BCE, since the cultures of these places in those periods probably had nothing whatsoever to do with developments in Sumer itself. (S. J. Liebermann gives a full critique.)

However, Schmandt-Besserat's general idea is not unacceptable, provided it is handled more methodically, by studying, not the entire collection of tokens in existence, but each subset of them in the context of its particular culture, in its specific location, at a particular period.

There must be major reservations about the overall conclusion concerning the origins of Sumerian writing. If it is ever demonstrated satisfactorily that there was a proper "system" of three-dimensional representation in these early Middle Eastern cultures, we will certainly find not one, but many different "systems" in the area. If we do establish a derivation from one such system to Sumerian writing, we are unlikely to establish it for more than a small number of individual signs.

All the same, these three-dimensional tokens must have meant *something* for their inventors and users, even if they did not form part of a system

| Token shape | Sumerian sign | Known meaning of the sign |
|---|---|---|
| | | jar, pot, vase |
| | | oil, grease, fat |
| | | sheep |
| | | bread, food |
| | | leather |
| | | clothing |

FIG. 10.3. *Comparison of tokens and their allegedly corresponding pictograms in early Sumerian writing (from D. Schmandt-Besserat, 1977)*

in the proper sense. They are obviously connected to ancient practices of symbolisation, which we can see in use on painted ceramics and in glyptics. One conclusion that might well come out of these speculations if sufficient evidence is found is that these tokens represent perhaps the final intermediate stage in the evolution of purely symbolic expressions of thought into formal notations of articulated language.

## MULTIFUNCTION OBJECTS

The variety of the tokens is so great, their geographical locations so diverse, and their chronological origins are so widely separated that they could not possibly have belonged to a single system.

Even within a single period and place, they did not all serve the same purposes.

We can all the same make a few plausible guesses about their meaning if we bear in mind the specific nature of the cultures to which they belong. Some of the tokens, for example, that have holes in the middle and were found threaded on string, were probably objects of personal decoration. Such "necklaces" may also have served as counting beads, much like rosaries, allowing priests to count out gods or prayers. Other tokens decorated with the heads of animals may have been amulets, invoking the spirits of the animals represented, in terms of superstitions about the protective values of the different species (warding off the evil eye, illnesses, accidents, etc.). And since clay was plentiful and easy to shape, we can suppose that a fair number of these tokens were *playing pieces*, for ancient games like fives, draughts, chess, and so on.

## FROM TOKENS TO *CALCULI*

However, the most interesting tokens from our point of view, and whose function is not in any doubt, are those small clay objects of varying shapes and sizes found inside the hollow clay balls called *bullae*. They were in use in Sumer and Elam (a region contiguous to Mesopotamia, covering the western part of the Iranian plateau and the plain to the east of Mesopotamia proper) from the second half of the fourth millennium BCE, and they served both as concrete accounting tools and also, as we shall see, as *calculi* which permitted the performance of the various arithmetical operations of addition, subtraction, multiplication, and even division. The Assyrians and the Babylonians called these counting tokens *abnū* (plural: *abnāti*), a word used to mean: 1. *stone*, 2. *stone object*, 3. *stone* (of a fruit), 4. *hailstone*, 5. *coin* [from R. Labat (1976) item 229]. Long before that, the Sumerians had called them *imna*, meaning "clay stone" (S. J. Liebermann in *AJA*). We will call them *calculi*, remembering that by this term we refer exclusively to the tokens found inside or close to hollow clay balls, the *bullae*.

## THE FORMAL ORIGINS OF SUMERIAN NUMERALS

Archaic Sumerian numerals suggest very strongly at first glance that they derive from a pre-existing concrete number- and counting system, but they also seem to have obviously formal origins. The various symbols used (Fig. 8.2, repeated in Fig. 10.4 below) look very much like some of the *calculi* "copied" onto clay tablets once writing had been invented: specifically, the little cone, the pellet, the large cone, the perforated large cone, the sphere, and the perforated sphere. To put things the other way round (Fig. 10.4):

- the fine line representing the unit in archaic Sumerian numerals looks a two-dimensional representation of the small cone token;
- the small circular imprint representing the tens looks like a pellet-shaped token;
- the thick indentation for 60 looks like a large cone;
- the thick dotted indentation for 600 looks like a large perforated cone;
- the large circular imprint (3,600) looks like a sphere;
- the large dotted circular imprint (36,000) looks like a perforated sphere.

These resemblances are so obvious that the relationship would have to be accepted even if there were no other proof. But as we shall see, the archaeological record contains more than adequate confirmation of these identifications.

| | SPOKEN NUMERALS | CALCULI | | WRITTEN NUMERALS | | |
|---|---|---|---|---|---|---|
| | Number-names | | | Archaic | Cuneiform | Mathematical structure |
| 1 | *geš* | | small cone | | | 1 |
| 10 | *u* | | pellet | | | 10 |
| 60 | *geš* | | large cone | | | 10.6 (= 60) |
| 600 | *gešu* | | perforated large cone | | | 10.6.10 (= 60.10) |
| 3,600 | *šàr* | | sphere | | | 10.6.10.6 (= $60^2$) |
| 36,000 | *šàr-u* | | perforated sphere | | | 10.6.10.6.10 (= $60^2$.10) |
| 216,000 | *šàrgal* | ? | | ? | | 10.6.10.6.10.6 (= $60^3$) |
| Archaeological date (BCE) | | From mid-4th millennium | | From c. 3200 | From c. 2650 | |

FIG. 10.4. *Number-names, numerals and* calculi *of Sumerian civilisation. The* calculi *come from several Mesopotamian sites (Uruk, Nineveh, Jemdet Nasr, Kiš, Ur, Tello, Šurrupak, etc.*

## THE HOLLOW CLAY BALLS FROM THE PALACE OF NUZI

It was in 1928–29 that Mesopotamian *calculi* were first properly identified, when the American archaeologists from the Oriental Research Institute in Baghdad excavating the Palace of Nuzi (a second-millennium BCE site near Kirkuk, in Iraq) came across a hollow clay ball clearly containing "something else", inscribed with cuneiform writing in Akkadian (Fig. 10.5) which in translation reads as follows:

*Abnāti* ("things") about sheep and goats:

| | | | |
|---|---|---|---|
| 21 | ewes which have lambed | 6 | she-goats that have had kids |
| 6 | female lambs | 1 | he-goat |
| 8 | adult rams | [2] | kids |
| 4 | male lambs | | |

The sum of the count is 48 animals. When the clay ball was opened, it was found to contain precisely 48 small, pellet-shaped, unbaked clay objects (which were subsequently mislaid). It seemed logical to assume that these tokens had previously been used to count out the livestock, despite the difficulty of distinguishing between the different categories by this system of reckoning.

FIG. 10.5. *Hollow clay ball or* bulla *found at the Palace of Nuzi, 48mm × 62mm x 50mm. Fifteenth century BCE. From the Harvard Semitic Museum, Cambridge, MA (inventory no. SMN 1854)*

The archaeologists might have thought nothing of their discovery without a chance occurrence that suddenly explained the original purpose of the find. One of the expedition porters had been sent to market to buy chickens, and by mistake he let them loose in the yard before they had been counted. Since he was uneducated and did not know how to count, the porter could not say how many chickens he had bought, and it would have been impossible to know how much to pay him for his purchases had he not come up with a bunch of pebbles, which he had set aside, he said, "one for each chicken". So an uneducated local hand had, without knowing it, repeated the very same procedures that herdsman had used at the same site over 3,500 years before.

Thirty years later, A. L. Oppenheim at the University of Chicago carried out a detailed study of all the archaeological finds at Nuzi, and discovered that the Palace kept a double system of accounting. The cuneiform tablets of the Palace revealed the existence of various objects called *abnu* ("stones") that were used to make calculations and to keep a record of the results. The texts written on the tablets make clear reference to the "deposit" of *abnu*, to "transfers" of the same, and to "withdrawals". The meticulous cuneiform accounts made by the Palace scribes were "doubled", as Schmandt-Besserat explains, by a tangible or concrete system. One set of *calculi* may for instance have represented the palace livestock. In spring, the season of lambing, the appropriate number of new *calculi* would have been added; *calculi* representing dead animals would have been withdrawn; perhaps *calculi* were even moved from one shelf to another when animals were moved between flocks, or when flocks moved to new pasture, or when they were shorn.

The hollow clay ball was therefore probably made by a Palace accountant for recording how many head of livestock had been taken to pasture by local shepherds. The shepherds were illiterate, to be sure, but the accountant must have known how to count, read, and write: he was probably a priest, as he possessed the great privilege of Knowledge, and must have been one of the managers of the Nuzi Palace's goods and chattels. The proof of this lies in the Akkadian word *sangu*, which means both "priest" and "manager of the Temple's wealth"; it is written in cuneiform in exactly the same way as the verb *manû*, which means "to count".

When shepherds left for pasture, the functionary would make as many unbaked clay pellets as there were sheep, and then put them inside the clay "purse". Then he would seal the purse and mark on it, in cuneiform, an account of the size of the flock, which he then signed with his mark. When the shepherd came back the purse could be broken open and the flock checked off against the pellets inside. There could be no disputing the numbers, since the signed account on the outside certified the size of the flock as far as the masters of the Palace were concerned, and the *calculi* provided the shepherd with his own kind of certified account.

The later discovery of an oblong accounting tablet shaped like the base of the hollow clay ball in the ruins of the same palace, but from a higher (and therefore more recent) stratum, gave further support to Oppenheim's views.

The story now moves to Paris, where, at the Musée du Louvre, there are about sixty of these hollow clay balls brought back c. 1880 by the French Archaeological Mission to Iran, which had been excavating the city of Susa (about 300 km east of Sumer, in present-day southwestern Iran, Susa was the capital of Elam and then of the Persian Empire under Darius). Up until recently the only interest that had been shown in them concerned the imprints of cylinder-seals with which most of them are decorated (Fig. 10.10). Several of the *bullae* had been broken during shipment to Paris, other had been found broken. All the same, some of them were intact, and sounded like rattles when shaken. X-ray photography showed that they contained *calculi* – but not all of the same uniform type. When some of them were very carefully opened, they were found to contain clay discs, cones, pellets, and sticks (Fig. 10.6)

As P. Amiet then argued, these "documents", since they came from a site dated about 3300 BCE, proved that Elam had an accounting system far more elaborate than that of Nuzi with its plain "unit counters", and had it 2,000 years earlier. In other words, this counting system had survived for two millennia, but had regressed over that period, losing the use of a base, and retreating to a rudimentary and purely cardinal method.

It was therefore correctly assumed that the counting system of Susa consisted of giving tangible form to numbers by the means of various *calculi* which symbolised numerical values both by their own number and by their respective shapes and sizes, which corresponded to some order of magnitude within a given number-system (for example, a stick was a unit of the first order of magnitude, a pellet for a unit of the second order, a disc for a unit of the third order, and so on).

More recent finds in Iran (Tepe Yahya, Chogha Miš, Tall-i-Malyan, Šahdad, etc.), in Iraq (Uruk, Nineveh, Jemdet Nasr, Kiš, Tello, Fara, etc.), and in Syria (Habuba Kabira) have proved Oppenheim and Amiet to be correct. What they have also shown is that the system was not restricted to Elam, but that similar accounting methods were used throughout the neighbouring region, including Mesopotamia. These methods are thus even more ancient than the one used for the accounting tablets of the Uruk period.

## FROM CLAY BALLS TO ACCOUNTING TABLETS

It then seemed very likely that the archaic accounting tablets of Sumer were directly descended from the clay *calculi*-and-*bulla* accounting system. The archaic Sumerian figures obviously were related to the *calculi*; and, unlike the later, perfectly rectangular tablets that were made to a standard pattern, the archaic counting tablets are just crude oblong or roughly oval slabs (Fig. 8.1 C above). So there really had been a point in time when the stones were supplanted by their own images in two-dimensional form, and the hollow clay balls replaced by these flat clay slabs. But this remained only a conjecture in the absence of all the archaeological evidence needed to reconstitute the intermediate stages of the supposed development and of evidence to allow firm datings.

In the 1970s, the French Archaeological Delegation to Iran (DAFI), under the direction of Alain Le Brun, excavated the Acropolis of Susa, and established a far more accurate and substantiated stratigraphy of Elamite civilisation than had previously been possible, and, in 1977–78, important finds were made which make the transition comprehensible in archaeological terms. A word of warning, however: the development we describe below is attested only at Susa. Nonetheless there are good reasons for believing that much the same thing happened at Sumer.

FIG. 10.6. *Sketch of the contents of an unbroken* bulla, *as revealed by X-rays*

The first reason is that Elamite civilisation is pretty much contemporary with Sumer, and flourished in very similar fashion in precisely similar circumstances in the second half of the fourth millennium BCE. For that reason various aspects of Elamite civilisation are used as reference points (or as potentially applicable models) for the civilisation of Uruk. All the same the Elamites retained many features that are distinct from those of their Mesopotamian neighbours.

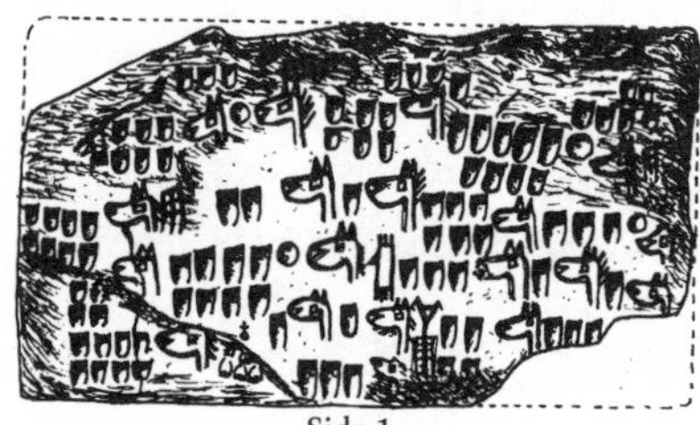
Side 1

Side 2

FIG. 10.7. *Proto-Elamite tablet (Susa, level unknown), c. 3000 BCE. J. Schell (1905) identified this as an inventory of stallions (erect manes), mares (flat manes), and colts (no manes), with the numbers of each indicated by various indentation-marks. Side 2 bears the imprint of a cylinder-seal representing standing and resting goats. Paris, Musée du Louvre, Sb 6310*

The second reason is that the Elamites, like the Sumerians, were fully conversant with the use of clay for expressing human thought visually and symbolically, and later on in using it to represent articulated language.

For we know that the Elamites acquired a writing system around 3000 BCE, the earliest traces of which are the clay "tablets" (Fig. 10.7) found at several Iranian sites, mainly at Susa, at archaeological level XVI. Like archaic Sumerian tablets, they bear on one side (sometimes both sides) a number of numerical signs alongside more or less schematic drawings, and occasionally the imprint of a cylinder-seal.

And finally, as we have seen, the system of *calculi* and *bullae* was used in Elam as well as Sumer since at least 3500–3300 BCE.

Such manifest analogies between the two civilisations lead us to hope that new archaeological finds at Sumerian sites will one day establish once and for all the relationship between Sumer and Elam.

## WHO WERE THE ELAMITES?

The oldest Iranian civilisation arose in the area now called Khuzestan. Its people called themselves *Haltami*, which the Bible transforms into *Elam*. The origins of Elam are as ill understood as its language, despite the efforts of many linguists to decipher it. We know only that the name of Elam means "land of God". Elamite appears to be an agglutinative language, like Sumerian and other Asianic languages; some linguists think it belongs to the Dravidian group (southern India) and is related to Brahaoui, which is currently spoken in Baluchistan. It should be noted that from the beginning of the third millennium BCE there appear to have been close relations between Elam and Tepe Yahya (Kirman), which is located on a possible migration route from India. The Elamite tablets found there have been dated as late fourth millennium BCE.

It seems most likely that the Elamites arrived and settled in the area that was to bear their name in the fifth millennium BCE, joining a farming culture of which the earliest traces date from the eighth millennium BCE. The earliest pieces of Susan art are decorated ceramics, showing archers and beasts of prey (Tepe Djowzi), and horned snakes (Tepe Bouhallan), and Susa, which became a full-blown city in the fourth millennium, seems to have been the most important Elamite town. Painted ceramics were abandoned during what Amiet calls the earlier period of "proto-urban" Elamite civilisation.

Throughout its history, Mesopotamia had relations with Elam, from which it imported wood, copper, lead, silver, tin, building stone, and rare stones such as alabaster, diorite, and obsidian, but from the start of the third millennium BCE relations were intense. The periods are divided as follows: from 3000 to 2800 BCE, the palaeo-Elamite period; from 2800 to 2500 BCE, the Sumero-Elamite period (subdivided into early and late, during which Sumerian influence is very noticeable); from 2500 to 1850 BCE, the Awan Dynasty, interrupted by an Akkadian conquest, was replaced by the dynasty of Shimash.

Susa became the central city in the second millennium BCE, and Elamite civilisation reaches its apogee in the middle of the thirteenth century BCE under the reign of Untash Gal who built Tchoga-Zanbil. During the first millennium BCE, Elam is closely connected to the Kingdom of Anshan which, from the sixth century BCE, became one of the key points in the Achaemenian Persian Empire.

## THE STAGES OF ELAMITE ACCOUNTING

With the help of the latest discoveries made by DAFI, we can now reconstruct the stages in the development of accounting systems in Elam. We begin in the second half of the fourth millennium BCE, in an advanced urban society where trading is increasing every day. And with an active economy, there is a pressing need to keep durable records of sales and purchases, stock lists and tallies, income and expenditure . . .

*First stage: 3500 – 3300 BCE*
Levels: Susa XVIII; Uruk IVb. For sources, see Fig. 10.4, 10.8, and 10.10

Susan officials have an accounting system through which they can represent any given number (for example, a price or a cost) by a given number of unbaked clay *calculi* each of which is associated with an order of magnitude according to the following conventions:

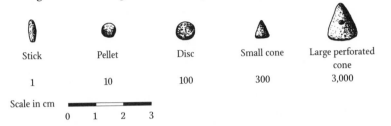

| Stick | Pellet | Disc | Small cone | Large perforated cone |
|---|---|---|---|---|
| 1 | 10 | 100 | 300 | 3,000 |

Scale in cm
0 1 2 3

FIG. 10.8. *The only* calculi *found in or very near hollow clay balls at the Acropolis of Susa. The values shown derive from the decipherment explained in Chapter 11 below. From DAFI 8, plate 1 (Susa, level XVIII)*

Intermediate numbers are represented by using as many of each type of *calculus* as required. For example, the number 297 calls for 2 discs, 9 pellets, and 7 sticks:

FIG. 10.9.

FIG. 10.10. *Exterior of a* bulla *marked with a cylinder-seal. Susa, c. 3300 BCE. From the Musée du Louvre, item Sb 1943*

You then place these objects with conventional values (whose use is not entirely dissimilar to our current use of coins or standard weights) into a hollow ball, spherical or ovoid in shape (Fig. 10.10), the outside of which is then marked by a cylinder-seal, so as to authenticate its origin and to guarantee its accuracy. For in Elam, as in Sumer, men of substance each had their own individual seal – a kind of tube of more or less precious stone on which a reversed symbolic image was carved. The cylinder-seal, invented around 3500 BCE, was its owner's representative mark. The owner used it to mark any clay object as his own, or to confer his authority on it, by rolling the cylinder on its axis over the still-soft surface (Fig. 10.11).

Let us imagine we are at the Elamite capital of Susa. A shepherd is about to set off for a few months to a distant pasture to graze a flock of 297 sheep that a wealthy local owner has entrusted to him. The shepherd and the owner call on one of the city's counting men to record the size of the flock. After checking the actual number of sheep, the counting master makes a hollow clay ball with his hands, about 7 cm in diameter, that is to say hardly bigger than a tennis ball. Then, through the thumb-hole left in the ball, he puts inside it 2 clay discs each standing for 100 sheep, 9 pellets that each stand for 10 sheep, and 7 little sticks, each one representing a single animal. Total contents: 297 heads (Fig. 10.9).

When that is done, the official closes up the thumb-hole, and, to certify the authenticity of the item he has just made up, rolls the owner's cylinder-seal over the outside of the ball, making it into the Elamite equivalent of a signed document. Then to guarantee the whole thing he rolls his own cylinder-seal over the ball. This makes it unique and entirely distinct from all other similar-looking objects.

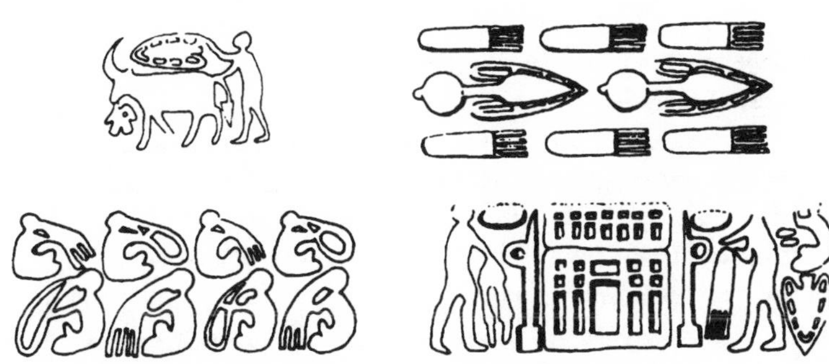

FIG. 10.11. *Cylinder-seal imprints from accounting documents found at Susa*

The counting master then lets the *bulla* dry and stores it with other documents of the same kind. With its tokens or *calculi* inside it, the *bulla* is now the official certification of the count of sheep that has taken place, and serves as a record for both the shepherd and the owner. On the shepherd's return from the pastures, they will both be able to check whether or not the right number of sheep have come back – all they need to do is break open the ball, and check off the returning sheep against the tokens that it contains.

At about the same period, the Sumerians used a very similar system: hollow clay balls have been found at Warka at the level of Uruk IVb, at Nineveh and Habuba Kabira (Fig. 10.4). The Sumerians, however, were accustomed to counting to base 60, using tens only as a supplementary system to reduce the need for memorisation (Fig. 8.5, 8.6 above), and the tokens that they used were also shaped rather differently. At Sumer, the small cone stood for 1, the pellet for 10, the large cone for 60, the perforated large cone stood for 600, the sphere represented 3,600, and the perforated sphere meant 36,000 (Fig. 10.4).

It was a sophisticated system for the period, since values were regularly multiplied by 10 by means of the perforation. By pushing a small circular stylus through the cone signifying 60, or through the sphere signifying 3,600, the values of 600 (60 × 10) and 36,000 (3,600 × 10) were obtained. The hole or circle was thus already a virtual graphic sign for the pellet, with a value of 10.

Let us now imagine ourselves in the market of the royal city of Uruk, capital of Sumer. A cattle farmer and an arable farmer have made a deal to exchange 15 head of cattle against 795 bags of wheat. However, the livestock dealer has only got 8 head of cattle at the market, and the grain seller has only 500 bags of wheat immediately available. The deal is done nonetheless, but to keep things above board there has to be a contract. The cattle man agrees to deliver a further 7 cattle by the end of the month, and the arable farmer promises to supply 295 bags of grain after that year's harvest. To make a firm record of the agreement, the cattle man makes a clay ball and puts in 7 small cones, one for each beast due, then closes the ball and marks its surface with his own cylinder-seal, as a signature. The arable farmer, for his part, makes another clay ball and puts in it 4 large cones, each one standing for 60 bags of wheat, 5 pellets each standing for 10 bags, and 5 small cones for the 5 remaining bags due, then seals and signs the clay ball in like manner. Then a witness puts his own "signature" on the two documents, to guarantee the completeness and accuracy of the transaction. Finally, the two traders exchange their *bullae* and go their separate ways.

So although this remains an illiterate society, it possesses a means of recording transactions that has exactly the same force and value as written contracts do for us today.

At a time when cities were still relatively small, and where trade was still relatively simple, business relations were conducted by people who knew each other, and whose cylinder-seals were unambiguously identifiable. For that reason, the nature of a transaction recorded in a *bulla* is implicit in the identity of the seal(s) upon it: the symbolic shapes on the outside of the clay ball tell you whether you are dealing with this farmer or that miller, with a particular craftsman or a specific potter. As for the numbers involved in the transaction, they are unambiguously recorded by the nature and number of the *calculi* inside.

Cheating is therefore ruled out. Each party to the deal possesses the record of what his partner owes him, a record certified by his business partner's own identity, in the form of his seal.

*Second stage: 3300 BCE*
Level: Susa XVIII. For sources see Fig. 10.13

The great defect of the system in place was that the hollow clay balls had to be broken in order to verify that settlements conformed to the contracts. To overcome this, the idea arose of making various imprints on the outer surface of the *bullae* (alongside the imprints of the necessary cylinder-seals) to symbolise the various tokens or *calculi* that are inside them. Technically, the device harks back to the more ancient practice of notching, but it is quite altered in its significance by the new context.

The corresponding marks are: a long, narrow notch, made by a stylus with its point held sideways on to the surface, to represent the stick; a small circular imprint, made by the same stylus pressed in vertically, to represent the pellet; a large circular imprint, made by a larger stylus or just by pressing in a finger-tip, to represent a disc; a thick notch, made by a large stylus held obliquely, to represent a cone; and a thick notch with a circular imprint to represent a perforated cone.

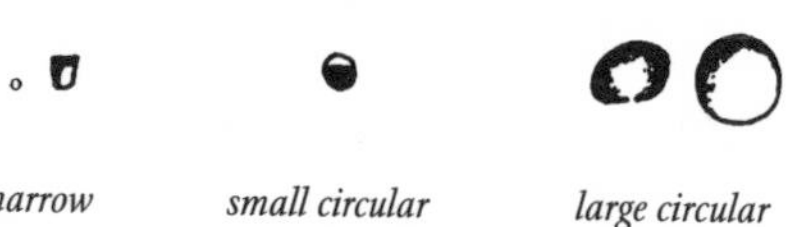

FIG. 10.12. *Numerical markings on* bullae *found at Susa*

This constitutes a kind of résumé of the contract, or rather a graphic symbolisation of the contents of each accounting "document".

Henceforth, an Elamite *bulla* containing (let us say) 3 discs and 4 sticks (making a total of 3 × 100 + 4 = 304 units) carries on its outer face, alongside the cylinder-seal imprints, 3 large circular indentations and 4 narrow lines. No longer is it necessary to break open the clay balls simply to check a sum or to make an inventory – because the information can now be "read" on the outside of the *bullae*.

The cylinder-seal imprint or imprints show the *bulla's* origin and guarantee it as a genuine document, and the indentations specify the quantities of beings or things involved in the accounting operation.

*Third stage: c. 3250 BCE*
Level: Susa XVIII. See Fig. 10.15 below

These indentations thus constitute real numerical symbols, since each of them is a graphic sign representing a number. Together they make up a genuine numbering system (Fig. 10.14). So why carry on using *calculi* and putting them in *bullae*, when it's much simpler to represent the corresponding values by making indentations on slabs of clay? Mesopotamian and Elamite accountants very quickly realised that of the two available systems, one was redundant, and the *calculi* were rapidly abandoned. The spherical or ovoid *bullae* came to be replaced by crudely rounded or oblong clay slabs, on which the same information as was formerly put on the casing of the *bullae* was recorded, but on one side only.

The cylinder-seal imprint remained the mark of authenticity on these new types of accounting records, whose shape, at the start, roughly imitates that of a *bulla*. The sums involved in the transaction are represented on the soft clay by graphic images of the *calculi* that would previously have been enclosed in a *bulla*. This stage therefore marks the appearance of the first "accounting tablets" in Elam.

It should be noted that the three stages laid out above occurred in a relatively short period of time, since all the evidence for them is attested at the same archaeological level (Susa XVIII), in the same room, and on the same floor level. The imprint of the same cylinder-seal on one *bulla* and two tablets (see *bulla* C in Fig. 10.13 and tablet B in Fig. 10.15 below, for example) seems to confirm that both systems existed side by side at least for a time.

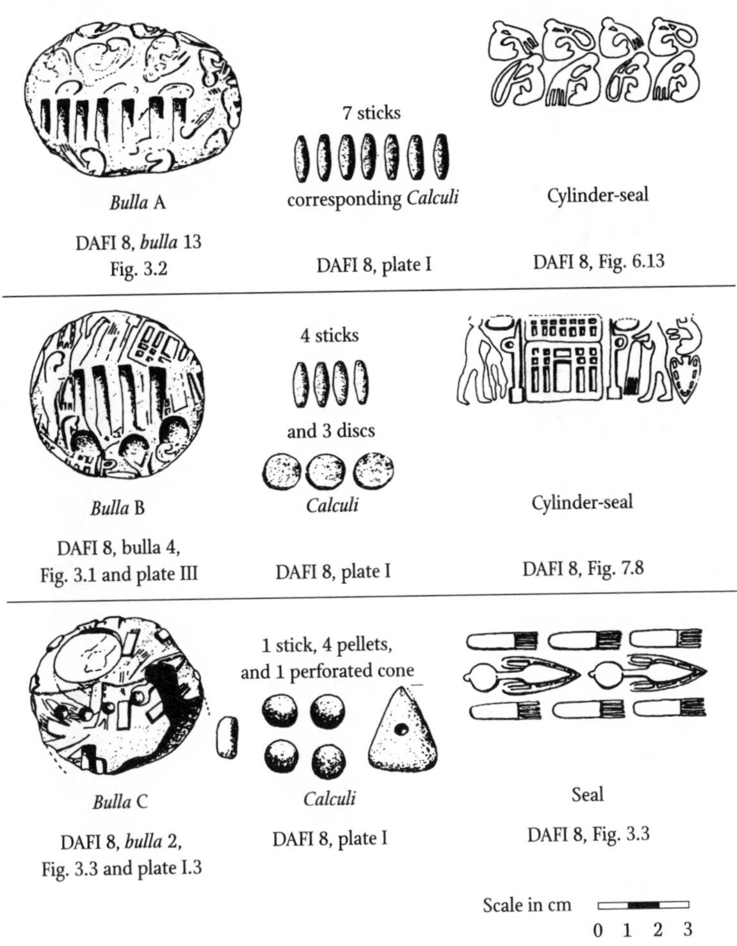

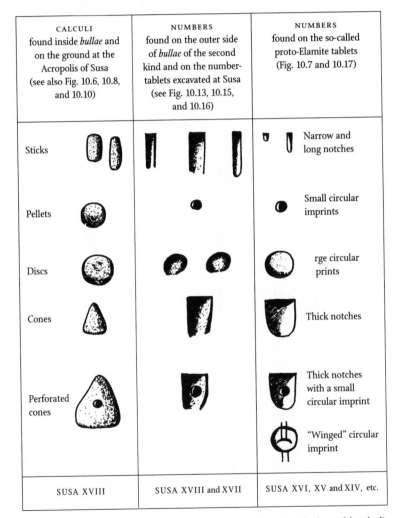


Fig. 10.13. Bullae *containing the same number of* calculi *as are symbolised on the outer surface by indentations next to the cylinder-seal imprints. Susa, level XVIII (approx. BCE 3300), excavated by DAFI in 1977–1978. Similar* bullae *have been found at Tepe Yahya and Habuba Kabira, but none so far at Uruk.*

| CALCULI found inside *bullae* and on the ground at the Acropolis of Susa (see also Fig. 10.6, 10.8, and 10.10) | NUMBERS found on the outer side of *bullae* of the second kind and on the number-tablets excavated at Susa (see Fig. 10.13, 10.15, and 10.16) | NUMBERS found on the so-called proto-Elamite tablets (Fig. 10.7 and 10.17) |
|---|---|---|
| Sticks | | Narrow and long notches |
| Pellets | | Small circular imprints |
| Discs | | rge circular prints |
| Cones | | Thick notches |
| Perforated cones | | Thick notches with a small circular imprint<br>"Winged" circular imprint |
| SUSA XVIII | SUSA XVIII and XVII | SUSA XVI, XV and XIV, etc. |

Fig. 10.14. *The indentations made on the outer side of the* bullae *imitate the shape of the* calculi *that are enclosed. Moreover, these marks resemble not only the number-tablets found at Susa but also the figures on the proto-Elamite tablets of later periods.*

*Fourth stage: 3200–3000 BCE*

Levels: Susa XVII; Uruk IVa. See Fig. 8.1 above and 10.16 below

This stage sees only a slow refinement of the system in place already: exactly the same types of information are included on the accounting tablets of the fourth period as on those of the third. However, the tablets themselves become less crudely shaped, the numbers are less deeply indented in the clay, and their shapes become more regular. In addition, the cylinder-seals are now imprinted on both sides of the tablet, and not just on the "top".

However, like the earlier *bullae* and crude tablets, this stage of development is still not "writing" in the proper sense. The notation records only numerical and symbolic information, and the things involved are described only in terms of their quantity, not by signs specifying their nature. Nor is the nature of the operation indicated by any of these documents: we have no idea if they are records of a sale, a purchase, or an allocation, nor can we know the names, the numbers, the functions, or the locations of any of the parties to the transaction. We have already made the assumption that the cylinder-seals, since they indicate the identities of the contracting parties, would also have indicated the type of transaction in a society where

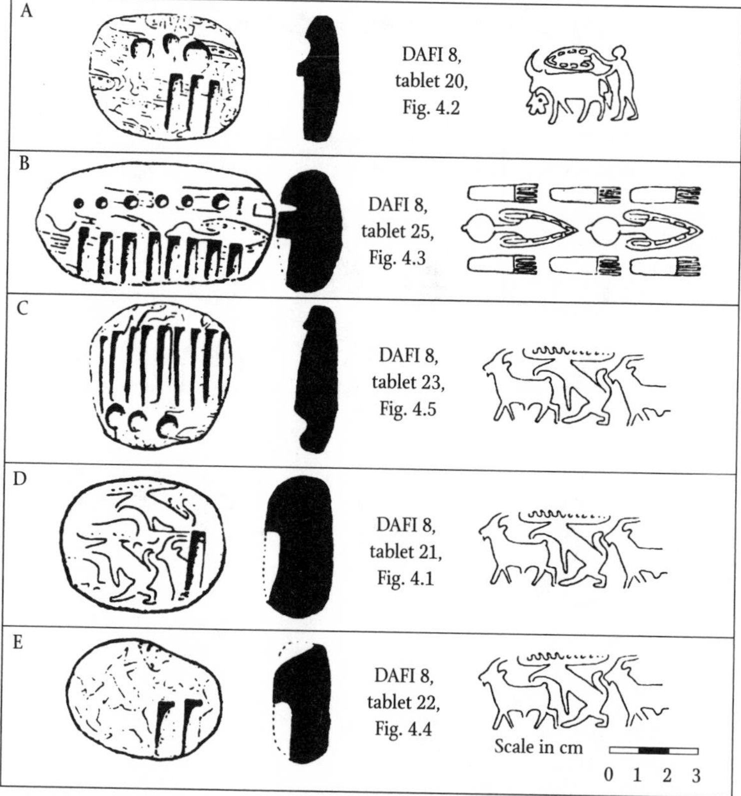


FIG. 10.15. *Roughly circular or oblong tablets containing indented numerical marks (similar to those found on* bullae*) alongside one or two cylinder-seal imprints. Items dated c. 3250 BCE, from Susa level XVIII, excavated dy DAFI in 1977–1978.*

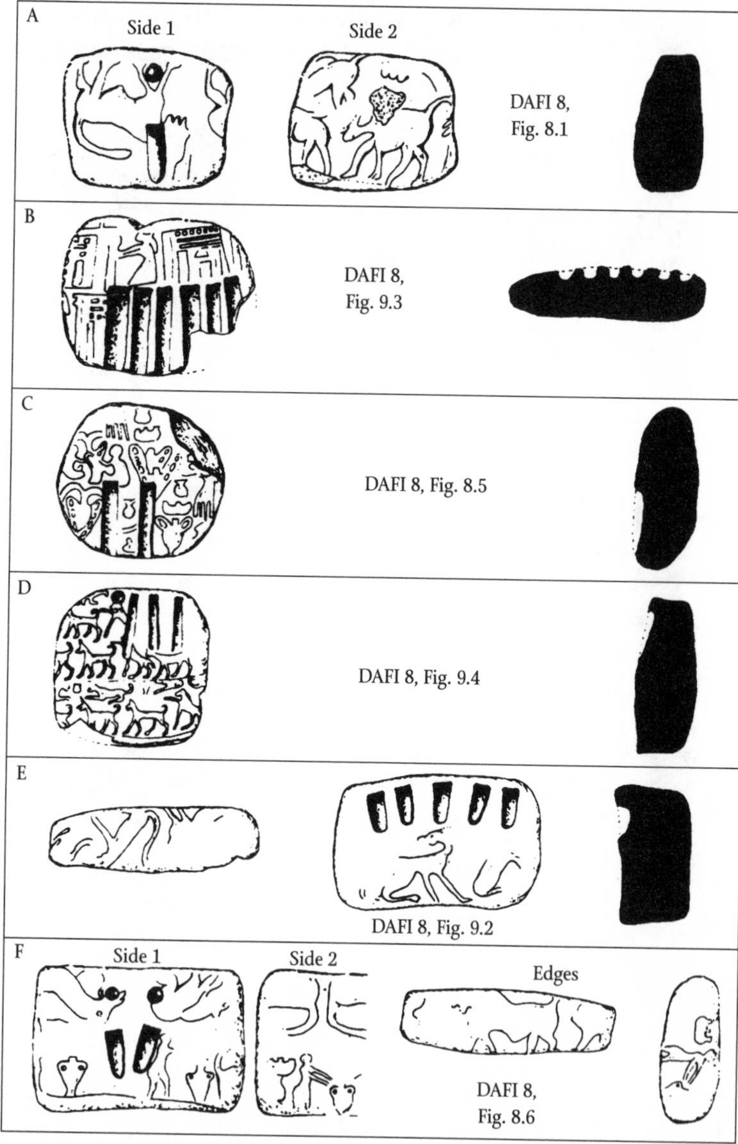


FIG. 10.16. *Numerical tablets from Susa level XVII , c. 3200– 3000 BCE, excavated by DAFI in 1972*

people were known to one another. This makes very clear just how concise, but also how imprecise are the purely symbolic visual notations of these documents, which constitute the trace of the very last stage in the prehistory of writing. Cylinder-seal imprints do in fact disappear from the tablets as soon as pictograms and ideograms make their appearance.

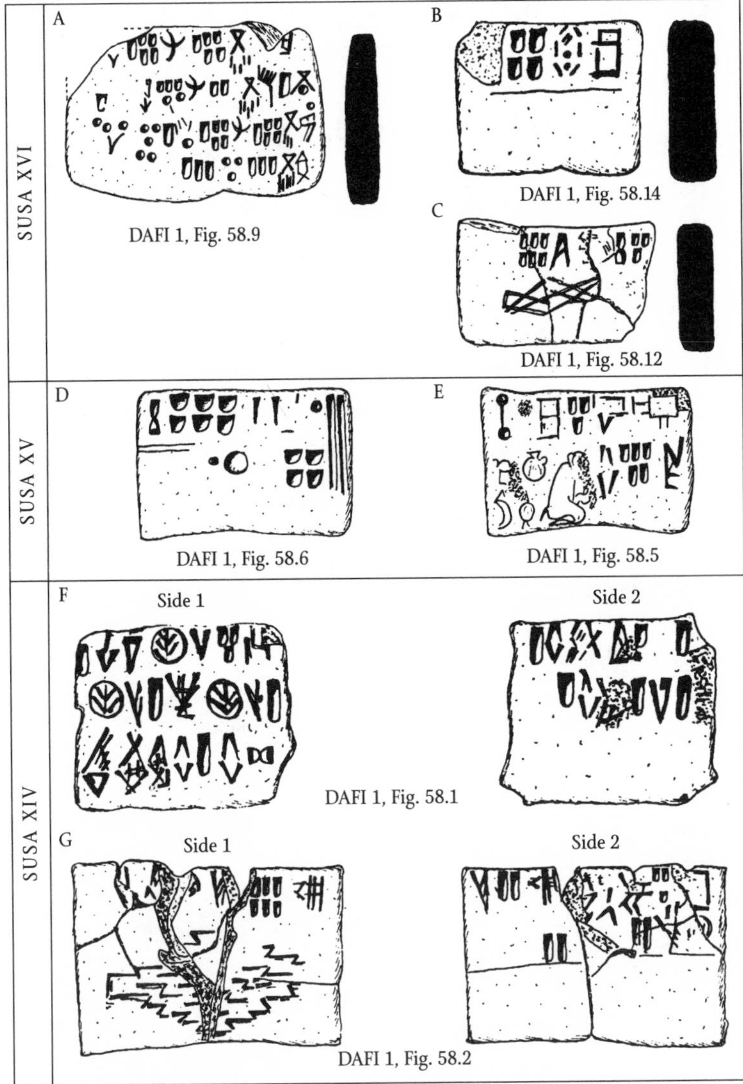


FIG. 10.17. *The first proto-Elamite tablets. They are less crude, rectangular tablets giving written name-signs alongside the corresponding numbers. From Susa, c. 3000–2800 BCE; excavated by DAFI in 1969–1971. (Cf. A. Le Brun)*

At Sumer, writing emerged at the same time as these regular tablets from Elam. The first Uruk tablets date from 3200–3100 BCE (Fig. 8.1 above) and, although they remain exclusively economic documents, they use a notation (archaic Sumerian numerals) which is founded not on making a "picture" of a vague idea, but on something much more precise, analytical and articulated. In tablet E of Fig. 8.1, for example, you can see how the document is divided into horizontal and vertical lines, marking out squares in which pictograms are placed beside groups of numbers. Sumerian tablets are thus ahead of the Susan ones of the same period: Sumer has something like writing, and Susa has only symbols.

### *Fifth stage: 3200–2900 BCE*

Level: Susa XVI. See also Fig. 10.17, tablets A, B, C

The tablets from this period are thinner and more regularly rectangular (standardised), but most significantly they carry the first signs of "proto-Elamite" script alongside numerical indentations. The purpose of the signs is to specify the nature of the objects involved in the transaction associated with the tablet. On several tablets found at Susa XVI, there are no cylinder-seal imprints.

### *Sixth stage: 2900–2800 BCE*

Level: Susa XV and XIV. See Fig. 10.17, D, E and F

In this period, the proto-Elamite script on the tablets grows to cover more of the surface than the number-signs. Could this mean that the script might hold the key to the grammar of the language? Is proto-Elamite the earliest alphabetic script? We do not know, as it remains to be deciphered.

## THE PROBLEMS OF SO-CALLED PROTO-ELAMITE SCRIPT

This script appeared at the dawn of the third millennium BCE and spread from the area around Susa to the centre of the Iranian plateau. It remained in use in Elam until around 2500 BCE, when it was supplanted by cuneiform writing systems from Mesopotamia, whence derived Elamite script proper, whose final form was neo-Elamite.

How did proto-Elamite arise? Some scholars believe the Elamites invented it, independently of the Sumerians. This presupposes that it resulted from a similar set of steps, starting from identical circumstances, and following the same generic idea based on earlier rudimentary trials in the area. That is not implausible, especially in the light of the developments we have just charted.

Other scholars take the opposite view, namely that proto-Elamite script was inspired by Sumerian. This is also quite plausible, even if the nature of the "inspiration" must have been quite a distant one. Some of the proto-Elamite signs look as if they might be related to specific Sumerian pictograms and ideograms, but most of the signs are too different to allow any systematic comparison of the two scripts. On the other hand, it may well be that the Sumerian invention of writing inspired their neighbours the Elamites (Uruk and Susa are less than two hundred miles apart) to invent a writing of their own. Sumerian accounting tablets are one or two centuries older than their Elamite equivalents, and there is no doubt in which direction the invention flowed.

It seems probable that writing would have been invented in Susa even without the example or inspiration of Sumer, since all the social and economic dynamics that led to the invention of writing elsewhere were present amongst the Elamites. For as the history of numbers shows, people in similar circumstances and faced with similar needs often do make very similar inventions, even when separated by centuries and continents.

Be that as it may, proto-Elamite script remains a mystery. The signs almost certainly represented beings and things of various kinds, but the forms used are simplified and conventionalised to a point where guessing their meaning is impossible. We also know next to nothing about the language which this script represents.

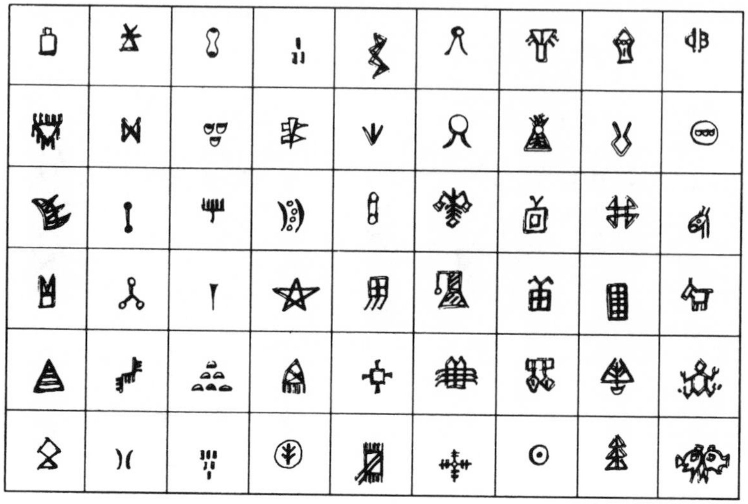

FIG. 10.18. *The signs of proto-Elamite script. References: Mecquenen; Scheil; Meriggi*

CHAPTER 11

# THE DECIPHERMENT OF A FIVE-THOUSAND-YEAR-OLD SYSTEM

In 1981, when I published the first edition of *The Universal History of Numbers*, the number-signs in the proto-Elamite script (Fig. 11.1) still presented major problems.

A table drawn up by W. C. Brice (1962), and later also referred to by A. Le Brun and F. Vallat (1978), clearly shows how these number-symbols received very varied, indeed contradictory, interpretations over the years on the part of the majority of epigraphists and specialists in these questions.

Despite the great difficulties, I decided to apply myself to the task. In 1979 I began my researches which, one year later, culminated in the complete decipherment of these number-signs, after close examination of a large number of invoice tablets which had been discovered by the French Archaeological Mission to Iran at the end of the last century. These documents may be found in the collections of the Louvre and the Museum of Teheran.

We shall come shortly to the method which I followed. But, in order to appreciate it, we must first make yet another visit to the land of Sumer . . .

## THE INVENTION OF THE BALANCE SHEET IN SUMERIA

The period from 3200 to 3100 BCE saw, as we have observed, the beginnings of written business accounts.

At first, however, the system was primitive. The documents held only one kind of numerical record at a time: one tablet for 691 jugs, for example (Fig. 8.1 C above), another tablet for 120 cattle (Fig. 8.1 D), another for 567 sacks of corn, another for 23 chickens, yet another for 89 female slaves imported from abroad, and so on.

But from around 3100 BCE as business transactions and distributions of goods became increasingly numerous and varied, the inventories and the accounts for each transaction also grew more complex and voluminous, and the accountants found they had to cut down on the cost of clay. From this time on the pictures and the numbers took up increasing amounts of space on the tablets. Onto a single rectangular sheet of clay, divided into boxes by horizontal and vertical lines, were recorded inventories of livestock in all their different kinds (sheep, fat sheep, lambs, lambkins, ewes, goats, kids male and female or half-grown, etc.) in all necessary detail. A single tablet, too, was used to summarise an agricultural audit in which all the different kinds of species were distinguished.

| A | B | C | D |
|---|---|---|---|
| E | F | G | H |
| I | J | K | L |
| M | N | O | P |

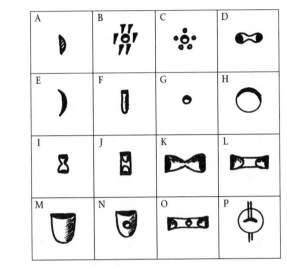

Fig. 11.1. *The proto-Elamite number-signs*

| | F | G | H | M | N |
|---|---|---|---|---|---|
| System proposed by Scheil<br>See MDP VI (1905) | 1 | 10 | 100 | 1,000 | 10,000 |
| System proposed by Scheil<br>See MDP XVII (1923) | 1 | 10 | 100 | 60 | 600 |
| System proposed by Scheil<br>See MDP XVII (1923) | 1 | 10 | 100 | 600 | 6,000 |
| System proposed by Langdon<br>See JRAS (1925) | 1 | ? | 100 | 1,000 | 10,000 |
| System proposed by Scheil<br>See MDP XXVI (1935) | 1 | 10 | 100 | 1,000 | 10,000 |
| System proposed by de Mecquenem<br>See MDP XXXI (1949) | 1 | 10 | 100 | 300 | 1,000 |

Fig. 11.2. *Various contradictory conclusions drawn over the years concerning the values of the proto-Elamite number-signs*

But then: the *balance sheet* was invented. Now people wrote on both sides of the tablet: the "recto" side bore the details of a transaction, the "verso" the totals under the various headings.

The idea took hold, and with refinement proved to be of the greatest usefulness. At Uruk, in 2850 BCE, a proposal of marriage has been made. The girl's father and the father of the future spouse have just agreed on the "bride price". When the ceremony has taken place, the bride's father will receive from the other 15 sacks of barley, 30 sacks of corn, 60 sacks of beans, 40 sacks of lentils, and 15 hens. But, in view of the frailties of human memory and in order to avoid any quarrels later, the two men betake themselves to one of the religious leaders of the town in order to draw up the contract in due form and give the force of law to the engagement.

Having taken note of all the elements of the marriage contract, the notary then fashions a roughly rectangular tablet of clay, and takes up his "tracing tools".

For writing, he uses two ivory sticks of different cross-section, pointed at one end and, at the other, fashioned into a kind of cylindrical stylus (Fig. 8.10 above). The pointed ends are used to draw lines or to trace pictograms on the soft clay (Fig. 8.11 above), and the cylindrical styluses are used to mark numbers by pressing at a given angle on the surface of the tablet. According to the angle between stylus and the tablet, the impression made on the soft clay will be either a notch or a circular imprint, whose size will depend on the diameter of the stylus which is used. As in Fig. 8.12 above, this will be a narrow or a wide notch, according as the wide or narrow stylus is used, if the angle is 30°–45°; or it will be a circular imprint of small or large diameter, according to the stylus, if it is applied perpendicular to the surface of the tablet.

Then, holding the tablet with its long side horizontal, the scribe draws four vertical lines, thereby dividing it into five sections, one for each item in the contract. At the bottom of the rightmost division he draws a "sack of barley", in the next a "sack of corn", then a "sack of beans", then a "sack of lentils", then finally in the leftmost division he draws a "hen". Then he places the corresponding numerical quantities: in the first division, a small circular imprint for the number 10, and 5 small notches each worth 1, thus making up the total of 15 sacks of barley; in the second, three imprints of 10 for the number 30; in the third, he marks the number 60 with a large imprint, and so on.

On the back of the tablet, he makes the summary, that is, the totals of the inventory according to the numbers on the front, namely "145 sacks (various)" and "15 hens".

This done, the two men append their signatures to the bottom of the tablet, but not as used to be done by rolling a cylinder-seal over it. Instead, they use the pointed end of the stylus to trace conventional signs which represent them. Then, having given the document into the safekeeping of the notary, they part.

### HOW THE SUMERIAN NUMBERS WERE DECIPHERED

The story reconstituted in the preceding section was not imaginary: it was achieved on the basis of the document shown in Fig. 11.3, which provides detailed evidence of how the Sumerian scribes used to note on one side of the tablets the details of the accounting, and on the other side a kind of summary of the transaction in the form of totals under different headings.

| Side 1 | Side 2 | Translation: Side 1 | Translation: Side 2 |
|---|---|---|---|
| | | 15 sacks of barley | 145 sacks (various) |
| | | 30 sacksof corn | |
| | | 60 sacks of ? | 15 hens |
| | | 40 sacks of ? | signature (?) |
| | | 15 hens | (?) |
| | | | (?) |

FIG. 11.3. *Sumerian "invoice" discovered at Uruk, said to be from the Jemdet Nasr era (c. 2850 BCE). Iraqi Museum, Baghdad. ATU 637*

But it is precisely this feature which has enabled the experts to decipher various ancient number-systems such as Sumerian, hieroglyphic or linear Cretan, and so on. The values of the numbers could therefore be determined with certainty by virtue of applying a large number of checks and verifications to these totals.

Observing, for example, that on the front of some tablet there were ten narrow notches here and there, while on the back there was a single small circular imprint, and then finding this correspondence confirmed in a sufficient number of similar cases, they can conclude that the narrow notch denotes unity and the small circular imprint denotes 10.

= 1   = 10

Now suppose that we are trying to discover the unknown value, which we shall denote by $x$, of the wide notch:

= $x$ ?

Of course, lacking any other indication, and in the absence of a bilingual "parallel text" (linguistic or mathematical), the value of this number would have long remained a mystery. But a happy chance has placed into our hands the tablet shown in Fig. 11.3, which bears the three numbers described above of which two have already been deciphered, which will indeed be our "Rosetta Stone".

We begin, of course, by ignoring the count of the 15 hens (one small circular imprint and 5 narrow notches, together with the pictogram of the bird), since this is reproduced exactly on the reverse of the document. So we shall only bother with the details of the inventory of sacks (goods denoted by the same writing sign throughout). Adding up the numbers on side 1, we therefore obtain

$$10 + 5 + 30 + x + 40 = x + 85$$

while on side 2 we find

$$2x + 20 + 5 = 2x + 25$$

On equating these two results, we obtain the equation

$$x + 85 = 2x + 25$$

which, on reduction, finally gives the result we are seeking, namely

$$= x = 60$$

However, we are only entitled to draw this conclusion as to the value of the sign in question if the value so determined gives consistent results for several other tablets of similar kind. And this turns out to be the case.

## SIMILAR PRACTICE OF THE ELAMITE SCRIBES

It was precisely by observing similar practice on the part of the Elamite scribes, and carrying out systematic verifications of the same kind on a multitude of proto-Elamite tablets (some of the most important of which will be shown below) that I was able, myself, to arrive at the solution of this thorny problem.

Some of these tablets can lead us to it, even though the values of the proto-Elamite numbers may remain unknown. Consider for example the tablet in Fig. 11.4 A which refers to a similar accounting operation. The goods in question are represented by writing signs (whose meaning, in many cases, still eludes us). But the numbers associated with the various goods are clearly indicated by groups of number-signs. The subsequent diagram (Fig. 11.4 B) shows what we shall from now on call the "rationalised transcription" of the original tablet.

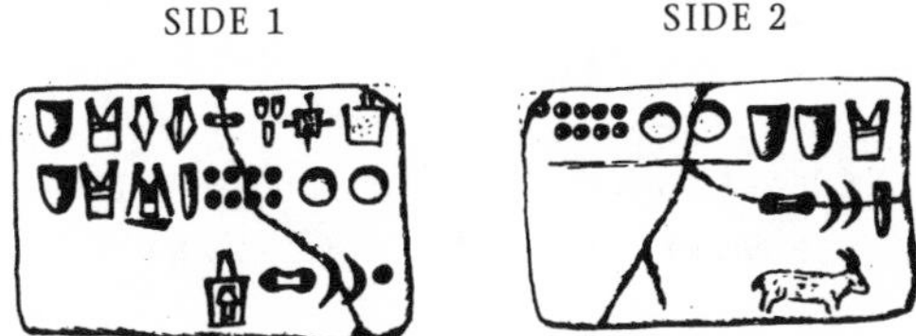


FIG. 11.4A. *Accounting tablet from Susa. Louvre. See MDP, VI, diagram 358*

| NUMBERS | WRITING | |
|---|---|---|
| | | SIDE 1 |
| | | SIDE 2 |

FIG. 11.4B.

Now we see, on the front of the tablet:

- the wide notch twice;
- the large circular impression twice;
- the small circular impression 9 times;
- the narrow, lengthened notch once;
- a circular arc twice;
- and a peculiar number (Fig. 11.1 D) once only.

This, moreover, is exactly what we also find on the reverse of the tablet. The number which is shown on side B therefore corresponds to the grand total of the inventory on the front.

In the same way, on the tablet shown in Fig. 11.5, the front and the reverse both show six narrow notches.

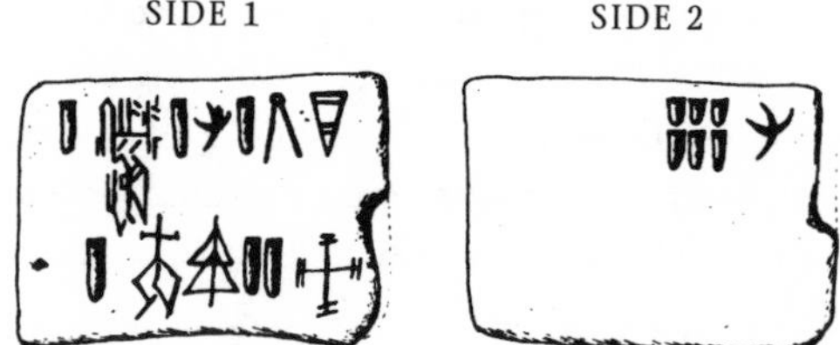


FIG. 11.5. *Tablets from Susa. Teheran Museum. See MDP, XXVI, diagram 437*

## DETERMINING THE VALUES OF THE PROTO-ELAMITE NUMBERS

Now consider the tablet shown in Fig. 11.6. In the present state of the tablet, on the front side the narrow notch occurs only 18 times, and the smaller circular impression occurs 3 times, while on the reverse the narrow notch occurs 9 times and the circular impression 4 times.

If we proceed by analogy with the Sumerian numbers of similar form, attributing value 1 to the narrow notch and value 10 to the circular imprint, then the total from the front of the tablet ($18 + 3 \times 10 = 48$) and the total from the reverse ($9 + 4 \times 10 = 49$) differ by 1. We may conjecture that this difference is the result of a missing piece broken off from its left-hand side, which would have damaged the numerical representation in the last line of the top face.

Since, moreover, there are similar tablets* on which we find exactly equal totals on the two sides, we may conclude that this explanation for the discrepancy is in fact correct.

Therefore we may definitively fix the value of the narrow notch as 1, and the value of the small circular impression as 10.

SIDE 1

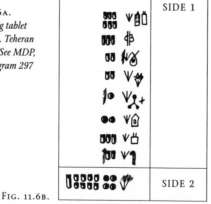

FIG. 11.6A. *Accounting tablet from Susa. Teheran Museum. See MDP, XXVI, diagram 297*

SIDE 2

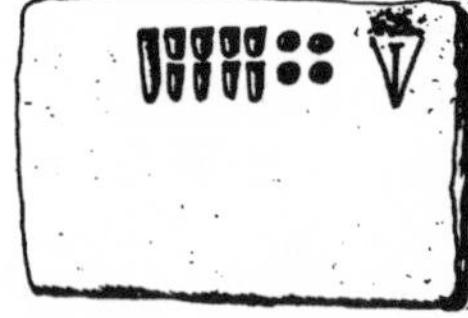

SIDE 1

SIDE 2

FIG. 11.6B.

* See, for example, tablet 353 of *MDP*, VI (Louvre; Sb 3046).

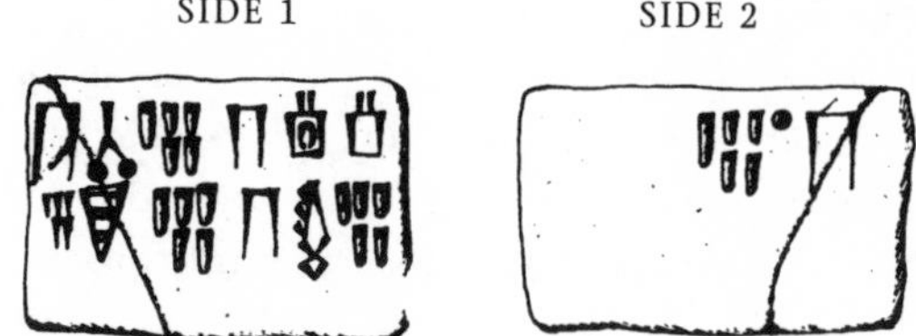


FIG. 11.7. *Accounting tablet from Susa. Louvre. See MDP, XVII, diagram 3*

Now we must take account of the fact that the Elamites set their numbers down from right to left (in the same direction as their writing), starting with the highest-order units and proceeding left towards the lower-order units. Furthermore, close examination of the tablets shows that the Elamite scribes used two different systems for writing numbers, both of which were based on the notion of juxtaposition to represent addition. These two systems made use, in general, of different symbols (Fig. 11.10 and 11.11).

For the first of these two proto-Elamite systems, it is pretty clear that the number-signs were always written in the following order, from right to left and from highest value to lowest value (Fig. 11.8).

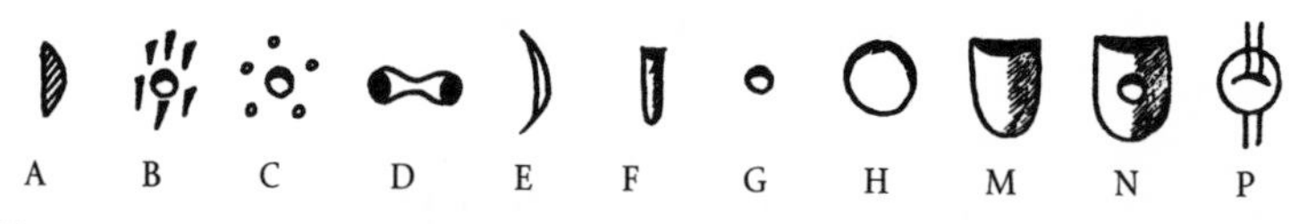


FIG. 11.8.

The number-signs of the second system always occur as follows, again from right to left and in decreasing order (Fig. 11.9).

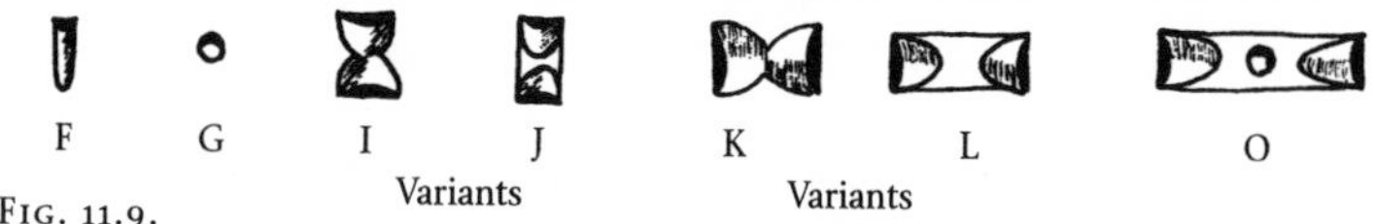


FIG. 11.9.

The above shows, therefore, that

- on the one hand, the numbers labelled A, B, C, D, and E (which always occur to the left of the narrow notch which represents 1) correspond to orders of magnitude below 1, that is to say to fractions;
- on the other hand, H, M, N, and P, and also I (or J), K (or L) and O correspond to orders of magnitude above 10 (since they always occur to the right of the small circular impression representing 10) (Fig. 11.10 and 11.11).

In the end, therefore, by working out the totals on many other tablets, I was able to obtain the following results which, as we shall see below, can be confirmed in other ways.

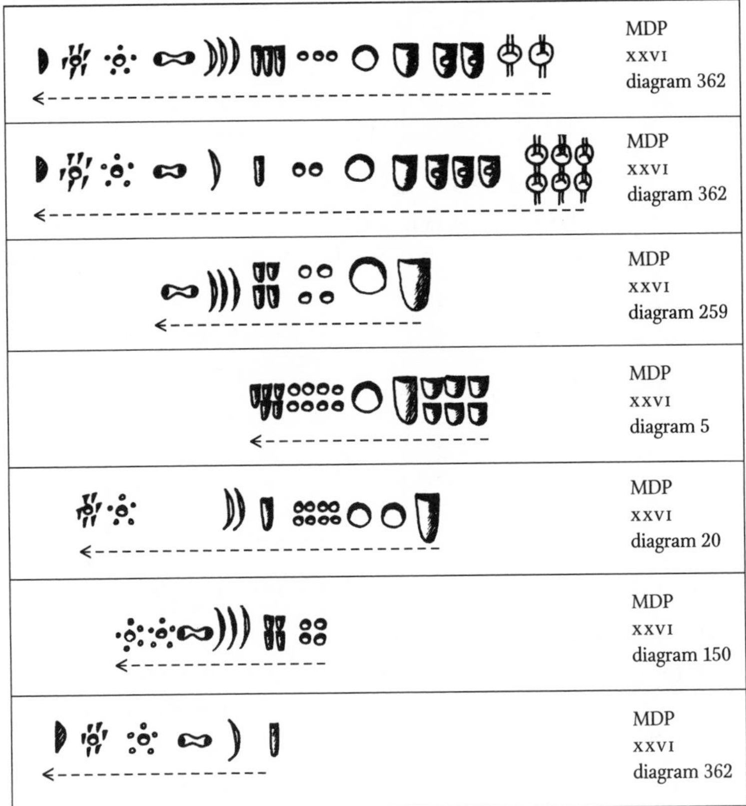


FIG. 11.10. *Instances of number taken from accounting tablets, which show how the earliest proto-Elamite number-system worked*

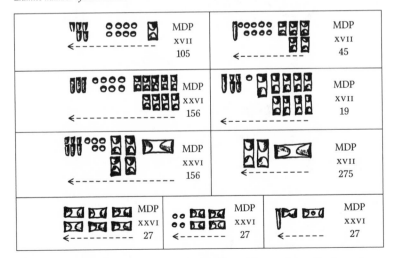

| | |
|---|---|
| MDP XVII 105 | MDP XVII 45 |
| MDP XXVI 156 | MDP XVII 19 |
| MDP XXVI 156 | MDP XVII 275 |
| MDP XXVI 27 / MDP XXVI 27 | MDP XXVI 27 |

FIG. 11.11. *Instances from accounting tablets which illustrate the second proto-Elamite number-system*

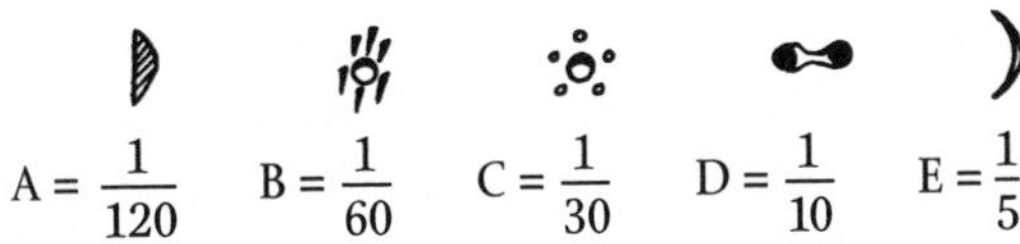

$$A = \frac{1}{120} \quad B = \frac{1}{60} \quad C = \frac{1}{30} \quad D = \frac{1}{10} \quad E = \frac{1}{5}$$

For the number E (the circular arc), for example, I considered the tablet shown in Fig. 11.12 which, as can be seen from its rationalised transcription, bears two kinds of inventory:

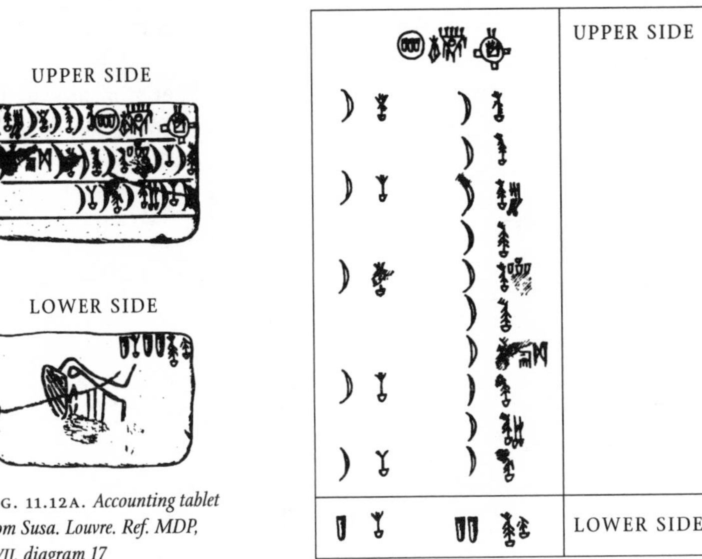


FIG. 11.12A. *Accounting tablet from Susa. Louvre. Ref. MDP, XVII, diagram 17*

FIG. 11.12B.

- one, associated with the script character [sign], which has 10 circular arcs on the top face and 2 narrow notches on the reverse;
- the other, associated with the ideogram [sign], which has 5 circular arcs on the top face and 1 narrow notch on the reverse.

Therefore, denoting by $x$ the unknown value of the number (E) in question, these two inventories give, according to the totals of the two sides, the two equations

$$x + x + x + x + x + x + x + x + x + x = 2$$
$$x + x + x + x + x = 1$$

namely

$$10x = 2$$
$$5x = 1$$

which is precisely how it was possible to determine the value $\frac{1}{5}$ for the circular arc.

Now let us try to evaluate the large circular imprint and the wide notch (H and M in Fig. 11.1). Because they look just like the Sumerian signs associated with 60 and 3,600 respectively (Fig. 8.7 and 9.15 above), we are at first tempted to conclude that the same values should be attributed to them in the present case. But when we examine the proto-Elamite tablets we find that this cannot be true. As we have seen, the Elamites set their numbers down from right to left, in decreasing order of magnitude and always commencing with the highest. Therefore, if these signs had the Sumerian values, the large circular impression should come before the wide notch in writing numbers. But this is not the case, as can be seen from Fig. 11.10 for example.

The document shown in Fig. 11.13 leads without difficulty to the ascertainment of the value of the proto-Elamite large circular impression.

UPPER SIDE

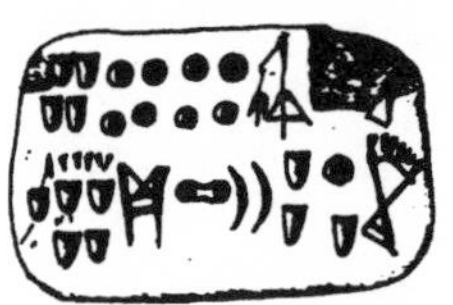

LOWER SIDE

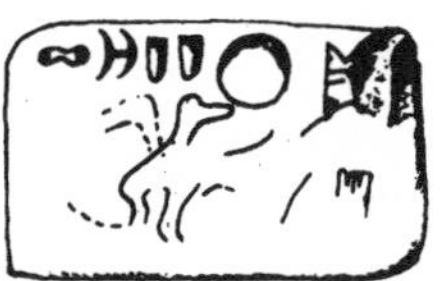

FIG. 11.13A. *Accounting tablet from Susa. Teheran Museum. Ref. MDP, XXXI, diagram 3*

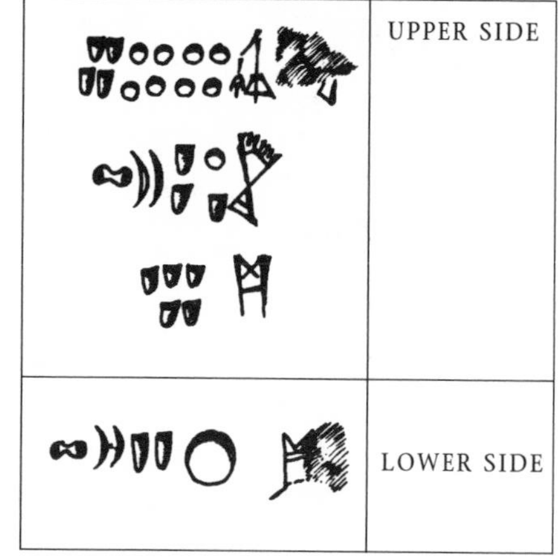


FIG. 11.13B.

Ignoring the two circular arcs and the doubled round imprint which are on both sides of the tablet, we find

- 9 small circular impressions and 12 narrow notches on the upper face;
- 1 large circular impression and 2 narrow notches on the lower.

Therefore, if we now evaluate these numerical elements on the two faces of the tablet, bearing in mind what we have already found out, we obtain the following:

| | |
|---|---|
| Upper | $9 \times 10 + 12 = 102$ |
| Lower | $1 \times x + 2 = x + 2$ |

Since these must be equal, we find the equation $x + 2 = 102$, whose solution is that $x = 100$.

Now consider the tablet shown in Fig. 11.14, on which we find

- 20 small circular impressions, and 2 large ones, on the upper face;
- 1 wide notch and one large circular impression, on the lower.

Let us now give the value 100 to the large circular impression, as we have just determined, and denote by $y$ the value of the wide notch. We then obtain the following totals:

| | |
|---|---|
| Upper | $20 \times 10 + 2 \times 100 = 400$ |
| Lower | $1 \times y + 100 = y + 100$ |

Since these also, as before, must be equal, we obtain the equation $y + 100 = 400$, whose solution is that $y = 300$.

From the preceding arguments, therefore, we attribute the value 100 to the large circular impression, and the value 300 to the wide notch.

UPPER SIDE LOWER SIDE

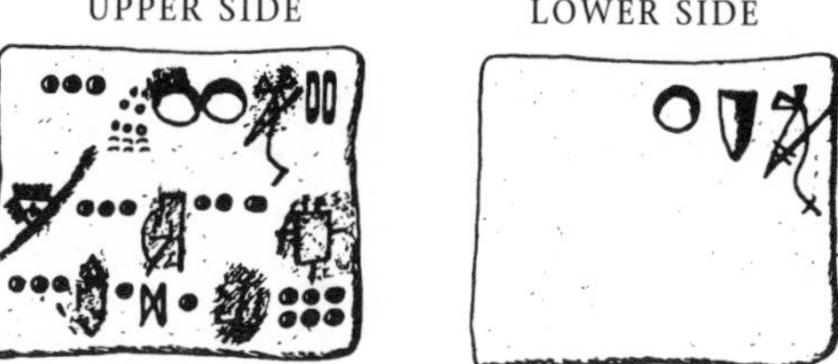

FIG. 11.14. *Tablet from Susa. Teheran Museum. Ref. MDP, XXVI, digram 118*

Of course, this would not allow us to conclude that these values correspond to a general reality unless we also find at least one other tablet which gives completely concordant results. This is, however, precisely the case for the tablets shown in Fig. 11.15 and 11.16.

UPPER SIDE LOWER SIDE

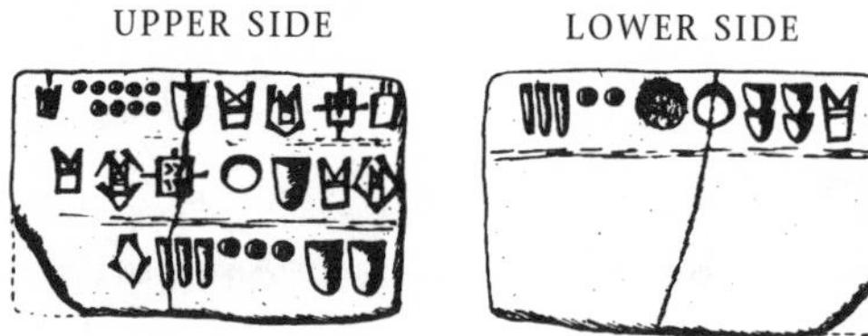

FIG. 11.15A. *Tablet from Susa. Louvre, Ref. MDP, VI, diagram 220*

| | |
|---|---|
| UPPER SIDE | $300 + 9 \times 10$ . . . . . . . . . . . . . . . . . 390 |
| | $300 + 100$ . . . . . . . . . . . . . . . . . . . . 400 |
| | $2 \times 300 + 3 \times 10 + 3$ . . . . . . . . . . . 633 |
| | 1,423 |
| LOWER SIDE | $4 \times 300 + 2 \times 100 + 2 \times 10 + 3$ . . . 1,423 |

FIG. 11.15B.

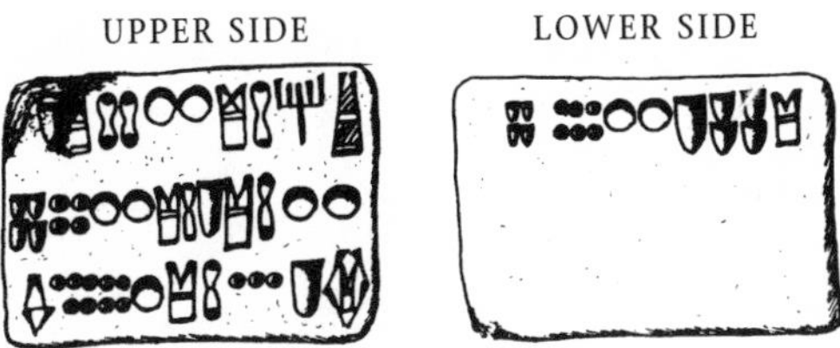


FIG. 11.16A. *Tablet from Susa. Teheran Museum, Ref. MDP, XXVI, diagram 439*

| | |
|---|---|
| | 2 × 100 ........................ 200 |
| | 300 + 2 × 100 .................. 500 |
| | 300 ........................... 300 |
| | 2 × 100 + 4 × 10 + 4 ............ 244 |
| | 300 + 3 × 10 .................. 330 |
| | 100 + 9 × 10 .................. 190 |
| | 1,764 |
| LOWER SIDE | 5 × 300 + 2 × 100 + 6 × 10 + 4 . . . 1,764 |

FIG. 11.16B.

In conclusion, the results established so far (which from now on will be considered definitive) are the following:

| | | | | | | | | |
|---|---|---|---|---|---|---|---|---|
| $\frac{1}{120}$ | $\frac{1}{60}$ | $\frac{1}{30}$ | $\frac{1}{10}$ | $\frac{1}{5}$ | 1 | 10 | 100 | 300 |

FIG. 11.17.

Therefore, of the eleven number-signs of the proto-Elamite system, nine have been deciphered.

Now let us consider the delicate problem of the following two number-signs:

FIG. 11.18.

As we have already shown in Fig. 11.2, these two numbers have been interpreted in the most diverse ways since the beginning of this century (the number labelled N, for example, has been assigned to 600, to 6,000, to 10,000, or even to 1,000). To try to have a better understanding of the situation, we shall consider the tablet shown in Fig. 11.19 A. According to V. Scheil, this is "an important example of an exercise in agricultural accounting". As far as I know, this is the only preserved intact proto-Elamite document which contains both the entire set of number-signs of the first system and also a grand summary total.

On this tablet, we find:

- on the top face, a series of twenty numerical entries (corresponding to an inventory of twenty lots of the same kind denoted, it would seem, by the script character at the right of the top line);
- on the reverse, the corresponding grand total (itself preceded by the same written character).

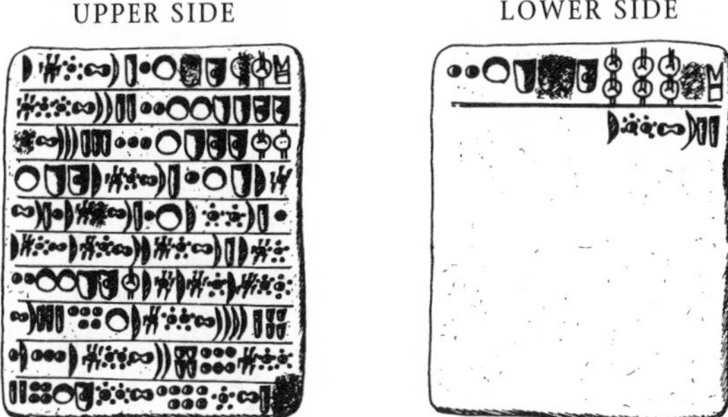


FIG. 11.19A. *Accounting tablet from Susa. Ref. MDP, XXVI, diagram 362*

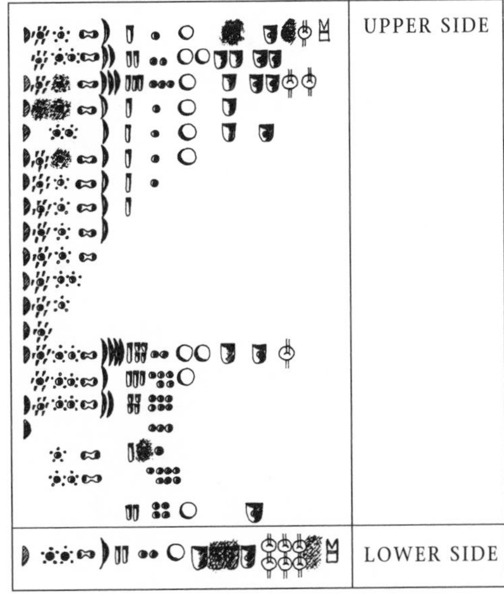


FIG. 11.19B.

Considering the results we have already obtained, we shall make various attempts to reconcile the totals of the numbers on this tablet, by trying various different possible values for the numbers labelled N and P, and making use of the numbers of occurrences of the different signs as shown in Fig. 11.19 C.

| | | | | | | | | | | | N | P |
|---|---|---|---|---|---|---|---|---|---|---|---|---|
| Number of times each sign occurs | on the upper side | 15 | 15 | 24 | 14 | 19 | 26 | 39 | 11 | 7 | 8 | 5 |
| | on the lower side | 1 | 0 | 2 | 1 | 1 | 2 | 2 | 1 | 1 | 3 | 6 |

FIG. 11.19C. *Complete listing of all the numerical signs on the tablet*

**First attempt:**

Following Scheil (1935, see MDP, XXVI), let us assign the value 10,000 to the wide notch with the circular impression (N), and the value 100,000 to the circle with the little wings (P). On the upper face of the tablet, we then obtain the following total for the numbers which appear there (Fig. 11.19 C):

$$15 \times \frac{1}{120} + 15 \times \frac{1}{60} + 24 \times \frac{1}{30} + 14 \times \frac{1}{10} + 19 \times \frac{1}{5}$$
$$+ 26 + 39 \times 10 + 11 \times 100 + 7 \times 300 + 8 \times 10{,}000 + 5 \times 100{,}000$$

namely $583{,}622 + \frac{45}{120}$

On the lower, similarly (Fig. 11.19 C):

$$1 \times \frac{1}{120} + 0 \times \frac{1}{60} + 2 \times \frac{1}{30} + 1 \times \frac{1}{10} + 1 \times \frac{1}{5}$$
$$+ 2 + 2 \times 10 + 1 \times 100 + 1 \times 300 + 3 \times 10{,}000 + 6 \times 100{,}000$$

namely $630{,}422 + \frac{45}{120}$

The difference between these two results is 46,800, far too great to allow this attempt to be considered correct, if we attribute the discrepancy to an error on the part of the scribe.

**Second attempt:**

Now consider the possibilities of assigning the values:

N = 6,000 [V. Scheil (1923)], P = 100,000 [V. Scheil (1935)]

By a similar calculation, we obtain (Fig. 11.19 C):

Upper side $551{,}622 + \frac{45}{120}$ Lower side B $618{,}000 + \frac{45}{120}$

This attempt also must be considered to fail, since the discrepancy between the two faces is again too large.

**Third attempt:**

Now let us try:

N = 6,000 [V. Scheil (1923)], P = 10,000 [S. Langdon (1925)]

This again fails, since we obtain (Fig. 11.19 C):

Upper side $101{,}622 + \frac{45}{120}$ Lower side $78{,}422 + \frac{45}{120}$

**Fourth attempt:**

Now let us consider the values proposed by R. de Mecquenem in 1949:

N = 1,000, and P = 10,000

Again from Fig. 11.19 C, we obtain the results

Upper side $61{,}622 + \frac{45}{120}$ Lower side $63{,}422 + \frac{45}{120}$

This possibility seemed to me for a long time to be the most likely solution. The results it gives are relatively satisfactory, since the discrepancy between the totals for the two faces of the tablet is only 1,800. On this belief, I had therefore supposed that the scribe had made some error in calculation, or had omitted to inscribe on the tablet the numbers corresponding to this difference. This, after all, could be likely enough, considering the many number-signs crowded onto the tablet – *errare humanum est*! Let us not forget that, just as in our own day, the scribes of old were capable of making mistakes in arithmetic.

Nonetheless, on reflection, it seemed to me that there was something illogical in attributing the value 1,000 to the number N, for two reasons.

Consider, first of all, the following two numerical entries taken from proto-Elamite tablets:

= $\mathscr{A}$ = $\mathscr{B}$

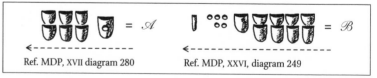

Ref. MDP, XVII diagram 280 Ref. MDP, XXVI, diagram 249

FIG. 11.20.

On Mecquenem's hypothesis, these would respectively have values

$$\mathscr{A} = 1 \times 1{,}000 + 6 \times 300 = 2{,}800$$
$$\mathscr{B} = 9 \times 300 + 5 \times 10 + 1 = 2{,}751$$

Now, still adopting this hypothesis, the following numbers would be units of consecutive orders of magnitude:

1 10 100 300 1,000 10,000

Therefore, in the first place, the question arises: if the notch with the circular impression really corresponded to the value 1,000, why should the scribes have adopted the above representations of the numbers 2,800 and 2,751, and not the more regular forms in Fig. 11.21 following?

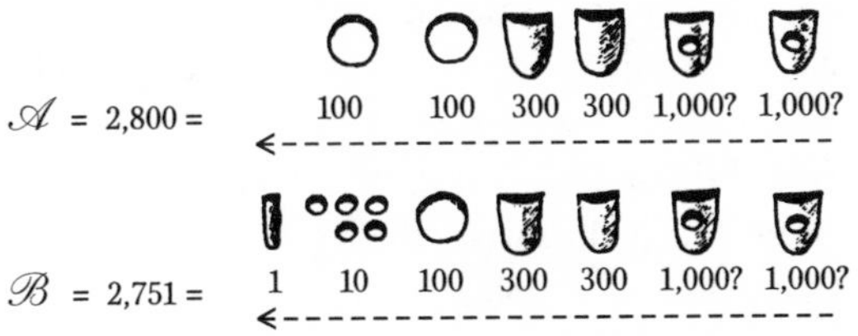

$\mathscr{A}$ = 2,800 = 100 100 300 300 1,000? 1,000?

$\mathscr{B}$ = 2,751 = 1 10 100 300 300 1,000? 1,000?

FIG. 11.21.

On the other hand, we know that for the Sumerians the small circular impression had value 10, the wide notch 60, and the combination of the latter including the former had value 600:

10 60 $60 \times 10 = 600$

FIG. 11.22.

in other words, that the last figure follows the multiplicative principle.

But for the Elamites the small circular impression had value 10 while the wide notch had value 300. By analogy with the Sumerian system, the value $300 \times 10 = 3{,}000$ should be assigned to the wide notch compounded with the small circle:

10 300 $300 \times 10 = 3{,}000?$

FIG. 11.23.

For these reasons I was led to reject Mecquenem's hypothesis.

**Fifth attempt:**

We are therefore now led to consider the proposed values:

N = 3,000 and P = 10,000

[the latter from S. Langdon (1925) and R. de Mecquenem (1949), the former from the above reasoning]. Again comparing the totals from the two faces of the tablet, this hypothesis gives the following results:

Upper side $77{,}622 + \frac{45}{120}$ Lower side B $69{,}422 + \frac{45}{120}$

This hypothesis therefore does not work either. But, if we wish to keep the value of 3,000 for the number N, we must seek a different value for the number P.

Now, close examination of the mathematical structure which can be inferred from the values so far determined in the proto-Elamite number-system caused me to suppose that the following three values could be possible for the number P:

9,000, 18,000 and 36,000

I was led to this supposition by postulating that the proto-Elamite system of fractions was developed on the same lines as the notation for the whole numbers, namely that there had to be a certain correspondence between a scale of increasing values, and a scale of decreasing values, relative to a given base number.

This, however, is exactly what one observes if one expresses the different values determined so far in terms of the number M = 300 (Fig. 11.24).

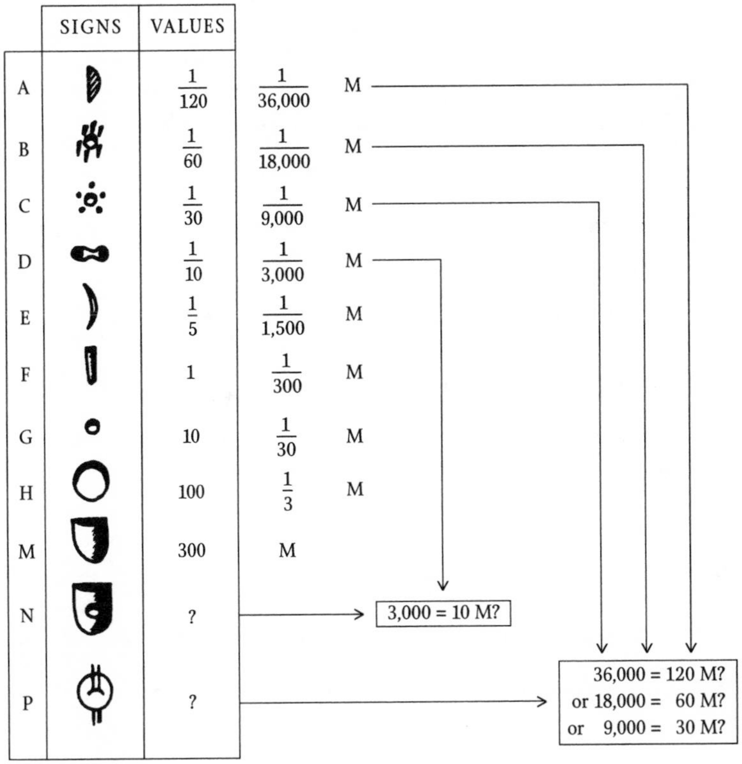

| | SIGNS | VALUES | | |
|---|---|---|---|---|
| A | | $\frac{1}{120}$ | $\frac{1}{36{,}000}$ M | |
| B | | $\frac{1}{60}$ | $\frac{1}{18{,}000}$ M | |
| C | | $\frac{1}{30}$ | $\frac{1}{9{,}000}$ M | |
| D | | $\frac{1}{10}$ | $\frac{1}{3{,}000}$ M | |
| E | | $\frac{1}{5}$ | $\frac{1}{1{,}500}$ M | |
| F | | 1 | $\frac{1}{300}$ M | |
| G | | 10 | $\frac{1}{30}$ M | |
| H | | 100 | $\frac{1}{3}$ M | |
| M | | 300 | M | |
| N | | ? | | 3,000 = 10 M? |
| P | | ? | | 36,000 = 120 M?<br>or 18,000 = 60 M?<br>or 9,000 = 30 M? |

FIG. 11.24.

**Sixth attempt:**

This now leads us to contemplate the possibilities based on these three possible values for P, of which the first is (Fig. 11.19 C):

$$N = 3{,}000,\ P = 9{,}000$$

But on comparing the totals which result, we find a serious discrepancy:

$$\text{Upper side } 72{,}622 + \frac{45}{120} \qquad \text{Lower side } 63{,}422 + \frac{45}{120}$$

Difference 9,200

Therefore this suggestion must be rejected.

**Seventh attempt:**
The same results from trying the second possibility inferred above, since the values:

$$N = 3{,}000,\ P = 36{,}000$$

also lead to implausible results (Fig. 11.19 C):

$$\text{Upper side } 207{,}622 + \frac{45}{120} \qquad \text{Lower side } 225{,}422 + \frac{45}{120}$$

Difference 17,800

**Final attempt, and the solution of the problem:**
Now consider the final possibility, with the following values:

$$N = 3{,}000,\ P = 18{,}000$$

This system, which is compatible with a coherent mathematical structure, also gives satisfyingly close agreement:

$$\text{Upper side } 117{,}622 + \frac{45}{120} \quad \left(117{,}622 + \frac{1}{5} + \frac{1}{10} + \frac{2}{30} + \frac{1}{120}\right)$$

$$\text{Lower side } 117{,}422 + \frac{45}{120} \quad \left(117{,}422 + \frac{1}{5} + \frac{1}{10} + \frac{2}{30} + \frac{1}{120}\right)$$

Whence, however, comes this discrepancy of 200 which exists between the two faces if we adopt this hypothesis? Quite simply, I believe, from a "typographical error".

Instead of inscribing on the lower side the grand total corresponding to the inventory on upper side, which should be in the form:

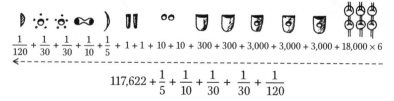

$$\frac{1}{120} + \frac{1}{30} + \frac{1}{30} + \frac{1}{10} + \frac{1}{5} + 1 + 1 + 10 + 10 + 300 + 300 + 3{,}000 + 3{,}000 + 3{,}000 + 18{,}000 \times 6$$

<---------------------------------------------------

$$117{,}622 + \frac{1}{5} + \frac{1}{10} + \frac{1}{30} + \frac{1}{30} + \frac{1}{120}$$

FIG. 11.25A.

the scribe in fact made a large circular impression in the place of one of the two wide notches:

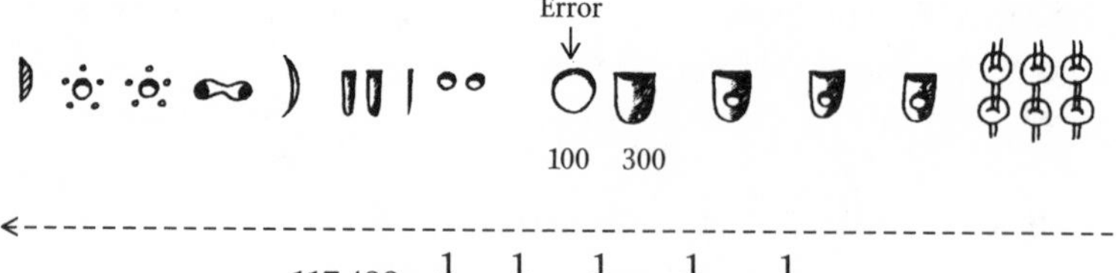


$$117{,}422 + \frac{1}{5} + \frac{1}{10} + \frac{1}{30} + \frac{1}{30} + \frac{1}{120}$$

FIG. 11.25B.

It is easy to see how this could happen. The scribe held his stylus with large circular cross-section in the wrong position (See Fig. 8.10 and 8.12 above): instead of pressing the stylus at an angle of 30°–45° to the surface of the soft clay, which would have given him a wedge, he held it perpendicular to the surface thereby obtaining the circle.

That is, instead of doing this:

FIG. 11.26A.

he did this:

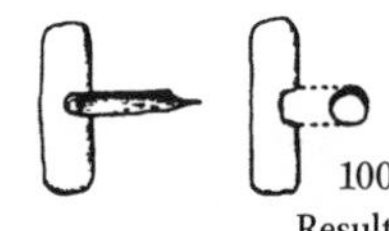


FIG. 11.26B.

Therefore, in all probability, we may conclude that the wide notch with a small circular imprint corresponds to the value 3,000, and the circle with the little wings corresponds to the value 18,000.

All the numbers in the proto-Elamite system have, therefore, been definitively deciphered.

We have good reason to suppose that this system is the more ancient of the two since the following numerals appear on the proto-Elamite accounting tablets from the archaic epoch onwards.

FIG. 11.27.

The same set of numerals appears on the earliest numerical tablets, as well as on the outside of the counting balls recently discovered on the site of the Acropolis of Susa. Finally, the numerals also are those which, according to their respective shapes, correspond to the archaic *calculi* which were formerly enclosed in the counting balls, in fact to the number-tokens of various shapes and sizes which stood for these numbers (and whose values, in turn, have themselves now been determined as a result of the decipherment described above; see also Fig. 10.8 and 10.14 above):

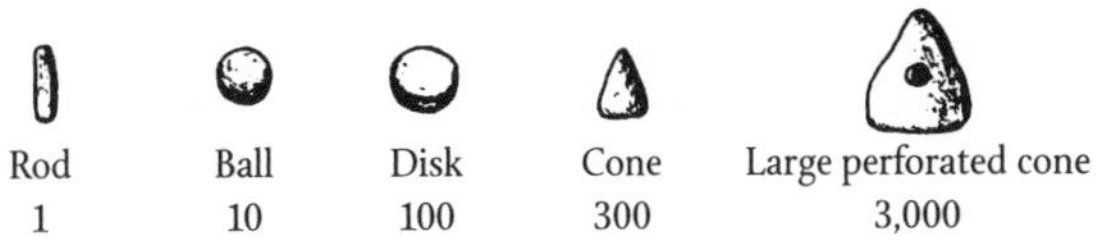

| Rod | Ball | Disk | Cone | Large perforated cone |
|---|---|---|---|---|
| 1 | 10 | 100 | 300 | 3,000 |

FIG. 11.28.

As to the second system of writing numbers, I believe that the Elamites constructed it – maybe in a relatively recent era – for the purpose of recording quantities of objects or of goods, or magnitudes, of a different kind from those for which the symbols of the first system were used.

I base this hypothesis on an analogy with Sumerian usage. During the third millennium BCE, the scribes of Lower Mesopotamia in fact used three different numerical notations:

- the first, the commonest and oldest, which we have studied in Chapter 8, was used for numbers of men, beasts, or objects, or for expressing measures of weight and length;
- the second was used for measures of volume;
- the third was used for measures of area.

This hypothesis is in fact confirmed by the tablet shown in Fig. 11.29, which carries two inventories which have been very clearly differentiated.

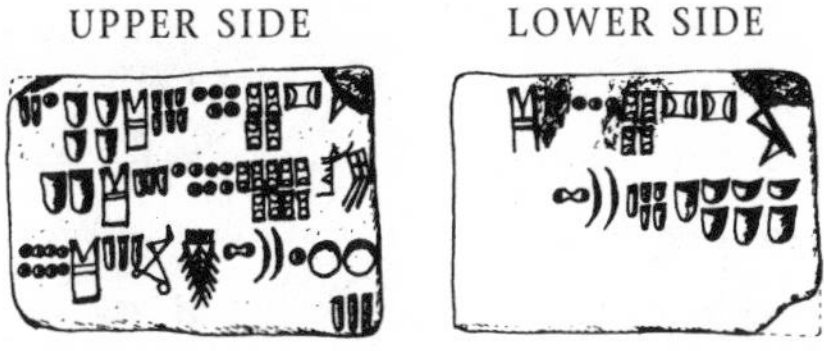


FIG. 11.29A. *Accounting tablet from Susa. Teheran Museum. Ref. MDP, XXVI, diagram 156*

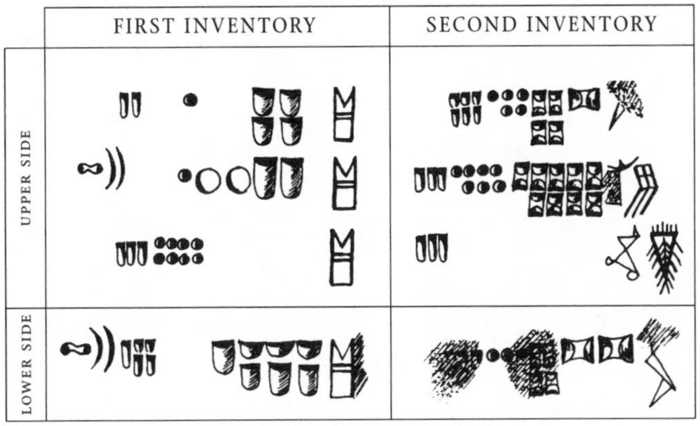

| | FIRST INVENTORY | SECOND INVENTORY |
|---|---|---|
| UPPER SIDE | | |
| LOWER SIDE | | |

FIG. 11.29B.

The first of these inventories is indicated by a characteristic script character, and the corresponding quantities are expressed in the numerals of the first proto-Elamite system (Fig. 11.29 B). The second inventory is indicated by the signs (which have not yet been deciphered):

and the corresponding quantities are expressed in the numerals of the second proto-Elamite system (Fig. 11.9).

The numbers given on the reverse of this tablet correspond respectively to the total of the first inventory and to the total of the second. Using the values we have already obtained, we can make the totals for the first inventory:

a) on upper side:

$$6 \times 300 + 2 \times 100 + 10 \times 10 + 5 + \frac{2}{5} + \frac{1}{10} = 2{,}105 + \frac{2}{5} + \frac{1}{10}$$

b) on lower side:

$$7 \times 300 + 5 + \frac{2}{5} + \frac{1}{10} = 2{,}105 + \frac{2}{5} + \frac{1}{10}$$

(which, by the way, is a further confirmation of the validity of our earlier result).

Now let us consider the different numerals on the second inventory, and let us give value 1 to the narrow notch, 10 to the small circular impression, 100 to the double vertical notch and 1,000 to the double horizontal notch. Then the totals come out as follows:

a) on upper side:

$$1{,}000 + 13 \times 100 + 12 \times 10 + 12 = 2{,}432$$

b) on lower side:

$$2 \times 1{,}000 + 4 \times 100 + 3 \times 10 + 2 = 2{,}432$$

We may therefore fix the values of the following numerals as shown:

FIG. 11.30.

(where the former of these values, for example, is confirmed by the tablet in Fig. 11.31, since the totals come to 591 on both sides).

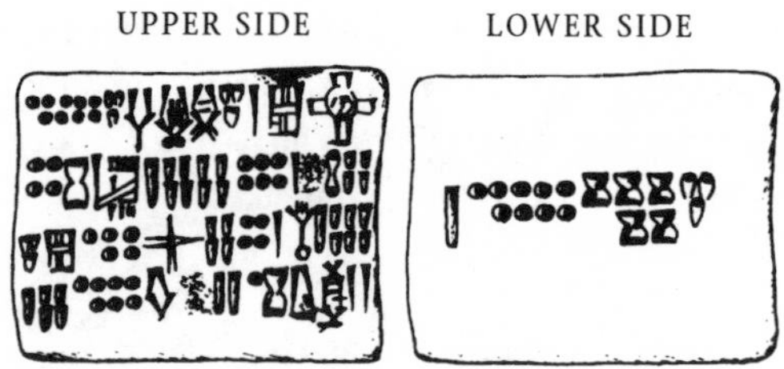


FIG. 11.31A. *Accounting tablet from Susa. Louvre. Ref. MDP, XVII, diagram 45*

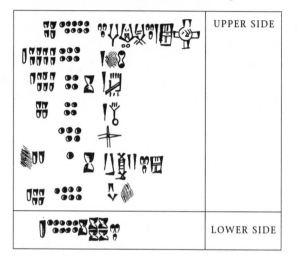

UPPER SIDE

LOWER SIDE

FIG. 11.31B.

So there we see pretty well all of the proto-Elamite numerals deciphered. At the same time, we have discovered that at Susa two different number-writing systems were in use, probably corresponding to two different systems of expressing numbers:

- one, strictly decimal* (Fig. 11.32);
- the other, visibly "contaminated" by the base 60 (Fig. 11.33).

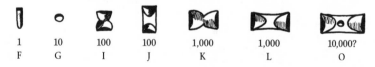

| 1 | 10 | 100 | 100 | 1,000 | 1,000 | 10,000? |
|---|---|---|---|---|---|---|
| F | G | I | J | K | L | O |

FIG. 11.32. *The values of the number-signs of the second proto-Elamite number-system*

* A question remains for the numeral formed from a double horizontal notch with a small circular impression in its centre (Fig. 11.32, sign O). Is this the numeral representing 10,000 = 1,000 × 10? It seems likely. But this could not be stated with certainty, since we lack documents better preserved than those we have at present, relevant to this numeral.

We may suppose that the first may have been used for counting such things as people, animals or things, while the second may have been used to express different measures in a system of measurement units (volumes and areas, for example).

| | SIGNS | X | Y | VALUES |
|---|---|---|---|---|
| A | | $\frac{1}{36,000}$ M | $\frac{1}{2}$ B | $\frac{1}{120}$ |
| B | | $\frac{1}{18,000}$ M | B | $\frac{1}{60}$ |
| C | | $\frac{1}{9,000}$ M | 2 B | $\frac{1}{30}$ |
| D | | $\frac{1}{3,000}$ M | 6 B | $\frac{1}{10}$ |
| E | | $\frac{1}{1,500}$ M | 12 B | $\frac{1}{5}$ |
| F | | $\frac{1}{300}$ M | 60 B | 1 |
| G | | $\frac{1}{30}$ M | 600 B | 10 |
| H | | $\frac{1}{3}$ M | 6,000 B | 100 |
| M | | M | 18,000 B = 300 × 60 B | 300 |
| N | | 10 M | 180,000 B = 300 × 600 B | ? |
| P | | 60 M | 1,800,000 B = 300 × 6,000 B | ? |

FIG. 11.33. *The mathematical structure of the first proto-Elamite number-system*

These are of course only hypotheses, but the above results lend confirmation to the existence of cultural and economic relations between Elam and Sumer, at any rate from the end of the fourth millennium BCE, and to the influence exerted by the Sumerians upon Elamite civilisation.

## CHAPTER 12

# HOW THE SUMERIANS DID THEIR SUMS

The arithmetical problems which the Sumerians had to deal with were quite complicated, as is shown by the many monetary documents which they have bequeathed to us. The question which we shall now address is to find out what methods they used in order to carry out additions, multiplications, and divisions. First of all, however, let us have a look at one very interesting document.

### A FOUR-THOUSAND-YEAR-OLD DIVISION SUM

The tablet shown in Fig. 12.1 is from the Iraqi site of Fara (Šuruppak), and it dates from around 2650 BCE.

We shall present its complete decipherment according to A. Deimel's *Sumerisches Lexikon* (1947). This document provides us with the most valuable information on Sumerian mathematics in the pre-Sargonic era (the first half of the third millennium BCE). It shows the high intellectual level attained by the arithmeticians of Sumer, probably since the most archaic era.

The tablet is divided into two columns, each subdivided into several boxes.

From top to bottom, in the first box of the left-hand column is a narrow notch, followed by a cuneiform group (*še-gur*$_7$), which signifies "granary of barley".

In the box beneath is a representation of the number 7, preceded by a sign which is to be read *sìla*.

In the third box, the numeral 1 is followed by the sign for "man" (*lú*); below this is a group which is to be read *šu-ba-ti* (the word *šu* means "hand") and which might be translated as "given in the hand".

Finally, at the very bottom of the left-hand column is the sign for "man" again, above which is the character *bi* which is simply the indicative "these".

The literal translation of this column therefore is: "1 granary of barley; 7 *sìla*; each man, given in the hand; these men."

In the first box of the right-hand column, we can recognise the representation of 164,571 in the archaic numerals (see Fig. 8.20 above), and in the box below a succession of signs which represent the phrase "granary of barley, there remains: 3".

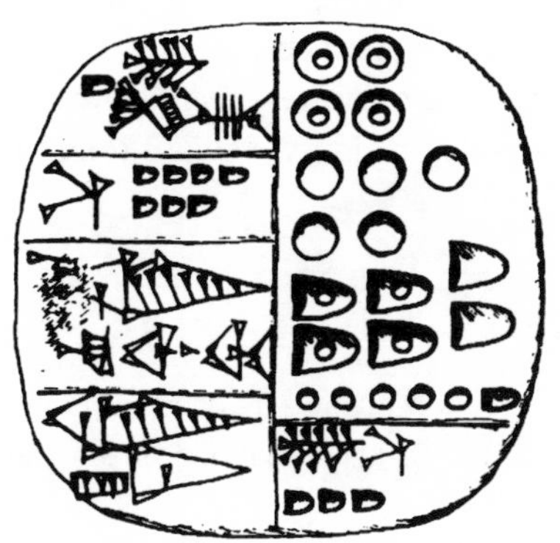

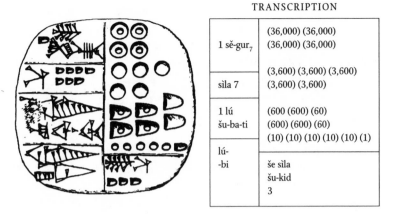

TRANSCRIPTION

| | |
|---|---|
| 1 sě-gur$_7$ | (36,000) (36,000)<br>(36,000) (36,000) |
| sìla 7 | (3,600) (3,600) (3,600)<br>(3,600) (3,600) |
| 1 lú<br>šu-ba-ti | (600 (600) (60)<br>(600) (600) (60)<br>(10) (10) (10) (10) (10) (1) |
| lú-<br>-bi | še sìla<br>šu-kid<br>3 |

LITERAL TRANSLATION

| Left-hand register | Right-hand register |
|---|---|
| 1 "granary of barley" | |
| 7 *sìla* (of barley) | 164,571 |
| Each man<br>in his hand receives | |
| Men<br>these | *sìla* of barley<br>remain<br>3 |

FIG. 12.1. *Sumerian tablet from Šuruppak (Fara). Date: c. 2650 BCE. Istanbul Museum. Ref. Jestin (1937), plate XXI, diagram 50 FS*

This tablet, which no doubt describes a distribution of grain, shows all the formal elements of arithmetical *division*: we have a *dividend*, a *divisor*, a *quotient*, and even, to an astonishing precision for the time, a *remainder*.

The *sìla* and the *še-gur*$_7$ ("granary of barley") are units of measurement of volume. At that time, the former contained the equivalent of 0.842 of our litre, while the latter came to about 969,984 litres, namely 1,152,000 *sìla* [see M. A. Powell (1972)]:

1 *še-gur*$_7$ (1 granary of barley) = 1,152,000 *sìla*

Thus this distribution involved the division of 1,152,000 *sìla* of barley between a certain number of people, each of whom is to receive 7 *sìla*.

Now let us do the calculation. 1,152,000 divided by 7 is 164,571, exactly the number in the first box of the right-hand column; and the remainder is 3, exactly the information given at the bottom of this column.

There is no doubt about it: you have before your very eyes the written testimony of the oldest known division sum in history – quite a complex one; and as old as Noah.

## OFFICIAL DOCUMENT OR LEARNER'S EXERCISE?

One may suppose that this tablet was probably an official document in the archives of the ancient Sumerian city of Šuruppak, unless it happened to be an exercise for apprentice calculators.

On the first supposition, then its translation into plain language is as follows:

> We have divided 1 granary of barley between a certain number of people, giving 7 *sìla* to each one. These men were 164,571 in number, and at the end of the distribution there were 3 *sìla* remaining.

On the other hand, if it was really an exercise for learners, then the appropriate translation would be:

| STATEMENT OF THE PROBLEM | SOLUTION OF THE PROBLEM |
|---|---|
| Given that a granary of barley has been divided between several men so that each man received 7 *sìla*, find the number of men. | The number of men was 164,571 and 3 *sìla* were left over after the distribution. |

For convenience of exposition, we shall adopt the latter interpretation in what follows.

There is no indication whatever in the document as to the method of calculation to be used to obtain the result. Nor do we yet know of any formal description. One thing however is certain, and that is that the calculation was not carried out by means of Sumerian numerals, which do not encapsulate an operational capability in the way that our own numerals do.

Nonetheless the results of the previous chapter give us some basis for supposition as to what the means of calculation may have been. The Sumerians most probably made use of the *calculi* (the very ones shown in Fig. 10.4), as much prior to the emergence of their numerical notation as subsequently, since we find these tokens in various archaeological sites of the third millennium BCE, that is to say, at a time when *bullae* had almost entirely been displaced by clay tablets (see Fig. 10.2 above).

We shall now put forward a speculative but entirely plausible reconstruction of the technique of calculation which was most probably used.

## CALCULATION WITH PELLETS, CONES, AND SPHERES

Let us imagine we are in the year 2650 BCE, in the Sumerian city of Šuruppak. We are in the school where scribes and accountants learn their skills, and the teacher has given a lesson on how to do a division. Now he begins the practical class, and sets the problem of dividing one granary of barley according to the conditions given.

The problem is therefore to divide 1,152,000 *sìla* of barley between a certain number of persons (to be determined) so that each one gets 7 *sìla* of barley, which comes down to dividing the first number by 7.

At this time, additions, multiplications, and divisions are carried out by means of the *calculi*, those good old *imnu* of former times which, in their several shapes and sizes, symbolise the different orders of magnitude of the units in the Sumerian number-system. Although their use has long disappeared from accounting practice, they are still the means that everyone uses for calculation. This has never worried any of the generations of scribes since the day when one of them thought of making replicas of the various *calculi* on clay tablets, to serve as numerical notations – a narrow notch for the small cone, a small hole for the pellet, a wide notch for the large cone, and so on (see Fig. 10.4 above).

Generally, the procedure for performing a division brings in successively: pierced spheres (= 36,000), plain spheres (= 3,600), large pierced cones (= 600), large plain cones (= 60), and so on. At each stage, the pieces are converted into their equivalents as multiples of smaller units whenever they are fewer than the size of the divisor.

Practically speaking, therefore, the above example proceeds as follows.

In Sumerian, the dividend 1,152,000 is expressed in words (see Fig. 8.5 above) as

*šàrgal-iá šàr-u-min*

which corresponds to the decomposition

$$216{,}000 \times 5 + 36{,}000 \times 2 = 5 \times 60^3 + 2 \times (10 \times 60^2)$$

The largest unit of the written numerals at this time, however, is only 36,000 (see Fig. 10.4 above), which is also the value of the largest of the *calculi*. Therefore the dividend must first be expressed in multiples of this smaller unit, therefore by 32 pierced spheres each of which stands for 36,000 units:

$$1{,}152{,}000 = 32 \times 36{,}000$$

But we are to divide this by 7, so we arrange these as best we can in groups of 7:

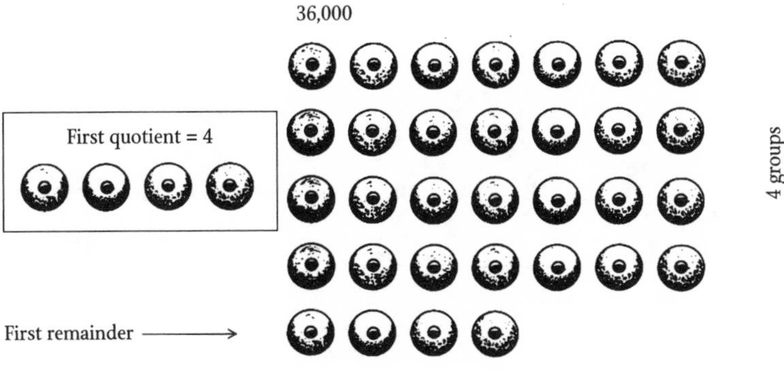


FIG. 12.2A.

The number of groups, each with 7 pierced spheres, in this first arrangement is 4, which is the quotient from this first partial division. This, in the context of the problem, is also equal to the number ( 4 × 36,000) of the first group of people who will receive 7 *sìla* of barley each. In order not to lose track of this partial result, we shall put 4 pierced spheres on one side to represent it.

After this, we have 4 pierced spheres left over. We therefore must divide these 4 × 36,000 *sìla*. But when it is expressed in pierced spheres, worth 36,000 each, we find that 4 cannot be divided by 7. At this point, therefore, we convert each one of these into its equivalent number of the next lower order of magnitude.

Each pierced sphere (36,000) is equivalent to 10 plain spheres, each worth 3,600. The 4 pierced spheres therefore become 4 × 10 = 40 plain spheres, which we once again arrange in groups of 7:

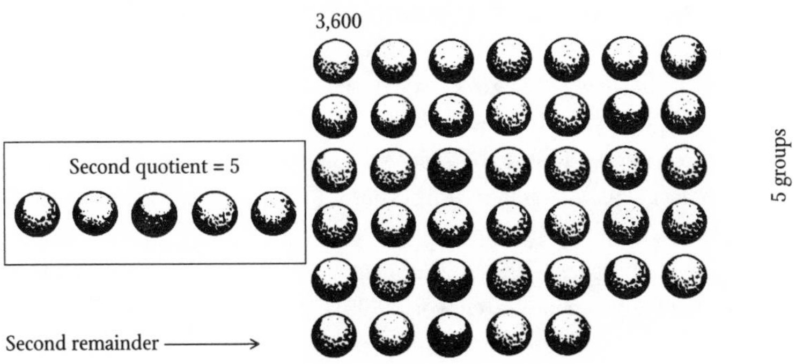


FIG. 12.2B.

Now we find that there are 5 complete groups of 7 plain spheres, so we put on one side 5 plain spheres (corresponding to the second group, 5 × 3,600, of people who will each receive 7 *sìla* of barley).

But we find that there are 5 plain spheres left over at the end of this second division, and 5 is not divisible by 7, so we now replace each plain sphere by its equivalent number of pieces of the next lower order of magnitude.

Each "3,600" sphere is equivalent to 6 large pierced cones worth 600 each, so we convert the 5 pierced spheres left over into 5 × 6 = 30 large pierced cones which we again arrange in groups of 7:

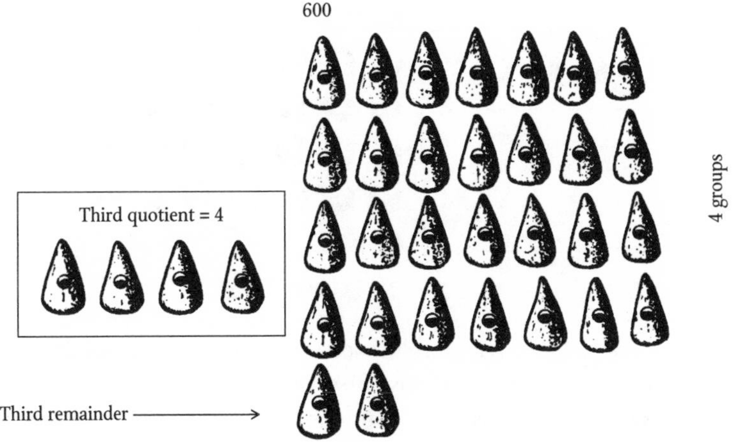


FIG. 12.2C.

Since we have 4 full groups of 7 pierced cones each, we therefore put aside 4 large pierced cones, corresponding to the third part of the men who will receive 7 *sìla* of barley each (4 × 600).

However, we now have 2 large pierced cones left over, so we still have to divide 2 × 600 *sìla* of barley.

Each "600" cone is equivalent to 10 large plain cones worth 60 each, so we convert the two large pierced cones left over into 2 × 10 = 20 large plain cones and we arrange these in groups of 7.

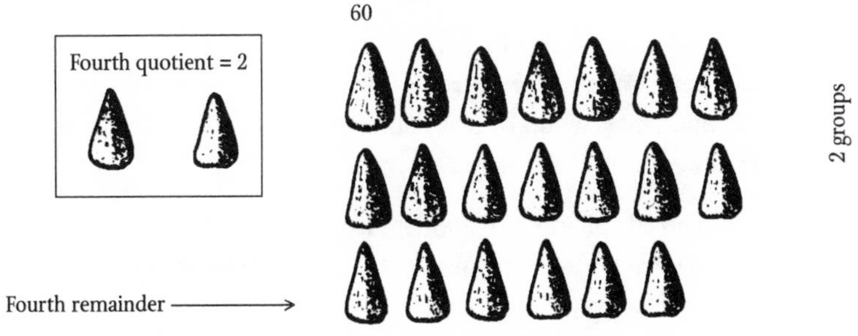


FIG. 12.2D.

We can form 2 complete groups of 7, with 6 large plain cones left over. As before, we put aside 2 cones to note the number of complete groups, corresponding to the 2 × 60 men who will each get 7 *sìla* of barley at this fourth stage of the distribution.

Now we convert the 6 large plain cones left over, worth 60 each, into their equivalent in pellets worth 10 each, therefore into 6 × 6 = 36 pellets, and we arrange these into groups of 7, with 1 left over:

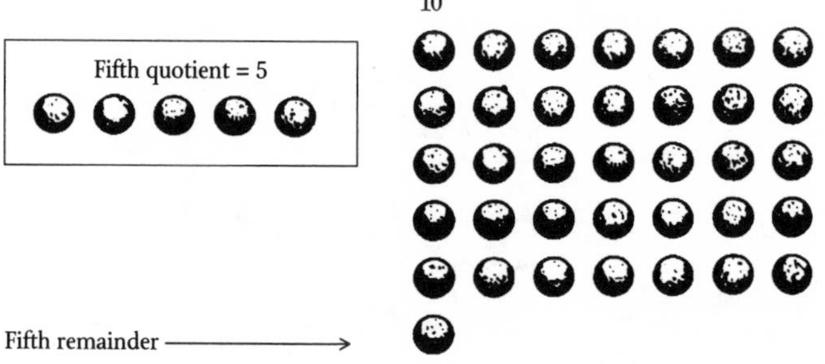


FIG. 12.2E.

Once again, we put aside 5 pellets corresponding to the 5 × 10 men who will each get 7 *sìla* of barley at this fifth stage of the distribution.

The single pellet left over, worth 10, is now converted into 10 small cones each worth 1. This makes one complete group of 7, with 3 left over.

FIG. 12.2F.

To note the one complete row, we put aside 1 small cone, and this corresponds to the number (10) of men who will each get 7 *sìla* of barley at this sixth stage of the distribution. Since the number corresponding to the left-over cones is 3, and this is less than the divisor, we can proceed no further in the division of the original number into whole units, and we have finished.

The final quotient can now be easily obtained by totalling the values of the pieces which we successively set aside in the course of the division, as follows:

| | |
|---|---|
| 4 pierced spheres | (quotient from the first division) |
| 5 plain spheres | (quotient from the second division) |
| 4 large pierced cones | (quotient from the third division) |
| 2 large plain cones | (quotient from the fourth division) |
| 5 pellets | (quotient from the fifth division) |
| *and* | |
| 1 small cone | (quotient from the sixth division) |

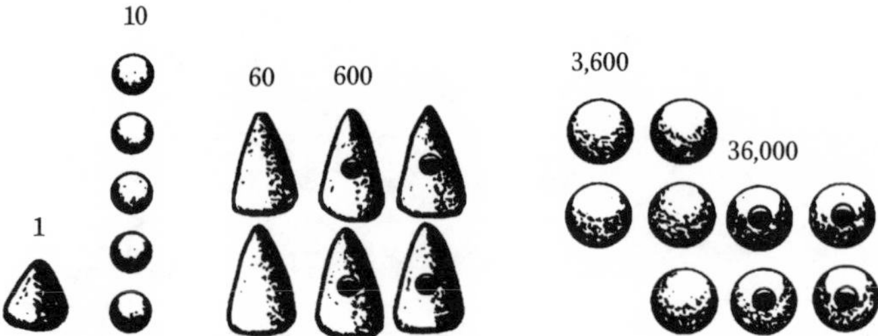


FIG. 12.2G. *Result of the division*

In other words, the total number of people to whom the barley will be distributed is

$$4 \times 36{,}000 + 5 \times 3{,}600 + 4 \times 600 + 2 \times 60 + 5 \times 10 + 1 = 164{,}571$$

Back at the school of arithmetic, one student raises his hand and gives his answer, in Sumerian words pronounced in the following order:

| | | | | |
|---|---|---|---|---|
| *šàr-u-limmu* | = | (3,600 × 10) × 4 | = | 4 pierced spheres |
| *šàr-iá* | = | 3,600 × 5 | = | 5 spheres |
| *geš-u-limmu* | = | (60 × 10) × 4 | = | 4 large pierced cones |
| *geš-min* | = | 60 × 2 | = | 2 large cones |
| *ninnû* | = | 50 | = | 5 pellets |
| *geš* | = | 1 | = | 1 small cone |

Not forgetting to add, of course

*še sìla šu-kid eš* ("and there are 3 *sìla* of barley left over")

Another of the students, however, shows up his work to the teacher as he has traced it onto his clay tablet, which he has divided into boxes and filled up with Sumerian script. In the top right-hand box, in archaic numerals, he has written the answer (164,571) exactly as shown in Fig. 8.20 above:

- 4 large circular impressions with small circular impressions within (a direct representation of 4 pierced spheres, each worth 36,000);
- 5 large plain circular impressions (a direct representation of 5 spheres, each worth 3,600);
- 4 wide notches with small circular impressions within (recalling the 4 large pierced cones, each worth 600);
- 2 plain large notches (for the 2 plain large cones, each worth 60);
- 5 small circular imprints (for the 5 pellets each worth 10); and
- 1 narrow notch (for the small cone representing 1).

And, since the spoken word vanishes into thin air, while what is written remains, it is thanks to the latter that the division sum from Šuruppak has survived for the thousands of years since the students who solved it vanished from the face of the earth.

## THE DISAPPEARANCE OF THE *CALCULI* IN MESOPOTAMIA

We can infer that Sumerian arithmetic was done in this kind of way from the most archaic times down to the pre-Sargonic era. The tablet shown in Fig. 12.1 is one piece of evidence, and the *calculi* from this epoch found in those regions provide another; but the most solid proof is the reconstruction of the method which we have shown, for it can be easily demonstrated that the same principles may be applied equally well to multiplication, addition, and subtraction.

Nonetheless, the historical problems of Mesopotamian arithmetic have not been completely solved.

At the time at which the tablet we have been examining was made (around 2650 BCE), the *calculi* were still in use throughout the region, and in appearance they remained close to the archaic, or *curviform*, numerals which had then come into use. These numerals, however, while still present at the time of Sargon I (around 2350 BCE), gradually disappeared during the second half of this millennium. Finally, at the time of the dynasty of Ur III (around 2000 BCE) they had been replaced by the cuneiform numerals. Correspondingly, the *calculi* themselves are no longer found in the majority of the archaeological sites of Mesopotamia dating from this period or later (Fig. 10.2 above).

While undergoing this transformation from archaic curviform to cuneiform aspect, the written numerals lost all resemblance to the *calculi* which were their concrete ancestors. The Sumerian written number-system, moreover, was essentially a *static* tool with respect to arithmetic, since it was not adapted to manipulation for calculations: the numerals, whether curviform or cuneiform, instead of having inherent potential to take part in arithmetical processes, were graphical objects conceived for the purpose of expressing in writing, and solely for the sake of recall, the results of calculations which had already been done by other means.

Therefore the calculators of Sumer, at a certain point in time, faced the necessity of replacing their old methods with new in order to continue to function. They therefore substituted for the old system of the *calculi* a new "instrument" which I shall shortly describe. Meanwhile, we make a detour to prepare the ground.

## FROM PEBBLES TO ABACUS

Only a few generations ago, natives in Madagascar had a very practical way of counting men, things, or animals. A soldier, for instance, would make his men pass in single file though a narrow passage. As each one emerged he would drop a pebble into a furrow cut into the earth. After the tenth had passed, the 10 pebbles would be taken out, and 1 pebble added to a parallel furrow reserved for tens. Further pebbles were then placed into the first furrow until the twentieth man had passed, then these 10 would be taken out and another added to the second furrow. When the second furrow had accumulated 10 pebbles, these in turn were taken out and 1 pebble was added to a third furrow, reserved for hundreds. And so on until the last man had emerged. So a troop of 456 men would leave 6 pebbles in the first furrow, 5 in the second, and 4 in the third.

Each furrow therefore corresponded to a power of 10: the ones, the tens, the hundreds, and so on. The Malagasies had unwittingly invented the abacus.

This was not unique to them, however. Very similar means have been devised since the dawn of time by peoples in every part of the earth, and the form of the instrument has also varied.

Some African societies used sticks onto which they slid pierced stones, each stick corresponding to an order of magnitude.

Amongst other peoples (the Apache, Maidu, Miwok, Walapai, or Havasupai tribes of North America, or the people of Hawaii and many Pacific islands) the practice was to thread pearls or shells onto threads of different colours.

Others, like the Incas of South America, placed pebbles or beans or grains of maize into compartments on a kind of tray made of stone, terracotta, or wood, or even constructed on the ground.

The Greeks, the Etruscans, and the Romans placed little counters of bone, ivory, or metal onto tables or boards, made of wood or marble, on which divisions had been ruled.

Other civilisations produced better implementations of the idea, by using parallel grooves or rods, with buttons or pierced pellets which could be slid along these. This is how the famous *suan pan* or Chinese Abacus came about, a most practical and formidable instrument which is still in common use throughout the Far East.

But before they used their abacus, the Chinese had for centuries used little ivory or bamboo sticks, called *chóu* (literally, "calculating sticks") which they arranged on the squares of a tiled floor, or on a table made like a chessboard.

The abacus did not evolve solely in form and construction. Far greater changes took place in the manner of its use.

The Madagascar natives, who did not profit fully from their great discovery, no doubt never understood that this way of representing numbers would give them the means to carry out complex calculations. So in order to add 456 persons to 328 persons, they would wait out the passage of the 456, and then of the 328 others, in order to finally observe the pebbles which gave the result.

Their use of the abacus was, therefore, purely for counting. Many other peoples were no doubt in the same state in the beginning. But, in seeking a practical approach to making calculations which were becoming ever more complex, they were able to develop procedures for the device by conceiving of a subtle game in which the pebbles were added or removed, or moved from one row to another.

To add one number to a number already represented on a decimal device, all they had to do was to represent the new number also on the abacus, as before, and then – after performing the relevant reductions – to read off the result. If there were more than 10 pebbles in a column, then 10 of these would be removed and 1 added to the next, starting with the lowest-order column. Subtraction can be done in a similar way, but by taking out pebbles rather than putting them in. Multiplication can be carried out by adding the results of several partial products.

The "heap of pebbles" approach to arithmetic, indeed the manipulation of various kinds of object for this purpose, thus once again is central in the history of arithmetic. These methods are at the very origin of the calculating devices which people have used throughout history, at times when the numerals did not lend themselves to the processes of calculation, and when the written arithmetic which we can achieve with the aid of "Arabic" numerals did not yet exist.

## THE SUMERIAN ABACUS RECONSTRUCTED

It is logical, therefore, to suppose that Sumerian calculators themselves made use of some sort of abacus, at any rate once their *calculi* had disappeared from use.

Archaeological investigation in the land of Sumer has failed so far to yield anything of this kind, nor has any text been discovered which precisely describes it as well as its principles and its structure. Nonetheless, we can with the greatest plausibility reconstruct it precisely.

We may in the first place suppose that the instrument was based on a large board of wood or clay. It may equally well have been on bricks or on the floor.

The abacus consists of a table of columns, traced out beforehand, corresponding to the different orders of magnitude of the sexagesimal system.

We may likewise suppose that the tokens which were used in the device were small clay pellets or little sticks of wood or of reed, which each had a simple unit value (unlike the archaic system of the *calculi*, whose pieces stood variously for the different orders of magnitude of the same number-system).

We may determine the mathematical principles of the Sumerian abacus by appealing to their number-system itself.

Their number-system, as we have seen, used base 60. This theoretically requires memorisation of 60 different words or symbols, but the spacing between successive unit magnitudes was so great that in practice an intermediate unit was introduced to lighten the load on the memory. In this way, the unit of tens was introduced as a stepping stone between the sexagesimal orders of magnitude. The system was therefore based on a kind of compromise, alternating between 10 and 6, themselves factors of 60. In other words, the successive orders of magnitude of the sexagesimal system were arranged as follows:

| | | | | | | |
|---|---|---|---|---|---|---|
| first order | first unit | 1 | = | 1 | = | 1 |
| of magnitude | second unit | 10 | = | 10 | = | 10 |
| second order | first unit | 60 | = | 60 | = | 10.6 |
| of magnitude | second unit | 600 | = | 10.60 | = | 10.6.10 |
| third order | first unit | 3,600 | = | $60^2$ | = | 10.6.10.6 |
| of magnitude | second unit | 36,000 | = | $10.60^2$ | = | 10.6.10.6.10 |
| fourth order | first unit | 216,000 | = | $60^3$ | = | 10.6.10.6.10.6 |
| of magnitude | second unit | 2,160,000 | = | $10.60^3$ | = | 10.6.10.6.10.6.10 |

On this basis, therefore, we can lay out the names of the numbers in a tableau as in Fig. 12.3. There are nine different units, five different tens, nine different sixties, and so on. From this table, therefore, we can clearly see that ten units of the first order are equivalent to one unit of the second, that six of the second are equivalent to one of the third, that ten of the third are equivalent to one of the fourth, and so on, alternating between bases of 10 and 6.

If, therefore, we accept that the Sumerians had an abacus, it must have been laid out as in Fig. 12.4.

Each column of the abacus therefore corresponded to one of the two sub-units of a sexagesimal order of magnitude. Since, moreover, the cuneiform notation of the numerals was written from left to right, in decreasing order of magnitude starting from the greatest, we may therefore reconstruct this subdivision in the following manner.

Proceeding from right to left, the first column is for the ones, the second for the tens, the third for the sixties, the fourth for the multiples of 600, the fifth for the multiples of 3,600, and so on (Fig. 12.4). To represent a given number on this abacus, therefore, one simply places in each column the number of counters (clay pellets, sticks, etc.) equal to the number of units of the corresponding order of magnitude.

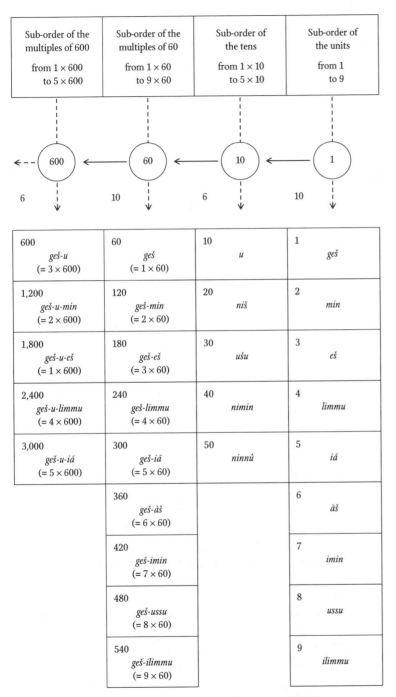

| SECOND SEXAGESIMAL ORDER | | FIRST SEXAGESIMAL ORDER | |
|---|---|---|---|
| Sub-order of the multiples of 600<br>from 1 × 600 to 5 × 600 | Sub-order of the multiples of 60<br>from 1 × 60 to 9 × 60 | Sub-order of the tens<br>from 1 × 10 to 5 × 10 | Sub-order of the units<br>from 1 to 9 |
| ← 600 (6) | ← 60 (10) | ← 10 (6) | ← 1 (10) |

| | | | |
|---|---|---|---|
| 600<br>*geš-u*<br>(= 3 × 600) | 60<br>*geš*<br>(= 1 × 60) | 10<br>*u* | 1<br>*geš* |
| 1,200<br>*geš-u-min*<br>(= 2 × 600) | 120<br>*geš-min*<br>(= 2 × 60) | 20<br>*niš* | 2<br>*min* |
| 1,800<br>*geš-u-eš*<br>(= 1 × 600) | 180<br>*geš-eš*<br>(= 3 × 60) | 30<br>*ušu* | 3<br>*eš* |
| 2,400<br>*geš-u-limmu*<br>(= 4 × 600) | 240<br>*geš-limmu*<br>(= 4 × 60) | 40<br>*nimin* | 4<br>*limmu* |
| 3,000<br>*geš-u-iá*<br>(= 5 × 600) | 300<br>*geš-iá*<br>(= 5 × 60) | 50<br>*ninnû* | 5<br>*iá* |
| | 360<br>*geš-àš*<br>(= 6 × 60) | | 6<br>*àš* |
| | 420<br>*geš-imin*<br>(= 7 × 60) | | 7<br>*imin* |
| | 480<br>*geš-ussu*<br>(= 8 × 60) | | 8<br>*ussu* |
| | 540<br>*geš-ilimmu*<br>(= 9 × 60) | | 9<br>*ilimmu* |

Fig. 12.3. *Structure of the Sumerian number-system (see also Fig. 8.6 and 10.4)*

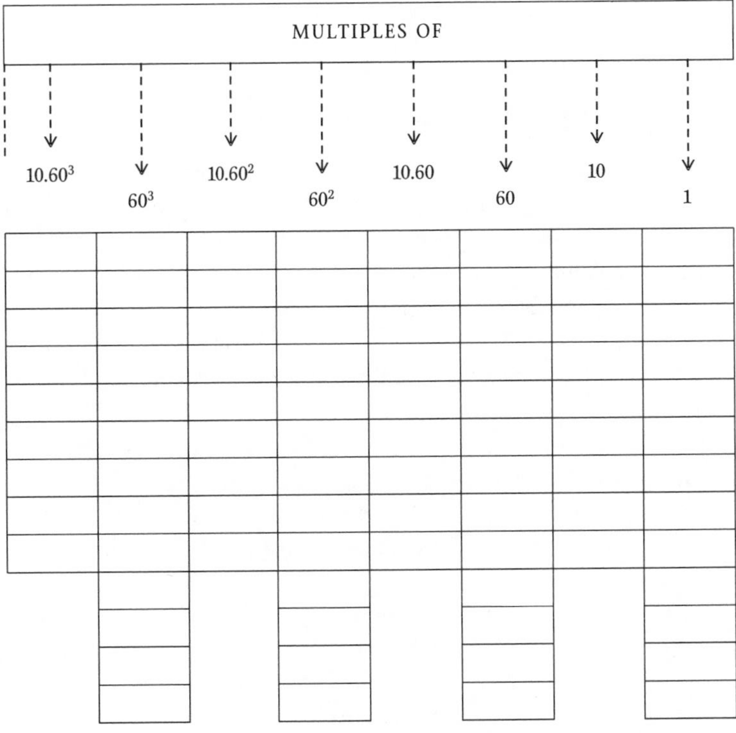


Fig. 12.4. *Form and structure of the divisions of the Sumerian abacus*

## CALCULATION ON THE SUMERIAN ABACUS

Suppose one number is already laid out on the abacus, and we wish to add another number to it. To do this, lay out the second number on the abacus as well. Then, if there are 10 or more counters in the first column, replace each 10 by a single counter added to the second. Then replace each 6 in the second column by 1 added to the third, then each 10 in the third by 1 added to the fourth, and so on, alternating between 10 and 6. When the left-hand column has been reached, the result of the addition can be read off. Subtractions proceed in an analogous way, and multiplication and division are done by repeated additions or subtractions.

Let us return to the problem in the tablet shown in Fig. 12.1, and try to solve it on the abacus. We want to divide 1,152,000 by 7. We shall proceed by means of a series of partial divisions, each one on a single order of magnitude and beginning with the greatest.

### First stage

In Sumerian terms, we are to divide by 7 the number whose expression, in number-names, is *šàrgal-iá šàr-u-min,*

which breaks down mathematically to:

$$5 \times 60^3 + 2 \times (10.60^2) = 5 \times 216{,}000 + 2 \times 36{,}000.$$

In the dividend there are therefore 5 units of order 216,000, and 2 of order 36,000. But, since the highest is present only five-fold and 5 is not divisible by 7, these units will be converted into multiples of the next lower order of magnitude, replacing the 5 counters in the highest order by the corresponding number of counters on the next.

One unit of order 216,000 is equal to 6 units of order 36,000, so we take 5 × 6 = 30 counters and add these to the 2 already there. There are, therefore, 32 counters on the board.

Now, 32 divided by 7 is 4, with remainder 4. I therefore place 4 counters (for the remainder) above the next column down (the 3,600 column) so as not to forget this remainder. Then the 4 counters (for the quotient) are placed in the 36,000 column. Then I remove the remaining counters.

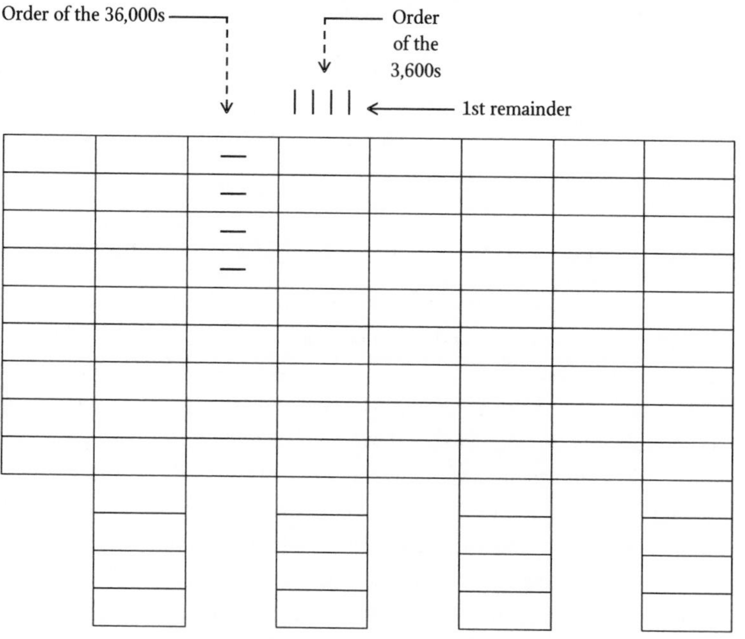


FIG. 12.5A.

**Second stage**

Now I convert the 4 counters of the preceding remainder into units of order 36,000.

One unit of 36,000 is 10 units of 3, 600, so I take 10 × 4 = 40 counters. But 40 divided by 7 is 5, with remainder 5. Therefore I now place 5 counters (for this remainder) above the next column down (600) so as not to forget it.

Then I place the 5 counters (for the quotient) in the 3,600 column, and remove the remaining counters.

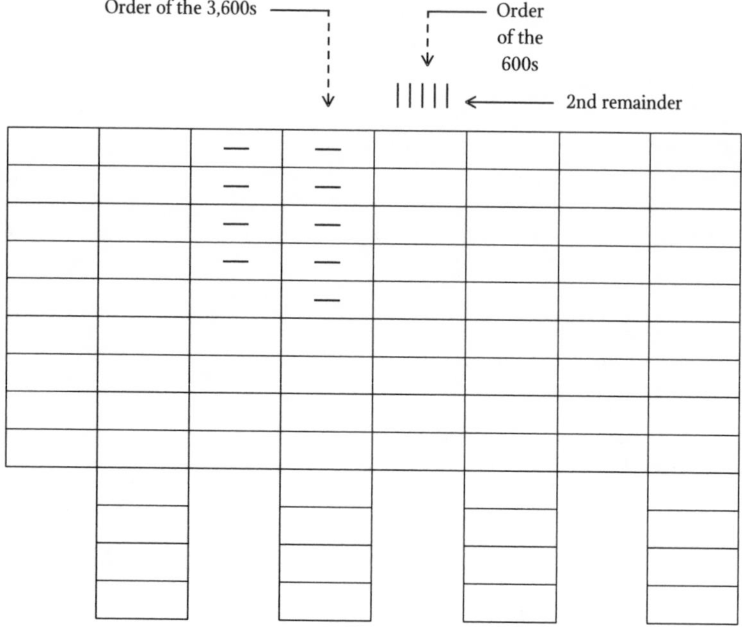


FIG. 12.5B.

**Third stage**

Now I convert the 5 counters for the preceding remainder into units of order 600. One unit of 3,600 is 6 units of 600, so I take 5 × 6 = 30 counters.

But 30 divided by 7 is 4, with remainder 2. I place 2 counters (for the remainder) above the next column down (60), as before.

Then I place 4 counters (for the preceding quotient) in the 600 column, and finally I remove the remaining counters (Fig. 12.5C, opposite).

**Fourth stage**

Now I convert the 2 counters for the preceding remainder into units of order 60. One unit of 600 is 10 units of 60, so I take 2 × 10 = 20 counters.

Now 20 divided by 7 is 2, with remainder 6. So I place 6 counters (for the remainder) above the next column down (10). Then I place 2 counters (for the preceding quotient) in the 60 column. Then I remove the remaining counters (Fig. 12.5D, opposite).

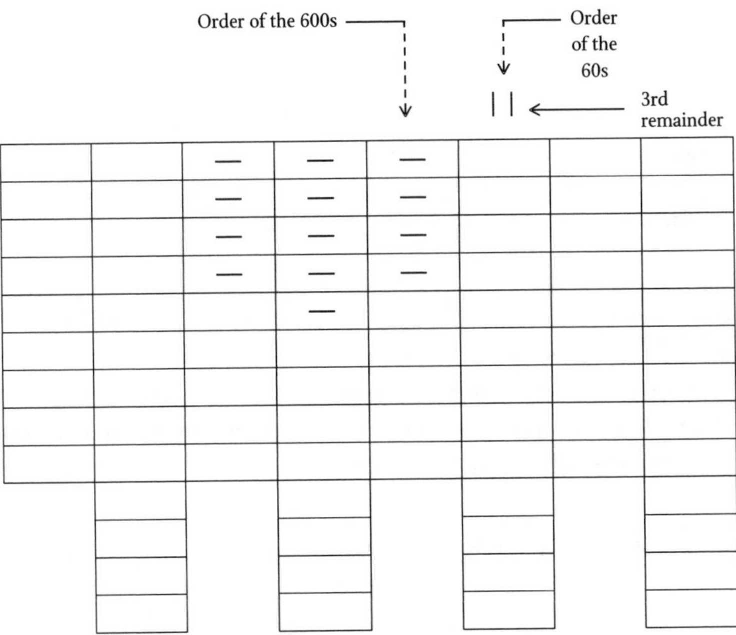


FIG. 12.5C.

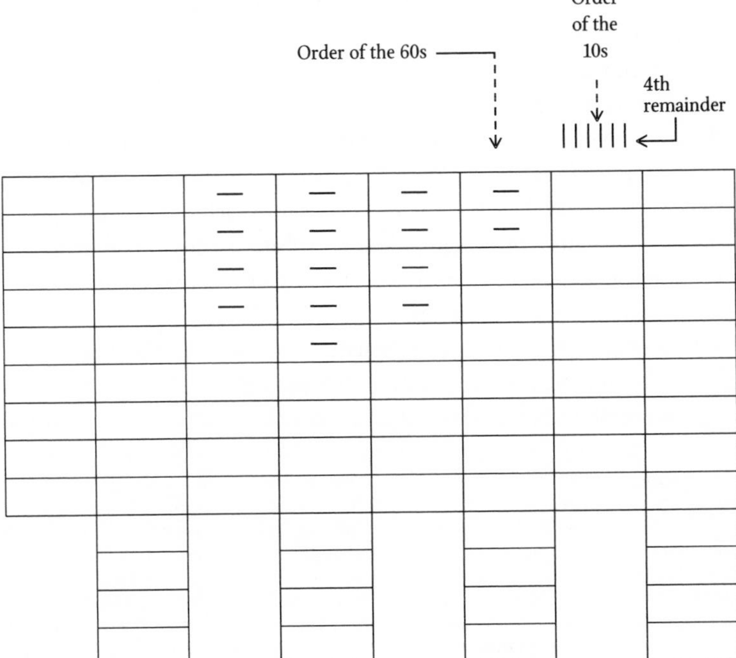


FIG. 12.5D.

**Fifth stage**

I convert the 6 counters for the preceding remainder into units of order 10. One unit of 60 is 6 units of 10, so I take $6 \times 6 = 36$ counters.

But 36 divided by 7 is 5 with remainder 1. So I place 1 counter (for the remainder) above the next column (units) and then 5 counters (for the preceding quotient) into the tens column. Then I remove the remaining counters.

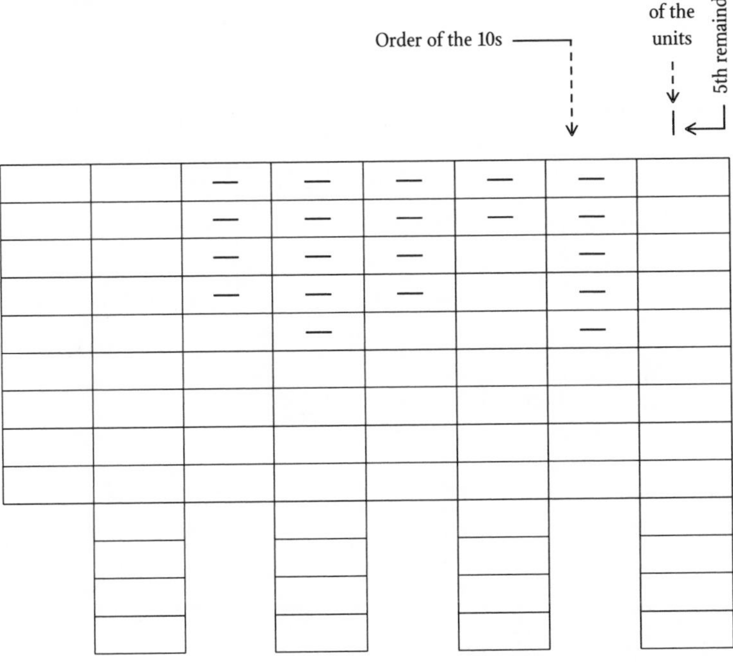


FIG. 12.5E.

**Sixth and final stage**

Now I convert the single counter for the preceding remainder into simple units. One unit of 10 is 10 simple units, so I take 10 counters.

But 10 divided by 7 is 1, with remainder 3. So I place 3 counters (for the remainder) to the right of the units column. Then I place 1 counter (for the preceding quotient) into the units column, and I remove the remaining counters.

Since I have now arrived at the final column, of simple units, the procedure is finished. To obtain the final result, I simply read off from the abacus to obtain the quotient (Fig. 12.5 F):

$$4 \times 36{,}000 + 5 \times 3{,}600 + 4 \times 600 + 2 \times 60 + 5 \times 10 + 1$$

and the 3 counters which I placed at the right of the last column give me the remainder.

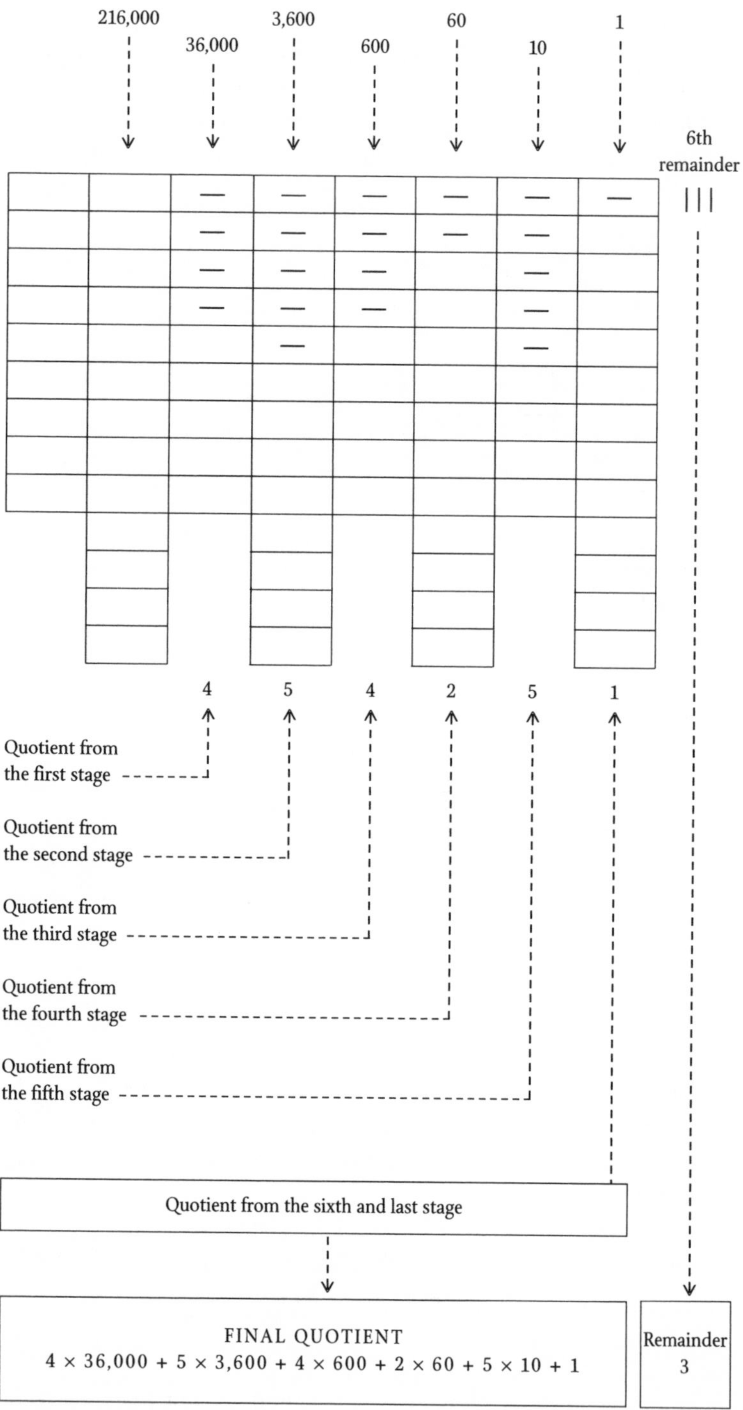


FIG. 12.5F.

On the abacus, therefore, the procedures for calculation were much simpler than for the much more ancient methods using the *calculi* of old. Undoubtedly both methods were in use together for a certain time, the more traditionally-minded tending to stay with the methods of their predecessors. These too were probably the same who continued to use the curviform notation of times past up until the end of the third millennium by which time the use of cuneiform notation had spread throughout Mesopotamia. We may therefore imagine the disputes between "calculists" and "abacists", the former standing to the defence of calculating by means of objects of different sizes and shapes, the latter attempting to demonstrate the many advantages of the new method.

What I have just said about the quarrel between the specialists is plausible, but it is merely a figment of my imagination. The rest of what I have been saying, though, is much more than merely probable.

## CONFIRMATION OF THE SUMERIAN ABACUS AND ABACISTS

The reconstructions described above have in fact received confirmation as a result of recent discoveries.

I am referring to Sumero-Akkadian texts on cuneiform tablets dating from the beginning of the second millennium BCE, from various Sumerian archaeological sites (including Nippur), which have been meticulously collated, translated and interpreted by Liebermann (in *AJA*). These texts are all reports, and detailed analyses, in two languages (Sumerian and Ancient Babylonian) of various professions exercised at the time in Lower Mesopotamia. They are, in a way, "yearbooks" for these professions, and were made in several copies. The reports refer to each profession by giving a description of its representative, and a brief title of the kind "man of . . .", but at the same time in each case they clearly specify the nature of any tools or devices used in each profession.*

Among all the many sorts of information in these texts, we find precisely the professions which are of prime interest for us. The lists give with great precision not only their official designation but also their tools and

* The bilingual texts from which have been taken the names given in Fig. 12.6 A–L occur mainly on tablets with the following museum references: – 3 NT 297, 3 NT 301 (cf. *Field Numbers of Tablets excavated at Nippur*); – IM 58433, IM 58496 (cf. *Tablets in the Collections of the Iraqi Museum of Baghdad*); – NBC 9830 (cf. *Tablets in the Babylonian Collection of the Yale University Library*, New Haven, Conn.); – MLC 653 and 1856 (cf. *Tablets in the Collection of the J. P. Morgan Library*, currently housed in the Babylonian Collection of the Yale University Library, New Haven). The article by S. J. Liebermann, of which the principal results are summarised here in a more accessible form and with some supplementary detail, provides the expert with all necessary philological information and correspondences, and all necessary bibliographical references, including those referring to the important publication by B. Landsberger (cf. *Materialen zum Sumerischen Lexikon*, Rome, 1937).

instruments, down to the very detail of their shape and material, and even which component goes with which instrument.

This is therefore a sufficiently significant discovery to justify a detailed philological explanation. The results will be displayed in successive diagrams each with three columns. On the left we shall place the Sumerian name (in capital letters); in the centre we shall place the Ancient Babylonian name (in italics); and, on the right, the equivalent English translation.

First we encounter a word which expresses the verb "to count":

| ŠID | *ma-nū* | to count |
|---|---|---|

FIG. 12.6A.

Remarkably, the Sumerian graphical etymology of this verb displays in itself evidence of the existence of the abacus. Originally, this verb was represented by the following pictogram (Fig. 12.6 B). Here we see a hand, or at any rate an extreme idealisation of one, doing something with a board in the shape of a frame or a tray and divided into rows and columns. Somewhat later on, the same verb was represented by a cuneiform ideogram, where we seem to see a frame divided into several columns and intersected by a vertical wedge resembling the figure for unity:

| MOST ANCIENT FORM OF THE SIGN *(archaic Sumerian of the Uruk period)* | ARCHAIC CUNEIFORM SIGN *(Sumerian of the epoch of Jemdet Nasr)* | MORE RECENT SIGN *(classical Sumerian)* |
|---|---|---|
| |  | |

FIG. 12.6B. *Sumerian notations of the verb "to count"* (šid). *Deimel (1947) no. 314*

Considering how ancient this sign for unity is (3000–2850 BCE), we are led to believe that the Sumerian abacus goes back to an even more ancient period than we had previously supposed. To come back to the "professional yearbooks", we find here also a clear reference to the system of *calculi*, which are referred to by a word which strictly means "small clay object":

| IMNA (IMNA$_4$NA or NA$_4$IM) | *abnū* | *calculus, calculi* (small clay object) |
|---|---|---|

FIG. 12.6C.

The "accounting" itself is denoted by a combination of the verb ŠID ("to count") and the word NIG̃ (total, sum):

| NIG̃$_2$-ŠID | *nik-kàs-si* | accounting (making the total) |
|---|---|---|

FIG. 12.6D.

Here again, the Sumerian etymology traces this to a suggestive origin, since the signs for the word NIGI (or NIGIN), meaning "total", "sum", "to collect together", clearly suggest the successive sections of the abacus:

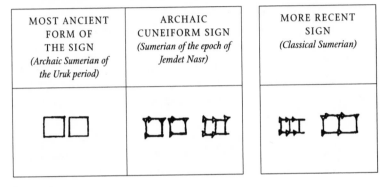

| MOST ANCIENT FORM OF THE SIGN *(Archaic Sumerian of the Uruk period)* | ARCHAIC CUNEIFORM SIGN *(Sumerian of the epoch of Jemdet Nasr)* | MORE RECENT SIGN *(Classical Sumerian)* |
|---|---|---|
| | | |

FIG. 12.6E.

Next we find the word for the expert in weights and measures, in a way the metrologist of that time and place, the "man of the stones":

| LÚ NA$_4$NA | *ša abnē e* | man of the stones |
|---|---|---|

FIG. 12.6F.

This designation obviously would not be confused with that of the calculator using the method of the *calculi*; this man is distinctly denoted in these texts by the terms:*

| LÚ IMNA$_4$NA<br>LÚ NA$_4$IM NA | *ša . . .*(?) | man of the *calculi* (the man with the small clay objects) |
|---|---|---|

FIG. 12.6G.

* The texts have been damaged by time at this point, and we cannot make out the corresponding Babylonian term. We see only its beginning, *ša*, which tells us little since *ša* is simply the Sumerian translation of the word for "man". But, following the work of A. L. Oppenheim (1959), we know that the Sumerian word for *calculi* was *abnū* (plural: *abnāti* or *abnē*, meaning literally "stone", "stone object", "kernel" or "hailstone". So we may suppose that the complete term was *ša abnāti-i*, where the scribe would use the second form of the plural in order to avoid confusion with *ša abnē-e*, unless he simply used the Sumerian word IMNA (*calculi*) in order to coin a term similar to *ša imnaki* (or *ša imnake*).

What follows next is even more thrilling, since it reveals not only the name but also the material used for the counter employed by the abacus-users of the period (the word "G̃EŠ" means "wood"):

| G̃EŠ-ŠID-MA<br>G̃EŠ-NIG$_2$-ŠID | *iṣ-ṣi mi-nu-ti*<br>*iṣ-ṣi nik-kàs-si* | counting stick;<br>stick for accounting |
|---|---|---|

FIG. 12.6H.

Moreover, it yields not only what the counter is made of; we also learn its shape, since the profession which made use of it comes under the heading of "the men of the small wooden sticks".

As we have supposed above, they indeed made use of rods to perform their operations on the abacus (Fig. 12.5).

As to the abacus itself, we find that the texts refer to it clearly, using a figurative expression. To understand what this means, let us first note that, in Sumerian, "tablet" is said as DAB$_4$ and, in the absence of any further details, this word is always understood to mean "clay tablet", the dominant medium of the region for writing on. But in this case the material of the tablet is specified by the word G̃EŠ, meaning "wood". The word G̃EŠDAB$_4$ therefore means "wooden tablet" and in this context is therefore quite other than the "paper of Mesopotamia".

Another word which enters into the makeup of the Sumerian term for the abacus is DÍM. As a verb, this means "to fashion", "to form", "to model in clay", "to construct" and so on, and hence, by association of ideas, "to elaborate", "to perfect", "to create", "to invent". As a noun, it means "fashion", "form", "construction", and, by extension, "perfecting", "formation", "elaboration", "creation", or "invention" [A. Deimel (1947), no. 440].

Now we understand that the word DÍM frequently, by association of ideas, referred to the activities associated with Mesopotamian accounting, not only modelling and moulding clay (to make the *calculi* and the tablets) but also, and above all, to perfect, to elaborate results, and consequently to create and to invent something which nature did not provide in the raw state. Moreover, calculation is essential for shaping and fashioning objects, and also to architects for whom it is a vital necessity in their constructions.

Bringing all these terms together into a logical compound, by composing the expression G̃EŠDAB$_4$-DÍM to designate the instrument in question, the scribes must have had several simultaneous meanings present in their minds:

1. "wooden tablet" meaning perfecting;
2. "wooden tablet" meaning elaboration;
3. "wooden tablet" meaning creation;
4. "wooden tablet" meaning invention;
5. "wooden tablet" endowed with form (namely the tablet);
6. "wooden tablet" endowed with forms (the columns);
7. "wooden tablet" meaning the accounts;
8. "wooden tablet" meaning accounting; and so on.

Here we have therefore the characteristics and the many purposes of the abacus. The word G̃EŠDAB$_4$-DÍM can have but one translation.

| G̃EŠDAB$_4$-DÍM | *g̃ešdab$_4$-dím mu* | abacus |
|---|---|---|

FIG. 12.6I.

Even more significant is this other designation of the instrument of calculation:

| G̃EŠŠU-ME-GE | *šu-me-ek-ku-ú* | abacus |
|---|---|---|

FIG. 12.6J.

The word ŠU which is one component of this expression literally means "hand". In certain contexts, however, it also means "total", "totality", alluding to the hand which assembles and totalises [A. Deimel (1947), no. 354].

The word ME, for its part, means "rite", "prescription"; in other words "the determination of that which must be done according to the rules", or "an action which is performed according to a precise order as well as a prescribed order" [A. Deimel (1947), no. 532].

It would not in fact be at all surprising if the practice of calculating on the abacus corresponded to a genuine ritual, since the knowledge of abstract numbers and, even more, skill in calculation were not within the grasp of everyone as they are today. Those who knew how to calculate were rare indeed.

With all the peoples of the earth, calculation did not merely evoke admiration for those skilled in the art: they were feared, and regarded as magicians endowed with supernatural powers. This naturally gave rise to a certain element of sacred ritual in their activities, not to mention the numerous privileges which kings and princes often granted them.

In any case, in a context such as the present one, the word ME must be understood as "the determination of that which must be done in the precise order prescribed by the rules of calculation". This is something like what computer scientists of modern times would mean by "algorithm".

The term GE (or GI) is the word for a reed and stands for all the names of objects which can be made from this material [A. Deimel (1947), no. 85].

When we put them all together, these terms give the expression G̃EŠŠU-ME-GE, which corresponds to one or other of the following literal translations:

1. A hand (ŠU), a reed (GE), the rules (of arithmetic) (ME) and wood (G̃EŠ) ("of the tablet" understood);
2. The wood (of the tablet), a reed (GE), the rules (of arithmetic) (ME) and a total (i.e. provided by the hand) (ŠU).

In plain language, the expression G̃EŠŠU-ME-GE clearly refers to the abacus.

Lastly, for the "professional calculator", the texts use one or other of the following expressions:

| | | |
|---|---|---|
| LÚ G̃EŠ DAB$_4$-DÍM<br>LÚ G̃EŠDAB$_4$ | *ša da-ab-di-mi* | abacist |

FIG. 12.6K.

The first of these means literally "the man (LÚ) of the wooden tablets for accounting (G̃EŠ DAB$_4$ DIM)", and the second simply means "the man (LÚ) of the wooden tablets (G̃EŠDAB$_4$)", there being no confusion possible about the material support.

We also find one or other of the two names

| | | |
|---|---|---|
| LÚ G̃EŠ ŠUMUN-GE<br>LÚ ŠUMUN-GI$_4$ | *ša šu-ma-ki-i* | abacist |

FIG. 12.6L.

The first of these means literally a man (LÚ) who manipulates the rules (MUN) with a reed (GE) on the wood (G̃EŠ) ("of the tablet" understood), while the second corresponds to a symbolic variant of the first which could be translated as a man (LÚ) who finds the total (ŠU) with a reed (GI) according to the rules (MUN).

There is at this time no doubt that the abacus indeed existed in Mesopotamia, and even coexisted with the archaic system of *calculi*, most probably throughout the third millennium BCE.

This abacus consisted of a tablet of wood, on which were traced beforehand lines of division which exactly corresponded to the Sumerian sexagesimal system (Fig. 12.5) and therefore delimited, column by column, each of the order of unity of this numerical system (1, 10, 10.6, 10.6.10, 10.6.10.6, 10.6.10.6.10, and so on).

The counting tokens themselves were thin rods of wood or of reed, given the value of a simple unit, such that their subtle arrangement over the columns of the abacus allowed all the operations of arithmetic to be carried out. (No doubt it is as a result of their perishable nature that archaeologists have never brought any of these to light. Another reason may be that, as we may well suppose, whenever one of these experts did not have a "calculating board" to hand, he could simply draw the "tablet" on the loose soil.)

Lastly, as with writing but perhaps more so, the use of the abacus gave rise to a guild, perhaps even with the privileges of a special caste, so much would its complex rules and practices have been inaccessible to ordinary mortals: this was the caste of the professional abacists, who no doubt jealously preserved the secrets of their art.

CHAPTER 13

# MESOPOTAMIAN NUMBERING AFTER THE ECLIPSE OF SUMER

## THE SURVIVAL OF SUMERIAN NUMBERS IN BABYLONIAN MESOPOTAMIA

For some time after the decline of Sumerian civilisation, the sexagesimal system remained in use in Mesopotamia. Just as many French people still use "old francs" in their everyday reckonings even though "new francs" replaced them officially as long ago as 1960, so the inhabitants of Mesopotamia continued to use the "old counting" based on multiples and powers of 60.

The following examples come from an accounting tablet excavated at Larsa (near Uruk) and probably dating from the reign of Rîm Sîn (1822–1763 BCE). They are characteristic examples of the everyday reckonings that constitute the city archives, and give an account of sheep with the following numerical values:

| | | | |
|---|---|---|---|
| 61 (ewes) | 60 1 | 96 (ewes) | 60 30 6 |
| 84 (rams) | 60 20 4 | 105 (rams) | 60 40 5 |
| 145 (sheep) | 120 20 5 | 201 (sheep) | 180 20 1 |

FIG. 13.1. *Birot, tablet 42, p. 85, plate XXIV*

The numerals used are indeed those of the old Sumerian cuneiform system, with its characteristic difficulty in the representation of numbers such as 61, the vertical wedge signifying 60 being almost indistinguishable from the wedge used for 1 – and this is certainly the reason why the scribe leaves a large space between the two symbols, so as to avoid confusion with the number 2.

There is nothing at all surprising in the persistence of the old system in Lower Mesopotamia, since that is where the system first arose, in the lands of Sumer. What is less obvious is why the sexagesimal system survived for so long in the lands to the north, that is to say in Akkadian areas. However, the evidence is indisputable. This is an example from a tablet written in Ancient Babylonian, dating from the thirty-first year of the reign of Ammiditana of Babylon (1683–1647 BCE). It provides an inventory of calves and cows in the following manner:

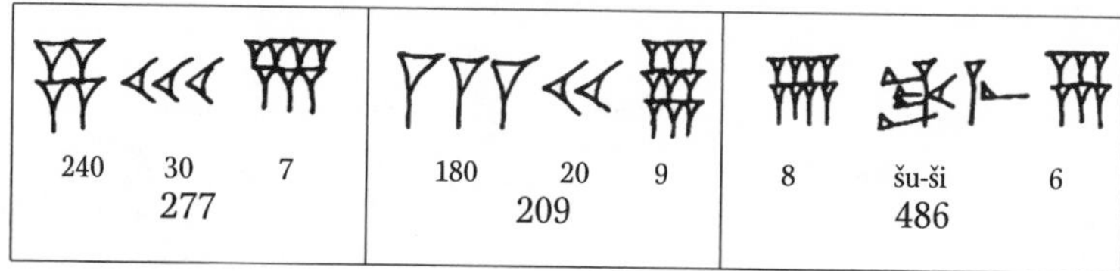


FIG. 13.2. *Finkelstein, tablet 348, plate CXIV, ll. 8–10*

The number 486 (the sum of 277 and 209) is represented not just by 8 large wedges each standing for 60 and 6 small wedges each standing for 1. The scribe has chosen to provide an additional phonetic confirmation of the number by putting the word *shu-shi* (the name of the number 60 in Akkadian) after the larger expression, rather like the way in which we write out cheques with a numerical and a literal expression of the sum involved.

All the same, these are just about the last traces of the unmodified system in use in Mesopotamia. Sumerian numbering was abandoned for good around the time that the first Babylonian Dynasty disappeared, in the fifteenth century BCE. By then, of course, modern Mesopotamian numerals, of Semitic origin, had been current for some time already.

## WHO WERE THE SEMITES?

The term "Semite" derives from the passage in the Old Testament (Genesis 10) where the tribes of Eber (the Hebrews), Elam, Asshur, Aram, Arphaxad, and Lud are said to be the descendants of *Shem*, one of Noah's three sons, the brother of Ham and Japheth. However, though it may have represented a real political situation in the first millennium BCE, the biblical map of the nations of the Middle East makes the Elamites, who spoke an Asianic language, cousins to the Hebrews, Assyrians, and Aramaeans, whose languages belong to the Semitic group.

"Asianic" is the term used for the earlier inhabitants of the Asian mainland whose languages, mostly of the agglutinative kind, were neither Indo-European nor Semitic. It is generally believed that Mesopotamia was originally inhabited by Asianic peoples, prior to the arrival of Sumerians. It is thought that Semitic-speaking populations came in a second wave, and that Akkadian civilisation constitutes the earliest Semitic nation in the area. However, significant Semitic elements are to be found in the cultures of Mari and Kiš at the beginning of the third millennium BCE, and it is even possible that the people of El Obeid were of Semitic origin themselves, though absorbed and assimilated by the Sumerians. The discovery of the Ebla tablets revealed the existence of a state speaking a language of the Semitic family in the mid-third millennium BCE, and so it becomes ever less certain that the "cradle" of the Semitic languages was the Arabian peninsula, as was long held to be true. Nonetheless, Arabic is probably the closest to the proto-Semitic stem-language, which began to differentiate into numerous branches (Ancient Egyptian, some aspects of the Hamitic languages of eastern Africa, and possibly even Berber, spoken in Algeria and Morocco) as early as the Mesolithic era, that is to say (for the Middle East) in the tenth to eighth millennia BCE. That is too far back in time for it to be possible to say exactly where Semitic languages first arose or who the people were who brought them to different civilisations in the Middle East. Like the term "Indo-European", "Semitic" does not designate any ethnic or cultural entity, but serves only to define a broad family of languages.

There was no single "Semitic civilisation", just as there was never such a thing as an Aryan or an Indo-European culture. Each of the main Semitic-speaking civilisations of antiquity developed its own specific culture, even if there are some features common to several or all of them. It is therefore important to distinguish amongst the Semitic cultures those of the *Akkadians*, the *Babylonians*, the *Assyrians*, the *Phoenicians*, the *Hebrews*, the *Nabataeans*, the *Aramaeans*, the various peoples of *Arabia*, *Ethiopia*, and so on. (See Guy Rachet, *Dictionnaire de l'archéologie*, for further details.)

## A BRIEF HISTORY OF BABYLON

At the beginning of the third millennium BCE Sumerians dominated the southern Mesopotamian basin, both numerically and culturally. To the north of them, between the Euphrates and the Tigris and on the northern and eastern edges of the Syro-Arabian desert, lived tribes of semi-nomadic pastoralists who spoke a Semitic language, called *Akkadian*. The Akkadian king, Sargon I The Elder, founded the first Semitic state when he defeated the Sumerians in c. 2350 BCE. His empire stretched over the whole of Mesopotamia and parts of Syria and Asia Minor. Its capital was at Agade (or *Akkad*), and, for one hundred and fifty years, it was the centre of the entire Middle East. As a result, Akkadian became the language of Mesopotamia and gradually pushed aside the unrelated language of Sumer. Assyrian and Babylonian are both descended from Akkadian and are thus Semitic, not Asianic, languages.

The Akkadian empire collapsed around 2150 BCE and for a relatively brief time thereafter Sumerians reasserted their control of the area. But that was the final period of Sumerian domination, for around 2000 BCE, the third empire of Ur collapsed under the simultaneous onslaughts of the Elamites (from the east) and the Amorites (from the west). Sumerian civilisation disappeared with it for ever, and in its place arose a new culture, that of the *Assyro-Babylonians*.

The Amorites, a Semitic people from the west, settled in Lower Mesopotamia and founded the city of Babylon, which would become and remain for many centuries the capital of the country known as *Sumer and Akkad*. The famous king and law-maker Hammurabi (1792–1750 BCE) was one of the outstanding figures of the first Babylonian dynasty established by the Semites, who became masters of the region. Hammurabi extended Babylonian territory by conquest over the whole of Mesopotamia and as far as the eastern parts of Syria.

This huge and powerful kingdom was nonetheless seriously weakened, from the seventeenth century BCE, by the Kassites, Iranian highlanders who made frequent raids, and it finally surrendered in 1594 BCE to the Hittites, who came from Anatolia.

Babylon then remained under foreign domination until the twelfth century BCE, when another Semitic people, the Assyrians, from the hilly slopes between the left bank of the Tigris and the Zagros mountain range, entered the concert of nations. The Assyrians were bearers of a version of Sumerian culture, which they developed most fully in military conquest, establishing an empire which stretched out in all directions and which was one of the most fearsome and feared military powers in the ancient world, until in 612 BCE, Nineveh, the Assyrians' capital, was destroyed in its turn.

The Babylonians, although dominated by the Assyrians from the ninth to the seventh centuries BCE, nonetheless retained their own distinctive culture throughout this period. However, the fall of Nineveh (and with it of the whole Assyrian Empire) in 612 BCE allowed a great flowering of Babylonian culture, which was the prime force in the Middle East for over a century, most especially under the reign of Nebuchadnezzar II (604–562 BCE). But that was Babylon's last glory: it was conquered in 539 BCE by Cyrus of Persia, then in 311 BCE by Alexander the Great, and finally expired completely shortly before the beginning of the Common Era.

## THE AKKADIANS, INHERITORS OF SUMERIAN CIVILISATION

In the Akkadian period (second half of the third millennium BCE) the Semites, who were now the masters of Mesopotamia, emerged as the preponderant cultural influence in the region. They naturally sought to impose their own language, and also to give it a written form. To do this they borrowed the cuneiform system of their predecessors, and adapted it progressively to their language and traditions.

By the time Sumerian cuneiform was adopted by the Akkadians, the writing system was already several centuries old. The ideas originally signified by the ideograms were mostly forgotten, and the signs were now purely symbolic. What the Akkadians found was a basically ideographic writing system with an already-established drift towards a phonetic system – a drift which the Akkadians accelerated, whilst retaining the ideographic meaning of some of the signs. They did so partly because their own language was less well suited to ideograms than Sumerian, and also because the signs which represented words for Sumerians represented only sounds to Akkadian ears.

The adaptation of cuneiform writing was however not a smooth or easy process. For one thing, Akkadian had sounds not present in Sumerian, and vice versa. The two ethnic groups of Akkadians (Babylonians and Assyrians) proceeded independently in this development, despite the numerous contacts between them. But by adopting the Sumerian cultural heritage, the Akkadians gave it its greatest flowering, leading it away from its origins in mnemotechnics and ultimately towards the creation of a true literary tradition.

## THE NUMBERING TRADITIONS OF SEMITIC PEOPLES

The spoken numbering system of the Semites was very different from the way Sumerians expressed numbers orally – not just linguistically, but also mathematically, since Semitic numbering was, and remains, strictly decimal.

However, Semitic numbering has one small grammatical oddity, in terms of the decimal numbering systems to which we are now accustomed.

Hebrew and Arabic numbering (see Fig. 13.3 below) provide characteristic examples.

In Hebrew as in Arabic, spoken numerals have feminine and masculine forms, according to the grammatical gender of the noun to which they are attached. For instance, the name of the number 1, treated as if it were an adjective, has one form if the noun it qualifies is masculine, and a different form if the noun is feminine. Similarly, the name of the number 2 agrees in gender with its noun. However, what is unusual is that for all numbers from 3, the number-adjective is feminine if the noun is masculine, and masculine if the qualified noun is feminine. In Hebrew, for example, where "men" is *anoshim* and "three" is *shalosh* (masculine) or *shloshah* (feminine), the expression "three men" is translated by *shloshah anoshim*, not, as you might expect from Latin or French grammar, by *shalosh anoshim*.

| | HEBREW | | ARABIC | |
|---|---|---|---|---|
| | Feminine | Masculine | Feminine | Masculine |
| 1 | *'eḥad* | *'aḥat* | *'aḥadun* | *'iḥda* |
| 2 | *shnaim* | *shtei* | *'iṯnān* | *'iṯnatāni* |
| 3 | *shloshah* | *shalosh* | *ṯalāṯun* | *ṯalāṯatun* |
| 4 | *'arba 'ah* | *'arba'* | *'arba'un* | *'arba'atun* |
| 5 | *ḥamishah* | *ḥamesh* | *khamsun* | *khamsatun* |
| 6 | *shishah* | *shesh* | *situn* | *sitatun* |
| 7 | *shib'ah* | *sheba'* | *sab'un* | *sab'atun* |
| 8 | *shmonah* | *shmoneh* | *ṯamāny* | *ṯamānyatun* |
| 9 | *tishah* | *tesha'* | *tis'un* | *tis'atun* |
| 10 | *'asarah* | *'eser* | *'ashrun* | *'asharatun* |

FIG. 13.3.

Numbers from 11 to 19 are formed by the name of the unit followed by the word for 10, each having masculine and feminine forms, used according to the previous rule:

| | HEBREW | | ARABIC | |
|---|---|---|---|---|
| | Feminine | Masculine | Feminine | Masculine |
| 11 | *'aḥad 'asar* | *'aḥat 'esreh* | *'aḥad 'ashara* | *'iḥda 'ashrata* |
| 12 | *shnaim 'asar* | *shtei 'esreh* | *'iṯnā 'ashara* | *'itnāta 'ashrata* |
| 13 | *shloshah 'asar* | *shlosh 'esreh* | *ṯalāṯut 'ashara* | *ṯalāṯa 'ashrata* |
| 14 | *'arba'ah 'asar* | *'arba 'esreh* | *'arba'ata 'ashara* | *'arba'a 'ashrata* |
| 15 | *ḥamishah 'asar* | *ḥamesh 'esreh* | *khamsata 'ashara* | *khamsa 'ashrata* |
| 16 | *shishah 'asar* | *shesh 'esreh* | *sitata 'ashara* | *sita 'ashrata* |

FIG. 13.4.

Apart from the number 20, which is derived from the dual form of the word for 10, the tens are derived from the name of the corresponding unit, with an ending that is derived from the customary mark of the plural:

| | HEBREW | ARABIC | |
|---|---|---|---|
| 20 | *'eshrim* | *'isrūn* | derived from dual of 10 |
| 30 | *shloshim* | *ṯalāṯūna* | plural of name of 3 |
| 40 | *'arba'im* | *'arba'ūna* | plural of name of 4 |
| 50 | *ḥamishim* | *khamsūna* | plural of name of 5 |
| 60 | *shishim* | *sitūna* | plural of name of 6 |
| 70 | *shibim* | *sib'ūna* | plural of name of 7 |
| 80 | *shmonim* | *ṯamānūna* | plural of name of 8 |
| 90 | *tishim* | *tis'ūna* | plural of name of 9 |

FIG. 13.5.

The system has special names for 100 and 1,000, and proceeds thereafter by multiplication for multiples of each of these powers of the base:

| | HEBREW | ARABIC | |
|---|---|---|---|
| 100 | *me'ah* | *mi'ātun* | |
| 200 | *ma'taim* | *mi'atāny* | dual of 100 |
| 300 | *shlosh me'ōt* | *ṯalāṯu mi'ātin* | (3 × 100) |
| 1,000 | *'elef* | *'alfun* | |
| 2,000 | *'alpaim* | *'alfāny* | dual of 1,000 |
| 3,000 | *shloshet 'alafim* | *ṯalāṯu 'alāf* | (3 × 1,000) |
| 10,000 | *'aseret 'alafim* | *'asharat 'alāf* | (10 × 1,000) |
| 20,000 | *'eshrim 'elef* | *'ishrūnat 'alāf* | (20 × 1,000) |
| 30,000 | *shloshim 'elef* | *ṯalāṯūnat 'alāf* | (30 × 1,000) |

*Note*: Classical Hebrew also has the word *ribō* ("multitude") to designate 10,000, together with its multiples: *shtei ribot* for 20,000, *shlosh ribōt* for 30,000, etc. Similar words exist in other ancient Semitic languages: *ribab* (Elamite), *ribbatum* (Mari), *r(b)bt* (Ugaritic).

FIG. 13.6.

NUMBER-NAMES IN ASSYRO-BABYLONIAN

| | |
|---|---|
| 1 | *ishtên* |
| 2 | *sita, sinâ* |
| 3 | *shalâshu* |
| 4 | *erbettu* |
| 5 | *khamshu* |
| 6 | *sheshshu* |
| 7 | *sîbu* |
| 8 | *shamânu* |
| 9 | *têshu* |

| | |
|---|---|
| 10 | *eshru, esheret* |
| 20 | *eshrâ* |
| 30 | *shalâshâ* |
| 40 | *arbâ* |
| 50 | *khamshâ* |
| 60 | *shushshu, shushi* |
| 70 | * |
| 80 | * |
| 90 | * |

* The pronunciation of these numbers is not known

| | | |
|---|---|---|
| 100 | *me'atu, me'at* | $= 10^2$ |
| 200 | *sita metin* | = 2 × 100 |
| 300 | *shalâsh me'at* | = 3 × 100 |
| 1,000 | *lim* | $= 10^3$ |
| 2,000 | *sinâ lim* | = 2 × 1,000 |
| 3,000 | *shalâshat limi* | = 3 × 1,000 |
| 10,000 | *esheret lim* | = 10 × 1,000 |
| 20,000 | *eshrâ lim* | = 20 × 1,000 |
| 100,000 | *me'at lim* | = 100 × 1,000 |
| 200,000 | *sita metin lim* | = 2 × 100 × 1,000 |

FIG. 13.7.

For intermediate numbers, addition and multiplication are used in conjunction. In Arabic, it should be noted, the units are always put before the tens: 57, for example, is *sab'un wa khamsūna* ("seven and fifty"), as in German (*siebenundfünfzig*).

The same order of expression is found in Ugaritic texts (Ugarit was a Semitic culture that flourished at Ras Shamra, in northern Syria, around the fourteenth century BCE) and in biblical Hebrew, most frequently in the Pentateuch and the Book of Esther. According to Meyer Lambert, this order of numbers is the archaic form.

However, the inverse order (hundreds followed by tens followed by units) is also found in the Hebrew Bible, and this is the commonest form in the first Books of the Prophets, and in most of the books written after the Exile (Haggai, Zechariah, Daniel, Ezra, Nehemiah, Chronicles). Modern Hebrew (Ivrit) also uses this order (except for numbers between 11 and 19), which is also the most frequent structure in Semitic languages as a whole (Assyro-Babylonian, Phoenician, Aramaic, Ethiopian, etc.).

All these numbering systems therefore demonstrate that they have a common origin, which gives all Semitic numbering its characteristic mark. It will now be easier to grasp how the Mesopotamian Semites radically transformed the cuneiform numerals of the Sumerians, and to understand the method that the western Semites (Phoenicians, Aramaeans, Nabataeans, Palmyreneans, Syriacs, the people of Khatra, etc.) invented to put their numbers in writing other than by spelling them out. (See Chapter 18 below, pp.227–32

## THE SUMERO-AKKADIAN SYNTHESIS

When the Akkadians took over cuneiform sexagesimal numbering, they were naturally hampered by a written system whose organisation differed entirely from the strictly decimal base of their own long-standing oral number-name system. The cuneiform numerals had a sign for 1 (the vertical wedge) and for 10 (the chevron) – but, since there was no sign for 100 or for 1,000, it occurred to them to write out the names of these numbers phonetically. "Hundred" and "thousand" were respectively *me'at*

"6,657" IN ANCIENT & MODERN SEMITIC LANGUAGES

| | | | | | | | | |
|---|---|---|---|---|---|---|---|---|
| ARABIC | *sitatunat 'alāf*<br>six thousand<br>6 × 1,000 | + | *sitatu mi'ātin*<br>six hundred<br>6 × 100 | + | *sab'un*<br>seven<br>7 | *wa*<br>&<br>+ | *khamsūna*<br>fifty<br>50 |
| UGARITIC | *ṯiṯ 'alpin*<br>six thousand<br>6 × 1,000 | + | *ṯiṯ mat*<br>six hundred<br>6 × 100 | + | *sab'a*<br>seven<br>7 | *l*<br>&<br>+ | *khamishuma*<br>fifty<br>50 |
| CLASSICAL HEBREW | *sheshet 'alafim*<br>six thousand<br>6 × 1,000 | + | *sesh me'ōt*<br>six hundred<br>6 × 100 | + | *shib'ah*<br>seven<br>7 | *we*<br>&<br>+ | *khamishim*<br>fifty<br>50 |
| CLASSICAL & MODERN HEBREW | *sheshet 'alafim*<br>six thousand<br>6 × 1,000 | + | *sesh me'ōt*<br>six hundred<br>6 × 100 | + | *khamishim*<br>fifty<br>50 | *we*<br>&<br>+ | *shib'ah*<br>seven<br>7 |
| ASSYRO-BABYLONIAN | *sheshshu limi*<br>six thousand<br>6 × 1,000 | + | *seshshu me'at*<br>six hundred<br>6 × 100 | + | *khamsha*<br>fifty<br>50 | <br><br>+ | *sîbu*<br>seven<br>7 |
| ETHIOPIAN | *sassā ma'āt*<br>sixty hundred<br>60 × 100 | + | *sadastū ma'āt*<br>six hundred<br>6 × 100 | + | *khamsā*<br>fifty<br>50 | *wa*<br>&<br>+ | *sab'atū*<br>seven<br>7 |

FIG. 13.8.

and *lim* in Akkadian, so they represented these numbers as words, using the Sumerian cuneiform signs for ME and AT, on the one hand, and for LI and IM on the other – rather as if we made puzzle-pictures of "Hun" and "Dread" to represent the sound and thus the number "hundred":

ME – AT
100

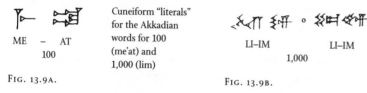

Cuneiform "literals" for the Akkadian words for 100 (me'at) and 1,000 (lim)

FIG. 13.9A.

LI–IM LI–IM
1,000

FIG. 13.9B.

However, they did not stop at the "writing out" stage, they also created genuine numerals, even if these were derived from the phonetic notation of the number-names. The symbols chosen were of course no more than sound-signs from their point of view, since they had lost the meanings that they had had in Sumerian. The symbol for 100 was soon shortened to its first syllable, ME, and for 1,000 the Akkadians used the chevron (= 10) followed by the sign for ME (1,000 = 10 ME = 10 × 100). And since this was the sign for the *word* meaning "thousand", pronounced *lim*, the cuneiform chevron followed by ME came to have the phonetic value of the sound LIM and to be used in all Akkadian words containing the sound LIM.

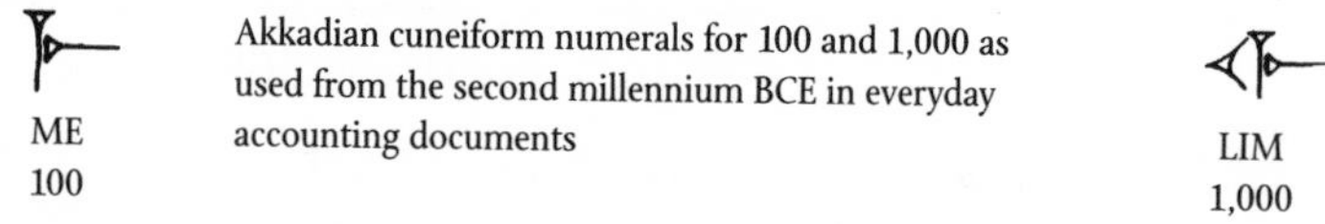

ME
100
Akkadian cuneiform numerals for 100 and 1,000 as used from the second millennium BCE in everyday accounting documents
LIM
1,000


FIG. 13.10A. FIG. 13.10B.

Because of the standard Semitic custom of counting orally in hundreds and thousands, the Akkadians therefore introduced *strictly decimal notations* into the sexagesimal numerals that they had adopted from the Sumerians. The result was a thoroughly mixed Akkadian number-writing system containing special signs for decimal and sexagesimal units, in the following manner:

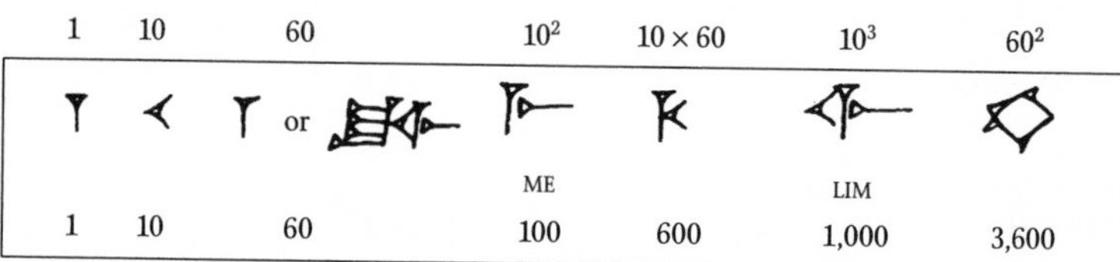


FIG. 13.11.

Let us look at a few characteristic examples. Those shown in Fig. 13.12 (M. J. E. Gautier, 1908, plates XVII, XLII and XLIII) come from clay tablets found at Dilbat, a small town in Babylonian territory that flourished in the nineteenth century BCE. Most of the tablets refer to the main events in the lives of members of a single family, and constitute as it were the family record.

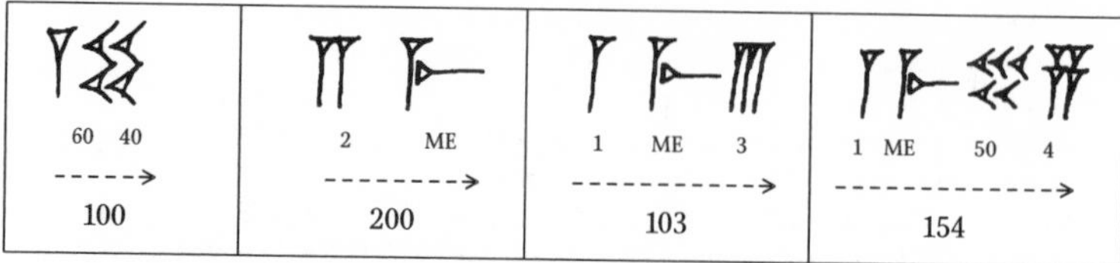


FIG. 13.12. *See Gautier, plates XVII, XLII and XLIII*

The next figure is a transcription of a tally of cattle found in northern Babylon (M. Birot, 1970, tab. 33, plate XVIII) dating from the seventeenth year of the reign of Ami-Shaduqa of Babylon (1646–1626 BCE):

| 1 SHU-SHI 3 | 60 10 3 | 60 20 5 | 1 ME 1 SHU-SHI 8 |
|---|---|---|---|
| 63 | 73 | 85 | 168 |

FIG. 13.13. *See Birot, tablet 33, plate XVIII*

These examples show how in this period the Akkadians did not seek to overturn the sexagesimal system that was deeply rooted in local tradition.

However, for the numbers 60 and 61, and in many cases for multiples of 60, the Semites coped with the corresponding difficulties of the notation system rather better than had the Sumerians. It occurred to them to represent the number 60 by the sound-group *shu-shi*, which was how they pronounced the number in Akkadian (see Fig. 13.7 above) or, in abbreviated form, as *shu* (see Fig. 13.2, 13.13 and 13.14).

| 1 SHU-SHI 1 | 1 SHU-SHII 2 | 1 SHU-SHI 6 | 3 SHU-SHI | 5 SHU-SHI |
|---|---|---|---|---|
| 61 | 62 | 66 | 180 | 300 |

FIG. 13.14.

In short, up to the middle of the second millennium BCE, Mesopotamian scribes of public, private, economic, juridical, and administrative tablets had recourse *either* to sexagesimal Sumerian numbering, *or* to decimal Semitic numbering, *or* finally to a system constituted by a kind of *interference between the two bases*.

## MESOPOTAMIAN DECIMALS

When Akkadian speech and writing finally supplanted their Sumerian counterparts in Mesopotamia, strictly decimal numbering became the norm in daily use. The ancient signs for 60, 600, 3,600, 36,000, and 216,000 progressively disappeared, and only the symbols ME (= 100) and LIM (= 1,000) remained, to provide the bases for the entire system of numerals.

As in classical Sumerian, units were represented by vertical wedges, repeated once for each unit, but whereas the Sumerians had grouped the wedges on a dyadic principle, the Akkadians put them in three groups:

| 1 | 2 | 3 | 4 | 5 | 6 | 7 | 8 | 9 |
|---|---|---|---|---|---|---|---|---|

FIG. 13.15.

The tens were also usually represented by repetition of the chevron (= 10), but here again the layout or grouping of the repeated symbols was quite distinct from older Sumerian patterns:

| 10 | 20 | 30 | 40 | 50 | 60 | 70 | 80 | 90 |
|---|---|---|---|---|---|---|---|---|
| | | | | | or 60 | 60 + 10 | 60 + 20 | 60 + 30 |

FIG. 13.16.

As for the hundreds and thousands, they were symbolised by notations based on multiplication, that is to say in accordance with the analytical combinations that existed in the spoken language of the Akkadians:

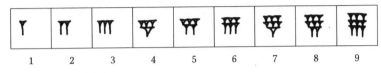

| | | | | | |
|---|---|---|---|---|---|
| 100 | 1 100 | 400 | 4 100 | 2,000 | 2 1,000 |
| 200 | 2 100 | 500 | 5 100 | 3,000 | 3 1,000 |
| 300 | 3 100 | 1,000 | 1 1,000 | 4,000 | 4 1,000 |

FIG. 13.17.

The following examples show just how radical the transformation of Sumerian cuneiform numerals was. The numbers shown relate to the booty taken during Sargon II's eighth campaign against Urartu (Armenia) in 714 BCE:

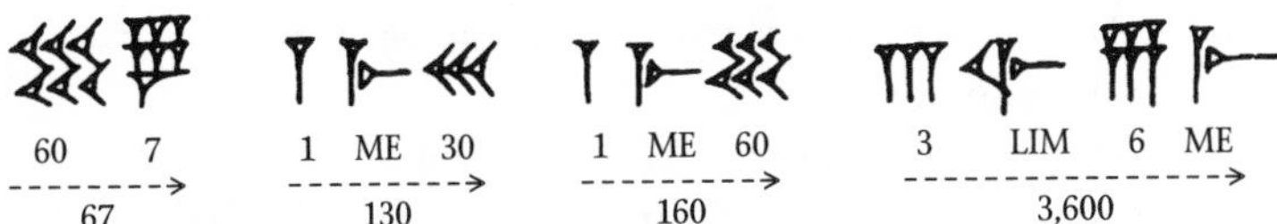


FIG. 13.18. *See Thureau-Dangin, lines. 380, 366 and 369*

As can be seen, 60 is now represented by six chevrons instead of the vertical wedge that formerly had this numerical value, and numbers such as 130, 160 and 3,600 are given strictly decimal representations.

We can also see that by grounding their written numerals on their spoken number-names, the Assyrians and Babylonians extended the arithmetical scope of their numeral system whilst restricting its basic figures to 100 and 1,000. All they needed to do was to combine these symbols with the multiplication principle, to produce expressions of the type 10,000 = 10 × 1,000, 40,000 = 40 × 1,000, 400,000 = 400 × 1,000, and so on. So Sargon II's scribe wrote out the number 305,412 in the following manner:

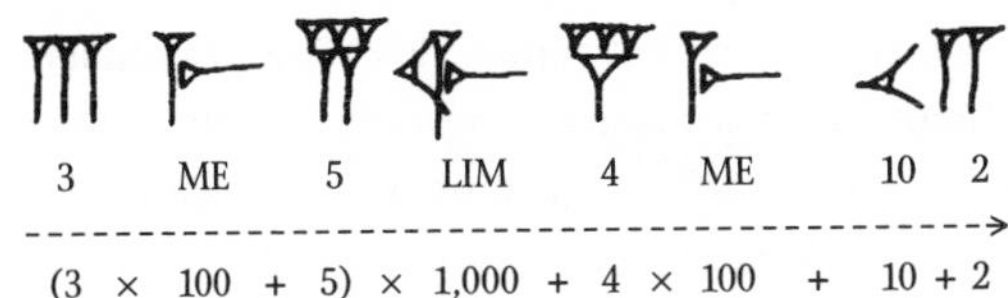


FIG. 13.19. *See Thureau-Dangin, l. 394*

## RECONSTRUCTING THE DECIMAL ABACUS

The Akkadians must surely have possessed a calculating device, for they could not otherwise have performed their complex arithmetical operations save by the archaic device of *calculi*, of which barely a handful have been found in archaeological levels of the second millennium BCE. Indeed, as we also saw in Chapter 12, the Sumerians themselves must have had a kind of abacus, which we reconstructed in its most probable form along with the rules and procedures for its use. Furthermore, the Akkadians, at least in the Babylonian period, had specific terms for referring not only to the instrument and the tokens which went with it, but also to the operator of the abacus.

In Ancient Babylonian (see Fig. 12.6H above), the arithmetical "token", which must have been a stick of wood or a swatch of reed stems, was called either

- *iṣ-ṣi mi-nu-ti* ("wood-for-counting"); or
- *iṣ-ṣi nik-kàs-si* ("wood-for-accounts").

As for the abacus itself, it was referred to by one of the two following loan-words borrowed from the corresponding Sumerian terms (see Fig. 12.6I and 12.6J above):

- *g̃ešdab-dim mu* ("wooden-tablet-for-accounts");
- *šu-me-ek-ku-ú* (literally, from the corresponding Sumerian word G̃EŠŠUMEGE, "wood (i.e. of the tablet), hand, rule, reed" or alternatively "wood, sum, rule, reed".

The abacus operator or abacist had two official names (see Fig. 12.6K and 12. 6L above):

- *ša da-ab-di-mi* ("the man for the tablet for accounts");
- *ša šu-ma-ki-i* ("the man for the abacus").

Our knowledge of these terms comes from various bilingual tablets dating from the beginning of the second millennium BCE, which provide a kind of "Yellow Pages" in both Sumerian and Ancient Babylonian, each entry consisting of a brief description of a representative of a profession ("the man for . . ."), followed by the name of the tools associated with the profession. (See Chapter 12 above, and for references to original sources, see S. J. Liebermann, in *AJA* 84.)

In view of all this, we have to suppose that the Akkadians first used the sexagesimal Sumerian abacus for as long as their arithmetic was dependent on Sumerian notation, but had to construct sexagesimal-decimal conversion tables for the requirements of their own decimal arithmetic during the long "transitional period" that lasted until the end of the first Babylonian dynasty, around the middle of the second millennium BCE. However, when Akkadian culture itself came to hold sway in Mesopotamia, the situation changed completely. The mathematical structure of the abacus had to be radically altered to adapt it to the modified cuneiform notation that was then used for strictly decimal arithmetic.

Indeed, the Assyro-Babylonian numeral system used base 10 and allowed all numbers up to one million to be represented by combinations of just these four signs:

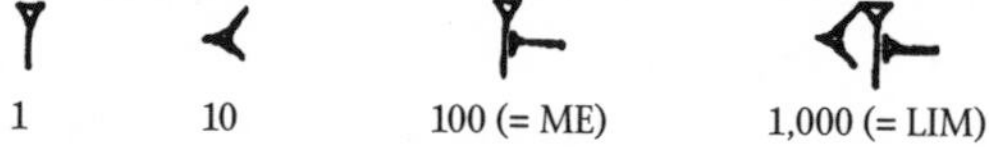


FIG. 13.20.

For numbers above 1,000, the system used analytical combinations of the given signs, that is to say it used the principle of multiplication to designate 10,000, 100,000, and 1,000,000, as follows:

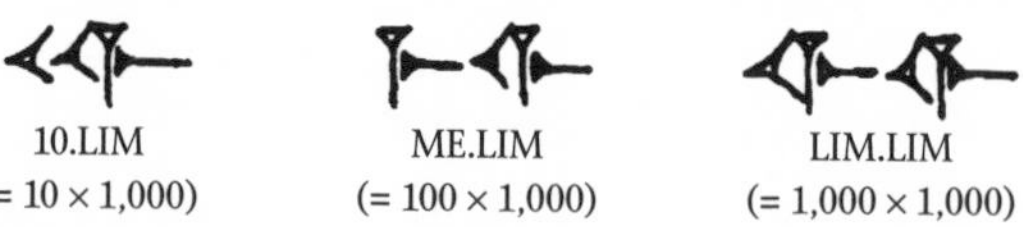


FIG. 13.21.

As we showed for the Sumerian abacus in Chapter 12 above, we can here show quite easily the most probable form of the Assyro-Babylonian abacus as it was used by "ordinary" counters (there are good reasons for thinking that there were two types of arithmeticians – the "ordinaries", whose arithmetic was exclusively decimal, and the "learned", who continued to use the sexagesimal system for mathematical and astronomical purposes). As for the way the abacus was used, it must have been very similar to the rules for the sexagesimal system, simply adapted to base 10:

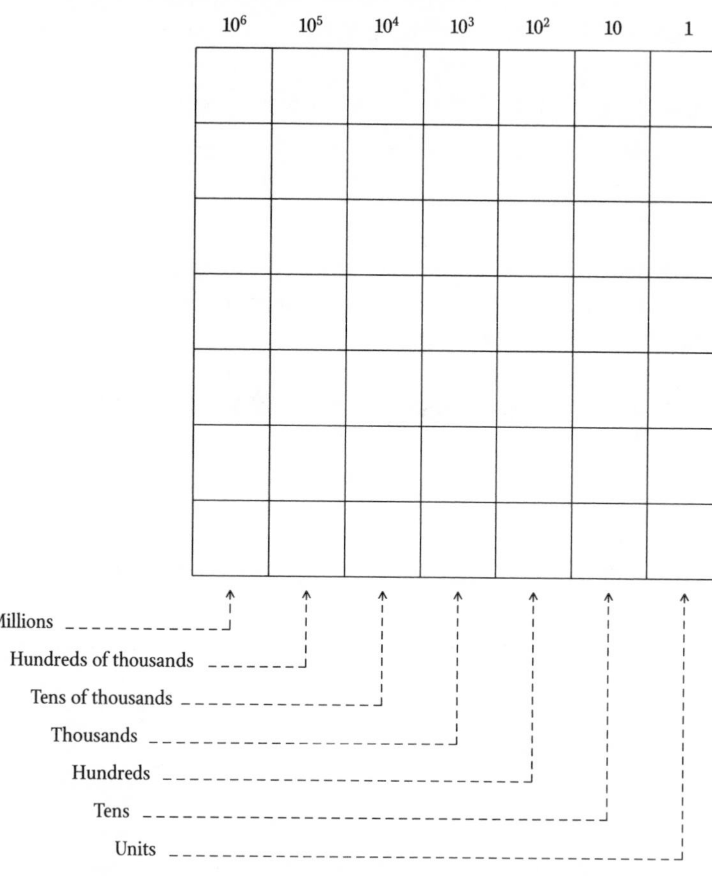


FIG. 13.22. *Reconstructed Assyro-Babylonian decimal abacus*

It should be noted that a brick marked with rows and columns as in Fig. 13.22 was discovered in the 1970s by the French Archaeological Delegation to Iran (DAFI) during the dig at the Acropolis of Susa, and that a few similar pieces were found in the same area during the Second World War. Up to now these objects have been taken as game-boards. *We suggest that they should rather be seen as arithmetical abaci*. Let us hope that further archaeological discoveries will provide suffficient evidence to confirm this interpretation. What we can be sure of, all the same, is that Susan accountants (and the Elamites in general) also used arithmetical tools, of which the first were of course the *calculi*. And there are very good reasons for thinking that the tools they used were similar to those of the Mesopotamians, for their operations were presumably just as complex as those being carried out a few hundred miles away by their Sumerian and Assyro-Babylonian counterparts.

## THE LAST TRACES OF SUMERIAN ARITHMETIC IN THE ASSYRO-BABYLONIAN DECIMAL SYSTEM

In the hands of the Semites, cuneiform numerals and Mesopotamian arithmetic were gradually adapted and finally transformed into a system with a different base working on quite different principles. All the same, base 60 did not disappear entirely, and even continued to play a major role as "big unit" in "ordinary" Mesopotamian accounting. Although it was often represented (at least, from the start of the first millennium BCE) by the decimal expression [cuneiform] Assyrians and Babylonians alike continued to represent the number 60 also by "spelling it out" inside numerical expressions, using either the sign for the sound *shu-shi* (which is how "sixty" was said in Semitic languages)

1 shu-shi

or in abbreviated form as *shu* (the first syllable of the word for "sixty")

1 shu

Above all, they went on figuring the numbers 70, 80, and 90 in the "old manner", that is to say in a way that carries the trace of the obsolete 60-based arithmetic of the Sumerians (just as, nowadays, the French words for 80 and 90 (*quatre-vingts, quatre-vingt-dix*) carry the trace of a vanished vigesimal arithmetic):

60 10 ------> 70

60 20 ------> 80

60 30 ------> 90

FIG. 13.23.

The signs for the old base units of 600 and 3,600 never disappeared entirely either. They continue to crop up in contracts and financial statements, in auguries and in historical and commemorative texts. In these later usages, the sign for 3,600 underwent a graphic development in line with the evolution of Mesopotamian cuneiform writing:

| CLASSICAL SUMERIAN | ASSYRIAN | | |
|---|---|---|---|
| | Ancient | Middle | Late |
| | BABYLONIAN | | |
| | Ancient | Middle | Late |

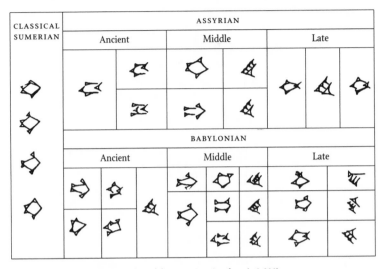

FIG. 13.24. *Evolution and stability of the Sumerian sign* shar *(= 3,600)*

For example, when Sargon II of Assyria inscribed the dimensions of the walls of his fortress at Khorsabad – 16,280 cubits* – he had the figure written not in what had by then become the standard notation:

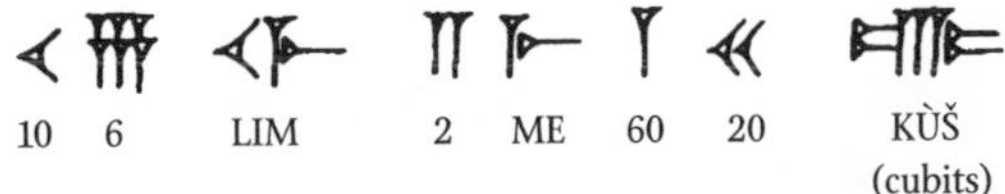


FIG. 13.25.

but in this arithmetically more archaic manner:

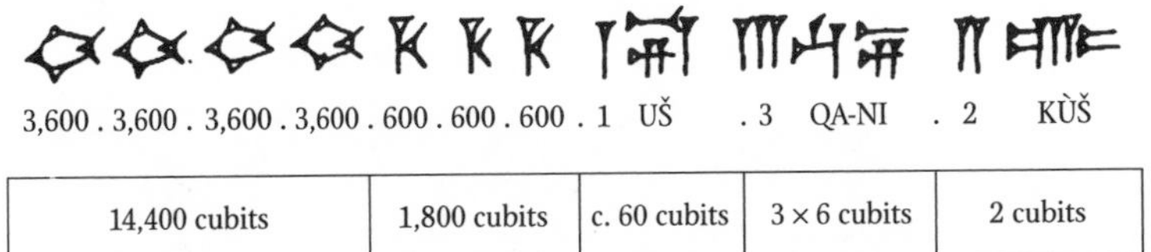


| 14,400 cubits | 1,800 cubits | c. 60 cubits | 3 × 6 cubits | 2 cubits |
|---|---|---|---|---|

FIG. 13.26. *See Lyon, p. 10, l. 65*

But such traces of the old system were mere relics, and had no influence at all on the strictly decimal arithmetic that the Assyro-Babylonians used throughout their history for everyday reckoning.

* The cubit (*kùš*) is a measure of length of approx. 50 cm. Six cubits make a *qânum*, and sixty cubits make an *uš*.

## RECAPITULATION: FROM SUMERIAN TO ASSYRO-BABYLONIAN NUMBERING

There were, in brief, three main stages in Mesopotamian culture after the establishment of the Akkadian Empire:

- in the first, the Semites assimilated the cultural heritage of their Sumerian predecessors in the region;
- the second is an intermediate period;
- the third is the period of Semitic predominance in Mesopotamian culture.

As far as numbers and arithmetic are concerned, these periods correspond respectively to: pure and simple borrowing of Sumerian sexagesimal numbering; the emergence of a mixed system using a combination of decimal and sexagesimal signs; and the development of a strictly decimal system. This profound transformation of cuneiform numbers occurred under the pressure of oral number-names, whose strictly decimal structure is a common feature of all Semitic languages (see Fig. 13.7 and 13.19 above). But this is not where the development came to a full stop: as we shall see, the scribes of the city of Mari evolved their own unique version of a decimal numeral system.

| | SUMERIAN SYSTEM (base 60 with 10 and 6 as auxiliary bases) | SUMERIAN-AKKADIAN SYNTHESIS (compromise between base 10 and base 60) | ORDINARY ASSYRO-BABYLONIAN SYSTEM (Strictly decimal base) |
|---|---|---|---|
| 1 | | | |
| 10 | | | |
| 60 | | 1 ŠU-ŠI or 1 ŠU | |
| 70 | 60 10 | | |
| 80 | 60 20 | | |
| 90 | 60 30 | | |
| 100 | 60 40 | or 1 ME | 1 ME |
| 120 | 60 60 | 2 ŠU-ŠI or 1 ME 20 | 1 ME 20 |
| 600 | | 6 ME | 6 ME |
| 1,000 | 600 360 40 | 1 LI-MI or 1 LIM | 1 LIM |
| 3,600 | | 3 LIM 6 ME | 3 LIM 6 ME |

FIG. 13.27. *Evolution of popular Mesopotamian numerals before and after the eclipse of Sumerian civilisation (see also Fig. 18.9 below)*

## THE ANCIENT SYRO-MESOPOTAMIAN CITY OF MARI

Various texts refer to the Sumero-Semitic city of Mari as an important place in the Mesopotamian world, but it was not until 1933 that André Parrot, led on by the suggestions of W. F. Albright and by the chance discovery of a statue, began to excavate at Tel-Hariri, on the border of Syria and Iraq. Over the following forty years, Parrot's team conducted a score of excavations and laid bare a whole civilisation.

The earliest traces of habitation at Mari date from the fourth millennium BCE, and by the first half of the third millennium it was already highly urbanised, with a ziggurat and a number of temples decorated with statuary and painted walls. The art and culture of Mari in this period resemble those of Sumer, but the facial types represented, as well as the names and the gods mentioned, are Semitic.

Mari became part of the Akkadian Empire, but regained some independence around the twenty-second century BCE. From the twentieth to the eighteenth century BCE Mari flourished as an independent and expanding city-state, but it was defeated and destroyed by Hammurabi around 1755 BCE. Though it continued to exist as a town, Mari never again regained any power or influence.

It was in the early eighteenth century BCE, under Zimri-Lim, that Mari built its most remarkable structures, including a 300-room palace occupying a ground area of 200 m × 120 m and in part of which were stored more than 20,000 cuneiform tablets, giving us a unique insight into the political, administrative, diplomatic, economic, and juridical affairs of a Mesopotamian state. The tablets include long lists of the palace's requirements (food, drink, etc.), and many letters written by women, which suggests that they played an important role in the life of the city.

### WHAT IS THE RULE OF POSITION?

Just as an alphabet allows all the words of a language to be written by different arrangements of a very limited set of signs, so our current numerals allow us to represent all the integers by different arrangements of a set of only ten different signs. From an intellectual point of view, this system is therefore far superior to most numerical systems of the ancient world. However, that superiority does not derive from the use of base 10, since bases such as 2, 8, 12, 20, or 60 can produce the same advantages and be used in exactly the same way as our current decimal positional system. As we have already seen, moreover, 10 is by far the most widespread numerical base in virtue not of any mathematical properties, but of a particularity of human physiology.

What makes our written numeral system ingenious and superior to others is the principle that *the value of a sign depends on the position it occupies in a string of signs*. Any given numeral is associated with units, tens, hundreds, or thousands depending on whether it occupies the first, second, third, or fourth place in a numerical expression (counting the places from right to left).

These reminders allow us to understand fully the numbering system of Mari and of the learned men of Babylon . . .

### THE MARI SYSTEM

It has recently come to light that the scribes of Mari used, alongside "classical" Mesopotamian number-notation, a system of numerals quite different to all that had preceded it.

As in previous systems, the first nine units were represented by an equivalent number of *vertical wedges*:

| 1 | 2 | 3 | 4 | 5 | 6 | 7 | 8 | 9 |
|---|---|---|---|---|---|---|---|---|

FIG. 13.28.

Similarly, the representation of the tens was in line with previous traditions, since it was based on the use of an equivalent number of *chevrons*. However, unlike the Assyro-Babylonians, the scribes of Mari did not use the old sexagesimal character for 60, but carried on multiplying chevrons for the numbers 60, 70, 80, and 90:

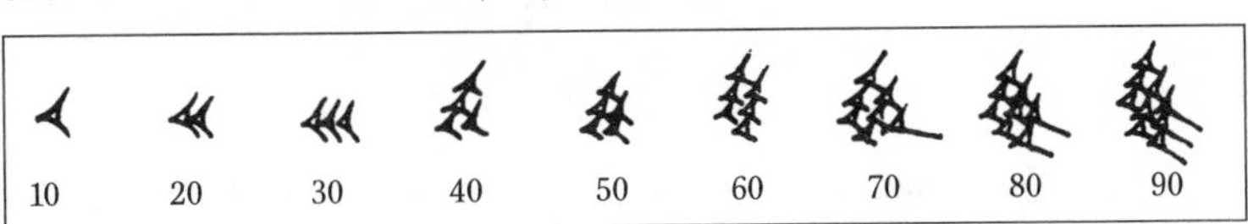


FIG. 13.29.

For 100, they did not use the old system of a wedge plus the sign for the word for 100 (ME), with the meaning 1 × "hundred": what they used was just the single vertical wedge. The number 200 was figured by two vertical wedges, 300 by three, and so on.

NOTATIONS OF THE HUNDREDS BY THE SCRIBES OF MARI

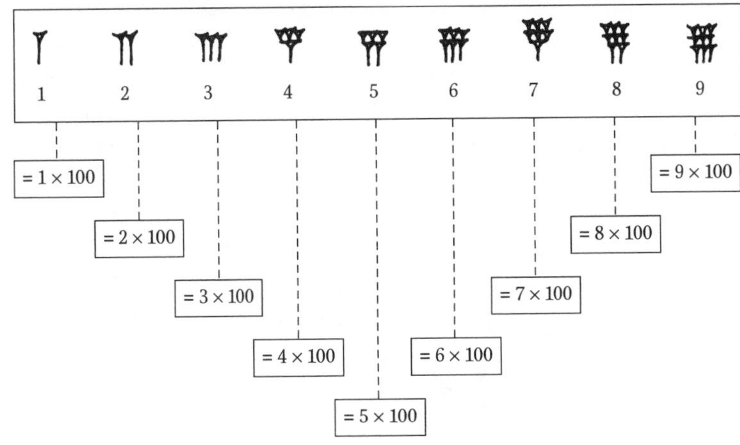


FIG. 13.30.

So a wedge represented either a unit or a hundred depending on where it came in the numerical expression.

For instance, to write "120", "130," and so on, the scribes of Mari put down 1 vertical wedge followed by 2, 3, etc. chevrons. And to represent a number such as 698, all that was needed was a representation of 6 followed by a representation of 98 (9 chevrons and 8 wedges):

| [1; 10] | [1; 20] | [1; 30] | [6; 98] |
|---|---|---|---|
| = 1 × 100 + 10<br>= 110 | = 1 × 100 + 20<br>= 120 | = 1 × 100 + 30<br>= 130 | = 6 × 100 + 98<br>= 698 |

FIG. 13.31.

It is clear that the scribes of Mari knew both the classical Mesopotamian decimal notation and also the positional sexagesimal system of the scholars (see below). When they drew up their tablets in Akkadian (a language which they handled with ease), they used the former for "current business" such as economic and legal documents, and the latter for "scientific" matters (tables, mathematical problems, and so on). In fact, the system we are now considering never was the official numeral-system of the city: for it is only found in quite particular places on the tablets (on the edges, on the reverse side, and in the margins) and, in most cases, the numbers worked out in the new system were written out again in one or the other of the two standard systems.

In other words, the new system seems to have served only as an aide-mémoire and checking device, to make doubly sure that the results written out in the traditional way were in fact correct. What we see is a kind of mathematical bilingualism, in which matching results reached by two separate notations resolve doubts about the correctness of the sums. And it is of course only because of the role that the system played, and because of the position of the new-style numerals on the tablets, that modern scholars have been able to read and interpret them.

The following examples come from the Royal Archives of Mari, as quoted, translated and decoded by D. Soubeyran. The first gives the last column of a tally of people, showing the totals for rows identified by the words in brackets, which refer to the categories of people counted:

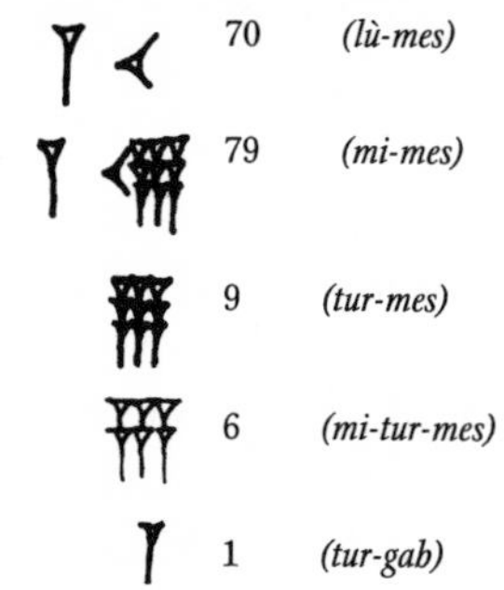


FIG. 13.32.

These numbers are written in classical Akkadian manner, so they represent: 70 + 79 + 9 + 6 + 1 = 165. However, after a space and before the title of the tablet comes the following expression:

FIG. 13.33A. 1 65

If this were a non-positional expression, its value would either be 1 + 65 = 66, or (allowing the vertical wedge to mean 60, as it often did in Akkadian arithmetic) 60 + 65 = 125. In neither case could it be a running total for the column which it follows after a space. However, if the wedge is given the value 100, then we do indeed get the running total of 165, in the following manner:

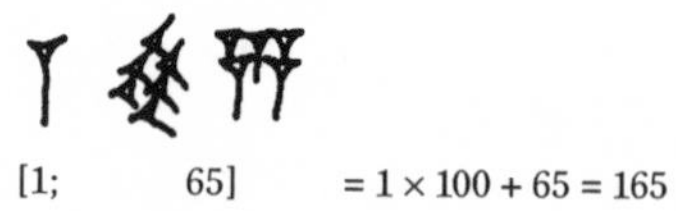


FIG. 13.33B.

The second example is also a list of people, perhaps of nobles. Each entry is accompanied by a number, which perhaps indicates the number of servants owned. There is a running total of 183 brought forward, which is written in classical decimal form as

1 ME-AT 83 ("1 hundred 83")

A second subtotal gives the figure of 26 servants, and, as you might expect, the grand total comes to 209, which is expressed in the same way as:

2 ME-TIM 9 (2 × 100 + 9 = 209)

However, the side-edge of the tablet has the following expression:

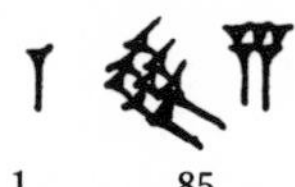

FIG. 13.34A. 1 85

If this were taken in the classical (non-positional) way, then it would mean either 1 + 85 (= 86) or 60 + 85 (= 145), and the totals would not match at all.

However, if the figures on the edge are taken as a centesimal positional expression, then the sum is 185, which is roughly the same as the first count of servants, 183.

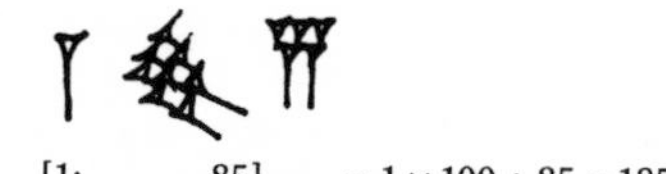


FIG. 13.34B.

The last of the three tablets details a sequence of deliveries of copper scythes, with a running total, written in the standard way, of 471 scythes. But between the markings for the month and the year, there is this:

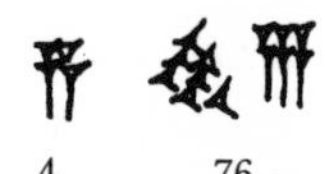

FIG. 13.35A. 4 76

Once again, this expression would not have much meaning if it were read in the classical manner, but, taking it as an expression in the positional system of the scribes of Mari, it would give 476, a good approximation of the previous running total (471):

FIG. 13.35B.

According to Soubeyran, the minor discrepancies between these figures and the totals, as well as their position on the tablets and the rough and ready way they are written, shows that they are rough drafts or workings-out, intended to check figures before they were inscribed on the tablets in a formal way. That makes it all the more interesting to see the scribes of Mari *thinking* in a positional, centesimal-decimal system, before converting their results into sexagesimal notation.

For numbers between 100 and 1,000, the Mari system used the *rule of position*, and its base was not 10, but 100: the first "large" unit was the hundred, with the ten playing the role of auxiliary base. On the other hand, the system did not have a zero. If it had had such a thing, then it would have served to mark the absence of units in a given order. In other words, if there had been a zero in the Mari system, then the multiples of the base would have been written in the same way as we write multiples of our base (20, 30, 40, etc.), with a zero indicating the absence of units of the first order.

All the same, the scribes of Mari were perfectly aware that the value of the numerals they wrote down depended on their position in a specific numerical expression. This is all the more noteworthy because very few civilisations have ever reached such a degree of simplification in written numerals, and by the same token discovered the rule of position. This development took place very early on: the tablets that bear the trace of the rule of position are not later than the eighteenth century BCE.

However, the system was not strictly or consistently positional. Had that been the case, then 1,000 (= 10 × 100, or ten units of the second centesimal order) would have been represented by a chevron, 2,000 by two chevrons, and so on. As for 10,000, the square of the base of the second centesimal order, it would have been represented by a vertical wedge (had there been a zero, it would have been figured in the form [1; 0; 0], the first zero signifying the absence of any units of the first order (numbers between 1 and 99), the second the absence of units of the second order (multiples of 100 by a number between 1 and 99)). And since 200 = 2 × 100, represented by two vertical wedges, so 20,000 = 2 × 10,000 would similarly have been represented by two vertical wedges.

But it was not so: the Mari system had special signs for 1,000 and for 10,000. However, the "Mari thousand" was rather different from the classical numeral, and it was combined with a multiplier to make numbers like 2,000, 3,000, etc.:

| LI-IM | 2 LI-IM |
|---|---|
| 1,000 | 2,000 |

FIG. 13.36.

This adds up to a mixed system, using simultaneously all the basic rules, of addition (for the total), of multiplication (for the thousands), and of position (for numbers less than 1,000).

The Mari scribes used a figure derived from the thousand overlaid with a chevron (= 10) to represent 10,000 (which was then combined with units for multiples of 10,000):

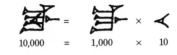

10,000 = 1,000 × 10

FIG. 13.37.

This is the only example amongst the decimal numerations of the whole Mesopotamian region where 10,000 is not written as an analytical combination of the numerals 10 and 1,000, and it is yet another way in which the Mari system is quite unique.

The Marian cuneiform sign for 10,000 (found not just in economic tablets, but in fields as diverse as tallies of bricks, of land areas, and of livestock) is related to the Sumerian ideogram GAL, which meant "large", and was pronounced *ribbatum* in the language of Mari, with the literal meaning of "multitude", whence "large number". So that was the *name* of the number 10,000, and it is clearly the same name as the one found at Ebla (*ri-bab*) in the twenty-fourth century BCE, at Ugarit (*r(b)bt*) in the fifteenth century BCE, and then in Syria (*ri-ib-ba-at*), and in Hebrew (*ribō*, pl. *ribōt*).

The following two examples from tablets found at Mari give a fuller view of how the system worked:

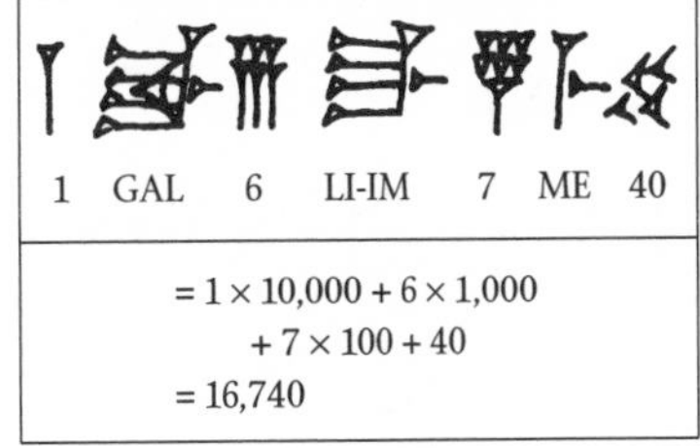


FIG. 13.38. *See Durand (1987)*

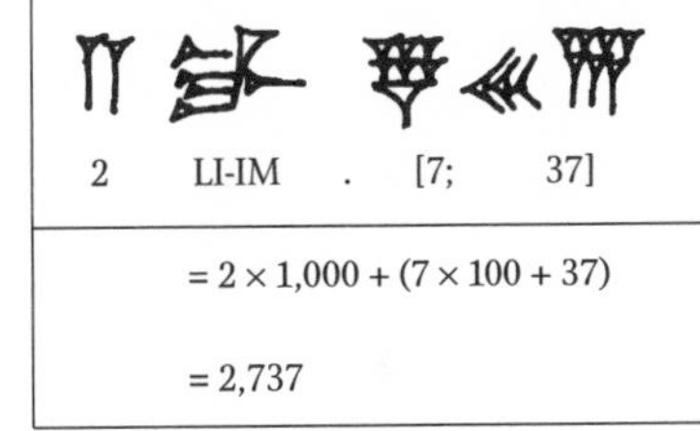


FIG. 13.39. *See Soubeyran (1984)*

In etymological and graphological terms, 10,000 was the "biggest number" in the system of Mari. (The scribes could of course represent far larger numbers by using the multiplication principle, even if no really large numbers have yet been found in the tablets.) It was a quite unique centesimal numeral system, found exclusively in this one city on the common border of Syria and Mesopotamia, at the time of the patriarch Abraham. It might have developed into a fully positional system had the Babylonian king Hammurabi not razed the city to the ground in 1755 BCE, and buried with it a very large part of Mari's culture. Ironically, it was the Babylonians themselves who actually devised the world's first true positional system – but it was neither a variant of Akkadian decimal arithmetic, nor a centesimal system like that of Mari. Used for mathematical and astronomical reckonings right down to the dawn of the Common Era, the "learned" numerals of Babylon were a direct inheritance of Sumer, whose memory they have perpetuated, directly and indirectly, right down to the present day.

## THE POSITIONAL SEXAGESIMAL SYSTEM OF THE LEARNED MEN OF MESOPOTAMIA

Although we cannot be sure about the exact date, the first real idea of a positional numeral system arose amongst the mathematicians and astronomers of Babylon in or around the nineteenth century BCE.

The Mesopotamian scholars' abstract numerals were derived from the ancient Sumerians' sexagesimal figures, but constituted a system far superior to anything else in the ancient world, anticipating modern notation in all respects save for the different base and the actual shapes used for the numerals.

Unlike the "ordinary" Assyro-Babylonian notation used for everyday business needs, the learned system used base 60 and was strictly positional. Thus a group of figures such as

[3; 1; 2]

which in modern decimal positional notation would express:

$$3 \times 10^2 + 1 \times 10 + 2$$

signified to Babylonian mathematicians and astronomers:

$$3 \times 60^2 + 1 \times 60 + 2$$

Similarly, the sequence [1; 1; 1; 1] which in our system would mean $1 \times 10^3 + 1 \times 10^2 + 1 \times 10 + 1$ (or 1,000 + 100 + 10 + 1) signified in the Babylonian system $1 \times 60^3 + 1 \times 60^2 + 1 \times 60 + 1$ (or 216,000 + 3,600 + 60 +1).

Instances of this system of numerals have been known since the very dawn of Assyriology, in the mid-nineteenth century, and, thanks to excavations made throughout Mesopotamia and Iraq at that time, many examples have come to rest in the great European museums (Louvre, British Museum, Berlin) and in the university collections at Yale, Columbia, Pennsylvania, etc. The types of document in which the learned system is used (and which come from Elam and Mari, as well as from Nineveh, Larsa, and other Mesopotamian cities) are for the most part as follows: tables intended to assist numerical calculation (e.g. multiplication tables, division tables, reciprocals, squares, square roots, cubes, cube roots, etc.); astronomical tables; collections of practical arithmetical and elementary geometrical exercises; lists of more or less complex mathematical problems.

The system is sexagesimal, which is to say that 60 units of one order of magnitude constitute one unit of the next (higher) order of magnitude. The numbers 1 to 59 constitute the units of the first order, multiples of 60 constitute the second order, multiples of 3,600 (sixty sixties) constitute the third order, multiples of 216,000 (the cube of 60) constitute the fourth order, and so on.

In fact, there were really only two signs in the system: a vertical wedge representing a unit, and a chevron representing 10:

1 10

Numbers from 1 to 59 inclusive were built on the principle of addition, by an appropriate number of repetitions of the two signs. Thus the numbers 19 and 58 were written

(1 chevron + 9 wedges) or (5 chevrons + 8 wedges)

So far the system is exactly the same as its predecessors. However, beyond 60, the learned system became strictly positional. The number 69, for instance, was not written

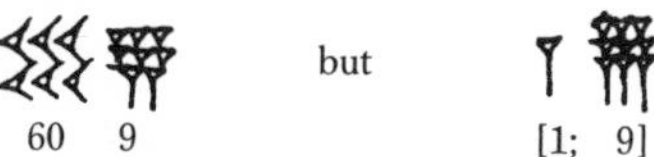


For example, this is how Asarhaddon, king of Assyria from 680 to 669 BCE, justified his decision to rebuild Babylon (wrecked by his father Sennacherib in 689 BCE) rather sooner than the holy writ prescribed:

> After inscribing the number 70 for the years of Babylon's desertion on the Tablet of Fate, the God Marduk, in his pity, changed his mind. He turned the figures round and thus resolved that the city would be reoccupied after only eleven years. [From *The Black Stone*, trans. J. Nougayrol]

The anecdote takes on its full meaning only in the light of Babylonian sexagesimal numbering. To begin with, Marduk, chief amongst the gods in the Babylonian pantheon, decides that the city will remain uninhabited for 70 years, and, to give full force to his decision, inscribes on the Tablet of Fate the signs:

Fig. 13.40A. [1; 10] ([1; 10] = 1 × 60 + 10)

Thereafter, feeling compassion for the Babylonians, Marduk inverts the order of the signs in the expression, thus:

10 . 1 (= 10 + 1)

FIG. 13.40B.

Since the new expression represents the number 11, Marduk decreed that the city would remain uninhabited only for that length of time, and could be rebuilt thereafter. The anecdote shows that the Mesopotamian public in general was at least aware of the rule of position as applied to base 60.

In the Babylonian system, therefore, the value of a sign varied according to its position in a numerical expression. The figure for 1 could for instance express

- a unit in first position from the right,
- a sixty in the second position,
- sixty sixties or $60^2$ in third position,

and so on.

| | | | |
|---|---|---|---|
| 1 | | 11 | |
| 2 | | 16 | |
| 3 | | 25 | |
| 4 | | 27 | |
| 5 | | 32 | |
| 6 | | 39 | or |
| 7 | | 41 | |
| 8 | | 46 | |
| 9 | or or * | 52 | |
| 10 | | 55 | |
| 20 | | 59 | |
| 30 | | | |
| 40 | or | | |
| 50 | or | | |

* Abbreviations used in the later period

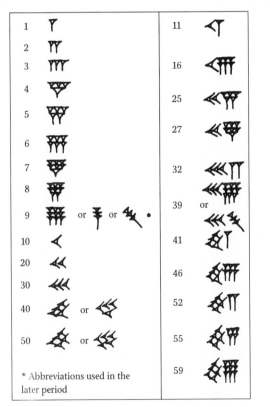

FIG. 13.41. *Representations of the fifty-nine significant units of the learned Mesopotamian numeral system*

For instance, to write the number 75 (one sixty and fifteen units) you put a "15" in first position and a "1" in second position, thus:

[1 ; 15] (= 1 × 60 + 15 = 75)

FIG. 13.42.

And to write 1,000 (16 sixties and 40 units) you put a "40" in first position and a "16" in second position, thus:

[16 ; 40] (= 16 × 60 + 40 = 1,000)

FIG. 13.43.

Conversely, an expression such as

[48 ; 20 ; 12]

FIG. 13.44.

expresses the number:

$$48 \times 60^2 + 20 \times 60 + 12 = 48 \times 3{,}600 + 20 \times 60 + 12 = 174{,}012$$

in exactly the same way as we would express "174,012 seconds" as:

48 h 20m 12s

Similarly, an expression such as

[1 ; 50 + 7 ; 30 + 6 ; 10 + 5] or [1 ; 57 ; 36 ; 15]

FIG. 13.45.

symbolises, in the minds of the Babylonian scholars, the number:

$$1 \times 60^3 + 57 \times 60^2 + 36 \times 60 + 15 \ (= 423{,}375)$$

The next examples come from one of the most ancient Babylonian mathematical tablets known (British Museum, BM 13901, dating from the period of the first kings of the Babylonian Dynasty), a collection of problems relating the solution of the equation of the second degree:

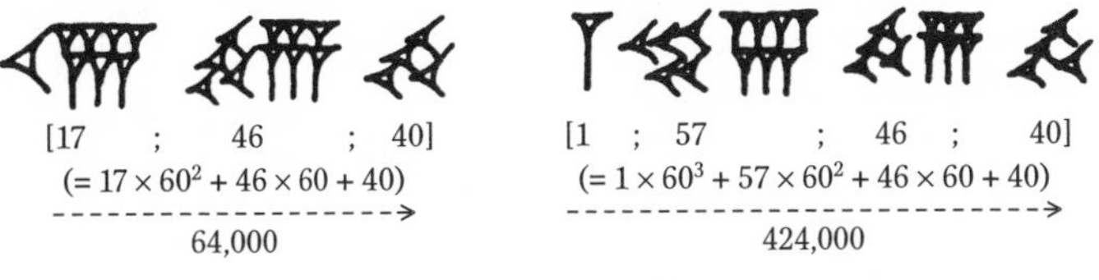


FIG. 13.46.

FIG. 13.47.

The difference between Sumerian numbers and the Babylonian "learned" system was simply this: the Sumerians relied on addition, the Babylonians on the rule of position. This can easily be seen by comparing the Sumerian and Babylonian expressions for the two numbers 1,859 and 4,818:

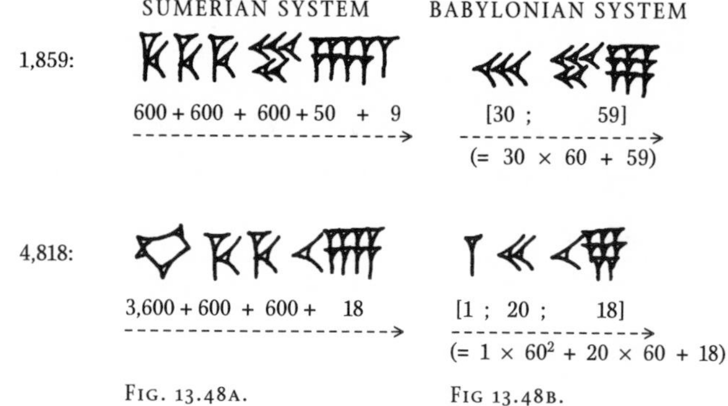


FIG. 13.48A. FIG 13.48B.

## THE TRANSITION FROM SUMERIAN TO LEARNED BABYLONIAN NUMERALS

One of the reasons for the "invention" of the learned Babylonian system is easy to understand – it was the "accident" which gave 1 and 60 the same written sign in Sumerian, and which originally constituted the main difficulty of using Sumerian numerals for arithmetical operations.

Moreover, the path to the discovery of positionality had been laid out in the very earliest traces of Sumerian civilisation. The two basic units were represented, first of all, by the same name, *geš* (see Fig. 8.5A and 8.5B above); then, in the second half of the fourth millennium BCE, they were represented by objects of the same shape (the small and large cone) (see Fig. 10.4 above); then, from 3200–3100 BCE to the end of the third millennium, by two figures of the same general shape, the narrow notch and the thick notch (see Fig. 8.9 above); then, from around the twenty-seventh century BCE, by cuneiform marks of the same type, distinguished only by their respective sizes; and, finally, from the third dynasty of Ur onwards (twenty-second to twentieth century BCE), especially in the writings of Akkadian scribes, by the same vertical wedge.

In other words, as we can see from Asarhaddon's story in *The Black Stone*, and in the Assyro-Babylonian representations of the numbers 70, 80 and 90 (see Fig. 13. 23 above), the large wedge meaning 60 had evolved in line with the general evolution of cuneiform writing so as to be indistinguishable from the small wedge meaning 1.

In everyday usage, that evolution was seen as a problem, which was got round by "spelling out" 60 as *shu-shi* in numbers such as 61, 62, 63, where the confusion was potentially greatest (see Fig. 13.14 above), and eventually by replacing the sexagesimal unit with a multiple of a decimal one (Fig. 13.18 above).

But in the usage of the learned men of Mesopotamia, the graphical equivalence of the signs for 1 and 60 gave rise (at least for numbers with two orders of magnitude) to a true rule of position. As the following notations show:

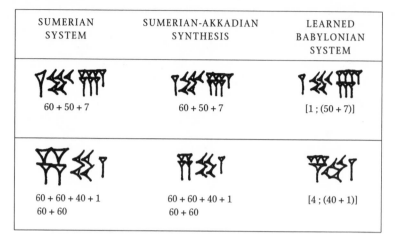

| SUMERIAN SYSTEM | SUMERIAN-AKKADIAN SYNTHESIS | LEARNED BABYLONIAN SYSTEM |
|---|---|---|
| 60 + 50 + 7 | 60 + 50 + 7 | [1 ; (50 + 7)] |
| 60 + 60 + 40 + 1<br>60 + 60 | 60 + 60 + 40 + 1<br>60 + 60 | [4 ; (40 + 1)] |

FIG. 13.49.

Babylonian scholars realised therefore that the rule or principle could be generalised to represent all integers, provided that the old Sumerian signs for the multiples and powers of 60 were abandoned. The first to go was the 600 (= 60 × 10), for which was substituted as many chevrons (= 10) as there were 60s in the number represented. Then the sign for 3,600 (the square of 60) was dropped, and, since this number was a unit of the third sexagesimal order, it was henceforth represented by a single vertical wedge. Subsequently the sign for 36,000 was eliminated, and replaced by the sign for 10 in the position reserved for the third sexagesimal order, and so on.

For instance, instead of representing the number 1,859 by three signs for 600 followed by the notation of the number 59 (1,859 = 3 × 600 + 59), Babylonian scholars now used [30; 59] (= 30 × 60 + 59), as shown in Fig. 13.48 above, which also gives the example of the "old" and "new" representations of 4,818.

The vertical wedge thus came to represent not only the unit, but any and all powers of 60. In other words, 1 was henceforth figured by the same wedge that signified 60, 3,600, 216,000, and so on, and all 10-multiples of the base (600, 36,000, 2,160,000, etc.) by the chevron.

The discovery was extremely fruitful in itself, but, because of the very circumstances in which it arose, it gave rise to many difficulties.

## THE DIFFICULTIES OF THE BABYLONIAN SYSTEM

Despite their strictly positional nature and their sexagesimal base, learned Babylonian numerals remained decimal and additive within each order of magnitude. This naturally created many ambiguous expressions and was thus the source of many errors. For example, in a mathematical text from Susa, a number [10; 15] (that is to say, $10 \times 60 + 15$, or 615) is written thus:

[10 ; 15]

FIG. 13.50A.

However, this expression could also just as easily be read as

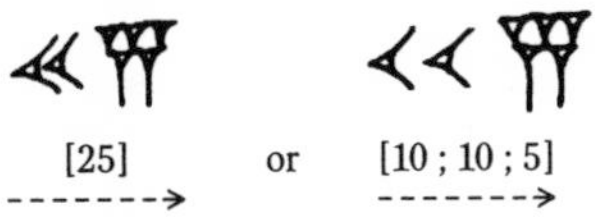

[25] or [10 ; 10 ; 5] $(= 10 \times 60^2 + 10 \times 60 + 5)$

FIG. 13.50B.

It is rather as if the Romans had adopted the rule of position and base 60, and had then represented expressions such as "10° 3′ 1″" (= 36,181″) by the Roman numerals X III I, which they could easily have confused with XI II I (11° 2′ 1″), X I III (10° 1′ 3″) , and so on. Scribes in Babylon and Susa were well aware of the problem and tried to avoid it by leaving a clear space between one sexagesimal order and the next. So in the same text as the one from which Fig. 13.50 is transcribed, we find the number [10; 10] $(= 10 \times 60 + 10)$, represented as:

[10 ; 10]

FIG. 13.51.

The clear separation of the two chevrons eliminates any ambiguity with the representation of the number 20.

In another tablet from Susa the number [1; 1; 12] $( = 1 \times 60^2 + 1 \times 60 + 12)$ is written

[1 ; 1 : 12]

FIG. 13.52A.

in which the clear separation of the leftmost wedge serves to distinguish the expression from

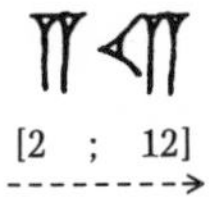

[2 ; 12] $(= 2 \times 60 + 12)$

FIG. 13.52B.

In some instances scribes used special signs to mark the separation of the orders of magnitude. We find double oblique wedges, or twin chevrons one on top of the other, fulfilling this role of "order separator"*:

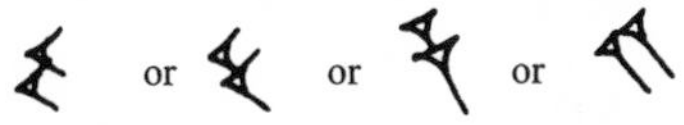

or or or

FIG. 13.53.

Here are some examples from a mathematical tablet excavated at Susa:

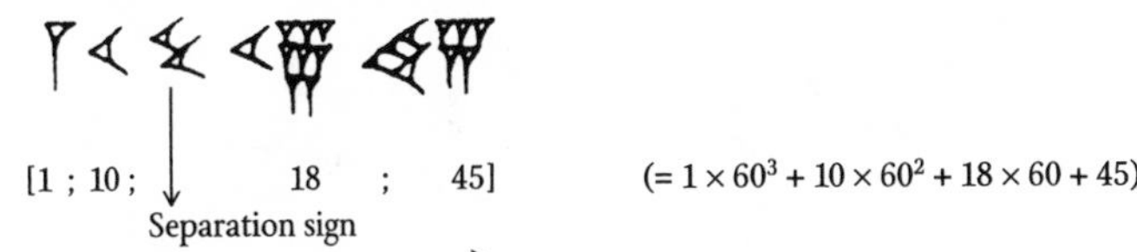

$(= 1 \times 60^3 + 10 \times 60^2 + 18 \times 60 + 45)$

FIG. 13.54A.

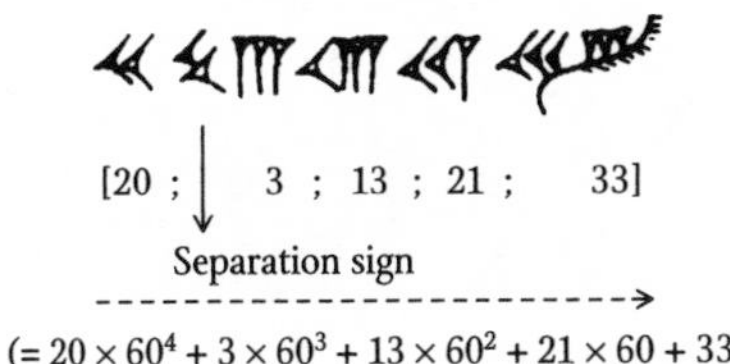

$(= 20 \times 60^4 + 3 \times 60^3 + 13 \times 60^2 + 21 \times 60 + 33)$

FIG. 13.54B.

The sign of separation makes the first number above quite distinct from the representation of [1; 10 + 18; 45] $(= 1 \times 60^2 + 28 \times 60 + 45)$; and for the same reason the second number above cannot be mistaken for [20 + 3; 13; 21; 33] $(= 23 \times 60^3 + 13 \times 60^2 + 21 \times 60 + 33)$.

This difficulty actually masked a much more serious deficiency of the system – the *absence of a zero*. For more than fifteen centuries, Babylonian mathematicians and astronomers worked without a concept of or sign for zero, and that must have hampered them a great deal.

In any numeral system using the rule of position, there comes a point where a special sign is needed to represent units that are missing from the number to be represented. For instance, in order to write the number *ten* using (as we now do) a decimal positional notation, it is easy enough to place the sign for 1 in second position, so as to make it signify one unit of the higher (decimal) order – but how do we signify that this sign is indeed

* In commentaries on literary texts, the same sign was used to separate head-words from their explications; in multilingual texts, the sign was used to mark the switch from one language to another; and in lists of prophecies, the sign was used to separate formulae and to mark the start of an utterance.

in second position if we have nothing to write down to mean that there is nothing in the first position? *Twelve* is easy – you put "1" in second position, and "2" in first position, itself the guarantee that the "1" is indeed in second position. But if all you have for ten is a "1" and then nothing . . . The problem is obviously acute. Similarly, to write a number like "seven hundred and two" in a decimal positional system, you can easily put a "7" in third position and a "2" in first position, but it's not easy to tell that there's an arithmetical "nothing" between them if there is indeed *no thing* to put between them.

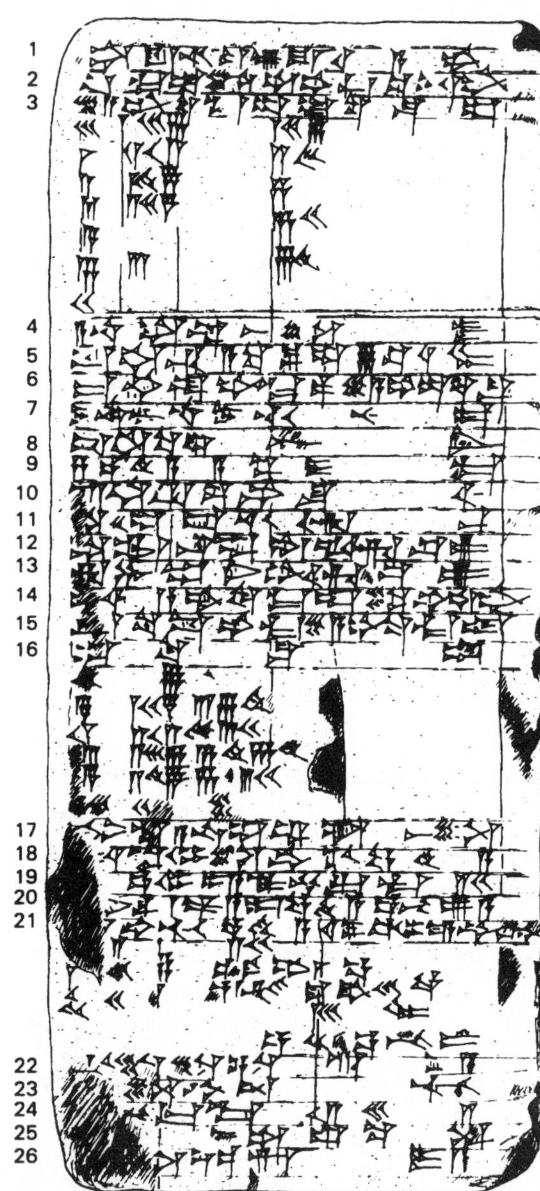

FIG. 13.55. *Important mathematical text from Larsa (Senkereh), dating from the period of the First Babylonian Dynasty (Louvre, AO 8862, side IV). See Neugebauer, tablet 38. Beneath line 16, note the representation of the number 18,144,220 as [1; 24; blank space; 3; 40].*

It became clear in the long run that such a *nothing* had to be represented by *something* if confusion in numerical calculation was to be avoided. The something that means nothing, or rather the sign that signifies the absence of units in a given order of magnitude, is, or would one day be represented by, zero.

The learned men of Babylon had no concept of zero around 1200 BCE. The proof can be seen on a tablet from Uruk (Louvre AO 17264) which gives the following solution:

*"Calculate the square of* 𒐖 𒌍𒐝 *and you get* 𒐚 𒐝*"*

In decimal numbers using the rule of position, the first of these expressions (2 × 60 + 27) is equal to 147, and the square of 147 is 21,609. This latter number can be expressed in sexagesimal arithmetic as 6 × 3,600 + 0 × 60 + 9, and should therefore be written in learned Babylonian cuneiform numbers with a "9" in first position, a "6" in third position, and "nothing" in second position. If the scribe had had a concept of zero he would surely have avoided writing the square of [2; 27] as the expression [6; 9] which we see on the tablet – since the simplest resolution of [6; 9] is 6 × 60 + 9 = 369, which is not the square of 147 at all!

Another example of the same kind can be found on a Babylonian mathematical tablet from around 1700 BCE (Berlin Archaeological Museum, VAT 8528), where the numbers [2; 0; 20] (= $2 \times 60^2 + 0 \times 60 + 20$ = 7,220) and [1; 0; 10] (= $1 \times 60^2 + 0 \times 60 + 10$ = 3,610) are represented by

FIG. 13.56.

These notations are manifestly ambiguous, since they could represent, respectively, [2; 20] (= 2 × 60 + 20 = 140) and [1; 10] (= 1 × 60 + 10 = 70).

To overcome this difficulty, Babylonian scribes sometimes left a blank space in the position where there was no unit of a given order of magnitude. Here are some examples from tablets excavated at Susa (examples A, B, C) and from Fig. 13.58 below (example D, line 15). Our interpretations are not speculative, since the values given correspond to mathematical relations that are unambiguous in context:

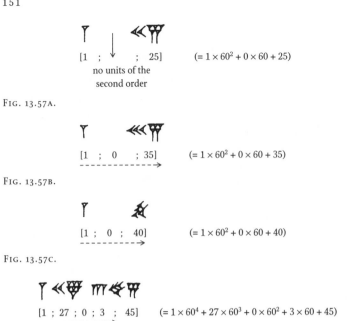

[1 ; ↓ ; 25] (= $1 \times 60^2 + 0 \times 60 + 25$)

no units of the second order

FIG. 13.57A.

[1 ; 0 ; 35] (= $1 \times 60^2 + 0 \times 60 + 35$)

FIG. 13.57B.

[1 ; 0 ; 40] (= $1 \times 60^2 + 0 \times 60 + 40$)

FIG. 13.57C.

[1 ; 27 ; 0 ; 3 ; 45] (= $1 \times 60^4 + 27 \times 60^3 + 0 \times 60^2 + 3 \times 60 + 45$)

FIG. 13.57D.

However, this did not solve the problem entirely. For a start, scribes often made mistakes or did not bother to leave the space. Secondly, the device did not allow a distinction to be made between the absence of units in one order of magnitude, and the absence of units in two or more orders of magnitude, since two spaces look much the same as one space. And finally, since the figure for 4, for instance, could mean $4 \times 60$, $4 \times 60^2$, $4 \times 60^3$, or $4 \times 60^4$, how could you know which order of magnitude was meant by a single expression?

These difficulties were compounded by fractions. Whereas their predecessors had given each fraction a specific sign (see Fig. 10.32 above for an example from Elam), the Babylonians used the rule of position for fractions whose denominator was a power of 60. In other words, positional sexagesimal notation was extended to what we would now call the negative powers of 60 ($60^{-1} = 1/60$, $60^{-2} = 1/60^2 = 1/3{,}600$, $60^{-3} = 1/60^3 = 1/216{,}000$, etc.). So the vertical wedge came to signify not just 1, 60, $60^2$, etc., but also 1/60, 1/3,600, and so on. Two wedges could mean 2 or 120 or 1/30 or 1/1,800; the figure signifying 15 could also signify 1/4 (= 15/60), and the number 30 might just as easily mean 1/2.

Numerals were written from right to left in ascending order of the powers of 60, and from left to right in ascending negative powers of 60, exactly as we now do with our decimal positional numbering – except that in Babylon there was nothing equivalent to the decimal point that we now use to separate the integer from the fraction.

TRANSCRIPTION

| LINE | | | | | |
|---|---|---|---|---|---|
| 1 | TA-KI-IL-TI ṢI-LI-IP-TIM | ÍB-SÁ SAG | ÍB-SÁ ṢI-LI-IP-TIM | MU-BI-IM | |
| 2 | ŠA IN NA-AS-SÀ-ḪU-Ú-[M]A SAG-I | ...-Ú | | | |
| 3 | 1, 59, *, 15 | 1; 59 | 2, 49 | KI | 1 |
| 4 | 1, 56, 56, 58, 14, 50, 6, 15 | 56, 7 | 3, 12, 1 | KI | 2 |
| 5 | 1, 55, 7, 41, 15, 33, 45 | 1, 16, 41 | 1, 50, 49 | KI | 3 |
| 6 | 1, 53, 10, *, 29, 32, 52, 16 | 3, 31, 49 | 5, 9, 1 | KI | 4 |
| 7 | 1, 48, 54, 1, 40 | 1, 5 | 1, 37 | KI | |
| 8 | 1, 47, 6, 41, 40 | 5, 19 | 8, 1 | KI | |
| 9 | 1, 43, 11, 56, 28, 26, 40 | 38, 11 | 59, 1 | KI | 7 |
| 10 | 1, 41, 33, 59, 3, 45 | 13, 19 | 20, 49 | KI | 8 |
| 11 | 1, 38, 33, 36, 36 | 9, 1 | 12, 49 | KI | 9 |
| 12 | 1, 35, 10, 2, 28, 27, 24, 26, 40 | 1, 22, 41 | 2, 16, 1 | KI | 10 |
| 13 | 1, 33, 45 | 45 | 1, 15 | KI | 11 |
| 14 | 1, 29, 21, 54, 2, 15 | 27, 59 | 48, 49 | KI | 12 |
| 15 | 1, 27, *, 3, 45 | 7, 12, 1 | 4, 49 | KI | 13 |
| 16 | 1, 25, 48, 51, 35, 6, 40 | 29, 31 | 53, 49 | KI | 14 |
| 17 | 1, 23, 13, 46, 40 | 56 | 53 | KI | 15 |

* Blank space indicating the absence of units in a given order of magnitude

FIG. 13.58. *Mathematical tablet, 1800–1700 BCE, showing that Babylonian mathematicians were already aware of the properties of right-angled triangles (Pythagoras' theorem). If we take the numbers in the leftmost column A, the second column B, and the third column C, we find that the numbers obey the relationship*

$$A = \frac{a^2}{c^2};\ B = b;\ C = c,\ \text{and}\ a^2 = b^2 + c^2$$

*This expresses the relationship by which in a right-angled triangle (with sides b and c and hypotenuse a) the square of the hypotenuse is equal to the sum of the squares on the other two sides. Columbia University, Plimpton 322. Author's own transcription*

Naturally enough, this led to enormous difficulties, such as are suggested by the following three interpretations (out of many others possible) of a single expression:

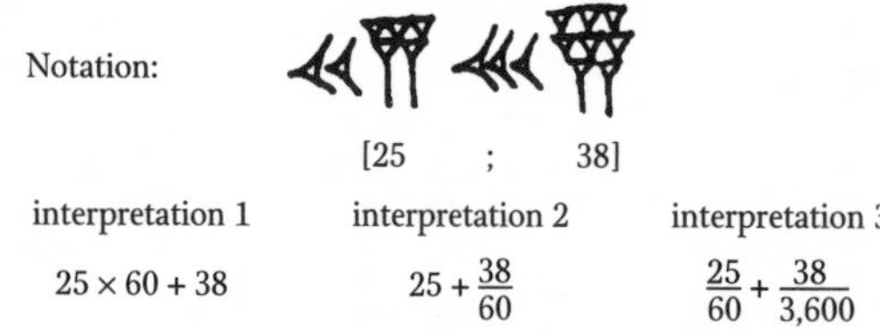


FIG. 13.59.

All the same, Babylonian mathematicians and astronomers managed to perform quite sophisticated operations for over a thousand years despite the imperfections of their numeral system. Of course, they had the orders of magnitude present in their minds, and the ambiguities of the notation were resolved by the context (that is to say, the premises of the problem being tackled) or by the commentary of the teacher, who must presumably have indicated the magnitudes involved.

FIG. 13.60. *Mathematical tablet from Uruk, late third or early second century BCE, containing one of the earliest known instances of the Babylonian zero. Louvre, AO 6484 side B. See Thureau-Dangin, tablet 33, side B, plate LXII.*

## THE BIRTH OF THE BABYLONIAN ZERO

At some point, probably prior to the arrival of the Seleucid Turks in 311 BCE, Babylonian astronomers and mathematicians devised a true zero, to indicate the absence of units of a given order of magnitude. They began to use, instead of a blank space, an actual sign wherever there was a missing order of the powers of 60, and the sign they used was a variant of the old "separator" sign discussed above (see Fig. 13.53):

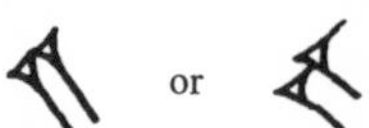


FIG. 13.61.

So, in an astronomical tablet from Uruk (now in the Louvre, AO 6456) from the Seleucid period, we can read:

[2; 0; 25; 38; 4]
$$(= 2 \times 60^4 + 0 \times 60^3 + 25 \times 60^2 + 38 \times 60 + 4)$$

written on the back of the tablet in the form:

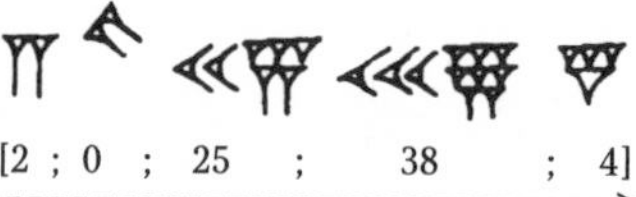


FIG. 13.62.

The diagonal double wedge thus marks the absence of any sexagesimal units of the fourth order of magnitude.

On lines 10, 14 and 24 of the tablet reproduced in Fig. 13.60 above, we can read:

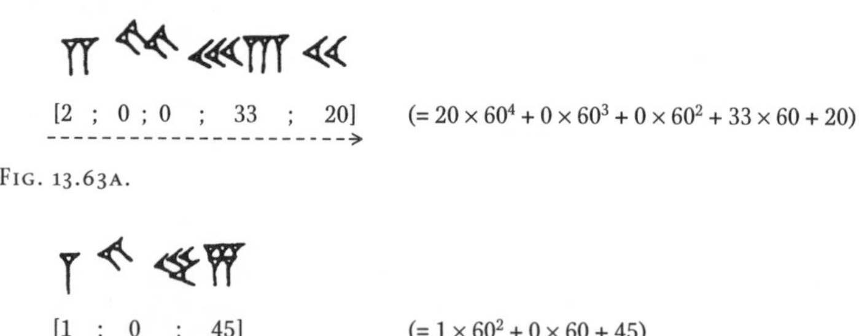

[2 ; 0 ; 0 ; 33 ; 20]
$(= 20 \times 60^4 + 0 \times 60^3 + 0 \times 60^2 + 33 \times 60 + 20)$
FIG. 13.63A.
[1 ; 0 ; 45]
$(= 1 \times 60^2 + 0 \times 60 + 45)$


FIG. 13.63B.

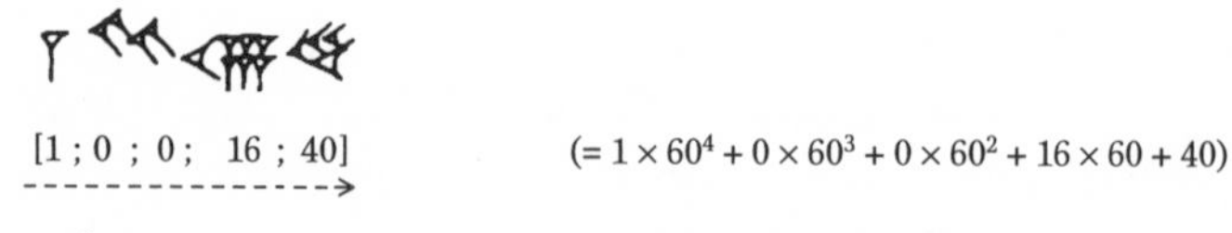

[1 ; 0 ; 0 ; 16 ; 40]
$(= 1 \times 60^4 + 0 \times 60^3 + 0 \times 60^2 + 16 \times 60 + 40)$


FIG. 13.63C.

[1 ; 0 ; 7 ; 30] (= 1 × 60³ + 0 × 60² + 7 × 60 + 30)

FIG. 13.63D.

The Babylonian mathematical documents that have been published to date show the zero only in median positions. For that reason, some historians of science have inferred that Mesopotamian scholars only ever used their zero in intermediate positions and that it would therefore be unwise to treat their zero as identical with ours. They argue that although Mesopotamians wrote expressions such as [1; 0; 3] or [12; 0; 5; 0; 33], they would never have thought of expressions of the form [5; 0] or [17; 3; 0; 0]. More recently, however, O. Neugebauer has shown that Babylonian astronomers used the zero not only in median, but also in initial and terminal positions. For instance, in an astronomical tablet from Babylon (Seleucid period), we find 60 written thus:

[1 ; 0] (= 1 × 60 + 0)

FIG. 13.64A.

Here the double slant chevron is used not as a separator, but to mark the absence of units of the first order. On the back of the same tablet, we also find 180 represented in the same way:

[3 ; 0] (= 3 × 60 + 0)

FIG. 13.64B.

And in another astronomical tablet from Babylon of the same period (British Museum, BM 34581), the number

$$[2; 11; 46; 0]\ (= 2 \times 60^3 + 11 \times 60^2 + 46 \times 60 + 0)$$

is represented as:

[2 ; 11 ; 46 ; 0]

FIG. 13.65.

The final zero in this last example is written in a rather special way, like a "10" with an elongated tail. Has the upper chevron of the zero just been omitted? Is it a scribal decoration? Or just a sign of haste? The latter seems the most likely, since there are other examples in tablets from the same period and the same astronomical source. The following example comes from one such tablet on which the zero is also represented several times in the normal manner:

[3 ; 0 ; 18] (= 3 × 60² + 0 × 60 + 18)

FIG. 13.66.

The double oblique wedge or chevron in initial position also allowed Babylonian astronomers to represent sexagesimal fractions unambiguously. Here are some examples from the tablet previously quoted:

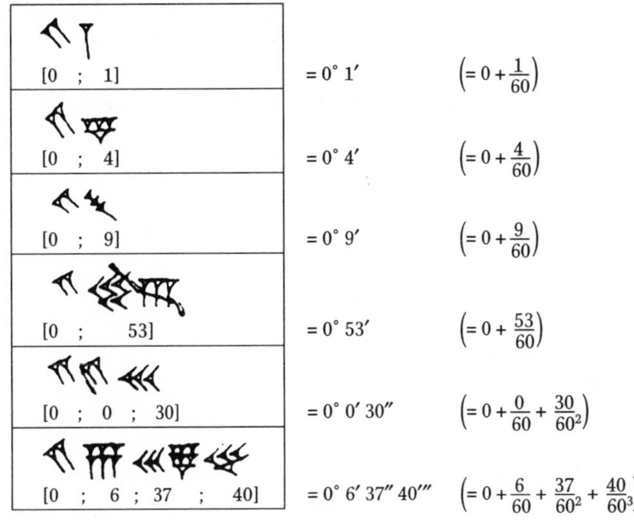

[0 ; 1] = 0° 1′ (= 0 + 1/60)
[0 ; 4] = 0° 4′ (= 0 + 4/60)
[0 ; 9] = 0° 9′ (= 0 + 9/60)
[0 ; 53] = 0° 53′ (= 0 + 53/60)
[0 ; 0 ; 30] = 0° 0′ 30″ (= 0 + 0/60 + 30/60²)
[0 ; 6 ; 37 ; 40] = 0° 6′ 37″ 40‴ (= 0 + 6/60 + 37/60² + 40/60³)


FIG. 13.67.

To summarise: the learned men of Mesopotamia perfected an abstract, strictly positional system of numerals at the latest around the middle of the second millennium BCE, a system far superior to any other in the Ancient World. At a much later date, they also invented zero, the oldest zero in history. Mathematicians seem only to have used it in median position; but the astronomers used it not only in the middle, but also in the final and initial positions of numerical expressions.

## THE DATING OF THE EARLIEST ZERO IN HISTORY

As we have seen, there is no zero in scientific texts of the First Babylonian Dynasty, and the figure is hardly attested in any texts prior to the third century BCE. Does that mean that the Mesopotamians only invented the zero in the Seleucid period? That cannot be so easily said, for there are distinctions to be made between the presumed date of an invention, the period of its propagation, and the dates of its first occurrence in texts that have come down to us. It is perfectly possible for an invention to have been made several generations before its use became widespread, just as it is possible for the "oldest documents known to bear a trace" to be several centuries later than the invention itself – either because the earlier documents have perished, or because they have not yet been discovered.

It is therefore legitimate to believe that the Babylonian zero arose several centuries before the third century BCE. This supposition is all the more plausible because we now know that the literary tablets of the Seleucid period are actually copies of much earlier documents (see H. Hunger, 1976): mathematical tablets of the Seleucid period may therefore not all be contemporary texts.

But these are only suppositions. Only further archaeological discoveries can provide definite proof.

## HOW WAS ZERO CONCEIVED?

The double wedge or double chevron had the meaning of "void", or rather of the "empty place" in the middle of a numerical expression, but it does not appear to have been imagined as "nothing", that is to say as the result of the operation 10 – 10.

In a mathematical tablet from Susa a scribe tried to explain the result of such an operation, thus:

*20 minus 20 comes to . . . you see?*

Similarly, in another mathematical text from Susa, at the end of an operation (referring to the distribution of grain) where you would expect the sum of 0 to occur, the scribe writes simply that "the grain is finished".

The concepts of "void" and "nothing" both certainly existed. But they were not yet seen as synonyms.

## HOW DID BABYLONIAN SCIENTISTS DO THEIR SUMS?

There are no known accounts of the computational methods used by Babylonian mathematicians and astronomers. These methods can nonetheless be reconstructed from the numerous mathematical texts that have been found and deciphered.

Although the rule of position had been adopted, learned Babylonian numerals remained close to their Sumerian roots in the sense that they remained sexagesimal, with 10 serving as an auxiliary base within each order of magnitude. Now, given that we have proved the existence of a Sumerian abacus and shown what shape it must have had, we can assume fairly safely that the tool was handed down to Babylonian scholars as part of their Sumerian heritage, and used by them for the same purposes. That is very probably how things happened, at least at the beginning of this story.

But there are very good reasons for believing that the rules and the shape of the abacus changed very quickly, and that the method became simpler as the centuries passed.

The simplification of the abacus counting method must have required as its counterpart the memorisation of "tables" for the numbers between 1 and 60 – these tables constituting the necessary mental "baggage" to be able to use the abacus for arithmetical operations.

In fact, the Babylonians never bothered to learn such number-tables by heart: they wrote them out once and for all, and handed the tablets down from generation to generation. Consequently, the mathematical tablets that have been discovered include a great number of multiplication tables.

Fig. 13.68 below is a typical example. The transcription can easily be followed by looking at the face of a clock or watch, and imagining the units of the first order as minutes, and the units of the second order as hours. It can then be seen that the tablet on the left gives the numbers from 1 to 20 followed by 30, 40, and 50, and on the right gives the result of multiplying those numbers by 25. It is therefore a 25-times table, completely analogous to one we could construct using our current decimal system:

| | | |
|---|---|---|
| 1 | (times 25 equals) | 25 |
| 2 | (times 25 equals) | 50 |
| 3 | (times 25 equals) | 75 |
| 4 | (times 25 equals) | 100 |
| 5 | (times 25 equals) | 125 |
| 6 | (times 25 equals) | 150 |
| 7 | (times 25 equals) | 175 |
| etc. | | |

Generally speaking, the multiplication tables give the products of a number $n$ (smaller than 60) of the first twenty integers, then of the numbers 30, 40, and 50. This clearly suffices to provide the product of $n$ multiplied by any number between 1 and 60.

With such tables in support, multiplications could be done fairly easily on an abacus.

The rule of position must have led rather quickly to the realisation that wooden tablets of the sort shown in Fig. 12.4 above were no longer necessary, and that the divisions of the Sumerian abacus did not have to be reproduced. All that was now needed was to draw parallel lines to create vertical columns, one for each of the magnitudes of the sexagesimal system. Since clay is easier to work than wood, we can surmise that the columns

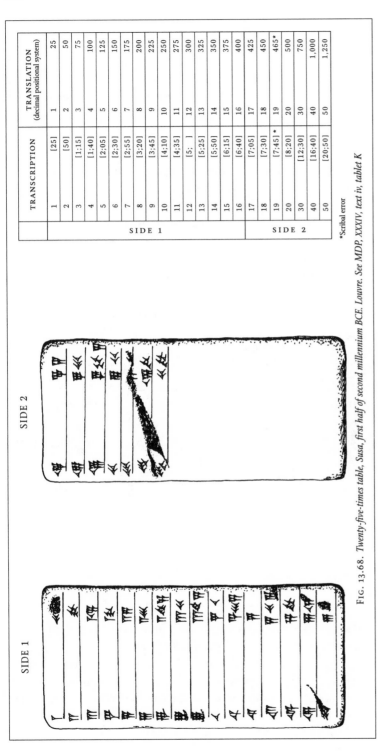

| | TRANSCRIPTION | | TRANSLATION (decimal positional system) | |
|---|---|---|---|---|
| SIDE 1 | 1 | [25] | 1 | 25 |
| | 2 | [50] | 2 | 50 |
| | 3 | [1;15] | 3 | 75 |
| | 4 | [1;40] | 4 | 100 |
| | 5 | [2;05] | 5 | 125 |
| | 6 | [2;30] | 6 | 150 |
| | 7 | [2;55] | 7 | 175 |
| | 8 | [3;20] | 8 | 200 |
| | 9 | [3;45] | 9 | 225 |
| | 10 | [4;10] | 10 | 250 |
| | 11 | [4;35] | 11 | 275 |
| | 12 | [5; ] | 12 | 300 |
| | 13 | [5;25] | 13 | 325 |
| | 14 | [5;50] | 14 | 350 |
| | 15 | [6;15] | 15 | 375 |
| | 16 | [6;40] | 16 | 400 |
| SIDE 2 | 17 | [7;05] | 17 | 425 |
| | 18 | [7;30] | 18 | 450 |
| | 19 | [7;45] * | 19 | 465* |
| | 20 | [8;20] | 20 | 500 |
| | 30 | [12;30] | 30 | 750 |
| | 40 | [16;40] | 40 | 1,000 |
| | 50 | [20;50] | 50 | 1,250 |

*Scribal error

SIDE 2

SIDE 1

FIG. 13.68. *Twenty-five-times table, Susa, first half of second millennium BCE. Louvre. See MDP, XXXIV, text iv, tablet K*

were drawn onto the wet clay of a tablet made afresh for each calculation. Sticks and tokens would not have been needed any longer, since the numbers involved could be drawn straight onto the clay in the relevant columns and wiped or scored out as the calculation proceeded. This reconstruction of the Mesopotamian abacus is of course only a speculation, but it is in our view a highly plausible one.

Here is an example of how it might have worked, using the multiplication table shown in Fig. 13.68.

The task is to multiply 692 by 25, or, in Babylonian terms, to multiply [11; 32] (= 11 × 60 + 32) by 25.

Let us begin by scoring the first three columns onto the wet clay tablet, in which the result will be entered in the three orders of magnitude, starting from the right (numbers from 1 to 59 will be entered in the rightmost column, multiples of 60 from 1 to 59 in the middle column, and multiples of 3,600 from 1 to 59 in the leftmost column).

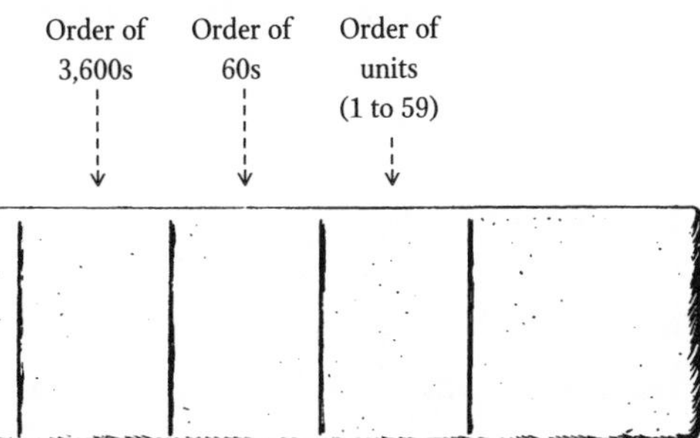


FIG. 13.69A.

To the right of the columns, let us inscribe the multiplicand [11; 32] (= 692) in cuneiform notation:

FIG. 13.69B.

Using the 25 × multiplication table, we look for the product of 2; finding 50, we enter that number in cuneiform notation in the units column of the abacus tablet:

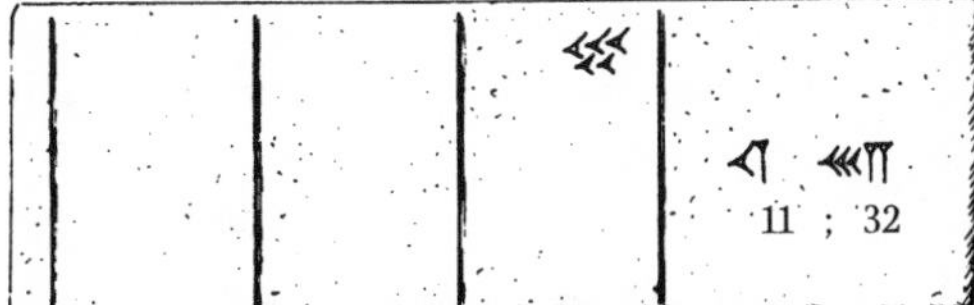


FIG. 13.69C.

We can now rub out the 2 from the multiplicand on the right of the tablet, and proceed to look up 30 in the 25 × multiplication table. The product supplied is [12; 30], so we enter 30 in the rightmost column of the units on our abacus, and 12 in the middle column, reserved for multiples of 60.

FIG. 13.69D.

So we rub out the 30 from the multiplicand on the right of the tablet, and proceed to look up 11 in the 25 × multiplication table. The product supplied is [4; 35], so we enter 35 in the middle column (since we have changed our order of magnitudes) and 4 in the leftmost column, the one reserved for multiples of 3,600.

So we can now rub out the 11 from the multiplicand, and find that there is nothing left on the right of the tablet. The first stage of the operation is complete.

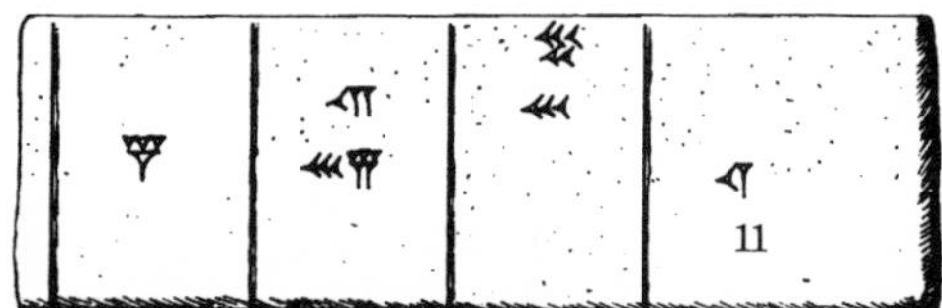


FIG. 13.69E.

The rightmost column now has 8 chevrons in it. Since this is more than the 6 chevrons which make a unit of the next order, we rub out 6 of them and "carry" them into a wedge which we enter in the middle column, leaving 2 chevrons in the units column.

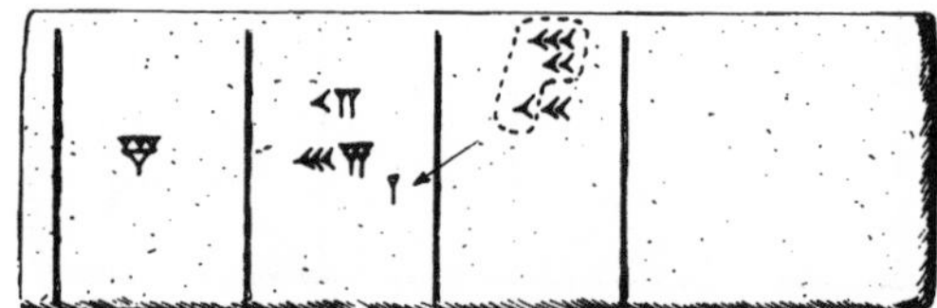

FIG. 13.69F.

So we now have 4 chevrons and 8 wedges altogether in the 60s column. The sum of these being not greater than 60, we simply rub out the numerals in the column and replace them with the numeral signifying 48, the sum of 4 chevrons (4 × 10) and 8 wedges (= 8). And as there is only a 4 in the column of the third order, the result of the multiplication is now fully entered on the abacus:

[11; 32] × 25 = [4; 48; 20]

(= 4 × 3,600 + 48 × 60 + 20 = 17,300)

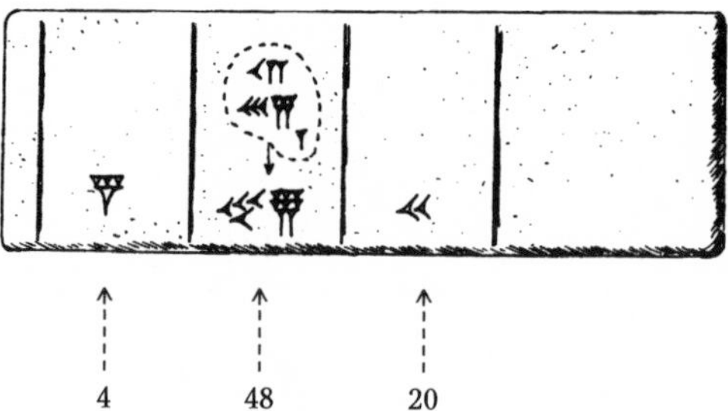

FIG. 13.69G.

The Babylonians also had tables of squares, square roots (Fig. 13.70), cube roots, reciprocals, exponentials, etc., for all numbers from 1 to 59, which enabled far more complex calculations to be performed. For instance, division was done by using the reciprocal table, i.e. to divide one number by another, you multiplied it by its reciprocal.

All this goes to show the great intellectual sophistication of the mathematicians and astronomers of Mesopotamia from the beginning of the second millennium BCE.

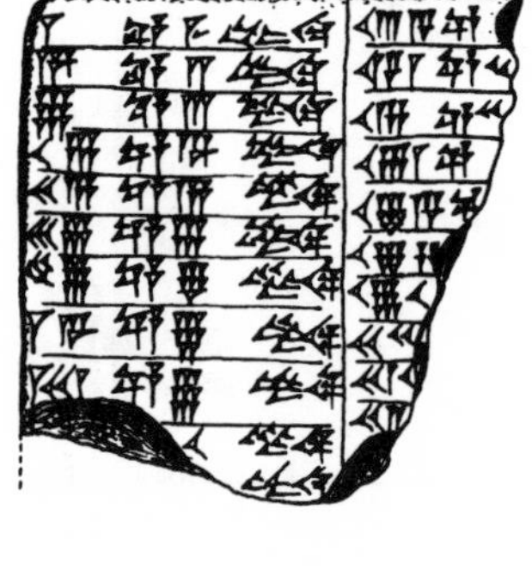

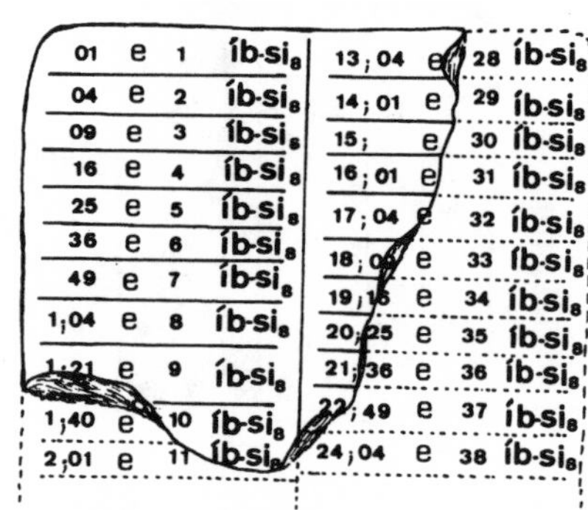

TRANSCRIPTION AND RECONSTRUCTION
01 e 1 íb-si8
04 e 2 íb-si8
09 e 3 íb-si8
16 e 4 íb-si8
25 e 5 íb-si8
36 e 6 íb-si8
49 e 7 íb-si8
1;04 e 8 íb-si8
1;21 e 9 íb-si8
1;40 e 10 íb-si8
2;01 e 11 íb-si8
13 ; 04 e 28 íb-si8
14 ; 01 e 29 íb-si8
15 ; e 30 íb-si8
16 ; 01 e 31 íb-si8
17 ; 04 e 32 íb-si8
18 ; e 33 íb-si8
19 ; e 34 íb-si8
20 ; 25 e 35 íb-si8
21 ; 36 e 36 íb-si8
;49 e 37 íb-si8
24 ; 04 e 38 íb-si8


FIG. 13.70. *Fragment of a table of square roots, c. 1800 BCE, from Nippur (100 miles SE of Baghdad). University of Pennsylvania, Babylonian section, CBS 14233 side 2.*

## THE BABYLONIAN LEGACY

The abstract system of the learned men of Babylon has had a powerful influence over the scientific world from antiquity down to the present.

From at least the second century BCE, Greek astronomers used the Babylonian system for expressing the negative powers of 60. However, instead of using cuneiform numerals, the Greeks used an adapted version of their own alphabetic numerals. For example, they wrote expressions like 0° 28′ 35″ and 0° 17′ 49″ in the following way:

ō KH ΛE $\left(= 0 + \frac{28}{60} + \frac{35}{60^2}\right)$
[0 ; 28 ; 35]

ō IZ MΘ $\left(= 0 + \frac{17}{60} + \frac{49}{60^2}\right)$
[0 ; 17 ; 49]

Fig. 13.71.

Fig. 13.72. *Greek astronomical papyrus, second century CE (after 109). Copied from Neugebauer, plate 2*

TRANSCRIPTION

| | | | | | | | | | | |
|---|---|---|---|---|---|---|---|---|---|---|
| | IΘ | | . . . | . . | | . . . . | IB | | . . . . . | KE] |
| | K | | . . . | . . | | . . . . | IΓ | | . . . . . | Kϛ] |
| | KA | | . . . | . . | | . . . . | IΔ | | . . . . . | KZ] |
| | KB | | . . . | A | | [· · · · · ·] | IE | | . . . . . | [KH] |
| | KΓ | | B MΓ | B | | KΔ Λϛ | Iϛ | | NB ME | KΘ |
| | KΔ | | Δ Λ | Γ | | Kϛ Θ | IZ | | NΔ Iϛ | Λ |
| ]. | KE | | Θ Kϛ | Δ | | KZ MA | IH | | TAYPOY | ΔIΔYM |
| ϛ | Kϛ | | ϛ MZ | E | | KΘ IΓ | IΘ | | ō KΘ Nϛ | ōK[·] |
| ]. | KZ | | H Θ | ϛ | | Λ Λϛ | K | | ō NΘ NB | ōM[·] |
| | KH | | I NB | H | | ΛB IH | KA | | ō KB MH | A |
| | KΘ | | IB Γ | Θ | | ΛΓ N | KB | | B NΔ Λϛ | B |
| | Λ | | IΓ ΛE | I | | ΛE KB | KΓ | | Δ KΘ | Γ |
| | | | IΔ Nϛ | IA | | ΛE NE | KΔ | | E NΘ | Δ |

TRANSLATION

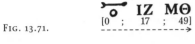

| | | | | | | | | | | |
|---|---|---|---|---|---|---|---|---|---|---|
| | 19 | | . . . . | . . | | . . . . | 12 | | . . . . | 25] |
| | 20 | | . . . . | . . | | . . . . | 13 | | . . . . | 26] |
| | 21 | | . . . . | . . | | . . . . | 14 | | . . . . | 27] |
| | 22 | | . . . . | 1 | | . . . . | 15 | | . . . . | [28] |
| | 23 | | 2 43 | 2 | | 24 36 | 16 | | 52 45 | 29 |
| | 24 | | 4 30 | 3 | | 26 9 | 17 | | 54 16 | 30 |
| ]. | 25 | | 9 26 | 4 | | 27 41 | 18 | | BULLS | GEMINI |
| 6 | 26 | | 6 47 | 5 | | 29 13 | 19 | | 0 29 56 | 0 20[.] |
| ]. | 27 | | 8 9 | 6 | | 30 36 | 20 | | 0 59 52 | 0 40[.] |
| | 28 | | 10 52 | 8 | | 32 18 | 21 | | 0 22 48 | 1 |
| | 29 | | 12 3 | 9 | | 34 50 | 22 | | 2 54 36 | 2 |
| | 30 | | 13 35 | 10 | | 35 22 | 23 | | 4 29 | 3 |
| | | | 14 56 | 11 | | 35 55 | 24 | | 5 59 | 4 |

Fig. 13.73. *Transcription and translation of a Greek astronomical table, from a third-century papyrus. University of Michigan Papyrus Collection, Inv. 924. See J. Garett Winter, pp. 118–20*

GREEK PAPYRI

| first century | after + 109 | 2nd century | 467 CE |
|---|---|---|---|
| Pap. Aberdeen | Pap. Lund | Pap. London | Pap. Michigan |
| No. 128 | Inv. 35A | No. 1278 | Inv. 1454 |

Fig. 13.74a. *The "sexagesimal" zero of Greek astronomers*

ARABO-PERSIAN MANUSCRIPTS

| + 1082 | +1436 | + 1680 | + 1788 |
|---|---|---|---|
| Bodleian Library, | Leyden Univ. Lib, | Princeton, Firestone Library, | |
| Ms Or. 516 | Cod. Or. 187 B | ELS 147 | ELS 1203 |

Fig. 13.74b. *Scribal variants of the "sexagesimal zero" in Arabic and Jewish astronomical texts*

Arab and Jewish astronomers also followed the Greeks' borrowing of the Babylonian system, which they "translated" into their own alphabetic numerals, giving the following forms for the illustrative numbers shown in Fig. 13.71 above:

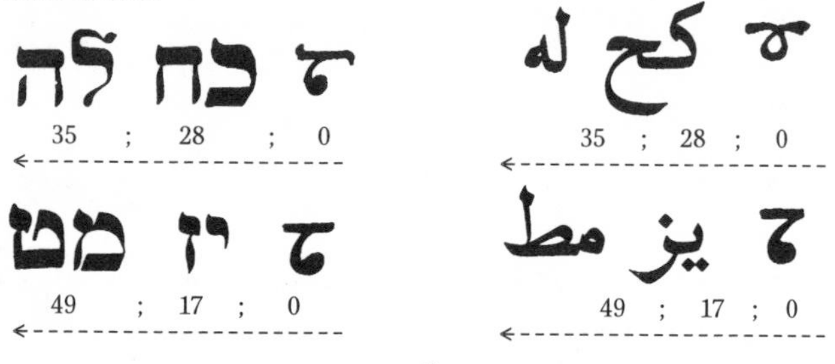


FIG. 13.75A.

35 ; 28 ; 0

49 ; 17 ; 0

FIG. 13.75B.

Thus the learned Babylonian system has come down to us and is perpetuated in the way we express measures of time in hours, minutes and seconds and in the way we count arcs and angles, despite the strictly decimal nature of the rest of our numerals and metric weights and measures. It is largely due to the Arabs that the system was transmitted to modern times.

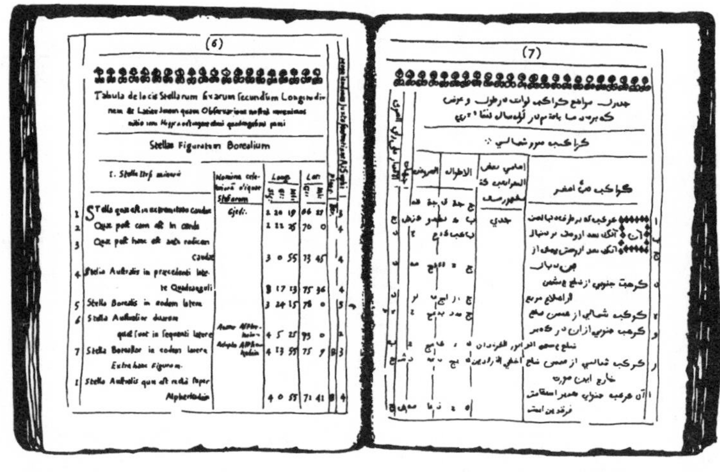


FIG. 13.76. *Bilingual (Latin-Persian) astronomical table, transcribed by Thomas Hyde, 1665. British Library 757 cc 11 (1), pp. 6–7*

FIG. 13.77. *Astronomical table by Levi Ben Gerson, a French-Jewish savant, 1288–1344 CE. British Museum, Add. 26 921, folio 20b. Transcribed by B. R. Goldstein, table 36.1*

## CODES AND CIPHERS IN CUNEIFORM NUMERALS

In some periods and in some fields, the scribes of Susa and Babylon were much given to playing cryptic games with numerals. Some of these games involved numerical transposition, that is to say the use of numerical expressions in lieu of words or ideograms, generally based on some coherent system of "coding", or on complex numerological symbolism.

One of the inscriptions of the name of King Sargon II of Assyria (722–705 BCE) provides an example of numerical transposition. Recording the construction of the great fortress of Khorsabad (Dur Šarukin), Sargon says:

> I gave its wall the dimensions of (3,600 + 3,600 + 3,600 + 3,600 + 600 + 600 + 600 + 60 + 3 × 6 + 2) cubits [i.e. 16,280 cubits] corresponding to the sound of my name. [Cylinder-inscription, line 65]

However, this assertion has not yet yielded all its secrets: we cannot reconstitute the coding system by which the name was transposed into numerals from this single example alone.

Another type of number-name game is shown in a tablet from Uruk of the Seleucid period. At the end of the *Exaltation of Ishtar* (published by F. Thureau-Dangin in 1914) the scribe indicates that the tablet belongs to someone called

21 35 35 26 44 son of 21 11 20 42

FIG. 13.78.

But who is he? The last line gives his name and the name of his father, but both names are written in numerals. The scribe gives us a puzzle without giving us the key [Thureau-Dangin (1914)].

Numerical cryptograms were also widely used for *haruspicy*, the "secret science" of divination or fortune-telling. Seers and fortune-tellers used several different numerical combinations for mystifying the profane and for ensuring that their magical texts remained impenetrable to the uninitiated (Fig. 13.79). Commenting on the *Esagil* tablet, which gives the dimensions of the great temple of Marduk at Babylon and of the tower of Babel, G. Contenau wrote:

> This difficult text looks on first reading like a bland statement of the dimensions of yards and terraces – a mere sequence of numbers, as on a stock list, with all it has to say stated plainly. However, the scribe has peppered his account with the intercalated formula so often found in hieratic texts:
>
> *May the initiated explain this to the initiated*
> *And the uninitiated see it not!*
>
> We should not forget the significant role played by the oral teaching of the pupil by the master which accompanied the lessons of the invariably summary texts themselves. Even texts which appear to be utterly ordinary hid esoteric meanings which we cannot imagine.

The scribes of Susa and Babylon also used cryptograms for word-games, or rather, writing-games, which are worth some attention. For instance, the

FIG. 13.79. *Astrological table with cryptograms in a code that remains to be deciphered. (Line 5, for instance, reads: 3; 5; 2; 1; 12; 4; 31). British Museum, 92685 side 1. Copy made by H. Hunger*

combination "3; 20" is often found used as an ideogram for the word meaning "king", which was pronounced *šàr* or *šarru* in Akkadian. An inscription on a brick from the reign of Šušinak-Šár-Ilâni, king of Susa (twelfth–eleventh century BCE) bears this formulation of the king's name:

ŠUŠINAK – [3 . 20] – ILÂNI [3 . 20] SUSI

("Shushinak-Shar-Ilâni, king of Susa")

Now, why is the numerical combination "3; 20" a logogram for "king"? In Akkadian, the word for "king" was *šár*, pronounced more or less exactly the same way as *šàr*, the name of the "higher" sexagesimal unit of counting in the Sumero-Babylonian system, that is to say 3,600. Elamite scribes thus seem to have made a pun by replacing the word "king" by a numerical combination [3; 20] which represented 3,600 according to a specific rule of interpretation.

But what was the rule? It clearly has nothing to do with the positional sexagesimal system of the learned men of Mesopotamia, since in that way of reckoning [3; 20] = 3 × 60 + 20 = 200, which is not the right answer. On the other hand, we could decompose *šàr* into a kind of "literal" numerical expression that would be represented by "sixty sixties", or, in cuneiform:

60 SHU-SHI

Punning scribes could have written this out, as a game, in this alternative way:

20 20 20 SHU-SHI

or finally as

3 × 20 SHU-SHI

So we can see that Susan scribes regarded the sequence [3; 20] as expressing the product of 3 × 20 (implicitly, × 60), that is to say 3,600, making a pun on *šár*, "the king".

Assyro-Babylonian scribes also used the combination [3; 20] to refer to the king, but they sometimes added a chevron, making [3; 30]. This latter variant cannot be accounted for by 3 × 60 + 30 = 210, nor by (3 × 30) × 60 = 5,400. However, if the addition of a chevron (= 10) to the expression [3; 20] is understood as the mark of a *multiplication* of [3; 20] by 10, then the symbol can be understood as:

[3; 30] = [3; 20] × 10 = 3,600 × 10 = 36,000

This gives the number called *šàr-u* in Sumerian, written in that older system as a 3,600 with an additional chevron in the middle:

3,600

3,600 × 10

The word *šàr-u* (which means "ten *šàr*" or "the great *šàr*") is thus what is meant by the numerical expression

3 ; 30 = 3 ; 20 × 10
ŠÀR-U

– because this Sumerian number-name has exactly the same sound as the Akkadian word *sharru*, meaning king. So when a scribe referred to the king by writing [3; 30], we can deduce that he meant to say "the great king".

There are many other Babylonian cryptograms which remain unsolved, however. For instance, we have no idea why the concepts of "right" and "left" came to be written by the cuneiform numerals [15] and [2; 30] respectively, nor why [1; 20] was used as an ideogram for "throne", nor, finally, why the vertical wedge, the sign for the unit in the numerical system, had the role of the determiner ("the man who . . . ") in the names of the main male functions.

SIDE 1

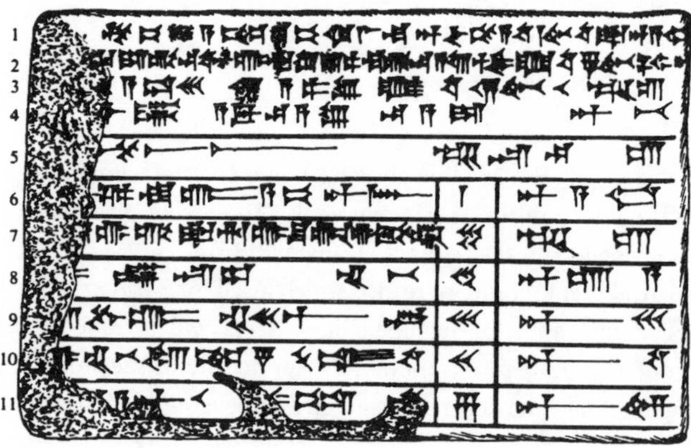

SIDE 2

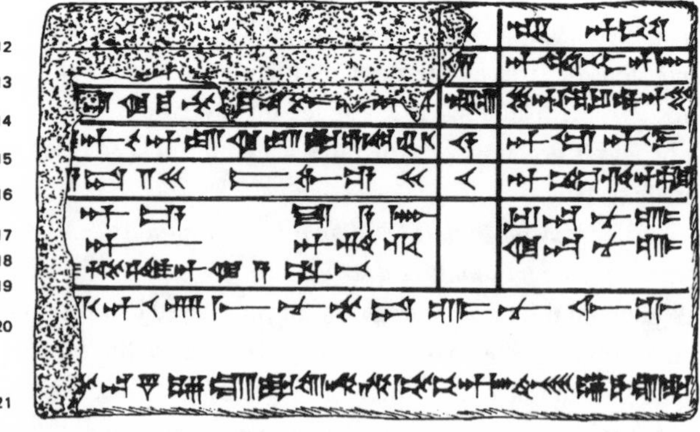

Fig. 13.80a. *Cuneiform tablet listing names of gods and their corresponding numbers. Seventh century BCE, from the "library" of Assurbanipal. British Museum K 170. Trans. J. Bottero*

## CODED CRYPTOGRAMS AND MYSTICAL NUMEROLOGY

Coded cryptograms were also used in theological speculation, for Mesopotamian scribes accorded great weight to the numerical transposition of the names of the gods. Indeed, the religions of the Assyro-Babylonians assumed that the celestial world was a "numerologically harmonious" one, in which the numerical value of a name was an essential attribute of the individual to which it belonged. For this reason, from the early part of the second millennium BCE and consistently throughout the first millennium, some of the Babylonian gods were represented by cuneiform numerals. Fig. 13.80 reproduces a tablet from the seventh century BCE which gives the names of the gods and for each one a number which could be used as that god's ideogram. These are the main points made on the tablet:

- **Anu**, god of heaven, is attributed the number 60, the higher unit of the sexagesimal Sumerian and Babylonian system, and considered to be the number of perfection, because, the scribe says, "Anu is the first god and the father of all the other gods";
- **Enlíl**, god of the earth, is represented by 50;
- **Ea**, god of water, is represented by 40 (elsewhere, she is sometimes ascribed the number 60);
- **Sîn**, the lunar god, corresponds to 30, because, the scribe says (line 9 , column 1, side 1), "he is the lord of the decision of the month", or, in other words, Sîn is the god who regulates the 30 days of the month
- **Shamash**, the sun-god, is worth 20;
- **Adad** here has the number 6 (more frequently, he has the number 10);
- **Ishtar**, daughter of Anu lord of the heavens and held to be the "queen of the heavens", has the number 15;
- **Ninurta**, son of Enlíl, has the same number as his father, 50;
- **Nergal** has the number 14;
- **Gibil** and **Nusku** are both represented by 10 because, according to the scribe (line 16, column 1, side B), "they are the companions of god 20 ( = Shamash): $2 \times 10 = 20$".

The numerological values of the gods of Babylon had all sorts of consequences. For example, the Babylonian *Creation Epic* concludes with a list of the "names" of **Marduk**, a series of epithets defining his virtues and powers and intended to demonstrate that he is truly the supreme god and the most godly of all. First comes a list of ten names, because Marduk's "number" is 10, then a second group of forty names, because Marduk is the son of Eá, whose number is 40; which adds up to fifty names, because 50 is the number of Enlíl, and the main point of the epic is to show how Marduk replaced Enlíl at the head of the universe of gods and men.

TRANSCRIPTION AND TRANSLATION OF THE TWO RIGHTMOST COLUMNS

| | Line | | | Number | Translation |
|---|---|---|---|---|---|
| SIDE 1 | LINE 6 | | | 1 or 60 | $^{d}$A-num |
| | 7 | | | 50 | $^{d}$En-lîl |
| | 8 | | | 40 | $^{d}$E-a |
| | 9 | | | 30 | $^{d}$Sîn (name written as 30) |
| | 10 | | | 20 | $^{d}$Shamash |
| | 11 | | | 6 | $^{d}$Adad |
| SIDE 2 | LINE 12 | | | 10 | $^{d}$Bêl $^{d}$Marduk (the Lord Marduk) |
| | 13 | | | 15 | $^{d}$Ishtar be-lit ilî (Ishtar queen of the heavens) |
| | 14 | | | En-lîl | 50 $^{d}$Nin-urta, mâr 50 (50 Nin-urta, son of the god Enlîl) (written as 50) |
| | 15 | | | 14 | $^{d}$U + gur, $^{d}$Nergal |
| | 16 | | | 10 | $^{d}$Gibil, $^{d}$Nusku |

Fig. 13.80b.

CHAPTER 14

# THE NUMBERS OF ANCIENT EGYPT

Egypt in the time of the Pharaohs had writing and written numerals. They first arose around 3000 BCE, that is to say, at about the same time that words and numbers were first written down in Mesopotamia.

We now know that there were regular contacts between Egypt and Mesopotamia before the end of the third millennium BCE. However, that does not mean that the Egyptians derived their writing or their counting from Sumerian models. Egyptian hieroglyphs, Jacques Vercoutter explains, use signs derived exclusively from the flora and fauna of the Nile basin; moreover, the tools used for making written signs existed in Egypt from the fourth millennium BCE.

The pictograms of Egyptian hieroglyphic writing are very different from Sumerian ideograms, even when we compare signs intended to represent the same idea or object; the shapes of such signs also seem quite unrelated. The media of the two systems likewise have little in common. As we saw, the Sumerians only ever wrote words and numbers by scoring clay tablets with a stylus, or else by pressing a shaped instrument onto wet clay; whereas the Egyptians carved their numerals and hieroglyphs in stone, with hammer and chisel, or else used the bruised tip of a reed to paint them on shards of stone or earthenware, or onto sheets made by flattening the dried-out, fibrous, and fragile stems of the papyrus reed.

Egyptian numerals are also quite different from Sumerian ones from a mathematical point of view. As we have seen, Sumer used a sexagesimal base; whereas the system of Ancient Egypt was strictly decimal.

So if there was something borrowed by Egypt from Sumer, it could only have been the idea of writing down numbers in the first place, and not any part of the way it was done.

Peoples very distant from each other in time and in place but facing similar situations and needs have discovered quite independently some of the same paths to follow, and have arrived at similar, if not identical, results. The Indus civilisation, the Chinese, and the pre-Columbian populations of Central America (Zapotecs, Maya, etc.), were all faced with situations probably very similar to those of the Sumerians, and made much the same mathematical discoveries for themselves. So it seems most sensible to suppose that around the dawn of the third millennium BCE the social, psychological, and economic conditions of Ancient Egypt were such that the invention of writing and of written numerals arose there of its own accord.

In fact, Egyptian society was already advanced, urbanised, and expanding rapidly long before 3000 BCE. Administrative and commercial logic led to the slow realisation that human memory could no longer suffice to fill all the needs of the state without some material support; oral culture must have come up against its natural limits. We must then suppose that the Egyptians felt an increasing need to record and thus to retain thoughts and words, and to fix in a durable form the accounts and inventories of their commercial activities. And, since necessity is the mother of invention, the Egyptians overcame the limits of oral culture by devising a system for writing down words and numbers.

## WHAT ARE EGYPTIAN HIEROGLYPHS?

Although they no longer knew how to read them, the Ancient Greeks recognised the signs carved on the many monuments of the Nile Valley (temples, obelisks, tombs, and funeral *stelae*) as "sacred signs" and thus called them *grammata hiera*, or, more precisely, *grammata hierogluphika* ("carved sacred signs"), whence our word, *hieroglyph*. It is from these carved signs, which the Ancient Egyptians considered to be "the expression of the words of the gods", that we have derived our knowledge of the spoken language of Ancient Egypt. The basic writing system for the representation of this language was designed for and was used for the most part only on stone monuments, and it is this writing system (rather than the language it represents) that we commonly call *hieroglyphs*.

## HOW TO READ HIEROGLYPHS

Hieroglyphs are very detailed pictograms representing humans, in various positions, all kinds of animals, buildings, monuments, sacred and profane objects, stars, plants, and so on.

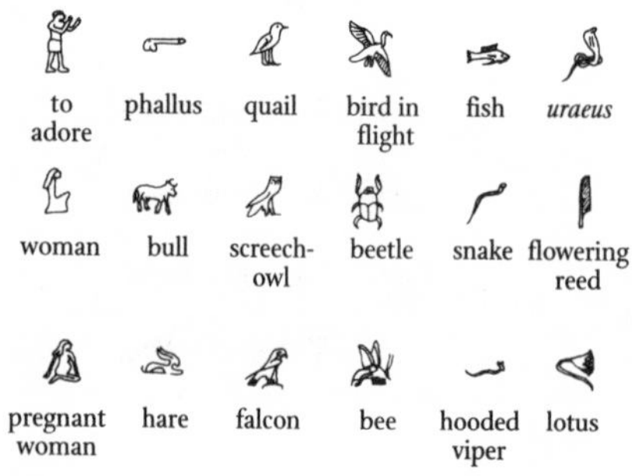

FIG. 14.1. *Some hieroglyphs*

Hieroglyphs may be written in lines from left to right or right to left, or in columns from top to bottom or from bottom to top. The direction of reading is indicated by the orientation of the animate figures (humans or animals) – they are "turned" so as to face the start of the line. So they look left in a text written/read from left to right:

Fig. 14.2.

and they look right in a text written/read from right to left:

Fig. 14.3.

Hieroglyphic signs could be used and understood, first of all, as "integral picture-signs" or *pictograms*: pictures that "meant" what they showed. In the second place, they could also be used as *ideograms*: that is to say, signs meaning something more than, or something connected to, what they showed. For example, an image-sign of a human leg could mean, first, "leg", as a pictogram, but also, as an ideogram, the related ideas or actions of "walking", "running", "running away", etc. Similarly the image of the sun's orb could mean "day", "heat", "light", or else refer to the sun-god. The ideogrammatic interpretation of a sign did not supplant its pictogrammatic meaning, but coexisted alongside it. The interpretation of a hieroglyphic sign is therefore open to infinite subjective variation.

Pictograms and ideograms cannot easily cope with every nuance of language. How can such a system represent actions such as *wishing, desiring, seeking, deserving*, and so on? Or abstract notions like *thought, luck, fear*, or *love*? Moreover, pictograms cannot represent the articulations of spoken language, and are independent of any particular language spoken.

To overcome these limitations, the Ancient Egyptians used their signs in a third way, quite at variance with their pictogrammatic and ideogrammatic values. A sign could also represent the *sound* of the *name* of the thing represented pictographically, and then be used in combination with other *sounds* represented by the ideograms of other words, to make a kind of visual pun or *rebus*. For instance, let us suppose that in Britain today we had only a hieroglyphic writing system and in that system had no pictogram for the things we call "carpets"; but did have a conventional pictogram for "car": ; and, given that we are a nation of dog-lovers, represented the general idea of "pet" by the ideogram: . Were we to proceed with the system as the Ancient Egyptians did, we would not invent a new pictogram for "carpet" but would create a compound picture-pun

Such a system has a built-in propensity towards ambiguity. This is not just because ideograms by their very nature have variable interpretations, but also and most especially because a rebus may make a sense in more than one reading of the phonetic value of the ideograms. To take an equally fictitious example from hieroglyphs to be realised by speakers of English, where the pictogram has the full pictorial meaning of "fir" and the broader meaning "wood" when taken as an ideogram, and the ideogram has the meanings of "house", "inn" or "home", the expression (read left to right)

could be realised phonetically as INN-FIR, with the punning meaning of the verb *infer*; or else, read right to left as HOME-WOOD, with the punning meaning of *homeward*. In order to reduce the number of total misapprehensions of that sort, Egyptian hieroglyphs therefore needed an additional sign in each compound expression, a kind of ideogrammatic hint or determiner that showed which way the sound-signs were to be taken. To continue our example, the determiner when added to would ensure that it was taken in the directional sense. So

would indeed be read as INFER, whereas

would be read as HOMEWARD.

That is roughly how Egyptian hieroglyphs evolved from pictorial evocations of things to phonetic representations of words. For example: the Ancient Egyptian for "quail chick" was pronounced Wa; the sign depicting a quail chick signified a quail chick, but also represented the sound Wa. Similarly, "seat" was pronounced Pe, and the drawing of a seat came to represent the sound Pe; "mouth" was eR, and a drawing of a mouth meant the syllable eR; a picture of a hare (WeN) stood for the sound WeN, a picture of a beetle (KhePeR) made the sound KhePeR, and so on.

Fig. 14.4.

Like Hebrew and other Semitic scripts, Egyptian hieroglyphs are consonantal, that is to say they represented only the consonants, leaving the vowels to be "understood" by convention and habit. (Where vowels are put in in modern transcriptions of the language, they are hypothetical and conventional: there is in fact no way of knowing how Ancient Egyptian was actually vocalised.) Since words in Ancient Egyptian contained either one, two, or three consonants, hieroglyphs used as sound-signs also belonged to one of three classes: uniliteral (representing a single consonantal sound), biliteral (representing two sounds), or triliteral (representing a group of three consonants). With their signs used simultaneously as pictograms, as ideograms, and with syllabic value, the Ancient Egyptians were thus able to represent all the words of their language.

An early example is Narmer's Palette (c. 3000–2850 BCE), which commemorates the victory of King Narmer over his enemies in Lower Egypt.

FIG. 14.5. *King Narmer's Palette, from Hieraconpolis, c. 3000–2850 BCE. Cairo Museum*

The king can be seen in the centre of the panel, wielding his club over a captive. The king's name, written in the cartouche above his regal headgear, is composed of the hieroglyphs "fish" and "scissor". The word meaning "fish" was pronounced N'R, and the word meaning "scissor" was pronounced M'R: the two together thus make N'RMR, or Narmer.

FIG. 14.6.

In similar fashion, the word for "woman", pronounced SeT, was represented by the image of a bolt (the word for "bolt" being a uniliteral with value S) and an image of a piece of bread ("piece of bread" being also a uniliteral, with value T). However, to ensure that S + T was read in the right way, a pictogram of a woman (unrealised in speech) was added as a determiner:

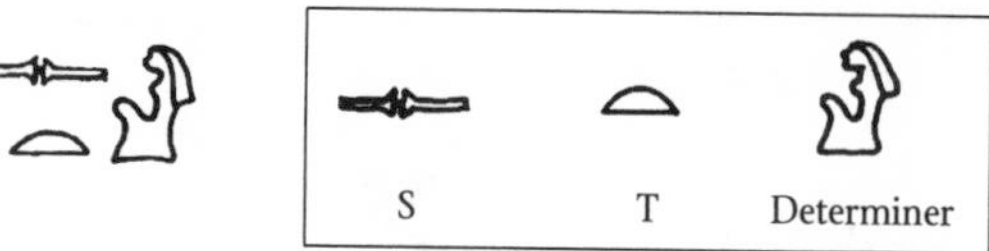


FIG. 14.7.

Likewise the vulture, NeReT in Ancient Egyptian, was represented by N ("stream of water"), R ("mouth") and T ("piece of bread"), plus the determiner, "bird", to ensure that the sound-signs were read as a word belonging to the class of birds.

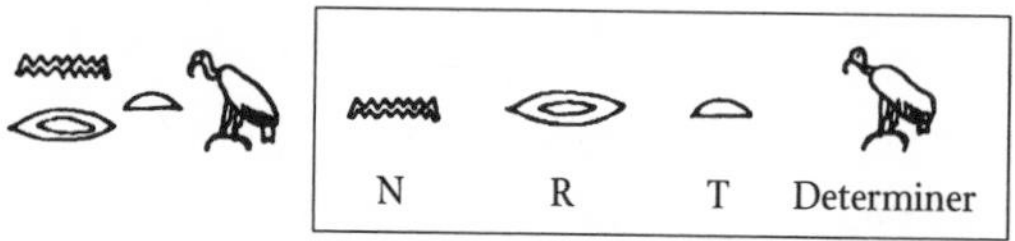


FIG. 14.8.

Hieroglyphic writing did not use only these kinds of determiners, however. In many cases, biliteral and triliteral signs are disambiguated by a phonetic "complement" which gives a supplementary clue as to how to read the sign. For instance, the hieroglyph of "hare", a word pronounced WeN, would be "confirmed" as meaning the biliteral sound WeN by the addition of the sign for "stream of water", a uniliteral sound pronounced N, as follows:

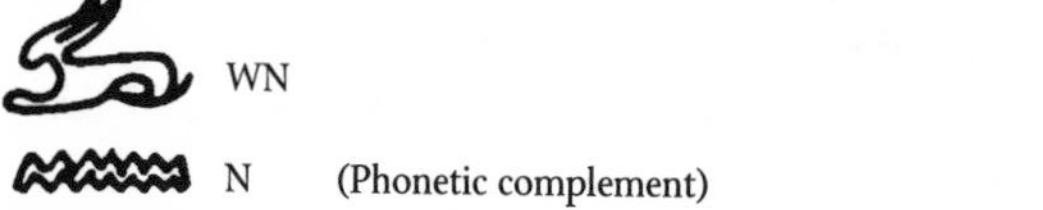


FIG. 14.9.

It is as if in our imaginary English hieroglyphs we added to the sign to ensure that

was recognised as a syllable containing the uniliteral consonant T (as in "cup of *tea*") and thus pronounced PET, and not seen as a pictogram meaning (for example) "Labrador".

In Ancient Egyptian the name of the god Amon was represented by the signs whose pronunciation was í ("reed in flower") and mn ("crenellation"), supplemented by a determiner (the ideogram signifying the class of gods) *plus* a phonetic complement, the sign for "stream of water", pronounced N, whose sole function was to confirm that the syllable was to be read in a way that made it include the sound n.

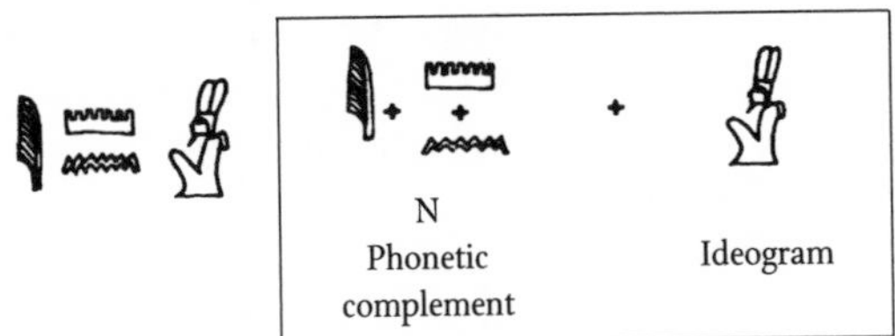


FIG. 14.10.

## HIEROGLYPHIC NUMERALS

Written Egyptian numerals from their first appearance were able to represent numbers up to and beyond one million, for the system contained specific hieroglyphs for the unit and for each of the following powers of 10: 10, 100 (= $10^2$), 1,000 (= $10^3$), 10,000 (= $10^4$), 100,000 (= $10^5$), and 1,000,000 (= $10^6$).

The unit is represented by a small vertical line. Tens are signified by a sign shaped like a handle or a horseshoe or an upturned letter "U". The hundreds are symbolised by a more or less closed spiral, like a rolled-up piece of string. Thousands are represented by a lotus flower on its stem, and ten thousands by a slightly bent raised finger. The hundred thousand has the form of a frog, or a tadpole with a visible tail, and the million is depicted by a kneeling man raising his arms to the heavens.

| | READING RIGHT TO LEFT | READING LEFT TO RIGHT |
|---|---|---|
| 1 | | |
| 10 | | |
| 100 | | |
| 1,000 | | |
| 10,000 | | |
| 100,000 | | |
| 1,000,000 | | |

FIG. 14.11. *The basic figures of hieroglyphic numerals with their main variants in stone inscriptions. Note that the signs change orientation depending on which way the line is to be read: the tadpole (100,000) and the kneeling man (1,000,000) must always face the start of the line.*

One of the oldest examples that we have of Egyptian writing and numerals is the inscription on the handle of the club of King Narmer, who united Upper and Lower Egypt around 3000–2900 BCE.

FIG. 14.12. *Tracing of the knob of King Narmer's club, early third millennium BCE*

Apart from King Narmer's name, written phonetically, the inscription on the club also provides a tally of the booty taken during the king's victorious expedition, consisting of so many head of cattle and so many prisoners brought back. The tally is represented as follows:

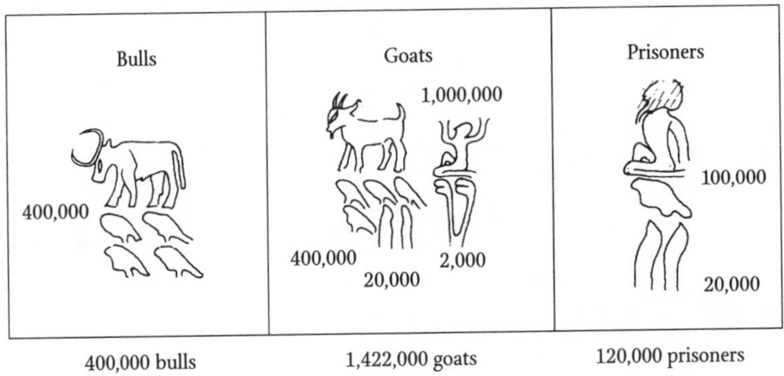


FIG. 14.13.

Are these real numbers, or are they purely imaginary figures whose sole aim is to glorify King Narmer? Scholars disagree. But we should note that the livestock tallies found on the *mastabas* of the Old Kingdom also often give very high numbers for individual owners, and that here we are dealing with the looting of an entire country.

Another example of high numbers can be found on a statue from Hieraconpolis, dating from c. 2800 BCE, where the number of enemies slain by a king called KhaSeKhem are shown as 47,209 by the following signs:

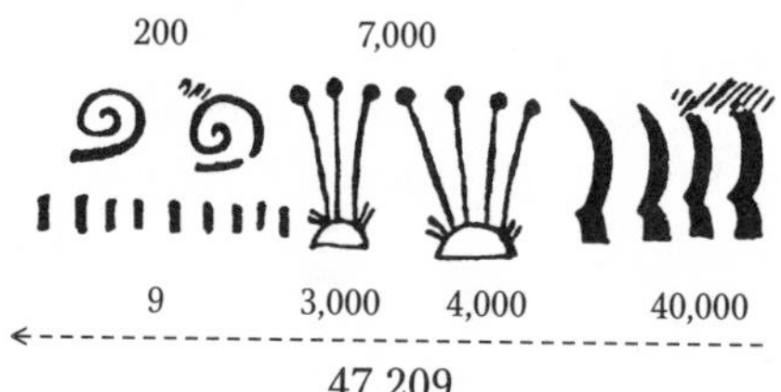


FIG. 14.14.

To represent a given number, then, the Egyptians simply repeated the numeral for a given order of decimal magnitude as many times as necessary, starting with the highest and proceeding along the line to the lowest order of magnitude (thousands before hundreds before tens, etc.).

Early examples show rather irregular outlines and groupings of the signs. In Fig. 14.13 above, for example, the number of goats (1,422,000) is written in a way that is contrary to the rules that were later laid down by Egyptian stone-cutters, since the figure for the million is placed to the right of the beast and on the same line, whilst the remainder of the number-signs are inscribed on the line below. The normal rule was for the signs to go from right to left in descending order of magnitude on the line below the sign for the object being counted, thus:

FIG. 14.15.

Similarly, Figure 14.14 shows rather primitive features in the representation of the finger (= 10,000), the grouping of the thousands (lotus flowers) into two distinct sets, and the relatively poor alignment of the unit signs. However, from the twenty-seventh century BCE, the execution of hieroglyphic numerals became more detailed and more regular. Also, to avoid making lines of numerals over-long, the custom emerged of grouping signs for the same order of magnitude onto two or three lines, which made them easier to add up by eye:

| I | II | III | II<br>II | III<br>II | III<br>III | IIII<br>III | IIII<br>IIII | III<br>III<br>III |
|---|---|---|---|---|---|---|---|---|
| 1 | 2 | 3 | 4 | 5 | 6 | 7 | 8 | 9 |

| ∩ | ∩∩ | ∩∩∩ | ∩∩<br>∩∩ | ∩∩∩<br>∩∩ | ∩∩∩<br>∩∩∩ | ∩∩∩∩<br>∩∩∩ | ∩∩∩∩<br>∩∩∩∩ | ∩∩∩<br>∩∩∩<br>∩∩∩ |
|---|---|---|---|---|---|---|---|---|
| 10 | 20 | 30 | 40 | 50 | 60 | 70 | 80 | 90 |

FIG. 14.16.

The evolution of Egyptian numerals can be traced as follows:

1: Old Kingdom period: funerary inscriptions of Sakhu-Rê, a Pharaoh of the Fifth Dynasty, who lived at the time of the building of the pyramids, around the twenty-fourth century BCE:

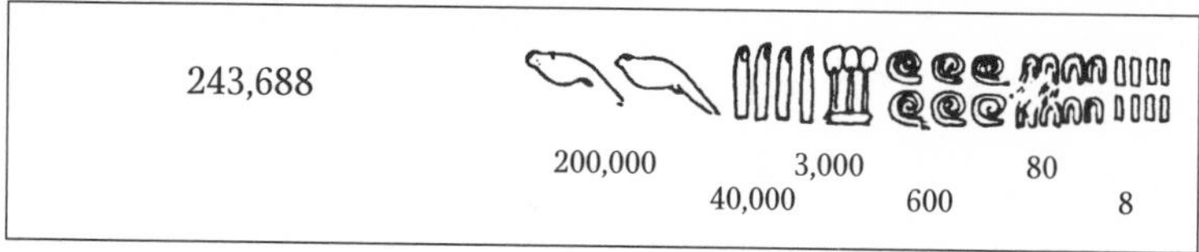


FIG. 14.17.

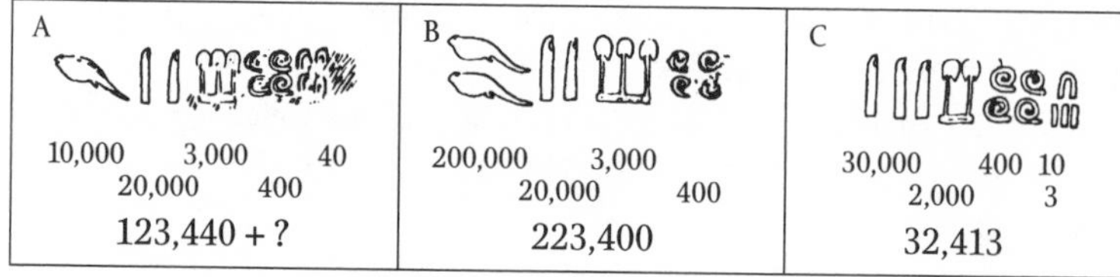


FIG. 14.18.

Although some parts of them have deteriorated somewhat from age, the hieroglyphic numerals are entirely recognisable. The tadpoles are all facing left, and thus these numerical expressions are read from left to right (see Fig. 14.11 above). In Fig. 14.17, the number 200,000 has been written along the line, unlike example B in figure 14.18, where the two tadpoles are put one above the other. The thousands are represented by lotus flowers connected at the base, a custom which disappeared by the end of the Old Kingdom period.

2: End of the First Intermediate period (end of third millennium BCE), from a tomb at Meir:

| A | B | C | D |
|---|---|---|---|
| 77 | 700 | 7,000 | 760,000 |

FIG. 14.19.

3: From the Annals of Thutmosis (1490–1436 BCE), a list of the plunder of the twenty-ninth year of the Pharaoh's reign (see Fig. 14.21):

The numerals can be transcribed as:

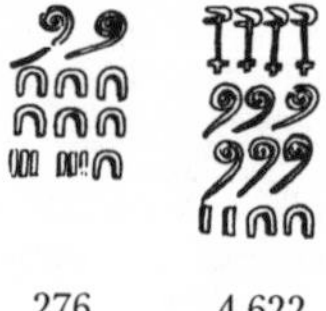


FIG. 14.20.

Fig. 14.21. *Stone bas-relief from Karnak. Louvre*

4: Numerical expression from the *stela* of Ptolemy V at Pithom, 282–246 BCE:

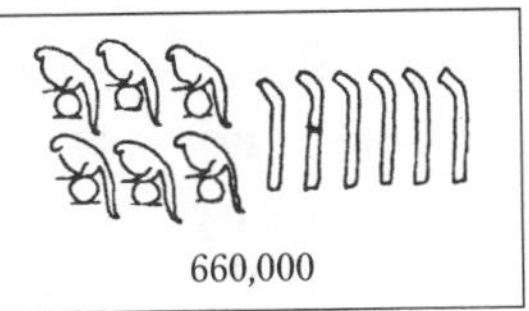



Fig. 14.22.

## THE ORIGINS OF EGYPTIAN NUMERALS

The numerical notation of Ancient Egypt was in essence a written-down trace of a concrete enumeration method that was probably used in earlier periods. The method was to represent any given number by setting out in a line or piling up into a heap the corresponding number of standard objects or tokens (pebbles, shells, pellets, sticks, discs, rings, etc.), each of which was associated with a unit of a given order of magnitude.

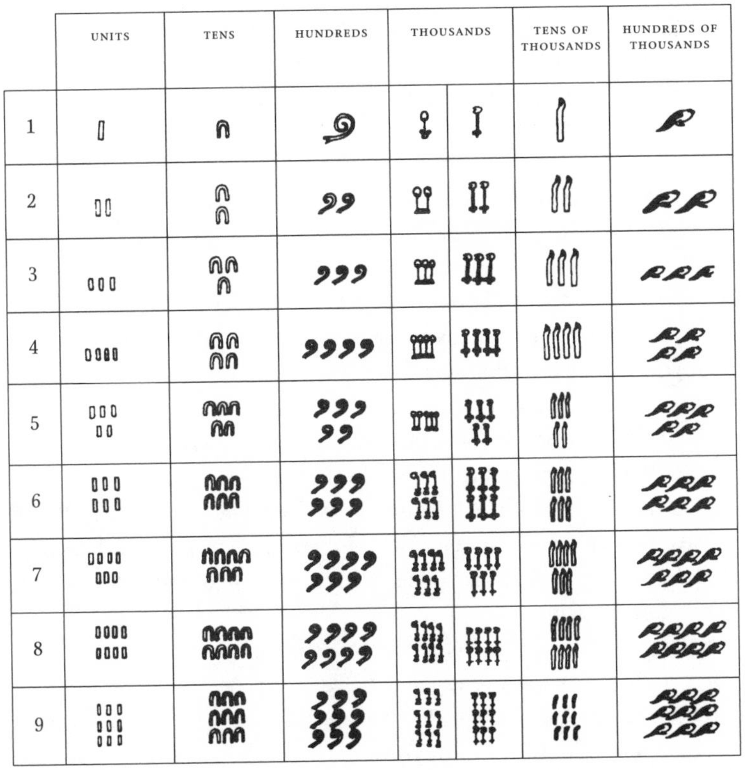

| | UNITS | TENS | HUNDREDS | THOUSANDS | | TENS OF THOUSANDS | HUNDREDS OF THOUSANDS |
|---|---|---|---|---|---|---|---|
| 1 | 𓏺 | 𓎆 | 𓍢 | 𓆼 | 𓆼 | 𓂭 | 𓆐 |
| 2 | 𓏻 | 𓎇 | 𓍣 | 𓆽 | 𓆽 | 𓂮 | 𓆐𓆐 |
| 3 | 𓏼 | 𓎈 | 𓍤 | 𓆾 | 𓆾 | 𓂯 | 𓆐𓆐𓆐 |
| 4 | 𓏽 | 𓎉 | 𓍥 | 𓆿 | 𓆿 | 𓂰 | 𓆐𓆐𓆐𓆐 |
| 5 | 𓏾 | 𓎊 | 𓍦 | 𓇀 | 𓇀 | 𓂱 | 𓆐𓆐𓆐𓆐𓆐 |
| 6 | 𓏿 | 𓎋 | 𓍧 | 𓇁 | 𓇁 | 𓂲 | 𓆐𓆐𓆐𓆐𓆐𓆐 |
| 7 | 𓐀 | 𓎌 | 𓍨 | 𓇂 | 𓇂 | 𓂳 | 𓆐𓆐𓆐𓆐𓆐𓆐𓆐 |
| 8 | 𓐁 | 𓎍 | 𓍩 | 𓇃 | 𓇃 | 𓂴 | 𓆐𓆐𓆐𓆐𓆐𓆐𓆐𓆐 |
| 9 | 𓐂 | 𓎎 | 𓍪 | 𓇄 | 𓇄 | 𓂵 | 𓆐𓆐𓆐𓆐𓆐𓆐𓆐𓆐𓆐 |

Fig. 14.23. *Hieroglyphic representations of the consecutive units in each decimal order*

Unlike Sumerian numerals, however, the hieroglyphs give no clue as to the nature of the tokens used in concrete reckoning prior to the invention of writing. It seems pretty unlikely that lotus flowers (1,000) or tadpoles (100,000), were ever practical counting tokens at any period of time. The spiral, the finger, and the kneeling man with upraised arms pose just as awkward and still unanswered questions.

It seems most likely to me that the origins of Egyptian numerals are much more complex than the origins of the written numbers of Sumer and Elam, and that their inventors used not one but several different principles at the same time. What follows are no more than plausible hypotheses about the origins of hieroglyphic numerals, unconfirmed by any hard evidence.

The origin of the numeral 1 could have been "natural" – the vertical line is just about the most elementary symbol that humans have ever invented for representing a single object. It was used by prehistoric peoples from over 30,000 years ago when they scored notches on bone, and as we have seen a whole multitude of different civilisations have given the line or notch the same unitary value over the ages.

In addition, the line (for 1) and the horseshoe (for 10) could well be the last traces in hieroglyphic numerals of the archaic system of concrete numeration. The line could have stood for the little sticks used with a value of 1, and the horseshoe might in fact have been at the start a drawing of the piece of string with which bundles of ten sticks were tied to make a unit of the next order.

As for the spiral and the lotus, they most probably arose through phonetic borrowing. We could imagine that the original Egyptian words for "hundred" and "thousand" were complete or partial homophones of the words for "lotus" and "spiral"; and that to represent the numbers, the Egyptians used the pictograms which represented words which had exactly or approximately the same sound, irrespective of their semantic meaning, as they did for many other words in their language and writing.

Parallels for such procedures exist in many other civilisations. In classical Chinese writing, for instance, the numeral 1,000 was written with the same character as the word "man", because "man" and "thousand" are reckoned to have had the same pronunciation in the archaic form of the language.

On the other hand, the Egyptian hieroglyph for 10,000, the slightly bent raised finger, seems to be a reminiscence of the old system of finger-counting which the Egyptians probably used. The system relies on various finger positions to make tallies up to 9,999.

The hieroglyphic sign for 100,000 may derive from a more strictly symbolic kind of thinking: the myriads of tadpoles in the waters of the Nile, the vast multiplication of frogspawn in the spring . . .

The hieroglyphic numeral for 1,000,000 might more plausibly be ascribed a psychological origin. The Egyptologists who first interpreted this sign thought that it expressed the awe of a man confronted with such a large number. In fact, later research showed that the sign (which also means "a million years" and hence "eternity") represented in the eyes of the Ancient Egyptians a *genie holding up the vault of heaven*. The pictogram's distant origin lies perhaps in some priest or astronomer looking up to the night sky and taking stock of the vast multitude of its stars.

## SPOKEN NUMBERS IN ANCIENT EGYPTIAN

The spoken numbers of Egyptian have been reconstructed from its modern descendant, Coptic, together with the phonetic transcriptions of numerical expressions found in hieroglyphic texts on the pyramids. Here are their syllabic transcriptions with their approximate phonetic realisations:

| | | | | | |
|---|---|---|---|---|---|
| 1 | *w'* | [wa'] | 10 | *mḏ* | [medj] |
| 2 | *snw* | [senu] | 20 | *dwty* | [dwetye] |
| 3 | *khmt* | [khemet] | 30 | *m'b'* | [m'aba'] |
| 4 | *fdw* | [fedu] | 40 | *khm* | [khem] |
| 5 | *díw* | [diwu] | 50 | *díyw* | [diyu] |
| 6 | *srsw* | [sersu] | 60 | *sí* | |
| 7 | *sfkh* | [sefekh] | 70 | *sfkh* | [sefekh] |
| 8 | *khmn* | [khemen] | 80 | *khmn* | [khemen] |
| 9 | *psḏ* | [pesedj] | 90 | *psḏ* | [pesedj] |

| *št* [shet] | *kh'* [kha'] | *ḏb'* [djebe'] | *ḥfn* [ḥefen] | *ḥḥ* [ḥeḥ] |
|---|---|---|---|---|
| 100 | 1,000 | 10,000 | 100,000 | 1,000,000 |

Note that 7, 8, and 9 have the same consonantal structure as 70, 80, and 90 respectively. The Egyptians may well have pronounced them slightly differently in order to avoid confusion: for instance, *sefekh* for 7 and *sefakh* for 70, *khemen* for 8 and *kheman* for 80, etc.

The spoken numerals, as can be seen, were strictly decimal. Compound numbers were expressed along the lines of the following example:

4,326:

| *fdw* | *kh* | *khmt* | *sht* | *dwty* | *srsw* |
|---|---|---|---|---|---|
| "four | thousand | three | hundred | twenty | six" |

## FRACTIONS AND THE DISMEMBERED GOD

Fractions were mostly expressed in Ancient Egyptian writing by placing the hieroglyph "mouth", pronounced eR and having in this context the specific sense of "part", over the numerical expression of the denominator, thus:

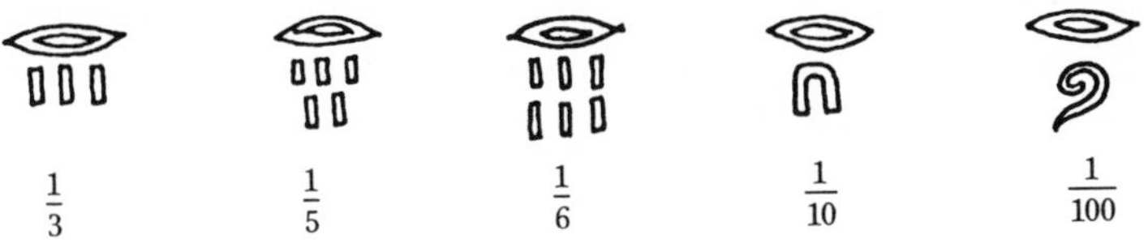


FIG. 14.24.

When the denominator was too large to go entirely beneath the eR sign, the remainder of it was placed to the right, thus:

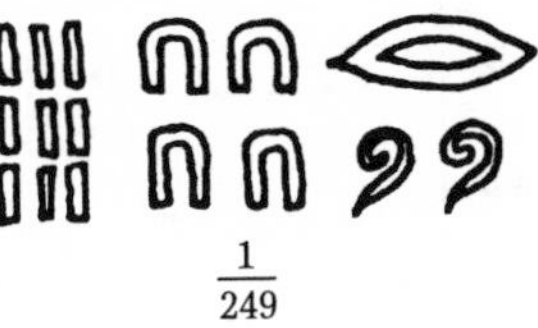


FIG. 14.25.

There were special signs for some fractions:

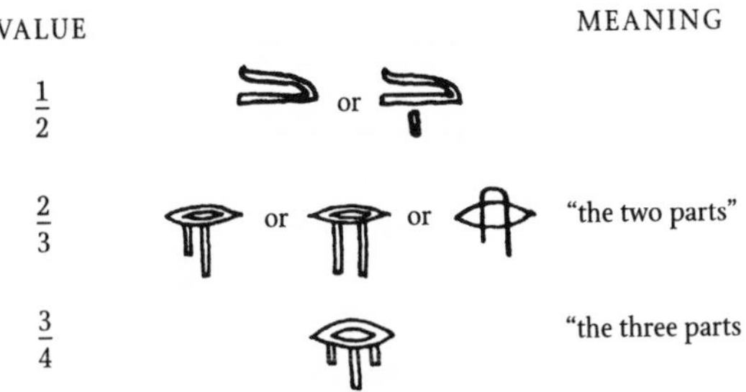

VALUE
MEANING
1/2
or
2/3
or
or
"the two parts"
3/4
"the three parts


FIG. 14.26.

Save for the last two expressions in Fig. 14.26, the only numerator used in Egyptian fractions was the unit. So to express (for instance) the equivalent of what we write as $\frac{3}{5}$, they did not write $\frac{1}{5}+\frac{1}{5}+\frac{1}{5}$ but decomposed the number into a sum of fractions with numerator 1.

$$=\frac{1}{2}+\frac{1}{10}=\frac{3}{5} \qquad =\frac{1}{3}+\frac{1}{4}+\frac{1}{5}=\frac{47}{60}$$

FIG. 14.27.

Measures of volume (dry and liquid) had their own curious system of notation which gave fractions of the *ḥeqat*, generally reckoned to have been equivalent to 4.785 litres. These volumetric signs used "fractions" of the hieroglyph representing the painted eye of the falcon-god Horus:

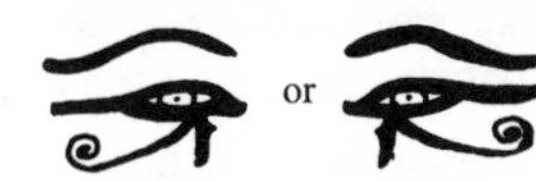


FIG. 14.28.

The name of Horus's eye was *oudjat*, written phonetically in hieroglyphs as follows:

FIG. 14.29.

The *oudjat* was simultaneously a human and a falcon's eye, and thus contained both parts of the cornea, the iris and the eyebrow of the human eye, as well as the two coloured flashes beneath the eye characteristic of the falcon. Since the most common fractions of the *ḥeqat* were the half, the quarter, the eighth, the sixteenth, the thirty-second and the sixty-fourth, the notation of volumetric fractions attributed to each of the parts or strokes in the *oudjat* sign the value of one of these fractions, as laid out in Fig. 14.30 below.

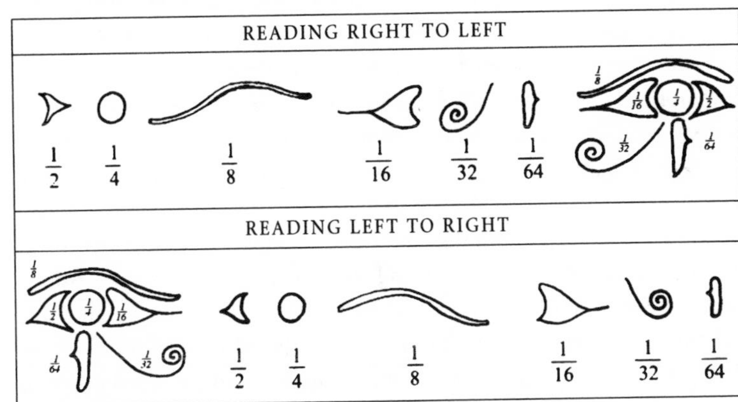


FIG. 14.30. *The fractions of the* ḥeqat

Horus was the son of Isis and Osiris, the god murdered and cut up into thirteen pieces by his brother Seth. When he grew up, Horus devoted himself to avenging his father, and his battles with his uncle Seth were long and bloody. In one of these combats, Seth ripped out Horus's eye, tore it into six pieces and dispersed the pieces around Egypt. Horus gave as good as he got, and castrated Seth. In the end, according to legend, the assembly of the gods intervened and put a stop to the fighting. Horus became king of Egypt and then the tutelary god of the Pharaohs, the guarantor of the legitimacy of the throne. Seth became the cursed god of the Barbarians and the Lord of Evil. The assembly of the gods instructed Thot, the god of learning and magic, to find and to reassemble Horus's eye and to make it healthy again. The *oudjat* thus became a talisman symbolising the wholeness of the body, physical health, clear vision, abundance and fertility; and so the scribes (whose tutelary god was Thot) used the *oudjat* to symbolise the fractions of the *ḥeqat*, specifically for measures of grain and of liquids.

An apprentice scribe one day observed to his master that the total of the fractions of the *oudjat* came to less than 1:

$$\frac{1}{2}+\frac{1}{4}+\frac{1}{8}+\frac{1}{16}+\frac{1}{32}+\frac{1}{64}=\frac{63}{64}$$

His master replied that the missing $\frac{1}{64}$ would be made up by Thot to any scribe who sought and accepted his protection.

## HIERATIC SCRIPT AND CURSIVE NUMERALS IN ANCIENT EGYPT

With its minutely complex and decorative signs, the hieroglyphic system of writing words and numbers was only really suitable for memorial inscriptions, and was used mainly, if not quite exclusively, on stone monuments such as tombs, funeral *stelae*, obelisks, palace and temple walls, etc. When Ancient Egyptians needed to note down or record accounts, censuses, inventories, reports, or wills, for example, or when they penned administrative, legal, economic, literary, magical, mathematical, or astronomical works, they had far more frequent recourse to a script that was easier to handle at speed, namely hieratic script.

Hieratic script uses signs that are simplifications and schematisations of the corresponding hieroglyphs, with fewer details and with shapes reduced to skeleton forms. In some cases, the hieratic versions can be recognised as variants of the original sign; but most often the relationship between the "cursive" and the "monumental" form is impossible to guess and has to be learned sign by sign.

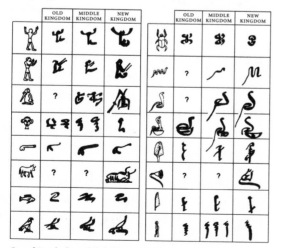

| | OLD KINGDOM | MIDDLE KINGDOM | NEW KINGDOM | | OLD KINGDOM | MIDDLE KINGDOM | NEW KINGDOM |
|---|---|---|---|---|---|---|---|
| | | | | | | | |
| | | | | | ? | | |
| | ? | | | | ? | | |
| | | | | | | | |
| | | | | | | | |
| | ? | ? | | | ? | ? | |
| | | | | | | | |
| | | | | | | | |

FIG. 14.31. *Some hieroglyphs and their hieratic equivalents*

There were also hieratic versions of the hieroglyphic numerals. These are the numerical signs found in the Harris Papyrus (British Museum), dating from the Twentieth Dynasty, which gives the possessions of the temples at the death of Ramses III (1192–1153 BCE):

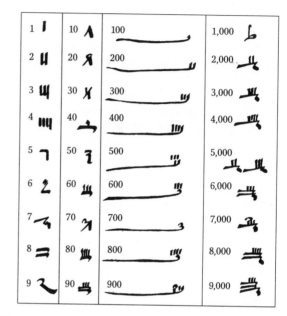

| | | | |
|---|---|---|---|
| 1 | 10 | 100 | 1,000 |
| 2 | 20 | 200 | 2,000 |
| 3 | 30 | 300 | 3,000 |
| 4 | 40 | 400 | 4,000 |
| 5 | 50 | 500 | 5,000 |
| 6 | 60 | 600 | 6,000 |
| 7 | 70 | 700 | 7,000 |
| 8 | 80 | 800 | 8,000 |
| 9 | 90 | 900 | 9,000 |

FIG. 14.32.

As can be seen, the hieratic numerals are for the most part visually quite unrelated to their equivalent hieroglyphs. Although the signs for the first four units are fairly self-explanatory ideograms, all the other numerals seem quite devoid of visually intuitive meaning.

So do hieratic numerals constitute a genuinely independent numbering system? Should we consider the numerals found in the Harris Papyrus as an arbitrary shorthand, invented by scribes for jotting down numbers intended to be written quite differently on stone monuments?

In fact, hieratic numerals, like the syllabic signs of this script, are developed from the corresponding hieroglyphs, and do not constitute an independent system. However, the changes in the shapes of the signs were very considerable, imposed in part by the characteristics of the reed-brushes used for hieratic characters (which, unlike hieroglyphs, were always written from right to left) and in part by a tendency to use ligatures, that is to say to run several signs together to produce single compounds. That is why the groups of five, six, seven, eight, and nine vertical lines became single signs devoid of any intuitive meaning:

| | |
|---|---|
| 5 | |
| 6 | |
| 7 | |
| 8 | |
| 9 | |

FIG. 14.33.

The relationship between hieratic numerals and hieroglyphs is difficult to see, but it was probably no more difficult for an Ancient Egyptian than it is for us to see the equivalence between the following ways of writing our own letters:

A B C D E F K R S

𝒜 ℬ 𝒞 𝒟 ℰ ℱ 𝒦 ℛ 𝒮

*a b c d e f k r s*

Imagine how hard it would be for a speaker of Chinese or Arabic, for example, with no knowledge of the Latin alphabet, to work out that the signs on the second and third lines have exactly the same value as the signs in the corresponding position on the first line!

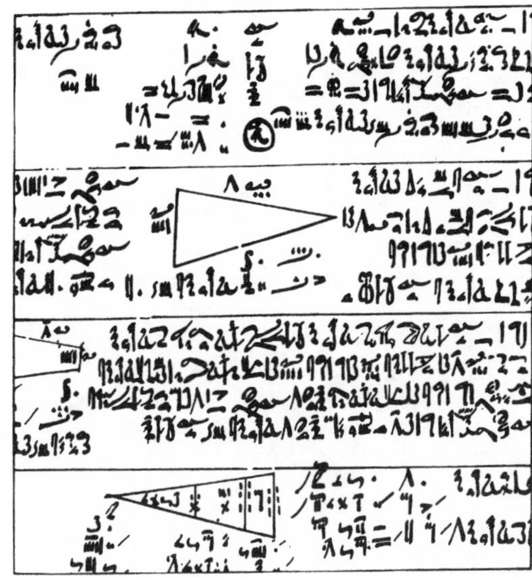

FIG. 14.34. *Detail from the Rhind Mathematical Papyrus (RMP), an important mathematical document written in hieratic script. From the Hyksos (Shepherd Kings) period (c. seventeenth century BCE), the RMP is a copy of an earlier document probably going back to the Twelfth Dynasty (1991–1786 BCE). The RMP is in the British Museum.*

Hieratic script was therefore not a form of "shorthand", in the sense that modern shorthand consists of purely arbitrary signs visually unrelated to the letters of the alphabet which they represent. Hieratic signs were indeed derived from hieroglyphs and represent the terminus of a long but specifically graphical evolution. Hieratic script never replaced the monumental script used for inscriptions on stone, and never had much impact on the shape of the hieroglyphs. The two systems were used in parallel for nearly 2,000 years, from the third to the first millennium BCE, and throughout this period hieratic script, despite its apparent difficulty, provided a perfectly serviceable tool for all administrative, legal, educational, magical, literary, scientific, and private purposes.

Hieratic script was gradually displaced from about the twelfth century BCE by a different cursive writing, called demotic. It survived in specific uses – notably in religious texts and in sacred funeral books – until the third century CE, which is why the Greeks called it *hieratikos*, meaning "sacred", whence our term "hieratic".

## FROM HIEROGLYPHIC TO HIERATIC NUMERALS

Hieratic numerals of the third millennium BCE are still fairly close to their hieroglyphic models; but over the centuries, the use of ligatures and the introduction of diacritics turn them little by little into apparently quite different signs with no intuitive resemblance to the original hieroglyphs. The end result was a set of numerals with distinctive signs for each of the following numbers:

| | | | | | | | | |
|---|---|---|---|---|---|---|---|---|
| 1 | 2 | 3 | 4 | 5 | 6 | 7 | 8 | 9 |
| 10 | 20 | 30 | 40 | 50 | 60 | 70 | 80 | 90 |
| 100 | 200 | 300 | 400 | 500 | 600 | 700 | 800 | 900 |
| 1,000 | 2,000 | 3,000 | 4,000 | 5,000 | 6,000 | 7,000 | 8,000 | 9,000 |

So though they began with a very basic additive numeration, the Egyptians developed a rapid notation system that was quite strikingly simple, requiring (for example) only four signs to represent the number 3,577, whereas in hieroglyphs it takes no fewer than 22 signs:

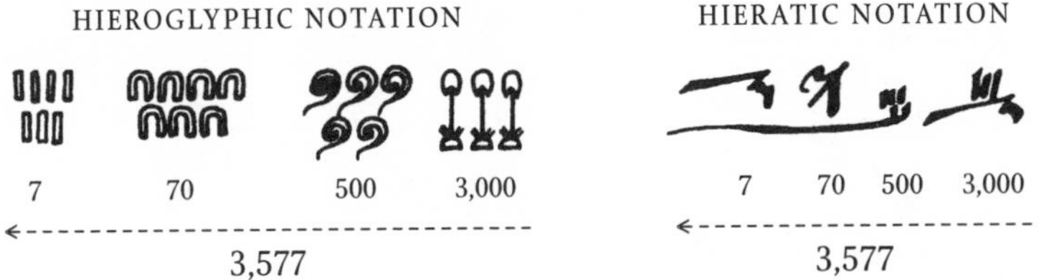


FIG. 14.35.

The main disadvantage of the hieratic system was of course that it required its users to memorise a very large number of distinct signs, and was thus quite impenetrable to all but the initiate. Here are the shapes that a hieratic mathematician had to know as well as we know 1 to 9:

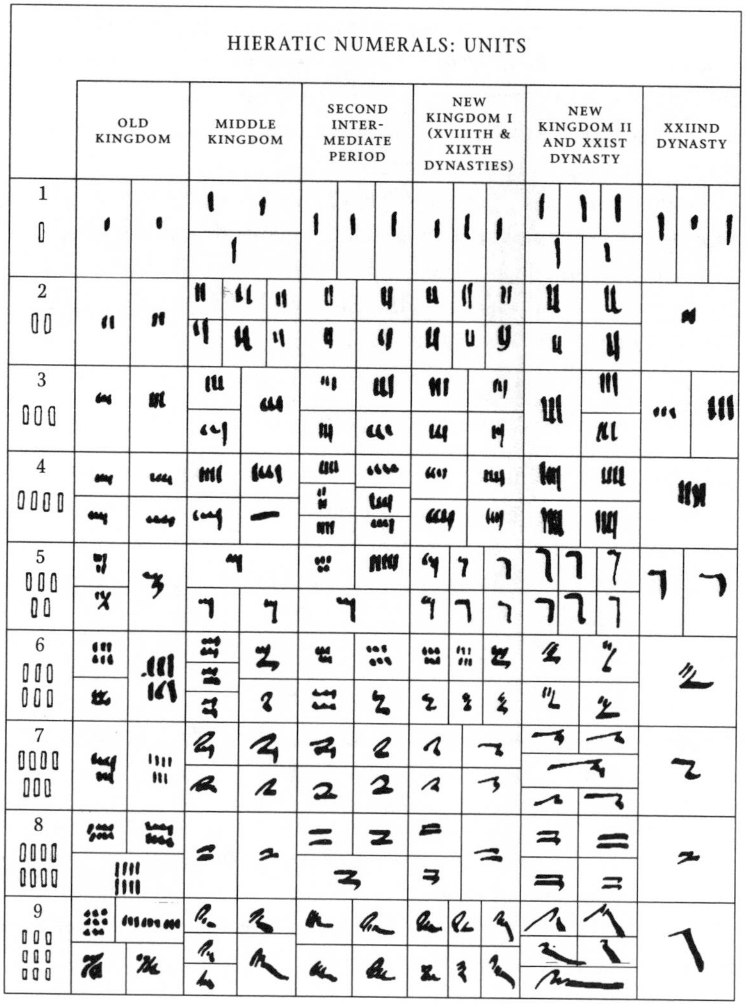

HIERATIC NUMERALS: UNITS

| | OLD KINGDOM | MIDDLE KINGDOM | SECOND INTER-MEDIATE PERIOD | NEW KINGDOM I (XVIIITH & XIXTH DYNASTIES) | NEW KINGDOM II AND XXIST DYNASTY | XXIIND DYNASTY |
|---|---|---|---|---|---|---|
| 1 | | | | | | |
| 2 | | | | | | |
| 3 | | | | | | |
| 4 | | | | | | |
| 5 | | | | | | |
| 6 | | | | | | |
| 7 | | | | | | |
| 8 | | | | | | |
| 9 | | | | | | |

FIG. 14.36A.

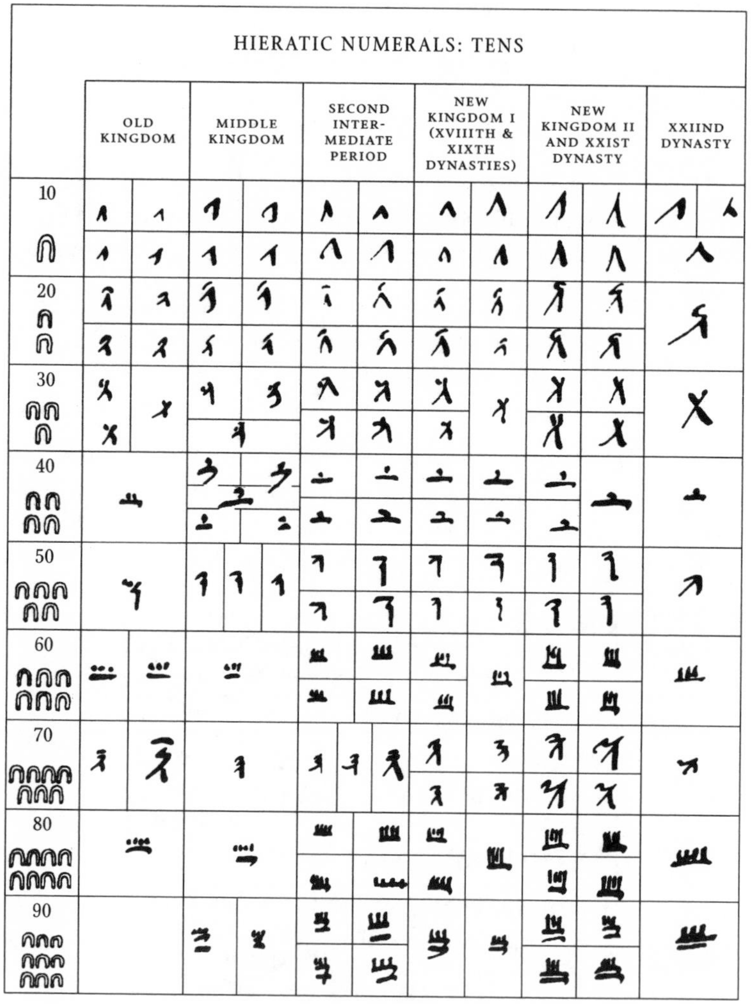

HIERATIC NUMERALS: TENS

| | OLD KINGDOM | MIDDLE KINGDOM | SECOND INTER-MEDIATE PERIOD | NEW KINGDOM I (XVIIITH & XIXTH DYNASTIES) | NEW KINGDOM II AND XXIST DYNASTY | XXIIND DYNASTY |
|---|---|---|---|---|---|---|
| 10 | | | | | | |
| 20 | | | | | | |
| 30 | | | | | | |
| 40 | | | | | | |
| 50 | | | | | | |
| 60 | | | | | | |
| 70 | | | | | | |
| 80 | | | | | | |
| 90 | | | | | | |

FIG. 14.36B.

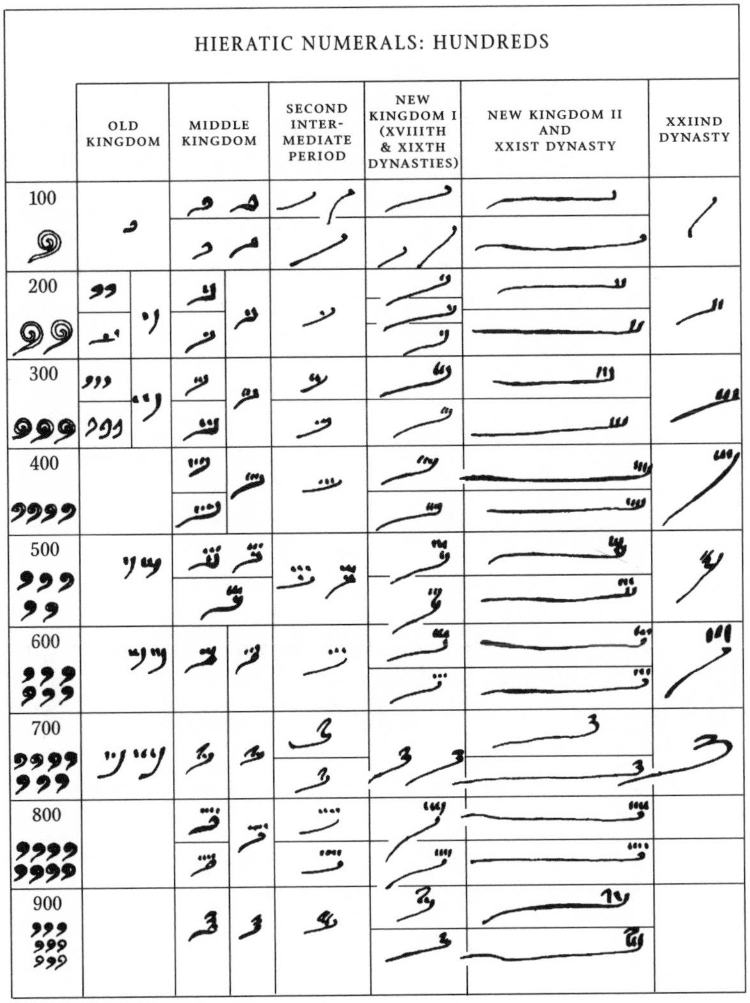

HIERATIC NUMERALS: HUNDREDS

| | OLD KINGDOM | MIDDLE KINGDOM | SECOND INTERMEDIATE PERIOD | NEW KINGDOM I (XVIIITH & XIXTH DYNASTIES) | NEW KINGDOM II AND XXIST DYNASTY | XXIIND DYNASTY |
|---|---|---|---|---|---|---|
| 100 | | | | | | |
| 200 | | | | | | |
| 300 | | | | | | |
| 400 | | | | | | |
| 500 | | | | | | |
| 600 | | | | | | |
| 700 | | | | | | |
| 800 | | | | | | |
| 900 | | | | | | |

Fig. 14.36c.

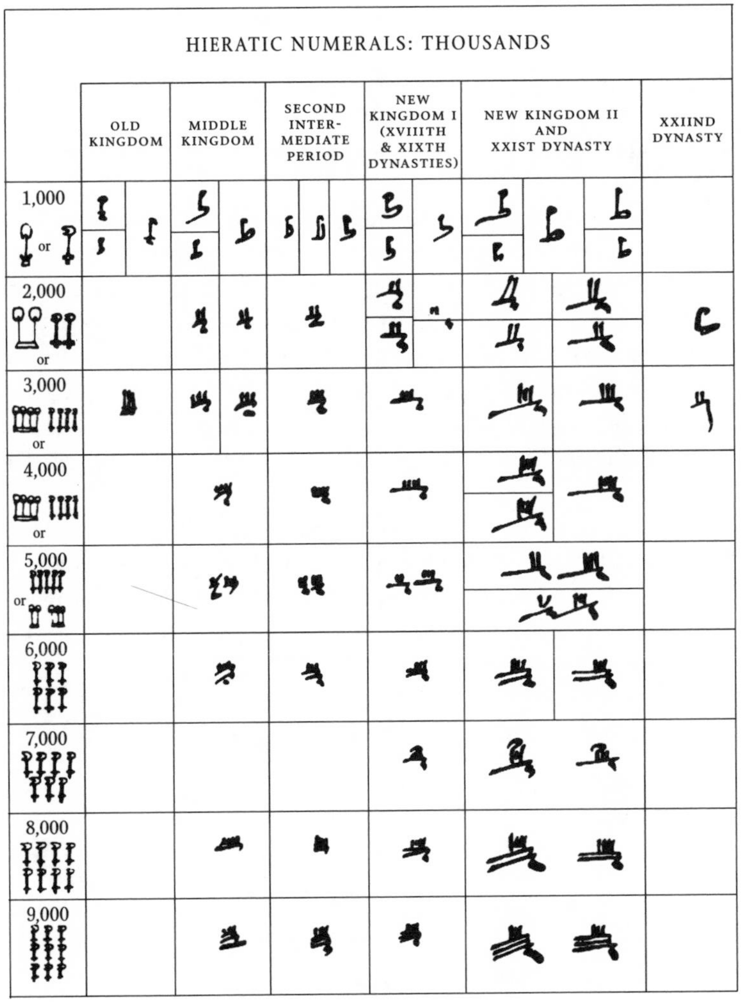

HIERATIC NUMERALS: THOUSANDS

| | OLD KINGDOM | MIDDLE KINGDOM | SECOND INTERMEDIATE PERIOD | NEW KINGDOM I (XVIIITH & XIXTH DYNASTIES) | NEW KINGDOM II AND XXIST DYNASTY | XXIIND DYNASTY |
|---|---|---|---|---|---|---|
| 1,000 | | | | | | |
| 2,000 | | | | | | |
| 3,000 | | | | | | |
| 4,000 | | | | | | |
| 5,000 | | | | | | |
| 6,000 | | | | | | |
| 7,000 | | | | | | |
| 8,000 | | | | | | |
| 9,000 | | | | | | |

Fig. 14.36d.

## DOING SUMS IN ANCIENT EGYPT

Let us imagine we're at a farm near Memphis, in the autumn of the year 2000 BCE. The harvest is in, and an inspector is here to make an assessment on which the annual tax will be calculated. So he orders some of the farm workers to measure the grain by the bushel and to put it into sacks of equal size.

This year's harvest includes white wheat, einkorn, and barley. So as to keep track of the different varieties of grain, the workers stack the white wheat in rows of 12 sacks, the einkorn in rows of 15 sacks, and the barley in rows of 19 sacks, and for each the total number of rows are respectively 128, 84, and 369.

When this is done, the inspector takes a piece of slate to use as a "notepad" and starts to do some sums on it in hieroglyphic numerals. For despite the primitive nature of their numerals, the Egyptians have known for centuries how to do arithmetic with them.

Adding and subtracting are quite straightforward. To add up, all you do is to place the numbers to be summed one above the other (or one alongside the other), then to make mental groups of the identical symbols and to replace each ten of one set of signs by one sign of the next higher decimal order.

For instance, to add 1,729 and 696, you first place (as in Fig. 14.37 below) 1,729 above 696. You then make mental groupings respectively of the vertical lines, the handles, the spirals, and the lotus flowers. By reducing them in packets of 10 to the sign of the next higher order, you get the correct result of the addition:

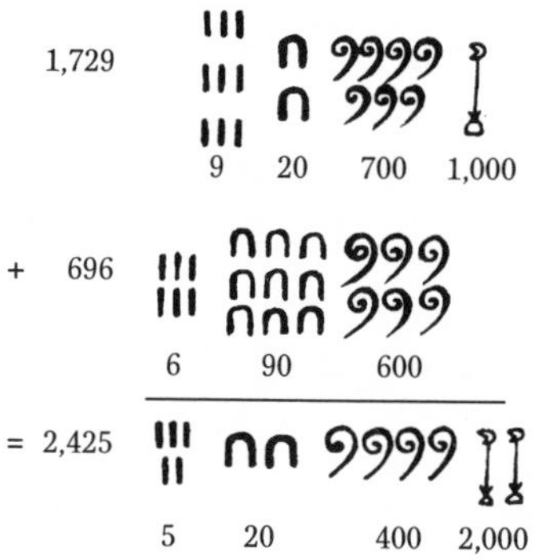

1,729 — 9 20 700 1,000

\+ 696 — 6 90 600

= 2,425 — 5 20 400 2,000

FIG. 14.37.

It is also quite easy to multiply and to divide by 10 in Egyptian hieroglyphics: to multiply, you replace each sign in the given number by the sign for the next higher order of decimal magnitude (or the next lower, for division by 10). So to multiply 1,464 by 10 you take:

4 60 400 1,000

FIG. 14.38.

and by following the regular procedure it becomes:

40 600 4,000 10,000

FIG. 14.39.

However, to multiply and to divide by any other factor, the Egyptians went about it quite differently. They knew only their two times table, and so they proceeded by a sequence of *duplications*.

To come back now to the tax-collector who needs to know the total number of sacks of white wheat in this year's harvest, and therefore needs to multiply 12 by 128. He goes about it like this:

| | |
|---|---|
| 1 | 12 |
| 2 | 24 |
| 4 | 48 |
| 8 | 96 |
| 16 | 192 |
| 32 | 384 |
| 64 | 768 |
| 128 | 1,536 |

That is to say, he writes the multiplier 12 in the right-hand column of his slate, and opposite it, in the left-hand column, he writes the number 1. He then doubles each of the two numbers in successive rows until the multiplier 128 appears on the left. As the number 1,536 appears on the right in the row where the left column shows 128, this is the result of the operation: 12 × 128 = 1,536.

To discover how many sacks of einkorn there are, he now has to multiply 84 by 15. His "doubling table" would look like this:

| | |
|---|---|
| 1 | 15 |
| 2 | 30 |
| 4 | 60 |
| 8 | 120 |
| 16 | 240 |
| 32 | 480 |
| 64 | 960 |

As the next doubling would take the multiplier beyond the required figure of 84, he stops there, and looks down the left-hand column to see which of the multipliers entered would sum to 84. He finds that he can reach 84 with just three of the multiplications already computed, and he checks the left-hand column numbers by making a little mark next to them, and putting an oblique stroke beside their right-hand column products, thus:

| | |
|---|---|
| 1 | 15 |
| 2 | 30 |
| –4 | 60/ |
| 8 | 120 |
| –16 | 240/ |
| 32 | 480 |
| –64 | 960/ |

He can then add up the numbers with the oblique check-mark and arrive at the result:

$$84 \times 15 = 960 + 240 + 60 = 1{,}260$$

To compute the number of sacks of barley, the inspector now has to multiply 369 by 19. He goes about it in exactly the same way, putting 1 in the left-hand column of his slate and 19 in the right-hand column, and then doubling the two terms successively as he goes down the rows. He stops when the left-hand column reaches 256, since the next step would give a multiplier of 512, which is higher than the required figure of 369:

| | |
|---|---|
| –1 | 19/ |
| 2 | 38 |
| 4 | 76 |
| 8 | 152 |
| –16 | 304/ |
| –32 | 608/ |
| –64 | 1,216/ |
| 128 | 2,432 |
| –256 | 4,864/ |

Then he looks down the left-hand column to find those numbers whose sum is 369, finds that they are 256, 64, 32, 16, and 1, and thus adds up the corresponding right-hand figures to arrive at his total:

$$369 \times 19 = 4{,}864 + 1{,}216 + 608 + 304 + 19 = 7{,}011$$

So the harvest adds up to 1,536 sacks of white wheat, 1,260 sacks of einkorn, and 7,011 sacks of barley. And since the Pharaoh's share of that is one tenth, the inspector can easily calculate the tax payable as 153 sacks of white wheat, 126 sacks of einkorn, and 701 sacks of barley.



So multiplication in the Egyptian manner is really quite simple and can be done without any multiplication tables other than the table of 2. Division is done similarly by successive duplication, but in reverse, as we shall see.

Let us suppose that in the time of Ramses II (1290–1224 BCE) robbers have just stripped the tomb of one of the Pharaohs of the preceding dynasty. They have stolen diadems, ear-rings, daggers, breast-plates, pendants – a whole mass of precious jewellery decorated with gold leaf and glass beads. Altogether there are 1,476 items in the robbers' haul, and the leader of the gang proposes to divide them equally amongst his eleven men and himself. So he has to divide 1,476 by 12. He goes about it just as if he were doing a multiplication, putting 12 in the right-hand column, and stopping when the right-hand figure reaches 768 since the next step would take the sequence beyond the total number of items to be shared:

| | |
|---|---|
| /1 | 12– |
| /2 | 24– |
| 4 | 48 |
| /8 | 96– |
| /16 | 192– |
| /32 | 384– |
| /64 | 768– |

He now has to find which of the numbers in the *right*-hand column total 1,476 and after various attempts to make the total he finds that 768, 384, 192, 96, 24, and 12 come out exactly right. So he makes a little mark against these figures in the right-hand column and puts an oblique against their corresponding numbers in the left-hand column. So he can now add up the checked numbers on the left to come out with the exact answer to the question: how many twelves go into 1,476?

$$1{,}476/12 = 64 + 32 + 16 + 8 + 2 + 1 = 123$$

So each of the robbers takes 123 pieces from the haul, and off they go with their fair shares.

This method of division only works when there is no remainder; where the dividend is not a multiple of the divisor, the Egyptians had a much more complicated method involving the use of fractions, which will not be explained here.*

The arithmetical methods of Pharaonic Egypt did not therefore require any great powers of memorisation, since, to multiply and to divide, all that you needed to know by heart was your two times table. Compared

*The method is explained in Richard J. Gillings, *Mathematics in the Time of the Pharaohs* (Cambridge, MA: MIT Press, 1972).

to modern arithmetic, however, Egyptian procedures were slow and very cumbersome.

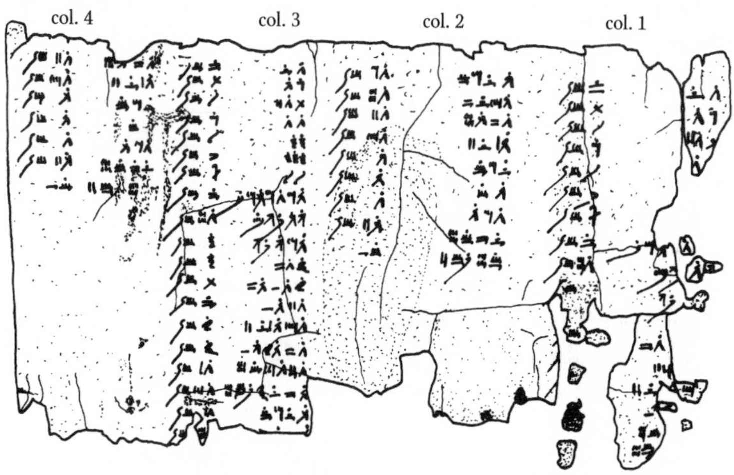


FIG. 14.40. *The Egyptian Mathematical Leather Roll (known as EMLR) in the British Museum. It contains, in hieratic notation, and in duplicate, twenty-six additions done in unit fractions and was probably used as a conversion table. [See Gillings (1972), pp. 89–103]*

## ANCIENT EGYPTIAN NUMBER-PUZZLES

Egyptian carvers, especially in the later periods, indulged in all sorts of puns and learned word-games, most notably in the inscriptions on the temples of Edfu and Dendara. Some of these word-games involve the names of the numbers, and the following tables (based on the work of P. Barguet, H. W. Fairman, J. C. Goyon, and C. de Wit) give a small sample of the innumerable curious scribal inventions for the representation of the numbers in hieroglyphs. The references are to Chassinat's transcription of the inscriptions on the walls of the temples of Edfu ("E") and Dendara ("D").

| VALUE | SIGN & MEANING | EXPLANATION | REFERENCE |
|---|---|---|---|
| 1 | harpoon | Homophony: "one" and "harpoon" are both pronounced *wa'* | E.VII, 18, 10 |
| 1 | sun | Because there is only one sun | E.IV, 6, 4 |
| 1 | moon | Because there is only one moon | E.IV, 6, 4 |
| 1 | fraction 1/30 | Only used in the expression "one day" or "the first day": 1/30 of a month is 1 day | E.IV, 8, 4; E.IV, 7, 1 |
| 2 | | Two × harpoon = 2 × 1 | E.IV, 14, 4 |
| 2 | | Sun + moon = 1 + 1 | E.VI, 7, 5 |
| 3 | | Three × harpoon = 3 × 1 | E.VII, 248, 10 |
| 4 | *jubilaeum* | No known explanation | E.IV, 6, 5; E.IV, 6, 6; E.VII, 15, 1 |
| 5 | 5-pointed star | Self-evident | E.IV, 6, 3; E.IV, 6, 5; E.VII, 6, 4 |
| 6 | | Standard sign for 1 + star = 1 + 5 | E.IV, 5, 4 |
| 7 | human head | The head has seven orifices: two eyes, two nostrils, two ears, mouth | E.IV, 4, 4; E.V, 305, 1 |
| 7 | | Standard sign for 2 + star = 2 + 5 | E.IV, 6, 5 |
| 7 | $\frac{1}{5}+\frac{1}{30}$ | Only in the expression "seven days": 1/5 of a month = 6 days + 1/30 = 1 day | E.IV, 8, 4; E.IV, 7, 1 |
| 8 | ibis | The sacred ibis was the incarnation of the god Thot, the principal divinity of the city of Hermopolis, formerly *Khmnw* or *Khemenu*, meaning "the city of eight" | E.III, 77, 17; E.VII, 13, 4; E. VII, 14, 2 |
| 8 | | A curious "re-formation" in hieroglyphics of the hieratic numeral 8 | E.VI, 92, 13 |
| 8 | | Standard notation of 3 + star = 3 + 5 = 8 | E.IV, 5, 2 |
| 8 | | Moon + head = 1 + 7 = 8 | E.IV, 6, 4 |
| 8 | | Standard notation of 1 + head = 1 + 7 = 8 | E.IV, 9, 3 |

| VALUE | SIGN & MEANING | EXPLANATION | REFERENCE |
|---|---|---|---|
| 9 | | Homophony: "nine" and "shine" are both pronounced *psḏ* | E.IV, 8, 2; E.VII, 8, 8 |
| 9 | scythe | Based on the fact that in hieratic the numeral 9 and the sign for scythe were identical | E.VII, 15, 3; E.VII, 15, 9; E.VII, 17, 3 |
| 9 | | Standard notation of 4 + star = 4 + 5 = 9 | E.IV, 6, 1 |
| 9 | | Standard notation of 2 + head = 2 + 7 = 9 | D.II, 47, 3 |
| 10 | falcon | The falcon-god Horus was the first to be added to the original nine divinities of Heliopolis, and thus represents 10 | E.V, 6, 5 |
| 14 | | falcon + *jubilaeum* = 10 + 4 = 14 | E.V, 6, 5 |
| 15 | fraction ½ | Only in the expression "15 days" or "fortnight": 1/2 of a month = 15 days | E.VII, 7, 6 |
| 17 | | Standard notation of 10 + head = 10 + 7 = 17 | E.VII, 248, 9 |
| 18 | $\frac{1}{2} + \frac{1}{10}$ | Only in the expressions "18 days" or "the 18th day": 1/2 month + 1/10 month = 15 + 3 = 18 | E.IV, 9, 1; E.VII, 7, 6; E.VII, 6, 1 |
| 19 | | Standard notation of 10 + scythe = 10 + 9 = 19 | E.VII, 248, 4 |
| 20 | | Two falcons = 2 × 10 = 20 | E.VII, 11, 8 |
| 107 | | Standard notation of 100 + head = 100 + 7 = 107 | E.VII, 248, 11 |

CHAPTER 15

# COUNTING IN THE TIMES OF THE CRETAN AND HITTITE KINGS

## THE NUMBERS OF CRETE

Between 2200 and 1400 BCE, the island of Crete was the centre of a very advanced culture: Minoan civilisation, as it is called, from the name of the legendary priest-king Minos who, according to Greek mythology, was one of the first rulers of Knossos, the ancient Cretan capital near the modern port of Heraklion (Candia).

The very existence of Minoan civilisation was almost completely unknown until the end of the last century, and it is only relatively recently that archaeologists have uncovered a brilliant and original culture which was, in many respects, the precursor of Greek civilisation.

When Minoan civilisation fell, around 1400 BCE, probably as a result of some natural disaster or of the invasion of the island by the Mycenaeans (of Greek origin), it disappeared almost without trace save for what was preserved in the fables and legends of the Ancient Greeks.

We owe the most spectacular discoveries – such as the famous Palace of Knossos – to the indefatigable enthusiasm and energy of the British archaeologist Sir Arthur Evans (1851–1941). He was the first to show that the Greek legends had a historical basis, and constituted a living trace of one of the oldest known European civilisations.

Since the end of the last century, archaeological investigations carried out mainly on the sites of Knossos and Mallia have brought to light a large number of documents whose analysis has revealed the existence of a "hieroglyphic" script between 2000 and 1660 BCE.

Cretan hieroglyphics have still not been deciphered, and these documents remain enigmatic. Nevertheless they show evidence of an accounting system adapted to a "bureaucracy" no doubt born within the earliest palaces of Minoan civilisation. In proof of this we find clay blocks and tablets covered with figures and hieroglyphic signs, which are more or less schematic drawings of all kinds of objects. These appear to be accounts giving details of inventories, supplies, deliveries, or exchanges. The purpose of the symbols was to note down the quantities of the different kinds of goods.

The numerical notation of Crete was strictly decimal, and was based on the additive principle. Unity was represented by a short slightly oblique stroke, or by a small circular arc which could be oriented anyhow. Cretan

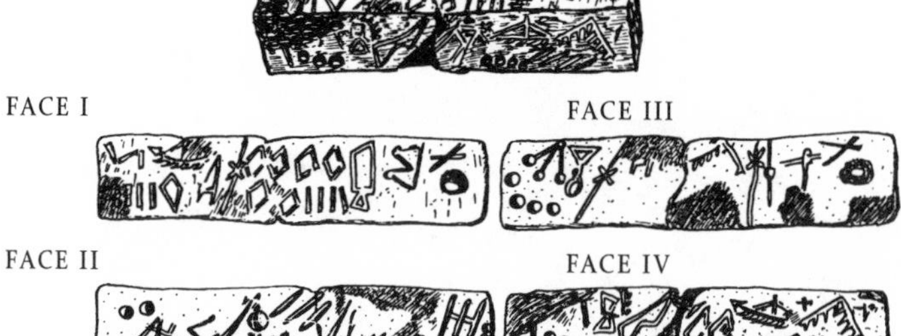


FIG. 15.1. *Inscriptions on bars of clay, showing Cretan hieroglyphic signs and numerals. Palace of Knossos, 2000–1500 BCE. [Evans (1909), Doc. P 100]*

hieroglyphic writing went from left to right and from right to left, in *boustrophedon* (as a ploughman ploughs a field from side to side). 10 was represented by a circle (or, on clay, by a small circular imprint as would be made by the pressure of a round-tipped stylus held perpendicularly to the surface of the clay). 100 was represented by a large oblique bar (distinctly different from the small stroke of unity), and 1,000 was represented by a kind of lozenge.

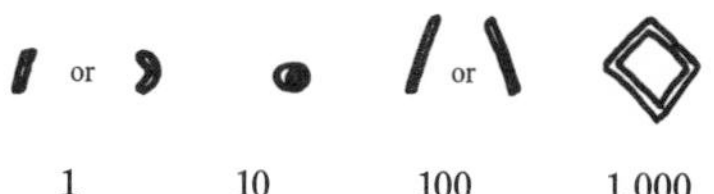

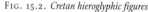

FIG. 15.2. *Cretan hieroglyphic figures*

With these as starting points, the Cretans represented other numbers by repeating each one as many times as required. The hieroglyphic figures were not, however, the only forms used. Other excavations have revealed a second script, no doubt derived from the hieroglyphic, in which the picture symbols give way to schematic drawings which, often, we cannot identify now. Analysis of these documents led Evans to distinguish two variants of this kind of writing, which he called "Linear A" and "Linear B".

The system known as "Linear A" is the older. It was in use from the start of the second millennium BCE up to around 1400 BCE, that is to say at about the same time as the hieroglyphic script.

The sites which have yielded documents in Linear A are several, notably Haghia Triada, Mallia, Phaestos, and Knossos. From Haghia Triada we have a large collection of accounting tablets, unfortunately in a somewhat sloppy script [Fig. 15.4]. These are, therefore, inventories, with ideograms and numbers; the tablets are in the format of small pages. But Linear A can be found as well on a wide variety of objects: vases (with inscriptions cut, painted, or written in ink), seals, stamps, and labels of clay; ritual objects (libation tables); large copper ingots; and so on. This writing may therefore be very widely found, not only in administrative environments but also in holy places and probably also in people's homes. [O. Masson (1963)]

| | | |
|---|---|---|
| man | ox | mountain |
| man crouching | ship | tree |
| eye | morning star | goat |
| axe | plant | wheat grain |
| plough | crescent moon | double axe |
| palace | bee | crossed arms |

FIG. 15.3. *A selection of Cretan hieroglyphics [after Evans]*

FIG. 15.4. *Cretan tablet with signs and numerals from the "Linear A" script. Haghia Triada, sixteenth century BCE. [GORILA (1976), HT 13, p. 26]*

The script known as "Linear B" is the more recent, and the best known, of the Cretan scripts. It is usually dated to the period between 1350 BCE and 1200 BCE. At this time, the Mycenaeans had conquered Crete, and ancient Minoan civilisation had spread onto the Greek mainland, especially in the region of Mycenae and Tyrinth.

The signs of this script were engraved on clay tablets, which were first unearthed in 1900. Since then, 5,000 tablets have been found in Crete (at Knossos only, but in large numbers) and on mainland Greece (mainly at Pylos and Mycenae). Linear B, therefore, may be found outside Crete. We may also note that this script, apparently derived by modification of Linear A, was used to record an archaic Greek dialect, as demonstrated by Michael Ventris, the English scholar who first deciphered it. It is the only Creto-Minoan script to have been deciphered to date (Linear A and the hieroglyphic script correspond to a language which still remains largely unknown).

A

B

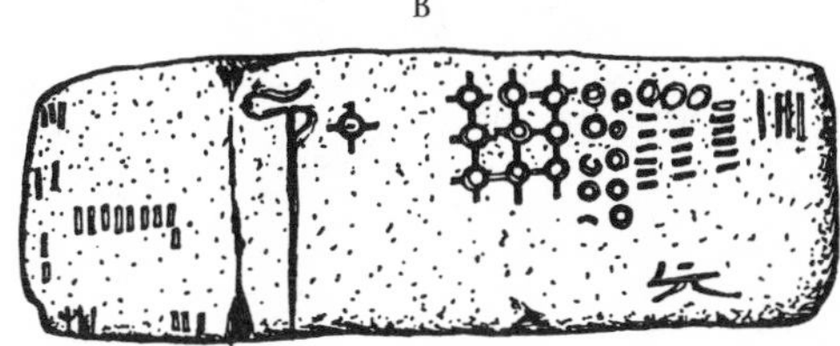

FIG. 15.5. *Cretan tablets with signs and numerals from the "Linear B" script, fourteenth or thirteenth century BCE. [Evans and Myres (1952)]*

Both Linear A and Linear B used practically the same number-signs (Fig. 15.6). These were:

- a vertical stroke for unity;
- a horizontal stroke (or, solely in Linear A, sometimes a small circular imprint) for 10;
- a circle for 100;
- a circular figure with excrescences for 1,000;
- the same, with a small horizontal stroke inside, for 10,000 (found only in Linear B inscriptions: Fig. 15.6, last line).

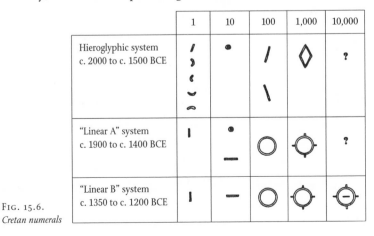

| | 1 | 10 | 100 | 1,000 | 10,000 |
|---|---|---|---|---|---|
| Hieroglyphic system c. 2000 to c. 1500 BCE | | | | | ? |
| "Linear A" system c. 1900 to c. 1400 BCE | | | | | ? |
| "Linear B" system c. 1350 to c. 1200 BCE | | | | | |

FIG. 15.6. *Cretan numerals*

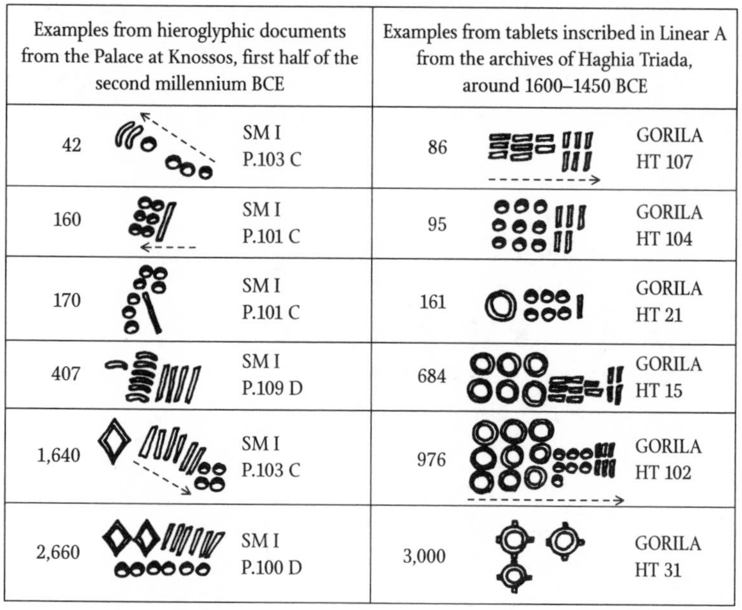

| Examples from hieroglyphic documents from the Palace at Knossos, first half of the second millennium BCE | | Examples from tablets inscribed in Linear A from the archives of Haghia Triada, around 1600–1450 BCE | |
|---|---|---|---|
| 42 | SM I P.103 C | 86 | GORILA HT 107 |
| 160 | SM I P.101 C | 95 | GORILA HT 104 |
| 170 | SM I P.101 C | 161 | GORILA HT 21 |
| 407 | SM I P.109 D | 684 | GORILA HT 15 |
| 1,640 | SM I P.103 C | 976 | GORILA HT 102 |
| 2,660 | SM I P.100 D | 3,000 | GORILA HT 31 |

FIG. 15.7. *The principle of the Cretan numerals*

To represent a given number, it was enough to repeat each of the above as many times as needed (Fig. 15.7).

The number-systems used in Crete in the second millennium BCE (hieroglyphic, Linear A, and Linear B) had, therefore, exactly the same intellectual basis as the Egyptian hieroglyphic notation and, for the whole time they were in use, underwent no modification of principle. (Similarly, the drawing of signs and numbers on clay did not give way to a cuneiform system, as happened in Mesopotamia). As in the monumental Egyptian system, these number-systems were founded on base 10 and used the principle of juxtaposition to represent addition. Moreover, the only numbers to which each system gave a special sign were unity, and the successive powers of 10.

The number 10,000 (found only in Linear B inscriptions) is derived from the number 1,000 by adding a horizontal bar in the interior of the latter. By all appearances, therefore, a multiplicative principle has been used (10,000 = 1,000 × 10), since the horizontal bar is simply the symbol for 10 in this system (Fig. 15.6).

## THE HITTITE HIEROGLYPHIC NUMBER-SYSTEM

From the beginning of the second millennium BCE the Hittites (a people of Indo-European origin) settled progressively in Asia Minor, no doubt by a process of slow immigration. Between the eighteenth and the sixteenth centuries BCE, they there established a great imperial power of which there were two principal phases: the Ancient Empire (pre-1600 to around 1450 BCE) and the New Empire (1450–1200 BCE).

In the course of the imperial era, the Hittites, with many successes and failures, pursued a policy of conquest in central Anatolia and northern Syria. But at the start of the thirteenth century BCE, no doubt under attack from the "Peoples of the Sea", this powerful empire abruptly collapsed. A renaissance, however, ensued from the ninth century BCE in the north of Syria where several small Hittite states maintained elements of the imperial tradition in the midst of mixed populations. This was the beginning of what is called the "neo-Hittite" phase of the civilisation. Finally, however, in the seventh century BCE, all these small states were absorbed by the Assyrian Empire.

The Hittites had two writing systems. One was a hieroglyphic system which seems to have been of their own creation, of which the earliest known evidence is from the fifteenth century BCE. The other was a cuneiform system borrowed from Assyro-Babylonian civilisation whose introduction dates from around the seventeenth century BCE.*

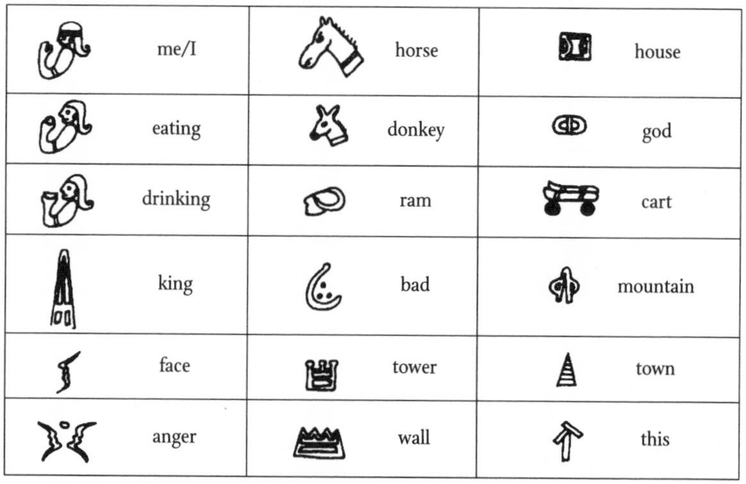

| | | |
|---|---|---|
| me/I | horse | house |
| eating | donkey | god |
| drinking | ram | cart |
| king | bad | mountain |
| face | tower | town |
| anger | wall | this |

FIG. 15.8. *The meanings of some of the Hittite hieroglyphics [after Laroche (1960)]*

Thus, for at least three centuries (1500–1200 BCE) the hieroglyphic lived alongside the cuneiform in Anatolia, and they constituted the dual medium of expression of the Hittite state. For a people to practise

* The cuneiform system, of Assyro-Babylonian origin, was adapted into at least three Hittite dialects: Nesitic, spoken in the capital of the empire; Louvitic, employed in southern Anatolia, and Palaitic in the north. Cuneiform characters were used for the numerous tablets making up the royal archives of the town of Hattuša, capital of the Hittite Empire, at the place which is now Boğazköy in Turkey, about 150 km east of Ankara; thanks to these documents, the history and language of the Hittites have been partially reconstructed.

two writing systems at the same time is not a frequent phenomenon. We are now able to perceive the reasons which induced the Hittites into this paradoxical situation. The scribes of Hattuša, who were the keepers of the Babylonian tradition, were a small and privileged group who had sole access to their literature and to the documents on clay. The establishment of a library answered a need, and the use of the cuneiform ensured that the kingdom could maintain communication with its representatives abroad. But the tablet was, in effect, a banned document: it made no public proclamation of the sublimity of the god, nor of the grandeur of the king. Without doubt the Hittites felt that these imprinted cuneiform characters, mechanical and lacking expression, should take second place to a different writing more visual, more monumental, more apt for writing of divine effigies and royal profiles. . . . The hieroglyphs are made to be gazed upon, and contemplated upon walls of rock: they give life to a name just as a relief brings the whole person to life. [E. Laroche (1960)]

All the same, hieroglyphic writing survived the cuneiform after the destruction of the Hittite Empire around 1200 BCE. It served not only for religious and dedicatory purposes, but also, and perhaps above all, for lay purposes in business documents.

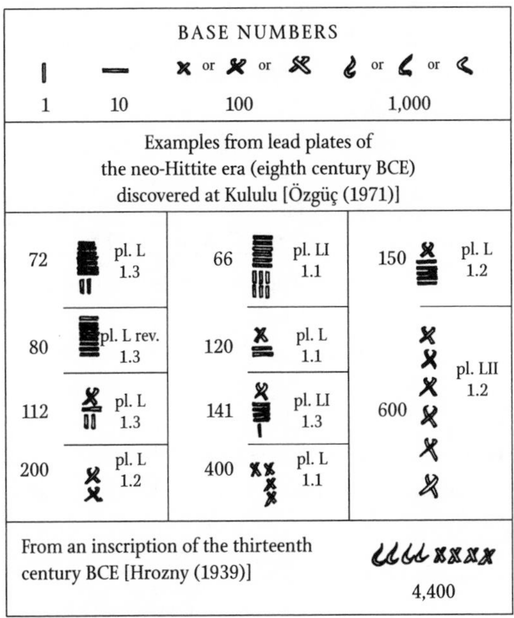


FIG. 15.9. *The Hittite hieroglyphic number-system*

In the Hittite hieroglyphic number-system, a vertical stroke represented unity. For the successive integers, small groups of two, three, four or five strokes were used to allow the eye to grasp the total sum of the units. The number 10 was represented by a horizontal stroke, a 100 by a kind of Saint Andrew's cross, and 1,000 by a sign which looked like a fish-hook (Figure 15.9). On this basis, the representation of intermediate numbers presented no difficulty, since it was sufficient to repeat each sign as many times as required.

The Hittite hieroglyphic number-system was, after the fashion of the Egyptian, strictly decimal and additive, since the only numbers to have specific signs were unity and the successive powers of 10.

## CHAPTER 16

# GREEK AND ROMAN NUMERALS

### THE GREEK ACROPHONIC NUMBER-SYSTEM

Let us now visit the world of the Ancient Greeks, and look at the number-systems used in the monumental inscriptions of the first millennium BCE.

The Attic system, which was used by the Athenians, assigns a specific sign to each of the numbers

1 5 10 50 100 500 1,000 5,000 10,000 50,000

and is based above all on the additive principle (Fig. 16.1).

| | | | | | |
|---|---|---|---|---|---|
| 1 | I | 100 | H | 10,000 | M |
| 2 | II | 200 | HH | 20,000 | MM |
| 3 | III | 300 | HHH | 30,000 | MMM |
| 4 | IIII | 400 | HHHH | 40,000 | MMMM |
| 5 | 𐅃 | 500 | 𐅅 | 50,000 | 𐅇 |
| 6 | 𐅃I | 600 | 𐅅H | 60,000 | 𐅇M |
| 7 | 𐅃II | 700 | 𐅅HH | 70,000 | 𐅇MM |
| 8 | 𐅃III | 800 | 𐅅HHH | 80,000 | 𐅇MMM |
| 9 | 𐅃IIII | 900 | 𐅅HHHH | 90,000 | 𐅇MMMM |
| 10 | Δ | 1,000 | X | | |
| 20 | ΔΔ | 2,000 | XX | | |
| 30 | ΔΔΔ | 3,000 | XXX | | |
| 40 | ΔΔΔΔ | 4,000 | XXXX | | |
| 50 | 𐅄 | 5,000 | 𐅆 | | |
| 60 | 𐅄Δ | 6,000 | 𐅆X | | |
| 70 | 𐅄ΔΔ | 7,000 | 𐅆XX | | |
| 80 | 𐅄ΔΔΔ | 8,000 | 𐅆XXX | | |
| 90 | 𐅄ΔΔΔΔ | 9,000 | 𐅆XXXX | | |

Fig. 16.1. *System of numerical annotation found in Attic inscriptions from around the fifth century BCE until the start of the Common Era. [Franz (1840); Guarducci (1967); Guitel; Gundermann (1899); Larfeld (1902–7); Reinach (1885); Tod]*

The Attic system has an interesting feature: with the exception of the vertical bar representing 1, the figures are simply the initial letters of the Greek names of the corresponding number, or are combinations of these: this is what is meant by an *acrophonic* number-system.

To show this:

| THE SIGN | WHICH IS THE SAME AS THE LETTER | WHOSE VALUE IS | IS THE FIRST LETTER OF THE WORD | WHICH IS THE GREEK NAME OF THE NUMBER |
|---|---|---|---|---|
| Γ | PI (the archaic form of the letter Π | 5 | Πεντε (Pente) | Five |
| Δ | DELTA | 10 | Δεκα (Deka) | Ten |
| H | ETA | 100 | Hεκατον (Hekaton) | Hundred |
| X | KHI | 1,000 | Χιλιοι (Khilioi) | Thousand |
| M | MU | 10,000 | Μυριοι (Murioi) | Ten thousand |

Fig. 16.2.

The signs for the numbers 50, 500, 5,000, and 50,000 are, as can be seen, made up by combining the preceding signs according to the multiplicative principle:

| | | | |
|---|---|---|---|
| 50 | 𐅄 = Γ . Δ | 5 × | 10 |
| 500 | 𐅅 = Γ . H | 5 × | 100 |
| 5,000 | 𐅆 = Γ . X | 5 × | 1,000 |
| 50,000 | 𐅇 = Γ . M | 5 × | 10,000 |

Fig. 16.3.

In other words, in the Attic system, in order to multiply the value of one of the alphabetic numerals Δ, H, X and M by 5, it is placed inside the letter Γ = 5.

This system, which in fact only recorded cardinal numbers, was used in metrology (to record weights, measure, etc.) and for sums of money. We shall later see it used for the Greek abacus.

Originally, ordinal numbers were spelled out in full, but from the fourth century BCE (probably, indeed, from the fifth) a different system was used to write these numbers, which we shall study later.

To write down a sum expressed in *drachmas*, the Athenians made use of these figures, repeating each one as often as required to add up to the quantity; each occurrence of the vertical bar for "1" was replaced by the symbol ⊢ which stood for "drachma":

XXX 𐅅 H ΔΔΔ ⊢⊢⊢

3,000 500 100 30 3

-------------------------->

3,633 drachmas

Fig. 16.4.

For multiples of the *talent*, which was worth 6,000 *drachmas*, they used the same number of signs but with T (the first letter of TALANTON) instead of ⊢:

𐅅 𐅄 ΔΔΔΔ 𐅃 TTT

500 50 40 5 3

-------------------------->

598 talents

Fig. 16.5.

For divisions of the *drachma* (the *obol*, the half- and the quarter-*obol*, and the *chalkos*) special signs were used:

| | | |
|---|---|---|
| 1 CHALKOS (or 1/8 of an obol) | X | X: initial letter of ΧΑΛΚΟΥΣ |
| 1 QUARTER-OBOL | Ɔ or T | T: initial letter of ΤΕΤΑΡΤΗΜΟΡΙΟΝ |
| 1 HALF-OBOL | C | |
| 1 OBOL (1/6 of a drachma) | I or O | O: initial letter of ΟΒΟΛΙΟΝ |

Fig. 16.6.

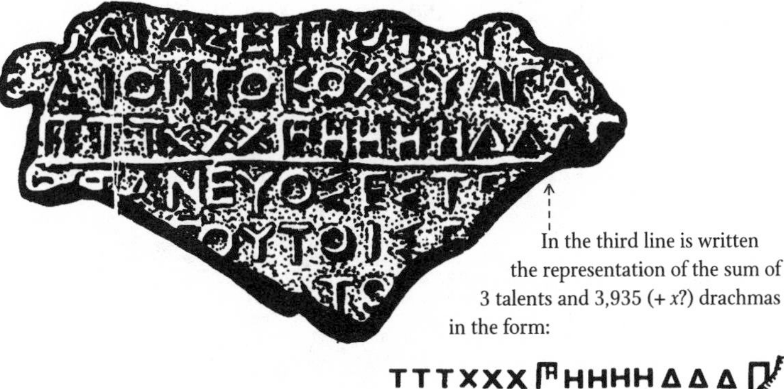

In the third line is written the representation of the sum of 3 talents and 3,935 (+ *x*?) drachmas in the form:

FIG. 16.7. *Greek inscription (fragment) from Athens dating from the fifth century BCE. (Museum of Epigraphy, Athens. Inv. Em12 355)*

By the use of these signs, the Athenians were able to write easily those sums of money which were of relatively frequent occurrence. The following examples give the idea. (A quite similar system was also used for weights and measures such as the *drachma*, *mina*, and *stater*.)

| Notation | Values | Reading |
|---|---|---|
| ΔΔ ⊢⊢⊢ III C T | 20, 3 (drachmas); 3, ½, ¼ (obols) | 23 drachmas and (3 + 1/2 + 1/4) obols |
| ⊢ΔΔΔΔ IIII | 40, 4 | *read:* 40 drachmas and 4 obols |
| XX 𐅅 HΔΔΔII | 2,000 500 100 30 2 | *read:* 2,630 drachmas and 2 obols |
| XXX(T)HH𐅄(T) ΔTTT — XX 𐅅ΔΔΔΔ⊢⊢⊢⊢ — IIIII | talents: 3,000 200 50 10 3; drachmas: 2,000 500 40 4; obols: 5 | 3,263 talents; 2,544 drachmas and 5 obols |

FIG. 16.8.

In the other states of the Ancient Greek world, the citizens also used similar acrophonic symbols in their various monumental inscriptions during the latter half of the first millennium BCE (Fig. 16.9 and 16.10). The Attic system itself, which is the oldest known of the Greek acrophonic systems, became more widespread at the time of Pericles, when the city of Athens was the capital of a number of Greek republics.

However it would be wrong to think that these different number-systems were all strictly identical to the Athenian one. Each had features which distinguished it from the others. We should not forget that each Greek state had its own system of weights and its own system of coinage (by this period the use of money was widespread throughout the Mediterranean). Furthermore the very notion of a unified metric system, on the lines of an international monetary system, was foreign to the Greek spirit.*

| | | | |
|---|---|---|---|
| 1 ⊢ (1 drachma) | 10 Δ | 100 H | 1,000 X |
| 2 ⊢⊢ | 20 ΔΔ | 200 HH | 2,000 XX |
| 3 ⊢⊢⊢ | 30 ΔΔΔ | 300 HHH | 3,000 XXX |
| 4 ⊢⊢⊢⊢ | 40 ΔΔΔΔ | 400 HHHH | 4,000 XXXX |
| 5 ⊢⊢⊢⊢⊢ | 50 𐅄 | 500 𐅅 | 5,000 𐅆 |
| 6 ⊢⊢⊢⊢⊢⊢ | 60 𐅄Δ | 600 𐅅H | 6,000 𐅆X |
| 7 ⊢⊢⊢⊢⊢⊢⊢ | 70 𐅄ΔΔ | 700 𐅅HH | 7,000 𐅆XX |
| 8 ⊢⊢⊢⊢⊢⊢⊢⊢ | 80 𐅄ΔΔΔ | 800 𐅅HHH | 8,000 𐅆XXX |
| 9 ⊢⊢⊢⊢⊢⊢⊢⊢⊢ | 90 𐅄ΔΔΔΔ | 900 𐅅HHHH | 9,000 𐅆XXXX |

Example: 𐅆 𐅅 HH ΔΔΔΔ ⊢⊢⊢⊢⊢⊢⊢⊢⊢
5,000 500 200 40 9
5,749 drachmas

FIG. 16.9. *Numerical notation in Greek inscriptions from the island of Cos (third century BCE). [Tod]*

| | | |
|---|---|---|
| 1 drachma | ⊢* or I** | |
| 5 | Γ* | Π: first letter of Πεντε, "five" |
| 10 | ▷** or Δ* | Δ: first letter of Δεκα, "ten" |
| 50 | ΓΕ or Γ^Ε* | ΠΕ: abbreviation of Πεντεδεκα, "fifty" |
| 100 | ⊢Ε | HE: abbreviation of Ηεκατον, "hundred" |
| 300 | ΤΕ* | T.HE: abbreviation of Τριακοσιοι, "three hundred" |
| 500 | ΓΗΕ or ΓΗΕ | Π.HE: abbreviation of Πεντακοσιοι, "five hundred" |
| 1,000 | Ψ | Ancient Boeotian form of the letter X: first letter of Χιλιοι, "thousand" |
| 5,000 | Γ^Ψ | Π.X: abbreviation of Πενταχιλιοι, "five thousand" |
| 10,000 | M | Letter M: first letter of Μυριοι, "ten thousand" |

* Found only at THESPIAE
** Found only at ORCHOMENOS

FIG. 16.10. *Numerical notation in Greek inscriptions from Orchomenos and from Thespiae (third century BCE) [Tod]*

* As P. Devambez (1966) explains: "Money was in the first place defined in terms of weight. Each state chose from its system of weights one unit to be the standard, and the others were multiples or sub-multiples of this. For instance, at Aegina in the Peloponnese, the standard unit of weight for commerce was the *mina* which weighed 628 gm. The unit of money was chosen to be one hundredth of this, the *drachma*, which therefore weighed 6.28 gm. The *didrachma* or *stater* was about twice this (12.57 gm). The sub-unit, the *obol*, weighing 1.04 gm, was a sixth of the *drachma*. At Euboea and in Attica, where the *mina* weighed 436 gm, the *drachma* was 4.36 gm; its multiples, the *didrachma* and *tetradrachma*, weighed twice and four times this, or 8.73 gm and 17.46 gm respectively; the *obol*, a sixth of the unit, weighed 0.73 gm."

| Drachmas | Signs | Places |
|---|---|---|
| 1 | 1 2 3 4 5 | 1 Epidaurus, Argos, Nemea<br>2 Karystos, Orchomenos<br>3 Attica, Cos, Naxos, Nesos, Imbros, Thespiae<br>4 Corcyra (Corfu), Hermione (Kastri)<br>5 Troezen, Chersonesus Taurica (Korsun), Chalcidice |
| 5 | 6 7 8 9 10 | 6 Epidaurus<br>7 Thera<br>8 Troezen<br>9 Attica, Corcyra, Naxos, Karystos, Nesos, Thebes, Thespiae, Chersonesus Taurica<br>10 Chalcidice, Imbros |
| 10 | 11 12 13 14<br>15 15 16 17 | 11 Argos<br>12 Nemea<br>13 Epidaurus, Karystos<br>14 Troezen<br>15 Corcyra, Hermione<br>16 Attica, Cos, Naxos, Nesos, Mytilene, Imbros, Chersonesus Taurica, Chalcidice, Thespiae<br>17 Orchomenos, Hermione |
| 50 | 18 19 20 21<br>22 23 24 24 | 18 Argos<br>19 Epidaurus, Troezen, Cos, Naxos, Karystos<br>20 Nemea, Cos, Nesos, Attica, Thebes<br>21 Imbros<br>22 Troezen<br>23 Chersonesus Taurica<br>24 Thespiae, Orchomenos |
| 100 | 25 26 27<br>28 29 30 | 25 Epidaurus, Argos, Nemea, Troezen<br>26 Attica, Thebes, Cos, Epidaurus, Corcyra, Naxos, Chalcidice, Imbros<br>27 Thespiae, Orchomenos 28 Karystos<br>29 Chersonesus Taurica<br>30 Chersonesus Taurica, Chios, Nesos, Mytilene |
| 500 | 31 32 33 34<br>35 36 37 38 | 31 Troezen<br>32 Epidaurus<br>33 Karystos<br>34 Cos<br>35 Naxos<br>36 Epidaurus<br>37 Epidaurus, Troezen, Imbros, Thebes, Attica<br>38 Thespiae, Orchomenos |
| 1,000 | 39 40 | 39 Attica, Thebes, Epidaurus, Argos, Cos, Naxos, Troezen, Karystos, Nesos, Mytilene, Imbros, Chalcidice, Chersonesus Taurica<br>40 Thespiae, Orchomenos |
| 5,000 | 41 42 | 41 Attica, Cos, Thebes, Epidaurus, Troezen, Chalcidice, Imbros<br>42 Thespiae, Orchomenos |
| 10,000 | 43 43 44 | 43 Attica, Epidaurus, Chalcidice, Imbros, Thespiae, Orchomenos<br>44 Attica |
| 50,000 | 45 46 | 45 Attica<br>46 Imbros |

FIG. 16.11. *Table of the numerical signs found in various Greek inscriptions of the period 1500–1000 BCE, used to express sums of money (in general, the numbers shown here refer to amounts in drachmas). When they are collected together as here we can see the common origin of all of the Greek acrophonic numerals which were in use at this time. [Tod]*

Bringing together all the different systems, we can observe their common origin (Fig. 16.11A and B).

FIG. 16.12. *The Ancient Greek world*

Looking now at Fig. 16.14, 16.15, and 16.16, we can see that the original number-systems were quite similar to the Egyptian hieroglyphic system and to the Cretan and Hittite systems.

The inconvenient feature of this kind of notation, in that it required multiple repetitions of identical symbols, led the Greeks to seek a simplification by assigning a specific sign to each of the numbers:

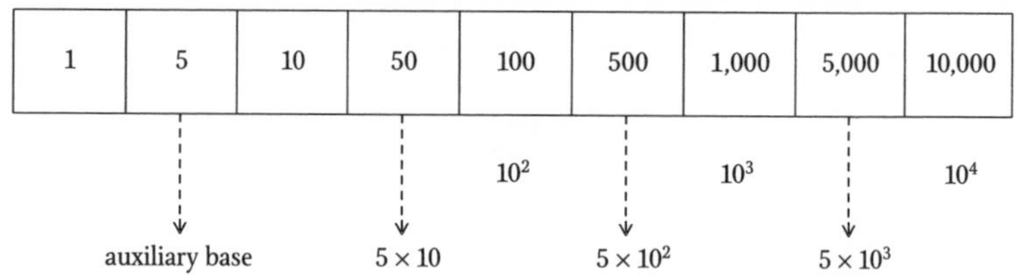


FIG. 16.13.

| •<br>1 drachma | –<br>10 drachmas | H<br>100 drachmas | X<br>1,000 drachmas |
|---|---|---|---|
| 2 •• | 20 = | 200 HH | 2,000 XX |
| 3 ••• | 30 =– | 300 HHH | 3,000 XXX |
| 4 •••• | 40 == | 400 HHHH | 4,000 XXXX |

FIG. 16.14A.

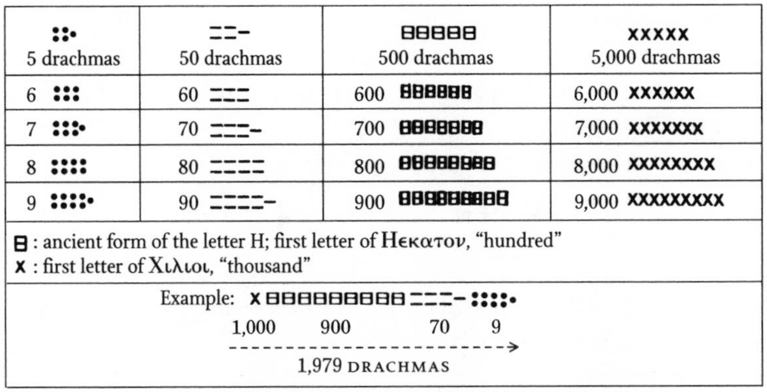

| ::• 5 drachmas | ==- 50 drachmas | ⊟⊟⊟⊟⊟ 500 drachmas | XXXXX 5,000 drachmas |
|---|---|---|---|
| 6 ::: | 60 === | 600 ⊟⊟⊟⊟⊟⊟ | 6,000 XXXXXX |
| 7 :::• | 70 ===- | 700 ⊟⊟⊟⊟⊟⊟⊟ | 7,000 XXXXXXX |
| 8 :::: | 80 ==== | 800 ⊟⊟⊟⊟⊟⊟⊟⊟ | 8,000 XXXXXXXX |
| 9 ::::• | 90 ====- | 900 ⊟⊟⊟⊟⊟⊟⊟⊟⊟ | 9,000 XXXXXXXXX |

⊟ : ancient form of the letter H; first letter of Ηεκατον, "hundred"
X : first letter of Χιλιοι, "thousand"

Example: X ⊟⊟⊟⊟⊟⊟⊟⊟⊟ ===- ::::•
1,000 900 70 9
1,979 DRACHMAS

Fig. 16.14b. *System of numerical notation in Ancient Greek inscriptions from Epidaurus (beginning of the fourth century BCE). This system, which is based on exactly the same principle as the Cretan number-systems, and is acrophonic for the numbers 100 and 1,000 only, has no symbols for 5, 50, 500, or 5,000, [Tod]*

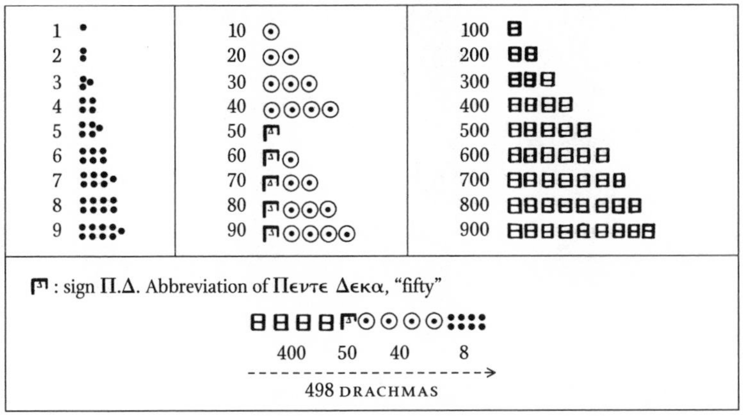

| | | |
|---|---|---|
| 1 • | 10 ⊙ | 100 ⊟ |
| 2 : | 20 ⊙⊙ | 200 ⊟⊟ |
| 3 :• | 30 ⊙⊙⊙ | 300 ⊟⊟⊟ |
| 4 :: | 40 ⊙⊙⊙⊙ | 400 ⊟⊟⊟⊟ |
| 5 ::• | 50 𐅄 | 500 ⊟⊟⊟⊟⊟ |
| 6 ::: | 60 𐅄⊙ | 600 ⊟⊟⊟⊟⊟⊟ |
| 7 :::• | 70 𐅄⊙⊙ | 700 ⊟⊟⊟⊟⊟⊟⊟ |
| 8 :::: | 80 𐅄⊙⊙⊙ | 800 ⊟⊟⊟⊟⊟⊟⊟⊟ |
| 9 ::::• | 90 𐅄⊙⊙⊙⊙ | 900 ⊟⊟⊟⊟⊟⊟⊟⊟⊟ |

𐅄 : sign Π.Δ. Abbreviation of Πεντε Δεκα, "fifty"

⊟⊟⊟⊟𐅄⊙⊙⊙⊙::::
400 50 40 8
498 DRACHMAS

Fig. 16.15. *System of numerical notation in Greek inscriptions from Nemaea (fourth century BCE): a decimal system with a supplementary sign for 50 only [Tod]*

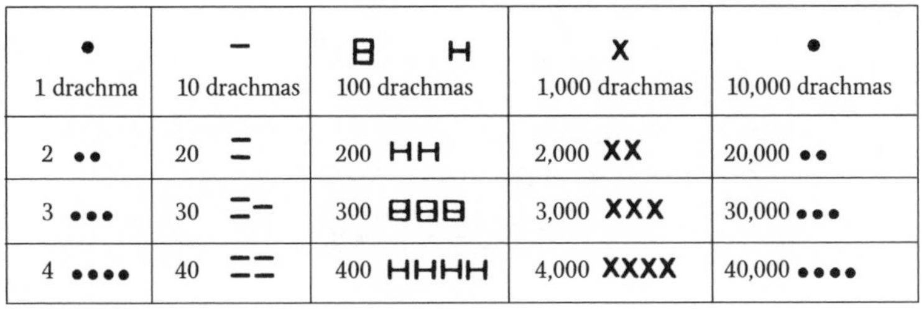

| • 1 drachma | – 10 drachmas | ⊟ H 100 drachmas | X 1,000 drachmas | ● 10,000 drachmas |
|---|---|---|---|---|
| 2 •• | 20 = | 200 HH | 2,000 XX | 20,000 ●● |
| 3 ••• | 30 =– | 300 ⊟⊟⊟ | 3,000 XXX | 30,000 ●●● |
| 4 •••• | 40 == | 400 HHHH | 4,000 XXXX | 40,000 ●●●● |

Fig. 16.16a.

| Γ• 5 drachmas | 𐅄 or 𐅄 50 drachmas | 𐅅 or 𐅅 or 𐅅 500 drachmas | 𐅆 5,000 drachmas | ? ● |
|---|---|---|---|---|
| 6 Γ•• | 60 𐅄– | 600 𐅅⊟ | 6,000 𐅆X | |
| 7 Γ••• | 70 𐅄= | 700 𐅅HH | 7,000 𐅆XX | |
| 8 Γ•••• | 80 𐅄=– | 800 𐅅⊟⊟⊟ | 8,000 𐅆XXX | |
| 9 Γ••••• | 90 𐅄== | 900 𐅅HHHH | 9,000 𐅆XXXX | |

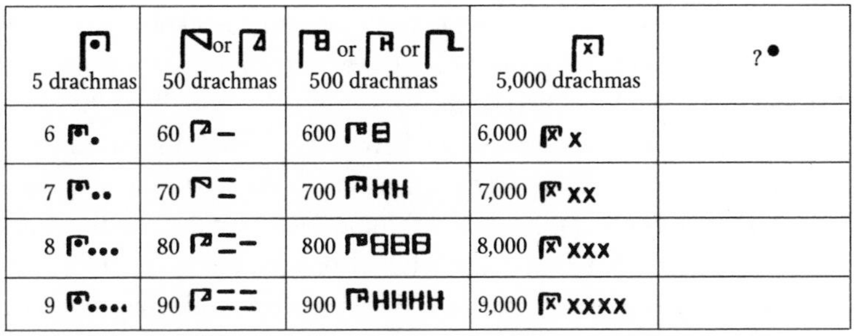

Fig. 16.16b. *Numerals in late inscriptions from Epidaurus (end of fourth to middle of third centuries BCE) [Tod]*

They therefore arrived at a mathematical system equivalent to the one used by the South Arabs and the Romans.

Thereafter, no more than fifteen different signs were required in order to represent the number 7,699 for example, instead of the thirty-one that were needed in the Cretan and the archaic Greek system.

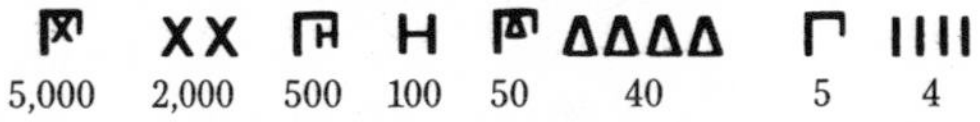

| 𐅆 | XX | 𐅅 | H | 𐅄 | ΔΔΔΔ | Γ | IIII |
|---|---|---|---|---|---|---|---|
| 5,000 | 2,000 | 500 | 100 | 50 | 40 | 5 | 4 |

Fig. 16.17.

Nevertheless, this advance in notation was a step backwards in the evolution of arithmetic itself. In the beginning, the Greeks had assigned specific symbols only to unity and to each power of the base, and they were able to do written arithmetic after the fashion of the Egyptians. But once they had introduced supplementary figures into their initial set, the Greeks deprived it of all operational capability. As result, the Greek calculators thenceforth had to resort to "counting tables".

## THE NUMBERS OF THE KINGDOM OF SHEBA

We now consider the numerical notation used by the ancient people of South Arabia, especially the Minaeans and the Shebans who shared what is now Yemen during the first millennium BCE. [M. Cohen (1958); J. C. Février (1959); M. Höfner (1943)]

The inscriptions which have come down to us from these peoples concern the most varied subjects: buildings constructed on several floors, irrigation systems retained by large dikes, offerings to the astral gods, animal sacrifices, tales of conquest, inventories of booty, and so forth. The writing, in which were written the neighbouring Semitic Arab languages, was no doubt derived (with some major changes) from the ancient Phoenician writing and had twenty-nine consonants represented by characters of geometric form, almost all of the same size. [M. Cohen (1958); J. G. Février (1959); M. Rodinson (1963)]

The system used by these people was based on the additive principle. A distinct symbol was assigned to each power of 10, and also to the number five and to the number fifty (Fig. 16.19).

Like the Greek systems which we have just analysed, this system was acrophonic in nature. Except for the signs for 1 and 50, all the others are letters of the alphabet, and are in fact the initial letters of the Semitic names of the numbers 5, 10, 100, and 1,000. (Quite possibly the South Arabs were influenced in this respect by the Greeks. This is conjectural, though we do in fact know from other studies that there were contacts between the Greeks, the Shebans, and the Minaeans.)

| | | |
|---|---|---|
| 1 | | Simple vertical bar |
| 5 | (a) or (b) | Letter ḪA: first letter of ḪAMSAT, Southern Arabic word for "five" |
| 10 | | Letter 'AYIN: first letter of the word 'ASARAT, "ten" |
| 50 | (a) or (b) | Half of the sign for 100 |
| 100 | (a) or (b) | Letter MIM: first letter of the word MI'AT, "hundred" |
| 1,000 | (a) or (b) | Letter 'ALIF, first letter of the word 'ALF, "thousand" |
| (a) reading from left to right | | (b) reading from right to left |

FIG. 16.18.

In the Minaean and Sheban inscriptions, numerals are usually enclosed between a pair of signs and in order to avoid confusion between letters representing numbers and letters standing for themselves (Fig. 16.22 and 16.24). It often happens, also, that the figures change orientation within the same inscription, since the South Arab writing was in *boustrophedon* (alternately from right to left and from left to right).

| | | | |
|---|---|---|---|
| 1 | 10 | 100 | 1,000 |
| 2 | 20 | 200 | 2,000 |
| 3 | 30 | 300 | 3,000 |
| 4 | 40 | 400 | 4,000 |
| 5 | 50 | 500 | 5,000 |
| 6 | 60 | 600 | 6,000 |
| 7 | 70 | 700 | 7,000 |
| 8 | 80 | 800 | 8,000 |
| 9 | 90 | 900 | 9,000 |

FIG. 16.19. *The symbols, and the principle, of the Southern Arabian number-system. This system is known only from the period from the fifth to the second or first centuries BCE. On inscriptions dating from after the beginning of the Common Era, it seems, numbers are spelled out in full.*

There is one interesting and important difference between the number-system of the South Arabs – at any rate those of Sheba – and the otherwise similar Greek system, in that the Arab system incorporated a rudimentary principle of position.

In fact, when one of the figures

10    50 (or)    100 (or)

FIG. 16.20.

is placed to the right of the sign for 1,000 (when reading from right to left), then this figure is (mentally) multiplied by 1,000. In the following, for example:

2,000   30   50   200 ←

FIG. 16.21.

we would at first be inclined to read the value

$$200 + 50 + 30 + 2{,}000 = 2{,}280$$

according to the traditional usage of the additive principle, whereas in fact it represented, in Sheba, the value

$$(200 + 50 + 30) \times 1{,}000 + 2{,}000 = 282{,}000$$

| | | | |
|---|---|---|---|
| CIS IV: inscr. 924 | ← | 5 | Notice the irregularity |
| RES: inscr. 2740, 1.7 | ← | 50 | |
| RES: inscr. 2868, 1.4 | ← | 60 | |
| RES: inscr. 2743, 1.10 | ← | 63 | |
| RES: inscr. 2774, 1.4 | ← | 47 | |
| RES: inscr. 2965, 1.4 | ← | 180 | Note the unusual manner of writing the number 30: instead of ○○○ |

FIG. 16.22. *Examples taken from Minaean inscriptions (third to first century BCE). The numbers shown above refer to the volume capacity of certain recipients offered to the astral gods of ancient Southern Arabia, or to lists of offerings to these gods, or to animals which have been sacrificed. [C. Robin (personal communication)].*

Similarly, when reading from left to right, the same effect is produced by placing the figure to the left of the sign for 1,000. Thus

200 50   30   2,000 →

FIG. 16.23.

gives 282,000 (and not 2,280!).

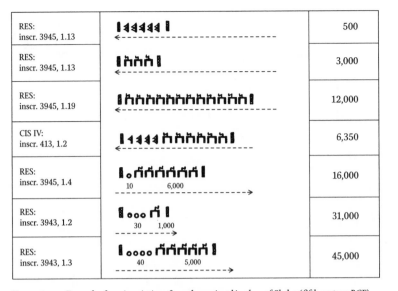

| Source | Notation | Value |
|---|---|---|
| RES: inscr. 3945, 1.13 | ← | 500 |
| RES: inscr. 3945, 1.13 | ← | 3,000 |
| RES: inscr. 3945, 1.19 | ← | 12,000 |
| CIS IV: inscr. 413, 1.2 | ← | 6,350 |
| RES: inscr. 3945, 1.4 | 10 6,000 → | 16,000 |
| RES: inscr. 3943, 1.2 | 30 1,000 → | 31,000 |
| RES: inscr. 3943, 1.3 | 40 5,000 → | 45,000 |

FIG. 16.24. *Examples from inscriptions from the ancient kingdom of Sheba (fifth century BCE). These inscriptions, principally from the site of Sirwah, tell of military conquests and give various inventories: numbers of soldiers, material resources, booty, prisoners, and so on. [C. Robin, personal communication]*

However, this practice must have surely given rise to confusion among the readers, and the Sheban stone-cutters therefore took the precaution of also writing out in words the number represented by the figures.

A lucky precaution, for it has enabled us today to arrive at an unambiguous interpretation of this number-system!

## ROMAN NUMERALS

Like the preceding systems, the Roman numerals allowed arithmetical calculation only with the greatest difficulty.

To be convinced of this, let us try to do an addition in these figures. Without translating into our own system, it is very difficult, if not impossible, to succeed.

The example which is most often cited is the following:

| | | | |
|---|---|---|---|
| | CCXXXII | | 232 |
| + | CCCCXIII | + | 413 |
| + | MCCXXXI | + | 1,231 |
| + | MDCCCLII | + | 1,852 |
| = | MMMDCCXXVIII | | 3,728 |

Roman numerals, in fact, were not signs which supported arithmetic operations, but simply abbreviations for writing down and recording numbers. This is why Roman accountants, and the calculators of the Middle Ages after them, always used the abacus with counters for arithmetical work.

As with the majority of the systems of antiquity, Roman numerals were primarily governed by the principle of addition. The figures (I= 1, V= 5, X= 10, L = 50, C = 100, D = 500 and M = 1,000) being independent of each other, placing them side by side implied, generally, addition of their values:

CLXXXVII = 100 + 50 + 10 + 10 + 10 + 5 + 1 + 1 = 187
MDCXXVI = 1,000 + 500 + 100 + 10 + 10 + 5 + 1 = 1,626

The Romans proceeded to complicate their system by introducing a rule according to which every numerical sign placed to the left of a sign of higher value is to be subtracted from the latter.

Thus the numbers 4, 9, 19, 40, 90, 400, and 900, for example, were often written in the forms

IV (= 5 – 1) instead of IIII
IX (= 10 – 1) instead of VIIII
XIX (= 10 + 10 –1) instead of XVIIII
XL (= 50 – 10) instead of XXXX
XC (= 100 – 10) instead of LXXXX
CD (= 500 – 100) instead of CCCC
CM (= 1,000 – 100) instead of DCCCC

It is remarkable that a people who, in the course of a few centuries, attained a very high technical level, should have preserved throughout that time a system which was needlessly complicated, unusable, and downright obsolete in concept.

In fact, the writing of the Roman numerals as well as its simultaneous use of the contradictory principles of addition and subtraction, are the vestiges of a distant past before logical thought was fully developed.

Roman numerals as we know them today seem at first sight to have been modelled on the letters of the Latin alphabet:

| I | V | X | L | C | D | M |
|---|---|---|---|---|---|---|
| 1 | 5 | 10 | 50 | 100 | 500 | 1,000 |

FIG. 16.25.

However, as T. Mommsen (1840) and E. Hübner (1885) have remarked, these graphic signs are not the first forms of the figures in this system. They were in fact preceded by much older forms which had nothing to do with letters of an alphabet. They are late modifications of much older forms.*

* The oldest known instances of the use of the letters L, D and M as numerals do not go back earlier than the first century BCE. As far as we know, the earliest Roman inscription which uses the letter L for 50 dates only from 44 BCE (CIL, I, inscr. 594). The earliest known use of the numerals M and D is in a Latin inscription which dates from 89 BCE, in which the number 1,500 is written as MD (CIL, IV, inscr. 590).

Originally, 1 was represented by a vertical line, the number 5 by a drawing of an acute angle, 10 by a cross, 50 by an acute angle with an additional vertical line, 500 by a semi-circle at an angle, and 1,000 by a circle with a superimposed cross (of which the *denarii* figure for 500 is geometrically one half):

| | I | V | X | ⩛ | ✱ | ◬ | ⊠ |
|---|---|---|---|---|---|---|---|
| FIG. 16.26. | 1 | 5 | 10 | 50 | 100 | 500 | 1,000 |

In an obvious way, the original figures for 1, 5 and 10 were assimilated to the letters I, V and X.

The original figure for 50 (which can still be found as late as the reign of Augustus, 27 BCE – 14 CE*) evolved progressively as shown below, finally merging with the letter L around the first century BCE:

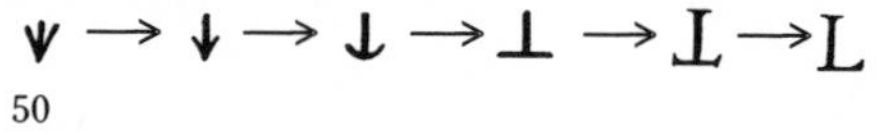

⩛ → ↓ → ⊥ → ⊥ → ⊥ → L

FIG. 16.27. 50

The original figure for 100 initially evolved in a similar way towards a more rounded form: ƆK and then, for the sake of abbreviation, was split into one or other of the forms Ɔ or C. By similarity of shape, and under the influence of the initial letter of the Latin word *centum* ("one hundred"), it was finally assimilated to the letter C.

The original figure for 500 first of all underwent an anticlockwise rotation of 45°. It then evolved towards the sign Ɓ (these signs can still be found on texts from the Imperial period†) and finally was assimilated to the letter D:

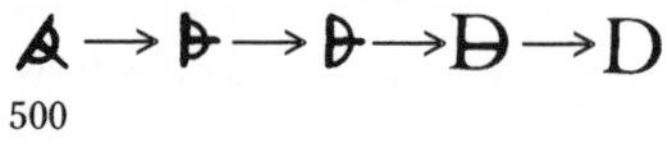

◬ → Ɓ → Ɓ → Ɓ → D

FIG. 16.28. 500

The figure for 1,000 first of all evolved towards the form ↀ. This gave rise to the many variant forms shown below for which, progressively, the letter M came to be substituted, from the first century BCE, under the influence of the first letter of the Latin word *mille*:

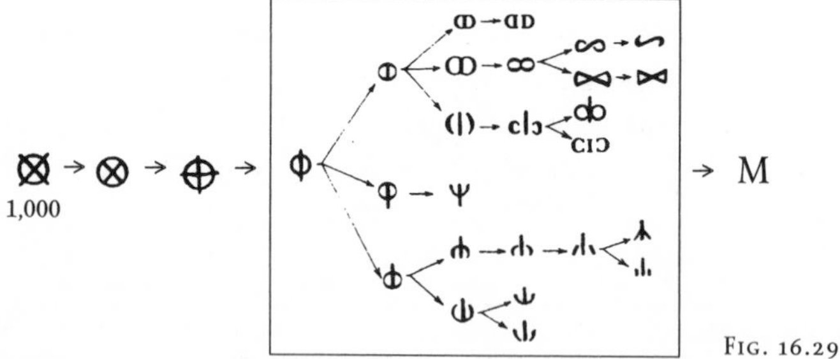


FIG. 16.29.

| | | |
|---|---|---|
| 1 | I | CIL I 638, 1449 |
| 2 | II | CIL I 638, 744 |
| 3 | III | CIL I 1471 |
| 4 | IIII | CIL I 638, 587, 594 |
| 5 | Λ or V | CIL I 1449 |
| | | CIL I 590, 809, 1449, 1479, 1853 |
| 6 | VI | CIL I 618 |
| 7 | VII | CIL I 638 |
| 8 | VIII | CIL I 698, 1471 |
| 9 | VIIII | CIL I 594, 590 |
| 10 | X | CIL I 638, 594, 809, 1449 |
| 14 | XIIII | CIL I 594 |
| 15 | XV | CIL I 1479 |
| 19 | XVIIII | CIL I 809 |
| 20 | XX | CIL I 638 |
| 24 | XXIIII | CIL I 1319 |
| 40 | XXXX | CIL I 594 |
| 50 | ↓ | CIL I 214, 411 and 450 |
| | or ⊥ | CIL I 1471, 638, 1996 |
| | or ⊥ | CIL I 617, 1853 |
| | or ⊥ | CIL I 744, 1853 |
| | or L | CIL I 594, 1479 and 1492 |
| 51 | ⊥ | CIL I 638 |
| 74 | ⊥ XXIIII | CIL I 638 |
| 95 | LXXXXV | CIL I 1479 |
| 100 | C | CIL I 638, 594, 25, 1853 |
| 100 | Ɔ | CIL VIII 21, 701 |
| 300 | CCC | CIL I 1853 |
| 400 | CCCC | CIL I 638 |
| 500 | Ɓ | CIL I 638, 1533 and 1853 |
| | or D | CIL I 590 |
| 837 | ƁCCCXXXVII | CIL I 638 |
| 1,000 | ↀ | CIL I 1533, 1578, 1853 and 2172 |
| | or ↀ | CIL X 39 |
| | or ∞ | CIL I 594 and 1853 |
| | or ⋈ | CIL X 1019 |
| | or ⋈ | CIL VI 1251a |
| | or M | CIL I 593 |
| | M | CIL I 590 |
| 1,200 | ∞ CC | CIL I 594 |
| 1,500 | MD | CIL I 590 |
| 2,000 | ∞∞ | CIL I 594 |
| 2,320 | ∞∞ CCCXX | CIL I 1853 |
| 3,700 | ↀↀↀ ƁCC | CIL I 25 |
| 5,000 | ↁ | CIL X 817 |
| | or ↁ | CIL I 1853 and 1533 |
| | or ↁ | CIL I 2172 |
| 5,000 | IƆƆ | CIL I 590, 594 |
| 7,000 | ↁ ↀↀ | CIL I 2172 |
| 8,670 | ↁ ↀↀↀ DCLXX | CIL I 1853 |
| 10,000 | (\|) | CIL I 1252, 198 |
| | or ↂ | CIL I 583 |
| | or CCIƆƆ | CIL I 1474 |
| | or ↂ | CIL I 744 |
| | or ↂ | CIL I 1724 |
| 12,000 | ↂ ↀↀ | CIL I 1578 |
| 21,072 | ↂↂↀ LXXII | CIL I 744 |
| 30,000 | CCIƆƆ CCI ƆƆ CCIƆƆ | CIL I 1474 |
| 30,000 | ↂↂↂ | CIL I 1724 |
| 50,000 | IƆƆƆ | CIL I 593 |
| 100,000 | | CIL I 801 |
| | or | CIL I 801 |
| | or CCCIƆƆƆ | CIL I 594 |

FIG. 16.30. *Written numbers from monumental Latin inscriptions, dating from the Republican and early Imperial periods*

* CIL, IV, inscr. 9934

† CIL, VIII, inscr. 2557

The various forms associated with the number 1,000 in Fig. 16.29 were mainly used during the period of the Republic, but they can also be found in some texts of the Imperial period.* A few of them even survived long after the fall of the Roman civilisation, since they can be found in quite a few printed works from the seventeenth century (Fig. 16.69 and 16.70).

| 28 | XXIIX | CIL I 1319 | 140 | CXL | CIL I 1492 |
|---|---|---|---|---|---|
| 45 | X⊥V | CIL I 1996 | 268 | C⊥XIIX | CIL I 617 |
| 69 | LXIX | CIL I 594 | 286 | CCXXCVI | CIL I 618 |
| 74 | LXXIV | CIL I 594 | 340 | CCCX⊥ | CIL I 1529 |
| 78 | LXXIIX | CIL I 594 | 345 | CCCX⊥V | CIL I 1853 |
| 79 | LXXIX | CIL I 594 | 1,290 | ∞ CCXC | CIL I 1853 |

FIG. 16.31. *Latin inscriptions from the Republican era showing the use of the principle of subtraction. Use of this principle (which undoubtedly reflects the influence of the popular system on the monumental system) was nevertheless unusual on well-styled inscriptions.*

VIAM·FECEI·AB·REGIO· AD·CAPVAM·ET
IN·EA·VIA·PONTEIS·OMNEIS·MILIARIOS
TABELARIOSQVE·POSEIVEI·HINCE·IS·VNT
NOVCERIAM·MEILIA·LI CAPVAM·XXCIIII
MVRANVM·↓XXIIII·COSENTIAM·CXXIII
VALENTIAM·C↓XXX AD·FRETVM·AD
STATVAM CXXXI· REGIVM·CCXXXVII
SVMA·AF·CAPVA·REGIVM·MEILIA·CCC
ET·EIDEM·PRAETOR·IN XXI
SICILIA·FVGITEIVOS·ITALICORVM
CONQVAEISIVEI·REDIDEIQVE
HOMINES·ÐCCCCXVII· EIDEMQVE
PRIMVS·FECEI·VT·DE·AGRO·POPLICO
ARATORIBVS·CEDERENT·PAASTORES
FORVM·AEDISQVE·POPLICAS·HEIC·FECEI

FIG. 16.32A. *Milestone engraving found at the Forum Popilii in Lucania (southern Italy), and made by C. Popilius Laenas, Consul in 172 BCE and 158 BCE. Now in the Museo della Civiltà Romana, Rome. [CIL, I, 638]*

| line 4 | ↓I | 51 | line 7 | CCXXXI | 231 |
|---|---|---|---|---|---|
| line 4 | XXCIIII | 84 | line 7 | CCXXXVII | 237 |
| line 5 | ↓ XXIIII | 74 | line 8 | CCCXXI | 321 |
| line 5 | CXXIII | 123 | line 12 | ÐCCCCXVII | 917 |
| line 6 | C ↓ XXX | 180 | | | |

FIG. 16.32B. *Written numbers on the inscription shown in Fig. 16.32 A*

* CIL, IV, inscr. 1251; CIL, X, inscr. 39 and 1019; CIL, IL, inscr. 4397: etc.

FIG. 16.33. Elogium *of Duilius, who conquered the Carthaginians at the battle of Mylae, 260 BCE. The inscription was re-cut at the start of the Imperial period, during the reign of Claudius (41–54 CE), in the style of the third century BCE. Found in the Roman Forum at the place of the rostra* (columna rostrata)*, and now in the Palazzo dei Conservatori in Rome [CIL, I, 195].*

*In lines 15 and 16, the figure for 100,000 is repeated at least 23 times (and at most 33, according to the restoration by the* Corpus*). On line 13, the number 3,700 is written in the form:*

ↀↀↀ DCC

*Note: The capital letters (in upright characters in the figure) correspond to that part of the inscription which remains intact. The italic letters correspond to the restoration (by the* Corpus*) of the part which has been damaged.*

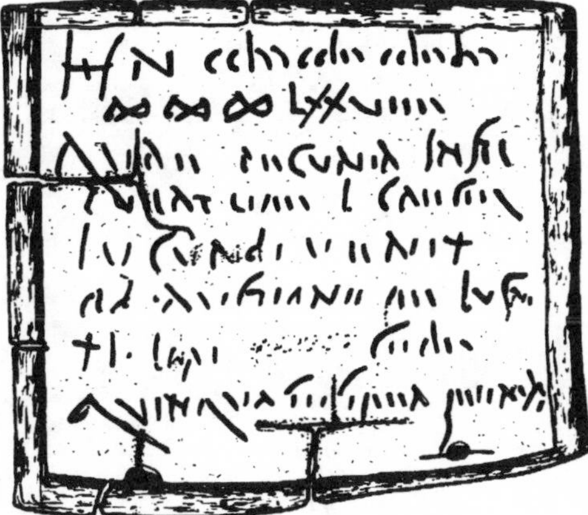

TRANSCRIPTION

HS n. ccIↄↄ ccIↄↄ ccIↄↄ Iↄↄ

∞ ∞ ∞ LXXVIII*

quæ pecunia in stipulatum
L. Caecili Iucundi venit ob
auctione (m) M. Lucreti Leri
[mer] cede quinquagesima
minu [s]

* See Fig. 16.29 and 16.30

FIG. 16.34. *Second panel of a triptych found at Pompeii, therefore prior to 79 CE (the year the city was destroyed)*

## ETRUSCAN NUMERALS

Roman numerals reached their standardised form, identical to letters of the Latin alphabet, late in the history of Rome; but in reality they began life many hundreds of years, maybe even thousands, before Roman civilisation, and they were invented by others.

The Etruscans, a people whose origins and language both remain largely unknown, dominated the Italian peninsula from the seventh to the fourth century BCE, from the plain of the Po in the north to the Campania region, near Naples, in the south. They vanished as a distinct people at the time of the Roman Empire, becoming assimilated into the population of their conquerors.

Several centuries before Julius Caesar the Etruscans, and the other Italic peoples (the Oscans, the Aequians, the Umbrians, etc.), had in fact invented numerals with form and structure identical to those of the archaic Roman numerals.

| | | |
|---|---|---|
| 1 | I | CIE 5710 |
| 2 | II | CIE 5708<br>TLE 26 |
| 3 | III | CIE 5741 |
| 4 | IIII | CIE 5748 |
| 5 | Λ or ∩ | CIE 5705, 5706, 5683, 5677 and 5741<br>ACII, Table IV 114 |
| 6 | IΛ | CIE 5700 |
| 7 | IIΛ | CIE 5635 |
| 8 | III∩ | ACII, Table IV 114 |
| 9 | IIIIΛ | CIE 5673 |
| 10 | X or X or + | CIE 5683, 5741, 5710 5748, 5695, 5763, 5797 5707, 5711 and 5834<br>CIE 5689 and 5677<br>TLE 126 |
| 19 | XIX | CIE 5797 |
| 36 | IΛXXX | CIE 5683 |
| 38 | IIIΛXXX | CIE 5741 |
| 38 | XIIXXX | CIE 5707 |
| 42 | IIX XXX | CIE 5710 |
| 44 | IIIIXXXX | CIE 5748 |
| 50 | ↑ or ↑ or ↑ | CIE 5708, 5695, 5705, 5706, 5677 and 5763<br>Buonamici, p. 245 |
| 52 | II↑ | CIE 5708 |
| 55 | Λ↑ | CIE 5705 and 5706 |
| 60 | X↑ | CIE 5695 |
| 75 | ΛX·X↑ | CIE 5677 |
| 82 | II+++↑ | TLE 26 |
| 86 | IIIIIIXXX↑ | CIE 5763 |
| 100 | ✱ or ✱ or ✱ | ACII, Table IV 114<br>Buonamici, p. 473 |
| 106 | IΛ✱ | SE, XXIII, series II (1965), p. 473 |

FIG. 16.35. *Written numbers from Etruscan inscriptions*

FIG. 16.36. *Etruscan coins dating from the fifth century BCE bearing the numbers*
Λ and X
5 10
*Collection of the Landes museum, Darmstadt [Menninger (1957) vol. II, p. 48]*

For many centuries, they used these figures according to the principles of addition and subtraction simultaneously. This is evidenced by several Etruscan inscriptions of the sixth century BCE, where the numbers 19 and 38 are written on the subtractive principle as 10 + (10 – 1) and 10 + 10 + 10 + (10 – 2) (Fig. 16.35).

FIG. 16.37. *Fragments of an Etruscan inscription bearing the numbers:*

X↑X 160 IIIΛXX 208 ΛX 15

[ACII table IV 114]

## A QUESTIONABLE HYPOTHESIS

A hypothesis commonly accepted nowadays asserts that all of these numerals derived from Etruscan numerals, themselves of Greek origin.

We should recall that Latin writing derived from Etruscan writing, and that this comes directly from Greek writing. The Greek alphabets fall into two groups: the Western type which (like the Chalcidean alphabet, for example) assigned the sound "kh" to the letter Ψ or ↓ or Ѵ; the Eastern type which (like the alphabet of Miletus or Corinth, for example) assigned to this symbol the sound "ps", while the sound "kh" is represented by the letter + or x. Etruscan writing, for several reasons, is associated with the Western type.

Therefore it has come to be supposed that the Etruscan alphabet "was borrowed from a Greek alphabet of Western type on the land of Italy itself, since the oldest of the Greek colonies which had such an alphabet, that of Kumi, dates from 750 BCE, and its establishment precedes the birth of the Tuscan civilisation by half a century." [R. Bloch (1963)]

On this basis, having compared the forms of the letters, many specialists in the Roman numbering system have therefore inferred that the ancient Latin signs for the numbers 50, 100 and 1,000 come respectively from the following letters, which belong to the Chalcidean alphabet (a Greek alphabet of Western type used, as it happens, in the Greek colonies in Sicily). These letters represented sounds which did not occur in Etruscan or in Latin, and later became assimilated to the Latin forms which we know.

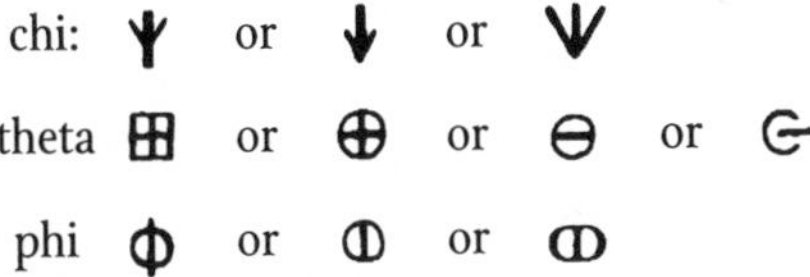

According to this hypothesis, the Greek letter *theta* Θ (originally ⊞ or ⊕) gradually turned into C, under the influence of the initial letter of the Latin word *centum*.

This explanation (which many Hellenists, epigraphers, and historians of science now hold as dogma) is seductive, but it cannot be accepted.

Why, in fact, should three particular foreign characters be introduced into the Roman number-system, and three only? And why should they be letters of the alphabet? No doubt, one may reply, because the Greeks themselves had often used letters of their alphabet as number-signs.

In antiquity, it is true, the Hellenes used two different systems of written numerals whose figures were in fact the letters of their alphabet. One of these used the initial letters of the names of the numbers. The other made use of all the letters of the alphabet (see Fig. 17.27 below):

| | | | | | | | | |
|---|---|---|---|---|---|---|---|---|
| A | Alpha | 1 | I | Iota | 10 | P | Rho | 100 |
| B | Beta | 2 | K | Kappa | 20 | Σ | Sigma | 200 |
| Γ | Gamma | 3 | Λ | Lambda | 30 | T | Tau | 300 |
| Δ | Delta | 4 | M | Mu | 40 | Y | Upsilon | 400 |
| E | Epsilon | 5 | N | Nu | 50 | Φ | Phi | 500 |
| . . . | . . . | . . . | Ξ | Xi | 60 | X | Chi | 600 |
| H | Eta | 8 | O | Omicron | 70 | Ψ | Psi | 700 |
| Θ | Theta | 9 | . . . | . . . | . . . | Ω | Omega | 800 |

Now the letter *chi*, which was supposed to be borrowed for the number 50 in Latin, has value 1,000 in the first of these systems, and 600 in the second; the letter *theta*, "borrowed" for the number 100 in Latin, has value 9 in the second Greek version; and the letter *phi*, supposed to have been borrowed for the Roman numeral for 1,000, is worth 500 in the second system. Why the differences?

If the Romans had borrowed the following Greek signs for the numbers 50 and 100:

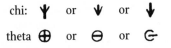

chi: Ψ or ↓ or ↓

theta ⊕ or ⊖ or ⊖

then the same would probably have been borrowed by the Etruscans as well. How then can we explain that, for the same values, the Etruscans in fact used quite different figures, namely (see Fig. 16.35):

↑ or ↑ for 50

and ✱ or ✱ for 100

One can see that the hypothesis is not very sound. The error is due to the fact that specialists have believed through many generations that Roman numerals are the children of Etruscan numerals, whereas in fact they are cousins.

## THE ORIGIN OF ROMAN NUMERALS

Though long obscure, the question is no longer in doubt. The signs I, V and X are by far the oldest in the series. Older than any kind of writing, older therefore than any alphabet, these figures, and their corresponding values, come naturally to the human mind under certain conditions. In other words, the Roman and Etruscan numerals are real prehistoric fossils: they are descended directly from the principle of the notched stick for counting, a primitive arithmetic, performed by cutting notches on a fragment of bone or on a wooden stick, which anyone can use in order to establish a one-to-one correspondence between the objects to be counted and the objects used to count them.

Let us imagine a herdsman who is in the habit of noting the number of his beasts using this simple prehistoric method.

Up to now, he has always counted as his forebears did, cutting in a completely regular manner as many notches as there are beasts in his herd. This is not very useful, however, because whenever he wants to know how many beasts he has, he has to count every notch on his stick, all over again.

The human eye is not a particularly good measuring instrument. Its capacity to perceive a number directly does not go beyond the number 4. Just like everyone else, our herdsman can easily recognise at a glance, without counting, one, two, three, or even four parallel cuts. But his intuitive perception of number stops there for, beyond four, the separate notches will be muddled in his mind, and he will have to resort to a procedure of abstract counting in order to learn the exact number.

Our herdsman, who has perceived the problem, is beginning to look for a way round it. One day, he has an idea.

As always, he makes his beasts pass by one by one. As each one passes, he makes a fresh notch on his tally stick. But this time, once he has made four marks he cuts the next one, the fifth, differently, so that it can be recognised at a glance. So at the number 5, therefore, he creates a new unit of counting which of course is quite familiar to him since it is the number of fingers on one hand.

For any individual, cutting into wood or bone presents the same problems, and will lead to the same solutions, whether in Africa or Asia, in Oceania, in Europe or in America.

Our herdsman only has a limited number of options. To distinguish the fifth notch from the first four, the first idea he has is simply to change the direction of cut. He therefore sets this one very oblique to the other four, and thereby obtains a representation all the more intuitive in that it reflects the angle that the thumb makes with the other four fingers.

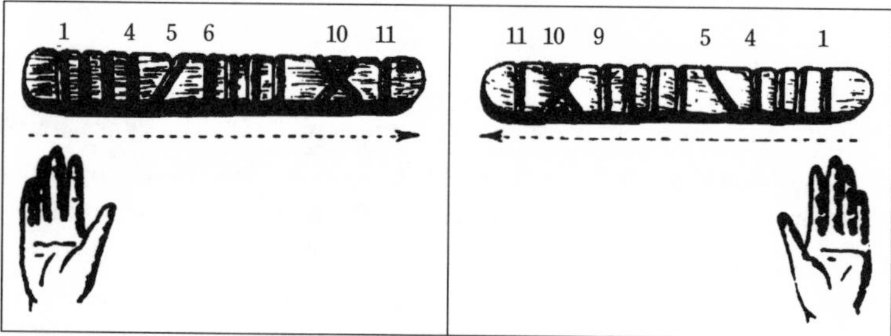


FIG. 16.38.

Another idea is to augment the fifth notch by adding a small supplementary notch (oblique or horizontal), so that the result is a distinctive sign in the form of a "t", a "Y" or a "V", variously oriented:

V Λ < > Y Y ⊣ ⊢ Y ⋋ ⋌

He resumes cutting notches in the same way as the first four, counting his beasts up to the ninth. But, at the tenth, he finds he must once again modify the notch so that it can be recognised at a glance. Since this is the total number of fingers on the two hands together, he therefore thinks of a mark which shall be some kind of double of the first. And so, as in all the numeral systems, he comes to make a mark in the form of an "X" or a cross:

X ✗ ✗ +

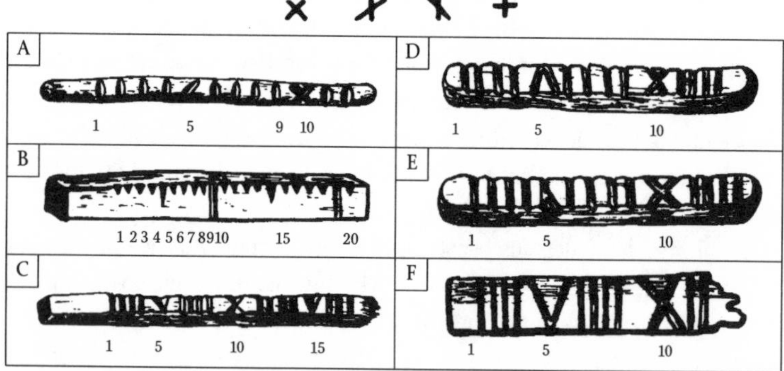


FIG. 16.39. *Anyone who counts by cutting notches on sticks will come to represent the numbers 1, 5, 10, 15, and so on in one of the above ways.*

So he has now created another numerical unit, the ten, and counting on the tally stick henceforth agrees with basic finger-counting.

Reverting to his simple notches, the herdsman continues to count beasts until the fourteenth and then, to help the eye to distinguish the fifteenth from the preceding ones, he again gives it a different form. But this time he does not create a new symbol. He simply gives it the same form as the "figure" 5, since it is like "one hand after the two hands together".

He carries on as before up to 19, and then he makes the twentieth the same as the tenth. Then again up to 24 with the ordinary notches, and the twenty-fifth is marked with the figure 5. And so on up to $9 + 4 \times 10 = 49$.

This time, however, he must once more imagine a new sign to mark the number 50, because he is not able to visually recognise more than four signs representing 10.

This is naturally done by adding a third cut to his notch, so he naturally chooses one of the following which can be made by adding one notch to one of the representations of the number 5:

Ѱ ⩚ И И K Ӿ 人 Y 人 Y ɦ h

Having done this, he can now proceed in the same way until he has gone through all the numbers from 50 to $50 + 49 = 99$.

At the hundredth, our herdsman once again faces the problem of making a distinct new mark. So equally naturally he will choose one of the following which can be made either by adding a further notch to one of the representations of 10, or by making a double of one of the representations of 50:

⁎ ⋈ Ӎ Ӎ Σ H Ħ K

Again as before, he continues counting up to $100 + 49 = 149$. For the next number, he re-uses the sign for 50 and then continues in the same way up to $150 + 49 = 199$.

At 200, he re-uses the figure for 100 and continues up to $200 + 49 = 249$. And so on until he reaches $99 + 4 \times 100 = 499$.

Now he creates a new sign for 500 and continues as before until $500 + 499 = 999$. Then another new sign for 1,000 which will allow him to continue the numbers up to 4,999 ($= 999 + 4 \times 1{,}000$), and so on.

And so, despite not being able to perceive visually a series of more than four similar signs, our herdsman, thanks to some well-thought-out notch-cutting, can now nonetheless perceive numbers such as 50, 100, 500, or 1,000, without having to count all the notches one by one. And if he runs out of space on his tally stick and cannot reach one of these numbers, then all he needs to do is to make as many more tally sticks as he needs.

When the notches are cut in a structured way like this, it is possible to go up to quite large numbers, as large as are likely to be needed in practice, without ever having to take account of any series of more than four signs of the same kind. Such a technique is therefore like a lever, the mechanical instrument which allows someone to raise loads whose weight far exceeds his raw physical strength.

The procedure also defines a written number-system which gives a distinct figure to each of the terms of the series

$$
\begin{aligned}
1 & \\
5 & \\
10 &= 5 \times 2 \\
50 &= 5 \times 2 \times 5 \\
100 &= 5 \times 2 \times 5 \times 2 \\
500 &= 5 \times 2 \times 5 \times 2 \times 5 \\
1{,}000 &= 5 \times 2 \times 5 \times 2 \times 5 \times 2 \\
5{,}000 &= 5 \times 2 \times 5 \times 2 \times 5 \times 2 \times 5
\end{aligned}
$$

Our herdsman's approach to cutting notches on sticks therefore gives rise to a decimal system in which the number 5 is an auxiliary base (and the numbers 2 and 5 are alternating bases), and its successive orders of magnitude are exactly the same as in the Roman system; furthermore, it will naturally give rise to graphical forms for the figures which are closely comparable with those in the archaic Roman and Etruscan systems.

Again, the use at the same time of both the additive and the subtractive principles in the Etruscan and Roman systems is yet another relic of this ancient procedure.

To return to our herdsman. Now that he has counted his various beasts under various categories, he wants to transcribe the results of this breakdown onto a wooden board. In total 144, his beasts are distributed as:

26 dairy cows
35 sterile cows
39 steers
44 bulls

In order to write down one of these numbers, say the steers, the first idea which occurs to him is to mark these by simply copying the marks of the tally stick onto the board:

| IIII | V | IIII | X | IIII | V | IIII | X | IIII | V | IIII | X | IIII | V | IIII |
|---|---|---|---|---|---|---|---|---|---|---|---|---|---|---|
| 1 | 5 | | 10 | | 15 | | 20 | | 25 | | 30 | | 35 | 39 |

FIG. 16.40.

But he soon becomes aware that such a *cardinal notation* is very tedious, because it brings in all of the successive marks made on the stick. To get around this difficulty, he therefore thinks of an ordinal kind of representation, much more abridged and convenient than the preceding one. For the numbers from 1 to 4, he at first adopts a cardinal notation writing them successively as

I II III IIII

He can hardly do otherwise for, to indicate that one of the lines is the third in the series, he must mark two others before it, in order that it shall be clear that it is indeed the third.

He does not do the same for the number 5, however, since this already has its own sign ("V", say), which distinguishes it from the preceding four. Therefore this "V" is sufficient in itself and dispenses with the need to transcribe the four notches that precede it on the tally stick. Instead of transcribing this number as IIIIV, all he needs to do is to write V.

Starting from this point, the number 6 (the next notch after the V) can be written simply VI, and not IIIIVI; the number 7 can be written as VII, and so on up to VIIII (= 9).

In turn, the sign in the shape of an "X" can represent the tenth mark in the series all on its own, and renders the nine preceding signs superfluous. On the same principle, the numbers 11, 12, 13, and 14 can be written as XI, XII, XIII, and XIIII (and not IIIIVIIIIXI, etc.). Now the number 15 can be written simply XV (and not IIIIVIIIIXIIIIV nor XIIIIV): each X can erase the nine preceding marks, and the last V the four preceding marks. The numbers from 16 to 19 can be written XVI, XVII, XVIII, XVIIII. Then, for the number 20, which corresponds to the second "X" in the series, we can write XX. And so on.

When he has counted his animals by means of the notches on his sticks, our herdsman can now transcribe the breakdown onto his wooden board:

| | | |
|---|---|---|
| XXVI | (= 26) | for the dairy cows |
| XXXV | (= 35) | for the sterile cows |
| XXXVIIII | (= 39) | for the steers |
| XXXXIIII | (= 44) | for the bulls |

However, looking for ways of shortening the work, our herdsman comes up with another idea. Instead of writing the number 4 using four lines (IIII), he writes it as IV, which is a way of marking the "I" as the fourth in the series on the stick, since this is the one that comes before the "V":

IIII → (III)IV → IV

In this way he cuts down on the number of symbols to write, saving 2. In the same way, instead of writing the number 9 as VIIII, he writes it as IX since this likewise marks the "I" as the ninth mark in the series on the notched stick:

IIIIVIIII → (IIIIVIII)IX → IX

He again cuts down on the number of symbols, saving 3. He does likewise for the numbers 14, 19, 24, and so on.

This is how one can explain why the Roman and Etruscan number-systems use forms such as IV, IX, XIV, XIX, etc., as well as IIII, VIIII, XIIII, XVIIII.

We can now conceive that all of the peoples who, for long ages, had been using the principle of notches on sticks for the purpose of counting should, in the course of time, with exactly the same motives as our herdsman and quite independently of any influence from the Romans or Etruscans themselves, be led to invent number-systems which are graphically and mathematically equivalent to the Roman and Etruscan systems.

This hypothesis seems so obvious that it could be accepted even if there were no concrete evidence for it. But such evidence exists, and in plenty.

## A REVEALING ETYMOLOGY

It is hardly an accident that the Latin for "counting" should refer to the practicalities of this primitive method of doing it.

In Latin, "to count" is *rationem putare*.* As M. Yon (cited by L. Gerschel) points out, the term *ratio* not only refers to counting,† but also has a meaning of "relationship" or "proportion between things".‡

Surely this is because, for the Romans, this word referred originally to the practice of notching, since counting, in a notch-based system, is a matter of establishing a correspondence, or one-to-one relationship, between a set of things and a series of notches. Gerschel has demonstrated this with a large number of examples.

As for the word *putare*:

> This strictly means to remove, to cut out from something what is superfluous, what is not indispensable, or what is damaging or foreign to that thing, leaving only what appears to be useful and without flaw. In everyday life it was employed above all to refer to cutting back a tree, to pruning. [L. Gerschel (1960)]

To sum up:

> In the method of counting described by the expression *rationem putare*, if the term *ratio* means representing each thing counted by a corresponding mark, then the action denoted by *putare* consists of cutting into a stick with a knife in order to create this mark: as many as there are things to count, so many are the notches on the stick, made by cutting out from the wood a small superfluous portion, as in the definition of *putare*. In a way, *ratio* is the mind which sees each object in relation to a mark; *putare* is the hand which cuts the mark in the wood. [L. Gerschel (1960)]

* To one of his contemporaries, Plautus wrote: *Postquam comedit rem, post rationem putat*. "It is now that he has consumed his resources that he counts the cost!" (*Trin*. 417).

† We find an example of this use in Cicero (*Flacc*. XXVII, 69): *Auri ratio constat; aurum in aerario est*. "The count of the gold is correct; the gold is in the public treasury."

‡ Cato (in *Agr*. I, 5) uses the expression *pro ratione* to mean "in proportion", giving *ratio* an arithmetical sense; Vitruvius (III, 3, 7) also uses the expression to mean "architectural proportion".

## FURTHER CONFIRMATION

A different confirmation from the preceding is given by F. Škarpa (1934) in a detailed study of the different kinds of notches used since time immemorial by herdsmen of Dalmatia (in the former Yugoslavia). In one of these, the number 1 is represented by a small line, the number 5 by a slightly longer one, and the number 10 by a line which is much longer than the others (which is very reminiscent of the measuring scales on rulers and on thermometers).

Another type of marking used by the Dalmatian herdsmen represented the number 1 by a vertical stroke, the number 5 by an oblique stroke, and the number 10 by a cross. In a third type, for the numbers 1, 5 and 10 we find:

I Λ X

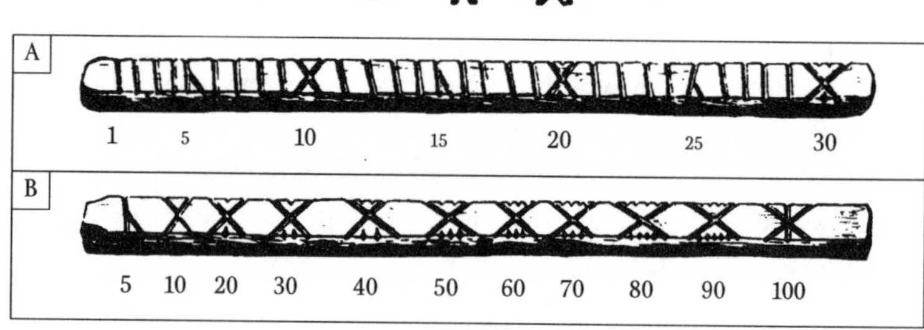


FIG. 16.41. *Herdsmen's tally sticks from Dalmatia [Škarpa (1934), Table II]*

Does this not rather closely resemble the Roman and Etruscan numerals? This is all the more striking in that this same type of tally (Fig. 16.41) shows that the sign for 100, as below, is identical to the Etruscan figure for 100:

FIG. 16.42. ✱

We may well ask why these people used such a figure for 100, but not the "half-figure" for 50 as the Etruscans did (Fig. 16.35). But close examination of the tally stick in question tells us why. We find that the marks for the tens, from 20 to 90, differ from the others by having very small notches on the edge of the stick, above and below, and the number of these small notches gives the corresponding number of tens.

This method of marking can dispense with a separate sign for 50. Suppose that a herdsman wants to keep a record of the fact that he has 83 dairy cows and 77 sterile cows, having already counted them as above. All he needs to do is write the results as shown below on a separate piece of wood which he will keep by him:

Fig. 16.43.

So this herdsman has no need for a special symbol for 50, since he has executed the idea described above as shown in Fig. 16.44:

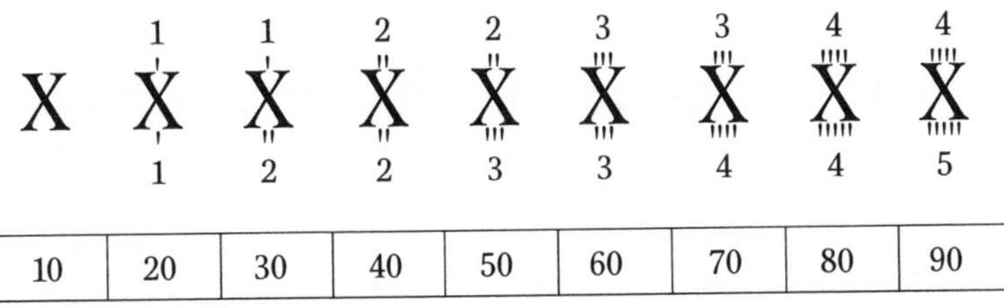


Fig. 16.44.

A final type of tally known from Dalmatia gives the following figures:

Fig. 16.45.

The presence of the sign N for 50 seems fairly natural since it can be made by adding a vertical bar to the figure for 5, just as the sign for 100 is made by adding a vertical bar to the sign X (Fig. 16.41).

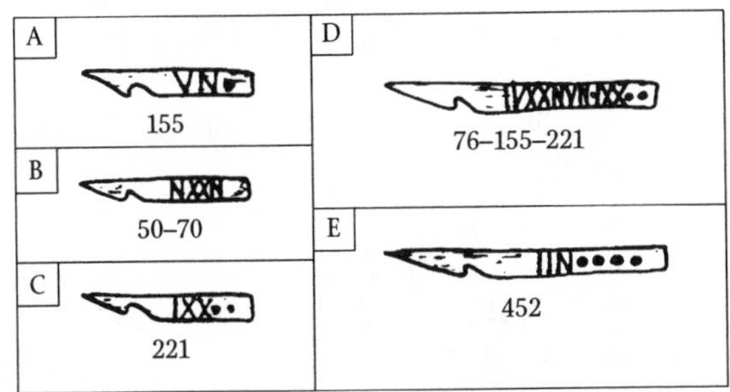


Fig. 16.46. *Herdsmen's tally sticks from Dalmatia [Škarpa(1934), Table IV]*

Very similar series are to be found in the Tyrol and in the Swiss Alps. They are found at Saanen on peasant double-entry tallies, at Ulrichen on milk-measuring sticks, as well as at Visperterminen on the famous *tallies of capital*, where sums of money lent by the commune or by religious foundations to the townsfolk were noted using the figures:

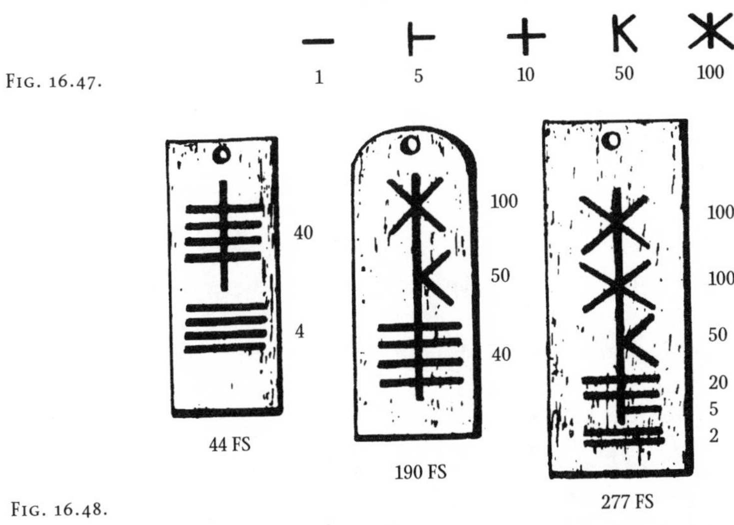


Fig. 16.47.

Fig. 16.48.

Further evidence can be found in the *calendrical ciphers*, strange numerical signs on the calendrical boards and sticks which were in use from the end of the Middle Ages up to the seventeenth century in the Anglo-Saxon and West Germanic world, from Austria to Scandinavia (Fig. 16.52 to 16.54).

| 1 2 3 4 | 5 | 6 7 8 9 | 10 | 11 12 13 14 | 15 | 16 17 18 19 |
|---|---|---|---|---|---|---|

Fig. 16.49.

These wooden almanacs give the Golden Number of the nineteen-year Metonic cycle in the graphical variants of the numerals shown on Fig. 16.52 to 16.54. See E. Schnippel (1926).

The figures used in the English clog almanacs of the Renaissance (see Fig. 16.54) are the following:

| 1 2 3 4 | 5 | 6 7 8 9 | 10 | 11 12 13 14 | 15 | 16 17 18 19 | 20 | 21 22 |
|---|---|---|---|---|---|---|---|---|

Fig. 16.50.

and those used in Scandinavian runic almanacs have the prototype:

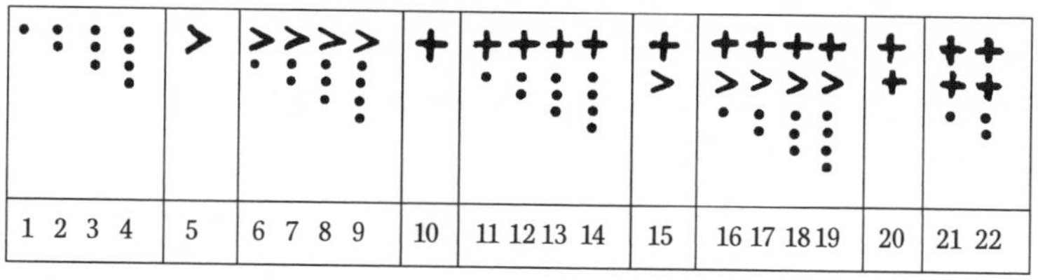

| · : ⁝ ⁞ | > | > > > ><br>· : ⁝ ⁞ | + | + + + +<br>· : ⁝ ⁞ | +<br>> | + + + +<br>> > > ><br>· : ⁝ ⁞ | +<br>+ | + +<br>+ +<br>· : |
|---|---|---|---|---|---|---|---|---|
| 1 2 3 4 | 5 | 6 7 8 9 | 10 | 11 12 13 14 | 15 | 16 17 18 19 | 20 | 21 22 |

FIG. 16.51.

In all of these notations, which appear dissimilar at first sight but which on close examination prove to originate with the tally-stick principle, the signs given to the numbers 1, 5 and 10 are unmistakably similar to the Roman numerals I, V and X and to the Etruscan numerals I, Λ, and + or X.

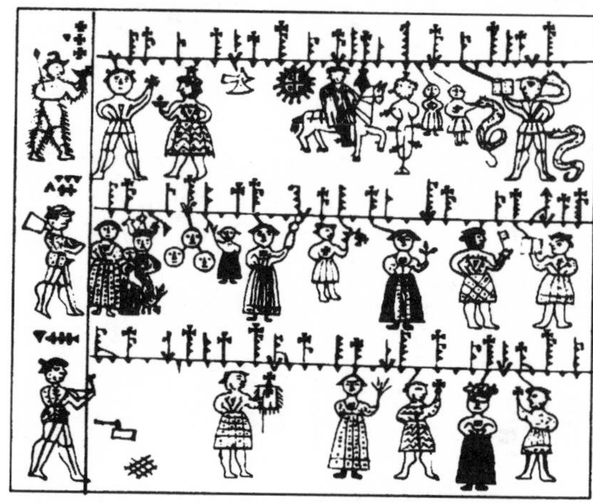

FIG. 16.52. *"Page" from a wooden almanac (Figdorschen Collection, Vienna, no. 799) [Riegl (1888), Table I]*

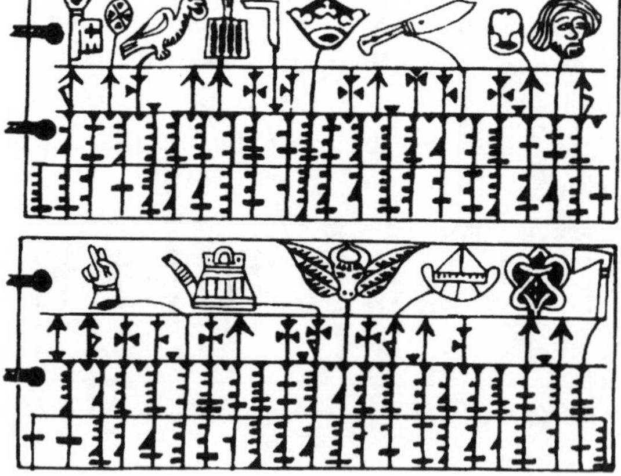

FIG. 16.53. *Two "pages" from a wooden almanac from the Tyrol (15th century) (Figdorschen Collection, Vienna, no. 800) [Riegl (1888), Table V]*

FIG. 16.54. *English clog almanac from the Renaissance (Ashmolean Museum, Oxford, Clog C) [Schnippel (1926), Table IIIa]*

Even better, in the nineteenth century the Zuñis (Pueblo Indians of North America living in New Mexico, at the Arizona frontier, whose traditions go back 2,000 years) still used "irrigation sticks" [F. H. Cushing (1920)] inscribed with numerals that were just as close to Roman figures:

- a simple notch for the number 1;
- a deeper notch, or an oblique one, for 5;
- a sign in the shape of an X for 10.

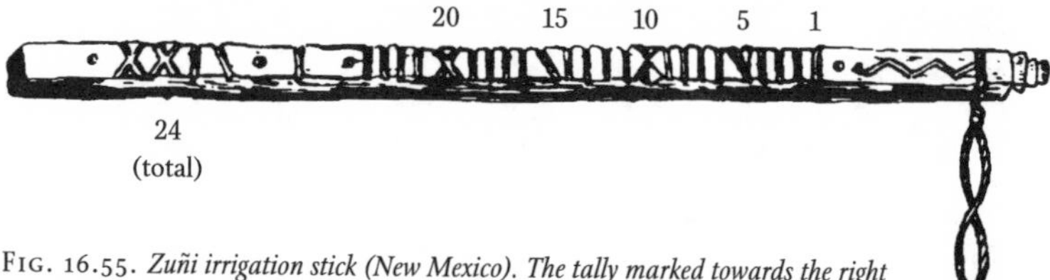


FIG. 16.55. *Zuñi irrigation stick (New Mexico). The tally marked towards the right of the stick totals 24, which is marked as a number at the left-hand end in the form XXI\, which is reminiscent of the Roman representation XXIV where the principle of subtraction has been applied. [Cushing (1920)]*

There can now be no possible doubt: Roman and Etruscan numerals derive directly from counting on tally sticks.

We are now in a position to put forward the following explanation of the genesis of such numerals.

Pastoral peoples, who lived in Italy long before the Etruscans and the Romans, since earliest antiquity (and possibly even in prehistoric times) counted by the method of tally sticks, and the Dalmatian herdsmen or the Zuñis, for example, independently discovered the same for their own use. In a quite natural way, all came to make use of the following signs:

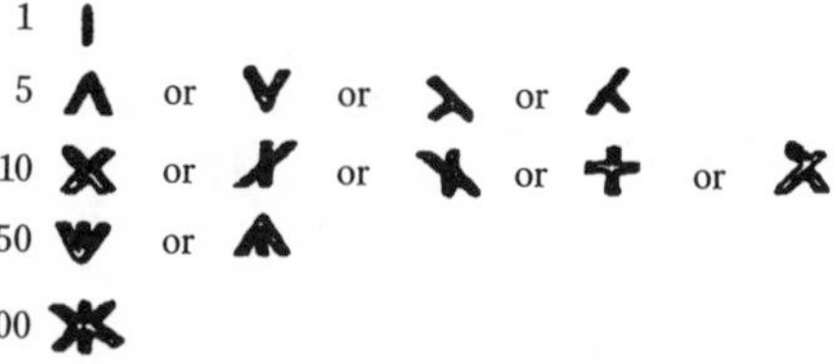


FIG. 16.56.

Inheriting this ancient tradition, the Etruscans and the Romans who came after them retained from these only the following:

| ETRUSCAN | ROMAN |
|---|---|
| 1 | 1 |
| 5 | 5 |
| 10 or or | 10 |
| 50 | 50 |
| 100 | 100 |

FIG. 16.57A.

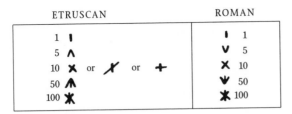

The Romans then completed the series by adding a sign for 500, and another for 1,000 (the former was the right-hand half of the latter, itself generated by drawing a circle on top of the figure for 10 (see Fig. 16.29, 16.30 and 16.35):

FIG. 16.57B. Figures for 1,000 ⊗ or ⊕

In their hands, these signs changed form over the centuries until they were replaced by the alphabetic numerals which we know.

This, therefore, is the most plausible explanation of the origins of the Roman and Etruscan numerals. The following does not gainsay it. A. P. Ninni (1899) reports that the Tuscan peasants and herdsmen were still using, in the last century, in preference to Arabic numerals, the following signs which they call *cifre chioggiotti*:

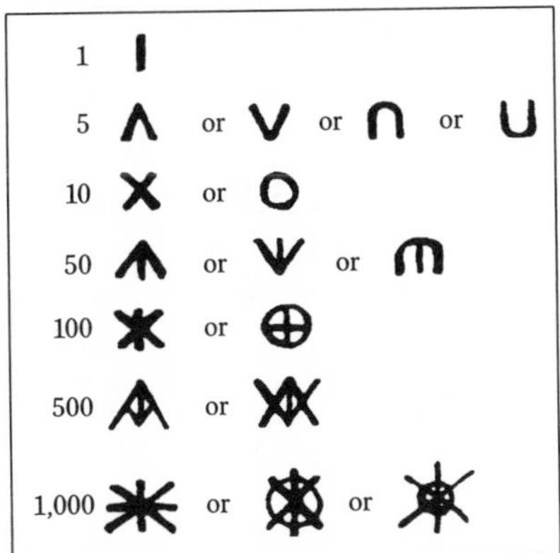


FIG. 16.58.

G. Buonamici (1932) saw these as descending from Etruscan or Roman numerals; may we not with more reason see them as a survival of the ancient practice of counting by cutting notches, a practice older than any writing and one which is to be found in every rural community on earth?

## ROMAN NUMERALS FOR LARGE NUMBERS

The largest number for which Roman numerals as we know them (and still sometimes use them) had a separate symbol is 1,000. The simple application of the additive principle to the seven basic figures of this system would only take us up to 5,000. Therefore, when we come to make use of these numerals, we find it effectively impossible to write large numbers. How do we represent, say, 87,000, except by writing down 87 copies of the letter M?

The ancients had some trouble getting round the problem, and adopted a variety of conventions for writing large numbers. The difficulties which they encountered, as did their successors in the European Middle Ages, deserve special consideration.

In the Republican period, the Romans had a simple graphical procedure by which they could assign a special notation to the numbers 5,000, 10,000, 50,000, and 100,000. The principal ones (found sporadically as late as the Renaissance) are the following:

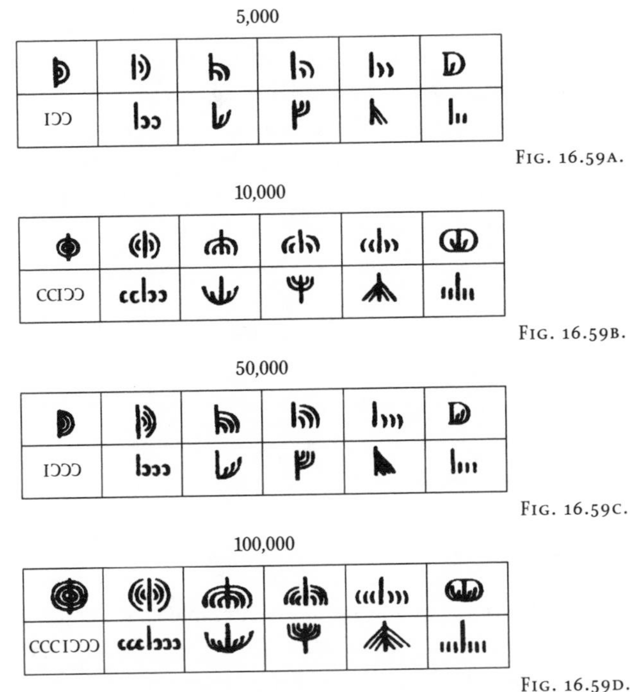


FIG. 16.59A.

FIG. 16.59B.

FIG. 16.59C.

FIG. 16.59D.

Comparing these with each other, and with the various ancient forms of the symbol for 1,000 (Fig. 16.30), we realise that they have a common origin. In fact, they are simply stylisations (more or less recognisable) of the original five signs.

The idea governing the formation of four of these consists of an extremely simple geometrical procedure. Taking as a starting point the primitive Roman sign for 1,000 (originally a circle divided in two by a vertical line), the signs for 10,000 and 100,000 were made by drawing one or two circles, respectively, around it; and the signs for 5,000 and for 50,000 were made by using the right-hand halves of these (Fig. 16.62):

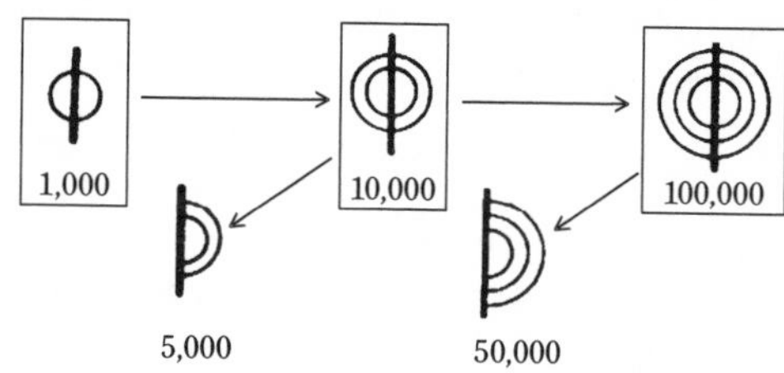


FIG. 16.59E.

Following the same principle, the Romans were able to write the numbers 500,000, 1,000,000, 5,000,000, etc., in the following forms:

FIG. 16.60. 500,000: or IↃↃↃↃ   1,000,000: or CCCCIↃↃↃↃ, etc.

But this kind of graphical representation is complicated, and it is difficult to recognise numbers above 100,000 at a glance; the Romans do not seem to have taken it any further. An additional possible reason is that there is no special word in Latin for numbers greater than 100,000: for example, Pliny (*Natural History*, XXXIII, 133) notes that in his time, the Romans were unable to name the powers of 10 above 100,000. For a million, for example, they said *decies centena milia*, "ten hundred thousand".

Nevertheless such a representation may be found, for numbers up to one million, in a work published in 1582 by a Swiss writer called Freigius (Fig. 16.61, 16.62 and 16.70):

| I | V | X | L | C | IↃ | CIↃ | IↃↃ | CCIↃↃ | IↃↃↃ | CCCIↃↃↃ | IↃↃↃↃ | CCCCIↃↃↃↃ |
|---|---|---|---|---|---|---|---|---|---|---|---|---|
| 1 | 5 | 10 | 5 × 10 | $10^2$ | $5 \times 10^2$ | $10^3$ | $5 \times 10^3$ | $10^4$ | $5 \times 10^4$ | $10^5$ | $5 \times 10^5$ | $10^6$ |

FIG. 16.61.

Other conventions were frequently used by the Romans, and may be found in use in the Middle Ages, which simplified the notation of numbers above 1,000 and allowed considerably larger numbers to be reached.

In one of these, a horizontal bar placed above the representation of a number meant that that number was to be multiplied by 1,000. In this way all numbers from 1,000 to 5,000,000 could be easily written.

It should however be noted that this convention could sometimes cause confusion with another older convention, in which, in order to distinguish between letters used to denote numbers from those used to write words, the Romans were in the habit of putting a line above the letters being used as numerals, as can be found in certain Latin abbreviations such as

$\overline{\text{II}}$VIR = duumvir; $\overline{\text{III}}$VIR = triumvir

| | 1,000 | 5,000 | 10,000 | 50,000 | 100,000 |
|---|---|---|---|---|---|
| Basic graphic form | | | | | |
| 1st stylisation | | | | | |
| 2nd stylisation | | | | | |
| 3rd stylisation | | | | | |
| 4th stylisation | | | | | |
| 5th stylisation | | | | | |
| 6th stylisation | | | | | |
| 7th stylisation | | | | | |
| 8th stylisation | | | | | |
| 9th stylisation | | | | | |
| 10th stylisation | | | | | |
| 11th stylisation | | | | | |
| 12th stylisation | c\|ɔ | \|ɔɔ | cc\|ɔɔ | \|ɔɔɔ | ccc\|ɔɔɔ |
| 13th stylisation | CIↃ | IↃↃ | CCIↃↃ | IↃↃↃ | CCCIↃↃↃ |

FIG. 16.62.

| | | |
|---|---|---|
| $\overline{\text{V}}$ | = 5,000 = 5 × 1,000 | CIL, VIII, 1577 |
| $\overline{\text{X}}$ | = 10,000 = 10 × 1,000 | CIL, VIII, 98 |
| $\overline{\text{LXXXIII}}$ | = 83,000 = 83 × 1,000 | CIL, I, 1757 |

*Examples from Latin inscriptions of which the oldest date from the end of the Republic*

FIG. 16.63.

| | | |
|---|---|---|
| | $\overline{\text{IIII}}$DCCCLXXII | 4,872 |
| | $\overline{\text{V}}$DLXVIII | 5,568 |
| | $\overline{\text{V}}$DCCCCXVI | 5,916 |
| | $\overline{\text{VI}}$CCLXIIII | 6,264 |

*Examples from a Latin astronomical manuscript of the 11th or 12th centuries CE (Bibliothèque nationale, Paris, Ms. lat. 14069, folio 19)*

FIG. 16.64.

Probably it is for this reason that at the time of the Emperor Hadrian (second century CE) the multiplication by 1,000 was indicated by placing a vertical bar at either side, as well as the horizontal line on top.

*Reconstructed examples*:

| | | |
|---|---|---|
| 35,000 | ⌈XXXV⌉ <br> 35 × 1,000 | |
| 557,274 | ⌈DLVII⌉ <br> 557 × 1,000 | CCLXXIV <br> + 274 |

FIG. 16.65.

However, this notation was generally reserved for a quite different purpose.

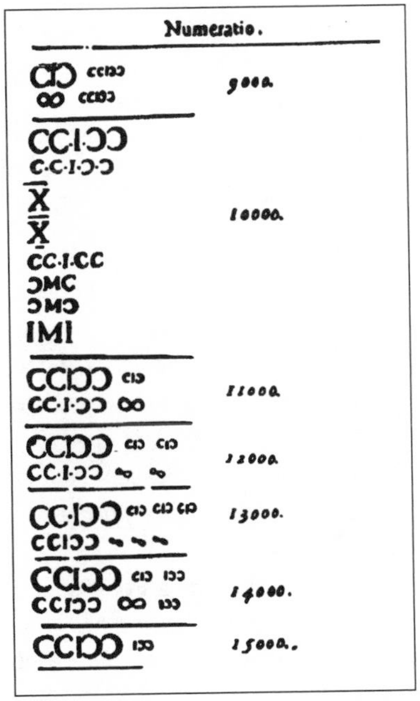

**Numeratio.**

| | |
|---|---|
| CIↃ CCIↃↃ <br> ∞ CCIↃↃ | 9000. |
| CC·I·ↃↃ <br> C·C·I·Ↄ·Ↄ <br> $\overline{X}$ <br> $\overline{X}$ <br> CC·I·CC <br> ↃMC <br> ↃMↃ <br> IMI | 10000. |
| CCIↃↃ CIↃ <br> CC·I·ↃↃ ∞ | 11000. |
| CCIↃↃ CIↃ CIↃ <br> CC·I·ↃↃ ∞ ∞ | 12000. |
| CC·IↃↃ CIↃ CIↃ CIↃ <br> CCIↃↃ ∞ ∞ ∞ | 13000. |
| CCIↃↃ CIↃ IↃↃ <br> CCIↃↃ ∞ IↃↃ | 14000. |
| CCIↃↃ IↃↃ | 15000. |

FIG. 16.66. *The archaic Roman numerals being used in a work by Petrus Bungus on the mystical significance of numbers* (Mysticae numerorum significationes opus . . .) *published at Bergamo in 1584–1585. (Bibliothèque nationale, Paris [R. 7489])*

Every Roman numeral enclosed in a kind of incomplete rectangle was, in fact, usually supposed to be multiplied by 100,000, which allowed the representation of all numbers between 1,000 and 500,000,000.

Examples from Latin inscriptions from the Imperial period in Rome:

| | | |
|---|---|---|
| ⌈XII⌉ [a] | 1,200,000 | 12 × 100,000 |
| ⌈XIII⌉ [b] | 1,300,000 | 13 × 100,000 |
| ⌈∞ ∞⌉ [c] | 200,000,000 | 2,000 × 100,000 |

a. Cf. CIL, I, 1409
b. Cf. CIL, VIII, 1641
c. Inscription from Ephesus, 103 CE [Cagnat (1899)]

FIG. 16.67.

According to some authors, the logical continuation of the convention of placing a line above the number was to place a double line to represent multiplication by 1,000,000, thus allowing the representation of numbers up to 5,000,000,000:

| | | |
|---|---|---|
| 1,000,000,000 | $\overline{\overline{\text{M}}}$ | = 1,000 × 1,000,000 |
| 2,300,000,000 | $\overline{\overline{\text{MMCCC}}}$ | = 2,300 × 1,000,000 |

However no evidence of this in currently known Roman inscriptions has been found.

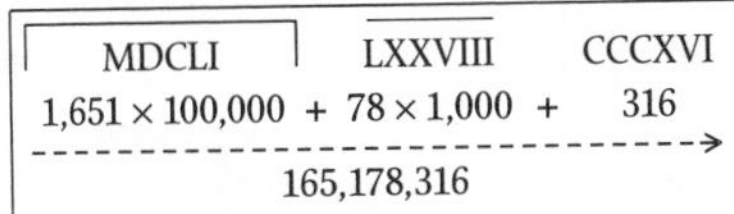

| ⌈MDCLI⌉ | $\overline{\text{LXXVIII}}$ | CCCXVI |
|---|---|---|
| 1,651 × 100,000 + | 78 × 1,000 + | 316 |

165,178,316

FIG. 16.68.

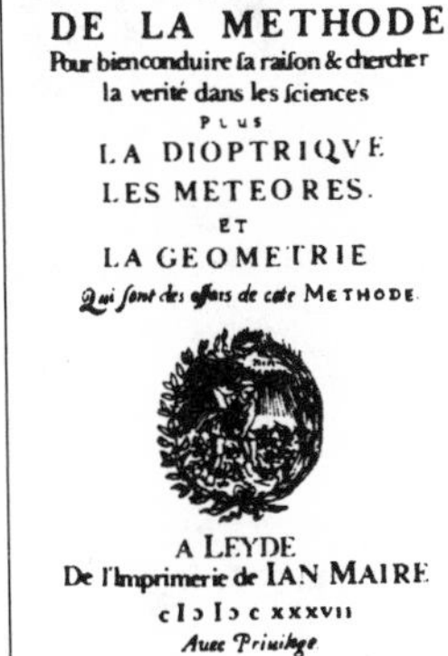

DISCOURS
DE LA METHODE
Pour bien conduire sa raison, & chercher
la verité dans les sciences
PLUS
LA DIOPTRIQVE
LES METEORES.
ET
LA GEOMETRIE
Qui sont des essais de cete METHODE.

A LEYDE
De l'Imprimerie de IAN MAIRE.
CIↃ IↃ C XXXVII.
Avec Privilege

B. D. S.

OPERA

POSTHUMA,

Quorum series post Præfationem exhibetur.

CIↃ IↃ C LXXVII.

FIG. 16.69. *Frontispieces from works by Descartes (published 1637) and Spinoza (published 1677). The dates are written in the archaic Roman numerals.*

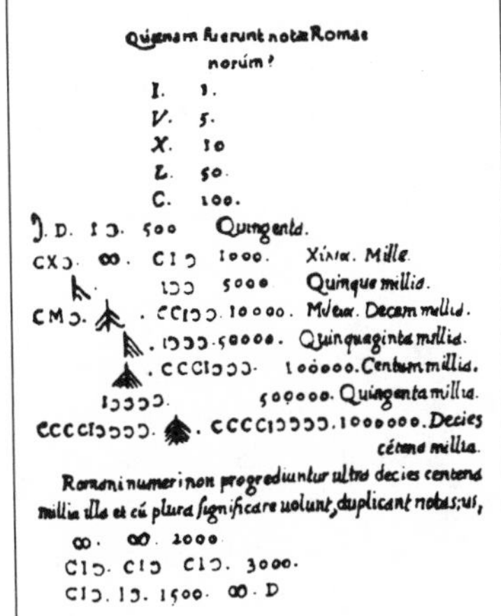

Quænam fuerunt notæ Romanorum?
I. 1.
V. 5.
X. 10
L. 50.
C. 100.
D. IↃ. 500 Quingenta.
CXↃ. ∞. CIↃ 1000. Χίλια. Mille.
IↃↃ 5000 Quinque millia.
CMↃ. CCIↃↃ. 10000. Decem millia.
IↃↃↃ. 50000. Quinquaginta millia.
CCCIↃↃↃ. 100000. Centum millia.
IↃↃↃↃ. 500000. Quingenta millia.
CCCCIↃↃↃↃ. CCCCIↃↃↃↃ. 1000000. Decies centena millia.
Romani numeri non progrediuntur ultra decies centena millia: illa et cum plura significare volunt, duplicant notas: ut,
∞. ∞. 2000
CIↃ. CIↃ CIↃ. 3000.
CIↃ. IↃ. 1500. ∞. D

FIG. 16.70. *Archaic Roman numerals in a work by Freigius, published in 1582 [Smith, D.E (1958)]*

FIG. 16.71. *Detail of a page of a Portuguese manuscript of 1200 CE, referring to the Venerable Bede's method of calculation (Lisbon Public Library, MS Alcobaça 394 (426), folio 252) [Burnham in PIB, plate XV]*

Figures on the drawings: $\overline{\text{XL}}$ XXX = 40,030
$\overline{\text{L}}$ XC = 50,090
$\overline{\text{LX}}$ = 60,000

But these kinds of notations could only cause confusion and errors of interpretation – as a future Roman Emperor learnt to his cost, according to Seneca (*Galba*, 5).

On succeeding to his mother Livia, Emperor Tiberius had to pay large sums of money to her legatees. Tiberius's mother had written the amount of her legacy to young Galba in the form: $\overline{\text{CCCCC}}$.

But Galba had not taken the precaution of checking that the amount was written out in words. So when he presented himself to Tiberius, Galba thought that the five Cs had been enclosed in vertical lines, and that therefore the sum due to him was

$$500 \times 100{,}000 = 50{,}000{,}000 \text{ sesterces.}$$

But Tiberius took advantage of the fact that the two sides-bars were very short, and claimed that this representation was a simple line above the five Cs. "My mother should have written them as $\overline{|\text{CCCCC}|}$ if you were to be right," he said. Since the simple line only represented multiplication by 1,000, Galba only received from Tiberius the sum of

$$500 \times 1{,}000 = 500{,}000 \textit{ sesterces}$$

Which goes to show that an unstable notation system can turn a large fortune into a mere pittance!

The Romans also devised other conventions. Instead of repeating the letters C and M for successive multiples of 100 or 1,000, they first wrote the number of hundreds or thousands they wanted, and then placed the letter C or M either as a coefficient or as a superscript index:

| | | | | | | | |
|---|---|---|---|---|---|---|---|
| 200: | II.C | or | $\text{II}^{\text{C}}$ | 2,000: | II.M | or | $\text{II}^{\text{M}}$ |
| 300: | III.C | or | $\text{III}^{\text{C}}$ | 3,000: | III.M | or | $\text{III}^{\text{M}}$ |

However, instead of simplifying the system, these various conventions only complicated it, since the principle of addition was completely subverted by the search for economy of symbols.

We therefore see the complexity and the inadequacy of the Roman number-system. *Ad hoc* conventions based on principles of quite different kinds made it incoherent and inoperable. There is no doubt that Roman numerals constituted a long step backwards in the history of number-systems.

## THE GREEK AND ROMAN ABACUSES

Given such a poor system of numerals, the Greeks, Etruscans and Romans did not use written numbers when they needed to do sums: they used abacuses.

The Greek historian Polybius (c. 210–128 BCE) was no doubt referring to one of these when he put the following words into the mouth of Solon (late seventh century to early sixth century BCE).

> Those who live in the courts of the kings are exactly like counters on the counting table. It is the will of the calculator which gives them their value, either a *chalkos* or a *talent*. (*History*, V, 26)

We can all the better understand the allusion when we know that the *talent* and the *chalkos* were respectively the greatest and the least valuable of the ancient Greek coins, and they were represented by the leftmost and rightmost columns of the abacus.

FIG. 16.72. *Detail of the Darius Vase from Canossa, c. 350 BCE (Museo Archeologico Nazionale, Naples)*

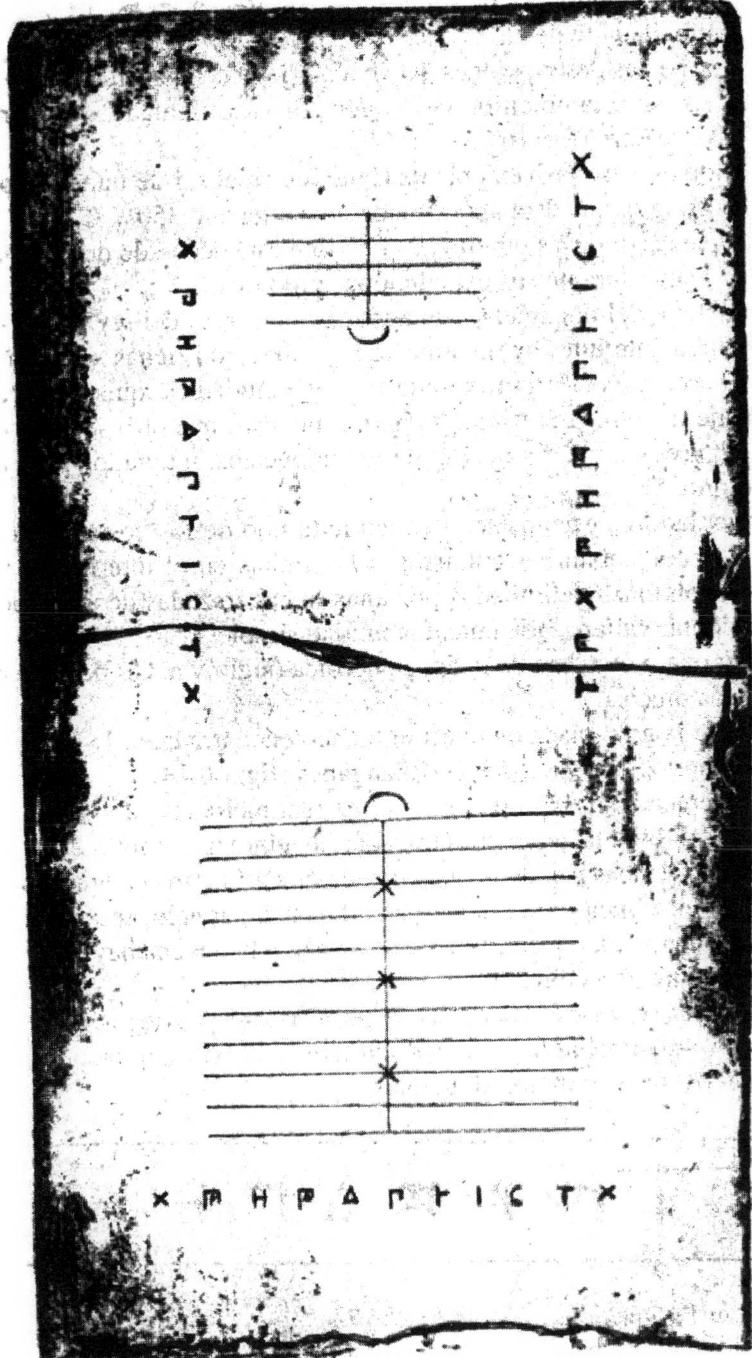

FIG. 16.73. *The Table of Salamis, originally considered to be a gaming table, which is in fact a calculating apparatus. Date uncertain (fifth or fourth century BCE). (National Museum of Epigraphy, Athens)*

The writings of many other Greek authors from Herodotus to Lysias also bear witness to the existence and use of the abacus.

Descriptions of the Greek abacus are not only to be found in literary text, but also in images. The "Darius Vase" is the most famous example (Fig. 16.72). It is a painted vase from Canossa in southern Italy (formerly a Greek colony) and dates from around 350 BCE. The various scenes painted on it are supposed to describe the activities of Darius during his military expeditions.

In one detail of the vase, we can see the King of Persia's treasurer using counters on an abacus to calculate the tribute to be levied from a conquered city. In front of him, a personage hands him the tribute, while another begs the treasurer to allow a reduction of taxes which are too heavy for the city he represents.

The Greek calculators stood by one of the sides of the horizontal table and placed pebbles or counters on it, within a certain number of columns marked by ruled lines. The counters or pebbles each had the value of 1.

A document from the Heroic Age (fifth century BCE) gives us a more detailed idea. It is a large slab of white marble, found on the island of Salamis by Rhangabes, in 1846 (Fig. 16.73).

It consists of a rectangular table 149 cm long, 75 cm wide and 4.5 cm thick, on which are traced, 25 cm from one of the sides, five parallel lines; and, 50 cm from the last of these lines, eleven other lines, also parallel, and divided into two by a line perpendicular to them: the third, sixth and ninth of these lines are marked with a cross at the point of intersection.

Furthermore, three almost identical series of Greek letters or signs are arranged in the same order along three of the sides of the table. The most complete of the series has the following thirteen symbols in it:

T 𐅇 X 𐅆 H 𐅅 Δ 𐅃 ⊢ I C T X

FIG. 16.74.

As we saw at the beginning of this chapter, these in fact correspond to the numerical symbols of the acrophonic number system (Fig. 16.1), and they serve here to represent monetary sums expressed in *talents*, *drachmas*, *obols*, and *chalkoi*, that is to say in multiples and sub-multiples of the *drachma*.

These symbols represented, from left to right in the order shown, 1 *talent* or 6,000 *drachmas*, then 5,000, 1,000, 500, 100, 50, 10, 5 and 1 *drachmas*, then 1 *obol* or one sixth of a *drachma*, 1 *demi-obol* or one twelfth of a *drachma*, 1 *quarter-obol* or one twenty-fourth of a *drachma*, and finally 1 *chalkos* (one eighth of an *obol* or one forty-eighth of a *drachma*). (Fig. 16.75)

| | | |
|---|---|---|
| T | 1 talent | First letter of TALANTON, "talent" |
| 𐅇 | 5,000 drachmas | |
| X | 1,000 drachmas | First letter of CHILIOI, "thousand" (drachmas) |
| 𐅆 | 500 drachmas | |
| H | 100 drachmas | First letter of HEKATON, "hundred" (drachmas) |
| 𐅅 | 50 drachmas | |
| Δ | 10 drachmas | First letter of DEKA, "ten" (drachmas) |
| Γ | 5 drachmas | First letter of PENTE, "five" (drachmas) |
| ⊢ | 1 drachma | |
| I | 1 obol | Unit mark for counting obols |
| C | 1/2 obol | Half of the letter O, first letter of OBOLION |
| T | 1/4 obol | First letter of TETARTHMORION |
| X | 1 chalkos | First letter of CHALKOUS |
| | 1 talent = 6,000 drachmas | |
| | 1 drachma = 6 obols | |
| | 1 obol = 8 chalkos | |

FIG. 16.75.

In the abacus of Salamis, each column was associated with a numerical order of magnitude.

The pebbles or counters disposed on the abacus changed value according to the position they occupied (see Fig. 16.76).

The four columns at the extreme right were reserved for fractions of a *drachma*, the one on the extreme right being for the *chalkos*, the next for the *quarter-obol*, the third for the *demi-obol*, and the last for the *obol*.

The next five columns (to the right of the central cross on Fig. 16.75) were associated with multiples of the *drachma*, the first on the right being for the units, the next for the tens, the third for the hundreds, and so on. In the bottom half of each column, one counter represented one unit of the value of the column. In the upper half, one counter represented five units of the value of the column.

The last five columns (to the left of the central cross in Fig. 16.76) were associated with *talents*, tens of *talents*, hundreds, and so on. One *talent* being worth 6,000 *drachmas*, the calculator would replace counters corresponding to 6,000 by one counter in the *talents* columns (sixth from the right).

As a result of this method of dividing up the table, additions, subtractions and multiplications could be done (Fig. 16.77 and 16.78).

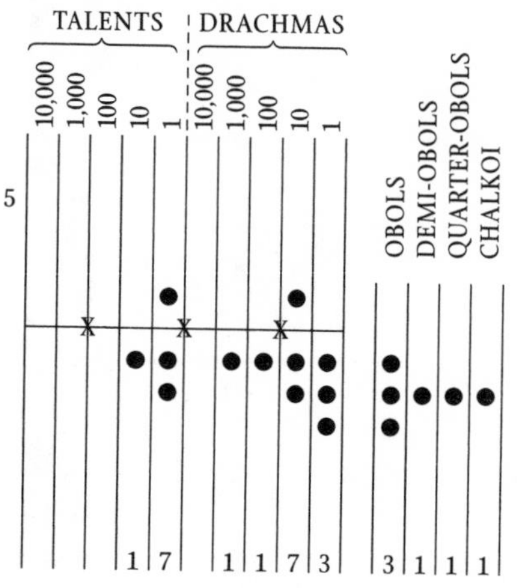


FIG. 16.76. *The principle of the Greek abacus from Salamis, showing the representation of the sum "17 talents, 1,173 drachmas, 3 obols, 1 demi-obol, 1 quarter-obol, and 1* chalkos*". (*Chalkoi *is the plural of* chalkos*.)*

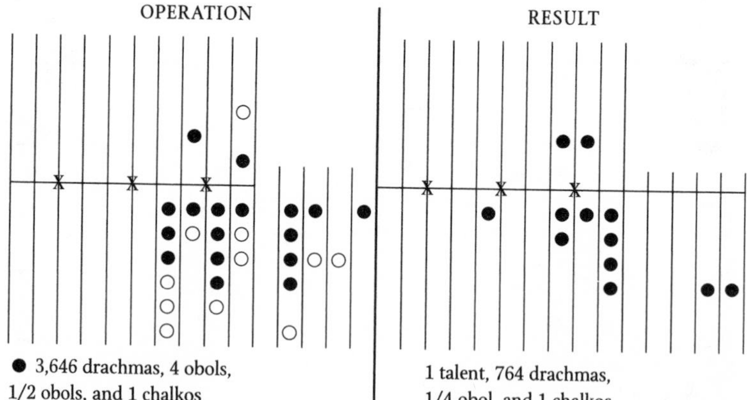


FIG. 16.77. *The method of addition on the Salamis abacus, showing the addition of "3,646* drachmas*, 4* obols*, 1/2* obol *and 1* chalkos*" (shown in black) and "3,117* drachmas*, 1* obol*, 1/2* obol *and 1/45* obol*" (shown in white). By reducing the counters according to the rules, the result is obtained as "1* talent*, 764* drachmas*, 1/4* obol*, and 1* chalkos*".*

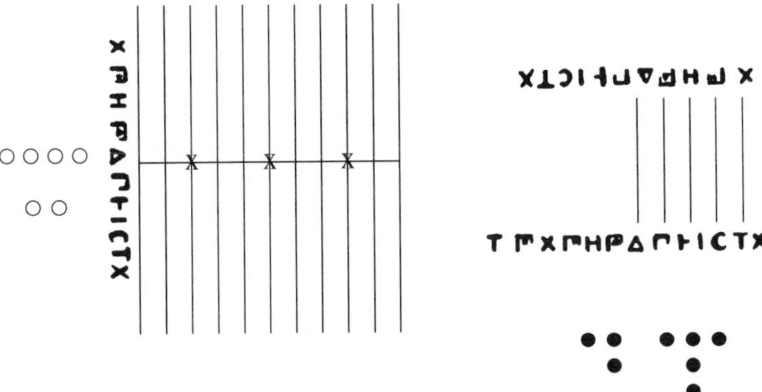

Fig. 16.78. *To multiply "121* drachmas, *3 obols, 1/2* obol, *and 1* chalkos*" by 42, for example, we start by placing the multiplier 42 on the abacus, by laying out the corresponding counters under the appropriate number-signs on the left of the table. Then the multiplicand, the sum of money, is laid out under the number-signs of one of the two series on the right (black circles). Then by manoeuvring the counters the result is obtained (see a similar method in Fig. 16.84).*

The Etruscans and their Roman successors also employed abacuses with counters. In Fig. 16.79 we reproduce an Etruscan medallion, a carved stone which shows a man calculating by means of counters on an abacus, noting his results on a wooden tablet on which Etruscan numerals can be seen (Fig. 16.35).

Fig. 16.79. *The medallion with the Etruscan calculator (date uncertain). (Coin Room, Bibliothèque nationale, Paris. Intaille 1898)*

ny Roman texts mention it:

*Coponem laniumque balneumque, tonsorem tabulamque calculosque et paucos . . . haec praesta mihi, Rufe . . .*

An innkeeper, a butcher, baths, a barber, a calculating table (= *tabulamque calculosque*) with its counters . . . fetch me all that, Rufus . . . *

*Computat, et cevet. Ponatur calculus, adsint cum tabula pueri; numeras sestertia quinque omnibus in rebus; numerentur deinde labores.*

He calculates, and he wriggles his rear. Let the counter (= *calculus*) be placed, let the slaves bring the (calculating) table: you find five thousand sesterces in all; now make the total of my works.†

At Rome, the abacus with counters was a table, on which parallel lines separated the different numerical orders of magnitude of the Roman number-system. The Latin word *abacus* denotes a number of devices with a flat surface which serve for various games, or for arithmetic (Fig. 16.80).

Fig. 16.80. *A Roman abacus with* calculi *(reconstruction)*

Each column generally symbolised a power of 10. From right to left, the first was associated with the number 1, the next with the tens, the third with the hundreds, the fourth with the thousands, and so on. To represent a number, as many pebbles or counters were placed as required. The Greeks called these counters *psephoi*, ("pebble" or "number") and the Romans called them *calculi* (singular: *calculus*). Certain authors (notably Cicero, *Philosophica Fragmenta*, V, 59) called them *aera* ("bronze"), alluding to the material they were often made of after the Imperial epoch (Fig. 16.81).

Fig. 16.81. *Roman calculating counters. After the originals in the Städtisches Museum, Wels, Germany*

To represent the number 6,021 on the columns of the abacus we therefore place one counter in the first column, two in the second, none in the third, and six in the fourth.

For 5,673 we place three in the first, seven in the second, six in the third, and five in the fourth (Fig. 16.82).

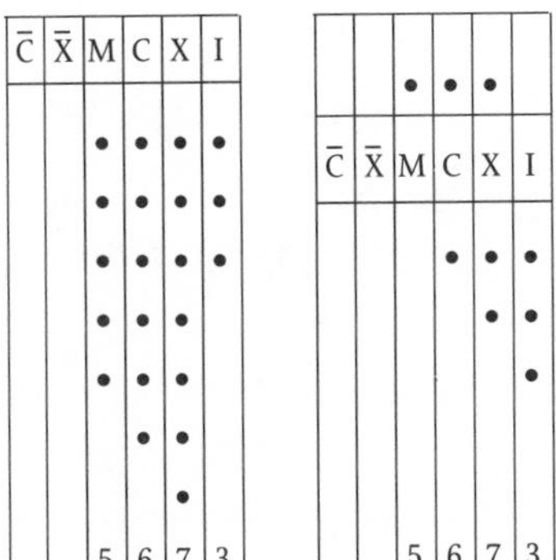


Fig. 16.82. *The principle of the Roman abacus with* calculi

Fig. 16.83. *Simplification of the principle of the Roman abacus with* calculi

* Martial, *Epigrams*, Vol. 1, book 2, 48

† Juvenal, *Satires*, IX, 40–43

To simplify calculation, each column is divided into an upper and a lower part. A counter in the lower half represents one unit of the value of the column, and a counter in the upper half represents half of one unit of the value of the next column (or five times the value of the column it is in). For the upper halves we therefore have five for the first column, fifty for the second, 500 for the third, and so on (Fig. 16.83).

By cleverly moving the counters between these divisions (adding to and taking away from the counters in each division) it is possible to calculate.

To add a number to a number which has already been set up on the abacus, it is set up in turn, and then the result is read off after the various manipulations have been performed. In a given column, if ten or more counters are present at any time then ten of these are removed and one is placed in the next higher column (to the left) (Fig. 16.82). On the simplified abacus, this procedure is somewhat modified. If there are five or more in the lower half, then five are removed and one is placed in the upper half; while if two or more are present in the upper half then two are removed, and one is placed in the lower half of the next column, to the left (Fig. 16.83). Subtraction is carried out in a similar way, and multiplication is done by addition of partial products.

For example, to multiply 720 by 62, we start setting up the numbers 720 and 62 as shown in Fig. 16.84A. Then the 7 of 720 (worth 700) and the 6 of 62 (worth 60) are multiplied, to give 42 (worth 42,000). Therefore two counters are placed in the fourth column and four counters are placed in the fifth.

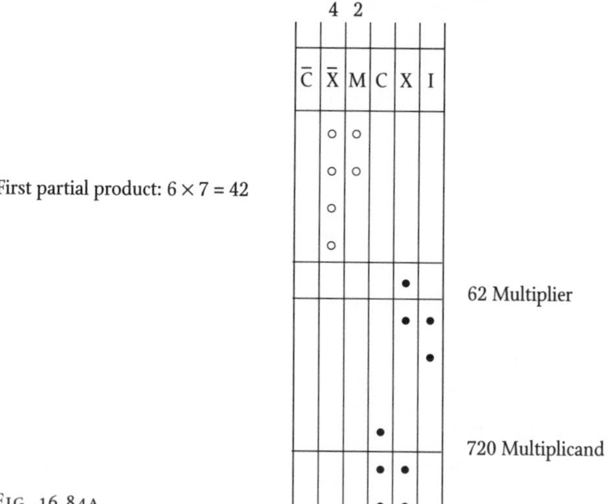


FIG. 16.84A.

Then the 7 of 720 (worth 700) and the 2 of 62 (worth 2) are multiplied to give 14 (worth 1,400), and four counters are placed into the third column and one is placed into the fourth.

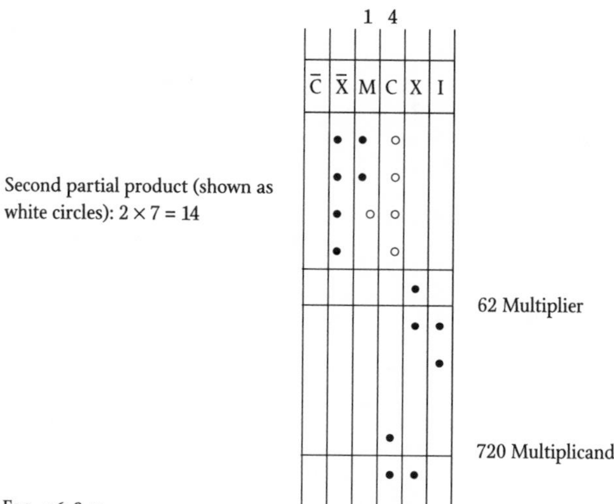


FIG. 16.84B.

Now the 7 of 720 has done its work, and can be removed. Next we multiply to 2 of 720 (worth 20) by the 6 of 62 (worth 60) to get 12 (worth 1,200), and so two counters are placed in the third column and one is placed in the fourth.

Finally, the 2 of 720 (worth 20) and the 2 of 62 (worth 2) are multiplied to get 40. Therefore four counters are placed in the second column.

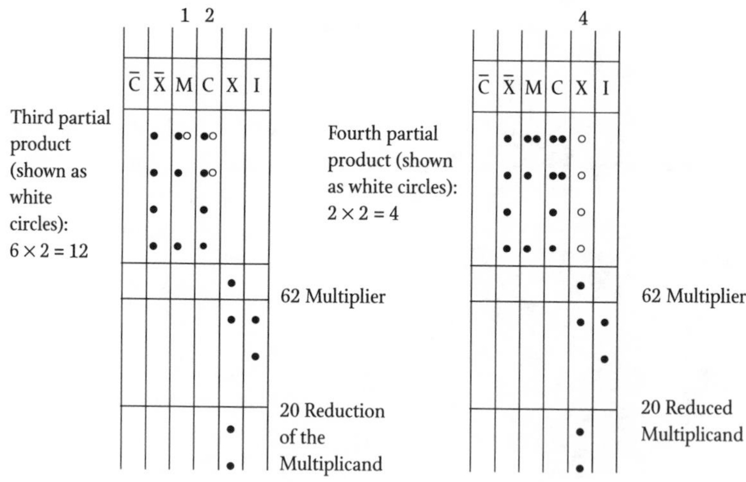


FIG. 16.84C.

FIG. 16.84D.

Now the various counters on the table are reduced as explained above to give the required result of the multiplication:

$$720 \times 62 = 44{,}640$$

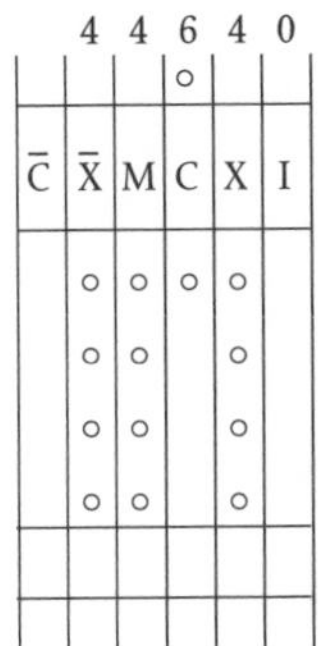


Result

FIG. 16.84E.

Calculating on the abacus with counters was therefore a protracted and difficult procedure, and its practitioners required long and laborious training. It is obvious why it remained the preserve of a privileged caste of specialists.

But traditions live on, and for centuries these methods of calculation remained extant in the West, deeply attached to Roman numerals and their attendant arithmetic. They even enjoyed considerable favour in Christian countries from the Middle Ages up to relatively recent times.

All the administrations, all the traders and all the bankers, the lords and the princes, all had their calculating tables* and struck their counters from base metal, from silver or from gold, according to their importance, their wealth, or their social standing. "I am brass, not silver!" was said at the time to express that one was neither rich nor noble. The clerks of the British Treasury, until the end of the eighteenth century, used these methods to calculate taxes, employing exchequers, or *checkerboards* (because of the way they were divided up). This is why the British Minister of Finance is still called "Chancellor of the Exchequer".

FIG. 16.85A. *The use of abacuses with counters continued in Europe until the Renaissance (and in some places until the French Revolution). Here we see an expert calculator in a German illustration from the start of the sixteenth century. [Treatise on Arithmetic by Köbel, published at Augsburg in 1514]*

* The existence of large numbers of treatises on practical arithmetic which mention these procedures throughout Europe in the sixteenth, seventeenth, and eighteenth centuries gives an idea of how widespread these practices were before the French Revolution.

FIG. 16.85B. *"Madame Arithmetic" teaching young noblemen the art of calculation on the abacus (sixteenth-century French tapestry). (Cluny Museum)*

FIG. 16.86. *Calculating counter bearing the arms of Montaigne (and surrounded with the chain of the order of Saint Michel de Montaigne). This counter was found earlier, in the ruins of the Château de Montaigne, though its original diestamp was found in the nineteenth century. [Brieux (1957)]*

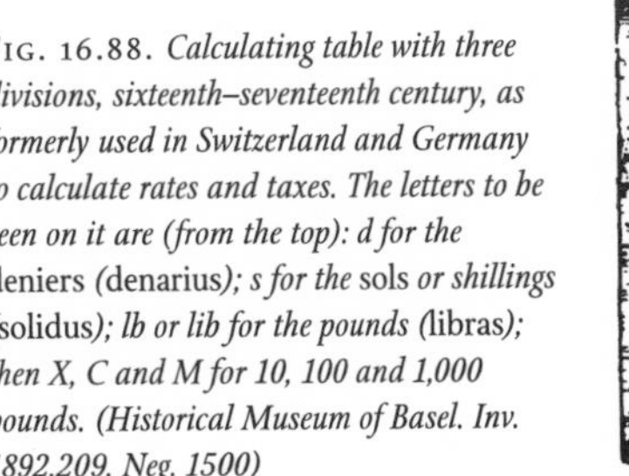

FIG. 16.87. *Fifteenth-century calculating table: one of the rare known abacuses from this period. (Historical Museum of Dinkelsbühl, Germany)*

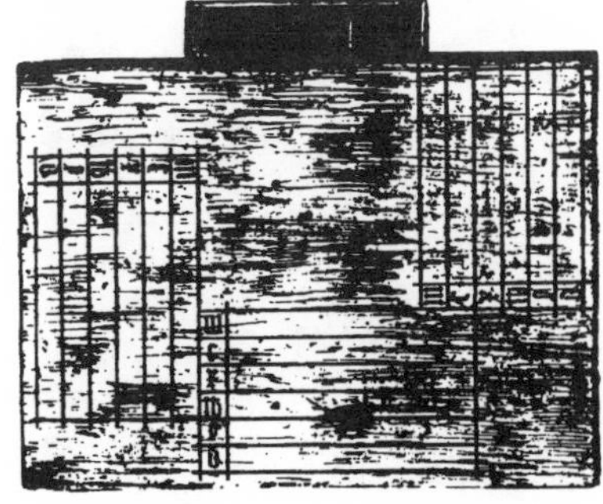

FIG. 16.88. *Calculating table with three divisions, sixteenth–seventeenth century, as formerly used in Switzerland and Germany to calculate rates and taxes. The letters to be seen on it are (from the top): d for the* deniers (denarius)*; s for the* sols *or shillings* (solidus)*; lb or lib for the pounds* (libras)*; then X, C and M for 10, 100 and 1,000 pounds. (Historical Museum of Basel. Inv. 1892.209. Neg. 1500)*

At the time of the Renaissance, many writers make reference to this. Thus Montaigne (1533–1592):

> We judge him, not according to his worth, but from the style of his counters, according to the prerogatives of his rank. (*Essays*, Book III, Bordeaux edition, 192, I, 17)

Likewise Georges de Brébeuf (1618–1661), adapting the formula of Polybius:

> Courtesans are counters;
> Their value depends on their place;
> If in favour, why then it's millions,
> But zero if they're in disgrace.

Again Fénelon (1651 – 1715), who makes Solon say:

> The people of the Court are like the counters used for reckoning: they are worth more or less depending on the whim of the Prince.

And Boursault (1638–1701):

> Never forget, if I may have your grace,
> Whatever more power either of us might have had
> We are still but counters stamped with value by the King.

Finally, Madame de Sévigné, who sent these words to her daughter in 1671:

> We have found, thanks to these excellent counters, that I would have had five hundred and thirty thousand pounds if I counted all my little successions.

The abacus of this period also consisted of a table marked out into divisions corresponding to the different orders of magnitude (Fig. 16.87 and 16.88). Numbers were set up on the table with counters (made of the most diverse materials), whose values depended on where they were placed. Placed on successive lines, from bottom to top, a counter would be worth 1, 10, 100, 1,000, and so on. Between successive lines, a counter was worth five of the value of the line below it (Fig. 16.89 and 16.90).

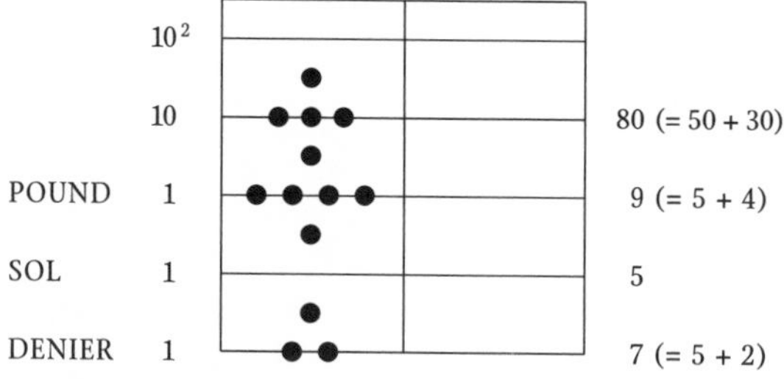


FIG. 16.89. *The layout of the sum "89 pounds, 5* sols *and 7* deniers*" on the French calculating table (sixteenth–eighteenth century).*

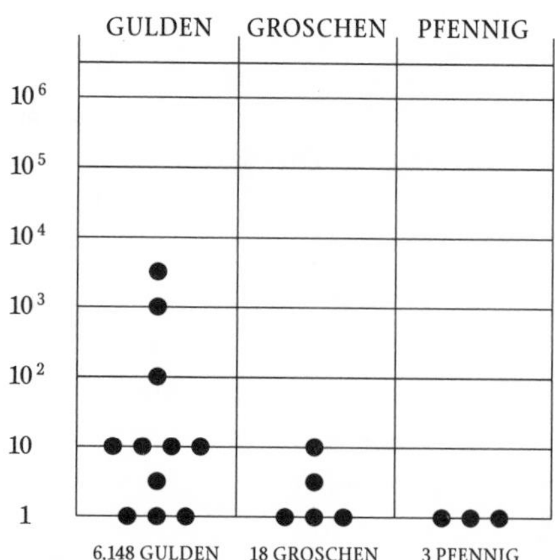


FIG. 16.90. *Representation of the sum of 6,148* gulden, *18* groschen *and 3* pfennigs *on the German calculating table (sixteenth–eighteenth century).*

The counting tables facilitated addition or subtraction, but lent themselves with difficulty to multiplication or division and even less well to more complex operations.

Arithmetical operations practised by this means had little in common with the operations of modern arithmetic with the same names. Multiplication, for example, was reduced to a sum of partial products or to a series of duplications. Division was reduced to a succession of separation into equal parts.

Such difficulties were at the origin of the fierce polemic which, from the beginning of the sixteenth century, ranged the *abacists* on one side, clinging to their counters and to archaic number-systems like the Greek and the Roman, against the *algorists* on the other, who vigorously defended calculation with pen and paper, the ancestor of modern methods.

Here, for example, is what Simon Jacob (who died in Frankfurt in 1564) had to say about the abacus:

> It is true that it seems to have some use in domestic calculations, where it is often necessary to total, subtract, or add, but in serious calculations, which are more complicated, it is often an embarrassment. I do not say that it is impossible to do these on the lines of the abacus, but every advantage that a man walking free and unladen has over he who stumbles under a heavy load, the figures have over the lines.

Pen and paper soon gained the day amongst mathematicians and astronomers. The abacus was in any case used almost exclusively in finance and in commerce. Only with the French Revolution would the use of the abacus finally be banished from schools and government offices.

## ABACUS IN WAX AND ABACUS IN SAND

The Latin word *abacus* derives from the Greek *abax* or *abakon* signifying "tray", "table" or "tablet", which possibly in turn derives from the Semitic word *abq*, "sand", "dust".

It is true that the "abacus in sand" is part of these oriental traditions, but it is mentioned also in the Graeco-Roman West, along with the abacus with counters, especially by Plutarch and by Apuleus. It consisted of a table with a raised border which was filled with fine sand on which the sections were marked off by tracing the dividing lines with the fingers or with a point. (Fig. 16.91).

FIG. 16.91. *Mosaic showing Archimedes (287?–212 BCE) calculating on an abacus with numerals (sand or wax), at the moment when a Roman soldier was about to assassinate him (eighteenth century). (Städtische Galerie/Liebieghaus, Frankfurt)*

Another type of calculating instrument used in Rome was the abacus in wax. It was a true portable calculator which was carried hanging from the shoulder, and it consisted of a small board of wood or of bone coated with a thin layer of black wax; the columns were marked by tracing in the wax with a pointed iron stylus (whose other end, being flat, was used to erase marks by pressing on the surface of the wax).



A specimen from Rome, dating from the sixth century, has been described by D. E. Smith. It is in the collections of the John Rylands Library in Manchester. It is made of bone, and consists of two rectangular iron plates joined by an iron hinge, with three iron styluses.

Horace (65–8 BCE) was perhaps alluding to this instrument in this passage from the first book of *Satires*:*

> . . . *causa fuit pater his, qui macro pauper agello noluit in Flavi ludum me mittere, magni quo pueri magnis e centurionibus orti laevo suspensi loculos tabulanque lacerte ibant octonos referentes Idibus aeris* . . .
>
> I owe this to my father who, poor and with meagre possessions, did not wish to send me to the school of Flavius, where the noble sons of noble centurions went, their box and their board (*tabulanque*) hanging from their left shoulder, paying at the Ides their eight bronze coins. . . .

The Europeans of the Middle Ages probably also used one or other of these, as well as the abacus with counters.

In his *Vocabularium* (1053), Papias (who may be considered one of the authorities on the knowledge of his time) also talks of the abacus as "a table covered with green sand", which is exactly what can be found in Rémy d'Auxerre in his commentary on the *Arithmetic* of Martianus Capella (c. 420–490 CE) where he describes it as "a table sprinkled with a blue or green sand, where the *figures* [the numbers] are drawn with a rod".

As for the abacus in wax, Adelard of Bath (c. 1095–c. 1160) alludes to it as follows [B. Boncompagni (1857)]:

> *Vocatur (Abacus) etiam radius geometricus, quia cum ad multa pertineat, maxime per hoc geometricae subtilitates nobilis illuminantur.*
>
> (The abacus) is also called the "geometrical radius" since it permits so many operations. In particular, thanks to it the subtleties of geometry become perfectly clear and comprehensible.

Finally, it is perfectly possible that Radulph de Laon (c. 1125) was thinking of one or other of these in writing [D. E. Smith & L. C. Karpinski (1911)]:

> . . . *ad arithmaticae speculationis investigandas rationes, et ad eos qui musices modulationibus deserviunt numeros, necnon et ad ea quae astrologorum sollerti industria de variis errantium siderum cursibus* . . . *Abacus valde necessarius inveniatur.*
>
> For the examination of the rules of mathematical thought and of the numbers which are at the base of musical modulations, and for the calculations which, thanks to the skilful industry of the astrologers, explain the various trajectories of the moving stars, the abacus shows itself absolutely indispensable.

* *Satires*, I, VI, 70–75

These authors do not however say what kinds of numeral were used with the abacuses of these two types, though especially at the time of Papias, Adelard and Radulph the Arab numerals were used and were already well known in Europe. But the Greek numerals were used also (from $\alpha = 1$ to $\theta = 9$) which had been much better known before this time, as well as the Roman numerals which were in a way the "official" numerals of mediaeval Europe.

In any case, which figures were used is not of great importance with instruments of this type for, by reason of its structure (which assigns variable values to the symbols according to their positions), the columns of the abacus in sand or the abacus in wax can render even the most primitive figures operational. Of which the proof follows, for the Roman numerals.

Let us again take up the multiplication of 720 by 62, and try to do it with Roman numerals on a tablet covered with sand or with wax.

The technique works for any decimal number-system whatever, provided the figures greater than or equal to 10 are not used. We start by writing the 720 and the 62 in the bottom lines. (Fig. 16.92A)

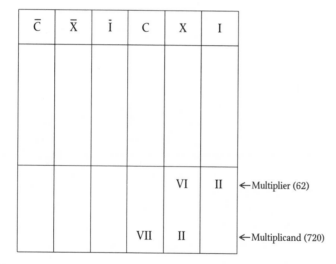

| $\bar{C}$ | $\bar{X}$ | $\bar{I}$ | C | X | I | |
|---|---|---|---|---|---|---|
| | | | | | | |
| | | | | VI | II | ←Multiplier (62) |
| | | | VII | II | | ←Multiplicand (720) |

FIG. 16.92A.

Now we multiply the 7 (700) of 720 by the 6 (60) of 62 and get 42 (42,000). Therefore we write this result at the top, the 2 in the fourth column and the 4 in the fifth. (Fig. 16.92B)

Then we multiply the 7 (700) of 720 by the 2 of 62 and get 14 (1,400), and we write this result at the top below the last one, with a 4 in the third column and a 1 in the fourth. (Fig. 16.92C)

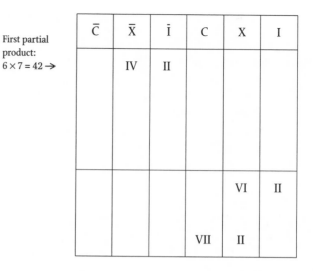

| | $\bar{C}$ | $\bar{X}$ | $\bar{I}$ | C | X | I |
|---|---|---|---|---|---|---|
| First partial product: 6 × 7 = 42 → | | IV | II | | | |
| | | | | | VI | II |
| | | | | VII | II | |

FIG. 16.92B.

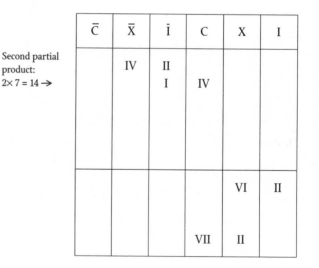

| | $\bar{C}$ | $\bar{X}$ | $\bar{I}$ | C | X | I |
|---|---|---|---|---|---|---|
| | | IV | II | | | |
| Second partial product: 2× 7 = 14 → | | | I | IV | | |
| | | | | | VI | II |
| | | | | VII | II | |

FIG. 16.92C.

Now we can forget the 7 of 762, and multiply the 2 (20) of 720 by the 6 (60) of 62, and get 12 (1,200) which we again write at the top below the last result: 2 in the third column and 1 in the fourth. (Fig. 16.92D)

Third partial product: 6 × 2 = 12 →

| $\bar{C}$ | $\bar{X}$ | $\bar{I}$ | C | X | I |
|---|---|---|---|---|---|
| | IV | II | | | |
| | | I | IV | | |
| | | I | II | | |
| | | | | VI | II |
| | | | | II | |

FIG. 16.92D.

Finally we multiply the 2 (20) of 720 by the simple 2 of 62 and get 4 (40), so we write a 4 in the second column.

Fourth partial product: 2 × 2 = 4 →

| $\bar{C}$ | $\bar{X}$ | $\bar{I}$ | C | X | I |
|---|---|---|---|---|---|
| | IV | II | | | |
| | | I | IV | | |
| | | I | II | | |
| | | | | IV | |
| | | | | VI | II |
| | | | | II | |

FIG. 16.92E.

We can now erase the 720 and the 62, and proceed to reduce the figures which remain. Here we can start with the second column.

Since this figure is less than 10, we pass immediately to the next column, the third. We add 4 and 2 and get 6, which is less than 10, we erase the two figures 4 and 2 and write in 6.

Then we pass to the fourth column, where we add 2, 1 and 1 to get 4, which is less than 10, so we erase the three figures and write a 4 in the fourth column.

The fifth column will remain unchanged since the single figure in it is less than 10.

It only remains to read the result directly off the columns:

$$720 \times 62 = 44{,}640$$

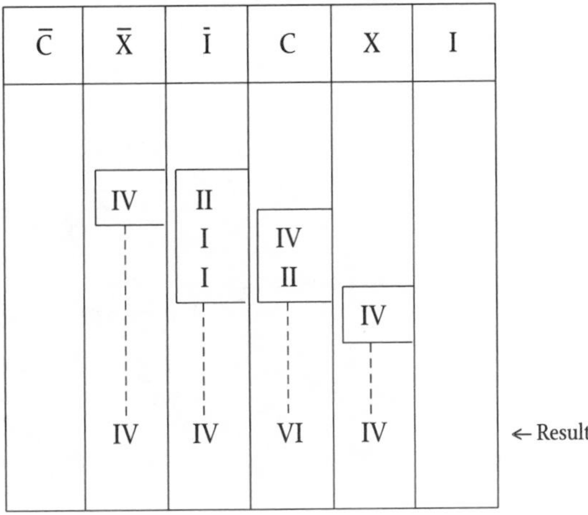


FIG. 16.92F.

## THE FIRST POCKET CALCULATOR

As well as the "desk-top models for school and office", some of the Roman accountants used a real "pocket calculator" whose invention undoubtedly predates our era. The proof of this is a bas-relief on a Roman sarcophagus of the first century, which shows a young calculator* standing before his master, doing arithmetic with the aid of an instrument of this type (Fig. 16.96). This instrument consisted of a small metal plate, with a certain number of parallel slots (usually nine). Each slot was associated with an order of magnitude, and mobile beads could slide along them.†

Ignoring for the moment the two rightmost slots, the remaining seven are divided into two distinct segments, a lower and an upper. The lower one

* Among the Romans, the word calculator meant, on the one hand, a "master of calculation" whose principal task was to teach the art of calculation to young people using a portable abacus or an abacus with counters; and, on the other hand, the keeper of the accounts or the intendant in the important houses of the patricians, where he was also called *dispensator*. If these were slaves, they were called *calculones*; but if they were free men then they were called *calculatores* or *numerarii*.

† As well as the abacus shown in Fig. 16.94, we know of at least two other examples. One is in the British Museum in London, and the other in the Museum of the Thermae in Rome.

contains four sliding beads, and the upper one, which is the shorter, contains only one.

In the space between these two rows of slots a series of signs is inscribed, one for each slot. These are figures expressing the different powers of 10 according to the classical Roman number-system which the bankers and the publicans used to count by *as*, by *sestertii*, and by *denarii** (Fig. 16.62 and 16.67 above):

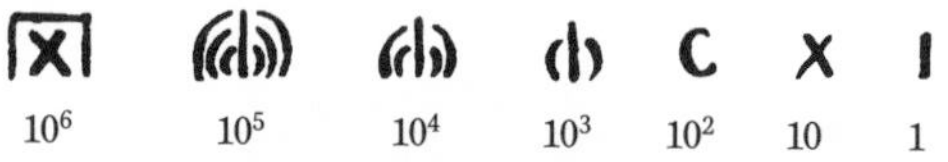

$10^6$ $10^5$ $10^4$ $10^3$ $10^2$ 10 1

FIG. 16.93.

FIG. 16.94. *Roman "pocket abacus" (in bronze), beginning of Common Era. (Cabinet des médailles, Bibliothèque nationale, Paris.) (br. 1925)*

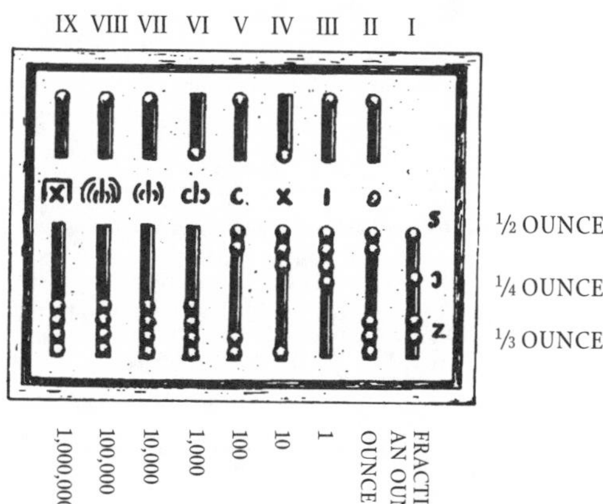


FIG. 16.95. *The principle of the portable Roman abacus. This specimen belonged to the German Jesuit Athanasius Kircher (1601–1680). (Museum of the Thermae, Rome)*

Each of these seven slots was therefore associated with a power of 10. From right to left, the third corresponded to the number 1, the fourth to the tens, the fifth to the hundreds and so on (Fig. 16.95).

If the number of units of a power of 10 did not exceed 4, it was indicated in the lower slot by pushing the same number of beads upwards. When it exceeded 4, the beads in the upper slot was pulled down towards the centre, and 5 units were removed from the number and this was represented in the lower slot.

If we are considering a calculation in *denarii*, the number represented on the abacus in Fig. 16.95 corresponds (leaving aside the first two slots on the right) to the sum of 5,284 *denarii*: 4 beads up in the lower slot III means 4 ones or 4 *denarii*; the upper bead down and 3 beads up in the lower slot IV means (5 + 3) tens or 80 *denarii*; two beads up in the lower slot V means two hundreds or 200 *denarii*; and finally the upper bead down in slot VI means five thousands or 5,000 *denarii*.

FIG. 16.96. *Bas-relief on a sarcophagus from a Roman tomb dating from the first century CE. (Capitoline Museum, Rome.)*

The first two slots on the right were used to note divisions of the *as*.* The second slot, marked with a single O, has an upper part with a single bead, and a lower which has not four, but five beads: it was used to represent multiples of the *uncia* (ounce) or twelfths of the *as*, each lower bead being worth one ounce and the upper bead being worth six ounces, which

* The unit of the Roman monetary system was the *as* of bronze. Its weight continually diminished, from the origin of the monetary system around the fourth century BCE until the Empire. It successively weighed 273 gm, 109 gm, 27 gm, 9gm and finally 2.3 gm. Its multiples were the *sestertius* (first silver, later bronze, then brass), the *denarius* (silver), and, from the time of Caesar, the *aureus* (gold). In the third century BCE, 1 *denarius* was 2.5 *as*, 1 *sestertius* was 4 *denarii* or 10 *as*. From the second century BCE, after a general monetary reform, 1 *sestertius* was 4 *as*, 1 *denarius* was 4 *sestertii* or 16 *as*, and 1 *aureus* was 25 *denarii* or 400 *as*.

* In Roman commercial arithmetic, fractions of a monetary unit were always expressed in terms of the *as*, the basic unit of money which was divided into twelve equal parts called *unciae* ("ounces") – which gave its name to the corresponding unit of value whatever its nature. Each multiple or sub-multiple of the *as* (or of the unit which the *as* represented) was then given a particular name. For example, for the sub-multiples we have 1/2: *as semis*; 1/3: *as triens*; 1/4: *as quadrans*; 1/5: *as quincunx*; 1/6: *as sextans*; 1/7: *as septunx*; 1/8: *as octans*; 1/9: *as dodrans*; 1/10: *as dextans*; 1/11: *as deunx*; 1/12: *as uncia*; 1/24: *as semuncia*; 1/48: *as sicilicus*; 1/72: *as sextula*.

allows counting up to 11/12 of an *as*. The first slot, divided into three and carrying four sliding beads, was used for the half ounce, the quarter ounce and the *duella*, or third part of the ounce. The upper bead was was worth 1/2 ounce or 1/24 *as* if it was placed at the level of the sign: S or Ƨ or Ƹ : the sign of *as semuncia*, 1/24 of an *as*.

The middle bead was worth 1/4 ounce or 1/48 *as* if it was placed at the level of the sign: Ɔ or ) or 7: the sign of *as sicilicus*, 1/48 of an *as*.

Finally, either of the two beads at the bottom of the slot was worth 1/3 ounce or 2/72 *as* if it was placed at the level of the sign: Z or 2 or ƻ: the sign of *as duae sextulae*, 2/72 of an *as* or *duella*.

The four beads of the first slot probably had different colours (one for the half ounce, one for the quarter ounce, and one for the third of an ounce) in case the three should find themselves on top of each other (as in Fig. 16.95). In certain abacuses these three beads ran in three separate slots.

Therefore we have here a calculating instrument very much the same as the famous Chinese abacus which still occupies an important place in the Far East and in certain East European countries.

With a highly elaborate finger technique executed according to precise rules, this "pocket calculator" (one of the first in all history) allowed those who knew how to use it to rapidly and easily carry out many arithmetic calculations.

Why did Western Europeans of the Middle Ages – the direct heirs of Roman civilisation – carry on using ancient calculating tables in preference to this more refined, better conceived, and far more useful instrument? We still do not know. Perhaps the invention belonged to one particular school of arithmeticians, which disappeared along with its tools at the fall of the Roman Empire.

CHAPTER 17

# LETTERS AND NUMBERS

## THE INVENTION OF THE ALPHABET

The invention of the alphabet was a huge step in the history of human civilisation. It constituted a far better way of representing speech in any articulated language, for it allowed all the words of a given language to be fixed in written form with only a small set of phonetic signs called *letters*.

This fundamental development was made by northwestern Semites living near the Syrian-Palestinian coast around the fifteenth century BCE. The Phoenicians were bold sailors and intrepid traders: once they had broken with the complex writing systems of the Egyptians and Babylonians by inventing their simpler method for recording speech, they took it with them to the four corners of the Mediterranean world. In the Middle East, they brought the idea of an alphabet to their immediate neighbours, the Moabites, the Edomites, the Ammonites, the Hebrews, and the Aramaeans. These latter were nomads and traders too, and thus spread alphabetic writing to all the cultures of the Middle East, from Egypt to Syria and the Arabian peninsula, from Mesopotamia to the confines of the Indian subcontinent. From the ninth century BCE, alphabetic writing of the Phoenician type also began to spread around the Mediterranean shores, and was gradually adopted by speakers of Western languages, who adapted it to their particular needs by modifying or adding some characters.

The twenty-two letters of the Phoenician alphabet thus gave rise directly to Palaeo-Hebraic writing (in the era of the Kings of Israel and Judaea), whence came the modern alphabet of the Samaritans, who have maintained ancient Jewish traditions. Aramaic script developed a little later, whence came the "square" or black-letter Hebrew alphabet, as well as Palmyrenean, Nabataean, Syriac, Arabic, and Indian writing systems. At the same time, Phoenician letters gave birth to the Greek alphabet, the first one to include full and rigorous representation of the vowel sounds. From Greek came the Italic alphabets (Oscan, Umbrian and Etruscan as well as Latin), and at a later stage the alphabets used for Gothic, Armenian, Georgian, and Russian (the Cyrillic alphabet). In brief, almost all alphabets in use in the world today are descended directly or indirectly from what the Phoenicians first invented.

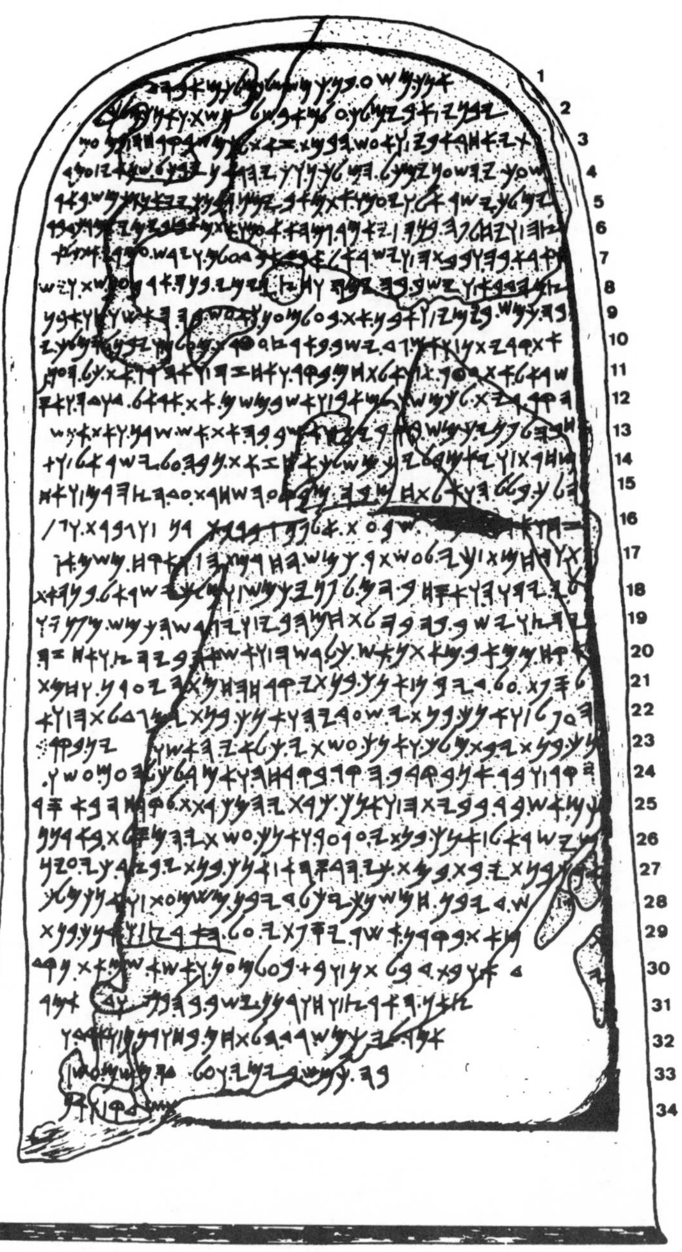

FIG. 17.1. Stela *of the Moabite king Mesha, a contemporary of the Jewish kings Ahab (874–853 BCE) and Joram (851–842 BCE), in the Louvre (M. Lidzbarski, vol. II, tab. 1). This is one of the oldest examples of palaeo-Hebrew script (used here to write in Moabitic, a dialect of Canaan close to Hebrew and Phoenician). This stela, put up in 842 BCE at Dibon-Gad, the Moabite capital, gives several clues to the relations that existed between Moab and Israel at that time; it is also the only document of the period found so far outside of Palestine in which the name of the God Yahweh is explicitly mentioned.*

## LETTERS AND ALPHABETIC NUMBERING

It is a remarkable fact that the names and the order of the twenty-two letters of the original Phoenician alphabet have been maintained more or less intact by almost all derivative alphabets, from Hebrew to Aramaic, from Etruscan to ancient Arabic, from Greek to Syriac. According to J. G. Février (1959), we can be sure of the order of the Phoenician letters because there are alphabetic primers in Etruscan dating from 700 BCE the order of whose letters is the same as the one encoded in many acrostics in biblical Hebrew (the lines of Psalms 9, 10, 25, 34, 111, 112, etc. begin with each of the letters of the Hebrew alphabet, in alphabetic order). In fact, the same order of the letters is even found in Ugaritic primers, dating from the fourteenth century BCE. These primers contain thirty letters written in cuneiform: however, as M. Sznycer has shown, the eight "extra" Ugaritic signs, intercalated or appended to the original twenty-two, do not alter the fundamental Phoenician order of the letters.

It is because the order of the ABC . . . is so ancient and so fixed that letters were able to play an important role in numbering systems.

| | ARCHAIC PHOENICIAN | | PALAEO-HEBREW SCRIPT | | | | | HEBREW | | |
|---|---|---|---|---|---|---|---|---|---|---|
| | Inscription of Akhiram, 11th century BCE | Inscription of Yehimilk, 10th century BCE | Stela of Mesha, 842 BCE | Samarian Ostraca, 8th century BCE | Arad Ostraca, 7th century BCE | Lakhish Ostraca, 6th century BCE | Aramaic cursive, Elephantine, 5th century BCE | Dead Sea Scrolls | Rabbinical cursive | Black-letter Hebrew |
| aleph | 𐤀 | 𐤀 | 𐤀 | 𐤀 | 𐤀 | 𐤀 | 𐡀 | א | א | א |
| bet | 𐤁 | 𐤁 | 𐤁 | 𐤁 | 𐤁 | 𐤁 | 𐡁 | ב | ב | ב |
| gimmel | 𐤂 | 𐤂 | 𐤂 | 𐤂 | 𐤂 | 𐤂 | 𐡂 | ג | ג | ג |
| dalet | 𐤃 | 𐤃 | 𐤃 | 𐤃 | 𐤃 | 𐤃 | 𐡃 | ד | ד | ד |
| he | 𐤄 | 𐤄 | 𐤄 | 𐤄 | 𐤄 | 𐤄 | 𐡄 | ה | ה | ה |
| vov | 𐤅 | 𐤅 | 𐤅 | 𐤅 | 𐤅 | 𐤅 | 𐡅 | ו | ו | ו |
| zayin | 𐤆 | 𐤆 | 𐤆 | 𐤆 | | 𐤆 | 𐡆 | ז | ז | ז |
| het | 𐤇 | 𐤇 𐤇 | 𐤇 | 𐤇 | 𐤇 | 𐤇 | 𐡇 | ח | ח | ח |
| tet | 𐤈 | | 𐤈 | | | 𐤈 | 𐡈 | ט | ט | ט |
| yod | 𐤉 | 𐤉 𐤉 | 𐤉 | 𐤉 | 𐤉 | 𐤉 | 𐡉 | י | י | י |
| kof | 𐤊 | 𐤊 | 𐤊 | 𐤊 | 𐤊 | 𐤊 | 𐡊 | כ | כ | כ |
| lamed | 𐤋 | 𐤋 | 𐤋 | 𐤋 | 𐤋 | 𐤋 | 𐡋 | ל | ל | ל |
| mem | 𐤌 | 𐤌 | 𐤌 | 𐤌 | 𐤌 | 𐤌 | 𐡌 | מ | מ | מ |
| nun | 𐤍 | 𐤍 | 𐤍 | 𐤍 | 𐤍 | 𐤍 | 𐡍 | נ | נ | נ |
| samekh | 𐤎 | | 𐤎 | 𐤎 | 𐤎 | 𐤎 | 𐡎 | ס | ס | ס |
| ayin | 𐤏 | 𐤏 | 𐤏 | 𐤏 | 𐤏 | 𐤏 | 𐡏 | ע | ע | ע |
| pe | 𐤐 | 𐤐 | 𐤐 | 𐤐 | | 𐤐 | 𐡐 | פ | פ | פ |
| tsade | [𐤑] | 𐤑 | 𐤑 | 𐤑 | 𐤑 | 𐤑 | 𐡑 | צ | צ | צ |
| quf | [𐤒] | 𐤒 | 𐤒 | 𐤒 | 𐤒 | 𐤒 | 𐡒 | ק | ק | ק |
| resh | 𐤓 | 𐤓 | 𐤓 | 𐤓 | 𐤓 | 𐤓 | 𐡓 | ר | ר | ר |
| shin | 𐤔 | 𐤔 | 𐤔 | 𐤔 | 𐤔 | 𐤔 | 𐡔 | ש ש | ש | ש |
| tav | 𐤕 𐤕 | 𐤕 | 𐤕 | 𐤕 | 𐤕 | 𐤕 | 𐡕 | ת | ת | ת |

FIG. 17.2. *Western Semitic alphabets*

| PHOENICIAN 10th to 6th centuries BCE | | HEBREW Early | HEBREW Modern | ANCIENT GREEK 5th century BCE | ITALIC ALPHABETS Oscan | Umbrian | Etruscan | |
|---|---|---|---|---|---|---|---|---|
| aleph | 𐤀 | א | א | Α alpha | 𐌀 | 𐌀 | 𐌀 | a |
| bet | 𐤁 | ב | ב | Β beta | 𐌁 | 𐌁 | 𐌁 | b |
| gimmel | 𐤂 | ג | ג | Γ gamma | 𐌂 | | 𐌂 | g |
| dalet | 𐤃 | ד | ד | Δ delta | 𐌃 | | 𐌃 | d |
| he | 𐤄 | ה | ה | Ε epsilon | 𐌄 | 𐌄 | 𐌄 | e |
| vov | 𐤅 | ו | ו | Ϝ digamma | 𐌅 | 𐌅 | 𐌅 | v |
| zayin | 𐤆 | ז | ז | zeta | 𐌆 | 𐌆 | 𐌆 | z |
| het | 𐤇 | ח | ח | Η eta | 𐌇 | 𐌇 | 𐌇 | h |
| tet | 𐤈 | ט | ט | Θ theta | | | 𐌈 | th |
| yod | 𐤉 | י | י | Ι iota | 𐌉 | 𐌉 | 𐌉 | i |
| kof | 𐤊 | כ | כ | Κ kappa | 𐌊 | 𐌊 | 𐌊 | k |
| lamed | 𐤋 | ל | ל | Λ lambda | 𐌋 | 𐌋 | 𐌋 | l |
| mem | 𐤌 | מ | מ | Μ mu | 𐌌 | 𐌌 | 𐌌 | m |
| nun | 𐤍 | נ | נ | Ν nu | 𐌍 | 𐌍 | 𐌍 | n |
| samekh | 𐤎 | ס | ס | Ξ xi | | | 𐌎 | s? |
| ayin | 𐤏 | ע | ע | Ο omicron | 𐌞 | | 𐌏 | o |
| pe | 𐤐 | פ | פ | Π pi | 𐌐 | 𐌐 | 𐌐 | p |
| tsade | 𐤑 | צ | צ | Ϻ san | | | 𐌑 | ṣ |
| quf | 𐤒 | ק | ק | Ϙ koppa | | | 𐌒 | q |
| resh | 𐤓 | ר | ר | Ρ rho | 𐌓 | 𐌛 | 𐌓 | r |
| shin | 𐤔 | ש | ש | Σ sigma | 𐌔 | 𐌔 | 𐌔 | s |
| tav | 𐤕 | ת | ת | Τ tau | 𐌕 | 𐌕 | 𐌕 | t |
| | | | | Υ upsilon | 𐌖 | 𐌖 | 𐌖 | u |
| | | | | phi | 𐌚 | 𐌚 | 𐌘 | f |
| | | | | chi | | | 𐌙 | kh |
| | | | | psi | | 𐌜 | | dh |
| | | | | omega | | 𐌛 | | c |

FIG. 17.3. *Phoenician and Hebrew alphabets compared to Greek and Italic*

| PHOENICIAN et ARAMAEAN | HEBREW | SYRIAC | ANCIENT ARABIC | GREEK |
|---|---|---|---|---|
| 'aleph (') | aleph (') | olap (') | alif (') | alpha (a) |
| bet (b) | bet (b, v) | bet (b) | ba (b) | beta (b) |
| gimmel (g) | gimmel (g) | gomal (g) | jim (j) | gamma (g) |
| dalet (d) | dalet (d) | dolat (d) | dal (d) | delta (d) |
| he (h) | he (h) | he (h) | ha (h) | epsilon (e) |
| waw (w) | vov (v) | waw (w) | wa (w) | faw* (f) |
| zayin (z) | zayin (z) | zayin (z) | zay (z) | zeta (z) |
| ḥet (ḥ) | het (h) | ḥet (ḥ) | ḥa (ḥ) | eta (e) |
| ṭet (ṭ) | tet (t) | ṭet (ṭ) | ṭa (ṭ) | theta (th) |
| yod (y) | yod (y) | yud (y) | ya (y) | iota (i) |
| kaf (k) | kof (k, kh) | kop (k) | kaf (k) | kappa (k) |
| lamed (l) | lamed (l) | lomad (l) | lam (l) | lambda (l) |
| mem (m) | mem (m) | mim (m) | mim (m) | mu (m) |
| nun (n) | nun (n) | nun (n) | nun (n) | nu (n) |
| samekh (s) | samekh (s) | semkat (s) | sin (s) | ksi (ks) |
| 'ayin (') | ayin (') | 'e (') | 'ayin (') | omicron (o) |
| pe (p) | pe (p, f) | pe (p, f) | fa (f) | pi (p) |
| ṣade (ṣ) | tsade (ts) | ṣode (ṣ) | ṣad (ṣ) | *san* (*s*) |
| qof (q) | quf (q) | quf (q) | qaf (q) | *qoppa* (*q*) |
| resh (r) | resh (r) | rish (r) | ra (r) | rho (r) |
| shin (s, sh) | shin (s, sh) | shin (sh) | shin (sh) | sigma (s) |
| taw (t) | tav (t) | taw (t) | ta (t) | tau (t) |
| | | | *tha* (*th*) | *upsilon* (*u*) |
| | | | *kha* (*kh*) | *phi* (*ph*) |
| | | | . . . . . . | . . . . . . |

* or digamma, subsequently dropped

FIG. 17.4. *The order of the twenty-two Phoenician letters has in most cases been preserved unaltered. The names here given to the Phoenician letters are only confirmed from the sixth century BCE, but their order and phonetic values go back much further, to at least the fourteenth century BCE.*

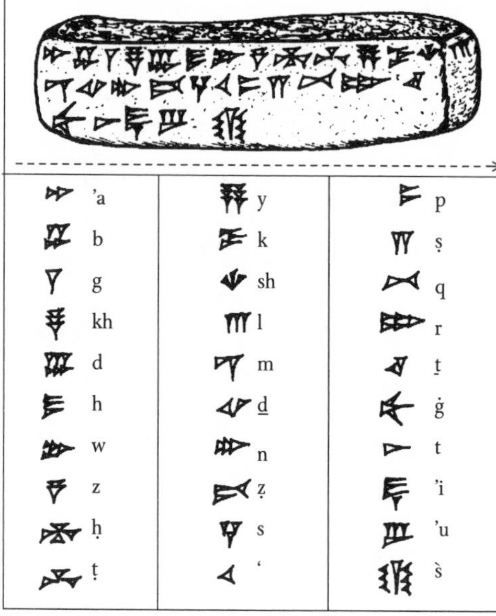


Fig. 17.5. *Ugaritic alphabet primer, fourteenth century BCE, found in 1948 at Ras Shamra. Damascus Museum. Transcription made by the author from a cast. See PRU II (1957), p. 199, document 184 A*

## SILENT NUMBERS

North African shepherds used to count their flock by reciting a text that they knew by heart: "Praise be to Allah, the merciful, the kind . . .". Instead of using the fixed order of the number-names (*one, two, three* . . . ), they would use the fixed order of the words of the prayer as a "counting machine". When the last of the sheep was in the pen, the shepherd would simply retain the last word that he had said of the prayer as the name of the number of his flock.

This custom corresponds to an ancient superstition in this and many other cultures that counting aloud is, if not a sin, then a hostage to the forces of evil. In this view, numbers do not just express arithmetical quantities, but are endowed with ideas and forces that are sometimes benign and sometimes malign, flowing under the surface of mortal things like an underground river. People who hold such a belief may count things that are not close (such as people or possessions belonging to others), but must not count aloud their own loved ones or possessions, for *to name an entity is to limit it*. So you must never say how many brothers, wives or children you have, never name the number of your cattle, sheep or dwellings, or state your age or your total wealth. For the forces of evil could capture the hidden power of the number if it were stated aloud, and thus dispose of the people or things numbered.

The North African shepherd using the prayer as a counting device was therefore doing so not only to invoke the protection of Allah, but also to avoid using the actual names of numbers. In that sense his custom is similar to the use of counting-rhymes by children – fixed rhythmic sequences of words which when recited determine whose go it is at a game. In Britain, for instance, children chant as they point round at each other in a circle: *eeny meeny miny mo – catch a blackman by his toe – if he hollers let him go – eeny meeny miny mo!* The child whose "go" it is is the one to whom the finger is pointing when the reciter reaches *mo!*

The use of a fixed sequence like this is reminiscent of the archaic counting methods of pre-numerate peoples, for whom points of the body functioned much like a counting rhyme. Similarly, disturbed children (and sometimes quite normal ones) invent their own counting sequences: one boy I got to know counted *André, Jacques, Paul, Alain, Georges, Jean, François, Gérard, Robert,* (for 1, 2 . . . 9) in virtue of the position of his dorm-mates' bunks with respect to his own; and G. Guitel (1975) reports the case of a girl who counted things as *January, February, March* . . . etc.

The girl could of course have used instead the invariable order of the letters of the alphabet (A, B, C, D, E . . . ), for any sequence of symbols can be used as a counting model – provided that the order of its elements is immutable, as it is with the alphabet. And for that reason many civilisations have thought of representing numbers with the letters of their alphabet, still set in the order given them by the Phoenicians.

From the sixth century BCE, the Greeks developed a written numbering system from 1 to 24 by means of alphabetic letters, known as acrophonics:

| | | | | | |
|---|---|---|---|---|---|
| Α | 1 | Ι | 9 | Ρ | 17 |
| Β | 2 | Κ | 10 | Σ | 18 |
| Γ | 3 | Λ | 11 | Τ | 19 |
| Δ | 4 | Μ | 12 | Υ | 20 |
| Ε | 5 | Ν | 13 | Φ | 21 |
| Ζ | 6 | Ξ | 14 | Χ | 22 |
| Η | 7 | Ο | 15 | Ψ | 23 |
| Θ | 8 | Π | 16 | Ω | 24 |

Fig. 17.6.

The tablets of Heliastes, like the twenty-four songs of the *Iliad* and the *Odyssey*, used this kind of numbering, which is also found on funerary inscriptions of the Lower Period. However, what we have here is really only a simple substitution of letters for numbers, not a proper alphabetic number-system which, as we will now see, calls for a much more elaborate structure.

## HEBREW NUMERALS

Jews still use a numbering system whose signs are the letters of the alphabet, for expressing the date by the Hebrew calendar, for chapters and verses of the Torah, and sometimes for the page numbers of books printed in Hebrew.

Hebrew characters, in common with most Semitic scripts, are written right to left and, a little like capital letters in the Latin alphabet, are clearly separated from each other. Most of them have the same shape wherever they come in a word: the five exceptions are the final *kof*, *mem*, *nun*, *pe*, and *tsade* (respectively, the Hebrew equivalents of our K, M, N, P, and a special letter for the sound TS):

| | *kof* | *mem* | *nun* | *pe* | *tsade* |
|---|---|---|---|---|---|
| Regular form | כ | מ | נ | פ | צ |
| Final form | ך | ם | ן | ף | ץ |

FIG. 17.7.

Black-letter ("square") Hebrew script is relatively simple and well-balanced, but care has to be taken with those letters that have quite similar graphical forms and which can mislead the unwary beginner:

| ב כ פ<br>b k p | ד ר ך<br>d r k final | ם ט<br>m t | ו ז<br>v z |
|---|---|---|---|
| ג נ<br>g n | ה ח ת<br>h kh t | ס ם<br>s m final | ע צ<br>(guttural) ts |

FIG. 17.8.

Hebrew numerals use the twenty-two letters of the alphabet, in the same order as those of the Phoenician alphabet from which they derive, to represent (from *aleph* to *tet*) the first nine units, then from *yod* to *tsade*, the nine "tens", and finally from *kof* to *tav*, the first four hundreds (see Fig. 17.10).

| | | | | |
|---|---|---|---|---|
| א aleph | ו vov | כ kof | ע ayin | ש shin |
| ב bet | ז zayin | ל lamed | פ pe | ת tav |
| ג gimmel | ח het | מ mem | צ tsade | |
| ד daleth | ט tet | נ nun | ק kuf | |
| ה he | י yod | ס samekh | ר resh | |

FIG. 17.9. *The Modern Hebrew alphabet*

| Letter | Name | Sound | Value |
|---|---|---|---|
| א | aleph | (h) a | 1 |
| ב | bet | b | 2 |
| ג | gimmel | g | 3 |
| ד | daleth | d | 4 |
| ה | he | h | 5 |
| ו | vov | v | 6 |
| ז | zayin | z | 7 |
| ח | het | kh | 8 |
| ט | tet | t | 9 |
| י | yod | y | 10 |
| כ | kof | k | 20 |

| Letter | Name | Sound | Value |
|---|---|---|---|
| ל | lamed | l | 30 |
| מ | mem | m | 40 |
| נ | nun | n | 50 |
| ס | samekh | s | 60 |
| ע | ayin | guttural | 70 |
| פ | pe | p | 80 |
| צ | tsade | ts | 90 |
| ק | kuf | k | 100 |
| ר | resh | r | 200 |
| ש | shin | sh | 300 |
| ת | tav | t | 400 |

FIG. 17.10. *Hebrew numerals*

Compound numbers are written in this system, from right to left, by juxtaposing the letters corresponding to the orders of magnitude in descending order (i.e., starting with the highest). Numbers thus fit quite easily in Hebrew manuscripts and inscriptions. But when letters are used as numbers, how do you distinguish numbers from "ordinary" letters?

THIS IS THE MONUMENT OF ESTHER DAUGHTER OF ADAIO, WHO DIED IN THE MONTH OF SHIVRAT OF YEAR 3 (ג) OF THE "SHEMITA". YEAR THREE HUNDRED AND 46 (שמו) AFTER THE DESTRUCTION OF THE TEMPLE (OF JERUSALEM)* PEACE! PEACE BE WITH HER!

* Year 346 of the Shemita +70 = 416 CE

FIG. 17.11. *Jewish gravestone (written in Aramaic), dated 416 CE. From the southwest shore of the Dead Sea. Amman Museum (Jordan). See IR, inscription 174*

HERE LIES AN
INTELLIGENT WOMAN
QUICK TO GRASP ALL
THE PRECEPTS OF FAITH
AND WHO FOUND THE
FACE OF GOD THE
MERCIFUL AT THE
TIME THAT COUNTS (?)
WHEN HANNA DEPARTED
SHE WAS 56 YEARS OLD

נ ו = נ'ו'
6 50
←------
56

Fig. 17.12. *Part of a bilingual (Hebrew-Latin) inscription carved on a soft limestone funeral* stela *found at Ora (southern Italy), seventh or eighth century CE. See CII, inscription 634 (vol. 1, p. 452).*

Numbers that are represented by a single letter are usually distinguished by a small slanted stroke over the upper left-hand corner of the character, thus:

| ש' | פ' | ל' | ג' | א' |
|---|---|---|---|---|
| 300 | 80 | 30 | 3 | 1 |

Fig. 17.13.

When the number is represented by two or more letters, the stroke is usually doubled and placed between the last two letters to the left of the expression (Fig. 17.14). But as these accent-strokes were also used as abbreviation signs, scribes and stone-cutters sometimes used other types of punctuation or "pointing" to distinguish numbers from letters (Fig. 17.15):

| שנ"ב | ל"ה |
|---|---|
| 2 + 50 + 300 | 5 + 30 |
| ←------------ | ←------ |

Fig. 17.14.

| | | |
|---|---|---|
| 150 | קנ | IHE, 183<br>date: 1389-90 CE |
| 175 | קע"ה | IHE, 100<br>date: 1415 CE |
| 196 | קצו | IHE, 201<br>date: 1436-37 CE |
| 219 | רי"ט | BM Add. 27 106<br>date: 1459 CE |
| 312 | שי"ב | BM Add. 27 146<br>date: 1552 CE |

| | | |
|---|---|---|
| 25 | כ"ה | IHE, 26<br>date: 1239 CE |
| 27 | כ"ז | IHE, 27<br>date: 1240 CE |
| 28 | כ"ח | IHE, 45<br>date: 1349 CE |
| 32 | לב | IHE, 110<br>date: 1271–72 CE |
| 44 | מ́ד́ | IHE, 139<br>date: 1283–84 CE |

Fig. 17.15. *Numerical expressions found in mediaeval Hebrew manuscripts and inscriptions. See Cantera and Millas*

The highest Hebrew letter-numeral is only 400, so this is how higher numbers were expressed:

| תתק | תת | תש | תר | תק |
|---|---|---|---|---|
| 100 400 400 | 400 400 | 300 400 | 200 400 | 100 400 |
| 900 | 800 | 700 | 600 | 500 |

Fig. 17.16.

So for numbers from 500 to 900, the customary solution was to combine the letter *tav* (= 400) with the letters expressing the complement in hundreds. Compound numbers in this range were written as follows:

| | |
|---|---|
| תתס"ט<br>9 60 400 400<br>←------------<br>869 | IHE, 102<br>date: 1108 CE |
| תתק"ם<br>40 100 400 400<br>←------------<br>940 | IHE, 107<br>date: 1180 CE |
| תתקמג<br>3 40 100 400 400<br>←------------<br>943 | IHE, 108<br>date: 1183 CE |

Fig. 17.17. *Expressions found on Jewish gravestones in Spain*

The numbers 500, 600, 700, 800, and 900 could also be represented by the final forms of the letters *kof, mem, nun, pe,* and *tsade* (see Fig. 17.7 above). However, this notation, which is found, for example, in the Oxford manuscript 1822 quoted by Gershon Scholem, was adopted only in Cabbalistic calculations. So in ordinary use, these final forms of the letters simply had the numerical value of the corresponding non-final forms of the letters.

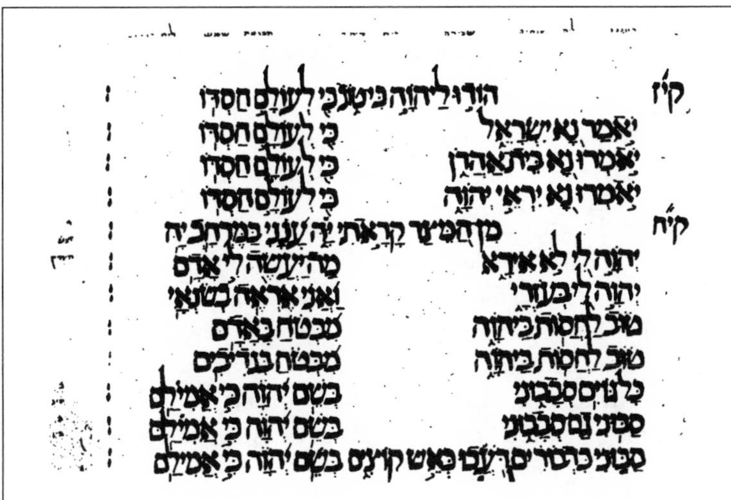

FIG. 17.18. *Page from a Hebrew codex, 1311 CE, giving Psalms 117 and 118. The numbers can be seen in the right-hand margin, in Hebrew letter-numbers. (Vatican Library, Cod. Vat. ebr. 12, fol. 58)*

To represent the thousands, the custom is to put two points over the corresponding unit, ten, or hundred character. In other words, when a character has two points over it, its numerical value is multiplied by 1,000.

| א --> א̈ | ב --> ב̈ | מ --> מ̈ | צ --> צ̈ |
|---|---|---|---|
| 1 1,000 | 2 2,000 | 40 40,000 | 90 90,000 |

FIG. 17.19.

The Hebrew calendar in its present form was fixed in the fourth century CE. Since then, the months of the Jewish year begin at a theoretical, calculated date and not, as previously, at the sighting of the new moon. The foundation point for the calculation was the *neomenia* (new moon) of Monday, 24 September 344 CE, fixed as 1 Tishri in the Hebrew calendar, that is to say New Year's Day. As it was accepted that 216 Metonic cycles, in other words 4,400 years, sufficed at that point to contain the entire Jewish past, the chronologists calculated that the first *neomenia* of creation took place on Monday, 7 October 3761 BCE. As a result, the Jewish year 5739, for example, corresponds to the period from 2, October 1978 to 21 September 1979, and it is expressed on Jewish calendars (of the kind you can find in any kosher grocer's or corner shop) as:

ה̈ ת ש ל״ט

9 30 300 400 5000

<--------------------------

FIG. 17.20.

Jewish scribes and stone-carvers did not always follow this rule, but exploited an opportunity to simplify numerical expressions that was implicit in the system itself. Consider the following expression found on a gravestone in Barcelona: it gives the year 5060 of the Jewish calendar (1299–1300 CE) in this manner:

ה̇ ס̇ (= 5 × 1,000 + 60)

60 5

<---------

FIG. 17.21.

Here, the points simply signify that the letters are to be read as numbers, not letters. But the expression appears to break one of the cardinal rules of Hebrew numerals – that the highest number always comes first, counting from right to left, which is the direction of writing in the Hebrew alphabet. So in any regular numerical expression, the letter to the right has a higher value than the one to its left. For that reason, the expression on the Barcelona gravestone is entirely unambiguous. Since the letter *he* can only have two values – 5 and 5,000, and the letter *samekh* counts for 60, the character to the right, despite not having its double point, must mean 5,000.

...שמואל ב״ר חלאבו...
...בשנת ת̇ת̇ד̇
804
"... SAMUEL SON OF KHALABU ...
... IN THE YEAR 804"

FIG. 17.22. *Fragment of a Jewish gravestone from Barcelona. The date is given as 804, for 4804 (4804 – 3760 = 1044 CE).*

Here are some other examples:

| | | |
|---|---|---|
| 5,109 | הׇ קׇ טׇ<br>9 100 5 | Toledo, 1349 CE; IHE, no. 85 |
| 5,156 | ה׳ק׳נ׳ו׳<br>6 50 100 5 | MS dated 1396; BM Add. 2806, fol. 11a |

There is an even more interesting "irregularity" in the way some mediaeval Jewish scholars wrote down the total number of verses in the Torah [see G. H. F. Nesselmann (1842), p. 484]. The figure, 5,845, was written by using only the letters for the corresponding units, thus:

החט״ה

| He | Mem | Het | He |
|---|---|---|---|
| 5 | 40 | 8 | 5 |

←-------------------

Because of the rule that we laid out above, this expression is not ambiguous. The letter *het*, for example, whose normal value is 8, cannot have a lower value than *mem*, to its left, and whose value is 40; nor can it be 8,000, since it is itself to the left of *he*, whose value must be larger. For that reason the *het* can only mean 800.

It is not difficult to account for this particular variant of Hebrew numerals. In speech, the number 5,845 is expressed by:

| KHAMISHAT | ALAFIM | SHMONEH | ME'OT | ARBA'IM | VE | KHAMISHA |
|---|---|---|---|---|---|---|
| "Five | thousand | eight | hundred | forty | & | five" |

The names of the numbers thus make the arithmetical structure of the number apparent:

$$5 \times 1{,}000 + 8 \times 100 + 40 + 5$$

This could be transposed into English as "five thousand eight hundred forty (&) five", or in Hebrew as:

ה׳ אלפים ח׳ טאות ט״ה

5 40 hundred 8 thousand 5

←-------------------------------------------------

FIG. 17.23.

"Mixed" formulations like these, combining words and numerals, are found on Hispano-Judaic tombstones (IHE, no. 61) and in some mediaeval manuscripts (for example, BM Add. 26 984, folio 143b). It is easy to see how such expressions can safely be abbreviated by leaving out the words for "hundred" and "thousand".

Another particularity arises in Hebrew numerals with the numbers 15 and 16. The regular forms would be:

יה (5 10) יו (6 10)

←------ ←-----

FIG. 17.24.

However, the letter-values of these numbers spell out parts of the name of *Yahweh* – and it is forbidden, in Jewish tradition, to write the name of the Lord, even if its literal form of four letters (the "divine tetragrammaton", יהוה "yahve") is perfectly well-known. To avoid writing the tetragrammaton, various abbreviations were devised (יהו, הו, יו, יה) but these two were covered by the prohibition on writing the name of God. So the regular forms of the numbers 15 and 16 could not be used, and were replaced by the expressions 9 + 6 and 9 + 7 respectively:

טו (6 9) טז (7 9)

←----- ←-----

FIG. 17.25.

These are the main features of Hebrew numerals. It was by no means the only one to use the letters of the alphabet for expressing numbers. Let us now look at the Greek system of alphabetic numbering.

## GREEK ALPHABETICAL NUMERALS

The Greek alphabet is absolutely fundamental for the history of writing and for Western civilisation as a whole. As C. Higounet (1969) explains, the Greek alphabet, quite apart from its having served to transmit one of the richest languages and cultures of the ancient world, forms the "bridge" between Semitic and Latin scripts. Historically, geographically, and also graphically, it was an intermediary between East and West; even more importantly, it was a structural intermediary too, in the sense that it first introduced regular and complete representations of the vowel sounds.

There is no question but that the Greeks borrowed their alphabet from the Phoenicians. Herodotus called the letters *phoinikeia grammatika*, "Phoenician writing"; and the early forms of almost all the Greek letters as well as their order in the alphabet and their names support this tradition. According to the Greeks themselves, Cadmos, the legendary founder of Thebes, brought in the first sixteen letters from Phoenicia; Palamedes was supposed to have added four more during the Trojan War; and four more were introduced later on by a poet, Simonides of Ceos.

| ARCHAIC PHOENICIAN ALPHABET | GREEK ALPHABETS: ARCHAIC (THERA) | GREEK ALPHABETS: EASTERN (MILETUS, CORINTH) | GREEK ALPHABETS: WESTERN (BOEOTIA) | CLASSICAL GREEK ALPHABET | |
|---|---|---|---|---|---|
| aleph | | | | Α α | alpha |
| bet | | | | Β β | beta |
| gimmel | | | | Γ γ | gamma |
| daleth | | | | Δ δ | delta |
| he | | | | Ε ϵ | epsilon |
| vov | | | | Ϝ Ϛ | digamma* |
| zayin | | | | Ζ ζ | zeta |
| het | | | | Η η | eta |
| tet | | | | Θ θ | theta |
| yod | | | | Ι ι | iota |
| kof | | | | Κ κ | kappa |
| lamed | | | | Λ λ | lambda |
| mem | | | | Μ μ | mu |
| nun | | | | Ν ν | nu |
| samekh | | | | Ξ ξ | ksi |
| ayin | | | | Ο ο | omicron |
| pe | | | | Π π | pi |
| tsade | | | | Ϻ | san* |
| kuf | | | | Ϙ ϙ | koppa* |
| resh | | | | Ρ ρ | rho |
| shin | | | | Σ σ | sigma |
| tav | | | | Τ τ | tau |
| | | | | Υ υ | upsilon |
| | | | | Φ φ | phi |
| | | | | Χ χ | chi |
| | | | | Ψ ψ | psi |
| | | | | Ω ω | omega |

*Greek letters that were eventually dropped from the alphabet

FIG. 17.26. *Greek alphabets compared to the archaic Phoenician script. See Février (1959) and Jensen (1969)*

The oldest extant pieces of writing in Greek date from the seventh century BCE. Some scholars believe that the original borrowing from Phoenician occurred as early as 1500 BCE, others think it did not happen until the eighth century BCE: but it seems most reasonable to suppose that it happened around the end of the second millennium or at the start of the first. At any rate, the Greek alphabet did not arise in its final form at all quickly. There was a whole series of regional variations in the slow adaptation of Phoenician letters to the Greek language, and these non-standard forms are generally categorised under the following headings: archaic alphabets (as found at Thera and Melos), Eastern alphabets (Asia Minor and its coastal archipelagos, the Cyclades, Attica, Corinth, Argos, and the Ionian colonies in Sicily and southern Italy), and Western alphabets (Eubeus, the Greek mainland, and non-Ionian colonies). Unification and standardisation did not occur until the fourth century BCE, following the decision of Athens to replace its local script with the so-called Ionian writing of Miletus, itself an Eastern form of the alphabet.

Early Greek writing was done right to left, or else in alternating lines (*boustrophedon*), but it settled down to left-to-right around 500 BCE. Since letters are formed from the direction of writing, this change of orientation has to be taken into account when we compare Greek characters to their Semitic counterparts.

The names of the original Greek letters are:

> *alpha, beta, gamma, delta, epsilon, digamma,*
> *zeta, eta, theta, iota, kappa, lambda, mu, nu,*
> *ksi, omicron, pi, san, koppa, rho, sigma, tau.*

Of these, the *digamma* was lost early on, and the *san* and *koppa* were also subsequently abandoned. However, a different form of the Semitic *vov* provided the *upsilon*, and three new signs, *phi*, *chi*, and *psi*, were added to represent sounds that do not occur in Semitic languages. Finally, *omega* was invented to distinguish the long *o* from the *omicron*. So the classical Greek alphabet, from the fourth century BCE, ended up having twenty-four letters, including vowels as well as consonants.

Semitic languages can be written down without representing the vowels because the position of a word in a sentence determines its meaning and also the vowel sounds in it, which change with different functions. In Greek, however, the inflections (word-endings) alone determine the function of a word in a sentence, and the vowel sounds cannot be guessed unless the endings are fully represented. The Phoenician alphabet had letters for guttural sounds that do not exist in Greek; Greek, for its part, had aspirated consonants with no equivalents in Semitic languages. So the

Greeks converted the Semitic guttural letters, for which they had no use, into vowels, which they needed. The "soft breathing sound" *aleph* became the Greek *alpha*, the sound of *a*; the Semitic letter *he* was changed into *epsilon* (*e*), and the *vov* first became *digamma* then *upsilon* (*u*); the Hebrew *yod* was converted into *iota* (*i*); and the "hard breathing sound" *ayin* became an *omicron* (*o*). For the aspirated consonants, the Greeks simply created new letters, the *phi*, *chi* and *psi*. In brief, the Greeks adapted the Semitic system to the particularities of their own language. But despite all that is clear and obvious about this process, the actual origin of the idea of representing the vowel sounds by letters remains obscure.

This survey of the development of the Greek alphabet allows us now to look at the principles of Greek numbering, often called a "learned" system, but which is in fact entirely parallel to Hebrew letter-numbers.

We can get a first insight into the system by looking at a papyrus (now in the Cairo Museum, Inv. 65 445) from the third quarter of the third century BCE (Fig. 17.31).

O. Guéraud and P. Jouguet (1938) explain that this papyrus is a "kind of exercise book or primer, allowing a child to practise reading and counting, and containing in addition various edifying ideas . . . As he learned to read, the child also became familiar with numbers. The place that this primer gives to the sequence of the numbers is quite natural, coming as it does after the table of syllables, because the Greek letters also had numerical values. It was logical to give the child first the combination of letters into syllables, and then the combinations of letters into numbers."

The numeral system the papyrus gives uses the twenty-four letters of the classical Greek alphabet, plus the three obsolete letters, *digamma*, *koppa* and *san* (see Fig. 17.26 above). These twenty-seven signs are divided into three classes. The first, giving the units 1 to 9, uses the first eight letters of the classical alphabet, plus *digamma* (the old Semitic *vov*), inserted in the sequence to represent the number 6. The second contains the eight following letters, plus the obsolete *koppa* (the old *quf*), to give the sequence of the tens, from 10 to 90. And the third class gives the hundreds from 100 to 900, using the last eight letters of the classical alphabet plus the *san* (the Semitic *tsade*) (for the value of 900) (see Fig. 17.27).

Intermediate numbers are produced by additive combinations. For 11 to 19, for instance, you use *iota*, representing 10, with the appropriate letter to its right representing the unit to be added. To distinguish the letters used as numerals from "ordinary" letters, a small stroke is placed over them. (The modern printing convention of placing an accent mark to the top right of the letter is not used in most Greek manuscripts.)

| UNITS | | | TENS | | | HUNDREDS | | |
|---|---|---|---|---|---|---|---|---|
| Α α | alpha | 1 | Ι ι | iota | 10 | Ρ ρ | rho | 100 |
| Β β | beta | 2 | Κ κ | kappa | 20 | Σ σ | sigma | 200 |
| Γ γ | gamma | 3 | Λ λ | lambda | 30 | Τ τ | tau | 300 |
| Δ δ | delta | 4 | Μ μ | mu | 40 | Υ υ | upsilon | 400 |
| Ε ϵ | epsilon | 5 | Ν ν | nu | 50 | Φ ϕ | phi | 500 |
| Ϛ ϛ | digamma* | 6 | Ξ ξ | ksi | 60 | Χ χ | chi | 600 |
| Ζ ζ | zeta | 7 | Ο ο | omicron | 70 | Ψ ψ | psi | 700 |
| Η η | eta | 8 | Π π | pi | 80 | Ω ω | omega | 800 |
| Θ ϑ | theta | 9 | Ϙ ϙ | koppa | 90 | Ϡ ϡ | san (sampi) | 900 |

*In manuscripts from Byzantium, 6 is written στ (sigma+tau). In Modern Greek, where alphabetic numerals are still used for specific purposes (rather like Roman numerals in our culture), this sign is called a *stigma*.

FIG. 17.27. *Greek alphabetic numerals*

The beginning of the primer scroll has the remnants of the number sequence up to 25:

| | | | |
|---|---|---|---|
| $\overline{\mathrm{H}}$ | 8 | $\overline{\mathrm{K}}$ | 20 |
| $\overline{\Theta}$ | 9 | $\overline{\mathrm{KA}}$ | 21 |
| $\overline{\mathrm{I}}$ | 10 | $\overline{\mathrm{KB}}$ | 22 |
| $\overline{\mathrm{IA}}$ | 11 | $\overline{\mathrm{K}\Gamma}$ | 23 |
| $\overline{\mathrm{IB}}$ | 12 | $\overline{\mathrm{K}\Delta}$ | 24 |
| $\overline{\mathrm{I}\Gamma}$ | 13 | $\overline{\mathrm{K}\epsilon}$ | 25 |

FIG. 17.28.

Guéraud & Jouguet (1938) note that the list is an elementary one, and does not even include all the symbols the pupil would need to understand the table of squares given at the end of the primer (Fig. 17.31). However, the table of squares itself, besides giving the young reader some basic ideas of arithmetic, also served to show the sequence of numbers beyond those given at the start of the scroll and to familiarise the learner with the principles of Greek numbering from 1 to 640,000, and that may have been its real purpose.

How could the scribe represent numbers from 1 to 640,000 when the highest numeral in the alphabet was only 900? For numbers up to 9,000, he just added a distinctive sign to the letters representing the units, thus*:

| 'Α | 'Β | 'Γ | 'Δ | 'Ε | 'Ϛ | 'Ζ | 'Η | 'Θ |
|---|---|---|---|---|---|---|---|---|
| 1,000 | 2,000 | 3,000 | 4,000 | 5,000 | 6,000 | 7,000 | 8,000 | 9,000 |

FIG. 17.29.

* Printed Greek usually puts the distinctive sign (a kind of *iota*) as a subscript, to the lower left corner of the character.

When he got to 10,000, otherwise called the myriad (Μυριοι)*, the second "base" of Greek numerals, he put an M (the first letter of the Greek word for "ten thousand") with a small *alpha* over the top. All following multiples of the myriad could therefore be written in the following way:

$$\overset{\alpha}{M} \quad \overset{\beta}{M} \quad \overset{\gamma}{M} \quad \overset{\delta}{M} \quad \overset{\epsilon}{M} \ldots \overset{\iota\alpha}{M} \quad \overset{\iota\beta}{M} \ldots \overset{\chi\xi\theta}{M} \ldots$$

10,000 20,000 30,000 40,000 50,000 110,000 120,000 6,690,000

FIG. 17.30.

As he gave these numbers in the form 1 myriad, 2 myriads, 3 myriads, etc. the scribe could reach 640,000 without any difficulty. He could obviously have continued the sequence up to the 9,999th myriad, which he would have written thus:

$$\overset{'\theta ϡ ϙ \theta}{M} \quad (\overset{9999}{M} = 99{,}990{,}000)$$

TRANSCRIPTION TRANSLATION

Left-hand column

| | | |
|---|---|---|
| | 4 | |
| | 5 | 25 |
| 6 | 6 | 36 |
| 7 | 7 | 79 |
| 8 | 8 | 64 |
| 9 | 9 | 81 |
| 10 | 10 | 100 |
| 20 | 20 | 400 |
| 30 | 30 | 900 |
| 40 | 40 | 1,600 |

Right-hand column

| | | |
|---|---|---|
| | 90 | 8,100 |
| 100 | 100 | 10,000 |
| 200 | 200 | 40,000 |
| 300 | 300 | 90,000 |
| 400 | 400 | 160,000 |
| 500 | 500 | 250,000 |
| 600 | 600 | 360,000 |
| 700 | 700 | 490,000 |
| 800 | 800 | 640,000 |

FIG. 17.31. *Fragment of a Greek papyrus, third quarter of the the third century BCE (Cairo Museum, inv. 65 445). See Guéraud & Jouguet (1938), plate X. The papyrus gives a table of squares, from 1 to 10 and then in tens to 40 (left-hand column), and from 50 to 800 (right-hand column). The squares of 1, 2 and 3 are missing from the start of the table.*

*When the accent is on the first syllable, the word means "ten thousand"; when the accent is on the second syllable, it has the meaning "a very large number".

These kinds of notation for very large numbers were frequently used by Greek mathematicians. For example, Aristarch of Samos (?310–?230 BCE) wrote the number 71,755,875 in the following way, according to P. Dedron & J. Itard (1959), p. 278:

'ζ ρ ο ε Μ 'ε ω ο ε

7,175 × 10,000 + 5,875

FIG. 17.32.

We find a different system in Diophantes of Alexandria (c. 250 CE): he separates the myriads from the thousands by a single point. So for him the following expression meant 4,372 myriads and 8,097 units, or 43,728,097 [from C. Daremberg & E. Saglio (1873), p. 426]:

δτοβ'η ϙζ

4,372 × 10,000 + 8,097

FIG. 17.33.

The mathematician and astronomer Apollonius of Perga (c. 262–c. 180 BCE) used a different method of representing very large numbers, and it has reached us through the works of Pappus of Alexandria (third century CE). This system was based on the powers of the myriad and used the principle of dividing numbers into "classes". The first class, called the elementary class, contained all the numbers up to 9,999, that is to say all numbers less than the myriad. The second class, called the class of *primary myriads*, contained the multiples of the myriad by all numbers up to 9,999 (that is to say the numbers 10,000, 20,000, 30,000, and so on up to 9,999 × 10,000 = 99,990,000). To represent a number in this class, the number of myriads in the number is written after the sign $\overset{\alpha}{M}$. A reconstructed example:

$$\overset{\alpha}{M}\chi\xi\delta$$

664 ------> means 664 × 10,000 = 6,640,000

FIG. 17.34.

Next comes the class of *secondary myriads*, which contains the multiples of a myriad myriads by all the numbers between 1 and 9,999 (that is to say, the numbers 100,000,000, 200,000,000, 300,000,000, and so on up to

9,999 × 100,000,000 = 999,900,000,000. A number in this range is expressed by writing *beta* over *M* before the number (written in the classical letter-number system) of one hundred millions that it contains. A reconstructed example:

$$\overset{\beta}{\mathrm{M}}\,{}'\epsilon\omega\xi\gamma$$

- - - - - - - - - - ->
5,863

FIG. 17.35.

This notation thus means: 5,863 × 100,000,000 = 586,300,000,000, and is "read" as 5,863 *secondary myriads*.

Next come the tertiary myriads, signalled by *gamma* over *M*, which begin at 100,000,000 × 10,000 = 1,000,000,000,000; then the quaternary myriads (signalled by *delta* over *M*), and so on.

The difference between the system used in the papyrus of Fig. 17.31 and the system of Apollonius is that whereas for the papyrus the superscribed letter over M is a *multiple* of 10,000, for Apollonius the superscript represents a *power* of 10,000.

In the Apollonian system, intermediate numbers can be expressed by breaking them down into a sum of numbers of the consecutive classes. Pappus of Alexandria [as quoted in P. Dédron & J. Itard (1959) p. 279] gave the example of the number 5,462,360,064,000,000, expressed as 5,462 tertiary myriads, 3,600 secondary myriads, and 6,400 primary myriads (in which the Greek word και can be taken to mean "plus"):

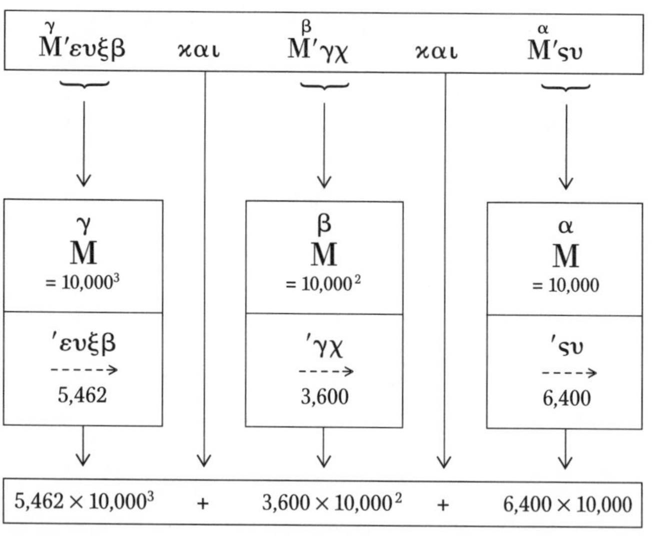


17.36.

Archimedes (?287– 212 BCE) proposed an even more elaborate system for expressing even higher magnitudes, and laid it out in an essay on the number of grains of sand that would fill a sphere whose diameter was equal to the distance from the earth to the fixed stars. Since he had to work with numbers larger than a myriad myriads, he imagined a "doubled class" of numbers containing eight digits instead of the four allowed for by the classical letter-number system, that is to say octets. The first octet would contain numbers between 1 and 99,999,999; the second octet, numbers starting at 100,000,000; and so on. The numbers belonged to the first, second, etc. class depending on whether they figure in the first, second, etc. octet.

As C. E. Ruelle points out in DAGR (pp. 425–31), this example suffices to show just how far Greek mathematicians developed the study and applications of arithmetic. Archimedes's conclusion was that the number of grains of sand it would take to fill the sphere of the world was smaller than the eighth term of the eighth octet, that is to say the sixty-fourth power of 10 (1 followed by 64 zeros). However, Archimedes's system, whose purpose was in any case theoretical, never caught on amongst Greek mathematicians, who it seems preferred Apollonius's notation of large numbers.

From classical times to the late Middle Ages, Greek alphabetic numerals played almost as great a role in the Middle East and the eastern part of the Mediterranean basin as did Roman numerals in Western Europe.

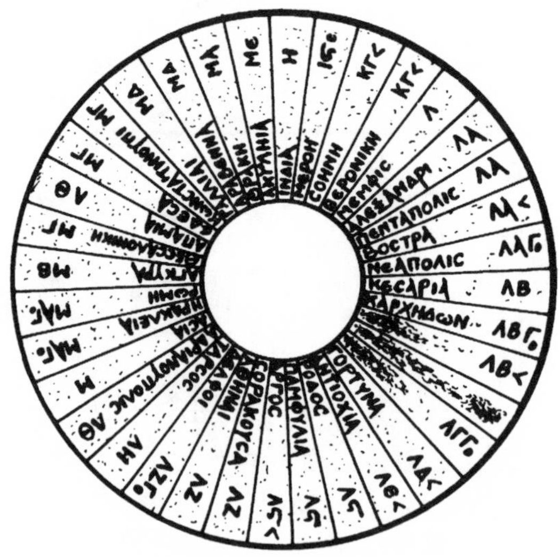

FIG. 17.37A. *Part of a portable sundial from the Byzantine era (Hermitage Museum, St Petersburg). This disc gives the names of the regions where it can be used, with latitudes indicated in Greek alphabetic numerals in ascending clockwise order.*

| TRANSCRIPTION | | TRANSLATION | |
|---|---|---|---|
| ΙΝΔΙΑ | Η | India | 8 |
| ΜΕΡΟΗ | Ις< | Meroe | 16 ½ |
| CΟΗΝΗ | ΚΓ< | Syena | 23 ½ |
| ΒΕΡΟΝΙΚΗ | ΚΓ< | Beronika | 23 ½ |
| ΜΕΜΦΙC | Λ | Memphis | 30 |
| ΑΛΕΞΑΝΔΡΙ | ΛΑ | Alexandria | 31 |
| ΠΕΝΤΑΠΟΛΙC | ΛΑ | Pentapolis | 31 |
| ΒΟCΤΡΑ | ΛΑ< | Bostra | 31 ½ |
| ΝΕΑΠΟΛΙC | ΛΑ Γο | Neapolis | 31 ⅔ |
| ΚΕCΑRΙΑ | ΛΒ | Caesaria | 32 |
| ΚΑΡΧΗΔΩΝ | ΛΒ Γο | Carthage | 32 ⅔ |
| . . . . . . . . . | ΛΒ< | . . . . . . . | 32 ½ |
| . . . . . . . . . | . . . | . . . . . . . | |
| . . . . . . . . . | ΛΓ Γο | . . . . . . . | 33 ⅔ |
| ΓΟΡΤΥΝΑ | ΛΔ< | Gortuna | 34 ½ |
| ΑΝΤΙΟΧΙΑ | ΛΕ< | Antioch | 35 ½ |
| ΡΟΔΟC | Λς | Rhodes | 36 |
| ΠΑΜΦΥΛΙΑ | Λς | Pamphilia | 36 |
| ΑΡΓΟC | Λς< | Argos | 36 ½ |
| CΟΡΑΚΟΥCΑ | ΛΖ | Syracuse | 37 |
| ΑΘΗΝΑΙ | ΛΖ | Athens | 37 |
| ΔΕΛΦΟΙ | ΛΖ Γο | Delphi | 37 ⅔ |
| ΤΑΡCΟC | ΛΗ | Tarsus | 38 |
| ΑΔΡΙΑΝΟΥΠΟΛΙC | ΛΘ | Adrianopolis | 39 |
| ΑCΙΑ | Μ | Asia | 40 |
| ΗΡΑΚΛΕΙΑ | ΜΑ Γο | Heraklion | 41 ⅔ |
| ΡΩΜΗ | ΜΑ Γο | Rome | 41 ⅔ |
| ΑΓΚΥΡΑ | ΜΒ | Ankara | 42 |
| ΘΕCCΑΛΟΝΙΚΗ | ΜΓ | Thessalonika | 43 |
| ΑΠΑΜΙΑ | ΛΘ | Apamea | 39 |
| ΕΔΕCΑ | ΜΓ | Edessa | 43 |
| ΚΩΝCΤΑʹΤΙΝΟΥΠΙ | ΜΓ | Constantinople | 43 |
| ΓΑΛΛΙΑΙ | ΜΔ | Gaul | 44 |
| ΑΡΑΒΕΝΝΑ | ΜΔ | Aravenna | 44 |
| ΘΡΑΚΗ | ΜΑ | Thrace | 41 (?44) |
| ΑΚΥΛΗΙΑ | ΜΕ | Aquileia | 45 |

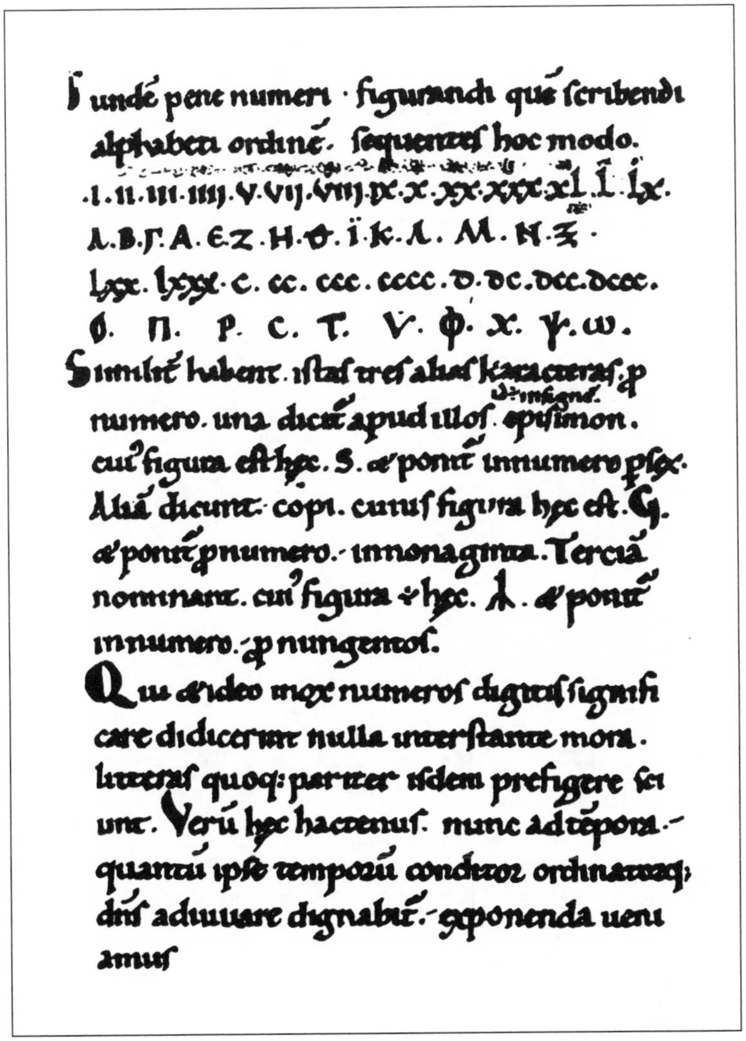

< = ½   Γο = ⅔

FIG. 17.37B.

*FIG. 17.38. Fragment of a Spanish manuscript concerning the Venerable Bede's finger-counting system, copied in c. 1130 CE, probably at Santa Maria de Ripoll (Catalonia). Madrid, National Library, Cod. A 19 folio 2 (top left). To explain the finger diagrams given on the following pages, the scribe uses two different numerical notations – Roman numerals and the Greek alphabetic system, with their correspondence.*

| | | | | | | | | |
|---|---|---|---|---|---|---|---|---|
| ⲁ̅ 1 | ⲃ̅ 2 | ⲅ̅ 3 | ⲇ̅ 4 | ⲉ̅ 5 | ⲋ̅ 6 | ⲍ̅ 7 | ⲏ̅ 8 | ⲑ̅ 9 |
| ⲓ̅ 10 | ⲕ̅ 20 | ⲗ̅ 30 | ⲙ̅ 40 | ⲛ̅ 50 | ⲝ̅ 60 | ⲟ̅ 70 | ⲡ̅ 80 | ϥ̅ 90 |
| ⲣ̅ 100 | ⲥ̅ 200 | ⲧ̅ 300 | ⲩ̅ 400 | ⲫ̅ 500 | ⲭ̅ 600 | ⲯ̅ 700 | ⲱ̅ 800 | ⳁ̅ 900 |
| ⲁ̿ 1,000 | ⲃ̿ 2,000 | ⲅ̿ 3,000 | ⲇ̿ 4,000 | ⲉ̿ 5,000 | ⲋ̿ 6,000 | ⲍ̿ 7,000 | ⲏ̿ 8,000 | ⲑ̿ 9,000 |
| ⲓ̿ 10,000 | ⲕ̿ 20,000 | ⲗ̿ 30,000 | ⲙ̿ 40,000 | ⲛ̿ 50,000 | ⲝ̿ 60,000 | ⲟ̿ 70,000 | ⲡ̿ 80,000 | ϥ̿ 90,000 |
| ⲣ̿ 100,000 | ⲥ̿ 200,000 | ⲧ̿ 300,000 | ⲩ̿ 400,000 | ⲫ̿ 500,000 | ⲭ̿ 600,000 | ⲯ̿ 700,000 | ⲱ̿ 800,000 | ⳁ̿ 900,000 |

FIG. 17.39. *Coptic numerals. [From Mallon, (1956); Till, (1955)]. The script of Egyptian Christians has 31 letters, of which 24 derive directly from Greek, and the others from demotic Egyptian writing. However, Coptic numerals use the same signs as the Greek system (that is to say, the 24 signs of the classical alphabet plus the three obsolete letters,* digamma, koppa *and* san, *with the same values as in Greek). In Coptic, letters used as numbers have a single superscripted line up to 999, and a double superscript for 1,000 and above.*

| ARMENIAN LETTERS | | NAMES OF THE LETTERS | SOUNDS | | NUMERICAL VALUES |
|---|---|---|---|---|---|
| UPPER-CASE | LOWER-CASE | | WESTERN ARMENIAN | EASTERN ARMENIAN | |
| Ա | ա | ayp/ayb | a | a | 1 |
| Բ | բ | pén/bén | p | b | 2 |
| Գ | գ | kim/gim | k | g | 3 |
| Դ | դ | ta/da | t | d | 4 |
| Ե | ե | yétch | é | ye/e | 5 |
| Զ | զ | za | z | z | 6 |
| Է | է | é | é | é | 7 |
| Ը | ը | et | e | e | 8 |
| Թ | թ | to | t | t/th | 9 |
| Ժ | ժ | jé | j | j | 10 |
| Ի | ի | ini | i | i | 20 |
| Լ | լ | lyoun | l | l | 30 |
| Խ | խ | khé | kh | kh | 40 |
| Ծ | ծ | dza/tsa | dz | ts | 50 |
| Կ | կ | gen/ken | g | k | 60 |
| Հ | հ | ho | h | h | 70 |
| Ձ | ձ | tsa/dza | tz | dz | 80 |
| Ղ | ղ | ghad | gh | gh | 90 |

FIG. 17.40. *Armenian numerals. Armenian uses an alphabet of 32 consonants and 6 vowels, designed specifically for this language in the fifth century CE by the priest Mesrop Machtots (c.362–440CE). The alphabet was based on Greek and Hebrew.*

| ARMENIAN LETTERS | | NAMES OF THE LETTERS | SOUNDS | | NUMERICAL VALUES |
|---|---|---|---|---|---|
| UPPER-CASE | LOWER-CASE | | WESTERN ARMENIAN | EASTERN ARMENIAN | |
| Ճ | ճ | djé/tché | dj | tch | 100 |
| Մ | մ | mén | m | m | 200 |
| Յ | յ | hi | y | y/h | 300 |
| Ն | ն | nou | h | n | 400 |
| Շ | շ | cha | ch | ch | 500 |
| Ո | ո | vo | o | o | 600 |
| Չ | չ | tcha | tch | tch | 700 |
| Պ | պ | bé/pé | b | p | 800 |
| Ջ | ջ | tché/djé | tch | dj | 900 |
| Ռ | ռ | ra | r | rr | 1,000 |
| Ս | ս | sé | s | s | 2,000 |
| Վ | վ | vév | v | v | 3,000 |
| Տ | տ | dyoun/tyoun | d | t | 4,000 |
| Ր | ր | ré | r | r | 5,000 |
| Ց | ց | tso | ts | ts | 6,000 |
| Ւ | ւ | hyoun | u | iu | 7,000 |
| Փ | փ | pyour | p | p | 8,000 |
| Ք | ք | ké | k | k | 9,000 |
| Օ | օ | o | ô | o | |
| Ֆ | ֆ | fé | f | f | |

FIG. 17.40*(continued). Like Greek, Armenian uses the first 9 letters to represent the units, the second 9 for the tens, the third 9 for the hundreds. However, as it has more letters than Greek, it can use the fourth set of 9 letters to represent the thousands. Note than only 36 of the 38 letters are used for numerical purposes.*

| GEORGIAN LETTERS | | VALUES | | GEORGIAN LETTERS | | VALUES | |
|---|---|---|---|---|---|---|---|
| UPPER-CASE | LOWER-CASE | PHONETIC | NUMERICAL | UPPER-CASE | LOWER-CASE | PHONETIC | NUMERICAL |
| Ⴀ | ⴀ | a | 1 | Ⴐ | ⴐ | r | 100 |
| Ⴁ | ⴁ | b | 2 | Ⴑ | ⴑ | s | 200 |
| Ⴂ | ⴂ | g | 3 | Ⴒ | ⴒ | t | 300 |
| Ⴃ | ⴃ | d | 4 | Ⴓ | ⴓ | u | 400 |
| Ⴄ | ⴄ | e | 5 | Ⴣ | ⴣ | vi | 500 |
| Ⴅ | ⴅ | v | 6 | Ⴔ | ⴔ | p' | 600 |
| Ⴆ | ⴆ | z | 7 | Ⴕ | ⴕ | k' | 700 |
| Ⴡ | ⴡ | h | 8 | Ⴖ | ⴖ | *v* | 800 |
| Ⴇ | ⴇ | t' | 9 | Ⴗ | ⴗ | q | 900 |
| Ⴈ | ⴈ | i | 10 | Ⴘ | ⴘ | š | 1,000 |
| Ⴉ | ⴉ | k | 20 | Ⴙ | ⴙ | tš | 2,000 |
| Ⴊ | ⴊ | l | 30 | Ⴚ | ⴚ | ts | 3,000 |
| Ⴋ | ⴋ | m | 40 | Ⴛ | ⴛ | dz | 4,000 |
| Ⴌ | ⴌ | n | 50 | Ⴜ | ⴜ | ts' | 5,000 |
| Ⴢ | ⴢ | ï | 60 | Ⴝ | ⴝ | tš' | 6,000 |
| Ⴍ | ⴍ | o | 70 | Ⴞ | ⴞ | ḫ | 7,000 |
| Ⴎ | ⴎ | p | 80 | Ⴤ | ⴤ | ẖ | 8,000 |
| Ⴏ | ⴏ | ž | 90 | Ⴟ | ⴟ | dž | 9,000 |
| | | | | Ⴠ | ⴠ | h | 10,000 |

FIG. 17.41. *Georgian alphabetic numerals. An example of a script and numeral system influenced by Greek in the Christian era. There are two distinct styles of writing the Georgian alphabet: the "priestly" script, or* khoutsouri, *reproduced above, and the "military", or* mkhedrouli. *Both have 38 letters.*

| GOTHIC LETTERS | VALUES | | GOTHIC LETTERS | VALUES | | GOTHIC LETTERS | VALUES | |
|---|---|---|---|---|---|---|---|---|
| | PHONETIC | NUMERICAL | | PHONETIC | NUMERICAL | | PHONETIC | NUMERICAL |
| 𐌰 | a | 1 | 𐌹 | i | 10 | 𐍂 | r | 100 |
| 𐌱 | b | 2 | 𐌺 | k | 20 | 𐍃 | s | 200 |
| 𐌲 | g | 3 | 𐌻 | l | 30 | 𐍄 | t | 300 |
| 𐌳 | d | 4 | 𐌼 | m | 40 | 𐍅 | w | 400 |
| 𐌴 | e | 5 | 𐌽 | n | 50 | 𐍆 | f | 500 |
| 𐌵 | q | 6 | 𐌾 | y | 60 | 𐍇 | ch | 600 |
| 𐌶 | z | 7 | 𐌿 | u | 70 | 𐍈 | hw | 700 |
| 𐌷 | h | 8 | 𐍀 | p | 80 | 𐍉 | o | 800 |
| 𐌸 | th | 9 | 𐍁 | | 90 | 𐍊 | | 900 |

FIG. 17.42. *Gothic: Another alphabetical numeral system influenced by Greek in the Christian era. The Goths – a Germanic people living on the northeastern confines of the Roman Empire, were Christianised by Eastern (Greek-speaking) priests in the second and third centuries CE. Wulfila (311–384 CE), a Christianised Goth who became a bishop, translated the Bible into his own tongue, and invented the Gothic alphabet, based on Greek together with some additional characters, in order to do this. The Goths eventually merged into other peoples, from Crimea to North Africa, and disappeared, leaving only the term "Gothic" with its various acquired meanings.*

| | | | | | |
|---|---|---|---|---|---|
| A | 1 | K | 10 | T | 100 |
| B | 2 | L | 20 | V | 200 |
| C | 3 | M | 30 | X | 300 |
| D | 4 | N | 40 | Y | 400 |
| E | 5 | O | 50 | Z | 500 |
| F | 6 | P | 60 | | |
| G | 7 | Q | 70 | | |
| H | 8 | R | 80 | | |
| I | 9 | S | 90 | | |

FIG. 17.43. *Numeral alphabet used by some mediaeval and Renaissance mystics. This adaptation of the Greek system to the Latin alphabet is described by A. Kircher in* Oedipi Aegyptiaci, *vol. II/1, p. 488 (1653).*

CHAPTER 18

# THE INVENTION OF ALPHABETIC NUMERALS

Greek alphabetic numerals were, as we have seen, pretty much identical to the system of Hebrew numerals, save for a few details. The similarity is such as to prompt the question: which came first?

What follows is an attempt to answer the question on the basis of what is currently known.

First of all, though, we have to clear away a myth that has been handed down uncritically as the truth for more than a hundred years.

## THE MYTH OF PHOENICIAN LETTER-NUMBERS

It has long been asserted that, long before the Jews and the Greeks, the Phoenicians first assigned numerical values to their alphabetic signs and thus created the first alphabetic numerals in history.

However, this assumption rests on no evidence at all. No trace has yet been discovered of the use of such a system by the Phoenicians, nor by their cultural heirs, the Aramaeans.

The idea is in fact but a conjecture, devoid of proof or even indirect evidence, based solely on the fact that the Phoenicians managed to simplify the business of writing down spoken language by inventing an alphabet.

As we shall see, Phoenician and Aramaic inscriptions that have come to light so far, including the most recent, show only one type of numerical notation – which is quite unrelated to alphabetic numerals.

In the present state of our knowledge, therefore, we can consider only the Greeks and the Jews as contenders for the original invention of letter-numerals.

## THE NUMERALS OF THE NORTHWESTERN SEMITES

The numerical notations used during the first millennium BCE by the various northwestern Semitic peoples (Phoenicians, Aramaeans, Palmyreneans, Nabataeans, etc.) are very similar to each other, and manifestly derive from a common source.

Leaving aside the cases of Hebrew and Ugaritic, the earliest instance of "numerals" found amongst the northwestern Semites dates from no earlier than the second half of the eighth century BCE. It is in an inscription on a monumental statue of a king called Panamu, presumed to have come from Mount Gercin, seven km northeast of Zencirli, Syria (not far from the border with Turkey). Semites generally liked to "write out" numbers, that is to say to spell out number-names, and this tradition, which continued for many centuries, no doubt explains why specific number-signs made such a late appearance. But that does not mean to say that their system of numerals is at all obscure.

The Aramaeans were traders who, from the end of the second millennium BCE, spread all across the Middle East; their language and culture were adopted in cities and ports from Palestine to the borders of India, from Anatolia to the Nile basin, and of course in Mesopotamia and Persia, over a stretch of time that goes from the Assyrian Empire to the rise of Islam. Thanks to the economic and legal papyri that constitute the archives of an Aramaic-speaking Jewish military colony established in the fifth century BCE at Elephantine in Egypt, we can easily reconstruct the Aramaeans' numeral system.

Aramaic numerals were initially very simple, using a single vertical bar to represent the unit, and going up to 9 by repetition of the strokes. To make each numeral recognisable at a glance, the strokes were generally written in groups of three (Fig. 18.1 A). A special sign was used for 10, and also (oddly enough) for 20 (Fig. 18.1 B and 18.1 C), whereas all other numbers from 1 to 99 were represented by the repetition of the basic signs. Aramaic numerals to 99 were thus based on the principle that any number of signs juxtaposed represented the sum of the values of those signs. As we shall see (Fig. 18.2), Aramaic numerals up to this point were thus identical to those of all other western Semitic dialects, namely:

| Sources | | |
|---|---|---|
| S 18 | | 1 |
| S 61 | | 2 |
| S 8 | | 3 |
| S 19 | | 4 |
| S 61 | | 5 |
| S 19 | | 6 |
| S 61 | | 7 |
| CIS. II[1] 147 | | 8 |
| S 62 | | 9 |

FIG. 18.1A. *Aramaic figures for the numbers 1 to 9. Copied from Sachau (1911), abbreviated as S, from fifth century BCE papyri from Elephantine (same source for Fig. 18.1 B – E)*

SIGNS FOR THE NUMBER 10

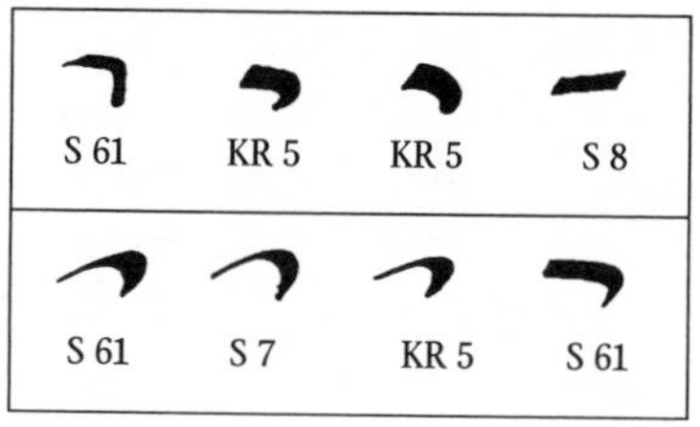

| | | | |
|---|---|---|---|
| S 61 | KR 5 | KR 5 | S 8 |
| S 61 | S 7 | KR 5 | S 61 |

FIG. 18.1B.

SIGNS FOR THE NUMBER 20

| | | | |
|---|---|---|---|
| S 18 | S 18 | S 25 | S 18 |
| S 19 | S 61 | S 15 | S 7 |

FIG. 18.1C.

REPRESENTATIONS OF THE TENS

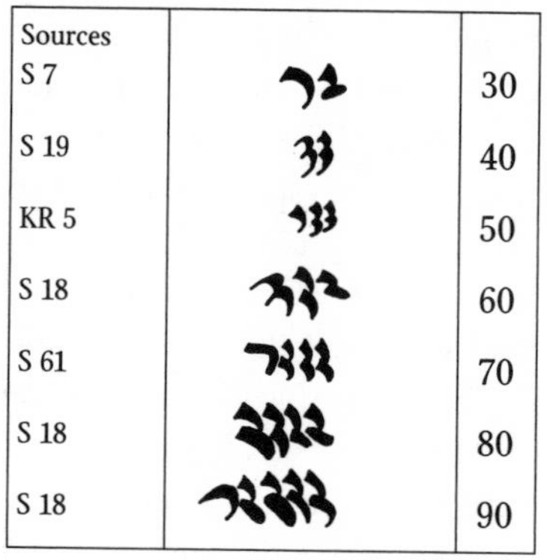

| Sources | | |
|---|---|---|
| S 7 | | 30 |
| S 19 | | 40 |
| KR 5 | | 50 |
| S 18 | | 60 |
| S 61 | | 70 |
| S 18 | | 80 |
| S 18 | | 90 |

FIG. 18.1D.

NUMBERS BELOW 100

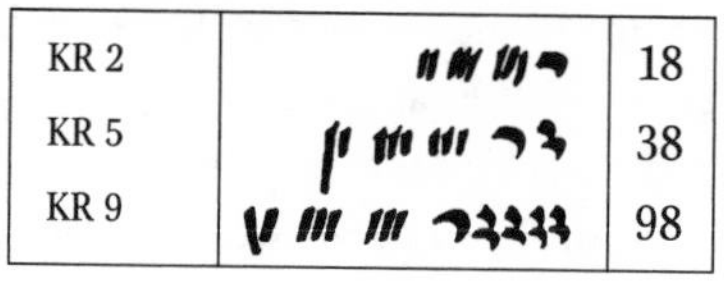

| | | |
|---|---|---|
| KR 2 | | 18 |
| KR 5 | | 38 |
| KR 9 | | 98 |

FIG. 18.1E.

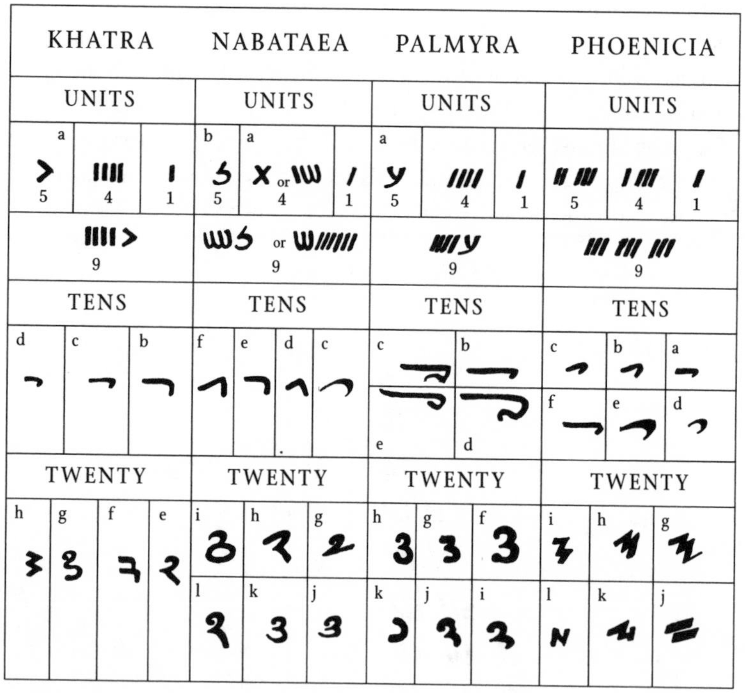

| KHATRA | NABATAEA | PALMYRA | PHOENICIA |
|---|---|---|---|
| References: B. Aggoula (1972); Milik (1972); Naveh (1972) | References: G. Cantineau (1930) | References: M. Lidzbarski (1962) | References: M. Lidzbarski (1962) |
| a Khatra no. 65 | a CIS II$^{1}$, 161 | a CIS II$^{3}$, 3 913 | a CIS I$^{1}$, 165 |
| b Khatra no. 65 | b CIS II$^{1}$, 212 | b CIS II$^{3}$, 3 952 | b CIS I$^{1}$, 165 |
| c Khatra no. 62 | c CIS II$^{1}$, 158 | c CIS II$^{3}$, 4 036 | c CIS I$^{1}$, 93 |
| d Abrat As-Saghira | d CIS II$^{1}$, 147 B | d CIS II$^{3}$, 3 937 | d CIS I$^{1}$, 88 |
| e Abrat As-Saghira | e CIS II$^{1}$, 349 | e CIS II$^{3}$, 3 915 | e CIS I$^{1}$, 165 |
| f Khatra no. 62 | f CIS II$^{1}$, 163 D | f CIS II$^{3}$, 3 937 | f CIS I$^{1}$, 3 A |
| g Abrat As-Saghira | g CIS II$^{1}$, 354 | g CIS II$^{3}$, 4 032 | g CIS I$^{1}$, 87 |
| h Khatra nos. 34, 65, 80 | h CIS II$^{1}$, 211 | h CIS II$^{3}$, 3 915 | h CIS I$^{1}$, 93 |
| i Doura-Europos | i CIS II$^{1}$, 161 | i CIS II$^{3}$, 3 969 | i CIS I$^{1}$, 7 |
| j Ashoka | j CIS II$^{1}$, 213 | j CIS II$^{3}$, 3 969 | j CIS I$^{1}$, 86 B |
| k Ostraca nos. 74 & 113 from Nisa | k CIS II$^{1}$, 204 | k CIS II$^{3}$, 3 935 | k CIS I$^{1}$, 13 |
| l Khatra nos. 62 & 65 | l CIS II$^{1}$, 204 | l CIS II$^{3}$, 3 915 | l CIS I$^{1}$, 165 |
| | m N, II, 12 | m CIS II$^{3}$, 3 917 | m CIS I$^{1}$, 143 |
| | n CIS II$^{1}$, 163D | | n CIS I$^{1}$, 65 |
| | o CIS II$^{1}$, 161 | | o IS I$^{1}$, 7 |
| | | | p CIS I$^{1}$, 217 |

| KHATRA | NABATAEA | PALMYRA | PHOENICIA |
|---|---|---|---|
| UNITS | UNITS | UNITS | UNITS |
| a: 5, 4, 1 | b, a: 5, 4 (or), 1 | a: 5, 4, 1 | 5, 4, 1 |
| 9 | 9 (or) | 9 | 9 |
| TENS | TENS | TENS | TENS |
| d, c, b | f, e, d, c | c, b, e, d | c, b, a, f, e, d |
| TWENTY | TWENTY | TWENTY | TWENTY |
| h, g, f, e | i, h, g, l, k, j | h, g, f, k, j, i | i, h, g, l, k, j |

FIG. 18.2.

- Phoenician, the language of a people of traders and sailors who settled, from the third millennium BCE, in Canaan (on the Mediterranean shore of Syria and Palestine); but Phoenician numerals are not found earlier than the sixth century BCE;
- Nabataean, spoken by people who, from the fourth century BCE, were settled at Petra, a city (now in Jordan) at the crossroads of trails leading from Egypt and Arabia to Syria and Palestine, and whose numeral system is attested from the second century BCE;
- Palmyrenean, spoken at Palmyra (east of Homs, in the Syrian desert), from around the beginning of the Common Era;
- Syriac, in use from the beginning of the Common Era;
- the dialect of Khatra, spoken in the early centuries of the Common Era by the inhabitants of the city of Khatra, in upper Mesopotamia, southwest of Mosul;
- Indo-Aramaic, a numeral system found in Kharoshthi inscriptions in the former province of Gandhara (on the borders of present-day Afghanistan and the Punjab), from the fourth century BCE to the third century CE;
- Pre-Islamic Arabic, in the fifth and sixth centuries CE.

However, despite affirmations to the contrary, the existence in these systems of a special sign for 20 is not a trace of an underlying vigesimal system borrowed by the Semites from a prior civilisation. The Semitic sign for 10 was originally a horizontal stroke or bar, and the tens were represented by repetitions of these bars, two by two:

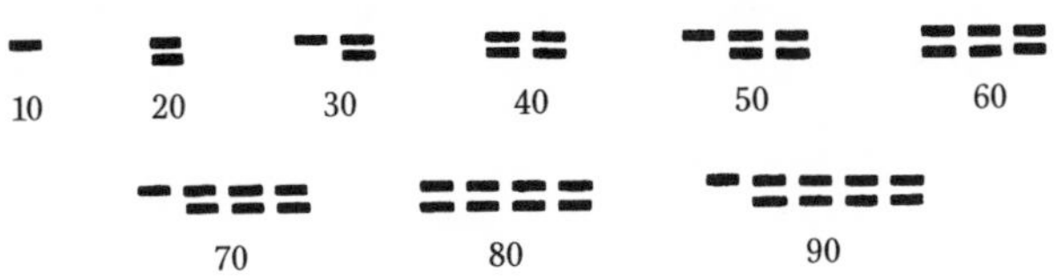


FIG. 18.3. *Figures for the tens on the Aramaic inscription at Zencirli (eighth century BCE). Donner & Röllig, Inscr. 215*

By a natural process of graphical development, which is found in all cursive scripts written with a reed brush on papyrus or parchment, the stroke became a line rounded off to the right. The double stroke for the number 20 developed into a ligature in rapid notation, and that "joined-up" form then gave rise to a whole variety of shapes, all deriving simply from writing two strokes without raising the reed brush.

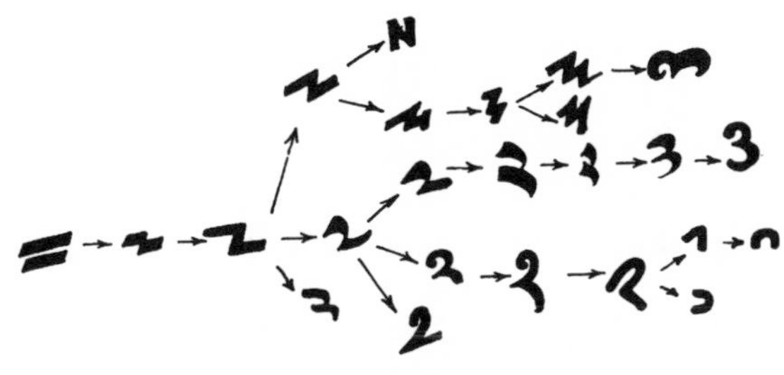

FIG. 18.4. *Origin and development of the figure for 20*

Aramaic numerals are thus strictly decimal, and do not have any trace of a vigesimal base. It was identical in principle to the Cretan Linear system for numbers below 100 – but that does not mean that it was a "primitive" form of number-writing nor that it lacked ways of coping with numbers above the square of its base. In fact, the system had a very interesting device for representing higher numbers which makes it significantly more sophisticated than many numeral systems of the Ancient World.

The Elephantine papyrus shows that Semitic numbering possessed distinctive signs for 100, 1,000 and 10,000 (though this last is not found on Phoenician or Palmyrenean inscriptions). What is more, the system did not require these higher signs to be repeated on the additive principle, but put unit expressions to the right of the higher numeral, that is to say used the multiplicative principle for the expression of large numbers (see Fig. 18.7 and 18.8).

## SOURCES

| | | | |
|---|---|---|---|
| a CIS II 147 | h Khatra | o CIS II 4 021 | u CIS II 147 |
| b S 19 | i S 15 | p CIS II 3 935 | v CIS I 7 |
| c S 61 | j KR 4 | q Sumatar Harabesi | w Assur |
| d Sari inscription | k S 61 | r CIS II 161 | x En-Namara (Cantineau, p. 49) |
| e Nisa ostracon 113 | l CIS I 165 | | y Bühler, p. 77 |
| f Qabr Abu Nayf | m CIS I 143 | s CIS II 163 D | |
| g Khatra | n CIS II 3 999 | t CIS II 3 915 | |

FIG. 18.5. *Variant forms of the Semitic numeral 100*

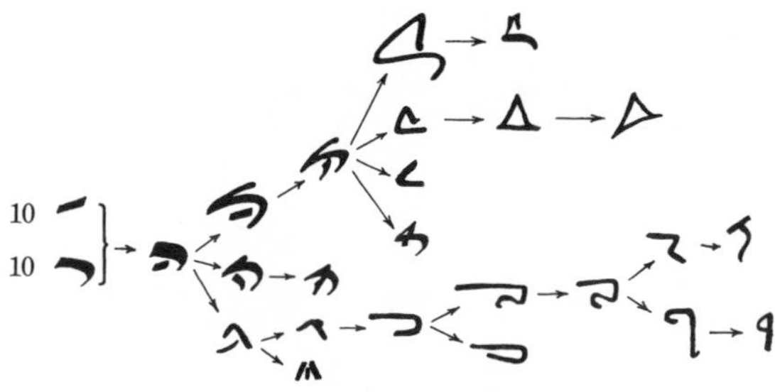


FIG. 18.6. *Origin and development of the figure 100. All these signs derive from placing two variants of the sign for 10 one above the other. This multiplicative combination has a kind of additional superscript to avoid confusing it with the sign for 20, and produced widely different graphical representations of the number 100.*

ARAMAIC (ELEPHANTINE PAPYRI)

| Source | Sign | Value | Source | Sign | Value |
|---|---|---|---|---|---|
| S 61 | 100 × 5 | 500 | S 19 | 100 × 1 | 100 |
| CIS II[1] | 100 × 8 | 800 | S fragm. 3 | 100 × 2 | 200 |
| S 61 | 100 × 9 | 900 | S 19 | 100 × 4 | 400 |

| KHATRA | | | NABATAEA | PALMYRA | PHOENICIA | | |
|---|---|---|---|---|---|---|---|
| k 100 × 1 | j 100 × 1 | i 100 × 1 | 100 × 1 | 100 × 1 | o 100 × 1 | n 100 × 1 | m 100 × 1 |
| k 100 × 2 | | | m 100 × 2 | 100 × 2 | 100 × 2 | | 100 × 2 |
| j 100 × 3 | | | n 100 × 3 | l 100 × 3 | 100 × 3 | | 100 × 3 |
| l 100 × 4 | | | o 100 × 4 | m 100 × 4 | p 100 × 4 | | 100 × 4 |

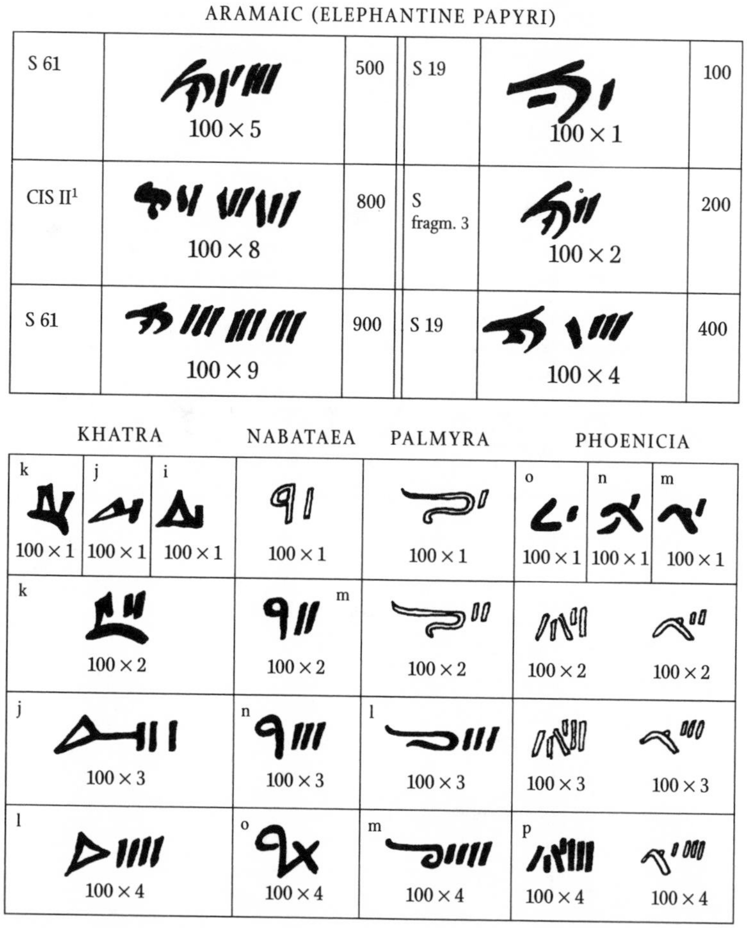

FIG. 18.7. *Semitic representations of the number 100. Attested examples are given in solid lines; reconstructed examples in outline. For sources, see list of references in Fig. 18.2 and 18.5.*

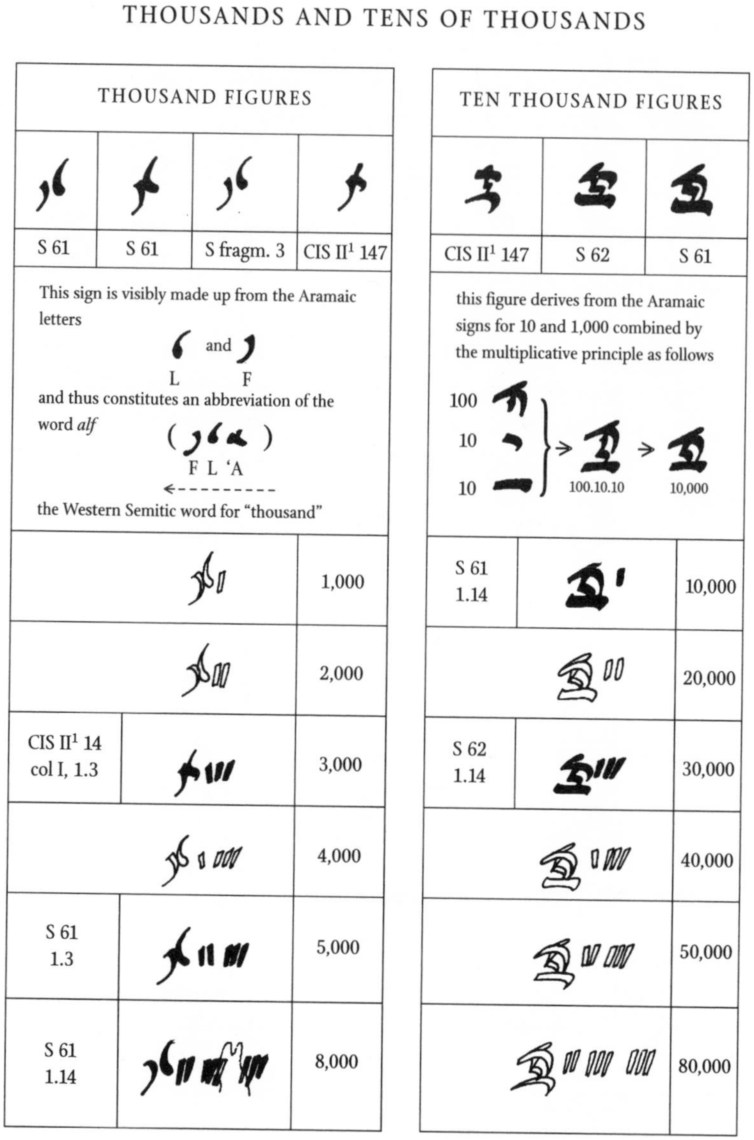

THOUSANDS AND TENS OF THOUSANDS

| THOUSAND FIGURES | | | |
|---|---|---|---|
| S 61 | S 61 | S fragm. 3 | CIS II[1] 147 |

This sign is visibly made up from the Aramaic letters L and F and thus constitutes an abbreviation of the word *alf* (F L 'A) the Western Semitic word for "thousand"

| Source | Sign | Value |
|---|---|---|
| | | 1,000 |
| | | 2,000 |
| CIS II[1] 14 col I, 1.3 | | 3,000 |
| | | 4,000 |
| S 61 1.3 | | 5,000 |
| S 61 1.14 | | 8,000 |

| TEN THOUSAND FIGURES | | |
|---|---|---|
| CIS II[1] 147 | S 62 | S 61 |

this figure derives from the Aramaic signs for 10 and 1,000 combined by the multiplicative principle as follows

100
10
10 → 100.10.10 → 10,000

| Source | Sign | Value |
|---|---|---|
| S 61 1.14 | | 10,000 |
| | | 20,000 |
| S 62 1.14 | | 30,000 |
| | | 40,000 |
| | | 50,000 |
| | | 80,000 |

FIG. 18.8. *Aramaic representations of the numbers 1,000 and over. Figures for these numbers have not been found in other northwestern Semitic numeral systems.*

In other words, the Semites used the additive principle for numbers from 1 to 99, but for multiples of 100, 1,000 and 10,000, they adopted the multiplicative principle by writing the numbers in the form 1 x 100, 2 x 100, 3 × 100, etc.; 1 × 1,000, 2 × 1,000, 3 × 1,000, etc. So for intermediate numbers above 100 they used a combination of the additive and multiplicative principles.

This corresponds with the general traditions of numbering amongst Semitic peoples. It is found amongst all the northwestern Semites (Phoenicians, Palmyreneans, Nabataeans, etc.) who used, as we have seen, numerical notations of the same kind as the Aramaic system of Elephantine. But it is also found amongst the so-called eastern Semites. The Assyrians and the Babylonians certainly inherited the additive sexagesimal system of the Sumerians, but they modified it completely even whilst adopting the cuneiform script for writing it down. Precisely because of their tradition of counting in hundreds and thousands, and finding no numeral for 100 or 1,000 in the Sumerian system, their scribes wrote those two numbers in phonetic script and represented their multiples not by addition of a sequence of signs, but by multiplication (Fig. 18.9).

So we can say that with the obvious exception of late Hebrew, none of the Semitic numeral systems had anything to do with the use of letters as numbers.

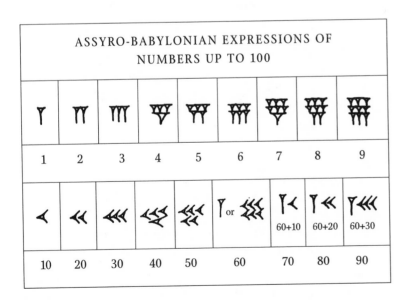

ASSYRO-BABYLONIAN EXPRESSIONS OF NUMBERS UP TO 100

| | | | | | | | | |
|---|---|---|---|---|---|---|---|---|
| 1 | 2 | 3 | 4 | 5 | 6 | 7 | 8 | 9 |
| | | | | | or | 60+10 | 60+20 | 60+30 |
| 10 | 20 | 30 | 40 | 50 | 60 | 70 | 80 | 90 |

FIG. 18.9. *Assyro-Babylonian "ordinary" numerals – an adaptation of Sumerian numerals to Semitic numbering traditions*

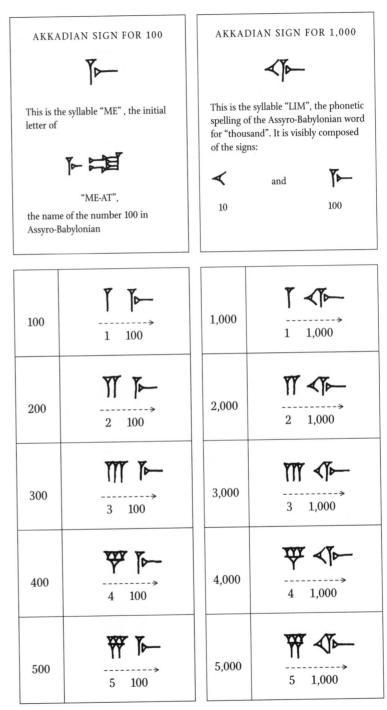

AKKADIAN SIGN FOR 100

This is the syllable "ME" , the initial letter of

"ME-AT",

the name of the number 100 in Assyro-Babylonian

AKKADIAN SIGN FOR 1,000

This is the syllable "LIM", the phonetic spelling of the Assyro-Babylonian word for "thousand". It is visibly composed of the signs:

10 and 100

| | |
|---|---|
| 100 | 1 100 |
| 200 | 2 100 |
| 300 | 3 100 |
| 400 | 4 100 |
| 500 | 5 100 |

| | |
|---|---|
| 1,000 | 1 1,000 |
| 2,000 | 2 1,000 |
| 3,000 | 3 1,000 |
| 4,000 | 4 1,000 |
| 5,000 | 5 1,000 |

FIG. 18.9*(Continued).*

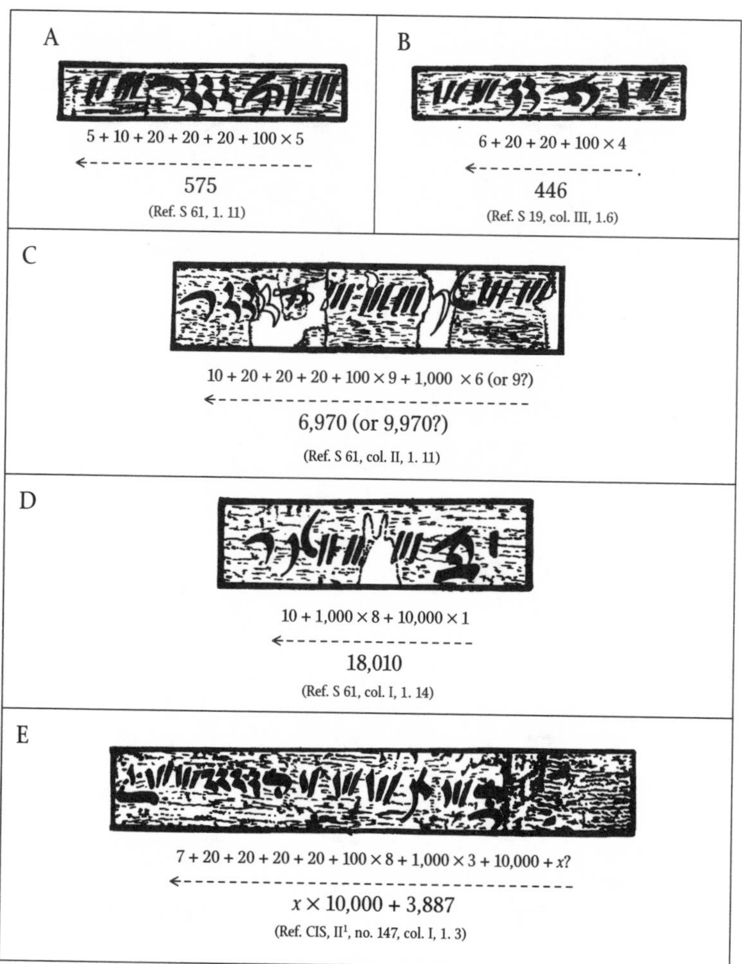


FIG. 18.10. *Facsimile and interpretation of numerical expressions in the Elephantine papyrus*

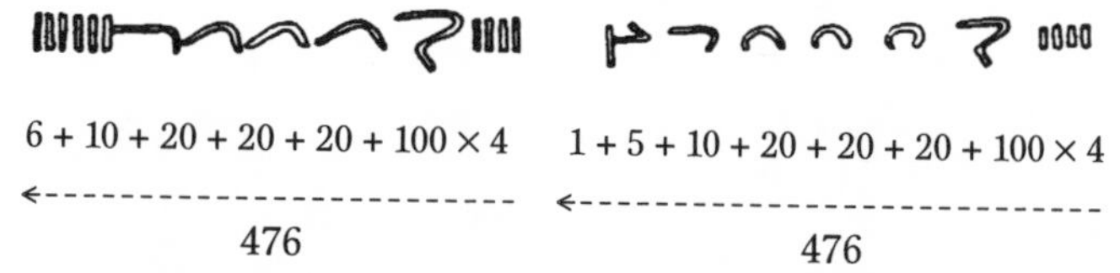


FIG. 18.11. *Tracing and interpretation of two examples from Syriac inscriptions at Sumatar Harabesi, dated 476 of the Seleucid era (165–166 CE). Source: Naveh*

FIG. 18.12. *Phoenician inscription, fifth century BCE. Source: CIS I', 7*

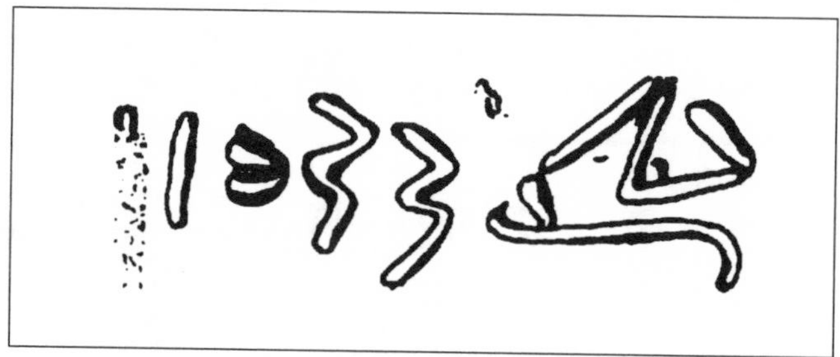

FIG. 18.13. *The number 547 on a Syriac inscription at Sari. Source: Naveh*

## THE OLDEST ARCHAEOLOGICAL EVIDENCE OF GREEK ALPHABETIC NUMERALS

Amongst the oldest known uses of Greek alphabetical numerals are those to be found on coins minted in the reign of Ptolemy II (286–246 BCE), the second of the Macedonian kings who ruled over Egypt after the death of Alexander the Great (Fig. 18.14).

| Coin inventory numbers | Date symbols | Transcription and translation | |
|---|---|---|---|
| CGC 44 | K | K | 20 |
| CGC 45 | KA | KA | 21 |
| CGC 46 | [illegible] | KA | 21 |
| CGC 48 | K B | KB | 22 |
| CGC 49 | KΓ | KΓ | 23 |
| CGC 50 | KΔ | KΔ | 24 |
| CGC 53 | KE | KE | 25 |
| CGC 57 | KZ | KZ | 27 |
| CGC 50 | KH | KH | 28 |

| Coin inventory numbers | Date symbols | Transcription and translation | |
|---|---|---|---|
| CGC 61 | Λ | Λ | 30 |
| CGC 63 | ΛA | ΛA | 31 |
| CGC 68 | ΛB | ΛB | 32 |
| CGC 70 | ΛΓ | ΛΓ | 33 |
| CGC 73 | ΛΔ | ΛΔ | 34 |
| CGC 99 | ΛE | ΛE | 35 |
| CGC 100 | ΛC | Λ | 36 |
| CGC 101 | ΛZ | ΛZ | 37 |
| CGC 77 | ΛH | ΛH | 38 |

FIG. 18.14. *Coins from the British Museum, catalogued by R. S. Poole*

Even earlier, in a Greek papyrus from Elephantine, we find a marriage contract that states that it was drawn up in the seventh year of the reign of Alexander IV (323–311 BCE), that is to say in 317–316 BCE, in which the dowry is expressed as *alpha drachma*, thus:

(transcription: ⊦ A
translation: *drachma* A)

FIG. 18.15.

The alphabetic numeral *alpha* probably means 1,000 in this case, unless the father of the bride was a real miser, since *alpha* could either mean 1,000 – or 1!

It therefore seems that the use of Greek alphabetic numerals was common by the end of the fourth century BCE.

Moreover, relatively recent excavations of the agora and north slope of the Acropolis in Athens prove that the system arose even earlier, in the fifth century BCE, since it is found on an inscription on the Acropolis that is assumed to date from the time of Pericles (see N. M. Tod, in ABSA, 45/1950).

## THE OLDEST ARCHAEOLOGICAL EVIDENCE OF HEBREW ALPHABETIC NUMERALS

Amongst the earliest instances of Hebrew alphabetic numerals are those found on coins struck in the second century CE by Simon Bar Kokhba, who seized Jerusalem in the Second Jewish Revolt (132–134 CE). The *shekel* coin shown in Fig. 18.16 bears an inscription in what were already the obsolete forms of the palaeo-Hebraic alphabet* that gives the date as *bet*, that is to say "Year 2", in alphabetic numerals, which corresponds (as Year 2 of the Liberation of Israel) to 133 CE.

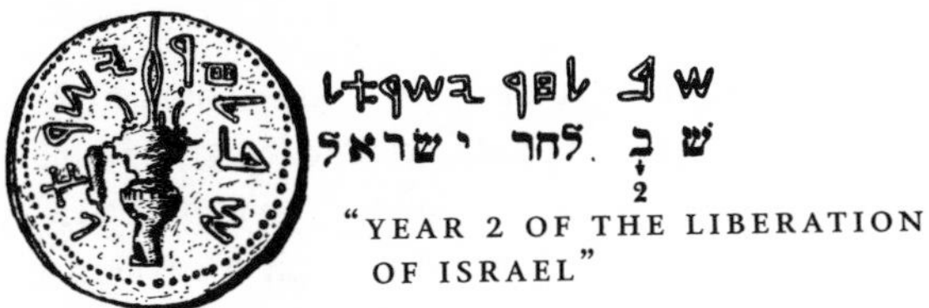


FIG. 18.16. *Coin from the Second Jewish Revolt (132–134 CE). Kadman Numismatic Museum, Israel*

Other earlier instances are found on coins from the First Jewish Revolt in 66–73 CE (Fig. 18.17), and Hasmonaean coins dating from the end of the first century CE. These inscriptions, such as the one reproduced as Fig. 18.18 (from a coin minted in 78 CE), are in the Aramaic language but written in palaeo-Hebraic script.

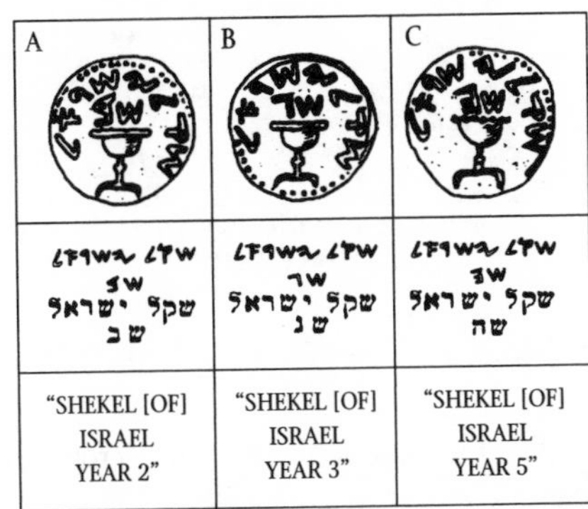

A
שקל ישראל
שב
"SHEKEL [OF] ISRAEL YEAR 2"
B
שקל ישראל
שג
"SHEKEL [OF] ISRAEL YEAR 3"
C
שקל ישראל
שה
"SHEKEL [OF] ISRAEL YEAR 5"


FIG. 18.17. *Coins struck during the First Jewish Revolt (66–73 CE):* shekels *dated Year 2 (A: 67 CE), Year 3 (B: 68 CE), and Year 5 (C: 70 CE) with alphabetic numerals in palaeo-Hebraic script. Kadman Numismatic Museum, Israel. See Kadman (1960), plates I–III.*

* Palaeo-Hebraic letters are close to Phoenician script. They were replaced by Aramaic script (which gave rise to modern square-letter Hebrew around the beginning of the CE) in the fifth century BCE (see Fig. 17.2 above). However, the archaic forms of the letters continued to be used sporadically up to the second century CE, most particularly by the leaders of the two Jewish revolts, to signify a return to the "true traditions of Israel". The alphabet of the present-day Samaritans is derived directly from palaeo-Hebraic script.

FIG. 18.18. *Coins struck in 78 BCE under Alexander Janneus. Kadman Numismatic Museum, Israel. See Naveh (1968), plate 2 (nos. 10 & 12) and plate 3 (no. 14).*

We must also mention a clay seal in the Jerusalem Archaeological Museum which must have originally served to fix a string around a papyrus scroll (Fig. 18.19). The seal bears an inscription in palaeo-Hebraic characters which can be translated as: "Jonathan, High Priest, Jerusalem, M". The letter *mem* at the end is still a puzzle, but it could be a numeral, with a value (= 40) referring to the reign of Simon Maccabeus, recognised by Demetrius II in 142 BCE as the "High Priest, leader and ruler of the Jews". If this were so, then the seal would date from 103 BCE (the "fortieth year" of Simon Maccabeus) and thus constitute the oldest known document showing the use of Hebrew alphabetic numerals.

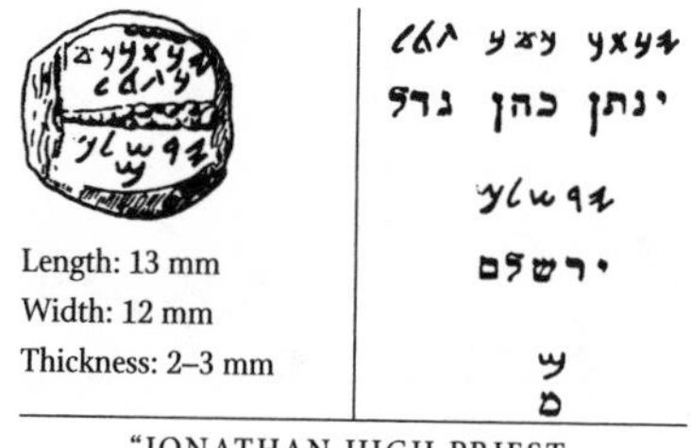


FIG. 18.19. Bulla *of* … *aean period (second century BCE). Israel Museum, Jerusalem, item 75.35. See Avigad (1975), Fig. 1 and Plate I-A.*

Finally, there is this fragment of a parchment scroll from Qumran (one of the "Dead Sea Scrolls"):

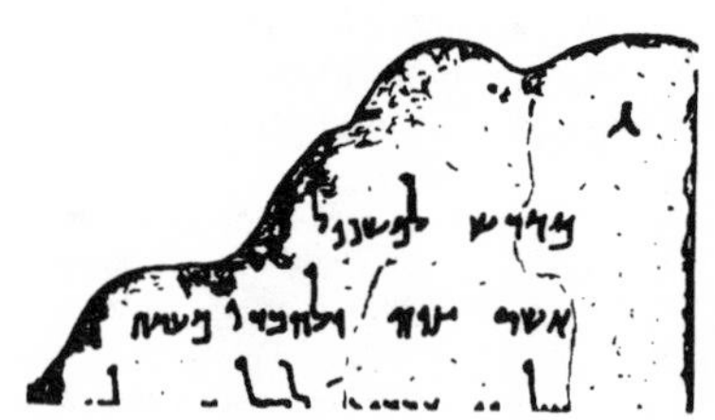

FIG. 18.20. *Fragment of a parchment scroll, recently found at Khirbet Qumran. Scroll 4QSd, no. 4Q 259. See Milik (1977).*

The scroll contains a copy of the Rule of the Essene community, written in square-letter Hebrew of a style that dates from the first century BCE at the earliest. The fragment comes from the first column of the third sheet of the scroll as it was found in the caves at Qumran. In the top right-hand corner there is a letter, *gimmel*: since this is the third letter of the Hebrew alphabet, people have assumed that the letter gives the sheet number, 3. However, the *gimmel* was not written by the same hand as the rest of the scroll; J. T. Milik has explained that the page-numbering was probably the work of an apprentice, using what was then a novel procedure for numbering manuscripts by the letters of the alphabet, whereas the main scribe used an older form of writing.

## JEWISH NUMERALS FROM THE PERSIAN TO THE HELLENISTIC PERIOD

The preceding section shows that in Palestine Hebrew letters were only just beginning to be used as numerals at the start of the Common Era.

This is confirmed by the discovery, in the same caves at Qumran, of several economic documents belonging to the Essene sect and dating from the first century BCE. One of them, a brass cylinder-scroll (Fig. 18.21), uses number-signs that are quite different from Hebrew alphabetic numerals.

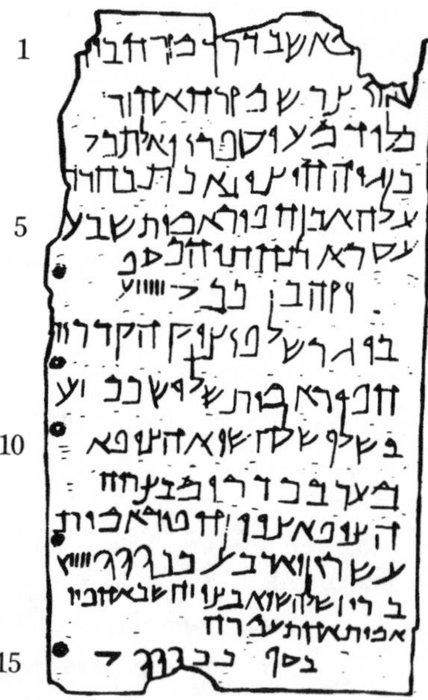

FIG. 18.21A. *Fragment of a brass cylinder-scroll, first century BCE, from the third of the Qumran caves. See DJD III, 3Q, plate LXII, column VIII.*

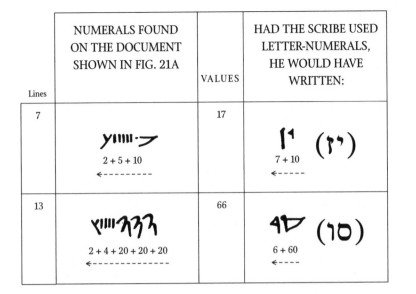

| Lines | NUMERALS FOUND ON THE DOCUMENT SHOWN IN FIG. 21A | VALUES | HAD THE SCRIBE USED LETTER-NUMERALS, HE WOULD HAVE WRITTEN: |
|---|---|---|---|
| 7 | 2 + 5 + 10 ←--------- | 17 | 7 + 10 ←----- (יז) |
| 13 | 2 + 4 + 20 + 20 + 20 ←-------------- | 66 | 6 + 60 ←---- (סו) |

FIG. 18.21B.

Further confirmation is provided by the many papyri from the fifth century BCE left by the Jewish military colony at Elephantine (near Aswan and the first cataract of the Nile). These consist of deeds of sale, marriage contracts, wills, and loan agreements, and they use numerals that are identical to those of the Essene scroll. For example, one such papyrus [E. Sachau (1911), papyrus no. 18] uses the following representations of 80 and 90, which are obviously unrelated to the Semitic letter-numbers *pe* (for 80) and *tsade* (for 90).

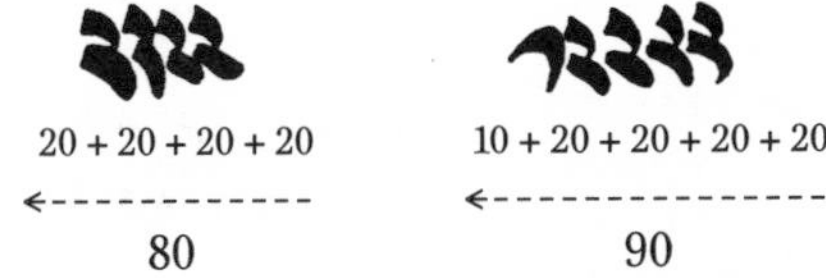


FIG. 18.22.

An even more definitive piece of evidence comes from the archaeological site of Khirbet el Kom, not far from Hebron, on the West Bank (Israel). It is a flat piece of stone that was used, at some point in the third century BCE, for writing a receipt for the sum of 32 *drachmas* loaned by a Semite called Qos Yada to a Greek by the name of Nikeratos – and is thus written in both Aramaic and Greek.

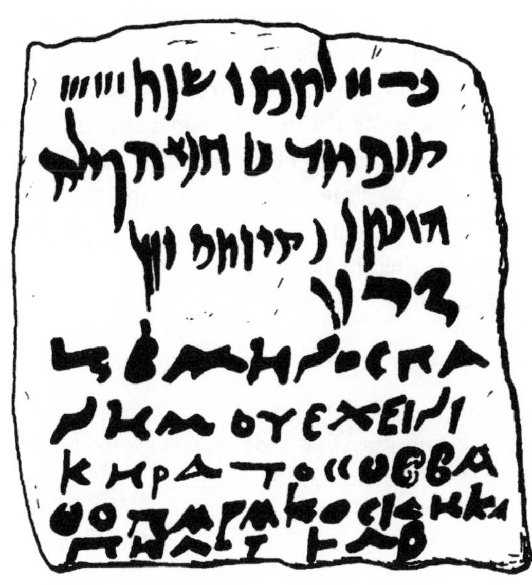

TRANSCRIPTION

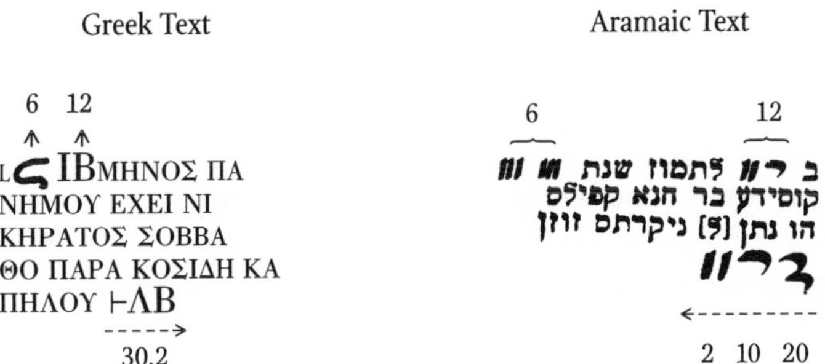


TRANSLATION

6th year, the 12th of the month of Panemos, Nikeratos, son of Sobbathos, received from Koside the moneylender [the sum of] 32 drachma

The 12th [of the month] of Tammuz [of] the 6th year Qos Yada son of Khanna the trader gave Nikeratos in "Zuz": 32.

FIG. 18.23. *Bilingual* ostracon *from Khirbet el Kom (Israel), probably dating from 277 BCE (Year 6 of Ptolemy II). See Geraty (1975), Skaist (1978).*

Close scrutiny of the inscription shows first of all that the two languages are written by different hands: probably the moneylender wrote the Aramaic and the borrower wrote the Greek. Moreover, we can see that Nikeratos the Greek wrote the sum he had borrowed and the date of the loan ("6th year, on the 12th of the month of Panemos") using Greek alphabetic numerals: ϛ *digamma* ( = 6), ιβ *iota-beta* ( = 12), and λβ *lambda-beta* (=32). On the other hand, Qos Yada the Semite wrote the sum of the loan (32 *zuz*) using the numeral system we have seen on the Essene scroll above, broken down as:

20 + 10 + 1 + 1

It seems indisputable that if Hebrew alphabetic numerals had been in use in Palestine at this time, then Qos Yada would have used them, and written the number 32 much more simply as

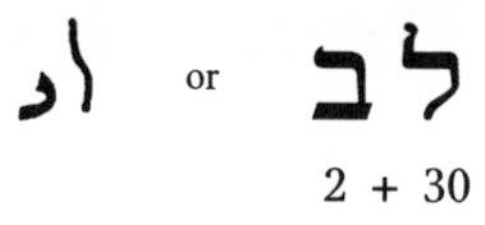

לב or לב

2 + 30

←------

FIG. 18.24.

We can therefore conclude that in all probability the inhabitants of Judaea did not use alphabetic numerals in ordinary transactions until the dawn of the Common Era.

The numeral system we have found in use amongst Jews from the Persian to the Hellenistic period (fifth to second centuries BCE) is in fact nothing other than the old western Semitic system, borrowed by the Hebrews from the Aramaeans together with their language (Aramaic) and script. Because the Aramaeans were very active in trade and commerce – their role across the land-mass of the Middle East was similar to that of the Phoenicians around the shores of the Mediterranean Sea – Aramaic script spread more or less everywhere. It finally killed off the cuneiform writing of the Assyro-Babylonians, and became the normal means of international correspondence.

## ACCOUNTING IN THE TIME OF THE KINGS OF ISRAEL

How did the Jews do their accounting in the age of the Kings, roughly from the tenth to the fifth centuries BCE? In the absence of archaeological evidence, it was long thought that numbers were simply written out as words, for the numeral system explained below remained undiscovered until less than a hundred years ago.

That was when excavations in Samaria uncovered a hoard of *ostraca* in palaeo-Hebraic script in the storerooms of the palace of King Omri. An *ostracon* is a flat piece of rock, stone or earthenware used as a writing surface. (The use of *ostraca* as "scribble-pads" for current accounts, lists of workers, messages and notes of every kind was very common amongst the Ancient Egyptians, the Phoenicians, the Aramaeans, and the Hebrews.) The Samarian *ostraca* consist of bills and receipts for payments in kind to the stewards of the King of Israel, and reveal that the Jews wrote out their numbers as words and also used a real system of numerals.

Subsequent discoveries confirmed the existence of these ancient Hebrew numerals. They have been found on a hoard of about a hundred *ostraca* unearthed at a site at Arad (in the Negev Desert, on the trail from Judaea to Edom); on another score of *ostraca* found at Lakhish in 1935, which contain messages from a Jewish military commander to his subordinates, written in the months prior to the fall of Lakhish to Nebuchadnezzar II in 587 BCE; numerous Jewish weights and measures; and on various similar discoveries made at the Ophel in Jerusalem, at Murabba'at and at Tell Qudeirat.

Although it took a long time to decipher these inscriptions, there is no longer any doubt (Fig. 18.26) but that these number-signs are Egyptian hieratic numerals in their fully developed form from the New Empire (shown in Fig. 14.39 and 14.46 above). This incidentally provides additional confirmation of the significant cultural relations between Egypt and Palestine which historians have revealed in other ways. In other words, in the period of the Kings of Israel, the Jews were influenced by the civilisation of the Pharaohs to the extent of adopting from it Egyptian cursive hieratic numerals (Fig. 18.25 and 18.27).

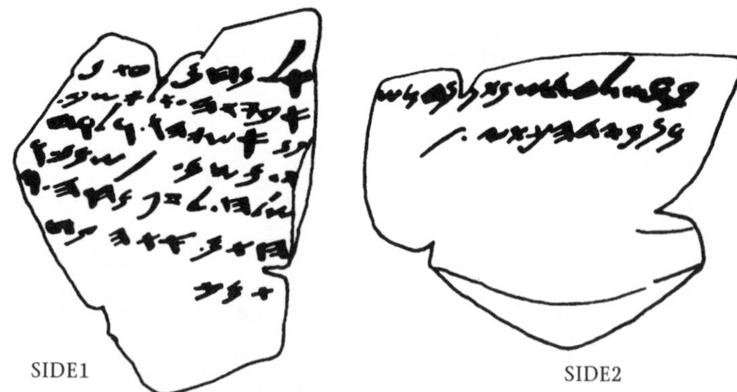


FIG. 18.25. *Hebrew* ostracon *from Arad, sixth century BCE (ostracon no 17). Written in palaeo-Hebraic script, side 2 has the number 24 written as: See Aharoni (1966).*

4 20

←------

| DATES BCE | SOURCES | | 1 | 2 | 3 | 4 | 5 | 6 | 7 | 8 | 9 | 10 | 20 | 30 | 40 | 50 | 60 | 70 | 80 | 90 | 100 | 200 | 300 |
|---|---|---|---|---|---|---|---|---|---|---|---|---|---|---|---|---|---|---|---|---|---|---|---|
| 9TH C | ARAD | Ostracon no. 72 | | | | | | | | | | | | | | | | | | | | | |
| 8TH C | SAMARIA | Ostraca published in 1910 | | | | | | | | | | | | | | | | | | | | | |
| | | Ostracon C 1101 | | | | | | | | | | | | | | | | | | | | | |
| 8TH – 7TH C | Inscribed Jewish weights | | | | | | | | | | | | | | | | | | | | | | |
| LATE 8TH C | Jerusalem Ophel | Ostr. no. 2 | | | | | | | | | | | | | | | | | | | | | |
| | | Ostr. no. 3 | | | | | | | | | | | | | | | | | | | | | |
| | | Ostr. no. 4 | | | | | | | | | | | | | | | | | | | | | |
| EARLY 7TH C | ARAD | Ostraca no. 33–36 | | | | | | | | | | | | | | | | | | | | | |
| 7TH C | MESHAD HASHAVYAHU | Ostr. 6 | | | | | | | | | | | | | | | | | | | | | |
| | MURABBA'AT | Papyrus no. 18 | | | | | | | | | | | | | | | | | | | | | |
| | ARAD | Ostracon no. 34 | | | | | | | | | | | | | | | | | | | | | |
| 6TH C | LAKHISH | Ostr. no. 9 | | | | | | | | | | | | | | | | | | | | | |
| | | Ostr. no. 19 | | | | | | | | | | | | | | | | | | | | | |
| | ARAD | Ostr. no. 1–4 | | | | | | | | | | | | | | | | | | | | | |
| | | Ostr. no. 16–18 | | | | | | | | | | | | | | | | | | | | | |
| | | Ostr. no. 24–29 | | | | | | | | | | | | | | | | | | | | | |

EGYPTIAN HIERATIC NUMERALS (NEW KINGDOM, CURSIVE). FROM MÖLLER (1911).

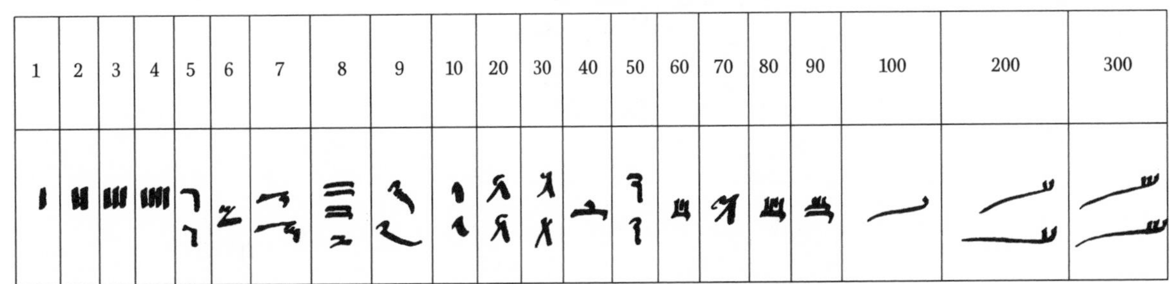


FIG. 18.26. *Table showing the identity of numerals used in Palestine under the Jewish Kings with Egyptian hieratic numerals*

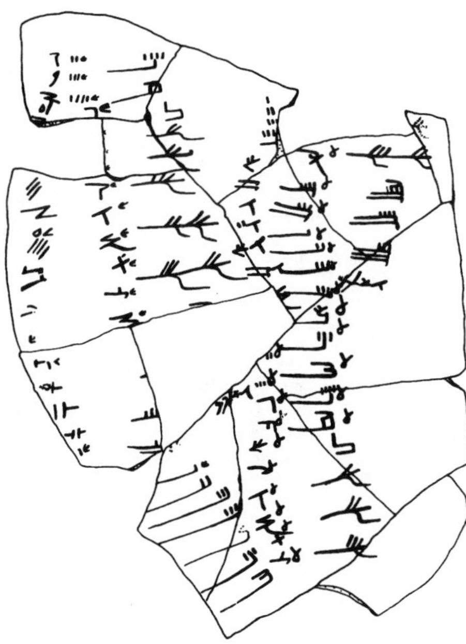

FIG. 18.27. Ostracon *no. 6 from Tell Qudeirat, late seventh century BCE, the largest known palaeo-Hebraic* ostracon, *found by R. Cohen in 1979. This text confirms the results of Fig. 18.26, since it gives almost the whole series of the hundreds and thousands in Egyptian hieratic script.*

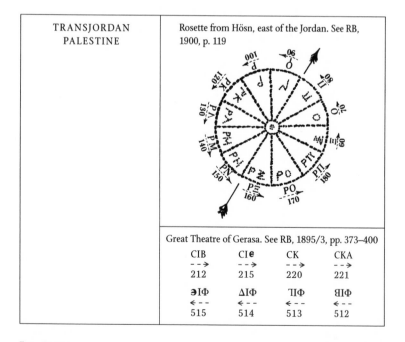

| | |
|---|---|
| TRANSJORDAN PALESTINE | Rosette from Hösn, east of the Jordan. See RB, 1900, p. 119<br>90 100 80 120 70 130 60 140 150 180 160 170<br>PK PΛ PM PN PΞ PO PΠ |
| | Great Theatre of Gerasa. See RB, 1895/3, pp. 373–400<br>CIB ---> 212 · CIЄ ---> 215 · CK ---> 220 · CKA ---> 221<br>϶IΦ <--- 515 · ΔIΦ <--- 514 · ΓIΦ <--- 513 · ΒIΦ <--- 512 |

FIG. 18.28A.

| | |
|---|---|
| EGYPT | *Coptic* inscription concerning Luke and two of his works. See ASAE, X, 1909, p.51<br>KH ----> 28 · KΔ ----> 24 · KZ ----> 27 |
| | *Jewish* funerary *stelae* from Tell el Yahudieh (10 km north of Cairo), dating from the first century CE. See CII 1454, 1458 and 1460<br>IB ---> 12 · IΓ ---> 13 · KΓ ---> 23 · ΛЄ ---> 35 · N ---> 50 · PB ---> 102 |
| PHRYGIA | *Jewish* inscription dated 253–254 CE. See CII 773<br>TΛH -----> 338 |
| ETHIOPIA | Aksum inscription, third century CE. See DAE 3 and 4<br>KΔ ----> 24 · ΓPIB ----> 3112 · ϚCKΔ --------> 6224 |
| LATIUM | *Jewish* catacombs on the Via Nomentana, Via Labicana and Via Appia Pignatelli. See CII 44, 78, 79<br>ΛΓ ----> 33 · KA ----> 21 · ΞΘ ----> 69 |
| NORTHERN SYRIA | Synagogue mosaic. *Jewish* inscription dated 392 CE. See CII 805<br>ΨΓ ----> 703 |
| SOUTH OF THE DEAD SEA | *Jewish* grave marking dated 389–390 CE. See CII 1209<br>ΠΠ -----> 86 · CΠΓ ------> 283 |

FIG. 18.28B.

## JEWISH LAPIDARY NUMERALS AT THE DAWN OF THE COMMON ERA

There is a final curiosity to add to this story. From the first century BCE to the seventh century CE, the use of Hebrew alphabetic numerals grew ever more common amongst Jews all over the Mediterranean basin, from Italy to Palestine and northern Syria, from Phrygia to Egypt and even Ethiopia. However, during this period, Jewish stone-carvers, who could write just as well in Hebrew as in Greek or Latin, most often put dates and numbers not in Hebrew, but in Greek alphabetic numerals, as the examples reproduced in Fig. 18.28 show.

## THE JEWS: NATIONAL IDENTITY AND CULTURAL COMPLEXITY

The people of Israel certainly played a major role in the history of the world's religions; but at the same time, Jewish culture has, throughout its history, accepted and adopted influences of the most diverse kinds.

The most notable of these "foreign influences" include:

- the adoption of the Phoenician alphabet in the period of the Kings;
- the adoption of the Assyro-Babylonian sexagesimal system for weights and measures (see Ezekiel XLV:12, where the *talent* is set at 60 *maneh*, and the *maneh* at 60 *shekels*);
- the presumed adoption of the Canaanites' calendar, in which each month starts with the appearance of the new moon;
- the borrowing of the names of the months from the ancient calendar of Nippur, used throughout Mesopotamia from the time of Hammurabi (*Nisān, Ayar, Siwan, Tammuz, Ab, Elul, Teshrêt, Arashamna, Kisilimmu, Tebet, Shebat,* and *Adar*). In Modern Hebrew, the names are still almost identical;
- the adoption of Aramaic and its script (the only ones in general use in Judaea at the time of Jesus).

What is remarkable about Jewish culture is that despite these numerous borrowings it retained a separate identity. Since the expulsion of the Jews from Palestine in the first century CE, and for the following 1,800 years, it has not ceased to adapt itself to the most diverse situations and to incorporate new elements, whilst also exercising a determining influence over developments in Western and Islamic culture. As Jacques Soustelle sees it, this long history of a cultural identity within a complex of cultural influences is what accounts for the successful re-founding of a Jewish nation-state in the twentieth century: Israel today is made of more than a score of distinct ethnic groups with many different mother-tongues, but sharing a common cultural identity.

## SUMMARY

From the tenth to the sixth century BCE (the era of the Kingdom of Israel), the Hebrews used Egyptian hieratic numerals; from the fifth to the second century BCE, they used Aramaic numerals; and from around the start of the common era, many Jews used Greek alphabetical numerals.

In the present state of knowledge, it seems that Greek alphabetic numerals go back at least as far as the fifth century BCE; whereas Hebrew alphabetic numerals are not found before the second century BCE.

Does that mean to say that the Greeks invented the idea of representing numbers by the letters of their alphabet, and that the Jews copied it during the Hellenistic period? It seems very likely, and all the more plausible in the light of the Jews' adoption of numerous other "outside" influences.

However, this is not the only possible conclusion. Many passages in the Torah (the Old Testament) suggest very strongly that the scribes or authors of these ancient texts were familiar with the art of coding words according to the numerical value of the letters used (see further explanations in Chapter 20 below). It is currently reckoned that the oldest biblical texts were composed in the reign of Jeroboam II (eighth century BCE) and that the definitive redaction of the main books of the Torah took place in the sixth century BCE, around the time of the Babylonian exile.

Do Hebrew alphabetic numerals go so far back in time? Or are the passages showing letter-number coding later additions?

If the system is as old as it seems, and which would imply that Hebrew letter-numbers were invented independently of the Greek model, we would still have to explain why they had no use in everyday life until the Common Era. One plausible answer to that question would be that since the letters of the Hebrew alphabet acquired a sacred character very early on, the Jews avoided using sacred devices for profane purposes.

In conclusion, let us say that the "Greek hypothesis" seems to have most of the actual evidence on its side; but that the possibility of an independent origin for Hebrew alphabetic numerals and of their restriction over several centuries to religious texts alone is not to be rejected out of hand.

# CHAPTER 19

# OTHER ALPHABETIC NUMBER-SYSTEMS

## SYRIAC LETTER-NUMERALS

The Arabic-speaking Christians of the Maronite sect have maintained, mainly for liturgical use, a relatively ancient writing system which is known as *serto* or *Jacobite* script.

Christians of the Nestorian sect, who are found mainly in the region of Lake Urmia (near the common frontier between Iraq, Turkey, Iran and the former Soviet Union), still speak a dialect of Aramaic which they write in a graphical system called *Nestorian* writing.

Each of these two writing systems has an alphabet of twenty-two letters, and is derived from a much older script called *estranghelo,* formerly used to write Syriac, a ancient Semitic language related to Aramaic.

Graphically, the Nestorian form, which is more rounded than the *estranghelo*, is intermediate between this and *serto* which in turn has a more developed and cursive form (Fig. 19.1). The letters themselves are written from right to left, are joined up, and, as in the writing of Arabic, undergo various modifications according to their position within a word, i.e. according to whether they stand alone or are in the initial, medial, or final position (Fig. 19.1 only shows the independent forms of Syriac letters).

The oldest known Syriac inscriptions seem to date from the first century BCE. *Estranghelo* writing seems to have been used only up to the sixth or seventh century. As used by the Nestorian Christians, fairly numerous in Persia in the period of the Sassanid Dynasty (226–651 CE), it gradually evolved until, around the ninth century, it attained its canonical Nestorian form. With the Jacobites, who mainly lived in the Byzantine Empire, it seems to have evolved more rapidly towards the *serto* form, since it was gradually replaced by this after the seventh or eighth century.

Finally, *estranghelo* (which is simply a variant of Aramaic script and therefore ultimately derives from the Phoenician alphabet) has preserved in its entirety the order of the original twenty-two Phoenician letters (the same order which is to be found with all the western Semites).

In *serto*, however, as in Nestorian, letters have been used (and still are used) as number-signs. This is confirmed by the fact that in all Syriac manuscripts (at least those later than the ninth century), codices are made up of serially numbered quires, ensuring the correct order of composition of the bound book. (The manuscript folios, however, were only numbered later, often using Arabic numerals.)

The numerical values of the Syriac letters are assigned as follows. The first nine letters are assigned to the units, the next nine letters to the tens, and the remaining four are assigned to the first four hundreds. Also, as in Hebrew, the numbers from 500 to 900 are written as additive combinations of the sign for 400 with the signs for the other hundreds, according to the schema:

$$500 = 400 + 100$$
$$600 = 400 + 200$$
$$700 = 400 + 300$$
$$800 = 400 + 400$$
$$900 = 400 + 400 + 100$$

The thousands are represented by a kind of accent mark placed beneath the letters representing the units, and the tens of thousands by a short horizontal mark beneath these same letters:

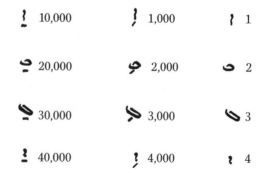

| | | |
|---|---|---|
| 10,000 | 1,000 | 1 |
| 20,000 | 2,000 | 2 |
| 30,000 | 3,000 | 3 |
| 40,000 | 4,000 | 4 |

Similar conventions allowed the Maronites to represent numbers greater than the tens of thousands. With a few exceptions, this number-system is quite analogous to that of the Hebrew letter-numerals. It is however a relatively late development in Syriac writing, since the oldest documents show that it does not go back earlier than the sixth or seventh centuries. Older Syriac inscriptions only reveal a single kind of numerical notation related to the "classical" Aramaic system.

| HEBRAIC LETTERS | | ARCHAIC PHOENICIAN | PALMYRENEAN | ESTRANGHELO | NESTORIAN | SERTO | NAMES TRANSCRIPTIONS AND NUMERICAL VALUES OF SYRIAC LETTERS | | |
|---|---|---|---|---|---|---|---|---|---|
| Aleph | א | 𐤀 | 𐡠 | ܐ | ܐ | ܐ | Ōlap | ʾ | 1 |
| Bet | ב | 𐤁 | 𐡡 | ܒ | ܒ | ܒ | Bēt | b bh | 2 |
| Gimmel | ג | 𐤂 | 𐡢 | ܓ | ܓ | ܓ | Gōmal | g gh | 3 |
| Dalet | ד | 𐤃 | 𐡣 | ܕ | ܕ | ܕ | Dōlat | d dh | 4 |
| He | ה | 𐤄 | 𐡤 | ܗ | ܗ | ܗ | Hē | h | 5 |
| Vov | ו | 𐤅 | 𐡥 | ܘ | ܘ | ܘ | Waw | w | 6 |
| Zayin | ז | 𐤆 | 𐡦 | ܙ | ܙ | ܙ | Zayn | z | 7 |
| Het | ח | 𐤇 | 𐡧 | ܚ | ܚ | ܚ | Ḥēt | ḥ | 8 |
| Tet | ט | 𐤈 | 𐡨 | ܛ | ܛ | ܛ | Ṭēt | ṭ | 9 |
| Yod | י | 𐤉 | 𐡩 | ܝ | ܝ | ܝ | Yud | y | 10 |
| Kof | ך כ | 𐤊 | 𐡪 | ܟ | ܟ | ܟ | Kōp | k kh | 20 |
| Lamed | ל | 𐤋 | 𐡫 | ܠ | ܠ | ܠ | Lōmad | l | 30 |
| Mem | ם מ | 𐤌 | 𐡬 | ܡ | ܡ | ܡ | Mim | m | 40 |
| Nun | ן נ | 𐤍 | 𐡭 | ܢ | ܢ | ܢ | Nun | n | 50 |
| Samekh | ס | 𐤎 | 𐡯 | ܣ | ܣ | ܣ | Semkat | ṣ | 60 |
| Ayin | ע | 𐤏 | 𐡰 | ܥ | ܥ | ܥ | ʿE | ʿ | 70 |
| Pé | ף פ | 𐤐 | 𐡱 | ܦ | ܦ | ܦ | Pē | p ph | 80 |
| Tsade | ץ צ | 𐤑 | 𐡲 | ܨ | ܨ | ܨ | Ṣōdē | ṣ | 90 |
| Quf | ק | 𐤒 | 𐡳 | ܩ | ܩ | ܩ | Quf | q | 100 |
| Resh | ר | 𐤓 | 𐡴 | ܪ | ܪ | ܪ | Rish | r | 200 |
| Shin | ש | 𐤔 | 𐡵 | ܫ | ܫ | ܫ | Shin | sh | 300 |
| Tav | ת | 𐤕 | 𐡶 | ܬ | ܬ | ܬ | Taw | t | 400 |

Fig. 19.1. *Syriac alphabets compared with Phoenician, Aramaean (from Palmyra) and Hebraic alphabets. The use of Syriac letters as number-signs is attested notably in a manuscript in the British Museum (Add. 14 620) which features the above order. (See M. Cohn, Costaz, Duval, Février, Hatch, Pihand, W. Wright)*

When did letter-numerals in Syriac writing first arise? In the absence of documents, it is hard to say. But there are several good reasons to suppose that the introduction of this system owed much to Jewish influence on the Christian and Gnostic communities of Syria and Palestine.

One final question: a Syriac manuscript, now in the British Museum (reference Add. 14 603), which probably dates from the seventh or eighth century [W. Wright (1870), p. 587a], reveals some interesting information. Its quires are numbered in the usual way, with Syriac letters according to their numerical values; but these have alongside them the corresponding older number-signs. Should we conclude that, at the date of this manuscript, the system of letter-numerals had not been universally adopted? Or, taking the question in the other sense, should we conclude that at that time the use of the old system was already a traditional but archaic usage, and the letter-numerals were by then not only widespread but considered by the majority of Syrians to be the only normal and official system of notation? The documentation which we have to hand does not give us an answer.

## ARABIC LETTER-NUMERALS

Arabic has a number-system modelled not only on the Hebrew system, but also on the Greek system of letter-numerals. But first we need to look at a curious problem.

The order of the twenty-eight letters of the Arabic alphabet, in its Eastern usage, is quite different from the order of the letters in the Phoenician, Aramaic or Hebrew alphabets.

A glance at the names of the first eight Arabic letters compared with the first eight Hebrew letters shows this straight away:

| ARABIC | HEBREW |
|---|---|
| *'alif* | *'aleph* |
| *ba* | *bet* |
| *ta* | *gimmel* |
| *tha* | *dalet* |
| *jim* | *he* |
| *ḥa* | *vov* |
| *kha* | *zayin* |
| *dal* | *het* |

We would expect to find the twenty-two western Semitic letters in the Arabic alphabet, and in the same order, since Arabic script derives from archaic Aramaic script. So how did the traditional order of the Semitic letters get changed in Arabic? The answer lies in the history of their system for writing numbers.

The Arabs have frequently used a system of numerical notation in which each letter of their own alphabet has a specific numerical value (Fig. 19.3); according to F. Woepke, they "seem to have considered [this system] as uniquely and by preference their own".

They call this *ḥurūf al jumal*, which means something like "totals by means of letters".

But, if we look closely at the numerical value which this system assigns to each letter, we are bound to note that the method used by the Arabs of the East is not quite the same as the one adopted, later, by western (North African) Arabs, since the values for six of the letters differ in the two systems.

| LETTER | | ITS VALUE | |
|---|---|---|---|
| | | IN THE MAGHREB | IN THE EAST |
| س | sin | 300 | 60 |
| ص | ṣad | 60 | 90 |
| ش | shin | 1,000 | 300 |
| ض | ḍad | 90 | 800 |
| ظ | ḍha | 800 | 900 |
| غ | ghayin | 900 | 1,000 |

FIG. 19.2.

Now, let us first note that the numerical values of the Arabic letters can be arranged into a regular series, as follows:

1; 2; 3; 4; . . . 10; 20; 30; 40; . . . 100; 200; 300; 400; . . . ; 1,000,

and if we set out, according to this sequence, the letter-numerals of the eastern Arabic system (the more ancient of the two) we obtain the order of the western Semitic letters of which we have just written (Fig. 17.2 and 17.4 above). Furthermore, if we tabulate the letter-numerals of the Arabic system (as in Fig. 19.4) and compare this with the Hebrew letter-numerals (Fig. 17.10) and also with the Syriac system of alphabetic numbering (Fig. 19.1), then it is easy to see that for the numbers below 400 all three systems agree perfectly. This shows that "in the initial system of numeration, the order of the northern Semitic alphabet was preserved, and additional letters from the Arabic alphabet were added later in order to go up to 1,000" [M. Cohen (1958)].

| LETTERS | | | | | | NUMERICAL VALUES | |
|---|---|---|---|---|---|---|---|
| LETTERS ON THEIR OWN | LETTER-NAMES | PHONETIC VALUES OF LETTERS | LETTERS IN INITIAL POSITION | LETTERS IN MEDIAN POSITION | LETTERS IN END POSITION | IN THE EAST | IN THE MAGHREB |
| ا | 'Alif | ' | ا | ـا | ـا | 1 | 1 |
| ب | Ba | b | بـ | ـبـ | ـب | 2 | 2 |
| ت | Ta | t | تـ | ـتـ | ـت | 400 | 400 |
| ث | Tha | th | ثـ | ـثـ | ـث | 500 | 500 |
| ج | Jim | j | جـ | ـجـ | ـج | 3 | 3 |
| ح | Ḥa | ḥ | حـ | ـحـ | ـح | 8 | 8 |
| خ | Kha | ḫ | خـ | ـخـ | ـخ | 600 | 600 |
| د | Dal | d | د | ـد | ـد | 4 | 4 |
| ذ | Dhal | dh | ذ | ـذ | ـذ | 700 | 700 |
| ر | Ra | r | ر | ـر | ـر | 200 | 200 |
| ز | Zay | z | ز | ـز | ـز | 7 | 7 |
| س | Sin | s | سـ | ـسـ | ـس | 60 | 300 |
| ش | Shin | sh | شـ | ـشـ | ـش | 300 | 1,000 |
| ص | Ṣad | ṣ | صـ | ـصـ | ـص | 90 | 60 |
| ض | Ḍad | ḍ | ضـ | ـضـ | ـض | 800 | 90 |
| ط | Ṭa | ṭ | طـ | ـطـ | ـط | 9 | 9 |
| ظ | Ḍha | ḍh | ظـ | ـظـ | ـظ | 900 | 800 |
| ع | 'Ayin | ' | عـ | ـعـ | ـع | 70 | 70 |
| غ | Ghayin | gh | غـ | ـغـ | ـغ | 1,000 | 900 |
| ف | Fa | f | فـ | ـفـ | ـف | 80 | 80 |
| ق | Qaf | q | قـ | ـقـ | ـق | 100 | 100 |
| ك | Kaf | k | كـ | ـكـ | ـك | 20 | 20 |
| ل | Lam | l | لـ | ـلـ | ـل | 30 | 30 |
| م | Mim | m | مـ | ـمـ | ـم | 40 | 40 |
| ن | Nun | n | نـ | ـنـ | ـن | 50 | 50 |
| ه | Ha | h | هـ | ـهـ | ـه | 5 | 5 |
| و | Wa | w | و | ـو | ـو | 6 | 6 |
| ي | Ya | y | يـ | ـيـ | ـي | 10 | 10 |

FIG. 19.3. *The Arabic alphabet, in its modern representation*

We may therefore conclude that the use of alphabetic numerals by the Arabs was introduced in imitation of the Jews and the Christians of Syria for the first twenty-two letters (numbers below 400), and according to the example of the Greeks for the remaining six (values from 400 to 1,000).

| | | | | | | | |
|---|---|---|---|---|---|---|---|
| ا | 'Alif | ' | 1 | س | Sin | s | 60 |
| ب | Ba | b | 2 | ع | ʿAyin | ʿ | 70 |
| ج | Jim | j | 3 | ف | Fa | f | 80 |
| د | Dal | d | 4 | ص | Ṣad | ṣ | 90 |
| ه | Ha | h | 5 | ق | Qaf | q | 100 |
| و | Wa | w | 6 | ر | Ra | r | 200 |
| ز | Zay | z | 7 | ش | Shin | sh | 300 |
| ح | Ḥa | ḥ | 8 | ت | Ta | t | 400 |
| ط | Ṭa | ṭ | 9 | ث | Tha* | th | 500 |
| ي | Ya | y | 10 | خ | Kha* | kh | 600 |
| ك | Kaf | k | 20 | ذ | Dhal* | dh | 700 |
| ل | Lam | l | 30 | ض | Ḍad* | ḍ | 800 |
| م | Mim | m | 40 | ظ | Ḍha* | ḍh | 900 |
| ن | Nun | n | 50 | غ | Ghayin* | gh | 1,000 |
| | | | | * subsequently added | | | |

FIG. 19.4. *The order of Arabic letters as ordained according to the regular development of the values of the alphabetic number-system of eastern Arabs*

In fact, "following the conquest of Egypt, Syria and Mesopotamia, numbers were habitually written, in Arabic texts, either spelled out in full or by means of characters borrowed from the Greek alphabet" [A. P. Youschkevitch (1976)].

Thus we find in an Arabic translation of the Gospels, the manuscript verses have been numbered with Greek letters:

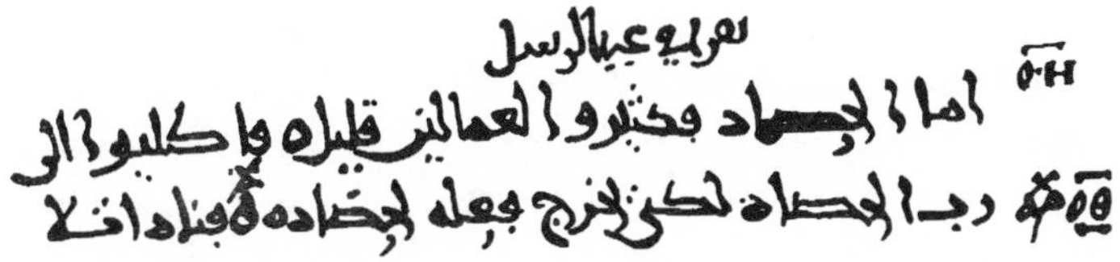

FIG. 19.5. *Excerpt from a Christian ninth-century manuscript. In this manuscript, which gives a translation from the Gospels, the corresponding verses have been numbered by reference to Greek letter-numbers, (first line, right: OH = v. 78; second line, right: Oθ = v. 79. Vatican Library, Codex Borghesiano arabo 95, folio 173. (See E. Tisserant, pl. 55)*

Similarly, in a financial papyrus written in Arabic and dating from the year 248 of the Hegira (862–863 CE), the sums were written exclusively according to the Greek system. [This document is, along with others of similar kind, in the Egyptian Library, inventory number 283; cf. A. Grohmann (1962)].

This usage persisted in Arabic documents for several centuries, but disappeared completely in the twelfth century. For all that, we should not conclude that Arabic letter-numerals were introduced only at that time. The system certainly first arose before the ninth century. We have, in fact, a mathematical manuscript copied at Shiraz between the years 358 and 361 of the Hegira (969–971 CE) in which all of the Arabic alphabetic numerals are used according to the Eastern system.*

Likewise, there is an astrolabe† dating from year 315 of the Hegira (927–928 CE) where this date is expressed in a palaeographic style known as Kufic script (Fig. 19.10). Other older documents indicate that the introduction of this system to the Arabs occurred as early as the eighth century, or, at the earliest, the end of the seventh.

From then on, all becomes clear. After adding six letters to the western Semitic alphabet which they inherited, and having established their system of alphabetic numerals preserving the traditional order of the letters, the Arab grammarians of the seventh or eighth century, apparently for pedagogical reasons, completely changed the original order of the letters by bringing together letters which had much the same graphical forms. At that time these grammarians "worked mainly in Mesopotamia where Jewish and Christian studies flourished, with Greek influences" (M. Cohen).

Thus it was that, from that time on, letters such as *ba*, *ta*, and *tha*, or *jim*, *ḥa*, and *kha* were placed in sequence in the Arabic alphabet (Fig. 19.3).

| خ | ح | ج | ث | ت | ب |
|---|---|---|---|---|---|
| kha | ḥa | jim | tha | ta | ba |
| 600 | 8 | 3 | 500 | 400 | 2 |

←-------------------------------------------------

FIG. 19.6.

* "Treatise by Ibrahim ibn Sinan on the Methods of Analysis and Synthesis in Problems of Geometry", a tenth-century copy of fifty-one works on mathematics (BN Ms.arab. 2457; see for example ff. 53v and 88).

† A scientific instrument for observing the position of stars and their height above the horizon. It was used in particular by Arabian astrologers, but some examples have been found from the Graeco-Byzantine era.

The better to establish the order of the alphabetic numerals, the eastern Arabs invented eight mnemonics which every user had to learn by heart in order to be able to recall the number-letters according to their regular arithmetic sequence (Fig. 19.7).

This clearly shows that the "ABC" order, pronounced *Abajad* (or *Abjad, Abujad, Aboujed*, etc. depending on accent), which sometimes governs the order of letters in the Arabic alphabet, does not correspond to their phonetic value nor to their graphical form, but to their respective numerical values according to the eastern Arabian system (Fig. 19.4).

In the usage of the Maghreb, it should be noted, the numerical values given to six of the twenty-eight Arabic letters are different from those in the Eastern system; also, the grouping of the number-letters is different, being done according to nine mnemonics which yield the following groups of values: (1; 10; 100; 1,000); (2; 20; 200); (3; 30; 300); etc. (Fig. 19.11).

| MNEMONIC WORDS | | DECOMPOSED AS | |
|---|---|---|---|
| Abajad | ابجد ← | ا ب ج د<br>d j b 'a ← | 4. 3. 2. 1 ← |
| Hawazin | هوز ← | ه و ز<br>z w h ← | 7. 6. 5. ← |
| Ḥuṭiya | حطي ← | ح ط ي<br>y ṭ ḥ ← | 10. 9. 8 ← |
| Kalamuna | كلمن ← | ك ل م ن<br>n m l k ← | 50. 40. 30. 20 ← |
| Sa'faṣ | سعفص ← | س ع ف ص<br>ṣ f ʿ f ← | 90. 80. 70. 60 ← |
| Qurshat | قرشت ← | ق ر ش ت<br>t sh r q ← | 400. 300. 200. 100. ← |
| Thakhudh | ثخذ ← | ث خ ذ<br>dh kh th ← | 700. 600. 500. ← |
| Ḍaḍhugh | ضظغ ← | ض ظ غ<br>gh ḍh ḍ ← | 1,000. 900. 800 ← |

Fig. 19.7. *Mnemonic words enabling eastern practitioners to find the order of numerical values associated with Arabic letters*

| | | | | | |
|---|---|---|---|---|---|
| 604 | خ د<br>4 600 ← | خد | 12 | ي ب<br>2 10 ← | يب |
| 472 | ت ع ب<br>2 70 400 ← | تعب | 58 | ن ح<br>8 50 ← | نح |
| 1,283 | غ ر ف ج<br>3 80 200 1,000 ← | غرفج | 96 | ص و<br>6 90 ← | صو |
| 1,631 | غ خ ل ا<br>1 30 600 1,000 ← | غخلا | 169 | ق س ط<br>9 60 100 ← | قسط |
| 1,629 | غ خ ك ط<br>9 20 600 1,000 ← | غخكط | 315 | ش ي ه<br>5 10 300 ← | شيه |

Fig. 19.8. *The writing of numbers by reference to the number-letters of the eastern Arabic system (transcribed into current characters) is always from right to left in descending order of values, starting with the highest order. Moreover, these number-signs (as with ordinary Arabic letters) have an inter-relationship generally by undergoing slight graphic modifications according to the position they occupy within the body of the number- or word-combinations. Examples reconstituted from an Arabic manuscript copied at Shiraz c. 970. Paris: Bibliothèque nationale, Ms. ar. 2457*

On the other hand, this same order occurs not only with the Jews, but also with all the northwestern Semites, as well as the Greeks, the Etruscans and the Armenians, to cite but a few. It is a very ancient ordering since, more than twenty centuries earlier than the Arabs, the inhabitants of Ugarit were familiar with it.

> Nonetheless the Arabs, lacking knowledge of the other Semitic languages . . . sought other explanations for the mnemonics *abjad*, etc. which had come down to them by tradition but which they found incomprehensible. The best that they could propose on this subject, interesting though it is, is pure fable. According to some, six kings of Madyan arranged the Arabic letters according to their own names. According to a different tradition, the first six mnemonics were the names of six demons. According to a third, it was the names of the days of the week. . . . We may none the less discern an interesting detail amongst these fables. One of the six kings of Madyan had supremacy over the others (*ra'īsuhum*): this was Kalaman, whose name bears perhaps some relation with the Latin *elementa**. In North Africa, the adjective *bujadi* is still used to mean beginner, novice literally, someone who is still on his ABC. [G. S. Colin]

* According to M. Cohen (p. 137), the Latin word *elementum* goes back to an earlier alphabet that began in the middle, with the letters L, M, N. So giving the LMN (*elemen-tum*) of a matter was the same as "saying the ABC of it all".

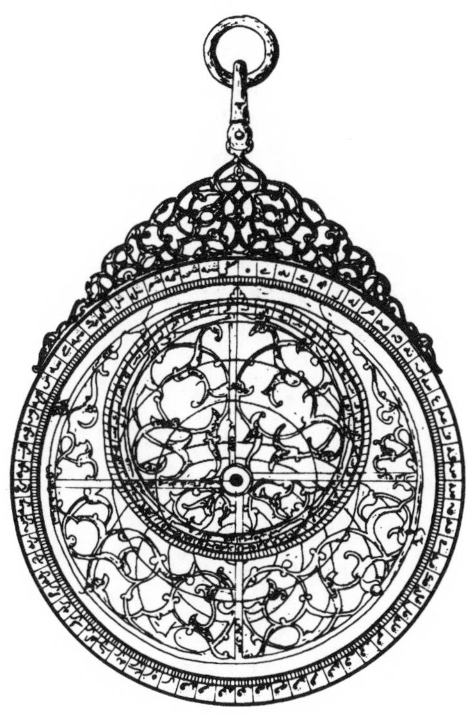

FIG. 19.9. *Seventeenth-century Persian astrolabe inscribed by Mohannad Muqim (Delhi, Red Fort, Isa 8). Note that the rim is marked in fives to 360 degrees by means of Arabic letter-numbers. (See B. von Dorn)*

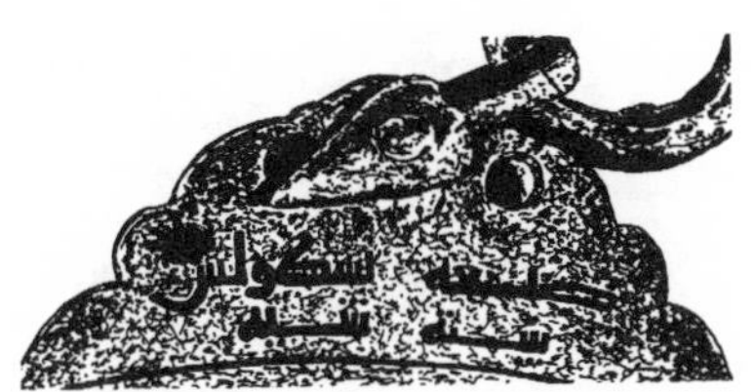

"WORK OF BASTULUS
YEAR 315"

FIG. 19.10. *Detail from an early oriental astrolabe, ostensibly once the property of King Farouk of Egypt, inscribed by Bastulus and dating from 315 of the Hegira (927–928 CE). The date is expressed by means of letter-numbers from the eastern number-system ("Kufic" characters with diacritics). (Personal communication from Alain Brieux)*

| | | | | | | | | | | | | | MNEMONIC WORDS RETAINED BY THIS USAGE |
|---|---|---|---|---|---|---|---|---|---|---|---|---|---|
| 1 | ا | 'alif | 10 | ي | ya | 100 | ق | qaf | 1,000 | ش | shin | Ayqash | ايقش ← |
| 2 | ب | ba | 20 | ك | kaf | 200 | ر | ra | | | | Bakar | بكر ← |
| 3 | ج | jim | 30 | ل | lam | 300 | س | sin | | | | Jalas | جلس ← |
| 4 | د | dal | 40 | م | mim | 400 | ت | ta | | | | Damat | دمت ← |
| 5 | ه | Ha | 50 | ن | Nun | 500 | ث | Tha | | | | Hanath | هنث ← |
| 6 | و | Wa | 60 | ص | Ṣad | 600 | خ | Kha | | | | Waṣakh | وصخ ← |
| 7 | ز | Zay | 70 | ع | 'Ayin | 700 | ذ | Dhal | | | | Za'adh | زعذ ← |
| 8 | ح | Ḥa | 80 | ف | Fa | 800 | ظ | Dha | | | | Ḥafaḍh | حفظ ← |
| 9 | ط | Ṭa | 90 | ض | Ḍad | 900 | غ | Ghayin | | | | Ṭaḍugh | طضغ ← |

FIG. 19.11. *Numeral alphabet used by African Arabs. (For mnemonic words see Fig. 19. 7 and footnote [same page])*

The eastern Arabs represented thousands, tens of thousands and hundreds of thousands by the multiplicative method. For this purpose, they adopted the convention of putting the letter associated with the corresponding numbers of units, of tens or of hundreds to the right of the Arab letter *ghayin*, whose value was 1,000 (Fig. 19.12).

| Arabic letter-number attributed to 1,000* | | | | | |
|---|---|---|---|---|---|
| | غ<br>isolated form | | | ـغ<br>final form | |
| 1,000 × 8<br>←-------- | حغ<br>gh Ḥ | 8,000 | 1,000 × 2<br>←-------- | بغ<br>gh B | 2,000 |
| 1,000 × 9<br>←-------- | طغ<br>gh Ṭ | 9,000 | 1,000 × 3<br>←-------- | جغ<br>gh J | 3,000 |
| 1,000 × 10<br>←--------- | يغ<br>gh Y | 10,000 | 1,000 × 4<br>←-------- | دغ<br>gh D | 4,000 |
| 1,000 × 20<br>←--------- | كغ<br>gh K | 20,000 | 1,000 × 5<br>←-------- | هغ<br>gh H | 5,000 |
| 1,000 × 30<br>←--------- | لغ<br>gh L | 30,000 | 1,000 × 6<br>←-------- | وغ<br>gh W | 6,000 |
| 1,000 × 40<br>←--------- | مغ<br>gh M | 40,000 | 1,000 × 7<br>←-------- | زغ<br>gh Z | 7,000 |

* i.e. the letter *ghayin*, twenty-eighth in the Abjad system (Fig. 19.4)

FIG. 19.12. *Eastern Arabic notation for numbers above 1,000*

## THE ETHIOPIAN NUMBER-SYSTEM

The Ethiopians borrowed the Greek alphabetical numbering system during the fourth century CE, no doubt under the influence of Christian missionaries who came from Egypt, Syria and Palestine.*

Starting, however, with 100, they radically altered the Greek system. Having adopted the first nineteen Greek alphabetic numerals to represent the first hundred whole numbers, they decided to indicate the hundreds and thousands by putting the letters for the units and tens to the left of the sign P (Greek *rho*) whose value was 100. That is to say that instead of representing the numbers 200, 300, . . . 9,000 after the Greek fashion:

| Σ | Τ | Υ | . . . | ϡ | ʹΑ | ʹΒ | . . . | ʹΘ |
|---|---|---|---|---|---|---|---|---|
| 200 | 300 | 400 | | 900 | 1,000 | 2,000 | | 9,000 |

they expressed them as follows (see Fig. 19. 13 A):

| ΒΡ | . . . | ΗΡ | . . . | ΚΡ | . . . | ΠΡ |
|---|---|---|---|---|---|---|
| 2 × 100<br>------→<br>200 | | 8 × 100<br>-------→<br>800 | | 20 × 100<br>-------→<br>2,000 | | 80 × 100<br>-------→<br>8,000 |

They denoted 10,000 by marking a ligature between two identical P signs (making a composite sign equivalent to multiplying 100 by itself, which we shall transcribe as P̵P. Then multiples of 10,000 were expressed by placing the symbol for the multiplier to the left of this symbol P̵P for 10,000.

| ΒP̵P | . . . | ΗP̵P | . . . | ΚP̵P | . . . | ΠP̵P |
|---|---|---|---|---|---|---|
| 2 × 10,000<br>---------→<br>20,000 | | 8 × 10,000<br>---------→<br>80,000 | | 20 × 10,000<br>---------→<br>200,000 | | 80 × 10,000<br>---------→<br>800,000 |

* The numerals that Ethiopians still sometimes use today are actually much more rounded stylisations of the numerical signs found on the Aksum inscriptions (Aksum, near the modern port of Adowa, was the capital of the Kingdom of Abyssinia from the fourth century CE). The modern signs follow the same principles as the ancient ones, which are themselves derived from the first nineteen letter-numerals of the Greek alphabet. Since the fifteenth century Ethiopian numerals have always been written inside two parallel bars with a curlicue at either end, signifying that they are to be taken as numbers, not as letters.

| VALUES | GREEK LETTER-NUMBERS | ETHIOPIAN INSCRIPTIONS AT AKSUM (4th century CE) DAE no. 7, 10, 11 | MODERN ETHIOPIAN NUMBERS |
|---|---|---|---|
| 1 | Α | ο | ፩ |
| 2 | Β | Β Β | ፪ |
| 3 | Γ | Γ Γ | ፫ |
| 4 | Δ | ∇ | ፬ |
| 5 | Ε | Ε Ε | ፭ |
| 6 | ϛ | ϛ ϛ ϛ | ፮ |
| 7 | Ζ | Ζ Ζ | ፯ |
| 8 | Η | Ι | ፰ |
| 9 | Θ | Θ Θ Θ | ፱ |
| 10 | Ι | ι | ፲ |
| 20 | Κ | Κ Κ | ፳ |
| 30 | Λ | Λ Λ | ፴ |
| 40 | Μ | Μ | ፵ |
| 50 | Ν | Ν Ν Ν | ፶ |
| 60 | Ξ | Ξ | ፷ |
| 70 | Ο | | ፸ |
| 80 | Π | Π | ፹ |
| 90 | Ϙ | | ፺ |
| 100 | Ρ | Ρ Υ Υ | ፻ |

Fig. 19.13a *Ethiopian numbering*

| VALUES AND GREEK LETTER-NUMBERS | | ETHIOPIAN INSCRIPTIONS AT ASKUM (4th century CE) DAE no. 7, 10, 11 | MODERN ETHIOPIAN NUMBERS AND ARITHMETICAL TRANSLATIONS | |
|---|---|---|---|---|
| 200 | Σ | ΒΥ | ፪፻ | $2 \times 100$ |
| 300 | Τ | ΓΥ | ፫፻ | $3 \times 100$ |
| 400 | Υ | ∇Υ | ፬፻ | $4 \times 100$ |
| 500 | Φ | ΕΥ | ፭፻ | $5 \times 100$ |
| 600 | Χ | ϛΥ | ፮፻ | $6 \times 100$ |
| 700 | Ψ | ΖΥ | ፯፻ | $7 \times 100$ |
| 800 | Ω | ΙΥ | ፰፻ | $8 \times 100$ |
| 900 | ϡ | ΘΥ | ፱፻ | $9 \times 100$ |
| 1,000 | ʹΑ | ιΥ | ፲፻ | $10 \times 100$ |
| 2,000 | ʹΒ | ΚΥ | ፳፻ | $20 \times 100$ |
| 3,000 | ʹΓ | ΛΥ | ፴፻ | $30 \times 100$ |
| 4,000 | ʹΔ | ΜΥ | ፵፻ | $40 \times 100$ |
| 5,000 | ʹΕ | ΝΥ | ፶፻ | $50 \times 100$ |
| 6,000 | ʹϛ | ΞΥ | ፷፻ | $60 \times 100$ |
| 8,000 | ʹΗ | ΠΥ | ፹፻ | $80 \times 100$ |
| 10,000 | $\overset{\alpha}{\mathrm{M}}$ | ΥΥ | ፼ | $100 \times 100$ |
| 20,000 | $\overset{\beta}{\mathrm{M}}$ | ΒΥΥ | ፪፼ | $2 \times 10,000$ |
| 31,900 | | ΓΥΥ ΙΥ ΘΥ | ፫፼ ፲፻ ፱፻ | |
| | | $3 \times 100 \times 100 + 10 \times 100 + 9 \times 100$ | | |
| 25,140 | | ΒΥΥ ΝΥ ΡΜ | ፪፼ ፶፻ ፻ ፵ | |
| | | $2 \times 100 \times 100 + 50 \times 100 + 100 + 40$ | | |

Fig. 19.13b *Ethiopian numbering for numbers above 100*

CHAPTER 20

# MAGIC, MYSTERY, DIVINATION, AND OTHER SECRETS

## SECRET WRITING AND SECRET NUMBERS IN THE OTTOMAN EMPIRE

We shall close our account of alphabetic numerals with an examination of the secret writing and secret numerals used until recently in the Middle East and, especially, in the official services of the Ottoman Empire.*

> The Turks used cryptography with abandon. Documents on Mathematics, Medicine and the occult, written or translated by the Turks, teem with secret alphabets and numerals, and they made use of every alphabet they knew. Usually they adopted such alphabets in the form in which they came across them, but sometimes they changed them; either deliberately, or as a result of the mutations which attend repeated copying. [M. J. A. Decourdemanche (1899)]

Fig. 20.1 shows secret numerals which were used for a long time in Egypt, Syria, North Africa, and Turkey. At first sight these would seem to have been made up throughout. However, if they are put alongside the Arabic letters which have the same numerical values, and then we put alongside these the corresponding Hebrew and Palmyrenean characters, we can at once see that the figures of these secret numerals are simply survivals of the ancient Aramaic characters in their traditional *Abjad* order (Fig. 20.2; see also Fig. 17.2, 17.4, 17.10 and 19.4).

* Such esoteric writing was used in a great variety of contexts: occultism, divination, science, diplomacy, military reports, business letters, administrative circulars, etc. Until the beginning of this century, the Turkish and Persian offices of the Ministry of Finance used a system of numerals known as *Siyaq*, whose figures were used in balance sheets and business correspondence. These figures were abbreviations of the Arabic names of the numbers, and their purpose was both to keep the sums of money secret from the public and also to prevent fraudulent alteration (see Chapter 25).

| | | | | | | | | | |
|---|---|---|---|---|---|---|---|---|---|
| | 9 | 8 | 7 | 6 | 5 | 4 | 3 | 2 | 1 |
| | or 90 | or 80 | 70 | 60 | or 50 | or 40 | 30 | or 20 | 10 |
| 1,000 | 900 | 800 | 700 | 600 | 500 | 400 | 300 | 200 | 100 |

FIG. 20.1.

Among these secret numerals there were alternative forms for the values 20, 40, 50, 80 and 90. These are in fact the final forms of the Hebrew and Palmyrenean letters *kof, mem, nun, pe* and *tsade*. The correspondences noted here are confirmed by treatises on arithmetic. The Egyptian treatises refer to this system as *al Shāmisī* ("sunlit"), which was used in those parts to designate things related to Syria. The Syrian documents themselves called it *al Tadmurī* ("from Tadmor"), which was the former Semitic name of Palmyra, an ancient city on the road linking Mesopotamia to the Mediterranean via Damascus to the south and via Homs to the north.

The people who had devised these secret writings had therefore taken the twenty-two Aramaic letters as they found them and (as has been explicitly mentioned by Turkish writers) they added six further conventional signs in order to complete a correspondence with the Arabic alphabet and to achieve a system of numerals which was complete from 1 to 1,000. This system was used until recent times, not only for writing numbers, but also as secret writing:

> In 1869, in order to draw up for French military officers a comparison between the abortive expedition of Charles III of Spain against Algiers, and the French expedition of 1830, the Ministry of War brought to Paris the original military report on the expedition of Charles III, which had been written in Turkish by the Algerian Regency at the Porte. This document was given to a military interpreter to be summarised. The manuscript, which I have seen, carried the stamp of a library in Algiers. After a whole wad of financial accounts came the report from the Regency. Following this came a series of annexes amongst which is an espionage report written as a long letter in the Hebraic script called *Khat al barāwāt*.

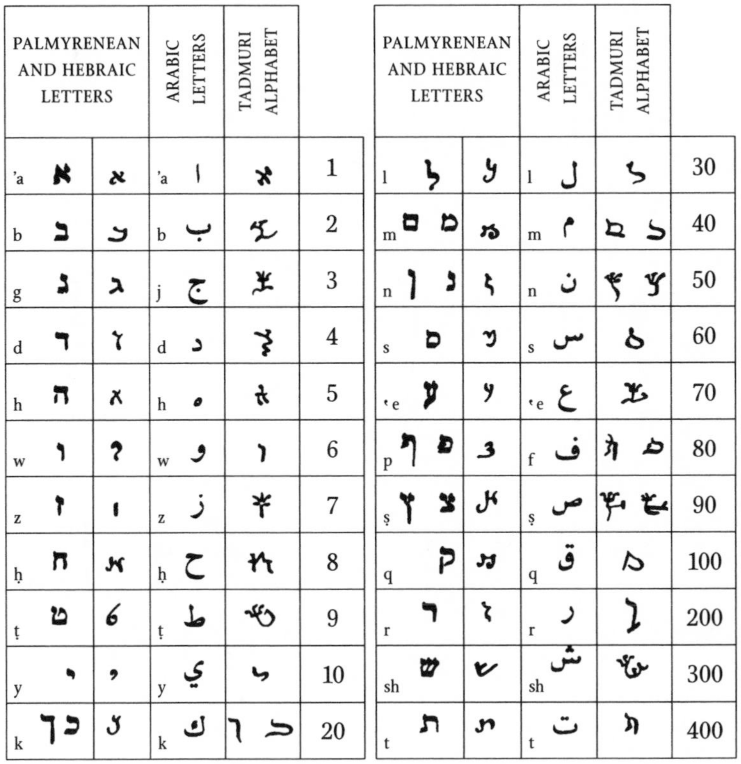

| PALMYRENEAN AND HEBRAIC LETTERS | | | ARABIC LETTERS | | TADMURI ALPHABET | |
|---|---|---|---|---|---|---|
| 'a | א | | 'a | ا | | 1 |
| b | ב | | b | ب | | 2 |
| g | ג | | j | ج | | 3 |
| d | ד | | d | د | | 4 |
| h | ה | | h | ه | | 5 |
| w | ו | | w | و | | 6 |
| z | ז | | z | ز | | 7 |
| ḥ | ח | | ḥ | ح | | 8 |
| ṭ | ט | | ṭ | ط | | 9 |
| y | י | | y | ي | | 10 |
| k | ך כ | | k | ك | | 20 |

| PALMYRENEAN AND HEBRAIC LETTERS | | | ARABIC LETTERS | | TADMURI ALPHABET | |
|---|---|---|---|---|---|---|
| l | ל | | l | ل | | 30 |
| m | ם מ | | m | م | | 40 |
| n | ן נ | | n | ن | | 50 |
| s | ס | | s | س | | 60 |
| 'e | ע | | 'e | ع | | 70 |
| p | ף פ | | f | ف | | 80 |
| ṣ | ץ צ | | ṣ | ص | | 90 |
| q | ק | | q | ق | | 100 |
| r | ר | | r | ر | | 200 |
| sh | ש | | sh | ش | | 300 |
| t | ת | | t | ت | | 400 |

FIG. 20.2. *Secret alphabet (still used in Turkey, Egypt, and Syria in the nineteenth century) compared with the Arabic, Palmyrenean, and Hebraic alphabets*

The signature was written in Tadmuri characters, not Latin: *Felipe, rabbina Yusuf ben Ezer, nacido en Granada.*

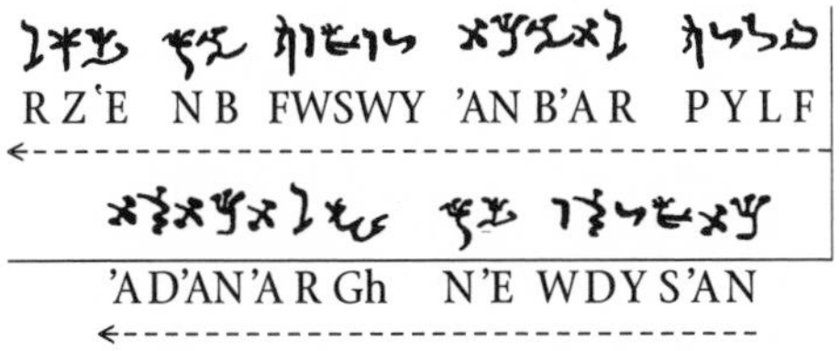

R Z'E N B FWSWY 'AN B'A R P Y L F

'AD'AN'A R Gh N'E W D Y S'AN

FIG. 20.3.

Then, on exactly the same kind of paper as the letter, is a detailed analysis of the Spanish land and sea forces, again written in Tadmuri characters. Since this analysis is also reproduced line for line in normal Turkish characters in the Regency report, it was easy for me to discern the value of each of the Tadmuri signs.

As an example, here in reproduction is the first line of the analysis, possibly for the army, possibly for the navy:

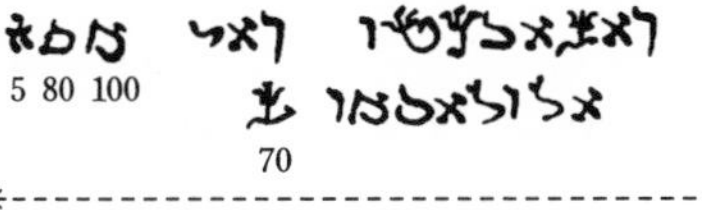


FIG. 20.4.

in which the following Spanish expressions are written in Tadmuri script:

| | | |
|---|---|---|
| *Regimento (del) Rey,* | 185 | (hombres) |
| "King's Army" | 185 | men |
| *El Velasco,* | 70 | (cañones) |
| "Navy" | 70 | guns |

[M. J. A. Decourdemanche (1899)]

We have no intention of presenting a general survey of the very many clandestine systems of the East; nonetheless we shall discuss two other systems of secret numerals which were used until recent times in the Ottoman army.

We begin with the simplest case. This is a system of numerals used in Turkish military inventories of provisions, supplies, equipment, and so on.

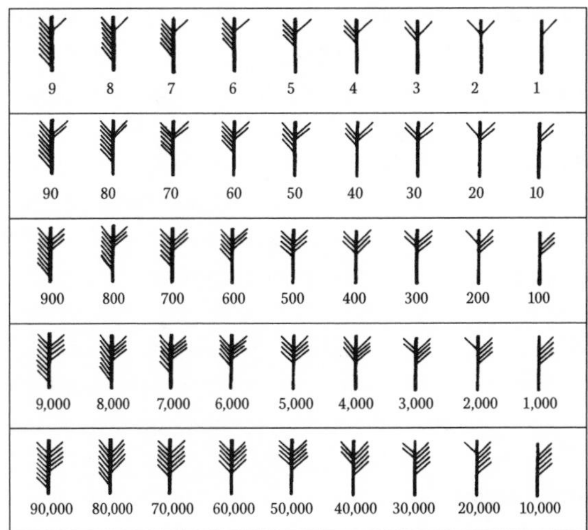


FIG. 20.5.

Here the numbers 1, 10, 100, 1,000 and 10,000 are represented by a vertical stroke with, on the right, one, two, three, four, or five upward oblique strokes. Adding one upward oblique stroke on the left of each of these gives the figures for 2, 20, 200, 2,000, and 20,000; two strokes on the left gives the figures for 3, 30, 300, 3,000, and 30,000; and so on, until with eight oblique strokes on the left we have the figures for 9, 90, 900, 9,000, and 90,000.

The above system is very straightforward, which is not the case for the next one. This was used in the Turkish army for recording the strengths of their units.

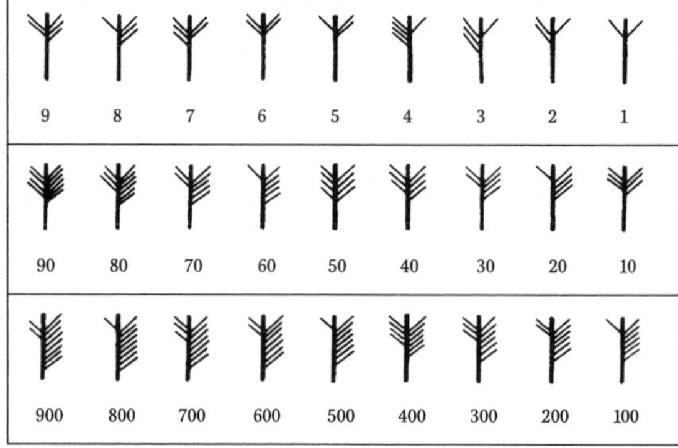


FIG. 20.6.

To the uninitiated, this system follows no obvious pattern. However, it was used both for writing numbers and also as a means of secret writing, which leads us to suppose that each of these signs corresponded to the Arabic letter corresponding to the numeral in question.

Proceeding as we did before, placing each of these numerals beside the Arabic letter corresponding to the same numerical value (see Fig. 19.4 above), we now consider the eight mnemonics for the letters of the Arabic numerals (Fig. 19.7), and it becomes clear how the figures of this system were formed.

For the numbers 1, 2, 3, 4 (corresponding to the first mnemonic, *ABJaD*), we take a vertical stroke with one oblique upward stroke on its right, and adjoin successively one, two, three, or four oblique upward strokes on its left.

Then, for the second mnemonic, *HaWaZin*, we take a vertical stroke with two upward oblique strokes on the right, and add successively one, two, or three upward oblique strokes on the left, and so on (Fig. 20.7).

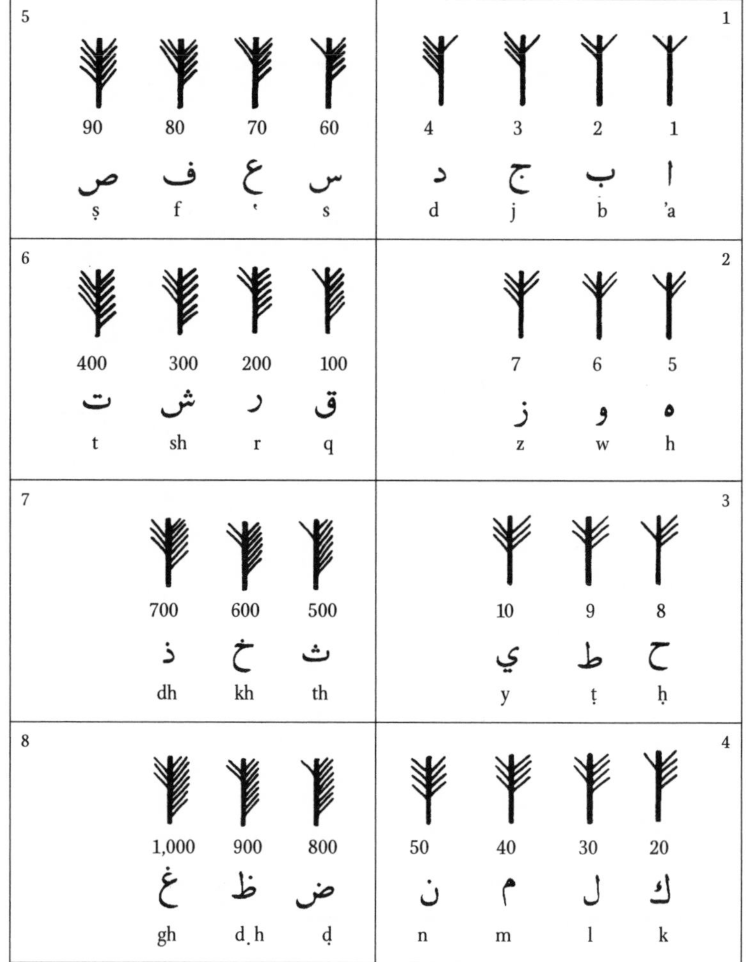


FIG. 20.7. *Secret numerical notation based on the succession of eight mnemonic words in the eastern Arab alphabetical numbering*

## THE ART OF CHRONOGRAMS

Jewish and Muslim writings since the Middle Ages abound in what are called "chronograms": these correspond to a method of writing dates, but – like calligraphy or poetry – are an art form in themselves.

This is the *Ramz* of the Arab poets, historians and stone-carvers in North Africa and in Spain, the *Tarikh* of the Turkish and Persian writers, which "consists of grouping, into one meaningful and characteristic word or short phrase, letters whose numerical values, when totalled, give the year of a past or future event." [G. S. Colin]

The following example occurs on a Jewish tombstone in Toledo [IHE, inscr. 43]:

שנת אגלי טל על חמשת אלפם
THOUSAND FIVE ON DEW DROP YEAR
YEAR: "ONE DROP OF DEW ON FIVE THOUSAND"


FIG. 20.8.

If we take it literally, the phrase is meaningless. But if we add up the numerical values of the letters in the phrase translated as "drop of dew", we discover that this phrase represents, according to the Hebrew calendar, the date of death of the person buried here:

"ONE DROP OF DEW" א ג ל י ט ל

30 9 10 30 3 1

83

FIG. 20.9.

This person died, in fact, in the year "eighty-three [= *drop of dew*] on five thousand," or, in plain language, in the year 5083 of the Hebrew era, i.e. 1322–1323 CE.

In the following two further examples from the Jewish cemetery in Toledo we find the years 5144 (Fig. 20.10) and 5109 (Fig. 20.11, in two different forms) shown in the chronograms: but note that the "5000" is not indicated, since it would have been implicitly understood, much as we understand "1974" when someone says "I was born in seventy-four". Also, note that in these examples the words whose letters represent numerals have been marked with three dots.

שנת הייֹנה אֹיֹן אֹב

2 1 | 50 1 10 | 5 10 5 50 10 | YEAR

YEAR: "WE HAVE BEEN MADE FATHERLESS"

144

FIG. 20.10.

לֹ ו הֹחיים | מנ וֹחה

40 10 10 8 5 6 30 | 5 8 6 50 40

"FOR LIFE" 109 | "REST" 109

FIG. 20.11.

The same procedure is found in Islamic countries, especially Turkey, Iraq, Persia, and Bīhar (in northwest India); but, like the oriental art of calligraphy, it seems to go no further back than the eleventh century.* The dreadful death of King Sher of Bīhar in an explosion occurred in the year 952 of the Hegira (1545 CE), which is recorded in the following chronogram [CAPIB, vol. X, p. 368]:

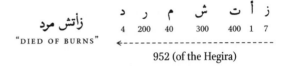

زآتش مرد "DIED OF BURNS"

ز أ ت ش م ر د

4 200 40 300 400 1 7

952 (of the Hegira)

FIG. 20.12.

Another interesting chronogram was made by the historian, mathematician and astronomer Al Biruni (born 973 CE at Khiva, died 1048 at Ghazni) in his celebrated *Tarikh ul Hind*. This learned man accused the Jews of deliberately changing their calendar so as to diminish the number of years elapsed since the Creation, in order that the date of birth of Christ should no longer agree with the prophecies of the coming of the Messiah; he boldly asserted that the Jews awaited the Messiah for the year 1335 of the Seleucid era (1024 CE), and he wrote this date in the following form:

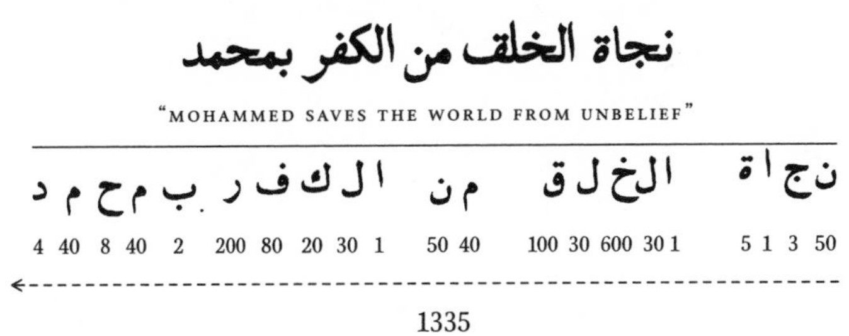


FIG. 20.13.

* In Persian and in Turkish certain letters have exactly the same numerical values as the equivalent Arabic letters according to the Eastern usage. For instance:

the letter پ, or P, has the same value as ب, or B;

the letter چ, or Ch, has the same value as ج, or J;

the letter گ, or G, has the same value as ک, or K.

Chronograms were also common in Morocco, but only from the seventeenth century CE (possibly the sixteenth, or earlier, according to recent documentation). They were often used in verse inscriptions commemorating events or foundations, and by writers, poets, historians and biographers, including the secretary and court poet Muhammad Ben Ahmad al Maklati (died 1630), and also the poets Muhammed al Mudara (died 1734) and 'Abd al Wahab Adaraq (died 1746) who both composed instructional historical synopses on the basis of chronograms, which in one case referred to the notabilities of Fez, and in the other to the saints of Maknez.*

The following example comes from an Arabic inscription discovered by Colin in the Kasbah of Tangier over fifty years ago, in the south chamber of the building known as *Qubbat al Bukhari*, in the old Sultan's Palace. We make a brief detour in time so as to stand in the period when this building was constructed.

The inscription was written to the glory of Ahmad ibn 'Ali ibn 'Abdallah. This notable person was:

> the son of the famous 'Ali ibn 'Abdallah, governor (*qā'id*) of Tetuan and chief of the Rif contingents destined for holy war (*mujāhidīn*) who, after a long siege, entered Tangier in 1095 of the Hegira (1684 CE) after its English occupiers had abandoned it . . .
>
> When *Qa'id* 'Ali ibn 'Abdallah, commandant (*amir*) of all the people of the Rif, died in year 1103 of the Hegira (1691–1692), Sultan Isma'il gave to them as chief the dead man's son, *basa* Ahmad ibn 'Ali; henceforth, almost all the history of northwest Morocco can be found in this man's biography . . . After 1139 of the Hegira (1726–1727), following the death of Sultan Isma'il, he took the opportunity provided by the weakness of his successor, Ahmad ad Dahabi, to try to seize Tetuan which was administered by another, almost independent, governor (*amīr*), Muhammed al Waqqas, but he was repulsed with loss.
>
> In 1140 of the Hegira (1727–1728), when Sultan Ahmad ad Dahabi (who had been overturned by his brother 'Abd al Malik) was restored to the throne, Ahmad ibn 'Ali refused to recognise him and declined to send him a deputation (a snub which was imitated by the town of Fez). The enmity between the Rif chieftain and the 'Alawite kings waxed from then on, and an impolitic gesture by Sultan 'Abdallah, successor of Ahmad ad Dahabi, transformed this into overt hostility . . .
>
> In 1145 of the Hegira, when a delegation of 350 holy warriors from the Rif came from Tangier to Sultan 'Abdallah to try to resolve the differences between him and *basa* Ahmad ibn 'Ali, he had them killed. The Rif chieftain distanced himself from the King and came closer to his brother and rival Al Mustadi. Thenceforth, until his unfortunate death in 1156 of the Hegira (1743), he did not cease from fighting with 'Abdallah, son of Sultan Isma'il, and to support his rivals against him. [G. S. Colin]

Returning now to our inscription, the date 1145 is given in the following verse (in which the numerical values have been calculated from the Arabic alphabetic numerals according to the Maghreb usage; see Fig. 19.11).

عام"حل بيت السعد بدر جمالي"

YEAR: "THE FULL MOON OF MY BEAUTY HAS ENTERED THE CHAMBER OF HAPPINESS"

| ج م ا ل ي | ا ل س ع د ب د ر | ح ل ب ي ت |
|---|---|---|
| 10 30 1 40 3 | 200 4 2 4 70 300 30 1 | 400 10 2 30 8 |

← 1145

FIG. 20.14.

In other words, the *Qubbat al Bukhari* in the Kasbah of Tangier was constructed in the year 1145 of the Hegira, the very time when *basa* Ahmad ibn 'Ali broke away from Sultan 'Abdallah.

We find in this chronogram, therefore, testimony to an art in which one's whole imagination is deployed to create a phrase which is both eloquent and, at the same time, has a numerical value that reveals the date of an event which one wishes to commemorate.

## GNOSTICS, CABBALISTS, MAGICIANS, AND SOOTHSAYERS

Once the letters of an alphabet have numerical values, the way is open to some strange procedures. Take the values of the letters of a word or phrase and make a number from these. Then this number may furnish an interpretation of the word, or another word with the same or a related numerical value may do so. The Jewish *gematria*,* the Greek *isopsephy* and the Muslim *khisab al jumal* ("calculating the total") are examples of this kind of activity.

* In epigraphic texts, chronograms were often written in a contrasting colour, and sometimes also in manuscripts where, however, we also find them written with thicker strokes. Arab chronograms, like those in Hebraic inscriptions, were always preceded by the preposition *fi* or by *Sanat 'ama* in the year, etc.

* Possibly a corruption of the Greek *geometrikos arithmos*, geometrical number

Especially among the Jews, these calculations enriched their sermons with every kind of interpretation, and also gave rise to speculations and divinations. They are of common occurrence in Rabbinic literature, especially the Talmud* and the Midrash.† But it is chiefly found in esoteric writings, where these cabbalistic procedures yielded hidden meanings for the purposes of religious dialectic.

Though not adept in the matter, we would here like to describe some examples of religious, soothsaying or literary practices which derive from such procedures.

The two Hebrew words *Yayin*, meaning "wine", and *Sod*, meaning "secret", both have the number 70 in the normal Hebrew alphabetic numerals (Fig. 20.15), and for this reason some rabbis bring these words together: *Nichnas Yayin Yatsa Sod*: "the secret comes out of the wine" (Latin: *in vino veritas*, the drunken man tells all).

FIG. 20.15.

In *Pardes Rimonim*, Moses Cordovero gives an example which relates *gevurah* ("force") to *arieh* ("lion"), which both have value 216. The lion, traditionally, is the symbol of divine majesty, of the power of *Yahweh*, while *gevurah* is one of the Attributes of God.

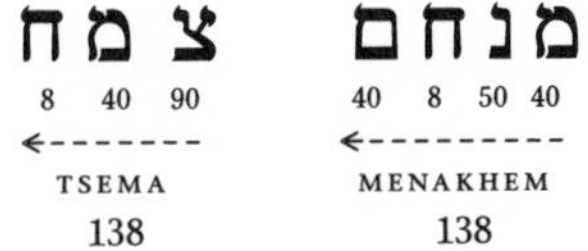


FIG. 20.16.

The Messiah is often called *Shema*, "seed", or *Menakhem*, "consoler", since these two words have the same value:



FIG. 20.17.

* The Rabbinic compilation of Jewish laws, customs, traditions and opinions which forms the code of Jewish civil and canon law

† Hebrew commentaries on the Old Testament

The letters of *Mashiyakh*, "Messiah", and of *Nakhash*, "serpent", give the same value:

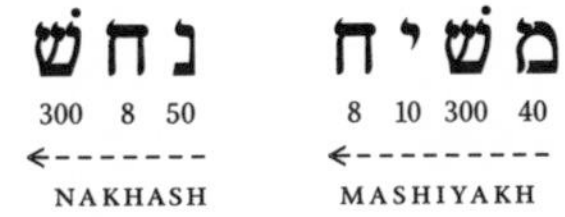


FIG. 20.18.

and this gives rise to the conclusion that "When the Messiah comes upon earth, he shall measure himself against Satan and shall overcome him."

We may also conclude that the world was created at the beginning of the Jewish civil year, from the fact that the two first words of the Torah (*Bereshit Bara*, "in the beginning [God] created") have the same value as *Berosh Hashanah Nibra*, "it was created at the beginning of the year":

FIG. 20.19.

In Genesis XXXII:4, Jacob says "I have sojourned with Laban" (in Hebrew, *'Im Laban Garti*). According to the commentary by Rashi* on this phrase (*Bereshit Rabbati*, 145), this means that "during his sojourn with Laban the impious, Jacob did not follow his bad example but followed the 613 commandments of the Jewish religion"; for, as he explains, *Garti* ("I have sojourned") has the value 613:

FIG. 20.20.

Genesis recounts elsewhere (XIV:12–14) how, in the battle of the kings of the East in the Valley of Siddim, Lot of Sodom, the kinsman of Abraham, was captured by his enemies: "When Abraham heard that his brother was taken captive, he armed his trained servants, born in his own house, three hundred and eighteen, and pursued them unto Dan", where he smote his adversaries with the help of "God Most High" (XIV:20). Then he addresses God in these words: "Lord GOD [*Yahweh*], what wilt thou give me, seeing I go childless and the steward of my house is this Eliezer of Damascus?" (Genesis XV:2).

* Rabbenu Shelomoh Yishakhi (1040–1105)

The *barayta* of the thirty-two Haggadic rules (for the interpretation of the Torah) gives the following interpretation (rule 29): the 318 servants are none other than the person of Eliezer himself. In other words, Abraham smote his enemies with the help of Eliezer alone, his trusted servant who was to be his heir; and whose name in Hebrew means "My God is help". The argument put forward for this brings together the two verses

> his trained servants, born in his own house, *three hundred and eighteen*

and

> the steward of my house is this *Eliezer of Damascus*

and the fact that the numerical value of the name *Eliezer* is 318:

אליעזר
200 7 70 10 30 1
ELI'EZER
318

Fig. 20.21.

Another concordance which the exegetes have achieved brings *Ahavah* ("Love") together with *Ekhad* ("One"):

אחד
4 8 1
EKHAD
13

אהבה
5 2 5 1
AHAVAH
13

Fig. 20.22.

As well as their numerical equivalence, it is explained that these two terms correspond to the central concept of the biblical ethic, that "God is Love", since on the one hand "One" represents the One God of Israel and, on the other hand, "Love" is supposed to be at the very basis of the conception of the Universe (Deuteronomy: V 6–7; Leviticus XIX:18). At the same time, the sum of their values is 26, which is the number of the name *Yahweh* itself:

יהוה
5 6 5 10
YHWH
26

Fig. 20.23.

The common Semitic word for "God" is *El*, but in the Old Testament this only occurs in compounds (*Israel, Ismael, Eliezer*, etc.). To refer to God, the Torah uses *Elohim* (which in fact is plural), and is the word which is supposed to express all the force and supernatural power of God. The Torah refers also to the attributes of God, such as *khay* ("living"), *Shadai* ("all-powerful"), *El Ilyion* ("God Most High") and so on. But *YHWH*, "Yahweh", is the only true Name of God: it is the *Divine Tetragram*. It is supposed to incorporate the eternal nature of God since it embraces the three Hebrew tenses of the verb "to be", namely:

היה
HaYaH "He was"

הוה
HoWeH "He is"

יהיה
YiHYeH "He shall be"

Fig. 20.24.

To invoke God by this name is therefore to appeal to His intervention and His concern for all things. But this name may be neither written nor spoken casually, and in order not to violate what is holy and incommunicable, in common use it must be read as *Adonai* ("My Lord").

Every kind of speculation has been founded on the numerical value of 26 which the Tetragram assumes according to the classical system of alphabetic numerals. Some adept writers have thereby been led to point out that in Genesis I:26, God says: "Let us make man in our image"; that 26 generations separate Adam and Moses; that 26 descendants are listed in the genealogy of Shem, and the number of persons named in this is a multiple of 26; and so on. According to them, the fact that God fashioned Eve from a rib taken from Adam is to be found in the numerical difference (= 26) between the name of Adam (= 45) and the name of Eve (= 19):

Fig. 20.25.

The usual alphabetic numerals were not the only basis adopted by the rabbis and Cabbalists for this kind of interpretation. A manuscript in the Bodleian Library at Oxford (Ms. Hebr. 1822) lists more than seventy different systems of *gematria*.

One of these involves assigning to each letter the number which gives its position in the Hebrew alphabet but with reduction of numbers above 9, that is to say with the same units figure as in the usual method, but ignoring tens and hundreds. The letter מ (*mem*), for example, which traditionally has the value 40, is given the value 4 in this system.* Similarly, the

* This can be found by the alternative method of noting that *Mem* is in the thirteenth place, so its value is equal to 1 + 3 = 4.

letter ש (*shin*), whose usual value is 300, has value 3 in this system.*From this, some have concluded that the name *Yahweh* can be equated to the divine attribute *Tov* ("Good"):

יחוה    טוב

5 6 5 1 ← YHWH 17    2 6 9 ← TOV "Good" 17

FIG. 20.26.

Another method gives to the letters values equal to the squares of their usual values, so that *gimmel*, for example, which usually has value 3, is here assigned the value 9 (Fig. 20.29, column B). According to a further system, the value 1 is assigned to the first letter, the sum (3) of the first two to the second letter, the sum (6) of the first three to the third, and so on. The letter *yod* , which is in the tenth position, therefore has a value equal to the sum of the first ten natural numbers: 1 + 2 + 3 + . . . + 9 + 10 = 55 (Fig. 20.29, column C).

Yet another system assigns to each letter the numerical value of the word which is the name of the letter. Thus *aleph* has the value 1 + 30 + 80 = 111:

טוב

80 30 1 ← 111

FIG. 20.27.

With these starting points, one can make a concordance between two words by evaluating them numerically according to either the same numerical system, or two different numerical systems. For instance, the word *Maqom* ("place"), which is another of the names of God, can be equated to *Yahweh* because in the traditional system the word *Maqom* has value 186, and *Yahweh* also has value 186 if we use the system which gives each letter the square of its usual value:

מקום    יהוה

40 6 100 40 ← MAQOM 186    $5^2$ $6^2$ $5^2$ $10^2$ ← YHWH 186

FIG. 20.28.

* *Shin* is in the twenty-first place, so its value is 2 + 1 = 3.

| Order number and normal values of the letters | | | A | B | C | D |
|---|---|---|---|---|---|---|
| 1 | א | 1 | 1 | $1^2$ | 1 | 111 value of אלף ALEPH |
| 2 | ב | 2 | 2 | $2^2$ | 1 + 2 | 412 " בית BET |
| 3 | ג | 3 | 3 | $3^2$ | 1 + 2 + 3 | 73 " גמל GIMMEL |
| 4 | ד | 4 | 4 | $4^2$ | 1 + 2 + 3 + 4 | 434 " דלת DALET |
| 5 | ה | 5 | 5 | $5^2$ | 1 + 2 + 3 + 4 + 5 | 6 " הא HE |
| 6 | ו | 6 | 6 | $6^2$ | 1 + 2 + 3 + 4 + 5 + 6 | 12 " וז VOV |
| 7 | ז | 7 | 7 | $7^2$ | 1 + 2 + 3 + 4 + 5 . . . + 7 | 67 " זין ZAYIN |
| 8 | ח | 8 | 8 | $8^2$ | 1 + 2 + 3 + 4 + 5 . . . + 8 | 418 " חית ḤET |
| 9 | ט | 9 | 9 | $9^2$ | 1 + 2 + 3 + 4 + 5 . . . + 9 | 419 " טית TET |
| 10 | י | 10 | 1 | $10^2$ | 1 + 2 + 3 + 4 + 5 . . . + 10 | 20 " יוד YOD |
| 11 | כ | 20 | 2 | $20^2$ | 1 + 2 + 3 + 4 + 5 . . . + 11 | 100 " כף KOF |
| 12 | ל | 30 | 3 | $30^2$ | 1 + 2 + 3 + 4 + 5 . . . + 12 | 74 " למד LAMED |
| 13 | מ | 40 | 4 | $40^2$ | 1 + 2 + 3 + 4 + 5 . . . + 13 | 90 " מים MEM |
| 14 | נ | 50 | 5 | $50^2$ | 1 + 2 + 3 + 4 + 5 . . . + 14 | 110 " נון NUN |
| 15 | ס | 60 | 6 | $60^2$ | 1 + 2 + 3 + 4 + 5 . . . + 15 | 120 " סמך SAMEKH |
| 16 | ע | 70 | 7 | $70^2$ | 1 + 2 + 3 + 4 + 5 . . . + 16 | 130 " עין AYIN |
| 17 | פ | 80 | 8 | $80^2$ | 1 + 2 + 3 + 4 + 5 . . . + 17 | 85 " פה PE |
| 18 | צ | 90 | 9 | $90^2$ | 1 + 2 + 3 + 4 + 5 . . . + 18 | 104 " צדי TSADE |
| 19 | ק | 100 | 1 | $100^2$ | 1 + 2 + 3 + 4 + 5 . . . + 19 | 104 " קוף QUF |
| 20 | ר | 200 | 2 | $200^2$ | 1 + 2 + 3 + 4 + 5 . . . + 20 | 510 " ריש RESH |
| 21 | ש | 300 | 3 | $300^2$ | 1 + 2 + 3 + 4 + 5 . . . + 21 | 360 " שין SHIN |
| 22 | ת | 400 | 4 | $400^2$ | 1 + 2 + 3 + 4 + 5 . . . + 22 | 406 " תו TAV |

FIG. 20.29. *Some of the many systems for the numerical evaluation of Hebraic letters. They are used by rabbis and Cabbalists for the interpretation of their homilies.*

This, it is emphasised, is confirmed by Micah I:3.

> For, behold, the LORD [*Yahweh*] cometh forth out of his place [*Maqom*].

This selection of examples – which could easily be much extended – gives a good idea of the complexities of Cabbalistic calculations and investigations which the exegetes went into, not only for the purpose of interpreting certain passages of the Torah but for all kinds of speculations.*

The Greeks also used similar procedures. Certain Greek poets, such as Leonidas of Alexandria (who lived at the time of the Emperor Nero), used them to create distichs and epigrams with the special characteristic of being *isopsephs*. A distich (consisting of two lines or two verses) is an isopseph if the numerical value of the first (calculated from the sum of the values of its letters) is equal to that of the second. An epigram (a short poem which might, for example, express an amorous idea) is an isopseph if all of its distichs are isopsephs, with the same value for each.

More generally, isopsephy consists of determining the numerical value of a word or a group of letters, and relating it to another word by means of this value.

At Pergamon, isopseph inscriptions have been found which, it is believed, were composed by the father of the great physician and mathematician Galen, who, according to his son, "had mastered all there was to know about geometry and the science of numbers."

At Pompeii an inscription was found which can be read as "I love her whose number is 545", and where a certain Amerimnus praises the mistress of his thoughts whose "honourable name is 45."

In the *Pseudo-Callisthenes*† (I, 33) it is written that the Egyptian god Sarapis (whose worship was initiated by Ptolemy I) revealed his name to Alexander the Great in the following words:

> Take two hundred and one, then a hundred and one, four times twenty, and ten. Then place the first of these numbers in the last place, and you will know which god I am.

Taking the words of the god literally, we obtain

200 1 100 1 80 10 200

which corresponds to the Greek name

Fig. 20.30.

In recalling the murder of Agrippina, Suetonius (*Nero*, 39) relates the name of Nero, written in Greek, to the words *Idian Metera apekteine* ("he killed his own mother"), since the two have exactly the same value according to the Greek number-system:

ΝΕΡΩΝ ΙΔΙΑΝ ΜΗΤΕΡΑ ΑΠΕΚΤΕΙΝΕ

50 5 100 800 50 10 4 10 1 50 40 8 300 5 100 1 1 80 5 20 300 5 10 50 5

"NERO" 1005 "HE KILLED HIS OWN MOTHER" 1005

Fig. 20.31.

The Greeks apparently came rather late to the practice of speculating with the numerical values of letters. This seems to have occurred when Greek culture came into contact with Jewish culture. The famous passage in the Apocalypse of Saint John clearly shows how familiar the Jews were with these mystic calculations, long before the time of their Cabbalists and the *Gematria*. Both Jews and Greeks were remarkably gifted for arithmetical calculation and also for transcendental speculation; every form of subtlety was apt to their taste, and number-mysticism appealed to both predilections at the same time. The Pythagorean school, the most superstitious of the Greek philosophical sects, and the most infiltrated by Eastern influence, was already addicted to number-mysticism. In the last age of the ancient world, this form of mysticism experienced an astonishing expansion.

It gave rise to arithmomancy; it inspired the Sybillines, the seers and soothsayers, the pagan *Theologoi*; it troubled the Fathers of the Church, who were not always immune to its fascination. Isopsephy is one of its methods. [P. Perdrizet (1904)]

Father Theophanus Kerameus, in his *Homily* (XLIV) asserts the numerical equivalence between *Theos* ("God"), *Hagios* ("holy") and *Agathos* ("good") as follows:

* We claim no competence to make the slightest commentary on these matters, neither on the delicate questions of the historical origins of *Gematria* in the Hebrew texts, nor on its evolution, nor on the extent to which it was regarded (or discredited) in Rabbinic and Cabbalistic writings throughout the centuries and in various countries. The reader who is interested in these questions may consult F. Dornseiff (1925) or G. Scholem.

† A spurious work associated with the name of Callisthenes, companion of Alexander in his Asiatic expedition.

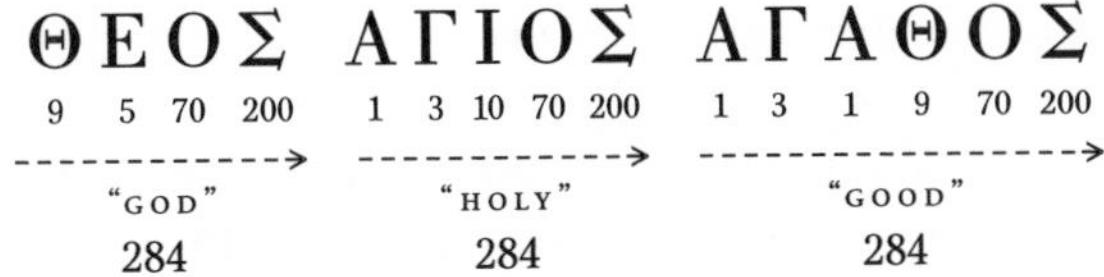


FIG. 20.32.

He likewise saw in the name *Rebecca* (wife of Isaac and mother of the twins Jacob and Esau) a figure of the Universal Church. According to him, the number (153) of great fish caught in the "miraculous draught of fishes" is the same as the numerical value of the name Rebecca in Greek (*Homily* XXXVI; John XXI).

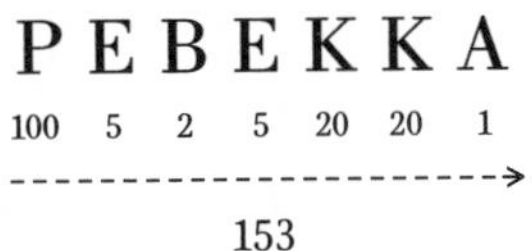


FIG. 20.33.

In the New Testament, the phrase *Alpha and Omega* (Apocalypse XXII: 13) is a symbolic designation of God: formed from the first and last letters of the Greek alphabet, in the Gnostic and Christian theologies it corresponded to the "Key of the Universe and of Knowledge" and to "Existence and the Totality of Space and Time". When Jesus declares that he is *the Alpha and the Omega*, he therefore declares that he is the beginning and the end of all things. He identifies himself with the "Holy Ghost" and therefore, according to Christian doctrine, with God Himself. According to Matthew III:16, the Holy Ghost appeared to Jesus at the moment of his birth in the form of a dove; the Greek word *Peristera* for "dove" has the value 801; and this is also the value of the letters of the phrase "Alpha and Omega" which, therefore, is no other than a mystical affirmation of the Christian doctrine of the Trinity.

A and Ω — ΠΕΡΙΣΤΕΡΑ

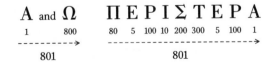

1 800 — 80 5 100 10 200 300 5 100 1

801 — 801

FIG. 20.34.



In another conception much exercised in the Middle Ages, numbers were given a supernatural quality according to the graphical shape of their symbols.

In a manuscript which is in the Bibliothèque nationale in Paris (Ms. lat. 2583, folio 30), Thibaut of Langres wrote as follows, about the number 300 represented by the Greek letter T (*tau*), which is also the sign of the Cross:

> The number is a secret guarded by writing, which represents it in two ways: by the letter and by its pronunciation. By the letter, it is represented in three ways: shape, order, and secret. By shape, it is like the 300 who, from the Creation of the World, were to find faith in the image of the Crucifix since, to the Greeks, these are represented by the letter T which has the form of a cross.

Which is why, according to Thibaut, Gideon conquered Oreb, Zeeb, Zebah, and Zalmunna with only the three hundred men who had drunk water "as a dog lappeth" (Judges VII:5).

A similar Christian interpretation is to be seen in the *Epistle of Barnabas*. In the patriarch Abraham's victory over his enemies with the help of 318 circumcised men, Barnabas finds a reference to the cross and to the two first letters of the name of Jesus (Ιησους)

T + IH = 318

300 10 + 8

FIG. 20.35.

He considers that the number 318 means that these men would be saved by the crucifixion of Jesus.

In the same fashion, according to Cyprian (*De pascha computus*, 20), the number 365 is sacred because it is the sum of 300 (T, the symbol of the cross), 18 (IH, the two first letters of the name of Jesus), 31 (the number of years Christ is supposed to have lived, in Cyprian's opinion) and 16 (the number of years in the reign of Tiberius, within which Jesus was crucified). This may well also explain why certain heretics believed that the End of the World would occur in the year 365 of the Christian era.*

* "But because this sentence is in the Gospel, it is no wonder that the worshippers of the many and false gods . . . invented I know not what Greek verses, . . . but add that Peter by enchantments brought it about that the name of Christ should be worshipped for three hundred and sixty-five years, and, after the completion of that number of years, should at once take end. Oh the hearts of learned men!" [Augustine, *The City of God*, Book 18, Chapter 53]

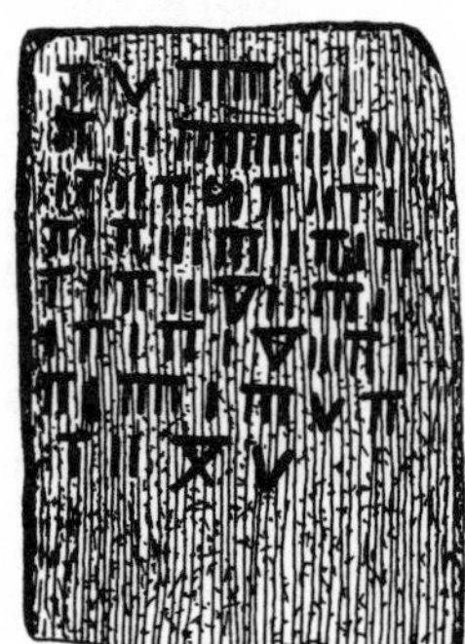

TRANSCRIPTION

I V $\overline{\mathrm{IIIIII}}$ V I
$\overline{\mathrm{II}}$ III $\overline{\mathrm{IIIIIIII}}$ III II
I $\overline{\mathrm{I}}$ II $\overline{\mathrm{II}}$ VI $\overline{\mathrm{II}}$ II $\overline{\mathrm{I}}$ I
$\overline{\mathrm{II}}$ I $\overline{\mathrm{II}}$ III $\overline{\mathrm{III}}$ II $\overline{\mathrm{II}}$ I $\overline{\mathrm{II}}$
$\overline{\mathrm{I}}$ I $\overline{\mathrm{II}}$ III $\overline{\mathrm{V}}$ II $\overline{\mathrm{III}}$ I
I $\overline{\mathrm{II}}$ I $\overline{\mathrm{II}}$ I $\overline{\mathrm{V}}$ III $\overline{\mathrm{II}}$ I
$\overline{\mathrm{II}}$ I $\overline{\mathrm{IIII}}$ I $\overline{\mathrm{III}}$ V $\overline{\mathrm{II}}$
$\overline{\mathrm{I}}$ II $\overline{\mathrm{X}}$ V

FIG. 20.36. *Wooden tablet found in North Africa, dating from the late fifth century CE. Note that on each line the Roman numerals total 18 (the overline denotes a part-total). It is not known whether this is a mathematical (indeed a teaching) document or a "magic" tablet relating to speculations on the numerical value of Greek or Hebraic letters. (See TA, act XXXIV, tabl. 3a)*

Clearly, all possible resources have been exploited for these purposes.

The Christian mystics, who wished to support the affirmation that Jesus was the Son of God, often equated the Hebrew phrase *'Ab Qal* which Isaiah used to mean "the swift cloud" on which "the Lord rideth" (XIX: 1) and the word *Bar* ("son"):

| עב קל | בר |
|---|---|
| 30 100 2 70 | 200 2 |
| ←----------- | ←----- |
| 202 | 202 |

FIG. 20.37.

For their part, the Gnostics* were able to draw almost miraculous consequences from the practice of isopsephy. P. Perdrizet (1904) explains:

> A text, which is probably by Hippolytus, says that in certain Gnostic sects isopsephy was a normal form of symbolism and catechesis. It did not serve only to wrap a revelation in a mystery: if in certain cases it served to conceal, in others it served to reveal, throwing light on things which otherwise would never have been understood . . .

Gnosticism seems loaded with a huge burden of Egyptian superstitions. It purported to rise to knowledge of the Universal Principle; in fact it was preoccupied with the quest to know the name of God and thence, with the aid of magic (the ancient magic of Isis), the means to induce God to allow Man to raise himself to God's own level. The name, like the shadow or the breath, is a part of the person: more, it is identical with the person, it is the person himself.

To know the name of God, therefore, was the problem which Gnosticism addressed. At first it seems insoluble: how can we know the Ineffable? The Gnostics did not pretend to know the name of God, but they believed it possible to learn its formula; and for them this was sufficient, since for them the formula of the divine name contained its complete magical virtue: and this formula was the number of the name of God.

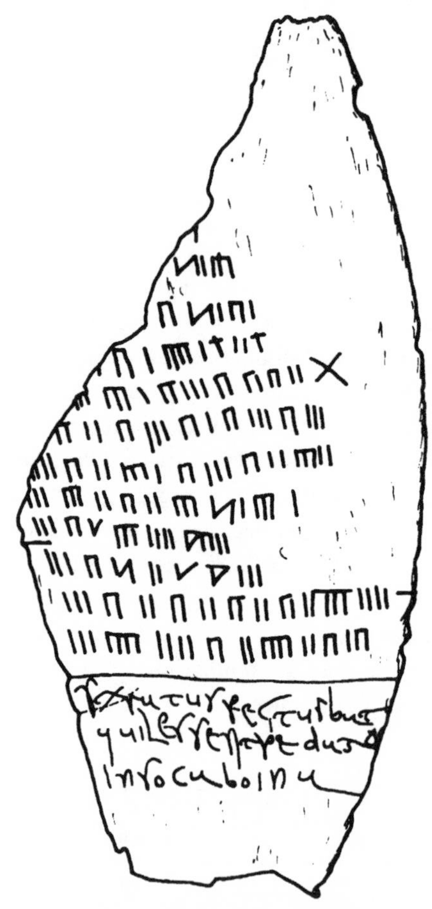

FIG. 20.38. *One of many slates found in the region of Salamanca. This one was discovered at Santibanez de la Sierra and dates from about the sixth century. It is a document similar to the previous one; each line that has remained intact shows a total count of 26. (See G. Gomez-Moreno, pp. 24, 117)*

* Gnosticism (from the Greek *gnosis*, "knowledge") is a religious doctrine which appeared in the early centuries of our era in Judaeo-Christian circles, but was violently opposed by rabbis and by the New Testament apostles. It is based essentially on the hope that salvation may be attained through an esoteric knowledge of the divine, as transmitted through initiation.

The supreme God of the Gnostics united in himself, according to Basilides the Gnostic, the 365 minor gods who preside over the days of the year . . . and so the Gnostics referred to God as "He whose number is 365" (ουεστιν η ψηαος ΤΞΕ). From God, on the other hand, proceeded the magical power of the seven vowels, the seven notes of the musical scale, the seven planets, the seven metals (gold, silver, tin, copper, iron, lead, and mercury); and of the four weeks of the lunar month. Whatever was the name of the Ineffable, the Gnostic was sure it involved the magic numbers 7 and 365. We may not know the unknowable name of God, so instead we seek a designation which would serve as its formula, and we only have to combine the mystic numbers 7 and 365. Thus Basilides created the name *Abrasax*, which has seven letters whose values add up to 365:

| Α | Β | Ρ | Α | Σ | Α | Ξ |
|---|---|---|---|---|---|---|
| 1 | 2 | 100 | 1 | 200 | 1 | 60 |

------------------------>
365

FIG. 20.39.

God, or the name of God (for they are the same) has first the character of holiness. Αγιος ο Θεος (*Hagios o Theos*) says the seraphic hymn; "hallowed be thy name" says the Lord's Prayer, that is "let the holiness of God be proclaimed."

Though the name of God remained unknown, it was known that it had the character to be the ideal holy name. Nothing therefore better became the designation of the Ineffable than the locution *Hagion Onoma* ("Holy Name") which the Gnostics indeed frequently employed. But this was not only for the above metaphysical or theological reason, nor because they had borrowed this same appellation from the Jews, but for a more potent mystical reason peculiar to them. By a coincidence of which Gnosticism had seen a revelation, the biblical phrase *Hagion Onoma* had the same number (365) as *Abrasax*.

| Α | Γ | Ι | Ο | Ν | | Ο | Ν | Ο | Μ | Α |
|---|---|---|---|---|---|---|---|---|---|---|
| 1 | 3 | 10 | 70 | 50 | | 70 | 50 | 70 | 40 | 1 |

------------------------------------->
365

FIG. 20.40.

Once embarked on this path, Gnosticism made other discoveries no less gripping.

Mingled as it was with magic, Gnosticism had a fatal tendency to syncretism. In isopsephy it had the means to identify with its own supreme God the national god of Egypt. The Nile, which for the Egyptians was the same as Osiris, was a god of the year, for the regularity of its floods followed the regular course of the years; and now, the number of the name of the Nile, *Neilos*, is 365:

| Ν | Ε | Ι | Λ | Ο | Σ |
|---|---|---|---|---|---|
| 50 | 5 | 10 | 30 | 70 | 200 |

------------------>
365

FIG. 20.41.

By isopsephy, Gnosticism achieved another no less interesting syncretism. The Mazdean cult of Mithras underwent a prodigious spread in the second and third centuries of our era. The Gnostics noticed that *Mithras*, written ΜΕΙΘΡΑΣ, has the value

| Μ | Ε | Ι | Θ | Ρ | Α | Σ |
|---|---|---|---|---|---|---|
| 40 | 5 | 10 | 9 | 100 | 1 | 200 |

--------------------->
365

FIG. 20.42.

Therefore the Sun God of Persia was the same as the "Lord of the 365 Days".

As Perdrizet says, the Christians often put new wine in old bottles, and they found that this kind of practice offered ample scope for fantasy. When the scribes and stone-carvers wished to preserve the secret of a name, they wrote only its number instead.

In Greek and Coptic Christian inscriptions, following an imprecation or an exhortation to praise, we sometimes come across the sign ϟΘ made up of the letters *Koppa* and *Theta*. This cryptogram remained obscure until the end of the nineteenth century, when J. E. Wessely (1887) showed that it was simply a mystical representation of *Amen* ('Αμην), since both have numerical value 99:

| Α | Μ | Η | Ν | | ϟ | Θ |
|---|---|---|---|---|---|---|
| 1 | 40 | 8 | 50 | | 90 | 9 |

------------> ---->
99 99

FIG. 20.43.

Similarly, the dedication of a mosaic in the convent of Khoziba near Jericho begins:

Φ Λ Ε ΜΝΗΣΦΗΤΙ ΤΟΥ ΔΟΥΛΟΥΣΟΥ

ΦΛΕ REMEMBER YOUR SERVANT

FIG. 20.44.

What does the group *Phi-Lambda-Epsilon* stand for? The problem was solved by W. D. Smirnoff (1902). These letters correspond to the Greek word for "Lord", **Κυριε**, whose numerical value is 535:

| Φ | Λ | Ε | | Κ | Υ | Ρ | Ι | Ε |
|---|---|---|---|---|---|---|---|---|
| 500 | 30 | 5 | | 20 | 400 | 100 | 10 | 5 |
| 535 | | | | 535 | | | | |

FIG. 20.45.

Much more significant are the speculations of the Christian mystics surrounding the number 666, which the apostle John ascribed to the *Beast of the Apocalypse*, a monster identified as the Antichrist, who shortly before the end of time would come on Earth to commit innumerable crimes, to spread terror amongst men, and raise people up against each other. He would be brought down by Christ himself on his return to Earth.

> 16 And he shall make all, both little and great, rich and poor, freemen and bondsmen, to have a character in their right hand, or on their foreheads.
>
> 17 And that no man might buy or sell, but he that hath the character, or the name of the beast, or the number of his name.
>
> 18 Here is wisdom. He that hath understanding, let him count the number of the beast. For it is the number of a man: and the number of him is six hundred and sixty-six. [Apocalypse, XIII:16–18]

We clearly see an allusion to isopsephy here, but the system to be used is not stated. This is why the name of the Beast has excited, and continues to excite, the wits of interpreters, and many are the solutions which have been put forward.

Taking 666 to be "the number of a man", some have searched amongst the names of historical figures whose names give the number 666. Thus Nero, the first Roman emperor to persecute the Christians, has been identified as the Beast of the Apocalypse since the number of his name, accompanied by the title "Caesar", makes 666 in the Hebraic system:

קסר נרון

50 6 200 50 200 60 100

<-----------------------

QSAR NERO

666

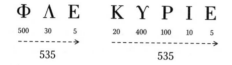

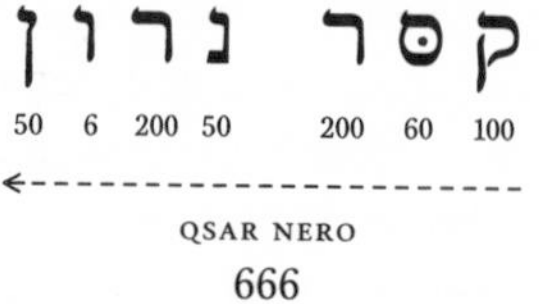

FIG. 20.46.

On the same lines, others have found that the name of the Emperor Diocletian (whose religious policies included the violent persecution of Christians), when only the letters that are Roman numerals are used, also gives the number of the Beast:

(Diocletian Augustus)

DIoCLEs AVGVsTVs

500 1 100 50 5 5 5

----------------------------------------->

666

FIG. 20.47.

Yet others, reading the text as "the number of a *type of man*", saw in 666 the designation of the Latins in general since the Greek word *Lateinos* gives this value:

| Λ | Α | Τ | Ε | Ι | Ν | Ο | Σ |
|---|---|---|---|---|---|---|---|
| 30 | 1 | 300 | 5 | 10 | 50 | 70 | 200 |

------------------------------->

666

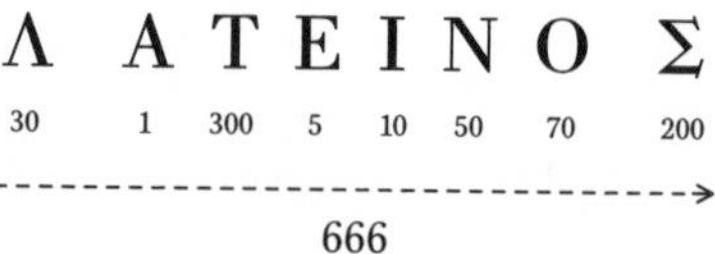

FIG. 20.48.

Much later, at the time of the Wars of Religion, a Catholic mystic called Petrus Bungus, in a work published in 1584–1585 at Bergamo, claimed to have demonstrated that the German reformer Luther was none other than the Antichrist since his name, in Roman numerals, gives the number 666:

FIG. 20.49.

But the disciples of Luther, who considered the Church of Rome as the direct heir of the Empire of the Caesars, lost no time in responding. They took the Roman numerals contained in the phrase *VICARIUS FILII DEI* ("Vicar of the Son of God") which is on the papal tiara, and drew the conclusion that one might expect:

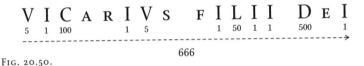

V I C A R I V S F I L I I D E I
5 1 100 1 5 1 50 1 1 500 1

666

FIG. 20.50.

The numerical evaluation of names was also used in times of war by Muslim soothsayers, under the name of *khisab al nim*, to predict which side would win. This process was described as follows by Ibn Khaldun in his "Prolegomena" (*Muqāddimah*, I):

> Here is how it is done. The values of the letters in the name of each king are added up, according to the values of the letters of the alphabet; these go from one to 1,000 by units, tens, hundreds and thousands. When this is done, the number nine is subtracted from each as many times as required until what is left is less than nine. The two remainders are compared: if one is greater than the other, and if both are even numbers or both odd, the king whose name has the smaller number will win. If one is even and the other odd, the king with the larger number will win. If both are equal and both are even numbers, it is the king who has been attacked who will win; if they are equal and odd, the attacking king will win.

Since each Arabic letter is the first letter of one of the attributes of Allah (*Alif*, the first letter of *Allah*; *Ba*, first letter of *Baqi*, "He who remains", and so on), the use of the Arabic alphabet led to a "Most Secret" system. In this, each letter is assigned, not its usual value, but instead the number of the divine attribute of which it is the first letter. For instance, the letter *Alif*, whose usual value is 1, is given the value 66 which is the number of the name of Allah calculated according to the *Abjad* system. This is the system used in the symbolic theology called *da'wa*, "invocation", which allowed mystics and soothsayers to make forecasts and to speculate on the past, the present and the future.

| LETTERS | | VALUES | ASSOCIATED DIVINE ATTRIBUTES | | | VALUES |
|---|---|---|---|---|---|---|
| | | | NAMES | | MEANING | |
| ا | 'alif | 1 | الله | ALLAH | Allah | 66 |
| ب | ba | 2 | باقي | BĀQĪ | He who remains | 113 |
| ج | jim | 3 | جامع | JĀMIʿ | He who collects | 114 |
| د | dal | 4 | ديّان | DAYĀN | Judge | 65 |
| ه | ha | 5 | هادي | HĀDĪ | Guide | 20 |
| و | wa | 6 | ولي | WALĪ | Master | 46 |
| ز | zay | 7 | زكي | ZAKĪ | Purifier | 37 |
| ح | ḥa | 8 | حق | ḤAQ | Truth | 108 |
| ط | ṭa | 9 | طاهر | ṬĀHIR | Saint | 215 |
| ي | ya | 10 | يسين | YASSĪN | Chief | 130 |
| ك | kaf | 20 | كافي | KĀFĪ | Sufficient | 111 |
| ل | lam | 30 | لطيف | LAṬĪF | Benevolent | 129 |
| م | mim | 40 | ملك | MALIK | King | 90 |
| ن | nūn | 50 | نور | NŪR | Light | 256 |
| س | sin | 60 | سميع | SAMĪʿ | Listener | 180 |
| ع | ʿayin | 70 | علي | ʿALĪ | Raised up | 110 |
| ف | fa | 80 | فتاح | FATĀḤ | Who opens | 489 |
| ص | ṣad | 90 | صمد | ṢAMAD | Eternal | 134 |
| ق | qaf | 100 | قادر | QĀDIR | Powerful | 305 |
| ر | ra | 200 | رب | RAB | Lord | 202 |
| ش | shin | 300 | شفيع | SHAFĪʿ | Who accepts | 460 |
| ت | ta | 400 | توب | TAWAB | Who restores to the good | 408 |
| ث | tha | 500 | ثابت | THĀBIT | Stable | 903 |
| خ | kha | 600 | خالق | KHĀLIQ | Creator | 731 |
| ذ | dhal | 700 | ذاكر | DHĀKIR | Who remembers | 921 |
| ض | ḍad | 800 | ضار | ḌĀR | Chastiser | 1,001 |
| ظ | dḥa | 900 | ظاهر | DḤĀHIR | Apparent | 1,106 |
| غ | gha | 1,000 | غفور | GHAFŪR | Indulgent | 1,285 |

FIG. 20.51. *The* Daʿwa *system, after the tabulation made by Sheikh Abu'l Muwwayid of Gujarat in* Jawahiru'l Khamsah

The same type of procedure allowed magicians to contrive their talismans, and to indulge in the most varied practices. In order to give their co-religionists the means to get rich quickly, to preserve themselves from evil and to draw down on themselves every grace of God, some *tolba* of North Africa offered their clients a *kherz* ("talisman") containing:

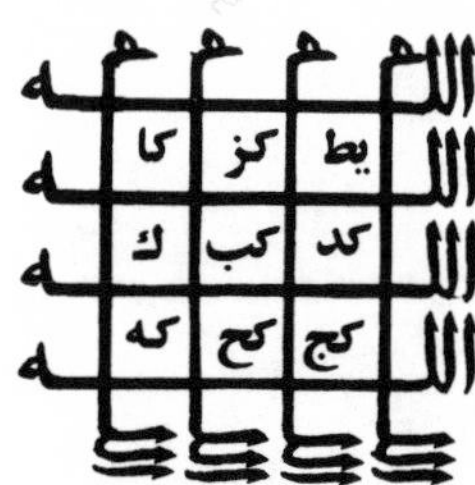

Fig. 20.52a.

This is a "magic square" whose value is 66, which can be obtained as the sum of every row, of every column, and of each diagonal:

| | | |
|---|---|---|
| 21 | 26 | 19 |
| 20 | 22 | 24 |
| 25 | 18 | 23 |

Fig. 20.52b.

and is itself the number of the name of Allah according to the *Abjad*:

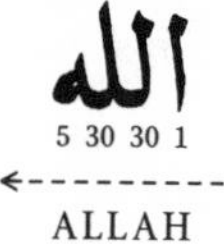


Fig. 20.53.

We can see, therefore, to what lengths the soothsayers, seers and other numerologists were prepared to go in applying these principles of number to the enrichment of their dialectic.

CHAPTER 21

# NUMBERS IN CHINESE CIVILISATION

## THE THIRTEEN FIGURES OF THE TRADITIONAL CHINESE NUMBER-SYSTEM*

The Chinese have traditionally used a decimal number-system, with thirteen basic signs denoting the numbers 1 to 9 and the first four powers of 10 (10, 100, 1,000, and 10,000). Fig. 21.1 shows the simplest representations of these, which is the one most commonly used nowadays.

| | | | |
|---|---|---|---|
| 1 | 一 | 10 | 十 |
| 2 | 二 | 100 | 百 |
| 3 | 三 | 1,000 | 千 |
| 4 | 四 | 10,000 | 萬 |
| 5 | 五 | | |
| 6 | 六 | | |
| 7 | 七 | | |
| 8 | 八 | | |
| 9 | 九 | | |

Fig. 21.1.

To an even greater extent than in the ancient Semitic world, this written number-system corresponds to the true type of "hybrid" number-system, since the tens, the thousands, and the tens of thousands are expressed according to the multiplicative principle (Fig. 21.2).

* I wish to express here my deep gratitude to my friends Alain Briot, Louis Frédéric and Léon Vandermeersch for their valuable contributions, and for their willing labour in reading this entire chapter.

| TENS | | HUNDREDS | | THOUSANDS | | TENS OF THOUSANDS | |
|---|---|---|---|---|---|---|---|
| 10 | 一十 1 × 10 | 100 | 一百 1 × 100 | 1,000 | 一千 1 × 1,000 | 10,000 | 一萬 1 × 10,000 |
| 20 | 二十 2 × 10 | 200 | 二百 2 × 100 | 2,000 | 二千 2 × 1,000 | 20,000 | 二萬 2 × 10,000 |
| 30 | 三十 3 × 10 | 300 | 三百 3 × 100 | 3,000 | 三千 3 × 1,000 | 30,000 | 三萬 3 × 10,000 |
| 40 | 四十 4 × 10 | 400 | 四百 4 × 100 | 4,000 | 四千 4 × 1,000 | 40,000 | 四萬 4 × 10,000 |
| 50 | 五十 5 × 10 | 500 | 五百 5 × 100 | 5,000 | 五千 5 × 1,000 | 50,000 | 五萬 5 × 10,000 |
| 60 | 六十 6 × 10 | 600 | 六百 6 × 100 | 6,000 | 六千 6 × 1,000 | 60,000 | 六萬 6 × 10,000 |
| 70 | 七十 7 × 10 | 700 | 七百 7 × 100 | 7,000 | 七千 7 × 1,000 | 70,000 | 七萬 7 × 10,000 |
| 80 | 八十 8 × 10 | 800 | 八百 8 × 100 | 8,000 | 八千 8 × 1,000 | 80,000 | 八萬 8 × 10,000 |

Fig. 21.2. *The modern Chinese notation for consecutive multiples of the first four powers of 10.*

For intermediate numbers, the Chinese used a combination of addition and multiplication, so that the number 79,564, for example, is decomposed as:

七 萬 九 千 五 百 六 十 四

←-------------------------------------------

7 × 10,000 + 9 × 1,000 + 5 × 100 + 6 × 10 + 4

79,564

Fig. 21.3.

NUMERALS OCCURRING IN THE DOCUMENT IN FIGURE 21.5

| Col. VIII | Col. VII | Col. IV | Col. I |
|---|---|---|---|
| 一百六十一<br>1 × 100 + 6 × 10 + 1<br>161 | 三百四十五<br>3 × 100 + 4 × 10 + 5<br>345 | 二百四十<br>2 × 100 + 4 × 10<br>240 | 一萬六千三百四十三<br>1 × 10,000 + 6 × 1,000 + 3 × 100 + 4 × 10 + 3<br>16,343 |
| 三十二<br>3 × 10 + 2<br>32 | 一十二<br>1 × 10 + 2<br>12 | 一千三百二十八<br>1 × 1,000 + 3 × 100 + 2 × 10 + 8<br>1,328 | |

Traditionally, these numbers, as in Chinese generally, would be placed vertically from top to bottom of columns which would be placed from right to left. However, in the People's Republic of China it is now preferred to write them horizontally from left to right.

FIG. 21.4. *Examples of numbers written with Chinese numerals*

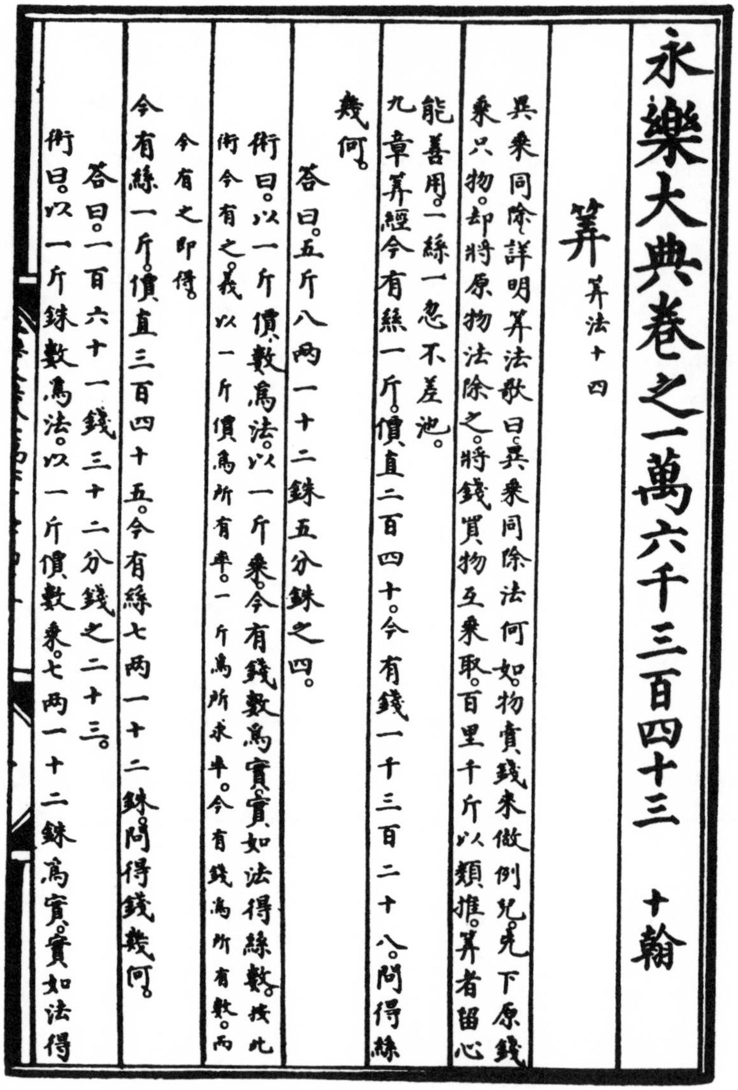

永樂大典卷之一萬六千三百四十三　十翰

筭　筭法十四

異乘同除詳明筭法歌曰。異乘同除法何如。物貨錢來做例兒。先下原錢乘只物。却將原物法除之。將錢買物互乘取。百里千斤以類推。筭者留心能善用。一絲一忽不差池。

九章筭經今有絲一斤。價直二百四十。今有錢一千三百二十八。問得絲幾何。

荅曰。五斤八兩一十二銖五分銖之四。

術曰。以一斤價數爲法。以一斤乘今有錢數爲實。實如法得絲數。按此術今有之義。以一斤價爲所有率。一斤爲所求率。今有錢爲所有數。而今有之即得。

今有絲一斤。價直三百四十五。今有絲七兩一十二銖。問得錢幾何。

荅曰。一百六十一錢三十二分錢之二十三。

術曰。以一斤銖數爲法。以一斤價數乘七兩一十二銖爲實。實如法得

FIG. 21.5. *A page from a Chinese mathematical document dating from the beginning of the fifteenth century. Cambridge University Library [Ms. Yong-le da dian, chapter 16 343, introductory page. From Needham (1959), III, Fig. 54].*

## TRANSCRIPTION OF CHINESE CHARACTERS

To transcribe Chinese characters into the Latin alphabet, we shall adopt the so-called *Pinyin* system in what follows. This has been the official system of the People's Republic of China since 1958. "This transcription", according to D. Lombard (1967), "was developed by Chinese linguists for use by the Chinese people and especially to assist schoolchildren to learn the language and its characters, and it is based mainly on phonological principles. The majority of Western Chinese scholars nowadays tend to abandon the older transcription systems (which sought in vain to represent pronunciation in terms of the spelling conventions of various European languages) in favour of this one. The reader is therefore no longer obliged to remember any spelling conventions, but instead must try to remember certain equivalences between sound and letter (as in beginning the study of German or Italian)."

Since the *Pinyin* system was not conceived with European readers in mind, it is natural that the values of its letters do not always coincide with English pronunciation. Here is a list of the most important aspects from the point of view of the English reader.

- b corresponds to our letter "p"
- c corresponds to our "ts"
- d corresponds to our "t"
- g corresponds to our "k"
- u corresponds to the standard English pronunciation of "u" as in "bull" (except after j, q or x)
- | corresponds to the pronunciation of "u" as, for instance, in Scotland or in French
- z corresponds to our "dz"
- zh corresponds to "j" as in "join"
- ch corresponds to "ch" as in "church"
- h in initial position, corresponds to the hard German "ch" (as in "Bach")
- x in initial position, corresponds to the soft "ch" (as in German "Ich")
- i corresponds to our "i" (as in "pin"); but, following z, c, s, sh, sh or r it is pronounced like "e" (in "pen") or like "u" in "fur"; following a or u, it is pronounced like the "ei" in "reign".
- q stands for a complex sound consisting of "ts" with drawing-in of breath
- r in initial position is like the "s" in pleasure; in other cases it is like the "el" in "channel".

## THE CHINESE ORAL NUMERAL SYSTEM

The number-signs shown above are in fact ordinary characters of Chinese writing. They are therefore subject to the same rules as govern the other Chinese characters. These are, in fact, "word-signs" which express in graphical form the ideographic and phonetic values of the corresponding numbers. In other words, they constitute one of the graphical representations of the thirteen monosyllabic words which the Chinese language possesses to denote the numbers from 1 to 9 and the first four powers of 10.

Having a decimal base, the oral Chinese number-system gives a separate name to each of the first ten integers:

| *yī* | *èr* | *sān* | *sì* | *wǔ* | *liù* | *qī* | *bā* | *jiǔ* | *shí* |
|---|---|---|---|---|---|---|---|---|---|
| 1 | 2 | 3 | 4 | 5 | 6 | 7 | 8 | 9 | 10 |

The numbers from 11 to 19 are represented according to the additive principle:

| | | | |
|---|---|---|---|
| 11 | *shí yī* | ten-one | = 10 + 1 |
| 12 | *shí èr* | ten-two | = 10 + 2 |
| 13 | *shí sān* | ten-three | = 10 + 3 |
| 14 | *shí sì* | ten-four | = 10 + 4 |
| ... | ....... | .......... | ......... |

The tens are represented according to the multiplicative principle:

| | | | |
|---|---|---|---|
| 20 | *èr shí* | two-ten | $= 2 \times 10$ |
| 30 | *sān shí* | three-ten | $= 3 \times 10$ |
| 40 | *sì shí* | four-ten | $= 4 \times 10$ |
| 50 | *wǔ shí* | five-ten | $= 5 \times 10$ |
| 60 | *liù shí* | six-ten | $= 6 \times 10$ |
| ... | ....... | .......... | ......... |

For 100 ($= 10^2$), 1,000 ($= 10^3$) and 10,000 ($= 10^4$), the words *bǎi*, *qiān* and *wàn* are used; for the various multiples of these the multiplicative principle is used:

| | | | |
|---|---|---|---|
| 100 | *yī bǎi* | one-hundred | |
| 200 | *èr bǎi* | two-hundred | $= 2 \times 100$ |
| 300 | *sān bǎi* | three-hundred | $= 3 \times 100$ |
| 400 | *sì bǎi* | four-hundred | $= 4 \times 100$ |
| ....... | ........ | ................. | ............... |
| 1,000 | *yī qiān* | one-thousand | |
| 2,000 | *èr qiān* | two-thousand | $= 2 \times 1{,}000$ |
| 3,000 | *sān qiān* | three-thousand | $= 3 \times 1{,}000$ |
| 4,000 | *sì qiān* | four-thousand | $= 4 \times 1{,}000$ |
| ....... | ........ | ................. | ............... |
| 10,000 | *yī wàn* | one-myriad | |
| 20,000 | *èr wàn* | two-myriad | $= 2 \times 10{,}000$ |
| 30,000 | *sān wàn* | three-myriad | $= 3 \times 10{,}000$ |
| 40,000 | *sì wàn* | four-myriad | $= 4 \times 10{,}000$ |
| ....... | ........ | ................. | ............... |

Starting with these, intermediate numbers can be represented very straightforwardly:

| 53,781 | *wǔ wàn* | *sān qiān* | *qī bǎi* | *bā shí* | *yī* |
|---|---|---|---|---|---|
| | ( five-myriad | three-thousand | seven-hundred | eight-ten | one ) |
| | (= $5 \times 10{,}000$ + | $3 \times 1{,}000$ + | $7 \times 100$ + | $8 \times 10$ + | 1 ) |

Thus the Chinese number-signs are a very simple way of writing out the corresponding numbers "word for word".

Finally, note that such a system has no need of a zero. For the numbers 504, 1,058, or 2,003, for example, one simply writes (or says):

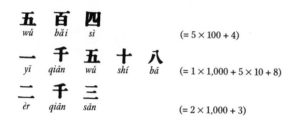

五 百 四
*wǔ bǎi sì* (= 5 × 100 + 4)

一 千 五 十 八
*yī qiān wǔ shí bā* (= 1 × 1,000 + 5 × 10 + 8)

二 千 三
*èr qiān sān* (= 2 × 1,000 + 3)

FIG. 21.6.

Note, however, that in current usage the word 零, *ling* (which means "zero"), is mentioned whenever any power of 10 is not represented in the expression of the number. This is done in order to avoid any ambiguity. But this usage was only established late in the development of the Chinese number-system.

504 五 百 零 四
5 100 0 4
*wǔ bǎi ling sì*
("five hundred zero four")

1,058 一 千 零 五 十 八
1 1,000 0 5 10 8
*yī qiān ling wǔ shí bā*
("one thousand zero five ten eight")

2,003 二 千 零 三
2 1,000 0 3
*èr qiān ling sān*
("two thousand zero three")

FIG. 21.7.



## CHINESE NUMERALS ARE DRAWN IN MANY WAYS

Even today, the thirteen basic number-signs are drawn in several different ways. Obviously they are spoken in the same way, but are a result of the many different ways of writing Chinese itself.

The forms we have considered so far, which may be called "classical", is the one in common use nowadays, especially in printed matter. It is also the simplest. Some of these signs are among the "keys" of Chinese writing: they are used in the elementary teaching of Chinese, at the stage of learning the Chinese characters.

They are part of the now standard *kǎishū* notation, a plain style in which the line segments making up each character are basically straight, but of varying lengths and orientations; they are to be drawn in a strict order, according to definite rules (Fig. 21.8).

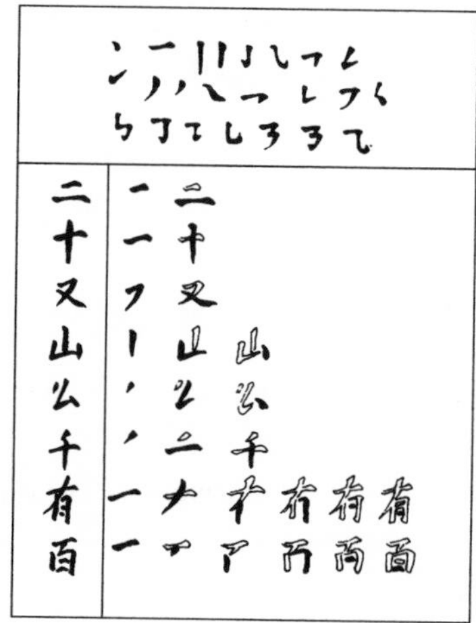

FIG. 21.8. *The basic strokes of Chinese writing in the standard style called* kǎishū, *and the order in which they are to be written in composing certain characters*

It is also the oldest of the common contemporary forms, having been used as early as the fourth century CE, and it is derived from the ancient writing called *lìshū** ("the writing of clerks") which was used in the Han Dynasty (Fig. 21.9).

* The *lìshū* style of notation is the earliest of the modern forms: it is the first "line writing" in Chinese history. However, "in seeking the maximum enhancement of the precision of the *lìshū* an even more geometrical style resulted, the inflexibly regular *kǎishū*." [V. Alleton (1970)]. This regular style became fixed as the standard for Chinese writing in the earliest centuries of the current era: administrative documents, official and scientific writings, were usually written in this style from that time on, when most such works were printed and the fonts for the characters had been made. When, below, we refer to "Chinese writing" without further qualification, it is this style which is meant.

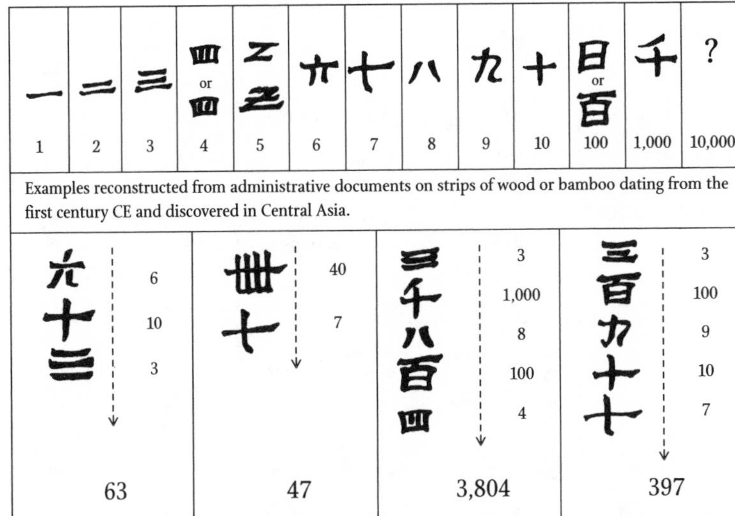

| 一 | 二 | 三 | 亖 or [illegible] | [illegible] | 六 | 七 | 八 | 九 | 十 | 日 or 百 | 千 | ? |
|---|---|---|---|---|---|---|---|---|---|---|---|---|
| 1 | 2 | 3 | 4 | 5 | 6 | 7 | 8 | 9 | 10 | 100 | 1,000 | 10,000 |

Examples reconstructed from administrative documents on strips of wood or bamboo dating from the first century CE and discovered in Central Asia.

| 63 | 47 | 3,804 | 397 |
|---|---|---|---|
| 六 6 / 十 10 / 三 3 | 卌 40 / 七 7 | 三 3 / 千 1,000 / 八 8 / 百 100 / 四 4 | 三 3 / 百 100 / 九 9 / 十 10 / 七 7 |

Fig. 21.9. *The earliest of the modern Chinese numerical notations. This is of the* lìshū *type and was in use during the Han Dynasty (206 BCE to 220 CE). The documents used for this diagram were written by scribes of the first century CE. [See de Chavannes (1913); Maspéro; Guitel]*

The second form of the Chinese numerals is called *guān zí* ("official writing"). It is used mainly in public documents, in bills of sale, and to write the sums of money on cheques, receipts or bills. Although still written like the classic *kăishū*, it is somewhat more complicated, having been made more elaborate in order to avoid fraudulent amendments in financial transactions (Fig. 21.10).

| Classical notation | 一 萬 三 千 六 百 八 十 四 |
|---|---|
| *Guàn zí* notation | 壹 萬 參 仟 陸 佰 捌 拾 肆 |
| | yī wàn sān qiān liù băi bā shí sì<br>1 × 10,000 + 3 × 1,000 + 6 × 100 + 8 × 10 + 4 |

Fig. 21.10.

The third style of writing the numerals is a cursive form of the classical numerals, which is routinely used in handwritten letters, personal notes, drafts, and so on. It belongs to the *xíngshū* style of writing, a cursive style which was developed to meet the need for abbreviation without detracting from the structure of the characters; the changes lay in the manner of drawing the characters more rapidly and flexibly using upward and downward brushstrokes. (Fig. 21.11).

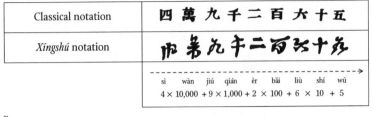

| Classical notation | 四 萬 九 千 二 百 六 十 五 |
|---|---|
| *Xíngshū* notation | [illegible] |
| | sì wàn jiú qiān èr băi liù shí wú<br>4 × 10,000 + 9 × 1,000 + 2 × 100 + 6 × 10 + 5 |

Fig. 21.11.

A combination of exaggerated abbreviation with virtuosity and imagination on the part of calligraphers rapidly brought these cursive forms, which still resembled the classical style, into an exaggeratedly simplified style which the Chinese call *căoshū* (literally, "plant-shaped"). It can only be deciphered by initiates, with the result that nowadays it is used only in painting and in calligraphy* (Fig. 21.12 and 21.13).

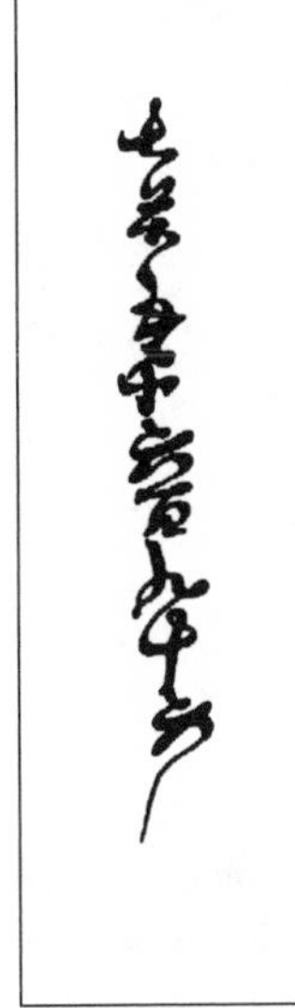

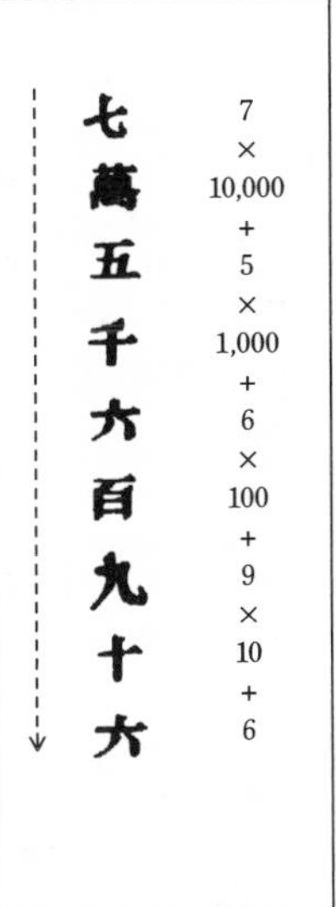


Fig. 21.12. *Example of 75,696*

* "Chinese writing underwent two transformations in the *căoshū*:

a. Lines and elements of characters were suppressed; save for characters with a small number of strokes, almost all elements are represented by symbols, leading to a kind of "writing of writing".

b. The strokes lose their individuality and join up: eventually a character is written in one movement; then the characters themselves join up, and even a whole column may be written without lifting brush from paper." [V. G. Alleton (1970)]

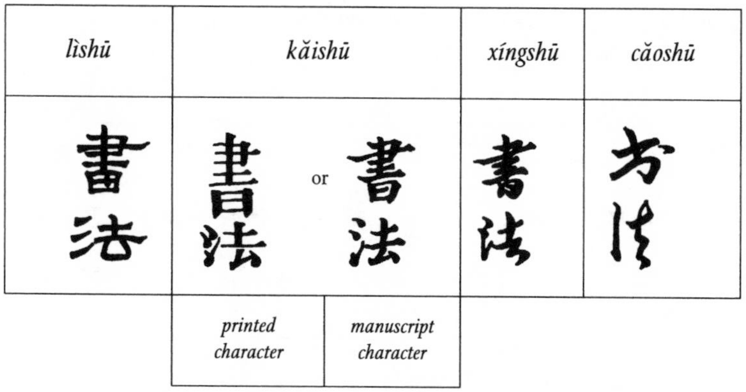

| *lìshū* | *kǎishū* | | | *xíngshū* | *cǎoshū* |
|---|---|---|---|---|---|
| 書法 | 書法 | or | 書法 | 書法 | 書法 |
| | *printed character* | | *manuscript character* | | |

FIG. 21.13. *The difference between the principal styles of modern Chinese writing, as shown in writing the word* shūfǎ *("calligraphy") in the styles of* lìshū *("official writing", used in the Han period),* kǎishū *("standard style", which replaced the* lìshū *and has been used since the fourth century CE),* xíngshū *(the current cursive style) and* cǎoshū *(a cursive style which has been reduced to maximum abbreviation and is now used only in calligraphy). [Alleton (1970)]*

Yet another form corresponds to a curiously geometrical way of drawing the numerals and characters, called *sháng fāng dà zhuàn*, which is still employed on seals and signatures (Fig. 21. 14).

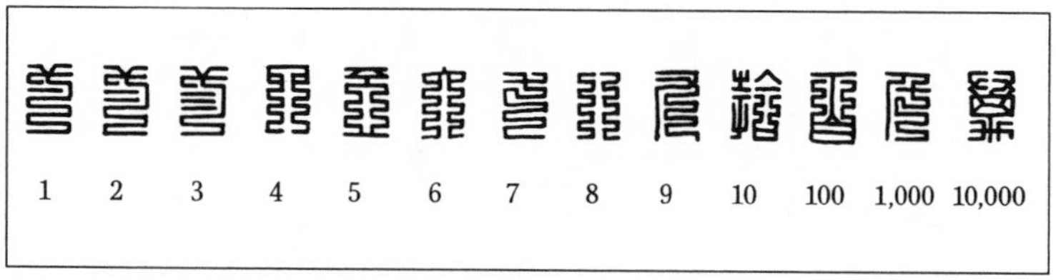

| 一 | 二 | 三 | 四 | 五 | 六 | 七 | 八 | 九 | 十 | 百 | 千 | 萬 |
|---|---|---|---|---|---|---|---|---|---|---|---|---|
| 1 | 2 | 3 | 4 | 5 | 6 | 7 | 8 | 9 | 10 | 100 | 1,000 | 10,000 |

FIG. 21.14. *Example of the singular* sháng fāng dà zhuàn *calligraphy as used for the thirteen basic characters of the Chinese number-system on seals and in signatures. [See Perny (1873); Pihan (1860)]*

As well as the forms already mentioned, there is the form used by traders to display the prices of goods. This is called *gán mà zí* ("secret marks"). Anyone who travels to the interior of China should be sure of knowing these numerals by heart, if he wishes to understand his restaurant bill (Fig. 21.15).

There are so many different styles for writing numerals in China that we should stop at this point, having described the important ones; to describe them all would be self-indulgent, and little to our purpose.

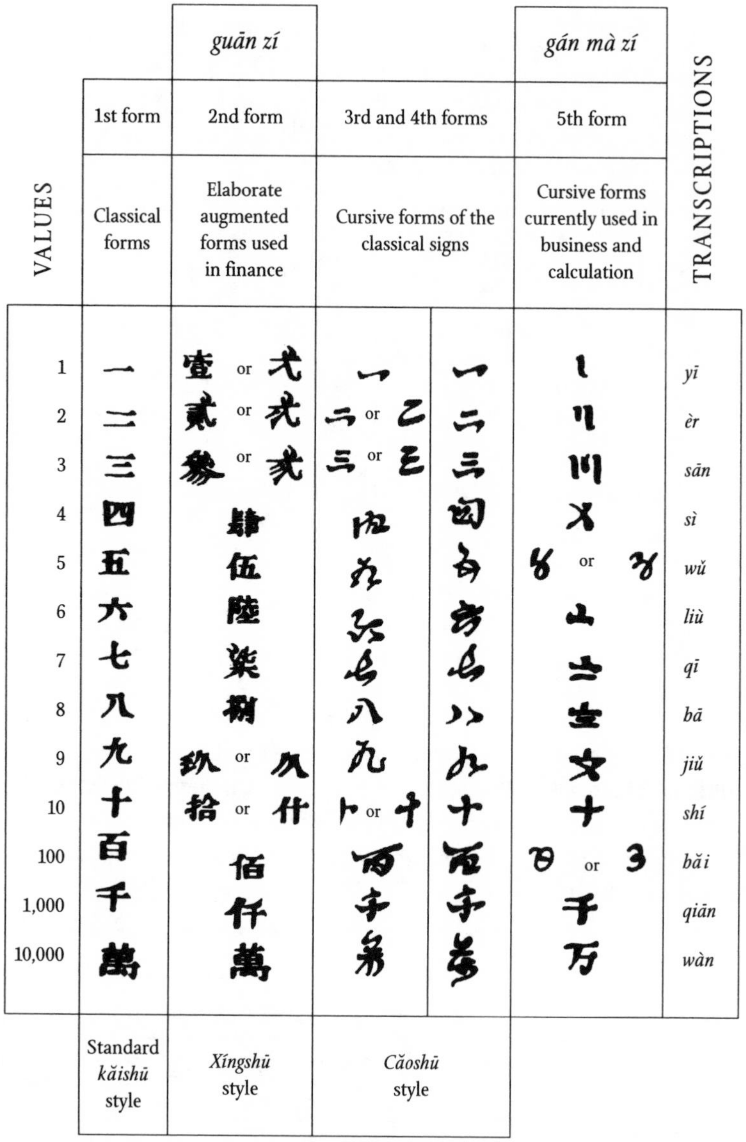

| | | *guān zí* | | | *gán mà zí* | |
|---|---|---|---|---|---|---|
| | 1st form | 2nd form | 3rd and 4th forms | | 5th form | |
| VALUES | Classical forms | Elaborate augmented forms used in finance | Cursive forms of the classical signs | | Cursive forms currently used in business and calculation | TRANSCRIPTIONS |
| 1 | 一 | 壹 or 弌 | 一 | 一 | 〡 | *yī* |
| 2 | 二 | 貳 or 弍 | 二 or 二 | 二 | 〢 | *èr* |
| 3 | 三 | 參 or 弎 | 三 or 三 | 三 | 〣 | *sān* |
| 4 | 四 | 肆 | 四 | 四 | 〤 | *sì* |
| 5 | 五 | 伍 | 五 | 五 | 〥 or 〥 | *wǔ* |
| 6 | 六 | 陸 | 六 | 六 | 〦 | *liù* |
| 7 | 七 | 柒 | 七 | 七 | 〧 | *qī* |
| 8 | 八 | 捌 | 八 | 八 | 〨 | *bā* |
| 9 | 九 | 玖 or 玖 | 九 | 九 | 〩 | *jiǔ* |
| 10 | 十 | 拾 or 什 | 十 or 十 | 十 | 十 | *shí* |
| 100 | 百 | 佰 | 百 | 百 | 百 or 3 | *bǎi* |
| 1,000 | 千 | 仟 | 千 | 千 | 千 | *qiān* |
| 10,000 | 萬 | 萬 | 萬 | 萬 | 万 | *wàn* |
| | Standard *kǎishū* style | *Xíngshū* style | *Cǎoshū* style | | | |

FIG. 21.15. *The principal graphic styles for the thirteen basic signs of the modern Chinese number-system. [Giles (1912); Mathews (1931); Needham (1959); Perny (1873); Pihan (1860)]*

## THE ORIGINS OF THE CHINESE NUMBER-SYSTEM

Several thousand bones and tortoise shells: these are the most ancient evidence we have of Chinese writing and numerals. They have for the most part been found since the end of the nineteenth century at the archaeological site of Xiao dun;* called *jiaguwen* ("oracular bones"), they date from around the Yin period (fourteenth–eleventh centuries BCE). On one side they bear inscriptions graven with a pointed instrument, on the other the surface is a maze of cracks due to heat. They would once have belonged to soothsayer-priests attached to the court of the Shang kings (seventeenth–eleventh centuries BCE) and would have been used in divination by fire.†

The writing on them is probably pictographic in origin, and seems to have reached a well-developed stage since it is no longer purely pictographic nor purely ideographic. The basis of the ancient Chinese writing in fact consists of a few hundred basic symbols which represent ideas or simple objects, and also of a certain number of more complicated symbols composed of two elements, of which one relates to the spoken form of a name and the other is visual or symbolic.‡ It represents a rather advanced stage of graphical representation (Fig. 21.16). "The stylisation and the economy of means are so far advanced in the oldest known Chinese writings that the symbols are more letters than drawings" [J. Gernet (1970), p. 31].

FIG. 21.16. *Some archaic Chinese characters*

* Village in the northwest of the An'yang district in the province of Henan

† According to H. Maspéro, this ritual took place as follows. Ancestor worship was of great importance in Chinese religion, and the priests consulted the royal ancestors on a great diversity of subjects. They first inscribed their questions on the ventral side of a tortoise shell which had been previously blessed (or on one side of the split shoulder-blade of a stag, of an ox or of a sheep). They then brought the other side towards the fire and the result of the divination was supposed to be decipherable from the patterns of cracks produced by the fire.

‡ "The peculiarities of the Chinese language may possibly explain the creation and persistence of this very complicated writing system. In ancient times, the language seems to have consisted of monosyllables of great phonemic variety, which did not allow the sounds of the language to be analysed into constituents, so Chinese writing could not evolve towards a syllabic notation, still less towards an alphabetic one. Each written sign could correspond to a single monosyllable and a single linguistic unit." (J. Gernet).

Gernet continues: "Moreover, this writing abounds in its very constitution with abstract elements (symbols reflected or rotated, strokes that mark this or that part of a symbol, representations of gestures, etc.) and with compounds of simpler signs with which new symbols are created."

The numerals, in particular, seem to have already embarked on the road towards abstract notation and appear to reflect a relatively advanced intellectual perspective.

In this system, unity is represented by a horizontal line, and 10 by a vertical line. Their origin is clear enough, since they reflect the operation of the human mind in given conditions: we know, for instance, that the people of the ancient Greek city of Karystos, and the Cretans, the Hittites and the Phoenicians, all used the same kind of signs for these two numerals. A hundred is denoted by what Joseph Needham called a "pine cone", and a thousand by a special character which closely resembles the character for "man" in the corresponding writing.

The figures 2, 3 and 4 are represented each by a corresponding number of horizontal strokes: an old ideographic system which is not used for the figures from 5 onwards. Like all the peoples who have used a similar numerical notation, the Chinese also stopped at 4; in fact few people can at a glance (and therefore without consciously counting) recognise a series of more than four things in a row. The Egyptians continued the series from 4 by using parallel rows, and the Babylonians and Phoenicians had a ternary system, but the Chinese introduced five distinct symbols for each of the five successive numbers: symbols, apparently, devoid of any intuitive suggestion. The number 5 was represented by a kind of X closed above and below by strokes; the number 6, by a kind of inverted V or by a design resembling a pagoda; 7, by a cross; 8, by two small circular arcs back to back; and the number 9 by a sign like a fish-hook (Fig. 21.17).

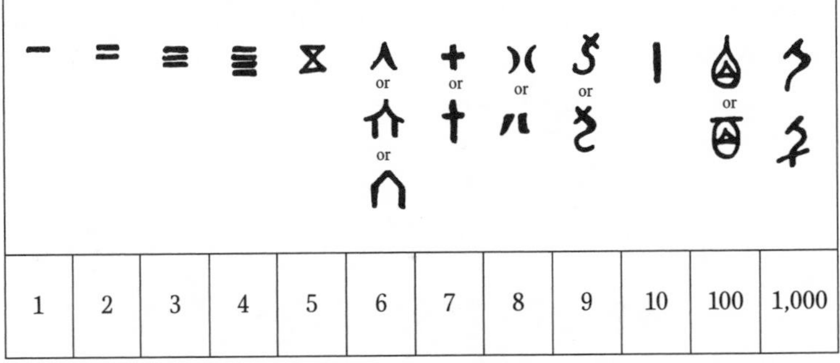


FIG. 21.17. *The basic signs of archaic Chinese numerals. They have been found on divinatory bones and shells from the Yin period (fourteenth to eleventh centuries BCE), and also on bronzes from the Zhou period (tenth to sixth centuries BCE). [Chalfant (1906); Needham (1959); Rong Gen (1959); Wieger (1963)]*

Now, did these number-signs evolve graphically from forms which originally consisted of groupings of corresponding numbers of identical elements? Or are they original creations? The history of Chinese writing leads us to form two hypotheses about these questions, both of them plausible, and not incompatible with each other.

We may in fact suppose that, for some of these numbers, their signs were, more or less, "phonetic symbols" which were used for the sake of the sounds they stood for, independently of their original meaning just as, indeed, was the case for Chinese writing. Such, for example, may well be why the number 1,000 has the same representation as "man", since the two words were probably pronounced in the same way at the time in question.

Another possible explanation may be of religious or magical origin, and may have determined the choice of the other symbols. Gernet (*EPP*) writes: "From the period of the inscriptions on bones and tortoiseshells at the end of the Shang Dynasty until the seventh century BCE, writing remained the preserve of colleges of scribes, adepts in the arts of divination and, by the same token, adepts also in certain techniques which depended on number, who served the princes in their religious ceremonies. Writing was therefore primarily a means of communication with the world of gods and spirits, and endowed its practitioners with the formidable power, and the respect mingled with dread, which they enjoyed. In a society so enthralled to ritual in behaviour and in thought, its mystical power must have preserved writing from profane use for a very long period."

Therefore it is by no means impossible that certain of the Chinese number-signs may have had essentially magical or religious roots, and were directly related to an ancient Chinese number-mysticism. Each number-sign, according to its graphical form, would have represented the "reality" of the corresponding number-form.

Whatever the case may be, the system of numerals which may be seen in the divinatory inscriptions on the bones and tortoise shells from the middle of the second millennium BCE is, intellectually speaking, already well on the way to the modern Chinese number-notation.

FIG. 21.18A. *Copy of a divinatory inscription on the ventral surface of a tortoise shell discovered at Xiao dun, which dates from the Yin period (fourteenth to eleventh centuries BCE). [Diringer (1968) plate 6-4: Yi 2908, translated and interpreted by L. Vandermeersch]*

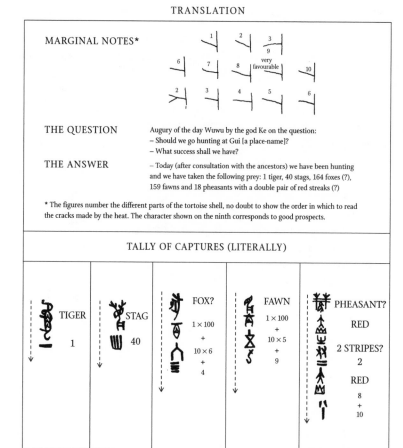

TRANSLATION

MARGINAL NOTES*

1 2 3 9

6 7 8 very favourable 10

2 3 4 5 6

THE QUESTION

Augury of the day Wuwu by the god Ke on the question:
– Should we go hunting at Gui [a place-name]?
– What success shall we have?

THE ANSWER

– Today (after consultation with the ancestors) we have been hunting and we have taken the following prey: 1 tiger, 40 stags, 164 foxes (?), 159 fawns and 18 pheasants with a double pair of red streaks (?)

* The figures number the different parts of the tortoise shell, no doubt to show the order in which to read the cracks made by the heat. The character shown on the ninth corresponds to good prospects.

TALLY OF CAPTURES (LITERALLY)

| | | | | |
|---|---|---|---|---|
| TIGER 1 | STAG 40 | FOX? 1 × 100 + 10 × 6 + 4 | FAWN 1 × 100 + 10 × 5 + 9 | PHEASANT? RED 2 STRIPES? 2 RED 8 + 10 |
| 1 Tiger | 40 Stag | 164 Fox | 159 Fawn | 18 Pheasant with a double pair of red stripes (?) |

FIG. 21.18B.

Leaving aside the numbers 20, 30 and 40 (to which we shall shortly return), the tens, hundreds and thousands are in fact represented according to the multiplicative principle by combining the signs corresponding to the units associated with them: in other words, the numbers from 50 to 90, for instance, are represented by superpositions according to the principle:

$$10 \times 5 \quad 10 \times 6 \quad 10 \times 7 \quad 10 \times 8 \quad 10 \times 9$$

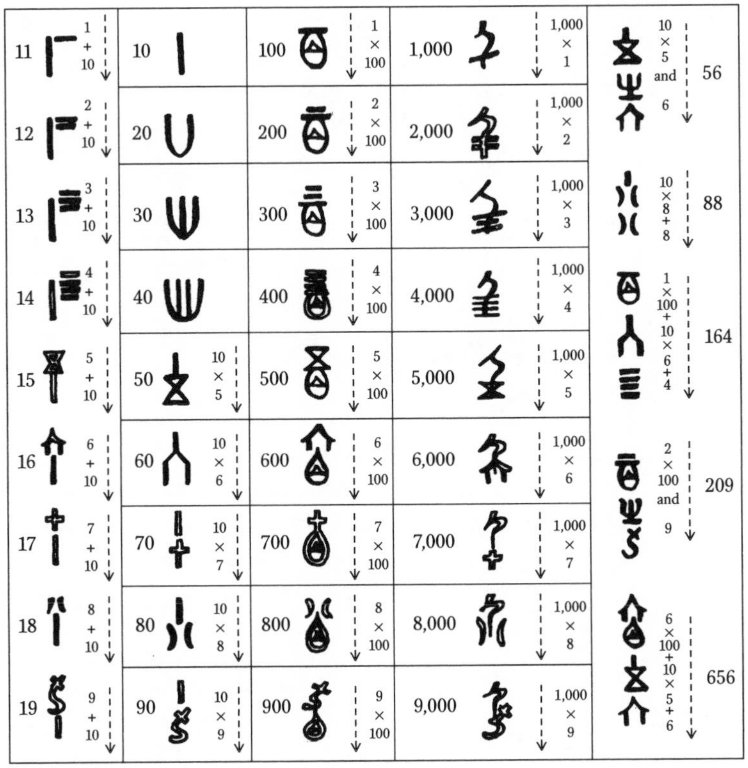

| | | | | |
|---|---|---|---|---|
| 11 (1 + 10) | 10 | 100 (1 × 100) | 1,000 (1,000 × 1) | 56 (10 × 5 and 6) |
| 12 (2 + 10) | 20 | 200 (2 × 100) | 2,000 (1,000 × 2) | |
| 13 (3 + 10) | 30 | 300 (3 × 100) | 3,000 (1,000 × 3) | 88 (10 × 8 + 8) |
| 14 (4 + 10) | 40 | 400 (4 × 100) | 4,000 (1,000 × 4) | 164 (1 × 100 + 10 × 6 + 4) |
| 15 (5 + 10) | 50 (10 × 5) | 500 (5 × 100) | 5,000 (1,000 × 5) | |
| 16 (6 + 10) | 60 (10 × 6) | 600 (6 × 100) | 6,000 (1,000 × 6) | 209 (2 × 100 and 9) |
| 17 (7 + 10) | 70 (10 × 7) | 700 (7 × 100) | 7,000 (1,000 × 7) | |
| 18 (8 + 10) | 80 (10 × 8) | 800 (8 × 100) | 8,000 (1,000 × 8) | 656 (6 × 100 + 10 × 5 + 6) |
| 19 (9 + 10) | 90 (10 × 9) | 900 (9 × 100) | 9,000 (1,000 × 9) | |

FIG. 21.19. *Principle of archaic Chinese numbering*

This representation should not be confused with the one used for the numbers 15 to 19, which was:

| 5 | 6 | 7 | 8 | 9 |
|---|---|---|---|---|
| + | + | + | + | + |
| 10 | 10 | 10 | 10 | 10 |

The numbers from 100 to 900 were written by placing the symbols for the successive units above the symbol for 100, and the thousands were written in a similar way to the tens (Fig. 21.19). Intermediate numbers were usually written by combining the additive and multiplicative methods.

We therefore see that, since the time of the very earliest known examples, the Chinese system was founded on a "hybrid" principle. That the numbers 20, 30 and 40 were often written as requisite repetitions of the symbol for 10 is quite simply due to the fact that the use of the multiplicative method would not have made the result any simpler. This kind of ideographic notation, natural though it was, was nevertheless limited, for psychological reasons, to a maximum of four identical elements.

The structure of the Chinese numerals stayed basically the same throughout its long history, even though the arrangement of the signs changed somewhat and their graphical forms underwent some variations (see Fig. 21.17, 21.21, then 21.9 and finally 21.15).

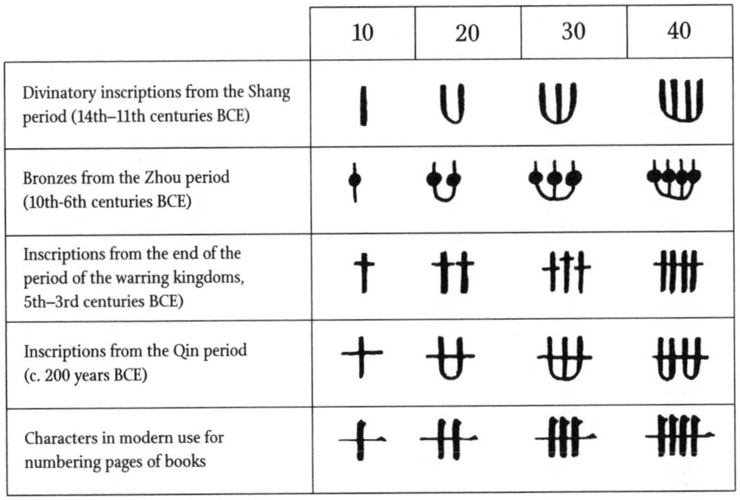

| | 10 | 20 | 30 | 40 |
|---|---|---|---|---|
| Divinatory inscriptions from the Shang period (14th–11th centuries BCE) | | | | |
| Bronzes from the Zhou period (10th-6th centuries BCE) | | | | |
| Inscriptions from the end of the period of the warring kingdoms, 5th–3rd centuries BCE) | | | | |
| Inscriptions from the Qin period (c. 200 years BCE) | | | | |
| Characters in modern use for numbering pages of books | | | | |

FIG. 21.20. *The durability of the ideographic forms of the first four numbers, as seen throughout the history of Chinese numerals*

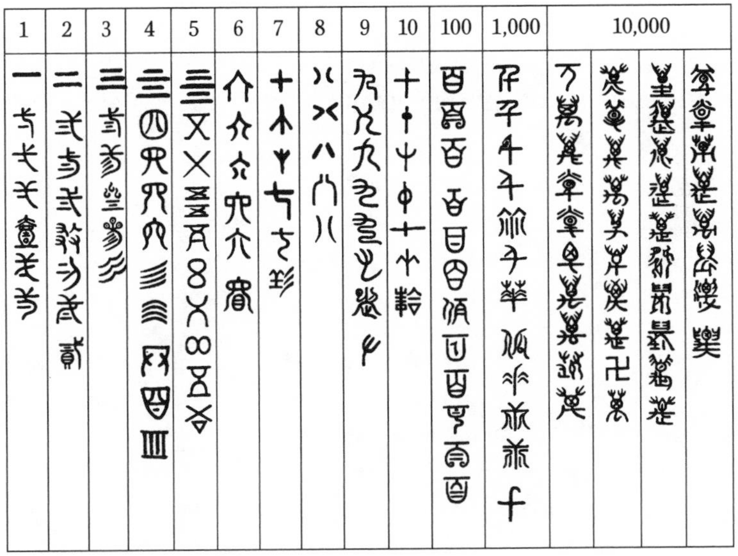

| 1 | 2 | 3 | 4 | 5 | 6 | 7 | 8 | 9 | 10 | 100 | 1,000 | 10,000 |
|---|---|---|---|---|---|---|---|---|---|---|---|---|

FIG. 21.21. *Variations in the graphical forms of the Chinese numerals, as found on inscriptions from the end of the period of the warring kingdoms (fifth to third centuries BCE). [Perny (1873); Pihan (1860)]*

## THE SPREAD OF WRITING THROUGHOUT THE FAR EAST

Over all the centuries, the structure of the Chinese characters has not fundamentally changed at all. The Chinese language is split into many regional dialects, and the characters are pronounced differently by the people of Manchuria, of Hunan, of Peking, of Canton, or of Singapore. Everywhere, however, the characters have kept the same meanings and everyone can understand them.

> For example, the word for "eat" is pronounced *chi* in Mandarin and is written with a character which we shall denote by "A". In Cantonese, this character is pronounced like *hek* but the Cantonese word for "eat" is pronounced *sik* and itself is represented by a character which we shall denote by "B". Nevertheless, all educated Chinese – even if in their dialect the word for "eat" is pronounced neither *chi* nor *sik* – readily understand the characters "A" and "B", which both mean "eat". [V. Alleton (1970)]

Chinese writing is therefore, in the words of B. Karlgren, a visual Esperanto: "The fact that people who are unable to communicate by the spoken word can understand each other when each writes his own language in Chinese characters has always been seen as one of the most remarkable features of this graphical system." [V. Alleton (1970)] We can easily understand why it is that some of China's neighbours have adopted this writing system for their own languages.

## NUMERALS OF THE FORMER KINGDOM OF ANNAM

This last was especially the case for the literate people of Annam (now Vietnam). They considered that the Chinese language was superior to their own, richer and more complete, and they adopted the Chinese characters as they stood but pronounced them in their own way (called "Sino-Annamite"). This gave rise to the Vietnamese writing called *chữ' nôm* (meaning "letter writing").

The Chinese numerals were also borrowed at the same time, and were read as follows in the Sino-Annamite pronunciation (*sô dêm táu*) which derived from an ancient Chinese dialect (Fig. 21.22).

| 一 | 二 | 三 | 四 | 五 | 六 | 七 | 八 | 九 | 十 | 百 | 千 | 萬 |
|---|---|---|---|---|---|---|---|---|---|---|---|---|
| *nhât* | *nhi* | *tam* | *tír* | *ngũ* | *luc* | *thàt* | *bát* | *ciru* | *thập* | *bách* | *thiên* | *vạn* |
| 1 | 2 | 3 | 4 | 5 | 6 | 7 | 8 | 9 | 10 | 100 | 1,000 | 10,000 |

FIG. 21.22. *The Chinese numerals and the Sino-Annamite names of the numbers [Dumoutier (1888)]*

In the present day, and since a date usually taken as the end of the thirteenth century CE, for most purposes (including letters, contracts, deeds and popular literature) the numerals are made in the *chữ' nôm* writing which is perfectly adapted to the Annamite number-names (the *sô dêm annam* system) (Fig. 21.23).

| 𠬠 | 𠄩 | 𠀧 | 𦊚 | 𠄼 | 𦒹 | 𦉱 | 𠔭 | 𠃩 | 𨒒 | 𤾓 | 𠦳 | 𣹓 |
|---|---|---|---|---|---|---|---|---|---|---|---|---|
| *một* | *hai* | *ba* | *bòn* | *năm* | *sáu* | *bảy* | *tám* | *chín* | *muòi* | *trăm* | *nghìn* | *muôn* |
| 1 | 2 | 3 | 4 | 5 | 6 | 7 | 8 | 9 | 10 | 100 | 1,000 | 10,000 |

FIG. 21.23. Chữ nôm *numerals and the Annamite names of the numbers. [Dumoutier (1888); Fossey (1948)]*

Although they look different from their Chinese prototypes, these numerals are in fact made up by combining a Chinese character (generally one of the Chinese numerals) as an ideogram, with some element of a character (or the whole character) chosen to represent the pronunciation of the pure Annamite number which is to be written (Fig. 21.24).

| figures | 2 | 3 | 4 | 5 | 6 | 7 | 8 | 9 | 10 | 100 | 1,000 | 10,000 |
|---|---|---|---|---|---|---|---|---|---|---|---|---|
| Chinese | 二 | 三 | 四 | 五 | 六 | 七 | 八 | 九 | 十 | 百 | 千 | 万 |
| *chữ' nôm* | 𠄩 | 𠀧 | 𦊚 | 𠄼 | 𦒹 | 𦉱 | 𠔭 | 𠃩 | 𨒒 | 𤾓 | 𠦳 | 𣹓 |

FIG. 21.24.

This changed nothing in the number-system itself, which continued to follow the Chinese rule of alternating digit and decimal order of magnitude, as in Fig. 21.25.

| | | |
|---|---|---|
| 𦒹 | *sáu* | 6 |
| | | × |
| 𠦳 | *nghìn* | 1,000 |
| | | + |
| 𦊚 | *bòn* | 4 |
| | | × |
| 𤾓 | *trăm* | 100 |
| | | + |
| 𠃩 | *chín* | 9 |
| | | × |
| 𨒒 | *muòi* | 10 |
| | | + |
| 𠔭 | *tám* | 8 |

FIG. 21.25.

Chinese characters were however abandoned in Vietnam at the start of the twentieth century in favour of an alphabetic system of Latin origin. The Annamite number-names (which are the only ones in current use) are either spelled out using Latin letters or are represented by Arabic numerals.

## JAPANESE NUMERALS

The Japanese also borrowed Chinese writing. However, according to M. Malherbe (1995), this was ill-adapted to the multiple grammatical suffixes of Japanese which are intrinsically incapable of ideographic representation. Therefore the Japanese early adopted (around the ninth century) a mixed system based on the following principle:

> Whatever corresponds to an idea is rendered by one of the Chinese *kanji* ideograms [the *kanji* system has been simplified to the point that there now remain only 1,945 official *kanji* characters, plus 166 for personal names, of which 996 are considered essential and are taught as part of primary education]. The more complicated ideograms have fallen into disuse and have been replaced by the *hiragana* characters.
>
> *Hiragana* is a syllabary: there are fifty-one signs, each of which represents a syllable, and not a letter as in the case of our alphabet. This can represent all the grammatical inflections and endings and, indeed, anything which cannot be written using ideograms.
>
> *Katakana* is a syllabary which exactly matches the *hiragana* but is used for recently imported foreign words, geographical names, foreign proper names, and so on.
>
> Finally, the *rômaji*, that is to say our own Western alphabet, is used in certain cases where using the other systems would be too complicated. For example, in a dictionary it is much more convenient to arrange the Japanese words according to the alphabetical order of their transcriptions into Latin characters.
>
> This writing system, which is the most complicated in the world, is regarded as inviolable by the Japanese who would consider themselves cut off from their culture if they gave themselves over to the use of *rômaji*, even though this would cause no practical difficulties nor inconvenience. [M. Malherbe (1995)]

The traditional Japanese numerals continue to be used despite the growing importance of Arabic numerals; they are the same as the Chinese numerals, in all their diverse forms (classical, cursive, commercial, etc.).

However, they are not pronounced as in Chinese. There are two different pronunciations: one is the "Sino-Japanese" which is derived from their Chinese pronunciation at the time when these characters were borrowed into Japanese; the other is "Pure Japanese".

The Japanese language therefore has two completely different series of number-names which still exist side by side.

The "Pure Japanese" system is a vestige of the ancient indigenous number-system. It consists of an incomplete list of names, which have short forms and complete forms (Fig. 21.26).

| | Short forms | | | Full forms | |
|---|---|---|---|---|---|
| 1 | *hi-* | or | *hito-* | *hitotsu*[a] | *hitori*[b] |
| 2 | *fu-* | or | *futa-* | *futatsu*[a] | *futari*[b] |
| 3 | *mi-* | | | *mitsu*[a] | *mitari*[b] |
| 4 | *yo-* | | | *yotsu*[a] | *yotari*[b] |
| 5 | *itsu-* | | | *itsutsu* | |
| 6 | *mu-* | | | *mutsu* | |
| 7 | *nana-* | | | *nanatsu* | |
| 8 | *ya-* | | | *yatsu* | |
| 9 | *kokono-* | | | *kokonotsu* | |
| 10 | *tō* | | | | |

*a. The number-names ending in -tsu are only used to refer to objects*
*b. The number-names ending in -tari are only used to refer to persons*

FIG. 21.26. *The Pure Japanese names of numbers. [Frédéric (1994 and 1977–87); Haguenauer (1951); Miller (1967); Plaut (1936)]*

Only the first four number-names have the ending *-tari* when applied to persons. From five persons upwards the base forms are used, which have neither inflection nor gender. This provides another instance of the psychological phenomenon described in Chapter 1, that only four items can be directly perceived.

The name of the number 8 also means "big number" and occurs in numerous locutions which express great multiplicity. So, where we for instance would say "break into a thousand pieces", the Japanese say

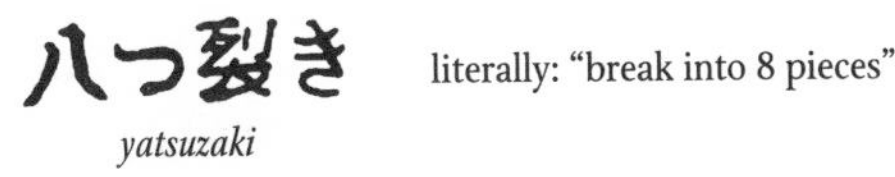


FIG. 21.27.

A market greengrocer – who sells every kind of fruit and vegetable – is likewise called

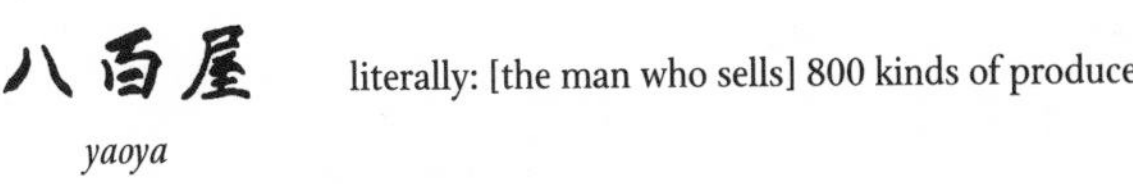

八百屋
yaoya
literally: [the man who sells] 800 kinds of produce


FIG. 21.28.

The city of Tokyo, which is of enormous extent, used to be called

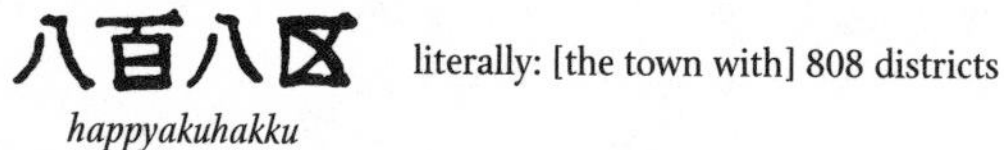

literally: [the town with] 808 districts

*happyakuhakku*

FIG. 21.29.

And to indicate the innumerable gods of their Shintô religion, the Japanese say

八百万の神 literally: 8 million gods

*happyakuman no kami*

FIG. 21.30.

As C. Haguenauer (1951) points out for the Pure Japanese number-names, there is a clear relation between the odd forms and the even forms, in the series "one–two" [*hito–futa*] and "three–six" [*mi–mu*], and an equally clear one between the even numbers four and eight [*yo–ya*]. The even numbers 2 and 6 have been obtained from the corresponding odd numbers by simple sound changes. In the latter case, a mere change of vowel makes the difference between "four" [*yo*] and "eight" [*ya*]. At first sight, only *i.tsu*, "five", and *tô*, "ten" are exceptions – as well, of course, as the odd numbers greater than 5. (Fig. 21.31)

| | | | | |
|---|---|---|---|---|
| 1 | *hito* ≈ *hi* | ←------→ $2 \times 1$ | 2 | *futa* ≈ *fu* |
| 3 | *mi* | ←------→ $2 \times 3$ | 6 | *mu* |
| 4 | *yo* | ←------→ $2 \times 4$ | 8 | *ya* |

FIG. 21.31.

This could indicate that long ago, among the indigenous peoples of Japan, the series of numbers came to a second break at 8 (the sequence 1, 2, 3, 4 being extended up to 8 by the additive principle: 5 = 3 + 2, 6 = 3 + 3, 7 = 4 + 3, 8 = 4 + 4).

In the aboriginal Japanese number-system there were also special names for some orders of magnitude above 10: a word for 20 (whose root is *hat'*) and individual names for 100 (*momo*), 1,000 (*chi*) and 10,000 (*yorozu*).

Nowadays, however, this system has been reduced to the barest minimum and is only now used for numbers between 1 and 10. The words for higher numbers have mostly fallen out of use except for the word for 20 (still used for lengths of time) and the word for 10,000 (sometimes used for the number itself, but most often simply to mean a boundless number).

The second of the Japanese number-systems has considerably greater capability than the one we have just looked at. It has a complete set of names for numbers, as follows:

| | | | | |
|---|---|---|---|---|
| 1 | *ichi* | 10 | | *jû* |
| 2 | *ni* | 100 | $(= 10^2)$ | *hyaku* |
| 3 | *san* | 1,000 | $(= 10^3)$ | *sen* |
| 4 | *shi* | 10,000 | $(= 10^4)$ | *man* |
| 5 | *go* | | | |
| 6 | *roku* | | | |
| 7 | *shichi* | | | |
| 8 | *hachi* | | | |
| 9 | *ku* | | | |

FIG. 21.32. *The Sino-Japanese number-names. [Haguenauer (1951); Miller (1967); Plaut (1936)]*

The numbers from 11 to 19 are represented according to the additive principle:

| | | | |
|---|---|---|---|
| 11 | *jû.ichi* | ten-one | $= 10 + 1$ |
| 12 | *jû.ni* | ten-two | $= 10 + 2$ |
| 13 | *jû.san* | ten-three | $= 10 + 3$ |

For the tens, hundreds and thousands, and so on, it used the multiplicative principle:

| | | | |
|---|---|---|---|
| 20 | *ni.jû* | two-ten | $= 2 \times 10$ |
| 30 | *san.jû* | three-ten | $= 3 \times 10$ |
| ... | | | |
| 100 | *hyaku* | hundred | $= 10^2$ |
| 200 | *ni.hyaku* | two-hundred | $= 2 \times 100$ |
| 300 | *san.hyaku* | three-hundred | $= 3 \times 100$ |
| ... | | | |
| 1,000 | *sen* | thousand | $= 10^3$ |
| 2,000 | *ni.sen* | two-thousand | $= 2 \times 1{,}000$ |
| 3,000 | *san.sen* | three-thousand | $= 3 \times 1{,}000$ |
| ... | | | |
| 10,000 | *ichi.man* | myriad | $= 10^4$ |
| 20,000 | *ni.man* | two-myriad | $= 2 \times 10{,}000$ |

| 五 萬 | | 三 千 | | 六 百 | | 八 十 | | 一 |
|---|---|---|---|---|---|---|---|---|
| *go.man* | | *san.sen* | | *roku.hyaku* | | *hachi.jû* | | *ichi* |
| ("five-myriad | | three-thousand | | six-hundred | | eight-ten | | one") |
| (= 5 × 10,000 | + | 3 × 1,000 | + | 6 × 100 | + | 8 × 10 | + | 1) |
| | | | | 53,681 | | | | |

FIG. 21.33.

The word for 10,000 in Sino-Japanese is *man*. Previously, *ban* was also used but nowadays it is only used in the sense of "unlimited number" or, rather, "maximum". While *sen.man* means "a thousand myriad", namely 10,000,000, its obsolete homologue *sen.ban* nowadays means "in the highest degree" or "extremely". The famous Japanese war-cry *banzai*, "long life (to) . . ." (to the Emperor, is understood), is made up of *ban*, "10,000", and *zai*, a modification of *sai*, "life". On its own, the word also means "bravo", in the sense that "for what you are doing you deserve to live ten thousand years!"

This oral number-system is of Chinese origin and so it is called the Sino-Japanese system. It long ago displaced the old Pure Japanese system whose structure was rather complicated. The changeover took place under the influence of Chinese culture and manifested itself not only in the disappearance of the number-names for the indigenous numbers above 10, but also by the adoption of the Chinese characters which express the names of these numbers; these characters are, of course, pronounced in the Japanese way. This is the reason why there are two systems in use together.

Two parallel systems are also used in Korea. In the aboriginal, true Korean system it is only possible to count up to 99, and it is only written in *hangŭl* (a Korean alphabet which has nothing to do with Chinese or Japanese writing and was created in 1443 by King Sezhong of the Yi Dynasty). The second, Sino-Korean system was derived from Chinese and allows arbitrarily large numbers; it is written with characters of Chinese origin or by means of Arabic numerals [see J. M. Li (1987)].

## CUSTOM AND SUPERSTITION: LINGUISTIC TABOOS

For numbers from 1 to 10, the Sino-Japanese system is only used in special circumstances, but is used without exception for larger numbers. In conversation, however, the Japanese often use both systems at the same time.

The main reason for this is the speaker's desire to make sure that the listener does not misunderstand. Since different words often sound alike in Japanese, ambiguity can only be avoided by careful choice of words.

This can be seen in the following examples (Fig. 21.34 and 21.35).

The word for "evening" is *ban*. For "one evening" one would say *hito.ban* and not *ichi.ban* since the latter spoken words may also mean "ordinal number" or "first number".

Similarly, *jû.nana* (combining the Sino-Japanese for 10 with the Pure Japanese for 7) can be heard more clearly than *jû.shichi* (in which both elements are Sino-Japanese) and so is more commonly used for 17; and for the same reason 70 is pronounced *nana.jû* and not *shichi.jû*. For 4,000, the indigenous word *yon* for 4 is combined with the Sino-Japanese *sen* for 1,000 in saying *yon.sen* rather than *shi.sen*. C. Haguenauer (1951) also gives the following examples:

| To say: | A Japanese would never use the form: | He would rather use the word or expression: |
|---|---|---|
| 4 | shi | yo |
| 7 | shichi | nana |
| 9 | ku | kokono |
| 14 | jûshi | jû.yon |
| 17 | jûshichi | jû.nana |
| 40 | shi.jû | yon.jû |
| 42 | shi.jûni | yon.jû.ni |
| 47 | shi.jûshichi | yon.jû.nana |
| 70 | shichi.jû | nana.jû |
| 400 | shi.hyaku | yon.hyaku |
| 4,000 | shi.sen | yon.sen |
| 7,000 | shichi.sen | nana.sen |

FIG. 21.34.

However, concern for clarity is not the whole story. Another reason is that the Japanese have always had scrupulous respect for certain linguistic taboos imposed by mystical fears.

In Japan, a "name" (in the widest sense of the term) has a very special significance. The sound of the name, it is held, is produced by the action of motive forces which, indeed, are the very essence of the name, so to pronounce a name is not merely to utter some expression but also – and above all – is to set in motion forces which may have malign powers. This is an ancient and universal belief: to name a being or a thing is to assume power over it; to pronounce a name, or even to utter a sound resembling the name of some malevolent spirit, is to risk awakening its powers and suffering their evil effects. We can therefore understand why the Japanese have attached such importance to precision of utterance and why they take such trouble to avoid using a name which might resemble the sound of a name of evil import.

In addition to this, there are mystical reasons. Numbers, in Japan as elsewhere, have hidden meanings. The Japanese even today still have a degree of numerical superstition, manifest as a respect or even an instinctive fear for certain numbers such as 4 or 9. Try to park your car in bays 4, 9, 14, 19, or 24 of a Tokyo car park: you may locate these places, perhaps, if the secret of perpetual motion is ever discovered. Seat number 4 in a plane of Japanese Airlines, rooms 304 or 309 of a hotel – these can hardly ever be found (still less in a hospital!). Simply because the number in "Renault 4" has always been one of the most menacing, the Japanese launch of this car failed miserably.

This superstition originates in an unfortunate coincidence of sound (resulting from the adoption of the Chinese number-system and its development according to the rules for reading and writing Sino-Japanese). In the Sino-Japanese system, the word for 4 is *shi* which has the same sound as the word for death. Therefore the Japanese recoil from using the Sino-Japanese word for 4, usually using the Pure Japanese word *yo-*. For 9, the Sino-Japanese word is *ku*, with the same sound as the word for pain. Throughout the Far East, including Japan, the ills of the human race are popularly attributed to Spirits of Evil which breathe their poisoned breath all round. Always meticulous about their health, the Japanese therefore sought to avoid attracting the malign attention of these spirits by avoiding the use of this word for 9, using instead the indigenous word *kokono-*.

For exactly the same reason, 4,000 is spoken as *yon.sen* rather than *shi.sen* which has the same sound as the expression for "deadly line"; for "four men" they say *yo.nin* and not *shi.nin* which also means "death" or "corpse". The indigenous word *nana* for "seven" is preferred to the numeral *shichi* (7) because the latter might be mistaken for *shitou* which means death, or loss. Finally, 42 is never spoken as *shi.ni* (a simplified expression: "four-two") nor *shi.jû.ni* (= 4 × 10 + 2), because of the dread presence of " death" in the name of the number 4 as *shi* in each case. There is a further reason: in the first form, the listener may hear *shin.i* – "occurrence of death"; in the second form we also have the name of "42 years of age" which is held to be an especially dangerous age for a man. This number is therefore usually expressed as *yon.jû.ni*.

It is a strange paradox that a civilisation which is at the forefront of science and technology has preserved the fears and superstitions of thousands of years ago, and that there is no thought that these should be overturned.

| | NUMERALS OF CHINESE ORIGIN | | | | READ AS: | | |
|---|---|---|---|---|---|---|---|
| | Standard forms | Cursive forms | Calligraphic forms | Commercial forms | Sino-Japanese | Pure Japanese short | Pure Japanese complete |
| 1 | 一 | 一 | 一 | 〡 | *ichi* | *hi-, hito-* | *hitotsu* |
| 2 | 二 | 二 or [illegible] | 二 | 〢 | *ni* | *fu-, futa-* | *futatsu* |
| 3 | 三 | 三 or [illegible] | 三 | 〣 | *san* | *mi-* | *mitsu* |
| 4 | 四 | 四 | 四 | 〤 | *shi* | *yo-* | *yotsu* |
| 5 | 五 | 五 | 五 | 〥 or 〥 | *go* | *itsu-* | *itsutsu* |
| 6 | 六 | 六 | 六 | 〦 | *roku* | *mu-* | *mutsu* |
| 7 | 七 | 七 | 七 | 〧 | *shichi* | *nana-* | *nanatsu* |
| 8 | 八 | 八 | 八 | 〨 | *hachi* | *ya-* | *yatsu* |
| 9 | 九 | 九 | 九 | 〩 | *ku* | *kokono-* | *kokonotsu* |
| 10 | 十 | 十 or 十 | 十 | 十 | *jû* | *tō* | |
| 100 | 百 | 百 | 百 | [illegible] or [illegible] | *hyaku* | | |
| 1,000 | 千 | 千 | 千 | 千 | *sen* | | |
| 10,000 | 萬 or 万 | 萬 | 萬 | 万 | *man* | | |

FIG. 21.35. *Number-names and numerals in current use in Japan*

## WRITING LARGE NUMBERS

In everyday use, neither the Chinese nor the Japanese have need of special signs for very large numbers. Using only the thirteen basic characters of their present-day number-system they can write down any number, up to at least a hundred billion ($10^{11}$).

Although usually only used for numbers up to $10^8$, the method they use is a simple extension of their ordinary number-system, namely introducing ten thousand ($10^4$) as an additional counting unit. The following shows how the Chinese represent consecutive powers of 10 (Fig. 21.36):

| | | |
|---|---|---|
| 10,000 : | yī wàn (= | 1 × 10,000) |
| 100,000 : | shí wàn (= | 10 × 10,000) |
| 1,000,000 : | yī bǎi wàn (= | 1 × 100 × 10,000) |
| 10,000,000 : | yī qiān wàn (= | 1 × 1,000 × 10,000) |
| 100,000,000 : | yī wàn wàn (= | 1 × 10,000 × 10,000) |
| 1,000,000,000 : | shí wàn wàn (= | 10 × 10,000 × 10,000) |
| 10,000,000,000 : | yī bǎi wàn wàn (= | 1 × 100 × 10,000 × 10,000) |
| 100,000,000,000 : | yī qiān wàn wàn (= | 1 × 1,000 × 10,000 × 10,000) |

FIG. 21.36A. *The usual Chinese notation for the successive powers of 10. [Guitel; Menninger (1957); Ore (1948); Tchen Yon-Sun (1958)]*

| | | | | | |
|---|---|---|---|---|---|
| $10^4$ | 一萬 *yī wàn* | $1 \times 10^4$ | $10^8$ | 一萬萬 *yī wàn wàn* | $1 \times 10^4 \times 10^4$ |
| $10^5$ | 十萬 *shí wàn* | $10 \times 10^4$ | $10^9$ | 十萬萬 *shí wàn wàn* | $10 \times 10^4 \times 10^4$ |
| $10^6$ | 一百萬 *yī bǎi wàn* | $1 \times 10^2 \times 10^4$ | $10^{10}$ | 一百萬萬 *yī bǎi wàn wàn* | $1 \times 10^2 \times 10^4 \times 10^4$ |
| $10^7$ | 一千萬 *yī qiān wàn* | $1 \times 10^3 \times 10^4$ | $10^{11}$ | 一千萬萬 *yī qiān wàn wàn* | $1 \times 10^3 \times 10^4 \times 10^4$ |

FIG. 21.36B.

For a very large number such as 487,390,629, therefore, they would write:

四 萬 八 千 七 百 三 十 九 萬 六 百 二 十 九

*sí wàn bā qiān qī bǎi sān shí jiǔ wàn liù bǎi èr shí jiǔ*

$(4 \times 10^4 + 8 \times 10^3 + 7 \times 10^2 + 3 \times 10 + 9) \times 10^4 + (6 \times 10^2 + 2 \times 10 + 9)$

FIG. 21.37.

decomposing it as

$(4 \times 10{,}000 + 8 \times 1{,}000 + 7 \times 100 + 3 \times 10 + 9) \times 10{,}000 + 6 \times 100 + 2 \times 10 + 9$

or $48{,}739 \times 10{,}000 + 629$.

The system just described is in practice the only one used for ordinary purposes. However, though only in scientific and especially astronomical texts, one may encounter special characters for higher orders than $10^4$ which can therefore be used to express much larger numbers than are possible with the usual system. However, the signs used have meanings which vary according to which of three value conventions is being used. Each sign may have one of three different values depending on whether it is used on the *xià deng* system ("lower degree"), the *zhōng deng* system ("middle degree") or the *shàng deng* system ("higher degree").

The character 兆, *zháo*, therefore, may represent a million ($10^6$) in the lower degree, a thousand billion ($10^{12}$) in the middle degree, and $10^{16}$ in the higher degree.

In the lower degree (*xià deng*) the system is a direct continuation of the ordinary number-system since the ten successive additional characters are simply the ten consecutive powers of 10 following $10^4$, namely

$$10^5, 10^6, 10^7, 10^8 \ldots, 10^{13}, 10^{14}$$

which are represented by the characters

*yì, zháo, jing, gai, . . . , zheng, zài.*

So, written in the lower degree, one million and three million would be written as follows:

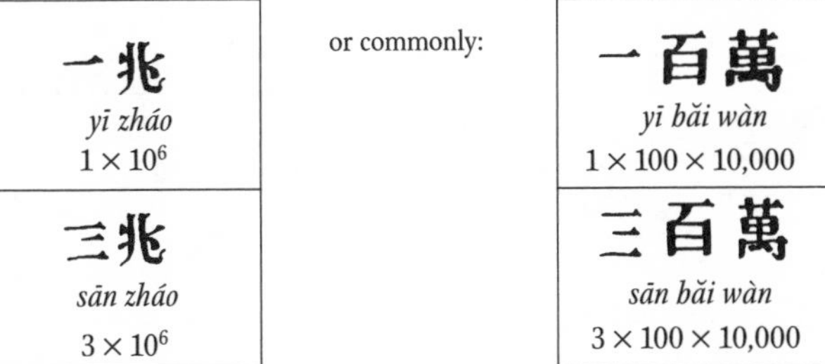

一兆
yī zháo
1 × 10⁶
三兆
sān zháo
3 × 10⁶
or commonly:
一百萬
yī bǎi wàn
1 × 100 × 10,000
三百萬
sān bǎi wàn
3 × 100 × 10,000


FIG. 21.38.

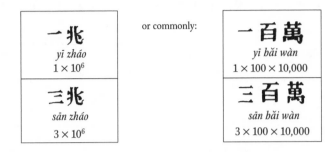

The *xià deng* system therefore allows any number less than $10^{15}$ to be written down straightforwardly. For example, the number 530,010,702,000,000 would be written as

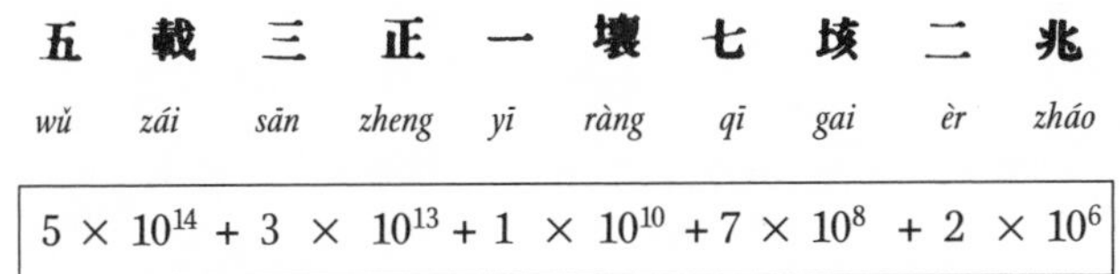


FIG. 21.39.

In the middle system the same ten consecutive characters represent increasing powers of 10 greater than $10^4$, but they now increase, not by a factor of 10 each time, but by a factor of 10,000, namely

$$10^8, 10^{12}, 10^{16}, \ldots, 10^{40}, 10^{44} \text{ (Fig. 21.42).}$$

With the convention that two of these characters should never occur consecutively, this system can be used to represent all the numbers less than $10^{48}$. For example:

三 百 五 十 壤 七 千 三 百 兆 二 十 六 億

*sān bǎi wǔ shí ràng qī qiān sān bǎi zháo èr shí liù yì*

| $(3 \times 10^2 + 5 \times 10).\ 10^{28} + (7 \times 10^3 + 3 \times 10^2).\ 10^{12} + (2 \times 10^6 + 6).\ 10^8$ |
|---|
| 3,500,000,000,000,007,300,002,600,000,000 |

FIG. 21.40.

In the higher degree system, only the first three of these ten characters are used, namely *yì*, *zháo* and *jing*. These are given the values $10^8$, $10^{16}$ and $10^{32}$ respectively. With these, it is possible to represent all numbers less than $10^{64}$. For example:

三京五千三百一億二百七萬六千一百八十五兆三億一萬

*sān jing wǔ qiān sān bǎi yī yì èr bǎi qī wàn liù qiān yī bǎi bā shí wǔ zháo sān yì yī wàn*

| $(3 \times 10^{32} + [[5 \times 10^3 + 3 \times 10^2 + 1] . 10^8 + [2 \times 10^2 + 7] \cdot 10^4 + 6 \times 10^3 + 1 \times 10^2 + 8 \times 10 + 5] \cdot 10^{16} + 3 \times 10^8 + 1 \times 10^4$ |
|---|
| 300,005,301,020,761,850,000,000,300,010,000 |

FIG. 21.41.

| | *Xià deng* LOWER DEGREE SYSTEM | *Zhōng deng* MIDDLE DEGREE SYSTEM | *Shàng deng* HIGHER DEGREE SYSTEM | |
|---|---|---|---|---|
| 萬 *wàn* | $10^4$ | $10^4$ | $10^4$ | |
| 億 *yì* [a] | $10^5$ | $10^8$ | $10^8$ | |
| 兆 *zháo* | $10^6$ | $10^{12}$ | $10^{16}$ | |
| 京 *jing* | $10^7$ | $10^{16}$ | $10^{32}$ | |
| 垓 *gai* | $10^8$ | $10^{20}$ | $10^{64}$ | THEORETICAL VALUES |
| 補 *bù* [b] | $10^9$ | $10^{24}$ | $10^{128}$ | |
| 壤 *ràng* | $10^{10}$ | $10^{28}$ | $10^{256}$ | |
| 溝 *gou* [c] | $10^{11}$ | $10^{32}$ | $10^{512}$ | |
| 澗 *jián* | $10^{12}$ | $10^{36}$ | $10^{1024}$ | |
| 正 *zheng* | $10^{13}$ | $10^{40}$ | $10^{2048}$ | |
| 載 *zái* | $10^{14}$ | $10^{44}$ | $10^{4096}$ | |

[a] Graphical variant 亿 [b] Equivalent word 溝 [c] Graphical variant 柫

FIG. 21.42. *Chinese scientific notation for large numbers [Giles (1912); Mathews (1931); Needham (1959)]*

Such very large numbers are, however, very infrequently used: "in mathematics, business or economics numbers greater than $10^{14}$ are very rare; only in connection with astronomy or the calendar do we sometimes find larger numbers" [R. Schrimpf (1963–64)].

Finally, let us draw attention to a very interesting notation which Chinese and Japanese scientists have used to express negative powers of 10:

$$10^{-1} = 1/10,\ 10^{-2} = 1/100,\ 10^{-3} = 1/1{,}000,\ 10^{-4} = 10{,}000, \text{ etc.}$$

They especially find mention in the arithmetical treatise *Jinkoki* published in 1627 by the Japanese mathematician Yoshida Mitsuyoshi (Fig. 21.43).

| | | |
|---|---|---|
| 分 | *fēn* | $10^{-1}$ |
| 厘 | *lí* | $10^{-2}$ |
| 毛 | *máo* | $10^{-3}$ |
| 糸 | *mi* | $10^{-4}$ |
| 忽 | *hū* | $10^{-5}$ |
| 微 | *wēi* | $10^{-6}$ |
| 纖 | *xiān* | $10^{-7}$ |
| 沙 | *shā* | $10^{-8}$ |
| 塵 | *chén* | $10^{-9}$ |
| 埃 | *āi* | $10^{-10}$ |

FIG. 21.43. *Sino-Japanese scientific notation for negative powers of 10 [Yamamoto (1985)]*

## THE CHINESE SCIENTIFIC POSITIONAL SYSTEM

Further evidence of advanced intellectual development in the Far East comes from the written positional notation formerly used by Chinese, Japanese, and Korean mathematicians.

Though we only know examples of this system dating back to the second century BCE, it seems probable that it goes back much further.

Known by the Chinese name *suan zí* (literally, "calculation with rods"), and by the Japanese name *sangi*, this system is similar to our modern number-system not only by virtue of its decimal base, but also because the values of the numerals are determined by the position they occupy. It is therefore a strictly positional decimal number-system.

However, whereas our system uses nine numerals whose forms carry no intrinsic suggestion of value, this system of numerals makes use of systematic combinations of horizontal and vertical bars to represent the first nine units. The symbols for 1 to 5 use a corresponding number of vertical strokes, side by side, and the symbols for 6, 7, 8, and 9 show a horizontal bar capping 1, 2, 3, or 4 vertical strokes:

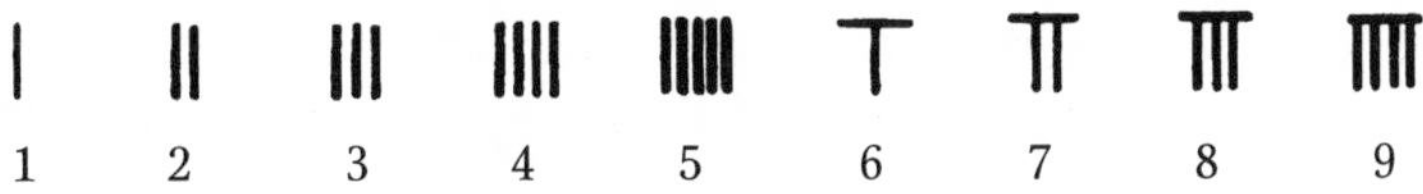


Fig. 21.44.

Examples of numbers written in this system are given by Cai Jiu Feng, a Chinese philosopher of the Song era who died in 1230 [in *Huang ji*, in the chapter *Hong fan* of his "Book of Annals", cited by A.Vissière (1892)]. Example:

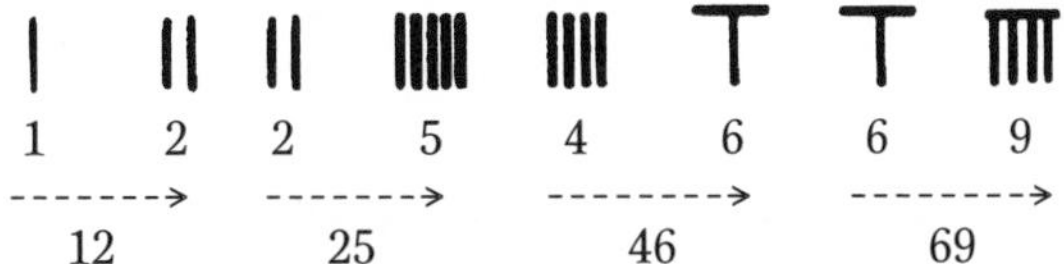


Fig. 21.45.

Ingenious as it was, this system lent itself to ambiguity.

For one thing, people writing in this system tended to place the vertical bars for the different orders of magnitude side by side. So the notation for the number 12 could be confused with that for 3 or for 21; 25 could be confused with 7, 34, 43, 52, 214, or 223, and so on (Fig. 21.45).

However, the Chinese found a way round the problem, by introducing a second system for the units, analogous to the first but made up of horizontal bars rather than vertical. The first five digits were represented by as many horizontal bars, and the numbers 6, 7, 8, 9 by erecting a vertical bar (with symbolic value 5) on top of one, two, three, or four horizontal bars:

Fig. 21.46.

Then, to distinguish between one order of magnitude and the next, they alternated figures from one series with figures from the other, therefore alternately vertical and horizontal. The units, hundreds, tens of thousands, millions, and so on (of odd rank) were drawn with "vertical" symbols (Fig. 21.44), whereas the tens, thousands, hundreds of thousands, tens of millions, etc. (of even rank) were drawn with "horizontal" symbols (Fig. 21.46), by which means the ambiguities were elegantly resolved (Fig. 21.48).

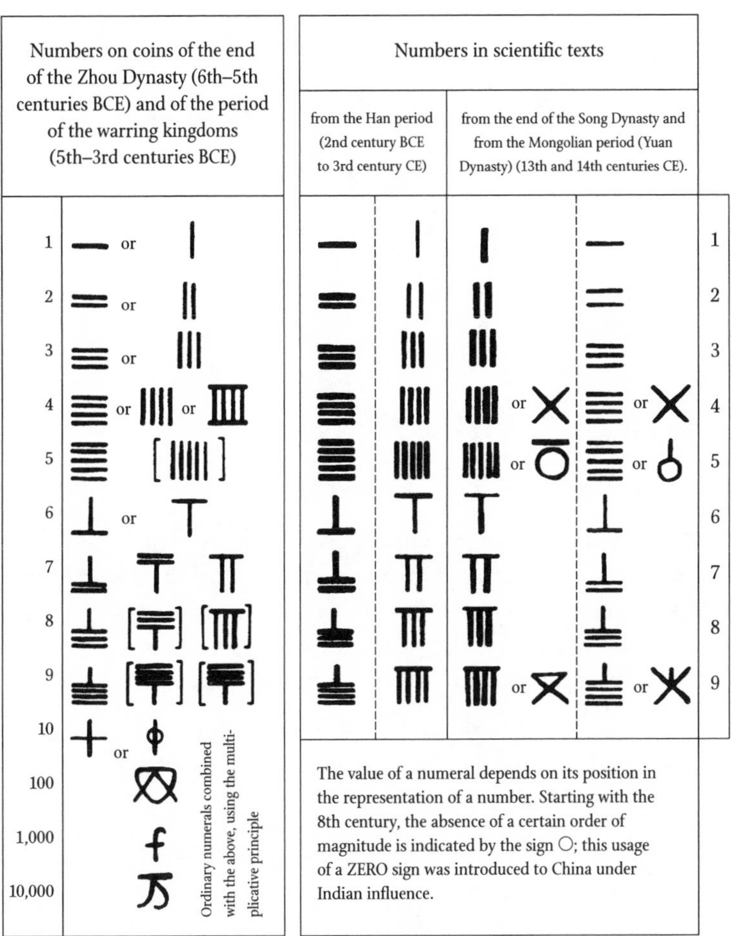


Fig. 21.47. *Chinese bar numerals through the ages [Needham (1959)]*

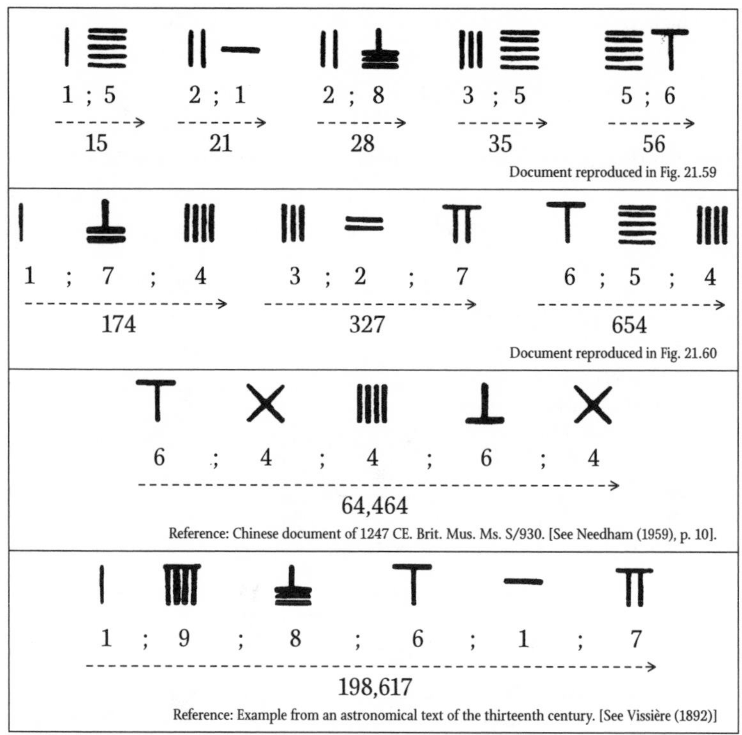


FIG. 21.48. *Examples of numbers written in the Chinese bar notation* (suan zí)

This step was taken at the time of the Han Dynasty (second century BCE to third century CE). This did not solve all the problems there and then, however, since the Chinese mathematicians were to remain unaware of zero for several centuries yet. The following riddle bears witness to this, in the words of the mathematician Mei Wen Ding (1631–1721):

> The character *hai* 亥 has 2 for its head and 6 for its body. Lower the head to the level of the body, and you will find the age of the Old Man of Jiangxian.

In the above, the character playing the main role in the riddle has been written in the *kǎishū* style:

FIG. 21.49. hai

and the riddle remains obscure since the modern character is not the same shape as it was before. According to Chinese sources, however, the riddle dates from long before the Common Era, originating in the middle of the Zhou era (seventh to sixth centuries BCE; see Needham (1959), p. 8). And since at that time Chinese characters were drawn in the *dà zhuàn* ("great seal") style, we must therefore see the character in question drawn in this style if we are to solve the riddle.

In this style, the word was written:

FIG. 21.50. hai

Its "head", therefore, is indeed the figure 2 ⚌, and its lower part is a "body" consisting of three identical signs each of which resembles the "vertical" symbol for the figure 6 (Fig. 21.47). Arrange the two horizontal lines of the head vertically and on the left-hand side of the body, and you find

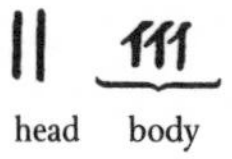

head body

FIG. 21.51.

or, nearly enough,

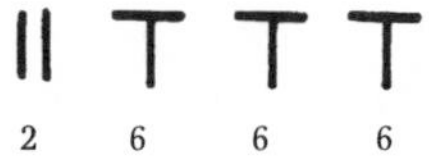

2 6 6 6

FIG. 21.52.

The Chinese system being decimal and strictly positional, this represents the number

$$2 \times 1,000 + 6 \times 100 + 6 \times 10 + 6 = 2,666$$

so the solution of the riddle is the number 2,666. But this cannot be an age in years, unless the Old Man of Jiangxian was a Chinese Methuselah. To consider them as 2,666 days would give an absurd answer, since the "Old Man" would then only be seven and a half years old. In fact, this number system had no zero until much later, so the answer can only be one of the numbers 26,660, 266,600, 2,666,000, etc. But since 266,600 or any higher number is out of the question, we are left with 26,660 days. In the riddle, the number sought does not represent days but tens of days: the Old Man of Jiangxian had lived 2,666 tens of days, or about 73 years.

The lack of a sign to represent missing digits also gave rise to confusion. In the first place, a blank space was left where there was no digit, but this was inadequate since numbers like 764, 7,064, 70,640 and 76,400 could easily be confused:

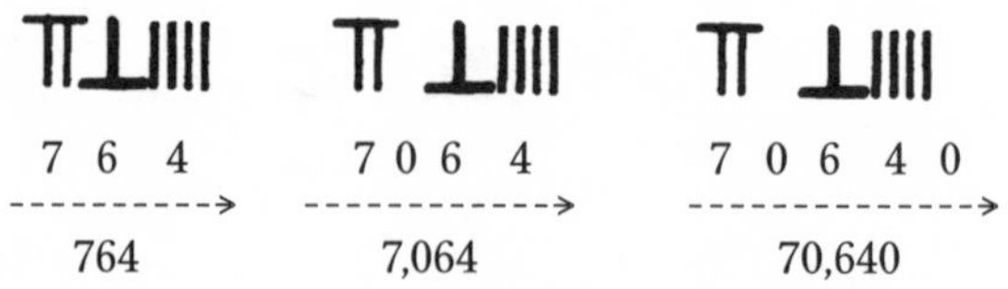

7 6 4 → 764

7 0 6 4 → 7,064

7 0 6 4 0 → 70,640

FIG. 21.53.

To avoid such ambiguities, some used signs indicating different powers of 10 from the traditional number-system, so that numbers such as 70,640 and 76,400 would be written as:

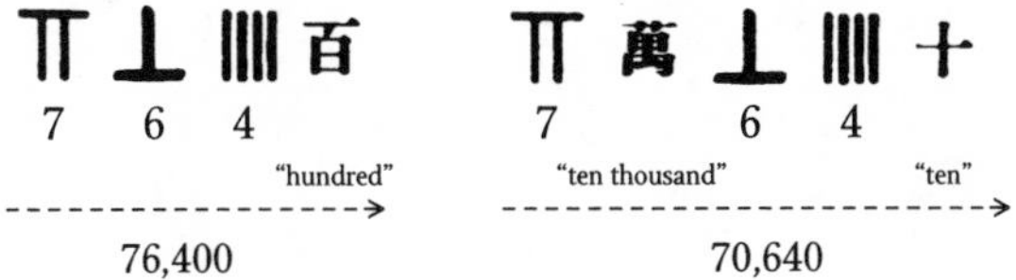


FIG. 21.54.

Others used the traditional expression, therefore writing out in full:

FIG. 21.55.

Yet others placed their numerals in the squares of a grid, leaving an empty square for each missing digit:

76,400

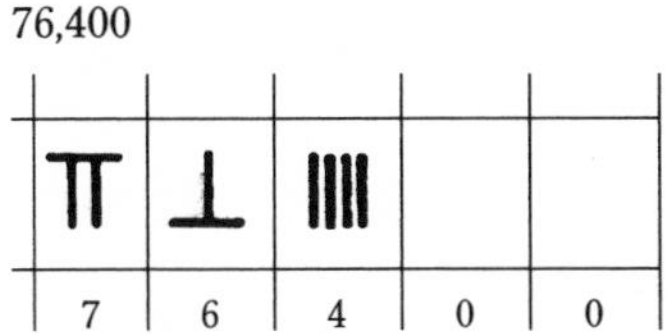


70,064

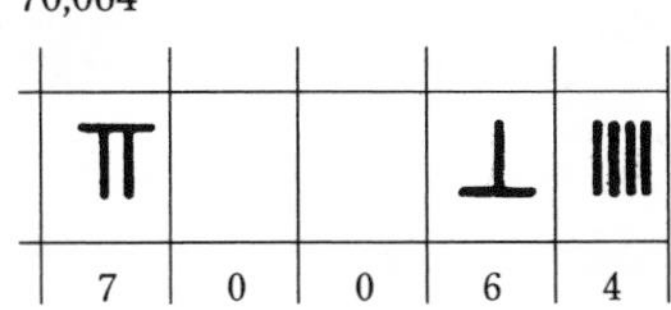


FIG. 21.56.

Only since the eighth century CE did the Chinese begin to introduce a special positional sign (drawn as a small circle) to mark a missing digit (Fig. 21.57); this idea no doubt reached them through the influence of Indian civilisation.

Once this had been achieved, all of the rules of arithmetic and algebra were brought to a degree of perfection similar to ours of the present day.

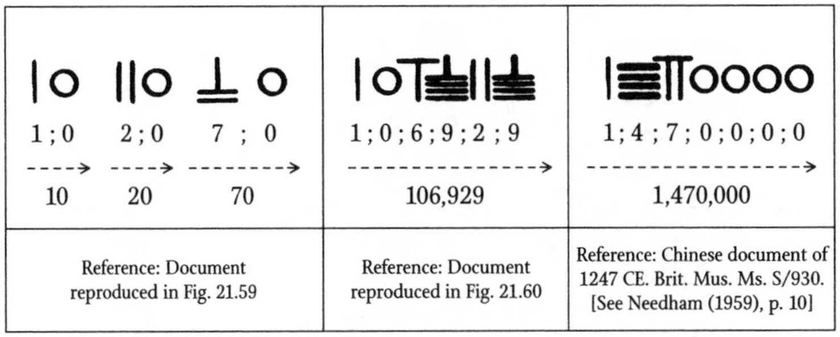

| 1 ; 0 → 10   2 ; 0 → 20   7 ; 0 → 70 | 1 ; 0 ; 6 ; 9 ; 2 ; 9 → 106,929 | 1 ; 4 ; 7 ; 0 ; 0 ; 0 ; 0 → 1,470,000 |
|---|---|---|
| Reference: Document reproduced in Fig. 21.59 | Reference: Document reproduced in Fig. 21.60 | Reference: Chinese document of 1247 CE. Brit. Mus. Ms. S/930. [See Needham (1959), p. 10] |

FIG. 21.57. *The use of zero in the Chinese bar numerals*

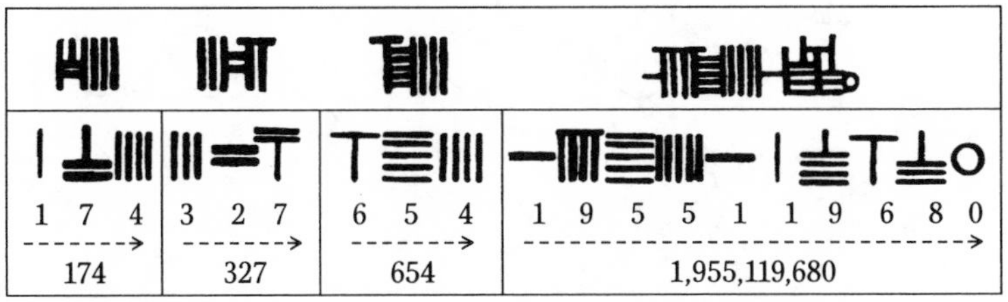

| 1 7 4 | 3 2 7 | 6 5 4 | 1 9 5 5 1 1 9 6 8 0 |
|---|---|---|---|
| 174 | 327 | 654 | 1,955,119,680 |

FIG. 21.58. *As a rule, in Chinese manuscripts or printed documents, numbers written in the bar notation are written as monograms, i.e. in a condensed form in which the horizontal strokes are joined to the vertical ones. (Examples taken from the document reproduced in Fig. 21.60)*

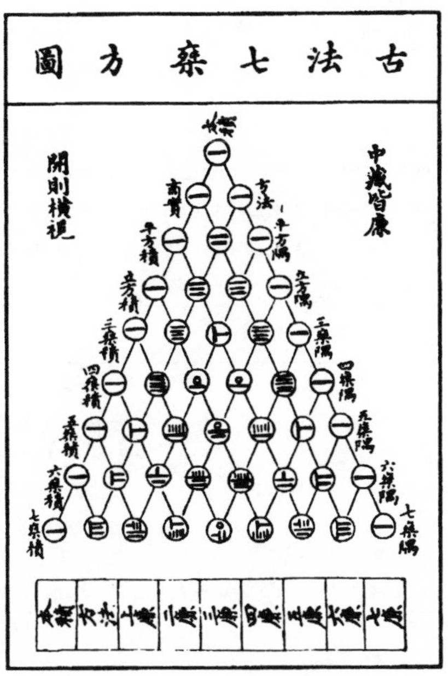


FIG. 21.59A. *Page from a text entitled* Su Yuan Yu Zhian, *published in 1303 by the Chinese mathematician Zhu Shi Jie (see the commentary in the text). [Reproduced from Needham (1959), III, p. 135, Fig. 80]*

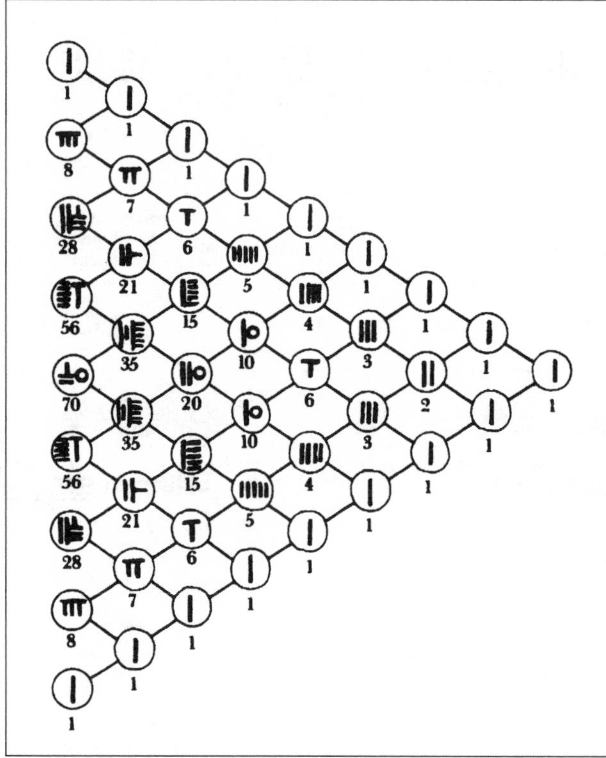


Fig. 21.59B.

Blaise Pascal was long believed in the West to have been the first to discover the famous "Pascal triangle" which gives the numerical coefficients in the expansion of $(a + b)^m$, where m is zero or a positive integer:

| BINOMIAL EXPANSIONS | PASCAL'S TRIANGLE |
|---|---|
| $(a+b)^0 = 1$ | 1 |
| $(a+b)^1 = a + b$ | 1 1 |
| $(a+b)^2 = a^2 + 2ab + b^2$ | 1 2 1 |
| $(a+b)^3 = a^3 + 3a^2b + 3ab^2 + b^3$ | 1 3 3 1 |
| $(a+b)^4 = a^4 + 4a^3b + 6a^2b^2 + 4ab^3 + b^4$ | 1 4 6 4 1 |
| $(a+b)^5 = a^5 + 5a^4b + 10a^3b^2 + 10a^2b^3 + 5ab^4 + b^5$ | 1 5 10 10 5 1 |
| $(a+b)^6 = a^6 + 6a^5b + 15a^4b^2 + 20a^3b^3 + 15a^2b^4 + 6ab^5 + b^6$ | 1 6 15 20 15 6 1 |
| ------------------------------------> | -----------------------> |

In fact, as we can see from Fig. 21.59A, which is schematically redrawn on its side in Fig. 21.59B (to be read from right to left), the Chinese had known of this triangle long before the famous French mathematician.

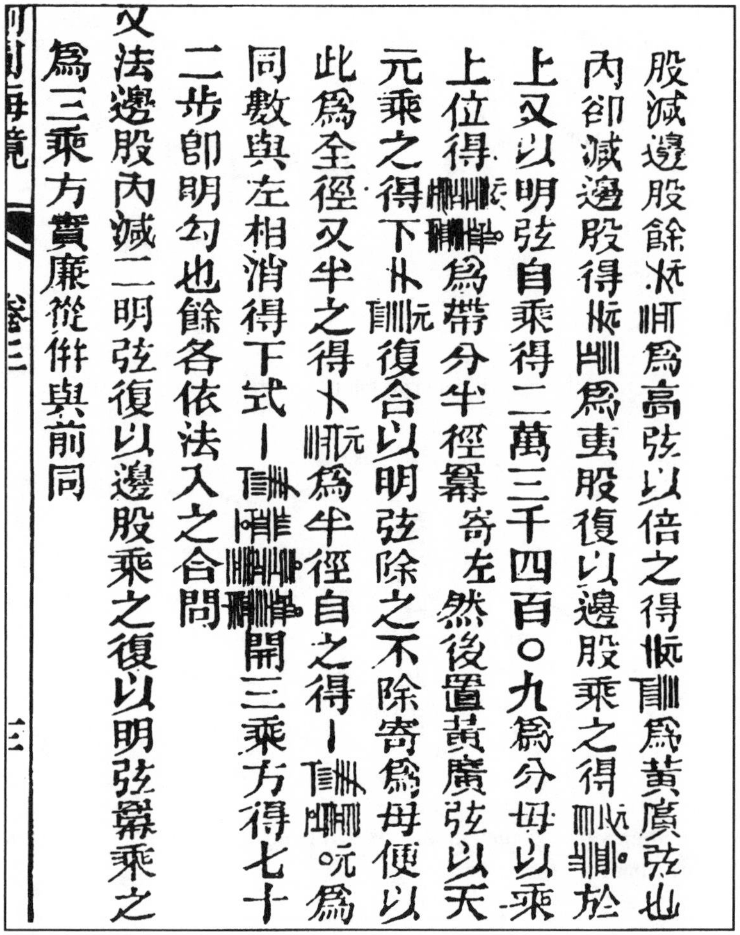

Fig. 21.60. *Extract from* Ce Yuan Hai Jing, *published in 1248 by the mathematician Li Ye. [Reproduced from Needham (1959), III, page 132, Fig. 79]*

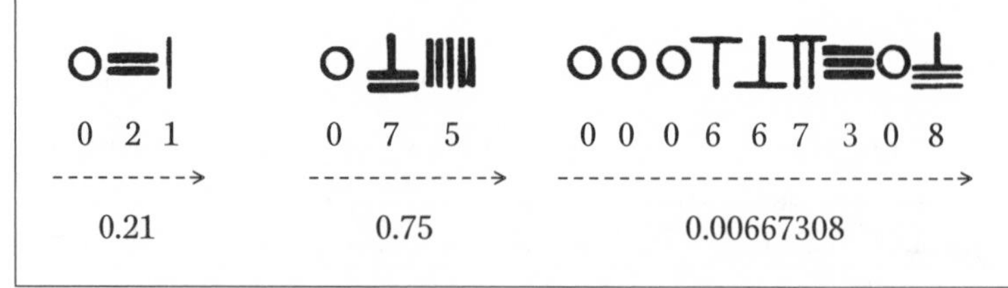


Fig. 21.61. *How Chinese mathematicians extended their positional notation to decimal fractions. Reconstructed examples based on a text from the Mongol period: Biot (1839)*

| EXAMPLES FROM A 13TH-CENTURY CHINESE TREATISE (cf. Fig. 21.60) | | | | EXAMPLES FROM AN 18TH-CENTURY JAPANESE TEXT |
|---|---|---|---|---|
| | | | | |
| | 654 | 1360 | 1536 | 152710100928 |
| | ------> | -----> | -------> | -------------> |
| −2 | −654 | −1,360 | −1,536 | −152,710,100,928 |

FIG. 21.62A. *Extension of scientific numerical notation to negative numbers. To indicate a negative number, the Chinese and Japanese mathematicians often drew an oblique stroke through the rightmost symbol of the written number. [Menninger (1957); Needham (1959)]*

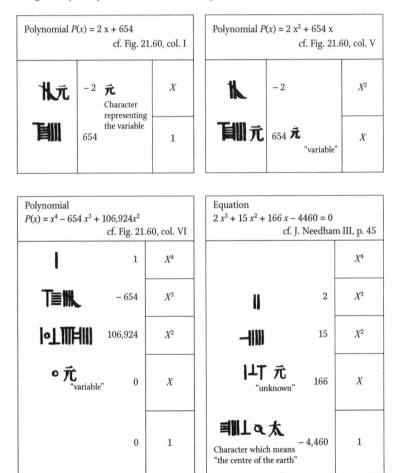

| Polynomial $P(x) = 2x + 654$ cf. Fig. 21.60, col. I | | |
|---|---|---|
| 元 | −2 元 Character representing the variable | $X$ |
| | 654 | 1 |

| Polynomial $P(x) = 2x^2 + 654x$ cf. Fig. 21.60, col. V | | |
|---|---|---|
| | −2 | $X^2$ |
| 元 | 654 元 "variable" | $X$ |

| Polynomial $P(x) = x^4 - 654x^3 + 106{,}924x^2$ cf. Fig. 21.60, col. VI | | |
|---|---|---|
| | 1 | $X^4$ |
| | −654 | $X^3$ |
| | 106,924 | $X^2$ |
| ○ 元 "variable" | 0 | $X$ |
| | 0 | 1 |

| Equation $2x^3 + 15x^2 + 166x - 4460 = 0$ cf. J. Needham III, p. 45 | | |
|---|---|---|
| | | $X^4$ |
| | 2 | $X^3$ |
| | 15 | $X^2$ |
| 元 "unknown" | 166 | $X$ |
| 太 Character which means "the centre of the earth" | −4,460 | 1 |

FIG. 21.62B. *Notation for polynomials and for equations in one unknown, used by Li Ye (1178–1265)*

## THE CHINESE VERSION OF THE RODS ON THE CHECKERBOARD

Although the numerals discussed above served for writing, they were not used for calculation. For arithmetical calculation, the Chinese used little rods made of ivory or bamboo which were called *chóu* ("calculating rods") which were placed on the squares of a tiled surface or a table ruled like a checkerboard.

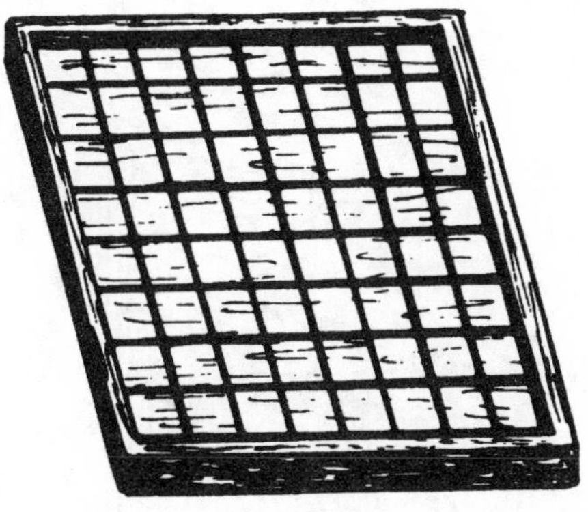

FIG. 21.63. *Model of a Chinese checkerboard used for calculation*

The following story from the ninth century CE is evidence in point. It tells how the Emperor Yang Sun selected his officials for their skill and rapidity in calculation.

> Once two clerks, of the same rank, in the same service, and with the same commendations and criticisms in their records, were candidates for the same position. Unable to decide which one to promote, the superior officer called upon Yang Sun, who had the candidates brought before him and announced: Junior clerks must know how to calculate at speed. Let the two candidates listen to my question. The one who solves it first will have the promotion. Here is the problem:
>
> > A man walking in the woods heard thieves arguing over the division of rolls of cloth which they had stolen. They said that, if each took six rolls there would be five left over; but if each took seven rolls, they would be eight short. How many thieves were there, and how many rolls of cloth?
>
> Yang Sun asked the candidates to perform the calculation with rods upon the tiled floor of the vestibule. After a brief moment, one of the clerks gave the right answer and was given the promotion, and all then departed without complaining about the decision. (See J. Needham in HGS 1, pp. 188–92).

FIG. 21.64. *A Chinese Master teaches the arts of calculation to two young pupils, using an abacus with rods. Reproduced from the* Suan Fa Tong Zong, *published in 1593 in China: [Needham (1959) III, p. 70]*

FIG. 21.65. *An accountant using the arithmetic checkerboard with rods. Reproduced from the Japanese* Shojutsu Sangaka Zue *of Miyake Kenriyû, 1795: (D. E. Smith)*

On an abacus of this kind, each column corresponds to one of the decimal orders of magnitude: from right to left, the first is for the units, the second for the tens, the third for the hundreds, and so on. A given number, therefore, is represented by placing in each column, along a chosen line, a number of rods equal to the multiplicity of the corresponding decimal order of magnitude. For the number 2,645, for example, there would be 5 rods in the first column, 4 in the second, 6 in the third and 2 in the fourth.

For the sake of simplicity, Chinese calculators adopted the following convention (in the words of the old Chinese textbooks of arithmetic): "Let the units lie lengthways and the tens crosswise; let the hundreds be upright and the thousands laid down; let the thousands and the hundreds be face to face, and let the tens of thousands and the hundreds correspond."

The mathematician Mei Wen Ding explains that there was a fear that the different groups might get muddled because there were so many of them. Numbers such as 22 or 33 were therefore represented by two groups of rods, one horizontal and the other vertical, which allowed them to be differentiated. To prevent errors of interpretation, the rods were laid down vertically in the odd-numbered columns (counting from the right), and horizontally in the even-numbered columns (Fig. 21.67).

| 1 | 2 | 3 | 4 | 5 | 6 | 7 | 8 | 9 |
|---|---|---|---|---|---|---|---|---|
| 10 | 20 | 30 | 40 | 50 | 60 | 70 | 80 | 90 |

FIG. 21.66. *How the units and tens are represented by rods on the arithmetical checkerboard*

| | UNITS OF ODD ORDER (columns for even powers of 10) | UNITS OF EVEN ORDER (columns for odd powers of 10) |
|---|---|---|
| 1 | 𝍠 | 𝍩 |
| 2 | 𝍡 | 𝍪 |
| 3 | 𝍢 | 𝍫 |
| 4 | 𝍣 | 𝍬 |
| 5 | 𝍤 | 𝍭 |
| 6 | 𝍥 | 𝍮 |
| 7 | 𝍦 | 𝍯 |
| 8 | 𝍧 | 𝍰 |
| 9 | 𝍨 | 𝍱 |

FIG. 21.67. *The rods are laid vertically for the units, the hundreds, the tens of thousands, and so on; they are laid horizontally for the tens, the thousands, the hundreds of thousands, and so on.*

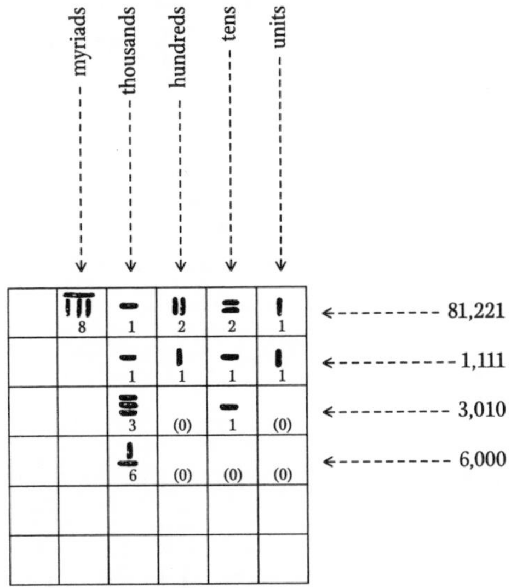


FIG. 21.68. *How certain numbers are represented by laying rods on the checkerboard*

From antiquity until recent times, the Chinese were able to perform every kind of arithmetical operation by means of this device: addition, subtraction, multiplication, division, raising to a power, extraction of square and cube roots, and so on.

The methods used for addition and subtraction were straightforward. The numbers to be added or subtracted were represented in the squares, and rods were added or removed column by column. Multiplication was almost as simple: the multiplier was placed at the top of the board, with the number to be multiplied placed a few rows lower down. The partial products were then set out on an intermediate line and added in as they were obtained.

For example, to work out the product $736 \times 247$ (as set out by Yang Hui in the thirteenth century), first of all the two numbers are set out on the board as follows, keeping two empty squares at the right of the multiplier:

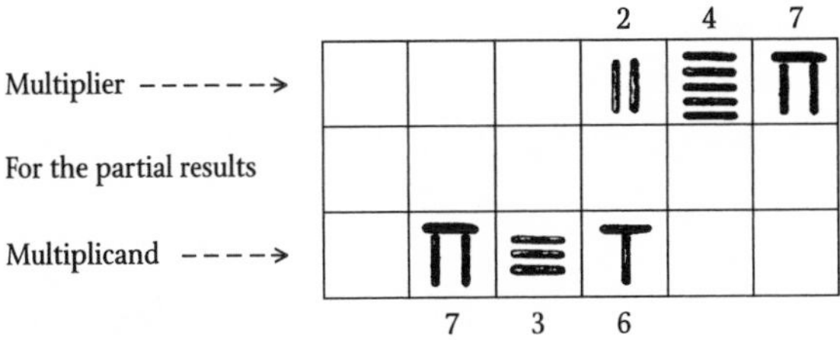


FIG. 21.69A.

Since the multiplier contains three figures, the method proceeds in three stages.

*First stage: multiplying 736 by 200*

Mentally multiply the 2 of the multiplier by the 7 of the multiplicand, and place the result 14 (in fact 140,000) in the middle line, taking care to place the units of the result above the hundreds of the multiplicand:

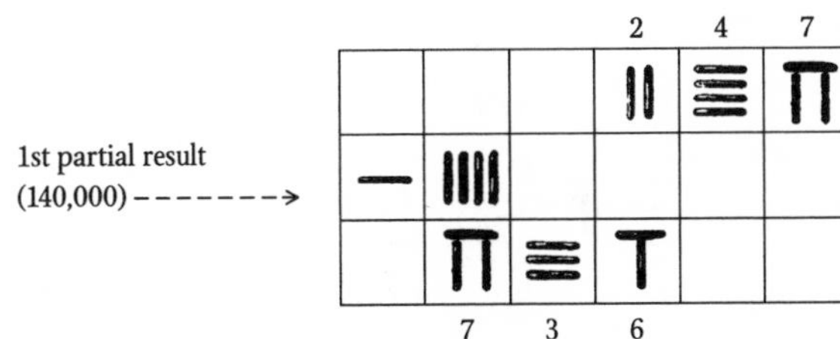


FIG. 21.69B.

Then multiply the 2 of the multiplier by the 3 of the multiplicand, and add the result 6 (in fact 6,000) to the partial result already obtained, placing it on the square to the right of the 4 in 14:

2nd partial result
(140,000 + 6,000 =
146,000)

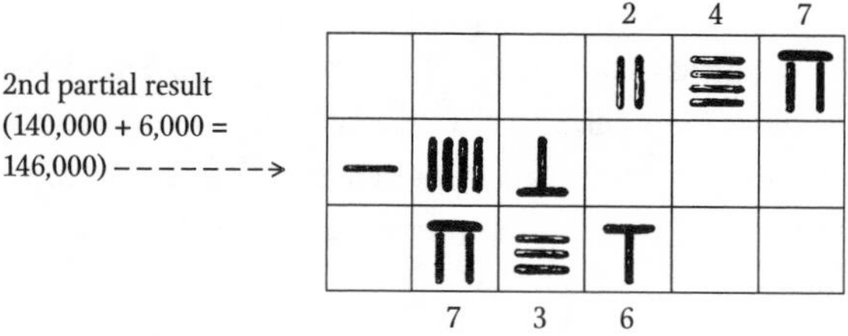


FIG. 21.69C.

Then multiply the 2 of 247 by the 6 of 736, and add this result 12 (in fact 1,200) to the partial result already obtained: in this case, the 2 is placed on the square to the right of the 6 from the preceding stage, and the 1 is placed on the next square to the left thereby being added to the number already there:

3rd partial result
(146,000 + 1,200 =
147,200)

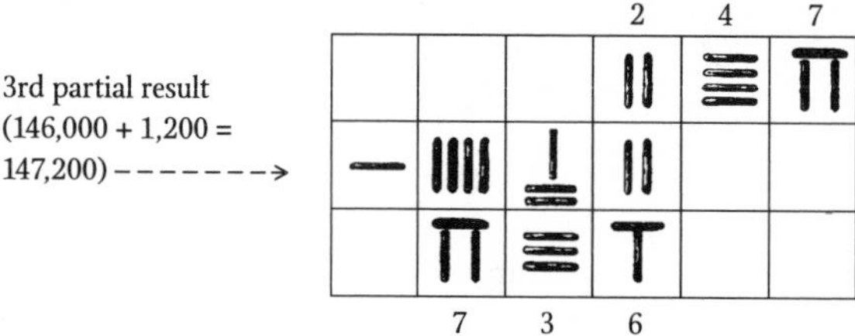


FIG. 21.69D.

*Second stage: multiplying 736 by 40*

The 2 of the multiplier has now done its work, so it is removed, and the multiplicand is moved bodily one square to the right:

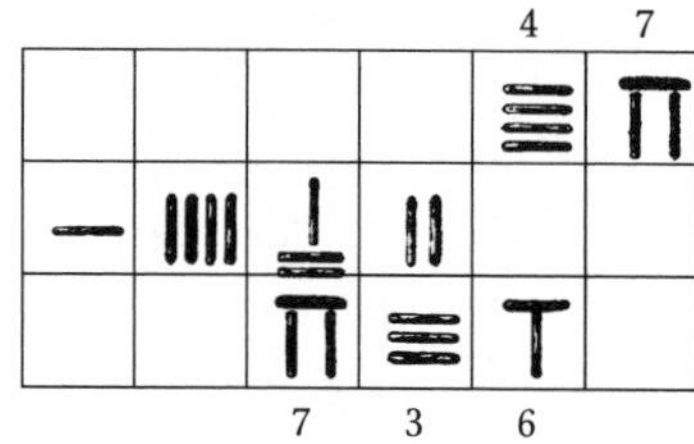


FIG. 21.69E.

Now multiply the 4 of the multiplier by the 7 of the multiplicand, place the result 28 (in fact 28,000) to the partial result in the middle row, and complete the addition:

4th partial result
(147,200 + 28,000 =
175,200)

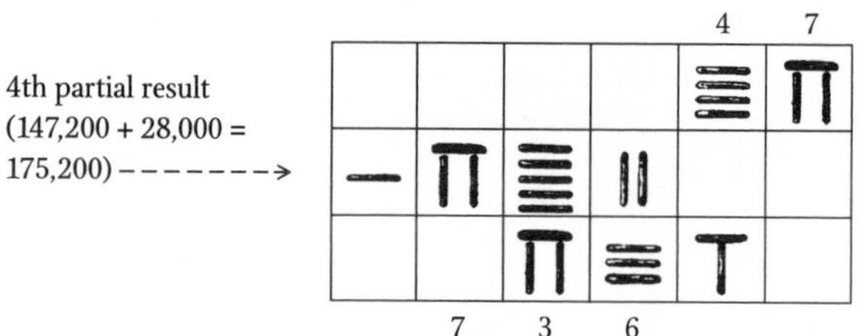


FIG. 21.69F.

Now multiply the 4 by the 3 of 736, and add the result 12 (in fact 1,200) to the middle line:

5th partial result
(175,200 + 1,200 =
176,400)

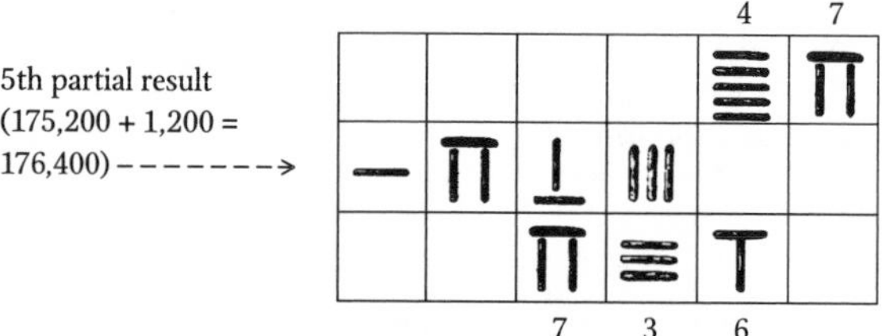


FIG. 21.69G.

Now multiply the 4 by the 6 of 736 and add the result 24 (in fact 240) to the middle line:

6th partial result
(176,400 + 240 =
176,640)

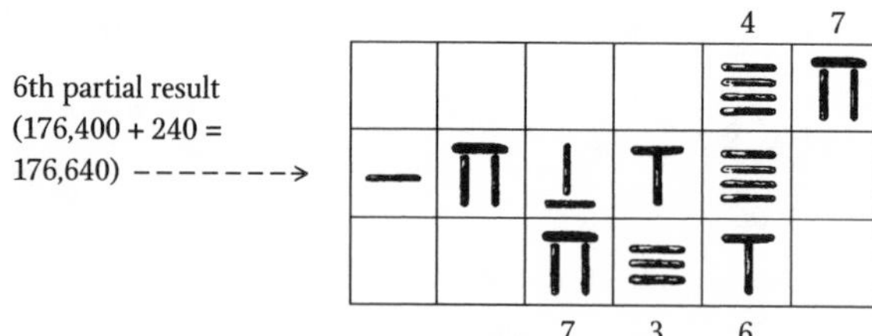


FIG. 21.69H.

*Third stage: multiplying 736 by 7*

The 4 of the multiplier has done its work and it too is now removed, and the multiplicand again moved bodily one square to the right:

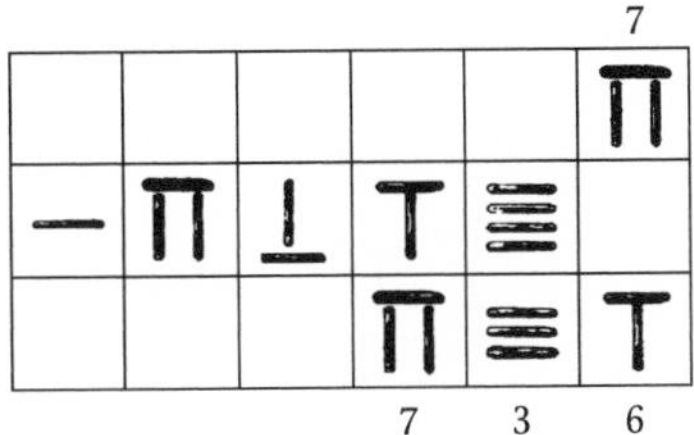


Fig. 21.69I.

The remaining 7 of the multiplier is now multiplied by the 7 of the multiplicand, and the result 49 (in fact 4,900) is added to the middle line:

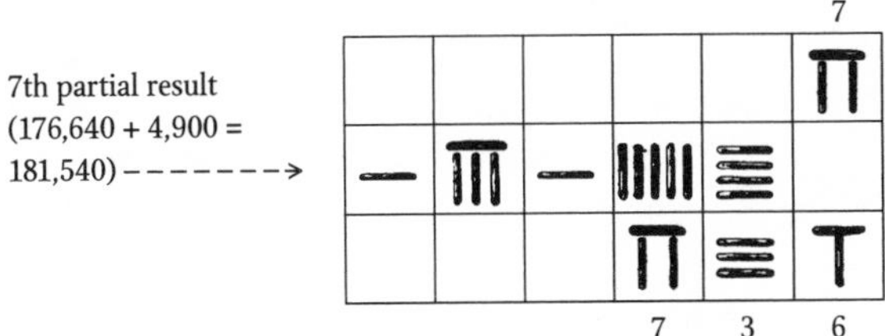


Fig. 21.69J.

Now multiply the 7 by the 3 of 736, and add the result 21 (in fact 210) to the middle line:

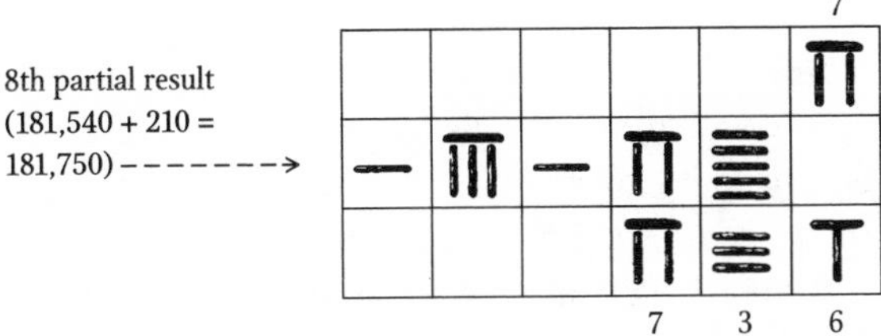


Fig. 21.69K.

Finally multiply the 7 by the 6 of 736, and add the result 42 to the middle line. This gives the following tableau, where the middle line shows the result of the multiplication (736 × 247 = 181,792):

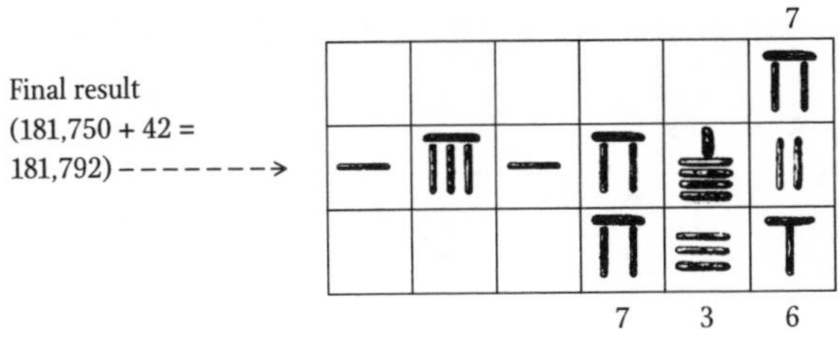


Fig. 21.69L.

Division was carried out by placing the divisor at the bottom and the dividend on the middle line. The quotient, which was placed at the top, was built up by successively removing partial products from the dividend.

On this numerical checkerboard it was also possible to solve equations, and systems of algebraic equations in several unknowns. The *Jiu Zhang Suan Shu* ("Art of calculation in nine chapters"), an anonymous work compiled during the Han Dynasty (206 BCE to 220 CE), gives much detail about the latter. Each vertical column is associated with one of the equations, and each horizontal row is associated with one of the unknowns, with the co-efficient of an unknown in an equation being placed in the square where the row intersects the column. Also, for this purpose, as well as the ordinary rods (reserved for "true" (*zheng*) numbers, i.e. positive numbers), black rods were used for negative numbers (*fu*: "false" numbers). A system of equations such as the following, for example:

$$2x - 3y + 8z = 32$$
$$6x - 2y - z = 62$$
$$3x + 21y - 3z = 0$$

was therefore represented as:

*The representation of a system of three equations in three unknowns on the arithmetical checkerboard. (From a treatise on mathematics of the Han period: 206 BCE to 220 CE): The first column on the left represents $2x - 3y + 8z = 32$; the second column represents $6x - 2y - z = 62$; the third column represents $3x + 21y - 3z = 0$.*

Fig. 21.70A.

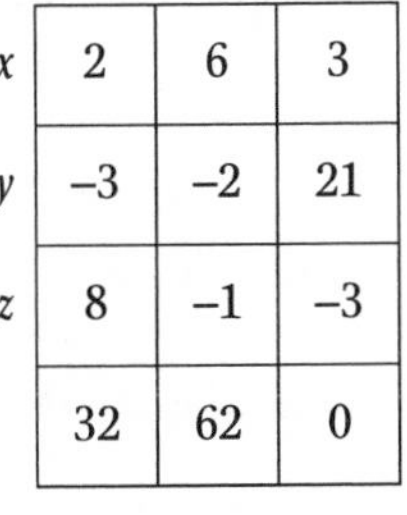

| | | | |
|---|---|---|---|
| *x* | 2 | 6 | 3 |
| *y* | −3 | −2 | 21 |
| *z* | 8 | −1 | −3 |
| | 32 | 62 | 0 |

Fig. 21.70B.

It could be solved quite easily by skilful manipulation of the rods.

This system of numerals is of particular interest for the history of numerical notation, since it is what led to the discovery of the principle of position by the Chinese.

Their system of writing numbers with vertical and horizontal strokes was simply the written copy of the way numbers were represented by rods on the abacus, where the different decimal orders of magnitude progressed in decreasing order from left to right. Once a calculation had been completed on the abacus by manipulation of the rods, their disposition on the abacus could be copied in writing, ignoring the lines dividing the abacus into squares. However, the rods were arranged on the abacus according to the principle of position, for the purposes of calculation, and so this principle was carried over into the written copy.

REPRESENTATION OF THE NUMBER 3,764

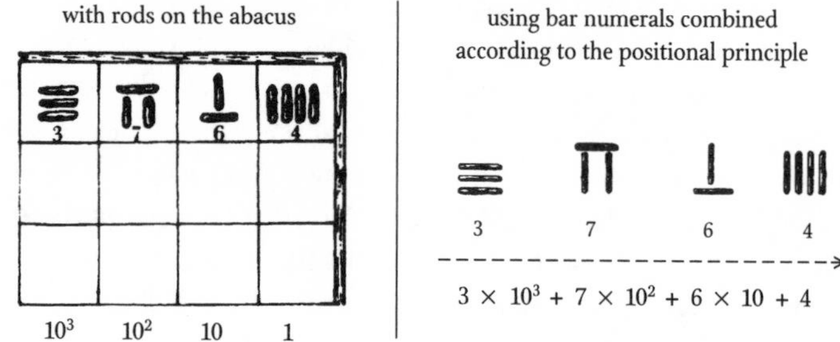


FIG. 21.71. *Origin of the Chinese bar numerals: how a manual calculating aid led to a written positional number-system*

The system of rods on the abacus was the practical means of performing arithmetic calculations, and the *suan zí* notation was used to transcribe the results into their mathematical texts.

The earliest known examples of the use of this abacus date from the second century BCE, but it is very likely that it goes much further back in time.

In any case, the characters used today for the Chinese word *suan*, which means "calculation", have a suggestive etymology. This word may be written using three apparently quite different characters, namely:

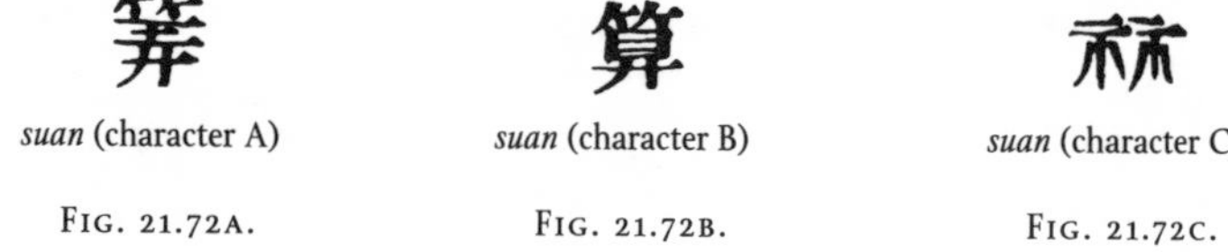

*suan* (character A) — FIG. 21.72A.

*suan* (character B) — FIG. 21.72B.

*suan* (character C) — FIG. 21.72C.

Derived from the following archaic form A′, the first character is an ideogram expressing two hands, a ruled table and a bamboo rod:

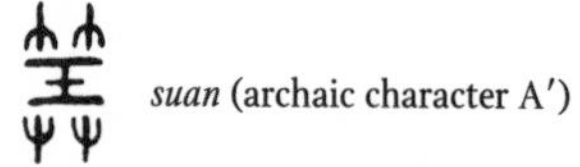

*suan* (archaic character A′)

FIG. 21.73A.

The second character is derived from the following archaic form B′ which expresses two hands and a ruled table:

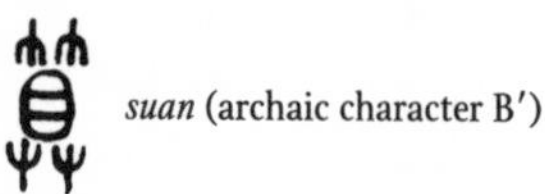

*suan* (archaic character B′)

FIG. 21.73B.

and the third comes from the following ancient form C′ which clearly evokes the representation of numbers on the checkerboard by means of rods vertically and horizontally oriented:

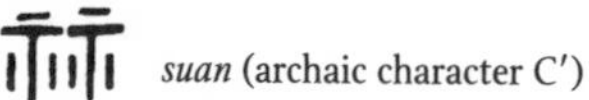

*suan* (archaic character C′)

FIG. 21.73C.

## THE CHINESE ABACUS: THE CALCULATOR OF MODERN CHINA

The celebrated "Chinese abacus" is, therefore. neither the first nor the only calculating device which has been used in China in the course of her long history. It is in fact of relatively recent creation, the earliest known examples being not older than the fourteenth century CE.

Amongst all the calculating devices which the Chinese have used, however, the *suan pan* (meaning "calculating board") is the only one with which all the arithmetical procedures can be performed simply and quickly. In fact almost everyone in China uses it: illiterate trader or accountant, banker, hotelier, mathematician, or astronomer. The most Westernised Chinese or Vietnamese, whether in Bangkok, Singapore, Taiwan, Polynesia, Europe, or the United States, carry out every kind of calculation using the abacus despite having ready access to electronic calculators, so deeply ingrained in their culture is its use. Even the Japanese, major world manufacturers of pocket calculators, still consider the *soroban* (the Japanese word for the abacus) as the principal calculating device and the one item that every schoolchild, businessman, peddler or office-worker should carry with them.

Likewise in the former Soviet Union the *schoty* (счёты), as the abacus is called, may be seen alongside the cash register and will be used to calculate the bill, in boutiques and hotels, department stores and banks.

A friend of mine, on a visit to the former Soviet Union, changed some French francs into roubles. The cashier first worked out the amount on an electronic calculator, and then checked the result on his abacus.

FIG. 21.74. *A Chinese shopkeeper doing his accounts with an abacus. (Reproduced from an illustration in the Palais de la Découverte in Paris)*

FIG. 21.75. *A Japanese accountant working with a* soroban. *From an eighteenth-century Japanese book,* Kanjô Otogi Zōshi *by Nakane Genjun, 1741: [Smith and Mikami (1914)]*

Westerners are invariably astonished at the speed and dexterity with which the most complicated calculations can be done on an abacus. Once, in Japan, there was even a contest between the Japanese Kiyoshi Matzusaki (*soroban* champion of the Post Office Savings Bank – a significant title, given what it means to be champion of anything in Japan) and the American Thomas Nathan Woods, Private Second Class in the 240th Financial Section of US Army HQ in Japan, the acknowledged "most expert electric calculator operator of the American forces in Japan". It took place in November 1945, just after the end of the Second World War, and the men of General MacArthur's army were eager to show the Japanese the superiority of modern Western methods.

The match took place over five rounds involving increasingly complicated calculations. And who won, four rounds out of five with numerous mistakes on the part of the loser? Why, the Japanese with the abacus! (Fig. 21.76)

RESULTS OF THE MATCH

KIYOSHI MATSUZAKI versus THOMAS NATHAN WOODS

*Soroban* champion of the Japanese Post Office Savings Bank

Private 2nd class in the 240th financial section of the US Forces HQ in Japan. The "top expert with the calculator in Japan"

Contested on 12 November 1945 under the auspices of the US Army daily *Stars and Stripes*

| 1st round | 2nd round | 3rd round | 4th round | Composite round |
|---|---|---|---|---|
| Additions of numbers with 3 to 6 figures | Subtractions of numbers with 6 to 8 figures | Multiplications of numbers with 5 to 12 figures | Divisions of numbers with 5 to 12 figures | 30 additions<br>3 subtractions<br>3 multiplications<br>3 divisions<br>(Numbers with from 6 to 12 figures) |
| Matsuzaki beat Woods | Matsuzaki beat Woods | Woods beat Matsuzaki | Matsuzaki beat Woods | Matsuzaki beat Woods |
| 1'14"8 / 2'00"2<br>1'16"0 / 1'53"0 | 1'04"0 / 1'20"0<br>1'00"8 / 1'36"0<br>1'00"0 / 1'22"0<br>(with mistakes) | (with mistakes by the loser) | 1'36"6 / 1'48"0<br>1'23"4 / 1'19"0<br>1'21"0 / 1'26"6 | 1'21"0 / 1'26"6<br>(with mistakes by the loser) |

Overall: Woods on the calculator is beaten 4 to 1 by Matsuzaki on the *soroban*

FIG. 21.76. Reader's Digest *no. 50, March 1947, p. 47*

The match, contested on 12 November 1945 under the auspices of the American Army daily *Stars and Stripes*, was a sensation. Their reporter wrote that: "Machinery suffered a setback yester-day in the Ernie Pyle theatre in Tokyo, when an abacus of centuries-old design crushed the most modern electrical equipment of the United States Government." The *Nippon Times* was exultant at this modest intellectual revenge for military defeat: "In the dawn of the atomic age, civilisation reeled under the blows of the 2,000-year-old *soroban*." An exaggeration, of course – above all concerning the age of the *soroban* – but one which must be viewed in the context of a Japan which, less than three months earlier, had seen two of its greatest cities destroyed by unprecedented military force. But anyone who has watched a Japanese of any competence operate the abacus would have no doubt that the same result could be obtained even today, with electronic instead of electrical calculators, at any rate for additions and subtractions. The keyboard speeds of most of us would be no match for the dexterity of the *soroban* operator. (*Science et Vie*, no. 734, November 1978, pp. 46–53).

The Chinese form of the instrument has a hardwood frame which holds a number of metal rods upon each of which slide wooden (or plastic) beads which may be of somewhat flattened shape. The beads are on either side of a wooden partition, two beads above and five below, and the beads may be slid towards the partition. Each of the metal rods corresponds to one of the decimal orders of magnitude, the value of a bead increasing by a factor of 10 as one moves from one rod to the rod on its left. (In theory, a base different from 10 may be used – 12 or 20 for example – provided each rod carries a sufficient number of beads.)

The normal abacus will have between eight and twelve rods, but the number may be fifteen, twenty, thirty, or even more, according to need. The more rods there are, the larger the numbers that the abacus can handle. With fifteen rods, for example, it can handle up to $10^{15}–1$ (a thousand million million, minus one!)

As a rule, the first two rods on the right are reserved for decimal fractions of first and second order, i.e. for the first two decimal places, and it is the third rod which is used for the units, the fourth for the tens, the fifth for the hundreds, and so on.

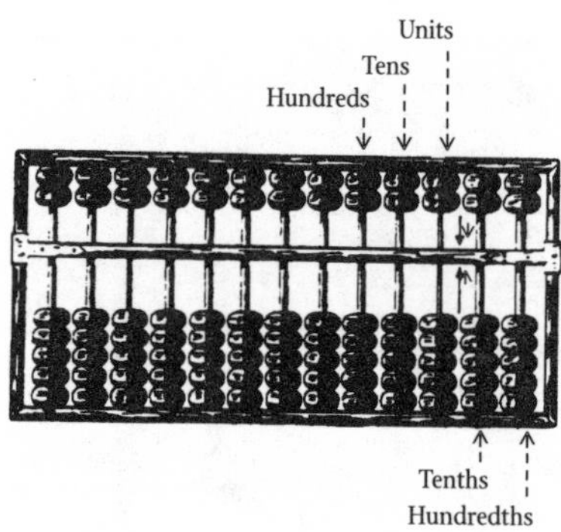


FIG. 21.77. *The representation of numbers on the Chinese* suan pan

The Russian abacus is somewhat different in design from the Chinese *suan pan* (Fig. 21.78). It has ten beads on each rod, of which two (the fifth and the sixth) are usually of a different colour, which makes it easier for the eye to recognise the numbers from 1 to 10. To represent a number the corresponding number of beads are slid towards the top of the frame.

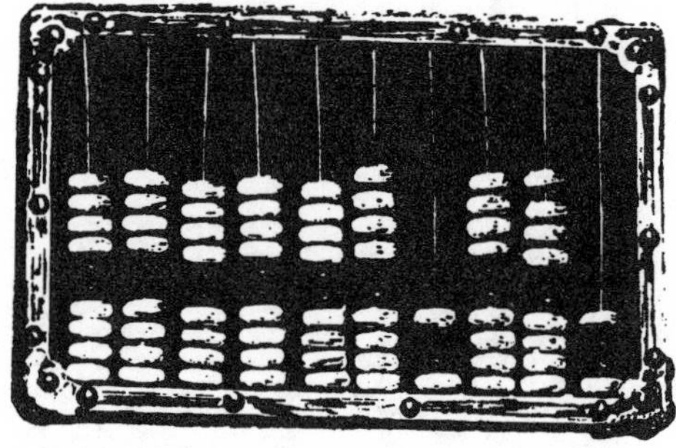

FIG. 21.78. *Russian abacus* (schoty). *It generally has four white beads, then two black and then four white. This type of instrument is still in use in Iran, Afghanistan, Armenia and Turkey*

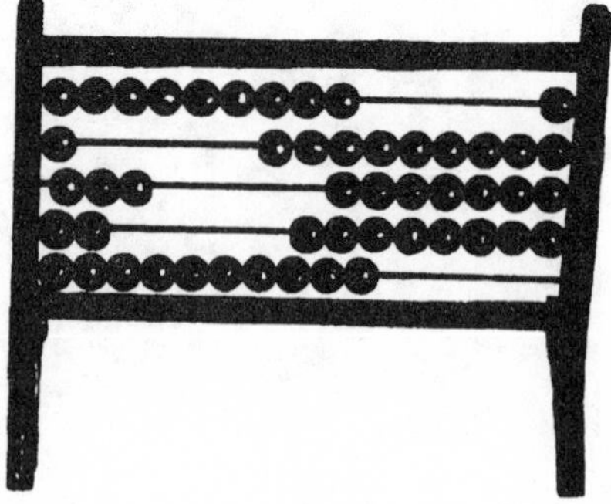

FIG. 21.79. *French abacus used for teaching arithmetic in municipal schools in the nineteenth century*

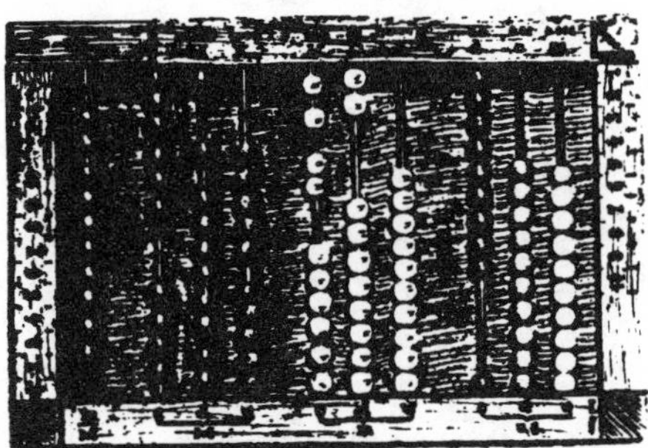

FIG. 21.80. *Abacus marketed by Fernand Nathan at the beginning of the twentieth century as a teaching aid*

On the Chinese abacus, each of the five beads on the lower part is worth one unit, and each of the two on the upper part is worth five. Arithmetical operations involve sliding beads from either side towards the central partition.

To place the number 3 on the abacus, slide three of the five beads on the lower part of the first rod upwards towards the partition. To place the number 9, slide four of the five lower beads upwards towards the partition, and one of the two upper beads downwards towards the partition:

FIG. 21.81.

For a larger number such as 4,561,280, the same principle is adopted for each digit: since the first digit is zero, the beads on the first rod are not displaced (denoting absence of number in this position), giving the result shown:

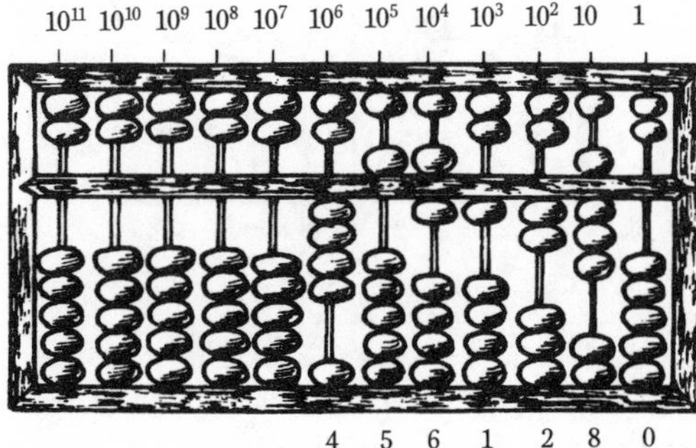


FIG. 21.82.

To place the number 57.39, which has a decimal fraction part, the same principle is used for the hundredths, then the tenths, and then the units, tens and hundreds (Fig. 21.83):

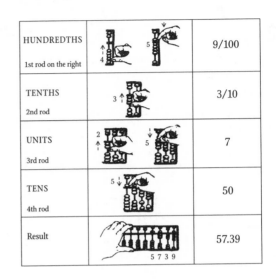

| | | |
|---|---|---|
| HUNDREDTHS<br>1st rod on the right | 4 ↑ 5 ↓ | 9/100 |
| TENTHS<br>2nd rod | 3 ↑ | 3/10 |
| UNITS<br>3rd rod | 2 ↑ 5 ↓ | 7 |
| TENS<br>4th rod | 5 ↓ | 50 |
| Result | 5 7 3 9 | 57.39 |

FIG. 21.83.

It is therefore a very simple matter to enter a number onto the Chinese abacus. Actual arithmetic is hardly any more complicated, provided one has learned the addition and multiplication tables by heart for the numbers from 1 to 9.

For convenience of exposition, we shall only consider whole numbers, and therefore we can allocate the first rod to the units, the second rod to the tens, and so on. Now consider addition of the three numbers 234, 432 and 567.

First of all we "clear" the abacus by sliding all the beads to the top and bottom extremities of the rods, leaving the central partition clear. To enter the number 234, first on the third rod from the right (for the hundreds), we slide two beads upwards; then, on the second rod (for the tens) three beads upwards; and finally, on the first rod (for the units), four beads upwards:

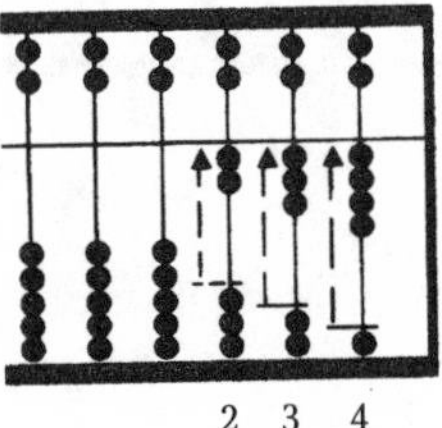


FIG. 21.84A.

Next, to add to this the number 432, we move the corresponding number of beads towards the centre in a similar way. However, on the hundreds rod there are already two beads touching the partition so we do not have four beads available to slide; but we can bring down one bead (representing 5) from the top against the partition and slide one of the lower beads back down away from the centre, since 5 – 1 = 4. On the tens rod, where three beads have already been moved upwards leaving two, in order to add in the 3 of 432 we again slide down one of the upper beads (for 5) and retract two of the lower beads (since 5 – 2 = 3). Finally, on the units rod, we slide down one of the upper beads (for 5) and retract three of the lower beads (since 5 – 3 = 2):

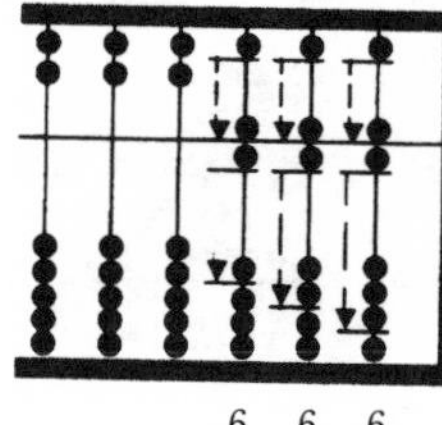


FIG. 21.84B.

As the third and final stage, to add the number 567 to this result, we start by sliding one of the upper beads (for 5) downwards on the hundreds rod. Then, on the tens rod, we slide down one of the upper beads (for 5) and we slide up one of the lower beads (for 1), since 5 + 1 = 6. Finally, on the units rod, we slide downwards one of the upper beads (for 5) and we slide upwards two of the lower beads (for 2), since 5 + 2 = 7. Our abacus now looks like the following:

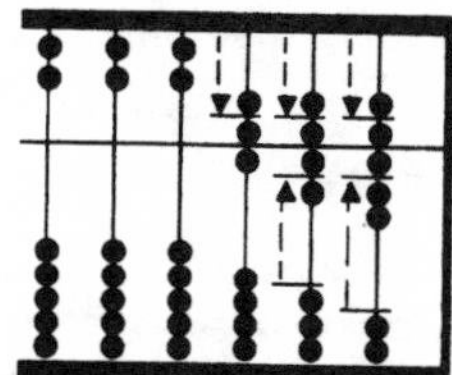

FIG. 21.84C.

But it is not yet all over: what is represented on each rod is no longer a decimal digit, and some further reduction is required before the result can be announced. Therefore, on the third (hundreds) rod, we slide the two upper beads away upwards: each counts for five hundreds, and so we then slide one lower bead (for 1) of the thousands rod towards the centre. Next, in a similar way, the two upper beads of the tens rods are slid upwards away from the centre and one lower bead of the hundreds rod is slid towards the centre; and, finally, the two upper beads of the units rod are replaced by a single bead on the tens rod. When this has been done, the abacus looks like the following, and the result can be read off from it: 234 + 432 + 567 = 1,233.

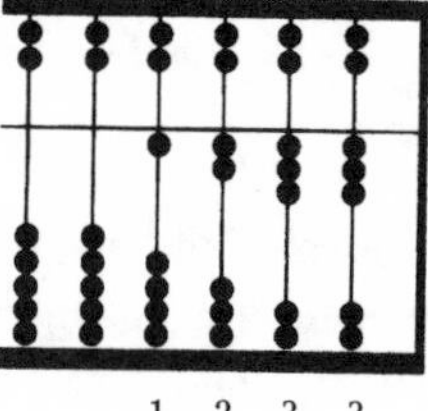


FIG. 21.84D.

Subtraction is carried out by the reverse process, multiplication by repeated addition of the multiplicand for as many times as each digit in the multiplier, and division by repeated subtraction of the divisor from the dividend as many times as possible, this number then being the quotient.

Suppose we want to evaluate the product 24 × 7.

We first note that the method is independent of the overall order of magnitude of the result: technically, the procedure is identical whether we want 24 × 7, 24,000 × 7, 24 × 700, 0.24 × 7 or 24 × 0.007, and the digits in the result will be the same; to get the correct result it is enough to keep the order of magnitude in mind.

To work out the above calculation, we start by placing the multiplier (7) on a rod at the left, and the multiplicand (24) towards the right, making sure to leave a few empty rods between them.

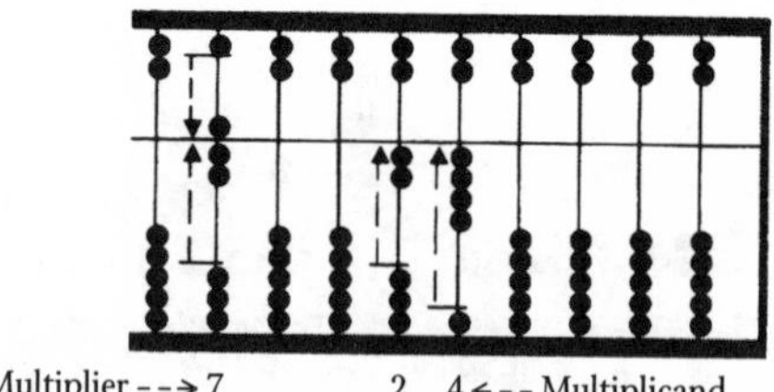


FIG. 21.85A.

Now we mentally multiply 7 by 4, getting 28, and we place this result immediately to the right of the multiplicand:

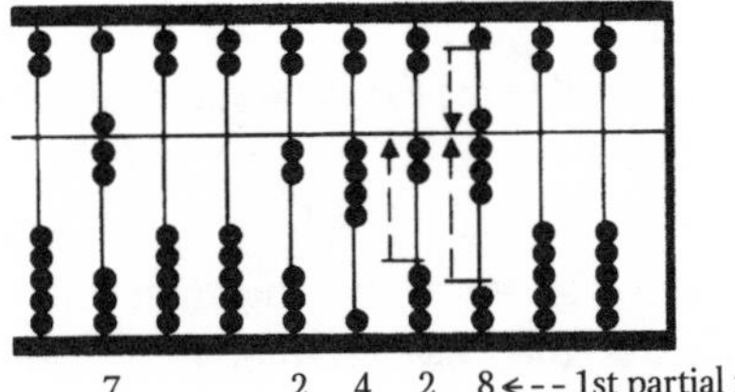


Fig. 21.85b.

Now the 4 of the multiplicand is eliminated by sliding its four units beads back downwards:

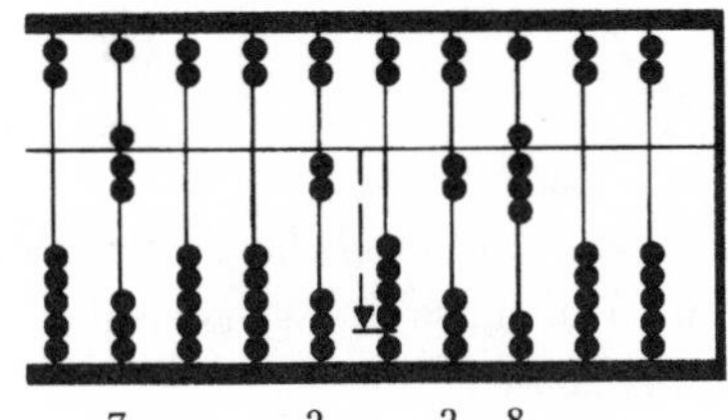


Fig. 21.85c.

Next we mentally multiply 7 by 2, getting 14, and we now enter this result as before but at one place further left. Adding it to what is already there, we therefore slide one lower bead upwards on the hundreds rod, and on the tens rod we slide one upper bead downwards and one lower bead upwards:

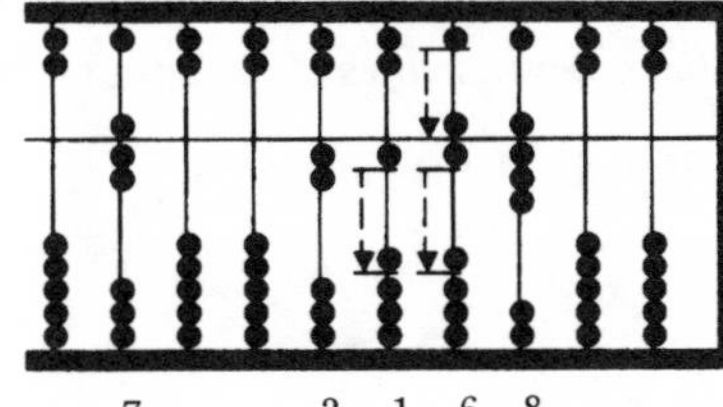


Fig. 21.85d.

The 2 of the multiplicand is now eliminated, and the multiplier also, and all that remains is to read off the result (168):

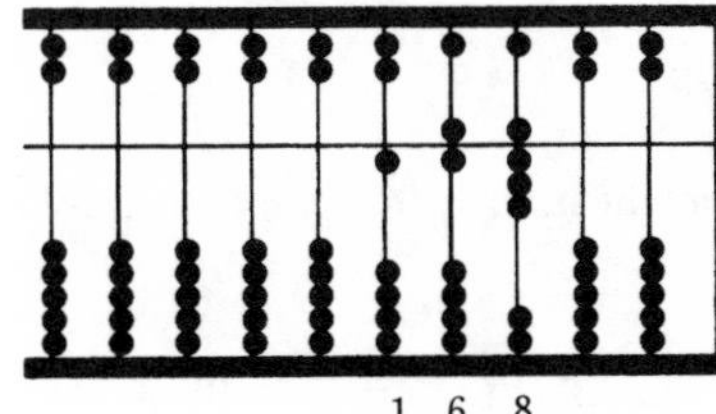


Fig. 21.85e.

So it is not very complicated to do arithmetic on the Chinese abacus. Even square roots or cube roots, or more complicated problems still, can be worked out by operators who know how to use it well. (Our intention here is only to give a general idea of how to use the abacus; we therefore abstain from describing the detailed technique for manipulating it, and we do not discuss its general arithmetical or algebraic applications.)

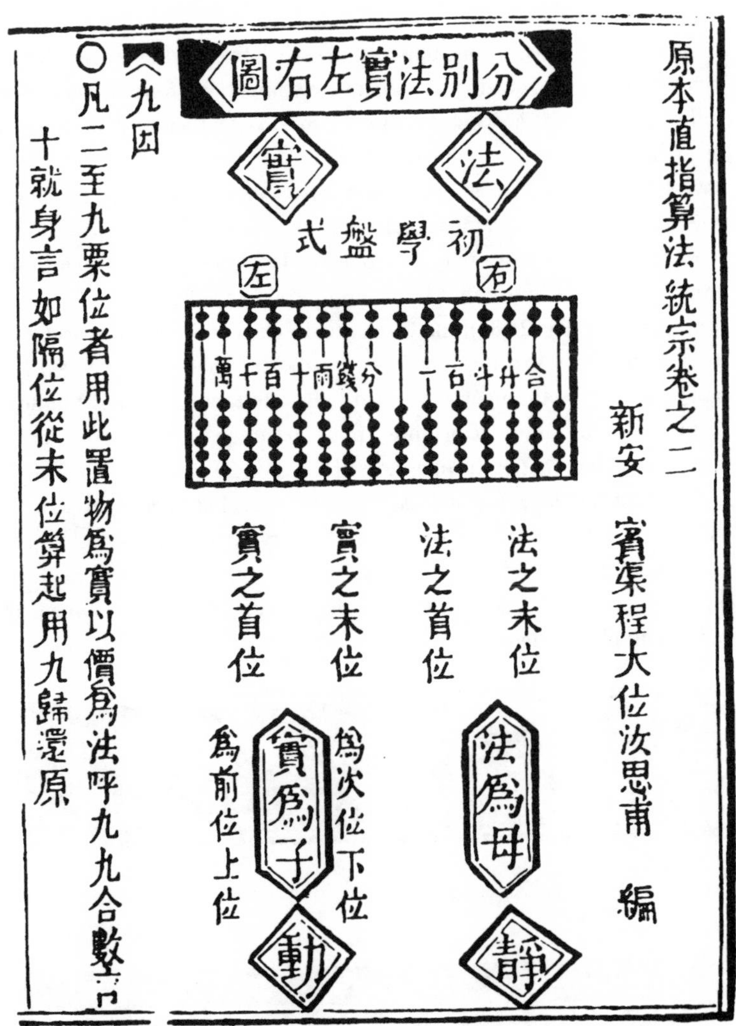


Fig. 21.86. *Instructions for using the* suan pan *in the Chinese* Suan Fa Tong Zong *printed in 1593. [Reproduced from Needham (1959), III, p. 76]*

For all its convenience, this aid to calculation has a number of disadvantages. It takes a long time, and thorough training, to learn how to use it. The finger-work must be extremely accurate, and the abacus must rest on a very solid support. Moreover, if one single error is made the whole procedure must be restarted from scratch, since the intermediate results (partial products, etc.) disappear from the scene once they have been used. None of this, however, detracts from the ingenious simplicity of the device.

After a little thought, however, we are led to ask a question touching on the basic concept of the Chinese abacus. We have seen that on each rod nine units are represented by one upper bead (worth five) and four lower beads (worth one each). Therefore five beads (one upper and four lower) always suffice to represent any number from 1 to 9. Why, therefore, do we find seven beads, whose total value is 15? The answer lies in the fact that (as we have seen in some of the above examples), it is often useful to represent on one rod, temporarily, an intermediate result whose value exceeds 9.

In this connection we may note that the Japanese *soroban* began to do away with the second upper bead, from around the middle of the nineteenth century (Fig. 21.87), and that since the end of the Second World War it has definitively lost the fifth lower bead. This change has obliged the Japanese abacists to undergo an even longer and more arduous training, and it has obliged them to acquire a finger technique even more elaborate and precise than that of the operators of the Chinese *suan pan* (Fig. 21.88).

The post-war Japanese abacus is therefore the fully perfected state of the instrument and marks the close of an evolution in the techniques of calculation which derive from arithmetical manipulations of pebbles, an evolution which has largely been independent of the development of written number-systems.

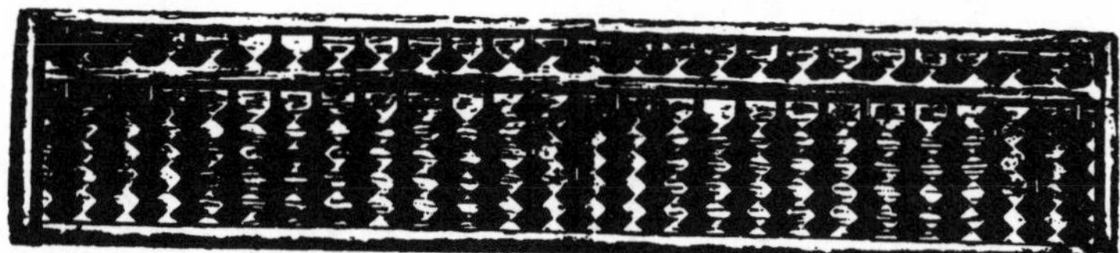

FIG. 21.87. *Pre-war Japanese* soroban *with a single upper bead and five lower beads*

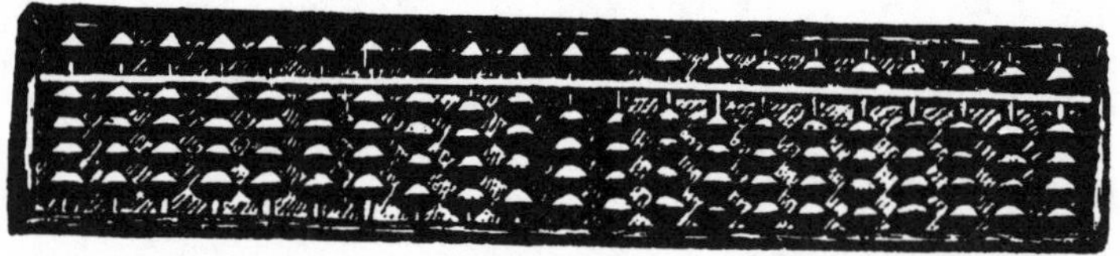

FIG. 21.88. *Post-war Japanese* soroban *with a single upper bead and four lower beads*

## NUMBER-GAMES AND WORD-PLAYS

We should not bid farewell to the Far Eastern civilisations without enjoying some examples of their wit.

Both the Chinese and the Japanese have always had a great weakness for plays on words and characters. Since their numerals correspond both to words and to characters, they have taken every opportunity to indulge it. Here are some examples.

The first example (noted by Mannen Veda) bears on the character for the figure 8. For the age of a 16-year-old girl, the Chinese use the expression *pogua*, which literally means "to cut the watermelon in two":

| 破 | 瓜 |
|---|---|
| *po* ("*cut into two*") | *gua* ("*watermelon*") |

This is a number-play on the form of the character *gua* ("watermelon") which seems to be composed of two characters identical to the figure 8 side by side, representing an addition:

瓜＝八＋八 $= 8 + 8 = 16$

Furthermore, the pun involves the fact that "watermelon" can also mean virginity (much as we use the word "flower"), which means that *pogua* is also an erotic image of the "defloration" of the young girl.

Other examples (noted by Masahiro Yamamoto) concern the names given to the various major anniversaries of old age in Japan.

1. The 77th birthday is the "happy anniversary". In Japanese it is called *kiju* and written

*kiju*

Graphically this yields the 77, since the word *ki* ("happy") is written, in the cursive style, as

*ki*

namely as a character which can be decomposed as follows:

七 十 七 $= 7 \times 10 + 7 = 77$

2. The 88th birthday is the "rice anniversary". In Japanese it is called *beiju* and written

米寿

*beiju*

Graphically this yields the 88, since the word *bei* ("rice") is written using a character which can be decomposed as follows:

米 = 八十八 $= 8 \times 10 + 8 = 88$

3. The 90th birthday is the "accomplished anniversary". In Japanese it is called *sotsuju* and is written

卒寿

*sotsuju*

Graphically this yields the 90, since the word *sotsu* ("accomplished") is written using a character which in turn may be replaced by an abbreviation which itself may be decomposed as follows:

卒 (*sotsu*) = 卆 (*sotsu*) = 九十 $= 9 \times 10 = 90$

4. The 99th birthday is the "white-haired anniversary". In Japanese it is called *hakuju* and is written

白寿

*hakuju*

Graphically this yields the 99, since the word *haku* ("white") is written using a character which is none other than the character for 100 from which one unit (the horizontal line) has been removed:

白 (*haku*) = 百 (*hyaku*) − 一 (*ichi*) $= 100 - 1 = 99$

5. Finally, the 108th birthday is the "tea anniversary". In Japanese it is called *chaju* and is written

茶寿

*chaju*

Graphically this yields the 108, since the word *cha* ("tea") is written using a character which can be decomposed as

茶 = 十 (10) 十 (10) 八 (8) 十 (10) 八 (8) $= 10 + 10 + (8 \times 10 + 8) = 108$

We may also note the strange number-names used by Zen monks to express sums of money in the Edo period (eighteenth century). For these monks, anything to do with money was considered vulgar and not to be mentioned directly. Therefore, to express numerical sums of money, they euphemistically made use of plays on characters (Fig. 21.89).

| | ZEN NUMERALS | LITERAL MEANING | EXPLANATION OF NUMERICAL INTERPRETATION |
|---|---|---|---|
| 1 | 大無人 *dai ni jin nashi* | "size without man" | = 大 without 人 ---> 一 = 1 |
| 2 | 天無人 *ten ni jin nashi* | "heaven without man" | = 天 without 人 ---> 二 = 2 |
| 3 | 王無中 *ô ni chû nashi* | "king without centre" | = 王 without 丨 ---> 三 = 3 |
| 4 | 罪無非 *zai ni hi nashi* | "fault without evil" | = 罪 without 非 ---> 四 = 4 |
| 5 | 吾無口 *go ni kuchi nashi* | "myself without mouth" | = 吾 without 口 ---> 五 = 5 |
| 6 | 交無人 *kô ni jin nashi* | "exchange without man" | = 交 without 人 ---> 六 = 6 |
| 7 | 切無刀 *setsu ni to nashi* | "cutting without a knife" | = 切 without 刀 ---> 七 = 7 |
| 8 | 分無刀 *bun ni to nashi* | "dividing without a knife" | = 分 without 刀 ---> 八 = 8 |
| 9 | 丸無点 *gan ni chu nashi* | "circle without accent" | = 丸 without ㇏ ---> 九 = 9 |
| 10 | 針無金 *shin ni kin nashi* | "needle without metal" | = 針 without 金 ---> 十 = 10 |

FIG. 21.89. *Esoteric numerals of the Zen monks (eighteenth century) (M. Yamamoto. Personal communication from Alain Birot)*

We close with the following Japanese verses, attributed to Kôbô Daishi (775–835)*:

いろはにほへとちりぬるを
わかよたれそつねならむ
うゐのおくやま
けふこえてあさき
ゆめみしゑひもせすん

TRANSCRIPTION

*I-ro-ha-ni-ho-he-to-*
*Chi-ri-nu-ru-wo*
*Wa-ka-yo-ta-re-so*
*Tsu-ne-na-ra-mu*
*U-i-no-o-ku-ya-ma*
*Ke-fu-ko-e-te*
*A-sa-ki-yu-me-mi-shi*
*E-hi-mo-se-su-n'*

TRANSLATION

*Though pretty be its colour,*
*The flower alas will fade;*
*What is there in this world*
*That can forever stay?*
*As I go forward from today,*
*To the end of the visible world,*
*I shall see no more dreams drift by*
*And I shall not be fooled by them.*

This poem contains every sound of the Japanese language with no repetitions. It is therefore often used in teaching Japanese.

However, number has never been far from poetry in the oriental cultures: these same syllables which have been so to speak frozen into a given order by this poem, have finally acquired numerical values. Which is why the Japanese often count using the syllables of the poem:

| I- | ro- | ha- | ni- | ho- | he- | to- | chi- | ri- | . . . |
|---|---|---|---|---|---|---|---|---|---|
| 1 | 2 | 3 | 4 | 5 | 6 | 7 | 8 | 9 | . . . |

* See L. Frédéric, *Encyclopaedia of Asian Civilisation*, J.-M. Place, Paris, 1977–87, vol. III.

CHAPTER 22

# THE AMAZING ACHIEVEMENTS OF THE MAYA

The civilisation of the Maya was without question the most glorious of all the pre-Columbian cultures of Central America. Its influence over the others, particularly over Aztec culture, can be likened to the influence of Greece over Rome in European antiquity.

## SIX CENTURIES OF INTELLECTUAL AND ARTISTIC CREATION

In the course of the first millennium CE the Maya people produced art, sculpture, and architecture of the highest quality and made great strides in education, trade, mathematics, astronomy, etc.

Maya builders discovered cement, learned how to make arches, built roads, and, of course, they put up vast and complex cities whose buildings were heavily decorated with sculpture and painting. Surprisingly, all this was done with tools that had not developed since the Stone Age: the Maya did not discover the wheel, nor use draught animals, nor any metals. The Mayas' true glory rests on their abstract, intellectual achievements.

They were, in the first place, astronomers of far greater precision than their European contemporaries. As C. Gallenkamp (1979) tells us, the Maya used measured sight-lines, or alignments of buildings that served the same purpose, to make meticulous records of the movements of the sun, the moon, and the planet Venus. (They may also have observed the movements of Mars, Jupiter and Mercury.) They studied solar eclipses in sufficient detail to be able to predict their recurrence. They were acutely aware that apparently small errors could lead in time to major discrepancies; the care they took with their observations allowed them to reduce margins of error to almost nothing. For example, the Maya calculation of the synodic revolution of Venus was 584 days, compared to the modern calculation of 583.92.

FIG. 22.1. *The Great Jaguar Temple at Tikal, constructed in c. 702 CE. Copy by the author from Gendrop (1978), p. 72*

The Maya also made their own very accurate measurement of the solar year, putting it at 365.242 days.* The latest computations give us the figure of 365.242198: so the Maya were actually far nearer the true figure than the current Western calendar of 365 days (which, with leap years, gives a true average of 365.2425).

They were no less precise in their measurement of the lunar cycle. Modern measuring devices of the most sophisticated kind allow us to fix the average length of a lunar cycle at 29.53059 days. Using only their eyes and their brains, the Mayan astronomers of Copán found that 149 new moons occurred in 4,400 days, which gives an average for each lunar month of 29.5302. At Palenque, the same calculation was made over 81 new moons and produced the even more accurate figure of 2,392 days, or 29.53086 per cycle.

* The Maya did not express the figure in this way of course, since they could only operate arithmetically in integers.

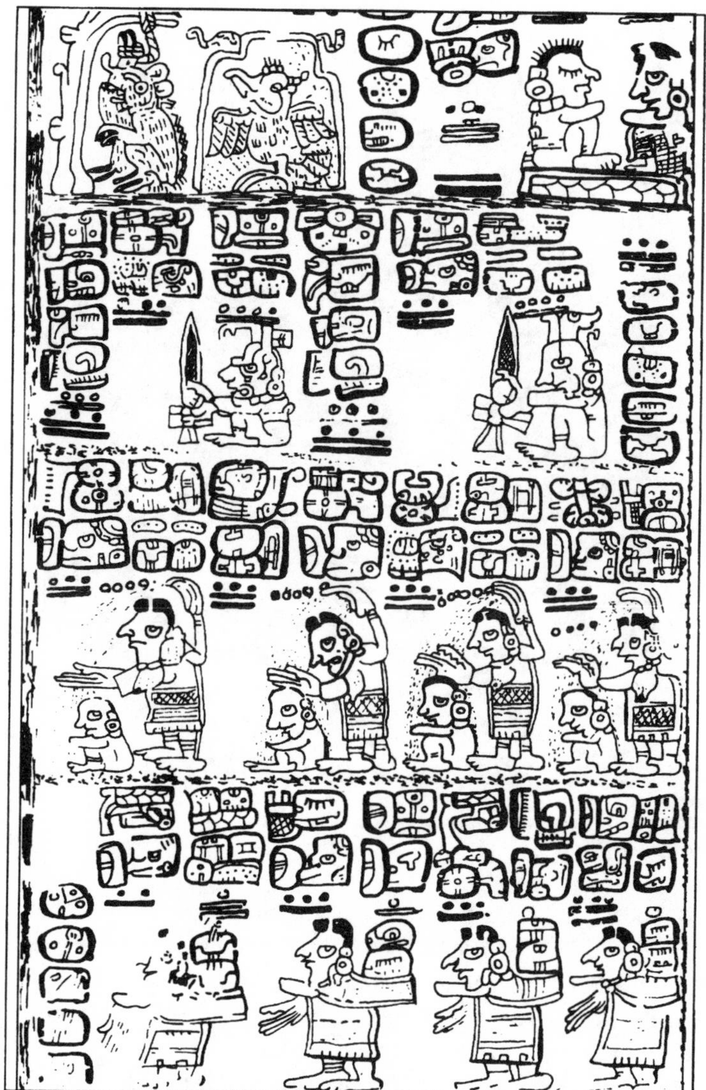

FIG. 22.2. *Extract from a Maya manuscript (lower part of p. 93 of the Codex Tro-Cortesianus, from the American Museum, Madrid). It shows a kind of memorandum for prophet-priests, part of a treatise on ritual magic which includes some astronomical observations.*

Even more fascinating is the Mayas' use of very high numbers for the measurement of time. On a *stela* at Quiriguà, for instance, there is an inscription that mentions the last 5 *alautun*, a period of no less than 300,000,000 years, and gives the precise start and end of the period according to the ritual calendar. Why did they count in terms so far beyond any human experience of life? Perhaps that will always remain a mystery; but it suggests that the Maya had a concept if not of infinity, then of a boundless, unending stretch of time.

FIG. 22.3. *Alone in the darkness of the night, a Maya astronomer observes the stars. Detail from the Codex Tro-Cortesianus. Copied from Gendrop (1978), p. 41, Fig. 2*

It is even more puzzling that the Maya measurements were done without any tools to speak of. They had not discovered glass, so there were no optical instruments. They had no clockwork, no hour-glasses, no idea of water-clocks (*clepsydras*), no means at all of measuring time in units less than a day (such as hours, minutes, seconds, etc.); nor did they have any concept of fractions. It is hard to imagine how to measure time without at least basic measuring devices.

The tool that the Maya used for measuring the true solar day was the very simple but utterly reliable device called a *gnomon*. It consists of a rigid stick or post fixed at the centre of a perfectly flat area. The stick's shadow alters as the day progresses. When the shadow is at its shortest, then the sun is at its meridian: that is to say, the sun has reached its highest point above the horizon, and it is "true noon".

As for astronomical observations, according to P. Ivanoff (1975), these were done by means of a jadeite tube placed over a wooden cross-bar, as shown in codices, thus:

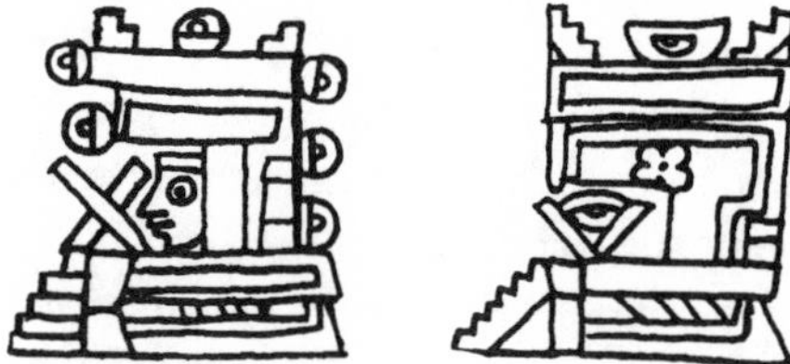

FIG. 22.4. *Astronomical observations, as shown in the Mexican manuscripts, Codex Nuttall and Codex Selden. Copied from Morley (1915). In the left-hand drawing, an astronomer seen in profile watches the sky through a wooden X; the right-hand drawing shows an eye looking through the angle of the X.*

The Maya also developed an elaborate writing system, consisting of intricate signs known as glyphs. These include numerals (as we shall see below) and many names or "emblem glyphs" associated with the main cities in the central Mayan area. The decipherment of Maya glyphs is currently the subject of intense and recently successful research.*

MAYA GODS

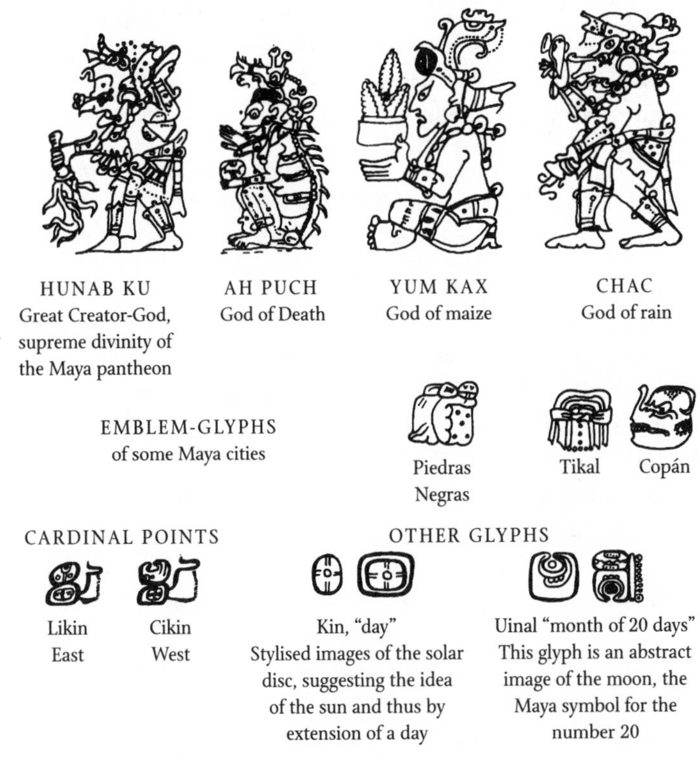


Fig. 22.5. *Some of the Maya hieroglyphs deciphered to date*

## MAYA CIVILISATION

Several dozen abandoned cities buried in the tropical jungles and savannah of Central America bear witness to one of the most mysterious episodes of human history.

With their stately temples perched atop pyramids up to 170 feet high, with their intricately carved pillars and altars and brightly painted earthenware vessels, these forgotten cities are all that is left of a sophisticated civilisation that is thought to have begun in the jungles of Peten. At the height of its glory, Maya civilisation covered the area shown in Fig. 22.6, and included:

- the present-day Mexican provinces of Tabasco, Campeche, and Yucatan, the region of Quintana Roo and a part of Chiapas province;
- the Peten region and almost all the uplands of present-day Guatemala;
- the whole of Belize (formerly British Honduras);
- parts of Honduras;
- the western half of Salvador;

making an area of about 325,000 km$^2$.

There are reckoned to be about two million direct descendants of the Maya alive today, most of whom are in Guatemala, and the remainder spread around Honduras and the Mexican provinces of Yucutan, Tabasco, and Chiapas.

Maya civilisation was fully developed at least as early as the third century CE and reached its greatest heights of artistic and intellectual creation long before the discovery of the New World by Christopher Columbus.

Fig. 22.6. *Map by the author, after P. Ivanoff*

* See Michael D. Coe, *Breaking the Maya Code* (London: Thames and Hudson, 1992), for a fascinating account of recent breakthroughs.

It is widely assumed that there was an early period of Maya civilisation dating from about the fifth century BCE, during which the Maya differentiated themselves from other Amerindian cultures; but of this era of formation, there remain few traces apart from shards of pottery, and little can be known of it.

The period from the third to the tenth century CE is the "classical" period of Maya civilisation, and it is in these centuries that the Maya developed their arts and sciences to their highest point. But at some point in the ninth or tenth centuries there occurred an unexpected and mysterious event which Americanists have not yet fully explained: the Maya began to abandon their ritual centres and cities in the central area of the "Old Empire". Their departure was so sudden in some places that buildings were left half-finished.

It was long thought that what had happened was an exodus of the entire population, but recent excavations have shown this not to be true. Various theories have been put forward to explain this resettlement of the Maya to the north and the south – epidemics, earthquakes, climate change, invasion, perhaps even their priests' interpretation of the wishes of the gods. The most plausible of these hypotheses are those that see the main cause of the exodus in the exhaustion of the soil. Mayan agriculture was based on the use of burnt clearings, which created ever more extensive infertile areas. In addition, there may well have been a peasant revolt, provoked by the vast inequality between the classes of Maya society.

Whatever the real cause, large sections of the Maya people left the central area, leaving a much reduced population which gave up the traditional rituals in the cities and allowed the religious monuments to fall into decay.

There was also an invasion of a different people, from the west. To judge by the ruins of Chichén Itzá (Yucatan), these invaders were probably Toltecs, who came from an area north of present-day Mexico City. After the "interregnum" (925–975 CE), the period following the fall of classical Maya culture is called the "Mexican period", and it lasted until 1200 CE.

The Maya accepted Toltec domination and adopted some of the Mexican gods, including Quetzalcoatl, the plumed serpent. The Maya also became more warlike, in line with the traditions of the Mexicans, whose gods required countless human sacrifices. However, even if the Maya of the Mexican period tore the hearts out of their human sacrificial victims, they were never as bloodthirsty as their neighbours, the Aztecs, whose religious rituals were frenetically violent.

Toltec and Maya civilisations gradually merged into one. The language, religion and even the physical characteristics of the Maya changed so much that it is hard to compare Maya civilisation before and after the Mexican invasion.

Between 1200 and 1540, the course of Maya history changed completely once again. Mexican civilisation was rejected, and the invaders adopted Maya customs. This period is called the age of "Mexican absorption". Maya civilisation continued to decline, as can be seen in the art and architecture of the period. Wars of annihilation broke out, and Maya civilisation soon came to an end. Only a small group from Chichén Itzá managed to escape and resettle on the island of Tayasal, in Lake Peten, where they maintained their independence until 1697.

## THE DOCUMENTARY SOURCES OF MAYA HISTORY

The first light to be shed on the civilisation of the Maya was the work of the famous American diplomat and traveller, John Lloyd Stephens, who explored the jungles of Guatemala and southern Mexico with the English artist Frederick Catherwood in 1839. A more detailed survey of Maya sites and buildings was carried out from 1881 by Alfred Maudslay, which marked the true beginning of scholarly research on the world of the Maya. But most of the knowledge we now have of this lost civilisation has been gained in the last few decades.

When the Spaniards conquered Central America in the sixteenth century, Maya civilisation had been all but extinct for several generations, and most of its magnificent cities were but inaccessible ruins in the midst of the jungle. This explains why the early Spanish chroniclers were bedazzled by the Aztecs and hardly mentioned the Maya at all.

Pre-Columbian cultures, moreover, were systematically suppressed by the conquistadors. Deeply shocked by the bloodthirstiness of Aztec and Maya rituals, and believing that their mission was to convert the natives to Christianity, the Spaniards sought to eradicate all traces of the devilish practices that they came across. In order to ensure that such abominable religions would never re-emerge, they burnt everything they could find in *autos-da-fé*.

Nonetheless it is to a Spaniard that we owe a significant part of our present knowledge of the history, customs and institutions of the Maya. In 1869, the colourful and indefatigable French monk Brasseur de Bourbourg unearthed in the Royal Library of Madrid a manuscript entitled *Relación de las Cosas de Yucatán* by the first bishop of Merida (Yucatan), Diego de Landa. Written shortly after the Spanish conquest, the *Relación* is full of priceless ethnographic information, including descriptions and drawings of the glyphs used by the indigenous population of Yucatan in the sixteenth century. Ironically, Landa was proud of having burned all the texts using this writing, the better to bring the natives into the embrace of the Catholic

Church. He wrote his chronicle in order to explain why he had destroyed all those precious painted codices – but thereby unwittingly preserved the basic elements of one of the most important pre-Columbian civilisations of the Americas.

The discovery of this sixteenth-century manuscript aroused great interest, because the glyphs copied down by Landa were similar to the carved shapes on the ruins found in the virgin jungle of Central America by Stephens and later explorers. It provided solid evidence of the cultural connection between the sixteenth-century population of the Yucatan peninsula and the builders of the lost cities of the jungle, both in Yucatan and further south.

Landa's manuscript is a major source for the history of the Maya, but it is not the only one. Much was also written down by the natives themselves, who were taught by Spanish missionaries to read and write in the Latin alphabet, which they then also used for writing in their own tongue. Although the teaching was intended to support the spread of Christianity, it was also used – inevitably – to set down the fast-disappearing oral traditions of the local populations.

A good number of anonymous accounts of this kind have survived, and give a reflection of the history, traditions and customs of the indigenous peoples of Spanish Central America. From the Guatemalan uplands comes the manuscript known as *Popol Vuh*, which contains fragments of the mythology, cosmology and religious beliefs of the Quiché Maya; and it was in the same area that the *Annals of the Cakchiquels* were found, which provide in addition the story of the tribe of that name during the Spanish conquest. The *Books of Chilám Balám* are a collection of native chronicles from Yucatan, and are named after a class of "Jaguar Priests", famed for their prophetic powers and their mastery of the supernatural. Fourteen of these manuscripts go a long way back in history; though they deal mostly with traditions, calendars, astrology, and medicine, three of them mention historical events that can be precisely situated in the year 1000 CE. Some parts of the *Chilám Balám* may even have been copied directly from ancient codices.

The ancient Maya codices used parchment, tree bark, or mashed vegetable fibres strengthened with glue to provide a writing surface. The glyphs were written with a brush pen dipped in wood ash, and then coloured with dyes from various animal and vegetable sources. The pages were glued together, then folded like a concertina and bound between wood or leather covers, much like a book. Three of them miraculously escaped the attention of the conquistadors, and found their way back to Europe, where they are now known by the names of the cities where they are kept: the Dresden Codex (in the Sächsische Landesbibliothek, Dresden, Germany) is an eleventh-century copy of an original text drafted in the classical period, and deals with astronomy and divination; the Codex Tro-Cortesianus (American Museum, Madrid) is less elaborate and was probably composed no earlier than the fifteenth century; and the Paris Codex (Bibliothèque nationale, Paris), likewise from the late period, gives illustrations of ceremonies and prophecies.

Despite these various documentary sources, much of Maya civilisation remains mysterious and unexplained to this day.

## AZTEC CIVILISATION

The legendary homeland of the Aztecs, according to the few manuscripts that have survived and the tales of Spanish conquerors, was called Aztlan and was located somewhere in northwestern Mexico, maybe in Michoacan. In a cave in Aztlan they are supposed to have found the "colibri wizard", Huitzilopochtli, who gave such good advice that he became the Aztecs' tribal god. Then began their long migration, by way of Tula and Zumpango (on the high plateau), and the Chapultepec, where they lived peaceably for more than a generation. Thereafter, they were defeated in battle and exiled to the infertile lands of Tizapan, infested with poisonous snakes and insects. A group of rebels took refuge on the islands in Lake Texcoco, where, in 1325 CE (or 1370, according to more recent calculations), they founded the city of Tenochtitlán, which has become present-day Mexico City.

Within a century Tenochtitlán became the centre of a vast empire. The Aztec King Itzcoatl subdued and enslaved most of the tribes in the valley; then under Motecuhzoma I (1440–1472) they battled on into the Puebla region in the south. Axayacatl, son of Motecuhzoma, led the Aztec armies even further south, as far as Oaxaca; he also attacked, but failed to conquer the Matlazinca and Tarasques in the west.

By the time the Spaniards arrived in 1519, the Aztecs possessed most of Mexico, and their language and religion held sway over a vast territory stretching from the Atlantic to the Pacific Oceans and from the northern plains to Guatemala. The name of the king, Motecuhzoma (Europeanised as "Montezuma") struck fear from one end to another of the empire; Aztec traders, with great caravans of porters, scoured the entire kingdom; and taxes were levied everywhere by the king's administrators. It was a relatively recent civilisation, at the height of its wealth and glory.

FIG. 22.7. *Page 1 of the Codex Mendoza (post-conquest). Through a number of Aztec hieroglyphs, this illustration sums up Aztec history and relates the founding of the city of Tenochtitlán.*

It was also a very violent civilisation. The continual military campaigns were for the most part undertaken in the service of the Aztec gods – for every aspect of Aztec history, culture, and society can only be understood in terms of a tyrannical religion which left no space for anything resembling hope or even virtue in the Christian sense. The main purpose of war-making was to seize prisoners who could be used in the ritual sacrifices. About 20,000 people were thus slaughtered every year in the service of magic. The Aztecs believed that the Sun and the Earth (both considered gods) required constant replenishment with human blood, or else the world's mechanism would cease to function. The slaughter also had a straightforward nutritional use, for only the victims' hearts were reserved for the gods' consumption. Human legs, arms and rumps were treated much as we treat butcher's meat, and sold retail at Aztec markets, for ordinary consumption.

Beside the priestly and the warrior castes, there were also castes of artisans and traders, organised into a set of guilds. The main market of the empire was at Tlatelolco, Tenochtitlán's twin town, founded in 1358, where merchandise of every sort, brought from the four corners of the Aztec empire, was traded. The records of the taxes levied by the imperial administrators of the Tlatelolco market have survived, and give a good picture of the wealth and variety of trade in the Aztec empire: gold, silver, jade, shells, feathers for ceremonial wear, ceremonial garb, shields, raw cotton for spinning, cocoa beans, coats, blankets, embroidered cloth, etc.

The empire and the whole of Aztec civilisation collapsed in the early sixteenth century. "Stout" Cortez, accompanied by a mere handful of men armed with guns, landed at Vera Cruz and marched towards the highlands. He gained the support of tribes that were the Aztecs' enemies or their subjects, and from them acquired supplies and reinforcements. After a violent struggle, Cortez seized Tenochtitlán on 13 August 1521, and destroyed Aztec civilisation for ever.

## AZTEC WRITING

At the time of the Spanish Conquest, Mexican script was a mixture of ideographic and phonetic representation, with some more or less "pictorial" signs designating directly the beings, objects or ideas that they resembled, and others (including the same ones) standing for the sound of the thing that they represented. Names of people and places were written in the manner of a rebus or puzzle (rather approximate ones, in fact, since the writing took no account of case-endings). For example, the name of the city of Coatlan (literal meaning: "snake-place") was represented by the drawing of a snake (= *coatl*) together with the sign for "teeth", pronounced *tlan*. The name of the city of Coatepec (literal meaning: "snake-mountain-place") was represented similarly by a snake (= *coatl*) together with the sign for "mountain" (*tepetl*).

FIG. 22.8. *Examples of Aztec names written in the form of a rebus*

Aztec script is used in a number of Mexican documents written just before and just after the Spanish conquest. Some of these deal with matters of religion, ritual, prophecy, and magic; others are narratives of real or mythical history (tribal migrations, foundations of cities, the origins and history of different dynasties, etc.); and others are registers of the vast taxes paid in kind (goods, food supplies, and men) by the subject cities to the lords of Tenochtitlán.

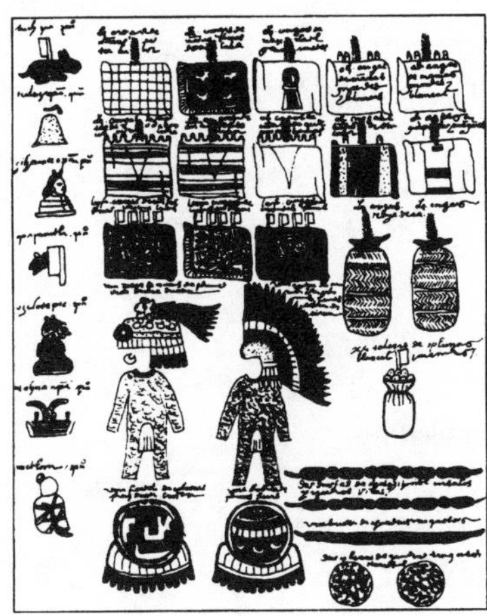

FIG. 22.9. *Codex Mendoza (folio 52 r), showing the tributes to be paid by seven Mexican cities to Tenochtitlán*

The most important by far of these Aztec documents in the Codex Mendoza, drawn up by order of Don Antonio de Mendoza, the first Viceroy of New Spain, and sent to the court of Spain. It contains three parts, dealing respectively with the conquests of the Aztecs, the taxes that they levied on each of the conquered towns, and with the life-cycle of an Aztec, from birth through education, punishment, recreation, military insignia, battles, the genealogy of the royal family, and even the ground-plan of Motecuhzoma's palace . . . It was written in a period of ten days (since the fleet was about to put to sea) in the native language and script, but with a simultaneous commentary on the meaning of every detail in Spanish. And it is largely thanks to the Spanish commentary that we can now seek to understand Aztec numerals . . .



## HOW THE MAYA DID THEIR SUMS

Most of what could be "read" in Maya texts and inscriptions until very recently consists of numerical, astronomical and calendrical information. However, before we can approach Mayan arithmetic, we need to know what their oral numbering system was.

Like all the other peoples of pre-Columbian Central America, the Maya counted not to base 10, but to base 20. As we now know, this was due to their ancestors' habits of using their toes as well as their fingers as a model set.

The language of the Maya and various dialects of it are still in use nowadays in the Mexican states of Yucatan, Campeche, and Tabasco, in a part of the Chiapas and the region of Quintana Roo, in most of Guatemala and in parts of Salvador and Honduras. The names of the numbers are as follows:

| | | | | |
|---|---|---|---|---|
| 1 | *hun* | 11 | *buluc* | |
| 2 | *ca* | 12 | *lahca* | (*lahun* + *ca* = 10 + 2) |
| 3 | *ox* | 13 | *ox-lahun* | (3 + 10) |
| 4 | *can* | 14 | *can-lahun* | (4 + 10) |
| 5 | *ho* | 15 | *ho-lahun* | (5 + 10) |
| 6 | *uac* | 16 | *uac-lahun* | (6 + 10) |
| 7 | *uuc* | 17 | *uuc-lahun* | (7 + 10) |
| 8 | *uaxac* | 18 | *uaxac-lahun* | (8 + 10) |
| 9 | *bolon* | 19 | *bolon-lahun* | (9 + 10) |
| 10 | *lahun* | | | |

FIG. 22.10A.

The units up to and including 10 thus have their own separate names, and above that number are made of additive compounds that rely on 10 as an auxiliary base. The one exception is the name of the number 11, *buluc,* which was probably invented to avoid confusion of a regular form *hun-lahun,* "one + ten" with *hun-lahun,* in the meaning "a ten".

Numbers from 20 to 39 are expressed as follows:

| | | |
|---|---|---|
| 20 | *kun kal* | score (*hun uinic*, "one man", in some dialects) |
| 21 | *hun tu-kal* | one (after) twentieth |
| 22 | *ca tu-kal* | two (after) twentieth |
| 23 | *ox tu-kal* | three (after) twentieth |
| 24 | *can tu-kal* | four (after) twentieth |
| 25 | *ho tu-kal* | five (after) twentieth |
| 26 | *uac tu-kal* | six (after) twentieth |
| 27 | *uuc tu-kal* | seven (after) twentieth |
| 28 | *uaxac tu-kal* | eight (after) twentieth |
| 29 | *bolon tu-kal* | nine (after) twentieth |
| 30 | *lahun ca kal* | ten-two-twenty |
| 31 | *buluc tu-kal* | eleven (after) twentieth |
| 32 | *lahca tu-kal* | twelve (after) twentieth |
| 33 | *ox-lahun tu-kal* | thirteen (after) twentieth |

| | | |
|---|---|---|
| 34 | *can-lahun tu-kal* | fourteen (after) twentieth |
| 35 | *holhu ca kal* | fifteen-two-twenty |
| 36 | *uac-lahun tu-kal* | sixteen (after) twentieth |
| 37 | *uuc-lahun tu-kal* | seventeen (after) twentieth |
| 38 | *uaxac-lahun tu-kal* | eighteen (after) twentieth |
| 39 | *bolon-lahun tu-kal* | nineteen (after) twentieth |

FIG. 22.10B.

So, as a general rule, these numbers are formed by inserting the ordinal prefix *tu* between the name of the unit and the name of the base, 20. But there are two exceptions:

30 "ten-two-twenty", instead of "ten (after) twentieth"
35 "fifteen-two-twenty", instead of "fifteen (after) twentieth"

These two anomalies cannot be explained by addition or by subtraction, since 35 is neither 15 + (2 × 20) nor (2 × 20) – 15. Moreover, the irregularity is repeated in the next sequence of numbers, which begin "one-three-twenty", "two-three-twenty" and so on.

| | | |
|---|---|---|
| 40 | *ca kal* | two score |
| 41 | *hun tu-y-ox kal* | one – third score |
| 42 | *ca tu-y-ox kal* | two – third score |
| 43 | *ox tu-y-ox kal* | three – third score |
| 44 | *can tu-y-ox kal* | four – third score |
| | ........................ | |
| 58 | *uaxac-lahun tu-y-ox kal* | eighteen – third score |
| 59 | *bolon-lahun tu-y-ox kal* | nineteen – third score |
| 60 | *ox kal* | three score |
| 61 | *hun tu-y-can kal* | one – fourth score |
| 62 | *ca tu-y-can kal* | two – fourth score |
| | ........................ | |
| 78 | *uaxac-lahun tu-y-can kal* | eighteen – fourth score |
| 79 | *bolon-lahun tu-y-can kal* | nineteen – fourth score |
| 80 | *can kal* | four score |
| 81 | *hun tu-y-ho-kal* | one – fifth score |
| 82 | *ca tu-y-ho-kal* | two – fifth score |
| | ........................ | |
| 98 | *uaxac-lahun tu-y-ho-kal* | eighteen – fifth score |
| 99 | *bolon-lahun tu-y-ho-kal* | nineteen – fifth score |
| 100 | *ho kal* | five score |
| | ........................ | |
| 400 | *hun bak* | one four-hundreder |
| 8,000 | *hun pic* | one eight-thousander |
| 160,000 | *hun calab* | one hundred-and-sixty-thousander |

FIG. 22.10C.

To work out how such a numbering system might have come into being, we have to imagine something like the following scene taking place several thousand years ago somewhere in Central America.

FIG. 22.11.

As they prepare to set off to fight a skirmish, warriors line up a few men to serve as "counting machines" or model sets, and one of the men proceeds to check off the number of warriors in the group. As the first one files past, the checker touches the first finger of the first "counting machine", then for the second he touches the second finger, and so on up to the tenth. The "accountant" then moves on to the toes of the first model set, up to the tenth, which matches the twentieth warrior that has filed past. For the next man, the accountant proceeds in exactly the same way using the second of the "counting machines", and when he gets to the last toe of the second man, he will have checked off forty warriors. He moves on to the third man, which would take him up to sixty, and so on until the count is finished.

Let us suppose that there are 53 men in the group. The accountant will reach the third toe of the first foot of the third man, and will announce the result of the count in something like the following manner: "There are as many warriors as make three toes on the first foot of the third man". But the result could also be expressed as: "Two hands and three toes of the third man" or even "ten-and-three of the third twenty". If applied to English, such a system would produce a set of number-names of the following sort:

| | | | |
|---|---|---|---|
| 1 | one | 11 | ten-one |
| 2 | two | 12 | ten-two |
| 3 | three | 13 | ten-three |
| 4 | four | 14 | ten-four |
| 5 | five | 15 | ten-five |
| 6 | six | 16 | ten-six |
| 7 | seven | 17 | ten-seven |
| 8 | eight | 18 | ten-eight |
| 9 | nine | 19 | ten-nine |
| 10 | ten | 20 | one man |

| Style A | | Style B |
|---|---|---|
| one after the first man | 21 | one of the second man |
| two after the first man | 22 | two of the second man |
| three after the first man | 23 | three of the second man |
| four after the first man | 24 | four of the second man |
| five after the first man | 25 | five of the second man |
| six after the first man | 26 | six of the second man |
| seven after the first man | 27 | seven of the second man |
| eight after the first man | 28 | eight of the second man |
| nine after the first man | 29 | nine of the second man |
| ten after the first man | 30 | ten of the second man |
| ten-one after the first man | 31 | ten-one of the second man |
| ten-two after the first man | 32 | ten-two of the second man |
| ten-three after the first man | 33 | ten-three of the second man |
| ten-four after the first man | 34 | ten-four of the second man |
| ten-five after the first man | 35 | ten-five of the second man |
| ... | ... | ... |
| ten-nine after the first man | 39 | ten-nine of the second man |
| two men | 40 | two men |
| one after the second man | 41 | one of the third man |
| two after the second man | 42 | two of the third man |
| three after the second man | 43 | three of the third man |
| ... | ... | ... |
| ten-one after the second man | 51 | ten-one of the third man |
| ten-two after the second man | 52 | ten-two of the third man |
| ten-three after the second man | 53 | ten-three of the third man |
| ... | ... | ... |
| ten-nine after the second man | 59 | ten-nine of the third man |
| three men | 60 | three men |
| one after the third man | 61 | one of the fourth man |
| two after the third man | 62 | two of the fourth man |
| ... | ... | ... |
| nineteen after the third man | 79 | nineteen of the fourth man |
| four men | 80 | four men |

FIG. 22.12.

It is now easy to see how the irregularities of the Maya number-names arose. The numbers 21 to 39 (except 30 and 35) are expressed in terms of Style A: 21 = *hun tu-kal* = "one (after) the twentieth" or "one (after the) first twenty", 39 = *bolon-lahun tu-kal* = "nine-ten (after the) twentieth" or "nine-ten (after the) first twenty"; whereas the numbers from 41 to 59, 61 to 79, etc. as well as the numbers 30 and 35, are expressed in terms of Style B: 30 = *lahun ca kal* = "ten-two-twenty" or "ten of the second twenty", and so forth.

The Maya were not alone in counting in this way. The number 53, for instance, is expressed as follows:

- by the Inuit of Greenland, as *inûp pingajugsane arkanek pingasut*, literally, "of the third man, three on the first foot";
- by the Ainu of Japan and Sakhalin, as *wan-re wan-e-re-hotne*, literally "three and ten of the third twenty" [see K. C. Kyosuke and C. Mashio (1936)];
- by the Yoruba (Senegal and Guinea) as *eeta laa din ogota*, literally "ten and three before three times twenty" [see C. Zaslavsky (1973)];
- and other instances of similar systems can be found amongst the Yedo (Benin) and the Tamanas of the Orinoco (Venezuela).

## THE "ORDINARY" NUMBERS OF THE MAYA

Now that we can see the reasons for the irregularities of the Maya number-name system, we can try to grasp their written numerals. Or rather, we would have been able to, had the Spanish Inquisition not stupidly destroyed almost every trace of it. So we are forced to take a step backwards.

Amongst the cultures of pre-Columbian Central America there are four main types of writing system: Maya, Zapotec (in the Oaxaca Valley), Mixtec (southwest Mexico), and Aztec (around Mexico City). Zapotec is the oldest, probably dating from the sixth century BCE, and Aztec is the most recent (see above). Now, although these scripts served to represent languages belonging to quite different linguistic families, they possess a number of graphical features in common, including (as far as Aztec, Mixtec and Zapotec are concerned) the basic features of numerical notation.

In Aztec vigesimal numerals, for instance, the unity was represented by a dot or circle, the base by a hatchet, the square of the base (20 × 20 = 400) by a sign resembling a feather, the cube of the base (20 × 20 × 20 = 8,000) by a design symbolising a purse.

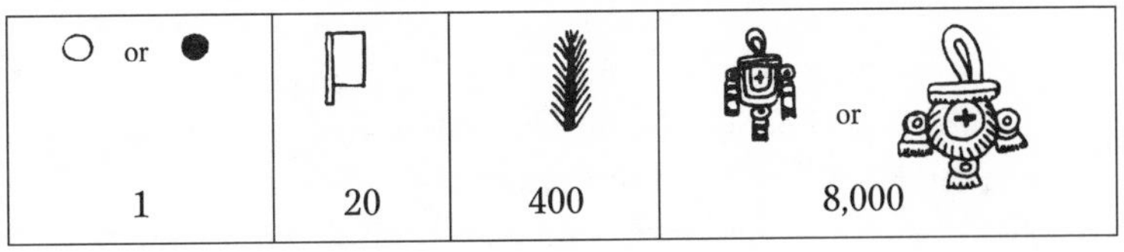

| or | | | or |
|---|---|---|---|
| 1 | 20 | 400 | 8,000 |

FIG. 22.13. *Aztec numerals*

The numeral system relied on addition: that is to say, numbers were expressed by repeating the component figures as many times as necessary. To express 20 shields, 100 sacks of cocoa beans, or 200 pots of honey, for example, one, five or ten "hatchets" were attached to the pictogram for the relevant object:

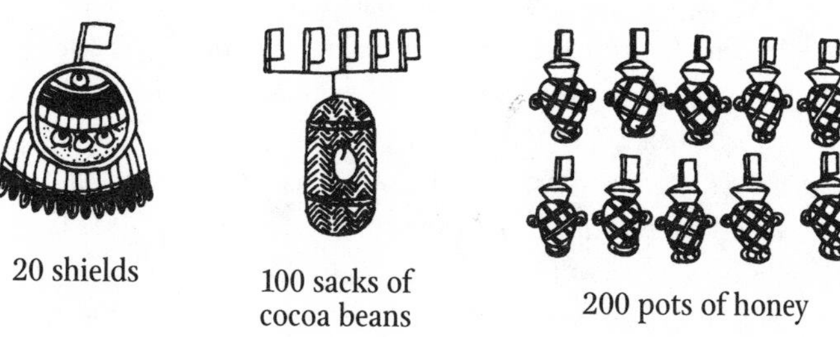


FIG. 22.14.

To record 400 embroidered cloaks, 800 deerskins or 1,600 cocoa bean-pods, one, two or four "feather" signs were similarly attached to the respective object-sign:

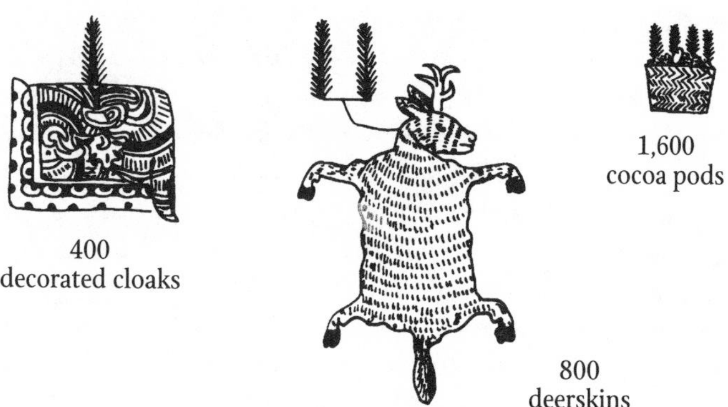


FIG. 22.15.

This was the way that the scribe of the Codex Mendoza recorded the taxes that were paid once, twice or four times a year by the subject-cities to the Aztec lords of Tenochtitlán. The page shown in Fig. 22.9 above gives the taxes due from seven cities in one province, and expresses them as follows:

1. *Left column*: the names of the seven cities, expressed by combinations of signs in the manner of a rebus:

FIG. 22.16A.

2. *Line 1, horizontally:*

FIG. 22.16B.

- 400 cloaks of black-and-white chequered cloth
- 400 cloaks of red-and-white embroidered cloth (worn by the lords of Tenochtitlán)
- 400 loincloths
- 2 sets of 400 white cloaks, size 4 *braza* (a unit of length indicated by the finger-sign)

3. *Line 2*

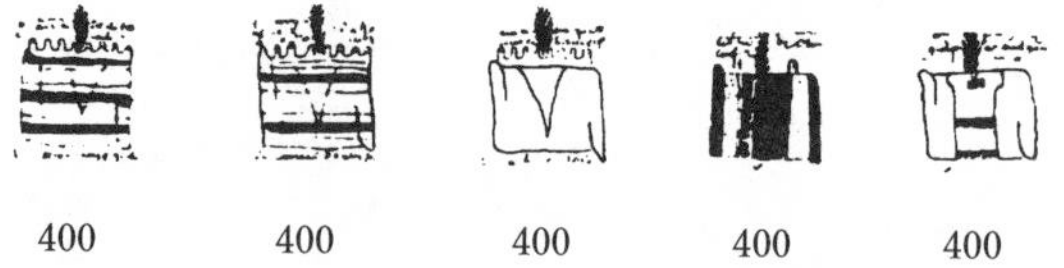


FIG. 22.16C.

- 2 sets of 400 orange-and-white-striped cloaks, size 8 *braza*
- 400 white cloaks, size 8 *braza*
- 400 polychrome cloaks, size 2 *braza*
- 400 women's skirts and tunics

4. *Line 3*

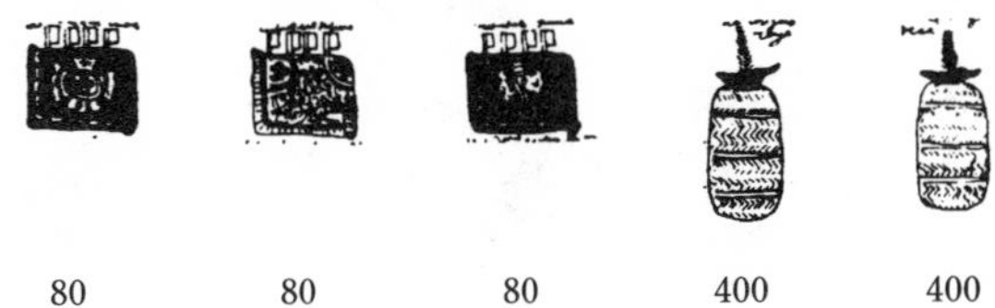


FIG. 22.16D.

- 3 sets of 80 coloured and embroidered cloaks (as worn by the leading figures of the capital)
- 2 sets of 400 bundles of dried peppers (used amongst other things to punish young people for breaking rules)

5. *Line 4*

FIG. 22.16E.

- 2 ceremonial costumes, 20 sacks of down, and 2 strings of jade pearls

6. *Last line*

FIG. 22.16F.

- 2 shields, a string of turquoise, and 2 plates with turquoise incrustation

The Codex Telleriano Remensis, another post-conquest document in Aztec script, also provides examples of numerals:

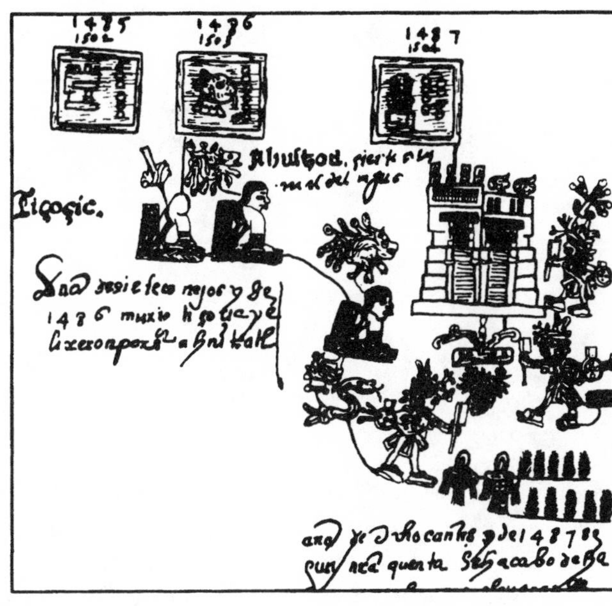

FIG. 22.17. *Detail from a page of the Aztec Codex Telleriano Remensis*

What this page says in effect is that 20,000 men from the subject provinces were sacrificed in 1487 CE to consecrate a new building. The number was written by the native scribe thus:

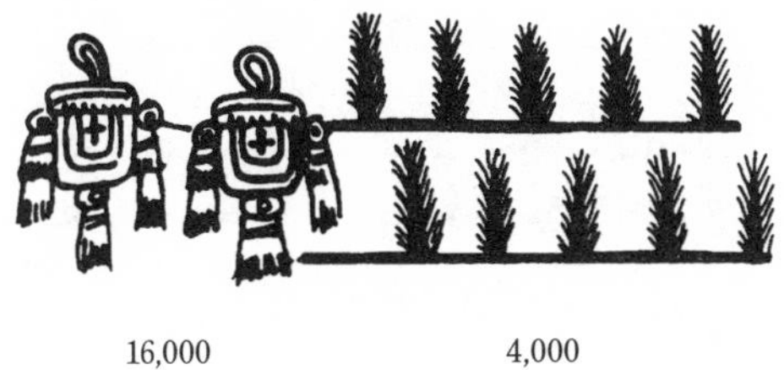


FIG. 22.18.

The Spanish annotator, however, made a mistake in transcribing this number: as he did not know the meaning of the two purses worth 8,000 each, he "translated" only the ten feathers, giving a total of 4,000.

Aztec numerals were identical to those of the Zapotecs and Mixtecs, as the following painting shows. It was done in Zapotec by order of the Spanish colonial authorities in Mexico in 1540 CE and shows the numbering conventions common to Zapotec, Mixtec and Aztec cultures:

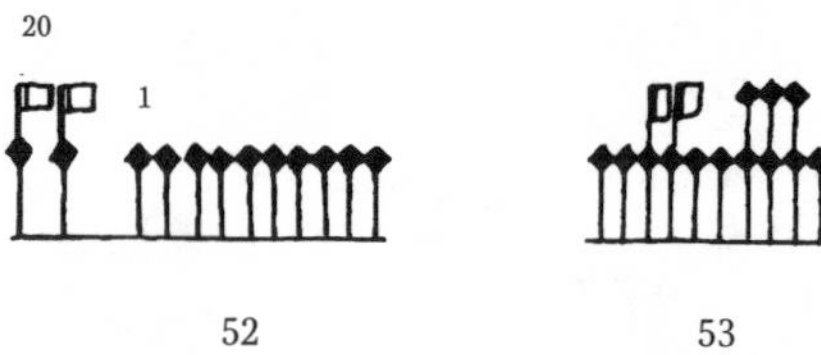


FIG. 22.19. *Numerical representations from a Zapotec painting made by order of the Spanish colonial authorities in 1540. It shows graphical conventions common to Zapotec, Mixtec, and Aztec numeral systems.*

So it seems certain that "ordinary" Maya numerals must also have been strictly vigesimal and based on the additive principle. It can be safely assumed that a circle or dot was used to represent the unity (the sign is common to all Central American cultures, and derives from the use of the cocoa bean as the unit of currency), that there was a special sign, maybe similar to the "hatchet" used by other Central American cultures, for the base (20), and other specific signs for the square of the base (400) and the cube (8,000), etc.

As we shall see below, it is also quite probable that, like the Zapotecs, the Maya introduced an additional sign for 5, in the form of a horizontal line or bar.

Even though no trace of it remains, we can reasonably assume that the Maya had a numeral system of this kind, and that intermediate numbers were figured by repeating the signs as many times as was needed. But that kind of numeral system, even if it works perfectly well as a recording device, is of no use at all for arithmetical operations. So we must assume that the Maya and other Central American civilisations had an instrument similar to the abacus for carrying out their calculations.

The Inca of South America certainly did have a real abacus, as shown in Fig. 22.20. The Spaniards were amazed at the speed with which Inca accountants could resolve complex calculations by shifting ears of maize, beans or pebbles around twenty "cups" (in five rows of four) in a tray or table, which could be made of stone, earthenware or wood, or even just laid out in the ground. Inca civilisation was obviously quite different from the Maya world, but it did have one thing in common: a method of recording numbers and tallies (the *quipus*, or knotted string) that was entirely unsuitable for performing arithmetical operations. For that reason the Inca were obliged to devise a different kind of operating tool.

FIG. 22.20. *Document proving the use of the abacus amongst the Peruvian and Ecuadorian Incas. It shows a* quipucamayoc *manipulating a* quipu *and on his right a counting table. From the Peruvian Codex of Guaman Poma de Ayala (16th century), Royal Library, Copenhagen*

## THE PLACE-VALUE SYSTEM OF "LEARNED" MAYA NUMERALS

The only numerical expressions of the Maya that have survived are in fact not of the ordinary or practical kind, but astronomical and calendrical calculations. They are to be found in the very few Maya manuscripts that exist, and most notably in the Dresden Codex, an astronomical treatise copied in the eleventh century CE from an original that must have been three or four centuries older.

What is quite remarkable is that Maya priests and astronomers used a numeral system with base 20 which possessed a true zero and gave a specific value to numerical signs according to their position in the written expression. The nineteen first-order units of this vigesimal system were represented by very simple signs made of dots and lines: one, two, three and four dots for the numbers 1 to 4; a line for 5, one, two, three and four dots next to the line for 6 to 9; two lines for 10, and so on up to 19:

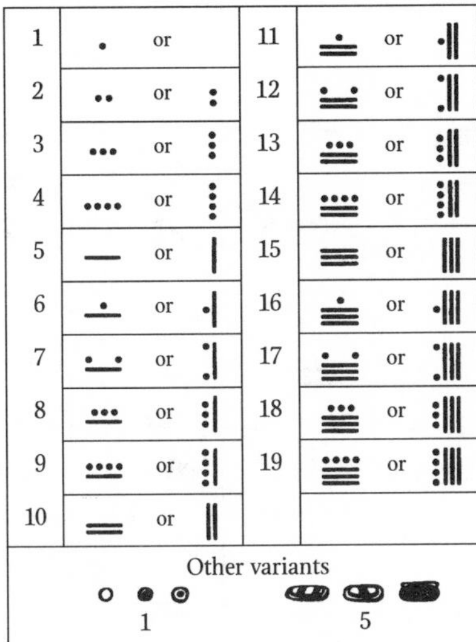


FIG. 22.21. *The first nineteen units in the numeral system of the Maya priests*

Numbers above 20 were laid out vertically, with as many "floors" as there were orders of magnitude in the number represented. So for a number involving two orders, the first order-units were expressed on the first or "bottom floor" of the column, and the second-order units on the "second floor". The numbers 21 (= 1 × 20 + 1) and 79 (3 × 20 + 19) were written thus:

FIG. 22.22. FIG. 22.23.

The "third floor" should have been used for values twenty times as great as the "second floor" in a regular vigesimal system. Just as in our decimal system the third rank (from the right) is reserved for the hundreds (10 × 10 = 100), so in Maya numbering the third level should have counted the "four hundreds" (20 × 20 = 400). However, in a curious irregularity that we will explain below, the third floor of Mayan astronomical numerals actually represented multiples of 360, not 400. The following expression:

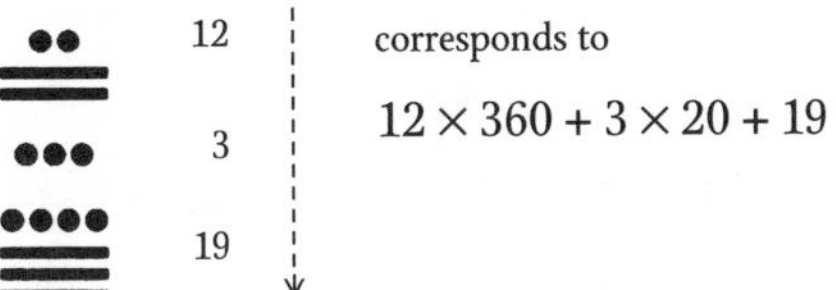


FIG. 22.24.

actually meant 12 × 360 + 3 × 20 + 19 = 4,399, and not 12 × 400 + 3 × 20 + 19 = 4,879!

Despite this, higher floors in the column of numbers were strictly vigesimal, that is to say represented numbers twenty times as great as the immediately preceding floor. Because of the irregularity of the third position, the fourth position gave multiples of 7,200 (360 × 20) and the fifth gave multiples of 144,000 (20 × 7,200) – and not of 8,000 and 160,000.

A four-place expression can thus be resolved by means of three multiplications and one addition, thus:

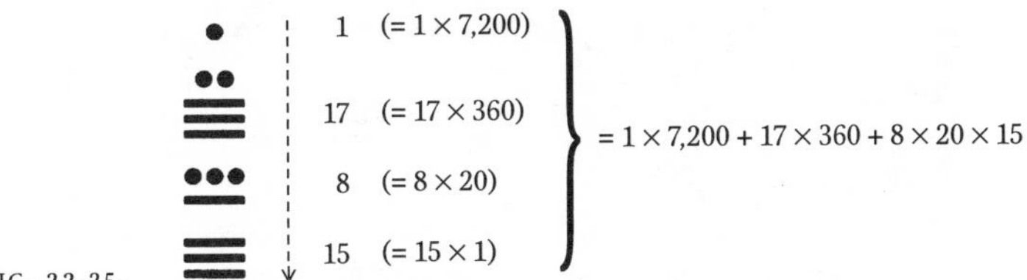


FIG. 22.25.

So that each numeral would be in its right place even when there were no units to insert in one or another of the "floors", Mayan astronomers invented a zero, a concept which they represented (for reasons we cannot pierce) by a sign resembling a snail-shell or sea-shell.

For instance, a number which we write as 1,087,200 in our decimal place-value system and which corresponds in Mayan orders of magnitude to 7 × 144,000 + 11 × 7,200 and no units of any of the lower orders of 360, 20 or 1, would be written in Maya notation thus:

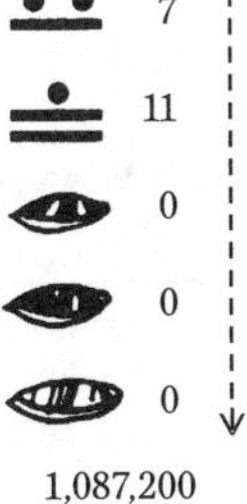


FIG. 22.26.

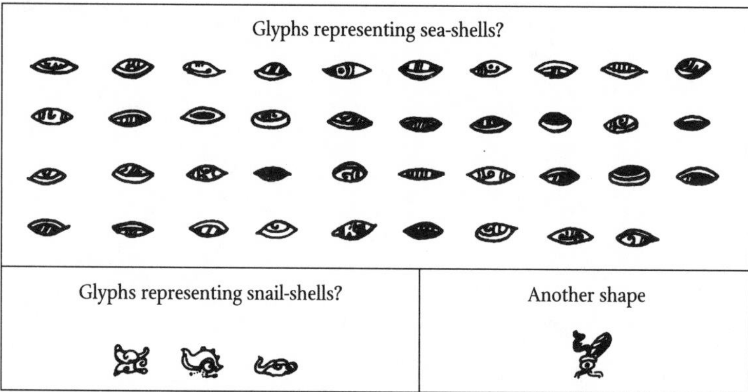


FIG. 22.27.

We can see the system in operation in these very interesting numerical expressions in the Dresden Codex:

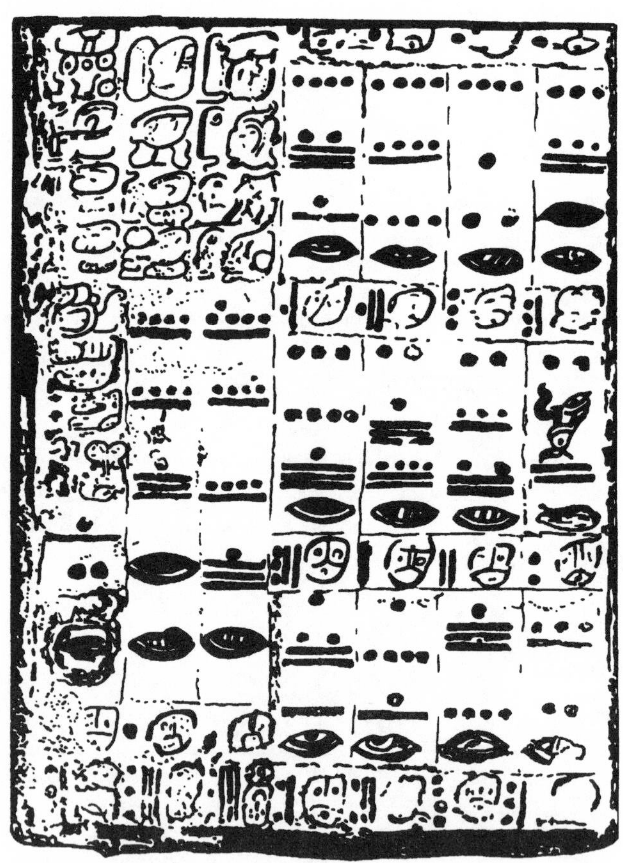

FIG. 22.28. *The Dresden Codex, p. 24 (part). Sächsische Landesbibliothek, Dresden*

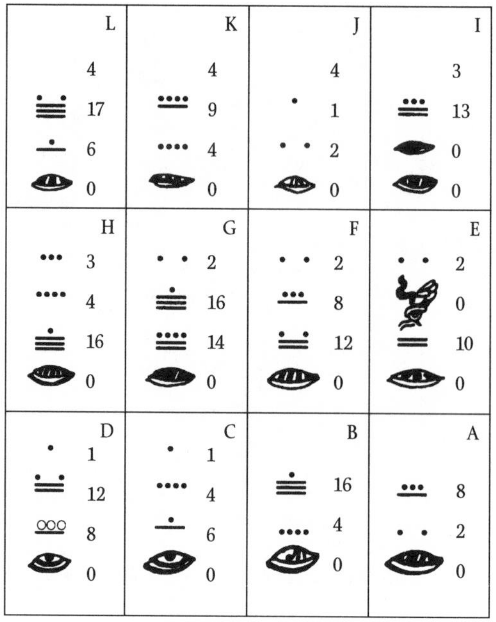

| L | K | J | I |
|---|---|---|---|
| 4<br>17<br>6<br>0 | 4<br>9<br>4<br>0 | 4<br>1<br>2<br>0 | 3<br>13<br>0<br>0 |
| **H** | **G** | **F** | **E** |
| 3<br>4<br>16<br>0 | 2<br>16<br>14<br>0 | 2<br>8<br>12<br>0 | 2<br>0<br>10<br>0 |
| **D** | **C** | **B** | **A** |
| 1<br>12<br>8<br>0 | 1<br>4<br>6<br>0 | 16<br>4<br>0 | 8<br>2<br>0 |

FIG. 22.29. *Transcriptions of the numerals on the right-hand side of Fig. 22.28*

Each of these expressions in Mayan astronomical notation refers to a number of days (we know this from the context) and gives the following set of equivalences:

| | | | | |
|---|---|---|---|---|
| A = | [8; 2; 0] = | 2,920 = | 1 × 2,920 = | 5 × 584 |
| B = | [16; 4; 0] = | 5,840 = | 2 × 2,920 = | 10 × 584 |
| C = | [1; 4; 6; 0] = | 8,760 = | 3 × 2,920 = | 15 × 584 |
| D = | [1; 12; 8; 0] = | 11,680 = | 4 × 2,920 = | 20 × 584 |
| E = | [2; 0; 10; 0] = | 14,600 = | 5 × 2,920 = | 25 × 584 |
| F = | [2; 8; 12; 0] = | 17,520 = | 6 × 2,920 = | 30 × 584 |
| G = | [2; 16; 14; 0] = | 20,440 = | 7 × 2,920 = | 35 × 584 |
| H = | [3; 4; 16; 0] = | 23,360 = | 8 × 2,920 = | 40 × 584 |
| I = | [3; 13; 0; 0] = | 26,280 = | 9 × 2,920 = | 45 × 584 |
| J = | [4; 1; 2; 0] = | 29,200 = | 10 × 2,920 = | 50 × 584 |
| K = | [4; 9; 4; 0] = | 32,120 = | 11 × 2,920 = | 55 × 584 |
| L = | [4; 17; 6; 0] = | 35,040 = | 12 × 2,920 = | 60 × 584 |

So this series is nothing other than a table of the synodic revolutions of Venus, calculated by Mayan astronomers as 584 days.

This gives us two indisputable proofs of the mathematical genius of Maya civilisation:

- it shows that they really did invent a place-value system;
- it shows that they really did invent zero.

These are two fundamental disoveries that most civilisations failed to make, including especially Western European civilisation, which had to wait until the Middle Ages for these ideas to reach it from the Arabic world, which had itself acquired them from India.

One problem remains: why was this system not strictly vigesimal, like the Mayas' oral numbering? For instead of using the successive powers of 20 (1, 20, 400, 8,000, etc.), it used orders of magnitude of 1, 20, $18 \times 20 = 360$, $18 \times 20 \times 20 = 7{,}200$, etc. In short, why was the third "floor" of the system occupied by the irregular number 360?

If Maya numerals had been strictly vigesimal, then its zero would have acquired operational power: that is to say, adding a zero at the end of a numerical string would have multiplied its value by the base. That is how it works in our system, where the zero is a true operational sign. For instance, the number 460 represents the product of 46 multiplied by the base. For the Maya, however, [1; 0; 0] is not the product of [1; 0] multiplied by the base, as the first floor gives units, the second floor gives twenties, but the third floor gives 360s. [1; 0] means precisely 20; but [1; 0; 0] is not 400 ($20 \times 20 + 0 +$ ), but 360. The number 400 had to be written as [1; 2; 0] or ($1 \times 360 + 2 \times 20 + 0$):

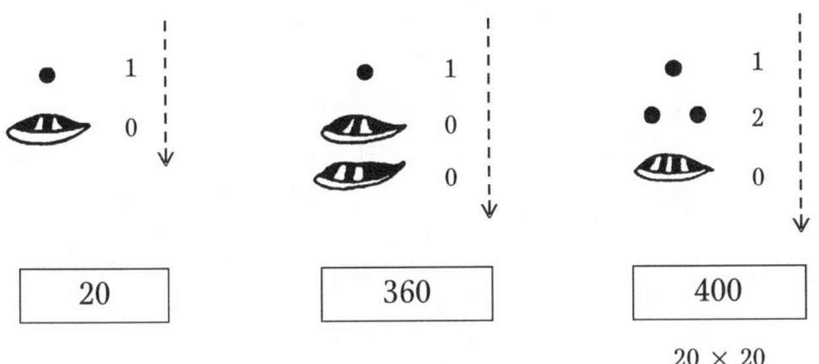


Fig. 22.30.

This anomaly deprived the Maya zero of any operational value, and prevented Mayan astronomers from exploiting their discovery to the full. We must therefore not confuse the Maya zero with our own, for it does not fulfil the same role at all.

## A SCIENCE OF THE HIGH TEMPLES

To understand the odd anomaly of the third position in the Maya place-value system we have to delve deep into the very sources of Maya mathematics, and make a long but fascinating detour into Maya mysticism and its reckoning of time.

Maya learned numerals were not invented to deal with the practicalities of everyday reckoning – the business of traders and mere mortals – but to meet the needs of astronomical observation and the reckoning of time. These numerals were the exclusive property of priests, for Maya civilisation made the passing of time the central matter of the gods.

Maya science was practised in the high temples: astronomy was what the priests did. Mayan achievements in astronomy, including the invention of one of the best calendars the world has ever seen, were part and parcel of their mystical and religious beliefs.

The Maya did not think of time as a purely abstract means of ordering events into a methodical sequence. Rather, they viewed it as a supernatural phenomenon laden with all-powerful forces of creation and destruction, directly influenced by gods with alternately kindly and wicked intentions. These gods were associated with specific numbers, and took on shapes which allowed them to be represented as hieroglyphs. Each division of the Maya calendar (days, months, years, or longer periods) was thought of as a "burden" borne on the back of one or another of the divine guardians of time. At the end of each cycle, the "burden" of the next period of time was taken over by the god associated with the next number. If the coming cycle fell to a wicked god, then things would get worse until such time as a kindly god was due to take over. These curious beliefs supported the popular conviction that survival was impossible without learned mediators who could interpret the intentions of the irascible gods of time. The astronomer-priests alone could recognise the attributes of the gods, plot their paths across time and space, and thus determine times that would be controlled by kindly gods, or (as was more frequent) times when the number of kindly gods would exceed that of evil gods. It was an obsession for calculating periods of luck and good fortune over long time-scales, in the hope that such foreknowledge would enable people to turn circumstances to their advantage. [See C. Gallenkamp (1979)]

FIG. 22.31. *The cyclical conception of events in the Mayas' mystical thinking. The inexorable cycle of Chac, god of rain, planting a tree, followed by Ah Puch, god of death, who destroys it, and by Yum Kax, god of maize and of agriculture, who restores it. From the Codex Tro-Cortesianus, copy from Girard (1972), p. 241, Fig. 61*

The priests were thus the possessors of the arcana of time and of the foretelling of the gods bearing the burden of particular times. Mysticism, religion and astronomy formed a single, unitary sphere which gave the priestly caste enormous power over the people, who needed priestly mediation in order to learn of the mood of the gods at any given moment. So despite its amazing scientific insights, Mayan astronomy was very different from what we now imagine science to be: as Girard puts it [R. Girard (1972)], its main purpose was to give mythical interpretations of the magical powers that rule the Universe.

## THE MAYA CALENDAR

The Maya had two calendars, which they used simultaneously: the *Tzolkin* – the "sacred almanac" or "magical calendar" or "ritual calendar", used for religious purposes; and the *Haab*, which was a solar calendar.

The religious year of the Maya consisted of twenty cycles of thirteen days, making 260 days in all. It had a basic sequence of twenty named days in fixed order:

| | | | |
|---|---|---|---|
| *Imix* | *Cimi* | *Chuen* | *Cib* |
| *Ik* | *Manik* | *Eb* | *Caban* |
| *Akbal* | *Lamat* | *Ben* | *Etznab* |
| *Kan* | *Muluc* | *Ix* | *Cauac* |
| *Chicchan* | *Oc* | *Men* | *Ahau* |

Each day had its distinct hieroglyph, which also represented directly the corresponding deity or sacred animal or object. As J. E. Thompson explains, prayers were addressed to the days, each of which was the incarnation of a divinity, such as the sun, the moon, the god of maize, the god of death, the Jaguar, etc.

FIG. 22.32. *Hieroglyphs for the twenty days of the Maya calendar, with their names in the Yucatec language. [See Gallenkamp (1979), Fig. 9; Peterson (1961), Fig. 55]*

Each of the days was also associated with a number-sign, in the range 1 to 13 (itself associated with thirteen Maya gods of the "upper world" or *Oxlahuntiku*).

In the first cycle, the first day was associated with the number 1, the second day with the number 2, and so on to the thirteenth day. The numbering then started over, so that the fourteenth day was associated with the number 1, the fifteenth with the number 2, and the last day of the first cycle had number 7.

The second cycle thus began with 8 and reached 13 with the sixth day, so that the numbering began again at 1 with the seventh day of the second cycle.

Thus it took thirteen cycles for the numbering to come back to where it started, with day one counting once again as 1. As there are $13 \times 20$ possible pairings of the sets 1–13 and 1–20, the whole series of cycles lasted 260 days.

Each day of the religious year therefore had a unique name consisting of its hieroglyph together with its number resulting from the cyclical recurrence explained above. So a day-hieroglyph plus number gives an unambiguous identification of any day in the religious year. The following expressions, for instance:

13 CHUEN

4 IMIX

FIG. 22.33.

specify the 91st and 121st days of a religious year that begins on *1 Imix*. (Fig. 22.34 below shows the whole cycle.)

| | I | II | III | IV | V | VI | VII | VIII | IX | X | XI | XII | XIII |
|---|---|---|---|---|---|---|---|---|---|---|---|---|---|
| IMIX | 1 | 8 | 2 | 9 | 3 | 10 | 4 | 11 | 5 | 12 | 6 | 13 | 7 |
| IK | 2 | 9 | 3 | 10 | 4 | 11 | 5 | 12 | 6 | 13 | 7 | 1 | 8 |
| AKBAL | 3 | 10 | 4 | 11 | 5 | 12 | 6 | 13 | 7 | 1 | 8 | 2 | 9 |
| KAN | 4 | 11 | 5 | 12 | 6 | 13 | 7 | 1 | 8 | 2 | 9 | 3 | 10 |
| CHICCHAN | 5 | 12 | 6 | 13 | 7 | 1 | 8 | 2 | 9 | 3 | 10 | 4 | 11 |
| CIMI | 6 | 13 | 7 | 1 | 8 | 2 | 9 | 3 | 10 | 4 | 11 | 5 | 12 |
| MANIK | 7 | 1 | 8 | 2 | 9 | 3 | 10 | 4 | 11 | 5 | 12 | 6 | 13 |
| LAMAT | 8 | 2 | 9 | 3 | 10 | 4 | 11 | 5 | 12 | 6 | 13 | 7 | 1 |
| MULUC | 9 | 3 | 10 | 4 | 11 | 5 | 12 | 6 | 13 | 7 | 1 | 8 | 2 |
| OC | 10 | 4 | 11 | 5 | 12 | 6 | 13 | 7 | 1 | 8 | 2 | 9 | 3 |
| CHUEN | 11 | 5 | 12 | 6 | 13 | 7 | 1 | 8 | 2 | 9 | 3 | 10 | 4 |
| EB | 12 | 6 | 13 | 7 | 1 | 8 | 2 | 9 | 3 | 10 | 4 | 11 | 5 |
| BEN | 13 | 7 | 1 | 8 | 2 | 9 | 3 | 10 | 4 | 11 | 5 | 12 | 6 |
| IX | 1 | 8 | 2 | 9 | 3 | 10 | 4 | 11 | 5 | 12 | 6 | 13 | 7 |
| MEN | 2 | 9 | 3 | 10 | 4 | 11 | 5 | 12 | 6 | 13 | 7 | 1 | 8 |
| CIB | 3 | 10 | 4 | 11 | 5 | 12 | 6 | 13 | 7 | 1 | 8 | 2 | 9 |
| CABAN | 4 | 11 | 5 | 12 | 6 | 13 | 7 | 1 | 8 | 2 | 9 | 3 | 10 |
| ETZNAB | 5 | 12 | 6 | 13 | 7 | 1 | 8 | 2 | 9 | 3 | 10 | 4 | 11 |
| CAUAC | 6 | 13 | 7 | 1 | 8 | 2 | 9 | 3 | 10 | 4 | 11 | 5 | 12 |
| AHAU | 7 | 1 | 8 | 2 | 9 | 3 | 10 | 4 | 11 | 5 | 12 | 6 | 13 |

FIG. 22.34. *The 260 consecutive days of the Maya liturgical year*

Each day of the religious year had its own specific character. Some were propitious for marriages or military expeditions, others ruled out such events. More generally, an individual's character and prospects were indissolubly linked to the character of the day of his birth, a belief that is still held by many Central American peoples, notably in the Guatemalan uplands.

Why did the pre-Columbian civilisations of Central America choose 260 as the number of days in their liturgical calendars? F. A. Peterson (1961) pointed out that the difference between the religious year (260) and the solar year (365) is 105 days. Moreover, between the tropic of Cancer and the tropic of Capricorn, the sun is at the zenith twice in every year, at intervals of 105 and 260 days precisely. At Copán, an ancient Maya city in Honduras, the relevant dates for the sun passing through its zenith are 13 August and 30 April. The rainy season begins straight after the sun passes through its "spring" zenith; 105 days later, the sun passes through its "autumn" zenith. So the year could be divided into a period of planting and growth that lasted 105 days, and then a period of harvesting and religious feasts that lasted 260.

This astronomical observation, even if it has not been accepted as the ultimate source of the Maya calendar, is certainly very interesting. Unfortunately the correlation of the sun's zenith with the rainy season only fits at Copán, which is on the fringes of the Maya area.

Other scholars have pointed out that 260 must be thought of as the product of 13 and 20, the divine (since there are 13 divinities in the "upper world" of the Maya) and the human (since Maya numbering is vigesimal, and the name of the number 20, *uinic*, means " a man").

Alongside the ritual calendar, the Maya used a solar-year calendar called the *Haab*, and referred to as the "secular" or "civil" or "approximate" calendar. It had a year of 365 days divided into eighteen *uinal* (twenty-day periods), plus a short "extra" period of five days added at the end of the eighteenth *uinal*. The names of the Maya twenty-day "months" were:

| | | |
|---|---|---|
| *Pop* | *Yaxkin* | *Mac* |
| *Uo* | *Mol* | *Kankin* |
| *Zip* | *Chen* | *Muan* |
| *Zotz* | *Yax* | *Pax* |
| *Tzec* | *Zac* | *Kayab* |
| *Xul* | *Ceh* | *Cumku* |

These names referred to various agricultural or religious events, and they were represented by the hieroglyphs of the tutelary god or animal-spirit associated with the event.

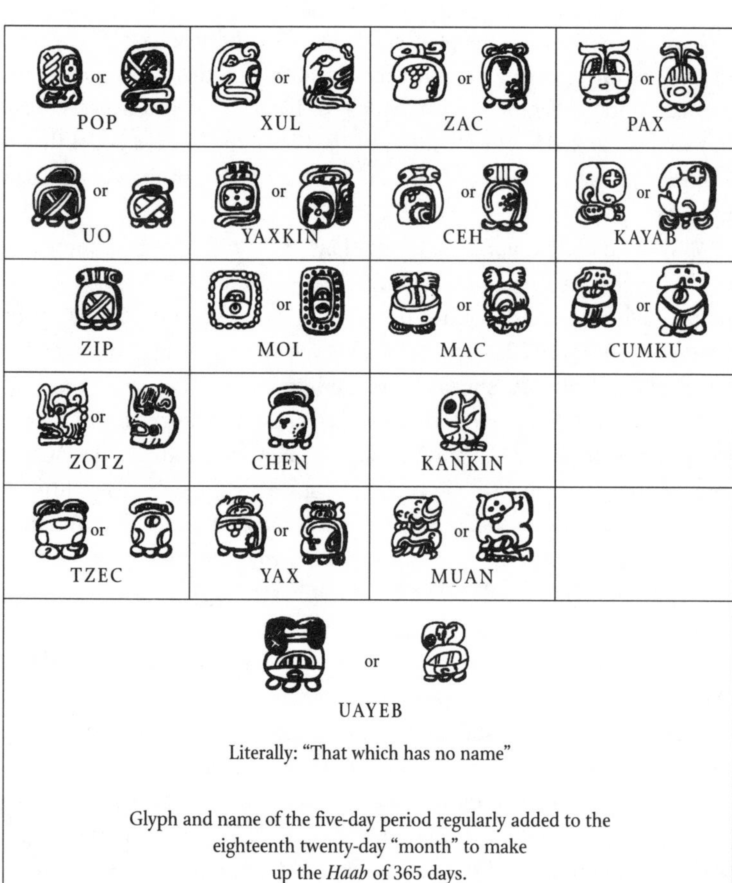


FIG. 22.35. *Glyphs and names of the eighteen 20-day "months" of the Maya solar calendar. [See Gallenkamp (1979), p. 80; Peterson (1961), p. 225]*

The "extra" five-day period was called *Uayeb*, meaning "The one that has no name", and it was represented by a glyph associated with the idea of chaos, disaster and corruption. They were thought of as "ghost" days, and considered empty, sad and hostile to human life. Anyone born during *Uayeb* was destined to have bad luck and to remain poor and miserable all his life long. Peterson quotes Diego de Landa, who reported that during *Uayeb* the Maya never washed, combed their hair, or picked their nits; they did no regular or demanding work, for fear that something untoward would happen to them.

The first day of each "month", including *Uayeb*, was represented by the glyph for the "month", that is to say the sign for its tutelary divinity, together with a special sign:

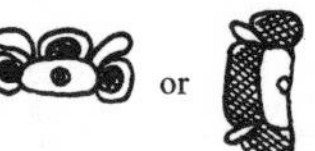


FIG. 22.36.

This sign, which is usually translated by specialists as "0", signified that the god who had carried the burden of time up to that point was passing it on to the following month-god. So since *Zip* and *Zotz* are the names of two consecutive "months" in the "approximate" Maya calendar, the hieroglyph:

FIG. 22.37.

meant that *Zip* was handing over the weight of time to *Zotz*.

As a result the remaining days of each "month", including *Uayeb*, were numbered from 1 to 19, with the second day having the number 1, the third day the number 2, and so on (see Fig. 22.40 below). As a result, the following "date" in the secular or civil calendar:

FIG. 22.38.

signified not the fourth, but the fifth day in the twenty-day "month" of *Xul*!

Each of the twenty days of the basic series (laid out in Fig. 22.32 above) kept exactly the same rank-number in each of the eighteen "months" of the civil or secular year. If the "zero day" of the first month of the year was *Eb*, for example, then the "zero day" of the following seventeen months was also *Eb*. But because of the extra five days added on in each annual cycle, the day-names stepped back by five positions each year. So, for example, if *Ahau* was day 8 in year N, it became day 3 in year N + 1, day 18 in year N + 2, day 13 in year N + 3, and day 8 again in year N + 4. The full cycle thus took four years to complete, and only in the fifth year did the correspondence between the names and the numbers of the days of the "months" return to its starting position.

Within the system, there were only four day-names from the basic series that could correspond to the calendrical expression:

FIG. 22.39.

These "new year days" were *Eb*, *Caban*, *Ik* and *Manik*, and they were referred to as the "year-bearer" days.

| POP | UO | ZIP | ZOTZ | TZEC | XUL | YAXKIN | MOL | CHEN | YAX | ZAC | CEH | MAC | KANKIN | MUAN | PAX | KAYAB | CUMKU | UAYEB |
|---|---|---|---|---|---|---|---|---|---|---|---|---|---|---|---|---|---|---|
| 0 | 0 | 0 | 0 | 0 | 0 | 0 | 0 | 0 | 0 | 0 | 0 | 0 | 0 | 0 | 0 | 0 | 0 | 0 |
| 1 | 1 | 1 | 1 | 1 | 1 | 1 | 1 | 1 | 1 | 1 | 1 | 1 | 1 | 1 | 1 | 1 | 1 | 1 |
| 2 | 2 | 2 | 2 | 2 | 2 | 2 | 2 | 2 | 2 | 2 | 2 | 2 | 2 | 2 | 2 | 2 | 2 | 2 |
| 3 | 3 | 3 | 3 | 3 | 3 | 3 | 3 | 3 | 3 | 3 | 3 | 3 | 3 | 3 | 3 | 3 | 3 | 3 |
| 4 | 4 | 4 | 4 | 4 | 4 | 4 | 4 | 4 | 4 | 4 | 4 | 4 | 4 | 4 | 4 | 4 | 4 | 4 |
| 5 | 5 | 5 | 5 | 5 | 5 | 5 | 5 | 5 | 5 | 5 | 5 | 5 | 5 | 5 | 5 | 5 | 5 | |
| 6 | 6 | 6 | 6 | 6 | 6 | 6 | 6 | 6 | 6 | 6 | 6 | 6 | 6 | 6 | 6 | 6 | 6 | |
| 7 | 7 | 7 | 7 | 7 | 7 | 7 | 7 | 7 | 7 | 7 | 7 | 7 | 7 | 7 | 7 | 7 | 7 | |
| 8 | 8 | 8 | 8 | 8 | 8 | 8 | 8 | 8 | 8 | 8 | 8 | 8 | 8 | 8 | 8 | 8 | 8 | |
| 9 | 9 | 9 | 9 | 9 | 9 | 9 | 9 | 9 | 9 | 9 | 9 | 9 | 9 | 9 | 9 | 9 | 9 | |
| 10 | 10 | 10 | 10 | 10 | 10 | 10 | 10 | 10 | 10 | 10 | 10 | 10 | 10 | 10 | 10 | 10 | 10 | |
| 11 | 11 | 11 | 11 | 11 | 11 | 11 | 11 | 11 | 11 | 11 | 11 | 11 | 11 | 11 | 11 | 11 | 11 | |
| 12 | 12 | 12 | 12 | 12 | 12 | 12 | 12 | 12 | 12 | 12 | 12 | 12 | 12 | 12 | 12 | 12 | 12 | |
| 13 | 13 | 13 | 13 | 13 | 13 | 13 | 13 | 13 | 13 | 13 | 13 | 13 | 13 | 13 | 13 | 13 | 13 | |
| 14 | 14 | 14 | 14 | 14 | 14 | 14 | 14 | 14 | 14 | 14 | 14 | 14 | 14 | 14 | 14 | 14 | 14 | |
| 15 | 15 | 15 | 15 | 15 | 15 | 15 | 15 | 15 | 15 | 15 | 15 | 15 | 15 | 15 | 15 | 15 | 15 | |
| 16 | 16 | 16 | 16 | 16 | 16 | 16 | 16 | 16 | 16 | 16 | 16 | 16 | 16 | 16 | 16 | 16 | 16 | |
| 17 | 17 | 17 | 17 | 17 | 17 | 17 | 17 | 17 | 17 | 17 | 17 | 17 | 17 | 17 | 17 | 17 | 17 | |
| 18 | 18 | 18 | 18 | 18 | 18 | 18 | 18 | 18 | 18 | 18 | 18 | 18 | 18 | 18 | 18 | 18 | 18 | |
| 19 | 19 | 19 | 19 | 19 | 19 | 19 | 19 | 19 | 19 | 19 | 19 | 19 | 19 | 19 | 19 | 19 | 19 | |

FIG. 22.40. *The 365 consecutive days of the Maya "civil" year*

| List of the 20 basic days | 1st year *U* | UAYEB | 2nd year *U* | | 3rd year *U* | | 4th year *U* | | 5th year *U* | UAYEB |
|---|---|---|---|---|---|---|---|---|---|---|
| Eb | 0 | 0 | . 15 | | . 10 | | . 5 | | 0 | 0 |
| Ben | 1 | 1 | . 16 | | . 11 | | . 6 | | 1 | 1 |
| Ix | 2 | 2 | . 17 | | . 12 | | . 7 | | 2 | 2 |
| Men | 3 | 3 | . 18 | | . 13 | | . 8 | | 3 | 3 |
| Cib | 4 | 4 | . 19 | UAYEB | . 14 | | . 9 | | 4 | 4 |
| Caban | 5 | | 0 . | 0 | . 15 | | . 10 | | 5 | |
| Etznab | 6 | | 1 . | 1 | . 16 | | . 11 | | 6 | |
| Cauac | 7 | | 2 . | 2 | . 17 | | . 12 | | 7 | |
| Ahau | 8 | | 3 . | 3 | . 18 | | . 13 | | 8 | |
| Imix | 9 | | 4 . | 4 | . 19 | UAYEB | . 14 | | 9 | |
| Ik | 10 | | 5 . | | 0 . | 0 | . 15 | | 10 | |
| Akbal | 11 | | 6 . | | 1 . | 1 | . 16 | | 11 | |
| Kan | 12 | | 7 . | | 2 . | 2 | . 17 | | 12 | |
| Chicchan | 13 | | 8 . | | 3 . | 3 | . 18 | | 13 | |
| Cimi | 14 | | 9 . | | 4 . | 4 | . 19 | UAYEB | 14 | |
| Manik | 15 | | 10 . | | 5 . | | 0 . | 0 | 15 | |
| Lamat | 16 | | 11 . | | 6 . | | 1 . | 1 | 16 | |
| Muluc | 17 | | 12 . | | 7 . | | 2 . | 2 | 17 | |
| Oc | 18 | | 13 . | | 8 . | | 3 . | 3 | 18 | |
| Chuen | 19 | | 14 . | | 9 . | | 4 . | 4 | 19 | |

*U: any given month of the 18-month year*

FIG. 22.41. *Successive positions of the twenty basic days in the Maya "civil" calendar*

## THE SACRED CYCLE OF MESO-AMERICAN CULTURES

The Maya, as we have seen, used two different calendars simultaneously, the *Tzolkin*, or religious calendar, of 260 days, and the *Haab*, or civil calendar, of 365 days. So to express a date in full, they combined the signs of its place in the religious calendar with the signs of its place in the civil year, thus:

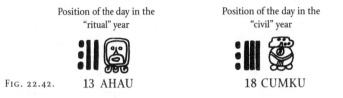

Position of the day in the "ritual" year

13 AHAU

Position of the day in the "civil" year

18 CUMKU

FIG. 22.42.

Since both these cycles permuted the days in regular recurrent order, the correspondence between the two calendars returned to its starting positions after a fixed period of time, which elementary arithmetic shows must be 18,980 days, or 52 "approximate" or civil years. In other words, the amount of time required for a given date in the civil calendar to match a given date in the religious calendar a second time round was equal to 52 years of 365 days or 72 years of 260 days.

You can imagine how this worked by thinking of a huge bicycle, with a chain wheel of 365 numbered teeth pulling round a sprocket with 260 numbered teeth. For the same chain link at the front sitting on tooth 1 to match the same chain link at the back also sitting on tooth 1, the pedals will have to turn 52 times, or (which is necessarily the same thing) the back wheel will have to go round 73 times.

The number of days in this cycle is equal to the lowest common multiple of 260 and 365. Since both these numbers are divisible by 5 and since 5 is moreover the highest common factor of 260 and 365, the number sought is

$$\frac{260 \times 365}{5} = 18{,}980 = 52 \text{ civil years} = 73 \text{ religious years}$$

That is the origin of the celebrated *sacred cycle of fifty-two years*, otherwise known as the Calendar Round, which played such an important role in Maya and Aztec religious life. (The Aztecs, for example, believed that the end of each Round would be greeted by innumerable cataclysms and catastrophes; so at the approach of the fateful date, they sought to appease the gods by making huge human sacrifices to them, in the hope of being allowed to live on through another cycle.)

We must mention, finally, that Maya astronomers also took the Venusian calendar into consideration. They had observed that after each period of 65

Venusian years, the start of the solar year, of the religious year and of the Venusian year all coincided precisely with the start of a new sacred cycle of 52 "civil" years. Such a remarkable occurrence was celebrated with enormous festivities.

## TIME AND NUMBERS ON MAYA *STELAE*

Alongside their two calendars, the Maya also used a third and rather amazing way of calculating the passage of time on their *stelae* or ceremonial columns. This "Long Count", as it is called by Americanists, began at zero at the date of 13 *baktun*, 4 *ahau*, 8 *cumku*, corresponding quite precisely, according to the concordance established by J. E. Thompson (1935), to 12 August 3113 BCE in the Gregorian calendar. It is generally assumed that this date corresponded to the Mayas' calculation of the creation of the world or of the birth of their gods [S. G. Morley (1915)]. However, this kind of reckoning did not use solar years, nor lunar years, nor even the revolutions of Venus, but multiples of recurrent cycles.

Its basic unit was the "day" and an approximate "year" of 360 days. Time elapsed since the start of the Mayan era was reckoned in *kin* ("day"), *uinal* (20-day "month"), *tun* (360-day "year"), *katun* (20-"year" period), *baktun* (400-"year" period), *pictun* (8,000-"year" cycle), and so on as laid out in Fig. 22.43.

The *katun* (= 20 *tun*) obviously did not correspond exactly to twenty years as we reckon them, but to 20 years less 104.842 days; similarly, the *baktun* (= 20 *katun* = 400 *tun*) was not exactly 400 years, but 400 years less 2,096.84 days. However, Mayan astronomers were perfectly aware of the discrepancies and of the corrections needed to the "Long Count" to make it correspond properly to actual solar years.

| Order of magnitude | Names and definitions | | Equivalences | Number of days |
|---|---|---|---|---|
| First | *kin* | DAY | | 1 |
| Second | *uinal* | "MONTH" OF 20 DAYS | 20 *kin* | 20 |
| Third | *tun* | "YEAR" OF 18 "MONTHS" | 18 *uinal* | 360 |
| Fourth | *katun* | CYCLE OF 20 "YEARS" | 20 *tun* | 7,200 |
| Fifth | *baktun* | CYCLE OF 400 "YEARS" | 20 *katun* | 144,000 |
| Sixth | *pictun* | CYCLE OF 8,000 "YEARS" | 20 *baktun* | 2,880,000 |
| Seventh | *calabtun* | CYCLE OF 160,000 "YEARS" | 20 *pictun* | 57,600,000 |
| Eighth | *kinchiltun* | CYCLE OF 3,200,000 "YEARS" | 20 *calabtun* | 1,152,000,000 |
| Ninth | *alautun* | CYCLE OF 64,000,000 "YEARS" | 20 *kinchiltun* | 23,040,000,000 |

FIG. 22.43. *The units of computation of time used in Maya calendrical inscriptions (the "Long Count" system)*

As we have seen, when counting people, animals or objects, the Maya used a strictly vigesimal system (see Fig. 22.10 above); but their time-counting method had an irregularity at the level of the third order of magnitude, which made the whole system cease to be vigesimal:

| | | | |
|---|---|---|---|
| 1 *kin* | | 1 = | 1 day |
| 1 *uinal* | = 20 *kin* | 20 = | 20 days |
| 1 *tun* | = 18 *uinal* | 18 × 20 = | 360 days |
| 1 *katun* | = 20 *tun* | 20 × 18 × 20 = | 7,200 days |
| 1 *baktun* | = 20 *katun* | 20 × 20 × 18 × 20 = | 144,000 days |
| 1 *pictun* | = 20 *baktun* | 20 × 20 × 20 × 18 × 20 = | 2,880,000 days |

If they had used a *tun* of 20 instead of 18 *uinal*, that is to say, using a truly vigesimal system, then their "year" would have had 400 days, and would have thus been even further "out" from the true solar year than was the 360-day *tun* of their calendrical computations.

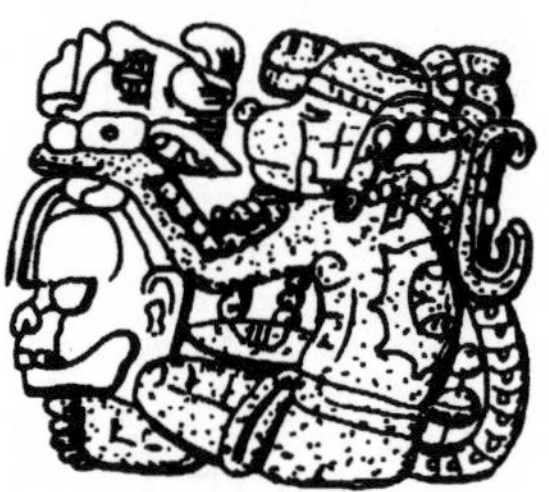

FIG. 22.44. *Detail of lintel 48 from Yaxchilán showing a bizarre representation of the expression "16* kin*" ("16 days"): a squatting monkey (a zoomorphic glyph sometimes associated with the word* kin*) holding the head of the god 6 in his hands and, in his legs, the death's-head which represents the number 10*

Each of these units of time had a special sign, which, like most Mayan hieroglyphs, had at least two different realisations, depending on whether it was being written with some kind of ink or paint on a codex, or carved in stone on a monument or ceremonial column. In other words, each of these units of time could be figured :

- by a relatively simple graphical sign, which could be more or less motivated by what it represented, or else an abstract geometrical shape;
- by the head of a god, a man, or an animal – otherwise called cephalomorphic glyphs, which were used for carved inscriptions;
- exceptionally, at Quiriguà and Palenque, by anthropomorphic glyphs, that is to say, by a god, man, or animal drawn in full.

| SYMBOLIC | CEPHALOMORPHIC | ANTHROPOMORPHIC |
|---|---|---|
| | | |

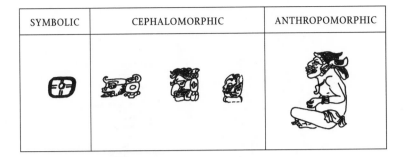

FIG. 22.45. *Various hieroglyphs for* kin, *"day"*

To represent the numerical coefficients of the units of time in the "Long Count", Maya scribes and sculptors used numerals which, like the unit-signs themselves, had more than one visual realisation.

FIG. 22.46. *Hieroglyphs for the units of time (from the Quiriguà* stelae*).*

Method One for showing the numbers was to use the cephalomorphic signs for the thirteen gods of the upper world (the set of gods and signs known as the *Oxlahuntiku*) for numbers 1 to 13. The maize-god, for instance, was associated with and therefore represented the number 5, and the god of death represented number 10.

FIG. 22.47. *Maya cephalomorphic numerals 1 to 19 (found on pieces of pottery and sculpture, on* stelae *J and F at Quiriguà, and on the "hieroglyphic staircase" at Palenque). [See Peterson (1961), p. 220, Fig. 52; Thompson (1960), p. 173, Fig. 13]*

For the numbers 14 to 19, however, the system used the numbers 4 to 9 with a modification that can be seen in the following figure:



VARIANTS OF THE GLYPH FOR "9"

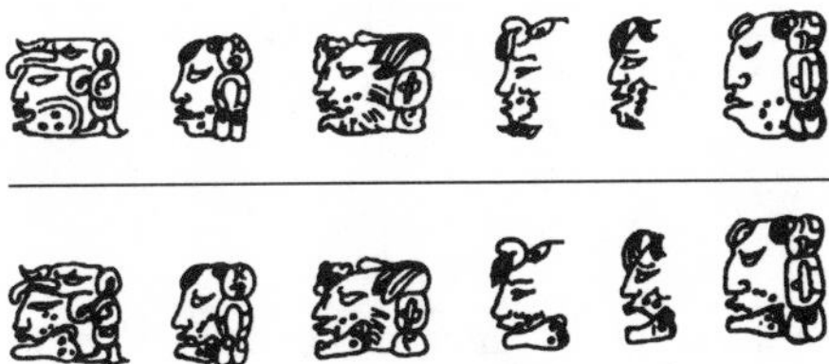

VARIANTS OF THE GLYPH FOR "19"

FIG. 22.48.

If you look closely you can see that the jawbone of the "nine-god" has been removed to make the glyph represent the number 19. Arithmetically, this is elementary, because, as the jawbone symbolised the god of death, it enabled an "extra ten" to be shown in the sign:

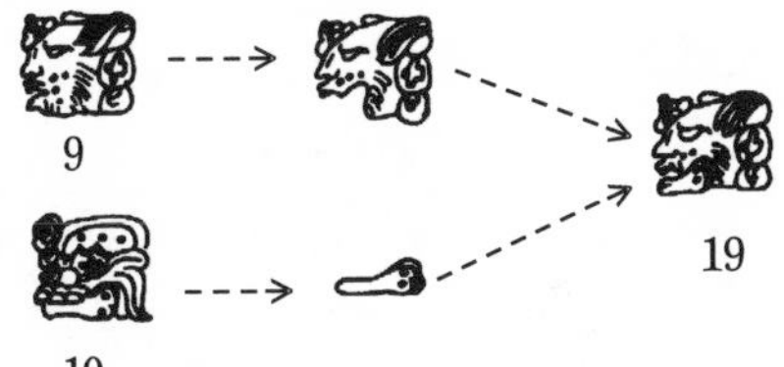


FIG. 22.49.

Method One was not used very often; more frequently, even in calendrical inscriptions, the dot-and-line system (see above, Fig. 22.21) is found.

In any case, dates and lengths of time could be expressed fairly simply within the systems explained so far. Americanists call these expressions "initial series". Our first example of an initial series comes from the "hieroglyphic staircase" of Palenque, where the numbers are represented by heads of the divinities, as shown in Fig. 22.50.

FIG. 22.50. *Initial series on the "hieroglyphic staircase" at Palenque. The date is given in cephalomorphic figures [From Peterson (1961), p. 232, Fig. 58]*

The inscription begins with the glyph called the "initial series start sign", or POP:

POP

Fig. 22.51.

This sign corresponds to the name of the divinity "responsible" for the "month" of the "civil" calendar on the day that the inscription was carved (or, to be more precise, the name of the month in the "secular" calendar in which the last day of the inscribed date falls).

Then, at the foot of the inscription, we can read the position of the date with respect to the "civil" and to the "religious" year, thus:

8 AHAU 13 POP

Fig. 22.52.

As for the number of days elapsed since the initial date of the Mayan era, it is expressed in the "Long Count" as follows:

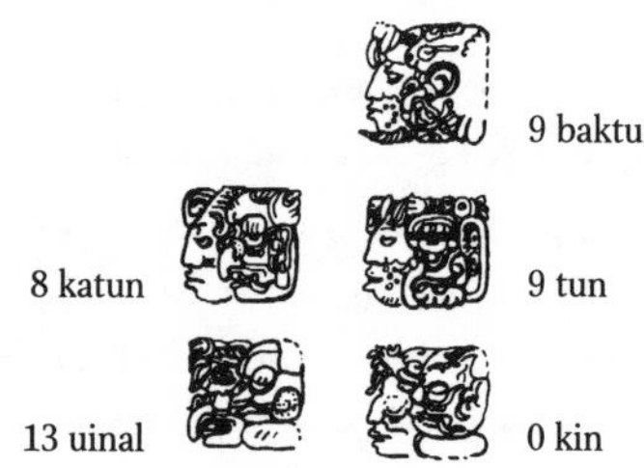


Fig. 22.53.

The date is read from top to bottom, and in descending order of magnitude of the counting units of the Maya calendar. It can be transcribed as follows:

| | | | |
|---|---|---|---|
| 9 *baktun* | = | 9 × 144,000 days ................ | 1,296,000 |
| 8 *katun* | = | 8 × 7,200 days ................. | 57,600 |
| 9 *tun* | = | 9 × 360 days ................... | 3,240 |
| 13 *uinal* | = | 13 × 20 days .................... | 260 |
| 0 *kin* | = | 0 × 1 day ...................... | 0 |
| Total | | | 1,357,100 days |

A fairly simple calculation reveals this to be the year 603 CE.

The Leyden Plate provides another example:

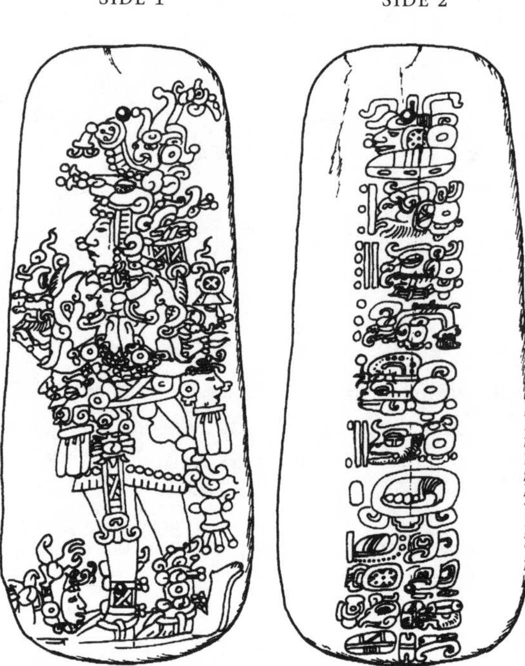


Fig. 22.54. *The Leyden Plate. This thin jade pendant, 21.5 cm high, was found in Guatemala, near Puerto Barrios, and is thought to have been carved at Tikal. On side 1 it shows a richly-clad Maya (probably a god) trampling a prisoner, and, on side 2, a date corresponding to 320 CE. Rijksmuseum voor Volkenkunde, Leyden, Holland*

As in the illustration from Palenque, this expression begins with an "introductory glyph", in this case the name of the god whose "burden" it was to carry the "month" of YAXKIN, during which the building on which this inscription was carved was completed:

YAXKIN

Fig. 22.55.

The date of completion is also expressed in terms of its position in the civil and religious calendars, thus:

1 EB 0 YAXKIN

Fig. 22.56.

As for the corresponding date in the "Long Count" system, it is given in this form:

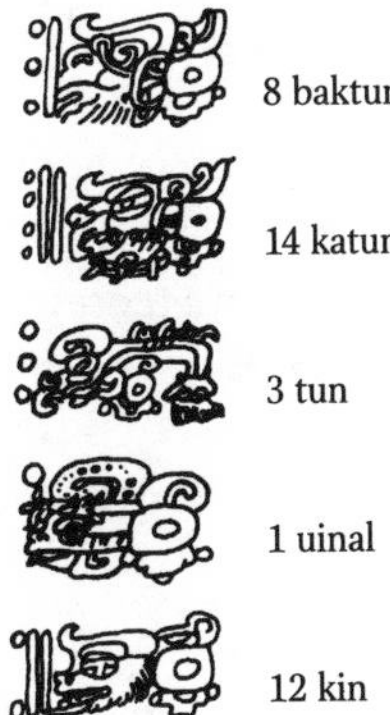


FIG. 22.57.

This date is also to be read from top to bottom in descending order of magnitudes, and from left to right within each glyph, and produces the following numbers:

| | | | |
|---|---|---|---|
| 8 *baktun* | = | 8 × 144,000 days ................ | 1,152,000 |
| 14 *katun* | = | 14 × 7,200 days ................. | 100,800 |
| 3 *tun* | = | 3 × 360 days ................... | 1,080 |
| 1 *uinal* | = | 1 × 20 days .................... | 20 |
| 12 *kin* | = | 12 × 1 day ..................... | 12 |
| Total | | | 1,253,912 days |

Once again, a simple calculation reveals that in view of the number of days since the beginning of the Mayan era this inscription was carved in the year 320 CE.

It was long thought that the Leyden Plate was the oldest dated artefact from Maya civilisation. However, in 1959, archaeological excavations in the ruins of the city of Tikal, in Guatemala, turned up an even older dated inscription. *Stela* no. 29 carries an inscription which can be translated as:

| | | | |
|---|---|---|---|
| 8 *baktun* | = | 8 × 144,000 days ............... | 1,152,000 |
| 12 *katun* | = | 12 × 7,200 days ................. | 86,400 |
| 14 *tun* | = | 14 × 360 days .................. | 5,040 |
| 8 *uinal* | = | 8 × 20 days .................... | 160 |
| 0 *kin* | = | 0 × 1 day ..................... | 0 |
| Total | | | 1,243,600 days |

which works out at the year 292 CE.

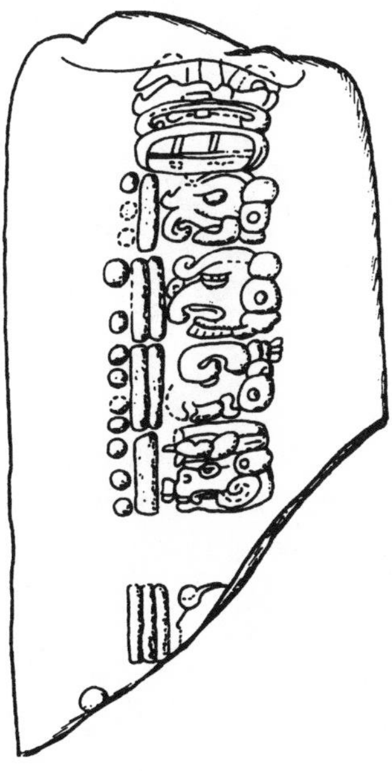

FIG. 22.58. *Side 2 of* stela *29 from Tikal (Guatemala), the oldest dated Mayan inscription found so far. The date written on it – usually transcribed as 8.12.14.8.0 – matches the year 292 CE. [See Shook (1960), p. 33]*

There are many other examples of calendrical inscriptions on the numerous *stelae* of the Maya, each one teeming with fantastical and elaborate signs. To conclude this section, let us look at one date found on *stela E* from Quiriguà.

The date of the *stela*'s erection begins on the top line with two glyphs: the first, on the left, is composed of the figure 9 with the head of the god representing *baktun*, and the other of the figure 17 with the head of the god representing *katun*. It then goes on, on the next line, with two compound glyphs signifying "0 *tun*" and "zero *uinal*" respectively; and, on the bottom line, the date ends with a sign meaning "0 *kin*".

9 BAKTUN
9 × 144,000
(= 1,296,000 days)

17 KATUN
17 × 7,200
(= 122,400 days)

0 TUN
0 × 360
(= 0 days)

0 UINAL
0 × 20
(= 0 days)

0 KIN
0 × 1
(= 0 days)

FIG. 22.59.

So the people who put up this column expressed the number of days elapsed since the start of the Mayan era up to the date on which they made this inscription, which is tantamount to expressing the latter date as:

| | | | |
|---|---|---|---|
| 9 *baktun* | = | 9 × 144,000 days ............... | 1,296,000 |
| 17 *katun* | = | 17 × 7,200 days ................. | 122,400 |
| 0 *tun* | = | 0 × 360 days .................. | 0 |
| 0 *uinal* | = | 0 × 20 days ................... | 0 |
| 0 *kin* | = | 0 × 1 day .................... | 0 |
| Total | | | 1,418,400 days |

So one million four hundred and eighteen thousand and four hundred days had passed since the "beginning of time" and, given that we know what the start-date was, we can calculate fairly easily that *stela E* at Quiriguà was completed on 24 January 771 CE.

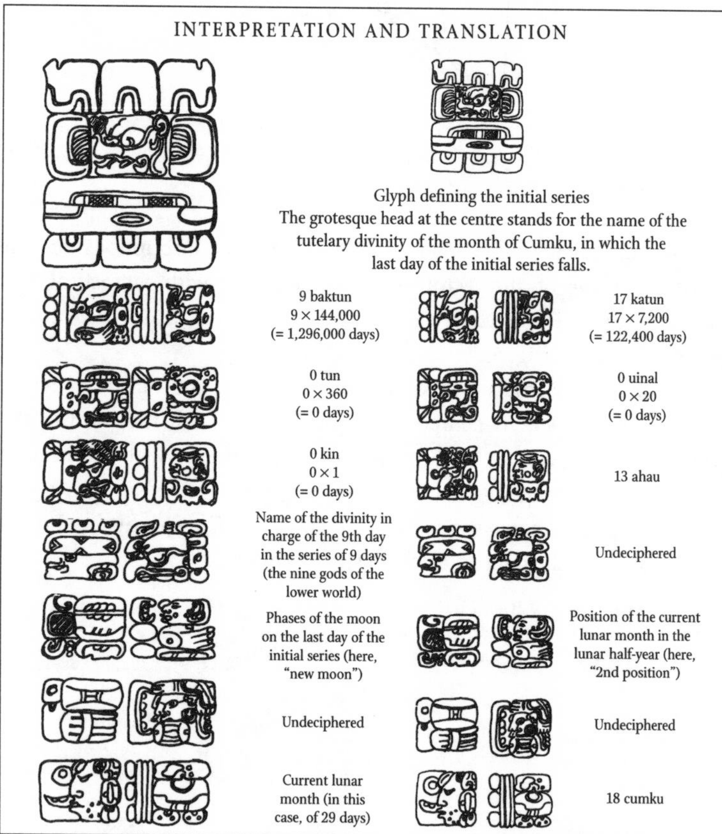


FIG. 22.60. *Detail from* stela E *at Quiriguà, giving an initial series together with a complementary series that provides other details on the date of the* stela*'s erection. The date is 9.17.0.0.0 and 13* ahau, *18* cumku, *which matches 24 January 771 CE, in the Gregorian calendar. [See Morley (1915), Fig. 25]*

We should note that these *stelae* contain some of the most interesting Mayan inscriptions that have been found. If we compare the oldest and newest dates found in particular places, we can get an idea of the duration of the great Maya cities. For example, at Tikal, the oldest date found is 292 CE and the latest is 869 CE; at Uaxactún, the limit-dates are 328 CE and 889 CE; at Copán, the relevant inscriptions are of 469 CE and 800 CE;* and so on. The important point in this long digression is to note that in their calendrical inscriptions the Maya represented the "zero", that is to say the absence of units in any one order, by glyphs and signs of the most diverse kinds.

FIG. 22.61. *Hieroglyphs for "zero" found on various Maya* stelae *and sculptures. Left to right: the first six, the commonest, are symbolic notations; the seventh and eighth are cephalomorphic, and the last is anthropomorphic. [See Peterson (1961), Fig. 51; Thompson (1960). Fig. 13]*

FIG. 22.62. *Detail of a plaque found at the Palenque Palace: an unusual anthropomorphic representation of the expression "0 kin" ("no days"). From Peterson (1961), fig. 14, p. 72.*

* See M. D. Coe, *op.cit.*, p. 68

## MAYA MATHEMATICS: A SCIENCE IN THE SERVICE OF ASTRONOMY AND MYSTICISM

The Maya system for counting time and for expressing the date did not really require a zero: the date expressed in Fig. 22.60 above, for example, could have been represented just as easily and just as unambiguously by:

9 *baktun*, 17 *katun*

as by the glyphs we actually have, which say

9 *baktun*, 17 *katun*, 0 *tun*, 0 *uinal*, 0 *kin*

So why did Maya calendrical computation bother to invent a zero?

The answers have to do with the religious, aesthetic and graphical ideas and customs of the Maya.

In religious terms, each of the time-units was imagined as a burden carried by one of the gods, the "tutelary god" of that cycle of time. At the end of the relevant cycle, the god passed on the burden of time to the god designated by the calendar as his successor.

On the date of "9 *baktun*, 11 *katun*, 7 *tun*, 5 *uinal*, and 2 *kin*", for instance, the god of the "days" carried the number 2, the god of the "months" carried the number 5, the god of the "years" carried the number 7, and so on, in this manner:

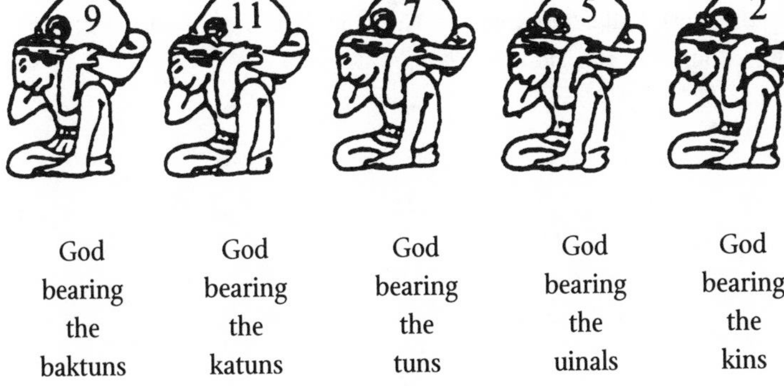


FIG. 22.63.

FIG. 22.64. Stela A *at Quiriguà. Erected in 775 CE, this column has gods carved on its front and back, and calendrical, astronomical and other glyphs carved on its other two sides. From Thompson (1960), Fig. 11, p. 163*

If we were to transpose this system to our own Gregorian calendar, we would need six gods to carry the "burden" of the date "31 December 1899". One god – the "day-god" – would "carry" the number 31; the second would bear the number 12, for the months; the third would carry the number 9, for the years; the fourth would "carry" the decades; and we would need two more, for the centuries and the millennia. At the end of the day of the 31 December 1899, these gods would have rested for a moment before setting off on a new cycle. The day-god would resume his burden, but with the number 1, and similarly for the month-god. But as the decade and the century would change (to 1900), the year-god and the decade-god would be released from their burdens for a period of time, the century-god would now bear the burden of the number 8, and the millennium-god would carry on with his 1 as he had been doing for the previous 900 years.

In Maya mystical thought, the fact that the gods occasionally had a rest from their burdens did not justify simply eradicating them from the representation of the task of carrying the burdens of time. Failing to put them in their right places in the inscription might have angered them! It would also have destroyed the absolute regularity of the system, in which calendrical expressions always ran from top to bottom in descending orders of magnitude. The aesthetically pleasing sequence of symbols in an unchanging order would have been altered if there had been no sign for zero. So we can say, in conclusion, that the demands of the writing system itself, the aesthetic appearance of inscriptions intended to be ceremonial, and a set of religious beliefs made the invention of a "zero-count" an absolute necessity (see Fig. 22.65).

Nonetheless, the calendrical system of the Maya is also part of a long and slow evolution leading towards the discovery of a place-value system. The Mayan units of time were always placed in precisely the same position in an inscription, with the same regularity as the tokens in an abacus or "counting table". And Mayan astronomer-priests did not fail to notice the arithmetical potential of their system.

When writing manuscripts, as opposed to inscriptions carved in stone, they eventually came to omit the glyphs representing the units of time (or the gods that were responsible for them), and wrote down only the corresponding numerical coefficients, since the order of the magnitudes was firm and fixed. So dates in the manuscripts are expressed just by numbers. For example, instead of writing the date "8 *baktun*, 11 *katun*, 0 *tun*, 14 *uinal*, 0 *kin*" as follows:

Fig. 22.65a.

they wrote (top to bottom, and with the numerical expressions rotated through 90°) simply:

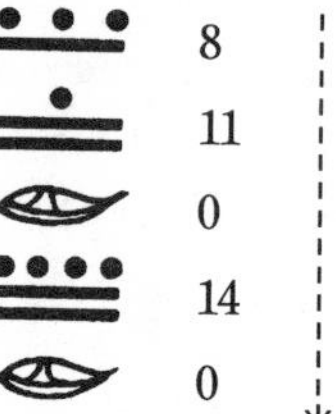


Fig. 22.65b.

Omitting the glyphs of the tutelary gods must have had less religious consequence in manuscripts than in the ceremonial and sacred *stelae*.

Taken outside of the context of mysticism and theology, the Maya system constitutes a remarkable written numeral system, incorporating both a true zero and the place-value principle. However, since it had been developed exclusively to express dates and to serve astronomical and calendrical computation, the system retained an irregular value in its third position, which, as we may recall, was $20 \times 18 = 360$, and not $20 \times 20 = 400$. This flaw made the system unsuitable for arithmetical operations and blocked any further mathematical development.

It is true that Maya scholars were concerned above all with matters religious and prophetic; but have not astrology and religion opened the path to philosophical and scientific developments in many places in the world? So we must pay homage to the generations of brilliant Mayan astronomer-priests who, without any Western influence at all, developed concepts as sophisticated as zero and positionality, and, despite having only the most rudimentary equipment, made astronomical calculations of quite astounding precision.

# CHAPTER 23

# THE FINAL STAGE OF NUMERICAL NOTATION

## THE LEGEND OF SESSA

In Arabic and Persian literature it is often written that the Indian world may glory in three achievements:

- the positional decimal notation and methods of calculation;
- the tales of the *Panchatantra* (from which probably came the well-known fable of *Kalila wa Dimna*);
- the *Shaturanja*, the ancestor of chess, about which a famous legend (adapted into modern terms) will give us an apt introduction to this very important chapter.

In order to prove to his contemporaries that a monarch, no matter how great his power, was as nothing without his subjects, an Indian Brahmin of the name of Sessa one day invented the game of *Shaturanja*.

This game is played between four players on an eight by eight chessboard, with eight pieces (King, Elephant, Horse, Chariot, and four Soldiers), which are moved according to points scored by rolling dice.

When the game was shown to the King of India, he was so amazed by the ingenuity of the game and by the myriad variety of its possible plays that he summoned the Brahmin, that he might reward him in person.

"For your extraordinary invention," said the King, "I wish to make you a gift. Choose your reward yourself and you shall receive it forthwith. I am so rich and so powerful that I can fulfil your wildest desire."

The Brahmin reflected on his reply, and then astonished everyone by the modesty of his request.

"My good Lord," he replied, "I wish that you would grant me as many grains of wheat as will fill the squares on the board: one grain for the first square, two for the second, four for the third, eight for the fourth, sixteen for the fifth, and so on, putting into each square double the number of grains that were put in the square before."

"Are you mad to suggest so modest a demand?" exclaimed the astonished King. "You could offend me with a request so unworthy of my generosity, and so trivial compared with all that I could offer you. But let it be! Since that is your wish, my servant will bring you your bag of wheat before nightfall."

The Brahmin made the merest smile, and withdrew from the Palace. That evening, the King remembered his promise and asked his Minister if the madman Sessa had received his meagre reward. "Lord," replied the Minister, "your orders are being carried out. The mathematicians of your august Court are at this moment working out the number of grains to give to the Brahmin."

The King's brow darkened. He was not used to such delay in obeying his orders. Before retiring to bed, the King asked once more whether the Brahmin had received his bag of grain.

"O King," replied the Minister, hesitating, "Your mathematicians have still not completed their calculations. They are working at it unceasingly, and they hope to finish before dawn."

The calculations proved to take far longer than had been expected. But the King, who did not wish to hear about the details, ordered that the problem should be solved before he awoke.

The next morning, however, his order of the night before remained unfulfilled, and the monarch, incensed, dismissed the calculators who had been working at the task.

"O good Lord," said one of his Counsellors, "you were right to dismiss these incompetents. They were using ancient methods! They are still counting on their fingers and moving counters on an abacus. I permit myself to suggest that the calculators of the central province of your Kingdom have for generations already been using a method far better and more rapid than theirs. They say it is the most expeditious, and the easiest to remember. Calculations which your mathematicians would need days of hard work to complete would trouble those of whom I speak for no more than a brief moment of time."

On this advice, one of these ingenious arithmeticians was brought to the Palace. He solved the problem in record time, and came to present his result to the King.

"The quantity of wheat which has been asked of you is enormous," he said in a grave voice. But the King replied that, no matter how huge the amount, it would not empty his granaries.

He therefore listened with amazement to the words of the sage.

"O Lord, despite all your great power and riches, it is not within your means to provide so great a quantity of grain. This is far beyond what we know of numbers. Know that, even if every granary in your Kingdom were emptied, you would still only have a negligible part of this huge quantity. Indeed, so great a quantity cannot be found in all the granaries of all the kingdoms of the Earth. If you desire absolutely to give this reward, you should begin by emptying all the rivers, all the lakes, all the seas and the oceans, melting the snows which cover the mountains and all the regions of the world, and turning all this into fields of corn. And then, when you have sown and reaped 73 times over this whole area, you will finally be quit of this huge debt. In fact, so huge a quantity of grain would have to be stored in a volume of twelve billion and three thousand million cubic metres, and require a granary 5 metres wide, 10 metres long and 300 million kilometres high (twice the distance from the Earth to the Sun)!"

The calculator revealed to the King the characteristics of the revolutionary method of numeration of his native region.

"The method of representing numbers traditionally used in your Kingdom is very complicated, since it is encumbered with a panoply of different signs for the units from 10 upwards. It is limited, since its largest number is no greater than 100,000. It is also totally unworkable, since no arithmetical operation can be carried out in this representation. On the other hand, the system which we use in our province is of the utmost simplicity and of unequalled efficacity. We use the nine figures 1, 2, 3, 4, 5, 6, 7, 8, and 9, which stand for the nine simple units, but which have different values according to the position in which they are written in the representation of a number, and we use also a tenth figure, 0, which means "null" and stands for units which are not present. With this system we can easily represent any number whatever, however large it may be. And this same simplicity is what makes it so superior, along with the ease which it brings to every arithmetical operation."

With these words, he then taught the King the principal methods of the calculation of the reward, and explained his operations as follows.

According to the demand of the Brahmin, we must place

1 grain of corn on the first square;
2 grains on the second square;
4 $(2 \times 2)$ on the third;
8 $(2 \times 2 \times 2)$ on the fourth;
16 $(2 \times 2 \times 2 \times 2)$ on the fifth;

and so on, doubling each time from one square to the next. On the sixty-fourth square, therefore, must be placed as many grains as there are units in the result of 63 multiplications by 2 (namely $2^{63}$ grains). So the quantity the Brahmin demanded is equal to the sum of these 64 numbers, namely

$$1 + 2 + 2^2 + 2^3 + \ldots + 2^{63}.$$

"If you add one grain to the first square," explained the calculator, "you would have two grains there, therefore $2 \times 2$ in the first two squares. By the third square you would then find a total of $2 \times 2 + 2 \times 2$ grains, or $2 \times 2 \times 2$ in all. By the fourth the total would be $2 \times 2 \times 2 + 2 \times 2 \times 2$, or $2 \times 2 \times 2 \times 2$ in all. Proceeding in this way, you can see that by the time you reach the last square of the board the total would be equal to the result of 64 multiplications by 2, or $2^{64}$. Now, this number is equal to the six-fold product of 10 successive multiplications by 2, further multiplied by the number 16:

$$\begin{aligned} 2^{64} &= 2^{10} \times 2^{10} \times 2^{10} \times 2^{10} \times 2^{10} \times 2^{10} \times 2^4 \\ &= 1{,}024 \times 1{,}024 \times 1{,}024 \times 1{,}024 \times 1{,}024 \times 1{,}024 \times 16 \end{aligned}$$

"And so," he concluded, "since this number has been obtained by adding one to the quantity sought, the total number of grains is equal to this number diminished by one grain. By completing these calculations in the way I have shown you, you may satisfy yourself, O Lord, that the number of grains demanded is exactly eighteen quadrillion, four hundred and forty-six trillion, seven hundred and forty-four billion, seventy-three thousand seven hundred and nine million, five hundred and fifty-one thousand, six hundred and fifteen (18,446,744,073,709,551,615)!"

"Upon my word!" replied the King, very impressed, "the game this Brahmin has invented is as ingenious as his demand is subtle. As for his methods of calculation, their simplicity is equalled only by their efficiency! Tell me now, my wise man, what must I do to be quit of this huge debt?" The Minister reflected a moment, and said:

"Catch this clever Brahmin in his own trap! Tell him to come here and count for himself, grain by grain, the total quantity of wheat which he has been so bold to demand. Even if he works without a break, day and night, one grain every second, he will gather up just one cubic metre in six months, some 20 cubic metres in ten years, and, indeed, a totally insignificant part of the whole during the remainder of his life!"

## THE MODERN NUMBER-SYSTEM: AN IMPORTANT DISCOVERY

The legend of Sessa thus attributes to Indian civilisation the honour of making this fundamental realisation which we may call the modern number-system. We shall see in due course that, despite the mythical character of the story, this fact is completely true.

But first we must weigh the importance of this written number-system, which nowadays is so commonplace and familiar that we have come to forget its depth and qualities.

Anyone who reflects on the universal history of written number-systems cannot but be struck by the ingeniousness of this system, since the concept of zero, and the positional value attached to each figure in the representation of a number, give it a huge advantage over all other systems thought up by people through the ages.

To understand this, we shall go back to the beginning of this history. But instead of following its different stages purely chronologically, and according to the various civilisations involved, we shall for the moment let ourselves be guided by a kind of logic of time, the regulator of historic data, which has made of human culture a profound unity.

## THE EARLIEST NUMERICAL RULE: ADDITION

This story begins about five thousand years ago in Mesopotamia and in Egypt, in advanced societies in full expansion, where it was required to determine economic operations far too varied and numerous to be entrusted to the limited capabilities of human memory. Making use of archaic concrete methods, and feeling the need to preserve permanently the results of their accounts and inventories, the leaders of these societies understood that some completely new approach was required.

To overcome the difficulty, they had the idea of representing numbers by graphic signs, traced on the ground or on tablets of clay, on stone, on sheets of papyrus, or on fragments of pottery. And so were born the earliest number-systems of history.

Independently or not, several other peoples embarked on this road during the millennia which followed. And it all worked out as though, over the ages and across civilisations, the human race had experimented with the different possible solutions of the problem of representing and manipulating numbers, until finally they settled on the one which finally appeared the most abstract, the most perfected and the most effective of all.

To begin with, written number-systems rested on the *additive principle*, the rule according to which the value of a numerical representation is obtained by adding up the values of all the figures it contains. They were therefore very primitive. Their basic figures were totally independent of each other (each one having only one absolute value), and had to be duplicated as many times as required.

The Egyptian hieroglyphic number-system, for example, assigned a special sign to unity and to each power of 10: a vertical stroke for 1, a sign like an upside-down U for 10, a spiral for 100, a lotus flower for 1,000, a raised finger for 10,000, a tadpole for 100,000, and a kneeling man with arms outstretched to the sky for 1,000,000. To write the number 7,659 required 7 lotus flowers, 6 spirals, 5 signs for 10 and 9 vertical strokes of unity, all of which required a total of 27 distinct figures (Fig. 23.1).

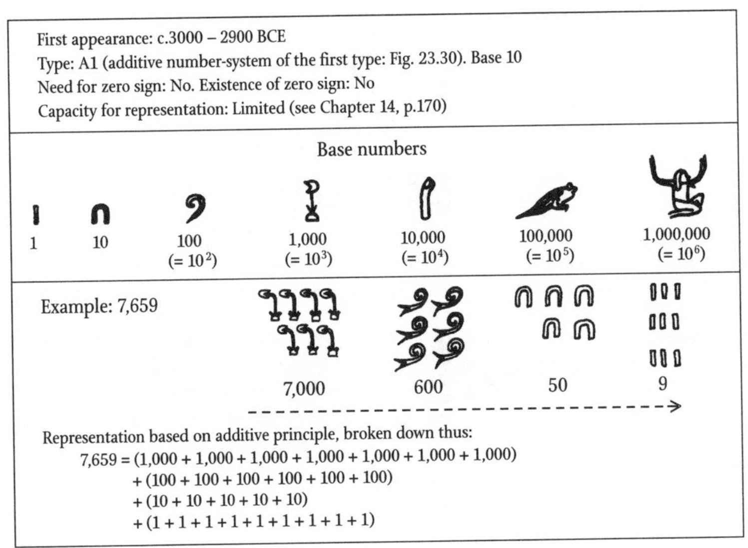


FIG. 23.1. *Egyptian hieroglyphic number-system*

The Sumerian number-system (which used base 60, with 10 as auxiliary base) gave a separate sign to each of the following numbers, in the order of their successive unit orders of magnitude:

| 1 | 10 | 60 | 600 | 3,600 | 36,000 | 216,000 |
|---|---|---|---|---|---|---|
| | | | $= 10 \times 60$ | $= 60^2$ | $= 10 \times 60^2$ | $= 60^3$ |

But it too was limited to repeating the figures as many times as required to make up the number. The number 7,659 was therefore represented according to the following arithmetical decomposition, which twice repeats the sign for 3,600, seven times the sign for 60, three times that for 10 and nine times the sign for unity, so that 21 distinct signs are required to represent this number (Fig. 23.2):

$$\begin{aligned}7{,}659 = &(3{,}600 + 3{,}600)\\ &+ (60 + 60 + 60 + 60 + 60 + 60 + 60)\\ &+ (10 + 10 + 10)\\ &+ (1 + 1 + 1 + 1 + 1 + 1 + 1 + 1 + 1)\end{aligned}$$

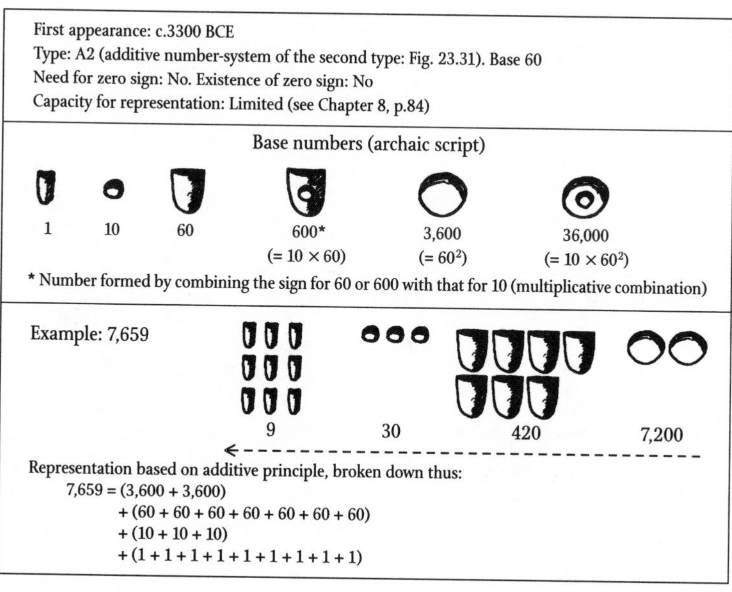


FIG. 23.2. *Sumerian number-system*

Other similar notations include the Proto-Elamite, the Cretan systems (Hieroglyphic and Linear A and Linear B), the Hittite hieroglyphic system, and even the Aztec number-system (which differed from the others only in that it used a base of 20) (Fig. 23.3 to 23.6).

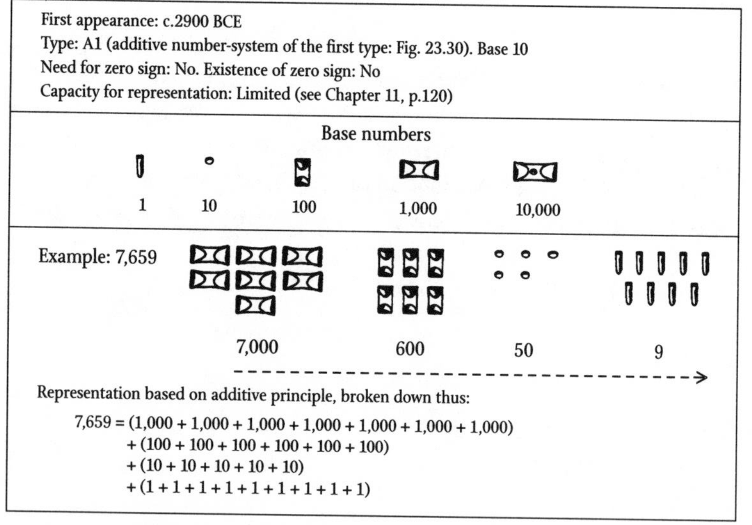


FIG. 23.3. *Proto-Elamite number-system*

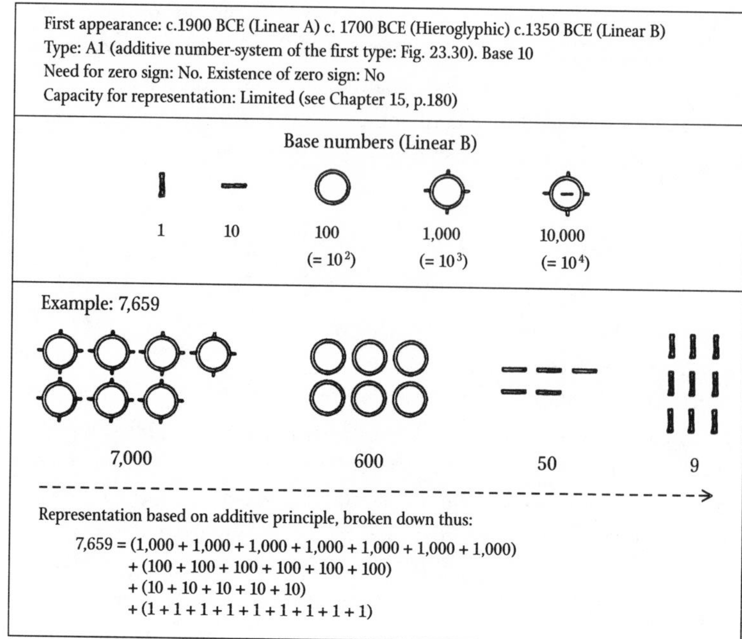


FIG. 23.4. *Cretan number-system (hieroglyphic and Linear A and B)*

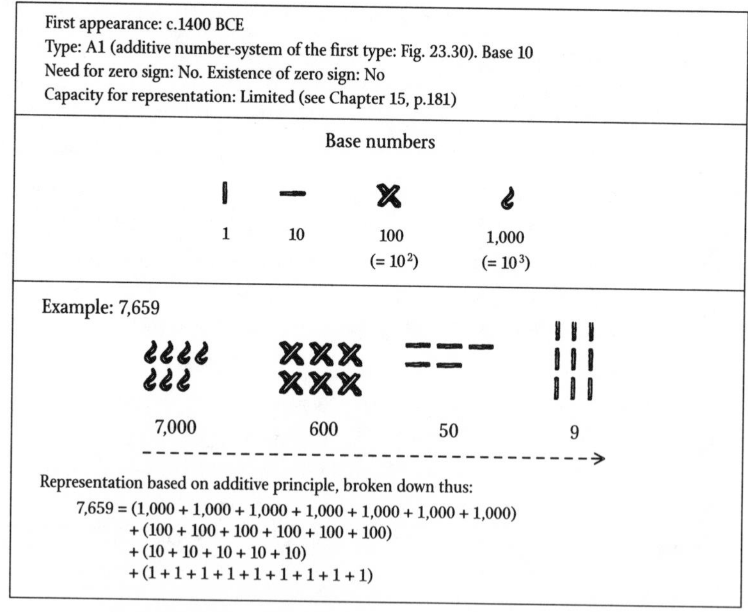


FIG. 23.5. *Hittite hieroglyphic number-system*

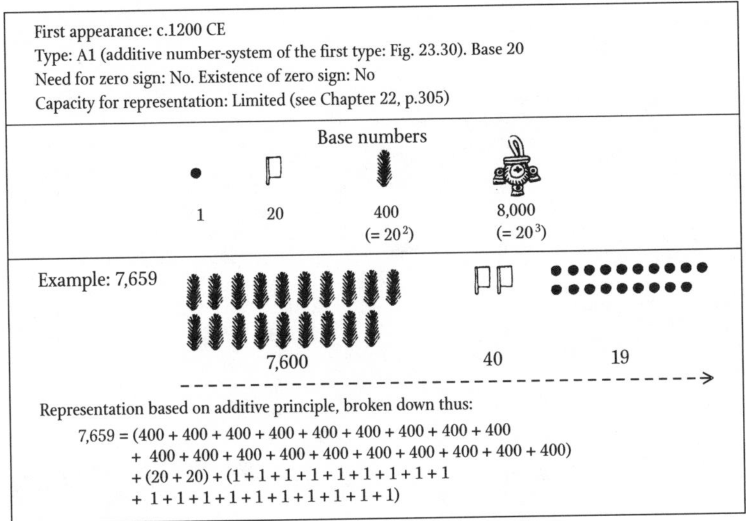

First appearance: c.1200 CE
Type: A1 (additive number-system of the first type: Fig. 23.30). Base 20
Need for zero sign: No. Existence of zero sign: No
Capacity for representation: Limited (see Chapter 22, p.305)
Base numbers
1
20
400 (= 20²)
8,000 (= 20³)
Example: 7,659
7,600
40
19
Representation based on additive principle, broken down thus:
7,659 = (400 + 400 + 400 + 400 + 400 + 400 + 400 + 400 + 400
+ 400 + 400 + 400 + 400 + 400 + 400 + 400 + 400 + 400 + 400)
+ (20 + 20) + (1 + 1 + 1 + 1 + 1 + 1 + 1 + 1 + 1 + 1
+ 1 + 1 + 1 + 1 + 1 + 1 + 1 + 1 + 1)


Fig. 23.6. *Aztec number-system*

To avoid the encumbrance of such a multitude of symbols, certain peoples introduced supplementary signs which corresponded to inter-mediate units. Such was the case for the Greeks, the Shebans, the Etruscans, and the Romans, who assigned a separate symbol to each of the numbers 5, 50, 500, 5,000, and so on, in addition to those they already had for the different powers of 10 (Fig. 23.7 and 23.8).

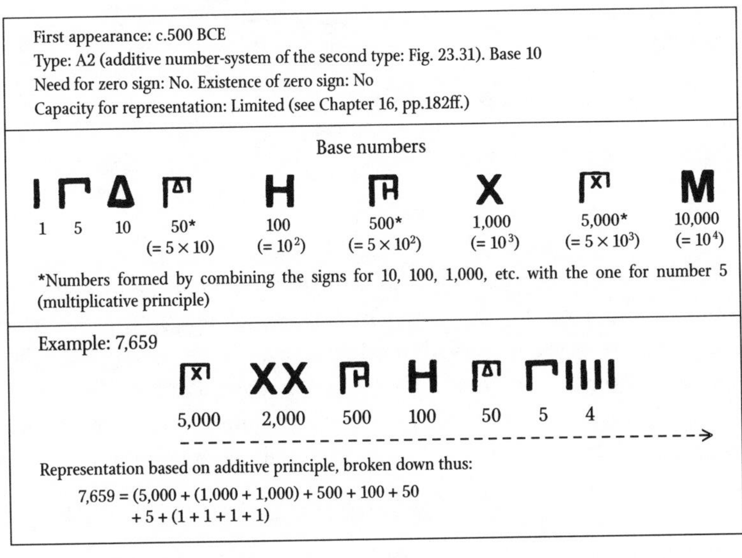


Fig. 23.7. *Greek acrophonic number-system*

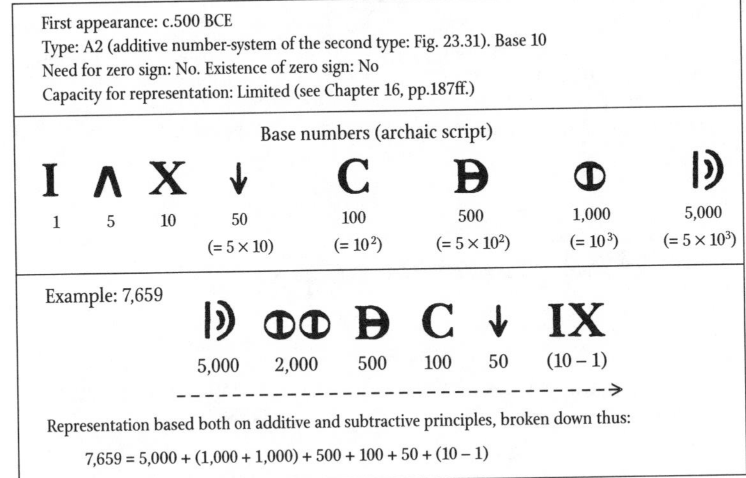

First appearance: c.500 BCE
Type: A2 (additive number-system of the second type: Fig. 23.31). Base 10
Need for zero sign: No. Existence of zero sign: No
Capacity for representation: Limited (see Chapter 16, pp.187ff.)
Base numbers (archaic script)
1
5
10
50 (= 5 × 10)
100 (= 10²)
500 (= 5 × 10²)
1,000 (= 10³)
5,000 (= 5 × 10³)
Example: 7,659
IX
5,000
2,000
500
100
50
(10 – 1)
Representation based both on additive and subtractive principles, broken down thus:
7,659 = 5,000 + (1,000 + 1,000) + 500 + 100 + 50 + (10 – 1)


Fig. 23.8. *Roman number-system*

LARGE ROMAN NUMBERS

To note down large numbers the Romans and the Latin peoples of the Middle Ages developed various conventions. Here are the principal ones (see Chapter 16, pp. 197ff.):

1. *Overline rule*
This consisted in multiplying by 1,000 every number surmounted by a horizontal bar:

$\overline{\mathrm{X}}$ = 10,000   $\overline{\mathrm{C}}$ = 100,000   $\overline{\mathrm{CXVII}}$ = 127 × 1,000 = 127,000

2. *Framing rule*
This consisted in multiplying by 100,000 every number enclosed in a sort of open rectangle:

|X| = 1,000,000   |CCLXIV| = 264 × 100,000 = 26,400,000

3. *Rule for multiplicative combinations*
The rule is occasionally found in Latin manuscripts in the early centuries CE, but most often in European mediaeval accounting documents. To indicate multiples of 100 and 1,000, first the number of hundreds and thousands to be entered are noted down, then the appropriate letter (C or M) is placed as a coefficient or superscript indication:

| | | | | | | | |
|---|---|---|---|---|---|---|---|
| 100 | C | | | 1,000 | M | | |
| 200 | II.C | or | $\mathrm{II^c}$ | 2,000 | II.M | or | $\mathrm{II^m}$ |
| 300 | III.C | or | $\mathrm{III^c}$ | 3,000 | III.M | or | $\mathrm{III^m}$ |
| ..... | | | | ..... | | | |
| 900 | VIIII.C | or | $\mathrm{VIIII^c}$ | 9,000 | VIIII.M | or | $\mathrm{VIIII^m}$ |

Examples taken from Pliny the Elder's *Natural History*, first century CE (VI, 26; XXXIII, 3).

LXXXIII.M for 83,000
CX.M for 110,000

Fig. 23.9a. *Latin notation of large numbers (late period)*

The same system is to be found in the Middle Ages, notably in King Philip le Bel's Treasury Rolls, one of the oldest surviving Treasury registers. In this book, dated 1299, we find what is reproduced here below, drawn up in Latin (from Registre du Trésor de Philippe le Bel, BN, Paris, Ms. lat. 9783, fo. 3v, col.1, line 22):

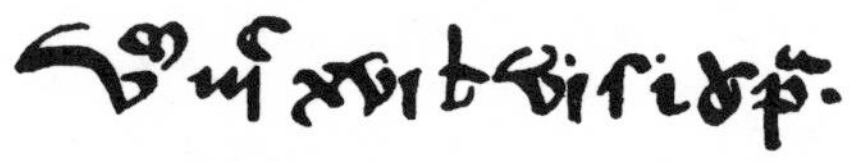

V$^{m}$. IIIe.XVI.l(ibras). VI.s(olidos) I. d(enarios). p(arisiensium)

"5,316 livres, 6 sols & 1 denier parisis"

FIG. 23.9B.

But this was out of the frying pan into the fire, for such systems required even more tedious repetitions of identical signs. In the Roman system, the conventions for writing numbers proliferated so much that the system finally lost coherence (Fig. 23.8 and 23.9). Furthermore, since it made use at the same time of two logically incompatible principles (the additive and the subtractive), this system finally represented a regression with respect to the other historic systems of number representation.

The first notable advance in this respect is in fact due to the scribes of Egypt who, seeking means for rapid writing, early sought to simplify both the graphics and the structure of their basic system. Starting from excessively complicated hieroglyphic signs, they strove to devise extremely schematic signs which could be written in a continuous trace, without interruption, such as are obtained by small rapid movements and often by a single stroke of the brush. Great changes thus occurred in the forms of the hieroglyphic numbers, so that the later forms had only a vague resemblance to their prototypes. This finally resulted in a very abbreviated numerical notation, as in the Egyptian hieratic number-system, giving a separate sign to each of the following numbers (Fig. 23.10):

| | | | | | | | | |
|---|---|---|---|---|---|---|---|---|
| 1 | 2 | 3 | 4 | 5 | 6 | 7 | 8 | 9 |
| 10 | 20 | 30 | 40 | 50 | 60 | 70 | 80 | 90 |
| 100 | 200 | 300 | 400 | 500 | 600 | 700 | 800 | 900 |
| 1,000 | 2,000 | 3,000 | 4,000 | 5,000 | 6,000 | 7,000 | 8,000 | 9,000 |

It was a cursive notation, and was succeeded by an even more abbreviated one, known as the demotic number-system.

First appearance: c.2500 BCE
Type: A3 (additive number-system of the third type: Fig. 23.32). Base 10
Need for zero sign: No. Existence of zero sign: No
Capacity for representation: Limited (see Chapter 14, pp.171ff.)

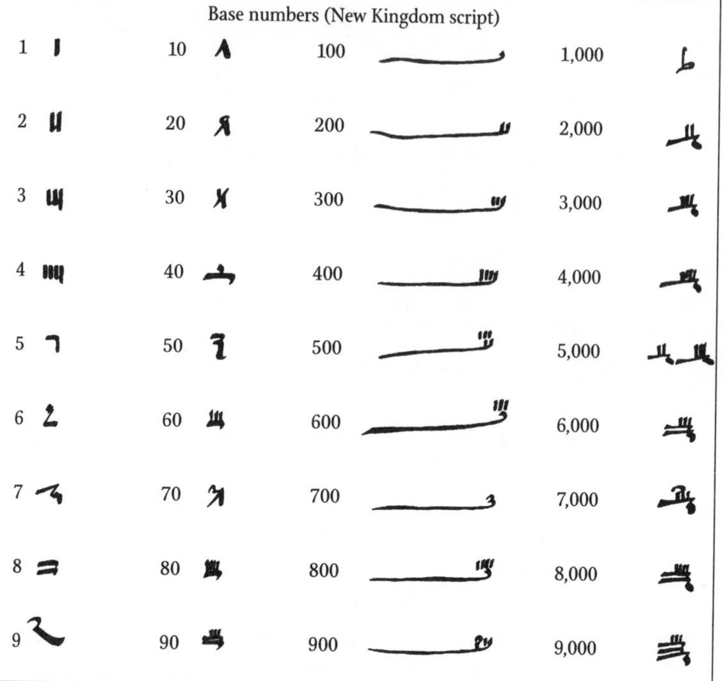

Base numbers (New Kingdom script)

| | | | | | | | |
|---|---|---|---|---|---|---|---|
| 1 | | 10 | | 100 | | 1,000 | |
| 2 | | 20 | | 200 | | 2,000 | |
| 3 | | 30 | | 300 | | 3,000 | |
| 4 | | 40 | | 400 | | 4,000 | |
| 5 | | 50 | | 500 | | 5,000 | |
| 6 | | 60 | | 600 | | 6,000 | |
| 7 | | 70 | | 700 | | 7,000 | |
| 8 | | 80 | | 800 | | 8,000 | |
| 9 | | 90 | | 900 | | 9,000 | |

Example: 7,659

9 50 600 7,000

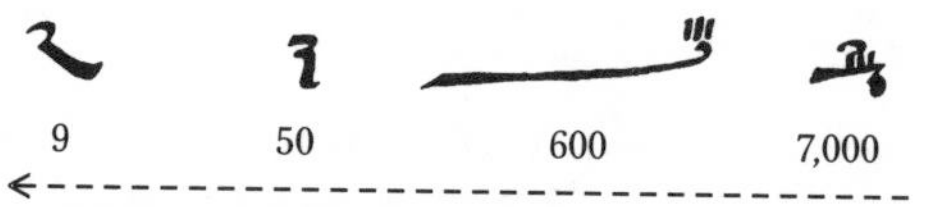

Representation based on additive principle, broken down thus:
7,659 = 7,000 + 600 + 50 + 9

FIG. 23.10. *Egyptian hieratic number-system*

In both cases, there were nine special signs for the units, nine more for the tens, nine more for the hundreds, and so on. Such systems allowed numbers to be represented with much greater economy of symbols. The number 7,659 now only needed four signs (as opposed to the 27 required by the hieroglyphic system), since it only requires writing down the symbols for 7,000, 600, 50, and 9 according to the decomposition

$$7{,}659 = 7{,}000 + 600 + 50 + 9.$$

The inconvenience of such a notation is, of course, the burden on the memory of retaining all the different symbols of the system.

The Greeks and the Jews, and later the Syriacs, the Armenians and the Arabs, used notations which are mathematically equivalent to this system (Fig. 23.11 to 23.13, and Fig. 19.4 above). But, instead of proceeding as the Egyptians had done to the progressive refinement of the forms of their figures, they constructed their systems on the basis of the letters of their alphabets. Taking these letters in their usual order (the Phoenician "ABC") associates the first nine letters with the nine units, the next nine with the nine tens, and so on.

First appearance: c. fourth century BCE
Type: A3 (additive number-system of the third type: Fig. 23.32). Base 10
Need for zero sign: No. Existence of zero sign: No
Capacity for representation: Limited (see Chapter 17, p.220)

Base numbers

| | | | | | | | | |
|---|---|---|---|---|---|---|---|---|
| Α | Β | Γ | Δ | Ε | Ϛ | Ζ | Η | Θ |
| 1 | 2 | 3 | 4 | 5 | 6 | 7 | 8 | 9 |
| Ι | Κ | Λ | Μ | Ν | Ξ | Ο | Π | Ϙ |
| 10 | 20 | 30 | 40 | 50 | 60 | 70 | 80 | 90 |
| Ρ | Σ | Τ | Υ | Φ | Χ | Ψ | Ω | ϡ |
| 100 | 200 | 300 | 400 | 500 | 600 | 700 | 800 | 900 |

Example: 7,659

| ʹΖ | Χ | Ν | Θ |
|---|---|---|---|
| 7,000 | 600 | 50 | 9 |

Representation based both on additive principle, broken down thus:

7,659 = 7,000 + 600 + 50 + 9

(The notation for the number 7,000 has been derived from that for 7, applying to this a small distinctive sign upper left.)

FIG. 23.11. *Greek alphabetic number-system*

First appearance: c. second century BCE
Type: A3 (additive number-system of the third type: Fig. 23.32). Base 10
Need for zero sign: No. Existence of zero sign: No
Capacity for representation: Limited (see Chapter 17, p.215)

Base numbers

| | | | | | | | | |
|---|---|---|---|---|---|---|---|---|
| א | ב | ג | ד | ה | ו | ז | ח | ט |
| 1 | 2 | 3 | 4 | 5 | 6 | 7 | 8 | 9 |
| י | כ | ל | מ | נ | ס | ע | פ | צ |
| 10 | 20 | 30 | 40 | 50 | 60 | 70 | 80 | 90 |
| ק | ר | ש | ת | | | | | |
| 100 | 200 | 300 | 400 | | | | | |

Example: 7,659

| ט | נ | ר | ת | ז̈ |
|---|---|---|---|---|
| 9 | 50 | 200 | 400 | 7,000 |

Representation based both on additive principle, broken down thus:

7,659 = 7,000 + 400 + 200 + 50 + 9

(The notation for the number 7,000 has been derived from that for 7, placing two dots above this.)

FIG. 23.12. *Hebraic alphabetic number-system*

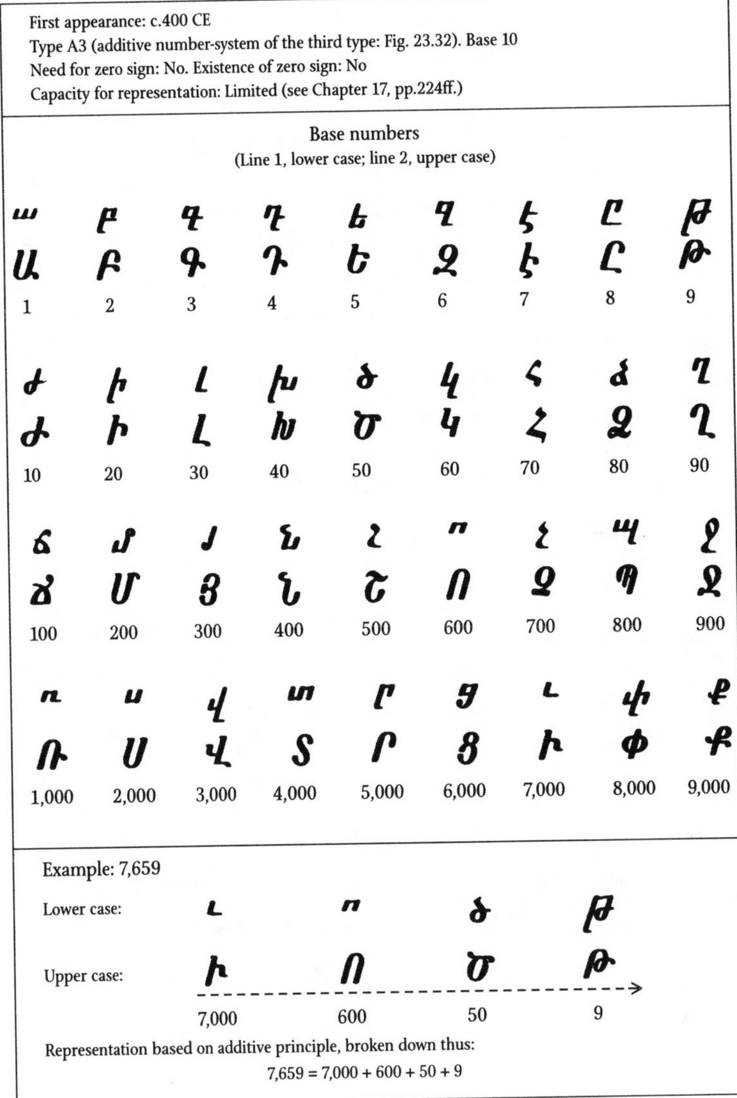


FIG. 23.13. *Armenian alphabetic number-system*

Such procedures allow the words of the language to be converted into numbers, which provides ample raw material for every kind of speculation, occultist fantasy or magical imagining, and for superstitious beliefs and practices. But, leaving aside this inconvenient by-product, the procedure gives a more or less acceptable solution to the problem according to the needs of the time. As with the Egyptian hieratic and demotic systems, the number 7,659 requires only four signs to be written down.

## THE DISCOVERY OF THE MULTIPLICATIVE PRINCIPLE

There was still a long road ahead before people could arrive at a system so well perfected as our own. Means for a numeric notation were still limited. Various peoples, it must be said, remained deeply attached to the old additive principle and were therefore in a blind alley. One major reason for this blockage concerned the problem of representing large numbers, which lie beyond the capability of the imagination when one is restricted solely to the additive principle. For this reason, some peoples made a radical change in their number-systems by adopting a hybrid principle which involved both multiplication and addition.

This change took place in two stages. The introduction of the new principle at first served only to extend the capabilities of number-systems which had been very primitive (Fig. 23.14 and 23.15).

*The Sumerians*

From c.3300 BCE the Sumerians tended to represent the units of different orders in their number-system by means of objects of conventional size and shape.

They had begun by using *calculi* to symbolise 1, 10, 60, and $60^2$ (see Chapter 10, p.100)

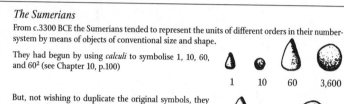

1 | 10 | 60 | 3,600

But, not wishing to duplicate the original symbols, they invoked the multiplicative principle to represent the order of 600 and of 36,000:

600 (= 10 × 60) | 36,000 (= 10 × $60^2$)

They had thus come up with the idea (very abstract for the time) of symbolising × 10 by making in the soft clay a small circular impression ("written" symbol for the pebble representing 10) within the large cone representing the value 60 or within the sphere representing the value 3,600.

And they used the same idea in representing these same numbers when they embarked on a written number-system in archaic script as well as in cuneiform (see Chapter 8, p.84):

Curviform number-symbols

Cuneiform number-symbols

600 (= 10 × 60) | 36,000 (= 10 × $60^2$)

*The Cretans* (second millennium BCE):

The Cretans introduced the number for 10,000 by combining the horizontal stroke of 10 with the sign for 1,000 (see Chapter 15, p.180):

10,000 (= 1,000 × 10)

*The Greeks* (from the fifth century BCE):

The Greeks invoked the same principle, completing their acrophonic number-system by introducing a notation with its own traits for each of the numbers 5, 50, 500, and 5,000 (see Chapter 16, pp.182ff.):

5 | 50 (= 5 × 10) | 500 (= 5 × $10^2$) | 5,000 (= 5 × $10^3$)

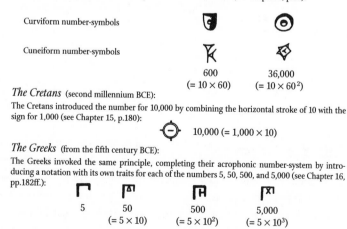

FIG. 23.14. *First emergence of the multiplicative principle*

Thus from the beginning of history, people have sometimes introduced the multiplication rule into systems essentially based on the additive principle. But during this first stage, the habit was confined to certain particular cases and the rule served only to form a few new symbols.

But in the subsequent stage, it gradually became clear that the rule could be applied to avoid not only the awkward repetition of identical signs, but also the unbridled introduction of new symbols (which always ends up requiring considerable efforts of memory).

And that is how certain notations that were rudimentary to begin with were often found to be extensible to large numbers.

*The Greeks*

This idea was exploited by ancient Greek mathematicians whose "instrument" was their alphabetic number-system: in order to set down numbers superior to 10,000, they invoked the multiplicative rule, placing a sign over the letter M (initial of the Greek word for 10,000, μυριοι) to indicate the number of 10,000s (see Chapter 17, p.220):

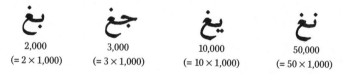

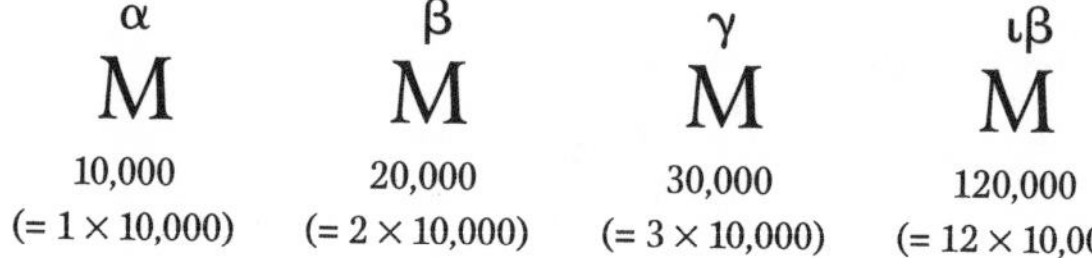


10,000 (= 1 × 10,000) | 20,000 (= 2 × 10,000) | 30,000 (= 3 × 10,000) | 120,000 (= 12 × 10,000)

*The Arabs*

Using the twenty-eight letters of their number-alphabet the Arabs proceeded likewise, but on a smaller scale: to note down the numbers beyond 1,000, all they had to do was to place beside the letter *ghayin* (worth 1,000 and corresponding to the largest base number in their system) the one representing the corresponding number of units, tens or hundreds (see Chapter 19, p.246):

بغ | جغ | يغ | نغ
---|---|---|---
2,000 (= 2 × 1,000) | 3,000 (= 3 × 1,000) | 10,000 (= 10 × 1,000) | 50,000 (= 50 × 1,000)

*The ancient Indians*

The same idea was invoked by the Indians from the time of Emperor Asoka until the beginning of the Common Era in the numerical notation that related to Brâhmi script (see Chapter 24, pp.378ff.). To write down multiples of 100, they used the multiplicative principle, placing to the right of the sign for 100 the sign for the corresponding units. For numbers beyond 1,000 they wrote to the right of the sign for 1,000 the sign for the corresponding units or tens:

400 (= 100 × 4) | 4,000 (= 1,000 × 4) | 6,000 (= 1,000 × 6) | 10,000 (= 1,000 × 10)

FIG. 23.15A. *First extension of the multiplicative principle*

*The Egyptian hieroglyphic system* (late period)

In Egyptian monumental inscriptions we find (at least from the beginning of the New Kingdom) a remarkable diversion from the "classical" system: when a tadpole (hieroglyphic sign for 100,000) was placed over a lower number-sign, it behaved as a multiplicator. In other words, by placing a tadpole over the sign for 18, for instance, the number 100,018 (= 100,000 + 18) was no longer being expressed, but rather the number $100{,}000 \times 18 = 1{,}800{,}000$ (a number which in the classical system would have been expressed by setting eight tadpoles adjacent to the hieroglyphic for 1,000,000).

Example, 27,000,000
Expressed in the form:
$100{,}000 \times 270$

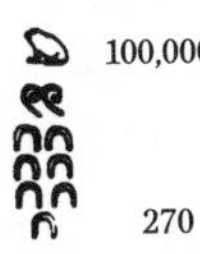


Taken from a Ptolemaic hieroglyphic inscription (third – first century BCE)

*The Egyptian hieratic system*

But the preceding irregularity was actually the result of the way the hieroglyphic system was influenced by hieratic notation: this used a more systematic method to note down numbers above 10,000 according to the rule in question. (See Chapter 14, pp. 171ff.)

| | Early Kingdom | Middle Kingdom | New Kingdom |
|---|---|---|---|
| 70,000 | [illegible] | [illegible] | [illegible] |
| 90,000 | | [illegible] | |
| 200,000 | [illegible] | [illegible] | |
| 700,000 | [illegible] | [illegible] | [illegible] |
| 1,000,000 | [illegible] | [illegible] | [illegible] |
| 2,000,000 | [illegible] | [illegible] | [illegible] |
| 10,000,000 | | | [illegible] |

Example: The number 494,800

(From the Great Harris Papyrus; 73, line 3. New Kingdom)

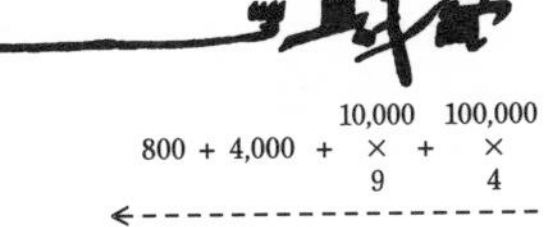


FIG. 23.15B.

The Assyro-Babylonians and the Aramaeans provide a case in point. They had a separate symbol for each of the numbers 1, 10, 100 and 1,000, but instead of representing the hundreds or thousands by separate signs or by repeating the 100 or 1,000 symbol as often as required, they had the idea of placing the signs for 100 or 1,000 side by side with the symbols for the units, thereby arriving at a multiplicative principle representing arithmetical combinations such as

| | |
|---|---|
| $1 \times 100$ | $1 \times 1{,}000$ |
| $2 \times 100$ | $2 \times 1{,}000$ |
| $3 \times 100$ | $3 \times 1{,}000$ |
| $4 \times 100$ | $4 \times 1{,}000$ |
| $5 \times 100$ | $5 \times 1{,}000$ |
| ... | ... |
| $9 \times 100$ | $9 \times 1{,}000$ |

However, they continued to write numbers below 100 according to the old additive method, repeating the sign for 1 or for 10 as often as required. The number 7,659, for example, was written according to the following decomposition (Fig. 23.16 and 23.17):

$$\begin{aligned}7{,}659 = &(1+1+1+1+1+1+1) \times 1{,}000\\ &+ (1+1+1+1+1+1) \times 100\\ &+ (10+10+10+10+10)\\ &+ (1+1+1+1+1+1+1+1+1)\end{aligned}$$

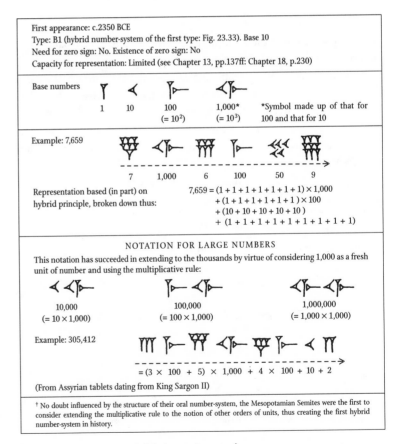

First appearance: c.2350 BCE
Type: B1 (hybrid number-system of the first type: Fig. 23.33). Base 10
Need for zero sign: No. Existence of zero sign: No
Capacity for representation: Limited (see Chapter 13, pp.137ff: Chapter 18, p.230)

Base numbers: 1; 10; 100 (= $10^2$); 1,000* (= $10^3$)
*Symbol made up of that for 100 and that for 10

Example: 7,659 — 7, 1,000, 6, 100, 50, 9

Representation based (in part) on hybrid principle, broken down thus:

$$\begin{aligned}7{,}659 = &(1+1+1+1+1+1+1) \times 1{,}000\\ &+ (1+1+1+1+1+1) \times 100\\ &+ (10+10+10+10+10)\\ &+ (1+1+1+1+1+1+1+1+1)\end{aligned}$$

NOTATION FOR LARGE NUMBERS

This notation has succeeded in extending to the thousands by virtue of considering 1,000 as a fresh unit of number and using the multiplicative rule:

10,000 (= $10 \times 1{,}000$) 100,000 (= $100 \times 1{,}000$) 1,000,000 (= $1{,}000 \times 1{,}000$)

Example: 305,412

$= (3 \times 100 + 5) \times 1{,}000 + 4 \times 100 + 10 + 2$

(From Assyrian tablets dating from King Sargon II)

† No doubt influenced by the structure of their oral number-system, the Mesopotamian Semites were the first to consider extending the multiplicative rule to the notion of other orders of units, thus creating the first hybrid number-system in history.

FIG. 23.16. *Common Assyro-Babylonian number-system*†

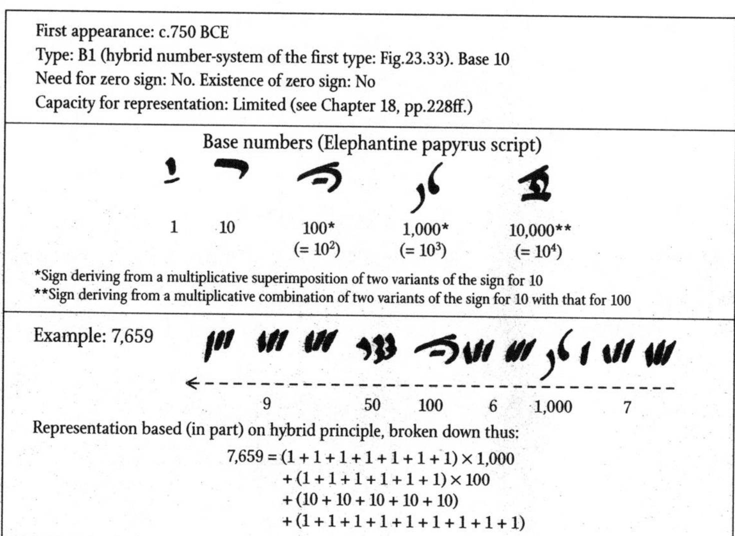

First appearance: c.750 BCE
Type: B1 (hybrid number-system of the first type: Fig.23.33). Base 10
Need for zero sign: No. Existence of zero sign: No
Capacity for representation: Limited (see Chapter 18, pp.228ff.)
Base numbers (Elephantine papyrus script)
1 10 100* (= 10²) 1,000* (= 10³) 10,000** (= 10⁴)
*Sign deriving from a multiplicative superimposition of two variants of the sign for 10
**Sign deriving from a multiplicative combination of two variants of the sign for 10 with that for 100
Example: 7,659
9 50 100 6 1,000 7
Representation based (in part) on hybrid principle, broken down thus:
7,659 = (1 + 1 + 1 + 1 + 1 + 1 + 1) × 1,000
+ (1 + 1 + 1 + 1 + 1 + 1) × 100
+ (10 + 10 + 10 + 10 + 10)
+ (1 + 1 + 1 + 1 + 1 + 1 + 1 + 1 + 1)


FIG. 23.17. *Aramaean number-system*

By such partial use of the multiplicative principle, the Assyro-Babylonian number-system was therefore of the "partial hybrid" type.

At a later period, the inhabitants of Ceylon went through the same change, but starting from a much better system than those above. They assigned a separate sign not only to every power of 10, but also to each of the nine units and to each of the nine tens, and then applied the same principle as above. In this way, the number 7,659 can be broken down (Fig. 23.18) as

$$7 \times 1{,}000 + 6 \times 100 + 50 + 9.$$

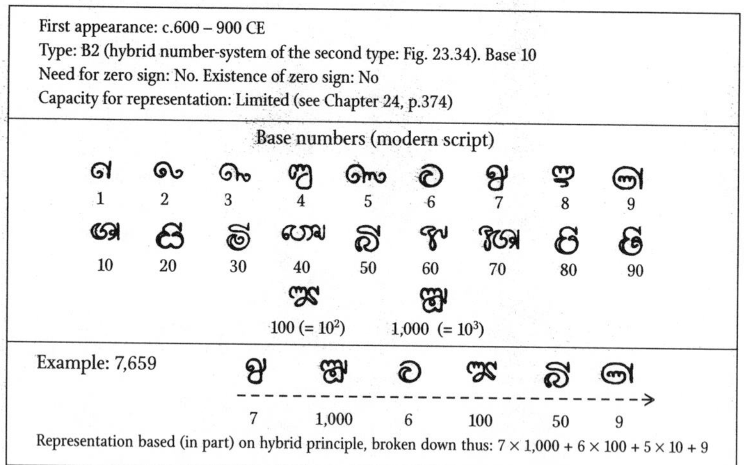


FIG. 23.18. *Singhalese number-system*

However, it was the Chinese, and the Tamils and Malayalams of southern India, who made the best use of this approach. They too had special signs for the numbers 1, 2, 3, 4, 5, 6, 7, 8, 9, 10, 100, 1,000, 10,000 but, instead of representing the tens by special signs, they had the idea of extending the multiplicative principle to all the orders of magnitude of their system, from the unit upwards. For intermediate numbers, they placed the sign for 10 between the sign for the number of units and the sign for the number of hundreds, the sign for 100 between the sign for the number of hundreds and the sign for the number of thousands, and so on. For the number 7,659 this gave rise to a decomposition of the type

$$7{,}659 = 7 \times 1{,}000 + 6 \times 100 + 5 \times 10 + 9.$$

Such systems are of "complete hybrid" type, in which the representation of a number resembles a polynomial whose variable is the base of the number-system (Fig. 23.19 to 23.21).

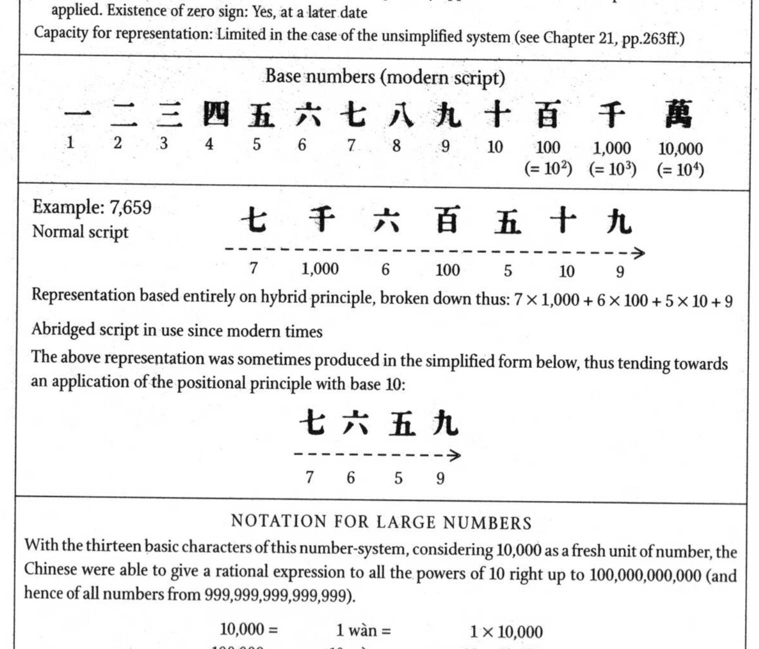

First appearance: c.1450 BCE
Type: B5 (hybrid number-system of the fifth type: Fig. 23.37). Base 10
Need for zero sign: No, when the hybrid principle is rigorously applied. Yes, when the simplified rule below is applied. Existence of zero sign: Yes, at a later date
Capacity for representation: Limited in the case of the unsimplified system (see Chapter 21, pp.263ff.)
Base numbers (modern script)
一 二 三 四 五 六 七 八 九 十 百 千 萬
1 2 3 4 5 6 7 8 9 10 100 (= 10²) 1,000 (= 10³) 10,000 (= 10⁴)
Example: 7,659
Normal script
七 千 六 百 五 十 九
7 1,000 6 100 5 10 9
Representation based entirely on hybrid principle, broken down thus: 7 × 1,000 + 6 × 100 + 5 × 10 + 9
Abridged script in use since modern times
The above representation was sometimes produced in the simplified form below, thus tending towards an application of the positional principle with base 10:
七 六 五 九
7 6 5 9
NOTATION FOR LARGE NUMBERS
With the thirteen basic characters of this number-system, considering 10,000 as a fresh unit of number, the Chinese were able to give a rational expression to all the powers of 10 right up to 100,000,000,000 (and hence of all numbers from 999,999,999,999,999).
10,000 = 1 wàn = 1 × 10,000
100,000 = 10 wàn = 10 × 10,000, etc
Example: 487,390,629
四 萬 八 千 七 百 三 十 九 萬 六 百 二 十 九
(4 × 10⁴ + 8 × 10³ + 7 × 10² + 3 × 10 + 9) × 10⁴ + (6 × 10² + 2 × 10 + 9)


FIG. 23.19. *Common Chinese number-system*

The discovery of such hybrid principles was a great step forward, in the context of the needs of the time, since it not only avoided tedious repetitions of identical signs but also lightened the burden on the memory, no longer required to retain a large number of different signs.

By the same token, the written representation of numbers could be brought into line with their verbal expression (the linguistic structure of the majority of spoken numbers had conformed, since the earliest times, to this kind of mixed rule).

The principal benefit, however, of this procedure was greatly to extend the range of numbers that could be represented (Fig. 23.15,16 and 19).

## THE DIFFICULTIES OF THE PRECEDING SYSTEMS

Despite the considerable advance which these changes represent, the capabilities of numerical notation remained very limited.

By making use of certain conventions of writing, the Greek mathematicians managed to extend their alphabetic notation to cope with larger numbers. Archimedes provides an important example. In his short arithmetical treatise *The Psammites*, he conceived a rule which would allow him to express very large numbers by means of the numeric letters of the Greek alphabet, such as the number of grains of sand which would be contained in the Sphere of the World (whose diameter is the distance from the earth to the nearest fixed stars). In our modern notation, this number would be expressed as a 1 followed by 64 zeros.

Chinese mathematicians also succeeded in extending their number-system to accommodate numbers which could exceed $10^{4096}$, a number which is far beyond any quantity that could be physically realised.

None of these systems, however, succeeded in achieving a rational notation for all numbers, since they did not have the unlimited capacity for representation which our own system has. The greater the order of magnitude required, the more special symbols must be invented, or further conventions of writing imposed.

We can therefore appreciate the undoubted superiority of our modern system of numerical notation, which is one of the foundations of the intellectual equipment of modern humankind. With the aid of a very small number of basic symbols, any number whatever, no matter how large, may be represented in a simple, unified and rational manner without the need for any further artifice.

Yet another reason for the superiority of our system is that it is directly adapted to the written performance of arithmetic.

It is precisely this fact which underlies the difficulty, or even impossibility, of doing arithmetic with the ancient number-systems, which remained blocked in this respect for as long as they were in use.

For example, let us try to perform an addition using Roman numerals:

```
    CCLXVI
+      DCL
+    MLXXX
+ MDCCCVII
----------
= ??????
```

Clearly, unless we translate this into our modern notation, this would be very hard:

```
    266
+   650
+ 1,080
+ 1,807
-------
= 3,803
```

But this is a mere addition! What about multiplication or division?

In systems of this kind, we are barely able to do arithmetic. This is due to the static nature of the number-signs, which have no operational significance but are more like abbreviations which can be used to write down the results of calculations performed by other means.

In order to do arithmetical calculations, the ancients generally made use of auxiliary aids such as the abacus or a table with counters. This requires long and difficult training and practice, and remains beyond the reach of ordinary mortals. It therefore remained the preserve of a privileged caste of specialist professional calculators. This is not to say, however, that such systems did not allow any written calculation.

The above addition can be carried out in the Roman system. This involves proceeding by stages, by counting and then reducing the results from each order of magnitude (five "I" replaced by one "V", two "V" by one "X", five "X" by one "L", two "L" by one "C", and so on):

| | | | CC | L | X | V | I |
|---|---|---|---|---|---|---|---|
| + | M | D | CCC | | | V | II |
| + | | D | C | L | | | |
| + | M | | | L | XXX | | |
| | MMM | D | CCC | | | | III |

The Romans probably did use such a method. But since it is at bottom a reduction to written form of operations performed on an abacus, they probably preferred to continue to use that instrument whose counters, for all their inconvenience, were nonetheless easier to manipulate than the symbols in their primitive representation of numbers.

We know also that, despite its very primitive character, the Egyptian number-system allowed arithmetical calculations. The methods certainly had the advantage of not obliging calculators to rely on memory. To multiply or to divide, it was in fact enough to know how to multiply or to divide by 2. Their methods, however, were slow and complicated compared with our modern ones. Worse, though, they lacked flexibility, unification and coherence.

On the other hand, the Graeco-Byzantine mathematicians certainly succeeded in devising various rules for multiplication and division in terms of the number-letters of their alphabet. There again, however, their procedures were much more complicated, and above all far more artificial and less coherent than ours.

These are all, therefore, mere attempts to invent rules of calculation during the ancient times. But, "The fact is that the difficulties encountered in former times were inherent in the very nature of the number-systems themselves, which did not lend themselves to simple straightforward rules" [T. Dantzig (1967)].

Therefore it was the discovery of our modern number-system, and above all its popularisation, which allowed the human race to overcome the obstacles and to dispense with all auxiliary aids to calculation such as we have been considering.

## DECISIVE FIRST STEP: THE PRINCIPLE OF POSITION

In order to achieve a system as ingenious as our own, it is first necessary to discover the *principle of position*. According to this, the value of a figure varies according to the position in which it occurs, in the representation of a number. In our modern decimal notation, a "3" has value 3 units, 3 tens or 3 hundreds depending on whether it is in the first, second or third position. To write seven thousand, six hundred and fifty-nine, all we have to do is to write down the figures 7, 6, 5, and 9 in that order, since according to the rule the representation 7,659 denotes the value

$$7 \times 1{,}000 + 6 \times 100 + 5 \times 10 + 9.$$

Because of this fundamental convention, only the coefficients of the powers of the base, into which the number has been decomposed, need appear.

This, therefore, is the principle of position. Apparently as simple as Columbus's egg; but it had to be thought of in the first place!

Nowadays, this principle seems to us to have such an obvious simplicity that we forget how the human race has stammered, hesitated and groped through thousands of years before discovering it, and that civilisations as advanced as the Greek and the Egyptian completely failed to notice it.

## SYSTEMS WHICH COULD HAVE BEEN POSITIONAL

For all that, even in the earliest times a goodly number of different number-systems could have led on to the discovery of the principle of position.

Consider for example the Tamil and Malayalam systems from south India. According to the hybrid principle, the figure representing the number of tens was placed to the left of the symbol for 10, the one representing the number of hundreds to the left of the symbols for 100, and so on (Fig. 23.20 and 23.21).

First appearance: c.600 – 900 CE
Type: B5 (hybrid number-system of the fifth type: Fig.23.37). Base 10
Need for zero sign: No, when the hybrid principle is rigorously applied. Yes, when the simplified rule below is applied.
Existence of zero sign: Not before the modern era
Capacity for representation: Limited in the case of the unsimplified system (see Chapter 24, p.372)
System used among the Tamils (southern India)

Base numbers (modern script)

| ௧ | ௨ | ௩ | ௪ | ௫ | ௬ | ௭ | ௮ | ௯ |
|---|---|---|---|---|---|---|---|---|
| 1 | 2 | 3 | 4 | 5 | 6 | 7 | 8 | 9 |

| ௰ | ௱ | ௲ |
|---|---|---|
| 10 | 100 (= $10^2$) | 1,000 (=$10^3$) |

Example: 7,659
Normal script

| ௭ | ௲ | ௬ | ௱ | ௫ | ௰ | ௯ |
|---|---|---|---|---|---|---|
| 7 | 1,000 | 6 | 100 | 5 | 10 | 9 |

Representation based entirely on hybrid principle, broken down thus:

$$7 \times 1{,}000 + 6 \times 100 + 5 \times 10 + 9$$

Abridged script in use since modern times

The above representation was sometimes produced in the simplified form below, thus tending towards an application of the positional principle with base 10:

| ௭ | ௬ | ௫ | ௯ |
|---|---|---|---|
| 7 | 6 | 5 | 9 |

FIG. 23.20. *Tamil number-system*

First appearance: c.600 – 900 CE
Type: B5 (hybrid number-system of the fifth type: Fig. 23.37). Base 10
Need for zero sign: No, when the hybrid principle is rigorously applied. Yes, when the simplified rule below is applied.
Existence of zero sign: Not before the modern era
Capacity for representation: Limited in the case of the unsimplified system (see Chapter 24, p.373)
System used among the Malayalam (southern India, Malabar coast)

Base numbers (modern script)

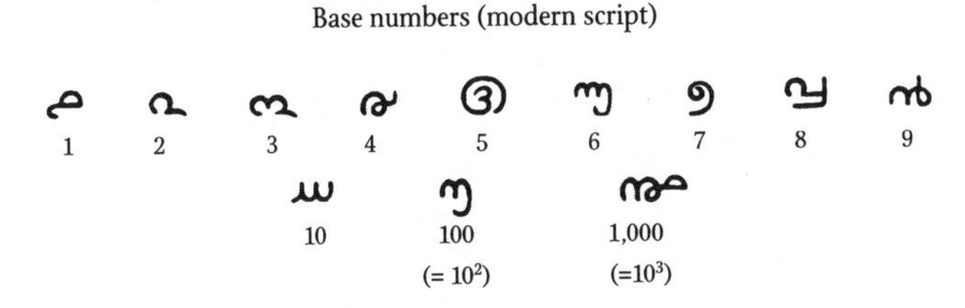


Example: 7,659
Normal script

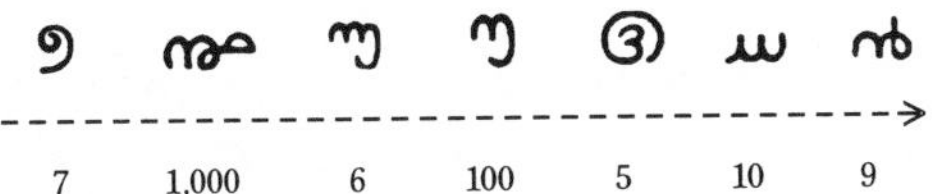


Representation based entirely on hybrid principle, broken down thus:

$$7 \times 1{,}000 + 6 \times 100 + 5 \times 10 + 9$$

Abridged script in use since modern times

The above representation was sometimes produced in the simplified form below, thus tending towards an application of the positional principle with base 10:

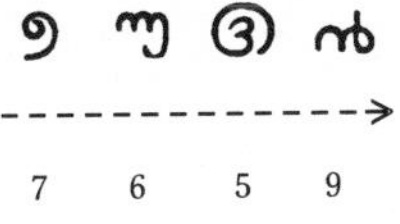


FIG. 23.21. *Malayalam number-system*

In this way, the number 6,657, for example, would usually be written as follows:

6 1,000 6 100 5 10 7 ---> Tamil

6 1,000 6 100 5 10 7 ---> Malayalam

which corresponded to the decomposition

$$6 \times 1{,}000 + 6 \times 100 + 5 \times 10 + 7.$$

Now, when we look at certain Tamil or Malayalam writings, we find that the symbols for 10, 100, and 1,000 have in many cases been suppressed [L. Renou and J. Filliozat (1953)]. The number 6,657 would then appear in the abbreviated notation

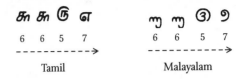

6 6 5 7 ---> Tamil

6 6 5 7 ---> Malayalam

The result of this simplification is that the figures 6, 6, 5, and 7 have been assigned values as follows:

- seven units to the figure 7 in the first place;
- five tens to the figure 5 in the second place;
- six hundreds to the figure 6 in the third place;
- six thousands to the figure 6 in the fourth place.

Thus the Tamil and Malayalam figures could be assigned values which depended on where they occurred in the representation of a number.

This remarkable potential for evolution towards a positional number-system is characteristic of hybrid numbering systems.

In such systems, in fact, the signs which indicate the powers of the base (10, 100, 1,000) are always written in the same order, either increasing or decreasing. Therefore it is natural that the people who used these systems would be led, for the sake of abbreviation, to suppress these signs leaving only the figures representing their coefficients.

This is what led certain Aramaic stone-cutters of the beginning of our era to sometimes leave out the sign for 100 in their numeric inscriptions.

The inscription of Sa'ddiyat is a remarkable piece of evidence for this. We know that in this region a hybrid system was used, whose basic signs had the following forms and values:

| 1 | 5 | 10 | 20 | 100 |
|---|---|---|---|---|

But we see in this inscription (which dates from the 436th year of the Seleucid era, or 124–125 CE) that the number 436 is written in the form [B. Aggoula (1972), plate II]

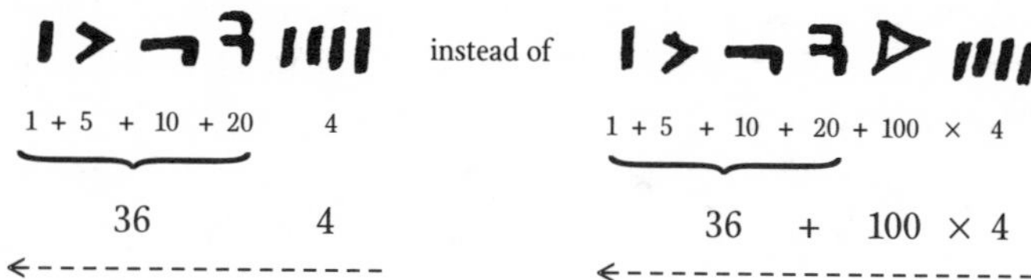


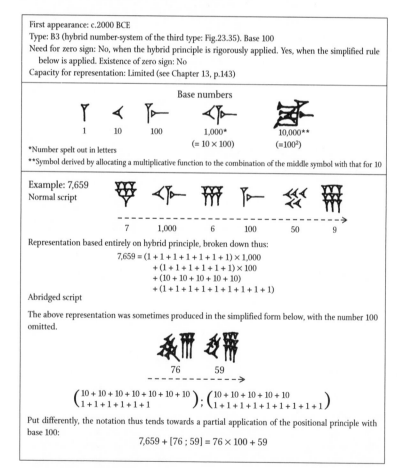

For the same reason, the scribes of Mari often left out the cuneiform figure for 100. This is all the more remarkable since the Mari system, uniquely among Mesopotamian systems, was in use around the nineteenth century BCE, therefore earlier than the period in which the Babylonian positional system appeared (J.-M. Durand).

First appearance: c.2000 BCE
Type: B3 (hybrid number-system of the third type: Fig.23.35). Base 100
Need for zero sign: No, when the hybrid principle is rigorously applied. Yes, when the simplified rule below is applied. Existence of zero sign: No
Capacity for representation: Limited (see Chapter 13, p.143)

Base numbers

1 10 100 1,000* (= 10 × 100) 10,000** ($=100^2$)

*Number spelt out in letters
**Symbol derived by allocating a multiplicative function to the combination of the middle symbol with that for 10

Example: 7,659
Normal script

7 1,000 6 100 50 9

Representation based entirely on hybrid principle, broken down thus:

$$7{,}659 = (1+1+1+1+1+1+1)\times 1{,}000 + (1+1+1+1+1+1)\times 100 + (10+10+10+10+10) + (1+1+1+1+1+1+1+1+1)$$

Abridged script

The above representation was sometimes produced in the simplified form below, with the number 100 omitted.

76 59

$$\begin{pmatrix}10+10+10+10+10+10+10\\1+1+1+1+1+1\end{pmatrix};\begin{pmatrix}10+10+10+10+10\\1+1+1+1+1+1+1+1+1\end{pmatrix}$$

Put differently, the notation thus tends towards a partial application of the positional principle with base 100:

$$7{,}659 + [76\,;\,59] = 76 \times 100 + 59$$

FIG. 23.22. *Mari number-system*

At Mari they used a hybrid system whose basic signs had the following forms and values (Fig. 23.22):

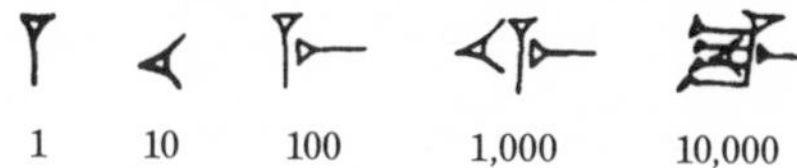


The number 476 would therefore be represented as:

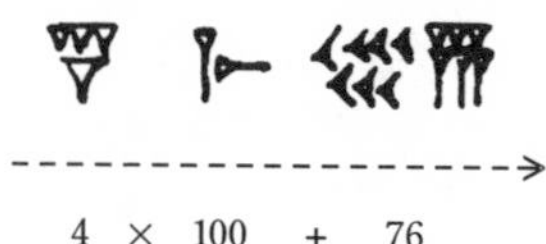


At any rate, that is the normal representation of this number. But, as we have only recently discovered, the Mari gave an abbreviated representation to this number [D. Soubeyran (1984), tablet ARM, XXII 26]:

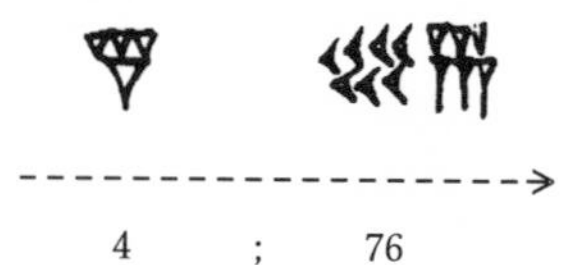


This simplification was, nevertheless, only made for the hundreds figure, not for all the powers of the base. For this reason, the Mari system never became positional in the full sense. This system of notation remained strongly bound to the methods of the old additive principle, and was therefore held back from taking the one vital further step forward from this significant advance.

A similar simplification can be found in certain Chinese writers, who also simplified their writing of numbers by suppressing the signs indicating the tens, hundreds, thousands, etc. (see Fig. 23.19 above). For the number 67,859 we therefore find [E. Biot (1839); K. Menninger (1969)]:

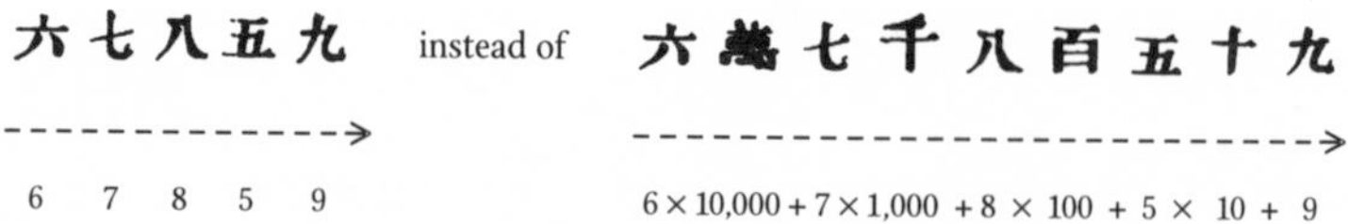

六七八五九
instead of
六萬七千八百五十九
6 7 8 5 9
6 × 10,000 + 7 × 1,000 + 8 × 100 + 5 × 10 + 9

Finally, consider the Maya priests and astronomers. In order to simplify the "Long Count" of their representation of dates, they too were led to suppress all indications of the glyphs for their units of time, leaving only the series of corresponding coefficients.

Let us take, for example, the Maya period of time expressed, in days, as $5 \times 144{,}000 + 17 \times 7{,}200 + 6 \times 360 + 11 \times 20 + 19$. This would usually be shows on the *stelae* as:

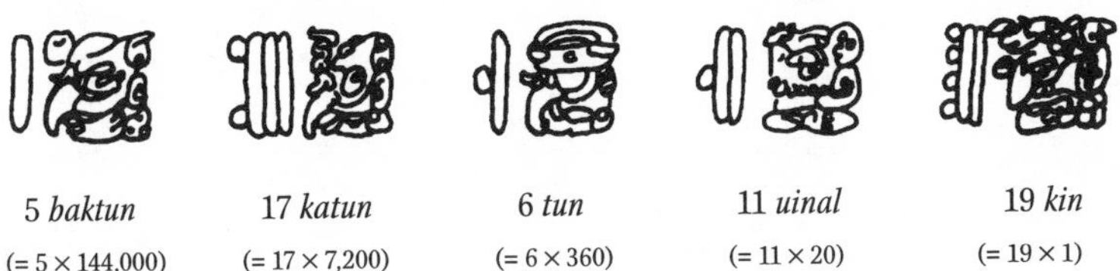


But, in their manuscripts, these astronomer-priests often preferred the following form in which appear only the numerical coefficients associated with the different time periods *kin* (days), *uinal* (periods of 20 days), *tun* (periods of 360 days), *katun* (periods of 7,200 days), etc. This gives a strictly positional representation:

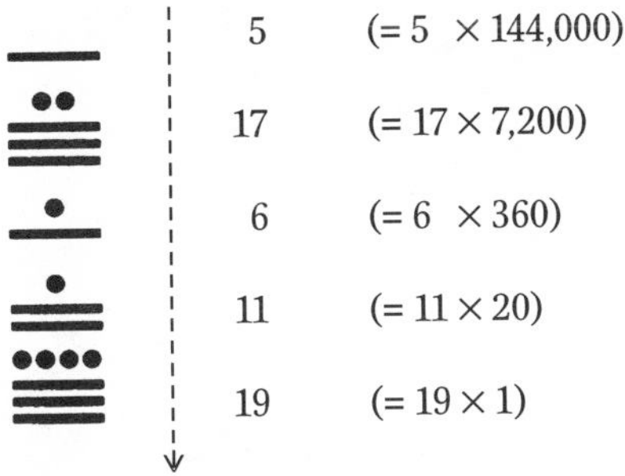


This proves clearly that hybrid numbering systems had the potential to lead to the discovery of the principle of position. However, a simplification of a partial hybrid system could only lead to an incomplete implementation of the rule of position, whereas the simplification of a fully hybrid system was capable of leading to its complete implementation.

The simplification of the Maya notation for "Long Count" dates did give rise to a positional system, as also did changes in certain other systems (but these were belated and therefore of no consequence for the universal history); apart from these marginal exceptions it must be said that none of these earlier systems arrived at the level of a truly positional numbering system.

We therefore see yet again how people who have been widely separated in time or space have, by their tentative researches, been led to very similar if not identical results.

In some cases, the explanation for this may be found in contacts and influences between different groups of people. But it would not be correct to suppose that the Maya were in a position to copy the ideas of the people of the Ancient World. The true explanation lies in what we have previously referred to as the profound unity of human culture: the intelligence of *homo sapiens* is universal, and its potential is remarkably uniform in all parts of the world. The Maya simply found themselves in favourable conditions, strictly identical to those of others who obtained the same results.

## THE EARLIEST POSITIONAL NUMBER-SYSTEMS OF HISTORY

The civilisation which developed the basis of our modern number system was therefore neither the first nor the only one to discover the principle of position.

In fact, three peoples came to its full discovery earlier, and independently. The numerical rule which is the basis of the positional system was created:

- for the first time, some 2,000 years BCE, by the Babylonians;
- for the second time, slightly before the Common Era, by Chinese mathematicians;
- for the third time, between the fourth and the ninth century CE, by the Mayan astronomer-priests.

The Babylonian sexagesimal system represented a number such as 392 by writing the number 6 in the second (sixties) place, and the number 32 in the first place, corresponding to a notation which might be transcribed (Fig. 23.23) as [6; 32] ($= 6 \times 60 + 32$).

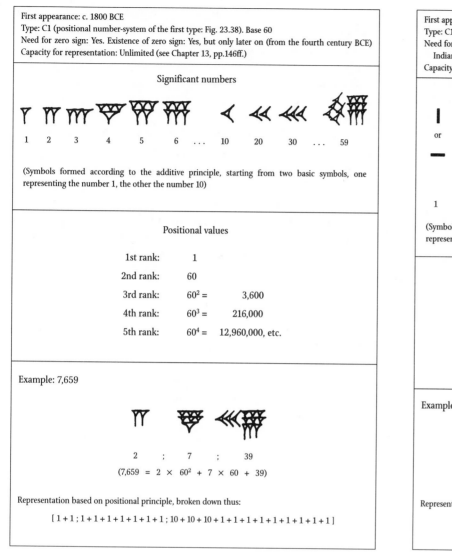

First appearance: c. 1800 BCE
Type: C1 (positional number-system of the first type: Fig. 23.38). Base 60
Need for zero sign: Yes. Existence of zero sign: Yes, but only later on (from the fourth century BCE)
Capacity for representation: Unlimited (see Chapter 13, pp.146ff.)

**Significant numbers**

(Symbols formed according to the additive principle, starting from two basic symbols, one representing the number 1, the other the number 10)

**Positional values**

| | | |
|---|---|---|
| 1st rank: | 1 | |
| 2nd rank: | 60 | |
| 3rd rank: | $60^2$ = | 3,600 |
| 4th rank: | $60^3$ = | 216,000 |
| 5th rank: | $60^4$ = | 12,960,000, etc. |

Example: 7,659

($7{,}659 = 2 \times 60^2 + 7 \times 60 + 39$)

Representation based on positional principle, broken down thus:

[ 1 + 1 ; 1 + 1 + 1 + 1 + 1 + 1 + 1 ; 10 + 10 + 10 + 1 + 1 + 1 + 1 + 1 + 1 + 1 + 1 + 1 ]

FIG. 23.23. *Learned Babylonian number-system (the first positional number-system in history)*

First appearance: c. 200 BCE
Type: C1 (positional number-system of the first type: Fig. 23.38). Base 10
Need for zero sign: Yes. Existence of zero sign: Yes, but only later on (from the eighth century, under Indian influence)
Capacity for representation: Unlimited

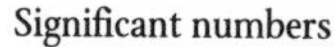

**Significant numbers**

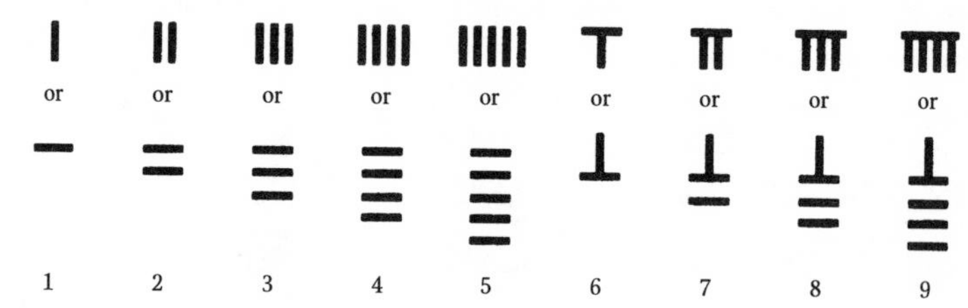


(Symbols formed according to the additive principle, starting from two basic symbols, one representing the number 1, the other the number 5)

**Positional values**

| | | |
|---|---|---|
| 1st rank: | 1 | |
| 2nd rank: | 10 | |
| 3rd rank: | $10^2$ = | 100 |
| 4th rank: | $10^3$ = | 1,000 |
| 5th rank: | $10^4$ = | 10,000, etc. |

Example: 7,659

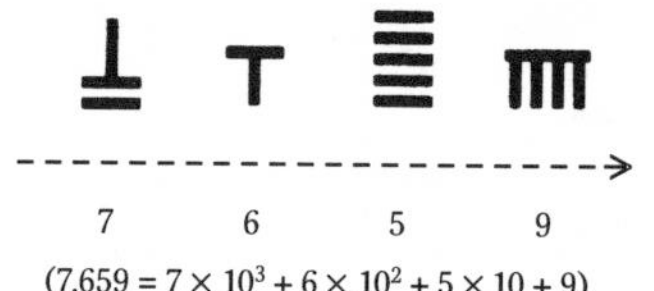


($7{,}659 = 7 \times 10^3 + 6 \times 10^2 + 5 \times 10 + 9$)

Representation based on positional principle, broken down thus:

[ 5 + 1 + 1 ; 5 + 1 ; 1 + 1 + 1 + 1 + 1 ; 5 + 1 + 1 + 1 + 1 ]

FIG. 23.24. *Learned Chinese number-system*

This is very much as we might today write $392' = 6 \times 60' + 32'$ in the form $6° 32'$ (6 degrees, 32 minutes).

The Chinese system was based on the same principle, with the difference that the base of the number-system was decimal instead of being equal to 60. To write 392 in this system, we therefore place the figures 3, 9 and 2 in this order in a notation which we may (Fig. 23.24) transcribe as [3; 9; 2] $(= 3 \times 100 + 9 \times 10 + 2)$.

In the Maya system with base 20, we may write (Fig. 23.25) [19; 12] $(= 19 \times 20 + 12)$. These Babylonian, Chinese and Maya systems were, therefore, the earliest positional number-systems of history.

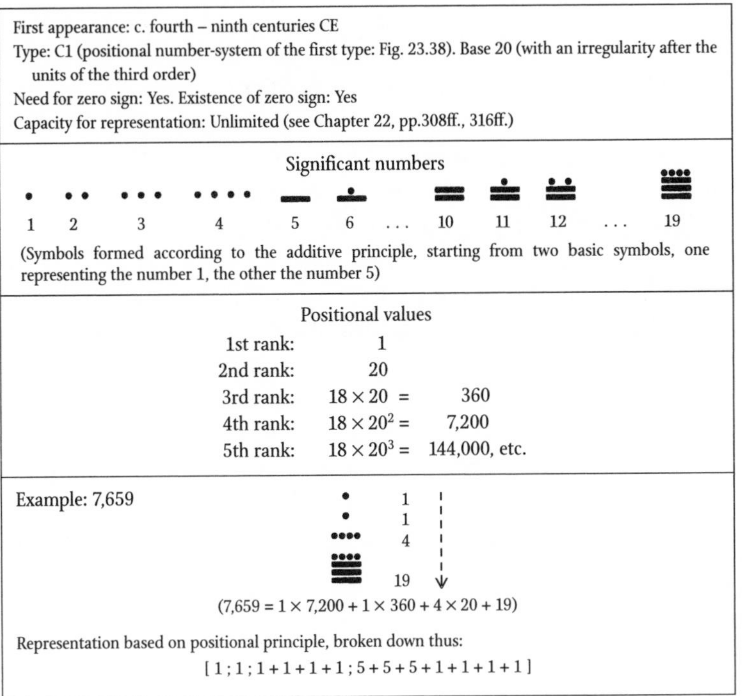


FIG. 23.25. *Learned Maya number-system*

## SYSTEMS WHICH DID NOT SUCCEED

Having made this fundamental and essential discovery, the way was in fact open to each of these three peoples to represent any number whatever, no matter how large, by means of a small set of basic signs. But none of these three succeeded in taking advantage of their discovery.

The Babylonians indeed discovered the principle of position and applied it to base 60. But it never occurred to them, for more than two thousand years, to attach a particular symbol to each unit in their sexagesimal system. Instead of fifty-nine different figures, they in reality had only two: one for unity, and one for 10. All the rest had to be composed by duplicating these as many times as necessary up to 59 (Fig. 23.23).

The Chinese also discovered the principle of position and applied it to base 10. But they did no better, for, instead of assigning a different sign to each of the nine units, they preserved their ideographic system, in which the number 8 was represented by the symbol for 5 with three copies of the symbol for unity (Fig. 23.24).

Likewise the Maya system used the principle of position applied to base 20. But they again had only two distinct figures, one for unity and the other for 5, instead of the nineteen which are required for full dynamic notation in base 20 (Fig. 23.25).

For each of these three, it is somewhat as if the Romans had applied the rule of position to their first few number-signs, for example writing 324 in the form III II IIII, which would surely have led to confusion with:

| | |
|---|---|
| I IIII IIII | (144) |
| II III IIII | (234) |
| II IIII III | (243) |
| III III III | (333) |
| III IIII II | (342) |
| IIII I IIII | (414) |
| IIII II III | (423) etc. |

The Maya system had another source of difficulty inherent in its very structure. The rule of position was not applied to the powers of the base, but to values which were in fact adapted to the requirements of the calendar and of astronomy.

Each number greater than 20 was written in a vertical column with as many levels as there were orders of magnitude: the units were at the bottom level, the twenties on the second level, and so on.

This system therefore became irregular from the third level onwards, and was not rigorously founded on base 20. Instead of giving the multiples of $20^2 = 400$, $20^3 = 8,000$, and so on, the different levels from the third upwards in fact indicated multiples of $360 = 18 \times 20$, $7,200 = 18 \times 20^2$, and so on.

But there was no such problem with the Babylonian and Chinese systems, whose positional values corresponded exactly to the progression of the values of their base:

| Units of | Learned Babylonian system (base 60) | Learned Chinese system (base 10) | Regular positional system (base 20) | Learned Maya system (irregular use of base 20) |
|---|---|---|---|---|
| 1st order | 1 | 1 | 1 | 1 |
| 2nd order | 60 | 10 | 20 | 20 |
| 3rd order | $60^2$ | $10^2$ | $20^2$ | $18 \times 20$ |
| 4th order | $60^3$ | $10^3$ | $20^3$ | $18 \times 20^2$ |
| 5th order | $60^4$ | $10^4$ | $20^4$ | $18 \times 20^3$ |
| 6th order | $60^5$ | $10^5$ | $20^5$ | $18 \times 20^4$ |

If the Maya positional system had been constructed regularly on base 20, the expression [7; 9; 3] would surely have signified

$$7 \times 20^2 + 9 \times 20 + 3 = 7 \times 400 + 9 \times 20 + 3.$$

But for the Maya priests this corresponded to $7 \times 360 + 9 \times 20 + 3$.

This is one of the reasons why their system remained unsuited to practical written calculation.

## A MAJOR SECOND STEP: DEVELOPMENT OF A DYNAMIC NOTATION FOR THE UNITS

From what we have seen so far, it is clear that for a numerical notation to be well adapted to written calculation, it must not only be based on the principle of position but must also have distinct symbols corresponding to graphic characters which have no intuitive visual meaning.

Otherwise put, the graphical structure of the number-signs must be like that of our modern written numbers, in that "9", for example, is not composed of nine points nor of nine bars, but is a purely conventional symbol with no ideographic significance (Fig. 23.26):

1 2 3 4 5 6 7 8 9

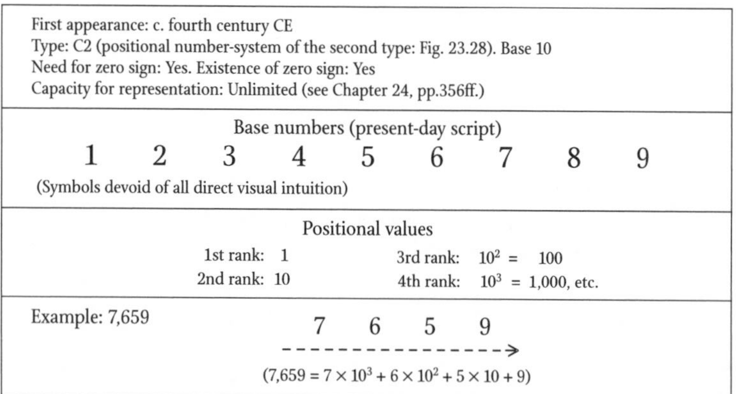

First appearance: c. fourth century CE
Type: C2 (positional number-system of the second type: Fig. 23.28). Base 10
Need for zero sign: Yes. Existence of zero sign: Yes
Capacity for representation: Unlimited (see Chapter 24, pp.356ff.)

Base numbers (present-day script)

1 2 3 4 5 6 7 8 9

(Symbols devoid of all direct visual intuition)

Positional values

1st rank: 1
2nd rank: 10
3rd rank: $10^2$ = 100
4th rank: $10^3$ = 1,000, etc.

Example: 7,659

7 6 5 9

- - - - - - - - - - - - - - - - →

$(7{,}659 = 7 \times 10^3 + 6 \times 10^2 + 5 \times 10 + 9)$

FIG. 23.26. *Modern number-system*

## THE FINAL FUNDAMENTAL DISCOVERY: ZERO

A no less fundamental condition for any number-system to be as well developed and as effective as our own is that it must possess a *zero*.

For so long as people used non-positional notations, the necessity of this concept did not make itself felt. The fact that there were signs for values greater than the base of the system meant that these systems could avoid the stumbling block which occurs whenever units of a certain order of magnitude are absent. To write, for instance, 2,004 in Egyptian hieroglyphics, it was sufficient to put two lotus flowers (for the thousands) and four vertical bars (for the units), the total of the values thus being

$$1{,}000 + 1{,}000 + 1 + 1 + 1 + 1 = 2{,}004.$$

In the Roman numerals, this number would be written MMIIII, and there was no need to have a special symbol to show that there were no hundreds and no tens. In the Chinese system, they would represent this number in the hybrid system, as a "2" followed by the symbol for 1,000 followed by a "4", corresponding to the decomposition $2{,}004 = 2 \times 1{,}000 + 4$.

On the other hand, once one has begun to apply place values on a regular basis, it is not long before one faces the requirement to indicate that tens, or hundreds, etc., may be missing. The discovery of zero was therefore a necessity for the strict and regular use of the rule of position, and it was therefore a decisive stage in an evolution without which the progress of modern mathematics, science and technology would be unimaginable.

Consider our decimal system. To write thirty, we have to place "3" in the second position, to have the value of three tens. But how do we show that it is in the second position if there is nothing at all in the first position? Therefore it is essential to have a special sign whose purpose is to indicate the absence of anything in a particular position. This thing which signifies nothing, or rather empty space, is in fact the zero. To arrive at the realisation that empty space may and must be replaced by a sign whose purpose is precisely to indicate that it is empty space: this is the ultimate abstraction, which required much time, much imagination, and beyond doubt great maturity of mind.

In the beginning, this concept was simply synonymous with empty space thus filled. But it was gradually perceived that "empty" and "nothing", originally thought of as distinct, are in reality two aspects of one and the same thing. Thus it is that the zero sign originally introduced to mark empty space finally symbolises in our eyes the value of the null number, a concept at the heart of algebra and modern mathematics.

Nowadays this is so familiar that we are no longer aware of the difficulties which its lack caused to the early users of positional number-systems.

Its discovery was far from a foregone conclusion, for apart from India, Mesopotamia and the Maya civilisation, no other culture throughout history came to it by itself. We can gain some idea of its importance when we recall that it escaped the eyes of the Chinese mathematicians, who nonetheless succeeded in discovering the principle of position. Only since the eighth century of our era, under the influence of our modern number-system, did this concept finally appear in Chinese scientific writings.

The Babylonians themselves were unaware of it for more than a thousand years, leading as one can imagine to numerous errors and confusions.

They certainly tried to get round the difficulty by leaving empty space where the missing units of a certain order would normally be found. Therefore they wrote much as if we wrote the number one hundred and six as 1. .6. But this was not enough to solve the problem in practice, since scribes could easily overlook it in copying, through fatigue or carelessness. Moreover it was difficult to indicate precisely the absence of two or more consecutive orders of magnitude, since one empty space beside another empty space is not easily distinguished from a single empty space.

It was therefore necessary to await the fourth century BCE to see the introduction of a special sign dedicated to this purpose. This was a cuneiform sign, which looked like a double oblique chevron, which was used not only in the medial and final positions but also in the initial position to indicate sexagesimal fractions of unity.

Medial: $[3; 0; 9; 2] = 3 \times 60^3 + 0 \times 60^2 + 9 \times 60 + 2$

$[3; 0; 0; 2] = 3 \times 60^3 + 0 \times 60^2 + 0 \times 60 + 2$

Final: $[3; 1; 5; 0] = 3 \times 60^3 + 1 \times 60^2 + 5 \times 60 + 0$

$[3; 1; 0; 0] = 3 \times 60^3 + 1 \times 60^2 + 0 \times 60 + 0$

Initial: $[0; 3; 4; 2] = 0 + 3 \times \frac{1}{60} + 4 \times \frac{1}{60^2} + 2 \times \frac{1}{60^3}$

This epoch, late in the history of Mesopotamia, saw the emergence of an eminently abstract concept, the Babylonian zero, the first zero of all time, to be followed some centuries later by the Maya zero.

## IMPERFECT ZEROS

Even so, neither the Babylonians nor the Maya managed to get the most from their prime discovery (Fig. 23.27).

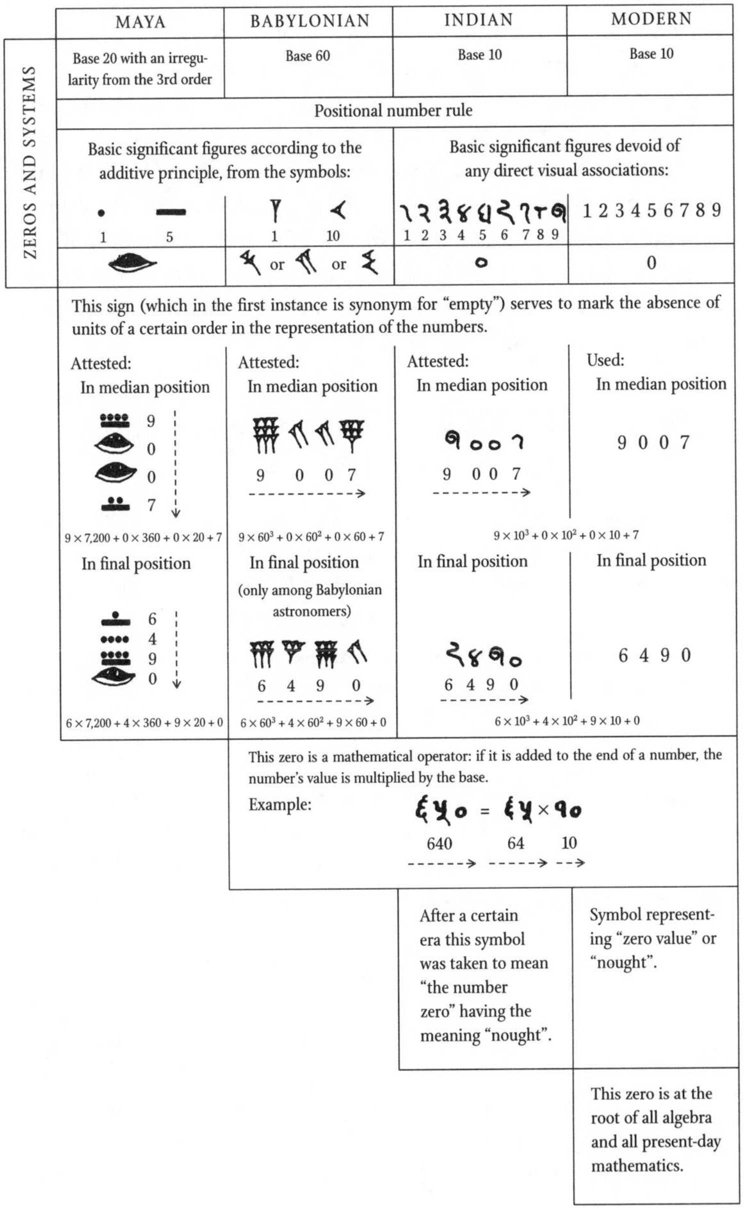

| ZEROS AND SYSTEMS | MAYA | BABYLONIAN | INDIAN | MODERN |
|---|---|---|---|---|
| | Base 20 with an irregularity from the 3rd order | Base 60 | Base 10 | Base 10 |
| | Positional number rule | | | |
| | Basic significant figures according to the additive principle, from the symbols: | | Basic significant figures devoid of any direct visual associations: | |
| | 1 5 | 1 10 | 1 2 3 4 5 6 7 8 9 | 1 2 3 4 5 6 7 8 9 |
| | | or or | | 0 |

This sign (which in the first instance is synonym for "empty") serves to mark the absence of units of a certain order in the representation of the numbers.

| MAYA | BABYLONIAN | INDIAN | MODERN |
|---|---|---|---|
| Attested: In median position | Attested: In median position | Attested: In median position | Used: In median position |
| 9 0 0 7 | 9 0 0 7 | 9 0 0 7 | 9 0 0 7 |
| $9 \times 7{,}200 + 0 \times 360 + 0 \times 20 + 7$ | $9 \times 60^3 + 0 \times 60^2 + 0 \times 60 + 7$ | $9 \times 10^3 + 0 \times 10^2 + 0 \times 10 + 7$ | |
| In final position | In final position (only among Babylonian astronomers) | In final position | In final position |
| 6 4 9 0 | 6 4 9 0 | 6 4 9 0 | 6 4 9 0 |
| $6 \times 7{,}200 + 4 \times 360 + 9 \times 20 + 0$ | $6 \times 60^3 + 4 \times 60^2 + 9 \times 60 + 0$ | $6 \times 10^3 + 4 \times 10^2 + 9 \times 10 + 0$ | |

This zero is a mathematical operator: if it is added to the end of a number, the number's value is multiplied by the base.

Example: 640 = 64 × 10

After a certain era this symbol was taken to mean "the number zero" having the meaning "nought".

Symbol representing "zero value" or "nought".

This zero is at the root of all algebra and all present-day mathematics.

FIG. 23.27. *Classification of zeros*

The Maya of course understood that it was a genuine zero sign, since they used it in medial as well as in final position. But, because of the anomalous progression they introduced at the third position of their positional system, this concept lost all operational usability.

The Babylonian zero not only had this possibility, it even filled the role of an arithmetical operator, at least in the hands of the astronomers (adjoining the zero sign at the end of a representation multiplied the number represented by sixty, i.e. by the value of the base). But it was never understood as a number synonymous with "empty", and never corresponded to the meaning of "null quantity" (Fig.23.27).

Despite these fundamental discoveries, therefore, none of these peoples was able to take the decisive step which would result in the ultimate perfection of numerical notation. Because of these imperfections, neither the Babylonian nor the Chinese nor the Maya positional system ever became adapted to arithmetical calculation, nor could ever give rise to mathematical developments such as our own.

## NUMBER-SYSTEMS WHICH COULD HAVE BECOME DYNAMIC

We saw above how the complete adaptation of modern numerical notation to practical arithmetic comes not only from the principle of position and from the zero, but also from the fact that its figures correspond to graphic signs which have no direct intuitive visual meaning.

Once again, the inventors of this system have neither the privilege nor the honour of priority, since certain other systems had already enjoyed this property since the earliest times.

With the Egyptians, as we have seen, the transition from hieroglyphic to hieratic, and then to demotic script, radically changed the notation for the first whole numbers. Starting with groupings of identical strokes representing the nine units, in the end we find cursive signs, independent of each other, with no apparent intuitive meaning [G. Möller (1911–12); R. W. Erichsen]:

| | 1 | 2 | 3 | 4 | 5 | 6 | 7 | 8 | 9 |
|---|---|---|---|---|---|---|---|---|---|
| Hieroglyphic notation | I | II | III | II II | III II | III III | IIII III | IIII IIII | III III III |
| Hieratic notation | [illegible] | [illegible] | [illegible] | [illegible] | [illegible] | [illegible] | [illegible] | [illegible] | [illegible] |
| Demotic notation | [illegible] | [illegible] | [illegible] | [illegible] | [illegible] | [illegible] | [illegible] | [illegible] | [illegible] |

The Egyptian cursive notations could therefore have risen to the status of a number-system mathematically equivalent to our modern one if they had only eliminated all the signs for numbers greater than or equal to 10, replacing their additive principle by a principle of position which would then have been applied to the signs for the first nine units. However, this did not take place, since the Egyptian scribes remained profoundly attached to their old and traditional method.

The same characteristic was present in yet another number-system, the Singhalese, whose first nine number-signs certainly correspond to independent graphics stripped of any capacity to directly and visually evoke the corresponding units (Fig. 23.18):

| | 1 | 2 | 3 | 4 | 5 | 6 | 7 | 8 | 9 |
|---|---|---|---|---|---|---|---|---|---|
| Singhalese notation | 𑇡 | 𑇢 | 𑇣 | 𑇤 | 𑇥 | 𑇦 | 𑇧 | 𑇨 | 𑇩 |

But this system too preserved its initial hybrid principle, and therefore remained stuck throughout its existence.

Why therefore did not well-conceived systems like the Tamil or the Malayalam take this decisive step, and why did they not become positional number-systems worthy of the name?

This is all the more surprising since both underwent simplification conducive to such an end, since they had distinct signs for the nine units which had no immediate visual associations as we have seen (Fig. 23.20 and 23.21):

| | 1 | 2 | 3 | 4 | 5 | 6 | 7 | 8 | 9 |
|---|---|---|---|---|---|---|---|---|---|
| Tamil notation | ௧ | ௨ | ௩ | ௪ | ௫ | ௬ | ௭ | ௮ | ௯ |
| Malayalam notation | ൧ | ൨ | ൩ | ൪ | ൫ | ൬ | ൭ | ൮ | ൯ |

The reason is that this simplification was not extended to all the numbers. The largest order of magnitude represented in these systems was 1,000. Numbers greater than or equal to 10,000 were either spelled out in full, or else they used the hybrid principle with the signs for 10, 100 and 1,000. These systems therefore remained firmly attached to their original principle, and for this reason they too remained blocked.

Furthermore, because there was no zero, the rule of simplification would only work on condition that every missing power of the base was followed by the sign for the order of magnitude immediately below.

In order to avoid confusion between the abbreviated Tamil notation for

3,605, and the number 365, it was necessary to keep the indicator for the hundreds in the representation of the former:

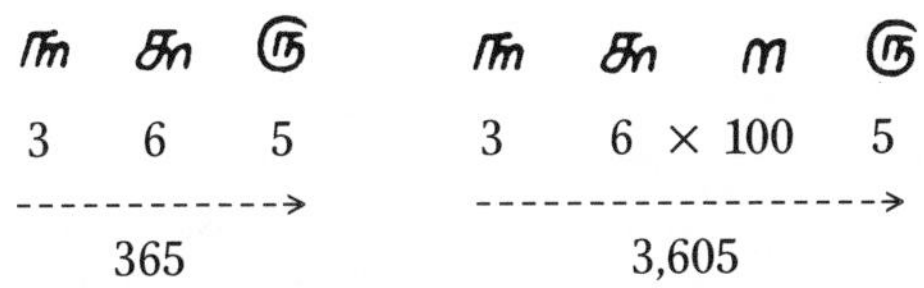

| 3 | 6 | 5 | | 3 | 6 × 100 | 5 |
|---|---|---|---|---|---|---|
| 365 | | | | 3,605 | | |

These systems were surely capable of rising to the level of our own if only they had eliminated the signs for the numbers greater than or equal to 10, and if the principle of position had been rigorously applied to the remaining figures. For a while there would have been difficulties due to the absence of zero, but, as necessity is the mother of invention, these would have been overcome by the invention of zero.

The common Chinese system of numeration (which, as we have seen, is in the same category as the two above) indeed went through this change.

In a table of logarithms, which is part of a collection of mathematical works put together on the orders of the Emperor Kangshi (1662–1722 CE) and published in 1713, we see the number 9,420,279,060 written in the form [K. Menninger (1957), II, pp. 278–279]:

九四二〇二七九〇六〇

9 4 2 0 2 7 9 0 6 0

By fully suppressing the classical signs for 10, 100, 1,000, and 10,000, by systematising the rule of position for all numbers, and by introducing a sign in the form of a circle to signify absence of an order of magnitude, the ordinary Chinese notation has been transformed into a number system equipped with a structure which is strictly identical to our own (Fig. 23.19). These number representations are perfectly adapted to arithmetical calculation.

The following example is taken from a work entitled *Ding zhu suan fa* ("Ding zhu's Method of Calculation"), published in 1335. It gives a table showing the multiplication of 3,069 by 45 laid out in a way which no one will have any difficulty in recognising [K. Menninger (1957), II, p. 300]:

| | TRANSLATION |
|---:|---:|
| 三〇六九 | 3069 |
| 四五 | 45 |
| 一五三四五 | 15345 |
| 一二二七六 | 12276 |
| 一三八一〇五 | 138105 |

This change only took place very late in the history of number-systems, however; the "push in the right direction" to the traditional Chinese system in fact came from the influence of the modern number-system.

## THE "INVENTION" OF THE MODERN SYSTEM: AN IMPROBABLE CONJUNCTION OF THREE GREAT IDEAS

This fundamental "discovery" did not, therefore, appear all at once like the fully formed act of a god or a hero, or single act of an imaginative genius. These pages show clearly that it had an origin and a very long history. Fruit of a veritable cascade of inventions and innovations, it emerged little by little, following thousands of years during which an extraordinary profusion of trials and errors, of sudden breakthroughs and of standstills, regressions and revolutions occurred.

The discovery is the "fruit of slow maturation of primitive systems, initially well conceived, and patiently perfected through long ages. With the passage of time, some scholars succeeded when the circumstances were right in perfecting the primitive instrument they had inherited from their ancestors. Their motive for this effort was the passion they had to be able to express large numbers. Other scholars, coming after them, realistic and persistent, managed to get this revolutionary novelty accepted by the calculators of their time. We inherit from both" (G. Guitel).

Finally it all came to pass as though, across the ages and the civilisations, the human mind had tried all the possible solutions to the problem of writing numbers, before universally adopting the one which seemed the most abstract, the most perfected, and the most effective of all (Fig. 23.26, 23.27, 23.28, and 23.29).

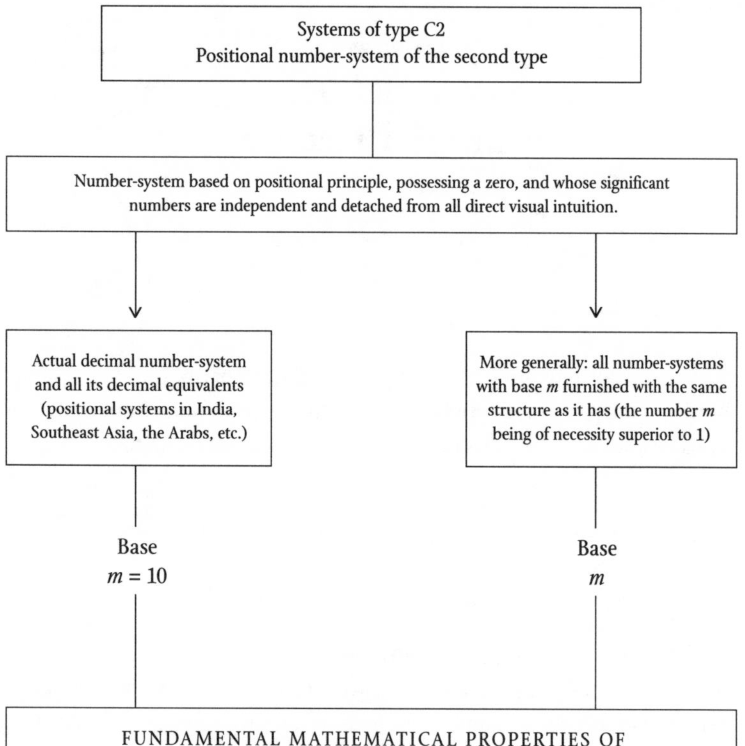

Systems of type C2
Positional number-system of the second type
Number-system based on positional principle, possessing a zero, and whose significant numbers are independent and detached from all direct visual intuition.
Actual decimal number-system and all its decimal equivalents (positional systems in India, Southeast Asia, the Arabs, etc.)
More generally: all number-systems with base m furnished with the same structure as it has (the number m being of necessity superior to 1)
Base
m = 10
Base
m


### FUNDAMENTAL MATHEMATICAL PROPERTIES OF POSITIONAL NUMBER-SYSTEMS WITH BASE $m$

1. The number of digits (including zero) is equal to $m$.
2. Every integer $x$ may be broken down in one single manner in the form of a polynomial in the degree $k - 1$, with base $m$ as a variable, and with coefficients all smaller than $m$. In other words, any number $x$ may be written in one way in the form:

$$x = u_k\, m^{k-1} + u_{k-1}\, m^{k-2} + \ldots + u_4\, m^3 + u_3\, m^2 + u_2\, m + u_1$$

where the integers $u_k, u_{k-1}, \ldots u_2, u_1$, all inferior to $m$, are symbolised by numbers in the system under consideration. One may agree to write the number $x$ in the following manner (where the horizontal dash serves to avoid any confusion with the product $u_k\, u_{k-1} \ldots u_4\, u_3\, u_2\, u_1$):

$$x = \overline{u_k\, u_{k-1} \ldots u_4\, u_3\, u_2\, u_1}$$

3. The four fundamental arithmetical operations (addition, subtraction, multiplication and division) are easily carried out in such a system, according to simple rules entirely independent of the base $m$ envisaged.

4. This positional notation may be extended easily to fractions with a base power for denominator, and thus to a simple and coherent notation for all the other numbers, rational and irrational, by dint of a point, following developments in positive and negative powers of $m$, thus analogous to decimal numbers.

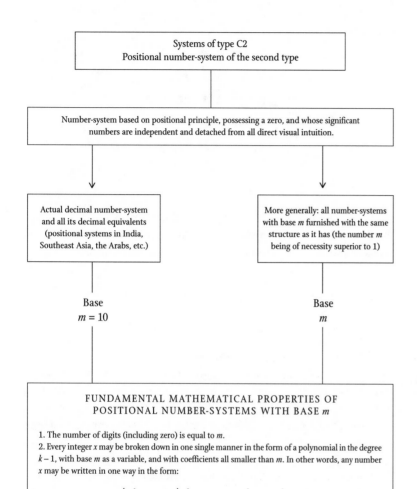

FIG. 23.28. *Classification of positional number-system (Type C2)*

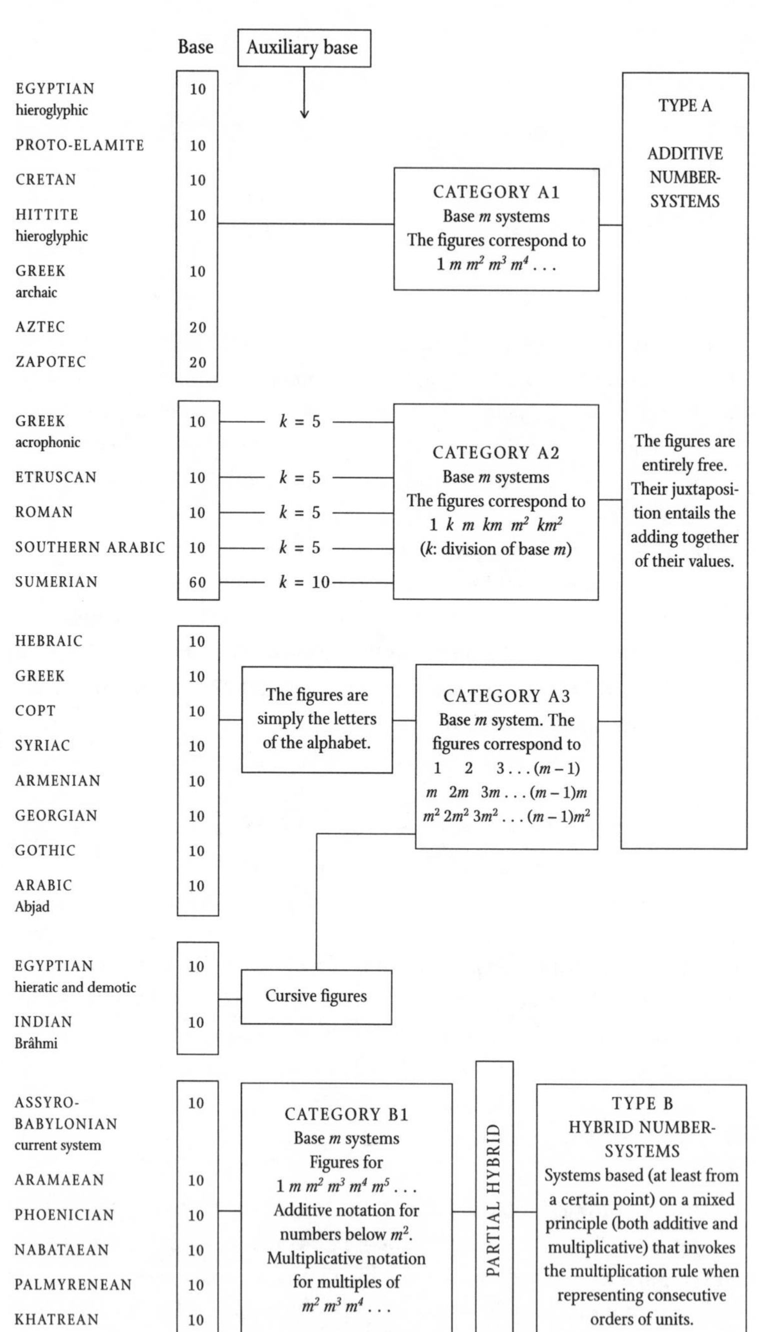


FIG. 23.29A. *Classification of written number-systems*

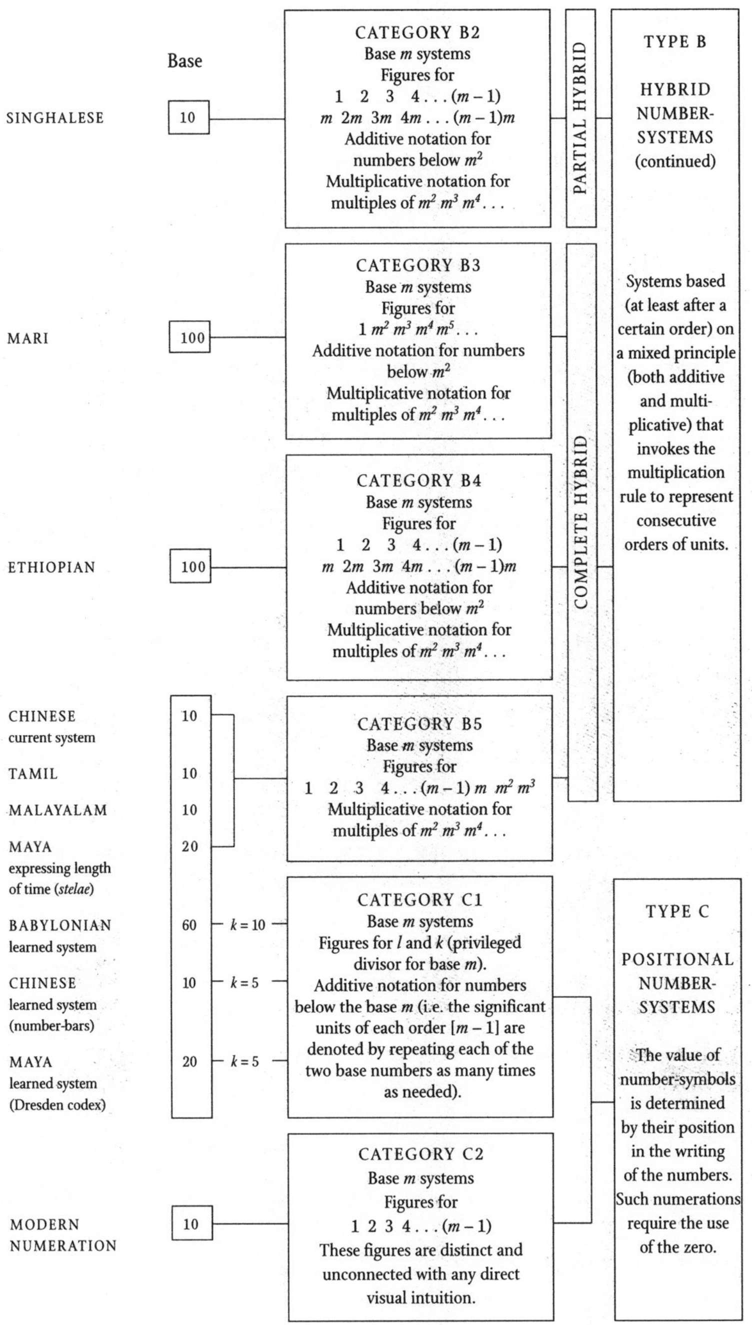


FIG. 23.29B. *Classification of written number-systems*

The story begins with primitive systems whose structure was based on the realities encountered in the course of accounting operations in ancient times. A certain amount of progress in the right direction was made, resulting in the creation of number-systems distinctly superior to the incoherent Roman numerals. But the paths which were taken led to dead ends, because these procedures incorporated only addition.

The awkwardnesses of these representations, together with the need for rapid writing, then brought about the development of hybrid systems, very conveniently mirroring spoken language, of which they can be seen as a more or less faithful transcription, sometimes showing a polynomial structure identical to that of the counting table, and at the same time extending considerably the power to express large numbers. Here too, however, the road was blocked. The principle they incorporated was inappropriate for arithmetical calculation, allowing addition and subtraction at best though at the cost of complicated manoeuvres, but useless for multiplication or division. In short, these systems were really only adequate for noting and recording numbers.

The decisive step in the adoption of systems of numerical notation with unlimited capacity, simple, rational, and immediately useable for calculation, could only be taken by inventing a well-conceived positional notation. This step was finally taken by simplifying hybrid notation, or by abbreviating systems for transferring numbers to the abacus, by the suppression of the signs indicating the powers of the base or by eliminating the columns of the abacus itself.

On the other hand this progress demanded a much higher level of abstraction, and the most delicate concept of the whole story: the zero. This was the supreme and belated discovery of the mathematicians who soon would come to extend it, from its first role of representing empty space, to embrace the truly numeric meaning of a null quantity (Fig. 23.27).

## THE KEYSTONE OF OUR MODERN NUMBER-SYSTEM

Number and culture are one, for "to know how a people counts is to know what kind of people it is" (to adapt Charles Morazé). At least from this point of view, the degree of civilisation of a people becomes something measurable.

Thus it now appears to us indisputable that the Babylonians, the Chinese and the Maya were superior to the Egyptians, the Hebrews and the Greeks. For, while the former took the lead with their fundamental discoveries of the principle of position and the zero, the others remained locked up for centuries with number-systems which were primitive, incoherent, and unuseable for practically any purpose save writing numbers down.

The measure of the genius of Indian civilisation, to which we owe our modern system, is all the greater in that it was the only one in all history to have achieved this triumph.

Some cultures succeeded, earlier than the Indian, in discovering one or at best two of the characteristics of this intellectual feat. But none of them managed to bring together into a complete and coherent system the necessary and sufficient conditions for a number-system with the same potential as our own.

We shall see in Chapter 24 that this system began in India more than fifteen centuries ago, with the improbable conjunction of three great ideas (Fig. 23.26), namely:

- the idea of attaching to each basic figure graphical signs which were removed from all intuitive associations, and did not visually evoke the units they represented;
- the idea of adopting the principle according to which the basic figures have a value which depends on the position they occupy in the representation of a number;
- finally, the idea of a fully operational zero, filling the empty spaces of missing units and at the same time having the meaning of a null number.

This fundamental realisation therefore profoundly changed human existence, by bringing a simple and perfectly coherent notation for all numbers and allowing anyone, even those most resistant to elementary arithmetic, the means to easily perform all sorts of calculations; also by henceforth making it possible to carry out operations which previously, since the dawn of time, had been inconceivable; and opening up thereby the path which led to the development of mathematics, science and technology.

It is also the ultimate perfection of numerical notation, as we shall see in the classification of the numerical notations of history to follow. In other words, no further improvement of numerical notation is necessary, or even possible, once this perfect number-system has been invented. Once this discovery had been made, the only possible changes remaining could only affect

- the choice of base (which could be 2, 8, 12, or any other number greater than 2);
- the graphical form of the figures.

But no further change is possible in the essential structure of the system, now once and for all unchangeable by virtue of its mathematical perfection.

Apart from the base (which is only a matter of how things are to be grouped, and therefore of the number of different basic figures for the units), a number-system structurally identical to ours is completely independent of its symbolism. It does not matter if the symbols are conventional graphic signs, letters of the alphabet, or even spoken words, provided it rests strictly and rigorously on the principle of position and it incorporates the full concept of the symbol for zero.

Here is an instructive example. It concerns the great Jewish scholar Rabbi Abraham Ben Meir ibn Ezra of Spain, better known as Rabbi Ben Ezra. He was born at Toledo around 1092, and in 1139 undertook a long journey to the East, which he completed after passing some years in Italy. Then he lived in the South of France, before emigrating to England where he died in 1167. No doubt influenced by his encounters while travelling, he instructed himself in the methods of calculation which had come out of India (precursors of our own). He then set out the principal rules of these in a work in Hebrew entitled *Sefer ha mispar* ("The Book of Number") [M. Silberberg (1895); M. Steinschneider (1893)].

Instead of conforming strictly to the graphics of the original Indian figures, he preferred to represent the first nine whole numbers by the first nine letters of the Hebrew alphabet (which, of course, he knew well since childhood). And, instead of adopting the old additive principle, on which the alphabetic Hebrew number-system had always been based (Fig. 23.12), he eliminated from his own system every letter which had a value greater than or equal to 10. He kept only the following nine, to which he applied the principle of position, and he augmented the series with a supplementary sign in the shape of a circle, which he called either *sifra* (from the Arab word for "empty") or *galgal* (the Hebrew word for "wheel"):

| א | ב | ג | ד | ה | ו | ז | ח | ט |
|---|---|---|---|---|---|---|---|---|
| 1 | 2 | 3 | 4 | 5 | 6 | 7 | 8 | 9 |
| aleph | bet | gimmel | dalet | he | vov | zayin | het | tet |

Thus, instead of representing the number 200,733 in the traditional Hebrew form (below, on the right), he wrote it as follows (below, left):

| ג | ג | ז | 0 | 0 | ב | instead of | ג | ל | ש | ת | ר̈ |
|---|---|---|---|---|---|---|---|---|---|---|---|
| 3 | 3 | 7 | 0 | 0 | 2 | | 3 | 30 | 300 | 400 | 200,000 |

Thus it was that the Hebrew number-system, in his hands, changed from a very primitive static decimal notation, by becoming adapted to the principle of position and the concept of zero, into a system with a structure rigorously identical to our own and, therefore, infinitely more dynamic.

However, this remarkable transformation seems not to have been followed by anyone other than Rabbi Ben Ezra himself, a unique case, it would seem, in the history of this system.

This isolated case, nonetheless, provides us with a model for a situation which must have come about many times following the invention and propagation of the positional system originating in India, mother of the modern system and of all those influenced by it. This is the situation in which scholars and calculators making contact, individually or in groups, with Indian civilisation and then, becoming aware of the ingenuity and many merits of their positional number-system, decide either to adopt it (individually or collectively) in its entirety or else to borrow its structure in order to perfect their own traditional systems.

Now that we can stand back from the story, the birth of our modern number-system seems a colossal event in the history of humanity, as momentous as the mastery of fire, the development of agriculture, or the invention of writing, of the wheel, or of the steam engine.

## THE CLASSIFICATION OF THE WRITTEN NUMBER-SYSTEMS OF HISTORY

With this survey we shall close our chapter. Its aim is to systematise the various comparisons we have made up to this point in a more formal and mathematical manner.

Before I enter into the heart of the matter, I wish at this point to render special homage to Geneviève Guitel, whose remarkable *Classification hiérarchisée des numérations écrites* has, for the first time, permitted me to bring together, intellectually speaking, systems which distance and time have separated almost totally.

This classification was published in her monumental *Histoire comparée des numérations écrites*, which has been an essential contribution to my understanding of this field.

Prior to her, as Charles Morazé has emphasised, there were certainly other histories of the number-systems, but none has attributed such importance to the comparisons which she has established on the basis of a principle of classification "which has the double merit of being both mathematically rigorous and remarkably relevant to the historical data which were to be put in order".

This classification, which I take up in my turn (while presenting it under a new light and amending certain details, resulting especially from the most recent archaeological discoveries), reveals that the numerical notations devised over five thousand years of history and evolution were not of unlimited variety. They may in fact be divided into three main types, of which each may be subdivided into various categories (Fig. 23.29):

- *additive systems*, which fundamentally are simply transcriptions of even more ancient concrete methods of counting (Fig. 23.30 to 23.32);
- *hybrid systems*, which were merely written transcriptions of more or less organised verbal expressions of number (Fig. 23.33 to 23.37);
- *positional systems*, which exhibit the ultimate degree of abstraction and therefore represent the ultimate perfection of numerical notation (Fig. 23.28 and 23.38).

### NUMBER-SYSTEMS OF THE ADDITIVE TYPE

These are the ones based on the *principle of addition*, where each figure has a characteristic value independent of its position in a representation. Number-systems of this type in turn fall into three categories.

*Additive number-systems of the first kind*

Our model for this is the Egyptian hieroglyphic system, which assigns a separate symbol to unity and to each power of 10, and uses repetitions of these signs to denote other numbers (Fig. 23.1).

CLASSIFICATION OF ADDITIVE NUMBER-SYSTEMS

Number-systems of this type fall into three kinds whose mathematical characteristics are summed up in Fig. 23.30 to 23.32: they require the adoption of a new writing convention based on a certain order of magnitude in order to note down high numbers.

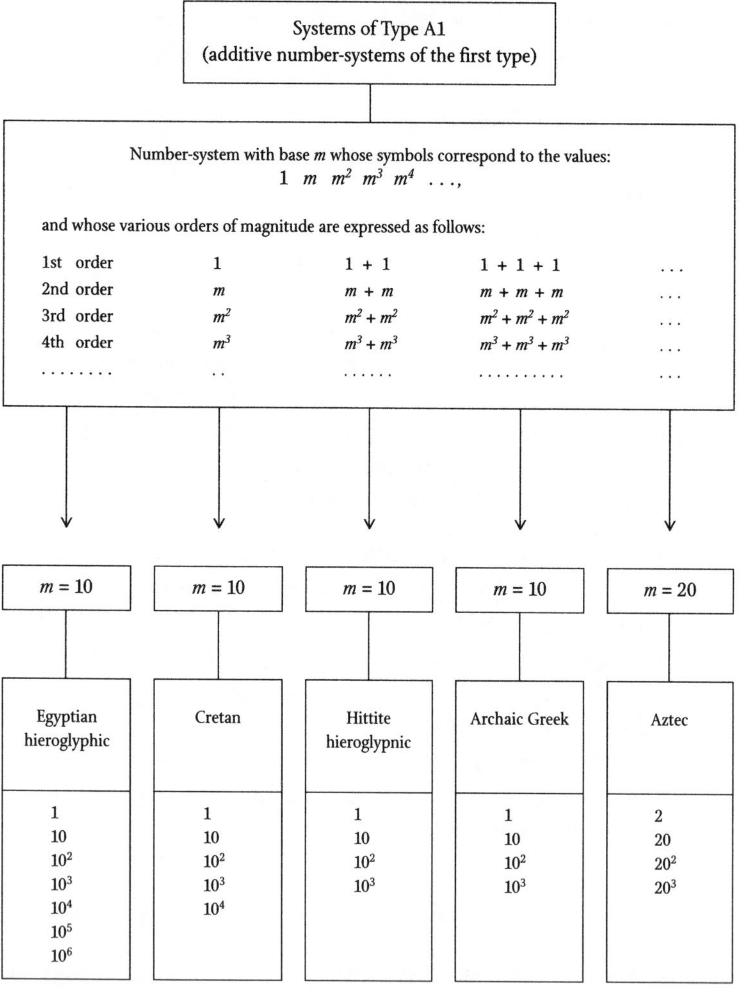


FIG. 23.30. *Classification of additive number-systems (Type A1)*

This is exactly what also happens in the Cretan number-system and in the Hittite hieroglyphic and archaic Greek systems. All of these systems are therefore strictly identical, and they differ only in the written forms of their respective figures (Fig. 23.3 to 23.5).

When they are in base 10, the additive systems of the first kind are therefore characterised by a notation which is based on arithmetical decompositions of the type:

Table 1

| 1st decimal order (units) | 2nd decimal order (tens) | 3rd decimal order (hundreds) | 4th decimal order (thousands) |
|---|---|---|---|
| 1 | 10 | $10^2$ | $10^3$ |
| 1 + 1 | 10 + 10 | $10^2 + 10^2$ | $10^3 + 10^3$ |
| 1 + 1 + 1 | 10 + 10 + 10 | $10^2 + 10^2 + 10^2$ | $10^3 + 10^3 + 10^3$ |
| Special notation for 1, 10, $10^2$, $10^3$, etc.<br>Additive notation for all other numbers. | | | |

Now if we consider the Aztec number-system, we find that even though it uses a different base (base 20), still like the others it assigns a special symbol only to unity and to the powers of the base (Fig. 23.6):

| Base | *Aztec system* | | | | | |
|---|---|---|---|---|---|---|
| 20 | 1 | 20 | $20^2$ | $20^3$ | $20^4$ | ... |
| $m$ | 1 | $m$ | $m^2$ | $m^3$ | $m^4$ | ... |
| 10 | 1 | 10 | $10^2$ | $10^3$ | $10^4$ | ... |

*Egyptian hieroglyphic system*

Since this is an additive system and proceeds by repetition of identical signs, it is characterised by a notation which depends on arithmetical decompositions of the type:

Table 2

| 1st vigesimal order (units) | 2nd vigesimal order (twenties) | 3rd vigesimal order (four hundreds) | 4th vigesimal order (eight thousands) |
|---|---|---|---|
| 1 | 20 | $20^2$ | $20^3$ |
| 1 + 1 | 20 + 20 | $20^2 + 20^2$ | $20^3 + 20^3$ |
| 1 + 1 + 1 | 20 + 20 + 20 | $20^2 + 20^2 + 20^2$ | $20^3 + 20^3 + 20^3$ |
| Special notation for 1, 20, $20^2$, $20^3$, etc.<br>Additive notation for all other numbers. | | | |

The Aztec system, therefore, is intellectually related to the preceding ones, and differs only in having base 20 instead of base 10.

All of these notations therefore belong to the same type (Fig. 23.30).

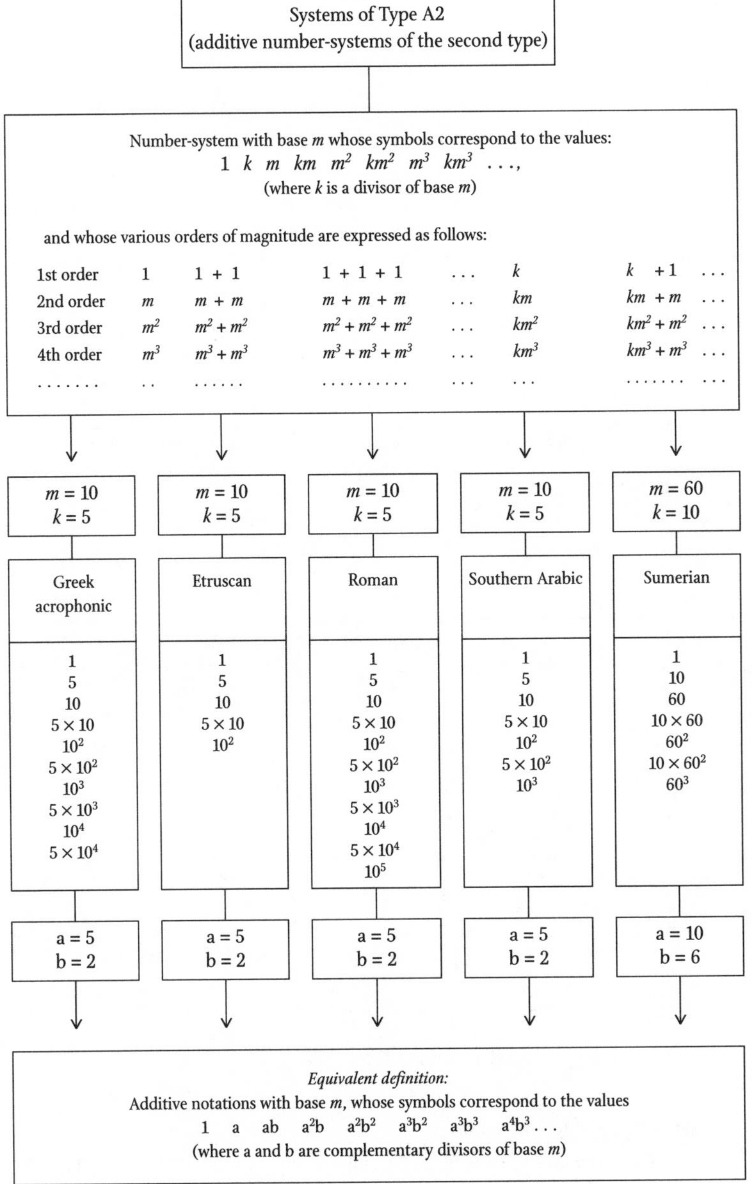


FIG. 23.31. *Classification of additive number-systems (type A2)*

*Additive systems of the second kind*

A characteristic example is the Greek acrophonic system. It is in base 10, and adopts the principle of addition assigning a special symbol to each of the numbers 1, 10, 100, 1,000, etc., as well as to each of the following: 5, 50, 500, 5,000, and so on (Fig. 23.7). Intellectually, therefore, it is of the same kind as the Southern Arabic system, the Etruscan, and the Roman, characterised by arithmetical decompositions of the type (Fig. 16.18, 16.35 and 23.8):

Table 3

| 1st decimal order (units) | 2nd decimal order (tens) | 3rd decimal order (hundreds) | 4th decimal order (thousands) |
|---|---|---|---|
| 1 | 10 | $10^2$ | $10^3$ |
| 1 + 1 | 10 + 10 | $10^2 + 10^2$ | $10^3 + 10^3$ |
| 1 + 1 + 1 | 10 + 10 + 10 | $10^2 + 10^2 + 10^2$ | $10^3 + 10^3 + 10^3$ |
| . . . | . . . | . . . | . . . |
| 5 | $5 \times 10$ | $5 \times 10^2$ | $5 \times 10^3$ |
| 5 + 1 | $5 \times 10 + 10$ | $5 \times 10^2 + 10^2$ | $5 \times 10^3 + 10^3$ |
| 5 + 1 + 1 | $5 \times 10 + 10 + 10$ | $5 \times 10^2 + 10^2 + 10^2$ | $5 \times 10^3 + 10^3 + 10^3$ |
| Special notation for 1, 5, 10, $5 \times 10$, $10^2$, $5 \times 10^2$, etc. Additive notation for all other numbers. | | | |

Denoting by $k$ the divisor of the base $m$ which thus acts as auxiliary base (here, $m = 10$ and $k = 5$), we see that these systems assign a special symbol not only to each power of the base ($1, m, m^2, m^3, \ldots$) but also to the product of each of these with $k$ ($k, km, km^2, km^3, \ldots$). As the following table shows, this is exactly the structure which can be discerned in the regular progression of the Sumerian number-system (Fig. 23.2):

*Sumerian system* (where $m$=60 and $k$=10)

| 1 | 10 | 60 | $10 \times 60$ | $60^2$ | $10 \times 60^2$ | $60^3$ | $10 \times 60^3$ |
|---|---|---|---|---|---|---|---|
| 1 | $k$ | $m$ | $km$ | $m^2$ | $km^2$ | $m^3$ | $km^3$ |
| 1 | 5 | 10 | $5 \times 10$ | $10^2$ | $5 \times 10^2$ | $10^3$ | $5 \times 10^3$ |

*Greek acrophonic numerals* (where $m = 10$ and $k = 5$)

Looking at it from another point of view, the succession of numbers receiving a particular sign in the Sumerian system may be expressed as:

1st order 1 <· · · · · · · · · · · · · · · · · · · · ·> 1
10 <· · · · · · · · · · · · · · · · · · · · ·> 10
2nd order 60 <· · · · · · · · · · · · · · · · · · · · ·> 10.6
$10 \times 60$ <· · · · · · · · · · · · · · · · · · · · ·> 10.6.10
3rd order $60^2$ <· · · · · · · · · · · · · · · · · · · · ·> 10.6.10.6
$10 \times 60^2$ <· · · · · · · · · · · · · · · · · · · · ·> 10.6.10.6.10
4th order $60^3$ <· · · · · · · · · · · · · · · · · · · · ·> 10.6.10.6.10.6
$10 \times 60^3$ <· · · · · · · · · · · · · · · · · · · · ·> 10.6.10.6.10.6.10

and so on, alternating the numbers 10 and 6.

Let $a$ and $b$ denote the divisors of m which act as alternating auxiliary bases (where, in the Sumerian case, we have $m = 60$, $a = 10$ and $b = 6$). This succession therefore exactly corresponds to that of the Greek acrophonic system (where $m = 10$, $a = 5$ and $b = 2$):

Table 4

| Sumerian | | Mathematical Characterisation | | | | Greek |
|---|---|---|---|---|---|---|
| 1 | <· · · ·> | 1 | | | <· · · ·> | 1 |
| 10 | <· · · ·> | a | | | <· · · ·> | 5 |
| 10.6 | <· · · ·> | a.b | | | <· · · ·> | 5.2 |
| 10.6.10 | <· · · ·> | $a^2b$ | = | a.b.a | <· · · ·> | 5.2.5 |
| 10.6.10.6 | <· · · ·> | $a^2b^2$ | = | a.b.a.b | <· · · ·> | 5.2.5.2 |
| 10.6.10.6.10 | <· · · ·> | $a^3b^2$ | = | a.b.a.b.a | <· · · ·> | 5.2.5.2.5 |
| 10.6.10.6.10.6 | <· · · ·> | $a^3b^3$ | = | a.b.a.b.a.b | <· · · ·> | 5.2.5.2.5.2 |
| ⋮ · · · · · · · · · · · · · · | ·> | a = 10<br>b = 6 | | a = 5<br>b = 2 | <· · · · · · · · · · · · · · | ⋮ |

The Greek structure is thus mathematically identical to that of the Sumerian, corresponding to arithmetical decompositions of the type:

| 1st sexagesimal order (units) | 2nd sexagesimal order (sixties) | 3rd sexagesimal order (multiples of 60) | 4th sexagesimal order (multiples of 60) |
|---|---|---|---|
| 1 | 60 | $60^2$ | $60^3$ |
| 1 + 1 | 60 + 60 | $60^2 + 60^2$ | $60^3 + 60^3$ |
| 1 + 1 + 1 | 60 + 60 + 60 | $60^2 + 60^2 + 60^2$ | $60^3 + 60^3 + 60^3$ |
| . . . | . . . | . . . | . . . |
| 10 | $10 \times 60$ | $10 \times 60^2$ | $10 \times 60^3$ |
| 10 + 1 | $10 \times 60 + 60$ | $10 \times 60^2 + 60^2$ | $10 \times 60^3 + 60^3$ |
| 10 + 10 | $10 \times 60 + 10 \times 60$ | $10 \times 60^2 + 10 \times 60^2$ | $10 \times 60^3 + 10 \times 60^3$ |
| Special notation for 1, 10, 60, $10 \times 60$, $60^2$, $10 \times 60^2$, etc.<br>Additive notation for all other numbers. | | | |

All these systems therefore belong to the same category (Fig. 23.31).

CLASSIFICATION OF ADDITIVE NUMBER-SYSTEMS (concluded)

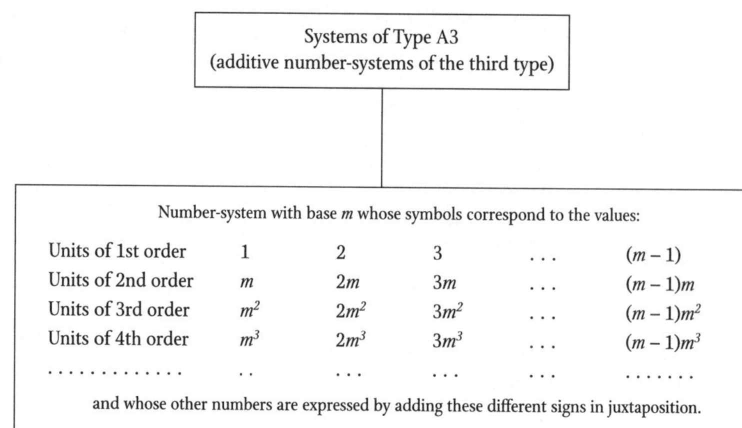


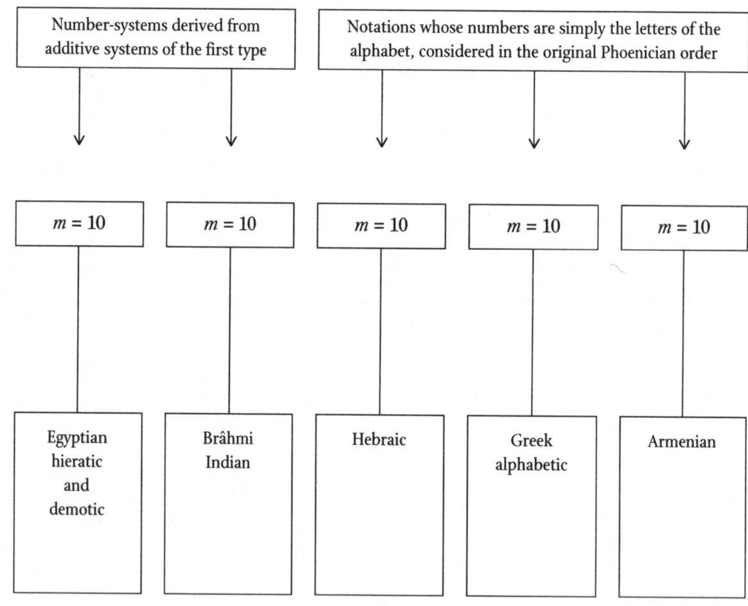


FIG. 23.32. *Classification of additive number-systems (Type A3)*

*Additive systems of the third kind*

The Egyptian hieratic system and the Greek alphabetic system are typical examples of this type. Intellectually, they correspond to the following characterisation (Fig. 23.10 to 23.13, and 23.32):

| 1st decimal order (units) | 2nd decimal order (tens) | 3rd decimal order (hundreds) | 4th decimal order (thousands) |
|---|---|---|---|
| 1 | 10 | 100 | 1,000 |
| 2 | 20 | 200 | 2,000 |
| 3 | 30 | 300 | 3,000 etc. |

Special notation for each unit of each number.

| | | | | | | | | |
|---|---|---|---|---|---|---|---|---|
| 1 | 2 | 3 | 4 | 5 | 6 | 7 | 8 | 9 |
| 10 | 2.10 | 3.10 | 4.10 | 5.10 | 6.10 | 7.10 | 8.10 | 9.10 |
| $10^2$ | $2.10^2$ | $3.10^2$ | $4.10^2$ | $5.10^2$ | $6.10^2$ | $7.10^2$ | $8.10^2$ | $9.10^2$ |
| $10^3$ | $2.10^3$ | $3.10^3$ | $4.10^3$ | $5.10^3$ | $6.10^3$ | $7.10^3$ | $8.10^3$ | $9.10^3$ |
| $10^4$ | $2.10^4$ | $3.10^4$ | $4.10^4$ | $5.10^4$ | $6.10^4$ | $7.10^4$ | $8.10^4$ | $9.10^4$ |
| ..... | ..... | ..... | ..... | ..... | ..... | ..... | ..... | ..... |

Additive notation for all other numbers.

## SYSTEMS OF HYBRID TYPE

These are founded on a mixed system in which both addition and multiplication are involved. On this basis, the multiples of the powers of the base are, from a certain order of magnitude onwards, expressed multiplicatively. This type of system can be divided into five categories.

### CLASSIFICATION OF HYBRID NUMBER-SYSTEMS

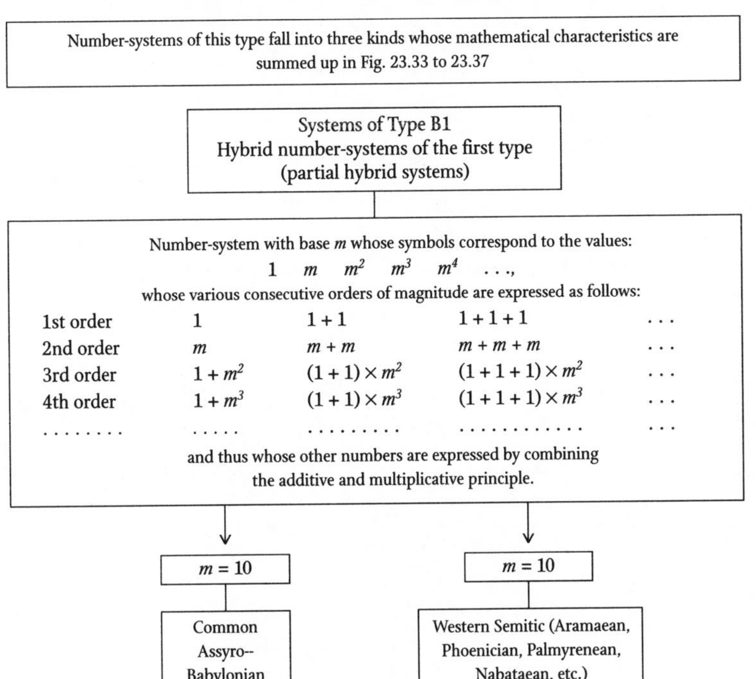
Number-systems of this type fall into three kinds whose mathematical characteristics are summed up in Fig. 23.33 to 23.37
Systems of Type B1
Hybrid number-systems of the first type
(partial hybrid systems)
Number-system with base $m$ whose symbols correspond to the values:
$1 \quad m \quad m^2 \quad m^3 \quad m^4 \quad \ldots,$
whose various consecutive orders of magnitude are expressed as follows:
1st order $1 \quad 1+1 \quad 1+1+1 \quad \ldots$
2nd order $m \quad m+m \quad m+m+m \quad \ldots$
3rd order $1+m^2 \quad (1+1)\times m^2 \quad (1+1+1)\times m^2 \quad \ldots$
4th order $1+m^3 \quad (1+1)\times m^3 \quad (1+1+1)\times m^3 \quad \ldots$
and thus whose other numbers are expressed by combining the additive and multiplicative principle.
$m = 10$
$m = 10$
Common Assyro-Babylonian
Western Semitic (Aramaean, Phoenician, Palmyrenean, Nabataean, etc.)

FIG. 23.33. *Classification of hybrid number-systems (Type B1, partial hybrid)*

### *Hybrid systems of the first kind*

The common Assyro-Babylonian system and that of the western Semitic peoples (Aramaeans, Phoenicians, etc.) are typical examples of this type. They have base 10, and assign a special symbol to each of the numbers 1, 10, 100, 1,000, etc., and use multiplicative notation for consecutive multiples of each of these powers of 10. At the same time, the units and the tens are still represented according to the old principle of additive juxtaposition.

When in base 10, hybrid systems of the first kind are characterised by arithmetical decompositions of the type (Fig. 23.16, 23.17, and 23.33):

Table 5

| 1st order (units) | 2nd order (tens) | 3rd order (hundreds) | 4th order (thousands) |
|---|---|---|---|
| 1 | 10 | $1 \times 10^2$ | $1 \times 10^3$ |
| 1 + 1 | 10 + 10 | $(1 + 1) \times 10^2$ | $(1 + 1) \times 10^3$ |
| 1 + 1 + 1 | 10 + 10 + 10 | $(1 + 1 + 1) \times 10^2$ | $(1 + 1 + 1) \times 10^3$ |

Special notation for 1, 10, $10^2$, $10^3$, etc. Additive notation for the numbers 1 to 99.
Multiplicative notation for the multiples of the powers of 10, starting with 100.
A notation involving both addition and multiplication for other numbers

### CLASSIFICATION OF HYBRID NUMBER-SYSTEMS (continued)

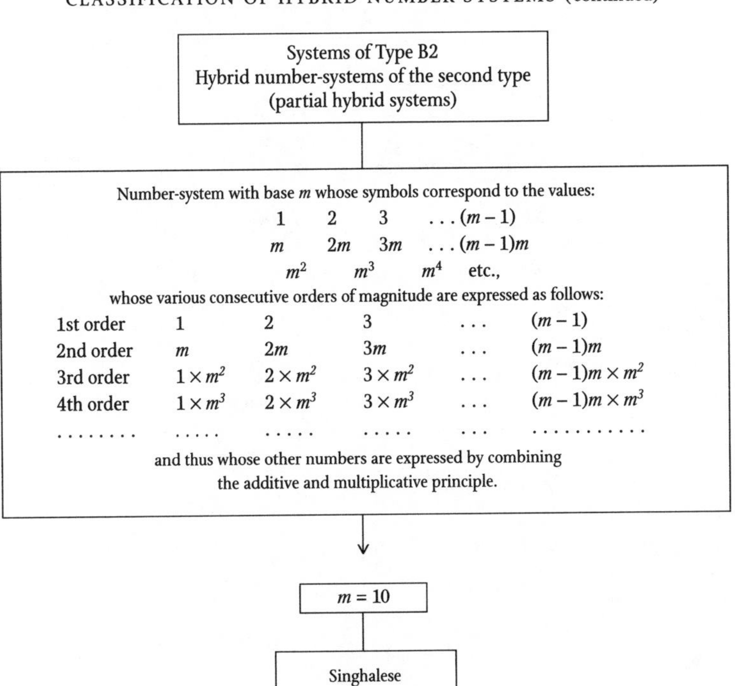
Systems of Type B2
Hybrid number-systems of the second type
(partial hybrid systems)
Number-system with base $m$ whose symbols correspond to the values:
$1 \quad 2 \quad 3 \quad \ldots (m-1)$
$m \quad 2m \quad 3m \quad \ldots (m-1)m$
$m^2 \quad m^3 \quad m^4$ etc.,
whose various consecutive orders of magnitude are expressed as follows:
1st order $1 \quad 2 \quad 3 \quad \ldots \quad (m-1)$
2nd order $m \quad 2m \quad 3m \quad \ldots \quad (m-1)m$
3rd order $1\times m^2 \quad 2\times m^2 \quad 3\times m^2 \quad \ldots \quad (m-1)m\times m^2$
4th order $1\times m^3 \quad 2\times m^3 \quad 3\times m^3 \quad \ldots \quad (m-1)m\times m^3$
and thus whose other numbers are expressed by combining the additive and multiplicative principle.
$m = 10$
Singhalese

FIG. 23.34. *Classification of hybrid number-systems (Type B3, complete hybrid)*

*Hybrid systems of the second kind*

The model for this type is the Singhalese system. It has base 10, and assigns a special symbol to each unit, to each of the tens, and to each of the powers of 10. The notation for the hundreds, thousands, etc. follows the multiplicative rule (Fig. 23.18).

When in base 10, hybrid systems of this kind are characterised by a notation which is based on arithmetical decompositions of the type (Fig. 23.34):

| 1st order (units) | 2nd order (tens) | 3rd order (hundreds) | 4th order (thousands) |
|---|---|---|---|
| 1 | 10 | $1 \times 10^2$ | $1 \times 10^3$ |
| 2 | 20 | $2 \times 10^2$ | $2 \times 10^3$ |
| 3 | 30 | $3 \times 10^2$ | $3 \times 10^3$ |
| Special notation for units, tens, $10^2$, $10^3$, etc. Additive notation for the numbers 1 to 99. Multiplicative notation for the multiples of the powers of 10, starting with 100. A notation involving both addition and multiplication for other numbers | | | |

CLASSIFICATION OF HYBRID NUMBER-SYSTEMS (continued)

Systems of Type B3
Hybrid number-systems of the third type
(complete hybrid systems)

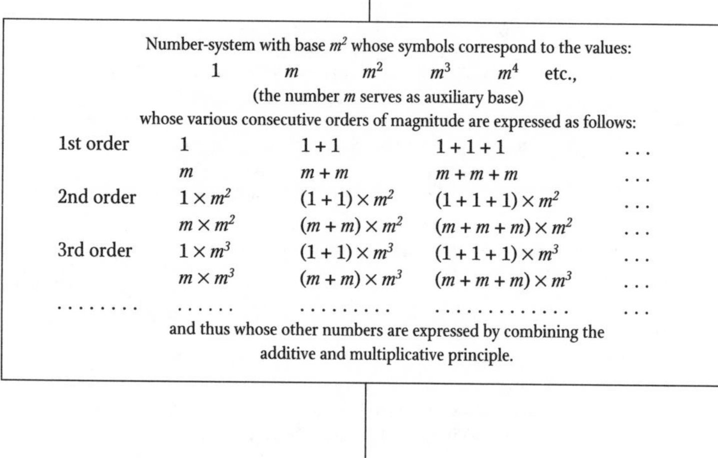

Number-system with base $m^2$ whose symbols correspond to the values:
1 $m$ $m^2$ $m^3$ $m^4$ etc.,
(the number $m$ serves as auxiliary base)
whose various consecutive orders of magnitude are expressed as follows:
1st order 1 1 + 1 1 + 1 + 1 . . .
$m$ $m + m$ $m + m + m$ . . .
2nd order $1 \times m^2$ $(1 + 1) \times m^2$ $(1 + 1 + 1) \times m^2$ . . .
$m \times m^2$ $(m + m) \times m^2$ $(m + m + m) \times m^2$ . . .
3rd order $1 \times m^3$ $(1 + 1) \times m^3$ $(1 + 1 + 1) \times m^3$ . . .
$m \times m^3$ $(m + m) \times m^3$ $(m + m + m) \times m^3$ . . .
and thus whose other numbers are expressed by combining the additive and multiplicative principle.


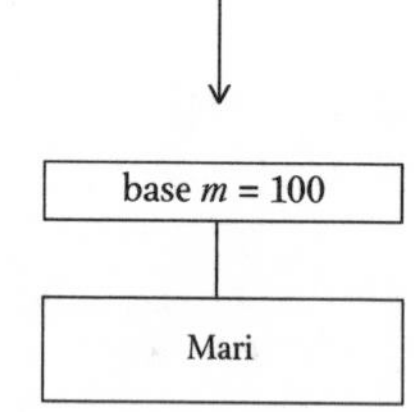


FIG. 23.35.

*Hybrid systems of the third kind*

The model for this type is the Mari system. It uses base 100, and gives a special symbol for each unit, for 10, and for each power of 100. The notation for the hundreds, the ten thousands, etc. uses the multiplicative rule. The system is characterised by a notation based on arithmetical decompositions of the type (Fig. 23.22 and 23.35):

| 1st centennial order | | 2nd centennial order | |
|---|---|---|---|
| units | tens | hundreds | thousands |
| 1 | 10 | $1 \times 10^2$ | $1 \times 10^3$ |
| 1 + 1 | 10 + 10 | $(1 + 1) \times 10^2$ | $(1 + 1) \times 10^3$ |
| 1 + 1 + 1 | 10 + 10 + 10 | $(1 + 1 + 1) \times 10^2$ | $(1 + 1 + 1) \times 10^3$ |
| . . . | . . . | . . . | . . . |
| Special notation for 1, 10, $10^2$, $10^3$, etc. Additive notation for the numbers from 1 to 99. Additive notation for the numbers 1 to 99. Multiplicative notation for multiples of the powers of $10^2$, starting with the first (100). A notation involving both addition and multiplication for other numbers. | | | |

CLASSIFICATION OF HYBRID NUMBER-SYSTEMS (continued)

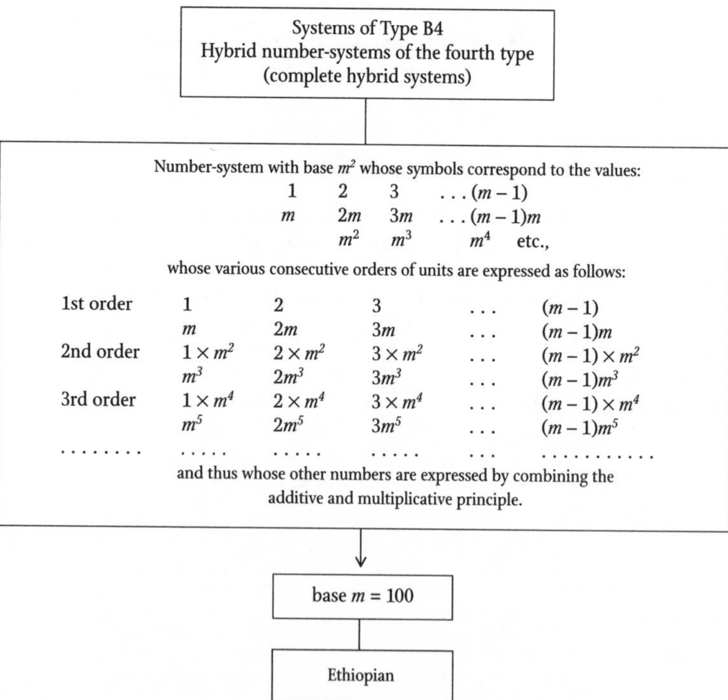


FIG. 23.36. *Classification of hybrid number-systems (Type B4, complete hybrid)*

*Hybrid systems of the fourth kind*

The model for this type is the Ethiopian system. It has base 100, and assigns a special sign to each unit and to each of the tens, and also to each power of 100. The notation for the hundreds, the ten thousands, etc. uses a multiplicative rule applied to these figures. The system is characterised by a notation based on arithmetical decompositions of the type (Fig. 23.36):

| 1st centennial order | | 2nd centennial order | |
|---|---|---|---|
| units | tens | hundreds | thousands |
| 1 | 10 | $1 \times 10^2$ | $1 \times 10^3$ |
| 2 | 20 | $2 \times 10^2$ | $2 \times 10^3$ |
| 3 | 30 | $3 \times 10^2$ | $3 \times 10^3$ |
| 4 | 40 | $4 \times 10^2$ | $4 \times 10^3$ |
| 5 | 50 | $5 \times 10^2$ | $5 \times 10^3$ |
| 6 | 60 | $6 \times 10^2$ | $6 \times 10^3$ |
| 7 | 70 | $7 \times 10^2$ | $7 \times 10^3$ |
| 8 | 80 | $8 \times 10^2$ | $8 \times 10^3$ |
| 9 | 90 | $9 \times 10^2$ | $9 \times 10^3$ |

Special notation for each unit, each ten and for each of $10^2$, $10^3$, etc. Additive notation for the numbers from 1 to 99. Multiplicative notation for multiples of the powers of $10^2$, starting with the first (100). A notation involving both addition and multiplication for other numbers.

CLASSIFICATION OF HYBRID NUMBER-SYSTEMS (concluded)

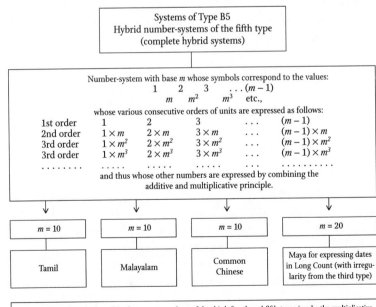

Systems of Type B5
Hybrid number-systems of the fifth type
(complete hybrid systems)

Number-system with base $m$ whose symbols correspond to the values:

$1 \quad 2 \quad 3 \quad \ldots (m-1)$
$m \quad m^2 \quad m^3$ etc.,

whose various consecutive orders of units are expressed as follows:

| | | | | | |
|---|---|---|---|---|---|
| 1st order | 1 | 2 | 3 | ... | $(m-1)$ |
| 2nd order | $1 \times m$ | $2 \times m$ | $3 \times m$ | ... | $(m-1) \times m$ |
| 3rd order | $1 \times m^2$ | $2 \times m^2$ | $3 \times m^2$ | ... | $(m-1) \times m^2$ |
| 3rd order | $1 \times m^3$ | $2 \times m^3$ | $3 \times m^3$ | ... | $(m-1) \times m^3$ |
| ........ | ..... | ..... | ..... | ... | .......... |

and thus whose other numbers are expressed by combining the additive and multiplicative principle.

| $m = 10$ | $m = 10$ | $m = 10$ | $m = 20$ |
|---|---|---|---|
| Tamil | Malayalam | Common Chinese | Maya for expressing dates in Long Count (with irregularity from the third type) |

Unlike the hybrid numbering of the first two types, those of the third, fourth and fifth types invoke the multiplicative principle in the notation of all units superior or equal to the base. Hence they are termed "complete hybrid". Thus the representation of the numbers is made here by expressing the various numerical values of a polynomial with the corresponding base $m$ as a variable (apart, that is, from the Maya system which embraces an irregularity in the values of its consecutive orders of units).

FIG. 23.37. *Classification of hybrid number-systems (Type B5, complete hybrid)*

*Hybrid systems of the fifth kind*

The model for this type is the common Chinese system, as well as the Tamil and Malayalam systems. These systems have base 10, and assign a special symbol to each unit and to each power of 10. The notation for the tens, the hundreds, the thousands, etc. uses the multiplicative principle.

When in base 10, hybrid systems of the fifth kind are characterised by a notation based on arithmetical decompositions of the following type (Fig. 23.37):

| 1st order (units) | 2nd order (tens) | 3rd order (hundreds) | 4th order (thousands) |
|---|---|---|---|
| 1 | $1 \times 10$ | $1 \times 10^2$ | $1 \times 10^3$ |
| 2 | $2 \times 10$ | $2 \times 10^2$ | $2 \times 10^3$ |
| 3 | $3 \times 10$ | $3 \times 10^2$ | $3 \times 10^3$ |
| 4 | $4 \times 10$ | $4 \times 10^2$ | $4 \times 10^3$ |
| 5 | $5 \times 10$ | $5 \times 10^2$ | $5 \times 10^3$ |
| 6 | $6 \times 10$ | $6 \times 10^2$ | $6 \times 10^3$ |
| 7 | $7 \times 10$ | $7 \times 10^2$ | $7 \times 10^3$ |
| 8 | $8 \times 10$ | $8 \times 10^2$ | $8 \times 10^3$ |
| 9 | $9 \times 10$ | $9 \times 10^2$ | $9 \times 10^3$ |

Special notation for each unit of the first order and for each of the numbers 10, $10^2$, $10^3$. Multiplicative notation for multiples of powers of the base, starting with 10. Notation involving both addition and multiplication for other numbers.

Unlike hybrid systems of the first kind, which only partially use the multiplicative principle, those of types 3, 4 and 5 bring the principle into play in the notation for all the orders of magnitude greater than or equal to the base. Additionally, the representation of other numbers is based on the coefficients of a polynomial whose variable is the base. For these reasons systems of this type are also called complete hybrid systems.

## POSITIONAL SYSTEMS

The systems are based on the principle that the value of the figures is determined by their position in the representation of a number.

Historically, there have been only four originally created positional systems:

- the system of the Babylonian scholars;
- the system of the Chinese scholars;
- the system of the Mayan astronomer-priests;
- and finally our modern system which, as we shall see in the next chapter, originated in India.

These systems (which require the use of a zero) may be divided into two categories.

CLASSIFICATION OF POSITIONAL NUMBER-SYSTEMS

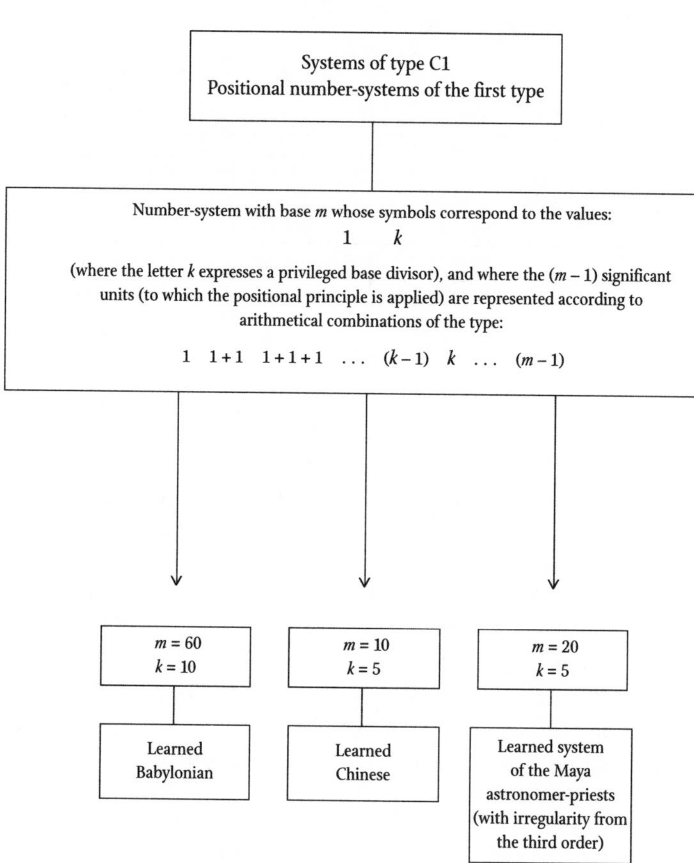


FIG. 23.38. *Classification of positional number-systems (Type C1)*

*Positional systems of the first kind*

This type includes:

1. – The *system of the Babylonian scholars*

This has base 60. The notation for the units of the first order (from 1 to 59) corresponds to arithmetical decompositions of the following type for the two basic figures, of which one represents unity and the other 10 (Fig. 23.23):

1 1 + 1 1 + 1 + 1 1 + 1 + 1 + 1 ...

10 10 + 10 10 + 10 + 10 10 + 10 + 10 + 10 ...

10 + 10 + 10 + 10 + 10 + 1 + 1 + 1 + 1 + 1 + 1 + 1 + 1 +1

2. – The *system of the Chinese scholars*

This has base 10. The notation for the units of the first order (from 1 to 9) corresponds to arithmetical decompositions of the following type for the two basic figures, of which one represents unity and the other 5 (Fig. 23.24):

1 1+1 1+1+1 1+1+1+1 5

5+1 5+1+1 5+1+1+1 5+1+1+1+1

3. – The *system of the Maya scholars*

This has base 20. The notation for the units of the first order (from 1 to 19) corresponds to arithmetical decompositions of the following type using two base figures, one representing unity and the other the number 5. In addition there is an irregularity starting with the third order in the succession of positional values (Fig. 23.25):

1 1+1 1+1+1 1+1+1+1 5

5+1 5+1+1 5+1+1+1 5+1+1+1+1 5+5

5+5+1 5+5+1+1 5+5+1+1+1 5+5+1+1+1+1 5+5+5

5+5+5+1 5+5+5+1+1 5+5+5+1+1+1 5+5+5+1+1+1+1

Systems of this type with base *m* use the principle of position, but they only possess two digits in the strict sense: one for unity, and the other for a particular divisor of the base, here denoted by *k*. The *m*–1 units are represented according to the additive principle (Fig. 23.38).

All of these systems clearly require a zero, and in the end have come to possess one, independently or not of outside influence.

*Positional systems of the second kind*

This category includes our own modern decimal notation, whose nine units are represented by figures (Fig. 23.26):

1 2 3 4 5 6 7 8 9

augmented by a tenth sign, written 0. Known as *zero*, this is used to mark the absence of units of a given rank, and at the same time enjoys a true numerical meaning, that of null number.

The fundamental characteristic of this system is that its conventions can be extended into a notation both simple and completely consistent for all numbers: integers, fractions, and irrationals (whether these be transcendental or not). In other words, the discovery of this system enables us to write down, in a simple and rational way, and using a completely natural extension of the principle of position and of the zero, not only fractions but entities such as $\sqrt{2}$, $\sqrt{3}$ or $\Pi$.

A decimal fraction is a fraction of which the denominator is equal to 10 or to a power of 10. 3/10, 1/100, 251/10,000 are therefore decimal fractions.

Now, the sequence of decimal fractions of unity (those which have numerator 1 and denominator a power of 10) has its terms called successively one tenth (or decimal unit of the first order), one hundredth (or decimal unit of the second order), one thousandth (or decimal unit of the third order), and so on:

$$\frac{1}{10} \quad \frac{1}{10^2} \quad \frac{1}{10^3} \quad \frac{1}{10^4} \quad \frac{1}{10^5} \quad \frac{1}{10^6} \quad \text{etc.}$$

Thus we have a sequence where each term is the product of its predecessor by 1/10, which means that the convention of our decimal notation applies here also, ten units of any order being equal to one unity of the order immediately above. These decimal units may therefore be unambiguously represented by a convention which extends the convention which applies to the integers, so that we may represent them in the form:

$$\begin{matrix} 0.1 & 0.01 & 0.001 & 0.0001, \text{ etc} \\ (= 10^{-1}) & (= 10^{-2}) & (= 10^{-3}) & (= 10^{-4}) \end{matrix}$$

If we now consider any decimal fraction, for example 39,654/1,000, we find its arithmetical decomposition according to the positive and negative powers of 10:

$$\frac{39{,}654}{1{,}000} = \frac{39{,}000}{1{,}000} + \frac{600}{1{,}000} + \frac{50}{1{,}000} + \frac{4}{1{,}000}$$

We observe therefore that this may be written in the form:

$$\frac{39{,}654}{1{,}000} = 39 + \frac{600}{1{,}000} + \frac{50}{1{,}000} + \frac{4}{1{,}000}$$

or, in accordance with the preceding convention:

$$\begin{aligned} \frac{39{,}654}{1{,}000} &= 39 + 0.6 + 0.05 + 0.004 \\ &= 39 + 6 \times 10^{-1} + 5 \times 10^{-2} + 4 \times 10^{-3} \end{aligned}$$

This number is therefore composed of 39 units, 6 tenths, 5 hundredths and 4 thousandths. Adopting the convention for the representation of the integers, one may make the convention of separating the integer units from the decimal units by a point, so that the fraction in question may now be put in the form:

$$\frac{39{,}654}{1{,}000} = 39.654$$

It is therefore expressed as a *decimal number* which can be read as 39 units and 654 thousandths.

Thus we see how the principle of position allows us to extend its application to decimal numbers.

One can also show that any number whatever can be expressed as a decimal number whose development may be finite or infinite (i.e. having a finite or an infinite number of figures following the decimal point).

One can therefore see the many mathematical advantages which flow from the discovery of our number-system.

But, clearly, this system is only a special case of the systems in this category. These are nowadays known as *systems* with base $m$, the number $m$ being at least equal to 2 ($m>1$). Historically speaking, these are simply positional systems with base $m$ furnished with a fully operational zero, whose ($m-1$) figures are independent of each other and without any direct visual significance (Fig. 23.28).

The written positional systems of the second kind are therefore the most advanced of all history. They allow the simple and completely rational representation of any number, no matter how large. Above all, they bring within the reach of everyone a simple method for arithmetical operations. And all this is independent of the choice of base (Fig. 23.29). It is precisely in these respects that our modern written number-system (or any one of its equivalents) is one of the foundations of the intellectual equipment of the modern human being.

# CHAPTER 24

# INDIAN CIVILISATION

## THE CRADLE OF MODERN NUMERALS

As G. Beaujouan (1950) has said, "the origin of the so-called 'Arabic' numerals has been written about so often that every view on the question seems plausible, and the only way of choosing between them is by personal conviction." Most of the literature (much of which is indeed of great value and has been used in the following pages) deals with one particular discipline from the many that are relevant to this tricky question of the origin of Arabic numerals. The few comprehensive works on the subject (Cajori, Datta and Singh, Guitel, Menninger, Pihan, Smith and Karpinski, or Woepcke) are now several decades old, and many discoveries have been made in more recent years. Since the beginning of the twentieth century, a wealth of reliable information has been compiled from the various specialised fields, and the findings all point to the fact that the number-system that we use today is of Indian origin. But no collective work has been produced that contained rigorous reasoning or an entirely satisfactory methodology. Moreover, the problem has been tackled in a somewhat loose manner in the past and was seen from a more limited and biased perspective than it is today. So it is well worth while going back to square one and looking at the question from a completely new angle, not only in the light of the results seen in the previous chapter and those of certain recent developments, but also, and most importantly, using a multidisciplinary process which takes into account the events of Indian civilisation.*

First, however, it is necessary (in order to eliminate them once and for all) to remember some of the main and rather unlikely theories which are still in circulation today on this subject.

## FANCIFUL EXPLANATIONS FOR THE ORIGIN OF "ARABIC" NUMERALS

According to a popular tradition that still persists in Egypt and northern Africa, the "Arabic" numerals were the invention of a glassmaker-geometer from the Maghreb who came up with the idea of giving each of the nine numerals a shape, the number of angles each one possessed being equal to the number it denoted: *one angle* represented the number 1, *two angles* the number 2, *three angles* the number 3, and so on (Fig. 24.1A).

FIG. 24.1A. *The first unlikely hypothesis on the origin of our numerals: the number of angles each numeral possesses*

At the end of the nineteenth century, P. Voizot, a Frenchman, put forward the same theory, apparently influenced by a Genoese author. But he also thought that it was "equally probable" that the numerals were formed by certain numbers of lines (Fig. 24.1B).

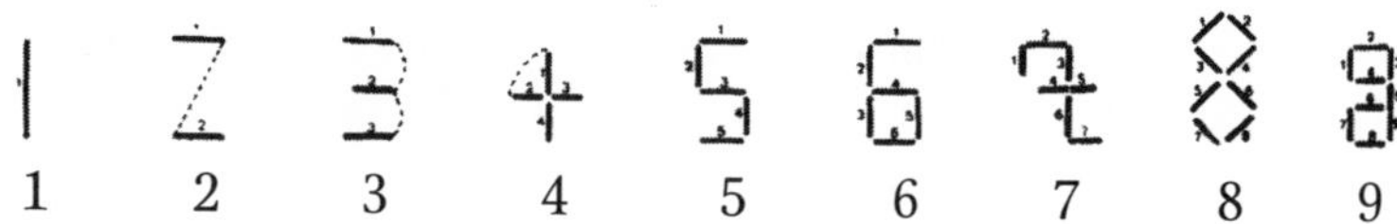


FIG. 24.1B. *Second unlikely hypothesis: the number of lines each numeral possesses*

Another similar hypothesis was put forward in 1642 by the Italian Jesuit Mario Bettini, which was taken up in 1651 by the German Georg Philip Harsdörffer. This time the idea was that the ideographical representation of the nine units would have been based on a number of points which were joined up to form the nine signs (Fig. 24.1C). In 1890, the Frenchman Georges Dumesnil also adopted this theory, believing the system to be of Greek invention: attributing the form of our present-day numerals to the Pythagoreans, his argument stated that the joining of the points to form the geometrical representations of the whole numbers played an important role for the members of this group.

FIG. 24.1C. *Third unlikely hypothesis: the number of points*

A corresponding theory was put forward by Wiedler in 1737 which he attributed to the tenth-century astrologer Abenragel: according to him, the invention of numerals was the result of the division into parts of the shape which is formed by a circle and two of its diameters. In other words, according to Wiedler, all the figures could be made from this one geometrical shape "as if they were inside a shell": thus the vertical diameter would

* Due to the complex nature of this civilisation, a "Dictionary of the numerical symbols of Indian Civilisation" has been compiled (see the end of the present chapter), which acts as both a thematic index and a glossary of the many notions which it is necessary to understand in order to grasp the ideas introduced in the following pages. In the present chapter, each word (whether in Sanskrit or in English) which is also found in the dictionary, is accompanied by an asterisk. (Examples: **anka*, **ankakramena*, **Ashvin*, **Indian Astronomy*, **Infinity*, **Numeral*, **sthâna*, **Symbols*, **yuga*, **zero*, etc.)

have formed 1; the same diameter plus two arcs at either end formed 2; a semi-circle plus a median horizontal radius made 3, and so on until zero, which was said to be formed by the complete circle (Fig. 24.1D).

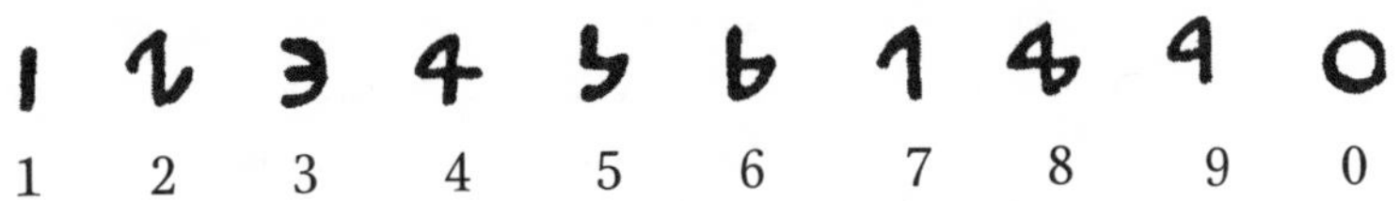


FIG. 24.1D. *Fourth unlikely hypothesis: the shapes formed by a circle and its diameters*

It is also worthwhile mentioning the theories of the Spaniard Carlos Le Maur (1778), who believed that the signs in question acquired their shape from a particular arrangement of counting stones (Fig. 24.1C) or from the number of angles that can be obtained from certain shapes formed by a rectangle, its diagonals, its medians, etc. (Fig. 24.1E).

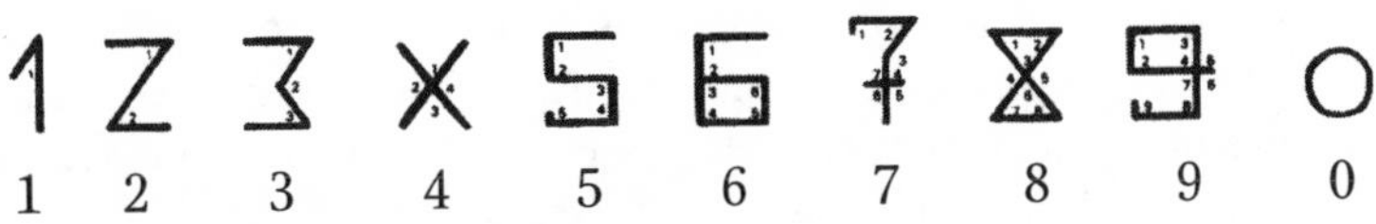


FIG. 24.1E. *Fifth unlikely hypothesis: a variation of Fig. 24.1A*

Finally, Jacob Leupold, in 1727, offered an "explanation" which goes by the name of the legend of Solomon's ring. According to this theory, the numerals were formed successively by the ring inscribing a square and its diagonals (Fig. 24.1F).

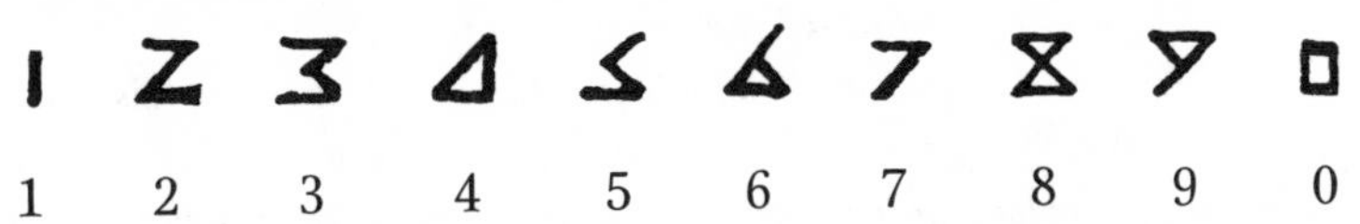


FIG. 24.1F. *Sixth unlikely hypothesis: the numerals come from a square (legend of Solomon's ring)*

If we were to believe any of these theories, it would mean that the appearance of the numerals that we use today would have to have been the fruit of one isolated individual's imagination. An individual who would have given each number a specific shape through a system based either on the use of different numbers of lines, angles or dots to add up to the amount of units the sign represents, or through the use of geometrical representations such as a triangle, rectangle, square or circle, which would mean that the signs were created according to a simple process of geometrical ordering.

These theories, then, all have one thing in common: their "explanation" for the appearance of our numerals is that these figures were the result of some kind of spontaneous generation; their shape, right from the outset, being perfectly logical. In fact, as F. Cajori (1928) explains, "the validity of any hypothesis depends upon the way in which the established facts are presented and the extent to which it opens the door to new research." In other words, a hypothesis can only acquire "scientific" value if it has the potential to broaden our knowledge of a given subject.

The hypotheses that have been mentioned so far in this chapter are basically sterile. None of them offers any explanation for the fact that the nine figures have appeared in an immense variety of shapes and forms over the centuries and in different parts of the world. Their approach is merely to consider the final product, in other words the numerals that we use today (as they appear in print), which fails to take account of the fact that these figures appear at the end of a very long story and have slowly evolved over several millennia.

These a posteriori hypotheses are flawed because they are the fruit of the pseudo-scientific imaginations of men who are fooled by appearances and who jump to conclusions which completely contradict both historical facts and the results of epigraphic and palaeographic research.*

It is still widely believed that the number-system that we use today was invented by the Arabs.

However, it definitely was not the Arabs who invented what we know as "Arabic" numerals. Historians have known for some time now that the name was coined as the result of a serious historical error. Significantly, and curiously, no trace of this belief is to be found in actual Arabic documents.

In fact, many Arabic works that concern mathematics and arithmetic reveal that Muslim Arabic authors, without the slightest hint of prejudice or complex, have always acknowledged that they were not the ones that made the discovery. But whilst it is incorrect, the name which was given to our numerals is not totally unfounded. There is always some basis for an historical error, no matter how widespread or long-standing it is, and this one is no exception, especially considering the fact that we are dealing with a broad geographical area and a duration of many centuries.

The belief that our numerals were invented by the Arabs is only found in Europe and probably originated in the late Middle Ages. This theory was only really voiced by mathematicians or arithmeticians who, in order to

* Moreover, this book demonstrates quite clearly that despite the significance and vast number of inventions that have punctuated the history of numerals as a whole, the findings have always been anonymous. Men would work for and in groups and gain no qualifications for their work. Certain documents made of stone, papyrus, paper and fabric immortalising the names of men who are sometimes associated with numbers mean nothing to us. Names of those who made use of and who reported numbers and counting systems are also known. But those of the inventors themselves are irretrievably lost, perhaps because their discoveries were made so long ago, or even because these brilliant inventions belonged to relatively humble men whose names were not deemed worthy of recording. It is also possible that the discoveries could be a result of the work of a team of men and so they could not really be attributed to a specific person. The "inventor" of zero, a meticulous scribe and arithmetician, whose main concern was to define a specific point in a series of numbers ruled by the place-value system, was probably never aware of the revolution that he had made possible. All of this proves the absurdity of the preceding hypotheses.

distinguish themselves from the masses, wanted to fill what they perceived as a void with random hypotheses based on preconceived ideas, and thus sacrificing historical truth to satisfy the whims of their own individual inspiration. To the uninitiated, the writings of these mathematicians would have seemed to constitute the linchpin of a doctrine that was sure to survive for many centuries. This is due to the fact that numerals and calculation have always been considered (rightly or wrongly) to be the very essence of mathematical science. The cause of the error is more easily understood now that it is known that the numerals in question arrived in the West at the end of the tenth century via the Arabs. At that time, the Arabs were relatively superior to Western civilisations in terms of both culture and science. Therefore the figures were given the name "Arabic".

This theory, however, was just one of the many explanations offered.

As the following evidence shows, European Renaissance authors offered many similar and equally unreliable theories, attributing the invention of our numerals to the Egyptians, the Phoenicians, the Chaldaeans and the Hebrews alike, all of whom are totally unconnected to this discovery.

It is interesting to note that even in the twentieth century certain authors, whilst being known for the quality of their work, have fallen into the trap of supporting unsatisfactory explanations and taking things solely at face value. At the turn of the century, historical scientists (G. R. Kaye, N. Bubnov, and B. Carra de Vaux, etc., who strongly opposed the idea that our number-system could be of Indian origin) alleged that our numerals were developed in Ancient Greece [see JPAS 8 (1907), pp. 475 ff.; N. Bubnov (1908); *SC* 21 (1917), pp. 273 ff.].

These men believed that the system originated in Neo-Pythagorean circles shortly before the birth of Christianity. They claimed that the system came to Rome from the port of Alexandria and soon after made its way to India via the trade route; it also travelled from Rome to Spain and the North African provinces, where it was discovered some centuries later by the Muslim Arabic conquerors. As for Middle Eastern Arabs, they picked up the system from Indian merchants. According to this view of things, European and North African numerals were formed by the "Western" transmission, and the radically different Indian and Eastern Arabic figures emerged by the "Eastern" route.

This tempting explanation is in fact an amalgam of the speculations of the early humanists, as we can see from the list of quotations that follows:

1. Köbel, *Rechenbiechlin*, first published in 1514: *Vom welchen Arabischen auch disz Kunst entsprungen ist*: "This art was also invented by the Arabs". [Köbel (1531), f° 13]

2. N. Tartaglia, *General trattato di numeri et misuri* ("General treatise of numbers and measures") first published in 1556: . . . *& que esto fu trouato di fare da gli Arabi con diece figure*: " . . . and this is what the Arabs did with ten figures [ten numerals]." [Tartaglia (1592), f° 9]

3. Robert Recorde, *The Grounde of Artes*: *In that thinge all men do agree, that the Chaldays, whiche fyrste inuented thys arte, did set these figures as thei set all their letters. For they wryte backwarde as you tearme it, and so doo they reade. And that may appeare in all Hebrew, Chaldaye and Arabike bookes . . . where as the Greekes, Latines, and all nations of Europe, do wryte and reade from the lefte hand towarde the ryghte.* [Recorde (1558), f° C, 5]

4. Peletarius, *Commentaire sur l'Arithmétique de Gemma Frisius* ("Commentary on Arithmetic by Gemma Frisius") first published in 1563: *La valeur des Figures commence au coste dextre vers le coste senestre: au rebours de notre maniere d'escrire par ce que la premiere prattique est venue des Chaldees: ou des Phéniciens, qui ont été les premiers traffiquers de marchandise*: "The figures read in ascending order from right to left which is the opposite of our way of writing. This is because the former practice comes from the Chaldaeans: or the Phoenicians, who were the first to trade their merchandise." [Peletarius, f° 77]

5. Ramus, *Arithmetic*, published in 1569: *Alii referunt ad Phoenices inventores arithmeticae, propter eandem commerciorum caussam: Alii ad Indos: Ioannes de Sacrobosco, cujus sepulchrum est Lutetiae . . . , refert ad Arabes:* "Others attribute the invention of arithmetic to the Phoenicians, for the same commercial reasons; others credit the Indians. Jean de Sacrobosco, whose tomb is in Paris . . . attributes the discovery to the Arabs." [Ramus (1569), p. 112]

6. Conrad Dasypodius, *Institutionum Mathematicarum*, published in 1593-1596: *Qui est harum Cyphrarum auctor? A quibus hae usitatae syphrarum notae sint inventae: hactenus incertum fuit: meo tamen iudicio, quod exiguum esse fateor: a Graecis librarijs (quorum olim magna fuit copia) literae Graecorum quibus veteres Graeci tamquam numerorum notis usunt usu fuerunt corruptae, vt ex his licet videre. Graecorum Literae corruptae.*

1 2 3 4 5 6 7 8 9

*Sed qua ratione graecorum literae ita fuerunt corruptae? Finxerunt has corruptas Graecorum literarum notas: vel abiectione vt in nota binarij numeri, vel inuersione vt in septenarij numeri nota nostrae notae, quibus hodie utimur, ab his sola differunt elegantia, vt apparet:* "Who invented these signs that are used as numerals? Until now no one has really known;

however, as far as I know (and I admit I know little), the letters that Ancient Greeks used to denote numbers were distorted and transformed through their use by Greek scribes (of which there were many), as one can see below.This is how the distorted letters look:

1 2 3 4 5 6 7 8 9

(as shown above).

"But how were these letters corrupted? The sign for number two has been reversed, the sign for the number seven has been inverted. The only difference between the signs that we use today and the Greek signs is that our signs are more elegant in appearance." [Dasypodius, quoted in Bayer]

7. Erpenius, *Grammatica Arabica,* published in 1613: "Arabic" numerals are "actually the figures used by Toledo's men of law", which he believes would have been transmitted to them by the Pythagoreans of Ancient Greece. But Golius, who published the book after the death of the author, realised that Erpenius had been mistaken, and suppressed that particular passage in the 1636 edition. [Erpenius]

8. Laurembergus, *The Mathematical Institution,* first published in 1636: *Supersunt volgares illi characteres Barbari, quibus hodie utitur universus fere orbis. Suntque universum novem: 1, 2, 3, 4, 5, 6, 7, 8, 9, queis additur o cyphra: seu figura nihili, Nulla, Zero Arabibus. Nonnullorum sententia est, primos harum figurarum inventores fuisse Arabes (alii Phoenices malunt; alii Indos) quae sane opinio non est a veritate aliena. Nam sicut Arabes olim totius fere orbis potiti sunt, ita credibile est, scientiarum quoque fuisse propagatores. Quicunque sit Inventor maxima sane illi debetur gratia*: "These ordinary, barbaric characters have survived the ages and are used throughout most of the modern world. There are nine altogether: 1 2 3 4 5 6 7 8 9, to which the figure 0 can be added, which denotes "nothing", the Arabic zero. Some think that it was the Arabs who originally invented these signs (whilst others believe it was the Phoenicians or even the Indians), and this is highly probable; the Arabs once dominated most of the world and it is likely that they invented the sciences. Whoever is to thank for the existence of our numerals deserves the highest recognition." [Laurembergus, p. 20, 1. 14; p. 21, 1. 2]

9. I. Vossius, *De Universae mathesos Natura et constitutione* (c.1604): "Arabic" numerals passed from "the Hindus or Persians to the Arabs, then to the Moors in Spain, then finally to the Spanish and the rest of Europe". His theory that the series was originally passed from the Greeks to the Hindus is without foundation. [Vossius (1660), pp. 39–40]

10. Nottnagelus, *The Mathematical Institution,* first published in 1645: *Computatores autem ob majorem supputandi commoditatem peculiares sibi finxerunt notas (quarum quidem inventionem nonnulli Phoenicibus adscribunt, quidam, ut Valla et Cardanus, Indis assignant, plerique vero Arabibus et Saracenis acceptam referunt) quas tamen alii ab antiqua vel potius corrupta Graecarum literarum forma, nonnulli vero aliunde derivatas autumant. Atque his posterioribus hodierni quoque utuntur Arithmetici*: "To facilitate calculation, arithmeticians invented their own unique signs (some believe it was the Phoenicians who invented them, others, such as Valla and Cardanus, believe it was the Indians; most people attribute the invention to the Arabs or Saracens); however, others claim that the numerals originated from the ancient, or rather, distorted shape of Greek letters; some even suggest another origin. The signs are still used by arithmeticians today." [Nottnagelus, p. 185]

11. Theophanes, *Chronicle,* first published in 1655: *Hinc numerorum notas et characteres, cifras vulgo dictos, Arabicum inventum aut Arabicos nulla ratione vocandos, qui haec legerit, mecum contendet . . .*: "The reader can appreciate that I can find no reason why the signs and characters that express numbers – which we vulgarly refer to as figures – are an Arabic invention . . ."

The following is an extract from a note written by Father Goar, which comments on the above passage: *Notas itaque characteresque, quibus numeros summatim exaramus, 1, 2, 3, 4, 5, 6, 7, 8, 9, ab Indis et Chaldaeis usque ad nos venisse scite magis advocat Glareanus in Arithmaticae praeludiis.* "In his Preludes to Arithmetic, Glareanus claims that the signs and characters which we use to write the numbers in an abbreviated form (1, 2, 3, etc.) actually came from the Indians and the Chaldaeans." [*Theophanis Chronographia*, p. 616, 2nd col., and p. 314]

12. P. D. Huet, Bishop of Avranches in his *Demonstratio Evangelica ad serenissimum Delphinum* claims that mediaeval European numerals were invented by the Pythagoreans. [Huet (1690)]

13. Dom Calmet (1707) upholds the theory of the Greek origin of our numerals [Calmet], as does J. F. Weidler, in *Spicilegium observationum ad historiam notarum numeralium pertinentium.* [Weidler (1755)]

14. C. Levias (1905), a contributor to the *Jewish Encyclopaedia,* states that our numerals were invented by the people of Israel and were introduced in Islamic countries around 800 CE by the Jewish scholar Mashallah. [Levias (1905), IX, p. 348]

15. Levi della Vida upholds the theory of the Greek origin of our numerals. [Levi della Vida (1933)]

16. M. Destombes says that European numerals are derived from the following letters of the Graeco-Byzantine alphabet I, θ, H, Z, …., Γ, B, by reversing the series of letters: B, Γ, … Z, H, θ, I, written in capitals and graphically adapted to the "shapes of the Visigothic letters from the third quarter of the tenth century CE". [Destombes (1962)]

The basis for all these hypotheses is, of course, invalid, because no evidence has ever been found to support the theory that the Greeks used a similar system to our own. However, rather than admit defeat in the face of solid counter-arguments firmly based in reality, the authors of these hypotheses persevered stubbornly, using all their imagination to come up with something resembling proof or confirmation of their unlikely theories.

As A. Bouché-Leclerq (1879) remarks, "it is almost tempting to admire the cunning way in which an unshakeable belief can transform into proof the very objections which threaten to destroy it, and nothing better demonstrates the psychological history of humanity than the irresistible prestige of the preconceived idea." Bouché-Leclerq is actually denouncing certain charlatans of Ancient Greece who mastered the art of exploiting trusting souls through the use of divinatory practices that were based on the interpretation of numerological dreams using the numeral letters of the Greek alphabet: "Perhaps the most embarrassing case", he explains, "was one which involved a dream which promised an elderly man a number of years that was too high to be added on to his current age and too low to represent his life-span as a whole. The charlatan, however, found a way to overcome such a dilemma. If a man of seventy heard someone say, 'You will live for fifty years,' he would live for another thirteen years. He has already lived for over fifty years and it is impossible that he will live for another fifty years, being seventy already. So the man will live for another thirteen years (according to the charlatans) because the letter *Nu* (N), whilst representing the number fifty, comes thirteenth in the Greek alphabet!"

It is likely that the same author would have also condemned the methods of historical scientists, who have been known to be somewhat economical with the truth. No doubt he would have said something similar about them if he had heard one particular historian's rather flimsy "explanation" which was soon adopted by all of his peers. When a shrewd man asked him why the Greeks had left no written trace of zero or of decimal place-value numeration, the historian in question, not to be deterred, replied: "That is because of the level of importance that they placed in oral tradition and also the great secrecy with which the Neo-Pythagoreans surrounded their knowledge"! If everyone reasoned in this way, history would amount to little more than a fairy tale.

Bearing in mind the fact that these authors were ardent admirers of Hellenistic civilisation, it is easy to understand why their theory was supported solely by claims that were unaccompanied by any shred of evidence, their main aim being to glorify the famous "Greek miracle".

The admiration that these authors display for Greek civilisation is, of course, perfectly justified. The Greeks were responsible for innovations in such varied fields as art, literature, philosophy, medicine, mathematics, astronomy, the sciences and engineering; their enormous contribution to our sciences and culture is undeniable. The paradox lies in the fact that the very men who wished to add to these achievements that are already acknowledged by the rest of the world, were unaware of the real story surrounding the scholars and mathematicians upon whom they wanted to bestow this undeserved honour. This clearly demonstrates narrow-mindedness on their part, attributing the development of our place-value notation solely to the origin of the graphical representation of the nine numerals in question.

J. F. Montucla (1798) quite rightly points out that "if the characters originate from Greek letters, they have drastically changed somewhere along the way. In fact, these letters could only resemble our numerals if they were shortened and turned about in a very odd fashion.* Moreover, the appearance of these characters is much less important than the ingenious way in which they are used; using only ten characters, it is possible to express absolutely any number. The Greeks were a highly intelligent race, and if this had been their invention, or even if they had simply got wind of it, they certainly would have made use of it."

Ancient Greece only had two systems of numerical notation: the first was the mathematical equivalent of the Roman system and the other was alphabetical, like the one used by the Hebrews. With a few exceptions towards the end of the era, neither of these systems were based on the rule of position, nor did they possess zero. Therefore the systems were not really of much practical use when it came to mathematical calculations, which were generally carried out using abacuses, upon which there were different columns for each decimal order.

Considering that the Greeks had invented such an instrument, the next logical step would have been their discovery of the place-value system and zero, through eliminating the columns of the instrument.

* Using such methods, it is always possible to find a way of promoting a theory: it is easy to manipulate the nine characters in order to "prove" that our nine numerals originate from them. This is precisely how certain extravagant comtemporary authors, ignoring not only the history of mathematical notation and writing, but also and above all the laws of palaeography, have come to "demonstrate" that these numerals derive from the first nine Hebrew letters, or even from the graphical representations for the twelve signs of the Zodiac. This goes to show that you can put the words of a song to any tune you like; in other words, appearances can be deceptive.

This would have provided them with the fully operational counting system that we use today.

However, the Greeks did not bother themselves with such practical concerns.

## INDIA: THE TRUE BIRTHPLACE OF OUR NUMERALS

The real inventors of this fundamental discovery, which is no less important than such feats as the mastery of fire, the development of agriculture, or the invention of the wheel, writing or the steam engine, were the mathematicians and astronomers of Indian civilisation: scholars who, unlike the Greeks, were concerned with practical applications and who were motivated by a kind of passion for both numbers and numerical calculations.

There is a great deal of evidence to support this fact, and even the Arabo-Muslim scholars themselves have often voiced their agreement.

## EVIDENCE FROM EUROPE WHICH SUPPORTS THE CLAIM THAT MODERN NUMERATION ORIGINATED IN INDIA

The following is a succession of historical accounts in favour of this theory, given in chronological order, beginning with the most recent.

1. P. S. Laplace (1814): "The ingenious method of expressing every possible number using a set of ten symbols (each symbol having a place value and an absolute value) emerged in India. The idea seems so simple nowadays that its significance and profound importance is no longer appreciated. Its simplicity lies in the way it facilitated calculation and placed arithmetic foremost amongst useful inventions. The importance of this invention is more readily appreciated when one considers that it was beyond the two greatest men of Antiquity, Archimedes and Apollonius." [Dantzig, p. 26]

2. J. F. Montucla (1798): "The ingenious number-system, which serves as the basis for modern arithmetic, was used by the Arabs long before it reached Europe. It would be a mistake, however, to believe that this invention is Arabic. There is a great deal of evidence, much of it provided by the Arabs themselves, that this arithmetic originated in India." [Montucla, I, p. 375]

3. John Wallis (1616–1703) referred to the nine numerals as *Indian figures* [Wallis (1695), p. 10]

4. Cataneo (1546) *le noue figure de gli Indi*, "the nine figures from India". [Smith and Karpinski (1911), p. 3]

5. Willichius (1540) talks of *Zyphrae Indicae*, "Indian figures". [Smith and Karpinski (1911) p. 3]

6. *The Crafte of Nombrynge* (c. 1350), the oldest known English arithmetical tract: II *fforthermore ye most vndirstonde that in this craft ben vsed teen figurys, as here bene writen for esampul 0 9 8 ʌ 6 5 4 3 2 1 . . . in the quych we vse teen figurys of Inde. Questio.* II *why ten figurys of Inde? Solucio. For as I have sayd afore thei were fonde fyrst in Inde.* [D. E. Smith (1909)]

7. Petrus of Dacia (1291) wrote a commentary on a work entitled *Algorismus* by Sacrobosco (John of Halifax, c. 1240), in which he says the following (which contains a mathematical error): *Non enim omnis numerus per quascumque figuras Indorum repraesentatur* . . .: "Not every number can be represented in Indian figures". [Curtze (1897), p. 25]

8. Around the year 1252, Byzantine monk Maximus Planudes (1260–1310) composed a work entitled *Logistike Indike* ("Indian Arithmetic") in Greek, or even *Psephophoria kata Indos* ("The Indian way of counting"), where he explains the following: "There are only nine figures. These are:

1 2 3 4 5 6 7 8 9

[figures given in their Eastern Arabic form].

"A sign known as *tziphra* can be added to these, which, according to the Indians, means 'nothing'. The nine figures themselves are Indian, and *tziphra* is written thus: 0". [B. N., Paris. *Ancien Fonds grec*, Ms 2428, f° 186 r°]

9. Around 1240, Alexandre de Ville-Dieu composed a manual in verse on written calculation (algorism). Its title was *Carmen de Algorismo*, and it began with the following two lines: *Haec algorismus ars praesens dicitur, in qua Talibus Indorum fruimur bis quinque figuris*: "*Algorism* is the art by which at present we use those Indian figures, which number two times five". [Smith and Karpinski (1911), p. 11]

10. In 1202, Leonard of Pisa (known as Fibonacci), after voyages that took him to the Near East and Northern Africa, and in particular to Bejaia (now in Algeria), wrote a tract on arithmetic entitled *Liber Abaci* ("a tract about the abacus"), in which he explains the following: *Cum genitor meus a patria publicus scriba in duana bugee pro pisanis mercatoribus ad eam confluentibus preesset, me in pueritia mea ad se uenire faciens, inspecta utilitate et commoditate futura, ibi me studio abaci per aliquot dies stare uoluit et doceri. Vbi ex mirabili magisterio in arte per nouem figuras Indorum introductus . . . Novem figurae Indorum hae sunt:*

9 8 7 6 5 4 3 2 1

*cum his itaque novem figuris, et cum hoc signo o. Quod arabice zephirum appellatur, scribitur qui libet numerus*: "My father was a public scribe of Bejaia, where he worked for his country in Customs, defending the interests of Pisan merchants who made their fortune there. He made me learn how to use the abacus when I was still a child because he saw how I would benefit from this in later life. In this way I learned the art of counting using the nine Indian figures . . .

The nine Indian figures are as follows:

9 8 7 6 5 4 3 2 1

[figures given in contemporary European cursive form].

"That is why, with these nine numerals, and with this sign 0, called *zephirum* in Arab, one writes all the numbers one wishes."
[Boncompagni (1857), vol.I]

11. C. 1150, Rabbi Abraham Ben Meïr Ben Ezra (1092–1167), after a long voyage to the East and a period spent in Italy, wrote a work in Hebrew entitled: *Sefer ha mispar* ("Number Book"), where he explains the basic rules of written calculation.

He uses the first nine letters of the Hebrew alphabet to represent the nine units. He represents zero by a little circle and gives it the Hebrew name of *galgal* ("wheel"), or, more frequently, *sifra* ("void") from the corresponding Arabic word.

However, all he did was adapt the Indian system to the first nine Hebrew letters (which he naturally had used since his childhood).

In the introduction, he provides some graphic variations of the figures, making it clear that they are of Indian origin, after having explained the place-value system: "That is how the learned men of India were able to represent any number using nine shapes which they fashioned themselves specifically to symbolise the nine units." [Silberberg (1895), p. 2; Smith and Ginsburg (1918); Steinschneider (1893)]

12. Around the same time, John of Seville began his *Liber algoarismi de practica arismetrice* ("Book of Algoarismi on practical arithmetic") with the following:

*Numerus est unitatum collectio, quae quia in infinitum progreditur (multitudo enim crescit in infinitum), ideo a peritissimis Indis sub quibusdam regulis et certis limitibus infinita numerositas coarcatur, ut de infinitis difinita disciplina traderetur et fuga subtilium rerum sub alicuius artis certissima lege teneretur:* "A number is a collection of units, and because the collection is infinite (for multiplication can continue indefinitely), the Indians ingeniously enclosed this infinite multiplicity within certain rules and limits so that infinity could be scientifically defined; these strict rules enabled them to pin down this subtle concept.
[B. N., Paris, Ms. lat. 16 202, f° 51; Boncompagni (1857), vol. I, p. 26]

13. C. 1143, Robert of Chester wrote a work entitled: *Algoritmi de numero Indorum* ("Algoritmi: Indian figures"), which is simply a translation of an Arabic work about Indian arithmetic. [Karpinski (1915); Wallis (1685), p. 12]

14. C. 1140, Bishop Raimundo of Toledo gave his patronage to a work written by the converted Jew Juan de Luna and archdeacon Domingo Gondisalvo: the *Liber Algorismi de numero Indorum* ("Book of Algorismi of Indian figures) which is simply a translation into a Spanish and Latin version of an Arabic tract on Indian arithmetic. [Boncompagni (1857), vol. I]

15. C. 1130, Adelard of Bath wrote a work entitled: *Algoritmi de numero Indorum* ("Algoritmi: of Indian figures"), which is simply a translation of an Arabic tract about Indian calculation. [Boncompagni (1857), vol. I]

16. C. 1125, The Benedictine chronicler William of Malmesbury wrote *De gestis regum Anglorum*, in which he related that the Arabs adopted the Indian figures and transported them to the countries they conquered, particularly Spain. He goes on to explain that the monk Gerbert of Aurillac, who was to become Pope Sylvester II (who died in 1003) and who was immortalised for restoring sciences in Europe, studied in either Seville or Cordoba, where he learned about Indian figures and their uses and later contributed to their circulation in the Christian countries of the West. [Malmesbury (1596), f° 36 r°; Woepcke (1857), p. 35]

17. Written in 976 in the convent of Albelda (near the town of Logroño, in the north of Spain) by a monk named Vigila, the *Codex Vigilanus* contains the nine numerals in question, but not zero. The scribe clearly indicates in the text that the figures are of Indian origin: *Item de figuris aritmetice. Scire debemus Indos subtilissimum ingenium habere et ceteras gentes eis in arithmetica et geometrica et ceteris liberalibus disciplinis concedere. Et hoc manifestum est in novem figuris, quibus quibus designant unum quenque gradum cuiuslibet gradus. Quarum hec sunt forma:*

9 8 7 6 5 4 3 2 1.

"The same applies to arithmetical figures. It should be noted that the Indians have an extremely subtle intelligence, and when it comes to arithmetic, geometry and other such advanced disciplines, other ideas must make way for theirs. The best proof of this is the nine figures with which they represent each number no matter how high. This is how the figures look:

9 8 7 6 5 4 3 2 1."

(In the original, the figures are presented in a style very close to the North African Arabic written form.) [Bibl. San Lorenzo del Escorial, Ms. lat. d.I.2, f° 9v°; Burnam (1912), II, pl. XXIII; Ewald (1883)]

## EVIDENCE FROM ARABIC SOURCES WHICH SUGGESTS THAT MODERN NUMERATION ORIGINATED IN INDIA

The following evidence proves that for over a thousand years, Arabo-Muslim authors never ceased to proclaim, in a praiseworthy spirit of openness, that the discovery of the decimal place-value system was made by the Indians.*

1. In *Khulasat al hisab* ("Essence of Calculation"), written c. 1600, Beha' ad din al 'Amuli, in reference to the figures in question, remarks that: "It was actually the Indians who invented the nine characters." [Marre (1864), p. 266]

2. C. 1470, in a commentary on an arithmetical tract, Abu'l Hasan al Qalasadi (d. 1486) wrote the following in reference to the nine figures used in Muslim Spain and Northern Africa: "Their origin is traditionally attributed to an Indian." [Woepcke (1863), p. 59]

3. In "Prolegomena" (*Muqqadimah*), written c. 1390, Abd ar Rahman ibn Khaldun (1332–1406) says that the Arabs first learned about science from the Indians along with their figures and methods of calculation in the year 156 of the Hegira (= 776 CE). [Ibn Khaldun, vol. III, p. 300]

4. In *Talkhis fi a 'mal al hisab* ("Brief guide to mathematical operations") written c. 1300, Abu'l 'abbas ahmad ibn al Banna al Marrakushi (1256–1321) makes a direct reference to the Indian origin of the figures and counting techniques. [Marre (1865); Suter (1900), p. 162]

5. C. 1230, Muwaffaq al din Abu Muhammad al Baghdadi wrote a tract entitled *Hisab al hindi* ("Indian Arithmetic"). [Suter (1900), p. 138]

6. C. 1194, Persian encyclopaedist Fakhr ad din al Razi (1149–1206) wrote a work entitled *Hada'iq al anwar*, which included a chapter called *Hisab al hindi* ("Indian Calculation"). [B. N., Anc. Fds pers., Ms. 213, f° 173r]

7. C. 1174, mathematician As Samaw'al ibn Yahya ibn 'abbas al Maghribi al andalusi, a Jew converted to Islam, wrote a work entitled *Al bahir fi 'ilm al hisab* ("The lucid book of arithmetic"), in which a direct reference is also made concerning the Indian origin of the figures and the methods of calculation. [Suter (1900), p. 124; Rashed and Ahmed (1972)]

* Henceforth, the scientific transcription of Arabic words will not be scrupulously adhered to. "Kh", "gh" and "sh" will be used in the place of h, g, and s to facilitate the reading of Arabic for those who are not specialists.

8. In 1172 Mahmud ibn qa'id al 'Amuni Saraf ad din al Meqi wrote a tract entitled *Fi'l handasa wa'l arqam al hindi* ("Indian geometry and figures"). [Suter (1900), p. 126]

9. C. 1048, 'Ali ibn Abi'l Rijal abu'l Hasan, alias Abenragel, in a preface to a treatise on astronomy, wrote that "the invention of arithmetic using the nine figures belongs to the Indian philosophers". [Suter (1900), p. 100]

10. C. 1030, Abu'l Hasan 'Ali ibn Ahmad an Nisawi wrote a work entitled *al muqni 'fi'l hisab al hindi* ("Complete guide to Indian arithmetic"). [Suter (1900), p. 96]

11. Between 1020 and 1030, in his autobiography, Al Husayn ibn Sina (Avicenna) tells of how, when he was very young, he heard conversations between his father and his brother which were often about Indian philosophy, geometry and calculation, and when he was ten (in the year 990), his father sent him to a merchant who was well-versed in numerical matters to learn the art of Indian calculation.

In his tract on speculative arithmetic, Ibn Sina writes the following: "As for the verification of squares using the Indian method (*fi'l tariq al hindasi*) . . . One of the properties of a cube consists of the way of verifying it using the methods of Indian calculation (*al hisab al hindasi*) . . ." [Woepcke (1863), pp. 490, 491, 502, 504; Leiden Univ. Lib., Ms. legs Warnerien, no. 84]

12. C. 1020, Abu'l Hasan Kushiyar ibn Labban al-Gili (971–1029) wrote a work which carries the Arabic title, *Fi usu'l hisab al hind* ("Elements of Indian calculation"), the opening words of which being: "This [tract] of calculations [written] in Indian [figures] is formed by . . ." [Library of Aya Sofia. Istanbul. Ms 4,857, f° 267 r; Mazaheri (1975)]

13. In roughly the same year, mathematician Abu Ali al Hasan ibn al Hasan ibn al Haytham, from Basra, wrote *Maqalat fi 'ala 'l hisab al hind* ("Principles of Indian calculation"). [Woepcke (1863), p. 489]

14. Astronomer and mathematician Muhammad ibn Ahmad Abu'l Rayhan al Biruni (973–1048), after living in India for thirty years, and having been introduced to Indian sciences, wrote a number of works between 1010 and 1030, including *Kitab al arqam* ("Book of figures"), and *Tazkira fi'l hisab wa'l mad bi'l arqam al sind wa'l hind* ("Arithmetic and counting using Sind and Indian figures").

In his work entitled *Kitab fi tahqiq i ma li'l hind* (which is one of the most important works about India to be written at that time), in which he mentions the diversity of the graphical forms of the figures used in India, and insists that the figures used by the Arabs originated in India, he makes the following remark: "Like us, the Indians use these numerical signs in their arithmetic. I have written a tract which shows, in as much detail as possible, how much more advanced the Indians are than we are in this field."

And in *Athar wu 'l baqiya* ("Vestiges of the past", or "Chronology of ancient nations"), he calls the nine figures *arqam al hind* ("Indian figures"), and demonstrates both how they differ from the sexagesimal system (which is Babylonian in origin), and their superiority over the Arab system of numeral letters. [Al-Biruni (1879) and (1910); Smith and Karpinski (1911), pp. 6–7; Datta and Singh (1938), pp. 98–9; Woepcke (1900), pp. 275–6]

15. Curiously, in his "Book of creation and history" (c. 1000), Mutahar ibn Tahir gives, in the *Nâgarî* form of the figures, the decimal positional expression of a number which the Indians believed represented the age of the planet. [Smith and Karpinski (1911), p. 7]

16. In 987, historian and biographer Ya 'qub ibn al Nadim of Baghdad wrote one of the most important works on the history of Arabic Islamic people and literature: the *Al Kitab al Fihrist al 'ulum* ("Book and index of the sciences"), in which he particularly refers to the work of the great Arabic Muslim astronomers and mathematicians of his time, and in which he constantly refers to methods of calculation as *hisab al hindi* ("Indian calculation"). [Dodge (1970); Suter (1892) and (1900); Karpinski (1915)]

17. Before 987, Sinan ibn al Fath min ahl al Harran (quoted in *Fihrist* by Ibn al Nadim) wrote a work entitled *Kitab al takht fi'l hisab al hindi* ("Tract on the wooden tablets used in Indian calculation"). [Suter (1892), pp. 37–8; Woepcke (1863), p. 490]

18. Also before 987, Ahmad Ben 'Umar al Karabisi (quoted in Ibn al Nadim's *Fihrist*) wrote *Kitab al hisab al hindi* ("A tract on Indian calculation"). [Suter (1900), p. 63; Woepcke (1863), p. 493]

19. Before 987 again, 'Ali Ben Ahmad Abu'l Qasim al Mujitabi al Antaki al Mu'aliwi (who died in 987) wrote a tract entitled *Kitab al takht al kabir fi'l hisab al hindi* ("Book of wooden tablets relating to Indian calculation"). [Suter (1900), p. 63; Woepcke (1863), p. 493]

20. Before 986, Al Sufi (who died in 986) wrote a work entitled *Kitab al hisab al hind* "Treatise on Indian calculation". [Smith and Karpinski (1911)]

21. C. 982, Abu Nasr Muhammad Ben 'Abdallah al Kalwadzani wrote *Kitab al takht fi'l hisab al hindi* ("Treatise on the tablet relative to Indian calculation"), quoted in *Fihrist* by Ibn al Nadim. [Suter (1900), p. 74; Woepcke (1863), p. 493]

22. C. 952, Abu'l Hasan Ahmad ibn Ibrahim al Uqlidisi wrote a work entitled: *Kitab al fusul fi'l hisab al hind* ("Treatise on Indian arithmetic"). [Saidan (1966)]

23. In 950, Abu Sahl ibn Tamim, a native of Kairwan (now Tunisia), wrote a commentary on *Sefer Yestsirah* (a Hebrew work concerning Cabbala) in which he explains the following: "The Indians invented the nine signs which denote units. I have already spoken about these at great length in a book which I wrote on Indian mathematics [he uses the expression *hisab al hindi*], known as *hisab al ghubar* ("calculations in the dust"). [Reinaud, p. 399; Datta and Singh (1938), p. 98]

24. C. 900, arithmetician Abu Kamil Shuja' ibn Aslam ibn Muhammad al Hasib al Misri (his last two names meaning "the Egyptian arithmetician") wrote an arithmetical work using the rule of the two false positions, which he attributed to the Indians. This work, which is only found in Latin translation, is called: "Book of enlargement and reduction, entitled 'the calculation of conjecture', after the achievements of the wise men of India and the information that Abraham[?] compiled according to the 'Indian' volume". [Suter, BM3; Folge, 3 (1902)]

25. Before 873, Abu Yusuf Ya 'qub ibn Ishaq al Kindi wrote *Kitab risalat fi isti mal 'l hisab al hindi arba' maqalatan* ("Thesis on the use of Indian calculation, in four volumes"), quoted in *Fihrist* by Ibn al Nadim. [Woepcke (1900), p. 403]

26. C. 850, the Arabic philosopher Al Jahiz (who died in 868) refers to the figures as *arqam al hind* ("figures from India") and remarks that "high numbers can be represented easily [using the Indian system]", even though the author expresses contempt for the Indian system. He asks the following question: "Who invented Indian figures . . . and calculation using the figures?" [Carra de Vaux (1917); Datta and Singh (1938), p. 97]

27. C. 820, Sanad Ben 'Ali, a Jewish mathematician who was converted to Islam, and who was one of Caliph al Ma'mun's astronomers, wrote a tract entitled: *Kitab al hisab al hindi* ("A treatise on Indian calculation") quoted by Ibn al Nadim in *Fihrist*. [Smith and Karpinski (1911), p. 10; Woepcke (1900), p. 490]

28. C. 810, Abu Ja 'far Muhammad ibn Musa al Khuwarizmi wrote: *Kitab al jam' wa'l tafriq bi hisab al hind* ("Indian technique of addition and subtraction"), of which there are Latin translations dating from the twelfth century. The tract begins thus:

" . . . we have decided to explain Indian calculating techniques using the nine characters and to show how, because of their simplicity and conciseness, these characters are capable of expressing any number."

He goes on to give a detailed explanation of the positional principle of decimal numeration, with reference to the Indian origin of the nine numerical symbols and of "the tenth figure in the shape of a circle" (zero), which he advises be used "so as not to confuse the positions". [Allard (1975); Boncompagni (1857)]; Vogel (1963); Youschkevitch (1976)]

## HOW RELIABLE IS THIS EVIDENCE?

All the above evidence points to the same conclusion: the numerical symbols that are used in the modern world were created in India.

However, there still remains the task of judging how reliable this evidence is. According to E. Claparède (1937), "reliable evidence is not the rule but the exception". This idea is perhaps best expressed by Charles Péguy, through the character of Clio, Muse of History (*Oeuvres complètes*, VIII, 301–302): "Humankind lies most when giving evidence (because the testimony becomes part of history), and . . . people lie even more when giving formal evidence. In everyday life, it is important to be truthful. When giving evidence, it is necessary to be twice as truthful. It is a well-known fact, however, that people lie all the time, but people lie less when not testifying than when they are testifying."

Etymologically, "testimony" derives from the Latin *testis* ("witness"), from which we get the verbs "to attest", "to contest", etc. Thus "testimony" means "the written or verbal declaration with which a person certifies the reality of a fact of which they have had direct knowledge" (P. Foulquié, 1982).

Often, however, the fact in question is certified by an anterior declaration given by an eye-witness, as if one was testifying to a scene which a friend had seen and then recounted.

This is precisely the conditions in which nearly all the above declarations were written.

By its very nature, a testimony is never objective:

> It is always marred by the subjectivity of its author, the unreliability of his memory, as well as gaps in perception and the unavoidable distortions of human memory (it is estimated that these errors increase at a rate of 0.33 per cent per day). Swiss psychologist Édouard Claparède and Belgian criminologist L. Vervaeck, using their pupils as subjects, found that correct testimonies were rare (only 5 per cent) and that the feeling of certainty increased with time . . . at the same rate as the increase in errors! [N. Sillamy (1967)].

It is because of its capital role in courtroom cases that the study of testimony plays such a major part in the applications of judicial psychology (see H. Piéron, 1979). The courtroom saying, *testis unus, testis nullus* (one sole witness is as useful as no witness at all) does not apply here because the origin of the numerals has been mentioned many times in the space of more than a thousand years. This case would in fact seem highly plausible.

But are all these accounts really completely independent of one another? If all these concurring pieces of evidence originate from one single source, then the proof might as well not exist at all.

The following example, taken from M. Bloch (1949), illustrates this point very clearly:

> Two contemporaries of Marbot – the Count of Ségur and General Pelet – gave accounts of Marbot's alleged crossing of the Danube which were analogous to Marbot's own account. Ségur's evidence came after Pelet's: he read the latter's account and did little more than copy it. It made no difference if Pelet wrote his account before Marbot; he was Marbot's friend and there is no doubt that Pelet had often heard Marbot recount his fictitious heroic deeds. This leaves Marbot as the only witness because his would-be guarantors both based their accounts on what he himself had related about the event.

In this kind of situation there is quite literally no witness at all.

However, Planudes, Fibonacci, Ibn Khaldun, Avicenna, al-Biruni, al-Khwarizmi and others, of whom many were actual eye-witnesses to the event, are neither Pelets, nor Ségurs, and certainly not Marbots. Their evidence and their accounts, as will be seen later, are firmly rooted in reality. These men are all in agreement, but this stems from neither a similar state of mind nor a phenomenon of collective psychology.

Despite the basic unreality of memory and the gaps and distortions which characterise the evidence given by any member of the human race, these accounts as a whole might still be an important item to add to the file for this investigation.

## EVIDENCE FROM PRE-ISLAMIC SYRIA

The Arabs and the Europeans were not the first to offer evidence about the origin of our digits. There were others; people who were around long before and who lived far beyond the frontiers of Islam. Proof is to be found in the Middle East, at a time when Muslim religion was only just beginning to emerge, shortly after the first Ommayad caliph came to power in Damascus.

At that time there lived a Syrian bishop named Severus Sebokt. He studied philosophy, mathematics and astronomy at the monastery of Keneshre on the banks of the Euphrates: a place that was exposed to a great wealth of knowledge because of its situation at the crossroads of Greek, Mesopotamian and Indian learning.

Severus Sebokt, then, knew Greek and Babylonian sciences as well as Indian science. Irritated by the belief that Greek learning was superior to that of other civilisations, he wrote a short article in the hope of bringing the Greeks down a peg or two.

Nau, who wrote a commentary on and published this manuscript, explains the circumstances under which it was written:

> In the Greek year 973 (662 in our calendar), Severus Sebokt, clearly offended by Greek pride, reclaimed the invention of astronomy for the Syrians. He explained that the Greeks had gleaned their knowledge from the Chaldaeans and the Babylonians, who he claimed were in fact Syrians. He quite rightly concludes that science belongs to everyone and that it is accessible to any race or individual who takes the trouble to understand it; it is not the property of the Greeks [F. Nau (1910)].

It is in order to reinforce this point that Severus uses the Indians as an example:

> The Hindus, who are not even Syrians, have made subtle discoveries in the field of astronomy which are even more ingenious than those of the Greeks [sic] and the Babylonians; as for their skilful methods of calculation and their computing which belies description, they use only nine figures. If those who think they are the sole pioneers of science, simply because they speak Greek, had known of these innovations, they would have realised (albeit a little late) that there are others who speak different languages who are also knowledgeable.

This piece of evidence is indispensable. The "computing that belies description which uses only nine figures" is, to Sebokt's mind, infinitely superior to spoken numeration: it is not possible to express all numbers using the latter method (because, like most oral methods of numeration, it involves a hybrid principle, using addition and multiplication of the names of the basic numbers); the Indian system makes it possible to write any number using only the nine figures.

In other words, the Indian system, as described by Severus Sebokt, has an unlimited capacity for representation because it has positional numeration.

This numeration is decimal because it uses nine digits.

It might seem curious that Sebokt does not mention the use of zero, but this is probably because he only had an abacus upon which to carry out his mathematical operations. It is likely that his "abacus" was a board sprinkled with sand or dust upon which he would write numbers using the nine Indian symbols within various columns corresponding to the consecutive decimal denominations. Therefore, zero was not physically represented: the absence of a unit in a given column was communicated by means of an empty space.

Sebokt's evidence proves that the Indian counting system was known and esteemed outside India by the middle of the seventh century CE.

## FROM THE EVIDENCE TO THE ACTUAL EVENT

The above evidence proves that all the preceding accounts are independent of each other but, however reliable these accounts are, they merely serve as confirmation of the truth. Alone, they do not constitute what is known as "historical truth". As F. de Coulanges said, "History is a science: it is a product of observation, not imagination; in order for the observation to be accurate, authentic documentation is needed."

A. Cuvillier (1954) explains that history, in the scientific sense of the word, is

> the study of human facts through time. So defined, historical facts are distinguished from those that are the subject of other sciences by their unique nature . . . Suspended in time, historical facts are, as a rule, in the past. Even when dealing with contemporary facts, the historian is still only personally privy to a very small percentage of the facts. The first task of a historian is to establish the facts through the use of documents, in other words the traces of these facts which still remain in the present.

Sociologist F. Simiand said that history is "information gleaned from left-over traces". The "traces" which are of interest here are the surviving written documents from Indian civilisation or from any culture connected to it.

Of course, it is essential to ensure that these documents are authentic. The traces in question came from an area of incredible diversity which, whilst proving the wonderful fertility of Indian civilisation, also shows an infinite complexity, with an added difficulty (to name but one): the considerable number of fakes produced by members of this same civilisation.

This, then, is the terrain the historian must embark upon; one of undeniable cultural wealth, even exuberance, yet it is crucial to remain extremely cautious when faced with documentation which is often tricky to date and which has to be closely examined in order to separate the genuine from the counterfeit, the ancient from the modern, the collective work from the individual work, a commentary from a copy of the original, etc.*

However, the vital work of historians from India and Southeast Asia must not be forgotten. For over a century, they have been separating the authentic from the fake, establishing the source and the date of a great many documents (even if this chronology is only approximate), restoring documents which had been damaged by the passage of time to their original state, studying the content and the allusions made in each work, and carrying out many other indispensable tasks.

All these results were collected in random order. To paraphrase H. Poincaré (1902), the science of history is built out of bricks; but an accumulation of historical facts is no more a science than a pile of bricks is a house.

## PROOF OF THE EVENT

In the previous chapter we offered a classification of written numbering systems that are historically attested, and through it we drew out a genuine chronological logic: the guiding thread, leading through centuries and civilisations, taking the human mind from the most rudimentary systems to the most evolved. It enabled us to identify the foundation stone (and, more generally, the abstract structure) of the contemporary written numeral system, the most perfect and efficient of all time. And it is precisely this chronological logic of the mind which shows us the path to follow in order to arrive at a historical synthesis. A synthesis intended to show just how the invention of numerals actually "worked", and to place it in its overall context, in terms of period, sequence of events, influences, etc.

Using this approach, we will be able to tell the story much more rigorously and to track the invention of the Indian system very closely indeed.

Drawing on all the available evidence to prove that India really was the cradle of modern numeration, the problem will be divided into the following subsections:

* Indian history is a constantly shifting terrain, where "forgeries" or "modern documents presented as ancient ones" abound in great quantities. It is an area where even documents that are believed to be authentic could quite possibly have been the fruit of several successive corrections or re-workings and the result of some apparently homogenous fusion of various commentaries, even commentaries on the commentaries themselves, so that the seemingly authentic document might have absolutely nothing in common with what the author to whom the work is attributed orginally intended. It is a field where certain specialists, who have not always been as rigorous as they might have been, have confused the issue by supporting their arguments with documents that have no historical worth whatsoever. This would appear to explain why the origin of the decimal point system was such an enigma for so long.

To untangle this apparently inextricable knot was no simple task because it involved the elimination of all unreliable sources (which are still used in a great many scientific publications) in order to include, as far as possible, nothing but trustworthy sources, from the most ancient documents on Indian civilisation.

1. To show that this civilisation discovered, and put into practice, the place-value system;
2. To prove that this same civilisation invented the concept of zero, which the Indian mathematicians knew could represent both the idea of an "empty space" and that of a "zero number";
3. To establish that the Indians formed their basic figures in the absence of any direct visual intuition;
4. To show that the early form of their symbols prefigured not only all the varieties currently in use in India and in Central and Southeast Asia, but also the respective shapes of Eastern and Western Arabic figures as well as the appearance of those figures used today and their various European predecessors of the same kind;
5. To prove that the learned men of that civilisation perfected the modern system of numeration for integers;
6. Finally, to establish once and for all that these discoveries took place in India, independent of any outside influence.

Historical reality, it can be seen, is not as simple as is generally thought: it is in any case not as simple as what an expression like "the invention of Arabic numerals", so cherished by the general public, seems to signify. For in terms of "invention" there would have to have been not only quite an exceptional combination of circumstances but also and above all an improbable conjunction of several great thoughts, created over fifteen centuries ago thanks to the genius of Indian scholars.

This would have taken exceptional powers of reflection, guided over a long period of time, not by logic or conscience, but by chance and necessity; chance discoveries and the need to remedy the problems engendered.

A. Vandel said, "A new idea is never the result of conscious or logical work. It emerges one day, fully formed, after a long gestation period which takes place within the subconscious."

It is true, as J. Duclaux says, that "the essential characteristic of scientific discoveries is that they cannot be made to order", because "the mind only makes discoveries when it is thinking of nothing".

## INDIAN NUMERICAL NOTATION

With the aim of establishing the Indian origin of modern numerical figures, the following is a review of the numerical notations in common use in India before and since this colossal event, beginning with the symbols currently in use in this particular part of the world.*

* Henceforth, the references given relate to the works which write out each of the styles in question. As for the geographical location of the regions concerned, these are taken mainly from L. Frédéric's *Dictionnaire de la civilisation indienne*.

It should be made clear straight away that the modern figures 1, 2, 3, 4, 5, 6, 7, 8, 9, 0 acquired their present form in the fifteenth century in the West, modelled on specific prototypes and adopted permanently when the printing press was "invented" in Europe. Today they are used all over the world, thus constituting a kind of universal language which can be understood by East and West alike.

However, this form is not the only one which can express the decimal positional system. Particular symbols representing the same numbers still coexist with the figures that we all know in several oriental countries.

From the Near East and the Middle East to Muslim India, Indonesia and Malaysia, the following symbols are preferred:

| 1 | 2 | 3 | 4 | 5 | 6 | 7 | 8 | 9 | 0 | Ref. |
|---|---|---|---|---|---|---|---|---|---|---|
| ١ | ٢ | ٣ | ۴ | ۵ | ٦ | ٧ | ٨ | ٩ | ٠ | EIS<br>Peignot and Adamoff<br>Pihan<br>Smith and Karpinski |
| | | | ٤ | ٥ | | | | | | |

Geographical area (see Fig. 25. 3):
Used in Libya, Egypt, Jordan, Syria, Saudi Arabia, Yemen, the Lebanon, Syria, Iraq, Iran, etc., as well as in Afghanistan, Pakistan, Muslim India, Indonesia Malaysia and formerly in Madagascar.

FIG. 24.2. *Current Eastern Arabic numerals (known as "Hindi" numerals)*

This is also the case in non-Muslim India, Central and Southeast Asia.

In these countries, symbols are still used that are graphically different from our own, and whose cursive form varies considerably from one region to another, according to the local style of writing.

Of course, this diversity dates back to ancient times, as the following pages will prove.

## *Nâgarî figures*

In his *Kitab fi tahqiq i ma li'l hind*, (an account of what he had witnessed in India, written around 1030) al-Biruni, the Muslim astronomer of Persian origin, after having lived in India and Sind for nearly thirty years, described the great diversity of the graphical forms of figures in common use at that time in different regions of India; his commentary begins thus [see al-Biruni (1910); Woepcke (1863), pp. 275–6]:

> Whilst we use letters for calculation according to their numerical value, the Indians do not use their letters at all for arithmetic.
>
> And just as the shape of the letters [that they use for writing] is different in [different regions of] their country, so the numerical symbols [vary].
>
> These are called **anka*.
>
> What we [the Arabs] use [for figures] is a selection of the best [and most regular] figures in India.
>
> Their shapes are not important, however, as long as their meaning is understood.
>
> The Kashmiris number their pages using figures which resemble ornamental drawings or letters [= characters used for writing] invented by the Chinese, which take a long time and a lot of effort to learn, but which are not used in calculation [which is carried out] in the dust (*hisab 'ala 't turab*).

Amongst the figures which were used long ago and are still used today most commonly in the various regions of India, the most regular are *Nâgarî*, which are also called *Devanâgarî*, from the name of the superb writing which they belong to (the words literally means "writing of the gods" in Sanskrit) (Fig. 24.3).

Al-Biruni (who mastered written and spoken Sanskrit), was alluding to precisely these figures when he said that the Arabs, in adopting the place-value system from India, had taken, as a means of notation for the nine units, "the best and most regular figures".

| 1 | 2 | 3 | 4 | 5 | 6 | 7 | 8 | 9 | 0 | Ref. |
|---|---|---|---|---|---|---|---|---|---|---|
| १ | २ | ३ | ४ | ५ | ६ | ७ | ८ | ९ | ० | Desgranges<br>Frédéric, DCI<br>Pihan<br>Renou and Filliozat |
| १ | २ | ३ | ४ | ५ | ६ | ७ | ८ | ९ | ० | |
| १ | २ | ३ | ४ | | ६ | | ८ | ९ | ० | |
| १ | २ | ३ | ४ | ५ | ६ | ७ | ८ | ९ | ० | |
| १ | | | ४ | ५ | ६ | | ८ | ९ | ० | |
| | | | | | | | ८ | ९ | | |

Geographical area (Fig. 24. 27 and 24. 53):
Used in the Indian states of Madhya Pradesh (Central Province), Uttar Pradesh (Northern Province), Rajasthan, Haryana, Himachal Pradesh (the Himalayas) and Delhi.

FIG. 24.3. *Modern Nâgarî (or Devanâgarî) numerals*

This point will be confirmed later in a palaeographical study, where it will be shown how these figures, or at least their ancestors, were over the years transformed by the hands of Arabic Muslim scribes to provide:

- in the Near East, the forms of the symbols in Fig. 24.2;
- in Northwestern Africa, other graphical representations, which would gradually be transformed, this time by European scribes, into the figures that we use today.

Furthermore, a striking resemblance still persists between the first three and the last of these signs and our own numerals 1, 2, 3 and 0.

*Marâthî figures*

These figures are used in the west of India, in the state-province of Maharashtra (capital, Bombay). They are, as a rule, the cursive form of their corresponding *Nâgarî*, except for a slight variation in the shape of the 5 and the 6 (Fig. 24.3). There is a resemblance between these symbols for 2, 3, and 0 and our own, and the *Marâthî* nine is symmetrical to the European nine.

| 1 | 2 | 3 | 4 | 5 | 6 | 7 | 8 | 9 | 0 | Ref. |
|---|---|---|---|---|---|---|---|---|---|---|
| १ | २ | ३ | ४ | ५ | ६ | ७ | ८ | ९ | ० | Drummond<br>Frédéric, DCI<br>Pihan |
| १ | २ | ३ | ४ | ५ | ६ | ७ | ८ | ९ | ० | |

Geographical area (Fig. 24. 27 and 24. 53) :
Used in the area bordered in the west by the coasts of Konkan and Daman, and in the north by Gujarat and Madhya Pradesh, in the south by Karnataka and in the southeast by Andhra Pradesh.

FIG. 24.4. *Modern Marâthî numerals*

*Punjabi figures*

Used in the state of Punjab (capital, Chandigarh), in the northwest of India, bordering Pakistan. These are the same as the corresponding *Nâgarî* figures, except for the shape of the 7 (Fig. 24.3). There are similarities between these symbols and our figures 2, 3, 7 and 0:

| 1 | 2 | 3 | 4 | 5 | 6 | 7 | 8 | 9 | 0 | Ref. |
|---|---|---|---|---|---|---|---|---|---|---|
| ੧ | ੨ | ੩ | ੪ | ੫ | ੬ | ੭ | ੮ | ੯ | ੦ | Pihan |

Geographical area (Fig. 24. 27 and 24. 53) :
Used in the northwest of India bordering Pakistan where the Indus, the Chenab, the Jhelam, the Ravi and the Satlej rivers meet; as well as in the states of Himachal Pradesh and Haryana.

FIG. 24.5. *Modern Punjabi numerals*

*Sindhî figures*

These are symbols used in Sind, whose name derives from that of the river Sindh (the Indus). These signs are more or less identical to their corresponding *Nagârî*, but their shape is generally more cursive than the latter (Fig. 24.3). The figures 2, 3 and 0 are similar to our own, and the *Sindhî* 5 is rather like a symmetrical version of the European 4.

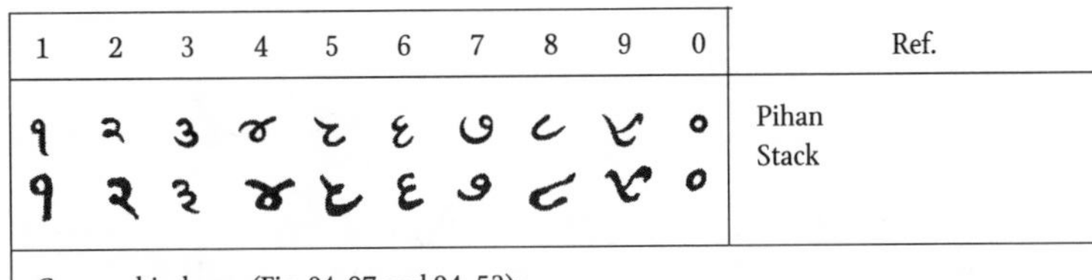

| 1 | 2 | 3 | 4 | 5 | 6 | 7 | 8 | 9 | 0 | Ref. |
|---|---|---|---|---|---|---|---|---|---|---|
| | | | | | | | | | | Pihan<br>Stack |

Geographical area (Fig. 24. 27 and 24. 53):
Used south of Punjab, on the lower banks of the Indus, in a region bordered in the south by the Gulf of Oman and in the west by the Thar desert.

FIG. 24.6. *Modern Sindhî numerals*

*Gurûmukhî figures*

In the city of Hyderabad (on the River Indus, to the east of Karachi, not to be confused with the other Hyderabad, capital of Andhra Pradesh), the merchants used to use a slight variant of the preceding figures, known as *Khudawadî*.

The traders of Shikarpur and Sukkur, on the other hand, sometimes used *Sindhî* or *Punjâbî* figures, sometimes eastern Arabic figures and sometimes *Gurûmukhî* figures, which are a mixture of *Sindhî* and *Punjâbî* styles (Fig. 24.5 and 24.6):

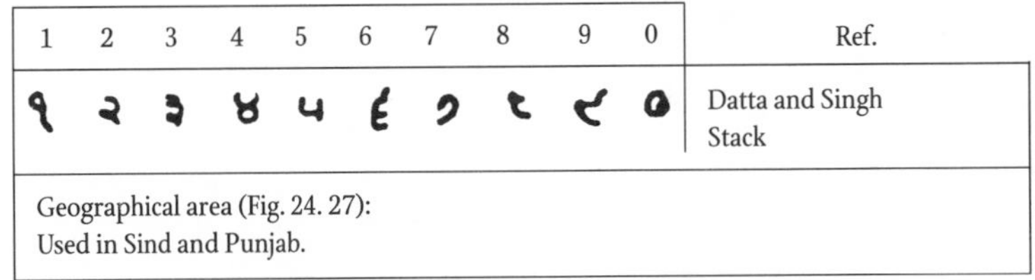

| 1 | 2 | 3 | 4 | 5 | 6 | 7 | 8 | 9 | 0 | Ref. |
|---|---|---|---|---|---|---|---|---|---|---|
| | | | | | | | | | | Datta and Singh<br>Stack |

Geographical area (Fig. 24. 27):
Used in Sind and Punjab.

FIG. 24.7. *Gurûmukhî numerals*

*Gujarâtî figures*

These are used in Gujarat State (capital, Ahmadabad), on the edge of the Indian Ocean, between Bombay and the border of Pakistan. Again, these are derived from *Nâgarî* figures, but they are more cursive in form, particularly the 6 (Fig. 24.3). There are similarities between the *Gujarâtî* figures 2, 3 and 0 and our own numerals, as well as the figure 6.

| 1 | 2 | 3 | 4 | 5 | 6 | 7 | 8 | 9 | 0 | Ref. |
|---|---|---|---|---|---|---|---|---|---|---|
| | | | | | | | | | | Drummond<br>Forbes<br>Frédéric, DCI<br>Pihan |
| Geographical area (Fig. 24. 27 and 24. 53):<br>Used in the west of India, bordering the Indian Ocean,<br>between Bombay and the border with Pakistan, on the Gulf of Cambay. | | | | | | | | | | |

FIG. 24.8. *Modern Gujarâtî numerals*

### *Kaîthî figures*

Used mainly in Bihar State, in Eastern India, and sometimes in Gujarat State. They evidently derive from *Nâgarî* figures and are similar in form to *Gujarâtî* figures (Fig. 24.3 and 24.8):

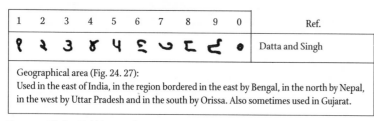

| 1 | 2 | 3 | 4 | 5 | 6 | 7 | 8 | 9 | 0 | Ref. |
|---|---|---|---|---|---|---|---|---|---|---|
| | | | | | | | | | | Datta and Singh |
| Geographical area (Fig. 24. 27):<br>Used in the east of India, in the region bordered in the east by Bengal, in the north by Nepal, in the west by Uttar Pradesh and in the south by Orissa. Also sometimes used in Gujarat. | | | | | | | | | | |

FIG. 24.9. *Modern Kaîthî numerals*

### *Bengâlî figures*

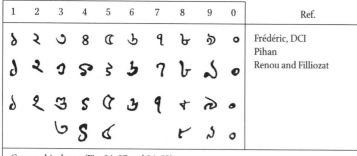

| 1 | 2 | 3 | 4 | 5 | 6 | 7 | 8 | 9 | 0 | Ref. |
|---|---|---|---|---|---|---|---|---|---|---|
| | | | | | | | | | | Frédéric, DCI<br>Pihan<br>Renou and Filliozat |
| Geographical area (Fig. 24. 27 and 24. 53):<br>Used in the regions in the northwest of the Indian sub-continent, between Bihar, Nepal, Assam, Sikkim, Bhutan, and the Bay of Bengal. Also widely used in Assam (along the Brahmaputra). | | | | | | | | | | |

FIG. 24.10. *Modern Bengâlî numerals*

Used in the northeast of the Indian sub-continent in Bangladesh (capital, Dacca), in the Indian state of West Bengal (capital, Calcutta), and in much of central Assam (along the Brahmaputra River).

Of all the *Bengâlî* figures, there are four which resemble *Nâgarî* figures: 2, 4, 7 and 0 (Fig. 24.3). The others, however, are very different from those used in other parts of India. In one of the following variants, our figures 2, 3, 7 and 0 are recognisable; one of the variants of 8 also constitutes a sort of prefiguration of our 8.

### *Maithilî figures*

Used mainly in the north of Bihar State, these derive mainly from *Bengâlî* figures (Fig. 24.10):

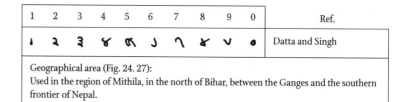

| 1 | 2 | 3 | 4 | 5 | 6 | 7 | 8 | 9 | 0 | Ref. |
|---|---|---|---|---|---|---|---|---|---|---|
| | | | | | | | | | | Datta and Singh |
| Geographical area (Fig. 24. 27):<br>Used in the region of Mithila, in the north of Bihar, between the Ganges and the southern frontier of Nepal. | | | | | | | | | | |

FIG. 24.11. *Modern Maithilî numerals*

### *Oriya figures*

Used mainly in Orissa State (capital, Bhubaneswar), these are also known as *Orissî* figures. Although they derive from the same source as *Nâgarî* figures, they present significant differences (Fig. 24.3):

| 1 | 2 | 3 | 4 | 5 | 6 | 7 | 8 | 9 | 0 | Ref. |
|---|---|---|---|---|---|---|---|---|---|---|
| | | | | | | | | | | Frédéric, DCI<br>Pihan<br>Renou and Filliozat<br>Sutton |
| Geographical area (Fig. 24. 27 and 24. 53):<br>Used in the region to the south of the eastern coast of Deccan, bordered in the north by Bengal and Bihar, in the west by Madhya Pradesh and in the south by Andhra Pradesh. | | | | | | | | | | |

FIG. 24.12. *Modern Oriyâ (or Orissî) numerals*

### *Tâkarî figures*

In everyday use in Kashmir, alongside eastern Arabic figures. They are also called *Tankrî* figures, of which a variant, *Dogrî*, is used in the Indian part of Jammu (in southwestern Kashmir):

| 1 | 2 | 3 | 4 | 5 | 6 | 7 | 8 | 9 | 0 | Ref. |
|---|---|---|---|---|---|---|---|---|---|---|
| | | | | | | | | | | Datta and Singh |

Geographical area (Fig. 24. 27):
Used in the region in the extreme northwest of the Indian sub-continent, currently divided by the Indian-Pakistani border, joining the country of Jammu to the north of Himachal Pradesh, the plain of Kashmir in the high basin of the Jhelum, the valley of Zaskar in the north of the Himalayas and that of Ladakh, adjoining Tibet and China.

FIG. 24.13. *Modern Tâkarî (or Tankrî) numerals*

### *Shâradâ figures*

The figures that were used for many centuries in Kashmir and Punjab, from which, among others, *Dogrî* and *Tâkarî* figures derived (Fig. 24.13).

| 1 | 2 | 3 | 4 | 5 | 6 | 7 | 8 | 9 | 0 | Ref. |
|---|---|---|---|---|---|---|---|---|---|---|
| | | | | | | | | | | Pihan<br>Renou and Filliozat<br>Smith and Karpinski |

Geographical area (Fig. 24. 27 and 24. 53):
Formerly used in Kashmir and Punjab (before the sixteenth century).

FIG. 24.14. *Shâradâ numerals (relatively recent forms)*

These figures are connected to *Shâradâ* writing, which was used in the region at least since the ninth century, before it was replaced, relatively recently, by the Persian Arabic characters that are used for writing.

This notation (even in its most recent form) deserves special attention, because instead of representing zero with an oval or a small circle, it uses a dot, the circle being used to denote the number 1 (the shape was slightly modified according to the base).

The *Shâradâ* 2 is like the *Nâgarî* 3, except that the lower appendage is absent from the *Shâradâ* figure.

To the untrained eye, it should be pointed out, the figures 2 and 3 are not sufficiently distinct from one another, although the top of the 3 differs from that of the 2 because it is long and snaking.

The 6 is symmetrical to the European 6, whilst the 8 is very similar to the hand-written form of our 3.

As for the 7 and the 9, they are, respectively, almost identical to the figures 1 and 7 of *Nâgarî* notation (Fig. 24.3).

### *Nepâlî figures*

Used mainly in the independent state of Nepal (capital, Kathmandu), these are also called *Gurkhalî* figures.

In one of the following variations, our 1, 2, 3, 4, 7 and 0 can be recognised, as well as our 8 to a certain extent (first set of figures, Fig. 24.15).

| 1 | 2 | 3 | 4 | 5 | 6 | 7 | 8 | 9 | 0 | Ref. |
|---|---|---|---|---|---|---|---|---|---|---|
| | | | | | | | | | | Datta and Singh<br>Renou and Filliozat |

Geographical area (Fig. 24. 27 and 24. 53):
Formerly used in Kashmir and Punjab (before the sixteenth century).

FIG. 24.15. *Current Nepâlî numerals*

There is an obvious similarity between these figures and the *Nâgarî* and *Shâradâ* figures, with which they share a common source (Fig. 24.3 and 24.14).

### *Tibetan figures*

These are the figures used in Tibet. They are similar to *Devanâgarî* figures (Tibetan writing comes from the same source as *Nâgarî*, introduced to the region in the seventh century CE at the same time as Buddhism). The 2, 3, the 9 (written backwards) and the 0 are alike.

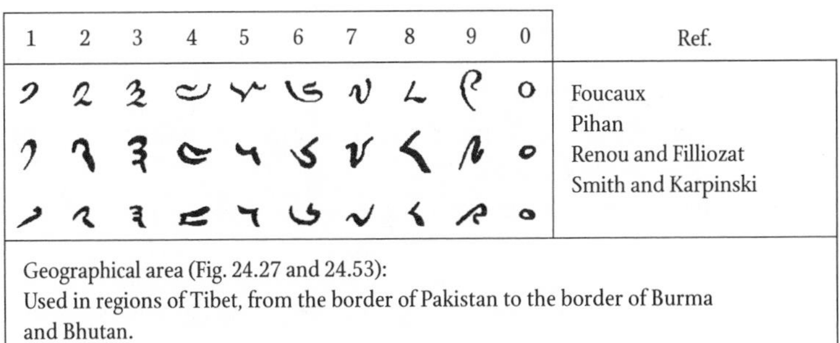

| 1 | 2 | 3 | 4 | 5 | 6 | 7 | 8 | 9 | 0 | Ref. |
|---|---|---|---|---|---|---|---|---|---|---|
| | | | | | | | | | | Foucaux<br>Pihan<br>Renou and Filliozat<br>Smith and Karpinski |

Geographical area (Fig. 24.27 and 24.53):
Used in regions of Tibet, from the border of Pakistan to the border of Burma and Bhutan.

FIG. 24.16. *Tibetan numerals*

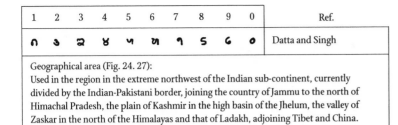

### *Tamil figures*

Unlike Northern and Central India, in Southern India, namely Tamil Nadu, Karnataka, Andhra Pradesh and Kerala, the Dravidian people do not speak Indo-European languages.

Tamil figures are used in Southeast India, in Tamil Nadu state (capital, Madras):

| 1 | 2 | 3 | 4 | 5 | 6 | 7 | 8 | 9 | 0 | Ref. |
|---|---|---|---|---|---|---|---|---|---|---|
| ௧ | ௨ | ௩ | ௪ | ௫ | ௬ | ௭ | ௮ | ௯ | | Frédéric, DCI |
| ௧ | ௨ | ௩ | ௪ | ௫ | ௬ | ௭ | ௮ | ௯ | | Pihan |
| ௧ | ௨ | ௩ | ௪ | ௫ | ௬ | ௭ | ௮ | ௯ | | Renou and Filliozat |

Geographical area (Fig. 24. 27 and 24. 53):
Used in the region on the eastern coast of the Indian peninsula, from the north of Madras to the tip of Cape Comorin (Kanya Kumari) and bordered in the east by the Bay of Bengal, in the west by Kerala, in the northwest by Karnataka and in the north by Andhra Pradesh. Also used in the north and northwest of Sri Lanka.

FIG. 24.17. *Current Tamil numerals (or "Tamoul" numerals, according to an erroneous transcription)*

It should be noted, however, that the Tamils do not use zero in this system, which is only vaguely based on the place-value system.

Along with the signs for nine units, their system actually possesses a specific sign for 10, 100 and 1,000. To express multiples of 10, or hundreds or thousands, the sign for 10, 100 or 1,000 is proceeded by that of the corresponding units, which thus play the part of multiplier.

In other words, the *Tamil* system is based upon a principle which is at once additive and multiplicative, known as the *hybrid principle* and which has been used in many systems since early antiquity (see Fig. 23.20).

Equally, in terms of their appearance, these figures have nothing in common with the preceding notations.

For these reasons, it was believed that the Tamil figures were an original creation of the Dravidians, after they came up with the idea of using certain letters of their alphabet as signs for counting with.

It is true that there is a degree of resemblance between the first ten figures and what might constitute the corresponding letters of the Tamil alphabet, although the correspondence is not always very rigorous:

*Comparison between the numeral and the letter*

| | | | |
|---|---|---|---|
| 1 | ௧ | க | ka, ga |
| 2 | ௨ | உ | u |
| 3 | ௩ | ஞ | ña |
| 4 | ௪ | ஷ | sha |
| 5 | ௫ | ரு | ra |
| 7 | ௬ | ச | cha |
| 7 | ௭ | எ | ê |
| 8 | ௮ | அ | a |
| 9 | ௯ | கூ | kû, gû |

*Tamil name for the corresponding number*

| | |
|---|---|
| ûru | 1 |
| irandu | 2 |
| mûnru | 3 |
| nâlu, nângu | 4 |
| aïndu, andju | 5 |
| âru | 6 |
| êrla, êzha | 7 |
| ettu | 8 |
| onbadu | 9 |

FIG. 24.18.

There is one question that cries out to be asked: if the theory is correct, why were these particular letters used to denote these numerical values? The obvious answer would be that the initials of the Tamil names for the numbers were used, but this is not the case, as the preceding table clearly demonstrates.

Then why were these letters singled out to represent numbers? Why did these people not give a numerical value to all the Tamil letters, as the Greeks and the Jews did with their respective alphabets when they created their systems of numeral letters?

This theory is rather far-fetched; it is merely a coincidence that these figures resemble the above Tamil letters. Moreover, the correspondence can only be established using the modern forms of the letters.

In fact, Tamil letters and figures are connected to all the other systems used in India: they all derive from the same source. Tamil writing, however, evolved in an entirely different manner from the others, both in terms of appearance and linguistic structure, introducing innovations which gave it its distinctive character. In particular, the characters and numerical symbols are considerably more rounded, with curves and volutes. It is not impossible that the material on which the characters were written played a role in this evolution, if it did not actually cause it.

In other words, the first nine Tamil figures are from the same family as the other corresponding Indian numerical symbols, the difference lying in their style and their adaptation to the unique shape of Tamil writing.

### *Malayâlam figures*

These figures are used by the Dravidian people of Kerala State, on the ancient coast of Malabar, in the southwest of India. They have the same name as the form of writing used in the area.

| 1 | 2 | 3 | 4 | 5 | 6 | 7 | 8 | 9 | 0 | Ref. |
|---|---|---|---|---|---|---|---|---|---|---|
| ൧ | ൨ | ൩ | ൪ | ൫ | ൬ | ൭ | ൮ | ൯ | | Drummond |
| ൧ | ൨ | ൩ | ൪ | ൫ | ൬ | ൭ | ൮ | ൯ | | Frédéric, DCI |
| ൧ | ൨ | ൩ | ൪ | ൫ | ൬ | ൭ | ൮ | ൯ | | Peet, J. |
| | | | | | | | | | | Pihan |
| | | | | | | | | | | Renouand Fillozat |

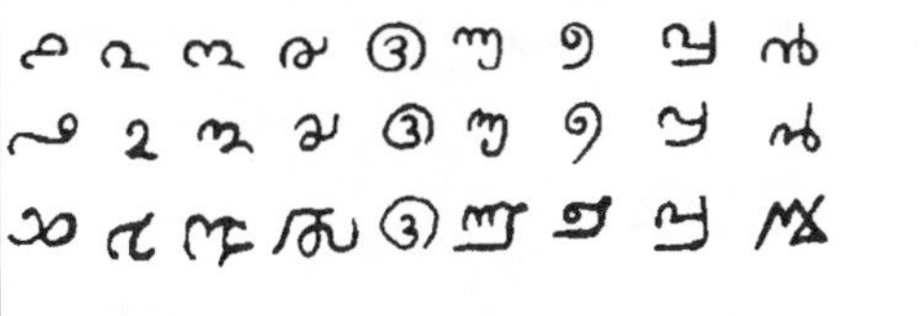

Geographical area (Fig. 24. 27 and 24. 53):
Used in the region stretching the length of the southeast coast of India, from Mangalore in the north to the southernmost point of India, and which is made up of a long coastal strip stretching from the coast of Malabar and by the Ghats encompassing the peaks of the Cardamoms.

FIG. 24.19. *Current Malayâlam numerals*

Like the Tamils, the people of Kerala did not use zero in their notation system for many centuries: *Malayâlam* figures are not based on the place-value system, and there are specific figures for 10, 100 and 1,000. It was only since the middle of the nineteenth century, under the influence of Europe, that zero was introduced and combined with the symbols for the nine units according to the positional principle.

Thus the Tamil and *Malayâlam* figures were the only ones in India that did not include zero and were not based on the positional principle until relatively recently.

However, it should be noted that Tamil figures, a few centuries ago, before they evolved into their current forms, closely resembled their *Malayâlam* cousins which have conserved a style close to the original.

The graphical link with the numerical signs of other regions of India is more easily seen through examining the original appearance of the Tamil figures than through looking at their modern form (Fig. 24.17 and 24.19):

The *Nâgarî* 1 is easily recognised, whose former shape was almost horizontal (Fig. 24.39) and which evolved in Tibet into a form constituting a sort of intermediate with the *Malayâlam* 1 (Fig. 24.16).

The *Nâgarî* 2 is also recognisable, although the "head" of the sign is very neatly rounded at the bottom.

On the other hand, the *Malayâlam* 3 is much closer to the corresponding *Oriyâ* figure (Fig. 24.12), with an extra "tail" which the *Nâgarî* 3 also has (Fig. 24.3).

The 4 is similar to its *Sindhi* equivalent except for the characteristic curve on the left (Fig. 24.6).

The 5 is very similar to one of the corresponding *Bengâlî* figure (Fig. 24.10) and is reminiscent of the *Malayâlam* style.

The 6 resembles its *Sindhî* counterpart (Fig. 24.6), but it has an extra loop on the top, the whole figure being in a position which is obtained by rotating it through 90° anti-clockwise.

The 7 resembles its *Marâthî, Gujarâtî* and *Oriyâ* equivalents (Fig. 24.4, 24.8 and 24.12), whose prototype is found in the ancient *Nâgarî* style (Fig. 24.39).

The 8 is the symmetrical equivalent of the *Gujarâtî* 8 (Fig. 24.4).

As for the 9, it particularly resembles the *Nâgarî* style of the ninth century CE.

There can be no doubt: the Dravidian figures for the nine units have the same origin as all the others; the similarities found scattered amongst these diverse figures could not possibly be the product of chance.

The following two varieties of Dravidian figures serve as confirmation of this fact.

### *Telugu figures*

These are the numerical symbols used by Dravidian people of the former Telingana, the Indian state of Andhra Pradesh (capital, Hyderabad). They are also called *Telinga* figures (Fig. 24.20).

| 1 | 2 | 3 | 4 | 5 | 6 | 7 | 8 | 9 | 0 | Ref. |
|---|---|---|---|---|---|---|---|---|---|---|
| ౧ | ౨ | ౩ | ౪ | ౫ | ౬ | ౭ | ౮ | ౯ | ౦ | Burnell |
| ౧ | ౨ | ౩ | ౪ | ౫ | ౬ | ౭ | ౮ | ౯ | ౦ | Campbell |
| ౧ | ౨ | ౩ | ౪ | ౫ | ౬ | ౭ | ౮ | ౯ | ౦ | Datta and Singh |
| ౧ | ౨ | ౩ | ౪ | ౫ | ౬ | ౭ | ౮ | ౯ | ౦ | Pihan |
| | | | | | | | | | | Renou and Filliozat |
| | | | | | | | | | | Smith and Karpinski |

Geographical area (Fig. 24. 27 and 24, 53):
Used in the southeast of India, bordered in the southeast by the Bay of Bengal, in the north by the States of Orissa and Madhya Pradesh, in the northwest by Maharashtra, in the west by Karnataka and in the south by Tamil Nadu.

FIG. 24.20. *Modern Telugu (or Telinga) numerals*

## *Kannara figures*

Used by the Dravidian people of central Deccan, including the state of Karnataka (capital, Bangalore) and part of Andhra Pradesh:

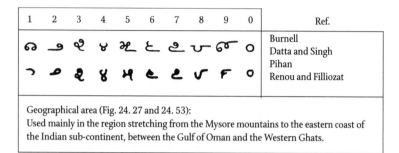

| 1 | 2 | 3 | 4 | 5 | 6 | 7 | 8 | 9 | 0 | Ref. |
|---|---|---|---|---|---|---|---|---|---|---|
| ೧ | ೨ | ೩ | ೪ | ೫ | ೬ | ೭ | ೮ | ೯ | ೦ | Burnell<br>Datta and Singh<br>Pihan<br>Renou and Filliozat |
| ೧ | ೨ | ೩ | ೪ | ೫ | ೬ | ೭ | ೮ | ೯ | ೦ | |

Geographical area (Fig. 24. 27 and 24. 53):
Used mainly in the region stretching from the Mysore mountains to the eastern coast of the Indian sub-continent, between the Gulf of Oman and the Western Ghats.

FIG. 24.21. *Modern Kannara (or Kannada or Karnata) numerals*

## *Sinhalese figures*

Used mainly in Sri Lanka and in the Maldives as well as in the islands to the north of the latter. (In the north and northwest of Sri Lanka, Tamil figures are also used due to the high number of Tamil people who live in these areas of the island.)

| 1 | 2 | 3 | 4 | 5 | 6 | 7 | 8 | 9 | 0 | Ref. |
|---|---|---|---|---|---|---|---|---|---|---|
| 𑇡 | 𑇢 | 𑇣 | 𑇤 | 𑇥 | 𑇦 | 𑇧 | 𑇨 | 𑇩 | | Alwis (de)<br>Charter<br>Frédéric, DCI<br>Pihan<br>Renou and Filliozat |
| 𑇡 | 𑇢 | 𑇣 | 𑇤 | 𑇥 | 𑇦 | 𑇧 | 𑇨 | 𑇩 | | |
| 𑇡 | 𑇢 | 𑇣 | 𑇤 | 𑇥 | 𑇦 | 𑇧 | 𑇨 | 𑇩 | | |

Geographical area (Fig. 24. 27 and 24. 53):
Used in Sri Lanka, in the Maldives, as well as in the islands to the north of the Maldives.

FIG. 24.22. *Current Sinhalese (or Sinhala) numerals*

It should be noted that although Sinhalese writing is linked to Dravidian forms of writing (even though it is more stylish, striving as it does towards an ornamental effect), the language of this writing is not Dravidian. Sinhalese is an Indo-European language: "it is a language that belongs to Prakrit (dialects) of 'Middle Indian', as several inscriptions written in *Brâhmî* dating from around the second century BCE show. However, after the fifth century CE, the Sinhalese language, separated from India's Indo-European languages by the Tamil area, developed in an individual style, as did its writing. The two seem to have changed little since 1250" (L. Frédéric).

There are twenty Sinhalese figures. This number of numerical signs is due to the absence of zero and the fact that the system, which is not based upon the place-value system, uses a specific figure for every ten units, as well as special figures that represent 10, 100 and 1,000 (see Fig. 23.18).

## *Burmese figures*

Used in Burma. Formerly used in the kingdom of Magadha, these were once known as *cha lum* figures, they are part of Burmese writing, which itself derives from the former *Pâlî* alphabet, introduced to the region by Buddhists (Fig. 24.23).

| 1 | 2 | 3 | 4 | 5 | 6 | 7 | 8 | 9 | 0 | Ref. |
|---|---|---|---|---|---|---|---|---|---|---|
| ၁ | ၂ | ၃ | ၄ | ၅ | ၆ | ၇ | ၈ | ၉ | ၀ | Carey<br>Datta and Singh<br>Latter<br>Pihan |
| ၁ | ၂ | ၃ | ၄ | ၅ | ၆ | ၇ | ၈ | ၉ | ၀ | |
| ၁ | ၂ | ၃ | ၄ | ၅ | ၆ | ၇ | ၈ | ၉ | ၀ | |
| ၁ | ၂ | ၃ | ၄ | ၅ | ၆ | ၇ | ၈ | ၉ | ၀ | |
| ၁ | ၂ | ၃ | ၄ | ၅ | ၆ | ၇ | ၈ | ၉ | ၀ | |

Geographical area (Fig. 24.27 and 24.53):
Used in the region stretching from Laos to the Bay of Bengal, and from Manipur to Pegu; also, in a slightly modified form, around Tenasserim and along the coast from Chittagong.

FIG. 24.23. *Modern Burmese numerals*

In modern Burmese writing, the principal element of the shape of the letters is a little circle, the value of which varies according to the breaks, juxtapositions or appendages.

The same applies to the figures, or at least to three of them, whose shapes should not be confused.

These are:

- the 1, formed by a circle, a quarter open on the left;
- the 8, which is a circle that is a quarter open at the bottom;
- and the 0 which is a whole circle.

The 3 is an open circle like the 1, with an appendage which slants towards the right, and the 4 is formed by the mirror image of the 3.
As for the 9, it is the 6 turned upside-down.

However this graphical rationalisation is relatively recent: the Indian origin (via former *Pâlî* figures) of the Burmese figures was still unknown in the seventeenth century.

### *Thai-Khmer figures*

These are the official numerical symbols of Thailand, Laos and Cambodia. They also belong to the family of numerical signs that are of Indian origin, actually belonging to the former *Pâlî* style.

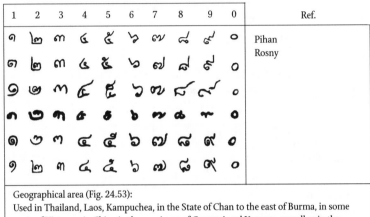

| 1 | 2 | 3 | 4 | 5 | 6 | 7 | 8 | 9 | 0 | Ref. |
|---|---|---|---|---|---|---|---|---|---|---|
| ๑ | ๒ | ๓ | ๔ | ๕ | ๖ | ๗ | ๘ | ๙ | ๐ | Pihan<br>Rosny |
| ๑ | ๒ | ๓ | ๔ | ๕ | ๖ | ๗ | ๘ | ๙ | ๐ | |
| ๑ | ๒ | ๓ | ๔ | ๕ | ๖ | ๗ | ๘ | ๙ | ๐ | |
| ๑ | ๒ | ๓ | ๔ | ๕ | ๖ | ๗ | ๘ | ๙ | ๐ | |
| ១ | ២ | ៣ | ៤ | ៥ | ៦ | ៧ | ៨ | ៩ | ០ | |
| ១ | ២ | ៣ | ៤ | ៥ | ៦ | ៧ | ៨ | ៩ | ០ | |
| Geographical area (Fig. 24.53):<br>Used in Thailand, Laos, Kampuchea, in the State of Chan to the east of Burma, in some parts of Vietnam, in China in the provinces of Guangxi and Yunnan, as well as in the Nicobar islands. | | | | | | | | | | |

FIG. 24.24. *Modern Thai-Khmer (known as "Siamese") numerals*

Some of these figures look so alike that they are easily confused. Unlike the various "true" Indian figures, the Thai-Khmer 2 is more complicated than the 3. The 5 only differs from the 4 because it has an extra loop at the top. The 8 is more or less symmetrical to the 6, and the figure 7 is easily confused with the 9.

### *Balinese figures*

These are from Bali, and also developed from the *Pâlî* figures.

| 1 | 2 | 3 | 4 | 5 | 6 | 7 | 8 | 9 | 0 | Ref. |
|---|---|---|---|---|---|---|---|---|---|---|
| ᭑ | ᭒ | ᭓ | ᭔ | ᭕ | ᭖ | ᭗ | ᭘ | ᭙ | ᭐ | Renou and Filliozat |
| Geographical area (Fig. 24.53):<br>Used in Bali, Borneo and the Celebes islands. | | | | | | | | | | |

FIG. 24.25. *Modern Balinese numerals*

### *Javanese figures*

The final figures in this list of numerical symbols currently in use in Asia are those from the island of Java:

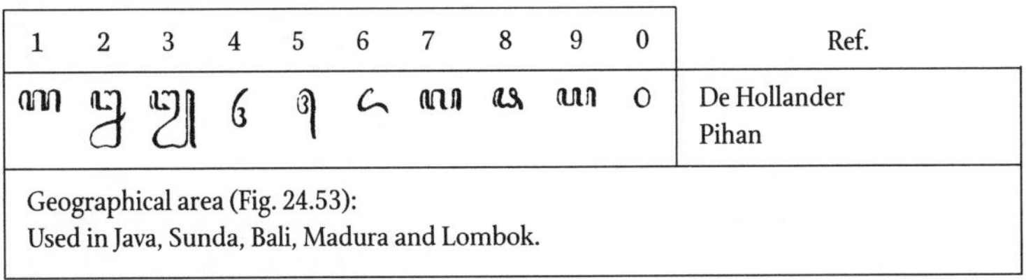

| 1 | 2 | 3 | 4 | 5 | 6 | 7 | 8 | 9 | 0 | Ref. |
|---|---|---|---|---|---|---|---|---|---|---|
| ꧑ | ꧒ | ꧓ | ꧔ | ꧕ | ꧖ | ꧗ | ꧘ | ꧙ | ꧐ | De Hollander<br>Pihan |
| Geographical area (Fig. 24.53):<br>Used in Java, Sunda, Bali, Madura and Lombok. | | | | | | | | | | |

FIG. 24.26. *Modern Javanese numerals*

Apart from the figures 0 and 5 (whose Indian origin is obvious), this notation actually corresponds to a relatively recent artificial innovation, the appearance of the figures curiously having been made to resemble the shape of certain letters of the current Javanese alphabet. Before this, however, the Javanese people used a notation which belonged to the *Pâlî* group of the family of Indian figures: the notation known as *Kawi* (attested since the seventh century CE), which belongs to the writing of the same name (from which the current Javanese alphabet derives).

### *Brâhmî, "mother" of all Indian writing*

Despite the high number of graphical representations of the nine units, there is no doubt as to their common origin.

Leaving European and Arabic numerals on one side for a moment, each of the preceding styles were graphically connected to one of the various styles of writing belonging to either India, Central or Southeast Asia: it is clear from extensive palaeographical research that they all derive, directly or indirectly, from the same source.

Therefore, it is worthwhile saying a few words about the history of the styles of writing of this region.

The oldest known writing of the sub-continent of India appeared on the stamps and plaques of the civilisation of the Indus (c. 2500 – 1500 BCE), discovered mainly in the ruins of the ancient cities of Mohenjo-daro and Harappâ. However, as this writing has not yet been deciphered, the corresponding language remains unknown; therefore there is a large gulf separating these inscriptions of the first known texts in Indian writing and the language, assuming that a link exists between the two systems.

In fact, the history of Indian writing begins with the inscriptions of Asoka, third emperor of the dynasty of the Mauryas of the Magadha, who reigned in India from c. 273 to 235 BCE, whose empire stretched from

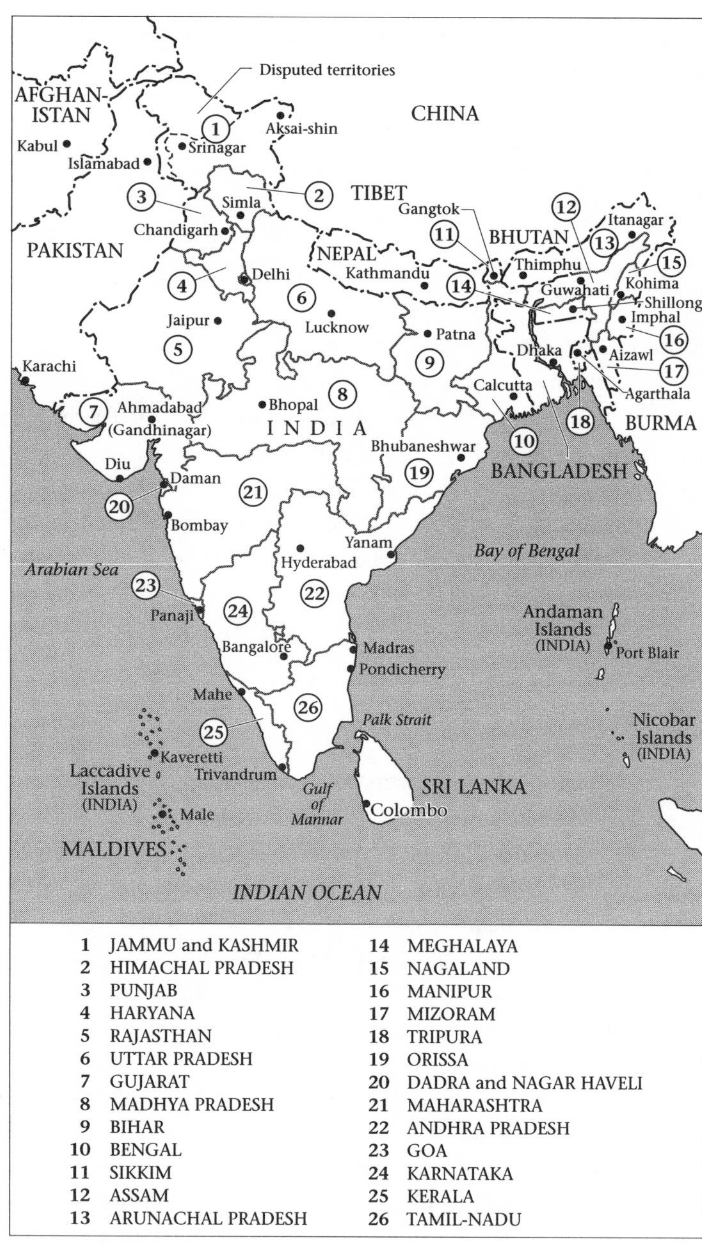


FIG. 24.27. *The states of present-day India*

Afghanistan to Bengal and from Nepal to the south of Deccan [see L. Frédéric (1987)]. These inscriptions are mainly edicts carved on rocks or columns for which diverse styles of writing were used: Greek and Aramaean in Kandahar and Jalalabad in Afghanistan; the *Kharoshthî* system in Manshera and Shahbasgarhi to the north of the Indus; and *Brâhmî* writing in all the other regions of the Empire.

*Kharoshthî* comes directly from the old Aramaean alphabet and is similarly written from right to left. This is why it is also labelled "Aramaeo-Indian" writing. Probably introduced in the fourth century BCE, it remained in use in the northwest of India until the end of the fourth century CE.

As for the written form of *Brâhmî,* it was written from left to right and was used to note the sounds of Sanskrit.

The origin of this writing is still not known. Attempts have been made to prove that it comes from *Kharoshthî* writing, but the explanation for this is far from convincing. *Brâhmî* certainly derives from the Western Semitic world, doubtless via some other variety of Aramaean, of which specimens have not yet been found [see M. Cohen (1958); J. G. Février (1959)].

Since the first millennium BCE, India was already open to outside influences, due to long-established ties with the Persians and Aramaean merchants who used the routes which went from Syria and Mesopotamia to the valley of the Indus.

However, the appearance of *Brâhmî* probably pre-dates Emperor Asoka, by whose time it was in widespread use in the different regions of the subcontinent of India.

This language outlived all the others, becoming the unique source of all the forms of writing that later emerged in India and her neighbouring countries. It was given the name *Brâhmî,* in Hindu religion one of the names of the seven **mâtrikâ* or "mothers of the world": one of the feminine energies (**shakti*) supposed to represent the Hindu divinities. Represented as sitting on a goose, her power was equal to that of Brahma, the "Immeasurable", god of the Sky and the horizons, who "endlessly gives birth to the Creation" and who one day invented *Brâhmî* writing for the well-being and diversity of humankind.

According to the edicts of Asoka, *Brâhmî* appeared, in a slightly modified form, in contemporary inscriptions of the Shunga Dynasty (185 – c. 75 BCE on the Magadha, in the present Bihar state, south of the Ganges, then in those of the Kanva Dynasty (who succeeded the former from 73 to c. 30 BCE).

The following is a more developed exploration of *Brâhmî*, first through the inscriptions of the Shaka Dynasty (Scythians, who reigned over Kabul in Afghanistan, Taxilâ in Punjab and Mathura, from the second century BCE to the first century CE) and through the coins embossed with the sovereigns of the Shaka Dynasty who reigned from the second to the fourth century CE in Maharashtra (under the name of *Kshatrapa*, "Satraps").

*Brâhmî* evolved a little more in the writing of the Andhra and Satavahana Dynasties which reigned during the first two centuries CE in the northwest of Deccan.

Then the system appeared, in an even more developed form, in the inscriptions of the Kushan emperors (who reigned from the first to the third century CE, and who, at first based in Gandhara and Transoxiana attempted to conquer Northwestern India).

Thus through numerous successive and perceptible modifications, *Brâhmî* gave birth to many highly individual styles of writing; styles which constitute the main groups currently in use (Fig. 24.28):

1. the group of types of writing in Northern and Central India and in Central Asia (Tibet and East Turkestan);
2. the writing of Southern India;
3. oriental writing (Southeast Asia).

The apparently considerable differences between the forms of writing of these various groups is ultimately due either to the specific character of the language and traditions to which they have been adapted, or to the techniques of the scribes of each region and the nature of the material they used.

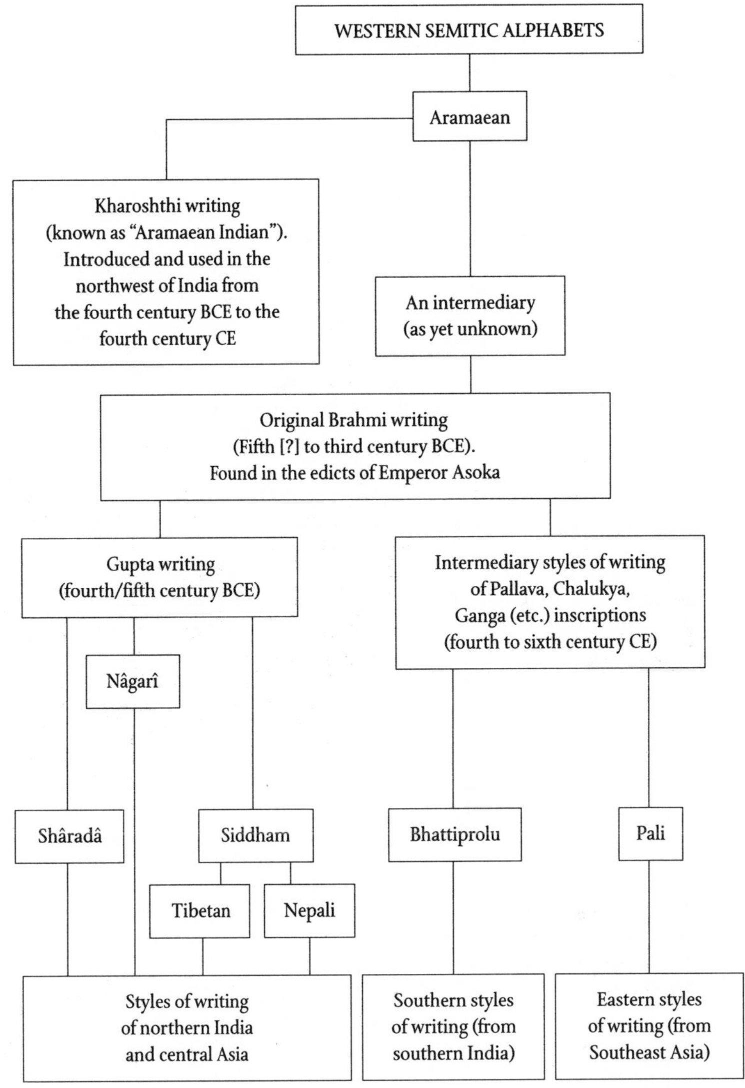


FIG. 24.28. *Indian styles of writing*

### *A parallel evolution: Indian figures*

In this context, everything becomes clear: in India and the surrounding regions, the notation of the nine units evolved in much the same way as the styles of writing that were born out of *Brâhmî*. In other words, in the same way as the writing they belong to, the various series of 1 to 9 formerly or currently in use in India, Central and Southeast Asia all derive more or less directly from the *Brâhmî* notation for the corresponding numbers.

### *The numerical symbols of the original Brâhmî notation*

This notation appeared for the first time in the middle of the third century BCE in edicts written in both *Ardha-Mâgadhî* and *Brâhmî* which the emperor Asoka had engraved on rocks, polished sandstone columns and temples hewn out of the rock, in diverse regions of his empire.

But the numerical notation that is found within these edicts is fragmentary, only giving the representations for the numbers 1, 2, 4, and 6:

| 1 | 2 | 3 | 4 | 5 | 6 | 7 | 8 | 9 | 0 | Ref. |
|---|---|---|---|---|---|---|---|---|---|---|
| 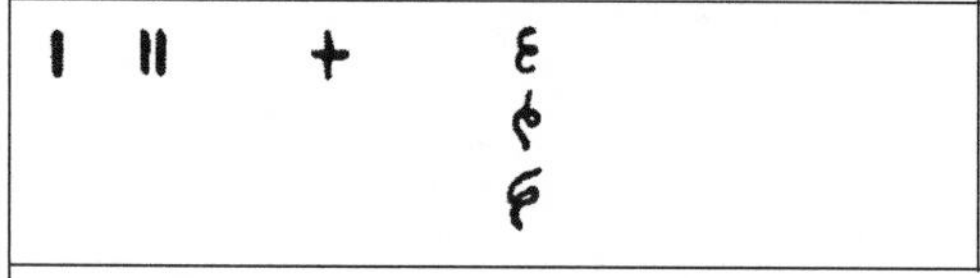 | | | | | | | | | | EI, III p. 134<br>IA, VI, pp. 155 ff.<br>IA, X, pp. 106 ff.<br>Indrâji, JBRAS XII |

Date: third century BCE.
Source: edicts of Asoka written in Brahmi, in various regions of the Empire of the Mauryas, from the regions of Shahbazgarhi, Manshera, Kalsi, Girnar and Sopara (north of Bombay) to Tosali and Jaugada in Kallinga (Orissa), Yerragudi in Kannara, Rampurwa and Lauriya-Araraj in the north of Bihar, Toprah and Mirath north of Delhi, and Rummindei and Nigliva in Nepal (Fig. 24.27).

FIG. 24.29. *Numerals of the original Brâhmî style of writing: our present-day 6 is already recognisable*

### *The numerical symbols of intermediate notations*

The same system appears in the documents of the eras which followed and this gives a much more precise idea of how *Brâhmî* figures looked.

The following figures appeared at the beginning of the Shunga and Magadha dynasties in the Buddhist inscriptions which adorn the walls of the grottoes of Nana Ghat:

| 1 | 2 | 3 | 4 | 5 | 6 | 7 | 8 | 9 | 0 | Ref. |
|---|---|---|---|---|---|---|---|---|---|---|
| 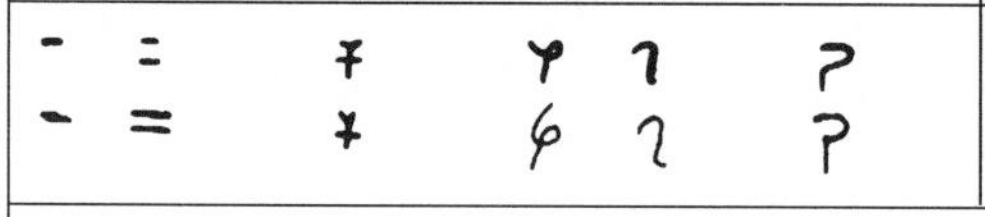 | | | | | | | | | | Datta and Singh<br>Indraji, JBRAS XII<br>Smith and Karpinski |

Date: second century BCE.
Source: the caves of Nana Ghat (central India, Maharashtra, c. 150 km from Poona), Buddhist inscriptions written for a sovereign named Vedishri which mainly concern various presents offered during religious ceremonies.

FIG. 24.30. *Numerals of the intermediary notation of the Shunga: we can already see the prefiguration of our numerals 4, 6, 7 and 9.*

The same series appeared a little later, but in a much more complete form, in the first or second century CE, in the inscriptions of the Buddhist grottoes of Nasik (Fig. 24.31).

*Brâhmî* figures are also found, in more and more varied forms, in Mathuran inscriptions (Fig. 24.32), Kushana and Andhran inscriptions (Fig. 24.33 and 24.34), western Satrap coins (Fig. 24.35), the inscriptions of Jaggayyapeta (Fig. 24.36), and of the Pallava Dynasty (Fig. 24.37).

As these numerals derive from *Brâhmî* figures and consequently serve as a go-between with the later forms of the numerals, they shall henceforth be referred to as the *numerical symbols of the intermediate notations.*

| 1 | 2 | 3 | 4 | 5 | 6 | 7 | 8 | 9 | 0 | Ref. |
|---|---|---|---|---|---|---|---|---|---|---|
| 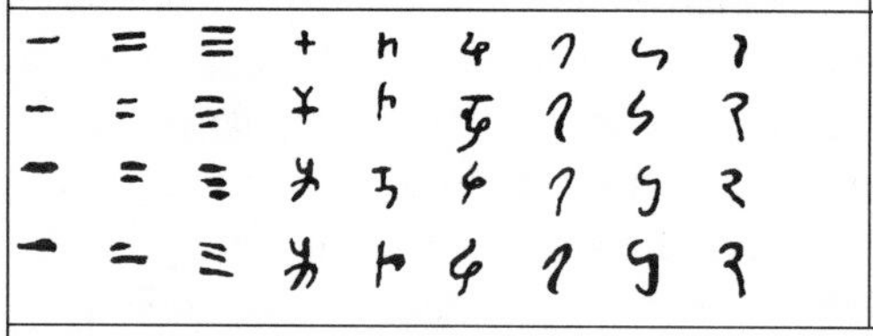 | | | | | | | | | | EI, VIII, pp. 59–96<br>EI, VII, pp. 47–74<br>Bühler<br>Datta and Singh<br>Renou and Filliozat<br>Smith and Karpinski |

Date: first or second century CE.
Source: Buddhist caves of Nasik (in Maharashtra, at least 200 km north of Bombay).

FIG. 24.31. *Numerals of the intermediary system of Nasik: we can see the prefiguration of our numerals 4, 5, 6, 7, 8 and 9.*

| 1 | 2 | 3 | 4 | 5 | 6 | 7 | 8 | 9 | 0 | Ref. |
|---|---|---|---|---|---|---|---|---|---|---|
| 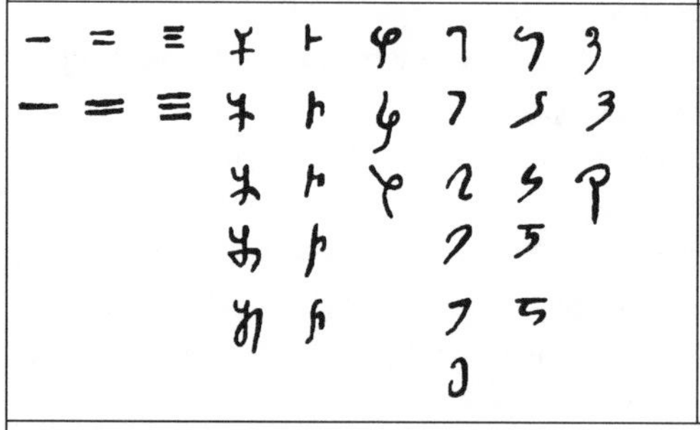 | | | | | | | | | | Bühler<br>Datta and Singh<br>Ojha |

Date: first – third century CE.
Source: inscriptions of Mathura (town of Uttar Pradesh, on the banks of the Yamuna 60 km northwest of Agra), contemporary with a Shaka dynasty.

FIG. 24.32. *Numerals of the intermediary system of Mathura*

| 1 | 2 | 3 | 4 | 5 | 6 | 7 | 8 | 9 | 0 | Ref. |
|---|---|---|---|---|---|---|---|---|---|---|
| 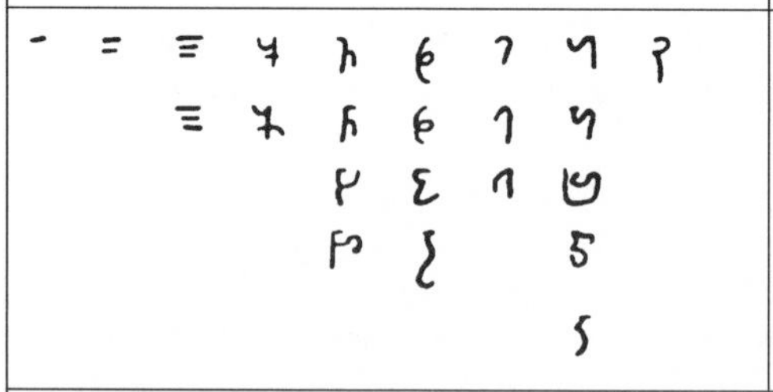 | | | | | | | | | | EI, I, p. 381<br>EI, II, p. 201<br>Bühler<br>Datta and Singh<br>Ojha<br>Smith and Karpinski |

Date: first – second century CE.
Source: contemporary inscriptions of the Kushana dynasty.

FIG. 24.33. *Numerals of the intermediary system of the Kushana*

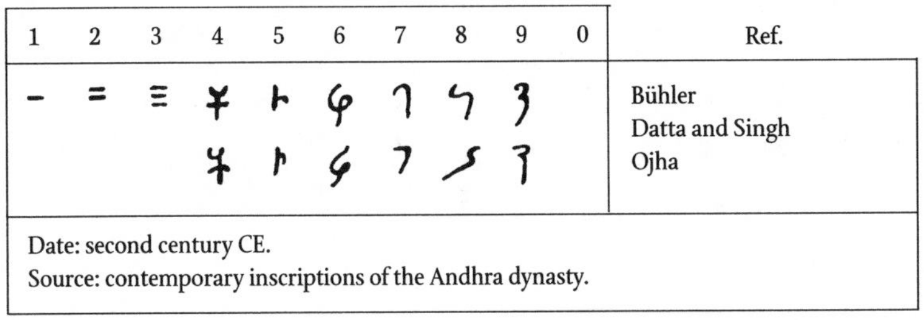

| 1 | 2 | 3 | 4 | 5 | 6 | 7 | 8 | 9 | 0 | Ref. |
|---|---|---|---|---|---|---|---|---|---|---|
| | | | | | | | | | | Bühler<br>Datta and Singh<br>Ojha |
| Date: second century CE.<br>Source: contemporary inscriptions of the Andhra dynasty. | | | | | | | | | | |

FIG. 24.34. *Numerals of the intermediary notation of the Andhra*

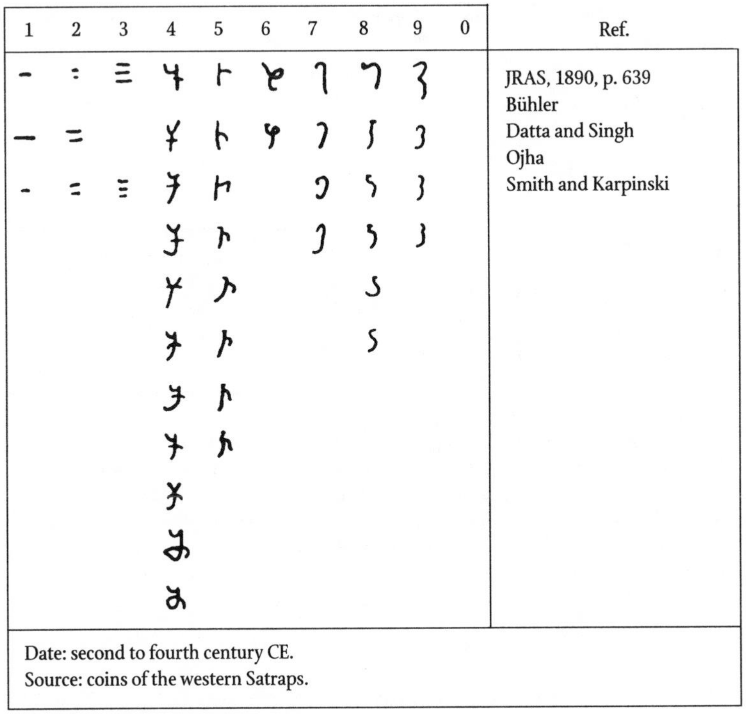

| 1 | 2 | 3 | 4 | 5 | 6 | 7 | 8 | 9 | 0 | Ref. |
|---|---|---|---|---|---|---|---|---|---|---|
| | | | | | | | | | | JRAS, 1890, p. 639<br>Bühler<br>Datta and Singh<br>Ojha<br>Smith and Karpinski |
| Date: second to fourth century CE.<br>Source: coins of the western Satraps. | | | | | | | | | | |

FIG. 24.35. *Numerals of the intermediary notation of the western Satraps*

These intermediate notations spread over the various regions of India and the neighbouring areas, as did the letters of the corresponding writing, and, over the centuries, they underwent graphical modifications, finally to acquire extremely varied cursive forms, each with a regional style.

### *The origin of the notations of Northern and Central India*

One of the first individual notations to appear was *Gupta* notation, used during the dynasty of the same name (its sovereigns reigned over the Ganges and its tributaries from c. 240 to 535 CE) (Fig. 24.38).

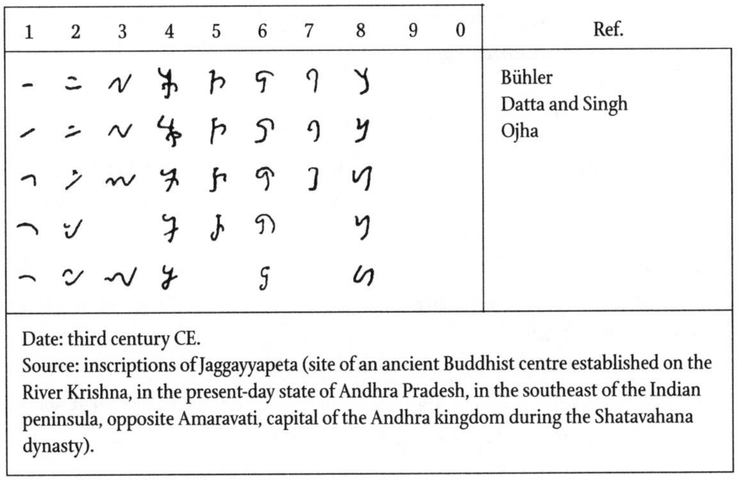

| 1 | 2 | 3 | 4 | 5 | 6 | 7 | 8 | 9 | 0 | Ref. |
|---|---|---|---|---|---|---|---|---|---|---|
| | | | | | | | | | | Bühler<br>Datta and Singh<br>Ojha |
| Date: third century CE.<br>Source: inscriptions of Jaggayyapeta (site of an ancient Buddhist centre established on the River Krishna, in the present-day state of Andhra Pradesh, in the southeast of the Indian peninsula, opposite Amaravati, capital of the Andhra kingdom during the Shatavahana dynasty). | | | | | | | | | | |

FIG. 24.36. *Numerals of the intermediary notation of Jaggayyapeta*

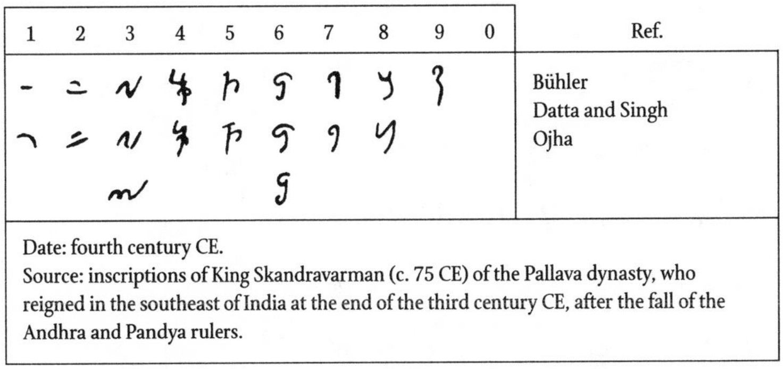

| 1 | 2 | 3 | 4 | 5 | 6 | 7 | 8 | 9 | 0 | Ref. |
|---|---|---|---|---|---|---|---|---|---|---|
| | | | | | | | | | | Bühler<br>Datta and Singh<br>Ojha |
| Date: fourth century CE.<br>Source: inscriptions of King Skandravarman (c. 75 CE) of the Pallava dynasty, who reigned in the southeast of India at the end of the third century CE, after the fall of the Andhra and Pandya rulers. | | | | | | | | | | |

FIG. 24.37. *Numerals of the first intermediary notation of the Pallava*

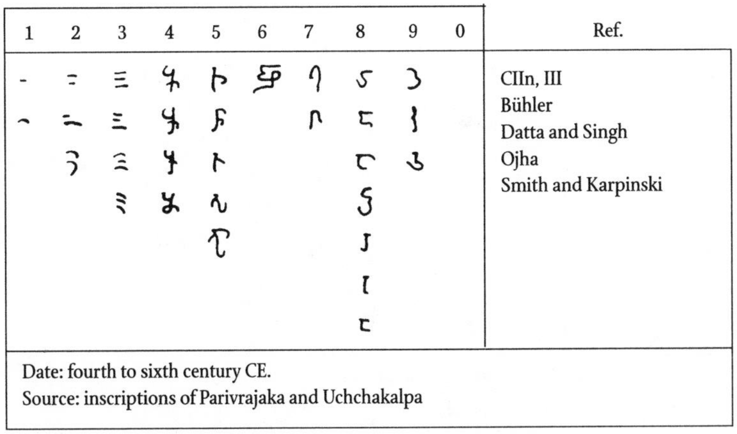

| 1 | 2 | 3 | 4 | 5 | 6 | 7 | 8 | 9 | 0 | Ref. |
|---|---|---|---|---|---|---|---|---|---|---|
| | | | | | | | | | | CIIn, III<br>Bühler<br>Datta and Singh<br>Ojha<br>Smith and Karpinski |
| Date: fourth to sixth century CE.<br>Source: inscriptions of Parivrajaka and Uchchakalpa | | | | | | | | | | |

FIG. 24.38. *Gupta numerals*

This notation was the origin of all the series of figures in common use in Northern India and Central Asia.

### *The first developments in Nâgarî notation*

As *Gupta* writing became more refined, it gave birth to *Nâgarî* notation (or "urban" writing, the magnificent regularity of which gave it the name of *Devanâgarî*, or "Nâgarî of the gods").

This writing soon acquired great importance, becoming not only the main writing of the Sanskrit language, but also of Hindi, the great language of modern Central India.

As numerical notation experienced a parallel evolution, so *Nâgarî* figures were born out of *Gupta* figures, which later led to the emergence of modern *Nâgarî* figures (see also Fig. 24.3 above):

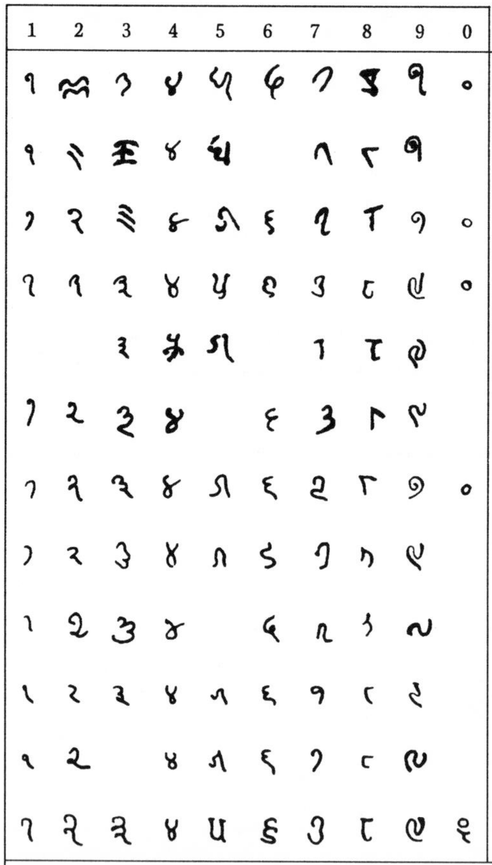

| 1 | 2 | 3 | 4 | 5 | 6 | 7 | 8 | 9 | 0 | Ref. |
|---|---|---|---|---|---|---|---|---|---|---|
| | | | | | | | | | | EI, I, p. 122<br>EI, I, P. 162<br>EI, I, p. 186<br>EI, II, p. 19<br>EI, III, p. 133<br>EI, IV, p. 309<br>EI, IX, p. 1<br>EI, IX, p. 41<br>EI, IX, p. 197<br>EI, IX, p. 198<br>EI, IX, p. 277<br>EI, XVIII, p. 87 |
| | | | | | | | | | | JA, 1863, p. 392 |
| | | | | | | | | | | IA, VIII, p. 133<br>IA, XI, p. 108<br>IA, XII, p. 155<br>IA, XII, p. 249<br>IA, XII, p. 263<br>IA, XIII, p. 250<br>IA, XIV, p. 351<br>IA, XXV, p. 177 |
| | | | | | | | | | | Bühler<br>Datta and Singh<br>Ojha |

Date: seventh to twelfth century CE (Fig. 24.75).
Source: various inscriptions on copper from Northern and Central India.

FIG. 24.39A. *Ancient Nâgarî numerals*

| 1 | 2 | 3 | 4 | 5 | 6 | 7 | 8 | 9 | 0 | Ref. |
|---|---|---|---|---|---|---|---|---|---|---|
| | | | | | | | | | | Datta and Singh<br>Ojha<br>Smith and Karpinski |

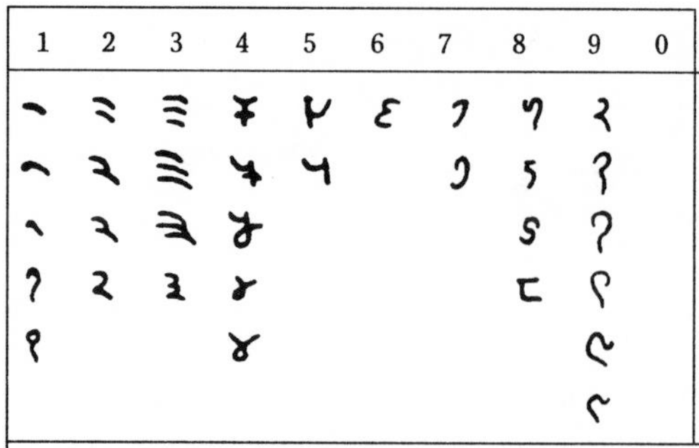

Date: eighth to twelfth century CE (Fig. 24.3).
Source: various manuscripts from northern and central India (which use neither zero nor the place-value system).

FIG. 24.39B. *Ancient Nâgarî numerals*

| 1 | 2 | 3 | 4 | 5 | 6 | 7 | 8 | 9 | 0 | Ref. |
|---|---|---|---|---|---|---|---|---|---|---|
| | | | | | | | | | | ASI, Rep. 1903–1904, pl. 72<br>EI, 1/1892, pp. 155–62<br>Datta and Singh<br>Guitel |

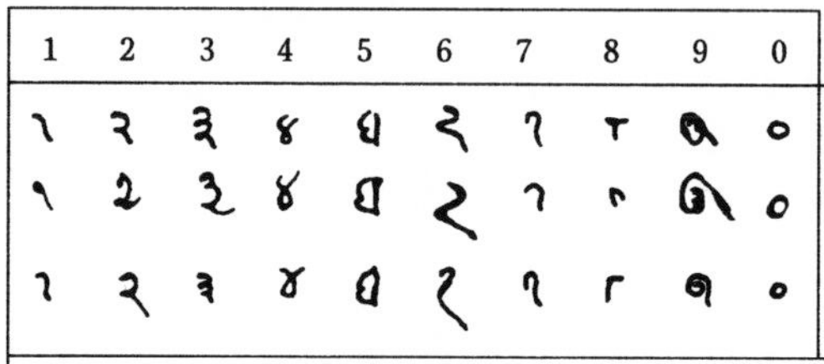

Date: 875 to 876 CE (Fig. 24.73).
Source: inscriptions of Gwalior (capital of the ancient princely state of Madhyabharat, situated between the present-day states of Madhya Pradesh and Rajasthan, c.120 km from Agra and over 300 km south of Delhi). The two Sanskrit inscriptions are from the temple of Vaillabhatta-svamim dedicated to Vishnu, and are from the time of the reign of Bhojadeva, dated 932 and 933 of the Vikrama Samvat era, or 875 and 876 CE.

FIG. 24.39C. *Ancient Nâgarî numerals*

These are the forms that the Arabs used when they adopted Indian numeration: the proof of this will be seen later on; moreover, in the following tables it can be seen that these figures, if not identical, are very similar to the numerical symbols that we use today.

### *Notations which are derived from Nâgarî*

In Maharashtra, via a southern variant, *Nâgarî* gave birth to *Mahârâshtrî*, which gradually evolved into modern *Marâthî* writing, of which there are currently two forms: *Bâlbodh* (or "academic" writing), used to write Sanskrit, and *Modî*, which is more cursive in form, and is only used to write *Marâthî*. A similar evolution took place for the notation of the nine units (Fig. 24.4 above).

In the state of Rajasthan (bordering Pakistan in the west, Punjab, Haryana and Uttar Pradesh in the north, Madhya Pradesh in the east and Gujarat in the south) *Nâgarî* evolved into *Râjasthanî*. In the northwest of India, however, between the Aravalli Range and the Thar Desert, *Nâgarî* diversified into the cursive forms of *Mârwarî* and *Mahâjanî*, mainly used for commercial purposes.

After the end of the eleventh century, a notation called *Kutilâ* (or "Proto-Bengali") was also born out of *Nâgarî*, from which, in turn, modern *Bengâlî* evolved, sometime after the beginning of the seventeenth century (Fig. 24.10), to which *Oriyâ* (Fig. 24.12), *Gujarâtî* (Fig. 24.8), *Kaîthî* (Fig. 24.9), *Maithilî* (Fig. 24.11) and *Manipurî* can be linked.

### *The development of Shâradâ notation*

After the beginning of the ninth century in Kashmir and Punjab, a northern variant of *Gupta* led to *Shâradâ* notation, which was used in the above parts of India until the fifteenth century at least (Fig. 24.14).

| 1 | 2 | 3 | 4 | 5 | 6 | 7 | 8 | 9 | 0 | Ref. |
|---|---|---|---|---|---|---|---|---|---|---|
| | | | | | | | | | | IA, XVII, pp. 34–48<br>Datta and Singh<br>Kaye: *Bakhshâlî manuscript*<br>Smith and Karpinski |

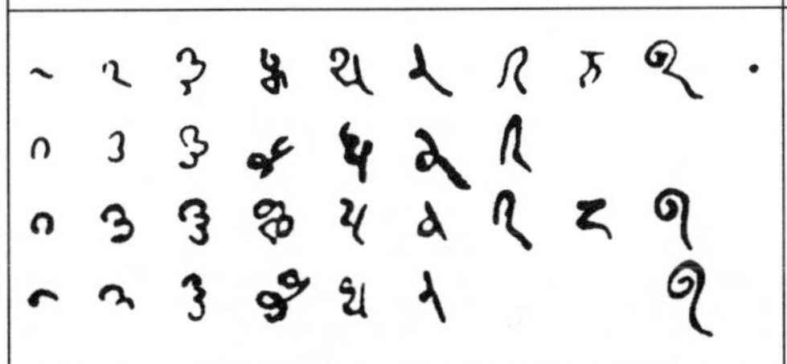

Date: between the ninth and twelfth century CE. (Fig. 24.14).
Source: Manuscript from Bakshali (a village in Gandhara, near Peshawar, in present-day Pakistan, where it was discovered in 1881). The manuscript is written entirely in the *Shâradâ* style, in the Sanskrit language, in both verse and prose, by an anonymous author. It deals with algebraic problems, the numbers being expressed in *Shâradâ* numerals using the place-value system, zero being written as a dot (*bindu*). This manuscript could not have been written earlier than the ninth century CE or later than the twelfth century, but it is possible that it is a copy of – or a commentary on – an earlier document.

FIG. 24.40A. *Ancient Shâradâ numerals*

### *Notations derived from Shâradâ*

It is from this notation that *Tâkarî* (Fig. 24.13), *Dogrî, Chameali, Mandealî, Kuluî, Sirmaurî, Jaunsarî, Kochî, Landa, Multânî, Sindhî* (Fig. 24.6), *Khudawadî, Gurûmukhî* (Fig. 24.7), *Punjâbî* (Fig. 24.5), etc., originated.

| 1 | 2 | 3 | 4 | 5 | 6 | 7 | 8 | 9 | 0 | Ref. |
|---|---|---|---|---|---|---|---|---|---|---|
| | | | | | | | | | | KAV<br>Smith and Karpinski |

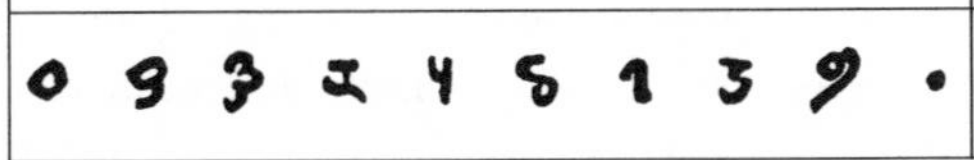

Date: the fifteenth century CE (approximately).
Source: A Kashmiri document which reproduces the Vedi hymns and texts of the *Atharvaveda* in *Shâradâ* characters (the document is preserved at Tübingen University).

FIG. 24.40B. *Shâradâ numerals (most recent style)*

### *Nepalese notations*

| 1 | 2 | 3 | 4 | 5 | 6 | 7 | 8 | 9 | 0 | Ref. |
|---|---|---|---|---|---|---|---|---|---|---|
| | | | | | | | | | | Bendall<br>Bühler<br>Datta and Singh<br>Ojha<br>Smith and Karpinski |

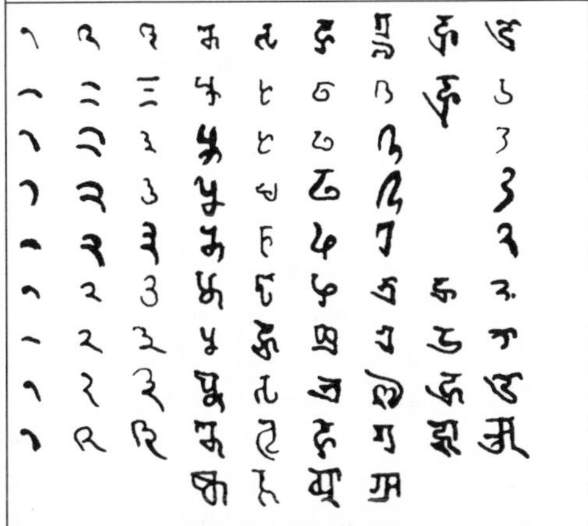

Date: eighth to twelfth century CE (Fig. 24.15).
Source: inscriptions from Nepal and various Buddhist manuscripts from Nepal.

FIG. 24.41. *Ancient Nepali numerals*

Many other systems originated from *Gupta*. After the fifth century CE, one variation evolved into *Siddhamatrikâ* (or Siddam) writing which was used mainly in China and Japan for Sanskrit notation. During its development, some time after the beginning of the ninth century, it gave birth to *Limbu* and modern Nepali (also called *Gurkhalî*), specific notations of Nepal whose numerical symbols underwent a parallel evolution (Fig. 24.41).

### *Notations which originated in India and Central Asia*

From the time of the Kushana Empire (first to third century CE) until the Empire of the Guptas, Indian civilisation, along with Buddhism, stretched to Chinese Turkestan, as well as towards northern Afghanistan and Tibet.

Thus one of the notations to be born out of Gupta reached these regions.

Without any radical change, this notation evolved into the writings of Chinese Turkestan, which were used to write Agnean, Kutchean and Khotanese. Each style would have possessed its own figures.

On the other hand, in the various regions of Tibet, the high valleys of the Himalayas and the neighbouring areas of Burma, *Gupta* underwent quite drastic changes to enable spoken languages with very different inflexions to be written down. This is how the Tibetan alphabet came about, the Guptan numerical symbols also being adapted to this graphical style (Fig. 24.16).

*Mongolian figures*

When the great conqueror Genghis Khan died in 1227, the Mongolian Empire stretched from the Pacific to the Caspian Sea.

J. G. Février (1959) claims that "the Mongolians did not possess any form of writing and that all their conventions were oral; their 'contracts' were alleged to be certain signs carved onto wooden tablets."

But by conquering nearly all of Asia, these half-savage hordes could no longer be contented with such rudimentary methods; so they decided to adopt the writing of the Uighur people of Turfar after they defeated them (the Uighur alphabet constituting a type of Syriac writing, imported by Nestorian monks).

The Mongolians then decided that they wanted an alphabet that was more appropriate for writing their language, mainly because of pressure from the propagators of Buddhism to have their own specific instrument for translating their texts. Their alphabet was created with the collaboration of the Uighurs. They wrote in vertical columns which read from left to right.

However, instead of adopting the non-positional system of the aforementioned region, the Mongolians preferred to use Tibetan figures, after the contact that they had had with the latter. Thus "Mongolian" figures were born:

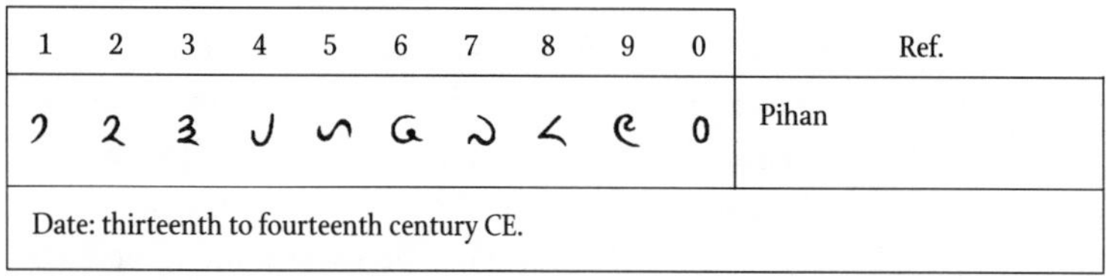

| 1 | 2 | 3 | 4 | 5 | 6 | 7 | 8 | 9 | 0 | Ref. |
|---|---|---|---|---|---|---|---|---|---|---|
| | | | | | | | | | | Pihan |

Date: thirteenth to fourteenth century CE.

Fig. 24.42. *Mongolian numerals: the numerals 2, 3, 6 and 0 are recognisable, as well as 9 (or rather its mirror image).*

*An evolution from the South to the East*

Like *Gupta*, there is another style of writing to come out of *Brâhmî* that is very different from its origins.

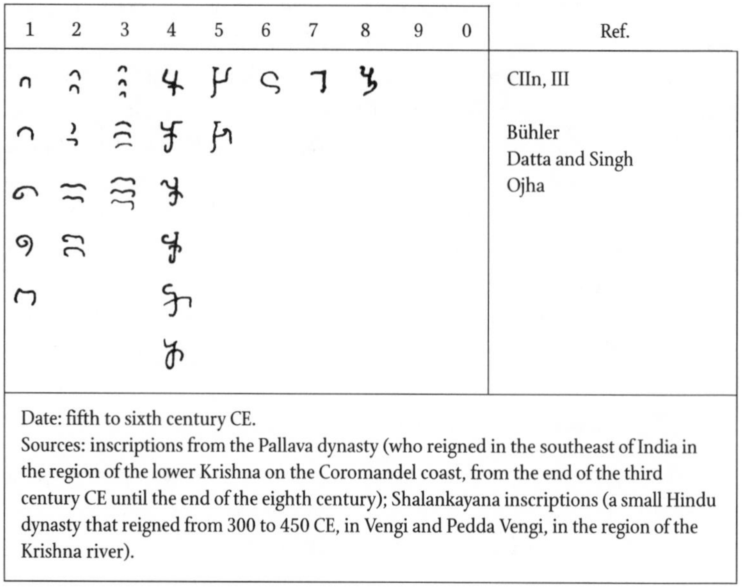

| 1 | 2 | 3 | 4 | 5 | 6 | 7 | 8 | 9 | 0 | Ref. |
|---|---|---|---|---|---|---|---|---|---|---|
| | | | | | | | | | | CIIn, III<br><br>Bühler<br>Datta and Singh<br>Ojha |

Date: fifth to sixth century CE.
Sources: inscriptions from the Pallava dynasty (who reigned in the southeast of India in the region of the lower Krishna on the Coromandel coast, from the end of the third century CE until the end of the eighth century); Shalankayana inscriptions (a small Hindu dynasty that reigned from 300 to 450 CE, in Vengi and Pedda Vengi, in the region of the Krishna river).

Fig. 24.43. *Numerals of the intermediary counting system of the Pallava*

This is the writing used in inscriptions in Pallava, Shâlankâyana and Valabhî (Fig. 24.43 and 24.44) and the more individualised style of Chalukya and Deccan (Fig. 24.45) and Ganga and Mysore (Fig. 24.46):

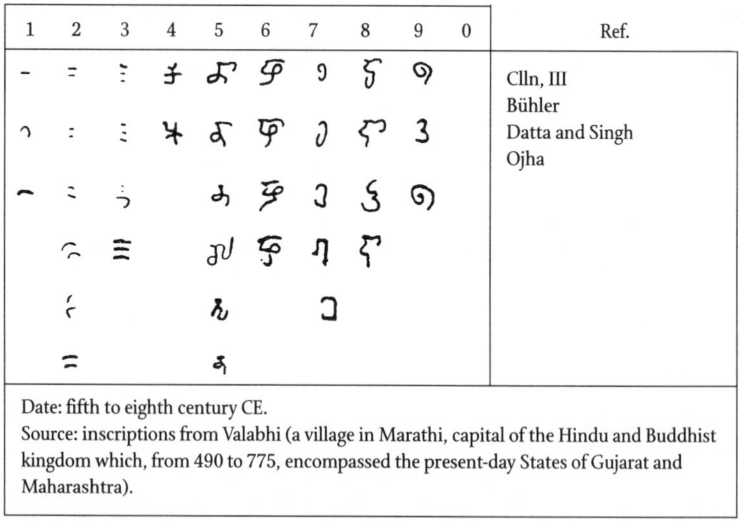

| 1 | 2 | 3 | 4 | 5 | 6 | 7 | 8 | 9 | 0 | Ref. |
|---|---|---|---|---|---|---|---|---|---|---|
| | | | | | | | | | | CIln, III<br>Bühler<br>Datta and Singh<br>Ojha |

Date: fifth to eighth century CE.
Source: inscriptions from Valabhi (a village in Marathi, capital of the Hindu and Buddhist kingdom which, from 490 to 775, encompassed the present-day States of Gujarat and Maharashtra).

Fig. 24.44. *Numerals of the intermediary counting system of Valabhî*

This is the common basis which would lead progressively on the one hand to the formation of the southern Indian style (attached to Dravidian

| 1 | 2 | 3 | 4 | 5 | 6 | 7 | 8 | 9 | 0 | Ref. |
|---|---|---|---|---|---|---|---|---|---|---|
| | | | | | | | | | | CIIn, III<br>Bühler<br>Datta and Singh<br>Ojha |

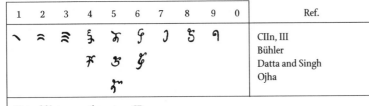

Date: fifth to seventh century CE.
Source: inscriptions of the oldest branch of the Chalukya dynasty of Deccan (known as "de Vatapi", who lived in Badami, in the present-day district of Bijapur, during the sixth century CE).

Fig. 24.45. *Numerals of the intermediary counting system of the Chalukya of Deccan*

| 1 | 2 | 3 | 4 | 5 | 6 | 7 | 8 | 9 | 0 | Ref. |
|---|---|---|---|---|---|---|---|---|---|---|
| | | | | | | | | | | CIIn, III<br>Bühler<br>Datta and Singh<br>Ojha |

Date: sixth to eighth century CE.
Source: inscriptions of the Ganga dynasty of Mysore (who ruled over a substantial part of the present-day State of Karnataka, from the fifth to the sixteenth century).

Fig. 24.46. *Numerals of the intermediary counting system of the Ganga of Mysore*

styles of writing), and on the other hand to the development of *Pâlî* styles, attached to eastern styles of writing.

*Southern (or Dravidian) styles*

From one of these systems was derived *Bhattiprolu* writing.

In Telingana, to the east of Andhra Pradesh and the south of Orissa, this gradually became *Telugu* (Fig. 24.20):

In the centre of Deccan, in Karnataka and Andhra Pradesh, it became *Kannara* (Fig. 24.21).

| 1 | 2 | 3 | 4 | 5 | 6 | 7 | 8 | 9 | 0 | Ref. |
|---|---|---|---|---|---|---|---|---|---|---|
| | | | | | | | | | | Burnell<br>Renou and Filliozat |

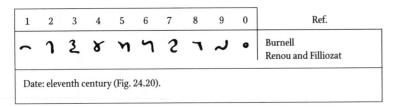

Date: eleventh century (Fig. 24.20).

Fig. 24.47. *Ancient Telugu numerals*

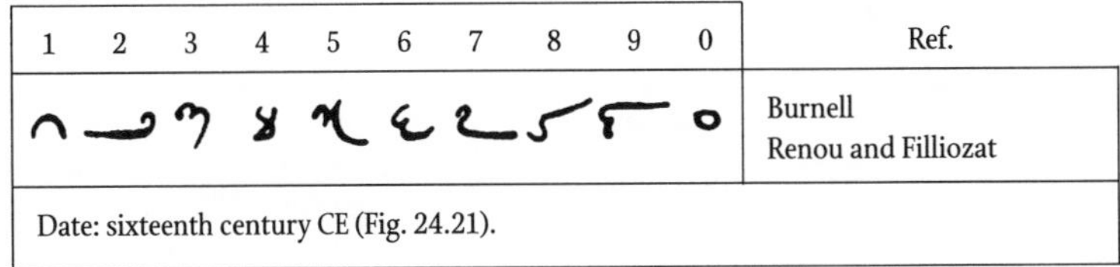

| 1 | 2 | 3 | 4 | 5 | 6 | 7 | 8 | 9 | 0 | Ref. |
|---|---|---|---|---|---|---|---|---|---|---|
| | | | | | | | | | | Burnell<br>Renou and Filliozat |

Date: sixteenth century CE (Fig. 24.21).

Fig. 24.48. *Ancient Kannara numerals*

To the east of the more southern regions, this became *Grantha* and *Tamil* (Fig. 24.17), as well as *Vatteluttu* (used primarily on the coast of Malabar from the eighth to the sixteenth century), whilst in the west this became the styles known as *Tulu* and *Malayâlam* (Fig. 24.19).

| 1 | 2 | 3 | 4 | 5 | 6 | 7 | 8 | 9 | 0 | Ref. |
|---|---|---|---|---|---|---|---|---|---|---|
| | | | | | | | | | | Burnell<br>Pihan<br>Renou and Filliozat |

Date: sixteenth century (Fig. 24.19).

Fig. 24.49. *Ancient Tamil numerals*

Finally, in the extreme south, primarily in Sri Lanka, *Sinhalese* was derived (Fig. 24.22).

*The styles of writing of Southeast Asia*

At the same time, to the east of India, another variety of intermediate systems developed to lead to the first forms of *Pâlî*. Attached to the ancient writing *Ardha-Mâgadhî* (the ancient language spoken in Magadha), these diversified, and led to the characteristic forms of writing used (and still used today) to the east of India and in Southeast Asia.

From this system was derived: *Old Khmer* (developed some time after the beginning of the sixth century CE); *Cham*, used in part of Vietnam, from the seventh century to some time around the thirteenth century; *Kawi* in Java, Bali and Borneo (Fig. 24.50), which dates back to the end of the seventh century, but which has now fallen into disuse; modern *Thai* writing (Shan, Siamese, Laotian, etc.), whose first developments date back to the thirteenth century (Fig. 24.24); *Burmese* (Fig. 24.24 and 24.51), which derived from *Môn* in the eleventh century, used by populations of Pegu before the Burmese invasion; *Old Malay* (Fig. 24.51), from which *Batak* (in the central region of the island of Sumatra), *Redjang* and *Lampong* (in the southeast of the same island), *Tagala* and *Bisaya* in the Philippines, as well as *Macassar* and *Bugi* (from Sulawezi) derived.

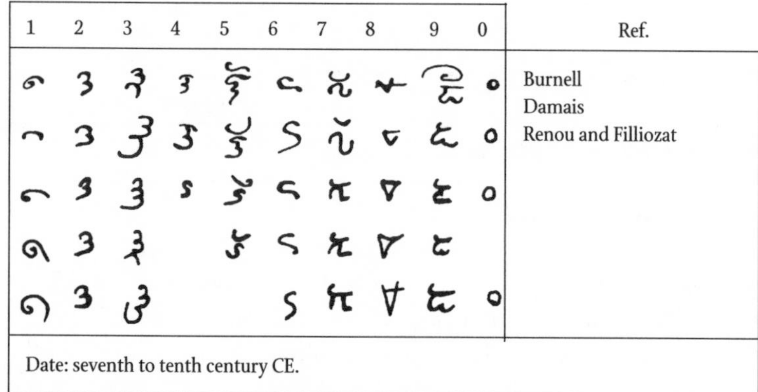

| 1 | 2 | 3 | 4 | 5 | 6 | 7 | 8 | 9 | 0 | Ref. |
|---|---|---|---|---|---|---|---|---|---|---|
| | | | | | | | | | | Burnell<br>Damais<br>Renou and Filliozat |
| Date: seventh to tenth century CE. | | | | | | | | | | |

FIG. 24.50. *Kawi (ancient Javanese and Balinese) numerals*

| 1 | 2 | 3 | 4 | 5 | 6 | 7 | 8 | 9 | 0 | Ref. |
|---|---|---|---|---|---|---|---|---|---|---|
| | | | | | | | | | | Latter<br>Smith and Karpinski |
| Date: seventeenth century CE (approx.). | | | | | | | | | | |

FIG. 24.51. *Ancient Burmese numerals*

## *Types of numerals that derive from Brâhmî*

These fall into three categories, like the types of writing of the same names (Fig. 24.52 and 24.53):

I. The family of writing styles from Northern and Central India and Central Asia, which developed from *Gupta* writing:

1. Forms of writing derived from *Nâgarî*:
   a. Mahârâshtrî numerals;
   b. Marâthî, Modî, Râjasthânî, Mârwarî and Mahâjani (derived from Mahârâshtrî) numerals;
   c. Kutilâ numerals;
   d. Bengalî, Oriyâ, Gujarâtî, Kaîthî, Maithilî and Manipurî (derived from Kutilâ) numerals.
2. Forms of writing derived from *Shâradâ*:
   a. Tâkarî and Dogrî numerals;
   b. Chamealî, Mandealî, Kuluî, Sirmaurî and Jaunsarî numerals;
   c. Sindhî, Khudawadî, Gurûmukhî, Punjâbî (etc.) numerals;
   d. Kochî, Landa, Multânî (etc.) numerals.
3. Types of *Nepalese* writing:
   a. Siddham numerals (influenced by the Nâgarî style);
   b. Modern Nepali numerals (derived from Siddham numerals).

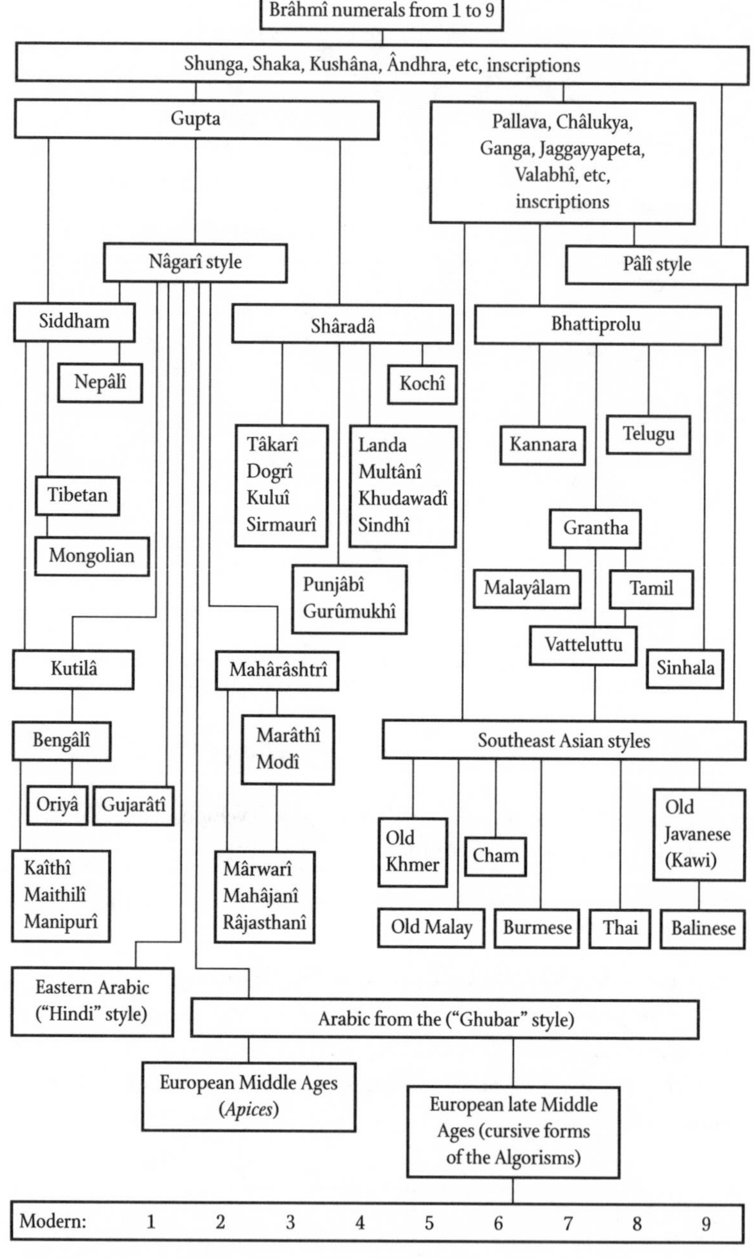


FIG. 24.52. *Numerals which derived from Brâhmî numerals*

4. Types of *Tibetan* writing:
   a. Tibetan numerals (derived from Siddham numerals);
   b. Mongolian numerals (derived from Tibetan numerals).
5. Types of writing from Chinese Turkestan (derived from *Siddham* numerals).

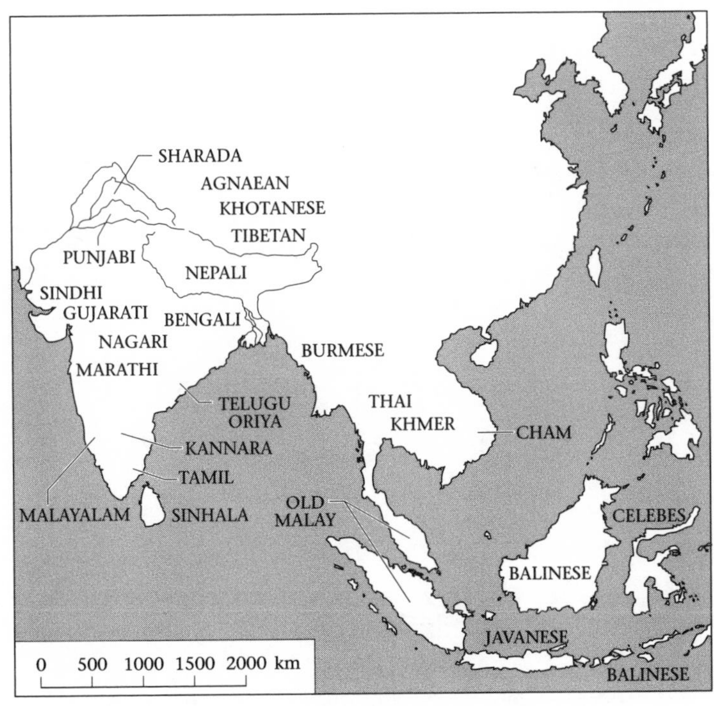


FIG. 24.53. *Geographical areas where writing styles of Indian origin are used*

II. The family of writing styles from southern India, which developed from *Bhattiprolu*, a distant cousin of Gupta:
1. Telugu and Kannara numerals;
2. Grantha, Tamil and Vatteluttu numerals;
3. Tulu and Malayâlam numerals;
4. Sinhalese (Singhalese) numerals.

III. The family of "oriental" writing styles, which developed from *Pâlî* writing, which itself derives from the same source as Bhattiprolu:
1. Old Khmer numerals;
2. Cham numerals;
3. Old Malay numerals;
4. Kawi (old Javanese and Balinese) numerals;
5. Modern Thai-Khmer (Shan, Lao and Siamese) numerals;
6. Burmese numerals;
7. Balinese, Buginese, Tagalog, Bisaya and Batak numerals.

As we will see later, Arabic numerals, East and North African alike, derive from the Indian *Nâgarî* numerals, and the numerals that we use today come from the *Ghubar* numerals of the Maghreb. Thus these diverse signs can be placed in the first category of group I.

### *The mystery of the origin of Brâhmî numerals*

Having demonstrated how the above types of numerals all derived from *Brâhmî*, it is now time to explain the origin of the *Brâhmî* numerals themselves.

For some time now, this writing has conserved an ideographical representation for the first three units: the corresponding number of horizontal lines. However, since their emergence, the numerals 4 to 9 have offered no visual key to the numbers that they represent (for example, the 9 was not formed by nine lines, nine dots or nine identical signs; rather, it was represented by a conventional graphic). This is a significant characteristic which has yet to be satisfactorily explained. To try and understand this enigma, let us now examine the principle hypotheses that have been put forward on this subject over the last century.

### *First hypothesis: The numerals originated in the Indus Valley*

S. Langdon (1925) believed that Indian styles of writing and numerals derived from the Indus Valley culture (2500 – 1800 BCE).

The first objection to this theory concerns the claim that there is a link between Indian letters and Proto-Indian pictographic writing.

We have just seen that *Brâhmî* writing actually developed from the ancient alphabets of the western Semitic world through the intermediate of a variety of Aramaean: this link has been satisfactorily established, even if samples of this intermediate writing have not yet been found (Fig. 24.28). Documents from this civilisation are separated from the first texts written in *Brâhmî* and in a purely Indian language by the space of two thousand years. However, the fact that the writing of Indus civilisation has not yet been deciphered does not concern us here.

It is not known whether the ancient civilisations of Mohenjo-daro and Harappâ still existed when India was invaded by the Aryans, or if their writing had developed during this interval.

Moreover, no mention is made of this link in Indian literature, and with good cause: the invaders probably found writing repugnant because, like all Indian European peoples, they attached great importance to oral tradition [see T. V. Gamkrelidze and V. V. Ivanov (1987); A. Martinet (1986)].

It is almost certain that when the Aryans arrived in India they brought no form of writing with them, as happened in Greece and the rest of Europe, whilst various Indo-European peoples came in successive waves to conquer the continent. Their intellectual and spiritual leaders would certainly have had "a knowledge of the great religious poems ; but it seems that their literature was written at a later date, and then the literate men would doubtless have preferred to keep the oral tradition going as long as possible to perpetuate their prestige and their privileges" [M. Cohen (1958)].

Therefore, Langdon's hypothesis has no foundation, because we do not know if any link exists between numerals used by the Indus civilisation and "official" Indian numerals. The theory becomes even more unlikely when one considers that the documentation which survived from the Indus Valley is very fragmentary and does not provide enough information for us to reconstruct the system as a whole.

### *Second hypothesis: Brâhmî numerals derive from "Aramaean-Indian" numerals*

Since Indian letters derive from the Aramaean alphabet, would it not be natural to presume that *Brâhmî* numerals were similarly the offshoot of one of the ancient systems of numerical notation of the western Semitic world? At first glance, this hypothesis seems plausible, since a numerical notation which derives directly from Aramaean, Palmyrenean, Nabataean, etc., can be found in many inscriptions from Punjab and Gandhara. This style is known as "Aramaean-Indian", and is related to *Karoshthî* writing (see Fig. 24.54 and Fig. 18.1 to 12).

However, once we have looked at *Brâhmî* numerals for numbers higher than the first nine units, we will see that this system is too different from Aramaean-Indian for this hypothesis to be taken into consideration.

On the one hand, Aramaean-Indian reads from right to left, whilst *Brâhmî* (and nearly all the styles of writing related to it) reads from left to right.

On the other hand, in the *Karoshthî* system, the numbers 4 to 9 are generally represented by the appropriate number of vertical lines, whilst the *Brâhmî* system gives them independent signs which give no direct visual indication as to their meaning.

Moreover, the original *Brâhmî* system possesses specific numerals for 10, 20, 30, 40, 50, 60, 70, 80, 90, 100, 200, etc., whilst the *Karoshthî* system only has specific figures for 1, 10, 20 and 100.

Finally, the initial Indian system is essentially based on the principle of addition, whilst Aramaean-Indian is based on a hybrid principle combining addition and multiplication.

Thus this hypothesis must be discarded.

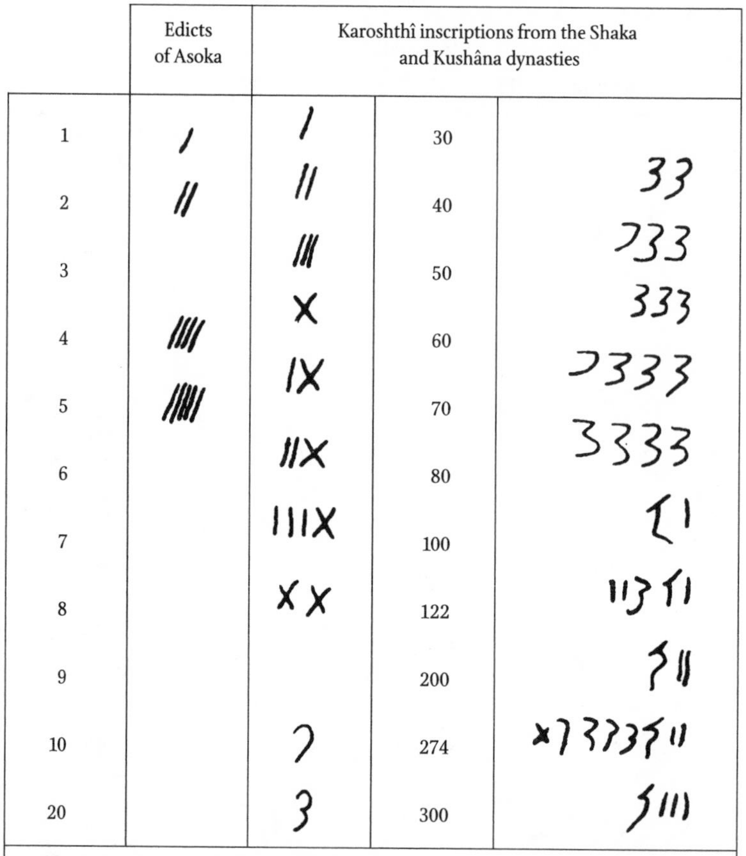

| | Edicts of Asoka | Karoshthî inscriptions from the Shaka and Kushâna dynasties | | |
|---|---|---|---|---|
| 1 | / | / | 30 | [illegible] |
| 2 | // | // | 40 | [illegible] |
| 3 | | /// | 50 | [illegible] |
| 4 | //// | X | 60 | [illegible] |
| 5 | ///// | IX | 70 | [illegible] |
| 6 | | IIX | 80 | [illegible] |
| 7 | | IIIX | 100 | [illegible] |
| 8 | | XX | 122 | [illegible] |
| 9 | | | 200 | [illegible] |
| 10 | | [illegible] | 274 | [illegible] |
| 20 | | [illegible] | 300 | [illegible] |

Date: third century BCE to fourth century CE.
Sources: inscriptions written in Karoshthî from the edicts of Asoka (3rd c. BCE), where the numerals are partially attested; and Karoshthî inscriptions (2nd c. BCE – 4th c. CE) from the north of Punjab and the former province of Gandhara (region in the north-west of India, the extreme north of Pakistan and the northeast of Afghanistan, which was part of the Persian Empire, before it was conquered in 326 BCE by Alexander the Great), where these numerals are more fully attested.

Fig. 24.54. *"Aramaean-Indian" numerical notations*

### *Third hypothesis: Brâhmî numerals derive from the Karoshthî alphabet*

Another hypothesis (suggested by Cunningham, and later shared by Bayley and Taylor), proposes that *Brâhmî* numerals derived from the letters of the *Brâhmî* alphabet, used as the initials of the Sanskrit names of the corresponding numbers. The following table demonstrates this theory:

| Forms given to Karoshthî letters by supporters of this theory | | Brâhmî numerals: forms found in Asoka's edicts, and the inscriptions of Nana Ghat and Nasik | | Names of numbers in Sanskrit | | Karoshthî letters: forms found in Asoka's edicts | |
|---|---|---|---|---|---|---|---|
| cha | | 4 | | 4 | chatur | | cha |
| pa | | 5 | | 5 | pañcha | | pa |
| sha | | 6 | | 6 | shat | | sha |
| sa | | 7 | | 7 | sapta | | sa |
| na | | 9 | | 9 | nava | | na |

FIG. 24.55.

The link that has been established here, however, is too tenuous, for at least three reasons.

The first is that the forms given by the supporters of this theory actually come from inscriptions from different eras, most often from later eras, therefore holding little significance for the problem in question, which concerns a graphical system. This is how such evolved forms like those of Kushâna inscriptions in the northwest of Punjab (second to fourth century CE) came to be confused with more ancient styles such as inscriptions from the Shaka era (second century BCE to first century CE) or those from Asoka's time (third century BCE).

The second reason is that the signs which are given for the presumed phonetic values are very similar (if not identical) to letters which are known to represent other numerical values.

Thirdly, the supporters of this hypothesis allowed themselves to get so carried away that they themselves actually added the final touch to the Aramaean-Indian letters which was needed in order to prove their theory.

There is another even more fundamental reason, however, why the above two theories must be rejected: they presume that *Karoshthî* pre-dates *Brâhmî*, whilst today's specialists believe precisely the opposite.

It is certain that *Karoshthî* writing derives from the Aramaean alphabet, because several of its characters are identical (in form and structure) to their Aramaean equivalents; and, like Semitic writing, it reads from right to left. *Karoshthî* remains very different from the latter style of writing, however, because it was adapted to the sounds and inflexions of Indian-European languages. It was probably introduced to the northwest of India in Alexander the Great's time (c. 326 BCE), and was used there until the fourth century CE, and until a slightly later date in Central Asia.

Nevertheless, *Brâhmî* does not derive from this writing. *Brâhmî* stems from another variant of Aramaean, whose characters were adapted to Indian languages whilst the direction of the writing was changed so that it read from left to right.

It is highly probable that *Brâhmî* had been around long before Asoka's time, because by then it was not only fully established, but also and above all it was in use in all of the Indian sub-continent. Therefore, it is very likely that *Karoshthî* was not used in other parts of India except for the regions of Gandhara and Punjab because, as J. G. Février (1959) has already pointed out, it emerged when an Indian style of writing already existed, namely *Brâhmî*, which was in use since roughly the fifth century BCE.

Thus it would seem unlikely that *Karoshthî* could have influenced the formation of *Brâhmî* letters and numerals.

### *Fourth hypothesis: Brâhmî numerals derive from the Brâhmî alphabet*

This hypothesis would initially seem quite feasible when one considers that many kinds of numerals have developed in this way.

The Greeks and the southern Arabic people, for example, gave, as a numerical symbol, the initial letter from their respective alphabets of the name of the number.

We also know that the Assyro-Babylonians, who had no numeral for 100 in their Sumerian cuneiform system, decided to use acrophonics; thus, the syllable *me* was used to denote this number, the initial of the word *me'at*, which means "hundred" in Akkadian.

Ethiopian numerals, which now appear to be completely independent from Ethiopian writing, actually derive from the first nineteen letters of the Greek numeral alphabet; this dates back to the fourth century CE, when the town of Aksum (not far from the modern town of Adoua) was the capital of the ancient kingdom of Abyssinia.

Thus the theory that the numerals of a given civilisation derive from its own alphabet is quite feasible.

In other words, the Indians could quite possibly have used a certain number of the letters of the *Brâhmî* alphabet to create a corresponding numbering system. This is the substance of J. Prinsep's hypothesis (1838); he believed that the prototypes of the Indian numerals constituted the initial, in *Brâhmî* characters, of the Sanskrit names for the corresponding numbers.

However, as this hypothesis has never been confirmed, it remains in the realm of conjecture. Moreover, the author also mixed archaic styles with later ones, and "customised" the characters in question to make his theory appear to hold water.

*Fifth hypothesis: Brâhmî numerals derive from a previous numeral alphabet*

B. Indrâji (1876) put forward the theory that *Brâhmî* numerals derived from an alphabetical numeral system that was in use in India before Asoka's time.*

If we compare the shapes of the numerals with the letters that appear in Asoka's *Brâhmî* inscriptions of Nana Ghat and Nasik, we can see that there are quite obvious similarities. The numeral for the number 4, a kind of "+" in Asoka's edicts, is identical to the sign used to write the syllable *ka*. Likewise, the 6 is very similar to the syllable (Fig. 24.29). The 7 resembles the syllable *kha*, whilst the 5 has the same appearance as the *ia*, *ña*, etc. (Fig. 24.30).

However, this link which has been established between the original Indian numerals and the letters of the *Brâhmî* alphabet is not convincing.

First, the *Brâhmî* numerals for 1, 2 and 3 do not resemble any letter: they are formed respectively by one, two and three horizontal lines (Fig. 24.29 to 35). Moreover, no phonetic value was assigned to the ancient forms of the *Brâhmî* numeral which represented multiples of 10 (see Fig. 24.70). Even where this relationship has been established, there is too much variation in the attribution of phonetic values to the signs in question. Thus, whilst the numeral 4 has been connected to the syllable *ka*, in its diverse forms it can equally be said to resemble the letter *pka* or the syllables *pna, lka, tka* or *pkr*. Similarly, the shape of the numeral representing the number 5 resembles the syllable *tr* as well as the following: *ta, tâ, pu, hu, ru, tr, trâ, nâ, hr, hra or ha* [see B. Datta and A. N. Singh (1938), p. 34].

In other words, if such a system did exist in Asoka's time, it is impossible to discover the principle by which it might have functioned.

* Along with the various styles of numerals, the Indians have long known and used a system for representing numbers which involves vocalised consonants of the Indian alphabet which, in regular order, each have numerical value. These are known as **Varnasankhyâ* in Sanskrit, the system of "letter-numbers". The system varies according to the era and region but always follows the method of attributing numerical values to Indian letters, sometimes even following the principle used in numerical representations (the place-value system or the principle of addition). Included in the numerous systems of this kind is the one which the famous astronomer Âryabhata (c. 510 CE) used to record his astronomical data; there is also the system called *Katapayâdi* used (amongst others) by the ninth-century astronomers Haridatta and Shankaranârâyana, as well as *Âksharapallî* frequently used in *Jaina manuscripts. These systems are still in common use today in various regions of India, from Maharashtra, Bengal, Nepal and Orissa to Tamil Nadu, Kerala and Karnataka. They are also used by the Sinhalese, the Burmese, the Khmer, the Siamese and the Japanese, as well as the Tibetans, who often use their letters as numerical signs, mainly to number their registers and manuscripts. Details can be found under the entries **Varnasankhyâ*, **Aksharapallî*, *Numeral Alphabet, **Âryabhata* and **Katapayâdi* numeration of the Dictionary.

Despite the shakiness of their explanations, the supporters of Indrâji's theory conjectured that the idea of assigning numerical values to letters of the alphabet dated back to the most ancient of times, their reasoning being that "Indian, Hindu, Jaina and Buddhist traditions attribute the invention of *Brâhmî* writing and its corresponding numeral system to Brahma, the god of creation."

(Of course, such an argument cannot be taken seriously, especially in the case of Indian civilisation, where such traditions were actually only developed relatively recently and are due to two basic traits of the Indian mentality. First, there is the desire of some of these theorists to make such concepts hold more water in the eyes of their readers, disciples or speakers, in attributing their invention to Brahma. There are also those who, convinced of the innate character of the Indian letters and numerals, do not even consider it necessary to give a historical explanation for their existence. In the first instance, the motive was to make these concepts sacred, and in the second, to make them timeless. The latter conveys India's fundamental psychological character; an obsession with the past which always involves wiping out historical events and replacing fact with religious history, which takes no account of archaeology, palaeography or, most importantly, history.)

The pioneer of the above theory even went so far as to claim that the first Indian numeral alphabet dates back to the eighth century BCE. According to Indrâji, it was Pânini (c. 700 BCE) who first had the idea of using the consonants and vowels of the Indian alphabet to represent numbers.

| | | | | |
|---|---|---|---|---|
| क<br>ka | ख<br>kha | ग<br>ga | घ<br>gha | ङ<br>ṅa |
| च<br>cha | छ<br>chha | ज<br>ja | झ<br>jha | ञ<br>ña |
| ट<br>ṭa | ठ<br>ṭha | ड<br>ḍa | ढ<br>ḍha | ण<br>ṇa |
| त<br>ta | थ<br>tha | द<br>da | ध<br>dha | न<br>na |
| प<br>pa | फ<br>pha | ब<br>ba | भ<br>bha | म<br>ma |
| य<br>ya | र<br>ra | ल<br>la | व<br>va | |
| श<br>śha | ष<br>sha | स<br>sa | | |
| ह<br>ha | | | | |

Fig. 24.56. *Consonants of the Devanagarî (or Nâgarî) alphabet*

Pânini is the famous grammarian of the Sanskrit alphabet: born in Shalatula (near to Attock on the Indus, in the present-day Pakistan), he is considered to be the founder of Sanskrit language and literature; his work, the *Ashtâdhyâyî* (also known as *Pâninîyam*), remains the most famous work on Sanskrit grammar [see L. Frédéric (1987)]. We have no exact dates in Pânini's life, and there is much doubt surrounding the work which is attributed to him.

In other words, the date suggested by Indrâji for the invention of the first Indian alphabetical numeral system has no foundation at all, especially when one considers that there is no known written document, nor specimen of true Indian writing, which dates so far back in Indian history.

It goes without saying, then, that this hypothesis must also be rejected.

### *The origin of Indian alphabetical numerals*

So where does the idea of writing numbers using the Indian alphabet come from?

It must be made clear straight away that the idea did not come from Aramaean merchants, who brought their own writing system to India (Fig. 24.28). With a few later exceptions, the northwestern Semites never used their letters for counting; their numerals were of the same kind as the *Karoshthî* (or Aramaean-Indian) system (see Fig. 24.54 and Fig. 18. 1 to 12).

One could attribute the idea to a Greek influence, in the light of Alexander the Great's conquest of the Indus in 326 BCE, and moreover because this kind of system was in use in Greece since the fourth century BCE. However, this hypothesis is not plausible, because no Indian inscriptions, during or after Asoka's reign, show any evidence of alphabetical numeration.

In fact, the first numeral system of this kind was invented in the Indian sub-continent by the famous mathematician and astronomer, *Âryabhata. This system was undeniably unique compared to all the other previous and contemporary systems; not only has it been quoted in numerous works and commentaries over the centuries, but it has also inspired a considerable diversity of authors, commentators and transcribers, in various eras, to draw comparisons with it and both similar and very different systems (see **Âryabhata* and **Katapayâdi* in the Dictionary).

### *Sixth Hypothesis: Brâhmî numerals came from Egypt*

Here are some other hypotheses put forward as to the origin of *Brâhmî* numerals.

Bühler (1896) and A. C. Burnell (1878) believed that the Indians owed their *Brâhmî* writing to Pharaonic Egypt. Bühler claimed that it derived from hieratic writing (see Fig. 23.10), and Burnell believed that *Brâhmî* writing derived from demotic writing.

Bühler's theory is not totally unfounded, because there is a much stronger similarity between *Brâhmî* writing and Egyptian hieratic writing than there is between the former and the demotic writing of the same civilisation. However, is this partial resemblance significant enough to suggest that Egypt had such an influence on India's distant past?

Arabia, the legendary land of "Pount", was a staging post for Egyptian trade. Ships sailed to the Red Sea along the eastern Nile delta and along a canal first to the Bitter Lakes, then to the Gulf of Suez. It is possible that these same merchant ships, in their quest for eastern goods, travelled further than Arabia, not only to the areas around the Persian Gulf, but also as far as the mouth of the Indus [see A. Aymard and J. Auboyer (1961)].

Conversely, during Alexander the Great's time, India communicated with the Caspian and the Black Sea by river navigation, notably along the Amu Darya; overland routes also led from Europe to India through Bactria, Gandhara and the Punjab, giving access to ports on the western coast of India. Commercial relations became more and more firmly established between Egypt and India, and ships even sailed as far as the coast of Malabar, in particular to the port of Muziris (now the town of Canganore) [see Aymard and Auboyer (1961)].

These relations, however, occurred at a relatively late time and do not really prove anything in terms of the transmission of Egyptian hieratic numerals: the amount of time which separates these numerals from their *Brâhmî* counterparts is too great to allow this hypothesis to be taken into consideration.

(It must be remembered that hieratic numerals were almost obsolete in Egypt by the eighth century BCE; therefore, if this system was transmitted to India, the transmission cannot have taken place any later than this time. As we possess no information about India at this time, this hypothesis cannot be proved.)

Moreover, the above comparison only concerns units; there is a clear difference between the other symbols (numerals representing 10 or above, which have not yet been discussed: see Fig. 24.70). Therefore the comparison only concerns a small percentage of the numerals.

### *The origin of the first nine Indian numerals*

There is another hypothesis which seems much more plausible, even in the absence of any documentation.

Basically, we have already proved this hypothesis: different civilisations have had the same needs, living under the same social, psychological, intellectual and material conditions. Independently of one another, they have followed identical paths to arrive at similar, if not identical results.

This explains the existence of certain numerals of other civilisations which resemble and often represent the same numbers as *Brâhmî* numerals, and which generally date back several centuries before Asoka's time.

On consulting Figs. 24.57 and 24.29 to 35, we can see signs which are not Indian, yet which are very similar to the various ways of writing the numbers 1, 2 and 3 in Indian civilisations. We can also see the evident similarity between the Nabataean or Palmyrenean "5" and that of ancient India, as well as the physical resemblance between the Egyptian hieratic or demotic "7" and "9" and their Indian counterparts.

What these analogies actually prove is not the unlikely theory that the first nine Indian numerals came from one of the other civilisations, but rather that there are universal constants caused by the fundamental rules of history and palaeography. These similarities occur because other civilisations used similar writing materials to those of ancient India, for example the calamus (a type of reed whose blunted end was dipped in a coloured substance), and which was used by Egyptian and western Semitic scribes (Aramaeans, Nabataeans, Palmyreneans, etc.) to write on papyrus or parchment, which was long used on tree bark or palm leaf in Bengal, Nepal, the Himalayas and in all of the north and northwest of India.

We know to what extent the nature of the instrument influenced, on the one hand, Egyptian manuscript writing, and on the other hand, the writings of the ancient Semitic world.

In the first case, the use of the calamus turned the hieroglyphics of monumental Egyptian writing into cursive hieratic signs, changing the detailed, pictorial symbols into a shortened, more simplified form, perfectly adapted to the needs of manuscript writing and rapid notation.

In the latter case, the same writing apparatus was used to transform the rigid and angular shape of Phoenician writing into the rounder, more cursive and fluid forms, like that of Elephantine Aramaean scribes.

Thus the superposition of two or three horizontal lines, first transformed into one complete sign by a ligature, gave birth to the same forms as the Indian numerals for 2 and 3, whose palaeographical styles vary considerably according to the era, the region and the habits of the scribe (Fig. 24.58).

This explanation relies on the assumption that horizontal lines formed the first three of the ancient ideographical Indian numerals, and this is what *Brâhmî* inscriptions written after the third century BCE (Shunga, Shaka, Kushâna, Mathurâ, Kshatrapa, etc.) would suggest (Fig. 24.30 to 38). This figurative representation was still in use during the time of the Gupta Dynasty (third to sixth century CE), and even persisted in some areas until the eighth century.

| 5 | 6 | 7 | 8 | 9 |
|---|---|---|---|---|
| a | c | f | k | p |
| b | d | g | l | q |
| x | e | h | m | r |
| y | w | i | n | s |
| | | j | o | t |
| | | | | u |
| | | | | v |

- Egyptian numerals: a (HP I, 618, Abusir); b (HP I, 618, Elephantine); c (HP II, 619, Louvre 3226); d (HP II, Louvre 3226); e (HP II, 619, Gurob); f (HP I, 620, Illahun); g (HP I, 620, Bulaq 18); h (HP II, 620, Louvre 3226); i (HP II, 620, P. Rollin); j (HP III, 620, Takelothis); k (HP I, 621, Elephantine); l (HP I, 621, Illahun); m (HP I, 621, Math); n (HP I, 621, Ebers); o (HP III, 621, Takelothis); p (HP I, 622, Abusir); q (HP I, 622, Illahun); r (HP I, 622, Illahun); s (HP I, 622, Bulaq, Harris); t (HP II, 622, P. Rollin); u (HP II, 622, Gurob); v (HP II, 622, Harris); w (DG, 697, Ptol.); Nabataean numerals: x (CIS, II1, 212); Palmyrenean numerals: y (CIS, II3, 3913).

FIG. 24.57. *Numerals which have the same appearance and numerical value as their Brâhmî equivalents. [Ref. Möller (1911–12); Cantineau; Lidzbarski (1962)]*

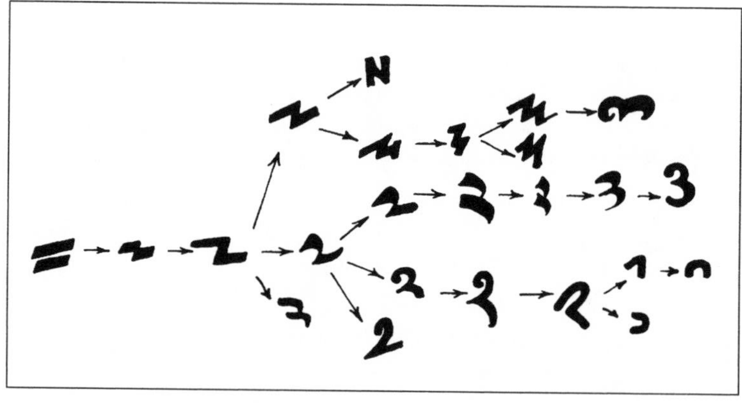

FIG. 24.58. *Evolution of Indian numerals 2 and 3*

However, if we examine Asoka's edicts (c. 260 BCE), we can see that throughout the Mauryan empire, the numbers 1, 2 and 3 were not represented by superposed horizontal lines, but by one, two or three vertical lines (Fig. 24.29).

Why did this change of direction occur? And why did it happen between the third and second century BCE, when the representations had been horizontal since the time of the Buddhist inscriptions of Nana Ghat (Fig. 24.30)?

The second question is difficult to answer, as no documentation has been found from that time on this subject (if indeed anyone took the trouble to write about something which must have seemed so insignificant). However, this is of little importance; we are only interested here in how such a change came about.

Could it have occurred due to aesthetic reasons? This is as unlikely as the possibility that the new notation evolved for practical reasons. To draw a line one, two or three times, whether vertically or horizontally, has no aesthetic value, and involves practically the same amount of exertion, unless it goes against what one is accustomed to writing.

On the other hand, this modification could have been due to the realisation that a vertical representation of the first three units was likely to be confused with the *danda*. This is a punctuation mark in the form of a small vertical line ( | ), which the Indians have long used in their Sanskrit texts to mark the end of a line or of part of a sentence, which they double ( || ) to indicate the end of a sentence, couplet or strophe.

The invention of the *danda* in the second century BCE could be responsible for the change in direction of the lines representing the numbers 1 to 3. This is mere conjecture which for the moment remains without proof or confirmation.

Here is another question: why did the Indians conserve these representations of the first three units for so long, when the numerals for 4 to 9 had already graphically evolved into independent numerals, which offered no visual clue as to the numbers they represented (Fig. 24.29 to 38)? This is not only true of the Indians: many other civilisations have offered us similar puzzles over the ages, notably those of China and Egypt.

The explanation lies in a basic human psychological trait, which was discussed in Chapter 1. Whilst it was necessary to have other signs than four or five to nine lines for the numbers 4 or 5 to 9, it was not judged necessary to change the signs for units which were lower or equal to 4; this was not only because these symbols could be drawn quickly and easily, but also and above all because without needing to count, the eye can easily distinguish between a number of lines when they number four or less. Four is the limit, beyond which the human mind has to begin to count in order to determine the exact quantity of a given number of elements.

So what was the reasoning behind the formation of the other six *Brâhmî* numerals? Are they purely conventional signs, created artificially to supply a need? Probably not. Taking the universal laws of palaeography into account, and the evidence surrounding the formation of similar numerals in other cultures, it is more likely that the numerals were born out of prototypes formed by the primitive grouping of a number of lines representing the value of the unit.

In other words, to all appearances, the *Brâhmî* numerals of Asoka's edicts were to their ideographical prototypes as Egyptian hieratic numerals were to their corresponding hieroglyphic numerals.

As the lines representing the numbers 1 to 3 were vertical before they were horizontal, one could reasonably presume that the first nine *Brâhmî* numerals constituted the vestiges of an old indigenous numerical notation, where the nine numerals were represented by the corresponding number of vertical lines; a notation, doubtless older than *Brâhmî* itself*, where, like the Egyptian hieroglyphic system, the Cretan or the Hittite system for example, the vertical lines were set out as in Figure 24.59.

FIG. 24.59. *A plausible reconstruction of the original Indian ideographical notation : the starting point of the evolution which led to the Brâhmî numerals for 4 to 9 (those for 1 to 3 retaining their ideographical form for many centuries, although represented horizontally rather than vertically)*

To enable the numerals to be written rapidly, in order to save time, these groups of lines evolved in much the same manner as those of old Egyptian Pharaonic numerals. Taking into account the kind of material that was written on in India over the centuries (tree bark or palm leaves) and the limitations of the tools used for writing (calamus or brush), the shape of the numerals became more and more complicated with the numerous ligatures (Fig. 24.60), until the numerals no longer bore any resemblance to the original prototypes. Thus a primitive numbering system became one of numerals of distinct forms which gave no visual indication as to their numerical value: the *Brâhmî* numerals of the first three centuries BCE.

Taking into consideration the universal constants of both psychology and palaeography, this is currently the most plausible explanation of the origin of the nine Indian numerals. The summary at the end of this chapter demonstrates the likely stages of the development of *Brâhmî* numerals.

Therefore, it appears that *Brâhmî* numerals were autochthonous, that is to say, their formation was not due to any outside influence. In all probability, they were created in India, and were the product of Indian civilisation alone.

In other words, one could say that the problem of the origin of our present-day numerals has been satisfactorily solved. This is also demonstrated in

*This is not at all surprising if one considers that, on the one hand, the ancient civilisation of the Indus which preceded the Aryans used exactly this type of notation (Fig. 1.14), and that on the other hand, Sumerian civilisation developed a numeral system even before creating its own writing system.

the palaeographical tables of Fig. 24.61 to 69, which constitute the complete historical synthesis of the question, and which have been set out taking into account all the details proved both previously and subsequently.

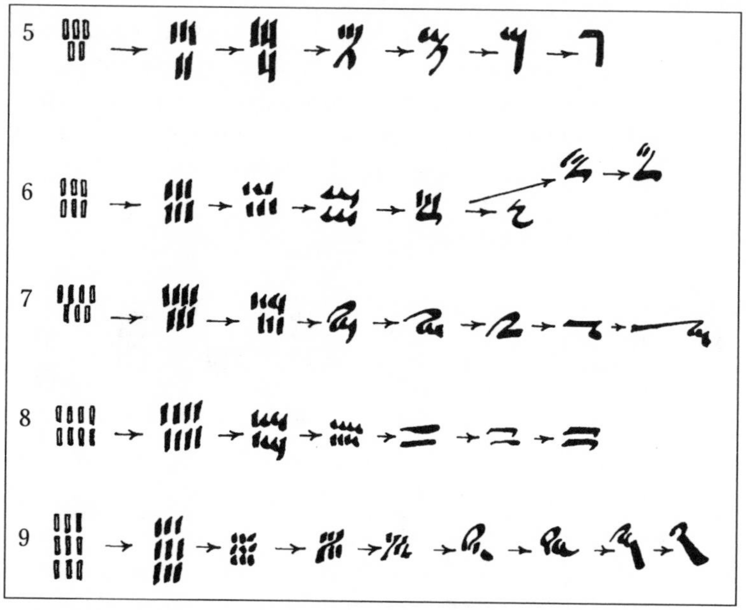

FIG. 24.60. *Results of the graphical evolution of the signs which were originally formed by the juxtaposition or superposition of several vertical lines, these lines being drawn on a smooth surface and written with a calamus with a blunt tip dipped into a coloured substance. This evolution took place among the Egyptians.*

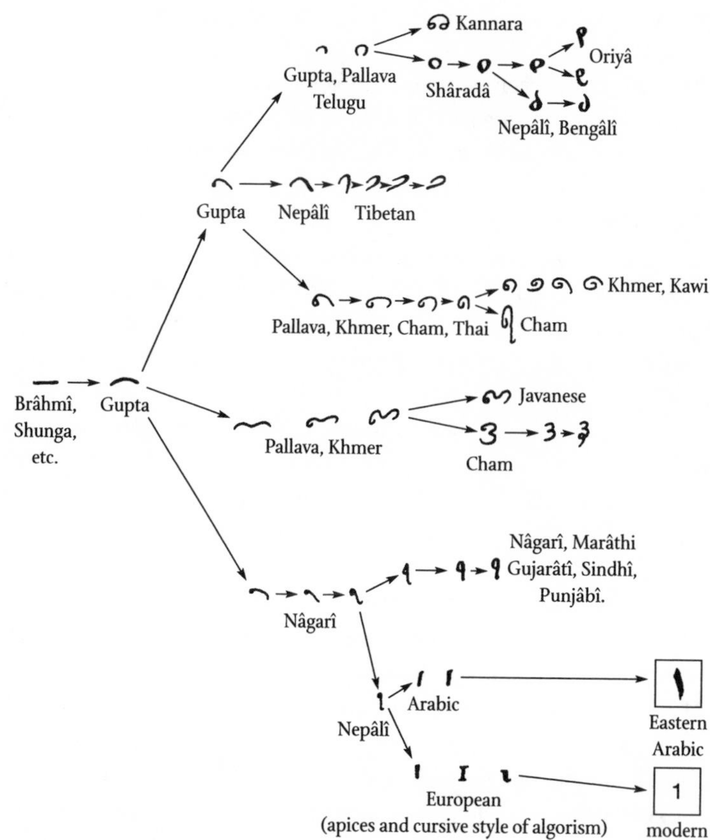


FIG. 24.61. *Origin and evolution of the numeral 1. (For Arabic and European numerals, see Chapters 25 and 26.)*

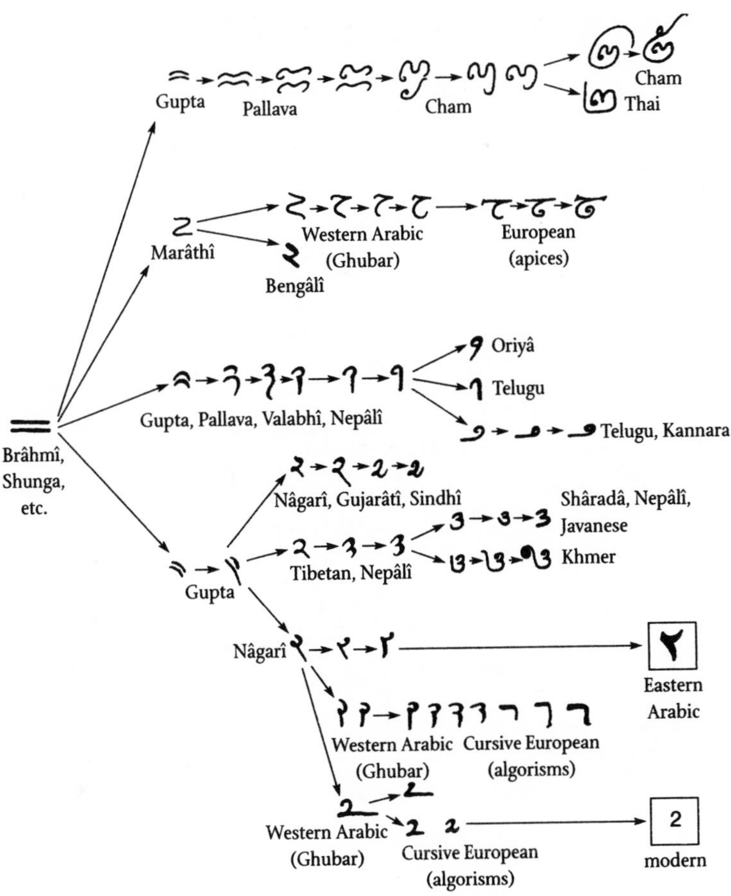


Fig. 24.62. *Origin and evolution of the numeral 2. (For Arabic and European numerals, see Chapters 25 and 26.)*

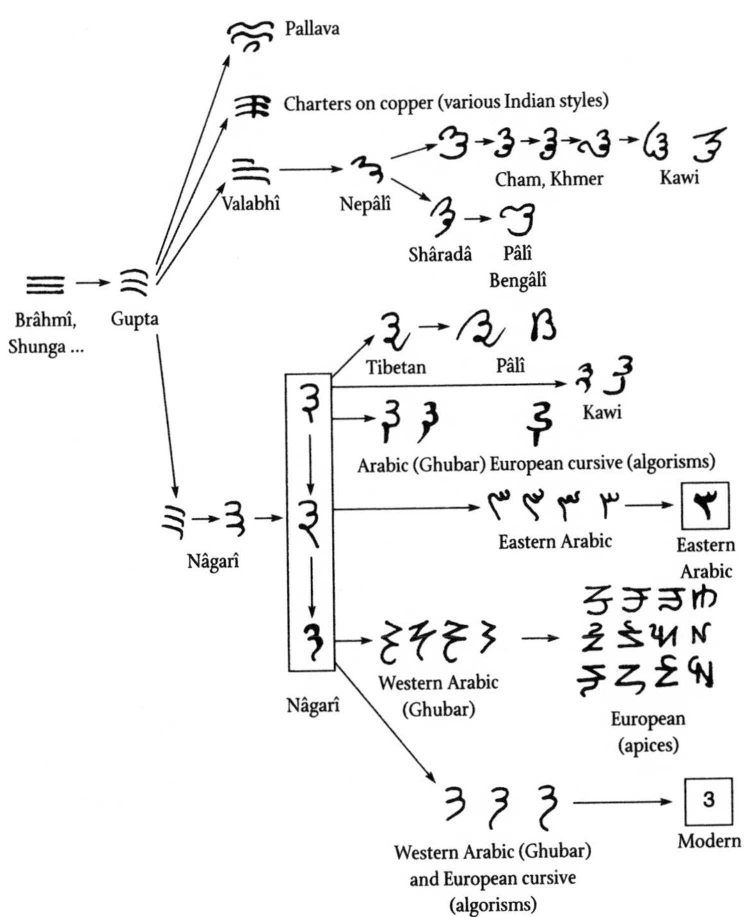


Fig. 24.63. *Origin and evolution of the numeral 3. (For Arabic and European numerals, see Chapters 25 and 26.)*

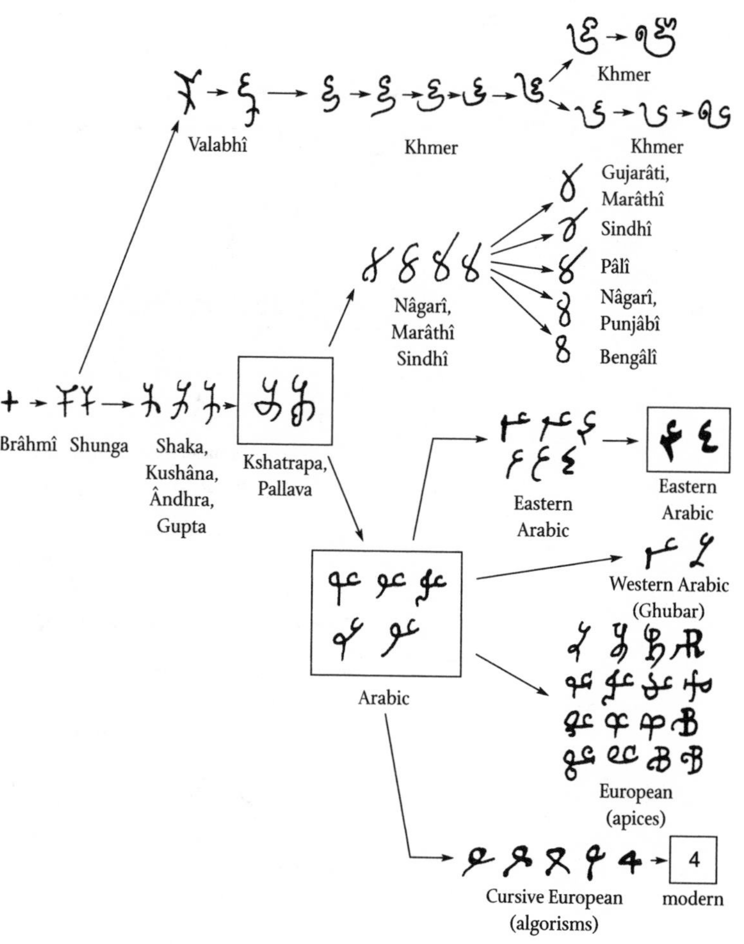


FIG. 24.64. *Origin and evolution of the numeral 4. (For Arabic and European numerals, see Chapters 25 and 26.)*

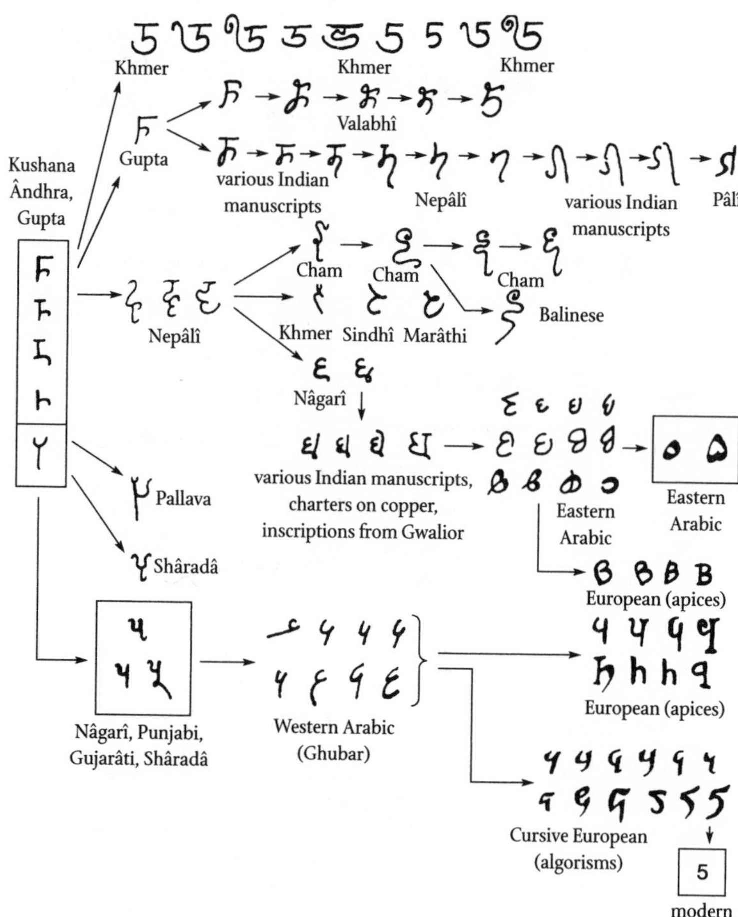


FIG. 24.65. *Origin and evolution of the numeral 5. (For Arabic and European numerals, see Chapters 25 and 26)*

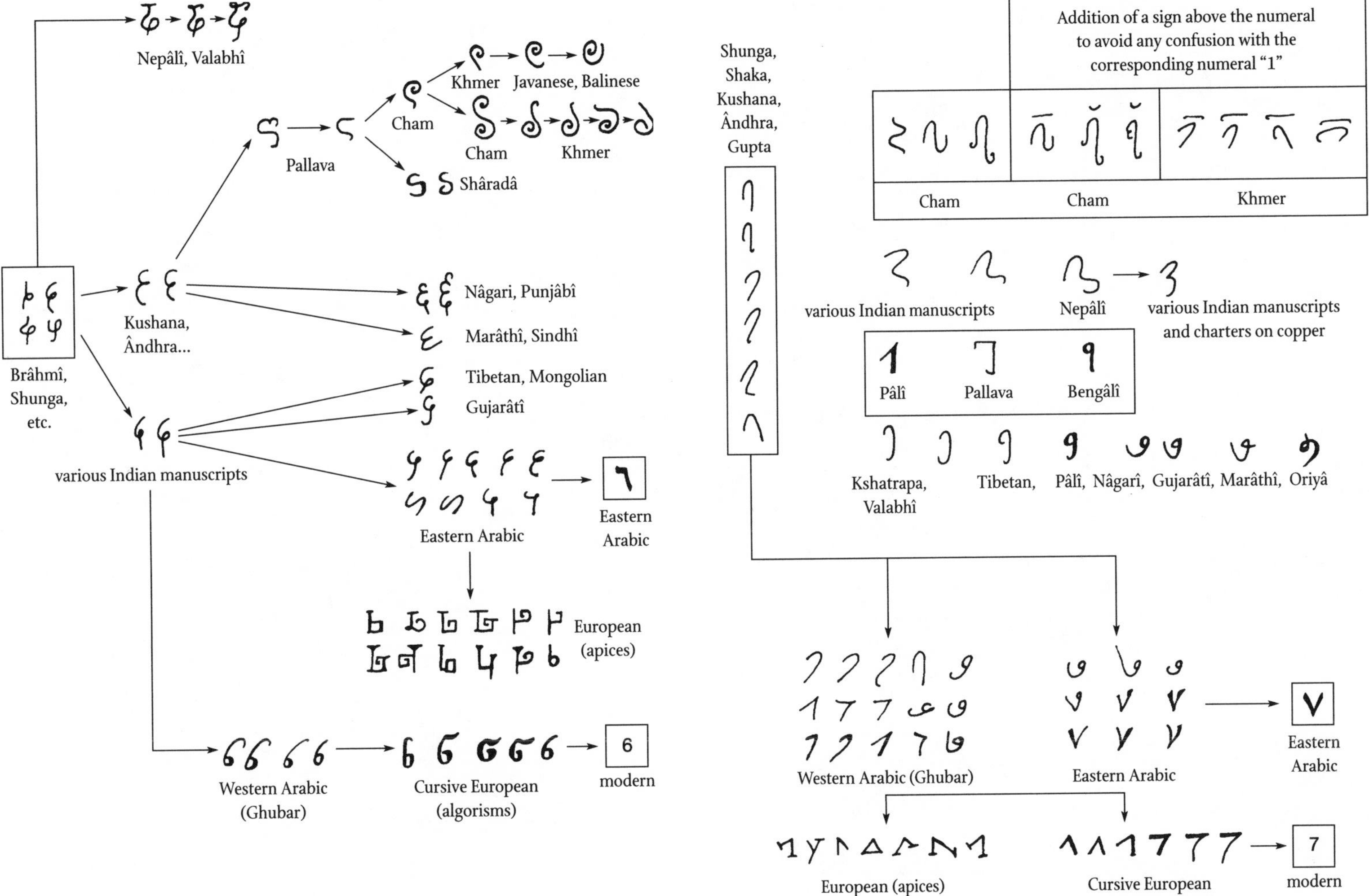


FIG. 24.66. *Origin and evolution of the numeral 6. (For Arabic and European numerals, see Chapters 25 and 26.)*

FIG. 24.67. *Origin and evolution of the numeral 7. (For Arabic and European numerals, see Chapters 25 and 26.)*

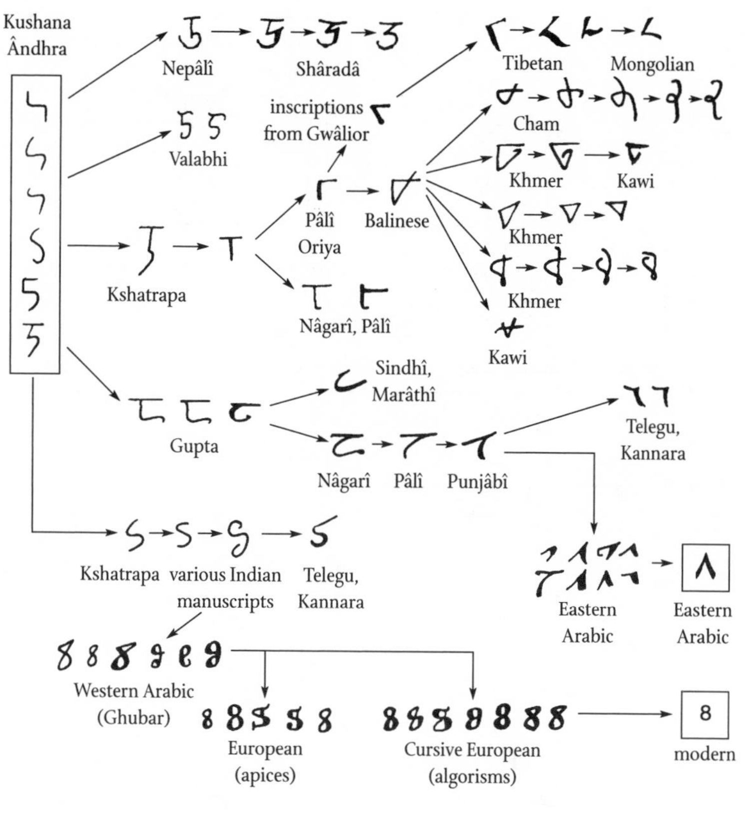


Fig. 24.68. *Origin and evolution of the numeral 8. (For Arabic and European numerals, see Chapters 25 and 26.)*

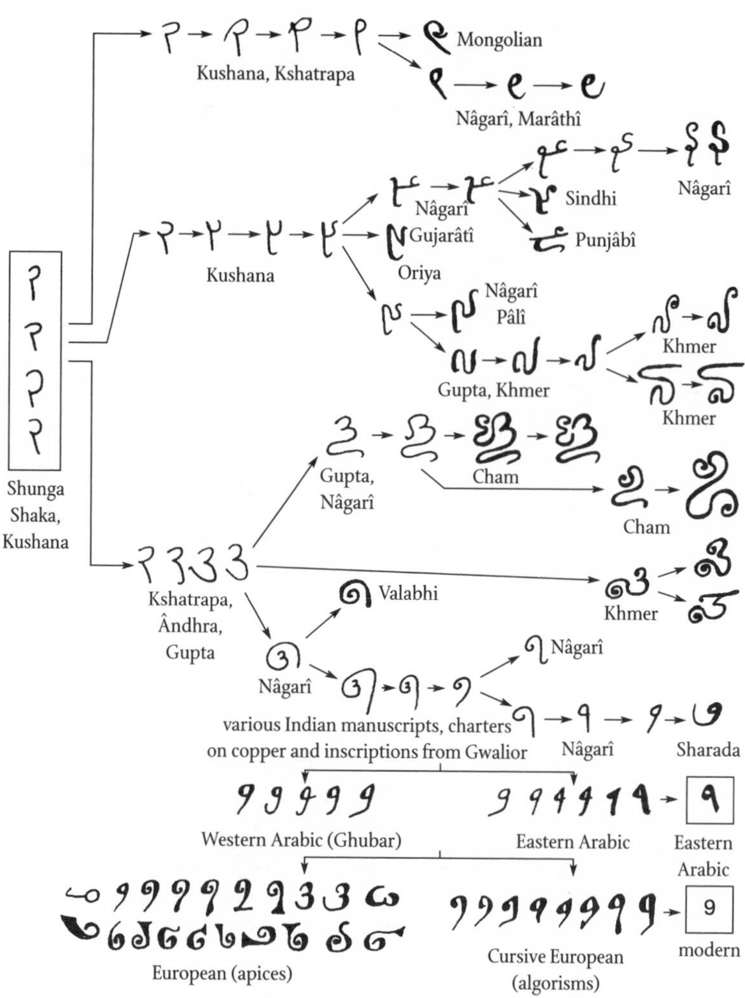


Fig. 24.69. *Origin and evolution of the numeral 9. (For Arabic and European numerals, see Chapters 25 and 26.)*

## OLD INDIAN NUMERALS: A VERY BASIC SYSTEM

As the preceding diagrams have shown, Indian numerals, even in their earliest forms, were the forerunners of the nine basic numerals of our present-day system. In other words, it was from these signs that, some centuries later, the numerals that we wrongly call "Arabic" were derived. As we shall see later, modern numerals are the descendants of North African numerals, which themselves are cousins of the eastern Arabic numerals, which in turn are linked to *Nâgarî* numerals, which belong to the family of decimal numeral systems currently in use in India and Southeast and central Asia.

From a graphical point of view, then, the first nine *Brâhmî* numerals shared one of the fundamental characteristics of our present-day numerals. This, however, was the only aspect which they originally had in common.

If we examine *Brâhmî* inscriptions or intermediate Indian inscriptions, from Asoka's edicts to those of the Shungas, Shakas, Kushânas, Ândhras, Kshatrapas, Guptas, Pallavas or even the Châlukyas, that is to say from the third century BCE to the sixth and seventh centuries CE, we can see that the corresponding principle of numerical notation is very rudimentary.

For a decimal base, this system relied largely upon the principle of addition, attributing a specific sign to each of the following numbers (Fig. 24.70):

| | | | | | | | | |
|---|---|---|---|---|---|---|---|---|
| 1 | 2 | 3 | 4 | 5 | 6 | 7 | 8 | 9 |
| 10 | 20 | 30 | 40 | 50 | 60 | 70 | 80 | 90 |
| 100 | 200 | 300 | 400 | 500 | 600 | 700 | 800 | 900 |
| 1,000 | 2,000 | 3,000 | 4,000 | 5,000 | 6,000 | 7,000 | 8,000 | 9,000 |
| 10,000 | 20,000 | 30,000 | 40,000 | 50,000 | 60,000 | 70,000 | 80,000 | 90,000 |

This written numeral system had special numerals, not only for each basic unit, but also for every ten, hundred, thousand and ten thousand units. To represent a number such as 24,400, one needed only to juxtapose, in this order, the numerals 20,000, 4,000 and 400 (Fig. 24.71):

20,000 4,000 400

Of course, if these numerals had belonged to a place-value system, the number in question would have been written in the following way, using the style of numerals in use at that time, the zero being represented by a little circle, as it appears in later Indian inscriptions:

2 4 4 0 0

UNITS

| | 1 | 2 | 3 | 4 | 5 | 6 | 7 | 8 | 9 |
|---|---|---|---|---|---|---|---|---|---|
| Third century BCE: Brahmi of Asoka | I | II | | + | | [illegible] | | | |
| Second century BCE: Inscriptions of Nana Ghat | — | = | | [illegible] | | [illegible] | [illegible] | | [illegible] |
| First or second century CE: Inscriptions of Nasik | — | = | ≡ | [illegible] | [illegible] | [illegible] | [illegible] | [illegible] | [illegible] |
| First to second century CE: Kushana inscriptions | — | = | ≡ | [illegible] | [illegible] | [illegible] | [illegible] | [illegible] | [illegible] |
| First to third century CE: Andhra, Mathura and Kshatrapa inscriptions | — | = | ≡ | [illegible] | [illegible] | [illegible] | [illegible] | [illegible] | [illegible] |
| Fourth to sixth century CE: Gupta inscriptions | [illegible] | [illegible] | [illegible] | [illegible] | [illegible] | [illegible] | [illegible] | [illegible] | [illegible] |
| Sixth to ninth century CE: Inscriptions of Nepal | [illegible] | [illegible] | [illegible] | [illegible] | [illegible] | [illegible] | [illegible] | [illegible] | [illegible] |
| Fifth to sixth century CE: Pallava inscriptions | [illegible] | [illegible] | [illegible] | [illegible] | [illegible] | [illegible] | [illegible] | [illegible] | |
| Sixth to seventh century CE: Valabhi inscriptions | [illegible] | [illegible] | [illegible] | [illegible] | [illegible] | [illegible] | [illegible] | [illegible] | [illegible] |
| Various Indian manuscripts | [illegible] | [illegible] | [illegible] | [illegible] | [illegible] | [illegible] | [illegible] | [illegible] | [illegible] |

FIG. 24.70A *Numerical notation linked to Brâhmî writing and its immediate derivatives. There is evidence of the signs formed by straight lines; the others are reconstructions based on a comparative study of forms. (For references, see Fig. 24. 29 to 38 and 24.41 to 46.)*

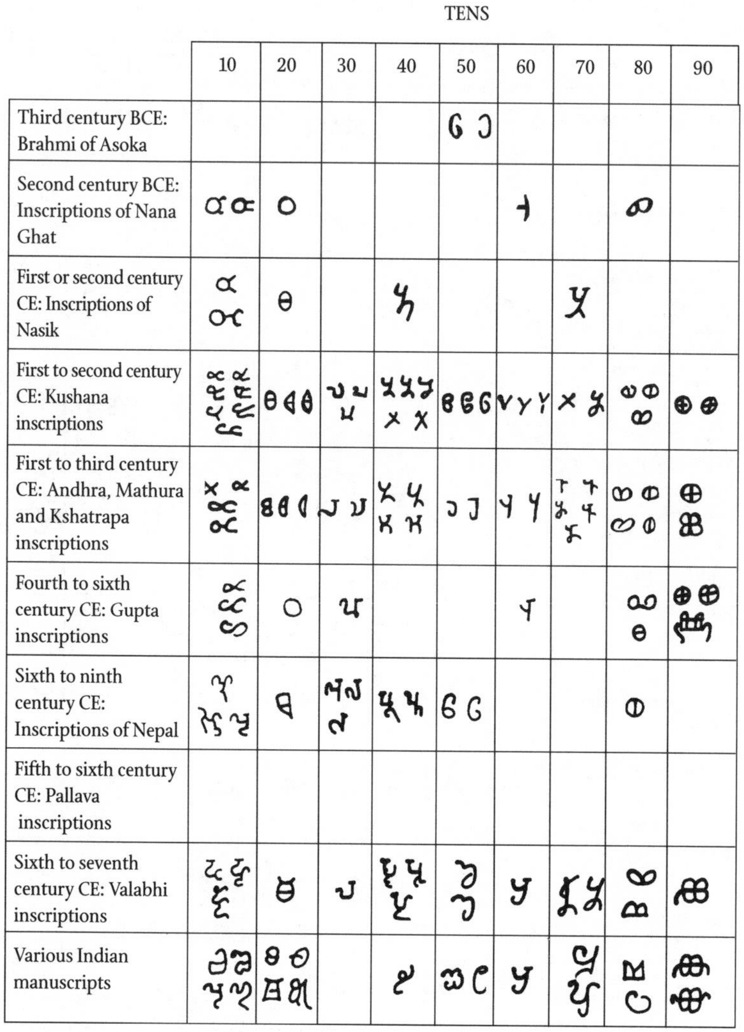

TENS

| | 10 | 20 | 30 | 40 | 50 | 60 | 70 | 80 | 90 |
|---|---|---|---|---|---|---|---|---|---|
| Third century BCE: Brahmi of Asoka | | | | | [illegible] | | | | |
| Second century BCE: Inscriptions of Nana Ghat | [illegible] | [illegible] | | | | [illegible] | | [illegible] | |
| First or second century CE: Inscriptions of Nasik | [illegible] | [illegible] | | [illegible] | | | [illegible] | | |
| First to second century CE: Kushana inscriptions | [illegible] | [illegible] | [illegible] | [illegible] | [illegible] | [illegible] | [illegible] | [illegible] | [illegible] |
| First to third century CE: Andhra, Mathura and Kshatrapa inscriptions | [illegible] | [illegible] | [illegible] | [illegible] | [illegible] | [illegible] | [illegible] | [illegible] | [illegible] |
| Fourth to sixth century CE: Gupta inscriptions | [illegible] | [illegible] | [illegible] | | | [illegible] | | [illegible] | [illegible] |
| Sixth to ninth century CE: Inscriptions of Nepal | [illegible] | [illegible] | [illegible] | [illegible] | [illegible] | | | [illegible] | |
| Fifth to sixth century CE: Pallava inscriptions | | | | | | | | | |
| Sixth to seventh century CE: Valabhi inscriptions | [illegible] | [illegible] | [illegible] | [illegible] | [illegible] | [illegible] | [illegible] | [illegible] | [illegible] |
| Various Indian manuscripts | [illegible] | [illegible] | | [illegible] | [illegible] | [illegible] | [illegible] | [illegible] | [illegible] |

FIG. 24.70B.

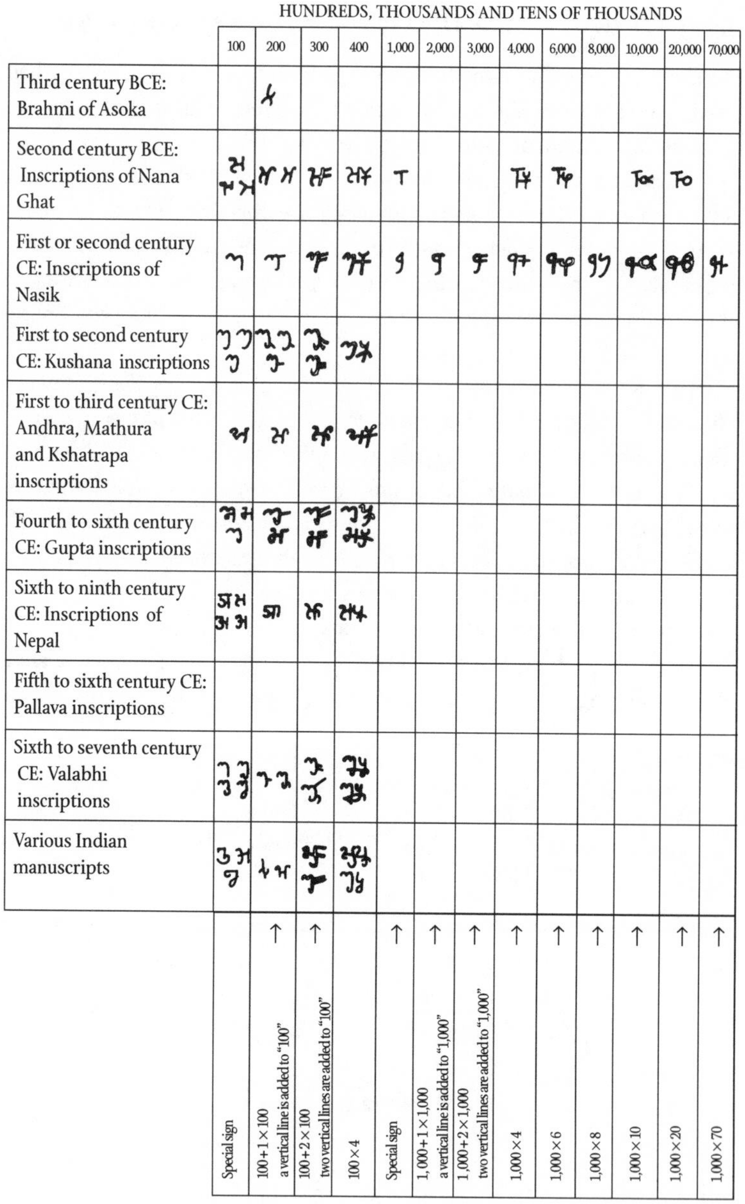

HUNDREDS, THOUSANDS AND TENS OF THOUSANDS

| | 100 | 200 | 300 | 400 | 1,000 | 2,000 | 3,000 | 4,000 | 6,000 | 8,000 | 10,000 | 20,000 | 70,000 |
|---|---|---|---|---|---|---|---|---|---|---|---|---|---|
| Third century BCE: Brahmi of Asoka | | [illegible] | | | | | | | | | | | |
| Second century BCE: Inscriptions of Nana Ghat | [illegible] | [illegible] | [illegible] | [illegible] | [illegible] | | | [illegible] | [illegible] | | [illegible] | [illegible] | |
| First or second century CE: Inscriptions of Nasik | [illegible] | [illegible] | [illegible] | [illegible] | [illegible] | [illegible] | [illegible] | [illegible] | [illegible] | [illegible] | [illegible] | [illegible] | [illegible] |
| First to second century CE: Kushana inscriptions | [illegible] | [illegible] | [illegible] | [illegible] | | | | | | | | | |
| First to third century CE: Andhra, Mathura and Kshatrapa inscriptions | [illegible] | [illegible] | [illegible] | [illegible] | | | | | | | | | |
| Fourth to sixth century CE: Gupta inscriptions | [illegible] | [illegible] | [illegible] | [illegible] | | | | | | | | | |
| Sixth to ninth century CE: Inscriptions of Nepal | [illegible] | [illegible] | [illegible] | [illegible] | | | | | | | | | |
| Fifth to sixth century CE: Pallava inscriptions | | | | | | | | | | | | | |
| Sixth to seventh century CE: Valabhi inscriptions | [illegible] | [illegible] | [illegible] | [illegible] | | | | | | | | | |
| Various Indian manuscripts | [illegible] | [illegible] | [illegible] | [illegible] | | | | | | | | | |
| | Special sign | ↑ 100+1 × 100 a vertical line is added to "100" | ↑ 100+2 × 100 two vertical lines are added to "100" | 100 × 4 | ↑ Special sign | ↑ 1,000+1 × 1,000 a vertical line is added to "1,000" | ↑ 1,000+2 × 1,000 two vertical lines are added to "1,000" | ↑ 1,000 × 4 | ↑ 1,000 × 6 | ↑ 1,000 × 8 | ↑ 1,000 × 10 | ↑ 1,000 × 20 | ↑ 1,000 × 70 |

FIG. 24.70C.

Like certain other systems of the ancient world, this numeration was very limited. Arithmetical operations, even simple addition, were virtually impossible. Moreover, the highest numeral represented 90,000: therefore such a system could not be used to record very high numbers.

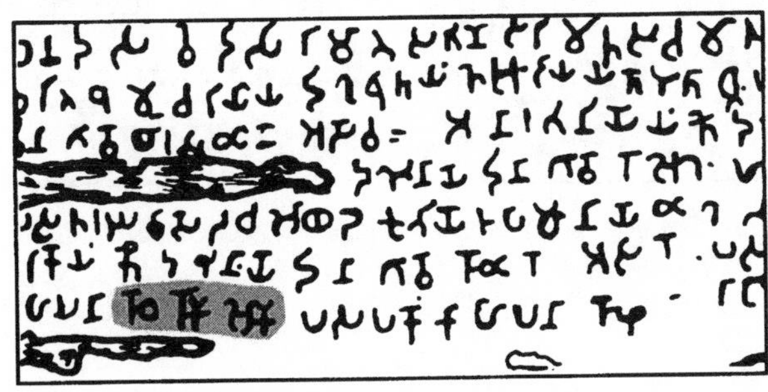

FIG. 24.71. *Detail of a Buddhist inscription in Brâhmî characters adorning one of the walls of the cave at Nana Ghat (second century BCE). The shaded section shows the Brâhmî notation for the number 24,400. [Ref. Smith and Karpinski (1911), p. 24]*

## THE PROBLEM OF THE DISCOVERY OF THE INDIAN PLACE-VALUE SYSTEM

Thus the ancestors of our numerals remained static for a long time before acquiring the dynamic and manageable character that they have today thanks to the place-value system.

This leads us to ask two fundamental questions, which we will tackle through an archaeological, epigraphic and philological examination of the mathematical, astrological and astronomical texts of India: *When and how did the first nine numerals of this rudimentary system come to be governed by this essential rule? And when was zero first used?*

### *The first significant clues*

Before we look at archaeology and epigraphy, it is worthwhile investigating whether some clues about zero and the place-value system can be discovered in Indian Sanskrit mathematical literature.

Here, for example, is an extract from the *Ganitasârasamgraha* (Chapter 1, line 27) by the mathematician Mahâvîrâchârya who, giving 12345654321 as the result of a previous calculation, defines this number in the following way [see B. Datta and A. N. Singh (1938)]:

*ekâdishadantâni kramena hînâni*

which means the quantity "beginning with one [which then grows] until it reaches six, then decreases in reverse order".

This comment is significant because it constitutes a numerical "palindrome": the number reads the same from left to right or from right to left, which is only possible if we are dealing with the place-value system:

12345654321
<—— · ——>

It should be noted that these types of numbers possess unusual properties; take the following, for example:

$$1^2 = 1$$
$$11^2 = 121$$
$$111^2 = 12321$$
$$1111^2 = 1234321$$
$$11111^2 = 123454321$$
$$111111^2 = 12345654321$$
$$1111111^2 = 1234567654321$$
$$11111111^2 = 123456787654321$$
$$111111111^2 = 12345678987654321$$

These are properties that could not have been worked out using a non-positional system, due to its inconsistencies and the rules that would have governed it.

In other words these types of numbers could only have been discovered after the place-value system was invented. As we know that the *Ganitasârasamgraha* is dated c. 850 CE, we can infer that the place-value system was discovered before the middle of the ninth century.

Here is another piece of evidence which places the discovery of the place-value system at an earlier date: the arithmetician Jinabhadra Gani, who lived at the end of the sixth century, gave to the number 224,400,000,000 the following Sanskrit expression, in his *Brihatkshetrasamâsa* I, 69 (see Datta and Singh, p. 79):

*dvi vimshati cha chatur chatvârimshati cha ashta shûnyâni*
"twenty-two and forty-four and eight zeros" (=224400000000).

This proves that the Indians knew of zero and the place-value system in the sixth century.

The preceding examples do not constitute "proof" in the strictest sense of the word, but they show that the place-value system must have been in use for some time if its subtleties were understood and appreciated by the contemporary public.

### *Evidence found in Indian epigraphy*

The first known Indian lapidary documents to bear witness to the use of zero and the decimal place-value system actually only date back to the second half of the ninth century CE.

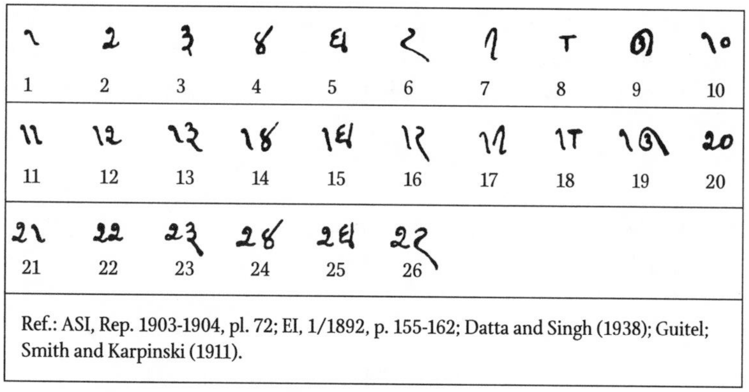

| 1 | 2 | 3 | 4 | 5 | 6 | 7 | 8 | 9 | 10 |
|---|---|---|---|---|---|---|---|---|---|
| 11 | 12 | 13 | 14 | 15 | 16 | 17 | 18 | 19 | 20 |
| 21 | 22 | 23 | 24 | 25 | 26 | | | | |

Ref.: ASI, Rep. 1903-1904, pl. 72; EI, 1/1892, p. 155-162; Datta and Singh (1938); Guitel; Smith and Karpinski (1911).

FIG. 24.72. *Numerals from the first inscription of Gwalior*

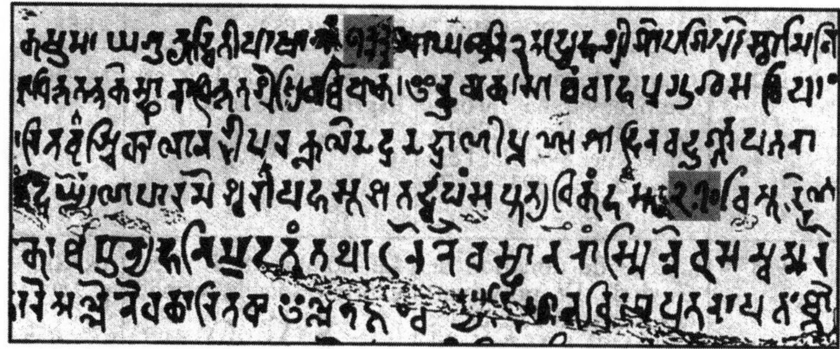

FIG. 24.73. *Detail from the second inscription of Gwalior (876 CE). The shaded section shows the representation of the numbers 933 and 270. [Ref. : EI, I, p. 160]*

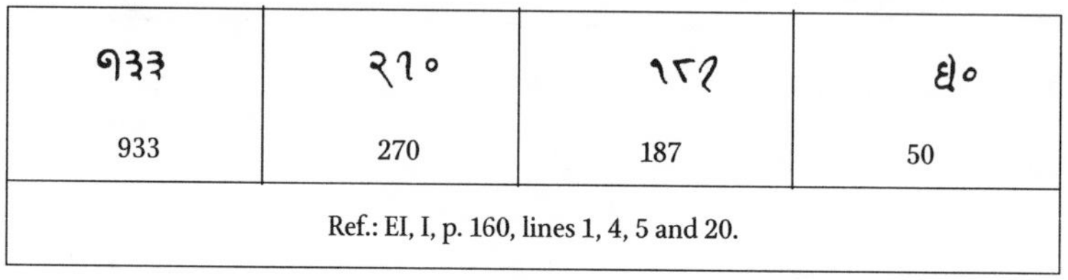

| ९३३ | २७० | १८७ | ५० |
|---|---|---|---|
| 933 | 270 | 187 | 50 |
| Ref.: EI, I, p. 160, lines 1, 4, 5 and 20. | | | |

FIG. 24.74. *Numerals from the second inscription of Gwalior*

These are two stone inscriptions from Bhojadeva's reign, discovered in the nineteenth century in the temple of Vâillabhattasvâmin, dedicated to Vishnu, near the town of Gwalior (capital of the ancient princely state of Madhyabharat, situated approximately 120 kilometres from Agra and a little over 300 kilometres south of Delhi).

The first inscription is quite clearly dated 932 in the *Vikrama* calendar (932 – 57 = 875 CE, see **Vikrama*, Dictionary). It is in Sanskrit, and consists of twenty-six stanzas, which are numbered in the following manner using *Nâgarî* numerals (the signs for the numbers 1, 2, 3, 7, 9 and 0 already strongly resemble their modern equivalents) (Fig. 24.72).

The second inscription is dated (in numerals) the year 933 in the *Vikrama* calendar ( = 876 CE). Written in Sanskrit prose, it gives an account of the offerings the inhabitants of Gwalior made to Vishnu. It tells mainly of the offering of a piece of land 270 × 187 *hasta*, which was to be turned into a flower garden, and of fifty garlands of seasonal flowers which the gardeners of Gwalior were to bring to the temple as a daily offering. The number denoting the date (933), as well as the three other numbers mentioned, are represented by *Nâgarî* numerals as they appear in Fig. 24.74.

There is no question as to the authenticity of these two inscriptions, and they clearly demonstrate the extent to which the inhabitants of the region were familiar with zero and the place-value system during the second half of the ninth century.

The inscriptions from Gwalior are not the oldest documents to contain evidence of the use of this system. Of the numerous other examples, of which there follows a list in ascending chronological order, there are documents engraved on copper which come from diverse regions of central and western India and date back to the era between the end of the sixth century and the tenth century CE.

These documents are legal charters written in Sanskrit and engraved in ancient Indian characters. They record donations given by kings or wealthy personages to the Brahmans. Each one contains details of the religious occasion when the donation(s) was (or were) offered and gives the name of the donor, the number of gifts plus a description of them, as well as a date which corresponds to one of the Indian calendars (**Chhedi*, **Shaka*, **Vikrama*, etc.; see Dictionary).

These dates are usually expressed in both letters and numerals, with the basic numerals, written in various Indian styles, varying in value according to their decimal position (Fig. 24.75 and 76).

The preceding evidence led historians, in the nineteenth century, to conclude that our present-day numerals were of Indian origin, and that they had been in use at least since the end of the sixth century CE (Fig. 24.75).

The foundations of this evidence seemed to crumble, however, at the beginning of the twentieth century, when three science historians, G. R. Kaye, N. Bubnov and B. Carra de Vaux, who were among those the most opposed to the idea that our numerals originated in India, questioned the authenticity of the copper inscriptions. They claimed that these documents had been re-written, altered or falsified at a much later date than the years given in the lists. It was concluded that, of all the texts which had been thought to be of Indian origin, only the inscriptions carved in stone could be regarded as proof of the existence of the system in question.

| | DOCUMENTS AND SOURCES | | |
|---|---|---|---|
| 972 | Donation charter of Amoghavarsha of the Râshtrakûtas, dated 894 in the *Shaka calendar (= 894 + 78 = 972 CE). | The number 894 is expressed: | IA, XII, p. 263 |
| 933 | Donation charter of Govinda IV of the Râshtrakûtas, dated 855 in the *Shaka Samvat calendar (= 855 + 78 = 933 CE). | The number 855 is expressed: | IA, XII, p. 249 |
| 917 | Donation charter of Mahîpâla, dated 974 in the *Vikrama Samvat calendar (= 974 – 57 = 917 CE). | The numbers 974 and 500 are expressed: | IA, XVI, p. 174 |
| 837 | Bâuka inscription. Dated 894 in the *Vikrama Samvat calendar (= 894 – 57 = 837 CE). | The number 894 is expressed: | EI, XVIII, p. 87 |
| 815 | Donation charter of Nâgbhata of Buchkalâ. Dated 872 in the *Vikrama Samvat calendar (= 872 – 57 = 815 CE). | The number 872 is expressed: | EI, IX, p. 198 |
| 793 | Donation charter of Shankaragana of Daulatâbâd. Dated 715 in the *Shaka calendar (= 715 + 78 = 793 CE). | The number 715 is expressed: | EI, IX, p. 197 |
| 753 | Donation charter of Dantidurga of the Râshtrakûtas. Dated 675 in the *Shaka Samvat calendar (= 675 + 78 = 753 CE). | The number 675 is expressed: | IA, XI, p. 108 |
| 753 | Inscription of Devendravarman. Dated 675 in the *Shaka calendar (= 675 + 78 = 753 CE). | The number 20 is expressed: | EI, III, p. 133 |
| 737 | Donation charter of Dhiniki. Dated 794 in the *Vikrama Samvat calendar (= 794 –57 = 737 CE). | The number 794 is expressed: | IA, XII, p. 155 |
| 594 | Donation charter of Dadda III, of Sankheda in Gujarat (Bharukachcha region). Dated 346 in the *Chhedi calendar (= 346 + 248 = 594 CE). | The number 346 is expressed: (see Fig. 24.76) | EI, II, p. 19 |

FIG. 24.75.

Since the inscriptions of Gwalior (875/876 CE) constituted the first evidence of this kind, these authors surmised that in India, zero and the place-value system could not have been used much before the second half of the ninth century CE.

It is true that amongst the charters recorded copper found in India, the authenticity of a certain number of them has been questioned, and quite rightly so, by Indianists (including Torkhede, Kanheri and Belhari, dated respectively 813, 674 and 646 CE [EI, III, p. 53; IA, XXV, p. 345; JA, 1863, p. 392]. Therefore, we have eliminated them from our investigation. As for the other documents of this kind, their authenticity has never been questioned by anyone except for Kaye and others who shared his motives.

The evidence was questioned in the hope of proving that Greek mathematicians were the "real" inventors of our numeral system, and that historians had been mistaken in attributing the invention to the Indians. However, as we have already seen, this hypothesis had no historical foundation, it was simply concocted in order to extend the tradition of the "Greek miracle".

The questioning of the authenticity of the Indian charters has never been satisfactorily justified.

The authors of the controversy would have it that these documents were "fabricated" at a later date, when the opportunity presented itself to a group of dishonest people who wished to take possession of the wealth which had long belonged to religious institutions and which the local authorities had confiscated or requisitioned some time before.

This explanation sounds feasible; however, there is no evidence to prove it, and the event was given a totally arbitrary date (some time during the eleventh century).

It was alleged that on the oldest known dated charter (Fig. 24.76), the numerals 3, 4 and 6, which come at the end of the inscription and which denote the **Chhedi* year 346 (594 CE), were added at a later date.

If this were true, then why is the numeral 3 written as three horizontal lines? At the end of the sixth century (which corresponds with the date on the document), this way of writing the number was still used, although it was already becoming obsolete. It had disappeared completely by the next epoch, to be replaced by the non-ideographical sign belonging to the same style as the 4 or 6 which appear in the same document.

Of course, it could be argued that the forger (if there had been a forger) could have studied the palaeography of Indian numerals before imitating the style in question. The date on the legal document which tells of offerings made by someone's ancestor (authentic or not) would have been important to the descendant or person claiming to be so in order for them to prove that they were the rightful owners of the property mentioned on the charter.

But why would someone go to so much trouble, when the date is already given in the text in the form of the names of the numbers in Sanskrit? At that time, this indication was quite acceptable on its own; it was even more reliable than the numerals, whose appearance was susceptible to so many alterations in the hands of scribes and engravers.

What would have been the point of such an addition? And why would the date have been written in keeping with the place-value system when non-positional notation derived from the *Brâhmî* system was still frequently used (at least by the lay person) to write this type of legal document (Fig. 24.70)?

In other words, if the document was forged, why was the place-value system favoured rather than the old non-positional system?

No acceptable answers to these questions have been provided by those who put forward the theory of a forgery. On the other hand, to support his theory, Kaye did not hesitate to cite the charters inscribed on copper containing dates written using the old system and dating back to the era between the end of the sixth century and the ninth century (source: IA, VI, p. 19 ; EI, III, p. 133, etc.).

The most amusing part of this story is that the above dispute only centred on the oldest copper charters containing examples of the use of the place-value system, and not on the numerous other documents of the same nature which were written after or at the same time as the Gwalior inscriptions (876 CE). As for those containing examples of numbers written in the old non-positional Indian numerals whose date oscillates between the sixth and eighth century, their arithmetic was never questioned by Kaye. Thus we can see that these authors had worked out their conclusions far better than their arguments.

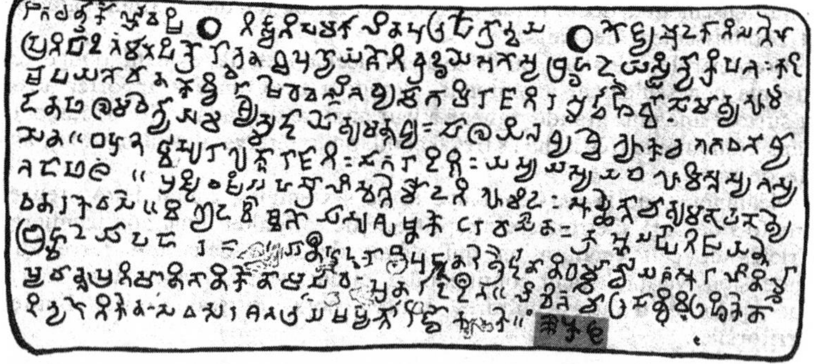

Fig. 24.76. *Donation charter of Dadda III, from Sankheda in Gujarat (region of Bharukachcha). Dated 346 in the *Chhedi calendar (= 346 + 248 = 594 CE), this document is the oldest known formal evidence of the use of the place-value system in India (in the shaded section, the number 346 is expressed according to this system). [Ref: EI, II, p. 19]*

We must be careful, however, because there is no way of ascertaining whether or not any of these copper charters are authentic; it is easy to make forgeries with copper, and we are dealing with a region of the world where counterfeiters, since time immemorial, have been masters at their craft.

The preceding counter-arguments would seem to suggest that these charters could well be authentic.

For the benefit of the doubt, however, we will not use these documents as evidence in our investigation, even though, from a purely graphical point of view, the letters and numerals they contain are indisputably authentic, unless the "forgers" pushed their talents to the limit to make exact copies of the contemporary and regional styles for each of the charters in question.

The fact remains that the history of the Indian decimal place-value system owes much to men such as Kaye. They proved that the subject was a lot more complicated than it seemed at first, and that all the documentation must be scrutinised very closely in establishing the facts. The controversy obliged scholars to go back to square one and apply stricter rules to their analysis of the facts and documents in this very rich and fertile field where they had not always exercised the correct degree of caution. On the other hand, men such as Kaye displayed a certain narrow-mindedness in limiting themselves to the literal frontiers of this civilisation which spread across a geographical area of truly continental dimensions, and which influenced and was witness to the flourishing of many other cultures which were situated beyond the limits of its own territory.

The following demonstrates that there are a great many other (unquestionably authentic) documents, which prove that zero and the place-value system are truly and exclusively Indian inventions, and that their discovery dates back even further than the oldest known inscription on copper.

### *Proof found in epigraphy from Southeast Asia*

The texts that we will consider now are of considerable value to this investigation, for at least two reasons: first, they are all carved in stone, which means that there can be no doubt as to their authenticity; secondly, they are extracts from dated inscriptions, the oldest of which date back into the distant past.

These inscriptions are written either in Sanskrit or in vernacular language, that is to say in the regional language, be it Old Khmer, in Old Malay, in *Cham*, in Old Javanese, etc. Many of them record offerings to temples, their interest being an indication of the date (the year in which the inscription was written) and a detailed description of the offerings.

The way in which the corresponding numbers are expressed gives us the most significant indication of the use of the Indian place-value system.

If we only look at the indigenous inscriptions for the moment (those which are unique to each of the civilisations in question), we can see a very interesting particularity: the commonly-used numbers are not expressed in the same manner as the dates.

For the common numbers (expressing units of length, surface areas or capacities; the number of slaves, objects or animals; the quantity of gifts offered to the divinities and temples, etc.), the engravers usually simply expressed them in the letters of their vernacular language.

However, Cambodia is an exception to this rule; the Khmer engravers often preferred to use their local numeral system, which is immediately identifiable due to its undeniably primitive character (Fig. 24.77). This system uses one, two, three or four vertical lines to represent the first four units, although the fourth is often represented by a sign which gives no ideographical clue to the number it represents. As for the units 5 to 9, these are also represented by independent signs. This system also has a particular sign for 10, 20 and 100. As the system relies on the additive principle to represent numbers below 100, the multiples of 10, from 30 to 90, are expressed by combinations of the numerals for 20 and 10, according to the following rule:

| | |
|---|---|
| 30 = 20 + 10 | Juxtaposition of the signs 20 and 10 |
| 40 = 20 + 1 × 20 | A vertical line is added to the sign for 20 |
| 50 = 40 + 10 | Juxtaposition of the signs 40 and 10 |
| 60 = 20 + 2 × 20 | Two vertical lines are added to 20 |
| 70 = 60 + 10 | Juxtaposition of the signs 60 and 10 |
| 80 = 20 + 3 × 20 | Three vertical lines are added to 20 |
| 90 = 80 + 10 | Juxtaposition of the signs 80 and 10 |

The multiples of 100 are expressed in much the same manner, the numeral 100 being accompanied by the corresponding unit:

| | |
|---|---|
| 200 = 100 + 1 × 100 | A vertical line is added to the sign 100 |
| 300 = 100 + 2 × 100 | Two vertical lines are added to 100 |

The system seems to be limited to numbers below 400: there is no example of a higher number than this; above this quantity the Khmers wrote the names of the numbers in the letters of their language.

Thus, in terms of graphical representation, the ancient Khmer vernacular numeral system derived from the old *Brâhmî* system, as can be seen in the above table.

On the other hand, the structure of the system stems from the counting system of the Old Khmer language, for which we know the base was 20

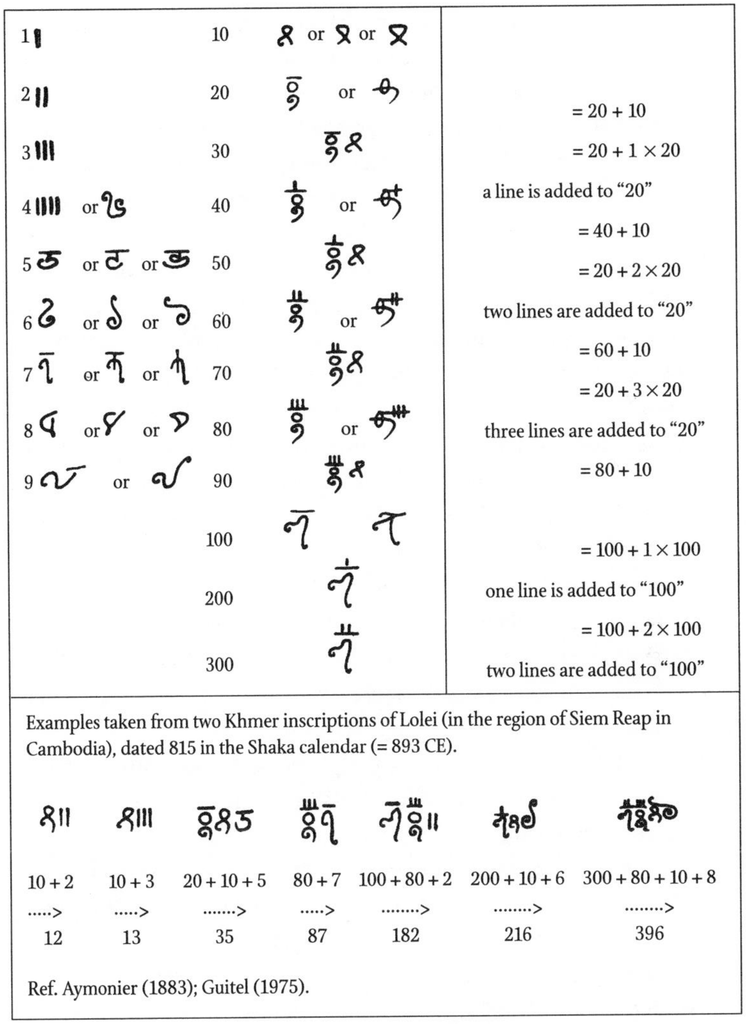

| Units | Tens and hundreds | Signs | Explanation |
|---|---|---|---|
| 1 | 10 | or or | |
| 2 | 20 | or | |
| 3 | 30 | | = 20 + 10 |
| 4 or | 40 | or | = 20 + 1 × 20<br>a line is added to "20" |
| 5 or or | 50 | | = 40 + 10 |
| 6 or or | 60 | or | = 20 + 2 × 20<br>two lines are added to "20" |
| 7 or or | 70 | | = 60 + 10 |
| 8 or or | 80 | or | = 20 + 3 × 20<br>three lines are added to "20" |
| 9 or | 90 | | = 80 + 10 |
| | 100 | | |
| | 200 | | = 100 + 1 × 100<br>one line is added to "100" |
| | 300 | | = 100 + 2 × 100<br>two lines are added to "100" |

Examples taken from two Khmer inscriptions of Lolei (in the region of Siem Reap in Cambodia), dated 815 in the Shaka calendar (= 893 CE).

| 10 + 2 | 10 + 3 | 20 + 10 + 5 | 80 + 7 | 100 + 80 + 2 | 200 + 10 + 6 | 300 + 80 + 10 + 8 |
|---|---|---|---|---|---|---|
| .....> | .....> | .......> | .....> | ........> | ........> | ........> |
| 12 | 13 | 35 | 87 | 182 | 216 | 396 |

Ref. Aymonier (1883); Guitel (1975).

FIG. 24.77. *The written numeration of the ancient Khmers: a system which uses the additive principle and which contains a curious trace of base 20. Used until the thirteenth century CE in vernacular inscriptions of Cambodia to express everyday numbers.*

(which explains the presence of a special sign for 20 and its multiples). As Coedès observes:

> The numeral system was not decimal, and today, despite the fact that Siamese numerals are used to represent multiples of ten above thirty, and likewise for 100, 1,000, etc., it still is not completely decimal: the names of the numbers from six to nine are expressed as five-one, five-

two, five-three, five-four, and special names for four and many of the multiples of twenty are still in common use. In ancient times, the Khmer people used no more than the names for one, two, three, four, five, ten, twenty and some multiples of twenty to express numbers, no matter how high the numbers were, and they used the Sanskrit word **shata* for "hundred", to which the term *slika* was added, which they also used to express the number 400 (= $20^2$).

In other words, the spoken Khmer numeral system constituted a kind of compromise between Indian decimal numeration and a very old and far more primitive indigenous system, based both on 4 and 5 [see BEFEO, XXIV (1924) 3–4, pp. 347–8; JA, CCLXII (1974) 1–2, pp. 176–91].

On the other hand, in order to express dates, the stone-carvers of the diverse civilisations of Southeast Asia never used their vernacular numeral system or wrote the numbers in word form in their own language; as we will see, this fact is of great significance.

They only recorded dates using one of the two following methods: either the names of the numbers in Sanskrit, or, more frequently after a certain date, a decimal numeral system using nine numerals and a zero in the form of a dot or a little circle, strictly adhering to the place-value system (Fig. 24.50 and 24.78 to 80).

There is evidence of this use of the place-value system until the thirteenth century, from the ninth, eighth and even the seventh century CE, depending on the region.

In Champa, it was used consistently, at least since the *Shaka* year 735 (813 CE), which is the date of the oldest known *Cham* inscription of Po Nagar (Fig. 24.80).

In the Indian islands, however, the system appeared much earlier:

- at the end of the eighth century in Java; the oldest vernacular inscription (in *Kawi* writing) to bear witness to the use of the place-value system on this island is from Dinaya, dated the **Shaka* year 682 (760 CE) (Fig. 24.80);
- at the end of the seventh century at Banka; the most ancient vernacular inscription (in Old Malay) which attests its use in this island is that of Kota Kapur, dated 608 Shaka (686 CE) (Fig. 24.80);
- at the end of the seventh century in Sumatra; the oldest vernacular inscriptions (in Old Malay) to bear witness to its use in the region come from Talang Tuwo and Kedukan Bukit in Palembang, dated the respective *Shaka* years 606 and 605 (684 and 683 CE) (Fig. 24.80);
- and also at the end of the seventh century in Cambodia; the oldest vernacular inscription (in Old Khmer) to bear witness to its use in this country is from Trapeang Prei, in the province of Sambor, the respective *Shaka* year 605 (683 CE) (Fig. 24.80).

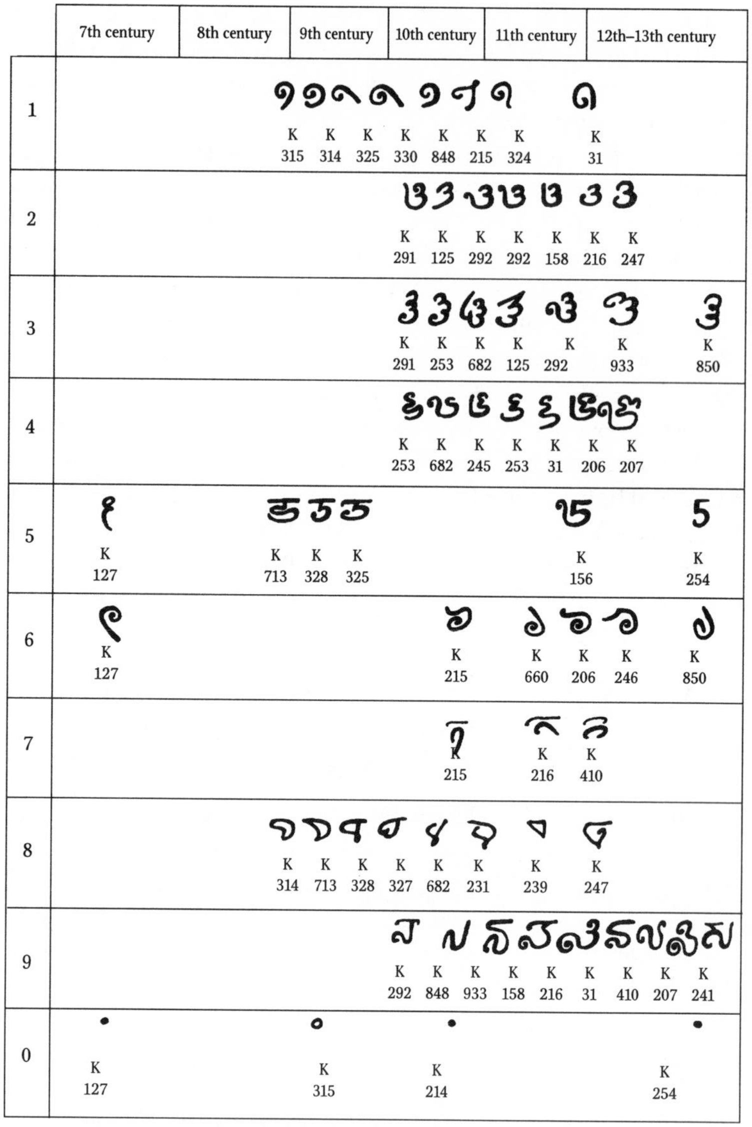

| | 7th century | 8th century | 9th century | 10th century | 11th century | 12th–13th century |
|---|---|---|---|---|---|---|
| 1 | | | K 315, K 314, K 325 | K 330, K 848, K 215 | K 324 | K 31 |
| 2 | | | | K 291, K 125, K 292 | K 292, K 158 | K 216, K 247 |
| 3 | | | | K 291, K 253, K 682 | K 125, K 292 | K 933, K 850 |
| 4 | | | | K 253, K 682, K 245 | K 253, K 31 | K 206, K 207 |
| 5 | K 127 | K 713, K 328 | K 325 | | K 156 | K 254 |
| 6 | K 127 | | | K 215 | K 660, K 206 | K 246, K 850 |
| 7 | | | | K 215 | K 216 | K 410 |
| 8 | | K 314 | K 713, K 328 | K 327, K 682, K 231 | K 239 | K 247 |
| 9 | | | | K 292, K 848, K 933 | K 158, K 216 | K 31, K 410, K 207, K 241 |
| 0 | K 127 | | K 315 | K 214 | | K 254 |

FIG. 24.78. *A selection of palaeographical variants (dated) of numerals of the place-value system which, in vernacular inscriptions of Cambodia written in Old Khmer, were exclusively used to express dates of the Shaka calendar. (For the K references, see IMCC)*

In Cambodia, however, this is not the oldest existing dated vernacular inscription. There is one which dates back even further, the earliest possible inscription to contain a date; it is from Prah Kuha Luhon, dated the *Shaka* year 596 (674 CE), this date being written in letters using the

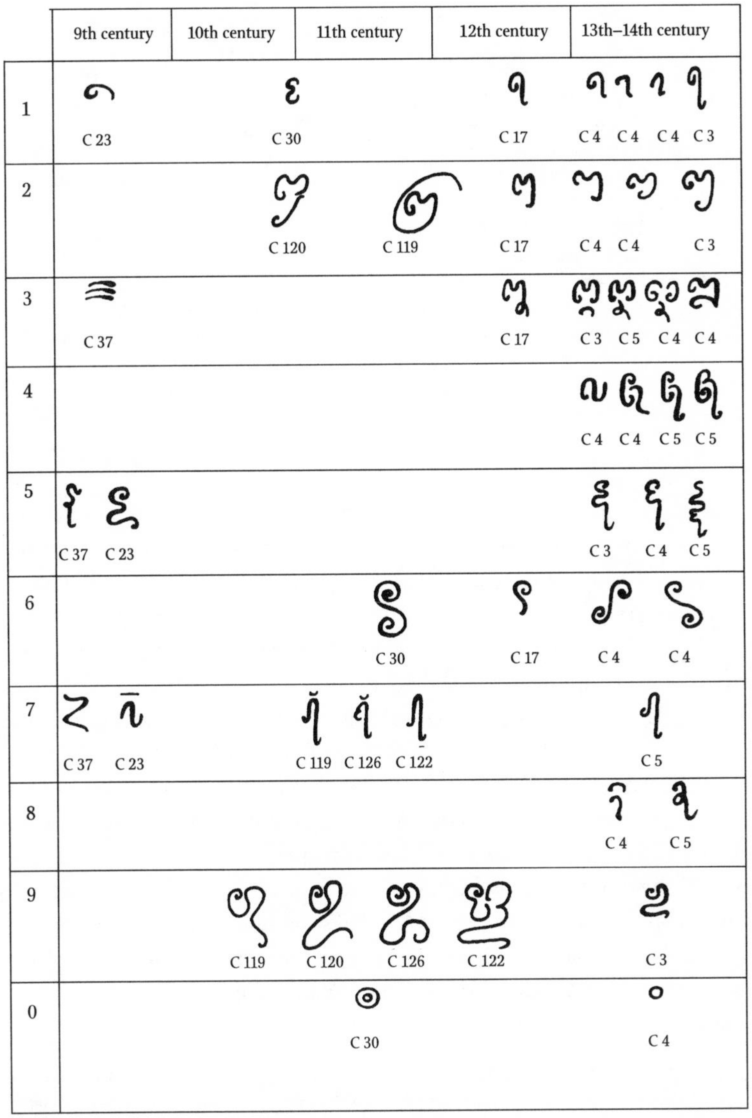

| | 9th century | 10th century | 11th century | 12th century | 13th–14th century |
|---|---|---|---|---|---|
| 1 | C 23 | C 30 | | C 17 | C 4 C 4 C 4 C 3 |
| 2 | | C 120 | C 119 | C 17 | C 4 C 4 C 3 |
| 3 | C 37 | | | C 17 | C 3 C 5 C 4 C 4 |
| 4 | | | | | C 4 C 4 C 5 C 5 |
| 5 | C 37 C 23 | | | | C 3 C 4 C 5 |
| 6 | | | C 30 | C 17 | C 4 C 4 |
| 7 | C 37 C 23 | | C 119 C 126 C 122 | | C 5 |
| 8 | | | | | C 4 C 5 |
| 9 | | C 119 | C 120 C 126 | C 122 | C 3 |
| 0 | | | C 30 | | C 4 |

Fig. 24.79. *A selection of palaeographical variants (dated) of numerals of the place-value system which, in (Cham) vernacular inscriptions of Champa, were exclusively used to express dates of the Shaka calendar. (For the C references, see IMCC)*

| | DOCUMENTS AND SOURCES | HOW THE DATE IS RECORDED IN THE *SHAKA* CALENDAR | |
|---|---|---|---|
| 1084 | Cham inscription of Pô Nagar, northern tower. Dated 1006 in the *Shaka* calendar (= 1006 + 78 = 1084 CE). |  1 0 0 6 | IMCC, C 30<br>BEFEO, XV, 2, p. 48 |
| 1055 | Cham inscription of Lai Cham, region of Hanoi. Dated 977 in the *Shaka* calendar (= 977 + 78 = 1055 CE). |  9 7 7 | IMCC, C 126 BEFEO, XV, 2, pp. 42–3 |
| 1055 | Cham inscription of Phú-qui, province of Phanrang. Dated 977 in the *Shaka* calendar (= 977 + 78 = 1055 CE). | 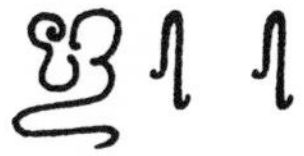 9 7 7 | IMCC, C 122<br>BEFEO, XV, 2, p. 41<br>BEFEO, XII, 8, p. 17 |
| 1050 | Cham inscription of Pô Klaun Garai (first inscription). Dated 972 in the *Shaka* calendar (= 972 + 78 = 1050 CE). | 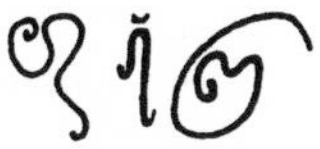 9 7 2 | IMCC, C 120 BEFEO, XV, 2, p. 40 |
| 1050 | Cham inscription of Pô Klaun Garai (second inscription). Dated 972 in the *Shaka* calendar (= 972 + 78 = 1050 CE). | 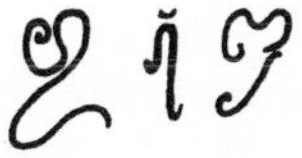 9 7 2 | IMCC, C 120 BEFEO, XV, 2, p. 40 |
| 1007 | Khmer inscription of Phnom Práh Nét Práh (foot of southern tower). Dated 929 in the *Shaka* calendar (= 929 + 78 = 1007 CE). | 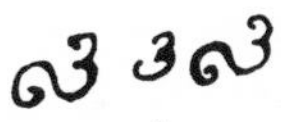 9 2 9 | IMCC, K 216<br>BEFEO, XXXIV, p. 423 |
| 1005 | Khmer inscription of Phnom Práh Nét Práh (foot of southern tower). Dated 927 in the *Shaka* calendar (= 927 + 78 = 1005 CE). |  9 2 7 | IMCC, K 216<br>BEFEO, XXXV, p. 201 |
| 880 | Balinese inscription of Taragal. Dated 802 in the *Shaka* calendar (= 802 + 78 = 880 CE). | 8 0 2 | Damais, p. 148, g. |

Fig. 24.80a.

Sanskrit names for the numbers (see IMCC, K 44, 1. 6; CIC, IV):

*shannavatyuttarapañchashata Shakaparigraha*
"the *Shaka* [year] numbering five hundred and ninety-six"

Thus in the vernacular inscriptions of Southeast Asia, the everyday numbers were always expressed through the names of the numbers in the

| | DOCUMENTS AND SOURCES | HOW THE DATE IS RECORDED IN THE *SHAKA* CALENDAR | |
|---|---|---|---|
| 878 | Balinese inscription of Mamali. Dated 800 in the *Shaka* calendar (= 800 + 78 = 878 CE). | 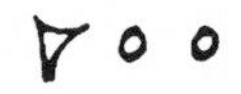 8 0 0 | Damais, p. 148, f. |
| 877 | Balinese inscription of Haliwanghang. Dated 799 in the *Shaka* calendar (= 799 + 78 = 877 CE). |  7 9 9 | Damais, p. 148, f. |
| 829 | Cham inscription of Bakul. Dated 751 in the *Shaka* calendar (= 751 + 78 = 829). | 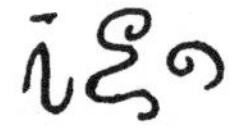 7 5 1 | IMCC, C 23<br>ISCC, p. 238<br>BEFEO, XV, 2, p. 47 |
| 813 | Cham inscription of Pô Nagar, northwest tower. Dated 735 in the *Shaka* calendar (= 735 + 78 = 813 CE). *This is the first Cham inscription to contain the date written in numerals.* |  7 3 5 | IMCC, C 37<br>JA 1891, i, p. 24<br>BEFEO, XV, 2, p. 47 |
| 686 | Malaysian inscription of Kota Kapur (isle of Banka). Dated 608 in the *Shaka* calendar (= 608 + 78 = 686 CE). | 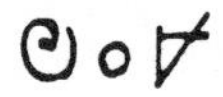 6 0 8 | BEFEO, XXX, pp. 29ff.<br>Kern VIII, p. 207 |
| 684 | Malaysian inscription of Talang Tuwo, Palembang (Sumatra). Dated 606 in the *Shaka* calendar (= 606 + 78 = 684 CE). |  6 0 6 | BEFEO, XXX, pp. 29ff. ACOR, II, p. 19 |
| 683 | Malaysian inscription of Kedukan Bukit, Palembang (Sumatra). Dated 605 in the *Shaka* calendar (= 605 + 78 = 683). *This is the oldest inscription in Old Malay to be dated in numerals.* |  6 0 5 | BEFEO, XXX, pp. 29ff. ACOR, II, p. 13 |

Fig. 24.80b.

| | DOCUMENTS AND SOURCES | HOW THE DATE IS RECORDED IN THE *SAKA* CALENDAR | |
|---|---|---|---|
| 683 | Khmer inscription of Trapeang Prei, province of Sambór. Dated 605 in the *Shaka* calendar (= 605 + 78 = 683 CE). *This is the first Khmer inscription dated in numerals.* |  6 0 5 | IMCC, K 127 CIC, XLVII |

Note: In Java, the oldest Kawi inscription (written in Old Javanese) to be dated in numerals is of Dinaya, which bears the date 682 in the *Shaka* calendar = 682 + 78 = 760 CE. [Ref: Tijdschrift, LVII, (1976), p. 411; LXIV, (1924), p. 227].

Fig. 24.80c.

indigenous language or its very rudimentary numerals. For the dates of the *Shaka* calendar, however, either the Sanskrit word, or, more commonly, the decimal place-value system was used, from no later than the end of the seventh century CE.

As we have seen, the different numerals used throughout Southeast Asia were actually nothing more than palaeographical variations of Indian numerals, which themselves derived from the *Brâhmî* form of the first nine units (Fig. 24.52, 24.53, and 24.61 to 69). The only difference between these diverse systems is their complete transformation into local cursive forms (Khmer, Javanese, *Cham*, Malay, Balinese, etc.), according to the habits of the scribes and engravers of the region.*

On the other hand, as well as the use of Sanskrit (the learned language of Indian civilisation) to record the dates, all the vernacular inscriptions reveal that the dates were written exclusively, for many centuries, according to a system whose Indian origin is indisputable: the *Shaka* era of the Indian astronomers [see R. Billard (1971)].

* It should be noted that the interpretation of the *Cham* numerals posed considerable difficulties because of mistakes made as to their values when they were first deciphered. 7 was mistaken for 1, the (more recent form of) 1 was mistaken for the very ancient form of 5, 7 for 9, etc. This caused even graver errors in the nineteenth century when it came to dating the inscriptions, which in turn led to mistakes in the interpretation of historical events and chronology. Thus it appeared that inscriptions of the same person, referring to the same event, were written at completely different points in time. The inscriptions of King Parameshvaravarman, for example, gave the *Shaka* date 972 (1050) in Sanskrit chronograms, whilst other inscriptions, or the same one, gave the date as the *Shaka* year 788 (866). Another example is an inscription of Mî-s'on (BEFEO, IV, 970, 24) which lists a series of religious foundations set up by King Jaya Indravarman of Gramapura; logically, these dates should have been listed in chronological order; however, when the values that were believed to be correct were applied, the dates emerged in the following incoherent manner: 1095, 1096, 1098, 1097, 1070 and 1072. There were many similar enigmas, which seemed to have no solution and were distorting all the acquired data, until L. Finot (BEFEO, XV, 2, 1915, pp. 39–52) discovered the true origin of the *Cham* numerals and with it the solution which had eluded his predecessors. The inaccurate interpretations of the *Cham* numerals were due to the very unusual variations that they had undergone over the centuries because of the whims and aesthetical preoccupations of the corresponding engravers.

These facts are even more significant because they concern the ancient civilisations of Indo-China and Indonesia (Cambodia, Champa, Java, Bali, Malaysia), which were strongly influenced by India in the early centuries CE, partly because of the widespread nature of Shivaism and Buddhism, and also because of the important intermediate role they played in the trading of spices, silk and ivory between India and China [see G. Coedès (1931), (1964)].

Champa is the ancient kingdom that stretched along the southeast coast of what is now Vietnam with the region of Hue at its centre. The native inhabitants of Champa had become Hindu by religion due to their frequent contact with Indian traders. Champa first became a powerful nation at the beginning of the fifth century CE, under the rule of Bhadravarman, who dedicated the shrine of Mi-so'n (which would remain the religious centre of the kingdom) to Shiva, one of the greatest divinities of the Indian Brahmanic pantheon.

Not far from Champa is Cambodia, which previously belonged to the Hindu kingdom of Fu Nan from the first to the sixth century CE, before becoming the centre of Khmer civilisation, which, conserving the foundations of an Indian culture, flourished until the fourteenth century.

In Java, too, which entered into relations with India at the beginning of the second century CE, and all of ancient Indonesia, early developments owe much to Indian civilisation through the influence of Buddhism and Brahmanism.

Thus we can see the great influence that Indian astronomers and mathematicians once exercised over the various cultures of Southeast Asia. All the preceding facts are highly significant for they show how the Khmer, the Cham, the Malaysian, the Javanese, the Balinese, and other races, were profoundly influenced by Indian culture, and borrowed elements of Indian astronomy, in particular the *Shaka* calendar, and conformed to the corresponding arithmetical rules [see F. G. Faraut (1910)].

In these regions, the appearance of zero and the place-value system coincides directly with the appearance of the dates of the *Shaka* calendar:

> The place-value system was used in Indo-China and Indonesia from the seventh century CE, in other words at least two centuries before Kaye claimed to find evidence of its use in India itself. However, unless zero and the Arabic numerals came from the Far East [which is precisely the opposite of what did happen] evidence of their use in the Indian colonies would suggest that they were in use in India at an even earlier date [G. Coedès (1931)].

Thus we have confirmed the words of the Syrian Severus Sebokt, who wrote, in the seventh century, that the Indian place-value system was already known and held in high esteem beyond the borders of India.

## THE INDIAN PLACE-VALUE SYSTEM: OUTSIDE INFLUENCE OR INDIAN INVENTION?

The place-value system is unquestionably of Indian origin, and its discovery doubtless dates back much further than the seventh century CE. The question we must now ask is whether this concept was inspired by an outside influence or whether it was a purely Indian discovery.

We know that during the course of history the Babylonians, the Chinese, the Maya and, of course, the Indians, succeeded in inventing a place-value system. If the Indians were influenced by any other civilisation, it would have to have been one of the other three we have just mentioned, either directly or through an intermediary.

Putting to one side the Maya civilisation, which apparently had no contact with the ancient world, this leaves us with the Chinese and Babylonian.

### *The possibility of Babylonian influence*

If the Indian place-value system was derived from that of the Babylonians, it might have been through the intermediary of Greek civilisation.

In 326 BCE, Alexander the Great took possession of the land of the Indus and of the ancient province of Gandhara, from the northeast of Afghanistan and the extreme north of present-day Pakistan to the northwest of India, before these regions were governed by the "Indian-Greek" Satraps c. 30 BCE. We know that many elements and methods of Babylonian astronomy were introduced into India shortly before the beginning of the first millennium CE, probably through the northwest of the Indian sub-continent; no doubt this took place in the eastern part of the present-day state of Gujarat, probably in the region of the port of Bharukachcha, which saw the development of both cultural and maritime activities and trade with the West during the first centuries CE. Thus, as R. Billard (1971) stresses, the period between the third century BCE and the first century CE is characterised by the appearance of the *tithi*, a unit of time used in the Babylonian tablets and corresponding to the thirtieth of a synodic revolution of the Moon, more or less the equivalent of a day or nychthemer: elements which are known to have been transmitted to the Greek astronomers by their Mesopotamian colleagues no later than the Hellenistic era.

We know that Babylonian scholars had invented and used a place-value system with 60 as a base since the nineteenth century BCE, and they had used zero since the fourth or third century BCE. As the sexagesimal system was "one of the elements that the Greeks had acquired from Babylonian astronomy, the mother and wet-nurse of their own astronomy" [F. Thureau-Dangin (1929)], it is possible to suppose that the idea of the place-value system arrived in India at the same time as Babylonian astronomy.

Although this hypothesis cannot be ruled out, we can nevertheless raise one serious objection to it. The Greek astronomers only used the Babylonian sexagesimal system to write the negative values of 60, in other words the sexagesimal fractions of the unit, whilst the system was originally developed in order to express whole numbers as well as fractions. Thus, if a similar influence was exercised over the Indians by the Greeks (for if the Indians were influenced by the Babylonian system it could only have been via the Greeks), how could an incomplete system, only used to record fractions, moreover with a base of 60, have influenced the invention of a decimal place-value system which was originally invented to record whole numbers? This is an obvious flaw, which makes this hypothesis appear rather paradoxical.

*The possibility of a Chinese influence*

Therefore, at first glance, a Chinese influence would seem more plausible. We know that since the time of the Han (206 BCE to 220 CE), Chinese scholars used a decimal place-value system known as *suan zi* ("calculation using rods"). A regular system which combined horizontal and vertical lines was used to represent the nine basic units, constituting a written transcription of a concrete counting system which used reeds, standing on one end or placed horizontally on a counting board like abacuses in columns.

Thus one could be forgiven for assuming that following the links established between China and India at the beginning of the first millennium CE, Indian scholars were influenced by Chinese mathematicians to create their own system in an imitation of the Chinese counting method.

However, this hypothesis is contradicted by the fact that zero only appeared in the *suan zi* system relatively late. The Chinese scholars overcame the difficulties this caused by expressing a number such as 1,270,000 either in the characters of their ordinary counting system (a non-positional system which did not require the use of zero) or by placing their rod numerals in a series of squares, the missing units being represented by empty squares:

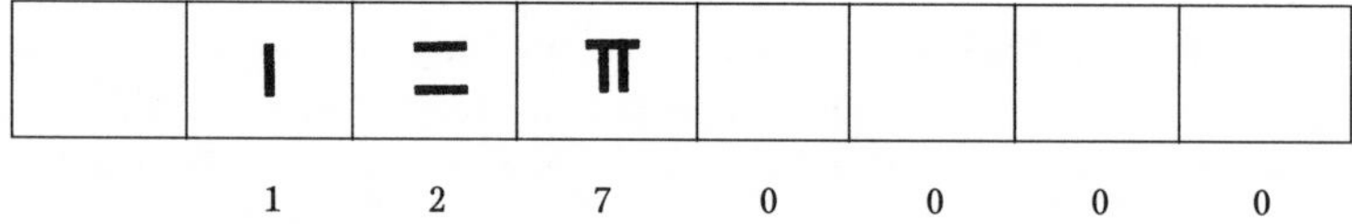

1 2 7 0 0 0 0

It was only after the eighth century CE, and doubtless due to the influence of the Indian Buddhist missionaries, that Chinese mathematicians introduced the use of zero in the form of a little circle or dot (signs that originated in India), thus representing the preceding number in the following manner:

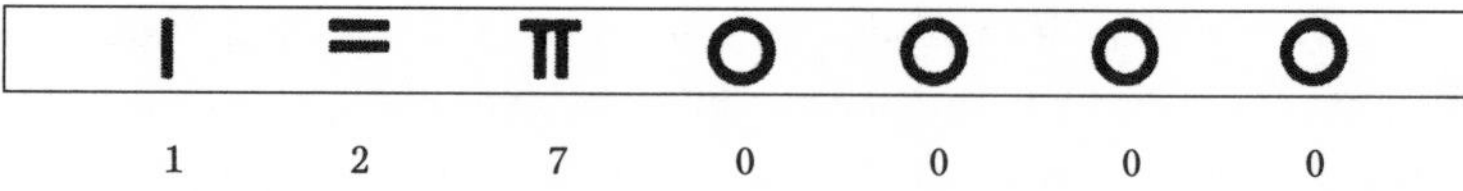

1 2 7 0 0 0 0

> A symbol for zero is mentioned in the *Kai yun zhan jing*, a major work on astronomy and astrology published by *Qutan Xida between 718 and 729 CE. The chapter of this work devoted to the *jui zhi* calendar of 718 CE contains a section on Indian calculating techniques [J. Needham (1959)].

After saying that the (Indian) figures are all written in the cursive form in just one stroke, Qutan Xida continues:

> When one or another of the nine numerals has to be used to express the number ten [literally: "when it reaches ten"], it is then written in a preceding column [before the numerals for the units] (*qian wei*). And each time an empty space appears in a column, a dot is always written [to convey the empty space] (*mei gong wei qu heng an yi dian*).

The author of this book on astronomy was not Chinese: *Qutan Xida was actually an adaptation of the Indian name *Gautama Siddhânta, the famous Indian Buddhist mathematician and astronomer living in China and the head of a school of astronomy at Chang'an since approximately 708 CE. According to L. Frédéric (1987), he was the one to introduce the notion of zero in China as well as the division of a circle into sixty sections.

This remarkable account confirms the influence, which has already been proved, of the rapid expansion of the Buddhist movement which accompanied the propagation of Indian science in the Far East. It also adds an important piece of evidence to our investigation of the origin of our modern numeral system:

> Living in China, doubtless knowing all the subtleties of the Chinese language, Qutan Xida insists not only on the fact that Indian numerals were written in a cursive form, but also that each one was written in just one stroke [G. Guitel, (1975)].

In the Chinese place-value system (the "learned" system), the units were written by juxtaposing or superposing one or more vertical or horizontal lines:

1 2 3 4 5 6 7 8 9

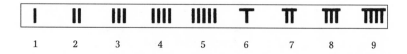

The Chinese numerals in common usage are formed by lines, in various positions, and written in a strict order, lifting the writing tool several times, the symbol for 2 being formed by two lines, 4 by six lines, 6 by four lines, and so on (Figure 21. 1 above). This will become clearer later, when we see the succession of lines that forms the Chinese numeral for 100 (see Fig. 21. 8 above):

> This could only have surprised the learned men amongst which Gautama Siddhânta (alias Qutan Xida) lived, because Chinese words are grouped according to the number of lines their drawing requires; for each character, one is taught the order in which the successive lines must be drawn [G. Guitel (1975)].

The nine numerals (of Indian origin) that we use today, on the other hand, are drawn in just one stroke of a pen or pencil. This is one of the characteristics of our numeral system, whose remarkable simplicity we forget because we have been using it all our lives.

This evidence proves that at the beginning of the eighth century, zero and the place-value system had spread as far as China. At the same time, it almost completely rules out any possibility of a Chinese influence over the development of our present-day numerals.

## THE AUTONOMY OF THE INDIAN DISCOVERIES

Thus it would seem highly probable under the circumstances that the discovery of zero and the place-value system were inventions unique to Indian civilisation. As the *Brâhmî* notation of the first nine whole numbers (incontestably the graphical origin of our present-day numerals and of all the decimal numeral systems in use in India, Southeast and Central Asia and the Near East) was autochthonous and free of any outside influence, there can be no doubt that our decimal place-value system was born in India and was the product of Indian civilisation alone.

## THE NUMERICAL SYMBOLS OF THE INDIAN ASTRONOMERS

We are now going to look at a truly remarkable method of expressing numbers which is frequently found on mathematical and astronomical texts written in Sanskrit; there is no doubt that these texts are of Indian origin.

It is to curious to note that historians of science have not always accorded it the importance it deserves. It constitutes the main piece of evidence of our investigation: added to all the other evidence, it allows us not only to prove beyond doubt that our present-day numeration is of Indian origin, and Indian alone, but also and above all to date the discovery even earlier than the seventh century CE. Moreover, it is even more significant when we consider that the nature of this system is unique in the history of numerals.

By way of introduction, here is a passage from the first modern Indian historian, the Persian astronomer al-Biruni, who wrote the following c. 1010, in his famous work on India [see al-Biruni (1910); F. Woepcke (1863), pp. 283–90]:

> When [the Indian scholars] needed to express a number composed of many orders of units in their astronomical tables, they used certain words for each number composed of one or two orders. For each number, however, they used a certain number of words, so that, if it was difficult to place one word in a certain place, they could choose another from "amongst its sisters" [amongst those which denoted the same number]. Brahmagupta said: If you want to write one, express it through a word which denotes something unique, like the Earth or the Moon; likewise, you can express two with any words which come in pairs, like black and white [this is probably an allusion to the "half black" and "half white" of the month, which corresponds to a division used by the Indians], three by things that come in threes, zero with the names for the sky, and twelve by the names of the sun . . . [Such is the way the system works] as I have understood it. It is an important element in the analysis of their [the Indians'] astronomical tables . . .

Instead of the word **eka*, which means "one", the Indian astronomers used names such as **âdi* (the "beginning"), **tanu* ("the body"), or **pitâmaha* ("the Ancestor", which alludes to *Brahma, considered to be the creator of the universe).

Instead of **dvi*, which means "two", they used all the words which express ideas, things or people which come in pairs: **Ashvin* ("the twin gods"), **Yama* ("the Primordial Couple"), **netra* ("the eyes"), **bâhu* ("the arms"), **paksha* ("the wings"), etc.

In other words, rather than using the ordinary Sanskrit names for the numbers 1 to 9 (**eka, *dvi, *tri, *chatur, *pañcha, *shat, *sapta, *ashta, *nava*), the Indian scholars expressed them by names which had symbolical value. For each number, there was a wide choice of words, whose literal translation evoked the numerical value they denoted in the reader's mind.

It is difficult to give an exhaustive list of these diverse symbolic words, there being an abundant, if not infinite, number of synonyms. However, the reader will get some idea of the variety of these words from the following examples:

ONE

| | |
|---|---|
| *eka*: | Ordinary name for the number 1 |
| *pitâmaha*: | First father |
| *âdi*: | Beginning |
| *tanu*: | Body |
| *kshiti, go . . .*: | Words meaning "Earth" |
| *abja, indu, soma . . .*: | Words meaning "Moon" |

TWO

| | |
|---|---|
| *dvi*: | Ordinary name for the number 2 |
| *Ashvin*: | Horsemen |
| *Yama*: | Primordial Couple |
| *yamala, yugala . . .*: | Words meaning twins or couples |
| *netra*: | Eyes |
| *bâhu:* | Arms |
| *gulphau*: | Ankles |
| *paksha*: | Wings |

THREE

| | |
|---|---|
| *tri:* | Ordinary name for the number 3 |
| *guna*: | Primordial properties |
| *loka*: | [Three] worlds |
| *kâla*: | Time |
| *agni, vahni . . .*: | Fire |
| *Haranetra*: | "Eyes of Hara" |

FOUR

| | |
|---|---|
| *chatur*: | Ordinary name for the number 4 |
| *dish*: | The [four] cardinal points |
| *abdhi, sindhu . . .*: | The [four] oceans |
| *yuga*: | The [four] cosmic cycles |
| *iryâ*: | The positions [of the human body] |
| *Haribâhu*: | The arms of Vishnu |
| *brahmâsya*: | The faces of Brahma |

FIVE

| | |
|---|---|
| *pañcha*: | Ordinary name for the number 5 |
| *bâna, ishu . . .*: | Arrows |
| *indriya*: | The [five] senses |
| *rudrâsya*: | The [five] faces of Rudra |
| *bhûta*: | The elements |
| *mahâyajña*: | The sacrifices |

SIX

| | |
|---|---|
| *shat:* | Ordinary name for the number 6 |
| *rasa*: | The senses |
| *anga*: | The [six] limbs [of the human body] |
| *shanmukha*: | The [six] faces of Kumara |

SEVEN

| | |
|---|---|
| *sapta*: | Ordinary name for the number 7 |
| *ashva*: | Horses |
| *naga*: | Mountains |
| *rishi*: | The [seven] sages |
| *svara*: | The vowels |
| *sâgara*: | The [seven] oceans |
| *dvîpa*: | The island-continents |

EIGHT

| | |
|---|---|
| *ashta*: | Ordinary name for the number 8 |
| *gaja*: | The [eight] elephants |
| *nâga*: | Word meaning "serpent" |
| *mûrti*: | Forms |

NINE

| | |
|---|---|
| *nava*: | Ordinary name for the number 9 |
| *anka*: | Numerals |
| *graha*: | Planets |
| *chhidra*: | The orifices [of the human body] |

ZERO

| | |
|---|---|
| *shûnya*: | Ordinary name for 0 |
| *bindu*: | The point or dot |
| *kha, gagana . . .*: | Words meaning "sky" |
| *âkâsha*: | Ether |
| *ambara, vyoman . . .*: | Atmosphere |

The Sanskrit language, which is very learned and rich, lends itself admirably to this system, as it does to poetry and the Indian way of thinking.

These symbols are all taken from nature, human morphology, animal or plant representations, everyday life, legends, traditions, religions, attributes of the divinities of the Vedic, Brahman, Hindu, Jaina or Buddhist pantheons, as well as from the associations of traditional or mythological ideas or from diverse social conventions of Indian civilisation.

With this unique system of numerical notation, we have now entered into the world of symbols of Indian civilisation.

To give the reader a better understanding of the characteristic way of thinking of Indian philosophers, astrologers, cosmographers, astronomers and mathematicians, (the true "inventors" of our present-day counting system), we have included the "Dictionary of Numerical Symbols of Indian Civilisation" at the end of this chapter, the necessity and usefulness of which will become clear in the course of the following pages. *

## THE PLACE-VALUE SYSTEM OF THE INDIAN NUMERICAL SYMBOLS

To give us some idea of the principle this system was based on, here is a literal translation of a Sanskrit verse taken from a work on astronomy entitled *Sûrya Siddhânta* (or "Astronomical canon of the Sun"; [see Anon. (1955), I, 33; Burgess and Whitney (1860)]:

> *Chandrochchasyâgnishûnyâshvivasusarpârnavâ yuge*
> *Vâmam pâtasya vasvagniyamâshvishikhidasrakâh*
> "The apsids of the moon in a yuga
> Fire. Vacuum. Horsemen. *Vasu*. Serpent. Ocean,
> and of its waning node
> *Vasu*. Fire. Primordial Couple. Horsemen. Fire. Twins"

This verse is incomprehensible to a reader who does not know that the words "Fire. Void. Horseman. Vasu. Serpent. Ocean" (*âgnishûnyâshvivasusarpârnavâ*) and "Vasu. Fire. Primordial Couple. Horseman. Fire. Horseman" (*vasvagniyamâshvishikhidasra*), in the minds of the Indian astronomers, represented the numbers 488,203 and 232,238 respectively.

Here is a comprehensible translation of the verse:

"[The number of revolutions] of the apsids of the moon in a yuga [is]: 488,203, and [of] its waning node: 232,238."

* For each of the word-symbols in question (**Ashvin, *Graha, *Kha*, etc.), the reader might find it interesting to consult the corresponding rubric, where the symbol is denoted by [S], then defined in terms of its numerical value and its literal meaning in Sanskrit, before, as far as possible, its implied symbolism is explained. To find the list of Sanskrit word-symbols used (in their abundant synonymy) for a given number, one only need consult the corresponding English word in the Dictionary (**One, *Two, *Zero*, etc.).

Thus the author of this text expressed, in his own way, two pieces of astronomical numerical data, concerning a **yuga* or "cosmic cycle" (in this case a cosmic cycle named **Mahâyuga* and corresponding to a period of time of 4,320,000 years).

The key to the system lies in knowing that, in a number-system which has 10 as its base, the first nine whole numbers, 10 and each multiple of 10 have a specific name; thus one expresses a given number by placing the name for "ten" between that of the units of the first order and that of the units of the second order, then the name for "hundred" between those of the second and third orders, and so on, respecting a previously agreed method of reading.

The number 8,237, for example, might be expressed in the following manner: "eight thousand, two hundred, three times ten and seven", according to this mathematical breakdown of the components:

$$8 \times 10^3 + 2 \times 10^2 + 3 \times 10 + 7 = 8{,}237.$$

As well as writing the number in terms of decreasing powers of ten, it can also be written in the opposite order, in increasing powers of ten, starting with the smallest unit, for example:

"Seven, three times ten, two hundred, eight thousand".

This is exactly how the Indian astronomers expressed numbers when they used the Sanskrit names of the numbers. Thus the preceding number can be mathematically broken down in the following way:

$$7 + 3 \times 10 + 2 \times 10^2 + 8 \times 10^3 = 8{,}237.$$

The method of expressing numbers that we are interested in here is the "oral" method, because it uses Sanskrit *words*, the difference being that it simply gives a succession of the corresponding names of the units, in keeping with the method of representation that we have just seen. In other words, there is no mention of the names which indicate the base and its various powers ("ten", "hundred", "thousand", etc.) Thus the preceding number would be expressed in the following manner:

Seven. Three. Two. Eight.

In the same way, two.eight.nine.three corresponds to the value:

$$2 + 8 \times 10 + 9 \times 10^2 + 3 \times 10^3 = 3{,}982.$$

In other words, the Sanskrit names for the numbers 1 to 10 had a varying value according to their position in the description of numbers of several orders of units. In saying *one, three, nine* for 931 for example, the word one is given the simple value of *one* unit, *three* is given the power of ten and *nine* the value of a multiple of one hundred.

Thus there can be no doubt that we are dealing with a decimal place-value system. This seems even more remarkable when we consider that the Indian scholars were the only ones to invent a system of this kind.

The above example, however, poses a fundamental question. We have just seen that in this system, a number such as 931 can be expressed relatively easily, by writing *one, three, nine*. On the other hand, it is difficult to express a number such as 901, where there is an empty space, if you like, in the decimal order (the "ten" column). To write this number, one could obviously not simply write *one, nine*, because this would convey the number 91 ($= 1 + 9 \times 10$), and not 901. How, then, do we communicate that there is nothing in the decimal order?

In other words, when one rigorously applies the place-value system to the nine simple units, the use of a special terminology is indispensable to indicate the absence of units in a certain order.

The Indian astronomers overcame this obstacle by using the Sanskrit word **shûnya* meaning "void" and by extension "zero". Thus they were able to express the number 901 in words which can be translated in the following way:

One. Zero. Nine ($= 1 + 0 \times 10 + 9 \times 10^2 = 901$).

The word *shûnya* ("zero") actually became the concept it signified; it played the role of zero in the place-value system, and thus prevented any confusion as to the value of the number expressed.

If we return to the verse quoted above, the Sanskrit numerical expression *agnishûnyâshvivasusarpârnava* (which represents the number 488,203) can be broken down as follows:

*agni.shûnya.ashvi.vasu.sarpa.arnava*

The words which act as components of this expression, however, are not the ordinary Sanskrit names of numbers. They are word-symbols, the literal translation of which, due to the association of ideas which characterises the Indian way of thinking, evoked a numerical value, rather like the way that the words pair and triad evoke the numbers two and three in our minds, except that the Sanskrit language had a greater choice of synonyms. Indian astronomers nearly always chose to express their numerical data using this almost infinite synonymy.

In order to represent the above number, the word-symbols appeared with the value indicated below:

| | | |
|---|---|---|
| **agni* | = "fire" | = 3 |
| **shûnya* | = "void" | = 0 |
| *ashvi* ( = **Ashvin*) | = "horsemen" | = 2 |
| **vasu* | | = 8 |
| **sarpa* | = "serpent" | = 8 |
| **arnava* | = "ocean" | = 4 |

Thus one can translate the above expression in the following manner:

| Fire. | Void. | Horsemen. | *Vasu*. | Serpent. | Ocean. |
|---|---|---|---|---|---|
| 3 | 0 | 2 | 8 | 8 | 4 |

Remembering the earlier explanation of the system, we can see that the number represented is:

$$3 + 0 \times 10 + 2 \times 10^2 + 8 \times 10^3 + 8 \times 10^4 + 4 \times 10^5 = 488{,}203.$$

The second numerical expression that appears in the verse is *vasvagniyamâshvishikhidasra*, which can also be broken down in the following way:

*vasv.agni.yama.ashvi.shikhi.dasra*

These are also word-symbols possessing the following numerical values:

| | | |
|---|---|---|
| *Vasv* ( = **Vasu*) | | = 8 |
| **agni* | = "fire" | = 3 |
| **yama* | = "Primordial Couple" | = 2 |
| *ashvi* (= **Ashvin*) | = "Horsemen" | = 2 |
| *shiki* ( = **Shikhin*) | = "fire" | = 3 |
| **dasra* | = "(one of the) Twins" | = 2 |

Which is interpreted as:

| *Vasu*. | Fire. | Primordial Couple. | Horsemen. | Fire. | Twins. |
|---|---|---|---|---|---|
| 8 | 3 | 2 | 2 | 3 | 2 |

This corresponds to the number:

$$8 + 3 \times 10 + 2 \times 10^2 + 2 \times 10^3 + 3 \times 10^4 + 2 \times 10^5 = 232{,}238.$$

This method of expressing numbers shows a perfect understanding of zero and the place-value system using 10 as a base.

It is a type of symbolic representation subject to many variations, yet the numerical symbols were always perfectly comprehensible to the Indian astronomers. Even if the value of certain words could vary according to the author, region or the time when they were written, the context always confirmed the intended numerical value.

### *Dating the Indian numerical word-symbols*

When were these word-symbols first used? The answer is highly significant because the concept of zero and the place-value system in India are at least as old as this method of expressing numbers.

### *Dates on Sanskrit inscriptions from Southeast Asia*

In India itself, as well as outside India, many documents exist which prove that this method of counting was, for a great many years, the privileged system of the Indian scholars, from the end of the sixth century at least until a relatively recent date.

The dated Sanskrit inscriptions of Southeast Asia figure very prominently amongst these documents.

It is important to make a clear distinction between the vernacular inscriptions and those written in Sanskrit. Both, however, date back to the *Shaka* era of the Indian astronomers. Primarily, in both types of inscriptions, the dates were recorded in words using the Sanskrit names for the numbers.

In the vernacular inscriptions (according to the region, written in Old Khmer, Old Javanese, Cham, etc.), these dates were then expressed using the nine numerals and zero of the Indian place-value system (Fig. 24.80).

In the Sanskrit inscriptions, however, the dates were recorded exclusively in the Indian word-symbols observing the place-value system and using 10 as the base. Here are some examples, taken from the oldest documents found in each of the regions in question.

The oldest dated Sanskrit inscription from Java is the *Stela* of Changal, the *Shaka* date of which is expressed in the following way [see H. Kern, VII, 118]:

*shrutîndriyarasair*

This can be broken down into separate words:

*shruti.indriya.rasair*

On consulting the Dictionary, under the headings **shruti*, **indriya* and **rasa* ( = *rasair*), the following meanings are obtained:

*shruti = Veda = 4
*indriya = properties = 5
*rasair = senses = 6

Bearing in mind that the numbers are always written according to the decimal place-value system, beginning with the smallest unit, in ascending powers of ten (which the Indian astronomers called **ankânâm vâmato gatih*, or the principle of the "movement of the numbers [the numerical symbols] from the right to the left"), we can see that the date in question can be interpreted as:

Veda.Properties.Senses.
4 5 6

This corresponds to the number:

$$4 + 5 \times 10 + 6 \times 10^2 = 654.$$

Thus the inscription in question dates back to the *Shaka* year 654 + 78 (732 CE).

The oldest dated inscription from Champa is the *Stela* of Mî-so'n, the Shaka date of which is written in the following numerical symbols [see G. Coedès and H. Parmentier (1923), C 74 B; BEFEO, XI, p. 266):

*ânandâmvarashatshata*

which can be translated as follows (bearing in mind that *ânanda* means the "(nine) Nanda"; *amvara* = **ambara* = "space" = 0; and *shatshata* = "six hundreds"):

*Nanda*.Space.Six.Hundred.
9 0 6 × 100

which corresponds to the date: $9 + 0 \times 10 + 6 \times 100 = 609 + 78$ *Shaka* (687 CE).

The use of the term *shatshata* to denote six hundred shows a certain inexperience in the writing of numerical symbols, because the number 609 can be written *ânandâmvarashat*, which places the symbols for nine (*ânanda*), zero (*amvara*) and six (*shat*) in order.

The oldest dated Sanskrit inscription from Cambodia is that of Prasat Roban Romas, in the province of Kompon Thom. This is also the oldest dated Sanskrit inscription in the whole of Southeast Asia. It contains the following *Shaka* date [see Coedès and Parmentier (1923), K 151; BEFEO, XLIII, 5, p. 6]:

*khadvishara*

Here is the literal translation (where **kha* = "space" = 0; **dvi* = "two" = 2; and **shara* = arrows = 5):

Space.Two.Arrows.
0 2 5

which corresponds to the date: $0 + 2 \times 10 + 5 \times 10^2 = 520 + 78$ *Shaka* (598 CE).

This proves that the use of Sanskrit word-symbols to express numbers was already widespread in Indo-China and Indonesia at the end of the sixth century CE.

As the civilisations were greatly influenced by Indian astronomers and astrologers, we can quite rightly presume that Indian scholars were using this technique at an even earlier date.

### *Evidence from the astronomers and mathematicians of India*

There is a great deal of evidence pointing to the fact that the system was used by Indian scholars from the sixth century CE until a relatively recent date, as the following (non-exhaustive) list of Indian texts (containing many examples of the word-symbols) shows. The list is written in reverse chronological order (after R. Billard, 1971):

1. *Trishatikâ* by Shrîdharâchârya (date unknown) [B. Datta and A. N. Singh (1938) p. 59]
2. *Karanapaddhati* by Putumanasomayâjin (eighteenth century CE) [K. S. Sastri (1937)]
3. *Siddhântatattvaviveka* by Kamâlakara (seventeenth century CE) [S. Dvivedi (1935)]
4. *Siddhântadarpana* by Nîlakanthaso-mayâjin (1500 CE) [K. V. Sarma (undated)]
5. *Drigganita* by Parameshvara (1431 CE) [ Sarma (1963)]
6. *Vâkyapañchâdhyâyi* (Anon., fourteenth century CE) [Sarma and Sastri (1962)]
7. *Siddhântashiromani* by Bhâskarâchârya (1150 CE) [B. D. Sastri (1929)]
8. *Râjamrigânka* by Bhoja (1042 CE) [Billard (1971), p. 10]
9. *Siddhântashekhara* by Shrîpati (1039 CE) [Billard (1971), p. 10]
10. *Shishyadhîvrddhidatantra* by Lalla (tenth century CE) [Billard (1971), p. 10]
11. *Laghubhâskarîyavivarana* by Shankaranârâyana (869 CE) [Billard (1971), p. 8]
12. *Ganitasârasamgraha* by Mahâvîrâchâryâ (850 CE) [M. Rangacarya (1912)]
13. *Grahachâranibandhana* by Haridatta (c. 850 CE) [Sarma (1954)]
14. *Bhâskarîyabhâsya* by Govindasvâmin (c. 830 CE) [Billard (1971), p. 8.]
15. Commentary on the *Âryabhatîya* by Bhâskara (629 CE) [K. S. Shukla and K. V. Sarma (1976)]
16. *Brahmasphutasiddhânta* by Brahmagupta (628 CE) [S. Dvivedi (1902)]
17. *Pañchasiddhântikâ* by Varâhamihîra (575 CE) [O. Neugebauer and D. Pingree (1970)]

*Examples taken from the work of Bhâskara I*

We will now look at some examples in their original form, taken from some of the oldest texts, which give a clearer indication than the above table of the earliest uses of this system in India.

The first concerns an example of how the number of years (4,320,000) that make up a **chaturyuga* (see also **yuga*) was expressed in word-symbols. It is an extract from the commentary which Bhâskara I wrote in 629 CE on the **Âryabhatîya* [see Shukla and Sarma (1976) p. 197]:

*viyadambarâkâshashûnyayamarâmaveda*

This can be broken down in the following manner:

*viyad. ambara. âkâsha. shûnya. yama. râma. veda*

On consulting the Dictionary, the following meanings are obtained:

| | | |
|---|---|---|
| **viyat* (here written *viyad*) | = "sky" | = 0 |
| **ambara* | = "atmosphere" | = 0 |
| **âkâsha* | = "ether" | = 0 |
| **shûnya* | = "void" | = 0 |
| **yama* | = "(the) Primordial Couple" | = 2 |
| **râma* | = "(the) Râma" | = 3 |
| **veda* | = "(the) Veda" | = 4 |

This gives the following translation, with the corresponding mathematical breakdown:

Sky.Atmosphere.Ether.Void.Primordial Couple.Râma.Veda.
0 0 0 0 2 3 4

$$= 0 + 0 \times 10 + 0 \times 10^2 + 0 \times 10^3 + 2 \times 10^4 + 3 \times 10^5 + 4 \times 10^6 = 4{,}320{,}000.$$

Here are three lines from the same work by Bhâskara (Commentary on the *Âryabhatîya*, manuscript R 14850 of the Government Oriental Manuscript Library, Madras, *Dashagîtika*, [see R. Billard (1971), pp.105–6], in Sanskrit, with the corresponding translation (the numerical word-symbols are underlined to distinguish them from the rest of the text):

*tadânayanam idânîm kalpâder adyanirodhâd ayam abdarashir itîritah*
*khâgnyadrirâmârkarasavasurandhrendavah*
*te chânkair api 1986123730.*

Before we look at the translation, it should be noted that the above word-symbols can be broken down in the following way:

*kha.agny.adri.râma.arka.rasa.vasu.randhra.indavah*

The Dictionary gives the following meanings for these words:

| | | |
|---|---|---|
| **kha* | = "space" | = 0 |
| *agny* (= **agni*) | = "fire" | = 3 |
| **adri* | = "mountains" | = 7 |
| **râma* | = "(the) Râma" | = 3 |
| **arka* | = "sun" | = 12 |
| **rasa* | = "senses" | = 6 |
| **vasu* | | = 8 |
| **randhra* | = "orifices" | = 9 |
| *indavah* (= **indu*) | = "moon" | = 1 |

Thus the following translation is obtained for the preceding extract from the Sanskrit text:

"In order to carry out the translation, here are the number of years which have transpired since the beginning of the [current] **kalpa* until the present day:

Space.Fire.Mountain.*Râma*.Sun.Sense.*Vasu*.Orifice.Moon.

"In figures this reads (*te chânkair api*): 1986123730".

As with the above example, here is the meaning of the word-symbols:

Space.Fire.Mountain.*Râma*.Sun.Sense.*Vasu*.Orifice.Moon.
0 3 7 3 12 6 8 9 1

This corresponds to the following number:

$$0 + 3 \times 10 + 7 \times 10^2 + 3 \times 10^3 + 12 \times 10^4 + 6 \times 10^6 + 8 \times 10^7 + 9 \times 10^8 + 1 \times 10^9 = 1{,}986{,}123{,}730.$$

One might be surprised to find, in a place-value system, word-symbols denoting values higher than or equal to ten, such as the word **arka* (= "sun" = 12) which is used here to express a number which contains two orders of units. Later, however, we will see why this symbol is used here, which does not constitute an exception to the rule of position where 10 is the base. If in this example, the word *arka*, on its own, expresses the number 12, it only acquires the value of 120,000 (= $12 \times 10^4$) because of the place it occupies in the above expression.

Moreover, the value (1,986,123,730) of the preceding word-symbols is clearly indicated "in figures" according to the place-value system, in the third line of the Sanskrit text, accompanied by the words "in figures this reads . . .", evidently in order to prevent any ambiguity as to the intended value. Thus we have a bilingual text of sorts which reinforces the above explanations.

This is not the only instance where Bhâskara felt the need to give the number in its corresponding numerals (using the place-value system of nine units and zero) as well as in astronomical word-symbols. Here is another example, this time involving a much higher number than the previous one [see Shukla and Sarma (1976), pp. 155]:

*shûnyâmbarodadhiviyadagniyamâkâshasharasharâdri-*
*shûnyendurasâmbarângânkâdrishvarendu*
*ankair api 1779606107550230400.*

As in the previous example, this compound word can be literally translated in the following way (given that: **shûnya* = 0, **ambara* = 0, **udadhi* [= *dadhi*] = 4, *viyad* (= **vyant*) = 0, **agni* = 3, **yama* = 2, **âkâsha* = 0, **shara* = 5, **shara* = 5, **adri* = 7, **shûnya* = 0, **indu* = 1, **rasa* = 6, **ambara* = 0, **anga* = 6, **anka* = 9, **adri* = 7, **Ashva* = 7 and **indu* = 1), where the following two consecutive expressions constitute two ways of writing the same number according to the same principle:

Void.Sky.Ocean.Sky.Fire.Couple.Space.Arrow.Arrow.Mountain.
0 0 4 0 3 2 0 5 5 7 →

Void.Moon.Sense.Atmosphere.Limb.Numeral.Mount.Horse.Moon
0 1 6 0 6 9 7 7 1

"In figures this reads: 1,779,606,107,550,230,400."

The number expressed in word symbols is the one expressed "in figures"; according to the text itself:

1,779,606,107,550,230,400.

Here Bhâskara uses the Sanskrit word *anka*, the "numerals", not only to indicate the equivalent of the number concerned in the place-value system using nine numerals (*ankair api*, "in figures this reads . . ."), but also to designate the number 9. This is of great importance, because the basic meaning of *anka* is "a mark" or "a sign", which by extension can mean "numeral", although there is no connection between its other meanings and the number 9. Therefore, Bhâskara's use of *anka* to represent the number 9 proves that nine numerals and the place-value system were already being used to write numbers in India when the commentary was written.

Bhâskara gives the number "in figures" as well as in word-symbols, and this leaves no doubt that he was alluding to the nine basic numerals of the decimal place-value system which was invented in India: which, along with zero, enabled the Indian astronomers not only to represent any number, however high it might have been, but also and above all to carry out any mathematical operation with the minimum of complication.

Thus, in 629, the methods of expressing numbers either in numerals or in word-symbols were widely recognised by the learned men of India.

*Examples found in the work of Varâhamihîra*

Here are some more examples from the *Pañchasiddhântikâ*, the astronomical work of Varâhamihîra (VIII, lines 2,4 and 5). [See S. Dvivedi and G. Thibaut; O. Neugebauer and D. Pingree] [Personal communication of Billard]:

1) How the number 110 is expressed:
*shûnyaikaika* = **shûnya.*eka.eka*
= void. one.one
0 1 1
$= 0 + 1 \times 10 + 1 \times 10^2 = 110.$

2) How the number 150 is expressed:
*khatithi* = **kha*.**tithi*
= space.day
0 15
$= 0 + 15 \times 10 = 150$.

3) How the number 38,100 is expressed:
*khakharûpâshtaguna* = **kha*. **kha*.**rûpa*.**ashta*.**guna*
= space.space.shape.eight.quality
= 0 0 1 8 3
$= 0 + 0 \times 10 + 1 \times 10^2 + 8 \times 10^3 + 3 \times 10^4 =$
38,100.

This astronomical text was written c. 575 CE. This proves that zero and the place-value system were already in use in India in the second half of the sixth century CE.

## THE EARLIEST KNOWN EVIDENCE OF THE INDIAN PLACE-VALUE SYSTEM

We will now look at the most important source of evidence relative to the history of the place-value system: the **Lokavibhâga* (or *The Parts of the Universe*), a work on *Jaina cosmology which constitutes the oldest known use of word-symbols.

Besides the fact that the "minus one" is expressed by *rûponaka* (literally: "diminished form", *rûpo* = **rûpa* = "shape" or "form" = 1) and that the concept of zero is expressed by **shûnya* (void) or by words such as **kha*, **gagana* or **ambara* ("sky", "atmosphere", "space", etc.), we find the following expression used for the number 14,236,713 [source: Anon. (1962), Chapter III, line 69, p. 70] [Personal communication of Billard]:

*trîny ekam sapta shat trîni dve chatvâry ekakam*

As the words used here are all names of numbers, they can be translated as follows (given that *eka* = 1 [= *ekaka*, the suffix *ka* here being a device used to regulate the metre of the line]; *dve* = 2; *trîni* = 3; *chatvâry* = 4; *shat* = 6; *sapta* = 7):

Three.One.Seven.Six.Three.Two.Four.One
3 1 7 6 3 2 4 1

$(= 3 + 1 \times 10 + 7 \times 10^2 + 6 \times 10^3 + 3 \times 10^4 + 2 \times 10^5 + 4 \times 10^6 + 1 \times 10^7 =$ 14,236,713).

The author of this text seems generally to have avoided the abundant synonyms for the numerals and chosen to almost exclusively use the ordinary Sanskrit names of the numbers (*eka, dvi, tri, chatur, pañcha,* etc.).

The reason for this is, perhaps, that the word-symbols were not sufficiently well-known outside "learned" circles. However, there is another probable reason: the author wanted to make his work accessible in order to promote the merits of the philosophy of his religion and the superiority of Jaina science to the public at large, and therefore avoided technical terms.

Nevertheless, at times the author does use certain word-symbols, as in this expression of the number 13,107,200,000 [see Anon. (1962), Chapter IV, line 56, p. 79]:

*pañchabhyah khalu shûnyebhyah param dve sapta*
*châmbaram ekam trîni cha rûpam cha . . .*

five voids, then two and seven, the sky, one and three and the form
00000 2 7 0 1 3 1

$(= 0 + 0 \times 10 + 0 \times 10^2 + 0 \times 10^3 + 0 \times 10^4 + 2 \times 10^5 + 7 \times 10^6 + 0 \times 10^7 + 1 \times 10^8 + 3 \times 10^9 + 1 \times 10^{10} =$ 13,107,200,000).

However, each time the author uses one of these expressions, careful not to confuse his readers, he feels obliged to:

- either be more precise by adding:
  **kramât*, "in order",
  or **sthânakramâd*, "in positional order (**sthâna*)"
- or, which is even more remarkable, to add the following explanation:
  **ankakramena*, in the order of the numerals (**anka*)".

In other words, the concept of zero and the place-value system was widespread in India in the fifth century CE and had probably already been known for some time in "learned" circles.

In fact, the **Lokavibhâga* is the oldest known authentic Indian document to contain the use of zero and decimal numeration. As we shall see, it dates back to the middle of the fifth century CE.

We even know the exact year of the document thanks to the following verses [see Anon. (1962), Chapter XI, lines 50–54, pp. 224ff.] [Personal communication of Billard]:

*vaishve sthite ravisute vrshabhe cha jîve*
*râjottareshu sitapaksham upetya chandre*
*grâme cha pâtalikanâmani pânarâshtre*
*shâstram purâ likhitavân munisarvanandî* (verse 52)

*samvatsare tu dvâvimshe kâñchîshah simhavarmanah*
*ashîtyagre shakâbdânâm siddham etach chhatatraye* (verse 53).

Here is the translation:

Verse 52: "This work was written long ago by the *Muni* Sarvanandin, in the town called Pâtalika, in the kingdom of Pâna, when Saturn was in

*Vaishva*, Jupiter in Taurus, the Moon in *Râjottara*, on the first day of the light fortnight."

Verse 53: "Year twenty-two [of the reign] of Simhavarman, king of Kânchî, three hundred and eighty *Shaka* years."

In verse 52, we are told that when the text (or the copy of it) was written, the Moon was in *Râjottara*. This word means the *nakshatra** called *Uttaraphalgunî*: one of the twenty-seven constellations of the sidereal sphere, divided according to the sidereal revolution of the Moon. As it is the tenth constellation which is referred to here, this position corresponds (according to reliable mathematical calculations) to the interval between 146° 40' and 160° of sidereal longitude. We are also told that the Moon was in its phase corresponding to the first day of the "light fortnight": the first half of the month. We can determine that the work was written in the *Shaka* year 380, the corresponding date being written "entirely in letters" using the ordinary Sanskrit names for the numbers.

Looking at the information given in the verses, which has been interpreted according to the elements of Indian history and astronomy, we have:

- the year, namely the *Shaka* year 380;
- the day of the month, in other words, the Moon is in the first day of the first fortnight of the month;
- and the position of the Moon: 146° 40' / 160° of sidereal longitude, which allows us to determine the month.

Without going into too much detail about the methodology used to determine the dates and to study the astronomical data, suffice to say that the information given leaves us in no doubt as to the date expressed here; the date, in the Julian calendar, corresponds exactly to:

Monday, 25 August, 458 CE.

This is the precise date of the Jaina cosmological text, **Lokavibhâga* (or *The Parts of the Universe*").†

We can now add the other two pieces of information given in verse 52: the planet Jupiter was in Taurus, in the second sign of the zodiac, thus occupying a position of 30° to 60° of sidereal longitude; at the same time, Saturn was in *Vaishva*, the **nakshatra* called *Uttarâshâdha* (the nineteenth constellation of the sidereal revolution), therefore between 266° 40' and 280° of sidereal longitude. As this data agrees with the preceding date, the date is astronomically confirmed.

Whilst this information allows us to date the *Lokavibhâga* with precision, it also irrefutably proves the authenticity of the document, due to the very nature of one of the preceding pieces of astronomical data.

Because Jupiter is situated in the text according to its position in a zodiacal sign, we can also find, for astrological reasons, the position of Saturn in the **nakshatra* system.

This is an irrefutable archaism characterised by the very history of Indian astrology. After this time, there are no more examples where the positions of the planets (with the exception of the Moon) are described in**nakshatra*, they are only expressed in relation to the position of the twelve signs of the zodiac (previously unpublished information given by Billard).

The very existence of this archaism and its almost total disappearance from later Indian texts prove the complete authenticity of its usage, of the document, and all the information it gives us. Moreover, the *Lokavibhâga* as a whole, from an astronomical and cosmological point of view, is undeniably archaic in character in comparison with later texts of the same genre.

Let us now look even more closely at the problem in hand. This text was "written" long before by a *Muni* named Sarvanandin, but the word "written" is ambiguous because in Sanskrit it can mean "copied" as well as "written". The *Lokavibhâga* appears to be the Sanskrit translation of an earlier work written in Prakrit (probably in a Jaina dialect), judging from the translation of verse 51:

> The **Rishi* Simhasûra translated into the Language [= Sanskrit] that which the uninterrupted line of doctors had transmitted [in dialect], which the revered *arhant* Vardhamâna [= the *Jina] delivered to the saints during the grand assembly of the gods and men, namely all that [the disciples of Jina such as] the Sudharma know about the creation of the universe. Let him be praised by all ascetics.

This could and very probably does mean that the current version of the *Lokavibhâga* is an exact reproduction of an original which was written before 458 CE.

Of course we must be wary of relatively recent Indian texts which are frequently attributed to the **Rishi*, the "Sages" of the Vedic era (twelfth to eighth century BCE) who are said to have received the great "Revelation" from the divinities.

The *Lokavibhâga*, however, is much more modest, as it attributes its writing to a *Muni*. This *Muni* could well have lived one or two generations before the above date.

* Here, this word is used to explain the "lunar mansions" in equal divisions. See **Nakshatra*.

† We also see in verse 53 that this text is dated the 22nd year of the rule of Simhavarman, king of Kâñchî (the "Golden Town", sacred place of the Hindus, in Tamil Nadu, approximately 60 km southwest of Madras). According to Frédéric, DCI (1987), pp. 819–20, this king, the son of Skandavarman II, issued from ones of the lines of the Pallava Dynasty, reigned from 436. As this was the 22nd year of his reign, this date corresponds to 436 + 22 = 458 CE. We do not know, however, if the chronology of this sovereign was established by specialists using the text of the **Lokavibhâga*. If this is the case, then this information is of no interest to us. On the other hand, if this is not the case, if the dates of the reign of Simhavarman were established from another inscription, then we have real confirmation of the date we have just determined using the astronomical data in the text in question.

This seems even more likely when we consider that, on the one hand, the numbers which appear in this text conform totally to the rules of the decimal place-value system, and on the other hand, the care that the author took to popularise the text. As we have already seen, when this text was written, Indian scholars were already familiar with the place-value system.

Who, then, is a *Muni*? The answer to this question is in the text itself, in verse 50:

> *Muni* is he who achieves perfection, and, displaying [the] strength of a lion, escapes the terrible [cycle of renaissance], through obeying the decree of respect to all animal life, the exercises of piety such as the vow of honesty, the holiness which conquers all false doctrine and all futility, dominates the empire of the senses, and even defeats the eternal *Karma* through the fire of fervent austerities.

That, in a nutshell, is the doctrine of Jaina, as well as what became of the *Muni* Sarvanandin to whom the writing of the *Lokavibhâga* is attributed.

When did this *Muni* live? A hundred or two hundred years previously? We will never know. What we do know for certain is that the discovery of our present-day numeral system was made well before that famous Monday 25 August, 458 CE.

## HIGHLY CONSISTENT EVIDENCE

Considering the quantity and extreme diversity of the information contained in this chapter, it would seem appropriate to present a summary of all the historical facts which have been established concerning the discovery of zero and the place-value system. The following is a list in reverse chronological order, with references to the Dictionary for those wishing to know more details.

### *Summary of the historical facts relating to the place-value system*

1150 CE. The Indian mathematician *Bhâskarâchârya (known as Bhâskara II) mentions a tradition, according to which zero and the place-value system were invented by the god Brahma. In other words, these notions were so well established in Indian thought and tradition that at this time they were considered to have always been used by humans, and thus to have constituted a "revelation" of the divinities. See *Place-value system.

1010–1030 CE. Date of evidence given by the Muslim scholar of Persian origin, *al-Biruni, about India and in particular her place-value system and methods of calculation; a highly documented piece of evidence to add to the others from the Arabic-Muslim world and the Christian West.

End of the ninth century CE. The philosopher *Shankarâchârya makes a direct reference to the Indian place-value system.

875–876 CE. The dates of the inscriptions of Gwalior: the oldest known "real" Indian inscriptions in stone to use zero (in the form of a little circle) and the nine numerals (in *Nâgarî*) according to the place-value system. See **Nâgarî* numerals, and Figs. 24.72 to 74.

869 CE. The Indian astronomer *Shankaranârâyana frequently uses the place-value system with word-symbols.

c. 850 CE. The Indian astronomer *Haridatta invents a system of numerical notation using letters of the Indian alphabet and based on the place-value system using zero (randomly represented by two different letters): this is the first example of a place-value system which uses letters of the alphabet. See **Katapayâdi* numeration.

850 CE. The Indian mathematician *Mahâvîrâchârya frequently uses the place-value system with the nine numerals or with Sanskrit numerical symbols [M. Rangacarya (1912)].

c. 830 CE. The Indian astronomer *Govindasvâmin frequently uses the place-value system [R. Billard (1971), p. 8].

813 CE. This is the date of the oldest known vernacular inscription of Champa (Indianised civilisation of Southeast Asia), the *Shaka* date of which is indicated using the nine Indian numerals and zero. See **Cham* numerals, and Fig. 24.80.

760 CE. The date of the oldest known vernacular inscription of Java, the *Shaka* date of which is expressed using the nine numerals and zero from India. See **Kawi* numerals, and Fig. 24.80.

732 CE. Date of the oldest known Sanskrit inscription from Java, the *Shaka* date of which is expressed using the place-value system and word-symbols of the Indian astronomers [H. Kern (1913–1929)].

718–729 CE. Date of the *Kai yuan zhan jing*, a work on astronomy and astrology by the Chinese Buddhist *Qutan Xida, who was in fact of Indian origin, real name *Gautama Siddhânta, who lived in China from c. 708 CE, and who, in his work, describes zero, the place-value system of the nine numerals and the Indian methods of calculation.

Seventh century CE. The poet *Subandhu makes direct references to the Indian zero (in the form of a dot) as a mathematical processing device. Thus zero and the place-value system were so well-established in India that the poet could use such subtleties with his metaphors. See *Zero and Sanskrit poetry.

687 CE. Date of the oldest known Sanskrit inscription of Champa, the *Shaka* date of which is expressed using the place-value system and the word-symbols of the Indian astronomers [G. Coedès and H. Parmentier, C 74 B;BEFEO, XI, p. 266].

683 CE. The date of the oldest known vernacular inscription from Malaysia, the *Shaka* date of which is written in the Indian numerals (including zero). See Fig. 24.80.

683 CE. Date of the oldest known vernacular inscription of Cambodia, the *Shaka* date of which is written in Indian numerals (including zero). See *Old Khmer numerals and Fig. 24.80.

662 CE. Syrian bishop Severus Sebokt writes of the nine numerals and Indian methods of calculation.

629 CE. Indian mathematician and astronomer *Bhâskara I frequently uses the place-value system with the word-symbols, often also expressing the number using the nine numerals and zero [K. S. Shukla and K. V. Sarma (1976)].

628 CE. Indian astronomer and mathematician Brahmagupta frequently uses the place-value system with the nine numerals as well as with the word-symbols. He also describes methods of calculation using the nine numerals and zero (very similar to the methods we still use today). He also provides fundamental rules of algebra, where zero is present as a mathematical concept (the number nought), and talks of infinity, defining it as the opposite of zero. See *Zero. *Infinity. **Khachheda*.

598 CE. Date of the oldest known Sanskrit inscription of Cambodia, the *Shaka* date of which is written in word-symbols according to the place-value system [Coedès and Parmentier, K 151; BEFEO, XLIII, 5, p. 6].

594 CE. Date of the donation charter engraved on copper of Dadda III of Sankheda, in Gujarat. This is the oldest known Indian text to bear witness to the use of the nine numerals according to the place-value system (see Fig. 24.75). As we saw earlier, there can be no doubt as to the authenticity of this document.

End of the sixth century CE. The arithmetician *Jinabhadra Gani gives several numerical expressions which prove that he was well acquainted with zero and the place-value system [Datta and Singh (1938)].

c. 575 CE. Indian astronomer and astrologer *Varâhamihîra makes frequent use of the place-value system with Sanskrit numerals. See *Indian astrology.

c. 510 CE. *Âryabhata invented a unique method of recording numbers which required perfect understanding of zero and the place-value system. Moreover, he used a remarkable process of calculating square and cube roots, which would have been impossible without the place-value system, using nine different numerals and a tenth sign which performed the functions of zero. See *Âryabhata (Numerical notations of), *Âryabhata's numeration, *Square roots (How Âryabhata calculated his).

(Monday 25 August) 458 CE. The exact date of the **Lokavibhâga*, (*The Parts of the Universe*), the Jaina cosmological text: the oldest known Indian text to use zero and the place-value system with word-symbols.

Thus one can see the impressive amount of evidence proving that our modern number-system is of Indian origin, and that it was invented long before the sixth century CE. All the evidence points to the fact that this invention is entirely Indian, and born out of a very specific context.

Moreover, we are not dealing with one isolated piece of evidence, or even a limited number of documents, but a huge collection of proofs from all the disciplines, dating from the most significant eras, which have been situated through the study of the palaeography, epigraphy and philology of Indian civilisations both within and outside India.

### THE MOST LIKELY TIME OF THESE DISCOVERIES

It is most likely that the place-value system and zero were discovered in the middle of the reign of the Gupta Dynasty, whose empire stretched the whole length of the Ganges Valley and its tributaries from 240 to approximately 535, known as the "classic" period.

This period saw the highest forms of Indian art (sculpture, painting, in the caves of Ajanta for example, etc.) reach maturity. It was also a classic period because, as Coomaraswamy said, "almost everything that belongs to the Asian spiritual conscience is of Indian origin and dates back to the Gupta Dynasty."

This era coincides with a kind of rebirth of Brahmanism, before it evolved in the wider sense of Hinduism in the following centuries.

Trade was also flourishing at this time, with the Near East, via Persia, and across the sea with the Roman Empire, particularly through Lâta or the eastern area of the present-day state of Gujarat.

Medicine was also developing at this time, particularly dissection.

In the field of literature, Sanskrit, previously the official language of the court and of Brahmanism, was adopted by the Jainas and Buddhists, who did much for the development of the language. And it was probably in this period too that Sanskrit grew to be a much richer language than it had been in the time of the Vedas. This time also saw the beginnings of the **Mahâbhârata*, one of the greatest Indian epic poems, and of the *Dharmashâstra*, collections of texts, mainly about customs, laws and castes.

It was during this time that the **Darshana* – the six systems of Indian philosophy – were developed.

The stories and fables, such as those of the **Pañchatantra*, (the main source of inspiration for the Persian fable *Kalila wa Dimna*), also appeared for the first time, whilst the theatre knew its first blossoming with the poet Kâlidâsa, considered to be one of the greatest dramatists of Indian history, and the **Navaratna* or "Nine Jewels" of Indian tradition.

As for Indian writing, Gupta constitutes the first notation to be individualised in relation to its *Brâhmî* ancestor. As it became more refined, it gave birth to *Nâgarî* (or *Devanâgarî*), in the seventh century CE, which became the principal style in which Sanskrit and then Hindi were written. From *Nâgarî* came the various styles of northern and central India. Another, more northern variant of *Gupta*, evolved into *Shâradâ* of Kashmir, or its derivatives, and also diversified into *Siddham*, from which the script of Nepal, Chinese Turkestan and Tibet would be derived (Fig. 24.52). (See *Indian numerals).

Thus the Gupta period saw the most spectacular progress in almost all the fields of learning, and was a veritable "explosion" of Indian culture.

This was also the time when **Lalitavistara Sûtra* was written, which tells the legend of Buddha and mentions numbers of the highest orders, following very surprising numerical speculation; speculation which grows rapidly after this period, but for which there is no evidence before this time.

It is doubtless no coincidence that the Gupta era saw the first blossoming of *Indian mathematics.

This was also the time of the first developments of trigonometrical astronomy and "Greek" astrology, which was very different from that which existed previously in India, both in terms of claims and material, and which, being in appearance very systematic, already had the scientific foundations of what would soon become Indian astronomy.

Moreover, this was the time when Âryabhata lived. His work would soon lead to a decisive about-turn in Indian astronomy, breaking once and for all with the old Greek-Babylonian traditions and developing the cosmic cycles called **yuga*, devoid of physical value but nonetheless based on a series of unique observations which were more or less precise.

The *Lokavibhâga* is dated 458 CE. This being the oldest known testimony of the use of zero and of the Indian decimal place-value system, the latest possible date of this discovery has to be the middle of the Gupta era. Documents written earlier or at the same time show use of either the ordinary system of the Sanskrit names of the numbers or, as we shall see in the following chronology, that of the old non-positional system derived from the *Brâhmî* system (Fig. 24.70).

Thus the earliest possible date of this discovery is the beginning of the Gupta Dynasty. We must take into consideration, however, the fact that documents bearing witness to the use of word-symbols or the decimal place-value system are only found in abundance after the beginning of the sixth century.

Bearing in mind, on the one hand, the perfect understanding of the place-value system displayed in the *Lokavibhâga* and the clear desire to popularise the text, and on the other hand the fact that the text was more than likely a Sanskrit translation of an earlier document (no doubt written in a Jaina dialect), it would not be unreasonable to suggest *the fourth century CE as the date of the discovery of zero and the place-value system*.

*Third to second century BCE*. First appearances of *Brâhmî* numerals in the edicts of Emperor Asoka and the inscriptions of Nana Ghat. These are very rudimentary. But the first nine figures already constitute the prefiguration of the nine numerals that we use today (Indian, then Arabic and European).

Sanskrit numerals are already worked out and there are particular names for the ascending powers of ten up to $10^8$ (= 100,000,000) at least.

*First century BCE to third century CE*. The numerals found in many inscriptions are derived from *Brâhmî* numerals and constitute a sort of intermediary between *Brâhmî* numerals and later styles, but the place-value system is not yet in use.

The Sanskrit system is extended to include powers of ten up to $10^{12}$ (= 1,000,000,000,000). [See *Names of numbers]

*Fourth to fifth century CE*. The numerals derived from *Brâhmî* numerals begin to diversify into specific styles (*Gupta, Pali, Pallava, Châlukya,* etc.)

The Sanskrit system is capable of expressing and using powers of ten up to $10^{421}$ and above, as we see in the **Lalitavistara* (before 308 CE).

## Discovery of zero and the place-value system

458 CE. (*To this day, no document has been found to prove that the nine units were used at this date according to the place-value system.*)

*The names of the first nine numbers are used according to the place-value system*, as we shall see in the **Lokavibhâga*, dated 458 CE, where the names of the numbers are sometimes replaced by word-symbols and the word **shûnya* ("void") and its synonyms are used as zeros.

*From the sixth century onwards*. The use of the place-value system and zero begin to appear frequently in documents from India and Southeast Asia (the following list is non-exhaustive):

594 CE. Sankheda charter on copper
628 CE. *Brâhmasputasiddhânta* by Brahmagupta
629 CE. Commentary on the *Âryabhatîya* by Bhâskara
683 CE. Khmer inscription of Trapeang Prei
683 CE. Malaysian inscription of Kedukan Bukit
684 CE. Malaysian inscription of Talang Tuwo
686 CE. Malaysian inscription of Kota Kapur
737 CE. Charter of Dhiniki on copper
753 CE. Inscriptions of Devendravarmana
760 CE. Javanese inscription of Dinaya
793 CE. Charter of Râshtrakûta on copper
813 CE. Cham inscription of Pô Nagar
815 CE. Charter of Buchkalâ on copper

829 CE. Cham inscription of Bakul
837 CE. Inscription of Bauka
850 CE. *Ganitasâramgraha* of Mahâvîrâchârya
862 CE. Inscription of Deogarh
875 CE. Inscriptions of Gwalior
877 CE. Balinese inscription of Haliwanghang
878 CE. Balinese inscription of Mamali
880 CE. Balinese inscription of Taragal
917 CE. Charter on copper of Mahipala, etc.

*Seventh century CE*. Gupta notation gave birth to *Nâgarî* numerals, which in turn were the forerunners of the numerals of northern and central India (*Bengalî, Gujâratî, Oriyâ, Kaîthî, Maithilî, Manipurî, Marâthî, Mârwarî*, etc.).

*Eighth century CE*. First appearance of the stylised numerals of Southeast Asia (*Khmer, Cham, Kawi*, etc.).

*Ninth century CE*. A northern variant of *Gupta* led to the *Shâradâ* numerals of Kashmir, ancestors of the numerals of northwest India (*Dogrî, Tâkârî, Multânî, Sindhî, Punjabî, Gurûmukhî*, etc.).

*Eleventh century CE*. The first appearances of *Telugu* numerals (southern India).

## A CULTURE WITH A PASSION FOR HIGH NUMBERS

The early passion which Indian civilisation had for high numbers was a significant factor contributing to the discovery of the place-value system, and not only offered the Indians the incentive to go beyond the "calculable", physical world, but also led to an understanding (much earlier than in our civilisation) of the notion of mathematical infinity itself.

The Indian love for high numbers can be seen in the **Lalitavistara Sûtra* or *Development of the Games* [of Buddha] (a Sanskrit text of the Buddhism of Mahâyâna, written in verse and prose, about the life of Buddha, the "Saint of the Shaka family", as he is said to have told his disciples), where high numbers are constantly evoked:

> Choosing a few random examples, we find in this text a meeting of ten thousand monks, eighty-four million Apsarâs, thirty-two thousand Bodhisattvas, sixty-eight thousand Brahmas, a million Shakras, a hundred thousand gods, hundreds of millions of divinities, five hundred Pratyeka-Buddhas, eighty-four thousand sons of gods, then thirty-two thousand and thirty-six million other sons of gods, sixty-eight thousand **kotis* [= 680,000,000,000] sons of gods and Bodhisattva, eighty-four hundred thousand **niyuta kotis* [= 8,400,000,000,000,000,000,000] of divinities.
>
> The principal signs of Buddha are given the number thirty-two, secondary signs eighty, signs of his mother thirty-two, those of the dwelling-place and the family where he is said to have been born eight and sixty-four. The queen Mâyâ-Devî is served by ten thousand women; the ornaments of the throne of Buddha are enumerated in hundreds of thousands; the hundreds of thousands of divinities and hundred thousand millions of Bodhisattvas and Buddhas pay homage to this throne which is the result of merits accumulated over one hundred thousand million **kalpas*, one *kalpa* being the equivalent of four billion, three hundred and twenty million years. The lotus flower that blossomed the night that Buddha was conceived has a diameter of sixty-eight million *yojana*. Two hundred thousand treasures appeared when Buddha was born; this filled the three thousand great hosts of worlds, and living creatures came to pay homage to his mother, the queen Mâyâ-Devî, in throngs of eighty-four thousand and sixty thousand [F. Woepcke (1863)].

Likewise, in *The Light of Asia*, Edwin Arnold reproduces this passage from the *Lalitavistara Sûtra*, about the education of Buddha as a child, aged eight, by the Sage Vishvâmitra, who explains, in another passage, that numeration, numbers and arithmetic constitute the most important discipline among the seventy-two arts and sciences that the Bodhisattva must acquire:

> And Vishvâmitra said: That's enough [now],
> Let us turn to Numbers. Count after me
> Until you reach **lakh* (= one hundred thousand):
> One, two, three, four, up to ten,
> Then in tens, up to hundreds and thousands.
> After which, the child named the numbers,
> [Then] the decades and the centuries, without stopping.
> [And once] he reached *lakh*, [which] he whispered in silence,
> Then came **koti*, **nahut*, **ninnahut*, **khamba*,
> **viskhamba*, **abab*, **attata*,
> Up to **kumud*, **gundhika*, and **utpala*
> [Ending] with **pundarîka* [leading]
> Towards **paduma*, making it possible to count
> Up to the last grain of the finest sand
> Heaped up in mountainous heights.

Let us interrupt the master for a moment to clarify the numerical values mentioned in the passage:

| | | |
|---|---|---|
| *lakh* | is worth | 100,000 = $10^5$ |
| *koti* | is worth | 10,000,000 = $10^7$ |
| *nahut* | is worth | 1,000,000,000 = $10^9$ |
| *ninnahut* | is worth | 100,000,000,000 = $10^{11}$ |
| *khamba* | is worth | 10,000,000,000,000 = $10^{13}$ |
| *viskhamba* | is worth | 1,000,000,000,000,000 = $10^{15}$ |
| *abab* | is worth | 100,000,000,000,000,000 = $10^{17}$ |
| *attata* | is worth | 10,000,000,000,000,000,000 = $10^{19}$ |
| *kumud* | is worth | 1,000,000,000,000,000,000,000 = $10^{21}$ |
| *gundhika* | is worth | 100,000,000,000,000,000,000,000 = $10^{23}$ |
| *utpala* | is worth | 10,000,000,000,000,000,000,000,000 = $10^{25}$ |
| *pundarîka* | is worth | 1,000,000,000,000,000,000,000,000,000 = $10^{27}$ |
| *paduma* | is worth | 100,000,000,000,000,000,000,000,000,000 = $10^{29}$ |

Thus we are dealing with a centesimal scale, the value of each name being one hundred times bigger than the one preceding it.

But beyond this counting system,
There is the *kâtha* which is used to count the stars in the night sky,
The *kôti-kâtha* for [enumerating] the drops of the ocean,
*Ingga*, to calculate the circular [movements],
*Sarvanikchepa*, with which it is possible to calculate
All the sand of a *Gunga*,
Until we reach *antahkapa*,
Which is [made up of ten] *Gungas*.
[And] if a more intelligible scale is required,
The mathematical ascensions, through the **asankhya*, which is the sum
Of all the drops of rain which, in ten thousand years,
Would fall each day on all the worlds,
Lead [the arithmetician] to the **mahâkalpa*,
Which the gods use to calculate their future and their past.

## THE LIMITATIONS OF THE (INDIAN) "INCALCULABLE"

The *asankhya* or **asankhyeya*, which was poetically defined as "the sum of all the drops of rain which, in ten thousand years, would fall each day on all the worlds", is actually none other than the Sanskrit term meaning "*incalculable*". It literally means: "number which is impossible to count" (from **sankhya* or *sankhyeya*, "number", accompanied by the privative "a").

This word is used in Brahman cosmogony, where it is sometimes used to denote the length of the "*day of Brahma", in other words 4,320,000,000 human years.

In **Bhagavad Gîtâ*, however, "incalculable" corresponds to the entire length of Brahma's life, which is 311,040,000,000,000 human years. In one of the commentaries on the work, it is pointed out that "this incredible longevity, for us infinite, represents no more than zero in the stream of eternity."

Naturally, the value given to "the incalculable" varies considerably according to the text, the author, the region and the era. Thus, the **Sankhyâyana Shrauta Sûtra* fixes this limit at 10,000,000,000,000 giving this number the name **ananta*, signifying "infinity" [see Datta and Singh (1938) p. 10)]. The Tibetans and the Sinhalese pushed the limit of **asankhyeya* much further in giving it a value of one followed by ninety-seven zeros. In the *Pâlî* Grammar of Kâchchâyana, the same concept is given a value of $10^{140}$ (ten million to the power of twenty), placing this term at the end of this very impressive nomenclature the scale of which is tens of millions [see JA 6/17 (1871), p. 411, lines 51–2)]:

| | | |
|---|---|---|
| A hundred times a hundred times a thousand makes a | *koti* | = $10^7$ |
| A hundred times a hundred times a thousand *koti* makes a | *pakoti* | = $10^{14}$ |
| A hundred times a hundred times a thousand *pakoti* | *kotippakoti* | = $10^{21}$ |
| A hundred times a hundred times a thousand *kotippakoti* | *nahuta* | = $10^{28}$ |
| A hundred times a hundred times a thousand *nahuta* | *ninnahuta* | = $10^{35}$ |
| A hundred times a hundred times a thousand *ninnahuta* | *akkhobhini* | = $10^{42}$ |
| A hundred times a hundred times a thousand *akkhobhini* | *bindu* | = $10^{49}$ |
| A hundred times a hundred times a thousand *bindu* | *abbuda* | = $10^{56}$ |
| A hundred times a hundred times a thousand *abbuda* | *nirabbuda* | = $10^{63}$ |
| A hundred times a hundred times a thousand *nirabbuda* | *ahaha* | = $10^{70}$ |
| A hundred times a hundred times a thousand *ahaha* | *ababa* | = $10^{77}$ |
| A hundred times a hundred times a thousand *ababa* | *atata* | = $10^{84}$ |
| A hundred times a hundred times a thousand *atata* | *sogandhika* | = $10^{91}$ |
| A hundred times a hundred times a thousand *sogandhika* | *uppala* | = $10^{98}$ |
| A hundred times a hundred times a thousand *uppala* | *kumuda* | = $10^{105}$ |
| A hundred times a hundred times a thousand *kumuda* | *pundarîka* | = $10^{112}$ |
| A hundred times a hundred times a thousand *pundarîka* | *paduma* | = $10^{119}$ |
| A hundred times a hundred times a thousand *paduma* | *kathâna* | = $10^{126}$ |
| A hundred times a hundred times a thousand *kathâna* | *mahâkathâna* | = $10^{133}$ |
| A hundred times a hundred times a thousand *mahâkathâna* | *asankhyeya* | = $10^{140}$ |

### *The extravagant numbers of the legend of Buddha*

Thus we can see the extent to which the Indians took their naming of numbers.

We can get an even clearer idea of this if we return to the **Lalitavistara Sûtra*, where Bodhisattva (Buddha), now an adult, is almost forced to take part in a competition:

When Bodhisattva reached a marriageable age, he was betrothed to Gopâ, the daughter of Shâkya Dandapâni. But Dandapâni refused to let him marry his daughter, unless the son of the king Shuddhodana [Bodhisattva] made a public show of his mastery of the arts. Thus a type of contest, the winner of which would be given Gopâ's hand in marriage, took place between Bodhisattva and five hundred other young Shakyas. This contest included writing, arithmetic, wrestling and archery [F. Woepcke (1863)].

After easily beating all the young Shakyas, Bodhisattva was invited by his father to pit his wits against the great mathematician Arjuna, who had judged the contest:

"Young man," said Arjuna, "do you know how we express numbers that are higher than a hundred **koti*?"

Bodhisattva nodded, but Arjuna impatiently continued:

"So how do we count beyond a hundred **koti* in hundreds?"

Here is Bodhisattva's reply, bearing in mind that one **koti* is the equivalent of ten million (= $10^7$):

"One hundred *koti* are called an **ayuta*, a hundred *ayuta* make a **niyuta*, a hundred *niyuta* make a **kankara*, a hundred *kankara* make a **vivara*, a hundred *vivara* are a **kshobhya*, a hundred *kshobhya* make a **vivaha*, a hundred *vivaha* make a **utsanga*, a hundred *utsanga* make a **bahula*, a hundred *bahula* make a **nâgabala*, a hundred *nâgabala* make a **titilambha*, a hundred *titilambha* make a **vyavasthânaprajñapati*, a hundred *vyavasthânaprajñapati* make a **hetuhila*, a hundred *hetuhila* make a **karahu*, a hundred *karahu* make a **hetvindriya*, a hundred *hetvindriya* make a **samâptalambha*, a hundred *samâptalambha* make a **gananâgati*, a hundred *gananâgati* make a **niravadya*, a hundred *niravadya* make a **mudrâbala*, a hundred *mudrâbala* make a **sarvabala*, a hundred *sarvabala* make a **visamjñagati* , a hundred *visamjñagati* make a **sarvajña* , a hundred *sarvajña* make a **vibhutangamâ*, a hundred *vibhutangamâ* make a **tallakshana*."

Thus, in his reply, Bodhisattva had given the following table:

| | | |
|---|---|---|
| 1 *ayuta* | = 100 *koti* | $= 10^{9}$ |
| 1 *niyuta* | = 100 *ayuta* | $= 10^{11}$ |
| 1 *kankara* | = 100 *niyuta* | $= 10^{13}$ |
| 1 *vivara* | = 100 *kankara* | $= 10^{15}$ |
| 1 *kshobhya* | = 100 *vivara* | $= 10^{17}$ |
| 1 *vivaha* | = 100 *kshobhya* | $= 10^{19}$ |
| 1 *utsanga* | = 100 *vivaha* | $= 10^{21}$ |
| 1 *bahula* | = 100 *utsanga* | $= 10^{23}$ |
| 1 *nâgabala* | = 100 *bahula* | $= 10^{25}$ |
| 1 *titilambha* | = 100 *nâgabala* | $= 10^{27}$ |
| 1 *vyavasthânaprajñapati* | = 100 *titlambha* | $= 10^{29}$ |
| 1 *hetuhila* | = 100 *vyavasthânaprajñapati* | $= 10^{31}$ |
| 1 *karahu* | = 100 *hetuhila* | $= 10^{33}$ |
| 1 *hetvindriya* | = 100 *karahu* | $= 10^{35}$ |
| 1 *samâptalambha* | = 100 *hetvindriya* | $= 10^{37}$ |
| 1 *gananâgati* | = 100 *samâptalambha* | $= 10^{39}$ |
| 1 *niravadya* | = 100 *gananâgati* | $= 10^{41}$ |
| 1 *mudrâbala* | = 100 *niravadya* | $= 10^{43}$ |
| 1 *sarvabala* | = 100 *mudrâbala* | $= 10^{45}$ |
| 1 *visamjñagati* | = 100 *sarvabala* | $= 10^{47}$ |
| 1 *sarvajña* | = 100 *visamjñagati* | $= 10^{49}$ |
| 1 *vibhutangamâ* | = 100 *sarvajña* | $= 10^{51}$ |
| 1 *tallakshana* | = 100 *vibhutangamâ* | $= 10^{53}$ |

In modern terms, the value of the *tallakshana* corresponds to the following formula:

$$1\ {}^*\textit{tallakshana} = (10^7) \times (10^2)^{23} = 10^{7+46\times 1} = 10^{53}.$$

"Having thus reached the **tallakshana*, which we would write today as 1 followed by fifty-three zeros, Bodhisattva added that this whole table forms only one counting system, the **tallakshana* counting system, [from the name of its last term]; but there is, above this system, that of **dhvajâgravati*; beyond that, the counting system **dhvajâgranishâmani*, and beyond that again, six other systems for which he gave the respective names" [Woepcke (1863)].

The **dhvajâgravati* system is also made up of twenty-four terms, and its first term is the **tallakshana* (the largest number in the preceding system, that is $10^{53}$). Since its progression increases geometrically by a ratio equivalent to one hundred, its final term therefore has the value:

$$1\ \textit{dhvajâgravati} = (10^{7+46\times 1}) \times (10^2)^{23} = 10^{7+46\times 2} = 10^{99}.$$

As this is the last term in the preceding system, it becomes the first in the following one, that is to say the third system, the *dhvajâgranishâmani*, the final number of which being equal to:

$$1\ \textit{dhvajâgranishâmani} = (10^{7+46\times 2}) \times (10^2)^{23} = 10^{7+46\times 3} = 10^{145}.$$

Step by step, we thus arrive at the ninth counting system, of which the name of the last term has the value:

$$(10^{7+46\times 8}) \times (10^2)^{23} = 10^{7+46\times 9} = 10^{421}.$$

(We write this number as 1 followed by 421 zeros).

Arjuna, full of admiration for the superiority of Buddha's knowledge, and wanting nothing more than to learn from him, asked him to explain how one enters into "the counting system which extends to the particles of the first atoms (**Paramânu*)" (literally: "first-atom-particle-penetration-enumeration") and to teach him and the young Shakyas how many first atoms there were in a *yojana* (a unit of length).

Here is Buddha's reply:

> If you want to know this number, use the scale that takes you from the *yojana* to four *krosha* of Mâgadha, from the *krosha* of Mâgadha to a thousand arcs (*dhanu*), from the arc to four cubits (*hasta*), from the cubit to two spans (*vitasti*), from the span to twelve phalanges of fingers (*angulî parva*), from the phalanx of the finger to seven grains of barley (*yava*), from the grain of barley to seven mustard seeds (*sarshapa*), from the mustard seed to seven poppy seeds (*likshâ râja*), from the poppy seed to seven particles of dust stirred up by a cow (*go râja*), from the particle of dust stirred up by a cow to seven specks of dust stirred up by a ram (*edaka râja*), from the speck of dust disturbed by a ram to seven specks of dust stirred up by a hare (*shasha râja*), from the speck of dust stirred up by a hare to seven specks of dust carried off by the wind (*vâyâyana râja*), from the speck of dust carried away by the wind to seven tiny specks of dust (*truti*), from a tiny speck of dust to seven minute specks of dust (*renu*), and from the minute speck of dust to seven particles of the first atoms (*paramânu râja*).

In other words, if we use the modern notation of the exponents and if we use the letter "p" to denote these "first atoms" (*paramânu*), this "scale" can be written in the following manner, starting with the smallest and finishing with the largest quantity:

1 minute speck of dust
= 7 particles of dust of the first atoms .................... 7 p

1 tiny speck of dust
= 7 minute specks of dust .................... $7^2$ p

1 speck of dust carried away by the wind
= 7 tiny specks of dust .................... $7^3$ p

1 speck of dust stirred up by a hare
= 7 specks of dust carried away by the wind .................... $7^4$ p

1 speck of dust stirred up by a ram
= 7 specks of dust stirred up by a hare .................... $7^5$ p

1 speck of dust stirred up by a cow
= 7 specks of dust stirred up by a ram .................... $7^6$ p

1 poppy seed
= 7 specks of dust stirred up by a cow .................... $7^7$ p

1 mustard seed
= 7 poppy seeds .................... $7^8$ p

1 grain of barley
= 7 mustard seeds .................... $7^9$ p

1 phalanx of a finger
= 7 grains of barley .................... $7^{10}$ p

1 span
= 12 phalanges of fingers .................... $12 \times 7^{10}$ p

1 cubit
= 2 spans .................... $2 \times 12 \times 7^{10}$ p

1 arc
= 4 cubits .................... $8 \times 12 \times 7^{10}$ p

1 *krosha* from Mâgadha
= 1,000 arcs .................... $1{,}000 \times 8 \times 12 \times 7^{10}$ p

1 *yojana*
= 4 *krosha* from Mâgadha .................... $4 \times 1{,}000 \times 8 \times 12 \times 7^{10}$ p

Carrying out the multiplication $4 \times 1{,}000 \times 8 \times 12 \times 7^{10}$ which is denoted by the last term in this scale, Buddha gives the sum by expressing in words the number of first atoms contained in the "length" of a *yojana*, namely:

108,470,495,616,000.

### *From very high numbers to very small numbers*

Using the corresponding Sanskrit terms and taking the scale in reverse order, we have, using the preceding data, the following table which begins with the phalanges of the digits (*angulî parva*) and ends with the atoms (*paramânu râja*):

| | |
|---|---|
| 1 *angulî parva* | = 7 *yava* |
| 1 *yava* | = 7 *sarshapa* |
| 1 *sarshapa* | = 7 *likshâ râja* |
| 1 *likshâ râja* | = 7 *go râja* |
| 1 *go râja* | = 7 *edaka râja* |
| 1 *edaka râja* | = 7 *shasha râja* |
| 1 *shasha râja* | = 7 *vâtayana râja* |
| 1 *vâtayana râja* | = 7 *truti* |
| 1 *truti* | = 7 *renu* |
| 1 *renu* | = 7 *paramânu râja*. |

Thus:

$$1 \; \textit{angulî parva} = 7^{10} \; \textit{paramânu râja}.$$

The **paramânu* or "highest atom" constitutes, in Indian thought, the smallest indivisible material particle, which has a taste, a smell and a colour.*

In terms of weight, a **paramânu* is the equivalent of one seventh of an "atom" (**anu*).

As an **anu* is approximately equal to 1/2,707,200 of a *tola*, which is itself equal to 11.644 grams, the **paramânu* weighs the equivalent of 1/18, 950,400 of 11.644 g; thus:

$$1 \; \textit{paramânu} = 0.000000614 \text{ g} = 6.14 \times 10^{-7} \text{ g}.$$

We will now look at the calculation from another angle.

According to the above table, a phalanx of a finger (*angulî parva*) corresponds to $7^{10}$ "specks of dust of a supreme atom" (**paramânu râja*) ; thus:

$$1 \; \textit{*paramânu râja} = 7^{-10} \; \textit{angulî parva}.$$

Three phalanges of the fingers make an "inch"; therefore a *paramânu râja* is equal to $3.7^{-10}$ inches. As an inch is equal to 27.06995 mm, we have:

$$1 \; \textit{*paramânu râja} = 0.000000287 \text{ mm} = 2.87 \times 10^{-7} \text{ mm}.$$

These constitute the smallest units of weight and length in India in the early centuries CE.

Thus we have seen how the Indians could easily deal with both "very high" and "very small" numbers.

## THE BEGINNINGS OF THESE NUMERICAL SPECULATIONS

The *Shâkyamuni* or "Sage of the Shâkyas", the Indian prince named Gautama Siddhârtha, better known as Buddha, is said to have lived during the fifth century BCE. Does this mean that the Indian passion for high numbers began at this time? We do not know, because no work by Buddha himself has ever been found.

The **Lalitavistara Sûtra* is a collection of stories and ancient legends which was actually only compiled relatively recently.

However, a passage of the **Vâjasaneyî Samhitâ* enumerates the stones needed to construct the sacred altar of fire using the following words [see Weber, in: JSO, XV, pp. 132–40)]:

**ayuta* = 10,000
**niyuta* = 100,000
**prayuta* = 1,000,000
**arbuda* = 10,000,000
**nyarbuda* = 100,000,000
**samudra* = 1,000,000,000
**madhya* = 10,000,000,000
**anta* = 100,000,000,000
**parârdha* = 1,000,000,000,000.

This example, like many others of the same genre, comes from a text belonging to Vedic literature. We know that the texts of the **Vedas* and most of the literature which derives from this civilisation date far back in terms of Indian history, but it is impossible to give an exact date to this era; the texts were first transmitted orally before being transcribed at a later date. As Frédéric explains, "the only chronological order we can give them is a purely internal one. The *Samhitâ* (the three Vedas: *Rigveda*, *Yajurveda*, and *Sâmaveda*) seem to have been composed first; next we have the fourth Veda or *Atharvaveda*, followed by the *Brâhmanas*, the *Kalpasûtras* and finally the *Aranyakas* and the *Upanishads*." What we can say with some certainty is that most of these texts were already in their finished form in the early centuries CE.

The numerical speculation contained in the legend of Buddha cannot have appeared later than the beginning of the fourth century CE, as the **Lalitavistara Sûtra* was translated into Chinese by Dharmarâksha in the year 308 CE.

Thus it would not be unreasonable to place the date of the first developments of these impressive numerical speculations around the third century CE.

### *The incredible speculations of the Jainas*

The members of the Jaina religious movement figure first and foremost amongst the Indian scholars to be well acquainted with such numerical speculations.

There are many examples in the text **Anuyogadvâra Sûtra*, where the sum of the human beings of the creation is given as $2^{96}$.

There are other, even older Jaina texts, where numbers containing eighty or even a hundred orders of units are described as "minuscule" in comparison with those under speculation: these numbers are as high as, or greater than *ten to the power of 250*, which we would write today as 1 followed by at least two hundred and fifty zeros.

There is also a period of time called **Shîrshaprahelikâ*, mentioned in several Jaina texts on cosmology, and expressed, according to Hema Chandra

*The **paramânu* bears no relation to our present-day concept of the "atom", but is more akin to what we would call a molecule: the smallest particle which constitutes a quantity of a compound body.

(1089 CE), as "196 positions of numerals of the decimal place-value system", and which corresponds, according to the same source, "to the product of 84,000,000 multiplied by itself twenty-eight times". Thus:

the *Shîrshaprahelikâ* = $(84{,}000{,}000)^{28} \approx (8^7)^{28} = 8^{7 \times 28} = 8^{196}$.

As for the ages of the world, the Jainas used the Brahman classification. Thus the fifth age (which we live in) would have begun in 523 BCE and would be characterised by pain. It would be followed by the sixth and last "age" of 21,000 years, at the end of which the human race would undergo horrific mutations, without the world actually coming to an end. According to Jaina doctrine, the universe is indestructible; this is because it is infinite in terms of both time and space. It was in order to define their vision of this impalpable universe, which is both eternal and limitless, that the Jainas undertook their impressive speculations on gigantic numbers and thus created a "science" which was characteristic of their way of thinking.

Their discovery of *infinity was doubtless due to the fact that they were constantly pushing the limits of the **asamkhyeya* (the "impossible to count", the "innumerable", the "number impossible to conceive") further and further.

## THE BIRTH OF MODERN NUMERALS

We can only admire the perfect ease with which the authors and readers of the texts we have just seen were able to write and pronounce these high numbers without ever being struck by a feeling of vertigo at the enormous quantities they were dealing with.

Sanskrit notation had an excellent conceptual quality. It was easy to use and moreover it facilitated the conception of the highest imaginable numbers. This is why it was so well suited to the most exuberant numerical or arithmetical-cosmogonic speculations of Indian culture.

This spoken counting system had a special name for each of the nine simple units:

| **eka* | **dvi* | **tri* | * **chatur* | **pañcha* | **shat* | **sapta* | **ashta* | **nava* |
|---|---|---|---|---|---|---|---|---|
| 1 | 2 | 3 | 4 | 5 | 6 | 7 | 8 | 9 |

There was an independent name for ten, and for each of its multiples, which were used alongside other words in the form of analytical combinations to express intermediate numbers. Like all Indo-European spoken counting systems, the numbers were often expressed – at least in everyday use – in descending order, from the highest to the smallest units.

However, around the dawn of the Common Era (probably from the second century BCE), this order was reversed, particularly in learned and official texts, the numbers being expressed in ascending order, from the smallest to the highest units. (It has been suggested that this radical transformation was due to the intervention of another civilisation. This idea is totally without foundation: why and how could this change in direction be due to an outside influence, bearing in mind that none of the known civilisations, Greek, Babylonian and Chinese included, had reached the same level as the Indians in terms of numerical concepts and expression? As we shall see later, the reason for this change has absolutely nothing to do with any outside influence.)

Where we would say "three thousand seven hundred and fifty-nine", Indian arithmeticians would have said:

*nava pañchâshat sapta shata cha trisahasra*
"nine, fifty, seven hundred and three thousand".

Apart from saying the numbers in the opposite order, the way that numbers were said in Sanskrit and the way in which we say them are very similar.

However, there is one fundamental difference. When we say the numbers 10,000, 100,000, 10,000,000, 100,000,000, etc., we say *ten thousand, a hundred thousand, ten million, a hundred million*, etc. In other words, *thousand, million*, etc., play the role of auxiliary bases.

There are no such auxiliary bases in the Sanskrit system, at least none which were used by learned men; each power of ten had a particular name which was completely independent of all the others.

These names are discussed in c. 1000 by the Muslim astronomer of Persian origin, al-Biruni, in his *Kitab fi tahqiq i ma li'l hind* (the book relating to his experiences of Indian civilisation):

> One thing that all nations agree on when it comes to calculations is the proportionality of the knots of calculation* according to the ratio of ten [= the decimal base]. This means that there is no order in which the unit is not worth one tenth of the unit which appears in the following order and ten times the value of the unit of the preceding order. I carefully researched the names of the different orders of numbers used in different languages to the best of my capabilities. I found that the same names are repeated once the numbers reach the thousands, as was the case with the Arabic system, which is the most appropriate method, and the most fitting to the nature of the subject in question. I have also written a whole dissertation on this subject. However, the Indians go beyond the thousands in their nomenclature, but not in a uniform manner; some use improvised names, others use names which derive from specific etymologies; others even mix both these types of names. This

* According to the contemporary Arabic terminology, the *knot* of calculation is the constituent of a given "order of units"; thus the knots of units are 1, 2, 3, 4, 5, 6, 7, 8, 9; the knots of tens 10, 20, 30, 40, 50, 60, 70, 80, 90; the knots of hundreds 100, 200, 300, 400, 500, 600, 700, 800, 900; and so on.

| Order of unit | Corresponding name | Numerical value | Power of ten |
|---|---|---|---|
| 1 | Atmosphere | 1 | 1 |
| 2 | Ether | 10 | 10 |
| 3 | Atmosphere | 100 | $10^2$ |
| 4 | Immensity of space | 1,000 | $10^3$ |
| 5 | Atmosphere | 10,000 | $10^4$ |
| 6 | Point (or Dot) | 100,000 | $10^5$ |
| 7 | Canopy of heaven | 1,000,000 | $10^6$ |
| 8 | Voyage on water | 10,000,000 | $10^7$ |
| 9 | Sky, Atmosphere | 100,000,000 | $10^8$ |
| 10 | Sky, Atmosphere | 1,000,000,000 | $10^9$ |
| 11 | Entire, Complete | 10,000,000,000 | $10^{10}$ |
| 12 | Hole | 100,000,000,000 | $10^{11}$ |
| 13 | Void | 1,000,000,000,000 | $10^{12}$ |
| 14 | Point (or Dot) | 10,000,000,000,000 | $10^{13}$ |
| 15 | Foot of Vishnu | 100,000,000,000,000 | $10^{14}$ |
| 16 | Sky | 1,000,000,000,000,000 | $10^{15}$ |
| 17 | Sky, Space | 10,000,000,000,000,000 | $10^{16}$ |
| 18 | Path of the gods | 100,000,000,000,000,000 | $10^{17}$ |

FIG. 24.81. *List of Sanskrit names (translated) for powers of ten according to al-Biruni*

naming reaches as far as the eighteenth order due to certain subtleties which were suggested to the people who use these names, by lexicographers, through the etymologies of these names. I will now describe the differences [which exist in the Indians' usage of these names]. One difference is that some people claim that after the **parârdha* [the name of the eighteenth order of units] there is a nineteenth order, which is called *bhûri*, and that beyond that there is no more need for calculation. But if calculation stops at a certain point, and there is a limit to the order of numbers used, this is only a convention; because this could only occur if one understood nothing besides the names used in the calculations. We also know [according to the same people] that a unit of this order [the nineteenth] is one fifth of the biggest nychthemeron. However, in terms of this method, no mention is made of the influence of any tradition in the work of those who share this opinion. But traditions do exist which shall be explained which mention periods made up of the largest nychthemeron. Adding a nineteenth order is taking the matter to extremes. Another difference lies in the fact that some people claim that the furthest limit of calculation is in the **koti* [$10^7$] and that beyond this order we return to multiples of tens, hundreds and thousands because the number of the divinities (*Deva*) is included in this order. These people say that the number of divinities is thirty-three **koti* [= 330,000,000], and that on each of the three [gods] Brahma, Nârâyana and Mahâdeva depend eleven *koti* [= 110,000,000] [of these divinities]. As for the names which come after the eighth order, they were created by the grammarians for reasons we shall give below. A further difference is due to the fact that in everyday usage, the Indians use **dasha sahasra* ["ten thousand"] for the fifth order, and **dasha laksha* for the seventh [the tens of millions], because the names of these two orders are hardly ever used. In the work entitled *Arjabhad* [the Arabic name for **Âryabhata*] from the town of Kusumapura, the names of the orders, from the tens of thousands to the tens of **kotis*, are as follows: **ayuta*, **niyuta*, **prayuta*, **koti*, **padma*, *parapadma*. Yet another difference lies in the fact that some people create names out of pairs. Thus they call the sixth order **niyuta* to follow the name of the fifth [**ayuta*], and they call the eighth **arbuda* so that the ninth order [**vyarbuda*] can follow on, as the twelfth [**nikharva*] follows the eleventh [**kharva*]. They also call the thirteenth order **shankha*, and the fourteenth **mahâshankha* [the "big *shankha*"]; and according to this rule the **mahâpadma* [the thirteenth order] was preceded by the **padma* [twelfth]. These are the differences it is worthwhile knowing. But there are many more which are of no use to us, and only exist because the numbers are taught without the slightest regard for their proper order, or because some people [use them but] claim that [they do not] know [their exact meaning]. This [knowing the precise meanings of all the names] would be difficult for tradesmen. According to the *Pulisha Siddhanta*, after **sahasra*, which is the fourth order, the fifth is **ayuta*, the sixth **niyuta*, the seventh **prayuta*, the eighth **koti*, the ninth **arbuda*, the tenth **kharva*. The names which follow are the same as the ones above [in Fig. 24.81].

These differences apart, the Sanskrit spoken counting system shows the remarkable spirit of organisation of the Indian scholars who, being the good arithmeticians and lexicographers that they were, sought, at an early stage, to give this system an impeccably ordered structure.

This fact is even more remarkable given that the Greeks got no further than ten thousand. As for the Romans, they only had specific names for numbers up to a hundred thousand. In his *Natural History* (XXXIII, 133), Pliny explains that the Romans, scarcely able to name the powers of ten superior to a hundred thousand, contented themselves with expressing "million" as: *decies centena milia*, "ten hundred thousand".

The French had to wait until the thirteenth century for the introduction of the word *million* in their vocabulary which took place c. 1270 [O. Bloch and W. von Wartburg (1968)], and until the end of the fifteenth century for the names of numbers higher than that.

In 1484, Nicolas Chuquet invented the very first set of names for high numbers above a million, using the million $10^6$ as the multiplier: "byllion" = $10^{12}$, "tryllion" = $10^{18}$, "quadrillion" = $10^{24}$, "quyllion" = $10^{30}$, "sixlion" = $10^{36}$, "septyllion" = $10^{42}$, "octyllion" = $10^{48}$, and "nonyllion" = $10^{54}$. Chuquet's work was never published, so that it was not until the middle of the seventeenth century that words like billion, trillion, etc. were found at all commonly. Nowadays, US English has the most regular naming system, using $10^3$ as the multiplier, as follows: $10^6$ million, $10^9$ billion, $10^{12}$ trillion, $10^{15}$ quadrillion. $10^{18}$ quintillion, $10^{21}$ sextillion, $10^{24}$ septillion.

In British English, however, the term "billion" is used for $10^{12}$ ($10^9$ being just "a thousand million"), and the multiplier used remains $10^6$, so that trillion = $10^{18}$ and quadrillion = $10^{24}$. Despite this, the American sense of billion is now used in all financial calculations, and is rapidly displacing the dictionary meaning in British English. French officially uses the same system as the US, except that the older term "milliard" is commonly used for $10^9$; "billion", officially given the value of $10^{12}$ in 1948, is rarely used, and $10^{12}$ is most often expressed (as in US English) by "trillion".

A comparison between the Arabic, Greek, Chinese and current British systems of expressing high numbers will give a better idea of the impressive conceptual quality of the Sanskrit system.

To make this even clearer, we will use the following number, which will be expressed successively according to the above systems:

523 622 198 443 682 439.

As we know, in their nomenclature of the powers of ten, the ancient Arabs always stopped at one thousand, then superposed thousand upon thousand, whilst still using the names of the inferior powers of ten. In other words, in their language, the above number would be expressed rather like this:

Five hundred *thousand thousand thousand thousand thousand* and twenty-three *thousand thousand thousand thousand thousand* and six hundred *thousand thousand thousand thousand* and twenty-two *thousand thousand thousand thousand* and a hundred *thousand thousand thousand* and ninety-eight *thousand thousand thousand* and four hundred *thousand thousand* and forty-three *thousand thousand* and six hundred *thousand* and eighty-two *thousand* and four hundred and thirty-nine.

Equally, in their nomenclature of powers of ten, the ancient Greeks and the Chinese always stopped at the *myriad* (ten thousand); from there, they superposed myriads on top of myriads, mixed with the names of the inferior powers of ten. In other words, in these languages, the above number would have been expressed rather like this [see Daremberg and Saglio (1873); Dedron and Itard (1974); Guitel (1975); Menninger (1957); Ore (1948); Woepcke (1863)]:

Fifty-two *myriads of myriads of myriads of myriads* and three thousand six hundred and twenty-two *myriads of myriads of myriads* and one thousand nine hundred and eighty-four *myriads of myriads* and four thousand three hundred and sixty-eight *myriads* and two thousand four hundred and thirty-nine.

In the United States this would be expressed as:
Five hundred and twenty-three quadrillion, six hundred and twenty-two trillion, one hundred and ninety-eight billion, four hundred and forty-three million, six hundred and eighty-two thousand, four hundred and thirty-nine.

In British English, this number would be expressed as:
Five hundred and twenty-three thousand six hundred and twenty-two billion, one hundred and ninety-eight thousand four hundred and forty-three million, six hundred and eighty-two thousand four hundred and thirty-nine.

All the above methods are rather complicated, and it is difficult to get a clear idea of the positional value of the number.

Since around the time of the *Vedas, the Sanskrit system was much clearer; it possessed names for all the powers of ten up to $10^8$ (= 100,000,000). Later, this was extended to $10^{12}$ (1,000,000,000,000) (probably at the start of the first millennium CE). When the powers of ten were named up to $10^{17}$ (and sometimes even further, as we saw in the *Jaina texts and in the *legend of Buddha) around 300 CE, it is likely that this was due to the development of the language itself.

Thus the following would have sufficed to express the above number in Sanskrit, using as an example for the base the nomenclature reported by al-Biruni (Fig. 24.81):

*nava cha trimshati cha chaturshata cha dvisahasra cha ashtâyuta cha shatlaksha cha triprayuta cha chaturkoti cha chaturvyarbuda cha ashtapadma cha navakharva cha ekanikharva cha dvimahâpadma cha dveshañka cha shatsamudra cha trimadhya cha dvântya cha pañchaparârdha.*

In semi-translation, the number reads something like this:

> Nine and three *dasha* and four *shata* and two *sahasra* and eight *ayuta* and six *laksha* and three *prayuta* and four *koti* and four *vyarbuda* and eight *padma* and nine *kharva* and one *nikharva* and two *mahâpadma* and two *shankha* and six *samudra* and three *madhya* and two *antya* and five *parârdha*.

In giving each power of ten an individual name, the Sanskrit system gave no special importance to any number.

Thus the Sanskrit system is obviously superior to that of the Arabs (for whom the thousand was the limit), or of the Greeks and the Chinese (whose limit was ten thousand) and even to our own system (where the names thousand, million, etc. continue to act as auxiliary bases).

Instead of naming the numbers in groups of three, four or eight orders of units, the Indians, from a very early date, expressed them taking the powers of ten and the names of the first nine units individually. In other words, to express a given number, one only had to place the name indicating the order of units between the name of the order of units that was immediately below it and the one immediately above it.

That is exactly what is required in order to gain a precise idea of the place-value system, the rule being presented in a natural way and thus appearing self-explanatory. To put it plainly, *the Sanskrit numeral system contained the very key to the discovery of the place-value system*.

> In order to grasp this idea, the names of the powers of ten need not always be the same.
>
> In fact, if the mathematical genius of the Indians could embrace variations on the names of the numbers whilst maintaining a clear idea of the series of the ascending powers of ten, this only made it more disposed to understanding the place-value system.
>
> These names need not necessarily have been in everyday use in India. They need only have been familiar to those who were capable of developing the potential ideas behind them, namely to learned men.
>
> We can understand al-Biruni's surprise at seeing grammarians creating these names, and being practically the only ones to use them, for, in the scientific development of Arabic civilization, grammar, lexicography and literature were completely separate movements from the mathematical, medical and philosophical sciences [F. Woepcke (1863)].

However, grammar and interpretation in ancient India were closely linked to the handling of high numbers. Studies relating to poetry and metrics initiated "scientists" to both arithmetic and grammar, and grammarians were just as competent at calculations as the professional mathematicians.

Thus we can see the importance of the role of Indian "scientists", philosophers and cosmographers who, in order to develop their arithmetical-metaphysical and arithmetical-cosmogonical speculations concerning ever higher numbers, became at once arithmeticians, grammarians and poets, and gave their spoken counting system a truly mathematical structure which had the potential to lead them directly to the discovery of the decimal place-value system.

In fact, since a time which was undoubtedly earlier than the middle of the fifth century CE, all mention of the names indicating the base and its diverse powers was suppressed in the body of the numerical expressions expressed by the names of the numbers.

In other words, the Indian scholars quite naturally arrived at the idea of writing numbers without the names **dasha* (= 10), **shata* (= $10^2$), **sahasra* (= $10^3$), **ayuta* (= $10^4$), **laksha* (= $10^5$), **prayuta* (= $10^6$), **koti* (= $10^7$), **vyarbuda* (= $10^8$), **padma* (= $10^9$), **kharva* (= $10^{10}$), **nikharva* (= $10^{11}$), **mahâpadma* (= $10^{12}$), **shankha* (= $10^{13}$), **samudra* (= $10^{14}$), **madhya* (= $10^{15}$), **antya* (= $10^{16}$), **parârdha* (= $10^{17}$), etc. From that time on, they simply wrote, in strict order, the names of the units which acted as multiplying coefficients in their numerical expression, according to the order of the ascending powers of ten. Thus they expressed numbers using nothing more than the names of the units.

Instead of writing the number 523 622 198 443 682 439 using the names of the numbers according to the ordinary principle of the Sanskrit language (the complete form of the *Sanskrit numeral system), they only retained the names of the units forming the coefficients of the diverse consecutive powers (abridged form, characteristic of the *simplified Sanskrit numeral system):

*Complete form*

Nine and three *dasha* and four *shata* and two *sahasra* and eight *ayuta* and six *laksha* and three *prayuta* and four *koti* and four *vyarbuda* and eight *padma* and nine *kharva* and one *nikharva* and two *mahâpadma* and two *shanka* and six *samudra* and three *madhya* and two *antya* and five *parârdha*.

*Mathematical breakdown*

$$\begin{aligned} &= 9 + 3 \times 10 \quad + 4 \times 10^2 \quad + 2 \times 10^3 \quad + 8 \times 10^4 \quad + 6 \times 10^5 \\ &\quad + 3 \times 10^6 \quad + 4 \times 10^7 \quad + 4 \times 10^8 \quad + 8 \times 10^9 \quad + 9 \times 10^{10} \\ &\quad + 1 \times 10^{11} + 2 \times 10^{12} + 2 \times 10^{13} + 6 \times 10^{14} + 3 \times 10^{15} \\ &\quad + 2 \times 10^{16} + 5 \times 10^{17} \\ &= 523{,}622{,}198{,}443{,}682{,}439 \end{aligned}$$

*Abridged form*

Nine.three.four.two.eight.six.three.four.four.
eight.nine.one.two.two.six.three.two.five

The numbers in the **Jaina* text, the **Lokavibhâga*, (the first document that we know of to make regular use of the place-value system) were expressed in a very similar manner.

In other words, the Indian system of numerical symbols (or at least the ancestor of this unique system) was born out of a simplification of the Sanskrit spoken numeral system.

Such a simplification is not at all surprising when we consider the consistency and potential of the human mind, as well as humankind's intelligence, actions and thoughts upon such matters. When two human beings or two cultures have the same needs and methods due to identical basic (social, psychological, intellectual and material) conditions, they inevitably follow the same paths to arrive at similar, if not identical results.

This is exactly what happened amongst the priest-astronomers of Maya civilisation. Due to a need to abbreviate increasingly high numerical expressions, and also because the units in their system of expressing lengths of time were presented in an impeccable order which was always rigorously followed, they discovered a place-value system, to which they added a sign which performed the function of zero.

As with the Maya, this simplification held no ambiguity for the Indians.

The fact that the successive names of the powers of ten had always followed an invariable order which was firmly established in the mind made the simplification even more comprehensible.

The actual reason for the simplification was doubtless a need for abbreviation. This would have become increasingly necessary as the Indian mathematicians gradually dealt with higher and higher numbers.

To write numbers containing dozens of orders of units according to their names would have taken up whole "pages" of writing. Even expressing one single number could take up several "sheets".

The scholars would have also wanted to be economical with their writing materials. They had to go and pick palm leaves themselves which they used for writing upon. These had to be picked just at the right time, before they opened out entirely, then dried and smoothed out in order to make them fit for the writing of manuscripts. (See *Indian styles of writing). The scholars wanted to give themselves as much time as possible to devote to the more noble task of contemplation, for example studying sacred texts or practising the physical, spiritual and moral exercises of yoga.

The simplification brought about an authentic place-value system which had the Sanskrit names of the nine units as its base symbols. Their value varied according to their relative position in a numerical expression.

Thus *three, two, one* gave the value of simple units to *three*, the value of a multiple of ten to *two* and the value of a multiple of a hundred to *one*.

However, as we can see in the following example, this method could present certain difficulties.

Given that the Sanskrit name for the number three is *tri*, in order to express the number 33333333333, it would be written thus:

*tri.tri.tri.tri.tri.tri.tri.tri.tri.tri.tri*
Three.Three.Three.Three.Three.Three.Three.Three.Three.Three.Three
3 3 3 3 3 3 3 3 3 3 3

$$(= 3 + 3 \times 10 + 3 \times 10^2 + 3 \times 10^3 + 3 \times 10^4 + 3 \times 10^5 + 3 \times 10^6 + 3 \times 10^7 + 3 \times 10^8 + 3 \times 10^9 + 3 \times 10^{10}).$$

This expression, which involves the repetition of *tri* eleven times, neither sounds pleasant nor is conducive to the memorisation of the number in question.

Moreover, this number only has eleven orders of units. It would be much worse if it had thirty or a hundred, or even two hundred orders of units.

To avoid this repetition of the same word, the Indian astronomers used synonyms for the Sanskrit names of the numbers. They used all kinds of ideas from traditions, mythology, philosophy, customs and other characteristics of Indian culture in general. This is how they gradually replaced the ordinary Sanskrit names of numbers with an almost infinite synonymy.

Thus the above number would have been expressed by the following kind of symbolic expression:

*agni.mûrti.guna.loka.jagat.dahana.kâla.hotri.vâchana.Râma.vahni*
Fire.Shape.Quality.World.World.Fire.Time.Fire.Voice.*Râma*.Fire.
3 3 3 3 3 3 3 3 3 3 3

This substitution of the ordinary names of numbers marked the birth of the representation of numbers by *numerical symbols in the place-value system.

Why did Indian astronomers favour this use of numerical symbols over the nine numerals and the sign for zero?

The fact is that they had excellent reasons for this choice.

First and foremost, the concept of zero and the decimal place-value system is totally independent of the chosen style of expressing the numbers (be it conventional graphics, letters of the alphabet or words with or without evocative meaning). All that matters is that there is no ambiguity and that the chosen system of representation contains a perfect concept of zero and the place-value system.

There are other reasons which are specific to the field of Indian astronomy and mathematics, which were generally written in Sanskrit, as were all important erudite Indian texts. The first thing to remember is that in India and Southeast Asia Sanskrit played, and still does play, a role comparable with that of Latin or Greek in Western Europe, with the added virtue of being the only language capable of translating, at the time of the meditations, the mystical transcendental truths said to have been revealed to the Rishi of Vedic times. Bearing in mind the power given to the language (and thus to its written expression), Sanskrit is considered to be the "language of the gods"; whoever masters the language is said to possess divine consciousness and the divine language (see *Mysticism of letters). This explains why the Sanskrit inscriptions of Cambodia, Champa and other indianised civilisations of Southeast Asia do not contain "numerals" for the expression of the *Shaka* dates. These inscriptions were nearly always in verse. As far as the stone-carvers of these regions were concerned, the introduction of numerical signs (considered "vulgar") in Sanskrit texts in verse would have constituted a sort of heresy, not only from an aesthetic point of view, but also and moreover in terms of mysticism and religion. This is why the dates were firstly written in the names of the numbers and then usually in numerical symbols as well. Moreover, the actual name "Sankskrit" is rather significant in this respect, as the word **samskrita* (Sanskrit) means "complete", "perfect" and "definitive". In fact, this language is extremely elaborate, almost artificial, and is capable of describing multiple levels of meditation, states of consciousness and psychic, spiritual and even intellectual processes. As for vocabulary, its richness is considerable and highly diversified [see L. Renou (1959)]. Sanskrit has for centuries lent itself admirably to the diverse rules of prosody and versification. Thus we can see why poetry has played such a preponderant role in all of Indian culture and Sanskrit literature.

This explains why the Indian astronomers preferred to use numerical symbols instead of numerals.

Numerical tables, Indian astronomical and mathematical texts, as well as mystical, theological, legendary and cosmological works were nearly always written in verse:

*Whilst making love a necklace broke.*
*A row of pearls mislaid.*
*One sixth fell to the floor.*
*One fifth upon the bed.*
*The young woman saved one third of them.*
*One tenth were caught by her lover.*
*If six pearls remained upon the string*
*How many pearls were there altogether?*

This is a mathematical problem posed in the **Lîlâvatî*, a famous mathematical work in the form of poems, written by *Bhâskarâchârya (in 1150), the title of which is the name of the daughter of the mathematician. Here is another example:

*Of a cluster of lotus flowers,*
*A third were offered to *Shiva,*
*One fifth to *Vishnu,*
*One sixth to *Sûrya.*
*A quarter were presented to Bhâvanî.*
*The six remaining flowers*
*Were given to the venerable tutor.*
*How many flowers were there altogether?*

From this type of game, the Indian scholars went on to use imagery to express numbers; the choice of synonyms was almost infinite and these were used in keeping with the rules of Sanskrit versification to achieve the required effect. Thus the transcription of a numerical table or of the most arid of mathematical formulae resembled an epic poem.

We need only look at the following lines from a text recording astronomical data to see how poetic and elliptical such documentation could be:

*The apsids of the moon in a yuga*
*fire. void. horsemen. Vasu. serpent. ocean*
*and of its waning node*
*Vasu. fire. primordial couple. horsemen. fire. horsemen.*

However, aesthetic refinement was not the only motive. This method also offered enormous practical advantages. Billard provides us with the precise fundamental reasons as to why the Indian astronomers chose to use word-symbols to express numbers:

> Indian astronomical texts were always written in Sanskrit. They contained little historical information, were totally devoid of discussions and demonstrations and of the kind of observations which we recognise the value of today – except for the occasional commentary, which was always written in prose – yet they possessed remarkable, even exceptional qualities. *The astronomical data is not only always explicit, but moreover the numerical values are still perfectly conserved after all this time and after so many copies have been made.* Although expressed in a very elliptical manner in the text, where the tradition of versification, used here for mnemonic purposes, led to a synonymy which was often infinite within the technical language – a rather unusual occurrence in the history of astronomy – the astronomical data is very precise and unrivalled in terms of reliability.

The importance of *numerical data* in the Indian astronomical texts is so great because the texts contain so little direct information. All we know of their *astronomical canons, for example, is the terminology by which they were described (average elements, apsids, nodes, eccentricities, exact longitudes, average longitudes, etc.), the terminology being the word-symbols.

It is precisely the study of the numerical data which led to the finding of a given canon in various different texts from very different eras, as well as facilitating the distinction between different canons (see *Indian astronomy).

Thus we can appreciate just how reliable this numerical data had to be in order for it to be transmitted from one generation to the next.

Although initially it might seem puerile, the use of word-symbols was in fact extremely efficient in conserving the exact value of the numbers they expressed, and it was doubtless to this end that the word-symbols were invented. "This conservation of the value of numbers in Sanskrit texts", writes Billard, "is even more impressive when one considers that Indian manuscripts, in material terms, generally do not survive more than two or three centuries [due to climate and above all vermin, which render the conservation of manuscripts extremely difficult], and had the numerical data been recorded in numerals, it would no doubt have reached us in an unusable state."

And Guitel observes that "from a purely mathematical point of view, the use of many different words to express each of the numbers presented no ambiguity; a text written in word-numbers could easily be translated into numerals [and vice versa]. All one would have needed was a glossary of all the words possessing a numerical value, which could be used like a dictionary of rhymes." Whatever the benefits of this system, however, it could not be used to carry out calculations.

The reason for this is obvious: numbers could be expressed using the place-value system with nine numerals and zero, and this system was doubtless invented at the same time as the positional system of the word-symbols.

However, no one would have dreamt of adding **fire*, **arrows*, **planets* and **serpents*, or of subtracting **oceans*, **orifices* or **nâga* from **elephants*, or multiplying the **faces of Kumâra* by the **eyes of Shiva* or dividing the **arms of Vishnu* by the **great sacrifices*!

Since no later than the fifth century, Indian arithmeticians used the place-value system with the nine numerals and zero to carry out complicated mathematical operations. They avoided the use of numerals for recording numerical data, but used them in their rough calculations.

On the other hand, it was very difficult for the Indian astronomers to express their numerical data in numerals, because numerals were far less reliable than the word-symbols. This is because, graphically speaking, the numerals varied according to the style of writing of each region (Fig. 24.3 to 52), and also according to the era and the author or transcriber. A shape which represented the number 2 to one person might well have represented a 3, a 7, or even a 9 to another.

The situation is completely different for us in the twentieth century, because the shapes of the numerals we use and their respective values are the same the world over. For the Indian astronomers, however, the use of numerals could cause confusion. The use of verse and word symbols, on the other hand, was very reliable, because the slightest error could break the rhythm of the verse or verses in question, and therefore would not escape notice. This is why Indian astronomers favoured word-symbols for many centuries.

There is also another, equally fundamental reason. As we have seen, Indian astronomical texts were always in verse: a prosody of long or short syllables was used, as in Graeco-Latin metrics, except that the metre and the number of syllables used in the Indian texts were always perfectly clear and very systematic. Thus the word-symbols not only guaranteed the conservation of the values of the numbers expressed, but also served a mnemonic function. "As well as allowing the writer to find a synonym which gave the required scansion, the word-symbol formed part of the metre, and the number that it expressed was at once firmly established in the text and in the mathematician's memory, who recited the verses as he worked out his calculations" (Billard). The method facilitated and reinforced the Indian scholars' memory: it allowed them to make the best use of their memory through associations of ideas or images contained within rhythms determined by the metre which was dictated by the rules of Sanskrit versification.

When we consider the above conditions, we can understand why the Indian astronomers developed the Sanskrit word-symbols, and continued to use them for such a prolonged length of time.

The same conditions led the astronomer *Âryabhata to develop his famous alphabetical numeral system. He was no doubt familiar with the Sanskrit word-symbols, but needed a system which was more concise whilst meeting the requirements of certain versified Sanskrit compositions. It is likely that he found the word-symbols to be lacking in brevity and perhaps also precision, especially when he wrote his famous sine table.

Similar reasons led astronomers such as *Haridatta or *Shankaranârâyana, at a later date, to use an alphabetical numeral system which was even more efficient then the **katapaya* system.

> The coexistence of different methods of achieving the same goals is one of the characteristics particular to the highly inventive genius of the Indians, which enjoyed both the finest distinctions and determinations, and the fluctuating wave of an abundant production, and was little inclined towards that precise and rather dry sobriety of the ancient Semites [F. Woepcke (1863)].

The discovery of the place-value system required another, equally basic progression. As soon as the place-value system was rigorously applied to the nine simple units, the use of a special terminology was indispensable to indicate the absence of units in a given order.

The Sanskrit language already possessed the word **shûnya* to express "void" or "absence". Synonymous with "vacuity", this word had for several centuries constituted the central element of a mystical and religious philosophy which had become a way of thinking.

Thus there was no need to invent a new terminology to express this new mathematical notion: the term **shûnya* ("void") could be used. This is how the word finally came to perform the function of zero as part of this exceptional counting system.

A number such as 301 could now be expressed:

*eka.shûnya.tri*

one. void. three.

1 0 3

The Sanskrit language, however, being an unrivalled literary instrument in terms of wealth, possessed more than just one word to express a void: there was a whole range of words with more or less the same meaning: words whose literal meaning was connected, directly or indirectly, with the world of symbols of Indian civilisation.

Thus words such as **abhra*, **ambara*, **antariksha*, **gagana*, **kha*, **nabha*, **vyant* or **vyoman*, which literally meant the sky, space, the atmosphere, the firmament or the canopy of heaven, came to signify not only a void, but also zero. There was also the word **âkâshâ*, the principal meaning of which was "ether", the last and the most subtle of the "five elements" of Hindu philosophy, the essence of all that is believed to be uncreated and eternal, the element which penetrates everything, the immensity of space, even space itself.

To the Indian mind, space was the "void" which had no contact with material objects, and was an unchanging and eternal element which defied description; thus the association between the elusive character and very different nature of zero (as regards numerals and ordinary numbers) and the concept of space was immediately obvious to the Indian scholars. The association between ether and "void" is also obvious because *âkâsha* (to the Indian mind) is devoid of all substance, being considered the condition of all corporal extension and the receptacle of all substance formed by one of the other four elements (earth, water, fire and air). In other words, once zero had been invented and put into use, it brought about the realisation that, in terms of existence, *âkâsha* played a role comparable with the one which zero performed in the place-value system, in calculations, in mathematics, and the sciences.

The following are other Indian numerical symbols for zero: **bindu*, "point"; **ananta*, "infinity"; **jaladharapatha*, "voyage on water"; **vishnupada*, "foot of Vishnu"; **pûrna*, "fullness, wholeness, integrity, completeness"; etc. (See also **Zero*.)

The use of one of these words prevented any misunderstanding. Later than the Babylonians, and most likely before the Maya, the Indian scholars invented zero, although for the time being it was little more than a simple word which formed part of everyday vocabulary.

So just how did the place-value system come to be applied to the nine Indian numerals?

Let us now go back to the numeral system of ancient India: the *Brâhmî* system, which, as we have already seen, constituted the prefiguration of the nine basic numerals that we use today (Fig. 24.29 to 52 and 24.61 to 69).

Current documentation suggests that the history of truly Indian numerals began with the *Brâhmî* inscriptions of Emperor Asoka (in the middle of the third century BCE). But the numerals were invented before the Maurya Dynasty, by which time the numerals were highly developed graphically speaking, and widespread throughout the Indian territory.

In fact, as we have already seen, the first nine *Brâhmî* numerals which appear in Asoka's edicts are probably vestiges of an old indigenous system (no doubt older than *Brâhmî* writing itself), where the nine units were represented by the necessary number of vertical lines, similar to the arrangements in Fig. 24.59.

We will now sum up the evidence we have compiled in this chapter on the early stages in the history of these numerals.

Like all the other civilisations of the world, the Indians initially used the required number of vertical lines to write the first nine numbers. However, as a row of vertical lines was not conducive to rapid reading and comprehension, this system was gradually abandoned, at least for the numbers 4 to 9. To overcome the problem, the lines representing the units were split into two groups (two lines on top of two others for 4, three lines on top of two others for 5, etc.; see Fig. 1.26), or a ternary arrangement was used (three lines on top of one line for 4, three above two for 5, etc.; see

Fig. 1. 27). This was how the Sumerians, Cretans and Urarteans proceeded, as well as the Egyptians, Assyrians, Phoenicians, Aramaeans and Lydians. In the long run, however, such groupings of lines did not allow for rapid writing, or time-saving, which was the main preoccupation of the scribes. Thus – due to a combination of circumstances imposed by the very nature of the writing materials used (the scribes wrote upon tree bark or palm leaves with a brush or calamus) depending on the region – a numeral system evolved which was unique to each civilisation and the numerals no longer visually represented their respective values. In each civilisation the change was brought about by both the nature of the writing materials and the desire to save time. Signs were invented which could be written in one single stroke or in short, quick strokes. Ligatures were exploited wherever possible, so that the brush need not be lifted, allowing several lines to be grouped together in one single sign. The initial representations of the numbers were radically modified, as we can see with the *Brâhmî* numerals for the numbers four to nine (Fig. 24.57, 58 and 60).

At the outset, these nine signs were not used in conjunction with the place-value system: the *Brâhmî* system relied on the principle of addition and a specific sign was given to each of the nine units in each decimal order, up to and including tens of thousands (Fig. 24.70).

Mathematical operations, even simple addition, were almost impossible without the invention of a device. The ancestors of our modern numerals remained static for some time before acquiring the dynamic and workable nature of the current numerals. Like certain other systems of the ancient world, this system was also rudimentary whilst it was only used to express numbers.

Mathematicians, philosophers, cosmographers and all others who, for one reason or another, were handling high numbers at that time, resorted to using the Sanskrit names of the numbers.

However, like all the mathematicians of the ancient world, Indian arithmeticians, before discovering the place-value system, used their fingers or, more often, concrete mathematical devices.

It seems that the most common was the abacus: from right to left, the first column represented the units, the next the tens, the third the hundreds, and so on.

Unlike the Greeks, Romans or Chinese, however, who then went on to use pebbles, tokens or reeds, the Indian mathematicians had the idea of using the first nine numerals of their counting system, tracing them in fine sand or dust, inside the column of the corresponding decimal order.*

* This information was obtained from descriptions given by various Indian authors, and later accounts from many Arabic, Persian, European and even Chinese authors [see Allard (1992); Cajori (1980); Datta and Singh (1938); Iyer (1954); Kaye (1908); Levey and Petruck (1965); Waeschke (1878)].

Thus a number such as 7,629 would have been represented in the following manner, with nine in the units column, two in the tens column, six in the hundreds and seven in the thousands:

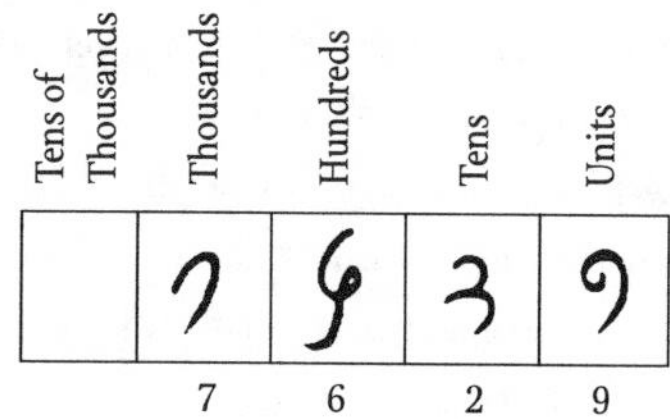


Of course, when a unit within an order of units was missing, one only needed to leave the appropriate column empty; thus the representation of 10,267,000:

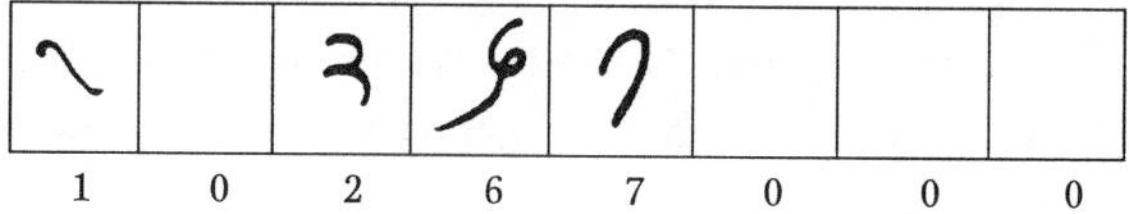


The mathematical operation would be carried out by successively erasing the results of the intermediary calculations. (There is a simple example of this in Chapter 25.)

Like us today, the "Pythagorean" tables had to be known by heart, which give the results of the four elementary operations of the nine significant numerals.

Before the beginning of the fifth century BCE, then, all the necessary "ingredients" for the creation of the written place-value system had been amassed by the Indian mathematicians:

- the units one to nine could be expressed by distinct numerals, whose forms were unrelated to the number they represented, namely the first nine *Brâhmî* numerals;
- they had discovered the place-value system;
- they had invented the concept of zero.

A few stages, however, were still lacking before the system could attain perfection:

- the nine numerals were only used in accordance with the additional principle for analytical combinations using numerals higher than or equal to ten, and the notation was very basic and limited to numbers below 100,000;
- the place-value system was only used with Sanskrit names for numbers;
- and zero was only used orally.

In order for the "miracle" to take place, the three above ideas had to be combined.

By using the nine *Brâhmî* numerals in the appropriate columns of the "dust" abacus, the Indian mathematicians had already reached the stage which would inevitably lead them to this major discovery.

This becomes clear when we imagine the Indian mathematicians at work, recording the result of a calculation they had carried out by drawing their abacus in the dust, bearing in mind that they had two methods of expressing numbers: the *Brâhmî* numerals and the Sanskrit names of the numbers.

In the abacus, they would have drawn the numerals in a contemporary style (those from the inscriptions of Nana Ghat, for example, dating from the second century BCE; see Fig. 24.30 and 71), and a calculation might have given the following result:

The figure obtained is 4,769.

As we know, from this time on, the numbers were expressed in their Sanskrit names in the order of ascending powers of ten, from the smallest to the highest.

Therefore this result would have been expressed as follows:

*nava shashti saptashata cha chatursahasra*
"nine sixty seven hundred and four thousand".

In numerals, however, the numbers would have been written in the opposite order, reading from left to right:

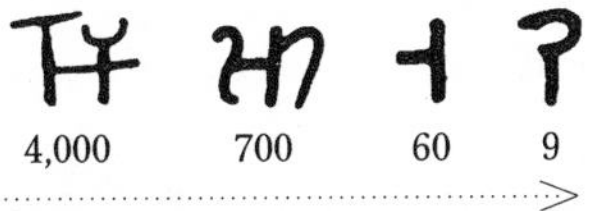


We have evidence of these methods of expressing figures in Indian inscriptions, since the third century BCE, from those of Asoka, Nana Ghat and Nasik to those of the Shunga, Shaka, Kusana, Gupta and Pallava dynasties. The corresponding numerical notations, all issuing from the original *Brâhmî* system, possess a different numeral for each unit of each decimal order (Fig. 24.70 and 71).

When we examine the signs used, we discover that these numerals are not all independent of one another; the only numerals which are really unique are the following:

| 1 | 2 | 3 | 4 | 5 | 6 | 7 | 8 | 9 |
|---|---|---|---|---|---|---|---|---|
| 10 | 20 | 30 | 40 | 50 | 60 | 70 | 80 | 90 |
| | | 100 | | | 1,000 | | | |

The numerals for 200 and 300, as well as those for 2,000 and 3,000, derive from the sign for 100 and 1,000 respectively, with the simple addition of one or two horizontal lines (Fig. 24.70 C).

In other words, the four numerals in question conformed to the following mathematical rules:

$$200 = 100 + 1 \times 100 \qquad 2{,}000 = 1{,}000 + 1 \times 1{,}000$$
$$300 = 100 + 2 \times 100 \qquad 3{,}000 = 1{,}000 + 2 \times 1{,}000$$

As for the remaining multiples of one hundred and one thousand, they were represented using the principle of multiplication and placing the numeral for the corresponding unit to the right of the sign for one hundred or one thousand:

$$400 = 100 \times 4 \qquad 4{,}000 = 1{,}000 \times 4$$
$$500 = 100 \times 5 \qquad 5{,}000 = 1{,}000 \times 5$$
$$600 = 100 \times 6 \qquad 6{,}000 = 1{,}000 \times 6$$
$$700 = 100 \times 7 \qquad 7{,}000 = 1{,}000 \times 7$$
$$800 = 100 \times 8 \qquad 8{,}000 = 1{,}000 \times 8$$
$$900 = 100 \times 9 \qquad 9{,}000 = 1{,}000 \times 9$$

It is visibly evident that this rule also applied to the notation of tens of thousands by placing the corresponding number of tens next to the sign for a thousand:

$$10{,}000 = 1{,}000 \times 10$$
$$20{,}000 = 1{,}000 \times 20$$
$$30{,}000 = 1{,}000 \times 30$$
$$40{,}000 = 1{,}000 \times 40$$

Thus, the number 4,769 was written:

1,000 × 4 + 100 × 7 + 60 + 9

This corresponds exactly, but in the opposite order, to the above Sanskrit expression of the figure:

*nava shashti saptashata cha chatur sahasra*
9 + 60 + 7 × 100 + 4 × 1,000

If we look at either way of expressing the sum in the opposite direction from the way it would have been spoken or written, we obtain the arithmetical breakdown of the other. This is what the Indian arithmeticians expressed in the phrase **ankânâm vâmato gatih*, which literally means the "principle of the movement of numerals from the right to the left", which applies to the reading of numbers from the smallest unit to the highest in ascending order of powers of ten.

The inscriptions of Nana Ghat provide the earliest known significant evidence of this principle. Thus we know that from at least as early as the second century BCE, the numbers were expressed in ascending powers of ten in *Brâhmî* numerals; in other words, they were expressed in the opposite order than from left to right. This means that the structure of *Brâhmî* notation had been copied exactly from the Sanskrit system.

Since the highest *Brâhmî* numeral expressed the number 90,000, it was impossible to use this system to express a number that was higher than 99,999.

As the *Brâhmî* numerals constituted an abbreviated written form of the spoken numeration, it had been developed to avoid having to express frequently used numbers through the long-winded Sanskrit names of the numbers.

In other words, the result of a calculation which was equal to or higher than 100,000 could only be written down in the Sanskrit names of the numbers.

The abacus traced in the dust could be used to carry out calculations involving extremely high numbers: each column represented a power of ten, and there was no limit to the number of columns which could be drawn.

Thus there was a very close link between the ability to carry out calculations on this abacus and the level of conception of high numbers and the capacity to express them orally or through writing.

In Indian calculation, the successive columns of the abacus always rigorously corresponded to the consecutive powers of ten. As the Sanskrit counting system possessed the same mathematical structure, these columns corresponded exactly to the impeccable succession of names which the Sanskrit system possessed for the various powers of ten. Thus each system constituted the mirror image of the other.

This is exactly where the potential to discover the place-value system of the nine numerals lay. As with the Sanskrit names of the numbers, the structure of the abacus contained the key to the discovery of the decimal place-value system. This is why the Sanskrit notation was perfect for recording the results of the calculations which were carried out on the abacus.

This becomes even clearer when we take the number 523,622,198,443, 682,439, as it would have appeared when written on the abacus using the nine *Brâhmî* numerals (see adjacent column).

We can see how the close relationship between the representation of the numbers on the abacus and the Sanskrit system led to the change in direction, before the second century BCE, of numerical expressions given using the Sanskrit names of the numbers.

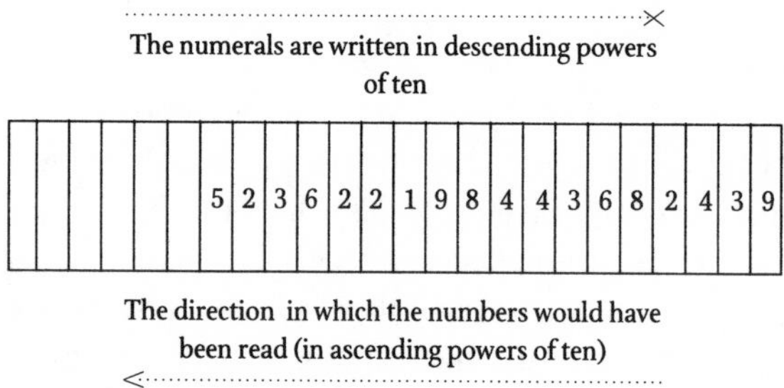
The numerals are written in descending powers of ten
5 2 3 6 2 2 1 9 8 4 4 3 6 8 2 4 3 9
The direction in which the numbers would have been read (in ascending powers of ten)

Whilst the numerals read from left to right on the abacus in descending powers of ten, they came to be read from right to left, from the smallest number to the highest.

Bearing in mind the conditions imposed by the very nature of the calculating instrument, the Indian mathematicians had no other choice but to adopt the expression of numbers in the direction described by **ankânâm vâmato gatih*: "the principle of the movement of the numerals from the right to the left". How could they know how to write a high number on the abacus if they began with the highest order? They would have had to work out which column each order had to be placed in by counting each corresponding column beginning with the column for the simple units. This would have wasted time. Thus the best solution was to always start with the column for the simple units.

Thus the old system was abandoned. By beginning with the highest power of ten, the arithmetician immediately knew the size of the number he was dealing with, but this did not facilitate the drawing, on the columns of the abacus, the successive numerals of a number which possessed more than four or five orders of units.

This is why the opposite direction was adopted, the advantage of which being that, no matter how high a number might be, there could be no mistake as to which column each numeral must be written in. It was for the same reason that this direction of expressing the numbers was conserved later on when the positional notation was invented using numerical symbols:

> We must not forget that the numbers which appear in the scientific poems [such as the numerical data given by the Indian astronomers] were destined for mathematical use. Certain lists contain numbers proportional to the differences of sines of angles which ascend in mathematical order; these enabled, with the aid of additions, an almost instant reconstruction of the numbers proportional to the sines of these angles.

The *pandit* dictated the poetic text which the scribe wrote in numerals. How could an addition be carried out if the data consisted of *wing* (= 2) and *fire* (= 3)? If only one number had to be reproduced, even a very high number, it would have been easy to translate it directly onto "paper", but if a series of numbers was involved, how could they be correctly placed on the counting table if they were read out in descending orders of units?

It would have been impossible to transcribe a number in this way unless it was known in advance. The only solution was to read out the numbers in ascending orders of units.

However, when the *pandit* was reading out a high number, the scribe needed to know its highest order; this is why the *pandit* started with the highest powers of the base; this is not possible if one uses the spoken positional numeration, yet this system did enable one to place the number correctly on the counting table, and then one could plainly see the powers of the base which had been recorded [G. Guitel, (1966)].

If we look again at the representation of the number given above, on the "dust" abacus, its mathematical breakdown according to **ankânâm vâmato gatih* was as follows:

$$\begin{aligned} &= 9 + 3 \times 10 + 4 \times 10^2 + 2 \times 10^3 + 8 \times 10^4 + 6 \times 10^5 \\ &\quad + 3 \times 10^6 + 4 \times 10^7 + 4 \times 10^8 + 8 \times 10^9 + 9 \times 10^{10} \\ &\quad + 1 \times 10^{11} + 2 \times 10^{12} + 2 \times 10^{13} + 6 \times 10^{14} + 3 \times 10^{15} \\ &\quad + 2 \times 10^{16} + 5 \times 10^{17} \\ &= 523{,}622{,}198{,}443{,}682{,}439 \end{aligned}$$

This corresponds exactly to the following Sanskrit expression:

"Nine and three *dasha* and four *shata* and two *sahasra* and eight *ayuta* and six *laksha* and three *prayuta* and four *koti* and four *vyarbuda* and eight *padma* and nine *kharva* and one *nikharva* and two *mahâpadma* and two *shankha* and six *samudra* and three *madhya* and two *antya* and five *parârdha*."

Once the Sanskrit numeration was simplified, this number could be expressed in the following manner:

nine.three.four.two.eight.six.three.four.four.
eight.nine.one.two.two.six.three.two.five

Why was the number written in the names of the numbers instead of in *Brâhmî* numerals, to which the place-value system could have been applied?

This is surely what the Indian mathematicians asked themselves one day, in their continuing desire to economise with time and materials.

They carried out calculations involving high numbers, for which even the intermediate results constituted very high numbers, and which could be difficult to memorise. The results were first recorded in rough. As they needed to be sparing with materials and time, the Indian mathematicians sought a way to write faster and in a more abridged form than the Sanskrit system, even in its simplified form. They realised that the nine *Brâhmî* numerals could provide the "stenography" that they required.

However, bearing in mind the position of the numerals on the abacus, it was necessary to revert to the descending order, thus going from the highest orders of units to the smallest, so as not to cause confusion in the numeral representations; for the results of calculations carried out on the abacus, the numerals had to be placed in the columns in the same positions as they appeared when written in rough.

Thus the number in question acquired the following notation, the numbers reading from left to right in descending order of powers of ten, constituting a faithful reproduction, minus the columns, of its representation on the abacus, as well as the reflection of the abridged form of the corresponding Sanskrit expression:

5 2 3 6 2 2 1 9 8 4 4 3 6 8 2 4 3 9.

Whence came the decimal position values which were given to the first nine numerals of the old notation which originated at the time of the reign of Emperor Asoka.

This was the birth of modern numerals, which signalled the death of the abacus and its columns.

The introduction of a new symbol proved indispensable in order to convey the absence of units in a given decimal order; whilst this sign was not needed when the abacus was used, it was of utmost necessity in the new positional numeral notation.

The Indians, never lacking in resources in these matters, again turned to their unique symbolism.

As we saw earlier, the word-symbols that the Sanskrit language used to express the concept of zero conveyed concepts such as the sky, space, the atmosphere or the firmament.

In drawings and pictograms, the canopy of heaven is universally represented either by a semi-circle or by a circular diagram or by a whole circle. The circle has always been regarded as the representation of the sky and of the Milky Way as it symbolizes both activity and cyclic movements [see J. Chevalier and A. Gheerbrant (1982)].

Thus the little circle, through a simple transposition and association of ideas, came to symbolise the concept of zero for the Indians.

Another Sanskrit term which came to mean zero was the word **bindu*, which literally means "point".

The point is the most insignificant geometrical figure, constituting as it does the circle reduced to its simplest expression, its centre.

For the Hindus, however, the *bindu* represents the universe in its non-manifest form, the universe before it was transformed into the world of appearances (*rûpadhâtu*). According to Indian philosophy, this uncreated universe possessed a creative energy, capable of generating everything and anything: it was the causal point.

The most elementary of all geometrical figures, which is capable of creating all possible lines and shapes (*rûpa*) was thus associated with zero, which is not only the most negligible of quantities, but also and above all the most fundamental of all abstract mathematics.

The point was thus used to represent zero, most notably in the *Shâradâ* system of Kashmir, and in the vernacular notations of Southeast Asia (Fig. 24.82).

From the fifth century CE, the Indian zero, in its various forms, already surpassed the heterogeneous notions of vacuity, nihilism, nullity, insignificance, absence and non-being of Greek-Latin philosophies. **Shûnya* embraced all these concepts, following a perfect homogeneity: it signified not only void, space, atmosphere and "ether", but also the non-created, the non-produced, non-being, non-existence, the unformed, the unthought, the non-present, the absent, nothingness, non-substantiality, nothing much, insignificance, the negligible, the insignificant, nothing, nil, nullity, unproductiveness, of little value and worthlessness (see **Shûnyatâ*, **Zero*, and Fig. 24.D10 and D11 of the latter entry in the Dictionary).

It was also, and above all, an eminently abstract concept: in the simplified Sanskrit system, as well as in the positional system of the numerical symbols, the word **shûnya* and its various synonyms served to mark the absence of units within a given decimal order, in a medial position as well as in an initial or final position; the point or the little circle were used in the same way.

This zero was also a mathematical operator: if it was added to the end of a numerical representation, the value of the representation was multiplied by ten.

By freeing the nine basic numerals from the abacus and inventing a sign for zero, the Indian scholars made significant progress, primarily simplifying quite considerably the rules of a technique which would lead to the birth of our modern written calculation.

The Indian people were the only civilisation to take the decisive step towards the perfection of numerical notation. We owe the discovery of modern numeration and the elaboration of the very foundations of written calculations to India alone.

It is very likely that this important historical event took place around the fourth century CE. It is thanks to the genius of the Indian arithmeticians that three significant ideas were combined:

LIST OF SANSKRIT WORDS FOR ZERO

| SYMBOLS | THEIR MEANINGS |
|---|---|
| **Abhra* | Atmosphere |
| **Akâśha* | Ether |
| **Ambara* | Atmosphere |
| **Ananta* | Immensity of space |
| **Antariksha* | Atmosphere |
| **Bindu* | Point (or Dot) |
| **Gagana* | Canopy of heaven |
| **Jaladharapatha* | Voyage on water |
| **Kha* | Space |
| **Nabha* | Sky, Atmosphere |
| **Nabhas* | Sky, Atmosphere |
| **Pûrna* | Entire, Complete |
| **Randhra* (rare) | Hole |
| *´*Shûnya* | Void |
| **Vindu* | Point (or Dot) |
| **Vishnupada* | Foot of Vishnu |
| **Vyant* | Sky |
| **Vyoman* | Sky, Space |

Ref.: Al-Biruni; Bühler; Burnell; Datta and Singh; Fleet, in: CIIN, III; Jacquet, in: JA, XVI, 1935; Renou and Filliozat; Sircar; Woepcke (1863).

GRAPHICAL SIGNS FOR ZERO

First sign: the little circle ○

Nowadays used in nearly all the notations of India and Southeast Asia (*Nâgarî, Marâthî, Punjabî, Sindhî, Gujarati, Bengâlî, Oriyâ, Nepâlî, Telugu, Kannara, Thai, Burmese, Javanese, etc.*). There is evidence of the use of this sign which dates back many centuries for nearly all these systems.

Second sign: the point, or dot ●

Formerly used in the regions of Kashmir (*Shâradâ* numerals). There is also evidence of the use of this sign in the Khmer inscriptions of ancient Cambodia and the vernacular inscriptions of Southeast Asia.

Fig. 24.3 to 51, 24.78 to 80. See also *Numerals "0", *Zero and Fig. 24.D11.

FIG. 24.82. *The various representations of the Indian zero**

- nine numerals which gave no visual clue as to the numbers they represented and which constituted the prefiguration of our modern numerals;
- the discovery of the place-value system, which was applied to these nine numerals, making them dynamic numerical signs;
- the invention of zero and its enormous operational potential.

Thus we can see that the Indian contribution was essential because it united calculation and numerical notation, enabling the democratisation of calculation. For thousands of years this field had only been accessible to the privileged few (professional mathematicians). These discoveries made the domain of arithmetic accessible to anyone.

It still remained for the Indian scholars to perfect the concept of zero and enrich its numerical significance.

Beforehand, the **shûnya* had only served to mark the absence of units in a given order. The Indian scholars, however, soon filled in the gap. Thus, in

* The Arabs acquired their signs for zero, as well as the place-value system, from the Indians. This is why we find the point and the little circle used to express zero in Arabic texts. The circle was the sign to prevail in the West, after the Arabs transmitted it to the Europeans some time after the beginning of the twelfth century.

a short space of time, the concept became synonymous with what we now refer to as the "number zero" or the "zero quantity".

The **shûnya* was placed amongst the **Samkhyâ*, which means it was given the status of a "number".

In c. 575, astronomer *Varâhamihira, in *Pañchasiddhântika,* mentioned the use of zero in mathematical operations, as did *Bhâskhara in 629 in his commentary on the *Âryabhatîya*.

In 628, in *Brâhmasphutasiddhânta*, *Brahmagupta defined zero as the result of the subtraction of a number by itself (a – a = 0), and described its properties in the following terms:

> When zero (**shûnya*) is added to a number or subtracted from a number, the number remains unchanged; and a number multiplied by zero becomes zero.

Moreover, in the same text, Brahmagupta gives the following rules concerning operations carried out on what he calls "fortunes" (*dhana*), "debts" (*rina*) and "nothing" (*kha*) [see S. Dvivedi (1902), pp. 309–10, rules 31–5)]:

> A debt minus zero is a debt.
> A fortune minus zero is a fortune.
> Zero (**shunya*) minus zero is nothing (**kha*).
> A debt subtracted from zero is a fortune.
> So a fortune subtracted from zero is a debt.
> The product of zero multiplied by a debt or a fortune is zero.
> The product of zero multiplied by itself is nothing.
> The product or the quotient of two fortunes is one fortune.
> The product or the quotient of two debts is one debt.
> The product or the quotient of a debt multiplied by a fortune is a debt.
> The product or the quotient of a fortune multiplied by a debt is a debt.

Modern algebra was born, and the mathematician had thus formulated the basic rules: by replacing "fortune" and "debt" respectively with "positive number" and "negative number", we can see that at that time the Indian mathematicians knew the famous "rule of signs" as well as all the fundamental rules of algebra.

It is clear how much we owe to this brilliant civilisation, and not only in the field of arithmetic; by opening the way to the generalisation of the concept of the number, the Indian scholars enabled the rapid development of algebra, and thus played an essential part in the development of mathematics and exact sciences.

The discoveries of these men doubtless required much time and imagination, and above all a great ability for abstract thinking. The reader will not be surprised to learn that these major discoveries took place within an environment which was at once mystical, philosophical, religious, cosmological, mythological and metaphysical.

*Sarasvati, goddess of knowledge and music. From Moor's* Hindu Pantheon

CHAPTER 24 PART II

# DICTIONARY OF THE NUMERICAL SYMBOLS OF INDIAN CIVILISATION

As we have seen in the course of this chapter, India has always dominated the world in the field of arithmetic, and indeed the art of numbers plays a leading role in Indian culture.

It is precisely this skill in the field of mathematics which led Indian scholars, at a very early stage, to develop their astonishing **arithmetical speculations* which could involve numbers comprising hundreds of decimal places, whilst maintaining a clear idea of the order of the ascending powers of ten within a nomenclature which contained a highly diversified terminology, based as it was upon both specific etymologies and improvised terms born out of a highly creative symbolical imagination.

This same arithmetical genius led to their inevitable discovery of zero and the place-value system and even enabled them to come within touching distance of the concept of mathematical infinity.

Therefore one should not be surprised to learn that, a thousand years earlier than the Europeans, Indian mathematicians already knew that zero and infinity were inverse concepts. They realised that when any number is divided by zero the result is infinity: $a / 0 = \infty$, this "quantity" undergoing no change if it is added to or subtracted from a finite number.*

We should not forget that these crucial discoveries were not the fruit of just one genius's individual inspiration nor even that of a group of "mathematicians" as we understand the term today. They were, of course, learned Indians. However, there should be no ambiguity about the meaning of this term: to be termed as learned at that time meant that one was a thinker, a little like the scholarly gentlemen of sixteenth-century Europe, except that an Indian's way of thinking would have been very different from that of the European scholars. The Indian scholar would have been a man of constant reflection whose studies covered the most diverse topics. Moreover, mystical, symbolical, metaphysical and even religious considerations came first and foremost in his learning:

> As India knew nothing of the work of either Aristotle or Descartes, and was ignorant of Jewish or Christian ethics, it would be futile to attempt to draw a parallel between these vastly different civilisations. The foundations are completely different, as are the customs and ways of thinking. It is impossible to make any comparisons, even if some aspects of Indian culture do seem to coincide with our own. [L. Frédéric, *Dictionnaire de la civilisation indienne* (1987)]

It would also be futile to try and make any comparisons between Indian mathematics and modern mathematics, modern mathematics being the very refined product of contemporary western civilisation: a highly abstract science that has been stripped of any mystical, philosophical or religious influence.

Moreover, the following pages prove that the main preoccupations of Indian scholars had nothing to do with what we in the West today refer to as "hard sciences". In fact, we will see that these major discoveries stem from the incessant study of astronomy, poetry, metric theory, literature, phonetics, grammar, philosophy or mysticism, and even astrology, cosmology and mythology all at once.

In India, an aptitude for the study of numbers and arithmetical research was often combined with a surprising tendency towards metaphysical abstractions: in fact, the latter is so deeply ingrained in Indian thought and tradition that one meets it in all fields of study, from the most advanced mathematical ideas to disciplines completely unrelated to 'exact' sciences.

In short, Indian science was born out of a mystical and religious culture and the etymology of the Sanskrit words used to describe numbers and the science of numbers bears witness to this fact.† Together, the discoveries in question represent the culmination of the uniqueness, wealth and incredible diversity of Indian culture.

To give the reader a clearer idea of the circumstances and conditions under which these major discoveries were made, it seems useful now to reiterate the principal notions that have already been explained in this chapter in the form of a Sanskrit and English dictionary and so to define (if need be) these ideas in a more analytical form. This dictionary can serve as a glossary to the numerous ideas which have been covered, each term being marked with an asterisk.

* In other words, Indian scholars knew the following properties: $(a/0) \pm k = k \pm (a/0) = (a/0)$, which is to say: $\infty \pm k = k \pm \infty = \infty$.

† The term for the "Science of numbers" in Sanskrit is **samkhyâna* (also spelt *sankhyâna*) which means "arithmetic": and, by extension, "astronomy" (from the time when the science of the stars was not considered to be a separate discipline from arithmetic). The word is frequently used in this sense in **Jaina* literature, where the science of numbers was considered to be one of the fundamental requirements for the full development of a priest; likewise in later Buddhist literature it was considered to be the most noble of the arts. the word "number" itself is **samkhya* or **samkhyeya*. One should note that this term not only applies to the concept of number but also to the actual numerical symbol. Arithmetician or mathematician is denoted by **sâmkhya*. But **sâmkhya* is also the term used for one of six orthodox (and most ancient) systems of the Hindu philosophy of the six **darshana* ("contemplations"), which teaches "number" as a way of thinking which is connected to the liberation of the soul, and according to which the universe was born out of the union of **prakriti* (nature) and *purusha* (the conscience). It is significant that the word **sâmkhyâ*, which also means a follower of this philosophy, is the term used to refer to the "calculator", but in a mystical sense in this context.

Thus the dictionary can help the reader through the maze of obscure rubric of the Sanskrit language and the complex concepts of Indian science and philosophy.

The dictionary is not only aimed at specialists: the entries, recorded with careful clarity and precision, are concise and easily accessible to the layman. It is not even necessary to have read Chapter 24 (or the preceding chapters) to grasp the concepts it explains.

Its entries are recorded alphabetically, regardless of whether the terminology is in English or Sanskrit.

This dictionary also serves as a thematic index which can clarify the ideas presented in this chapter through an effective reference system of general or specific rubric, giving not only references to Chapter 24 but also to those of the forthcoming Chapters 25 and 26. For example, the reader has only to turn to the entries **Chhedi*, **Shaka* or **Vikrama* to find out about each era.

Under **Asankhyeya*, the Sanskrit for "incalculable", we learn that the same term was also sometimes used to express the rather more modest sum of ten to the power 140.

Similarly, the term **Padma*, or **Paduma*, reveals that the poetic name "pink lotus" was used to denote the number ten to the power of 14 (or 29) as well as ten to the power of 119. The lotus flower was used to represent various numbers in Indian mathematics, the values of which depended on the colour, the number of petals and whether the flower was open, just flowering or still in bud. Thus **white lotus* came to mean ten to the power 27 or ten to the power 112, **pink lotus* ten to the power 21 or 105, and **blue lotus (half-open)* ten to the power 25 or 98. (See **Utpala*, **Pundarika*, **Kumud* and **Kumuda* that can all mean "lotus", according to slight characteristic differences of the flower.)

Under entries entitled **High numbers*, there are numerous other examples of the unique symbolism which without a doubt was the source of inspiration for the names of these large figures. The same entries also demonstrate how in ancient India, grammar and interpretation were inextricably linked to the handling of high numbers to the extent to which the study of poetry and the Sanskrit metric system inevitably initiated the Indian scholars into the art of arithmetic as well as grammar. Consequently poets, grammarians and astronomers, in fact all learned men, were as skilled at calculation as the arithmeticians and the teachers themselves.

Under the entries **Ananta*, we see that the Sanskrit name for infinity was not only used to denote the sum of ten thousand million (ten billion in US English) but also, curiously, it was used as the symbol for zero.

The entries **Infinity* and **Serpent* will allow the reader to understand the relationship between infinity as we understand the term, and the mythological world of Hinduism, and that **Ananta*, often represented as coiled up in a sort of sleeping "8" (like our symbol ∞), is none other than the immense serpent of infinity and eternity, which is linked to the ancient myths of the original serpent.

This dictionary provides an insight into the circumstances under which the Indian scientists invented zero and the place-value system. See **Names of numbers*, **Sanskrit*, **Place-value system*, **Position of numerals*, **Numerical symbols (Principle of the numeration of)*, **Shûnya* and **Zero*.

Under the entries **Shûnya* and **Shûnyatâ*, the Indian philosophical notions of "void" or "emptiness" are very briefly explained, and we can see how these early Indian concepts went far beyond corresponding but very heterogeneous notions of contemporary Graeco-Latin philosophies.

It also shows how, from a very early stage, **shûnya* meant zero as well as emptiness, the central element of the deeply religious and mystical philosophy, **shûnyatâ*. The word came to represent zero when the place-value system was discovered, the two concepts fitting together naturally. The dictionary also explains how, through the use of symbolism, this concept finally came to be graphically represented as the little circle that we all recognise as zero.

The entries **Yuga* and **Kalpa* tell of Indian cosmic cycles, and the speculations developed about them, both in Indian cosmogony and in the learned astronomy introduced by *Âryabhata at the beginning of the sixth century. See also **Âryabhata*, **Cosmic cycles*, **Day of Brahma* and **Indian astronomy*.

The entry **Âryabhata's number-system* serves as further proof that it was the Indians who discovered zero and who are responsible for our current written number-system. In fact, we will see that this scholar, whilst constructing his own numerical notation (a very clever alphabetical number-system), could not have failed to have known of zero and the place-value system, given the very structure of the system.

If we look at the entries beginning with the expression **numeral alphabet*, we can find information about Indian alphabetical numbering systems. This will also give the reader some idea of the practices which were quite naturally born out of their usage: the composition of chronograms, the invention of secret codes, the preparation of talismans (closely linked to the Indian mysticism of letters and numbers), the development of homiletic or symbolic interpretations, predictive calculations and magical and divinatory practices.

Under the appropriate headings, one can similarly find short biographical notes about the great Indian scientists such as *Âryabhata, *Bhâskara, *Bhâskarâchârya, *Brahmagupta or *Varâhamihîra, often accompanied by very precise accounts of the numerical notations they adopted (including bibliographical references).

If we consult the entries **Brâhmî numerals*, **Gupta numerals*, **Nâgarî numerals*, **Shâradâ numerals*, etc., we can also find out all about each style and see the impressive diversity of **Indian written numeral systems*. Alternatively, the reader can consult **Indian numerals*.

Extra details can be found about **Indian arithmetic*, the different **ages* of the sub-continent and **Indian astrology*, **Indian astronomy* and the **Indian mathematics* of this civilisation.

However, it is not necessary to look up all of the references given here: entries are accompanied by references to similar terms.

Entries such as **Algebra*, **Arithmetic*, **High numbers*, **Names of numbers*, **Numerical notation* etc., include either an alphabetical or a numerical list of terms relating to each of the ideas in question.

As for references which seem to have little to do with arithmetic, see: **Astronomical speculations*, **Buddhism*, **Brahmanism*, **Cosmogonic speculations*, **Hinduism*, **Indian cosmogonies and cosmologies*, **Indian divinities*, **Indian mythologies*, **Indian thought*, **Indian religions and philosophies*, **Jaina*, etc.

The main aim behind creating this dictionary has been, however, to give the reader a better idea of the subtle and complicated world of Indian numbers: a world which is largely unknown to Western readers and which is closely linked to Indian legends and cosmogonies. Its symbols, rather than being ordinary graphic signs and names of numbers, derive from a huge wealth of synonyms inspired by nature, human morphology, everyday activities, social conventions and traditions, legends, religion, philosophy, literature, poetry, the attributes of the divinities and by traditional and mythological ideas.

Thus, depending on the context, the idea of **wind* can evoke the number 5, the number 7 or the number 49. This demonstrates a subtlety which Westerners would not grasp if it was not shown from the correct perspective. The reasoning behind these diverse meanings offers a fascinating example of a logic and a way of thinking which is highly characteristic of the Indian mentality, and will help the reader to understand Cartesianism, which can often, due to the very nature of its rationalism, seem in total contradiction. To find out about the logic behind the above values of **wind*, see **Vâyu*, **Pâvana* and **Mount Meru*.

Other examples include: the term **anga*, which literally means "limbs" or "parts" and is often used as the numerical symbol for the number six; the word **rasa*, "sensations", is frequently used to denote the same number; the name of **Rudra*, the ancient Vedic god of the Sky, was used as the numerical symbol for 11, etc. Similarly, the **God of carnal love*, whose name is **Kâma*, was a symbol for the number 13; the **God of water and oceans*, **Varuna*, was the symbol for the 4; **Agni*, the **God of sacrificial fires* meant "three", etc.

These examples (along with many others) show the subtlety of the Indian symbolic system as well as demonstrating one of the most characteristic traits of the Indian cast of mind.

This dictionary contains a considerable amount of symbolism. A term with a symbolic meaning is denoted by an [S], an abbreviation of the actual Indian numerical symbol, and is defined firstly by its numerical value and literal meaning in Sanskrit and then, where possible, its symbolism is explained.

To gain a better understanding of the world in which the Indians created such symbols, the reader might find it useful to read the entries **Symbols* and **Numerical symbols*.

To find the Indian numerical symbol for a given number, one only has to look up the English (or Sanskrit) equivalent: **One*, **Two*, **Three*, etc. (or **Eka*, **Dva*, **Tri*, etc.) Under **Ashta*, for example, the normal Sanskrit word for the number 8, there is a list of terms in which the word appears because of a direct link with the idea of the number 8 (for example **ashtadiggaja*, the "eight elephants", guardians of the eight horizons of Hindu cosmogony). But if the reader wishes to know about words that have a more symbolical relationship with the same number, he should refer to the entry **Eight*, where there is not only a list of numerical symbols which are synonymous with this number, but also a summary explanation of its different symbolical meanings: the serpent (**Ahi*, **Naga*, **Sarpa*), the serpent of the deep (**Ahi*), the elephant (**Dantin*, **Dvipa*, **Gaja*), the eight elephants (**Diggaja*), a sign that augurs well (**Mangala*), etc.

Of course, one could also consult the more detailed rubric either in Sanskrit: **Hastin*, **Lokapâla*, **Murti*, **Tanu*, etc., or the English translation of the concepts behind Indian numerical symbols, such as: **Sky*, **Space*, **Elephant*, **Moon*, **Earth*, **Sun*, **Zodiac*, **Serpent*, etc.

The entry **Numerical symbols (general alphabetic list)* contains all the word-symbols of the Sanskrit language that are included in this dictionary, whilst the entry **Symbolism of words with a numerical value* gives an alphabetical list of English words which correspond with associated ideas contained within the Sanskrit symbols.

The entry **Symbolism of numbers* provides a list of ideas (in arithmetical order) found in the symbolism of ordinary numbers, in high numbers and in the concept of infinity or zero.

This dictionary is the first of its kind to attempt to understand the thought process of the symbolic mind that characterises Indian numerical thinking.

Through a multidisciplinary process, mainly concerned with numbers and the symbols which represent them, the following is a kind of "vertical reading" of literary, philosophical, religious, mystical, mythological, cosmological, astronomical, and of course mathematical elements of this incredibly rich and subtle civilisation: elements which can be found in "horizontal" presentation in a great many wide-ranging publications in the most specialised of libraries.

This dictionary could be said to complement *L'Inde Classique* by L. Renou and J. Filliozat (1953), and also the monumental *Dictionnaire de la civilisation indienne* by L. Frédéric (1987) (the first of its kind to condense, in a remarkably simple yet well-documented manner, the essential facts about the India of both yesterday and today from a historical, geographical, ethnographic, religious, philosophical, literary and linguistic perspective). It also supplements the *Dictionnaire de la sagesse orientale* by K. Friedrichs, I. Fischer-Schreiber, F. K. Erhard and M. S. Diener (1989), which constitutes a vast yet accessible range of references and a very enriching insight to those who are interested in philosophy, mysticism and meditation, or in a general introduction to the doctrines of Hinduism, Buddhism, Taoism and the religion of Zen. It is also the perfect companion to the *Dictionnaire des symboles* by J. Chevalier and A. Gheerbrant (1982), which not only explains the history of symbolic language through the ages and in different civilisations and the indelible yet hidden imprint it has left in our minds, but also opens the doors of the imagination and invites the reader to meditate on the symbols, just as Gaston Bachelard invited us to muse on our dreams in order to discover within them the taste and feel of a living reality. These works have all influenced the writing of this book; without them the following dictionary could not have been compiled because the required research would have taken an inordinate amount of time and would have involved reading analytical works which are inaccessible to the public, and which, moreover, are devoid of any synthesis.

The author warmly thanks Billard, Frédéric, Chevalier and Jacques for their invaluable personal correspondence, especially Billard and Frédéric for reading the rubrics of this dictionary and who offered pertinent and constructive remarks.

The writing of this dictionary had to be handled with utmost caution (especially considering that this field of study was completely new to the author) for several reasons:

- The author was in danger of being carried away by his own enthusiasm; a justified enthusiasm, yet capable of leading to erroneous interpretations.
- The vertiginous world of Indian symbols is highly complex.
- Moreover, the culture in question (whose diverse aspects were studied, notably the countless symbols which are often multivalent) is not only incredibly complex but is also full of pitfalls. See **Indian documentation* (*Pitfalls of*).
- Finally, Indian astronomy has played a significant role in this historiography. (The available documentation only offered a relatively simple insight into the literature of Indian astronomy. However, C. Jacques obligingly recommended Billard's *Astronomie Indienne* (1971) which is an unprecedented publication on the subject.)

Suffice to say that embarking upon this domain was rather like coming face to face with one of the many-headed dragons of the legends of Indian mythology: it was merciless and threatened to devour the author at any moment, as it had those who had set foot in this territory before without the necessary amount of precaution and vigilance. Now tamed, however, the appeased monster offers the reader all the delights of Eastern subtlety, and the wonder of this ingenious civilisation and its inestimable contributions.

*Shiva's dance of the creation and destruction of the world. After Moor's* Hindu Pantheon

# A

**ABAB**. Name given to the number ten to the power seventeen (= a hundred trillion). See ***Abhabâgamana***, **Names of numbers**, **High numbers**.
Source: **Lalitavistara Sûtra* (before 308 CE).

**ABABA**. Name given to the number ten to the power seventy-seven. See ***Abhabâgamana***, **Names of numbers**, **High numbers**.
Source: **Vyâkarana* (Pâlî grammar) by Kâchchâyana (eleventh century CE).

**ABBUDA**. Name given to the number ten to the power fifty-six. See **Names of numbers**, **High numbers**.
Source: **Vyâkarana* (Pâlî grammar) by Kâchchâyana (eleventh century CE).

**ABDHI**. [S]. Value = 4. "Sea". Four seas were said to surround **Jambudvîpa* (India). See ***Sâgara***, **Four**. See also **Ocean**.

**ABHABÂGAMANA**. "Beyond reach". Sanskrit term used to express the uncountable and unlimited. It is possible that the words **abab* (ten to the power seventeen) and **ababa* (ten to the power seventy-seven) were abbreviations of this word. See **Names of numbers**, **High numbers**, **Infinity**. See also ***Asamkhyeya***.

**ABHRA**. [S]. Value = 0. "Atmosphere". The atmosphere represents "emptiness". See ***Shûnga***, **Zero**.

**ABJA**. Literally: "Moon". Name given to the number ten to the power nine (= a thousand million). See **Names of numbers**. For an explanation of this symbolism: see **High numbers (Symbolic meaning of)**.
Source: **Lîlâvatî* by Bhâskarâchârya (1150 CE); **Trishatikâ* by Shrîdharâchârya (date uncertain).

**ABJA**. [S]. Value = 1. "Moon". The moon is unique. Another reason for this symbolism could be that in Indian tradition the moon is considered to be the source and symbol of fertility. It is likened to the primordial waters from whence came the revelation: it is the receptacle of seeds of the cycle of rebirth. It is thus the unity as well as the starting point. See **One**.

**ABLAZE**. [S]. Value = 3. See ***Shikhin*** and **Three**.

**ABSENCE, ABSENT**. See ***Shûnyatâ***, **Zero**.

**ABSOLUTE**. As a symbol representing a large quantity. See **High numbers (Symbolic meaning of)**, **Infinity**.

**ADDITION**. [Arithmetic]. **Samkalita* in Sanskrit.

**ÂDI**. [S]. Value = 1. "Commencement, primordial principle". In Hindu and Brahman philosophy, this principle is said to be found in all things before the creation; thus it is the unity as well as the starting point. See **One**.

**ÂDITYA**. [S]. Value = 12. "Children of Âditi". In Brahman and Vedic cosmogony, *Âditi* is the infinite sky, the original space. The *Âditya* are its children. In Vedic times, there were five, then they became seven, and finally twelve and were consequently identified by the twelve months of the year and the course of the sun during this period of time. The same word also signifies Sûrya, the sun god of the **Vedas*. As Sûrya = 12, *the children of Âditi* = 12. See ***Sûrya***.

**ADRI**. [S]. Value = 7. "Mountain". Allusion to *Mount Meru, sacred mountain which, according to ancient Indian cosmological representation, was situated at the centre of the universe and constituted a meeting and resting place for the gods: a representation where we know seven played a significant role. See **Seven**.

**AGA**. [S]. Value = 7. "Mountain". See ***Adri***. **Seven**.

**AGES (The four)**. See ***Chaturyuga***.

**AGNI**. [S]. Value = 3. "Fire". In Brahman mythology, Agni is the god of sacrificial fire (the three Vedic fires), which is represented as a man with three bearded heads who appears in three different forms: in the sky as the sun, in the air as lightning and on the earth as fire. Hence: "fire" = 3. See **Fire**, **Three**.

**AGNIPURÂNA**. See ***Purâna* and positional numeration**.

**AHAHA**. Name given to the number ten to the power seventy. See **Names of numbers**, **High numbers**.
Source: **Vyâkarana* (Pâlî grammar) by Kâchchâyana (eleventh century CE).

**AHAR**. [S]. Value = 15. "Day". See ***Tithi***, **Fifteen**.

**AHI**. [S]. Value = 8. Probably an allusion to Ahirbudhnya (or Ahi Budhnya) who, in Vedic mythology, designates the serpent of the depths of the ocean, born of dark waters. Thus: *Ahi* = 8, because the serpent corresponds symbolically to the number eight. See ***Nâga***, **Eight**. See also **Serpent (Symbolism of)**.

**AHIRBUDHNYA**. See ***Ahi***.

**ÂKÂSHA**. [S]. Value = 0. "Ether", the "element which penetrates everything", "space". It was considered as emptiness which could not mix with material things, immobile and eternal, beyond description. The association of ideas with the "void" or "emptiness" (*shûnya*) was established well before *shûnya* was identified with the concept of zero. In Indian thought ether was not only the void; it was also and above all the most subtle of the five elements of the revelation. It is certainly devoid of substance, but *âkâsha* is regarded as the condition of all corporeal extension and the receptacle of all matter formed by one of the other four elements (earth, water, fire or air). The association of ideas with zero became even more evident when this fundamental discovery was made: zero not only signified a void and "that which has no meaning", but also played an important role in the place-value system, and in terms of an abstract number, an equally essential role in mathematics and all the other sciences. Hence the symbolism: *âkâsha* = "space" = "void" = "ether" = "element which penetrates everywhere" = 0. See ***Shûnya***, ***Shûnyatâ***, **Zero**.

**AKKHOBHINI**. Name given to ten to the power forty-two. See **Names of numbers**. **High numbers**.
Source: **Vyâkarana* (Pâlî grammar) by Kâchchâyana (eleventh century CE).

**AKRITI**. [S]. Value = 22. In terms of the poetry of Sanskrit expression, Akriti means the metre of four times twenty-two syllables per verse. See **Indian metric**.

**ÂKSHARA**. [S]. Value = 1. "Indestructible". A Sanskrit word which, in Hindu philosophy, denotes the "undying" part of the vocal sound corresponding to the revelation of the Brahman. This is a direct reference to the word **ekâkshara*, the "Unique and undying" which is often expressed by the Sacred Syllable **AUM*. See ***Trivarna***, **Mysticism of letters**, **One**.

**AKSHARAPALLÎ**. Prakrit word meaning "letter-phoneme, syllable". It denotes a numerical notation of the alphabetical type frequently used in *Jaina manuscripts. See **Numeral alphabet**.

**AKSHITI**. Name given to the number ten to the power fifteen (= trillion). See **Names of numbers**, **High numbers**.
Source: **Pañchavimsha Brâhmana* (date uncertain).

**AL-BIRUNI** (Muhammad ibn Ahmad Abu'l Rayhan) (973–1048). Muslim astronomer and mathematician of Persian origin. After having lived in India for nearly thirty years, and having been initiated into the Indian sciences, he wrote many works, including *Kitab al arqam* ("Book of numerals"), *Tazkira fi'l hisab wa'l mad bi'l arqam al Sind wa'l hind* ("Arithmetic and counting systems using numerals in Sind and the Indias"), and above all *Kitab fi tahqiq i ma li'l hind* which constitutes one of the most important accounts of India in mediaeval times. Al-Biruni described the system of Sanskrit numerical symbols in minute detail, and stressed the importance of the place-value system and zero. He also went into much detail about the Sanskrit counting system, attaching particular importance to the Indian nomenclature of high numbers (see Fig. 24. 81). Here is a list of the principal names of numbers mentioned in *Kitab fi tahqiq i ma li'l hind* [see Woepcke (1863) p. 279]:**Eka* (= 1). **Dasha* (= 10). **Shata* (= $10^2$). **Sahasra* (= $10^3$). **Ayuta* (= $10^4$). **Laksha* (= $10^5$). **Prayuta* (= $10^6$). **Koti* (= $10^7$). **Vyarbuda* (= $10^8$). **Padma* (= $10^9$). **Kharva* (= $10^{10}$). **Nikharva* (= $10^{11}$). **Mahâpâdma* (= $10^{12}$). **Shankha* (= $10^{13}$). **Samudra* (= $10^{14}$). **Madhya* (= $10^{15}$). **Antya* (= $10^{16}$). **Parârdha* (= $10^{17}$). See **Indian numerals**, **Nâgarî numerals**, **Names of numbers**, **High numbers**, **Sanskrit**, **Numerical symbols**.

**ALGEBRA**. Alphabetical list of the words relating to this discipline, to which a rubric is dedicated in this dictionary: **Avyaktaganita*. **Bîja*. **Bîjaganita*. **Ghana*. *Indian mathematics (History of). **Samîkarana*. **Varga*. **Varga-Varga*. **Varna*. **Vyavahâra*. **Yâvattâvat*.

**ALPHABETICAL NUMERATION**. See ***Aksharapallî***, **Âryabhata's numeration**, **Katapayâdi numeration**, **Numeral alphabet**, ***Varnasamjña*** and ***Varnasankhya***.

**AMARA**. [S]. Value = 33. "Immortal". Allusion to the thirty-three gods. See ***Deva***, **Thirty-three**.

**AMBARA**. [S]. Value = 0. "Atmosphere". See ***Abhra***, **Zero**.

**AMBHODHA (AMBHODHI)**. [S]. Value = 4. "Sea". It was said that four seas surrounded **Jambudvîpa* (India). See ***Sâgara***, **Four**. See also **Ocean**.

**AMBHONIDHI**. [S]. Value = 4. "Sea". See ***Sâgara***, **Four**. See also ***Jala***.

**AMBODHA (AMBODHI, AMBUDHI)**. [S]. Value = 4. "Sea". See ***Sâgara***. **Four**. See also **Ocean**.

**AMBURÂSHI**. [S]. Value = 4. "Sea". See ***Sâgara***. **Four**. See also **Ocean**.

**AMRITA**. Nectar of "Immortality". See ***Soma***, **Serpent (Symbolism of the)**.

**ANALA**. [S]. Value = 3. "Worlds". See **Loka**. **Three**.

**ANANTA**. Literally "Infinity". Name given to the number ten to the power thirteen (= ten billion). See ***Asamkhyeya***, **Names of numbers**. For an explanation of this symbolism see **High numbers (symbolic meaning of)**.
Source: **Sankhyâyana Shrauta Sûtra* (date uncertain), which defines this number as the "limit of the calculable".

**ANANTA.** Word which literally means "infinity". In Hindu mythology, the *ananta* denotes a huge serpent representing eternity and the immensity of space. It is shown resting on the primordial waters of original chaos (Fig. D. 1). Vishnu is lying on the serpent, between two creations of the world, floating on the "ocean of unconsciousness". The serpent is always represented as coiled up, in a sort of figure eight on its side (like the symbol ∞), and theoretically has a thousand heads. It is considered to be the great king of the **nâgas* and lord of hell (**pâtâla*). Each time the serpent opens its mouth it produces an earthquake because there is a belief that the serpent also supported the world on its back. It is the serpent that at the end of each **kalpa*, spits the destructive fire over the whole of creation. See **Infinity**. See also **Serpent**.

**ANANTA.** [S]. Value = 0. "Infinity". It seems paradoxical, yet this symbolism comes from the association of *Ananta*, the serpent of infinity, with the immensity of space. As "space" = 0, the name of the serpent became a synonym of zero. See ***Ananta*** (the second of the above entries). **Zero.**

**ANCESTOR.** [S]. Value = 1. See **Pitâmaha. One.**

**ÂNDHRA NUMERALS.** Signs derived from *Brâhmî numerals, through the intermediary of Shunga, Shaka and Kushâna numerals, used in the contemporary inscriptions of the Ândhra dynasty (second – third century CE). These signs are notably found in the inscriptions of Jaggayapeta. The corresponding system did not function according to the place-value system and moreover did not possess zero. See **Indian written numeral systems (Classification of)**. See also Fig. 24.34, 36, 24.52, and 24.61 to 70.

**ANGA.** [S]. Value = 6. "Limb". The human body consists of six "limbs", or members: the head, the trunk, two arms and two legs. This is not the only reason, however, for this symbolism: there are six appended texts of the *Veda* (a group of Vedic texts called *Vedânga* which deal mainly with Vedic rituals, of their conservation and their perfect transmission). As *Vedânga* means the "members of Veda", we can see how the idea of "member" or "limb" came to signify the number six. See ***Veda*, *Vedanga*, Six.**

**ANGULI.** [S]. Value = 10. "Digit", because we have ten fingers. See **Ten.**

**ANGULI.** [S]. Value = 20. "Digit", because we have ten fingers and toes. See **Twenty.**

**ANKA.** Literally "mark, sign". The term means "numeral", "sign of numeration". See ***Anka*** [S]. See also all entries beginning with **Numeral.**

**ANKA.** [S]. Value = 9. "Numerals". Allusion to the nine significant numerals of the Indian place-value system. This symbol was in use no later than the time of *Bhâskara I (629 CE). See ***Anka. Ankasthâna.* Nine.**

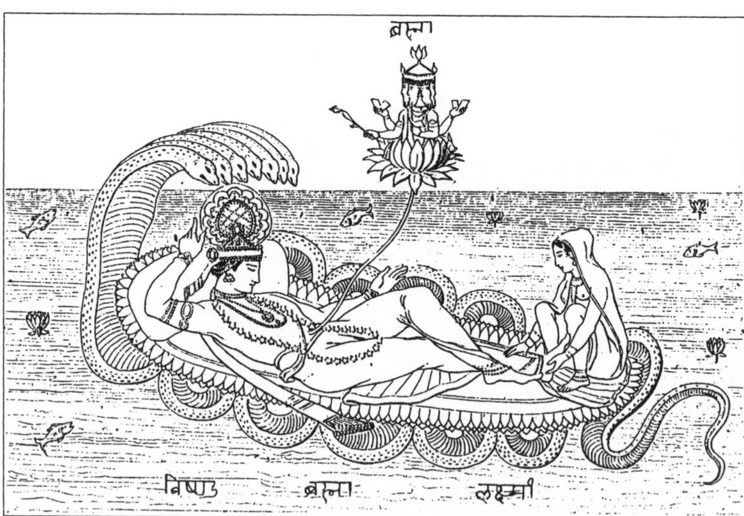


FIG. 24D.1. *Vishnu with Lakshmi and the serpent Ananta and Brahma sitting on a lotus flower which grows out of Vishnu's navel. From Dubois de Jancigny,* L'Univers pittoresque, *Hachette, Paris, 1846*

**ANKAKRAMENA.** Expression which literally means "in the order of the numerals", and alludes to the principle which the numerals are subjected to in the place-value system. See ***Anka. Sthâna.***

Source: **Lokavibhâga* (458 CE).

**ANKÂNÂM VÂMATO GATIH.** Expression which means "principle of the movement of the figures from the right to the left". The numbers were read out in ascending order, from the smallest units to the highest multiple of ten. This was the reverse of how the numbers were presented in Indian numerical notations (from left to right).

**ANKAPALLÎ.** Prakrit term which literally means "numerals, representation". It is applied to any system of representing numbers using actual numerals. Thus it denotes "numerical notation".

**ANKASTHÂNA.** Literally "Numerals in position". The Sanskrit name for positional numeration.

Source: **Lokavibhâga* (458 CE).

**ANKLE.** [S]. Value = 2. See ***Gulpha*** and **Two.**

**ANTA.** Name given to the the number ten to the power eleven (= a hundred thousand million). See **Names of numbers, High numbers.**

Sources: **Vâjasaneyî Samhitâ*, **Taittirîya Samhitâ* and **Kâthaka Samhitâ* (from the start of the first millennium CE); **Pañchavimsha Brâhmana* (date uncertain).

**ANTARIKSHA.** [S]. Value = 0. "Atmosphere". See ***Abhra.* Zero.**

**ANTYA.** Literally "(the) last". Name given to the number ten to the power twelve (= a billion). See **Names of numbers, High numbers.**

Source: **Sânkhyâyana Shrauta Sûtra* (date uncertain). An allusion is made to the highest order of units of the ancient Sanskrit numeration at the time of the **Vâjasaneyî Samhitâ*, **Taittirîya Samhitâ* and **Kâthaka Samhitâ* (from the start of the first millennium CE), where the nomenclature stopped at ten to the power twelve.

**ANTYA.** Literally "(the) last". Name given to the number ten to the power fifteen (= a trillion). See **Names of numbers, High numbers.**

Sources: **Lîlâvatî* by Bhâskarâchârya (1150 CE); **Ganitakaumudî* by Nârâyana (1350 CE), **Trishatikâ* by Shrîdharâchârya (date uncertain). At this later date, when the Sanskrit names for numbers by far surpassed the simple power of fifteen, the name of this number still retained a vestige of the limitation of the spoken numeration of ancient times. See the first entry under ***Antya***.

**ANTYA.** Literally "(the) last". Name given to the number ten to the power sixteen (ten trillion). See ***Antya*** (the above entries, Sources), **Names of numbers, High numbers.**

Source: **Kitab fi tahqiq i ma li'l hind* by al-Biruni (c. 1030).

**ANU.** Sanskrit name for "atom". See ***Paramânu.***

**ANUSHTUBH.** [S]. Value = 8. Name given to certain groups of verses of Vedic poetry. This is an allusion to the eight syllables which made up each one of the four elements which constituted the stanzas which were called *anushtubh*. See **Eight, Indian metric.**

**ANUYOGADVÂRA SÛTRA.** Title of a Jaina cosmological text giving countless examples of extremely high numbers, the corresponding speculations reaching (and even surpassing) easily as high as ten to the power two hundred (one followed by two hundred zeros). Thus, the figure said to express the total number of human beings of the creation is described as the "quantity obtained by multiplying the sixth power of the square of two $[ = (2^2)^6 = 2^{12}]$ by the third power of two $[ = 2^3 = 8]$, which is equal to the number which can be divided by 2 ninety-six times $[ = (2^{12})^8 = 2^{12 \times 8} = 2^{96}]$" [see Datta and Singh (1938), p.12]. There is also the period of time called **Shîrshaprahelikâ*, expressed, according to Hema Chandra (1089 CE), by "196 places of the place-value system" and which corresponds approximately "to the product of 84,000,000 multiplied by itself twenty-eight times" (see Datta and Singh, *op.cit*). This text, amongst many others, shows that the Jainas were amongst the Indian scholars who were most familiar with such arithmetical-cosmogonical speculations. See **Names of numbers, High numbers, Infinity.**

**ÂPA.** Sanskrit term meaning "water". See ***Jala.***

**APHORISM.** [S]. Value = 3. See ***Vâchana.* Three.**

**ÂPTYA.** [S]. Value = 3. "Spirit of the Waters". Allusion to the Vedic divinity named *Trita Âptya*, the "Third Spirit of the Waters", who killed Vishvarûpa, the three-headed demon. See **Three.**

**ARABIC NUMERATION (Positional systems of Indian origin).** See **"Hindi" numerals** and **Ghubar numerals.**

**ARAMAEAN-INDIAN NUMERALS.** See **Kharoshthî numerals.**

**ARAMAEAN-INDIAN NUMERATION.** See **Kharoshthî numerals.**

**ARBUDA.** Name given to the number ten to the power seven (= ten million). See **Names of numbers. High numbers.**

Sources: **Vâjasaneyî Samhitâ* (beginning of Common Era); **Taittirîya Samhitâ* (beginning of Common Era); **Kâthaka Samhitâ* (beginning of Common Era); **Pañchavimsha Brâhmana* (date unknown); **Sankhyâyana Shrauta Sûtra* (date unknown); **Âryabhatîya* (510 CE).

**ARBUDA.** Name given to the number ten to the power eight (= one hundred million). See **Names of numbers. High numbers.**

Source: *Lîlâvatî* by Bhâskarâchârya (1150 CE); **Ganitakaumudî* by Nârâyana (1350 CE); **Trishatikâ* by Shrîdharâchârya (date unknown).

**ARBUDA**. Name given to the number ten to the power ten (= ten billion). See **Names of numbers, High numbers.**

Source: **Ganitasârasamgraha* by Mahâvîrâchârya (850 CE).

**ARITHMETIC**. Here is an alphabetical list of terms relating to this discipline which appear as headings in this dictionary: **Abhabâgamana*, *Addition, *Algebra, **Anka*, **Ankakramena*, **Ankasthâna*, *Arithmetical operations, *Âryabhata, *Âryabhata (Numerical notations of), *Âryabhata's numeration, **Asamkhyeya*, *Base 10, *Base of one hundred, *Bhâskara, *Bhâskarâchârya, **Bîja*, **Bîjaganita*, *Brahmagupta, *Buddha (Legend of), *Calculation (The science of), *Calculation (Methods of), *Calculation on the abacus, *Calculator, *Cube, *Cube root, **Dashaguna*, **Dashagunâsamjñâ*, *Day of Brahma, **Dhûlikarma*, *Digital calculation, *Divi-dend, *Division, *Divisor, *Equation, *Fractions, *High numbers, *Indeterminate equation, *Indian mathematics (The history of), *Infinity, **Kaliyuga*, **Kalpa*, **Katapayâdi* numeration, **Khachheda*, **Khahâra*, **Mahâvîrâchârya*, *Mathematician, *Mathematics, *Mental arithmetic, *Multiplication, *Names of numbers, *Nârâyana, *Numeral alphabet, *Numerals, *Numerical symbols, *Numerical symbols (Principle of the numeration of), **Pâtî*, *Quotient, *Remainder, *Rule of five, *Rule of eleven, *Rule of nine, *Rule of seven, *Rule of three, *Sanskrit, **Shatottaragananâ*, **Shatottaraguna*, **Shatottarasamjñâ*, *Shrîdharâchârya, *Square root, **Sthâna*, *Subtraction, *Total, **Yuga*, *Zero.

**ARITHMETICAL OPERATIONS**. See **Calculation, *Dhûlikarma*, Indian methods of calculation, *Parikarma*, *Pâtî*, *Pâtîganita*, Square roots (how Âryabhata calculated his).**

**ARITHMETICAL SPECULATIONS**. See ***Anuyogadvâra Sûtra*, *Asamkhyeya*, Calculation, Day of Brahma, *Yuga*, *Kalpa*, Jaina, Names of numbers, High numbers**, and **Infinity**.

**ARITHMETICAL-COSMOGONICAL SPECULATIONS**. See ***Anuyogadvâra Sûtra*, *Asamkhyeya*, Calculation, Cosmic cycles, Day of Brahma, *Yuga* (Definition), *Yuga* (Systems of calculation of), *Yuga* (Cosmogonic speculations on), *Kalpa*, Jaina, Names of numbers, High numbers, Infinity.**

**ARJUNÂKARA**. [S]. Value = 1,000. "Hands of Arjuna". Allusion to the mythical sovereign Arjunakârtavîrya, leader of the Haihayas and king of the "seven isles", who, according to one of the legends of the *Mahâbhârata*, had a thousand arms. See **Thousand.**

**ARKA**. [S]. Value = 12. "Bright". An epithet given to Sûrya, the sun god, who, symbolically, represents the number twelve. See **Sûrya. Twelve.**

**ARMS**. [S]. Value = Two. See ***Bâhu*** and **Two.**

**ARMS OF ARJUNA**. [S]. Value = 1,000. See ***Arjunâkara*** and **Thousand.**

**ARMS OF KÂRTTIKEYA**. [S]. Value = Twelve. See ***Shanmukhabâhu*** and **Twelve.**

**ARMS OF RÂVANA**. [S]. Value = Twenty. See ***Râvanabhuja*** and **Twenty.**

**ARMS OF VISHNU**. [S]. Value = Four. See ***Haribâhu*** and **Four.**

**ARNAVA**. [S]. Value = 4. "Sea". Four seas were said to surround **Jambudvîpa* (India). See ***Sâgara*. Four.** See also **Ocean.**

**ARROW**. [S]. Value = 5. See ***Bâna*, *Ishu*, *Kalamba*, *Mârgana*, *Sâyaka*, *Shara*, *Vishikha*** and **Five.**

**ÂRYABHATA**. A veritable pioneer of Indian astronomy, Âryabhata is without doubt one of the most original, significant and prolific scholars in the history of Indian science. He was long known by Arabic Muslim scholars as *Arjabhad*, and later in Europe in the Middle Ages by the Latinised name of *Ardubarius*. He lived at the end of the fifth century and the beginning of the sixth century CE, in the town of Kusumapura, near to Pataliputra (now Patna, in Bihar). His work, known as *Âryabhatîya* was written c. 510 CE. It is the first Indian text to record the most advanced astronomy in the history of ancient Indian astronomy. The work also involves trigonometry and gives a summary of the main mathematical knowledge in India at the beginning of the sixth century, bearing witness to the high level of understanding that had been reached in this field at this time. The following rapturous declamation by *Bhâskara (one of Âryabhata's disciples and most fervent of admirers), taken from the Commentary which he wrote on the *Âryabhatîya* in 629 CE gives some indication of the high level of abstract thought achieved by the scholar way ahead of his time [see Billard, IJHS. XII,2, p. 111]: "Âryabhata is the master who, after reaching the furthest shores and plumbing the inmost depths of the sea of ultimate knowledge of mathematics, kinematics and spherics, handed over the three [sciences] to the learned world."

See **Indian astronomy (History of). Indian mathematics (History of).**

**ÂRYABHATA. (Numerical notations of).** When referring to numerical data, Âryabhata often used the Sanskrit names of the numbers: at least this is the impression we get if we look at the sections of his work respectively entitled *Ganitapâda*, (which deals with "mathematics"), *Kâlakriyâ* (which talks of "movements", in particular his system of exact longitudes in his **Astronomical canon*) and *Golapâda* (which relates to "spherics" and other three-dimensional problems). Here is a list of the principal names of numbers mentioned in the *Âryabhatîya* [see *Ârya*, II, 2]:

**Eka* (= 1). **Dasha* (= 10). **Shata* (= $10^2$). **Sahastra* (= $10^3$). **Ayuta* (= $10^4$). **Niyuta* (= $10^5$). **Prayuta* (= $10^6$). **Koti* (= $10^7$). **Arbuda* (= $10^8$). **Vrindâ* (= $10^9$).

See **Names of numbers** and **High numbers.**

Âryabhata also used a method of recording numbers which he invented himself: it was a clever (if not terribly practical) alphabetical system. However, he certainly knew the system of *numerical symbols, as we can see if we look at the *Ganitapâda*, which contains two examples of numbers expressed in this way [see *Ârya*, II, line 20; Billard, p. 88]:

*sarûpa*, "added to the form", and: *râshiguna*, "multiplied by the zodiac".

**Samkalita* means addition (literally: "put together") and **gunana* means multiplication. These words can be abbreviated to *sa* ("plus") and *guna* ("multiplied by"). **Rûpa* and **râshi* are the respective numerical symbols for "shape" (or "form") and "zodiac", the numerical values for which are one and twelve. Thus the two above expressions can be translated as follows: *sarûpa*, "added to one", and: *râshiguna*, "multiplied by twelve". This is concrete proof that Âryabhata was familiar with the method of recording numbers using the numerical symbols. These are the only two examples that have been found in his work; however, Billard (pp. 88-89) shows that the *Âryabhatîya*, in its present state, is in fact two different works put together, or rather the result of reorganisation carried out on the original version. Some parts were left unaltered, some were slightly modified, and others still were radically changed in terms of numerical data, basic constants and metre. The text we have today consists of nothing more than the reworked parts, as the original has not been found. It is probable that Âryabhata used the numerical symbols in the first version of his work and later changed his method of representing numbers as he reorganised his work. Finally, it is extremely likely that Âryabhata knew the sign for zero and the numerals of the place-value system. This supposition is based on the following two facts: first, the invention of his alphabetical counting system would have been impossible without zero or the place-value system; secondly, he carries out calculations on square and cubic roots which are impossible if the numbers in question are not written according to the place-value system and zero. See **Indian written numeration (Classification of), Sanskrit numeration, Numerical symbols (Principle of the numeration of), Âryabhata's numeration.** See also **Square roots (How Âryabhata calculated his).**

**ÂRYABHATA'S NUMBER-SYSTEM**. This is an alphabetical numerical notation invented by the astronomer Âryabhata c. 510 CE. It is a system which uses thirty-three letters of the Indian alphabet and is capable of representing all the numbers from 1 to $10^{18}$.

Âryabhata, it appears, was the first man in India to invent a numerical alphabet. He developed the system in order to express the constants of his **astronomical canons*, and his surprising astronomical speculations about **yugas*, and the system is more elegant and also shorter than that which uses numerical symbols. See **Âryabhata (Numerical notations of), Numeral alphabet, Numerical symbols, Numeration of numerical symbols, *Yuga* (Systems of calculating)** and ***Yuga* (Astronomical speculations about).**

The use of this notation is found throughout his work entitled *Dashagîtikapâda*, where he describes it in the following way: *Vargâksharâni varge 'varge 'vargâsha râni kât nmau yah Khadvinavake svarâ nava varge "varge navântya varge vâ.*

Translation: "The letters which are [said to be] classed (*varga*) [are], from [the letter] *ka*, [those which are placed] in odd rows (*varga*); the letters which are [called] unclassed (*avarga*) [are those which are placed] in even rows (*avarga*); [thus, one] *ya* is equal to *nmau* (= *na* + *ma*]; the nine vowels [are used to record] the nine pairs of places (*kha*) in even or odd [rows]. The same [procedure] can be repeated after the last of the nine even rows".

Ref.: JA, 1880, II, p. 440; JRAS, 1863, p. 380; TLSM, I, 1827, p. 54; ZKM, IV, p. 81; Datta and Singh, (1938), p. 65; Shukla and Sarma, *Ganita Section*, (1976), pp. 3ff.

To put it plainly, Âryabhata's method consists of assigning a numerical value to the consonants of the Indian alphabet in the following manner:

1. For the first twenty-five consonants, the order of normal succession is followed for whole numbers starting with 1.
2. The twenty-sixth represents five units more than the twenty-fifth.
3. For the remaining seven, the progression grows by tens.
4. The last consonant of the alphabet receives the value of one hundred.

Thus this notation contains a number of peculiarities which are unique to Indian syllables (which are transcribed below in modern *Nâgarî* characters). For a better understanding of this principle, it should be remembered that this writing uses thirty-three different characters, which represent the consonants, and many other signs representing the vowels in an isolated position (*a, â, i, î, u, û, ri, rî, l, e, o, ai, au*).

An isolated consonant is always pronounced with a short *a*, but when it is combined with another vowel, a special sign is added which graphically has nothing in common with the sign representing this vowel in an isolated position (a vertical line to the right of the consonant, a line above, a loop below the letter, a horizontal line above the letter, with a loop, and so on).

As for the writing of the word, it is done with a continuous horizontal line called *mâtrâ* (see Fig. D. 2).

It must not be forgotten that the essential phonetic elements of this syllable system are constituted by the association of a consonant with a following vowel ( which is either short like *a, i, u, ri, la*, or long like *â, î, û, rî*), or a diphthong (*e, o, ai, au*), which is always long by definition. Thus to a given consonant, a short vowel *a*, or a long vowel *â*, or even any one of remaining vowel or diphthongs can be joined (*i, u, ri, la, î, û, rî, lâ, e, o, ai, au*). Therefore, the following syllables correspond to the consonant *ma* (m) [for example]: *ma, mâ, mi, mî, mu, mû, mri, mrî, mla, mlâ, me, mo, mai.*

Conversely, the thirty-three consonants can be joined to any vowel. Take, for example, *a*:

5 gutturals *ka kha ga gha na*
5 palatals *cha chha ja jha ña*
5 cerebrals *ṭa ṭha ḍa ḍha ṇa*
5 dentals *ta tha da dha na*
5 labials *pa pha ba bha ma*
4 semivowels *ya ra la va*
3 sibilants *sha sha sa*
1 aspirated *ha*

The alphabetical notation of numbers invented by this astronomer/phonetician/mathematician is based upon precisely this structure. Starting from the first vowel (*a*):

- the five guttural consonants (*ka, kha, ga, gha, na*) receive the values 1 to 5;
- the five palatals (*cha chha ja jha h*) those of 6–10;
- the 5 cerebrals (*ta tha da dha na*) those of 11 to 15;
- the 5 dentals (*ta ths da dha na*) those of 16 to 20; the five labials (*pa pha ba bha ma*) those of 21 to 25
- the 4 semivowels (*ya ra la va*) receive the values 30, 40, 50 and 60;
- the 3 sibilants (*sha sha sa*) those of 70, 80 and 90;
- and the last letter of the alphabet (the aspirated *ha*) that of 100.

However, if a vertical line is added to the right of a *devanâgarî* consonant (thus vocalising the consonant with a long *a*), the value remains the same: *kâ* = ka; *khâ* = kha; *tâ* = ta, etc.

In other words, this numeration does not distinguish between long and short vowels when attributing numerical values to the letters (*kâ* = ka, *mî* = mi, *ñû* = ñu, *prî* = pri, etc.).

Thus, from this point on, to avoid confusion, only the consonants accompanied by a short vowel will be referred to.

To record numbers above one hundred, Âryabhata came up with the idea of using the rules of the vocalisation of consonants.In keeping with the order of the letters of the alphabet, the first thirty-three consonants with the vowel *a* represent the numbers from 1 to 100 according to the above rule. But if these consonants are vocalised using an *i* or an *î*, ( which follow *a* in the Indian syllable system), the value of each is multiplied by a hundred (Fig. D. 3). If they are then accompanied by a *u* or a *û*, their initial values are multiplied by 10,000 (= $10^4$). Thus, when either *ri* or *rî* are attached to the successive consonants, they represent the initial

| | | | | | |
|---|---|---|---|---|---|
| gutturals | क<br>*ka* = 1 | ख<br>*kha* = 2 | ग<br>*ga* = 3 | घ<br>*gha* = 4 | ङ<br>*ṅa* = 5 |
| palatals | च<br>*cha* = 6 | छ<br>*chha* = 7 | ज<br>*ja* = 8 | झ<br>*jha* = 9 | ञ<br>*ña* = 10 |
| cerebrals | ट<br>*ṭa* = 11 | ठ<br>*ṭha* = 12 | ड<br>*ḍa* = 13 | ढ<br>*ḍha* = 14 | ण<br>*ṇa* = 15 |
| dentals | त<br>*ta* = 16 | थ<br>*tha* = 17 | द<br>*da* = 18 | ध<br>*dha* = 19 | न<br>*na* = 20 |
| labials | प<br>*pa* = 21 | फ<br>*pha* = 22 | ब<br>*ba* = 23 | भ<br>*bha* = 24 | म<br>*ma* = 25 |
| semivowels | य<br>*ya* = 30 | र<br>*ra* = 40 | ल<br>*la* = 50 | व<br>*va* = 60 | |
| sibilants | श<br>*śha* = 70 | ष<br>*sha* = 80 | स<br>*sa* = 90 | | |
| aspirates | ह<br>*ha* = 100 | | | | |

FIG.24D.2. *Alphabetical numeration of Âryabhata: numerical value of consonants in isolated position (vocalisation using a short "a"). Ref. : NCEAM, p. 257. Datta and Singh (1937); Guitel (1966); Jacquet (1835); Pihan (1860); Rodet*

| | | | | |
|---|---|---|---|---|
| कि<br>*ki* = 100 | खि<br>*khi* = 200 | गि<br>*gi* = 300 | घि<br>*ghi* = 400 | ङि<br>*ṅi* = 500 |
| चि<br>*chi* = 600 | छि<br>*chhi* = 700 | जि<br>*ji* = 800 | झि<br>*jhi* = 900 | ञि<br>*ñi* = 1,000 |
| टि<br>*ṭi* = 1,100 | ठि<br>*ṭhi* = 1,200 | डि<br>*ḍi* = 1,300 | ढि<br>*ḍhi* = 1,400 | णि<br>*ṇi* = 1,500 |
| ति<br>*ti* = 1,600 | थि<br>*thi* = 1,700 | दि<br>*di* = 1,800 | धि<br>*dhi* = 1,900 | नि<br>*ni* = 2,000 |
| पि<br>*pi* = 2,100 | फि<br>*phi* = 2,200 | बि<br>*bi* = 2,300 | भि<br>*bhi* = 2,400 | मि<br>*mi* = 2,500 |
| यि<br>*yi* = 3,000 | रि<br>*ri* = 4,000 | लि<br>*li* = 5,000 | वि<br>*vi* = 6,000 | |
| शि<br>*shi* = 7,000 | षि<br>*shi* = 8,000 | सि<br>*si* = 9,000 | | |
| हि<br>*hi* = 10,000 | | | | |

FIG.24D.3. *Alphabetical numeration of Âryabhata: numerical value of consonants vocalised by a short "i")*

values multiplied by 1,000,000 (= $10^6$). And so on with each of the consecutive vowels of the alphabet, multiplying by successive powers of 100 ($10^2$, $10^4$, $10^6$, etc.). Using all the possible phonemes, this rule enables numbers to be expressed up to the value of the number that we recognise today in the form of a 1 followed by eighteen zeros ($10^{18}$).

If we look at the question from another angle, Âryabhata's alphabetical notation follows the successive powers of a hundred: it is thus an additional numeration with a base of 100, where the units and tens (units of the first centesimal order) are expressed by the first thirty-three successive consonants in an isolated position, according to the vocalisation with an *a* (long or short). The units of the second centesimal order (units and tens, multiplied by a hundred = $10^2$) are expressed by the same consonants, this time vocalised by an *i* (short or long). Those of the third centesimal order (units and tens multiplied by ten thousand = $10^4$) are expressed by the consonants accompanied by the vowel *u* (or *û*). And so on until the units of the ninth centesimal order (units and tens multiplied by $10^{16}$) using the thirty-three consonants with *au* (which corresponds to the last vowel. This is how the values for the consonant *ka* are obtained, using the successive vocalisations (Fig. D. 4). The first four orders of Âryabhata's numeration are presented in Fig. D. 5.

| Centesimal order of units | 1st | 2nd | 3rd | 4th | 5th | 6th | 7th | 8th | 9th |
|---|---|---|---|---|---|---|---|---|---|
| Syllable | क *ka* | कि *ki* | कु *ku* | कृ *kri* | कॢ *kli* | के *ke* | कै *kai* | को *ko* | कौ *kau* |
| Value | 1 | $10^2$ | $10^4$ | $10^6$ | $10^8$ | $10^{10}$ | $10^{12}$ | $10^{14}$ | $10^{16}$ |

Fig.24D.4. *Consecutive orders of units in Âryabhata's alphabetical numeration (successive values of syllables formed beginning with "ka").*

| Vocalisation | with an *a* | | with an *i* | | with a *u* | | with an *r* | |
|---|---|---|---|---|---|---|---|---|
| Associated centesimal order | 1st | | 2nd | | 3rd | | 4th | |
| Power of ten | 1 | | $10^2$ | | $10^4$ | | $10^6$ | |
| Row of syllable | odd | even | odd | even | odd | even | odd | even |

Fig.24D.5.

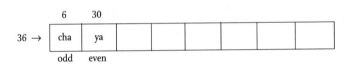

As the numbers were set out according to the ascending powers of a hundred, starting with the smallest units, the representation was carried out – at least theoretically – within a rectangle subdivided into several successive rectangles, where the syllables were written from left to right according to the centesimal order (Fig. D. 6). The number 57,753,336 corresponds to the number of synodic revolutions of the moon during a **chaturyuga*.

In Âryabhata's language (Sanskrit), this number is expressed as follows (see *Ârya*, II, 2): *Shat thimshati trishata trisahasra pañchâyuta saptaniyuta saptaprayuta pañchakoti*.

This can be translated as follows, where the numbers are expressed in ascending order, starting with the smallest unit:

"Six [= *shat*],
three tens [= *trimshati*],
three hundreds [= *trishata*],
three thousands [= *trisahasra*],
five myriads [= *pañchâyuta*],
seven hundred thousand [= *saptaniyuta*],
seven millions [= *saptaprayuta*],
five tens of millions [= *pañchakoti*]".

See **Âryabhata (Numerical notations of)**, ***Ankânâm vâmato gatih,*** **Names of numbers, Sanskrit.**

Âryabhata's notation conformed rigorously to this order, the only difference being that it functioned according to a base of 100 and not

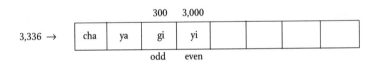

| Centesimal order | ←1st→ | | ←2nd→ | | ←3rd→ | | ←4th→ | |
|---|---|---|---|---|---|---|---|---|
| Syllable | | | | | | | | |
| Row | odd | even | odd | even | odd | even | odd | even |

Fig. 24D.6.

of 10. Thus it was necessary to break down the expression of the number in question (at least mentally) into sections of two decimal orders, as follows:

1st centesimal order: *six, three* tens
2nd centesimal order: *three* hundreds, *three* thousand,
3rd centesimal order: *five* myriads, *seven* hundred thousand,
4th centesimal order: *seven* million, *five* tens of millions.

For the first centesimal order, it was necessary to take the consonants, vocalised by *a*, which correspond respectively to the values 6 and 30 (six and three tens), which gives the syllables *cha* and *ya* (see Fig. D. 6A).

For the second centesimal order, the consonants were vocalised by *i*, corresponding respectively to the values 300 and 3,000, the syllables being *gi* and *yi* (see Fig. D. 6B).

For the third centesimal order, the consonants were vocalised by *u*, and corresponded respectively to the values 50,000 and 700,000, the syllables being *nu* and *shu* (see Fig. D.6C).

Finally, for the fourth centesimal order, the consonants were vocalised with *ri*, corresponding respectively to the values 7,000,000 and 50,000,000, the syllables being *chhri* and *lri* (see Fig. D.6D).

| | 6 | 30 | | | | | | |
|---|---|---|---|---|---|---|---|---|
| 36 → | cha | ya | | | | | | |
| | odd | even | | | | | | |

Fig. 24D.6A.

| | | | 300 | 3,000 | | | | |
|---|---|---|---|---|---|---|---|---|
| 3,336 → | cha | ya | gi | yi | | | | |
| | | | odd | even | | | | |

Fig.24D.6B.

| | | | | | 50,000 | 700,000 | | |
|---|---|---|---|---|---|---|---|---|
| 753,336 → | cha | ya | gi | yi | nu | shu | | |
| | | | | | odd | even | | |

Fig.24D.6C.

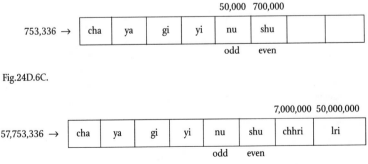

| | | | | | | | 7,000,000 | 50,000,000 |
|---|---|---|---|---|---|---|---|---|
| 57,753,336 → | cha | ya | gi | yi | nu | shu | chhri | lri |
| | | | | | odd | even | | |

Fig.24D.6D.

Thus, the notation for the number in question would be: *chayagiyinushuchhrilri.*

Fig. D. 6E shows the main breakdowns for this value (where the value of a syllable is called *absolute* when the vocalisation accompanying the consonant is ignored, and it is called *relative* where the opposite is the case).

The number 4,320,000 is expressed in the same manner, this figure corresponding to the total number of years in a **chaturyuga* (Fig. D. 6F): *khuyughri*. This notation is proof that inventive genius does not always go hand in hand with simplicity.

Thus, contrary to the opinion of several authors, this notation was not based on the place-value system, and certainly did not use zero. It is in fact an additional numeration of the third category of the classification given in Chapter 23.

However, it is very likely that Âryabhata knew about zero and the place-value system.It is precisely because he already knew about these concepts that he was able to achieve the degree of abstraction that was needed to develop a numerical notation such as this, which is unique in the whole history of written numerations. Whilst this system is additional in principle, its mathematical structure is full of the purest concepts of zero and the place-value system. This is made clear in Fig. D. 4. The consonant *ka* is the chosen graphical sign upon which everything else is based. Going from left to right, the rule of numerical vocalisation invented by Âryabhata can be resumed as follows: by adding *i* to this sign, in reality two zeros are being added to the decimal representation of the value of the letter *ka* (in other words, the unit); but by adding a *u*, a *ri*, a *li*, an *e*, an *ai*, an *o*, or an *au*, four, six, eight, ten, twelve, fourteen or sixteen zeros would be added.

The Jews, the Syrians and the Greeks certainly used similar conventions, but only for highly specialised usage and without perceiving them from the same angle as Âryabhata. By adding an accent, a dot or even one or two suffixes to a letter, they multiplied its value by 100 or 1,000, but they never managed to generalise their convention from such an abstract angle as Âryabhata.

| Syllables | *cha* | *ya* | *gi* | *yi* | *nu* | *shu* | *chhri* | *lri* |
|---|---|---|---|---|---|---|---|---|
| Absolute values | 6 | 30 | 3 | 30 | 5 | 70 | 7 | 50 |
| Relative total values for each column | 36 | | $33 \times 10^2$ | | $75 \times 10^4$ | | $57 \times 10^6$ | |
| Total value | $36 + 33 \times 10^2 + 75 \times 10^4 + 57 \times 10^6$ | | | | | | | |
| 57,753,336 → | *cha* | *ya* | *gi* | *yi* | *nu* | *shu* | *chhri* | *lri* |

Fig. 24D.6E.

| Syllables | | | | | *khu* | *yu* | *ghri* | |
|---|---|---|---|---|---|---|---|---|
| Absolute values | | | | | 2 | 30 | 4 | |
| Relative total values for each column | | | | | $32 \times 10^4$ | | $4 \times 10^6$ | |
| Total value | | | | | $32 \times 10^4 + 4 \times 10^6$ | | | |
| 57,753,336 → | | | | | *khu* | *yu* | *ghri* | |

Fig. 24D.6F.

Âryabhata had an advantage, because the phonetic structure which characterises the Indian syllable system is almost mathematical itself. This is confirmed by *Bhâskara I, a faithful disciple separated by a century from Âryabhata. In his Commentary on the *Âryabhatîya* (629 CE), he gives this brief explanation of the rule in question: *nyâsashcha sthânânâm* oooooooooo. Translation: "By writing in the places (**sthâna*), we have: oooooooooo [= ten zeros]".

[See: commentary on the *Ganitapâda*, line 2; Shukla and Sarma, (1976), pp. 32–4; Datta and Singh, (1938), pp. 64–7].

In his text, the commentator uses not only the word *sthâna* which means "place" (which the Indian scholars often used in the sense of "positional principle"), but also and above all the little circle, which is the numeral "zero" of the Indian place-value system. See ***Sthâna*, Numeral 0** and **Zero**.

Later on in his commentary, Bhâskara I writes the following: *khadvinavake svarâ nava varge 'varge khâni shûnyâni, khânâm dvinavakam khadvinavakam tasmin khadvinavake ashtâdashasu shûnyopalakshiteshu* . . . Translation: "The nine vowels (*nava varge*) [are used to note] the nine pairs of zeros (*khadvinavake*); [because] **kha* means zero (**shûnya*). In the nine pairs of places, that is, in the eighteen (*ashtâdashasu*) [places] marked by zeros (*shûnyopalakshiteshu*) . . ."

The use of the term **kha* as one of the designations of zero is explained as one of the synonyms of **shûnya*, a word meaning "void" which Indian mathematicians and astronomers have always used, since at least the fifth century CE, in the sense of zero.

This leaves no doubt: even if the master was not very loquacious on the subject, his disciple and commentator explains Âryabhata's system and uses the Indian symbol for zero (the little circle), and also the three fundamental Sankrit terms (**sthâna*, **kha*, **shûnya*). See **Zero**.

The Sanskrit term **kha*, literally "space", signifies "sky" and "void", and thus by extension "zero" in its mathematical sense. As for "place" (**sthâna*), Âryabhata gave it the meaning of the place occupied by a given syllable; thus, by extension, it meant "order of unit" in his alphabetical numeration. This is due to the "row" occupied by the syllable in a square which is formed by the structure of his notation system (Fig. D. 5). To his mind, it was a completely separate "place", one for the even row, and one for the odd row, within a unit of the centesimal order (the odd row having a value of a simple unit and the even row a value of a multiple of ten). It is due to the fact that such a "place" can be "emptied" if the units or tens of the corresponding "order" are absent that "place" came to mean both "position" and void. As for the expression *khadvinavake*, for Âryabhata this meant the "nine pairs of *zeros*", the eighteen zeros added to the decimal positional representation of the initial value of a given consonant, at the end of the successive vocalisation operations.

Moreover, in *Golapâda*, Âryabhata alludes to the essential component of our place-value system when he states that "from place to place (**sthâna*), each [of the numerals] is ten times [greater] than the preceding one" [see Clark (1930), p. 28]. What is more, in the chapter of the *Ganitapâda* on arithmetic and methods of calculation, he gives rules for operations in decimal base for the extraction of square roots and cube roots. Neither of these two operations can be carried out if the numbers are not expressed *in writing*, using the place-value system with nine distinct numerals and a tenth sign which performs the function of zero. See ***Pâtîganita*, Indian methods of calculation** and **Square roots (How Âryabhata calculated his)**.

This is mathematical proof that, at the beginning of the sixth century CE, Âryabhata had perfect knowledge of zero and the place-value system, which he used to carry out calculations. The question remains: why did he invent such a complicated system when he could have used a much simpler one? It seems that the alphabet offered an almost inexhaustible supply when it came to creating mnemonic words, especially for complicated numbers, and this facilitated the readers' memorising of them. He always wrote in Sanskrit verse, and thus he had found a very convenient way not only to write numbers in a condensed form, but also to meet the demands of the metre and versification of the Sanskrit language.

Luckily, Âryabhata was the only one to make use of the system that he had invented. His successors, including those that referred to his work, generally adopted the method of **numerical symbols*. Even those that opted for an alphabetical numerical notation did not choose to use his system: they used a radically transformed form which was much simpler. See **Katapayâdi numeration**.

When Âryabhata's alphabetical numeration became widespread in the field of Indian astronomy, it no doubt was disastrous for the preservation of mathematical data. Worse yet, it caused the Indian discoveries of the place-value system and zero, which took place before Âryabhata's time, to be irretrievably lost to history.

**ÂRYABHATÎYA**. Title given to Âryabhata's work by his successors.

**ASAMKHYEYA** (or **ASANKHYEYA**). Literally: "number impossible to count" (from **samkhyeya* or **sankhyeya*, "number", the "highest number imaginable". See **High numbers** and **Infinity**.

**ASANKHYEYA**. Literally: "non-number". Term designating the "incalculable". See **Asamkhyeya**.

**ASANKHYEYA**. Literally: "impossible to count". Name given to the number ten to the power 140. See **Names of numbers**. For an explanation of this symbolism, see **High numbers (Symbolic meaning of)**.

Source: **Vyâkarana* (Pâlî grammar) by Kâchchâyana (eleventh century CE).

**ÂSHÂ**. [S]. Value = 10. "Horizons". See ***Dish***, **Ten**.

**ASHÎTI**. Ordinary Sanskrit name for the number eighty.

**ASHTA** (or **ASHTAN**). Ordinary Sanskrit name for the number eight. It is used in the composition of several words which have a direct relationship with the idea of this number. Examples: **Ashtadanda*, **Ashtadiggaja*, **Ashtamangala*, **Ashtamûrti*, **Ashtânga* and **Ashtavimoksha*.

For words which have a more symbolic relationship with this number, see **Eight** and **Symbolism of numbers**.

**ASHTACHATVÂRIMSHATI**. Ordinary Sanskrit name for the number forty-eight. For words which have a more symbolic relationship with this number, see **Forty-eight** and **Symbolism of numbers**.

**ASHTADANDA**. "Eight parts". These are the eight parts of the body that we use to conduct a profound veneration by stretching out face down on the ground. See ***Ashtânga***.

**ASHTADASHA**. Ordinary Sanskrit name for the number eighteen. For words which have a more symbolical relationship with this number, see **Eighteen** and **Symbolism of numbers**.

**ASHTADIGGAJA**. "Eight elephants". Collective name for the guardians of the eight "horizons" of Hindu cosmogony (these being: *Airâvata*, **Pundarîka*, *Vâmana*, **Kumuda*, *Anjana*, *Pushpadanta*, *Sarvabhauma* and *Supratîka*). See ***Diggaja***.

**ASHTAMANGALA**. "Eight things that augur well". This concerns the "eight jewels" that Buddhism considers as the witnesses of the veneration of Buddha. See ***Mangala***.

**ASHTAMÛRTI**. "Eight shapes (or forms)". The name of the most important forms of Shiva. See ***Ashta*** and ***Mûrti***.

**ASHTAN**. A synonym of **Ashta*.

**ASHTÂNGA**. "Eight limbs (or members)". This term denotes the eight limbs of the human body which are used in prostration (the head, the chest, the two hands, the two feet and the two knees).

**ASHTAVIMOKSHA**. "Eight liberations". This refers to a Buddhist meditation exercise, which has eight successive stages of concentration, the aim of which being to liberate the individual from all corporeal and incorporeal attachments.

**ASHTI**. [S]. Value = 16. In terms of Sanskrit poetry, this refers to the metre of four times sixteen syllables per line. See **Sixteen** and **Indian metric**.

**ASHVA**. [S]. Value = 7. "Horse". Allusion to the seven horses (or horse with seven heads) of the chariot upon which **Sûrya*, the Brahmanic god of the sun, raced across the sky. See **Seven**.

**ASHVIN**. [S]. Value = 2. "Horsemen". Name of the twin gods Saranyû and Vivashvant (also called *Dasra and *Nâsatya) of the Hindu pantheon. They symbolise the nervous and vital forces, and are supposed to respectively represent the morning star and the evening star. They are the offspring of horses, hence their name (from **Ashva*). These divinities are considered as the "Primordial couple" who appeared in the sky before dawn in a horse-drawn golden chariot. See **Two**.

**ASHVINA**. [S]. Value = 2. "Horsemen". See **Ashvin** and **Two**.

**ASHVINAU**. [S]. Value = 2."Horsemen". See **Ashvin** and **Two**.

**ASTRONOMICAL CANON**. A group of elements conceived as a whole by the author of an astronomical text. These elements are always presented together in a text, commentary or quotation, being effectively (astronomically), or supposedly, interdependent. Often, however, except for historical information, and even though complete, the canons are only in the form of the game of **bîja*. Thus, mathematically, a given canon can be placed in any era or represent any unit of time. See **Indian astronomy (The history of)**.

**ASTRONOMICAL SPECULATIONS**. See ***Yuga*** **(Astronomical speculation on)**.

**ASURA**. "Anti-god". Name given to the Titans of Indian mythology.

**ATATA**. Name given to the number ten to the power eighty-four. See **Names of numbers** and **High numbers**.

Source: **Vyâkarana* (Pâlî grammar) by Kâchchâyana (eleventh century CE).

**ATIDHRITI**. [S]. Value = 19. The metre of four times nineteen syllables per verse. See **Indian metric**.

**ÂTMAN**. [S]. Value = 1. In Hindu philosophy, this term describes the "Self", the "Individual soul", the "Ultimate reality", even the *Brahman himself, who is said to possess all the corresponding characteristics. The uniqueness of the "Self" and above all the first character of the Brahman as the "great ancestor" explain the symbolism. See ****Pitâmaha*** and **One**.

**ATMOSPHERE**. [S]. Value = 0. See **Infinity**, ***Shûnya*** and **Zero**.

**ATRI**. [S]. Value = 7. Proper noun designating the seventh of the *Saptarishi* (the "Seven Great Sages" of Vedic India), considered to be the founder of Indian medecine. See ***Rishi*** and **Seven**.

**ATRINAYANAJA**. [S]. Value = 1. "Moon". See ***Abja*** and **One**.

**ATTATA**. Name given to the number ten to the power nineteen (= ten British trillions). See **Names of numbers** and **High numbers**.

Source: **Lalitavistara Sûtra* (before 308 CE).

**ATYASHTI**. [S]. Value = 17. The metre of four times seventeen syllables per line in Sanskrit poetry. See **Indian metric**.

**AUM**. Sacred symbol of the Hindus. See **Mysticism of letters** and ***Ekâkshara***.

**AVANI**. [S]. Value = 1. "Earth". See ***Prithivî*** and **One**.

**AVARAHAKHA**. Generic name of the five elements of the revelation. See ***Bhûta*** and ***Mahâbhûta***.

**AVATÂRA**. [S]. Value = 10. "Descent". The incarnation of a Brahmanic divinity, birth through transformation, the aim being to carry out a terrestrial task in order to save humanity from grave danger. The allusion here is made to the *Dashâvatâra*, the "ten *Avatâra*", or major incarnations of *Vishnu, attributed to the four "ages" of the world (**yugas*), according to Hindu cosmogony. See ***Dashâvatâra*** and **Ten**.

**AVYAKTAGANITA**. Name given to algebra (literally: "science of calculating the unknown"), as opposed to arithmetic, called *vyaktaganita*. See ***Vyaktaganita***, **Algebra**, **Arithmetic**.

**AYUTA**. Name for the number ten to the power four (= ten thousand). See **Names of numbers** and **High numbers**.

Source: **Vâjasaneyî Samhitâ* (beginning of Common era); **Taittirîya Samhitâ* (beginning of Common era); **Kâthaka Samhitâ* (beginning of Common era); **Pañchavimsha Brâhmana* (date uncertain); **Sankhyâyana Shrauta Sûtra* (date uncertain); **Âryabhatîya* (510 CE); **Kitab fi tahqiq i ma li'l hind* by al-Biruni (1030 CE); **Lîlâvatî* by Bhâskarâchârya (1150 CE); **Ganitakaumudî* by Nârâyana (1350 CE); **Trishatikâ* by Shrîdharâchârya (date uncertain).

**AYUTA**. Name given to the number ten to the power nine (= a thousand million). See **Names of numbers** and **High numbers**.

Source: **Lalitavistara Sûtra* (before 308 CE).

# B

**BÂHU**. [S]. Value = 2. "Arms", due to the symmetry of the two arms. See **Two**.

**BAHULA**. Name given to the number ten to the power twenty-three (= a hundred thousand [British] trillion). See **Names of numbers** and **High numbers**.

Source: **Lalitavistara Sûtra* (before 308 CE).

**BAKSHÂLÎ'S MANUSCRIPT**. See **Indian documentation (Pitfalls of)**.

**BALINESE NUMERALS**. Signs derived from *Brâhmî numerals, through the intermediary of Shunga, Shaka, Kushâna, Ândhra, Pallava, Châlukya, Ganga, Valabhî, "Pâlî", Vatteluttu and Kawi numerals. Currently in use in Bali, Borneo and the Celebes islands. The corresponding system functions according to the place-value system and possesses zero (in the form of a little circle). For ancient numerals, see Fig. 24.50 and 80. For modern numerals, see Fig. 24.25. See **Indian written numeral systems (Classification of)**. See also Fig. 24.52 and 24.61 to 69.

**BÂNA**. [S]. Value = 5. "Arrow". See ***Shara*** and **Five**. See also ***Pañchabâna***.

**BASE OF ONE HUNDRED**. See ***Shatottaraganaṇâ***, ***Shatottaraguna*** and ***Shatottarasamjñâ***.

**BASE TEN**. See ***Dashaguna*** and ***Dashagunâsamjñâ***.

**BEARER.** [S]. Value = 1. See ***Dharanî*** and **One.**

**BEGINNING.** [S]. Value = 1. See ***Âdi*** and **One.**

**BENGALI NUMERALS.** Signs derived from *Brâhmî numerals, through the intermediary of Shunga, Shaka, Kushâna, Andhra, Gupta, Nâgarî and Kutila numerals. Currently used in the northeast of India, in Bangladesh, Bengal and in much of the centre of Assam (along the Brahmaputra river). The corresponding system functions according to the place-value system and possesses zero (in the form of a little circle). See **Indian written numeral systems (Classification of).** See also Fig. 24.10, 52 and 24.61 to 69.

**BENGALÎ SÂL (Calendar).** See ***Bengalî San.***

**BENGALÎ SAN (Calendar).** The solar era beginning in the year 593 CE. It is still used today in Bengal. To obtain the corresponding date in Common years, add 593 to a date expressed in this calendar. It is also called *Bengalî Sâl*. See **Indian calendars.**

**BHA.** [S]. Value = 27. "Star". Allusion to the twenty-seven **nakshatra*. See **Twenty-seven.**

**BHÂGAHÂRA.** Term used in arithmetic to denote division, although the word is most often used to denote the divisor (which is also called *bhâjaka*). See ***Chhedana.***

**BHAGAVAD GÎTÂ.** "Song of the Lord". A long philosophical Sanskrit poem containing the essence of **Vedânta* philosophy, explained by Krishna and Arjuna in a dialogue about action, discrimination and knowledge. It is a relatively recent text (c. fourth century CE) and is found in the sixth book of the **Mahâbhârata* [see Frédéric; *Dictionnaire* (1987)].

**BHÂJAKA.** Term used in arithmetic to denote the divisor. See ***Bhâgahâra.***

**BHÂJYA.** Term used in arithmetic to denote the dividend. Synonym: *hârya*. See ***Bhâgahâra*** and ***Chhedana.***

**BHÂNU.** [S]. Value = 12. An epithet of **Sûrya*, the Sun-god. *Bhânu* = *Sûrya* = twelve. See **Twelve.**

**BHARGA.** [S]. Value = 11. One of the names of *Rudra. See **Rudra-Shiva** and **Eleven.**

**BHÂSKHARA.** Indian mathematician, disciple of **Âryabhata* (a century after his death). He was born in the first half of the seventh century. He is known mainly for his Commentary on the *Âryabhatîya, in which examples of the use of the place-value system expressed by means of the Sanskrit numerical symbols are found in abundance. The translation of the numbers expressed in this manner is often given using the nine numerals and zero (also according to the rules of the place-value system) [see Shukla and Sarma (1976)]. He is usually called "Bhâskara I" so that he is not confused with another mathematician of the same name (*Bhâskarâchârya). See **Âryabhata (Numerical notations of), Âryabhata's numeration, Numerical symbols, Numerical symbols (Principle of the numeration of),** and **Indian mathematics (The history of).**

**BHÂSKARA I.** See **Bhâskara.**

**BHÂSKARA II.** See **Bhâskarâchârya.**

**BHÂSKARÂCHÂRYA.** Indian mathematician, astronomer and mechanic, usually referred to as Bhâskara II. He lived in the second half of the twelfth century CE. He is famous for his work, the *Siddhântashiromani*, an astronomical text accompanied by appendices relating to mathematics, amongst which we find the **Lîlâvâtî* (the "Player"), which contains a whole collection of problems written in verse. He frequently uses zero and the place-value system, which are expressed in the form of *numerical symbols. He also describes methods of calculation which are very similar to our own and are carried out using the nine numerals and zero. Moreover, he explains the fundamental rules of algebra where the zero is presented as a mathematical concept, and defines *Infinity* as the inverse of zero [see Sastri (1929)].

Here is a list of the main names of numbers given in the *Lîlâvâtî* (Lîl, p. 2) [see Datta and Singh (1938), p.13]:

**Eka* (= 1). **Dasha* (= 10). **Shata* (= $10^2$). **Sahasra* (= $10^3$). **Ayuta* (= $10^4$). **Laksha* (= $10^5$). **Prayuta* (= $10^6$). **Koti* (= $10^7$). **Arbuda* (= $10^8$). **Abja* (= $10^9$). **Kharva* (= $10^{10}$). **Nikharva* (= $10^{11}$). **Mahâpadma* (= $10^{12}$). **Shanku* (= $10^{13}$). **Jaladhi* (= $10^{14}$). **Antya* (= $10^{15}$). **Madhya* (= $10^{16}$). **Parârdha* (= $10^{17}$).

See **Names of numbers, High numbers, Positional numeration, Numerical symbols (Principle of the numeration of), Zero, Infinity, Arithmetic, Algebra,** and **Indian mathematics (The history of).**

**BHÂSKARÎYABHÂSYA.** See *Govindasvâmin.*

**BHATTIPROLU NUMERALS.** Signs derived from *Brâhmî numerals, through the intermediary of Shunga, Shaka, Kushâna, Ândhra, Pallava, Châlukya, Ganga and Valabhî numerals. Used since the eighth century CE by the Dravidians in southern India. Kannara, Telugu, Grantha, Malayalam Tamil, Sinhalese, etc. numerals derived from these numerals. The corresponding system does not use the place-value system or zero. See **Indian written numeral systems (Classification of).** See also Fig. 24.52 and 24.61 to 69.

**BHAVA.** [S]. Value = 11. "Water". One of the names of *Rudra, the etymological meaning of which is related to the tears. See **Rudra, Rudra-Shiva** and **Eleven.**

**BHAVISHYAPURÂNA.** See ***Purâna.***

**BHINNA.** [Arithmetic]. Sanskrit term used to denote "fractions" in general (literally "broken up"). It is synonymous with *bhâga, amsha*, etc. (literally "portion", "part", etc.).

**BHOJA.** Indian astronomer who lived in the eleventh century CE. He is known as the author of a text entitled *Râjamrigânka*, in which there are many examples of the place-value system expressed through Sanskrit numerical symbols [see Billard (1971), p. 10]. See **Numerical symbols, Numerical symbols (Principle of the numeration of),** and **Indian mathematics (The history of).**

**BHÛ.** [S]. Value = 1. "Earth". See ***Prithivî*** and **One.**

**BHÛBHRIT.** [S]. Value = 7. "Mountain". Allusion to *Mount Meru. See ***Adri*** and **Seven.**

**BHÛDHARA.** [S]. Value = 7. "Mountain". Allusion to *Mount Meru. See ***Adri*** and **Seven.**

**BHÛMI.** [S]. Value = 1. "Earth". See ***Prithivî*** and **One.**

**BHÛPA.** [S]. Value = 16. "King". See ***Nripa*** and **Sixteen.**

**BHÛTA.** [S]. Value = 5. "Element". In Brahman and Hindu philosophy, there are five elements (or states) in the manifestation: air (**vâyu*), fire (**agni*), earth (**prithivî*), water (**âpa*) and ether (**âkâsha*). See ***Pañchabhûta*** and **Five.** See also ***Jala.***

**BHUVANA.** [S]. Value = 3. "World". The "three worlds" (**triloka*). See ***Loka, Triloka,*** and **Three.**

**BHUVANA.**[S]. Value = 14. "World". According to Mahâyâna Buddhism, the thirteen "countries of election" or "heavens" of Jina and Bodhisattva, to which was added **Vaikuntha*. See **Fourteen.**

**BÎJA.** Word meaning "letters" in terms of *mathematical symbols* (letters used to express unknown values). In algebra, the word is also used in the sense of "element" or even "analysis". See **Algebra** and ***Bîjaganita.***

**BÎJA.** Word meaning "letters" in terms of *religious symbols* (which generally represent the divinities of the Brahman pantheon or the Buddhist tantric) and esoteric symbols (according to a power which is believed to be creative or evocative). See **Mysticism of letters.**

**BÎJA.** Word used in astronomy to denote corrective terms expressed numerically and applying to the elements of a given text, modifying those of the corresponding *astronomical canon. See **Indian astronomy (The history of).**

**BÎJAGANITA.** Word denoting algebraic science or science of analytical arithmetic and the calculation of elements (from **bîja*: "letter-symbol", "element", "analysis" and from **ganita*: "science of calculation"). The word was used in this sense since Brahmagupta's time (628 CE). However, Indian mathematicians only ever used the first syllable of the word denoting a given operation as their algebraic symbols. See **Indian mathematics (The history of).**

**BILLION.** (= ten to the power twelve. US, ten to power of nine). See ***Antya, Kharva, Mahâbja, Mahâpadma, Mahâsaroja, Parârdha,*** and ***Shankha.***

**BINDU.** Literally "point". This is the name given to the number ten to the power forty-nine. See **Names of numbers.** For an explanation of this symbolism, see **High numbers (The symbolic meaning of).**

Source: **Vyâkarana* (Pâlî grammar) by Kâchchâyana (eleventh century CE).

**BINDU.** [S]. Value = 0. This word literally means "point". This is the symbol of the universe in its non-manifest form, before its transformation into the world of appearances (*rûpadhâtu*). The comparison between the uncreated universe and the point is due to the fact that this is the most elementary mathematical symbol of all, yet it is capable of generating all possible lines and shapes (**rûpa*). Thus the association of ideas with "zero", which is not only considered to be the most negligible quantity, but also and above all it is the most fundamental of mathematical concepts and the basis for all abstract sciences. See **Zero.**

**BIRTH.** [S]. Value = 4. See ***Gati, Yoni*** and **Four.**

**BLIND KING.** [S]. Value = 100. See ***Dhârtarâshtra*** and **Hundred.**

**BLUE LOTUS (half-open).** This has represented the number ten to the power twenty-five. See ***Utpala*** and **High numbers (The symbolic meaning of)**.

BLUE LOTUS **(half-open)**. This has represented the number ten to the power ninety-eight. See ***Uppala*** and **High numbers (The symbolic meaning of)**.

BODY. [S]. Value = 1. See ***Tanu*** and **One**.

BODY. [S]. Value = 6. See ***Kâya*** and **Six**.

BODY. [S]. Value = 8. See ***Tanu*** and **Eight**.

BORN TWICE. [S]. Value = 2. See ***Dvija*** and **Two**.

BOW WITH FIVE FLOWERS. See ***Pañchabâna***.

BRAHMA. Name of the "Universal creator", the first of the three major divinities of the Brahman pantheon (Brahma, *Vishnu, *Shiva). See ***Pitâmaha***, ***Âtman*** and ***Parabrahman***.

BRAHMA. [S]. Value = 1. See ***Âtman***, ***Pitâmaha***, **Parabrahman** and **One**.

BRAHMAGUPTA. Indian astronomer who lived in the first half of the seventh century CE. His best-known works are *Brahmasphutasiddhânta* and *Khandakhâdyaka*, where there are many examples of the place-value system using the nine numerals and zero, as well as the *Sanskrit numerical symbols. He also describes methods of calculation which are very similar to our own using the nine numerals and zero. Moreover, he gives basic rules of algebra where zero is presented as a mathematical concept, and he defines *Infinity* as the number whose denominator is zero [see Dvivedi (1902)]. See **Numerical symbols (Principle of the numeration of)**, **Zero**, **Infinity**, **Arithmetic**, **Algebra**, and **Indian mathematics (The history of)**.

BRAHMAN. See ***Âtman***, ***Pitâmaha***, ***Parabrahman***, ***Paramâtman***, **Day of Brahma** and ***High numbers*** (***The symbolic meaning of***).

BRAHMANICAL RELIGION. See **Indian philosophies and religions**.

BRAHMANISM. See **Indian religions and philosophies**.

BRAHMASPHUTASIDDHÂNTA. See **Brahmagupta** and **Indian mathematics (The history of)**.

BRAHMÂSYA. [S]. Value = 4. "Faces of *Brahma". In representations, this god generally has four faces. He also has four arms and he is often depicted holding one of the four **Vedas* in each hand. See **Four**.

BRÂHMÎ ALPHABET. See Fig. 24. 28.

BRÂHMÎ NUMERALS. The numerals from which all the other series of 1 to 9 in India Central and Southeast Asia are derived. These are found notably in Asoka's edicts and in the Buddhist inscriptions of Nana Ghat and Nasik. The corresponding system does not function according to the place-value system, nor does it possess zero. See Fig. 24.29 to 31 and 70. For notations derived from Brâhmî, see Fig. 24.52. For their graphic evolution, see Fig. 24.61 to 69. See **Indian written numeral systems (Classification of)**.

BREATH. [S]. Value = 5. See ***Prâna***, ***Pâvana*** and **Five**.

BRILLIANT. [S]. Value = 12. See ***Arka*** and **Twelve**.

BUDDHA **(The legend of)**. Legend recounted in the **Lalitavistara Sûtra*, which is full of examples of immense numbers. See **Indian mathematics (The history of)**.

BUDDHASHAKARÂJA **(Calendar)**. Buddhist calendar which is hardly used outside of Sri Lanka and the Buddhist countries of Southeast Asia. It generally begins in 543 BCE, thus by adding 543 to a date in this calendar we obtain the corresponding date in our own calendar. See **Indian calendars**.

BUDDHISM. Here is an alphabetical list of all the terms related to Buddhism which can be found as headings in this dictionary: **Ashtamangala*, **Ashtavimoksha*, **Bhuvana*, *Buddha (Legend of), **Chaturmukha*, **Chaturyoni*, **Dashabala*, **Dashabûmi*, **Dharma*, **Dvâdashadvârashâstra*, **Dvâtrimshadvaralakshana*, **Gati*, *High numbers (The symbolic meaning of), **Indriya*, **Jagat*, **Kâya*, **Loka*, **Mangala*, **Pañchâbhijñâ*, **Pañcha Indryâni*, *Pañchachaksus*, **Pañchaklesha*, **Pañchânantara*, **Ratna*, **Saptabuddha*, **Shûnya*, **Shûnyatâ*, **Tallakshana*, **Trikâya*, **Tripitaka*, **Vajra* and *Zero.

BUDDHIST RELIGION. See **Buddhism** and **Indian religions and philosophies**.

BURMESE NUMERALS. Signs derived from *Brâhmî numerals, through the intermediary of Shunga, Shaka, Kushâna, Ândhra, Pallava, Châlukya, Ganga, Valabhî, "Pâlî", Vatteluttu and Môn numerals. Used since the eleventh century CE by the people of Burma. The corresponding system uses the place-value system and zero (in the form of a little circle). For ancient numerals, see Fig. 24.51. For modern numerals, see Fig. 24.23. See **Indian written numeral systems (Classification of)**. See also Fig. 24.52 and 24.61 to 69.

## C

CALCULATING BOARD. See ***Pâtî***, ***Pâtîganita***.

CALCULATING SLATE. See ***Pâtîganita***.

CALCULATION **(Methods of)**. See ***Dhûlîkarma***, ***Pâtî***, ***Pâtîganita*** and **Indian methods of calculation**.

CALCULATION (The science of). See ***Ganita***.

CALCULATION ON THE ABACUS. See ***Dhûlîkarma***.

CALCULATOR. [Arithmetic]. See ***Samkhya***.

CANOPY OF HEAVEN. [S]. Value = 0. See **Zero**, **Zero (Indian concepts of)** and **Zero and Infinity**.

CARDINAL POINT. [S]. Value = 4. See ***Dish*** and **Four**.

CAUSAL POINT. See ***Paramabindu*** and **Indian atomism**.

CELESTIAL YEAR. See ***Divyavarsha***.

CENTESIMAL NUMERATION. See ***Shatottaragananâ***, ***Shatottaraguna*** and ***Shatottarasamjñâ***.

CHAITRA. Lunar-solar month corresponding to March / April.

CHAITRÂDI. "The beginning of *Chaitra*". This is the name of the year beginning in spring with the month of **Chaitra*, the first lunar-solar month.

CHAKRA. [S]. Value = 12. "Wheel". This refers to the zodiac wheel. See ***Râshi*** and **Twelve**.

CHAKSHUS. [S]. Value = 2. "Eye". See ***Netra*** and **Two**.

CHÂLUKYA **(Calendar)**. Calendar of the dynasty of the eastern Châlukyas, beginning in the year 1075 CE. This calendar was used until the middle of the twelfth century (until c. 1162). To obtain the corresponding date in our own calendar, add 1075 to a date expressed in this calendar. See **Indian calendars**.

CHÂLUKYA NUMERALS. Signs derived from *Brâhmî numerals, through the intermediary of Shunga, Shaka, Kushâna, Ândhra and Pallava numerals. Contemporaries of the "Vatapi" dynasty of the Châlukyas of Deccan (fifth to seventh century CE). The corresponding system does not use the place-value system or zero. See Fig. 24.45 and 70. See **Indian written numeral systems (Classification of)**. See also Fig. 24.52 and 24.61 to 69.

CHAM NUMERALS. Signs derived from *Brahmi numerals, through the intermediary of Shunga, Shaka, Kushâna, Ândhra, Pallava, Châlukya, Ganga, Valabhi, "Pâlî" and Vatteluttu numerals. Used from the eighth to the thirteenth century CE to express dates of the Shaka calendar in the vernacular inscriptions of Champa (in part of Vietnam). The corresponding system uses the place-value system and zero. See **Indian written numeral systems (Classification of)**. See Fig. 24.79 and 80. See also Fig. 24.52 and 24.61 to 69.

CHANDRA. [S]. Value = 1. "Luminous". An attribute of the *Moon as a (male) divinity of the Brahmanic pantheon. See ***Abja***, ***Soma*** and **One**.

CHARACTERISTIC. [S]. Value = 5. See ***Purânalakshana*** and **Five**.

CHATUR. Ordinary Sanskrit name for the number four, which forms part of the composition of many words which have a direct relationship with the idea of this number.

Examples: **Chaturânanavadana*, **Chaturyuga*, **Chaturanga*, **Chaturangabalakâya*, **Chaturâshrama*, **Chaturdvîpa*, **Chaturmahârâja*, **Chaturmâsya*, **Chaturmukha*, **Chaturvarga*, **Chaturyoni*. For words which have a more symbolic relationship with the number four, see **Four** and **Symbolism of numbers**.

CHATURÂNANAVÂDANA. [S]. Value = 4. The "four oceans". See ***Chatur***, ***Sâgara***, **Four**. See also **Ocean**.

CHATURANGA. "Four parts". Name given to an ancient Indian game, the ancestor of chess: there were four players and the board consisted of eight by eight squares and eight counters (the king, the elephant, the horse, the chariot and four soldiers). See ***Chatur***.

CHATURANGABALAKÂYA. "Four corps". Name given to the ancient organisation of the Indian army, which consisted of elephants (*hastikâya*), the cavalry (*ashvakâya*), the chariots (*râthakâya*) and the infantry (*pattikâya*). See ***Chatur***.

CHATURÂSHRAMA. "Four stages". According to Hindu philosophy, there were four stages to a man's life, in keeping with the Vedic concept: in the first (called *brahmâchârya*), intellectual capacities are developed, profane and religious instruction are received and the virtues of spiritual life are cultivated; in the second (*grihastha*), marriage and home-making take place; in the third (*hânaprastha*), having fulfilled his role as master of the house and having served his community, he goes alone into the forest to devote himself to intensive meditation, philosophical studies and the Scriptures; finally, in the fourth stage (*sannyâsa*), he gives up all his possessions and no longer cares for earthly things. See ***Chatur***.

**CHATURDASHA.** Ordinary Sanskrit name for the number fourteen. For words with a symbolic relationship with this number, see: **Fourteen** and **Symbolism of numbers**.

**CHATURDVÎPA.** "Four Islands". In Brahmanic mythology and Hindu cosmology this is the name given to the four island-continents said to surround India (*Jambudvîpa*). See ***Chatur***. For an explanation of this choice of number, see **Ocean**, which gives the same explanation about the four seas (**chatursâgara*).

**CHATURMAHÂRÂJA.** "Four great kings". These are the four guardian divinities of the cardinal points, who are said to live on the peaks of *Mount Meru (Vaishravana in the North, Virûpâksha in the West, Virûdhaka in the South and Dhritarâshtra in the East). See ***Chatur***.

**CHATURMÂSYA.** "Four months". Name of an Indian ritual which takes place every four months, once at the start of spring, once at the start of the rain season, and once at the start of autumn. See ***Chatur***.

**CHATURMUKHA.** "Four faces". Name given to all the Brahmanic or Buddhist divinities who are represented as having four faces (*Brahma, *Shiva, etc.). See ***Chatur***.

**CHATURSÂGARA.** "Four oceans". These are the four seas said to surround India (*Jambudvîpa*). See ***Sâgara***. For an explanation of this choice of number, see **Ocean**.

**CHATURVARGA.** "Four aims". These are the **trivarga* of Hindu philosophy (the three objectives of human existence), namely: *artha*, (material wealth), **kâma* (passionate love), and **dharma* (duty), to which sometimes a fourth is added, *moskha*, the liberation of the soul. See ***Chatur***.

**CHATURVIMSHATI.** Ordinary Sanskrit name for the number twenty-four. For words which have a symbolic relationship with this number, see: **Twenty-four** and **Symbolism of numbers**.

**CHATURYONI.** The "four types of reincarnation". According to Hindus and Buddhists, there are four different ways to enter the cycle of rebirth (**samsâra*): either through a viviparous birth (*jarâyuva*), in the form of a human being or mammal; or an oviparous birth (*andaja*), in the form of a bird, insect or reptile; or by being born in water and humidity (*samsvedaja*), in the form of a fish or a worm; or even through metamorphosis (*aupapâduka*), which means there is no "mother" involved, just the force of *Karma* [see Frédéric (1994)]. See ***Chatur***.

**CHATURYUGA.** The "four periods". Cosmic cycle of 4,320,000 human years, subdivided into four periods. Synonymous with **mahâyuga*. See ***Chatur*** and ***Yuga* (Definition)**.

**CHATURYUGA.** [Astronomy]. According to speculations about **yugas*, the *chaturyuga*, or cycle of 4,320,000 years, is defined as the period at the beginning and end of which the nine elements (namely the Sun, the Moon, their apsis and node and the planets) are in average perfect conjunction at the starting point of the longitudes. See ***Chaturyuga*** (previous entry) and ***Yuga* (Astronomical speculation on)**.

**CHATVÂRIMSHATI.** Ordinary Sanskrit name for the number *forty.

**CHHEDANA.** [Arithmetic]. Term meaning division (literally: "to break into many pieces"). Synonyms: *bhâgahâra*, *bhâjana*, etc.

**CHHEDI (Calendar).** Calendar beginning 5 September, 248 CE, which was used in the region of Malva and in Madhya Pradesh. To obtain the corresponding date in our own calendar, add 248 to a given *Chhedi* date. Sometimes called *kalachurî*, it was in use until the eighteenth century CE. See **Indian calendars**.

**CHHIDRA.** [S]. Value = 9. "Orifice". The nine orifices of the human body (the mouth, the two eyes, the two nostrils, the two ears, the anus and the sexual orifice). See **Nine**.

**CHRONOGRAM.** A short phrase or sentence whose numerical value amounted to the date of a given event. There are many methods of composing chronograms in India.

**CHRONOGRAMS (Systems of letter numerals).** One of the processes of composing chronograms involves the use of a *numeral alphabet. The hidden date is revealed by evaluating the various letters of each word of the sentence in question, then totalling the value of each word. This requires a mixture of mathematical and poetical skill, using the imagination to create sentences which have both literal and mathematical meaning. These types of chronograms (for which the system of evaluation clearly varies according to the system of attribution of numerical values to the letters of the alphabet) were not only written in Sanskrit, but also in various *Prakrits (local dialects). Many examples have been found throughout India, from Maharashtra, Bengal, Nepal or Orissa to Tamil Nadu, Kerala or Karnataka. They were also used by the Sinhalese, the Burmese, the Khmers, and in Thailand, Java and Tibet. Many other examples also exist in Muslim India and in Pakistan, but these are many chronograms which employ numeral letters of the Arabic-Persian alphabet using a process called *Abjad*. See **Numeral alphabet and composition of chronograms**.

**CHRONOGRAMS (Systems of numerical symbols).** Another method of composing chronograms is only used for expressing the dates of the Shaka calendar: the language used is always Sanskrit and the dates are always expressed metaphorically, using Indian *numerical symbols ruled by the place-value system. This process was used for many centuries in India and in all the Indianised civilisations of Southeast Asia (Khmer, Cham, Javanese, etc., kingdoms). See **Numerical symbols, Numerical symbols (principle of the numeration of)**.

**CIRCLE.** As a symbolic representation of the sky. See **Serpent (Symbolism of the)**.

**CIRCLE.** As the graphic representation of zero. See ***Shûnya-chakra*, The numeral 0, Zero**.

**CITY-FORTRESS.** [S]. Value = 3. See ***Pura***, ***Tripura*** and **Three**.

**COBRA (Cult and symbolism of).** See **Serpent (Symbolism of)** and ***Nâga***.

**CODE** (secret writing and numeration). See **Numeral alphabet and secret writing**.

**COLOUR.** [S]. Value = 6. See ***Râga*** and **Six**.

**COMPLETE.** As a synonym of a large quantity. See **High numbers (Symbolic meaning of)**.

**COMPLETE.** As a synonym of zero. See ***Pûrna***.

**CONSTELLATION.** [S]. Value = 27. See ***Nakshatra*** and **Twenty-seven**.

**CONTEMPLATION.** [S]. Value = 6. See ***Darshana*** and **Six**.

**COSMIC CYCLES.** The division and length of cosmic cycles has always been of great importance in terms of Brahmanism: These periods represented the successive sections of cosmic life, conceived as cyclical and eternally revolving. The divisions of time were naturally the key elements of these cycles. The temporal dimension was meant to correspond to the duration of the creative and animating power of the cosmos, the "Word" (**vâchana*), which was uttered by the supreme progenitor of the world, Brahman-Prajâpati, and that which assimilates "knowledge" *par excellence*, the Veda. Thus the progenitor resembles the year which is taken as a unit of measurement of its cyclical activity, and the **Veda*, a collection of lines, is divided into as many metric elements as there are moments in the "year" (see HGS, I, pp. 157–8). Of course, the "year" in question here is a "divine" year as opposed to a human year. See **Divine Year**, ***Yuga* (Definition)**, ***Yuga* (Systems of calculation of)**, ***Yuga* (Cosmogonic speculations on)**, ***Kalpa***, **Day of Brahma**.

**COSMIC ERAS.** See **Cosmic cycles** and ***Yuga* (Definition)**.

**COSMOGONIC SPECULATIONS.** See ***Yuga* (Cosmogonic speculations on), Kalpa, Jaina.**

**COURAGE.** [S]. Value = 14. See ***Indra*** and **Fourteen**.

**COW.** [S]. Value = 1. See ***Go*** and **One**.

**COW.** [S]. Value = 9. See ***Go*** and **Nine**.

**CUBE ROOT.** [Arithmetic]. See ***Ghanamûla***.

**CUBE.** [Arithmetic]. See ***Ghana***.

## D

**DAHANA.** [S]. Value = 3. "Fire". See ***Agni*** and **Three**.

**DANTA.** [S]. Value = 32. "Teeth". Humans have thirty-two teeth. See **Thirty-two**.

**DANTIN.** [S]. Value = 8. "Elephant". See ***Diggaja*** and **Eight**.

**DARSHANA.** [S]. Value = 6. "Vision", "contemplation", "system", and by extension "demonstration" and "philosophical point of view". This concerns the six principal systems of Hindu philosophy: mental research (*mîmâmsâ*); method (*nyâya*); the study and description of nature (*vaisheshika*); number as a way of thinking applied to the liberation of the soul (**sâmkhya*); the philosophies and practices of the liberation of the spirit from material ties (*yoga*); and studies based on the *Vedânta Sûtras* which deal with the basic identity of the soul and the *Brahman (*vedânta*). See ***Shaddarshana*** and **Six**.

**DASHA (or DASHAN).** Ordinary Sanskrit name for the number ten, which appears in the composition of many words which have a direct relationship with the idea of this number. Examples: **Dashabala*, **Dashabhûmi*, **Dashaguna*, **Dashagunâsamjñâ*, **Dashagunottarasamjñâ*, **Dashaharâ*, **Dashâvatâra*.

For words which have a more symbolic relationship with this number, see **Ten** and **Symbolism of numbers**.

**DASHABALA.** "Ten powers". This refers to the ten faculties possessed by a Buddha, which give him ten powers, namely: the intuitive knowledge of the possible and the impossible, whatever the situation; the development of actions; the superior and inferior faculties of living beings; the diverse elements of the world; the paths which lead to purity or impurity; contemplation, concentration, meditation and the three deliverances; death; and the purification of all imperfections .

**DASHABHÛMI**. "Ten lands", "ten paradises". This refers to the "ten stages" of the Buddha Shâkyamuni.

**DASHAGUNA**. "Ten, primordial property". Sanskrit name for the decimal base. This word can be found in such works as the **Trishatikâ* by Shrîdharâchârya [see TsT, R. 2 – 3] and in the **Lîlavâtî* by Bhâskarâchârya [see Lîl, p. 2].

**DASHAGUNÂSAMJÑÂ**. "Words representing powers of ten". Term which applies to names of numbers of the Sanskrit numeration, distributed according to a decimal scale (base 10). See ***Dashaguna***, **Names of numbers** and **High numbers**. This word can be found in such works as the **Trishatikâ* by Shrîdharâchârya [see TsT, R. 2 – 3].

**DASHAGUNOTTARASAMJÑÂ**. "Words representing powers of ten". Term which applies to names of numbers of the Sanskrit numeration, distributed according to a decimal scale (base 10). The contrast is made here with the word *shatottarasmjñâ* which applies to the centesimal scale (base 100). See **Dashaguna**, **Names of numbers**, **High numbers** and ***Shatottarasamjñâ***.

**DASHAHARÂ**. Name of the Feast of the tenth day. See ***Dasha*** and ***Durgâ***.

**DASHAKOTI**. Literally "ten **kotis*". Name given to the number ten to the power eight ( = a hundred million). See **Names of numbers** and **High numbers**.

Source: **Ganitasârasamgraha* by Mahâvîrâchârya (850 CE).

**DASHALAKSHA**. Literally "ten **lakshas*". This is the name given to the number ten to the power six (one million). See **Names of numbers** and **High numbers**.

Source: **Ganitasârasamgraha* by Mahâvîrâchârya (850 CE).

**DASHAN**. Ordinary Sanskrit name for the number ten. See ***Dasha***.

**DASHASAHASRA**. Literally "ten **sahastras*". Name given to the number ten to the power four (ten thousand). See **Names of numbers** and **High numbers**.

Source: **Ganitasârasamgraha* by Mahâvîrâchârya (850 CE).

**DASHÂVATÂRA**. Name of the "ten major incarnations" of *Vishnu, which are as follows: *Matsya* (incarnation as a fish); *Kûrma* (incarnation as a tortoise); *Varâha* (as a boar); *Narasimha* (as a lion-man); *Vâmana* (as a dwarf); *Parashu-Râma* (as Râma of the axe); **Râma* (the hero of *Râmâyana*); **Krishna* (the god); *Budha* (the god); and *Kalki*. See ***Avatâra***.

**DASRA**. [S]. Value = 2. Name of one of the two twin gods Saranyû and Vivashvant of the Hindu pantheon (also called Dasra and Nâsatya). Symbolism through association of ideas with the "Horsemen". See ***Ashvin*** and **Two**.

**DAY**. [S]. Value = 15. See ***Tithi***, ***Ahar*** and **Fifteen**.

**DAY OF BRAHMA (Arithmo-cosmogonical speculations about the)**. According to Brahman cosmogony, the lifespan of the material universe is limited, and it manifests itself by **kalpa* cycles: "All the planets of the universe, from the most evolved to the most base, are places of suffering, where birth and death take place. But for the soul that reaches my Kingdom, O son of Kunti, there is no more reincarnation. One day of *Brahma is worth a thousand of the ages [**yuga*] known to humankind; as is each night" (**Bhagavad Gîtâ*, VIII, lines 16 and 17). Thus each *kalpa* is worth one day in the life of Brahma, the god of creation. In other words, the four ages of a **mahâyuga* must be repeated a thousand times to make one "day of Brahma", a unit of time which is the equivalent of 4,320,000,000 human years.

According to this cosmogony, this is the total length of one created universe. The *kalpa* or "day of Brahma" is meant to correspond to the appearance, evolution and disappearance of a world, and this cycle is followed by a period of "cosmic repose" of equal length, which is followed by a new *kalpa*, and so on indefinitely. In other words, each *kalpa* ends with the total destruction (*pralaya*) of the universe which is followed by a period of reabsorption which is equivalent to a "night of Brahma", of equal length to the corresponding "day", before life is breathed into a new universe. It is precisely during this period of non-creation that *Vishnu, lying on *Ananta, the serpent of Infinity and Eternity, rests while he waits for Brahma to accomplish his work of Creation. This philosophy was developed as far as to speculate on the "length of the life of the god Brahma". A Commentary on the **Bhagavad Gîtâ* says: ". . . nothing in the material universe, not even Brahma can escape birth, ageing and death . . . the Causal Ocean contains countless Brahmas, who, being in a constant state of flux, appear and disappear like bubbles of air".

Here are some calculations relating to this this subject. Given that one whole "twenty-four hour day" in this god's life is the sum of one of his "days" and one of his "nights", "twenty-four hours in the life of Brahma" corresponds to: 4,320,000,000 + 4,320,000,000 = 8,640,000,000 (= eight thousand, six hundred and forty million) human years. One "year of Brahma" is made up of 360 of these "days". Thus it corresponds to 8,640,000,000 × 360 = 3,110,400,000,000 (= three billion, one hundred and ten thousand, four hundred million) human years. As this god is said to live for one hundred of his "years", the total length of his existence is equal to: 3,110,400,000,000 × 100 = 311,040,000,000,000 ( = three hundred and eleven billion, forty thousand million) human years. According to certain traditions reported notably by al-Biruni, the "day of Brahma" does not correspond to a simple *kalpa*, but to a **parârdha* of kalpas, which is the length of a kalpa multiplied by *ten to the power seventeen*.

Thus: 1 "day of Brahma" = 100,000,000, 000, 000, 000 (= one hundred trillion) *kalpas*. As one *kalpa* is 4,320,000,000 years long, one "day" of this god corresponds to: 432,000,000, 000,000,000,000,000,000 (= four hundred and thirty-two sextillions) human years. Thus the complete "day" = 864,000,000,000,000,000, 000,000,000 (= eight hundred and sixty-four sextillions) human years. If we multiply this number by 36,000, the "life of Brahma" lasts thirty-one octillion and one hundred and four septillion human years. Childish at first sight, such speculations are very revealing of the Indian tendency towards metaphysical abstraction and of the high conceptual level achieved at an early stage by this civilisation. See ***Ananta***, ***Asamkhyeya***, **Calculation**, **High numbers**, **Infinity**, **Speculative arithmetic**, **Sanskrit**, ***Sheshashîrsha***, **Indian mathematics (The history of)** and ***Yuga*** **(Cosmogonical speculations on)**.

**DAY OF BRAHMA**. Cosmic period corresponding to the total length of one creation of the universe. According to Brahman cosmogony, this "day" is equal to 12,000,000 divine years (**divyavarsha*); and as one divine year is equal to 360 human years, the "day of Brahma" is equal to 4,320,000,000 human years. See ***Divyavarsha***, ***Mahâyuga*** and ***Yuga***.

**DAY OF THE WEEK**. [S]. Value = 7. See ***Vâra*** and **Seven**.

**DECIMAL NUMERATION**. See ***Dashaguna*** and ***Dashagunâsamjñâ***.

**DELECTATION**. [S]. Value = 6. See ***Rasa*** and **Six**.

**DEMONSTRATION**. [S]. Value = 6. See ***Darshana*** and **Six**.

**DESCENT**. [S]. Value = 10. See ***Avatâra*** and **Ten**.

**DEVA**. [S]. Value = 33. "Gods". This is the general name given to all the divinities of the Hindu, Brahmanic, Vedic and Buddhist pantheons. These are the inhabitants of Mount Meru (mythical mountain, situated at the axis of the universe), who are ruled by a god. Unlike the great divinities (*Mahâdeva*) such as *Brahma, *Vishnu and *Shiva, these divinities have neither strength nor creative power. Theoretically numbering thirty-three million, they are reduced to thirty-three in Hindu cosmogony, which also gives their group the name **Trâiyastrimsha* ("thirty-three"). See **Thirty-three**. See also **Mount Meru**.

**DEVANAGARI NUMERALS**. See **Nagari numerals**.

**DEVAPÂRVATA**. "Mountain of the gods". One of the names of Mount Meru, the home of the gods in Brahmanic mythology and Hindu cosmology. See **Mount Meru**, ***Adri***, ***Dvîpa***, ***Pûrna***, ***Pâtâla***, ***Sâgara***, ***Pushkara***, ***Pâvana*** and ***Vâyu***.

**DHARÂ**. [S]. Value = 1. "Earth". See ***Prithivî*** and **One**.

**DHARANÎ**. [S]. Value = 1. Literally "Bearer". This is synonymous here with the "earth", in the sense of "the bearer". See ***Prithivî*** and **One**.

**DHARMA**. In Indian philosophies, the *Dharma* is the general law, the Duty, the thing which is permanently fixed, the ensemble of rules and natural phenomena which rule the order of things and of men. In Buddhist philosophy in particular, the *dharma* is considered to be one of the three Treasures (**Triratna*) and one of the three refuges of the faithful. It is thus the social duty of the disciple. It represents the ultimate Only Reality, Virtue, Natural Order of all that exists, the Doctrine of Buddha as well as all the perceptions (ideas) hidden in the Manas [see Frédéric, *Dictionnaire*]. See ***Shûnyatâ***.

**DHÂRTARÂSHTRA**. [S]. Value = 100. There is a legend related in the *Mahâbhârata* about the blind king Dhritarâshtra, son of Ambîkâ and the king Vichitravîrya, who married Gandhârî, with whom he had a hundred sons, called *Dhârtarâshtra*. During the Great Battle against the sons of Pându, the latter were all killed and became demons [see Frédéric, *Dictionnaire*]. See ***Pândava*** and **Hundred**.

**DHÂTRÎ**. [S]. Value = 1. "Earth". See ***Prithivî*** and **One**.

**DHRITI**. [S]. Value = 18. This refers to the metre of four times eighteen syllables per verse in Sanskrit poetry. See **Indian metric**.

**DHRUVA**. [S]. Value = 1. In Hindu mythology, this was the son of a king called Uttânapâda and his queen Sunîtî, who, through the power of his will, became the ***Sudrishti***, the "divinity who never moves": the Pole star, whose uniqueness and fixedness are doubtless at the root of this symbolism. See **One**.

**DHÛLÎKARMA**. Literally "work on dust" (from *Dhûlî*, "sand", "dust", and *karma*, "act"). Term used in ancient Sanskrit literature to denote the "act of carrying out mathematical operations", in allusion to the ancient Indian practice of carrying out calculations on a board covered in sand. Today, the word is only used in the abstract sense of "superior mathematics". See **Calculation (Methods of)**.

**DHVAJÂGRANISHÂMANI**. Name given to the number ten to the power 145. See **Names of numbers** and **High numbers**.

Source: **Lalitavistara Sûtra* (before 308 CE).

**DHVAJÂGRAVATI**. Name given to the number ten to the power ninety-nine. See **Names of numbers** and **High numbers**.

Source: **Lalitavistara Sûtra* (before 308 CE).

**DIAMOND**. A representation of the number ten to the power thirteen. See ***Shanku***.

**DIGGAJA**. [S]. Value = 8. In Hindu cosmogony, the collective name given to the *Ashtadiggaja*, the "eight Elephants", who are said to guard the eight horizons of space. See ***Dish*** and **Eight**.

**DIGITAL CALCULATION**. See ***Mudrâ***.

**DIKPÂLA**. [S]. Value = 8. "Guardian of the points of the compass". In Hindu cosmogony, this is the collective name given to the eight divinities considered to be the guardians of the horizons and the points of the compass (*Indra in the east, *Agni in the southeast, *Yama in the south, Nirritî in the southwest, *Varuna in the west, Kuvera in the north, *Vâyu in the northwest and Îshâna in the northeast). See ***Diggaja***, ***Dish***, ***Lokapâla*** and **Eight**.

**DISH**. [S]. Value = 4. "Horizon". The four cardinal points (north, south, east and west). See **Four**.

**DISH**. [S]. Value = 8. "Horizon". The horizons corresponding to the eight points of the compass: the north, the northwest, the west, the southwest, the southeast, the south, the east and the northeast. See **Eight**.

**DISH**. [S]. Value = 10. "Horizon". The ten horizons of space: the eight normal horizons, plus the *nadir* and the *zenith*. See **Ten**.

**DISHA**. [S]. Value = 4. "Horizon". See ***Dish*** and **Four**.

**DISHÂ**. [S]. Value = 10. "Horizon". See ***Dish*** and **Ten**.

**DIVÂKARA**. [S]. Value = 12. "Sun". See ***Sûrya*** and **Twelve**.

**DIVIDEND**. [Arithmetic]. See ***Bhâjya***.

**DIVINATION**. See **Numeral alphabet, magic, mysticism and divination**, **Indian astrology**, and **Indian astronomy (The history of)**.

**DIVINE MOTHER**. [S]. Value = 7. See ***Mâtrikâ*** and **Seven**.

**DIVINE PERFECTION**. As a symbol for a large quantity. See **High numbers (Symbolic meaning of)**.

**DIVINE YEAR**. See ***Divyavarsha***.

**DIVISION**. [Arithmetic]. See ***Chhedana***, ***Bhâgahâra***, ***Labdha***, ***Shesha*** and ***Bhâjya***.

**DIVISOR**. [Arithmetic]. See ***Bhâgahâra***.

**DIVYAVARSHA**. "Celestial or divine year". To convert a number of divine years into human years, it must be multiplied by 360.

**DOGRI NUMERALS**. Signs derived from *Brâhmî numerals, through the intermediary of Shunga, Shaka, Kushâna, Andhra, Gupta and Sharada numerals, and constituting a variation of Takari numerals. These are currently used in the Indian part of Jammu (in the southwest of Kashmir). The corresponding system uses the place-value system and possesses zero (in the form of a little circle). See **Indian written numeral systems (Classification of)**. See Fig. 24.13, 52 and 24.61 to 69.

**DOT**. A graphical sign representing zero. see **Numeral 0**, ***Bindu***, ***Shûnya-bindu***, **Zero**.

**DOT**. A name for ten to the power forty-nine. See ***Bindu***, **High numbers**, **High numbers (Symbolic meaning of)**.

**DOT**. [S]. Value = 0. See ***Bindu***, **Indian atomism** and **Zero**.

**DRAVIDIAN NUMERALS**. Numerals used in the southern regions of India, namely Tamil Nadu, Karnataka, Andhra Pradesh and Kerala, where the people are referred to as "Dravidian", and who, unlike the people from northern and central India, do not speak Indo-European languages. These signs are derived from *Brâhmî numerals, through the intermediary of Shunga, Shaka, Kushâna, Andhra, Pallava, Chalukya, Ganga, Valabhi and Bhattiprolu numerals. The corresponding system has not always used the place-value system or possessed zero. See **Tamil numerals**, **Malayalam numerals**, **Telugu numerals**, **Kannara numerals** and **Indian written numeral systems (Classification of)**.

**DRAVYA** [S]. Value = 6. "Substances". The six "bodies", or "substances" which make up existence according to *Jaina philosophy (these are: *dharmâshtikâya*, which constitutes the means of movement; *adharmâshtikâya*, which allows the animate to become inanimate; *akshatikâya*, which creates the space in which the animate and the inanimate live; *pudgalâshtikâya*, which enables the very existance of matter; *jîvâshtikâya*, which allows the mind to exist through inferences; and *kâla*, which is nothing other than time [see Frédéric (1987)]. See **Six**. This symbol is found in **Ganitasârasamgraha* by Mahâvîrâchârya [see Datta and Singh (1938), p. 55].

**DRIGGANITA**. See ***Parameshvara***.

**DRISHTI**. [S]. Value = 2. This term is generally used in the sense of "vision", "contemplation", "revelation", "conception of the world" and "theory". Its primary sense, however, is "eye"; hence *drishti* = 2. See ***Netra*** and **Two**.

**DROP**. [S]. Value = 1. See ***Indu*** and **One**.

**DUALITY**. See ***Dvaita***.

**DURGÂ**. [S]. Value = 9. "Inaccessible". This is the name of a Hindu divinity, the "Divine Mother", wife of Shiva, who is worshipped during the "Feast of the nine days" (*navarâtrî*), which is celebrated at the end of the rain season in the month of *Ashvina* (September – October). The association of ideas which led to *Durgâ* becoming the numerical symbol equivalent to nine is obvious, but the choice of this value for the number of days of the feast is difficult to explain. This divinity, who is said to possess great powers, is often represented as having ten arms; moreover, she is depicted standing on a lion, which symbolises her power. The "Feast of nine days", which marks the end of the monsoon, ends on the tenth day with the grand feast of the *dashaharâ* (from *dasha*, "ten"), which is dedicated to Durgâ. The Hindus commemorate the victory of their divinities over the forces of Evil.

In this double religious symbolism, it is possible that, in accordance with the characteristic Indian way of thinking, these nine days were associated with the nine numerals of the place-value system, with which it is possible to write all numbers. The tenth day might then be associated with the tenth sign in this system: the zero, which corresponds to the most elusive, "inaccessible" and abstract concept; a concept whose invention is attributed to *Brahma, and which certainly constituted a great victory over the difficulties presented by numerical calculation. As for the tenth whole number, which in this system is written using a 1 and a 0, this would have corresponded, in the Indian symbolic mind, to an achievement, followed by the return to the unit at the end of the development of the cycle of the first nine numbers. However, this is mere conjecture for which there is no proof or foundation, it is simply based on one of the possible attitudes which characterise Indianity so well. See ***Shûnya***, ***Shûnyatâ***, **Zero** and **Nine**.

**DUST BOARD**. See ***Pâtî***, ***Pâtîganita***.

**DVA (or DVE, DVI)**. Ordinary Sanskrit name for the number two, which forms a component of many words which have a direct relationship with the concept of duality, opposition, complementarity, etc. Examples: **Dvaipâyanayuga*, **Dvaita*, **Dvandva*, **Dvandvamoha*, **Dvâparayuga*, **Dvaya*, **Dvija*, **Dvivâchana*.

For words which have a more symbolic link with this number, see **Two** and **Symbolism of numbers**.

**DVÂDASHA**. Literally "twelve". This term is used symbolically in the *Rigveda* to mean "year", in allusion to the twelve months of the year.

Ref: *Rigveda*, VII, 103, 1; Datta and Singh (1938), p. 57.

**DVÂDASHA**. Ordinary Sanskrit name for the number twelve. For words which have a symbolic link with this number, see **Symbolism of numbers**.

**DVÂDASHADVÂRASHÂSTRA**. "Tract of the twelve doors". Title of a work by Nâgârjuna, one of the principal Buddhist philosophers, founder of the school of **Mâdhyamika*. See ***Dvâdasha*** and ***Shûnyatâ***.

**DVAIPÂYANAYUGA**. Synonym of **dvâparayuga*.

**DVAITA**. "Duality". Term applied to a dualist philosophy, according to which a human creature is different from the *Brahman, its creator. This philosophy opposes the pure doctrine of the *Vedântas*, which is monistic (*Advaitavedânta*, "non dualist *Vedânta*").

**DVANDVA** [S]. Value = 2. "Couple, contrast". The symbolism is self-explanatory. See **Two**.

**DVANDAMOHA**. From *dvandva*, "couple, contrast", and *moha*, "illusion". This is the name given by the Hindus to what they consider to be the illusory impression that couples composed of opposites exist, such as shadow and light, joy and pain, etc.

**DVAPARAYUGA** (or **DVAIPÂYANAYUGA**). Name of the third of the four cosmic ages which make up a **mahâyuga*. This cycle, which is meant to be the equivalent of 864,000 human years, is regarded as the age during which humans have only lived for half of their lives, and where the forces of good have balanced out those of evil. See ***Yuga* (Definition)**.

**DVÂTRIMSHADVARALAKSHANA**. "Thirty-two distinctive signs of perfection". According to Buddhism, these are the signs which allow Buddha to differentiate between ordinary humans from a moral, physical or spiritual perspective. See ***Dvatrimshati***.

**DVATRIMSHATI** (or **DVITRIMSHATI**). Ordinary Sanskrit name for the number thirty-two. For words which have a symbolic connection with this number, see **Thirty-two** and **Symbolism of numbers**.

**DVAVIMSHATI** (or **DVIVIMSHATI**). Ordinary Sanskrit name for the number twenty-two. For words which have a symbolic link with this number, see **Twenty-two** and **Symbolism of numbers**.

**DVAYA**. [S]. Value = 2. Word meaning "pair". The symbolism is self-explanatory. See **Two**.

**DVE**. Ordinary Sanskrit name for the number two. See **Dva**.

**DVI**. Ordinary Sanskrit name for the number two. See **Dva**.

**DVIJA**. [S]. Value = 2. "Twice born". Epithet given to people belonging to the first three Brahmanic casts having the right to wear the sacred sash and who, during the ceremony of the handing over of the sash, are considered to be beginning a second life, this time of a spiritual nature [see Frédéric (1987)]. See ***Dva*** and **Two**.

**DVÎPA**. [S]. Value = 7. "Island-continent". Allusion to the seven island-continents which, in Hindu cosmology, are meant to radiate out from *Mount Meru. See ***Adri*** and **Seven**. See also ***Sapta Dvîpa***.

**DVIPA**. [S]. Value = 8. "Elephant". See ***Diggaja*** and **Eight**.

**DVITRIMSHATI**. Synonym of **dvatrimshati*.

**DVIVÂCHANA**. Name of the dual of Sanskrit verbs.

**DVIVIMSHATI**. Synonym of **dvavimshati*.

**DYUMANI**. [S]. Value = 12. "Sun". See ***Sûrya*** and **Twelve**.

# E

**EARTH**. As a mystical symbol for the number four. See ***Nâga***, ***Jala***, **Ocean**, **Serpent (Symbolism of the)**.

**EARTH**. As a name for the number ten to the power sixteen, ten to the power seventeen, ten to the power twenty, ten to the power twenty-one. See ***Kshiti***, ***Kshoni***, ***Mahâkshiti***, ***Mahâkshoni*** and **High numbers**.

**EARTH**. [S]. Value = 1. See ***Avani***, ***Bhû***, ***Bhûmi***, ***Dharâ***, ***Dharanî***, ***Dhâtrî***, ***Go***, ***Jagatî***, ***Kshaunî***, ***Kshemâ***, ***Kshiti***, ***Kshoni***, ***Ku***, ***Mahî***, ***Prithivî***, ***Vasudhâ***, ***Vasundharâ*** and **One**.

**EARTH**. [S]. Value = 9. See ***Go*** and **Nine**.

**EASTERN ARABIC NUMERALS**. Signs derived from *Brâhmî numerals, through the intermediary of Shunga, Shaka, Kushâna, Andhra, Gupta and Nâgarî numerals. Currently in use in Near and Middle East and in Muslim India, Malaysia and Indonesia. The corresponding system functions according to the place-value system and possesses zero (formerly either in the form of a little circle or dot but today exclusively represented by a dot). See **Indian written numeral systems (Classification of)**. See Fig. 24.2, 24.52 and 24.61 to 24.69.

**EIGHT**. Ordinary Sanskrit names for the number eight: **ashta*, **ashtan*. Here is a list of the corresponding numerical symbols: **Ahi*, *Anîka*, **Anushtubh*, *Bhûti*, **Dantin*, **Diggaja*, *Dik*, **Dikpâla*, **Dish*, *Durita*, **Dvipa*, *Dvirada*, **Gaja*, **Hastin*, *Ibha*, *Karman*, **Kuñjara*, **Lokapâla*, *Mada*, **Mangala*, **Matanga*, **Mûrti*, **Nâga*, *Pushkarin*, **Sarpa*, **Siddhi*, *Sindhura*, **Takshan*, **Tanu*, **Vasu* and *Yâma*.

These words can either be translated by the following words or have a symbolic relationship with them: 1. The serpent (*Ahi*, *Nâga*, *Sarpa*). 2. The serpent of the deep (*Ahi*). 3. The elephant (*Dantin*, *Dvipa*, *Diggaja*). 4. The eight elephants (*Diggaja*). 5. That which augurs well (*Mangala*). 6. The jewel (*Mangala*). 7. The shapes, or forms (*Mûrti*). 8. The horizons (*Dish*). 9. The guardians of the horizons and of the points of the compass (*Lokapâla*). 10. The guardians of time (*Dikpâla*). 11. Supernatural powers (*Siddhi*). 12. Certain groups of lines of Vedic poetry (*Anushtubh*). 13. A group of eight divinities (*Vasu*). 14. The spheres of existence of *Adibhautika* (*Vasu*). 15. The "acts" (*Karman*) (only in *Jaina philosophy). 16. The "body" (*Tanu*).

See **Numerical symbols**.

**EIGHTEEN**. Ordinary Sanskrit name: **ashtadasha*. The corresponding numerical symbol is **Dhriti*.

**EIGHTY**. See ***Ashîti***.

**EKA**. Ordinary Sanskrit word for the number one, which appears in the composition of many words which have a direct relationship with the concept of unity. Examples: **Ekachakra*, **Ekadanta*, **Ekâgratâ*, *Ekâkshara*, **Ekântika*, **Ekatva*.

For words which have a symbolic connection with the concept of this number, see **One** and **Symbolism of numbers**.

**EKACHAKRA**. "Who has only one wheel". Attribute of *Sûrya (the Sun-god).

**EKADANTA**. "Who has only one tooth". Attribute of *Ganesha, son of *Shiva and Pârvati, who is represented as having the body of a man and the head of an elephant, endowed with a unique defence. He is the Hindu divinity of wisdom, guaranteeing success in terrestrial existence and spiritual life.

**EKADASHA**. The ordinary Sanskrit name for the number eleven. For words which have a symbolic link with this number, see **Eleven** and **Symbolism of numbers**.

**EKADASHARÂSHIKA**. [Arithmetic]. Sanskrit name for the Rule of Eleven.

**EKÂDASHÎ**. Name of the eleventh day after the new moon, which orthodox Hindus spend fasting and meditating.

**EKÂGRATÂ**. In Hindu philosophy, a term which denotes a particular type of esoteric yoga, consisting in concentrating all of one's attention on a single point or object, which allows one to achieve *dhyâna* or "active contemplation".

**EKÂKSHARA**. "Unique and indestructible". Name of the sacred syllable of the Hindus (**AUM*).

**EKÂNNACHATVÂRIMSHATI**. "One away from forty". The ancient form of the Sanskrit name for the number thirty-nine (in Vedic times). See **Names of numbers**.

**EKÂNNATRIMSHATI**. "One away from thirty". The ancient form of the Sanskrit name for the number twenty-nine (in Vedic times). See **Names of numbers**.

**EKÂNNAVIMSHATI**. "One away from twenty". The ancient from of the Sanskrit name for the number *nineteen (during Vedic times). See **Names of numbers**.

**EKÂNTIKA**. Name of the monotheistic doctrine of the Vishnuite tradition.

**EKATVA**. In Hindu philosophical systems, a term denoting Unity, the contemplation of Everything. This is the ability to see the Self or the Divine in everything, and everything in the Self or the Divine.

**EKAVIMSHATI**. Ordinary Sanskrit name for the number twenty-one. For words which have a symbolic connection with this number, see **Twenty-one** and **Symbolism of numbers**.

**ELEMENT**. [S]. Value = 5. See ***Bhûta***, Five and ***Pañchabhûta***.

**ELEMENTS OF THE REVELATION**. See ***Bhûta***, ***Pañchabhûta***, ***Jala***, **Five**, **Numeral alphabet, magic, mysticism and divination** and **Ocean**.

**ELEPHANT**. A symbol for ten to the power twenty-one, ten to the power twenty-seven, ten to the power 105 or ten to the power 112. See ***Kumud***, ***Kumuda***, ***Pundarîka*** and **High numbers**.

**ELEPHANT**. [S]. Value = 8. See ***Dantin***, ***Diggaja***, ***Gaja***, ***Hastin***, ***Kuñjar***, **Eight** and ***Ashtadiggaja***.

**ELEVEN**. Ordinary Sanskrit name: **ekâdasha*. Here is a list of corresponding numerical symbols: *Akshauhinî*, **Bharga*, **Bhava*, **Hara*, **Îsha*, **Îshvara*, *Lâbha*, **Mahâdeva*, **Rudra*, **Shiva*, **Shûlin*, *Trishtubh*.

These words have the following translation or symbolic meaning: 1. A name or attribute of Rudra-Shiva (*Bharga*, *Bhava*, *Hara*, *Îsha*, *Îshvara*, *Mahâdeva*, *Rudra*, *Shiva*, *Shûlin*). 2. The "Supreme Divinity" (*Îshvara*). 3. The "Lord of the Universe" (*Îshvara*). 4. The "Great God" (*Mahâdeva*). 5. "Grumbling" (*Rudra*). 6. The "Lord of tears" (*Rudra*). 7. "Violent" (*Rudra*). 8. The "Master of the animals" (*Shûlin*). See **Numerical symbols**.

**ENERGY (feminine)**. [S]. Value = 3. See ***Shakti*** and **Three**.

**EQUATION**. [Algebra]. See ***Ghana***, ***Varga***, ***Vargavarga***, ***Samîkarana***, ***Vyavahâra***, ***Yâvattâvat*** and **Indian mathematics (The history of)**.

**ERAS (of Southeast Asia)**. See **Shaka**, ***Buddhashakarâja*** and **Indian calendars**.

**ESOTERICISM**. See ***Âkshara***, **Numeral alphabet and secret writing**, **Numeral alphabet, magic, mysticism and divination**, ***Âtman***,

*AUM*, ***Bîja***, ***Ekâgratâ***, ***Ekâkshara***, ***Kavacha***, **Mantra**, ***Trivarna***, ***Vâchana*** and **Serpent**.

**ETERNITY**. See ***Ananta*** and **Infinity**.

**ETHER**. [S]. Value = 0. See ***Âkâsha***, ***Shûnya*** and **Zero**.

**EUROPEAN NUMERALS (Algorisms)**. Numerals used after the twelfth century by European mathematicians (written calculation). The corresponding system functioned according to the place-value system and possessed a zero (in the form of a little circle). These signs derived from *Brâhmî numerals, firstly through the intermediary of types of Indian numerals such as Shunga, Shaka, Kushâna, Andhra, Gupta and Nâgarî, and then via the numerals used by the Arabs. The appearance of the numerals varied greatly from one school to another. Some styles derived from "Hindi" numerals, but most came from Arabic numerals. One such style, standardised due to the requirements of typography, became the origin of the numerals we use today: 1 2 3 4 5 6 7 8 9 0. **See Indian written numeral systems (Classification of)**. See also Fig. 24.52 and 24.61 to 69.

**EUROPEAN NUMERALS (Apices of the Middle Ages)**. Numerals used by European mathematicians in the Middle Ages (who carried out their calculations on an abacus). They derive from *Brâhmî numerals, first through the intermediary of types of Indian numerals such as Shunga, Shaka, Kushâna, Andhra, Gupta and Nâgarî, and then via Ghubar numerals of North African Arabs. The appearance of the numerals varied greatly from one school to another. The corresponding system did not possess zero because calculations were carried out on the abacus. See **Indian written numeral systems (Classification of)**. See also Fig. 24.52 and 24.61 to 69.

**EYE**. [S]. Value = 2. See ***Netra***, ***Drishti*** and **Two**.

**EYE**. [S]. Value 3. See ***Netra*** and **Three**.

**EYE OF SHUKRA**. [S]. Value = 1. See ***Shukranetra*** and **One**.

**EYES**. [S]. Value = 2. See ***Lochana*** and **Two**.

**EYES OF INDRA**. [S]. Value = 1,000. See ***Indradrishti*** and **Thousand**.

**EYES OF SENÂNÎ**. [S]. Value = 12. See ***Senânînetra*** and **Twelve**.

**EYES OF SHIVA**. [S]. Value = 3. See ***Haranetra*** and **Three**.

# F

**FACE**. [S]. Value = 4. See ***Mukha*** and **Four**.

**FACES OF BRAHMA**. [S]. Value = 4. See ***Brahmâsya*** and **Four**.

**FACES OF KÂRTTIKEYA**. [S]. Value = 6. See ***Kârttikeyâsya*** and **Six**.

**FACES OF KUMÂRA**. [S]. Value = 6. See ***Kumâravadana*** and **Six**.

**FACES OF RUDRA**. [S]. Value = 5. See ***Rudrâsya*** and **Five**.

**FACULTY**. [S]. Value = 5. See ***Indriya*** and **Five**.

**FIFTEEN**. Ordinary Sanskrit name: **pañchadasha*. Here is a list of corresponding numerical symbols: **Ahar*, *Dina*, *Ghasra*, **Paksha*, **Tithi*. These words have the following translation or symbolic meaning: 1. "Wing", in allusion to the number of days in one of the two "wings" of the month (*Paksha*). 2. "Day", in allusion to the number of days in one of the two "wings" of the month (*Ahar*, *Tithi*). See **Numerical symbols**.

**FIFTY**. See ***Pañchâshat*** and **Names of numbers**.

**FINGER**. [S]. Value = 10. See ***Anguli*** and **Ten**.

**FINGER (or Digit)**. [S]. Value = 20. See ***Anguli*** and **Twenty**.

**FINITE (Number)**. See **Infinity** and **Indian mathematics (The history of)**.

**FIRE**. [S]. Value = 3. See ***Agni***, ***Anala***, ***Dahana***, ***Hotri***, ***Hutâshana***, ***Jvalana***, ***Krishânu***, ***Pâvaka***, ***Shikhin***, ***Tapana***, ***Udarchis***, ***Vahni***, ***Vaishvânara*** and **Three**.

**FIRE** [S]. Value = 12. See ***Tapana***. **Twelve**.

**FIRMAMENT**. [S]. Value = 0. See ***Shûnya***, **Zero** and **Infinity**.

**FIRST FATHER**. [S]. Value = 1. See ***Pitâmaha*** and **One**.

**FIVE**. Ordinary Sanskrit name: **pañcha*. Here is a list of corresponding numerical symbols: *Artha*, **Bâna*, *Bhâva*, **Bhûta*, **Gavyâ*, **Indriya*, **Ishu*, **Kalamba*, **Karanîya*, *Kshâra*, *Lavana*, **Mahâbhûta*, **Mahâpâpa*, **Mahâyjña*, **Mârgana*, *Pallava*, **Pândava*, *Parva*, *Parvan*, **Pâtaka*, **Pâvana*, **Prâna*, **Purânalakshana*, **Putra*, **Ratna*, **Rudrâsya*, **Sâyaka*, **Shara*, *Shastra*, **Suta*, *Tanmâtra*, *Tata*, **Tattva*, **Tryakshamukha*, **Vishaya*, **Vishikha*.

The translation, or symbolic meaning of these words is as follows: 1. Arrows (**Bâna*, **Ishu*, **Kalamba*, **Mârgana*, **Sâyaka*, **Shara*, **Vishikha*). 2. Statistics (**Purânalakshana*). 3. "That which must be done" (**Karanîya*). 4. Purification (**Pâvana*). 5. The gifts of the Cow (**Gavya*). 6. The elements, in allusion to the five elements of the revelation (**Bhûta*). 7. The Great Elements, in allusion to the five elements of the revelation (**Mahâbhûta*). 8. The faculties (**Indriya*). 9. The worst sins (**Mahâpâpa*). 10. The great sacrifices (**Mahâyajñe*). 11. The main observances (**Karanîya*). 12. The fundamental principles, realities, truths, the "true natures" (*Tattva*). 13. The Jewels (**Ratna*). 14. The breaths (**Prâna*). 15. The senses, or the sense organs (**Vishaya*). 16. The Sons of Pându (**Pândava*). 17. The Sons, in allusion to the sons of Pându (**Putra*). 18. The faces of Rudra (**Rudrâsya*, **Tryakshamukha*). See **Numerical symbols**.

**FIVE ELEMENTS (philosophy of the)**. See ***Bhûta***, ***Pañchabhûta***, ***Jala***, **Five**, **Numeral alphabet**, **Magic**, **Mysticism and Divination**.

**FIVE SUPERNATURAL POWERS**. See ***Pañchabhijñâ***.

**FIVE VISIONS OF BUDDHA**. See ***Pañchachakshus***.

**FORM**. [S]. Value = 1. See ***Rûpa*** and **One**.

**FORM**. [S]. Value = 3. See ***Mûrti***, ***Trimûrti*** and **Three**.

**FORM**. [S]. Value = 8. See ***Mûrti*** and **Eight**.

**FORTY**. Ordinary Sanskrit name: **chatvârimshati*. Corresponding numerical symbol: *Naraka*.

**FORTY-EIGHT**. Ordinary Sanskrit name: **ashtachatvârimshati*. Corresponding numerical symbol: **Jagatî*.

**FORTY-NINE**. Ordinary Sanskrit name: **navachatvârimshati*. Corresponding numerical symbols: **Tâna* and **Vâyu*.

**FOUR**. Ordinary Sanskrit name for this number: **chatur*. Here is a list of the corresponding numerical symbols: **Abdhi*, **Ambhodha*, *Ambhodhi*, **Ambhonidhi*, *Ambudhi*, **Amburâshi*, **Arnava*, *Âshrama*, *Aya*, *Âya*, *Bandhu*, **Brahmâsya*, **Chaturânanavadana*, *Dadhi*, **Dish*, **Disha*, **Gati*, *Gostana*, **Haribâhu*, **Îryâ*, **Jala*, **Jaladhi*, **Jalanidhi*, *Jalâshaya*, *Kashâya*, *Kendra*, *Khatvâpâda*, *Koshtha*, **Krita*, **Mukha*, *Payodhi*, *Payonidhi*, *Purushârtha*, **Sâgara*, *Salilâkara*, **Samudra*, *Senânga*, **Shruti*, **Sindhu*, **Turîya*, **Udadhi*, *Vanadhi*, **Varidhi*, **Vârinidhi*, **Veda*, *Vishanidhi*, *Vyûha*, **Yoni*, **Yuga*.

These words have the following translation or symbolic meaning: 1. Water (*Jala*). 2. Sea or ocean (**Abdhi*, **Ambhonidhi*, **Ambudhi*, **Amburâshi*, **Arnava*, **Jaladhi*, **Jalanidhi*, **Jalâshaya*, **Sâgara*, **Samudra*, **Sindhu*, **Udadhi*, **Vâridhi*, **Vârinidhi*). 3. The four oceans (*Chaturânanavadana*). 4. The "horizons", in the sense of the cardinal points (*Dish*, *Disha*). 5. The conditions of existence (*Gati*). 6. The "Fourth" as an epithet of the Brahman (*Turîya*). 7. The "revelations" (*Shruti*). 8. The "positions" (*Îryâ*). 9. The arms of Vishnu (*Haribâhu*). 10. The births (*Gati*, *Yoni*). 11. The vulva (*Yoni*). 12. The *Vedas* (*Veda*). 13. The faces of Brahma (*Brahmâsya*). 14. The "faces" (*Mukha*). 15. The four ages of a *mahâyuga* (*Yuga*). 16. The last of the four ages of a *mahâyuga* (*Krita*). See **Numerical symbols**. See also **Ocean**.

**FOUR CARDINAL POINTS**. [S]. Value = 4. See ***Dish*** and **Four**.

**FOUR ISLAND-CONTINENTS**. See ***Chaturdvîpa*** and **Ocean**.

**FOUR OCEANS (or FOUR SEAS)**. See ***Chatursâgara***, ***Sâgara*** (= 4) and **Ocean**.

**FOUR STAGES**. See ***Chaturâshrama***.

**FOURTEEN**. Ordinary Sanskrit name: **chaturdasha*. Here is a list of corresponding numerical symbols: *Bhuvana*, **Indra*, **Jagat*, **Loka*, **Manu*, *Pûrva*, **Ratna*, **Shakra*, **Vidyâ*.

These words have the following translation or symbolic meaning: 1. The god Indra (*Indra*). 2. "Courage", "strength", "power" (*Indra*). 3. Powerful (*Shakra*). 4. "Human", in the sense of progenitor of the human race (*Manu*). 5. The worlds (*Bhuvana*, *Jagat*, *Loka*). 6. The Jewels (*Ratna*). See **Numerical symbols**.

**FOURTH (The)**. [S]. Value = 4. Word used as an epithet for *Brahma. See ***Turîya*** and **Four**.

**FRACTIONS**. [Arithmetic]. See ***Bhinna***, ***Kalâvarna***, ***Pañcha Jâti***.

**FUNDAMENTAL PRINCIPLE**. [S]. Value = 1. See ***Âdi*** and **One**.

**FUNDAMENTAL PRINCIPLE**. [S]. Value = 5. See ***Tattva*** and **Five**.

**FUNDAMENTAL PRINCIPLE**. [S]. Value = 7. See ***Tattva*** and **Seven**.

**FUNDAMENTAL PRINCIPLE**. [S]. Value = 25. See ***Tattva*** and **Twenty-five**.

# G

**GAGANA**. [S]. Value = 0. Word meaning "the canopy of heaven", "firmament". This symbolism is explained by the fact that the sky is nothing but a "void". See **Zero** and ***Shûnya***.

**GAJA**. [S]. Value = 8. "Elephant". See ***Diggaja*** and **Eight**.

**GAME OF CHESS**. See **Chaturanga**.

**GANANÂ**. Word meaning "arithmetic" in ancient Buddhist literature. More commonly, however, it has been used in the sense of "mental arithmetic" (which was and still is particularly developed in the art of Indian calculation).

**GANANÂGATI**. From **ganană*, "arithmetic", and **gati*, "condition of existence". Name given to the number ten to the power thirty-nine. See **Names of numbers**. For an explanation of this symbolism, see **High numbers (Symbolic meaning of)**.

Source: **Lalitavistara Sûtra* (before 308 CE).

**GANESHA**. Hindu divinity of wisdom, also called **Ekadanta*. See ***Eka***.

**GANESHA**. Indian mathematician who lived around the middle of the sixteenth century. Notably his works include a work entitled *Ganitamañjarî*.

**GANGA NUMERALS**. Signs derived from *Brâhmî numerals, through the intermediary of Shunga, Shaka, Kushâna, Ândhra, Pallava and Chalukya numerals. These were contemporaries of the beginnings of the dynasty of the Gangas of Mysore (sixth to eighth century CE) The corresponding system did not use the place-value system or zero. See **Indian written numeral systems (Classification of)**. See Fig. 24.46, 52 and 24.61 to 69.

**GANITA**. Sanskrit name for mathematics. In Vedic literature, this word is used to mean "the science of calculation", which is no doubt its original meaning. By extension, this word later acquired the meaning "science of measuring". In ancient Buddhist literature, there are three types of *ganita*: **mudrâ* or "manual arithmetic"; **gananâ* or "mental arithmetic"; and **samkhyâna* or "high arithmetic". Note that the word *ganita* was often used in ancient times to mean astronomy and even geometry (*kshetraganita*). See **Arithmetic**, **Calculation** and **Indian mathematics (The history of)**.

**GANITAKAUMUDÎ**. See **Nârâyana**.

**GANITÂNUYOGA**. Word meaning "explanation of mathematical principles". Term used mainly in *Jaina texts.

**GANITASÂRASAMGRAHA**. See **Mahâvîrâchârya**.

**GATI**. [S]. Value = 4. Literally "condition of existence". This word denotes the different forms of existence that reincarnation can assume (**samsâra*). The word became the numerical symbol for 4, synonymous with **yoni*, "birth" [see Frédéric (1994)]. See ***Chaturyoni*** and **Four**.

**GAUTAMA SIDDHÂNTA**. (Not to be confused with Gautama Siddhârtha, the Buddha). Chinese Buddhist astronomer of Indian origin, author of a work on astronomy and astrology entitled *Kai yuan zhan jing* (718 – 729 CE), where he describes zero, the place-value system and Indian methods of calculation. See **Place-value system**, and **Zero**.

**GAVYÂ**. [S]. Value = 5. "Gifts of the Cow". These are the **Pañchagavyâ*, the "five gifts of the Cow" (namely: milk, curds, dung, *ghî* and urine), which make up the sacred drink *gavyâ*, used by certain *samnyâsin* ascetics for its supposedly curative and purifying properties [see Frédéric (1994)]. See **Five**.

**GÂYATRÎ**. [S]. Value = 24. In expressive Sanskrit poetry, this is a stanza composed of three times eight syllables. See **Indian metric**.

**GEOMETRY**. See ***Kshetraganita*** and **Indian mathematics (The history of)**.

**GHANA**. "Cube". Sanskrit term used in arithmetic and algebra to denote the operation of cubing a number.

**GHANA**. Word used in algebra to denote "cube", in allusion to the third degree of equations of this order. See ***Varga***, ***Varga-Varga*** and ***Yâvattâvat***.

**GHANAMÛLA**. Sanskrit term used in arithmetic and algebra to denote the operation of the extraction of the cubic root.

**GHUBAR NUMERALS**. Signs derived from *Brâhmî numerals, through the intermediary of Shunga, Shaka, Kushâna, Ândhra, Gupta and Nâgarî numerals. Formerly used by the Arabic mathematicians of North Africa (for calculations carried out on the "dust" abacus). The corresponding system did not always possess zero. See **Indian written numeral systems (Classification of)**. See Fig. 24.52 and 24.61 to 69.

**GIFTS OF THE COW**. [S]. Value = 5. See ***Gavyâ*** and **Five**.

**GIRI**. [S]. Value = 7. "Mountain, hill". See ***Adri*** and **Seven**.

**GO**. [S]. Value = 1. "Cow", "Earth". This is the name of the sacred cow worshipped by the Hindus. This cow is said to have been created by *Brahma on the first day of the month of *Vaishâkha* (April–May). The word forms part of the composition of the name *Govinda* ("Cowherd") attributed to *Vishnu as "Saviour of the earth". This is also an allusion to the fact that the earth (**Prithivî*) is often symbolically associated with a cow named *Prishnî*. This relationship (which also explains the veneration of the cow in Hindu religion) stems from the fact that the cow, like the earth, gives life [see Frédéric (1994)]. See **One**.

**GO**. [S]. Value = 9. "Cow, Earth". Another meaning of this word is "radiance", and by extension "star". This is why the word became synonymous with **graha*, "planets" (in the sense of **navagraha*, the "nine planets of the Hindu cosmological system"). Thus *Go* = 9. See **Nine**.

**GOAL (The three)**. See ***Trivarga***.

**GOAL (The four)**. See ***Chaturvarga***.

**GOD OF CARNAL LOVE**. [S]. Value = 13. See ***Kâma*** and **Thirteen**.

**GOD OF COSMIC DESIRE**. [S]. Value = 13. See ***Kâma*** and **Thirteen**.

**GOD OF SACRIFICIAL FIRES**. [S]. Value = 3. See ***Agni*** and **Three**.

**GOD OF WATER AND OCEANS**. See ***Varuna***.

**GODS**. [S]. Value = 33. See ***Deva*** and **Thirty-three**.

**GOOGOL**. This term is of English origin. It was invented by the American mathematician Edward Kastner in the 1940s. It denotes the number ten to the power 100. This number, which no longer represents anything palpable, surpasses all that is possible to count or measure in the physical world. See **Infinity** and **High numbers**.

**GOVINDASVÂMIN**. Indian astronomer c. 830 CE. Notably, his works include *Bhâskarîyabhâsya*, in which there are many examples of the use of the place-value system using Sanskrit numerical symbols [see Billard (1971), p. 8]. See **Numerical symbols**, and **Numeration of numerical symbols**.

**GRAHA**. [S]. Value = 9. "Planet". This alludes to the **navagrahas*, the "nine planets" of the Hindu cosmological system (namely: **Sûrya*, the Sun; **Chandra*, the Moon; *Angâraka*, Mars; *Budha*, Mercury; *Brihaspati*, Jupiter; *Shukra*, Venus; *Shani*, Saturn; and the two demons of the eclipses **Râhu* and *Ketu*. See ***Paksha*** and **Nine**.

**GRAHA**. "Planet". See previous entry, ***Saptagraha*** and ***Navagraha***.

**GRAHACHÂRANIBANDHANA**. See ***Haridatta***.

**GRAHÂDHÂRA**. "Axis of the planets". Name given to the Pole star. See **Dhruva** and ***Sudrishti***.

**GRAHAGANITA**. Name given to astronomy by Brahmagupta (628 CE). Literally: "calculation of the planets", and, by extension, "mathematics of the stars". See **Indian astronomy (The history of)** and ***Ganita***.

**GRAHAPATI**. "Master of the planets". Name sometimes given to **Sûrya*, the Sun-god. See ***Graha***.

**GRAHARÂJA**. "King of the planets". Name sometimes given to **Sûrya*, the Sun-god. See ***Graha***.

**GRANTHA NUMERALS**. Symbols derived from *Brâhmî numerals and influenced by Shunga, Shaka, Kushana, Andhra, Pallava, Chalukya, Ganga, Valabhi and Bhattiprolu numerals. Formerly used by the Dravidian peoples of Kerala and Tamil Nadu. The symbols corresponded to a mathematical system that was not based on place-values and therefore did not possess a zero. See: **Indian written numeral systems (Classification of)**. See also Fig. 24.52 and 24.61 to 69.

**GREAT ANCESTOR**. [S]. Value = 1. See ***Pitâmaha*** and **One**.

**GREAT ELEMENT**. See ***Mahâbhûta***. Value = 5.

**GREAT GOD**. [S]. Value = 11. See ***Mahâdeva***, **Eleven** and ***Rudra-Shiva***.

**GREAT KINGS** (The four). See ***Chaturmahârâja***.

**GREAT SACRIFICE**. [S]. Value = 5. See ***Mahâyajña*** and **Five**.

**GREAT SIN**. [S]. Value = 5. See ***Mahâpâpa*** and **Five**.

**GREAT YEAR OF BEROSSUS**. Cosmic period mentioned in the work of the Babylonian astronomer Berossus (fourth – third century BCE), 432,000 years long. There is an "arithmetical" relationship between this "Great year" and the Indian cosmic cycles called **yugas*, because it corresponds: to a **kaliyuga*, to 1/10 of a **mahâyuga*, and to 2/5 of a **yugapâda*. However, it is not known if there is a historical link between this "year" and the Indian **yugas*. See **Great year of Heraclitus** and ***Yuga* (Astronomical speculations)**.

**GREAT YEAR OF HERACLITUS**. Cosmic period of the ancient Mediterranean world which, according to Censorinus, is 10,800 years long. There is a mathematical relationship between this "Great year" and the Indian cosmic cycles known as *yugas*, because it corresponds: to 1/40 of a **kaliyuga*, to 1/100 of a **yugapâda* and to 1/400 of a **mahâyuga*. However, it is not known if there is a historical link between this "year" and the Indian **yugas*. See **Great year of Berossus**.

**GUARDIAN OF THE HORIZONS**. [S]. Value = 8. See ***Lokapâla*** and **Eight**.

**GUARDIAN OF THE POINTS OF THE COMPASS**. [S]. Value = 8. See ***Lokapâla***, ***Dikpâla*** and **Eight**.

**GUJARATI NUMERALS**. Signs derived from *Brâhmî numerals, through the intermediary of Shunga, Shaka, Kushâna, Ândhra, Gupta, Nâgarî and Kûtilâ numerals. Currently in use in Gujarat State, on the Indian Ocean, between Bombay and the border of Pakistan. The corresponding system functions according to the place-value system and possesses zero (in the form of a little circle). See **Indian written numeral systems (Classification of)**. See also Fig. 24.8, 52 and 24.61 to 69.

**GULPHA**. [S]. Value = 2. "Ankle". This symbolism is due to the symmetry of this part of the body. See **Two**.

**GUNA**. [S]. Value = 3. "Merit", "Quality", "primordial property". Philosophically, the *gunas* are the qualities or conditions of existence which make up Nature. They are in a state of rest when the qualities are in perfect equilibrium, and in a state of evolution when one or more of them prevail over the others. According to the philosophy of the **Sâmkhya*, these qualities are composed of three natural "materials": *Sattva* (representing kindness, the pure essence of things), *Rajas* (active energy, passion), and *Tamas* (passivity, apathy). Here the word is synonymous with *Triguna*, "three qualities", "three primordial properties" [see Frédéric, *Dictionnaire* (1987)]. See ***Triguna*** and **Three**.

**GUNA**. [S]. Value = 6. "Merit", "quality", "primordial property". The allusion here is to **shadâyatana*, the "six *gunas*" of Buddhist philosophy. This value was only acquired relatively recently. See ***Shâdayatana*** and **Six**.

**GUNANA**. Term used in arithmetic to mean multiplication. Other synonyms: *hanana*, *vadha*, *kshaya*, etc. (which literally mean: "destroy", "kill", etc., in allusion to the successive erasing of the results of the partial products whilst carrying out calculations on sand or using chalk on a board). See **Calculation**, ***Pâtîganita***, and **Indian methods of calculation**. See also Chapter 25.

**GUNDHIKA**. Name given to the number ten to the power twenty-three. See **Names of numbers** and **High numbers**.

Source: **Lalitavistara Sûtra* (before 308 CE).

**GUPTA (Calendar)**. A calendar (with normal years) established by Chandragupta I beginning in 320 CE. To find the date in the universal calendar which corresponds to one expressed in Gupta years, add 320 to the Gupta date. Sometimes the first year of this calendar is given as 318 or 319. It was used during the Gupta dynasty. In Central India and Nepal, it persisted until the thirteenth century. See **Indian calendars**.

**GUPTA NUMERALS**. Signs derived from *Brâhmî numerals, through the intermediary of Shunga, Shaka, Kushâna and Ândhra numerals. Contemporaries of the Gupta dynasty (inscriptions of Parivrajaka and Uchchakalpa). The corresponding system does not use the place-value system or zero. These numerals were the ancestors of Nâgarî, Shâradâ and Siddham notations. See Fig. 24.38 and 24.70. For notations derived from Gupta numerals, see Fig. 24.52. For their graphical evolution, see Fig. 24.61 to 69. See **Indian written numeral systems (Classification of)**.

**GURKHALI NUMERALS**. See **Nepali numerals**.

**GURUMUKHI NUMERALS**. Signs derived from *Brâhmî numerals, through the intermediary of Shunga, Shaka, Kushâna, Ândhra, Gupta and Shâradâ numerals, and constituting a sort of mixture of Sindhi and Punjabi numerals. Once used by the merchants of Shikarpur and Sukkur. (These merchants also used Sindhi or Punjabi numerals, as well as the eastern Arabic "Hindi" numerals.) The corresponding system functions according to the place-value system and possesses zero (in the form of a little circle). See Fig. 24.7. See also **Indian written numeral systems (Classification of)** and Fig. 24.52 and 24.61 to 69.

## H

**HALF OF THE BEYOND**. As a representation of the numbers ten to the power twelve, ten to the power seventeen and ten to the power eighteen. See ***Parârdha***.

**HALF OF THE MONTH**. [S]. Value = 2. See ***Paksha***.

**HALF OF THE MONTH**. [S]. Value = 15. See ***Paksha***.

**HAND**. [S]. Value = 2. See ***Kara*** and **Two**.

**HARA**. [S]. Value = 11. One of the names of *Shiva who is an emanation of *Rudra, the symbolic value of which is eleven. See ***Rudra-Shiva*** and **Eleven**.

**HARANAYANA**. [S]. Value = 3. The "eyes of *Hara". See ***Haranetra***.

**HARANETRA**. [S]. Value = 3. "Eyes of *Hara". *Shiva, who has a multitude of names and attributes, one of which is **Hara*, often represented with a third eye in his forehead, which is meant to symbolise perfect knowledge. From which: *Haranetra* = 3. See **Three**.

**HARIBÂHU**. [S]. Value = 4. "Arms of Hari". *Hari* (literally "he who removes sin") is one of the names for *Vishnu, who is always represented as having four arms.

**HARIDATTA**. Indian astronomer c. 850 CE. Notably, his works include *Grahachâranibandhana*, in which he tells of the fruit of his invention: a system of numerical notation which uses the letters of the Indian alphabet. This is based on the place-value system and a zero (always expressed by one of two letters). This system is called *katapayâdi*: the first ever alphabetical positional number system [see Sarma (1954)]. See **Katapayâdi numeration**, and **Indian Mathematics (The history of)**.

**HARSHAKÂLA (Calendar)**. Calendar beginning in the year 606 CE, created by Harshavardhana, King of Kanauj and Thaneshvar. To find the date in the universal calendar which corresponds to one expressed in Harshakâla years, add 606 to the Harshakâla date. This calendar was only used during the reign of Harshavardhana and for a short time afterwards in Nepal. See **Indian calendars**.

**HÂRYA**. "Dividend" (in the mathematical sense). See ***Bhâjya***.

**HASTIN**. [S]. Value = 8. "Elephant". See ***Diggaja*** and **Eight**.

**HEADS OF RÂVANA**. [S]. Value = 10. See ***Râvanashiras*** and **Twenty**.

**HEADS OF RUDRA**. See ***Rudrâsya***.

**HEGIRA (Calendar of the)**. See ***Hijra***.

**HELL**. Value = 7. See ***Pâtâla***.

**HEMADRÎ**. One of the names of *Mount Meru.

**HETUHILA**. Name given to the number ten to the power thirty-one. See **Names of numbers** and **High numbers**.

Source: **Lalitavistara Sûtra* (before 308 CE).

**HETVINDRIYA**. Name given to the number ten to the power thirty-five. See **Names of numbers** and **High numbers**.

Source: **Lalitavistara Sûtra* (before 308 CE).

**HIGH NUMBERS**. Early in Indian civilisation, there was a sort of "craze" for high numbers. *Sanskrit numeration lent itself admirably to the expression of high numbers because it possessed a specific name for each power of ten. There are numerous examples to be found, not only in works on mathematics, but also in those concerning astronomy, cosmology, grammar, religion, legends and mythology. This proves that these names were not in everyday use in India, but rather they were familiar in learned circles, at least as early as the beginning of the Common Era. See **Names of numbers**.

In the naming of high numbers, these texts generally reached the highest numbers that were used in calculations. Thus each of the ascending powers of ten up to a *billion (ten to the power twelve), or even up to *quadrillion (ten to the power 18) were named. In cosmological texts, however (especially those developed by members of the religious cult of *Jaina, such as the *Anuyogadvâra Sûtra*), this limit was pushed much further, bearing witness to the extraordinary fertility of Indian imagination. The Jainas attempted to define their vision of an eternal and infinite universe; thus they undertook impressive arithmetical speculations, which always involve extremely high numbers, equal to or higher than numbers such as ten to the power 190 or ten to the power 250.

This obsession with high numbers is also found in **Vyâkarana*, a famous Pali grammar of Kâchchâyana, and in the legend of Buddha, related in the **Lalitavistara Sûtra*, which juggles with numbers as high as ten to the power 421. At first glance childish, this passion for high numbers can tell us something about the high conceptual level achieved early on by Indian arithmeticians. It led the Indians not only to expand the limits of the "calculable", physical world, but also and above all to conceive of the notion of infinity, long before the Western world. See **Googol** and all other entries entitled **High numbers** as well as those entitled **Infinity**.

**HIGH NUMBERS.** Here is a (non-exhaustive) alphabetical list of Sanskrit words which represent high numbers:*Abab (= $10^{17}$), *Ababa (= $10^{77}$), *Abbuda (= $10^{56}$), *Abja (= $10^{9}$), *Ahaha (= $10^{70}$), *Akkhobhini (= $10^{42}$), *Akshiti (= $10^{15}$), *Ananta (= $10^{13}$), *Anta (= $10^{11}$), *Antya (= $10^{12}$), *Antya (= $10^{15}$), *Antya (= $10^{16}$), *Arbuda (= $10^{7}$), *Arbuda(= $10^{8}$), *Arbuda (= $10^{10}$), *Asankhyeya (= $10^{140}$), *Atata (= $10^{84}$), *Attata (= $10^{19}$), *Ayuta (= $10^{4}$), *Ayuta (= $10^{9}$), *Bahula (= $10^{23}$), *Bindu (= $10^{49}$), *Dashakoti (= $10^{8}$), *Dashalaksha (= $10^{6}$), *Dashasahasra (= $10^{4}$), *Dhvajâgravati (= $10^{99}$), *Dhvajâgranshâmani (= $10^{145}$), *Gananâgati (= $10^{39}$), *Gundhika (= $10^{23}$), *Hetuhila (= $10^{31}$), *Hetvindriya (= $10^{35}$), *Jaladhi (= $10^{14}$), *Kankara (= $10^{13}$), *Karahu (= $10^{33}$), *Kathâna (= $10^{119}$), *Khamba (= $10^{13}$), *Kharva (= $10^{10}$), *Kharva (=$10^{12}$), *Kharva (= $10^{39}$), *Koti (= $10^{7}$), *Kotippakoti (= $10^{21}$), *Kshiti (= $10^{20}$), *Kshobha (= $10^{22}$) *Kshobhya (= $10^{17}$), *Kshoni (= $10^{16}$), *Kumud (=$10^{21}$), *Kumuda (= $10^{105}$), *Lakh (= $10^{5}$), *Lakkha (= $10^{5}$), *Laksha (= $10^{5}$), *Madhya (= $10^{10}$), *Madhya (= $10^{11}$), *Madhya (= $10^{15}$), *Madhya (= $10^{16}$), *Mahâbja (= $10^{12}$), *Mahâkathâna (= $10^{126}$), *Mahâkharva (= $10^{13}$), *Mahâkshiti (= $10^{21}$), *Mahâkshobha (= $10^{23}$), *Mahâkshoni (= $10^{17}$), *Mahâpadma (= $10^{12}$), *Mahâpadma (=$10^{15}$), *Mahâpadma (= $10^{34}$), *Mahâsaroja (= $10^{12}$), *Mahâshankha (= $10^{19}$), *Mahâvrindâ (= $10^{22}$), *Mudrâbala (= $10^{43}$), *Nâgabala (= $10^{25}$), *Nahut (= $10^{9}$), *Nahuta (= $10^{28}$), *Nikharva (= $10^{9}$), *Nikharva (= $10^{11}$), *Nikharva (= $10^{13}$), *Ninnahut (= $10^{11}$), *Ninnahuta (= $10^{35}$), *Nirabbuda (= $10^{63}$), *Niravadya (= $10^{41}$), *Niyuta (= $10^{5}$), *Niyuta (= $10^{6}$), *Niyuta (= $10^{11}$), *Nyarbuda ($10^{8}$), *Nyarbuda (= $10^{11}$), *Padma (= $10^{9}$), *Padma (= $10^{14}$), *Padma (= $10^{29}$), *Paduma (= $10^{29}$), *Paduma (= $10^{119}$), *Pakoti (= $10^{14}$), *Parârdha (= $10^{12}$), *Parârdha (= $10^{17}$), *Pârâvâra (= $10^{14}$), *Prayuta (= $10^{5}$), *Prayuta (= $10^{6}$), *Pundarîka (= $10^{27}$), *Pundarîka (= $10^{112}$), *Salila (= $10^{11}$), *Samâptalambha (= $10^{37}$), *Samudra (= $10^{9}$), *Samudra (= $10^{10}$), *Samudra (= $10^{14}$), *Saritâpati (= $10^{14}$), *Saroja (= $10^{9}$), *Sarvabala (= $10^{45}$), *Sarvajña (= $10^{49}$), *Shankha (= $10^{12}$), *Shankha (= $10^{13}$), *Shankha (= $10^{18}$), *Shanku (= $10^{13}$), *Shatakoti (= $10^{9}$), *Sogandhika (= $10^{91}$), *Tallakshana (= $10^{53}$), *Titilambha (= $10^{27}$), *Uppala (= $10^{98}$), *Utpala (= $10^{25}$), *Utsanga (= $10^{21}$), *Vâdava (= $10^{9}$), *Vâdava (= $10^{14}$), *Vibhutangamâ (= $10^{51}$), *Visamjñagati (= $10^{47}$), *Viskhamba (= $10^{15}$), *Vivaha (= $10^{19}$), *Vivara (= $10^{15}$), *Vrindâ (= $10^{9}$), *Vrindâ (= $10^{17}$), *Vyarbuda (= $10^{8}$), *Vyavasthânaprajñapati (= $10^{29}$).

**HIGH NUMBERS (SYMBOLIC MEANING OF).** The preceding list is enough to give the reader some idea of the arithmetical genius of the Indian scholars. However, it only gives the mathematical value of the words in question, and neither their literal nor their symbolic meaning. The following (summary) explanations should give the reader a precise idea of the associations of ideas and the symbolism which is implied in this unique terminology.

Firstly, the word *padma* (which represents the number ten to the power nine, ten to the power fourteen or ten to the power twenty-nine) literally means "*lotus". However, there is another word *paduma* (which can represent ten to the power twenty-nine as well as ten to the power 119), as well as the terms *utpala* (ten to the power twenty-five), *uppala* (ten to the power ninety-eight), *pundarîka* (ten to the power twenty-seven or ten to the power 112), *kumud* (ten to the power twenty-one) and *kumuda* (ten to the power 105), which also mean "lotus". The reasoning behind this metaphor lies in the fact that the lotus flower is the best-known and most widespread symbol in the whole of Asia. "Born of miry waters, this flower maintains a flat and immaculate purity above the water in all its splendour. Thus it became the symbol of a pure spirit leaving the impure matter of the body. Nearly all Indian philosophies and religions adopted the flower as their symbol, and its diffusion throughout Asia took place due to the spread of Buddhism, even though it is almost certain that the lotus flower was already used as a symbol by many peoples before the advent of Buddhism. Indian philosophers saw it as the very image of divinity, which remains intact and is never soiled by the troubled waters of this world. The closed flower of the lotus, in the shape of an egg, represents the seed of creation which rose out of the primordial waters, and as it opens all the latent possibilities contained within the seed develop in the light. This is why, in Hindu iconography, *Brahma is seen to be born from a lotus flower growing out of the navel of *Vishnu who lies upon the serpent *Ananta who is coiled up on the primordial waters which represent infinity [see Fig. D. 1, p. 446, of the entry entitled **Ananta**]. Similarly, this flower is the 'throne' of Buddha and most of his manifestations: here the lotus represents the *bodhi*, the 'nature of Buddha' which remains pure when it leaves the *samsâra*, the cycle of rebirth of this world. A whole symbolic system developed around the lotus flower, which takes into account its colour, the number of petals it possesses, and whether it is open, half-closed or in bud. In the *Kundalinî Yoga*, it is the stem of the lotus which forms the spiritual axis of the world and upon which the lotus flowers become steadily more fully open and the number of petals becomes greater and greater up until the highest illumination where the corolla, which has become divine and of unequalled brilliance, possesses a thousand petals," [see Frédéric, *Le Lotus*, (1987)].

Indian art also seized upon this flower and it has been widely represented in painting as well as in sculpture. We can appreciate why Indian mathematicians, with their perfect command of symbolism, also adopted the lotus flower and all its corresponding mysticism in order to express in Sanskrit gigantic quantities. The *padma* (or *paduma*), is the pink lotus. As well as the purity which it represents, this flower, to the Indian mind, symbolises the highest divinity, as well as innate reason. Written *Padmâ* (with a long a), the pink lotus flower figures amongst the names of Lakshmî as feminine energy (*shakti*) of *Vishnu. In the word *sahasrapadma* (the "thousand-petalled lotus"), it represents the "third eye", that of perfect Knowledge; it also represents the superior illumination and the divine corolla, of unequalled brilliance, flowering on the axis of the spirituality of the world as a thousand-petalled pink lotus [see Frédéric, *Le Lotus* (1987).

It is probably the idea of absolute and divine perfection which gave *padma* a value as elevated as ten to the power 119. However, it did not initially represent such a quantity. Initially, as the Indian mathematicians were gradually becoming more accustomed to dealing with large quantities, and with the idea of perfection and absolution, they probably gave the lotus the value of ten to the power nine. Its value gradually increased as it was successively attributed the values of ten to the power fourteen, then ten to the power twenty-nine and finally ten to the power 119. The flower in question here possesses a thousand petals.

*Padmâ* is also the name of one of the branches of the Ganges at its mouth. It is interesting to note that this swampy delta with branches radiating from it, like the petals of a lotus flower, is often referred to as *jahnavîvaktra*, literally the "mouths of *Jâhnavî* (the Ganges)". The name, as an ordinary numerical symbol, corresponds to the number one *thousand, precisely because of the hundreds of branches which characterise it. Moreover *Vishnu is associated with the thousand-petalled lotus and has a thousand attributes, amongst which are: *Sahasranâma*, "the thousand names (of Vishnu)".What is more, in Hindu mythology, it is from the "feet of *Vishnu" (*Vishnupada) that the celestial *Gangâ* (the Ganges) sprang, considered to be the "divine mother of India". Thus this flower was associated with both the name and the concept of thousand (*sahasra*). However, the terminology which was used had recourse to a secondary symbolism: *thousand was no longer really a numerical concept; its figurative sense was the idea of plurality, of "vast number". Like the word *padma* which initially only meant ten to the power nine, the name of this vast number grew to have the value of ten to the power twelve; which then gave ten to the power fifteen the name *mahâpadma*, which means "great lotus". Through a similar association of ideas, the word *shankha*, which means "sea conch", was assigned to the numbers ten to the power thirteen and ten to the power eighteen. This symbolises certain Buddhist or Hindu divinities (such as *Vishnu or *Varuna for example). In India, the conch represents riches, good fortune and beauty. This can be associated with the image of a diamond which is pure and beautiful in equal measures. As the diamond is everlasting and shines with a thousand fires, the beauty represented by the conch can be compared to this precious stone. Thus, for some Indian arithmeticians, *shanku* ("diamond") is equal to ten to the power thirteen.

Returning to the conch, one of the attributes of Vishnu is expressed by the Sanskrit word for conch (*shankha*), which symbolises the conservative principle of the revelation, due to the fact that the sound and the pearl are conserved within the shell. The conch is also the symbol of abundance, fertility and fecundity, which are precisely the characteristics of the sea from which the shell comes. The shell is also related to the water. This explains the connection with *Varuna, the lord of the Waters. Here there is also the connection with the lotus, which also symbolises not only abundance, but also and above all, in the eyes of humans, a limitless expanse. This is why the word *samudra*, which means "ocean", came to mean, in this symbolic terminology, the number ten to the power nine, ten to the power ten or ten to the power fourteen. This is the reason why *Bhâskarâchârya used the word *jaladhi*, which also means *ocean, to denote the number ten to the power fourteen. The mathematician must have chosen this word because he was writing in verse and in Sanskrit and he chose his words in order to achieve the desired effect, using an almost limitless choice of synonyms, following the exacting rules of Sanskrit versification. See **Poetry and writing of numbers**.

The Indians also see **samudra* as the waves of superior consciousness which bring immortality to mere mortals; eternal existence and infinity. This explains the connection with **soma*, which is the **amrita*, the "drink of immortality". *Soma* can also mean *moon, which became a metaphor for a goblet full of the heady brew. Thus it was quite natural that this star should also be associated with incalculably vast quantity. So **abja* ("moon") and **mahâbja* ("great moon"), came to represent numbers such as ten to the power nine or a billion (ten to the power twelve). As well as being connected to water, the conch is symbolically related to the moon, as it is white, the colour of the full moon. This gives double justification to this association of ideas. The apparently limitless expanse of the sea is the most immense thing in the "terrestial world". As the *earth is called **kshiti* or **kshoni*, (also referred to as **mahâkshiti* and **mahâkshoni*, meaning "great earth") we can see how these words came to represent such immense values as ten to the power sixteen, ten to the power seventeen, ten to the power twenty or ten to the power twenty-one.

The Sanskrit word **abhabâgamana* means "the unachievable". The term **ababa* is used in the **Lalitavistara Sûtra* (before 308 CE), to express the quantity ten to the power seventy, and it is possible that this is an abbreviation of **abhabâgamana*. The word **ahaha*, used in the same text to express ten to the power seventy-seven, is almost definitely an abbreviation of the word *abaharaka*, which means extravagant and is similar to our word "abracadabra".

**Pundarîka* means white lotus with eight petals and is the symbol of spiritual and mental perfection. The term is generally reserved for esoteric divinities, and was dedicated to Shikhin, the second Buddha of the past. This lotus has the same number of petals as the eight directions of space, the eight points of the compass and the eight elephants (**diggaja*) of Hindu cosmogony. Amongst these elephants figures **Pundarîka* who guards the "southeastern horizon" of the universe for the god of fire **Agni*. The "southwestern horizon" is guarded by the elephant **Kumuda*, whose real name also means "lotus", but this time refers to the white-pink flower. The sun is not far from this lotus, as it is situated at the axis of the eight horizons. The elephant *Kumuda* also symbolises the Sun-god *Sûrya, who is often denoted by names which evoke the idea of a thousand or the lotus flower: **Sahasrâmshu* ("Thousand of the Shining", in allusion to its rays), **Sahasrakirana* ("Thousand rays"), **Sahasrabhûja* ("Thousand arms") and *Padmapâni* ("Lotus carrier"). Thus the Indians expressed the omniscience of this god and his incalculable powers.

If the sun is a source of light, warmth and life, then like the petals of a lotus, its rays must also contain the spiritual influences received by all things on earth. This is why the names of **Pundarîka*, **Kumud* and **Kumuda* came to represent such vast quantities as ten to the power twenty-one, ten to the power 105 and ten to the power 112. Indian mathematicians soon took the step from the Sun to the canopy of heaven. **Parârdha*, one of the names attributed to ten to the power twelve or ten to the power seventeen, comes from *para*, "beyond", and from *ardha*, "half of beyond". Due to a similar association of ideas, **Madhya*, "middle" (representing the "middle of the beyond") was used to express such numbers as ten to the power ten, ten to the power eleven, ten to the power fifteen or ten to the power sixteen.

According to the Indians, *parârdha* is the spiritual half of the path which leads to death. It is the same as *devayâna*, the "path of the gods", which, according to the *Vedas, was one of the two possibilities offered to human souls after death, *parârdha* leading to deliverance from **samsâra* or cycle of reincarnation. On reaching the sky, one cannot fail to achieve divine transcendence, power, durability and sanctity, thus touching upon the incalculable in terms of mental and physical perfection, represented by the word **pundarîka*. Intelligence, wisdom and the triumph of the mind over the senses is represented by **utpala*, the blue, half open lotus. This is why these words came to be worth such quantities as ten to the power twenty-five, ten to the power twenty-seven, ten to the power ninety-eight or even ten to the power 112.

No living being can attain divine transcendence, which is conveyed by the Inaccessible, the Absolute, and the Ineffable. This is similar to the "incalculable", **Asamkhyeya* (or **Asankhyeya*), "that which cannot be counted". According to the **Lalitavistara Sûtra*, this word means "the sum of all the drops of rain which, in ten thousand years, would fall each day on all the worlds". In other words, this is the "highest number imaginable". **Asankhyeya* is the term used to express the number ten to the power 140. The terminology used here deals symbolically with the notion of eternity. This is explained by al-Biruni in the thirty-third chapter of his work on India, where he gives this word the value of ten to the power seventeen (Fig. 24. 81):

"The name of the eighteenth order is **parârdha*, which means half of the sky, or more precisely, half of what is above. The *reason* for this is that if a period of time is made up of **kalpas* (cycles of 4,320,000,000 years), a unit of this order is a day of the *purusha* (= one day of the Supreme Being, namely *Brahma). As there is nothing beyond the sky, this is the largest body. Half of the biggest nychthemer (= the longest possible day) is similar to the other; in doubling it, we obtain a "night" with a "day", and thus complete the biggest nychtemer. It is certain that *parârdha* is from *para*, which means the whole sky". Ref: *Kitab fi tahqiq i ma li'l hind* (1030 CE).

Al-Biruni also tells that "according to some, the day of *purusha* (the day of Brahma) is made up of a *parârdha* and a *kalpa*". As a *kalpa* is 4,320,000,000,000 years, this "day" corresponds to: 432,000,000,000,000,000,000,000,000 (four hundred and thiry-two sextillion) years. See **Day of Brahma**.

Traditional brahmanic cosmogony more modestly attributes the length of 4,320,000,000 human years to the "*day of Brahma". This is also what it refers to as **asankhyeya*, the "incalculable". The *Bhagavad Gîtâ* assigns 311,040,000,000,000 years to this word. In a commentary on the text is written: "This formidable longevity, to us infinite, only represents a zero in the tide of eternity." The word *Padmabhûta*, "born from the lotus (with a thousand petals)" is an attribute of *Brahma. Brahma is said to have been born from the lotus which grew out of Vishnu's navel as he lay on the serpent *Ananta floating on the primordial waters (see Fig. D. 1 and **Ananta**). Another attribute of Brahma is *Padmanâbha*, which means "the one whose navel is the lotus (Vishnu)". This is why the word *ananta*, which means "infinity" and "eternity", has sometimes been used to express the number ten to the power thirteen, in memory of distant times when the Sanskrit names for numbers went no further than ten to the power twelve. See **Antya** (first entry, note in the reference).

Ananta is another name for *Shesha, the king of the **nâga* who resides in the lower regions of the **Pâtâla*. It is an immense serpent with a thousand heads, who serves as a seat for *Vishnu as he rests amongst creatures between two periods of creation. At the end of each **kalpa*, he spits the fire which destroys creation. Considered as the "Remainder" (**Shesha*), the "Vestige" of destroyed universes and as the seed of all future creations, he represents immensity and space, and infinity and eternity all at once. See **Serpent (Symbolism of the)**.

The words *eternal* and *infinity* mean "that which has no end, that which never ends, that which can never be reached". This leads to ideas of absoluteness and totality, in the strongest sense of the terms. Words such as **Sarvabala* and **Sarvajña*, formed with the Sanskrit adjective *sarva*, meaning "everything" or "totality", have respectively been associated with numbers as high as ten to the power forty-five or ten to the power forty-nine. Moreover, *Sarvajñata* expresses omniscience in Buddhism, the knowledge of Buddha, one of his fundamental attributes. In the Buddhism of Mahâyâna, this word has even acquired the meaning "the knowledge of all the **dharmas* and of their true nature"; a nature which, in essence, is **shûnyatâ*, vacuity. According to the Indians, vacuity materialises in the centre of the **vajra*, the "diamond", symbol of what remains once appearances have disappeared. The *vajra* is also the projectile "of a thousand points", reputed to never miss its mark, and made of bronze by Tvashtri for *Indra; but this is above all a religious instrument, symbol of the *linga* and divine power, indicating the stability and resoluteness of mind as well as its indestructible character. And as vacuity also means the void for Indians (also caused as much by nothingness, absence or insignificance as by the unthought, immateriality, insubstantiality and non-being), this explains why the **bindu*, the "point" (destined to become a numerical symbol and a graphical representation for zero), represented, for Indian arithmetician-grammarians, a number as high as ten to the power forty-nine.

Before the discovery of infinity or zero, the *bindu* (the "point"), was the Indian symbol for the universe in its non-manifest form, thus that of a universe before its transformation into *rûpadhâtu*, the "world of appearances". For Indian scholars, "nothing" could be united with "everything", even before mathematics made these two concepts inverse notions of one another. See **Zero**, **Infinity** and **Indian Mathematics (The history of)**. See **Names of numbers**, **Numerical symbols**, **Arithmetical speculations**, **Cosmogonical speculations**, ***Sheshashîrsha***,

***Shûnya*, *Shûnyatâ*, Indian atomism**. See also **Serpent (Symbolism of the)**.

**HIJRA (Calendar)**. Arabic name for the Islamic calendar, which, according to tradition, begins on the 16 July 622 CE, day of the "Escape" or "Flight" (*hijra*, "Hegira") of the prophet Mohammed of Mecca to Medina. As the Muslim year has twelve lunar months each twenty-nine or thirty days long, making up a year of 354 days, this calendar must be rectified by the addition of eleven intercalary years of 355 days every thirty years to catch up with normal solar years. To obtain a date in the universal calendar from one in the Hegira, multiply the latter by 0.97 and add 625.5 to the result. For example: the start of the year of the Hegira 1130 corresponds to July 1677:

1130 *Hegira* = (1130 × 0.97) + 625.5 = 1721.6. Inversely, to find a date of the Hegira from a date in the universal calendar, subtract 625.5 from the latter, then multiply the result by 1.0307 and add 0.46. If there are decimals remaining, add a unit.

For example, to convert the year 1982 into the Hegira calendar, proceed as follows: 1st stage: 1982 – 625.5 = 1356.5.

2nd stage: 1356.5 × 1.0307 = 1398.14.

3rd stage: 1398.14 + 0.46 = 1398.6.

In rounding off this result, the year of the Hegira 1399 is obtained [see Frédéric, *Dictionnaire* (1987)]. See **Indian calendars**.

**HINDI NUMERALS**. See **Eastern Arabic numerals**.

**HINDU RELIGION**. See **Indian religions and philosophies**.

**HINDUISM**. See **Indian religions and philosophies**.

**HOLE**. [S]. Value = 0. See ***Randhra***.

**HOLE**. [S]. Value = 9. See ***Chhidra*, *Randhra*** and **Nine**.

**HORIZON**. [S]. Value = 4. See ***Dish*** and **Four**.

**HORIZON**. [S]. Value = 8. See ***Dish*** and **Eight**.

**HORIZON**. [S]. Value = 10. See ***Dish*** and **Ten**.

**HORSE**. [S]. Value = 7. See ***Ashva*** and **Seven**.

**HORSEMEN**. [S]. Value = 2. See ***Ashvin*** and **Two**.

**HOTRI**. [S]. Value = 3. "Fire". See ***Agni*** and **Three**.

**HUMAN**. [S]. Value = 14. In the sense of progenitor of the human race. See ***Manu*** and **Fourteen**.

**HUNDRED**. Ordinary Sanskrit name: **shata*. Corresponding numerical symbols: *Abjadala*, **Dhârtarâshtra*, *Purushâyus* and *Shakrayajña*. See **Numerical Symbols**.

**HUNDRED BILLION** (= ten to the power fourteen). See ***Jaladhi*, *Padma*, *Pakoti*, *Pârâvâra*, *Samudra*, *Saritâpati*, *Vâdava***. See also **Names of numbers**.

**HUNDRED MILLION** (= ten to the power eight). See ***Arbuda*, *Dashakoti*, *Nyarbuda*, *Vyarbuda***. See also **Names of numbers**.

**HUNDRED THOUSAND BILLION (UK)** (= ten to the power seventeen). See ***Abab*, *Kshobhya*, *Mahâkshoni*, *Parârdha*, *Vrindâ***. See also **Names of numbers**.

**HUNDRED THOUSAND MILLION (UK)** (= ten to the power eleven). See ***Anta*, *Madhya*, *Nikharva*, *Ninnahut*, *Niyuta*, *Nyarbuda*, *Salila***. See also **Names of numbers**.

**HUNDRED THOUSAND TRILLION (UK)** (ten to the power twenty-three). See ***Bahula*, *Gundhika*, *Mahâkshobha***. See also **Names of numbers**.

**HUNDRED TRILLION (UK)** (= ten to the power twenty). See ***Kshiti***. See also **Names of numbers**.

**HUTÂSHANA**. [S]. Value = 3. "Fire". See ***Agni*** and **Three**.

# I

**IMMENSE**. [S]. Value = 1. See ***Prithivî*** and **One**.

**IMPURITIES (The five)**. See ***Pañchaklesha***.

**IMRAJÎ (Calendar)**. See ***Kristâbda***.

**INACCESSIBLE**. [S]. Value = 9. See ***Durgâ*** and **Nine**.

**INCARNATION**. [S]. Value = 10. See ***Avatâra*** and **Ten**.

**INDESTRUCTIBLE**. [S]. Value = 1. See ***Âkshara*** and **One**.

**INDETERMINATE EQUATION (Analysis of an)**. See ***Kuttaganita*** and **Indian mathematics (History of)**.

**INDETERMINATE**. See **Infinity**.

**INDIA (States of the present-day Indian union)**. See Fig. 24. 27.

**INDIAN ARITHMETIC**. Alphabetical list of words related to this discipline which appear as entries in this dictionary:*Arithmetic. **Bhâgahâra*, **Bhâjya*, **Bhinna*, **Chhedana*, **Dashaguna*, **Dashagunâsamjña*, **Dhûlikarma*, **Ekadasharâshika*, **Gananâ*, **Ganita*, **Ghana*, **Ghanamûla*, **Gunana*, *High numbers, *Indian mathematics (history of), *Indian methods of calculation, *Infinity, **Kalâsavarnana*, **Labhda*, **Mudrâ*, *Names of numbers, **Navarâshika*, **Pañcharâshika*, **Pañcha jâti*, **Parikarma*, **Pâtî*, **Pâtîganita*, **Râshi*, **Râshividyâ*, **Samkalita*, **Samkhyâna*, **Samkhyeya*, **Saptarâshika*, **Sarvadhana*, **Shatottaragananâ*, **Shatottaraguna*, **Shatottarasamjñâ*, **Shesha*, *Square roots (How Âryabhata calculated his), **Trairâshika*, **Varga*, **Vargamûla*, **Vyastrairâshika*, **Vyavakalita*.

**INDIAN ASTROLOGY**. This discipline used to go by the name of **Jyotisha*, which literally means the "science of the stars". But today this term is more commonly used to describe astronomy. This naming dates back to the time when astronomy was not yet considered to be a separate discipline from arithmetic and calculation.

Until early in the Common Era, astrology was often confused with astronomy, the latter at that time having no other objective than to serve the former. Knowledge of the stars and their movements was a method of predicting the future and determining the favourable dates and times of any given human action: consecration of a ritual sacrifice, commercial transactions, setting out on a voyage, etc. [see Frédéric (1994)]. Thus, when an individual was born, the astrologers, having determined the exact time and the position of the planets and the sun, established the horoscope of the newly-born infant, which was considered indispensable in ascertaining the child's birth chart.

*Varâhamihîra stands out as one of the most famous astrologers of Indian history. He lived in the sixth century CE, and is known principally for his work, *Pañchasiddhântikâ* (the "*Five *Astronomical Canons*), which is dated c.575 CE. But he also wrote many works on astrology, divination and practical knowledge. The most important of these is *Brihatsamhitâ* (the "great compilation") which covers many subjects: descriptions of heavenly bodies, their movements and conjunctions, meteorological phenomena, indications of the omens these movements, conjunctions and phenomena represent, what action to take and operations to accomplish, signs to look for in humans, animals, precious stones, etc. [see Filliozat, in: HGS, I, pp. 167–8]. See **Indian astronomy (The history of)**, ***Ganita*, *Râshi*, *Tanu*** and ***Yuga***.

**INDIAN ASTRONOMY (The history of)**. Here is an alphabetical list of terms related to this discipline which appear in this dictionary: *Âryabhata, *Astronomical canon, *Bhâskara, **Bhâskarâchârya*, *Bhoja, **Bîja*, *Brahmagupta, **Chaturyuga*, **Ganita*, *Govindasvâmin, *Great year of Berossus, *Great year of Heraclitus, *Haridatta, *Indian astrology, *Indianity (fundamental mechanisms of), *Indian mathematics (The history of), **Jyotisha*, **Kaliyuga*, **Kalpa*, *Kamâlakara, **Karana*, *Lalla, **Mahâyuga*, **Nakshatra*, **Nakshatravidyâ*, *Nîlakanthasomayâjin, *Numerical symbols, *Parameshvara, *Putumanasomayâjin, **Samkhyâna*, *Shankaranârâyana, *Shrîpati, **Siddhânta*, **Tithi*, **Varâmihîra*, **Yuga* (definition), **Yuga* (systems of calculating), **Yuga*, (astronomical speculation on), **Yugapâda*.

**INDIAN ASTRONOMY (The history of)**. If we take the word "astronomy" in its wider sense, we can traditionally distinguish three principal periods. The first corresponds to the astronomy of era and ritual: a lunar-solar era devoid of any time-scale or era. The corresponding "material" is characterised by the **nakshatra*, division of the sidereal sphere in twenty-seven or twenty-eight constellations or asterisms according to the twenty-seven or twenty-eight days of the sidereal revolution of the Moon. The planets (it is unlikely that they had all been discovered at this time) only played a very small part in divination. Between the third century BCE and the first century CE, elements and procedures of Babylonian astronomy appeared in Indian astronomy. A unity of time appeared called the **tithi*, which is approximately the length of a day or nychthemer, and corresponds to one thirtieth of the synodic revolution of the Moon. It was at this time that the planets came to the fore in divination and were subjected to arithmetical calculations, based on their synodic revolutions. However, it was at the beginning of the sixth century CE that Indian astronomy underwent its most spectacular developments: *scientific astronomy* was inaugurated by the work of *Âryabhata, which dates back to c. 510 CE. From the outset, this astronomy was based on an astonishing speculation about the cosmic cycles called **yugas*, of a very different nature from arithmetical cosmogonic speculations. This speculation involves astronomical elements, where the **mahâyuga* (or **chaturyuga*), a cycle of 4,320,000 years is presented as the period at the beginning and the end of which the nine elements (the sun, the moon, their apsis and node, and the planets) should be found in average perfect conjunction with the starting point of the longitudes. Thus the length of the revolutions, which had hitherto been considered commensurable, were subjected to common multiplication and general conjunctions.

It is precisely this which makes the speculation so surprising and audacious, because this fact is totally devoid of any physical value. These supposed general conjunctions confer absurd values upon average elements even by the approximative standards of ancient astronomy. However, thanks to a veritable paradox, it is this strange coupling of speculation and reality that enabled Billard to develop a powerful method of determining (with precision) hitherto unknown facts and a chronology which had been despaired of due to the unique conditions of the Indian astronomical text. It is interesting to note that the speculative elements of this astronomy have been as useful to contemporary historical science as they were once harmful. For more details, see: ***Yuga* (cosmogonical speculation on), Indian astrology, Yuga (astronomical speculation about), Indian mathematics (The history of), Indianity (Fundamental mechanisms of)**. [See Billard (1971)].

**INDIAN ATOMISM.** See ***Paramânu, Paramânu Raja*** and ***Paramabindu***.

**INDIAN CALENDARS.** India (which only adopted the universal calendar in 1947) has known a great many different eras during the course of its history. Certain eras, mythical or local, have existed on a very limited scale. Others, however, have become so widely used that they still exist today. The (non-exhaustive) list in the following entry gives an idea of the vast number of eras which have been used, and also allow for a better understanding of the elements of Chapter 24, where the documentation is often dated in one of these eras.

**INDIAN CALENDARS.** An alphabetical list of terms relating to these eras which can be found in this dictionary: *Bengalî San*, **Buddhashakarâja*, **Châlukya*, **Chhedi*, **Gupta*, **Harshakâla*, **Hijra*, **Kaliyuga*, **Kollam*, **Kristabda*, **Lakshamana*, **Laukikasamvat*, **Marâtha*, **Nepâlî*, *Parthian, **Samvat*, **Seleucid*, **Shaka*, **Simhasamvat*, **Thâkurî*, **Vikrama*, **Vilâyatî*, **Vîrasamvat*. [See Cunningham (1913); Frédéric, *Dictionnaire*, (1987); Renou and Filliozat (1953)].

**INDIAN COSMOGONIES AND COSMOLOGIES.** Here is an alphabetical list of terms relating to these subjects which appear as entries in this dictionary: *Âditya*, **Adri*, **Anuyogadvâra Sûtra*, *Arithmetical-cosmogonic speculations, **Ashtadiggaja*, **Avatâra*, **Bhuvana*, **Chaturânanavâdana*, **Chaturdvîpa*, **Chaturmahârâja*, **Chaturyuga*, *Cosmognic speculations, **Dantin*, *Day of Brahma, **Diggaja*, **Dikpâla*, **Dvaparayuga*, **Dvîpa*, **Gaja*, **Go*, **Graha*, **Hastin*, *High numbers (the symbolic meaning of), **Indra*, **Jaina*, **Kala*, **Kaliyuga*, **Kalpa*, **Krita*, **Kritayuga*, **Kuñjara*, **Lokapâla*, **Mahâkalpa*, **Mahâyuga*, **Manu*, *Mount Meru, *Ocean, **Paksha*, **Râhu*, **Sâgara*, *Serpent (Symbolism of the), **Shîrshaprahelikâ*, **Takshan*, **Tretâyuga*, **Tribhuvana*, **Triloka*, **Vaikuntha*, **Vasu*, **Vishnupada*, **Vishva*, **Vishvadeva*, **Yuga*.

**INDIAN DIVINITIES.** Here is an alphabetical list of terms relating to this theme, which can be found as entries in this dictionary:**Âditya*, **Agni*, **Amara*, **Âptya*, **Arka*, **Ashtamangala*, **Ashtamûrti*, **Ashva*, **Ashvin*, **Âtman*, **Avatâra*, **Bhânu*, **Bharga*, **Bhava*, **Bhuvana*, **Bîja*, **Brahma*, **Brahmâsya*, *Buddha (Legend of), **Chandra*, **Chaturmahârâja*, **Chaturmukha*, **Dashabala*, **Dashabhûmi*, **Deva*, **Dhruva*, **Dikpâla*, **Divâkara*, **Durgâ*, **Dvâtrimshadvaralakshana*, **Dyumani*, **Ekachakra*, **Ekadanta*, **Ganesha*, **Go*, *High numbers (Symbolic meaning of), **Hara*, **Haranayana*, **Haranetra*, **Haribâhu*, **Indra*, *Infinity (Indian mythological representation of), **Îsha*, **Îshadrish*, **Îshvara*, **Jaina*, **Kâma*, **Kârttikeya*, **Kârttikeyâsya*, **Kâya*, **Keshava*, **Krishna*, **Kumârâsya*, **Kumâravadana*, **Lokapâla*, *Lotus, **Mahâdeva*, **Mârtanda*, **Netra*, **Pañchabâna*, **Pañchânana*, **Parabrahman*, **Pâvana*, **Pinâkanayana*, **Pitâmaha*, **Ravi*, **Ravibâna*, **Rudra*, **Rudra-Shiva*, **Rudrâsya*, **Sâgara*, **Sahasrâmshu*, **Sahasrakirana*, **Sahasrâksha*, *Sahasranâma, **Shakra*, **Shakti*, **Shatarûpâ*, **Shikhin*, **Shiva*, **Shukranetra*, **Shûla*, **Shûlin*, **Sura*, **Sûrya*, **Tapana*, **Triambaka*, **Tribhuvaneshvara*, **Tripurasundarî*, **Trishûlâ*, **Trivarna*, **Tryakshamukha*, **Tryambaka*, **Turîya*, **Vaikuntha*, **Vaishvanara*, **Varuna*, **Vasu*, **Vâyu*, **Vrindâ*, **Yama*.

**INDIAN DOCUMENTATION (Pitfalls of).** The purpose of this entry is to warn readers about texts which have absolutely no historical worth whatsoever, which contemporary Indianists – doubtless through bias towards material or excessive admiration of Indian culture – have put forward, due to shocking carelessness and lack of objectivity, in order to claim that the invention of zero and Indian positional numeration date back to the most ancient times. These documents are either fakes, or works resulting from recent compilations, or even ancient texts which successive generations reproduced whilst constantly correcting and revising them over the course of time.

Amongst these documents figures the manuscript of Bakshâlî, discovered in 1881 in the village of Bakshâlî in Gandhara, near Peshawar, in present-day Pakistan. The author of this mathematical text is unknown. It is written in Sanskrit (in verse and prose) and is mainly concerned with algebraic problems. This document is interesting from the point of view of the history of Indian numeration because it contains many examples of numbers written using the sign zero and the *place-value system, as well as several numerical entries expressed in *numerical symbols. According to certain historians of mathematics, this manuscript dates back to "the fourth, or possibly even the second century CE". This document undeniably constitutes an invaluable source of information about *Indian mathematics, but the manuscript itself, in its present form, could not possibly be as old as is claimed. The reason for this is that the numerals, like the characters used for writing, are written in the *Shâradâ* style, of which we know both the origin and the date of its first developments. See **Indian styles of writing** and **Shâradâ numerals**. See also Fig. 24. 38 and 40A.

To give the second or fourth century CE as the date of this document would be an evident contradiction; it would mean that a northern derivative of Gupta writing had been developed two or three centuries before Gupta writing itself appeared. Gupta only began to evolve into the Shâradâ style around the ninth century CE. In other words, the Bakshâlî manuscript cannot have been written earlier than the ninth century CE. However, in the light of certain characteristic indications, it could not have been written any later than the twelfth century CE. Nevertheless, when certain details are considered, which probably reveal archaisms of styles, terminologies and mathematical formulations, it seems likely that the manuscript in its present-day form constitutes the commentary or the copy of an anterior mathematical work. See **Shâradâ numerals, Indian styles of writing** and **Indian mathematics (The history of)**.

Other so-called "proofs" put forward to demonstrate that zero and the place-value system were discovered well before the Common Era include the texts of the **Purânas* (particularly *Agnipurâna*, *Shivapurâna* and *Bhavishyapurâna*).

In the *Purânas*, great importance is placed in decimal numeration. Thus, in *Agnipurâna*, the eighth text, during an explanation of the place-value system, it is written that "after the place of the units, the value of each place (**sthana*) is ten times that of the preceding place". Similarly, in the *Shivapurâna*, it is explained that usually "there are eighteen positions (**sthana*) for calculation", the text also pointing out that "the Sages say that in this way, the number of places can also be equal to hundreds". These cosmological-legendary texts have often been dated from the fourth century BCE, and some have even been dated as far back as 2000 years BCE. These dates, however, are totally unrealistic, because these texts are from diverse sources and they are the fruit of constant reworkings carried out within an interval of time oscillating between the sixth and the twelfth centuries CE.

In fact, the *Purânas* only seem to have become part of traditional Indian writings after the twelfth century CE. This is a characteristic trait of the Indian mentality which, in order to give more weight to explanations of mythology and legends and to support its tendency to sanctify, immortalise and distort certain elements of knowledge, often invokes an authority from scripture which assumes antiquity. See **Indianity**. Of course these texts do stem from a relatively ancient source; but this source, which has accrued diverse rubrics in quite recent times, has been steadily and constantly reworked.

Here is a typical passage: *ravivâre cha sande cha phâlgune chaiva pharvarî shashtish cha siksatî jñeyâ tad udâhâram îdrisham*

Translation: Namely, for example, that *ravivâre* (Sunday) means *sande*, [the month of] *phâlguna* (February) *pharvâri*, and sixty *siksatî*. Ref: *Bhavishyapurâna* (III, *Pratisargaparvan*, I, line 37); ed. Shrivenkateshvar, Bombay, 1959, p. 423) (Personal communication of Billard). The text refers to a "barbaric language" which is none other than Old English! This would suggest that the English race already existed four thousand years ago, and were contemporaries of the Sumerians. This demonstrates just how far biased authors can go with their unreliable dating of documents. The above line (which was doubtless added at the time of the British domination of India) is clearly out of context. Thus we can see the difficulties we are faced with when dealing with Indian documentation.

[See Datta, in BCMS, XXI, pp. 21ff.; IA, XVII, pp. 33–48; Datta and Singh (1938); Kaye, *Bakhshâlî* (1924); Smith and Karpinski (1911)].

**INDIAN MANUSCRIPTS (First material of).** See ***Pâtîganita*** and **Indian styles of writing (Material of).**

**INDIAN MATHEMATICS (The history of).** What we know of this discipline only really dates back to the beginning of the Common Era, as no documents written in Vedic times survive. Of course this does not mean that Indian mathematical activities only commenced at this time. However, vital information about geometry can be found from this time in the *kalpasûtra*, a collection of Brahmanic rites including the *Shulvasûtra* (or "Aphorisms on lines"), dedicated to the description of the rules of construction for altars and the measurements of sacrificial altars.

Three versions of these texts exist: these are called *Baudhâyana*, *Apastamba* and *Kâtyâyana*. The best known is the *Apastamba* version, in which a similar statement to Pythagoras's theorem can be found as well as some problems similar to those in *Elements* by Euclid. Thus, to build altars of a predetermined size, a square equal to the sum of the difference of two others had to be built. The altars, which were constructed out of bricks, had to be constructed in certain dimensions and with a determined number of bricks, and in certain cases had to undergo transformations to increase their surface area by a quantity specified in advance.

Some historians think that Indian mathematics of this era only constituted a utilitarian science. However, there is no evidence to prove this theory. As documentation currently stands, only the obtained mathematical results are known. The concise and essentially technical style of the texts in question did not leave room for even a summary description of the corresponding reasoning and methodology.

During the "classic" era (third to sixth century CE), there was a veritable renaissance in all fields of learning, especially in arithmetic and calculation, which underwent rapid expansion at this time. Moreover, it is probably at the beginning of this period that the impressive Indian speculations on high numbers were developed.

In Vedic literature, there is already evidence of skilful handling of quantities as large as ten to the power seven or ten to the power ten, the **Vedas* mentioning names of numbers from **eka* (= 1) to **arbuda* (= ten million). In the texts **Vâjasaneyî Samhitâ*, **Taittirîya Samhitâ* and **Kâthaka Samhitâ* (written at the start of the Common Era), there are numbers as high as **parârdha*, which, according to contemporary values, represents a billion.

Before the third century CE, however, there was no known text as long as the **Lalitavistara Sûtra*. It is a text about the life of the prince Gautama Siddhârtha (founder of Buddhist doctrine and thereafter named Buddha), which tells of how Buddha, whilst still a boy, becomes master of all sciences. It tells of the evaluation of the number of grains of sand in a mountain, which evokes the famous problem of Archimedes's *Sand-Reckoner*. What is significant is that the speculation goes way beyond the limits of numbers considered by Greek mathematicians: in one passage, when Buddha arrives at the number which today we write as "1" followed by fifty-three zeros, he adds that the scale in question is only one counting system, and that beyond it there are many other counting systems, and cites all their names without exception. See **Buddha (Legend of)**, **Names of numbers** and **High numbers**.

When the **Lalitavistara Sûtra* was written (before 308 CE), Indian arithmetical speculation had reached and surpassed the number ten to the power 421! After this time, equally vast quantities are found in *Jaina cosmological texts, which, speculating on the dimensions of a universe believed to be infinite in terms of both time and space, easily reach and surpass numbers equivalent to ten to the power 200. See **Jaina**, ***Shîrshaprahelikâ***, ***Anuyogadvâra Sûtra***. This means that, since the third century CE, the Indian mind had an extraordinary penchant for calculation and handling numbers which no other civilisation possessed to the same degree. See **Arithmetic**, **Calculation**, **Arithmetical speculations** and **Arithmetical-cosmogonical speculations.**

In fact, long before the *Lalitavistara Sûtra* and Jaina speculations, astrological and astronomical considerations had led the Indians to be deeply interested in mathematics. This took place between the third century BCE and the first century CE, under the influence of Greek astronomers and after the deployment of India's cultural, maritime and commercial activities with the West during this period. This was the time when astronomical procedures of Babylonian origin were introduced to Indian astronomy: a period characterised notably by the appearance of the unit of time called **tithi*, of similar length to the nychthemer (the "day") and consequently corresponding to 1/30 of the synodic revolution of the moon. It was also at this time that planets came into divination and became subjects for arithmetical calculation, based on their apparent synodic revolutions. This is how mathematics, which was essentially applied to religion, came to be used in astronomy at the time of the Gupta dynasty.

The beginning of the third period of this history roughly corresponds to the end of the "classic" era, around the end of the fifth century CE and the beginning of the sixth century, thus coinciding with the epoch of the **Âryabhatîya*. The *Âryabhatîya* is the name of the work by the mathematician and astronomer *Âryabhata, one of the most original, productive and significant scholars in the history of Indian science. This work (written c. 510 CE) is the first Indian text to display deep knowledge of astronomy, and is arguably the most advanced in the history of ancient Indian astronomy, which at this time developed amazing speculations about cosmic cycles called **yugas*. See ***Yuga*** **(Astronomical speculations about)** and **Indian astronomy (The history of).**

This work also deals with trigonometry and gives a summary of the principal Indian mathematical knowledge at the beginning of the sixth century: rules for working out square and cubic roots; rules of measurement; elements and formulas of geometry (triangle, circle, etc.); rules of arithmetical progressions; etc. Here is another important particularity of the *Âryabhatîya*: whilst Ptolemy's trigonometry was based principally on a relationship between the chords of a circle and the angle at the centre which subtends each one of them, Âryabhata's trigonometry established a relationship of a different nature between the chord and the arc of the centre, which is the *sine function* as a trigonomic ratio. That is one of the fundamental contributions of Âryabhata. The work also gives a sine table with the "approximate value" (*âsanna*) of the number $\pi$ (*pi*):$\pi \approx 62{,}832/20{,}000 \approx 3.1416$ [see Shukla and Sarma, (1976) and Sarma (1976)]. See **Âryabhata.**

Âryabhata invented a unique numerical notation, the conception of which required a perfect knowledge of zero and the place-value system. He also used a remarkable procedure for calculating square and cube roots, which would be impossible to carry out if the envisaged numbers were not expressed in written form, according to the place-value system, using nine distinct numerals and a unique tenth sign performing the function of zero. See **Âryabhata (Numerical notations of)**, **Âryabhata's numeration**, ***Pâtîganita***, **Indian methods of calculation** and **Square roots (How Âryabhata calculated).**

Of course, Âryabhata was not the first to use zero and the place-value system: the **Lokavibhâga*, or "Parts of the universe", contains numerous examples of them more than fifty years before the *Âryabhatîya* was written: it is a *Jaina cosmological work, which is very precisely dated Monday 25 August 458 CE in the Julian calendar. Moreover, this is the oldest known Indian text to use zero and the place-value system, except that the text only uses the system of *Sanskrit numerical symbols. See **Numerical symbols (Principle of the numeration of).** However, bearing in mind the perfect conception of the examples taken from the **Lokavibhâga* and the concern about vulgarisation which is clearly expressed, and moreover taking into consideration the fact that this text was probably the Sanskrit translation of an earlier document (most likely written in a Jaina dialect), it seems very likely that these major discoveries were made in the fourth century CE. This system had become widespread amongst the learned in India by the end of the sixth century. After the beginning of the seventh century, it had gone beyond the frontiers of India into the Indian civilisations of Asia.

As a consequence, calculation and the science of mathematics made substantial progress, as did astronomy, the most spectacular developments of which took place after the start of the sixth century CE. See **Indian astronomy (The history of).** Amongst the many successors of Âryabhata was Bhâskara I, his faithful disciple and fervent admirer, who wrote a Commentary on the *Âryabhatîya* in 629 CE. This gives invaluable information about the events which took place during the century which separated him from his preceptor. Moreover, this work reveals that Bhâskara had fully mastered mathematical operations which employed the nine numerals and zero using, for example, the Rule of Three. He dealt with arithmetical fractions with ease, expressing them in a very similar manner to our own, although he did not use the horizontal line (which was introduced by the Arabic-Muslim mathematicians). Bhâskhara's work also gives many clues about the development of algebra during that time.

*Brahmagupta was a contemporary of Bhâskhara. In 628 CE he wrote an astronomical text called *Brahmasphutasiddhânta* ("Revised system of Brahma"). In 664 CE he wrote a text on astronomical calculation called *Khandakhâdyaka*. He contradicted some of

Âryabhata's accurate ideas about the rotation of the earth, thus delaying the progress of certain aspects of astronomy. However, his work marked progress in fields such as algebra. He is without a doubt the greatest astronomer and mathematician of the seventh century CE. He made frequent use of the place-value system, and described methods of calculation which are very similar to our own using nine numerals and zero. Amongst his most important contributions are his systematisation of the science of negative numbers, his generalisation of Hero of Alexandria's formula for calculating the area of a quadrilateral, as well as his explanations of general solutions of quadratic equations. The progress of Indian mathematics was stimulated by the development of astronomy initiated by Âryabhata. Indian astronomers used all sorts of mathematical techniques and theories in this discipline. See **Indian astronomy (The history of)** and ***Yuga* (Astronomical speculations on)**.

Through resolving indeterminate equations, which depend entirely upon knowing the properties of whole numbers, the Indians arrived at discoveries which went far beyond those of other races of Antiquity or the Middle Ages, and which modern science only arrived at through the efforts of Euler [Woepcke (1863)].

Indian algebra never took the decisive step which would have elevated it to the same level as modern algebra. Instead of using symbols such as *a*, *b*, *x*, *y*, etc., which are independent of the real quantities that they represent, it never occurred to the Indian mathematicians to use symbols other than the first syllables of words which denoted the operations concerned. Moreover, the presentation of and solutions to their various mathematical problems were usually written in verse and consequently subjected to the rules of Sanskrit metric. This explains why their algebraic symbols remained for so long wrapped up in verbiage which was subject to diverse interpretations. See **Sanskrit** and **Poetry and the writing of numbers.**

Brahmagupta's successors included the *Jaina mathematician *Mahâvîrâchârya c. 850 CE, in Kannara in southern India. He wrote a work entitled *Ganitasârasamgraha*: "This work deals with the teachings of Brahmagupta but contains both simplifications and additional information. First he explains the mathematical terminology that he uses, then he deals with arithmetical operations, fractions, the Rule of Three, areas, volumes and in particular calculations with practical applications. He gives examples of solutions to problems. Although like all Indian versified texts it is extremely condensed, this work, from a pedagogical point of view, has a significant advantage over earlier texts" [Filliozat (1957–64)].

Other astronomers or mathematicians who corrected or significantly improved the work of their predecessors include *Govindasvâmin (c. 830), *Shankaranârâyana (c. 869), *Lalla (ninth century CE), *Shrîpati (c. 1039), *Bhoja (c. 1042), *Nârâyana (c. 1356), *Parameshvara (c. 1431), *Nîlakanthasomayâjin (c. 1500) and *Shrîdharâchârya (date uncertain). After Âryabhata and Brahmagupta, however, one of the greatest Indian mathematicians of the Middle Ages was without a doubt *Bhâskarâchârya, who is usually referred to as Bhâskara II, to avoid confusion with Bhâskara I. Born in 1114, the son of Chûdâmani Maheshvar, the astronomer in charge of the observatory of *Ujjain, he finished writing his *Siddhântashiromani* in 1150. This work is divided into four parts, the first two being devoted to mathematics and the second two to astronomy. These are respectively: the **Lîlâvatî* (named after his daughter), in which he explains the principle rules of arithmetic; the **Bîjaganita* (**bîja* means "letter" or "symbol", and **ganita* means "calculation"), which is about algebra; *Grahganita*, or "Calculation of the Planets"; and finally the *Golâdhyâya*, or "Book of the Spheres". In the field of astronomy, Bhâskarâchârya "repeats his predecessors but criticises them, even Brahmagupta, who he agrees with most often . . . he compares the gravitational forces of the stars to the winds, in distinguishing these winds from the atmosphere and its deplacements. Mathematically, he accounts for the movements by a developed theory of epicycles and eccentrics. One of the most interesting aspects of his work is that he analyses the movement of the sun, not only in considering the difference of the longitudes from one day to another, but even dividing the day into several intervals, and considers the movement in each one of them to be uniform" [Filliozat in: HGS, I]. The mathematical sections are mainly the study of linear or quadratic equations, indeterminate or otherwise; measurements, arithmetical and geometrical progressions, irrational numbers, and many other arithmetical questions of an algebraic, trigonomic or geometric nature.

Thus we have the names and the principle contributions of some of the most renowned figures in the history of Indian science. See **Infinity**, **Infinity (Indian concepts of)** and **Infinity (Indian mythological representation of)**.

**INDIAN METHODS OF CALCULATION.** When they first started out, Indian mathematicians carried out their arithmetical operations by drawing the nine numerals of their old *Brahmi numeration on the soft soil inside a series of columns of an abacus drawn in advance with a pointed stick. If a certain order lacked units the corresponding column was simply left empty. See ***Dhûlîkarma***. This archaic method was used later by the Arabic arithmeticians, particularly those of the Maghreb and Andalusia, who had adopted the nine Indian numerals but who did not tend to use zero or carry out their calculations without the aid of columns. However, these mathematicians did not only write out their calculations on the ground: they normally used a wooden board covered in dust, fine sand, flour or any kind of powder, and wrote with the point of a stylet, the flat end of which was used to erase mistakes. This board might be placed on the ground, on a stool or a table, or sometimes the board was equipped with its own legs, like the "counting tables" which were later used in Arabic, Turkish and Persian administrations. Sometimes this board constituted part of a type of kit, and was thus smaller and could be carried in a case. See ***Pâti*** and ***Pâtîganita***.

In the fifth century CE, the first nine Indian numerals taken from the *Brâhmî* notation began to be used with the place-value system and were completed by a sign in the form of a little circle or dot which constituted zero: this system was to be the ancestor of our modern written numeration. See **The place-value system**, **Position**, **Zero (Indian concepts of)**.

Thus the Indian mathematicians radically transformed their traditional methods of calculation by suppressing the columns of their old abacus and using the place-value system with their first nine numerals whilst continuing to use the board covered in dust. This step thus marked the birth of our modern written calculation.

To start with, although the corresponding techniques had been liberated from the abacus, they were still a faithful reproduction of the old methods of calculation: they were still carried out, as always, by successive corrections, continually erasing the results at each stage of the calculation, and this limited human memory whilst also preventing them from finding out the errors they had committed on the way to arriving at the final result. This method was used with various variations by *Mahâvîrâchârya (850 CE), *Shrîpati (1039 CE), *Bhâskarâchârya (1150 CE) and even by *Nârâyana (1356 CE). Alongside this technique, the Indian mathematicians (and the Arabic arithmeticians after them) developed a way of carrying out operations without any erasing, by writing the intermediary results above the final result. This, of course, was a great advantage because they could see if they had made a mistake in their calculations if the final result turned out to be wrong, yet brought with it the inconvenience of a lot more writing and more difficulty in deciphering the results, and this is why this method of calculation using nine numerals and zero remained beyond the understanding of the layman for so long.

Moreover, it was impossible for these methods to progress further without a radical transformation of the writing materials which were being used. The use of chalk and blackboard, long before the use of pen and paper became widespread, made the task much less onerous because the intermediary results could either be preserved or rubbed out with a cloth.*Bhâskarâchârya (1150 CE) used his **pâti* to work out highly advanced methods of calculation (notably multiplication which he referred to as *sthânakkhanda*, which literally means: the procedure of "separating the positions"). Even before him, the mathematician *Brahmagupta, in his *Brahmasphutasiddhânta* (628 CE), had described four methods of multiplication which were even more advanced and are more or less identical to those we use today. See also **Square roots (How Âryabhata calculated his)**.

**INDIAN METRIC.** Here is an alphabetical list of terms related to this discipline, which appear as entries in this dictionary:**Akriti*, **Anushtubh*, **Ashti*, **Atidhriti*, **Atyashti*, **Dhriti*, **Gâyatrî*, **Jagatî*, **Kriti*, *Poetry and writing of numbers, **Prakriti*, *Sanskrit, *Serpent (Symbolism of the), *Numerical symbols, **Vikriti*.

**INDIAN MYTHOLOGIES.** Here is an alphabetical list of words relating to this theme, which appear as article headings in this dictionary: **Agni*, **Ahi*, **Ananta*, **Âptya*, **Arjunâkara*, **Ashva*, **Ashtadiggaja*, **Ashvin*, **Atri*, **Avatâra*, *Buddha (Legend of), **Brahmâsya*, **Chaturdvîpa*, **Chaturmukha*, **Dantin*, **Dasra*, **Dhârtarâshtra*, **Dhruva*, **Diggaja*, **Dvîpa*, **Gaja*, *High numbers (The symbolic meaning of), **Haribâhu*, **Hastin*, **Indrarishti*, **Indu*, *Infinity (Indian mythological representation of), *Jagat, *Jaladharapatha, *Kâla, *Kârttikeya, *Kârttikeyâsya, *Kumârâsya, *Kumâravadana, *Kumud, *Kumuda, *Kuñjara, *Lokapâla, *Manu, *Mount Meru, *Mukha, *Muni, *Mûrti, *Naga, *Nâga, *Nâsatya, *Nripa, *Paksha, *Pândava, *Pâtâla, *Pâvana, *Pundarîka, *Pûrna, *Pushkara, *Putra, *Râhu, *Râvana, *Râvanabhuja, *Râvanashiras, *Rishi, *Sahasrârjuna, *Saptarishi, *Senâninetra, *Shanmukhabâhu, *Shatarûpâ, *Sheshashîrsha, *Shukranetra, *Soma, *Trimûrti, *Tripura, *Trishiras, *Uchchaishravas, *Vaikuntha, *Vasu, *Vâyu, *Vishnupada.

**INDIAN NUMERALS (or numerals of Indian origin).** List of the principal series of numerals that originated in India (graphical signs which derived from Brahmi numerals): *Agni, *Andhra, Balbodh, *Balinese, Batak, *Bengali, *Bhattiprolu, Bisaya, *Brahmi, Bugi, *Burmese, *Chalukya, *Cham, Chameali, *Dogri, *Dravidian, *Eastern Arabic, *European (*apices*), *European (*algorisms*), *Ganga, *Ghubar, *Grantha, *Gujarati, *Gupta, *Gurumukhi, Jaunsari, *(Ancient) Javanese, *Kaithi, *Kannara, *Kawi, *Khudawadi, Khutanese, Kochi, *Kshatrapa, Kului, *Kushâna, Kutchean, *Kutila, Landa, (Ancient) Laotian, Mahajani, *Maharashtri, *Maharashtri-Jaina, *Maithili, *Malayalam, Mandeali, *Manipuri, *Marathi, *Marwari, *Mathura, Modi, *Mon, *Mongol, Multani, *Nagari, *Nepali, *Old Khmer, Old Malay, *Oriya, *"Pali", *Pallava, *Punjabi, *Rajasthani, *Shaka, Shan, *Sharada, *Shunga, Siamese, *Siddham, *Sindhi, *Sinhala, Sirmauri, Tagala, *Takari, *Tamil, *Telugu, *Thai-Khmer, *Tibetan, Tulu, *Valabhi, and *Vatteluttu numerals.

For origins and graphical evolution, see Fig. 24.61 to 69. For genealogy, classification and geographical distribution, see Fig. 24.52 and 53. For all numerical notations of both Ancient and Modern India, see **Numerical notation**. See also **Indian styles of writing** and **Indian written numeral systems (Classification of)**.

**INDIAN RELIGIONS AND PHILOSOPHIES.** Here is an alphabetical list of terms related to this theme which appear as entries in this dictionary: *Abhra, *Abja, *Âdi, *Âditya, *Agni, *Âkâsha, *Amara, *Anala, *Âptya, *Arka, *Ashtadiggaja, *Ashtamangala, *Ashtamûrti, *Ashtavimoksha, *Ashva, *Ashvin, *Âtman, *Avatâra, *Bhânu, *Bharga, *Bhava, *Bhûta, *Bhuvana, *Bîja, *Bindu, *Brahma, *Brahmâsya, *Chandra, *Chaturâshrama, *Chaturmahârâja, *Chaturmukha, *Chaturvarga, *Chaturyoni, *Dahana, *Dantin, *Darshana, *Dashabala, *Dashabhûmi, *Dashaharâ, *Dashâhavatâra, *Dasra, *Deva, *Dharma, *Dhruva, *Diggaja, *Dikpâla, *Dish, *Divâkara, *Divyavarsha, *Dravya, *Drishti, *Durgâ, *Dvaita, *Dvandvamoha, *Dvâtrimshadvaralakshana, *Dvija, *Dvîpa, *Dyumani, *Ekachakra, *Ekadanta, *Ekâdashî, *Ekâgratâ, *Ekâkshara, *Ekântika, *Ekatattvâbhyâsa, *Ekatva, *Eleven, *Gagana, *Gaja, *Ganesha, *Gati, *Go, *Guna, *Hara, *Haranayana, *Haranetra, *Haribâhu, *Hastin, *High numbers (The symbolic meaning of), *Hotri, *Hutâshana, *Indian atomism, *Indra, *Indradrishti, *Indriya, *Infinity, *Infinity (Indian concepts of), *Infinity (Indian mythological representations of), *Îsha, *Îshadrish, *Ishvara, *Jagat, *Jaina, *Jala, *Jvalana, *Kâla, *Kâma, *Karanîya, Karttikeya, Karttikeyâsya, *Kârttikeya, *Kârttikeyâsya, *Kâya, *Keshava, *Kha, *Krishânu, *Kumârâsya, *Kumâravadana, *Kumud, *Kumuda, *Kuñjara, *Loka, *Lokapâla, *Lotus, *Mahâbhûta, *Mahâdeva *Mahâpâpa, *Mahâyajña, *Mantra, *Manu, *Mârtanda, *Mâtrikâ, *Nâsatya, *Netra, *Pañchabâna, *Pañchâbhijña, *Pañchabhûta *Pañchachakshus, *Pañcha Indriyâni, *Pañchaklesha, Pañchalakshana, *Pañchânana, *Parabrahman, *Paramânu, *Pâtâla, *Pâvaka, *Pâvana, *Pinâkanayana, *Pitâmaha, *Prâna, *Prithivî, *Pundarîka, *Purâ, *Pûrna, *Purânalakshana, *Pushkara, *Râga, *Râma, *Rasa, *Ratna, *Ravi, *Ravibâna, *Ravichandra, *Rudra, *Rudra-Shiva, *Rudrâsya, *Sâgara, *Sahasrakirana, *Sahasrâksha, *Sahasrâmshu, *Sahasranâma, *Samkhya, *Samkh yâ, *Sâmkhya, *Sâmkhyâ, *Samkhyeya, *Samsâra, *Saptamâtrika, *Shâdanga, *Shadâyatana, *Shaddarshana, *Shakra, *Shakti, *Shatarûpâ, *Shatapathabrâhmana, *Shatkasampatti, *Shikhin, *Shiva, *Shruti, *Shukranetra, *Shûla, *Shûlin, *Shûnya, *Shûnyatâ, *Siddha, *Siddhi, *Soma, *Sura, *Sûrya, *Tallakshana, *Tapana, *Tattva, *Triambaka, *Tribhûvaneshvara, *Trichîvara, *Triguna, *Tripurâ, *Tripurasundarî, *Triratna, *Trishûla, *Trivarga, *Trivarna, *Trividyâ, *Tryakshamukha, *Turîya, *Udarchis, *Vahni, *Vaikuntha, *Vaishvanara, *Vajra, *Varuna, *Vasu, *Vishaya, *Vrindâ, *Yama, *Yoni.

**INDIAN STYLES OF WRITING (The materials of).** The Indians have used various materials in the history of their writing, starting with stone, which, like nearly all other civilisations, has served for the writing of official inscriptions and important commemorative texts. Stone has often been replaced, at later times, with copper and other metals. Parchment has also been used, but only really in central Asia, religious reasons probably preventing its use in India. Tree-bark was used, mainly in Assam and southern regions, upon which scribes wrote in ink.

In Kashmir and the whole northwest of India, as well as in the regions of the Himalayas, ink and brush were (and still are) used on birch-bark. This manuscript writing was called *bhûrjapattra*, and its use was mentioned by Quintus Curtius in this region at the time of Alexander the Great: *Libri arborum teneri, haud secus quam chartae, litterarum notas capiunt* ("The tender part of the bark of trees can be written on, like papyrus") [quoted in Février (1959)]. Wooden boards were also used, upon which characters were not carved, but written in ink. Cotton was another writing support, as reported in the same region by Nearchus, Alexander's admiral.

As for manuscripts (the oldest known examples dating back to the first century CE), palm-leaves were the most popular supports in India and Southeast Asia. These were used since ancient times, in regions of Nepal, Burma, Bengal, as well as in southern India, Ceylon, Siam, Cambodia and Java. Its popularity was due to its availability and the ease with which it could be used: "The leaves chosen for writing were picked young, when they had not yet unfurled. The middle vein was removed and they were left to dry out under pressure. To join them together, they were placed between two boards or between two big dried nervures. Then a thread passed through all the leaves to join them together. Only one instrument was needed to pierce, slice, prepare, join and write a book. The extraordinary simplicity of such material certainly played an important role in the diffusion of Indian culture" [Février, (1959)]. One of the ways of writing on a palm leaf was to engrave the characters using a pointed instrument: "It is undeniable that the characters traced with a point appear pale and unclear, but when sprinkled with dust, they become black, and the dust does not stick to the rest of the leaf because its surface is naturally smooth." Another writing instrument is the calamus, a type of reed whose blunted end is dipped in a type of dye; this has been used since time immemorial in Bengal, Nepal and all southern regions of India. Thus the writing materials used in India and Southeast and Central Asia are as varied as the styles of writing themselves. These conditions account for the great diversity of India writing styles: this diversity has not come about by chance, as the nature of the writing materials has had a profound influence over the appearance of the corresponding styles of writing. **See Indian styles of writing.**

**INDIAN STYLES OF WRITING.** The various styles of writing which are currently in use in India, Central and Southeast Asia all derive more or less directly from the ancient *Brâhmî* writing, as it is found in the edicts of Emperor Asoka and in a whole series of inscriptions which are contemporaries of the Shunga, Kushâna, Ândhra, Kshatrapa, Gupta, Pallava, etc., dynasties. This writing underwent many successive and relatively subtle modifications over the course of the centuries, which led to the development of various completely individual styles of writing. The apparently considerable differences are due to either the specific character of the languages and traditions to which they have been adapted, or the regional customs and the writing materials used. **See Indian styles of writing (The materials of the).**

These styles of writing can be put into three groups (Fig. 24. 28): the group of styles of writing of northern and central India and of Central Asia (Tibet and Chinese Turkestan): the group of styles of writing of southern India; and finally the group of styles of writing known as "oriental". Naturally, the writing of the first nine numbers has undergone a similar evolution over the centuries: all the series of numerals from 1 to 9 currently in use in India and Central and Southeast Asia derive from the ancient *Brâhmî* notation for the corresponding numbers and can be placed in the same groups as those for the styles of writing (Fig. 24. 52). For all the corresponding varieties, see **Indian numerals.**

**INDIAN THOUGHT.** Here is an alphabetical list of words related to this theme, which appear as entries in this dictionary: *Abhra, *Abja, *Âdi, *Âditya, *Adri, *Agni, *Ahi, *Âkshara, *Amara, *Anala, *Ananta, *Anga, *Anuyogadvâra Sûtra, *Âptya, *Arithmetical-cosmogonical speculations, *Arjunâkara, *Arka, *Asamkhyeya, *Âshâ, *Ashtadanda, *Ashtadiggaja, *Ashtamangala, *Ashtamûrti, *Ashtânga, *Ashtavimoksha, *Ashva, *Ashvin, *Âtman, *Atrinayanaja, *AUM, *Avatâra, *Bâna, *Bhânu, *Bharga, *Bhava, *Bhûta, *Bhuvana, *Bîja, *Bindu, *Brahma, *Brahmâsya, *Calculation, *Chakskhus, *Chandra, *Chaturânanavâdana, *Chaturdvîpa, *Chaturmahârâja, *Chaturmukha, *Chaturyoni,

*Chaturyuga, *Cosmogonic speculations, *Dahana, *Dantin, *Darshana, *Dashabala, *Dashabhûmi, *Day of Brahma, *Deva, *Dharanî, *Dharma, *Dhruva, *Diggaja, *Dikpâla, *Dish, *Divâkara, *Dravya, *Drishti, *Durgâ, *Dvâdashadvârashâstra, *Dvaparayuga, *Dvâtrimshadvaralakshana, *Dvija, *Dvipa, *Dvîpa, *Dyumani, *Eight, *Eka, *Ekachakra, *Ekadanta, *Ekâdashî, *Ekâgratâ, *Ekâkshara, *Ekatva, Ekavâchana, *Eleven, *Fifteen, *Five, *Four, *Fourteen, *Gagana, *Gaja, *Ganesha, *Gati, *Gavyâ, *Go, *Graha, *Hara, *Haranayana, *Haranetra, *Haribâhu, *Hastin, *High numbers (The symbolic meaning of), *Hotri, *Hutâshana, *Indian astrology, *Indian atomism, *Indian documentation (Pitfalls of), *Indianity (Fundamental mechanisms of), *Indra, *Indradrishti, *Indriya, *Indu, *Infinity, *Infinity (Indian concepts of), *Infinity (Indian mythological representation of), *Îsha, *Îshadrish, *Ishu, *Îshvara, *Jagat, *Jaladharapatha, *Jvalana, *Jyotisha, Kabubh, *Kâla, *Kalamba, *Kaliyuga, *Kalpa, *Kâma, *Karanîya, *Kârttikeya, *Kârttikeyâsya, *Kavacha, *Kâya, *Keshava, *Kha, *Krishânu, *Krita, *Kritayuga, *Kshapeshvara, *Kumârâsya, *Kumâravadana, *Kumud, *Kumuda, *Kuñujara, *Loka, *Lokapâla, *Lotus, *Mahâdeva, *Mahâkalpa, *Mahâyuga, *Mangala, *Mantra, *Manu, *Mârgana, *Mârtanda, *Mrigânka, *Mukha, *Mûrti, *Mysticism of infinity, *Mysticism of zero, *Nâga, *Netra, *Nine, *Numeral alphabet, magic, mysticism and divination, *Numerical symbols, *Ocean, *One, *Paksha, *Pañchabâna, *Pañchâbhijña, *Pañchachakshus, *Pañcha Indriyâni, *Pañchaklesha, *Pañchalakshana, *Pañchânana, *Parabrahman, *Pâtâla, *Pâvaka, *Pâvana, *Pinâkanayana, *Pitâmaha, *Pundarîka, *Purâ, *Pûrna, *Putra, *Râga, *Râhu, *Rasa, *Ratna, *Râvanabhuja, *Ravi, *Ravibâna, *Rudra, *Rudra-Shiva, *Rudrâsya, *Sâgara, *Sahasrakirana, *Sahasrâksha, *Sahasrâmshu, *Sahasranâma, *Samkhya, *Samkhyâ, *Sâmkhya, *Sâmkhyâ, *Samkhyâna, *Samkhyeya, *Sanctification of a concept, *Sanskrit, *Saptabuddha, *Sarpa, *Sâyaka, *Seven, *Shakra, *Shakti, *Shankha, *Shanku, *Shanmukhabâhu, *Shara, *Shashadhara, *Shashanka, *Shashin, *Shatarûpâ, *Shikhin, *Shîrshaprahelikâ, *Shîtâmshu, *Shîtarashmi, *Shiva, *Shukranetra, *Shûla, *Shûlin, *Shûnya, *Shûnyatâ, *Six, *Sixteen, *Soma, *Sudhâmshu, *Sura, *Sûrya, *Symbolism of numbers, *Symbols, *Takshan, *Tallakshana, *Tapana, *Ten, *Thirteen, *Thirty-three, *Thousand, *Three, *Tretâyuga, *Triambaka, *Tribhuvana, *Tribhûvaneshvara, *Trikâya, *Triloka, *Trimûrti, *Tripitaka, *Tripurâ, *Tripurasundarî, *Trishûla, *Trivarna, *Tryakshamukha, *Tryambaka, *Turîya, *Twelve, *Twenty, *Twenty-five, *Two, *Udarchis, *Uppala, *Utpala, *Vâchana, *Vahni, *Vaikuntha, *Vaishvanara, *Vajra, *Varuna, *Vasu, *Vidhu, *Vishika, *Vishnupada, *Vishva, *Vishvadeva, *Vrindâ, *Yama, *Yuga, *Yuga (Astronomical speculation on), *Yuga (Cosmogonical speculations on), *Zero.

**INDIAN WRITTEN NUMERAL SYSTEMS (Graphical classification of).** The aim of this article is to give a quick recapitulation of the various numerical notations formerly or currently used in the Indian sub-continent, in order to identify the palaeographic type of each one. The following references to figures are mainly the ones which can be found in Chapter 24. More or less all of the numerical notations which are currently in use in India, Central Asia and Southeast Asia (see Fig. 24. 61 to 69) derive from the ancient *Brâhmî* notation (Fig. 24. 29 to 31 and 70), which is found in the edicts of Emperor Asoka and a whole series of contemporary inscriptions of the Shunga, Kushâna, Ândhra, Kshatrapa, Gupta, Pallava, etc. dynasties (Fig. 24. 29 t 38 and 70). This original notation (which surely derives from an earlier ideographical notation) has undergone several subtle graphical modifications over the course of the centuries (Fig. 24. 70), which led to the development of various types of notations which are all highly individual like *Gupta* (Fig. 24. 38), Bhattiprolu and *"Pali"*. See also **Ândhra numerals**, **Bhattiprolu numerals**, **Brâhmî numerals**, **Châlukya numerals**, **Ganga numerals**, **Gupta numerals**, **Kshatrapa numerals**, **Kushâna numerals**, **Mathurâ numerals**, **Pâlî numerals**, **Pallava numerals**, **Shaka numerals**, **Shunga numerals** and **Valabhî numerals**.

For the graphical origin of *Brâhmî* numerals, see Fig. 24. 57 to 24. 59. For notations derived from *Brâhmî*, see Fig. 24. 52. For their graphical evolution, see Fig. 24. 61 to 24. 69.

The apparently considerable differences between these notations are due to either the specific character of the languages and traditions to which they belong to which the corresponding writing would have been adapted, or to the regional habits of the scribes and the nature of the writing material used. See **Indian styles of writing.**

The notations can be divided into three broad groups (see Fig. 24. 52):

1. – The group of notations from Central India, from Northern India, from Tibet and Chinese Turkestan. These notations are the ones which come from Gupta writing. These can be divided in turn into five sub-groups:

1. 1. – *The sub-group of notations derived from Nâgarî*. This group is made up of notations issued from Nâgarî numerals (Fig. 24.3, 39 and 72 to 74), amongst which are: Mahârâshtrâ; Marâthî (Fig. 24.4); Balbodh; Modi; Rajasthani; Mârwarî; Mahajani; Kutilâ; Bengali (Fig. 24.10); Oriya (Fig. 24.12); Gujarati (Fig. 24.8); Maithili (Fig. 24.11); Manipuri; Kaithi (Fig. 24. 9); etc.

The Arabic notations *"Hindi"* and *Ghubar* also belong to this sub-group, as well as the European *Apices* and *Algorisms*: Arabic numerals both from the East and the Maghreb (Fig. 25.3 and 25.5), derive more or less directly from Nâgarî numerals. The numerals that we use today, and the European numerals of the Middle Ages (Fig. 26.3 and 10), derive from the *Ghubar* numerals of the Maghreb (Fig. 25.5). See **Eastern Arabic numerals**, **Bengali numerals**, **European numerals (Apices)**, **European numerals (Algorisms)**, **Ghubar numerals**, **Gujaratî numerals**, **Kâithî numerals**, **Kutila numerals**, **Mahârâshtrî numerals**, **Mahârâshtrîjaina numerals**, **Maithilî numerals**, **Manipuri numerals**, **Marâthî numerals**, **Mârwarî numerals**, **Nâgarî numerals**, **Oriya numerals and Rajasthani numerals.**

1.2. – *The sub-group of notations derived from Sharada writing*. This is composed of notations derived from the numerals of the same name (Fig. 24. 14 and 40), including: Tâkarî (Fig. 24. 13); Dogri (Fig. 24. 13); Chameali ; Mandeali; Kului; Sirmauri; Jaunsari; Sindhi (Fig. 24. 6); Khudawadi (Fig. 24. 6); Gurumukhi (Fig. 24. 7); Punjabi (Fig. 24. 5); Kochi; Landa; Multani; etc. See **Dogrî numerals**, **Gurumukhi numerals**, **Khudawadî numerals**, **Punjabi numerals**, **Shâradâ numerals**, **Sindhi numerals**, **Sirmauri numerals and Tâkarî numerals.**

1.3. – *The sub-group of notations from Nepal*. This includes modern Nepali (Fig. 24. 15), which derives from the ancient Siddham notation (Fig. 24. 42) which itself comes from Gupta but under the influnce of Nagari. See Nepali numerals and Siddham numerals.

1.4. – *The sub-group of Tibetan notations*. This contains Tibetan notations (Fig. 24. 16), which all derive from *Siddham*, and which are notably related to ancient Mongol writing (Fig. 24.42). See **Tibetan numerals and Mongol numerals.**

1.5. – *The sub-group of notations from Central Asia*. This contains notations of Chinese Turkestan, which also all derive from *Siddham*.

2. – The group of notations from Southern India. These are notations which come from *Bhattiprolu* (Fig. 24. 43 to 24. 46), distant cousin of Gupta. They can be subdivided into four groups:

2. 1. – *The sub-group of Telugu notations*. This is made up of Telugu and Kannara notations (Fig. 24. 20, 21, 47 and 48).

2.2. – *The sub-group of Grantha notations*. This contains Grantha, Tamil and Vatteluttu notations (Fig. 24. 17 and 24. 49).

2.3. – *The sub-group of Tulu notations*. This contains Tulu and Malayalam notations (Fig. 224. 19).

2.4. – *The sub-group of Sinhalese notations*. In this group Sinhala notation can be found (Fig. 24. 22).

See **Dravidian numerals**, **Grantha numerals**, **Kannara numerals**, **Malayalam numerals**, **Sinhala numerals**, **Tamil numerals**, **Telugu numerals** and **Vatteluttu numerals.**

3. – The group of eastern notations. These are the notations of Southeast Asia, which are all derived from "Pali" writing, which itself comes from the same source as Gupta and Bhattiprolu (Fig. 24. 43 to 46). These in turn can be subdivided into seven groups:

3.1. – *The sub-group of Burmese notations*. This contains ancient and modern Burmese notations (Fig. 24. 23).

3.2. – *The sub-group of Old Khmer notations*. In this group there is the ancient notation of Cambodia (Fig. 24. 77, 78 and 80).

3.3. – *The sub-group of Cham notations*. This contains the notation of Champa (Fig. 24. 79 and 80).

3.4. – *The sub-group of Old Malay notations*. This group contains the writing style once used in Malaysia (Fig. 24. 80).

3.5. – *The sub-group of Old Javanese notations*. This group contains "Kawi" writing which was once used in Java and Bali (Fig. 24. 50 and 24. 80).

3.6. – *The sub-group of present-day Thai-Khmer writing*. This includes Shan, Laotian and Siamese, as well as the notation which is currently used in Cambodia, Laos and Thailand (Fig. 24. 24).

3. 7. – *The sub-group of current Balinese notations*. This sub-group is made up of the current

Balinese, Buginese, Tagala, Bisaya and Batak notations (Fig. 24.25). See **Balinese numerals, Burmese numerals, Cham numerals, Ancient Javanese numerals, Kawi numerals, Thai-Khmer numerals and Old Khmer numerals.** For an overview of all these notations, see Fig. 24.52. For their geography, see Fig. 24.27 and 53. For their mathematical classification, see **Indian written numeral systems (The mathematical classification of).**

**INDIAN WRITTEN NUMERAL SYSTEMS (The mathematical classification of).** Here is a quick summary of the mathematical structure of the various notations which were once used, or are still in use, in the Indian sub-continent. The numerical notations which derive from *Brâhmî* (see **Indian written numeral systems (Graphical classification of)**) are not the only ones to be used in the Indian sub-continent. In northwest India, after Asoka's time until the sixth or seventh century CE, a style of writing was used which was imported by Aramaean traders. This was known as *Karoshthî* (Fig. 24. 54). See **Karoshthî numerals.** There is also the system which was found in Mohenjo-daro and Harappa (in present-day Pakistan), which was used from c. 2500 to 1500 BCE by the ancient Indus civilisation, long before the Aryans arrived on Indian soil. See **Proto-Indian numerals.**

From a mathematical point of view, according to the classification of numerations in Chapter 23, these different systems (which generally have a decimal base) can be divided into three broad categories:

A. – The category of additional numerations. These are systems which are based upon the additional principle, each numeral possessing its own value, independent of its position in numerical representations. This category can be subdivided into three types:

A.1. – *The first type of additional numerations*. These are numerations which (like the Egyptian hieroglyphic system for example) attribute a particular numeral to each of the numbers 1, 10, 100, 1,000, 10,000, etc., and which repeat these signs as many times as necessary to record other numbers (Fig. 23.30). The ancient *Indusian numeration no doubt belonged to this type.

A.2. – *The second type of additional numeration*. These are numerations which (like the Roman system for example) attribute a specific numeral to each of the numbers 1, 10, 100, 1,000, etc., as well as to 5, 50, 500, etc., and which repeat these signs as many times as necessary to record other numbers (Fig. 23.31). There is no known example of this type in India.

A. 3. – *The third type of additional numeration*. These are numerations which (like the Egyptian hieratic system for example) attribute a particular sign to each unit of each decimal order (1, 2, 3, . . . 10, 20, 30, . . . 100, 200, 300, . . ., etc.) and which use combinations of these different signs to write other numbers (Fig. 23.32). This is the type that all notations derived from Brâhmî belong to, at least initially (Fig. 24.70). Thus the following notations belong to this sub-category: Ândhra notation (Fig. 24.34 and 36); Bhattiprolu notation; Chalukya notation (Fig. 24.45); Ganga notation (Fig. 24.46); Gupta notation (Fig. 24.38); Kshatrapa notation (Fig. 24.35); Kushâna notation (Fig. 24.33); Mathura notation (Fig. 24.32); Ancient Nâgarî notation (Fig. 24.39B); Ancient Nepali notation (Fig. 24.41); Pallava notation (Fig. 24.34 and 24.36); Valabhî notation (Fig. 24.44); etc. Alphabetical notations also fall into this category (which use vocalised consonants of the Indian alphabet, to which a numerical value is assigned in a regular order, and which are still used today in various regions of India, from Tibet, Nepal, Bengal or Orissa to Maharashtra, Tamil Nadu, Kerala, Karnataka and Sri Lanka, and from Burma to Cambodia, in Thailand and in Java); notably that of Âryabhata (the difference being that the latter had a centesimal base, not a decimal one). One exception is *Katapayâdi* numeration (which seems to have been invented by Haridatta), which was alphabetical but based on the place-value system. See **Numeral alphabet, Âryabhata's numeration** and **Katapayâdi numeration.**

B. – The category of hybrid numerations. These are numerations which use both multiplication and addition in their representations of numbers. This category can be divided into five types:

B.1. – *The first type of hybrid numeration*. These are numerations which (like the Babylonian system) attribute a particular numeral to each of the numbers 1, 10, 100, 1,000, etc., using an additive notation for numbers inferior to one hundred and a multiplicative notation for the hundreds, the thousands, etc., and which represents other numbers through combinations which use both the additive and multiplicative principles (Fig. 23.33). Aramaean numeration belongs to this group (Fig. 23.17) as well as *Kharoshthî numeration which is derived from the former (Fig. 24.54).

B.2. – *The second type of hybrid numeration*. These are numerations which function exactly like the Sinhalese system (Fig. 23.18 and 24.22): a particular numeral is given to each simple unit, as well as to each power of ten (10, 100, 1,000, etc.), and the notation of hundreds, thousands, etc., follows the multiplicative rule (Fig.23.34).

B.3. – *The third type of hybrid numeration*. These are Mari numerations (Fig. 23.22), which do not seem to exist in India.

B.4. – *The fourth type of hybrid numeration*. These are Ethiopian numerations (Fig. 23.36), which do not seem to exist in India.

B.5. – *The fifth type of hybrid numeration*. These are the numerations for which Tamil and Malayâlam numerations provide the models (Fig. 23.20 and 21); these give a specific numeral to each simple unit (1, 2, 3 . . .), as well as to diverse multiples of each power of ten (10, 20, 30, . . . 100, 200, 300, . . ., etc.), and where the notation of tens, hundreds, thousands, etc., is carried out using the multiplicative principle (Fig. 23.37).

C. – The category of positional numerations. These are numerations founded on the principle according to which the basic value of numerals is determined by their position in the writing of the numbers, and which thus requires the use of a zero (Fig. 23.27). This category can be subdivided into two types:

C.1. – *The first type of positional numerations*. These are Babylonian, Chinese or Maya (Fig. 23. 23, 24, 25 and 38), which are not found in India.

C.2. – *The second type of positional numerations*. These are the numerations (Fig. 23.28), which belong to the one which was invented in India over fifteen centuries ago and which is the origin of all decimal positional notations which are currently in use (Fig. 24.3 to 16 and 20 to 26), including our own (Fig. 23.26) and the one which is still used in Arabic countries (Fig. 24.3). This system has a decimal base, and nine distinct numerals which give no visual indication as to their value, which represent the nine significant units (from which our signs 1, 2, 3, 4, 5, 6, 7, 8, 9 are) derived; it also possesses a tenth sign, called **shûnya* (zero), which is written as a little dot or circle (Fig. 24.82 and **Zero*, Fig. D. 11), and is the ancestor of our zero, whose function is to mark the absence of units in any given order, and which possesses a veritable numerical value: that of "nil" (Fig. 23.27). The fundamental characteristic of this system is that it can express all numbers in a simple and coherent way, whether they are whole, fractional, irrational or transcendental (Fig. 23.28). Thus the Indian place-value system (for that is what it is) is the first of the category of the most evolved written numerations in history (Fig. 28.29). The following are the notations which belong to this category:

Modern Nâgarî (Fig. 24.3, 39 A and 39 C); Marathî (Fig. 24.4); Punjabi (Fig. 24.5); Sindhi (Fig. 24.6); Gurumukhi (Fig. 24.7); Gujarati (Fig. 24.8); Kâithî (Fig. 24.9); Bengali (Fig. 24.10); Maithilî (Fig. 24.11); Oriya (Fig. 24.12); Tâkarî (Fig. 24.13); Shâradâ (Fig. 24.14 and 40); modern Nepali (Fig. 24.15); Tibetan (Fig. 24.16); Telugu (Fig. 24.20 and 47); Kannara (Fig. 24.21 and 48); Burmese (Fig. 24.23 and 51); Thai-Khmer (Fig. 24.24); Balinese (Fig. 24.25); modern Javanese (Fig. 24.26); ancient Javanese (Fig. 24. 50); Mongol (Fig. 24.42); the "Hindi" form of Arabic writing (Fig. 24.3 and 25.5); the "Ghubar" form of Arabic writing, whilst it was used to represent numbers with zero, without the columns of the abacus drawn in the dust (Fig. 25.3); the "Algorism" form of European writing (Fig. 26.10); etc.

Thus the discovery of Indian positional numeration not only allowed the simple and perfectly rational representation of absolutely any number (however large or complex), but also and above all a very easy way of carrying out mathematical operation; this discovery made it possible for anyone to do sums. The Indian contribution to the history of mathematics was essential, because it united calculation with numerical notation, thus enabling the democratisation of the art of calculation.

For the graphical classification of the various numerical notations, see *Indian written numerations (The graphical classification of). For the Sanskrit names, usage, conditions and discovery of positional numeration, see: ***Anka, Sthâna, Ankakramena, Ankasthâna, Sthânakramâd*, Names of numbers, High numbers, Sanskrit, Numerical symbols, Numeration of numerical symbols, Katapayâdi numeration, Âryabhata's numeration.**

For zero and its graphic or symbolic representations, see **Zero, *Shûnya*, Numeral 0.**

For corresponding methods of calculation, see ***Patîganita*, Indian methods of calculation.**

For the subtleties relating to zero and the place-value system in Sanskrit poetry, see **Poetry, zero and positional numeration.**

**INDIVIDUAL SOUL.** [S]. Value = 1. See ***Âtman***. One.

**INDO-EUROPEAN NAMES OF NUMBERS.** See Chapter 2, especially Fig. 2. 4 A to 4 J and 2. 5, where the Sanskrit names of numbers are compared to those of other languages of Indo-European origin. See Fig. D. 2 of the entry entitled **Âryabhata's numeration.**

**INDRA.** [S]. Value = 14. "Strength", "Courage", "Power". The name of one of the principal gods of Vedic times and of the Brahm anic pantheon. He represents the source of cosmic life that he transmits to the earth through the intermediary of rain. His strength lies in the seminal fluid of all beings, this god being said to be "made of all the gods put together". He is eternally young, because he rejuvenates himself at the start of each *manvantara*, which means each of the fourteen "ages" of our world which make up a **kalpa*. Thus *Indra* = 14. See ***Yuga, Manu*** and **Fourteen.**

**INDRADRISHTI.** [S]. Value = 1,000. "Eyes of Indra". One of this god's attributes is **Sahastâksha*, "of the thousand eyes". See **Indra, Thousand.**

**INDRIYA** [S]. Value = 5. "Power". This is due to the Buddhist physical and mental powers, which are divided into five groups: the foundations (*âyatana*); the natures (*bhâva*); the senses (*vedâna*); the spiritual powers (*balâ*); and the supramundane powers. The same word also means the "five roots" (**pañcha indriya*), which, as positive agents, enable a person to lead a moral life: faith (*shraddendriya*), energy (*vîyendriya*), memory (*smritîndriya*), meditation (*samâdhîndriya*), and wisdom (*prâjnendriya*) [see Frédéric, *Dictionnaire* (1987)]. See **Five.**

**INDU.** [S]. Value = 1. "Drop". This represents the moon, and alludes to the "dew" (*chandrakanta*), the mythical pearl said to have been made from concentrated moonbeams. The moon being worth 1, this symbolism is self-explanatory. This word should not be confused with **bindu* ("point") which is a synonym for zero. See **One.**

**INDUSIAN NUMERALS.** See **Proto-Indian numerals.**

**INDUSIAN NUMERATION.** See **Proto-Indian numerals.**

**INFERIOR WORLD.** [S]. Value = 7. See ***Pâtâla*** and **Seven.**

**INFINITELY BIG.** See **High numbers, *Asamkhyeya*, High numbers (Symbolic meaning of).**

**INFINITELY SMALL.** See **Low numbers, *Paramânu, Shûnya, Shûnyatâ*, Zero** and **Infinity.**

**INFINITY (Indian concepts of).** Amongst the Sanskrit words for zero is **ananta*, which literally means "infinity": Ananta is an immense serpent, who, in Indian cosmology and mythology, represents the serpent of infinity, eternity and the immensity of space. *Vishnu is said to rest on the serpent in between creations. See **Serpent (Symbolism of the), High numbers** and **Infinity (Indian mythological representation of).**

In Indian mysticism, the concept of zero and that of infinity are very closely linked. Thus words such as **ambara*, **kha*, **gagana*, etc., meaning "space", "sky" or the "canopy of heaven" came to represent zero. See **Zero, *Shûnya, Âkâsha, Vishnupada*** and ***Pûrna***.

Of course, Indian mathematicians knew perfectly well how to distinguish between these two notions, which are the inverse of one another, for to their mind, division by zero was equal to infinity. This was the case at least since the time of *Brahmagupta (628 CE), who defined infinity with the term **khachheda*, literally "the quantity whose denominator is zero" [see Datta and Singh (1938), pp. 238–44]. In **Lîlâvatî*, *Bhâskarâchârya wrote the following about the same concept, which he refers to as **khahara*, which literally means "division by zero" [see Datta and Singh (1938), pp. 238–44]: "In this quantity which has zero as divisor, there is no [possible] modification, even though several [quantities] can be extracted or introduced; in the same way, no changes can be carried out on the constant and infinite God [*Brahma] during the period of the destruction or creation of worlds, however many living species are projected forward or are absorbed." This is what Ganesha wrote on the subject in *Grahalâghava* (c. 1558 CE): "The *Khachheda* is an indefinite quantity, unlimited and infinite; it is impossible to know how high this quantity is. It can be modified by neither the addition nor the subtraction of limited [= finite] quantities, for in the preliminary operation of reducing all the fractional expressions to the same denominator, which it is necessary to do beforehand in order to be able to calculate their sum or their difference, the numerator and the denominator of the finite quantity both disappear." So Indian scholars, at least since Brahmagupta's time, knew that division by zero equalled infinity:

$$a/0 = \infty.$$

To their mind, this "quantity" remained unchanged if a finite number was added to it or subtracted from it; thus:

$$a/0 + k = k + a/0 = a/0$$

and

$$a/0 - k = k - a/0 = a/0.$$

This means that the Indians, at least as early as the beginning of the seventh century CE, knew these mathematical formulas that we use today:

$$\infty + k = k + \infty = \infty$$

and

$$\infty - k = k - \infty = \infty.$$

Brahmagupta, however, (and several of his successors) committed the error of thinking that when zero was divided by itself the result was zero, when in reality the result is an "indefinite quantity". Bhâskarâchârya, who made the necessary corrections to the erroneous assertions of his predecessors, and who quite rightly affirmed that a number other than zero divided by zero gives an infinite quotient, himself committed an error by saying that the product of infinity multiplied by zero is a finite number. However, this in no way detracts from the merits of Indian civilisation which was so advanced in comparison with all the other civilisations of Antiquity and the Middle Ages. See **Infinity, Infinity (Indian mythological representation of)** and **Indian mathematics (The history of).**

**INFINITY (Indian mythological representation of).** It seems that the lemniscate which today represents the concept of infinity, was introduced for the first time in 1655 by the English mathematician John Wallis. Hindu mythological iconography contains a very similar symbol representing the same idea, although it seems that it was never used in the domain of mathematics. This symbol is that of Ananta, the famous serpent of infinity and eternity, which is always represented coiled up in a sort of figure of eight on its side like the symbol ∞. See ***Ananta*** (in particular Fig. D. 1), ***Pûrna*** and ***Vishnupada***.

This begs the following question: Did Wallis know of the Indian mythological symbol when he introduced this sign into the list of mathematical conventions? The answer is no; this graphism and its many variants (∞, 8, S, etc.) can be found in diverse civilisations and many different epochs and parts of the world, and the symbolism is similar, if not identical, to that of the Indian mythological representations. This symbolism can be found in many ancient astrological, magical, mystical and divinatory representations, for example in ancient and mediaeval talismans, both Eastern and Western, where the S is very common and is meant to express, for the wearer of the amulet, a sign favourable to *eternal* union and *infinite* happiness. The sign which looks like a figure of eight lying on its side can be found painted on the walls of masonic lodges or embroidered upon clothing. It is not there for merely decorative purposes; it symbolises the bonds which unite the members of a social body: the interlacing expresses the sentiment *united until death* [see Chevalier and Gheerbrant (1982)]. This symbol can also be found in the manuscripts of mediaeval alchemists, where three Ss signify the *abundance* of rain water, as well as its *constance*. The S can also be connected to the *celestial wheel* of the Romanised god of Ancient Gaul, and to talismans which have *celestial* meaning in Greek-Roman magical traditions [see Marquès-Rivière (1972)]. For the Assyrians, *hawu* was also in the form of an S, like the serpent of *eternal life*. This symbol was later used by the Hebrews to represent the "bronze serpent" before it was destroyed by Hezekiah (2 Kings XVIII, 4). This is the serpent made by Moses to save the Israelites who had spoken against the Lord, and who had been bitten by the fiery serpents sent by Yahweh (Numbers, XXI, 6)[see GLE, IX, p. 770].

The interlace is often a symbol for water or for the vibration of the air. In many cosmogonies, the interlace symbolises the very nature of creation, energy and all existence. In Celtic art, it symbolises the notion of *ourobouros*: the endless movement of evolution and involution through the muddle of cosmic and human facts. The *ourobouros* is the serpent which bites its own tail ; this symbolises self-evolution, or self-fertilisation and, consequentially, eternal return. This evokes the **samsâra* (or the Indian cycle of "rebirth"), which is an *indefinite* circle of rebirths, of *continual* repetition. Thus the serpent gradually came to be represented by a circular graphism. Sometimes this circle has been dissected by two perpendicular diameters in order to show the inter-relationship between the sky and the earth. The sign which looks like an X or a cross symbolises the earth with its four horizons. Thus the circle dissected by the cross is none other than the celestial-terrestrial opposition of the mysticism of the serpent. See **Serpent (Symbolism of the)**.

Palaeography proves that this dissected circle is, cursively speaking, the S or 8 sign denoting a vast quantity or eternity. This is very significant when we look at the shapes of Roman numerals. The Roman numerals that we know today look like Roman letters: **I** (1), **V** (5), **X** (10), **L** (50), **C** (100), **D** (500), **M** (1,000).

In reality, however, these symbols are not the original ones used to write the numbers. In fact, Roman numerals derive from the ancient practice of counting using a "tally" system which led to numbers being represented by the following symbols:

1 5 10 50 100

Originally, the unit was represented by a vertical line, the number 5 by an acute angle, the number 10 by a cross, 50 by an acute angle dissected by a vertical line and 100 was a cross dissected by a vertical line. We can easily see how the primitive signs for 1, 5 and 10 became the letters I, V and X. The sign for 50 originally looked like an arrow pointing downwards. This evolved into what looked like a T on its head before finally being mistaken for the letter L. As for the representation of 100, this initially evolved into a sign which looked like this:

Then, in order to save time, this symbol was cut in half so that it looked like the letter C, or its mirror image. This is also the initial of the Latin word for hundred, *Centum*. To create a sign for 1,000, the Romans decided to use the symbol for 10 (the cross) and draw a circle around it. Then, for 500, they cut the sign for a thousand in half: ↁ. This sign would later be mistaken for the letter D. The circle dissected by a cross (1,000) evolved into various shapes, which were replaced by the M due to the Latin word *Mile* (see Fig. 16. 26 to 34): Thus we can see how, graphically, the circle dissected by a cross became a sign which was shaped like an S or an 8 on its side. In Latin, the term *Mile*

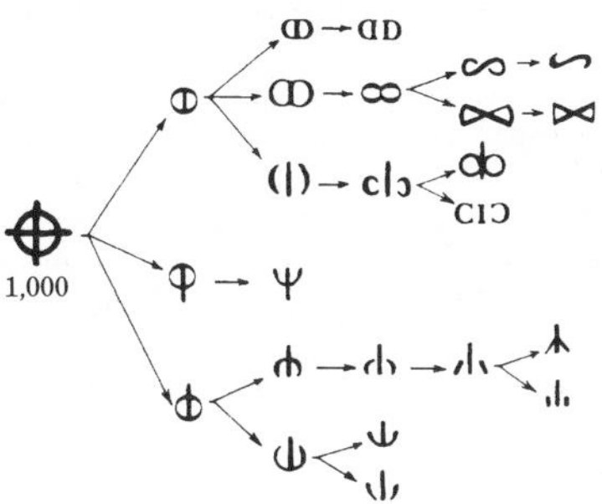


corresponded to the highest number in spoken numeration and, by extension, in everyday language, it meant "vast number" and "the incalculable". In his *Natural History* (XXXIII, 133), Pliny wrote that the Romans had no names for powers of ten superior to a hundred thousand, and so referred to a million as *decies centena milia* ("ten hundred thousand"). The snake as the sign for infinity, in its various forms, has been connected to ideas such as the sky, the universe, the axis of the world, the night of beginnings, the primordial substance, the vital principle, life, eternal life, sexual energy, spiritual energy, vestiges of the past, the seed of times to come, cyclical development and resorption, longevity, extreme fertility, the incalculable quantity, abundance, immensity, totality, absolute stability, endless movement, etc. See **High numbers (The symbolic meaning of)**.

**INFINITY AND MYSTICISM.** See **Infinity (Indian mythological representation of)** and **Serpent (Symbolism of the)**.

**INFINITY.** All confusion must be avoided between *infinity* and *indefinite*. *Indefinite* comes from the Latin *indefinitus*, signifying "vague". This word also has other possible meanings. The first is the opposite of "defined", that which is "unspecified", which remains "undetermined", like death for example: "That which is certain in death is softened a little by that which is uncertain; it is an indeterminate length of time which has something of infinity and of what we call eternity." The second meaning expresses the opposite of that which is "finite"; it is a quantity which, whilst remaining finite, is susceptible to unlimited expansion or growth. This is the meaning of indefinite progress. The third sense can be found in this extract from Descartes: "Each body can be divided into infinitely small parts. I do not know if their number is infinite or not, but certainly, to the best of our knowledge, it is indefinite" [*Oeuvres*, XI, 12]. This is the opposite of that which is infinite: here the indefinite is "that which is only infinite from a certain point of view, because we cannot see its end" [Foulquié (1982)].

On the other hand, a fourth meaning makes this word a synonym of infinity. This is *potential infinity*, illustrated by this quote from Pascal: "The eternal silence of these infinite spaces fill me with fear" [*Pensées*, 428]. This extract inspired the following commentary from Paul Valéry: "This phrase, which is so powerful and magnificent that it is one of the most famous ever to have been uttered, is a *Poem*, and by no means a thought. *Eternal* and *Infinite* are symbols of non-thought. Their value is entirely emotional. They act on our sensitivity. They provoke the peculiar sensation of the inability to imagine" [*Variété, La Pléiade*, I, 458]. Thus potential infinity is "that which, being effectively finite, has the potential for limitless growth" [Foulquié]. In terms of either potential or reality, infinity has posed one of the most serious problems to the human mind in all the history of civilisation. Confronting infinity has been a little like meeting Cerberus at the entrance to the Underworld. *There is one, final number, but it is beyond the power of mortals to reach it; this power belongs only to the gods who are the only ones who may count the stars and the firmament*. Such is the leitmotiv of both ancient and modern religions. It bears witness to humanity's constant obsession with this concept. It demonstrates not only our ability to count numbers "to the end", but also to learn the true meaning of that which conceals the rather vague notion of the unlimited: "We imagine some kind of finite range, then we disregard the limits of the range, and we have the idea of an infinite range. In this way, and perhaps in this way alone, we are able to conceive of an infinite number, an infinite duration, etc. Through this definition, or rather this analysis, we can see to what extent our notion of *infinity* is vague and imperfect; it is only really the notion of the *indefinite*, if we understand by this word a vague quantity to which no limits have been assigned, and not, as one could understand by another meaning of this word, a quantity for which there are limits, yet these limits have not been specified" [D'Alembert, *Essai sur les éléments de philosophie*, Éclairc., XIV].

This explains why comparisons are sometimes made which are reminiscent of religious metaphors and parables. The grains of sand of the desert, the drops of water of the ocean or the stars in the sky are evoked, without the realisation that such comparisons are puerile, as they only involve the domain of the finite. In everyday use, infinity is only understood by its negation. In fact, the word "infinity" derives from the Latin *infinitus*, "that which has no end", "that which never ends". It is the negation of the finite, in the sense that infinity is "that which can never be reached".

[See Blaise (1954); Bloch and von Wartburg (1968); Chantraine (1970); Du Cange (1678); Ernout and Meillet (1959); Estienne (1573); Gaffiot; GLF (1971); Littré (1971); Robert (1985)].

It is precisely this limited conception which prevented the Greek mathematicians from making progress in this domain. Historically, it was in Greece, after Pythagoras's discoveries, that the evolution of this concept began with the undisputed statement that "infinity is something which cannot be measured". The problem, according to Bertrand Russell, represented "in one form or another, the basis of nearly all the theories of space, time and of infinity which persisted from that time up until the present day".

Descartes was one of the first European scholars to establish infinity as a fundamental reality. This notion later became a perfectly precise, objective concept, presenting no basic problems such as those often conferred upon it by the profane. The symbol for infinity ($\infty$) seems to have appeared for the first time in 1655 in a list of mathematical signs compiled by the English mathematician John Wallis.

Mathematically, infinity is that which is bigger than any other quantity, and no finite number can be added to it. Fléchier compared infinity to God, as God is "infinitely powerful and thus infinitely free". Zero is the opposite of infinity: it is infinitely small, the variable quantity which is inferior to all positive numbers, however small they might be. Infinity, or the impossibility of counting all the numbers, remains a *mathematical hypothesis*; it is one of the fundamental axioms upon which contemporary mathematics is based. See **Infinity (Indian concepts of)** and **Infinity (Indian mythological representations of)**.

**INFINITY.** Term used as a synonym for "potential infinity". See **Infinity**. See also **Indian mathematics (The history of)**.

**INFINITY**. Term used as a synonym for the "indeterminable". See **Infinity** and **High numbers**. See also **Serpent (Symbolism of the)**.

**INFINITY**. Term used as a synonym of the "unlimited". See **Infinity**, **High numbers** and **High numbers (The symbolic meaning of)**. See also **Serpent (Symbolism of the)**.

**INFINITY**. Term used as a synonym of the eternity and immensity of space. See ***Ananta***, **Infinity (The Indian mythological representation of)**, **High numbers (The symbolic meaning of)** and **Infinity**.

**INFINITY**. Term used to represent the number ten to the power fourteen. See ***Ananta*** and **High numbers (The symbolic meaning of)**.

**INFINITY**. [S]. Value = 0. See **Infinity**, ***Âkâsha***, ***Ananta***, ***Vishnupada***, ***Shûnya*** and **Zero**.

**INNATE REASON**. As a symbol for a large quantity. See **High numbers (Symbolic meaning of)**.

**INNUMERABLE**. See ***Abhabâgamana***, ***Asamkhyeya***, **High numbers** and **Infinity**.

**INNUMERABLE**. Term used as the name for the number ten to the power 140. See ***Asankhyeya***.

**INSIGNIFICANCE**. See **Low numbers**, ***Shûnyatâ*** and **Zero**.

**INTERLACING**. See **Infinity (Indian mythological representation of)** and **Serpent (Symbolism of the)**.

**IRYÂ**. [S]. Value = 4. "Position". Allusion to the four principal positions of the human body (positions: lying flat on one's stomach, lying flat on one's back, standing up or sitting down). See **Four**.

**ÎSHA**. [S]. Value = 11. This is the shortened form of *Îshâna*, one of the names of *Rudra, the symbolic value of which is eleven. See **Rudra-Shiva** and **Eleven**.

**ÎSHADRISH**. [S]. Value = 3. The "eyes of Hara". See ***Îsha***, ***Haranetra*** and **Three**.

**ISHU**. [S]. Value = 5. "Arrow". See ***Shara*** and **Five**.

**ÎSHVARA**. [S]. Value = 11. "Lord of the universe", "Supreme divinity". One of the attributes of *Shiva, who is an emanation of *Rudra, whose name has the symbolic value of eleven. See **Rudra-Shiva** and **Eleven**.

**ISLAND-CONTINENT**. [S]. Value = 7. See ***Dvîpa*** and **Seven**.

**ISLAND-CONTINENTS (The four)**. See ***Chaturdvîpa***.

**ISLAND-CONTINENTS (The seven)**. See ***Sapta*** and ***Dvîpa***.

**ÎSVÎ (Calendar)**. See ***Kristâbda***.

# J

**JAGAT**. [S]. Value = 3. "Universe", "Phenomenal world". Here this word is taken in the sense of three "worlds". See ***Loka***, ***Triloka*** and Three.

**JAGAT**. [S]. Value = 14. "World". Here the word is taken in the sense of the fourteen chosen lands of the Buddhism of the Mahâyana (including **Vaikuntha*). See ***Bhuvana*** and **Fourteen**.

**JAGATÎ**. [S]. Value = 1. "Earth". See ***Prithivî*** and **One**.

**JAGATÎ**. [S]. Value = 12. In Sanskrit poetry, this is the metre which is made up of a verse of four times twelve syllables. See **Indian metric** and **Twelve**.

**JAGATÎ**. [S]. Value = 48. In Sanskrit poetry, this is the metre which is made up of a verse of four times twelve syllables. See **Indian metric**.

**JÂHNAVÎVAKTRA**. [S]. Value 1,000. The "mouths of Jâhnavî". The name Jâhnavî denotes the river Ganges (*Gangâ*), considered to be the daughter of Jâhnu. According to legend, Jâhnu drank the river because it disturbed his prayer, but the water came out of his ears. The Sanskrit name for "thousand" (**sahasra*) often means "multiplicity" and "multitude". The swampy delta of this river is divided into many hundreds of branches, and so these "mouths" came to represent the quantity *thousand* because they are so numerous. See **Thousand** and **High numbers (Symbolic meaning of)**.

**JAINA RELIGION**. See **Indian religions and philosophies**.

**JAINA**. This is the name of an Indian religious sect. This religion seems to have been founded around the sixth century BCE by a "sage" (*muni*) named Vardhamâna, better known as *Jina*. Jaina philosophy and logic is accompanied by a very strict moral doctrine, born out of several concepts including *nayavâda* (a highly developed science of the knowledge of the real from its most diverse aspects) and *syâdvâda* (which consists of a relativist vision which is meant to adjust the affirmation and negation of things to their moving reality). Nature is divided into "categories", which are classed in different orders depending on the point of view from which they are considered.

In one of these "categories", there are "principles" and "masses of beings", the most important of which are the soul, matter, the cause of movement, the cause of the halting of movement and space (**âkâsha*). Matter is of atomic structure. Each "atom" of corpororal nature is uncreated, indivisible and indestructible, whilst possessing particular tastes, smells and colours. As for time, it is considered a substance without space, yet according to Jaina philosophy, it is made up of an infinite number of "temporal atoms" (**kâlânu*). These diverse theories are accompanied by a highly developed cosmological vision of the universe (**loka*) in which the universe is represented as a man made up of three worlds, his head forming the superior world, his body the middle world and his legs the inferior world. These three worlds are surrounded by a triple atmospheric cover, made up of air, vapour and ether (**âkâsha*), beyond which is nothing but empty space (**shûnyatâ*). This universe is organised around a hollow vertical axis, inside which live all "mobile" living beings.

Each world is divided into several stages: the inferior world; the middle world, which includes our world and the island-continents (**dvîpa*, **chaturdvîpa*); and finally the superior world, situated above *Mount Meru, the mythical mountain of Hindu and Brahmanic cosmology, which is said to be the centre of the universe where the gods live. The summit, which constitutes the "chignon" of the cosmic man, is said to be occupied by liberated souls. As for the ages of the world, the Jainas accept Brahmanic classification. Thus the fifth age (the age which we are living in) would have begun in 523 BCE and be characterised by pain. This would be followed by a sixth and last "age", 21,000 years long, at the end of which the human race would undergo terrible mutations.

However, the world would not disappear, for, according to Jaina doctrine, the universe is indestructible. This is because it is infinite, in terms of both time and space. It was in order to define their vision of this impalpable universe, situated in the unlimited and the eternal, that the Jainas began their impressive numerical speculations and thus created a science which was characteristic of their way of thinking: a "science" which, by using incredibly high numbers and constantly expanding the limits of **asamkhyeya* (the "incalculable", the "impossible to count") finally allowed them to get within reach of the world of infinite numbers. [See Frédéric, *Dictionnaire* (1987)]. See ***Anuyogadvâra Sûtra***, **Names of numbers**, **High numbers** and **Infinity**.

**JALA**. [S]. Value = 4. Synonymous with **âpa*, "water". This symbolism is explained by the Brahmanic doctrine of the "elements of the revelation" (**bhûta*). According to this philosophy, the universe is the result of the interaction of five "powers" (*nritya*) personified respectively by *Brahma, *Vishnu, *Rudra, Maheshvara and *Shiva. These powers are: creation (*shrishti*), conservation (*stithi*), creative emotion (*tirobhava*), destruction (*shangara*) and rest (*anugraha*). On account of these five "powers", the universe is thus the result of the transformation and interaction of the "five elements" (**pañchabhûta*). These elements are respectively: ether (**âkâsha*), water (**âpa*), air (**vâyu*), fire (**agni*) and water (**prithivî*). Ether, the most subtle of the five elements, is considered to be the condition of all corporal extension and the receptacle of all matter which manifests itself in the form of any one of the other four elements. Ether is thus space, the "element which penetrates everything", the **shûnya*, the "void". In other words, according to this philosophy, **âkâsha* (ether) is the immobile and eternal element which is the essential condition for all manifestation, but which, by its very essence, is indescribable, and cannot be mixed with any material thing. Thus this element is not meant to participate directly with the "material order of nature", which comes from **prakriti* (the supposed original material substance of the universe).

Hence we are dealing with "natural order" which is very similar to the doctrine of the great philosophers of Ancient Greece (Pythagoras, Plato, Aristotle, etc.). This doctrine states that: the various phenomena of life can be attributed to the manifestations of the elements which determine the essence of the forces of Nature, who carries out her work of generation and destruction using these elements: *water*, *air*, *fire*, and *earth*. Each one of these elements is created by the combination of two primordial constituents: water comes from coldness and humidity, air comes from humidity and heat, fire is made by heat and dryness, and earth comes from dryness and cold. Each one of these is representative of a state, liquid, gas, igneous and solid. In each of these groups is a collection of fixed conditions of life, and the groups together form a cycle, which begins with the first element (water) and ends with the last (earth), after passing through the intermediary stages (air and fire). This gives a *quaternary order of nature*, which corresponds to both the human temperaments and to the stages of human life: winter, spring, summer, autumn; midnight to dawn, dawn to midday, midday to dusk, dusk to midnight; phlegmatic, sanguine, bilious and choleric; childhood, youth, maturity and old age; learning, blossoming, culminating, declining; etc. [Chevalier and Gheerbrant (1982)]. It is thus on this basis that water (*jala*) came to symbolise the number four in Sanskrit. This quaternary symbolism is also responsible for the fact that the value of four has often been attributed to the word for "ocean" (**sâgara*). See **Four** and ***Sâgara***.

**JALADHARAPATHA.** [S]. Value = 0. "Voyage on the water". Allusion to *Ananta, the serpent with a thousand heads, who floats on the primordial waters, or the "ocean of unconsciousness", during the space of time which separates two succesive creations of the world. This symbolism thus corresponds to the identification of infinity with zero, because Ananta is none other than the serpent of infinity and eternity. See **Zero**.

**JALADHI.** [S]. Value = 4. "Sea". See **Sâgara, Four, Ocean.**

**JALADHI** Name given to the number ten to the power fourteen (= a hundred billion). See **Names of numbers.** For an explanation of this symbolism, see **High numbers (Symbolic meaning of).**

Source: **Lîlâvatî* by Bhâskarâchârya (1150 CE).

**JALANIDHI.** [S]. Value = 4. "Sea". See ***Sâgara*, Four.** See also **Ocean.**

**JAMBUDVÎPA.** "Isle of the Jambu tree". Name in Hindu cosmology for the whole of the Indian subcontinent, which is situated (according to a characterised representation of the structure of the universe) to the south of *Mount Meru.

**JAVANESE NUMERALS (Ancient).** See **Kawi numerals.**

**JAVANESE NUMERALS (Modern).** Currently in use in the island of Java, in Bali, Madura and Lombok, as well as in the Sounda islands. The corresponding system functions according to the place-value system and possesses zero (in the form of a little circle). Apart from the numerals 0 and 5 (whose graphical origin is evident), this notation actually corresponds to a relatively recent graphical creation, the shape of the numerals having (curiously) become similar in appearance to some of the letters of the contemporary Javanese alphabet. The Javanese people formerly used a notation which derived from Brâhmî numerals, which belongs to the group of numerals known as "Pali". See Fig. 24.26 and 52. See also **Indian written numeral systems (Classification of)** and **Kawi numerals.**

**JEWEL.** [S]. Value = 3. See ***Ratna*** and **Three.**

**JEWEL.** [S]. Value = 5. See ***Ratna*** and **Five.**

**JEWEL.** [S]. Value = 8. See ***Mangala*** and **Eight.**

**JEWEL.** [S]. Value = 9. See ***Ratna*** and **Nine.**

**JEWEL.** [S]. Value = 14. See ***Ratna*** and **Fourteen.**

**JINA.** Name of the founder of the religious sect of the *Jainas.

**JINABHADRA GANI.** Indian arithmetician who lived at the end of the sixth century. His works notably include *Brihatkshetrasâmâsa*, where he gives an expression for the number 224,400,000,000 in the simplified Sanskrit system using the place-value system [see Datta and Singh (1938) p. 79]. See **Indian mathematics (The history of).**

**JVALANA.** [S]. Value = 3. "Fire". See ***Agni*** and **Three.**

**JYOTISHA.** Sanskrit name attributed to astronomy, once it was considered to be a separate discipline from arithmetic and calculation. This name, however, (which literally means "science of the stars") was long attributed to astrology. See **Indian astrology, *Ganita*** and **Indian astronomy (The history of).**

**JYOTISHAVEDÂNGA.** "Astronomical Element of Knowledge". Name of an ancient text on astrology, notably concerning the determination of the exact dates of the sacrifices of the Brahman cult [see Billard (1971)]. See **Jyotisha, Indian astrology** and **Indian astronomy (The history of).**

# K

**KÂCHCHÂYANA.** Grammarian from Sri Lanka who is believed to have written the *Vyâkarana*, a Pali grammar divided into eight parts. He probably lived during the eleventh century CE. Here is a list of the principal names of numbers mentioned in this work: **Koti* ( = $10^7$), **Pakoti* (= $10^{14}$), **Kotippakoti* (= $10^{21}$), **Nahuta* (= $10^{28}$), **Ninnahuta* (= $10^{35}$), **Akkhobhini* (= $10^{42}$), **Bindu* (= $10^{49}$), **Abbuda* (= $10^{56}$), **Nirabbuda* (= $10^{63}$), **Ahaha* (= $10^{70}$), **Ababa* (=$10^{77}$), **Atata* (= $10^{84}$), **Sogandhika* (= $10^{91}$), **Uppala* (= $10^{98}$), **Kumuda* (= $10^{105}$), **Pundarîka* (= $10^{112}$), **Paduma* (= $10^{119}$), **Kathâna* (= $10^{126}$), **Mahâkathâna* (= $10^{133}$), **Asankhyeya* (= $10^{140}$).

See **Names of numbers** and **High numbers**

Source: *Vyâkarana* by Kâchchâyana [see JA, 6th Series, XVII, 1871, p. 411, line 51–52].

**KAÎTHÎ NUMERALS.** Signs derived from *Brâhmî numerals, through the intermediary of Shunga, Shaka, Kushâna, Ândhra, Gupta, Nâgarî, Kutila and Bengali numerals. Currently in use in Bihar state, in the east of India, and sometimes used in Gujarat state. The corresponding system functions according to the place-value system and possesses zero (in the form of a little circle). See also Fig. 24.9, 52 and 61 to 69. See **Indian written numeral systems (Classification of).**

**KAKUBH.** [S]. Value = 10. "Horizon". See ***Dish*** and **Ten.**

**KÂLA.** [S]. Value = 3. "Time". In Brahman mythology, time is personified by the terrible Kâla, Lord of Creation and Destruction. He is often identified as Shiva holding his Trident (**trishûlâ*), which symbolises the three aspects of the revelation (creation, preservation, destruction), as well as the three primordial properties (**guna*) and the three states of consciousness. Here, the word is synonymous with **trikâla*, "three times". See ***Guna, Shûla, Triguna*** and **Three.**

**KALACHURI (Calendar).** See ***Chhedi.***

**KALAMBA.** [S]. Value = 5. "Arrow". See ***Shara*** and **Five.**

**KÂLÂNU.** "Temporal atom". In *Jaina philosophy, time (**kâla*) is made up of an infinite number of temporal atoms (atom = **anu*).

**KALÂSAVARNA.** Word used in arithmetic to denote "fundamental operations" carried out on fractions. See ***Parikarma.***

**KALIYUGA (Calendar).** Calendar of fictitious times, which is sometimes referred to in Hindu religious texts and Indian astronomical texts. It begins on the 18 February of the year 3101 BCE. Characteristically, the beginning of this calendar is traditionally related to a theoretical starting point of celestial revolutions corresponding to a supposed general conjunction in average longitude with the starting point of the sidereal longitudes of the sun, the moon and the planets (the ascending apogees and node of the moon being respectively at 90° and 180° of these longitudes). To find the corresponding date in our calendar, simply subtract 3,101 from a date in the *Kaliyuga* calendar. See ***Kaliyuga*, Indian calendars** and ***Yuga* (Astronomical speculation).**

**KALIYUGA.** Name of the last of the four cosmic calendars which make up a **mahâyuga*. This cycle, said to be 432,000 human years long, is the "iron age", during which living things only live for a quarter of their existence and the forces of evil triumph over good: we are living in this age now, and it is meant to end with a *pralaya* (destruction by fire and water). See ***Yuga* (Definition), *Yuga* (Systems of calculating)** and ***Yuga* (Cosmogonical speculations about).**

**KALPA.** Unit of cosmic time which, according to Indian speculations, corresponds to the length of 1,000 **mahâyugas*. Thus one Kalpa corresponds to 4,320,000,000 human years. See ***Yuga* (Definition).**

**KALPA.** [Astronomy]. According to Brahmagupta (628 CE), the *kalpa* cycle, or period of 4,320,000,000 years, is delimited by two perfect conjunctions in real longitude of the totality of elements, each one accompanied by a total eclipse of the sun at exactly six o'clock in *Ujjayinî. See ***Kalpa*** (first entry above) and ***Yuga* (Astronomical speculations about).**

**KALPAS (Cosmogonical speculations about).** In Buddhist cosmogony, the term *kalpa* denotes an infinite length of time. The *kalpa* is made up of four periods: the creation of worlds, the lifespan of existing worlds, the destruction of worlds and the duration of the existence of chaos. During the period of creation the different universes are formed with their living beings. The second period sees the appearance of the sun and the moon, the differentiation between the sexes and the development of social life. During the phase of destruction, fire, water and wind destroy everything apart from the fourth *dhyâna*. Chaos represents total annihilation. These four phases make up one "big" *kalpa* (**mahâkalpa*); Each one of them is made up of twenty "little" *kalpas*, which themselves are broken down into fire, bronze, silver and golden ages. During the entire creation phase of a "little" *kalpa*, the life expectancy of humans increases by one year per century until it reaches 84,000 years. In a parallel fashion, the human body grows to a height of 84,000 feet. During the "little" *kalpa's* period of disappearance, which is made up of successive phases of plague, war and famine, human life is shortened to ten years and the human body returns to the height of one foot.

[This article is taken from the *Dictionnaire de la sagesse orientale*, Friedrichs, Fischer-Schreiber, Erhard and Diener (1989)]. See ***Kalpa*** (first entry) and **Day of Brahma**.

**KÂMA**. [S]. Value = 13. Name of the Hindu divinity of Cosmic Desire and Carnal Love whose action decides the laws of the reincarnation of living beings (**samsâra*). Kâma presides over the thirteenth lunar day. See **Thirteen** and ***Pañchabâna***.

**KAMÂLAKARA**. Indian astronomer of the seventeenth century CE. His works notably include *Siddhântatattvaviveka*, in which the place-value system with Sanskrit numerical symbols is frequently used [see Dvivedi (1935)]. See **Numerical symbols**, **Numerical symbols (Principle of the numeration of)**, and **Indian mathematics (The history of)**.

**KANKARA**. Name given to the number ten to the power thirteen (= ten billion). See **Names of numbers** and **High numbers**.

**KANNARA NUMERALS**. Signs derived from *Brâhmî numerals, through the intermediary of Shunga, Shaka, Kushâna, Ândhra, Pallava, Chalukya, Ganga, Valabhî and Bhattiprolu numerals. Currently used by the Dravidians of Karnataka state and part of Andhra Pradesh. They are also called Kannada (or even Karnata) numerals. The corresponding system today uses the place-value system and zero (in the from of a little circle). For ancient numerals, see Fig. 24.48. For modern numerals, see Fig. 24.21. See also Fig. 24.52 and 24.61 to 69. See **Indian written numeral systems (Classification of)**.

**KARA**. [S]. Value = 2. "Hand". This is because of the symmetry of the two hands. See **Two**.

**KARAHU**. Name given to the number ten to the power thirty-three. See **Names of numbers** and **High numbers**.

Source: **Lalitavistara Sûtra* (before 308 CE).

**KARANA**. Name of the astronomical formula employing, for example, in the workings of real longitudes, the interpolation – generally linear – of tabulated values. **See Indian astronomy (The history of)** and ***Yuga*** **(Astronomical speculation on)**.

**KARANAPADDHATI**. See ***Putumanasomayâjin***.

**KARANÎYA**. [S]. Value = 5. "That which must be done". This refers to the five major observances of *Jaina religion, which constitute the basic rules of their philosophy: not to harm living beings (*ahimsâ*); not to be false (*sunrita*); not to steal (*asteya*); carnal discipline (*brahmâchârya*); and detachment from earthly possessions (*aparigraha*). See **Five**.

**KARNATA NUMERALS**. See **Kannara numerals**.

**KARNIKÂCHALA**. One of the names of *Mount Meru. See ***Adri***, ***Dvîpa***, ***Pûrna***, ***Pâtâla***, ***Sâgara***, ***Pushkara***, ***Pâvana*** and ***Vâyu***.

**KÂRTTIKEYA**. Hindu divinity of war and the planet Mars, son of Shiva, often likened to *Kumâra.

**KÂRTTIKEYÂSYA**. [S]. Value = 6. "Faces of *Kârttikeya". Allusion to the six heads of this divinity. See **Six**.

**KATAPAYA (Spoken numeration)**. See **Katapayâdi numeration**.

**KATAPAYÂDI NUMERATION**. Method of writing numbers using the letters of the Indian alphabet. In this system, the numerical attribution of of syllables corresponds to the following rule, according to the regular order of succession of the letters of the Indian alphabet (see Fig. 24. 56): the first nine letters (*ka*, *kha*, *ga*, *gha*, *na*, *cha*, *chha*, *ja* and *jha*) represent the numbers 1 to 9, whilst the tenth (*ña*) corresponds to zero; the following nine letters (*ta*, *tha*, *da*, *dha*, *na*, *ta*, *tha*, *da*, *dha*) also receive the values 1 to 9, whilst the following letter (*na*) has the value of 0; the next five (*pa*, *pha*, *ba*, *bha*, *ma*) represent the first five units; and the last eight (*ya*, *ra*, *la*, *va*, *sha*, *sha*, *sa* and *ha*) represent the numbers 1 to 8.

Thus each simple unit is represented by two, three or four different letters: numeral 1 by one of the letters *ka*, *ta*, *pa* or *ya* (hence *katapaya* is the name of the system); 2 by *kha*, *tha*, *pha* or *ra*; 3 by *ga*, *da*, *ba*, *la*; 4 by *gha*, *dha*, *bha* or *va*; 5 by *na*, *na*, *ma* or *sha*; 6 by *cha*, *ta* or *sha*; 7 by *chha*, *tha* or *sa*; 8 by *ja*, *da* or *ha*; 9 by *jha* or *dha*; and 0 by *ña* or *na*. This system is infinitely simpler than Âryabhata's.

The complete key is given in the following lines, which are an extract from *Sadratnamâlâ* by Shankaravarman: *Nañâvachashacha shûyâni Samkhyâ katapayâdayah*

*Mishre tûpânta hal samkhya Na cha chintyohalasvarah*

Translation: "[The letter] *na* and *a*, as well as the vowels, are zero. [The letters] starting with *ka*, *ta*, *pa*, *ya*, represent the numbers [from 1 to 9]. When two consonants are joined, only the last one corresponds to a number. And a consonant which is not joined to a vowel is insignificant." [See EI, VI, p. 121; Datta and Singh, p. 70]. In other words, in this system, the vowels and the consonants which are not vocalised have no numerical value; and groups such as *ksha*, *tva*, *ktya*, etc., often considered as unitary in Indian alphabets, receive respectively the same value as the letters *sha*, *va*, *ya*, etc. The letters *ña* and *na*, represent zero. Thus the vocalised consonants are the only "numerals" in the system, their numerical value being entirely independent of the vocalisations in question. This means that, unlike Âryabhata's system, there is no difference between syllables such as *sa*, *si*, *su*, *se*, *so*, *sai*, etc. In fact, this system constitutes a simplification of Âryabhata's alphabetical numeration. See **Âryabhata** and **Âryabhata's numeration**.

Historically, the first author who is known to have used this system employing the name of *katapayâdi* is the astronomer Shankaranârâyana, author of a work entitled *Laghubhahâskarîyavivarana* written in 869 CE. This date is given by the author himself, and is expressed as the **Shaka* year 791, which is 791 + 78 = 869 CE.

However, the latter did not invent *katapayâdi*, because the system had already appeared, under the name of *varnasamjñâ*, "from syllables", in *Grahachâranibandhana* by the astronomer Haridatta, for which there is overwhelming evidence to suggest that he was the inventor of this system. First, there is no mention is made of the system by his predecessors; secondly, in his work (where he makes frequent reference to Âryabhata), he takes the trouble to give all the details (like the inventor of a new system who feels obliged to explain it to readers who are used to using a different method); finally, it is Haridatta who is the first and last person to explain the system, which suggests that afterwards it became common knowledge. [Personal communication of Billard]. According to a tradition in Kerala, Haridatta wrote his text in 684 CE [see Sarma (1954), p. v]. However, this date does not seem to correspond to a significant piece of evidence found in the work of astronomer Shankaranârâyana, where he is paying homage

| | LETTERS USED | | | |
|---|---|---|---|---|
| for the numeral 1 | क *ka* | ट *ta* | प *pa* | य *ya* |
| for the numeral 2 | ख *kha* | ठ *tha* | फ *pha* | र *ra* |
| for the numeral 3 | ग *ga* | ड *da* | ब *ba* | ल *la* |
| for the numeral 4 | घ *gha* | ढ *dha* | भ *bha* | व *va* |
| for the numeral 5 | ङ *ṅa* | ण *ṇa* | म *ma* | श *sha* |
| for the numeral 6 | च *cha* | त *ta* | | ष *sha* |
| for the numeral 7 | छ *chha* | थ *tha* | | स *sa* |
| for the numeral 8 | ज *ja* | द *da* | | ह *ha* |
| for the numeral 9 | झ *jha* | ध *dha* | | |
| for the numeral 0 | ञ *ña* | न *na* | | |

FIG. 24D.7 *Letter-numerals of the "Katapayâdi system". Ref. : Datta and Singh (1938); Jacquet (1835); Pihan (1860); Renou and Filliozat (1953); Sarma (1954)*

to his illustrious predecessors, and quotes their respective names, using the word **kramâd*, which means "in the order":

1. *Âryabhata* [c. 510 CE]
2. *Varâhamihira* [c. 575 CE]
3. *Bhâskara* [c. 629 CE]
4. *Govindasvâmin* [c. 830 CE]
5. *Haridatta*.

Thus Haridatta is placed after Govindasvâmin, of whom Shankaranârâyana was a disciple. Such a list is very rare for an Indian scholar; chronology was not generally of much interest to them. It seems even more remarkable in light of the fact that Indian astronomical texts are usually very sparing with historical facts, and it is very rare to find a reference to another text. If mention is made of earlier authors, the list is usually in order, to aid the rhythm of the versification. This is the only known example of such a list accompanied by a chronological indication. In short, if Haridatta's work was written before 869 (the date of Shankaranârâyana's text), it probably dates back to c. 850 CE. (Personal communication of Billard.) This means that *katapayâdi* numeration was not created until the middle of the ninth century, three centuries after Âryabhata. Through abandoning the method of successively vocalising the consonants of the Indian alphabet, and replacing each value which was equal to or higher than ten with a zero or one of the nine numerals, the inventor of the *katapayâdi* system transformed Âryabhata's system into a place-value system equipped with a zero.

The proof of this is in the following mention in an anonymous text from the tenth century, where there is frequent use of the *katapayâdi* notation: *vibhâvonashakâbdam . . .*

"The *Shaka* date decreased by 444 . . ." Ref.: *Grahachâranibandhanasamgraha*, line 17; Billard (1971), p. 142.

This mention contains the expression *vibhâvona* (= *vibhâva* + *ûna*) which means "444 (= *vibhâva*) decreased by". Bearing in mind the principle of this notation, where the value of a consonant is independent of its vocalisation, and where the numbers are expressed in ascending order starting with the smallest units, the number 444 is written as follows:

*vibhâva* (= *va.bha.va*).

According to the values of the numeral letters in the *katapayâdi* system, this gives the following (Fig. D. 7):

(*va*) (*bhâ*) (*va*) $= 4 + 4 \times 10 + 4 \times 10^2$
4 4 4 = 444.

Thus the numeral letters are combined and are never modified by vowels; these can be inserted wherever necessary, as they have no numerical value. As for the principle of the notation, which is the rule of position applied to any of the nine letter-units and the two letter-zeros, it follows the ascending powers of ten starting with the smallest unit, as it does with the numerical symbols. Here is another example found in an astronomical table of Haridatta's *Grahachâranibandhana* (II, 14), giving a trigonometric function for Saturn [see Sarma (1954), p. 12]:

| *dhanâdhya* | *dha.nâ.dhya* |
|---|---|
| = *dhîrâdhya* | = *dhî.râ.dhya* |
| = *dha.nâ.ya* | = *dha.ra.ya* |
| 9 0 1 | 9 2 1 |
| = 109 | = 129 |

This is more proof that Âryabhata was fully aware of zero and the place-value system, but by confining himself as he did to his system of vocalisation, he made it impossible for his numeration to be positional. (See **Ankânâm vâmato gatih**.)

It is surprising to note the numbers of letters that could be used to record the same numeral. In fact, this system, like the notation which inspired it, offered many possibilities to the mnemonics of numbers. Moreover, it was perfectly capable of meeting the needs of the rules of versification or prosody, with the advantage of being especially useful when reproducing abundant tables of trigonometric functions in a much shorter form than the system of *numerical symbols. Added to the possibility of expressing a given numeral with many different letters was the ability to vocalise these letters without changing the values they expressed. Thus it was always possible to find several intelligible words to express a number. This is doubtless the reason why this system came to be used, in various forms, in southern India (the notation in this case being applied to letters of the Grantha, Tulu, Telugu (etc.) alphabets).

**KÂTHAKA SAMHITÂ.** Text derived from the *Yajurveda* "black". It figures amongst the texts of Vedic literature. Passed down through oral transmission since ancient times, it only found its definitive form at the beginning of Christianity. See ***Veda***. Here is a list of the main names of numbers mentioned in this text:

**Eka* (= 1), **Dasha* (= 10), **Sata* (= $10^2$), **Sahasra* (= $10^3$), **Ayuta* (= $10^4$), **Prayuta* (= $10^5$), **Niyuta* (= $10^6$), **Arbuda* (= $10^7$), **Nyarbuda* (= $10^8$), **Samudra* (= $10^9$), **Vâdava* (= $10^9$), **Madhya* (= $10^{10}$), **Anta* (= $10^{11}$), **Parârdha* (=$10^{12}$).

See **Names of numbers** and **High numbers**.

Ref.: *Kâthaka Samhitâ*, XVII, 10 [see Datta and Singh (1938), p. 10].

**KATHÂNA.** Name given to the number ten to the power 119 See **Names of numbers** and **High numbers.**

Ref.: **Vyâkarana* (Pali grammar) by Kâchchâyana (c. eleventh century).

| | | | | | |
|---|---|---|---|---|---|
| Gutturals | क *ka* | ख *kha* | ग *ga* | घ *gha* | ङ *ṅa* |
| Âryabhata's system[1] | 1 | 2 | 3 | 4 | 5 |
| Katapayâdi system[2] | 1 | 2 | 3 | 4 | 5 |
| Palatals | च *cha* | छ *chha* | ज *ja* | झ *jha* | ञ *ña* |
| Âryabhata's system | 6 | 7 | 8 | 9 | 10 |
| Katapayâdi system | 6 | 7 | 8 | 9 | 0 |
| Cerebrals | ट *ta* | ठ *tha* | ड *da* | ढ *dha* | ण *ṇa* |
| Âryabhata's system | 11 | 12 | 13 | 14 | 15 |
| Katapayâdi system | 1 | 2 | 3 | 4 | 5 |
| Dentals | त *ta* | थ *tha* | द *da* | ध *dha* | न *na* |
| Âryabhata's system | 16 | 17 | 18 | 19 | 20 |
| Katapayâdi system | 6 | 7 | 8 | 9 | 0 |
| Labials | प *pa* | फ *pha* | ब *ba* | भ *bha* | म *ma* |
| Âryabhata's system | 21 | 22 | 23 | 24 | 25 |
| Katapayâdi system | 1 | 2 | 3 | 4 | 5 |
| Semivowels | य *ya* | र *ra* | ल *la* | व *va* | |
| Âryabhata's system | 30 | 40 | 50 | 60 | |
| Katapayâdi system | 1 | 2 | 3 | 4 | |
| Sibilants | श *sha* | ष *sha* | स *sa* | | |
| Âryabhata's system | 70 | 80 | 90 | | |
| Katapayâdi system | 5 | 6 | 7 | | |
| Aspirates | ह *ha* | | | | |
| Âryabhata's system | 100 | | | | |
| Katapayâdi system | 8 | | | | |

1. See Fig. D.2, p. 448
2. See Fig. D.7, p. 474

FIG. 24D.8

**KAVACHA.** Literally "Charm, armour". This is the name for Tantric talismans and amulets. See **Numeral alphabet, magic, mysticism and divination.**

**KAWI NUMERALS.** Signs derived from *Brâhmî numerals, through the intermediary of Shunga, Shaka, Kushâna, Ândhra, Pallava, Chalukya, Ganga, Valabhî, "Pali" and Vatteluttu numerals. Formerly used (since the seventh century CE) in Java and Borneo. These are the numerals of Old Javanese writing. The corresponding system uses the place-value system and zero (in the form of a little circle). See Fig. 24.50, 52, 61 to 69 and 80. See also **Indian written numeral systems (Classification of).**

**KÂYA.** [S]. Value = 6. "Body". Allusion to the **trikâya*, the "three bodies" that a Buddha can assume simultaneously, and which are often associated with the "three spheres" of Buddhas' existence. Thus the symbolic sum: 3 + 3 = 6. See **Six.**

**KESHAVA.** [S]. Value = 9. This concerns one of the epithets of *Vishnu (and *Krishna). The symbolism is due to the fact that *keshava* is another name for the month of *mârgâshîrsha*, the ninth month of the **chaitrâdi* year. See **Nine.**

**KHA.** [S]. Value = 0. Word meaning "space". This symbolism is explained by the fact that space is nothing but a "void". See ***Shûnya*** and **Zero.**

**KHACHHEDA.** Sanskrit term used to denote infinity. Literally "divided by zero" (from **kha*, "space" as a symbol for zero, and *chheda*, "the act of breaking into many parts", "division"). Thus it is the "quantity whose denominator is zero". The term is used notably in this sense by *Brahmagupta in his *Brahmasphutasiddânta* (628 CE). See ***Chhedana***, **Infinity** (entries beginning with), **Zero** and **Indian mathematics (The history of).**

**KHAHÂRA.** Sanskrit word for infinity. Literally "division by zero". Notably used by *Bhâskarâchârya. See ***Khachheda.***

**KHAMBA.** Name given to the number ten to the power thirteen (= ten billion). See **Names of numbers** and **High numbers.**

Source: **Lalitavistara Sûtra* (before 308 CE).

**KHAROSHTHÎ ALPHABET.** See Fig. 24. 28.

**KHAROSHTHÎ NUMERALS.** Numerals derived from the numerical notations of western Semitic civilisations. This is attested notably in the edicts of Asoka written in Aramaean Indian. The corresponding system does not use the place-value system, nor does it possess zero. See **Indian written numeral systems (Classification of).** See Fig. 24.54.

**KHARVA.** Name given to the number ten to the power ten (ten thousand million). See **Names of numbers** and **High numbers.**

Sources: **Kitab fi tahqiq i ma li'l hind* by al-Biruni (c. 1030 CE); **Lîlâvatî* by Bhâskarâchârya (1150 CE); **Ganitakaumudî* by Nârâyana (1350 CE); **Trishatikâ* by Shrîdharâchârya (date unknown).

**KHARVA.** Name given to the number ten to the power twelve (= one billion). See **Names of numbers** and **High numbers.**

Source: **Ganitasârasamgraha* by Mahâvîrâchârya (850 CE).

**KHARVA.** Name given to the number ten to the power thirty-nine. See **Names of numbers** and **High numbers.**

Source: **Râmâyana* by Vâlmîki (in the first centuries of the Common Era).

**KHMER NUMERALS.** For modern numerals, see **Thai-Khmer numerals.** For ancient numerals, see **Old Khmer numerals.**

**KHUDAWADI NUMERALS.** Signs derived from *Brâhmî numerals, through the intermediary of Shunga, Shaka, Kushâna, Ândhra, Gupta and Shâradâ numerals, and constituting a slight variation of Sindhi numerals. Once used by the merchants of Hyderabad (a town of Sind, built on the delta of the Indus, to the east of Karachi, not to be confused with the other Hyderabad, capital of Andhra Pradesh). The corresponding system functions according to the place-value system and possesses zero (in the form of a little circle). See **Indian written numeral systems (Classification of)** and Fig. 24.6, 52 and 61 to 69.

**KING.** [S]. Value = 16. See ***Bhûpa***, ***Nripa*** and **Sixteen.**

**KITAB FI TAHQIQ I MA LI'L HIND.** Arabic work by al-Biruni, which constitutes one of the most important pieces of evidence about Indian civilisation at the beginning of the eleventh century. See **al-Biruni.**

**KOLLAM (Calendar).** Beginning in 825 CE, created by the sovereign of the town of the same name situated in Kerala near to Travancore, on the Malabar coast. To find the corresponding date in the Common Era, add 825 to a date expressed in Kollam years. This calendar is also called *Parashurâma*. It is rarely used. See **Indian calendars.**

**KOTI.** Name given to the number ten to the power seven (= ten million). See **Names of numbers** and **High numbers.**

Sources: **Râmâyana* by Vâlmîki (in the first centuries CE); **Lalitavistara Sûtra* (before 308 CE); **Âryabhatîya* (510 CE); **Ganitasârasamgraha* by Mahâvîrâchârya (850 CE); **Kitab fi tahqiq i ma li'l hind* by al-Biruni (c. 1030 CE); **Vyâkarana* (Pali grammar) by Kâchchâyana (eleventh century CE); **Lîlâvatî* by Bhâskarâchârya (1150 CE); **Ganitakaumudî* by Nârâyana (1350 CE); **Trishtikâ* by Shrîdharâchârya (date unknown).

**KOTIPPAKOTI.** Name given to the number ten to the power twenty-one (= quintillion). See **Names of numbers** and **High numbers.**

Source: **Vyâkarana* (Pali grammar) by Kâchchâyana (eleventh century CE).

**KRAMÂD (KRAMÂT).** Word meaning "in the order". See ***Sthâna***, ***Sthânakramâd*** and ***Ankakramena.***

**KRISHÂNU.** [S]. Value = 3. "Fire". See ***Agni*** and **Three.**

**KRISHNA.** See ***Avatâra.***

**KRISTÂBDA (Calendar).** Name given to the Christian calendar. It is also referred to as *îsvî* or *imrâjî*. See **Indian calendars.**

**KRITA.** [S]. Value = 4. The name of the first of four cosmic cycles (**kritayuga*) which make up a **chaturyuga* (or **mahâyuga*) in Brahman cosmogony. The symbolism is not due to the fact that the *kritayuga* was the "first age" of the world, but because it inaugurated a new *chaturyuga*. Thus it began a new cosmic cycle composed of four periods corresponding to the life of a universe. See ***Yuga*** and **Four.**

**KRITAYUGA.** Name of the first of the four cosmic eras which make up a **mahâyuga* (or **chaturyuga*). This cycle, said to last 1,728,000 human years, is regarded as the "golden age" during which humans have an extremely long life and everything is perfect. See ***Yuga.***

**KRITI.** [S]. Value = 20. In Sanskrit poetry, this is a metre of four times twenty syllables. See **Indian metric** and **Twenty.**

**KSHAPESHVARA.** [S]. Value = 1. "Moon". See ***Abja*** and **One.**

**KSHATRAPA NUMERALS.** Signs derived from *Brâhmî numerals, through the intermediary of Shunga, Shaka and Kushâna numerals. Contemporaries of the western Satraps (second to fourth century CE). The corresponding system did not function according to the place-value system and moreover did not possess zero. See **Indian written numeral systems (Classification of).** See also Fig. 24.35, 52, 24.61 to 69 and 70.

**KSHAUNÎ.** [S]. Value = 1. "Earth". See ***Prithivî*** and **One.**

**KSHEMÂ.** [S]. Value = 1. "Earth". See ***Prithivî*** and **One.**

**KSHETRAGANITA.** Term used in early times meaning geometry. See ***Ganita.***

**KSHITI.** Literally "earth". Name given to the number ten to the power twenty (= a hundred quadrillions). See **Names of numbers.** For an explanation of this symbolism, see **High numbers (Symbolic meaning of).**

Source: **Ganitasârasamgraha* by Mahâvîrâchârya (850 CE).

**KSHITI.** [S]. Value = 1. "Earth". See ***Prithivî*** and **One.**

**KSHOBHA.** Name given to the number ten to the power twenty-two (= ten quintillions). See **Names of numbers** and **High numbers.**

Source: **Ganitasârasamgraha* by Mahâvîrâchârya (850 CE).

**KSHOBHYA.** Literally "Movement". Name given to the number ten to the power seventeen. This name might have been attributed to such a high number because of the "endless movement" of the waves, since "ocean" (**samudra*, **jaladhi*) was also sometimes used to represent large quantities. See **Names of numbers** and **High numbers.**

Source: **Lalitavistara Sûtra* (before 308 CE).

**KSHONI.** Literally "earth". Name given to the number ten to the power sixteen. See **Names of numbers.** For an explanation of this symbolism, see **High numbers (Symbolic meaning of).**

Source: **Ganitasârasamgraha* by Mahâvîrâchârya (850 CE).

**KSHONI.** [S]. Value = 1. "Earth". See ***Prithivî*** and **One.**

**KU.** [S]. Value = 1. "Earth". See ***Prithivî*** and **One.**

**KUMÂRA.** See ***Kumârâsya***, ***Kumâravadana*** and ***Kârttikeya.***

**KUMÂRÂSYA.** [S]. Value = 6. "Faces of *Kumâra". Allusion to the six heads of *Kârttikeya. See ***Kumâra*** and **Six.**

**KUMÂRAVADANA.** [S]. Value = 6. "Faces of *Kumâra". Allusion to the six heads of *Kârttikeya. See ***Kumâra*** and **Six.**

**KUMUD.** Literally "(pink-white) lotus". Name given to the number ten to the power twenty-one (= quintillion). See **Names of numbers.** For an explanation of this symbolism, see **High numbers (Symbolic meaning of).**

Source: **Lalitavistara Sûtra* (before 308 CE).

**KUMUDA**. Literally "pink-white lotus". Name given to the number ten to the power 105. See **Names of numbers**. For an explanation of this symbolism, see **High numbers (Symbolic meaning of)**.

Source: **Vyâkarana* (Pali grammar) by Kâchchâyana (eleventh century CE).

**KUÑJARA**. [S]. Value = 8. "Elephant". See ***Diggaja*** and ***Eight***.

**KUSHÂNA NUMERALS**. Signs derived from *Brâhmî numerals, through the intermediary of Shunga and Shaka numerals. Contemporaries of the Kushâna dynasty (first to second century CE). The corresponding system did not function according to the place-value system and moreover did not possess zero. See **Indian written numeral systems (Classification of)**. See also Fig. 24.33, 52, 24.61 to 69 and 70.

**KUTILA NUMERALS**. Signs derived from *Brâhmî numerals, through the intermediary of Shunga, Shaka, Kushâna, Ândhra, Gupta and Nâgarî numerals. Formerly used in Bengal and Assam. The corresponding system was based on the place-value system and possessed zero (in the form of a little circle). These numerals were the ancestors of Bengali, Oriyâ, Gujarâtî, Kaîthî, Maithilî and Manipuri numerals. See **Indian written numeral systems (Classification of)**. See Fig. 24.52 and 24.61 to 69.

**KUTTAKAGANITA**. In algebra, the name given to the part related to the analysis of indeterminate equations of the first degree. See **Indian mathematics (The history of)**.

# L

**LABDHA**. Term used in arithmetic to denote the quotient of a division. Synonym: *labdhi*. See ***Bhâgahâra***, ***Chhedana*** and ***Shesha***.

**LAGHUBHÂSKARÎYAVIVARANA**. See **Shankaranârâyana**.

**LAKH**. Name given to the number ten to the power five (= a hundred thousand). See **Names of numbers** and **High numbers**.

Source: **Lalitavistara Sûtra* (before 308 CE).

**LAKKHA**. Name given to the number ten to the power five (= a hundred thousand). See **Names of numbers** and **High numbers**.

Source: **Vyâkarana* (Pali grammar) by Kâchchâyana (eleventh century CE).

**LAKSHA**. Name given to the number ten to the power five (= a hundred thousand). See **Names of numbers** and **High numbers**.

Source: **Ganitasârasamgraha* by Mahâvîrâchârya (850 CE); **Kitab fi tahqiq i ma li'l hind* by al-Biruni (c. 1030 CE); **Lîlâvatî* by Bhâskarâchârya (1150 CE); **Ganitakaumudî* by Nârâyana (1350 CE); **Trishtikâ* by Shrîdharâchârya (date unknown).

**LAKSHAMANA (Calendar)**. This calendar begins in the year 1118 CE. To find the corresponding date in the Common Era, add 1118 to the date expressed in the *Lakshamana* calendar. Formerly used in the region of Mithila (north of Bihar). See **Indian Calendars**.

**LALITAVISTARA SÛTRA**. "Development of games'. Sanskrit text on the Buddhism of the Mahâyâna, written in verse and prose, about the life of Buddha, as he is said to have recounted it to his own disciples, where there is constant reference to numbers of gigantic proportions. This text is in fact a relatively recent compilation of ancient stories and legends. It is clearly later than the **Vâjasaneyî Samhitâ* (written at the start of the Common Era) but not later than the beginning of the fourth century, because the *Lalitavistara Sûtra* was translated into Chinese by Dharmarâksha in the year 308 CE. Here is a list of some of the names of high numbers mentioned in the text:

*Lakh* (= $10^5$), **Koti* (=$10^7$), **Nahut* (= $10^9$), **Ninnahut* (= $10^{11}$), **Khamba* (= $10^{13}$), **Viskhamba* (= $10^{15}$), **Abab* (= $10^{17}$), **Attata* (= $10^{19}$), **Kumud* (= $10^{21}$), **Gundhika* (= $10^{23}$), **Utpala* (= $10^{25}$), **Pundarîka* (= $10^{27}$), **Paduma* (= $10^{29}$).

Here is another list of high numbers mentioned in the text (legend of Buddha):

**Ayuta* (= $10^9$), **Niyuta* (= $10^{11}$), **Kankara* (= $10^{13}$), **Vivara* (= $10^{15}$), **Kshobhya* (= $10^{17}$), **Vivaha* (= $10^{19}$), **Utsanga* (= $10^{21}$), **Bahula* (= $10^{23}$), **Nâgabala* (= $10^{25}$), **Titilambha* (= $10^{27}$), **Vyavasthânaprajñapati* (= $10^{29}$), **Hetuhila* (= $10^{31}$), **Karahu* (= $10^{33}$), **Hetvindriya* (= $10^{35}$), **Samâptalambha* (= $10^{37}$), **Gananâgati* (= $10^{39}$), **Niravadya* (= $10^{41}$), **Mudrâbala* (= $10^{43}$), **Sarvabala* (= $10^{45}$), **Visamjñagati* (= $10^{47}$), **Sarvajña* (= $10^{49}$), **Vibhutangamâ* (= $10^{51}$), **Tallakshana* (= $10^{53}$), **Dhvajâgravati* (= $10^{99}$), **Dhvajâgranishâmani* (= $10^{145}$), etc.

See **Names of numbers** and **High numbers**. [See Lal Mitra (1877); Datta and Singh (1938), pp. 10–11; Woepcke (1863)].

**LALLA**. Indian astronomer who lived in the ninth century CE. His works notably include an interesting astronomical text entitled *Shishyadhîvriddhidatantra*, in which there is abundant usage of the place-value system recorded by means of *Sanskrit numerical symbols [see Billard (1971), p. 10]. See **Numerical symbols (Principle of the numeration of)**, and **Indian mathematics (The history of)**.

**LAUKIKAKÂLA (Calendar)**. See **Laukikasamvat**.

**LAUKIKASAMVAT (Calendar)**. Beginning in 3076 BCE, this calendar was formerly used in Punjab and Kashmir. To find the corresponding date in the Common Era, take away 3076 from a date expressed in the *Laukikasamvat* calendar. This calendar also goes by the names of *Laukikakâla*, *Lokakâla*, *Saptarishikâl*a, etc. See **Indian calendars**.

**LEGEND OF BUDDHA**. See **Buddha (Legend of)** and ***Lalitavistara Sûtra***.

**LÎLÂVATÎ**. "The (female) player". Name of a mathematical work from the twelfth century CE written in a highly poetic style. See **Bhâskarâchârya** and **Indian mathematics (The history of)**.

**LINGA (LINGAM)**. Literally "sign". Erected stone, in the shape of a prism or cylinder, phallic in appearance, which, in Hinduism, represents the universe and fundamental nature, complement of the **yoni*, the "feminine vulva", which is symbolised by a stone lying on its side and represents manifest energy [see Frédéric, *Dictionnaire* (1987)].

**LOCHANA**. [S]. Value = 2. The (two) "eyes". See **Two**.

**LOKA**. [S]. Value = 3. "World". Division of the Hindu universe. There are three *loka*: the earth (*bhurloka*), the space between the earth and the sun (*bhuvarloka*), and the space between the sun and the pole star (*svarloka*). In Buddhism, there are also three *lokas* and these represent the "spheres" of existence which make up the universe: *kâmaloka* (the world of sensations), *rûpaloka* (world of shapes or forms), and *arûpaloka* (the formless, immaterial world) [see Frédéric *Dictionnaire* (1987)]. See ***Triloka*** and **Three**.

**LOKA**. [S]. Value = 7. "World". Here the allusion is to another classification which tells of the existence of seven superior worlds: *bhurloka* (the earth); *bhuvarloka* (the space between the earth and the sun, supposedly the home of the **muni*, the **siddha*, etc.); the *svarloka* (the sky of *Indra); *maharloka* (where Bhrigu and many other "saints" are said to reside); *janaloka* (the land of the three sons of *Brahma); *taparloka* (home of the *vairâja*); and *satyaloka* or *brahmaloka* (the domain of Brahma). These seven superior worlds defend themselves against seven **pâtâla* ("inferior worlds") [see Frédéric *Dictionnaire*, (1987)]. See **Seven**.

**LOKA**. [S]. Value = 14. "world". See ***Bhuvana*** and **Fourteen**.

**LOKAKÂLA (Calendar)**. See ***Laukikasamvat***.

**LOKAPÂLA**. [S]. Value = 8. "Guardian of the horizons". In Hindu mythology, this is the name of the eight divinities who are guardians of the eight "horizons" and the eight points of the compass, who are represented as warriors in armour riding elephants. See ***Diggaja***, ***Dikpâla***, ***Dish*** and **Eight**.

**LOKAVIBHÂGA**. "Parts of the Universe". A *Jaina cosmological text which possesses the very exact date of Monday, August 25th of the year 458 CE in the Julian calendar. It is the oldest Indian text known to be in existence which contains zero and the place-value system expressed in numerical symbols [see Anon. (1962), chapter IV, line 56, p. 79]. See ***Anka***, ***Ankakramena***, ***Sthâna***, ***Sthânakramâd*** and ***Ankasthâna***.

**LORD OF THE UNIVERSE**. [S]. Value = 11. See **Îshvara**, **Rudra-Shiva** and **Eleven**.

**LOTUS**. This flower is the most famous symbol in all of Asia. It symbolises the pure spirit leaving the impure vessel of the body. It is the very image of divinity, which remains intact and is never soiled by the troubled waters of this world. A whole symbolism has developed around the lotus, according to its colour, the number of petals, and whether it is open, fresh-blown or in bud [see Frédéric, *Le Lotus* (1987)]. Thus it is not surprising that Indian arithmetic is full of related vocabulary and that such symbolism was often used to express very high numbers. In many texts, the words **padma*, **paduma*, **utpala*, **pundarîka* (also spelt **pundarika*), **kumud* and **kumuda* (which all literally mean "lotus") express numbers such as: ten to the power four, ten to the power nine, ten to the power fourteen, ten to the power twenty-one, ten to the power twenty-five, ten to the power twenty-seven, ten to the power twenty-nine, ten to the power ninety-eight, ten to the power 105, ten to the power 112 or ten to the power 119. See **High numbers (The symbolic meanings of)**.

**LOW NUMBERS**. See ***Paramânu***.

**LUMINOUS**. [S]. Value = 1. See ***Chandra*** and **One**.

**LUNAR MANSION**. [S]. Value = 27. See ***Nakshatra*** and **Twenty-seven**.

# M

**MADHYA**. Literally "Milieu". Name given to the number ten to the power ten (= ten thousand million). See **Names of numbers.** For an explanation of this symbolism, see **High numbers (The symbolic meaning of).**

Sources: **Vâjasaneyî Samhitâ* (beginning of the Common Era); **Taitirîya Samhitâ* (beginning of the Common Era); **Pañchavimsha Brâhmana* (date unknown).

**MADHYA.** Literally "Milieu". Name given to the number ten to the power eleven (= thousand million). See ***Names of numbers.*** For an explanation of this symbolism, see **High numbers (The symbolic meaning of).**

Sources: **Kâthaka Samhitâ* (beginning of the Common Era).

**MADHYA.** Literally "Milieu". Name given to the number ten to the power fifteen (= trillion). See **Names of numbers.** For an explanation of this symbolism, see **High numbers (The symbolic meaning of).**

Source: **Kitab fi tahqiq i ma li'l hind* by al-Biruni (c. 1030 CE).

**MADHYA.** Literally "Milieu". Name given to the number ten to the power sixteen (= ten trillion). See **Names of numbers.** For an explanation of this symbolism, see **High numbers (The symbolic meaning of).**

Source: **Lîlâvatî* by Bhâskarâchârya (1150 CE); **Ganitakaumudî* by Narâyana (1350 CE); **Trishatikâ* by Shrîdharâchârya (date uncertain).

**MÂDHYAMIKA.** Name given to the adepts of the Buddhist doctrine called the "Middle Path". This doctrine does not separate the reality and non-reality of things, and even considers the latter as a type of "vacuity" (**shûnyatâ*). This is why its adepts are sometimes called the **shûnyavâdin*, the "vacuists". See ***Shûnya*** and **Zero.**

**MAGIC.** See **Numeral alphabet, magic, mysticism and divination.**

**MAHÂBHÂRATA.** Name of a great Indian epic. See ***Arjunakara, Dhârtarâshtra, Nripa*** and ***Vasu.***

**MAHÂBHÛTA.** [S]. Value = 5. "Great element". This term is used by Hindus to denote collectively the five elements of the revelation. It is thus synonymous with the word **bhûta*, which can denote any one of the elements. Another generic term for the five elements is *Avarahakha*, made up of the five letters which symbolise each one of them: *A* (earth); *Va* (water); *Ra* (fire); *Ha* (wind); and *Kha* (ether). See ***Pañchabhûta*** and **Five.**

**MAHÂBJA.** Literally "great moon". Name given to the number ten to the power twelve (= billion). See ***Abja*** and **Names of numbers.** For an explanation of this symbolism, see **High numbers (The symbolic meaning of).**

Source: **Ganitakaumudî* by Narâyana (1350 CE).

**MAHÂDEVA.** [S]. Value = 11. "Great god". One of the names for *Rudra, whose symbolic value is eleven. See ***Rudra, Shiva*** and **Eleven.**

**MAHÂKALPA.** "Great **kalpa*". According to arithmetical-cosmogonical speculations, this term denotes a unit of cosmic time which is even bigger than the *kalpa* (= 4,320,000,000 human years). It is the equivalent of twenty "little" *kalpa* or ordinary *kalpa*. Thus one *mahâkalpa* = 86,400,000,000 human years. See ***Yuga.***

**MAHÂKATHÂNA.** Literally "great **kathâna*". Name given to the number ten to the power 126. See **Names of numbers** and **High numbers.**

Source: **Vyâkarana* (Pali grammar) by Kâchchâyana (eleventh century CE).

**MAHÂKHARVA.** Literally "great **kharva*". Name given to the number ten to the power thirteen (= ten billion). See **Names of numbers** and **High numbers.**

Source: **Ganitasârasamgraha* by Mahâvîrâchârya (850 CE).

**MAHÂKSHITI.** Literally: "great earth". Name given to the number ten to the power twenty-one (= quintillion). See **Names of numbers.** For an explanation of the symbolism, see **High numbers (Symbolic meaning of).**

Source: **Ganitasârasamgraha* by Mahâvîrâchârya (850 CE).

**MAHÂKSHOBHA.** Literally "great earth". Name given to the number ten to the power twenty-three (= a hundred quintillions). See **Names of numbers** and **High numbers.**

Source: **Ganitasârasamgraha* by Mahâvîrâchârya (850 CE).

**MAHÂKSHONI.** Literally "great earth." Name given to the number ten to the power seventeen ( = a hundred trillion). See **Names of numbers** and **High numbers.**

Source: **Ganitasârasamgraha* by Mahâvîrâchârya (850 CE).

**MAHÂPADMA.** Literally "great (pink) *lotus". Name given to the number ten to the power twelve (= billion). See ***Padma,*** **High numbers** and **Names of numbers.**

Sources: *Kitab fi tahqiq i ma li'l hind* by al-Biruni (c. 1030); **Lîlâvatî* by Bhâskarâchârya (1150 CE).

**MAHÂPADMA.** Literally "great (pink) *lotus". Name given to the number ten to the power fifteen (= trillion). See ***Padma,*** **High numbers** and **Names of numbers.**

Source : **Ganitasârasamgraha* by Mahâvîrâchârya (850 CE).

**MAHÂPADMA.** Literally "great (pink) *lotus". Name given to the number ten to the power thirty-four. See ***Padma,*** **High numbers** and **Names of numbers.**

Source : **Râmâyana* by Vâlmîki (in the early centuries CE).

**MAHÂPÂPA.** [S]. Value = 5. "Great sin". Allusion to the **Pañchaklesha*, the "five impurities", which, in Hindu and Buddhist philosophies, constitute the five main obstacles denying the faithful the Way of Realisation (*bodhi*) : greed, anger, thoughtlessness, insolence and doubt. See **Five.**

**MAHARASHTRI NUMERALS.** Signs derived from *Brâhmî numerals, through the intermediary of Shunga, Shaka, Kushâna, Ândhra, Gupta and Nâgarî numerals. Formerly used in Maharashtra State. The corresponding system was based on the place-value system and possessed zero (in the form of a little circle). These numerals were the ancestors of Marathi, Modi, Marwari, Mahajani and Rajasthani numerals. See **Indian written numeral systems (Classification of).** See Fig. 24.52 and 24.61 to 69.

**MAHARASHTRI-JAINA NUMERALS.** Signs derived from *Brahmi numerals, through the intermediary of Shunga, Shaka, Kushâna, Ândhra, Gupta and Nâgarî numerals. Formerly used by the *Jainas (*Shvetâmbara*). The corresponding system was based on the place-value system and possessed zero (in the form of a little circle). See **Indian written numeral systems (Classification of).** See Fig. 24.52 and 24.61 to 69.

**MAHÂSAROJA.** Literally "great **saroja*". Name given to the number ten to the power twelve (= billion). See **Names of numbers** and **High numbers.**

Source : **Trishatikâ* by Shrîdharâchârya (date uncertain).

**MAHÂSHANKHA.** Literally "great conch". Name given to the number ten to the power nineteen (= ten quadrillions). See ***Shankha,*** **Names of numbers** and **High numbers.**

Source : **Ganitasârasamgraha* by Mâhavîrâchârya (850 CE).

**MAHÂVÎRÂCHÂRYA.** *Jaina mathematician who lived in the ninth century. His works notably include *Ganitasârasamgraha*, where there is frequent use of the place-value system, written not only in numerical symbols, but also with nine numerals and the sign for zero. Here is a list of the principle names for numbers mentioned in *Ganitasârasamgraha* :

**Eka* (= 1), **Dasha* (= 10), **Shata* (= $10^2$), *Sahasra* (= $10^3$), **Dashasahasra* (= $10^4$), **Laksha* (= $10^5$), **Dashalaksha* (= $10^6$), **Koti* (= $10^7$), **Dashakoti* (= $10^8$), **Shatakoti* (= $10^9$), **Arbuda* (= $10^{10}$), **Nyarbuda* (= $10^{11}$), **Kharva* (= $10^{12}$), **Mahâkharva* (= $10^{13}$), **Padma* (= $10^{14}$), **Mahâpadma* (= $10^{15}$), **Kshoni* (= $10^{16}$), **Mahâkshoni* (= $10^{17}$), **Shankha* (= $10^{18}$), **Mahâshankha* (= $10^{19}$), **Kshiti* (= $10^{20}$), **Mahâkshiti* (= $10^{21}$), **Kshobha* (= $10^{22}$), **Mahâkshobha* (= $10^{23}$).

See **Names of numbers, High numbers, Numeration of numerical symbols, Zero** and **Indian mathematics (The history of).**

Source : GtsS, I, p. 63–68 [See Datta and Singh (1938), p. 13; Rangacarya (1912)].

**MAHÂVRINDÂ.** Literally "great **vrindâ*". Name given to the number ten to the power twenty-two (= ten quintillions). See **Names of numbers,** ***Vrindâ*** and **High numbers (The symbolic meaning of).**

Source : **Râmâyana* by Vâlmîki (first centuries CE).

**MAHÂYAJÑA.** [S]. Value = 5. "Great sacrifice". This is the name for the five daily sacrifices that all orthodox Hindus must make : prayer and devotion (*hapûjâ*), the placing of offerings in various places (*baliharana*), offerings to the shades of ancestors (*pitriyajña*), the offering of a ritual meal (*manushyayajña*) and a sacrifice in honour of the fire which cooks the food [see Frédéric *Dictionnaire* (1987)]. See **Five.**

**MAHÂYUGA.** "Great period". This is the largest cosmic cycle of Indian speculations. Considered as the "Great age", this cycle is made up of four successive periods (**kritayuga*, **tretâyuga*, **dvâparayuga*, **kaliyuga*); this is why it is also called the **chaturyuga* ("four ages"). It is said to be made up of 4,320,000 human years. See ***Yuga.***

**MAHÎ.** [S]. Value = 1. This term (which is also the name of a river in Rajasthan) means "curds", the first product derived from milk. Milk itself is the first and most important of the "gifts of the Cow" (**gavyâ*), which is the first nourishment by which all others potentially exist. Thus this symbolism embraces the idea of the cow as a whole and even, in an esoterical sense, the sacred Cow of the Hindus, which is identified with the whole world, because the Cow dispenses life. As "Earth" is a numerical symbol for the value 1, *Mahi* = "curds" = 1. See ***Prithivî, Go*** and ***One.***

**MAHÎDHARA.** [S]. Value = 7. "Mountain". See **Mount Meru,** ***Adri*** and **Seven.**

**MAHORÂGA.** "Great serpents". Category of demons in the form of cobras. See **Serpent (Symbolism of the).**

**MAIN OBSERVANCE.** [S]. Value = 5. See ***Karanîya*** and **Five.**

**MAITHILI NUMERALS.** Signs derived from *Brâhmî numerals, through the intermediary of Shunga, Shaka, Kushâna, Ândhra, Gupta, Nâgarî and Bengali numerals. Currently found mainly in the north of Bihar State. The corresponding system is based on the place-value system and possesses zero (in the form of a little circle). See **Indian written numeral systems (Classification of)**. See Fig. 24.11, 52 and 24.61 to 69.

**MALAYALAM NUMERALS.** Signs derived from *Brâhmî numerals, through the intermediary of Shunga, Shaka, Kushâna, Ândhra, Pallava, Chalukya, Ganga, Valabhî, Bhattiprolu and Grantha numerals. Currently used by the Dravidians of Kerala State on the ancient coast of Malabar, to the southwest of India. The corresponding system is not based on the place-value system and has only possessed zero since a relatively recent date. See **Indian written numeral systems (Classification of)**. See also Fig. 24.19, 52 and 24.61 to 69.

**MANDARA.** One of the names for Mount Meru. See **Mount Meru**, ***Adri***, ***Dvîpa***, ***Pûrna***, ***Pâtâla***, ***Sâgara***, ***Pushkara***, ***Pâvana*** and ***Vâyu***.

**MANGALA.** [S]. Value = 8. "Jewel", "thing which augurs well". Here the allusion is to **ashtamangala*, the eight "things which augur well", Buddhist symbols which represent the veneration of the "Master of the world" (and, by extension, Buddha). These are: the parasol (symbol of royal dignity meant to protect against misfortune); the two fish (signs of the Indian master of the universe); the conch (symbol of victory in combat); the lotus flower (symbol of purity); the container of lustral water (filled with Amrita, the nectar of immortality); the rolled flag (sign of victorious faith); the knots of eternal life; and the wheel of the Doctrine (*Dharmachakra*). See **Eight**.

**MANIPURI NUMERALS.** Signs derived from *Brâhmî numerals, through the intermediary of Shunga, Shaka, Kushâna, Ândhra, Gupta, Nâgarî, Kutila and Bengali numerals. Currently in use in Manipur State, to the east of Assam and next to the border of Burma. The corresponding system functions according to the place-value system and possesses zero (in the form of a little circle). See **Indian written numeral systems (Classification of)**. See also Fig. 24.52 and 24.61 to 69.

**MANTRA.** Sacred formula which constitutes the digest, in a material from, of the divinity which it is meant to invoke. See **Numeral alphabet, magic, mysticism and divination** and **Mysticism of letters**.

**MANU.** [S]. Value = 14. Literally "human". This is the name given in traditional legends to the Progenitor of the human race as a symbol of the thinking being and considered as the intermediary between the Creator and the human race. According to the **Vedas*, the *manus* constituted the first divine legislators who fixed the rules of religious ceremonies and ritual sacrifices. According to the **purânas*, there were fourteen successive *manus*, sovereigns living in ethereal worlds where they are meant to direct the conscious life of humankind and its ability to think. Thus: *manu* = 14. (The *manu* of the present era is the seventh: named *Vaivashvata*, "Born of the Sun") See **Fourteen**.

**MANUAL ARITHMETIC.** See ***Mudrâ***.

**MANUSMRITI.** Important religious work considered to be the foundation of Hindu society.

**MARÂTHA (Calendar).** This calendar begins in the year 1673 CE, and was founded by Shivâjî. To find the corresponding date in the Common Era, add 1673 to a date expressed in the *Marâtha* calendar. Formerly used in Maharashtra. See **Indian calendars**.

**MARATHI NUMERALS.** Signs derived from *Brâhmî numerals, through the intermediary of Shunga, Shaka, Kushâna, Ândhra, Gupta, Nâgarî and Mahârâshtrî numerals. Currently used in the west of India, in the state-province of Maharashtra. The corresponding system is based on the place-value system and possesses zero (in the form of a little circle). See **Indian written numeral systems (Classification of)**. See Fig. 24.4, 52 and 24.61 to 69.

**MÂRGANA** [S]. Value = 5. "Arrow". See ***Shara*** and **Five**.

**MÂRTANDA.** [S]. Value = 12. One of the names of **Sûrya*. See **Twelve**.

**MARWARI NUMERALS.** Signs derived from *Brâhmî numerals, through the intermediary of Shunga, Shaka, Kushâna, Ândhra, Gupta, Nâgarî and Mahârâshtrî numerals. Currently used in the northwest of India (Rajasthan) and in the Aravalli mountains, and between the afore-mentioned mountains and the Thar desert (Mârusthali). The corresponding system is based on the place-value system and possesses zero (in the form of a little circle). See **Indian written numeral systems (Classification of)**. See Fig. 24.52 and 24.61 to 69.

**MÂSA.** [S]. Value = 12. "Month". Allusion to the twelve months of the year. See **Twelve**.

**MÂSÂRDHA.** [S]. Value = 5. "Season". See ***Ritu*** and **Five**.

**MATANGA.** [S]. Value = 8. "Elephant". See ***Diggaja*** and **Eight**.

**MATHEMATICIAN.** See ***Samkhyâ***.

**MATHEMATICS.** See ***Ganita***, ***Ganitânuyoga***, **Arithmetic**, **Calculation** and **Indian mathematics (The history of)**.

**MATHURA NUMERALS.** Signs derived from *Brâhmî numerals, through the intermediary of Shunga, Shaka, Kushâna and Ândhra numerals. Contemporaries of a Shaka dynasty (first to third century CE). These are attested mainly in the inscriptions of Mathura (in Uttar Pradesh). The corresponding system did not use the place-value system or zero. See Fig. 24.32, 52 and 24.61 to 69, and 70. See also **Indian written numeral systems (Classification of)**.

**MÂTRIKÂ.** [S]. Value = 7. "Divine Mother". Name given in Hinduism to the **saptamâtrikâ*, the seven aspects of *shakti*, "feminine energy" of the divinities : aspects which are considered to be the "mothers of the world". Thus: *mâtrikâ* = 7. See **Seven**.

**MATTER (Indian concept of).** See **Indian atomism**, **Jaina** and ***Jala***.

**MEMBER.** [S]. Value = 6. See ***Anga*** and **Six**.

**MENTAL ARITHMETIC.** See ***Gananâ***.

**MERIT.** [S]. Value = 3. See ***Guna***, ***Triguna*** and **Three**.

**MERIT.** [S]. Value = 6. See ***Guna***, ***Shadâyatana*** and **Six**.

**MILIEU.** As a name of a high number. See ***Madhya***.

**MILLION** (= ten to the power six). See ***Dashalaksha***, ***Niyuta***, ***Prayuta*** and **Names of numbers**.

**MON NUMERALS.** Signs derived from *Brâhmî numerals, through the intermediary of Shunga, Shaka, Kushâna, Ândhra, Pallava, Châlukya, Ganga, Valabhî, "Pâlî" and Vateluttu numerals. Formerly used by the people of Pegu before the Burmese invasion. The corresponding system was not based upon the place-value system and did not possess zero. See Fig. 24.52 and 24.61 to 69. See also **Indian written numeral systems (Classification of)**.

**MONGOL NUMERALS.** Signs derived from *Brâhmî numerals through the intermediary notations of Shunga, Shaka, Kushâna, Âdhra, Gupta, Siddham and Tibetan numerals. Used by the Mongols during the thirteenth and fourteenth centuries. The corresponding system functioned according to the place-value system and possessed zero (in the form of a little circle). See Fig. 24. 42, 52 and 24.61 to 69. See **Indian written numeral systems (Classification of)**.

**MONTH (Rite of the four).** See ***Chaturmâsya***.

**MONTH.** [S]. Value = 12. See ***Mâsa*** and **Twelve**.

**MOON.** Used as a name for ten to the power nine or ten to the power twelve. See ***Abja*** and ***Mahâbja***. See also **High numbers (The symbolic meaning of)**.

**MOON.** [S]. Value = 1. See ***Abja***, ***Atrinayanaja***, ***Chandra***, ***Indu***, ***Kshapeshvara***, ***Mriganka***, ***Shashadhara***, ***Shashanka***, ***Shashin***, ***Shîtâmshu***, ***Shîtarashmi***, ***Soma***, ***Sudhâmshu***, ***Vidhu*** and **One**.

**MORTAL SINS (The Five).** See ***Pañchânantarya***.

**MOUNT MERU.** Mythical mountain in Hindu cosmology and Brahman mythology. It has many Sanskrit names : *Ratnasanu*, *Sumeru*, *Hemadrî*, *Mandara*, *Karnikâchala*, *Devapârvata*, etc. Mount Meru was meant to be the place where the gods lived and met. It was said to be situated at the centre of the universe, under the Pole star, and also constituted the "axis of the world". *Indra lived at the summit, the head of the **deva*, whilst the slopes were peopled with the **Trâyastrimsha*, the thirty-three *deva* (gods).

Mount Meru plays an important role in mythology and Brahman and Hindu cosmological texts. Thus this mountain was said to act as a pivot between the **deva* and the **asura* ("anti-gods") during the churning of the sea of milk.

In corresponding representations, Mount Meru, and all that is connected to it, is always associated with the number seven. First there is the concept of the "mountain" and, by extension, that of "hill", which is generally symbolically connected with this number. There are also the "seven oceans" (**sapta sâgara*). Then there are the "island-continents" (**dvîpa*), each one flooded by one of the seven oceans, which surround Mount Meru. As for Mount Meru itself, it has seven faces, each facing one of the seven seas and one of the seven continents. It is above the **pâtâla*, the seven underworlds or "inferior worlds", where the **nâga* live, the master of whom is the king Muchalinda, the chthonian genie in the form of a cobra, depicted with seven heads.

Fig. 24D.9. *Mount Meru, centre of the universe in Hindu and Brahmanic cosmology. Ref.: Dubois de Jancigny,* L'Univers pittoresque, *Hachette, Paris, 1846*

Thus Mount Meru represents total stability and the absolute centre of the universe, around which the universe and the firmament revolve. This image of Mount Meru connects it to one of the universal images of the Pole star. According to the legend, Mount Meru is directly underneath the Pole star and is "on the same axis". The symbolic correspondence between Mount Meru and the number seven comes from the fact that the Pole star, the **Sudrishti*, "That which never moves", is the last of the seven stars of the constellation: the Bear. According to Indian tradition, this constellation is the personification of the seven "great Sages" of Vedic times, the **Saptarishi*, who are thought to be the authors of both the hymns and invocations of the *Rigveda*, and of the most important Vedic texts.

In Sanskrit, the word for "mountain" is **pârvata*, which appears in one of the names of Mount Meru : *Devapârvata*, "mountain of the gods". Because this sacred mountain was associated with the number seven, the mountain, daughter of Himâlaya, sister of Vishnu and wife of *Shiva also came to be synonymous with this number: she was *Kâlî*, the "Black", who represented the destructive power of time (**Kâla*) and was considered in the **Veda* to be the seventh tongue of Agni, "Fire". It is perhaps not by chance that Manasâ, the Hindu tantric divinity, who symbolises the destructive and regenerative aspects of Pârvati, has been considered as one of the sisters of Muchalinda (the king of the *nâga* with seven cobra heads) and as the **Pâtâla Kumârâ*, divinity of the serpents and "princess of the (seven) Underworlds".

Even *Sûrya, the Sun-god, traditionally associated with the number twelve in the system of *numerical symbols, has represented the number seven : he is often represented as a warrior flying through the sky on a chariot pulled either by seven horses (**ashva*), or by *Aruna*, the horse with seven heads. See ***Adri***, ***Dvîpa***, ***Sâgara***, ***Pâtâla***, ***Pushkara***, ***Pâvana***, ***Vâyu***, ***Loka*** (= 7) and **Seven**.

**MOUNTAIN.** [S]. Value = 7. See ***Adri***, ***Pârvata***, **Seven** and **Mount Meru**.

**MOUTHS OF JÂHNAVÎ.** [S]. Value = 1,000. "Mouths of the Ganges". See ***Jâhnavîvaktra*** and **Thousand**.

**MRIGÂNKA.** [S]. Value = 1. "Moon". See ***Abja*** and **One**.

**MUCHALINDA (MUCHILINDA).** Name of the king of the **nâga*. See **Serpent (Symbolism of the)**.

**MUDRÂ.** "Mark, sign". In Indian mysticism, mainly in esoteric Buddhism, this word denotes the gestures made by the hands and is meant to symbolise a mental attitude of the divinities. They are mainly used during ceremonies and prayers to invoke Buddha and the power of his divinites.

**MÛDRÂ.** Term denoting manual arithmetic and digital calcultaion in Ancient Sanskrit literature. See Chapter 3.

**MUDRÂBALA.** Literally : "Power of the **mudrâ*". Name given to the number ten to the power forty-three. To those using it, such a high number must have symbolically represented a quantity which was as incalculable as the powers concealed within the mystical gestures called *mudrâ*. See ***Mudrâ*** (first article), **Names of numbers** and **High numbers**.

Source : **Lalitavistara Sûtra* (before 308 CE).

**MUKHA.** [S]. Value = 4. "Face". Allusion to the **chaturmukha* ("Four Faces"), which refers to all the of the Brahmanic (or Buddhist) divinities who are represented as having four faces (*Brahma, *Shiva, etc.). See **Four**.

**MULTIPLICATION.** [Arithmetic]. See ***Gunana***, ***Pâtîganita*** and **Indian methods of calculation**.

**MUNI.** [S]. Value = 7. "Sage". This is an allusion to the seven mythical sages of Vedic times. Strictly speaking, the word *muni*, "sage", is much less strong than **Rishi*, which denotes the seven "Sages". But the name began to be used as a symbol for the number seven because of the desired effect in the versification of expressions using numerical symbols. See **Seven**, **Sanskrit** and **Poetry and the writing of numbers**.

**MÛRTI.** [S]. Value = 3. "Form". Allusion to the "three forms" of Hindu triads (**trimûrti*), constituted by either three different divinities (usually *Brahma, *Shiva and *Vishnu), or three aspects of one single divinity. See **Three**.

**MÛRTI.** [S]. Value = 8. "Form". Allusion to the **ashtamûrti*, the "eight" most important "forms" of *Shiva : **Rudra*, who represents the power of fire; **Bhava*, water; *Sharva*, the earth; *Ishâna*, the sun; *Pashupati*, sacrifice; *Bhîma*, the terrible; and *Ugra* and *Mahâdeva*. See ***Rudra-Shiva*** and **Eight**.

**MUSICAL MODE.** [S]. Value = 6. See ***Râga*** and **Six**.

**MUSICAL NOTE.** [S]. Value = 7. See ***Svara*** and **Seven**.

**MUSLIM INDIA.** See **Numeral alphabet and composition of chronograms** and **Eastern Arabic numerals**.

**MYSTICISM AND POSITIONAL NUMERATION.** See ***Durgâ***.

**MYSTICISM OF HIGH NUMBERS.** See **High numbers (The symbolic meaning of)**.

**MYSTICISM OF INFINITY.** See **Infinity (Mythological representation of)** and **Serpent (Symbolism of the)**.

**MYSTICISM OF LETTERS.** See ***Akshara***, **Numeral alphabet, magic, mysticism and divination**, ***Bîja***, **Mantra**, ***Trivarna***, ***Vâchana***.

**MYSTICISM OF NUMBERS.** See **Numerical symbols**, **Symbolism of words with a numerical value**, **Symbolism of numbers (Concept of large quantity)**, **Symbolism of zero and High numbers (The symbolic meaning of)**.

**MYSTICISM OF THE NUMBER FOUR.** See ***Nâga***, ***Jala***, **Ocean** and **Serpent (Symbolism of the)**.

**MYSTICISM OF THE NUMBER SEVEN.** See **Mount Meru** and **Ocean**.

**MYSTICISM OF ZERO.** See ***Shûnya***, ***Shûnyatâ***, **Zero**, **Zero (Indian concepts of)**, **Zero and Sanskrit poetry** and **Symbolism of Zero**.

**MYTHICAL PEARL.** [S]. Value = 1. See **Indu**.

## N

**NABHA.** [S]. Value = 0. "Sky, atmosphere". This symbolism is due to the fact that the sky is considered to be the "void". See **Zero** and ***Shûnya***.

**NABHAS.** [S]. Value = 0. "Sky, atmosphere". See ***Nabha***, **Zero** and ***Shûnya***.

**NÂDÎ.** Hindu word denoting the arteries of the human body. See **Numeral alphabet, magic, mysticism and divination**.

**NAGA.** [S]. Value = 7. "Mountain". Literally, "That which does not move". This is an allusion to *Mount Meru, the mythical mountain of Hindu cosmology and Brahman mythology, the dwelling and meeting place of the gods, which is said to be situated at the centre of the universe and thus constitute the axis of the world. This symbolism comes from the fact that the number seven plays an important role in mythological representations related to

Mount Meru, and because the Pole star, situated directly above this mountain, is the **Sudrishti*, the divinity "who never moves". See ***Adri*, Mount Meru, Seven** and ***Dhruva***.

**NÂGA**. [S]. Value = 8. "Serpent". This symbolism is due to the fact that the serpent (especially the *nâga*) is considered to be not only a sun genie who owns the earth and its treasures, but also an aquatic symbol. It is a "spirit of the waters" that lives in the **pâtâla* or "underworlds". In Sanskrit, water is **jala*, and this word is used as a numerical symbol for the number four. In their subterranean kingdom, the *nâga* reproduce in couples and evolve in the company of the *nâginî* (the females), so "water", in this case, has been symbolically multiplied by two, to give their generic name the symbolic value of : 4 × 2 = 8. In traditional Indian thought, the earth (to which the serpent is also associated) corresponds symbolically to the number four, being associated with the square and its four horizons (or cardinal points). As the *nâga* is also aquatic, water (= 4) has been symbolically added to give the serpent its generic designation as a numerical symbol with a value equal to eight (*nâga* = earth + water = 4 + 4 = 8). See **Eight, Serpent (Symbolism of the)** and **Infinity**.

For a documented example of this : see EI, XXXV, p. 140.

**NÂGABALA**. Literally, "Power of the **nâga*". Name given to the number ten to the power twenty-five. See **Names of numbers, High numbers** and **Serpent (Symbolism of the)**.

Source : **Lalitavistara Sûtra* (before 308 CE).

**NÂGARÎ ALPHABET**. See Fig. 24.56. See also **Âryabhata's numeration**.

**NÂGARÎ NUMERALS**. Signs derived from *Brâhmî numerals through the intermediary notations of Shunga, Shaka, Kushâna, Âdhra and Gupta. Today these are the most widely used numerals in India, from Madhya Pradesh (central province) to Uttar Pradesh (northern province), Rajasthan, Haryana, Himachal Pradesh (the Himalayas) and Delhi. These numerals are also called *Devanagari* because they are the most regular numerals of India. The corresponding system is based on the place-value system and possesses zero (in the form of a little circle). However, this was not always the case, as a considerable number of documents written before the eighth century CE prove. These signs were the ancestors of Siddham, Nepali, Tibetan, Mongol, Kutilâ, Bengali, Oriya, Kaîthî, Maithilî, Manipurî, Gujarâtî, Mahârâshtrî, Marâthî, Modî, Mârwarî, Mahâjani, Râjasthanî, etc. numerals, as well as the "Hindi" numerals of the eastern Arabs, the Ghubar numerals of North Africa, the *apices* and algorisms of mediaeval Europe, not to mention our own modern numerals. For ancient Nâgarî numerals recorded on copper charters, see Fig. 24.39 A and 75; for those recorded on manuscripts, see Fig. 24.39 B; for inscriptions of Gwalior, see Fig. 24.39 C and 24. 72 to 74. For modern Nâgarî numerals, see Fig. 24.3. For notations which derived from Nâgarî, see Fig. 24.52. For the corresponding graphical evolution, see Fig. 24.61 to 69. See also **Indian written numeral systems (Classification of)**.

**NÂGINÎ**. Female of the **nâga*. See ***Nâga*** and **Serpent (Symbolism of the)**.

**NAHUT**. Name given to the number ten to the power nine. See **Names of numbers** and **High numbers**.

Source : **Lalitavistara Sûtra* (before 308 CE).

**NAHUTA**. Name given to the number ten to the power twenty-eight. See **Names of numbers** and **High numbers**.

Source : **Vyâkarana* (Pâli grammar) by Kâchâyana (eleventh century CE).

**NAIL**. [S]. Value = 20. See ***Nakha*** and **Twenty**.

**NAKHA**. [S]. Value = 20. "Nail". This is because of the nails of the ten fingers and ten toes. See **Twenty**.

**NAKSHATRA**. [S]. Value = 27. "Lunar Mansion". This refers to the houses occupied successively by the moon in its monthly cycle, which in solar days lasts between twenty-seven and twenty-eight days. For the representation of the sidereal movements of the moon, however, Indian astronomers usually used the system of twenty-seven *nakshatra* marking twenty-seven ideal equal divisions of the ecliptic zone (each one equal to 13° 20'). This is why the word came to symbolically signify the number twenty-seven. See **Twenty-seven**.

**NAKSHATRAVIDYÂ**. Literally : "Knowledge of the **nakshatra*". Name given to "astronomy" in the *Chândogya Upanishad*.

**NAMES OF NUMBERS (up to thousand)**. Here is a list of ordinary Sanskrit names of numbers :

**Eka* (= 1); **Dva* (= 2); **Dve* (= 2); **Dvi* (= 2); **Trai* (= 3); **Traya* (= 3); **Tri* (= 3); **Chatur* (= 4); **Pañcha* (= 5); **Shad* (= 6); **Shash* (= 6); **Shat* (= 6); **Sapta* (= 7); **Saptan* (= 7); **Ashta* (= 8); **Ashtan* (= 8); **Nava* (= 9); **Navan* (= 9).

**Dasha* (= 10); **Dashan* (= 10); **Ekadasha* (= 11); **Dvâdasha* (= 12); **Trayodasha* (= 13); **Chaturdasha* (= 14); **Pañchadasha* (= 15); **Shaddasha* (= 16); **Saptadasha* (= 17); **Ashtadasha* (= 18); **Navadasha* (= 19).

**Vimshati* (= 20); **Ekavimshati* (= 21); **Dvavimshati* (= 22); *Trayavimshati* (= 23); **Chaturvimshati* (= 24); **Pañchavimshati* (= 25); **Shadvimshati* (= 26); **Saptavimshati* (= 27); *Ashtavimshati* (= 28); *Navavimshati* (= 29).

**Trimshat* (= 30); **Chatvarimshat* (= 40); **Pañchashat* (= 50); **Shashti* (= 60); **Saptati* (= 70); **Ashîti* (= 80); **Navati* (= 90).

At the start of the Common Era, the subtractive forms were also used for the numbers 19, 29, 39, 49, etc. : **ekânnavimshati* (= 20 – 1 = 19); **ekânnatrimshati* (= 30 – 1 = 29); etc.

[See **Taittirîya Samhitâ*, VII, 2, 11; Datta and Singh (1938), pp. 14–15].

**Shata* (= 100). This is the classical Sanksrit form of this number. However, at the beginning of the Common Era, the Indo-European form *Sata* was still used.

Ref.: There is evidence of the use of this form in **Vâjasaneyî Samhitâ*, **Taittirîya Samhitâ*, **Kâthaka Samhitâ*, **Pañchavimsha Brâhmana* and **Sankhyâyana Shrauta Sûtra*.

*Dvashata* (= 200); *Trishata* (= 300); *Chatuhshata* (= 400); etc. **Sahasra* ( = 1,000); *Dvasahasra* (= 2,000); *Trisahasra* (= 3,000); *Chatursahasra* (= 4,000); etc.

See **Sanskrit**.

**NAMES OF NUMBERS (Powers of ten above thousand)**. After ten thousand, Sanskrit spoken numeration assigns names to the various powers of ten which differ considerably from one author to another and from one era to another; thus the same word can have several numerical values depending on the source in question. The use of these names was not commonplace in India. However, they were very familiar to scholars, since the following terms are found in astronomical, mathematical, cosmological, grammatical and religious texts, as well as in legend and mythology.

In the following list, the letters in brackets indicate the source of each word in question; here are the letters, the sources they represent, and the era in which they were written :

(a) *Vâjasaneyî Samhitâ*. (b) *Taittirîya Samhitâ*. (c) *Kâthaka Samhitâ*. (d) *Râmâyana* by Vâlmîki. (e) *Lalitavistara Sûtra*. (f) *Pañchavimsha Brâhmana*. (g) *Sankhyâyana Shrauta Sûtra*. (h) *Âryabhatîya* by Âryabhata. (i) *Ganitasârasamgraha* by Mahâvîrâchârya. (j) *Kitab fi tahqiq i ma li'l hind* by al-Biruni. (k) *Vyâkarana*, Pâli grammar, by Kâchchâyana. (l) *Lîlâvatî* by Bhâskarâchârya. (m) *Ganitakaumudî* by Nârâyana. (n) *Trishatikâ* by Shrîdharâchârya.

(a, b, c : beginning of the Common Era; d : early centuries of the Common Era; e : before 308 CE; f, g : date uncertain; h : c. 510 CE; i : 850 CE; j : c. 1030 CE; k : eleventh century CE; l : 1150 CE; m : 1356 CE; n : date uncertain).

Here is a (non-exhaustive) arithmetical list of the Sanskrit names of high numbers :

TEN TO THE POWER 4: **Ayuta* (a, b, c, f, g, h, j, l, m, n); **Dashasahasra* (i).

TEN TO THE POWER 5: **Lakh* (e); **Lakkha* (k); **Laksha* (i, j, l, m, n); **Niyuta* (a, b, f, h); **Prayuta* (c).

TEN TO THE POWER 6: **Dashalaksha* (i); **Niyuta* (c); **Prayuta* (a, b, f, g, h, j, l, m, n).

TEN TO THE POWER 7: **Arbuda* (a, b, c, f, g, h); **Koti* (d, e, h, i, j, k, l, m, n).

TEN TO THE POWER 8: **Arbuda* (l, m, n); **Dashakoti* (i); **Nyarbuda* (a, b, c, f, g); **Vyarbuda* (j).

TEN TO THE POWER 9: **Abja* (l, n); **Ayuta* (e); **Nahut* (e); **Nikharva* (g); **Padma* (j); **Samudra* (a, b, c, f); **Saroja* (m); **Shatakoti* (i); **Vâdava* (c); **Vrindâ* (h).

TEN TO THE POWER 10: **Arbuda* (i); **Kharva* (j, l, m, n); **Madhya* (a, b, f); **Samudra* (g).

TEN TO THE POWER 11: **Anta* (a, b, c, f); **Madhya* (c); **Nikharva* (j, l, m, n); **Ninnahut* (e); **Niyuta* (e); **Nyarbuda* (i); **Salila* (g).

TEN TO THE POWER 12: **Antya* (g); **Kharva* (i); **Mahâbja* (m); **Mahâpadma* (j, l); *Mahâsaroja* (n); **Parârdha* (a, b, c, f); **Shankha* (d).

TEN TO THE POWER 13: **Ananta* (g); **Kankara* (e); **Khamba* (e); **Mahâkharva* (i); **Nikharva* (f); **Shankha* (j); **Shanku* (l, m, n).

TEN TO THE POWER 14: **Jaladhi* (l); **Padma* (i); **Pakoti* (k); **Pârâvâra* (m); **Samudra* (j); **Saritâpati* (n); **Vâdava* (f).

TEN TO THE POWER 15: **Akshiti* (f); **Antya* (l, m, n); **Madhya* (j); **Mahâpadma* (i); **Viskhamba* (e); **Vivara* (e).

TEN TO THE POWER 16: **Antya* (j); **Madhya* (l, m, n); **Kshoni* (i).

TEN TO THE POWER 17: **Abab* (e); **Kshobhya* (e); **Mahâkshoni* (i); **Parârdha* (j, l, m, n); **Vrindâ* (d).

TEN TO THE POWER 18: **Shankha* (i).

TEN TO THE POWER 19: **Attata* (e); **Mahâshankha* (i); **Vivaha* (e).

TEN TO THE POWER 20: **Kshiti* (i).

TEN TO THE POWER 21: **Kotippakoti* (k); **Kumud* (e); **Mahâkshiti* (i); **Utsanga* (e).

TEN TO THE POWER 22: **Kshobha* (i); **Mahâvrindâ* (d).

TEN TO THE POWER 23: *Bahula (e); *Gundhika (e); *Mahâkshobha (i).

TEN TO THE POWER 25: *Nâgabala (e); *Utpala (e).

TEN TO THE POWER 27: *Pundarîka (e); *Titilambha (e).

TEN TO THE POWER 28: *Nahuta (k).

TEN TO THE POWER 29 : *Padma (d); *Paduma (e); *Vyavasthânaprajñapati (e).

TEN TO THE POWER 31: *Hetuhila (e).

TEN TO THE POWER 33: *Karahu (e).

TEN TO THE POWER 34: *Mahâpadma (d).

TEN TO THE POWER 35: *Hetvindriya (e); *Ninnahuta (k).

TEN TO THE POWER 37: *Samâptalambha (e).

TEN TO THE POWER 39: *Gananâgati (e); *Kharva (d).

TEN TO THE POWER 41: *Niravadya (e).

TEN TO THE POWER 42: *Akkhobhini (k).

TEN TO THE POWER 43: *Mudrâbala (e).

TEN TO THE POWER 45: *Sarvabala (e).

TEN TO THE POWER 47: *Visamjñagati (e).

TEN TO THE POWER 49: *Bindu (k); *Sarvajña (e).

TEN TO THE POWER 51: *Vibhutangamâ (e).

TEN TO THE POWER 53: *Tallakshana (e).

TEN TO THE POWER 56: *Abbuda (k).

TEN TO THE POWER 63: *Nirabbuda (k).

TEN TO THE POWER 70: *Ahaha (k).

TEN TO THE POWER 77: *Ababa (k).

TEN TO THE POWER 84: *Atata (k).

TEN TO THE POWER 91: *Sogandhika (k).

TEN TO THE POWER 98: *Uppala (k).

TEN TO THE POWER 99: *Dhvajâgravati (e).

TEN TO THE POWER 105: *Kumuda (k).

TEN TO THE POWER 112: *Pundarîka (k).

TEN TO THE POWER 119: *Kathâna (k); *Paduma (k).

TEN TO THE POWER 126: *Mahâkathâna (k).

TEN TO THE POWER 140: *Asankhyeya (k).

TEN TO THE POWER 145: *Dhvajâgranishâmani (e). And so on until ten to the power 421 (e).

See **Sanskrit** and **Poetry and writing of numbers.**

Indian scholars did not specialise in just one field of study; they embraced diverse disciplines all at once, such as mathematics, astronomy, literature, poetry, phonetics or philosophy, and even mysticism, divination and astrology. Thus it is not surprising that in arithmetic, their fertile imaginations led them to use subtle symbolism to name high numbers. They gave a unique name to each power of ten up to at least as high as ten to the power 421. This is why their spoken numeration had a mathematical structure with the potential to lead them to the discovery of the place-value system and consequently the "invention" of zero. See **High numbers.** For an explanation of the symbolism of these diverse words, see: **High numbers (The symbolic meaning of), Zero, Numeration of numerical symbols** and **Numerical symbols (Principle of the numeration of).**

**NÂRÂYANA.** Indian mathematician c. 1356. His works notably include *Ganitakaumudî.*

Here is a list of the principal names of numbers mentioned in that work: *Eka (= 1),*Dasha (= 10), *Shata (= $10^2$), *Sahasra (= $10^3$), *Ayuta (= $10^4$), *Laksha (= $10^5$), *Prayuta (= $10^6$), *Koti (= $10^7$), *Arbuda (= $10^8$), *Saroja (= $10^9$), *Kharva (= $10^{10}$), *Nikharva (= $10^{11}$). *Mahâpadma (= $10^{12}$), *Shanku (= $10^{13}$). *Pârâvâra (= $10^{14}$), *Madhya (= $10^{15}$), *Antya (= $10^{16}$), *Parârdha (= $10^{17}$).

See **Names of numbers** and **High numbers.** [See Datta and Singh (1938), p. 13]

**NÂSATYA.** [S]. Value = 2. Name of one of the two twin gods Saranyû and Vivashvant of the Hindu pantheon (also called *Dasra and Nâsatya). The symbolism is through an association of ideas with the "Horsemen". See ***Ashvin*** and **Two.**

**NAVA (NAVAN).** Ordinary Sanskrit names for the number nine, which appear in the composition of many words which have a direct relationship with the concept of this number. Examples: *Navagraha, *Navaratna, *Navarâshika and *Navarâtrî. For words which have a more symbolic relationship with this number, see: **Nine** and **Symbolism of numbers.**

**NAVACHATVÂRIMSHATI.** Ordinary Sanskrit name for the number forty-nine. For words having a symbolic link to this number, see **Forty-nine** and **Symbolism of numbers.**

**NAVADASHA.** Ordinary Sanskrit name for the number nineteen. For words which have a symbolic link to this number, see : **Nineteen** and **Symbolism of numbers.**

**NAVAGRAHA.** Literally : "nine planets". This relates to the nine planets of the Hindu cosmological system: the seven planets (*saptagraha) plus the demons of the eclipses ***Râhu*** and ***Ketu.*** See ***Graha*** and ***Paksha.***

**NAVAN.** Ordinary Sanskrit name for the number nine. See ***Nava.***

**NAVARÂSHIKA.** [Arithmetic]. Sanskrit name for the Rule of Nine. See ***Nava.***

**NAVARATNA.** "Nine jewels", "Nine precious stones". Collective name given to the nine famous poets of Sanskrit expression who are said to have lived in the court of King Vikramâditya (namely : *Dhavantari*, the pearl; *Kshapanaka*, the ruby; *Amarasimah*, the topaz; *Shanku*, the diamond; *Vetâlabhatta*, the emerald; *Ghatakarpara*, the lapis-lazuli; *Kâlidâsa*, the coral; *Varâhamîhira*, the sapphire; and *Vararuchi*, not identified to any specific stone). See ***Nava*** and ***Ratna*** (= 9).

**NAVARÂTRÎ.** Name of the nine-day Feast. See ***Durgâ.***

**NAVATI.** Ordinary Sanskrit name for the number ninety.

**NAYANA.** [S]. Value = 2. "Eye". See ***Netra*** (= 2) and **Two.**

**NEPÂLÎ** (Calendar). Beginning in 879. To find the corresponding date in the Common Era, simply add 879 to a date expressed in this calendar. Still used occassionally in Nepal. Also called *Newârî*. See **Indian calendars.**

**NEPÂLÎ NUMERALS.** Signs derived from *Brahmi numerals through the intermediary notations of Shunga, Shaka, Kushâna, Adhra, Gupta, Nâgarî and Siddham numerals. Currently used mainly in the independent state of Nepal. They are also called *Gurkhali* numerals. The corresponding system is based on the place-value system and has a zero (in the form of a little circle). For ancient numerals, see Fig. 24.41. For modern numerals, see Fig. 24.15. See Fig. 24.52 and 24.61 to 69. See also **Indian written numeral systems (Classification of).**

**NETHER WORLD.** [S]. Value = 7. See ***Pâtâla.*** **Seven.**

**NETRA.** [S]. Value = 2. "Eye". See **Two.**

**NETRA.** [S]. Value 3. "Eye". Symbol used only in regions of Bengal, where this word is generally used to denote the three eyes of *Shiva. See **Three.**

**NEWARÎ (Calendar).** See ***Nepâlî.***

**NIHILISM.** See ***Shûnyatâ*** and **Zero.**

**NIKHARVA.** Name given to the number ten to the power nine. See **Names of numbers** and **High numbers.**

Source: *Sankhyâyana Shrauta Sûtra (date uncertain).

**NIKHARVA.** Name given to the number ten to the power eleven. See **Names of numbers** and **High numbers.**

Sources : *Kitab fi tahqiq i ma li'l hind by al-Biruni (c. 1030 CE); *Lîlâvatî by Bhâskarâchârya (1150 CE); *Ganitakaumudî by Nârâyana (1350 CE); *Trishatikâ by Shrîdharâchârya (date unknown).

**NIKHARVA.** Name given to the number ten to the power thirteen. See **Names of numbers** and **High numbers.**

Source: *Pañchavimsha Brâhmana (date uncertain).

**NIL, NULLITY.** See ***Shûnyatâ*** and **Zero.**

**NÎLAKANTHASOMAYÂJIN.** Indian astronomer c. 1500 CE. His works notably include *Siddhântadarpana*, in which the place-value system with Sanskrit numerical symbols is used frequently [see Sarma, *Siddhântadarpana*]. See **Numerical symbols, Numeration of numerical symbols** and **Indian mathematics (The history of).**

**NINE.** Ordinary Sanskrit names: *nava, *navan. Here is a list of the corresponding numerical symbols: *Abjagarbha, Aja,* *Anka, *Brihatî,* *Chhidra, *Durgâ, *Dvâra,* *Go, *Graha, *Keshava, *Khanda, Laddha, Labdhi, Nanda, Nidhâna, Nidhi, Padârtha,* *Randhra, *Ratna, *Târkshyadhvaja, Upendra, Varsha.* These words have the following literal or symbolic meaning: 1. The Brahman (*Abjagarbha, Aja*). 2. The name of the ninth month of the *chaitradi* year (*Keshava*). 3. The numerals of the place-value system (*Anka*). 4. The "Inaccessible", the "Divine Mother", in allusion to a divinity of the same name (*Durgâ*). 5. The Jewels (*Ratna*). 6. The holes, the orifices (*Chhidra, Randhra*). 7. The planets (*Graha*). 8. The radiance (*Go*). 9. The "Cow" to denote the earth (*Go*). See **Numerical symbols.**

**NINETEEN.** Ordinary Sanskrit name : *navadasha. The corresponding numerical symbol is *Atidhriti. Note that at the beginning of the Common Era, and probably since Vedic times, this number was also called *ekânnavimshati*, which literally means "one away from twenty" [see *Taittirîya Samhitâ*, VII, 2. 11]; but it is also used in its normal form from this time [See *Taittirîya Samhitâ*, XIV, 23; Datta and Singh (1938), pp. 14–15].

**NINETY.** See ***Navati.***

**NINNAHUT.** Name given to the number ten to the power eleven. See **Names of numbers** and **High numbers.**

Source : *Lalitavistara Sûtra (before 308 CE).

**NINNAHUTA.** Name given to the number ten to the power thirty-five. See **Names of numbers** and **High numbers.**

Source: *Vyâkarana (Pâlî grammar) by Kâchchâyana (eleventh century CE).

**NIRABBUDA.** Name given to the number ten to the power sixty-three. See **Names of numbers** and **High numbers.**

Source: *Vyâkarana (Pâlî grammar) by Kâchchâyana (eleventh century CE).

**NIRAVADYA.** Name given to the number ten to the power forty-one. See **Names of numbers** and **High numbers.**

Source : **Lalitavistara Sûtra* (before 308 CE).

**NÎRVANA.** According to Indian philosophers, this is the supreme state of non-existence, reincarnation and absorption of the being in the Brahman. See ***Shûnyatâ*** and **Zero.**

**NIYUTA.** Name given to the number ten to the power five. See **Names of numbers** and **High numbers.**

Sources : **Vâjasaneyî Samhitâ*, **Taittirîya Samhitâ* and **Kâthaka Samhitâ* (from the start of the first millennium CE); **Pañchavimsha Brâhmana* (date uncertain); **Âryabhatîya* (510 CE).

**NIYUTA.** Name given to the number ten to the power six. See **Names of numbers** and **High numbers.**

Source : **Kâthaka Samhitâ* (start of the Common Era).

**NIYUTA.** Name given to the number ten to the power eleven. See **Names of numbers** and **High numbers.**

Source : **Lalitavistara Sûtra* (before 308 CE).

**NON-BEING.** See ***Shûnyatâ*** and **Zero.**

**NON-EXISTENCE.** See ***Shûnyatâ*** and **Zero.**

**NON-PRESENT.** See ***Shûnyatâ*** and **Zero.**

**NON-PRODUCT.** See ***Shûnyatâ*** and **Zero.**

**NON-SUBSTANTIALITY.** See ***Shûnyatâ*** and **Zero.**

**NON-VALUE.** See ***Shûnyatâ*** and **Zero.**

**NOTHING.** See ***Shûnya*** and **Zero.**

**NOTHINGNESS.** See ***Shûnyatâ*** and **Zero.**

**NRIPA.** [S]. Value = 16. "King". This is an allusion to the sixteen kings of the epic poems of the **Mahâbhârata* (Brihadbala, king of Koshala; Chitrasena, king of the Gandharva; Dhritarâshtra, the blind king of Indraprastha; Drupada, king of the Panchala; Jayadrâtha, king of the Sindhu; Kartavîrya, king of the Haihaya; Kâshîpati, king of the Kâshî; Madreshvara, king of the Madra; king Pradîpa; Shatayûpa, ascetic king; Shishupâla, king of the Chedi; Subala, king of Gandhâra; Vajra, king of Indraprastha; Virâta, king of the Matsya; Yavanâdhipa, king of the Yavana; and Yudhisthira, king of Indraprastha). See **Sixteen.**

**NUMBERS (Philosophy and science of).** See ***Samkhya, Samkhyâ, Sâmkhya, Sâmkhyâ,*** **Numerical symbols, Symbolism of words with a numerical value, Symbolism of numbers, *Shûnya, Shûnyatâ,* Zero, Infinity** and **Mysticism of infinity.**

**NUMBERS (The science of).** See ***Samkhyâna.*** See also **Numbers (The philosophy and science of).**

**NUMERAL "0"** (in the form of a little circle). Currently the symbol used in nearly all the numerical notations of India (the following types of modern numerals : Nâgarî, Gujarâtî, Marâthî, Bengali, Oriyâ, Punjabi, Sindhi, Gurûmukhî, Kâithî, Maithilî, Tâkarî, Telugu, Kannara, etc.), of central Asia (Nepali and Tibetan numerals) and of Southeast Asia (Thai-Khmer, Balinese, Burmese, Javanese, etc. numerals). There is evidence of the use of this sign since the seventh century CE in the Indianised civilisations of Southeast Asia (Champa, Cambodia, Sumatra, Bali, etc.). See Fig. 24.3 to 13, 24.15, 16, 21, 24, 25, 26, 28, 39, 41, 42, 50, 51, 52, 78, 79 and 24.80. See **Indian written numeral systems (Classification of).** See also **Circle** and **Zero.**

**NUMERAL "0"** (in the shape of a point or dot). This was formerly in use in the regions of Kashmir and Punjab (Sharada numerals). There is evidence of the use of this sign since the seventh century CE in the Khmer inscriptions of ancient Cambodia. Today, this sign is still used in Muslim India in eastern Arabic numeration ("Hindi" numerals). See Fig. 24.2, 14, 40, 78 and 80. See **Indian written numeral systems (Classification of), Eastern Arabic numerals, Dot** and **Zero.**

**NUMERAL "1". (The origin and evolution of the).** See Fig. 24. 61.

**NUMERAL "2". (The origin and evolution of the).** See Fig. 24. 62.

**NUMERAL "3". (The origin and evolution of the).** See Fig. 24. 63.

**NUMERAL "4". (The origin and evolution of the).** See Fig. 24. 64.

**NUMERAL "5". (The origin and evolution of the).** See Fig. 24. 65.

**NUMERAL "6". (The origin and evolution of the).** See Fig. 24. 66.

**NUMERAL "7". (The origin and evolution of the).** See Fig. 24. 67.

**NUMERAL "8". (The origin and evolution of the).** See Fig. 24. 68.

**NUMERAL "9". (The origin and evolution of the).** See Fig. 24. 69.

**NUMERAL** (as a sign of written numeration). See ***Anka*** and **Signs of numeration.**

**NUMERAL ALPHABET AND COMPOSITION OF CHRONOGRAMS.** Chronograms can be found on certain monuments. These are short phrases written in Sanskrit (or Prakrit), the words of which, when evaluated then totalled according to the numerical value of their letters, give the date of an event which has already taken place or will take place in the future. In Muslim India, the same procedure was used frequently, this time using the numeral letters of the Arabic-Persian alphabet. They are commonly found on epitaphs to express the date of death of the person buried in the tomb. See **Numeral alphabet, Chronogram. Chronograms (System of letter numerals).**

**NUMERAL ALPHABET AND SECRET WRITING.** Like all those who have used an alphabetical numeration, the Indians, Sinhalese, Burmese, Khmers, Thais, Javanese and Tibetans alike have used it to write in a secret code. We still use such systems today to write information or incantatory or magic formulas. In this way numerical series are written to hide their meanings should they fall into the hands of the profane or uninitiated. Likewise, if the order of pages are numbered in this way, it prohibits the profane from reading the texts, thus keeping them secret in a coherent manner; the initiated only has to put the pages in the correct order before he reads the text. See **Numeral alphabet.**

**NUMERAL ALPHABET, MAGIC, MYSTICISM AND DIVINATION.** As with the Greeks, the Jews, the Syrians, the Arabs and the Persians, the Indian mystics, Magi and soothsayers used their numeral alphabets as the basic instruments of their magical, divinatory or numerological interpretations or practices. A whole mystical-religious practice, just like gnosis, Judaeo-Christian Cabbala or Muslim Sufism was created in this manner. This led to all kinds of homilectic and symbolic interpretations, to various predictive calculations and to the creation of certain **kavachas*, talismans curiously resembling Hebrew Cabbalistic pentacles and Muslim *herz* from North Africa. The practice was based on a doctrine of sound and the Sanskrit alphabet: **bîjas* or "letter-seeds", where each syllable of the alphabet characterised a divinity of the Brahmanic pantheon (or of the pantheon of tantric Buddhism in the schools in the North), whom it was believed that one could evoke just by pronouncing the letter. The sound, by definition, was considered to be the creative and evocative element *par excellence*. Hence the mystical value attached to each letter in association with the esoterical meaning of its numerical equivalent. The external sound of the voice is born in the secret centre of the person in the form of the essence of the sound, and passes through three vibratory processes before becoming audible: *parâ, pashyantî* and *madhyamâ*. Beginning subtly, the sound turns into one of the forty-six letters of the Sanskrit alphabet. As the sound is transmitted by the **nâdî*, it becomes one or another of the Sanskrit alphabetical letters. The matter, in Hindu cosmology, is divided into five states of manifestation : air, fire, earth, water, ether. Each state corresponds to a Sanskrit letter as is shown in the following table:

| | |
|---|---|
| Air | (**Vâyu*): *ka, kha, ga, gha, na, a, â, ri, ha, sha, ya.* |
| Fire | (**Agnî*): *cha, chha, ja, jha, ña, i, î, rî, ksa, ra.* |
| Earth | (**Prithivî*): *ta, tha, da, dha, na, u, û, li, sha, va, la.* |
| Water | (**Âpa*): *ta, tha, da, dha, na, ê, ai, lî, sa.* |
| Ether | (**Âkâsha*): *pa, pha, ba, bha, ma, o, au, am, ah.* |

We can now understand the principle of the creation of a **mantra*, which is a combination of sounds which have been carefully studied in terms of their secret values. It is not worth trying to make intelligible sense of a mantra because this is not its aim; just as certain numerical combinations enabled Cabbalists to invent ingenious secret names, names which are impossible to translate (they are artificial creations), the mantra is a precise combination of sounds created with some secret aim in mind [Marquès-Rivière (1972)] See ***Âkshara,*** **Numeral alphabet, *Bhûta, Mahâbhûta, Trivarna, Vâchana.*** See also Chapter 20, for similar practices in other cultures.

**NUMERAL ALPHABET.** This denotes any system of representing numbers which uses vocalised consonants of the Indian alphabet, to which a numerical value is assigned, in a predetermined, regular order. In keeping with their diverse systems of recording numbers (in numerals, in symbols or spoken), the Indians knew and used different systems of this kind. This is what is conveyed by the collective name **varnasankhya*, or systems of "letter-numbers".

The inventor of the first numerical alphabet in Indian history was the astronomer *Âryabhata, who, c. 510 CE, had the idea of using the thirty-three letters of the Indian alphabet to represent all the numbers from 1 to $10^{18}$. His aim in creating this system was to express the constants of his **astronomical canon*, as well as the numerical data of his diverse speculations on **yugas*. See **Âryabhata (Numerical notations of),**

**Âryabhata's numeration, Yuga (astronomical speculation about).**

After Âryabhata, many other numeral alphabets were invented using Indian letters. These vary both according to the numerical value of the letters and the period and region, and sometimes even the principle employed in the numerical representations.

One such system is the *katapayâdi* system, which is still called **varnasamjña* (or "proceeding from syllables"); it was almost certainly created by the astronomer *Haridatta in the ninth century CE. and later adopted by many astronomers, including Shankaranârâyana (c. 869 CE). It is a simplified version of Âryabhata's system; the successive vocalisations of the consonants of the Indian alphabet are suppressed. Each value which is superior or equal to ten is replaced with a zero or one of the first nine units. The author of the system thus transformed the earlier system into an alphabetical numeration which used the place-value system and zero. See **Katapayâdi numeration**.

Amongst the diverse alphabetical notations, it is also worth mentioning the **aksharapallî* system, which is frequently used in Jaina manuscripts. Such systems are still in use today in various regions of India, from Maharashtra, Bengal, Nepal and Orissa to Tamil Nadu, Kerala and Karnataka. They are also found amongst the Sinhalese, the Burmese, the Khmers, the Thais and the Javanese. They can also be found amongst the Tibetans, who have long used their letters as numerical signs, particularly when numbering their registers and the pages of their manuscripts. See Chapters 17 to 20 for similar uses in other cultures.

**NUMERAL.** [S]. Value = 9. See ***Anka*** and **Nine**.

**NUMERATION OF NUMERICAL SYMBOLS.** Name given here to the place-value system written using Sanskrit numerical symbols, used by Indian astronomers and mathematicians since at least the fifth century CE. In Sanskrit, this is often called **samkhya* (or **sankhya*). See **Numerical symbols (Principle of the numeration of)**.

**NUMERICAL NOTATION.** Here is an alphabetical list of terms relating to this notion, which appear as headings in this dictionary: **Aksharapallî*, **Ândhra numerals*, **Anka*, **Ankakramena*, **Ankânâm Vamâto Gatih*, **Ankapallî*, **Ankasthâna*, *Arabic numeration (Positional systems of Indian origin), *Âryabhata's numeration, *Brâhmî numerals, *Eastern Arabic numerals, *High numbers, *Indian numerals, *Indian written numeral systems (Classification of), *Indusian numeration, *Katapayâdi numeration, *Kharoshthî numeration, *Numeral alphabet, *Numeral 1, *Numeral 2, etc., *Numerical symbols (Principle of the numeration of), *Sanskrit *Sthâna, *Sthânakramâd, *Varnasamjnâ and *Zero.

**NUMERICAL SYMBOLS.** These are words which are given a numerical value depending what they represent. They can be taken from nature, the morphology of the human body, representations of animal or plants, acts of daily life, any types of tradition, philosophical, literary or religious elements, attributes and morphologies connected to the divinities of the Hindu, Jaina, Vedic, Brahmanic, Buddhist, etc. pantheons, legends, traditional associations of ideas, mythologies or social conventions of Indian culture. See **Symbols**. See also all entries entitled **Numerical symbols** or **Symbolism of numbers**.

**NUMERICAL SYMBOLS (General alphabetic list).** These are Sanskrit numerical symbols which are found in texts on mathematics or astronomy, as well as in various Indian epigraphic inscriptions (this list is not exhaustive):

**Abdhi* (= 4), **Abhra* (= 0), **Abja* (= 1), *Abjadala* (= 100), *Abjagarbha* (= 9), *Achala* (= 7), **Adi* (= 1), **Âditya* (= 12), **Adri* (= 7), **Aga* (= 7), *Aghosha* (= 13), **Agni* (= 3), **Ahar* (= 15), **Ahi* (= 8), *Airâvata* (= 1), *Aja* (= 9), **Âkashâ* (= 0), **Akriti* (= 22), **Âkshara* (= 1), *Akshauhinî* (= 11), *Akshi* (= 2), **Amara* (= 33), *Ambaka* (= 2), **Ambara* (= 0), **Ambhodha* (= 4), *Ambhodhi* (= 4), *Ambhonidhi* (= 4), **Ambodha* (= 4), *Ambodhi* (= 4), *Ambudhi* (= 4), **Amburâshi* (= 4), **Anala* (= 3), **Ananta* (= 0), **Anga* (= 6), **Anguli* (= 10), **Anguli* (= 20), *Anîka* (= 8), **Anka* (= 9), **Antariksha* (= 0), **Anushtubh* (= 8), **Âptya* (= 3), *Arhat* (= 24), *Ari* (= 6), **Arjunâkara* (= 1,000), **Arka* (= 12), **Arnava* (= 4), *Artha* (= 5), **Âshâ* (= 10), *Âshrama* (= 4), **Ashti* (= 16), **Ashva* (= 7), **Ashvin* (= 2), **Ashvina* (= 2), **Ashvinau* (= 2), **Atidhriti* (= 19), *Atijagatî* (= 13), **Âtman* (= 1), **Atri* (= 7), **Atrinayanaja* (= 1), **Atyashti* (= 17), **Avani* (= 1), **Avatâra* (= 10), *Aya* (= 4), *Âya* (= 4), *Ayana* (= 2).

**Bâhu* (= 2), **Bâna* (= 5), *Bandhu* (= 4), **Bha* (= 27), **Bhânu* (= 12), **Bharga* (= 11), *Bhâva* (= 5), **Bhava* (= 11), *Bhaya* (= 7), **Bhû* (= 1), **Bhûbrit* (= 7), **Bhûdhara* (= 7), **Bhûmi* (= 1), **Bhûpa* (= 16), **Bhûta* (= 5), *Bhûti* (= 8), **Bhuvana* (= 3), **Bhuvana* (= 14), **Bindu* (= 0), **Brahmâsya* (= 4), *Brihatî* (= 9).

**Chakra* (= 12), **Chakshus* (= 2), *Chandah* (= 7), *Chandas* (= 7), **Chandra* (= 1), **Chaturânanavâdana* (= 4), **Chhidra* (= 9).

*Dadhi* (= 4), **Dahana* (= 3), **Danta* (= 32), **Dantin* (= 8), **Darshana* (= 6), **Dasra* (= 2), **Deva* (= 33), **Dharâ* (= 1), **Dharanî* (= 1), **Dhârtarâshtra* (= 100), **Dhâtrî* (= 1), *Dhâtu* (= 7), *Dhî* (= 7), **Dhriti* (= 18), **Dhruva* (= 1), **Diggaja* (= 8), *Dik* (= 8), **Dikpâla* (= 8), *Dina* (= 15), **Dish* (= 4), **Dish* (= 8), **Disha* (= 10), **Dishâ* (= 4), * *Disha* (= 10), **Divâkara* (= 12), *Dosha* (= 3), **Dravya* (= 6), **Drishti* (= 2), **Durgâ* (= 9), *Durita* (= 8), **Dvandva* (= 2), *Dvâra* (= 9), **Dvaya* (= 2), **Dvija* (= 2), **Dvipa* (= 8), **Dvîpa* (= 7), *Dvirada* (= 8), *Dyumani* (= 12).

**Gagana* (= 0), **Gaja* (= 8), *Gangâmarga* (= 3), **Gati* (= 4), **Gavyâ* (= 5), **Gâyatrî* (= 24), *Ghasra* (= 15), **Giri* (= 7), **Go* (= 1), **Go* (= 9), *Gostana* (= 4), **Graha* (= 9), *Grahana* (= 2), **Gulpha* (= 2), **Guna* (= 3), **Guna* (= 6).

**Hara* (= 11), **Haranayana* (= 3), **Haranetra* (= 3), **Haribâhu* (= 4), **Hastin* (= 8), *Haya* (= 7), *Himagu* (= 1), *Himakara* (= 1), *Himâmshu* (= 1), **Hotri* (= 3), **Hutâshana* (= 3).

*Ibha* (= 8), *Îkshana* (= 2), *Ilâ* (= 1), **Indra* (= 14), **Indradrishti* (= 1,000), **Indriya* (= 5), **Indu* (= 1), **Iryâ* (= 4), **Îsha* (= 11), **Îshadrish* (= 3), **Ishu* (= 5), **Îshvara* (= 11).

**Jagat* (= 3), **Jagat* (= 14), **Jagatî* (= 1), **Jagatî* (= 12), **Jagatî* (= 48), **Jahnavîvaktra* (= 1,000), **Jala* (= 4), **Jaladharapatha* (= 0), **Jaladhi* (= 4), **Jalanidhi* (= 4), *Jalâshaya* (= 4), *Jana* (= 1), *Janghâ* (= 2), *Jânu* (= 2), *Jâti* (= 22), *Jina* (= 24), **Jvalana* (= 3).

**Kakubh* (= 10), **Kâla* (= 3), *Kalâ* (= 16), *Kalamba* (= 5), *Kalatra* (= 7), **Kâma* (= 13), **Kara* (= 2), *Kâraka* (= 6), **Karanîya* (= 5), *Karman* (= 8), *Karman* (= 10), *Karna* (= 2), **Kârttikeyâsya* (= 6), *Kashâya* (= 4), **Kâya* (= 6), *Kendra* (= 4), **Keshava* (= 9), **Kha* (= 0), *Khanda* (= 9), *Khara* (= 6), *Khatvâpâda* (= 4), *Koshtha* (= 4), **Krishânu* (= 3), **Krita* (= 4), **Kriti* (= 20), *Kritin* (= 22), *Kshapâkara* (= 1), **Kshapeshvara* (= 1), *Kshâra* (= 5), **Kshaunî* (= 1), **Kshemâ* (= 1), **Kshiti* (= 1), **Kshoni* (= 1), **Ku* (= 1), *Kucha* (= 2), **Kumârâsya* (= 6), **Kumâravadana* (= 6), **Kuñjara*= 6), (= 8), *Kutumba* (= 2).

*Labdha* (= 9), *Labdhi* (= 9), *Lâbha* (= 11), *Lakâra* (= 10), *Lavana* (= 5), *Lekhya* (= 6), **Loka* (= 3), **Loka* (= 7), **Loka* (= 14), **Lokapâla* (= 8), **Lochana* (= 2).

*Mada* (= 8), **Mahâbhûta* (= 5), **Mahâdeva* (= 11), **Mahâpâpa* (= 5), **Mahâyajña* (= 5), **Mahî* (= 1), **Mahîdhara* (= 7), *Mala* (= 6), **Mangala* (= 8), *Manmatha* (= 13), **Manu* (= 14), **Mârgana* (= 5), **Mârtanda* (= 12), **Mâsa* (= 12), **Mâsârdha* (= 6), **Matanga* (= 8), **Mâtrika* (= 7), **Mrigânka* (= 1), **Mukha* (= 4), *Mûlaprakriti* (= 1), **Muni* (= 7), **Mûrti* (= 3), **Mûrti* (= 8).

**Nabha* (= 0), **Nabhas* (= 0), *Nâdî* (= 3), *Nadîkûla* (= 2), **Naga* (= 7), **Nâga* (= 8), **Nakha* (= 20), **Nakshatra* (= 27), *Nanda* (= 9), *Naraka* (= 40), **Nâsatya* (= 2), *Naya* (= 2), *Nâyaka* (= 1), **Nayana* (= 2), **Netra* (= 2), **Netra* (= 3), *Nidhâna* (= 9), *Nidhi* (= 9), **Nripa* (= 16).

*Oshtha* (= 2).

*Padârtha* (= 9), **Paksha* (= 2), **Paksha* (= 15), *Pallava* (= 5), **Pândava* (= 5), *Pankti* (= 10), **Parabrahman* (= 1), *Parva* (= 5), *Parvan* (= 5), **Parvata* (= 7), **Pâtaka* (= 5), **Pâtâla* (= 7), **Pâvaka* (= 3), **Pâvana* (= 5), **Pâvana* (= 7), *Payodhi* (= 4), *Payonidhi* (= 4), **Pinâkanayana* (= 3), **Pitâmaha* (= 1), **Prakriti* (= 21), *Prâleyâmshu* (= 1), **Prâna* (= 5), **Prithivî* (= 1), **Pura* (= 3), **Purâ* (= 3), **Purânalakshana* (= 5), **Pûrna* (= 0), *Purushârtha* (= 4), *Purushâyus* (= 100), *Pûrva* (= 14), **Pushkara* (= 7), *Pushkarin* (= 8), **Putra* (= 5).

**Rada* (= 32), **Râga* (= 6), *Rajanîkara* (= 1), **Râma* (= 3), *Râmanandana* (= 2), **Randhra* (= 0), **Randhra* (= 9), **Rasa* (= 6), **Râshi* (= 12), *Rashmi* (= 1), **Ratna* (= 3), **Ratna* (= 5), **Ratna* (= 9), **Ratna* (= 14), **Râvanabhuja* (= 20), **Râvanashiras* (= 10), **Ravi* (= 12), **Ravibâna* (= 1,000), **Ravichandra* (= 2), *Ripu* (= 6), **Rishi* (= 7), **Ritu* (= 6), **Rudra* (= 11), **Rudrâsya* (= 5), **Rûpa* (= 1).

**Sâgara* (= 4), **Sâgara* (= 7), **Sahasrâmshu* (= 12), *Sahodarâh* (= 3), *Salilâkara* (= 4), **Samudra* (= 4), **Samudra* (= 7), *Sankrânti* (= 12), **Sarpa* (= 8), **Sâyaka* (= 5), *Senânga* (= 4), **Senânînetra* (= 12), **Shadâyatana* (= 6), **Shaddarshana* (= 6), **Shâdgunya* (= 6), **Shaila* (= 7), **Shakra* (= 14), *Shakrayajña* (= 100), **Shakti* (= 3), **Shankarâkshi* (= 3), **Shanmukha* (= 6), **Shanmukhabâhu* (= 12), **Shara* (= 5), **Shashadhara* (= 1), **Shashanka* (= 1), **Shashin* (= 1), *Shâstra* (= 5), *Shâstra* (= 6), **Sheshashîrsha* (= 1,000), **Shikhin* (= 3), **Shîtâmshu* (= 1), **Shîtarashmi* (= 1), **Shiva* (= 11), **Shruti* (= 4), **Shukranetra* (= 1), **Shûla* (= 3), **Shûlin* (= 11), **Shûnya* (= 0), *Shveta* (= 1), *Siddha* (= 24), **Siddhi* (= 8), **Sindhu* (= 4), *Sindhura* (= 8), **Soma* (= 1), **Sudhâmshu* (= 1), **Sura* (= 33), **Sûrya* (= 12), **Suta* (= 5), *Svagara* (= 21) **Svara* (= 7).

*Takshan (= 8), *Tâna (= 49), Tanmâtra (= 5), *Tanu (= 1), *Tanu (= 8), *Tapana (= 3), *Tapana (= 12), Tarka (= 6), Târkshadhvaj (= 9), Tata (= 5), *Tattva (= 5), *Tattva (= 7), *Tattva (= 25), *Tithi (= 15), *Trailokya (= 3), *Trayî (= 3), Tridasha (= 33), Trigata (= 3), *Triguna (= 3), *Trijagat (= 3), *Trikâla (= 3), *Trikâya (= 3), *Triloka (= 3), *Trimûrti (= 3), *Trinetra (= 3), *Tripurâ (= 3), *Triratna (= 3), *Trishiras (= 3), Trishtubh (= 11), *Trivarga (= 3), *Trivarna (= 3), *Tryakshamukha (= 5), *Tryambaka (= 3), *Turaga (= 7), *Turangama (= 7), *Turîya (= 4).

*Uchchaishravas (= 1), *Uda (= 27), *Udadhi (= 4), *Udarchis (= 3), Upendra (= 9), *Utkriti (= 26), * Urvarâ (= 1).

*Vâchana (= 3), *Vahni (= 3), *Vaishvânara (= 3), *Vâjin (= 7), Vanadhi (= 4), *Vâra (= 7), *Vâridhi (= 4), *Vârinidhi (= 4), Varsha (= 9), *Vasu (= 8), *Vasudhâ (= 1), *Vasundharâ (= 1), *Vâyu (= 49), *Veda (= 3), *Veda (= 4), *Vidhu (= 1), Vidyâ (= 14), *Vikriti (= 23), *Vindu (= 0), Vishanidhi (= 4), *Vishaya (= 5), *Vishikha (= 5), *Vishnupada (= 0), Vishtapa (= 3), Vishuvat (= 2), *Vishva (= 13), *Vishvadeva (= 13), Viyata (= 0), Vrata (= 5), *Vyant (= 0), Vyasana (= 7), Vyaya (= 12), *Vyoman (= 0), Vyûha (= 4).

*Yama (= 2), Yâma (= 8), *Yamala (= 2), *Yamau (= 2), Yati (= 7), *Yoni (= 4), *Yuga (= 2), *Yuga (= 4), *Yugala (= 2), *Yugma (= 2).

To gain an idea of the symbolism of these words, see **Symbolism of words with a numerical value** and **Symbolism of numbers**. The first of these two entries gives an alphabetical list of English terms which explain the various corresponding associations of ideas, and the second entry gives a list of the same associations of ideas, set out this time in numerical order (one, two, three, etc.). To understand the principle for using word-symbols to represent numbers, see **Numerical symbols (Principle of the numeration of)**.

Source: Bühler (1896), pp. 84ff; Burnell (1878); Datta and Singh (1938), pp. 54–7; Fleet, in : CIIn, VIII; Jaquet, in : JA, XVI, 1835; Renou and Filliozat (1953), p. 708–9; Sircar (1965), pp. 230–3; Woepcke (1863).

**NUMERICAL SYMBOLS (Principle of the numeration of).** Procedure used to record numbers by Indian scholars since at least as early as the fifth century CE. This is simply a series of Sanskrit word-symbols (which are used as names of units), which are written in conformity with the "principle of the movement of numerals from the right to the left" (*ankânâm vâmato gatih*). See **Sanskrit** and **Numerical symbols**.

In other words, in this system, numerical symbols have a variable value depending on their position when numbers are written down. The system possesses several different special terms which symbolise zero and which thus serve to mark the absence of units in any given decimal order in this positional notation (*shûnya, *âkâsha, *abhra, *ambara, *antariksha, *bindu, *gagana, *jaladharapatha, *kha, *nabha, *nabhas, etc.). An expression such as:

*agni. shûnya. ashvi. vasu.*

[literally : "fire (= 3). void (= 0). Horsemen (= 2). Vasu (= 8)"] corresponds to the numbers: $3 + 0 \times 10 + 2 \times 10^2 + 8 \times 10^3 = 8{,}203$.

This method of expressing numbers uses the place-value system and zero. What is remarkable about it is that Indian scholars are the only ones to have invented such a system. See **Position of numerals**, and **Zero**.

**NUMERICAL SYMBOLS (Sanskrit designation of).** The generic term for words used as numerical symbols is *samkhya*, which literally means "number". Also used to refer to the system as a whole, which is the place-value system expressed through numerical symbols.

**NUMEROLOGY.** See **Numeral alphabet, magic, mysticism and divination.**

**NYARBUDA.** Name given to the number ten to the power eight (= one hundred million). See **Names of numbers** and **High numbers.**

Sources: *Vâjasaneyî Samhitâ* (beginning of the Common Era); **Taittirîya Samhitâ* (beginning of the Common Era); **Kâthaka Samhitâ* (beginning of the Common Era); **Pañchavimsha Brâhmana* (date uncertain); **Sankhyâyana Shrauta Sûtra* (date uncertain).

**NYARBUDA.** Name given to the number ten to the power 11. See **Names of numbers** and **High numbers.**

Source: **Ganitasârasamgraha* by Mahâhavîrâchârya (850 CE).

# O

**OCEAN.** Name given to the number ten to the power four, ten to the power nine or ten to the power fourteen. See ***Jaladhi***, ***Samudra*** and **High numbers.**

**OCEAN.** [S]. The entries entitled **sâgara* or **samudra*, which, as numerical symbols, translate the idea of "sea" or "ocean", can have the value of either 4 or 7. The relation between *sâgara* and 4 can be explained through the allusion to the "four oceans" (**chatursâgara*) which, according to Hindu and Brahmanic mythologies, surround **Jambudvîpa*, (India). However, this explanation does not give the real reason for the choice of the number four for the oceans surrounding India. In reality, it is due to the fact that the mystical symbol for "water" (**jala*) is the number four. According to Brahmanic doctrine of the five elements of the manifestation (**bhûta*), water (which is also called **apa*) forms, along with earth (*prithivî*), air (*vâyu*) and fire (*agnî*), the ensemble of elements which are said to participate directly in the "material order of nature". This order is believed to be quaternary, and the diverse phenomena of life boil down to the manifestations of these four elements in the determination of the essence of the forces of nature as well as in the realisation of the latter in its work of generation and destruction. In traditional Indian thought (and even according to a universal constant), the earth itself corresponds symbolically to the number four, because it is associated with a square due to its four horizons (or cardinal points).

As for the relationship between **sâgara* and the number seven, this can be explained by direct reference to the seven mythical oceans (namely: The ocean of salt water, the ocean of sugar cane juice, the ocean of wine, the ocean of thinned butter, the ocean of whipped cheese, the ocean of milk and the ocean of soft water), which are meant to surround **Mount Meru*. See ***Sapta sâgara***.

Mount Meru is the mythical and sacred mountain of Brahman mythology and Hindu cosmology, which constitutes the meeting place and dwelling of the gods. Situated at the centre of the universe, this mountain is placed above seven hells (**pâtâla*), and has seven faces, each one looking at one of the seven "island-continents", themselves each in one of the seven oceans, etc. In this symbolism, Mount Meru represents the total fixedness and the absolute centre around which the firmament and the whole universe pivot in their eternal course. This image is connected to one of the universal symbolic representations of the *Pole star. Mount Meru is said to be situated directly underneath this star, and along exactly the same axis. This symbolic correspondence comes from the fact that the Pole star, the **Sudrishti*, "That which never moves", is the last of the seven stars of the Little Bear, which themselves are considered by Indian tradition to be the personification of the seven "great Sages" (in other words the **saptarishi* of Vedic times, believed to be the authors of hymns and invocations, as well as of the most important texts of the **Veda*). This is why the number seven came to play a preponderant symbolic role in the mythological representations associated with Mount Meru.

It is this symbolism which determined the number of cosmic oceans in the legends about the creation of the universe, and gave words expressing the idea of "ocean" a value of 7. In its representations, India, (*Jambudvîpa*) is considered to be the "centre of the earth", whilst Mount Meru was regarded as the centre of the universe. Ocean has two different numerical values in order to mark the opposition between the human character, essentially terrestrial, of the oceans surrounding India, and the divine character, essentially celestial, of the oceans surrounding Mount Meru. In spite of the apparent paradox, Indian scholars managed to avoid any confusion. The words **samudra* and **sâgara*, which both mean "ocean", were both sometimes used as symbols for the number four. But they were usually used (never simultaneously) to express the number seven, words such as **abdhi*, **ambhonidhi*, *ambudhi*, **amburâshi*, **jaladhi*, **jalanidhi*, **jalâshaya*, **sindhu*, **vâridhi* or **vârinidhi* being reserved for the number four, and which more modestly meant "sea".

**OCEAN.** [S]. Value = 4. See ***Abdhi***, ***Ambhonidhi***, ***Ambudhi***, ***Amburâshi***, ***Arnava***, ***Jaladhi***, ***Jalanidhi***, ***Sâgara***, ***Samudra***, ***Sindhu***, ***Udadhi***, ***Vâridhi*** and ***Vârinidhi***. See also **Four**, ***Jala***.

**OCEAN.** [S]. Value = 7. See ***Sâgara*** and ***Samudra***. See also **Seven, Mount Meru.**

**OLD KHMER NUMERALS.** Symbols derived from *Brâhmî numerals and influenced by Shunga, Shaka, Kushâna, Ândhra, Pallava, Châlukya, Ganga, Valabhî, "Pâli" and Vatteluttu numerals. Used from the seventh century CE in the ancient kingdom of Cambodia. The notation used for dates in the **Shaka* era were based on a place-value system and had a zero (a dot or small circle), whereas vernacular notation was very rudimentary. See: **Indian written numerals systems (Classification of)**. See Fig., 24.52, 61 to 69, 77, 78 and 80.

**ONE.** Ordinary Sanskrit name for this number: **Eka*. Here is a list of corresponding numerical symbols: **Abja*, **Âdi*, *Airâvata*, **Âkshara*, **Âtman*, **Atrinayanaja*, **Avani*, **Bhû*, **Bhûmi*, **Chandra*, **Dharâ*, **Dharanî*, **Dhâtrî*, **Dhruva*, **Go*, *Himagu*, *Himakara*, *Himâmshu*, *Ilâ*, **Indu*, **Jagatî*, *Jana*, *Kshapâkara*, **Kshapeshvara*, **Kshaunî*, **Kshemâ*, **Kshiti*, **Kshoni*,

*Ku, *Mahî, *Mrigânka, Mûlaprakriti, Nâyaka, *Parabrahman, *Pitâmaha, Prâleyâmshu, *Prithivî, Rajanîkara, Rashmi, *Rûpa, *Shashadhara, *Shashanka, *Shashin, Shveta, *Shîtâmshu, *Shîtarashmi, *Shukranetra, *Soma, *Sudhâmshu, *Tanu, *Uch-chaishravas, *Urvarâ, *Vasudha, *Vasundharâ, *Vidhu.* These words have the following translation or symbolic meaning: 1. The "Moon". (*Abja, Atrinayanaja, Chandra, Indu, Jagatî, Kshapeshvara, Mrigânka, Shashadhara, Shashanka, Shashin, Shîtâmshu, Shîtarashmi, Soma, Sudhâmshu, Vidhu*). 2. The drink of immortality (*Soma*). 3. The "Earth" (*Avani, Bhû, Bhûmi, Dharâ, Dharanî, Dhâtrî, Go, Jagatî, Kshaunî, Kshemâ, Kshiti, Kshoni, Ku, Mahî, Prithivî, Urvarâ, Vasudha, Vasundharâ*). 4. The "Ancestor", the "First Father", the "Great Ancestor" (*Pitâmaha*). 5. Individual soul, supreme soul, ultimate Reality, the Self (*Âtman*). 6. The Brahman (*Âtman, Pitâmaha, Parabrahman*). 7. The beginning (*Âdî*). 8. The body (*Tanu*). 9. The Pole star (*Dhruva*). 10. The form (*Rûpa*). 11. The "drop" (*Indu*). 12. The "immense" (*Prithivî*). 13. The "Indestructible" (*Akshara*). 14. The rabbit (*Shashin, Shashadhara*). 15. The "Luminous", in allusion to the moon as a masculine entity (*Chandra*). 16. The "cold Rays" of the moon (*Shîtamshu, Shîtarashmi*). 17. The terrestrial world (*Prithivî*). 18. The eye of Shukra (*Shukranetra*). 19. The "Bearer", in allusion to the earth (*Dharanî*). 20. The primordial principle (*Âdî*). 21. Rabbit figure (*Shashadhara*). 22. The Cow (*Go, Mahî*). 23. Curdled milk (*Mahî*). See **Numerical symbols**.

**OPINION.** [S]. Value = 6. See ***Darshana*** and **Six**.

**ORDERS OF BEINGS (The five).** See ***Pañchaparamesthin***.

**ORIFICE.** [S]. Value = 9. See Chhidra, Randhra and Nine.

**ORIGINAL SERPENT (Myth of the).** See **Infinity (Indian mythological representation of)** and **Serpent (Symbolism of the).**

**ORISSÎ NUMERALS.** See **Oriyâ Numerals**.

**ORIYÂ NUMERALS.** Symbols derived from *Brâhmî numerals and influenced by Shunga, Shaka, Kushâna, Ândhra, Gupta, Nâgarî, Kutilâ and Bengali. Now used mainly in the state of Orissâ. Also called Orissî numerals. The symbols correspond to a mathematical system that has place values and a zero (shaped like a small circle). See **Indian written numeral systems (Classification of)**. See Fig. 24.12, 52 and 24.61 to 69.

**OUROBOUROS.** See **Infinity (Indian mythological representation of)** and **Serpent (Symbolism of the).**

# P

**PADMA** (or **PADUMA**). This is the name for the pink lotus. As well as the purity it represents, to the Indian mind it symbolises the highest divinity as well as innate reason.

**PADMA.** Name given to the number ten to the power nine. See **Names of numbers**. See also **High numbers (The symbolic meaning of).**

Source: **Kitab fi tahqiq i ma li'l hind* by al-Biruni (c. 1030 CE).

**PADMA.** Name given to the number ten to the power fourteen. See **Names of numbers**. For an explanantion of this symbolism, see **Padma** (or **Paduma**). See also **High numbers (The symbolic meaning of).**

Source: **Ganitasârasamgraha* by Mahâvîrâchârya (850 CE).

**PADMA.** Name given to the number ten to the power twenty-nine. See **Names of numbers**. See also **High numbers (The symbolic meaning of).**

Source: **Râmâyana* by Vâlmîki (early centuries CE).

**PADUMA.** Literally, "(pink) lotus". Name given to the number ten to the power twenty-nine. See **Names of numbers**. See also **High numbers (The symbolic meaning of).**

Source: **Lalitavistara Sûtra* (before 308 CE).

**PADUMA.** Name given to the number ten to the power 119. See **Names of numbers**. See also **High numbers (The symbolic meaning of).**

Source: **Vyâkarana* (Pali grammar) by Kâchchâyana (eleventh century CE).

**PAIR.** [S]. Value = 2. See **Dvaya** and **Two**.

**PAKOTI.** Name given to the number ten to the power fourteen. See **Names of numbers** and **High numbers.**

Source: **Vyâkarana* (Pali grammar) by Kâchchâyana (eleventh century CE).

**PAKSHA.** [S]. Value = 2. "Wing". This is due to the symmetry of this organ. The word can also mean one of the two halves of a month. Thus it is sometimes also used to represent the number fifteen. This double symbolism can be explained by the division of the month (**mâsa*) into two periods of fifteen days called *paksha*, each one corresponding to one phase of the moon. The first, called "shining" (*shudi*), is progressive, and the second, called "shadow" (*badi*), is degressive. According to Hindu mythology and cosmogony, these two periods formed one whole being (before the churning of the sea of milk); this being was decapitated by Indra when he drank the **amrita* (the nectar of eternal life) that he had stolen. This created the "Cut in twos" (*Ashleshâbava*): two beings named *Râhu and *Ketu, who personify the ascending and descending nodes of the moon. See ***Mâsa***, ***Râhu*** and **Two**.

**PAKSHA.** [S]. Value = 15. See **Fifteen**.

**"PÂLÎ" NUMERALS.** Symbols derived from *Brâhmî numerals and influenced by Shunga, Shaka, Kushâna, Ândhra, Pallava, Châlukya, Ganga and Valabhî. Formerly used in Magadha (the ancient Hindu kingdon of present-day Bihar, south of the Ganges) from the Mauryan period. All the later numeral symbols of the eastern and southeast Asia (Mon, Burmese, Cham, Ancient Khmer, Thai-Khmer, Balinese, etc.) derive from Pâlî numerals. The symbols corresponded to a mathematical system that was not based on place values and therefore did not possess a zero. See: **Indian written numerals systems (Classification of)**. See Fig. 24.52 and 24.61 to 69.

**PALLAVA NUMERALS.** Symbols derived from •Brâhmî numerals and influenced by Shunga, Shaka, Kushâna and Ândhra, arising at the time of the Pallava dynasty (fourth to sixth centuries CE). The symbols correspond to a mathematical system that was not based on place values and therefore did not possess a zero. See: **Indian written numeral systems (Classification of)**. See Fig. 24.37, 24.61 to 24.69 and 24.70.

**PAÑCHA.** Ordinary Sanskrit term for the number five, which appears in many words which have a direct relationship with the idea of this number. Examples:

**Pañchabâna, *Pañchâbhijñâ, *Pañchabhûta, *Pañchachakshus, *Pañchadisha, *Pañchagavyâ, *Pañcha Indriyâni, *Pañcha Jâti, *Pañchaklesha, *Pañchânana, *Pañchanantarya, *Pañchaparameshtin, *Pañcharâshika, *Pañchatantra.*

For words which have a more symbolic relationship with this number, see **Five** and **Symbolism of numbers**.

**PAÑCHABÂNA.** "Bow of five flowers". This is one of the attributes of *Kâma, Hindu divinity of Cosmic Desire and Carnal Love, who is generally invoked in marriage ceremonies. Kâma is often represented as a young man armed with a bow of sugar cane and five arrows covered in or constituted by five flowers.

**PAÑCHÂBHIJÑÂ.** Name given by the Sinhalese to the "five supernatural powers" of Buddha. The Buddhists of Sri Lanka only recognise five of the six *Abhijña*, or "supernatural powers", which other Buddhist philosophies believe in.

**PAÑCHABHÛTA.** "Five elements". Collective name for the five elements of the manifestation of Brahman and Hindu philosophies. See ***Bhûta*** and ***Jala***.

**PAÑCHACHAKSHUS.** "Five visions of Buddha". According to Buddhists, Buddha possesses the five following types of visions: that of the body, of the divine form, wisdom, doctrine and of his eye.

**PAÑCHADASHA.** Ordinary Sanskrit name for the number fifteen. For words with a symbolic relationship with this number, see **Fifteen** and **Symbolism of numbers**.

**PAÑCHADISHA.** "Five horizons". These are the four cardinal points plus the zenith. See ***Dish***.

**PAÑCHAGAVYÂ.** "Five gifts of the Cow". See ***Gavyâ***.

**PAÑCHA INDRIYÂNI.** "Five faculties". These are the mental and physical faculties of Buddhist philosophy, which are divided into five groups. See ***Indriya***.

**PAÑCHA JÂTI.** Name of the five fundamental arithmetical rules of the reduction of fractions.

**PAÑCHAKLESHA.** "Five impurities". According to Hindu and Buddhist philosophies, these are the five major obstacles which keep the faithful off the Way of Realisation. See ***Mahâpâpa***.

**PAÑCHÂNANA.** Name of the five heads of *Rudra. See ***Rudrâsya***.

**PAÑCHÂNANTARYA.** "Five mortal sins" of Buddhism. These are the following sins: parricide; matricide; the killing of an *arhat* (a saint issued from *karma*); causing division in the Buddhist community (*sangham*); and wounding a Buddha.

**PAÑCHAPARAMESHTIN.** Name of the five orders of beings, considered to be the "five treasures" (*Pañcha Ratna*) of *Jaina religion.

**PAÑCHAPARÂSHIKA.** [Arithmetic]. Sanskrit name for the Rule of Five.

**PAÑCHÂSHAT.** Ordinary Sanskrit name for the number fifty.

**PAÑCHASIDDHÂNTIKA.** "Five astronomical canons". See ***Varâhamihîra*** and **Indian astrology.**

**PAÑCHATANTRA.** "Five books". Name of the famous collection of moralistic tales and fables, made up of five books. The fables of Aesop and La Fontaine are more or less directly inspired by this collection. See ***Pañcha***.

**PAÑCHAVIMSHA BRÂHMANA.** Text derived from the *Samaveda*, a text of Vedic literature. The contents were transmitted orally since ancient times, but were constantly re-worked and added to, and did not achieve their finished form until relatively recently. Date uncertain. See ***Veda***. Here is a list of the main names of numbers mentioned in the text [see Datta and Singh (1938), p. 10]:

**Eka* (= 1), **Dasha* (= 10), **Sata* (= $10^2$), **Sahasra* (= $10^3$), **Ayuta* (= $10^4$), **Niyuta* (= $10^5$), **Prayuta* (= $10^6$), **Arbuda* (= $10^7$), **Nyarbuda* (= $10^8$), **Samudra* (= $10^9$), **Madhya* (= $10^{10}$), **Anta* (= $10^{11}$), **Parârdha* (=$10^{12}$), **Nikharva* (= $10^{13}$), **Vâdava* (= $10^{14}$), **Akshiti* (= $10^{15}$). See **Names of numbers** and **High numbers.**

**PAÑCHAVIMSHATI.** Ordinary Sanskrit name for the number twenty-five. For words which are symbolically related to this number, see **Twenty-five** and **Symbolism of numbers.**

**PÂNDAVA.** [S]. Value = 5. "Son of Pându". This refers to one of the five brothers, semi-legendary heroes of the epic **Mahâbhârata* (namely: Yudishtira, Arjuna, Bhîma, Nakula, and Sahadeva), son of the king Pându of Hastinâpura. See **Five.**

**PAPER.** See **Pâtîganita.**

**PARÂ.** See **Numeral alphabet, magic, mysticism and divination.**

**PARABRAHMAN.** [S]. Value = 1. Literally, "Supreme Brahman". Expression synonymous with **Paramâtman*, in terms of "Supreme Soul", and an epithet given to Mahâpurusha (supreme entity of the global spirit of humanity), considered in Hindu philosophy to be the Absolute Lord of the universe and thus identified with the Brahman. See **Âtman, Pitâmaha** and **One.**

**PARADISE.** [S]. Value = 13. See ***Vishvadeva*** and **Thirteen.**

**PARADISE.** [S]. Value = 14. See ***Bhuvana*** and **Fourteen.**

**PARAMABINDU.** "Supreme Point". This is the supreme causal point, which, according to Buddhist philosophy, is both inexistent and identical to all the universe; it is also time considered as a point (****bindu***) which lasts no sequential time but gives the impression of having a duration [see Frédéric, *Dictionnaire* (1987)].

**PARAMÂNU.** "Supreme Atom". This is the smallest indivisible material particle, and has a taste, odour and colour. This is different to our notion of the "atom", and is more like what we call a "molecule", the smallest particle which constitutes part of a compound body. The *paramânu* and the **paramânu râja* (or "grain of dust of the first atoms") have long been the smallest units of length and weight in India. These are found notably in the Legend of Buddha, told in the **Lalitavistara Sûtra*, where the *paramânu* corresponds to 0.000000287 mm and the *paramânu râja* to 0.000000614 g.

**PARAMÂNU RÂJA.** "Grain of dust of the first atoms". Name of the smallest Indian unit of weight. At the time of the writing of the **Lalitavistara Sûtra* (before 308 CE), it corresponded to 0.000000614g. See ***Paramânu.***

**PARAMÂTMAN.** "Supreme Soul". Epithet given to the *Brahman. See ***Parabrahman.***

**PARAMESHVARA.** Indian astronomer c. 1431 CE. His works notably include the text entitled *Drigganita*, in which there is abundant use of the place-value system using Sanskrit numerical symbols [see Sarma (1963)]. See **Numerical symbols, Numeration of numerical symbols** and **Indian mathematics (History of).**

**PARÂRDHA.** From *para*, "beyond", and *ardha* "half". This is the spiritual half of the path which leads to death, identical to *devayâna*, the "way of the gods", which, according to the **Vedas*, is one of the two possibilities offered to human souls after death (this path being said to lead to the deliverance from **samsâra* or cycles of rebirth). The symbolism which has led to these words having such high numerical values as ten to the power twelve or ten to the power seventeen comes from an association of ideas, not only with the immeasurable immensity of the sky, but also with the eternity which it represents. For more details, see **High numbers (Symbolic meaning of).**

**PARÂRDHA.** Literally "half of the beyond". Name given to the number ten to the power twelve (= billion). See **Names of numbers.** For an explanation of this symbolism, see ***Parârdha*** (first entry) and **High numbers (Symbolic meaning of).**

Sources: **Vâjasaneyî Samhitâ*, **Taittirîya Samhitâ* and **Kâthaka Samhitâ* (from the start of the first millennium CE); **Pañchavimsha Brâhmana* (date uncertain).

**PARÂRDHA.** Literally "half of the beyond". Name given to the number ten to the power seventeen. See **Names of numbers.** For an explanation of this symbolism, see ***Parârdha*** (first entry) and **High numbers (The symbolic meaning of).**

Sources: **Kitab fi tahqiq i ma li'l hind* by al-Biruni (c. 1030 CE); **Lîlâvatî* by Bhâskarâchârya (1150 CE); **Ganitakaumudî* by Nârâyana (1350 CE); **Trishatikâ* by Shrîdharâchârya (date unknown).

**PARASHURÂMA (Calendar).** See ***Kollam.***

**PÂRÂVÂRA.** Name given to the number ten to the power fourteen. See **Names of numbers** and **High numbers.**

Source: **Ganitakaumudî* by Nârâyana (1350 CE).

**PARIKARMA.** Word used in arithmetic to denote "fundamental operations" carried out on whole numbers. See ***Kalâsavarna.***

**PART.** [S]. Value = 6. See ***Anga*** and **Six.**

**PARTHIAN (Calendar).** Calendar beginning in the year 248 BCE. Formerly used in the northwest of the Indian sub-continent. To find a corresponding date in the Common Era, subtract 248 from a date expressed in the Parthian calendar. See **Indian calendars.**

**PÂRVATA.** [S]. Value = 7. "Mountain". Clearly an allusion to the "Mountain of the gods" (**devapârvata*), one of the names for *Mount Meru, which is said to be the home of the gods. This numerical symbolism is due to the preponderance of the number seven in the mythological representations of Mount Meru. See ***Adri*** and **Seven.**

**PÂRVATÎ.** See **Mount Meru.**

**PASHYANTÎ.** See **Numeral alphabet, magic, mysticism and divination.**

**PASSION.** [S]. Value = 6. See ***Râga*** and **Six.**

**PÂTAKA.** [S]. Value = 5. "Great sin". See ***Mahâpâpa*** and **Five.**

**PÂTÂLA.** [S]. Value = 7. "Inferior world". This refers to one of the seven "hells" of Hindu and *Jaina mythology (namely: *Atâla, Vitâla, Nitâla, Gabhastimat, Mahâtâla, Sutâla* and *Pâtâla*). These inferior worlds are said to be situated one on top of the other underneath *Mount Meru. They are the dwelling place of the **nâga*, who are ruled by *Muchalinda, a chthonian genie in the form of a cobra, depicted as having seven heads. See **Seven.**

**PÂTÂLA KUMÂRA.** "Princess of the Underworlds". Name given to the daughter of Himalaya, sister of Vishnu and wife of Shiva. See ***Pârvatî.***

**PÂTÎ.** Literally "Board", "tablet". Term used for the calculating board or tablet, upon which Indian mathematicians carried out their calculations. See ***Pâtîganita*** and **Indian methods of calculation.**

**PÂTÎGANITA (or GANITAPÂTÎ).** In its most general sense, this word is used today to mean "abstract mathematics". In the past, however, it referred to "arithmetic" and to the "practice of calculation", and appeared in the titles of many works relating to this discipline, for example: *Pâtîsâra* by Munishvara (1658); *Ganitapâtîkaumudî* by *Nârâyana (1356), which deals notably with magic squares; and *Ganitatilaka* by *Shrîpati (1039), the sub-heading of which is *Pâtîganita*. See: Datta and Singh (1938); Kapadia (1935).

Moreover, in his *Brahmasphutasiddhânta* (628), *Brahmagupta describes the ensemble of basic arithmetical operations with the word *pâtîganita*. He writes: "Those that know the twenty logistic operations separately and individually, [these being] addition, multiplication, etc., as well as the eight [methods] of determination, including [in particular measurement by] shadow, is a [true] mathematician." See: BrSpSi; Datta and Singh (1938).

To Brahmagupta's mind, the eight fundamental operations of the Indian mathematicians were the same as the first eight operations of *pâtîganita* (namely: addition, subtraction, multiplication, division, the squaring or cubing of a number, the extraction of the square or cube root), to which the five fundamental rules of the reduction of fractions were added: the **trairâshika* or "Rule of Three", etc. This shows the high level that had been reached by the Indian mathematicians in their calculating techniques at the beginning of the seventh century CE. The methods of calculation which originated in India are known to us today not only because of the information provided by Arabic and European authors, but also by Indian authors themselves. See **Square roots (How Âryabhata calculated his).**

See: Allard (1981); Datta and Singh (1938); Iyer (1954); Kaye (1908); Waeschke (1878).

In some rural regions of India, these processes have been taught through the centuries with hardly any modifications, and calculations are still carried out on the *pâtî* (small board) [see Datta and Singh (1938)]. The word *pâtîganita* (or *ganitapâtî*) is composed of **ganita*, which means "calculation, arithmetic, science of calculation", and **pâtî*, synonymous with *Patta* in the sense of "board" or "tablet". See: AMM, XXXV, p. 526; Datta and Singh, pp. 7–8 and 123. This etymology dates back to the time when Indian mathematicians carried out their calculations on either a board or a tablet. Today, the most natural support for carrying out mathematical operations on is paper. Paper was invented in China, although the circumstances are not fully known. There are texts that attribute the invention of a type of paper made from the pulp created by removing the fibre from rags and fishing nets to Cai-Lun in the year 105 BCE. However, the ideogram used to write the word *paper* in Chinese contains the sign for silk. It seems that paper made from silk preceded paper made from vegetable fibres, the latter quickly replacing the former type because it was cheaper and more resistant. Cai-Lun and other paper makers then went on to use the pulp of vegetable matter, particularly the bark of the mulberry tree. It was probably in the tenth century that they began to use bamboo and, around the fourteenth century, straw. It would be a long time after the Chinese discovery before the West would know about paper. The production of paper began in Samarkand in 751 after the Chinese were taken prisoner by the Arabs at the battle of Talas. Paper was then made by Chinese workers in Baghdad (from 793) and Damascus, which for centuries remained the principal supplier to Europe. From there, methods of fabrication spread to Egypt (c. 900) and Morocco, from where the Arabic invaders introduced it to Spain [see Galiana, (1968)].

Paper was introduced to India by the Persians, who learned the methods of manufacture from the Arabs. It was not until the fourteenth century, however, that the Indians learned the secrets of paper-making. In other words, Indians almost never used paper to carry out their calculations, until very late on in their history. The Arabs and Persians never used paper for this purpose until the twelfth or thirteenth century, because it was such a rare and expensive commodity. The Indians could have used the same material as they used for their manuscripts, carving or writing on palm-leaves or tree-bark (see **Indian styles of writing, The materials of**). However, carrying out calculations was a completely different practice to writing manuscripts: working out sums was "rough work", whilst manuscripts were written to last, on durable material and in indelible ink. They used something much more economic than palm-leaves or tree-bark for their calculations: they used chalk and slate, just as most people in the Western world used at school until very recently (or chalk and a blackboard). The mathematician *Bhâskarâchârya (whose favourite instrument was the *pâtî*, or "board", which he wrote upon with a piece of chalk) refers to the use of these materials in his texts, notably in his **Lîlavâti*, where he writes the following:

*khatikâyâ rekhâ ucchâdya . . .*, "After having drawn the lines [of the numerals for the calculations] on the *pâtî* with chalk . . ."

[See: Datta and Singh (1938), p. 129; Dvivedi (1935), p. 41.]

In other words, the Indian mathematicians began, at some point to use if not slate, then at least a wooden board painted black, and chalk to write their numerals on and cross them out, and a rag to rub them out.

Just as the Arabs and Persians adopted the Indian numerals and methods of calculation, so they began to use the support upon which the Indians carried out their mathematical operations. They gave it the Arabic name *takht* or *luha* (especially in northern Africa) which means "table" or "board" (whether it is made of wood, leather, metal, earth, clay or even slate). As for "arithmetic", this was described by the expression *'ilm al hisab al takht* ("science of calculation on the board"). This support had the advantage of overcoming all the difficulties created when calculations were carried out on boards covered in dust. See **Indian methods of calculation.**

**PÂVAKA.** [S]. Value = 3. "Fire". See ***Agni*** and **Three.**

**PÂVANA.** "Purification", and by extension, "He who purifies". This is another name for *Vâyu, the ancient Brahmanic god of the wind. He is often represented on a mount in the form of an antelope or a deer and holding a fan, an arrow and a standard, respectively symbols of the air (*vâyu*), of speed and of the wind [see Frédéric, *Dictionnaire* (1987)].

**PÂVANA.** [S]. Value = 5. "Purification". This symbolism is due to this word being associated with one of the attributes of *Vâyu, god of the wind, because the wind itself in Indian cosmologies is considered to be the "cosmic breath". Vâyu is king of the subtle and intermediary domain between the sky and the earth who penetrates, breaks up and purifies. Vâyu is also known by the name *Anila*, which means "breath of life". Thus, according to the Hindus, he is the cosmic energy that penetrates and conserves the body and is manifested most clearly in the form of breath in creatures. Vâyu is also the **prâna*, the "breath" in terms of "vital respiration". Hinduism distinguishes between five types of breath: *prajña*, the very essence of breath, the pure vital force; *vyâna*, the regulator of the circulation of the blood; *samâna*, which regulates the process of absorption and assimilation of food and maintains the balance of the body by looking after the processes of feeding; *apâna*, which looks after secretion; and *udâna*, which acts on the upper part of the organism and facilitates spiritual development by creating a link between the physical part and the spiritual part of the being. Thus *pâvana* = 5. See ***Pâvana*** (previous entry), ***Prâna*** and **Five.**

The use of this numerical symbol can be found in Bhâskarâchârya [see SiShi, I, 27] and in Bhattotpala's Commentary on *Brihatsamhitâ* (chapter II). [See Datta and Singh (1938), p. 55].

**PÂVANA.** [S]. Value = 7. "Purification", "He who purifies". This is one of the attributes of *Vâyu, god of the wind (see previous article). To understand the reason for this symbolism, it is necessary to be familiar with the relevant episode in Brahmanic mythology. One day Vâyu revolted against the *deva*, or "gods", who live on *Mount Meru. He decided to destroy the mountain, and started a powerful hurricane. But the mountain was protected by the wing of Garuda, the bird-helper of Vishnu, which meant that the assaults of the wind had no effect. One day, however, when Garuda was absent, Vâyu cut off the peak of Mount Meru and threw it into the ocean. This is how Lankâ was born, the island of Sri Lanka. This mythological tale explains how the wind came to have this value. The mythical mountain, *Mount Meru, the living and meeting place of the gods, and centre of the universe, is said to be situated above the seven **pâtâla* (or "inferior worlds"), and has seven faces, each one turned towards one of the seven **dvîpa* (or "island-continents") and one of the seven **sâgara* (or "oceans"); when Vâyu attacked the mountain, he created seven strong winds, one for each face of Mount Meru. Thus: *pâvana* = 7. See **Seven.**

**PERFECT.** A synonym for a large quantity. See **High numbers (Symbolic meaning of).**

**PERFECT.** [S]. Value = 0. See ***Pûrna*** and **Zero.**

**PHENOMENAL WORLD.** [S]. Value = 3. See **Jagat, Loka, Three, Triloka.**

**PHILOSOPHICAL POINT OF VIEW.** [S]. Value = 6. See ***Darshana*** and **Six.**

**PHILOSOPHY OF VACUITY.** See ***Shûnya, Shûnyatâ.***

**PHILOSOPHY OF ZERO.** See **Symbolism of zero, *Shûnya, Shûnyatâ,*** and **Zero.**

**PINÂKANAYANA.** [S]. Value = 3. This is one of Shiva's names, the third divinity of the Hindu trinity, god of destruction and dissolution. He is often represented with a third eye on his forehead (which symbolises perfect Knowledge). Moreover, his emblem is the **trishûla*, or "trident", symbols of the three aspects of the manifestation (creation, preservation, destruction). See ***Haranetra*** and **Three.**

**PINK LOTUS.** As name of the numbers ten to power nine, ten to power fourteen and ten to power twenty-nine. See: **Padma, High Numbers (Symbolic Meaning of).**

**PINK LOTUS.** As name of the numbers ten to power twenty-nine, ten to power 119. See: **Padma, High Numbers (Symbolic Meaning of).**

**PITÂMAHA.** [S]. Value = 1. "Great ancestor", "grandfather", "first father". This is an allusion to the god Brahma, first divinity of the trinity of Hinduism; he is the "Director of the sky", the "Master of the horizons", the "One" amongst the diversity. See **One.**

**PLACE-VALUE SYSTEM.** The most common Sanskrit term for this is **sthâna*, which literally means "place". See ***Sthâna, Anka, Ankakramena, Ankasthâna, Sthânakramâd*** and **Indian written numeral systems (Classification of).**

**PLANET.** [S]. Value = 9. See ***Graha*** and **Nine.**

**PLANETS.** See ***Graha, Saptagraha*** and ***Navagraha.***

**PLENITUDE.** [S]. Value = 0. See ***Pûrna*** and **Zero.**

**POETRY.** See **Indian metric, Poetry and writing of numbers, *Nâga*, Serpent (Symbolism of the)** and **Poetry, zero and positional numeration.**

**POETRY AND WRITING OF NUMBERS.** Like all the scholars of India, astronomers and mathematicians of this civilisation usually wrote in Sanskrit, often writing their numerical tables and texts in verse. These scholars loved to play with and speculate with numbers, and their enjoyment can be seen in the form of their wording which, if not lyrical, is at least in verse. Thus numbers came to be written using words which were connected to them symbolically, and one such word could be chosen from an almost limitless selection of synonyms so that it would fit the rules of Sanskrit versification and give the desired effect. The transcription of a numerical table or of the most arid mathematical formula would often resemble an epic poem. Their language lent itself admirably to the rules of versification, thus giving poetry a significant role in Indian culture and Sanskrit literature. See **Sanskrit** and **Numerical symbols.**

**POETRY, ZERO AND POSITIONAL NUMERATION.** See **Zero and Sanskrit poetry.**

**POINT.** [S]. Value = 3. See ***Shûla*** and **Three.**

**POSITION.** [S]. Value = 4. See ***Îryâ*** and **Four.**

**POSITION OF NUMERALS.** See ***Sthâna, Sthânakramâd, Ankasthâna*** and ***Ankakramena.***

**POWER.** [S]. Value = 14. See ***Indra*** and **Fourteen.**

**POWERFUL.** [S]. Value = 14. See ***Shakra*** and **Fourteen.**

**POWERS OF TEN.** See **Ten, Hundred, Thousand, Ten thousand, Million, Ten million, Hundred million, Thousand million, Ten thousand million, Hundred thousand million, Billion, Ten billion, Hundred billion, Trillion, Ten trillion, Hundred trillion, Quadrillion, Quintillion, Names of numbers, High numbers** and **Infinity.**

**PRÂKRIT.** "Unrefined", "Basic". Generic name commonly used by Indians to refer to the numerous Indo-European dialects of the "Indo-Aryan" category.

**PRAKRITI.** "Nature, material". According to Indian philosophy, this is the original material that the universe was made from. It is the principal transcendental material, which is associated with terrestrial elements, as opposed to the principal spirit (which is represented by the skies).

**PRAKRITI.** [S]. Value = 21. In Sanskrit poetry, this is the metre of four times twenty-one syllables per verse. See **Indian metric.**

**PRALAYA.** Name of the total destruction of the universe in Hindu and Brahman cosmogonies. See **Day of Brahma, *Kalpa, Kaliyuga*** and ***Yuga.***

**PRÂNA.** [S]. Value = 5. "Breath". In Hindu philosophy, this describes the five breaths which are said to govern the vital functions of the human being (*prajña, apâna, vyâna, udâna* and *samâna*). This term not only applies to respiratory rhythms (like the *prânâyama* physical exercises, which are meant to control breathing and form part of the techniques of *hathayoga*), but also to "subtle breathing" identified with intelligence and wisdom (*prajña*) [see Frédéric, *Dictionnaire* (1987)]. See ***Pâvana*** and **Five.**

**PRAYUTA.** Name given to the number ten to the power five (a hundred thousand). See **Names of numbers** and **High numbers.**

Source: *Kâthaka Samhitâ* (beginning of the Common Era).

**PRAYUTA.** Name given to the number ten to the power six (= million). See **Names of numbers** and **High numbers.**

Sources: **Vâjasaneyî Samhitâ*, **Taittirîya Samhitâ* and **Kâthaka Samhitâ* (from the start of the first millennium CE); **Pañchavimsha Brâhmana* (date uncertain); **Âryabhatîya* (510 CE). **Lîlâvatî* by Bhâskarâchârya (1150 CE); **Ganitakaumudî* by Nârâyana (1350 CE); **Trishatikâ* by Shrîdharâchârya (date uncertain); **Kitab fi tahqiq i ma li'l hind* by al-Biruni (c. 1030 CE); **Sankhyâyana Shrauta Sûtra* (date unknown).

**PRECEPT.** [S]. Value = 6. See **Six.**

**PRIMORDIAL PRINCIPLE.** [S]. Value = 1. See ***Âdi.*** **One.**

**PRIMORDIAL PROPERTY.** [S]. Value = 3. See ***Guna*** and ***Triguna.***

**PRIMORDIAL PROPERTY.** [S]. Value = 6. See ***Guna*** and ***Shadâyatana.***

**PRINCIPLE OF THE ENUNCIATION OF NUMBERS.** See ***Ankânâm vâmato gatih*** and **Sanskrit.**

**PRINCIPLE OF POSITION.** The Sanskrit term usually designating it is **sthâna*, literally: "place". See ***Sthâna.***

**PRITHIVÎ.** [S]. Value = 1. "Immense", "Earth", "terrestrial world". This symbolism primarily refers to the unique nature of the earth, considered to be the spouse of the sky. However, this is also and above all an allusion to the fact that the earth, as principal transcendental material (**prakriti*), as opposed to the principal spirit (represented by the skies), is regarded as the mother of all things. See **One.**

**PROGENITOR OF THE HUMAN RACE.** [S]. Value = 14. See ***Manu*** and **Fourteen.**

**PROTO-INDIAN NUMERALS.** Symbols used from about 2500 to 1500 BCE by people of the Indus civilisation (Mohenjo-daro, Harappa, etc.) who preceded the Aryan settlement of the Indian sub-continent. It is not known how these very different symbols could have evolved into early Brâhmî numerals (nor if indeed there is a connection between them). Only the signs for the nine units have been identified so far; a full understanding of proto-Indian numerals must await further archaeological evidence. See Fig. 1.14.

**PUNDARÎKA.** Literally "(white) lotus". Name given to the number ten to the power twenty-seven. See **Names of numbers.** For an explanation of this symbolism, see: **High numbers (Symbolic meaning of).** Source: **Lalitavistara Sûtra* (before 308 CE).

**PUNDARÎKA.** Literally "(white) lotus". Name given to the number ten to the power 112. See **Names of numbers.** For an explanation of this symbolism, see: **High numbers (Symbolic meaning of).**

Source: **Vyâkarana* (Pali grammar) by Kâchchâyana (eleventh century CE).

**PUNJABI NUMERALS.** Symbols derived from *Brâhmî numerals and influenced by Shunga, Shaka, Kushâna, Ândhra, Gupta and Shârada. Currently used in the Punjab (Northwest India). The symbols correspond to a mathematical system that has place values and a zero (shaped like a small circle). See: **Indian written numeral systems (Classification of).** See Fig. 24.5, 52, and 24.61 to 69.

**PURA.** [S]. Value = 3. "City". Allusion to the **tripura*, the "three cities" of the **Asura* (or "anti-gods"), flying iron fortresses from which they directed the war they waged against the **deva.* See **Three.**

**PURÂ.** [S]. Value = 3. "State". Allusion to the three **tripurâ*, the "three states of consciousness" according to Hinduism (awake, asleep and dreaming). See **Three.**

**PURÂNA.** Literally: "Ancients". Traditional Sanskrit texts, dealing with highly diverse subjects, such as the creation of the world, mythology, legends, the genealogy of mythical sovereigns, castes, etc. These texts were written for ordinary people and those of "low caste". Analysis has shown that they are made up of documents written at various times and are from many different sources, and were compiled, revised, added to and corrected in an interval of time oscillating between the sixth and the twelfth century, some even being dated as nineteenth century. Thus the documentation that they contain should be treated with caution, as, from a purely historical point of view at least, they are of no interest. See **Indian documentation (Pitfalls of).**

**PURÂNA AND POSITIONAL NUMERATION.** See **Indian documentation (Pitfalls of).**

**PURÂNALAKSHANA.** [S]. Value = 5. (Late usage). Allusion to the texts of the **Purâna*, which tell of the *Pañchalakshana*, which, in Hindu philosophy, correspond to the "five characteristics" which are said to have defined history: creation (*sarga*), periodical creations (*pratisarga*), divine geneaologies (*vamsha*), the era of a **manu* (*manvantara*) and the genealogies of human sovereigns (*vamshânucharita*) [see Frédéric, *Dictionnaire* (1987)]. See **Five.**

**PURIFICATION.** [S]. Value = 5. See ***Pâvana*** and **Five.**

**PURIFICATION.** [S]. Value = 7. See ***Pâvana*** and **Seven.**

**PÛRNA.** [S]. Value = 0. Literally "full, fullness, fulfilled, perfect, finished". To a Western reader, this symbolism might seem paradoxical: how can a word that means "full" represent zero, the void? The allusion is to *Vishnu, the second great divinity of the Hindu and Brahman trinity, whose essential role is to preserve, and cause the evolution of, creation (*Brahma being the creator, *Vishnu the conserver and *Shiva the destroyer). Vishnu is considered to be the internal cause of existence and the guardian of **dharma*. Each time the world goes wrong, he hastens (incarnating himself in the form of **avatâra*) to show humanity new ways in which to develop. He is often represented as a handsome young man with four arms, holding a conch in the first hand, a bow in the second, a club in the third and a lotus flower in the fourth. The conch represents riches, fortune and beauty, which are the attributes of Vishnu as the principal conserver of the manifestation, because the sound and the pearl are conserved within the shell. As for the *lotus, it symbolises the highest divinity, innate reason and mental and spiritual perfection. It also symbolises the "third eye", that of perfect Knowledge; however, it is also the superior illumination and the divine corolla, the totality of revelation and illumination, as well as intelligence, wisdom and the victory of the mind over the senses. See **High numbers (Symbolic meaning of).**

Moreover, like the thousand-petalled lotus, Vishnu possesses a thousand attributes and qualities (**sahasranâma*). He represents the innumerable (*thousand* here being treated in its figurative sense). See **Thousand.** Thus Vishnu is associated with the idea of wholeness, integrity, completeness, absoluteness and perfection. The "foot of Vishnu" (****vishnupada***), is the "sky", the "zenith", the "place of the blessed" and the meeting place of the gods; it is, in Hindu cosmology, the summit of *Mount Meru, the mythical mountain situated at the centre of the universe, the source of the celestial *Gangâ* (the sacred Ganges). This makes it easier to understand how "full" came to mean infinity, eternity, and by extension completion and perfection. It is upon *Ananta, the serpent with a thousand heads floating on the primordial waters of the "ocean of unconsciousness", that Vishnu lies to rest during the time separating two creations of the universe. Ananta represents infinity, and has also often represented zero as a numerical symbol. Thus it is clear how *pûrna* came to mean zero. See ***Ananta, Jaladharapatha, Shûnya,*** **Zero.** See also **Infinity, Infinity (Indian mythological representation of)** and **Serpent (Symbolism of the).**

**PUSHKARA.** [S]. Value = 7. This is a surname attributed to Krishna and Shiva, as well as to

Dyaus (the Sky) considered to be a "reservoir of water". The allusion here is to *Pushkara*, one of the seven mythical continents that surround *Mount Meru. See ***Dvîpa*, *Sapta*, *Sâgara*, Ocean** and **Seven**.

**PUTRA**. [S]. Value = 5. "Son". In this symbolism, the word in question is a synonym of **Pândava*, which means the "sons of Pându". See **Five**.

**PUTUMANASOMAYÂJIN**. Indian astronomer of the eighteenth century. His works notably include a text entitled *Karanapaddhati*, in which there is frequent use of the place-value system written in the Sanskrit numerical symbols [see Sastri (1929)]. See **Numerical symbols, Numeration of numerical symbols,** and **Indian mathematics (History of)**.

## Q

**QUADRILLION** (= ten to the power eighteen). See ***Shankha*** and **Names of numbers**.

**QUALITY**. [S]. Value = 3. See ***Guna*, *Triguna*** and **Three**.

**QUALITY**. [S]. Value = 6. See ***Guna*, *Shadâyatana*** and **Six**.

**QUINTILLION** (= ten to the power twenty-one). See ***Kotippakoti*, *Kumud*, *Mahâkshiti*** and ***Utsanga***. See also **Names of numbers**.

**QUOTIENT**. [Arithmetic]. See ***Labdha***.

**QUTAN XIDA**. Chinese astronomer of Indian origin. Qutan Xida is none other than the Chinese rendering of the Indian name *Gautama Siddhânta.

## R

**RABBIT**. [S]. Value = 1. See **One**, ***Shashin*, *Shashadhara***.

**RADA**. [S]. Value = 32. "Tooth". See ***Danta*** and **Thirty-two**.

**RADIANCE**. [S]. Value = 9. See ***Go*** and **Nine**.

**RÂGA**. [S]. Value = 6. "Attraction, colour, passion, musical mode". This Sanskrit term describes the moments of emotion provoked by a piece of music (the modes and rhythms causing diverse sensations in the listener) or by a visual work of art. The instants of emotion, which can be provoked by the perception of an exterior agent such as the rain, the wind, a storm, etc., or even by an interior sentiment such as love, nostalgia, sadness, etc., combine with lines and colours to provoke diverse sensations within the spectator. In the symbolism in question, the allusion is to the *janaka râga*, the six "eastern *râga*", who are male, associated with their six *râgini* (or female *râga*), and with the six "sons" of the latter ones, each of these groups in turn being associated with the **shadâyatana* or "six **guna*" of Buddhist philosophy (in other words the six sense organs: eye, nose, ear, tongue, body and mind) [see Frédéric, *Dictionnaire* (1987)]. See ***Rasa***, **Six** and ***Nâga***.

**RÂHU**. Demon who, according to ancient Indian mythology and cosmology, caused eclipses by "devouring" the sun or the moon, due to a privilege conferred on him by *Brahma. See ***Paksha***.

**RÂJAMRIGÂNKA**. See ***Bhoja***.

**RÂJASTHANÎ NUMERALS**. Symbols derived from *Brâhmî numerals and influenced by Shunga, Shaka, Kushâna, Ândhra, Gupta, Nâgarî and Mahârâshtrî. Currently used in the state of Rajasthan in the west of the sub-continent (bordering on Pakistan, Punjab, Haryana, Uttar Pradesh, Madhya Pradesh and the Gujurat). Râjasthanî numerals are a variant of Mârwarî numerals. The symbols correspond to a mathematical system that has place values and a zero (shaped like a small circle). See: **Indian written numeral systems (Classification of)**. See Fig. 24.52 and 24.61 to 69.

**RÂMA**. [S]. Value = 3. Allusion to the three Râma of Indian tradition and philosophy: the first, also called Parashu-Râma, or "Râma of the axe", is the sixth incarnation of Vishnu, who came to crush the tyranny of the Kshatriyas, the caste of warriors; the second, called Râma-chandra, seventh incarnation of Vishnu, came to develop *sattva* in humankind, in other words uprightness, equilibrium, serenity and peacefulness; and the third, called simply Râma, was the famous hero of the epic poem **Râmâyana* [see Frédéric, *Dictionnaire* (1987)].

**RÂMÂYANA**. "The march of Râma". This is an epic Indian poem, written down in Sanskrit by the poet Vâlmîki. It is derived from very ancient legends, but did not find its definitive form until the early centuries of the Common Era. Here is a list of names of the high numbers mentioned in this text (from a passage where, in order to express the number of monkeys that made up Sugriva's army, the author gives the following names successively, which increase each time on a scale of one hundred thousand): **Koti* (= $10^7$), **Shanka* (= $10^{12}$), **Vrindâ* (= $10^{17}$), **Mahâvrinda* (= $10^{22}$), **Padma* (= $10^{29}$), **Mahâpadma* (= $10^{34}$), **Kharva* (= $10^{39}$).

See **Names of numbers** and **High numbers**.

[See Weber in: JSO, XV, pp. 132–40; Woepcke (1863)].

**RANDHRA**. [S]. Value = 0. "Hole". Numerical word-symbol used rarely and not until a relatively recent date. The origin of this association of ideas clearly comes from the lack of consideration attached to the anal orifice. See **Zero**.

**RANDHRA**. [S]. Value = 9. "Hole". This is an allusion to the nine orifices of the human body (the two eye sockets, the two ears, the two nostrils, the mouth, the navel and the anal orifice). See ***Chhidra*** and **Nine**.

**RASA**. [S]. Value = 6. "Sensation". In its most general meaning, this word denotes the sensation(s) that a *Shadâyatana* can experience, in other words the "six senses or sense organs" of Indian philosophy (which are: the eye, the nose, the ear, the tongue, the body and the mind). However, the explanation for this symbolism is much more subtle than that. It can only be understood with reference to the Indian aesthetic canons, where this word describes the emotional state of the spectator, listener or reader, in terms of the essence of the evocative power of the musical, pictorial, poetic, theatrical, (etc.) art. This aesthetic distinguishes between nine different types of *rasa*, including the least agreeable, namely: *shringâra* (love or erotic passion); *hâshya* (comedy and humour); *karunâ* (compassion); *vîra* (heroic sentiment); *adbhuta* (amazement); *shânta* (peace and serenity); *raudra* (anger and rage); *bhayânaka* (fear or anguish); and *vîbhatsa* (disgust or repulsion). The first six are the ones which enable enjoyment, and this is what *rasa* refers to in this symbolism: the idea of "savouring". Thus *rasa* = 6. See ***Shadâyatana*** and **Six**.

**RÂSHI**. "Rule". Often used in arithmetic to denote the "Rule of Three".

**RÂSHI**. [S]. Value = 12. "Zodiac". This, of course, refers to the twelve signs of the Indian zodiac: *Mesha* (Aries); *Vrishabha* (Taurus); *Mithûna* (Gemini); *Karka* (Cancer); *Simha* (Leo); *Kanyâ* (Virgo); *Tulâ* (Libra); *Vrishchika* (Scorpio); *Dhanus* (Sagittarius); *Makara* (Capricorn); *Kumbha* (Aquarius); *Mîna* (Pisces). See **Twelve**.

**RÂSHIVIDYÂ**. Name given to arithmetic in the *Chândogya Upanishad*. Literally: "Knowledge of the rules".

**RATNA**. [S]. Value = 3. "Jewel". This is probably an allusion to the **triratna*, the "three jewels" of Buddhism, namely: the Community (*sangha*), the Buddhist Law (**dharma*) and Buddha. These "jewels" are represented by a trident. See ***Dharma*, *Shûla*** and **Three**.

Note: this symbol is found very rarely representing this value, except for in the **Ganitasârasamgraha* by Mahâvîrâchârya [see Datta and Singh (1938), p. 55].

**RATNA**. [S]. Value = 5. "Jewel". This is the most frequent value that this word is used for as a numerical symbol. It is probably an allusion to the **pañchaparameshtin*, the "five orders of beings" considered to be the "five treasures" of *Jaina religion: the **siddha*, human beings who are omniscient and who became immortal after being freed from the bonds of *karma* and **samsâra*; the *arhat*, sages liberated from the bonds of *karma*, but still subject to the laws of **samsâra*; the *âchârya* or "great masters"; the *upadhya* or "masters"; and the ascetics (*sâdhu*) [see Frédéric, *Dictionnaire* (1987)]. See **Five**.

**RATNA**. [S]. Value = 9. "Jewel". This allusion could be to the **Navaratna*, "Nine Jewels", the collective name given to the nine famous poets who wrote in Sanskrit who lived in the court of the king Vikramâditya. See **Nine**.

**RATNA**. [S]. Value = 14. "Jewel". There is no concrete explanation for this symbolism. However, it could have some connection to the **saptaratna* or "seven jewels" of Buddhism, which constitute the seven attributes of the current Buddha ("Golden wheel"; *Chintâmani*, or miraculous pearl said to grant all wishes; "White horse"; "Noble woman"; "Elephant" carrying the sacred Scriptures; "Minister of Finances"; and "Head of war"); these are attributes that would have been associated symbolically with the **saptabuddha*, or seven Buddhas of the past (Vipashyin, Shikhin, Vishvabhû, Krakuchhanda, Kanakamuni and Kâshyapa), including the current Buddha (Shâkyamuni Siddhârtha Gautama); thus, by symbolic addition: *ratna* = 7 + 7 = 14. See **Fourteen**.

**RATNASANU**. One of the names for *Mount Meru. See ***Adri*, *Dvîpa*, *Pûrna*, *Pâtâla*, *Sâgara*, *Pushkara*, *Pâvana*** and ***Vayu***.

**RÂVANA**. Name of the king-demon Lankâ who, according to the legends of **Râmâyana*, usurped the throne of his half-brother Kuvera and stole his flying palace (*pushpaka*).

**RÂVANABHUJA**. [S]. Value = 20. "Arms of *Râvana". Allusion to the twenty arms of this king-demon. See **Twenty**.

**RÂVANASHIRAS**. [S]. Value = 10. "Heads of *Râvana". This king-demon is said to have had ten heads. See **Ten**.

**RAVI.** [S]. Value = 12. This is another name for *Sûrya, the divinity of the sun. See **Twelve.**

**RAVIBÂNA.** [S]. Value = 1,000. "Beams of Ravi". This refers to one of the attributes of *Ravi (= *Sûrya), the divinity of the sun, and expresses the **sahasrakirana* or "Thousand Rays" of the sun. See **Thousand.**

**RAVICHANDRA.** [S]. Value = 2. The couple uniting Ravi and Chandra (named Ravi after *Sûrya, the sun whose other attribute is **Ravi*, and *Soma, the moon, the masculine entity also called **Chandra*). See **Two.**

**REALITY.** [S]. Value = 5. See ***Tattva*** and **Five.**

**REALITY.** [S]. Value = 7. See ***Tattva*** and **Seven.**

**REALITY.** [S]. Value = 25. See ***Tattva*** and **Twenty-five.**

**REMAINDER.** [Arithmetic]. See ***Shesha.***

**RISHI.** [S]. Value = 7. "Sage". Allusion to the **Saptarishi*, the seven great mythical Sages of Vedic times (Gotama, Bharadvâja, Vishvamitra, Jamadagni, Vasishtha, Kashyapa and *Atri), created by *Brahma and said to be the authors of the hymns and invocations of the *Rigveda* and most of the other **Vedas*. They are said to form the seven stars of the Little Bear. See **Seven.**

**RITU.** [S]. Value = 6. "Season". Allusion to the six seasons, each lasting two "months" in the Hindu calendar: spring (*vasanta*); the hot season (*grishma*); the rain season (*varsha*); autumn (*sharada*); winter (*hemanta*) and the cold season (*shishira*). See **Six.**

**RUDRA-SHIVA (Attributes of).** [S]. Value = 11. See ***Bharga, Bhava, Hara, Îsha, Îshvara, Mahâdeva, Rudra, Shiva, Shûlin*** and **Eleven.**

**RUDRA.** [S]. Value = 11. "Rumbling", "Violent", "Lord of tears". This is the name of the ancient Vedic divinity of the tempest who, according to the **Vedas*, was the personification of the vital breaths, which came from *Brahma's forehead, of which there were eleven. Thus: *Rudra* = 11. See **Eleven.**

**RUDRÂSYA.** [S]. Value = 5. "Faces of *Rudra". This god is said to have had five heads. He is also lord of the "five elements", "the five sense organs", the five "human races" and the five points of the compass (if the zenith is included). See **Five.**

**RULE OF THREE.** [Arithmetic]. See ***Râshi, Trairâshika*** and ***Vyastatrairâshika.***

**RULE OF FIVE.** [Arithmetic]. See ***Pañchaparâshika.***

**RULE OF SEVEN.** [Arithmetic]. See ***Saptarâshika.***

**RULE OF NINE.** [Arithmetic]. See ***Navarâshika.***

**RULE OF ELEVEN.** [Arithmetic]. See ***Ekâdasharâshika.***

**RÛPA.** [S]. Value = 1. "Form", "Appearance". This word is synonymous here with "body" as a symbol for the number one. See ***Tanu*** and **One.**

# S

**SÂGARA.** [S]. Value = 4. "Sea, Ocean". This symbolism can be explained by an allusion to the four "oceans" (**chatursâgara*) which surround the four "island-continents" (**chaturdvîpa*) which, according to Hindu cosmology, surround *Jambudvîpa* (India). See **Four** and **Ocean.**

**SÂGARA.** [S]. Value = 7. "Sea, Ocean". This symbolism can be explained by an allusion to the seven "oceans" (**sapta Sâgara*) which, according to Hindu cosmology and Brahmanic mythology, surround *Mount Meru. See **Four** and **Ocean.**

**SAGE.** [S]. Value = 7. See ***Atri, Rishi, Saptarishi, Muni*** and **Seven.**

**SAHASRA.** Ordinary Sanskit name for the number *thousand, the consecutive multiples of which are formed by placing the word *sahasra* to the right of the name of the corresponding unit: *dvasahasra* (two thousand), *trisahasra* (three thousand), *chatursahasra* (four thousand), *pañchasahasra* (five thousand), etc. This name appears in many words which have a direct relationship with the idea of this number.

Examples: **Sahasrabhûja, *Sahasrakirana, *Sahasrâksha, *Sahasrâmshu, *Sahasranâma, *Sahasrapadma, *Sahasrârjuna.*

For words which have a more symbolic relationship with this number, see **Thousand** and **Symbolism of numbers.**

**SAHASRABHÛJA.** "Thousand arms". This is one of the names of the Sun-god *Sûrya (in allusion to his rays). In the schools of Buddhism of the North, this term refers to an ancient divinity whose thousand arms represented his multiple powers and omniscience.

**SAHASRAKIRANA.** "Thousand rays". One of the names of the Sun-god *Sûrya.

**SAHASRÂKSHA.** "Thousand eyes". One of the attributes of *Indra and *Vishnu. See ***Indradrishti*** and ***Sahasra.***

**SAHASRÂMSHU.** [S]. Value = 12. "Thousand (of the) Shining" (from *sahasra,* "thousand", and *âmshu,* "shining"). This is a metaphorical name for the Sun (the "thousand rays" of its "shining"), and the symbolism has nothing to do with the idea of a thousand, but with the name of the Sun-god as a numerical symbol equal to twelve. See ***Sûrya*** and **Twelve.**

**SAHASRANÂMA.** "Thousand names". One of the attributes of *Vishnu and *Shiva.

**SAHASRAPADMA.** "Lotus of a thousand petals". See **Lotus** and **High numbers (Symbolic meaning of).**

**SAHASRÂRJUNA.** "Arjuna's thousand". Name for the thousand arms of Arjunakârtavîrya, mythical sovereign of the *Mahâbhârata. See ***Arjunâkara.***

**SALILA.** Name given to the number ten to the power eleven. See **Names of numbers** and **High numbers.** Source: **Samkhyâyana Shrauta Sûtra* (date uncertain).

**SAMÂPTALAMBHA.** Name given to the number ten to the power thirty-seven. See **Names of numbers** and **High numbers.**

Source: **Lalitavistara Sûtra* (before 308 CE).

**SAMÎKARANA.** Term used to denote an "equation". Literally "to make equal" (from *sama* "equal", and *kara* "to make"). Synonyms: *samîkâra, sâmîkriyâ,* etc.

**SAMKALITA.** Sanskrit term denoting addition. Literally: "put together". Synonyms: *samkalana* (literally: "act of reuniting"); *mishrana* ("act of mixing"); *sammelana; prakshepana; samyojana,* etc.

**SAMKHYA (SANKHYA).** "Number". Term often used to describe the system of writing numbers using numerical symbols. See **Numerical symbols** and **Numeration of numerical symbols.**

**SÂMKHYÂ (SÂNKHYÂ).** Literally "calculator". This term describes the adept of the mystical philosophy of **sâmkhya*.

**SÂMKHYA (SÂNKHYA).** Literally "number". This denotes one of the six orthodox systems of Indian philosophies. See ***Darshana*** and ***Tattva.***

**SAMKHYÂ (SANKHYÂ).** Literally "number". Word used to denote "expert-calculator" and, by extension, the arithmetician and mathematician. See ***Darshana*** and ***Samkhyâna.***

**SAMKHYÂNA (SANKHYÂNA).** "Science of numbers", and by extension "arithmetic" and "astronomy". Word used in this sense in Buddhist and Jaina literature. This science was considered to be one of the fundamental conditions for the development of a Jaina priest. For the Buddhists, it was also considered (although somewhat later) to be the first and most noble of arts.

**SAMKHYEYA (SANKHYEYA).** "Number", in the operative and arithmetical sense of the word.

**SAMSÂRA.** Cycle of rebirth. See ***Gati, Kâma*** and ***Yoni.***

**SAMSKRITA.** "Complete", "perfect", "definitive". Term used to denote the Sanskrit language. See **Sanskrit.**

**SAMUDRA.** Literally "ocean". Name given to the number ten to the power nine. See **Names of numbers.** For an explanation of this symbolism, see **High numbers (Symbolic meaning of).**

Sources: **Vâjasaneyî Samhitâ, *Taittirîya Samhitâ* and **Kâthaka Samhitâ* (from the start of the first millennium CE); **Pañchavimsha Brâhmana* (date uncertain).

**SAMUDRA.** Literally "ocean". Name given to the number ten to the power ten. See **Names of numbers.** For an explanation of this symbolism, see **High numbers (Symbolic meaning of).**

Source: **Sankhyâyana Shrauta Sûtra* (date uncertain).

**SAMUDRA.** Literally "ocean". Name given to the number ten to the power fourteen. See **Names of numbers.** For an explanation of this symbolism, see **High numbers (Symbolic meaning of).**

Source: *Kitab fi tahqiq i ma li'l hind* by al-Biruni (c. 1030 CE).

**SAMUDRA.** [S]. Value = 4. "Ocean". This is because of the four oceans that are said to surround **Jambudvîpa* (India). See ***Sâgara,*** **Four** and **Ocean.**

**SAMUDRA.** [S]. Value = 7. "Ocean". This is because of the seven oceans that are said to surround *Mount Meru. See ***Sâgara,*** **Seven** and **Ocean.**

**SAMVAT (Calendar).** See ***Vikrama.***

**SANKHYA,** etc. See ***Samkhya,*** etc.

**SANKHYÂNA.** See ***Samkhyâna.***

**SANKHYÂYANA SHRAUTA SÛTRA.** Philosophical Sanskrit text (date uncertain). Here is a list of the principal names of numbers mentioned in the text [see Datta and Singh (1938), p. 10]:

**Eka* (= 1), **Dasha* (= 10), **Sata* (= $10^2$), **Sahasra* (= $10^3$), **Ayuta* (= $10^4$), **Niyuta* (= $10^5$), **Prayuta* (= $10^6$), **Arbuda* (= $10^7$), **Nyarbuda* (= $10^8$), **Nikharva* (= $10^9$), **Samudra* (= $10^{10}$), **Salila* (= $10^{11}$), **Antya* (= $10^{12}$), **Ananta* (= $10^{13}$). See **Names of numbers** and **High numbers.**

**SANKHYEYA.** See ***Samkhyeya.***

**SANSKRIT.** In India and Southeast Asia, Sanskrit has played, and still does play today, a role comparable with Greek and Latin in Western Europe. This language is capable of translating, through meditation, the mystical transcendental truths said to have been revealed to the **Rishi* in Vedic times. See ***Akshara, AUM, Trivarna, Vâchana*** and **Mysticism of letters.** Moreover, the name of the Sanskrit language itself is quite significant, because the word **samskrita* ("Sanskrit") means "complete", "perfect" and "definitive". The people who know this Sanskrit are said to speak the divine language and are thus gifted with divine knowledge. Bearing in mind the power accorded to the spoken word (and consequentially its written expression), Sanskrit is considered to be the "language of the gods".

In fact, this language is extremely elaborate, almost artificial. It is capable of describing multiple levels of meditations, states of consciousness and physical, spiritual and even intellectual processes. The inflection of nouns is richly articulated and there are numerous personal forms of the verb, even though the syntax is rudimentary. The vocabulary is very rich and highly diversified according to the means for which it is intended [see Renou (1930); see also Filliozat (1992)].

This shows how, over the centuries, Sanskrit has lent itself admirably to the rules of prosody and versification. This explains why poetry has always played such an important role in Indian culture and Sanskrit literature. It is clear why Indian astronomers favoured the use of Sanskrit numerical symbols, based on a complex symbolism which was extraordinarily fertile and sophisticated, possessing as it did an almost limitless choice of synonyms. See **Poetry and writing of numbers** and **Numerical symbols.**

**SAPTA (SAPTAN).** Ordinary Sanskrit name for the number seven, which forms part of the composition of many words directly related to the idea of this number. Examples: **Saptabuddha, *Saptagraha, *Saptamâtrikâ, *Saptapadî, *Saptarâshika, *Saptarishi, *Saptarishikâla, *Saptasindhava*. For words which have a more symbolic connection with this number, see **Seven** and **Symbolism of numbers.**

**SAPTA.** Literally "seven". Term used symbolically in the texts of the *Atharvaveda* as a synonym for each of the following ideas: "sage", "ocean", "mountain", "island-continent", etc. The allusion here is to the "Seven Sages" of Vedic times (**saptarishi*), to the seven cosmic oceans (**sapta sâgara*), to the seven peaks of Mount Meru, or to the seven "island continents" (*sapta dvîpa*) of Indian mythology and cosmology. See ***Saptarishi, Adri, Giri, Sâgara, Dvîpa,*** **Mount Meru** and **Ocean.**

For an example, see *Atharvaveda*, I, 1, 1; Datta and Singh (1938), p. 17.

**SAPTABUDDHA.** Name of the seven Buddhas. See ***Sapta*** and ***Ratna*** (= 14).

**SAPTADASHA.** Ordinary Sanskrit name for the number seventeen. For words which have a symbolic link with this number, see **Seventeen** and **Symbolism of numbers.**

**SAPTA DVÎPA.** "Seven islands". In Hindu cosmology and Brahmanic mythology, this is the name given to the seven island-continents said to surround *Mount Meru. See ***Sapta.*** For an explanantion of the symbolism and the choice of this number, see **Ocean.**

**SAPTAGRAHA.** Literally "seven planets". These are the following: **Sûrya* (the Sun); **Chandra* (the Moon); **Angâraka* (Mars); **Budha* (Mercury); **Brihaspati* (Jupiter); **Shukra* (Venus); and **Shani* (Saturn). See ***Graha*** and ***Paksha.***

**SAPTAMÂTRIKÂ.** Name for the seven "divine Mothers". See ***Mâtrikâ.***

**SAPTAN.** Ordinary name for the number seven. See ***Sapta.***

**SAPTAPADÎ.** "Seven paces". Name of a Hindu rite which forms part of the nuptial ceremonies, where the bride and groom must take seven paces around the sacred fire in order to consummate the union.

**SAPTARÂSHIKA.** [Arithmetic]. Sanskrit name for the Rule of Seven.

**SAPTARATNA.** Name of the "Seven Jewels of Buddhism". See ***Ratna*** (= 14).

**SAPTARISHI.** "Seven Sages". These are the seven **Rishi* of Vedic times, who are said to have resided in the seven stars of the Little Bear. See ***Atri*** and **Mount Meru.**

**SAPTARISHIKÂLA.** "Time of the seven **Rishi*". Name of an Indian calendar. See ***Saptarishi, Kâla*** and ***Laukikasamvat.***

**SAPTA SÂGARA.** "Seven oceans". These are the seven oceans which are said to surround *Mount Meru in Hindu cosmology and Brahmanic mythology: the ocean of salt water, the ocean of sugar cane juice, the ocean of wine, the ocean of thinned butter, the ocean of whipped cheese, the ocean of milk and the ocean of soft water). See ***Sâgara.*** For an explanation of the choice of this number, see **Ocean.**

**SAPTASINDHAVA.** "Seven rivers". This is one of the seven sacred rivers of ancient Brahmanism (*Gangâ, Yamunâ, Sarsvatî, Satlej, Parushni, Marurudvridhâ* and *Arjîkîyâ*).

**SAPTATI.** Ordinary Sanskrit name for the number seventy.

**SAPTAVIMSHATI.** Ordinary Sanskrit name for the number twenty-seven. For words which have a symbolic relationship with this number, see **Twenty-seven** and **Symbolism of numbers.**

**SARITÂPATI.** Name given to the number ten to the power fourteen (= hundred billion). See **Names of numbers** and **High numbers.**

Source: **Trishatikâ* by Shrîdharâchârya (date uncertain).

**SAROJA.** Name given to the number ten to the power nine. See **Names of numbers** and **High numbers.**

Source: **Ganitakaumudî* by Nârâyana (1350 CE).

**SARPA.** [S]. Value = 8. "Serpent". See **Nâga, Eight** and **Serpent (Symbolism of).**

**SARVABALA.** Name formed with the Sanskrit adjective *sarva*, which signifies "everything". It is given to the number ten to the power forty-five. See **Names of numbers.** For an explanation of this symbolism, see **High numbers (Symbolic meaning of).**

Source: **Lalitavistara Sûtra* (before 308 CE).

**SARVADHANA.** [Arithmetic]. Term denoting the "total", or the "whole".

**SARVAJÑA.** Name formed with the Sanskrit adjective *sarva*, which means "everything". Given to the number ten to the power forty-nine. See **Names of numbers.** For an explanation of this symbolism, see **High numbers (Symbolic meaning of).**

**SATA.** Ancient Sanskrit form of the name for hundred. See ***Shata*** and **Names of numbers.** Use of this word is notably found in **Vâjasaneyî Samhitâ, *Taittirîya Samhitâ* and **Kâthaka Samhitâ* (from the start of the first millennium CE); and in **Pañchavimsha Brâhmana* (date uncertain) and **Sankhyâyana Shrauta Sûtra* (date uncertain).

**SATYAYUGA.** Synonym of *Kritayuga. See **Yuga.**

**SÂYAKA.** [S]. Value = 5. "Arrow". Allusion to the *Pañchasâyaka*, the "five arrows" of *Kâma. See ***Bâna, Pañchabâna, Shara*** and **Five.**

**SEASON.** [S]. Value = 6. See ***Ritu*** and **Six.**

**SELEUCID (Calendar).** This calendar began in the year 311 BCE, and was used in the northwest of the Indian subcontinent. To find the corresponding date in the Common Era, subtract 311 from a date expressed in the Seleucid calendar. See **Indian calendars.**

**SELF (THE).** [S]. Value = 1. See ***Abja*** and **One.**

**SENÂNÎNETRA.** [S]. Value = 12. "Eyes of Senânî". This is one of the names of *Kârttikeya, who is often depicted as having six heads. Thus *Senânînetra* = 6 × 2 = 12 eyes. See ***Kârttikeyâsya*** and **Twelve.**

**SENSATION.** [S]. Value = 6. See ***Rasa*** and **Six.**

**SENSE.** [S]. Value = 5. See ***Vishaya*** and **Five.**

**SERPENT (Cult of the).** See **Serpent (Symbolism of the).** See also **Infinity (Indian mythological representation of).**

**SERPENT (Symbolism of the).** In India and all its neighbouring regions, since the dawn of Indian civilisation, the Serpent has been an object of veneration worshipped by the most diverse of religions. At the beginning of the rain season in Rajasthan, Bengal and Tamil Nadu, the serpent is worshipped through offerings of milk and food. In popular religion, the cobra is very highly considered and these snakes are to be found adorning stones called *Grâmadevata*, or "divinities of the village", which are placed under the banyans. Frédéric (1987) explains that the serpents, in most local religions, are genies of the ground, chthonian spirits who possess the earth and its treasures. The cobras are the most significant type of snake in Indian mythology; they are deified and have their own personality. They are often associated with the cult of *Shiva, and in some pictures of Shiva, he has a cobra wound one of his left arms. In these representations, cobras are actually **nâga*, chthonian divinities with the body of a serpent, considered to be the water spirits in all folklore of Asia, especially in the Far East where they are depicted as dragons.

In fact, in traditional Indian iconography, the **nâga* are usually represented as having the head of a human with a cobra's hood. They live in the **pâtâla*, the underworlds, and guard the treasure which is under the earth. They are said to live with the females, the *nâginî* (renowned for their beauty) and devote themselves to poetry. They are considered to be excellent poets, and are even called the princes of poetry:

first they mastered numbers, which led them naturally to becoming masters of the art of poetic metric. They are also princes of arithmetic because, according to legend, there are a *thousand of them. In other words, due to their considerable fertility, the *nâga* represent the incalculable. Just as metric involves the regulation of rhythm, so they are also sometimes associated with the rhythm of the seasons and the weather cycles.

Coming back to the cobra, this is a long snake which can measure between one metre and one and a half metres. Because of this, the Hindus classified them amongst the demons called **mahorâga* (or "large serpents"). It is the "royal" cobra, however, (which can be up to two metres in length) that was the logical choice of leader of the tribe. This snake, as king of the *nâga*, was given several different names: *Vâsuki, *Muchalinda, Muchilinda, Muchalinga, Takshasa, *Shesha, etc., and there are many corresponding myths. See **Vâsuki**. According to a Buddhist legend, the king Muchilinda protected the Buddha, who was in deep meditation, from the rain and floods, by making his coils into a high seat and sheltering him with the hoods of his seven heads. The name which is used most frequently, however, is *Shesha. He is sometimes depicted as a snake with seven heads, but he is usually represented as a serpent with a thousand heads. This is why the term **Sheshashîrsha* (literally "head of Shesha") often means "thousand" when it is used as a numerical word-symbol. Etymologically, the word *shesha* means "vestige", "that which remains". Shesha is also called *Âdi Shesha* (from **Âdi*, "beginning"). This is because Shesha is also and most significantly the "original serpent", born out of the union of Kashyapa and Kadru (Immortality). And because he married Anantashîrsha (the "head of *Ananta"), Shesha, according to Indian cosmology and mythology, became the son of immortality, the vestige of destroyed universes and the seed of all future creations all at once.

The king of the *nâga* thus represents primordial nature, the limitless length of eternity and the boundless limits of infinity. Thus Shesha is none other than *Ananta*: that immense serpent that floats on the primordial waters of original chaos and the "ocean of unconsciousness", and *Vishnu lies on his coils when he rests in between two creations of the world, during the birth of *Brahma who is born out of his navel (see Fig. D. 1 in the entry entitled **Ananta*). Ananta is also the great prince of darkness. Each time he opens his mouth, he causes an earthquake. At the end of each **kalpa* (cosmic cycle of 4,320,000,000 years), Ananta spits and causes the fire which destroys all creation in the universe. He is also **Ahirbudhnya* (or *Ahi Budhnya*), the famous serpent of the depths of the ocean who, according to Vedic mythology, is born out of dark waters. Thus, as well as being the genie of the ground and the chthonian spirit who owns the earth and its treasures, the serpent is also a "spirit of the waters" (**âptya*), who lives in the "inferior worlds" (**pâtâla*).

Some myths clearly indicate this ambivalence surrounding the nature of the reptile, for example the legend which tells the story of Kâliya, the king of the *nâga* of the Yamunâ river; this is a serpent with four heads of monstrous proportions, who defeated by *Krishna, who was then only five years old, went to hide in the depths of the ocean. In this myth, the *four heads* of the king of the *nâga* is significant, because when this serpent goes by the name of Muchalinda, it often has seven heads, or a thousand heads like *Ananta. The choice of these numerical attributions is not simply a question of chance. In fact, in these allegories, the seven heads of Muchalinda represent the subterranean kingdom of the *nâga*, each one being associated with one of the seven hells which constitute the "inferior worlds". These Hells are situated just below *Mount Meru, the centre of the universe, which itself has seven faces, each one facing one of the seven oceans (**sapta sâgara*) and one of the "island-continents" (**sapta dvîpa*). Muchalinda was the "original serpent" who created primordial nature. *Mount Meru, the sacred and mythical mountain of Indian religions, which is thus associated with the number seven, receives its light from the *Pole star, the last of the seven stars of the Little Bear, situated on exactly the same line as this "axis of the world".

On the other hand, the four heads of Kâliya represent the essentially terrestrial nature of the serpent, which crawls along the ground. In Indian mystical thought, earth corresponds symbolically to the number four, which is linked to the square, which in turn is associated with the four cardinal points. On the other hand, the thousand heads of Shesha-Ananta symbolise both the incalculable multitude and an eternal length of time. As for the battle mentioned above between *Krishna and the king of the *nâga*, this is the mystical expression of the rivalry between man and serpent. This man-snake duality is expressed in a very symbolic manner in Vedic literature (notably in the *Chhândogya Upanishad*), where Krishna, the "Black", before his deification, is a simple scholar or **asura* (an "anti-god"). After his victory over the snake, he becomes one of the divinities of the Hindu pantheon: he becomes the *eighth* "incarnation" (**avatâra*) of Vishnu, even before becoming the "beneficent protector of humanity".

This duality is also expressed numerically, because Krishna's position as an incarnation of Vishnu is equal to eight, which is exactly the mystical value of the *nâga*. The *nâga* is not only considered to be a genie of the ground, a chthonian spirit who owns the earth and its treasure, but also and above all an aquatic symbol; it is a "spirit of the waters" living in the underworlds. The symbolic value of the earth is 4. In Indian mystical thought, water (see **Jala*) also has the value 4, thus the ambivalence surrounding the serpent is expressed by the relation: *nâga* = earth + water = 4 + 4 = 8. This value is confirmed by the fact that the *nâga* reproduce in couples and always develop in the company of the *nâginî* (their females); this gives the number eight as the result of the symbolic multiplication of two (the *nâga* and his *nâginî*) by 4 (the earth or water). This is why the name of this species became a word-symbol for the numerical value of 8 (see the entry entitled ***Nâga**).

As well as its terrestrial character, the serpent symbolises primordial nature: "The underworlds and the oceans, the primordial water and the deep earth form one *materia prima*, a primordial substance, which is that of the serpent. He is spirit of the first water and spirit of all waters, be they below, on the surface of or above the earth. Thus the serpent is associated with the cold, sticky and subterranean night of the origins of life: All the serpents of creation together form a unique primordial mass, an *incalculable* primordial thing, which is constantly in the process of deteriorating, disappearing and being reborn." [Keyserling, quoted in Chevalier and Gheerbrant (1982)]. Thus the serpent symbolises life. The *primordial thing* is life in its latent form. Keyserling says that the Chaldaeans only had one word to express both "life" and "serpent". The symbolism of the serpent is linked to the very idea of life; in Arabic, *serpent* is *hayyah* and *life* is *hayat*. [Guénon, quoted in Chevalier and Gheerbrant (1982)]. The serpent is one of the most important archetypes of the human soul [Bachelard, quoted in Chevalier and Gheerbrant (1982)]. The same images are found in Indian cosmological and mythological representations. Thus in tantric doctrine, the *Kundalinî*, literally the "Serpent" of Shiva, source of all spiritual and sexual energies (energies = **shakti*) is said to be found coiled up at the base of the vertebral column, on the *chakra* of the state of sleep. And when he wakes up, "the serpent hisses and becomes tense, and the successive ascent of the *chakra* begins: this is the arousal of the libido, the renewed manifestation of life" [Frédéric, *Dictionnaire* (1987)]. Moreover, from a macroscopic point of view, the *Kundalinî* is the equivalent of the serpent *Ananta, who grasps in his coils the very base of the *axis of the universe*. He is associated with Vishnu and Shiva, and symbolises cyclical development and reabsorption, but, as guardian of the nadir, he is the *bearer of the world*, and ensures its continuity and stability. But Ananta is first and foremost the serpent of infinity, immensity and eternity. All these meanings are in fact various applications, depending on the field in question, of the myth of the original Serpent, which represents primordial indifferentiation. The serpent is considered to be both the beginning and the end of all creation. It is not by chance that the Sanskrit language uses the word *Shesha*, the "remainder", to denote the serpent Ananta; to the Indians, the *nâga* with a thousand heads represents the "vestige" of worlds which have disappeared as well as the seed of worlds yet to appear. This explains the importance which so many cosmologies and mythologies place on the eschatological symbolism of the serpent.

In summary, the snake has always been associated with ideas of the sky, celestial bodies, the universe, of the night of origins, *materia prima*, the axis of the world, primordial substance, the vital principle, life, eternal life and sexual or spiritual energy. It is also connected to ideas of the vestige of past creations and the seed of future creations, cyclical development and reabsorption, longevity, an innumerable quantity, abundance, fertility, immensity, wholeness, absolute stability, endless undulating movement, etc.

In other words, since time immemorial, and amongst all the races of the earth, the serpent, as well as being a symbol of the earth and water, personifies the notion of infinity and *eternity*. See **Infinity, Infinity (Indian concepts of)** and **Infinity (Indian mythological representation of)**.

**SERPENT OF INFINITY AND ETERNITY.** See **Ananta,** ***Sheshashîrsha*****, Infinity (Indian mythological representation of)** and **Serpent (Symbolism of the).**

**SERPENT OF THE DEEP.** [S]. Value = 8. See ***Ahi*, Eight** and **Serpent (Symbolism of the).**

**SERPENT WITH A THOUSAND HEADS.** [S]. Value = 1,000. See ***Sheshashîrsha*** and **Thousand.** See also **Infinity (Indian mythological representation of)** and **Serpent (Symbolism of the).**

**SERPENT.** [S]. Value = 8. See ***Nâga*, *Ahi*, *Sarpa*** and **Eight.**

**SEVEN.** The ordinary Sanskrit words for the number seven are **sapta* and **saptan*. Here is a list of corresponding numerical symbols: **Abdhi*, *Achala*, **Adri*, **Aga*, **Ashva*, **Atri*, *Bhaya*, **Bhûbhrit*, **Bhûdhara*, *Chandas*, *Dhâtu*, *Dhî*, **Dvîpa*, **Giri*, *Haya*, *Kalatra*, **Loka*, **Mahîdhara*, **Mâtrikâ*, **Muni*, **Naga*, **Pârvata*, **Pâtâla*, **Pâvana*, **Pushkara*, **Rishi*, **Sâgara*, **Sâgara*, **Samudra*, **Shaila*, **Svara*, **Tattva*, **Turaga*, **Turangama*, **Vâjin*, **Vâra*, **Vyasana* and *Yati*. These words have the following symbolic meaning or translation: 1. "Purification" and by extension "Purifier" (*Pâvana*). 2. The horses (*Ashva*, *Turaga*, *Turangama*, *Vâjin*). 3. The island-continents (*Dvîpa*). 4. The seas or oceans (*Sâgara*, *Samudra*). 5. The divine mothers (*Mâtrikâ*). 6. The worlds (*Loka*). 7. The inferior worlds (*Pâtâla*). 8. The mountains or hills (*Adri*, *Aga*, *Bhûbhrit*, *Bhûdhara*, *Giri*, *Mahîdhara*, *Naga*, *Pârvata*, *Shaila*). 9. The syllables (*Svara*). 10. The musical notes (*Svara*). 11. The "Sages" of Vedic times (*Muni*, *Rishi*). 12. The last of the seven Rishi (*Atri*). 13. The days of the week (*Vâra*). 14. "That which does not move" (*Naga*). 15. The seventh "island-continent" (*Pushkara*). 16. The fears (*Bhaya*) (only in *Jaina religion). 17. The winds (*Pâvana*).

See **Numerical symbols.**

**SEVENTEEN.** Ordinary Sanskrit name: **saptadasha*. Corresponding numerical symbol: **atyashti*.

**SEVENTY.** See ***Saptati*.**

**SEVERUS SEBOKT.** Syrian bishop of the seventh century CE. His works notably include a manuscript dated 662 CE, where he talks of the system of nine numerals and Indian methods of calculation.

**SHAD (SHASH, SHAT).** Ordinary Sanskrit name for the number six, this word forms part of the composition of many other words which are directly related to the idea of this number.

Examples: **Shâdanga*, **Shadâyatana*, **Shaddarshana*, **Shâdgunya*, **Shatkasampatti*. For words which are symbolically related to this number, see **Six** and **Symbolism of numbers.**

**SHÂDANGA.** "Six parts". This is the name for the six aesthetic rules of painting, which are described in a commentary on the *Kâmasutra* by Yashodhara (these six rules being as follows: *rûpabheda*, "shape"; *pramanam*, "size"; *bhava*, "sentiment"; *lavana*, "grace"; *sadrîshyam*, "comparison"; and *varnikabahanga*, "colour").

**SHADÂYATANA.** [S]. Value = 6. "Six **guna*". These are the "six bases", or "six categories". These are the six senses, objects or sense organs of Buddhist philosophy (namely: the eye, the nose, the ear, the tongue, the body and the mind). See **Six.**

**SHADDARSHANA.** [S]. Value = 6. "Six visions", "six contemplations", "six philosophical points of view". These are the six principal systems of Hindu philosophy. See ***Darshana*** and **Six.**

**SHADDASHA.** Ordinary Sanskrit name for the number sixteen. For words which are symbolically connected to this number see **Sixteen** and **Symbolism of numbers.**

**SHÂDGUNYA.** [S]. Value = 6. "six **guna*". This is a synonym of **shadâyatana*. See **Six.**

**SHADVIMSHATI.** Ordinary Sanskrit name for the number twenty-six. For words which are symbolically associated with this number see **Twenty-six** and **Symbolism of numbers.**

**SHAILA.** [S]. Value = 7. "Mountain". This concept is related to the myth of *Mount Meru, where the numbers seven plays a significant role. See ***Adri*** and **Seven.**

**SHAKA (Calendar).** This is the most widely used calendar in Hindu India, as well as in the parts of Southeast Asia influenced by India. It is also known as *Shakakâla*, *Shakarâja* or *Shakasamvat*. It began in the year 78 of the Common Era. According to certain traditions, this calendar was begun in the first century CE by a Satrap (Kshatrapa) king called Shâlivâhana (or Nahapâna), who then reigned over the city of *Ujjain. To find a corresponding date in the Common Era, add seventy-eight to a date expressed in *Shaka* years. See **Indian calendars.**

**SHAKA NUMERALS.** Symbols derived from *Brâhmî numerals and influenced by Shunga numerals, arising at the time of the Shunga dynasty (second to first centuries BCE). The symbols corresponded to a mathematical system that was not based on place values and therefore did not possess a zero. See: **Indian written numeral systems (Classification of).** See Fig. 24.52 and 24.61 to 69.

**SHAKASAMVAT (Calendar).** See ***Shaka*.**

**SHAKRA.** [S]. Value = 14. "Powerful". Allusion to the "strength" of the god *Indra, amongst whose attributes is *Shakradevendra*, "Powerful Indra". This explains the symbolism in question, becuase *Indra* = 14. See **Fourteen.**

**SHAKTI.** [S]. Value = 3. "Energy". In Brahmanism and Hinduism, this word denotes feminine energy or the active principle of all divinity. The allusion here is to the *shakti* of the most important divinities, namely those of the triad formed by *Brahma, *Vishnu and *Shiva. See **Three.**

For an example of the use of this word-symbol, see: EI, XIX, p. 166.

**SHANKARÂCHÂRYA.** Hindu philosopher of the late ninth century. His works notably include *Shârîrakamîmâmsâbhâshya* (great commentary on the *Vedântasûtra*), where there is a reference to the place-value system of the Indian numerals.

**SHANKARÂKSHI.** [S]. Value = 3. Synonym of **Haranetra*, "eyes of *Shiva". See **Three.**

**SHANKARANÂRÂYANA.** Indian astronomer c. 869 CE. His works notably include a text entitled *Laghubhâskarîyavivarana* in which the place-value system of Sanskrit numerical symbols is used frequently. He also uses the *katapayâdi* method invented by Haridatta [see Billard (1971), p. 8]. See **Numerical symbols, Katapayâdi numeration** and **Indian mathematics (History of).**

**SHANKHA.** Word which expresses the sea conch. It is a symbol of riches and of certain Hindu and Buddhist divinities (such as *Vishnu). It is a name given to the number ten to the power twelve. See **Names of numbers.** For an explanation of this symbolism, see **High numbers (Symbolic meaning of).**

Source: **Râmâyana* by Vâlmîki (early centuries CE).

**SHANKHA.** Word which expresses the sea conch. It is given to the number ten to the power thirteen (ten billion). See **Names of numbers.** For an explanation of this symbolism, see **High numbers (Symbolic meaning of).**

Source: **Kitab fi tahqiq i ma li'l hind* by al-Biruni (c. 1030 CE).

**SHANKHA.** Word meaning sea conch. It is given to the number ten to the power eighteen. See **Names of numbers.** For an explanation of this symbolism, see **High numbers (Symbolic meaning of).**

Source: **Ganitasârasamgraha* by Mâhavîrâchârya (850 CE).

**SHANKU.** Literally: "Diamond". Name given to the number ten to the power thirteen (ten billion). See **Names of numbers.** For an explanation of this symbolism, see **High numbers (Symbolic meaning of).**

Sources: **Lîlâvatî* by Bhâskarâchârya (1150 CE); **Ganitakaumudî* by Nârâyana (1350 CE), **Trishatikâ* by Shrîdharâchârya (date uncertain).

**SHANMUKHA.** [S]. Value = 6. Synonym of **Kumârâsya*, "Faces of **Kumâra* (= Shanmukha)". This is an allusion to the six heads of *Kârttikeya. See ***Kârttikeyâsya*** and **Six.**

**SHANMUKHABÂHU.** [S]. Value = 12. "Arms of *Shanmukha (= *Kumâra = *Kârttikeya)". Kârttikeya is said to have had twelve arms. See ***Kârttikeyâsya*** and **Twelve.**

**SHARA.** [S]. Value = 5. "Arrow". This is one of the attributes of *Kâma, Hindu divinity of Cosmic Desire and Carnal Love, who is generally invoked during wedding ceremonies, and whose action is said to determine the laws of **samsâra* for human beings. The symbolism in question is due to the fact that Kâma is often represented as a young man armed with a bow made of sugar cane shooting five arrows (**pañchabâna*) which are either flowers or adorned with flowers. See **Arrow** and **Five.**

**SHÂRÂDA NUMERALS.** Symbols derived from *Brâhmî numerals and influenced by Shunga, Shaka, Kushâna, Ândhra, and Gupta. Used in Kashmir and the Punjab from the ninth to the fifteenth centuries CE. The symbols correspond to a mathematical system that has place values and a zero (shaped like a dot). The more or less direct ancestor of Tâkarî, Dogrî, Kuluî, Sirmaurî, Kochî, Landa, Maltânî, Khudawadî, Sindhî, Punjabi, Gurûmukhî, etc. numerals. For historic symbols, see Fig. 24.40; for current symbols,see Fig. 24.14; for derived notations, see Fig. 24.52. For the corresponding graphical development, see Fig. 24.61 to 69. See: **Indian written numeral systems (Classification of).**

**SHASH.** Ordinary Sanskrit word for the number six. See ***Shad*.**

**SHASHADHARA.** [S]. Value = 1. "Which represents a rabbit". This is connected with an attribute of the moon. According to legend, a rabbit, who offered its own flesh to relieve the poor, was rewarded by having its own image impressed on the face of the moon. This explains the symbolism in question, because "Moon" = 1. See ***Abja*** and **One.**

**SHASHANKA**. [S]. Value = 1. "Moon". See ***Abja*** and **One**.

**SHASHIN**. [S]. Value = 1. "To the Rabbit". This is the rabbit which, according to legend, was drawn on the face of the moon. See ***Shashadhara, Abja*** and **One**.

**SHASHTI**. Ordinary Sanskrit name for the number sixty.

**SHAT**. Ordinary Sanskrit name for the number six. See ***Shad***.

**SHATA**. Ordinary Sanskrit name for the number one hundred. Its multiples are formed by placing it to the right of the names of the corresponding units: *dvashtat* (two hundred), *trishata* (three hundred), *chatushata* (four hundred), etc. This name forms part of the composition of several words which are symbolically associated with the idea of this number.

Examples: **Shatapathabrâhmana*, **Shatarudriya*, **Shatarûpâ*, **Shatottaragananâ*, **Shatottaraguna*, **Shatottarasamjnâ*. For words which have a more symbolic link with this number, see **Hundred** and **Symbolism of numbers**. See also ***Sata*** for an ancient form of this number.

**SHATAKOTI**. Literally: a hundred **koti*". This is the name given to the number ten to the power nine. See **Names of numbers** and **High numbers**.

Source: **Ganitasârasamgraha* by Mâhavîrâchârya (850 CE).

**SHATAPATHABRÂHMANA**. "Brâhmana of the Hundred ways". This is the title of an important work of Vedic literature, divided into a hundred *adhyâya* ("recitations").

**SHATARUDRIYA**. Name of a Sanskrit hymn which is part of the **Taittirîya Samhitâ* (*Yajurveda*), addressed to *Rudra considered from a hundred different perspectives.

**SHATARÛPÂ**. "Of a hundred forms". One of the names for the first woman, daughter and wife of *Brahma, who is said to have been gifted with a "hundred bodies". See ***Rûpa***.

**SHATKASAMPATTI**. Literally "six great victories" (from *shatka*, "made up of six", and *sampatti*, "to obtain, achieve, succeed"). In Hinduism, this refers to the "Six Great Victories" of Tattvabodha of Shankara, which constitutes the first of the four conditions that an adept of the philosophy of the *Vedânta* must fulfil.

**SHATOTTARAGANANÂ**. "Centesimal arithmetic". There is a reference to this in **Lalitavistara Sûtra* [see Datta and Singh (1938), p. 10].

**SHATOTTARAGUNA**. "Hundred, primordial property". Sanskrit name for the centesimal base. Reference to this is found in the **Lalitavistara Sûtra*.

**SHATOTTARASAMJÑÂ**. "Names of multiples of a hundred". This term applies to names of numbers in Sanskrit numeration in the centesimal base. There is a reference to this in **Lalitavistara Sûtra* [see Datta and Singh (1938), p. 10]. The equivalent of this word in terms of the decimal base is **Dashagunâsamjnâ*. See ***Shatottaraganana, Shatottaraguna*** and ***Dashagunâsamjña***.

**SHESHA**. "Vestige", "that which remains" or "he who remains". In Brahman and Hindu mythologies, this is the name of *Ananta, the king of the **nâga* and serpent of the infinity, eternity and immensity of space. See **Serpent (Symbolism of the)**.

**SHESHA**. [Arithmetic]. "Vestige". Term describing the remainder in division.

**SHESHASHÎRSHA**. [S]. Value = 1,000. Literally "heads of *Shesha". Shesha is the king of the **nâga* who lives in the inferior worlds (**pâtâla*) and who is considered to be the "Vestige" of destroyed universes as well as the seed of all future creation. This symbolism comes from the fact that Shesha is represented as a serpent with a thousand heads, the number *thousand* here meaning "multitude" and the "incalculable". See **Ananta, Thousand, High numbers (Symbolic meaning of)** and **Serpent (Symbolism of the)**.

**SHIKHIN**. [S]. Value = 3. "Ablaze". This is one of the names for *Agni, the god of sacrificial fire, whose name is equal to the number three. See **Three**.

**SHÎRSHAPRAHELIKÂ**. From *shîrsha*, "head", and *prahelikâ*, "awkward question, enigma". This term is used in the texts of *Jaina cosmology to denote a period of time which corresponds approximately to ten to the power 196. See ***Anuyogadvâra Sûtra***, **Names of numbers, High numbers** and **Infinity**.

**SHÎTÂMSHU**. [S]. Value = 1. "Of the cold rays". This is a synonym of "moon", the opposite of the warm rays of the sun. See ***Abja*** and **One**.

**SHÎTARASHMI**. [S]. Value = 1. "Of the cold rays". This is a synonym of "Moon", the opposite of the warm rays of the sun. See ***Abja*** and **One**.

**SHIVA**. [S]. Value = 11. One of the three main divinities of the Brahmanic pantheon (*Brahma, *Vishnu, *Shiva). There is no mention of Shiva in the **Veda*, and it would seem that Shiva did not become a god until relatively recently. The symbolism in question comes from the fact that Shiva is none other than an incarnation of *Rudra, ancient Vedic divinity of tempests and cosmic anger. As Rudra symbolises the number 11 (because of the eleven vital breaths, born from Brahma's forehead, of which he was the personification), the name of Shiva also came to represent this number. See ***Rudra-Shiva*** and **Eleven**.

**SHRÎDHARÂCHÂRYA**. Indian mathematician. The date of his birth is not known. His works notably include *Trishatikâ*. Here is a list of the principal names of numbers mentioned in this work:

**Eka* (= 1), **Dasha* (= 10), **Shata* (= $10^2$), **Sahasra* (= $10^3$), **Ayuta* (= $10^4$), **Laksha* (= $10^5$), **Prayuta* (= $10^6$), **Koti* (= $10^7$), **Arbuda* (= $10^8$), **Abja* (= $10^9$), **Kharva* (= $10^{10}$), **Nikharva* (= $10^{11}$), **Mahâsaroja* (= $10^{12}$), **Shanku* (= $10^{13}$), **Saritâpati*, (= $10^{14}$), **Antya* (= $10^{15}$), **Madhya* (= $10^{16}$), **Parârdha* (= $10^{17}$).

Ref.: TsT, R. 2–3 [see Datta and Singh (1938), p. 13].

See **Names of numbers, High numbers** and **Indian mathematics (History of)**.

**SHRÎPATI**. Indian astronomer c. 1039 CE. His works notably include a text entitled *Siddhântashekhara*, in which the place-value system of the Sanskrit numerical system is used frequently [see Billard (1987), p. 10]. See **Numerical symbols, Numeration of numerical symbols,** and **Indian mathematics (History of)**.

**SHRUTI**. [S]. Value = 4. "Recital". Name given to the ancient Brahmanic and Vedic religious texts, which are said to have been revealed by a divinity to one of the seven "Sages" (**rishî*), poets and soothsayers of Vedic times (**Saptarishi*). As this allusion primarily concerns the **Veda*, and there are four of them, *shruti* = 4. See **Four**.

**SHUKRANETRA**. [S]. Value = 1. The "Eye of Shukra". According to legend, this divinity had one eye destroyed by *Vishnu, thus the symbolism in question. See **One**.

**SHÛLA**. [S]. Value = 3. "Point". Allusion to the three points of *Shiva's Trident (**trishûlâ*), which symbolise the three aspects of the manifestation (creation, preservation, destruction), as well as the three primordial principles (**triguna*), and the three states of consciousness (**tripurâ*). See ***Guna*** and **Three**.

**SHÛLIN**. [S]. Value = 11. This is one of the attributes of *Rudra, who is invoked as "lord of the animals" in the *Shûlagava*, Brahmanic sacrifices of two-year-old calves with the aim of obtaining prosperity. Thus *Shûlin* = Rudra = 11. See ***Rudra-Shiva*** and **Eleven**.

**SHUNGA NUMERALS**. Symbols derived from *Brâhmî numerals, arising during the Shunga dynasty (second century BCE). The symbols corresponded to a mathematical system that was not based on place-values and therefore did not possess a zero. See: **Indian written numeral systems (Classification of)**. See Fig. 24.30, 24.52 and 24.61 to 69.

**SHÛNYA**. Literally "void". This is the principal Sanskrit term for "zero". However, the Sanskrit language (the excellent literary instrument of mathematicians, astronomers and all Indian scholars) has many synonyms for expressing this concept (**abhra*, **âkâsha*, **ambara*, **ananta*, **antariksha*, **bindu*, **gagana*, **jaladharapatha*, **kha*, **nabha*, **nabhas*, **pûrna*, **vishnupada*, **vindu*, **vyoman*, etc.). The words **kha*, **gagana*, etc., are used for "sky", "firmament", and the words **ambara*, **abhra*, **nabhas*, etc., signify "space", "atmosphere", etc. The word **âkâsha* means the fifth "element", "ether", the immensity of space, as well as the essence of all that is uncreated and eternal. There is also the word **bindu*, which means "dot" or "point". At least since the beginning of the Common Era, *shûnya* means not only void, space, atmosphere or ether, but also nothing, nothingness, negligible, insignificant, etc. In other words, the Indian concept of zero far surpassed the heterogeneous notions of vacuity, nihilism, nothingness, insignificance, absence and non-being of Greek and Latin philosophies. See ***Shûnyatâ***, **Numerical symbols, Zero, Zero (Graeco-Latin concepts of), Zero (Indian concepts of)** and **Indian atomism**.

**SHÛNYA-BINDU**. Literally: "void-dot". Name of the graphical representation of zero in the shape of a dot. See ***Shûya, Bindu***, **Numeral 0 (in the shape of a dot)** and **Zero**.

**SHÛNYA-CHAKRA**. Literally: "void-circle". Name of the graphical representation of zero in the shape of a little circle. See ***Shûnya***, **Numeral 0 (in the shape of a little circle)** and **Zero**.

**SHÛNYA-KHA**. Literally: "void-space". Name given to the function of zero in numerical representations: it is the *empty space* which marks the absence of units of a given order in positional numeration. See ***Kha, Shûnya*** and **Zero**.

**SHÛNYA-SAMKHYA.** Literally: "void-number". Name given to zero as a numerical symbol. It is also the "zero quantity" considered to be a whole number in itself. See ***Samkhya, Shûnya*** and **Zero.**

**SHÛNYATÂ.** In Sanskrit, the privileged term for the designation of zero is **shûnya*, which literally means "void". But this word existed long before the discovery of the place-value system. Since Antiquity, this word has constituted the central element of a mystical and religious philosophy, developed as a way of thinking and behaving, namely the philosophy of "vacuity" or *shûnyatâ*. See ***Shûnya.*** This doctrine is a fundamental concept of Buddhist philosophy and is called the "Middle Way" (*Mâdhyamaka*), which teaches that every thing in the world (*samskrita*) is empty (**shûnya)*, impermanent (*anitya*), impersonal (*anâtman*), painful (*dukha*) and without original nature. Thus this vision, which does not distinguish between the reality and non-reality of things, reduces these things to complete insubstantiality.

This philosophy is summed up in the following answer that the Buddha is said to have given to his disciple Shariputra, who wrongly identified the void (**shûnya*) with form (**rûpa*): "That is not right," said the Buddha, "in the *shûnya* there is no form, no sensation, there are no ideas, no volitions, and no consciousness. In the *shûnya,* there are no eyes, no ears, no nose, no tongue, no body, no mind. In the *shûnya*, there is no colour, no noise, no smell, no taste, no contact and no elements. In the *shûnya,* there is no ignorance, no knowledge, or even the end of ignorance. In the *shûnya*, there is no aging or death. In the *shûnya*, there is no knowledge, or even the acquisition of knowledge.

"Buddhists did not always use *shûnya* in this sense: in the ancient Buddhism of *Hînayâna* (known as the "Lesser Vehicle"), this notion only applied to the person, whereas in *Mahâyâna* Buddhism (of the schools of the North and known as the "Greater Vehicle"), the idea of vacuity stretched to include all things. To explain the difference between these two concepts, the Buddhists of the schools of the North make the following comparison: in the ancient vision, things were regarded as if they were empty shells, whereas in the *Mahâyâna* the very existence of the empty shells is denied. This concept of the whole of existence being a void should not lead to the conclusion that this is an attitude of nihilism. Far from meaning that things do not exist, this philosophy expresses that things are merely illusions. Through criticising the knowledge of things as being a pure illusion (*mâyâ*), it actually means that absolute truth is independent of the being and the non-being, because it is the *shûnyatâ* or "vacuity". The *shûnyatâ* has a real existence; it is composed of **âkâsha*, or "ether", the last and most subtle of the "five elements" (**pañchabhûta* or "ether) of Hindu and Buddhist philosophies, which is considered to be the essence of all that is uncreated and eternal, and the element which penetrates everything. The **âkâsha* has no substance, yet it is considered to be the condition of all corporeal extension and the receptacle of all matter which manifests itself in the form of one of the other four elements (earth, water, fire, air). According to this philosophy, the *shûnyatâ* is the ether-universe, the only "true universe".

This is why the being and the non-being are considered to be insignificant and even illusory compared to the *shûnyatâ,* which thus excludes any possible mixing with material things, and which, as an unchanging and eternal element, is impossible to describe. In **Mâdhyamika* Buddhism (the followers are still called **Shûnyavâdin* or "vacuitists"), the void has been identified with the absence of self and salvation. Both are meant to achieve redemption, which is only possible in the *shûnyatâ*. In other words, in order to be granted deliverance, vacuity must be achieved; for this, the mind must be purified of all affirmation and all negation at once.

This ontology is inextricably linked to the mysticism of universal vacuity, and represented the great philosophical revolution of Buddhism amongst the schools of the North, implemented by Nâgârjuna, the patriarch of this sect. The *Madhyamakashâstra*, the fundamental text which is traditionally attributed to Nâgârjuna, was translated into Chinese in the year 409 CE, when he had already achieved almost god-like status, and was renowned far beyond the frontiers of India. [See Bareau (1966), pp. 143ff.-; Frédéric (1987); Percheron (1956); Renou and Filliozat (1953)]. This proves that the fundamental concepts of this mysticism were already fully established at the beginning of the Common Era.

These concepts were pushed to such an extent that twenty-five types of *shûnya* were identified. Amongst these figured: the void of non-existence, of non-being, of the unformed, of the unborn, of the non-product, of the uncreated or the non-present; the void of the non-substance, of the unthought, of immateriality or insubstantiality; the void of non-value, of the absent, of the insignificant, of little value, of no value, of nothing, etc. This means that in the *shûnyatâvâda*, the philosophical notions of vacuity, nihilism, nullity, non-existence, insignificance and absence were conceived of very early and unified according to a perfect homogeneity under the unique label of *shûnyatâ* ("vacuity"). In this domain at least, India was very advanced in comparison with corresponding Graeco-Latin ideas. See **Zero (Graeco-Latin concepts of), Zero** and **Zero (Indian concepts of).**

**SHÛNYATÂVADÂ.** Name of the Buddhist doctrine of vacuity. See ***Shûnya*** and ***Shûnyatâ.***

**SHÛNYAVÂDIN.** "Vacuitist". Name given to the followers of the Buddhist philosophy of vacuity. See ***Shûnyatâ*** and ***Mâdhyamika.***

**SIDDHA.** In Hindu and Jaina philosophies, this is the name given to human beings that become immortals after having obtained liberation.

**SIDDHAM NUMERALS.** Symbols derived from *Brâhmî numerals and influenced by Shunga, Shaka, Kushâna, Ândhra, Gupta and Nâgarî. Used in Ancient Nepal (sixth to eighth centuries CE). The symbols corresponded to a mathematical system that was not based on place-values and therefore did not possess a zero. Ancestor of Nepali, Tibetan, Mongolian, etc. numerals. Siddham also influenced the shapes of Kutilâ numerals, whence came Bengali, Oriyâ, Kaîthî, Maithilî, Manipurî, etc. numerals. See Fig. 24.41. For systems derived from Siddham, see Fig. 24.52. For graphical development, see Figs. 24.61 to 69. **See: Indian written numeral systems (Classification of).**

**SIDDHÂNTA.** [Astronomy]. Generic name of the astronomical texts which describe such things as the calculation for an eclipse of the Moon or the Sun, and the procedures, methods and instruments of observation. Diverse parameters and data are supplied, as well as the procedure for trigonometric operations, etc. See **Indian astronomy (History of)** and **Yuga (Astronomical speculations on).**

**SIDDHÂNTADARPANA.** See **Nilakanthasomayâjin.**

**SIDDHÂNTASHEKHARA.** See **Shrîpati.**

**SIDDHÂNTASHIROMANI.** See **Bhâkarâchârya.**

**SIDDHÂNTATATTVAVIVEKA.** See **Kamâlakara.**

**SIDDHI.** [S]. Value = 8. "Supernatural power". This is an allusion to the *ashtasiddhi*, the eight major *siddhi*, or eight supernatural powers which the *siddha* and the *pûrnayogin* (perfect adepts of the techniques of yoga) are gifted with. See **Eight.**

**SIGNS IN THE FORM OF "S" OR "8".** See **Numeral 8, Serpent (Symbolism of the)** and **Infinity (Indian mythological representation of).**

**SIGNS OF NUMERATION.** See Fig. 24.61 to 69. See **Indian numerals**, which gives the complete list of signs of numeration, as well as **Numerical notation**, which gives a list of the main systems of numeration used in India since Antiquity. See also **Indian written numeral systems (Classification of)**, which recapitulates on all the numerical notations of the Indian sub-continent, from both a mathematical and a palaeographic point of view.

**SIMHASAMVAT (Calendar).** Calendar beginning in the year 1113 CE. Add 1113 to a date in this calendar to find the corresponding year in the Common Era. Formerly used in Gujarat. It was probably abandoned during the thirteenth century. See **Indian calendars.**

**SIMPLE YUGA (Non-speculative).** See **Yuga (Astronomical speculation on).**

**SINDHI NUMERALS.** Symbols derived from *Brâhmî numerals and influenced by Shunga, Shaka, Kushâna, Ândhra, Gupta and Shâradâ. Currently used in the region of Sindh (whose name derives from the river now called the Indus). The symbols correspond to a mathematical system that has place values and a zero (shaped like a small circle). **See: Indian written numeral systems (Classification of).** See Fig. 24.6, 24.52 and 24.61 to 69.

**SINDHU.** [S]. Value = 4. "Sea". See *Sâgara*, **Four** and **Ocean.**

**SINE (Function).** This is referred to as *ardhajyâ,* which literally means: "demi-chord". This is the name used since *Âryabhata (c. 510 CE) by Indian astronomers to denote this function of trigonometry.

**SINHALA (SINHALESE) NUMERALS.** Symbols derived from *Brâhmî numerals and influenced by Shunga, Shaka, Kushâna, Ândhra, Pallava, Châlukya, Ganga, Valabhî and Bhattiprolu numerals. Currently used mainly in Sri Lanka (Ceylon), in the Maldives and in other islands to the north of the Maldives. (Note that in the north and northwest of Sri Lanka, *Tamil numerals are used by the Tamil inhabitants.). The symbols correspond to a mathematical system that is not based on place values and therefore does not possess a zero. See: **Indian written numeral systems (Classification of)**. See Fig. 24.22, 24.52 and 24.61 to 69.

**SIX.** Ordinary Sanskrit names for this number: **shad, *shash, *shat.* Here is a list of corresponding numerical symbols: **Anga, Ari, *Darshana, *Dravya, *Guna, Karâka, *Kârttikeyâsya, *Kâya, Khara, *Kumârâsya, *Kumâravadana, Lekhya, Mala, *Mâsârdha, *Râga, *Rasa, Ripu, *Ritu, *Shâdâyatana, *Shaddarshana, *Shâdgunya, *Shanmukha, Shâstra, Tarka.*

These words have the following symbolic meanings or translations: 1. The philosophical points of view *(Darshana)* 2. The six philosophical points of view *(Shaddarshana)*, 3. The bodies *(Kâya)*, 4. The colours *(Râga)*, 5. The musical modes *(Râga)*, 6. The weapons *(Shâstra)*, 7. The limbs *(Anga)*, 8. The **Vedânga (Anga)*, 9. The merits, the qualities, the primordial properties *(Guna)*, 10. The six primordial properties, the six bases, the six categories *(Shadâyatana, Shâdgunya)*, 11. The seasons *(Mâsârdha, Ritu)*, 12. The substances *(Dravya)*, 13. The faces of Kârttikeya-Kumâra *(Kârttikeyâsya, Kumârâsya, Kumâravadana, Shanmukha)*, 14. The sensations, in the sense of "flavours" *(Rasa)*. See **Numerical symbols.**

**SIX AESTHETIC RULES OF PAINTING.** See ***Shâdanga.***

**SIXTEEN.** Ordinary Sanskrit name: **shaddasha.* Here is a list of corresponding numerical symbols: **Ashti, *Bhûpa, Kalâ* and **Nripa.* These words refer to or are related to the following: 1. A particular element of Indian metric (*Ashti*) 2. The sixteen kings of the legend of the **Mahâbhârata* (*Bhûpa* and *Nripa*) 3. The "fingers of the Moon" (*Kalâ*). See **Numerical symbols.**

**SIXTY.** See ***Shashti.***

**SMALLEST UNIT OF LENGTH.** See ***Paramânu.***

**SMALLEST UNIT OF WEIGHT.** See ***Paramânu râja.*** See also *Indian weights and measures.

**SOGANDHIKA.** Name given to the number ten to the power ninety-one. See **Names of numbers** and **High numbers.**

Source: **Vyâkarana* (Pâlî grammar) by Kâchchâyana (eleventh century CE).

**SOMA.** [S]. Value = 1. Name of an intoxicating drink, used in Vedic times for religious ceremonies and sacrifices: "It is a drink made from a climbing plant, with which an offering is made to the gods and which is drunk by Brahmanic priests. This drink plays an important role in the *Rigveda.* It is considered to be capable of conferring supernatural powers and is worshipped as though it were a god. The Hindus also call it the wine of immortality (**amrita*). It is the symbol of the transition from ordinary sensory pleasures to divine happiness *(ânanda).* K. Friedrichs, etc. "In Indian thought, *Soma* also represents the source of all life and symbolises fertility; thus it is the sperm, the receptacle of the seeds of cyclic rebirth. In this respect, the *soma* is connected to the symbolism of the moon. This is why the *soma* is also the lunar star, a masculine entity compared with a full goblet of the drink of immortality. Thus: *soma* = 1. See ***Abja*** and **One.**

Source: **Lalitavistara Sûtra* (before 308 CE).

**SPECULATIVE YUGA.** See ***Yuga*** **(Astronomical speculation on).**

**SQUARE ROOT.** [Arithmetic]. See ***Varganmûla.*** See also **Square roots (How Âryabhata calculated his).**

**SQUARE ROOTS (How Âryabhata calculated his).** In the chapter of *Ganitapâda* in *Âryabhatîya* devoted to arithmetic and methods of calculation, the astronomer *Âryabhata (c. 510 CE) described, amongst other operations, the rule for the extraction of square roots [see Arya, *Ganita,* line 4]:

*Always divide the even column by twice the square root. Then, after subtracting the square of the even column, put the quotient in the next place. This will give you the square root.*

The rule, thus formulated, is a typical example of Âryabhata's extremely concise style, only giving the essential information in his definitions, operations or concepts, any other information being deemed useless for reasons only known by the man himself. Here is the extract again, with the necessary information added for easy understanding:

[After subtracting the largest possible square from the figure found in the last uneven column, then having written the square root of the subtracted number in the line of the square root] *always divide the* [figure in the] *even column* [written on the right] *by twice the square root. Then, after subtracting the square* [of the quotient] *from the* [figure found in the] *even column* [written on the right], *place the quotient in the next place* [to the right of the figure which is already written down in the line of the square root]. *This will give you the square root* [desired]. [But if there are figures remaining on the right, repeat the process until there are no more of these figures].

[See: Datta and Singh (1938), pp. 169–75; Clark (1930), pp. 23ff; Shukla and Sarma (1976), pp. 36–7; Singh, in BCMS, 18, (1927)]. Here is the reproduction (with no theoretical justification) of the first of these rules, in order to calculate the square root of the number 55, 225, according to the information given notably by Bhâskara (in 629) in his Commentary on the *Âryabhatîya:* First, the number in question is written in the following manner, marking each uneven place with a vertical line and each even place with a horizontal line:

```
| – | – |
5 5 2 2 5
```

Then a horizontal line is drawn (to the right of the number in question), in order to write down the successive numbers of the square root:

```
| – | – |     ________
5 5 2 2 5
              line of the square root
```

By beginning the operation with the highest figure of the uneven column, the biggest square it contains is 4, thus the square root is equal to 2. Therefore a 4 is placed in a line underneath and a 2 on the line of square roots:

```
| – | – |         2
              ________
5 5 2 2 5
4             line of the square root
```

Then a line is drawn below the 4, which is subtracted from the preceding 5; the result is 1, and this figure is placed under the line in the even position of this first section, without forgetting to return the 4 to the extreme left of this lower line:

```
| – | – |         2
              ________
5 5 2 2 5
4             line of the square root
_______
4) 1
```

Next the figure in the even column written immediately to the right (the 5) is considered, and is placed below the bottom line, to the right of the 1:

```
| – | – |         2
              ________
5 5 2 2 5
4             line of the square root
_______
4) 1 5
```

Now the number 15 which has been obtained is divided by twice the square root that was previously found (2), in other words by 4; as the quotient found is 3, thus 3 is written on the line of square roots, to the right of the 2 that is already there, without forgetting to record the same figure on the extreme right of the line of the 15:

```
| – | – |         2 3
              ________
5 5 2 2 5
4             line of the square root
_______
4) 1 5   (3
```

The product of the numbers 4 and 3 (placed to the left and right of the line of 15) is 12, and this is placed on the line below 15:

```
| – | – |         2 3
              ________
5 5 2 2 5
4             line of the square root
_______
4) 1 5   (3
   1 2
```

Then 12 is subtracted from the above 15, and the result is placed on the line below, after drawing a line below the number 12:

```
| – | – |         2 3
              ________
5 5 2 2 5
4             line of the square root
_______
4) 1 5   (3
   1 2
_______
     3
```

Then the 2 from the following uneven column is placed next to the 3:

```
| – | – |         2 3
              ________
5 5 2 2 5
4             line of the square root
_______
4) 1 5   (3
   1 2
_______
     3 2
```

And a 9 (the square of the quotient 3 found above, indicated to the right of the line of 15) is placed in the line below the 32:

```
| – | – |        2 3
                 ─────
5 5 2 2 5
4                line of the square root
─────────
4) 1 5    (3
   1 2
─────────
     3 2
       9
```

A line is drawn and the 9 is subtracted from the 32, then the result is placed below this line:

```
| – | – |        2 3
                 ─────
5 5 2 2 5
4                line of the square root
─────────
4) 1 5    (3
   1 2
─────────
     3 2
       9
─────────
     2 3
```

Now the 2 is taken from the even column and placed to the right of the positions of 23:

```
| – | – |        2 3
                 ─────
5 5 2 2 5
4                line of the square root
─────────
4) 1 5    (3
   1 2
─────────
     3 2
       9
─────────
     2 3 2
```

Then the number 232 which has been thus obtained is divided by 46, which is double the square root found (23), and as the quotient is 5, the numbers 46 and 5 are written as follows (the divisor 46 on the left and the quotient 5 on the right), by placing a 5 on the line of square roots to the right of the 3:

```
| – | – |        2 3 5
                 ─────
5 5 2 2 5
4                line of the square root
─────────
4) 1 5    (3
   1 2
─────────
     3 2
       9
─────────
46)  2 3 2   (5
```

And as the product of 46 times 5 is 230, this number is placed below 232:

```
| – | – |        2 3 5
                 ─────
5 5 2 2 5
4                line of the square root
─────────
4) 1 5    (3
   1 2
─────────
     3 2
       9
─────────
46)  2 3 2   (5
     2 3 0
```

Another line is drawn, and the following subtraction is carried out:

```
| – | – |        2 3 5
                 ─────
5 5 2 2 5
4                line of the square root
─────────
4) 1 5    (3
   2 2
─────────
     3 2
       9
─────────
46)  2 3 2   (5
     2 3 0
─────────
         2
```

The last figure (5) is lowered and placed to the right of the 2:

```
| – | – |        2 3 5
                 ─────
5 5 2 2 5
4                line of the square root
─────────
4) 1 5    (3
   1 2
─────────
     3 2
       9
─────────
46)  2 3 2   (5
     2 3 0
─────────
         2 5
```

The last quotient is equal to 5, and the square of this number is taken (25) and subtracted from this last number. As the result is equal to zero, the operation is finished. It is clear that the operation has worked, because the square root of 55 225 is equal to 235.

```
| – | – |        2 3 5
                 ─────
5 5 2 2 5
4                line of the square root
─────────
4) 1 5    (3
   1 2
─────────
     3 2
       9
─────────
46)  2 3 2   (5
     2 3 0
─────────
       2 5
       2 5
─────────
         0
```

Thus it is clear that this procedure is not algebraic (contrary to Kaye's allegations, who gave the unwarranted affirmation that Âryabhata's method was identical to that of Theon of Alexandria), and it is also clear that it is impossible to use Âryabhata's method if the numbers in question are not expressed *in writing* using distinct numerals as a base for the calculations. In other words, the operations described by Âryabhata involve placing the numbers involved in the calculation in two or three blocks of numbers, according to whether it is the square root or the cube root that is being extracted. It can be proved mathematically that these operations could not be carried out using a written numeration that was not based upon the place-value system and did not have a zero.

**STHÂNA.** Sanskrit term meaning "place". Generally used by Indian scholars to express the place-value system. See ***Sthânakramâd, Ankakramena*** and ***Ankasthâna.***

Source: **Lokavibhâga* (458 CE).

**STHÂNAKRAMÂD.** Sanskrit term which literally means: "in the order of the position". Often used by Indian scholars in ancient times (fifth – seventh century CE) to indicate that a series of numbers or numerical word-symbols were written according to the place-value system. An example of this is found in the *Jaina cosmological text, the **Lokavibhâga* ("Parts of the universe"), which is the oldest known Indian text to contain an example of the place-value system written in numerical symbols. [See Anon. (1962), chap. IV, line 56, p. 79].

**SUBANDHU.** Indian poet from the beginning of the seventh century CE. His works notably include a love story entitled *Vâsavadattâ*, where there are precise references to zero written as a dot *(*shûnya-bindu)*. See **Zero.**

**SUBSTANCE.** [S]. Value = 6. See ***Dravya*** and **Six.**

**SUBTRACTION.** [Arithmetic]. See ***Vyavakalita.***

**SUDDHA SVARA.** These are the seven notes of the *sa-grâmma (Sa, Ri, Ga, Ma, Pa, Dha, Ni)*, the first scale in Indian music. The notes are represented by short syllables, each one corresponding to the initial of the Sanskrit name of the note *(Ni = Nishâda; Ga = Gandhâra*, etc.) [see Frédéric, *Dictionnaire*, (1987)].

**SUDHÂMSHU.** [S]. Value = 1. "Moon". See ***Abja*** and **One.**

**SUDRISHTI.** "That which is seen clearly". Name given to the Pole star, "the Star which never moves". See ***Dhruva, Grahâdhâra*** and **Mount Meru.**

**SUMERU.** One of the names of *Mount Meru. See ***Adri, Dvîpa, Pûrna, Pâtâla, Sâgara, Pushkara, Pâvana*** and ***Vâyu.***

**SUN.** As a concept associated with the number thousand. See ***Sahasrakirana, Sahasrâmshu*** and **High numbers.**

**SUN.** As a mystical value equal to 7. See **Mount Meru.**

**SUN.** [S]. Value = 12. See ***Bhânu, Divâkara, Dyumani, Mârtanda, Shasrâmshu, Sûrya*** and **Twelve.**

**SUN RAYS.** [S]. Value = 12. See ***Shasrâmshu*** and **Twelve.**

**SUN-MOON (The couple).** [S]. Value = 2. See ***Ravichandra*** and **Two.**

**SUPERNATURAL POWER.** [S]. Value = 8. See ***Pañchâbhijñâ, Siddhi*** and **Eight.**

**SURA.** [S]. Value = 33. "Gods". See ***Deva*** and **Thirty-three.**

**SÛRYA.** [S]. Value = 12. Name of the Brahmanic sun god. This symbolism is explained by the "course" of the sun during the twelve months of the year. See ***Râshi*** and **Twelve.**

**SUTA.** [S]. Value = 5. "Son". See ***Putra*** and **Five.**

**SVARA.** [S]. Value = 7. "Note", "syllable". This is probably an allusion to the **suddha svara*, the seven notes of the first scale in Indian music. See **Seven.**

**SYLLABLE.** [S]. Value = 7. See ***Svara*** and **Seven.**

**SYMBOLISM OF NUMBERS.** Here is a list of associations of ideas contained in Indian numerical symbolism, given in arithmetical order (list not exhaustive):

**Number One.** Concept often directly or symbolically related to: the god *Sûrya; the god *Ganesha; a type of deep concentration (**ekâgratâ*); the sacred Syllable of the Hindus (**ekâkshara*); a certain monotheist doctrine (**ekântika*); the study of the unique reality; the contemplation of Everything (**ekatva*); the *moon; the drink of immortality (**soma*); the *earth; the *Ancestor; the *Great Ancestor; the *First Father; the *beginning; the *body; the *Self; *Ultimate reality; the superior soul; the *individual soul; the *Brahman; the "*form"; the "*drop"; the "*immense"; the "*indestructible"; the *rabbit; the "*Luminous"; the *Pole star; the "Cold rays"; the *eye of Shukra; the *terrestrial world; the "*Bearer"; the *primordial principle; the *rabbit figure; the *Cow; the sour milk, etc. See ***Eka*** and **One.**

**Number Two.** Concept often directly or symbolically related to: *duality; the idea of couple, *pair, twins or contrast; the *symmetrical organs; *wings; the *hand; the *arms; the *eyes; *vision; the *ankles; the primordial couple; the couple; *Sun-Moon; the twin gods; the conception of the world; *contemplation; revelation; the *Horsemen; the epithet "Twice born"; the third age of a **mahâyuga* (**Yuga*); etc. See ***Dva*** and **Two.**

**Number Three.** Concept often directly or symbolically related to: the *three "classes" of beings; the *Triple science; the first three **Vedas*; the *eyes; the three eyes of *Shiva; the "*three worlds"; the god Shiva; the god Vishnu; the god Krishna; the ritual dress of Buddhist monks (**trichîvara*); the "three primary forces"; the "three eras"; the "*three bodies"; the "*three forms"; the "three baskets"; the *three city-fortresses; the *three states of consciousness; the "*three jewels"; the triple town-fortress (**tripura*); the town-fortress with the triple rampart (*tripura*); a demon with three heads (**trishiras*); Shiva's Trident; the principal castes of Brahmanism; the "*three aims"; the "*three letters"; the god *Agni; "fire"; the *god of sacrificial fires; the "three rivers; the *phenomenal world; the "*aphorism"; *Feminine Energies; the "*merits"; the "*qualities"; the Spirit of the waters; the *Eye; the *points; the *times; the "*three heads"; the *three *Râma; etc. See ***Traya, Vajra*** and **Three.**

**Number Four.** Concept often directly or symbolically related to: the "*four oceans"; the "*four stages"; the "*four island-continents"; the four "*great kings"; the "four *months"; the "four *faces"; the "four aims"; the "four *ages"; the "four *ways of rebirth"; water; sea; *ocean; "*horizons"; the *cardinal points; the *arms of Vishnu; the *positions; the *vulva; the *births; the "*Fourth" (as an epithet of Brahma); the conditions of existence; the **Vedas*; the *faces of Brahma; the four ages of a **mahâyuga*; "*faces"; etc. See ***Chatur*** and **Four.**

**Number Five.** This number is considered to be sacred and magic in India and all Indianised civilisations of Southeast Asia. It is often directly or symbolically related to: the *Bow with five flowers; the "*five supernatural powers"; the *five elements of the manifestation; the "five visions of Buddha"; the "five *horizons"; the *gifts of the Cow; the "five *faculties"; the "five *impurities"; the five *heads of Rudra (= Shiva); the "five *mortal sins"; the five *orders of beings; the "five treasures" of Jaina religion (**pañchaparameshtin*"); the "sons of Pându"; the *arrows; the characteristics; *Purification; the "*Great Elements"; the *great sacrifices; the *main observances; the *fundamental principles; the *realities; the *truths; the "*true natures"; the *Jewels; the *breaths; the *senses; the *winds; the sense organs; the *faces of Rudra; etc. See ***Pañcha;*** and **Five.**

**Number Six.** Concept often directly or symbolically related to: the "six *parts; the "six *bases"; the "six categories"; the six *philosophical points of view; the six aesthetic rules (**shâdanga*); the *bodies; the *colours; the *musical modes; the weapons; the limbs; the *merits; the *qualities; the *primordial properties; the *substances; the *seasons; the **Vedânga*; the *faces of Kârttikeya (= *Kumâra); the *sensations; the flavours; etc. See ***Shad*** and **Six.**

**Number Seven.** Concept often directly or symbolically related to: the seven Buddhas (**saptabuddha*); the seven *planets; the "seven paces" (**saptapâdi*); the "seven *Jewels" (*saptagraha*); the "seven *sages"; the **Rishi*; "*Purification"; the *horses; the "seven *divine mothers"; the "seven rivers" (**saptasindhava*); the seven *horses of Sûrya; the *island-continents; the *seas; the *oceans; the "worlds"; the seven *inferior worlds; the seven *hells; the *mountain; the seven *syllables; the seven *musical notes; the last of the seven *Rishi* (**Atri*); the seven *days of the *week; "That which never moves"; *blue lotus flower; the seven *winds; etc. See ***Sapta*** and **Seven.**

**Number Eight.** Concept often directly or symbolically related to: the "eight parts (*ashtasansa*)"; the "eight *horizons"; the "eight *forms"; the "eight *limbs" of prostrating oneself (**ashtânga*); the *serpent; the *serpent of the deep; the "eight liberations" (**ashtavimoksha*); the *elephant; the eight "things which augur well"; the "eight *elephants"; the *guardians of the horizons; the *guardians of the points of the compass; the "*jewel"; the "shapes"; the eight divinities (**Vasu*); the spheres of existence; the *supernatural powers; the acts; the "body" (**tanu*); etc. See ***Ashta*, Serpent (Symbolism of the)** and **Eight.**

**Number Nine.** Concept often directly or symbolically related to: the nine planets (**navagraha*); the "nine *Jewels"; the Feast of nine days; the "nine precious stones" (**navaratna*); the *Brahman; the ninth month of the year **chaitradi*; the numeral of the place-value system (**anka*); the "Unborn"; the "*Inaccessible"; the "*Divine Mother"; the divinity *Durgâ; the *holes; the *orifices; the *radiance; the "*Cow"; etc. See ***Nava*** and **Nine.**

**Number Ten.** Concept often directly or symbolically related to: the digits; the Feast of the tenth day; the ten powers of a Buddha (**dashabala*); the *descents; the "ten *earths"; the "ten paradises" (**dashabhûmi*); the "ten stages" of the Buddha (**dashabhûmi*); the *horizons; the *heads of Râvana; the ten *major incarnations of Vishnu (**dashâvatâra*); etc. See ***Dasha*, Ten,** and **Durgâ.**

**Number Eleven.** Concept often symbolically associated with: the god *Rudra (= *Shiva), who is often referred to by one of his attributes instead of by name ("Supreme Divinity", "*Great God", "*Lord of the universe", "*Lord of tears", "Rumbling", Lord of the animals", "Violent", etc.). See ***Ekadasha*** and **Eleven.**

**Number Twelve.** Concept often symbolically associated with: the "brilliant"; the sun; the Sun-god; the "solar fire"; the *sun rays; the "*months"; the *zodiac; the *arms of Kârttikeya; the "*wheel"; the *eyes of Senânî; the sons of Âditî; etc. See ***Dvâdasha*** and **Twelve.**

**Number Thirteen.** Concept often symbolically associated with the *god of carnal love and of cosmic desire (**Kâma*) and with the *universe formed by thirteen worlds. See ***Trayodasha*** and **Thirteen.**

**Number Fourteen.** Concept often symbolically associated with: the god *Indra, who is often referred to by one of his attributes instead of by name ("*Courage", "Strength", "*Power", "*Powerful", etc.); the "*human" (in the sense of the progenitor of the human race); the worlds; the fourteen universes (**bhuvana*); the "*Jewels"; etc. See ***Chaturdasha*** and **Fourteen.**

**Number Fifteen.** Concept often symbolically associated with: "*wing"; "*day"; etc. See ***Pañchadasha*** and **Fifteen.**

**Number Sixteen.** Concept often symbolically associated with: the sixteen *kings of the legend of the **Mahâbhârata*; the "fingers of the moon" (*kalâ*). See ***Shaddasha*** and **Sixteen.**

**Number Twenty.** Concept often directly or symbolically associated with: the digits; the *nails; the *arms of Râvana; etc. See ***Vimshati*** and **Twenty.**

**Number Twenty-five.** Concept often symbolically associated with: the *fundamental principles; the "*true natures"; the *truths; the *realities; etc. See ***Pañchavimshati*** and **Twenty-five.**

**Number Twenty-seven.** Concept often directly related to: the "stars"; "*lunar mansions"; the *constellations; etc. See ***Saptavimshati*** and **Twenty-seven.**

**Number Thirty-two.** Concept often directly related to: the teeth. See ***Dvatrimshati*** and **Thirty-two.**

**Number Thirty-three.** Concept symbolically associated with: the "*gods"; the "immortals". See ***Trâyastrimsha*** and **Thirty-three.**

**Number Forty-nine.** Concept often symbolically associated with the *winds. See ***Navachatvârimshati*** and **Forty-nine.**

**Number thousand.** Concept often interpreted in the sense of the multitude or the incalculable, often associated with: the attributes of many Hindu and Brahmanic divinities (the "Thousand arms", the "*Thousand rays" or the "Thousand of the Brilliant" all denote the Sun-god *Sûrya; the "Thousand names" denotes the gods *Vishnu and *Shiva; the "Thousand eyes" refers to the gods Vishnu and Indra; etc.); or mythological figures (such as the demon Arjuna, who is referred to by the name "Thousand arms of Arjuna"). This number is also associated with: the Mouths of the Ganges (**jâhnavîvaktra*); the Arrows of Ravi (= Sûrya); *Ananta (the serpent with a thousand heads); the *lotus with a thousand petals; etc. See ***Sahasra*** and **Thousand.**

**SYMBOLISM OF NUMBERS (Concept of a large quantity).** Here is an alphabetical list of English words which have a connection with Indian high numbers, and which can be found as entries in this dictionary: *Arithmetical speculations, *Astronomical speculations, *Billion, *Blue lotus, Conch, *Cosmic cycles,

*Day of Brahma, *Diamond, *Dot, *Earth, *High numbers, *High numbers (Symbolic meaning of), *Hundred billion, *Hundred million, Hundred quadrillion, Hundred quintillion, Hundred thousand, *Hundred thousand million, *Hundred trillion, Incalculable, *Indeterminate, *Infinity, *Kalpa, *Kalpa (Arithmetical-cosmogonical speculations on), *Lotus, *Million, *Moon, *Names of numbers, *Ocean, *Pink lotus, Pink-white lotus, *Powers of ten, *Quadrillion, *Quintillion, *Serpent with a thousand heads, *Serpent of infinity and eternity, Sky, *Ten billion, *Ten million, Ten quadrillion, Ten quintillion, *Ten thousand, *Ten thousand million, *Ten trillion, *Thousand (in the sense of "multitude"), *Thousand, *Thousand million, *Trillion, Unlimited, *White lotus, *Zero. See **High numbers**, which gives a list of the principal corresponding Sanskrit words, as well as all the necessary explanations.

**SYMBOLISM OF NUMBERS (Concept of Infinity).** Here is an alphabetical list of English words which are connected to the Indian idea of infinity, and which can be found as entries in this dictionary: *Arithmetical speculations, Arithmetical-cosmogonical speculations, *Blue lotus, Conch, *Cosmic cycles, *Cosmogonic speculations, *Day of Brahma, *Diamond, *Dot, *Earth, *Eternity, *High numbers, *High numbers (Symbolic meaning of), Incalculable, Indefinite, *Infinitely big, *Infinity, *Infinity (Indian concepts of), *Kalpa, *Kalpa (Arithmetical-cosmogonical speculations on), *Lotus, *Moon, *Names of numbers, *Ocean, *Pink lotus, Pink-white lotus, *Serpent of infinity and eternity, *Serpent (Symbolism of the), *Serpent with a thousand heads, Sky, *Thousand, Unlimited, *White lotus.

**SYMBOLISM OF NUMBERS (Concept of Zero).** Here is an alphabetical list of words which are connected to Indian notions of vacuity, the void and zero, and which appear as entries in this dictionary: *Sanskrit terms:* *Abhra, *Âkâsha, *Ambara, *Ananta, *Antariksha, *Bindu, *Gagana, *Jaladharapatha, *Kha, *Khachheda, *Khahâra, *Nabha, *Nabhas, *Pûrna, *Randhra, *Shûnya, *Shûnya-bindu, *Shûnya-chakra, *Shûnya-kha, *Shûnya-samkhya, *Shûnyatâ, *Shûnyavâdin, *Vindu, *Vishnupada, *Vyant, *Vyoman. *English terms:* *Absence, *Atmosphere, *Canopy of heaven, *Dot, *Ether, *Firmament, *Hole, *Indian atomism, *Infinitely small, *Infinity, *Insignificance, *Low numbers, Negligible, *Nihilism, *Non-being, *Non-existence, *Non-present, *Non-substantiality, *Nothing, *Nothingness, *Numeral 0, Sky, Space, Uncreated, Unformed, Unproduced, *Unthought, *Vacuity, *Void, *Zero, *Zero (Graeco-Latin concepts of), *Zero (Indian concepts of), *Zero and Sanskrit poetry. See also **Durgâ**.

**SYMBOLISM OF WORDS WITH A NUMERICAL VALUE.** Here is an alphabetical list of English words which correspond to the associations of ideas contained in Sanskrit numerical symbols, which appear as entries in this dictionary (the list is not exhaustive):

*Ablaze (= 3), *Ancestor (= 1), *Ankle (= 2), *Aphorism (= 3), *Arms (= 2), *Arms of Arjuna (= 1,000), *Arms of Kârttikeya (= 12), *Arms of Râvana (= 20), *Arms of Vishnu (= 4), *Arrow (= 5), *Arrows of Ravi (= 1,000), *Atmosphere (= 0). *Bearer (= 1), *Beginning (= 1), *Birth (= 4), *Blind king (= 100), *Body (= 1), *Body (= 6), *Body (= 8), *Brahma (= 1), *Breath (= 5), *Brilliant (= 12).

*Canopy of heaven (= 0), *Cardinal point (= 4), *Characteristic (= 5), *City-fortress (= 3), *Colour (= 6), Condition of existence (= 4), *Constellation (= 27), *Contemplation (= 6), *Courage (= 14), *Cow (= 1), *Cow (= 9).

*Day (= 15), *Day of the week (= 7), *Delectation (= 6), *Demonstration (= 6), *Descent (= 10), Digit (= 10), Digit (= 20), *Divine mother (= 7), *Dot (= 0), *Drop (= 1).

*Earth (= 1), *Earth (= 9), *Element (= 5), *Elephant (= 8), Energy (= 3), *Ether (= 0), *Eye (= 2), *Eye (= 3), *Eye of Shukra (= 1), *Eyes (= 2), *Eyes of Senânî (= 12), *Eyes of Shiva (= 3), *Eyes of Indra (= 1,000).

*Face (= 4), *Faces of Brahma (= 4), *Faces of Kârttikeya (= 6), *Faces of Kumâra (= 6), *Faces of Rudra (= 5), *Faculty (= 5), *Fire (= 3), *Fire (= 12), *Firmament (= 0), *First father (= 1), *Form (= 1), *Form (= 3), *Form (= 8), *Four cardinal points (= 4), *Fourth (= 4), *Fundamental principle (= 5), *Fundamental principle (= 7), *Fundamental principle (= 25).

*Ganges (= 1,000), *Gift of the Cow (= 5), *God of carnal love (= 13), *God of cosmic desire (= 13), *God of sacrificial fires (= 3), *Gods (= 33), *Great Ancestor (= 1), *Great god (= 11), *Great element (= 5), *Great sacrifice (= 5), *Great sin (= 5), *Guardian of the horizons (= 8), *Guardian of the points of the compass (= 8).

*Hand (= 2), He who has three heads (= 3), *Heads of Râvana (= 20), *Hell (= 7), *Hole (= 0), *Horizon (= 4, *Horizon (= 8), *Horizon (= 10), *Horse (= 7), *Horsemen (= 2), *Human (= 14).

*Immense (= 1), *Inaccessible (= 9), *Incarnation (= 10), Indestructible (= 1), *Individual soul (= 1), *Indra (= 14), *Inferior world (= 7), *Infinity (= 0), *Island-continent (= 4), *Island-continent (= 7).

*Jewel (= 8), *Jewel (= 5), *Jewel (= 9), *Jewel (= 14).

*King (= 16). Limb (= 6), *Lord of the universe (= 11), *Luminous (= 1), *Lunar mansion (= 27).

*Main observance (= 5), *Merit (= 6), *Merit (= 3), *Month (= 12), *Moon (= 1), *Mountain (= 7), *Mouths of Jâhnavî (= 1,000), *Musical mode (= 6), *Musical note (= 7).

*Nail (= 20), *Numeral (= 9).

*Ocean (= 4), *Ocean (= 7), *Opinion (= 6), *Orifice (= 9).

*Pair (= 2), *Paradise (= 13), *Paradise (= 14), *Part (= 6), *Passion, *Phenomenal world (= 3), *Philosophical point of view (= 6), *Planet (= 9), *Point (= 3), *Position (= 4), *Power (= 14), *Powerful (= 14), *Precept (= 6), Primordial couple (= 2), *Primordial principle (= 1), *Primordial property (= 3), *Primordial property (= 6), *Progenitor of the human race (= 14), *Purification (= 7).

*Quality (= 3), *Quality (= 6).

*Rabbit (= 1), Rabbit figure (= 1), *Radiance (= 9), *Reality (= 5), *Reality (= 7), *Reality (= 25), *Rudra-Shiva (= 11), Rumbler (= 11). *Sage (= 7), *Season (= 6), *Self (= 1), *Sensation (= 6), *Sense (= 5), *Sense organs (= 5), *Serpent (= 8), *Serpent of the deep (= 8), *Serpent with a thousand heads (= 1,000), Sky (= 0), Son (= 5), Sons of Âditî (= 12), *Sons of Pându (= 5), Sour milk (= 1), Space (= 0), Spirit of the waters (= 3), Star (= 27), State (= 3), State of the manifestation (= 5), Strength (= 14), *Substance (= 6), *Sun (= 12), *Sun (= 1,000), *Sun-god (= 12), *Sun-Moon (= 2), *Sun rays (= 12), *Supernatural power (= 8), Supreme Divinity (= 11), Supreme soul (= 1), *Syllable (= 7), *Symmetrical organs (= 2).

*Taste (= 6), *Terrestrial world (= 1), That which augurs well (= 8), That which must be done (= 5), That which belongs to all humans (= 3), *Thousand (= 12), *Thousand rays (= 12), *Three aims (= 3), *Three bodies (= 3), *Three city-fortresses (= 3), *Three classes of beings (= 3), *Three eyes (= 3), *Three forms (= 3), *Three fundamental properties (= 3), *Three heads (= 3), *Three jewels (= 3), *Three letters (= 3), *Three sacred syllables (= 3), *Three states (= 3), *Three times (= 3), *Three universes (= 3), *Three worlds (= 3), *Time (= 3), *Tone (= 49), Tooth (= 32), *Triple science (= 3), *True nature (= 7), *True nature (= 25), *Truth (= 5), *Truth (= 7), *Truth (= 25), Twice born (= 2), Twin gods (= 2), Twins, pairs or couples (= 2).

*Ultimate reality (= 1), *Universe (= 13).

*Veda (= 3), *Veda (= 4), *Vedânga (= 6), *Violent (= 11), *Vision (= 6), *Voice (= 3), *Void (= 0), *Vulva (= 4).

Water (= 4), *Week (= 7), *Wheel (= 12), *Wind (= 5), *Wind (= 7), *Wind (= 49), *Wing (= 2), *Wing = 15), *Word (= 3), *World (= 3), *World (= 7), *World (= 14).

*Yuga (= 2), *Yuga (= 4).

*Zenith (= 0), *Zodiac (= 12).

See **Symbols, Numerical symbols, One, Two, Three, Four, Five, Six, Seven, Eight, Nine, Ten, Eleven, . . . Zero** and **Names of numbers.**

**SYMBOLISATION OF THE CONCEPT OF INFINITY.** See **Infinity, Infinity (Indian concepts of), Infinity (Mythological representation of)** and **Serpent (Symbolism of the).**

**SYMBOLISATION OF THE CONCEPT OF ZERO.** See **Zero, Dot** and **Circle.**

**SYMBOLS.** In the Brahmanic religion, and other religions of the Indian sub-continent, symbols have always been very important. They are either visible and understood by everyone and resume a number of concepts which are difficult to write down (*stûpa,* for example), or they are invisible because they have a sense which the profane cannot see (such as the *bîja,* the *yantra,* the **mudrâ,* etc.).

The symbols are represented by numerous categories of beings (such as animals), objects or even plants. As with Mahâyâna Buddhism, each divinity of Brahmanism possesses a carrier-animal which symbolises the god himself: Garuda for *Vishnu, Nandin for *Shiva, etc.: they also have a *bîja* (a letter-symbol for the corresponding sound to invoke them), **mantras* (or sacred formulas), *yantras* (geometrical diagrams with symbolic meaning) and various "signs" or distinctive marks which allow the faithful to identify the representations of the gods immediately.

The combination of signs is also symbolic, and different from a sole, isolated symbol (like **vajra* and *ganthâ*). Some symbols are raw materials like the **linga* of Shiva or the *shâlagrâma* of Vishnu; others are constructions (such as *stûpas, chaityas,* temples and various sculptures).

As for the plant kingdom, many trees (pipal, banyan, etc.) plants (*tulasî*) and seeds (*rudrâksha*) constitute symbols to Hindus, Buddhists and followers of the Jaina religion. In India, all things are potentially symbolic, not only in philosophy and religion, but also in literature, art and music. The most significant symbols are the attributes of the divinities. The Trident (**trishûlâ*) belongs to Shiva, but like the serpent (**nâga*) or the elephant, it has other meanings. See **Serpent (Symbolism of the).** The club (*danda, gadâ*) is the sign of the guardians of the gate (*dvârapâla*), but also a symbol of solar energy. The lance (*shakti*) and other weapons: dagger (*kshurikâ*), axe (*parashu*), bow and arrow (*dhanus, *bâna*), shield (*khetaka*), sword (*khadga*), etc., are used to show the power of divinities.

*Lotus flowers are most important to Buddhism, but are also highly symbolic of the pure nature of Hindu divinities.

Other very common symbolic attributes include: musical instruments (the *vînâ* of Sarasvatî, the *damarû* of Shiva-Nâtarâja); the conch (**shankha*); the bell (*ganthâ*); everyday objects (the mirror of Mâyâ, *darpana*); the cord (*pâsha*) that joins the soul to matter; the book (*pushtaka*) which represents all the **Vedas*; etc.

The sun (*chakra*) and the moon (*kulikâ*), the symbols of constellations, all have specific meanings which are either obvious or hidden (esoteric or tantric). There is a lot of symbolism connected to the human body: nudity suggests detachment from contingencies; colour of skin means anger and fury or peace and joy. Hair (in a bun) symbolises Yogin; dishevelled hair represents the mobility of Mâyâ; frizzy, untidy hair means rage.

The number of arms and legs that a divinity possesses is also highly symbolic: the more arms, the more active the god is. When a god only has two arms, this represents "angelic", peaceful qualities. If a god has no attributes whatsoever, this represents neutrality, like the *Brahman. Jewels and ornaments also have precise meanings, which can vary according to era, beliefs and philosophies. [The information in this entry is taken from Frédéric, *Dictionnaire de la civilisation indienne* (1987)].

**SYMMETRICAL ORGANS.** As symbols for the number two. See ***Bâhu, Gulpha, Nayana, Netra, Paksha*** and **Two.**

# T

**TAITTIRÎYA SAMHITÂ.** Text derived from the *Yajurveda* "black", which figures amongst the texts of Vedic literature. It is the result of oral transmission since ancient times, and did not appear in its definitive form until the beginning of the Common Era. See **Veda.** Here is a list of the principal names of numbers mentioned in the text: **Eka* (= 1), **Dasha* (= 10), **Sata* (= $10^2$), **Sahasra* (= $10^3$), **Ayuta* (= $10^4$), **Niyuta* (= $10^5$), **Prayuta* (= $10^6$), **Arbuda* (= $10^7$), **Nyarbuda* (= $10^8$), **Samudra* (= $10^9$), **Madhya* (= $10^{10}$), **Anta* (= $10^{11}$), **Parârdha* (=$10^{12}$). [See **Names of numbers** and **High numbers.** See: **Taittirîya Samhitâ,* IV, 40. 11. 4; VII, 2. 20. 1; Datta and Singh (1938), p. 9; Weber, in: JSO, XV, p. 132-40].

**TÂKÂRI NUMERALS.** Symbols derived from *Brâhmî numerals and influenced by Shunga, Shaka, Kushâna, Ândhra, Gupta and Shâradâ numerals. Currently used in Kashmir alongside the so-called "Hindi" numerals of eastern Arabs. Also called Tankrî numerals. The symbols correspond to a mathematical system that has place values and a zero (shaped like a small circle). See **Indian written numeral systems (Classification of).** See Fig. 24.13, 52 and 24.61 to 69.

**TAKSHAN.** [S]. Value = 8. "Serpent". See ***Nâga,*** **Eight** and **Serpent (Symbolism of the).**

**TAKSHASA.** Name of the king of the **nâga.* See **Serpent (Symbolism).**

**TALLAKSHANA.** Name given to the number ten to the power fifty-three. According to the legend of Buddha, this number is the highest in the first of the ten numerations of high numbers defined by the Buddha child during a contest in which he competed against the great mathematician Arjuna. *Tallakshana* contains the word *lakshana,* which literally means "character", "mark", "distinguishing feature". In Buddhism, this word often expresses the "hundred and eight distinctive signs of perfection" which distinguish a Buddha from other human beings (108 being considered a magic and sacred number which symbolises perfection). See **Names of numbers** and **High numbers (Symbolic meaning of).**

Source: **Lalitavistara Sûtra* (before 308 CE).

**TAMIL NUMERALS.** Symbols derived from *Brâhmî numerals and influenced by Shunga, Shaka, Kushâna, Ândhra, Pallava, Châlukya, Ganga, Valabhî, Bhattiprolu and Grantha numerals. Currently in use by the Dravidian population of the state of Tamil nadu (Southeast India). The symbols correspond to a mathematical system that is not based on place values and therefore does not possess a zero. For contemporary symbols, see Fig. 24.17; for historical symbols, see Fig 24, 49. See **Indian written numeral systems (Classification of).** See also Fig. 24.52 and 24.61 to 69.

**TÂNA.** [S]. Value = 49. "Tone". In Indian music, this refers to the combinations of seven octaves of seven notes.

**TANKRÎ NUMERALS. See Tâkarî Numerals.**

**TANU.** [S]. Value = 1. "Body". This symbolism comes from astrology, where "house I" is that which refers to the person, and by extension the body (*tanu*) of the person, whose horoscope is being prepared. See **One.**

**TANU.** [S]. Value = 8. "Body". This is an allusion to the **âshtânga,* the "eight limbs" of the human body that are involved in the act of prostrating oneself. See ***Ashtânga*** and **Eight.**

**TAPANA.** [S]. Value = 3. "Fire". See ***Agni,*** **Three** and **Fire.**

**TAPANA.** [S]. Value = 12. The word means "fire", but here it is taken in the sense of "solar fire" and thus of the Sun-god *Sûrya. See ***Sûrya*** and **Twelve.**

**TASTE.** [S]. Value = 6. See ***Rasa*** and **Six.**

**TATTVA.** [S]. Value = 5. "Reality, truth, true nature, fundamental principle". Allusion to the five "fundamental principles" identified by Indian philosophers and considered to be the basis for thought. See **Five.**

**TATTVA.** [S]. Value = 7. "Reality, truth, true nature, fundamental principle". Allusion to the seven "fundamental principles" identified by Jaina philosophy and considered to be the basis of the system for thought. See **Seven.** This symbol is very rarely used to represent this value, except for in the *Ganitasârasamgraha* by the Jaina mathematician *Mahâvîrâchârya [see Datta and Singh (1938), p. 56].

**TATTVA.** [S]. Value = 25. "Reality, truth, true nature, fundamental principle". Allusion to the twenty-five "fundamental principles" identified by the orthodox philosophy of **Sâmkhya: avyakta* (the "non-manifest"); *buddhi* (intelligence); *ahamkâra* (Ego, the consciousness of the Me); the *tanmâtra* (or "original substances", five subtle elements from which the basic elements are said to derive); the *mahâbhûta* (the five elements of the manifestation); the *buddhîndriya* (the five "sense organs"); the *karmendriya* (the five organs of activity, namely: the tongue, the hands, the legs, the organs of evacuation, and the reproductive organs); *manas* (the "Ability for reflection"; and *purusha* (the Self, the Absolute, pure consciousness) See **Twenty-five.**

**TELINGA NUMERALS.** See **Telugu numerals.**

**TELUGU NUMERALS.** Symbols derived from *Brâhmî numerals and influenced by Shunga, Shaka, Kushâna, Ândhra, Pallava, Châlukya, Ganga, Valabhî and Bhattiprolu numerals. Currently in use amongst the Dravidian population of Andhra Pradesh (formerly Telingana). Also called *Telinga* numerals. The symbols correspond to a mathematical system that has place values and a zero (shaped like a small circle). For contemporary symbols, see Fig. 24.20; for historical symbols, see Fig. 24, 47. See: **Indian written numeral systems (Classification of).** See Fig. 24.13, 52 and 24.61 to 69.

**TEN.** Ordinary name in Sanskrit: *•dasha.* List of corresponding numerical symbols: **Anguli, *Âsha, *Avatâra, *Dish, *Dishâ, *Kakubh, Karman, Lakâra, Pankti, *Râvanshiras.*

These terms translate or designate symbolically: 1. Descendances and incarnations (*Avatâra*); 2. Fingers (*Anguli*); 3. Horizons (*Dish, Dishâ, Âshâ, Kakubh*); 4. The heads of Râvana (*Râvanshiras*). **See Numerical symbols.**

**TEN BILLION** ( = ten to power thirteen; in US expressed as "ten trillion"). See ***Ananta, Kankara, Khamba, Makâkharva, Nikharva, Shankha, Shangku.*** See also **Names of numbers.**

**TEN MILLION** ( = ten to power seven). See ***Arbuda. Koti.*** See also **Names of numbers.**

**TEN THOUSAND** ( = ten to power four). See ***Ayuta, Dashashasra.*** See also **Names of numbers.**

**TEN THOUSAND MILLION** ( = ten to power ten; in US expressed as "ten billion"). See ***Arbuda, Kharva, Madhya, Samudra.*** See also **Names of numbers.**

**TEN TRILLION** (in British sense of ten to power nineteen; otherwise called "ten quadrillion"). See ***Attata, Mahâshankha Vivaha.*** See also **Names of numbers.**

**TERRESTRIAL WORLD.** [S]. Value = 1. See **One, Prithivî.**

**THAI (THAI-KHMER) NUMERALS.** Symbols derived from *Brâhmî numerals and influenced by Shunga, Shaka, Kushâna, Ândhra, Pallava, Châlukya, Ganga, Valabhî,

"Pâlî" and Vatteluttu numerals. Currently used in Thailand, Laos and Cambodia (Kampuchea). The symbols correspond to a mathematical system that has place values and a zero (shaped like a small circle). See **Indian written numeral systems (Classification of)**. See Fig. 24.24, 52 and 24.61 to 69.

**THÂKURÎ (Calendar).** Calendar beginning in the year 595 CE. To find the corresponding date in the Common Era, add 595 to a date expressed in the *Thâkurî* calendar. Formerly used in Nepal. See **Indian calendars**.

**THIRTEEN.** Ordinary Sanskrit name: **trayodasha*. Here is a list of the corresponding numerical symbols: *Aghosha, Atijagatî, *Kâma, Manmatha, *Vishva, *Vishvadeva.*

These words have the following translation or symbolic meaning: 1. The god of carnal love and of cosmic desire (Kâma). 2. The universe comprised of thirteen worlds (*Vishva, Vishvadeva*).

See **Numerical symbols.**

**THIRTY.** Ordinary Sanskrit name: **trimshat.*

**THIRTY-TWO.** Ordinary Sanskrit name: **dvatrimshati.* The corresponding numerical symbols are: **Danta* and **Rada.* These words both mean "teeth". See **Numerical symbols.**

**THIRTY-THREE.** Ordinary Sanskrit word: **trâyastrimsha.* The corresponding numerical symbols are: **Amara, *Deva, *Sura, Tridasha.* These words have the following meaning: 1. The "gods" (*Amara, Deva, Sura*) 2. The "immortals", in allusion to the gods (*Amara*).

See **Numerical symbols.**

**THOUSAND.** Ordinary Sanskrit name: **Sahasra*. Corresponding numerical symbols: **Arjunâkara, *Indradrishti, *Jâhnavîvaktra, *Ravibâna, *Sheshashîrsha.*

These terms name or refer to: 1. The mouth of the Ganges or Jâhnavî (*Jâhnavîvaktra*). 2. The arms of Arjuna (*Arjunâkara*). 3. The arrows of Ravi (*Ravibâna*). 4. The thousand-headed serpent (*Sheshashîrsha*). 5. The eyes of Indra (*Indradrishti*). See **Numerical Symbols.**

**THOUSAND.** In the sense of "many, a multitude of. . .". See ***Jâhnavîvakta***. See also **High Numbers (Symbolic Meaning of).**

**THOUSAND.** In the sense of infinity and eternity. See ***Sheshashîrsha***.

**THOUSAND.** [S]. Value = 12. See ***Sahasrâmshu***, **Twelve.**

**THOUSAND MILLION.** ( = ten to power nine, known in US as "one billion"). See ***Abja, Ayuta, Nahut, Nikharva, Padma, Samudra, Saroja, Shatakoti, Vâdava, Vrindâ***. See also **Names of numbers.**

**THOUSAND RAYS.** [S]. Value = 12. See ***Sahasrâmshu***. **Twelve.**

**THREE.** The ordinary Sanskrit names for this number are: **traya, *trai* and **tri.* Here is a list of corresponding word-symbols:

**Agni, *Anala, *Âptya, *Bhuvana, *Dahana, Dosha, Gangâmarga, *Guna, *Haranayana, *Haranetra, *Hotri, *Hutâshana, *Îshadrish, *Jagat, *Jvalana, *Kâla, *Krishânu, *Loka, *Mûrti, Nâdî, *Netra, *Pâvaka, *Pinâkanayana, *Purâ, *Râma, *Ratna, Sahodara, *Shakti, *Shankarâkshi, *Shikhin, *Shûla, *Tapana, *Trailokya, *Trayî, Trigata, *Triguna, *Trijagat, *Trikâla, *Trikâya, *Triloka, *Trimûrti, *Trinetra, *Tripura, *Tripurâ, *Triratna, *Trishiras, *Trivarga, *Trivarna, *Tryambaka, *Udarchis, *Vâchana, *Vahni, *Vaishvânara, *Veda, Vishtapa.*

These words have the following translation or symbolic meaning: 1. The god of fire (*Agni*). 2. "Fire", in allusion to the god of sacrificial fire (*Agni, Anala, Dahana, Hotri, Hutâshana, Jvalana, Krishânu, Pâvaka, Shikhin, Tapana, Udarchis, Vahni, Vaishvânara*). 3. "That which belongs to all humans" (*Vaishvânara*). 4. Ablaze (*Shikhin*). 5. The worlds, the universe (*Bhuvana, Loka*). 6. The three worlds (*Triloka*). 7. The phenomenal worlds (*Jagat*). 8. The three phenomenal world (*Trijagat*). 9. The "three letters", in allusion to the three sacred syllables (*Trivarna*). 10. The "aphorism" (*Vâchana*). 11. Feminine energies (*Shakti*). 12. The City-Fortresses (*Pura*). 13. The Three City, Fortresses (*Tripura*). 14. The "States", in allusion to the States of consciousness (*Purâ*). 15. The Three states of consciousness (*Tripurâ*). 16. The "forms" (*Mûrti*). 17. The three forms (*Trimûrti*). 18. The Jewels (*Ratna*). 19. The three Jewels (*Triratna*). 20. The "qualities", the "primordial properties" (*Guna*). 21. The "three primordial properties" (*Triguna*). 22. The Eye, in allusion to the "three eyes" (*Netra*). 23. The three eyes (*Trinetra, Tryambaka*). 24. The *points* (*Shûla*). 25. Time, in allusion to the "three times" (*Kâla*). 26. The three times (*Trikâla*). 27. The triple science (*Trayî*). 28. The three aims (*Trivarga*). 29. The three classes of beings (*Trailokya*). 30. The three bodies (*Trikâya*). 31. The three states (*Tripurâ*). 32. The spirit of the waters (*Âptya*). 33. The eyes of Shiva (*Haranetra*). 34. The god Shiva (*Pinâkanayana*). 35. "The one with three heads" (*Trishiras*). 36. The three Râmas (*Râma*). See **Numerical symbols.**

**THREE AIMS.** [S]. Value = 3. See ***Trivarga*** and **Three.**

**THREE BODIES.** [S]. Value = 3. See ***Trikâya*** and **Three.**

**THREE CITY-FORTRESSES.** [S]. Value = 3. See ***Tripura*** and **Three.**

**THREE CLASSES OF BEINGS.** [S]. Value = 3. See ***Trailokya*** and **Three.**

**THREE EYES.** [S]. Value = 3. See ***Tryambaka*** and **Three.**

**THREE FORMS.** [S]. Value = 3. See ***Trimûrti*** and **Three.**

**THREE HEADS.** [S]. Value = 3. See ***Trishiras*** and **Three.**

**THREE JEWELS.** [S]. Value = 3. See ***Triratna*** and **Three.**

**THREE LETTERS.** [S]. Value = 3. See ***Trivarna*** and **Three.**

**THREE PRIMORDIAL PROPERTIES.** [S]. Value = 3. See ***Triguna*** and **Three.**

**THREE SACRED SYLLABLES.** [S]. Value = 3. See ***Trivarna*** and ***AUM.***

**THREE STATES.** [S]. Value = 3. See ***Tripurâ*** and **Three.**

**THREE TIMES.** [S]. Value = 3. See ***Trikâla*** and **Three.**

**THREE UNIVERSES** [S]. Value = 3. See ***Jagat, Loka, Trijagat*** and **Three.**

**THREE WORLDS.** [S]. Value = 3. See ***Triloka*** and **Three.**

**TIBETAN NUMERALS.** Symbols derived from *Brâhmî numerals and influenced by Shunga, Shaka, Kushâna, Ândhra, Gupta, Nâgarî and Siddham numerals. Used in areas of Tibet since the eleventh century CE. The symbols correspond to a mathematical system that has place values and a zero (shaped like a small circle). However, it was not always thus: many Tibetan manuscripts show that a structure identical to the archaic Brâhmî system was used in former times. See **Indian written numeral systems (Classification of).** See Fig. 24.16, 52 and 24.61 to 69.

**TILAKA.** "Sesame". Name given to the dot that Hindus stick to their foreheads whcih represents the third eye of *Shiva, the eye of knowledge. See **Poetry, zero and positional numeration.**

**TIME.** [S]. Value = 3. See ***Kâla, Trikâla*** and **Three.**

**TITHI.** Unit of time used in Babylonian tablets which corresponds to a thirtieth of a synodic revolution of the Moon. This length of time is approximately the same as a day or *nychthemer.* See **Indian astronomy (History of).**

**TITHI.** [S]. Value = "Day". 15. Allusion to the 15 days of each **paksha* of the month. See ***Tithi*** and **Fifteen.**

This symbol is notably found in **Varâhamihîra:* PnSi, VIII, line 4; Dvivedi and Thibaut (1889); Neugebauer and Pingree (1970–71).

**TITILAMBHA.** Name given to the number ten to the power twenty-seven. See **Names of numbers** and **High numbers.** Source: *Lalitavistara Sûtra (before 308 CE).

**TONE.** [S]. Value = 49. See ***Tâna.***

**TOTAL.** [Arithmetic]. See ***Sarvadhana.***

**TRAI. (TRAYA, TRI).** Ordinary Sanskrit terms for the number three which form part of several words which are directly related to the number in question.

Examples: **Trailokya, *Trairâshika, *Trayî, *Triambaka, *Tribhuvana, *Tribhûvaneshvara, *Trichîvara, *Triguna, *Trijagat, *Trikâla, *Trikâlajñâna, *Trikândî, *Trikâya, *Triloka, *Trimûrti, *Trinetra, *Tripitaka, *Tripura, *Tripurâ, *Tripurasundarî, *Triratna, *Trishiras, *Trishûlâ, *Trivamsha, *Trivarga, *Trivarna, *Trivenî, *Trividyâ, *Tryambaka.*

For words which are symbolically associated with this number, see **Three** and **Symbolism of numbers.**

**TRAILOKYA.** [S]. Value = 3. "Three classes". This name denotes the three classes of beings envisaged by Hindu and Buddhist philosophies: the *kâmadhâtu,* beings evolving in desire; the *rûpadhâtu,* those of the world of forms; and the *arûpadhâtu,* those of the world of the formless. See ***Trai*** and **Three.**

**TRAIRÂSHIKA.** [Arithmetic]. Sanskrit name for the Rule of Three. See ***Trai.***

**TRAYA.** Ordinary Sanskrit name for the number three. See ***Trai.***

**TRÂYASTRIMSHA.** Ordinary Sanskrit name for the number thirty-three. For words which are symbolically associated with this number, see **Thirty-three,** ***Deva*** and **Symbolism of numbers.**

**TRAYÎ.** [S]. Value = 3. "Triple science". Allusion to the *Samhitâ* (*Rigveda, Yajurveda, Sâmaveda*), who are the three first **Vedas.* See ***Trai, Veda*** and **Three.**

**TRAYODASHA.** Ordinary Sanskrit name for the number thirteen. For words which are symbolically associated with this number, see **Thirteen** and **Symbolism of numbers.**

**TRETÂYUGA.** Name of the second of the four cosmic eras which make up a **mahâyuga.* This cycle, which is said to be worth 1,296,000 human years, is regarded as the age during which human beings would live no more than three quarters of their life. See ***Mahâyuga*** and ***Yuga.***

**TRI.** Ordinary Sanskrit word for the number three. See Trai.

**TRIAMBAKA.** "With three eyes". See ***Tryambaka.***

**TRIBHUVANA.** Name of the "three worlds" of Hindu cosmogony: the skies (*svarga*), the earth (**bhûmi*) and the hells (**pâtâla*). See ***Trai.***

**TRIBHÛVANESHVARA.** "Lord of the three worlds". One of the titles attributed to *Shiva, *Vishnu and *Krishna. See ***Trai.***

**TRICHÎVARA.** "Three garments". Term denoting the ritual costume comprising the loincloth, sash and robe worn by Buddhist monks of the schools of the South (Hînayâna, Theravâda). See ***Trai.***

**TRIGUNA.** [S]. Value = 3. "Three primordial properties", "three primary forces". Symbolism which corresponds to the representation of the group Vishnu-Sattva, Brahma-Rajas and Rudra-Tamas, this group being thus composed of the energies which personify the three main divinities of the Brahmanic pantheon. See ***Guna,*** **Brahma, Vishnu, Shiva** and **Three.**

**TRIJAGAT.** [S]. Value = 3. "Three universes". See ***Jagat, Triloka*** and **Three.**

**TRIKÂLA.** [S]. Value = 3. "Three times". Allusion to the three divisions of time: the past, the present and the future. See ***Kâla*** and **Three.**

**TRIKÂLAJÑÂNA.** From **trikâla,* "three times", "three eras", and from *jñâna,* "knowledge". Name denoting the magic and occult power which is given to the **Siddhi* to enable them to know the past, the present and the future all at once. See ***Kâla, Trikâla*** and ***Trai.***

**TRIKÂNDÎ.** "Three chapters". This name is sometimes given to the *Vâkyapadîya* of Bhartrihari, famous text of "grammatical philosophy" divided into three *kânda* or "chapters". See ***Trai.***

**TRIKÂYA.** [S]. Value = 3. "Three bodies". Allusion to the three bodies that a Buddha may assume simultaneously: the "body of the Law" (*dharmakâya*), the "body of enjoyment" (*sambhogakâya*) and the "body of magical creation or transformation" (*nirmânakâya*). See **Three.**

**TRILLION.** See ***Akshiti, Antya, Madhya, Mahâpadma, Viskhamba, Vivara*** and **Names of numbers.**

**TRILOKA.** [S]. Value = 3. "Three worlds". In allusion to the worlds of Hindu cosmogony: the Skies (*svarga*), the earth (**bhûmi*) and the hells (**pâtâla*). See **Three.**

**TRIMSHAT.** Ordinary Sanskrit name for the number thirty.

**TRIMÛRTI.** [S]. Value = 3. "Three forms". See ***Mûrti*** and **Three.**

**TRINETRA.** [S]. Value = 3. "Three eyes". See ***Haranetra*** and **Three.**

**TRIPITAKA.** "Three baskets". Term denoting the Buddhist Law written in Sanskrit which constitutes the sacred Scriptures of this religion. The allusion is to the three different baskets into which the three principal compilers placed the three fundamental Buddhist texts: the *vinâyapitaka,* which deals with monastic discipline; the *sûtrapitaka* and the *abhidharmapitaka* which deals with Buddha's teaching [see K. Friedrichs, etc, (1989)]. See ***Trai.***

**TRIPLE SCIENCE.** [S]. Value = 3. See ***Trayî*** and **Three.**

**TRIPURA.** [S]. Value = 3. Literally: "Three City-fortress". Name of a triple fortress-town (or triple rampart) which was built by the **Asura* and destroyed by Shiva. See ***Pura*** and **Three.**

**TRIPURÂ.** [S]. Value = 3. Literally: "three states". Name which collectively denotes the three states of consciousness of Hinduism. See ***Purâ*** and **Three.**

**TRIPURASUNDARÎ.** "Beauty of the three cities". One of the names given to *Pârvatî, the "mountain dweller", daughter of Himâlaya, sister of *Vishnu and wife of *Shiva. See ***Trai.***

**TRIRATNA.** [S]. Value = 3. "Three jewels". See ***Ratna*** and **Three.**

**TRISHATIKÂ.** See ***Shrîdharâchârya.***

**TRISHIRAS.** [S]. Value = 3. "He who has three heads". This is the name of the demon with three heads, younger brother of *Râvana, who, according to the legend of **Râmâyana,* was killed by *Râma. See **Râvana** and **Three.**

**TRISHÛLA.** "Three points". Name of *Shiva's Trident. See ***Shûla*** and ***Trai.***

**TRIVAMSHA.** Name which collectively denotes the three principal castes of Brahmanism (namely: the Brahmans, the *kshatriya* and the *vaishya*). See ***Trai.***

**TRIVARGA.** [S]. Value = 3. "Three aims". This is an allusion to the three objectives of human existence according to Hindu philosophy, namely: material wealth (*artha*), love with desire (**kâma*) and duty (**dharma*). See **Three.**

**TRIVARNA.** [S]. Value = 3. "Three letters". This refers to the letters A, U and M of the Indian alphabet, which spell AUM, the sacred Syllable of the Hindus, which means something approximating "I bow". This represents all of the following at once: the divine Word in an audible form; the fullblown *Brahman; the Fire of the Sun; the Unity; the Cosmos; the Immensity of the Universe; the past; the present; the future; as well as all Knowledge. According to Hindu religion, AUM contains the very essence of all the sounds that have been, that are, and that will be made, and within it is reunited the three great powers of the three great divinities of the Brahmanic pantheon [see Frédéric (1987)]. See ***AUM, Akshara,*** **Mysticism of letters,** ***Trai*** and **Three.**

**TRIVENÎ.** "Three rivers". Name sometimes given to the town of Prayâga (now Allâhâbâd) where the following three rivers are said to meet: the Ganges, the Yamunâ and the mythical Sarasvatî. See ***Trai.***

**TRIVIDYÂ.** Name given to the "three axioms" of Buddhist philosophy: *anitya,* the impermanence of all things; *dukha,* pain, suffering; and *anâtmâ,* the non-reality of existence. See ***Trai.***

**TRIVIMSHATI.** Ordinary Sanskrit name for the number twenty-three. For words which are symbolically connected with this number, see **Twenty-three** and **Symbolism of numbers.**

**TRUE NATURE.** [S]. Value = 5. See ***Tattva*** and **Five.**

**TRUE NATURE.** [S]. Value = 7. See ***Tattva*** and **Seven.**

**TRUE NATURE.** [S]. Value = 25. See ***Tattva*** and **Twenty-five.**

**TRUTH.** [S]. Value = 5. See ***Tattva*** and **Five.**

**TRUTH.** [S]. Value = 7. See ***Tattva*** and **Seven.**

**TRUTH.** [S]. Value = 25. See ***Tattva*** and **Twenty-five.**

**TRYAKSHAMUKHA.** [S]. Value = 5. Synonymous with **Rudrâsya,* "faces of *Rudra". See **Five.**

**TRYAMBAKA.** [S]. Value = 3. "With three eyes", "with three sisters". Epithet given to many Hindu divinities, especially Shiva . See ***Haranetra, Traya*** and **Three.**

**TURAGA.** [S]. Value = 7. "Horse". See ***Ashva*** and **Seven.**

**TURANGAMA.** [S]. Value = 7. "Horse". See ***Ashva*** and **Seven.**

**TURÎYA.** [S]. Value = 4. "Fourth". Epithet occasionally given to the Brahman who transcends the three states of consciousness. See ***Tripurâ*** and **Four.**

**TWELVE.** Ordinary Sanskrit name: **dvâdasha.*

**TWENTY.** Ordinary Sanskrit name: **vimshati.* Here is a list of corresponding numerical symbols: **Anguli, *Kriti, *Nakha, *Râvanabhuja.* These words express: 1. The arms of Râvana (*Râvanabhuja*). 2. The fingers (*Anguli*). 3. The nails (*Nakha*). 4. An element of Indian metrication (*Kriti*). See **Numerical symbols.**

**TWENTY-ONE.** Ordinary Sanskrit name: **ekavimshati.* Corresponding numerical symbols: **Prakriti, Svaga* ("heaven"), *Utkriti.*

**TWENTY-TWO.** Ordinary Sanskrit name: **dva vimshati.* Corresponding numerical symbols: **Akriti, Jâti* ("Caste"), *Kritin.*

**TWENTY-THREE.** Ordinary Sanskrit name: **trayavimshati* (or *trivimshati*). Corresponding numerical symbol: **Vikriti.*

**TWENTY-FOUR.** Ordinary Sanskrit name: **chaturvimshati.* Corresponding numerical symbols: *Arhat, *Gâyatrî, Jina, Siddha.*

**TWENTY-FIVE.** Ordinary Sanskrit name: **pañchavimshati.* Corresponding numerical symbol: ***Tattva.*** This word expresses: 1. The fundamental principles. 2. The "true natures". 3. The realities. 4. The truths.

**TWENTY-SIX.** Ordinary Sanskrit name: **shadvimshati.* Corresponding numerical symbol: **Utkriti.*

**TWENTY-SEVEN.** Ordinary Sanskrit name: **saptavimshati.* Corresponding numerical symbols: **Bha, *Uda, *Nakshatra.* These words express or symbolise: 1. The "stars" (*Bha, Uda*). 2. The "lunar mansions" (*Nakshatra*). 3. The constellations (*Nakshatra*).

**TWO.** Ordinary Sanskrit names: **dva, dve, dvi.* Corresponding numerical symbols: *Akshi, Ambaka, *Ashvin, *Ashvina, *Ashivinau, Ayana, *Bâhu, *Chakshus, *Dasra, *Drishti, *Dvandva, *Dvaya, *Dvija, Grahana, *Gulpha, Îshana, Janghâ, Jânu, *Kara, Karna, Kucha, Kutumba, *Lochana, Nadîkulâ, *Nâsatya, Naya, *Nayana, *Netra, Oththa, *Paksha, Râmananddana, Ravichandra, Vishuvat, *Yama, *Yamala, *Yamau, *Yuga, *Yugala, *Yugma.*

These terms symbolically refer to or designate: 1. Twins. pairs or couples (*Ashvin, Ashvina, Ashvinau, Dasra, Dvandva, Dvaya, Dvija, Nâsatya, Ravichandra, Yama, Yamala, Yugala, Yugma*). 2. Symmetrical organs (*Bâhu,*

*Gulpha, Kara, Nayana, Netra, Paksha*). 3. Wings (*Paksha*). 4. Arms (*Bâhu*). 5. The Horsemen (*Ashvin, Ashvina, Ashvinau*). 6. Ankles (*Gulpha*). 7. The conception of the world, contemplation, revelation, theory (*Drishti*). 8. The primordial couple (*Yama*). 9. The epithet "twice born" (*Dvija*). 10. The twin gods (*Ashvin, Dasra, Nâsatya*). 11. The hand (*Kara*). 12. The pair (*Dvaya*). 13. The Sun-Moon couple (*Ravichandra*). 14. The eye (*Netra, Chakshus*). 15. Eyes (*Lochana*). 16. Vision (*Drishti*). 17. The third age of a *mahâyuga* (*Yuga*). **See Numerical symbols.**

# U

**UCHCHAISHRAVAS.** [S]. Value = 1. This is the name of a wonderful white horse which, according to Brahmani and Hindu mythology, came from the "churning of the sea of milk" and which Indra appropriated. He is considered to be the ancestor of all horses, thus the symbolism in question. See **One.**

**UDA.** [S]. Value = 27. "Star". This is an allusion to the twenty-seven **nakshatra*. See ***Nakshatra*** and **Twenty-seven.**

**UDADHI.** [S]. Value = 4. "Ocean". See ***Sâgara*, Four** and **Ocean.**

**UDARCHIS.** [S]. Value = 3. "Fire". See ***Agni*** and **Three.**

**UJJAYINÎ.** Town situated in the extreme west of what is now the state of Madhya Pradesh. It defines the first meridian of Indian astronomy. See **Indian astronomy (History of)** and **Yuga (Astronomical speculation on).**

**ULTIMATE REALITY.** [S]. Value = 1. See ***Âtman*** and **One.**

**UNIQUE REALITY.** [S]. Value = 1. See ***Âtman*** and **One.**

**UNIVERSE.** [S]. Value = 13. See ***Vishva, Vishvada*** and **Thirteen.**

**UPPALA.** Pali word which literally means: "(blue) lotus flower (half open)". Name given to the number ten to the power ninety-eight. See **Names of numbers.** For an explanation of this symbolism, see **Lotus** and **High numbers (Symbolic meaning of).**

Source: **Vyâkarana* (Pali grammar) by Kâchchâyana (eleventh century CE).

**URVARÂ.** [S]. Value = 1. "Earth". See ***Prithivî.***

**UTKRITI.** [S]. Value = 26. In Sanskrit poetry, this is a metre of four lines of twenty-six syllables per stanza. See **Indian metric.**

**UTPALA.** Literally: "(blue) lotus flower (half open)". In Hindu and Buddhist philosophies, this lotus (which is never represented in full bloom) notably symbolises the victory of the mind over the body. Name given to the number ten to the power twenty-five. See **Names of numbers.** For an explanation of this symbolism, see **Lotus** and **High numbers (Symbolic meaning of).**

Source: **Lalitavistara Sûtra* (before 308 CE).

**UTSANGA.** Name given to the number ten to the power twenty-one (= quintillion). See **Names of numbers** and **High numbers.**

Source: **Lalitavistara Sûtra* (before 308 CE).

# V

**VÂCHANA.** [S]. Value = 3. "Aphorism". From *vâch*, "voice", "speech", "spoken word", and form *annâ*, "nourishment". This is an allusion to the creative and evocative power of sound and acoustic resonance (especially through speech) and to its "indestructible and imperishable" nature, which correspond to the revelation of the *Brahman, which is said to be resumed in the three letters of the sacred Syllable *AUM. See ***Akshara, Trivarna*** and **Three.**

**VACUITY.** See ***Shûnya, Shûnyatâ*, Zero, Zero (Graeco-Latin concepts of), Zero (Indian concepts of)** and **Zero and Sanskrit poetry.**

**VÂDAVA.** Name given to the number ten to the power nine. See **Names of numbers** and **High numbers.**

Source: **Kâthaka Samhitâ* (from the start of the Common Era).

**VÂDAVA.** Name given to the number ten to the power fourteen. See **Names of numbers** and **High numbers.**

Source: **Pañchavimsha Brâhmana* (date uncertain).

**VAHNI.** [S]. Value = 3. "Fire". See ***Agni*** and **Three.**

**VAIKUNTHA.** Celestial home of *Vishnu and *Krishna. See ***Bhuvana.***

**VAISHESHIKA.** See ***Darshana.***

**VAISHVÂNARA.** [S]. Value = 3. "that which belongs to all humans". This is one of the Vedic names for *Agni (= 3), the god of sacrificial fire, who is said to possess the powers of fire, lightning and light. See ***Agni*** and **Three.**

**VÂJASANEYÎ SAMHITÂ.** This is a text which forms part of the *Yajurveda* "white", which is one of the oldest Vedic texts. Passed down through oral transmission since ancient times, it only found its definitive form at the beginning of Christianity. See **Veda.** Here is a list of the main names of numbers mentioned in this text:

**Eka* (= 1), **Dasha* (= 10), **Sata* (= $10^2$), **Sahasra* (= $10^3$), **Ayuta* (= $10^4$), **Niyuta* (= $10^5$), **Prayuta* (= $10^6$), **Arbuda* (= $10^7$), **Nyarbuda* (= $10^8$), **Samudra* (= $10^9$), **Madhya* (= $10^{10}$), **Anta* (= $10^{11}$), **Parârdha* (=$10^{12}$). See **Names of numbers** and **High numbers.**

[See: *Vâjasaneyî Samhitâ*, XVII, 2; Datta and Singh (1938), p. 9; Weber, in: JSO, XV, pp. 132–40; Woepcke (1863).]

**VÂJIN.** [S]. Value = 7. "Horse". See ***Ashva*** and **Seven.**

**VAJRA.** In Hindu and Buddhist philosophies, the *vajra* is the "diamond" that symbolises all that remains when appearances have disappeared. Thus it is the vacuity (**shûnyatâ*) that is as indestructible as a diamond. It is also the missile "with a thousand points", which is said to never miss its target, and made out of bronze by Tvashtri for *Indra. This weapon is a symbol of **linga* and divine power. It also indicates a strong, stable and indestructible mind. As a word-symbol, *vajra* has several meanings: the weapon is usually a short bronze baton, which has three, five, seven or nine points at each end. With three points, for example, *vajra* symbolises: the **triratna* (or "three jewels" of Buddhism); time in its three tenses (**trikâla*); the three aspects of the world (**tribhuvana*); etc. [see Frédéric, *Dictionnaire* (1987)]. See ***Shûnyatâ*** and **Symbols.**

**VALABHI NUMERALS.** Symbols derived from *Brâhmî numerals and influenced by Shunga, Shaka, Kushâna, Ândhra, Pallava, Châlukya, and Ganga numerals. The system arose at the time of the inscriptions of Valabhi, the capital city of a Hindu-Buddhist kingdom that ruled over present-day Gujurat and Maharastra. The symbols correspond to a mathematical system that is not based on place values and therefore does not possess a zero. See **Indian written numeral systems (Classification of).** See Fig. 24.44, 52 and 24.61 to 69.

**VÂRA.** [S]. Value = 7. "Day of the week". This is because of the seven days: *ravivâra* or *âdivâra* (Sunday), *induvâra* or *somavâra* (Monday), *mangalavâra* (Tuesday), *budhavâra* (Wednesday), *brihaspativâra* (Thursday), *shukravâra* (Friday), and *shanivâra* (Saturday). See **Seven.**

**VARÂHAMIHÎRA.** Indian astronomer and astrologer c. 575 CE. His works notably include *Pañchasiddhântikâ* (the "Five *Siddhântas*"), where there are many examples of the place-value system [see Neugebauer and Pingree (1970–71)]. See **Indian astrology, Indian astronomy (History of)** and **Indian mathematics (History of).**

**VARGA.** Word used in arithmetic to denote the squaring operation. Synonym: *kriti.* In algebra, the same word is used for the square, in allusion to cubic equations. See ***Ghana, Varga-Varga*** and ***Yâvattâvat.***

**VARGAMÛLA.** Word used in arithmetic to describe the extraction of the square root. See ***Pâtîganita,* Indian methods of calculation** and **Square roots (How Âryabhata calculated his).**

**VARGA-VARGA.** Algebraic word for quadratic equations.

**VÂRIDHI.** [S]. Value = 4. "Sea". See ***Sâgara,* Four** and **Ocean.**

**VÂRINIDHI.** [S]. Value = 4. "Sea". See ***Sâgara,* Four** and **Ocean.**

**VARNA.** Literally "letter", in mathematics "symbol". See ***AUM, Bîja*** and ***Bîjaganita.***

**VARNASAMJÑA.** "Syllable system". Name that Haridatta gave to the **katapaya* system.

**VARNASANKHYÂ.** Literally: "letter-numbers". This word denotes any system of representing numbers which uses the vocalised consonants of the Indian alphabet, each one being assigned a numerical value. See **Numeral alphabet.**

**VARUNA.** Vedic and Hindu divinity of the water, the sea and the oceans. See **High numbers (Symbolic meaning of).**

**VASU.** [S]. Value = 8. Name in the **Mahâbhârâta* which is given to a group of eight divinities, who are meant to correspond, philosophically speaking, to the eight "spheres of existence" of the *Adibhautika*, which in turn represent the visible forms of the laws of the universe. See **Eight.**

**VASUDHÂ.** [S]. Value = 1. "Earth". See ***Prithivî*** and **One.**

**VÂSUKI.** In Brahmanic mythology, this is the name given to the king of the **naga.* He is said to have been used by the **deva* (the gods) and the **asura* (the anti-gods) as a "rope" with which to spin *Mount Meru on its axis in order to churn the sea of milk and thus extract the "nectar of immortality" (**amrita*). See **Serpent (Symbolism of the).**

**VASUNDHARÂ.** [S]. Value = 1. "Earth". See ***Prithivî*** and **One**.

**VATTELUTTU NUMERALS.** Symbols derived from *Brâhmî numerals and influenced by Shunga, Shaka, Kushâna, Ândhra, Pallava, Châlukya, and Ganga, Valabhî, Bhattiprolu and Grantha numerals as well as by Ancient Tamil. Used from the eighth to the sixteenth centuries CE in the Dravidian areas of South India, particularly the Malabar coast. The symbols correspond to a mathematical system that is not based on place values and therefore does not possess a zero. See: **Indian written numeral systems (Classification of)**. See Fig. 24.52 and 24.61 to 69.

**VÂYU.** "Wind". This is a name for the god of the wind. Other names include: *Marut* ("Immortal"), *Anila* ("Breath of life"), *Vâta* ("Wandering", "He who is in perpetual movement") or **Pâvana* ("Purifier"). According to Brahmanic and Hindu cosmogonies, he is one of the eight **Dikpâla* (divinities who guard the horizons and points of the compass), whose task is to guard the northwest "horizon".

**VÂYU.** [S]. Value = 49. "Wind". This symbolism can be explained by reference to tales of Brahman mythology. One day Vâyu revolted against the **deva*, the "gods" who live on the peaks of *Mount Meru. He decided to destroy the mountain, and unleashed a powerful hurricane. However, the mountain was protected by the wings of Garuda, the carrier-bird of *Vishnu, which rendered all the assaults of the wind ineffectual. One day, in Garuda's absence, Vâyu chopped off the peak of *Mount Meru, and threw it into the ocean. That is how the island of Sri Lanka was created. Mount Meru was meant to be the place where the gods lived and met. It was said to be situated at the centre of the universe, above the seven **pâtâla* (or "inferior worlds"); it has seven faces, each one facing one of the seven **dvîpa* (or "island-continents) and the seven **sâgara* ("*oceans"). When Vâyu attacked the mountain, he created seven strong winds, one for each face. Once the summit of the sacred mountain had been rased, the seven winds, thus placed at the centre of the universe and no longer encountering any barrier, each went to one of the seven continents and the seven oceans. Thus: *Vâyu* = 7 × 7 = 49. See other entry entitled ***Vâyu***.

**VEDA.** Name of the oldest sacred texts of India, they are made up of four principal books (namely: the *Rigveda*, the *Yajurveda*, the *Sâmaveda*, and the *Atharvaveda*). These texts and those of derived literature probably date back to ancient times in the history of India. But it is impossible to date them exactly, because they were primarily transmitted orally before being transcribed at a later date. In fact, it is only possible to give them a chronological position in relation to each other. The three *Samhitâ* (the texts of the *Rigveda*, the *Yajurveda* and the *Sâmaveda*) seem to have been composed first. As for the fourth *Veda*, (the *Atharvaveda*), it was followed by the *Brâhmana*, the *Kalpasûtra*, and lastly by the *Âranyaka* and the *Upanishad* [see Frédéric, *Dictionnaire* (1987)].

**VEDA.** [S]. Value = 3. (Very rarely used as a numerical symbol). The allusion here is probably to the three *Samhitâ* (the *Rigveda*, the *Yajurveda* and the *Sâmaveda*), which constitute the first three texts of the *Veda*. See ***Trayî*** and **Three**.

**VEDA.** [S]. Value = 4. (The most frequent value of this word as a numerical symbol.) Here the allusion is to the four principal books of which the *Veda* is composed (the *Rigveda*, the *Yajurveda*, the *Sâmaveda*, and the *Atharvaveda*). See **Four**.

**VEDÂNGA.** [S]. Value = 6. "Members of the **Veda*". Group of six Vedic and Sanskrit texts dealing principally with the Vedic ritual, its conservation and its transmission. See ***Darshana***.

**VEDIC RELIGION.** See **Indian religions and philosophies.**

**VIBHUTANGAMÂ.** Name given to the number ten to the power fifty-one. See **Names of numbers** and **High numbers**.

Source: **Lalitavistara Sûtra* (before 308 CE).

**VIDHU.** [S]. Value = 1. "Moon". See ***Abja*** and **One**.

**VIKALPA.** Word used in mathematics since the eighth century to designate "permutations" and "combinations".

**VIKRAMA. (Calendar).** Formerly used in the centre, west and northwest of India. Also called *vikramâdityakâla, vikramasamvat*, or quite simply *samvat*. It began in the year 57 BCE. To find an approximate corresponding date in the Common Era, subtract 57 from a date in the *Vikrama* calendar.

**VIKRAMÂDITYAKÂLA (Calendar).** See ***Vikrama***.

**VIKRAMASAMVAT (Calendar).** See ***Vikrama***.

**VIKRITI.** [S]. Value = 23. In Sanskrit poetry, this is the metre of four times twenty-three syllables per stanza. See **Indian metric**.

**VILÂYATÎ (Calendar).** Solar calendar commencing in the year 592 CE. Used in Bengal and Orissa. To find a date in the Common Era, add 592 to a date expressed in the *Vilâyatî* calendar. See **Indian calendars**.

**VIMSHATI.** Ordinary Sankrit name for the number twenty. For words which have a symbolic relationship with this number, see **Twenty** and **Symbolism of numbers**.

**VINDU.** [S]. Value = 0. **Prâkrit* word which has the literal meaning and symbolism of **bindu*. See **Zero**.

**VIOLENT.** [S]. Value = 11. See **Rudra-Shiva** and **Eleven**.

**VÎRASAMVAT (Calendar).** Commencing in the year 527 BCE, it is only used in *Jaina texts. To find a corresponding date in the Common Era, subtract 527 from a date expressed in this calendar. See **Indian calendars**.

**VISAMJÑAGATI.** Name given to the number ten to the power forty-seven. See **Names of numbers** and **High numbers**.

Source: **Lalitavistara Sûtra* (before 308 CE).

**VISHAYA.** [S]. Value = 5. "Sense, sense organ". See ***Shara*** and **Five**.

**VISHIKHA.** [S]. Value = 5. "Arrow". See ***Shara*** and **Five**.

**VISHNU.** Name of one of the three major divinities of the Brahmanic and Hindu pantheon (*Brahma, *Vishnu, *Shiva). See ***Vishnupada***, ***Pûrna*** and **High numbers (Symbolic meaning of)**.

**VISHNUPADA.** [S]. Value = 0. Literally: "foot of Vishnu", and by extension (and characteristically of Indian thought): "zenith", "sky". This is an allusion to the "Supreme step of Vishnu", the zenith, which denotes *Mount Meru, home of the blessed. The symbolism in question also refers to the "Three Steps of Vishnu", symbols of the rising, apogee and setting of the sun, which allowed him to measure the universe. It is also from the "feet of Vishnu" that, according to Hindu mythology, the sacred *Gangâ* (the Ganges) springs and, before it divides into terrestrial rivers, has its source at the summit of Mount Meru (which is situated at the centre of the universe and over which are the heavens or "worlds of Vishnu"). Vishnu rests upon *Ananta, the serpent with a thousand heads who floats on the primordial waters and the "ocean of unconsciousness", during the time that separates two creations of the universe. Thus this symbolism corresponds to the connection in Indian philosophy between infinity and zero, because Ananta is the serpent of infinity, eternity and of the immensity of space. Space also means sky, which is considered to be the "void" which has no contact with material things. Thus Vishnu is identified with ether (**âkâsha*), an immobile, eternal and indescribable space. In other words, Vishnu is synonymous with vacuity (**shûnyatâ*). See ***Abhra***, ***Âkâsha***, ***Kha***, ***Ananta***, **Zero** and **Zero (Indian concepts of)**.

**VISHVA.** [S]. Value = 13. Contraction of the word *Vishvadeva* and a symbol for the number 13. See ***Vishvadeva*** and **Thirteen**.

**VISHVADEVA.** [S]. Value = 13. This is an allusion to the universe formed by thirteen paradises or chosen lands (**bhuvana*), and does not include the **vaikuntha*. See ***Bhuvana***, ***Vaikuntha*** and **Thirteen**.

**VISION.** [S]. Value = 2. See ***Drishti*** and **Two**.

**VISION.** [S]. Value = 6. See ***Darshana*** and **Six**.

**VISKHAMBA.** Name given to the number ten to the power fifteen. See **Names of numbers** and **High numbers**.

Source: **Lalitavistara Sûtra* (before 308 CE).

**VIVAHA.** Name given to the number ten to the power nineteen. See **Names of numbers** and **High numbers.**

Source: **Lalitavistara Sûtra* (before 308 CE).

**VIVARA.** Name given to the number ten to the power fifteen. See **Names of numbers** and **High numbers.**

Source: **Lalitavistara Sûtra* (before 308 CE).

**VOICE.** [S]. Value = 3. See ***Vâchana*** and **Three**.

**VOID.** [S]. Value = 0. See ***Shûnya***, ***Shûnyatâ*** and **Zero**.

**VRINDÂ.** Name given to a plant which is similar to basil, the leaves of which are said to have the power to purify the body and mind. It is believed to be an incarnation of Vishnu: according to the legend, Vrindâ was the wife of a Titan then was seduced by Vishnu. She cursed her husband and transformed him into a *shâlagrâma* stone before killing herself by throwing herself onto a fire of live coals; the plant (still called *tulasî*) was born out of the ashes. See ***Ananta***, ***Vishnupada***, ***Samudra***, **Names of numbers** and **High numbers**.

**VRINDÂ.** Name given to the number ten to the power nine. See **Names of numbers** and **High numbers**. For an explanation of this symbolism, see ***Vrindâ*** (first entry) and **High**

**numbers (Symbolic meaning of).**
Source: *Âryabhatîya* (510 CE).

**VRINDÂ.** Name given to the number ten to the power seventeen. See **Names of numbers** and **High numbers.** For an explanation of this symbolism, see ***Vrindâ*** (first entry) and **High numbers (symbolic meaning of).**
Source: **Kâmâyana* by Vâlmikî (early centuries CE).

**VULVA.** [S]. Value = 4. See ***Yoni*** and **Four.**

**VYÂKARANA.** See ***Kâchchâyana.***

**VYAKTAGANITA.** Name for arithmetic (literally: "science of calculating the known"), as opposed to algebra, which is called **Avyaktaganita.*

**VYANT.** [S]. Value = 0. "Sky". The symbolism can be explained by the fact that the sky (or heaven) is the "void" in Indian beliefs. See ***Shûnya*** and **Canopy of heaven.**

**VYARBUDA.** Name given to the number ten to the power eight (= hundred million). See **Names of numbers** and **High numbers.**
Source: *Kitab fi tahqiq i ma li'l hind* by al-Biruni (c. 1030 CE).

**VYASTATRAIRÂSHIKA.** [Arithmetic]. Name of the inverse of the Rule of Three. See ***Trairâshika.***

**VYAVAHÂRA.** Literally: "procedure". Term used in algebra (since the seventh century CE) to denote the solving of equations.

**VYAVAKALITA.** [Arithmetic]. Sanskrit term for subtraction. Literally: "taken away".

**VYAVASTHÂNAPRAJÑAPATI.** Name given to the number ten to the power twenty-nine. See **Names of numbers** and **High numbers.**
Source: **Lalitavistara Sûtra* (before 308 CE).

**VYOMAN.** [S]. Value = 0. Word meaning "sky", "space". See **Zero** and ***Shûnya.***

**VYUTTKALITA.** [Arithmetic]. Sanskrit term for subtraction. See ***Vyavakalita.***

## W

**WAYS OF REBIRTH (The four).** See ***Chaturyoni*** and ***Yoni.***

**WEEK.** [S]. Value = 7. See ***Vâra*** and **Seven.**

**WHEEL.** [S]. Value = 12. See ***Chakra, Râshi*** and **Twelve.**

**WHITE LOTUS.** As a representation of the numbers ten to power twenty-seven and ten to power 112. See **Pundarika, High Numbers (Symbolic meaning of).**

**WIND.** [S]. Value = 5. See ***Pâvana.***

**WIND.** [S]. Value = 7. See ***Pâvana.***

**WIND.** [S]. Value = 49. See ***Vâyu.***

**WING** [S]. Value = 15. See ***Paksha.*** **Fifteen.**

**WING.** [S]. Value = 2. See ***Paksha.*** **Two.**

**WORLD.** [S]. Value = 3. See **Bhuvana.**

**WORLD.** [S]. Value = 7. See **Loka.**

**WORLD.** [S]. Value = 14. See **Bhuvana.**

## Y

**YAMA.** [S]. Value = 2. "Primordial couple". Allusion to the couple in Hindu mythology, formed by *Yama* (the first mortal who became god of death) and *Yamî*, his twin sister, wife and his feminine energy (**shakti*). See **Two.**

**YAMALA.** [S]. Value = 2. Synonym of **Yama.* See **Two.**

**YAMAU.** [S]. Value = 2. Synonym of **Yama.* See **Two.**

**YÂVATTÂVAT.** Literally : "as many as". Term used in algebra to denote the "equation" in general.

**YONI.** [S]. Value = 4. "Vulva". Allusion to the four lips that form the entrance of the vulva. By extension, the word also means "birth". Here, the reference is to the **Chaturyoni* which, according to Hindus and Buddhists, correspond to the "four ways of rebirth". According to this philosophy, there are four different ways to enter the cycle of rebirth (**samsâra*) : either through a viviparous birth (*jarâyuva*), in the form of a human being or mammal; or an oviparous birth (*andaja*), in the form of a bird, insect or reptile; or by being born in water and humidity (*samsvedaja*), in the form of a fish or a worm; or even through metamorphosis (*aupapâduka*), which means there is no "mother" involved, just the force of *Karma*. See **Four.**

**YUGA (Definition).** "Period". Generic names for the cosmic cycles of Indian speculations which are either based upon Brahmanic cosmogny or the learned astronomy founded by *Âryabhata. The principal cycle is the **mahâyuga* (or "great period") made up of 4,320,000 human years. This is divided into four successive *yugas*. Thus the words **mahâyuga* and **chaturyuga* (literally : four periods) are treated as synonymous. These four successive ages are named respectively : **kritayuga* (or **satyayuga*), **tretâyuga*, **dvaipâyanayuga* (or **dvâparayuga*), and **kaliyuga*. The corresponding lengths can considered to be equal or unequal depending on which system of calculation is used. See other entries entitled ***Yuga.***

**YUGA (Astronomical speculation on).** Since its emergence at the start of the sixth century CE, *learned Indian astronomy* has been marked by its amazing speculation about the cosmic cycles (known as **yugas*), which is very different from the cosmogonical speculations. See ***Yuga*** **(Definition)** and ***Yuga*** **(Systems of calculating).**

According to this speculation, directly linked to astronomical elements, the **chaturyuga* or cycle of 4,320,000 years is the period at the beginning and end of which the nine elements (namely the sun, the moon, their apsis and node and the planets) are in mean perfect conjunction at the starting point of the longitudes. Thus the durations of the revolutions, previously considered to be the same lengths, are (in this astronomy) subjected to common multiples and general conjunctions. See **Indian astronomy (History of)** and **Indian mathematics (History of).** This speculation seems so audacious because it is obviously devoid of any physical meaning.

As for the cycle called **kalpa*, which constitutes an even longer period of time of 4,320,000,000 years, it is delimited, according to *Brahmagupta (628 CE), by two perfect conjunctions in true longitude of the totality of elements, themselves each matched by a total eclipse of the Sun on the stroke of six in the secular time in **Ujjayinî*. In practice, however, these fictional eras can be reduced to the age of the **kaliyuga*, the present age, which traditionally starts at a theoretical point of departure of the celestial revolutions corresponding to the 18 February 3101 BCE at zero hours. (This moment is fixed itself at the general conjunction in mean longitude at the starting point of the sidereal longitudes of the sun, the moon and the planets, the apogees and node ascending from the moon being respectively at 90° and 180° from these longitudes.) Literally, the word **yuga* signifies "yoke", "link". In ancient Indian astronomy, this term was employed in the very limited sense of the simple "cycle". Thus in the **Jyotishavedânga* ("Astronomic Element of Knowledge"), a *yuga* of five years is used, this being a period at the end of which the sun and the moon are considered to have each completed a whole number of revolutions. On the other hand, in the *Romakasiddhânta* (start of the fourth century CE), the *yuga* is a lunar-solar cycle, the length of which is 2,850 years. These cycles, however, do not constitute an "astronomical speculation" like the one that began to be developed in Âryabhata's time. No speculative system relating to *yugas* is found in the texts of the **Vedas*. This means that the *yuga* speculations were probably unknown in India during Vedic times and until the early centuries of the Common Era.

Nevertheless, purely arithmetical speculative calculations on these cycles appear in the **Manusmriti* (a significant religious work considered to form the basis of Hindu society), as well as in the much later texts of the *Yâjñavalkyasmriti* and the epic of the **Mahâbhârata*. It is difficult, however, if not impossible, to glean from this a chronology for the history of speculative *yugas*, since a great deal of uncertainty presides over the dates of these documents.

On the other hand, the work of Âryabhata, in which astronomical speculation of *yugas* appears for the first time, is dated in a rather precise manner, to within a few years of 510 CE.

In fact, as far as it is possible to tell, it was Âryabhata who, after the beginning of the sixth century CE, introduced speculative *yugas* into mathematical astronomy and made them generally known in India.

None of the Indian speculative canons (on *yugas*) that are known today is dated before Âryabhata's time. Âryabhata's astronomical speculations on *yugas* use basic numbers, some of which can be seen in the following calculations (**nakshatra* here denoting the twenty-seven lunar mansions divided into equal lengths) :

1 **mahâyuga* = 4,320,000 years = 12,000 (moments) × 360 = 27 (*nakshatra*) × 4 × 4 (phases) × 10,000 = "great period". 1 **yugapâda* = 1,080,000 years = 3,000 (moments) × 360 = 27 (*nakshatra*) × 4 (phases) × 10 000 = one quarter of a "great period". 1 **kaliyuga* = 432,000 years = 1,200 (moments) × 360 = 27 (*nakshatra*) × 4 × 4 (phases) × 1,000 = 1,200 ("divine years") × 360 = one tenth of a "great period".

According to Censorinus, Heraclitus's "great year" was 10,800 years long. On the other hand, the surviving fragments of the work of Babylonian astronomer Berossus (fourth–third century BCE) contain mention of a cosmic period 432,000 years long, which is also called "Great Year" :

1 **Great Year of Heraclitus* = 10,800 years = 30 (moments) × 360. 1 **Great Year of Berossus*= 432,000 years = 1,200 (moments) × 360. In other words, in all the cycles there is the following arithmetical relationship: 1 **yugapada* = 100 times the Great Year of

Heraclitus = 2.5 times the Great Year of Berossus. 1 **kaliyuga* = one Great Year of Berossus = 40 times the Great Year of Heraclitus. 1 **mahâyuga* = 400 times the Great Year of Heraclitus = 10 times the Great Year of Berossus.

From what is known today, it is impossible to establish whether there is any link between Âryabhata's *yugas* and the cosmic periods of the Mediterranean world. What *is* known is that Heraclitus belonged to the time when Persia dominated certain countries of the Greek world as well as part of India, whilst Berossus belonged to the end of the Persian rule and the beginning of the conquests of Alexander the Great . . . So why did Âryabhata develop his remarkable speculation? "As far as Âryabhata was concerned, speculation about *yugas* was just a theory. Convinced of the existence of common multiples of the different revolutions, he had set himself the task of researching the cycles of this astronomy, which was the most advanced of his time, and of which he was fully aware of the value. Whether it was a spontaneous idea, or drawn from a revival of the the *Great Year* 432,000 years long of the Babylonian astronomer Berossus, or even inspired by a wholly verbal, strictly arithmetical speculation, in any case Âryabhata drew out the constants of the mean movements in order to construct these common multiples and general conjunctions, from a single reality in time, that is to say the astronomical reality of 510 CE almost to the year. Of course, the theory was regrettable, but we must not forget the serious and extreme rigour he showed in undertaking such a work." (Billard)

**YUGA (Cosmogonical speculations on).** According to speculations developed by cosmogonies on what is referred to as the "decline of Proper moral and cosmic Order over the course of time", the **mahâyuga* corresponds to the appearance, evolution and disappearance of a world, and the whole cycle is followed by a new *mahâyuga*, and so on until the destruction of the universe. The four ages of this "great cycle" are considered to be unequal in terms of both length and worth. Qualitatively, this is how things are meant to unfold [see Friedrichs, etc (1989), *Dictionnaire de la sagesse orientale*] :

1. The **kritayuga*, the first of the four *yugas*, is the golden age, during which humans enjoy extremely long lives and everything is perfect. This is the ideal age, where virtue, wisdom and spirituality reign supreme, and there is a total absence of ignorance and vice. Hatred, jealousy, pain, fear and menace are unknown. There is only one god, one sole **Veda*, one law and one religion, each caste fulfilling its duties with the utmost selflessness. This age is said to have lasted 4,800 divine years (**divyavarsha*), which is equal to 1,728,000 human years.
2. The **tretâyuga* is the age during which humans are only believed to live three quarters of their lives. They are now marked by vice, there are the beginnings of laxity in their behaviour and the first ritual sacrifices are carried out. During this age, humans begin to act with intention and in self-interest. Rectitude diminishes by a quarter. The age is said to last 3,600 divine years, or 1,296,000 human years.
3. The **dvâparayuga* is said to be the age during which the forces of Evil equal those of Good, and where honest behaviour, virtue and spirituality are reduced by half. Illnesses have made their appearance and humans now only live half their lives. This age is said to last 2,400 divine years, or 864,000 human years.
4. Finally, the **kaliyuga* or "iron age" is the age we are living in now. "True virtue" is something which has all but disappeared and conflicts, ignorance, irreligion and vice have increased manifold. Illnesses, exhaustion, anger, hunger, fear and despair reign supreme. Living things only live for a quarter of their existence and the forces of evil triumph over good. Only a quarter of the original rectitude displayed by humans remains. This age is meant to have begun in the year 3101 BCE, and is meant to last 1,200 divine years, or 432,000 human years. It is meant to end with a *pralaya* (destruction by fire and water) before a new **chaturyuga* begins.

See ***Yuga* (Definition)**, ***Yugas* (Systems of calculating)** and any other entry entitled ***Yuga***.

**YUGA (Systems of calculating).** In the traditional system, the four ages of a **chaturyuga* are of unequal length, with the ratios of 4, 3, 2 and 1, from the **kritayuga* to the **kaliyuga* whose length is equal to 1/10 of the **mahâyuga*. In other words, the four successive ages of a *chaturyuga* are divided unequally as follows :

1 *mahâyuga* = 4/10 + 3/10 + 2/10 + 1/10. Thus the corresponding values in "divine" years: 1 **kritayuga* = 4,800 divine years (= 4/10 of *mahâyuga*); 1 **tretâyuga* = 3,600 divine years (= 3/10 of *mahâyuga*); 1 **dvâparayuga* = 2,400 divine years (= 2/10 of *mahâyuga*); 1 **kaliyuga* = 1,200 divine years (= 1/10 of *mahâyuga*); 1 *mahâyuga* = 12,000 divine years.

As one divine year is said to be equal to 360 human years, these cycles have the following durations in human time:

1 *kritayuga* = 1,728,000 human years; 1 *tretâyuga* = 1,296,000 human years; 1 *dvâparayuga* = 864,000 human years; 1 *kaliyuga* = 432,000 human years; 1 *mahâyuga* = 4,320,000 human years.

See ***Divyavarsha***. The system for calculating unequal *yugas* was used by a considerable number of Indian astronomers (including *Brahmagupta), as well as by a great many cosmographers, philosophers and authors of religious texts (traditional system).

However, the system used by *Âryabhata consists in dividing the *mahâyuga* in the following manner :

1 *kritayuga* = 1,080,000 human years; 1 *tretâyuga* = 1,080,000 human years; 1 *dvâparayuga* = 1,080,000 human years; 1 *kaliyuga* = 1,080,000 human years; 1 *mahâyuga* = 4,320,000 human years.

In other words, the four cycles of the *chaturyuga* are all considered to be equal here. This is the system of the **yugapâdas* or "quarters of a *yuga*". However, the *mahâyuga* is not the longest unit of cosmic time in these systems of calculation. There is also the cycle called a **kalpa*, which corresponds to 12,000,000 divine years:

1 *kalpa* = 12,000,000 divine years = 12,000,000 × 360 = 4,320,000,000 human years. Bearing in mind the length of the *mahâyuga* (= 12,000 divine years), this cycle is thus also defined by :

1 *kalpa* = 1,000 *mahâyuga* = 4,320,000 × 1,000 = 4,320,000,000 human years.

An even longer period exists, the **mahâkalpa*, or "great *kalpa*", which is the length of twenty "ordinary" *kalpas* (20,000 *mahâyugas*) :

1 *mahâkalpa* = 12,000,000 × 20 = 240,000,000 divine years = 240,000,000 × 360 = 86,400,000,000 years.

**YUGA.** [S]. Value = 2. (This symbol is very rarely used to represent this value.) The allusion here is to the cycle called **Dvâipayanayuga*, where, according to Brahmanic cosmogony, men have only lived half of their lives and where the forces of Evil are counteracted by the equal strength of the forces of Good. The duality (**dvaita*) between Good and Evil is at the root of this symbolism. See ***Yuga* (Definition)** and **Two**.

**YUGA.** [S]. Value = 4. The allusion here is to the most important of the cosmic cycles of this name : the **mahâyuga* (or "Great Age"), also called **chaturyuga* (or "four ages"). Composed of four successive "ages", in Hindu cosmogony the *mahâyuga* is said to correspond to the appearance, evolution and disappearance of a world. See **Yuga (Definition)** and **Four**.

**YUGALA.** [S]. Value = 2. Synonym of **Yama*. See **Two**.

**YUGAPÂDA.** "Quarter of a *yuga*". Name given to each of the four cycles of a **chaturyuga*, subdivided into four equal parts, according to the system of calculation used by *Âryabhata. A *yugapâda* thus corresponds to 1,080,000 human years. See ***Yuga* (Definition)** and ***Yuga* (Systems of calculating)**.

**YUGMA.** [S]. Value = 2. Synonym of **Yama*. See **Two**.

# Z

**ZENITH.** [S]. Value = 0. See ***Vishnupada*** and **Zero.**

**ZERO.** Ordinary Sanskrit name for zero : **shûnya*. Here is a list of corresponding numerical symbols: **Abhra*, **Âkâsha*, **Ambara*, **Ananta*, **Antariksha*, **Bindu*, **Gagana*, **Jaladharapatha*, **Kha*, **Nabha*, **Nabhas*, **Pûrna*, **Randhra*, **Vindu*, **Vishnupada*, **Vyant*, **Vyoman*. These words translate or symbolically express :

1. The void (*Shûnya*). 2. Absence (*Shûnya*). 3. Nothingness (*Shûnya*). 4. Nothing (*Shûnya*). 5. The insignificant (*Shûnya*). 6. The negligible quantity (*Shûnya*). 7. Nullity (*Shûnya*). 8. The "dot" (*Bindu, Vindu*). 9. The "hole" (*Randhra*). 10. Ether, or "element which penetrates everything" (*Âkâsha*). 11. The atmosphere (*Abhra, Ambara, Antariksha, Nabha, Nabhas*). 12. Sky (*Nabha, Nabhas, Vyant, Vyoman, Vishnupada*). 13. Space (*Âkâsha, Antariksha, Kha, Vyant, Vyoman*). 14. The firmament (*Gagana*). 15. The canopy of heaven (*Gagana*). 16. The immensity of space (*Ananta*). 17. The "voyage on water" (*Jaladharapatha*). 18. The "foot of Vishnu" (*Vishnupada*). 19. The zenith (*Vishnupada*). 20. The full, the fullness (*Pûrna*). 21. The state of that which is entire, complete or finished (*Pûrna*). 22. Totality (*Pûrna*). 23. Integrity (*Pûrna*). 24. Completion (*Pûrna*). 25. The serpent of eternity (*Ananta*). 26. The infinite (*Ananta, Vishnupada*). See **Numerical symbols.**

**ZERO (Graeco-Latin concepts of).** Western cultures have obviously had a concept of the void since Antiquity. To express it, the Greeks had the word *oudén* ("void"). As for the Romans, they used the term *vacuus* ("empty"), *vacare* ("to be empty"), and *vacuitas* ("emptiness"); they also had the words *absens*, *absentia*, and even *nihil* (nothing), *nullus* and *nullitas*. But these words actually corresponded to notions that were understood very distinctly from each other. With the help of some appropriate examples, an etymological approach will enable us hereafter to form quite a clear idea of the evolution of the concepts down the ages and to perceive better the essential difference which exists between these diverse notions and the Indian concept of the zero. "Presence" (from the Latin *praesens*, present participle of *praesse*, "to be before [*prae*]", "to be facing") is properly speaking the fact of being where one is. But the adjective present also means "what is there in the place of which one is speaking"; this meaning is applicable then both to an object and to a living being. In the figurative sense, applied to people, present means "that which is present in thought to the thing being spoken of" (to be present in thought at a ceremony, despite the physical absence); applied to things, however, it means "that which is there for the speaker, or for what he is aware of". It is thus a moral or deliberate presence. Another meaning of presence, in opposition this time to the past and the future, is "that which exists or is really happening, either at the moment of speaking or at the moment of which one is speaking". Consequently, this meaning corresponds to the present situation. Figuratively, it is rather a matter of "that which exists for the consciousness at the moment one is speaking", somewhat like a scene one witnessed and which remains present in one's mind.

This preamble allows a better understanding of "absence", since it is a term that is opposed to presence. The word comes from the Latin *absentia*, which derives from *abest*, "is far". Thus it expresses the character of "that which is far from". It is thus by definition the fact of not being present at a place where one is normally or one is expected. And the absent is the person or the thing which is lacking or missing. As for non-presence, it is simply the *void* left by an absence, since it is the space that is not occupied by any being or any thing. If it is an unoccupied place, it is this that is empty, whether it be a seat, an administrative post or even one of the "places" of the place-value system. By dint of thinking solely of the void, some thinkers have arrived at vacuism, a type of physics, according to which there exist spaces where all material reality is void of all existence. It was developed notably by the Epicureans, who conceded the existence of places where all matter, visible or invisible, was absent. Others opted rather for anti-vacuism, like Descartes, who considered an absolute void to be a contradictory notion. Nowadays, it is still sometimes said that nature abhors a vacuum. This aphorism was created by those who held to the physics of the ancient world in order to make sense of the existence of certain phenomena for which they were incapable of providing a satisfactory explanation. It was not until the experiments of the Italian physicist Torricelli on the laws of atmospheric weight, that the lie was given to this idea.

**ZERO (Indian concepts of).** In Sanskrit, the principal term for zero is *Shûnya*, which literally means "void"or "empty". But this word, which was certainly not invented for this particular circumstance, existed long before the discovery of the place-value system. For, in its meaning as "void", it constituted, from ancient times, the central element of a veritable mystical and religious philosophy, elevated into a way of thinking and existing. The fundamental concept in **shûnyatâvâda*, or philosophy of "vacuity", **shûnyatâ*, this doctrine is in fact that of the "Middle Way" (*Mâdhymakha*), which teaches in particular that every made thing (*samskrita*), is void (**shûnya*), impermanent (*anitya*), impersonal (*anâtman*), painful (*dukh*) and without original nature. Thus this vision, which does not distinguish between the reality and non-reality of things, reduces the same things to total insubstantiality. See ***Shûnya*** and ***Shûnyatâ***.

This is how the philosophical notions of "vacuity", nihilism, nullity, non-being, insignificance and absence, were conceived early in India (probably from the beginning of the Common Era), following a perfect homogeneity, contrary therefore to the Graeco-Latin peoples (and more generally to all people of Antiquity) who understood them in a disconnected and totally heterogeneous manner.

The concepts of this philosophy have been pushed to such an extreme that it has been possible to distinguish twenty-five types of *shûnya*, expressing thus different nuances, among which figure the void of non-existence, of non-being, of the unformed, of the unborn, of the non-product, of the uncreated or the

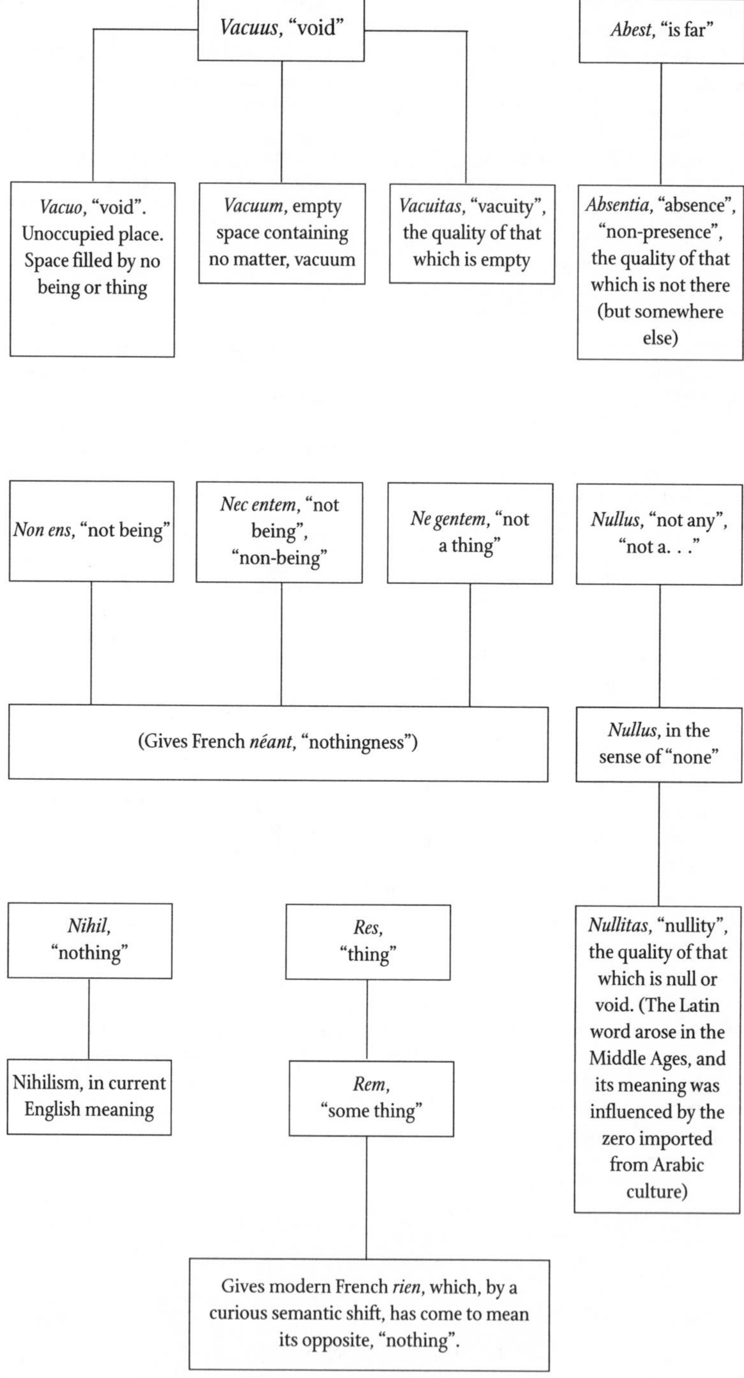


FIG.24D.10. *The Western concept of nought*

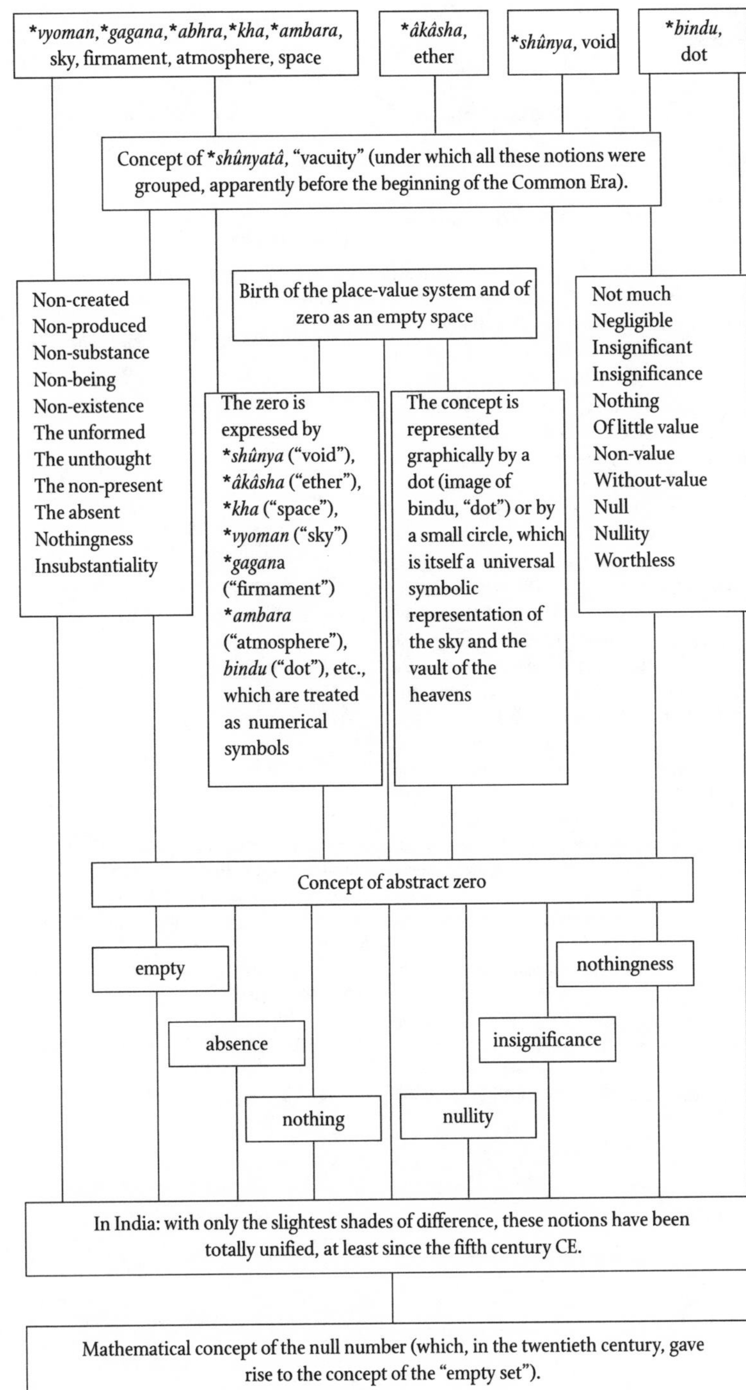


FIG.24D11. *The perfect homogeneity of the Indian zero*

non-present; the void of the non-substance, of the unthought, of immateriality or insubstantiality; the void of non-value, of the absent, of the insignificant, of little value, of no value, of nothing, etc. In brief, the zero could have hardly germinated in a more fertile ground than the Indian mind. Once the place-value system was born, the *shûnya*, as a symbol for the void and its various synonyms (absence, nothing, etc.), naturally came to mark the absence of units in a given order [see Fig. D. 11]. It is important to remember that the Indian place-value system was born out of a simplification of the **Sanskrit place-value system* as a consequence of the suppression of the word-symbols for the various powers of ten. This was a decimal positional numeration which used the nine ordinary names of numbers and the term *shûnya* ("void") as the word that performs the role of zero. Thus the Indian zero has meant from an early time not only the void or absence, but also heaven, space, the firmament, the canopy of heaven, the atmosphere and ether, as well as nothing, a negligible quantity, insignificant elements, the number nil, nullity and nothingness, etc. This means that the Indian concept of zero by far surpassed the heterogenous notions of vacuity, nihilism, nullity, insignificance, absence and non-being of all the contemporary philosophies. See all other entries entitled **Zero**.

The Sanskrit language, which is an incomparably rich literary instrument, possessed more than just one word to express "void". It possessed a whole panoply of words which have more or less the same meaning; these words are related, in a direct or indirect manner, to the universe of symbolism of Indian civilisation. See **Sanskrit**, **Numerical symbols**, **Numeration of numerical symbols**.

Thus words which literally meant the sky, space, the firmament or the canopy of heaven came to mean not only the void but also zero. See ***Abhra***, ***Âkâsha***, ***Ambara***, ***Antariksha***, ***Gagana***, ***Kha***, ***Nabha***, ***Vyoman*** and **Zero**. In Indian thought, space is considered as the void which excludes all mixing with material things, and, as an immutable and eternal element, is impossible to describe. Because of the elusive character and the very different nature of this concept as regards ordinary numbers and numerals, the association of ideas with zero was immediate. Other Indian numerical symbols used to mean zero were: **pûrna* "fullness", "totality", "wholeness", "completion"; **jaladharapatha*, "voyage on the water"; **vishnupada*, "foot of Vishnu; etc. To find out more about this symbolism, see the appropriate entries. Such a numerical symbolism has played a role that has been all the more important in the history of the place-value system because it is in fact at the very origin of a representation that we are very familiar with. The ideas of heaven, space, atmosphere, firmament, etc., used to express symbolically, as has just been seen, the concept of zero itself. And as the canopy of heaven is represented by human beings either by a semi-circle or a circular diagram or again by a complete circle, the little circle that we know has thus come, through simple transposition or association of ideas, to symbolise graphically, for the Indians, the idea of zero itself. It has always been true that "The circle is universally regarded as the very face of heaven and the Milky Way, whose activity and cyclical movements it indicates symbolically" [Chevalier and Gheerbrant (1982)]. And so it is that the little circle was put beside the nine basic numerals in the place-value system, to indicate the absence of units in a given order, thereafter acquiring its present function as arithmetical operator (that is to say that if it is added to the end of a numerical representation, the value is thus multiplied by ten). See **Numeral 0** (in the form of a little circle), ***Shûnya-chakra***.

The other Sanskrit term for zero is the word **bindu*, which literally means "dot". The dot, it is true, is the most elementary geometrical figure there is, constituting a circle reduced to its centre. For the Hindus, however, the **bindu* (in its supreme form of a **paramabindu*) symbolises the universe in its non-manifest form and consequently constitutes a representation of the universe before its transformation into the world of appearances (*rûpadhâtu*). According to Indian philosophies, this uncreated universe is endowed with a creative energy capable of engendering everything; it is thus in other terms the causal point whose nature is consequently *identical* to that of ""vacuity" (**shûnyatâ*). See ***Bindu***, ***Paramânu***, ***Paramabindu***, ***Âkâsha***, ***Shûnyatâ*** and **Zero**.

Thus this natural association of ideas with this geometrical figure, which is the most basic of them all, yet capable of engendering all possible lines and shapes (*rûpa*). It is the perfect symbol for zero, the most negligible quantity there is, yet also and above all the most basic concept of all abstract mathematics. Thus the dot came to be a representation of zero (particularly in the *Shâradâ* system of Kashmir and in the notations of Southeast Asia; see Fig. 24. 82) which possesses the same properties as the first symbol, the little circle. See **Numeral 0** (in the form of a dot) and ***Shûnya-bindu***.

This is the origin of the eastern Arabic zero in the form of a dot : when the Arabs acquired the Indian place-value system, they evidently acquired zero at the same time. This is why, in Arabic writings, sometimes the sign is given in the form of a dot, sometimes in the form of a small circle. It is the little circle that prevailed in the West, after the Arabs of the Maghreb transmitted it themselves to the Europeans after the beginning of the twelfth century. To return to India, this notion was gradually enriched to engender a highly abstract mathematical concept, which was perfected in *Brahmagupta's time (c. 628 CE); that of the "number zero" or "zero quantity". It is thus that the *shûnya* was classified henceforth in the category of the **samkhya*, that is to say the "numbers", so marking the birth of modern algebra. See ***Shûnya-samkhya***. So, from abstract zero to infinity was a single step which Indian scholars took early and nimbly. The most surprising thing is that amongst the Sanskrit words used to express zero, there is the term **ananta*, which literally means "Infinity". Ananta, according to Indian mythologies and cosmologies, is in fact the immense serpent upon which the god *Vishnu is said to rest between two creations; it represents infinity, eternity and the immensity of space all at once. Sky, space, the atmosphere, the canopy of heaven were, it is true, symbols for zero, and it is impossible not to draw a comparison in these conditions, between the void of the spaces of the cosmos with the multitude represented by the stars of the firmament, the immensity of space and the eternity of the celestial elements. As for the ether (**âkâsha*), this is said to be made up of an infinite number of atoms (**anu*, **paramânu*). This is why, from a mythological, cosmological and metaphysical point of view, the zero and infinity have come to be united, for the Indians, in both time and space. See ***Ananta***, ***Shesha***, ***Sheshashîrsha***, **Infinity (Indian mythological representation of)** and **Serpent (Symbolism of the)**.

But from a mathematical point of view, however, these two concepts have been very clearly distinguished, Indian scholars having known that one equalled the inverse of the other. See **Infinity**, **Infinity (Indian concepts of)** and **Indian mathematics (History of)**.

To sum up, the Indians, well before and much better than all other peoples, were able to unify the philosophical notions of void, vacuity, nothing, absence, nothingness, nullity, etc. They started by regrouping them (from the beginning of the Common Era) under the single heading **shûnyatâ* (vacuity), then (from at least the fifth century CE) under that of the **shûnyakha* (the sign zero as empty space left by the absence of units in a given order in the place-value system) before recategorising them (well before the start of the seventh century CE) under the heading of *shûnya-samkhya* (the "zero number") [see Fig. D. 11]. Once again, this indicates the great conceptual advance and the extraordinary powers of abstraction of the scholars and thinkers of Indian civilisation. The contribution of the Indian scholars is not limited to the domain of arithmetic; by opening the way to the generalising idea of number, they enabled the rapid development of algebra and consequently played an essential role in the development of mathematics and all the exact sciences. It is impossible to exaggerate the significance of the Indian discovery of zero. It constituted a natural extension of the notion of *vacuity, and gave the means of filling in the space left by the absence of an order of units. It provided not only a word or a sign, it also and above all became a numeral and a numerical element, a mathematical operator and a whole number in its own right, all at the point of convergence of all numbers, whole or not, fractional or irrational, positive or negative, algebraic or transcendental.

**ZERO AND INFINITY**. See **Zero**, **Infinity**, **Infinity (Indian concepts of)**, **Infinity (Mythological representations of)**, **Serpent (Symbolism of the)**, **Zero (Graeco-Latin concepts of)**, **Zero (Indian concepts of)** and **Indian mathematics (History of)**.

**ZERO AND SANSKRIT POETRY**. In India, the use of zero and the place-value system has been a part of the way of thinking for so long that people have gone as far as to use their principal characteristics in a subtle and very poetic form in a variety of Sanskrit verse. As proof, here is a quotation from the poet Bihârîlâl who, in his *Satsaî*, a famous collection of poems, pays homage to a very beautiful woman in these terms : "The dot [she has] on her forehead Increases her beauty tenfold, Just as a zero dot (**shûnya-bindu*) Increases a number tenfold. " [See Datta and Singh, in: AMM, XXXIII, (1926), pp. 220ff.].

First of all, it should be remembered that the dot that the woman has on her forehead is none other than the **tilaka* (literally: sesame), a mark representing for the Hindus the third eye of *Shiva, that is the eye of knowledge. While young girls put a black spot between their eyebrows by means of a non-indelible colouring matter, married women put a permanent red dot on their foreheads; it would seem then that the homage was being paid to a married woman. It is known that the dot (**bindu*) figures among the numerous numerical symbols with a value equal to zero, and is even used as one of the graphical representations of this concept. See **Zero**, ***Shûnya-bindu***, **Numeral 0** (in the form of a dot). This is a very clear allusion to the arithmetical operative property of zero in the place-value system, because if zero is added to the right of the representation of a given number, the value of the number is multiplied by ten (see Fig. 23.26 and 27). Another quotation, taken this time from the *Vâsavadattâ* by the poet *Subandhu (a long love story, written in an extremely elaborate language, swarming with word plays, implications and periphrases):

"And at the moment of the rising of the Moon
With the darkness of the falling night,
It was as if, with folded hands
Like closed blue lotus blossoms,
The stars had begun straightway
To shine in the firmament (**gagana*)...
Like zeros in the form of dots (**shûnya-bindu*),
Because of the emptiness (**shûnyatâ*) of the **samsâra*,
Disseminated in space (*kha*),
As if they had been [dispersed]
In the dark blue covering the skin of the Creator [= *Brahma],
Who had calculated their sum total
By means of a piece of Moon in the guise of chalk."

[See *Vâsavadattâ of *Subandhu*, Hall, Calcutta (1859), p. 182; Datta and Singh (1938), p. 81.]

Here too the metaphor used leaves the reader in no doubt; the void (**shûnya*) – which is placed in relation to the emptiness (**shûnyatâ*) of the cycle of rebirths (**samsâra*) – is also implied in its representation in the form of a dot (**shûnya-bindu*), as an operator in the art of written calculation. These concepts really had to have been part of the way of thinking for a long time for the subtleties used in this way to have been understood and appreciated by the wider public of the time.

**ZODIAC**. [S]. Value = 12. See ***Râshi*** and **Twelve**.

CHAPTER 25

# INDIAN NUMERALS AND CALCULATION IN THE ISLAMIC WORLD

As we saw in the previous chapter, it was indeed the Indians who invented zero and the place-value system, as well as the very foundations of written calculation as we know it today.

These highly significant inventions date back at least as far as the fifth century CE.

However, it was not until five centuries later that the nine basic numerals appeared in Christian Europe.

Another two or three centuries elapsed before zero was first used in Europe, along with the afore-mentioned methods of calculation, and it was later still that these revolutionary new ideas were propagated and fully accepted in the Western world.

Thus the Indian inventions were not transmitted directly to Europe: Arab-Muslim scholars (amongst their numerous fundamental roles) played an essential part as vehicles of Indian science, acting as "intermediaries" between the two worlds.*

Therefore, before we proceed with our history, it is worth knowing a little about the Arabs, in terms of their culture, their way of thinking, their own science and their fundamental contributions to the evolution of science the world over. This will give the reader a clearer idea of the conditions under which this transmission of ideas took place, which led to the internationalisation of Indian science and methods of calculation.

## THE SCIENTIFIC CONTRIBUTIONS OF ARAB-ISLAMIC CIVILISATION

In the century following the death of the prophet Mohammed the Islamicised Arabs built up an enormous empire through their conquests. At the beginning of the eighth century CE, the Empire stretched from the Pyrenees to the borders of China, and included Spain, southern Italy, Sicily, North Africa, Tripolitania, Egypt, Palestine, Syria, part of Asia Minor and Caucasia, Mesopotamia, Persia, Afghanistan and the Indus Valley.

* Words preceded by an asterisk have entries in the Dictionary (pp. 445–510).

FIG. 25.1. *Detail of a page from Al bahir fi 'ilm al hisab (The Lucid Book of Arithmetic) by As Samaw'al ibn Yahya al-Maghribi (died c. 1180 in Maragha), a Jewish mathematician, doctor and philosopher from the Maghreb, who converted to Islam [Istanbul, Aya Sofia Library, Ms. ar. 2,718. See Rashed and Ahmed 1972]. This document, which uses "Hindi" numerals to reproduce what is known as "Pascal's triangle", shows that Muslim mathematicians knew about the binomial expansion* (a + b)m, *where "m" is a positive integer, as early as the tenth century. The author admits that this triangle is not his, and attributes it to al-Karaji, who lived near the end of the tenth century [Anbouba; Rashed in DSB].*

Nevertheless, the advance came to a halt when it met with successful resistance: in 718 by the Byzantine army near Constantinople; in 732 by Charles Martel at Poitiers; and in 751 by the Chinese on the border of Sogdiana.

Once the political influence of the "Son of the Arabian Desert" fell into decline, the Empire was controlled for nearly a century by the caliphs of the Omayyad Dynasty (661–750), with Damascus as their capital. Power then went to the Abbasid caliphs (750–1258) who made Baghdad their capital in 772 and reigned over the empire for the next 500 years.

There followed a period of exaltation characteristic of expansion, and this was a highly fertile era of cultural assimilation and scientific development. Arabic culture dominated the world for several centuries, before Mongol invasions, the Crusades, the division of the Empire and the anarchy of internal wars brought it to an end in the thirteenth century.

### THE ASSIMILATION OF OTHER CULTURES

When the Arab nomads who had been converted to Islam left the desert to conquer this immense territory, they lived from trading spices, medicines, cosmetics and perfume. Their level of literacy and numeracy was very basic. The little that they knew of science was based on practical applications involving simple formulae, and was often tinged with arithmology, mysticism and all kinds of magical and divinatory practices.

Thus the first Islamicised Arabs initially possessed none of the intellectual means they would need to realise their desire to conquer other lands and to deal with the enormous revenue that taxation and capitation would soon bring to this vast new Empire.

However, through their conquests and trade relations, they found themselves increasingly in contact with people from different cultures: Syrians, Persians, Greek émigrés, Mesopotamians, Jews, Sabaeans, Turks, Andalusians, Berbers, peoples from Central Asia, inhabitants of the shores of the Caspian, Afghans, Indians, Chinese, etc. Thus they discovered cultures, sciences and technologies far superior to their own. They were quick to adapt and to get to grips with the new concepts and knowledge, which scientists, intellectuals and engineers from the conquered lands had accumulated over the ages, and in some cases had developed to quite an advanced level.

### THE METROPOLIS OF NEAR EASTERN SCIENCE BEFORE ISLAM

Long before the Arabic conquest, the philosophy of Aristotle and the sciences of nature, mathematics, astronomy and medicine, according to the doctrines of Hippocrates and Galen, were all taught in Syria and Mesopotamia, notably at the schools of Edesse, Nisibe and Keneshre.

At the same time, Persia constituted an important crossroads and centre of influence for the meeting of Greek, Syrian, Indian, Zoroastrian, Manichaean and Christian cultures.

The Persian King Khosroes Anûshîrwân (531–579) sent a cultural mission to India and brought many Indian scientists to Jundishapur. Nestorian Christians, who had been expelled from the school at Edesse by Byzantine orthodoxy, found refuge in the same town. This is also where the Neo-Platonist philosophers of Athens (such as Simplicius who wrote commentaries on the works of Aristotle and Euclid) were welcomed by King Anûshîrwân when their academy was closed in 529 under the orders of Emperor Justinian (527–565). It was at Edesse, Nisibe, Keneshre and Jundishapur that Greek works were first translated into Syrian or Persian, and that the first works in Sanskrit were discovered. The first translations into Arabic were undertaken at Jundishapur shortly after the constitution of the Islamic Empire [see L. Massignon and R. Arnaldez in HGS; A. P. Youschkevitch (1976)].

### BAGHDAD, FIRST ISLAMIC CENTRE OF SCIENTIFIC LEARNING

The importance of these cultural and scientific centres gradually declined during the Abbassid Dynasty, and so the town of Baghdad became the centre of intellectual activity in the Near East, thus playing a vital role in this history.

Founded in 762, then elevated to capital of the Arabic Empire in 772, Baghdad was initially the obvious centre for international trade. Then, owing to both its privileged location, and to the generous action of sovereign patrons, such as caliphs al-Mansur (754–775), Harun al-Rashid (786–809) and al-Ma'mun (813–833), whose subsidies contributed to the development of science and culture in Islam, Baghdad became the most vivacious intellectual centre of the East. This is where Arabic science truly began.

> If we put together the religious and social conditions, we shall understand the position of Islamic intellectuals and the fillip they gave to intellectuals of all creeds and races, by their mobilisation for a common labour in the Arabic tongue. For science is one of the Islamic city's institutions. Not only do patrons encourage it, but caliphs work to create and develop it. It is sufficent to cite Khalid, the "philosopher prince", whose actions were "perhaps legendary", or al-Mansur, the founder of Baghdad, and al-Ma'mun "who eagerly sent out emissaries to look for manuscripts and have them translated" [L. Massignon and R. Arnaldez].

## THE GOLDEN AGE OF ARABIC SCIENCE

One of the most outstanding periods in the history of science took place in Islam between the eighth and thirteenth centuries of the Common Era.

This was at a time when Western civilisation was devastated by epidemics, famine and war, and was in no position to relay the cultural heritage of Antiquity. The Arab-Muslim scholars were able to develop not only mathematics, astronomy and philosophy, but also medicine, pharmacy, zoology, botany, chemistry, mineralogy and mechanics.

Through a collective effort, the Muslims and the peoples conquered by Islam collected together all the Greek works that they could find on philosophy, literature, science and technology.

It is sufficient to cite the names of Euclid, Archimedes, Ptolemy, Aristotle, Plato, Galen, Hero of Alexandria, Apollonius, Menelaus, Philo of Byzantium, Plotinus and Diophantus to give an idea of the variety and richness of the works that were translated into Arabic.

These translations and collected works grew in number and circulation, as universities and libraries sprang up all over the Islamic world. Towns such as Damascus, Cairo, Kairouan, Fez, Granada, Cordoba, Bokhara, Chorem, Ghazni, Rey, Merv and Isfahan soon became intellectual and artistic centres which were centuries ahead of the Christian capitals.

## "ARABIC" OR "ISLAMIC" SCIENCE?

Arabic science is not necessarily the same thing as Muslim science. The Arabic language was a vehicle for science, which, during that long period of time, became the international language of the scholars of the Muslim world, and consequently the intellectual link between the different races.

Amongst the diverse cultures which were conquered or influenced by Islam was Persia, the birthplace of many brilliant minds, including al-Fazzari, al-Khuwarizmi, al-Razi, Avicenna, al-Biruni, Kushiyar ibn Labban al-Gili, Umar al-Khayyam, Nasir ad din at Tusi and Ghiyat ad din Ghamshid ibn Mas'ud al-Kashi.

During the assimilation of Indian science, the Arabs were helped by many Hindu Brahmans, who were often received at the court of Baghdad by enlightened caliphs.

They were assisted by Persians and Christians from Syria and Mesopotamia, who, being fervent admirers of Indian cultures, had gone so far as to learn Sanskrit.

The Buddhists also greatly contributed, especially those converted to Islam, such as Barmak who was sent to India to study astrology, medicine and pharmacy and who, on his return to Muslim territory, translated many Sanskrit texts into Arabic [see L. Frédéric (1989)].

There were also non-Muslim Arabic scholars, such as the Christian and Jewish intellectuals, who were often referred to as *ahl al kitab*, the "people of books", and whom the caliphs of Baghdad and Cordoba integrated to a certain extent amongst the members of their empires, sometimes allowing them the privileged right to hospitality which they called *dhimma*.

Often mistranslated as "tolerance", *dhimma* really means "right to hospitality", a "protection" that the caliphs sometimes gave to non-Islamic residents. They did also show a certain "tolerance" towards their conquered peoples, sometimes even "allowing" them to profess a different religion. But this tolerance had its limits. The expression of ideas contrary to official dogma, and even more, living by non-orthodox ideas, was vigorously repressed. Non-believers were often considered to be "internal emigrants" and not permitted to rise to the same rank as Muslims. The "Pact of 'Umar" even forced Jews and Christians to wear a "circlet": a round piece of cloth, yellow for the former and blue for the latter. Conversion to Islam offered a number of social, pecuniary and fiscal advantages.

Even the brilliant culture of the Kharezm Province was discriminated out of existence, as al-Biruni (a native of Kharezm) explained: "Thus Qutayba did away with those who knew the script of Kharezm, who understood the country's traditions and taught the knowledge of its inhabitants; he submitted them to tortures so that they were wrapped up in shadows and no one could know (even in Kharezm) what had (preceded) or followed the birth of Islam" [Youschkevitch].

The case of the Maghreb and especially that of Islamic Spain (before the virulence of the Almohads) do still prove that "tolerance" was practised for almost six centuries, in the sense of a greater liberty for Jews and for *Mozarabes* ("Arabic" Christians) who lived peacefully according to their own philosophies, organisations and traditions, with their synagogues, churches and convents [V. Monteil (1977)].

Thus the Christian scholars of the Arabic world often worked as "catalysts" by collecting, translating and commenting on, in Syriac or Arabic, many scientific and philosophical works of Greek or Indian origin. Amongst these men were: the astrologer Theophilus of Edesse, who translated many Greek medical texts into Syriac; the doctor Ibn Bakhtyashu, head of the Jundishapur hospital; the doctor Salmawayh ibn Bunan; the astronomer Yahya ibn Abi Mansur; the doctor Massawayh al-Mardini; the philosopher, doctor, physician, mathematician and translator Qusta ibn Luqa, from Baalbek; and the translators Yahya ibn Batriq, Hunayn ibn Ishaq, Matta ibn Yunus and Yahya ibn 'Adi.

As for Jewish intellectuals, or those issued from Judaism, it is worth mentioning the astronomer Ya'qub ibn Tariq, one of the first scholars of the Empire to study Indian astronomy, arithmetic and mathematics;

astronomers Marshallah and Sahl at Tabari; the astrologer Sahl ibn Bishr; the mathematician converted to Islam As Samaw'al ibn Yahya ibn 'Abbun al-Maghribi, who continued al-Karaji's work on algebra; and the converted doctor and historian Rashid ad din, who compiled a history of China.

There was also the philosopher-rabbi Abu 'Amran Musa ibn Maymun ibn 'Abdallah, better known as Rabbi Moshe Ben Maimon or Maimonides, whose encyclopaedic interests embraced not only philosophy, but also mathematics, astronomy and medicine. Born in Cordoba in 1135, he was initially one of the victims of religious persecution at the hands of the Almohad sovereigns, who forced him to proclaim himself a Muslim for sixteen years. The rabbi remained a Jew, and at the end of this period of time, he went first to Fez, then to Palestine, before settling in Egypt where he became a doctor at the court of the Fatimids in Cairo, until his death in 1204. He wrote many works on medicine (*Aphorisms of Medicine, Tract of Conservation and of the Regime of Health* and *Rules of Morals* being the only ones to have survived). These works were mainly concerned with:

> the treatment of haemorrhoids (a surgical operation which should only be carried out as a last resort), of asthma by a dry climate, nervous depression or "melancholy" through psychotherapy: recovery being seen as a return to equilibrium; and diets, all embraced by a global vision of the human being, always presented in a spirit of compassion and charity [V. Monteil (1977)].

He also wrote the famous *Moreh Nebukhim* (*Guide for the Lost*), in which his Aristotelian philosophy seeks to reconcile faith and reason, and he asserts himself as one of the first intermediaries between Aristotle and the doctors of scholasticism. Another of his fundamental contributions, this time to Judaism, was his *Commentary on Mishna* (1158–1165) and his *Second Law or the Strong Hand* (1170–1180). Before they were even translated into Hebrew or Latin, the medical and philosophical works of Maimonides were generally written in Arabic first. In other words, despite their profound attachment to Judaism, scholars such as Maimonides were authentic Arabic thinkers.

After the Abbasid school of Baghdad (ninth to eleventh century CE), there came the schools of Toledo and Seville, and Jewish scholars such as Yehuda Halevi, Salomon ibn Gabirol (Avicebron) and Abraham ibn Ezra or Abraham bar Hiyya (who would have spoken Hebrew, Arabic, Castilian and even Latin or Greek) acted as intermediaries between the Arabic and Christian worlds.

Of course, Arabic science was also and above all the creation of Muslim scholars. Amongst these men were: al-Fazzari, al-Kindi, al-Razzi, al-Khuwarizmi, Thabit ibn Qurra, al-Battani, Abu Kamil, al-Farabi, al-Mas'udi, Abu'l Wafa, al-Karayi, al-Biruni, Ibn Sina (Avicenna), Ibn al-Haytham, 'Umar al-Khayyam, Ibn Rushd (Averroes) and Ibn Khaldun (see the Chronology, pp. 519ff. for further information).

Islamic religion played an important role in the scientific discoveries of this civilisation. The Koran preaches humanism in the search for knowledge; one of the necessities imposed by the study of this holy book and of Islamic thought is "the development of scientific research where Revelation, Truth and History are considered in their dialectic relationship as structural terms of human existence" [M. Arkoun (1970)].

The Koran frequently encourages the faithful to look for signs of proof of their faith in the heavens and on Earth: "Search for science from the cradle to the grave, even if you have to travel as far as China . . . Those that follow the path of scientific research will be led by God on the path to Paradise" [L. Massignon and R. Arnaldez in HGS]. (We have not been able to find the source for this advice, which many attribute to Mohammed. But it would seem that it forms an integral part of Islamic culture, at least since the time of Ibn Rushd.)

It is true that the science in question here is knowledge of religious Law (*'ilm*), but in Islam this is not set apart from secular science. Thus we find a whole series of *hadith* about medicine, remedies and the legitimacy of their use. Moreover, scholars and philosophers have not hesitated to quote the texts in order to defend their activities.

Averroes wrote in his *Authoritative Text*: "It is clear in the Koran that the Law invites rational observation of living beings in the search for an understanding of these beings through reason."

This is the opinion of all Muslims who have accepted and cultivated science. It is because the Koran invites the faithful to contemplate the power of Allah in the organisation of the universe that astrology has always been considered the "highest, noblest and most beautiful of sciences" [al-Battani] in the Islamic world.

The patient assimilation of observations and calculations relative to the positions of the planets, the moon's phases, equinoxes, etc., were often directly related to the demands of Islamic religion: the calculation of the exact times of the ritual prayer of the *'asr*, the dates of religious ceremonies, the month of Ramadan, orientation towards Mecca, etc.

This is why, despite the considerations above, the science and culture of the Islamic world should be more accurately termed "Arab-Islamic".

## THE SPREAD OF SCIENTIFIC KNOWLEDGE: ANOTHER ACHIEVEMENT OF ISLAM

Other sciences existed before Islam (in Ancient Greece, Persia, India, China, etc.), but although these were all mainly concerned with the same problems, they all had their own unique way of dealing with them. In other words, before Islam, there was no universal science as we know it today.

In fact, different cultures sought to preserve their knowledge and keep it a secret from the outside world. An example of this is the Neo-Pythagoreans in Greece.

Part of the reason why the Muslims were responsible for the unification of science is their success in conquering other peoples.

International trade played an important role, as did the Arabic geographers, travellers and cosmographers, translators, philologists, lexicographers and writers of commentaries:

By describing different areas of the globe, those unusual men described the marvels of nature, products of the earth, fauna, agriculture and crafts. This was a considerable source of information. Some geographers were also great scholars, experts in all fields, such as the famous al-Biruni [Massignon and Arnaldez].

Another factor was the cultural assimilation by the Muslims of the most diverse of cultures: this began at the time of the caliphs of Damascus, but it was not until the time of the enlightened caliphs of Baghdad and Cordoba that the results were felt.

The "tolerance" of these caliphs towards other cultures, beliefs, customs and traditions for nearly six centuries was also an important factor.

The promotion of study and research in the Koran has already been mentioned in this chapter. This was not only a fundamental condition for the development of Arabic Islamic science, but also one of the main causes for Islam's ready acceptance of the most diverse of cultures. (But it should also be noted that Arab-Islamic science, despite its universal nature, was always oriented towards knowledge of divine Law. It is necessary to wait until the European Renaissance before science gains the non-religious character we now recognise.)

## THE DEVELOPMENT OF ARABIC ISLAMIC SCIENCES

The Islamic conquerors were not always in favour of science and culture.

Caliph 'Umar (634–644 CE) ordered the destruction of countless works seized in Persia. His argument was as follows: "If these books contain the key to the truth, Allah has given us a more reliable way to find it. And if these books contain certain falsehoods, they are useless" [see A. P. Youschkevitch (1976)].

There were certainly other similar cases of religious or xenophobic opposition, leading to great cultural losses. In the Islamic world, scholars suffered from the whims of totalitarian leaders. They had to avoid direct confrontation with official dogma if they did not want to lose their state subsidies and risk even greater repression. At the end of the eleventh century, the famous poet, astronomer and mathematician Omar Khayyam reported, in his *Mathematical Treatise*:

> We have witnessed a decline in scholarship, few scholars are left, and those who remain experience vexations. Their troubled times stop them from concentrating on deepening and bettering their knowledge. Most so-called scholars today mask the truth with lies.
>
> In science, they go no further than plagiarism and hypocrisy and use the little knowledge they have for vile material ends. And if they come across others who stand apart for their love of the truth and rejection of falsehood and hypocrisy, they attack them with insults and sarcasm.

But according to Youschkevitch, "this situation could not in the long term stop the triumph of scientific progress. Schools, libraries and observatories were built in the cities. To make a name for themselves, enlightened sovereigns set up academies similar to those founded by European monarchs in the seventeenth and eighteenth centuries. The transmission of knowledge was thus assured; but it was only later that the discovery of printing facilitated it."

However, such opinions were exceptional and not held by caliphs ruling later in Islam. In fact, the role of Islam and of Arabic scholars in the fields of science and culture can never be overstated.

The famous library of Alexandria, the most important one of Ancient Greece, was pillaged and destroyed twice: the first time in the fourth century CE by the Christian Vandals, and the second time (through a perverse paradox of history) by the Muslims in the seventh century. Many original manuscripts of inestimable value disappeared. Many Greek literary and scientific masterpieces would have been lost forever if they had not been collected and translated into Arabic.

It was thanks to the work of the North African philosopher Ibn Rushd (Averroes) that Saint Thomas Aquinas could study and understand the importance of Aristotle's work. Similarly, it is thanks to Avicenna that Albertus Magnus could develop the philosophy of universality. It is largely thanks to the work of Arabic translators that the works by Ancient Greeks are known to us today.

Moreover, the Arabs have never hesitated to underline the importance of Greek science and to express their admiration for it: "The language of the Hellenistic people is Greek; it is the most vast and the most robust of languages" [Sa'id al-Andalusi, *Tabaqat al umam*, in R. Taton (1957), I, p. 432].

Greek culture played a huge part in the development of Arabic science.

But it would be a mistake to believe that the latter was nothing more than the continuation of Greek science. This would be as far-fetched as the opinion that "India, and not Greece, formed the religious ideals of Arabia and inspired its art, literature and architecture".

Of course the framework of Arabic scientific thinking constituted an obvious extension of, and was largely based upon Hellenic science. However, the Arabs used the discoveries of Ancient Greece as a source of inspiration and actually expanded upon them. Moreover, Greece was not the only civilisation to inspire the Arabic scholars. They were also interested in oriental culture, from which they borrowed different elements which they adopted to suit their needs.

Thus their numeral alphabet was forged from a combination of Jewish, Greek and Syriac systems by adopting the corresponding principle to the twenty-eight letters of their own written alphabet.

Through the Christians of Syria and Mesopotamia they discovered the place-value system of the Babylonians, which they used in their tables and their astronomical texts to record sexagesimal fractions. Through trading with Persia and parts of the Indian sub-continent, they also came into contact with Indian civilisation. Thus they discovered Indian arithmetic, algebra and astronomy. Sa'id al-Andalusi (see above) expressed his admiration for Indian culture. He recognised its precedence over Islam and went as far as to call it "a mine of wisdom and the source of law and politics". He also wrote that "the Indian scientists devote themselves to the science of numbers (*'ilm al 'adad*), to the rules of geometry, to astronomy and generally to mathematics . . . they are unrivalled in medicine and the knowledge of treatments". His conclusion, however, is a little subjective. He claims that the intellectual talents and qualities of the Indians are nothing more than the product of "good fortune (*hazz*)", due to "astral influences" [R. Taton (1957), I, p. 432].

The Chinese were another foreign influence. After the battle of Talas in 751, the Arabs learned the secrets of making paper from linen or hemp from their Chinese prisoners. The first factory was built in Baghdad c. 800. It would be another four centuries before paper was used in Europe, through the intermediary of Spain.

At the beginning of the fourteenth century, Rashid ad din, grand vizier of the Mongolian sultan Ghazan Khan a Tabriz, and himself a converted Jew, compiled a library of 60,000 manuscripts, many of which came from Chinese and Indian sources.

In his *Universal History* (*Jami'at tawarikh*), he carefully described how Chinese characters were engraved on wood, and gave their transcription in Arabic. He translated extracts from the best known medical works of China and Mongolia into Arabic and Persian, including *Mejing*, a text on sphygmology (or science of the arterial pulse) by Wang Shuhe (265 – 317), which identifies four standard methods of medical examination, namely observation, auscultation, interrogation and palpation. These would not be studied in Western Europe until the eighteenth century [see V. Monteil (1977)].

However, the Arabs were not content merely to preserve the discoveries of Greece, Babylon, China and India. They wanted to make their own contribution to the world of science.

As they carefully collected, translated and studied works from the past, they added various commentaries, after mixing explanations with original developments, and always maintaining a critical perspective which rejected fixed dogmatism [see A. P. Youschkevitch (1976)].

Thus in mathematics, Greek methods were often mixed with Indian methods, sometimes with Babylonian ones or even, at a later date, with Chinese approaches.

The Arabs combined the strict systematisation of Greek mathematics and philosophy with the practicality of Indian science. This enabled them to make significant progress in the fields of arithmetic, algebra, geometry, trigonometry and astronomy. Through collecting, propagating and teaching the use of Indian numerals and calculation, and by pushing the study of certain remarkable properties of numbers towards the first seeds of a theory of numbers, the Arabs made considerable progress in the field of arithmetic.

Scholars such as al-Khuwarizmi, Abu Kamil, al-Karaji, As Samaw'al al-Maghribi, 'Umar al-Khayyam, al-Kashi and al-Qalasadi led arithmetic towards algebrised operations.

The Arabs (and more generally the Semites) "personalised" the number. Instead of an object which had various properties, it became an active being. They did not see numbers as being static and limited, as the Greeks did. The Arabs were interested in the ordinal, rather than the cardinal numbers: they were not deterred as the Greeks were by the irrational [see L. Massignon and R. Arnaldez].

The assimilation of the classical heritage allowed the mathematicians of Islamic countries to develop algorithms and corresponding problems, and thus achieve a higher level than that reached by Indian or Chinese mathematicians. It also enabled them to find more efficient ways to resolve and generalise these problems than the Chinese and the Indian methods. Where the latter were content to establish a specific rule of calculation, the mathematicians of Islam often managed to develop an entire theory [A. P. Youschkevitch (1976)].

In short, the work of the Arabic scholars involved objectivity, the questioning of the doctrines of the ancient scientists and systematic recourse to analysis, synthesis and experimentation.

The progress of sciences, in terms of knowledge, is dependent on the scientific mind . . . Perhaps some thought that all of science had already been discovered, and that all that remained to do was to assimilate all the information. But this gathering of knowledge was in fact an excellent prelude to methodical research and progress. The need for an inventory led to classifications of the sciences (such as those of al-Farabi or Avicenna) which was enough to cause an evolution of the concept of science. Under the influence of Plato and Aristotle, the Ancient Greeks classified the sciences according to their method, and the degree of intelligibility of their purpose. The Arabs took a more straightforward stance: the sciences exist and they must be ordered so that none is forgotten. The lack of conceptual analysis which characterises Arabic classification of the sciences was in fact an advantage from a purely scientific point of view. Knowledge itself promotes learning and marks out the direction to follow towards the acquisition of further knowledge [L. Massignon and R. Arnaldez].

For the Arabs, then, to know was not to contemplate, but to do; in other words to verify, challenge, experiment, observe, rethink, describe, identify, measure, correct, even complete and generalise. This is the Arab influence on science: it became an operating science following the development of "scientific reason". The Arabs had a great deal of curiosity, love and estimation for knowledge [L. Massignon and R. Arnaldez], which meant that they not only preserved and transmitted the science of Antiquity, but they transformed it and established it along new lines, giving it a new lease of life and originality.

## THE ARABIC LANGUAGE: THE AGENT AND VEHICLE OF ISLAMIC SCIENCE

Right from the start of the history of Arab-Islamic science, anything concerning the science had to be written in Arabic if it was to be of any consequence, this language having become the permanent intellectual link between the scholars of various origins during this long period of time.

For many philosophers, mathematicians, physicians, chemists, doctors and astronomers, this language was more than a mere obligation: it was a real passion. It was the preferred language for expressing science and knowledge.

For example, the Persian scientists Avicenna and al-Biruni, rather than writing in Turkish or Persian wrote in Arabic, despite having been born in Kharezm, to the north of Iran in what is now Uzbekistan (formerly in the USSR). Al-Biruni explains his preference for Arabic in his *Kitab as saydana* ("*Treatise on Drugs*" [see V. Monteil (1977), p. 7]:

> It is in Arabic that, through translation, the sciences of the world were transmitted [to us] and were embellished and found a place in our hearts. The beauty of the Arabic language has circulated with them in our arteries and veins. Of course, every nation has its own language, the one used for trading and talking to our friends and companions. But personally I feel that if a science found itself eternalised in my own mother-tongue, it would be as surprised as a camel finding itself in a gutter of Mecca or a giraffe in the body of a thoroughbred. When I compare Arabic with Persian (and I am equally competent in both languages) I admit that I prefer invective in Arabic to praise in Persian. You would agree with me if you saw what happens to a scientific work when it is translated into Persian: it loses all its brilliance and has less impact, it becomes muddled and quite useless. Persian is only good for transmitting historical stories about kings or telling tales at evening gatherings.

Of course, al-Biruni's description of Farsi is totally unjustified. Many Muslim scholars from Persia, Afghanistan and the Indus Valley also wrote in Arabic, although Persian is perfectly capable of expressing any concept, as well as the rigour, nuances and foundations of scientific thought. However, al-Biruni's preference for Arabic was not brought about by chance, and was certainly not due to a passing fad.

In terms of structure, Arabic became a much richer language and gradually acquired its scientific character in order to meet the demands of the translation of foreign works and the transposition of scientific texts.

When a scientific text is translated from its original language into an equally well-equipped language, there might be grammatical problems but there are no technical or conceptual difficulties. This was not the case when Greek was first translated into Arabic: vocabulary had to be created, or existing words adjusted to meet the needs of science. There was often an intermediate Syriac word which prepared the way for Arabic. The creation of the scientific Arabic language was not only philological, it also involved two scientific methods: the identification and verification of concepts [L. Massignon and R. Arnaldez].

It was in this scientific spirit that the lexicographers made an inventory of the Arabic language, as scholars had made an inventory of knowledge by attempting to classify different fields of learning through rethinking and evaluating concepts, then organising them in relation to one another. As for those who translated or commented on texts, they looked for Arabic equivalents for foreign terms in lexicons and in nature, and also in the different elements of knowledge, either to introduce new words and concepts, or even to express new ideas using the most ancient vocabulary.

This is how Arabic acquired its unique aptitude for expressing scientific thought, and for developing it in the service of the exact sciences.

This language, which was originally considered to be the language of the Revelation and the fundamental criterion for anyone wishing to belong to the Muslim religion and the Islamic community (*Umma*), became not only the vehicle of international science, but also and above all the essential agent of the Renaissance and the dominant factor in the Arabic scientific revolution.

## OTHER ARABIC CONTRIBUTIONS TO THE WORLD OF SCIENCE

The Arabs also contributed significantly in the field of technology, developing upon the knowledge passed down by the Ancient Greeks.

The Greek school (which had turned out such prestigious scholars as Archimedes, Ctesibios, Philon of Byzantium and Hero of Alexandria) had seen the discovery of quite advanced mechanical technology: the endless screw, the hollow screw, pulley blocks, mobile pulleys, levers and weapons; clepsydras (types of clocks activated by water); astrolabes (for observing the positions of the stars and determining their height above the horizon); the construction of automata (devices capable of moving by themselves); the use of the odometer (an instrument designed to measure distances, comprising a series of chains and endless screws, moved by chariot wheels and pulling a needle along a graduated scale which indicated the distance travelled); etc. [see A. Feldman and P. Fold (1979); B. Gille (1980, 1978); C. Singer (1955–7); D. de Solla Price (1975)].

The Greeks of Byzantium carried on the work of the Greeks of Alexandria, and, to a certain extent, were one of the transmission links with mediaeval Europe.

However, the handing-on of the Greek tradition was also and above all the work of the engineers of the Muslim world. Here again, the Arabs gathered all the information, then made improvements and even innovations. Under orders from the caliph Ahmad ibn Mu'tasim, Qusta ibn Luqa al-Ba'albakki translated Hero of Alexandria's work on the traction of heavy bodies into Arabic; others translated or were inspired by the work of Philo of Byzantium [see B. Gille (1978)]. The Arabs also distinguished themselves in the art of clock-making. They even created their own inventions, above all in the field of automata and astronomical clocks, this being not only the legacy of the Greeks but probably also the Chinese, who were likewise experts in this field.

The following were amongst the most famous of the Arabic-Muslim engineers: the Banu Musa ibn Shakir brothers, whose works notably include *Al'alat illati tuzammi bi nafsiha* (*The Instrument Which Plays Itself*, written c. 850), largely inspired by the ideas of Hero of Alexandria; Ibn Mu'adh Abu 'Abdallah al-Jayyani, who wrote *Kitab al asrar fi nata'ij al afkar*, which describes several water clocks (second half of the tenth century); Badi'al-Zaman al-Asturlabi, famous for the automata he built for the Seleucid monarchs (first half of the twelfth century); 'Abu Zakariyya Yahya al-Bayasi, known for his mechanical pipe organs (second half of the twelfth century); and Ridwan of Damascus, made famous by his automata activated by ball-cocks (1203).

The most famous and most productive of the Arabic engineers was Isma'il ibn al-Razzaz al-Jazzari, whose *Kitab fi ma'rifat al hiyat al handasiyya* (*Book of the Knowledge of Ingenious Mechanical Instruments*, 1206) shows a perfect knowledge of Greek traditions and records apparently hitherto unknown innovations. This work not only contains the plans for constructing perpetual flutes, water clocks, mechanical pump systems for fountains, automata activated by ball-cocks and movements transmitted by chains and cords, it also contains descriptions of sequential automata, mainly using camshafts, thus transforming the circular movement of a type of crankshaft into the alternating movement of distribution instruments.

As well as the diverse instruments, there is also the astrolabe which became known in the West (at the same time as the "Arabic" numerals) thanks to Pope Sylvester II (Gerbert of Aurillac), who acquired the astrolabe from the Arabs when he lived as a monk in Spain from 967 to 970 CE.

There was also the compass, that ingenious device which has a magnetic needle and made navigation possible. It was invented by the Chinese at the beginning of the Common Era and was retrieved by the Arabs (in all likelihood in 752 during the battle of Talas), who in turn passed it on to the Europeans during the Renaissance.

The scholars of Islam also made their mark on the science of optics, which led to the invention of the mirror:

> Optics was particularly studied by Ibn al-Haytham. His work included physiological optics and a philosophical discussion of the nature of light, but he is known above all for his research in the field of geometry. He knew about reflection and refraction; he experimented with different mirrors, planes, spheres, parabolas, cylinders, both concave and convex. He wrote a text about the measurement of the paraboloid of revolution. He embarked on actual physical research through his work on the light of the stars, the rainbow, the colours, shadows and darkness. For a scientist of this calibre there was no fixed distinction between mathematical sciences and natural sciences, and Ibn al-Haytham was always shifting between the two [L. Massignon and R. Arnaldez].

Alchemy, too, was a fanciful art, the aim of which was to find the so-called "philosopher's stone" from which could be extracted a miraculous property which would at once cure all illnesses, give eternal life and transform metals: it was a vain science whose basis was denounced by great minds such as al-Kindi, Avicenna and Ibn Khaldun. However, as Diderot pointed out, alchemy, in spite of its frivolous nature, "often led to the discovery of important truths on the great path of the imagination". By stripping it of some of its arithmology and magic, the early Arabian scholars began to prepare the way for the creation and expansion of modern chemistry.

## THE FORERUNNERS OF CONTEMPORARY SCIENCE?

Certain Arabic scholars, such as al-Biruni and Averroes, and doctors such as 'Ali Rabban at Tabari and Ibn Massawayh were well ahead of their time.

Perhaps the most significant contribution of the Arabic world, however, came from the historian 'Abd ar Rahman ibn Khaldun (who was born in Tunis in 1332 and died in Cairo in 1406), a visionary of modern science, who was gifted with truly extraordinary insight. One only need read this extract from his *Prolegomena* to appreciate his foresight: "The human world is the next step after the world of apes (*qirada*) where sagacity and wisdom may be found, but not reflection and thought. From this point of view, the first human level comes after the ape world: our observation ends here" [see *Muqaddimah*, p. 190; V. Monteil (1977), p. 101].

This is a surprising opinion for a time when such ideas were practically inconceivable. It would not be until 1859, in Darwin's *Origin of Species*, that these ideas would be presented and even then some time elapsed before they were accepted and developed in the Western world.

Thus, we can see how much Europe owes to this civilisation which is largely unknown or at least unrecognised by the Western public.

## SIGNIFICANT DATES IN THE HISTORY OF ARABIC-ISLAMIC CIVILISATION

The following chronology is divided into sections, each representing half a century in the golden age of Arab-Islamic civilisation. Its aim is to give an idea of cultural, literary, scientific and technical activity which ran parallel to military and religious events. The list (which, of course, is not exhaustive) is of scholars and intellectuals, including the most illustrious poets, writers, mathematicians, physicians, astronomers, geographers, engineers, chemists, naturalists and doctors of the Arab world. In some cases, a summary of their fundamental contributions is supplied [see A. A. al-Daffa (1977); M. Arkoun (1970); O. Becker and J. E. Hoffman (1951); E. Dermenghem (1955); EIS; O. Fayzoullaiev (1983); A. Feldman and P. Fold (1979); L. Frédéric (1987 and 1989); L. Gille (1978); C. Gillespie (1970–80); L. Massignon and R. Arnaldez in HGS; A. Mazaheri (1975); A. Mieli (1938); V. Monteil (1977); R. Rashed (1972); G. Sarton (1927); C. Singer (1955–7); H. Suter (1900–02); G. J. Toomer; K. Vogel (1963); H. J. J. Winter (1953); A. P. Youschkevitch (1976)].

*Second half of the sixth century*

c. 571 CE. "Year of the Elephant". Supposed birth-date of the prophet Mohammed.

*First half of the seventh century*

612. Year of the "Revelation", when Mohammed began his prophecy in Mecca.

622. Mohammed and the first followers of the new faith, the "Muslims" (*al muslimin*, from the Arabic word "believers") were expelled from Mecca. They found refuge in Yahtrib, which then became the "Town" of the Prophet or "Medina" (*madinah*). This year marked the beginning of the Muslim calendar, which is called the Hegira (from *hijra*, "expatriation").

624. Battle of Badr. The *qibla* is established, the symbol of the "new people of God". Beginning of the "Muslim institutions".

628. Mecca is seized by Mohammed and his followers.

632. The death of Mohammed.

632–661. Time of the "orthodox" caliphs (Abu Bakr, 'Umar, 'Uthman and 'Ali); capital: Medina.

632–634. Abu Bakr is caliph, the successor of Mohammed.

634. The conquest of Syria by the Arabs, who defeat Heraclius's Byzantine army near Jerusalem.

634–644. 'Umar (Omar) is caliph.

635. The Arabs take Damascus and overturn the Persian Empire.

637. Battle of Kadisiya, defeat of the Persians.

637–638. Founding of the towns Basra and Kufa. The writing of the Koran begins.

637–640. Conquest of Mesopotamia, Khuzistan, Azerbaidjan and Media.

638. Jerusalem is surrendered to Omar.

639. Arabs attack Armenia.

640. The conquest of Palestine.

641. Egypt is conquered by the Arabs.

642. Victory over the Persians.

642–646. The Arabs attack Armenia.

643. The Arabs complete their conquest of Persia and Tripolitania, and arrive in Sind (now Pakistan).
644–656. Rule of 'Uthman (Ottman).
647. Barka in Tripolitania is taken (now Libya).
649. Cyprus is conquered by the Arabs.

*Second half of the seventh century*

655. Battle of Lycia, where the Muslim fleet destroys the Byzantine fleet.
656–661. Rule of 'Ali, the son-in-law of the prophet.
657. Battles of Jamal and Siffin, where the followers of 'Ali (then considered to be the first man converted to Islam) fought the followers of Mu'awiyah (rival and hostile descendants of Mohammed's family).
661–750. The Omayyad Dynasty. Capital: Damascus. Rule henceforth becomes hereditary. Effort to centralise Omayyad administration.
665. First attacks in the Maghreb.
670. Successful campaigns in North Africa. Founding of the town of Qairawa (Kairouan, in present-day Tunisia). Appearance and beginning of Shiite and Kharajite movements.
673–678. Siege of Constantinople by the Arabs.
680. Death of Husayn in Kerbala. Martyrdom of the Shiites.
695. First use of coins by the Arabs.

*Culture, Science and Technology*
Period of:

- the poet Imru' al-Qays.
- the poet Yahya ibn Nawfal al-Yamani.
- Khalil ibn Ahmad (one of the founders of Arabic poetry).

*First half of the eighth-century*

707. Development of political, "courtly" and urban poetry. First theological-political discussions.
707–718. The Muslims seize the mouths of the River Indus (Sind) and part of the Punjab (India).
709. The Maghreb surrenders to Arabic domination.
711. Musa Ben Nusayir dispatches Tariq ibn Ziyad, who crosses the Straits of Gibraltar (called Jabal Tariq), then successively occupies Seville, Cordoba and Toledo, before continuing north.
712. Arabic conquest of Samarkand (now Uzbekistan).
715. The Arabic Empire extends its confines to China and the Pyrenees.
718. The Arabs meet resistance from the Byzantine army at Constantinople. Thus the Arabic advance comes to a halt at the Taurus mountains.
720. The Arabs cross the Pyrenees and penetrate the kingdom of the Franks. First Arabic colony in Sardinia.
732. The Arabs are defeated at Poitiers by Charles Martel; the end of the Arabic advance in Europe.

*Culture, Science and Technology*
Period of:

- the Christian doctor Yuhanna ibn Massawayh.
- the poets al-Farazdaq, al-Akhtal and Jarir.
- the mystic thinker Hasan al-Basri.
- the Arabic version of the *Kalila wa Dimna* fables by Ibn al-Muqafa' (ancient Persian tale inspired by the Indian **Pañchatantra*).
- the first paintings of Islamic art.

*Second half of the eighth century*

750. Abu'l 'Abbas founds the Dynasty of the same name.
750–1258. Abbasid Dynasty. Capital: Baghdad (from 772).
751. Battle of Talas in present-day ex-Soviet Kyrghyzstan, where the Chinese armies are defeated by Arab troops. But Chinese reprisals later stop the Arabic advance at the limits of Sogdania.
754–775. Reign of the Abassid caliph al-Mansur.
756–1031. Omayyad Dynasty in Spain. Capital: Cordoba.
760. Arabic expedition against Kabul (in Afghanistan).
761–911. Rustamid Dynasty in Tiaret.
762. Caliph al-Mansur founds the town of Baghdad.
768. Sind is governed by the Arabs.
786. The Arabs seize Kabul.
786–809. Reign of the Abassid caliph Harun al-Rashid.
786–922. Idrissid Dynasty in the Maghreb. Capital: Fez.
795. Disorder in Egypt.

*Culture, Science and Technology*
Period of:

- the introduction of Indian science, numerals and calculation to the Islam world.
- the Persian astronomers Abu Ishaq al-Fazzari, and Muhammad al-Fazzari (his son), and of Jewish astronomer Ya'qub ibn Tariq. These are the men who would translate the *Brahmasphutasiddhânta* by Brahmagupta and study, for the first time in Islam, Indian astronomy and mathematics.
- the Persian astrologer al-Nawabakht and his son al-Fadl, chief librarian of caliph Harun al-Rashid.
- the Jewish astronomer Mashallah.
- the Christian Abu Yahya, translator of *Tetrabiblos* by Ptolemy.

- the Persian Christian Ibn Bakhtyashu', first of a large family of doctors, head of the hospital at Jundishapur.
- the Sabaean alchemist Jabir ibn Hayyan (Gebir in mediaeval Latin) who studied chemical reactions and bonds between chemical bodies.
- the alchemist Abu Musa Ja'far al-Sufi who wrote that there are two types of distillation, depending on whether or not fire is used.
- the Christian astrologer Theophilus of Edesse, translator of Greek works.
- the philologist and naturalist al-Asma'i.
- Abu Nuwas, one of the greatest Arabic poets.
- the poet Abu al-'Atahiya.
- the mystic thinker Abu Shu'ayb al-Muqafa.
- and the first production of paper in Islam.

*End of the eighth century*

At this time, the provinces of Africa, the Maghreb and Spain freed themselves from the links with the caliph of Baghdad.

*Ninth – tenth century*

This was the time of the development of the Sunni (Hanbali, Maliki, Hanafi, Shafi, Mutazili, Zahiri, etc.) and Shiite (Immami, Zayidi, Ismaeli, etc.) religious movements and of the mystical philosophy of the Sufi; popular Islam prevailed over classical Islam, which was reduced to a few common cultural and religious signs. This time was also marked by the rapid development of Arab-Islamic civilisation in all fields. It was also the era when the *Alf layla wa layla*, the *Thousand and One Nights* was written (anonymous masterpiece of Arabic literature, a collection of tales and legends, such as those of Scheherazade, Ali Baba, Sinbad the sailor, the magic lamp, etc., which have become an integral part of universal mythology).

*First half of the ninth century*

800. Charlemagne is named Emperor of the West.
800–809. Aghlabite Dynasty in "Ifriqiya" (territory composed of present-day Tunisia and part of Algeria).
813–833. Reign of the Abassid caliph al-Ma'mun who, as a grand patron, would favour cultural and scientific translations.
820–999. Independent indigenous dynasties in eastern Persia: Tahirid (820–873), Saffarid (863–902), Samanid (874–999).
826. Crete is taken by the Arabs.
827–832. Sicily is taken.
846. Sacking of Rome by the Saracens.

*Culture, Science and Technology*

Period of:

- the founding of the "House of Wisdom" (*Bayt al-Hikma*) in Baghdad, a kind of academy of sciences, where the cultural heritage of Antiquity was welcomed with enthusiasm and where the development of Arab-Islamic science began.
- the Persian astronomer and mathematician al-Khuwarizmi. His work on the Indian place-value system and on algebra with quadratic equations contributed greatly to the knowledge and propagation of Indian numerals, methods of calculation and algebraic procedures, not only in the Muslim world but also in the Christian West. He also wrote an interesting series of problems with examples taken from the methods of merchants and executors which required a great deal of mathematical skill due to the complex structure of the legacies of the Koran.
- the mathematician 'Abd al-Hamid ibn Wasi ibn Turk.
- the Christian translator Yahya ibn Batriq.
- al-Hajjaj ibn Yusuf, translator of Euclid's *Elements*.
- the astronomer and mathematician al-Jauhari, who carried out some of the first work on the parallel postulate.
- the converted Jewish astronomer Sanad ibn 'Ali, who had the observatory built in Baghdad.
- the philosopher al-Nazzam.
- the great philosopher and physician al-Kindi, who was interested in logic and mathematics, and sought to analyse the essence of definition and demonstration; he also wrote about geometrical optics and physiology.
- the philosopher al-Jahiz, author of the famous *Book of Animals*.
- the Persian Christian astronomer Yahya ibn Abi Mansur, who drew up *Al zij al muntahan* (*Established Astronomical Tables*).
- the astronomer Abu Sa'id al-Darir, from the Caspian region, who wrote about the course of the meridian.
- the astronomer al-Abbas, who introduced the tangent function.
- the astronomer Ahmad al-Nahawandi of Jundishapur.
- the astronomer Hasbah al-Hasib, from Merv, who established a table of tangents.
- the astronomer al-Farghani, who wrote an Arabic version of Ptolemy's *Almagestus*.
- the astronomer al-Marwarradhi, from Khurasan.
- the astronomer 'Umar ibn al-Farrukhan, from Tabaristan.
- the Jewish astronomer Sahl at Tabari, from Khurasan.
- the Jewish astronomer Sahl ibn Bishr, from Khurasan.
- the astrologer Abu Ma'shar, from Balkh (Khurasan).
- Ali ibn 'Isa al-Asturlabi, famous maker of astronomical instruments.

- al-Himsi, who made the work of Apollonius known.
- the Banu Musa ibn Shakir brothers, translators, mathematicians and engineers, who wrote a work on automata.
- Ibn Sahda, who translated medical works.
- the Christian doctor Jibril ibn Bakhtyashu'.
- the Christian doctor Salmawayh ibn Bunan.
- the surgeon Abu'l Qasim az Zahrawi (Abulcassis in mediaeval Latin), from Cordoba.
- the Christian pharmacologist Ibn Massawayh, author of *Aphorisms*.
- the writer As Suli.
- the doctor and philosopher 'Ali Rabban at Tabari, author of *Paradise of Wisdom*, inspired by the *Aphorisms* of the Indian Brahman heretic Chanakya of the third century BCE.
- the mystical thinkers Dhu 'an Nun Misri, al-Muhasibi, Ibn Karram, Bistami.
- and the poets Abu Tammam and Buhturi.

*Second half of the ninth century*

868–905. Tulunid Dynasty in Egypt and Syria.
869. Malta is taken by the Arabs.
875–999. Samanid Dynasty in the north and east of present-day Iran, Tadjikistan and Afghanistan. Capital: Bokhara.
880. Italy is recaptured from the Arabs by Basil I.

*Culture, Science and Technology*
Period of:

- al-Mahani the geometer and astronomer from the region of Kirman, who studied the problems of the division of the sphere using the cubic equation which bears his name.
- al-Nayrizi (Anaritius in mediaeval Latin), astronomer and mathematician from the Shiraz region, who wrote commentaries on Euclid and Ptolemy.
- the Egyptian mathematician Ahmad ibn Yusuf, who wrote a work dealing with proportions.
- the mathematician Thabit ibn Qurra, who translated Archimedes's treatise on the sphere and the cylinder and who did important work on conic sections; he also produced a brilliantly clear proof of Pythagoras's theorem, the first general rule for obtaining pairs of amicable numbers* and a method for constructing magic squares.

* Two numbers are "amicable" if the sum of the distinct divisors of each one (including 1 but excluding the number itself) is equal to the other number. For instance, 220 has divisors 1, 2, 4, 5, 10, 11, 20, 22, 44, 55, 110, which add up to 284; while 284 has divisors 1, 2, 4, 71, 142, which add up to 220. The numbers 220 and 284 form an "amicable pair", and they are the smallest to do so.

- the mathematicians Abu Hanifa Ahmad and al-Kilwadhi.
- al-Battani (Albategnus in mediaeval Latin) the astronomer who accompanied his theory of planets with insights into trigonometry, which were later to be thoroughly investigated by Western astronomers; he determined the inclination of the ecliptic and the precession of the equinoxes with great accuracy using cotangents.
- the astronomer Hamid ibn 'Ali.
- the Persian astrologist Abu Bakr.
- Qusta ibn Luqa al-Ba'albakki, the Christian mathematician and engineer of Greek origin, who in particular translated Hero of Alexandria's *Mechanics* which deals with the traction of heavy objects, as well as works by Autolycus, Theodosius, Hypsicles and Diophantus.
- the Christian doctor Hunayn ibn Ishaq, who translated Greek medical works into Arabic, as well as works by Archimedes, Theodosius and Menelaus.
- the Christian Yahya ibn Sarafyun, who wrote a medical encyclopaedia in Syriac.
- the pharmacologist Sabur ibn Sahl, from Jundishapur, author of a book of antidotes.
- Muhammad Abu Bakr Ben Zakariyya al-Razi (Razhes in mediaeval Latin) the great Persian clinician, alchemist and physician who was thought to be the greatest doctor of his age; he first differentiated between German measles and measles; he described how to equip a chemical "laboratory" and his *Sirr al Asrar* (*The Secret of Secrets*), contains important work on distillation.
- the philosopher Abul Hasan 'Ali ibn Ismail al-Ash'ari, founder of Muslim scholasticism and of the Mutaqallimin school. He expounded a theological system based on an atomism similar to that of Epicurus.
- the geographer al-Ya'qubi.
- the Persian geographer Ibn Khurdadbeh, alias Ibn Hauqal, author of the *Book of Routes and Provinces*.
- the mystical thinker Tirmidhi, known as "the philosopher".
- the poets Mutanabbi and Ibn Sa'ad.

*First half of the tenth century*

905. End of Tulunid Dynasty in Egypt, power taken by the governors of the caliphs.
909. Beginning of the rule of the Fatimid caliphs in Ifriqiya.
932–1055. The Buyid Dynasty, unifying eastern Persia and Media.
935. Muhammad ibn Tughaj reconquers Alexandria and southern Syria.
943. The Caliphate of Baghdad confers the rule of Egypt to Ibn Tughaj for thirty years.

945. The Buyids enter Baghdad. The Caliphate is now no more than a "legal fiction".

*Culture, Science and Technology*
Period of:

- Abu Kamil, the great algebraist from Egypt, who continued the work of al-Khuwarizmi, and whose algebraic discoveries were to be used, c. 1206, by the Italian mathematician Leonard of Pisa (or "Fibonacci"); also devised interesting formulas related to the pentagon and decagon.
- the geometer Abu 'Uthman, translator of the tenth book of Euclid's *Elements* and of Pappus's *Commentary*.
- the Christian translators Matta ibn Yunus and Yahya ibn 'Adi.
- Sinan ibn Thabit, mathematician, physician, astronomer and doctor.
- the mathematician Ibrahim ibn Sinan ibn Thabit, who dealt with the problem of constructing conic sections, and who studied the surface of the parabola and conoids.
- the mathematician Abu Nasr Muhammad, who made an interesting discovery with his theorem of sines in plane and spherical trigonometry.
- the mathematician Abu Ja'far al-Khazini, from Khurasan, who worked on algebra and geometry, and who solved al-Mahani's cubic equation by using conic sections.
- the astronomer al-Husayn Ben Muhammad Ben Hamid, known as Ibn al-Adami.
- the astrologist and mathematician al-Imrani, who wrote a commentary on Abu Kamil's algebra.
- the arithmeticians Ali ibn Ahmad and Nazif ibn Yumn al-Qass.
- Bastulus, the famous maker of astronomical instruments.
- the great geographer and mathematician al-Mas'udi.
- the geographer Qudama.
- the geographer Abu Dulaf.
- the geographer Ibn Rusta of Isfahan.
- the Persian geographer Ibn al-Faqih, from Hamadan.
- the geographer Abu Zayd, from Siraf (Arabic-Persian gulf).
- the geographer al-Hamdani, from the Yemen.
- the philosopher al-Farabi (Alpharabius in mediaeval Latin), from Turkestan, who devised a metaphysics based on Aristotle, Plato and Plotinus and who, in his *Ihsa' al 'ulum*, came up with a "Classification of the Sciences" in five branches: linguistics and philology; logic; mathematical sciences, subdivided into arithmetic, geometry, perspective, astronomy, mechanics and gravitation; physics and metaphysics; and finally the political, legal and theological sciences.
- the alchemist and agronomist known as Ibn Wahshiya.
- the mystic thinkers Junayd and Abu Mansur ibn Husayn al-Hallaj.
- the poet Ibn Dawud.
- and the Persian poet Rudaki.

*Second half of the tenth century*

957. The Byzantines in northern Syria.
961–969. The Byzantines reconquer Crete and Cyprus, as well as Antioch and Aleppo (Syria).
962. A Turkish tribe conquers the Afghan kingdom of Ghazna.
969. The Fatimids of Tunisia occupy Egypt, which puts up no resistance, then settle there.
972–1152. Zirid and Hammadid Dynasties in Ifriqiya.
973. Foundation of *Al Kahira* (Cairo).
998–1030. Reign of Mahmud, or "the Ghaznavid" (because he settled in Ghazna), over what is now Afghanistan, Khurasan and various annexed regions in the north of India.

*Culture, Science and Technology*
Period of:

- the founding in Cairo of the *Dar al Hikma* (House of Wisdom), a sort of scientific academy similar to that of Baghdad.
- the founding of the al-Azhar university in Cairo.
- the blossoming of the sciences in the Caliphate of Cordoba, thanks to Caliph al-Hakam II, who put together an immense library.
- the mathematician Abu'l Wafa' al-Bujzani, from Quistan, who wrote commentaries on Euclid, Diophantus and al-Khuwarizmi. He introduced the tangent function, and his work on trigonometry led to great improvements in methods of solving spherical triangles where, instead of Ptolemy's formula (derived from Menelaus's theorem) which involved the six great-circle arcs of a quadrilateral, a formula involving the four arcs generated on a transversal of the figure composed of a spherical triangle and the perpendiculars dropped from two of its vertices to opposite sides is used. Thanks to him, the Arabs acquired Diophantus's *Arithmetica* with its studies of algebra and number theory.
- the mathematician al-Uqlidisi (whose name means "the Euclidean") who published an important study of decimal fractions.
- the mathematician Ibn Rustam al-Kuhi, from Tabaristan, who studied geometrical problems posed by Archimedes and Apollonius.
- the Persian mathematician/astronomer Abu'l Fath from Isfahan, who revised the Arabic translation of much of Apollonius's work.

- the mathematician al-Sijzi, from Sigistan, who studied problems of conic intersections and the trisecting of angles.
- the mathematician al-Khujandi, from the Sir Daria region, who established a proof concerning the sine in spherical triangles, and proved that the sum of two cubes cannot equal a cube.
- the mathematicians Sinan ibn al-Fath and Abu Nasr.
- the secret society of the Brothers of Purity (*Ikhwan al-Safa*), whose *Epistles* were a sort of encyclopaedia based on Pythagorean and Neo-Platonic mysticism, and which divided the sciences into four sorts: mathematics; science of physical bodies; science of rational souls; and science of divine laws.
- the Andalusian astronomer and mathematician Maslama ibn Ahmad, based in Cordoba.
- the Persian astronomer and mathematician 'Abd ar-Rahman al-Sufin, who drew up a catalogue of stars containing the first observation of the Andromeda nebula.
- the great doctor and surgeon Abu'l Qasim from Zakna, near Cordoba, author of the *Kitab al-Tasrif*, which deals with practical surgery, cauterising wounds, tying up arteries, operating on bones, the eyes, etc.
- 'Ali ibn 'Abbas, a doctor from southern Persia.
- Abu Mansur Muwaffak, a doctor who wrote an important medical treatise in Persian.
- the Andalusian doctor Ibn Juljul.
- the Persian geographer al-Istakhri, from near Persepolis.
- the geographer Buzurg ibn Shahriyyar, from Khuzistan.
- the Palestinian traveller and geographer al-Muqaddasi, from Jerusalem.
- the philosopher Abu Bakr Ahmad ibn 'Ali al-Baqilani.
- the historian Ya 'qub ibn al-Nadim, author of *Kitab al-Fihrist al 'ulum* (*The Book and Index of Sciences*) containing biographies of contemporary thinkers.
- the mechanical and hydraulic engineer Ibn Mu'adh Abu 'Abdallah al-Jayyani, author of an important treatise on water clocks.
- and the mystic thinker Tawhidi.

*First half of the eleventh century*

1000. First clashes in Khurasan between the Ghaznavids and the Seljuks (Turkomans pushed out of Central Asia by the Chinese).

1001–1018. Mahmud the Ghaznavid takes possession of Peshawar, crushes a Hindu coalition and sacks Muttra, one of India's holy cities.

1030. The Ghaznavids crushed by the Seljuks.

1030–1050. The Seljuks occupy various towns in eastern then western Persia (where they clash with the Buyids), before turning away towards Syria and Asia Minor.

*Culture, Science and Technology*

Period of:

- the mathematician, astronomer, physician and geographer al-Biruni, from Khiva in Kharezm, who travelled widely in India, where he learnt Sanskrit and became acquainted with Indian science; he later took back what he had learnt and wrote numerous works on astronomy, arithmetic and mathematics; he also made a new calculation of trigonometric tables based on Archimedes's premises, an equivalent of Ptolemy's theory.
- the mathematician al-Karaji, who did important work on the arithmetic of fractions; basing himself on the work of Diophantus and Abu Kamil, he devised an algebra in which, alongside the standard forms of second degree equations, he dealt with certain 2n degree equations; his work showed how a rigorous approach can, by using irrational numbers, arrive at forms that are more supple than those of Greek geometric algebra; this was, in fact, the start of a development which would lead to the elimination of geometrical representations in Arabic arithmetic and algebra thanks to the use of symbols.
- the mathematician Kushiyar ibn Labban al-Gili, from the south of the Caspian Sea, who worked on Indian arithmetic and sexagesimal calculations.
- the Persian mathematician An Nisawi, from Khurasan, who continued the work of al-Khuwarizmi in arithmetic and algebra.
- the mathematician Abu'l Ghud Muhammad ibn Layth.
- the mathematician Abu Ja'far Muhammad ibn al-Husayn.
- the astronomer Ibn Yunus, appointed to the Dar al-Hikma observatory in Cairo.
- the mathematician, physician and doctor Ibn al-Haytham (Alhazen in mediaeval Latin) whose *Book of Optics* contains important discoveries about eyesight, the theory of the reflection and refraction of light, the fundamental laws of which he established.
- the Andalusian mathematician al-Kirmani, from Cordoba.
- the Andalusian mathematician and astronomer Ibn al-Samh, from Granada.
- the mathematician and astronomer Ibn Abi'l Rijal (Abenragel in mediaeval Latin), from Cordoba but working in Tunis.
- the Andalusian mathematician and astronomer Ibn al Saffar, from Cordoba.

• the great philosopher Al Husayn ibn Sina (Avicenna), a universal mind as interested in medicine as in mathematics; based on Aristotle's ideas, his philosophy rejected mysticism and theology and dwelt instead on science and nature; his *Canon Medecinae* remained a textbook in Europe until the seventeenth century; in his *Aqsam al 'ulum al 'aqliyya* (or *Division of the Rational Sciences*) he drew up a consistent classification by means of an analytical division to allow a hierarchy of the sciences to be established.
• the Christian philosopher Miskawayh whose rational thought makes him one of Ibn Khaldun's precursors.
• the chemist al-Kathi.
• the Christian doctor Massawayh al-Mardini, settled in Cairo.
• the doctor 'Ali ibn Ridwan, from Cairo.
• the doctor Abu Sa'id 'Ubayd Allah.
• the Andalusian doctor Ibn al-Wafid, from Toledo.
• the Jewish doctor Ibn Janah, author of a treatise on medicinal herbs.
• the doctor Ibn Butlan.
• the oculist 'Ammar.
• the oculist 'Ali ibn 'Isa, author of an important treatise on ophthalmology.
• the jurist and poet Ibn Hazem.
• the atheist Syrian poet Abu'l 'ala al-Ma'ari.
• and the Persian poet Abu'l Qasim Firduzi, author of the famous *Book of Kings*.

*Second half of the eleventh century*

1050. Troubled times in Egypt. Order re-established by Badr al-Jamali who then ruled over Egypt until 1121 for the Fatimids.
1055. Tughril Beg, the Seljuk, enters Baghdad as the protector of the Empire of the Caliphs.
1055–1147. Almoravid Dynasty in the Maghreb.
1062. Yusuf Ben Tashfin (the true founder of modern Morocco) founds Marrakech, which becomes the Almoravids' capital.
1069. Yusuf Ben Tashfin takes Fez, an Arab-Islamic centre, then develops its intellectual, artistic and economic activities.
1076. The Seljuks take Damascus and Jerusalem.
1085. The Christians occupy Toledo.
1086. Faced with a dangerous situation in Andalusia, Yusuf Ben Tashfin declares a "Holy War" against Christian Spain, stops the activities of Alphonsus VI of Castile, then annexes the whole of the south of Spain, uniting it with the Maghreb and thus forming the Almoravid Empire.
1090. The Turks arrive between the Danube and the Balkans.
1096. Start of the First (People's) Crusade. The badly organised crusaders are massacred in Asia Minor.
1097–1099. The crusaders take Nicea, defeat the Turks at Dorylaeum then take Jerusalem.

*Culture, Science and Technology*
Period of:
• the great Persian poet and mathematician 'Umar al-Khayyam (Omar Khayyam), from Nishapur, author of the *Rubaiyat*, the famous collection of poems; he also wrote commentaries on Euclid's *Elements*, worked on the theory of proportions and studied third-degree equations, suggesting geometric solutions for some of them.
• the mathematician Yusuf al-Mu'tamin (the enlightened king of Saragossa).
• the mathematician Muhammad ibn 'Abd al-Baqi.
• the Andalusian astronomer al-Zarqali, from Cordoba, who reworked Ptolemy's *Planisphaerium*.
• the poet, philosopher, mathematician and astronomer al-Hajjami, who played a vital part in reforming the calendar; he also provided an overview of third-degree equations and made an important study of Euclid's postulates.
• the Persian oculist Zarrin Dast.
• the Andalusian geographer and chronicler al-Bakri, from Cordoba.
• the doctors Ibn Jazla and Sa'id ibn Hibat Allah.
• the Andalusian agronomist Abu 'Umar ibn Hajjaj, from Seville.
• the mystic philosopher Abu Hamid al-Ghazali (Algazel in mediaeval Latin), whose teachings stood against Islam's scientific progress.
• the "sociologist" al-Mawardi.
• and the Persian poet Anwari.

*First half of the twelfth century*

1100. Foundation of the Christian Kingdom of Jerusalem.
1125. Revolt of the Masmudas of the Atlas under Ibn Tumert, the inventor of the Almohad doctrine.
1136. Cordoba, Western Islam's cultural capital, is taken by Ferdinand III, king of Castile and Leon.
1147. 'Abd al-Mu'min, the successor of Ibn Tumert, destroys the power of the Almoravids and proclaims himself Caliph after taking Fez (in 1146) and Marrakech (in 1147). He then extends his conquests to Ifriqiya and reaches Spain.

1147–1269. The Almohad Dynasty in the Maghreb and Andalusia.
1148. The crusaders defeated at Damascus.

*Culture, Science and Technology*
Period of:

- the Andalusian mathematician Jabir ibn Aflah, from Seville, particularly famous for his work on trigonometry.
- the great Andalusian geographer al-Idrisi, who made important contributions to the development of mathematical cartography.
- the Jewish mathematician from Spain Abraham Ben Meïr ibn 'Ezra (better known as Rabbi Ben Ezra).
- the engineer Badi al-Zaman al-Asturlabi, famous for the automata he made for the Seljuk kings.
- the philosopher Ibn Badja (Avempace in mediaeval Latin and during the Renaissance).
- the philosopher and doctor Abu al-Barakat, author of the *Kitab al mu'tabar* (*Book of Personal Reflection*).
- and the Andalusian philosopher Ibn Zuhr (alias Avenzoar).

*Second half of the twelfth century*

1150. Allah ud din Husayn, Sultan of Ghur, destroys the Ghazni Empire.
1169–1171. Salah ad din (Saladin), a Muslim of Kurdish origins, succeeds his uncle as Vizier of Egypt then ends the reign of the Fatimids by recognising only the suzerainty of Nur ad din, the unifier of Syria, and the Abbasid Caliph of Baghdad.
1174. Salah ad din succeeds Nur ad din and founds the Ayyubid Dynasty which dominated Egypt and Syria thenceforth. Leaning on the Arab traditionalists, he declares "Holy War" against the Christians of the West, hence reinforcing the links between the eastern peoples and their Arab-Islamic traditions.
1187. Salah ad din takes back Jerusalem. Victory of the Almohads at Gafsa under the Maghribi Sultan Abu Yusuf Ya'qub al-Mansur.
1188. Genghis Khan unifies the Mongols.
1191. Under Muhammad of Ghur, the Islamic Afghan and Turkish tribes of Central Asia try to conquer the north of India, but are pushed back at the very gates of Delhi.
1192. Battle of Tarain: Muhammad of Ghur defeats Prithiviraj and takes Delhi.
1192–1526. Sultanate of Delhi.
1193. The Muslims take Bihar and Bengal.
1195. Victory of the Almohads at Alarcos.

*Culture, Science and Technology*
Period of:

- the mathematician al-Amuni Saraf ad din al-Meqi.
- the converted Jewish mathematician, philosopher and doctor As Samaw'al ibn Yahya al-Maghribi, from the Maghreb, who continued the work of al-Karaji.
- the Persian encylopaedist and mathematician Fakhar ad din al-Razi.
- the Persian engineer Abu Zakariyya Yahya al-Bayasi, famous for his mechanical pipe organs.
- the great Jewish philosopher Maimonides (Rabbi Moshe Ben Maimon), from Cordoba, whose encyclopaedic interests included astronomy, mathematics and medicine.
- the great Andalusian philosopher Ibn Rushd (Averroes), born in Cordoba and died in Marrakech, the finest flowering of Arab philosophy and a profound influence on the West.
- the Maghrebi philosopher Ibn Tufayl (Abubacer in mediaeval Latin and during the Renaissance).
- the mystical thinker Ruzbehan Baqli.
- the Persian mystical poet Nizami.
- and the Persian poet Khaqani.

*First half of the thirteenth century*

1202. The Muslims arrive on the banks of the Ganges, at Varanasi (Benares).
1203. Continuation of Muhammad of Ghur's conquest of northern India.
1206–1211. Reign of Qutb ud din (Sultanate of Delhi).
1208. The Albigensian Crusade.
1212. Defeat of the Almohads at Las Navas de Tolosa.
1211–1222. Under Genghis Khan, the Mongols invade China, Transoxiana and Persia, before continuing their migration under Hulagu Khan towards Mesopotamia and Syria.
1211–1227. Reign of Iltumish (Sultanate of Delhi) who obtains recognition of his authority over India from the Caliph of Baghdad. Under this domination, India will remain relatively stable until 1290.
1214–1244. The Banu Marin (Merinids) conquer the north of the Maghreb.
1221. The Mongols press against the borders of the Sultanate of Delhi, but are held back by Iltumish.
1227. Death of Genghis Khan, whose empire stretched from the Pacific to the Caspian Sea.

1248. The Christians take back Seville from the Muslims.

*Culture, Science and Technology*
Period of:

- the mathematician Muwaffaq al din Abu Muhammad al-Baghdadi.
- the leading court official and patron Abu'l Hasan al-Qifti, author of *Tarikh al huqama* (*Chronology of the Thinkers*).
- Muhammad ibn Abi Bakr, famous maker of astronomic instruments.
- the engineer Ridwan of Damascus, best known for his ball-cock automata.
- the great Persian engineer al-Jazzari, author of the *Book of the Knowledge of Ingenious Mechanical Instruments*, in which he provided plans for perpetual flutes, water clocks, and different sorts of sequential automata using ball-cocks and camshafts.
- Ya'qub ibn 'Abdallah ar Rumi, who produced an important encyclopaedia of Arab geography.
- the esoteric Muslim Ibn 'Arabi.
- and the poets Ibn al-Farid and Shushtari.

*Second half of the thirteenth century*

1250. The Mamelukes take power in Egypt. The Kingdom of Fez created by the Banu Marin (Merinids).
1254–1517. Reign of the Mamelukes in Egypt.
1258. Hulagu Khan's Mongols retake and sack Baghdad.
1259. Mongol invasion of Syria.
1260. Mongols crushed at the border of Egypt by the Mameluke monarch.
1261. Egypt becomes the centre of the Arab world and also, to a certain degree, of the Islamic world.
1269. In the Maghreb, the Banu Marin take Marrakech and found their own dynasty (the "Merinids").
1291. The Mamelukes take Acre and eliminate the Christians on the Syrian–Palestinian coast.
1297. 'Ala ud din Khalji (Sultanate of Delhi) defeats the Mongols then starts sacking Gujarat and Rajasthan.

*Culture, Science and Technology*
Period of:

- the mathematician and astronomer Nasir ad din at Tusi, from Tus in Khurasan, who did important work on arithmetic, algebra and geometry; his work undoubtedly marks the high point of Arabic trigonometry, dealing thoroughly with spherical right-angled triangles and successfully broaching the study of spherical triangles in general, even bringing in the polar triangle; in astronomy, he published his famous "Ikhanian" Tables; and, in geometry, he corrected the translations of Greek geometrical works and his discussion of Euclid's propositions was later to inspire the Italian mathematician Saccheri in his initial research in 1773 into non-Euclidean geometry.
- the doctor Ibn al-Nafis of Damascus, who wrote a commentary on Avicenna's *Canon*, with important developments concerning pulmonary circulation.
- the pharmacologist and botanist Ibn al-Baytar.
- the Persian mystical poet and hagiographer Farid ad din 'Attar.
- the Persian poet Sa'adi of Chiraz, author of the *Gulistan*.

*First half of the fourteenth century*

1306. The Sultans of Delhi repulse the Mongols once more.
1307–1325. The Sultans of Delhi attack the kingdoms of Deccan and reach the south of India, conquering the lands of the Maratha, Kakatiya and Hoysala.
1333. The Moors recapture Gibraltar from the Kingdom of Castile.

*Culture, Science and Technology*
Period of:

- the great Maghrebi arithmetician Ibn al-Banna al-Marrakushi.
- the converted Jewish doctor and historian Rashid ad din, author of a *Universal History*, in which he reproduced large extracts of the best-known medical works of China and Mongolia.
- the historian al-Umari.
- the moralist Ibn Taymiyya.
- the Andalusian mystic thinker Ibn Abbad of Ronda.
- the great Maghrebi traveller Ibn Battutua who, in thirty years, covered more than 120,000 kilometres in the Islamic world, from Northern Africa to China, via India.
- and the Persian poets Hamdallah al-Mustawfi and Tebrizi.

*Second half of the fourteenth century*

1356. India is "given" by the Caliph of Baghdad to Firuz Shah Tughluq.
1371. The Ottomans defeat the Serbs at Chirmen.
1389. The Ottomans crush the Serbs at Kosovo Polje.
1390. The Ottomans occupy the remaining territories of the Byzantine Empire in Asia Minor.

1392. The Ottomans arrive in the Balkans.
1398–1399. Timur (Tamerlane) sacks Delhi.

*Culture, Science and Technology*
Period of:

- the great thinker Ibn Khaldun, from Tunis, remarkable for his rationalism, his feeling for general laws and his extraordinarily acute scientific insights; in many ways a precursor of Auguste Comte.
- the writer Ibn al-Jazzari.
- the writer Taybugha.
- and the Persian poet Hafiz of Chiraz, author of *Bustan*.

*First half of the fifteenth century*

1400–1401. Incursion of Timur and sacking of Baghdad.
1405. Return to Baghdad of the Jalayrid leaders.
1422. The Ottomans besiege Constantinople.
1400–1468. Constant disputes between Turkomans and Mongols.
1444. The Viceroy of the Baghdad Timurid Dynasty founds his empire in Mesopotamia and Kurdistan.
1447. End of the empire of Timur, independence of Persia and of the Afghan and Indian regions.

*Culture, Science and Technology*
Period of:

- Ulugh Bek, the enlightened monarch of Samarkand, builder of an observatory equipped with the finest instruments of the age; author of trigonometric tables, among the most precise of the numeric tables produced by Islam's thinkers.
- the Persian mathematician Ghiyat ad din Ghamshid ibn Mas'ud al-Kashi, who did important work on algebra, sexagesimal calculations and arithmetic, especially on the binomial formula, decimal fractions, exponential powers of whole numbers, n roots, the theory of proportions and irrational numbers.
- the historian al-Maqrizi.

*Second half of the fifteenth century*

1453. Constantinople falls to Sultan Mehmet II. The beginning of the Ottoman Empire, which will later cover Anatolia, Rumelia, Bulgaria, Albania, Greece, the Crimea, Syria, Mesopotamia, Palestine, Egypt, Hejaz, Armenia, Kurdistan and Bessarabia and which, after 1520, will extend its frontiers as far as Hungary, southern Mesopotamia, the Yemen, Georgia, Azerbaidjan, with Tripoli and the whole of Ifriqiya as dependencies (excepting the Maghreb which managed to remain autonomous during this period).
1468. The Turkoman al-Koyunlu establishes his authority in Mesopotamia.
1492. The Catholics Ferdinand and Isabella retake Granada.
1499–1722. Reign of the Safavids in Persia; Shi'ism becomes the official religion.

*Culture, Science and Technology*
Period of:

- the mathematician al-Qalasadi, who did important work on arithmetic, especially algebra, greatly developing its symbols.
- and the Persian historian Mirkhond.

*The sixteenth century*

1508. The Safavids push the Turkomans out of Mesopotamia.
1516. Turkish corsairs establish themselves in Algiers.
1517. Ottoman conquest of Syria and Egypt, thus ending the Caliphate of Baghdad and bringing about the fall of the Mamelukes in Egypt.
1524. Babur, a descendant of Timur, invades the Punjab and takes Lahore.
1526. Babur kills the last Sultan of Delhi and takes the throne. The beginning of the Mogul Empire in India and Afghanistan (1526–1707).
1571. Turks defeated by Holy League in naval battle of Lepanto.
1574. The central and eastern regions of North Africa come under Ottoman control.
1578–1603. Beginning of the Sa'adian Dynasty with the reign of al-Mansur (Maghreb).

*Culture, Science and Technology*
Period of:

- the Turkish arithmetician Tashköprüzada.
- and the Turkish poets Baki and Fuzuli.

*The seventeenth century*

1672–1727. Beginnings of the 'Alawite Dynasty in the Maghreb with the reign of Mulay Ismail, contemporary of Louis XIV.

*Culture, Science and Technology*
Period of:

- the mathematician Beha ad din al-Amuli.
- the arithmetician and commentator al-Ansari.
- the writers Hajji Khalifa, 'Abd al-Qadir al-Baghdadi, 'Abd ar Rashid Ben 'Abd al-Ghafur and Ad Damamini.
- the encylopaedist Jamal ad din Husayn Indju.

- the Turkish poets Nefi, Nabî and Karaja Oghlan.
- and the Turkish traveller and writer Evliya Chelebi.

*The eighteenth century*

1799. Start of the *Nahda* ("Renaissance").

*Culture, Science and Technology*
Period of:

- the Turkish poet Nedim.
- the Turkish writer and historian Naima.

*The nineteenth century*

1804. The Wahhabis take Mecca and restore Hanbali Islam.
1805–1849. Reign of Muhammad 'Ali, Pasha of Egypt.
1811–1818. 'Ali defeats the Wahhabis.

*Culture, Science and Technology*
Period of:

- the Turkish thinkers Namik Kemal, Ziya Pasha, Ahmet Mithat, Chinassi and Avdülhak Hamit.

*Beginning of the twentieth century*

1918–1922. Reign of Sultan Mehmed IV (whom the Treaty of Sèvres obliged to accept the dismemberment of the Turkish Empire. Turkey was reduced to the landmass of Anatolia).
1922. Mehmed IV overturned by Mustafa Kemal, founder of modern, republican Turkey.
1924. Official end of the Ottoman Empire.

## THE ARRIVAL OF INDIAN NUMERALS IN THE ISLAMIC WORLD

How were Indian numerals and calculating methods introduced into Islam?

The Arabs possibly encountered them at the beginning of the eighth century CE, when Hajjaj sent out an army under Muhammad Ben al-Qasim to conquer the Indus Valley and the Punjab.

But it is far more likely that the army had nothing to do with it, and that it was necessary to wait for a delegation of scholars before Indian science was transmitted to the Islamic world.

This is, indeed, Ibn Khaldun's explanation, who says in his *Prolegomena* that the Arabs received science from the Indians, as well as their numerals and calculation methods, when a group of erudite Indian scholars came to the court of the caliph al-Mansur in year 156 of the Hegira (= 776 CE) [see *Muqaddimah*, trans. Slane, III, p. 300].

This is a late source, dating from about 1390. But Ibn Khaldun's version corresponds closely with earlier texts, especially with a tale told by the astronomer Ibn al-Adami in about 900, which is referred to by the court patron Hasan al-Qifti (1172–1288) in his *Chronology of the Scholars*:

> Al-Husayn Ben Muhammad Ben Hamid, known as Ibn al-Adami, tells in his Great Table, entitled *Necklace of Pearls*, that a person from India presented himself before the Caliph al-Mansur in the year 156 [of the Hegira = 776 CE] who was well versed in the *sindhind* method of calculation related to the movement of heavenly bodies, and having ways of calculating equations based on *kardaja* calculated in half-degrees, and what is more various techniques to determine solar and lunar eclipses, co-ascendants of ecliptic signs and other similar things. This is all contained in a work, bearing the name of Fighar, one of the kings of India, from which he claimed to have taken the *kardaja* calculated for one minute. Al-Mansur ordered this book to be translated into Arabic, and a work to be written, based on the translation, to give the Arabs a solid base for calculating the movements of the planets. This task was given to Muhammad Ben Ibrahim al-Fazzari who thus conceived a work known by astronomers as the *Great Sindhind*. In the Indian language *sindhind* means "eternal duration". The scholars of this period worked according to the theories explained in this book until the time of Caliph al-Ma'mun, for whom a summary of it was made by Abu Ja'far Muhammad Ben Musa al-Khuwarizmi, who also used it to compose tables that are now famous throughout the Islamic world [F. Woepcke (1863)].

Much can be learned from this. The repetition of the word *sindhind* is significant; it is the Arabic translation of the Sanskrit **siddhânta*, the general term for Indian astronomic treatises, which contained a complete set of instructions for calculating, for example, lunar or solar eclipses, including the trigonometric formulae for true longitude [see R. Billard in IJHS]. The "*sindhind*" method thus stands for the set of elements contained in such treatises. As for the word *kardaja*, which is also frequently used, it means "sine" and derives from an Arabic deformation of the Sanskrit *ardhajyâ* (literally "semi-chord") which Indian astronomers had used, from the time of *Âryabhata, for this trigonometric function which is the basis of all calculations in the Indian *siddhânta* system.

This method is presented in the mathematician and astronomer Brahmagupta's (628) *Brahmasphutasiddhânta* and the astrologer *Varâhamihîra's (575) *Pañchasiddhântikâ*. But it was explained long before these treatises in the astronomer *Âryabhata's *Âryabhatîya* (c. 510).

Now, apart from the *Âryabhatîya* (which uses a special form of alphabetic numeration), all Indian astronomers noted their numbers by using Sanskrit numerical symbols: this notation gave them a solid base for noting numeric data and was based on a decimal place-value system using zero. As for their calculations, they used a system quite similar to our own one with their nine numerals plus a tenth sign written as a circle or point and acting as a true zero (see *Zero, etc).

In other words, when the Arabs learnt Indian astronomy, they inevitably came up against Indian numerals and calculation methods, so that the arrival of the two branches of knowledge precisely coincided. This is confirmed by al-Biruni's *Kitab fi tahqiq i ma li'l hind* (c. 1030), which tells of his thirty-year stay in India.

We must now try to date this transmission.

Now, al-Qifti, Ibn al-Adami and other authors agree on the date mentioned in the quotation above; i.e. 156 of the Hegira, or 776 CE. Several facts about Arabic science make this date plausible. According to A. P. Youschkevitch:

> If the arrival of Indian scholars gave the astronomers of Baghdad the possibility of acquainting themselves with the astronomy of the *siddhânta*, there was already much interest in the subject. Three astronomers who worked during the reign of Caliph al-Mansur are known to us, thanks to al-Qifti: Abu Ishaq Ibrahim al-Fazzari (died c. 777) who first made Arabic astrolabes, his son Muhammad (died c. 800), and finally Ya'qub ibn Tariq (died c. 796), who wrote works dealing with spherical geometry and who also compiled various tables.

All we now have to discover is which of the Indian *siddhânta* was adapted by al-Fazzari during the reign of al-Mansur. Now, the Fighar who is mentioned in the text is none other than Vyâgramukha (abbreviated to Vyâgra then deformed into Fighar), an Indian sovereign of the Châpa Dynasty who, according to an inscription, was defeated by Pulakeshin II, king of the Deccan in about 634. His capital was Bhillamâla (now Bhinmal), in the southwest of what is now Rajasthan. And it was precisely under the reign of Vyâgramukha, in the year 550 of the **Shaka* era (i.e. 628 CE), that *Brahmagupta composed his *Brahmasphutasiddhânta* (*Brahma's Revised System*) at the age of thirty.

Thus, one or other of the Indian scholars who arrived in Baghdad in 773 probably gave the caliph a copy of the *Brahmasphutasiddhânta*, along with other Sanskrit works.

It thus seems quite likely that not only Indian astronomy, but mathematics too, were introduced to the Muslims through the work of Brahmagupta.*

What led these Indian scholars to give such a present to al-Mansur? They had been kept for some time in his palace, which gave that enlightened monarch, with his lifelong thirst for knowledge, the opportunity to learn some Indian astronomy and arithmetic. Thus it was that these Brahmans, as worthy representatives of Indian culture, were led to demonstrate to him what seemed to them to be most important, original and ingenious in their science. They then, quite probably, gave the caliph copies of Brahmagupta's *Brahmasphutasiddhânta* and *Khandakhâdyaka*, which contained not only the *siddhânta* method, but also the principle of the decimal place-value system, the zero, calculation methods and the basics of Indian algebra.

It is easy to imagine the enthusiasm of al-Khuwarizmi, Abu Kamil, al-Karaji, al-Biruni, An Nisawi and others, too, who could appreciate the superiority of the Indians' place-value system and methods of calculation.

In his *Chronology of the Scholars*, Abu'l Hasan al-Qifti speaks of their admiration:

> Among those parts of their sciences which came to us, [I must mention] the numerical calculation later developed by Abu Ja'far Muhammad Ben Musa al-Khuwarizmi; it is the swiftest and most complete method of calculation, the easiest to understand and the simplest to learn; it bears witness to the Indians' piercing intellect, fine creativity and their superior understanding and inventive genius [F. Woepcke (1863)].

We must, in passing, admire this author's objectivity and lack of chauvinism, his ability to recognise the superiority of a discovery made by foreigners and his praise for a civilisation which had produced such a superior system to his own culture's.

* Even if Brahmagupta made some mistakes (he argued against the rotation of the earth demonstrated by Âryabhata in 520, for example), he was incontestably the greatest mathematician of the seventh century – a reputation he would keep for several centuries among Indian mathematicians and astronomers, and also among many Arabic-Islamic scholars, such as al-Biruni. His work, first presented in his *Brahmasphutasiddhânta* (628) then expanded in his *Khandakhâdyaka* (664), made considerable progress compared to earlier work, including that of Âryabhata and Bhâskara, particularly in algebra, one of his main innovations. Among his fundamental contributions can be cited his own system of a negative or zero arithmetic (with a clear and accurate statement of the rules of algebraic symbols), and his presentation of general solutions to quadratic equations with positive, negative or zero roots.

This quotation also leads us to look at one of the Islamic world's most famous mathematicians: al-Khuwarizmi, who was born in 783 in Khiva (Kharezm) and died in Baghdad in about 850 [see O. Fayzoullaiev (1983); G. J. Toomer in DSB; K. Vogel (1963)]. Little is known about his life, except that he lived at the court of the Abbasid caliph al-Ma'mun, shortly after the time when Charlemagne was made Emperor of the West, and that he was one of the most important of the group of mathematicians and astronomers who worked at the "House of Wisdom" (*Bayt al-Hikma*), Baghdad's scientific academy.

His fame is due to two works which made significant contributions to the popularisation of Indian numerals, calculation methods and algebra in both the Islamic world and the Christian West. One of them, *Al jabr wa'l muqabala* (*Transposition and Reduction*), dealt with the basics of algebra. It has come down to us both in its original Arabic and in Geraldus Cremonensis's mediaeval Latin translation, entitled *Liber Maumeti filii Moysi Alchoarismi de algebra et almuchabala*. This book was extremely famous, to such an extent that we owe to it the term for that fundamental branch of mathematics, "algebra". The first word of its title stands for one of the two basic operations which must be made before solving any algebraic equation. *Al jabr* is the operation of transposing terms in an equation such that both sides become positive; later compressed into *aljabr*, it was translated into Latin as "algebra", giving us the term we know today. As for *Al-muqabala*, it stands for the operation consisting in the reduction of all similar terms in an equation.

According to Ibn al-Nadim's *Fihrist*, al-Khuwarizmi's other work was called *Kitab al jami' wa'l tafriq bi hisab al hind* (*Indian Technique of Addition and Subtraction*). The original has, unfortunately, been lost but several post-twelfth century Latin translations of it survive. It is the first known Arabic book in which the Indian decimal place-value system and calculation methods are explained in detail with numerous examples. Like his other book, it became so famous in Western Europe that the author's name became the general term for the system. Latinised, al-Khuwarizmi first became *Alchoarismi*, then *Algorismi*, *Algorismus*, *Algorisme* and finally *Algorithm*. This term originally stood for the Indian system of a zero with nine digits and their methods of calculation, before acquiring the more general and abstract sense it now has.

Unbeknown to him, al-Khuwarizmi provided the name for a fundamental branch of modern mathematics, and gave his own name to the science of algorithms, the basis for one of the practical and theoretical activities of computing. What more can be said about this great scholar's influence?

FIG. 25.2. *Muhammad Ben Musa al-Khuwarizmi (c. 783–850). Portrait on wood made in 1983 from a Persian illuminated manuscript for the 1200th anniversary of his birth. Museum of the Ulugh Begh Observatory, Urgentsch (Kharezm), Uzbekistan (ex USSR). By calling one of its fundamental practices and theoretical activities the "algorithm", computer science commemorates this great Muslim scholar.*

| | 1 | 2 | 3 | 4 | 5 | 6 | 7 | 8 | 9 | 0 |
|---|---|---|---|---|---|---|---|---|---|---|
| Mathematical treatise copied in Shiraz in 969 by the mathematician 'Abd Jalil al-Sijzi. Paris, BN, MS. ar. 2547, f° 85 v-86 | | | | | | | | | | |
| Astronomocal treatise by al-Biruni (*Al Qanun al Mas'udi*), copied in 1082. Oxford, Bodleian, Ms. Or. 516, f° 12 v | | | | | | | | | | |
| Eleventh-century astronomical treatise. Paris, BN, Ms. ar, 2511, f° v 10, 14,19 | | | | | | | | | | |
| Eleventh-century astronomical tables. Paris, BN, Ms. ar. 2495, f° 10 | | | | | | | | | | |
| Twelfth-century astronomical treatise. Paris, BN, Ms. ar. 2494, f° 10 | | | | | | | | | | |
| Thirteenth-century copy of a ninth-century manuscript. Paris, BN, Ms. ar. 4457 f° 20 v | | | | | | | | | | |
| Kushyar ibn Labban's astronomical treatise, copied in 1203 in Khurasan. University of Leyden, Ms. al madkhal | | | | | | | | | | |
| Thirteenth-century astronomical tables. Paris, BN, Ms. ar. 2513, f° 2 v | | | | | | | | | | |
| A 1470 manuscript, Paris, BN, MS. ar. 601, f° 1 v | | | | | | | | | | |
| A1507 manuscript. University of Leyden, Cod. OR. 204 (3) | | | | | | | | | | |
| A 1650 manuscript from Istanbul. Princeton University, ELS 373 | | | | | | | | | | |
| Seventeenth-century work of practical arithmetic. Paris, BN, Ms. ar. 2475, f° 25, 26, 53 v | | | | | | | | | | |
| Seventeenth-century manuscript. Paris, BN, Ms. ar. 2460, f° 6 v | | | | | | | | | | |
| Seventeenth-century manuscript, Paris, BN, Ms. 2475, f° 91–94 | | | | | | | | | | |
| Modern characters | ١ | ٢ | ٣ | ۴ ٤ | ۵ ٥ | ٦ | ٧ | ٨ | ٩ | ٠ |

FIG. 25.3. *The "Hindi" numerals, used by Eastern Arabs*

## THE GRAPHIC EVOLUTION OF INDIAN NUMERALS IN EASTERN ISLAMIC COUNTRIES

When the Arabs learnt this number-system, they quite simply copied it (Fig. 25.3).

In the middle of the ninth century, the Eastern Arabs' 1 ( ), 2 ( ), 3 ( ), 4 ( ), 5 ( ), 6 ( ) and 9 ( ) could easily be confused with their Indian *Nâgarî* prototypes, thus:

1 2 3 4 5 6 7 8 9

But Arabic scribes gradually modified them, until they no longer resembled their prototypes (Fig. 25.3).

Such a development was a normal adaptation of the Indian models to the style typical of Arab writing. In other words, as they became integral parts of the writing system and associated with its graphic style, the Indian numerals gradually changed until they looked like a set of original symbols.

But these stylistic changes cannot explain everything. A close examination of Arab manuscripts, dating from the early centuries of Islam, shows that the Indian numerals became inverted.

And thus, in Islamic countries of the Near East:

| | | | | |
|---|---|---|---|---|
| Indian 1 ( ) | became: | | | |
| Indian 2 ( ) | became: | then: | and finally: | ٢ |
| Indian 3 ( ) | became: | then: | and finally: | ٣ |
| Indian 4 ( ) | became: | then: | and finally: | ٤ |
| Indian 5 ( ) | became: | then: | and finally: | ٥ |
| Indian 6 ( ) | became: | then: | and finally: | ٦ |
| Indian 7 ( ) | became: | then: | and finally: | ٧ |
| Indian 8 ( ) | became: | then: | and finally: | ٨ |
| Indian 9 ( ) | became: | then: | and finally: | ٩ |

This inversion came about for practical, material reasons.

During the early centuries of the Hegira, eastern Arabic scribes used to write the characters of their cursive script from top to bottom, rather than from right to left, in successive lines from left to right. They wrote somewhat as follows:

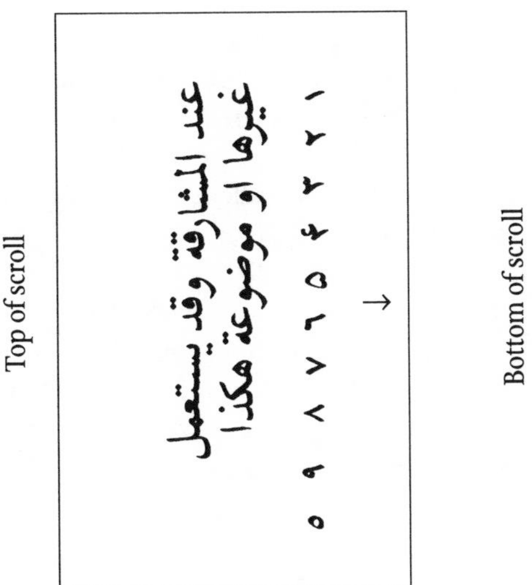


FIG. 25.4A.

Then to read, they turned their manuscript clockwise through 90°, so that the lines could be read from right to left:

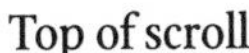


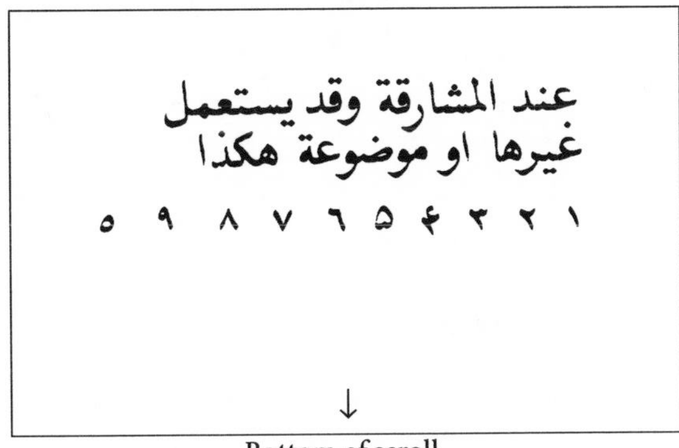


FIG. 25.4B.

This was the old custom of Aramaic scribes of the ancient city of Palmyra, perpetuated then transmitted to the Arabs by Syriac scribes [see M. Cohen (1958)].

It came about for the following reasons, essentially to do with manuscript writing on papyrus, which, until the ninth century, was widely used in the Islamic world.

First of all, stalks were cut into sections, the length of which determined the height of the sheet. The tissue was then cut open with a knife, hammered flat, then the strips thus obtained were laid side by side in two layers at right-angles to each other. They were then struck repeatedly. The finished sheets were glued along the longer sides so that the horizontal fibres were on one side (the facing page) and the vertical ones on the other. Once the horizontal fibres had been placed on the inside and the vertical ones on the outside, the sheet could be rolled up into a scroll [see L. Cottrell (1962)].

In order to write, Arabic scribes (like their Palmyrenean and Syriac predecessors) sat cross-legged, with their robe pulled up as a writing table.

Bearing in mind this position and the fragility of the sheet, it is easy to understand why scribes held their manuscripts lengthways, perpendicular to their bodies, with the head of the scroll to their left, thus writing their cursive script from top to bottom, in successive lines from left to right.

This explains the inversion of most Indian numerals in Arabic manuscripts dating from the early centuries of Islam.

As for zero, it was originally written as a "little circle resembling the letter 'O'," to borrow al-Khuwarizmi's explanation, who was referring to the Arabic letter *ha* ( ه ), shaped like a small circle [see A. Allard (1957); B. Boncompagni (1857); K. Vogel (1963); A. P. Youschkevitch (1976)].

Several Arabic manuscripts prove that this usage continued in certain places until the seventeenth century.

Here is a pun, typical of twelfth-century Arabic poetry. It occurs in two lines taken from the poem Khaqani composed in praise of Prince Ghiyat ad din Muhammad (c. 1155), to exhort him to free the province of Khurasan from its Oghuzz Turkoman invaders [see A. Mazaheri (1975)]:

Your enemy will be *mutawwaq* ("captured with a metal collar")
Like zero (*al sifr*) on the earthen tablet (*takht al turab*);
At his side will be the units ("of soldiers")
Like a sigh (*aah*) of regret.
It is true; among your subjects, your enemy is nothing.
If we did pay attention to him,
He would merely be a zero to the left of the figures (*arqam*).

The meaning of this fine passage is clearer if we consider that:

- the Arabic word for "sigh" is *aah*, composed of a double *alif* ( ا ) and a single *ha* ( ه );
- the first of these two letters looks like the vertical line representing the number 1, while the other resembles zero;
- the phrase "your enemy will be *mutawwaq*" means "your enemy will be captured with a metal collar, as the zero which is shaped like an O" the (hence, by extension: "your enemy will be imprisoned, then hanged").

The poet's metaphor thus plays on the graphic resemblance between the word *aah* (a sigh) and the numerical notation 011 to give the image of the leader of the opposing army being dragged by the neck (0) by the victorious troops (11):

| ه ا ا | ٥ ١ ١ |
|---|---|
| <······· | <······· |
| H A A | 0 1 1 |

These verses thus mean: "The Turkoman will have a chain round his neck, and be dragged by the troops in front of Sultan Muhammad."

This confirms that the small circle still stood for zero in the twelfth century in certain eastern provinces of the Muslim empire.

This is not surprising, for it is the *Shûnya-chakra* (the "zero-circle"), one of the Indian ways of depicting zero (see **Shûnya*; **Shûnya-chakra*; *Zero).

But, in the long term, this circle became so small that it was reduced to a point (Fig. 25.3).

The point is, in fact, the second way the Indians used to depict zero. It appeared at an early period in India and Southeast Asia (see **Shûnya*; **Bindu*; **Shûnya-bindu*; *Zero). Al-Biruni also speaks of this in his *Kitab fi tahqiq i ma li'l hind*, where he discusses Indian numerals and the Sanskrit numeric symbol system and lists the words symbolising zero: he cites the Sanskrit words **shûnya* ("vacuum", "zero") and **kha* ("space", "zero") before adding "*wa huma 'n naqta*" ("they mean 'point'") [see F. Woepcke (1863)].

To conclude, it was in this stylised and slightly modified form that the nine Indian numerals spread across the eastern provinces of Islam, in a fixed series that was only to be changed in insignificant ways throughout the succeeding centuries, particularly for the numbers 5 and 0 (Fig. 25.3). And these were what Arab authors have always referred to as *arqam al hindi* ("Indian numerals"):

| ١ | ٢ | ٣ | ۴ or ٤ | ٥ or ۵ | ٦ or ۶ | ٧ | ٨ | ٩ | ٠ |
|---|---|---|---|---|---|---|---|---|---|
| 1 | 2 | 3 | 4 | 5 | 6 | 7 | 8 | 9 | 0 |

These forms can be found in 'Abd Jalil al-Sijzi (951–1024), al-Biruni (c. 1000), Kushiyar ibn Labban al-Gili (c. 1020) and As Samaw'al al-Maghribi (c. 1160) (Fig. 25.1), and they are still used in all the Gulf countries, from Jordan and Syria to Saudi Arabia, the Yemen, Iraq, Egypt, Iran, Pakistan, Afghanistan, Muslim India, Malaya and Madagascar.

## THE WESTERN ARABS' "GHUBAR" NUMERALS

But this was not exactly the origin of our "Arabic" numerals. We inherited them from the Arabs, true enough, but from the Arabs of the West (the inhabitants of North Africa and Spain) and not from the Arabs of the Near East.

Before proceeding further, we should like to quote three revealing passages from manuscripts in the Bibliothèque nationale and translated by Woepcke [F. Woepke (1863), pp. 58–69].

They are three commentaries on mathematical works. In each of them, the commentator's explanations are mixed in with the original text, which is written in red ink to distinguish it from the commentary, which is written in black ink. Thus, in the following extracts, the original text is printed in italics and the commentary in Roman.

*First passage*

*The nine Indian numerals* [arqam al hindi] *are as follows*:

| 1 | 2 | 3 | 4 | 5 | 6 | 7 | 8 | 9 |
|---|---|---|---|---|---|---|---|---|
| ١ | ٢ | ٣ | ۴ | ۵ | ۶ | ٧ | ٨ | ٩ |

*Or like this:*

| 1 | 2 | 3 | 4 | 5 | 6 | 7 | 8 | 9 |
|---|---|---|---|---|---|---|---|---|

*. . . which are the "Ghubar" numerals.*

*Second passage*

The author says: *The first order goes from one to nine and is called the order of units.*

These nine symbols, called "ghubar" [ = "dust"] numerals, are widely used in the provinces of Andalusia and in the lands of the Maghreb and Ifriqiya. Their origin is said to have occurred when an Indian picked up some fine dust, spread it over a board (*luha*) made of wood, or of some other material, or else over any plane surface, on which he marked the multiplications, divisions or other operations he wanted to carry out. When he had finished his problem, he put it [the board] away in its case until he needed it again.

[In order to memorise their shapes] the following verses have been written about these numerals [in which the shapes of the letters, words and figures mentioned evoke the numerals being referred to]:

| | 1 | 2 | 3 | 4 | 5 | 6 | 7 | 8 | 9 | 0 |
|---|---|---|---|---|---|---|---|---|---|---|
| Practical arithmetical treatise by Ibn al-Banna al-Marrakushi. Fourteenth century. University of Tunis, Ms. 10 301, f° 25 v. CF. M. Souissi | | | | | | | | | | |
| *Guide to the Katib* (work which gives details of the various number-systems used by scribes, accountants, officials etc.) Manuscript dated to 1571–72 (see Fig. 25.10). Paris, BN, Ms. ar. 4 441, f° 22 | | | | | | | | | | |
| Sharishi, *Kashf al talkhis* ("Commentary on the Arithmetical Treatise. . .). Manuscript dated 1611. University of Tunis, M. 2 043, f° 16r | | | | | | | | | | |
| Bashlawi, *Risala fi'l hisab* ("Letter Concerning Arithmetic"). Seventeenth-century manuscript. University of Tunis, Ms. 2 043, f° 32 r. Cf. M. Souissi | | | | | | | | | | |
| Anonymous. Arithmetical treatise entitled *Fath al wahhab 'ala nuzhat al husab 'al ghubar* ("Guide to the Art of *Ghubar* Calculations"). Commentary by al-Ansari, written in 1620 and completed by1629. Paris, BN, Ms. ar. 2 475, f° 46 r, 152 v and 156 v | | | | | | | | | | |
| Copy of a treatise of practical arithmetic by Ibn al-Banna (*Talkhis a 'mal al hisab*, "Concise Summary of Arithmetical Operations") Seventeenth century. Paris, BN, Ms. ar. 2 464, f° 3v | | | | | | | | | | |
| As Sakhawi, *Mukhtasar Fi 'ilm al hisab* ("Summary of Arithmetic"). Eighteenth century. Paris, BN, Ms. ar. 2463, f° 79 v – 80 | | | | | | | | | | |

FIG. 25.5. *The Western Arabs' numerals ("Ghubar" script)*

These are an *alif* ( ) [for number 1],

And a *ya* ( ) [for 2],

Then the word *hijun* ( ) [for 3].

After that the word *'awun* (عو) [for 4];

And after *'awun*, one traces an *'ayin* (ع) [for 5].

Then a *ha* [final] ( ) [for 6].

And after the *ha*, appears a number [7], which,

When it is written, looks like an iron with a bent head ( ).

The eighth (of these signs is made of) two zeros [*sifran*]

[Connected by] an *alif* ( ). And the *waw* ( ) is the

Ninth, which completes the series.

The shape of the *ha* ( ) [sometimes given to number 2] is not pure. Here are the nine signs (which must be written so that) the one appears in the highest place, with the two below it, as follows:

| | | | | | | | | |
|---|---|---|---|---|---|---|---|---|
| 1 | 2 | 3 | 4 | 5 | 6 | 7 | 8 | 9 |

*Third passage*

*The preface deals with* the shape of the *Indian signs*, as they were drawn up by the Indian nation, *and these are*, i.e. the Indian signs, *nine figures* which must be formed *as follows*, that is: one, two, three, four, five, six, seven, eight, nine, with the following forms:

| ١ | ٢ | ٣ | ۴ | ۵ | ۶ | ٧ | ٨ | ٩ |
|---|---|---|---|---|---|---|---|---|
| 1 | 2 | 3 | 4 | 5 | 6 | 7 | 8 | 9 |

*which are most often used by us*, i.e. the Easterners, but others too are used. *Or*, they must be formed *as follows*:

1 2 3 4 5 6 7 8 9

*which are not much used* by us, while their use is widespread among the Western [Arabs].

Note. The author's meaning is clearly that both series come from India, which is true. The learned al-Shanshuri says in his commentary on the Murshidah: and they are called, i.e. the second way [of forming these signs], Indian, because they were devised by the Indian nation. End of quotation. But they are distinguished by different names, the former are called *Hindi* and the latter *Ghubar*, and they are termed *Ghubari* because people used to spread flour over their board and trace figures in it.

The following verses have been written about these signs . . . [the same as those quoted in the second passage above, with one slight difference which is described in Fig. 25.6].

(But) they have been brought together better in one single verse, as follows:

An *alif* (ا) (for number1),

a *ha* (ح) [for 2],

*hizun* (ح) (for 3),

'*awun* (عو) [for 4],

an '*ayin* (ع) (for 5),

a *ha* (final) ( ) [for 6],

an inverted *waw* ( ) (for 7),

two zeros [linked by an *alif*] ( ) (for 8),

and a *waw* (و) [for 9].

Certain points are worthy of note in these passages.

Firstly, we learn that the *Ghubar* numerals were used in the Maghreb (the western region of North Africa, between Constantine and the Atlantic), in Muslim Andalusia and in Ifriqiya (the eastern region of North Africa, between Tunis and Constantine). And it can be observed that they are written in a completely different way from the eastern provinces' *Hindi* numerals.

We have also learnt about a means of calculation: a sort of wooden board sprinkled with dust, the use of which was, as we shall see, linked with *Ghubar* numerals.

We can also see that the tradition of an Indian origin for these numerals had been transmitted by Arab and Maghrebi arithmeticians.

But the most important point concerns the verses written about *Ghubar* numerals, and which ingeniously fix their shapes. The stability of these verses from one manuscript to another is remarkable when one considers that they are not copies of the same source, but two completely independent manuscripts from different periods and locations.

They are an excellent way of memorising the nine numerals, by associating them with certain Arabic letters (or groups of letters), written in the typical style of the old Maghribi and Andalusian script. They were presumably composed to teach pupils how to write the nine Indian numerals in the style of their native province; it is rather as though we gave the shapes of the Roman letters O, I, Z etc. to our children for them to learn the numbers 0, 1, 2 etc.

Figure 25.6 contains further explanations of each line, as it appears in manuscript. The exact forms have been recreated, with reference to local scripts and drawing on parallels with the numerals contained in these manuscripts.

The two oldest known documents which refer to *Ghubar* numerals and calculation date back to 874 and 888 CE [see JASB 3/1907; SC XXIV/1918]. The shapes of the numerals they contain are close to those in Fig. 25.6 and, of course, to those described in the verses quoted above. And, as the most recent manuscript containing these verses comes from the beginning of the nineteenth century, it can be supposed that the forms of the *Ghubar* numerals were fixed centuries ago and passed down from generation to generation in this manner. In other words, an attempt was made to prevent the *Ghubar* numerals from being altered by scribes. These verses can also be found in numerous other arithmetical treatises.

The original forms of these numerals were conserved no doubt because the Maghrebi are attached to traditions coming from the Muslim conquest of Andalusia and North Africa. And that is when these numerals arrived in these regions and were then adapted to the local cursive scripts.

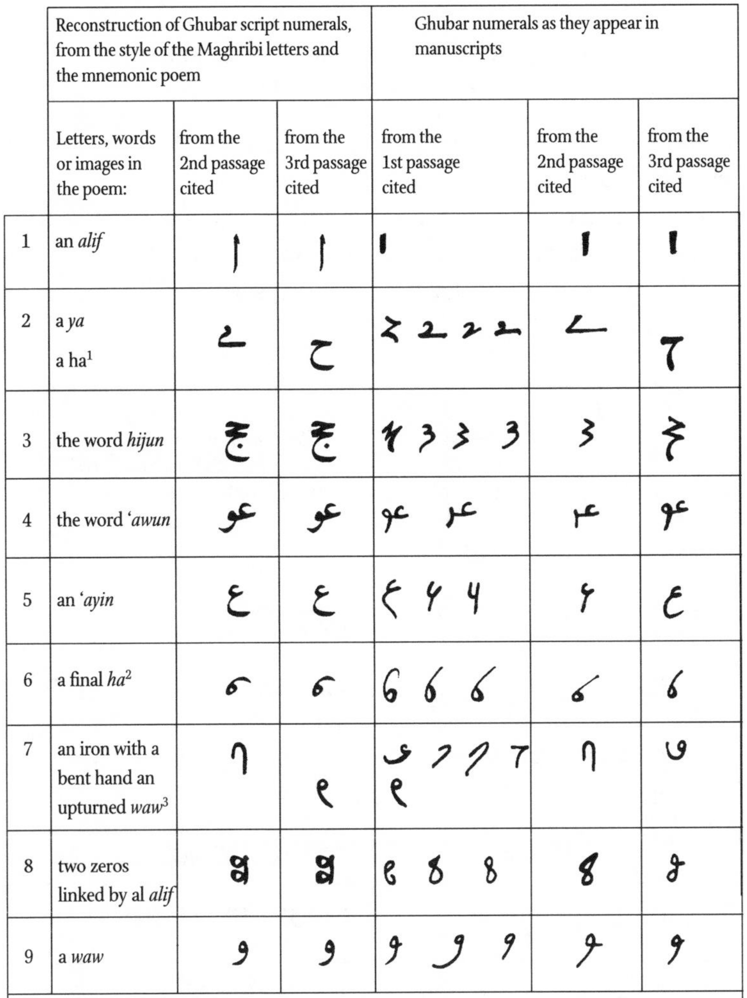

| | Reconstruction of Ghubar script numerals, from the style of the Maghribi letters and the mnemonic poem | | | Ghubar numerals as they appear in manuscripts | | |
|---|---|---|---|---|---|---|
| | Letters, words or images in the poem: | from the 2nd passage cited | from the 3rd passage cited | from the 1st passage cited | from the 2nd passage cited | from the 3rd passage cited |
| 1 | an *alif* | | | | | |
| 2 | a *ya*<br>a ha[1] | | | | | |
| 3 | the word *hijun* | | | | | |
| 4 | the word *'awun* | | | | | |
| 5 | an *'ayin* | | | | | |
| 6 | a final *ha*[2] | | | | | |
| 7 | an iron with a bent hand an upturned *waw*[3] | | | | | |
| 8 | two zeros linked by al *alif* | | | | | |
| 9 | a *waw* | | | | | |

1: The author of the second passage notes that the *ha* "is not pure". This remark, referring here of course to the number 2, seems to mean that the variant similar to this letter (which is also found in manuscripts) was not the original shape of 2 and that it had initially been more like the final form of *ya*, which is often written in this way in the Maghribi script (Fig. 25.8).

2: Such is, in fact, the final form of *ha*, as it occurs in Maghrebi and Andalusian manuscripts (Fig. 25.8A).

3: The existence of this variant of the number 7 (as an upturned *waw*) is confirmed in a marginal note which occurs in the manuscript of the first passage.

Fig. 25.6.

## THE TRANSMISSION OF INDIAN NUMERALS TO WESTERN ARABS

The question that now needs answering is how and when Indian arithmetic arrived in North Africa and Spain.

Woepcke provides us with part of the answer:

> Even though the unity of the caliphs came to an early end, pilgrimages to Mecca, flourishing trade, individual travels, migrations of entire populations and even wars kept up a constant communication between the various lands inhabited by Muslims. Once Indian arithmetic was known in the East, it inevitably became introduced into the West. A lack of precise information concerning this event in the history of science makes dating it impossible, but we are probably not far from the truth if we say that Indian arithmetic arrived in North Africa and Spain during the ninth century.

It is important to remember the special relationship the Caliphate of Cordoba had with Byzantium, which allowed the circulation of certain ancient texts. It can also be supposed that this facilitated contacts and meetings with representatives of Indian culture in the cosmopolitan world of Byzantium. But we should also bear in mind the contact that the Andalusians and Maghrebi must have had with their eastern cousins, without passing through Byzantium.

The arrival of Indian arithmetic in these regions could easily have come about either through texts written by eastern Arabs, or via more direct contacts with Indian scholars; thus in a similar way to what happened between India and the eastern Arabs.

But we must not overlook the vital role Jewish tradesmen and merchants probably played in this transmission. This is, in fact, suggested by Abu'l Qasim 'Ubadallah, a Persian geographer working in Baghdad. Better known as Ibn Khurdadbeh, he wrote as follows in his work entitled *Book of Routes and Provinces* (c. 850 CE):

> Jewish merchants speak Arabic as well as Persian, Greek, Latin and all other European languages. They travel constantly from the Orient to the Occident and from the West to the East, by both land and sea. They take ship from the land of the Latins [*franki*] by the western sea [the Mediterranean] and sail towards Farama; there, they unload their merchandise, place it in caravans and take the overland route to Colzom, on the edge of the eastern sea [the Red

Sea]. From there, they take ship again and sail towards Hejaz [Arabia] and Jidda, before moving on to Sind, India and China. Then they return, bringing with them goods from the east . . . These travels are also made by road. The merchants leave the land of the Latins, go towards Andalusia, cross the patch of sea [the Straits of Gibraltar] and travel across the Maghreb before reaching the African provinces and Egypt. They then travel towards Ramalla, Damascus, Kufa, Baghdad and Basra, before coming to Ahwaz, the Fars, Kerman, the Indus, India and China [quoted in Smith and Karpinski (1911)].

Similar information about these merchants can be found in this extract from the *Gulistân* (*Rose Garden*), written by the Persian poet Sa'adi in the first half of the thirteenth century [see E. Arnold (1899); D. E. Smith and L. C. Karpinski (1911)]:

I met a merchant who had a hundred and forty camels
And fifty porters and slaves . . .
He replied: I want to take Persian sulphur to China,
Which, from what I have heard,
Fetches a high price in that country;
Then procure goods made in China
And take them to Rome (*Rum*);
And from Rome load a boat with brocades for India;
And with that trade for Indian steel (*pûlab*) in Halib;
From Halib, I shall transport glass to the Yemen,
And take back Yemeni painted cloth to Persia.

Unlike Ibn Hauqal, the poet does not specify the origin of this travelling merchant, who may not be Jewish. The Jews have never had a monopoly over international trade. So Jewish traders were merely one of the numerous links in this chain of transmission.

Whether they were or were not Jewish, these tradesmen used numbers as often as they travelled or traded. And, like the various languages they learnt in their business, they must also have become acquainted with the different systems of arithmetic used by the peoples they encountered.

As India was part of their route, they must surely have been obliged to learn Indian numerals and arithmetic, and were thus one form of communication between India and the Maghreb.

## FROM HINDI NUMERALS TO GHUBAR NUMERALS – A SIMPLE QUESTION OF STYLE

To return to Arabic numerals, the Indian influence is evident, whether it be on the *Hindi* symbols, or the *Ghubar* (Fig. 25.3 and 6).

Even a rapid comparison between the Indian *Nâgarî* numerals and the *Ghubar* shows of course the presence of the Indian 1, but also 2, 3, 4 (with a slightly different orientation in Arabic), 6, 7, 9 and 0, and even 5 and 8 (Fig. 25.5 and 7).

| | The Arabic numerals below (attested in the early period of the Maghrebi and Andalusian provinces) | Correspond to the Indian numerals below (in a variety of styles, from Brâhmî to Nâgarî, including others attested from the beginning of the CE to the eighth century) | |
|---|---|---|---|
| 1 | | | 1 |
| 2 | | | 2 |
| 3 | | | 3 |
| 4 | | | 4 |
| 5 | | | 5 |
| 6 | | | 6 |
| 7 | | | 7 |
| 8 | | | 8 |
| 9 | | | 9 |

Fig. 25.7.

In palaeographic terms, there is thus no difference between the *Hindi* numerals of the Machreq and the *Ghubar* numerals of the Maghreb. Both come from the same source. Any differences between them simply derive from the habits of scribes and copyists in the two regions.

The history of Arabic writing styles helps us to understand these changes more clearly (Fig. 25.8). From the beginning of Islam, two distinct forms of writing evolved: a lapidary cursive style, derived from pre-Islamic inscriptions; and an even more cursive style, from the earliest written Arabic manuscripts, also dating to before the Hegira.

The lapidary cursive style produced the *Kufic* script, for inscriptions and manuscripts, with its characteristic horizontal base line on which the rigid, angular letters are set vertically. According to Ibn al-Nadim's *Fihrist* (987), this script derived from the early habits of the stone-carvers and scribes from Kufa on the Euphrates, hence its name. (Founded in 638 CE, Kufa was a centre of learning under the Omayyad caliphs until the foundation of Baghdad in 762.) This script was also used, during the first centuries of Islam, for legal and religious texts (in particular for the first copies of the Koran, in mosques and on tombstones), which explains its hieratic nature.

It was then gradually replaced by the *naskhi* script, generally used by copyists, and leading to the elegant calligraphy of the "Avicenna" Arabic alphabet which is most commonly used today. Derived from ancient cursive Arabic manuscripts, this style is marked by its smooth rounded forms, broken up into small curved elements. It is also the source of the *nastalik* script, used in Persia, Mesopotamia and Afghanistan, and the *sulus* script of the Turkish Ottoman Empire. With certain exceptions, the form of the letters remained very similar to *Naskhi*.

The difference between the two styles, at least at the beginning, was really due to what they were used for and the material they were written on. While the cursive manuscript style was used for everyday texts on papyrus or parchment, the other one was reserved for inscriptions on stone, wood or metal. The former was traced onto the papyrus, parchment or other smooth surface with a quill or a reed (the famous *qalam*, or "calamus") dipped in thick ink. But the latter was sculpted into stone, carved into wood or engraved into copper. This naturally explains the former's smooth rounded forms, contrasting with the latter's angular rigidity.

If we now return to the numerals and compare the signs contained in Fig. 25.3 and 5, we can see that the cursive *Hindi* numerals are far more rounded than those of the Maghreb, with the base line of the former breaking up into small curves. In other words, the eastern Arabs' numerals follow closely the rules of the *Naskhi* script.

On the other hand, the *Ghubar* numerals, while remaining cursive, are nevertheless obviously more angular, stiff and rigid. A closer look reveals that their curves, down-strokes and angles are absolutely identical to those used in the *Kufic* script. This is, at least, what is revealed in the original of *Kashf al asrar 'an 'ilm al gobar*, by the Andalusian mathematician al-Qalasadi. Its letters and *Ghubar* numerals are written in a way which reflects the pure *Kufic* tradition from the early centuries of Islam. This manuscript dates from the fifteenth century and the Institut des Langues Orientales in Paris possesses a copy of it from a lithograph made in Fez [see A. Mazaheri (1975)].

This is not surprising, for the *Maghribi* (or *African*) script which spread across North Africa, Sudan and Muslim Spain after the ninth century is in fact nothing more than a manuscript *Kufic*.

It should not be forgotten that the Maghrebi and Andalusians were extremely attached to ancient Islamic traditions. This is particularly true of the lapidary cursive style of the first conquerors of the region, the Abbasids of Samara, which gave the Maghribi script its stiffness and rigidity.

By fixing their forms by means of the verses quoted above (Fig. 25.6), they were made to adopt the characteristic shapes of Maghribi letters and thus follow its cursive rules.

To sum up, whatever differences there may be between *Hindi* and *Ghubar* numerals, their common source is demonstrably Indian.

But it was not in their *Hindi* form, but in the *Ghubar* style that Indian numerals migrated from Spain to the Christian peoples of Western Europe, before finally taking the shape they have today.

## ARAB RESISTANCE TO INDIAN NUMERALS

It is tempting to think that the Indian system spread through the Islamic world, replacing all other ways of representing numbers and, because of their ingenious simplicity, the corresponding calculation methods were rapidly accepted at all levels of Arab-Islamic society. The author humbly admits that he was wrong in the first edition of the present work in which he subscribed to that idea and neglected the following interesting details. Of course, certain scholars such as al-Khuwarizmi and An Nisawi were sufficiently astute to understand the superiority of this system. But there was an equal number of Muslims who were, sometimes violently, opposed to the use of numerals and even more so to their becoming generalised.

This means that, contrary to what is often believed, the domination of the Indian system was a long, difficult process. Many arithmetical treatises, for example, contain not a single Indian numeral, and sometimes no numerals at all, because the numbers in each line are expressed by their Arabic names. And if Indian numbers are to be found anywhere, then it is most probably, or even one would think inevitably, in arithmetical works.

| NAME OF LETTER | NUMERICAL VALUE | SHAPE OF LETTER in Naskhi Arabic | in Maghribi Arabic | in Persian, in the nasta'lik script |
|---|---|---|---|---|
| alif | 1 | ا | ا | ا |
| ba | 2 | ب | ب | ب |
| jim | 3 | ج | ج | ج |
| dal | 4 | د | د | د |
| ha | 5 | ه | ه | ه |
| wa | 6 | و | و | و |
| zay | 7 | ز | ز | ر or ر |
| ḥa | 8 | ح | ح | ح |
| ṭa | 9 | ط | ط | ط |
| ya | 10 | ى | ي or ے | ى |
| kaf | 20 | ك | ك | ک |
| lam | 30 | ل | ل | ل |
| mīm | 40 | م | م | م |
| nūn | 50 | ن | ن | ن |
| sin | 60 | س | س | س |
| 'ayin | 70 | ع | ع | ع |
| fa | 80 | ف | ڢ | ف |
| sad | 90 | ص | ص | ص |
| qaf | 100 | ق | ف | ق |
| ra | 200 | ر | ر | ر |
| shin | 300 | ش | ش | ش |
| ta | 400 | ت | ت | ت |
| tha | 500 | ث | ث | ث |
| kha | 600 | خ | خ | خ |
| dhal | 700 | ذ | ذ | ذ |
| dad | 800 | ض | ض | ض |
| dha | 900 | ظ | ظ | ظ |
| ghayin | 1,000 | غ | غ | غ |

FIG. 25.8A. *The Arabic alphabet in the Naskhi and Maghribi scripts*

NASKHI STYLE

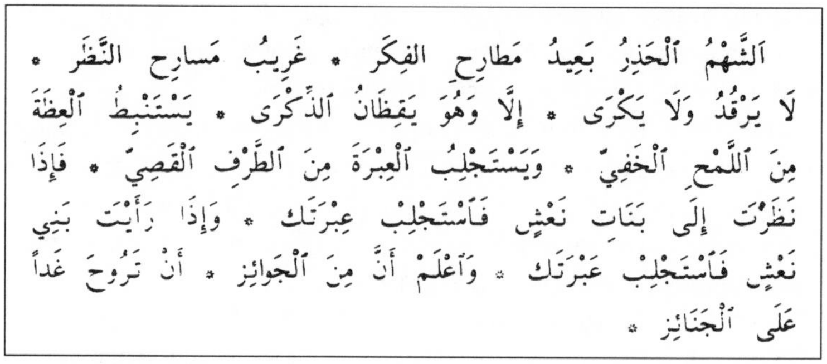

اَلشَّهْمُ ٱلْحَذِرُ بَعِيدُ مَطارِحِ الفِكَرِ ۞ غَرِيبُ مَسارِحِ النَّظَرِ ۞
لَا يَرْقُدُ وَلَا يَكْرَى ۞ إِلَّا وَهُوَ يَقِظَانُ ٱلذِّكْرَى ۞ يَسْتَنْبِطُ ٱلْعِظَةَ
مِنَ ٱللَّمْحِ ٱلْخَفِيِّ ۞ وَيَسْتَجْلِبُ ٱلْعِبْرَةَ مِنَ ٱلطَّرْفِ ٱلْقَصِيِّ ۞ فَإِذَا
نَظَرْتَ إِلَى بَنَاتِ نَعْشٍ فَٱسْتَجْلِبْ عِبْرَتَكَ ۞ وَإِذَا رَأَيْتَ بَنِي
نَعْشٍ فَٱسْتَجْلِبْ عَبْرَتَكَ ۞ وَٱعْلَمْ أَنَّ مِنَ ٱلْجَوَائِزِ ۞ أَنْ تَرُوحَ غَداً
عَلَى ٱلْجَنَائِزِ ۞

KUFIC STYLE

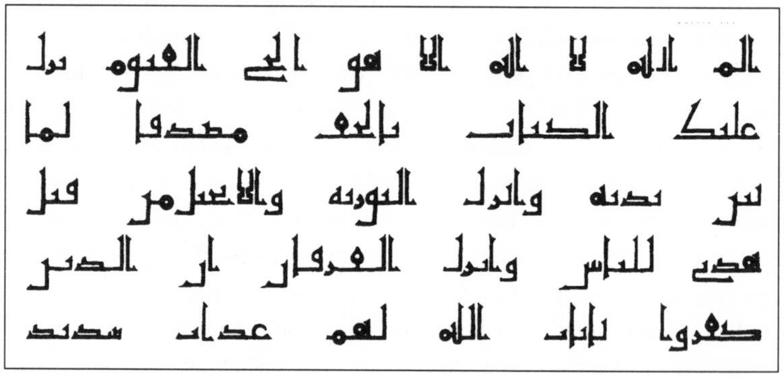

الم الله لا اله الا هو الحى القيوم نزل
عليك الكتاب بالحق مصدقا لما
بين يديه وانزل التوريه والانجيل من قبل
هدى للناس وانزل الفرقان ان الذين
كفروا بايات الله لهم عذاب شديد

MAGHRIBI STYLE

حكى الحارث بن همام قال لما اقتعدت غارب الاغتراب، وأنأتني المتربة
عن الأتراب، طوحت بي طوائح الزمن، الى صنعاء اليمن، فدخلتها
خاوي الوفاض، بادي الإنفاض، لا أملك بلغة، ولا أجد في جرابي
مضغة، فطفقت أجوب طرقاتها مثل الهائم، وأجول في حوماتها
جولان الحائم، وأرود في مسارح لمحاتي، ومسايح غدواتي وروحاتي،
كريما أخلق له ديباجتي، وأبوح اليه بحاجتي، أو أديبا تفرج رؤيته

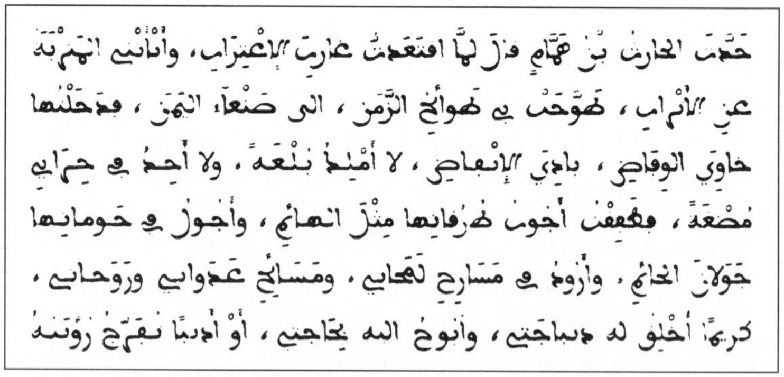

FIG 25.8B. *Different styles of written Arabic [CPIN; see also de Sacy; Sourdel in EIS]*

For Islam, like everywhere else, had its "traditionalists", bookkeepers and accountants who remained deeply attached to previous practices and vigorously opposed to scientific and technological innovations.

## THE CONSERVATISM OF ARAB SCRIBES AND OFFICIALS

One of the reasons for this opposition was the conservatism of Arab and Islamic scribes and officials, who long remained attached to their ancestral methods of counting and calculating on their fingers.

Thus, in his *Kitab al mu'allimin* (*Schoolmasters' Book*), al-Jahiz gives this advice, which provides a clear idea of the polemic that must have confronted the users of Indian numerals and the ardent defenders of traditional methods for several generations: "It seems better to teach pupils digital calculation and avoid Indian arithmetic (*hisab al hindi*), geometry and the delicate problems of land measurement." [British Museum Ms. 1129, f° 13r].

This author, who scorned Indian numerals and arithmetic, thus recommended teaching calculation using fingers and joints (*hisab al 'aqd*) as being, to his mind, more useful for the future official scribe of the period. Some accountants even preferred manual calculation to the Arabs' traditional means of calculation, the dust board.

This is, for example, revealed in *Kitab al hisab bila takht bal bi'l yad* ("Treatise on calculation without the board, but with [the fingers of] the hand"), written in 985 by al-Antaki [see A. Mazaheri (1975)].

In his *Adab al kutab*, destined for scribes and accountants, the Persian writer As Suli (died 946) gives the reason for this preference for manual calculation. After mentioning the "nine Indian characters" and "the great simplicity of this system" when expressing "large quantities", he then adds: "Official scribes nevertheless avoid using this system because it requires equipment [i.e. a counting board] and they consider that a system that requires nothing but the members of the body is more secure and more fitting to the dignity of a leader." As Suli then eulogises the official accountants of the Arab-Islamic world, with their supple joints and movements "as fast as the twinkling of an eye". He quotes a certain 'Abdullah ibn Ayub who "compares the jagged lightning fork with the rapidity of the accountant's hand, when he says: 'It seems that its flash [of lightning] in the sky is made up of a scribe's or accountant's two hands!'" Then he concludes: "That is why they content themselves with just the *iqd* [i.e. counting on the fingers] and the system of joints" [see J. G. Lemoine (1932)].

Officials always, of course, claim they are irreplaceable in order to keep their privileged positions. They are thus never happy to see a new simple system becoming generalised, which anyone can use without going through their difficult and mysterious apprenticeship.

This is a universal tendency, which can be witnessed throughout Antiquity, and in Western Europe from mediaeval times up until the French Revolution. If Arab-Islamic scribes and officials violently opposed the introduction of Indian numerals, it was because it could mean an end to their monopoly.

But this traditionalism does not explain everything. We must also consider the multiplicity and diversity of the peoples that made up the Muslim empire. The heterogeneous nature of the cultures and populations of this complex world, along with regional and individual habits, also played a part.

"Culture", as E. Herriot put it, "is what remains when all else has been forgotten." It is the form of knowledge which enables the mind to learn new things. Hence the idea of developing and enriching our various mental faculties by intellectual exercises such as study and research.

But "culture", in any given civilisation, is also the intellectual, scientific, technological and even spiritual inheritance of its people. It is thus the sum of knowledge, which its great minds have assimilated, and which greatly adds to its enrichment.

In this way, Arab-Islamic civilisation was exceptional for its originality, strong culture and deep insights of its thinkers, scholars, poets and artists.

And, to quote P. Foulquié, a culture is also the "collective way people think and feel, the set of customs, institutions and works which, in any given society, are at once the effect and the means of personal culture." Thus (to run Martin du Gard and Mead together), it is the set of virtues, preconceptions, individual habits and works which make up a given nation in its ways of behaving, acquired and transmitted by its members, who are accordingly united by a shared tradition.

Like any other, Arab culture was also composed of varied customs, countless details, endless habits and presumptions, characteristic of its daily existence. Great minds thus coexisted with lesser, more ignorant souls whose unthinking conservatism led them to clutch onto methods that had been useful to their distant ancestors, but which had long since stopped being appropriate to modern times and activities.

## TRADITIONAL ACCOUNTANTS VERSUS USERS OF OUTMODED SYSTEMS

When the Arab-Islamic civilisation found itself in contact with the Christian West, some Arab accountants had the curious idea of adopting Latin calculation methods using counters on a board, and thus set about turning the clock back. This was the case with certain Syrian and Egyptian accountants, presumably under the influence of their trading links with the Genoans and Byzantines.

This was severely criticised by the Persian historian Hamdullah who, in his 1339 *Nuzhat al qolub* (work of geography and chronology), says: "In the year 420 [of the Hegira, thus 1032 CE], Ibn Sina invented the 'calculation knots', thus freeing our accountants from the tedium of totting up counters [*mishsara shumari*] on instruments and boards, like the Latin abacus [*takhata yi frenki*] and suchlike" [see A. Mazaheri (1975)].

As an accountant, Hamdullah had certainly been deeply impressed by a calculation method called *'uqud al hisab* ("calculation knots"), recommended for accountancy two centuries before by the famous Ibn Sina (Avicenna), then the finance minister of Persia, under Buyid domination.

To gain a better understanding of this method, we must remember that a "knot" (in Arabic *'aqd* or *'uqda*, the singular of *'uqud* or *'uqad*) had at this time not only its primary meaning, but also signified "class of numbers corresponding to the successive products of the nine units and any power of ten". In other words, the "knot" stood for the decimal system. There was the units *knot*, the tens *knot*, the hundreds *knot* and so on. This same term can be found in al-Maradini [see S. Gandz (1930)] and in Ibn Khaldun's *Prolegomena* [see *Muqaddimah*, trans. Slane, I, pp. 243–4].

By extension, the expression *'uqud al hisab* came to mean "calculation knots", in reference to an ancient way of recording numbers on knotted cords, used by the Arabs in antiquity. The various places of consecutive digits were marked by knots tied in predetermined positions. This system was thus very similar to the South American Incas' *quipus* and the ancient Japanese *ketsujo*, used until recently in the Ryu-Kyu Islands (Fig. 25.9).

The Arabs (presumably before the advent of Islam) had long used these knotted cords as a way of noting numbers for administrative records. The numbers thus tied on the strings recorded accounts and various inventories. This is reminiscent of the tradition, reported by Ibn Sa'ad, according to which Fatima, Mohammed's daughter, counted the ninety-nine attributes of Allah, and the supererogatory eulogies which followed the compulsory prayers, on knotted cords, and not on a rosary. These cords were also used as receipts and contracts. This is shown by the fact that, in Arabic, the word *'aqd* means both "knot" and "contract".

To return to the "calculation knots" which Avicenna is supposed to have invented, it is highly probable that Hamdullah was referring to a means of manual calculation.

The common Arabic expression for "hand counting" is *hisab al yad* (from *hisab*, "counting, calculation", and *yad*, "hand"). It can be found, for example, in al-Antaki and As Suli (*op. cit.*), as well as al-Baghdadi in his *Khizanat al 'adab*.

But in many authors, the word *'aqd* or *'uqda* ("knot") also means the "join" between the finger and the hand, and by extension the "joints" of the finger. For, this *hisab al 'uqud* ("counting with knots") is in fact "counting on the joints of the fingers", by allusion to the "knot" of the joints and the "join" between the fingers and the hand.

There were several ways of counting on fingers in Islam. Although Hamdullah is vague about Avicenna's method, it is possible to work out what it was by elimination. To Hamdullah's mind, the word *'uqud* in the expression *'uqud al hisab* ("calculation knots") could in fact have meant the "order of units" in an enumeration. And, as this concerns a manual method, the "knots" in question could refer to units in a highly evolved decimal system. What comes to mind is that "dactylonomy", similar to deaf and dumb sign language, which was used by the Arabs and Persians for centuries, in which the units and tens were counted on the phalanxes and joints of one hand, while the other one was symmetrically used for the hundreds and thousands (see Chapter 3). This system was famously described in a poem written in rajaz metre, called *Urjuza fi hisab al 'uqud*, composed before 1559 by Ibn al-Harb and dealing with the science of "counting on phalanxes and joints" [see J. G. Lemoine (1932)].

But this cannot be the method referred to by Hamdullah. As Guyard explains: "the word *'uqud*, taken as a noun, stands for the shapes obtained by bending the fingers and, by extension, the numbers thus formed." That is why the units in the manual systems already alluded to were called "knots". But, this same word *'uqud*, taken as an action, means "bending the fingers" [see JA, 6th series, XVIII (1871), p. 109]. And, since he is discussing arithmetic, what Hamdullah is talking about is definitely an action, not a state. It is thus the science of calculating with what may be called "moving knots" which is in question. For the other systems were mere static ways of counting on the fingers and joints of the hand (just simple manual representations of numbers), whereas the technique being envisaged here allows calculations to be made by actively bending the fingers.

By opposing "calculation knots" to the Latin abacus, Hamdullah was thinking of "knots" as an action, bending certain fingers and straightening others, allowing arithmetical operations to be carried out in a much easier way than on the abacus. That is why, according to this admirer of Avicenna, these "moving knots" had freed "our accountants from the tedium of totting up counters" thrown down onto "the Latin abacus and suchlike."

But Hamdullah is guilty of making an historical mistake. The method he attributes to Avicenna had already existed in the Islamic world for a long time.

This is not our accountant historian's only slip. For the method recommended by the famous philosopher was only of use in operations on common numbers. Hence Hamdullah's error of judgment. He had not understood that the Latin abacus, primitive though it was, allowed numbers to be reached that are far higher than can be obtained by any form of manual calculation, no matter how elaborate. For the limits of the human hand set the limits of the method.

Thus it was that, through ignorance of basic practical arithmetic, or perhaps through sheer bloody-mindedness, users of a totally outmoded means of calculation attacked other accountants with methods as primitive as their own. The latter were, of course, to be upbraided for falling for a technique that came from a culture that was quite alien to Islam, and which the former presumably held in disdain.

In this context, it is easy to imagine how both camps violently opposed the introduction of Indian numerals and calculation methods, whose evident superiority over their archaic ways they would never admit.

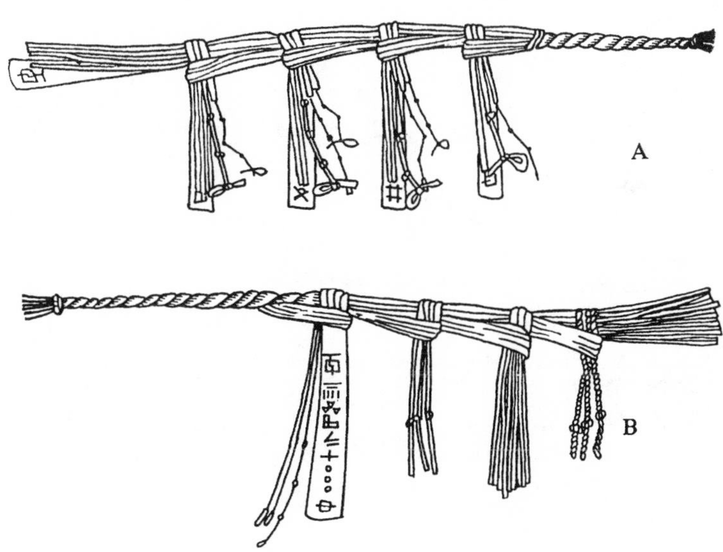


FIG. 25.9. *Japanese* ketsujo

*This was a concrete accountancy method, used in ancient Japan and analogous to the* quipus *of the Incas (Peru, Ecuador and Bolivia). Given the universal nature of this method, this Figure will provide a good idea of how Arabs used knotted cords in the pre-Islamic era and probably also in the early days of Islam (despite lack of evidence).*

*This* ketsujo *stands for(the knots represent sums of money, as used in the Ryu-Kyu Islands, particularly by workmen and tax collectors) [Frédéric 1985, 1986, 1977-1987, 1994]:*

*A – cloth account given to the State, or a temple, from left to right:*

*– Yoshimoto family: 1* jo, *8* shaku, *5* sun *and 7* bu*;*

*– 1* jo, *4* shaku, *3* sun *and 7* bu*;*

– Togei *family:* ibid.

*B – Horizontal strand: 20 households.*

*Others, from right to left: 3* hyo, *1* to, *3* shaku *and 2* sai.

## THE NUMERICAL NOTATION OF ISLAM'S OFFICIALS

In fact, the Indian system was introduced into the Islamic world in several steps. As operators, and thus as a means of calculation, the numerals were rapidly adopted by mathematicians and astronomers, soon followed by an ever-increasing number of intellectuals, mystics, magi and soothsayers. Meanwhile, others preferred to calculate by using the first nine letters of the Arabic alphabet (from *alef* to *ta*). But as a way of representing numbers (i.e. when noting numerical values and not making calculations), Indian numerals did not completely replace traditional notation until a relatively recent date.

Thus it was that Arab, Persian and Turkish officials continued to favour their own special notations, which had nothing to do with the Indian numerals in public use, for official and diplomatic documents, bills of exchange and administrative circulars until the nineteenth century.

This is shown in the *Guide to the Writer's Art* (1571–1572) [BN, Ms. ar. 4,441], which is a sort of handbook for professional writers. It gives a clear idea of the plurality of the numerical systems used by the scribes, officials and accountants of the Ottoman Empire at the end of the sixteenth century (Fig. 25.10).

Among these varied forms, let us mention the *Dewani* numerals used in Arab administrations, and the *Siyaq* numerals favoured by the accounts offices in the Ottoman Empire's finance ministry and in Persian administrations. These numerals were, originally, simply monograms or abbreviations of the names of the numbers in Arabic, written in an extremely cursive style. Later, they became so stylised and modified that their origins were scarcely recognisable. It is easy to understand how they were used to prevent fraudulent alterations to accounts, while at the same time leaving the general public in the dark as to what amounts were being described [see H. Kazem-Zadeh (1913); A. Chodzko (1852); L. Fekete (1955); A. P. Pihan (1860); C. Stewart (1825)].

We should also like to mention the *Coptic numerals*, used since antiquity by officials in the Arab administration of Egypt in their accounts, which were in fact slightly deformed letter numerals from the ancient alphabet of the Christian Copts of Egypt.

### *The Dewani numerals*

These numerals were used in Arab administrative offices (called *dewan*, hence their name).

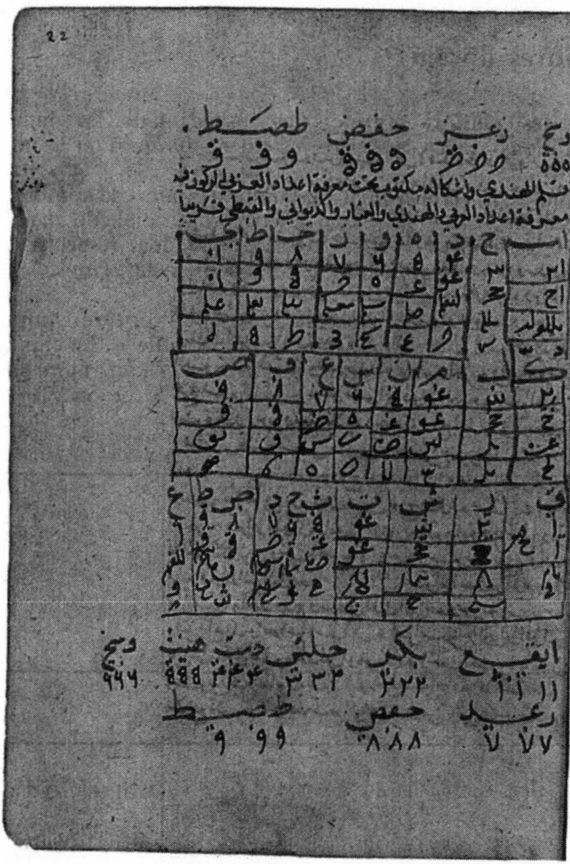

FIG. 25.10. *Page from an Arabic work, entitled* Murshida fi sana'at al katib *("Guide to the Writer's Art"). Dated 1571-1572, it is a sort of handbook for professional writers.*

*It gives a very clear idea of the numerous different ways Arab-Muslim scribes, accountants and officials wrote down their numbers at the end of the sixteenth century. It contains, counting from the top down: the Ghubar numerals (2nd line) (Fig. 25.5); the Arabic letter numerals (5th line); the Hindi numerals (6th line) (Fig. 25.3); then the Ghubar numerals again (7th line); the Dewani numerals (8th line); the Coptic numerals (9th line); the Arabic letter numerals (10th line); the Hindi numerals (11th line); the Ghubar numerals (12th line); two variants of the Coptic numerals (13th and 14th lines); etc. [BN Paris, Ms. ar. 4441, f° 22]*

They are abbreviations of the Arabic numerical nouns. Thus, number 1 is the letter *alif*, standing for *ahad*, "one". Similarly, numbers 5, 10 and 100 correspond to the letters *kha*, *'ayin* and *mim*, standing for *khamsa*, "five", *'ashara*, "ten", and *mi'at*, "hundred".

As for the number 1,000, it is a stylised form of the complete word *alf*, meaning "thousand". Number 10,000 corresponds to a monogram of *'asharat alaf*, "ten thousand" [A. P. Pihan (1860)].

*Units*

| | | | | | |
|---|---|---|---|---|---|
| 1 | ا | 4 | لىعا | 7 | ىعا |
| 2 | لا | 5 | حا | 8 | ىها |
| 3 | ٮـ or للا | 6 | سا | 9 | ىعا |

*Tens*

| | | | | | |
|---|---|---|---|---|---|
| 10 | عا | 40 | لىعا | 70 | ىعا |
| 20 | عو | 50 | حا | 80 | ىو |
| 30 | ىو | 60 | سا | 90 | ىعا |

*Hundreds*

| | | | | | |
|---|---|---|---|---|---|
| 100 | ٯ | 400 | لىعما | 700 | ىعما |
| 200 | ٯل | 500 | حما | 800 | ىط |
| 300 | سهما or ثلما | 600 | سما | 900 | ىعما |

*Thousands*

| | | | | | |
|---|---|---|---|---|---|
| 1,000 | ال or الف | 4,000 | لىعالف | 7,000 | ىعالف |
| 2,000 | الفى | 5,000 | حالف | 8,000 | ىهالف |
| 3,000 | سالف | 6,000 | سالف | 9,000 | ىعالف |

*Ten Thousands*

| | | | | | |
|---|---|---|---|---|---|
| 10,000 | عالف | 40,000 | لىعلا | 70,000 | ىعلا |
| 20,000 | عولا | 50,000 | حلا | 80,000 | ىلا |
| 30,000 | سلا | 60,000 | سلا | 90,000 | ىعلا |

*Hundred Thousands*

| | | | | | |
|---|---|---|---|---|---|
| 100,000 | طالف | 400,000 | لىعوطالف | 700,000 | ىعا طالف |
| 200,000 | لاطالف | 500,000 | حطالف | 800,000 | ىها طالف |
| 300,000 | ٮطالف | 600,000 | سطالف | 900,000 | ىعا طالف |

*Composite numbers*

Units are always placed before tens and between the hundreds and tens, as is done in spoken Arabic. Numerals are written from right to left, like the Arabic words they represent in this same order for composite numbers.

| | | | | | |
|---|---|---|---|---|---|
| 11 | [illegible] | 17 | [illegible] | 206 | [illegible] |
| 14 | [illegible] | 21 | [illegible] | 3,478 | [illegible] |
| 15 | [illegible] | 24 | [illegible] | 62,789 | [illegible] |

*The numerals of Egyptian Coptic officials*

The Arab administration of Egypt employed Christian Copts, who had their own special accountancy notation. These signs (which can be found in several Arabic manuscripts from this region) are cursive derivatives of the letter numerals of the Coptic alphabet, itself derived from Greek. Numbers up to nine thousand are reached by using the units and underlining them. For the ten thousands, the tens are underlined, as are the hundreds for the hundred thousands. Finally, composite numbers are always topped by a slightly curved line.

*Units*

| | | | | | |
|---|---|---|---|---|---|
| 1 | [illegible] | 4 | [illegible] | 7 | [illegible] |
| 2 | [illegible] | 5 | [illegible] | 8 | [illegible] |
| 3 | [illegible] | 6 | [illegible] | 9 | [illegible] |

*Tens*

| | | | | | |
|---|---|---|---|---|---|
| 10 | [illegible] | 40 | [illegible] | 70 | [illegible] |
| 20 | [illegible] | 50 | [illegible] | 80 | [illegible] |
| 30 | [illegible] | 60 | [illegible] | 90 | [illegible] |

*Hundreds*

| | | | | | |
|---|---|---|---|---|---|
| 100 | [illegible] | 400 | [illegible] | 700 | [illegible] |
| 200 | [illegible] | 500 | [illegible] | 800 | [illegible] |
| 300 | [illegible] | 600 | [illegible] | 900 | [illegible] |

*Thousands*

| | | | | | |
|---|---|---|---|---|---|
| 1,000 | [illegible] | 4,000 | [illegible] | 7,000 | [illegible] |
| 2,000 | [illegible] | 5,000 | [illegible] | 8,000 | [illegible] |
| 3,000 | [illegible] | 6,000 | [illegible] | 9,000 | [illegible] |

*Ten Thousands*

| | | | | | |
|---|---|---|---|---|---|
| 10,000 | [illegible] | 40,000 | [illegible] | 70,000 | [illegible] |
| 20,000 | [illegible] | 50,000 | [illegible] | 80,000 | [illegible] |
| 30,000 | [illegible] | 60,000 | [illegible] | 90,000 | [illegible] |

*Hundred Thousands*

| | | | | | |
|---|---|---|---|---|---|
| 100,000 | [illegible] | 400,000 | [illegible] | 700,000 | [illegible] |
| 200,000 | [illegible] | 500,000 | [illegible] | 800,000 | [illegible] |
| 300,000 | [illegible] | 600,000 | [illegible] | 900,000 | [illegible] |

*Composite numbers*

| | | | | | |
|---|---|---|---|---|---|
| 16 | [illegible] | 803 | [illegible] | 38,491 | [illegible] |
| 45 | [illegible] | 4,370 | [illegible] | 752,020 | [illegible] |

*The Persian Siyaq numerals*

These numbers were used in Persian administrations, and were also favoured by tradesmen and merchants. They are abbreviations of the words for the numbers in Arabic (and not in Persian). They are written from right to left, like the Arabic words they represent, as are the composite numbers [A. P. Pihan (1860); see also A. Chodzko (1852); H. Kazem-Zadeh (1913); C. Stewart (1825)].

*Units*

| | | | | | |
|---|---|---|---|---|---|
| 1 | [illegible] | 4 | [illegible] | 7 | [illegible] |
| 2 | [illegible] | 5 | [illegible] | 8 | [illegible] |
| 3 | [illegible] | 6 | [illegible] | 9 | [illegible] |

*Tens*

| | | | | | |
|---|---|---|---|---|---|
| 10 | [illegible] | 40 | [illegible] | 70 | [illegible] |
| 20 | [illegible] | 50 | [illegible] | 80 | [illegible] |
| 30 | [illegible] | 60 | [illegible] | 90 | [illegible] |

*Composite numbers from 11 to 18*

For these numbers, the final line of the units is rounded off and rises up towards the top of number ten:

| | | | | | |
|---|---|---|---|---|---|
| 11 | [illegible] | 14 | [illegible] | 17 | [illegible] |
| 12 | [illegible] | 15 | [illegible] | 18 | [illegible] |

*Composite numbers from 21 to 99*

The units and the other ten digits are linked together in the same way:

| | | | | | |
|---|---|---|---|---|---|
| 21 | [illegible] | 54 | [illegible] | 87 | [illegible] |
| 43 | [illegible] | 76 | [illegible] | 99 | [illegible] |

*Hundreds*

When written on their own, the hundreds have special signs, sometimes followed by a sort of upturned comma and full stop, which are always omitted in composite numbers. One sign calls for particular attention, because of possible errors ([illegible]). With a line before, it stands for 400, and with no line, 700. The same sign, with an additional curl to the right, stands for 900:

| | | | | | |
|---|---|---|---|---|---|
| 100 | [illegible] | 400 | [illegible] | 700 | [illegible] |
| 200 | [illegible] | 500 | [illegible] | 800 | [illegible] |
| 300 | [illegible] | 600 | [illegible] | 900 | [illegible] |

*Composite numbers from 101 to 999*

| | | | | | |
|---|---|---|---|---|---|
| 101 | [illegible] | 366 | [illegible] | 791 | [illegible] |
| 109 | [illegible] | 377 | [illegible] | 820 | [illegible] |
| 110 | [illegible] | 388 | [illegible] | 896 | [illegible] |
| 111 | [illegible] | 399 | [illegible] | 915 | [illegible] |
| 204 | [illegible] | 472 | [illegible] | 999 | [illegible] |

*Thousands*

To form the multiples of 1,000, the characteristic patterns of the units are used, with the final stroke lengthened from right to left. In this position, and with a pronounced broadening, it is enough to indicate the presence of the thousand in the combination:

| | | | | | |
|---|---|---|---|---|---|
| 1,000 | [illegible] | 4,000 | [illegible] | 7,000 | [illegible] |
| 2,000 | [illegible] | 5,000 | [illegible] | 8,000 | [illegible] |
| 3,000 | [illegible] | 6,000 | [illegible] | 9,000 | [illegible] |

*Composite numbers of four digits*

The group [illegible] stands for the thousand, but only that exact value. For, when followed by hundreds or tens, the group [illegible] is used (abbreviation of the Arabic word الف, *alf*, "thousand"):

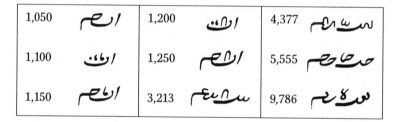

| | | | | | |
|---|---|---|---|---|---|
| 1,050 | [illegible] | 1,200 | [illegible] | 4,377 | [illegible] |
| 1,100 | [illegible] | 1,250 | [illegible] | 5,555 | [illegible] |
| 1,150 | [illegible] | 3,213 | [illegible] | 9,786 | [illegible] |

*Ten and Hundred Thousands*

After 10,000, the group [illegible] (abbreviation of the number 1,000) reappears, and the final stroke of the ten thousands is lengthened below the signs, instead of going down vertically:

| | | | |
|---|---|---|---|
| 10,000 | [illegible] | 99,112 | [illegible] |
| 25,072 | [illegible] | 110,100 | [illegible] |
| 34,683 | [illegible] | 245,123 | [illegible] |
| 45,071 | [illegible] | 300,000 | [illegible] |
| 50,008 | [illegible] | 456,789 | [illegible] |

*Other variants of the Persian Siyaq numerals*

| | Variants noted by Forbes | Variants noted by Stewart |
|---|---|---|
| 1 | [illegible] | [illegible] |
| 2 | [illegible] | [illegible] |
| 3 | [illegible] | [illegible] |
| 4 | [illegible] | [illegible] |
| 5 | [illegible] | [illegible] |
| 6 | [illegible] | [illegible] |
| 7 | [illegible] | [illegible] |
| 8 | [illegible] or [illegible] | [illegible] |
| 9 | [illegible] | [illegible] |
| 11 | [illegible] | [illegible] |
| 22 | [illegible] | [illegible] |
| 33 | [illegible] | [illegible] |
| 44 | [illegible] | [illegible] |
| 55 | [illegible] | [illegible] |
| 66 | [illegible] | [illegible] |
| 77 | [illegible] | [illegible] |
| 88 | [illegible] | [illegible] |
| 99 | [illegible] | [illegible] |

| | Variants noted by Forbes | Variants noted by Stewart |
|---|---|---|
| 100 | [illegible] | [illegible] |
| 200 | [illegible] | [illegible] |
| 300 | [illegible] or [illegible] | [illegible] |
| 400 | [illegible] or [illegible] | [illegible] |
| 500 | [illegible] | [illegible] |
| 600 | [illegible] | [illegible] |
| 700 | [illegible] or [illegible] | [illegible] |
| 800 | [illegible] or [illegible] | [illegible] |
| 900 | [illegible] or [illegible] | [illegible] |
| 1,000 | [illegible] | [illegible] |
| 2,000 | [illegible] | [illegible] |
| 3,000 | [illegible] | [illegible] |
| 40,000 | [illegible] | [illegible] |
| 50,000 | [illegible] | [illegible] |
| 100,000 | [illegible] | [illegible] |
| 200,000 | [illegible] | [illegible] |

Note that the number 100,000 is none other than the Sanskrit word *lakh* ( [illegible] ), used by the Indians for this amount.

## *The Siyaq numerals of the Ottoman Empire*

These numbers were favoured by the accounts offices in the Ottoman Empire's finance ministry.

They are abbreviations of the words for the numbers in Arabic (and not in Turkish). They are written from right to left, like the Arabic words they represent, as are the composite numbers. Also called *Siyaq*, they are analogous to the Persian numerals of the same name, even though they differ in certain respects.

Note that the point (which stands for 6) normally replaces the other sign ( [illegible] ) for the same value in composite numbers. But when this point is placed at the end of a number it is a mere punctuation mark, without any numerical value. Finally, in composite numbers made up of tens and units, the latter always come first, as in Arabic [A. P. Pihan (1860); see also L. Fekete (1955)].

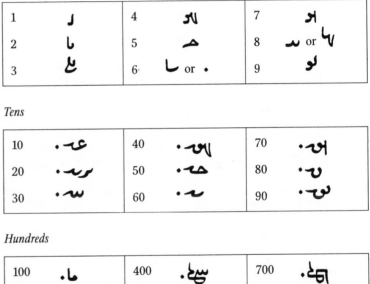

*Units*

| | | | | | |
|---|---|---|---|---|---|
| 1 | [illegible] | 4 | [illegible] | 7 | [illegible] |
| 2 | [illegible] | 5 | [illegible] | 8 | [illegible] or [illegible] |
| 3 | [illegible] | 6 | [illegible] or • | 9 | [illegible] |

*Tens*

| | | | | | |
|---|---|---|---|---|---|
| 10 | [illegible] | 40 | [illegible] | 70 | [illegible] |
| 20 | [illegible] | 50 | [illegible] | 80 | [illegible] |
| 30 | [illegible] | 60 | [illegible] | 90 | [illegible] |

*Hundreds*

| | | | | | |
|---|---|---|---|---|---|
| 100 | [illegible] | 400 | [illegible] | 700 | [illegible] |
| 200 | [illegible] | 500 | [illegible] | 800 | [illegible] |
| 300 | [illegible] | 600 | [illegible] | 900 | [illegible] |

*Thousands*

| | | | | | |
|---|---|---|---|---|---|
| 1,000 | [illegible] | 4,000 | [illegible] | 7,000 | [illegible] |
| 2,000 | [illegible] | 5,000 | [illegible] | 8,000 | [illegible] |
| 3,000 | [illegible] | 6,000 | [illegible] | 9,000 | [illegible] |

*Ten Thousands*

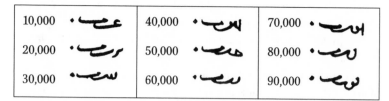

| | | |
|---|---|---|
| 10,000 | 40,000 | 70,000 |
| 20,000 | 50,000 | 80,000 |
| 30,000 | 60,000 | 90,000 |

*Composite numbers*

Note that for composite numbers containing several digits, the Turks generally used the letter س (*sin*), lengthening its horizontal stroke over the group. This letter stood for the word سياق (*siyaq*).

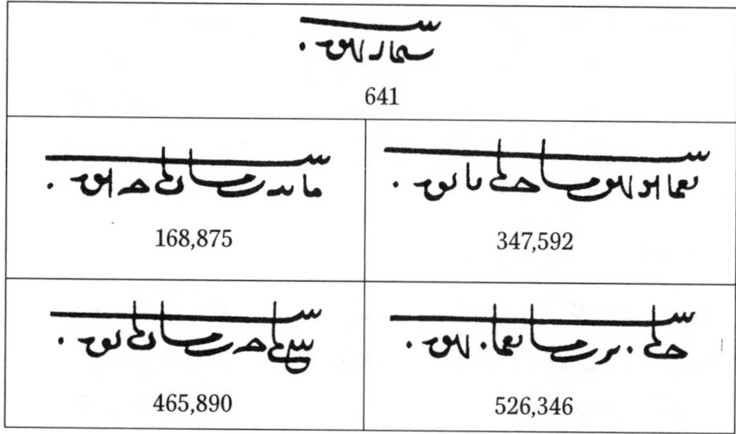


## INDIAN NUMERALS' MAIN ARAB RIVAL

Of all rival notations, with which Indian numerals were sometimes mixed in Arabic writings, the most important was certainly Arabic letter numerals. These were known as *huruf al jumal* (literally "letters [for calculating] series") and also as *Abjad* (from its first four letters), because it does not use the letters in the "dictionary" order, or *mu'jama*, (i.e. *alif, ba, ta, tha, jim, ha, kha, dal, dhal,* etc.) but in a special order, called *abajadi* beginning *alif, ba, jim, dal, ha, wa, zay, ḥa, ṭa etc.*, attributed as follows: *'a* = 1, *b* = 2, *j* = 3, *d* = 4, *h* = 5, *w* = 6, *z* = 7, *ḥ* = 8, *ṭ* = 9 etc. This is not, of course, a simple number-series (like one going from 1 to 26 by means of the Roman alphabet), but a true place-value system, the first nine letters being the units, the next nine the tens (*y* = 10, *k* = 20, *l* = 30, *m* = 40, *n* = 50, etc.), the following nine the hundreds (*q* = 100, *r* = 200, *sh* = 300, *ta* = 400, etc.) and, finally, the twenty-eighth letter standing for one thousand (*gh* = 1,000). Note that the *al abajadi* order is very close to Hebrew, Greek and Syriac letter numerals and is obviously the older order because it derives directly from the original Phoenician alphabet (see Fig. 25.8A). There were some differences between East and West. In the former *sin, sad, shin, dad, dha* and *ghayin* stood for 60, 90, 300, 800, 900 and 1,000, but in the latter 300, 60, 1,000, 90, 800 and 900 respectively.

Islamic scholars and writers often preferred to use this system. One example is the *Kitab fi ma yahtaju ilahyi al kuttab min 'ilm al hisab* (*Book of Arithmetic Needed by Scribes and Merchants*), written by the geometer and astronomer Abu'l Wafa al-Bujzani between 961 and 976.

The first two parts deal with calculating with whole numbers and fractions, the third with surfaces of plain figures, the volumes of solid bodies and the measurement of distances. The last four parts deal with various arithmetical problems, such as in business transactions, taxation, units of measurement, exchanges of currency, cereals and gold, paying and maintaining an army, constructing buildings, dams etc. [A. P. Youschkevitch (1976)]. Now, in this book, which was especially conceived for practical use, the Indian decimal place-value system is never used. All numbers are expressed by Arabic letter numerals.

A further significant example: the *Kitab al kafi fi'l hisab* (*Summary of the Science of Arithmetic*), written by the mathematician al-Karaji towards the end of the tenth century. It is rather similar to Abu'l Wafa's work and, like many later books, contains no mention of Indian numerals.

True, these works were especially for scribes, accountants and merchants, and we know that this form of arithmetic was favoured not only by scribes but also by officials and tradesmen. That is why this system stood up for so long against the new Indian way, which was supported by al-Khuwarizmi, An Nisawi and many others [A. P. Youschkevitch (1976)].

More surprisingly, the same phenomenon can be found in many Arabic works dealing with algebra, geometry and geography, which also contain only the letter system.

*Works on astronomy*

For astronomic treatises and tables, this was for a long time the only system the Arabs used.

It may be useful to remind ourselves of certain points concerning the sexagesimal system which the Arabs had inherited from the Babylonians, via Greek astronomers.

Babylonian scholars used a place-value system with base 60 and, from around the fourth century CE, they had a zero. These cuneiform numerals were the vertical wedge for units and the slanting wedge for tens (see Fig. 13.41). As for zero, it was represented either by a double oblique vertical, or by two superimposed slanting wedges.

This system was then adopted by Greek astronomers (at least from the second century BCE), but only to express the sexagesimal fractions of units (negative powers of 60). Otherwise, instead of using cuneiform signs, the Greeks had their own letter numerals, from α to θ for the first nine numbers, the next five (ι to ν) for the first five tens, with all the intermediate numbers expressed as simple combinations of these letters. Influenced by the Babylonians, they introduced a zero expressed either as sign written in various different ways (presumably the result of adapting old Mesopotamian cuneiform into a cursive script), or as a small circle topped by a horizontal stroke (probably the letter *omicron* (ο) the initial letter of *ouden*, "nothing", and topped with a stroke to avoid confusion with the letter o which stood for 70); or else as an upturned 2 (probably a cursive variant of the above) (see Fig. 13.74A and *Zero).

Arab astronomers also took over the Greek sexagesimal system, adapting it to their own alphabet. Note that to express zero in their sexagesimal calculations, the Arabs hardly ever used the Indian signs (the circle and the point). Instead they used a sign written in a variety of different ways (including the upturned 2 referred to above) which they had also inherited from the Greeks. Woepcke has this to say about the Arabs' sexagesimal system:

> Rather than [Indian] numerals, the Arabs preferred an alphabetic notation for their astronomic tables. They apparently found it more convenient. This use is confirmed in Arabic manuscripts containing astronomic tables, in which Indian numerals are rarely met with. The Arabs sometimes used them to express very large numbers, for example degrees over the circumference [see JA April–May (1860)]. However, this exception was unnecessary. Sexagesimal calculation, just as it had divided the degree into minutes, seconds, thirds etc., also had higher values, superior to the degree so that it was unnecessary to go higher than 59 in this notation. This is revealing about the relationship between sexagesimal calculation and alphabetic notation: it is after the number 60, useless in a rigorous sexagesimal system [i.e. based on place-value], that the divergence between the African and Asian alphabetic notations began [F. Woepcke (1857), p. 282].

The Arabs wrote an expression such as 0° 20' 35" as follows (reading from right to left):

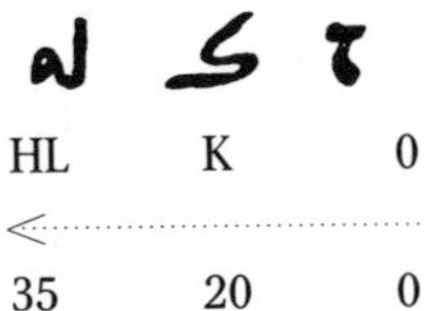


O being zero, K the letter *kaf* = 20 and the group LH, or *lam-ha*, the juxtaposition of *lam* ( = 30) and *ha* ( = 5).

To sum up, in their sexagesimal calculations and tables, Arab astronomers generally used their alphabet in the way described above (see Fig. 13.76). An exception to this rule was Abu'l Hasan Kushiyar ibn Labban al-Gili (971–1029), who wrote the *Maqalatan fi osu'l hisab al hind* (*Two Books Dealing with Calculations Using Indian Numerals*), the second book of which is concerned with base 60. The "tables of sixty" (*jadwal al sittini*) are expressed in the traditional Arabic letter numerals, but the operations are made using Indian numerals [Aya Sofia Library, Istanbul, Ms. 4857, f° 274 r and following; see A. Mazaheri (1975), pp. 96–141]. But, so far as I know, this is the only author to break the rule stated above.

### *Books of magic and divination*

The underlying reason for this preference is suggested in a work dealing with Arab astrology and white magic, dating from 1631 CE [BN Paris, Ms ar 2595, f° 1–308]. The author, a certain al-Gili (not to be confused with the mathematician cited above), uses a number of magic alphabets to name the spirits and the seven planets and shows how to make talismans "using Indian numerals", according to "the secret virtues of Arabic numerals". When drawing up "judgements of nativities" (i.e. horoscopes) the writer speculates about the numerals' "magical properties" in what he calls *hisab al jumal* (or "calculation of series") in which the letters of the Arabic alphabet are used, each with a number attached to it. He then draws up two lists of numbers, one called the *jumal al kabir* ("large series"), the other *jumal as saghir* ("little series") [see P. Casanova (1922); A. Winkler (1930)]. The author then explains how remarkable it is that "Arabic calculation (*bi hisab al 'arabi*) is always used for the little series, and Indian calculation (*bi hisab al hindi*) for the large series". In other words, the "large series" is always expressed in Indian numerals, the "little series" in Arabic letter numerals.

Why this difference? Large series were designed to give numerical values, true arithmetical numbers, while the little series was compared with it in order to give a name to each numerical value and determine its alleged secret virtues. For the author is of course referring to letter numerals when he mentions "the secret virtues of Arabic numerals". For him, the Indian numerals had no hidden powers.

Thus, the *Abjad* system (also called *Huruf al jumal*) was considered by the Arabs as "more their own than any other" [F. Woepcke (1857)]. They even gave their own name to it: *hisab al 'arabi* ("Arabic Calculation").

Arab magi and soothsayers presumably wanted to make a clear distinction between a system which they considered to be typically Arabic and part of Muslim traditions and practices, and another, the arithmetical superiority of which they were willing to recognise, but which remained in their eyes foreign and "not sacred".

### *A strange "machine" for thinking out events*

To gain a clearer idea of the Arabs' magical and divinatory practices, let us listen to Ibn Khaldun who, in his *Prolegomena*, describes that strange "machine" for thinking out events which is known as the *za'irja*. It inspired Ramon Lull (died 1315) in his famous *Ars Magna*, and, even at the end of the seventeenth century, Leibnitz was still one of its admirers.

> It is claimed that by using an artificial system, we can know about the contents of the invisible world. This is the *za'irjat al 'alam* ["circular chart of the universe"] supposedly invented by Abu'l 'Abbas as Sibti, from Ceuta, one of the most distinguished of the Maghribi *Sufi*. Near the end of the sixth century [of the Hegira = twelfth century CE], As Sibti was in the Maghreb while Ya'qub al-Mansur, the Almohad monarch, was on the throne.
>
> The construction of the *za'irja* ["circular chart"] is a wondrous piece of work. Many highly placed persons like to consult it to obtain useful knowledge from the invisible world. They try to use enigmatic procedures and sound out its mysteries in the hope of reaching their goals. What they use is a large circle, containing other concentric circles, some of which refer to the celestial spheres, and others to the elements, the sublunary world, spirits, all sorts of events and various forms of knowledge. The divisions of each circle are the same as the sphere they represent; and the signs of the zodiac, plus the four elements [air, earth, water and fire] are found within them. The lines which trace each division continue as far as the centre of the circle and are called "radii".
>
> On each radius appears a series of letters, each with a numerical value, some of which belong to the writing of records, that is to say to signs which Maghribi accountants and other officials still use for writing numbers. [The author is of course referring to the monograms and abbreviations of the Arabic names of the numbers, called *Dewani* numerals].
>
> There are also some *gobar* numerals [Fig. 25.5].
>
> Inside the *za'irja*, between the concentric circles, can be found the names of the sciences and various sorts of place name. On the other side of the chart of circles, there is a figure containing a large number of squares, separated by vertical and horizontal lines. This chart is fifty-five squares high, by one hundred and thirty-one squares across. [The author does not say that many of these squares are empty.] Some of these contain numbers [written in Indian numerals], and others letters. The rule which determines how the characters are placed in the squares is unknown to us, as is the principle that determines which squares are to be filled and which remain empty. Around the *za'irja* are found some lines of verse, written in the *tawil* metre, rhymed on the syllable *la*. This poem explains how to use the chart to obtain the answer to a question. But its lack of precision and vagueness mean that it is a veritable enigma.
>
> On one side of the chart is a line of verse written by Abu 'Abdallah Malik ibn Wuhaib [fl. 1122 CE], one of the West's most distinguished soothsayers. He lived under the Almoravid Dynasty and belonged to the *uleima* of Seville. This line is always used when consulting the *za'irja* in this way, or in any other way to obtain an answer to a question. To have an answer, the question must be written down, but with all the words split up into separate letters. Then, the sign of the zodiac is located [in the astronomical tables] and the degree of that sign as it rises above the horizon [i.e. its ascendant] coinciding with the moment of the operation. Then, on the *za'irja*, the radius is located which forms the initial boundary of the sign of the ascendant. This radius is followed to the centre of the circle, and thence to the circumference, opposite the place where the sign of the ascendant is indicated, and all the letters found on this radius, from beginning to end, are copied out.
>
> Also noted are the numerical signs [Indian numerals] written between the letters, which are then transformed into letters according to the *hisab al jumal* system [the "series calculation", used when replacing Indian numerals by letters and vice versa]. Sometimes units must be converted into tens, tens into hundreds, and vice versa, but always under the rules drawn up for the *za'ijra*. The result is placed next to the letters which make up the question. Then the radius which marks the third sign from the ascendant is examined. All the letters and numbers on this radius are written down, from its beginning to the centre, without going to the circumference. The numbers are then replaced by letters, according to the procedure already described, the letters being placed one beside the other.
>
> Then, the verse written by Malik ibn Wuhaib, the key for all operations is taken and, once it has been split into separate letters, it is put to one side. After that, the number of the degree of the ascendant is multiplied by what is called the sign's *'asas* [literally "base" or "foundation" an algebraic term for the index of a power, but here standing for

the number of degrees between the end of the last sign of the zodiac and the sign which is the ascendant at the time of the operation, the distance being taken in the opposite direction from the normal order of the signs].

To obtain this *'asas*, we count backwards, from the end of the series of signs; this is the opposite of the system used for ordinary calculations which starts at the beginning of the series. The product thus obtained is multiplied by a factor called the great *'asas* and the fundamental *dur* ["circuit", or "period", in astronomy used for the time it takes a point to make a complete orbit of the earth. A planet's *dur* is thus either its orbit, or the time taken to return to any given point in the heavens. But in the *za'irja, dur* also stands for certain numbers used for selecting the letters which will give the required answer].

The results are then applied to the squares on the chart, according to the rules governing the operation, and after using a certain number of *dur*. In this way, several letters are extracted [from the chart], some of which are eliminated, while the rest are placed opposite Ibn Wuhaib's verse.

Some of these letters are also placed among the letters forming the words of the question, which have already had others added to them. Letters of this series are eliminated when they occupy places indicated by the *dur* numbers. [Thus:] As many letters are counted as there are digits in the *dur*, when the last *dur* figure is arrived at, the corresponding letter is rejected; this operation is repeated until the series of letters is exhausted.

It is then repeated using other *dur*. The isolated letters remaining are put together and produce [the answer to the question asked by] a certain number of words forming a verse, in the same metre and rhyme as the key verse, composed by Malik ibn Wulaib. Many highly placed persons have become absorbed in this pursuit and eagerly use it in the hope of learning the secrets of the invisible world. They believed that the relevance of the answers showed that they were accurate. This belief is absolutely unfounded.

The reader will already have understood that the secrets of the invisible world cannot be discovered by such artificial means.

It is true that there is some connection between the questions and answers, in that the answers are intelligible and relevant, as in a conversation.

It is also true that the answers are obtained as follows: a selection is made between the letters in the question and on the radii of the chart. The products of certain factors are applied to the squares on the chart, whence some letters are extracted; certain letters are eliminated by several selections using the *dur*, and the rest are then placed opposite the letters making up the verse [of Malik ibn Wahaib].

Any intelligent person who examines the connections between the various steps in this operation will discover its secret. For these mutual connections give the mind the impression that it is in communication with the unknown, and also provide the way of going there. The faculty of noticing the connections between things is most often found in people used to spiritual exercises, and practice increases the power of reasoning and adds new strength to the faculty of reflection. This effect has already been explained on numerous occasions.

This idea has resulted in the fact that almost everybody has attributed the invention of the *za'irja* to people [the *Sufi*, Muslim esoterics], who had purified their souls by spiritual exercise.

Thus, the *za'irja* I have described is attributed to As Sibti [a *Sufi*]. I have seen another one, invented, it is said, by Sahl ibn 'Abdallah and must admit that it is an astounding work, a remarkable production of a profound spiritual application.

To explain why As Sibti's *za'irja* gives a versified answer, I tend to think that the use of Ibn Wuhaib's verse as a starting point influences the answer and gives it the same metre and rhyme.

To support this view, I have seen an operation made without this verse as a starting point, and the answer was not versified. We shall speak further of this later. Many people refuse to accept that this operation is serious and that it can answer one's questions. They deny that it is real and look on it as something suggested by fancy and imagination. If they are to be believed, people who use the *za'irja* take letters from a verse they have composed as they see fit and insert them among the letters making up the question and those from the radii. They then work by chance and without any rules; finally they produce the verse, pretending that it has been obtained by following a fixed procedure.

Such an operation would only be an ill-conceived game. No one using it would be capable of grasping the connections between beings and knowledge, or of seeing how different the operations of perception are from those of the intelligence. The observers would also be led to deny anything they do not perceive.

To answer those who call the *za'irja* a piece of juggling, suffice it to say that we have seen operations performed on it respecting the rules and, according to our considered opinion, they are always carried out in the same way and follow a genuine system of rules. Anybody possessed of a certain degree of penetration and attention would agree with this, once one of these operations has been witnessed.

Arithmetic, a science producing absolutely clear results, contains many problems which the intelligence cannot understand at once, because they include connections which are hard to grasp and elude observation.

How much more so, then, for the art of the *za'irja*, which is so extraordinary and whose connections with its subject are so obscure?

We shall cite one rather difficult problem here, to illustrate this point. Take several *dirhams* [silver coins] and, beside each coin, place three *fulus* [copper coins]. With the sum of the *fulus* you buy one bird, and with that of the *dirhams* several more at the same price. How many birds have you bought? The answer is nine. We know that there are twenty-four *fulus* to a *dirham*; so three fulus are the eighth of a *dirham*. Now, since each unit is made up of eight eighths, we can suppose that when making this purchase we have brought together the eighth of each *dirham* with the eighths of the other *dirhams*, and that each of these sums is the price of one bird. With the *dirhams* we have then bought just eight birds; the number of eighths in a unit; add to that the bird purchased with the *fulus* and we have nine birds in all, since the price in *dirhams* is the same as that in *fulus*.

This example shows us how the answer is hidden implicitly in the question and is arrived at by knowing the hidden connections between the quantities given in the problem.

The first time we encounter a question of this sort, we imagine that it belongs to a category that can be solved only by applying to the invisible world. But mutual connections allow us to extract the unknown from what is known. This is especially true of things in the sentient world and the sciences.

As for future events, they are secrets that cannot be known precisely because we are ignorant of their causes and have no certain knowledge of them.

From what we have explained, it can be seen how a procedure which, by using the *za'irja*, extracts an answer from the words of the question is a matter of making certain combinations of letters, which had initially been ordered to ask the question, appear in a different form.

For anyone who can see the connection between the letters of the question, and those of the answer, the mystery is now clear.

People capable of seeing these connections and using the rules we have explained can thus easily arrive at the solution they require.

Each of the *za'irja's* answers, seen under a different light, is like any other answer, according to the position and combination of its words; that is, it can either be negative or positive.

To return to the first point of view, the answer has another characteristic: its indications are in the class of predictions and their accordance with events [in other words, as Slane emphasises in modern terms, these indications are part of the category of agreements between discourse and the extrinsic].

But we shall never know [about future events] if we use procedures such as the one just described.

What is more, mankind is forbidden to use it for these ends. God communicates knowledge to whomsoever he wants; [for, as the Koran says (sura 2, verse 216) *God knows, but you know not.*] [See *Muqaddimah*, pp. 213–19; cf. Slane's translation, pp. 245–53.]

We must salute, in passing, Ibn Khaldun's eminently modern rationality, categorically rejecting the rather strange practices of Arab astrologers and soothsayers, which were in fact outlawed by Islam.

To this can be added the strange "revelation calculation" [*hisab 'an nim*], which soothsayers used in time of war to predict which of the two sovereigns would conquer or be conquered. Here is how Ibn Khaldun describes it in his *Prolegomena:*

> The numerical value of the letters in each sovereign's name was added up. Then each sum was reduced until it was under nine. The two remainders were compared. If one was higher than the other, and if both were odd or both even, then the king whose name had provided the lower figure would win. If one was even and the other odd, the king whose name had provided the higher figure would win. If both remainders were equal and even, then the king who had been attacked would vanquish. But if both were equal and odd, then the attacking king would be victorious [*Muqaddimah*, cf. Slane's translation, I, pp. 241–2].

### *The underlying reasons for the preponderance of Arabic letter numerals*

Thus, the system of Arabic letter numerals was favoured as a way of writing numbers not only by scholars, mathematicians, astronomers, physicians and geographers, but also by authors of religious works, mystics, alchemists, magicians, astrologers, soothsayers, scribes, officials and tradesmen, among both Arabs and Muslims.

The system was so common in the Islamic world, that Arab poets even invented a particular form of literary composition which used the letter numerals. These *ramz* were versified according to the arithmetic equalities or progressions of the numerical values of the letters in each line.

Even historians, and the lapidaries of North Africa, Spain, Turkey and Persia, were (at least in later periods) fond of a technique called *tarikh*, i.e. "chronograms", which consists in grouping a set of letters, the numerical value of which when added together produces the date of some past or future event, into one meaningful or significant word, or else into a short phrase.

This shows how the representation of numbers was of vital importance in the history of Islam. It was, of course, directly linked with both the meanings and the characters of Arabic writing, since the "numerals" were simply letters of the alphabet. This numerical notation was always written from right to left, like words, and, as for ordinary letters, the characters were generally joined up and slightly modified depending on whether they were isolated, initial, medial or final.

Thus, for poets enamoured of the *ramz*, these "letter numerals" or numerical letters were an integral part of their artistic expression, mirroring the beauty of the language. For artists, they also harmonised with the art of calligraphy, reflecting both their individual perspectives and the emotional state in which the work was created. And for those with a mystical bent, these same "numerals" allowed them to produce graphic or versified symbolic expressions, at once literal and numerical, of their quest for Allah.

Meanwhile, the scribes, who adhered to their characteristic embellishments, were able to give these numerical letters the same grace, balance and rhythm as the ordinary letters in their miniatures and illuminations.

All of which confirms the perfect continuity between this system and the purest of Arab and Islamic traditions, and the fundamental practices of Muslim mysticism.

It must not be forgotten that the Arabic script is considered to embody a Revelation and the spreading of the word of the Prophet; it is thus the basic criterion for belonging to the Islamic community (the *Umma*). It is this close connection between the Muslim religion and the Arabic script which gives the Arabic alphabet its privileged, almost fundamentally sacred, position. Tradition even has it that the reed pen, the famous *qalam* ("calamus"), was the first of all Allah's creations.

For the *Hurufi* ("Letterers"), sects based on beliefs attached to the symbolic meaning of Arabic letters, a name was the essence of the thing named.

And, as all names are supposed to be contained in the letters of the discourse, the entire universe was the product of Arabic letters. In other words, from these letters proceeded the universe. Hence the association between the "science of letters" (*'ilm al huruf*), the "science of words" (*'ilm al simiya*) and the "science of the universe" (*'ilm al 'alam*).

The mystic al-Buni was one follower of this belief. He established correspondences between the Arabic letters and what he thought were the elements of the visible world: the four elements (water, earth, air and fire), the celestial spheres, the planets and the signs of the zodiac. And, as there are twenty-eight letters, he associated them with the twenty-eight lunisolar mansions [see E. Doutté (1909)].

God is a force translated by the Word; he acts through his voice and so, by inference, through the very letters of the Arabic alphabet.

Thus the "sciences" of letters and words, once mastered, would reveal the attributes of Allah as they are manifested in nature through the Arabic letters. According to these doctrines:

> The Arabic letter symbolises the mystery of being, through its fundamental unity derived from the divine Word and its countless diversity resulting in virtually infinite combinations; it is the image of the multitude of creation, and even the very substance of the beings it names. Together, they are regarded as manifestations of the *Word Itself*, inseparable attributes of the Divine Essence, as indestructible as the Supreme Truth. Like the divine being, they are immanent in all things. They are merciful, noble and eternal. Each of them is invisible (hidden) in the Divine Essence [J. Chevalier and A. Gheerbrant (1982)].

This is why, according to the precepts of Islam, the Koran, as the Revelation of the Prophet Mohammed, cannot be read in another language than Arabic, nor can it be transcribed into a different script. For this Book, seen as one of the expressions of the Word of Allah, is identified with the Divine Essence. To quote Doutté:

> This conception takes us back to ancient times, when the Romans, by the word *litterae*, and the Nordic peoples, by the word "rune", meant the entirety of human knowledge. Nearer to the Arabs, in the Semitic world, the Talmud teaches that letters are the essence of things. God created the world by using two letters; Moses on going up to heaven met God who was weaving crowns with letters. Ibn Khaldun has much to say about these doctrines and gives a theory of written talismans [see *Muqaddimah*, trans. Slane, II, pp. 188–95]: as the letters composing them were formed from the elements which make up each being, they could act upon them.

Such is the basis of *'ilm al huruf* and *'ilm as simiya*, Islam's mystical "sciences" of letters and words.

One category of letters, whose magical powers have a religious origin and are thus characteristic of Arab magic, are those at the beginning of certain suras of the Koran, and whose meaning is totally unknown (or else jealously guarded by Muslim mystics). For example, sura II begins with *alif, lam, mim*; sura III with *alif, lam, mim, sad* etc. Orthodox Muslims call these letters *mustabih*, and say that their meaning is impenetrable for the human mind; thus, unsurprisingly, they have been adopted by magicians.

Al-Buni calls them *al huruf an nuraniya*; there are fourteen of them, exactly half the number (28) of lunar mansions, from which he draws further speculations. Each of them, he points out, is the initial of one of the names of God. Two of these groups, which contain five letters, have particularly attracted magicians. They are supposed to have extraordinary virtues and many *herz* ("talismans") have been made using them.

If letters have magical powers, then these powers are increased when they are written separately. In the Arabic script, individual letters are more perfectly formed than when they are joined up.

But the letters' most singular properties come from their numerical values. Two different words can have the same numerical total. The mysticism of letters then says that they are equivalent. In the Cabbala, this is the principle of "gematria". It is also a favourite of Muslim magic. Not only are words linked together by the numbers expressed by their letters, but these very letters can reveal their magical virtues through a numerical evaluation of the letters and words [E. Doutté (1909)].

In other words, Arabic words have a numerical value. A reciprocal logic even had it that numbers were charged with the semantic meaning of the word or words they corresponded to. Hence, as with the Cabbala, ciphered messages, "secret languages" and all sorts of speculations were cooked up by mystics, numerologists, alchemists, magi and soothsayers. Their aim was to stop laymen understanding and harmonising with these esoteric meanings, which supposedly held a hidden truth, or else to compose cryptographic texts wrapped in apparently indecipherable allegories and puzzles, or to use them for a variety of interpretations, conclusions, practices and predictions (see Chapter 20).

It can thus be seen how a numerical value was added to the letters' symbolic, magical and mystical powers, thus giving them the broadest and most effective range of meaning.

Words have always fascinated us, but numbers even more so. Since time immemorial, numbers have been the mystic's ideal tool. They do not express only arithmetical values but, inside what was considered to be their visible exteriors, numbers also contained magical and occult forces which ran on an unseen current, rather like an underground stream. Such ideas could be either for good or evil, depending on their inherent nature.

The magical and mystical character of numbers is a common human belief. Their importance in Mesopotamia, ancient Egypt, pre-Columbian America, China and Japan is beyond our scope. As are the theories and doctrines of the Pythagoreans and Neo-Platonists who, struck by the importance of numbers and their remarkable properties, made them into one of the bases of their metaphysics, believing that numbers were the principle, the source and the root of all beings and things. But what should be emphasised is the direct link between a belief in the magic of numbers and the fear of enumeration, present among the Hebrews (see for example Exodus, 30: 12 and II Samuel 24: 10), the Chinese and Japanese (who are particularly superstitious about the number four), and also among several African, Oceanian and American peoples, who find numbers repellent.

It should be said in passing, that the ancient fear of enumeration reveals the difficulties humans have always had in assimilating the concept of number, which they see, and rightly so, as highly abstract.

It is this very link between magic and the ancient fear of numbers which forbids, for example, North African Muslims from pronouncing numbers connected with people dear to them or personal possessions. For, according to this belief, giving the number of an entity allows it to be circumscribed. If you provide the number of your brothers, wives or children, your oxen, ewes or hens, the sum of your belongings, or even your age, you are giving Satan, who is ever on the lookout, the possibility to use the hidden power of these numbers. You thus allow him to act upon you and do evil to the people or things you so imprudently enumerated.

A sort of superstitious reciprocity led to the making of *herz* in the form of magic squares: talismans with alleged beneficent powers, such as curing female sterility, bringing happiness to a home or attracting material riches.

As a passing remark, Islamic religion and traditions see the number five as a good omen in, for example, the five *takbir* of the Muslim profession of faith, *Allah huwa akbar* ("God is Great"); the five daily prayers; the five days dedicated to 'Arafat; the five fundamental elements of the pilgrimage to Mecca, the five witnesses of the pact of the *Mubahala*; and the five keys to the mystery in the Koran (6: 59; 31: 34). There is also Thursday, called in Arabic *al khamis* ("the fifth"), which is a particularly sacred day. Then there are: the five goods given as a tithe; the five motives for ablution; the five sorts of fasting; the pentagram of the five senses and of marriage; the five generations that mark the end of tribal vengeance; and so on. Naturally, there are the five fingers of the hand, placed under special protection in memory of the five fingers of the "hand of Fatima", the daughter of Mohammed and Khadija, and wife of 'Ali, the Prophet's cousin [see J. Chevalier and A. Gheerbrant (1982); E. Doutté (1909); EIS; T. P. Hugues (1896)].

Even today if you foolishly ask Tunisians, Algerians or Moroccans how old they are, or how much money they have, they will cast off the evil eye by vaguely replying "a few" if they are polite, or else curtly say "five", or even brusquely slap the five fingers of their hand over your own "evil eye".

To sum up, each of the twenty-eight Arabic letters, as an ordinary letter, was supposed to have its own symbolic meaning, magical power and creative force. But as a numeral or written in cipher, each was linked with a number and, as such, was directly in touch with the supposed idea, power and force contained in that number. A name is the outward sign of the Word, considered to be one of the main magical and mystical forces. As it is made up of letters, and thus of the corresponding numbers too, it is easy to see why the Arabs' alphabetic numbering (as a particular case among their multiple ways of evaluating their letters and words) was for mystics, magi and soothsayers a product of sound, sign and number, and hence had powers that transcended the ordinary alphabet.

We can now see how important this system was at all levels of Islamic society. And we can also see why the Indian place-value system was considered by most authors to be something absolutely alien to their culture and traditions.

The direction it was written in added to its relative unpopularity. It ran from left to right (one hundred and twenty-seven, for example, being written as 127), the opposite way to Arabic script. And as the numerical letters were written from right to left, from the highest digit to the lowest, and obeyed the rules of the Arabic cursive script, they were favoured above any other system.

The direction of Indian numerals had been highly practical for Indian mathematicians and accountants, whose script went from left to right. But this fact (which caused obvious problems for people accustomed to writing from right to left) raised difficulties for Arab-Muslim scholars.

They would certainly have solved this problem if they had inverted the original order of the Indian decimal system, by writing something like this:

8 7 6 3 2

$\longleftarrow$

when an Indian would have written:

$$23{,}678\ (=\ 2\times10^4+3\times10^3+6\times10^2+7\times10+8).$$

They would thus have completely adapted the Indian system to their own script. But this idea apparently never occurred to the Arabs, or else they refused to break with the Indian tradition.

Another reason, the last we shall give here, for this opposition was as follows.

During their relations with India, the Arabs were in contact with the Hindî, but also with the Punjabî, the Sindhî, the Mahârâshtrî, the Manipurî, the Orissî, the Bihârî, the Multanî, the Bengalî, the Sirmaurî and even the Nepalî. A glance at Chapter 24 will confirm how much the writing of numerals in India varied, not only from one period to another, but also in different regions, and even with different scribes (Fig. 24.3 to 52). What was a 2 for some became something like a 3, 7, or 9 for others, for palaeographic reasons. In other words, a lack of standardisation meant that the written form of Indian numerals remained unstable. But for mathematicians and astronomers numbers had to remain the same and be absolutely consistent. How could one transmit the fundamental data of a work of astronomy, for example, if the numerical value of observations and results could be variously interpreted, depending on the time, place and habits of the user? What is more, if a scribe or copyist made a mistake, it might never be noticed. These numerals were therefore not sufficiently rigorous for works dealing with mathematics, geography or astronomy in which value was of prime importance. Hence the preference for numerical letters, which did not present such a problem.

Need we add that, if the so-called "Arabic" numerals had really been invented by the Arabs, then they would have been used more widely and adopted by Muslims much more rapidly? There is also a good chance that these numerals would have been written from right to left, like the Arabic script.

These important facts add to the indisputable evidence that our present number-system comes from India.

Among other imperishable merits, the Arab-Islamic civilisation did certainly transmit our modern numerals and methods of calculation to mediaeval Europe, which was at the time at a much lower scientific and cultural level. In gratitude for this basic contribution, Europe then named these numerals after the people who had provided them. But to say that Islam was the cradle of these numerals would be to fall into the trap laid by an erroneous term, which even Arab and Muslim scholars never used in their writings or vocabulary.

## DUST-BOARD CALCULATION

The time has come to discuss Indian calculation methods, which not only played an important role in the transmission of Indian numerals throughout the Islamic world, but also profoundly influenced how techniques evolved.

Many good reasons lead us to suppose that, from earliest times, Arab-Muslim arithmeticians in the East and the West made their calculations by sketching out the nine Indian numerals in loose soil, with a pointer, stick or just with a finger. This was known as *hisab al ghubar* ("calculating on dust") or *hisab 'ala at turab* ("calculating on sand").

But they did not always write on the ground; they also had other methods. Their most common tool seems to have been the counting board, what is called in the East *takht al turab* or *takht al ghubar* (from *takht*, "tablet" or "board", *turab*, "sand", and *ghubar*, "dust"), which was also known in the Maghreb and Andalusia as the *luhat al ghubar* (*luhat* being a synonym of *takht*).

Several Persian poets refer to it, at least from the twelfth century on, such as Khaqani, in his eulogy for Prince Ala al-Dawla Atsuz (1127–1157) [A. Mazaheri (1975)]:

> The seven climates tremble with quartan fever;
> And dust will cover the vaulted sky,
> Like the accountant's board (*takht*).

Or the mystical poet Nizami (died 1203) [Nizami (1313), cited by A. Mazaheri (1975)]:

> From the system of nine heavens
> [Marked] with nine figures,
> [God] cast the Indian numerals
> Onto the earth board.

This counting board was favoured not only by professional Arab accountants, mathematicians and astronomers, but also by magi, soothsayers and astrologers.

In about 1155, Nizami told this story, which features the philosopher al-Kindi (ninth century) [Nizami as above]: "Al Kindi asked for the dust board, got up and [with his astrolabe] read the height of the sun, the hour and traced the horoscope on the sand board (*takht al turab*) . . . "

It consisted of a board of wood, or of any other material, on which was scattered dust or fine sand, so that the Indian numerals could be traced out in it and calculations made. Powder, or sometimes even flour were also used, as our sources indicate. The word *ghubar* in fact means "powder" or "any powdery substance" as well as "dust".

This counting board was not unique to Arab arithmeticians. It was also used long before Islam by the Indians (see **Pâtîganita*).

## TRACES OF THE OLD PERSIAN ABACUS FROM THE TIME OF DARIUS

Old abacuses from time of Darius and Alexander were also used, at least in Persia during the first centuries of the Hegira. Calculations were made by throwing down pebbles or counters, and certain Persian accountants kept up this method (see Fig. 16.72 and 73).

The following is, of course, just a hypothesis, but it is supported by much of the evidence. The Persian verb "to count", "to calculate" is *endakhten*, which also means "to throw". At this time, arithmetical operations were carried out on tables or rugs, divided by horizontal and vertical lines, on which the counters were placed, their value changing as they moved from one column to another.

It is also interesting to note that the action which corresponds to the verb *endakhten* ("count", "calculate") is *endaza*, which means three things: "throwing", "counting" and "calculating". This is shown in this brief quotation from *Kalila wa Dimna*, a famous Persian fable, here in a twelfth-century version by Abu al Ma'ali [see A. Mazaheri (1975)]: "Having carefully listened to his mother's words / The lion threw them backwards (*baz endakht*) with his memory." This is so subtle that a commentary is necessary.

Even for a lion, "throwing words backwards" is meaningless. But if we take the verb to mean "to calculate" or, by extension, "to measure" we can then see that the king of the jungle had thought over, or "weighed", his mother's words.

But let us not take etymology too far in order to explain something which had already almost vanished from the old country of the Sassanids, for these words had lost their numerical meaning by the thirteenth century. And the instrument itself, rightly considered as cumbersome and impractical, had been rejected by the region's professional accountants at an early date. (Note also that they rejected the Chinese abacus, introduced by Mongol invaders during the thirteenth century; but the unpopularity of this excellent apparatus was due to the Persians' hatred of Genghis Khan and his successors.)

## THE BOARD AS A COLUMN ABACUS

To return to calculations made on the ground, or else in dust scattered over a flat surface or board, there were of course different ways of working. Here is the most rudimentary.

The arithmeticians began by tracing several parallel lines on the surface to be used, thus marking out a series of columns which corresponded to the places of the decimal system. Then they drew the nine numerals inside each one. In this way, they immediately acquired a place value.

The Arabs, like the ancient Indian arithmeticians, would write a number such as 4,769 by tracing the number 9 in the units column, the number 6 in the tens column, the number 7 in the hundreds column and the number 4 in the thousands column.

| *Ten thousands* | *Thousands* | *Hundreds* | *Tens* | *Units* |
|---|---|---|---|---|
| | 4 | 7 | 6 | 9 |

So there was no need for zero. It was sufficient just to leave the column empty, as in our next example which represents 57,040:

| *Ten thousands* | *Thousands* | *Hundreds* | *Tens* | *Units* |
|---|---|---|---|---|
| 5 | 7 | | 4 | |

As for the calculations, they were carried out in the dust, then erased.

There is a clear trace of this in the etymology of the Sanskrit words *gunara, hanana, vadha, kshayam*, etc., used by the Indians to mean "multiplication". Literally, they mean "to destroy" or "to kill", in allusion to the successive wiping out of intermediary products, as our example will now show.

Let us suppose that an accountant wants to multiply 325 by 28. The first thing to do is trace out the four columns required. Then, inside them, we place 325 and 28 as follows, with the highest place of the multiplicand in the same column as the lowest place of the multiplier.

| | | | |
|---|---|---|---|
| <br>2 | 3<br>8 | 2 | 5 |

We then multiply the upper 3 by the lower 2. As this equals 6, we place this figure to the left of the upper 3:

| ↓ | | | |
|---|---|---|---|
| 6<br>2 | 3<br>8 | 2 | 5 |

Then we multiply the upper 3 by the lower 8. As this equals 24, we wipe out the 3 and replace it with 4 (the unit column of 24, the partial product):

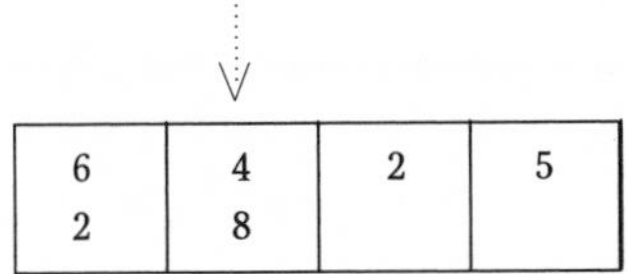

And, to the 6 we add 2 (the tens digit of 24):

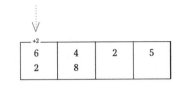

| ↓ +2 | | | |
|---|---|---|---|
| 6<br>2 | 4<br>8 | 2 | 5 |

Then, after erasing, we have:

| | | | |
|---|---|---|---|
| 8<br>2 | 4<br>8 | 2 | 5 |

The first step has now been carried out, both columns of the multiplier 28 having acted on the hundreds column of the multiplicand (the upper 3 of the initial layout).

We then proceed to the second step by moving all the numbers of the multiplier one place to the right:

| | | | |
|---|---|---|---|
| 8 | 4<br>2 | 2<br>8 | 5 |

——————>

Then, by using the tens digit of the multiplicand (the upper 2 of the initial layout), we multiply 2 by 2. As this equals 4, we then add 4 to the 4 which lies immediately to the left of upper 2:

| | ↓ +4 | | |
|---|---|---|---|
| 8 | 4<br>2 | 2<br>8 | 5 |

Then, after erasing, we have:

| | | | |
|---|---|---|---|
| 8 | 8<br>2 | 2<br>8 | 5 |

We then multiply the same upper 2 by the lower 8. This makes 16, so we replace, after erasing, the upper 2 with 6 (the units digit of the result):

| | | ↓ | |
|---|---|---|---|
| 8 | 8<br>2 | 6<br>8 | 5 |

We then add 1 (the tens digit of 16, as above) to the 8 just to the left of the new 6:

| 8 | 8 | 6 | 5 |
|---|---|---|---|
|  | 2 | 8 |  |

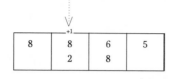

Then, after erasure, we have:

| 8 | 9 | 6 | 5 |
|---|---|---|---|
|  | 2 | 8 |  |

We have now finished the second step, since both digits of the multiplier 28 have operated on the tens digit of the multiplicand (the upper 2 of the initial layout).

We then begin the next step by moving the numbers of the multiplier one column to the right again:

| 8 | 9 | 6 | 5 |
|---|---|---|---|
|  |  | 2 | 8 |

This time we multiply the units digit of the multiplicand (the upper 5 of the initial layout) by the lower 2. This comes to 10, so we leave untouched the upper 6 (there being no unit digit in the number 10), but add 1 (the tens digit of 10) to the 9 immediately to the left of the 6:

| 8 | 9 | 6 | 5 |
|---|---|---|---|
|  |  | 2 | 8 |

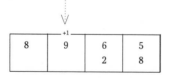

But as this makes 10 again, we wipe out the 9, leave the space empty (because of zero units in 10) and add 1 to the 8 just to the left of this blank column:

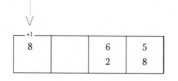

| 8 |  | 6 | 5 |
|---|---|---|---|
|  |  | 2 | 8 |

Then, after erasure, we have:

| 9 |  | 6 | 5 |
|---|---|---|---|
|  |  | 2 | 8 |

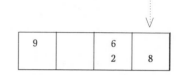

We then multiply the upper 5 by the lower 8. As this makes 40, we wipe out the upper 5, but leave the space empty because there is no unit in the product found:

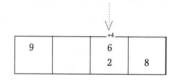

| 9 |  | 6 |  |
|---|---|---|---|
|  |  | 2 | 8 |

But we then add the 4 of the product to the upper 6:

| 9 |  | 6 |  |
|---|---|---|---|
|  |  | 2 | 8 |

As this again makes 10, we wipe out the 6, leave the space blank and add 1 to the number (zero) in the empty space immediately to the left:

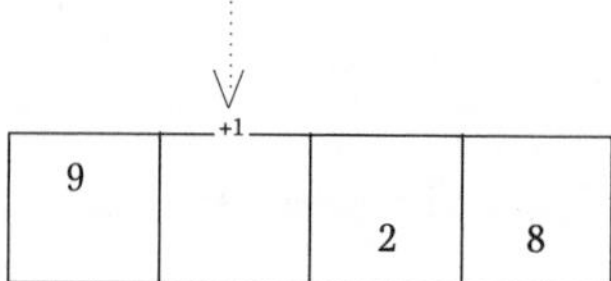

Then, after erasure, we have:

| 9 | 1 |  |  |
|---|---|---|---|
|  |  | 2 | 8 |

And, as the lowest place of the multiplier is now in the lowest place of the multiplicand (here, the units column of the abacus), we know that the multiplication of 325 by 28 has been completed.

All we have to do know is to read the number on the upper line, nine thousand, one hundred, no tens, no units; so the result is 9,100:

| 9 | 1 |  |  |
|---|---|---|---|
|  |  |  |  |

This method thus consists in carrying out a number of steps corresponding to the number of places in the multiplicand, each being subdivided into a series of products of one of the digits of the multiplicand successively operated on by all the digits of the multiplier.

In this case, the procedure (now called the operation's "algorithm") has three main phases, each subdivided into two simple steps consisting of calculating a partial product; hence six simple steps in all:

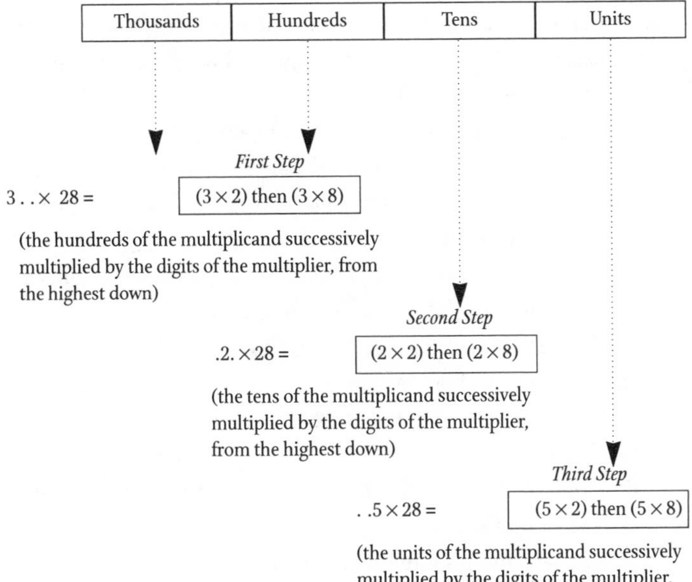
Thousands
Hundreds
Tens
Units
First Step
3 . . × 28 =
(3 × 2) then (3 × 8)
(the hundreds of the multiplicand successively multiplied by the digits of the multiplier, from the highest down)
Second Step
.2. × 28 =
(2 × 2) then (2 × 8)
(the tens of the multiplicand successively multiplied by the digits of the multiplier, from the highest down)
Third Step
. .5 × 28 =
(5 × 2) then (5 × 8)
(the units of the multiplicand successively multiplied by the digits of the multiplier, from the highest down)

In other words, this "algorithm" works according to the following formula:

$$
\begin{aligned}
325 \times 28 &= (3 \times 100 + 2 \times 10 + 5) \times (2 \times 10 + 8) \\
&= (3 \times 2) \times 1{,}000 + (3 \times 8) \times 100 \quad \textit{(first step)} \\
&\quad + (2 + 2) \times 100 + (2 \times 8) \times 10 \quad \textit{(second step)} \\
&\quad + (5 \times 2) \times 10 + 5 \times 8 \quad \textit{(third step)}
\end{aligned}
$$

This counting board thus allows us to carry out calculations without using zero, which explains why certain Arab manuscripts dealing with Indian numerals and methods of calculation make no mention of it.

In certain parts of North Africa, this method continued to be used until the end of the seventeenth century, which explains why the *Ghubar* numerals of the Maghreb generally come down to us in incomplete series, with the zero missing (Fig. 25.5).

But in the East, it gradually disappeared after the tenth or eleventh century and was replaced by more highly developed methods. It is true that this system is long, tiresome and requires considerable concentration and practice.

In fact, very little distinguishes it from methods used in Antiquity. The reason for this has less to do with the numerals themselves than with the method used. It makes no difference whether we trace out the nine Indian numerals, the first nine letters of the Greek or Arabic alphabet, or even the first nine Roman numerals. The principle would still remain virtually the same.

The Indians, as we have seen, certainly used such a system early in their history. But they abandoned it as soon as they had developed their own place-value system and their arithmetic allowed simpler rules to be found.

To carry out arithmetical operations, the early Indians used whatever was to hand. Like everybody else, they presumably began by using pebbles, or similar objects. Then, or perhaps at the same time, they carried out operations on their fingers. But during the next stage, when they developed their first written numerals, they conceived of the idea of drawing several parallel columns, putting the units in the first one, the tens in the second, the hundreds in the third, and so on. They thus invented the column abacus, as others did before and after them. But instead of using pebbles, counters or reeds, they preferred their own nine numerals, which they traced out in dust with a pointer inside the appropriate columns. This was the birth of their dust abacus, which they later improved by working on a table or board covered with sand or dust, instead of the ground.

But this system could not evolve further, so long as it continued to be a column abacus; this concept in fact trapped the human mind for centuries, preventing us from thinking out simpler and more practical rules.

This once again highlights the importance of the discovery of the place-value system. This principle had, of course, long been present in the way calculations were made, but without anybody noticing it. The creative genius of the Indians then brought together all the necessary ideas for discovering the perfect number-system. They had to:

- get rid of stones, reeds, knotted cords, manual techniques or, more generally, any concrete method;
- eliminate any notions of ideogrammatic representation (writing numerals as numbers of lines, points etc.), which certainly came later than the previous system, but was just as primitive;
- eliminate any notation of numbers higher than or equal to the base of the calculation system;

- keep only the nine numerals, in a decimal system, and apply place-value to them;
- replace all existing systems by this group of nine numerals, independent one from the other, and which visually represented only what they were supposed to represent;
- get rid of the abacus and its now useless columns, and apply the new principle to the numerals which were freed from any direct visual intuition;
- fill the gap now created by this method when a place was not filled by a numeral;
- think of replacing this gap by a written sign, acting as zero in the strict arithmetical and mathematical sense of the term.

To sum up, it was by rejecting the abacus that Indian scholars discovered the place-value system.

This raises a question concerning the arithmeticians of the Maghreb and Andalusia, who continued to use the dust abacus and its associated methods for several centuries: did the Western Arabs not know about zero and the place-value system? The answer is no, because these arithmeticians knew the *Hindi* numerals which, as we know, were based on the place-value system and included zero.

In other words, they were aware that the numerals they used could also be manipulated with zero and its associated rules. This is shown in certain Maghrebi manuscripts, in which zero is drawn as a circle (Fig. 25.5).

Why, then, did they not use them for "written calculation" instead of using a dust board? The answer seems to lie in the attachment the Maghrebi and Andalusians always felt for traditions coming from the time of the conquest of North Africa and Spain. Thus, the use of the dust board/abacus has the same traditionalist explanation as their cursive script, derived directly from the *Kufic*.

In fact, the Arabs inherited various arithmetical methods from the Indians, ranging from the most primitive to the most highly developed. In their thirst for knowledge, they presumably took from the Indians everything they could find in terms of calculation methods, without realising that certain things could well be left alone. We should not forget that India is a veritable sub-continent, cut up into regions, peoples, practices, customs and traditions, and it has always been difficult, if not impossible, to see it as a whole.

It is because they came into contact with people who used methods already abandoned by the scholars, and decided to uphold this tradition, that certain Arab-Muslim arithmeticians remained stuck in such a rudimentary rut for several generations.

## THE COLUMNLESS BOARD

But this was not, of course, the case for all the Arabs. Others were lucky or bright enough to take up the dust board freed of its columns.

Among them was al-Khuwarizmi. In his *Kitab al jama' wa'l tafriq bi hisab al hind* (*Book of Addition and Subtraction According to Indian Calculations*) he had not only explained the decimal place-value system when applied to Indian numerals, but also recommended "writing the zeros so as not to mix up the positions" [A. P. Youschkevitch (1976), p. 17]. There was also Abu'l Hasan 'Ali ibn Ahmad an Nisawi (died c. 1030), whose *Al muqni' fi'l hisab al hind* (*Complete Guide to Indian Arithmetic*) followed the same sources and methods as the previous work.

Abu'l Hasan Kushiyar ibn Labban al-Gili (971–1029) also deserves a mention. The first chapter of Book I of his *Maqalatan fi osu'l hisab al hind* (*Two Books Dealing with Calculations Using Indian Numerals*) begins as follows:

> The aim of any calculation is to find an unknown quantity. To do this, at least [one of these] three operations is necessary: multiplication [*al madrub*], division [*al qisma*] and [extraction of] the square root [*al jadr*] . . . There is also a fourth operation, less often used, which is the extraction of the side of a cube.
>
> But before learning how to carry out these operations, we must familiarise ourselves with each of the nine numerals [*huruf*], the position [*rutba*] of each in relation to the others in the [place-value] system [*al wad'*] . . .
>
> Here are the nine numerals [written in the *Hindi style*, but here updated]:
>
> 9 8 7 6 5 4 3 2 1.
>
> [Thus positioned], they represent a number and each stands in a position [*martaba*].
>
> The first is the image of one, the second of two, the third of three . . . and the last of nine. What is more, the first is in the position of the units, the second in the tens, the third in the hundreds, the fourth in the thousands . . .
>
> As for the number formed by these numerals, it must be read: nine hundred and eighty-seven million six hundred and fifty-four thousand three hundred and twenty-one.
>
> [When writing] a number [containing several place-values] we must put a zero [*sifr*, literally "void"] in each place where there is no numeral. For example, to write ten, we put a zero in the place of the units; to write a hundred, we put two zeros, one in the place of the missing tens and one in the place of the units.

Here are these two figures:

Ten: 10
Hundred: 100

There are no exceptions to this rule.

For any of the nine numerals under consideration, the one immediately to its left stands for tens, the next one to the left for hundreds, and the next one to the left for thousands.

In the same way, any of the nine numerals under consideration stands for the tens of the numeral immediately to its right, for the hundreds of the next numeral to its right, for the thousands for the following one, and so on [f° 267v and 268r; A. Mazaheri (1975), pp. 75–76].

These scholars had thus understood that the place-value system and zero removed the need for columns on a counting board.

So, like the Indians, they entered into the era of modern "written calculation".

But they now had to know off by heart the tables giving the results of the four basic operations on these numerals. This is what the Persian mathematician Ghiyat ad din Ghamshid ibn Mas 'ud al-Kashi explains in his *Miftah al hisab* (*Key to Calculation*), in which he reproduces one of these tables: "Here is the table for multiplying numbers inferior to ten. The arithmetician should learn it by heart and know it perfectly, for it can also be used for the multiplication of numbers superior to ten " [see A. Mazaheri (1975)].

*Calculating on a columnless board by erasing intermediate results*

Our first example of this method comes from the work of Kushiyar ibn Labban al-Gili, cited above [f° 269v to 270v]:

> We want to multiply three hundred and twenty-five by two hundred and forty-three.
>
> We put them on the board as follows:
>
> |   |   | 3 | 2 | 5 |
> |---|---|---|---|---|
> | 2 | 4 | 3 |   |   |
>
> the first numeral [on the right] of the bottom number being always under the last numeral [on the left] of the top number.
>
> We then multiply the upper three by the lower two; this makes six, which we place above the lower two, to the left of the upper three, thus:
>
> | 6 |   | 3 | 2 | 5 |
> |---|---|---|---|---|
> | 2 | 4 | 3 |   |   |
>
> If the six had contained tens, these would have been placed to its left.
>
> Then we multiply the upper three again by the lower four; this makes twelve, of which we place the two above the four and add the one [which represents the tens] to the six of sixty, obtaining seventy, thus:
>
> | 7 | 2 | 3 | 2 | 5 |
> |---|---|---|---|---|
> | 2 | 4 | 3 |   |   |
>
> Then we multiply the upper three by the lower three; this makes nine, which replaces the upper three:
>
> | 7 | 2 | 9 | 2 | 5 |
> |---|---|---|---|---|
> | 2 | 4 | 3 |   |   |
>
> We then advance the bottom number one place towards the right, thus:
>
> | 7 | 2 | 9 | 2 | 5 |
> |---|---|---|---|---|
> |   | 2 | 4 | 3 |   |
>
> And we multiply the two above the lower three by the lower two; this makes four which, added to the two above the lower two, makes six:
>
> | 7 | 6 | 9 | 2 | 5 |
> |---|---|---|---|---|
> |   | 2 | 4 | 3 |   |
>
> Then we multiply the upper two again by the lower four; this makes eight, which we add to the nine above the four:
>
> | 7 | 7 | 7 | 2 | 5 |
> |---|---|---|---|---|
> |   | 2 | 4 | 3 |   |
>
> Then we multiply the upper two again by the lower three; this makes six, which replaces the upper two above the lower three:
>
> | 7 | 7 | 7 | 6 | 5 |
> |---|---|---|---|---|
> |   | 2 | 4 | 3 |   |
>
> We then advance the bottom number one place [towards the right], thus:
>
> | 7 | 7 | 7 | 6 | 5 |
> |---|---|---|---|---|
> |   |   | 2 | 4 | 3 |

> Finally, we multiply the upper five by the lower two; this makes ten, which we thus add to the tens position above the lower two:
>
> | 7 | 8 | 7 | 6 | 5 |
> |---|---|---|---|---|
> | | | 2 | 4 | 3 |
>
> Then we multiply the five again by the lower four; this makes two [tens]. Added to the tens [in the position above] the four, [these two numbers] together make nine:
>
> | 7 | 8 | 9 | 6 | 5 |
> |---|---|---|---|---|
> | | | 2 | 4 | 3 |
>
> Finally, we multiply the five by the lower three; this makes fifteen, thus leaving the five alone, we just add one [the tens digit] to the tens, thus:
>
> | 7 | 8 | 9 | 7 | 5 |
> |---|---|---|---|---|
> | | | 2 | 4 | 3 |
>
> The [upper] number is the one we wanted to calculate.

This method thus consists in applying the same number of steps as there are places in the multiplicand, each being subdivided into as many products of one of its numbers and the successive digits of the multiplier.

The same method, with some variants, can be found in, for example al-Khuwarizmi and An Nisawi, as well as numerous Indian mathematicians such as Shrîdharâchârya (date uncertain), Nârâyana (1356), Bhâskarâchârya (1150), Shrîpati (1039), Mahâvîrâchârya (850), etc. [B. Misra (1932) XIII, 2; H. R. Kapadia (1935), 15; B. Datta and A. N. Singh (1938), pp. 137–43].

## THE DUST BOARD SMEARED WITH A TABLET OF MALLEABLE MATTER

Despite being freed of columns, this approach remained primitive. It was merely a written imitation of older methods and could hardly develop further because of the limitations imposed by the medium.

The dust board was certainly very practical for calculation methods with or without the abacus columns, and especially for the technique of wiping out intermediate results, as this passage from *Psephophoria kata Indos* shows (by Maximus Planudes (1260–1310), a Byzantine monk):

> It would perhaps not be superfluous to show another multiplication method. But it is extremely inconvenient when done with ink and paper, while it is suited for use on a board covered with sand. For it is necessary to wipe out certain numbers, then replace them with others; when using ink, this leads to much inextricable confusion, but with sand it is easy to wipe out a number with one's finger and replace it with others. This method of writing numbers in sand is especially useful, not only for multiplication, but for other operations as well . . . [BN Paris. Ancien Fonds grec, Ms 2381, f° 5v, ll. 30–35; Ms 2382, f° 9r, ll. 13–25; Ms 2509, f° 105v, ll. 2–10] [see A. Allard (1981); H. Waeschke (1878); F. Woepcke (1857), p. 240].

But the dust board became increasingly impractical as the numerals began to resemble one another more and more.

Just take a wooden board, sprinkle it with dust or flour, then draw numbers on it in the usual way. Then try to carry out one of the operations we have seen, following the same method. You will immediately see how hard it is to replace one number with another. If you sprinkle the number to be removed, or use a flat instrument to wipe it out, the very nature of the powdery matter means you risk wiping out all the adjacent numbers as well.

Attempts were made to get round this problem by leaving a large space between the different numbers. But there are limits to the size of the board, and this means that longer, more complicated calculations would require a larger space. What is more, by wiping out intermediate results, this method limits the contribution of the human memory and makes spotting intermediate mistakes extremely difficult. Hence an obvious block on finding out simpler and more practical methods.

It is possible to guess what replaced sand calculation and the use of the dust board in certain Islamic countries.

As we have seen, in Persian and Mesopotamian provinces, the preceding method of calculation was also called *takht al turab* (or in Persian *takhta-yi khak*), literally "board of sand". This expression is found, for example in the *Jami' al hisab bi't takht wa't turab*, by the mathematician and astronomer Nasir ad din at Tusi (1201–1274). This work's title can be translated literally as "Collection of arithmetic using a board and dust" [A. P. Youschkevitch (1976), p. 181, n. 71].

But the Arabic word *turab*, and its Persian equivalent *khak*, means not only "sand" or "dust", but also "earth", "clay" and "cement". Hence the difficulty in precisely translating this author's ideas: for Persian and Mesopotamian arithmeticians, did this word mean only "sand" and "dust", or did it also cover a wad of clay? We can, in fact, suppose that for reasons linked to climate and the nature of the soil in different regions, these arithmeticians were led to use clay for carrying out their calculations, rather than a board scattered with sand. This hypothesis is reasonable, given the limited number of material solutions. It becomes even more probable when we remember that, in these regions, clay tablets had been used for writing for thousands of years. It is sufficient to remember the Sumerians, the Elamites, the Babylonians, the Assyrians and the Acheminid Persians, the distant precursors of these Persian and Mesopotamian arithmeticians, to support the idea that, even under Islam, these peoples had not forgotten their ancient writing materials.

According to this hypothesis, these arithmeticians would then have smeared soft clay over their boards and traced numbers on them with a stylus, pointed at one end and flattened at the other. This is why the Arabic expression *takht al turab*, and its Persian equivalent *takhta-yi khak*, as in At Tusi's book cited above, could be translated by "board smeared with clay".

This hypothesis can be applied to the regions of Persia, Mesopotamia and Syria, but less so to other Muslim provinces.

If we return to the "board", the Arabic word *luha*, used by the Maghrebi and Andalusians for this article had, and always has had, as broad a range of meaning as its Eastern equivalent *takht* (which comes from the Persian *takhta*, itself derived from the Sassanid *takhtag*). Both words mean "table", but also "board", "plank", "tablet" and "plate" or "plaque", be it of wood, leather, metal, earth or even clay.

At a certain time, it is not impossible that wax came to replace the dust or flour used on the board in the Maghreb, and elsewhere. In other words, it can be supposed that the Maghrebi and other Islamic peoples calculated on tablets covered with wax, like those of the ancient Romans, using a stylus with a flattened tip for rubbing out.

All of these techniques perhaps coexisted, each being favoured at different times, in different regions and according to local customs. It is extremely unlikely that people living in such a vast and varied world as Islam would have all used the same method.

## CALCULATING WITHOUT INTERMEDIATE ERASURES

What is certain is that the Arab arithmeticians' next step was to "calculate without erasures, by crossing out and writing above their intermediate results".

This method is found, for example, in the *Kitab al fusul fi'l hisab al hind* (*Treatise on Indian Arithmetic*), written in Damascus in 952 (or 958) by Abu'l Hasan Ahmad ibn Ibrahim al-Uqlidisi. It can also be found in works by An Nisawi (1052), al-Hassar (c. 1175), al-Qalasadi (c. 1475), etc., in which it is described as the *a'mal al hindi* ("method of the Indians") or else as *tarik al hindi* (literally "way of the Indians") [see A. Allard (1976), pp. 87–100; A. Saidan (1966); H. Suter BMA, II, 3, pp. 16–17; F. Woepcke (1857), p. 407].

Here are the rules, applied to the product of 325 and 243:

As before, we begin by placing the multiplicand above the multiplier, thus:

| | | | | | |
|---|---|---|---|---|---|
| | | 3 | 2 | 5 | ← *Multiplicand* |
| 2 | 4 | 3 | | | ← *Multiplier* |

We then multiply the upper 3 by the lower 2; this makes 6, which we place on the line above the multiplicand, in the same column as the 2 of the multiplier:

| | | | | | |
|---|---|---|---|---|---|
| 6 | | | | | |
| | | 3 | 2 | 5 | ← *Multiplicand* |
| 2 | 4 | 3 | | | ← *Multiplier* |

And we cross out the 2 of the multiplier:

| | | | | | |
|---|---|---|---|---|---|
| 6 | | | | | |
| | | 3 | 2 | 5 | ← *Multiplicand* |
| ~~2~~ | 4 | 3 | | | ← *Multiplier* |

Then we multiply the upper 3 by the lower 4; this makes 12, we carry forward the 1 and place the 2 on the same line as the 6, above the 4:

| | | | | | |
|---|---|---|---|---|---|
| 6 | 2 | | | | |
| | | 3 | 2 | 5 | ← *Multiplicand* |
| ~~2~~ | 4 | 3 | | | ← *Multiplier* |

Then we add the carried-forward number to the 6; so we cross out 6 and write 7 on the line above, just over the crossed-out number:

| | | | | | |
|---|---|---|---|---|---|
| 7 | | | | | |
| ~~6~~ | 2 | | | | |
| | | 3 | 2 | 5 | ← *Multiplicand* |
| ~~2~~ | 4 | 3 | | | ← *Multiplier* |

And we cross out the 4 of the multiplier:

| | | | | | |
|---|---|---|---|---|---|
| 7 | | | | | |
| ~~6~~ | 2 | | | | |
| | | 3 | 2 | 5 | ← *Multiplicand* |
| ~~2~~ | ~~4~~ | 3 | | | ← *Multiplier* |

We then multiply the upper 3 by the lower 3; this makes 9, which we write in the same column as the 3 of the multiplier, but on the line above the multiplicand:

| | | | | | |
|---|---|---|---|---|---|
| 7 | | | | | |
| ~~6~~ | 2 | 9 | | | |
| | | 3 | 2 | 5 | ← *Multiplicand* |
| ~~2~~ | ~~4~~ | 3 | | | ← *Multiplier* |

And we cross out the 3 of the multiplier:

| | | | | | |
|---|---|---|---|---|---|
| 7 | | | | | |
| ~~6~~ | 2 | 9 | | | |
| | | 3 | 2 | 5 | ← *Multiplicand* |
| ~~2~~ | ~~4~~ | ~~3~~ | | | ← *Multiplier* |

The first step of the operation has now been completed, so we write the multiplier 243 again on the line below, but moving one column to the right, after having crossed out the 3 of the multiplicand:

| | | | | | |
|---|---|---|---|---|---|
| 7 | | | | | |
| ~~6~~ | 2 | 9 | | | |
| | | ~~3~~ | 2 | 5 | ← *Multiplicand* |
| ~~2~~ | ~~4~~ | ~~3~~ | | | |
| | 2 | 4 | 3 | | ← *Multiplier* |

Then we multiply the 2 of the multiplicand by the 2 of the multiplier; hence 4, which we add to the 2 to the right of the already crossed-out 6 on the line above the multiplicand; we thus cross out this 2, and write 6 on the line above, in the same column:

| | | | | | |
|---|---|---|---|---|---|
| 7 | 6 | | | | |
| ~~6~~ | ~~2~~ | ~~9~~ | | | |
| | | ~~3~~ | 2 | 5 | ← *Multiplicand* |
| ~~2~~ | ~~4~~ | ~~3~~ | | | |
| | 2 | 4 | 3 | | ← *Multiplier* |

And we cross out the 2 of the multiplier:

| | | | | | |
|---|---|---|---|---|---|
| 7 | 6 | | | | |
| ~~6~~ | ~~2~~ | ~~9~~ | | | |
| | | ~~3~~ | 2 | 5 | ← *Multiplicand* |
| ~~2~~ | ~~4~~ | ~~3~~ | | | |
| | ~~2~~ | 4 | 3 | | ← *Multiplier* |

We then multiply the 2 of the multiplicand by the 4 of the multiplier; this makes 8, which we add to the 9 in the same column in the line above the multiplicand; this makes 17, we carry forward 1 and place 7 on the line above (just over the 9), after crossing out the 9:

| | | | | | |
|---|---|---|---|---|---|
| 7 | 6 | 7 | | | |
| ~~6~~ | ~~2~~ | ~~9~~ | | | |
| | | ~~3~~ | 2 | 5 | ← *Multiplicand* |
| ~~2~~ | ~~4~~ | ~~3~~ | | | |
| | ~~2~~ | 4 | 3 | | ← *Multiplier* |

Then we add the carried-forward 1 to the 6 on the top line; we thus cross out this 6 and write a 7 on the line above, in the same column:

| | | | | | |
|---|---|---|---|---|---|
| | 7 | | | | |
| 7 | ~~6~~ | 7 | | | |
| ~~6~~ | ~~2~~ | ~~9~~ | | | |
| | | ~~3~~ | 2 | 5 | ← *Multiplicand* |
| ~~2~~ | ~~4~~ | ~~3~~ | | | |
| | ~~2~~ | 4 | 3 | | ← *Multiplier* |

And we cross out the 4 of the multiplier:

| | | | | | |
|---|---|---|---|---|---|
| | 7 | | | | |
| 7 | ~~6~~ | 7 | | | |
| ~~6~~ | ~~2~~ | ~~9~~ | | | |
| | | ~~3~~ | 2 | 5 | ← *Multiplicand* |
| ~~2~~ | ~~4~~ | ~~3~~ | | | |
| | ~~2~~ | ~~4~~ | 3 | | ← *Multiplier* |

Then we multiply the 2 of the multiplicand by the 3 of the multiplier; this makes 6, so we write 6 in the same column as the 2 in the line just above:

| | | | | | |
|---|---|---|---|---|---|
| | 7 | | | | |
| 7 | ~~6~~ | 7 | | | |
| ~~6~~ | ~~2~~ | ~~9~~ | 6 | | |
| | | ~~3~~ | 2 | 5 | ← *Multiplicand* |
| ~~2~~ | ~~4~~ | ~~3~~ | | | |
| | ~~2~~ | ~~4~~ | 3 | | ← *Multiplier* |

And we cross out the 3 of the multiplier:

| | | | | | |
|---|---|---|---|---|---|
| | 7 | | | | |
| 7 | ~~6~~ | 7 | | | |
| ~~6~~ | ~~2~~ | ~~9~~ | 6 | | |
| | | ~~3~~ | 2 | 5 | ← *Multiplicand* |
| ~~2~~ | ~~4~~ | ~~3~~ | | | |
| | ~~2~~ | ~~4~~ | ~~3~~ | | ← *Multiplier* |

The second step has now been completed, so we write the multiplier 243 once again on the line below, moving one column to the right, after having crossed out the 2 of the multiplicand:

```
    7
7   6̸   7
6̸   2̸   9̸   6
        3̸   2̸   5      ← Multiplicand
2̸   4̸   3̸
    2̸   4̸   3̸
        2   4   3      ← Multiplier
```

Then we multiply the 5 of the multiplicand by the 2 of the multiplier; this makes 10, we carry forward 1, but add nothing to the 7 in the same column as the 2 on the second line above the multiplicand. We then add the carried-forward 1 to the 7 on the top line; we cross out this 7 and write 8 on the line above:

```
    8
    7̸
7   6̸   7
6̸   2̸   9̸   6
        3̸   2̸   5      ← Multiplicand
2̸   4̸   3̸
    2̸   4̸   3̸
        2   4   3      ← Multiplier
```

And we cross out the 2 of the multiplier:

```
    8
    7̸
7   6̸   7
6̸   2̸   9̸   6
        3̸   2̸   5      ← Multiplicand
2̸   4̸   3̸
    2̸   4̸   3̸
        2̸   4   3      ← Multiplier
```

Then we multiply the 5 of the multiplicand by the 4 of the multiplier; this makes 20, we carry forward the 2, but add nothing to the 6 in the same column as the 2 on the line just above the multiplicand. Then we add the carried-forward 2 to the 7 in the column just to the left; we cross out this 7 and write 9 on the line above:

```
    8
    7̸   9
7   6̸   7̸
6̸   2̸   9̸   6
        3̸   2̸   5      ← Multiplicand
2̸   4̸   3̸
    2̸   4̸   3̸
        2̸   4   3      ← Multiplier
```

And we cross out the 4 of the multiplier:

```
    8
    7̸   9
7   6̸   7̸
6̸   2̸   9̸   6
        3̸   2̸   5      ← Multiplicand
2̸   4̸   3̸
    2̸   4̸   3̸
        2̸   4̸   3      ← Multiplier
```

Finally, we multiply the 5 of the multiplicand by the 3 of the multiplier; this makes 15, so we write a 5 above the 5 of the multiplicand:

```
    8
    7̸   9
7   6̸   7̸
6̸   2̸   9̸   6   5
        3̸   2̸   5      ← Multiplicand
2̸   4̸   3̸
    2̸   4̸   3̸
        2̸   4̸   3      ← Multiplier
```

Then we add the carried-forward 1 to the 6 immediately to the left on the same line; we cross out this 6 and write 7 on the line above:

```
    8
    7̸   9
7   6̸   7̸   7
6̸   2̸   9̸   6̸   5
        3̸   2̸   5      ← Multiplicand
2̸   4̸   3̸
    2̸   4̸   3̸
        2̸   4̸   3      ← Multiplier
```

And we cross out the 3 of the multiplier and the 5 of the multiplicand:

8
~~7~~ 9
7 ~~6~~ ~~7~~ 7
~~6~~ ~~2~~ ~~9~~ ~~6~~ 5
~~3~~ ~~2~~ ~~5~~ ← *Multiplicand*
~~2~~ ~~4~~ ~~3~~
~~2~~ ~~4~~ ~~3~~
~~2~~ ~~4~~ ~~3~~ ← *Multiplier*

As the operation has now been completed, all we have to do is read the uncrossed-out numerals, from left to right, to obtain the result:

8
~~7~~ 9
7 ~~6~~ ~~7~~ 7
~~6~~ ~~2~~ ~~9~~ ~~6~~ 5
~~3~~ ~~2~~ ~~5~~
~~2~~ ~~4~~ ~~3~~
~~2~~ ~~4~~ ~~3~~
~~2~~ ~~4~~ ~~3~~
↓ ↓ ↓ ↓ ↓
325 × 243 7 8 9 7 5

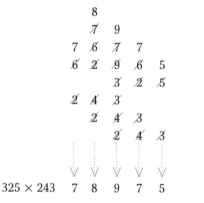

The advantage of this method over the preceding one is the possibility to check the operation and so spot any errors. This is why it was used by many Muslim arithmeticians for some time; and that is also why it survived in Europe until the end of the eighteenth century.

The disadvantage was to make the writing of calculations extremely crowded and their progression difficult to follow.

This can be seen in the following example of division "à la française", as explained in F. Le Gendre's *Arithmétique*:

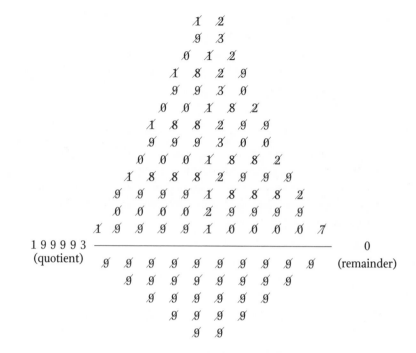

~~1~~ ~~2~~
~~9~~ ~~3~~
~~0~~ ~~1~~ ~~2~~
~~1~~ ~~8~~ ~~2~~ ~~9~~
~~9~~ ~~9~~ ~~3~~ ~~0~~
~~0~~ ~~0~~ ~~1~~ ~~8~~ ~~2~~
~~1~~ ~~8~~ ~~8~~ ~~2~~ ~~9~~ ~~9~~
~~9~~ ~~9~~ ~~9~~ ~~3~~ ~~0~~ ~~0~~
~~0~~ ~~0~~ ~~0~~ ~~1~~ ~~8~~ ~~8~~ ~~2~~
~~1~~ ~~8~~ ~~8~~ ~~8~~ ~~2~~ ~~9~~ ~~9~~ ~~9~~
~~9~~ ~~9~~ ~~9~~ ~~9~~ ~~1~~ ~~8~~ ~~8~~ ~~8~~ ~~2~~
~~0~~ ~~0~~ ~~0~~ ~~0~~ ~~2~~ ~~9~~ ~~9~~ ~~9~~ ~~9~~
~~1~~ ~~9~~ ~~9~~ ~~9~~ ~~9~~ ~~1~~ ~~0~~ ~~0~~ ~~0~~ ~~0~~ ~~7~~
1 9 9 9 9 3 (quotient) ——— 0 (remainder)
~~9~~ ~~9~~ ~~9~~ ~~9~~ ~~9~~ ~~9~~ ~~9~~ ~~9~~ ~~9~~ ~~9~~
~~9~~ ~~9~~ ~~9~~ ~~9~~ ~~9~~ ~~9~~ ~~9~~ ~~9~~
~~9~~ ~~9~~ ~~9~~ ~~9~~ ~~9~~ ~~9~~
~~9~~ ~~9~~ ~~9~~ ~~9~~
~~9~~ ~~9~~

We will not weary the reader by explaining this extraordinarily complex system. Suffice it to say that this represents 19,999,100,007 divided by 99,999 [Le Gendre, *Arithmétique en sa perfection* (Paris 1771), p. 54].

It can thus be understood that division, even when written down, long remained beyond the scope of the average person.

It is also true that this work was not meant for a large public. As the author himself makes clear, the "perfection" of this arithmetic was based on "the usage of financiers, experienced people, bankers and merchants".

## FROM THE WOODEN BOARD TO PAPER OR BLACKBOARD

However, long before Le Gendre, several Arab and Indian arithmeticians embarked on a far better way, omitting intermediate results and thus dropping the technique of constant erasure. But this method, and its consequent change of writing medium, called for a greater application of memory.

The changes in methods and the changes in materials affected each other, long before they resulted in our present day techniques. (See *Pâtî, *Pâtîganita.)

It can be supposed that, as in India, Muslim arithmeticians at some time started using a blackboard, or else a wooden board painted black, with chalk to write and cross out their numbers, and a cloth for erasing them. (See **Pâtîganita*.)

By using chalk and keeping or, even better, rubbing out intermediate results, certain Indian and Arab arithmeticians were able to free their imaginations and work towards the methods we now use.

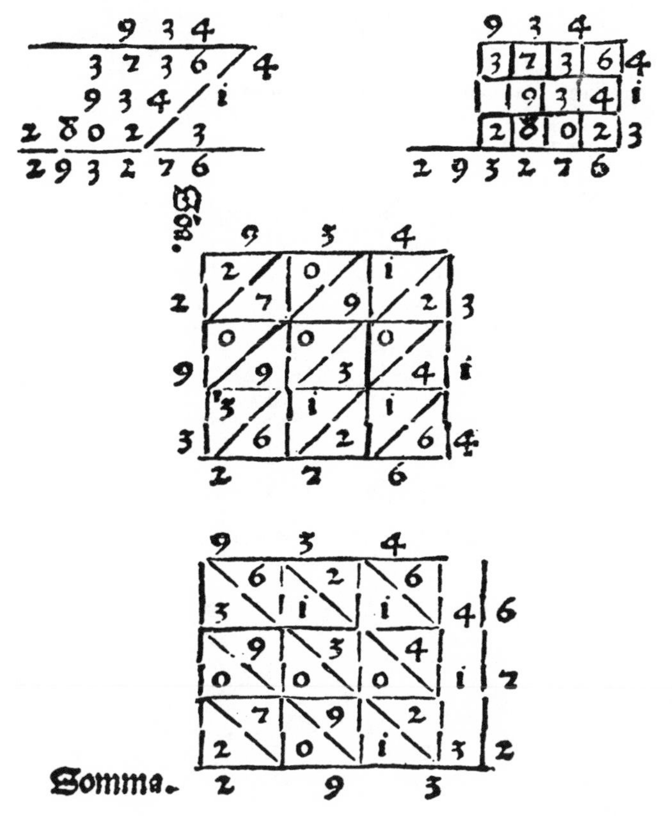

Uoglio pero che tu intendi che ſono altri modi de moltiplicare per ſcachiero:li quali laſſaro al ſtudio tuo:mettendo li exempli ſoi ſolamente in forma. come porai vedere qui ſotto
Or togli de fare lo predicto ſcachiero.3oe.3 i 4. fia.9 3 4.e nota de farlo per li quatro modi come qui da ſotto.

Somma.

Fig. 25.11. *Page of an anonymous Italian arithmetical treatise, published in Treviso in 1478. It contains various forms of "jealousy" multiplication* (per gelosia). *[Document in the Palais de la Découverte]*

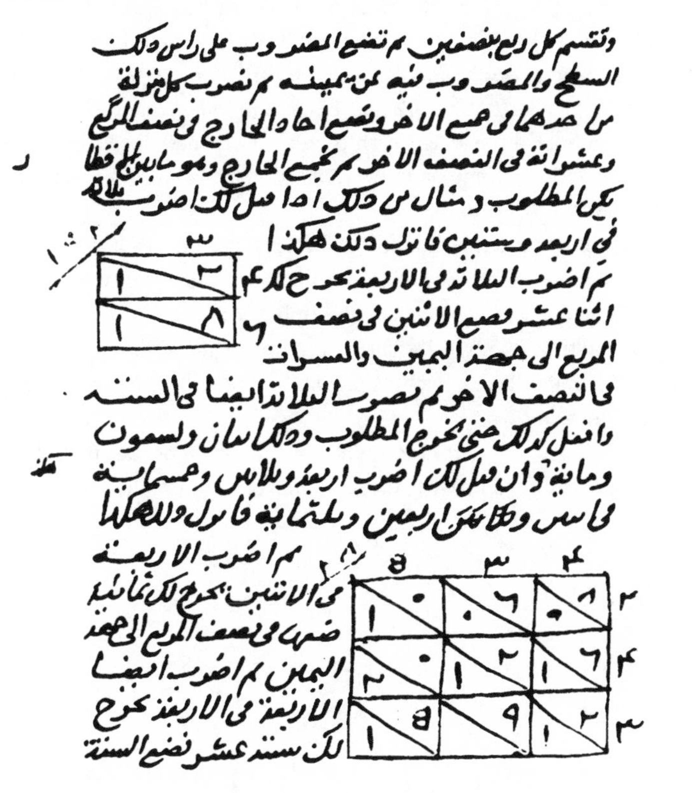

Fig. 25.12. *Page of an Arabic treatise dealing with written calculation using Indian numerals, explaining a method of multiplication "by a table" (known in the West as* per gelosia*). To the left we have the product of 3 and 64 and bottom right the product of 534 and 342. Sixteenth-century copy of* Kashf al mahjub min 'ilm al ghubar *(see below). Paris, BN, Ms. ar. 2473, f° 9)*

## "JEALOUSY" MULTIPLICATION

Here follows an example of a highly developed technique, which the Arabs must have invented in around the thirteenth century. At the end of the Middle Ages it was transmitted to Europe, where it was known as multiplication *per gelosia* ("by jealousy"), an allusion to the grid used in the operation which is reminiscent of the wooden or metal lattices through which jealous wives, and especially husbands, could see without being seen. It is described in an anonymous work published in Treviso in 1478 (Fig 25.11) and in the *Summa de arithmetica, geometria, proporzioni di proporzionalita* by Luca Pacioli, an Italian mathematician (Venice, 1494). The Arabs called this system "multiplication on a grid" (*al darb bi'l jadwal*) and it was described in about 1470 by Abu'l Hasan 'Ali ibn Muhammad al-Qalasadi, in his *Kashf al mahjub min 'ilm al ghubar* (*Revelation of the Secrets of the Science of Arithmetic*), the word *ghubar* here being used for "written arithmetic" in general and not in the original sense of "dust" (Fig. 25.12). But there is a much earlier version, from about 1299, by Abu'l 'Abbas Ahmad ibn Muhammad ibn al-Banna al-Marrakushi in his *Talkhis a'mal al hisab* (*Brief Summary of Arithmetical Operations*) [see A. Marre (1865); H. Suter (1900–02), p. 162]. But in India there is no trace of it before the middle of the seventeenth century. It was described there for the first time in 1658 by Ganesha in his *Ganitamañjarî* [see B. Datta and A. N. Singh (1938), pp. 144–5].

The layout is quite simple, and the final result is arrived at, rather as in our present-day system, by adding together the products of various numbers contained in the multiplier and multiplicand.

Let us multiply 325 by 243. There are three digits in the multiplier and three in the multiplicand, we thus draw a square grid with three columns and three lines.

Above the grid we write the numbers 3, 2 and 5 of the multiplicand from left to right; then we write the 2, 4 and 3 of the multiplier up the right-hand side of the grid:

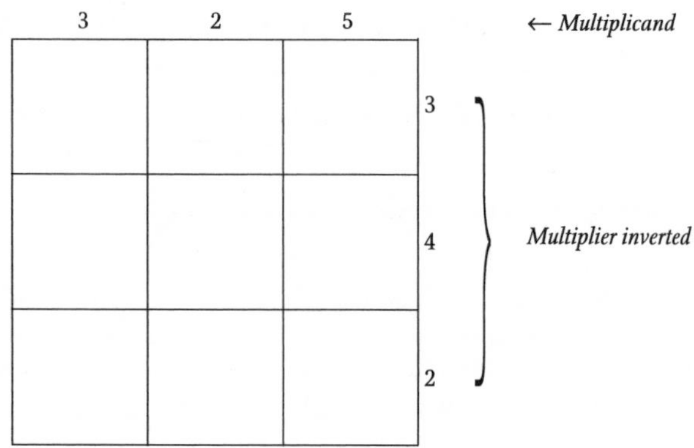


We then divide each square of the grid in half by drawing a diagonal from the top left-hand corner to the bottom right-hand corner. Then, in each square we write the product of the number on the same line to the right and the number in the same column at the top. This product must, of course, be inferior to 100.

We then write the tens digit in the left-hand triangle of the square and the units digit in the right-hand triangle. If either of these digits is missing, then we must write zero.

In the first upper right-hand square we thus write the product of 5 and 3, i.e. 15, by placing the 1 in the left-hand triangle and the 5 in the right-hand triangle.

And so on, thus:

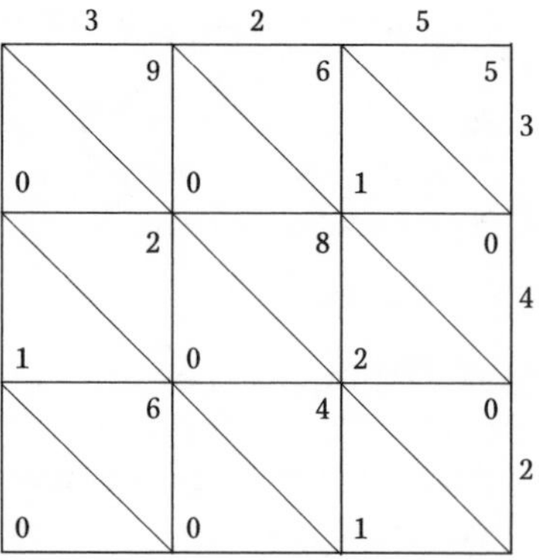


Outside the grid, we then add up the numbers contained in each oblique strip, beginning with the 5 in the top right-hand corner. We then proceed from right to left and from the top to the bottom. When necessary we carry forward any tens digits and add them onto the next strip and we thus obtain all the digits of the final result outside the grid, thus:

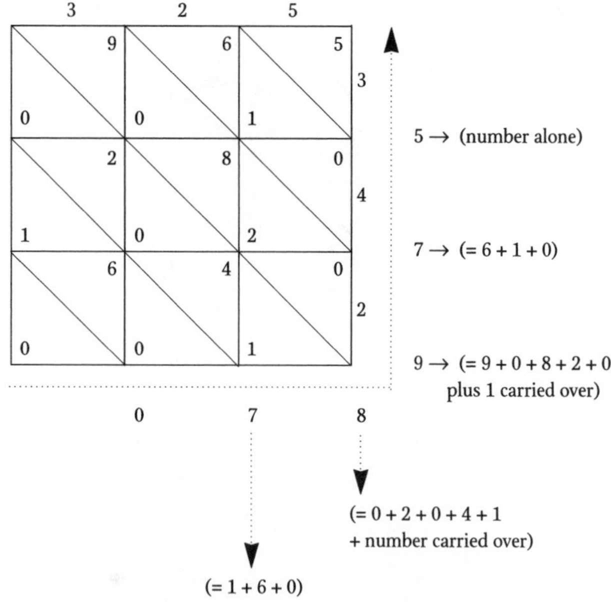


The result is then obviously read from left to right, following the arrow, therefore 78,975:

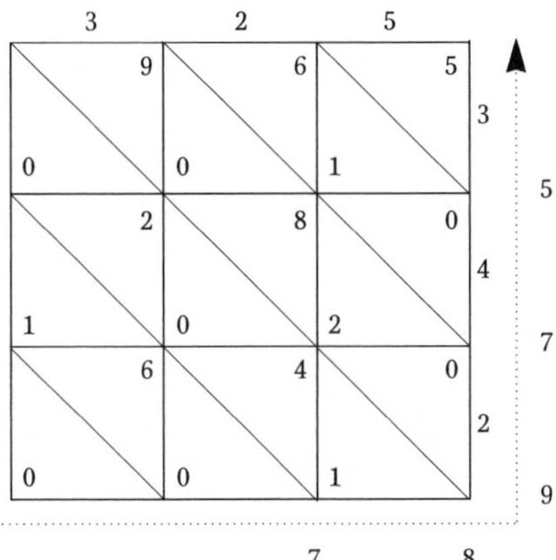


Note that the Arabs often wrote the resulting digits along an oblique segment, perpendicular to the main diagonal, to the left of the grid; the result, of course, still reads from left to right:

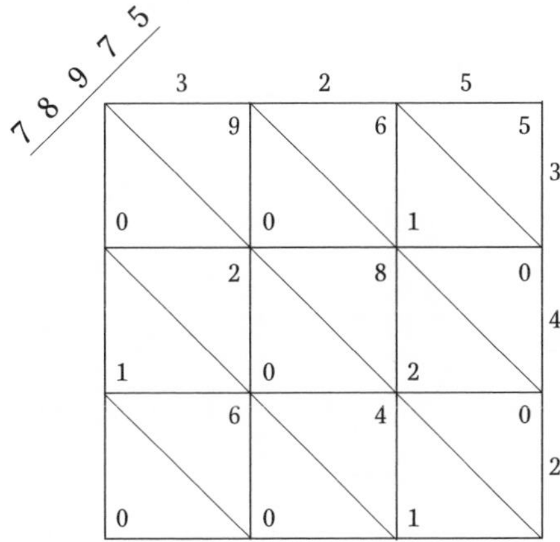


This method may seem long in comparison with the one we now use. But its advantage is that the final result is grouped together at the end whereas in our modern system it is produced gradually during the intermediate steps.

*Other ways of proceeding*

Instead of following a falling diagonal, we follow a rising one; and instead of writing the multiplier backwards, we write it the correct way round. Hence the following arrangement, which is used in the same way, except that the additions appear to the left of the grid:

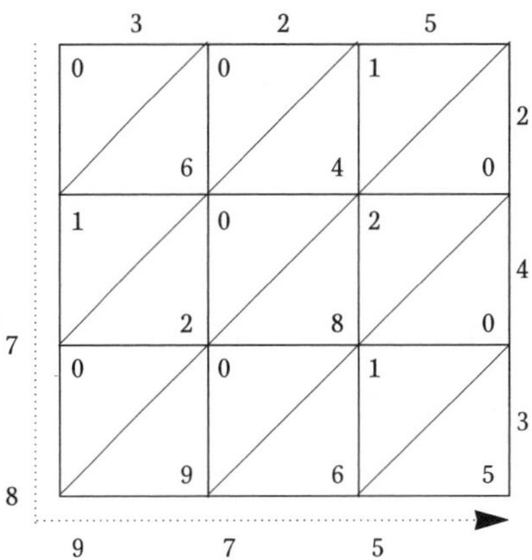


Other possibilities also exist, of course, by placing the digits of the multiplier to the left rather than to the right of the grid.

*Simplified techniques*

In the anonymous work cited above, we also find another layout alongside the preceding ones. Instead of noting down all the details of the operation, we simply give the results, which certainly requires a greater effort of memory, especially when it comes to carrying numbers forward during the intermediary steps (Fig. 25.11).

As both the multiplicand and multiplier have three digits, we draw a rectangular grid with four columns and three lines, the extra column being used for noting partial results with more digits than in the multiplicand.

Then we place the digits of the multiplicand and multiplier thus:

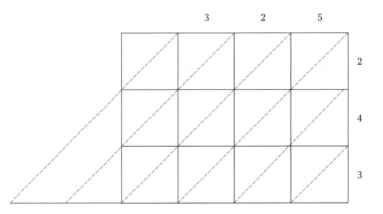

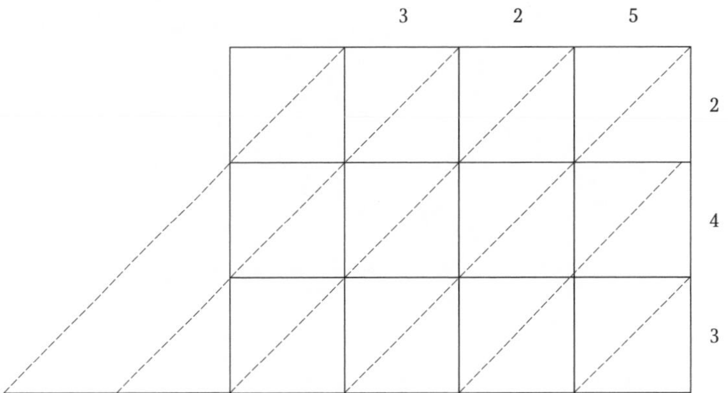

We then calculate the products at the intersections of the columns and lines. But, as there are here no diagonals, only one digit must be written in each square, with the tens digit being added onto the following square on the left.

On the first line, to the right, the first square thus gives 10; we note the 0 and carry forward the 1 to the next square on the left. Its own result is 4, to which the 1 is added, making 5; and so on:

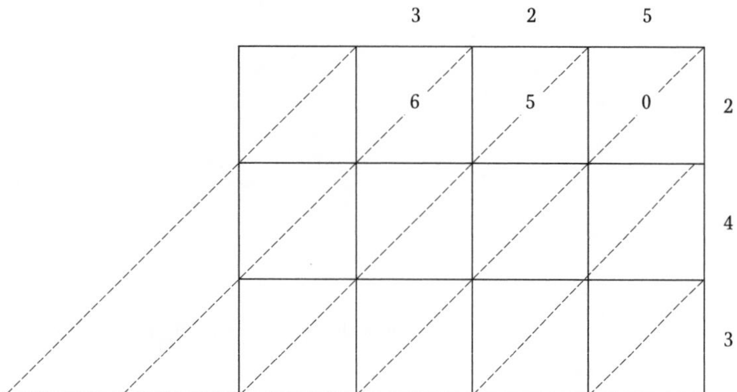


On the second line we thus obtain:

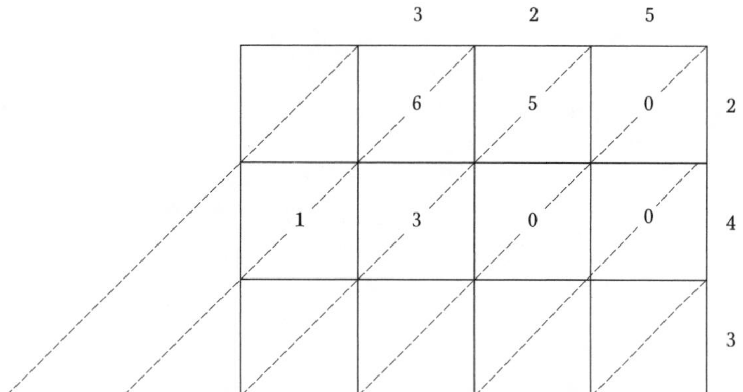


And, finally, on the third:

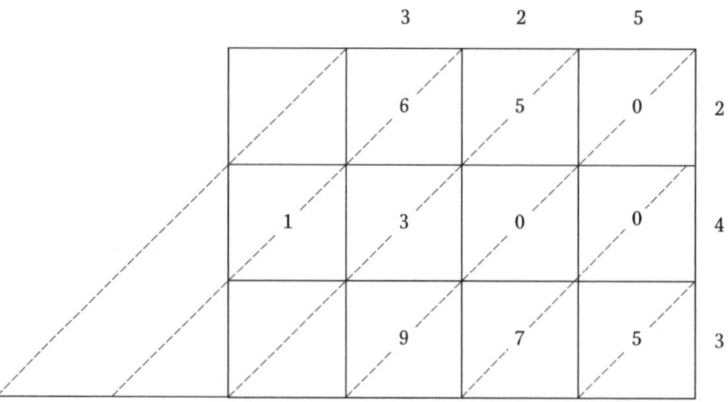


The result is then obtained by adding together the numbers along each line parallel to the rising diagonal from right to left; it is then read from left to right:

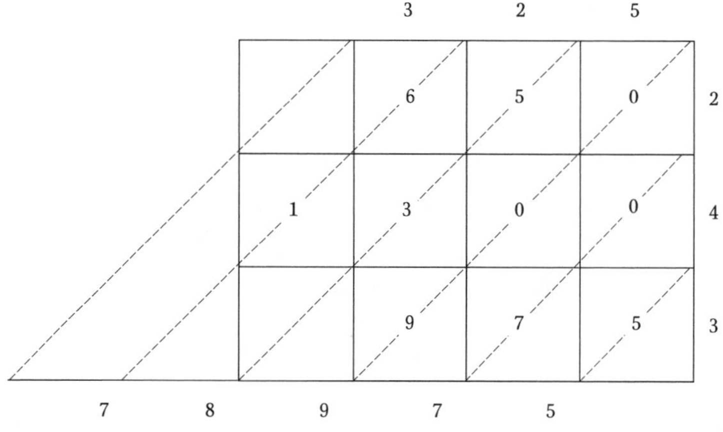


Note that we can work the other way round, by writing the digits of the multiplier backwards on the left. But we must then follow the falling diagonal from left to right to obtain the result:

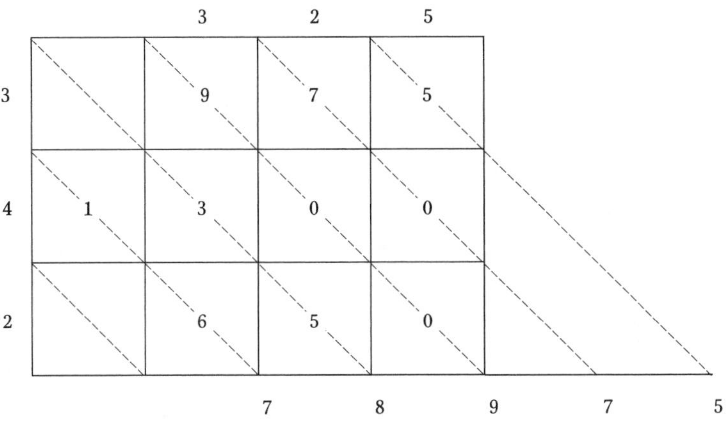


*An even simpler technique*

At the end of the fifteenth century, the following more highly developed variant could be found in Europe.

We draw a line and write the digits of the multiplicand above it then, below to the right, the digits of the multiplier obliquely rising from left to right:

| | | 3 | 2 | 5 | | | | |
|---|---|---|---|---|---|---|---|---|
| | | | | | | | | 3 |
| | | | | | | | 4 | |
| | | | | | | 2 | | |

To the left of the 3 of the multiplier, we then write its products with the digits of the multiplicand:

| | | 3 | 2 | 5 | | | | |
|---|---|---|---|---|---|---|---|---|
| | | 9 | 7 | 5 | | | | 3 |
| | | | | | | | 4 | |
| | | | | | | 2 | | |

Then we do the same with the 4:

| | | 3 | 2 | 5 | | | | |
|---|---|---|---|---|---|---|---|---|
| | | 9 | 7 | 5 | | | | 3 |
| 1 | 3 | 0 | 0 | | | | 4 | |
| | | | | | | 2 | | |

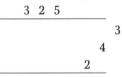

Then with the 2:

| | | 3 | 2 | 5 | | | | |
|---|---|---|---|---|---|---|---|---|
| | | 9 | 7 | 5 | | | | 3 |
| 1 | 3 | 0 | 0 | | | | 4 | |
| 6 | 5 | 0 | | | | 2 | | |

We then add up the products, which gives us:

| | | 3 | 2 | 5 | | | | |
|---|---|---|---|---|---|---|---|---|
| | | 9 | 7 | 5 | | | | 3 |
| 1 | 3 | 0 | 0 | | | | 4 | |
| 6 | 5 | 0 | | | | 2 | | |
| 7 | 8 | 9 | 7 | 5 | | | | |

This method is thus as highly developed as our own (which we append here to facilitate comparison), the only difference being the position of the multiplier:

| | | 3 | 2 | 5 |
|---|---|---|---|---|
| | | 2 | 4 | 3 |
| | | 9 | 7 | 5 |
| 1 | 3 | 0 | 0 | |
| 6 | 5 | 0 | | |
| 7 | 8 | 9 | 7 | 5 |

## NASIR AD DIN AT TUSI'S METHOD

Here now is a multiplication technique, already existing in the thirteenth century, particularly in the *Miftah al Hisab* (*Key to Calculation*) by Ghiyat ad din Ghamshid ibn Mas'ud al-Kashi, from the Persian town of Kashan; this work was completed in 1427 [see A. P. Youschkevitsch (1976), p. 181, n. 67].

But it was known and used two centuries earlier. It can be found, with a slight variation, in Nasir ad din at Tusi's *Jami' al hisab bi't takht wa't turab* (*Collection of Arithmetic Using a Board and Dust*), which dates back to 1265 and was copied by his disciple Hasan ibn Muhammad an Nayshaburi in 1283 [translated by S. A. Akhmedov and B. A. Rosenfeld; see A. P. Youschkevitsch 1976, p. 181, n. 71].

Let us multiply 325 by 243. The multiplicand and multiplier are placed as follows:

| 3 | 2 | 5 |
|---|---|---|
| 2 | 4 | 3 |

We multiply the 5 of the multiplicand by the 3 of the multiplier and place the result beneath the line, being careful to respect the place values:

```
    3 2 5
    2 4 3
---------
      1 5
```

Then we multiply the 3 of the multiplicand (not the 2, which is for the moment ignored) by the 3 of the multiplier and we place the product beneath the same line, on the left of the previous one:

```
    3 2 5
    2 4 3
---------
    9 1 5
```

We now return to the 2 of the multiplicand, which we multiply by 3 and this time place the result on the line below, one step to the left:

```
    3 2 5
    2 4 3
---------
    9 1 5
      6
```

We draw another horizontal below these results and then multiply the 5 of the multiplicand by the 4 of the multiplier, placing the result one step to the left:

```
    3 2 5
    2 4 3
---------
    9 1 5
      6
---------
    2 0
```

Then we multiply the 3 of the multiplicand by the 4 of the multiplier and write the product on the same line on the left of the preceding result:

```
    3 2 5
    2 4 3
---------
    9 1 5
      6
---------
1 2 2 0
```

Then we return to the 2 of the multiplicand, which we multiply by the 4 of the multiplier and place the result on the line below the preceding ones, one step to the left:

```
    3 2 5
    2 4 3
---------
    9 1 5
      6
---------
1 2 2 0
    8
```

Then we draw another line and carry out the preceding operations with the 2 of the multiplier, placing the first product one step towards the left. With the 5 of the multiplicand we obtain:

```
    3 2 5
    2 4 3
---------
    9 1 5
      6
---------
1 2 2 0
    8
---------
  1 0
```

Omitting the 2, with the 3 we obtain (on the same line):

```
    3 2 5
    2 4 3
---------
    9 1 5
      6
---------
1 2 2 0
    8
---------
6 1 0
```

And with the 2 of the multiplicand, on the line below and one step to the left, we have:

```
    3 2 5
    2 4 3
---------
    9 1 5
      6
---------
1 2 2 0
    8
---------
6 1 0
  4
```

The intermediate steps are thus over. All we have to do now is draw another line and add up the partial results, position after position, from the right to the left, and so easily obtain our result:

```
    3 2 5
    2 4 3
  ---------
    9 1 5
      6
  ---------
1 2 2 0
    8
  ---------
6 1 0
  4
  ---------
7 8 9 7 5
```

## THE INDIAN MATHEMATICIAN BHÂSKARÂCHÂRYA'S METHODS

In his *Lîlâvatî*, Bhâskharâchârya (c. 1150) often uses a more highly developed method than the preceding one, which he called, in Sanskrit, *sthânakhanda* (literally "separation of positions"). There are several variants, the main ones being as follows [see J. Taylor (1816), pp. 8–9; B. Datta and A. N. Singh 1938, p. 147]:

To multiply 325 by 243, we begin by setting out the operation like this, separating the three digits of the multiplier and copying the digits of the multiplicand three times:

```
2 4 3   2 4 3   2 4 3
  3       2         5
---------------------
```

We begin multiplying with the 5. First we take the product of 5 and 3 and write the full result below the line, without carrying anything forward; we then move to the product of 5 and 2 (skipping the product of 5 and 4) and write the result on the same line, again without carrying forward, just to the left of the first result:

```
2 4 3   2 4 3   2 4 3
  3       2         5
---------------------
              1 0 1 5
```

Then we take the product of 5 and the 4 we had omitted and write the result below the others, one step to the left:

```
2 4 3   2 4 3   2 4 3
  3       2         5
---------------------
              1 0 1 5
                2 0
```

We draw a line below these results and add them up:

```
2 4 3   2 4 3   2 4 3
  3       2         5
---------------------
              1 0 1 5
                2 0
---------------------
    1 2 1 5
```

We then multiply using the 2, in the same way, placing the sum of the partial results one step to the left from the first one:

```
2 4 3   2 4 3     2 4 3
  3       2           5
-----------------------
        4   6   1 0 1 5
          8       2 0
-----------------------
    1 2 1 5
    4 8 6
```

Then we multiply by 3 in the same way, placing the sum of the partial results one step to the left from the previous one:

```
2 4 3   2 4 3     2 4 3
  3       2           5
-----------------------
6   9   4   6   1 0 1 5
1 2       8       2 0
-----------------------
    1 2 1 5
    4 8 6
  7 2 9
```

We draw a final line, add up the totals and obtain the result:

```
2 4 3   2 4 3     2 4 3
  3       2           5
-----------------------
6   9   4   6   1 0 1 5
1 2       8       2 0
-----------------------
    1 2 1 5
    4 8 6
  7 2 9
-----------------------
  7 8 9 7 5
```

*Another method*

One variant of Bhâskarâchârya's method uses a layout like this:

```
  3 2 5
2 4 3
-----
```

We then multiply the 3 of the multiplicand (the highest place) by each of the numbers in the multiplier (this time from the lowest first):

```
  3 2 5
2 4 3
-----
7 2 9
```

Then we multiply the 2 of the multiplicand by each of the numbers of the multiplier, placing the result on the line below, one step to the right from the previous result:

```
  3 2 5
2 4 3
-----
7 2 9
  4 8 6
```

Finally, we carry out the same procedure with the 5 of the multiplicand and move one step more to the right to note the result:

```
  3 2 5
2 4 3
-----
7 2 9
  4 8 6
  1 2 1 5
```

We draw a line and add up the partial results to obtain the final answer:

```
  3 2 5
2 4 3
-----
7 2 9
  4 8 6
  1 2 1 5
---------
7 8 9 7 5
```

## THE INDIAN MATHEMATICIAN BRAHMAGUPTA'S METHODS

Long before Bhâskarâchârya, Brahmagupta, in his *Brahmasphutasiddhânta* (628 CE) described four even more highly developed methods, which he called *gomûtrikâ, khanda, bheda* and *isa* [S. Dvivedi (1902), p. 209; H. T. Colebrooke (1817); B. Datta and A. N. Singh (1938), p. 148].

Here, as an example, is the method called *gomûtrika* (which, in Sanskrit, literally means "like the trajectory of a cow's urine", an allusion to the zigzagging of the arithmetician's eyes as he carries out the operation).

To multiply 325 by 243, we begin with the following layout, copying the multiplicand onto three successive lines, moving one step to the right as we go down. We place the digits of the multiplier vertically from top to bottom starting on the top line:

```
2    3 2 5
4      3 2 5
3        3 2 5
     ---------
```

On the first line we then mentally multiply the 2 of the multiplier by the 5 (the lowest digit) of the multiplicand; this makes 10, we write 0 on a lower line in the same column as this 5 and carry forward the 1 which will be added to the next product:

```
2    3 2 5
4      3 2 5
3        3 2 5
     ---------
         0
```

Then we multiply the same 2 by the 2 of the multiplicand, which makes 4, and which is added to the carried-forward 1. The result is placed under the line, to the left of the 0:

```
2    3 2 5
4      3 2 5
3        3 2 5
     ---------
       5 0
```

Then we multiply the same 2 by the 3 of the multiplicand; this makes 6, which we place under the line to the left of the 5:

```
2    3 2 5
4      3 2 5
3        3 2 5
     ---------
     6 5 0
```

We then move to the line with the multiplier 4 and carry out the same steps, this time placing the results on a line below the 650, one step to the right, thus:
– with the product of 4 and 5:

```
2    3 2 5
4      3 2 5
3        3 2 5
     ---------
     6 5 0
           0
```

– then with 4 and 2 (adding the 2 carried forward):

```
2    3 2 5
4      3 2 5
3        3 2 5
     ---------
     6 5 0
         0 0
```

– and with 4 and 3 (adding the 1 carried forward):

```
2    3 2 5
4      3 2 5
3        3 2 5
     ---------
     6 5 0
     1 3 0 0
```

We then go down to the line with the multiplier 3, carrying out the same steps, this time with the partial results on a line below the 1,300, one step to the right, thus:

– with the product of 3 and 5:

```
2    3 2 5
4      3 2 5
3        3 2 5
     ---------
     6 5 0
     1 3 0 0
             5
```

– then with 3 and 2 (adding the 1 carried over):

```
2    3 2 5
4      3 2 5
3        3 2 5
     ---------
     6 5 0
     1 3 0 0
           7 5
```

– and finally with 3 and 3:

```
2    3 2 5
4      3 2 5
3        3 2 5
     ---------
     6 5 0
     1 3 0 0
         9 7 5
```

All we have to do now is add up the partial results to obtain the final answer:

```
2    3 2 5
4      3 2 5
3        3 2 5
     ---------
     6 5 0
     1 3 0 0
         9 7 5
     ---------
     7 8 9 7 5
```

*Other variants of this method*

Another layout Brahmagupta used was as follows, with the multiplicand copied three times on three successive lines, each moving one step to the left in comparison with the line above, and with the multiplier placed on the right, from the bottom to the top:

```
    3 2 5    3
  3 2 5      4
3 2 5        2
---------
    9 7 5
1 3 0 0
6 5 0
---------
7 8 9 7 5
```

Brahmagupta's method was, thus, highly developed. There was just one more small step to be taken for it to become as efficient as our present-day technique. This was, in fact, what happened as is shown in Brahmagupta's works, which contain the following extremely interesting variant.

Instead of copying the multiplicand three times, Brahmagupta wrote it just once in the layout below, in which the multiplier is written as in the preceding example, i.e. from the bottom to the top and below the initial line, each partial result being noted opposite and to the left of the number that produces it:

```
    3 2 5
  ---------
    9 7 5    3
  1 3 0 0    4
  6 5 0      2
  ---------
  7 8 9 7 5
```

This is exactly the same as the method which Italian mathematicians in the second half of the fifteenth century (Luca Pacioli, etc.) had deduced from simplifying the *per gelosia*, and laid out as follows (Fig. 25.11):

```
    3 2 5
  ---------
    9 7 5      3
  1 3 0 0    4
  6 5 0    2
  ---------
  7 8 9 7 5
```

In other words, from as early as the beginning of the seventh century, Indian mathematicians had a way of multiplying that was far simpler than the "jealousy" method; a procedure which, with a mere change in the layout of the numbers, was to give rise to our present-day technique.

It can now be seen just how advanced the Indians and their Arab successors were in this field.

CHAPTER 26

# THE SLOW PROGRESS OF INDO-ARABIC NUMERALS IN WESTERN EUROPE

All that is now left to tell is the story of how India's discoveries reached the Christian West through Arabic intermediaries. As is well known, this transmission did not happen in a day. Quite the contrary!

When they first encountered numeral systems and computational methods of Indian origin, Europeans proved so attached to their archaic customs, so extremely reluctant to engage in novel ideas, that many centuries passed before written arithmetic scored its decisive and total victory in the West.

## RENAISSANCE ARITHMETIC: AN OBSCURE AND COMPLEX ART

> I was borne and brought up in the Countrie, and amidst husbandry: I have since my predecessours quit me the place and possession of the goods I enjoy, both businesse and husbandry in hand. I cannot yet cast account either with penne or Counters [Montaigne, *Essays*, Vol. II (1588), p. 379].

These words were written by one of the most learned men of his day: Michel de Montaigne, born 1533, was educated by famous teachers at the College de Guyenne, in Bordeaux, travelled widely thereafter, and came to own a sumptuous library. He was a member of the *parlement* of Bordeaux and then mayor of that city, as well as a friend of the French kings François II and Charles IX. And he admits without the slightest embarrassment, that he cannot "cast account" – or, in modern language, do arithmetic!

Could he have been aware of the fabulous discoveries of Indian scholars, already over a thousand years old? Almost certainly not. Cultural contacts between Eastern and Western civilisations had been very limited ever since the collapse of the Roman Empire. Montaigne might have known of two ways, at most, of doing sums: with "Counters" on a ruled table or abacus; and using written Arabic numerals ("with penne"). The first operating method stands in the highly complicated tradition of Greece and Rome; the second, which Montaigne would no doubt have ascribed to the Arabs, was in fact the invention of Indian scholars. But no one had thought of teaching it to him; Montaigne, like most of his contemporaries, no doubt viewed it with mistrust and suspicion.

The following anecdote gives a good picture of the arithmetical state of Europe in the fifteenth and sixteenth centuries. A wealthy German merchant, seeking to provide his son with a good business education, consulted a learned man as to which European institution offered the best training. "If you only want him to be able to cope with addition and subtraction," the expert replied, "then any French or German university will do. But if you are intent on your son going on to multiplication and division – assuming that he has sufficient gifts – then you will have to send him to Italy."

It has to be said that arithmetical operations were not in everyone's grasp: they constituted an obscure and complex art, the specialist preserve of a privileged caste, whose members had been through a long and rigorous training which had allowed them to master the mysterious and infinitely complicated use of the classical (Roman) counter-abacus.

A student of those days needed several years of hard work as well as a long voyage to master the intricacies of multiplication and division – something not far short of a PhD curriculum, in today's terms.

The great respect in which such scholars were held provides a measure of the difficulty of the operational techniques. Specialists would take several hours of painstaking work to perform a multiplication which a child could now do in a few minutes. And tradesmen who wanted to know the total of the week's or the month's takings were obliged to employ the services of such counting specialists (Fig. 26.1).

FIG. 26.1. *Arithmetician performing a calculation on a counter-abacus. From a fifteenth-century European engraving, reproduced from Beauclair, 1968*

This situation did not alter in the conservative bureaucracies of the European nations throughout the seventeenth and eighteenth centuries. Samuel Pepys, for example, became a civil servant after taking a degree at Cambridge, and after a time in the Navy, became a clerk to the Admiralty. From 1662, he was in charge of naval procurement. Though thoroughly well-educated by the standards of the day, Samuel Pepys was nonetheless quite unable to make the necessary calculations for checking the purchases of timber made by the Admiralty. So he resolved to educate himself afresh:

> Up at 5 a-clock. . . By and by comes Mr Cooper, Mate of the *Royall Charles*, of whom I entend to learn Mathematiques; and so begin with him today. . . After an hour's being with him at Arithmetique, my first attempt being to learn the Multiplication table, then we parted till tomorrow; and so to my business at my office again. . . [Pepys, *Diary*, (1985), p. 212].

He eventually mastered the techniques, and was so proud of himself that he sought to teach his wife addition, subtraction and multiplication. But he didn't dare launch her into the subtleties of long division.

It is now perhaps easier to understand why skilled abacists were long regarded in Europe as magicians enjoying supernatural powers.

## THE EARLIEST INTRODUCTION OF "ARABIC" NUMERALS IN EUROPE

All the same, even before the Crusades, Westerners could have made full and profitable use of the Indian computational methods which the Arabs had brought to the threshold of Europe from the ninth century CE. A channel of transmission existed, and it was by no means a paltry one.

A French monk with a thirst for knowledge, named Gerbert of Aurillac, could indeed have played the same role in the West as had the learned Persian al-Khuwarizmi, in the Arab-Islamic world. In the closing stages of the tenth century CE, Gerbert – who was to become Pope in the year 1000 – could have broadcast in the West the discoveries of India which had reached North Africa and the Islamic province of Andalusia (Spain) some two centuries earlier. But he found no followers in this respect.

In order to understand the circumstances attendant on the first arrival of Indian numerals in Western Europe, we have to remember the long-drawn-out sequels of the collapse of the Roman Empire and the ensuing Barbarian invasions.

From the end of the Roman Empire in the fifth century until the end of the first millennium, Western Europe was continually laid waste by epidemics, by famine, and by warfare, and suffered centuries of political instability, economic recession, and profound obscurantism. The so-called "Carolingian renaissance" in the Benedictine monasteries of the ninth century may have revitalised the idea and structure of education in the era of Charlemagne and also laid the bases of mediaeval philosophy, but it actually brought only minor and temporary relief to the general situation.

Scientific knowledge available at that time was very elementary, if not entirely deficient. The few privileged men who received any "education" learned first to read and to write. They went on to grammar, dialectics and rhetoric, and sometimes also to the theory of music. Finally, they received basic instruction in astronomy, geometry and arithmetic.

"Theoretical" arithmetic in the High Middle Ages was drawn from a work attributed to the Latin mathematician Boethius (fifth century CE) who had himself drawn handsomely on a second-rate work by the Greek Nicomacchus of Gerasa (second century CE). As for "practical" arithmetic, it consisted mainly in the use of Roman numerals, and in operations with counters on the old abacus of the Romans; it also included the techniques of finger-counting transmitted by Isidore of Seville (died 636 CE) and by Bede (died 735 CE).

In these almost completely "dark ages", even the memory of human arts and sciences was almost lost. But a sudden reawakening occurred in the eleventh and twelfth centuries:

> A massive demographic explosion brought many consequences in its wake – the development of virgin lands, the growth of towns and of the monastic orders, the crusades, and the construction of ever larger churches. Prices rose, the circulation of money accelerated, and, as sovereign states began to control feudal anarchy, trade also began to prosper. An increase in international contacts created a favourable environment for the introduction of Arabic science in the West [G. Beaujouan (1947)].

Gerbert of Aurillac was certainly one of the most prominent scientific personalities of this whole period. Born in southwest France c. 945 CE, he took holy orders at the monastery of Saint-Géraud at Aurillac, where his sharp mind and passion for learning soon marked him out. He learned mathematics and astronomy from Atton, the Bishop of Vich, and then, probably as a result of a visit to Islamic Spain from 967 to 970, he absorbed the lessons of the Arabic school. He learned to use an astrolabe, he learned the Arabic numeral system, as well as the basic arithmetical operations in the Indian manner. From 972 to 987, Gerbert was in charge of the Diocesan school of Reims, and then, after a period as abbot of Bobbio (Italy), he became an adviser to Pope Gregory V and became successively Archbishop of Reims, Archbishop of Ravenna, and finally Pope Sylvester II, from 2 April 999 until his death on 12 May 1003.

Legend has it that Gerbert went as far as Seville, Fez and Cordoba to learn Indo-Arabic arithmetic and that he disguised himself as a Muslim pilgrim in order to gain entrance to Arab universities. Though that is not impossible, it is more likely that he remained in Christian Spain at the monastery of Santa Maria de Ripoll, a striking example, according to Beaujouan, of the hybridisation of Arabic and Isidorian traditions. The little town of Ripoll (near Barcelona, in Catalonia) had indeed long served as a meeting point for the Islamic and Christian worlds.

One thing is nonetheless quite certain: Gerbert brought back to France all the techniques necessary for modern arithmetic to exist. His teaching at the diocesan school at Reims was highly influential and did much to reawaken interest in mathematics in the West. And it was Gerbert who first introduced so-called Arabic numerals into Europe. Arabic numerals, indeed – but alas, only the first nine! He did not bring back the zero from his Spanish sojourn, nor did he include Indian arithmetical operations in his pack.

So what happened? Gerbert's initiative actually met fierce resistance: his Christian fellows clung with conservative fervour to the number-system and arithmetical techniques of the Roman past. Most clerics of the period, it has to be said, thought of themselves as the heirs of the "great tradition" of classical Rome, and could not easily countenance the superiority of any other system. The time was simply not ripe for a great revolution of the mind.

### *A Victorian howler*

The mediaeval forms of the Arabic numerals are found in a manuscript entitled *Geometria Euclidis* (*Euclidian Geometry*) which for a long time was attributed to Boethius, a Roman mathematician of the fifth century CE. The text itself claims that the nine numeral symbols shown and their use in a place-value system had been invented by Pythagoreans and derived from the use of the table-abacus in Ancient Greece. For this reason, the shapes of the nine Indo-Arabic numerals used in the Middle Ages came to be called "the *apices* of Boethius", even though, as we have seen, there is no possibility whatsoever that a Roman of the fifth century, let alone Greek Pythagoreans, could have known or understood Indo-Arabic arithmetic.

The solution of this conundrum is very simple. As modern analyses have shown, *Geometria Euclidis* was put together by an anonymous compiler in the eleventh century, and its attribution to Boethius is entirely apocryphal.

### *Early forms of Arabic numerals in the West*

The earliest actual appearances of Arabic numerals in the West are to be found in the *Codex Vigilanus* (copied by a monk named Vigila at the Monastery of Albelda, Spain, in the year 976) and the *Codex Aemilianensis*, copied directly from the *Vigilanus* in the year 992 at San Millan de la Cogolla, also in northern Spain (see Fig. 26.2).

These nine figures are clearly integrated in the cursive script of "full Visigothic of the Northern Spanish type" (in the terms of R. L. Burnam, 1912–25), but their Indian origin is nonetheless quite manifest. Both manuscripts give the numerals shapes that are very close to the *Ghubar* figures of the Western Arabs.

From the early eleventh century, the nine figures appear in a whole variety of shapes and sizes in a great number of mss copied in more or less every corner of the European continent. The variations in shape and style are the result of palaeographic modifications occurring in different periods and places, as can be seen in Figure 26.4.

However, contrary to first appearances, "Arabic" numerals did not first spread through the West by manuscript transmission, but through a piece of calculating technology called *Gerbert's abacus*. In other words, the numerals were disseminated not by writing but by the oral transmission of the knowledge necessary to learn how to operate the entirely new kind of abacus that Gerbert of Aurillac had promoted from around 1000 CE, and thereafter by his numerous disciples in Cologne, Chartres, Reims and Paris.

Let us recall that for many centuries the Christianised populations of Western Europe had expressed number almost exclusively through the medium of Roman numerals, a very rudimentary system of notation whose operational inefficacy lay at the root of all the difficulties experienced in

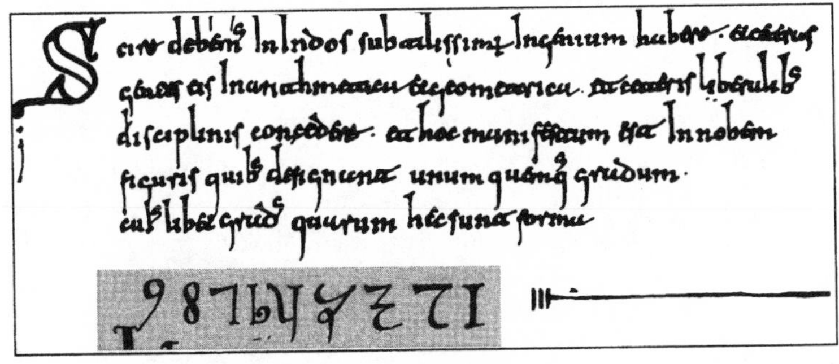

FIG. 26.2. *Detail from the* Codex Vigilanus *(976 CE, Northern Spain). The first known occurrence of the nine Indo-Arabic numerals in Western Europe. Escurial Library, Madrid, Ms. lat. d.I.2, f° 9v. See Burnam (1920), II, plate XXIII*

calculation throughout that long period of the "dark ages". First-millennium mediaeval arithmeticians made their calculations just like their Roman predecessors, through a complicated game of counters placed on tables marked out with rows and columns delimiting the different decimal orders.

On a Roman abacus, you place as many unit counters in a column for a specific decimal order as there are units in that order. But just before 1000 CE, it occurred to Gerbert of Aurillac, who had seen Arabic counting methods during his time in Andalusia, to reduce the number of counters used and so to simplify the material complexity of computation on an abacus.

Gerbert's system involved jettisoning multiple unit counters and replacing them with single counters in each decimal column, the new horn "singles" being marked with one of the nine numerals he had brought back from the Arabs. These number-tokens were called *apices* (*apex* in the singular) and were each dubbed with a specific name that has nothing to do with the number shown (though a few of them seem to hark back to Arabic and Hebrew number-names).

The *apex* for 1 was called *Igin*
for 2 was called *Andras*
for 3 was called *Ormis*
for 4 was called *Arbas*
for 5 was called *Quimas*
for 6 was called *Caltis*
for 7 was called *Zenis*
for 8 was called *Temenias*
for 9 was called *Celentis*

FIG. 26.3.

So the one, two, three, four, five, six, seven, eight or nine unit-counters in each column of the Roman abacus were replaced by a single *apex* bearing on it the corresponding numeral in "Arabic" script.

When a decimal order was "empty", the abacist simply put no *apex* in the corresponding column. So to represent the number 9,078, you put the apex for 8 in the unit column, the *apex* for 7 in the tens column, and the *apex* for 9 in the thousands column, leaving all the other columns empty.

| DATES | SOURCES | 1 | 2 | 3 | 4 | 5 | 6 | 7 | 8 | 9 | 0 |
|---|---|---|---|---|---|---|---|---|---|---|---|
| 976 | Spain: Codex *Vigilanus*. Escurial, Ms. lat. d.I.2, f° 9v | [illegible] | [illegible] | [illegible] | [illegible] | [illegible] | [illegible] | [illegible] | [illegible] | [illegible] | |
| 992 | Spain: Codex *Aemilianensis*. Escurial, Ms. lat. dI.1, f° 9v | [illegible] | [illegible] | [illegible] | [illegible] | [illegible] | [illegible] | [illegible] | [illegible] | [illegible] | |
| Before 1030 | France (Limoges). Paris, BN Ms. lat. 7231, f° 85v | [illegible] | [illegible] | [illegible] | [illegible] | [illegible] | [illegible] | [illegible] | [illegible] | [illegible] | |
| 1077 | Vatican Library, Ms. lat. 3101, f° 53v | [illegible] | [illegible] | [illegible] | [illegible] | [illegible] | [illegible] | [illegible] | [illegible] | [illegible] | |
| XIth C | Bernelinus, *Abacus*. Montpellier, Library of the Ecole de Médecine, Ms. 491, f° 79 | [illegible] | [illegible] | [illegible] | [illegible] | [illegible] | [illegible] | [illegible] | [illegible] | [illegible] | |
| 1049? | Erlangen, Ms. lat. 288, f° 4 | [illegible] | [illegible] | [illegible] | [illegible] | [illegible] | [illegible] | [illegible] | [illegible] | [illegible] | |
| XIth C | Montpellier, Library of the Ecole de Médecine, Ms. 491, f° 79 | [illegible] | [illegible] | [illegible] | [illegible] | [illegible] | [illegible] | [illegible] | [illegible] | [illegible] | [illegible] |
| XIth C | France: Gerbertus, *Raciones numerorum Abaci*. Paris, BN Ms. lat. 8663, f° 49v | [illegible] | [illegible] | [illegible] | [illegible] | [illegible] | [illegible] | [illegible] | [illegible] | [illegible] | |
| XIth/ XIIth C | Lorraine: Boecius, *Geometry*. Paris, BN, Ms. lat. 7377, f° 25v | [illegible] | [illegible] | [illegible] | [illegible] | [illegible] | [illegible] | [illegible] | [illegible] | [illegible] | |
| XIth C | Boecius, *Geometry*. London, BM, Ms. Harl. 3595, f° 62 | [illegible] | [illegible] | [illegible] | [illegible] | [illegible] | [illegible] | [illegible] | [illegible] | [illegible] | |
| XIth C | Germany (Regensburg). Munich, Bayerische Staatsbibliothek, Clm 12567, f° 8 | [illegible] | [illegible] | [illegible] | [illegible] | [illegible] | [illegible] | [illegible] | [illegible] | [illegible] | |
| XIth C | Boecius, *Geometry*. Chartres, Ms. 498. f° 160 | [illegible] | [illegible] | [illegible] | [illegible] | [illegible] | [illegible] | [illegible] | [illegible] | [illegible] | [illegible] |
| Early XIIth C | Bernelinus, *Abacus*. London, BM Add. Ms. 17808, f°, 57 | [illegible] | [illegible] | [illegible] | [illegible] | [illegible] | [illegible] | [illegible] | [illegible] | [illegible] | |
| Late XIth C | Bernelinus, *Abacus*. Paris, BN Ms. lat. 7193, f° 2 | [illegible] | [illegible] | [illegible] | [illegible] | [illegible] | [illegible] | [illegible] | [illegible] | [illegible] | |
| Late XIth C | France (Chartres?): Anon., Arithmetical tables. Paris, BN Ms. lat. 9377, f° 113 | [illegible] | [illegible] | [illegible] | [illegible] | [illegible] | [illegible] | [illegible] | [illegible] | [illegible] | |
| Late XIth C | Bernelinus, *Abacus*. Paris, BN Ms. lat. 7193, f° 2 | [illegible] | [illegible] | [illegible] | [illegible] | [illegible] | [illegible] | [illegible] | [illegible] | [illegible] | |
| XIIth C | Rome, Alessandrina Library, Ms. 171, f° 1 | [illegible] | [illegible] | [illegible] | [illegible] | [illegible] | [illegible] | [illegible] | [illegible] | [illegible] | [illegible] |
| XIIth C | Paris, Saint Victor. Gerlandus, *De Abaco*. Paris, BN, Ms. lat. 15119, f° 1 | [illegible] | [illegible] | [illegible] | [illegible] | [illegible] | [illegible] | [illegible] | [illegible] | [illegible] | |
| XIIth C | Boecius, *Geometry*. Paris, BN, Ms. lat. 7185, f° 70 | | [illegible] | [illegible] | [illegible] | [illegible] | [illegible] | [illegible] | [illegible] | [illegible] | |
| XIIth C | France, Chartres(?):Bernelinus, *Abacus*. Oxford, Bodley, Ms Auct. F. I. 9, f° 67v | [illegible] | [illegible] | [illegible] | [illegible] | [illegible] | [illegible] | [illegible] | [illegible] | [illegible] | |
| XIIth C | Gerlandus, *De Abaco*. London, BM Add. Ms. 22414, f° 5 | [illegible] | [illegible] | [illegible] | [illegible] | [illegible] | [illegible] | [illegible] | [illegible] | [illegible] | |
| XIIth C | Gerlandus, *De Abaco*. Paris, BN Ms. lat. 95, f° 150 | [illegible] | [illegible] | [illegible] | [illegible] | [illegible] | [illegible] | [illegible] | [illegible] | [illegible] | |
| Early XIIIth C | France (Chartres): Anon. Paris, BN Ms. lat. Fonds Saint-Victor, 533, f° 22v | [illegible] | [illegible] | [illegible] | [illegible] | [illegible] | [illegible] | [illegible] | [illegible] | [illegible] | |

FIG. 26.4. *Mediaeval* apices. *Sources: BSMF X (1877), p. 596; Burnam (1920), II, plates XXIII & XXIV; Folkerts (1970) Friedlein (1867) p. 397; Hill (1915) Smith and Karpinski (1911) p. 88*

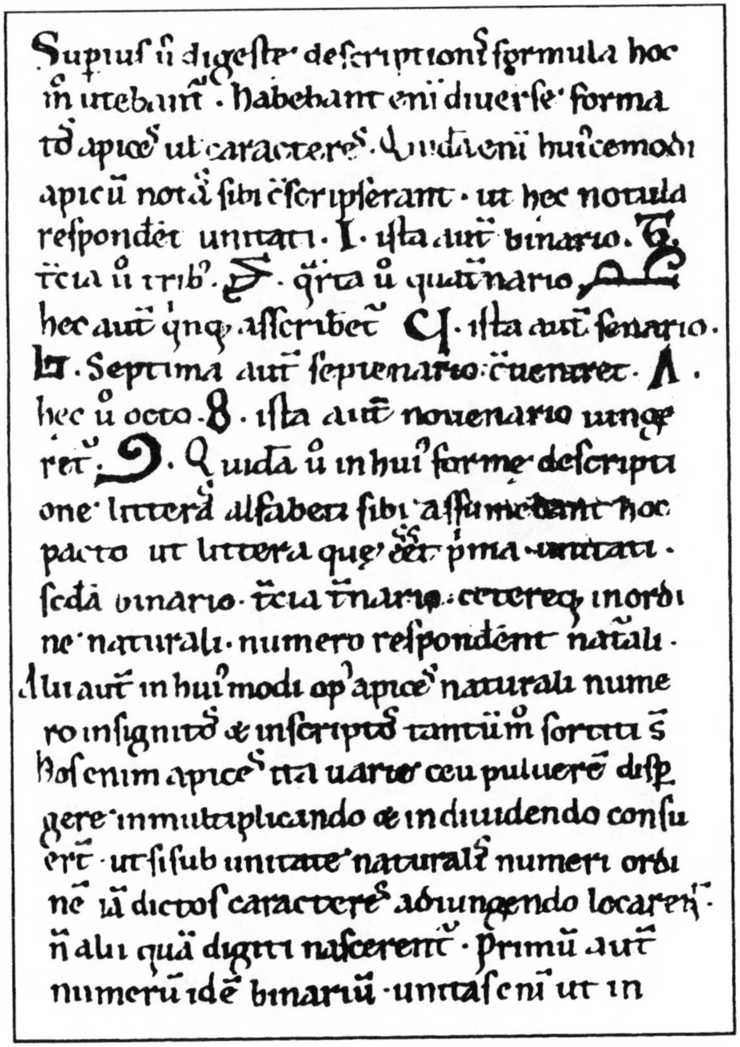

FIG. 26.5. Apices *in an eleventh-century Latin manuscript. Berlin, Ms. lat. 8° 162 (n), f° 74. From Folkerts (1970)*

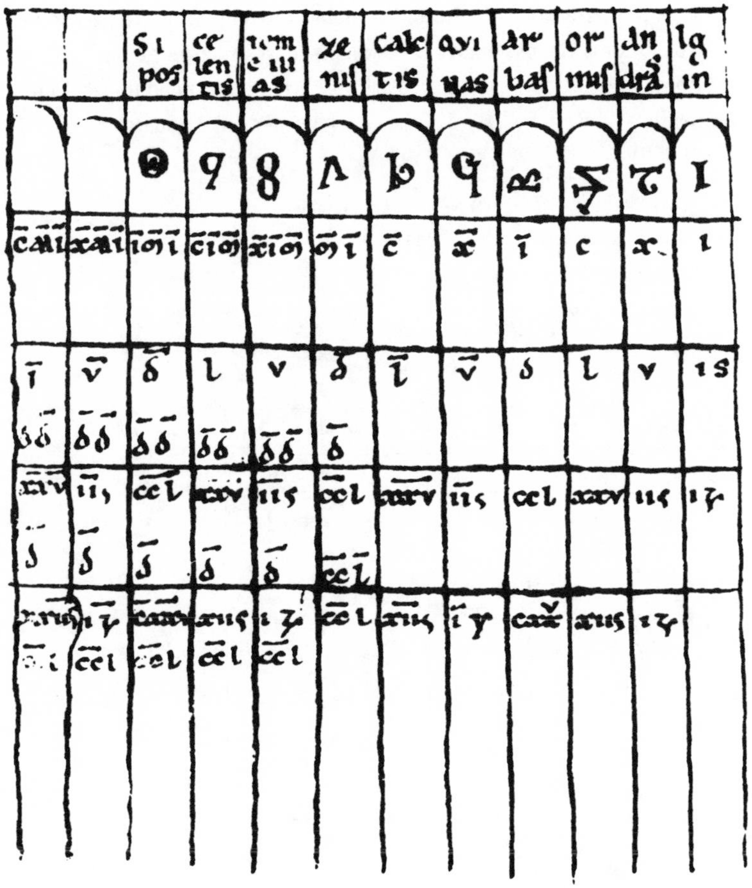

FIG. 26.6. Apices *and the columns of Gerbert's abacus in an eleventh-century Latin manuscript. Berlin, Ms. lat. 8° 162 (n), f° 73v. From Folkerts (1970)*

So at this early stage the Arabic numerals introduced by Gerbert served only to simplify the use of an abacus identical in structure to that of classical Rome. Indeed, some mediaeval arithmeticians continued to use Roman numerals – or even the letters of the Greek numeral alphabet ($\alpha = 1$, $\beta = 2$, . . . $\theta = 9$) – on their *apices*.

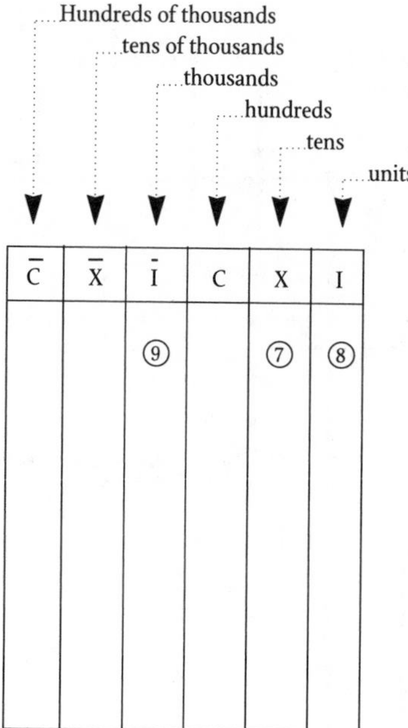


FIG. 26.7.

## ARITHMETICAL OPERATIONS ON GERBERT'S ABACUS

The following example shows how sums can be done on Gerbert's abacus without zero. The fact that it is possible to complete these operations correctly explains why mediaeval manuscripts of the eleventh and twelfth centuries contain no symbols for zero, nor ever even mention the concept. The nine symbols of Indian origin spread around Europe, but only in very restricted circles, since the whole business of counting was in the hands of a tiny elite of arithmeticians, appropriately called *abacists*.

*To multiply 325 by 28*

1. Place the *apices* for multiplicand (325) on the bottom row of the abacus, putting the "5" counter in the units column, the "2" counter in the tens column, and the "3" counter in the hundreds column:

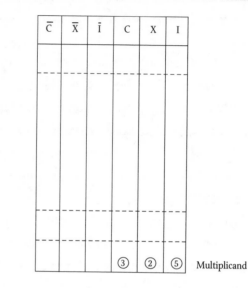

FIG. 26.8A.

2. Then place the *apices* for the multiplier (28) on the top row of the abacus, putting the "2" and "8" counters in the tens and units columns respectively:

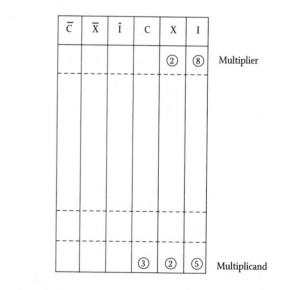

| C̄ | X̄ | Ī | C | X | I | |
|---|---|---|---|---|---|---|
| | | | | ② | ⑧ | Multiplier |
| | | | ③ | ② | ⑤ | Multiplicand |

FIG. 26.8B.

3. Now find the product of the 8 × 5 in the units column. Since the product is 40, place a "4" counter in the upper part of the central rows of the abacus, in the tens column, leaving the units column empty, thus:

| $\bar{C}$ | $\bar{X}$ | $\bar{I}$ | C | X | I | |
|---|---|---|---|---|---|---|
| | | | | ② | ⑧ | Multiplier |
| | | | | ④ | | |
| | | | | | | |
| | | | | | | |
| | | | ③ | ② | ⑤ | Multiplicand |

FIG. 26.8C.

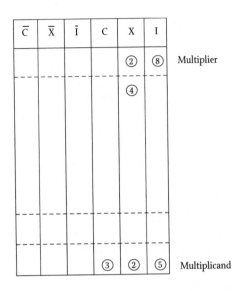

4. Now find the product of the counter in the units column by the one in the tens columns, in other words 2 × 8. The product being 16, place a "1" counter in the hundreds column and a "6" counter in the tens column, still leaving the units column empty, thus:

| $\bar{C}$ | $\bar{X}$ | $\bar{I}$ | C | X | I | |
|---|---|---|---|---|---|---|
| | | | | ② | ⑧ | Multiplier |
| | | | | ④ | | |
| | | | ① | ⑥ | | |
| | | | | | | |
| | | | | | | |
| | | | ③ | ② | ⑤ | Multiplicand |

FIG. 26.8D.

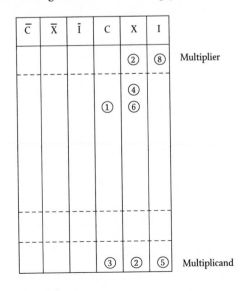

5. Now multiply the same unit 8 by the *apex* in the hundreds column, in other words 3 × 8. The product being 24, place a "2" counter in the thousands column and a "4" counter in the hundreds column:

| $\bar{C}$ | $\bar{X}$ | $\bar{I}$ | C | X | I | |
|---|---|---|---|---|---|---|
| | | | | ② | ⑧ | Multiplier |
| | | | | ④ | | |
| | | | ① | ⑥ | | |
| | | ② | ④ | | | |
| | | | | | | |
| | | | | | | |
| | | | ③ | ② | ⑤ | Multiplicand |

FIG. 26.8E.

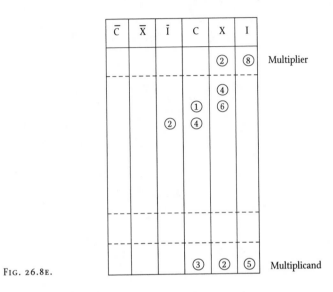

6. Since the multiplying of the "8" is now complete, remove the "8" token from the abacus before turning attention to the "2" in the multiplier:

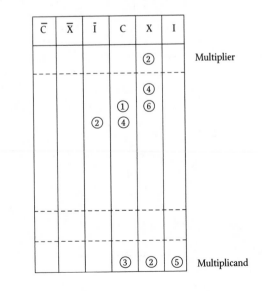

| $\bar{C}$ | $\bar{X}$ | $\bar{I}$ | C | X | I | |
|---|---|---|---|---|---|---|
| | | | | ② | | Multiplier |
| | | | | ④ | | |
| | | | ① | ⑥ | | |
| | | ② | ④ | | | |
| | | | | | | |
| | | | | | | |
| | | | ③ | ② | ⑤ | Multiplicand |

FIG. 26.8F.

7. Now multiply the 2 by the 5 in the multiplicand. Since the 2 is in the tens column, the product (10) requires us to place a "1" counter in the hundreds column, thus:

| C̄ | X̄ | Ī | C | X | I | |
|---|---|---|---|---|---|---|
| | | | | ② | | Multiplier |
| | | ② | ① ④ ① | ④ ⑥ | | |
| | | | | | | |
| | | | ③ | ② | ⑤ | Multiplicand |

FIG. 26.8G.

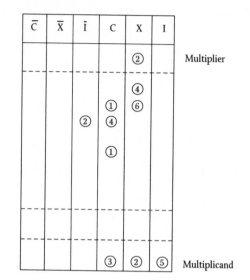

8. Now multiply the 2 of the multiplier by the 2 in the multiplicand, giving the answer 4. Since both factors are in the tens columns, the result (four tens of tens) is registered by placing a "4" in the hundreds column, thus:

| C̄ | X̄ | Ī | C | X | I | |
|---|---|---|---|---|---|---|
| | | | | ② | | Multiplier |
| | | ② | ① ④ ① ④ | ④ ⑥ | | |
| | | | | | | |
| | | | ③ | ② | ⑤ | Multiplicand |

FIG. 26.8H.

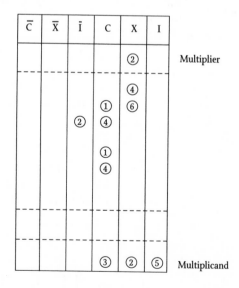

9 Now multiply the same 2 by the 3 in the hundreds column of the multiplicand. The product, 6, means six tens of hundreds, so we place a "6" counter in the thousands column, thus:

| C̄ | X̄ | Ī | C | X | I | |
|---|---|---|---|---|---|---|
| | | | | ② | | Multiplier |
| | | ② ⑥ | ① ④ ① ④ | ④ ⑥ | | |
| | | | | | | |
| | | | ③ | ② | ⑤ | Multiplicand |

FIG. 26.8I.

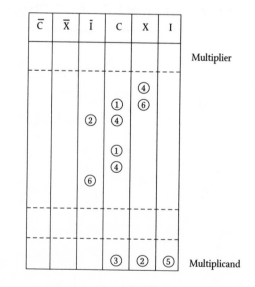

10. Since the multiplying of the 2 is now complete, remove the "2" token from the multiplier line of the abacus:

| C̄ | X̄ | Ī | C | X | I | |
|---|---|---|---|---|---|---|
| | | | | | | Multiplier |
| | | ② ⑥ | ① ④ ① ④ | ④ ⑥ | | |
| | | | | | | |
| | | | ③ | ② | ⑤ | Multiplicand |

FIG. 26.8J.

11. As all the multiplications of the highest number in the multiplier are now also complete, all that remains is to sum the partial products on the board, replacing counters whose total is more than 10 by a unit counter in the next-leftmost column. Since the "4" and "6" of the tens column total 10, they are taken off the board and replaced by a "1" counter in the hundreds column, thus:

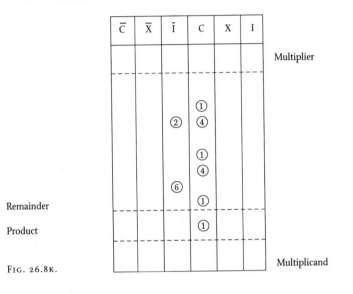

| $\bar{C}$ | $\bar{X}$ | $\bar{I}$ | C | X | I | |
|---|---|---|---|---|---|---|
| | | | | | | Multiplier |
| | | ② | ① ④ | | | |
| | | | ① ④ | | | |
| | | ⑥ | ① | | | Remainder |
| | | | ① | | | Product |
| | | | | | | Multiplicand |

FIG. 26.8K.

12. Now sum the tokens in the hundreds column. As they total 11, remove all counters bar the "1", and place a "1" in the thousands column, thus:

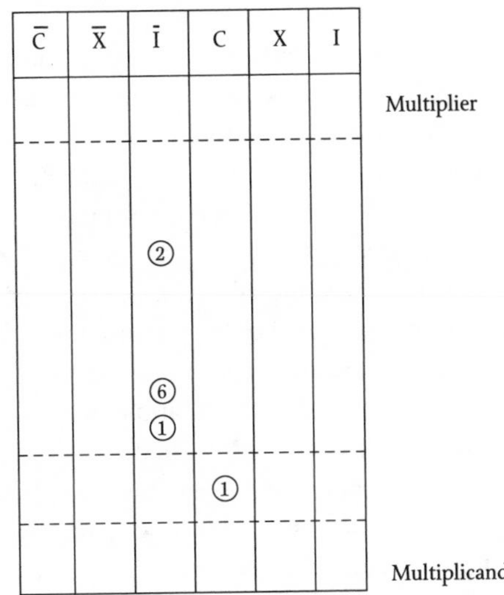


FIG. 26.8L.

13. Finally, sum the tokens in the thousands column, which gives 9, so remove all the tokens and replace them by a "9" in the thousands column, thus:

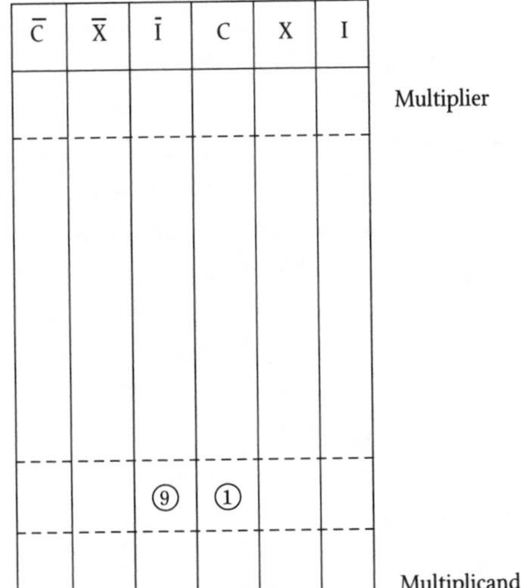


FIG. 26.8M.

14. The result of the operation is therefore 9,100 (since the tens and units columns are empty). This example shows how Gerbert's abacus made arithmetical operations long and complicated; its use presupposed lengthy training and a high degree of intelligence.

## FROM "ARABIC" NUMERALS TO EUROPEAN APICES

The shapes of the Arabic numerals brought back from Spain by Gerbert of Aurillac were represented with the most fantastical variations on European horn *apices*. Consider the following versions of "4" found over the first two hundred years of the second millennium CE:

| *Archetype* | | | | | | |
|---|---|---|---|---|---|---|
| *Archetype, Spain, Xth C* | *Limoges (France), XIth C* | *Fleury (France), XIth C* | *Lorraine (France), XIIth C* | *Auxerre (France), XIIth C* | *Regensburg (Bavaria), XIIth C* | *Chartres (France), XIIIth C* |

FIG. 26.9.

Styles obviously varied from one region to another, from one school to another, even from one engraver to another, in a period that had no concept of standardisation. Indeed, what we can see happening in these examples is the adaptation of the *Ghubar* forms of the Arabic numerals to the very different styles of writing practised in different parts of Europe. So in Italy we see numerals assimilated to the round shapes and wide openings of Italic script, in England to the narrower and more angular shapes of English script, in Germany to the thicker and squarer writing style of German script, and in France and Spain we see them being shaped in harmony with the dominant styles of Carolingian script.

A similar phenomenon has already been observed in India and in the Indic civilisations of Southeast Asia. Scribes and stone-carvers adapted the basic nine symbols to their own indigenous writing styles and applied their own aesthetic sense to the shapes, so that there quickly resulted widely differing sets of numerals that at first glance seem quite unrelated.

Similar diversity has been seen in the Arabic world too, where scribes and copyists adapted the same basic figures to the different scripts used in different areas of the Arabic-speaking world.

So there is no reason for the Western world not to have also generated a range of distinctive variations on the numeral set. However, as Beaujouan has pointed out, there was a supplementary factor in the West. All the different shapes found, he insists, are virtually superimposable on each other provided they are rotated by some degree. That is particularly noticeable for the 3, 4, 5, 6, 7 and 9 (see Fig. 26.4).

The reason is that the *apices* were often placed on the abacus without any particular regard for the original orientation of the shape. In some schools, for example, the *apices* were placed upside-down, so the 5 was sometimes found with its "tail" at the bottom. The 9 was sometimes placed on its right side, sometimes on its left side, and sometimes placed upside down so that it looked like today's 6.

Some scribes and stone-carvers simply replaced the original shape of the numeral with the shape that they had grown accustomed to, or which seemed more "logical" in their eyes. Confusion became generalised, and even mathematical course-books often taught the numbers upside-down and back-to-front.

The obvious solution would have been to mark the top or bottom of each horn *apex* with a dot, but people were content merely to distinguish the two figures that could most easily be confused by writing the 6 with sharp, angular lines and the 9 with curved and flowing lines.

However, mediaeval *apices* did not actually give rise directly to our current numerals. After the Crusades, these early forms of the numerals were simply abandoned, and shapes closer to the original Arabic forms were re-introduced – and it is these later arrivals, which eventually stabilised into standard forms, that ultimately gave rise to modern "Arabic" numerals.

## THE SECOND INTRODUCTION OF ARABIC NUMERALS IN EUROPE

We might have expected Pope Sylvester II to have opened the millennium onto a new era of progress in the West, thanks to the numerals and operational techniques he had brought back from the Arabic-Islamic world. But such expectations would be vain: the ignorance and conservatism of the Christian world blocked the way.

Although modern numerals and number-techniques were in fact available from the late tenth century, they were used only in the most rudimentary ways for over two hundred years. They served solely to simplify archaic counting methods and to give rise to rules of procedure which, according to William of Malmesbury, "perspiring abacists barely comprehended themselves".

Some arithmeticians even put up a solid resistance to the new-fangled figures from the East by inscribing their *apices* with the Greek letter-numerals from $\alpha = 1$ to $\theta = 9$, or the Roman figures I to IX. Anything was better than having recourse to the "diabolical signs" of the "satanic accomplices" that the Arabs were supposed to be!

Gerbert of Aurillac also suffered at the hands of the rearguard. It was rumoured that he was an alchemist and a sorcerer, and that he must have sold his soul to Lucifer when he went to taste of the knowledge of the Saracens. The accusation continued to circulate for centuries until finally, in 1648, papal authorities reopened the tomb of Sylvester II to make sure that it was not still infested by the devil!

The dawn of the modern age did not really occur until Richard Lionheart reached the walls of Jerusalem. From 1095 to 1270, Christian knights and princes tried to impose their religion and traditions on the Infidels of the Middle East. But what they actually achieved was to bring back to Europe the cultural riches they encountered in the Holy Land. It was these campaigns – or rather, their secondary consequences – that finally allowed the breakthrough which Gerbert of Aurillac, for all his knowledge and energy, had failed to achieve at the end of the tenth century. For the wars implied a whole range of contacts with the Islamic world, and a number of clerks travelling with the armies learned the written numerals and arithmetical methods of the Indo-Arabic school.

Gerbert's abacus thus slowly fell into disuse. Gradually, numerals written on sand or dust, instead of being engraved on horn-tipped *apices*, led to the disappearance of the columns on the abacus. This allowed much simpler, much faster and more elegant operations, which now came to be called *algorisms*, after al-Khuwarizmi, the first Islamic scholar who had generalised their application.

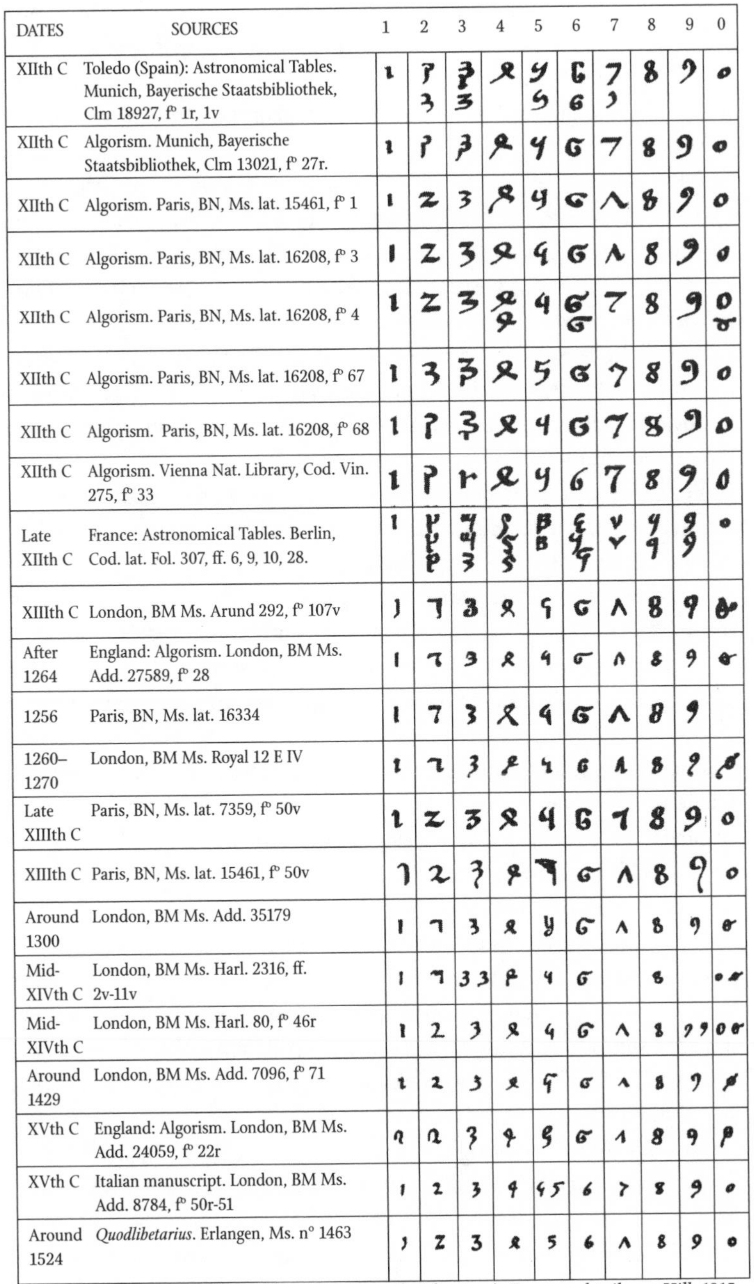

| DATES | SOURCES | 1 | 2 | 3 | 4 | 5 | 6 | 7 | 8 | 9 | 0 |
|---|---|---|---|---|---|---|---|---|---|---|---|
| XIIth C | Toledo (Spain): Astronomical Tables. Munich, Bayerische Staatsbibliothek, Clm 18927, f° 1r, 1v | | | | | | | | | | |
| XIIth C | Algorism. Munich, Bayerische Staatsbibliothek, Clm 13021, f° 27r. | | | | | | | | | | |
| XIIth C | Algorism. Paris, BN, Ms. lat. 15461, f° 1 | | | | | | | | | | |
| XIIth C | Algorism. Paris, BN, Ms. lat. 16208, f° 3 | | | | | | | | | | |
| XIIth C | Algorism. Paris, BN, Ms. lat. 16208, f° 4 | | | | | | | | | | |
| XIIth C | Algorism. Paris, BN, Ms. lat. 16208, f° 67 | | | | | | | | | | |
| XIIth C | Algorism. Paris, BN, Ms. lat. 16208, f° 68 | | | | | | | | | | |
| XIIth C | Algorism. Vienna Nat. Library, Cod. Vin. 275, f° 33 | | | | | | | | | | |
| Late XIIth C | France: Astronomical Tables. Berlin, Cod. lat. Fol. 307, ff. 6, 9, 10, 28. | | | | | | | | | | |
| XIIIth C | London, BM Ms. Arund 292, f° 107v | | | | | | | | | | |
| After 1264 | England: Algorism. London, BM Ms. Add. 27589, f° 28 | | | | | | | | | | |
| 1256 | Paris, BN, Ms. lat. 16334 | | | | | | | | | | |
| 1260–1270 | London, BM Ms. Royal 12 E IV | | | | | | | | | | |
| Late XIIIth C | Paris, BN, Ms. lat. 7359, f° 50v | | | | | | | | | | |
| XIIIth C | Paris, BN, Ms. lat. 15461, f° 50v | | | | | | | | | | |
| Around 1300 | London, BM Ms. Add. 35179 | | | | | | | | | | |
| Mid-XIVth C | London, BM Ms. Harl. 2316, ff. 2v-11v | | | | | | | | | | |
| Mid-XIVth C | London, BM Ms. Harl. 80, f° 46r | | | | | | | | | | |
| Around 1429 | London, BM Ms. Add. 7096, f° 71 | | | | | | | | | | |
| XVth C | England: Algorism. London, BM Ms. Add. 24059, f° 22r | | | | | | | | | | |
| XVth C | Italian manuscript. London, BM Ms. Add. 8784, f° 50r-51 | | | | | | | | | | |
| Around 1524 | *Quodlibetarius*. Erlangen, Ms. n° 1463 | | | | | | | | | | |

FIG. 26.10. *The second form of European numerals (algorisms). For more details, see Hill, 1915*

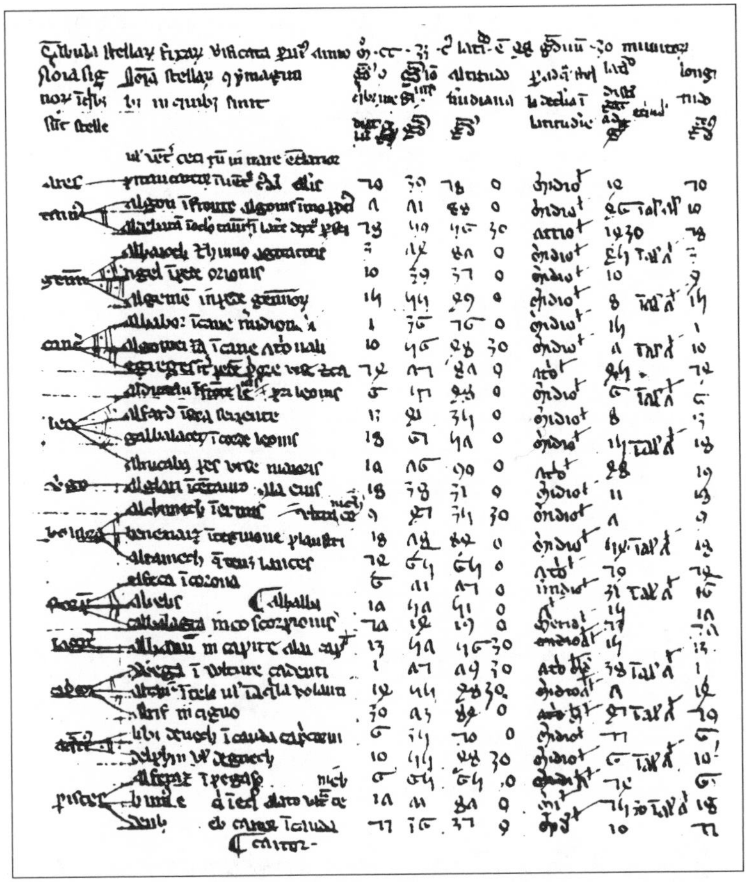

FIG. 26.11. *Numerals including zero in a thirteenth-century Latin manuscript. Paris, BN, Ms. lat. 7413, part II. Facsimile in the Ecole des Chartes, AF 1113*

So the first European "algorists" were born at the gates of Jerusalem. But unlike the "abacists", the new European counting experts were obliged to adopt the zero, to signify missing orders of magnitude, otherwise computations written in sand would lead to confusing representations of number and mistaken operations. At last, then, true "Arabic" numerals including zero, and the arithmetical tradition that had been born long before in India, were able to make their way into Europe.

There were of course other contacts with the Islamic world on the other side of the Mediterranean, by way of Sicily and most especially through Spain and North Africa. It was in Spain that a huge wave of translations began in the twelfth century, bringing into Latin works written in Arabic, and even more importantly Greek and Sanskrit texts already translated into Arabic. Thanks to translators like Adelard of Bath and to centres of scholarship at Cordoba and Toledo, the resources available for acquiring knowledge of arithmetic, mathematics, astronomy, natural sciences and philosophy swelled almost by the day; and it was by means of translation from the Arabic that the West eventually became familiar with the works of Euclid, Archimedes, Ptolemy, Aristotle, al-Khuwarizmi, al-Biruni, Ibn Sina, and many others.

Between them, the Crusaders at Jerusalem and the scholars of Toledo were ensuring the more or less rapid death of the abacus and of abacism.

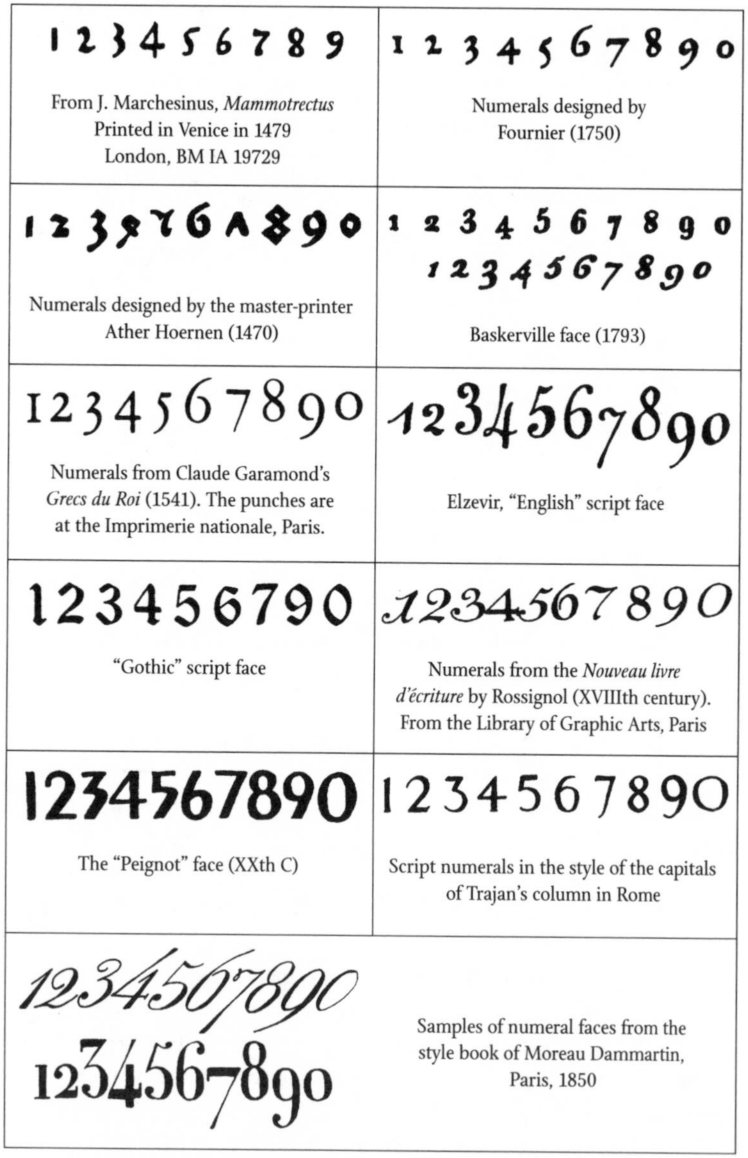

| | |
|---|---|
| 1 2 3 4 5 6 7 8 9<br>From J. Marchesinus, *Mammotrectus*<br>Printed in Venice in 1479<br>London, BM IA 19729 | 1 2 3 4 5 6 7 8 9 0<br>Numerals designed by<br>Fournier (1750) |
| 1 2 3 4 5 6 7 8 9 0<br>Numerals designed by the master-printer<br>Ather Hoernen (1470) | 1 2 3 4 5 6 7 8 9 0<br>1 2 3 4 5 6 7 8 9 0<br>Baskerville face (1793) |
| 1234567890<br>Numerals from Claude Garamond's<br>*Grecs du Roi* (1541). The punches are<br>at the Imprimerie nationale, Paris. | 1234567890<br>Elzevir, "English" script face |
| 123456790<br>"Gothic" script face | 1234567890<br>Numerals from the *Nouveau livre d'écriture* by Rossignol (XVIIIth century).<br>From the Library of Graphic Arts, Paris |
| 1234567890<br>The "Peignot" face (XXth C) | 1234567890<br>Script numerals in the style of the capitals<br>of Trajan's column in Rome |
| 1234567890<br>1234567890 | Samples of numeral faces from the<br>style book of Moreau Dammartin,<br>Paris, 1850 |

FIG. 26.12. *The development of printed numerals since the fifteenth century*

The spread of "algorism" was given renewed impetus from the start of the thirteenth century by a great Italian mathematician, Leonard of Pisa (c. 1170–1250), better known by the name of Fibonacci. He visited Islamic North Africa and also travelled to the Middle East. He met Arabic arithmeticians and learned from them their numeral system, the operational techniques, the rules of algebra and the fundamentals of geometry. This education was what underlay the treatise that he wrote in 1202 and which was to become the algorists' bible, the *Liber abaci* (*The Book of the Abacus*). Despite its title, Fibonacci's treatise (which assisted greatly the spread of Arabic numerals and the development of algebra in Western Europe) has no connection with Gerbert's abacus or the arithmetical course-books of that tradition – for it lays out the rules of written computation using both the zero and the rule of position. Presumably Fibonacci used "abacus" in his title in order to ward off attacks from the practical abacists who effectively monopolised the world of accounting and clung very much to their counters and ruled tables. At all events, from 1202 the trend began to swing in favour of the algorists, and we can thus mark the year as the beginning of the democratisation of number in Europe.

Resistance to the new methods was not easily overcome, however, and many conservative counting-masters continued to defend the archaic counter-abacus and its rudimentary arithmetical operations.

Professional arithmeticians, who practised their art on the abacus, constituted a powerful caste, enjoying the protection of the Church. They were inclined to keep the secrets of their art to themselves; they necessarily saw algorism, which brought arithmetic within everyone's grasp, as a threat to their livelihood.

Knowledge, though it may now seem rudimentary, brought power and privilege when it represented the state of the art, and the prospect of seeing it shared seemed fearful, perhaps even sacrilegious, for its practitioners. But there was another, more properly ideological reason for European resistance to Indo-Arabic numerals.

Even whilst learning was reborn in the West, the Church maintained a climate of dogmatism, of mysticism, and of submission to the holy scriptures, through doctrines of sin, hell and the salvation of the soul. Science and philosophy were under ecclesiastical control, were obliged to remain in accordance with religious dogma, and to support, not to contradict, theological teachings.

The control of knowledge served not to liberate the intellect, but to restrict its scope for several centuries, and was the cause of several tragedies. Some ecclesiastical authorities thus put it about that arithmetic in the Arabic manner, precisely because it was so easy and ingenious, reeked of magic and of the diabolical: it must have come from Satan himself! It was only a short step from there to sending over-keen algorists to the stake, along with witches and heretics. And many did indeed suffer that fate at the hands of the Inquisition.

The very etymology of the words "cypher" and "zero" provides evidence of this.

FIG. 26.13. *Written arithmetic using "Arabic" numerals. European engraving, sixteenth century. Paris, Palais de la Découverte*

When the Arabs adopted Indian numerals and the zero, they called the latter *sifr,* meaning "empty", a plain translation of the Sanskrit *shûnya. Sifr* is found in all Arabic manuscripts dealing with arithmetic and mathematics, and it refers unambiguously to the null figure in place-value numbering. (See for example the manuscripts in the Bibliothèque nationale, Paris, shelfmarks Ms. ar. 2457, f° 85v–86; Ms. ar. 2463, f° 79v–80; Ms. ar. 2464, f° 3v; Ms. ar. 2473, f° 9; Ms. ar. 2475, f° 45v–46r; and University of Tunis Ms. 10301, f° 25v; Ms. 2043, f°16v and 32v.) Etymologically, *sifr* means "empty" and also "emptiness" (the latter can also be expressed by *khalâ* or *farâgh*). The stem SFR can also be found in words meaning "to empty" (*asfara*), "to be empty" (*safir*) and "have-nothing" (*safr al yadyn*, literally "empty hands", that is to say, "he who has nothing in his hands".

When the concept of zero arrived in Europe, the Arabic word was assimilated to a near-homophone in Latin, *zephyrus*, meaning "the west wind" and, by rather convenient extension, a mere breath of wind, a light breeze, or – almost – nothing. In his *Liber Abaci*, Fibonacci (Leonard of Pisa) used the term *zephirum*, and the term remained in use in that form until the fifteenth century:

> The nine Indian figures [*figurae Indorum*] are the following: 9, 8, 7, 6, 5, 4, 3, 2, 1. This is why with these nine figures and the sign 0, called *zephirum* in Arabic, all the numbers you may wish can be written [Fibonacci, as reproduced by B. Boncompagni (1857)].

However, in his *Sefer ha mispar* (*Number Book*), Rabbi Ben Ezra (1092–1167) used the term *sifra* [see M. Silberberg (1895) p. 2; D. E. Smith and Y. Ginsburg (1918)]. In various spellings, the Arabic term *sifra* (*cifra, cyfra, cyphra, zyphra, tzyphra*. . .) continued to be used to mean "zero" by some mathematicians for many centuries: we find it in the *Psephophoria kata Indos* (*Methods of Reckoning of the Indians*) by the Byzantine monk Maximus Planudes (1260–1310) [A. L. Allard, (1981)], in the *Institutiones mathematicae* of Laurembergus, published in 1636, and even as late as 1801 in Karl Friedrich Gauss's *Disquisitiones arithmeticae* (Gauss must have been one of the very last scholars to write in Latin).

In popular language, words derived from *sifr* soon came to be associated not with figures in general but with "nothing" in particular: in thirteenth-century Paris, a "worthless fellow" was called a *cyfre d'angorisme* or a *cifre en algorisme*, i.e. "an arithmetical nothing".

However, it was Fibonacci's term, *zephirum*, which gave rise to the modern name of zero, by way of the Italian *zefiro* (*zero* is just a contraction of *zefiro*, in Venetian dialect). The first known occurrence of the modern form of the word occurs in *De arithmetica opusculum* by Philippi Calandri and which, despite its Latin title, was written in Italian, and published in Florence in 1491. There is absolutely no doubt that *zero* owes its spread to French (*zéro*) and Spanish (*cero*) (and later on to English and other languages) to the enormous prestige that Italian scholarship acquired in the sixteenth century.

Meanwhile, Arabic *sifr* had also developed into the French word *chiffre*, the English *cipher*, German *Ziffer*, Spanish *cifra*. To begin with, the Latin terms *figuris* and *numero* were used to refer to the set of number-symbols (in English they still are called *figures* or *numerals*, more or less interchangeably); but from about 1486 in French, we find *chiffre* being used not to mean zero, but to mean a figure or numeral; and a similar development can be found in sixteenth- and seventeenth-century mathematical texts written in Latin, such as those by Willichius (1540), Conrad Rauhfuss Dasypodius of Strasbourg (*Institutionum Mathematicarum*, 1593), and the *Chronicle of Theophanes* (1655).

FIG. 26.14. *The Quarrel of the Abacists (to the left) and the Algorists (to the right). Adapted from an illustration in Robert Recorde (1510-1558),* The Grounde of Artes *(1558)*

Why did the original name of zero come to be used for the whole set of Indo-Arabic numerals? The answer lies in the attitude of the Catholic authorities to the counting systems borrowed from the Islamic world. The Church effectively issued a veto, for it did not favour a democratisation of arithmetical calculation that would loosen its hold on education and thus weaken its power and influence; the corporation of accountants raised its own drawbridges against the "foreign" invasion; and in any case the Church preferred the abacists – who were most often clerics as well – to keep their monopoly on arithmetic. "Arabic" numerals and written calculation were thus for a long while almost underground activities. Algorists plied their skills in hiding, as if they were using a secret code.

All the same, written calculation (on sand or by pen and ink) spread amongst the people, who were keenly aware of the central role played by zero, then called *cifra*, or *chifre*, or *chiffre*, or *tziphra*, etc. By a very common form of linguistic development, known as synecdoche, the name of the part (in this case, zero) came to be used for the whole, as in a kind of shorthand, so that words derived from *sifr* came to mean the entire set of numerals or any one of them. Simultaneously, it also came to mean "a secret", or a secret code – a cipher.

So the history of words for zero also tell the history of our culture: each time we use the word "cipher", we are also reviving a linguistic memory of the time when a *zero* was a dangerous *secret* that could have got you burned at the stake.

It is now easier to understand why in the mid-sixteenth century Montaigne could not "cast account" either "with penne or with Counter". For even with the introduction of written arithmetic, multiplication and division long remained outside the grasp of ordinary mortals, given the complicated operating techniques that were used. It was not until the end of the eighteenth century that simpler techniques were generalised and brought basic arithmetical operations even to those with little taste for sums.

The quarrel between the *abacists* (the defenders of Roman numerals and of calculations done on ruled boards with counters) and the algorists, who supported the written calculation methods originally invented in India, actually lasted several centuries. And even after the latter's victory, the use of the abacus was still so firmly entrenched in people's habits that all written sums were double-checked on the old abacus, just to make sure.

Until relatively recently, the British Treasury still used the abacus to calculate taxes due. And because the reckoning-board was called an *exchequer* (related to the words for *chess* and *chess-board* in various European languages), the Finance Minister of the United Kingdom is still called the Chancellor of the Exchequer.

Even long after written arithmetic with Arabic numerals had become the sole tool of scientists and scholars, European businessmen, financiers, bankers and civil servants – all of whom turned out to be more conservative than men of learning – found it hard to abandon entirely the archaic methods of the bead and counter-abacus.*

Only the French Revolution had the strength to cut through the muddle and to implement what many could see quite clearly, that written arithmetic was to counting-tokens as walking on a well-paved road was to wading through a muddy stream. The use of the abacus was banned in schools and government offices from then on.

Calculation and science could thenceforth develop without hindrance. Their stubborn and fierce old enemy had finally been put to rest.

* Translator's note: my father was trained as an accountant in the City of London in the late 1920s. Although he had of course learned modern arithmetic at school, he was required to learn how to tally sums on a bead abacus before being allowed to draw a wage. (DB)

FIG. 26.15. *Wood-block engraving from Gregorius Reisch,* Margarita Philosophica *(Freiburg, 1503). Lady Arithmetic (standing in the centre) gives her judgment by smiling on the arithmetician (to our left, her right) working with Arabic numerals and the zero (the numerals also adorn her dress). The quarrel of the abacists and algorists is over, and the latter have won.*

CHAPTER 27

# BEYOND PERFECTION

That then was how numerical notation was brought to its full completion, democratised, and universalised: after a long history of twists and turns, with leaps forward and steps backward, ideas lost and found again, and with the friction between different systems used in conjunction ultimately generating the flash of genius on which it is all based: the decimal place-value system.

Is the story really at an end? After such a long and eventful history, could there not be more adventures to come? No, there could not. This really is the end. Our positional number-system is perfect and complete, because it is as economical in symbols as can be and can represent any number, however large. Also, as we have seen, it is the most efficacious in that it allows everyone to do arithmetic.

True, the development of computers and of electronic calculators with liquid crystal displays in the last half century has brought some changes in the graphical representation of the "Arabic" numerals. They have taken on more schematic shapes that would no doubt have horrified the scribes and calligraphers of yesteryear. In reality, however, these changes have had no effect whatever on the structure of the number-system itself. The numerals have been redesigned to meet the physical constraints of the display media, while also meeting the requirement to be readable both by machines and by the human eye.

Of course, as we have seen many times, a different base could have been used for our number-system. The base 12, for example, is in many ways more convenient than our decimal base; and the base 2 is well adapted to electronic computers which usually can recognise only two different states, symbolised by 0 and 1, of a physical system (perforation of a tape, or direction of magnetisation or of a current, etc.). But a change of base would change nothing in the structure of the number-system: this would continue to be a positional system and would continue to possess a zero, and its fundamental rules would be identical to those which we know already for our decimal system.

In short, the invention of our current number-system is the final stage in the development of numerical notation: once it was achieved, no further discoveries remained to be made in this domain.

The difficulties encountered on the road to a fully finished number-representation bear witness, on a limited front though one rich in possibilities, to true progress in human affairs.

From the beginning, human beings have shown the unique characteristic of harnessing the forces of nature to their development, their survival and their domination over other species, through discovering the laws of nature by means of observing the effects of their actions on their environment. Instead of following immutably programmed instinct, they act, seek to understand the "why" of things, ponder, and create.

In his novel *Les Animaux dénaturés*, Vercors recounts a telling story. A tribe of "primitive" people share a valley with a colony of beavers. The valley is swept by a flood. The beavers, driven by their hereditary instincts, build a dam and thus protect their dens. The humans, on the other hand, guided by their grand wizard, climb the sacred hill and meditate, begging mercy from their gods; this, however, does not prevent their village from being devastated by the flood.

At first sight, the behaviour of the humans seems stupid. But on reflection we see something really profound in it, for it is the germ of all future civilisation. They were certainly wrong to attribute the disaster to supernatural forces but, despite appearances, their reaction leapt beyond the mere instinct of the beavers, since they sought to understand the true cause of their misfortune. Humanity has surely passed through such phases: we know how far our tribulations have brought us.

This is not the place to retrace the evolution of the human race since the time of the first hominids. We must rather recall that human beings are characterised above all, not by what is innate and does not need to be learned, but by the predominance of what they can adjoin to their nature from learning, experience and education.

In other words, humankind is universally an intelligent social animal, and is differentiated from other higher animals by, above all, the predominance of what is acquired over what is inborn.

That fundamental truth has not always been, nor indeed yet is, obvious to everyone. For reasons ranging from the political to the criminal, this question has been subjected to systematic mystification in order that irrelevant criteria, such as the colour of the skin or the shape of the face, may be used to demonstrate the supposed superiority of one race over others.

The principal motivations and the basic ideas of racist and segregationist philosophies are directed towards maintaining great confusion between the notion of race and the ideas of a people, of a tribe, of an ethnicity and of a linguistic group, and towards cultivating a belief that there are so-called superior races who have a kind of natural right to exploit or even to suppress so-called inferior ones.

These indefensible racist mystifications, which the Nazis elevated to political ideology during Germany's Third Reich and which throughout the Second World War gave rise to the greatest barbarity of all time and led millions of innocents to slaughter, reflected an appalling eugenic mentality whose spirit still haunts the world decades after Nazism was crushed. All those who may have forgotten it, or who would wish that it should be forgotten, need to be reminded that "one man is not the same as another" but at the same time "one race is not unequal to another, still less is one people unequal to another" (J. Rostand, *Hérédité et Racisme*, p. 63).

As to the colour of the skin, this in fact (according to François Jacob) depends on the intensity of sunlight or, as the Arab philosopher Ibn Khaldun expressed it around 1390: "The climate gives the skin its colour. Black skin is the result of the greater heat of the South" [*Muqaddimah, Prolegomena*, p. 170; see V. Monteil (1977), p. 169].

The concept of race, in fact, is strictly biological, while that of people is historical. We talk, therefore, not of the French race but of the French people, which is made up of a mixture of several races. Nor is there a Breton race, but there is a Breton people; no Jewish race, but the Jewish people; no Arab race but Arab culture; no Latin race but Latin civilisation; and neither Semitic nor Aryan races, but Semitic and Aryan languages.

According to R. Hartweg (GLE Vol. 8, p. 976) the concept of race is "one of the categories of zoological classification. It denotes a relatively broad grouping within a species, a kind of sub-species, a collection of individuals of common origin which share a number of sufficiently meaningful biological characters." It therefore "rests on genetic, anatomical, physiological and pathological criteria. The difficulty with attempting to apply a racial classification to humankind therefore arises from: 1. the choice of criteria; 2. the fact that there are at present very few races which might be considered relatively 'pure', because of inter-breeding; 3. the transitory nature of the definition of any given race since races, like humanity itself, undergo continual evolution." D. L. Julia (1964) has the following view of this question:

> From the biological point of view, the notion of race as applied to humans is very imprecise. Features such as skin colour or facial structure are definite morphological characters, but they are biologically vague. Even if we suppose that different races exist, criteria such as physical strength, or intelligence (as measured by IQ tests), show no systematic variation. Though the people of industrialised nations may have weaker constitutions than those of African nations, for example, and although culture and education may seem less prevalent among the latter than among Western peoples, nonetheless this has no bearing on the physical potential of the former nor on the intellectual potential of the latter. On the other hand, differences of character – whereby we traditionally contrast the intellectual strictness of the "whites" with the intuitive mind and generous spirit of the "blacks", or the openness of both of these with the feline suppleness and deep capacity for dissimulation of the "yellow" peoples – bear no relationship to a scale of values. Differences of character should not be a source of conflict, but an occasion for learning and therefore of enrichment: in coming to understand other people, any persons of any race will come to better understand themselves as individuals, and learn wisdom for the conduct of their own lives.

In short, "racist theories are gratuitous constructs, based on tendentious and immature anthropological ideas" (J. Rostand, *Hérédité et Racisme*, p. 57). "The truth is, that there is no such thing as a pure race, and to base politics on ethnographic analysis is to base it on a chimera" (E. Renan, *Discours et Conférences*, pp. 93–4).

In the domain of the history of numbers, at least, we have seen that human intelligence is universal and that the progress has been achieved in the mental, cultural and collective endowment of the whole of humankind. From the Cro-Magnon to the modern period, no fundamental change in the human brain has in fact occurred: only cultural enrichment of mental furnishings. This means that all human beings, whether white, red, black or yellow, whether living in the town, the country or the bush, have without exception equal intellectual potential. Individuals will develop the possibilities of their intelligence, or not, according to their needs, their environments, their social circumstances, their cultural heritage and their diverse individual aptitudes. These strictly personal individual differences are what determine whether one mind will be more or less enlightened, more or less inventive, than another.

As was stated in the Preface, number and simple arithmetic nowadays seem so obvious that they often seem to us to be inborn aptitudes of the human brain.

This was no doubt why the great German mathematician Kronecker said "God created the integers, the rest is the work of Man", whereas in fact the whole is an invention, the pure creation of the human mind; as the German philosopher Lichtenberg said: "Mankind started from the principle that every magnitude is equal to itself, and has ended up able to weigh the sun and the stars."

And the invention is of purely human origin: no god, no Prometheus, no extra-terrestrial instructor, has given it to the human race.

The actual history of numbers serves also, incidentally, to refute all those popular stories of extra-terrestrials who came to Earth to civilise the human race. Had we been visited by a scientifically and technologically advanced civilisation from outer space, we would not first have learned from it mysterious methods of erecting megaliths, but a number-system based on the principle of position and endowed with a zero. There is abundant documentation which proves that these were of late appearance, and that historically there was a great variety of number-systems in use. Quite sufficient to disprove any extra-terrestrial source for arithmetic – and therefore for everything else.

This profoundly human invention is also the most universal of inventions. In more than one sense, it binds humanity together. There is no Tower of Babel for numbers: once grasped, they are everywhere understood in the same way. There are more than four thousand languages, of which several hundred are widespread; there are several dozen alphabets and writing systems to represent them; today, however, there is but one single system for writing numbers. The symbols of this system are a kind of visual Esperanto: Europeans, Asiatics, Africans, Americans or Oceanians, incapable of communicating by the spoken word, understand each other perfectly when they write numbers using the figures 0, 1, 2, 3, 4. . . , and this is one of the most notable features of our present number-system. In short, numbers are today the one true universal language. Anyone who thinks that number is inhuman would do well to reflect on this fact.

The invention and democratisation of our positional number-system has had immeasurable consequences for human society, since it facilitated the explosion of science, of mathematics and of technology.

This in its turn gave rise to the mechanisation of arithmetical and mathematical calculations.

Yet all the elements needed to construct a true calculating engine had already been in existence, known and utilised since ancient times by scholars and engineers such as Archimedes, Ctesibius or Hero of Alexandria – such devices as levers, the endless screw, gears, toothed wheels, etc. But when we look at the numerical notations which they used at the time we can see that it would have been out of the question for them to conceive of, let alone to construct, such machines.

Nor did the technology of the time permit their actual construction: not until the start of the seventeenth century, when clockwork mechanisms underwent enormous development, would the first implementations of such devices be seen. Without a positional number-system with a zero, Schickard and Pascal would have been unable to imagine the components of their calculating machines. Pascal, for example, would not have thought of the transferrer (a counter-balanced pawl which, when one counting-wheel advanced from "9" to "0" after completing a revolution, advanced the next wheel through one step), nor of the totalisator (a device which, for each power of ten, had a cylinder bearing two enumerations from "0" to "9", in opposite directions, one used for additions and the other for subtractions).

To sum up: if the positional number-system with a zero had not existed, the problem of mechanising the process of calculation would never have found a solution; still less would it have been conceivable to automate the process. This, however, is another story, the story of automatic calculation, which begins with the classical calculating or analytical engines, passes on to machines for sorting and classifying data, and culminates in the emergence of the computer.

These powerful developments would never have seen the light of day, had the Indian discovery of positional notation not influenced the art of calculation itself. Since, however, it evidently did, we are led to look far beyond the domain of mere figures into the universe of number itself.

Note first that, unlike almost all earlier systems, our modern number-system allows us to write out straightforwardly any number whatever, no matter how large it may be. But modern mathematicians have introduced a simplification in the representation of very large numbers by means of so-called "scientific" notation which makes use of the powers of ten. For example, 1,000 may be written as $10^3$, a million as $10^6$ and a billion as $10^9$, the small number in the exponent denoting the number of zeros in the standard representation of the number. For a billion, for example, we write down three figures instead of ten.

As it stands this is no more than an abbreviated notation, which effects no change in the number-system being used. Nonetheless, it is more than mere shorthand, since it lends itself to the procedure known as exponentiation ("raising to the power") which stands to multiplication as multiplication stands to addition, since we can write:

$$a^m \times a^n = a^{m+n};\ a^m/a^n = a^{m-n};\ (a^m)^n = a^{mn}$$

Using this notation, a very large number such as

72,965,000,000,000,000,000,000,000,000,000 (27 zeros)

can be written more economically as

$$72{,}965 \times 10^{27}$$

which simply indicates that by adjoining 27 zeros to 72,965 the complete representation of the number is obtained. We can also use "floating-point" notation, and express the first number as a decimal fraction followed by the appropriate power of ten, as in

$$7.2965 \times 10^{31}$$

which indicates that the decimal point is to be moved 31 places to the right in order to obtain the complete representation.

Most pocket calculators and electronic computers have a facility of this kind which allows them to show numerical results which exceed the decimal capacity of the display (or at least to show their approximate values).

The positional number-system gave rise to great advances in arithmetic, because it showed the properties of numbers themselves more clearly. It similarly enabled mathematicians of recent times to unify apparently distinct concepts, and to create theories which had previously been unthinkable.

Fractions, for example, had been known since ancient times, but owing to the lack of a good notation they were for long ages written using notations which were only loosely established, which were not uniform, and which were ill adapted to practical calculation.

Originally, remember, fractions were not considered to be numbers. They were conceived as relations between whole numbers. But, as methods of calculation and arithmetic developed, it was observed that fractions obeyed the same laws as integers, so that they could be considered as numbers (an integer, therefore, being a fraction whose denominator was unity). As a result, where numbers had previously served merely for counting, they now became "scales" which could be put to several uses. Thereafter, two magnitudes would no longer be compared "by eye"; they could be conceived as subdivided into parts equal to a magnitude of the same kind which served as a unit of reference. Despite this advance, however, the ancients, with their inadequate notations, were unable to unify the notion of fraction and failed to construct a coherent system for their units of measurement.

Using their positional notation with base 60, the Babylonians were the first to devise a rational notation for fractions. They expressed them as sexagesimal fractions (in which the denominator is a power of sixty) and wrote them much as we now write fractions of an hour in minutes and seconds:

33m 45s (= 33/60h + 45/3600h).

They did not, however, think of using a device such as the "decimal point" to distinguish between integers and sexagesimal fractions of unity, so that the combination [33; 45] could as well mean 33h 45m as 0h 33m 45s. They had, so to speak, a "floating notation" whose ambiguities could only be resolved by context.

The Greeks then tried to make a general notation for vulgar fractions, but their alphabetic numerals were ill-adapted for the purpose and so they abandoned the attempt. Instead, they adopted the Babylonian sexagesimal notation.

Our modern notation for vulgar fractions is due to the Indians who, using their decimal positional number-system, wrote a fraction such as 34/1,265 much as we do now:

34 (numerator)
1,265 (denominator).

This notation was adopted by the Arabs, who brought it into its modern form by introducing the horizontal bar between numerator and denominator.

Then, following the discovery of "decimal" fractions (in which the denominator is a power of ten), people gradually became aware of the importance of extending the positional system in the other direction, i.e. of representing numbers "after the decimal point", and this is what finally allowed all fractions to be written without difficulty, and which showed the integers to be a special kind of fraction, in which no figures appear after the decimal point.

The first European to make the decisive step towards our modern notation was the Belgian Simon Stevin. Where we would write 679.567, he wrote:

679(0) 5(1) 6(2) 7(3)

which stood for 679 integer units, 5 decimal units of the first order (tenths), 6 of the second order (hundredths) and 7 of the third (thousandths).

Later on, the Swiss Jost Bürgi simplified this notation by omitting the superfluous indication of decimal order, and by marking the digit representing the units with the sign°:

$67\overset{\circ}{9}567$

At the same time, the Italian Magini replaced the ring sign by a point placed between the units digit and the tenths digit, creating the decimal-point notation which is still the standard usage in English-speaking countries:

679.567

In continental Europe, a comma is commonly used instead of the point, and this was introduced at the start of the seventeenth century by the Dutchman Snellius:

679,567

This rationalisation of the concept and of the notation of fractions had immeasurable consequences in every domain. It led to the invention of the "metric system", built entirely on the base 10 and completely consistent: in 1792, the French Revolution offered "to all ages and to all peoples for their greater good" this system which replaced the old systems of arbitrary, inconsistent and variable units. We know full well the fantastic progress that this brought in every practical domain, by virtue of the enormous simplification of every kind of calculation.

Once established, positional decimal notation opened up the infinite complexity of the universe of number, and led to prodigious advances in mathematics.

In the sixth century BCE the Greek mathematicians, following Pythagoras, discovered that the diagonal of a square "has no common measure" with the side of the square. It can be observed by measurement, and deduced by reason, that the diagonal of a square whose side is one metre long has a length which is not a whole number of metres, nor of centimetres, nor millimetres. . . . In other words, the number $\sqrt{2}$ (which is its mathematical magnitude) is an "incommensurable" number. This was the moment of discovery of what we now call "irrational" numbers, which are neither integers nor fractions.

This discovery greatly perturbed the Pythagoreans, who believed that number ruled the Universe, by which they understood the integers and their simpler combinations, namely fractions. The new numbers were called "unmentionable", and the existence of these "monsters" was not to be divulged to the profane. According to the Pythagorean conception of the world, this inexplicable error on the part of the Supreme Architect must be kept secret, lest one incur the divine wrath.

But the secret soon became well known to right-thinking people who were prepared to mention the unmentionable, to name the unnameable, and who delivered it up to the profane world. That perfect harmony between arithmetic and geometry, which had been one of the fundamental tenets of the Pythagorean doctrine, was seen to be a vain mystification.

Once we are free of these mystical constraints, we can accept that there are numbers which are neither integers nor fractions. These are the "irrational" numbers, of which examples are $\sqrt{2}$ , $\sqrt{3}$, the cube root of 7 and of course the famous $\pi$.

Nevertheless, this class of numbers remained ill defined for many centuries, because the defective number-systems of earlier times did not allow such numbers to be represented in a consistent manner. They were in fact designated by words, or by approximate values which had no apparent relation to each other. Lacking the means to define them correctly, people were obliged to admit their existence but were unable to incorporate them into a general system.

Modern European mathematicians, with the benefits of effective numerical notation and continual advances in their science, finally succeeded where their predecessors had failed. They discovered that these irrational numbers could be identified as decimal numbers where the series of digits after the decimal point does not terminate, and does not eventually become a series of repetitions of the same sequence of digits. For example: $\sqrt{2}$ = 1.41421356237. . . This was a fundamental discovery: this property characterises the irrational numbers.

Of course, a fraction such as 8/7 also possesses a non-terminating decimal representation:.

$$8/7 = 1.142857142857142857...$$

but its representation is periodic: the sequence "142857" is indefinitely repeated, with nothing else intervening: we can therefore, for instance, easily determine that the 100th decimal digit will be "8" , since 16 repetitions will take us to the 96th place, and four more digits will give the digit "8".

On the other hand, the irrational numbers do not follow such a pattern. Their decimal expansion is not periodic, and there is no rule which allows us to determine easily what digit will be in any particular place. This is precisely the respect in which the vulgar fractions (what we today call "rational numbers" ) differ from the irrational numbers.

However, nowadays this is not how irrational numbers are defined. Instead, an algebraic criterion is used, according to which an irrational number is not the solution of any equation of the first degree with integer coefficients. The number 2, for example, is the solution of $x - 2 = 0$, and the fraction 2/3 is the solution of $3x - 2 = 0$. On the other hand, it can be proved that the number $\sqrt{2}$ cannot be the solution of any equation of this kind, and so it is irrational.

Nonetheless, the concept of such numbers would not have been fully understood without the introduction of a further extension of the notion of number: the " algebraic" numbers. This concept was discovered in the nineteenth century by the mathematicians Niels Henrik Abel of Norway, and Évariste Galois of France. An algebraic number is a solution of an algebraic equation with integer coefficients. Clearly this holds for any integer or fractional number, but it also holds for any irrational number which can be expressed by radicals. For example, $\sqrt{2}$ is a solution of the equation $x^2 - 2 = 0$, and the cube root of 5 is a solution of the equation $x^3 - 5 = 0$. The set of algebraic numbers, therefore, includes both the set of rational numbers (which itself includes the integers) and the set of all numbers that can be expressed by the use of radicals.

However, even this is inadequate to contain all numbers. After the discoveries of Liouville, Hermite, Lindemann and many others, we know that there are additional "real numbers", which are not integers, or fractions, or even algebraic irrational numbers. These are the so-called "transcendental" numbers, which cannot occur as a solution of any algebraic equation with integer or fractional coefficients. They are, of course, irrational; but they cannot be expressed by the use of radicals. There are infinitely many of them; examples include the number "π" (the area, and also half the circumference, of a circle with unit radius), the number "e" (the base of the system of natural logarithms invented in 1617 by the Scottish mathematician John Napier), the number "log 2" (the decimal logarithm of 2) and the number "cos 25°" (cosine of the angle whose measure is 25 degrees). However, we cannot here let ourselves be carried away into the further reaches of the theory of numbers.

Now, if it is possible as we have seen to write any number whatever in a simple and rational way, no matter how large or strange it may be, then we may well ask if there is a last number, greater than all the others. We can directly see from the positional notation that this cannot be so, since if we write down the decimal representation of an integer then all we have to do is to add a zero at the right-hand end, to multiply this number by 10. Proceeding indefinitely in this way, we readily see that the sequence of integers has no limit. All the more so for the fractions and the irrational numbers, for which we can demonstrate that there exist "several infinities" between any two consecutive integers.

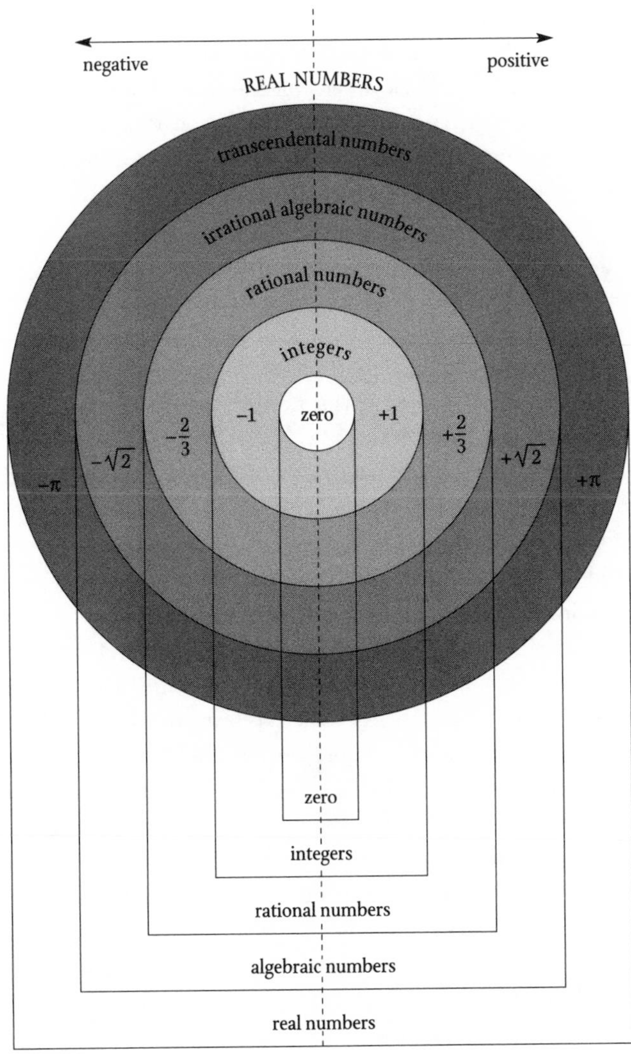


FIG. 27.1. *The successive algebraic extensions of the concept of number*

From the dawn of history, people came up against the dilemma of the infinite (see the article *Infinity, in the Dictionary). Since then, however, the concept of infinity has been made perfectly precise and objective, and presents no fundamental obscurity – at least, not such as the common mind attributes to it. Infinity has its own symbol: ∞, like a figure 8 on its side, called "lemniscate" by some and introduced quite recently into mathematical notation by the English mathematician John Wallis who first employed it in 1655. But we can hardly prove the existence of infinity – the impossibility of counting all numbers – since infinity, nowadays, is taken as an axiom, a mathematical hypothesis, on which the whole of contemporary mathematics is based.

It is but one step from infinity to zero, and it is a step which leads us on to algebra, since the null is the opposite of the unlimited.

For thousands of years, people stumbled along with inadequate and useless systems which lacked a symbol for "empty" or "nothing" . Similarly there was no way of conceiving of "negative" numbers (–1, –2, –3, etc.), such as we nowadays use routinely to express, for example, sub-zero temperatures or bank accounts in deficit. Therefore a subtraction such as "3 – 5" was for a long time considered to be impossible. We have seen how the discovery of zero swept away this obstacle so that ordinary ("natural") numbers were extended to include their "mirror images" with respect to zero.

That inspired and difficult invention, zero, gave rise to modern algebra and to all the branches of mathematics which have come about since the Renaissance (see the article *Zero in the Dictionary).

Algebra would not however have blossomed as it did if, as well as the zero, there had not also been another, equally important discovery made by Franciscus Vieta in 1591 and brought to perfection by René Descartes in 1637: this is the use of letters as mathematical symbols, which inaugurated a completely new era in the history of mathematics.

Algebra, in fact, is a generalisation of arithmetic. An $x$ or a $y$, or any other letter, is a new sort of "number": it stands for any number, whose value is unknown. One might say that it is a sign in wait for a number, holding the place for one or more figures yet to come, just as the zero sign itself filled the place of a digit corresponding to a missing decimal order of magnitude.

But this is no merely formal artifice. Using a letter to stand for a parameter or an unknown value finally freed algebra from enslavement to words, leading to the creation of a kind of "international language" which is understood unambiguously by mathematicians the world over.

In its turn, literal notation underwent a further liberation from certain restrictions acquired in its everyday usage. The symbol *x* or *y* did not simply represent a number: it could be considered in itself, independently of what kind or size of thing it represented. Thus the symbol itself transcends what it represents and becomes a mathematical object in its own right, obeying the laws of calculation. Mathematical arguments and calculations could therefore be abbreviated and systematised, and abstraction became directly accessible. Leibnitz wrote that "This method spares the work of the mind and the imagination, in which we must economise above all. It enables us to reason with small cost in effort, by using letters in place of things in order to lighten the load on the imagination." In turn, the spread of algebra throughout Europe brought about great scientific progress, and led to substantial refinement of operational symbolism in its widest sense.

Taking a very rapid overview of the history of mathematics, this science arose in Ancient Greece when her philosophers and mathematicians brought a decisive advance into human thought: that combination of abstraction, generalisation, synthesis and logical reasoning which had previously lain hid in shadow. The Greeks, however, were enamoured of what is beautiful and simple and, consequently, of what is divine. They thereby cut themselves off from the world of reality and therefore from applied mathematics. The epic Græco-Latin era was succeeded in the West by the long dark night of the Middle Ages, feebly lit up from time to time by a few individuals of no great stature.

It was the Arabs who took over. They were well placed to assimilate the whole of the Ancient Greek legacy, together with Indian science, saving the essentials from oblivion, and they developed and propagated it according to "scientific reasoning".

In due course, the first great European universities were founded and the pursuit of knowledge was resumed: the Western world once again awoke and initiated the study of nature based on independence of thought. This great dawning derives above all from the work of Fibonacci, *Liber abaci* (1202) which, over the next three centuries, was to prove a rich source of inspiration for the development of arithmetic and algebra in the West. But the West also established numerous contacts with Arabic and Islamic culture from the eleventh century onwards, whereby European mathematicians came to know not only the works of Archimedes, Euclid, Plato, Ptolemy, Aristotle and Diophantus, but also became acquainted with the work of Arab, Persian and Indian thinkers and learned the methods of calculation which had been invented in India.

The true renaissance – or rather the true awakening – of mathematics in Europe would not take place until the seventeenth century, first of all in the work of René Descartes who made full use of the new knowledge in his invention of algebraic and analytic geometry. Pascal later opened new questions in considering the problems of mathematical infinity, followed in this by Newton who also, with Leibnitz, ushered in the era of the infinitesimal calculus.

During the eighteenth century the spirits of Greek and of Cartesian mathematics were sustained together, leading on to a synthesis which, continuing into the nineteenth century, gave rise to the invention of determinants and matrices and the development of vector calculus.

In the nineteenth century, Gauss, Cauchy and Picard completed the Græco-Cartesian edifice. Lobachevsky questioned the foundations of Euclidean geometry and invented non-Euclidean geometry. On the last night of his all too short and dramatic life, the young Évariste Galois, a political revolutionary, left for the world his creation of the first abstract algebraic structures. George Boole laid the foundations of mathematical logic and Georg Cantor worked out the fundamentals of the theory of sets and of modern topology. The century closed with Hilbert's publication of his axiomatisation of geometry, which became the model for the modern axiomatic study of mathematics.

Since then, the explosion of modern mathematics has been characterised by an ever more pronounced algebraic approach: unlike the ancient mathematics which was based on very specific concepts of line and of magnitude, its basis is the universal and very abstract concept of a set. This recent unification in terms of logic and the theory of sets has made mathematics, for the first time in its history, an undivided subject.

And, finally, this unity in abstraction of modern mathematics laid the foundation of the computer science which is being developed today.

Therefore we must pay tribute to all the mathematicians, be they English, French, American, Italian, Russian, German, Japanese or any other, who have brought mathematics to its present extraordinary flowering, for which the words of Arthur Cayley in 1883 are still a beautifully apt description: "It is difficult to give an idea of the vast extent of modern mathematics. The word 'extent' is not the right one: I mean extent crowded with beautiful detail – not an extent of mere uniformity such as an objectless plain, but a tract of beautiful country to be rambled through and studied in every detail of hillside and valley, stream, rock, wood, and flower. But, as for everything else, so for mathematical beauty – beauty can be perceived but not explained." We may not, however, omit from this roll of honour to the glory of Western mathematics the Indian civilisation which invented the modern number-system in which the later great discoveries are rooted. Nor should we omit the Arabic and Islamic civilisation which carried the flame whilst the West slept.

There is a last great question. Could modern mathematics, in all its rigour, and in all its principles, with its theoretical extensions and practical applications which have revolutionised the way we live – could mathematics have possibly occurred in the absence of a positional numerical notation so perfect as the one we have? It seems incredibly unlikely. Modern science and technology may have their roots in antiquity, but they could only flourish as they have in the context of the modern era and in the framework of a number-system as revolutionary and efficient as our positional decimal system, which originated in India. To move mountains, the mind requires the simplest of tools.

And so our history of numbers is now completed. However, it is itself but a chapter in another history, the history of the representations of the world, and that history, beyond doubt, will never be completed.

# BIBLIOGRAPHY

ABBREVIATIONS USED ARE LISTED ON PP. VI–XIV ABOVE

*(Dates in brackets are the first edition when a more recent edition has been published)*

## 1. WORKS AVAILABLE IN ENGLISH


N. Abbot, "(No title)". *JRAS* : pp.277–80. London, 1938.

A. Aboul-Souf, "Tell-es-Sawwan. Excavations . . . Spring 1967". *SUM* 24: pp.3ff. Baghdad, 1968.

Y. Aharoni, "Hebrew ostraca from Tel Arad". *IEJ* 16: pp.1–10. Jerusalem, 1966.

Y. Aharoni, "The Use of Hieratic Numerals in Hebrew Ostraca and the Shekel". *BASOR* 184: pp.13–19. Ann Arbor, 1966.

Y. Aharoni, *Arad Inscriptions*. Jerusalem: Bialik Institute/Israel Exploration Society, 1975.

Q. Ahmad, "A Note on the Art of Composing Chronograms". *CAPIB* 10: pp.367–74. Patna, 1973.

Al-Biruni, *Chronology of Ancient Nations*. Transl. E. C. Sachau. London, 1879.

Al-Biruni (1888), *Alberuni's India, an account of the Religion, Philosophy, Literature, Geography, Chronology, Astronomy, Customs, Laws and Astrology of India about AD 1030*. Transl. E. C. Sachau. London: Paul, Trench & Trübner, 1910.

A. A. Al-daffa, *The Muslim Contribution to Mathematics*. Atlantic Highlands, NJ: Humanities Press, 1977.

E. Alföldi-Rosenbaum, "The Finger Calculus in Antiquity and the Middle Ages". *FMS* 5: pp.1–9. Berlin, 1971.

al-Jazzari (Ibu al-Razzaz), *The Book of Knowledge of Ingenious Mechanical Devices*. Transl. D.R.Hill. Dordrecht, 1974.

J. de Alwis, *A Grammar of the Singhalese Language*. Colombo, 1852.

A. Anbouba, "Al-Sammaw'al". In: C. Gillespie, *DSB* 12: pp.91–4. New York: Scribners, 1970.

Anonymous, *Sûryasiddhânta*. Bombay, 1955.

Anonymous, *Lokavibhaga*. Sholapur: P. Balchandra Siddhanta-Shastri, 1962.

F. Anton, *The Art of the Maya*. New York: Putnam, 1970.

E. Arnold, *Gulistân*. New York, 1899.

N. Avigad, "A Bulla of Jonathan the High Priest". *IEJ* 25: pp.8–12. Jerusalem, 1975.

L. M. Bagge, "The Early Numerals". *CR* 20: pp.259–67.

G. H. Baillie, *Clocks and Watches. An Historical Bibliography*. Vol. I. London: Holland Press, 1978.

R. Baker, *The Science Reference Library . . . Inventors and Inventions that have Changed the Modern World*. London: British Library, 1976.

W. Balasinski and W. Mrowka, "On Algorithms of Arithmetical Operations". *BAPS* 5: pp.803–4. Warsaw, 1957.

James Dyer Ball, *Things Chinese*. London: S. Low & Marston, 1900.

W. W. R. Ball, *A Short Account of the History of Mathematics*. London: Macmillan, 1888.

V. H. Barkley, *Historia Numorum. A Manual of Greek Numismatics*. Oxford: Clarendon Press, 1911.

Francis Pierrepont Barnard, *The Casting Counter and the Counting Board: a chapter in the history of numismatics and early arithmetic*. Oxford: Clarendon Press, 1916.

R. D. Barnett, "The Hieroglyphic Writing of Urartu". In: *Hommages à H. G. Güterbock*: pp.43ff. Istanbul: Anatolian Studies, 1974.

G. A. Barton, "Documents from the Temple Archives of Telloh". *HLCT*, Pt. I. New Haven, 1918.

E. C. Bayley, "On the Genealogy of Modern Numerals". *JRASI* 14: p.335. London, 1847.

E.A. Bechtel, "Finger-Counting among the Romans in the Fourth Century". *CPH* 4: pp.25–31. Chicago, 1909.

Eric Temple Bell (1937), *Men of Mathematics*. New York: Dover, 1965.

Eric Temple Bell, *The Development of Mathematics*. New York: McGraw Hill, 1940.

C. Bendall, "Table of Numerals". *CSKBM*. London, 1902.

P. K. Benedict, "Chinese and Thai Numeratives". *JAOS* 65. Baltimore, 1945.

S. R. Benedict, *A Comparative Study of the Early Treatises Introducing the Hindu Art of Reckoning into Europe*. PhD thesis, University of Michigan at Ann Arbor, 1914.

Ariel Bension, *The Zohar in Moslem and Christian Spain*. London: Routledge, 1922. New York: Hermon, 1974.

K. T. Bharat, *Vedic Mathematics*. Delhi, 1970.

Roger Billard, "Âryabhata and Indian Astronomy". *IJHS* 12/2.

S. Birch, *Facsimile of an Egyptian Hieratic Papyrus of the Reign of Ramses II*. London: British Museum, 1876.

A. N. Blackman, "The Rock Tombs of Meir". *ASE* 25. London, 1924.

J. Bonnycastle, *Introduction to Arithmetic*. London, 1810.

J. Bordaz, "The Suberde Excavations in Southwest Turkey. An Interim Report". *TAD* 17: pp.43–61. 1969.

J. Bottero, E. Cassin and J. Vercoutter, *The Near East: The Early Civilizations*. London: Weidenfeld & Nicolson, 1967.

J.G. Bourke, "Medicine Men of the Apache". *ARBE* 9: pp.555ff. Washington, DC, 1892.

Charles P. Bowditch, *The Numeration, Calendar Systems and Astronomical Knowledge of the Maya*. Cambridge: Cambridge University Press, 1910.

C. B. Boyer, "Fundamental Steps in the development of numeration". *IS* 35: pp.153–69. 1944.

C. B. Boyer (1968), *A History of Mathematics*. New York: Wiley. Revised by Uta C. Merzbach, new ed., 1991.

R. J. Braidwood and B. Howe, "Prehistoric Investigations in Iraqi Kurdistan". *SAOC* 3. Chicago, 1960.

P. N. Breton, *Illustrated History of Coins and Tokens relating to Canada*. Montreal: Breton, 1894.

W. C. Brice, "The Writing System of the Proto-Elamite Account Tablets". *BJRL* 45: pp.15–39. Manchester, 1962.

J. A. Brinkman, "Mesopotamian chronology of the historical period". *AM* : pp.335–7.

SIR JAMES BROOKE, *Ten Years in Sarawak*. Vol. I. pp. 139–40. [The author (1803–1868) was Rajah of Sarawak.]

BURGESS and WHITNEY, "Translation of the Sûrya-Siddhânta". *JAOS* 6: pp.141–498. Baltimore, 1860.

R. BURNAM, "A Group of Spanish Manuscripts". *BHI* 22: pp.229–33. Bordeaux, 1920.

A. C. BURNELL, *Elements of South Indian Paleography from the Fourth to the Seventeenth Centuries AD*. London, 1878.

C. CAHEN and R. S. SERJEANT, "A fiscal survey of medieval Yemen". *ARAB* 4: pp.23–32. Leyden, 1957.

FLORIAN CAJORI (1897), *A History of Mathematics*. New York, 1980.

FLORIAN CAJORI, "A Notable Case of Finger-Counting in America". *IS* 8: pp.325–7. 1926.

C. A. CAMPBELL, "A first season of excavations at the Urartian citadel of Kayalidere". *ANS* 16: pp.89ff. London, 1966.

A. D. CAMPBELL, *A Grammar of the Teloogoo Language*. Madras, 1820.

F. CAREY, *A Grammar of the Burman Language*. Serampore, 1814.

F. H. CHALFANT, "Chinese Numerals". *MCM* 4. Washington, DC, 1906.

J. CHALMERS, "Maipua and Namaua Numerals". *JAI* 27: pp.141ff. 1898.

B. H. CHAMBERLAIN, "A Quinary System of Notation Employed in Luchu". *JAI* 27: pp.338–95. 1898.

J. CHARTER, *A Grammar of the Cingalese Language*. Colombo, 1815.

A. B. CHASE, *The Rhind Mathematical Papyrus*. Oberlin, 1927.

D. C. CHENG, "The Use of Computing Rods in China". *AMM* 32: pp.492–9. 1925.

EDWARD CHIERA, *They Wrote on Clay*. Chicago: University of Chicago Press, 1938.

J. E. CIRLOT, *A Dictionary of Symbols*. Transl. Jack Sage. New York: Philosophical Library, 1962.

W. E. CLARCK, *The Âryabhatiya of Âryabhata*. Chicago, 1930.

CODRINGTON, *Melanesian Languages*. Pp.211–12. London, (no date).

MICHAEL D. COE, *The Maya Scribe and his World*. New York: Grolier, 1973.

MICHAEL D. COE, *Breaking the Maya Code*. New York: Thames and Hudson, 1992.

I. B. COHEN, *Album of Science from Leonardo to Lavoisier, 1450–1800*. New York: Scribner, 1980.

H. T. COLEBROOKE, *Algebra with Arithmetic and Mensuration from the Sanscrit*. London, 1817.

L.L.CONANT, *The Number Concept, its Origin and Development.* New York: Simon & Schuster, 1923.

L. L. CONANT, "Counting". *WM* 1: pp.432–41. 1956.

H. DE CONTENSON, "New Correlations between Ras Shamra and Al-'Amuq". *BASOR* 172: pp.35–40. Ann Arbor, 1963.

H. DE CONTENSON, "Tell Ramad". *ARCHN* 24: pp.278-85. New York, 1971.

C. S. COON, "Cave Exploration in Iran". *MMO* : 75. Philadelphia, 1951.

A. C. CROMBIE, *History of Science from St Augustine to Galileo*. London, 1961.

CUNNINGHAM, "Book of Indian Eras". *JRAS* : pp.627ff. London, 1913.

F. H. CUSHING, "Manual Concepts: A Study of the Influence of Hand-Usage on Culture-Growth". *AA* 5: pp.289–317. Menasha, Wis., 1892.

F. H. CUSHING, "Zuñi Breadstuffs". *INM* 8: pp.77ff. 1920.

TOBIAS DANTZIG (1930), *Number: The Language of Science*. New York: Macmillan, 1967.

B. DATTA, "The Scope and Development of Hindu Ganita". *IHQ* 5: pp.497–512. Calcutta, 1929.

B. DATTA and A.N. SINGH (1938), *History of Hindu Mathematics*. Bombay: Asia Publishing, 1962.

M. DAUMAS, Ed. (1969), *A History of Technology and Invention*. New York: Crown,1979.

P. DÉDRON and J. ITARD, *Mathematics and Mathematicians*. Transl. J. Field. Milton Keynes: Open University Press, 1974.

CHARLES DICKENS, Speech to the Administrative Reform Association. In: K. J. Fielding, *The Speeches of Charles Dickens*: pp.204–5. Hemel Hempstead: Harvester Wheatsheaf, 1855.

DAVID DIRINGER (1948), *The Alphabet. A Key to the History of Mankind*. London: Hutchinson, 1968. New York: Funk & Wagnalls, 1968.

R. B. DIXON, "The Northern Maidu". *BAMNH* 17: pp.228 & 271. New York, 1905.

B. DODGE, *The Fihrist of Al Nadim*. New York, 1970.

R. DRUMMOND, *Illustrations of the Grammatical Parts of the Guzerattee, Mahratta and English Languages*. Bombay, 1808.

S. DVIVEDI, *Trishâtika by Shrîdharâchârya*. Benares, 1899.

S. DVIVEDI, "Brâhmasphutasiddhânta, by Brahmagupta". *The Pandit, Vol.* 23–24. Benares, 1902.

S. DVIVEDI, *Lîlavâti by Bhâskarâ, II*. Benares, 1910.

S. DVIVEDI, *Siddhântatattvaviveka, by Kamâlakara*. Benares, 1935.

S. DVIVEDI and G. THIBAUT (1889), *The Pañchasiddhântikâ, the Astronomical Work of Varâha Mihira*. Lahore: Motilal Banarsi Dass, 1930.

W. C. EELS, "Number Systems of North American Indians". *AMM* 20: pp.293–8. 1913.

R. W. ERICHSEN, "Papyrus Harris I". *BAE* 5. Brussels, 1933.

R. W. ERICHSEN, Ed., *Relative Chronologies in Old World Archaeology*. Chicago, 1954.

SIR ARTHUR EVANS, *Scripta Minoa*. Vol. 1. Oxford: Clarendon Press, 1909.

SIR ARTHUR EVANS, *The Palace of Minos*. 5 vols. London: Macmillan, 1921–1936.

SIR ARTHUR EVANS and SIR JOHN MYRES, *Scripta Minoa*. Vol. 2. Oxford: Clarendon Press, 1952.

HOWARD EVES, *An Introduction to the History of Mathematics*. New York: Harcourt Brace, 1990.

H. W. FAIRMAN, "An introduction to the study of Ptolemaic signs and their values". *BIFAO* 43. Cairo (no date).

JOHN FAUVEL and JEREMY GRAY (1987), *The History of Mathematics. A Reader.* Contains many of the texts quoted by Ifrah, from Aristotle to Nicomacchus of Gerasa and Bacon. Milton Keynes: Open University, 1987.

P. L. FAYE, "Notes on the Southern Maidu". *UCAE* 20: pp.44ff. Berkeley, 1923.

A. FELDMAN and P. FOLD, *Scientists and Inventors. The People who Made Technology, from the earliest times to the present*. London, 1979.

L. N. G. FILON, "The Beginnings of Arithmetic". *MGA* 12: p.177. 1925.

J. J. FINKELSTEIN, "Late Old Babylonian Documents and Letters". *YOS* 13. New Haven, 1972.

J. C. FLEET, "The Last Words of Asoka". *JRAS* : pp.981–1016. London, 1909.

W. FORBES, *A Grammar of the Goojratee Language*. Bombay, 1829.

E. FÖRSTEMANN, "Commentary on the Maya manuscript in the Royal Library of Dresden". *PMAE* 6. Cambridge, 1906.

L. Frédéric, *Encyclopaedia of Asian Civilizations*. Paris, 1977–1987.

D. H. French, "Excavations at Can Hasan". *ANS* 20: p.27. London, 1970.

R. Fujisawa, "Note on the Mathematics of the Old Japanese School". In: Duporck, *op. cit.*: p. 384. 1902.

C. J. Gadd, "Omens expressed in numbers". *JCS* 21: pp.52ff. New Haven, 1967.

T.V. Gamkrelidze and V. V. Ivanov, *Indo-European and the Indo-Europeans*. In Russian. Tbilisi, 1984.

Solomon Gandz, "The Knots in Hebrew Literature". *IS* 14: pp.189ff. 1930.

Solomon Gandz, "The Origin of the Ghubar Numerals". *IS* 16: pp.393–424. 1931.

Solomon Gandz, *Studies in Hebrew Astronomy and Mathematics*. New York: Ktav, 1970.

N. R. Ganor, "The Lachich Letters". *PEQ* 99: pp.74–7. London, 1967.

A. Gardiner, *Egyptian Grammar*. Oxford: Oxford University Press, 1950.

J. Garett-Winter, *Papyri in the University of Michigan Collection*. (University of Michigan Studies, Vol. 3). Ann Arbor, 1936.

W. Gates, *Codex of Dresden*. Baltimore, 1932.

I. J. Gelb, *The Study of Writing. The Foundations of Grammatology*. Chicago: University of Chicago Press, 1963.

L.T. Geraty, "The Khirbet El-Kôm Bilingual Ostracon". *BASOR* 220: pp.55–61. Ann Arbor, 1975.

J. C. L. Gibson, *Textbook of Syrian Semitic Inscriptions*. Vol. 1. Oxford: Clarendon Press, 1971.

C. Gillespie, Ed., *Dictionary of Scientific Biography*. New York: Scribners, 1970–1980.

Richard J. Gillings (1972), *Mathematics in the Time of the Pharaohs*. New York: Dover Press, 1982.

S. R. K. Glanville, "The Mathematical Leather Roll in the British Museum". *JEA* 13: pp.232–9. London, 1927.

A. Glaser, *A History of Binary Numeration*. Philadelphia: Tomash, 1971.

B. R. Goldstein, "The Astronomical Tables of Levi Ben Gerson". *TCAS* 45. New Haven, 1974.

R. Gopalan, *History of the Pallava of Kâñchî*. Madras, 1928.

C. H. Gordon, "Ugaritic Textbook". *AOR* 38/7: pp.42–52. Rome, 1965.

J. Gouda, *Reflections on the Numerals "one" and "two" in Ancient Indo-European Language*. Utrecht, 1953.

J. Gow, *A Short History of Greek Mathematics*. New York: Chelsea, 1968.

F. L. Griffith, "The Rhind Mathematical Papyrus". *PSBA* 13: pp.328ff. See also Vol. 14, pp. 26ff.; Vol. 16, pp.164ff. London, 1891.

A. Grohmann, *Arabic Papyri in the Egyptian Library*. Vol. 4. Cairo, 1962.

B. Gunn, "Finger Numbering in the Pyramid Texts". *ZAS* 57: pp.283ff. Berlin, 1922.

B. Gunn, "The Rhind Mathematical Papyrus". *JEA* 12: p.123. London, 1926.

L. V. Gurjar, *Ancient Indian Mathematics and Vedha*. Poona, 1947.

A. C. Haddon, "The Ethnography of the Western Tribes of the Torres Straits". *JAI* 19: pp.305ff. 1890.

J. B. S. Haldane, *Science and Indian Culture*. Calcutta, 1966.

F. Hall, *The Sûrya-Siddhânta, or an Ancient System of Hindu Astronomy*. Amsterdam, 1975.

H. Hall, *The Antiquities of the Exchequer*. London: Elliott Stock, 1898.

R. T. Hallock (1969). *Persepolis Fortification Tablets*. Oriental Institute Publications, Vol. 42. Chicago: University of Chicago Oriental Institute, 1969.

G. B. Halsted, *On the Foundation and Technique of Arithmetic*. Chicago, 1912.

J. R. Harris, Ed., *The Legacy of Egypt*. Oxford: Oxford University Press, 1971.

A. P. Hatch, *An Album of Dated Syriac Manuscripts*. Boston, 1946.

E. C. Hawtrey, "The Lengua Indians of the Paraguayan Chaco". *JAI* 31: pp.296ff. 1902.

T. L. Heath, *A History of Greek Mathematics*. Oxford: Clarendon Press, 1921.

B. W. Henderson, *Selected Historical Documents of the Middle Ages*, pp.20ff. London, 1892.

T. Hideomi, *Historical Development of Science and Technology in Japan*. Tokyo, 1968.

D. R. Hill, *The Book of Knowledge of Ingenious Mechanical Devices*. Transl. D. R. Hill. Dordrecht: Reidel, 1974.

G. F. Hill, "On the early use of Arabic numerals in Europe". *ARCHL* 62: pp.137–90. London, 1910.

G. F. Hill, *The Development of Arabic Numerals in Europe*. Oxford: Clarendon Press, 1915.

E. Hincks, "On the Assyrian Mythology". *TRIA* 12/2: pp.405ff. 1855.

E. W. Hopkins, *The Religions of India*. Boston, 1898.

A. W. Howitt, "Australian Message Sticks and Messengers". *JAI* 18: pp.317–19. 1889.

T. P. Hugues, *Dictionary of Islam*. London, 1896.

G. Hunt, "Murray Island, Torres Straits". *JAI* 28: pp.13ff. 1899.

Georges Ifrah, *From One to Zero*. New York: Viking, 1985.

B. Indraji, "On Ancient Nagari Numeration". *JBRAS* 12: pp.404ff. Bombay, 1876.

B. Indraji, "The Western Kshatrapas". *JBRAS* 22: pp.639–62. Bombay, 1890.

R. A. K. Irani, "Arabic Numeral Forms". *CENT* 4: pp.1–13. Copenhagen, 1955.

R. V. Iyer, "The Hindu Abacus". *SMA* 20: pp.58–63. 1954.

J. Janssen, *Commodity Prices from the Rasmessid Period*. Leyden: E. J. Brill, 1975.

H. Jenkinson, "Exchequer Tallies". *ARCHL* 72: pp.367–80. London, 1911.

H. Jenkinson, "Medieval Tallies, public and private". *ARCHL* 74: pp.289–351. London, 1913.

L. Kadman, "The Coins of the Jewish War of 66-73 CE". *CNP* 2/3. Jerusalem, 1960.

H. Kalmus, "Animals as Mathematicians". *NA* 202: pp.1156–60. London, 1964.

H. J. Kantor, "The relative chronology of Egypt". In: Erichsen, *Relative Chronologies*: pp.10ff. Chicago, 1954.

H. R. Kapadia, *Ganitatilaka by Shrîpati*. (Gaikward Sanskrit Series). Baroda: Gaikward, 1935.

L. C. Karpinski, "Hindu numerals in the Fihrist". *BMA* 11: pp.121-4. 1911.

L. C. Karpinski, *Robert of Chester's Algoritmi de numero Indorum*. New York, 1915.

L. C. Karpinski, *The History of Arithmetic*. New York: Russell & Russell, 1965.

J. T. Kaufman, "New Evidence for Hieratic Numerals". *BASOR* 188: pp.67–9. Ann Arbor, 1967.

G. R. Kaye, "Notes on Indian Mathematics". *JPAS* 8: pp.475–508. Calcutta, 1907.

G. R. Kaye, "The Use of the Abacus in Ancient India". *JPAS* 4: pp.293–7. Calcutta, 1908.

G. R. Kaye, *Indian Mathematics*. Calcutta & Simla, 1915.

G. R. Kaye, *Bakhshâlî Manuscript. A Study in Mediaeval Mathematics*. Lahore, 1924.

G. R. Kaye, *Hindu Astronomy*. Calcutta, 1924.

R. G. Kent, "Old Persian Grammar, texts, lexicon". *AOS* . New Haven, 1953.

V. J. Kerkhof, "An Inscribed Stone Weight from Schechem". *BASOR* 184: pp.20ff. Ann Arbor, 1966.

E. K. Kingsborough, *Antiquities of Mexico*. 3 vols. London, 1831–1848.

G. Kleinwächter, "The Origin of the Arabic Numerals". *CHR* 11: pp.379–81. Cont. in Vol. 12: pp.28–30. 1882.

M. Kline, *Mathematical Thought from Ancient to Modern Times*. Oxford: Oxford University Press, 1972.

Y. V. Knorozov, "The Problem of the Study of Maya Hieroglyphic Writing". *AA*. Menasha, Wisc., 1946.

C. G. Knott, "The Abacus in its Historic and Scientific Aspects". *TASJ* 14. Yokohama, 1886.

O. Koehler, "The Ability of Birds to count". *WM* 1: pp.489ff. 1956.

E. Kraeling, *The Brooklyn Museum Aramaic Papyri*. New Haven, 1953.

R. Labat, "Elam and Western Persia, c. 1200–1000 BC". In: *CAH*: Cambridge, 1963

R. Labat, "Elam, c. 1600–1200 BC". In: *CAH*: Cambridge, 1963

A. Terrien de Lacouperie, "The Old Numerals, the Counting Rods and the Swan Pan in China". *NC* 3. London, 1888.

S. Langdon, *JRAS* : pp.169–73. London, 1925.

A. Langsdorf and D. F. McCown, "Tall-i-Bakun". *OIP* 59. Chicago, 1942.

T. Latter, *A Grammar of the Language of Burma*. Calcutta, 1845.

H. P. Lattin, "The Origins of our Present System of Notation according to the Theories of Nicolas Bubnov". *IS* 19. 1933.

J. D. Leechman and M. R. Harrington, "String Records of the Northwest". *INM* 9. 1921.

L. Legrain, *Historical Fragments*. (Publications of the Babylonian Section, Vol. 13). Philadelphia: University Museum, 1922.

R. Lemay, "The Hispanic Origin of our Present Numeral Forms". *VIAT* 8: pp.435–59. University of California Press, 1977.

J. Leslie, *The Philosophy of Arithmetic*. Edinburgh: Constable, 1817.

H. Levey and Petruck, *Principles of Hindu Reckoning*. Madison, 1965.

C. Levias, Ed., *Jewish Encyclopedia*. New York, 1905.

Lucien Lévy-Brühl (1922), *How Natives Think*. Transl. Lilian A. Clare. New York: Washington Square Press, 1966.

Saul Liebermann, *Hellenism in Jewish Palestine*. New York: Jewish Theological Seminary, 1950.

S. J. Liebermann, "Of Clay Pebbles, Hollow Clay Balls, and Writing". *AJA* 84: pp.339–58. New York (no date).

S. J. Liebermann, "A Mesopotamian Background for the so-called Aggadic 'Measures' of Biblical Hermeneutics?" *HUCA* 58: pp.157–223. 1987

A. Livingston, *Mystical and Mythological Explanatory Works of Babylonian and Assyrian Scholars*. Oxford: Clarendon Press, 1986.

L. Leland Locke, *The Ancient Quipo or the Peruvian Knot Record*. New York: American Museum of Natural History, 1923.

W. K. Loftus, *Travels and Researches in Chaldaea and Susiana*. London, 1857.

W. J. MacGee, "Primitive Numbers". *ARBE* 19: pp.821–51. Washington, DC, 1897.

E. J. H. MacKay, *Further Excavations at Mohenjo Daro*. Delhi, 1937–38.

G. Mallery, "Picture Writing of the American Indians". *ARBE* 10: pp.223ff. Washington, DC, 1893.

M. E. L. Mallowan, *The Development of Cities from Al Ubaid to the end of Uruk V*. Vol. 1. Cambridge: Cambridge University Press, 1967.

J. Marshall, *Mohenjo Daro and the Indus Civilisation*. London, 1931.

D. D. Mehta, *Positive Science in the Vedas*. New Delhi, 1974.

Karl W. Menninger (1957), *Number Words and Number Symbols: A cultural history of numbers*. Transl. P. Broneer. Boston: MIT Press, 1969.

B. D. Meritt, H. T. Wade-Gery and M. F. McGregor, *The Athenian Tribute Lists*. Cambridge: Harvard University Press, 1939.

H. Midonick, *The Treasury of Mathematics*. New York: Penguin, 1965.

R. A. Miller, *The Japanese Language*. Chicago, 1967.

Mingana, "Arabic Numerals". *JRAS* : pp.139–48. London, 1937.

B. Misra, *Siddhânta-Shiromani by Bhâskâra II*. Vol. 1. Calcutta, 1932.

R. Lal Mitra, *Lalitavistara Sûtra*. Calcutta, 1877.

P. Moon, *The Abacus*. New York: Gordon & Breach, 1971.

A. M. T. Moore, "The Excavations at Tell Abu Hureyra". *PPS* 41: pp.50–77. 1975.

A. de Morgan, *Arithmetical Books from the Invention of Printing up to the Present Time*. London: Taylor & Walton, 1847.

Sylvanus G. Morley, *An Introduction to the Study of Maya Hieroglyphs*. Washington, DC: Bureau of American Ethnology, 57. 1915.

Sylvanus G. Morley and F. R. Morley, "The Age and Provenance of the Leyden Plate". *CAAH* 5/509. Washington, DC, 1939.

P. Mortenson, "Excavations at Tepe Guran". *AAR* 34: pp.110–21. Copenhagen, 1964.

S. Moscati, Ed., *An Introduction to the Comparative Grammar of the Semitic Languages*. Wiesbaden, 1969.

J. V. Murra, *The Economic Organisation of the Inca State*. Chicago, 1956.

R. Van Name. "On the Abacus of China and Japan". *JAOS* X: p.cx. Baltimore (no date).

J. Naveh, "Dated Coins of Alexander Janneus". *IEJ* 18: pp.20–5. Jerusalem, 1968.

J. Naveh, "The North-Mesopotamian Aramaic Script-type". *IEJ* 2: pp.293–304. Jerusalem, 1972.

Joseph Needham, "Mathematics and the Science of the Heavens and the Earth". In: J. Needham, *Science and Civilization in China*, 3. Cambridge: Cambridge University Press, 1959.

Oscar Neugebauer, "The Rhind Mathematical Papyrus". *MTI* : pp.66ff. 1925.

Oscar Neugebauer (1952), *The Exact Sciences in Antiquity*. Princeton: Princeton University Press, 1969.

Oscar Neugebauer, *Astronomical Cuneiform Texts*. London, 1955.

Oscar Neugebauer, "Studies in Byzantine Astronomical Terminology". *TAPS* 50/2: p.5. 1960.

Oscar Neugebauer, *A History of Ancient Mathematical Astronomy*. Berlin & New York, 1975.

O. Neugebauer and D. Pingree, *The Pañcasiddhantika of Varahamihira*. Copenhagen: Munksgård, 1970–1971.

Oscar Neugebauer and A. Sachs, "Mathematical Cuneiform Texts". *AOS* 29. New Haven, 1945.

E. Nordenskjöld, "Calculation with years and months in the Peruvian Quipus". *CETS* 6/2. 1925.

E. Nordenskjöld, "The Secret of the Peruvian Quipus". *CETS* 6/1. 1925.

M. d' Ocagne (1893), *Simplified Calculation*. Transl. M. A. Williams & J. Howlett. Cambridge: MIT Press, 1986.

G. H. Ojha, *The Paleography of India*. Delhi, 1959.

A. L. Oppenheim, "An Operational Device in Mesopotamian Bureaucracy". *JNES* 18: pp.121–8. Chicago, 1959.

A. L. Oppenheim, *Ancient Mesopotamia. A Portrait of a Dead Civilization*. Chicago: University of Chicago Press, 1964.

O. Ore, *Number Theory and its History*. New York: McGraw-Hill, 1948.

T. Ozgüç, "Kültepe and its vicinity in the Iron Age". *TTKY* 5/29. Ankara, 1971.

R. A. Parker, *Demotic Mathematical Papyri*. London, 1972.

A. Parpola, S. Koskiennemi, S. Parpola and P. Aalto, *Decipherment of the Proto-Dravidian Inscriptions of the Indus Civilisation*. Copenhagen, 1969.

J. Peet, *A Grammar of the the Malayalim Language*. Cottayam, 1841.

T. E. Peet, *The Rhind Mathematical Papyrus*. London: British Museum, 1923.

Samuel Pepys, *The Shorter Pepys*. Ed. Robert Latham. Berkeley: University of California Press, 1985.

Frederick Peterson, *Ancient Mexico*. New York: Capricorn, 1962.

T. G. Pinches and J. N. Strasmaier, *Late Babylonian Astronomical and Related Texts*. Providence, 1965.

R. S. Poole, *A Catalogue of Greek Coins in the British Museum*. Bologna, 1963.

B. Porter and R. L. B. Moss, *Topographical Bibliography of Ancient Egyptian Hieroglyphs*. Oxford: Clarendon Press, 1927–1951.

M. A. Powell, "Sumerian Area Measures and the Alleged Decimal Substratum". *ZA* 62/2: pp.165–221. Berlin, 1972.

M. A. Powell, "The Antecedents of Old Babylonian Place Notation". *HMA* 3: pp.417–39. 1976.

M. A. Powell, "Three Problems in the History of Cuneiform Writing". *VL* 15/4: pp.419–40. 1981.

S. Powers, "Tribes in California". *CNAE* 3: pp.352ff. Washington, DC, 1877.

J. Prinsep, "On the Inscriptions of Piyadasi or Asoka". *JPAS*. Calcutta, 1838.

T. Prouskouriakoff, *An Album of Maya Architecture*. Vol. 558. Washington: Carnegie Institute, 1946.

T. Prouskouriakoff, "Sculpture and the Major Arts of the Maya Lowland". *HMAI* 2. Austin, 1965.

J. M. Pullam, *The History of the Abacus*. London: Hutchinson, 1968.

J. E. Quibbel, *Hierakonpolis*. London: Quaritch, 1900.

A. F. Rainey, "The Samaria Ostraca in the light of fresh evidence". *PEQ* 99: pp.32–41. London, 1967.

M. Rangacarya, *Ganitâsarasamgraha, by Mahâv"ra*. Madras, 1912.

R. Rashed, "Al-Karaji". In: C. Gillespie, *DSB*, 7: pp.240–6. New York, 1970.

C. Rawlinson, "Notes on the Early History of Babylonia". *JRAS* 15: pp.215–59. London, 1855.

Robert Recorde, *The Grounde of Artes*. London, 1558.

G. A. Reisner, *A History of the Giza Necropolis*. Vol. 1. London & Oxford, 1942.

C. T. E. Rhenius, *A Grammar of the Tamil Language*. Madras, 1846.

J. L. Richardson, "Digital Reckoning Used among the Ancients". *AMM* 23: pp.7–13. 1916.

J. T. Rogers, *The Story of Mathematics*. London: Hodder & Stoughton, 1966.

Bertrand Russell (1928), *Introduction to the Philosophy of Mathematics*. London: Allen & Unwin, 1967.

A. Saidan, "The Earliest Extant Arabic Arithmetic". *IS* 57. 1966.

S. Samburski, "On the Origin and Significance of the term Gematria". *JJS* 29: pp.35–8. London, 1978.

K. V. Sarma, *Grahachâranibandhana, by Haridatta*. Madras, 1954.

K. V. Sarma, *Drgganita, by Parameshvara*. Hoshiarpur, 1963.

K. V. Sarma, *Siddhântadarpana, by Nîlakanthasomayâjin*. Madras: Adyar Library and Research Centre

K. V. Sarma and K. S. Sastri, *Vâkyapañchâdhyayi*. Madras, 1962.

G. Sarton, *Introduction to the History of Science*. Baltimore: Johns Hopkins, 1927.

G. Sarton, "The First Explanation of Decimal Fractions in the West". *IS* 23: pp.153–245. 1935.

B. D. Sastri, *Siddhântashiromani, by Bhâskarâchârya*. Benares, 1929.

K. S. Sastri, *The Karanapadhati by Putumanasomayâjin*. Trivandrum, 1937.

L. Satterthwaite, *Concepts and Structures of Maya Calendrical Arithmetic*. Philadelphia, 1947.

F. C. Scesney, *The Chinese Abacus*. New York, 1944.

Denise Schmandt-Besserat, "The earliest precursor of writing". *SCAM (June)* : pp.38–47. New York, 1978.

Denise Schmandt-Besserat, "Reckoning before writing". *ARCHN* 32/3: pp.22–31. New York, 1979.

Denise Schmandt-Besserat, "The envelopes that bear the first writing". *TC* 21/3: pp.357–85. Austin, 1980.

Denise Schmandt-Besserat, "From Tokens to Tablets". *VL* 15/4: 321–44. 1981.

J. F. Scott, *A History of Mathematics*. London: Taylor & Francis, 1960.

R. B. Y. Scott, "The Scale-Weights from Ophel". *PEQ* 97: pp.128–39. London, 1965.

J. B. Segal, "Some Syriac Inscriptions of the 2nd–3rd century AD". *BSOAS* 16: pp.24–8. London, 1954.

A. Seidenberg, "The Ritual Origin of Counting". *AHES* 2: pp.1–40. 1962.

E. Senart, "The Inscriptions in the cave at Karle". *EI* 7: pp.47–74. Calcutta, 1902.

E. Senart, "The Inscriptions in the cave at Nasik". *EI* 8: pp.59–96. Calcutta, 1905.

R. Shafer, "Lycian Numerals". *AROR* 18: pp.250–61. Prague, 1950.

E. M. Shook, "Tikal Stela 29". *Expedition* 2: pp.29–35. Philadelphia, 1960

K. S. Shukla and K. V. Sarma, "Âryabhatiya of Âryabhata". *Indian National Science Academy* 77. Delhi, 1976.

L. G. Simons, "Two Reckoning Tables". *SMA* 1: pp.305–8. 1932.

C. Singer, *A History of Technology*. Oxford: Clarendon Press, 1955–1957.

D. C. Sircar, *Indian Epigraphy*. Delhi, 1965.

C. Sivaramamurti, "Indian Epigraphy and South Indian Scripts". *BMGM* 3/4. Madras, 1952.

A. Skaist, "A Note on the Bilingual Ostracon from Khirbet-el-Kom". *IEJ* 28: pp.106–8. Jerusalem, 1978.

D. SMELTZER, *Man and Number*. London: A. & C. Black, 1958.

DAVID EUGENE SMITH, "An Ancient English Algorism". In: *Festschrift Moritz Cantor*. Leipzig, 1909.

D. E. SMITH, *Rara Arithmetica*. Boston, 1909.

D. E. SMITH, "Computing jetons". *NNM* 9. New York, 1921.

D. E. SMITH, Ed., *A Source Book in Mathematics*. New York: McGraw-Hill, 1929.

D. E. SMITH, *History of Mathematics*. New York: Dover, 1958.

D. E. SMITH and JEKUTHIEL GINSBURG, "Rabbi Ben Ezra and the Hindu-Arabic Problem". *AMM* 25: pp.99–108. 1918.

D. E. SMITH AND L. C. KARPINSKI, *The Hindu-Arabic Numerals*. Boston: Ginn & Co, 1911.

D. E. SMITH AND Y. MIKAMI, *History of Japanese Mathematics*. Chicago: Open Court, 1914.

D. E. SMITH and M. SALIH, "The Dust Numerals amongst the Ancient Arabs". *AMM* 34: pp.258–60. 1927.

V. A. SMITH, "Asoka Notes". *IA* 37: pp.24ff. Also Vol. 38 (1909): pp.151–9. Bombay, 1908.

D. DE SOLLA-PRICE, "Portable Sundials in Antiquity". *CENT* 14: pp.242–66. Copenhagen, 1969

D. DE SOLLA-PRICE, "Gears from the Greeks. The Antikythera Mechanism". *TAPS*. November 1974

D. DE SOLLA-PRICE, *Gears from the Greeks*. New York: Science History Publications, 1975.

E. A. SPEISER, *Excavations at Tepe Gawra*. Vol. 1. Philadelphia: University of Pennsylvania Press, 1935.

H. J. SPINDEN, *Maya Art and Civilization*. Indian Hills, CO, 1959.

G. STACK, *A Grammar of the Sindhi Language*. Bombay, 1849.

C. STEWART, *Original Persian Letters and Other Documents*. London, 1825.

DIRK J. STRUIK, Gerbert d'Aurillac. In: C. Gillespie, *DSB*, 5: pp.364–6. New York, 1970.

DIRK J. STRUIK (1948), *A Concise History of Mathematics*. Mineola: Dover Books, 1987.

A. SUTTON, *An Introductory Grammar of the Oriya Language*. Calcutta, 1831.

SZEMERENYI, *Studies in the Indo-European System of Numerals*. Heidelberg, 1960.

C. TAYLOR, *The Alphabet*. London, 1883.

J. TAYLOR, *Lilawati*. Bombay, 1816.

J. E. TEEPLE, "Maya Astronomy". *CAA* 403/1: pp.29–116. Washington, DC, 1931.

J. ERIC THOMPSON, "Maya Chronology: the Correlation Question". *Publications of the Carnegie Institute of Washington* 456. Washington, DC, 1935.

J. ERIC THOMPSON, "Maya Arithmetic". *CAAH* 528/7–36. Washington, DC, 1942.

J. ERIC THOMPSON, *Maya Hieroglyphic Writing*. Washington, DC: Carnegie Institute, 1950

J. ERIC THOMPSON, *The Rise and Fall of Maya Civilization*. Norman: University of Oklahoma Press, 1954.

J. ERIC THOMPSON, "A Commentary on the Dresden Codex". *MAPS* 93. Philadelphia, 1972.

R. C. THOMPSON and M. E. L. MALLOWAN, "The British Museum Excavations at Nineveh". *LAA* 20: pp.71–186. Liverpool, 1933.

A. J. TOBLER, *Excavations at Tepe Gawra*. Philadelphia: University of Pennsylvania Press, 1950.

N. M. TOD, "The Greek Numeral Notation". *ABSA* 18: pp.98–132. London, 1911. Also Vol. 28: pp.141ff. 1926.

N. M. TOD, "Three Greek Numeral Systems". *JHS* 33: pp.27–34. London, 1913. Also Vol. 24: pp.54–67. 1919.

N. M. TOD (1918), "The Macedonian Era". *ABSA* 23: pp.206–17. London, 1918. Also Vol. 24 (1919): pp.54–67.

N. M. TOD, "The Greek Acrophonic Numerals". *ABSA* 37: pp.236–58. London, 1936. Also Vol. 24: pp.54–67. 1919.

N. M. TOD, "The Alphabetic Numeral System in Attica". *ABSA* 45: pp.126–39. London, 1950.

G. J. TOOMER, "Al Khowarizmi". In: C. Gillespie, *DSB*, 7: 358–65. New York, 1970.

E. B. TYLOR, *Primitive Culture*. London, 1871.

C. S. UPASAK, *The History and Paleography of Mauryan Brahmi Script*. Patna: Nalanda, 1960.

G. VAILLANT, *The Aztecs of Mexico*. London: Pelican, 1950.

MICHAEL VENTRIS and J. CHADWICK, *Documents in Mycenean Greek*. Cambridge: Cambridge University Press, 1956.

J. WALLIS, *A Treatise on Algebra, both Historical and Practical*. London, 1685.

A. N. WHITEHEAD and BERTRAND RUSSELL, *Principia Mathematica*. Cambridge, 1925–1927.

J. G. WILKINSON, *The Manner and Customs of the Ancient Egyptians*. London, 1837.

F. H. WILLIAMS, *How to Operate the Abacus*. London, 1941.

H. J. J. WINTER, "Formative Influences in Islamic Sciences". *AIHS* 6: pp.171–92. Paris, 1953.

C. E. WOODRUFF, "The Evolution of Modern Numerals from Tally-Marks". *AMM*. August 1909.

C. L. WOOLEY, *Ur Excavations: The Royal Cemetery*. London: Oxford University Press, 1934.

W. WRIGHT, *Catalogue of Syriac Manuscripts in the British Museum*. London: British Museum, 1870.

F. A. YELDHAM, *The Story of Reckoning in the Middle Ages*. London: Harrap, 1926.

Y. YOSHINO (1937), *The Japanese Abacus Explained*. Dover Books: New York, 1963.

C. ZASLAVSKY, "Black African Traditional Mathematics". *MT* 63: pp.345–52. 1970.

C. ZASLAVSKY, *Africa Counts*. Boston: Prindle Webster & Schmidt, 1973.

## 2. WORKS IN OTHER LANGUAGES

H. Adler, J. Bormann, W. Kammerer, I. O. Kerner and N. J. Lehmann. "Mathematische Maschinen". In: *EMDDR*. Berlin: VEB, 1974.

B. Aggoula, "Remarques sur les inscriptions hatréennes". *MUS* 47. Beirut, 1972.

E. Aghayan, "Mesrop Maschtots et la création de l'alphabet arménien". *ARM:* pp.10–13. January 1984.

A. Al-karmali, "Al 'uqad". *MACH* 3: pp.119–69. Baghdad, 1900.

A. Allard, *Les Plus Anciennes Versions latines du XIIe siècle issues de l'arithmétique d'Al Khwarizmi*. Louvain, 1975.

A. Allard, "Ouverture et résistance au calcul indien". In: *Colloque d'histoire des sciences: Résistance et ouverture aux découvertes*. Louvain, 1976.

A. Allard, *Le "Grand Calcul selon les Indiens" de Maxime Planudes: sa source anonyme de 1252*. Louvain, 1981.

A. Allard, *Le calcul indien (algorismus)*. Paris: Blanchard, 1992.

V. Alleton, *L'Ecriture chinoise*. Paris: PUF, 1970.

A. Amiet, *La Glyptique mésopotamienne archaïque*. Paris: Editions du CNRS, 1961.

P. Amiet, "Il y a cinq mille ans les Elamites inventaient l'écriture". *ARCH* 12: pp.20–2. Paris, 1966.

A. Amiet, *Elam*. Auvers-sur-Oise: Archée, 1966.

A. Amiet, "Glyptique susienne des origines à l'époque des Perses achéménides". *MDP* 43/2. 1967.

A. Amiet, *Bas-reliefs imaginaires de l'Ancien Orient d'après les cachets et les sceaux-cylindres.* Paris, 1973.

A. Amiet, *Les Civilisations antiques du Proche-Orient*. Paris: PUF, 1975.

A. Amiet, "Alternance et dualité. Essai d'interprétation de l'histoire Elamite". *AKK* 15: pp.2–22. Brussels, 1979.

A. Amiet, "Comptabilité et écriture à Suse". In: A. M. Christin, *ESIP*: pp.39–44. Paris: Le Sycomore, 1982

A. Apokin and L. E. Maistrov (1974). *Rasvitie Vyichistltelynih Masin.* Moscow: Nauka, 1974.

A. J. Arberry, *Le Soufisme*. Paris, 1952.

M. Arkoun, "Chronologie de l'Islam". In: Kasimirski, *Le Koran*. Paris: Garnier-Flammarion, 1970.

Arnauld and Nicole (1662). *Logique de Port-Royal, ou l'art de penser*. Paris: Hachette, 1854.

J. Auboyer, "La Monnaie en Inde". In: *DAT* 2: p.712. . Paris, Editions de l'Accueil, 1963.

A. Aymard and J. Auboyer, "L'Orient et la Grèce antique". In: A. Aymard and J. Auboyer, *Histoire générale des civilisations*, I. Paris: PUF, 1961.

E. Aymonier, "Quelques notions sur les inscriptions en vieux khmer". *JA*: pp.441–505. Paris, April 1883.

E. Babelon, "Moneta". In: C. Daremberg and E. Saglio, *DAGR*: pp.1963ff. Paris, 1873.

J. Babelon, *Mayas d'hier et d'aujourd'hui*. Paris, 1967.

J. Babelon, *ARCH* 22: p.21. Paris, 1968.

M. Baillet, J. T. Milik and R. De Vaux, "Les Petites Grottes de Qumran". *DJD* 3. Oxford, 1961.

L. Bakhtiar, *Le Soufisme, expression de la quête mystique*. Paris: Le Seuil, 1977.

A. Bareau, "Bouddhisme". In: *Les Religions de l'Inde*, 3. Paris, 1966.

P. Barguet, "Les Dimensions du temple d'Edfou et leurs significations". *BSFE* 72. Paris, March 1975.

E. Barnavi, *Histoire universelle des Juifs, de la Genèse à la fin du XXe siècle*. Paris: Hachette, 1992.

A. Barth and A. Bergaigne, "Inscriptions sanskrites du Champa et du Cambodge". *NEM* 27. Paris, 1885.

L. Baudin, "L'Empire socialiste des Inkas". *TMIE* 5. Paris, 1928.

M. Baudouin, *La Préhistoire par les étoiles*. Paris, 1926.

Bayer, *Historia regni Graecorum Bactriani*. Saint Petersburg, 1738.

W. de Beauclair, *Rechnen mit Maschinen. Eine Bildgeschichte der Rechnentechnik*. Braunschweig: Vieweg u. S., 1968.

G. Beaujouan, "Etude paléographique sur la rotation des chiffres et l'emploi des apices du Xe au XIIe siècle". *RHS* 1: pp.301–13. 1947.

G. Beaujouan, *Recherches sur l'histoire de l'arithmétique au Moyen Age*. Paris: Ecole des Chartes (unpublished thesis abstract), 1947.

G. Beaujouan, "Les soi-disant chiffres grecs ou chaldéens". *RHS* 3: pp.170–4. 1950.

G. Beaujouan, "La Science dans l'Occident médiéval chrétien". In: R. Taton, *HGS*, 1: pp.517–34. Paris: PUF, 1957.

G. Beaujouan, "L'enseignement du Quadrivium". In: *Settimane di studio del centro italiano di studi sull'alto medioevo*: pp.650ff. Spoleto, 1972.

O. Costa de Beauregard, *Le Second Principe de la Science du temps*. Paris: Le Seuil, 1963.

O. Becker and J. E. Hoffmann, *Geschichte der Mathematik*. Bonn: Athenäum Verlag, 1951.

Bede, The Venerable, "De Temporum ratione". In: Migne, *Patrologia cursus completus. Patres latini*: 90. Paris, 1850.

L. Benoist, *Signes, Symboles et Mythes*. Paris: PUF, 1977.

E. Benvéniste, *Problèmes de linguistique générale.* Paris: Gallimard, 1966, 1974.

D. Van Berchem, "Notes d'archéologie arabe. Monuments et inscriptions fatimides". *JA*: pp.80 & 392. Paris, 1891–1892.

D. Van Berchem, "Tessères ou calculi? Essai d'interprétation des jetons romains en plomb". *RN* 39: pp.297–315. Paris, 1936.

H. Berlin, "El glifo 'emblema' de las inscripciones mayas". *JSA* 47: pp.111–19. Paris, 1958.

J. J. G. Berthevin, *Eléments d'arithmétique complémentaire*. Paris, 1823.

M. Betini, *Apiaria universae philosophicae, mathematicae*. Vol. 2, 1642.

P. Beziau, *Origine des chiffres*. Angers, 1939.

O. Biermann, *Vorlesungen über mathematische Näherungsmethoden*. Braunschweig, 1905.

R. Billard, *L'Astronomie indienne. Investigation des textes sanskrits et des données numériques*. Paris: Ecole française d'Extrême-Orient, 1971.

E. Biot, "Notes sur la connaissance que les Chinois ont eue de la valeur de position des chiffres". *JA* December: pp.497–502. Paris, 1839.

M. Birot, *Tablettes d'époque babylonienne ancienne*. Paris, 1970.

K. Bittel, *Les Hittites*. Paris: Gallimard, 1976.

A. Blanchet and A. Dieudonné, *Manuel de numismatique*. Paris: Picard, 1930.

Marc Bloch, *Apologie pour l'histoire, ou le métier d'historien*. Paris: Armand Colin, 1949.

R. Bloch, "Etrusques et Romains: problèmes et histoire de l'écriture". In: *EPP:* pp.183–298. 1963.

R. Bloch, "La Monnaie à Rome". In: *DAT*, 2: pp.718ff. Paris, 1964.

Bobynin, *Enseignement mathématique*, p.362. Paris, 1904.

A. Boeckh, J. Franz, E. Curtius and A. Kirckhoff, *Corpus Inscriptionum Graecarum*. Berlin, 1828–1877.

M. Boll, *Histoire des mathématiques*. Paris: PUF, 1963.

B. Boncompagni, *Trattati d'aritmetica*. Rome, 1857.

B. Boncompagni, Ed., *Liber abaci. Scritti di Leonardo Pisano*. Contains the works of Fibonacci (Leonard of Pisa). Rome, 1857.

Ludwig Borchardt, "Das Grabdenkmal des Königs S'ahu-Re'". *ADOGA* 7. 1913.

Ludwig Borchardt, *Denkmäler des Alten Reiches im Museum von Kairo*. (General Catalogue of Egyptian Antiquities in the Cairo Museum). *DAR* Vol. 57. Berlin, 1937.

R. Borger, "Die Inschriften Asarhaddons, König von Assyrien". *AFO* 9. Graz, 1956.

J. Bottero, *La Religion babylonienne*, pp. 59–60. Paris: PUF, 1952.

J. Bottero, "Textes économiques et administratifs". *ARMA* 7: pp.326–41. Paris, 1957.

J. Bottero, "Symptômes, signes, écritures en Mésopotamie ancienne". In: *Divination et rationalité*: pp.159ff. Paris: Le Seuil, 1974.

J. Bottero, "De l'aide-mémoire à l'écriture". In: A.-M. Christin, *Ecritures. Systèmes idéographiques et pratiques expressives*: pp.13–37. Paris: Le Sycomore, 1982.

A. Bouché-Leclercq (1879), *Histoire de la divination dans l'antiquité*. Paris, 1963.

B. Bourdon, *La Perception visuelle de l'espace*. Paris: Costes, 1902.

P. Boutroux, *L'Idéal scientifique des mathématiciens*. Paris: Alcan, 1922.

J. Bouveresse, J. Itard and E. Sallé, *Histoire des mathématiques*. Paris: Larousse, 1977.

J.-P. Bouyou-Moreno, *Le Boulier chinois*. Paris: Self-published.

F. Braeunig, *Calcul mental*. Paris, 1882.

H. Breuil, "La formation de la science préhistorique". In: C. Zervos, *L'Art de l'époque du renne en France*: pp.13–20. Paris, 1959.

M. Brézillon, *Dictionnaire de la préhistoire*. Paris: Larousse, 1969.

H. Brugsch, *Thesaurus inscriptionum Aegyptiacarum*. Leipzig, 1883.

E. M. Bruins and M. Rutten, "Textes mathématiques de Suse". *MDP* 34. 1961.

H. Bruns, *Grundlinien des wissenschaftlichen Rechnens*. Leipzig, 1903.

L. Brunschvicg, *Les Etapes de la philosophie mathématique*. Paris: Alcan, 1912.

N. M. Bubnov, *Origin and History of Our Numbers*. In Russian. Kiev, 1908.

N. M. Bubnov, *Arithmetische Selbständigkeit der europäischen Kultur*. Berlin: Friedländer, 1914.

N. M. Bubnov (1899). *Gerberti opera mathematica*. Hildesheim, 1963.

G. Bühler, *Indische Paleographie*. Strasbourg, 1896.

G. Buonamici, *Epigrafia etrusca*. Florence, 1932.

A. Burloud, *La Pensée conceptuelle*. Paris: Alcan, 1927.

R. Burnam, *Paleographica Iberica. Facsimiles de manuscrits portugais et espagnols du IXe au XVIe siècle*. Paris, 1912–1925.

R. Cagnat, *Cours d'épigraphie latine*. Paris, 1890.

R. Cagnat, "Revue des publications épigraphiques relatives à l'antiquité romaine". *RAR* 3/34: pp.313–20. Paris, 1899.

P. Calandri, *De Arithmetica opusculum*. Florence, 1491.

A. Dom Calmet, "Recherches sur l'origine des chiffres de l'arithmétique". *MDT* pp.1620–35. September 1707.

C. D. du Cange (1678). *Glossarium ad scriptores mediae et infimae latinitatis*. Paris, 1937–1938.

F. Cantera and J. M. Millas, *Las Inscripciones hebraïcas de España*. Madrid: Consejo superior de Investigaciones cientificas, 1956.

G. Cantineau, *Le Nabatéen*. Vol. 1. Paris: Leroux, 1930.

M. Cantor, *Kulturleben der Völker*, pp.132–6. Halle, 1863.

M. Cantor, *Vorlesungen über die Geschichte der Mathematik*. Leipzig, 1880. See also third edition, 1: p.133. Leipzig, 1907.

P. Casanova, "Alphabets magiques arabes". *JA* 17/18: pp.37–55; 19/20: pp.250-67. Paris, 1921–1922.

G. Casaril, *Rabbi Simeon Bar Yochaï et la Cabbale*. Paris: Le Seuil, 1961.

Casiri, *Bibliotheca Arabo-Hispanica Escurialensis*. Madrid, 1760–1770.

A. Caso, *Calendario y escritura de las antiguas culturas de Monte-Albán*. Mexico, 1946.

A. Caso, *Las calendarios prehispánicos*. Mexico: Universidad Nacional Autónoma, 1967.

E. Cassin, "La monnaie en Asie occidentale". In: *DAT*, 2: pp.712–15. Paris: A. Colin, 1963.

Cataneo, *Le pratiche delle due prime mathematiche*. Venice, 1546.

J. M. A. Chabouillet, *Catalogue général et raisonné des camées et pierres gravées de la Bibliothèque impériale*. Paris, 1858.

C. Chadefaud, "De l'expression picturale à l'écriture égyptienne". In: *ESIP*: pp.81–99. Paris: Le Sycomore, 1982.

E. Chassinat, *Le Temple d'Edfou*. Cairo, 1897–1960.

E. Chassinat, *Le Temple de Dendara*. Cairo, 1934.

E. De Chavannes, *Les Documents chinois découverts par Aurel Stein*. Oxford, 1913.

Chen Tsu-Lung, "La Monnaie en Chine". *DAT* 2: pp.711–12. Paris: L'Accueil, 1963–1964.

J. Chevalier and A. Gheerbrant, *Dictionnaire des Symboles*. Paris: Laffont, 1982.

A. Chodzko, *Grammaire persane*. Paris: Imprimerie impériale, 1852.

A.-M. Christin, *Ecritures. Systèmes idéographiques et pratiques expressives*. Paris: Le Sycomore, 1982.

N. Chuquet (1484), *Triparty en la science des Nombres*. Ms, BN Paris, partly published by A. Marre, Rome, 1880–1881.

E. Claparède, In: *L'Invention*. Paris, Alcan/Centre international de synthèse, 1937.

G. Coedès and H. Parmentier, *Listes générales des inscriptions et des monuments du Champa et du Cambodge*. Hanoi: Ecole française d'Extrême-Orient, 1923.

Georges Coedès, *BEFEO* 24: pp.3–4. Paris, 1924

Georges Coedès, *Corpus des Inscriptions du Cambodge*. Paris: Paul Geuthner, 1926.

Georges Coedès, "Les Inscriptions malaises deçrivijaya". *BEFEO* 30: pp.29–80. Paris, 1930–1931.

Georges Coedès, "A propos de l'origine des chiffres arabes". *BSOAS* 6/2: pp.323–8. London, 1931.

Georges Coedès, *Les Etats hindouisés d'Indonésie et d'Indochine*. Paris: De Boccard, 1964.

Georges Coedès, *Inscriptions du Cambodge*. Paris: Ecole française d'Extrême-Orient, 1966.

D. Cohen, *Dictionnaire des racines sémitiques ou attestées dans les langues sémitiques*. The Hague: Mouton, 1970.

M. Cohen, "Traité de langue amharique". *TMIE* 24: pp.26ff. Paris, 1970.

M Cohen, *La Grande Invention de l'écriture et son évolution*. Paris: Klincksieck, 1958.

G. S. Colin, "Une nouvelle inscription arabe de Tanger". *HESP* 5: pp.93–9. 1924.

G. S. Colin, "De l'origine des chiffres de Fez". *JA* XIII/1: pp.193–5. Paris, 1933.

G. S. Colin, "Hisab al-Djummal". In: *EIS*: p.484. Paris, 1954.

G. S. Colin, "Abdjad". In: *EIS*: p.100. Paris, 1954.

J. -P. Collette, *Histoire des mathématiques*. Montreal: Renouveau pédagogique, 1973.

Combe, Sauvaget and Wiet, *Répertoire chronologique d'épigraphie arabe*. Cairo, 1931–1956.

A. Conrady, *De chinesischen Handschrift und sonstigen Kleinfunde Sven Hedins in Lou-Lan*. Stockholm, 1920.

G. Conteneau, "Notes d'iconographie religieuse assyrienne". *RA* 37: pp.154–65. Paris, 1940.

H. de Contenson, "Nouvelles données sur le néolithique précéramique . . . à Ghoraifé". *BSPF* 73/3: pp.80–2. Paris, 1976.

A. Cordoliani, "Contribution à la littérature ecclésiastique au Moyen-Age". *SME* 3/1: pp.107–37. Turin, 1960. See also Vol. 3/2: pp.169–208. Turin, 1961.

L. Costaz, *Grammaire syriaque*. Beirut, 1964.

Leonard Cottrell, Ed., *Dictionnaire encyclopédique d'archéologie*. Paris: SEDES, 1962.

P. Couderc, *Le Calendrier*. Paris: PUF, 1970.

C. Courtois, L. Leschi, C. Perrat and C. Saumagne, *Tablettes Albertini. Actes privés de l'époque vandale*. Paris: Gouvernement général de l'Algérie, 1952.

D. A. Crusius, *Anweisung zur Rechnenkunst*. Halle, 1746.

M. Curtze, *Petri Philomeni de Dacia in Algorismus Vulgarem Johannis de Sacrobosco commentarius, una cum Algorismo ipso*. Copenhagen, 1897.

A. Cuvillier, *Cours de philosophie*. Paris: Armand Colin, 1954.

A. Dahan-Dalmedico and J. Pfeiffer, *Histoire des mathématiques. Routes et dédales*. Montreal: Etudes vivantes, 1982.

L.C. Damais, "Etudes balinaises". *BEFEO* 50/1: pp.144–52. Paris, 1960.

P. Damerov and R. K. Englund, "Die Zahlzeichensysteme der archaischen Texte aus Uruk". *ATU* 2/5. Berlin, September 1985.

M. Danloux-Dumesnils, *Esprit et bon usage du système métrique*. Paris: Blanchard, 1965.

C. Daremberg and E. Saglio, *Dictionnaire des antiquités grecques et romaines*. Paris: Hachette, 1873–1919.

Dasypodius, *Institutionum Mathematicarum*. Strasbourg, 1593.

F. Daumas, *Les Dieux de l'Egypte*. Paris: PUF, 1970.

S. Debarbat and A. Ten, *Le Mètre et le système métrique*. Paris & Valence, 1993.

M. J. A. Decourdemanche, "Notes sur quatre systèmes turcs de notation numérique secrète". *JA* 9/14: pp.258ff. Paris, 1899.

J.J. DeHollander, *Handleiding bij de beoefening der Javaansche Taal en Letterkunde*. Breda, 1848.

A. Deimel, "Sumerische Grammatik der archaïschen Texte". *OR* 9/pp.43–44, 182–98. Rome, 1924.

A. Deimel, *Sumerisches Lexikon*. Scripta Pontificii Instituti Biblici. Rome, 1947.

H. Delacroix , *Le Langage et la pensée*. Paris: Alcan, 1922.

H. Delacroix, *Psychologie de l'art*. Paris: Alcan, 1927.

X. Delamarre, *Le Vocabulaire indo-européen*. Paris: Maisonneuve, 1984.

A. Délédicq, "Numérations et langues africaines". *BLPM* 27: pp.3–9. Paris, 1981.

M. Delphin, "L'astronomie au Maroc". *JA* 17. Paris, 1891.

P. Demiéville, *Matériaux pour l'enseignement du chinois*. Paris, 1953.

J. G. Dennler, "Los nombres indigenas en guarani". *PHYS* 16. Buenos Aires, 1939.

E. Dermenghem, *Mahomet et la tradition islamique*. Paris: Le Seuil, 1955.

M. Destombes, "Les chiffres coufiques des instruments astronomiques arabes". *RSS* 2/3: pp.197–210. Florence, 1960.

M. Destombes, "Un astrolabe carolingien et l'origine de nos chiffres arabes". *AIHS* 58/59: pp.3–45. Paris, 1962.

J. Dhombres, *Nombre, mesure et continu: épistémologie et histoire*. Paris: Nathan, 1978.

J. Dhombres, "La fin du siècle des lumières: un grand élan scientifique". In: G. Ifrah, *Deux Siècles de France*, 111: pp.5–12. Paris: Total Information, 1989.

J. Dhombres, *Une Ecole révolutionnaire en l'an III. Leçons de mathématiques*. Paris: Dunod, 1992.

J. Dhombres, "Résistance et adaptation du monde paysan au système métrique". In: A. Croix and J. Quéniart, *La Culture programme*, pp.128–42. Roanne, 1993.

Denis Diderot, "Arithmétique". In: D. Diderot and J. d'Alembert, *L'Encyclopédie*, I: pp.680–4. Paris, 1751.

E. Doblhoffer, *Le Déchiffrement des écritures*. Paris: Arthaud, 1959.

M. Dobrizhoffer, *Auskunft über die Abiponische Sprache*. Leipzig: Platzmann, 1902.

P. J Dols, "La vie chinoise dans la province de Kan-Su". *ANTH* 12-13: pp.964ff. Göteborg, 1917–1918.

E. Domluvil, "Die Kerbstöcke der Schafhirten in der mährischen Walachei". *ZOV* 10: pp.206–10. Vienna, 1904.

Dong-Zuobin, *Xiao dun yin xu wenzi: yi bián*. Taipei, 1949. In Chinese.

H. Donner and W. Röllig, *Kanaanaische und Aramäische Inschriften*. Wiesbaden, 1962.

B. von Dorn, "Über ein drittes in Russland befindliches Astrolabium mit morgenländischen Inschriften". *BSC* 9/5: pp.60–73. Paris, 1841.

F. Dornseiff(1925), *Das Alphabet in Mystik und Magie*. Leipzig/Berlin, 1977.

G. Dossin, "Correspondance de Iasmah-Addu". *ARMA* 5: Letter 20, pp. 36–7. Paris, 1952.

E. Doutte, *Magie et Religion dans l'Afrique du Nord*. Algiers, 1909.

A. Dragoni, *Sul metodo aritmetico degli antichi Romani*. Cremona, 1811.

G. Dumesnil, "Note sur la forme des chiffres usuels". *RAR* 16: pp.342–48. Paris, 1890.

P. Dupont-Sommier, "La Science hébraïque ancienne". In: R. Taton, *HGS*, 1: pp.141–51. Paris, 1957.

E. Duporck, Ed., *Comptes rendus du deuxième Congrès international de mathématiques de Paris*. Paris, 1902.

J.-M. Durand, "Espace et écriture en cunéiforme". In: Christin, *ESIP*: pp.51–62. Paris, 1982.

J.-M. Durand, "A propos du nom du nombre 10 000 à Mari". *MARI* 3: pp.278–9. Paris, 1984.

J.-M. Durand, "Questions de chiffres". *MARI* 5: pp.605–22. Paris, 1987.

R. Duval (1881), *Traité de grammaire syriaque*. Amsterdam, 1969.

W. Dyck, *Katalog matematische usw. Instrumente*, Munich: Wolf und Sohn, 1892.

A. Eisenlohr, *Ein mathematisches Handbuch der Alten Aegypter*. Leipzig, 1977.

Mircea Eliade, *Traité d'histoire des religions*. Paris, 1949.

R. W. Erichsen, *Demotisches Glossar*. Vol. 5. Copenhagen, 1954.

Erpenius, *Grammatica Arabica*. Leyden, 1613.

J. Essig, *Douze, notre dix futur*. Paris: Dunod, 1955.

J. Euting, *Nabatäische Inschriften aus Arabien*. Berlin, 1885.

P. Ewald, *Neues Archiv der Gesellschaft für ältere deutsche Geschichtskunde* 8: pp.258–357. 1883.

A. Falkenstein, "Archäische Texte aus Uruk". *ADFU*. Berlin, 1936.

A. Falkenstein, *Das Sumerische. Handbuch der Orientalistik*. Leyden: Brill, 1959.

F.-G. Faraut, *Astronomie cambodgienne*. Phnom Penh: F. H. Schneider, 1910.

O. Fayzoullaiev, *Mohammed al-Khorazmii*. Tashkent, 1983.

L. Fekete, *Die Siyaqat-Schrift in der türkischen Finanzverwaltung*. Vol. I. Budapest, 1955.

J. Feller, Ed., *Dictionnaire de la psychologie moderne*. Paris: Marabout, 1967.

E. Fettweis, "Wie man einstens rechnete". *MPB* 49. Leipzig, 1923.

J.-G. Février, *Histoire de l'écriture*. Paris: Payot, 1959.

J. Filliozat, "La Science indienne antique". In: R. Taton, *HGS*: pp.159ff. Paris, 1957.

P. S. Filliozat, *Le Sanskrit*. Paris: PUF, 1992.

O. Fine, *De arithmetica practica*, fo. 11–12. Paris, 1544.

L. Finot, "Note d'épigraphie: les inscriptions de Jaya Parameçvaravatman Ier, roi du Champa". *BEFEO* 15: pp.39–52. Paris, 1915.

H. Fleisch, *Traité de philologie arabe*. Beirut, 1961.

M. Folkerts, *Boethius' Geometrie, II. Ein Mathemathisches Lehrbuch des Mittelalters*. Wiesbaden: Franz Steiner, 1970.

Formaleoni, *Dei fonti degli errori nella cosmografia e geografia degli Antichi*. Venice, 1788.

C. Fossey, *Notices sur les caractères étrangers anciens et modernes*. Paris: Imprimerie nationale, 1948.

M. Foucaux, *Grammaire tibétaine*. Paris: Imprimerie nationale, 1858.

P. Foulquié, *Dictionnaire de la langue philosophique*. Paris: PUF, 1982.

J. Franz, *Elementa epigraphica graecae*. Berlin, 1840.

L. Frédéric, *Japon, l'Empire éternel*. Paris: Le Félin, 1985.

L. Frédéric, *Japon intime*. Paris: Le Félin, 1986.

L. Frédéric, *Dictionnaire de la civilisation indienne*. Paris: Laffont, 1987.

L. Fréréric, *Le Lotus*. Paris: Le Félin, 1987.

L. Frédéric, *Dictionnaire de Corée*. Paris: Le Félin, 1988.

L. Frédéric, *L'Inde de l'Islam*. Paris: Arthaud, 1989.

L. Frédéric, *Les Dieux du Bouddhisme*. Paris: Flammarion, 1992.

L. Frédéric, *L'Inde mystique et légendaire*. Paris: Rocher, 1994.

L. Frédéric, *Dictionnaire de la civilisation japonaise*. Paris: Laffont, 1994.

J.-C. Fredouille, *Dictionnaire de la civilisation romaine*. Paris: Larousse, 1968.

J. B. Frey, *Corpus Inscriptionum Iudaicorum*. Rome, 1936 – 1952.

G. Friedlein, *Boethii de Institutione arithmeticae*, pp. 373–428. Leipzig: Teubner, 1867.

G. Friedlein, *Die Zahlzeichen und das elementare Rechnen des Griechen und Römer und des Christlichen Abendlandes vom 7. bis 13. Jahrhundert*. Erlangen, 1869.

J. Friedrich, *Geschichte der Schrift*. Heidelberg, 1966.

K. Friedrichs, I. Fischer-Schreiber, F. K. Erhard and M. S. Diener, *Dictionnaire de la sagesse orientale*. Paris: Laffont, 1989.

W. Froehner, "Le Comput digital". *ASNA* 8: pp.232–8. Paris, 1884.

A. de La Fuye, J. T. Belaiew, R. de Mecquenem and J. M. Unvala, "Archéologie, Métrologie et Numismatique susiennes". *MDP* 25: pp.193ff. 1934.

A. M. von Gabain (1950). *Alttürkische Grammatik*. Leipzig: Harassowitz, 1950.

C. Gallenkamp, *Les Mayas. Découverte d'une civilisation perdue*. Paris: Payot, 1979.

G. F. Gamurrini, *Appendice al Corpus Inscriptionum Italicorum*. Florence, 1880.

Gaudefroy-Demombynes, *Grammaire de l'arabe classique*. Paris: Blachère, 1952.

M. J. E. Gautier, *Archives d'une famille de Dilbat*. Cairo, 1908.

R. Gemma-Frisius, *Arithmetica practicae methodus facilis*. Paris: Peletier, 1563.

Le Gendre, *Arithmétique en sa perfection*. Paris, 1771.

P. Gendrop, *Arte prehispánico en Mesoamerica*. Mexico City: Trillas, 1970.

H. de Genouillac, *Tablettes sumériennes archaiques*. Paris: Paul Geuthner, 1909.

H. de Genouillac, *Tablettes de Drehem*. Paris, 1911.

J. Gernet, "La Chine, aspects et fonctions psychologiques de l'écriture". In: *EPP*. Paris, 1963.

L. Gerschel, "Al Longue crôye. Autour des comptes à crédit". *EMW* 8: pp.256–92. 1959.

L. Gerschel, "Comment comptaient les anciens Romains". In *Latomus* 44: pp.386–97. 1960.

L. Gerschel, "La conquête du nombre". *AESC* 17: pp.691–714. Paris, 1962.

L. Gerschel, "L'Ogam et le nombre". *REC* 10/1: pp.127–66; 10/2: pp.516–77. Paris, 1962.

R. Ghirshman, *L'Iran, des origines à l'Islam*. Paris: Payot, 1951.

O. Gillain, *La Science égyptienne. L'Arithmétique au Moyen Empire*. Brussels: Editions de la Fondation Egyptologique, 1927.

B. Gille, *Les Ingénieurs de la Renaissance*. Paris: Le Seuil, 1978.

B. Gille, *Les Mécaniciens grecs*. Paris: Le Seuil, 1980.

R. Girard, *Les Popol-Vuh. Histoire culturelle des Mayas-Quichés*. Paris: Payot, 1972.

M. Gmür, "Schweizerische Bauermarken und Holzurkunden". *ASR*. Berne, 1917.

L. Godart and J.-P. Olivier, "Recueil des inscriptions en Linéaire A". *EC* 21. Paris, 1976.

G. Godron, "Deux notes d'épigraphie thinite". *RE* 8: pp.91ff. Paris, 1951.

I. Godziher, "Le Rosaire dans l'Islam". *RHR* 21: pp.295–300. Paris, 1890.

V. Goldschmidt, *Die Entstehung unseren Ziffern*. Heidelberg, 1932.

M. Gomez-Moreno, "Documentacion Goda en Pizarra". *BRAH*. Madrid, 1966.

J. Goschkewitsch, "Über das Chinesische Rechensbrett". *AKRG* 1: p.293. Berlin, 1858.

M. Granet, *La Pensée chinoise*. Paris: Albin Michel, 1988.

M. W. Green and H. J. Nissen, *Zeichenliste der archaischen Texte aus Uruk*. Vol. 2. Berlin: Max-Planck-Institut, 1985.

A. Grohmann, *Einführung und Chrestomathie zur arabischen Papyruskunde*. Prague, 1954.

H. Grotefend, *Taschenbuch der Zeitrechnung des deutschen Mittelalters*. Hanover/Leipzig, 1891.

M Guarducci, *Epigraphia greca*. Rome, 1967.

O. Guérard and P. Jouguet, *Un livre d'écolier du IIIe siècle avant J.-C.* Cairo: Société royale égyptienne de papyrologie, 1938.

E. Guillaume, "Abacus". In: Daremberg and Saglio, *Dictionnaire*: pp.1–5. Paris, 1873.

G. Guitel, "Comparaison entre les numérations aztèque et égyptienne". *AESC* 13: pp.687–705. Paris, 1958.

G. Guitel, "Signification mathématique d'une tablette sumérienne". *RA* 57: pp.145–50. Paris, 1963.

G. Guitel, "Classification hierarchisée des numérations écrites". *AESC* 21: pp.959–81. Paris, 1966.

G Guitel, *Histoire comparée des numérations écrites*. Paris: Flammarion, 1975.

G Gundermann, *Die Zahlzeichen*. Giessen, 1899.

S. Günther, "Die quadratischen Irrationalitäten der Alten". *AGM* 4. Leipzig, 1882.

G. Hager, *Memoria sulle cifre arabiche*. Milan, 1813.

L. Hambis, *Grammaire de la langue mongole écrite*. Paris: Maisonneuve, 1945.

L. Hambis, "La Monnaie en Asie centrale et en Haute Asie". In: *DAT*, 2: p.711. Paris, 1963.

E. T. Hamy, "Le Chimpu". *NAT* 21. Paris, 1892.

J. Harmand, "Le Laos et les populations sauvages de l'Indochine". *TDM* 38/2: pp.2–48. Also Vol. 39/1: pp.241–314. Paris, 1879.

J. Harmand, "Les races indochinoises". *MSA* 2/2: pp.338–9. Paris, 1882.

G. P. Harsdörffer, *Delitae Mathematicae et Physicae*. Nürnberg, 1651.

Charles Higounet, *L'Ecriture*. Paris: PUF, 1969.

M Höfner, *Altsüdarabische Grammatik*. Leipzig, 1943.

K. Homeyer, *Die Haus- und Hofmarken*. Berlin, 1890.

E. Hoppe, "Das Sexagesimalsystem und die Kreisteilung". *AMP* 3/15.4: pp.304–13. 1910.

E. Hoppe, "Die Entstehung des Sexagesimalsystems und der Kreisteilung". *ARCH* 8: pp.448–58. Paris, 1927.

B. Hrozny, "L'inscription hittite hiéroglyphe 'Messerschmidt' ". *AROR* 11: pp.1–16. Prague, 1939.

E Hübner, *Exempla scripturae epigraphicae Latinae e Caesaris dictatoris morte ad aetatem Justiniani*. Berlin, 1885.

Huet, *Demonstratio Evangelica ad serenissimum Delphinum*. Paris, 1690.

H. Hunger, "Kryptographische Astrologische Omina". *AOAT* 1: pp.133ff. 1969.

H. Hunger, "Spätbabylonische Texte aus Uruk". *ADFU* 1/9. Berlin, 1976.

M. Hyades, "Ethnographie des Fuégiens". *BSA* 3/10: pp.327ff. Paris, 1887.

C. Idoux, "Inscriptions lapidaires". *BCFM* 19: pp.24–35. Paris, 1959.

C. Idoux (1959). "Quand l'écriture naît sur l'argile". *BCFM* 19: pp.210ff. Paris, 1959.

P. Ivanoff, *Maya*. Paris: Nathan, 1975.

François Jacob, *Logique du vivant*. Paris: Gallimard, 1970.

H. von Jacobs (1896). *Das Volk der Siebener-Zähler*. Berlin, 1896.

E. Jacquet, "Mode d'expression symbolique des nombres employés par les Indiens, les Tibétains et les Javanais". *JA* 16: pp.118–21. Paris, 1835.

Jenner, in *JA* CCLXII/1–2: pp.176–91. Paris, 1974.

H. Jensen, *Die Schrift in der Vergangeneit und Gegenwart*. Berlin, 1969.

R. Jestin, *Tablettes sumériennes de Shuruppak*. Paris, 1937.

P. Joüon, "Sur les noms de nombre en sémitique". *MFO* 6: pp.133–9. Beirut, 1913.

D. Julia, *Dictionnaire de la philosophie*. Paris: Larousse, 1964.

P. Kaplony, "Die Inschriften der ägyptischen Frühzeit". *ÄGA* 8. Wiesbaden, 1963.

H. Kazem-Zadeh, "Les Chiffres Siyâk et la comptabilité persane". *RMM* 30: pp.1–51. 1913.

L. Keimer, "Bemerkungen zum Schiefertafel von Hierakonpolis". *AEG* 7. Milan, 1926.

H. Kern, *Verspreide Geschriften*. The Hague, 1913–1929.

G. Kewitsch, "Zweifel an der astronomischen und geometrischen Grundlage des 60-er Systems". *ZA* 18: pp.73–95. Berlin, 1904.

G. Kewitsch, "Zur Entstehung des 60-er Systems". *AMP* 3/18–2: pp.165ff. 1911.

G. Kewitsch, "Zur Entstehung des 60-er Systems". *ZA* 24: pp.265–83. Berlin, 1915.

Athanasius Kircher, *Oedipi Aegyptiaci*. Rome, 1653.

Köbel, *Ain new geordnet Rechenbiechlin*. Augsburg, 1514.

A. König, "Wesen und Verwendung der Rechenpfennige". *MM* 4. Frankfurt/Main, 1927.

M. Korostovtsev (1947). "L'hiéroglyphe pour 10 000". *BIFAO* 45: 81–88. 1947.

F. Kretzschmer and E.Heinsius, "Über eine Darstellung altrömischer Rechnenbretter". *TZG* 20: pp.96ff. 1951.

W. Kubitschek, "Die Salaminische Rechentafel". *NZ* 31: pp.393–8. Vienna, 1900.

A. Kuckuck, *Die Rechenkunst im sechzenten Jahrhundert*. Berlin, 1974.

R. Labat, "L'écriture cunéiforme et la civilisation mésopotamienne". In: *EPP*. Paris, 1963

R. Labat, "Jeux numériques de l'idéographie susienne". *ASS* 16: pp.257–60. Chicago, 1965.

R. Labat and F. Malbran-Labat, *Manuel d'épigraphie akkadienne*. Paris: Paul Geuthner, 1976.

G. Lafaye (1873). "Micatio". In: Daremberg and Saglio, *Dictionnaire*. Paris, 1889-1890.

Maurice Lambert, "La période présargonique. La vie économique à Shuruppak". *SUM* 9: pp.198–213. Baghdad, 1953.

Maurice Lambert, "La naissance de la bureaucratie". *RH* 224: pp.1–26. Paris, 1960.

Maurice Lambert, "Pourquoi l'écriture est née en Mésopotamie". *ARCH* 12: pp.24–31. Paris, 1966.

Maurice Lambert, *La Naissance de l' écriture en pays de Sumer*. Paris: Société des Antiquités nationales, 1976.

MAYER LAMBERT, *Traité de grammaire hébraïque*. Paris: E. Leroux, 1931.

W. LARFELD, *Handbuch der griechischen Epigraphik*. Leipzig, 1902–1907.

E. LAROCHE, *Les Hiéroglyphes hittites*. Vol. 1. Paris: CNRS, 1960.

PETRUS LAUREMBERGUS, *Pet. Laurembergi Rostochiensis Instituiones arithmeticae*. Hamburg, 1636.

A. LEBRUN, "Recherches stratigraphiques à l'acropole de Suse". *DAFI* 1: pp.163–214. Paris, 1971

A. LEBRUN and F. VALLAT, "L'origine de l'écriture à Suse". *DAFI* 8: pp.11–59. Paris, 1978.

G. LEFEBVRE, *Grammaire de l'égyptien classique*. Cairo: Institut français d'archéologie orientale, 1956.

J. LEFLON, *Gerbert*. Saint-Wandrille, 1946.

J. C. HOUZEAU DE LEHAIE, "Fragments sur le calcul numérique". *BAB* 2/39: pp. 487–548. Continued in Vol. 2/40:pp.74–139 & 455–524. Brussels, 1875.

H. LEHMANN, *Les Civilisations précolombiennes*. Paris: PUF, 1973.

A. LEMAIRE, *Inscriptions hébraïques*. Paris: Le Cerf, 1977.

A. LEMAIRE, "Les Ostraca paléo-hébreux des fouilles de l'Ophel". *LEV* 10: pp.156–61. 1978.

A. LEMAIRE and P. VERNUS, "Les Ostraca paléo-hébreux de Qadesh Barnéa". *OR* 49: pp.341–5. Rome, 1980.

A. LEMAIRE and P. VERNUS, L'Ostracon paléo-hébreu No. 6 de Tell Qudeirat. In: *FAP*: pp.302–27. 1983.

J. G. LEMOINE, "Les Anciens procédés de calcul sur les doigts en Orient et en Occident". *REI* 6: pp.1–58. Paris,1932.

K. LEPSIUS, *Denkmäler aus Aegypten und Aethiopien*. Berlin, 1845.

ANDRÉ LEROI-GOURHAN, *Le Geste et la Parole*. Paris: Albin Michel, 1964.

C. LEROY (1963). "La monnaie en Grèce". In: *DAT*, 2: pp.716–18. 1963.

J. LEUPOLD, *Theatrum Arithmetico-Geometricum*. Leipzig, 1727.

M. LIDZBARSKI (1898) *Handbuch der Nordsemitischen Epigraphik*. Hildesheim, 1962.

U. LINDGREN, *Gerbert von Aurillac und das Quadrivium*. Wiesbaden: Sudhoffs Archiv, 1976.

FRANÇOIS LE LIONNAIS, *Les Grands Courants de la Pensée mathématique*. Paris: Blanchard, 1962.

E. LITTMANN, "Sabäische, griechische und altabessinische Inschriften". *DAE* 4. Berlin, 1913.

A. LOEHR, "Méthodes de calcul du XVIe siècle". *SS* 9: p.8. Innsbruck, 1925.

E. LÖFFLER, "Die arithmetischen Kentnisse der Babyloner und das Sexagesimalsystem". *AMP* 17/2: pp.135–44. 1910.

D. LOMBARD, *La Chine impériale*. Paris: PUF, 1967.

M. G. LORIA, *Guida allo studio della storia delle Matematiche*. Milan: Hoepli, 1916.

J. LÜROTH, *Vorlesungen über numerisches Rechnen*. Leipzig, 1900.

D. G. LYON, "Keilschriften Sargons König von Assyrien". *ASB* 5. Leipzig, 1883.

E. J. H. MACKAY, *La Civilisation de l'Indus. Fouilles de Mohenjo-Dar et de Harappa*. Paris, 1936.

L. MALASSIS, E. LEMAIRE and R. GRELLET, "Bibliographie relative à l'arithmétique, au calcul simplifié et aux instruments à calculer". *BSEIN* 132: pp.739–57. Paris, 1920.

BERTIL MALMBERG, *Le Langage, signe de l'humain*. Paris: Picard, 1979.

WILLIAM OF MALMESBURY, *De gestis regum Anglorum libri*. London, 1596.

J. MARCUS, "L'écriture zapotèque". *PLS* 30: pp.48–63. Paris, 1980.

J. MARQUES-RIVIÈRE, *Amulettes, talismans et pentacles*. Paris: Payot, 1972.

A. MARRE, *Le Khulusat al hisab de Beha ad din al 'Amuli*. Rome, 1864.

A. MARRE, "Le Talkhys d'Ibn Albanna". *AANL* 17. Rome, 1865.

ANDRÉ MARTINET, *Des Steppes aux Océans: L'Indo-européen et les Indo-Européens*. Paris: Payot, 1986.

G. MASPÉRO, *Etudes de mythologie et d'archéologie égyptienne*. Paris, 1893–1916.

G. MASPÉRO, *Histoire ancienne des peuples de l'Orient classique*. Vol. 1: pp. 323–6. Paris, 1896.

H. MASPÉRO, *Les Documents chinois découverts par Aurel Stein*. Oxford, 1951.

H. MASPÉRO, *La Chine antique*. Paris, 1965.

L. MASSIGNON and R. ARNALDEZ, "La science arabe". In: R. Taton, *Histoire générale des sciences*: pp.431–71. Paris: PUF, 1957–1964.

O. MASSON, "La civilisation égéenne. Les écritures crétoises et mycéniennes". In: *EPP*: pp.93–9. Paris: A. Colin,1963.

C. LE MAUR, *Elementos de Mathematica pura*. Madrid, 1778.

A. MAZAHERI, "Les origines persanes de l'arithmétique". In *IDERIC* Vol. 8. Nice, 1975.

R. DE MECQUENEM, "Epigraphie proto-élamite". *MDP* 31. 1949.

ANTOINE MEILLET and MARCEL COHEN, *Les Langues du Monde*. Paris: CNRS, 1952.

G. MENENDEZ-PIDAL, "Los llamados numerales arabes en Occidente". *BRAH* 145: pp.179–208. Madrid, 1959.

P. MERIGGI, *La Scrittura proto-elamica*. Rome: Academia nazionale dei Lincei, 1971–1974.

A. MÉTRAUX, *Les Incas*. Paris: Le Seuil, 1961.

C. MEYER, *Die Historische Entwicklung der Handelsmarken in der Schweiz*. Bern, 1905.

C. B. MICHAELIS, *Grammatica syriaca*. Rome, 1829.

P. H. MICHEL, *Traité de l'astrolabe*. Paris, 1947.

A. MIELI, *La Science arabe et son rôle dans l'évolution scientifique mondiale*. Leyden, 1938.

A. MIELI, *Panorama general de la historia de la Ciencia*. Buenos Aires, 1946–1951.

J. T. MILIK, *Dédicaces faites par les dieux*. Vol. 1. Paris, 1972.

J.T. MILIK, "Numérotation des feuilles des rouleaux dans le Scriptorium de Qumran". *SEM* 27: pp.75–81. Paris, 1977.

G. MÖLLER, *Hieratische Paläographie*. Leipzig, 1911–1922.

T. MOMMSEN, *Die Unteritalischen Dialekte*. Leipzig, 1840.

V. MONTEIL, *La Pensée arabe*. Paris: Seghers, 1977.

J.-F. MONTUCLA (1798). *Histoire des mathématiques*. Paris: A. Blanchard, 1968.

GEORGES MOUNIN, *Histoire de la linguistique*. Paris: PUF, 1967.

C. MUGLER (1963). "Abaque". In: *DAT*, 1: 19–20.

A. MULLER (1971). *Les Ecritures secrètes*. Paris: PUF, 1971.

A. NAGL, "Der arithmetische Traktat von Radulph von Laon". *AGM* 5: pp.98–133. Leipzig, 1890.

A. NAGL, "Die Rechentafel der Alten". *SKAW* 177. Vienna, 1914.

A. NATUCCI, *Sviluppo storico dell'aritmetica generale e dell'algebra*. Naples, 1955.

F. NAU, "Notes d'astronomie indienne". *JA* 10/16: pp.209ff. Paris, 1910.

G. H. F. Nesselmann, *Die Algebra der Griechen*. Berlin, 1842.

Oscar Neugebauer, "Zur Entstehung des Sexagesimalsystems". *AGW* 13/1. Göttingen, 1927.

Oscar Neugebauer, "Zur Aegyptischen Buchrechnung". *ZÄS* 64: pp.44–8. Berlin, 1929.

Oscar Neugebauer, *Vorgriechische Mathematik*. Berlin, 1934.

K. Niebuhr, *Beschreibung von Arabien*. Copenhagen, 1772. French translation pub. Paris, 1779.

A. P. Ninni, *Sui segni prealfabeticci usati anche ora nella numerazione scritta dai pescatori Clodieusi*. Venice, 1889.

E. Nordenskjöld, "Le Quipu péruvien du musée du Trocadéro". *BMET*. Paris, January 1931.

M. Noth, "Exkurs über die Zahlzeichen auf den Ostraka". *ZDP* 50: pp.250ff. Leipzig, 1927.

Nottnagelus, *Christophori Nottnagelii Professoris Wittenbergensis Institutionum mathematicarum*. Wittenberg, 1645.

J. Nougayrol, "Textes hépatoscopiques d'époque ancienne". *RA* 60: pp.65ff. Paris, 1945.

A. Olleris, *Les Oeuvres de Gerbert*. Paris, 1867.

L. Pacioli, *Summa de arithmetica*. Venice, 1494.

G. A. Palencia, *Los Mozarabes de Toledo en los siglos XII e XIII*. Madrid, 1930.

A. Parrot, *Assur*. Paris: Gallimard, 1961.

H. Pedersen, *Vergleichende Grammatik der keltischen Sprachen*. Göttingen, 1909.

J. Peignot and G. Adamoff, *Le Chiffre*. Paris, 1969.

Peletarius, *Commentaire sur l'Arithmétique de Gemma Frisius*. Lyon, 1563.

M. Percheron, *Le Bouddha et le bouddhisme*. Paris: Le Seuil, 1956.

P. Perdrizet, "Isopséphie". *REG* 17: pp.350–60. Paris, 1904.

P. Perny, *Grammaire de la langue chinoise*. Paris: Maisonneuve & Leroux, 1873.

C. Perrat, "Paléographie médiévale". *HM* 11: pp.585–615. Paris, 1961.

J. Perrot, "Le gisement natoufien de Mallaha". *ANTHR* 70: pp.437–84. Paris, 1966.

J. Perrot, Ed., *Les Langues dans le monde ancien et moderne*. Paris: CNRS, 1981.

J. Piaget, *Le Langage et la pensée chez l'enfant*. Neuchâtel: Delachaux & Niestlé, 1923.

J. Piaget, *La Naissance de l'intelligence chez l'enfant*. Neuchâtel: Delachaux & Niestlé, 1936.

J. Piaget, *La Construction du réel chez l'enfant*. Neuchâtel: Delachaux & Niestlé, 1937.

J. Piaget, *La Représentation du monde chez l'enfant*. Paris: Alcan, 1938.

J. Piaget and Szeminska, *Le Genèse du nombre chez l'enfant*. Neuchâtel: Delachaux & Niestlé, 1941.

J. Piaget, *La Formation du symbole chez l'enfant*. Neuchâtel: Delachaux & Niestlé, 1945.

J. Piaget, *Psychologie de l'intelligence*. Paris: A. Colin, 1947.

J. Piaget, *La Représentation de l'espace chez l'enfant*. Paris: PUF, 1948.

J. Piaget and B. Inhelder, *La Psychologie de l'enfant*. Paris: PUF, 1966.

J. Piaget and B. Inhelder, *L'Image mentale chez l'enfant*. Paris: PUF, 1966.

Piccard, "Mémoire sur la forme et la provenance des chiffres". *SVSN* : pp.176 and 184. 1859.

H. Piéron, *Vocabulaire de la psychologie*. Paris: PUF, 1979.

A. P. Pihan, *Notice sur les divers genres d'écritures des Arabes, des Persans et des Turcs*. Paris, 1856.

A. P. Pihan, *Exposé des signes de numération usités chez les peuples orientaux anciens et modernes*. Paris, 1860.

B. Piotrovsky, *Ourartou*. Geneva & Paris, 1969.

A. Poebels, *Grundzüge der sumerischen Grammatik*. Rostock, 1923.

Henri Poincaré, *La Science et l'hypothèse*. Paris: Flammarion, 1902.

Henri Poincaré, *Science et méthode*. Paris: Flammarion, 1909.

G. Poma-de-Ayala, *Codex péruvien*. Paris: Institut d'ethnologie, 1963.

E. Porada, *Iran Ancien*. Paris, 1963.

F. A. Pott, *Die quinäre und vigesimale Zählenmethode bei Völkern aller Weltteile*. Halle, 1847.

J. Prätorius (1599), "Compendiosa multiplicatio". *ZMP* 40: p.7. 1895.

M. Prou, *Recueil de fac-similés d'écritures du Ve au XVIIe siècle*. Paris, 1904.

Guy Rachet, *Dictionnaire de l'archéologie*. Paris: Laffont, 1983.

Guy Rachet, *Civilisations et archéologie de la Grèce pré-hellénique*. Paris: Le Rocher, 1993.

Ramus, *Arithmeticae libri duo, geometriae septem et viginti*. Basel, 1569.

Ramus, *Scholarum mathematicarum libri unus et triginta*. Basel, 1569.

H. Ranke, "Das altägyptische Schlangenspiel". *SHAW* 4: pp.9–14. Heidelberg, 1920.

R. Rashed and S. Ahmed, *Al-Bahir en algèbre d'As-Samaw'al*. Damascus: University of Damascus, 1972.

R. Rashed, *Essai d'histoire des mathématiques par Jean Itard*. Paris: A. Blanchard, 1984.

P. Reichlen, "Abaque, Calcul". In: *DAT*, 1: pp.18 and 188. 1963.

W. J. Reichmann, *La Fascination des nombres*. Paris: Payot, 1959.

S. Reinach, *Traité d'épigraphie grecque*. Paris: E. Leroux, 1885.

Reinaud, *Mémoire sur l'Inde*. Paris, 1894.

Ernest Renan. "Discours et conférences".

L. Renou, *Grammaire sanscrite*. Paris, 1930.

L. Renou and J. Filliozat, *L'Inde classique. Manuel des études indiennes*. Hanoi, 1953.

L. Renou, L. Nitti and N. Stchoupak, *Dictionnaire Sanscrit-Français*. Paris, 1959.

L. Reti, *Léonard de Vinci*. Paris: Laffont, 1974.

J. B. Reveillaud, *Essai sur les chiffres arabes*. Paris, 1883.

A. Riegl, "Die Holzkalender des Mittelalters und der Renaissance". *MIOG* 9: pp.82–103. Innsbruck, 1888.

M. E. de Rivero and J. D. de Tschudi, *Antiquités péruviennes*, p.95. Paris, 1859.

P. Rivet, *Les Cités Mayas*. Paris: Payot, 1954.

J. Rivoire, *Histoire de la monnaie*. Paris: PUF, 1985.

L. Rodet. "Sur la véritable signification de la notation numérique inventée par Âryabhata". *JA* 16/7: pp.440–85. Paris (no date).

L. Rodet, "Les Souan-pan des Chinois et la banque des argentiers". *BSM* 8. Paris, 1880.

A. Rödiger, "Über die im Orient gebräuchliche Fingersprache für den Ausdruck der Zahlen". *ZDMG* : pp.112–29. Wiesbaden, 1845.

M. RODINSON, "Les Sémites et l'alphabet". In: *EPP*. Paris: Armand Colin, 1963.
A. ROHRBERG, "Das Rechnen auf dem chinesischen Rechenbrett". *UMN* 42: pp, 34–7. 1936.
RONG-GEN, *Jinwen-bián*. In Chinese. Beijing, 1959.
M. RÖSLER , "Das Vigesimalsystem im Romanischen". *ZRP* 26. Tübingen, 1910.
L. DE ROSNY, "Quelques observations sur la langue siamoise et son écriture". *JA*. Paris, 1855.
C. P. RUCHONNET, *Elements de calcul approximatif*. Lausanne, 1874.
J. RUSKA, "Zur ältesten arabischen Algebra und Rechenkunst". *SHAW* 2. Heidelberg, 1917.
J. RUSKA, "Arabische Texte über das Fingerrechnen". *DI* 10: pp.87–119. 1920.
J. RUSKA, "Zahl und Null bei Jabir ibn Hayyan". *AGMNT* IX. 1928.
M. RUTTEN, *La Science des Chaldéens*. Paris: PUF, 1970.
EDUARD SACHAU, *Aramäische Papyrus und Ostraka aus Elephantine*. Berlin: Königliches Museum, 1911.
J. A. SANCHEZ-PEREZ, *La Arithmetica en Roma, en India, y en Arabia*. Madrid, 1949.
S. SAUNERON, *L'Egyptologie*. Paris: PUF, 1968.
A. SCHAEFFER, *Le Palais royal d'Ugarit*. (Mission de Ras Shamra). Paris: Imprimerie nationale, 1955–1957.
V. SCHEIL, "Documents archaïques en écriture proto-élamite". *MDP* 6. 1905.
V. SCHEIL, "Notules". *RA* 12: pp.158–60. Paris, 1915
V. SCHEIL, "Textes de comptabilité proto-élamites". *MDP* 17–26. 1923–1935.
M. SCHMIDL, "Zahl und Zählen in Afrika". *MAGW* 45: pp.165–209. Vienna, 1915.
E. SCHNIPPEL, "Die Englischen Kalenderstäbe". *BEPH*, 5. Leipzig, 1926.
A. DE SCHODT, *Le jeton considéré comme instrument de calcul*. Brussels, 1873.
GERSHON SCHOLEM, *Les Grands Courants de la mystique juive*. Paris, 1950.
GERSHON SCHOLEM, *La Kabbale et sa symbolique*. Paris, 1966.
S. SCHOTT, "Eine ägyptische Schreibpalette als Rechenbrett". *NAWG* 1/5: pp.91–113. Göttingen, 1967.
L. VON SCHROEDER, *Pythagoras und die Inder*. Leipzig, 1884.
H. SCHUBERT, *Mathematische Mussestunden*. Leipzig, 1900.
W. VON SCHULENBURG, "Die Knotenzeichen der Müller". *ZE* 29: pp.491–4. Brunswick, 1897.
E. VON SCHULER, "Urartaïsche Inschriften aus Bastam". *AMI* 3: pp.93ff. (Also Vol. 5 [1972], pp.117ff.) Berlin, 1970.
F. SECRET, *Les Kabbalistes chrétiens de la Renaissance*. Paris, 1964.
SÉDILLOT, *Matériaux pour servir à l'histoire comparée des sciences mathématiques chez les Grecs et les Orientaux*. Paris, 1845–1849.
R. SEIDER, *Paläographie der griechischen Papyri*. Vol. 1. Stuttgart, 1967.
E. SELER, *Reproduccion fac similar del codice Borgia*. Mexico & Buenos Aires, 1963.
E. SENART, *Les Inscriptions de Piyadasi*. Paris, 1887.
H. SEROUYA, *La Kabbale, sa psychologie mystique, sa métaphysique*. Paris: Grasset, 1957.
H. SEROUYA, *Maimonide*. Paris: PUF, 1963.
H. SEROUYA, *La Kabbale*. Paris: PUF, 1972.
J. A. SERRET, *Traité d'arithmétique*. Paris, 1887.
K. SETHE, *Hieroglyphische Urkunden der griechisch-römischen Zeit*. Leipzig, 1904–1916.
K. SETHE and W. HELCK (1905–1908), *Urkunden der 18.ten Dynastie*. Leipzig, 1930.
K. SETHE, "Von Zahlen und Zahlworten bei den alten Aegyptern". *SWG* 25. 1916.
K. SETHE, "Eine altägyptische Fingerzahlreim". *ZAS* 54: pp.16–39. Berlin, 1916.
K. SETHE, *Urkunden des alten Reichs*. Leipzig, 1932–1933.
M. SILBERBERG, *Sefer Ha Mispar: Das Buch der Zahl, ein hebräisch-arithmetisches Werk des Rabbi Abraham Ibn Ezra*. Frankfurt/Main, 1895.
N. SILLAMY, *Dictionnaire de la psychologie*. Paris: Larousse, 1967.
E. SIMON, "Über Knotenschriften und ähnliche Kennzeichen der Riukiu-Inseln". *AMA* 1: pp.657–67. Leipzig/London, 1924.
M. SIMONI-ABBAT, *Les Aztèques*. Paris, 1976.
F. ŠKARPA, "Rabos u Dalmaciji". *ZNZ* 29/2: pp.169–83. Zagreb, 1934.
W. D. SMIRNOFF, "Sur une inscription du couvent de Saint-Georges de Khoziba". *CRSP* 12: pp.26–30. 1902.
A. SOGLIANO, "Isopsepha Pompeiana". *RRAL* 10: pp.7ff. Rome, 1901.
H. SOTTAS and E. DRIOTTON, *Introduction à l'étude des hiéroglyphes*. Paris, 1922.
D. SOUBEYRAN, "Textes mathématiques de Mari". *RA* 78: pp.19–48. Paris, 1984.
JACQUES SOUSTELLE, *La Pensée cosmologique des anciens Mexicains*. Paris, 1940.
JACQUES SOUSTELLE, *La Vie quotidienne des Aztèques*. Paris: Hachette, 1955.
JACQUES SOUSTELLE, *Les Quatre Soleils. Souvenirs et réflexions d'un ethnologue au Mexique*. Paris: Plon, 1967.
JACQUES SOUSTELLE, *Les Olmèques*. Paris, 1979.
F. G. STEBLER, "Die Tesseln im Oberwallis". *DS* 1: pp.461ff. 1897.
F. G. STEBLER, "Die Hauszeichen und Tesseln der Schweiz". *ASTP* 11: pp.165–205. 1907.
G. STEINDORFF, *Urkunden des Aegyptischen Altertums*. Leipzig, 1904–1916.
M. STEINSCHNEIDER, "Die Mathematik bei den Juden". *BMA* : pp.69ff. 1893.
M. STERNER, *Geschichte der Rechenkunst*. Munich & Leipzig, 1891.
J. STIENNON, *Paléographie du Moyen Age*. Paris, 1973.
H. STOY, *Zur Geschichte des Rechnenunterrichts*. Jena, 1876.
G. STRESSER-PÉAN (1957). "La science dans l'Amérique précolombienne". In: R. Taton, *HGS*, 1: pp. 419–29.
P. SUBTIL, "Numération égyptienne". *BLPM* 27: pp.11–23. Paris, 1981.
P. SUBTIL, "Numération babylonienne". *BLPM* 27: pp.25–38. Paris, 1981.
P. SUBTIL, "Numération chinoise". *BLPM* 27: p.50. Paris, 1981.
H. SUTER. "Das Rechenbuch des Abû Zakarija el-Hassar". *BMA* 2/3: p.15.
H. SUTER, "Das mathematiker-Verzeichnis im Fihrist des Ibn Abî Ja'kûb an Nadim". *ZMP* 37/Supp.: pp.1–88. 1892.
H. SUTER, *Die Mathematiker und Astronomen der Araber und ihre Werke*. Leipzig, 1900.
M. SZNYCER, "L'origine de l'alphabet sémitique". In: *L'Espace et la lettre*: pp.79–119. Paris: UGE, 1977.
P. TANNERY, "Sur l'étymologie du mot 'chiffre'". *RAR* . Paris, 1892.
P. TANNERY, *Leçons d'arithmétique théorique et pratique*. Paris, 1900.
P. TANNERY, "Prétendues notations pythagoriciennes sur l'origine de nos chiffres". *Mémoires scientifiques* 5: p.8. 1922.
N. TARTAGLIA, *General trattato di numeri e misuri*. Venice, 1556.
R. TATON, Ed. *Histoire générale des sciences*. Paris: PUF, 1957–1964
R. TATON, *Histoire du calcul*. Paris: PUF, 1969.

J.-B. Tavernier (1679), *Voyages en Turquie, en Perse et aux Indes*. Part II, pp.326–7. Paris, 1712.

Theophanes, *Theophanis Chronographia*. Paris, 1655.

F. Thureau-Dangin, *Une relation de la huitième campagne de Sargon*. Paris: Paul Geuthner, 1912.

F. Thureau-Dangin, "L'exaltation d'Ishtar". *RA* 11: pp.141–58. Paris, 1914.

F. Thureau-Dangin, *Tablettes d'Uruk*. Paris, 1922.

F. Thureau-Dangin, "L'origine du système sexagésimal". *RA* 26: p.43. Paris, 1929.

F. Thureau-Dangin, *Esquisse d'une histoire du système sexagésimal*. Paris: Paul Geuthner, 1932.

F. Thureau-Dangin, "Une nouvelle tablette mathématique de Warka". *RA* 31: pp.61–9. Paris, 1934.

F. Thureau-Dangin, "L'équation du deuxième degré dans la mathématique babylonienne". *RA* 33: pp.27–48. Paris, 1936.

F. Thureau-Dangin (1938). *Textes mathématiques de Babylone*. Leyden, 1938.

W. C. Till, *Koptische Grammatik*. Leipzig, 1955.

L. F. C. Tischendorff, *Notila editionis Codicis Bibliorum Sinaitici*. Leipzig, 1860.

E. Tisserant, *Specimena Codicum Orientalum*. Rome, 1914.

Dom Toustaint and Dom Tassin, "Chiffres". In: *Nouveau Traité de diplomatique*, 3: pp.508–33. Paris, 1757.

J. Trenchant, *L'arithmétique . . . avec l'art de calculer aux jetons*. See preface. Lyon: Michel, 1561.

F. Vallat, "Les tablettes proto-élamites de l'acropole de Suse". *DAFI* 3: pp.93–104. Paris, 1973.

J. M. Vallicrosa, *Assaig d'historia de les idees fisiques i matematiques a la Catalunya medieval*. Vol. 1. Barcelona, 1931.

J. M. Vallicrosa, *Estudios sobre historia de la Ciencia española*. Barcelona, 1949.

J. M. Vallicrosa, *Nuevos estudios sobre historia de la Ciencia española*. Barcelona, 1960.

B. Carra de Vaux, "Sur l'origine des chiffres". *SC* 21: pp.273–82. 1917.

B. Carra de Vaux, "Tar'ikh". In: H. A. R. Gibb, *Encyclopédie de l'Islam*: pp.705–6. Paris: Maisonneuve, 1970–1992.

R. de Vaux, "Titres et fonctionnaires égyptiens à la cour de David et de Salomon". *RB* 48: pp.394–405. Saint-Etienne, 1939.

Vercors, *Les Animaux dénaturés*. Paris: Albin Michel, 1952.

J. Vercoutter,"La Monnaie en Egypte". In: *Dictionnaire archéologique des techniques* 2: pp.715–16. Paris, 1963.

J. Vercoutter, *L'Égypte ancienne*. Paris: PUF, 1973.

J. Vézin, "Un nouveau manuscrit autographe d'Adhémar de Chabannes". *BSNAF* : pp.50–51. Paris, 1965.

G. Levi della Vida, *DI* 10: p.243. 1920.

G. Levi della Vida, "Appunti e quesiti di storia letteraria: numerali greci in documenti arabo-spagnoli". *RdSO* 14: pp.281–3. 1933.

A. Vieyra, *Les Assyriens*. Paris, 1961.

G. Ville, Ed., *Dictionnaire d'archéologie*. Paris: Larousse. 1968.

A. Vissière, "Recherches sur l'origine de l'abaque chinois". *BGHD* : pp.54–80. Paris, 1892.

K. Vogel, *Die Grundlagen der ägyptischen Arithmetik*. Munich, 1929.

K. Vogel, *Mohammed ibn Mussa Alchwarizmi's Algorismus*. Aalen, 1963.

H. Vogt, "Haben die alten Inder den Pythagoreischen Lehrsatz und das Irrationale gekannt?" *BMA* 7/3: pp.6–20.

P. Voizot, "Les chiffres arabes et leur origine". *NAT* 27: p.222. Paris, 1899.

I. Vossius (1658). *Pomponi Meloe libri tres de situ orbis*. Frankfurt, 1700.

I. Vossius, *De Universae mathesos Natura et constitutione*, pp.39–40. Amsterdam: J. Blaue, 1660.

H. Waeschke, *Rechnen nach der indischen Methode. Das Rechenbuch des Maximus Planudes*. Halle, 1878.

J. Wallis, *Opera Mathematica*. Oxford, 1695.

J. F. Weidler, *De characteribus numerorum vulgaribus*. Wittenberg, 1727.

J. F. Weidler, *Spicilegium observationum ad historiam notarum numeralium pertinentium*. Wittenberg, 1755.

H. Weissenborn, *Zur Geschichte der Einführung der jetzigen Ziffern in Europa durch Gerbert*. Berlin, 1892.

J. E. Wessely, "Die Zahl neunundneunzig". *MSPR* 1: pp.113–16. 1887.

L. Weyl-Kailey, *Victoire sur les maths*. Paris: Laffont, 1985.

Willichius, *Arithmeticae libri tres*. Strasbourg, 1540.

A. Winkler, "Siegel und Charakter in der muhammedanischen Zauberei". *SGKIO* 7. Berlin,1930

C. de Wit, "A propos des noms de nombre dans les textes d'Edfou". *CDE* 37: pp.272–90. Brussels, 1962.

F. Woepke, "Sur le mot *kardaga* et sur une méthode indienne pour calculer le sinus". *NAM* 13: pp.392ff. Paris, 1854.

F. Woepke, "Sur une donnée historique relative à l'emploi des chiffres indiens par les Arabes". *ASMF.* Rome, 1855.

F. Woepke, *Sur l'Introduction des de l'Arithmétique indienne en Occident*. Rome, 1857.

F. Woepke, "Mémoire sur la propagation des chiffres indiens". *JA* 6/1: pp.27–79, 234–90, 442–59. Paris, 1863.

Woisin, *De Graecorum notis numeralibus*. Leipzig, 1886.

H. Wuttke, *Geschichte der Schrift und des Schrifttums*. Vol. 1: pp.62ff. Leipzig, 1872.

A.P. Youschkevitch, *Geschichte der Mathematik im Mittelalter*. Leipzig, 1964.

A.P. Youschkevitch, *Les Mathématiques arabes*. Paris: Vrin, 1976.

K. Zangenmeister, *Die Entstehung der römischen Zahlzeichen*. Berlin, 1887.

Charles Zervos, *L'Art de l'époque du renne en France*. Paris, 1959.

# INDEX


Aba, tomb of 52
abacus 125–133, 207–211, 333–334, 366, 556–562; abacists and algorists 578, 590–591; Akkadian 139–141; Assyro-Babylonian 140; Chinese 125, 211, 283–294, 556; French 290; Gerbert's 579, 581–586, 588; Greek 200–203, 208; Inca 308; Indian 434, 559; Latin 542–543; *Liber Abaci* 361–362, 588–589; Mesopotamian tablet 155; multiplication 208–209, 557–559; Persian 556, 562–563; Roman 187, 202–207, 209–211, 577, 579–580, 582; Russian 290; suan pan 288–294; Sumerian sexagesimal 126–133, 140; Table of Salamis 201–203; wax or sand 207–209, 563
Ibn Abbad 527
al-Abbas 521
ibn 'Abbas, 'Ali 524
'Abbas, Caliph Abu'l 520
Abbasid caliphs 512, 514, 520
ibn 'Abdallah, Abu 'Amran Musa ibn Maymun *see* Maimonides
ibn 'Abdallah, Ahmed ibn 'Ali 252
ibn 'Abdallah, 'Ali 252
ibn 'Abdallah, Sahl 551
Abel, Niels Henrik 596
Abenragel 356, 363, 524
Abjad numerals (ABC) 244, 248, 250, 261–262, 548–555
aboriginals, Australian 6, 18
Abraham 73, 253–254, 257, 364
Abrasax 259
absolute quantity 21
abstraction 16; counting 10, 19–20, 76; numbers 5, 23 *see also* calculation; model collections; place-value system; zero
Abulcassis 522
Abusir 390
Abyssinia 96, 246, 387
Academy of Sciences (France) 42
accounting 101–120, 187, 541–543; balance sheet 109–111; Cretan 178; Elamite 102–107; Japan 288–289; Jews 236; Mayan 304–305; Mesopotamian 132; pocket calculator 209–211; Roman 187, 209–211; Sumerian 122–124, 131 *see also* bullae; calculi; quipus; tablet; tally sticks
Acor 406
acrophonic number systems 186, 214, 387
Adab 81
Adad 161
Adam (first man) 254
al-Adami, Ibn 523, 529–530
addition: abacus 127, 204–206, 285; calculi, Sumerian use 122; Egypt, Ancient 174; suan pan 291–292
additive principle 231, 325–329, 333–336, 347–351; Americas 306, 308; India 397, 434; Roman numerals 187; Sheban 186
Adelard of Bath 207, 362, 587
ibn 'Adi, Yahya 513, 523
Afghanistan 377, 386, 522–523, 528; and Arabs 512, 520, 523; counting 94, 290; numerals 228, 368, 534; writing 376, 539
Aflah, Jabir ibn 526
Africa: Arabian provinces 521; base five in 36; Central 5, 22; counting 10, 96, 125; East 72; Maghribi script 539; number mysticism 93–94, 554; South 5; West 19, 24–25, 70, 74
Africa, North: and Arab-Islamic world 520, 528, 587; calculation 559; counting 47, 214; Goths 226; Morra 51; number mysticism 248, 250, 262, 553; numerals 242, 244, 356, 534–537; writing 248, 539
Agade *see* Akkadian Empire
ages of the world 426
Aggoula, B. 335
Ah Puch, god of death 312
Aharoni, Y. 236
Ahmad, Abu Hanifa 522
Ahmad, Ali ibn 523
ibn Ahmad, Khalil 58, 520
ibn Ahmad, Maslama 524
Ahmed 363, 511, 519
Ainus 36, 305
Akhiram 213
al-Akhtal 520
Akkad 135
Akkadian Empire 81, 134–146; bullae 100–101; counting 139–141; Mari 74, 81, 134, 142–146, 336; number mysticism 93, 159–160; numbers 90, 134, 136–139, 142–146; writing 130–133, 135–136, 159–160 *see also* Assyrian; Babylonian civilisation
Aksharapallî numerals 388
Aksum 246, 387
al Shamishi system 248
al Tadmuri system 248
Albania 33–35, 528
Albategnus 522
Albright, W. F. 142
alchemy 518–519, 553–554
Alexander the Great 135, 256, 386, 407
Alexandria 515, 522
Algazel 525
algebra 588, 597–598; Arabs and 521, 524, 527–528, 531; Brahmagupta 419, 439, 530
algebraic numbers 596–597
Algeria 248, 521, 528; counting 49, 66, 555
algorists 587–590
algorithms 559, 587; al-Khuwarizmi 531, 587
Alhazen 524
'Ali, Abu'l Hassan 58
'Ali, caliph 519–520, 555
ibn 'Ali, Hamid 522
ibn 'Ali, Sanad 364, 521
Ali (language) 22
alien intervention theory 593–594
Allah 47, 59, 214, 514–515, 553; attributes of 11, 50–51, 71, 261–262, 542, 553
Allah, Abu Sa'id 'Ubayd 525
Allah, Sa'id ibn Hibat 525
Allard, A. L. 365, 434, 533, 562–563, 589
Alleton, V. G. 266–268, 272
almanacs 195–196
alphabet 212–214; Greek 190; Hebrew 215–218; palaeo-Hebrew 212, 233; Samaritans 212
alphabetic numerals 156–157, 212–262, 329, 483–484; Arabic 158, 241–246, 516, 548–555; Aryabhata's 432; Ethiopian 246–247; Greek 218–223, 227, 232–233, 238–239, 329, 333, 360; Hebrew 158, 227, 233–236, 238–239, 346, 362; Indian 389; Syriac 240–242, 329; Varnasankhyâ's 388 *see also* mysticism
Alpharabius 523
Alphonsus VI of Castile 525
Americans, native: counting 10, 64, 70, 72, 125, 196; number mysticism 93–94, 554; use base five 36 *see also* Maya
amicable numbers 522
Amiet, P. 80, 101–102
'Ammar 525
Ammonites 212
Amon 164
Amorites 39, 135
amp 43
al-'Amuli, Beha ad din 363, 528
Anaritius 522
Anatolia 75–76, 97–98 *see also* Hittite; Ottoman Empire; Turkey
Anbouba 511
al-Andalusi, Sa'id 515–516
Andalusia 525–526; abacus 556, 560, 563; and Arabs 512; numerals 534–539
Andhra numerals 397–398
Andromeda nebula 524
Anglo-French, word for money 72
Anglo-Saxon, number names 33–35
animals, counting abilities 3–4
anka (numerals) 368, 415–416
Annam 272–273 *see also* Vietnam
al-Ansari 528
al-Antaki 541–542
Antichrist 260
Anu 93–94, 161
Anûshîrwân, King Khosroes 512
*Anuyogadvâra Sûtra* 425
Anwari 58, 525
Api language 36
apices 580–586; of Boethius 579
Apollonius of Perga 221–222, 361, 513, 523
apostles, New Testament 258
Apuleus 55–56

Aquinas, St Thomas 515
ibn 'Arabi 527
Arabic numerals 25, 56, 392–396, 534–539, 592; in Europe 577–591; origins 356–359, 385
Arab-Islamic civilisation 52, 82, 157–158, 185–187, 389, 511–576
  language 135–137, 212–213, 513, 517–518
  number systems 58, 157, 228, 349, 438, 543–548, 595; alphabetic numerals 241–246, 516, 552–555; counting 39, 47, 49, 66, 70, 96, 428–429; Indian numerals 368, 511–541
  science 512, 514–515, 520 *see also* Baghdad; Muslims
Arad 213, 236
Aramaean Indian writing *see* Kharoshthî writing
Aramaeans 134, 236, 376; number system 39, 137, 227–231, 331, 335, 351
Aramaic script 212–213, 236, 240, 376–377, 387, 390; cryptography 248; Jews adopt 233, 239
Aranda people 5, 72
Arawak, base five in 36
archaic numerals 85, 87–90, 92–93; accounting tablets 107, 121; bullae 104; calculi 125; Sumerian 77–79, 83–84, 92–93, 99–100, 107, 117
Archimedes 207, 222, 361, 518, 522–524, 594, 598; Arabic translation 513, 588; high numbers 333
Ardha-Mâgadhî 383
are (unit of measurement) 42
Argos 219
Aristarch of Samos 221
Aristophanes 47
Aristotle 20, 512–515, 517, 588, 598
arithmetic 5, 10, 96, 206, 528; early 76, 96–97; during Renaissance 577–578; systems 185, 220–222, 248, 442
*Arithmetic* 207
Arithmetic, Lady 205, 591
*Arithmetical Introduction* 43
*Arjabhad* 427
Arjuna 423–424
Arkoun, M. 514, 519
Armenia 139, 290, 519, 528; alphabet 212; numerals 33–35, 224–225, 329
Arnaldez, R. 512, 514–519
Arnold, Edwin 421, 538
Artaxerxes, King 55
Aryabhata 388, 419–420, 427, 432, 447–451, 530
Aryans 385–386
as (Roman unit) 92, 210–211
Asankhyeya 451
Asarhaddon, King 146
al-Ash'ari, Abul Hasan 'Ali ibn Ismail 522
*Ashtâdhyâyî* 389
Asia 81, 402–407, 512; counting 36, 48–49; number systems 94, 412–413 *see also* China; India
Asia Minor 76, 81, 180, 219, 524
Asianics 134
al-Asma'i 521
Asoka, Emperor 375–377, 379, 386–387, 420, 433, 435; edicts 390–391, 397
Assemblée constituante 42
Assurbanipal, library of 160
Assyrian Empire 134–135, 180; counting 39, 99; language 135; number systems 92, 139, 141, 231
Assyro-Babylonian civilisation 135, 140; number system 9, 137, 141–142, 331–332, 351
astrology 159, 549, 553, 556; "Greek" 420; Indian 417, 463; and Koran 514
*Astronomie Indienne* 443
astronomy 92, 522, 524, 529, 551; Arabic 243, 530, 548–549; Babylonian 153, 156–158, 407; Chinese 277–278; Greek 156–157, 408, 549; Indian 409–411, 416–417, 431–432, 443, 463–464, 513–514; Mayan 297–298, 308–313, 315–316, 321–322; sexagesimal system 91–92, 95, 140, 157–158, 548–549; tables 146, 157–159, 198, 521; *Ikhanian Tables* 527; trigonometrical 420 *see also* lunar cycle
al-Asturlabi, Ali ibn 'Isa 521
al-Asturlabi, Badi al-Zaman 518, 526
al-Atahiya, Abu 521
Athens 182–183, 219, 233
'Attar, Farid ad din 527
Attica 183, 219
Atton, Bishop of Vich 578
Auboyer, J. 389
Augustine 257
aureus (Roman money) 210
Aurillac, Gerbert of 362, 518, 578–579, 581–586
Australia 5–6, 18, 72, 93–94
Austria 66
Autolycus 522
Avempace 526
Avenzoar 526
Averroes 514–515, 519, 526
Avestan 32–35
Avicebron 514
Avicenna 363, 513–515, 517, 519, 525, 527, 542; "Avicenna" Arabic alphabet 539
Avigad 234
Awan 81
Axayacatl 301
Ayala, Guaman Poma de 69, 308
Aymard, A. 389
Aymonier, E. 403
Azerbaidjan 519, 528
Aztecs 301–303, 315–316; monetary system 72–73, 302–303, 306; number system 36, 44, 47, 305–308, 326, 348–349; writing 302–303, 305–307
Aztlan 301

al-Ba'albakki, Qusta ibn Luqa 518, 522
Babel, tower of 159
Babur 528
Babylonian civilisation 81, 134–161, 180–181; arithmetic 40, 99, 139, 154–156; cryptograms 158–160; number system 92–93, 139–154, 231, 337–342, 345, 353–354, 407–408; writing 134, 138, 153, 158
Bachelard, Gaston 443
ibn Badja (Avempace) 526
Baghdad 527–528; House of Wisdom 512–514, 516, 520–523, 525–528, 530–531
al-Baghdadi, "Abd al-Qadir 528
al-Baghdadi, Muwaffaq al din Abu Muhammad 363, 527, 542
Bahrain 49
bakers, counting methods 65–66, 70
ibn Bakhtyashu', Jibril 513, 521–522
Baki 528
Bakr, Abu, caliph 519, 522
Bakr, Muhammad ibn Abi 527
al-Bakri 525
balance sheet 109–111
Bâlbodh writing 380
Bali 405–407, 421; numerals 375, 383–385
Balkans 528
Balmés, R. 21
Baltic 33
Bamouns, decimal counting 39
Banda 36, 44
Banka Island 404
banzai 275
Baoule 39
al-Baqi, Muhammad ibn "Abd 525
al-Baqilani, Abu Bakr Ahmad ibn "Ali 524
Baqli, Ruzbehan 526
al-Barakat, Abu 526
*barayta* 254
Barguet, P. 176
Barmak 513
Barnabus 257
barter 72–75
Barton, G. 88
Baru Musa ibn Shakir brothers 518, 522
base numbers 23–46, 96; auxilliary 426–429; eleven 41; five 36, 44–46, 62, 192–193; 'm' 355; six 142 *see also* binary; decimal; duodecimal; sexagesimal system
Basil I 522
Basilides the Gnostic 259
Baskerville face numerals 588
Basques 38–39
Basra 519
al-Basri, Hasan 520
Bastulus 523
Batak numerals 383, 385
Bath *see* Adelard
ibn Batriq, Yahya 513, 521

al-Battani (Albategnus) 514, 522
Ibn Battutua 527
Bavaria 585
al-Bayasi, 'Abu Zakariyya Yahya 518, 526
Bayer 359
Bayley, E. C. 61, 386–387
al-Baytar, Ibn 527
Beast, number of 260–261
Beauclair, W. 577
Beaujouan, G. 356, 578
Bebi-Hassan 52
Becker, O. 91, 519
Bede, Venerable 49–50, 52–56, 200, 223, 578
Beg, Tughril 525
Behar 251
Bek, Ulugh 528
Belgium 31, 65
Belhari 401
Belize 299
Bengal 49, 390, 526
Bengâlî numerals 370, 381, 384, 421, 438
Beni Hassan 51–52
Benin, Yedo 305
Béquignon sisters 26
Berbers 39, 512
*Bereshit Rabbati* 253
Bergamo, Gnosticism 260
Bernelinus 580
*Beschreibung von Arabien* 48
Bessarabia 528
Bete 36
Bettini, Mario 356
Bhadravarman, King 407
*Bhagavad Gîtâ* 422
Bhâskara 419, 452, 530; *Aryabhatîya, Commentary on* 414–415, 420, 439
Bhâskarâchârya (Bhâskara II) 414, 418, 431, 452, 562; multiplication method 573–574
Bhattiprolu writing 377, 383, 385
Bhoja 414
*Bible*
Old Testament 134, 253–254; Daniel 137; Deuteronomy 254; Esther, Book of 137; Exodus 73; Ezekial 239; Ezra 137; Genesis 134, 253–254; Leviticus 254; Nehemiah 137; Pentateuch 137; Prophets, Books of the 137; Psalms 213, 217; Samuel 73; Zechariah 137 *see also Torah*
New Testament 257; Gospels 243, 257; Matthew 257; Revelation (Apocalypse) 256–257, 260
Bihar 526
bijection 10
Billard, R. 406–407, 414–418, 431–432, 443, 529
billion 427–428
binary principle 6, 9, 89, 139, 166
binary system 40–41, 59, 592
binomial formula 528
Biot, E. 282, 336
Birman numerals 438
Birot, M. 134, 138
birth-date 313
al-Biruni, Muhammad ibn Ahmed Abu'l Rayhan 513–515, 519, 524, 588; Indian numerals 418, 438; *Kitab fi tahqiq i ma li'l hind* 363–365, 368, 409, 426–429, 530, 534; *Tarikh ul Hind* 251
Bisaya writing 383, 385
ibn Bishr, Sahl 514, 521
Bistami 522
biunivocal correspondence 10
*Black Stone, The* 146
blackboard 566–567
black-letter Hebrew (modern) 212–213, 215, 233
Bloch, O. 365, 427
Bloch, R. 190
boards: checkerboard 283–288; columnless 560–563; dust 555–563; wax 207–209, 563; wooden 64, 535–536 *see also* abacus; tally sticks
Bodhisattva *see* Buddha
body counting systems 5, 12–19, 23, 214; and base 44–46 *see also* finger counting
Boecius 580
Boethius 48, 578–579
Bokhara 513, 522
Bolivia 69–70, 543
Boncompagni, B. 207, 362, 365, 533
bones 62–63, 269 *see also* tally sticks
Bons, E. 518–519
*Book of Animals* 521
*Book of Kings* 525
Boole, George 598
Boorstin, D. J. 91
Borchardt 55
Borda 42–43
Borneo 375, 383
Botocoudos 5, 72
Bottero, J. 80–81, 160
Bouché-Leclerq, A. 360
Bourdin, P. 21
Boursault 206
boustrophedon writing 186, 219
Brahma 376, 418–419, 422, 427, 441
Brahmagupta 414, 419, 439, 453; *Brâhmasphutasiddhânta* 420, 439, 520, 530, 573–576; multiplication method 573–576
Brâhmî numerals 378–379, 382, 384–395, 402, 420, 433–436, 453
Brâhmî writing 375–378, 397–399
Brasseur de Bourbourg 300
Brazil *see* Botocoudos
de Brébeuf, Georges 206
Breton 33–35, 38
Brice, W. C. 109
bride, price of 72
Brieux 205
Britain, Great 92, 146, 170–171, 176, 214; Treasury 590
British Honduras 299
British New Guinea 13–14
Brockelmann, C. 589
Brooke 18–19
Brothers of Purity (Ikhwan al-safa) 524
Bruce Hannah, H. 26
Le Brun, Alain 101, 109
Bubnov, N. 358, 400
Buddha 71, 408–409, 418, 420–425, 428
Buddhism 11, 71, 407–408, 443, 513
Bugis 14, 383, 385
Bühler 386–387, 389, 438
Buhturi 522
al-Bujzani, Abu'l Wafa' 523, 548
Bulgaria 528
bullae 97, 99–105, 122, 234
ibn Bunan, Salmawayh 513, 522
Bungus, Petrus 199, 260
al-Buni 553–554
Buonamici, G. 190, 197
Bureau des Longitudes (Paris) 43
Burgess 411
Bürgi, Jost 595
Burma: numerals 374–375, 384–385, 388; writing 382–383
Burnam, R. L. 363, 579–580
Burnell, A. C. 389, 438
Burnham 57, 200
Bushmen 5, 72
Ibn Butlan 525
Byzantine Empire 222, 240, 360, 518, 523, 537; arithmetic 334

Cabbala 217, 554
Cadmos 219
Caesar, Julius 7
Cagnat 199
Cai Jiu Feng 279
La Caille 43
Cairo 513, 523
Cajori, F. 356–357, 434
*Cakchiquels, Annals of the* 301
calamus reed 539, 553
calculation 132, 541, 563–566; Babylonian 154–156; Egyptian 39, 174–176, 334; Mayan 303–305, 308, 321–322; North Africa 559; tables 127–130, 146, 203–206, 283–288, 555–563 *see also* abacus; body counting; calculi; notched bones; string; tally sticks
calculator, pocket, first 209–211
calculi 96–105, 118–119, 125–126, 139, 168; Elamites 103, 140–141; Roman abacus 203–205; Sumerian 121–124, 131
calculus 598

calendar 18, 50, 239, 525; ciphers 195–196; Hebrew 215, 217; lunar 19, 297, 407; Mayan 36, 297, 308, 311–322; Roman 7; Shaka 407
Callisthenes 256
Calmet, Dom 359
Cambodia 407, 419; inscriptions 404–405, 431; numerals 375, 403, 413, 438 *see also* Khmer
Cambridge Expeditions 12–14
Campeche 299, 303
Canaan 228, 239
candela 43
Canossa, Darius vase 200–201
Cantera 216
Canton 272
Cantor, Georg 598
Cantor, Moritz 91
căoshū writing 267–268
Capella, Martianus 207
cardinal numeration 20–22, 24–26, 193; Attic system 182–183; reckoning devices 15–19, 96; Yoruba 37
Carib 36
Carolinas Islands 70
Carolingian script 586
Carra de Vaux, B. 358, 364, 400
cartography 526
Casanova, P. 549
Catalonia 223
Cataneo 361
Catherwood, Frederick 300
Cato 194
cattle 72
Cauchy 598
Cayley, Arthur 598
*Ce Yuan Hai Jing* 282
Celebes Islands 375
Celtic numbers 38
censuses 68
centesimal-decimal system 144–145
Central America 300–302, 308; counting 10, 303; numbers 162, 313; trading methods 72–73 *see also* Maya
Ceylon 332 *see also* Vedda
cha lum numerals 374
Chalcidean alphabet 190
Chalfant 269
chalk 566–567
chalkos 182, 200–203
Chalmers, J. 14
Chamealî numerals 381, 384
Champa 404–407, 418; inscriptions 420–421, 431; numerals 383, 385, 413, 421
Chanakya 522
Chandra, Hema 425–426
Changal, Stela of 413
Chapultepec 301
Charlemagne 521
Charles III, king of Spain 248
Charpin, Dominique 88
Chassinat 176
de Chavannes 267
Chelebi, Evliya 529
Chermiss, tally sticks 66
chess 323–324
Chevalier, J. 437, 443, 553, 555
chevrons 148–149
Chhedi 454
Chiapas province 299, 303
Chichén Itzá 300
*Chilám Balám, Books of* 301
children 4–5, 10, 214
chimpu 70 *see also* string, knotted
China 51, 263–273, 276–296, 381; abacus 283–294, 556; counting 39, 49–50, 61, 66, 70, 343, 428; high numbers 276–278, 333, 429; monetary system 73, 75–76; number mysticism 554; number system 162, 168, 263–296, 332, 336–343, 353–354, 375; outside influences 408–409, 512, 516, 520, 526
Chinassi 529
Chinese Turkestan writing 382, 385, 420
Chodzko, A. 543, 545
Chogha Mish 101
Chorem 513
chóu 125, 283–288
Christ 251
Christianity 513; Arabic 240, 513; Central America 300–301; demonised Arabic numerals 588; isopsephy 259–261 *see also* Crusades
chronograms 250–252, 553 *see also* codes and ciphers
Chuquet, Nicolas 427–428
Chuvash 66
Cicero 47, 51, 194, 203
circle 92
*City of God, The* 257
Clandri, Philippi 589
Claparède, E. 365
classification of sciences 517, 523, 525
clay objects: accounting 78, 80, 109–111; tokens 96, 99 *see also* calculi; tablet
clock-making 518
Coatepec 302
Coatlan 302
*Code Napoléon* 66
codes and ciphers 158–161, 248–262, 553–554 *see also* mysticism; numerology
*Codex Aemilianensis* 579–580
*Codex Mendoza* 36, 302–303, 306
*Codex Morley* 298
*Codex Selden* 298
*Codex Telleriano Remensis* 307
*Codex Tro-Cortesianus* 298, 301, 312
*Codex Vigilanus* 362, 580
Codices, Hebrew 217
Codrington, M. 6, 19
Coe, M. D. 299, 320
Coedès, G. 403, 406–407, 413
Cohen, M. 185, 242–244, 376, 385–386, 533
Cohen, R. 238
coins 75–76, 183, 190, 520
Colebrooke, H. T. 573
Colin, G. S. 244, 250, 252
columnless board 562–563
Comte, Auguste 528
Conant, L. L. 19, 45
concrete numeration 21, 23, 167–168
La Condamine 42
Congo, early money 73
conic sections 522–523
Conrady, A. 66
Constantinople 520, 528
Contenau, G. 159
contracts 66, 70
Coomaraswamy 419
Copán 297, 313, 320
Coptic 168, 224
Copts 55
Cordoba 513, 523, 525, 587
Cordovero, Moses 253
Corinth 219
correspondence 21–23; biunivocal 10; one-for-one 10–12, 16–17, 19, 96, 191, 194
Cortez 302
Cos 183
Cottrell, L. 533
de Coulanges, F. 366
Coulomb 43
counting 10, 19–22, 76; cuneiform ideogram meaning 131; methods 62–63, 68–71, 99; rhymes 214; systems *see under* body counting; correspondence; mapping; *see also under specific race/country*
cowrie shells as currency 72
*Crafte of Nombrynge, The* 361
Creation 217, 251, 364; Mayan Long Count 316, 320
Cremonensis, Geraldus, *Liber Maumeti filii Moysi Alchoarismi de algebra et almuchabala* 531
Crete 178–180, 521, 523; Linear A and Linear B 229, 326; number system 9, 178–180, 326, 348
Crimea 226, 528
Cro-Magnon man 62–63
Crusades 525–526, 586–588
cryptography 158–161, 248–250, 259–261, 554 *see also* mysticism; numerology
Ctesibius 518, 594
cubes 363, 524; roots 285, 293, 596

cubit 141
cufik *see* kufic
cuneiform notation 87–88, 135–138, 142, 180–181; numerals 84, 89–90, 100, 125, 145; codes and ciphers 158–161; decimal 137, 139–140; script 107, 121, 148–149; tablets 130–134
Cunningham 386–387
Curr 6
currency 41, 72–76, 182–184, 308
Curtze 361
curviform notation 125, 130
Cushing, F. H. 15, 196
Cuvillier, A. 366
Cyclades 219
cylinders 522
cypher, etymology of 589–590
Cyprian 257
Cyprus 523
Cyril of Alexandria, Saint 56
Cyrillic alphabet 212
Cyrus of Persia 135
Czech, number names 33, 35

dà zhuàn writing style 280
Dadda III 419
al-Daffa, A. A. 519
DAFI (French Archeological Delegation in Iran) 101–103, 105–107, 109, 140, 248
*DAGR* 222
ad Dahabi, Ahmad 252
Dahomey 19
Daishi, Kôbô 296
Dalmatia 194–195
Damais 405–406
Damamini, Ad 528
Damascus 513, 519–520, 525–526
Damerov 92–93
Dammartin, Moreau 588
Dan 36, 253
Dantzig, T. 6, 22–23, 36, 46, 334
Daremberg, C. 221, 428
al-Darir, Abu Sa'id 521
Darius, King 70, 201
Darwin, Charles 519
Das, S. C. 26
Dast, Zarrin 525
Dasypodius, Conrad 358–359, 590
Datta, B. 356, 364, 386–388, 399–400, 414, 419, 422, 434, 438, 562, 568, 573
d'Auxerre, Rémy 207
da'wa 261–262
Dayak 18–19
*De pascha computus* 257
*De ratione temporum* 49–50, 52, 56
*Dead Sea Scrolls* 213, 234
decimal system 24–36, 39–44, 354–355, 542; Ben Ezra 346; Chinese 263, 278–283; counting 68–69, 96, 139–142, 192–193, 208–209; Cretan 178, 180; Egyptian 39, 162; fractions 282, 528, 595; hieroglyphs 167; Mesopotamia 138–146; metric system 42–43; proto-Elamite 120; Semitic 136; Sumerian 93–95
Decourdemanche, M. J. A. 248–249
Dedron, P. 48, 221–222, 428
Deimel, A. 82, 84, 89, 121, 131–132
Delambre 43
Delhi 526–528
Demetrius II 234
demotic writing 171
denarius 210
Dendara, temple of 176
Dene-Dindjie Indians 46
denier, French unit 92
Denmark 33, 38
Dermenghem, E. 519
Descartes, René 42, 199, 597–598
Destombes, M. 360
Devambez, P. 182–183
Devanagari numerals *see* Nâgarî numerals
Dewani numerals 543–545, 550
Dharmarâksha 425
*Dharmashâstra* 419
Dhombres, Jean 43
Dibon Gad 212
Dickens, Charles 65
dictionary of Indian numerical symbols 440–510
Diderot 519
Diener, M. S. 443
digital 59
Dilbat 138
*Ding zhu suan fa* 343
Diocletian 260
Diophantus of Alexandria 221, 513, 522–524, 598
Diringer 270
disability, spatio-temporal 5
divination 159, 269, 549–556 *see also* mysticism
Divine Tetragram 218, 254 *see also* Yahweh
division: à la française 566; abacus 127–130, 206, 285, 287; calculi 121–124; Egypt, Ancient 174–176
Dobrizhoffer, M. 6
Dodge 364
Dogon 72
Dogrî numerals 370–371, 384, 421
Dols, P. J. 49
Donner 229
Dornseiff, F. 256
Doutté, E. 553–555
dozen 41, 92, 95
drachma 182–183, 201–203
Dravidian numerals 373–374, 383
*Dresden Codex* 301, 308, 310
duality 32
Duclaux, J. 367
Duke of York's Island 6
Dulaf, Abu 523
Dumesnil, Georges 356
Dumoutier 272
dung (counting device) 12
duodecimal system 41–43, 92–95
duplications, abacus 206
Dupuy, Louis 42
Durand, J-M. 145, 336
dust-board calculation 555–563
Dutch, number names 33–35
ibn Duwad 523
Dvivedi, S. 414–415, 439, 573
dyadic principle *see* binary principle

e, number 597
Ea 161
Easter, determining 50
Ebla tablets 135, 145
eclipse 529
Ecuador 69, 543
Ed Dewachi, S. 519
Edesse 512
Edfu, temple of 176
Edomites 212
Egine 183
Egypt, Ancient 73–74, 162, 166, 236, 259, 389; and Arabs 519, 522–523, 525–529, 545; sign language 55
calculation 39, 174–176, 334; abacus 541–542; fingers 51–52, 61, 94
number system 9, 91, 162–177, 325–329, 342–350, 390; alphabetic numerals 232, 238, 243; Arabic numerals 356; Indian numerals 368, 534; number mysticism 554
writing 162, 212, 392; hieratic 170–171, 236–239; secret 248
Egyptian Mathematical Leather Roll (EMLR) 176
eight 34, 396; Chinese 269; Egypt 176; Hebrew 215; Indian 410; Japanese 273
eight hundred, Hebrew 216–217
eight thousand: Aztec 305; Mayan 308
eighteen, Egypt 177
eighty: cryptographic 248; Hebrew 215, 235
eighty-eight, Japanese 295
Eisenstein 96
El Obeid 134
El Salvador 299, 303
Elam 102, 134–135; accounting 99, 101–107, 111–120; calculi 140–141; cryptograms 159–160; numerals 9, 39, 96–120, 146; proto-Elamite script 107–120
Elema's body counting system 13–14
Elephantine 213, 227, 229–233, 235, 390

eleven, base 41
Eliezer of Damascus 253–254
Elogium of Duilius 189
Elzevir script face 588
EMLR (Egyptian Mathematical Leather Roll) 176
end of the world 426
engineering 518
England: clog almanacs 195–196; number names 31, 33–35, 428; score (twenty) 37–38; tally sticks 65
English, Old 72
English script 586
Englund 92–93
Enlíl 161
Ephesus 199
Ephron the Hittite 73
Epicurus 522
Epidaurus 185
epigraphy 399–400, 402–407, 419
*Epistle of Barnabas* 257
Equador 308
equations 283, 287, 511, 525, 530
equivalence between sets 3–4
Erhard, F. K. 443
Erichsen, R. W. 342
Erpenius 359
Erse, Old 33–35
Eskimos 36
Essene sect 234–235
Essig 41–42
estranghelo 240
Ethiopia, number system 137, 238, 246–247, 353
Ethiopian numerals 387
Etruscans: abacus 125, 203; alphabet 212–213; number system 9, 39, 189–190, 327, 349; Roman numerals 196–197
Eubeus 219
Euboea 183
Euclid 512–513, 522, 527, 588, 598; *Elements* 521, 523, 525
Euclidian geometry 598
Europe 42, 519–520, 571; Arabic numerals 577–582, 586–591
European numerals 392–396
Evans, Sir Arthur 178–179
Eve 254
evolution theory 519
Ewald 363
*Exaltation of Ishtar* 159
exponential powers 528, 594
Ezra, Rabbi Abraham Ben Meïr ibn 346, 362, 514, 526, 589

al-Fadl 520
*Fahangi Dijhangiri* 52
Fairman, H. W. 176
Falkenstein 81–82
al-Faqih, ibn 523
Far East 70, 272–276, 278–283, 294–296 *see also individual countries*
Fara 87, 101 *see also* Šuruppak
al-Farabi 514, 517, 523
Faraut, F. G. 407
al-Farazdaq 58, 520
al-Farghani 521
al-Farid, Ibn 527
al-Farrukhan, ʿUmar ibn 521
Farsi language 518
Fath, Abu'l 523
al-Fath, Sinan ibn 524
Fatima (Mohammed's daughter) 71, 542, 555
Fayzoullaiev, O. 519, 531
al-Fazzari, Abu Ishaq Ibrahim 513–514, 520, 530
al-Fazzari, Muhammad Ben Ibrahim 520, 529–530
feet and inches 92
Fekete, L. 543, 547
Feldman, A. 518–519
Fénelon 206
Ferdinand II 528
Février, J. G. 64, 66, 70, 79, 185, 213, 219, 376, 382, 387
Fez 252, 513, 520, 525, 527, 539
Fibonacci 365, 523, 588, 598; *Liber Abaci* 361–362, 588–589
fifteen 161, 177, 218
fifty 184, 186, 215; cryptogram 93, 161, 248; Roman numerals 188, 192
fifty thousand 184, 197–198
fifty-three 305
fifty-two 315–316
*Fihrist al alum, Al Kitab al* 364, 531, 539
Fijians 19
Filliozat, J. 335, 386–387, 431, 438, 443
finger counting 22, 28, 47–61, 168, 578; and base 44, 93–95 *see also* Bede; body counting
Finkelstein 134
Finot, L. 406
Firduzi, Abu'l Qasim 58, 525
Fischer-Schreiber, I. 443
five 34, 176, 194, 394, 442, 554–555; Attic 182; base 36, 44–46, 62, 192–193; Chinese 269; Greek 184; Hebrew 215; Indian 410; Mayan 308; Minaean 186; Roman 188, 192; rule of 9; Sheban 186
five hundred 184, 188, 216–217
five thousand 184, 197–198
Fleet 438
floating-point notation 594
Fold, P. 518–519
Folge 364
Folkerts, M. 580–581
Forbes, W. 547
Formaleoni 91
Formosa 73
fortune-telling *see* mysticism
forty 93, 161, 215, 248
forty-nine 442
forty-two 276
Fossey 272
Foulquié, P. 365, 541
four 33, 176, 215, 394, 410; base 94; limit of 7–9, 19, 22, 391; Chinese 269, 271; Japanese 273; mysticism 94, 276
four hundred 215, 305, 308
four thousand 276
Fournier 588
fourteen 161, 177
fractions 548–549, 594, 596; Babylonian 151, 153, 408; decimal 282, 528, 595; Egyptian 168–170; Indian 424–425; Maya 298
France 42–43, 51, 72, 577, 586; counting 32, 38, 65–66, 290; French Revolution 42, 206, 590, 595; number system 31, 33–35, 92, 427–428, 585; metric 42–43 *see also* DAFI
Franz J. 182
Frédéric, Louis 263, 273, 296, 367, 374, 376, 389, 408, 417, 425, 440, 443, 513, 519, 543
Freigius 198–199
French National Archives 43
Friedrichs, K. 443
Frieldlein 580
Frohner 55
Fuegians 5
Fulah 36
Fuzuli 528

von Gabain, A. M. 27
ibn Gabirol, Salomon (Avicebron) 514
Galba, Emperor 200
Galen 256, 512–513
Gallenkamp, C. 297, 311–312, 314
Galois, Évariste 596, 598
games 294–296
Gamkrelidze, T. V. 385
gán mà zí writing 268
Gandhara 228
Gandz, S. 542
Ganesha 568
Gani, Jinabhadra 399, 419
*Ganitasârasamgraha* 399, 421
Garamond, Claude 588
du Gard, Martin 541
Garett Winter, J. 157
Gauss, Karl-Friedrich 589, 598
Gautama Siddhânta *see* Buddha
Gautier, M. J. E. 138
Gebir 521
gematria 252–256, 554
Le Gendre, F. 566
Gendrop 297–298
Genjun, Nakane 289
de Genouillac 86, 88

geography 541, 555
*Geometria Euclidis* 579
geometry 92, 541, 548, 588, 598; base 91–92, 95; base 12 41; Non-Euclidian 527
Georgia 212, 225, 528
Geraty 235
Gerbert of Aurillac 362, 518, 578–579, 581–586
Germany: counting 65, 70–71, 205–206; language 28, 33–35, 72, 586; number system 31, 33–35
Gernet, J. 269–270
Gerschel, L. 6, 64, 66–67, 194
Gerson, Levi Ben 158
gestures, number 14–19, 58–59 *see also* body counting systems
Gettysburg Address 36
al-Ghafur, "Abd ar Rashid Ben "Abd 528
al-Ghazali, Abu Hamid 525
Ghaznavid 58, 523–524
Ghazni 513
Gheerbrant, J. 437, 443, 553, 555
ghubar numerals 385, 534–539, 550, 556, 559, 579, 585
Gibil 161
Gibraltar 520, 527
Gideon 257
Giles 268, 278
Gilgamesh 81
al-Gili 549
al-Gili, Abu'l Hasan Kushiyar ibn Labban 363, 513, 524, 534, 549, 560–562
Gill, Wyatt 12, 14
Gille, B. 518
Gille, L. 519
Gillespie, C. 519
Gillings, Richard J. 175–176
Ginsburg, Y. 199, 207, 284, 361–362, 589
Girard 312
Glareanus 359
glyphs *see* Maya
glyptics 81, 84
Gmür 64–66
gnomon 298
Gnosticism 258–259
Goar, Father 359
gobar numerals *see* ghubar numerals
Godart 179
Godrî numerals 381
gods: God (Judaeo-Christian) 258–259, 552; Mayan 300, 311–314; names and numbers 160–161, 258–259; and spirits 270 *see also* Allah
Godziher, I. 51
Goldstein, B. R. 158
Golius 359
Gondisalvo, Domingo 362
Goths 33–35, 212, 226; Gothic script face 588
Gourmanches 39
Govindasvâmin 414, 418
Goyon, J. C. 176
Granada 513
Grantha writing 383, 385
Greece, Ancient 182–191, 256; and Arabs 512, 515, 518, 528; currency 75–76, 183
science 515, 517–518; astronomy 82, 156–157, 408, 549; Greek Myth 360, 366, 401; isopsephy 252, 256–259, 360
Greeks, Ancient: abacus 200–202, 208; counting 39, 96, 125, 220, 427–428
number system 9, 33–35, 157, 327, 345, 348–350; acrophonic 182–187, 201–203, 214; alphabetic numerals 190–191, 218–223, 232–233, 238–239, 329; Arabic numerals 356, 358–361; fractions 595; high numbers 333, 429
writing 32, 162, 179, 376; alphabet 212–213, 219; papyri 157
Green 92
Greenland 36, 305
Gregory V, Pope 578
Griaule, M. 72
Grmek, M. D. 516
Grohmann, A. 243
gross 41, 92
guān zí writing 267
Guarani 36
Guarducci 182
Guatemala 313, 318–319 *see also* Maya civilisation
Guéraud, O. 220–221
*Guide to the Writer's Art* 543–544
Guitel, Geneviève 214, 343, 347, 408, 437
Guitel, R. L. 432
Guitel, G. 182, 267, 276, 356, 400, 403, 428
Gujarâtî numerals 369–370, 381, 384, 421, 438
Gundermann, G. 182
Gupta dynasty 419
Gupta numerals 378, 381–382, 394, 397–398, 421, 460
Gupta writing 377, 384, 420
Gurkhalî numerals *see* Nepali numerals
Gurûmukhî numerals 369, 381, 384, 421
Guyard 542
Gwalior 380, 394, 396, 400–401, 418, 421

Haab, Mayan solar calendar 312–313, 315
Habuba Kabira 101, 103
Haddon, A. C. 6, 14
*Hadiths* 47
Hafiz of Chiraz 528
*Haggai* 137
Haghia Triada 178, 180
Haguenauer, C. 273–275
Hajjaj, Abu 'Umar ibn 525, 529
al-Hajjami 525
al-Hakam II, caliph 523
Halevi, Yehuda 514
Halhed, N. 50
al-Hallaj, Abu Mansur ibn Husayn 523
Hambis, L. 27, 72
al-Hamdani 523
Hamdullah 542
Hamid, al-Husayn Ben Muhammad Ben 523
Hamit, Avdülhak 529
Hammurabi 81, 135, 142, 145; Code of 86
al Hanbali, Mawsili 55, 58
hand, counting with 47–61, 68 *see also* body counting; finger counting
hangŭ alphabet (Korea) 275
Hanoi 405
Harappâ 375, 385
al-Harb, *Urjuza fi hasab al "uqud* 542
Haridatta 388, 414, 418, 432
Harmand, J. 64, 66
al Harran, Sinan ibn al Fath min ahl 364
Harris Papyrus, The 170, 390
Harsdörffer, Georg Philip 356
Hartweg, R. 593
haruspicy *see* mysticism
Hasan, Ali ibn Abi'l Rijal abu'l (Abenragel) 363
al-Hasib, Hasbah 521
al-Hassar 563
Hassenfrantz 43
Hattuša 180–181
Ibn Hauqal 522, 538
Haüy 43
Havasupai 125
Havell, E. B. 516
Hawaii 70, 125
Hawtrey, E. C. 15
Hayes, J. R. 516, 519
al-Haytham, Abu Ali al Hasan ibn al Hasan ibn 363, 514, 518, 524
ibn Hayyan, Jabir 521
Ibn Hazem 525
Hebrew number system 39, 136–137, 145, 214–218, 345; accounting 236–238; alphabetic numerals 158, 215–218, 227, 233–236, 238–239, 241, 329, 346; and Arabic numerals 359; Ben Ezra 346, 362; mysticism 239, 250, 252–256, 554
Hebrews: calendar 215, 217; language 72, 137, 212–213, 215–218, 236 *see also* Israel; Jews
Hejaz 528
Helen of Troy 51
Heliastes, tablets of 214
Henan 269–270
heqat (Egyptian unit) 169–170
Heraclius 519
Heraklion 178
herdsmen *see* shepherds

Hermite 596
Hero of Alexandria 513, 518, 522, 594
Herodotus 70, 219
Herriot, E. 541
Hierakonpolis 164–165
hierarchy relation 20, 24
hieratic script 170–171, 236–239
hieroglyphs: Cretan 178–180; Egyptian 162–177; Hittite 180–181, 326; Mayan 298–301, 311–314, 316–322
high numbers 298, 333, 428–429, 594; China and Japan 276–278; India 421–429, 434, 440, 460–463; Roman 197–200
Higounet, C. 77, 86, 218
Hilbert 598
Hill 580, 587
Himalayas 390
al-Himsi 522
Hindi language 380
Hindi numerals 368, 511, 532, 536, 538–539, 560
Hinduism 376, 407, 419, 443; calendar 50
Hippocrates 512
Hippolytus 258
hiragana 273
Hisabal Jumal 252, 261–262
Hittite number system 33–35, 39, 180–181, 326, 348
Hiyya, Abraham bar 514
Hoernan, Ather 588
Hoffmann, J. E. 91, 519
Höfner, M. 185
Homer 72
Honduras 299, 303, 313
Hôpital des Quinze-Vingts 38
Hoppe, E. 91–92
Horace 207
horoscope 549
Horus 169, 177
Houailou 36
Hrozny 181
*Huang ji* 279
Hübner, E. 187
Huet, P. D. 359
Hugues, T. P. 555
Huitilopchtli 301
Hunan 272
hundred 25, 179, 194; Chinese 263, 265, 269; Greek 182, 184; Hebrew 215; hieroglyphic 165, 168, 178, 181, 325; Japanese 274; Mesopotamian 137–139, 142–144, 186, 229–231; Roman numerals 188, 192
hundred and eight 295
hundred and seven 177
hundred thousand 140, 165, 168, 197–198, 325
Hungary 528
Hunger, H. 154, 159
Hunt, G. 6
al-Husayn, Abu Ja'far Muhammad ibn 520
Husayn, Allah ud din 524, 526
Huygens, Christian 42
hybrid systems 330, 332, 334–335, 345; Aramaean-Indian 386; classification 351–353; Tamil 372
Hyde, Thomas 158
Hypsicles 522

*I Ching* 70
Icelandic, Old 33–35
ideographic representation 79–81, 98, 107, 136, 145, 163–164; Akkadian 159–160; Chinese 265, 271, 273; Linear A script 178 *see also* hieratic script; hieroglyphs
al-Idrisi 526
Ifrah, G. 137, 368 , 369–375
Ifriqiya 521–523, 525, 528; ghubari numerals 534, 536
*Ikhanian Tables* 527
*Iliad* 72, 214
Iltumish 526
Imperial measurements 92
al-Imrani 523
Inca civilisation 39, 68–69, 125, 308
incalculable, Indian 422
inch 92
India 356–439, 512, 520, 523, 526–528; astrology 417, 463; astronomy 409–411, 416–417, 431–432, 513–514; writing 212, 431–432
Indian number system 332, 341, 346–347, 361–439, 534; calculation 346, 435–437, 568; chronograms 251; counting 39, 49–50, 94, 559; dictionary of numerical symbols 440–510; fractions 424–425, 595; high numbers 421–426, 434, 440; Indian numerals 367–383, 389–399; in Islamic world 511–576; place-value system 334–335, 353, 399–409, 416–421
Indju, Jamal ad din Husayn 528
Indo-Aramaic 228
Indochina 49–50, 65–66, 407
Indo-Europeans 22, 29–32
Indonesia 368, 407
Indrâji, B. 388–389
Indus civilisation 39, 162, 385
infants *see* children
infinity 362, 419, 421–422, 426, 440, 470–472, 597–598
Intaille 203
integers 597; aspects of 21–22
International Standards system (IS) 43
Inuit 36, 44, 305
invoices 78, 110
Iran 81, 135, 522; accounting 97–99, 101–102; counting 94, 290; number system 368, 534 *see also* DAFI
Iraq 52; accounting 101, 121; counting 49, 94; number system 251, 368, 534 *see also* Sumer
Irish 33–35, 38
irrational numbers 528, 596–597
'Isa, 'Ali ibn 525
Isabella I 528
Isaiah 258
Isfahan 513
Ibn Ishaq, Hunayn 513, 522
Ishtar 161
Isidore of Seville 56, 578
Isis 169, 258
Iskhi-Addu, King 74
Islam *see* Muslims
Islamic world *see under* Arab-Islamic civilisation
Ismail, Mulay 528
Isma'il, Sultan 252
Isme-Dagan, King 74
isopsephy 252, 256–261
Israel 97–98, 212, 239 *see also* Hebrews; Jews
al-Istakhri 524
Italic script 212–213, 586
Italy 51, 522; number system 31, 238; writing 216, 219, 586
Itard, J. 48, 221–222, 428
Itzcoatl 301
Ivan IV Vassilievich 96
Ivanoff, P. 298–299
Ivanov, V. V. 385
Iyer 434

Jacob, François 593
Jacob, Simon 206
Jacobites 240
Jacques 443
Jacquet 438
Jaggayyapeta 378
Jaguar Priests 301
al-Jahiz 364, 521, 541
Jainas 425–426, 440; *Lokavibhâga* 416–420, 430
Jalalabad writing 376
al-Jamali, Badr 525
*Jami'at tawarikh* (Universal History) 516
Ibn Janah 525
de Jancigny, Dubois 444
Janus (god) 47
Japan 305, 381; games 294–296; mysticism 554; number system 36, 273–283, 289–290, 388, 542–543
Jarir 520
al-Jauhari 521
Jaunsarî numerals 381, 384
Java 406–407, 418, 420; Kawi writing 383–384, 404; Sanskrit 413
Javanese numerals 375, 392–393, 395, 438
al-Jayyani, Ibn Mu'adh Abu ''Abdallah 518, 524
Ibn Jazla 525
al-Jazzari, Isma'il ibn al-Razzaz 518, 527–528
jealousy multiplication 567–571, 576
Jefferson 42

Jelinek 62
Jemdet 101
Jemdet Nasr 81, 110
Jensen 219
Jerome, Saint 56
Jerusalem 233–234, 236, 525–526, 587–588
Jestin 89, 121
Jesus 257–258
Jews 134, 239, 256, 512–513, 537–538; mysticism 250, 252–256; number system 71, 157–158, 238 *see also* Hebrews; Israel
Jiangxian, Old Mann of 280
*Jinkoki* 278
*Jiu zhang suan shu* 287
John, St 256, 260
John of Halifax 361
John of Seville 362
Jonglet, René 65
Jordan 228, 368, 534
Jouguet, P. 220–221
Judaea 236, 239
Julia, D. L. 593
Julian calendar 50
Juljul, Ibn 524
Junayd 523
Jundishâpûr 512–513
Justinian, Emperor 512
Justus of Ghent 48
Juvenal 55, 203

Kabul 520
Kabyles 66
Kadman 233
*Kai yuan zhan jing* 408, 418
Kairouan 513
kăishū writing 267–268
Kaîthî numerals 370, 381, 384, 421
Kalaman, King 244
Kâlidâsa 419
*Kalila wa Dimna* 323, 419, 520, 556
kalpa 473–474
al Kalwadzani, Abu Nasr Muhammad Ben Abdullah 364
Kamalâkara 414
Kamil, 'Abu 514, 516, 523–524, 530
Kamilarai people 5
Kampuchea 375
Kandahar 376
Kangshi, Emperor 343
Kanheri 401
kanji ideograms 273
*Kanjô Otogi Zoshi* 289
Kannada numerals 374, 385
Kannara numerals 383, 438
Kapadia, H. R. 562
al Karabisi, Ahmad ben 'Umar 364
al-Karaji 511, 514, 516, 524, 526, 530, 548
Karlgren, B. 272
Karnata numerals 374
Karoshthi numerals 386–387
Karpinski, L. C. 207, 356, 361–362, 364, 381, 386–387, 399–400, 538, 580
ibn Karram 522
Karystos 269
al-Kashi 516
al-Kashi, Ghiyat ad din Ghamshid ibn Mas'ud 513, 528, 561, 571
Kashmir 368, 370–371, 381, 420–421, 438
katakana 273
Katapayâdi numerals 388
al-Kathi 525
Kawi writing 383–385, 404, 421
Kaye, G. R. 358, 400–402, 407, 434
Kazem-Zadeh, H. 543, 545
kelvin 43
Kemal, Mustafa 529
Kemal, Namik 529
Keneshre 366, 512
Kenriyû, Miyake 284
Kerameus, Father Theophanus 256
Kern, H. 406, 413, 418
ketsujo 542
Kewitsch, G. 91–93
ibn Khaldun, 'Abd ar Rahman 261, 363, 365, 514, 519, 525, 528, 553; *Prolegomena* 529, 542, 550–552, 593
Khalid 512
Khalifa, Hajji 528
Khaliji, 'Ala ud din 527
Khan, Genghis 382, 526, 556
Khan, Haluga 526–527
Khaqani 58–59, 526, 556
Kharezm Province 513
Kharoshthî writing 376–377, 386
Khas Boloven 65–66
KhaSeKhem, King 165
Khatra 228
Khayyam, Omar 513–516, 525
al-Khazini, Abu Ja'far 523
Khirbet el Kom 235
Khirbet Qumran 234
Khmer 383, 404–407, 420; number system 36–37, 388, 403–404
Khmer numerals 375, 385, 421, 438
Khorsabad 141, 159
khoutsouri 225
Khoziba 259
Khudawadî numerals 369, 381, 384
al-Khujani 524
Khurasan 523–524
ibn Khurdadbeh 522, 537–538
al-Khuwarizmi, Abu Ja'far Muhammad Ben Musa 364–365, 513–514, 516, 521, 523–524, 529–531, 533, 539, 548, 560, 562, 588; algorithms 531, 587
Khuzistan 519
kilogram 42–43
al-Kilwadhi 522
al-Kindi 364, 514, 519, 521, 556
king, ideogram for 159–160
Kircher, A. 210, 226
al-Kirmani 524
Kis 81, 101, 134
*Kitab al arqam* (Book of Figures) 363
*Kitab fi tahqiq i ma li'l hind* 363, 368, 426
Knossos 178–180
knot, meaning decimal system 542
Köbel 205, 358
Kochî numerals 381, 384
Kokhba, Simon Bar 233
*Koran* 514, 519, 521, 553–554
Korea 275, 278–283
Kota Kapur 404
al-Koyunlu 528
Kronecker 593
Kshatrapa numerals 397–398
Kufa, founded 519
Kufic script 243, 539-540
al-Kuhi, Ibn Rustam 523
Kulango 36
Kuluî numerals 381, 384
Kululu 181
Kumi 190
Kurdistan 528
Kushana numerals 397–398
Kutilâ numerals 381, 384
Kyosuke, K. C. 305

Labat, R. 84, 87, 99
Lafaye, G. 51
Lagaš 81, 93
Lagrange 42–43
Lakhish 213, 236
*Lalitavistara Sûtra* 420–425
Lalla 414
Lalou, M. 26
Lambert, Meyer 90, 137
Lampong writing 383
Landa, Diego de 300–301, 314
Landa numerals 381, 384
Landsberger, D. 130
Langdon, S. 109, 116–117, 385–386
Lao Tse 70
de Laon, Radulph 207
Laos 375, 383

Laplace, P. S. 42–43, 361
Larfeld 182
Laroche, E. 180–181
Larsa 146
laser 43
Latin 52, 72, 96, 194; alphabet 212; number names 7, 31, 33–35
Laurembergus 359, 589
Lavoisier 42
*Law of 10 Frimaire, Year VIII* 43
*Laws of 18 Germinal, Year III* 42–43
ibn Layth, Abu'l Ghud Muhammad 524
LCM (lowest common multiple) 93
Lebanon 368
Leclant 52
Lehmann-Haupt 91
Leibnitz 550, 598
Lemoine, J. G. 49–50, 56, 541–542
Lengua people 15
Lenoir 43
Leonard of Pisa *see* Fibonacci
Leonidas of Alexandria 256
Lepsius 75
letter numerals *see* Abjad numerals; alphabetic numerals
*Lettres* of Malherbe 51
Leupold, Jacob 57, 357
Levey 434
Levias, C. 359
Lévy-Bruhl, L. 5, 19, 45–46
Leydon Plaque 318–319
Li, J. M. 275
Li Ye 282–283
*Liber de Computo* 56
*Liber etymologiarum* 56
Libya 368, 520
Lichtenberg 593
Lidzbarski, M. 212, 390
Liebermann, S. J. 98–99, 130, 140
ligatures 170–171, 228–229, 246, 391, 434
*Light of Asia, The* 421
*Lîlâvatî* 431
Limbu numerals 381
Lincoln, Abraham 37
*L'Inde Classique* 443
Lindemann 596
Linear A script 178–180
Linear B script 178–180
lines, grouping of 433–434
Liouville 596
lìshū writing 266–268
Lithuania 33–35
*Lives of Famous Men* 55
Lobachevsky 598
Locke 68
Löfler 91
logarithms, natural 597
logic 598
*Lokavibhâga* 416–420, 430
Lombard, D. 265
Lombok 375
London, Royal Society of 42
Long Count 316–319
Lot of Sodom 253
Louis XI France 38
Louis XIV 528
Louvre 146
Lucania 189
Lull, Ramon 550
de Luna, Juan 362
lunar cycle 17–18, 50, 217; calendars 19, 297, 407; eclipse 529; mansions 554; and numerology 93
ibn Luqa, Qusta 513
Luther, Martin 260–261
Lutsu 64
Lycians 9, 39
Lydian civilisation 9
Lyon 141

al Ma'ali, Abu 556
al-Ma'ari, Abu'l 'ala 525
Macassar writing 383
Maccabeus, Simon 234
MacGregor, Sir William 14
Machtots, Mesrop 224
Madagascar 125, 368, 534
Madura 375
Magadha 383
Maghreb 244, 252, 513, 520–521, 525–528; calculation 556, 560, 563; numerals 356, 385, 534–539, 559; writing 539–540
al-Maghribi, As Samaw'al ibn Yahya 55, 363, 511, 514, 516, 526, 534
Maghribi script 539–540
magic 248–262, 298, 302, 549–556; talismans 262, 522, 554 *see also* mysticism
Magini 595
Magnus, Albertus 515
*Mahâbhârata* 419
Mahâjanî writing 381, 384
al-Mahani 522–523
Mahârâshtrî writing 380, 384
Mahâvîrâchâryâ 399, 414, 418, 421, 562
Mahmud 523
Mahommed *see* Mohammed
Maidu 125
Maimon, Rabbi Moshe Ben *see* Maimonides
Maimonides 526
Maithilî numerals 370, 381, 384, 421
*Majami* 55
al Maklati, Muhammad Ben Ahmed 252
Maknez, chronograms 252
Malagasy 39
Malay, Old 383, 385, 404, 406
Malaya 534
Malayâlam numerals 332, 334–335, 342, 353, 373, 383, 385
Malaysia 39, 368, 406–407, 418, 420
Maldives 374
Malherbe, M. 36, 51, 273
Mali 72
al Malik, 'Abd 252
Malinke 36, 44
Mallia 178
Mallon 224
Malta 522
al-Ma'mun, Caliph 512, 521, 529, 531
Manaeans 9
Manchuria 272
Manchus 39
Mandealî numerals 381, 384
Manipurî numerals 381, 384, 421
Mann 37
al-Mansur, Caliph 512, 520, 529–530
al-Mansur, Sultan Abu Yusuf Ya'qub 526, 528, 550
Mansur, Yahya ibn Abi 513, 521
many, concept of 5–6, 32, 94
mapping 10–12, 16–17, 21, 23
al-Maqrizi 528
al-Maradini 542
Marâthî numerals 369, 380, 384, 421, 438
Marchesinus, J. 588
al-Mardini, Massawayh 513, 525
Marduk 146–147, 159, 161
Mari 74, 81, 134, 142–146, 336, 352
Maronites 240
Marrakech 525, 527
al-Marrakushi, Abu'l Abbas Ahmad ibn al-Banna 363, 527, 568
Marre, A. 363, 568
Martel, Charles 512 , 520
Martial 203
Martinet, A. 385
Mârwarî numerals 381, 384, 421
al-Marwarradhi 521
ibn Masawayh, Yuhanna 519–520
Mashallah 359, 520
Ma'shar, Abu 521
Mashio, C. 305
Maspéro, H. 74, 267, 269
ibn Massawayh 522
Massignon, L. 512, 514–519
Masson, O. 179
al-Mas'udi 514, 523

*Materialen zum Sumerischen Lexikon* 130
*Mathematical Treatise* 515
*Mathematics in the Time of the Pharoahs* 175
Mathews 268, 278
Mathura numerals 397–398
Matlazinca 301
Matzusaki, Kiyoshi 289
Maudslay, Alfred 300
Le Maur, Carlos 357
al-Mawardi 525
Maximus, Claudius 56
Maya civilisation 72, 297–322; astronomy 315–316, 321–322; calculation 303–305, 321–322; calendars 36, 300, 311–319; mysticism 300, 311–314, 316–322; writing 298–301, 305, 311–314, 316–322
Mayan number system 9, 36, 44, 162, 303–312, 345; positional 322, 337, 339–340, 353–354, 430; zero 320–322, 341–342, 430
Mazaheri, A. 363, 519, 533, 539, 541–542, 549, 556, 561
Mead 541
measurement 82, 91, 153, 158
Mebaragesi 81
Mecca 519, 529, 537, 554
Méchain 43
de Mecquenem, R. 109, 116–117
Media 519, 522
mediating objects *see* model collections
Medina 519
Mediterranean 212, 222
Mehmed IV, Sultan 529
Mehmet II, Sultan 528
Mei Wen Ding 280, 284
*Mejing* 516
*Melanesian Languages* 6
Melos 219
Mendoza, Don Antonio de 303
Menelaus 513, 522–523
Menna, Prince 61
Menninger, K. 190, 276, 283, 336, 343, 356, 428
al-Meqi, al-Amuni Saraf ad din 363, 526
Merida, bishop of 300
meridian expedition 42–43
Merv 513
Mesha, King 212–213
Mesopotamia 94, 162, 239; Arabs 512, 519, 526, 528; Babylonian era 134–161; India 376, 513; Mari 142–146; mysticism 93–94, 554; writing 212, 539 *see also* Akkadian Empire; Elam; Semites; sexagesimal system; Sumerians
counting: abacus 130–133, 562–563; bullae 99, 101; calculi 97–98; clay tablets 84–89
number system 82, 96–108, 134–161, 325–329; Aramaic numerals 228; decimals 138–146; letter-numerals 243; zero 152–154, 341
Messiah (jewish) 253
*la Mesure de la Terre* 42
metal, as currency 73–74
*Metaphysics* 20
Metonic cycle 195
metric system 595; history 42–43
metrology 182
Mexico 36, 299–301, 303, 305, 307
Mexico City 301, 305
*Micah III* 256
Middle East: calculi 97–98; language 52, 212, 222, 248–250; Semites 134
*Midrash* 253
Mieli, A. 519
Mikame 289
Miletus 219
Milik, J. T. 234
Millas 216
Miller, J. 42, 273–274
milliard 428
million 140, 165, 168, 325, 427–428
Minaeans 185–186
Minoan civilisation 39, 178–180
Minos, King 178
minus (mathematical concept) 89
Mirkhond 528
Miskawayh 525
Mî-s'on 406–407, 413
Misra, B. 562
al Misri, Abu Kamil Shuja' ibn Aslam ibn Muhammad al Hasib 364
Misri, Dhu "an Nun 522
Mithat, Ahmet 529
Mithras 259
Mitsuyoshi, Yoshida 278
Miwok 125
Mixtecs 36, 305, 307–308
mkhedrouli 225
mnemonics 432, 537
Moab 212
model collections 10, 12, 17–18, 23
modern numerals 324–325, 343–347, 356–365, 368, 385, 426–439, 592–599
Modî numerals 380, 384
Mogul Empire 528
Mohammed 47, 50–51, 58–59, 514, 519, 553
Mohenjo-daro 375, 385
Mohini 50
mole 43
Molière 38
Möller, G. 342, 390
Mommsen, T. 187
Môn writing 194, 383
money 41, 72–76, 182–184, 308
Monge 42–43
Mongolia 22, 27, 39, 49, 51, 556
Mongolian Empire 382, 526–527
Mongolian numerals 382, 385, 395–396
monks 70–71; Zen 295
Montaigne, Michel de 205–206, 577, 590
Monteil, V. 513–514, 516, 518–519, 593
Montezuma 301, 303
Montucla, J. F. 360–361
moon 217, 239, 411 *see also* lunar cycle
Moor 439
Morazé, Charles 345, 347
*Moreh Nebukhim* (Guide for the Lost) 514
Morley, S. G. 298, 316, 320
Morocco 51, 252, 555
Morra 51–52
Moses 254 , 553
Moss 75
Mota 19
Motecuhzoma I 301, 303
Mouton, Abbé Gabriel 42
Moya, Juan Perez de 55
Mozarabes (Arabic Christians) 513
al Mu'aliwi, 'Ali Ben Ahmad Abu'l Qasim al Mujitabi al Antaki 364
Mu'awiyah 520
Mudara, Muhammedal 252
Muhammad, Abu Nasr 523
Muhammad of Ghur 526
al-Muhasibi 522
al Mulk, Nizam 58
Multânî numerals 381, 384, 421
multiplication methods: abacus 127, 204–206, 208–209, 285–287, 292, 557–559, 582–585; calculi 122; fingers 59–61; tables 154–156, 220, 561, 578; written 154–156, 174–176, 567–576
multiplicative principle 229, 231, 246, 263, 270, 330–334
al-Mu'min, 'Abd 525
al-Muqaddasi 524
*Muqaddimah* 261, 363, 519, 529, 542, 552–553, 593
al-Muqafa, Abu Shu'ayb 521
al-Muqafa', ibn 520
Murabba'at 236
Murray Islanders 5–6, 14
Muslims: finger gestures 47, 50, 52, 58–59; Hisabal Jumal 250, 252, 261–262; magic talismans 262, 522, 554; prayer 9, 50, 71 *see also* Arab-Islamic civilisation
al Mustadi 252
al-Mustawfi, Hamdallah 527
al-Mu'tamin 525
Mutanabbi 522
ibn Mu'tasin, Ahmad 518
Muwaffak, Abu Mansur 524
Mycenae 179

Myres 179
myriad 26, 221–222
*Mysticae numerorum signification esopus* 199
mysticism, number: Arabs 512, 553–555; China 270; fear of numbers 214, 275–276; India 431, 543; Mayan 321–322; sacred symbols 93–94, 162, 239; soothsayers 261–262, 269, 551–556 *see also* codes and ciphers; magic

'n' roots 528
Nabataean numerals 212, 227–228, 390
Nabî 529
al-Nadim, Ya'qub ibn 364, 524; Fihrist 364, 531, 539
al-Nafis, Ibn 527
Nâgarî numerals 364, 368–369, 384, 400, 421, 438, 481, 532, 538
Nâgarî writing 364, 377, 379–380, 388, 420
al-Nahawandi, Ahmad 521
Naima 529
Nakshatra 417
names of numbers 14–15, 19–23, 33–35, 136–137; games 159; gods' names used 95; Indian 481–482; Mayan 303–304; prayer words 214
Nânâ Ghât 379, 387–388, 391, 397–399, 420, 435–436
Napier, John 597
Nârâyana 562
Narmer, King 164–165
Nâsik 379, 387–388, 397–398, 435
Naskhi script 539–540
Nasr 101
Nasr, Abu 524
Nasr, S. H. 516, 519
nastalik script 539–540
Nathan, Ferdinand 291
*Natural History* 47, 198, 200, 427
Nau, F. 366
Naveh 232, 234
al-Nawabakht 520
al-Nayrizi (Anaritius) 522
Nayshaburi, Hasan ibn Muhammad an 571
al-Nazzam 521
Nebuchadnezzar II 135, 236
Nedim 529
Needham, Joseph 51, 264, 268–269, 278–284, 293, 408
Nefi 529
negative numbers 278, 283, 287, 597
Negev, A. 73
Nemea 185
Nepal 377, 384, 388, 390, 420
Nepali numerals 371, 381, 384, 392–398, 438
Nergal 161
Nero 256, 260
Nesselmann, G. H. F. 218
Nestor, King 55
Nestorian sect 240
Neugebauer, O. 91–92, 150, 153, 157, 414–415
New Guinea 13–14, 305
New Hebrides 36
New Mexico 196
Newberry 52
Newton, Isaac 42
Nichomachus of Gerasa 43–44
Nicobar Islands 375
Nicomacchus of Gerasa 578
Niehbuhr, Karsten 48–49
Nigeria 70
Nîlakanthaso-mayâjin 414
Nile 259
nine 35, 396; Chinese 269; Egyptian 177; Hebrew 215; Indian 410; Japanese 276
nine hundred 216–217
nineteen 177
ninety 215, 235, 248, 295
ninety-nine 295
ninety-three 58
Ninevah 101, 103, 135, 146
Ninni, A. P. 197
Ninurta 161
Nippur 81, 130, 239
Nisawi, Abu'l Hasan 'Ali ibn Ahmad an 363, 524, 530, 539, 548, 560, 562–563
Nisibe 512
Nissen 92
Nizami 526, 556
Nommo the Seventh 72
non-equivalence between sets 3–4
non-Euclidian geometry 598
notched bones 11 *see also* tally sticks
Nottnagelus 359
Nougayrot, J. 85, 146
nought *see* zero
Nubians 39
Numa, King 47
number systems: alphabetic numerals 156–157, 212–262, 483–484; Arab-Islamic 511–576; Chinese 263–296; Cretan 178–180; Egyptian 162–177; Europe 578–582, 586–591; Greek 182–187, 218–223, 232–233; Hebrew 214–218, 233–236; historical classification 347–355; Hittite 180–181; Indian 367–439; Dictionary 440–510; Mayan 297–322; Mesopotamian 96–108, 134–161; modern 324–325, 343–347, 356–367, 592–599; Roman 187–200 *see also* abacus; accounting; base numbers; body counting; calculation; decimal; mysticism; position, rule of; Sumerian; zero
numerology 93–94, 161, 360, 554 *see also* codes and ciphers; mysticism
Nur ad din 526
Nusayir, Musa Ben 520
Nusku 161
Nuwas, Abu 521
Nuzi, Palace of 100–101

Oaxaca Valley 301, 305
Oaxahunticu *see* Maya calendars
obols 182–183, 201–203
Oceania 10, 12–14, 36, 44, 554
*Odyssey* 214
*Oedipi Aegyptiaci* 226
Oghlan, Karaja 529
Ojha 386–387
Okinawa 70
Olivier 179
Omayyad dynasty 520
Omri, King 236
one 33, 194, 392; Aztec 305; Chinese 269; Greek 179, 182, 184, 186; Hebrew 215; hieroglyphic 165, 168, 176, 178, 181, 325; Indian 409–410; Maya 308; Roman numerals 188, 192; Sumerian 84, 148
one hundred and eight 71
one-for-one correspondence 10–12, 16–17, 19, 96, 191, 194
*Opera mathematica* 91
Ophel, accounting 236
Oppenheim, A. L. 100–101, 131
Ora 216
oral numeration 25–26, 265–266, 303
Orchomenos 183
order relation 20
ordinal numeration 20–22, 24, 182, 193
Ore 276, 428
Oriental Research Institute (Baghdad) 100
*Origin of Species* 519
Orissî numerals 370
Oriyâ numerals 370, 381, 384, 421, 438
Orontes 55
Oscan alphabet 212
Osiris 169, 259
ostraca 213, 236, 238
Otman, Khalif 58
Ottoman Empire 527–529, 543; secret writing 248–250; Siyaq numerals 547–548
oudjat 169–170
ounce 92
ownership, mark of 66
oxen 72
Özgüç 181

Pacific Islands 72, 125
Pacioli, Luca 57, 567, 576
pairing 6, 21
Pakistan 94, 386, 520; numerals 368, 381, 534; phalanx-counting 94
Palamedes 219
Palenque 297, 316–317, 320
palaeography 391, 401, 404–406, 419, 538–539, 579

palaeo-Hebraic alphabet 212–213, 233, 236, 238
Palestine 70, 236, 239, 519, 528; numerals 228, 236–238, 241, 246
Pâlî writing 374, 377, 383, 385
palindromes, numerical 399
Pallava dynasty 378; numerals 397–398
Palmyra 212, 227–228, 248, 533
Palmyrenean numerals 390
*Pañchasiddhântikâ* 414–416, 439
*Pañchatantra* 323, 419, 520
Pânini 388–389
*Pâninîyam* 389
*Pantagruel* 51
paper making 516, 521, 566–567
Papias 207
Pappus of Alexandria 221–222, 523
Papuans 13–14
papyrus 533
Paraguay 15
Parameshvara 414
parchment, Maya 301
*Pardes Rimonim* 253
Paris, B. N. 361–362
*Paris Codex* 301
Paris (of Athens) 51
parity, concept of 6
Parmentier, H. 413, 418–419
Parrot, André 142
Pascal, Blaise 282, 594, 598
Pascal's triangle 282, 511
Pasha, Ziya 529
pebbles 12, 15, 96–97, 125; counting 126
Péguy, Charles 365
Peignot 381
Peignot script face 588
Peking 272
Peletarius 358
Pellat, C. 55, 541–542
Peloponnese 75–76, 183
pendulums 42
Pepys, Samuel 578
perception, limits of 6–10
Perdrizet, P. 256, 258–259
Pergamon 256
Perny, P. 51, 268, 271
Persia 70, 259, 376, 512–528; abacus 556, 562–563; number system 39, 58, 250–251, 545–547, 553; writing 240, 248, 539
Persian Gulf *see* Sumerians
Peru 69–70, 308, 543
*Peruvian Codex* 308
Peten, Lake 299–300
Peter, Simon 257
Peterson, F. A. 312–314, 317, 320
Petitot 46
Petra 228
Petruck 434
Petrus of Dacia 361
Phaestos 178
phalanx-counting 94–95
Pheidon, King 75
Philippines 383
Phillipe, André 65
philology 419
Philo of Byzantium 513, 518
philosopher's stone 518–519
*Philosophica Fragmenta* 203
Phoenicians 359; alphabet 212–214, 219, 239; number system 9, 39, 137, 227–228, 351; writing 185, 232, 236, 390
phonograms 80, 136, 265 *see also* hieroglyphs
Phrygia 238
pi 596–597
Piaget 4–5
Picard 598
Picard, Abbé Jean 42
pictograms 78–81, 85, 97–99, 107–108, 306 *see also* hieroglyphs
Piéron, H. 365
Pihan, A. P. 268, 271, 356, 381, 543–545, 547
Pingree, D. 91–92, 150, 153, 157, 414–415
Pinyin system 265
Pisa, Leonard of *see* Fibonacci
Pizarro 68
place-value system 324, 559–560, 588; abacus 287–288, 434–437, 561; discovery 287–288, 337–339, 399–407, 416–421 *see also* position, rule of; positional systems
Planudes, Maximus 361, 365, 562, 589
plates, lead 181
Plato 512–513, 517, 523, 598
Plaut 273–274
Plautus 194
Pliny the Elder 47, 198, 427
Plotinus 513, 523
plurality 32
Plutarch 47, 55
Pô Nagar 404–406, 420
Poincaré, H. 367
Polish 33–35
Polybius 200
Polynesia 6, 72
polynomials 283
Pompeii 256
Popilius Laenas, C. 189
*Popol Vuh* 301
Porter 75
Portugal 33, 35, 51
Posener, G. 52, 533
position, rule of 24, 143, 145–155, 334–340, 345–346 *see also* place–value system
positional systems: Arabic 186; Babylonian 145–154; Chinese 278–283; historical classification 353–355; India 411–421; Mayan 308–312, 322
Pott, F. A. 36–37
Powell, M. A. 82, 121
powers: abacus 285; cubed 363; exponential 528, 594; negative 156, 278; squared 323–324, 363; ten 278, 426–429, 440, 594
Práh Kuhâ Lûhon 404
Prasat Roman Romas 413
prayer-beads 11, 50–51, 99
Pre-Sargonic era 81, 87, 89
Prescott, W. H. 69
priests, Mayan 311–312
primitive societies: barter in 72–73; counting 5, 10, 12–18, 46
Prinsep, J. 386–387
Prithiviraj 526
*Prolegomena* 261, 363, 519, 529, 542, 552–553, 593
Prophet, the *see* Mohammed
Proto-Elamite number system 326
Psammetichus, King 52
*Psammites, The* 333
*Pseudo-Callisthenes* 256
Ptolemy 513, 520–525, 588, 598
Ptolemy I 256
Ptolemy II 232
Ptolemy V 167
Pudentilla, Aemilia 55–56
Puebla region 301
*Pulisha Siddhanta* 427
Punjab 228
Punjabî numerals 369, 381, 384, 421, 438
Putumanasomayâjin 414
puzzles, number 176–177
Pygmies 5, 72
Pylos 179
Pythagoras 256, 515, 596
Pythagoras' theorem 151, 522

al-Qalasadi 539
al-Qalasadi, Abu'l Hasan 'Ali ibn Muhammad 363, 516, 528, 563, 567
Qasim, Abu'l 524
al-Qasim, Muhammad Ben 529
al-Qass, Nazif ibn Yumn 523
al-Qays, Imru' 520
al-Qifti, Abu'l Hasan 527, 529–530
quadrillion 427–428
Quahuacan 36
Quauhnahuac 36
Qubbut al Bukhari 252
Qudama 523
Quetzalcoatl 300
quinary systems 9, 44–46, 94–95

Quintana Roo 299, 303
Quintilian 47
quintillion 428
quipucamayoc 69, 308
quipus 64, 68–69, 308, 542–543 *see also* string, knotted
Quiriguà 298, 316–317, 319–321
ibn Qurra, Thabit 514
Qutan Xida *see* Buddha
Qutayba 513
Qutb ud din 526

Rabban, 'Ali 522
Rabelais 51
Rachet, Guy 81, 135
Raimundo of Toledo 362
Râjasthani numerals 381, 384
Ramus 358
Ramz 250
Rangacarya, M. 414, 418
rank-ordering 16
Ras Shamra 137, 214
Rashed, R. 363, 511, 519
Rashi 253
al-Rashid, Harun 512, 520
Rashid ad din 514, 516, 527
rational numbers 596–597
Razhes 522
al-Razi, Fakhar ad din 363, 513–514, 526
al-Razi, Muhammad Abu Bakr Ben Zakariyya (Razhes) 522
ready-made mappings 12, 17, 19
real numbers 597
Rebecca, wife of Isaac 257
rebus 302–303, 306–307
receipts 68, 70
Recorde, Robert 358, 590
recurrence 20
Red Sea 49
Redjang writing 383
Reinach 182
Reinaud 364
Reisch, Gregorius 591
*Relación de las Cosas de Yucatán* 300
Renaissance 529
Renou, L. 335, 386–387, 431, 438, 443
Rey 513
Reychman, J. 543, 547
Rhangabes 201
Rhind Mathematical Papyrus (RMD) 171
Richard Lionheart 586
Richer 42
Ridwan, 'Ali ibn 525
Ridwan of Damascus 518, 527
Riegl 196
Rif 252
right-angled triangles 151, 522
Rijal, ibn Abi'l 524
Rivero, Diego 47
de Rivero, M. E. 69
RMD (Rhind Mathematical Papyrus) 171
Robert of Chester 362
Robin, C. 186–187
Rodinson, M. 185
Röllig 229
rômaji 273
Roman Empire 7, 51, 92, 521, 577–578; calculation 39, 70, 96, 333–334, 427; abacus 125, 202–207, 209–211, 578–580, 582; currency 55, 76
Roman numerals 9, 187–200, 327–328, 349; used in Europe 208, 578–579
Romance languages 31–32
Romanian 33–35
Rong Gen 269
roots, square and cube 156, 285, 293, 419, 560, 596–597
rosaries 70–71
Rosenfeld, B. A. 571
Rostand, J. 593
Rudaki 523
Ruelle, C. E. 222
Rumelia 528
Rumi, Ya'qub ibn 'Abdallah ar 527
Ibn Rushd (Averroes) 514–515, 526
Russia 33–35, 66, 72, 212, 290
ibn Rusta of Isfahen 523
Rutten, M. 93
Ryu-Kyu islands 70, 542–543

Ibn Sa'ad 58, 71, 522, 542
Sa'adi of Chiraz 527, 538
Saanen 195
Sabaeans 9, 512
Saccheri 527
Sachau, E. 227, 235
sacred symbols 93–94, 162, 239 *see also* mysticism
de Sacrobosco, Jean 358, 361
Sa'ddiyat 335
Saffar, Ibn al 524
Saffarid dynasty 521
Saglio, E. 221, 428
ibn Sahda 522
Sahdad 101
ibn Sahl, Sabur 522
Saidan, A. 364, 563
Sakhalin, Ainu of 305
Saladin 526
Salamis, Table of 201–203
Samanid dynasty 521–522
Samaria 213, 236
Samaritans 212, 233
Samarkand 520, 528
al-Samh, Ibn 524
Sanayî, Abu'l Majîd 58
sangi 278–283
Sankheda 402
*Sankhyâyana Shrauta Sûtra* 422
Sanskrit 72, 433; number names 29, 32–35, 404–406, 411–420, 530; high numbers 427–429, 434; oral counting 426–431; Pânini 389; Shiddhamatrikâ 381 *see also* Brâhmî; Nâgarî writing
ibn Sarafyun, Yahya 522
Sarapis 256
Sarasvati (goddess) 439
Sardinia 520
Sargon I The Elder 135
Sargon II 139, 141, 159
Sari 232
Sarma, K. V. 414–415, 419
Sarton, G. 519
Sarvanandin 416–418
Sastri, B. D. 414
Sastri, K. S. 414
Satan 554, 588
*Satires* 207
Satraps 378, 407
Saudi Arabia 368, 534
Saul 73
Saxon, Old 33–35
Scandinavia 65, 196
Scheil, J. 102, 109, 116
Scheil, V. 115
Schickard 594
Schmandt-Besserat, Denise 97–100
Schnippel, E. 195–196
Scholem, Gershon 217, 256
Schopenhauer 20
Schrimpf, R. 278
science: classification 517, 523, 525; Koran 514
scientific notation 594
Scots Gaelic 33
Scott 52
Scythians 377
seals, cylinder 103–104, 106–107
Sebokt, Severus 366, 407, 419
secret writing 248–250 *see also* mysticism
*Sefer ha mispar* (Number Book) 362
Semites 81, 134–136; alphabet 212–213, 377; number system 22, 136–146, 227–232, 351 *see also particularly* Akkadian; Arab-Islamic; Assyrian; Babylonian; Hebrews; Phoenician cultures
Senart 386–387
Seneca 47, 200
Senegal 305
Sennacherib 146

separation sign 149
septillion 428
Serere 36
Sessa, legend of 323–324
sestertius 210
Seth 169
sets, theory of 598
seven 34, 395, 442; Chinese 269; Egyptian 176; Hebrew 215; Indian 410; Japanese 276
seven hundred 216–217
seventeen 177
seventy 215
seventy-seven 294
de Sévigné 206
Seville 514, 527
sexagesimal system 82–84, 90–95, 126, 139, 157; Akkadian 134, 138, 239; astronomy 91–92, 95, 140, 157–158, 548–549; Babylonian 134–161; calculation 126–133, 140, 528; proto-Elamite 120
sextillion 428
Sezhong, King 275
*Shah Nameh* 58
Shahadah, prayer of 47
ibn Shahriyyar, Buzurg 524
Shaka calendar 407, 494
Shamash 161
Shan 39
shàng deng number system 277–278
sháng fāng dà zhuàn writing 268
Shankarâchârya 418
Shankaranârâyana 388, 414, 418, 432
al-Shanshuri 536
Shâradâ numerals 371, 381, 384, 421, 438, 494
Shâradâ writing 371, 377, 420
Shaturanja (early chess) 323–324
Sheba 185–187, 327
shekel 73
shells 24–25, 37
Shem, son of Noah 134, 254
Shen Nong 70
shepherds, counting methods 47, 191–193, 214; and base 10 24–25; bullae 101, 103; pebbles 12; quipus 69; tally sticks 11, 64
Sher of Behar, King 251
Shiite Islam 521
Shiraz 243
Shivaism 407
*Shojutsu Sangaka Zue* 284
Shook 317
Shrîdhârâchârya 414, 562
Shrîpati 414, 562
Shukla, K. S. 414–415, 419
shûnya (zero) 412, 495–496
Shuri 70
Shushtari 527
Siamese numerals 375, 388, 403
Siamese writing 383
Siberia 71–72
Sibti, Abu'l 'Abbas as 550–551
Sicily 190, 219, 521, 587
Siddham numerals 384
Siddham writing 377, 381, 420
Siddhamatrikâ writing 381
siddhânta *see* India, astronomy
Siddim, Valley of 253
sign language 52–59
al-Sijzi, 'Abd Jalil 524, 534
Silberberg, M. 346, 362, 589
silent numbers 214
Sillamy, N. 4, 365
Simiand, F. 366
Simonides of Ceos 219
Simplicius 512
Sin 161
ibn Sina, Al Husayn *see* Avicenna
Sinan, Ibrahim ibn 243
Sindhî numerals 369, 381, 384, 421, 438
Singapore 272
Singer, C. 518–519
Singh, A. N. 356, 364, 386–388, 399–400, 414, 419, 422, 434, 438, 562, 568, 573
Singhalese numerals 342, 352, 374, 383, 385, 388
singularity 32
Sino-Annamite writing 272
Sino-Japanese numerals 273–276, 278
Sino-Korean number system 275
Sircar 438
Sirmaurî numerals 381, 384
Sitaq 248
Sivaramamurti 387
six 34, 161, 395; Chinese 269; Egyptian 176; Hebrew 215; Indian 410
six hundred 84, 216–217
six hundred and sixty-six 260–261
sixteen 218
sixty 91, 93; base 40, 82; Hebrew 215; Mesopotamian 84, 141–142, 148, 161
Siyaq numerals 545–548
Skaist 235
Skandravarman, King 378
Škarpa, F. 194–195
Slane, I. 542, 552–553
Slavonic Church 33–34
Smirnoff, W. D. 260
Smith, D. E. 199, 207, 284, 356, 361–362, 364, 381, 386–387, 399–400, 538, 580, 589
Smith, V. A. 386
Snellius 595
sol, French unit 92
solar cycle 49–50; calendar, Maya 297; eclipse 529
de Solla Price, D. 518
Solomon Islanders 19
Solomon's ring, legend of 357
Solon 200, 206
Sommerfelt, A. 5
soothsayers 261–262, 269, 551–556 *see also* mysticism
soroban 288–289, 294
Soubeyran, D. 144–145, 336
Sounda 375
Sourdel 540
Soustelle, Jacques 36, 72–73, 239
South America: counting 5, 10, 36, 125; Inca civilisation 68–69, 308
South Borneo 18–19
Spain 51, 250, 525, 553; and Arabs 248, 513, 520–521, 528, 587; Central America 300–303, 308; number system 31, 33–35, 359, 534–537, 585; Spanish Inquisition 588–589; writing 216, 539, 586
Spanish Inquisition 588–589
spatio-temporal disabilities 5
spheres 522
Spinoza 199
spirits, malign 275–276
square alphabet 212–213, 215, 233
square roots 156, 285, 293, 419, 560, 596–597
squares (power of two) 323–324, 363
Sri Lanka 5, 372, 374
*Stars and Stripes* 290
Steinschneider, M. 346, 362
Stele of the Vultures 86
Stephen, E. 19
Stephens, John Lloyd 300–301
sterling currency 41
Stevin, Simon 595
Stewart, C. 543, 545, 547
sticks as counting devices 15–16, 125 *see also* tally sticks
stone, as medium 162
string, knotted 64, 68–71
*Su Yuan Yu Zhian* 281
*Suan Fa Tong Zong* 61, 284, 293
suan pan 288–294 *see* abacus, Chinese
suan zí notation 278–283, 288, 408
Subandhu 418
subha (prayer) 50
subtraction 174; abacus 127, 204–206, 285, 292
subtractive principle 89, 328
succession 21–22
Sudan 97–98, 539
Suetonius 256
Sufi 521, 550–551
al-Sufi, Abu Musa Ja'far 364, 521
al-Sufin, 'Abd ar-Rahman 524

Sulawezi 383
Suli, As 522, 541–542
sulus script 539
Sumatar Harabesi 232
Sumatra 64, 383, 404
Sumer 101–102, 135
Sumerians 77–91; bullae 99, 103–104, 109–111; calculation 82–83, 94, 121–133, 140–142; number system 9, 77–95, 99–100, 109–120, 122, 139, 142, 147–148, 325–326, 349–350; Sumerian-Akkadian synthesis 137–138, 142, 148; writing 77–81, 86–90, 107 *see also* Mesopotamia
*Sumerisches Lexikon* 121
Sunni Islam 521
superstition *see* mysticism
Šuruppak 87–90, 121–122
*Sûrya Siddhânta* 411
Susa 101–107, 112–115, 119–120, 140, 149, 155, 158–160
Susinak-sar-Ilâni, King 159
Suter, H. 363–364, 519, 563, 568
Swedish 33–35
Switzerland 31, 65–66, 195, 205
Sylvester II *see* d'Aurillac, Gerbert
symbolism 78, 499–501
synonyms 409–421, 430–432, 438
Syria 52, 145, 526; Arabs 512, 519, 522–524, 526–528; Hittites 180; India 376, 513; writing 212–213, 232, 248 *see also* Ugaritic people
Syrian number system 227–228, 246; alphabetic numerals 238–243, 329; calculation 49, 94, 97–98, 101, 541–542, 563; Indian numerals 365–368, 534; Mari 142–146
Sznycer, M. 213

al-Tabari, Marshallah 514
al-Tabari, Sahl 514, 521
at Tabari, 'Ali Rabban 519
Tabasco 299, 303
tables: astronomical 146, 157–159, 198, 521, 527; mathematical 127–130, 146, 203–206, 283–288, 555–563; multiplication 154–156, 220, 561, 578
tablet: clay 77–80, 84–89, 92–93, 98–125, 132–135; accounting 79–80, 101–122, 134; Babylonian 134, 138, 140, 147–148, 159–160; calculation 122–125, 147–148, 151, 562–563; Cretan 178–179; Ebla tablets 135, 145; Heliastes 214; Hittite 181; proto-Elamite 102, 105; Sumerian 77–78, 98, 101–102, 107, 122, 134, 140, 562–563; Tablet of Fate 146; wooden (abacus) 132–133
Tabriz, Ghazan Khan a 516
Tadjikistan 522
Tadmor 248
Tafel 190
Tagala writing 383
Tagalog numerals 385
ibn Tahir, Mutahar 364
Tahirid dynasty 521
Tâkarî numerals 370–371, 381, 384, 421
talent (money) 182–183, 200–203
talismans 262, 522, 554
Talleyrand 42
Tall-i-Malyan 101
tally sticks 11–12, 16–18, 62–67, 191–197
*Talmud* 253
Tamanas 36, 44, 305
Tamil numerals 332, 334–335, 342, 353, 372–374, 383, 385
ibn Tamin, Abu Sahl 364
Tammam, Abu 522
Tangier 252
Tankrî numerals 370–371
*Tao Te Ching* 70
Taoism 443
al Tarabulusi, Ahmad al Barbir 58–59
Tarasques 301
Tarih 250
*Tarikh ul Hind* 251
ibn Tariq, Ya'qub 513, 520, 530
Tartaglia, N. 358
Tashfin, Yusuf Ben 525
Tashköprüzada 528
Taton, R. 515–516
Tavernier, J. B. 50
Tawhidi 524
tax collection 64–65, 68, 70, 302–303, 306
Tayasal 300
Taybugha 528
Taylor, C. 386–387
Taylor, J. 573
ibn Taymiyya 527
Tchen Yan-Sun 276, 428
Tebrizi 527
Tel-Hariri excavation 142
Telinga numerals 373
Tell Qudeirat 236, 238
Tello 101
Telugu numerals 373, 383, 385, 421, 438
ten 35; Chinese 263, 269; decimal system 24–32, 39–44; Greek 182, 184, 186; Hebrew 215; hieroglyphic 165, 168, 177–179, 181, 325; mysticism 43–44, 161; powers of 426–429, 440, 594; Roman numerals 188, 192; sexagesimal system 82–84, 93–95; tally sticks 194
ten thousand: Aramaic 230; Babylonian 140, 145; Chinese 263, 265; Greek 182, 184, 221–222; Hebrew 137; hieroglyphic 165, 168, 179–180, 325; Japanese 274–275; Roman numerals 197–198
*Ten Years in Sarawak* 19
Tenochtitlán 301–303, 306
Tepe Yahya 101–102
ternary principle 89, 139, 166, 227
Tertullian 47
*Tetrabiblos* 520
Tetuan 252
Texcoco, Lake 301
ibn Thabit, Ibrahim ibn Sinan 523
Thai numerals 375, 383, 385, 392–393, 438
Thebes 52, 219
Théodoret 55
Theodosius 522
Theon of Alexandria 91
Theophanes 359, 590
Theophilus of Edesse 513, 521
Thera 219
Thespiae 183
Thibaut, G. 415
Thibaut of Langres 257
thirty 93, 161, 215
thirty-six thousand 84
Thompson, J. E. 312, 316–317, 320–321
Thot 169–170, 176
thousand 25; Aramaic 230; Chinese 168, 263, 265, 269; Greek 182, 184, 186; hieroglyphic 165, 168, 178–179, 181, 325; Japanese 274; Mesopotamian 137–139, 142, 145, 231; Roman numerals 188–189, 192, 197–198
*Thousand and One Nights* 521
three 33, 393; base 40; Chinese 269; Egyptian 176; Hebrew 215; Indian 410; many as 4, 32, 94; ternary principle 9, 89
three hundred 215, 257
three hundred and sixty five 257–258
three thousand, six hundred 84, 93, 141–142, 148
Thureau-Dangin, F. 82, 91–92, 139, 152, 159, 407
Thutmosis 166
Tiberius, Emperor 200, 257
Tibet: counting 39, 70–71; number system 26–27, 371, 373, 388, 422; writing 377, 382, 420
Tibetan numerals 371, 385, 392–393, 395–396
Tijdschrift 406
Tikal 297, 318–320
time 17–19, 28, 49–50, 68, 298, 311; and base 41, 82, 158
Timur 528
Tiriqan, King 81
Tirmidhi 522
al Tirmidhî, Abù Dawud 51
Tizapan 301
Tlatelolco 302
Tod, N. M. 182–185, 233
tokens *see* abacus; calculi; currency; tally sticks
Tokharian language 32–35
Tokyo 274, 276, 289–290
Toledo 251, 514, 525, 587
Toltecs 300
Toluca 36
Toomer, G. J. 519, 531
topology 598
*Torah* 215, 218, 239, 253–254, 256 *see also Bible*
Torkhede 401

Torres Straits 6, 12, 14
Trajan's column 588
transcendental numbers 596–597
triangles, spherical 527
trigonometry 420, 523, 526–529
trillion 427–428
Tripoli 528
Tropfke, J. 537
Truffaut, François 4
Tschudi, J. D. 69
ibn Tufayl (Abubacer) 526
ibn Tughaj, Muhammad 522
Tughluq, Firaz Shah 527
Tula 301
Tulu numerals 383, 385
Tumert, Ibn 525
Tunisia 520–521, 523, 555
ibn Turk, Abu al-Hamid ibn Wasi 521
Turkestan, Chinese, writing 382, 385, 420
Turkey 512, 529; mysticism 248–251, 553; Russian abacus 290; writing 180, 248–250 *see also* Ottoman Empire
Turkish, Ancient 27–29
at Tusi, Nasir ad din 513, 527, 562, 571–573
twelve, base *see* duodecimal system
twenty 44; base *see* vigesimal system; Egyptian 177; Japanese 274; mysticism 93, 161, 248; Semitic 215, 228–229
twenty six 254
two 33, 393; base *see* binary system; Chinese 269; Egyptian 176; Hebrew 215; Indian 409–410
two hundred 215
Tyal tribe 73
Tylor, E. B. 5
Tyrol 195–196
*Tzolkin*, Mayan calendar 312, 315

Uaxactún 320
Uayeb 314
Ugaritic people 39, 137, 145, 214, 244
Ulrichen 195
al umam, Tabaqat 515
'Umar, caliph 515, 519
al-'Umari 527
'Umayyad dynasty 512
Umbrian alphabet 212
Umna 81
unciae (Roman ounce) 210
United Kingdom: Chancellor of the Exchequer 590 *see also* England; Scots Gaelic; Welsh
United States 92, 428
*Universal History* (Jami'at tawarikh) 516
Untash Gal 102
Upasak 387
al-Uqlidisi, Abu'l Hasan Ahmad ibn Ibrahim 364, 523, 563
Ur 81, 87, 90, 135
Urartu 9, 39, 139
Urmia, Lake 240
Uruk 81–81 , 86, 106, 110; clay tablets 77–78, 98, 101, 150, 152; number system 92–93, 101, 159
'Uthman, Abu 523
'Uthman, Caliph 519–520
Utu-Hegal, King 81
Uyghurs 28

*Vâjasaneyî Sâmhita* 425
*Vâkyapañchâdhyâyi* 414
Valabhi numerals 397–398
Vallat, F. 109
de La Vallée-Poussin, L. 530
value, concept of 72–76
Vandel, A. 367
Varâhamihîra 414–416, 419, 439, 504, 530
Varnasankhyâ numerals 388
Vatteluttu numerals 383, 385
Veda, Mannen 294
*Vedas* 29, 425
Vedda people 5, 72
Venezuela 36, 305
Ventris, Michael 179
Venus 297, 311, 315–316
Vera Cruz 302
Vercors 592
Vercoutter, Jacques 162
verse 431–432, 436–437
Vervaeck, L. 365
Vida, Levi della 58, 359
Vieta, Franciscus 597
Vietnam 272–273, 383, 407 *see also* Champa
vigesimal system 36–39, 44, 303–316; Aztec 306–308; Mayan 303–304, 306–311, 313, 316; Sumer 82
Vigila 362, 579
Vikrama 505
de Ville-Dieu, Alexandre 361
Vishmvâmitra 421
Vishnu 50, 444
Visperterminen 195
Vissière, A. 279
Vitruvius 194
*Vocabularium* 207
Vogel, K. 365, 519, 531, 533
Voizot, P. 356
von Wartburg, W. 427
Vossius, I. 359
vulgar fractions 595
Vyâgramukha (Fighar) 529–530

Waeschke, H. 434, 562
Wafa, Abu'l 514
al-Wafid, Ibn 525
al Wahab Adaraq, Abd 252
ibn Wahshiya 523
Walapai 125
Wallis, John 91, 361–362, 597
Wang Shuhe 516
Waqqas, Muhammedal 252
Warka 103
Warka, Lady of 81
wax calculating board 207–209, 563
Weber 425
wedge 148–149, 160
Weidler, J. F. 359
weights and measures 183, 239; International Bureau of 43
Welsh 33–35, 38
Wessely, J. E. 259
West Bank 97–98
Weyl-Kailey, L. 5
Whitney 411
Wiedler 356
Wieger 269
Wilkinson 52
William of Malmesbury 362, 586
Willichius 361, 590
wind, evokes numbers 442
Winkler, A. 549
Winter, H. J. J. 519
de Wit, C. 176
Woepcke, F. 242, 356, 362–364, 368, 409, 421, 423, 427–429, 433, 438, 529–530, 534, 537, 549, 562–563
Wolof 36
Woods, Thomas Nathan 289
Wright, W. 241
writing 81, 107–108, 272–275; styles 171, 186, 539 *see also under specific race/country*; mysticism
writing materials 85, 390–391, 430, 434; chalk 566–567; papyrus 301, 533; reeds 85–87, 539, 553
ibn Wuhaib, Abu ''Abdallah Malik 550
Wulfila 226
xià deng number system 277
Xiao dun 269–270
xíngshū writing 267–268

Yaeyama 70
Yahweh 71, 212, 218, 253–255
Yahya, Abu 520
Yamamoto, Masahiro 278, 294
al Yaman, Hudaifa ibn 58
al-Yamani, Yahya ibn Nawfal 58, 520
Yang Hui 285
Yang Sun 283
al-Ya'qubi 522
Yaxchilán 316
year, days of 91
Yebu 36–37, 70

Yedo 305
Yehimilk 213
Yemen 368, 528, 534 *see also* Sheba
Yishakhi, Rabbenu Shelomoh 253
Yong-le da dian 264
Yoruba 36–37, 305
Youschkevitch, A. P. 243, 365, 512–513, 515–516, 519, 530, 533, 548, 560, 562, 571
Yoyotte, J. 52, 533
Yucatan 299, 300–301, 303, 312
yuga (cosmic cycle) 411, 420, 506–507
Yum Kax 312
ibn Yunus, Matta 513, 523–524
ibn Yusuf, Ahmad 522
ibn Yusuf, al-Hajjaj 521

Zahrawi, Abu'l Qasim az (Abulcassis) 522
za'irja 550–552
Zajackowski, A. 543, 547
Zapotec 36, 162, 305, 307–308
al-Zarqali 525
Zaslavsky, C. 37, 44, 305
Zayd, Abu 523
Zen 295, 443
Zencirli, Aramaic numerals 229
zero 25, 324, 340–346, 354–357, 365, 416, 507–510, 587; abacus 366, 434–437, 559; absence 145, 149–151, 343, 366, 372–374, 559; Babylonian 152–154; Chinese 266, 280–281, 408; Europe 588–590; Greek 157; imperfect 341–342; India 371, 399, 410, 412, 415–416, 420, 433, 437–439; Islamic world 533–534; Mayan 308–311, 320–322
zhong deng system 277
Zhu Shi Jie 281
Zimri-Lim 142
ibn Ziyad, Tariq 520
zodiac 92, 549–551, 553
ibn Zuhr (Avenzoar) 526
Zulus 5
Zumpango 301
Zuñi 15, 196

# THE UNIVERSAL HISTORY OF NUMBERS

GEORGES IFRAH, now aged fifty, was the despair of his maths teachers at school – he lingered near the bottom of the class. Nevertheless he grew up to become a maths teacher himself and, in order to answer a pupil's question as to where numbers came from, he devoted some ten years to travelling the world in search of the answers, earning his keep as a night clerk, waiter, taxi-driver. Today he is a maths encyclopaedia on two legs, and his book has been translated into fourteen languages.

DAVID BELLOS is Professor of French at Princeton University and author of *Georges Perec: A Life in Words*. E. F. HARDING has taught at Aberdeen, Edinburgh and Cambridge and is a Director of the Statistical Advisory Unit at Manchester Institute of Science and Technology. SOPHIE WOOD is a specialist in technical translation from French and Spanish. IAN MONK, while skilled in technical translation, is better known for his translations of Georges Perec and Daniel Pennac.